ORGANIC
CHEMISTRY

Janice Gorzynski Smith

University of Hawai'i at Mānoa

Mc Graw Hill **Higher Education**

Boston   Burr Ridge, IL   Dubuque, IA   Madison, WI   New York   San Francisco   St. Louis
Bangkok   Bogotá   Caracas   Kuala Lumpur   Lisbon   London   Madrid   Mexico City
Milan   Montreal   New Delhi   Santiago   Seoul   Singapore   Sydney   Taipei   Toronto

Higher Education

ORGANIC CHEMISTRY

Published by McGraw-Hill, a business unit of The McGraw-Hill Companies, Inc., 1221 Avenue of the Americas, New York, NY 10020. Copyright © 2006 by The McGraw-Hill Companies, Inc. All rights reserved. No part of this publication may be reproduced or distributed in any form or by any means, or stored in a database or retrieval system, without the prior written consent of The McGraw-Hill Companies, Inc., including, but not limited to, in any network or other electronic storage or transmission, or broadcast for distance learning.

Some ancillaries, including electronic and print components, may not be available to customers outside the United States.

This book is printed on acid-free paper.

1 2 3 4 5 6 7 8 9 0 VNH/VNH 0 9 8 7 6 5 4

ISBN 0–07–111662–1

The credits section for this book begins on page C–1 and is considered an extension of the copyright page.

www.mhhe.com

For Megan Sarah.

Janice Gorzynski Smith was born in Schenectady, New York, and grew up following the Yankees, listening to the Beatles, and water skiing on Sacandaga Reservoir. She became interested in chemistry in high school, and went on to major in chemistry at Cornell University where she received an A.B. degree *summa cum laude*. Jan credits two truly gifted teachers—Sister Rita Boggs in high school and Professor Roald Hoffmann in general chemistry—for showing her the beauty and logic of the subject that has become her life's work. Jan earned a Ph.D. in Organic Chemistry from Harvard University under the direction of Nobel Laureate E.J. Corey, and she also spent a year as a National Science Foundation National Needs Postdoctoral Fellow at Harvard. During her tenure with the Corey group she completed the total synthesis of the plant growth hormone gibberellic acid.

Following her postdoctoral work Jan joined the faculty of Mount Holyoke College where she served on the faculty for 21 years. During this time she was active in teaching organic chemistry lecture and lab courses, conducting a research program in organic synthesis, and serving as department chair. Her organic chemistry class was named one of Mount Holyoke's "Don't-miss courses" in a survey by *Boston* magazine. After spending two sabbaticals amidst the natural beauty and diversity in Hawai'i in the 1990s, Jan and her family moved there permanently in 2000. She is currently a faculty member at the University of Hawai'i at Mānoa, where she teaches both the one- and two-semester organic chemistry lecture and lab courses, and serves as the faculty advisor to the student affiliate chapter of the American Chemical Society. In 2003, she received the Chancellor's Citation for Meritorious Teaching.

Jan resides in Hawai'i with her husband Dan, an emergency medicine physician. She has four children: Matthew and Zachary, age 9 (margin photo in Section 5.3); Jenna, a student at the University of Hawai'i; and Erin, who is a student at the Brown University School of Medicine and co-author of the *Student Study Guide/Solutions Manual* for this text. When not teaching, writing, or enjoying her family, she bikes, hikes, snorkels, and scuba dives in sunny Hawai'i, and time permitting, enjoys travel and Hawai'ian quilting.

Foreword by E. J. Corey

Students entering an introductory Organic Chemistry course often do so with mixed feelings. On the one hand, they are happy to have completed their first semester or year of studies in a science curriculum. On the other hand, they are fearful of what lies ahead in a course that is often considered to be one of the most challenging an undergraduate will take. Many of these students aspire to the practice of healthcare; others are primarily interested in scientific research. Even those who have performed well in the classroom up to this point in their academic careers may worry about what lies ahead in Organic Chemistry. As a person who has spent his career studying this science, I cannot resist the temptation to explain my love of the subject and to allay any such fears.

I must start with the fact that Organic Chemistry is the fundamental language of life, since all life involves organic molecules, their interactions and transformations. It provides a firm foundation for biology and medicine and the tools to develop new medicines and therapies. Although Organic Chemistry is rich in information, it also conforms to a set of principles and an inner logic that is stunningly beautiful to those who know the subject. The actors of organic chemistry (molecules) and their behavior (reactions) can be described by a powerful and beautiful graphical language, consisting of two- and three-dimensional chemical formulas that express the atoms, bonds, electrons, charges and geometries that characterize their structures. Indeed, this formula notation is so powerful that a trained organic chemist can tell at a glance much about the chemistry and reactivity of a molecule and even how it might be made from simpler substances.

Organic Chemistry appealed to me as an undergraduate because of the beauty of organic molecules and their relevance to human health. The logic and order in the behavior of these molecules stirred my curiosity. As a boy, I had dreamed of someday building things like radios or cars or airplanes. Organic Chemistry offered me the opportunity to construct equally amazing things, albeit on a much smaller scale, that could function to stop infections, cure diseases and make life better.

Over the years, I have worked with more than one hundred students pursuing Ph.D.'s in Organic Chemistry. The author of this text is one of them. Many of these students are now university professors who maintain their own research labs. An even larger number are research scientists working in the pharmaceutical and chemical industry. Dr. Smith, however, while trained to be a researcher in my lab, decided to direct her passion and energy to teaching. In her 25 years on the faculty, first at Mount Holyoke and now the University of Hawai'i at Mānoa, she has received numerous accolades for her outstanding teaching. This book is a compilation of the methods and notes that she has evolved over those years.

In reading a prepublication draft of *Organic Chemistry,* I was impressed by how clearly and efficiently she has presented the material. It is obvious that, through her interaction with thousands of students, she has developed an approach that provides just the right amount of direction needed to understand the concepts. She also has integrated biological applications in such a way that they are neither separate nor unrelated to the core topics. I believe that Dr. Smith's book will make Organic Chemistry more accessible to students and allow them to sense the many fascinating aspects of Organic Chemistry that caught my eye years ago.

I would like to congratulate Dr. Smith on this book. To cover in one concise text so many varied aspects of Organic Chemistry is a major achievement.

E. J. Corey
Harvard University

Contents in Brief

Contents

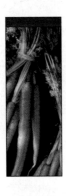

4 Alkanes 112

5 Stereochemistry 156

6 Understanding Organic Reactions 190

10   Alkenes 346

11   Alkynes 386

12   Oxidation and Reduction 412

13   Radical Reactions 446

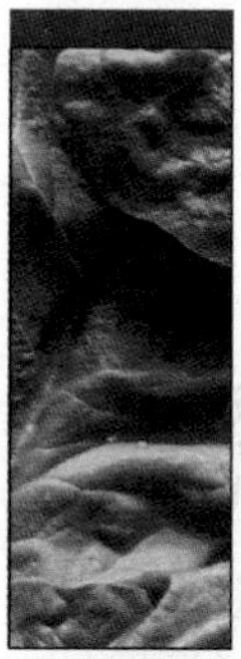

14   Mass Spectrometry and Infrared Spectroscopy 478

15   Nuclear Magnetic Resonance Spectroscopy 504

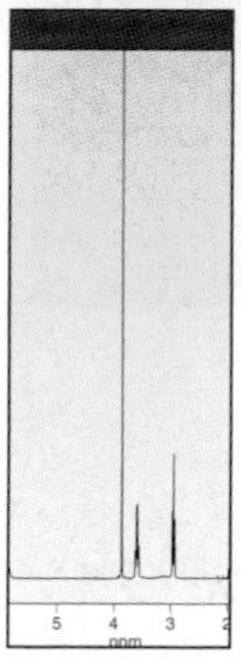

22   Carboxylic Acids and Their Derivatives—Nucleophilic Acyl Substitution 780

23   Substitution Reactions of Carbonyl Compounds at the α Carbon 832

24   Carbonyl Condensation Reactions 864

28   Amino Acids and Proteins 1012

List of *How To*'s

How To boxes provide detailed instructions for key procedures that students need to master. Below is a list of each *How To* and where it is presented in the text.

List of Mechanisms

Mechanisms are the key to understanding the reactions of organic chemistry. For this reason, great care has been given to present mechanisms in a detailed, step-by-step fashion. The list below indicates when each mechanism in the text is presented for the first time.

List of Selected Applications

Applications make any subject seem more relevant and interesting—for nonmajors and majors alike. The following is a list of some of the biological, medicinal, and environmental applications that have been integrated throughout *Organic Chemistry*. Each chapter opener showcases an interesting and current application relating to the chapter's topic.

	Description	Reference
Prologue	The structure of palytoxin, one of the most toxic substances known, is used to illustrate the conventions for drawing skeletal (line) structures.	Figure 1
Chapter 1	The structure of capsaicin, the spicy component of hot peppers, is used to apply the most important concepts from the chapter—how to determine the hybridization and geometry around an atom, how to label all polar and nonpolar bonds, and how to compare bond length and bond strength for different C−C bonds.	Section 1.13
Chapter 2	Aspirin must be in its neutral form to cross a cell membrane, so the different properties of acids and their conjugate bases can have important physiological consequences.	Section 2.7
Chapter 3	Because of their different functional groups, vitamin A is water insoluble, whereas vitamin C is water soluble. As a result, excess vitamin A is stored in the fat cells of tissue, but excess vitamin C is excreted.	Section 3.6
Chapter 4	Lipids, water-insoluble biological compounds, contain the same structural features as alkanes, the simplest organic molecules.	Section 4.15
Chapter 5	Starch and cellulose are used to illustrate how apparently minute differences in structure can result in vastly different properties.	Section 5.1
Chapter 6	The metabolism of glucose and the combustion of gasoline are two apparently different organic reactions that share a common thread—both reactions release a great deal of useful energy.	Chapter opener, Section 6.4
Chapter 7	Nitrosamines, one potential by-product of the sodium nitrite used to preserve meats, can be converted to unstable intermediates that react with biological nucleophiles (such as DNA or an enzyme). The end result can be cancer or death of the cell.	Section 7.16
Chapter 8	Ethylene, the simplest hydrocarbon containing a carbon–carbon double bond, is a hormone that regulates plant growth and fruit ripening.	Chapter opener
Chapter 9	New asthma drugs, such as zileuton, inhibit the enzyme responsible for the synthesis of leukotriene C_4 from arachidonic acid. By blocking the synthesis of a compound responsible for the disease, zileuton treats the cause of the asthma, not just its symptoms.	Section 9.16
Chapter 10	Fats are solids at room temperature, whereas oils are liquids, because oils have more Z double bonds, giving rise to more kinks in their hydrocarbon chains.	Chapter opener, Section 10.6
Chapter 11	Ethynylestradiol, a synthetic compound containing a C−C triple bond, is a component of several widely used oral contraceptives.	Chapter opener
Chapter 12	Antabuse, a drug given to alcoholics to prevent them from consuming alcoholic beverages, acts by interfering with the normal oxidation of ethanol. Instead, acetaldehyde builds up in the person's system, causing them to become violently ill.	Section 12.13
Chapter 13	Vitamin E is a fat-soluble vitamin that is thought to inhibit radical reactions that cause oxidative damage to the lipids in the cell membrane.	Chapter opener, Section 13.12

	Description	Reference
Chapter 14	Trace amounts of tetrahydrocannabinol (THC), the primary active constituent of marijuana, can be detected by modern instrumental methods, such as mass spectrometry and IR spectroscopy.	Chapter opener, Section 14.3
Chapter 15	Magnetic resonance imaging (MRI)—NMR spectroscopy in medicine—is a powerful diagnostic technique used routinely in hospitals today.	Section 15.12
Chapter 16	Commercial sunscreens contain conjugated compounds that absorb UV light, thus shielding your skin (for a time) from the harmful effects of UV radiation.	Section 16.15
Chapter 17	Zoloft, Viagra, and Claritin are just three of many drugs whose structures contain benzene rings.	Figure 17.5
Chapter 18	A key step in the laboratory synthesis of LSD involves carbon–carbon bond formation using electrophilic aromatic substitution.	Chapter opener, Figure 18.4
Chapter 19	Aspirin relieves pain and reduces inflammation by blocking the synthesis of prostaglandins, 20-carbon fatty acids with a five-membered ring that are responsible for pain, inflammation, and a wide variety of other biological functions.	Chapter opener, Section 19.6
Chapter 20	Juvenile hormones, the compounds that regulate the complex life cycles of some insects, are synthesized by reactions that form new carbon–carbon bonds.	Chapter opener, Figure 20.6
Chapter 21	11-*cis*-Retinal is the light-sensitive conjugated aldehyde that plays a key role in the complex chemistry of vision for all vertebrates, arthropods, and mollusks.	Chapter opener, Figure 21.9
Chapter 22	Penicillin and related β-lactams kill bacteria because they undergo a nucleophilic acyl substitution reaction that interferes with the synthesis of the bacterial cell wall.	Chapter opener, Section 22.14
Chapter 23	One method for synthesizing tamoxifen, a potent anticancer drug, forms a new carbon–carbon bond on the α carbon to a carbonyl group using an intermediate enolate.	Chapter opener, Section 23.8
Chapter 24	Gibberellic acid, a plant growth hormone, can be synthesized in the laboratory using two intramolecular aldol reactions to form two carbon–carbon bonds of the complex carbon skeleton.	Chapter opener
Chapter 25	Dopamine, a neurotransmitter, is a chemical messenger released by one nerve cell, which then binds to a receptor in a neighboring target cell. Proper dopamine levels are necessary to maintain an individual's mental and physical health. Parkinson's disease results when dopamine-producing neurons die and dopamine levels drop.	Figure 25.6
Chapter 26	Cholesterol, a lipid synthesized in the liver, is found in almost all body tissues. Even though it is a vital component for healthy cell membranes and it serves as the starting material for the synthesis of all other steroids, elevated cholesterol levels can lead to coronary artery disease.	Chapter opener, Section 26.8
Chapter 27	Lactose, a carbohydrate formed from glucose and galactose, is the principal sugar in dairy products. Individuals who lack adequate amounts of lactase, the enzyme necessary to digest and absorb lactose, suffer from a condition known as lactose intolerance.	Chapter opener, Section 27.12
Chapter 28	The two amino acids that comprise the sweetener aspartame are joined via the same peptide bond that is found in proteins and polypeptides in living organisms.	Chapter opener, Section 28.5

Preface

When I began this project, my goal was to write a text that showed students the beauty and logic of organic chemistry by giving them a book that they would *use*. Five years and three drafts later, I present to you *Organic Chemistry*, a text based on lecture notes and handouts that have been developed in my own organic chemistry courses over the years. I have followed two guiding principles: use relevant and interesting applications to illustrate chemical phenomena, and present the material in a student-friendly fashion using bulleted lists, solved problems, and extensive illustrations and summaries. *Organic Chemistry* is my attempt to simplify and clarify a course that intimidates many students—to make organic chemistry interesting, relevant, and accessible to *all* students, both chemistry majors and those interested in pursuing careers in biology, medicine, and other disciplines, without sacrificing the rigor they need to be successful in the future.

The Basic Features

◆ **Style** This text is different—by design. Today's students rely more heavily on visual imagery to learn than ever before. I have therefore written a text that uses less prose and more diagrams, equations, tables, and bulleted summaries to introduce and reinforce the major concepts and themes of organic chemistry.

◆ **Content** *Organic Chemistry* accents basic themes in an effort to keep memorization at a minimum. Relevant materials from everyday life are used to illustrate concepts, and this material is integrated throughout the chapter rather than confined to a boxed reading. Each topic is broken down into small chunks of information that are more manageable and easily learned. Sample Problems are used as a tool to illustrate stepwise problem solving. Exceptions to the rule and older, less useful reactions are omitted to focus attention on the basic themes.

◆ **Organization** *Organic Chemistry* uses functional groups as the framework within which chemical reactions are discussed. Thus, the emphasis is placed on the reactions that different functional groups undergo, not on the reactions that prepare them. Moreover, similar reactions are grouped together so that parallels can be emphasized. These include acid–base reactions (Chapter 2), oxidation and reduction (Chapters 12 and 20), radical reactions (Chapter 13), and reactions of organometallic reagents (Chapter 20).

By introducing one new concept at a time, keeping the basic themes in focus, and breaking complex problems down into small pieces, I have found that many students find organic chemistry an intense but learnable subject. Many, in fact, end the yearlong course surprised that they have actually *enjoyed* their organic chemistry experience.

Illustrations

The illustration program is a key component of the visual emphasis in *Organic Chemistry*. Besides traditional skeletal (line) structures and condensed formulas, there are numerous ball-and-stick molecular models and electrostatic potential maps to help students grasp the three-dimensional structure of molecules (including stereochemistry) and to better understand the distribution of electronic charge. Unique to *Organic Chemistry* are the micro-to-macro illustrations. These pieces combine line art and photos to place molecules in a broader context. Examples include starch and cellulose (Chapter 5), adrenaline (Chapter 7), and dopamine (Chapter 25).

Organization and Presentation

For the most part, the overall order of topics in the text is consistent with the way most instructors currently teach organic chemistry. There are, however, some important differences in the way topics are presented to make the material logical and more accessible. This can especially be seen in the following areas.

- **Review material**—Chapter 1 presents a healthy dose of review material covering Lewis structures, molecular geometry and hybridization, bond polarity, and types of bonding. While many of these topics are covered in general chemistry courses, they are presented here from an organic chemist's perspective. I have found that giving students a firm grasp of these fundamental concepts helps tremendously in their understanding of later material.

- **Acids and bases**—Chapter 2 on acids and bases serves two purposes. It gives students experience with curved arrow notation using some familiar proton transfer reactions. It also illustrates how some fundamental concepts in organic structure affect a reaction, in this case an acid–base reaction. Since many mechanisms involve one or more acid–base reactions, I emphasize proton transfer reactions early and come back to this topic often throughout the text.

- **Functional groups**—Chapter 3 uses the functional groups to introduce important properties of organic chemistry. Relevant examples—PCBs, vitamins, soap, and the cell membrane—illustrate basic solubility concepts. In this way, practical topics that are sometimes found in the last few chapters of an organic chemistry text (and thus often omitted because instructors run out of time) are introduced early so that students can better grasp why they are studying the discipline.

- **Stereochemistry**—Stereochemistry (the three-dimensional structure of molecules) is introduced early (Chapter 5) and reinforced often, so students have every opportunity to learn and understand a crucial concept in modern chemical research, drug design, and synthesis.

- **Modern reactions**—While there is no shortage of new chemical reactions to present in an organic chemistry text, I have chosen to concentrate on new methods that introduce a particular three-dimensional arrangement in a molecule, so-called asymmetric or enantioselective reactions. Examples include Sharpless epoxidation (Chapter 12), CBS reduction (Chapter 20), and enantioselective synthesis of amino acids (Chapter 28).

- **Grouping reactions**—Since certain types of reactions have their own unique characteristics and terminology that make them different from the basic organic reactions, I have grouped these reactions together in individual chapters. These include acid–base reactions (Chapter 2), oxidation and reduction (Chapters 12 and 20), radical reactions (Chapter 13), and reactions of organometallic reagents (Chapter 20). I have found that focusing on a group of reactions that share a common theme helps students to better see their similarities.

- **Synthesis**—Synthesis, one of the most difficult topics for a beginning organic student to master, is introduced in small doses, beginning in Chapter 7 and augmented with a detailed discussion of retrosynthetic analysis in Chapter 11. In later chapters, special attention is given to the retrosynthetic analysis of compounds prepared by carbon–carbon bond forming reactions (for example, Sections 20.11 and 21.10C).

- **Spectroscopy**—Since spectroscopy is such a powerful tool for structure determination, four methods are discussed over two chapters (Chapters 14 and 15).

- **Key concepts**—End-of-chapter summaries succinctly summarize the main concepts and themes of the chapter, making them ideal for review prior to working the end-of-chapter problems or taking an exam.

Acknowledgments

When I started working on this project in the fall of 1999, I had no sense of the magnitude of the task nor any idea of just how many people I would rely upon to complete it. Although mine is the only name that appears on the cover, this text is truly the result of extensive contributions from a number of individuals.

I must first thank two chemist–friends, Spencer Knapp and John Murdzek, for the roles they have played in seeing this project to its completion. I never imagined that Spencer Knapp, a former classmate from Cornell and co-worker from Harvard, would one day serve as my "right-hand person" for a number of aspects for this text. In addition to his excellent work in reviewing the manuscript and illustrations and overseeing the creation of the spectra, along with his colleagues Patrick J. O'Connor and Richard A. Huhn, his steady encouragement and keen insight throughout the writing process have been of enormous value.

John Murdzek helped develop every aspect of this project with uncanny attention to detail. He took a wordy and rambling first draft and helped me polish it into a cleaner and well-organized final manuscript. His thoughtful feedback has significantly improved the quality of this book. He is a true professional and I have been most fortunate to have had his assistance from first to final draft. I hope our collaboration continues for many future editions.

I have also benefited greatly from a team of advisors who have helped guide me through the many intricate aspects of the text and also provided great moral support. They include:

Robert S. Coleman *The Ohio State University*
Malcolm D.E. Forbes *University of North Carolina—Chapel Hill*
Susan M. King *University of California—Irvine*
Spencer Knapp *Rutgers University*
C. Peter Lillya *University of Massachusetts—Amherst*
Robert E. Maleczka, Jr. *Michigan State University*
Keith T. Mead *Mississippi State University*
Richard Pagni *University of Tennessee*
Michael Rathke *Michigan State University*
Ieva L. Reich *University of Wisconsin—Madison*
John W. Taylor *Rutgers University*
Laren M. Tolbert *Georgia Institute of Technology*
Jane E. Wissinger *University of Minnesota*

Having served as an accuracy checker for professional publications in the past, I know how important and challenging this can be for an organic chemistry text. I have had the good fortune to be supported by a great group of instructors who have helped insure that this text is as "error-free" as possible. They include:

Daniel A. Adsmond *Ferris State University*
Kay M. Brummond *University of Pittsburgh*
Donnie N. Byers *Johnson County Community College*
Todd S. Deal *Georgia Southern University*
Susan M. King *University of California—Irvine*
Kathleen V. Nolta *University of Michigan—Ann Arbor*
Catherine Woytowicz *The George Washington University*

Throughout the development of this first edition I have been impressed with the quality of the feedback I have received on how the manuscript could be improved. Some individuals have provided in-depth reviews, while others have agreed to class test sections of the manuscript. Still others have provided very candid feedback to me or my editors at focus groups and symposiums. Collectively, these individuals have made the most significant contributions to this project.

Reviewers

Christopher J. Abelt, *College of William & Mary*
Madeline Adamczeski, *San Jose City College*
Jennifer Adamski, *Old Dominion University*
Ruquia Ahmed-Schofield, *Xavier University of Louisiana*
Yu Mi Ahn, *Deciphera Pharmaceuticals LLC*
Igor Alabugin, *Florida State University*
Angela J. Allen, *University of Michigan—Dearborn*
Neil T. Allison, *University of Arkansas*
Eric Anslyn, *University of Texas—Austin*
Jon C. Antilla, *University of Mississippi*
Mark Arant, *University of Louisiana—Monroe*
Paramjit Arora, *New York University*
Arthur J. Ashe III, *University of Michigan—Ann Arbor*
Donald H. Aue, *University of California—Santa Barbara*
George C. Bandik, *University of Pittsburgh*
John Barbaro, *University of Florida*
Debra L. Bautista, *Eastern Kentucky University*
Byron L. Bennett, *University of Nevada—Las Vegas*
Stuart R. Berryhill, *California State University—Long Beach*
Steve Bertman, *Western Michigan University*
Robert S. Bly, *University of South Carolina*
Debra J. Boehmler, *University of Maryland—College Park*
Ned B. Bowden, *University of Iowa*
Stephen E. Branz, *San Jose State University*
Phil A. Brown, *North Carolina State University*
Kay M. Brummond, *University of Pittsburgh*
Christine Brzezowski, *University of Alberta*
Patrick E. Buick, *Florida Atlantic University—Boca Raton*
Arthur W. Bull, *Oakland University*
Paul Buonora, *California State University—Long Beach*
Kevin Burgess, *Texas A&M University—College Station*
Rachel L. Burroughs, *Columbus State Community College*
Donald J. Burton, *University of Iowa*
Sylvester Burton, *Southern University*

Donnie N. Byers, *Johnson County Community College*
Kevin C. Cannon, *Penn State Abington College*
Neal Castagnoli, *Virginia Tech*
Young-Tae Chang, *New York University*
Dana Chatellier, *University of Delaware*
Michelle Chatellier, *University of Delaware*
Clair J. Cheer, *San Jose State University*
Barry A. Coddens, *Northwestern University*
Randolph A. Coleman, *College of William & Mary*
David M. Collard, *Georgia Institute of Technology*
Gregory R. Cook, *North Dakota State University*
Scott W. Cowley, *Colorado School of Mines*
Sonja E. Davison, *Tarrant County College—North East Campus*
Todd S. Deal, *Georgia Southern University*
Mark R. DeCamp, *University of Michigan—Dearborn*
Roman Dembinski, *Oakland University*
Michael R. Detty, *SUNY—Buffalo*
Mark Distefano, *University of Minnesota*
Lisa M. Dollinger, *University of Arizona*
William A. Donaldson, *Marquette University*
Ralph Dougherty, *Florida State University*
Liangcheng Du, *University of Nebraska—Lincoln*
Norma Dunlap, *Middle Tennessee State University*
Ihsan Erden, *San Francisco State University*
Ken Feldman, *Penn State University*
Martin Feldman, *Howard University*
Gary B. Fisher, *De Anza College*
Steven A. Fleming, *Brigham Young University*
Paul Floreancig, *University of Pittsburgh*
David C. Forbes, *University of South Alabama*
Maryam Foroozesh, *Xavier University of Louisiana*
John W. Francis, *Columbus State Community College*
Natia L. Frank, *University of Washington*
Andreas H. Franz, *University of the Pacific*
Kevin P. Gable, *Oregon State University*

Warren P. Giering, *Boston University*

Timothy E. Glass, *University of Missouri—Columbia*

Galina Z. Goloverda, *Xavier University of Louisiana*

Geeta Govindarajoo, *Rutgers University*

Scott Gronert, *San Francisco State University*

Robert B. Grossman, *University of Kentucky*

Denise Guinn, *Regis University*

Robert D. Guthrie, *University of Kentucky*

Ronald L. Halterman, *University of Oklahoma*

Scott T. Handy, *SUNY—Binghamton*

Paul R. Hanson, *University of Kansas*

Aaron M. Hartel, *University of Georgia*

Frederick J. Heldrich, *College of Charleston*

Alvan C. Hengge, *Utah State University*

Gene Hiegel, *California State University—Fullerton*

Edwin F. Hilinski, *Florida State University*

Stephen Hixson, *University of Massachusetts—Amherst*

T. Keith Hollis, *University of California—Riverside*

Todd Houston, *Griffith University*

Christopher Ikediobi, *Florida A&M University*

Ralph Isovitsch, *Xavier University of Louisiana*

T.G. Jackson, *University of South Alabama*

Bruce B. Jarvis, *University of Maryland—College Park*

Anton W. Jensen, *Central Michigan University*

Leslie Jimenez, *Rutgers University*

David M. Johnson, *University of Texas—San Antonio*

Richard P. Johnson, *University of New Hampshire*

David C. Johnson II, *Indiana University—Bloomington*

Kyung Woon Jung, *University of South Florida*

Branko S. Jursic, *University of New Orleans*

Russell G. Kerr, *Florida Atlantic University—Boca Raton*

James J. Kiddle, *Western Michigan University*

C. A. Kingsbury, *University of Nebraska—Lincoln*

Rebecca M. Kissling, *SUNY—Binghamton*

Francis Klein, *Creighton University*

Michael R. Korn, *Southwest Texas State University*

Dalila G. Kovacs, *Michigan State University*

Bette A. Kreuz, *University of Michigan—Dearborn*

Paul J. Kropp, *University of North Carolina—Chapel Hill*

Grant Krow, *Temple University*

Dorothy B. Kurland, *West Virginia University*

Man Lung Kwan, *John Carroll University*

John Landrum, *Florida International University*

Thomas Lectka, *Johns Hopkins University*

Susan Z. Lever, *University of Missouri—Columbia*

Jack B. Levy, *University of North Carolina—Wilmington*

Harriet Lindsay, *Eastern Michigan University*

Robert S.H. Liu, *University of Hawai'i at Mānoa*

Robert Loeschen, *California State University—Long Beach*

Andrew B. Lowe, *University of Southern Mississippi*

Yun Lu, *Utah State University*

Hannia Lujan-Upton, *Long Island University*

Frederick A. Luzzio, *University of Louisville*

Ronald M. Magid, *University of Tennessee—Knoxville*

Michael Maguire, *Wayne State University*

Elahe Mahdavian, *Louisiana State University—Shreveport*

George Majetich, *University of Georgia*

Sanku Mallik, *North Dakota State University*

Przemyslaw Maslak, *Penn State University*

Jim Maxka, *Northern Arizona University*

Edward J. McIntee, *College of St. Benedict/St. John's University*

Lydia McKinstry, *University of Nevada—Las Vegas*

Miguel O. Mitchell, *Salisbury University*

Dillip K. Mohanty, *Central Michigan University*

Gary A. Molander, *University of Pennsylvania*

Andrew T. Morehead, Jr., *East Carolina University*

Richard W. Morrison, *University of Georgia*

Layne Morsch, *DePaul University*

Mark C. Morvant, *Texas A&M University—Corpus Christi*

Rabi Ann Musah, *State University of New York—Albany*

Jennifer Muzyka, *Centre College*

Donna Nelson, *University of Oklahoma*

Dallas New, *University of Central Oklahoma*

Elva Mae Nicholson, *Eastern Michigan University*

Kathleen V. Nolta, *University of Michigan—Ann Arbor*

Joseph M. O'Connor, *University of California—San Diego*

Patrick J. O'Connor, *Rutgers University*

Anne B. Padías, *University of Arizona*

Raj Pandian, *University of New Orleans*

Cyril Párkányi, *Florida Atlantic University—Boca Raton*

Keith O. Pascoe, *Georgia State University*

Timothy E. Patten, *University of California—Davis*

Robert Patterson, *University of Southern Mississippi*

John H. Penn, *West Virginia University*

James D. Pennington, *Texas A&M University*

Otto Phanstiel, *University of Central Florida*

Karen E. S. Phillips, *Hunter College of CUNY*

Eugene Pinkhassik, *University of Memphis*

James S. Poole, *Ball State University*

Daniel Quinn, *University of Iowa*

Morton Raban, *Wayne State University*

Jon D. Rainier, *University of Utah*

Rajendra Rathore, *Marquette University*

Jianhua Ren, *University of the Pacific*

Suzanne Ruder, *Virginia Commonwealth University*

Lev Ryzhkov, *Towson University*

Brian A. Salvatore, *Louisiana State University—Shreveport*

Robert Sammelson, *Ball State University*

Vyacheslav V. Samoshin, *University of the Pacific*

Paul Sampson, *Kent State University*

Linda S. Sapochak, *Pacific Northwest National Laboratory*

Tomikazu Sasaki, *University of Washington*

Chris Schaller, *College of St. Benedict/St. John's University*

Susan M. Schelble, *University of Colorado—Denver*

Peter R. Schreiner, *Institute of Organic Chemistry, Justus-Liebig-University Giessen*

David I. Schuster, *New York University*

Valerie Sheares, *University of North Carolina—Chapel Hill*

Jim Silliman, *Texas A&M University*

Virginia F. Smith, *United States Naval Academy*

Carl H. Snyder, *University of Miami*

Gary J. Snyder, *University of Massachusetts—Amherst*

Tami I. Spector, *University of San Francisco*

David Spurgeon, *University of Arizona*

Laurie S. Starkey, *Cal Poly—Pomona*

Richard Steiner, *University of Utah*

Chad E. Stephens, *Georgia State University*

J. William Suggs, *Brown University*

Sarah A. Tabacco, *Massachusetts Institute of Technology*

Jetze J. Tepe, *Michigan State University*

Eric S. Tillman, *Bucknell University*

Edward Turos, *University of South Florida*

Cyriacus Chris Uzomba, *Austin Community College*

M. Graca H. Vicente, *Louisiana State University*

Bruno M. Vittimberga, *University of Rhode Island*

Heidi R. Vollmer-Snarr, *Brigham Young University*

Thomas G. Waddell, *University of Tennessee—Chattanooga*

George H. Wahl, Jr., *North Carolina State University*

Philip Warner, *Northeastern University*

Sarah L. Weaver, *Southeastern Louisiana University*

Donald K. Wedegaertner, *University of the Pacific*

Matthew Weinschenk, *Emory University*

Barbara J. Whitlock, *University of Wisconsin—Madison*

Milton J. Wieder, *Metropolitan State College of Denver*

David F. Wiemer, *University of Iowa*

Laurie A. Witucki, *Grand Valley State University*

Stephen Woski, *University of Alabama*

Catherine Woytowicz, *The George Washington University*

Dennis L. Wright, *Dartmouth College*

Justin K. Wyatt, *College of Charleston*

Cong-Gui Zhao, *University of Texas—San Antonio*

Class Tests

Rutgers University
Instructors: Geeta Govindarajoo and John W. Taylor

Louisiana State University—Shreveport
Instructor: Brian A. Salvatore

University of Hawai'i at Mānoa
Instructor: Janice Gorzynski Smith

Organic chemistry is often regarded as one of the most difficult courses an undergraduate can take, and so it stands to reason that authoring a text for this course would be challenging. Fortunately I have had the steadfast support of a dedicated team of publishing professionals at McGraw-Hill.

I am especially grateful to Michael Lange, the Editor-In-Chief for the McGraw-Hill Science, Engineering and Math imprint, whose keen interest in my project kept me focused on its completion. Joyce Berendes has skillfully directed me through the complex production process with patience and a good sense of humor. Joan Weber, my developmental editor for the past three years, has served as the "point person" for the hundreds of reviewers, accuracy checkers, class testers, and advisors involved in the project. She helped me manage the multiple tasks required in developing a text like this one. Her professionalism and her attention to detail have been crucial to the success of this project. Gratitude is also due to Thomas Timp, Sponsoring Editor, who has come on to direct the last stages of this project; marketing managers Deb Hash and Tami Hodge whose insights have helped me see the project from an entirely different viewpoint; and Jeff Schmitt, who has led the development of the media supplements that accompany the text. Suzanne Guinn, my first developmental editor, guided me through the first chapters, and provided innumerable helpful suggestions that found their way into the body of the text.

I owe my biggest thanks to Kent Peterson, now Editorial Director, who has been my steady advocate since the beginning. Without his energy and his willingness to "think outside the box" this first edition would never have become a reality. He has become a friend and teacher to someone new to the publishing industry.

Finally, I especially thank my family. I was raised in a household of two girls, and so I learned at a very early age that "Girls can do anything," even chemistry. My parents taught me that hard work and a good education were keys to a bright future, and I am happy that my dad is alive and in good health to see the culmination of much hard work.

My immediate family has experienced the day-to-day demands of living with a busy author. My husband Dan read the entire manuscript at each stage, and his many suggestions and insights have made this a clearer and more accurate text. His willingness to assume added household responsibilities in the last year of this project was crucial in giving me the needed time to complete it in a timely fashion.

Thanks also go to my daughters, Erin and Jenna. Erin has co-authored the solutions manual with me, much of which she completed during her first two years of medical school. (And I thought I was the busy one!) Jenna, a third-year college student, prepared materials from my old exams that form much of the test bank of questions for instructors. In this way, this project has truly become a Smith family affair.

My twin sons were still in preschool when I began writing in the fall of 1999 and now they are independent fourth graders. Their soccer games and piano recitals kept me grounded in reality, and their "little boy humor" made me laugh when I most needed it.

Among the many others that go unnamed but who have profoundly affected this work are the thousands of students I have been lucky to teach over the last 25 years. I have learned so much from my daily interactions with them, and I hope that the wider chemistry community can benefit from this experience by the way I have presented the material in this text.

Although every effort has been made to make this text and its accompanying solutions manual as error-free as possible, some errors undoubtedly remain and for them, I am solely responsible. Please feel free to email me about any inaccuracies, so that subsequent editions may be further improved.

With much aloha,

Janice Gorzynski Smith
jgsmith@gold.chem.hawaii.edu

Illustrations

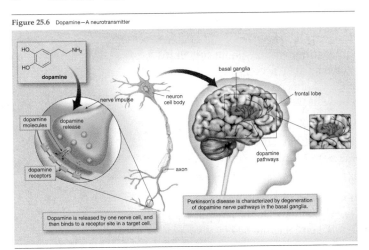

Figure 25.6   Dopamine—A neurotransmitter

Dopamine is released by one nerve cell, and then binds to a receptor site in a target cell.

Parkinson's disease is characterized by degeneration of dopamine nerve pathways in the basal ganglia.

Cocaine, amphetamines, and several other addicting drugs increase the level of dopamine in the brain, which results in a pleasurable "high." With time, the brain adapts to increased dopamine levels

affects brain processes that control movement and emotions, so proper dopamine levels are necessary to maintain an individual's mental and physical health. For example, when dopamine-producing neurons die, the level of dopamine drops, resulting in the loss of motor control symptomatic of Parkinson's disease.

Understanding the neurochemistry of these compounds has led to the synthesis and availability of several useful drugs. **Fentanyl** is a common narcotic pain reliever used in surgical procedures,

Micro-to-macro illustrations

At key points throughout the text, photos and illustrations are combined. These "micro-to-macro" illustrations reveal the underlying molecular structures that give rise to macroscopic properties of common phenomena.

"The artwork depicting intermolecular forces in Chapter 3 is the best that I have ever seen. I think these renderings will be a tremendous aid to the students' understanding of these often difficult to grasp concepts."

S. Todd Deal
Georgia Southern University

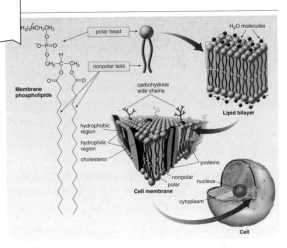

3.8   Application: The Cell Membrane   101

Phospholipids contain an ionic or polar head, and two long nonpolar hydrocarbon tails. In an aqueous environment, phospholipids form a lipid bilayer, with the polar heads oriented toward the aqueous exterior and the nonpolar tails forming a hydrophobic interior. Cell membranes are composed largely of this lipid bilayer.

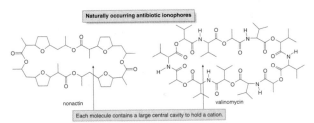

Naturally occurring antibiotic ionophores

Each molecule contains a large central cavity to hold a cation.

Several synthetic ionophores have also been prepared, including one group called **crown ethers.** *Crown ethers* **are cyclic ethers containing several oxygen atoms that bind specific cations depending on the size of their cavity.** Crown ethers are named according to the general format *x*-**crown**-*y*, where *x* is the total number of atoms in the ring and *y* is the number of oxygen atoms. For example, 18-crown-6 contains 18 atoms in the ring, including 6 O atoms. This crown ether binds potassium ions. Sodium ions are too small to form a tight complex with the O atoms, and larger cations do not fit in the cavity.

Spectra

Over 100 spectra created specifically for *Organic Chemistry,* are presented throughout the text. The spectra are color-coded by type and generously labeled.

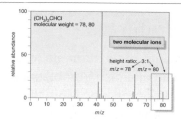

484   CHAPTER 14   Mass Spectrometry and Infrared Spectroscopy

Figure 14.3 Mass spectrum of 2-chloropropane [(CH₃)₂CHCl]

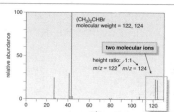

Figure 14.4 Mass spectrum of 2-bromopropane [(CH₃)₂CHBr]

When the molecular ion consists of two peaks (M and M + 2) in a 1:1 ratio, a Br atom is present in the molecule.

...molecular ions would you expect for compounds having each of the following molecular ...las: (a) C₄H₉Cl; (b) C₃H₇F; (c) C₆H₁₁Br; (d) C₄H₁₁N?

Mass Spectrometry

...ugh using the molecular ion to determine the molecular weight of an organic compound is ...d valuable, recent advances have greatly ex...ded the information obtained from mass

---

15.7   More Complex Examples of Splitting   523

Figure 15.6 The ¹H NMR spectrum of 2-bromopropane, [(CH₃)₂CHBr]

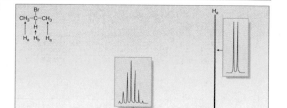

The ¹H NMR spectrum of 1-bromopropane (CH₃CH₂CH₂Br) illustrates a different result.

CH₃CH₂CH₂Br has three different types of protons—H$_a$, H$_b$, and H$_c$—so it exhibits three NMR signals (Figure 15.7). H$_a$ and H$_c$ are each triplets because they are adjacent to two H$_b$ protons. H$_b$ has protons on both adjacent carbons, but because H$_a$ and H$_c$ are *not equivalent to each other,* we cannot merely add them together and use the *n* + 1 rule.

Instead, to determine the splitting of H$_b$, we must consider the effect of the H$_a$ protons and the H$_c$ protons *separately*. The three H$_a$ protons split the H$_b$ signal into four peaks, and the two H$_c$ protons split each of these four peaks into three peaks—that is, the NMR signal due to H$_b$ consists of $4 \times 3 = \textbf{12 peaks}$. Figure 15.8 shows a splitting diagram illustrating how these 12 peaks arise. Often, an NMR signal with so many lines has several overlapping peaks, as is the case with the multiplet for H$_b$ in Figure 15.7.

Figure 15.7 The ¹H NMR   CH₃ CH₂ CH₂

Mechanisms

Curved arrow notation is used extensively to help students follow the movement of electrons in reactions. Where appropriate, mechanisms are presented in parts to promote a better conceptual understanding.

MECHANISM 13.1

Radical Halogenation of Alkanes

Initiation

Step [1] Bond cleavage forms two radicals.

- The reaction begins with homolysis of the weakest bond in the starting materials using energy from light or heat.
- Thus, the Cl−Cl bond (ΔH° = 58 kcal/mol), which is weaker than either the C−C or C−H bond in ethane (ΔH° = 88 and 98 kcal/mol, respectively), is broken to form two chlorine radicals.

Propagation

Steps [2] and [3] One radical reacts and a new radical is formed.

- The Cl· radicals are highly reactive (they lack an octet of electrons), so they abstract a hydrogen atom from ethane (Step [2]). This forms H−Cl and leaves one unpaired electron on carbon, generating the ethyl radical (CH₃CH₂·).
- CH₃CH₂· is highly reactive, so it can abstract a chlorine atom from Cl₂ (Step [3]), forming CH₃CH₂Cl and a new chlorine radical (Cl·).
- The Cl· radical formed in Step [3] is a reactant in Step [2], so Steps [2] and [3] can occur repeatedly without an additional initiation reaction (Step [1]).
- In each propagation step, one radical is consumed and one radical is formed. The two products—CH₃CH₂Cl and HCl—are formed during propagation.

Repeat Steps [2], [3], [2], [3], again and again.

Termination

Step [4] Two radicals react to form a σ bond.

- To terminate the chain, two radicals react with each other in one of three ways (Steps [4a, b, and c]). Because these reactions remove reactive radicals and form stable bonds, they prevent further propagation via Steps [2] and [3].

Problem Solving

Sample Problems

Sample Problems show students how to solve organic chemistry problems in a logical, stepwise manner. More than 800 follow-up problems are located throughout the chapters to test whether students understand concepts covered in the Sample Problems.

"Smith's experience in teaching has allowed her to design problems that seem to address students' most common misconceptions."

Clair J. Cheer
San Jose State University

How To's

How To's provide students with detailed instructions on how to work through key processes.

"Each type of problem is presented with a stepwise approach, which provides students with an expert's approach on how to solve the problem. This is invaluable and something that is missing from most organic texts."

Harriet Lindsay
Eastern Michigan University

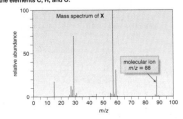

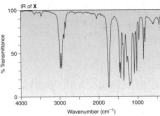

Applications and Summaries

Key Concept Summaries

Succinct summary tables reinforcing important principles and concepts are provided at the end of each chapter.

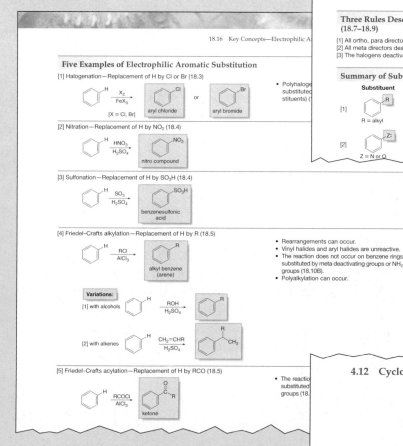

18.16 Key Concepts—Electrophilic Aromatic Substitution

Mechanism of Electrophilic Aromatic Substitution (18.2)

- Electrophilic aromatic substitution follows a two-step mechanism. Reaction of the aromatic ring with an electrophile forms a carbocation, and loss of a proton regenerates the aromatic ring.
- The first step is rate-determining.
- The intermediate carbocation is stabilized by resonance; a minimum of three resonance structures can be drawn. The positive charge is always located ortho or para to the new C–E bond.

    (+) ortho to E   (+) para to E   (+) ortho to E

Three Rules Describing the Reactivity and Directing Effects of Common Substituents (18.7–18.9)

[1] All ortho, para directors except the halogens activate the benzene ring.
[2] All meta directors deactivate the benzene ring.
[3] The halogens deactivate the benzene ring and direct ortho, para.

Summary of Substituent Effects in Electrophilic Aromatic Substitution (18.6–18.9)

	Substituent	Inductive effect	Resonance effect	Reactivity	Directing effect
[1]	R = alkyl	donating	none	activating	ortho, para
[2]	Z = N or O	withdrawing	donating	activating	ortho, para

Five Examples of Electrophilic Aromatic Substitution

[1] Halogenation—Replacement of H by Cl or Br (18.3)

$$\xrightarrow{\frac{X_2}{FeX_3}}$$ aryl chloride   or   aryl bromide

[X = Cl, Br]

- Polyhalogenated... stituents)...

[2] Nitration—Replacement of H by NO_2 (18.4)

$$\xrightarrow{\frac{HNO_3}{H_2SO_4}}$$ nitro compound

[3] Sulfonation—Replacement of H by SO_3H (18.4)

$$\xrightarrow{\frac{SO_3}{H_2SO_4}}$$ benzenesulfonic acid

[4] Friedel–Crafts alkylation—Replacement of H by R (18.5)

$$\xrightarrow{\frac{RCl}{AlCl_3}}$$ alkyl benzene (arene)

- Rearrangements can occur.
- Vinyl halides and aryl halides are unreactive.
- The reaction does not occur on benzene rings substituted by meta deactivating groups or NH_2 groups (18.10B).
- Polyalkylation can occur.

Variations:

[1] with alcohols $$\xrightarrow{\frac{ROH}{H_2SO_4}}$$

[2] with alkenes $$\xrightarrow{\frac{CH_2=CHR}{H_2SO_4}}$$

[5] Friedel–Crafts acylation—Replacement of H by RCO (18.5)

$$\xrightarrow{\frac{RCOCl}{AlCl_3}}$$ ketone

- The reactio... substituted... groups (18....

> *"The summaries are very effective. I think this is the strongest aspect of this text. The summaries will allow students a great way to review for exams (both during the course and when studying for MCATs, GREs, etc.)."*
>
> **Susan M. Schelble**
> **University of Colorado-Denver**

4.12 Cyclohexane

Let's now examine in detail the conformation of **cyclohexane**, the most common ring size in naturally occurring compounds.

4.12A The Chair Conformation

A planar cyclohexane ring would experience angle strain, because the internal bond angle between the carbon atoms would be 120°, and torsional strain, because all of the hydrogens on adjacent carbon atoms would be eclipsed.

If a cyclohexane ring were flat....

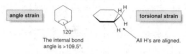

angle strain   120°   torsional strain

The internal bond angle is >109.5°.   All H's are aligned.

Visualizing the chair. If a cyclohexane conformer is tipped downward, we can more easily view it as a chair with a back, seat, and foot support.

In reality, cyclohexane adopts a puckered conformation, called the **chair** form, which is more stable than any other possible conformation.

The carbon skeleton of chair cyclohexane

The chair conformation is so stable because it eliminates angle strain (**all C–C–C bond angles are 109.5°**) and torsional strain (all hydrogens on adjacent carbon atoms are **staggered**, not eclipsed).

109.5°   All H's are **staggered**.

Margin Notes

Margin notes are placed carefully throughout the chapters, providing interesting information relating to topics covered in the text. Some margin notes are illustrated with photos to make the chemistry more relevant.

Supplements

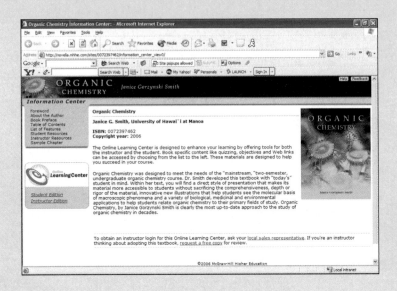

Online Learning Center

The Online Learning Center is a comprehensive, exclusive website that provides numerous electronic resources for both instructors and students. Access to this learning tool is FREE with the purchase of a new textbook. Log on at www.smithorganic.com

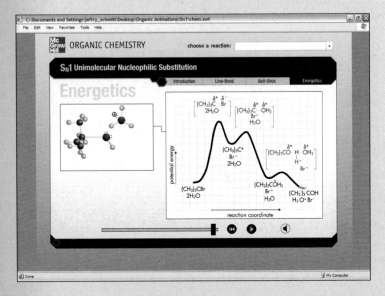

Digital Content Manager

This cross-platform CD-ROM is a collection of visual resources that allows instructors to create customized PowerPoint presentations from illustrations, figures, tables, and worked examples from the text. The CD also contains full-color animations, Active Art, and PowerPoint lecture outlines.

Student Study Guide/Solutions Manual

Written by Janice Gorzynski Smith and Erin R. Smith, the Student Study Guide/Solutions Manual provides step-by-step solutions to all in-chapter and end-of-chapter problems. Each chapter begins with an overview of key concepts and includes key rules and summary tables.

Transparency Acetates

This boxed set of over 200 full-color transparencies features key color images from the text for use in small and large classroom settings.

Prologue

O rganic chemistry. You might wonder how a discipline that conjures up images of eccentric old scientists working in basement laboratories is relevant to you, a student in the twenty-first century.

Consider for a moment the activities that occupied your past 24 hours. You likely showered with soap, drank a caffeinated beverage, ate at least one form of starch, took some medication, read a newspaper, listened to a CD, and traveled in a vehicle that had rubber tires and was powered by fossil fuels. If you did any *one* of these, your life was touched by organic chemistry.

What Is Organic Chemistry?

> ◆ Organic chemistry is the chemistry of compounds that contain the element carbon.

It is one branch in the entire field of chemistry, which encompasses many classical subdisciplines including inorganic, physical, and analytical chemistry, and newer fields such as bioinorganic chemistry, physical biochemistry, polymer chemistry, and materials science.

Organic chemistry was singled out as a separate discipline for historical reasons. Originally, it was thought that compounds in living things, termed *organic compounds,* were fundamentally different from those in nonliving things, called *inorganic compounds.* Although we have known for more than 150 years that this distinction is artificial, the name *organic* persists. Today the term refers to the study of the compounds that contain carbon, many of which, incidentally, are found in living organisms.

It may seem odd that a whole discipline is devoted to the study of a single element in the periodic table, when more than 100 elements exist. It turns out, though, that there are far more organic compounds than any other type. **Organic chemicals affect every aspect of our world, and for this reason, it is important and useful to know something about them.**

Clothes, foods, medicines, gasoline, refrigerants, and soaps are composed almost solely of organic molecules. Some, like cotton, wool, or silk are naturally occurring; that is, they can be isolated directly from natural sources. Others, such as nylon and polyester are synthetic, meaning they are produced by chemists in the laboratory. By studying the principles and concepts of organic chemistry, you can learn more about compounds such as these and how they affect the world around you.

Realize, too, what organic chemistry has done for us. Organic chemistry has made available both comforts and necessities that were previously nonexistent, or reserved for only the wealthy. We have seen an enormous increase in life span, from 47 years in 1900 to over 70 years currently. To a large extent this is due to the isolation and synthesis of new drugs, such as antibiotics and vaccines for childhood diseases. Chemistry has also given us the tools to control insect populations that spread disease, and there is more food for all because of fertilizers, pesticides, and herbicides. Organic chemistry has given us contraceptives, plastics, anesthetics, and synthetic heart valves. Our lives would be vastly different today without these products of organic chemistry.

Some Representative Organic Molecules

Perhaps the best way to appreciate the variety of organic molecules is to look at a few. Three simple organic compounds are **methane, ethanol,** and **trichlorofluoromethane.**

methane

◆ **Methane,** the simplest of all organic compounds, contains one carbon atom. Methane—the main component of natural gas—occurs widely in nature. Like other hydrocarbons—organic compounds that contain only carbon and hydrogen—**methane is combustible;** that is, it burns in the presence of oxygen. Methane is the product of the anaerobic (without air) decomposition of organic matter by bacteria. The natural gas we use today was formed by the decomposition of organic material millions of years ago. Hydrocarbons such as methane are discussed in Chapter 4.

ethanol

◆ **Ethanol,** the alcohol present in beer, wine, and other alcoholic beverages, is formed by the fermentation of sugar, quite possibly the oldest example of organic synthesis. Ethanol can also be made in the lab by a totally different process, but **the ethanol produced in the lab is identical to the ethanol produced by fermentation.** Alcohols including ethanol are discussed in Chapter 9.

trichlorofluoromethane

◆ **Trichlorofluoromethane** is a member of a class of molecules called **chlorofluorocarbons** or **CFCs,** which contain one or two carbon atoms and several halogens. Trichlorofluoromethane is an unusual organic molecule in that **it contains no hydrogen atoms.** Because it has a low molecular weight and is easily vaporized, trichlorofluoromethane has been used as an aerosol propellant and refrigerant. It and other CFCs have been implicated in the destruction of the stratospheric ozone layer, as is discussed in Chapter 13.

Because more complicated organic compounds contain many carbon atoms, organic chemists have devised a shorthand to draw them. Keep in mind the following when examining these structures:

◆ **Each solid line represents a two-electron covalent bond.**

◆ **When no atom is drawn at the corner of a ring, an organic chemist assumes it to be carbon.**

For example, in the six-membered ring drawn, there is one carbon atom at each corner of the hexagon.

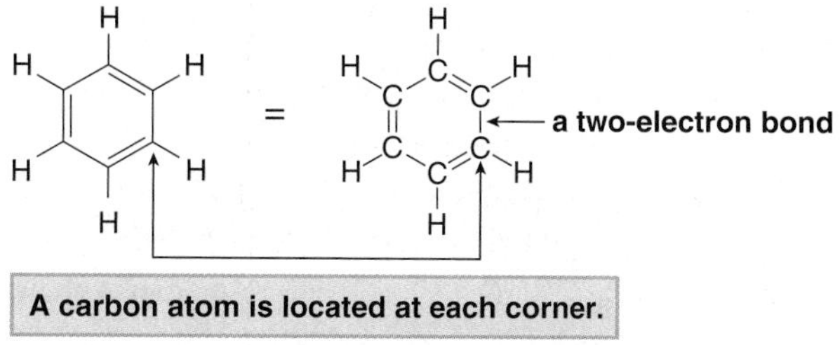

A carbon atom is located at each corner.

Three complex organic molecules that are important medications are **amoxicillin, fluoxetine,** and **AZT.**

◆ **Amoxicillin** is one of the most widely used antibiotics in the penicillin family. The discovery and synthesis of such antibiotics in the twentieth century have made routine the treatment of infections that were formerly fatal. You were likely given some amoxicillin to treat an ear infection when you were a child. The penicillin antibiotics are discussed in Chapter 22.

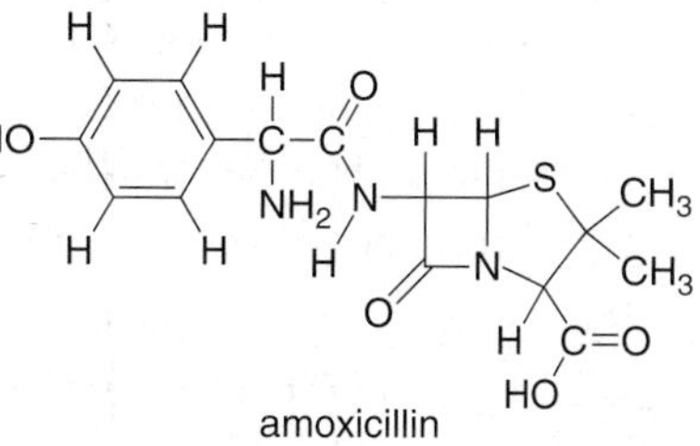

amoxicillin

◆ **Fluoxetine** is the generic name for the antidepressant **Prozac.** Prozac was designed and synthesized by chemists in the laboratory, and is now produced on a large scale in chemical factories. Because it is safe and highly effective in treating depression, Prozac is widely prescribed. An estimated 40 million individuals worldwide have used Prozac since 1986.

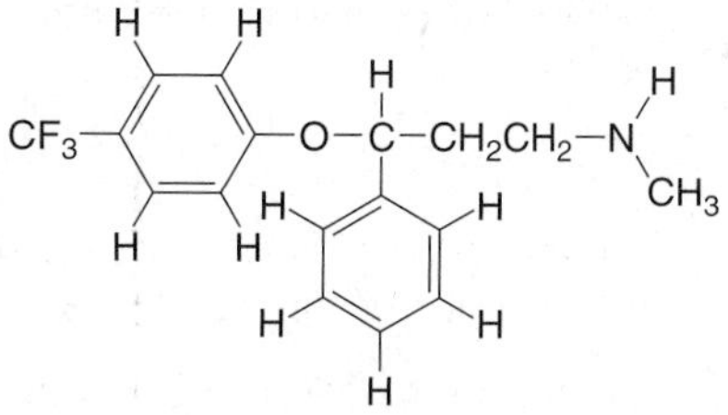

fluoxetine

◆ **AZT,** the abbreviation for **az**idodeoxy**t**hymidine, is a drug available to treat human immunodeficiency virus (HIV), the virus that causes acquired immune deficiency syndrome (AIDS). Also known by its generic name **zidovudine,** AZT represents a chemical success to a different challenge: synthesizing agents that combat viral infections.

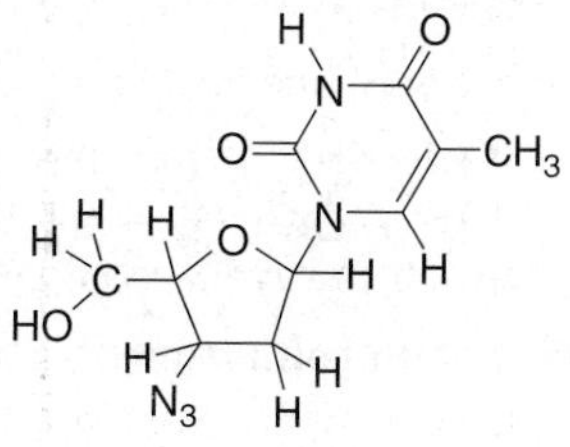

AZT

Other complex organic compounds having interesting properties are **capsaicin** and **DDT.**

◆ **Capsaicin,** one member of a group of compounds called *vanilloids,* is responsible for the characteristic spiciness of hot peppers. It is the active ingredient in pepper sprays used for personal defense. The properties of capsaicin are discussed in Chapter 1.

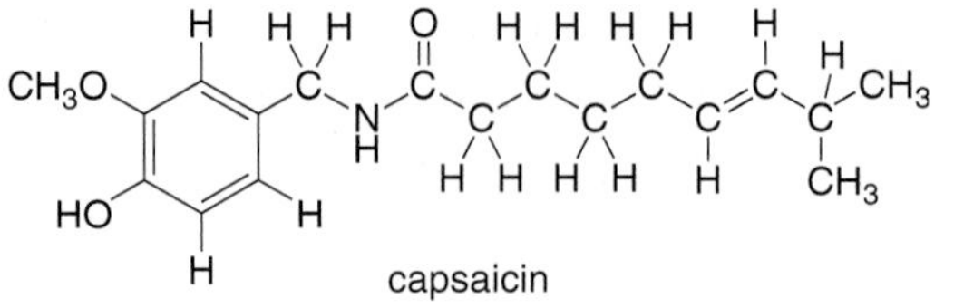

capsaicin

◆ **DDT,** the abbreviation for **d**ichloro**d**iphenyl**t**richloroethane, is a pesticide once called "miraculous" by Winston Churchill because of the many lives it saved by killing disease-carrying mosquitoes. DDT use is now banned in the United States and many developed countries because it is a nonspecific insecticide that persists in the environment.

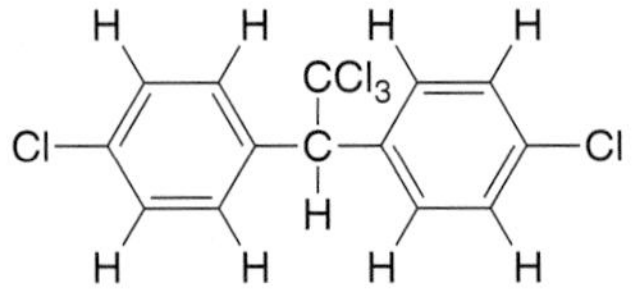

DDT

What are the common features of these organic compounds?

◆ **All organic compounds contain carbon atoms and most contain hydrogen atoms.**

◆ **All the carbon atoms have four bonds.** A stable carbon atom is said to be *tetravalent.*

◆ Other elements may also be present. **Any atom that is not carbon or hydrogen is called a *heteroatom.*** Common heteroatoms include N, O, S, P, and the halogens.

◆ Some compounds have **chains** of atoms and some compounds have **rings.**

These features explain why there are so many organic compounds: **Carbon forms four strong bonds with itself and other elements. Carbon atoms combine together to form rings and chains.**

Palytoxin—An Example of a Very Complex Organic Compound

Let's complete this discussion with **palytoxin** ($C_{129}H_{223}N_3O_{54}$), a complex organic molecule first isolated from marine soft corals of the genus *Palythoa* (Figure 1), and one of the most potent poisons known.

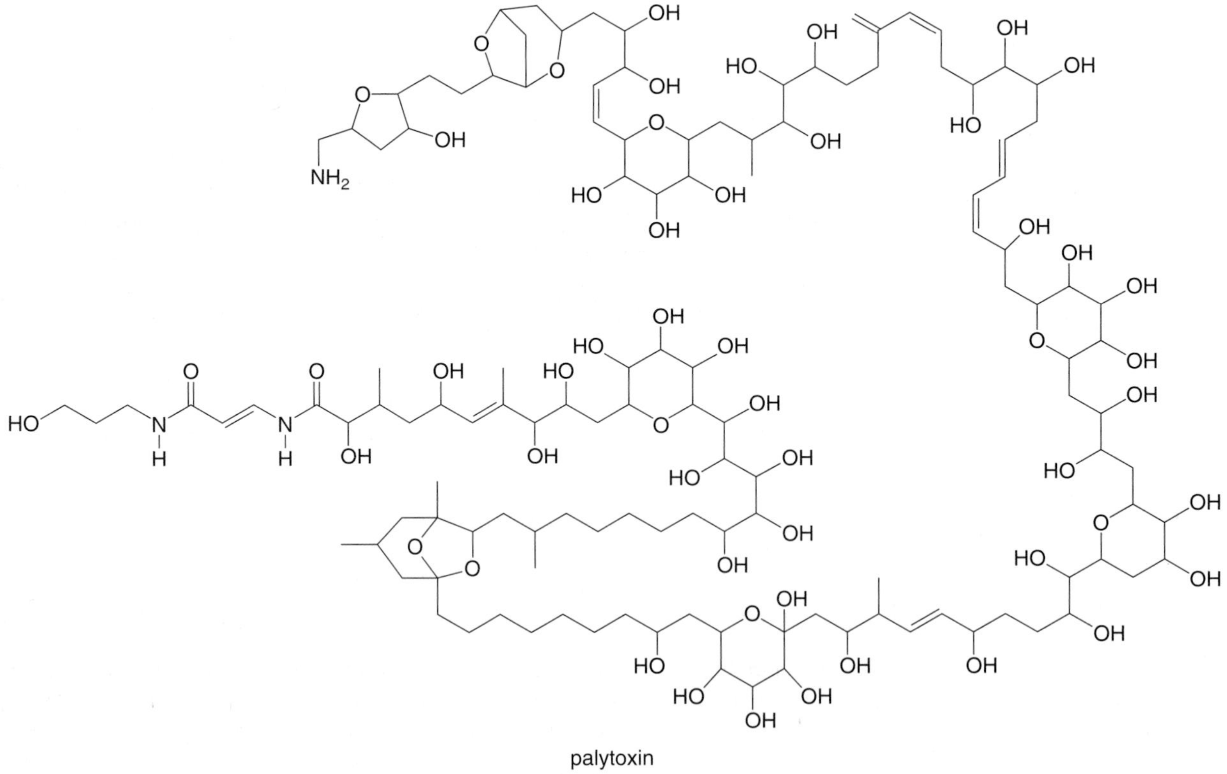

palytoxin

(a)

(b)

The modern story of palytoxin began in 1961 when a biologist from the Hawaiian Institute of Marine Biology noticed a reference in a Hawaiian-English dictionary for *limu-make-o-Hana,* which means *deadly seaweed of Hana.* Ancient Hawaiians once used this seaweed, in actuality a soft coral, to poison their spears.

Shortly thereafter, scientists began the tedious process of isolating the poisonous compound from this natural source. Palytoxin was finally isolated in pure form in 1971 at the University of Hawai'i at Mānoa, and its structure determined simultaneously by two different research groups in 1981. In 1994, Professor Yoshito Kishi and co-workers at Harvard University synthesized palytoxin in the laboratory. Some of the techniques used to determine the structure of palytoxin and to show that the compound synthesized in the lab was identical to the natural product, such as infrared and nuclear magnetic resonance spectroscopy, are discussed in Chapters 14 and 15.

To understand the structure of palytoxin we need to know more shorthand for drawing chains of organic molecules. The following conventions are often used.

◆ Assume there is a carbon atom at the junction of any two lines or at the end of any line.
◆ Assume there are enough hydrogens around each carbon to make it tetravalent.
◆ Draw in all heteroatoms and the hydrogens directly bonded to them.

A portion of the palytoxin chain is drawn here, both in shorthand and with all carbon and hydrogen atoms drawn in. Note how cluttered the second structure looks. Because you need to be able to interconvert shorthand structures with complete structures, this convention is discussed in greater detail in Chapter 1.

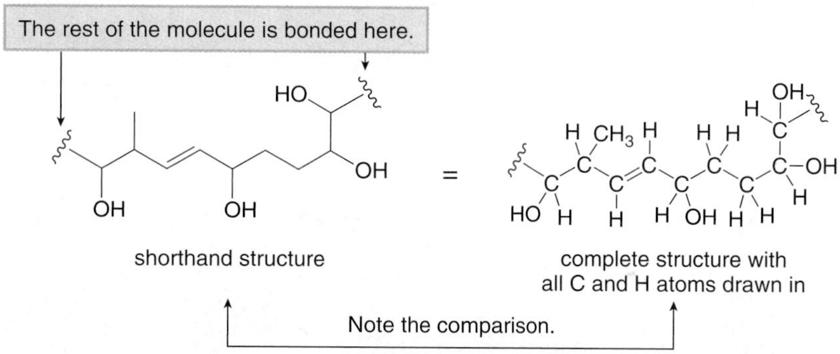

The rest of the molecule is bonded here.

shorthand structure = complete structure with all C and H atoms drawn in

Note the comparison.

In this introduction, we have seen a variety of molecules that have diverse structures. They represent a miniscule fraction of the organic compounds currently known and the many thousands that are newly discovered or synthesized each year. The principles you learn in organic chemistry will apply to all of these molecules, from simple ones like methane and ethanol, to complex ones like capsaicin and palytoxin. It is these beautiful molecules, their properties, and their reactions that we will study in organic chemistry.

WELCOME TO THE WORLD OF ORGANIC CHEMISTRY!

Structure and Bonding

Capsaicin is the compound responsible for the characteristic spicy flavor of jalapeño and habañero peppers. Although it first produces a burning sensation on contact with the mouth or skin, repeated application desensitizes the area to pain. This property has made it the active ingredient in several topical creams for treatment of chronic pain. Capsaicin has also been used as an animal deterrent in pepper sprays, and as an additive to make birdseed squirrel-proof. In Chapter 1, we discuss the structure, bonding, and properties of organic molecules like capsaicin.

Before examining organic molecules in detail, we must review some important features about structure and bonding learned in previous chemistry courses. We will discuss these concepts primarily from an organic chemist's perspective, and spend time on only the particulars needed to understand organic compounds.

Important topics in Chapter 1 include drawing Lewis structures, predicting the shape of molecules, determining what orbitals are used to form bonds, and how electronegativity affects bond polarity. Equally important is Section 1.7, on drawing organic molecules, both shorthand methods routinely used for simple and complex compounds, as well as three-dimensional representations that allow us to more clearly visualize them.

1.1   The Periodic Table

All matter is composed of the same building blocks called **atoms.** There are two main components of an atom.

- The **nucleus** contains positively charged **protons** and uncharged **neutrons.** Most of the mass of the atom is contained in the nucleus.
- The **electron cloud** is composed of negatively charged **electrons.** The electron cloud comprises most of the volume of the atom.

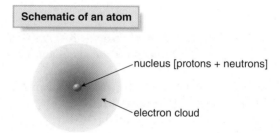

Schematic of an atom

nucleus [protons + neutrons]

electron cloud

The charge on a proton is equal in magnitude but opposite in sign to the charge on an electron. In a neutral atom, the **number of protons in the nucleus equals the number of electrons.** This quantity, called the **atomic number,** is unique to a particular element. For example, every neutral carbon atom has an atomic number of six, meaning it has six protons in its nucleus and six electrons surrounding the nucleus.

In addition to neutral atoms, we will also encounter **charged ions.**

- A *cation* is positively charged and has fewer electrons than its neutral form.
- An *anion* is negatively charged and has more electrons than its neutral form.

The number of neutrons in the nucleus of a particular element can vary. **Isotopes** are two atoms of the same element having a different number of neutrons. The **mass number** of an atom is the total number of protons and neutrons in the nucleus. Isotopes have different mass numbers.

Isotopes of carbon and hydrogen are sometimes used in organic chemistry, as we will see in Chapter 15.

- The most common isotope of hydrogen has one proton and no neutrons in the nucleus, but 0.02% of hydrogen atoms have one proton and one neutron. This isotope of hydrogen is called **deuterium,** and is symbolized by the letter **D.**
- Most carbon atoms have six protons and six neutrons in the nucleus, but 1.1% have six protons and seven neutrons.

The **atomic weight** is the weighted average of the mass of all isotopes of a particular element, reported in atomic mass units (amu).

Each atom is identified by a one- or two-letter abbreviation that is the characteristic symbol for that element. Carbon is identified by the single letter **C.** Sometimes the atomic number is indicated as a subscript to the left of the element symbol, and the mass number is indicated as a superscript, as shown in Figure 1.1.

Figure 1.1   A comparison of two isotopes of the element carbon

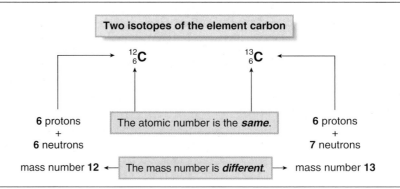

A **row** in the periodic table is also called a *period,* and a **column** is also called a *group.*

Long ago it was realized that groups of elements have similar properties, and that these atoms could be arranged in a schematic way called the **periodic table.** There are more than 100 known elements, arranged in the periodic table in order of increasing atomic number. A periodic table is located on the inside front cover for your reference.

The periodic table is arranged in rows and columns. The particular row and column tell us much about the properties of an element.

◆ Elements in the same row are similar in *size.*
◆ Elements in the same column have similar *electronic and chemical properties.*

Carbon's entry in the periodic table is as follows:

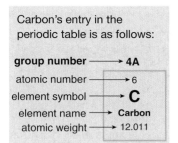

Each column in the periodic table is identified by a **group number,** an Arabic (1 to 8) or Roman (I to VIII) numeral followed by the letter A or B. For example, carbon is located in group **4A** in the periodic table in this text.

Although more than 100 elements exist, most are not common in organic compounds. Figure 1.2 contains a truncated periodic table, indicating the handful of elements that we will routinely deal with in this text. Notice that **most of these elements are located in the first and second rows of the periodic table.**

Although we don't need to go into a complete discussion of how the periodic table is arranged, remember that across each row, electrons are added to a particular shell of orbitals around the nucleus. The shells are numbered 1, 2, 3, and so on. Adding electrons to the first shell forms the first row. Adding electrons to the second shell forms the second row. **Electrons are first added to the shells closest to the nucleus.** These electrons are held most tightly.

Each shell contains a certain number of subshells called **orbitals.** An orbital is a region of space that is high in electron density. There are four different kinds of orbitals, called *s, p, d,* and *f.* The first shell has only one orbital, called an *s* orbital. The second shell has two kinds of orbitals, *s* and *p,* and so on. Each type of orbital occupies a certain space and has a particular shape.

Figure 1.2   A periodic table of the common elements seen in organic chemistry

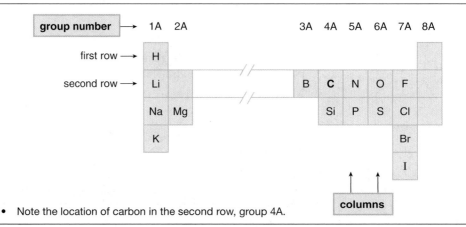

• Note the location of carbon in the second row, group 4A.

Because we will be dealing mainly with the first- and second-row elements, we will concentrate on only *s* **orbitals** and *p* **orbitals.**

◆ An *s* **orbital** has a **sphere of electron density.** It is *lower in energy* than other orbitals of the same shell, because electrons are kept close to the positively charged nucleus. An *s* orbital is filled with electrons before a *p* orbital in the same shell.

◆ A *p* **orbital** has a **dumbbell shape.** It contains a **node of electron density** at the nucleus. A node means there is no electron density in this region. A *p* orbital is *higher in energy* than an *s* orbital (in the same shell) because its electron density is farther away from the nucleus. A *p* orbital is filled with electrons only after an *s* orbital of the same shell is full.

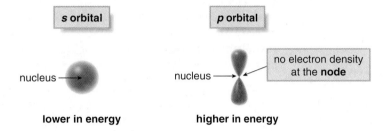

Let's now look at the elements in the first and second rows of the periodic table.

The First Row

The first row of the periodic table is formed by adding electrons to the first shell of orbitals around the nucleus. There is only one orbital in the first shell, called the **1*s* orbital.**

◆ Remember: Each orbital can have a maximum of two electrons.

This means that there are **two possible elements in the first row,** one having one electron added to the 1*s* orbital, and one having two. The element hydrogen (H) has what is called a $1s^1$ configuration with one electron in the 1*s* orbital, and helium (He) has a $1s^2$ configuration with two electrons in the 1*s* orbital.

The Second Row

Every element in the second row has a filled first shell of electrons. Thus, all second-row elements have a $1s^2$ configuration. These electrons in the inner shell of orbitals are called **core electrons,** and are not usually involved in the chemistry of a particular element.

Each of the elements in the second row of the periodic table has four orbitals available to accept additional electrons:

◆ **one 2*s* orbital,** the *s* orbital in the second shell

◆ **three 2*p* orbitals,** all dumbbell-shaped and perpendicular to each other along the *x, y,* and *z* axes

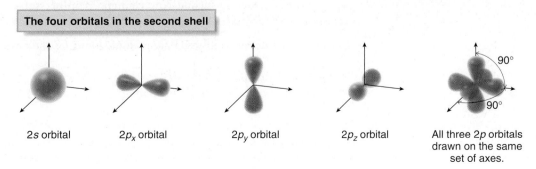

Because each of the four orbitals available in the second shell can hold two electrons, there is a *maximum capacity of eight electrons* for elements in the second row. The second row of the periodic table consists of eight elements, obtained by adding electrons to the 2*s* and three 2*p* orbitals.

The outermost electrons are called **valence electrons.** Valence electrons determine the properties of a given element. The valence electrons are more loosely held than the electrons closer to the nucleus, and as such, they participate in chemical reactions. **The group number of a second row element reveals its number of valence electrons.** For example, carbon in group 4A has four valence electrons, and oxygen in group 6A has six.

PROBLEM 1.1   The most common isotope of oxygen has a mass number of 16, but two other isotopes having mass numbers of 17 and 18 are also known. For each isotope, give the following information: (a) the number of protons; (b) the number of neutrons; (c) the number of electrons in the neutral atom; and (d) the group number.

1.2 Bonding

Until now our discussion has centered on individual atoms, but it is more common in nature to find two or more atoms joined together.

◆ *Bonding* is the joining of two atoms in a stable arrangement.

Bonding may occur between atoms of the same or different elements. It is a favorable process because it always leads to *lowered energy and increased stability.* Joining two or more elements forms **compounds.** Although only about 100 elements exist, more than 10 million compounds are currently known. Examples of compounds include hydrogen gas (H_2), formed by joining two hydrogen atoms; and methane (CH_4), the simplest organic compound, formed by joining a carbon atom with four hydrogen atoms.

One general rule governs the bonding process.

◆ Through bonding, atoms attain a complete outer shell of valence electrons.

Alternatively, because the noble gases in column 8A of the periodic table are especially stable as atoms having a filled shell of valence electrons, the general rule can be restated.

◆ Through bonding, atoms attain a stable noble gas configuration of electrons.

What does this mean for first- and second-row elements? **A first-row element like hydrogen can accommodate** *two electrons* **around it.** This would make it like the noble gas helium at the end of the same row. **A second-row element is most stable with** *eight valence electrons* **around it** like neon. Elements that behave in this manner are said to follow the **octet rule.**

There are two different kinds of bonding: **ionic bonding** and **covalent bonding.**

◆ *Ionic bonds* result from the *transfer* of electrons from one element to another.
◆ *Covalent bonds* result from the *sharing* of electrons between two nuclei.

The type of bonding is determined by the location of an element in the periodic table. An ionic bond generally occurs when elements on the **far left** side of the periodic table combine with elements on the **far right** side, ignoring the noble gases, which form bonds only rarely. **The resulting**

Atoms readily form ionic bonds when they can attain a noble gas configuration by gaining or losing just one or two electrons.

ions are held together by extremely strong electrostatic interactions. A positively charged cation formed from the element on the left side attracts a negatively charged anion formed from the element on the right side. The resulting **salts** are seen in many of the inorganic compounds with which you are familiar. Sodium chloride (**NaCl**) is common table salt, and potassium iodide (**KI**) is an essential nutrient added to make iodized salt.

Ionic compounds form extended crystal lattices that maximize the positive and negative electrostatic interactions. In NaCl, each positively charged Na^+ is surrounded by six negatively charged Cl^- ions, and each Cl^- ion is surrounded by six Na^+ ions.

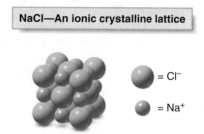

NaCl—An ionic crystalline lattice

$= Cl^-$

$= Na^+$

Lithium fluoride, LiF, is an example of an ionic compound.

- The element **lithium,** located in group 1A of the periodic table, has just one valence electron in its second shell. If this electron is lost, lithium forms the cation Li^+ having no electrons in the second shell. However, it will have a stable electronic arrangement with two electrons in the first shell like helium.

- The element **fluorine,** located in group 7A of the periodic table, has seven valence electrons. By gaining one it forms the anion F^-, which has a filled valence shell (an octet of electrons), like neon.

- Thus, lithium fluoride is a stable ionic compound.

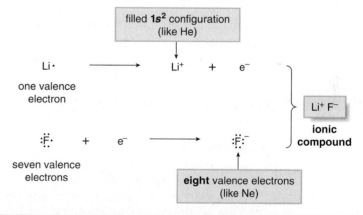

filled $1s^2$ configuration (like He)

Li·

one valence electron

Li^+ + e^-

:F̈· + e^-

:F̈:⁻

seven valence electrons

eight valence electrons (like Ne)

$Li^+ F^-$

ionic compound

- **The transfer of electrons forms stable salts composed of cations and anions.**

A **compound** may have either ionic or covalent bonds. A **molecule** has only covalent bonds.

The second type of bonding, **covalent bonding,** occurs with elements like carbon in the middle of the periodic table, which would otherwise have to gain or lose several electrons to form an ion with a complete valence shell. **A covalent bond is a two-electron bond,** and a compound with covalent bonds is called a **molecule.** Covalent bonds also form between two elements from the same side of the table, such as two hydrogen atoms or two chlorine atoms. H_2, Cl_2, and CH_4 are all examples of covalent molecules.

PROBLEM 1.2   Label each bond in the following compounds as ionic or covalent.
a. F_2 b. LiBr   c. CH_3CH_3 d. $NaNH_2$

How many covalent bonds will a particular atom typically form? As you might expect, it depends on the location of the atom in the periodic table. In the first row, **hydrogen forms one covalent bond** using its one valence electron. When two hydrogen atoms are joined in a bond, each has a filled valence shell of two electrons.

Second-row elements can have no more than eight valence electrons around them. For neutral molecules, two consequences result.

◆ **Atoms with one, two, or three valence electrons form one, two, or three bonds,** respectively, in neutral molecules.

Nonbonded pair of electrons = unshared pair of electrons = lone pair

◆ **Atoms with four or more valence electrons** form enough bonds to give an octet. This results in the following simple equation:

$$\text{predicted number of bonds} = 8 - \text{number of valence electrons}$$

The letter **X** is often used to represent one of the halogens in group 7A: F, Cl, Br, or I.

These guidelines are used in Figure 1.3 to summarize the usual number of bonds formed by the common atoms in organic compounds. Notice that when second-row elements form fewer than four bonds their octets consist of both **bonding (shared) electrons** and **nonbonding (unshared) electrons.** Unshared electrons are also called **lone pairs.**

SAMPLE PROBLEM 1.1   Without referring to Figure 1.3, how many covalent bonds are predicted for each atom?
a. B   b. N

SOLUTION

a. B has three valence electrons. Thus, it is expected to form three bonds.
b. N has five valence electrons. Because it contains more than four valence electrons, it is expected to form $8 - 5 = 3$ bonds.

PROBLEM 1.3   How many covalent bonds are predicted for each atom?
a. O   b. Al   c. Br

1.3 Lewis Structures

Lewis structures are electron dot representations for molecules. There are three general rules for drawing Lewis structures.

1. Draw only the valence electrons.
2. Give every second-row element an octet of electrons, if possible.
3. Give each hydrogen two electrons.

Drawing a Lewis structure for a diatomic molecule like **HF** is quite straightforward. Hydrogen has one valence electron and fluorine has seven. H and F each donate one electron to form a two-electron bond. The resulting molecule gives both H and F a filled valence shell.

Figure 1.3 Summary: The usual number of bonds of common *neutral* atoms

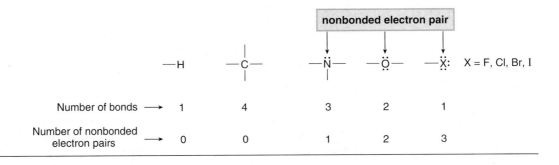

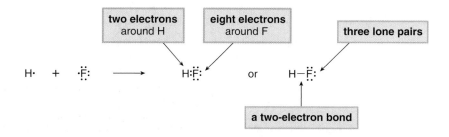

In a Lewis structure, **a *solid line* indicates a two-electron covalent bond.**

1.3A   A Procedure for Drawing Lewis Structures

Drawing a Lewis structure for larger molecules is easier if you follow a stepwise procedure.

How To   Draw a Lewis Structure

Step [1]   **Arrange atoms next to each other that you think are bonded together.**

This process may involve trial and error, but two guidelines help:

- Always place hydrogen atoms and halogen atoms on the periphery because H and X (X = F, Cl, Br, and I) form only one bond each.

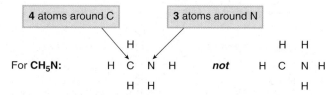

- As a first approximation, place no more atoms around an atom than the number of bonds it usually forms.

In truth, the proper arrangement of atoms may not be obvious, or more than one arrangement may be possible (see Sections 1.4A and 1.5). Even in many simple molecules, the connectivity between atoms must be determined experimentally. For this reason, atom arrangement will be specified for you in some problems.

Step [2]   **Count the electrons.**

- Count the number of valence electrons from all atoms.
- Add one electron for each negative charge.
- Subtract one electron for each positive charge.

This sum gives the total number of electrons that must be used in drawing the Lewis structure.

Step [3]   **Arrange the electrons around the atoms.**

- Place a bond between every two atoms, giving two electrons to each H and no more than eight to any second-row atom.
- Use all remaining electrons to fill octets with lone pairs.
- If all valence electrons are used and an atom does not have an octet, you need to form multiple bonds. Sample Problem 1.4 illustrates how to do this.

Step [4]   **Assign formal charges to all atoms.**

- Formal charges are discussed in Section 1.3C.

Sample Problems 1.2 and 1.3 illustrate how to draw Lewis structures in some simple organic molecules.

SAMPLE PROBLEM 1.2   Draw a Lewis structure for methane, CH_4.

SOLUTION

Step [1]   Arrange the atoms.

H
H C H
H

- Place C in the center and 4 H's on the periphery. In this case, only one arrangement is possible.
- Note that C is surrounded by four atoms, its usual number.

Step [2]   Count the electrons.

$1\ C \times 4\ e^- = 4\ e^-$
$4\ H \times 1\ e^- = 4\ e^-$
8 e^- total

Step [3]   Add the bonds and lone pairs.

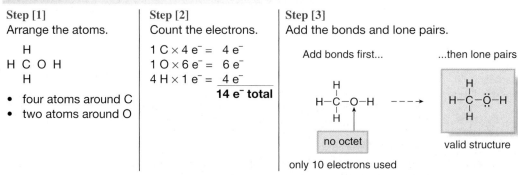

Adding four two-electron bonds around carbon uses all eight valence electrons in this example, and so there are no lone pairs. To check whether a Lewis structure is valid, we must be able to answer YES to the following three questions:

- Have all the electrons been used?
- Is each H surrounded by two electrons?
- Is each second-row element surrounded by no more than eight electrons?

The answer to all three questions is YES, so the Lewis structure drawn for CH_4 is valid.

SAMPLE PROBLEM 1.3   Draw a Lewis structure for methanol, a compound with molecular formula CH_4O.

SOLUTION

Step [1]	Step [2]	Step [3]
Arrange the atoms.	Count the electrons.	Add the bonds and lone pairs.

Step [1]
Arrange the atoms.

H
H C O H
H

- four atoms around C
- two atoms around O

Step [2]
Count the electrons.

$1\ C \times 4\ e^- =\ 4\ e^-$
$1\ O \times 6\ e^- =\ 6\ e^-$
$4\ H \times 1\ e^- =\ 4\ e^-$
14 e^- total

Step [3]
Add the bonds and lone pairs.

Add bonds first...   ...then lone pairs

H
H–C–O–H
H

no octet
only 10 electrons used

valid structure

H
H–C–Ö–H
H

In Step [3], placing bonds between all atoms uses only 10 electrons, and the O atom, with only four electrons, does not yet have a complete octet. To complete the structure we must give the O atom two nonbonded electron pairs. This uses all 14 electrons, giving every H two electrons and every second-row element eight. We have now drawn a valid Lewis structure.

PROBLEM 1.4   Draw a valid Lewis structure for each species:
a. CH_3CH_3 b. CH_5N c. CH_3^-

1.3B   Multiple Bonds

Sample Problem 1.4 illustrates two examples when multiple bonds are needed in Lewis structures.

SAMPLE PROBLEM 1.4   Draw a Lewis structure for each compound. Assume the atoms are arranged as follows:

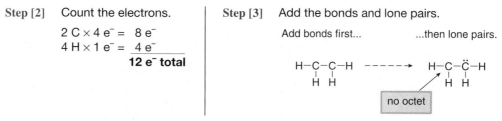

[a] ethylene, C_2H_4 [b] acetylene, C_2H_2

SOLUTION

[a] **Ethylene, C_2H_4:** Follow Steps 1 to 3 to draw a Lewis structure. After placing five bonds between the atoms and adding the two remaining electrons as a lone pair, one C still has no octet.

Step [2]   Count the electrons.

$2\,C \times 4\,e^- =\ 8\,e^-$
$4\,H \times 1\,e^- =\ \underline{4\,e^-}$
                    12 e^- total

Step [3]   Add the bonds and lone pairs.

Add bonds first...   ...then lone pairs.

H—C—C—H   ----→   H—C—C̈—H
  | | | |
  H H   H H
                         ↗
                    no octet

To give both C atoms an octet, **change *one* lone pair into *one* bonding pair of electrons between the two C atoms, forming a double bond.**

H—C‿C̈—H   -- Move a lone pair. -->   H—C=C—H   ← **Each C now has four bonds.**
  | | | |
  H H   H H

ethylene
a valid Lewis structure

This uses all 12 electrons, each C has an octet, and each H has two electrons. The Lewis structure is valid. **Ethylene contains a carbon–carbon double bond.**

[b] **Acetylene, C_2H_2:** A similar phenomenon occurs with acetylene. Placing the 10 valence electrons gives a Lewis structure in which one or both of the C atoms lacks an octet.

Step [2]   Count the electrons.

$2\,C \times 4\,e^- =\ 8\,e^-$
$2\,H \times 1\,e^- =\ \underline{2\,e^-}$
                    10 e^- total

Step [3]   Add the bonds and lone pairs.

Add bonds first...   ...then lone pairs.

H—C—C—H   ----→   H—C—C̈—H   or   H—C̈—C̈—H
                        ↑   ↑ ↑
                    no octet   no octets

Carbon always forms four bonds in stable organic molecules. Carbon forms single, double, and triple bonds to itself and other elements.

In this case, we must **change *two* lone pairs into *two* bonding pairs of electrons, forming a triple bond.**

For example:

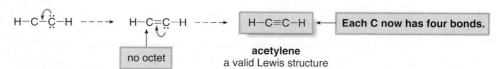

H—C‿C̈—H   --->   H—C=C—H   --->   H—C≡C—H   ← **Each C now has four bonds.**
                        ↑
                    no octet

acetylene
a valid Lewis structure

This uses all 10 electrons, each C has an octet, and each H has two electrons. The Lewis structure is valid. **Acetylene contains a carbon–carbon triple bond.**

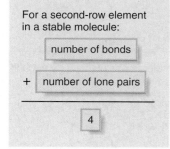

For a second-row element in a stable molecule:

number of bonds

+ number of lone pairs

―――――――――

4

◆ After placing all electrons in bonds and lone pairs, use a lone pair to form a multiple bond if an atom does not have an octet.

You must change *one* lone pair into *one* new bond for each *two* electrons needed to complete an octet. In acetylene, for example, four electrons were needed to complete an octet, so two lone pairs were used to form two new bonds, forming a triple bond. Double and triple bonds are very common in organic compounds.

PROBLEM 1.5   Draw an acceptable Lewis structure for each compound, assuming the atoms are connected as arranged.

a. HCN   H C N   b. H_2CO H C O   c. CH_3CO^+ H C C O
                                        H   H

1.3C   Formal Charge

To manage electron bookkeeping in the Lewis structure of a molecule, organic chemists use **formal charge.**

♦ *Formal charge* is the charge assigned to individual atoms in a Lewis structure.

By calculating formal charge, we determine how the number of electrons around a particular atom compares to its number of valence electrons. Formal charge is calculated as follows:

$$\boxed{\text{formal charge}} = \boxed{\begin{array}{c}\textbf{number of} \\ \textbf{valence electrons}\end{array}} - \boxed{\begin{array}{c}\textbf{number of electrons} \\ \textbf{an atom "owns"}\end{array}}$$

The number of electrons "owned" by an atom is determined by its number of bonds and lone pairs.

♦ An atom "owns" *all* of its unshared electrons and *half* its shared electrons.

The number of electrons "owned" by different carbon atoms is indicated in the following examples:

$$-\overset{|}{\underset{|}{C}}-\qquad\qquad -\overset{|}{C}=\overset{|}{C}-\qquad\qquad -\overset{|}{\underset{|}{C}}:$$

- • C shares eight electrons.
- • C "owns" **four** electrons.

- • Each C shares eight electrons.
- • Each C "owns" **four** electrons.

- • C shares six electrons.
- • C has two unshared electrons.
- • C "owns" **five** electrons.

Sample Problem 1.5 illustrates how formal charge is calculated on the atoms of a polyatomic ion. Note that the **sum of the formal charges on the individual atoms equals the net charge on the molecule or ion.**

SAMPLE PROBLEM 1.5   Determine the formal charge on each atom in the ion H_3O^+.

$$\left[\begin{array}{c} H-\overset{\cdot\cdot}{O}-H \\ | \\ H \end{array} \right]^+$$

SOLUTION
For each atom, two steps are needed:

Step [1]   **Determine the number of electrons an atom "owns."**

$$\boxed{\begin{array}{c}\text{number of} \\ \text{electrons owned}\end{array}} = \boxed{\begin{array}{c}\text{number of} \\ \text{unshared electrons}\end{array}} + \frac{1}{2}\left[\begin{array}{c}\text{number of} \\ \text{shared electrons}\end{array}\right]$$

Step [2]   **Subtract this sum from its number of valence electrons.**

O atom	H atoms
[1] number of electrons "owned" by O	[1] number of electrons "owned" by each H
$2 + \frac{1}{2}(6) = 5$	$0 + \frac{1}{2}(2) = 1$
[2] formal charge on O	[2] formal charge on each H
$6 - 5 = \boxed{+1}$	$1 - 1 = \boxed{0}$

ANSWER

The formal charge on each H is 0. The formal charge on oxygen is +1. The overall charge on the ion is the sum of all of the formal charges; 0 + 0 + 0 + 1 = **+1.**

PROBLEM 1.6   Calculate the formal charge on each second-row atom in the following species:

a. 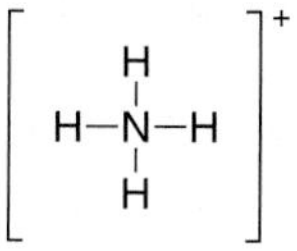 $\left[\begin{array}{c} H \\ | \\ H-N-H \\ | \\ H \end{array}\right]^{+}$ b. $CH_3-N\equiv C\!:$ c. $:\!\ddot{O}\!=\!\ddot{O}\!-\!\ddot{O}\!:$

PROBLEM 1.7   Draw a Lewis structure for each ion:

a. CH_3O^- b. HC_2^- c. $(CH_3NH_3)^+$

> Sometimes it is easier to count bonds, rather than shared electrons when determining formal charge.
> $\frac{1}{2}$**[number of shared electrons] = number of bonds**

When you first add formal charges to Lewis structures, you should use the procedure in Sample Problem 1.5 to determine them. With practice, you will notice that certain bonding patterns always result in the same formal charge. For example, any N atom with four bonds (and, thus no lone pairs) has a +1 formal charge. Table 1.1 lists the bonding patterns and resulting formal charges for carbon, nitrogen, and oxygen.

TABLE 1.1	**Formal Charge Observed with Common Bonding Patterns for C, N, and O**		
		Formal charge	
Atom	**Number of valence electrons**	**+1**	**0**	**−1**
C	4	$-\overset{+}{\underset{|}{C}}-$	$-\overset{|}{\underset{|}{C}}-$	$-\overset{..}{\underset{|}{C}}^{-}$
N	5	$-\overset{|}{\underset{|}{N}}^{+}-$	$-\overset{..}{\underset{|}{N}}-$	$-\overset{..}{\underset{..}{N}}^{-}$
O	6	$-\overset{..}{\underset{|}{O}}^{+}-$	$-\overset{..}{\underset{..}{O}}-$	$-\overset{..}{\underset{..}{O}}\!:^{-}$

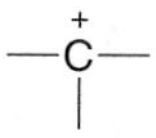

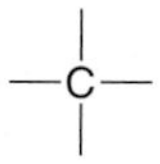

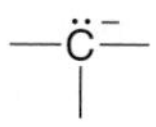

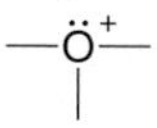

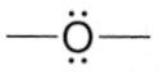

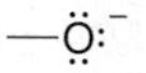

1.4   Lewis Structures Continued

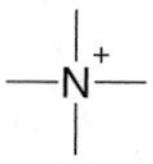

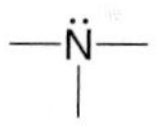

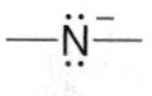

The discussion of Lewis structures concludes with the introduction of isomers and exceptions to the octet rule.

1.4A   Isomers

In drawing a Lewis structure for a molecule with several atoms, sometimes more than one arrangement of atoms is possible for a given molecular formula. For example, there are two acceptable arrangements of atoms for the molecular formula C_2H_6O.

$$\begin{array}{cc} \underset{\substack{|\\H}}{\overset{\substack{H\ H\\|\ |}}{H-C-C-\ddot{O}-H}} & \xleftarrow{\text{isomers}}\rightarrow & \underset{\substack{|\\H\ \ H}}{\overset{\substack{H\ \ \ H\\|\ \ \ |}}{H-C-\ddot{O}-C-H}} \end{array}$$

ethanol   dimethyl ether

same molecular formula
C_2H_6O

Both are valid Lewis structures, and both molecules exist. One is called ethanol, and the other, dimethyl ether. These two compounds are called **isomers.**

◆ *Isomers* are different molecules having the same molecular formula.

Ethanol and dimethyl ether are **constitutional isomers** because they have the same molecular formula, but **the *connectivity of their atoms is different.*** For example, ethanol has one $C-C$ bond and one $O-H$ bond, whereas dimethyl ether has two $C-O$ bonds. A second class of isomers, called **stereoisomers,** is introduced in Section 4.13B.

PROBLEM 1.8   Draw Lewis structures for each molecular formula.
a. $C_2H_4Cl_2$ (two isomers)   b. C_3H_8O (three isomers)   c. C_3H_6 (two isomers)

1.4B   Exceptions to the Octet Rule

DMSO is a widely used solvent in organic reactions.

Sulfuric acid is the most widely produced industrial chemical. It is also formed when sulfur oxides, emitted into the atmosphere by burning fossil fuels high in sulfur content, dissolve in water. This makes rain water acidic, forming acid rain.

Most of the common elements in organic compounds—**C, N, O, and the halogens**—follow the octet rule. Hydrogen is a notable exception, because it accommodates only two electrons in bonding. Additional exceptions include boron and beryllium (second-row elements in groups 2A and 3A, respectively), and elements in the third row (particularly phosphorus and sulfur).

Elements in Groups 2A and 3A

Elements in groups 2A and 3A of the periodic table, such as beryllium and boron, do not have enough valence electrons to form an octet in a neutral molecule. Lewis structures for BeH_2 and BF_3 show that these atoms have only four and six electrons, respectively, around the central atom. There is nothing we can do about this! There simply aren't enough electrons to form an octet.

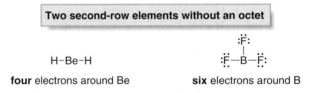

Two second-row elements without an octet

H—Be—H   :F̈—B—F̈:
                                        :F̈:

four electrons around Be   **six** electrons around B

Because the Be and B atoms each have less than an octet of electrons, these molecules are highly reactive.

Elements in the Third Row

Naled is a common insecticide found in pest strips and flea collars.

A second exception to the octet rule occurs with some elements located in the third row and later in the periodic table. These elements have empty *d* orbitals available to accept electrons, and thus they may have ***more than eight* electrons** around them. For organic chemists, the two most important elements in this category are phosphorus and sulfur, which can have 10 or even 12 electrons around them. Examples of these phosphorus and sulfur compounds include the following:

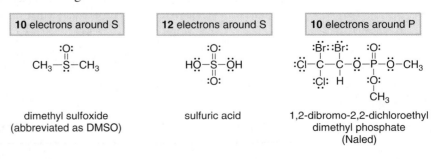

| **10** electrons around S | **12** electrons around S | **10** electrons around P |

dimethyl sulfoxide (abbreviated as DMSO)   sulfuric acid   1,2-dibromo-2,2-dichloroethyl dimethyl phosphate (Naled)

1.5 Resonance

Some molecules can't be adequately represented by a single Lewis structure. For example, two valid Lewis structures can be drawn for the anion (HCONH)⁻. One structure has a negatively charged N atom and a C−O double bond; the other has a negatively charged O atom and a C−N double bond. These structures are called **resonance structures** or **resonance forms**. A **double-headed arrow** is used to separate two resonance structures.

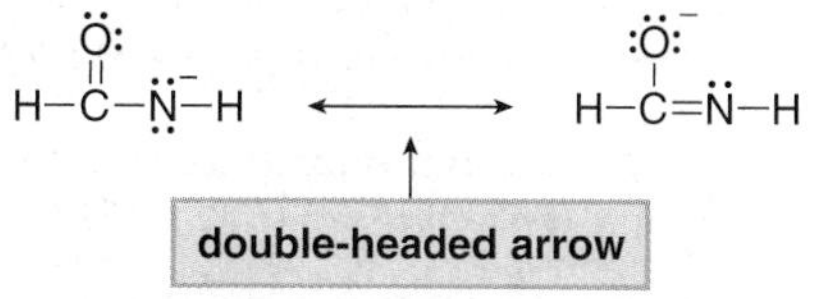

double-headed arrow

◆ *Resonance structures* are two Lewis structures having the *same* placement of atoms but a *different* arrangement of electrons.

Which resonance structure is an accurate representation for (HCONH)⁻? **The answer is *neither* of them.** The true structure is a composite of both resonance forms, and is called a **resonance hybrid.** The hybrid shows characteristics of **both** resonance structures.

Each resonance structure implies that electron pairs are localized in bonds or on atoms. In actuality, resonance allows certain electron pairs to be **delocalized** over several atoms, and this delocalization of electron density adds stability. **A molecule with two or more resonance structures is said to be *resonance stabilized.*** We will return to the resonance hybrid in Section 1.5C. First, however, we will concentrate on the general principles of resonance theory and learn how to interconvert two or more resonance structures.

1.5A An Introduction to Resonance Theory

Keep in mind the following basic principles of resonance theory.

◆ An individual resonance structure does not accurately represent the structure of a molecule or ion. Only the hybrid does.
◆ Resonance structures are *not* in equilibrium with each other. There is no movement of electrons from one form to another.
◆ Resonance structures are *not* isomers. Two isomers differ in the arrangement of *both* atoms and electrons, whereas resonance structures differ *only* in the *arrangement of electrons.*

SAMPLE PROBLEM 1.6 Label each pair of compounds as isomers or resonance structures.

a. CH₂=C−CH₃ (with Ö−H group) and CH₃−C−CH₃ (with :O: double bond)   b. CH₃−C≡Ö:⁺ and CH₃−C=Ö:

SOLUTION

an O−H bond

a. CH₂=C−CH₃ (with :Ö−H) and CH₃−C−CH₃ (with :O:)

one more C−H bond

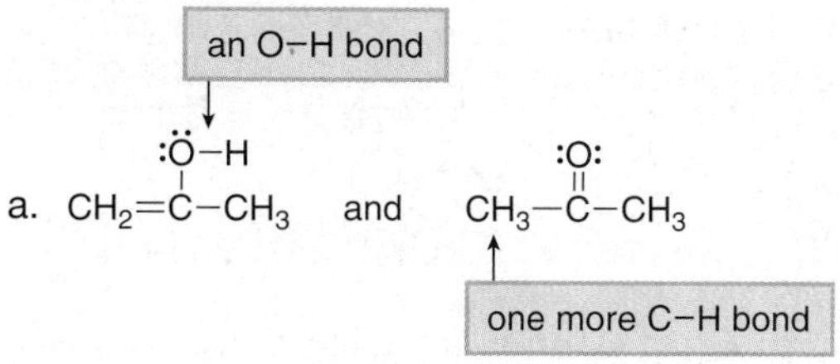

These compounds are **isomers** because the placement of atoms is different. One compound has an O−H bond and one compound has an additional C−H bond.

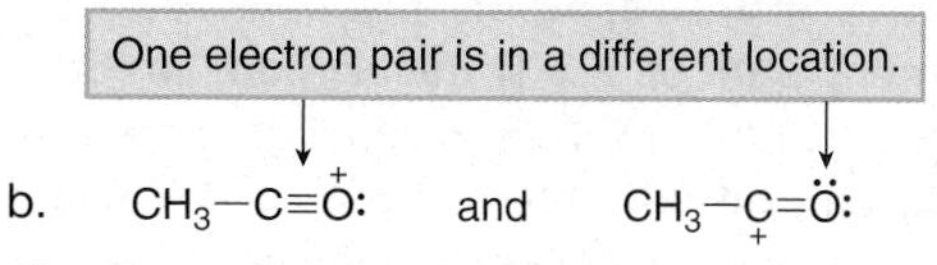

One electron pair is in a different location.

b. CH₃−C≡Ö:⁺ and CH₃−C=Ö:⁺

These compounds are **resonance structures.** The atom position is the same in both compounds, but the location of an electron pair is different.

Resonance structures are different ways of drawing the same compound. Two resonance structures are *not* different compounds.

PROBLEM 1.9   Classify each pair of compounds as isomers or resonance structures.

a. 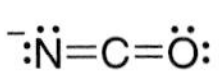 $:\!\overset{-}{N}\!=\!C\!=\!\overset{..}{O}\!:$ and $:\!C\!\equiv\!\overset{+}{N}\!-\!\overset{..}{\overset{-}{O}}\!:$ b. 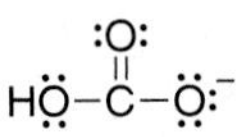 $H\overset{..}{\overset{}{O}}\!-\!\overset{\overset{\displaystyle :\!\overset{..}{O}\!:}{\|}}{C}\!-\!\overset{..}{\overset{-}{O}}\!:$ and $H\overset{..}{\overset{}{O}}\!-\!\overset{\overset{\displaystyle :\!\overset{..}{\overset{-}{O}}\!:}{|}}{C}\!=\!\overset{..}{O}\!:$

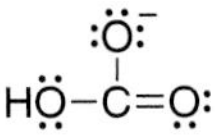

1.5B   Drawing Resonance Structures

To draw resonance structures, use the three rules that follow:

Rule [1]   **Two resonance structures differ in the position of multiple bonds and nonbonded electrons. The placement of atoms and single bonds always stays the same.**

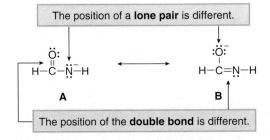

Rule [2]   **Two resonance structures must have the same number of unpaired electrons.**

two unpaired electrons

- **A** and **B** have no unpaired electrons.
- **C** is **not** a resonance structure of **A** and **B**.

Rule [3]   **Resonance structures must be valid Lewis structures. Hydrogen must have two electrons and no second-row element can have more than eight electrons.**

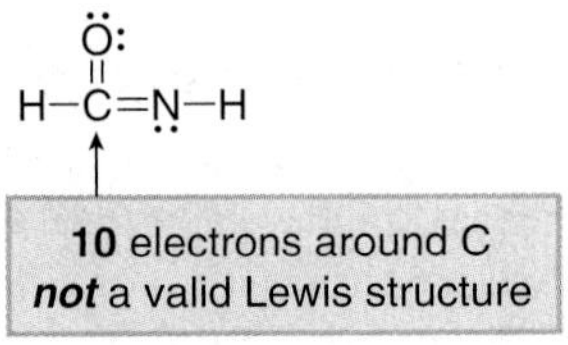

10 electrons around C
not a valid Lewis structure

Curved arrow notation is a convention that shows how electron position differs between the two resonance forms.

◆ *Curved arrow notation shows the movement of an electron pair. The tail of the arrow always begins at an electron pair, either in a bond or lone pair. The head points to where the electron pair "moves."*

A curved arrow always begins at an electron pair. It ends at an atom or a bond.

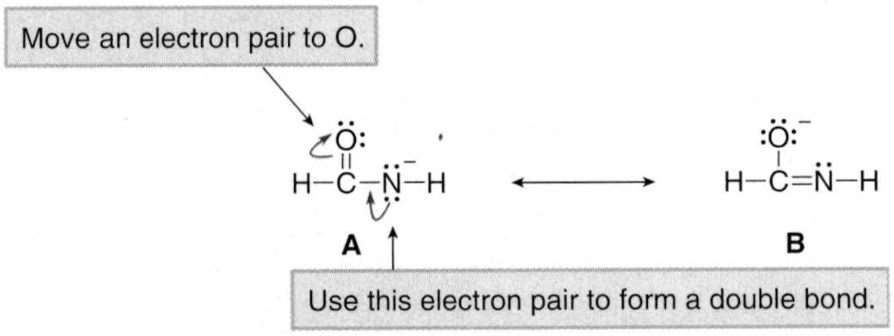

Move an electron pair to O.

Use this electron pair to form a double bond.

Resonance structures **A** and **B** differ in the location of two electron pairs, so two curved arrows are needed. To convert **A** to **B,** take the lone pair on N and form a double bond between C and N. Then, move an electron pair in the C–O double bond to form a lone pair on O. Curved arrows thus show how to reposition the electrons in converting one resonance form to another. **The electrons themselves do not actually move.** Sample Problem 1.7 illustrates the use of curved arrows to convert one resonance structure to another.

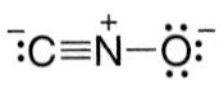

SAMPLE PROBLEM 1.7   Follow the curved arrows to draw a second resonance structure for each ion.

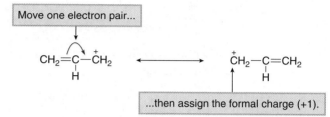

SOLUTION

[a] The curved arrow tells us to move **one** electron pair in the double bond to the adjacent C–C bond. We must then determine the formal charge on any atom whose bonding is different, making sure that the overall charge is the same in both resonance forms.

> Move one electron pair...

$$CH_2\!\!=\!\!\underset{H}{C}\!\!-\!\!\overset{+}{C}H_2 \longleftrightarrow \overset{+}{C}H_2\!\!-\!\!\underset{H}{C}\!\!=\!\!CH_2$$

> ...then assign the formal charge (+1).

Positively charged carbon atoms are called **carbocations.** Carbocations are unstable intermediates because they contain a carbon atom that is lacking an octet of electrons.

[b] **Two** curved arrows tell us to move **two** electron pairs. The second resonance structure has a formal charge of (–1) on O.

> Move two electron pairs...

$$H\!\!-\!\!\underset{H}{C}\!\!-\!\!\overset{\ddot{O}}{C}\!\!-\!\!CH_3 \longleftrightarrow H\!\!-\!\!\underset{H}{C}\!\!=\!\!\underset{}{C}\!\!-\!\!CH_3$$

> ...then calculate formal charges.

The arrows tell us precisely where to place the electrons in the second resonance structure. This type of resonance-stabilized anion is called an **enolate anion.** Enolates are important intermediates in many organic reactions, and all of Chapters 23 and 24 is devoted to their preparation and reactions.

PROBLEM 1.10   Follow the curved arrows to draw a second resonance structure for each species.

a. $H\!\!-\!\!\underset{H}{C}\!\!=\!\!\ddot{O}\!: \longleftrightarrow$

b. $CH_3\!\!-\!\!\underset{H}{\ddot{C}}\!\!-\!\!\underset{H}{C}\!\!=\!\!CH_2 \longleftrightarrow$

PROBLEM 1.11   Use curved arrow notation to show how the first resonance structure can be converted to the second.

a. $\overset{+}{C}H_2\!\!-\!\!\underset{H}{C}\!\!=\!\!\underset{H}{C}\!\!-\!\!CH_3 \longleftrightarrow CH_2\!\!=\!\!\underset{H}{C}\!\!-\!\!\underset{H}{\overset{+}{C}}\!\!-\!\!CH_3$

b. $:\!\ddot{O}\!\!=\!\!C\!\!-\!\!\ddot{O}\!:^{-} \longleftrightarrow {}^{-}\!:\!\ddot{O}\!\!-\!\!C\!\!-\!\!\ddot{O}\!:^{-}$

Two resonance structures can have exactly the same kinds of bonds, as they do in the carbocation in Sample Problem 1.7a, or they may have different types of bonds, as they do in the enolate in Sample Problem 1.7b. Either possibility is fine as long as the individual resonance structures are valid Lewis structures.

The ability to draw and manipulate resonance structures is an important skill that will be needed throughout your study of organic chemistry. With practice, you will begin to recognize certain common bonding patterns for which more than one Lewis structure can be drawn. For now, notice that two different resonance structures can be drawn in the following situations.

◆ **When a lone pair is located on an atom directly bonded to a multiple bond.**

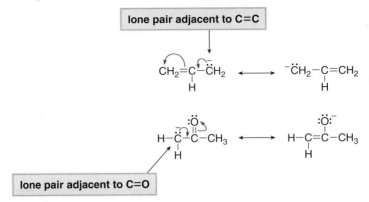

◆ **When an atom bearing a (+) charge is bonded either to a multiple bond or an atom with a lone pair.**

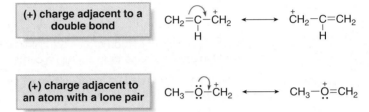

PROBLEM 1.12   Draw a second resonance structure for each species.

a. 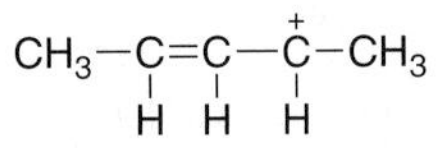 $CH_3-C=C-\overset{+}{C}-CH_3$
       $\;\;\;\;\;\;\;\;\;|\;\;|\;\;|$
       $\;\;\;\;\;\;\;\;\;H\;\;H\;\;H$

b. $CH_3-\overset{+}{C}-CH_3$
       $\;\;\;\;\;\;\;\;\;\;|$
       $\;\;\;\;\;\;\;\;\;:\overset{..}{\underset{..}{Cl}}:$

c. $H-C=C-\overset{..}{\underset{..}{Cl}}:$
       $\;\;\;\;\;\;|\;\;|$
       $\;\;\;\;\;\;H\;\;H$

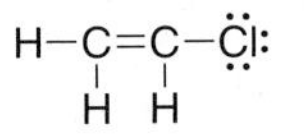

1.5C   The Resonance Hybrid

The resonance hybrid is the composite of all possible resonance structures. In the resonance hybrid, the electron pairs drawn in different locations in individual resonance structures are ***delocalized.***

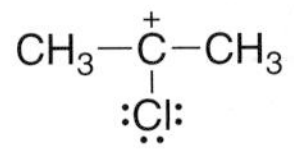

◆ **The resonance hybrid is more stable than any resonance structure because it delocalizes electron density over a larger volume.**

What does the hybrid look like? When all resonance forms are identical, as they were in the carbocation in Sample Problem 1.7a, each resonance form contributes **equally** to the hybrid.

When two resonance structures are different, the hybrid looks more like the "better" resonance structure. The "better" resonance structure is called the **major contributor** to the hybrid, and all others are **minor contributors.** The hybrid is the weighted average of the contributing resonance structures. What makes one resonance structure "better" than another? There are many factors, but for now, we will learn just two.

◆ **A "better" resonance structure is one that has *more bonds* and *fewer charges*.**

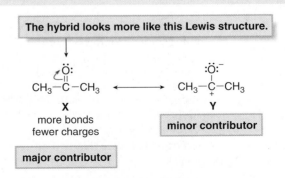

Comparing resonance structures **X** and **Y, X** is the major contributor because it has more bonds and fewer charges. Thus, the hybrid looks more like **X** than **Y**.

How can we draw a hybrid, which has delocalized electron density? First, we must determine what is different in the resonance structures. Two differences commonly seen are the **position of a multiple bond** and the **site of a charge**. The anion (HCONH)⁻ illustrates two common conventions for drawing resonance hybrids.

individual resonance structures

resonance *hybrid*

Appendix A lists common symbols and conventions used in organic chemistry.

◆ **Double bond position.** Structure **A** has a C−O double bond, whereas structure **B** has a C−N double bond. A dashed line in the hybrid indicates partial double bond character between these atoms.

◆ **Location of the charge.** A negative charge resides on different atoms in **A** and **B**. The symbol δ⁻ (for a partial negative charge) indicates that the charge is delocalized on the N and O atoms in the hybrid.

This discussion of resonance is meant to serve as an introduction only. You will learn many more facets of resonance theory in later chapters. In Chapter 2, for example, the enormous effect of resonance on acidity is discussed.

PROBLEM 1.13   Draw the hybrid for each ion in Sample Problem 1.7.

PROBLEM 1.14   Label the resonance structures in each pair as major, minor, or equal contributors to the hybrid. Then draw the hybrid.

1.6   Determining Molecular Shape

We can now use Lewis structures to determine the shape around a particular atom in a molecule. Consider the H_2O molecule. The Lewis structure tells us only which atoms are connected to each other, but it implies nothing about the geometry. What does the overall molecule look like? Is H_2O a bent or linear molecule?

A Lewis structure does **not** imply any geometry.

H−Ö−H

Two variables define a molecule's structure: **bond length** and **bond angle**.

We will use **angstroms** as the unit of measure for bond length because most bond lengths in the chemical literature are still reported in angstroms. The SI unit for bond length is the **picometer** (pm), where $1 \text{ pm} = 10^{-2} \text{ Å} = 10^{-12} \text{ m}$. Thus, 0.958 Å = 95.8 pm.

1.6A   Bond Length

Bond length **is defined as the average distance between the centers of two bonded nuclei.** Bond lengths are typically reported in angstroms, where $1 \text{ Å} = 10^{-10}$ m. For example, the O−H bond length in H_2O is 0.958 Å. Average bond lengths for common bonds are listed in Table 1.2.

◆ Bond length *decreases* across a row of the periodic table as the size of the atom *decreases.*

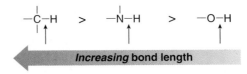

◆ Bond length *increases* down a column of the periodic table as the size of an atom *increases.*

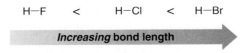

Learn these general trends. Often knowing such trends is more useful than learning a set of exact numbers, because we are usually interested in comparisons rather than absolute values.

TABLE 1.2   Average Bond Lengths

Bond	Length (Å)	Bond	Length (Å)	Bond	Length (Å)
H−H	0.74	H−F	0.92	C−F	1.33
C−H	1.09	H−Cl	1.27	C−Cl	1.77
N−H	1.01	H−Br	1.41	C−Br	1.94
O−H	0.96	H−I	1.61	C−I	2.13

1.6B   Bond Angle

Bond angle determines the shape around any atom bonded to two other atoms.

◆ The number of groups surrounding a particular atom determines its geometry. A group is either an atom or a lone pair of electrons.
◆ The most stable arrangement keeps these groups as far away from each other as possible. This is exemplified in Valence Shell Electron Pair Repulsion (VSEPR) theory.

A second-row element has only three possible arrangements, defined by the number of groups surrounding it.

To determine geometry: [1] Draw a valid Lewis structure; [2] count groups around a given atom.

Number of groups	Geometry	Bond angle
• two groups	linear	180°
• three groups	trigonal planar	120°
• four groups	tetrahedral	109.5°

Let's examine several molecules to illustrate this phenomenon. In each example, we first need a valid Lewis structure, and then we merely count groups around a given atom to predict its geometry.

Two Groups Around an Atom

Any atom surrounded by only two groups is linear and has a bond angle of 180°. Two examples illustrating this geometry are **BeH₂** (beryllium hydride) and **HC≡CH** (acetylene).

Acetylene has more than one second row element, so consider each carbon atom *separately.* Because each C is surrounded by two atoms and no lone pairs, each H−C−C bond angle in acetylene is 180°, and therefore all four atoms are linear.

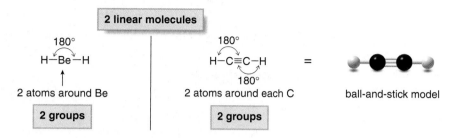

Acetylene illustrates another important feature: *ignore multiple bonds in predicting geometry.* **Count only atoms and lone pairs.**

We will begin to represent molecules with models having balls for atoms and sticks for bonds, as in the ball-and-stick model of acetylene just shown. These representations are analogous to a set of molecular models. Invest in a set of models *now.* Most students in organic chemistry find that building models helps them visualize the shape of molecules.

Three Groups Around an Atom

Trigonal = three-sided.

Any atom surrounded by three groups is trigonal planar and has bond angles of 120°. Two examples illustrating this geometry are **BF₃** (boron trifluoride) and **CH₂=CH₂** (ethylene). Note that *each* carbon atom of ethylene is surrounded by three atoms and no lone pairs, making *each* H−C−C bond angle 120°.

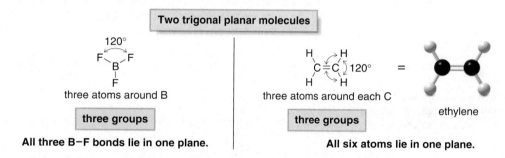

Four Groups Around an Atom

Any atom surrounded by four groups is tetrahedral and has bond angles of approximately 109.5°. For example, the simple organic compound methane, **CH₄,** has a central carbon atom with bonds to four hydrogen atoms, each pointing to a corner of a tetrahedron. This arrangement keeps four groups farther apart than a square planar arrangement in which all bond angles would be only 90°.

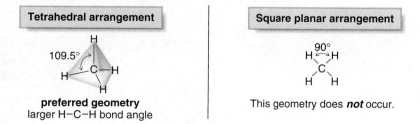

How can we represent the three-dimensional geometry of a tetrahedron on a two-dimensional piece of paper? **Place two of the bonds in the plane of the paper, one bond in front and one bond behind,** using the following conventions:

◆ A *solid line* is used for a bond *in* the plane.
◆ A *wedge* is used for a bond in *front* of the plane.
◆ A *dashed line* is used for a bond *behind* the plane.

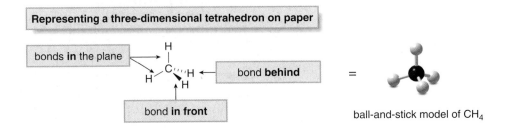

This is just one way to draw a tetrahedron for CH_4. We can turn the molecule in many different ways, generating many equivalent representations. For example, all of the following are acceptable drawings for CH_4.

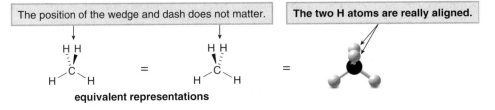

Finally, **wedges and dashes are used for groups that are really** *aligned one behind another.* It does not matter in the following two drawings whether the wedge or dash is skewed to the left or right, because the two H atoms are really aligned. To confirm that they are aligned, see the three-dimensional model, also provided.

All carbons in stable molecules are **tetravalent,** but the geometry varies with the number of groups around the particular carbon.

The position of the wedge and dash does not matter.

The two H atoms are really aligned.

equivalent representations

PROBLEM 1.15   Draw two different three-dimensional representations for CH_2Cl_2 (dichloromethane) using wedges and dashes.

Ammonia (NH_3) and water (H_2O) are molecules with central atoms surrounded by four groups, some of which are lone pairs. In **NH₃,** the three H atoms and one lone pair around N point to the corners of a tetrahedron. The H−N−H bond angle of 107° is close to the theoretical tetrahedral bond angle of 109.5°. This shape is referred to as a **trigonal pyramid,** because one of the groups around the N is a nonbonded electron pair, not another atom.

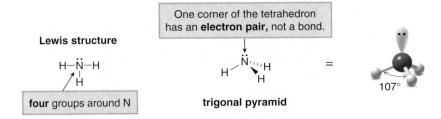

One corner of the tetrahedron has an **electron pair,** not a bond.

Lewis structure

four groups around N

trigonal pyramid

107°

In **H$_2$O,** the two H atoms and two lone pairs around O point to the corners of a tetrahedron. The H−O−H bond angle of 105° is close to the theoretical tetrahedral bond angle of 109.5°. Water has a **bent** shape, because two of the groups around oxygen are lone pairs of electrons.

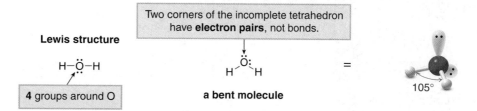

Lewis structure

H−Ö−H

4 groups around O

Two corners of the incomplete tetrahedron have **electron pairs**, not bonds.

Ö:
H   H

a bent molecule

=

105°

In both NH$_3$ and H$_2$O the bond angle is somewhat smaller than the theoretical tetrahedral bond angle because of repulsion of the lone pairs of electrons. The bonded atoms are thus compressed into a smaller space with a smaller bond angle.

Predicting geometry based on counting groups is summarized in Figure 1.4.

Figure 1.4 Summary: Determining geometry based on the number of groups

Number of groups around an atom	Geometry	Bond angle	Examples
2	linear	180°	BeH$_2$, HC≡CH
3	trigonal planar	120°	BF$_3$, CH$_2$=CH$_2$
4	tetrahedral	109.5°	CH$_4$, NH$_3$, H$_2$O

SAMPLE PROBLEM 1.8   Determine the geometry around the labeled atom in each species.

a. :Ö=C̈=Ö: b. H−N$^+$−H   (with H above and H below)

SOLUTION

a.   :Ö=C=Ö:
           180°
    two atoms around C
    no lone pairs

        two groups

         linear

b.   H−N$^+$−H   ⟶   N$^+$
                            109.5°
    four atoms around N
    no lone pairs

        four groups   ⟶   **tetrahedral**

PROBLEM 1.16   Determine the geometry around all second-row elements in each compound.

a. CH$_3$−C̈−CH$_3$ (with :O: double bonded above)   b. CH$_3$−Ö−CH$_3$ c. :NH$_2$ d. CH$_3$−C≡N:

PROBLEM 1.17   Predict the indicated bond angles in each compound.

a. CH$_3$−C≡C−Cl   b. CH$_2$=C−Cl   c. CH$_3$−C−Cl (with H above and H below)

PROBLEM 1.18   Using the principles of VSEPR theory, you can predict the geometry around any atom in any molecule, no matter how complex. Enanthotoxin is a poisonous compound isolated from a common variety of hemlock grown in England. Predict the geometry around the indicated atoms in enanthotoxin.

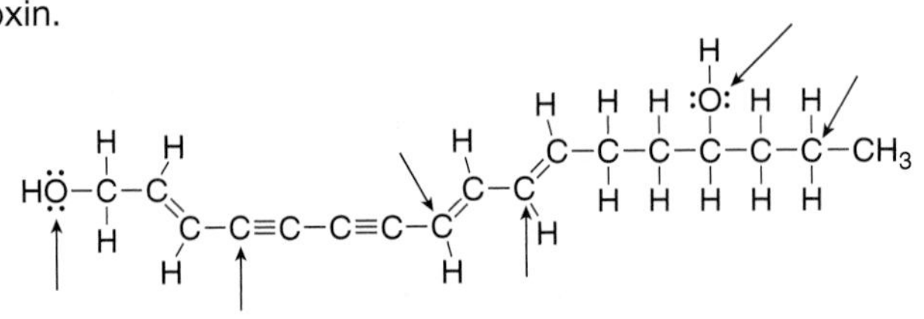

enanthotoxin

1.7   Drawing Organic Structures

Drawing organic molecules presents a special challenge. Because they often contain many atoms, we need shorthand methods to simplify their structures. You were briefly introduced to this topic in the prologue. The two main types of shorthand representations used for organic compounds are **condensed structures** and **skeletal structures.**

1.7A   Condensed Structures

Condensed structures are most often used for compounds having a chain of atoms bonded together, rather than a ring. The following conventions are used:

◆ All of the atoms are drawn in, but the two-electron bond lines are generally omitted.
◆ Atoms are usually drawn next to the atoms to which they are bonded.
◆ Parentheses are used around similar groups bonded to the same atom.
◆ Lone pairs are omitted.

To interpret a condensed formula, it is usually best to start at the **left side** of the molecule and remember that the **carbon atoms must be tetravalent.** A carbon bonded to three H atoms becomes CH_3; a carbon bonded to two H atoms becomes CH_2, and so forth. Two examples of condensed formulas for compounds having only carbon and hydrogen are given below.

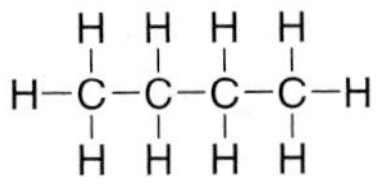

$$H-\overset{\overset{H}{|}}{\underset{\underset{H}{|}}{C}}-\overset{\overset{H}{|}}{\underset{\underset{H}{|}}{C}}-\overset{\overset{H}{|}}{\underset{\underset{H}{|}}{C}}-\overset{\overset{H}{|}}{\underset{\underset{H}{|}}{C}}-H \quad = \quad CH_3CH_2CH_2CH_3 \quad \text{or} \quad CH_3(CH_2)_2CH_3$$

• Draw a C with 3 H's as CH_3.
• Draw a C with 2 H's as CH_2.

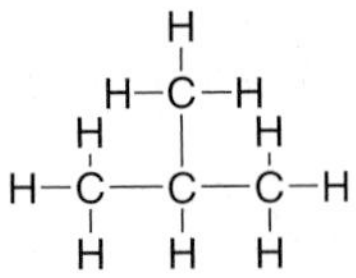

2 CH_2 groups bonded together

$$H-\overset{\overset{\overset{\overset{H}{|}}{\underset{}{C}}-H}{|}}{\underset{\underset{H}{|}}{C}}... \quad = \quad (CH_3)_3CH$$

• Draw 3 CH_3 groups bonded to 1 C.

> Recall from the prologue: A **heteroatom** is any atom that is not C or H.

Other examples of condensed structures with heteroatoms and carbon–carbon multiple bonds are given in Figure 1.5. You must learn how to convert a Lewis structure to a condensed structure, and vice versa.

Translating some condensed formulas is not obvious, and it will come only with practice. This is especially true for compounds containing a carbon–oxygen double bond. Some noteworthy examples in this category are given in Figure 1.6. Keep in mind that a **condensed formula must give rise to a valid Lewis structure.**

Figure 1.5 Examples of condensed structures

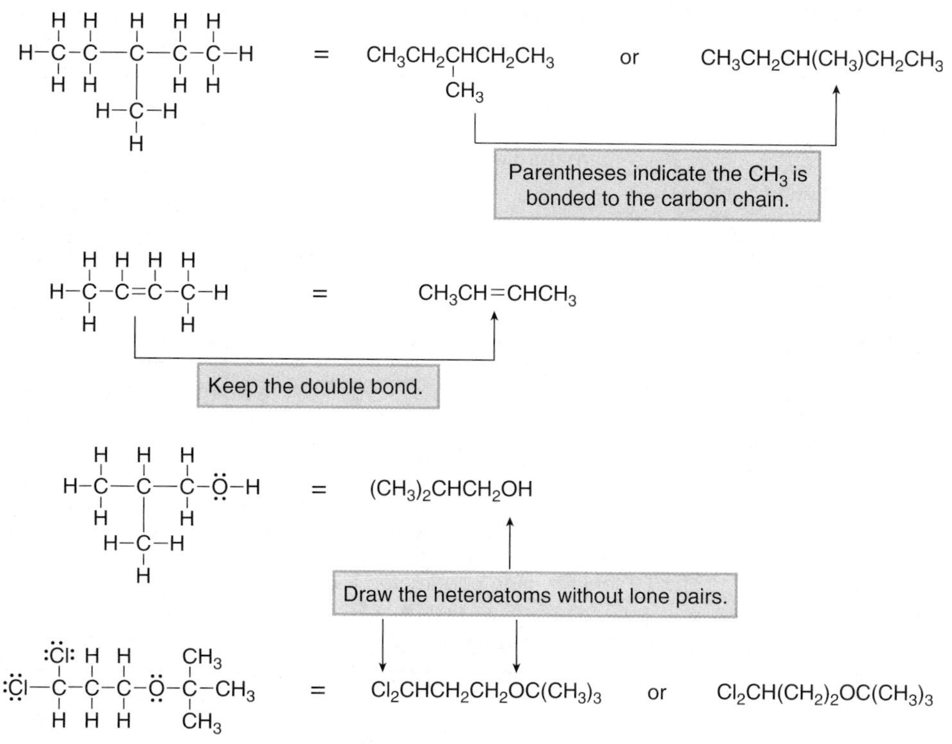

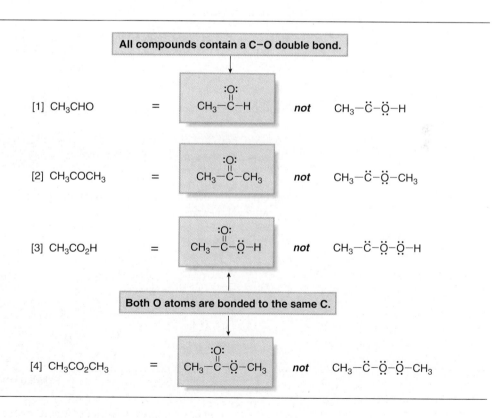

Figure 1.6 Condensed structures containing a C−O double bond

SAMPLE PROBLEM 1.9   Convert each condensed formula to a Lewis structure.
a. $(CH_3)_2CHOCH_2CH_2CH_2OH$ b. $CH_3(CH_2)_2CO_2C(CH_3)_3$

SOLUTION

Start at the left and proceed to right, making sure that each carbon has four bonds. Give any heteroatom enough lone pairs to have an octet.

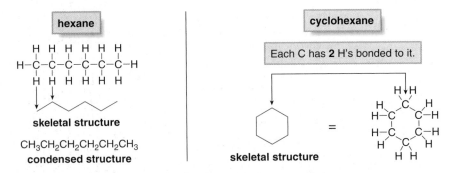

a. $(CH_3)_2CHOCH_2CH_2CH_2OH$

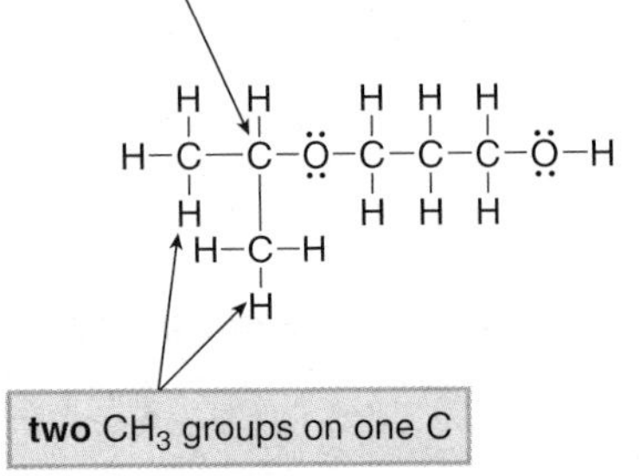

two CH₃ groups on one C

b. $CH_3(CH_2)_2CO_2C(CH_3)_3$

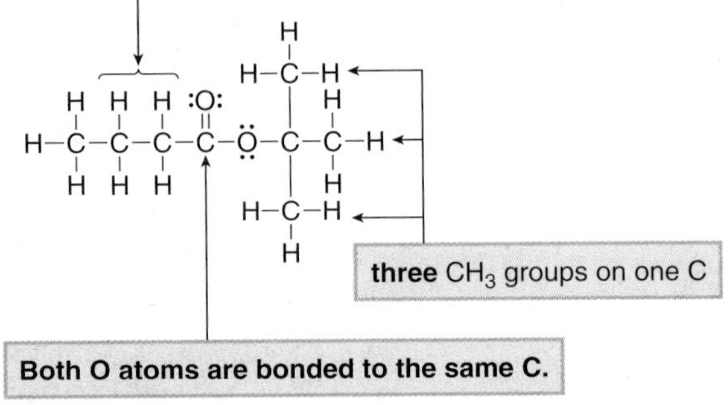

three CH₃ groups on one C

Both O atoms are bonded to the same C.

PROBLEM 1.19   Convert each condensed formula to a Lewis structure.
a. $CH_3(CH_2)_4CH(CH_3)_2$ c. $(CH_3)_3CCH(OH)CH_2CH_3$
b. $(CH_3)_2CHCH(CH_2CH_3)_2$ d. $(CH_3)_2CHCHO$

1.7B   Skeletal Structures

Skeletal structures were introduced in the prologue. Skeletal structures are used for organic compounds containing both rings and chains of atoms. Recall the three important rules used to draw them:

◆ Assume there is a carbon atom at the junction of any two lines or at the end of any line.
◆ Assume there are enough hydrogens around each carbon to make it tetravalent.
◆ Draw in all heteroatoms and the hydrogens directly bonded to them.

Carbon chains are drawn in a zigzag fashion, and rings are drawn as polygons, as shown for hexane and cyclohexane.

hexane

skeletal structure

$CH_3CH_2CH_2CH_2CH_2CH_3$
condensed structure

cyclohexane

Each C has **2** H's bonded to it.

skeletal structure

Figure 1.7 shows other examples of skeletal structures for a variety of Lewis structures. Keep in mind that **skeletal structures often leave out the lone pairs on heteroatoms.** *Don't forget about them.* They often participate in chemical reactions.

Figure 1.7 Examples of skeletal structures

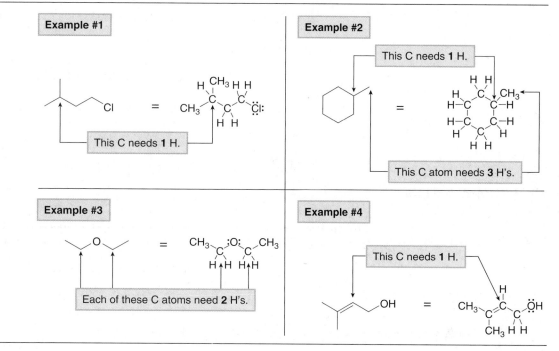

Example #1

This C needs **1** H.

Example #2

This C needs **1** H.

This C atom needs **3** H's.

Example #3

Each of these C atoms need **2** H's.

Example #4

This C needs **1** H.

PROBLEM 1.20

How many hydrogen atoms are present around each labeled carbon atom in the following molecules? Both compounds are active ingredients in some common sunscreens.

a.

octinoxate
(2-ethylhexyl 4-methoxycinnamate)

b.

avobenzone

PROBLEM 1.21

Convert each skeletal structure to a Lewis structure.

a.   b.   c.   d.

Take care in interpreting skeletal structures for positively and negatively charged carbon atoms, because *both* the hydrogen atoms *and* the lone pairs are omitted. Keep in mind the following:

◆ A charge on a carbon atom takes the place of one hydrogen atom.
◆ The charge determines the number of lone pairs. Negatively charged carbon atoms have one lone pair and positively charged carbon atoms have none.

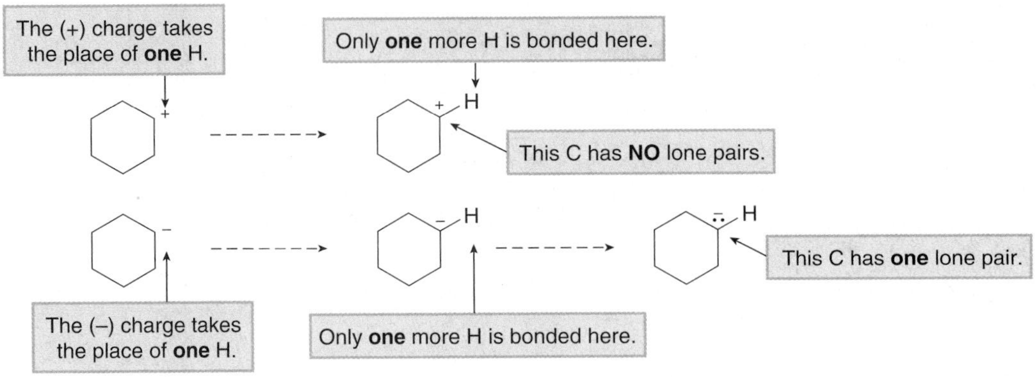

The (+) charge takes the place of **one** H.

Only **one** more H is bonded here.

This C has **NO** lone pairs.

This C has **one** lone pair.

Only **one** more H is bonded here.

The (−) charge takes the place of **one** H.

PROBLEM 1.22   Draw in all hydrogens and lone pairs on the charged carbons in each ion.

a. 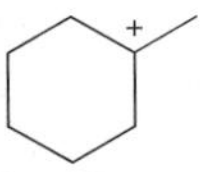   b. 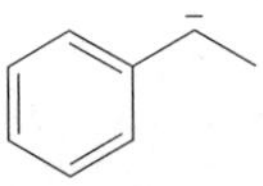   c. 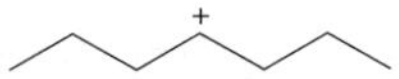   d.

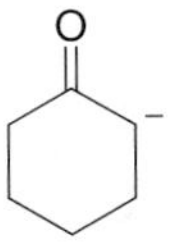

PROBLEM 1.23   Convert each molecule into the indicated representation.

a. Convert [structure with CO$_2$CH$_3$ and N] into a Lewis structure.

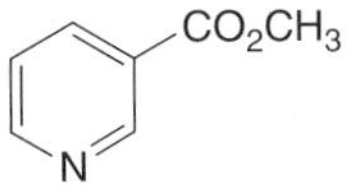

b. Convert [structure with OH] into a Lewis structure and a condensed structure.

c. Convert $CH_3-C=C-C-C-\ddot{O}-C-CH_3$ into a condensed structure and a skeletal structure.

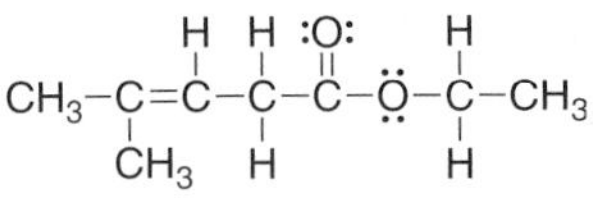

with H H :O: H above and CH$_3$ H H below

1.8 Hybridization

What orbitals do the first- and second-row atoms use to form bonds? Let's begin with hydrogen and then examine the orbitals used for bonding by atoms in the second row.

1.8A Hydrogen

Recall from Section 1.2 that two hydrogen atoms share each of their electrons to form H$_2$. Thus, the 1s orbital on one H overlaps with the 1s orbital on the other H to form a bond that concentrates electron density between the two nuclei. This type of bond, called a σ (sigma) **bond,** is cylindrically symmetrical because the electrons forming the bond are distributed symmetrically about an imaginary line connecting the two nuclei.

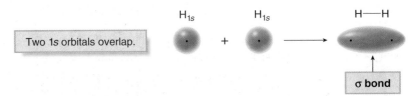

Two 1s orbitals overlap.

H$_{1s}$ +   H$_{1s}$ →   H—H

σ bond

◆ A σ bond concentrates electron density on the axis that joins two nuclei. All single bonds are σ bonds.

1.8B   Bonding in Methane

To account for the bonding patterns observed in more complex molecules, we must take a closer look at the 2s and 2p orbitals of atoms of the second row. Let's illustrate this with methane, CH_4.

Carbon has two core electrons, plus **four valence electrons.** To fill atomic orbitals in the most stable arrangement, electrons are placed in the orbitals of lowest energy. For carbon, this places two in the 2s orbital and one each in two 2p orbitals.

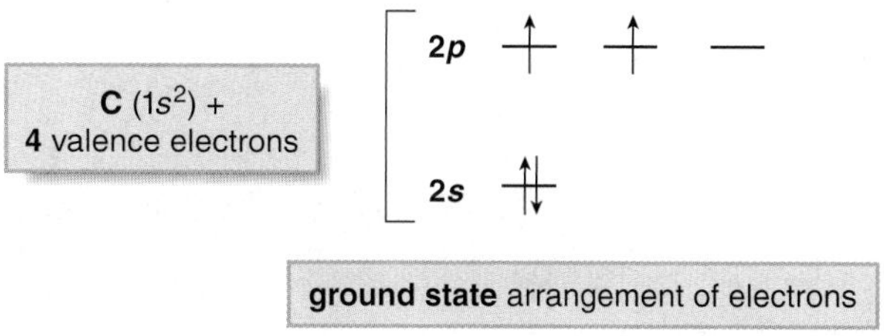

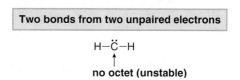

ground state arrangement of electrons

◆ **This lowest energy arrangement of electrons for an atom is called its *ground state.***

In this description, **carbon should form *only two bonds*** because it has only two unpaired valence electrons, and CH_2 should be a stable molecule. In reality, however, CH_2 is a highly reactive species that cannot be isolated under typical laboratory conditions. In CH_2, carbon would not have an octet of electrons.

Two bonds from two unpaired electrons

H–C̈–H

no octet (unstable)

There is a second possibility. Promotion of an electron from a 2s to a vacant 2p orbital would form **four** unpaired electrons for bonding. This process requires energy because it moves an electron to a higher energy orbital. This higher energy electron configuration is called an electronically **excited state.**

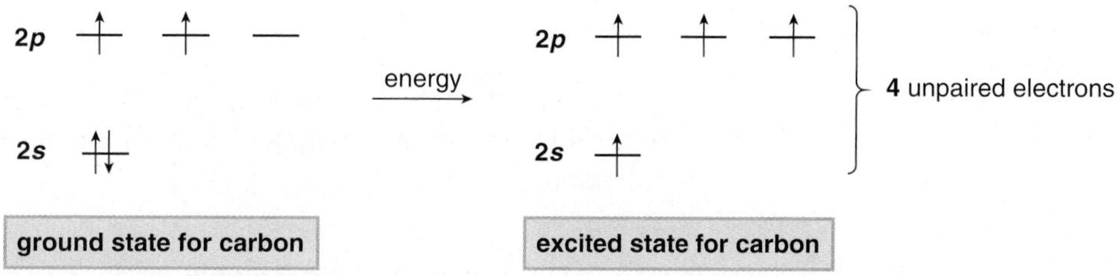

2p   ↑   ↑   —

2s   ↑↓

ground state for carbon

→ energy

2p   ↑   ↑   ↑

2s   ↑

excited state for carbon

4 unpaired electrons

This description is still not adequate. Carbon would form two different types of bonds: three with 2p orbitals and one with a 2s orbital. **But experimental evidence points to carbon forming *four identical bonds* in methane.**

To solve this dilemma, chemists have proposed that atoms like carbon do not use pure s and pure p orbitals in forming bonds. Instead, atoms use a set of new orbitals called **hybrid orbitals.** The mathematical process by which these orbitals are formed is called **hybridization.**

◆ *Hybridization* **is the combination of two or more atomic orbitals to form the same number of hybrid orbitals, each having the same shape and energy.**

Hybridization of one 2s orbital and three 2p orbitals for carbon is illustrated here. Each hybrid orbital has one electron. These new hybrid orbitals are intermediate in energy between the 2s and 2p orbitals.

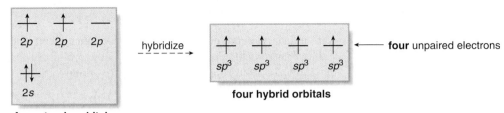

Forming four *sp*³ hybrid orbitals for carbon

four atomic orbitals

hybridize

four hybrid orbitals ← **four** unpaired electrons

◆ These hybrid orbitals are called *sp*³ *hybrids* because they are formed from *one s* orbital and *three p* orbitals.

What do these new hybrid orbitals look like? Mixing a spherical 2*s* orbital and three dumbbell-shaped 2*p* orbitals together produces four orbitals having one large lobe and one small lobe, oriented toward the corners of a tetrahedron. Each large lobe concentrates electron density in the bonding direction between two nuclei. This makes bonds formed from hybrid orbitals **stronger** than bonds formed from pure *p* orbitals. Figure 1.8 compares the shape of an individual *p* orbital and *sp*³ hybrid orbital.

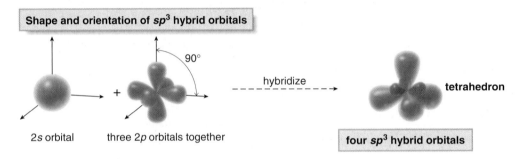

Shape and orientation of *sp*³ hybrid orbitals

2*s* orbital three 2*p* orbitals together 90° hybridize tetrahedron

four *sp*³ hybrid orbitals

The four hybrid orbitals form four equivalent bonds. We can now explain the observed bonding in CH_4.

◆ Each bond in CH_4 is formed by overlap of an *sp*³ hybrid orbital of carbon with a 1*s* orbital of hydrogen. These four bonds point to the corners of a tetrahedron.

All four C−H bonds in methane are **σ bonds,** because the electron density is concentrated on the axis joining C and H. An orbital picture of the bonding in CH_4 is given in Figure 1.9.

PROBLEM 1.24 What orbitals are used to form each of the C−C and C−H bonds in $CH_3CH_2CH_3$ (propane)? How many σ bonds are present in this molecule?

Figure 1.8 A comparison of a *p* orbital and an *sp*³ hybrid orbital

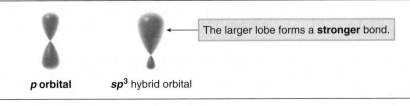

p orbital *sp*³ hybrid orbital The larger lobe forms a **stronger** bond.

Figure 1.9   Bonding in CH_4
using sp^3 hybrid orbitals

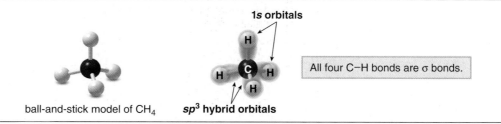

1s orbitals

All four C−H bonds are σ bonds.

ball-and-stick model of CH_4 sp^3 **hybrid orbitals**

1.8C   Other Hybridization Patterns—sp and sp^2 Hybrid Orbitals

Forming sp^3 hybrid orbitals is just one way that $2s$ and $2p$ orbitals can hybridize. In fact, three common modes of hybridization are seen in organic molecules. The number of orbitals is always conserved in hybridization; that is, a **given number of atomic orbitals hybridize to form an equivalent number of hybrid orbitals.**

- *One $2s$ orbital and three $2p$ orbitals form four sp^3 hybrid orbitals.*
- *One $2s$ orbital and two $2p$ orbitals form three sp^2 hybrid orbitals.*
- *One $2s$ orbital and one $2p$ orbital form two sp hybrid orbitals.*

The **superscripts** for hybrid orbitals correspond to the **number of atomic orbitals** used to form them. The number "1" is understood.

For example: $sp^3 = s^1p^3$

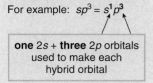

one $2s$ + **three** $2p$ orbitals used to make each hybrid orbital

We have already seen pictorially how four sp^3 hybrid orbitals are formed from one $2s$ and three $2p$ orbitals. Figures 1.10 and 1.11 illustrate the same process for sp and sp^2 hybrids. Each sp and sp^2 hybrid orbital has one large and one small lobe, much like an sp^3 hybrid orbital. Note, however, that both sp^2 and sp hybridization **leave one and two $2p$ orbitals *unhybridized*,** respectively, on each atom.

To determine the hybridization of an atom in a molecule, we count groups around the atom, just as we did in determining geometry. **The number of groups (atoms and nonbonded electron pairs) corresponds to the number of atomic orbitals that must be hybridized to form the hybrid orbitals.**

Figure 1.10   Forming two sp hybrid orbitals

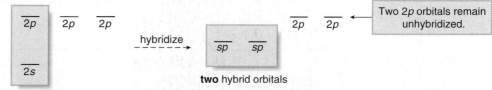

Two $2p$ orbitals remain unhybridized.

two atomic orbitals

- Forming **two sp hybrid orbitals** uses **one $2s$** and **one $2p$** orbital, leaving **two $2p$ orbitals *unhybridized*.**

Figure 1.11   Forming three sp^2 hybrid orbitals

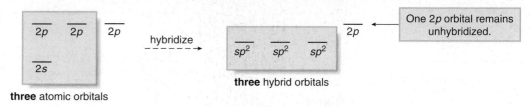

One $2p$ orbital remains unhybridized.

three atomic orbitals

- Forming **three sp^2 hybrid orbitals** uses **one $2s$** and **two $2p$** orbitals, leaving **one $2p$ orbital *unhybridized*.**

number of groups around an atom	number of orbitals used	type of hybrid orbital
2	2	two *sp* hybrid orbitals
3	3	three *sp²* hybrid orbitals
4	4	four *sp³* hybrid orbitals

Let's illustrate this phenomenon with BeH_2, BF_3, NH_3, and H_2O. We already determined the geometry in these molecules by counting groups in Section 1.6.

In **BeH₂**, the **Be atom is *sp* hybridized** because it is surrounded by two groups (two H atoms). Each Be–H bond is formed by overlap of an *sp* hybrid orbital from Be and a $1s$ orbital from H. The *sp* hybrid orbitals are oriented 180° away from each other.

In **BF₃**, the **B atom is *sp²* hybridized** because it is surrounded by three groups (three F atoms). Each B–F bond is formed by overlap of an *sp²* hybrid orbital from B and a $2p$ orbital from F. The *sp²* hybrid orbitals all lie in a plane, and are oriented 120° apart. The B atom also has a vacant unhybridized $2p$ orbital. This orbital is located ***above and below the plane*** of the BF_3 molecule.

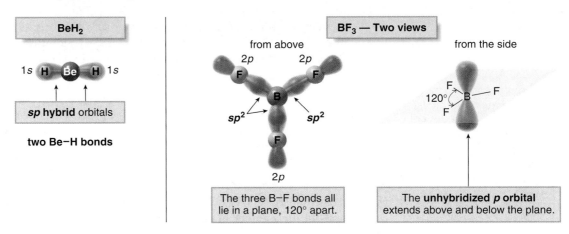

The N atom in **NH₃** and the O atom in **H₂O** are both surrounded by four groups, making them sp^3 hybridized. This means that each N–H and O–H bond in these molecules is formed by overlap of an sp^3 hybrid orbital with a $1s$ orbital from H. The lone pairs of electrons on N and O also occupy an sp^3 hybrid orbital. This is illustrated in Figure 1.12.

SAMPLE PROBLEM 1.10   What orbitals are used to form each bond in methanol, CH_3OH?

SOLUTION

To solve this problem we must draw a valid Lewis structure and then count groups around each atom. Then, use the rule to determine hybridization: two groups = sp, three groups = sp^2, and four groups = sp^3.

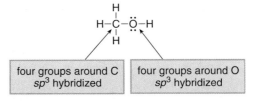

ANSWER
- All C–H bonds are formed from C_{sp^3}–H_{1s}.
- The C–O bond is formed from C_{sp^3}–O_{sp^3}.
- The O–H bond is formed from O_{sp^3}–H_{1s}.

Figure 1.12   Hybrid orbitals
of NH_3 and H_2O

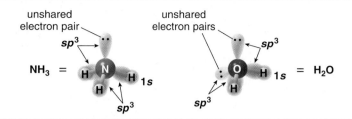

PROBLEM 1.25   What orbitals are used to form each bond in the following molecules?
a. CH_3BeH b. $(CH_3)_3B$ c. CH_3OCH_3

1.9   Ethane, Ethylene, and Acetylene

Let's now use the principles of hybridization to determine the type of bonds in **ethane, ethylene, and acetylene.**

ethane   ethylene   acetylene

1.9A   Ethane—CH_3CH_3

According to the Lewis structure for **ethane, CH_3CH_3,** each carbon atom is singly bonded to four other atoms. As a result:

- Each carbon is tetrahedral.
- Each carbon is sp^3 hybridized.

tetrahedral C atoms

All of the bonds in ethane are σ bonds. The C–H bonds are formed from the overlap of one of the three sp^3 hybrid orbitals on each carbon atom with the $1s$ orbital on hydrogen. The C–C bond is formed from the overlap of an sp^3 hybrid orbital on each carbon atom.

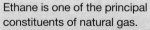

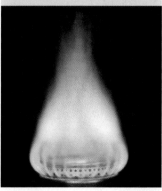

Orbital description

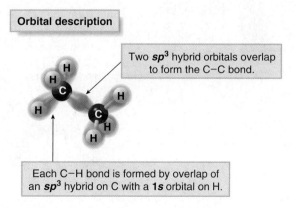

Two sp^3 hybrid orbitals overlap
to form the C–C bond.

Each C–H bond is formed by overlap of
an sp^3 hybrid on C with a $1s$ orbital on H.

Finally, making a model of ethane illustrates one additional feature about its structure. **Rotation occurs around the central C–C σ bond.** Note how the relative position of the H atoms on the adjacent CH_3 groups changes from one representation to another. This process is discussed in greater detail in Chapter 4.

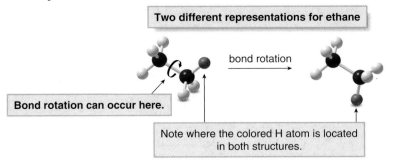

Two different representations for ethane

bond rotation

Bond rotation can occur here.

Note where the colored H atom is located in both structures.

1.9B   Ethylene—C_2H_4

Based on the Lewis structure of **ethylene, $CH_2 = CH_2$,** each carbon atom is singly bonded to two H atoms and doubly bonded to the other C atom, so each C is surrounded by three groups. As a result:

◆ Each carbon is trigonal planar (Section 1.6B).
◆ Each carbon is sp^2 hybridized.

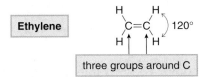

Ethylene

$120°$

three groups around C

What orbitals are used to form the two bonds of the C–C double bond? Recall from Section 1.8 that **sp^2 hybrid orbitals** are formed from **one $2s$ and two $2p$ orbitals,** leaving one $2p$ orbital **unhybridized.** Because carbon has four valence electrons, **each of these orbitals has one electron** that can be used to form a bond.

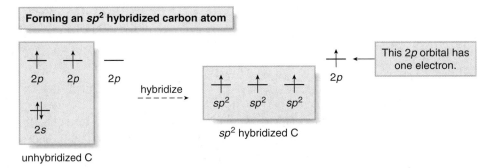

Forming an sp^2 hybridized carbon atom

$2p$ $2p$ $2p$

$2s$

hybridize

sp^2 sp^2 sp^2

$2p$

This $2p$ orbital has one electron.

unhybridized C

sp^2 hybridized C

Each C–H bond results from the end-on overlap of an sp^2 hybrid orbital on carbon and the $1s$ orbital on hydrogen. Similarly, one of the C–C bonds results from the end-on overlap of an sp^2 hybrid orbital on each carbon atom. Each of these bonds is a **σ bond.**

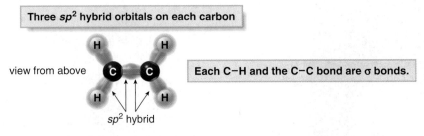

Three sp^2 hybrid orbitals on each carbon

view from above

Each C–H and the C–C bond are σ bonds.

sp^2 hybrid

The second C–C bond results from the side-by-side overlap of the $2p$ orbitals on each carbon. Side-by-side overlap creates an area of electron density above and below the plane containing the sp^2 hybrid orbitals (that is, the plane containing the six atoms in the σ bonding system).

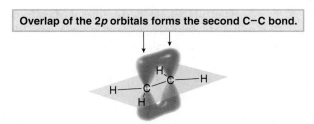

Overlap of the 2p orbitals forms the second C−C bond.

In this second bond, the electron density is **not** concentrated on the axis joining the two nuclei. This new type of bond is called a **π bond.** Because the electron density in a π bond is further from the two nuclei, **π bonds are usually weaker and therefore more easily broken than σ bonds.**

Thus, a carbon–carbon double bond has two components:

♦ a σ bond, formed by end-on overlap of two sp^2 hybrid orbitals;
♦ a π bond, formed by side-by-side overlap of two $2p$ orbitals.

Figure 1.13 summarizes the bonding observed in ethylene.

Unlike the C−C single bond in ethane, rotation about the C−C double bond in ethylene is **restricted.** It can only occur if the π bond first breaks and then reforms, a process that requires considerable energy.

All double bonds are composed of one σ and one π bond.

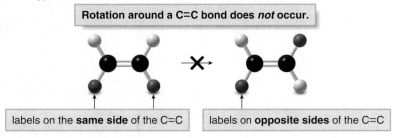

Rotation around a C=C bond does *not* occur.

labels on the **same side** of the C=C   labels on **opposite sides** of the C=C

1.9C   Acetylene—C₂H₂

Based on the Lewis structure of **acetylene, HC≡CH,** each carbon atom is singly bonded to one hydrogen atom and triply bonded to the other carbon atom, so each carbon atom is surrounded by two groups. As a result:

♦ Each carbon is linear (Section 1.6B).
♦ Each carbon is *sp* hybridized.

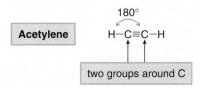

Acetylene   180°
                       H−C≡C−H

two groups around C

Figure 1.13   Summary: The σ and π bonds in ethylene

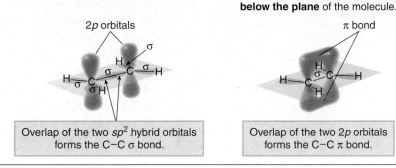

The five σ bonds are labeled.

$2p$ orbitals

Overlap of the two sp^2 hybrid orbitals forms the C−C σ bond.

The π bond extends above and below the plane of the molecule.

π bond

Overlap of the two $2p$ orbitals forms the C−C π bond.

Because acetylene produces a very hot flame on burning, it is often used in welding torches. The fire is very bright, too, so it was once used in the lamps worn by spelunkers—people who study and explore caves.

What orbitals are used to form the bonds of the C−C triple bond? Recall from Section 1.8 that *sp* **hybrid orbitals** are formed from **one 2s and one 2p orbital,** leaving **two 2p orbitals unhybridized.** Because carbon has four valence electrons, **each of these orbitals has one electron** that can be used to form a bond.

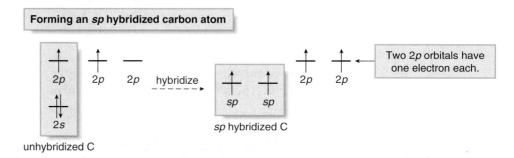

Forming an *sp* hybridized carbon atom

Each C−H bond results from the end-on overlap of an *sp* hybrid orbital on carbon and the 1s orbital on hydrogen. Similarly, one of the C−C bonds results from the end-on overlap of an *sp* hybrid orbital on each carbon atom. Each of these bonds is a **σ bond.**

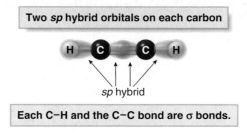

Each carbon atom also has two **unhybridized 2p orbitals** that are perpendicular to each other and to the *sp* hybrid orbitals. Side-by-side overlap between the two 2p orbitals on one carbon with the two 2p orbitals on the other carbon creates the second and third bonds of the C≡C triple bond. The electron density from one of these two bonds is above and below the axis joining the two nuclei, and the electron density from the second of these two bonds is in front of and behind the axis, so both of these bonds are **π bonds.**

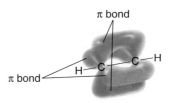

The side-by-side overlap of two p orbitals always forms a π bond.

Thus, a carbon–carbon triple bond has three components:

All triple bonds are composed of one σ and two π bonds.

◆ a σ bond, formed by end-on overlap of two *sp* hybrid orbitals;
◆ two π bonds, formed by side-by-side overlap of two sets of 2p orbitals.

Figure 1.14 summarizes the bonding observed in acetylene. Figure 1.15 summarizes the three possible types of bonding in carbon compounds.

SAMPLE PROBLEM 1.11   Answer each question for the molecule acetone, drawn below:

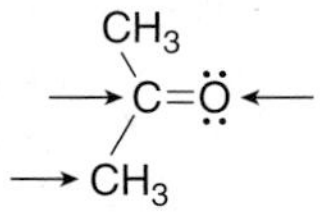

$$\begin{array}{c} CH_3 \\ \diagdown \\ \longrightarrow C = \ddot{O} \longleftarrow \\ \diagup \\ \longrightarrow CH_3 \end{array}$$
acetone

a. Determine the hybridization of the labeled atoms.
b. What orbitals are used to form the C−O double bond?
c. In what type of orbital does each lone pair reside?

SOLUTION

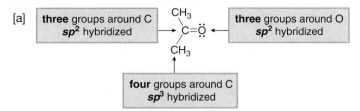

[a] **three** groups around C → **sp²** hybridized ← **three** groups around O **sp²** hybridized

four groups around C **sp³** hybridized

[b] • The σ bond is formed from the end-on overlap of $C_{sp^2} - O_{sp^2}$.
   • The π bond is formed from the side-by-side overlap of $C_{2p} - O_{2p}$.

[c] The O atom has three sp^2 hybrid orbitals.
   • One is used for the σ bond of the double bond.
   • The remaining two sp^2 hybrids are occupied by the lone pairs.

PROBLEM 1.26   Determine the hybridization around the labeled atoms in the following molecules:

a. $CH_3 - C \equiv CH$ b. (cyclohexane)$= \ddot{N} - CH_3$ c. $CH_2 = C = CH_2$

PROBLEM 1.27   Classify each bond in the following molecules as σ or π:

a. $CH_3 - \overset{\overset{\displaystyle O}{\|}}{C} - H$ b. $CH_3 - C \equiv N$ c. $H - \overset{\overset{\displaystyle O}{\|}}{C} - OCH_3$

Figure 1.14   Summary: The σ and π bonds in acetylene

The three σ bonds are labeled.

2p orbitals

2p orbitals

Overlap of the two sp hybrid orbitals forms the C–C σ bond.

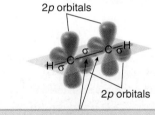

Two π bonds extend out from the axis of the linear molecule.

one π bond

second π bond

Overlap of two sets of two 2p orbitals forms two C–C π bonds.

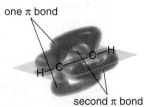

Figure 1.15   A summary of covalent bonding seen in carbon compounds

Number of groups bonded to C	Hybridization	Bond angle	Example	Observed bonding
4	sp^3	109.5°	**CH_3CH_3** ethane	one σ bond $C_{sp^3} - C_{sp^3}$
3	sp^2	120°	**$CH_2 = CH_2$** ethylene	one σ bond + one π bond $C_{sp^2} - C_{sp^2}$ $C_{2p} - C_{2p}$
2	sp	180°	**$HC \equiv CH$** acetylene	one σ bond + two π bonds $C_{sp} - C_{sp}$ $C_{2p} - C_{2p}$ $C_{2p} - C_{2p}$

1.10 Bond Length and Bond Strength

Let's now examine the relative bond length and bond strength of the C−C and C−H bonds in ethane, ethylene, and acetylene.

1.10A A Comparison of Carbon–Carbon Bonds

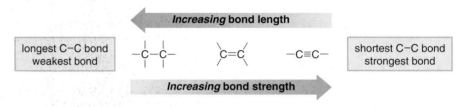

♦ As the number of electrons between two nuclei *increases,* bonds become shorter and stronger.

♦ Thus, triple bonds are shorter and stronger than double bonds, which are shorter and stronger than single bonds.

Although the SI unit of energy is the **joule** (J), organic chemists overwhelmingly report energy values in **calories** (cal). For this reason, energy values in the tables in this text are reported in calories, followed by the number of joules in parentheses. 1 cal = 4.18 J

Note the inverse relationship between bond length and bond strength. The shorter the bond, the closer the electron density is kept to the nucleus, and the harder the bond is to break. *Shorter* bonds are *stronger* bonds.

Values for bond lengths and bond strengths for CH_3CH_3, $CH_2=CH_2$ and $HC≡CH$ are listed in Table 1.3. Be careful not to confuse two related but different principles regarding multiple bonds such as C−C double bonds. **Double bonds, consisting of both a σ and a π bond, are *strong*. The π component** of the double bond, however, is usually much *weaker* **than the σ component.** This is a particularly important consideration when studying alkenes in Chapter 10.

TABLE 1.3	Bond Lengths and Bond Strengths for Ethane, Ethylene, and Acetylene	
Compound	**C−C bond length (Å)**	**Bond strength kcal/mol (kJ/mol)**
CH_3-CH_3	1.53	88 (368)
$CH_2=CH_2$	1.34	152 (635)
$HC≡CH$	1.21	200 (837)
Compound	**C−H bond length (Å)**	**Bond strength kcal/mol (kJ/mol)**
CH_3CH_2-H	1.11	98 (410)
$CH_2=C-H$ $\;\;\;\;|$ $\;\;\;\;H$	1.10	104 (435)
$HC≡C-H$	1.09	125 (523)

Increasing bond length (C−C table)
Increasing bond strength (C−C table)
Increasing bond length (C−H table)
Increasing bond strength (C−H table)

1.10B A Comparison of Carbon–Hydrogen Bonds

The length and strength of a C−H bond vary slightly depending on the hybridization of the carbon atom.

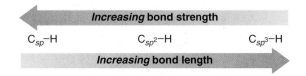

To understand why this is so, we must look at the atomic orbitals used to form each type of hybrid orbital. A single 2*s* orbital is always used, but the number of 2*p* orbitals varies with the type of hybridization. A quantity called **percent s-character** indicates the fraction of a hybrid orbital due to the 2*s* orbital used to form it.

$$\textit{sp} \text{ hybrid} \qquad \frac{\text{one } 2s \text{ orbital}}{\text{two hybrid orbitals}} = \textbf{50\% s-character}$$

$$\textit{sp}^2 \text{ hybrid} \qquad \frac{\text{one } 2s \text{ orbital}}{\text{three hybrid orbitals}} = \textbf{33\% s-character}$$

$$\textit{sp}^3 \text{ hybrid} \qquad \frac{\text{one } 2s \text{ orbital}}{\text{four hybrid orbitals}} = \textbf{25\% s-character}$$

Why should the percent *s*-character of a hybrid orbital affect the length of a C−H bond? A 2*s* orbital keeps electron density closer to a nucleus compared to a 2*p* orbital. As the **percent s-character increases,** a hybrid orbital holds its electrons closer to the nucleus, and the **bond becomes shorter and stronger.**

> **Increased percent s-character** − − → **Increased bond strength** − − → **Decreased bond length**

PROBLEM 1.28   Rank the labeled bonds in each compound in order of (a) increasing bond strength and (b) increasing bond length.

a.

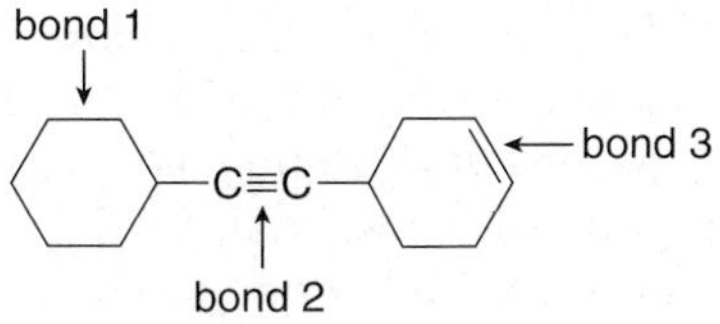

b.

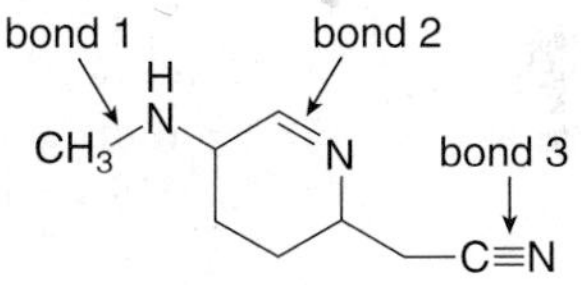

PROBLEM 1.29   Which of the labeled bonds in each pair of compounds is shorter?

a. 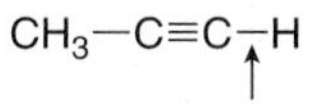 $CH_3\!-\!C\!\equiv\!C\!-\!H$ and $\underset{H}{\overset{CH_3}{C}}\!=\!CH_2$

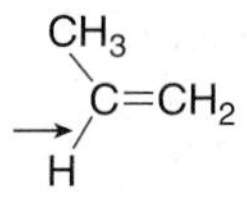

c. $CH_2\!=\!\overset{..}{N}\!-\!H$ and $CH_3\!-\!\underset{H}{\overset{..}{N}}\!-\!H$

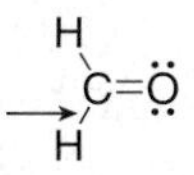

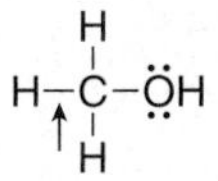

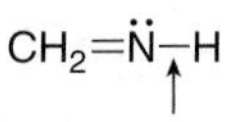

b. $C\!=\!\overset{..}{\underset{..}{O}}$ and $H\!-\!\overset{H}{\underset{H}{C}}\!-\!\overset{..}{\underset{..}{O}}H$

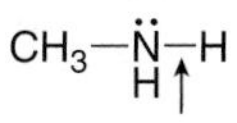

1.11   Electronegativity and Bond Polarity

Electronegativity **is a measure of an atom's attraction for electrons in a bond.** Thus, electronegativity indicates how much a particular atom "*wants*" electrons. The following trends in electronegativity are observed in the periodic table:

◆ **Electronegativity** *increases* **across a row** of the periodic table as the nuclear charge increases (excluding the noble gases).

◆ **Electronegativity** *decreases* **down a column** of the periodic table as the atomic radius increases, pushing the valence electrons further from the nucleus.

Figure 1.16
Electronegativity values
for some common elements

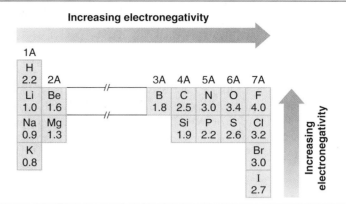

As a result, the most electronegative elements are located at the **upper right-hand corner** of the periodic table, and the least electronegative elements in the lower left-hand corner. A scale has been established to represent electronegativity values arbitrarily, from 0 to 4, as shown in Figure 1.16.

Electronegativity values are relative, so they can be used for comparison purposes only. When comparing two different elements, one is **more electronegative** than the other if it attracts electron density toward itself. One is less electronegative—**more electro*positive***—if it gives up electron density to the other element.

PROBLEM 1.30   Rank the following atoms in order of increasing electronegativity. Label the most electronegative and most electropositive atom in each group.
a. Li, Na, H   b. O, C, Be

Electronegativity values are used as a guideline to indicate whether the electrons in a bond are **equally shared** or **unequally shared** between two atoms. For example, whenever two identical atoms are bonded together, each atom attracts the electrons in the bond to the same extent. The electrons are equally shared, and the **bond is *nonpolar*.** Thus, a **carbon–carbon bond is nonpolar.** The same is true whenever two different atoms having similar electronegativities are bonded together. **C – H bonds are considered to be nonpolar,** because the electronegativity difference between C (2.5) and H (2.2) is small.

The small electronegativity difference
between C and H is ignored.

Bonding between atoms of different electronegativity values results in the **unequal sharing** of electrons. For example, in a C−O bond, the electrons are pulled away from C (2.5) toward O (3.4), the element of higher electronegativity. **The bond is *polar*, or *polar covalent*.** The bond is said to have a **dipole;** that is, **a separation of charge.**

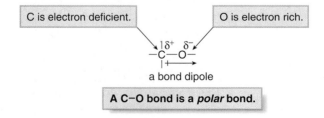

The direction of polarity in a bond is often indicated by an arrow, with the head of the arrow pointing toward the more electronegative element. The tail of the arrow, with a perpendicular line drawn through it, is drawn at the less electronegative element. Alternatively, the symbols δ^+ and δ^- indicate this unequal sharing of electron density.

> ◆ δ^+ means an atom is electron deficient (has a partial positive charge).
> ◆ δ^- means the atom is electron rich (has a partial negative charge).

PROBLEM 1.31   Show the direction of the dipole in each bond. Label the atoms with δ^+ and δ^-.

a. H—F   b. $-\overset{\displaystyle |}{\underset{\displaystyle |}{B}}-\overset{\displaystyle |}{\underset{\displaystyle |}{C}}-$ c. $-\overset{\displaystyle |}{\underset{\displaystyle |}{C}}-Li$ d. $-\overset{\displaystyle |}{\underset{\displaystyle |}{C}}-Cl$

Students often wonder how large an electronegativity difference must be to consider a bond polar. That's hard to say. We will set an arbitrary value for this difference and use it as an *approximation*. **Usually, a polar bond will be one in which the electronegativity difference between two atoms is ≥ 0.5 units.**

The distribution of electron density in a molecule can be shown using an **electrostatic potential map.** These maps are color coded to illustrate areas of high and low electron density. Electron-rich regions are indicated in red, and electron-deficient sites are indicated in blue. Regions of intermediate electron density are shown in orange, yellow, and green.

For example, an electrostatic potential map of CH_3Cl clearly indicates the polar nature of the C−Cl bond (Figure 1.17). The more electronegative Cl atom pulls electron density toward it, making it electron rich. This is indicated by the red around the Cl in the plot. The carbon is electron deficient, and this is shown with blue.

When comparing two maps, the comparison is useful only if they are plotted *using the same scale* of color gradation. For this reason, whenever we compare two plots in this text, they will be drawn side by side using the same scale. It will be difficult to compare two plots in different parts of the book, because the scale may be different. Despite this limitation, an electrostatic potential plot is a useful tool for visually evaluating the distribution of electron density in a molecule, and with care, comparing the electron density in two different molecules.

1.12   Polarity of Molecules

Thus far, we have been concerned with the polarity of one bond. To determine whether a molecule has a net dipole, we must use the following two-step procedure:

[1] Use electronegativity differences to **identify all of the polar bonds and the directions of the bond dipoles.**
[2] **Determine the geometry** around individual atoms by counting groups, and decide if individual dipoles **cancel** or **reinforce each other in space.**

Sample Problem 1.12 illustrates two different outcomes to this process.

Figure 1.17   Electrostatic potential plot of CH_3Cl

[a] Color scheme used for electron density

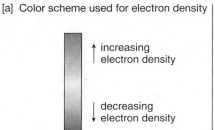

↑ increasing electron density

↓ decreasing electron density

[b] Electrostatic potential plot of CH_3Cl

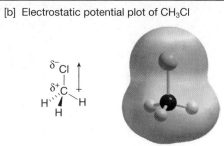

SAMPLE PROBLEM 1.12   Determine whether each of the following molecules is polar (i.e., has a *net* dipole): [a] H_2O; [b] CO_2.

SOLUTION

[a] **H_2O:** First identify polar bonds. Each O–H bond is polar because the electronegativity difference between O (3.4) and H (2.2) is large. Then, draw H_2O in three dimensions, as a bent molecule. Because the two dipoles **reinforce** (both point *up*), **H_2O has a net dipole.**

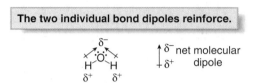

The two individual bond dipoles reinforce.

↑ δ^- net molecular
↓ δ^+ dipole

Do *not* draw H_2O as

H–Ö–H

The net dipole bisects the H–O–H bond angle.
The bent representation shows that the dipoles reinforce.

Answer: **H_2O is a polar molecule.**

A **polar molecule** has either one polar bond, or two or more bond dipoles that reinforce. A **nonpolar molecule** has either no polar bonds, or two or more bond dipoles that cancel.

Note: We must know the geometry to determine if two dipoles cancel or reinforce. For example, do *not* draw H_2O as a linear molecule, because you might think that the two dipoles cancel, when in reality, they reinforce.

[b] **CO_2:** First identify polar bonds. Each C–O bond is polar because the electronegativity difference between O (3.4) and C (2.5) is large. Then, draw CO_2 to illustrate its geometry. Two groups around C make it a linear molecule. In this case the two dipoles are equal and opposite in direction so they **cancel.** This results in a **nonpolar molecule** with **no net dipole.**

Whenever C or H is bonded to N, O, and all halogens, the bond is *polar.* Thus, the C–I bond is considered polar even though the electronegativity difference between C and I is small. Remember, electronegativity is just an approximation.

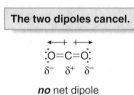

The two dipoles cancel.

:Ö=C=Ö:
δ^- δ^+ δ^-

no net dipole

Answer: **CO_2 is a nonpolar molecule.**

Electrostatic potential plots for H_2O and CO_2 appear in Figure 1.18. Additional examples of polar and nonpolar molecules are given in Figure 1.19.

Figure 1.18   Electrostatic potential plots for H_2O and CO_2

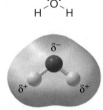

- The electron-rich (red) region is concentrated on the electronegative O atom. Both H atoms are electron deficient (blue-green).

- Both electronegative O atoms are electron rich (red) and the central C atom is electron deficient (blue).

Figure 1.19   More examples of polar and nonpolar molecules

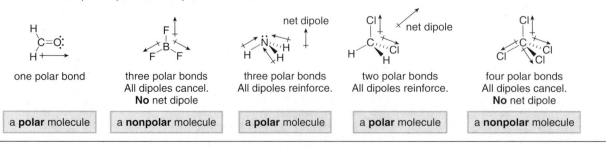

one polar bond	three polar bonds All dipoles cancel. **No** net dipole	three polar bonds All dipoles reinforce.	two polar bonds All dipoles reinforce.	four polar bonds All dipoles cancel. **No** net dipole
a **polar** molecule	a **nonpolar** molecule	a **polar** molecule	a **polar** molecule	a **nonpolar** molecule

PROBLEM 1.32   Indicate which of the following molecules is polar because it possesses a net dipole. Show the direction of the net dipole if one exists.

a. CH_3Br b. CH_2Br_2 c. CF_4 d. 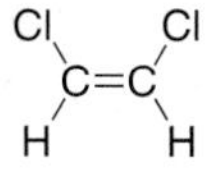 $\underset{H}{\overset{Cl}{C}}=\underset{H}{\overset{Cl}{C}}$ e. 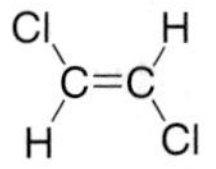 $\underset{H}{\overset{Cl}{C}}=\underset{Cl}{\overset{H}{C}}$

1.13   Capsaicin—A Representative Organic Molecule

The principles learned in this chapter apply to all organic molecules regardless of size or complexity. For example, we now know a great deal about the structure of **capsaicin,** the spicy component of hot peppers mentioned at the beginning of this chapter.

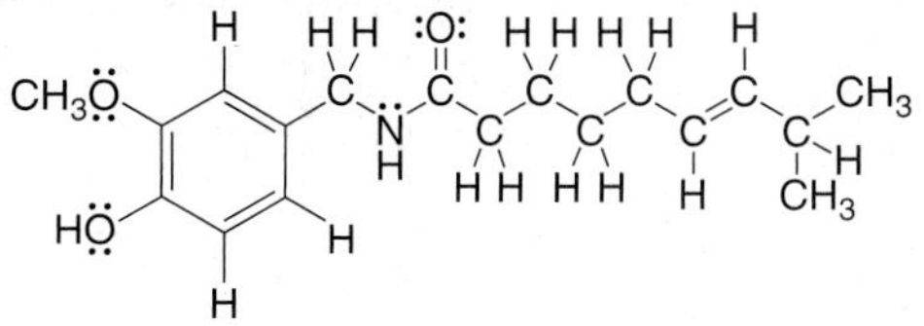

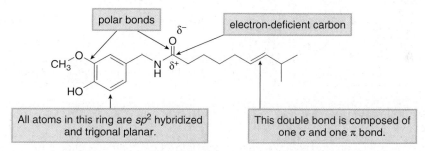

capsaicin

For example, you should be able to do all of the following:

- Convert this structure of capsaicin to a skeletal formula.
- Determine the hybridization and geometry around every atom.
- Label all polar and nonpolar bonds.
- Compare bond length and bond strength for certain C−C bonds.

Some of these concepts are illustrated in the skeletal structure for capsaicin. As we continue our study of organic chemistry you will see that these fundamental properties about a molecule determine its physical properties and its behavior in chemical reactions.

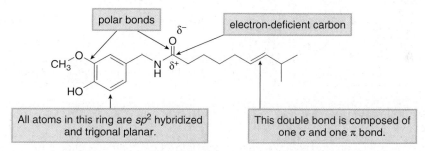

PROBLEM 1.33   Answer the following questions about capsaicin. In all questions, label different sites than those identified in the structure shown.
a. Label two polar and two nonpolar bonds.
b. Label all sp^3 hybridized C atoms.
c. Which two H atoms bear a partial positive charge (δ^+)?
d. Label three sp^2 hybridized C atoms and one sp^2 hybridized O atom.

1.14   Key Concepts—Structure and Bonding

Important Facts

- **The general rule of bonding.** Atoms "strive" to attain a complete outer shell of valence electrons (Section 1.2). H "wants" two electrons. Second-row elements "want" eight electrons.

- **Formal charge** is the difference between the number of valence electrons of an atom and the number of electrons it "owns" (Section 1.3C). See Sample Problem 1.5 for a stepwise example.
- **Curved arrow notation** shows the movement of an electron pair. The tail of the arrow always begins at an electron pair, either in a bond or a lone pair. The head points to where the electron pair "moves" (Section 1.5).
- **Electrostatic potential plots** are color-coded maps of electron density, indicating electron-rich and electron-deficient regions (Section 1.11).

The Importance of Lewis Structures (Sections 1.3, 1.4)

A properly drawn Lewis structure shows the number of bonds and lone pairs present around each atom in a molecule. In a valid Lewis structure, each H has two electrons, and each second-row element has no more than eight. This is the first step needed to determine many properties of a molecule.

	Geometry	[linear, trigonal planar, or tetrahedral] (Section 1.6)
Lewis structure	**Hybridization**	[sp, sp^2, or sp^3] (Section 1.8)
	Types of bonds	[single, double, or triple] (Sections 1.3, 1.9)

Resonance (Section 1.5)

The basic principles:
- Resonance exists when a compound cannot be represented by a single Lewis structure.
- Resonance structures differ only in the position of nonbonded electrons and π bonds, not atoms.
- The resonance hybrid is the only accurate representation for a resonance-stabilized compound. A hybrid represents the compound better than any single resonance structure because electron density is delocalized.

The difference between resonance structures and isomers:
- Two **isomers** differ in the arrangement of *both* atoms and electrons.
- **Resonance structures** differ *only* in the arrangement of electrons.

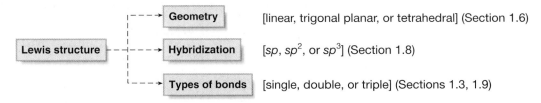

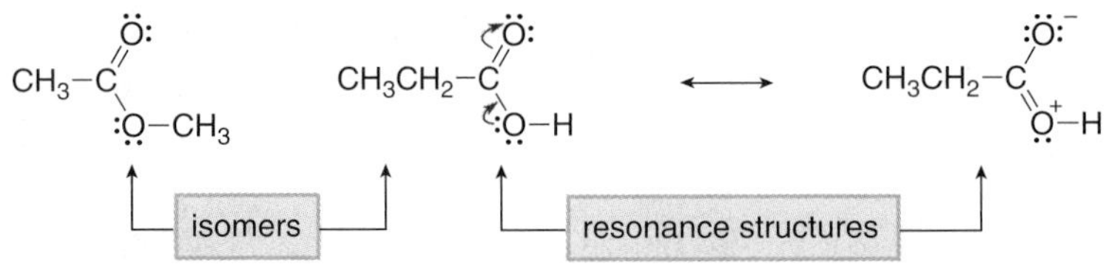

Geometry and Hybridization

The number of groups around an atom determines both its geometry (Section 1.6) and hybridization (Section 1.8).

Number of groups	Geometry	Bond angle (°)	Hybridization
2	linear	180	sp
3	trigonal planar	120	sp^2
4	tetrahedral	109.5	sp^3

Drawing Organic Molecules (Section 1.7)

- Shorthand methods are used to abbreviate the structure of organic molecules.

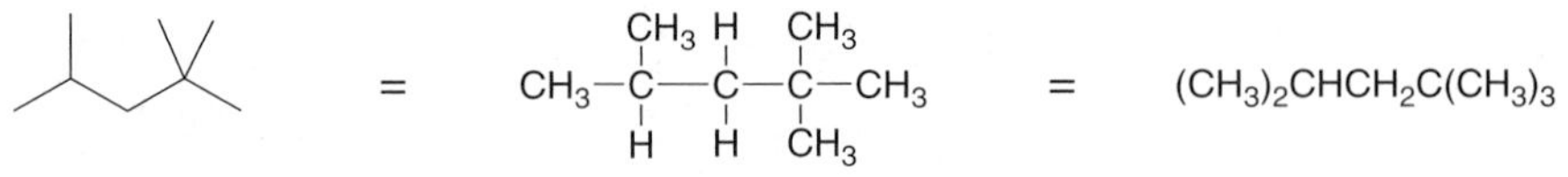

skeletal structure   =   isooctane   =   condensed structure

$(CH_3)_2CHCH_2C(CH_3)_3$

- A carbon bonded to four atoms is tetrahedral. The best way to represent a tetrahedron is to draw two bonds in the plane, one bond in front, and one bond behind.

Bond Length

- Bond length decreases across a row and increases down a column of the periodic table (Section 1.6A).
- Bond length decreases as the number of electrons between two nuclei increases (Section 1.10A).
- Bond length decreases as the percent s-character increases (Section 1.10B).
- Bond length and bond strength are inversely related. In general, shorter bonds are stronger bonds (Section 1.10).
- Sigma (σ) bonds are generally stronger than π bonds (Section 1.9).

Electronegativity and Polarity (Sections 1.11, 1.12)

- Electronegativity increases across a row and decreases down a column of the periodic table.
- A polar bond results when two atoms of different electronegativity values are bonded together. Whenever C or H is bonded to N, O, or any halogen, the bond is polar.
- A polar molecule has either one polar bond, or two or more bond dipoles that reinforce.

Problems

Ionic and Covalent Bonding

1.34 Label each bond in the following compounds as ionic or covalent.
     a. NaF   b. BrCl   c. HCl   d. CH_3NH_2 e. $NaOCH_3$

Lewis Structures and Formal Charge

1.35 Assign formal charges to each carbon atom in the given species. All lone pairs have been drawn in.

     a. $CH_2{=}\ddot{C}H$ b. H–$\ddot{\ddot{C}}$–H   c. H–$\dot{C}$–H   d. H–C–C

1.36 Assign formal charges to each N and O atom in the given molecules. All lone pairs have been drawn in.
     a. CH_3–$\ddot{\ddot{N}}$–CH_3 c. CH_3–N≡N: e. CH_3–$\ddot{O}$·

     b. :$\ddot{N}$=N=$\ddot{N}$:   d. CH_3–$\overset{\displaystyle :OH}{\overset{||}{C}}$–$CH_3$ f. CH_3–$\ddot{N}$=$\ddot{O}$

1.37 Draw one valid Lewis structure for each compound. Assume the atoms are arranged as drawn.

     a. CH_2N_2 H C N N   c. CH_3CNO H C C N O   e. HCO_3^- H O C O
                    H   H

     b. CH_3NO_2 H C N O   d. HCO_2^- O   f. $^-CH_2CN$ H C C N
                   H O   H C O   H

1.38 Draw a valid Lewis structure for each compound.
     a. N_2 b. $(CH_3OH_2)^+$ c. $(CH_3CH_2)^-$ d. HNNH   e. H_6BN

Isomers and Resonance Structures

1.39 Draw all possible isomers for each molecular formula.
     a. C_3H_7Cl (two isomers)   b. C_2H_4O (three isomers)   c. C_3H_9N (four isomers)

1.40 Draw Lewis structures for the nine isomers having molecular formula C_3H_6O, with all atoms having a zero formal charge.

1.41 With reference to compound **A** drawn below, label each compound as an isomer, a resonance structure, or neither.

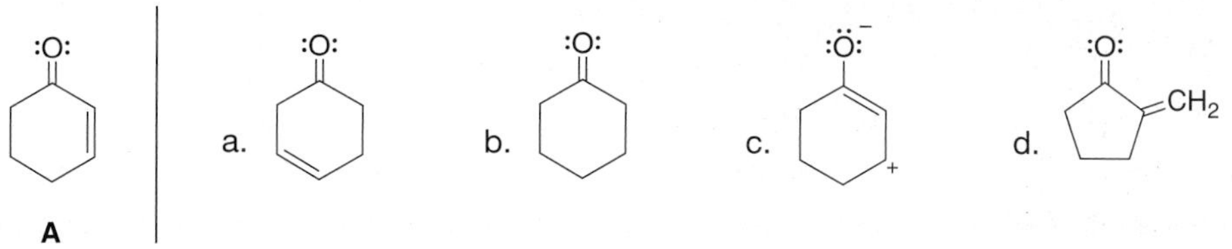

1.42 How are the molecules or ions in each pair related? Classify them as resonance structures, isomers, or neither.

a. $CH_3-\overset{..}{\underset{..}{O}}-CH_2CH_3$ and $CH_3-\overset{\underset{|}{CH_3}}{\underset{\overset{|}{H}}{C}}-\overset{..}{O}H$

c. $CH_2=CH-\overset{-}{\overset{..}{C}H}-CH_3$ and $\overset{-}{\overset{..}{C}H_2}-CH=CH-CH_3$

b. ☐ and $CH_3-\overset{}{\underset{\underset{H}{|}}{C}}=\overset{}{\underset{\underset{H}{|}}{C}}-CH_3$

d. $CH_3CH_2CH_3$ and $CH_3CH_2\overset{..}{C}H_2$

1.43 Add curved arrows to show how the first resonance structure can be converted into the second.

a. $CH_3-\overset{\overset{H}{|}}{\underset{+}{C}}-\overset{..}{N}-CH_3$ ⟷ $CH_3-\overset{\overset{H}{|}}{C}=\overset{+}{\underset{\underset{H}{|}}{N}}-CH_3$

c. (phenyl)–$\overset{..}{C}H_2$ ⟷ (cyclohexadienyl)–CH_2

b. $\overset{\overset{\displaystyle :O:}{||}}{H-C}-\overset{..}{N}H_2$ ⟷ $\overset{\overset{\displaystyle :\overset{-}{\overset{..}{O}}:}{|}}{H-C}=\overset{+}{N}H_2$

1.44 Follow the curved arrows to draw a second resonance structure for each species.

a. $CH_3-\overset{+}{\overset{..}{N}}\equiv N:$ ⟷

b. $CH_3-\overset{\overset{\displaystyle :\overset{-}{\overset{..}{O}}:}{|}}{C}=CH-\overset{+}{CH_2}$ ⟷

c. (methylcyclohexadienyl cation, CH_3) ⟷

d. (cyclohexadienyl with $\overset{+}{N}H_2$) ⟷

1.45 Draw a second resonance structure for each ion.

a. $CH_3-\overset{\overset{\displaystyle :O:}{||}}{C}-\overset{..}{\overset{-}{O}}:$

b. $CH_2=\overset{+}{N}H_2$

c. (cyclohexanone radical anion, :O: .⁻)

d. $H-\overset{\overset{\displaystyle +}{\overset{\displaystyle :O}{|}}H}{C}-H$

1.46 For each ion in Problem 1.45 draw the resonance hybrid.

1.47 Draw all reasonable resonance structures for each species.

a. O_3

c. N_3^-

e. (cyclohexadienyl cation, +)

g. (pyranyl cation :O: with +)

b. NO_3^- (a central N atom)

d. $CH_3-\overset{\overset{\displaystyle :O:}{||}}{C}-\overset{-}{\overset{..}{C}H}-\overset{\overset{\displaystyle :O:}{||}}{C}-CH_3$

f. $CH_2=CH-\overset{-}{\overset{..}{C}H}-CH=CH_2$

1.48 Draw four additional resonance structures for the following cation. Then draw the resonance hybrid.

(cyclohexadienyl cation with =CH_2, +)

1.49 Rank the resonance structures in each group in order of increasing contribution to the resonance hybrid. Label the resonance structure that contributes the most as **3** and the resonance structure that contributes the least as **1**. Label the intermediate contributor as **2**.

a. $CH_3-\overset{\overset{\displaystyle :O:}{||}}{C}-\overset{..}{O}-CH_3$ ⟷ $CH_3-\overset{\overset{\displaystyle :\overset{-}{\overset{..}{O}}:}{|}}{\underset{+}{C}}-\overset{..}{O}-CH_3$ ⟷ $CH_3-\overset{\overset{\displaystyle :\overset{-}{\overset{..}{O}}:}{|}}{C}=\overset{+}{\overset{..}{O}}-CH_3$

b. $CH_3-\overset{\overset{\displaystyle CH_3}{|}}{C}=\overset{..}{N}-\overset{..}{N}H_2$ ⟷ $CH_3-\overset{\overset{\displaystyle CH_3}{|}}{\overset{..}{C}}-\overset{..}{N}=\overset{+}{N}H_2$ ⟷ $CH_3-\overset{\overset{\displaystyle CH_3}{|}}{\underset{+}{C}}-\overset{..}{N}-\overset{..}{N}H_2$

1.50 Two compounds of molecular formula H_3NO exist. Draw Lewis structures for both compounds and decide which molecule is more stable.

Geometry

1.51 Predict all bond angles in each compound.

a. CH_3Cl b. NH_2OH c. $CH_2=NCH_3$ d. $HC\equiv CCH_2OH$ e. (chlorobenzene, Cl)

1.52 Predict the geometry around each indicated atom.

a. $CH_3CH_2\underset{\uparrow}{C}H_2CH_3$

c. (benzene, ←)

e. $CH_3-\overset{\overset{\displaystyle O}{||}}{\underset{\uparrow}{C}}-OH$

b. $(CH_3)_2\underset{\uparrow}{N}^-$

d. $\underset{\uparrow}{B}F_4^-$

f. $(CH_3)_3\underset{\uparrow}{N}$

1.53 Draw each compound in three dimensions, using solid lines, wedges, and dashes to illustrate the position of atoms.

a. $CH_3CH_2CH_3$ b. CH_3OH c. $(CH_3)_2NH$

Drawing Organic Molecules

1.54 How many hydrogens are present around each labeled atom?

a.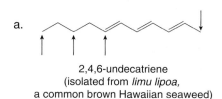

2,4,6-undecatriene
(isolated from *limu lipoa,*
a common brown Hawaiian seaweed)

b.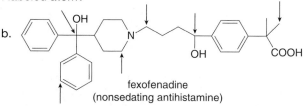

fexofenadine
(nonsedating antihistamine)

1.55 Draw in all the carbon and hydrogen atoms in each molecule.

a.   d.   g.

ethambutol
(drug used to treat tuberculosis)

b.   e.

menthol
(isolated from peppermint oil)

h.

c.   f.

myrcene
(isolated from bayberry)

estradiol
(a female sex hormone)

1.56 Convert each molecule into a skeletal structure.

a. (CH₃)₂CHCH₂CH₂CH(CH₃)₂   c. (CH₃)₃C(CH₂)₄CH₂CH₃   e.

limonene
(oil of lemon)

b. CH₃CH(Cl)CH(OH)CH₃   d.

f. CH₃(CH₂)₂C(CH₃)₂CH(CH₃)CH(CH₃)CH(Br)CH₃

1.57 Convert the following condensed formulas into Lewis structures.
a. CH₃CH₂COOH   c. CH₃COCH₂Br   e. (CH₃)₃CCHO
b. CH₃CONHCH₃   d. (CH₃)₃COH   f. CH₃COCl

1.58 Draw in all the hydrogen atoms and nonbonded electron pairs in each ion.

a.   b.   c.

Hybridization

1.59 Predict the hybridization and geometry around each indicated atom.
a. CH₃C̄H₂   c. (CH3)₃O⁺   e. CH₃–C≡C–H   g. CH₃CH=C=CH₂
   ↑   ↑   ↑   ↑ ↑

b.   d.   –CH₂Cl   f. CH₂=NOCH₃
   ←   ↑   ↑

1.60 What orbitals are used to form each indicated bond? For multiple bonds, indicate the orbitals used in individual bonds.

a.   b.   c.   d. H–C≡C–C=N–CH₃
                                                ↑ ↑ ↑ ↑
                                                        H

1.61 Citric acid and zingerone are two naturally occurring compounds. What orbitals are used to form each labeled bond?

a.

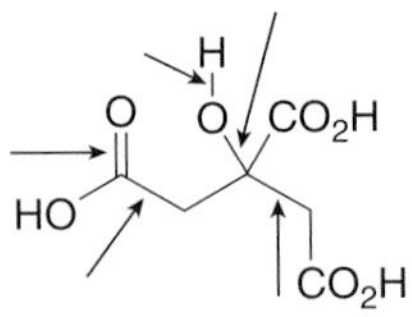

b.

citric acid
(responsible for the tartness
of citrus fruits)

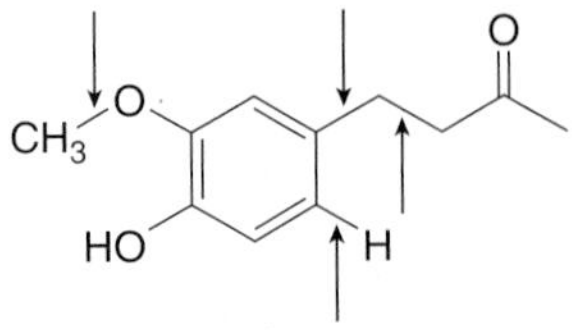

zingerone
(responsible for the pungent
taste of ginger)

1.62 Ketene, $CH_2=C=O$, is an unusual organic molecule that has a single carbon atom doubly bonded to two different atoms. Determine the hybridization of both C atoms and the O in ketene. Then, draw a diagram showing what orbitals are used to form each bond (similar to Figures 1.13 and 1.14).

1.63 Consider the unstable cation and anion drawn below.

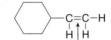   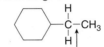

a. What is the hybridization of each carbon atom in these ions?
b. What orbitals are used to form both bonds in each carbon–carbon double bond?

Bond Length and Strength

1.64 Rank the following bonds in order of *increasing* bond length.

a.

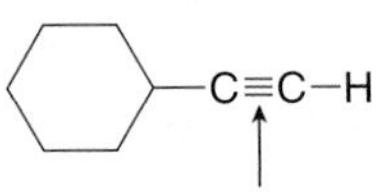

b. 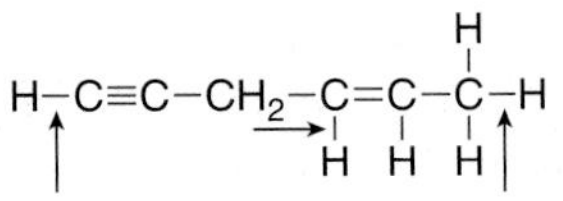 $H-C{\equiv}C-CH_2-C=C-C-H$

1.65 Indicate the longer labeled bond in each compound.

a. 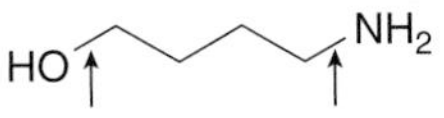 HO␣␣␣␣␣NH₂   b. 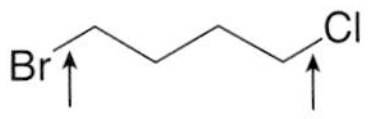 Br␣␣␣␣␣Cl   c.

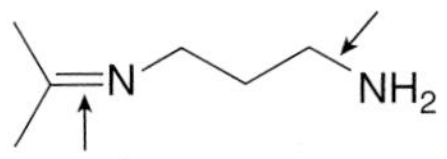

1.66 The labeled C–C bonds in the following compound are somewhat different in length. Predict which bond is shorter and explain why.

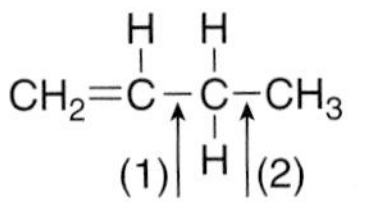

$$CH_2=C-C-CH_3$$
(1)   (2)

1.67 A σ bond formed from two sp^2 hybridized C atoms is stronger than a σ bond formed from two sp^3 hybridized C atoms. Explain.

1.68 Two useful organic compounds that contain Cl atoms are drawn below. Vinyl chloride is the starting material used to prepare poly(vinyl chloride), a plastic used in insulation, pipes, and bottles. Chloroethane (ethyl chloride) is a local anesthetic. Why is the C–Cl bond of vinyl chloride stronger than the C–Cl bond in chloroethane?

$CH_2=CHCl$ CH_3CH_2Cl

vinyl chloride   chloroethane
(ethyl chloride)

Bond Polarity

1.69 Use the symbols δ⁺ and δ⁻ to indicate the polarity of the labeled bonds.

a. Br–Cl   b. NH₂–OH   c. CH₃–NH₂   d. 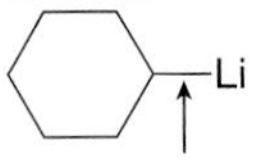   –Li

1.70 Label the polar bonds in each molecule. Indicate the direction of the net dipole (if there is one).

a. CHBr₃
b. CH₃CH₂OCH₂CH₃
c. CBr₄

d. 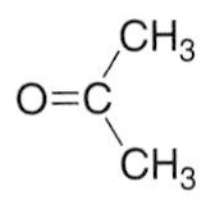 $O=C$ with CH_3

e.

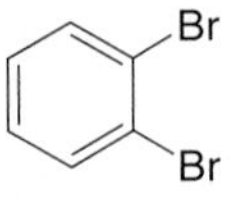

f.

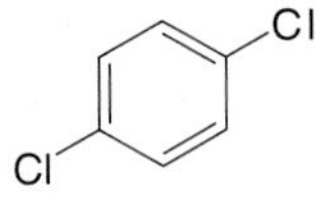

1.71 The net dipole of $CHCl_3$ is smaller than the net dipole of CH_2Cl_2. Explain why.

General Questions

1.72 Answer the following questions about acetonitrile ($CH_3C{\equiv}N{:}$).
a. Determine the hybridization of both C atoms and the N atom.
b. Label all bonds as σ or π.
c. In what type of orbital does the lone pair on N reside?

d. Label all bonds as polar or nonpolar.
e. Draw a diagram showing what orbitals are used to form each bond.

1.73 Benzene is the simplest member of a whole class of compounds called aromatic hydrocarbons.

 benzene

    a. How is each carbon atom hybridized?
    b. What is the geometry around each carbon atom? What is the overall geometry of the benzene ring?
    c. Draw a diagram showing the orbitals used to join the carbon atoms of the ring.
    d. Follow the indicated curved arrow notation to draw a second resonance structure.

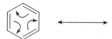

    e. Benzene and other aromatic hydrocarbons are shown in Chapter 17 to be very stable. Offer an explanation.

1.74 The principles of this chapter can be applied to organic molecules of any size. Answer the following questions about amoxicillin, an antibiotic from the penicillin family.

amoxicillin

    a. Predict the hybridization and geometry around each indicated atom.
    b. Label five polar bonds using the symbols δ^+ and δ^-.
    c. Draw a skeletal structure.
    d. How many π bonds does amoxicillin have? Label them.
    e. Find a C−H bond containing a carbon atom having a hybrid orbital with 33% s-character.

1.75

nicotine

    a. What is the hybridization of each N atom in nicotine?
    b. What is the geometry around each N atom?
    c. In what type of orbital does the lone pair on each N atom reside?

1.76 CH_3^+ and CH_3^- are two highly reactive carbon species.
    a. What is the predicted hybridization and geometry around each carbon atom?
    b. Two electrostatic potential plots are drawn for these species. Which ion corresponds to which diagram and why?

    A   **B**

Challenge Problems

1.77 When two carbons having different hybridization are bonded together, the C−C bond contains a slight dipole. In a C_{sp^2}−C_{sp^3} bond, what is the direction of the dipole? Which carbon is considered more electronegative?

1.78 Draw as many resonance structures as you can for the following compound.

O_2N—⟨benzene ring⟩—NH_2

1.79 Draw all possible isomers having molecular formula C_4H_8 that contain one π bond.

1.80 One of the longest C−C single bonds known is present in **A** below. Why is this bond so long?

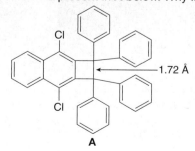

1.72 Å

A

Acids and Bases

Aspirin is one of the most widely used over-the-counter drugs. Whether you purchase Anacin, Bufferin, Bayer, or a generic, the active ingredient is the same: **acetylsalicylic acid.** Aspirin, a synthetic compound that does not occur in nature, was first marketed to the general public in 1899, and is now sold under more than 100 trade names throughout the world. Like many drugs, aspirin undergoes a proton transfer reaction after ingestion. In Chapter 2 we will learn about acidity and the role of acid–base reactions in aspirin's chemistry.

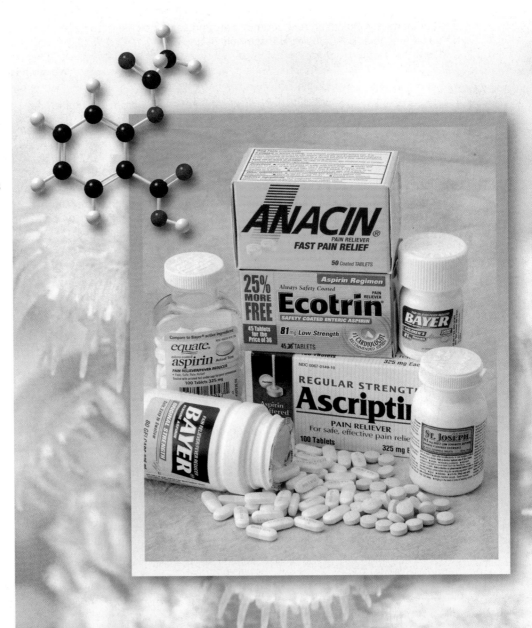

Chemical terms such as *anion* and *cation* may be unfamiliar to most nonscientists, but *acid* has found a place in everyday language. Commercials advertise the latest remedy for the heartburn caused by excess stomach *acid*. The nightly news may report the latest environmental impact of *acid* rain. Wine lovers know that wine sours because its alcohol has turned to *acid*. *Acid* comes from the Latin word *acidus,* meaning sour, because when tasting compounds was a routine method of identification, these compounds were sour.

In Chapter 2, we will concentrate on two definitions of acids and bases: the **Brønsted–Lowry** definition, which describes acids as ***proton donors*** and bases as ***proton acceptors,*** and the **Lewis definition,** which describes acids as ***electron pair acceptors*** and bases as ***electron pair donors.***

2.1 Brønsted–Lowry Acids and Bases

The general words "acid" and "base" usually mean a *Brønsted–Lowry* acid and *Brønsted–Lowry* base.

The Brønsted–Lowry definition describes acidity in terms of protons: positively charged **hydrogen ions, H$^+$.**

◆ A Brønsted–Lowry acid is a *proton donor.*
◆ A Brønsted–Lowry base is a *proton acceptor.*

H$^+$ = proton

A Brønsted–Lowry acid must contain a *hydrogen* atom, but its overall charge doesn't matter. This definition of an acid is often familiar to students, because many of the inorganic acids in general chemistry are Brønsted–Lowry acids. We will use the symbol **H–A** for a general Brønsted–Lowry acid.

H–A = general Brønsted–Lowry acid.
B: = general Brønsted–Lowry base.

A Brønsted–Lowry base must be able to form a bond to a proton. Because a proton has no electrons, **a base must contain an "available" electron pair** that can be easily donated to form a new bond. These include **lone pairs** or electron pairs in π **bonds.** We will use the symbol **B:** for a general Brønsted–Lowry base.

Some examples of Brønsted–Lowry acids and bases are given in Figure 2.1.

Charged species such as $^-$OH and $^-$NH$_2$ are used as **salts,** with cations such as Li$^+$, Na$^+$, or K$^+$ to balance the negative charge. These cations are called **counterions** or **spectator ions. The identity of the cation is usually inconsequential.** For this reason, the counterion is often omitted.

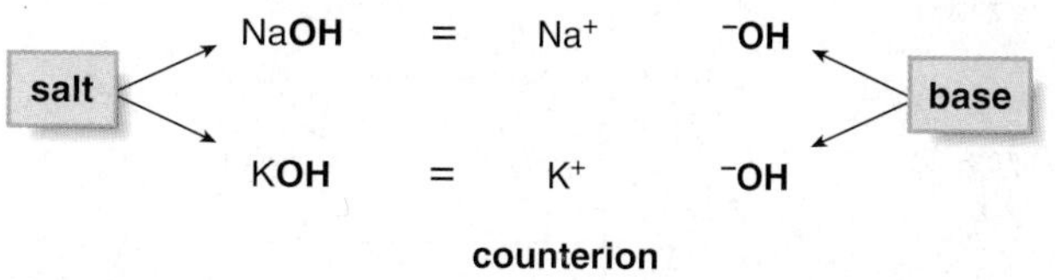

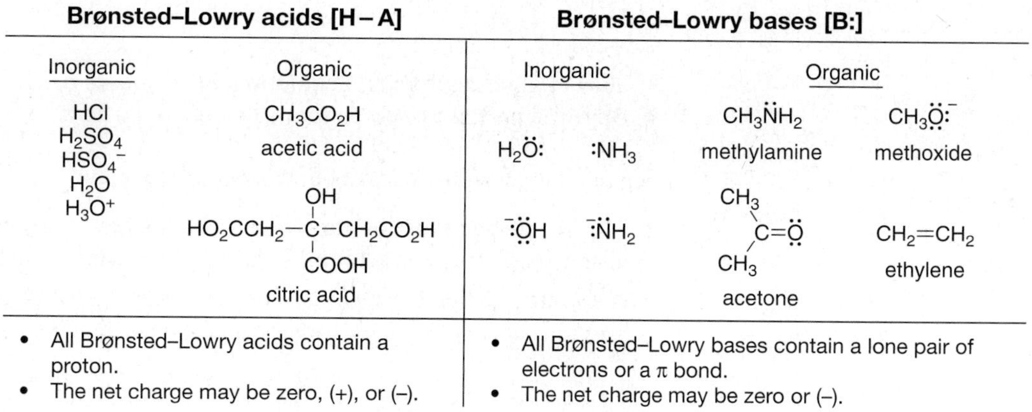

$$
\text{salt} \left\{
\begin{array}{l}
\text{NaOH} \quad = \quad \text{Na}^+ \quad {}^-\text{OH} \\
\text{KOH} \quad = \quad \text{K}^+ \quad {}^-\text{OH}
\end{array}
\right\} \text{base}
$$

counterion

Figure 2.1 Examples of Brønsted–Lowry acids and bases

Brønsted–Lowry acids [H–A]		Brønsted–Lowry bases [B:]	
Inorganic	Organic	Inorganic	Organic
HCl	CH$_3$CO$_2$H		CH$_3\ddot{\text{N}}$H$_2$ CH$_3\ddot{\text{O}}$:$^-$
H$_2$SO$_4$	acetic acid	H$_2\ddot{\text{O}}$: :NH$_3$	methylamine methoxide
HSO$_4^-$			
H$_2$O			CH$_3$
H$_3$O$^+$	OH	:$\ddot{\text{O}}$H :$\ddot{\text{N}}$H$_2$	\ C=$\ddot{\text{O}}$ CH$_2$=CH$_2$
	HO$_2$CCH$_2$—C—CH$_2$CO$_2$H		/ ethylene
	COOH		CH$_3$
	citric acid		acetone
• All Brønsted–Lowry acids contain a proton. • The net charge may be zero, (+), or (–).		• All Brønsted–Lowry bases contain a lone pair of electrons or a π bond. • The net charge may be zero or (–).	

Compounds like H_2O and CH_3OH that contain both hydrogen atoms and lone pairs may be either an acid or a base, depending on the particular reaction. These fundamental principles are true no matter how complex the compound. For example, the addictive pain reliever **morphine** is a Brønsted–Lowry acid because it contains many hydrogen atoms. It is also a Brønsted–Lowry base because it has lone pairs on O and N, and four π bonds.

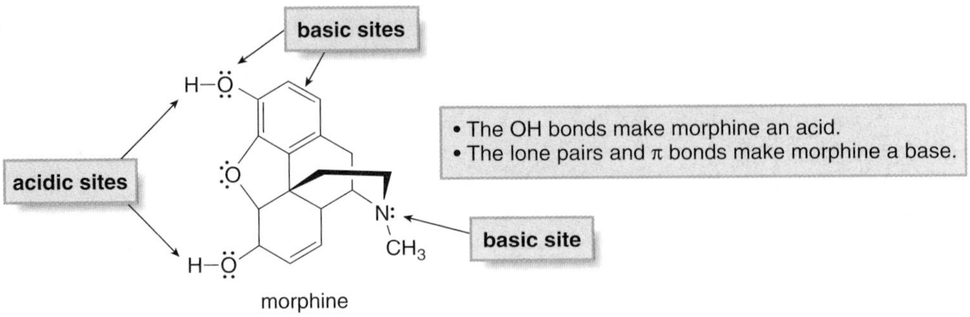

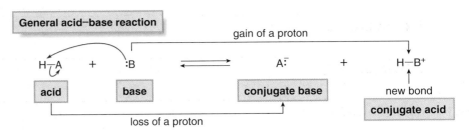

morphine

- The OH bonds make morphine an acid.
- The lone pairs and π bonds make morphine a base.

PROBLEM 2.1 a. Which compounds are Brønsted–Lowry acids: HBr, NH_3, CCl_4?
b. Which compounds are Brønsted–Lowry bases: CH_3CH_3, $(CH_3)_3CO^-$, $HC \equiv CH$?
c. Classify each compound as an acid, a base, or both: CH_3CH_2OH, $CH_3CH_2CH_2CH_3$, $CH_3CO_2CH_3$.

2.2 Reactions of Brønsted–Lowry Acids and Bases

A Brønsted–Lowry acid–base reaction results in transfer of a proton from an acid to a base. For this reason, these acid–base reactions are also called *proton transfer reactions.* Proton transfer reactions are fundamental to the study of organic chemistry.

Consider, for example, the reaction of the acid $H-A$ with the base :B. **In an acid–base reaction, one bond is broken and one is formed.**

♦ **The electron pair of the base B: forms a new bond to the proton of the acid.**

♦ **The acid $H-A$ loses a proton, leaving the electron pair in the $H-A$ bond on A.**

> Recall from Section 1.5 that a curved arrow shows the movement of an **electron pair.** The tail of the arrow always begins at an electron pair and the head points to where that electron pair "moves."

General acid–base reaction

gain of a proton

$$H{-}A \;+\; :B \;\rightleftharpoons\; A^{-} \;+\; H{-}B^{+}$$

acid   base   conjugate base   conjugate acid

new bond

loss of a proton

This "movement" of electrons in reactions can be illustrated using curved arrow notation. Because **two electron pairs** are involved in this reaction, **two curved arrows** are needed. Two products are formed.

♦ Loss of a proton from an acid forms its *conjugate base.*
♦ Gain of a proton by a base forms its *conjugate acid.*

Keep in mind two other facts about this general reaction:

♦ The **net charge must be the same** on both sides of any equation. In this example, the net charge on each side is zero. Individual charges can be calculated using formal charges.

♦ **A double reaction arrow** is used between starting materials and products to indicate that the reaction can proceed in the forward and reverse directions. These are **equilibrium arrows.**

Two examples of proton transfer reactions are drawn here with curved arrow notation.

A double reaction arrow indicates equilibrium.

equilibrium arrows

Remove H⁺ from an acid to form its conjugate base. Add H⁺ to a base to form its conjugate acid.

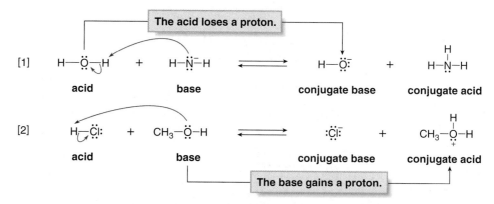

◆ **Brønsted–Lowry acid–base reactions always result in the transfer of a proton from an acid to a base.**

The ability to identify and draw a conjugate acid or base from a given starting material is a necessary skill, illustrated in Sample Problems 2.1 and 2.2.

SAMPLE PROBLEM 2.1   a. What is the conjugate acid of CH_3O^-?
b. What is the conjugate base of NH_3?

SOLUTION

[a] **Add H⁺ to CH_3O^- to form its conjugate acid.**

Add a proton to a lone pair.

$CH_3-\overset{..}{\underset{..}{O}}{:}^- \longrightarrow CH_3-\overset{..}{\underset{..}{O}}-H$

base   conjugate acid

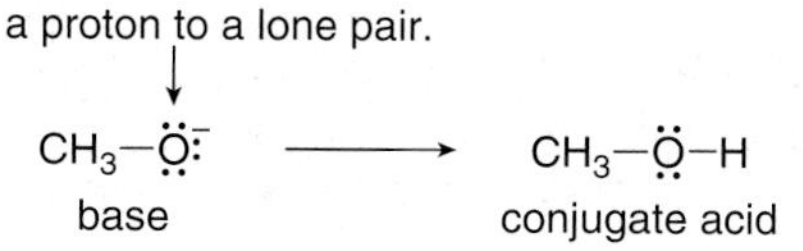

[b] **Remove H⁺ from NH_3 to form its conjugate base.**

Leave the electron pair on N.

$H-\overset{H}{\underset{H}{N}}-H \longrightarrow H-\overset{..}{\underset{H}{N}}{:}^-$

acid   conjugate base

Remove a proton.

PROBLEM 2.2   a. Draw the conjugate acid of each base: NH_3, Cl^-, $(CH_3)_2C{=}O$.
b. Draw the conjugate base of each acid: HBr, HSO_4^-, CH_3OH.

PROBLEM 2.3   The presence of a π bond also makes a compound a base. With this in mind, draw the conjugate acid of ethylene, $CH_2{=}CH_2$. What is ethylene's conjugate base?

SAMPLE PROBLEM 2.2   Label the acid and base, and the conjugate acid and base in the following reaction. Use curved arrow notation to show the movement of electron pairs.

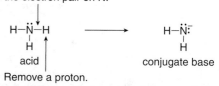

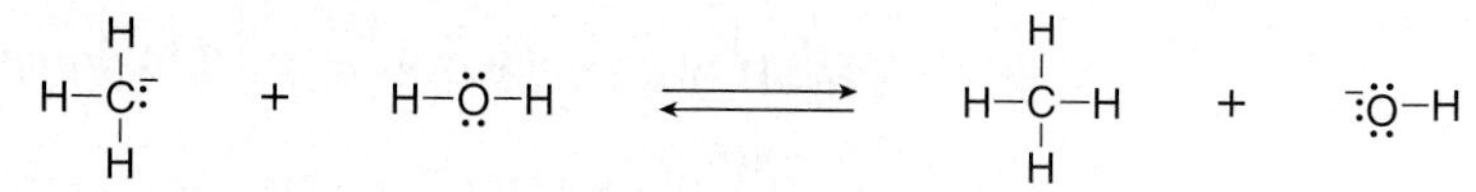

SOLUTION

In this example, CH_3^- is the base because it accepts a proton, forming its conjugate acid, CH_4. H_2O is the acid because it donates a proton, forming its conjugate base, ^-OH.

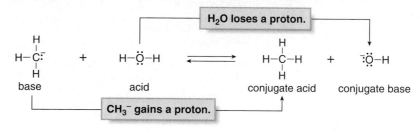

Two curved arrows are needed. One shows that the lone pair on CH_3^- bonds to a proton of H_2O, and the second shows that the electron pair in the $O-H$ bond remains on O.

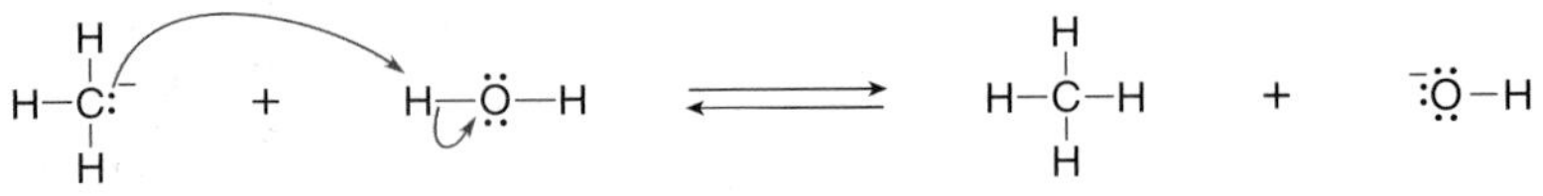

$$H-\overset{H}{\underset{H}{C}}{:}^- \quad + \quad H-\overset{\cdot\cdot}{\underset{\cdot\cdot}{O}}-H \quad \rightleftharpoons \quad H-\overset{H}{\underset{H}{C}}-H \quad + \quad {:}\overset{\cdot\cdot}{\underset{\cdot\cdot}{O}}-H$$

PROBLEM 2.4   Label the acid and base, and the conjugate acid and base in the following reactions. Use curved arrows to show the movement of electron pairs.

a. $HCl \quad + \quad H_2O \quad \rightleftharpoons \quad Cl^- \quad + \quad H_3O^+$

b. 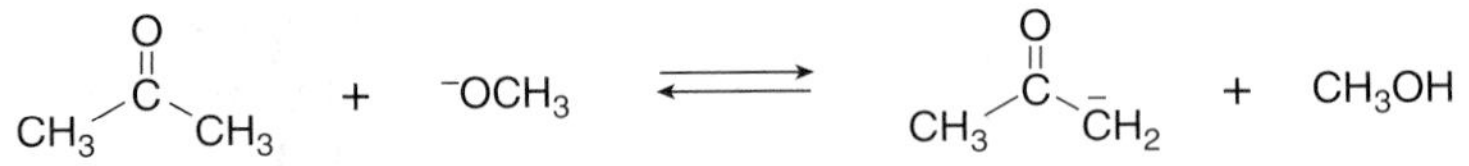 $\quad + \quad {}^-OCH_3 \quad \rightleftharpoons \quad$ $\quad + \quad CH_3OH$

In all proton transfer reactions, the **electron-rich base** donates an electron pair to the acid, which usually has a polar $H-A$ bond. Thus, the H of the acid bears a partial positive charge, making it **electron deficient.** This is the first example of a general pattern of reactivity.

◆ **Electron-rich species react with electron-deficient ones.**

Given two starting materials, how do you know which is the acid and which is the base in a proton transfer reaction? The following generalizations will help to decide this in many reactions:

[1] Common acids and bases introduced in general chemistry will often be used in the same way in organic reactions. HCl and H_2SO_4 are strong acids, and ^-OH is a strong base.
[2] When only one starting material contains a hydrogen, it must be the acid. If only one starting material has a lone pair or a π bond, it must be the base.
[3] A starting material with a net positive charge is usually the acid. A starting material with a negative charge is usually the base.

PROBLEM 2.5   Draw the products of each proton transfer reaction.

a. $CCl_3CO_2H \quad + \quad {}^-OCH_3 \rightleftharpoons$ c. $CH_3NH_2 \quad + \quad HCl \rightleftharpoons$

b. $H-C{\equiv}C-H \quad + \quad H{:}^- \rightleftharpoons$ d. $CH_3CH_2OH \quad + \quad H_2SO_4 \rightleftharpoons$

2.3   Acid Strength and pK_a

Acid strength is the tendency of an acid to donate a proton.

◆ **The more readily a compound donates a proton, the stronger the acid.**

Acidity is measured by an equilibrium constant. When a Brønsted–Lowry acid $H-A$ is dissolved in water, an acid–base reaction occurs, and an equilibrium constant K_{eq} can be written for the reaction.

Dissolving an acid in water	$H{-}A \quad + \quad H{-}\overset{\cdot\cdot}{\underset{\cdot\cdot}{O}}{-}H \quad \rightleftharpoons \quad A{:}^- \quad + \quad H{-}\overset{H}{\underset{+}{O}}{-}H$
	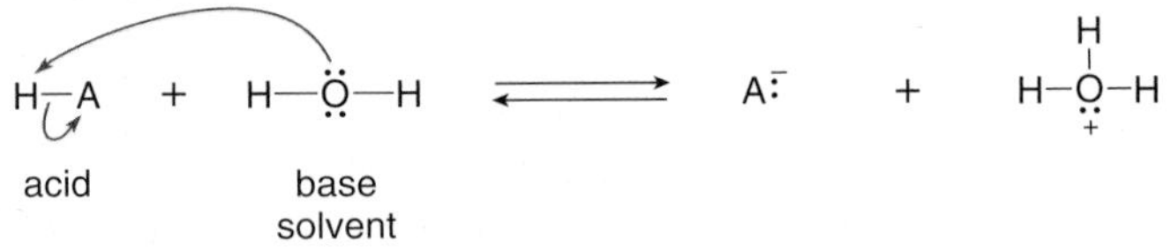 acid base solvent

Equilibrium constant	$K_{eq} = \dfrac{[\text{products}]}{[\text{starting materials}]} = \dfrac{[H_3O^+][A{:}^-]}{[H{-}A][H_2O]}$

Because the concentration of the solvent H_2O is essentially constant, the equation can be rearranged and a new equilibrium constant, called the **acidity constant, K_a,** can be defined.

$$\text{Acidity constant} = K_a = [H_2O]K_{eq} = \frac{[H_3O^+][A:^-]}{[H-A]}$$

How is the magnitude of K_a related to acid strength?

◆ The *stronger the acid,* the further the equilibrium lies to the right, and the *larger the K_a.*

For most organic compounds, K_a is small, typically 10^{-5} to 10^{-50}. This contrasts with the K_a values for many inorganic acids, which range from 10^0 to 10^{10}. Because dealing with exponents can be cumbersome, it is often more convenient to use pK_a values instead of K_a values.

Definition: pK_a = −log K_a

How does pK_a relate to acid strength?

K_a		pK_a = −log K_a	
K_a values of typical organic acids		**pK_a values of typical organic acids**	
10^{-5} to 10^{-50}		+5 to +50	
larger number stronger acid	smaller number weaker acid	smaller number stronger acid	larger number weaker acid

Recall that a **log** is an **exponent**; for example, log 10^{-5} = −5.

◆ The *smaller* the pK_a, the *stronger* the acid.

PROBLEM 2.6   Which compound in each pair is the stronger acid?

a. $CH_3CH_2CH_3$ and CH_3CH_2OH
   pK_a = 50   pK_a = 16

b.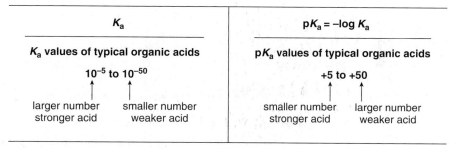

and

$K_a = 10^{-10}$ $K_a = 10^{-41}$

PROBLEM 2.7   Use a calculator when necessary to answer the following questions.
a. What is the pK_a for each K_a: 10^{-10}, 10^{-21}, and 5.2×10^{-5}?
b. What is the K_a for each pK_a: 7, 11, and 3.2?

We have thus far discussed acid strength only. What about base strength? **An inverse relationship exists between acidity and basicity.**

◆ A *strong acid* readily donates a proton forming a *weak conjugate base.*
◆ A *strong base* readily accepts a proton forming a *weak conjugate acid.*

Table 2.1 is a brief list of pK_a values for some common compounds, ranked in order of *increasing* pK_a and therefore *decreasing* acidity. Because strong acids form weak conjugate bases, this list also ranks their conjugate bases, in order of *increasing* basicity. For example, CH_4 is the weakest acid in the list, because it has the highest pK_a (50). Its conjugate base, CH_3^-, is therefore the strongest conjugate base. An extensive pK_a table is located in Appendix B.

Comparing pK_a values thus provides two useful bits of information: the **relative acidity of two acids,** and the **relative basicity of their conjugate bases,** as shown in Sample Problem 2.3.

TABLE 2.1   Selected pK_a Values

Acid	pK_a	Conjugate base
H–Cl	–7	Cl⁻
CH_3COO–H	4.8	CH_3COO^-
HO–H	15.7	HO^-
CH_3CH_2O–H	16	$CH_3CH_2O^-$
$HC\equiv CH$	25	$HC\equiv C^-$
H–H	35	H^-
H_2N–H	38	H_2N^-
$CH_2=CH_2$	44	$CH_2=\ddot{C}H$
CH_3–H	50	CH_3^-

Increasing acidity of the acid ↑ (left column)
Increasing basicity of the conjugate base ↓ (right column)

SAMPLE PROBLEM 2.3   Rank the following compounds in order of increasing acidity, and then rank their conjugate bases in order of increasing basicity.

$$CH_2=CH_2 \qquad HCl \qquad CH_3COOH$$

SOLUTION

The pK_a values in Table 2.1 allow us to rank these compounds in order of increasing acidity: **the lower the pK_a, the stronger the acid.**

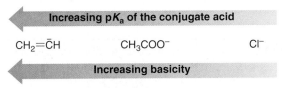

← Increasing pK_a		
$CH_2=CH_2$	CH_3COOH	HCl
$pK_a = 44$	$pK_a = 4.8$	$pK_a = -7$

Increasing acidity →

Because strong acids form weak conjugate bases, the **basicity of conjugate bases increases with increasing pK_a of their acids.**

← Increasing pK_a of the conjugate acid		
$CH_2=\ddot{C}H$	CH_3COO^-	Cl⁻

← Increasing basicity

PROBLEM 2.8   Rank the conjugate bases of each group of acids in order of increasing basicity.
a. NH_3, H_2O, CH_4 b. $CH_2=CH_2$, $HC\equiv CH$, CH_4

Note the large range in pK_a values (–7 to 50 in Table 2.1). The **pK_a scale is logarithmic** so a small difference in pK_a translates into a large numerical difference. For example, the difference between the pK_a of NH_3 (38) and $CH_2=CH_2$ (44) is six pK_a units. This means that NH_3 is 10^6 or **one million times more acidic** than $CH_2=CH_2$.

Although Table 2.1 is abbreviated, it gives pK_a values of many common compounds seen in organic chemistry. It is also a useful tool for *estimating* the pK_a of a compound similar though not identical to one in the table.

Suppose you are asked to estimate the pK_a of the N–H bond of CH_3NH_2. Although CH_3NH_2 is not listed in the table, we have enough information to *approximate* its pK_a. Because the pK_a of the N–H bond of NH_3 is 38, we can estimate the pK_a of the N–H bond of CH_3NH_2 to be 38. Its actual pK_a is 40, so this is a good first approximation.

PROBLEM 2.9   Estimate the pK_a of each of the indicated bonds.

a. [benzene ring]—N(H)—H   b. [benzene ring]—O—H   c. $BrCH_2COO$—H

2.4 Predicting the Outcome of Acid–Base Reactions

A proton transfer reaction represents an equilibrium. Because an acid donates a proton to a base, thus forming a conjugate acid and conjugate base, there are always two acids and two bases in the reaction mixture. Which pair of acids and bases is favored at equilibrium? **The position of the equilibrium depends on the relative strengths of the acids and bases.**

◆ **Equilibrium always favors formation of the weaker acid and base.**

> In a proton transfer reaction, the **stronger acid reacts with the stronger base** to form the weaker acid and the weaker base.

Because a strong acid readily donates a proton and a strong base readily accepts one, these two species react to form a weaker conjugate acid and base that do not donate or accept a proton as readily. Comparing pK_a values allows us to determine the position of equilibrium, as illustrated in Sample Problem 2.4.

SAMPLE PROBLEM 2.4   Determine the direction of equilibrium in the following proton transfer reaction.

$$H-C\equiv C-H + {}^-NH_2 \rightleftharpoons$$

SOLUTION

Follow three steps to determine the position of equilibrium:

Step [1]   Identify the acid and base in the starting materials.

$H-C\equiv C-H$ + $^-NH_2$ • Assume $^-NH_2$ is the base because it bears a net negative
     acid   base   charge.

Step [2]   Draw the products of proton transfer and identify the conjugate acid and base in the products.

$H-C\equiv C{-}H$ + $:\ddot{N}H_2$ $\rightleftharpoons$ $H-C\equiv C:^-$ + $:NH_3$
     acid   base   conjugate base   conjugate acid
$pK_a = 25$ $pK_a = 38$

Step [3]   Compare the pK_a values of the acid and the conjugate acid. Equilibrium favors formation of the weaker acid with the higher pK_a.

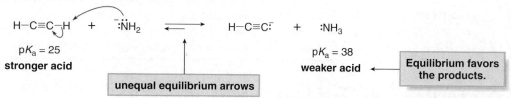

$H-C\equiv C{-}H$ + $:\ddot{N}H_2$ $\rightleftharpoons$ $H-C\equiv C:^-$ + $:NH_3$

$pK_a = 25$ $pK_a = 38$
stronger acid   **weaker acid** ← Equilibrium favors the products.

unequal equilibrium arrows

• Because the pK_a of the starting acid (25) is **lower** than the pK_a of the conjugate acid (38), $HC\equiv CH$ is a **stronger** acid and equilibrium favors the products.

PROBLEM 2.10   Draw the products of each reaction and determine the direction of equilibrium.

a. $H-C\equiv C-H$ + H^- $\rightleftharpoons$ c. CH_3COOH + $CH_3CH_2O^-$ $\rightleftharpoons$

b. CH_4 + ^-OH $\rightleftharpoons$ d. Cl^- + CH_3CH_2OH $\rightleftharpoons$

How can we know if a particular base is strong enough to deprotonate a given acid, so that the equilibrium lies to the right? The pK_a table readily gives us this information.

Compare any two consecutive entries in Table 2.1, such as ethanol (CH_3CH_2OH; $pK_a = 16$) and acetylene ($HC\equiv CH$; $pK_a = 25$), and their conjugate bases, ethoxide ($CH_3CH_2O^-$) and acetylide ($HC\equiv C^-$). Ethanol is a stronger acid than acetylene, so acetylide is a stronger base than ethoxide.

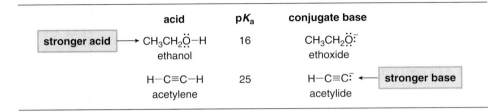

	acid	pK_a	conjugate base	
stronger acid	$CH_3CH_2\overset{..}{\underset{..}{O}}-H$ ethanol	16	$CH_3CH_2\overset{..}{\underset{..}{O}}{:}^-$ ethoxide	
	$H-C\equiv C-H$ acetylene	25	$H-C\equiv C{:}^-$ acetylide	stronger base

Because Table 2.1 is arranged from low to high pK_a, **an acid can be deprotonated by the conjugate base of any acid *below* it in the table.**

Two proton transfer reactions are possible.

[1] Reaction of acetylene with ethoxide forms acetylide and ethanol. Because the stronger acid is the product of the reaction, ***equilibrium favors the starting materials.*** **The base ethoxide is *not* strong enough to deprotonate acetylene.**

acid		base		conjugate base		conjugate acid	
$H-C\equiv C-H$ acetylene $pK_a = 25$	+	$\overset{..}{\underset{..}{:O}}CH_2CH_3$ ethoxide	⇌	${:}C\equiv C-H$ acetylide	+	$CH_3CH_2\overset{..}{\underset{..}{O}}-H$ ethanol $pK_a = 16$	stronger acid

Equilibrium favors the starting materials.

[2] Reaction of ethanol with acetylide forms ethoxide and acetylene. Because the weaker acid is the product of the reaction, ***equilibrium favors the products.*** Thus, the **base acetylide *is* strong enough to deprotonate ethanol.**

acid		base		conjugate base		conjugate acid	
$CH_3CH_2\overset{..}{\underset{..}{O}}-H$ ethanol $pK_a = 16$	+	${:}C\equiv C-H$ acetylide	⇌	$CH_3CH_2\overset{..}{\underset{..}{O}}{:}^-$ ethoxide	+	$H-C\equiv C-H$ acetylene $pK_a = 25$	weaker acid

Equilibrium favors the products.

Note that in the second reaction, **ethanol is deprotonated by acetylide, the conjugate base of an acid *weaker* than itself.** This is a specific example of a general fact.

◆ An acid can be deprotonated by the conjugate base of any acid having a *higher* pK_a.

PROBLEM 2.11 Answer the following questions by referring to Table 2.1:
a. Which of the following bases is strong enough to deprotonate CH_3COOH: H^-, $HC\equiv C^-$, and Cl^-?
b. List four bases that are strong enough to deprotonate $HC\equiv CH$.

2.5 Factors that Determine Acid Strength

We have already learned in Section 2.3 that a tremendous difference in acidity exists among compounds. HCl ($pK_a < 0$) is an extremely strong acid. Water ($pK_a = 15.7$) is moderate in acidity, and CH_4 ($pK_a = 50$) is an extremely weak acid. How are these differences explained? There is one general rule.

◆ Anything that stabilizes a conjugate base A:$^-$ makes the starting acid H—A more acidic.

For now we will concentrate on how structural differences between molecules can profoundly affect acidity. In Chapter 6, we will learn how to relate the stability of a species to its relative potential energy.

Four factors affect the acidity of H–A:

[1]   Element effects
[2]   Inductive effects
[3]   Resonance effects
[4]   Hybridization effects

No matter which factor is discussed, the same procedure is always followed. To compare the acidity of any two acids:

> ◆ **Always draw the conjugate bases.**
> ◆ **Determine which conjugate base is more stable.**
> ◆ **The *more stable* the conjugate base, the *more acidic* the acid.**

2.5A   Element Effects—Trends in the Periodic Table

Factors affecting the acidity of H–A:

◆ **Element effects**
◆ Inductive effects
◆ Resonance effects
◆ Hybridization effects

The most important factor determining the acidity of H–A is the location of A in the periodic table.

Comparing Elements in the Same Row of the Periodic Table

Examine acidity trends **across a row** of the periodic table by comparing CH_4 and H_2O, two compounds having H atoms bonded to a second-row element. We know from Table 2.1 that **H_2O has a much *lower* pK_a and therefore is much *more acidic* than CH_4,** but why is this the case?

To answer this question we must first draw both conjugate bases by removing a proton, and then determine which is more stable. Each conjugate base has a net negative charge, but the negative charge in ^-OH is on oxygen and in CH_3^- it is on carbon.

Because the oxygen atom is much **more electronegative** than carbon, oxygen more readily accepts a negative charge, making ^-OH much more stable than CH_3^-. **H_2O is a stronger acid than CH_4 because ^-OH is a more stable conjugate base than CH_3^-.**

This is a specific example of a general trend.

> ◆ Across a row of the periodic table, the acidity of H–A increases as the electronegativity of A increases.

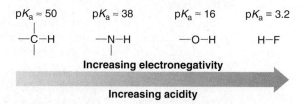

The enormity of this effect is evident by noting the approximate pK_a values for these bonds. **A C–H bond is approximately 10^{47} times *less acidic* than H–F.**

Comparing Elements Down a Column of the Periodic Table

Now examine acidity trends down a column of the periodic table by comparing H–F and H–Br. Once again we must first draw both conjugate bases by removing a proton, and then determine which is more stable. In this case, removal of a proton forms F^- and Br^-.

H—F $\longrightarrow$ F⁻ | H—Br $\longrightarrow$ Br⁻

acid   conjugate base   acid   conjugate base

There are two important differences between F⁻ and Br⁻—electronegativity and size. In this case, **size is more important than electronegativity.** The size of an atom or ion increases down a column of the periodic table, so Br⁻ is much larger than F⁻, and this stabilizes the negative charge.

◆ Positive or negative charge is stabilized when it is spread over a larger volume.

Because Br⁻ is larger than F⁻, Br⁻ is more stable than F⁻, and H–Br is a stronger acid than H–F.

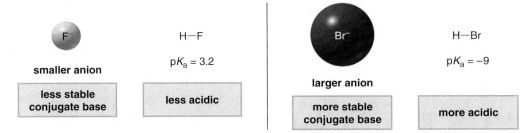

F⁻ smaller anion	H—F pKₐ = 3.2	Br⁻ larger anion	H—Br pKₐ = −9
less stable conjugate base	less acidic	more stable conjugate base	more acidic

This again is a specific example of a general trend.

◆ Down a column of the periodic table, the acidity of H–A increases as the size of A increases.

pKₐ = 3.2   pKₐ = −7   pKₐ = −9   pKₐ = −10

H—F   H—Cl   H—Br   H—I

Increasing size

Increasing acidity

This is *opposite* to what would be expected on the basis of electronegativity differences between F and Br, because F is more electronegative than Br. **Size and *not* electronegativity determines acidity down a column.**

Combining both trends together:

◆ The acidity of H–A increases both left-to-right across a row and down a column of the periodic table.

> Because of carbon's position in the periodic table (in the second row and to the left of O, N, and the halogens), **C–H bonds are usually the *least acidic* bonds in a molecule.**

SAMPLE PROBLEM 2.5   Without reference to a pKₐ table, decide which compound in each pair is the stronger acid: [a] H_2O and HF; [b] H_2S and H_2O.

SOLUTION

[a] H_2O and H–F both have H atoms bonded to a second-row element. Because the acidity of H–A increases across a row of the periodic table, the H–F bond is more acidic than the H–O bond. **HF is a stronger acid than H_2O.**

[b] H_2O and H_2S both have H atoms bonded to elements in the same column. Because the acidity of H–A increases down a column of the periodic table, the H–S bond is more acidic than the H–O bond. **H_2S is a stronger acid than H_2O.**

PROBLEM 2.12   Without reference to a pKₐ table, decide which compound in each pair is the stronger acid.
a. NH_3 and H_2O b. HBr and HCl   c. H_2S and HBr

When discussing acidity, the most acidic proton in a compound is the one removed first by a base. Although four factors determine the overall acidity of a particular hydrogen atom, **element effects—the identity of A—is the single most important factor in determining the acidity of the H–A bond.**

To decide which hydrogen is most acidic, **first determine what element each hydrogen is bonded to and then decide its acidity based on periodic trends.** This is illustrated in Sample Problem 2.6.

SAMPLE PROBLEM 2.6   Label the most acidic hydrogen in the following compound.

$$CH_3-N-CH_2CH_2CH_2CH_3$$
$$|$$
$$H$$

SOLUTION

This compound contains only $C-H$ bonds and $N-H$ bonds. Because the acidity of $H-A$ increases across a row of the periodic table, an $N-H$ bond is more acidic than a $C-H$ bond. The single H on N is the most acidic H in this compound.

$$CH_3-N-CH_2CH_2CH_2CH_3$$
$$|$$

most acidic hydrogen ⟶ H

PROBLEM 2.13   Which hydrogen in each molecule is most acidic?
a. $CH_3CH_2CH_2CH_2OH$ b. $HOCH_2CH_2CH_2NH_2$ c. $(CH_3)_2NCH_2CH_2CH_2NH_2$

2.5B   Inductive Effects

Factors affecting the acidity of $H-A$:
◆ Element effects
◆ **Inductive effects**
◆ Resonance effects
◆ Hybridization effects

A second factor affecting the acidity of $H-A$ is the presence of electronegative atoms. To illustrate this phenomenon, let's compare ethanol (CH_3CH_2OH) and 2,2,2-trifluoroethanol (CF_3CH_2OH), two different compounds containing $O-H$ bonds. The pK_a table in Appendix B indicates that CF_3CH_2OH is a stronger acid than CH_3CH_2OH. Because we are looking at the acidity of the $O-H$ bond in both compounds, what causes the difference?

CH_3CH_2O-H CF_3CH_2O-H

ethanol   2,2,2-trifluoroethanol

$pK_a = 16$ $pK_a = 12.4$ ⟵ **stronger acid**

Once again we must first draw both conjugate bases by removing a proton, and then determine which is more stable. Both bases have a negative charge on an electronegative oxygen, but the second anion has three additional electronegative fluorine atoms. These fluorine atoms withdraw electron density from the carbon to which they are bonded, making it electron deficient. Furthermore, this electron-deficient carbon pulls electron density through σ bonds from the negatively charged oxygen atom, stabilizing the negative charge. This is called an **inductive effect.**

$CH_3CH_2O^-$

| No additional electronegative atoms stabilize the conjugate base. | CF_3 withdraws electron density, stabilizing the conjugate base. |

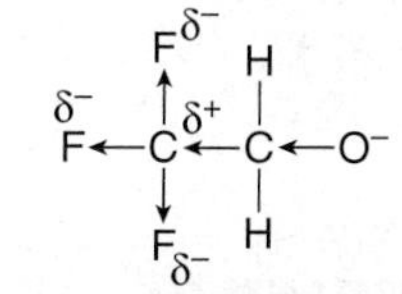

◆ An *inductive effect* is the pull of electron density through σ bonds caused by electronegativity differences of atoms.

In this case, the electron density is *pulled away* from the negative charge through σ bonds by the very electronegative fluorine atoms, and so it is called an **electron-*withdrawing* inductive effect. Thus, the three electronegative fluorine atoms stabilize the negatively charged conjugate base $CF_3CH_2O^-$, making CF_3CH_2OH a stronger acid than CH_3CH_2OH.**

We have learned two important principles from this discussion:

◆ More electronegative atoms stabilize regions of high electron density by an electron-withdrawing inductive effect.
◆ The acidity of $H-A$ increases with the presence of electron-withdrawing groups in A.

Figure 2.2 Electrostatic potential plots of $CH_3CH_2O^-$ and $CF_3CH_2O^-$

$CH_3CH_2O^-$

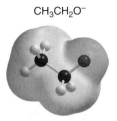

The dark red of the O atom indicates a region of high electron density.

$CF_3CH_2O^-$

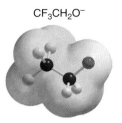

The O atom is yellow, indicating it is less electron rich.

Inductive effects result because an electronegative atom stabilizes the negative charge of the conjugate base. The *more electronegative* the atom and the *closer* **it is to the site of the negative charge, the greater the effect.** This effect is discussed in greater detail in Chapter 19.

Electrostatic potential plots in Figure 2.2 compare the electron density around the oxygen atoms in these conjugate bases. The darker red around the O atom of $CH_3CH_2O^-$ indicates a higher concentration of electron density compared to the O atom of $CF_3CH_2O^-$. This illustrates that the three F atoms pull electron density away from the O atom in $CF_3CH_2O^-$ by an electron-withdrawing inductive effect.

PROBLEM 2.14

Which compound in each pair is the stronger acid?
a. $ClCH_2COOH$ and FCH_2COOH
b. Cl_2CHCH_2OH and $Cl_2CHCH_2CH_2OH$
c. CH_3COOH and O_2NCH_2COOH

2.5C Resonance Effects

Factors affecting the acidity of H–A:

◆ Element effects
◆ Inductive effects
◆ **Resonance effects**
◆ Hybridization effects

A third factor that determines acidity is resonance. As we already learned in Section 1.5, resonance occurs whenever two or more different Lewis structures can be drawn for the same arrangement of atoms.

To illustrate this phenomenon, compare ethanol (CH_3CH_2OH) and acetic acid (CH_3COOH), two different compounds containing O–H bonds. Based on Table 2.1, CH_3COOH is a stronger acid than CH_3CH_2OH:

Recall that resonance structures are two Lewis structures having the same placement of atoms but a different arrangement of electrons.

CH_3CH_2O-H
ethanol
$pK_a = 16$

$CH_3-C\overset{O}{\underset{O-H}{\big\|}}$
acetic acid
$pK_a = 4.8$ ← **stronger acid**

Draw the conjugate bases of these acids to illustrate the importance of resonance. For ethoxide ($CH_3CH_2O^-$) the conjugate base of ethanol, only one Lewis structure can be drawn. The negative charge of this conjugate base is *localized* on the O atom.

CH_3CH_2O-H ⟶ $CH_3CH_2\ddot{O}\colon^-$ ← **The negative charge is localized on O.**
ethanol   ethoxide
acid   conjugate base

only **one** Lewis structure

With acetate (CH_3COO^-), however, two resonance structures can be drawn.

The negative charge is delocalized on two O atoms.

$$CH_3-C\underset{O-H}{\overset{O}{\Vert}} \longrightarrow CH_3-C\underset{:\overset{..}{O}:^-}{\overset{:\overset{..}{O}:}{\Vert}} \longleftrightarrow CH_3-C\underset{:\overset{..}{O}:}{\overset{:\overset{..}{O}:^-}{\Vert}} \quad\Big|\quad CH_3-C\underset{:\overset{..}{O}:_{\delta^-}}{\overset{\overset{..}{O}\,^{\delta^-}}{\Vert}}$$

acetic acid   acetate conjugate base   **hybrid**

two resonance structures   **resonance-stabilized conjugate base**

The difference in these two resonance structures is the **position of a π bond** and a **lone pair.** Curved arrow notation illustrates how one resonance structure can be converted to the other. Although each resonance structure of acetate implies that the negative charge is localized on an O atom, in actuality, charge is *delocalized* over both O atoms. **Delocalization of electron density stabilizes acetate, making it a weaker base.**

Remember that neither resonance form adequately represents acetate. The true structure is a **hybrid** of both structures. In the hybrid, the electron pairs drawn in different locations in individual resonance structures are *delocalized.* With acetate, we use a dashed line to show that each C–O bond has partial double bond character. The symbol δ^- (partial negative) indicates that the charge is delocalized on both O atoms in the hybrid.

Thus, **resonance delocalization makes CH_3COO^- more stable than $CH_3CH_2O^-$, so CH_3COOH is a stronger acid than CH_3CH_2OH.**

This is another example of a general rule.

◆ The acidity of H–A increases when the conjugate base A:⁻ is resonance stabilized.

Electrostatic potential plots of $CH_3CH_2O^-$ and CH_3COO^- in Figure 2.3 indicate that the negative charge is concentrated on a single O in $CH_3CH_2O^-$, but delocalized over the O atoms in CH_3COO^-.

PROBLEM 2.15   The C–H bond in acetone, $(CH_3)_2C=O$, has a pK_a of 19.2. Draw two resonance structures for its conjugate base. Then, explain why acetone is much more acidic than propane, $CH_3CH_2CH_3$ ($pK_a = 50$).

PROBLEM 2.16   Acetonitrile (CH_3CN) has a pK_a of 25, making it more acidic than many other compounds having only C–H bonds. Draw Lewis structures for acetonitrile and its conjugate base. Use resonance structures to account for the acidity of acetonitrile.

2.5D Hybridization Effects

The final factor affecting the acidity of H–A is the hybridization of A. To illustrate this phenomenon, compare ethane (CH_3CH_3), ethylene ($CH_2=CH_2$), and acetylene ($HC≡CH$), three

Figure 2.3 Electrostatic potential plots of $CH_3CH_2O^-$ and CH_3COO^-

$CH_3CH_2O^-$

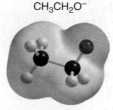

The negative charge is concentrated on the single oxygen atom, making this anion *less stable.*

CH_3COO^-

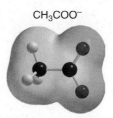

The negative charge is delocalized over both oxygen atoms, making this anion *more stable.*

different compounds containing C−H bonds. Appendix B indicates that there is a considerable difference in the pK_a values of these compounds.

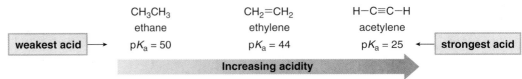

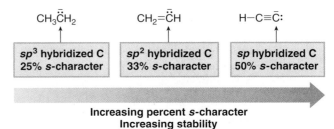

Increasing percent s-character
Increasing stability

Note, however, that the hybridization of the carbon bearing the negative charge is different in each anion, so the lone pair of electrons occupies an orbital with a different percent *s*-character in each case.

◆ **The *higher* the percent s-character of the hybrid orbital, the closer the lone pair is held to the nucleus, and the more stable the conjugate base.**

Thus, **acidity increases from CH_3CH_3 to $CH_2=CH_2$ to $HC\equiv CH$ as the negative charge of the conjugate base is stabilized by increasing percent *s*-character.**

Once again this is a specific example of a general trend.

◆ **The acidity of H−A increases as the percent s-character of the A:⁻ increases.**

Electrostatic potential plots of these carbanions appear in Figure 2.4.

Factors affecting the acidity of H−A:

◆ Element effects
◆ Inductive effects
◆ Resonance effects
◆ **Hybridization effects**

PROBLEM 2.17   For each pair of compounds: [1] Which indicated H is more acidic? [2] Draw the conjugate base of each acid. [3] Which conjugate base is stronger?

a. $CH_3CH_2-C\equiv C-H$ and $CH_3CH_2CH_2CH_2-H$ b.

Figure 2.4   Electrostatic potential plots of three carbanions

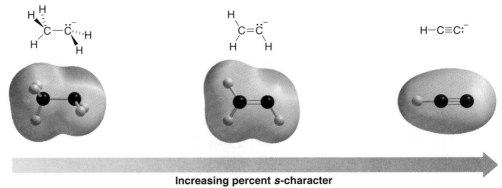

Increasing percent s-character
Increasing carbanion stability

• As the lone pair of electrons is pulled closer to the nucleus, the negatively charged carbon appears less intensely red.

Figure 2.5 Summary of the factors that determine acidity

Factor	Example
1. **Element effect:** The acidity of H−A increases both left-to-right across a row and down a column of the periodic table.	Increasing acidity → $-\overset{\mid}{\underset{\mid}{C}}-H$ $-N-H$ $-O-H$ $H-F$ $-S-H$ $H-Cl$ $H-Br$ $H-I$ (Increasing acidity ↓)
2. **Inductive effects:** The acidity of H−A increases with the presence of electron-withdrawing groups in A.	CH_3CH_2O-H CF_3CH_2O-H **more acidic**
3. **Resonance effects:** The acidity of H−A increases when the conjugate base A:⁻ is resonance stabilized.	CH_3CH_2O-H CH_3COO-H **more acidic**
4. **Hybridization effects:** The acidity of H−A increases as the percent s-character of A:⁻ increases.	CH_3CH_3 $CH_2{=}CH_2$ $H-C{\equiv}C-H$ **Increasing acidity** →

2.5E   Summary of Factors Determining Acid Strength

The ability to recognize the most acidic site in a molecule will be important throughout the study of organic chemistry. All the factors that determine acidity are therefore summarized in Figure 2.5.

The following two-step procedure shows how these four factors can be used to determine the relative acidity of protons.

How To   Determine Relative Acidity of Protons

Step [1]   **Identify the atoms bonded to hydrogen, and use periodic trends to assign relative acidity.**

- The most common H−A bonds in organic compounds are C−H, N−H, and O−H. Because acidity increases left-to-right across a row, the relative acidity of these bonds is C−H < N−H < O−H. Therefore, H atoms bonded to C atoms are usually *less acidic* than H atoms bonded to any heteroatom.

Step [2]   **If the two H atoms in question are bonded to the same element, draw the conjugate bases and look for other points of difference. Ask three questions:**

- Do electron-withdrawing groups stabilize the conjugate base?
- Is the conjugate base resonance stabilized?
- How is the conjugate base hybridized?

Sample Problem 2.7 shows how to apply this procedure to actual compounds.

SAMPLE PROBLEM 2.7   Rank the following compounds in order of increasing acidity of their most acidic hydrogen atom.

$$ClCH_2CH_2OH \qquad CH_3CH_2OH \qquad CH_3CH_2NH_2$$

A   **B**   **C**

SOLUTION

[1] Compounds **A, B,** and **C** contain C−H, N−H, and O−H bonds. Because acidity increases left-to-right across a row of the periodic table, the O−H bonds are most acidic. Compound **C** is thus the least acidic because it has *no* O−H bonds.

[2] The only difference between compounds **A** and **B** is the presence of an electronegative Cl in **A**. The Cl atom stabilizes the conjugate base of **A**, making it more acidic than **B**.

ANSWER

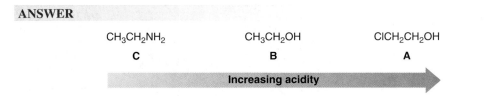

Increasing acidity

PROBLEM 2.18 Rank the compounds in each group in order of increasing acidity.
a. $CH_3CH_2CH_3$, CH_3CH_2OH, $CH_3CH_2NH_2$
b. $BrCH_2COOH$, CH_3CH_2COOH, $CH_3CH_2CH_2OH$
c. $CH_3CH_2NH_2$, $(CH_3)_3N$, CH_3CH_2OH

PROBLEM 2.19 Which proton in each of the following drugs is most acidic?

a.

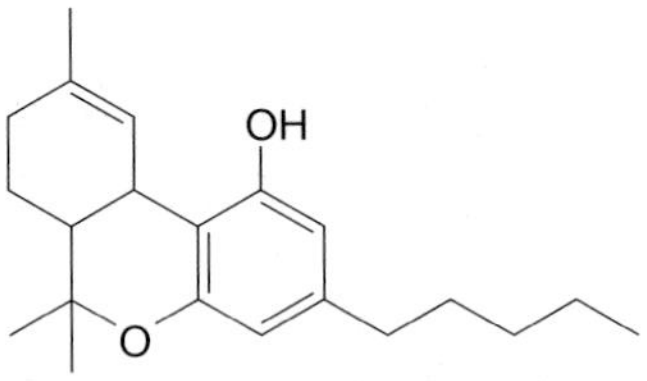

THC
tetrahydrocannabinol
(active component in marijuana)

b.

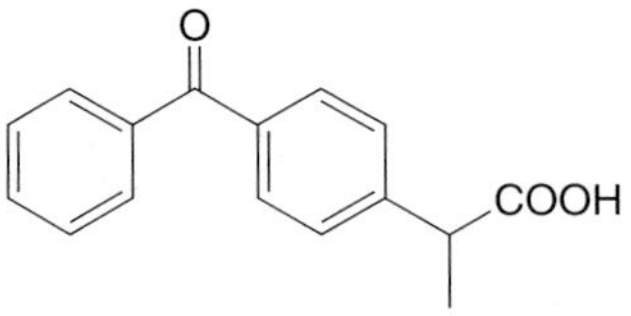

ketoprofen
(anti-inflammatory agent)

c.

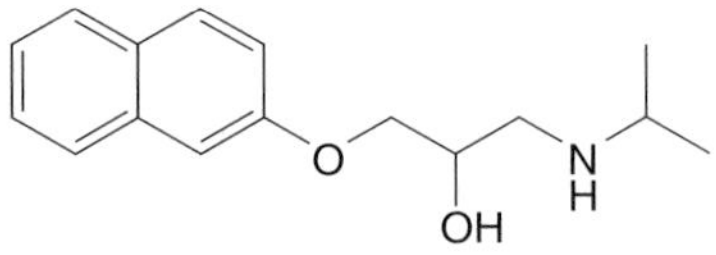

propranolol
(antihypertensive agent)

2.6 Common Acids and Bases

Many strong or moderately strong acids and bases are used as reagents in organic reactions.

2.6A Common Acids

Several organic reactions are carried out in the presence of strong inorganic acids, most commonly **HCl** and **H₂SO₄.** Less frequently **HNO₃** is used. All of these strong acids have **pK_a values ≤ 0.** These acids should be familiar from previous chemistry courses.

Two organic acids are also commonly used, namely **acetic acid** and **p-toluenesulfonic acid** (usually abbreviated as **TsOH**). Although acetic acid has a higher pK_a than the inorganic acids, making it a weaker acid, it is more acidic than most organic compounds. p-Toluenesulfonic acid is similar in acidity to the strong inorganic acids. Because it is a solid, small quantities can be easily weighed on a balance and then added to a reaction mixture.

CH_3COOH

acetic acid
pK_a = 4.8

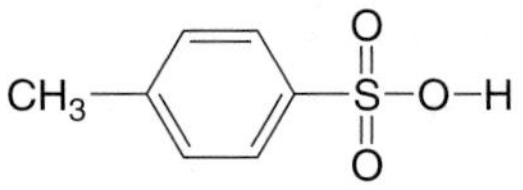

CH_3—⟨⟩—$\overset{\overset{O}{\|}}{\underset{\overset{\|}{O}}{S}}$—O–H   =   TsOH

p-toluenesulfonic acid
pK_a = −7

2.6B Common Bases

Common strong bases used in organic reactions are more varied in structure. Three common kinds of negatively charged bases include:

[1] Negatively charged oxygen bases: ⁻**OH** (hydroxide) and its organic derivatives
[2] Negatively charged nitrogen bases: ⁻**NH₂** (amide) and its organic derivatives
[3] Hydride (**H⁻**)

Figure 2.6 gives examples of these strong bases. Remember that each of these bases is used as a salt with a spectator ion (usually Li⁺, Na⁺, or K⁺) that serves to balance charge.

◆ **Strong bases have weak conjugate acids with high pK_a values, usually > 12.**

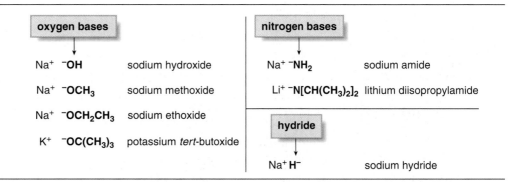

Figure 2.6 Some common negatively charged bases

oxygen bases

Na⁺ ⁻OH   sodium hydroxide

Na⁺ ⁻OCH₃   sodium methoxide

Na⁺ ⁻OCH₂CH₃   sodium ethoxide

K⁺ ⁻OC(CH₃)₃   potassium *tert*-butoxide

nitrogen bases

Na⁺ ⁻NH₂   sodium amide

Li⁺ ⁻N[CH(CH₃)₂]₂   lithium diisopropylamide

hydride

Na⁺ H⁻   sodium hydride

Strong bases have a net negative charge, but not all negatively charged species are strong bases. For example, none of the halides, F⁻, Cl⁻, Br⁻, or I⁻, is a strong base. These anions have very strong conjugate acids and have little affinity for donating their electron pairs to a proton.

Carbanions, negatively charged carbon atoms discussed in Section 2.5D, are especially strong bases. Perhaps the most common example is **butyllithium.** Butyllithium and related compounds are discussed in greater detail in Chapter 20.

very strong base

CH₃CH₂CH₂CH₂⁻ Li⁺

butyllithium

Two other weaker organic bases are **triethylamine** and **pyridine.** These compounds have a lone pair on nitrogen, making them basic, but they are considerably weaker than the amide bases because they are neutral, not negatively charged.

CH₃CH₂—N̈—CH₂CH₃
        |
       CH₂CH₃

triethylamine   pyridine

PROBLEM 2.20   Draw the products of the proton transfer reactions.

a. CH₃CH₂OH + NaH ⇌

b. CH₃COOH + NaOCH₂CH₃ ⇌

c. CH₃CH₂CH₂CH₂Li + H₂O ⇌

d. CH₃—⟨benzene ring⟩—SO₃H + N(CH₂CH₃)₃ ⇌

2.7 Aspirin

Aspirin, or acetylsalicylic acid, is the most well known member of a group of compounds called **salicylates.** Although aspirin was first used in medicine for its analgesic (pain-relieving), antipyretic (fever-reducing), and anti-inflammatory properties, today it is commonly used as an antiplatelet agent in the treatment and prevention of heart attacks and strokes. **Aspirin is a synthetic compound;** it does not occur in nature, though some related salicylates are found in willow bark and meadowsweet blossoms.

Like many drugs, aspirin is capable of undergoing a proton transfer reaction. Its most acidic proton is the H bonded to O, and in the presence of base, this H is readily removed.

The modern history of aspirin dates back to 1763 when Reverend Edmund Stone reported on the analgesic effect of chewing on the bark of the willow tree. Willow bark is now known to contain *salicin*, which is structurally related to aspirin.

salicin

willow tree

Why is this acid–base reaction important? After ingestion, aspirin first travels into the stomach and then the intestines. In the acidic environment of the stomach, aspirin remains in its neutral form, but in the basic environment of the small intestine, aspirin is deprotonated to form its conjugate base, an ion.

acetylsalicylic acid **neutral form**	conjugate base **ionic form**
This form exists in the **stomach**.	This form exists in the **intestines**.

Why is this important? **To be active, aspirin must cross a cell membrane, and to do so, it must be neutral, not ionic.** This means that aspirin crosses a cell membrane and is absorbed by the body in its neutral form in the stomach. Whether aspirin is present as its acid or its conjugate base is thus very important in determining whether it can permeate a cell. Aspirin's mechanism of action is discussed in greater detail in Chapter 19.

PROBLEM 2.21

amphetamine

Compounds like amphetamine that contain nitrogen atoms are protonated by the HCl in the gastric juices of the stomach, and the resulting salt is then deprotonated in the basic environment of the intestines to regenerate the neutral form. Write proton transfer reactions for both of these processes. Where is amphetamine likely to be absorbed by the body?

2.8 Lewis Acids and Bases

The Lewis definition of acids and bases is more general than the Brønsted–Lowry definition.

◆ A Lewis acid is an *electron pair acceptor.*
◆ A Lewis base is an *electron pair donor.*

Lewis bases are structurally the same as Brønsted–Lowry bases. Both have an **available electron pair**—a lone pair or an electron pair in a π bond. A Brønsted–Lowry base always donates this electron pair to a proton, but a Lewis base donates this electron pair to anything that is electron deficient.

All Brønsted–Lowry bases are Lewis bases.

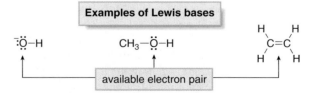

Examples of Lewis bases

available electron pair

A Lewis acid must be able to accept an electron pair, but there are many ways for this to occur. **All Brønsted–Lowry acids are also Lewis acids, but the reverse is not necessarily true.** Any species that is electron deficient and capable of accepting an electron pair is also a Lewis acid.

Common examples of Lewis acids (which are not Brønsted–Lowry acids) include BF_3 and $AlCl_3$. These compounds contain elements in group 3A of the periodic table that can accept an electron pair because they do not have filled valence shells of electrons.

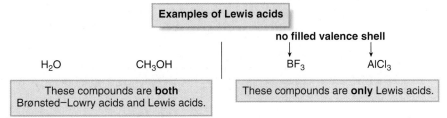

PROBLEM 2.22   Which compounds are Lewis bases?
a. NH_3 b. $CH_3CH_2CH_3$ c. H^- d. $H-C\equiv C-H$

PROBLEM 2.23   Which compounds are Lewis acids?
a. BBr_3 b. CH_3CH_2OH c. $(CH_3)_3C^+$ d. Br^-

In a Lewis acid–base reaction, a Lewis base donates an electron pair to a Lewis acid. Most of the reactions in organic chemistry involving movement of electron pairs can be classified as Lewis acid–base reactions. Lewis acid–base reactions illustrate a general pattern of reactivity in organic chemistry.

> Any reaction in which one species donates an electron pair to another species is a Lewis acid–base reaction.

◆ **Electron-rich species react with electron-poor species.**

In the simplest Lewis acid–base reaction one bond is formed and no bonds are broken. This is illustrated with the reaction of BF_3 with H_2O. BF_3 has only six electrons around B so it is the electron-deficient Lewis acid. H_2O has two lone pairs on O so it is the electron-rich Lewis base.

$$F-B(F)(F) + H-\overset{..}{\underset{..}{O}}-H \longrightarrow F-\overset{-}{\underset{F}{B}}(F)-\overset{+}{\underset{H}{\overset{..}{O}}}-H$$

Lewis acid   Lewis base   new bond

H_2O donates an electron pair to BF_3 to form one new bond. One curved arrow shows the movement of one electron pair. The electron pair in the new B–O bond comes from the oxygen atom, and a single product is formed. Both B and O bear formal charges in the product, but the overall product is neutral.

◆ A Lewis acid is also called an *electrophile.*
◆ When a Lewis base reacts with an electrophile other than a proton, the Lewis base is also called a *nucleophile.*

> Nucleophile = nucleus loving. Electrophile = electron loving.

In this Lewis acid–base reaction, **BF_3 is the electrophile and H_2O is the nucleophile.**

Two other examples are drawn. Note in each reaction that the **electron pair is not removed from the Lewis base;** instead, it is donated to an atom of the Lewis acid, and one new covalent bond is formed.

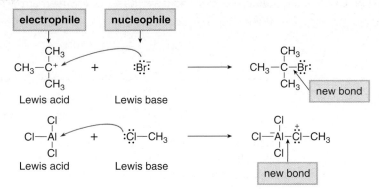

PROBLEM 2.24 For each reaction, label the Lewis acid and base. Use curved arrow notation to show the movement of electron pairs.

a. 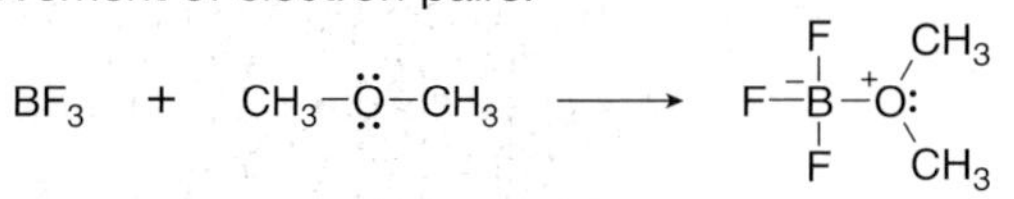 BF_3 + $CH_3-\overset{..}{\underset{..}{O}}-CH_3$ ⟶ F–B–O: (with F, F, CH_3, CH_3 substituents)

b. $(CH_3)_2\overset{+}{C}H$ + ^-OH ⟶ $(CH_3)_2CHOH$

PROBLEM 2.25 Draw the products of each reaction, and label the nucleophile and electrophile.

a. $CH_3CH_2-\overset{..}{\underset{..}{O}}-CH_2CH_3$ + BBr_3 ⟶

b. 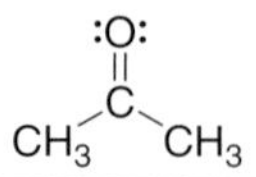 $\underset{CH_3}{}\overset{:O:}{\underset{}{\overset{||}{C}}}\underset{CH_3}{}$ + $AlCl_3$ ⟶

In some Lewis acid–base reactions, one bond is formed and one bond is broken. To draw the products of these reactions, keep the following steps in mind.

[1] **Always identify the Lewis acid and base first.**
[2] **Draw a curved arrow from the electron pair of the base to the electron-deficient atom of the acid.**
[3] **Count electron pairs and break a bond when needed to keep the correct number of valence electrons.**

For example, draw the Lewis acid–base reaction between cyclohexene and H–Cl. The Brønsted–Lowry acid HCl is also a Lewis acid, and cyclohexene, having a π bond, is the Lewis base.

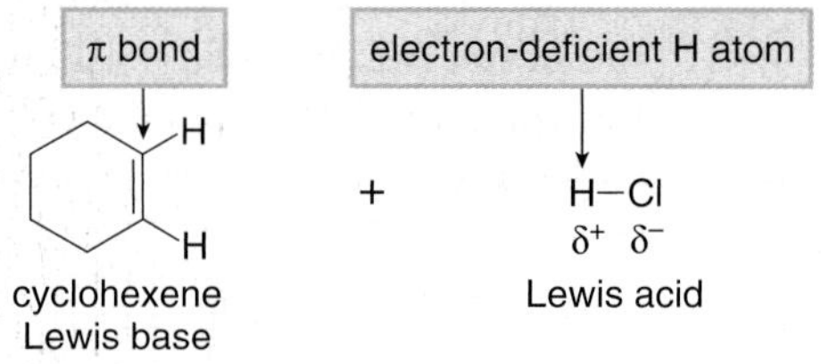

> Recall from Chapter 1 that a positively charged carbon atom is called a **carbocation.**

> In the reaction of cyclohexene with HCl, the new bond to H could form at **either carbon of the double bond,** because the same carbocation results.

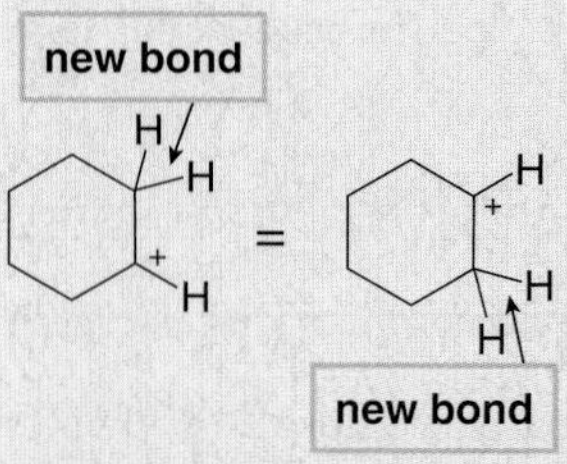

To draw the product of this reaction, the electron pair in the π bond of the Lewis base forms a new bond to the proton of the Lewis acid, forming a carbocation. The H–Cl bond must break, giving its two electrons to Cl, forming Cl^-. Because two electron pairs are involved, two curved arrows are needed.

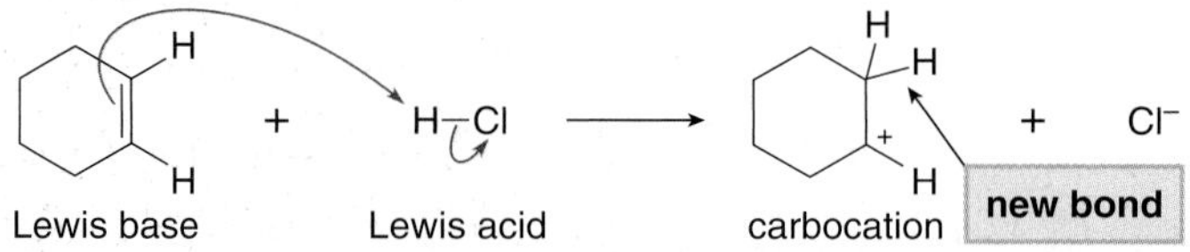

The Lewis acid–base reaction of cyclohexene with HCl is a specific example of a fundamental reaction of compounds containing C–C double bonds. The details of this reaction are discussed in Chapter 10.

PROBLEM 2.26 Label the Lewis acid and base. Use curved arrow notation to show the movement of electron pairs.

$CH_2=\underset{H}{\overset{CH_3}{C}}$ + H_3O^+ ⟶ $CH_3-\underset{H}{\overset{CH_3}{\overset{+}{C}}}$ + H_2O

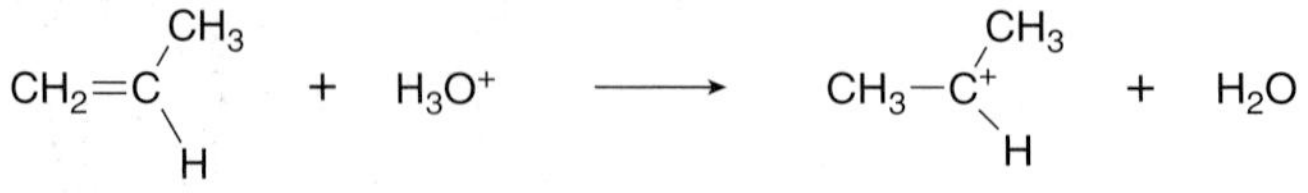

2.9   Key Concepts—Acids and Bases

A Comparison of Brønsted–Lowry and Lewis Acids and Bases

Type	Definition	Structural feature	Examples
Brønsted–Lowry acid (2.1)	proton donor	a proton	HCl, H_2SO_4, H_2O, CH_3COOH, TsOH
Brønsted–Lowry base (2.1)	proton acceptor	a lone pair *or* a π bond	^-OH, $^-OCH_3$, H^-, $^-NH_2$, NH_3, $CH_2{=}CH_2$
Lewis acid (2.8)	electron pair acceptor	a proton, *or* an unfilled valence shell, *or* a partial (+) charge	BF_3, $AlCl_3$, HCl, CH_3COOH, H_2O
Lewis base (2.8)	electron pair donor	a lone pair *or* a π bond	^-OH, $^-OCH_3$, H^-, $^-NH_2$, NH_3, $CH_2{=}CH_2$

Acid–Base Reactions

[1] A Brønsted–Lowry acid donates a proton to a Brønsted–Lowry base (2.2).

Example

H—Ö—H	+	H—N̈—H	⇌	H—Ö:	+	H—N—H (with H on top)
acid proton donor		**base** proton acceptor		**conjugate base**		**conjugate acid**

[2] A Lewis base donates an electron pair to a Lewis acid (2.8).

Example

$$CH_3\text{—}\overset{+}{\underset{CH_3}{\overset{CH_3}{C}}} \quad + \quad :\ddot{Br}:^- \quad \longrightarrow \quad CH_3\text{—}\underset{CH_3}{\overset{CH_3}{C}}\text{—}\ddot{Br}:$$

Lewis acid
electrophile   **Lewis base**
nucleophile

- Electron-rich species react with electron-poor ones.
- Nucleophiles react with electrophiles.

Important Facts

- Definition: **pK_a = −log K_a. The lower the pK_a, the stronger the acid** (2.3).
- The stronger the acid, the weaker the conjugate base (2.3).
- In proton transfer reactions, equilibrium favors the weaker acid and weaker base (2.4).
- An acid can be deprotonated by the conjugate base of any acid having a higher pK_a (2.4).

Periodic Trends in Acidity and Basicity (2.5A)

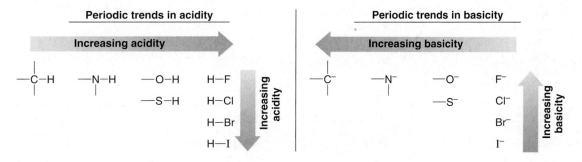

Factors that Determine Acidity (2.5)

[1] **Element effects** (2.5A)	The acidity of H—A increases both left-to-right across a row and down a column of the periodic table.	
[2] **Inductive effects** (2.5B)	The acidity of H—A increases with the presence of electron-withdrawing groups in A.	
[3] **Resonance effects** (2.5C)	The acidity of H—A increases when the conjugate base A:$^-$ is resonance stabilized.	
[4] **Hybridization effects** (2.5D)	The acidity of H—A increases as the percent s-character of the A:$^-$ increases.	

Problems

Brønsted–Lowry Acids and Bases

2.27 What is the conjugate acid of each base?

   a. H_2O b. $^-NH_2$ c. HCO_3^- d. $CH_3CH_2NHCH_3$ e. CH_3OCH_3 f. CH_3COO^-

2.28 What is the conjugate base of each acid?

   a. HCN   b. HCO_3^- c. $(CH_3)_2\overset{+}{N}H_2$ d. $HC{\equiv}CH$ e. CH_3CH_2COOH f. CH_3SO_3H

Reactions of Brønsted–Lowry Acids and Bases

2.29 Draw the products of each proton transfer reaction. Label the acid and base in the starting materials, and the conjugate acid and base in the products.

   a. $CH_3OH + {}^-NH_2 \rightleftharpoons$

   d. $(CH_3CH_2)_3N + H{-}Cl \rightleftharpoons$

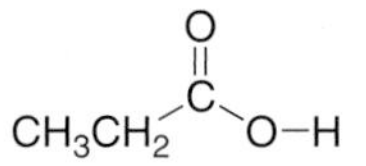

   b. $+ CH_3O^- \rightleftharpoons$

   e. $CH_3CH_2{-}O{-}H + H{-}Br \rightleftharpoons$

   c. $CH_3CH_2{-}C{\equiv}C{-}H + H^- \rightleftharpoons$

   f. $CH_3C{\equiv}C^- + H_2O \rightleftharpoons$

2.30 Each of the following reactions is written with an incorrect use of curved arrow notation. Rewrite each equation with correct curved arrows.

   a.

   b.

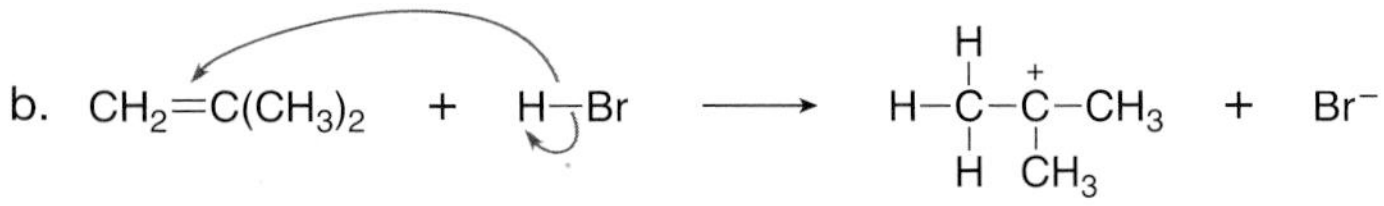

2.31 Draw the products of each acid–base reaction.

   a. 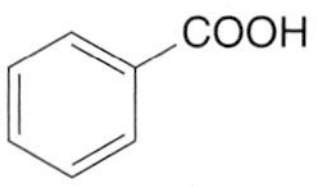 $+ NH_3 \rightleftharpoons$

   d. $CH_3CH_2{-}OH + KOH \rightleftharpoons$

   b. 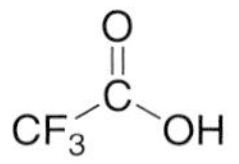 $+ NaHCO_3 \rightleftharpoons$

   e. $(CH_3)_2NH + $ $\rightleftharpoons$

   c. $CH_3{-}C{\equiv}C{-}H + NaNH_2 \rightleftharpoons$

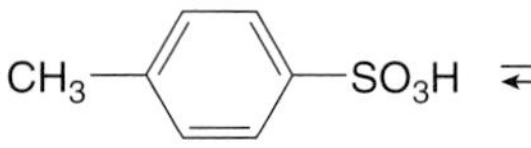

   f. 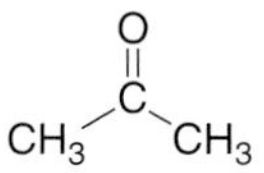 $+ H_2SO_4 \rightleftharpoons$

pK_a, K_a, and the Direction of Equilibrium

2.32 What is K_a for each compound? Use a calculator when necessary.

   a. H_2S
   $pK_a = 7.0$

   b. $ClCH_2COOH$
   $pK_a = 2.8$

   c.   HCN
   $pK_a = 9.1$

2.33 What is the pK_a for each compound? Use a calculator when necessary.

   a.

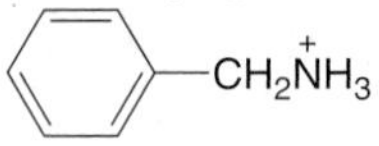

   b.

   c. CF_3COOH

   $K_a = 4.7 \times 10^{-10}$ $K_a = 2.3 \times 10^{-5}$ $K_a = 5.9 \times 10^{-1}$

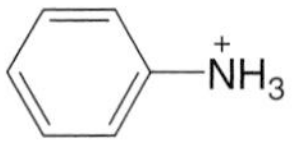

2.34 Which bases in Table 2.1 are strong enough to deprotonate each compound?

   a. H_2O b. NH_3 c. CH_4

2.35 Which compounds can be deprotonated by ^-OH, so that equilibrium favors the products? Refer to the pK_a table in Appendix B.

   a. HCOOH   b. H_2S c. 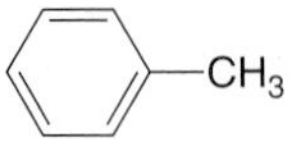 $-CH_3$ d. CH_3NH_2

2.36 Draw the products of each reaction. Use the pK_a table in Appendix B to decide if the equilibrium favors the starting materials or products.

a.
$CF_3C(=O)OH$ + $^-OCH_2CH_3$ ⇌

d.
phenol-OH + $NaHCO_3$ ⇌

b.
$CH_3CH_2C(=O)OH$ + $NaCl$ ⇌

e. $H-C\equiv C-H$ + $Li^+\ ^-CH_2CH_3$ ⇌

c. $(CH_3)_3COH$ + H_2SO_4 ⇌

f. CH_3NH_2 + H_2SO_4 ⇌

Relative Acid Strength

2.37 Rank the following compounds in order of increasing acidity.

a. NH_3, H_2O, HF
b. HBr, HCl, HF
c. H_2O, H_3O^+, HO^-
d. NH_3, H_2O, H_2S

e. CH_3OH, CH_3NH_2, CH_3CH_3
f. HCl, H_2O, H_2S
g. $CH_3CH_2CH_3$, $ClCH_2CH_2OH$, CH_3CH_2OH
h. $HC\equiv CCH_2CH_3$, $CH_3CH_2CH_2CH_3$, $CH_3CH=CHCH_3$

2.38 Rank the following ions in order of increasing basicity.

a. $CH_3\bar{C}H_2$, CH_3O^-, $CH_3\bar{N}H$

c. CH_3COO^-, $CH_3CH_2O^-$, $ClCH_2COO^-$

b. CH_3^-, HO^-, Br^-

d.
$\langle\text{cyclohexyl}\rangle-C\equiv C^-$, $\langle\text{cyclohexyl}\rangle-CH_2\bar{C}H_2$, $\langle\text{cyclohexyl}\rangle-CH=\bar{C}H$

2.39 The pK_a of three different C–H bonds is given below.

$CH_3CH_2CH_2\text{—}H$
$pK_a = 50$

$CH_2=C(CH_3)CH_2\text{—}H$
$pK_a = 43$

$CH_3C(=O)CH_2\text{—}H$
$pK_a = 19.2$

a. For each compound, draw the conjugate base, including all possible resonance structures.
b. Explain the observed trend in pK_a.

2.40 a. What is the conjugate acid of **A**?
b. What is the conjugate base of **A**?

H_2N—⟨cyclohexyl⟩—OH
A

2.41 Many drugs are Brønsted–Lowry acids or bases.
a. What is the most acidic proton in the analgesic ibuprofen? Draw the conjugate base.
b. What is the most basic electron pair in cocaine? Draw the conjugate acid.

ibuprofen

cocaine

2.42 The pK_a of CH_3NO_2 is 10, making its C–H bond more acidic than most C–H bonds. Explain.

2.43 The indicated hydrogen in 1,4-pentadiene is more acidic than the indicated hydrogen in pentane. Explain.

1,4-pentadiene   pentane

2.44 The base NaH reacts readily with CH_3CH_2OH but does not react readily with CH_3CH_3. Explain.

2.45 Dimethyl ether (CH_3OCH_3) and ethanol (CH_3CH_2OH) are isomers, but CH_3OCH_3 has a pK_a of 40 and CH_3CH_2OH has a pK_a of 16. Why are these pK_a values so different?

2.46 a. What is the hybridization of the N atom in each compound?
b. Which of these compounds is the most basic?

$CH_3C{\equiv}N{:}$ $\qquad$ $CH_2{=}\ddot{N}CH_3$ $\qquad$ $CH_3{-}\ddot{N}(H){-}CH_3$

Lewis Acids and Bases

2.47 Classify each compound as a Lewis base, a Brønsted–Lowry base, both, or neither.

a. b. $CH_3{-}Cl$ c. d.

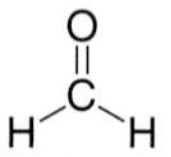

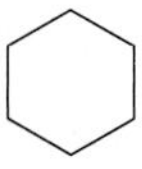

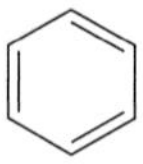

2.48 Classify each species as a Lewis acid, a Brønsted–Lowry acid, both, or neither.

a. H_3O^+ b. Cl_3C^+ c. BCl_3 d. BF_4^-

Lewis Acid–Base Reactions

2.49 Label the Lewis acid and Lewis base in each reaction. Use curved arrows to show the movement of electron pairs.

a.

c.

b.

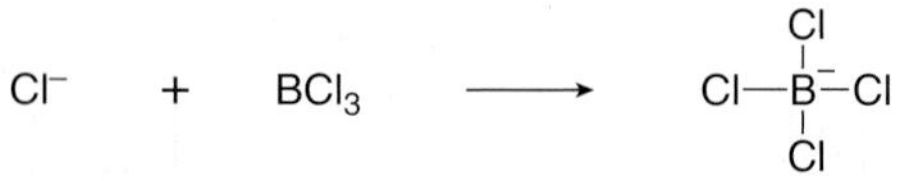

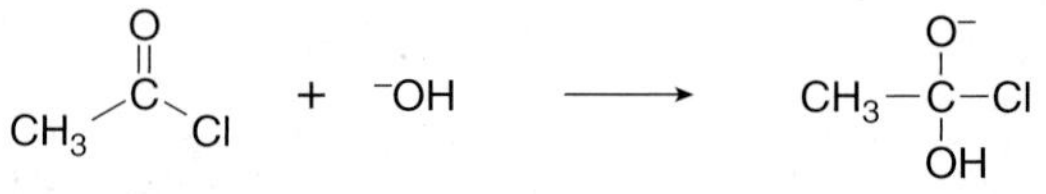

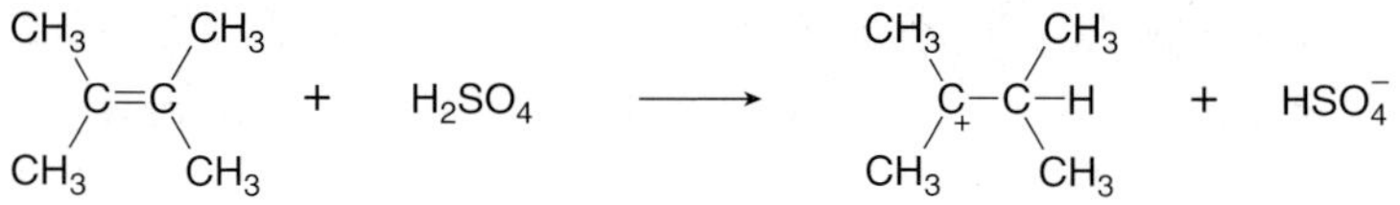

2.50 Draw the products of each Lewis acid–base reaction. Label the electrophile and nucleophile.

a. $CH_3CH_2OH + BF_3 \longrightarrow$

b. $CH_3SCH_3 + AlCl_3 \longrightarrow$

c.

d.

e. $Br_2 + FeBr_3 \longrightarrow$

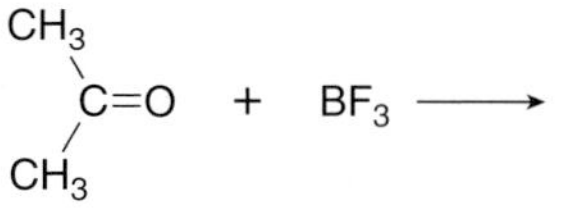

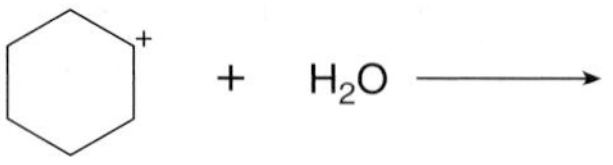

General

2.51 Classify each reaction as either a proton transfer reaction, or reaction of a nucleophile with an electrophile. Use curved arrows to show how the electron pairs move.

a.

b.

c.

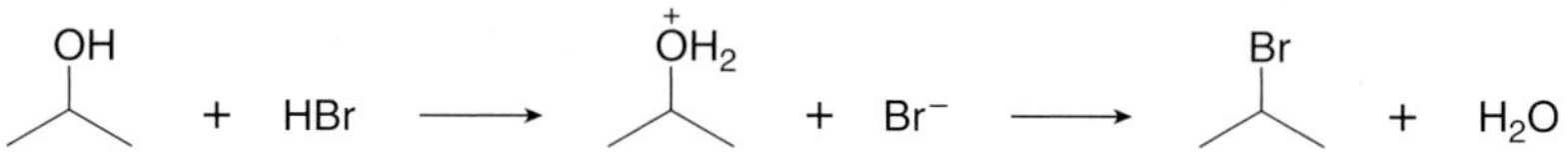

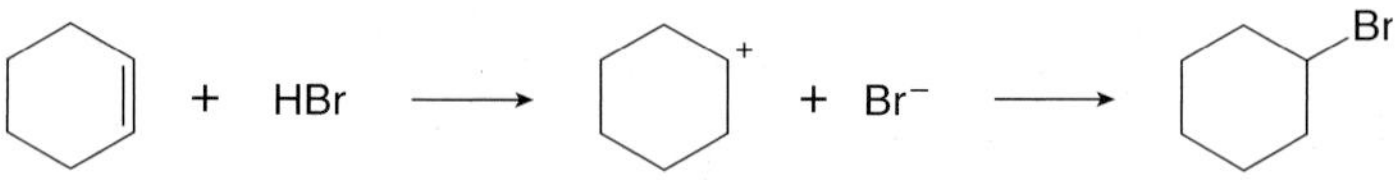

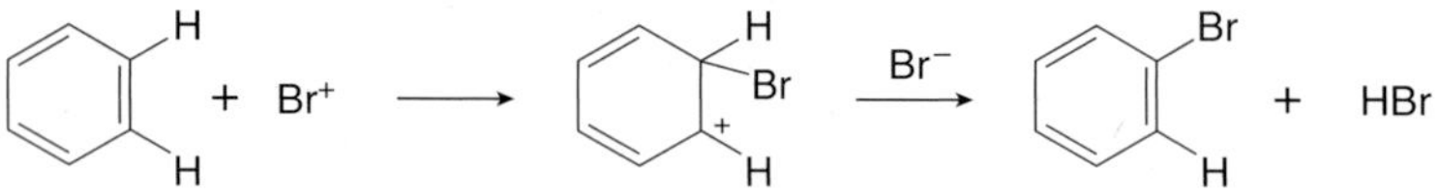

Challenge Problems

2.52 Molecules like acetamide can be protonated on either their O or N atoms when treated with a strong acid like HCl. Which site is more readily protonated and why?

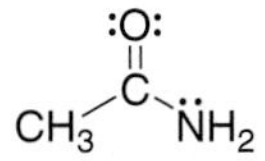

$$\underset{\text{acetamide}}{CH_3 - \overset{\overset{\displaystyle :O:}{\|}}{C} - \ddot{N}H_2}$$

2.53 Two pK_a values are reported for a compound like malonic acid that has two COOH groups.

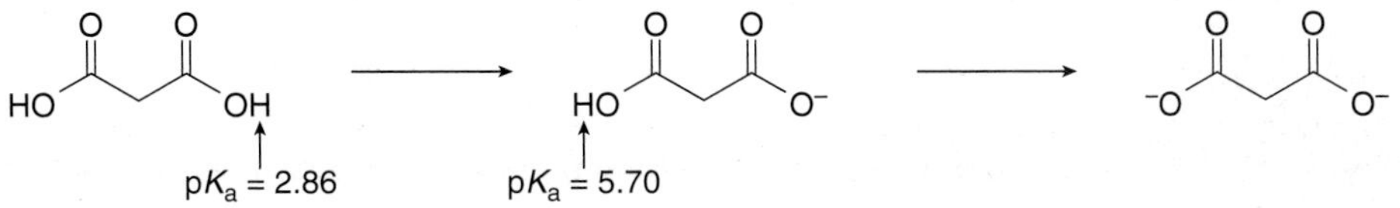

pK$_a$ = 2.86   pK$_a$ = 5.70

Explain why one pK_a is lower and one pK_a is higher than the pK_a of acetic acid (CH$_3$COOH, pK_a = 4.8).

2.54 Amino acids such as glycine are the building blocks of large molecules called proteins that give structure to muscle, tendon, hair, and nails.

$$\underset{\text{glycine}}{NH_2CH_2 - \overset{\overset{\displaystyle O}{\|}}{C} - OH} \qquad \underset{\text{zwitterion form}}{\overset{+}{N}H_3CH_2 - \overset{\overset{\displaystyle O}{\|}}{C} - O^-}$$

a. Explain why glycine does not actually exist in the form with all atoms uncharged, but actually exists as a salt called a zwitterion.
b. What product is formed when glycine is treated with concentrated HCl?
c. What product is formed when glycine is treated with NaOH?

2.55 Write a stepwise reaction sequence using proton transfer reactions to show how the following reaction occurs.

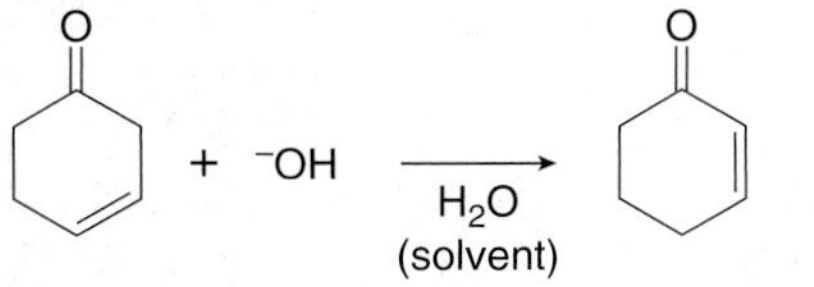

2.56 Which H atom in vitamin C (ascorbic acid) is most acidic?

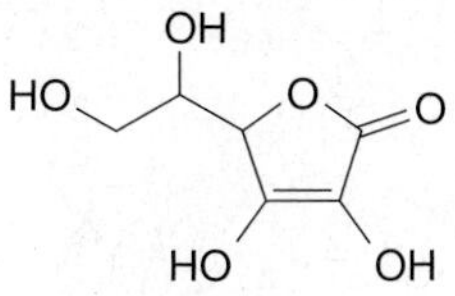

vitamin C
ascorbic acid

Introduction to Organic Molecules and Functional Groups

Vitamin A, or retinol, is a key component of the retina's visual receptors. It also helps to maintain the health of mucous membranes and the skin, so many anti-aging creams contain vitamin A. Vitamin A is synthesized in the body from β-carotene, the orange pigment found in carrots. In Chapter 3, we will learn why some vitamins like vitamin A can be stored in the fat cells in the body, whereas others like vitamin C are excreted in urine.

H aving learned some basic concepts about structure, bonding, and acid–base chemistry in Chapters 1 and 2, we will now concentrate on organic molecules.

    ◆ What are the characteristic features of an organic compound?

    ◆ What determines the properties of an organic compound?

After these questions are answered, we will use these principles to understand some important phenomena. For example, why do we store some vitamins in the body and readily excrete others? How does soap clean away dirt? We will also see that the properties of organic molecules explain some basic biological phenomena, such as the structure of cell membranes and the transport of species across these membranes. Finally, we will examine how the presence of certain structural features in a molecule allows us to predict the chemical reactions the molecule undergoes.

◆

3.1   Functional Groups

What are the characteristic features of an organic compound? Most organic molecules have C–C and C–H σ bonds. These bonds are strong, nonpolar, and not readily broken. Organic molecules may have the following structural features as well:

    ◆ *Heteroatoms—atoms other than carbon or hydrogen.* Common heteroatoms are nitrogen, oxygen, sulfur, phosphorus, and the halogens.

    ◆ *π Bonds.* The most common π bonds occur in C–C and C–O double bonds.

These structural features distinguish one organic molecule from another. They determine a molecule's geometry, physical properties, and reactivity, and comprise what is called a **functional group.**

> ◆ A *functional group* is an atom or a group of atoms with characteristic chemical and physical properties. It is the *reactive part* of the molecule.

Why do heteroatoms and π bonds confer reactivity on a particular molecule?

> ◆ Heteroatoms have lone pairs and create electron-deficient sites on carbon.
> ◆ π Bonds are easily broken in chemical reactions. A π bond makes a molecule a base and a nucleophile.

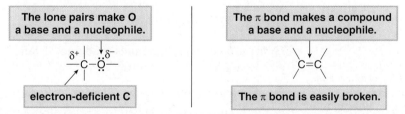

Don't think, though, that the C–C and C–H σ bonds are unimportant. They form the **carbon backbone** or **skeleton** to which the functional groups are bonded. A functional group usually behaves the same whether it is bonded to a carbon skeleton having as few as two or as many as twenty carbons. For this reason, we often abbreviate the carbon and hydrogen portion of the molecule by a capital letter **R,** and draw the **R** bonded to a particular functional group.

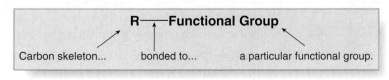

Ethane, for example, has only C−C and C−H σ bonds, so it has no functional group. Ethane has no polar bonds, no lone pairs, and no π bonds, so it has **no reactive sites.** Because of this, ethane and molecules like it are very unreactive.

Ethanol, on the other hand, has two carbons and five hydrogens in its carbon backbone, as well as an OH group, a functional group called a **hydroxy** group. Ethanol has lone pairs and polar bonds that make it reactive with a variety of reagents, including the acids and bases discussed in Chapter 2. The hydroxy group makes the properties of ethanol very different from the properties of ethane. Moreover, any organic molecule containing a hydroxy group has properties similar to ethanol.

Ethane
- all C−C and C−H σ bonds
- no functional group

Ethanol
carbon backbone
hydroxy group
- polar C−O and O−H bonds
- two lone pairs

Most organic compounds can be grouped into a relatively small number of categories, based on the structure of their functional group. Ethane, for example, is an **alkane,** whereas ethanol is a simple **alcohol.**

PROBLEM 3.1   What reaction occurs when CH_3CH_2OH is treated with (a) H_2SO_4? (b) With NaH? What happens when CH_3CH_3 is treated with these same reagents?

3.2 An Overview of Functional Groups

We can subdivide the most common functional groups into three types. A more complete list of functional groups is presented on the inside back cover.

◆ **Hydrocarbons**

◆ **Compounds containing a C−Z σ bond** where Z = an electronegative element

◆ **Compounds containing a C=O group**

3.2A Hydrocarbons

Hydrocarbons **are compounds made up of only the elements carbon and hydrogen.** They may be **aliphatic** or **aromatic.**

To review the structure and bonding of the simple aliphatic hydrocarbons, return to Section 1.9.

[1] **Aliphatic hydrocarbons.** Aliphatic hydrocarbons can be divided into three subgroups.

◆ *Alkanes* have only C−C σ bonds and no functional group. Ethane, **CH_3CH_3,** is a simple alkane.

The word *aliphatic* is derived from the Greek word *aleiphas* meaning *fat.* Aliphatic compounds have physical properties similar to fats.

◆ *Alkenes* have a C−C double bond as functional group. Ethylene, **$CH_2=CH_2$,** is a simple alkene.

◆ *Alkynes* have a C−C triple bond as functional group. Acetylene, **$HC≡CH$,** is a simple alkyne.

[2] **Aromatic hydrocarbons.** This class of hydrocarbons was so named because many of the earliest known aromatic compounds had strong, characteristic odors.

The simplest aromatic hydrocarbon is **benzene.** The six-membered ring and three π bonds of benzene comprise a *single* functional group.

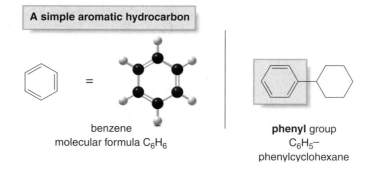

A simple aromatic hydrocarbon

benzene
molecular formula C_6H_6

phenyl group
C_6H_5-
phenylcyclohexane

When a benzene ring is bonded to another group, it is called a **phenyl group.** In phenylcyclohexane, for example, a phenyl group is bonded to the six-membered cyclohexane ring.

Table 3.1 summarizes the four different types of hydrocarbons.

TABLE 3.1	Hydrocarbons		
Type of compound	**General structure**	**Example**	**Functional group**
Alkane	R—H	CH_3CH_3	– –
Alkene	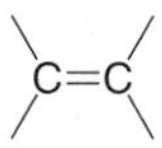 $\diagdown C=C \diagup$	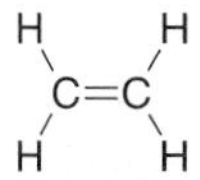 $\overset{H}{\underset{H}{\diagdown}} C = C \overset{H}{\underset{H}{\diagup}}$	double bond
Alkyne	—C≡C—	H—C≡C—H	triple bond
Aromatic compound	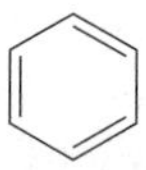 ⬡	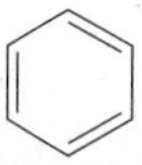 ⬡	phenyl group

Alkanes, which have no functional groups, are notoriously unreactive except under very drastic conditions. For example, **polyethylene** is a synthetic plastic and high molecular weight alkane, consisting of chains of $-CH_2-$ groups bonded together, hundreds or even thousands of atoms long. Because it is an alkane with no reactive sites, it is a very stable compound that does not readily degrade and thus persists for years in landfills.

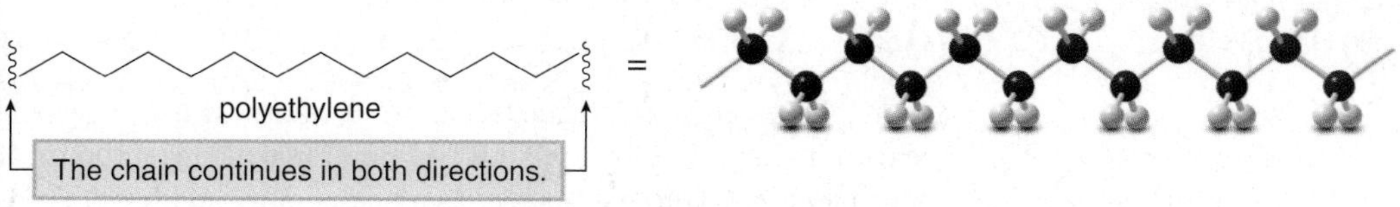

polyethylene

The chain continues in both directions.

3.2B Compounds Containing C–Z σ Bonds

Several types of functional groups that contain C–Z σ bonds are listed in Table 3.2. The electronegative heteroatom Z creates a polar bond, making carbon electron deficient. The lone pairs on Z are available for reaction with protons and other electrophiles, especially when Z = N or O.

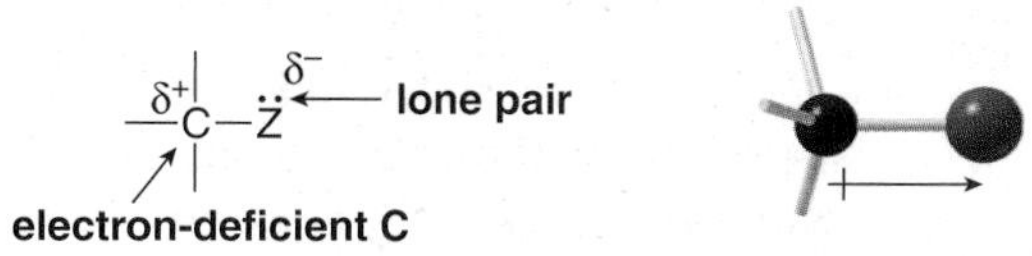

$$\overset{\delta^+}{\underset{|}{\overset{|}{C}}} - \overset{\delta^-}{\underset{\cdot\cdot}{Z}} \longleftarrow \text{lone pair}$$

electron-deficient C

TABLE 3.2   Compounds Containing C–Z σ Bonds

Type of compound	General structure	Example	3-D structure	Functional group
Alkyl halide	R—Ẍ: (X=F, Cl, Br, I)	CH_3—B̈r:	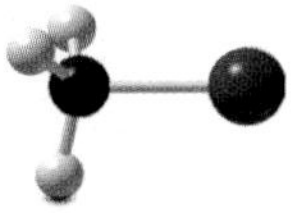	**–X** halo group
Alcohol	R—Ö H	CH_3—Ö H		**–OH** hydroxy group
Ether	R—Ö—R	CH_3—Ö—CH_3	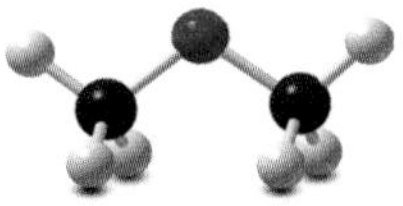	**–OR** alkoxy group
Amine	R—N̈H_2 or R_2N̈H or R_3N̈	CH_3—N̈H_2	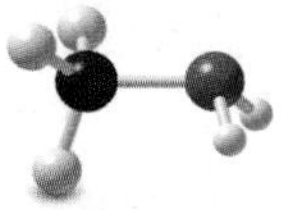	**–NH₂** amino group
Thiol	R—S̈ H	CH_3—S̈ H	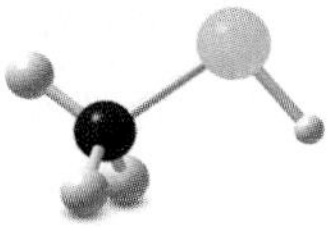	**–SH** mercapto group
Sulfide	R—S̈—R	CH_3—S̈—CH_3		**–SR** alkylthio group

Diethyl ether, the first ether used for general anesthesia, is discussed in greater detail in Chapter 9.

Hemibrevetoxin B is a neurotoxin produced by algae blooms referred to as "red tides," because of the color often seen in shallow ocean waters when these algae proliferate.

Molecules containing these functional groups may be simple or very complex. Diethyl ether, for example, is a simple ether because it contains a single O atom, depicted in red, bonded to two C atoms. Hemibrevetoxin B, on the other hand, is more complex. It contains four ether groups, in addition to other functional groups.

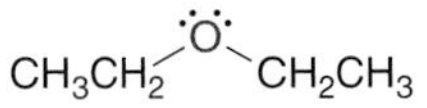

$$CH_3CH_2 \overset{\cdot\cdot}{\underset{\cdot\cdot}{O}} CH_2CH_3$$

diethyl ether

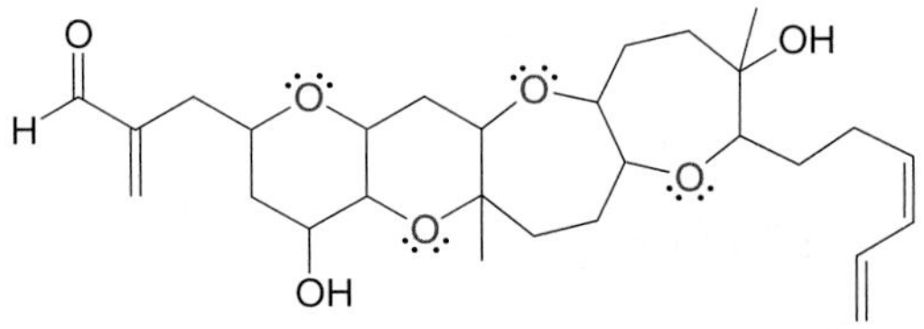

hemibrevetoxin B

3.2C   Compounds Containing a C=O Group

Many different types of functional groups possess a C–O double bond (a **carbonyl group**), as shown in Table 3.3. The polar C–O bond makes the carbonyl carbon an electrophile, while the lone pairs on O allow it to react as a nucleophile and base. The carbonyl group also contains a π bond that is more easily broken than a C–O σ bond.

Reactive features of a carbonyl group

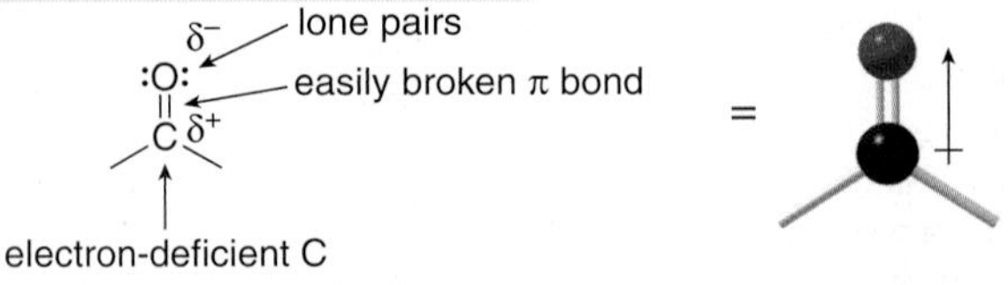

lone pairs

easily broken π bond

electron-deficient C

Atenolol and Viracept are examples of useful drugs that contain a variety of functional groups. **Atenolol is a β blocker,** a group of drugs used to treat hypertension. **Viracept is a protease inhibitor,** a class of drugs used to treat HIV.

TABLE 3.3 Compounds Containing a C=O Group

Type of compound	General structure	Example	3-D structure	Functional group
Aldehyde	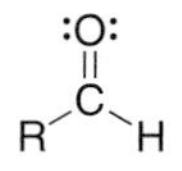 R–CHO (:O: on C, R and H)	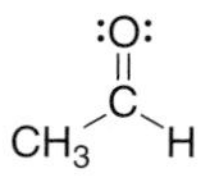 CH₃–CHO	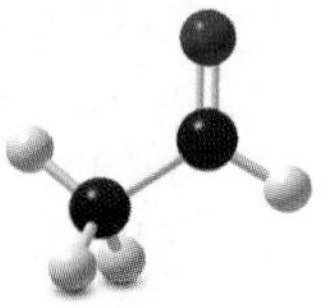	C=O carbonyl group
Ketone	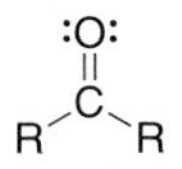 R–CO–R	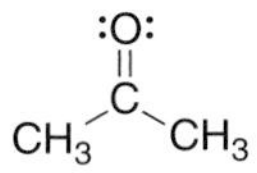 CH₃–CO–CH₃	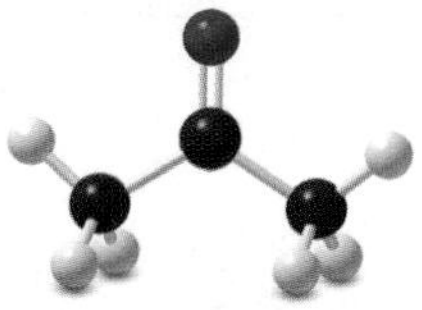	C=O carbonyl group
Carboxylic acid	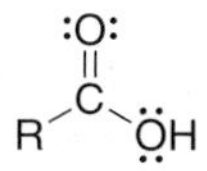 R–CO–OH	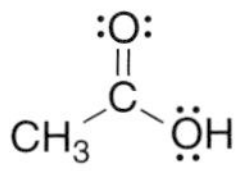 CH₃–CO–OH	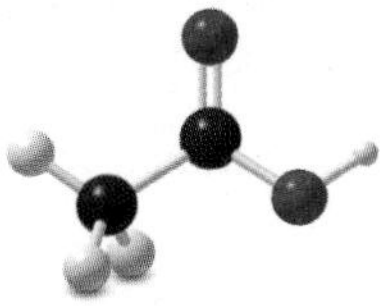	−COOH carboxy group
Ester	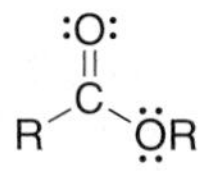 R–CO–OR	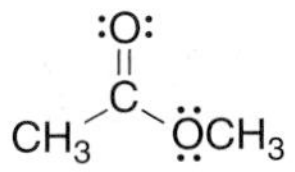 CH₃–CO–OCH₃	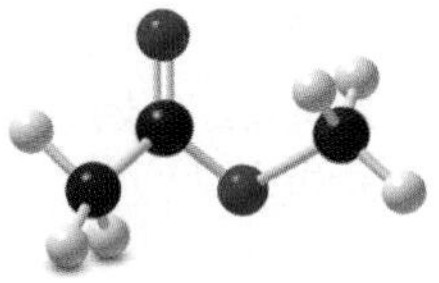	−COOR
Amide	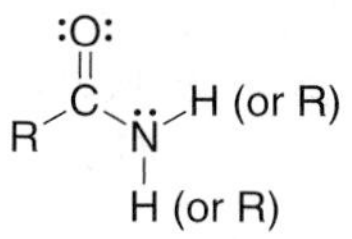 R–CO–N with H (or R) and H (or R)	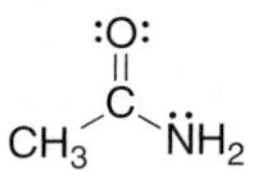 CH₃–CO–NH₂	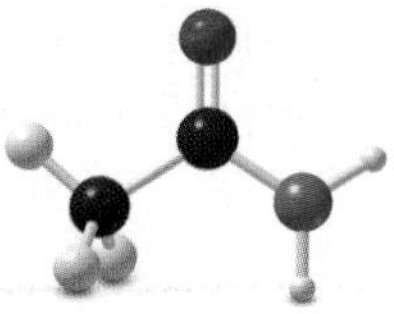	−CONH₂, −CONHR, or −CONR₂
Acid chloride	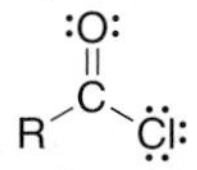 R–CO–Cl:	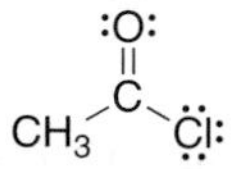 CH₃–CO–Cl:	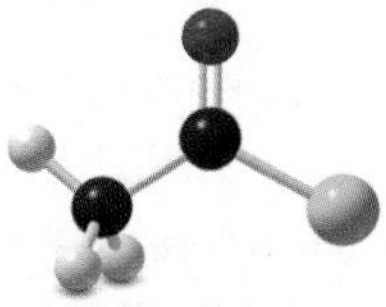	−COCl

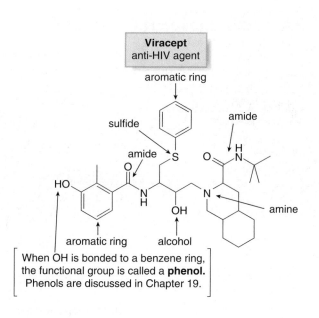

Viracept
anti-HIV agent

aromatic ring
sulfide
amide
amide
aromatic ring
alcohol
amide
amine

When OH is bonded to a benzene ring, the functional group is called a **phenol**. Phenols are discussed in Chapter 19.

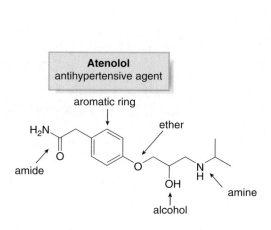

Atenolol
antihypertensive agent

aromatic ring
ether
amide
alcohol
amine

The importance of a functional group cannot be overstated. A functional group determines all the following properties of a molecule:

- bonding and shape
- type and strength of intermolecular forces
- physical properties
- nomenclature
- chemical reactivity

PROBLEM 3.2 Identify the functional groups in each of the following molecules:

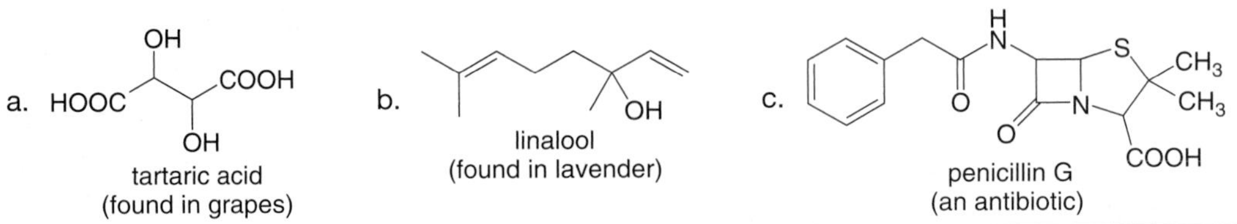
a. HOOC
OH
COOH
OH
tartaric acid
(found in grapes)
b.
OH
linalool
(found in lavender)
c.
H
N
S
CH3
CH3
O
N
O
COOH
penicillin G
(an antibiotic)

3.3 Biomolecules

***Biomolecules* are organic compounds found in biological systems.** Many are relatively small, with molecular weights of less than 1000 g/mol. There are four main families of these small molecules: amino acids, simple sugars, lipids, and nucleotides. A specific example of each of these families is illustrated in Figure 3.1. Some of these molecules are used to synthesize larger compounds. For example, simple sugars (such as glucose) combine to form complex carbohydrates (such as starch and cellulose). Some of these molecules also have special biological functions in the cell or are involved in reactions that build tissue or metabolize food.

- Unlike many simple organic molecules, **biomolecules often have several functional groups.** For example, the simple sugar glucose has five hydroxyl groups and an additional oxygen atom.
- **Biomolecules are organic compounds.** As such, their properties and chemistry are usually understandable in terms of basic organic chemistry.

Therefore, the fundamental properties of biomolecules will be integrated into our study of organic chemistry.

3.4 Intermolecular Forces

The types of interactions that exist *between* molecules are called **intermolecular forces.** A functional group determines the type and strength of these interactions.

Figure 3.1 Examples of small biomolecules

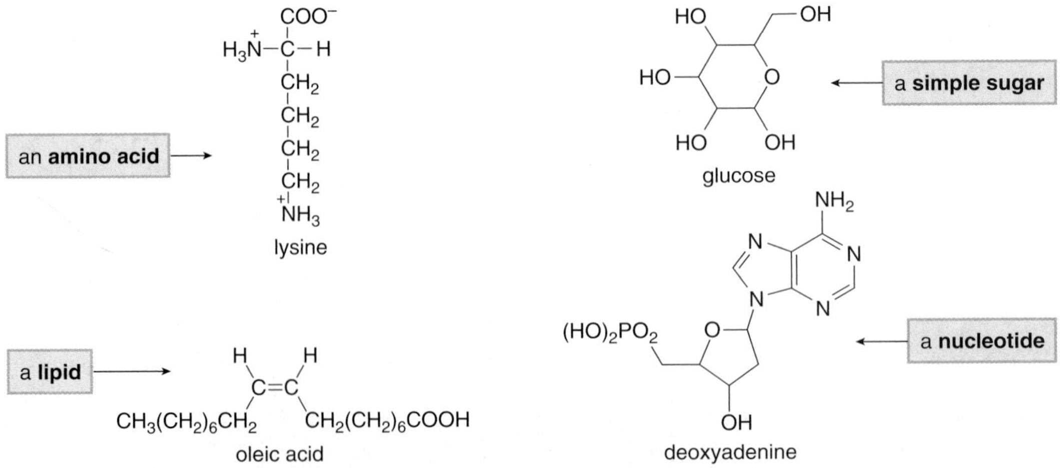
an amino acid
COO−
H3N−C−H
CH2
CH2
CH2
CH2
NH3
lysine
HO
OH
HO
O
HO
OH
glucose
a simple sugar
NH2
N
N
N
N
(HO)2PO2
O
OH
deoxyadenine
a nucleotide
a lipid
H
H
C=C
CH3(CH2)6CH2
CH2(CH2)6COOH
oleic acid

3.4A   Types of Intermolecular Forces

Intermolecular forces are also referred to as **noncovalent interactions** or **nonbonded interactions.**

Ionic compounds contain oppositely charged particles held together by **extremely strong electrostatic interactions.** These ionic interactions are much stronger than the intermolecular forces present between covalent molecules, so it takes a great deal of energy to separate oppositely charged ions from each other.

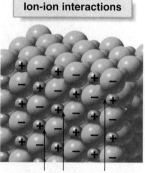

Ion–ion interactions

strong electrostatic interaction

Covalent compounds are composed of discrete molecules. The nature of the forces between the molecules depends on the functional group present. There are three different types of interactions, presented here in order of *increasing strength:*

- ◆ **van der Waals forces**
- ◆ **dipole–dipole interactions**
- ◆ **hydrogen bonding**

van der Waals Forces

van der Waals forces are very weak interactions caused by the **momentary changes in electron density in a molecule.** van der Waals forces are the only attractive forces present in nonpolar compounds.

For example, although a nonpolar CH_4 molecule has no net dipole, at any one instant its electron density may not be completely symmetrical, creating a *temporary* dipole. This can induce a temporary dipole in another CH_4 molecule, with the partial positive and negative charges arranged close to each other. **The weak interaction of these temporary dipoles constitutes van der Waals forces.** All compounds exhibit van der Waals forces.

van der Waals forces are also called **London forces.**

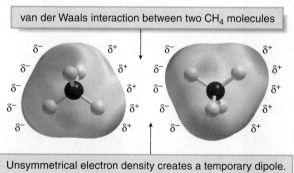

van der Waals interaction between two CH_4 molecules

Unsymmetrical electron density creates a temporary dipole.

The surface area of a molecule determines the strength of the van der Waals interactions between molecules. **The larger the surface area, the larger the attractive force between two molecules, and the stronger the intermolecular forces.** Long, sausage-shaped molecules such as $CH_3CH_2CH_2CH_2CH_3$ (pentane) have stronger van der Waals interactions than compact spherical ones like $C(CH_3)_4$ (neopentane), as shown in Figure 3.2.

Another factor affecting the strength of van der Waals forces is **polarizability.**

- ◆ *Polarizability* is a measure of how the electron cloud around an atom responds to changes in its electronic environment.

Figure 3.2 Surface area and van der Waals forces

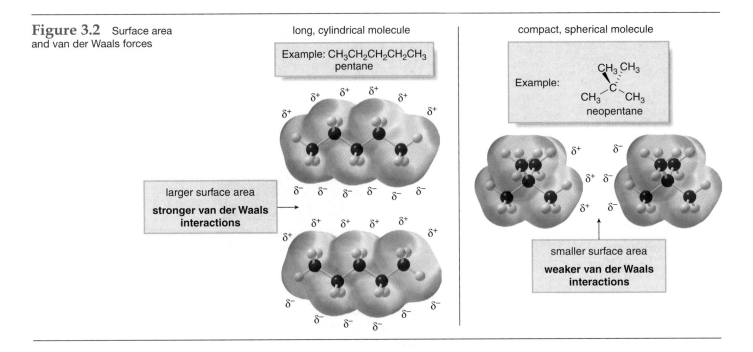

long, cylindrical molecule

Example: $CH_3CH_2CH_2CH_2CH_3$
pentane

larger surface area
stronger van der Waals interactions

compact, spherical molecule

Example:
neopentane

smaller surface area
weaker van der Waals interactions

Although any single van der Waals interaction is weak, a large number of van der Waals interactions creates a strong force. For example, geckos stick to walls and ceilings by van der Waals interactions of the surfaces with the 500,000 tiny hairs on each foot.

Larger atoms like iodine, which have more loosely held valence electrons, are more polarizable than smaller atoms like fluorine, which have more tightly held electrons. Because larger atoms have more easily induced dipoles, compounds containing them possess stronger intermolecular interactions.

Thus, two F_2 molecules have little force of attraction between them, because the electrons are held very tightly and temporary dipoles are difficult to induce. On the other hand, two I_2 molecules exhibit a much stronger force of attraction because the electrons are held much more loosely and temporary dipoles are easily induced.

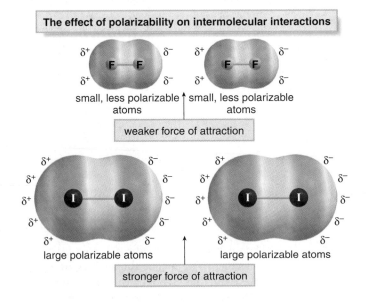

The effect of polarizability on intermolecular interactions

small, less polarizable atoms   small, less polarizable atoms

weaker force of attraction

large polarizable atoms   large polarizable atoms

stronger force of attraction

Dipole–Dipole Interactions

Dipole–dipole interactions **are the attractive forces between the permanent dipoles of two polar molecules.** In acetone, $(CH_3)_2C=O$, for example, the dipoles in adjacent molecules align so that the partial positive and partial negative charges are in close proximity. These attractive forces caused by permanent dipoles are much stronger than weak van der Waals forces.

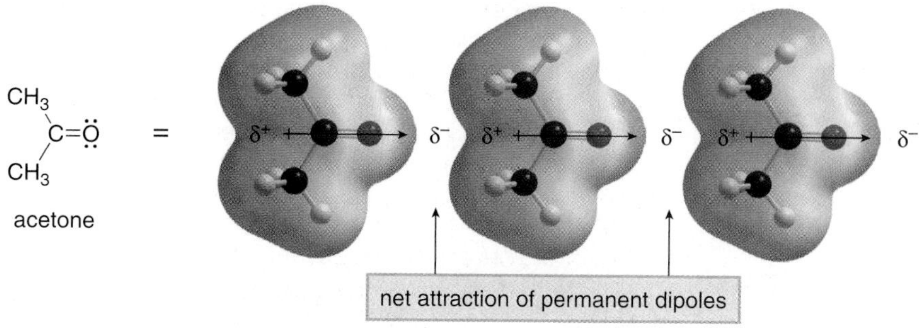

$$CH_3$$
$$\underset{CH_3}{\overset{CH_3}{C}}=\ddot{O}\quad =$$

acetone

net attraction of permanent dipoles

Hydrogen Bonding

Hydrogen bonding **typically occurs when a hydrogen atom bonded to O, N, or F, is electro-statically attracted to a lone pair of electrons on an O, N, or F atom in another molecule.** Thus, H_2O molecules can hydrogen bond to each other. When they do, an H atom covalently bonded to O in one water molecule is attracted to a lone pair of electrons on the O in another water molecule. Hydrogen bonds are the strongest of the three types of intermolecular forces, though they are still much weaker than any covalent bond.

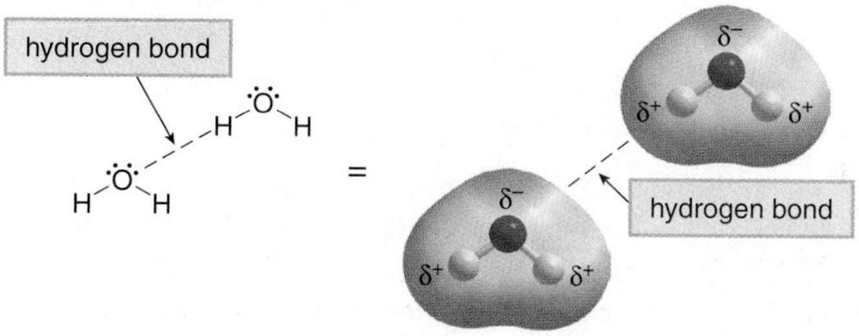

hydrogen bond

hydrogen bond

Table 3.4 summarizes the four types of interactions that affect the properties of all compounds. It is often necessary to determine the *relative* strength of intermolecular forces for a group of compounds. This is illustrated in Sample Problem 3.1.

TABLE 3.4	Summary of Types of Intermolecular Forces		
Type of force	**Relative strength**	**Exhibited by**	**Example**
van der Waals	weak	all molecules	$CH_3CH_2CH_2CH_2CH_3$ $CH_3CH_2CH_2CHO$ $CH_3CH_2CH_2CH_2OH$
dipole–dipole	moderate	molecules with a net dipole	$CH_3CH_2CH_2CHO$ $CH_3CH_2CH_2CH_2OH$
hydrogen bonding	strong	molecules with an O–H, N–H, or H–F bond	$CH_3CH_2CH_2CH_2OH$
ion–ion	very strong	ionic compounds	NaCl, LiF

SAMPLE PROBLEM 3.1

Rank the following compounds in order of increasing strength of intermolecular forces: $CH_3CH_2CH_2CH_2CH_3$ (pentane), $CH_3CH_2CH_2CH_2OH$ (1-butanol), and $CH_3CH_2CH_2CHO$ (butanal).

SOLUTION

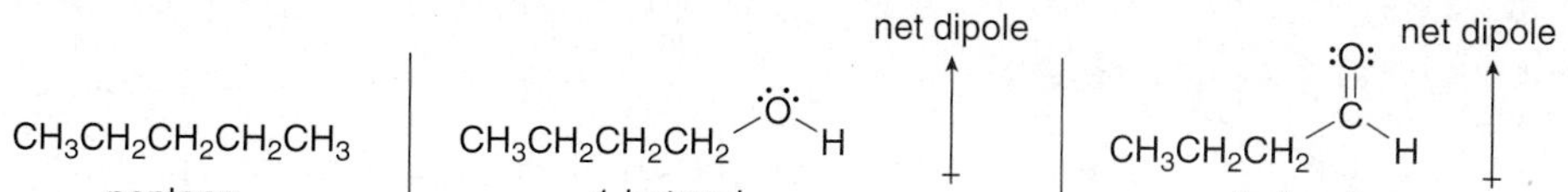

net dipole   net dipole

$$CH_3CH_2CH_2CH_2CH_3$$

pentane

• nonpolar molecule

$$CH_3CH_2CH_2CH_2\overset{\ddot{O}}{\diagdown}H$$

1-butanol

• bent molecule around O
• polar C–O and O–H bonds
• O–H bond for hydrogen bonding

$$CH_3CH_2CH_2\overset{:\ddot{O}:}{\diagdown}\overset{\parallel}{C}\diagdown H$$

butanal

• trigonal planar C
• polar C=O bond

• Pentane has only nonpolar C–C and C–H bonds, so its molecules are held together only by **van der Waals** forces.

- 1-Butanol is a polar bent molecule, so it can have **dipole–dipole** interactions in addition to **van der Waals** forces. Because it has an O–H bond, 1-butanol molecules are held together by intermolecular **hydrogen bonds** as well.
- Butanal has a trigonal planar carbon with a polar C=O bond, so it exhibits **dipole–dipole** interactions in addition to **van der Waals** forces. There is *no* H atom bonded to O, so two butanal molecules *cannot* hydrogen bond to each other.

ANSWER

$CH_3CH_2CH_2CH_2CH_3$ $CH_3CH_2CH_2CHO$ $CH_3CH_2CH_2CH_2OH$

Increasing strength of intermolecular forces →

PROBLEM 3.3   What types of intermolecular forces are present in each compound?

a. 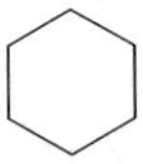 [hexagon structure]

c. $(CH_3CH_2)_3N$

e. $CH_3CH_2CH_2COOH$

b. [cyclic ether structure]

d. $CH_2=CHCl$

f. $CH_3-C\equiv C-CH_3$

3.4B   Application: Hydrogen Bonding in Nylon

Hydrogen bonding plays an important role in many organic molecules, including the synthetic fiber **nylon.**

Nylon is a *polymer,* **a large organic molecule composed of smaller units—called *monomers*— that are covalently bonded to each other in a repeating pattern.** It is also called a **polyamide,** because its carbon skeleton contains many amide units. A given molecule of nylon contains many thousands of atoms, so that only a portion is illustrated below.

Nylon was first synthesized at DuPont by Wallace Carothers in the 1930s. After Japanese control of silk production in the South Pacific made silk unavailable to the United States during World War II, nylon was used to replace silk in the manufacture of parachutes. The synthesis and properties of nylon are discussed in Chapter 22.

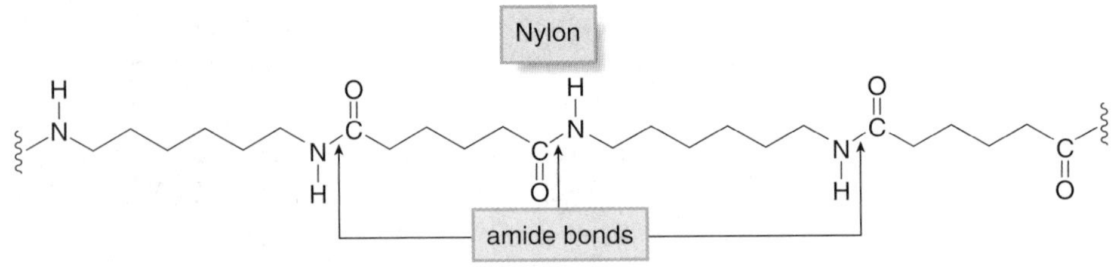

Nylon

amide bonds

The long chains of nylon are held together by hydrogen bonding interactions, with the N–H of one chain hydrogen bonded to an O atom in an adjacent chain. These intermolecular forces help strengthen nylon fibers.

Hydrogen bonding helps determine the three-dimensional shape of large biomolecules, too, such as carbohydrates and proteins. See Chapters 27 and 28 for details.

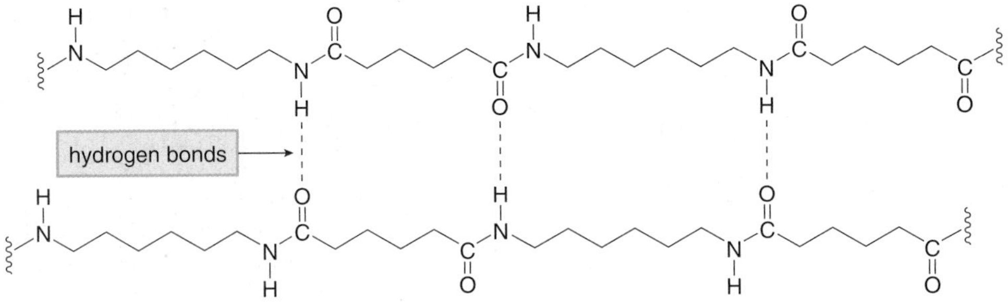

hydrogen bonds →

3.5   Physical Properties

The strength of a compound's intermolecular forces determines many of its physical properties, including its boiling point, melting point, and solubility.

3.5A   Boiling Point (bp)

The *boiling point* of a compound is the temperature at which a liquid is converted to a gas. In boiling, energy is needed to overcome the attractive forces in the more ordered liquid state.

Relative boiling points of compounds are understood by first knowing the strength of the forces that hold the liquid molecules together.

◆ **The stronger the intermolecular forces, the higher the boiling point.**

Because **ionic compounds** are held together by extremely strong interactions, they have **very high boiling points.** The boiling point of NaCl, for example, is 1413 °C. **With covalent molecules, the boiling point depends on the identity of the functional group.** For compounds of approximately the same molecular weight:

compounds with van der Waals forces	compounds with dipole−dipole interactions	compounds with hydrogen bonding

Increasing strength of intermolecular forces
Increasing boiling point

Recall from Sample Problem 3.1, for example, that the relative strength of the intermolecular forces increases from pentane to butanal to 1-butanol. The boiling points of these compounds increase in the same order.

$$CH_3CH_2CH_2CH_2CH_3 \qquad CH_3CH_2CH_2CHO \qquad CH_3CH_2CH_2CH_2OH$$
pentane   butanal   1-butanol
bp = 36 °C   bp = 76 °C   bp = 118 °C

Increasing strength of intermolecular forces
Increasing boiling point

Because surface area and polarizability affect the strength of intermolecular forces, they also affect boiling point. For two compounds with similar functional groups:

◆ **The larger the surface area, the higher the boiling point.**
◆ **The more polarizable the atoms, the higher the boiling point.**

Examples of each phenomenon are illustrated in Figure 3.3. In comparing two ketones having the same type of intermolecular forces but differing in size, 3-pentanone has a higher boiling point than acetone because it has a greater molecular weight and larger surface area. In comparing two alkyl halides having the same number of carbon atoms, CH_3I has a higher boiling point than CH_3F because I is more polarizable than F.

Figure 3.3 Effect of surface area and polarizability on boiling point

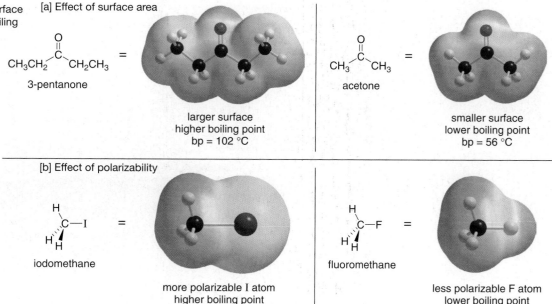

[a] Effect of surface area

CH_3CH_2—C(=O)—CH_2CH_3
3-pentanone

larger surface
higher boiling point
bp = 102 °C

CH_3—C(=O)—CH_3
acetone

smaller surface
lower boiling point
bp = 56 °C

[b] Effect of polarizability

H—C—I (iodomethane)

more polarizable I atom
higher boiling point
bp = 42 °C

H—C—F (fluoromethane)

less polarizable F atom
lower boiling point
bp = −78 °C

SAMPLE PROBLEM 3.2   Which compound in each pair has the higher boiling point?

a.   ⌇⌇⌇⌇   and   ⌇⌇⌇   b.   ⌇⌇⌇   and   ⌇⌇⌇ OH

A   B   C   D

SOLUTION

a. Isomers **A** and **B** have only nonpolar C–C and C–H bonds, so they exhibit only van der Waals forces. Because **B** is more compact, it has less surface area and a lower boiling point.

b. Compounds **C** and **D** have approximately the same molecular weight but different functional groups. **C** is a nonpolar alkane, exhibiting only van der Waals forces. **D** is an alcohol with an O–H group available for hydrogen bonding, so it has stronger intermolecular forces and a higher boiling point.

PROBLEM 3.4   Which compound in each pair has the higher boiling point?

a. $(CH_3)_2C=CH_2$ and $(CH_3)_2C=O$
b. CH_3CH_2COOH and CH_3COOCH_3
c. $CH_3(CH_2)_4CH_3$ and $CH_3(CH_2)_5CH_3$
d. $CH_2=CHCl$ and $CH_2=CHI$

Liquids having different boiling points can be separated in the laboratory using a *distillation* apparatus, which is pictured in Figure 3.4. When a mixture of two liquids is heated in the distilling flask, the lower boiling compound, called the **more volatile component,** distills first, followed by the **less volatile, higher boiling component.** By collecting the distillate in a series of receiver flasks, the two liquids can usually be separated from each other. The best separations are generally achieved, though, when the liquids in the mixture have widely different boiling points.

3.5B   Melting Point (mp)

The *melting point* is the temperature at which a solid is converted to its liquid phase. In melting, energy is needed to overcome the attractive forces in the more ordered crystalline solid. Two factors determine the melting point of a compound.

Figure 3.4   Schematic of a distillation apparatus

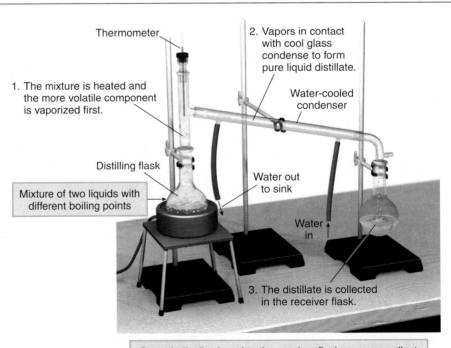

1. The mixture is heated and the more volatile component is vaporized first.

Thermometer

2. Vapors in contact with cool glass condense to form pure liquid distillate.

Water-cooled condenser

Distilling flask

Mixture of two liquids with different boiling points

Water out to sink

Water in

3. The distillate is collected in the receiver flask.

By periodically changing the receiver flask, one can collect compounds having different boiling points in separate flasks.

◆ The stronger the intermolecular forces, the higher the melting point.
◆ Given the same functional group, the more symmetrical the compound, the higher the melting point.

Because **ionic compounds** are held together by extremely strong interactions, they have **very high melting points.** For example, the melting point of NaCl is 801 °C. With covalent molecules, the melting point once again depends on the identity of the functional group. For compounds of approximately the same molecular weight:

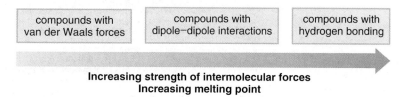

Increasing strength of intermolecular forces
Increasing melting point

The trend in the melting points of pentane, butanal, and 1-butanol parallels the trend observed in their boiling points.

$$CH_3CH_2CH_2CH_2CH_3 \qquad CH_3CH_2CH_2CHO \qquad CH_3CH_2CH_2CH_2OH$$
pentane   butanal   1-butanol
mp = −130 °C   mp = −96 °C   mp = −90 °C

Increasing strength of intermolecular forces
Increasing melting point

Symmetry also plays a role in determining the melting points of compounds having the same functional group and similar molecular weights, but very different shapes. A compact symmetrical molecule like neopentane packs well into a crystalline lattice whereas isopentane, which has a CH_3 group dangling from a four-carbon chain, does not. Thus, neopentane has a much higher melting point.

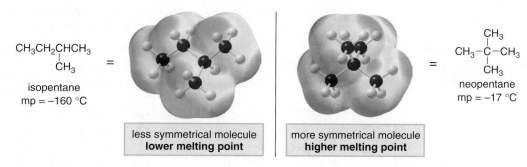

$CH_3CH_2CHCH_3$
|
CH_3
isopentane
mp = −160 °C

less symmetrical molecule
lower melting point

more symmetrical molecule
higher melting point

$CH_3-\overset{\overset{\displaystyle CH_3}{|}}{\underset{\underset{\displaystyle CH_3}{|}}{C}}-CH_3$
neopentane
mp = −17 °C

PROBLEM 3.5 Predict which compound in each pair has the higher melting point.

a. and NH_2 b. and

PROBLEM 3.6 Why do you suppose that symmetry affects the melting point of a compound but not its boiling point?

3.5C Solubility

Solubility **is the extent to which a compound, called the** *solute,* **dissolves in a liquid, called the** *solvent.* In dissolving a compound, the energy needed to break up the interactions between the molecules or ions of the solute comes from new interactions between the solute and the solvent.

Quantitatively, a compound may be considered soluble when 3 g of solute dissolves in 100 mL of solvent.

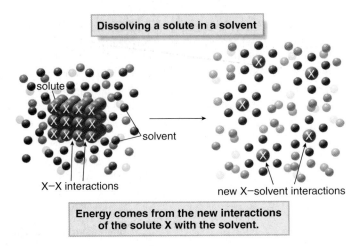

Dissolving a solute in a solvent

solute

solvent

X–X interactions

new X–solvent interactions

Energy comes from the new interactions of the solute X with the solvent.

Compounds dissolve in solvents having similar kinds of intermolecular forces.

◆ "Like dissolves like."
◆ Polar compounds dissolve in polar solvents. Nonpolar or weakly polar compounds dissolve in nonpolar or weakly polar solvents.

Water and **organic solvents** are two different kinds of solvents. Water is very polar because it is capable of hydrogen bonding with a solute. Many organic solvents are either nonpolar, like carbon tetrachloride (CCl_4) and hexane [$CH_3(CH_2)_4CH_3$], or weakly polar like diethyl ether ($CH_3CH_2OCH_2CH_3$).

Ionic compounds are held together by strong electrostatic forces, so they need very polar solvents to dissolve. **Most ionic compounds are soluble in water, but are insoluble in organic solvents.** To dissolve an ionic compound, the strong ion–ion interactions must be replaced by many weaker **ion–dipole interactions,** as illustrated in Figure 3.5.

Most organic compounds are soluble in organic solvents (remember, *like dissolves like*). **An organic compound is water soluble only if it contains one polar functional group capable of hydrogen bonding with the solvent for every five C atoms it contains.** In other words, a water-soluble organic compound has an O- or N-containing functional group that solubilizes its nonpolar carbon backbone.

Compare, for example, the solubility of butane and acetone in H_2O and CCl_4.

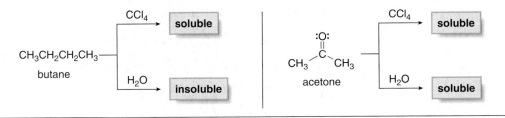

Figure 3.5   Dissolving an ionic compound in H_2O

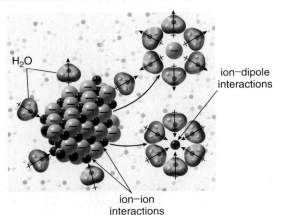

• When an ionic solid is dissolved in H_2O, the ion–ion interactions are replaced by ion–dipole interactions. Though these forces are weaker, there are so many of them that they compensate for the stronger ionic bonds.

Because butane and acetone are both organic compounds having a C–C and C–H backbone, they are soluble in the organic solvent CCl$_4$. Butane, a nonpolar molecule, is insoluble in the polar solvent H$_2$O. Acetone, however, is H$_2$O soluble because it contains only three C atoms and its O atom can hydrogen bond with an H atom of H$_2$O. In fact, acetone is so soluble in water that acetone and water are **miscible**—they form solutions in all proportions with each other.

> (CH$_3$)$_2$C=O molecules cannot hydrogen bond to each other because they have no OH group. However, (CH$_3$)$_2$C=O can hydrogen bond to H$_2$O because its O atom can hydrogen bond to one of the H atoms of H$_2$O.

Hydrogen bonding makes the small polar molecule acetone H$_2$O soluble.

> For an organic compound with one functional group, **a compound is water soluble only if it has ≤ five C atoms and contains an O or N atom.**

The size of an organic molecule with a polar functional group determines its water solubility. A low molecular weight alcohol like **ethanol is water soluble** because it has a small carbon skeleton (≤ five C atoms) compared to the size of its polar OH group. Cholesterol, on the other hand, has 27 carbon atoms and only one OH group. Its carbon skeleton is too large for the OH group to solubilize by hydrogen bonding, so **cholesterol is insoluble in water.**

> Hydrophobic = afraid of H$_2$O; hydrophilic = H$_2$O loving.

◆ The nonpolar part of a molecule that is not attracted to H$_2$O is said to be *hydrophobic.*
◆ The polar part of a molecule that can hydrogen bond to H$_2$O is said to be *hydrophilic.*

In cholesterol, for example, the **hydroxy group is hydrophilic,** whereas the **carbon skeleton is hydrophobic.**

MTBE (*tert*-butyl methyl ether) and 4,4′-dichlorobiphenyl (a polychlorinated biphenyl, abbreviated as PCB) demonstrate that solubility properties can help determine the fate of organic compounds in the environment.

MTBE
tert-butyl methyl ether

4,4′-dichlorobiphenyl
(a polychlorinated biphenyl, PCB)

MTBE is a high-octane additive used in unleaded gasoline that is now raising some concerns in the environment. Although MTBE is not toxic or carcinogenic, it has a distinctive, nauseating odor, and **it is water soluble.** Small amounts of MTBE have contaminated the drinking water in several California communities, making it unfit for consumption. For this reason, the use of MTBE in gasoline is now banned in several states.

4,4′-Dichlorobiphenyl is a polychlorinated biphenyl (**PCB**), a compound that contains two benzene rings joined by a C–C bond, and substituted by one or more chlorine atoms on each ring.

Figure 3.6 Solubility summary

Type of compound	Solubility in H$_2$O	Solubility in organic solvents (such as CCl$_4$)
• Ionic		
NaCl	**soluble**	**insoluble**
• Covalent		
CH$_3$CH$_2$CH$_2$CH$_3$	**insoluble** (no N or O atom to hydrogen bond to H$_2$O)	**soluble**
CH$_3$CH$_2$CH$_2$OH	**soluble** (≤ 5 C atoms and an O atom for hydrogen bonding to H$_2$O)	**soluble**
CH$_3$(CH$_2$)$_{10}$OH	**insoluble** (> 5 C atoms; too large to be soluble even though it has an O atom for hydrogen bonding to H$_2$O)	**soluble**

PCBs have been used as plasticizers in polystyrene coffee cups and coolants in transformers. They have been released into the environment during production, use, storage, and disposal, making them one of the most widespread organic pollutants. **PCBs are insoluble in H$_2$O, but very soluble in organic media,** so they are soluble in fatty tissue, including that found in all types of fish and birds around the world. Although PCBs are not acutely toxic, frequently ingesting large quantities of fish contaminated with PCBs has been shown to retard growth and memory retention in children.

Solubility properties of some representative compounds are summarized in Figure 3.6.

SAMPLE PROBLEM 3.3   Which compounds are water soluble?

Cl   and   OH

chlorocyclopentane   cyclopentanol
A   **B**

SOLUTION

A has five C atoms and a polar C−Cl bond, but it is incapable of hydrogen bonding with H$_2$O, making it H$_2$O insoluble. **B** has five C atoms and a polar OH group able to hydrogen bond to H$_2$O, making it H$_2$O soluble.

PROBLEM 3.7   Which compounds are water soluble?
a. CH$_3$CH$_2$OCH$_2$CH$_3$ b. CH$_3$CH$_2$CH$_2$CH$_2$CH$_3$ c. (CH$_3$CH$_2$CH$_2$CH$_2$)$_3$N

SAMPLE PROBLEM 3.4   (a) Which of the following molecules can hydrogen bond to another molecule like itself?
(b) Which of the following molecules can hydrogen bond with water?

CH$_3$CH$_2$—Ö—CH$_2$CH$_3$ and   CH$_3$CH$_2$—N$\overset{\cdots}{}$CH$_2$CH$_3$
                                                        |
                                                        H

diethyl ether   diethylamine
A   **B**

SOLUTION

a. Compounds **A** and **B** have polar C−O and C−N bonds but **A** does not have an O−H bond, so it cannot hydrogen bond to another **A** molecule. **B,** on the other hand, has an N−H bond, so two **B** molecules can hydrogen bond with each other.

b. Because **A** has an electronegative O atom and **B** has an electronegative N atom, both **A** and **B** can hydrogen bond with a hydrogen atom in water.

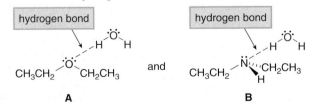

PROBLEM 3.8   Label the hydrophobic and hydrophilic portions of each molecule:

a.
norethindrone
(oral contraceptive component)

b.   COOH
arachidonic acid
(fatty acid)

c.
benzo[*a*]pyrene derivative
(carcinogenic component
derived from tobacco smoke)

3.6   Application: Vitamins

In Sections 3.6, 3.7, and 3.8, the principles of solubility are used to explain some common phenomena.

Vitamins **are organic compounds needed in small amounts for normal cell function.** Our bodies cannot synthesize these compounds in our cells, so they must be obtained in our diet. Most vitamins are identified by a letter, such as A, C, D, E, and K. There are several different B vitamins, though, so a subscript is added to distinguish them: for example, B_1, B_2, and B_{12}.

Vitamins are either **fat soluble** (they dissolve in organic media) or **water soluble.** You can memorize which are which, but it's usually easier to distinguish one from the other by applying the solubility principles discussed in Section 3.5C. Vitamins A and C illustrate the differences between fat-soluble and water-soluble vitamins.

3.6A   Vitamin A

Vitamin A, or **retinol,** is an essential component of the vision receptors in the eyes. A deficiency of this vitamin leads to a loss of night vision.

vitamin A

Vitamin A contains 20 carbons and a single OH group, making it **water insoluble.** Because it is organic, it is **soluble in any organic medium.** To understand the consequences of these solubility characteristics, we must learn about the chemical environment of the body.

About 70% of the body is composed of water. Fluids such as blood, gastric juices in the stomach, and urine are largely water with dissolved ions such as Na^+ and K^+. Vitamin A is insoluble in these fluids. There are also fat cells composed of organic compounds having C–C and C–H bonds. Vitamin A is soluble in this organic environment, and thus it is readily stored in these fat cells, particularly in the liver.

Vitamin A may be obtained directly from the diet. In addition, β-carotene, the orange pigment found in many plants including carrots, is readily converted to vitamin A in our bodies.

β-carotene
(orange pigment in carrots)

in the body

vitamin A

The name **vitamin** was first used in 1912 by the Polish chemist Casimir Funk, who called them *vitamines,* because he thought that they all contained an *amine* functional group. Later the word was shortened to vitamin, because some are amines but others, like vitamins A and C, are not.

Ingesting a moderate excess of vitamin A doesn't cause any harm, but a large excess causes headaches, loss of appetite, and even death. Early Arctic explorers who ate polar bear livers, which contain an unusually large amount of vitamin A, are thought to have died from consuming too much vitamin A.

Eating too many carrots does not result in an excess of stored vitamin A. If you consume more β-carotene than you need, your body stores this precursor until it needs more vitamin A. Some of the β-carotene reaches the surface tissues of the skin and eyes, giving them an orange color. This phenomenon may look odd, but it is harmless and reversible. When stored β-carotene is converted to vitamin A and is no longer in excess, these tissues will return to their normal hue.

3.6B   Vitamin C

Vitamin C, or **ascorbic acid,** is important in the formation of collagen, a protein that holds together the connective tissues of skin, muscle, and blood vessels. Vitamin C is obtained from eating citrus fruits. A deficiency of vitamin C causes scurvy, a common disease of sailors in the 1600s who had no access to fresh fruits on long voyages.

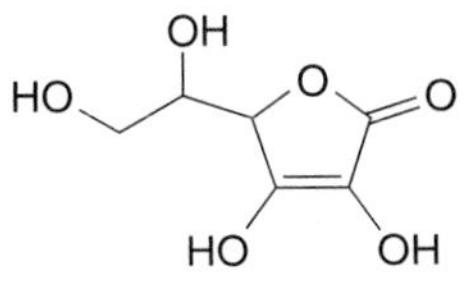

vitamin C
(ascorbic acid)

Vitamin C has six carbon atoms, each bonded to an oxygen atom that is capable of hydrogen bonding, making it **water soluble.** Vitamin C thus dissolves in urine. Although it has been acclaimed as a deterrent for all kinds of diseases, from the common cold to cancer, the consequences of taking large amounts of vitamin C are not really known, because any excess of the minimum daily requirement is excreted in the urine.

PROBLEM 3.9   Predict the water solubility of each of the following vitamins.

a.

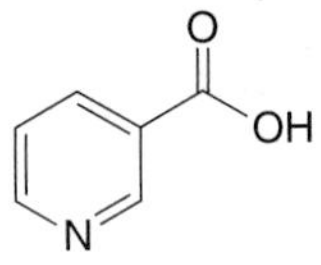

vitamin B_3
(niacin)

b.

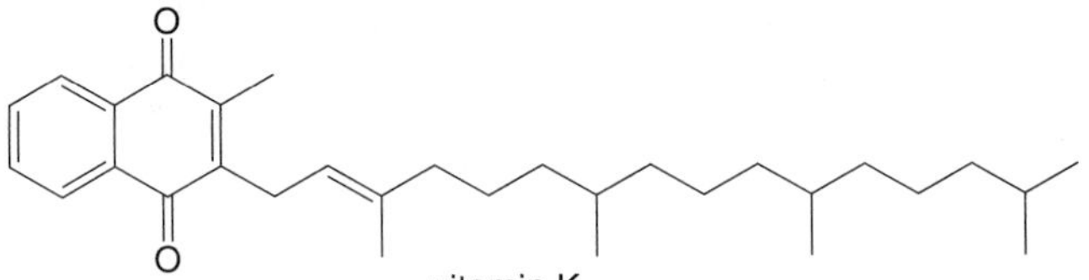

vitamin K_1
(phylloquinone)

3.7   Application of Solubility: Soap

Soap has been used by humankind for some two thousand years. Historical records describe its manufacture in the first century and document the presence of a soap factory in Pompeii. Before this time clothes were cleaned by rubbing them on rocks in water, or by forming soapy lathers from the roots, bark, and leaves of certain plants. These plants produced natural materials called *saponins,* which act in much the same way as modern soaps.

Soap molecules have two distinct parts:

◆ **a hydrophilic portion composed of ions called the *polar head***
◆ **a hydrophobic carbon chain of nonpolar C – C and C – H bonds, called the *nonpolar tail***

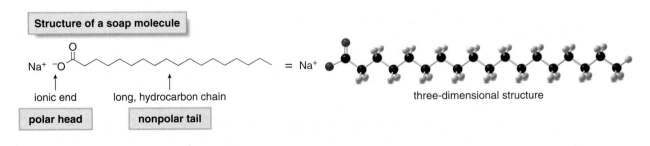

Structure of a soap molecule

Na^+ ionic end   long, hydrocarbon chain   = Na^+ three-dimensional structure

polar head   nonpolar tail

Figure 3.7   Dissolving soap in water

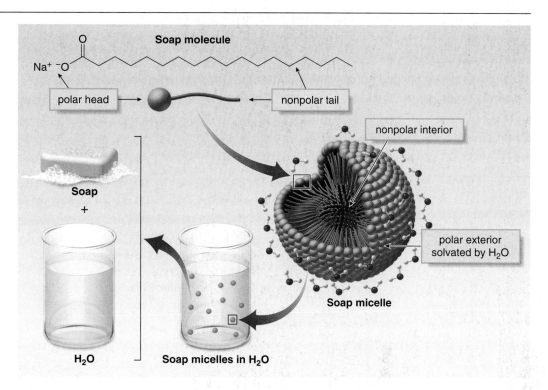

When soap is dissolved in H_2O, the molecules form micelles with the nonpolar tails in the interior and the polar heads on the surface. The polar heads are solvated by ion–dipole interactions with H_2O molecules.

Dissolving soap in water forms *micelles,* **spherical droplets having the ionic heads on the surface and the nonpolar tails packed together in the interior.** This is illustrated in Figure 3.7. In this arrangement, the ionic heads are solvated by the polar solvent water, thus solubilizing the nonpolar, "greasy" hydrocarbon portion of the soap.

How does soap dissolve grease and oil? Water alone cannot dissolve dirt, which is composed largely of nonpolar hydrocarbons. When soap is mixed with water, however, the nonpolar hydrocarbon tails dissolve the dirt in the interior of the micelle. The polar head of the soap remains on the surface of the micelle to interact with water. The nonpolar tails of the soap molecules are so well sealed off from the water by the polar head groups that the micelles are water soluble, allowing them to separate from the fibers of our clothes and be washed down the drain with water. In this way, soaps do a seemingly impossible task: they remove nonpolar hydrocarbon material from skin and clothes, by solubilizing it in the polar solvent water.

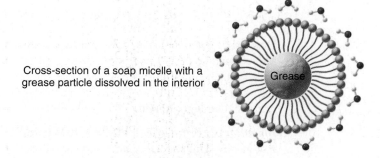

Cross-section of a soap micelle with a grease particle dissolved in the interior

PROBLEM 3.10   Today, synthetic detergents like the compound drawn here, not soaps, are used to clean clothes. Explain how this detergent cleans away dirt.

a detergent

3.8   Application: The Cell Membrane

The cell membrane is a beautifully complex example of how the principles of organic chemistry come into play in a biological system. We will examine both the structure of the cell membrane and transport of molecules and ions across the membrane.

3.8A   Structure of the Cell Membrane

The basic unit of living organisms is the **cell.** The cytoplasm is the aqueous medium inside the cell, separated from water outside the cell by the **cell membrane.** The cell membrane serves two apparently contradictory functions. It acts as a barrier to the passage of ions, water, and other molecules into and out of the cell, and it is also selectively permeable, letting nutrients in and waste out.

A major component of the cell membrane is a group of organic molecules called **phospholipids.** Like soap, they contain a hydrophilic ionic portion, and a hydrophobic hydrocarbon portion, in this case two long carbon chains composed of $C-C$ and $C-H$ bonds. **Phospholipids thus contain a polar head and** *two* **nonpolar tails.**

When phospholipids are mixed with water, they assemble in an arrangement called a **lipid bilayer,** with the ionic heads oriented on the outside and the nonpolar tails on the inside. The polar heads electrostatically interact with the polar solvent H_2O, while the nonpolar tails are held in close proximity by numerous van der Waals interactions. This is schematically illustrated in Figure 3.8.

Cell membranes are composed of these lipid bilayers. The charged heads of the phospholipids are oriented toward the aqueous interior and exterior of the cell. The nonpolar tails form the hydrophobic interior of the membrane, thus serving as an insoluble barrier that protects the cell from the outside.

3.8B   Transport Across a Cell Membrane

How does a polar molecule or ion in the water outside a cell pass through the nonpolar interior of the cell membrane and enter the cell? Some nonpolar molecules like O_2 are small enough to enter and exit the cell by diffusion. Polar molecules and ions, on the other hand, may be too large or too polar to diffuse efficiently, so they are transported across the membrane with the help of molecules called **ionophores.**

Ionophores **are organic molecules that complex cations.** They have a hydrophobic exterior that makes them soluble in the nonpolar interior of the cell membrane, and a central cavity with several oxygen atoms whose lone pairs complex with a given ion. The size of the cavity determines the identity of the cation with which the ionophore complexes. Two naturally occurring antibiotics that act as ionophores are **nonactin** and **valinomycin.**

Figure 3.8 The cell membrane

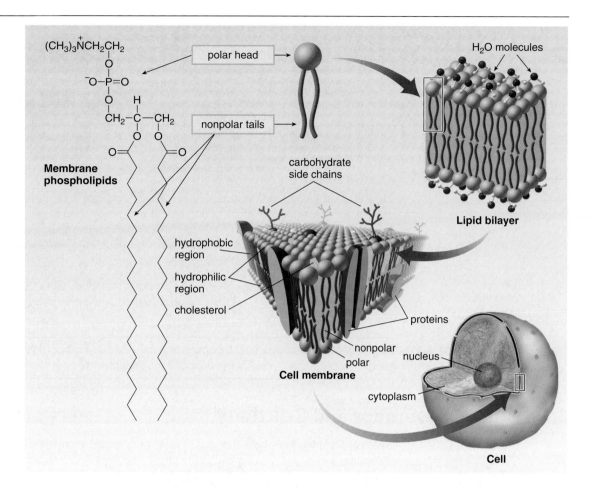

Phospholipids contain an ionic or polar head, and two long nonpolar hydrocarbon tails. In an aqueous environment, phospholipids form a lipid bilayer, with the polar heads oriented toward the aqueous exterior and the nonpolar tails forming a hydrophobic interior. Cell membranes are composed largely of this lipid bilayer.

Naturally occurring antibiotic ionophores

nonactin

valinomycin

Each molecule contains a large central cavity to hold a cation.

Several synthetic ionophores have also been prepared, including one group called **crown ethers**. ***Crown ethers* are cyclic ethers containing several oxygen atoms that bind specific cations depending on the size of their cavity.** Crown ethers are named according to the general format ***x-crown-y***, where ***x*** is the total number of atoms in the ring and ***y*** is the number of oxygen atoms. For example, 18-crown-6 contains 18 atoms in the ring, including 6 O atoms. This crown ether binds potassium ions. Sodium ions are too small to form a tight complex with the O atoms, and larger cations do not fit in the cavity.

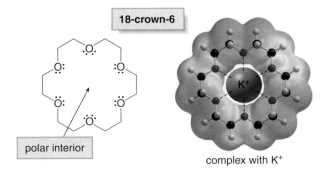

complex with K⁺

How does an ionophore transfer an ion across a membrane? The ionophore binds the ion on one side of the membrane in its polar interior. It can then move across the membrane because its hydrophobic exterior interacts with the hydrophobic tails of the phospholipid. The ionophore then releases the ion on the other side of the membrane. This ion-transfer role is essential for normal cell function. This process is illustrated in Figure 3.9.

In this manner, antibiotic ionophores like nonactin transport ions across a cell membrane of bacteria. This disrupts the normal ionic balance in their cells, thus interfering with cell function and causing the bacteria to die.

PROBLEM 3.11   Nonactin and valinomycin each contain only two different types of functional groups. What two functional groups are present in nonactin? In valinomycin?

3.9   Functional Groups and Reactivity

Much of Chapter 3 has been devoted to how a functional group determines the strength of intermolecular forces and, consequently, the physical properties of molecules. We'll conclude with a brief introduction to how a functional group also determines reactivity.

What type of reaction does a particular kind of organic compound undergo? Begin by recalling two fundamental concepts.

◆ Functional groups create reactive sites in molecules.
◆ Electron-rich sites react with electron-poor sites.

All functional groups contain a heteroatom, a π bond, or both, and these features make electron-deficient (or electrophilic) sites and electron-rich (or nucleophilic) sites in a molecule. Molecules react at these sites. To predict reactivity, we must first locate the functional group and then determine the resulting electron-rich or electron-deficient sites it creates. Keep three guidelines in mind.

Figure 3.9   Transport of ions across a cell membrane

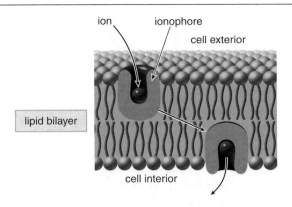

By binding an ion on one side of a lipid bilayer (where the concentration of the ion is high) and releasing it on the other side of the bilayer (where the concentration of the ion is low), an ionophore transports an ion across a cell membrane.

◆ **An electronegative heteroatom like N, O, or X makes a carbon atom *electrophilic*.**

CH₃CH₂—Cl   ⬡—OH
      δ⁺   δ⁺

electrophilic site

◆ **A lone pair on a heteroatom makes it basic and nucleophilic.**

$$CH_3-\ddot{O}-CH_3 \qquad CH_3-\overset{\underset{\displaystyle CH_3}{|}}{\underset{..}{N}}-CH_3$$

basic and nucleophilic site

◆ **π Bonds create *nucleophilic* sites and are more easily broken than σ bonds.**

$$\diagdown C = C \diagdown \qquad -C \equiv C-$$

π bonds

PROBLEM 3.12   Label the electrophilic and nucleophilic sites in each molecule.

a. ⬡—Br   b. H₂O   c. ⬡   d. ⬡=N—CH₃

Why is it so important to identify the nucleophilic and electrophilic sites in a compound? Once you do, you can begin to understand how it will react. In general, electron-rich sites react with electron-deficient sites:

◆ **Nucleophiles react with electrophiles.**

:Nu⁻ = a nucleophile; E⁺ = an electrophile.

An electron-deficient carbon atom reacts with a nucleophile, symbolized as :Nu⁻ in reactions. An electron-rich carbon reacts with an electrophile, symbolized as E⁺ in reactions.

At this point we don't know enough organic chemistry to draw the products of many reactions with confidence. We do know enough, however, to begin to predict if two compounds might react together based solely on electron density arguments, and at what atoms that reaction is most likely to occur.

For example, alkenes contain a C–C double bond, an electron-rich functional group with a nucleophilic π bond. Thus, alkenes react with electrophiles, E⁺, but *not* with other electron-rich species like ⁻OH or Br⁻.

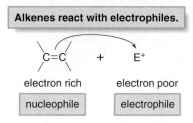

Alkenes react with electrophiles.

$$\diagdown C = C \diagdown \quad + \quad E^+$$

electron rich   electron poor

nucleophile   electrophile

On the other hand, alkyl halides possess an electrophilic carbon atom, so they react with electron-rich nucleophiles.

Alkyl halides react with nucleophiles.

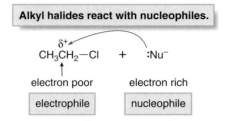

For now, you don't need to worry about the products of these reactions. At this point you should only be able to find reactive sites in molecules and begin to understand why a reaction might occur at these sites. After you learn more about the structure of organic molecules in Chapters 4 and 5, we will begin a detailed discussion of organic reactions in Chapter 6.

PROBLEM 3.13 Considering only electron density, state whether the following reactions will occur:

a. $CH_3CH_2-Br \;+\; {}^{-}OH \longrightarrow$

b. $CH_3-C{\equiv}C-CH_3 \;+\; Br^- \longrightarrow$

c. $CH_3C(=O)Cl \;+\; {}^{-}OCH_3 \longrightarrow$

d. $CH_3-C{\equiv}C-CH_3 \;+\; Br^+ \longrightarrow$

3.10 Key Concepts—Introduction to Organic Molecules and Functional Groups

Types of Intermolecular Forces (3.4A)

Increasing strength ↓

Type of force	Cause
van der Waals	Caused by the interaction of temporary dipoles • Larger surface area, stronger forces • Larger, more polarizable atoms, stronger forces
dipole–dipole	Caused by the interaction of permanent dipoles
hydrogen bonding	Caused by the electrostatic interaction of an H atom in an O–H, N–H, or H–F bond with the lone pair of another N, O, or F atom.
ion–ion	Caused by the charge attraction of two ions

Physical Properties

Property	Observation
Boiling point (3.5A)	• For compounds of comparable molecular weight, the stronger the intermolecular forces the higher the bp.

$CH_3CH_2CH_2CH_2CH_3$	$CH_3CH_2CH_2CHO$	$CH_3CH_2CH_2CH_2OH$
VDW	VDW, DD	VDW, DD, HB
MW = 72	MW = 72	MW = 74
bp = 36 °C	bp = 76 °C	bp = 118 °C

→

Increasing strength of intermolecular forces
Increasing boiling point

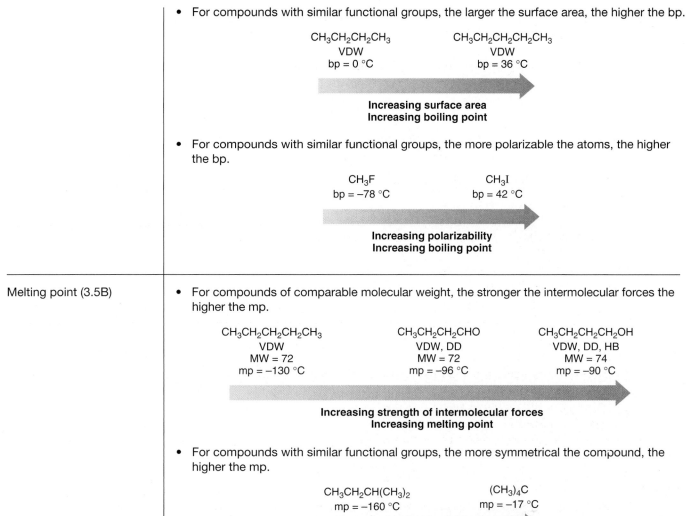

- For compounds with similar functional groups, the larger the surface area, the higher the bp.

$$CH_3CH_2CH_2CH_3$$
VDW
bp = 0 °C

$$CH_3CH_2CH_2CH_2CH_3$$
VDW
bp = 36 °C

Increasing surface area
Increasing boiling point

- For compounds with similar functional groups, the more polarizable the atoms, the higher the bp.

$$CH_3F$$
bp = −78 °C

$$CH_3I$$
bp = 42 °C

Increasing polarizability
Increasing boiling point

Melting point (3.5B)

- For compounds of comparable molecular weight, the stronger the intermolecular forces the higher the mp.

$$CH_3CH_2CH_2CH_2CH_3$$
VDW
MW = 72
mp = −130 °C

$$CH_3CH_2CH_2CHO$$
VDW, DD
MW = 72
mp = −96 °C

$$CH_3CH_2CH_2CH_2OH$$
VDW, DD, HB
MW = 74
mp = −90 °C

Increasing strength of intermolecular forces
Increasing melting point

- For compounds with similar functional groups, the more symmetrical the compound, the higher the mp.

$$CH_3CH_2CH(CH_3)_2$$
mp = −160 °C

$$(CH_3)_4C$$
mp = −17 °C

Increasing symmetry
Increasing melting point

Solubility (3.5C)

Types of H_2O soluble compounds:
- Ionic compounds
- Organic compounds having ≤ 5 C's, and an O or N atom for hydrogen bonding (for a compound with one functional group).

Types of compounds soluble in organic solvents:
- Organic compounds regardless of size or functional group.

Key: VDW = van der Waals, DD = dipole–dipole, HB = hydrogen bonding, MW = molecular weight

Reactivity (3.9)

- **Nucleophiles react with electrophiles.**
- Electronegative heteroatoms create electrophilic carbon atoms, which tend to react with nucleophiles.
- Lone pairs and π bonds are nucleophilic sites that tend to react with electrophiles.

Problems

Functional Groups

3.14 Identify the functional groups in each of the following molecules:

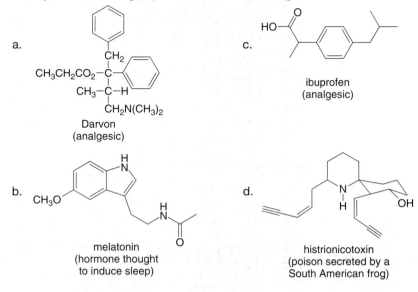

a.

CH₃CH₂CO₂—C
Darvon
(analgesic)

c.

ibuprofen
(analgesic)

b.

CH₃O

melatonin
(hormone thought
to induce sleep)

d.

histrionicotoxin
(poison secreted by a
South American frog)

3.15 Draw the seven constitutional isomers having molecular formula $C_4H_{10}O$. Identify the functional group in each isomer.

3.16 Draw the structure of a compound fitting each description:
   a. An aldehyde having molecular formula C_4H_8O
   b. A ketone having molecular formula C_4H_8O
   c. A carboxylic acid having molecular formula $C_4H_8O_2$
   d. An ester having molecular formula $C_4H_8O_2$

Intermolecular Forces

3.17 What types of intermolecular forces are exhibited by each compound with itself?

a. 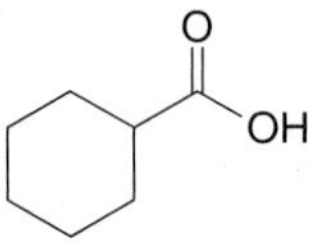   b. 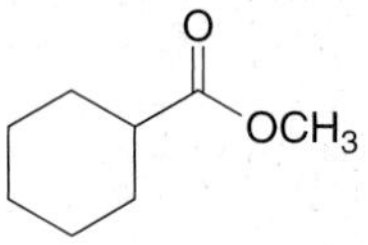   c. 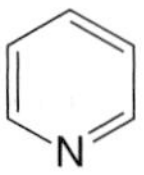   d.

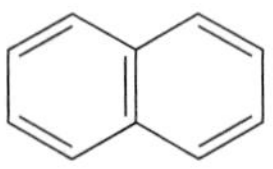

3.18 Rank the following compounds in order of increasing strength of intermolecular forces:
   a. CH_3NH_2, CH_3CH_3, CH_3Cl
   b. CH_3Br, CH_3I, CH_3Cl
   c. $(CH_3)_2C=C(CH_3)_2$, $(CH_3)_2CHCOOH$, $(CH_3)_2CHCOCH_3$
   d. $NaCl$, CH_3OH, CH_3Cl

3.19 Carboxylic acids (RCOOH) can exist as dimers in some situations, with two molecules held together by two intermolecular hydrogen bonds. Show how two molecules of acetic acid, the carboxylic acid present in vinegar, can hydrogen bond to each other.

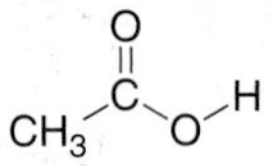

acetic acid

3.20 Intramolecular forces of attraction are often important in holding large molecules together. For example, some proteins fold into compact shapes, held together by attractive forces between nearby functional groups. A schematic of a folded protein is drawn here, with the protein backbone indicated by a blue-green ribbon, and various appendages drawn dangling from the chain. What types of intramolecular forces occur at each labeled site (**A–F**)?

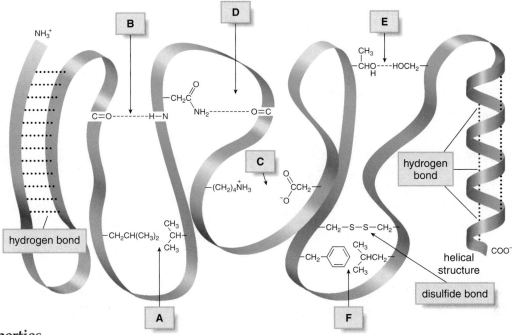

Physical Properties

3.21 Rank the compounds in each group in order of increasing boiling point.

  a. $CH_3(CH_2)_4-I$ $CH_3(CH_2)_5-I$ $CH_3(CH_2)_6-I$

  b. $CH_3CH_2CH_2NH_2$, $(CH_3)_3N$, $CH_3CH_2CH_2CH_3$

  c. $(CH_3)_3COC(CH_3)_3$, $CH_3(CH_2)_3O(CH_2)_3CH_3$, $CH_3(CH_2)_7OH$

  d.  ![structures]

  e. 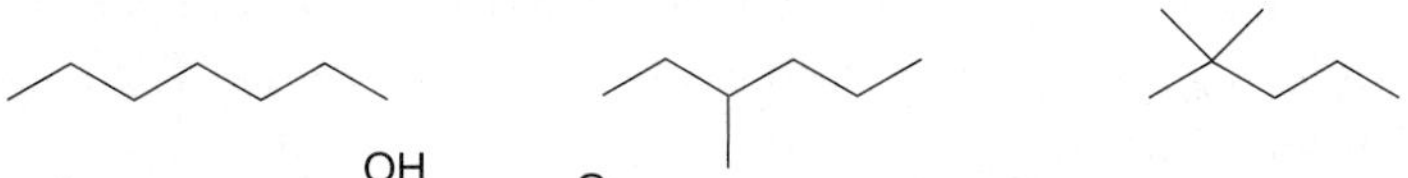 ![structures]

  f. 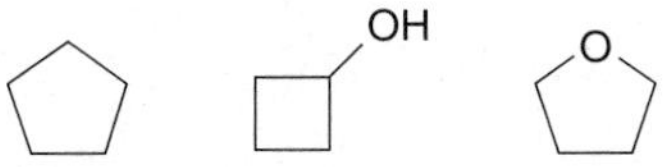 ![structures]

3.22 Explain why $CH_3CH_2NHCH_3$ has a higher boiling point than $(CH_3)_3N$, even though they have the same molecular weight.

3.23 Rank the compounds in each group in order of increasing melting point.

  a. $(CH_3)_2CHOH$, $(CH_3)_3CH$, $(CH_3)_2C=O$ c. ![cyclohexane with NH2, Cl, CH3]

  b. CH_3F, CH_3Cl, CH_3I

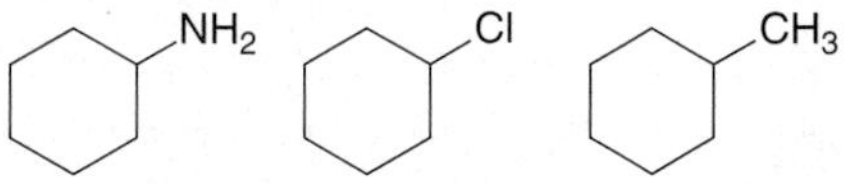

3.24 Explain the observed trend in the melting points for four isomers of molecular formula C_7H_{16}.

	mp (°C)
$CH_3CH_2CH_2CH(CH_3)CH_2CH_3$	−119
$CH_3CH_2CH_2CH_2CH(CH_3)_2$	−118
$CH_3CH_2CH_2CH_2CH_2CH_2CH_3$	−91
$(CH_3)_2CHC(CH_3)_3$	−25

3.25 Explain why benzene has a lower boiling point but much higher melting point than toluene.

  benzene
  bp = 80 °C
  mp = 5 °C

  and

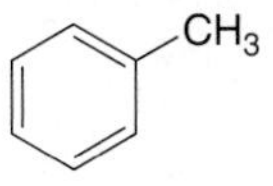

  toluene
  bp = 111 °C
  mp = −93 °C

3.26 Rank the compounds in each group in order of increasing water solubility.

  a. $(CH_3)_3CH$, $CH_3OCH_2CH_3$, $CH_3CH_2CH_2CH_3$, $CH_3CH_2CH_2OH$

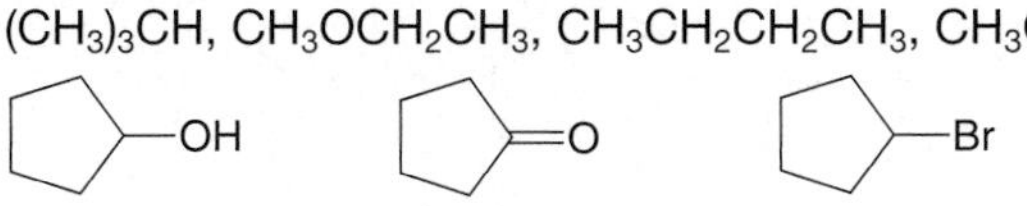

  b.

3.27 Which of the following molecules can hydrogen bond to another molecule of itself? Which can hydrogen bond with water?
   a. CBr_3CH_3 c. CH_3OCH_3 e. $CH_3CH_2CH_2CONH_2$ g. CH_3SOCH_3
   b. CH_3NH_2 d. $(CH_3CH_2)_3N$ f. $CH_3CH_2CH_2Cl$ h. $CH_3CH_2COOCH_3$

3.28 Explain why diethyl ether ($CH_3CH_2OCH_2CH_3$) and 1-butanol ($CH_3CH_2CH_2CH_2OH$) have similar solubility properties in water, but 1-butanol has a much higher boiling point.

3.29 Why is cyclohexanol more water soluble than 1-hexanol?

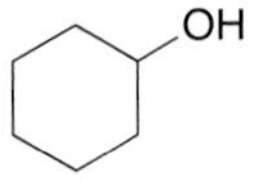

cyclohexanol 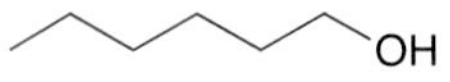   1-hexanol

3.30 Predict the water solubility of each of the following organic molecules:

a. 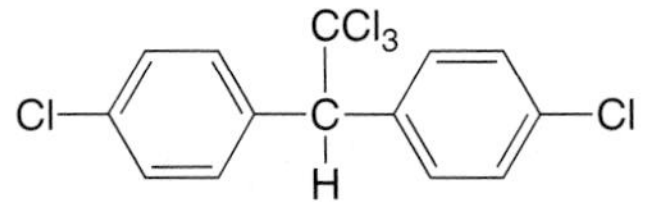
DDT
(nonbiodegradable pesticide)

c. 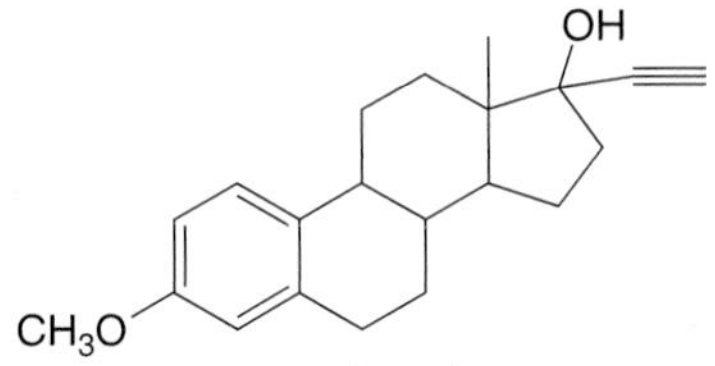
mestranol
(component in oral contraceptives)

b. 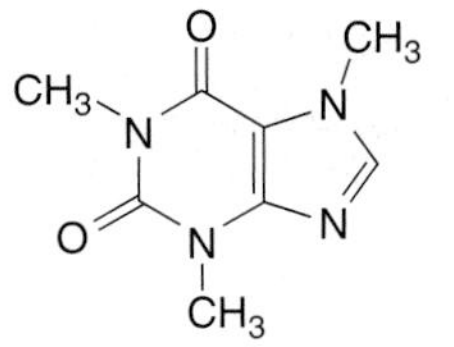
caffeine
(stimulant in coffee, tea,
and many soft drinks)

d. 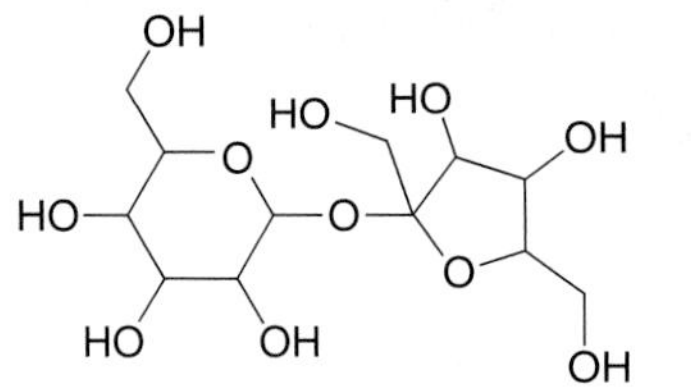
sucrose
(table sugar)

3.31 Polyfluorinated organic compounds often have much lower boiling points than might be expected. For example, although heptane ($CH_3CH_2CH_2CH_2CH_2CH_2CH_3$) has a boiling point of 90 °C, perfluoroheptane ($CF_3CF_2CF_2CF_2CF_2CF_2CF_3$) has a much higher molecular weight and a boiling point of 82 to 84 °C. Explain.

3.32 A mixture of four alkanes (**A–D**) was distilled and four fractions (1–4, in order of increasing boiling point) were collected. Which alkane was the principal component in each fraction?
   A. $CH_3(CH_2)_6CH_3$ **B.** $(CH_3)_2CHCH(CH_3)_2$ **C.** $CH_3(CH_2)_4CH_3$ **D.** $CH_3(CH_2)_5CH_3$

Applications

3.33 Predict the solubility of each of the following vitamins in water and in organic solvents:

a. 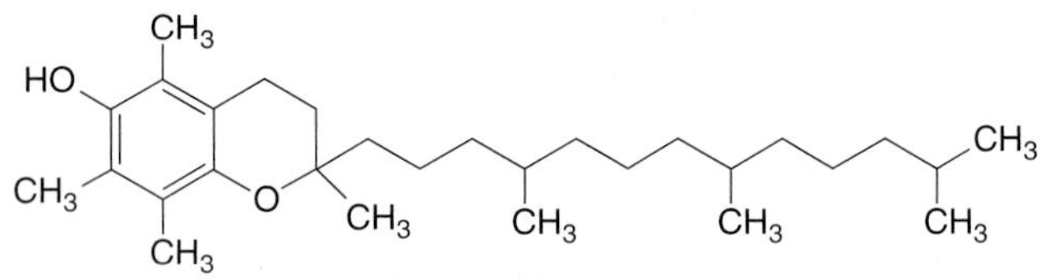
vitamin E

b. 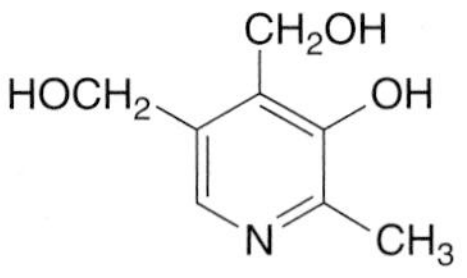
pyridoxine
vitamin B_6

3.34 THC is the active component in marijuana, and ethanol is the alcohol in alcoholic beverages. Explain why drug screenings are able to detect the presence of THC but not ethanol weeks after these substances have been introduced into the body.

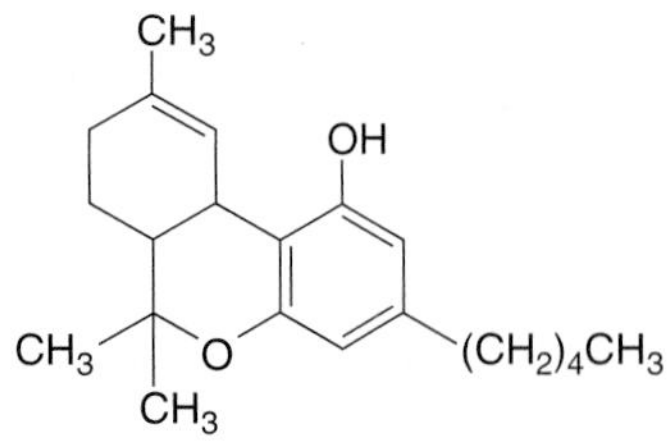

tetrahydrocannabinol
THC

CH_3CH_2—OH
ethanol

3.35 Cocaine is a widely abused, addicting drug. Cocaine is usually obtained as its hydrochloride salt (cocaine hydrochloride) but can be converted to crack (the neutral molecule) by treatment with base. Which of the two compounds here has a higher boiling point? Which is more soluble in water? How does the relative solubility explain why crack is usually smoked but cocaine hydrochloride is injected directly into the bloodstream?

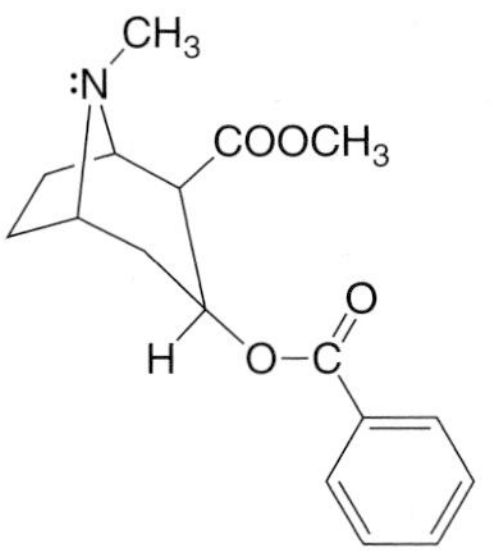

CH3
:N
COOCH3
O
H
O–C

cocaine (crack)
neutral organic molecule

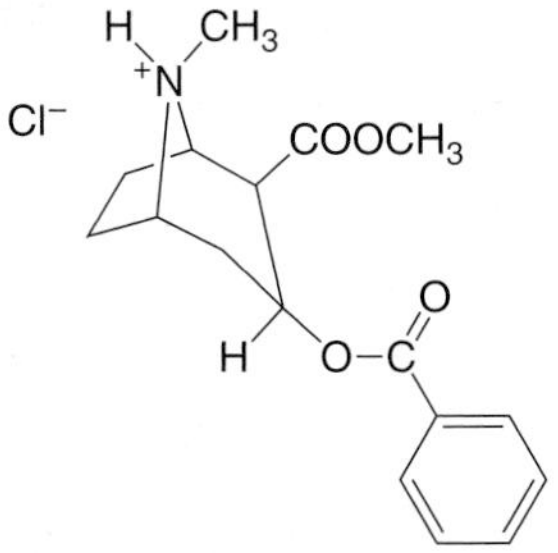

H
CH3
+N
Cl⁻
COOCH3
O
H
O–C

cocaine hydrochloride
a salt

3.36 Unlike soap, which is ionic, some liquid laundry detergents are neutral molecules. Explain how each of the following molecules behaves like soap and cleans away dirt.

a. 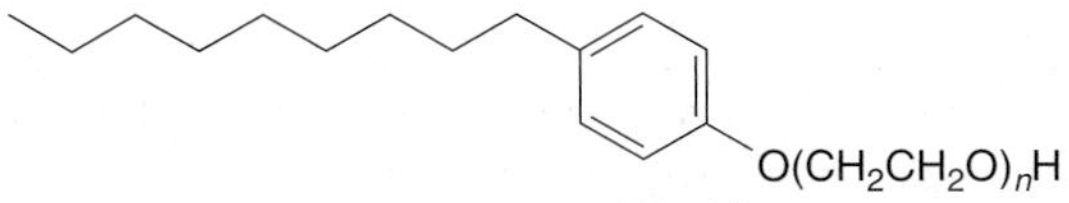

O(CH2CH2O)nH

$n = 7$ to 13

b. 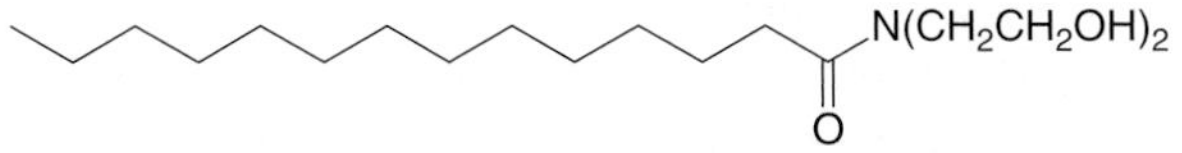

N(CH2CH2OH)2
O

3.37 Most mayonnaise recipes call for oil (nonpolar, mostly hydrocarbon), vinegar (H_2O and CH_3COOH), and egg yolk. The last ingredient is a source of phospholipids that act as emulsifying agents. Explain.

Reactivity of Organic Molecules

3.38 Label the electrophilic and nucleophilic sites in each of the following molecules:

a. 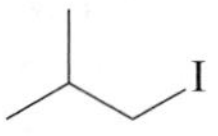

I

b. 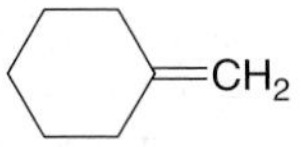

CH2

c. 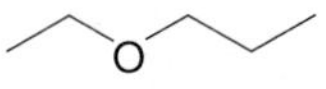

O

d. 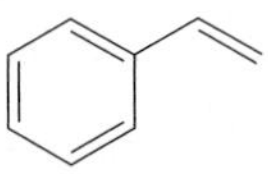

e. CH_3OH

f. 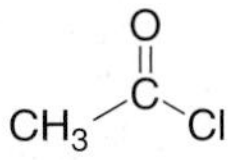

O
C
CH3
Cl

3.39 By using only electron density arguments, determine whether the following reactions will occur:

a. 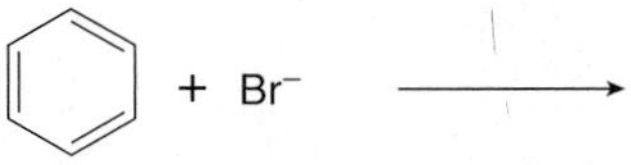

+ Br⁻

b. 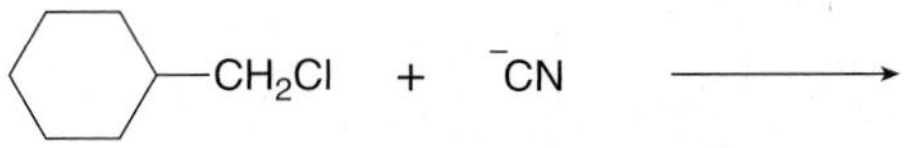

CH2Cl
+
⁻CN

c. 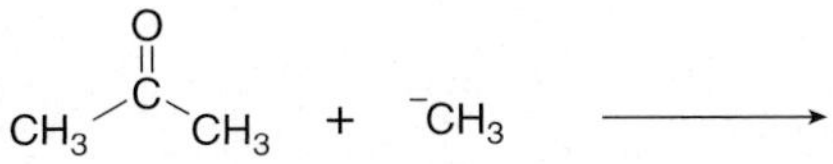

O
C
CH3
CH3
+
⁻CH3

d. 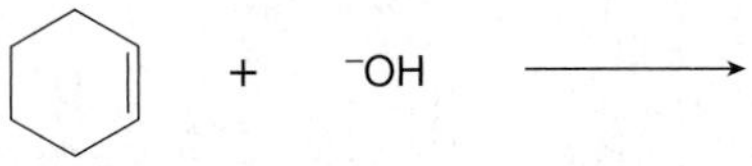

+
⁻OH

e. 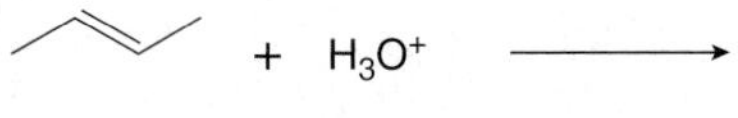

+ H3O+

Cell Membrane

3.40 The composition of a cell membrane is not uniform for all types of cells. Some cell membranes are more rigid than others. Rigidity is determined by a variety of factors, one of which is the structure of the carbon chains in the phospholipids that comprise the membrane. One example of a phospholipid was drawn in Section 3.8A, and another, having C−C double bonds in its carbon chains, is drawn here. Which phospholipid would be present in the more rigid cell membrane and why?

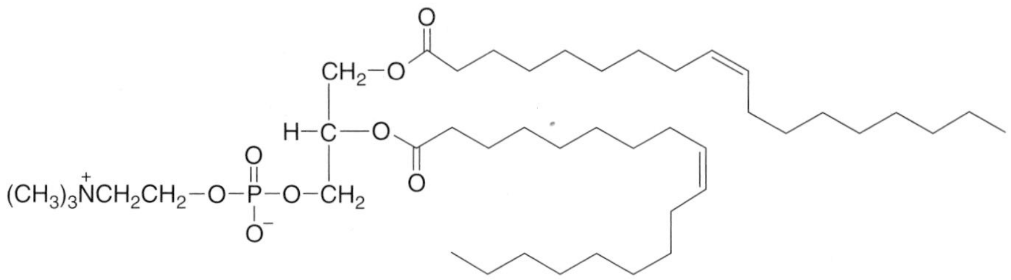

a phospholipid with π bonds in the long hydrocarbon chains

General Question

3.41 Vancomycin is an especially useful antibiotic for treating infections in cancer patients on chemotherapy and renal patients on dialysis. Unlike mammalian cells, bacterial cells are surrounded by a fairly rigid cell wall, which is crucial to the bacterium's survival. Vancomycin kills bacteria by interfering with their cell wall synthesis.

   a. How many amide functional groups are present in vancomycin?
   b. Which OH groups are bonded to sp^3 hybridized carbon atoms and which are bonded to sp^2 hybridized carbons?
   c. Would you expect vancomycin to be water soluble? Explain.
   d. Which proton is the most acidic?
   e. Label three *different* functional groups capable of hydrogen bonding.

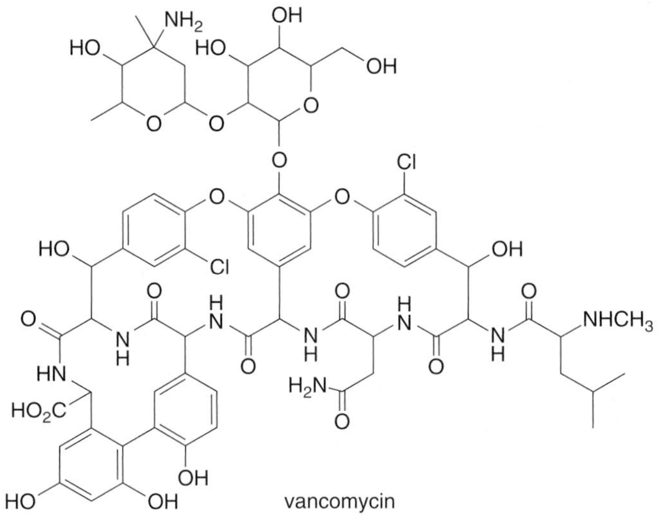

vancomycin

Challenge Problems

3.42 Explain why **A** is less water soluble than **B,** even though both compounds have the same functional groups.

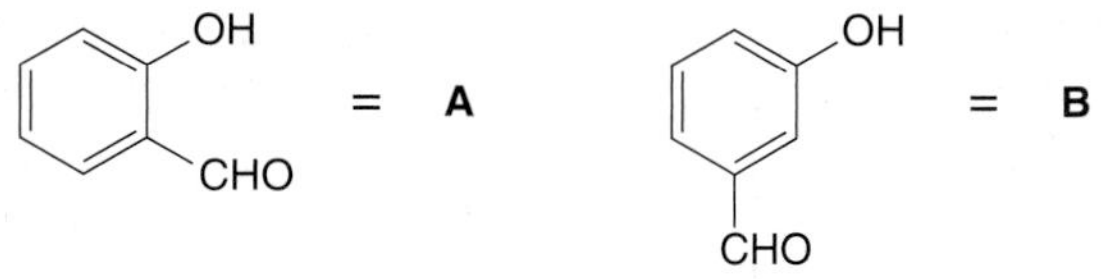

3.43 Recall from Section 1.9B that there is restricted rotation around carbon–carbon double bonds. Maleic acid and fumaric acid are two isomers with vastly different physical properties and K_a values for loss of both protons. Explain why each of these differences occurs.

	maleic acid	fumaric acid
mp (°C)	130	286
solubility (g/mol) in H_2O at 25 °C	780	7
K_{a1}	1.5×10^{-2}	1.0×10^{-3}
K_{a2}	2.6×10^{-7}	3×10^{-5}

Alkanes

Alkanes, the simplest hydrocarbons, are found in all shapes and sizes and occur widely in nature. They are the major constituents of the natural gas used to heat homes and the gasoline used to fuel automobiles. New and complex alkanes unknown in nature have also been synthesized in the laboratory. **Dodecahedrane,** a beautifully symmetrical compound composed of 12 five-membered rings joined together in a spherical structure, is one such alkane. In Chapter 4, we will learn about the properties of alkanes, how to name them (nomenclature), and oxidation—one of their important reactions.

In Chapter 4, we will apply the principles of bonding, shape, and reactivity discussed in Chapters 1 through 3 to our first family of organic compounds, the **alkanes.** Because alkanes have no functional group, they are much less reactive than other organic compounds, and for this reason, much of Chapter 4 is devoted to learning how to name and draw them, as well as to understanding what happens when they rotate about their carbon–carbon single bonds.

Studying alkanes also provides our first opportunity to learn about **lipids,** a group of biomolecules similar to alkanes, in that they are composed mainly of nonpolar carbon–carbon and carbon–hydrogen σ bonds. Section 4.15 serves as a brief introduction only, so we will return to lipids in Chapters 10 and 26.

4.1 Alkanes—An Introduction

Secretion of **undecane** by a cockroach causes other members of the species to aggregate. Undecane is a *pheromone,* **a chemical substance used for communication** in an animal species, most commonly an insect population. Minute amounts elicit an activity such as mating, aggregation, defense, and so forth. Because pheromones are biodegradable and highly specific, they can be used in some cases to control insect populations (Section 20.10).

Recall from Section 3.2 that *alkanes* **are aliphatic hydrocarbons having only C – C and C – H σ bonds.** Because their carbon atoms can be joined together in chains or rings, they can be categorized as acyclic or cyclic.

♦ **Acyclic** alkanes have the molecular formula C_nH_{2n+2} (where n = an integer) and contain only linear and branched chains of carbon atoms. They are also called **saturated hydrocarbons** because they have the maximum number of hydrogen atoms per carbon.

♦ **Cycloalkanes** contain carbons joined in one or more rings. Because their general formula is C_nH_{2n}, they have two fewer H atoms than an acyclic alkane with the same number of carbons.

Undecane and cyclohexane are two naturally occurring alkanes.

| **An acyclic alkane** | **A cycloalkane** |

$$CH_3(CH_2)_9CH_3$$

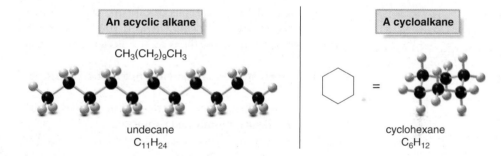

undecane
$C_{11}H_{24}$

cyclohexane
C_6H_{12}

4.1A Acyclic Alkanes Having One to Five C Atoms

We drew structures for the two simplest acyclic alkanes in Chapter 1. **Methane, CH_4,** has a single carbon atom, and **ethane, CH_3CH_3,** has two. All C atoms in an alkane are surrounded by four groups, making them *sp*³ **hybridized** and **tetrahedral,** and all bond angles are 109.5°.

Cyclohexane is one component of the mango, the most widely consumed fruit in the world.

CH_4
methane   =

H—C—H (Lewis structure)

3-D representation

1.09 Å   109.5°
ball-and-stick model

CH_3CH_3
ethane   =

H—C—C—H

3-D representation

1.53 Å

The three-carbon alkane **$CH_3CH_2CH_3$,** called **propane,** has molecular formula C_3H_8. Note in the three-dimensional drawing that each carbon atom has two bonds in the plane (solid lines), one bond in front (on a wedge), and one bond behind the plane (on a dashed line).

To draw the structure of an alkane, join the carbon atoms together with single bonds, and add enough H atoms to make each C tetravalent.

$CH_3CH_2CH_3$
propane
=

H—C—C—C—H (with H's above and below each C)

Lewis structure

3-D representation

ball-and-stick model

PROBLEM 4.1   Which molecular formulas correspond to an acyclic alkane?
a. $C_{12}H_{26}$ b. C_8H_{16} c. $C_{30}H_{64}$

The molecular formulas for methane, ethane, and propane fit into the general molecular formula for an alkane, C_nH_{2n+2}:

◆ Methane = CH_4 = $C_1H_{2(1)+2}$
◆ Ethane = C_2H_6 = $C_2H_{2(2)+2}$
◆ Propane = C_3H_8 = $C_3H_{2(3)+2}$

The three-dimensional representations and the ball-and-stick models for these alkanes indicate the tetrahedral geometry around each carbon atom. In contrast, **the Lewis structures are not meant to imply any three-dimensional arrangement.** Moreover, in propane and higher molecular weight alkanes, the carbon skeleton can be drawn in a variety of different ways and still represent the same molecule.

Two equivalent representations for propane

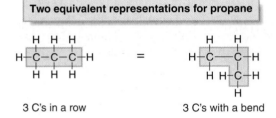

3 C's in a row   3 C's with a bend

For example, the three carbons of propane can be drawn in a horizontal row or with a bend. *These representations are equivalent.* If you follow the carbon chain from one end to the other, you move across the *same* three carbon atoms in both representations.

◆ **In a Lewis structure, the bends in a carbon chain don't matter.**

There are two different ways to arrange four carbons, giving two compounds with molecular formula C_4H_{10}, named **butane** and **isobutane.**

Two constitutional isomers having molecular formula C_4H_{10}

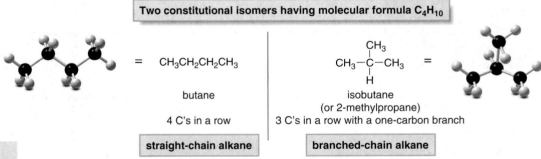

= $CH_3CH_2CH_2CH_3$ $CH_3—C—CH_3$ (with CH_3 above and H below)   =

butane   isobutane
                          (or 2-methylpropane)
4 C's in a row   3 C's in a row with a one-carbon branch

straight-chain alkane   **branched-chain alkane**

Constitutional isomers like butane and isobutane belong to the same family of compounds: they are both **alkanes.** In contrast, constitutional isomers like CH_3CH_2OH and CH_3OCH_3 have different functional groups and belong to different families: CH_3CH_2OH is an **alcohol** and CH_3OCH_3 is an **ether.**

Butane and isobutane are *isomers,* **two different compounds with the same molecular formula** (Section 1.4A). They belong to one of the two major classes of isomers called **constitutional** or **structural isomers.** The two isomers discussed in Section 1.4A, CH_3OCH_3 and CH_3CH_2OH, are also constitutional isomers. We will learn about the second major class of isomers, called **stereoisomers,** in Section 4.13B.

◆ *Constitutional isomers* differ in the way the atoms are connected to each other.

Butane, which has four carbons in a row, is a **straight-chain** or **normal alkane** (an *n*-alkane). Isobutane, on the other hand, is a **branched-chain** alkane.

PROBLEM 4.2   Label each representation as butane or isobutane.

a. H—C—C—C—H (with H H H on top, H H CH₃ on bottom)   b. (CH₃)₃CH   c. (structure)   d. (zigzag structure)   e. (branched structure)

With alkanes having more than four carbons, the names of the straight-chain isomers are systematic and derive from Greek roots: *pent*ane for five C atoms, *hex*ane for six, and so on. There are three constitutional isomers for the five-carbon alkane, each having molecular formula C₅H₁₂: **pentane, isopentane** (or 2-methylbutane), and **neopentane** (or 2,2-dimethylpropane).

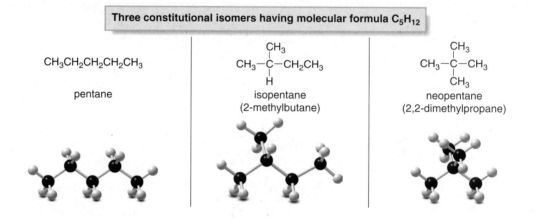

Three constitutional isomers having molecular formula C₅H₁₂

CH₃CH₂CH₂CH₂CH₃

pentane

CH₃—C—CH₂CH₃ (with CH₃ on top, H on bottom)

isopentane
(2-methylbutane)

CH₃—C—CH₃ (with CH₃ on top, CH₃ on bottom)

neopentane
(2,2-dimethylpropane)

PROBLEM 4.3   Which of the following is *not* another representation for isopentane?

a. CH₃CH₂—C—CH₃ (with H on top, CH₃ on bottom)

c. CH₃CH₂CH(CH₃)₂

e. (branched structure)

b. H—C—CH₃ (with H on top), CH₃—C—CH₃ (with H on bottom)

d. H—C——C—H (with H H on top), CH₃—C—H CH₃ (with H on bottom)

Carbon atoms in alkanes and other organic compounds are classified by the number of other carbons directly bonded to them.

◆ A *primary carbon* (**1° carbon**) is bonded to *one* other C atom.
◆ A *secondary carbon* (**2° carbon**) is bonded to *two* other C atoms.
◆ A *tertiary carbon* (**3° carbon**) is bonded to *three* other C atoms.
◆ A *quaternary carbon* (**4° carbon**) is bonded to *four* other C atoms.

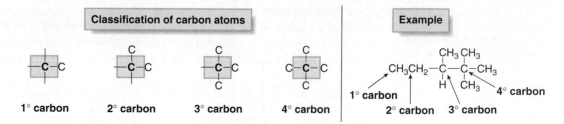

Classification of carbon atoms				Example

1° carbon   2° carbon   3° carbon   4° carbon

CH₃CH₂—C—C—CH₃ (with CH₃ CH₃ on top, H CH₃ on bottom)

1° carbon   2° carbon   3° carbon   4° carbon

Hydrogen atoms are classified as **primary (1°), secondary (2°),** or **tertiary (3°)** depending on the **type of carbon atom** to which they are bonded.

◆ A *primary hydrogen* is on a C bonded to one other C atom.
◆ A *secondary hydrogen* is on a C bonded to two other C atoms.
◆ A *tertiary hydrogen* is on a C bonded to three other C atoms.

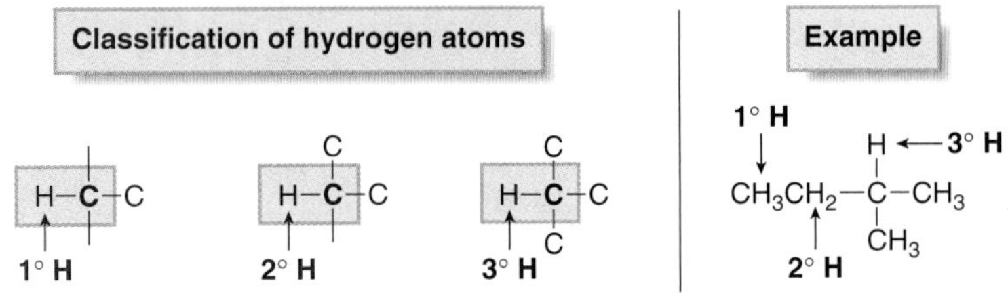

SAMPLE PROBLEM 4.1   (a) Classify the designated carbon atoms as 1°, 2°, 3°, or 4°. (b) Classify the designated hydrogen atoms as 1°, 2°, or 3°.

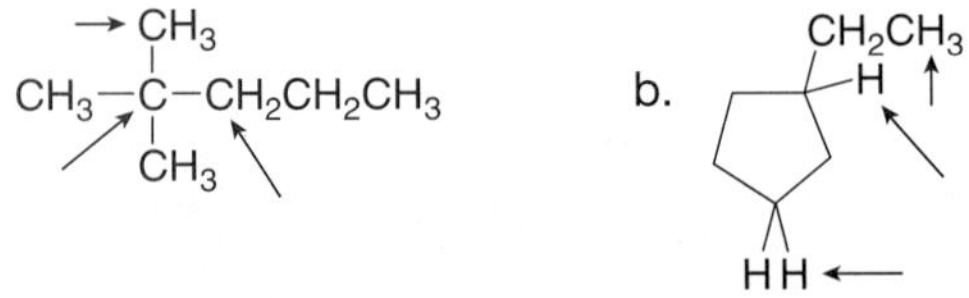

SOLUTION

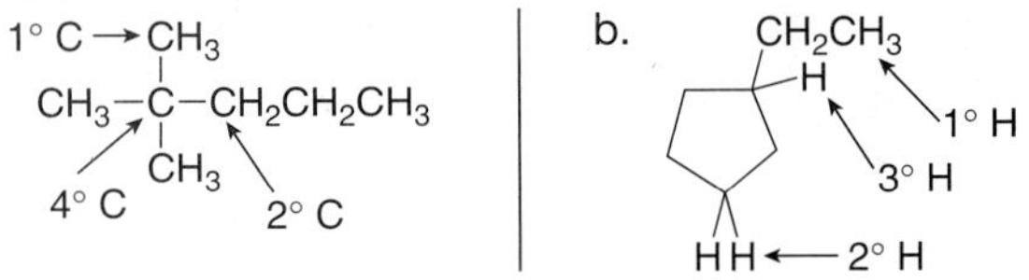

PROBLEM 4.4   Classify the carbon atoms in each compound as 1°, 2°, 3°, or 4°.

a. $CH_3CH_2CH_2CH_3$ b. $(CH_3)_3CH$ c. 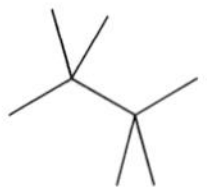   d.

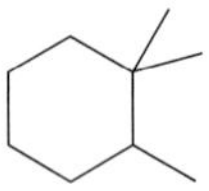

PROBLEM 4.5   Classify the hydrogen atoms in each compound as 1°, 2°, or 3°.

a. $CH_3CH_2CH_3$ b. $CH_3CH_2CH(CH_3)C(CH_3)_3$ c.

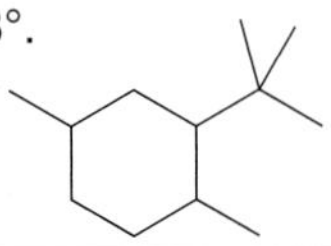

4.1B   Acyclic Alkanes Having More Than Five C Atoms

The maximum number of possible constitutional isomers increases dramatically as the number of carbon atoms in the alkane increases, as shown in Table 4.1. For example, there are 75 possible isomers for an alkane having 10 carbon atoms, but 366,319 possible isomers for one having 20 carbons.

Each entry in Table 4.1 is formed from the preceding entry by adding a CH_2 group. **A CH_2 group is called a *methylene group*. A group of compounds that differ by only a CH_2 group is called a *homologous series*.** Notice that the names of all alkanes end in the suffix *-ane,* and the syllable preceding the suffix identifies the number of carbon atoms in the chain.

The suffix *-ane* identifies a molecule as an alk*ane*.

PROBLEM 4.6   Draw the five constitutional isomers having molecular formula C_6H_{14}.

PROBLEM 4.7   Draw all constitutional isomers having molecular formula C_8H_{18} that contain seven carbons in the longest chain and a single CH_3 group bonded to the chain.

TABLE 4.1	Summary: Straight-Chain Alkanes		
Number of C atoms	Molecular formula	Name (*n*-alkane)	Number of constitutional isomers
1	CH_4	methane	—
2	C_2H_6	ethane	—
3	C_3H_8	propane	—
4	C_4H_{10}	butane	2
5	C_5H_{12}	pentane	3
6	C_6H_{14}	hexane	5
7	C_7H_{16}	heptane	9
8	C_8H_{18}	octane	18
9	C_9H_{20}	nonane	35
10	$C_{10}H_{22}$	decane	75
20	$C_{20}H_{42}$	eicosane	366,319

4.2   Cycloalkanes

Cycloalkanes have molecular formula C_nH_{2n} and contain carbon atoms arranged in a ring. Think of a cycloalkane as being formed by removing two H atoms from the end carbons of a chain, and then bonding the two carbons together. Simple cycloalkanes are named by adding the prefix **cyclo-** to the name of the acyclic alkane having the same number of carbons.

Cycloalkanes having three to six carbon atoms are shown in the accompanying figure. They are most often drawn in skeletal representations.

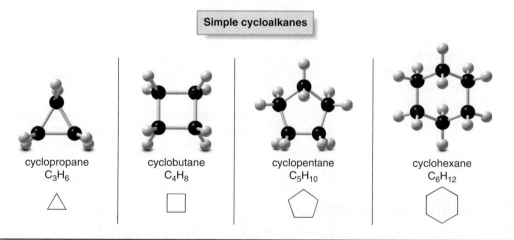

Simple cycloalkanes

cyclopropane
C_3H_6

cyclobutane
C_4H_8

cyclopentane
C_5H_{10}

cyclohexane
C_6H_{12}

PROBLEM 4.8   Draw the five constitutional isomers that have molecular formula C_5H_{10} and contain one ring.

PROBLEM 4.9   The largest known cycloalkane with a single ring has 288 carbon atoms. What is its molecular formula?

4.3   An Introduction to Nomenclature

How are organic compounds named? Long ago, the name of a compound was often based on the plant or animal source from which it was obtained. For example, the name for **formic acid,** a caustic compound isolated from certain ants, comes from the Latin word *formica,* meaning *ant;* and **allicin,** the pungent principal of garlic, is derived from the botanical name for garlic, *Allium sativum.*

Garlic has been a valued commodity throughout history. It has been used in Chinese herbal medicine for more than 4000 years, as a form of currency in Siberia, and as a repellent for witches by the Saxons. Today it is used as a dietary supplement because of its reported health benefits. **Allicin,** the molecule responsible for garlic's odor, is a rather unstable molecule that is not stored in the garlic bulb, but rather is produced by the action of enzymes when the bulb is crushed or bruised.

Other compounds were named by their discoverer for more personal reasons. Adolf von Baeyer supposedly named barbituric acid after a woman named Barbara, although speculation continues on Barbara's identity—a lover, a Munich waitress, or even St. Barbara.

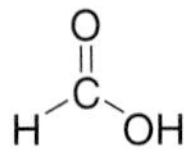

formic acid
(obtained from certain ants)

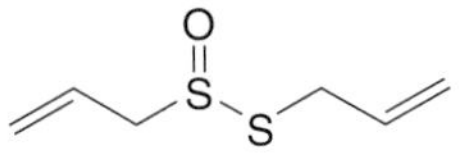

allicin
(odor of garlic)

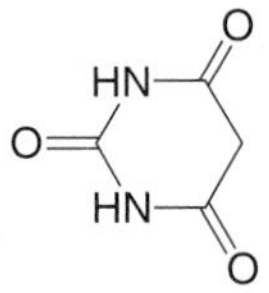

barbituric acid
(named for Barbara whom?)

With the isolation and preparation of thousands of new organic compounds it became clear that each organic compound must have an unambiguous name, derived from a set of easily remembered rules. A systematic method of naming compounds was developed by the *I*nternational *U*nion of *P*ure and *A*pplied *C*hemistry. It is referred to as the **IUPAC** system of nomenclature; how it can be used to name alkanes is explained in Sections 4.4 and 4.5.

Naming organic compounds has become big business for drug companies. The IUPAC name of an organic compound can be long and complex, and may be comprehensible only to a chemist. As a result, most drugs have three names:

◆ **Systematic:** The systematic name follows the accepted rules of nomenclature and indicates the compound's chemical structure; this is the IUPAC name.

◆ **Generic:** The generic name is the official, internationally approved name for the drug.

◆ **Trade:** The trade name for a drug is assigned by the company that manufactures it. Trade names are often "catchy" and easy to remember. Companies hope that the public will continue to purchase a drug with an easily recalled trade name long after a cheaper generic version becomes available.

The IUPAC system of nomenclature has been revised regularly since it was first adopted in 1892. The last extensive revision was done in 1993. In general, the latest IUPAC guidelines will be presented in this text, unless accepted practice follows earlier recommendations.

In the world of over-the-counter anti-inflammatory agents, the compound a chemist calls 2-[4-(2-methylpropyl)phenyl]propanoic acid has the generic name ibuprofen. It is marketed under a variety of trade names including Motrin and Advil.

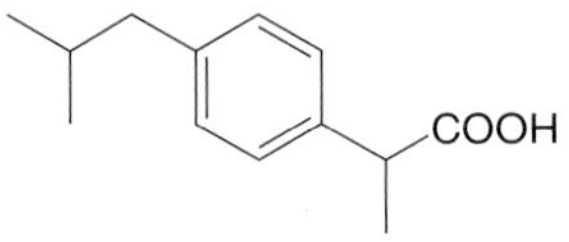

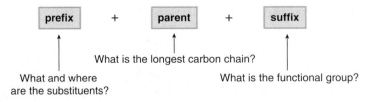

Systematic name: 2-[4-(2-methylpropyl)phenyl]propanoic acid
Generic name: ibuprofen
Trade name: Motrin *or* Advil

4.4   Naming Alkanes

The name of every organic molecule has three parts.

◆ The **parent name** indicates the number of carbons in the longest continuous carbon chain in the molecule.

◆ The **suffix** indicates what functional group is present.

◆ The **prefix** tells us the identity, location, and number of substituents attached to the carbon chain.

| prefix | + | parent | + | suffix |

What is the longest carbon chain?

What and where are the substituents?

What is the functional group?

The names listed in Table 4.1 of Section 4.1B for the simple *n*-alkanes consist of the parent name, which indicates the number of carbon atoms in the longest carbon chain, and the suffix **-ane,** which indicates that the compounds are alkanes. Notice that the parent name for **one carbon is *meth-*,** for **two carbons is *eth-*,** and so on. Thus, we are already familiar with two parts of the name of an organic compound.

To determine the third part of a name, the prefix, we must learn how to name the carbon groups or *substituents* that are bonded to the longest carbon chain.

4.4A   Naming Substituents

Carbon substituents bonded to a long carbon chain are called **alkyl groups.**

◆ An *alkyl group* is formed by removing one hydrogen from an alkane.

An alkyl group is a part of a molecule that is now able to bond to another atom or a functional group. **To name an alkyl group, change the *-ane* ending of the parent alkane to *-yl*.** Thus, methane (CH_4) becomes methyl (CH_3-) and ethane (CH_3CH_3) becomes ethyl (CH_3CH_2-).

Naming three- and four-carbon alkyl groups is more complicated because the parent hydrocarbons have more than one type of hydrogen atom. For example, propane has both 1° and 2° H atoms, and removal of each of these H atoms forms a different alkyl group with a different name, **propyl** or **isopropyl.**

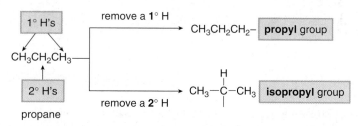

As we learned in Section 3.1, **R** denotes a general carbon group bonded to a functional group. **R** thus denotes any alkyl group.

Because there are two different butane isomers to begin with, each having two different kinds of H atoms, there are four possible alkyl groups containing four carbon atoms, each having a different name: **butyl,** *sec*-**butyl,** **isobutyl,** and *tert*-**butyl.**

The prefix **iso-** is part of the words *propyl* and *butyl,* forming a single word: **isopropyl** and **isobutyl.** The prefixes *sec*- and *tert*- are separated from the word *butyl* by a hyphen: *sec*-**butyl** and *tert*-**butyl.**

The prefix *sec*- is short for *secondary.* A *sec*-butyl group is formed by removal of a 2° **H.** The prefix *tert*- is short for *tertiary.* A *tert*-butyl group is formed by removal of a 3° **H.**

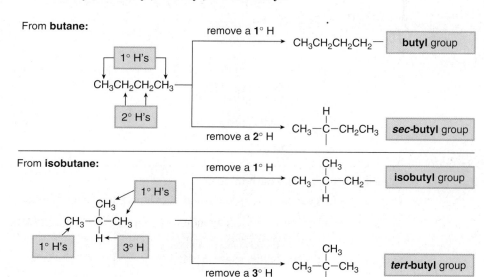

Abbreviations are sometimes used for certain common alkyl groups.

Alkyl group (abbreviation)

◆ methyl (Me)

◆ ethyl (Et)

◆ butyl (Bu)

◆ *tert*-butyl (*t*-Bu)

Now that we know how to name simple alkyl groups, we can learn the stepwise procedure for naming alkanes.

4.4B   Naming an Acyclic Alkane

Four steps are needed to name an alkane. In the following examples, only the C atoms of the carbon skeleton will be drawn. **Remember each C has enough H atoms to make it tetravalent.**

How To   Name an Alkane Using the IUPAC System

Step [1]   **Find the parent carbon chain and add the suffix.**

- Find the longest continuous carbon chain, and name the molecule by using the parent name for that number of carbons, given in Table 4.1. To the name of the parent, add the suffix *-ane* for an alkane. Each functional group has its own characteristic suffix.

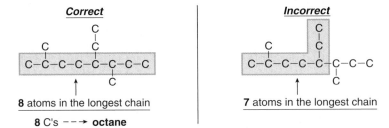

- Finding the longest chain is just a matter of trial and error. Place your pencil on one end of the chain, go to the other end without picking it up, and count carbons. Repeat this procedure until you have found the chain with the largest number of carbons.
- It does not matter if the chain is *straight* or has *bends.* All of the following representations are equivalent.

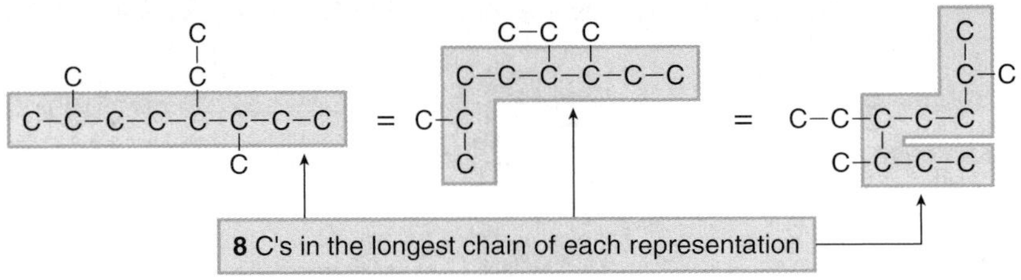

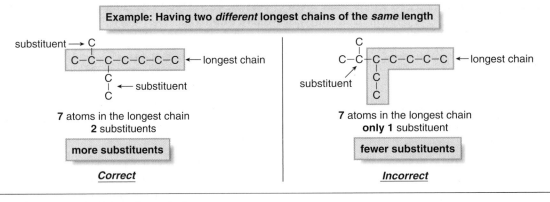

- If there are two chains of equal length, pick the chain with more substituents. In the following example, two different chains in the same alkane contain 7 C's, but the compound on the left has two alkyl groups attached to its long chain, whereas the compound to the right has only one.

Example: Having two *different* longest chains of the *same* length

substituent → C

C–C–C–C–C–C–C ← longest chain
        |
        C ← substituent
        |
        C

7 atoms in the longest chain
2 substituents

more substituents

Correct

C
|
C–C–C–C–C–C–C ← longest chain
    |
substituent   C
    |
    C

7 atoms in the longest chain
only 1 substituent

fewer substituents

Incorrect

Step [2]   **Number the atoms in the carbon chain.**

- Number the longest chain to give the *first* substituent the lower number.

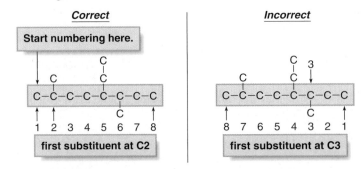

- If the first substituent is the same distance from both ends, number the chain to give the second substituent the lower number. Always look for the first point of difference in numbering from each end of the longest chain.

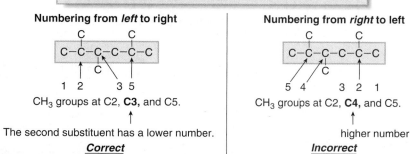

| Example: Giving a lower number to the *second* substituent |

Numbering from *left* to right

CH₃ groups at C2, **C3,** and C5.

The second substituent has a lower number.

Correct

Numbering from *right* to left

CH₃ groups at C2, **C4,** and C5.

higher number

Incorrect

- When numbering a carbon chain results in the same numbers from either end of the chain, assign the lower number *alphabetically* to the first substituent.

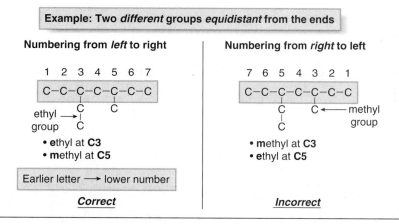

| Example: Two *different* groups *equidistant* from the ends |

Numbering from *left* to right

- **ethyl** at **C3**
- **methyl** at **C5**

Earlier letter ⟶ lower number

Correct

Numbering from *right* to left

methyl group

- **methyl** at **C3**
- **ethyl** at **C5**

Incorrect

Step [3]   Name and number the substituents.

- Name the substituents as alkyl groups, and use the numbers from Step 2 to designate their location.

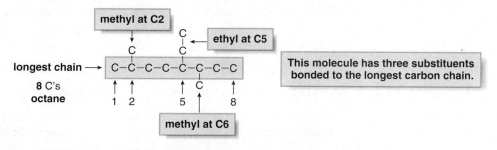

methyl at C2

ethyl at C5

longest chain ⟶

8 C's
octane

This molecule has three substituents bonded to the longest carbon chain.

methyl at C6

- Every carbon belongs to *either* the longest chain or a substituent, but *not both.*
- Each substituent needs its *own* number.
- If two or more identical substituents are bonded to the longest chain, use prefixes to indicate how many: **di-** for two groups, **tri-** for three groups, **tetra-** for four groups, and so forth. The preceding molecule has two methyl substituents, and so its name contains the prefix di- before the word methyl → *di*methyl.

Step [4]   Combine substituent names and numbers + parent + suffix.

- Precede the name of the parent by the names of the substituents.
- Alphabetize the names of the substituents, ignoring all prefixes except *iso,* as in isopropyl and isobutyl.
- Precede the name of each substituent by the number that indicates its location. There must be **one number for each substituent.**

- Separate numbers by commas and separate numbers from letters by hyphens. The name of an alkane is a single word, with no spaces after hyphens or commas.

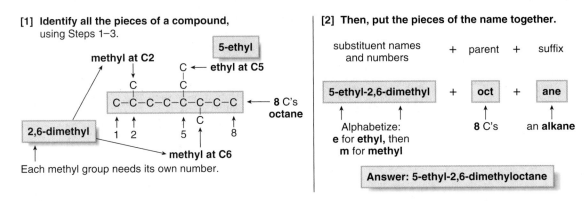

[1] Identify all the pieces of a compound, using Steps 1–3.

methyl at C2

5-ethyl

C ← ethyl at C5

C-C-C-C-C-C-C-C ← 8 C's octane

2,6-dimethyl

1 2   5   8

methyl at C6

Each methyl group needs its own number.

[2] Then, put the pieces of the name together.

substituent names and numbers + parent + suffix

5-ethyl-2,6-dimethyl + oct + ane

Alphabetize: e for ethyl, then m for methyl

8 C's   an alkane

Answer: 5-ethyl-2,6-dimethyloctane

Several additional examples of alkane nomenclature are given in Figure 4.1.

SAMPLE PROBLEM 4.2   Give the IUPAC name for the following compound.

$$CH_3-\underset{\underset{CH_3}{|}}{\overset{\overset{CH_3}{|}}{C}}-\underset{\underset{|}{|}}{\overset{\overset{H}{|}}{C}}-\underset{\underset{H}{|}}{\overset{\overset{H}{|}}{C}}-\underset{\underset{CH_2CH_2CH_2CH_3}{|}}{\overset{\overset{CH_2CH_3}{|}}{C}}-CH_3$$

SOLUTION

To help identify which carbons belong to the longest chain and which are substituents, always draw a box around the atoms of the long chain. Every other carbon atom then becomes a substituent that needs its own name as an alkyl group.

[H's on C's are omitted in the answer, for clarity.]

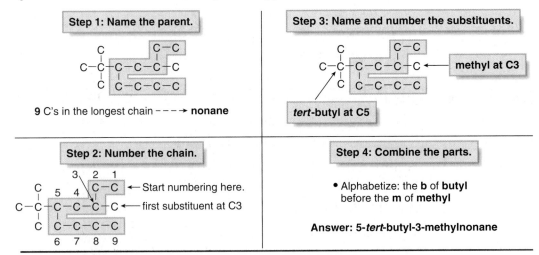

Step 1: Name the parent.

9 C's in the longest chain - - - → **nonane**

Step 2: Number the chain.

3 2 1

5 4

6 7 8 9

← Start numbering here.

← first substituent at C3

Step 3: Name and number the substituents.

← methyl at C3

tert-butyl at C5

Step 4: Combine the parts.

- Alphabetize: the **b** of **butyl** before the **m** of **methyl**

Answer: 5-*tert*-butyl-3-methylnonane

PROBLEM 4.10   Give the IUPAC name for each compound.

a.
$$\underset{\underset{CH_3}{|}}{\overset{\overset{CH_3}{|}}{CH_3CH_2CH_2-C-CH_2CH_2CH_2CH_3}}$$

b. $H-\underset{\underset{CH_3}{|}}{\overset{\overset{CH_2CH_3}{|}}{C}}-CH_2-\underset{\underset{CH_3}{|}}{CHCH_3}$

c.
$$CH_3CH_2CH_2-\underset{\underset{H}{|}}{\overset{\overset{H}{|}}{C}}-CH_2CH_2-\underset{\underset{H}{|}}{\overset{\overset{CH_2CH_3}{|}}{C}}-CH_3$$

Figure 4.1   Examples of alkane nomenclature

$$CH_3CH_2-\overset{\overset{\displaystyle CH_3}{|}}{\underset{\underset{\displaystyle H}{|}}{C}}-\overset{\overset{\displaystyle CH_3}{|}}{\underset{\underset{\displaystyle H}{|}}{C}}-CH_3$$

2,3-dimethylpentane

$$\left[\begin{array}{c}\text{Number to give the 1}^{\text{st}}\text{ methyl group}\\\text{the lower number.}\end{array}\right]$$

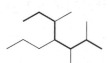

4-ethyl-5-methyloctane

$$\left[\begin{array}{c}\text{Assign the lower number to the 1}^{\text{st}}\text{ substituent}\\\text{alphabetically: the }\textbf{e}\text{ of }\textbf{e}\text{thyl before the }\textbf{m}\text{ of }\textbf{m}\text{ethyl.}\end{array}\right]$$

$$CH_3CH_2CH_2CH_2-\overset{\overset{\displaystyle CH_3}{|}}{\underset{\underset{\underset{\underset{\displaystyle CH_3}{|}}{CH_2}}{|}}{C}}-\overset{\overset{\displaystyle H}{|}}{C}-CH_2CH_3$$

4-ethyl-3,4-dimethyloctane

$$\left[\begin{array}{c}\text{Alphabetize the }\textbf{e}\text{ of }\textbf{e}\text{thyl}\\\text{before the }\textbf{m}\text{ of }\textbf{m}\text{ethyl.}\end{array}\right]$$

2,3,5-trimethyl-4-propylheptane

$$\left[\text{Pick the long chain with more substituents.}\right]$$

- The carbon atoms of each long chain are **drawn in red.**

PROBLEM 4.11   Give the IUPAC name for each compound.

a. $CH_3CH_2CH(CH_3)CH_2CH_3$

d.

f.

b. $(CH_3)_3CCH_2CH(CH_2CH_3)_2$

e.

c. $CH_3(CH_2)_3CH(CH_2CH_2CH_3)CH(CH_3)_2$

You must also know how to derive a structure from a given name. Sample Problem 4.3 illustrates a stepwise method.

SAMPLE PROBLEM 4.3   Give the structure corresponding to the following IUPAC name: 6-isopropyl-3,3,7-trimethyldecane.

SOLUTION
Follow three steps to derive a structure from a name.

Step [1]   Identify the parent name and functional group found at the *end* of the name.

        decane ---→ **10** C's ---→ C–C–C–C–C–C–C–C–C–C

Step [2]   Number the carbon skeleton in *either* direction.

        1 2 3 4 5 6 7 8 9 10
        C–C–C–C–C–C–C–C–C–C

Step [3]   Add the substituents at the appropriate carbons.

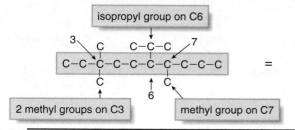

PROBLEM 4.12   Give the structure corresponding to each IUPAC name.
a. 3-methylhexane   c. 3,5,5-trimethyloctane   e. 3-ethyl-5-isobutylnonane
b. 3,3-dimethylpentane   d. 3-ethyl-4-methylhexane

PROBLEM 4.13   Give the IUPAC name for each of the five constitutional isomers of molecular formula C_6H_{14} in Problem 4.6.

4.5 Naming Cycloalkanes

Cycloalkanes are named by using similar rules, but the prefix **cyclo-** immediately precedes the name of the parent.

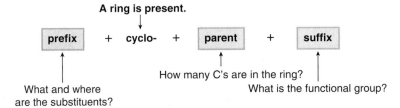

A ring is present.

prefix + cyclo- + parent + suffix

What and where are the substituents?

How many C's are in the ring?

What is the functional group?

| **How To** | Name a Cycloalkane Using the IUPAC System |

Step [1]   **Find the parent cycloalkane.**

- Count the number of carbon atoms in the ring and use the parent name for that number of carbons. Add the prefix **cyclo-** and the suffix **-ane** to the parent name.

6 C's in the ring ⟶ **cyclohexane**

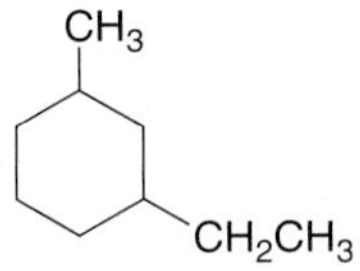

CH₃
CH₂CH₃

Step [2]   **Name and number the substituents.**

- No number is needed to indicate the location of a single substituent.

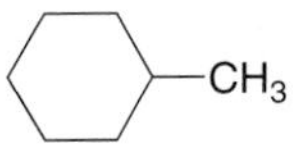

—CH₃

methylcyclohexane

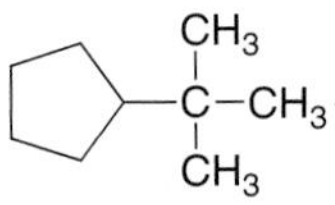

CH₃
|
—C—CH₃
|
CH₃

tert-butylcyclopentane

- For rings with more than one substituent, begin numbering at one substituent and proceed around the ring clockwise or counterclockwise to give the second substituent the lower number.

numbering clockwise

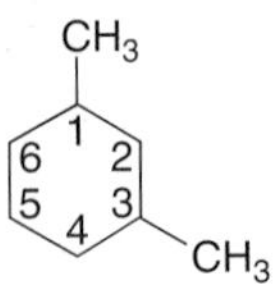

CH₃

6 1
5   2
  4 3
       CH₃

CH₃ groups at C1 and **C3**
The 2nd substituent has a lower number.

Correct: **1,3-dimethylcyclohexane**

numbering counterclockwise

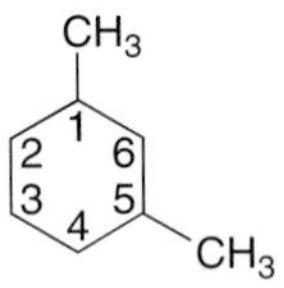

CH₃

2 1
3   6
  4 5
       CH₃

CH₃ groups at C1 and **C5**

Incorrect: 1,5-dimethylcyclohexane

- With two different substituents, number the ring to assign the lower number to the substituents *alphabetically.*

Begin numbering at the ethyl group.

CH₂CH₃

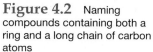

- **e**thyl group at **C1**
- **m**ethyl group at **C3**

$$\boxed{\text{earlier letter} \longrightarrow \text{lower number}}$$

Correct: **1-ethyl-3-methylcyclohexane**

Begin numbering at the methyl group.

CH₂CH₃

- **m**ethyl group at **C1**
- **e**thyl group at **C3**

Incorrect: 3-ethyl-1-methylcyclohexane

When an alkane is composed of both a ring and a long chain, what determines whether a compound is named as an acyclic alkane or a cycloalkane? If the number of carbons in the ring is greater than or equal to the number of carbons in the longest chain, the compound is named as a **cycloalkane.** Two examples are given in Figure 4.2.

Several additional examples of cycloalkane nomenclature are given in Figure 4.3.

Figure 4.2 Naming compounds containing both a ring and a long chain of carbon atoms

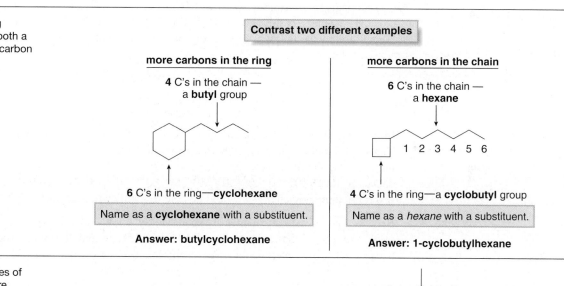

Contrast two different examples

more carbons in the ring

4 C's in the chain — a **butyl** group

6 C's in the ring—**cyclohexane**

Name as a **cyclohexane** with a substituent.

Answer: butylcyclohexane

more carbons in the chain

6 C's in the chain — a **hexane**

1 2 3 4 5 6

4 C's in the ring—a **cyclobutyl** group

Name as a *hexane* with a substituent.

Answer: 1-cyclobutylhexane

Figure 4.3 Examples of cycloalkane nomenclature

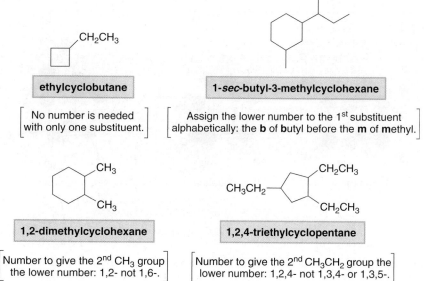

CH₂CH₃

ethylcyclobutane

$$\left[\begin{array}{l}\text{No number is needed} \\ \text{with only one substituent.}\end{array}\right]$$

1-*sec*-butyl-3-methylcyclohexane

$$\left[\begin{array}{l}\text{Assign the lower number to the 1}^{\text{st}}\text{ substituent} \\ \text{alphabetically: the }\textbf{b}\text{ of }\textbf{butyl}\text{ before the }\textbf{m}\text{ of }\textbf{methyl.}\end{array}\right]$$

CH₃

CH₃

1,2-dimethylcyclohexane

$$\left[\begin{array}{l}\text{Number to give the 2}^{\text{nd}}\text{ CH}_3\text{ group} \\ \text{the lower number: 1,2- not 1,6-.}\end{array}\right]$$

CH₂CH₃

CH₃CH₂—

CH₂CH₃

1,2,4-triethylcyclopentane

$$\left[\begin{array}{l}\text{Number to give the 2}^{\text{nd}}\text{ CH}_3\text{CH}_2\text{ group the} \\ \text{lower number: 1,2,4- not 1,3,4- or 1,3,5-.}\end{array}\right]$$

PROBLEM 4.14   Give the IUPAC name for each compound.

a.

c.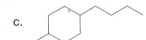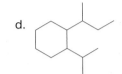

e.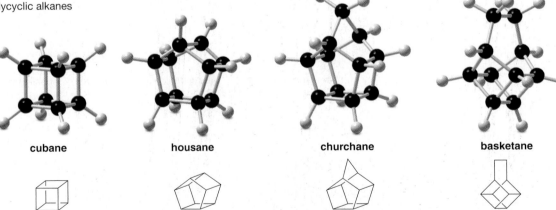

b.

d.

PROBLEM 4.15   Give the structure corresponding to each IUPAC name.

a. 1,2-dimethylcyclobutane

b. 1,1,2-trimethylcyclopropane

c. 4-ethyl-1,2-dimethylcyclohexane

d. 1-*sec*-butyl-3-isopropylcyclopentane

e. 1,1,2,3,4-pentamethylcycloheptane

PROBLEM 4.16   Give the IUPAC names for the five cyclic isomers of molecular formula C_5H_{10} in Problem 4.8.

4.6   Common Names

Some organic compounds are identified using **common names** that do not follow the IUPAC system of nomenclature. Many of these names were given to molecules long ago, before the IUPAC system was adopted. These names are still widely used. For example, isopentane, an older name for one of the C_5H_{12} isomers, is still allowed by IUPAC rules, although it can also be named 2-methylbutane. We will follow the IUPAC system except in cases in which a common name is widely accepted.

$$CH_3-\overset{\overset{\displaystyle CH_3}{|}}{\underset{\underset{\displaystyle H}{|}}{C}}-CH_2CH_3$$

isopentane or 2-methylbutane

In the past several years organic chemists have attempted to synthesize some unusual cycloalkanes not found in nature. **Dodecahedrane,** the molecule that introduced Chapter 4, is one such molecule, first prepared at Ohio State University in 1982. The IUPAC name for dodecahedrane is undecacyclo[9.9.9.0^{2,9}.0^{3,7}.0^{4,20}.0^{3,15}.0^{6,16}.0^{8,15}.0^{10,14}.0^{12,19}.0^{13,17}]eicosane, a name so complex that few trained organic chemists would be able to identify its structure.

Because these systematic names are so unwieldy, organic chemists often assign a name to a polycyclic compound that is more descriptive of its shape and structure. Dodecahedrane is named because its 12 five-membered rings resemble a dodecahedron. Figure 4.4 shows the names and

Figure 4.4   Common names for some polycyclic alkanes

cubane   housane   churchane   basketane

For a more comprehensive list of unusual polycyclic alkanes (including windowpane, davidane, catenane, propellane, and many others), see *Organic Chemistry: The Name Game* by Alex Nickon and Ernest Silversmith, Pergamon Press, 1987.

structures of several other cycloalkanes whose names were inspired by the shape of their carbon skeletons. Note that all the names end in the suffix *-ane,* indicating that they refer to alkanes.

4.7 Fossil Fuels

Natural gas is odorless. The smell observed in a gas leak is due to minute amounts of a sulfur additive such as methanethiol, CH_3SH, which provides an odor for easy detection.

Methane is formed and used in a variety of ways. The CH_4 released from decaying vegetable matter in New York City's main landfill is used for heating homes. CH_4 generators in China convert cow manure into energy in rural farming towns.

Only 19% of the distillate of a barrel of crude oil has the right alkane composition for gasoline, but the need for gasoline far exceeds this output. To increase the amount of gasoline from crude oil, two processes are employed. **Cracking** converts higher molecular weight compounds into lower molecular weight compounds, and **reforming** converts lower molecular weight compounds into higher molecular weight compounds. These two processes allow 47% of the distillate from petroleum refining to be converted into gasoline.

Products obtained from a barrel of crude oil; 1 barrel = 42 gallons.

Many alkanes occur in nature, primarily in natural gas and petroleum. Both of these fossil fuels serve as energy sources, formed from the degradation of organic material long ago.

Natural gas is composed largely of **methane** (60% to 80% depending on its source), with lesser amounts of ethane, propane, and butane. These organic compounds burn in the presence of oxygen, releasing energy for cooking and heating.

Petroleum is a complex mixture of compounds, most of which are hydrocarbons containing one to forty carbon atoms. Distilling crude petroleum, a process called **refining,** separates it into usable fractions that differ in boiling point (Figure 4.5). Most products of petroleum refining provide fuel for home heating, automobiles, diesel engines, and airplanes. Each fuel type has a different composition of hydrocarbons:

- **gasoline:** $C_5H_{12} - C_{12}H_{26}$
- **kerosene:** $C_{12}H_{26} - C_{16}H_{34}$
- **diesel fuel:** $C_{15}H_{32} - C_{18}H_{38}$

Petroleum provides more than fuel. About 3% of crude oil is used to make plastics and other synthetic compounds including drugs, fabrics, dyes, and pesticides. These products are responsible for many of the comforts we now take for granted in industrialized countries. Imagine what life would be like without air conditioning, refrigeration, anesthetics, and pain relievers, all products of the petroleum industry. Consider college students living without CDs and spandex!

Barrel of crude oil

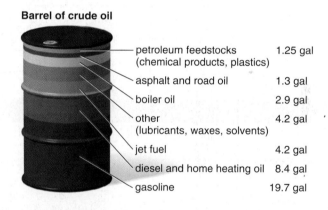

petroleum feedstocks (chemical products, plastics)	1.25 gal
asphalt and road oil	1.3 gal
boiler oil	2.9 gal
other (lubricants, waxes, solvents)	4.2 gal
jet fuel	4.2 gal
diesel and home heating oil	8.4 gal
gasoline	19.7 gal

Figure 4.5 Refining crude petroleum into usable fuel and other petroleum products.
(a) An oil refinery. At an oil refinery, crude petroleum is separated into fractions of similar boiling point by the process of **distillation.**
(b) Schematic of a refinery tower. As crude petroleum is heated, the lower-boiling, more volatile components distill first, followed by fractions of progressively higher boiling point.

(a)

(b)
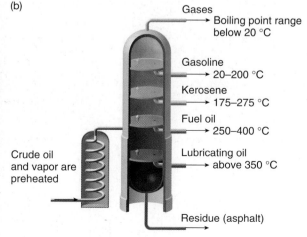

Gases → Boiling point range below 20 °C

Gasoline → 20–200 °C

Kerosene → 175–275 °C

Fuel oil → 250–400 °C

Lubricating oil → above 350 °C

Crude oil and vapor are preheated

Residue (asphalt)

Energy from petroleum is *nonrenewable.* Given the known petroleum reserves and the current rate of energy consumption, remaining oil reserves are estimated to last no more than 43 years, and this assumes no increase in the rate of energy consumption, which is highly unlikely. Given our dependence on petroleum, not only for fuel, but also for the many necessities of modern society, it becomes clear that we must both conserve what we have and find alternate energy sources.

4.8   Physical Properties of Alkanes

Alkanes contain only nonpolar C–C and C–H bonds, and as a result they exhibit only **weak van der Waals forces.** Table 4.2 summarizes how these intermolecular forces affect the physical properties of alkanes.

The gasoline industry exploits the dependence of boiling point and melting point on alkane size by seasonally changing the composition of gasoline in locations where it gets very hot in the summer and very cold in the winter. Gasoline is refined to contain a larger fraction of higher boiling hydrocarbons in warmer weather, so it evaporates less readily. In colder weather, it is refined to contain more lower boiling hydrocarbons, so it freezes less readily.

The mutual insolubility of nonpolar oil and very polar water leads to the common expression, "Oil and water don't mix."

Because nonpolar alkanes are not water soluble, crude petroleum spilled into the sea from a ruptured oil tanker creates an insoluble oil slick on the surface. The insoluble hydrocarbon oil poses a special threat to birds whose feathers are coated with natural nonpolar oils for insulation. Because these hydrophobic oils dissolve in the crude petroleum, birds lose their layer of natural protection and many die.

PROBLEM 4.17   Rank the following products of petroleum refining in order of increasing boiling point: diesel fuel, kerosene, and gasoline.

PROBLEM 4.18   Arrange the following compounds in order of increasing boiling point.
$CH_3(CH_2)_6CH_3$, $CH_3(CH_2)_5CH_3$, $CH_3CH_2CH_2CH_2CH(CH_3)_2$, $(CH_3)_3CCH(CH_3)_2$

TABLE 4.2   Physical Properties of Alkanes

Property	Observation
Boiling point	• Alkanes have low bp's compared to more polar compounds of comparable size.

$$CH_3CH_2CH_3 \qquad CH_3CHO \qquad CH_3CH_2OH$$
$$\text{VDW} \qquad\quad \text{VDW, DD} \qquad \text{VDW, DD, HB}$$
$$MW = 44 \qquad\quad MW = 44 \qquad\quad MW = 46$$

low bp → bp = –42 °C   bp = 21 °C   bp = 79 °C

Increasing strength of intermolecular forces
Increasing boiling point

• Bp increases as the number of carbons increases because of increased surface area.

$$CH_3CH_2CH_2CH_3 \qquad CH_3CH_2CH_2CH_2CH_3 \qquad CH_3CH_2CH_2CH_2CH_2CH_3$$
$$\text{bp} = 0\ °C \qquad\qquad \text{bp} = 36\ °C \qquad\qquad\qquad \text{bp} = 69\ °C$$

Increasing surface area
Increasing boiling point

TABLE 4.2   **Continued**

- The bp of isomers decreases with branching because of decreased surface area.

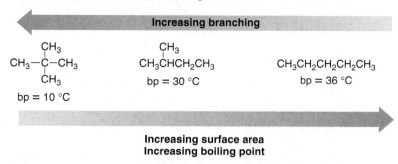

- Alkanes have low mp's compared to more polar compounds of comparable size.

Melting point

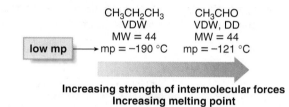

- Mp increases as the number of carbons increases because of increased surface area.

$$CH_3CH_2CH_2CH_3 \qquad CH_3CH_2CH_2CH_2CH_2CH_3$$
$$mp = -138\ °C \qquad\qquad mp = -95\ °C$$

Increasing surface area
Increasing melting point

- Mp increases with increased symmetry.

$$CH_3CH_2CH(CH_3)_2 \qquad\qquad (CH_3)_4C$$
$$mp = -160\ °C \qquad\qquad mp = -17\ °C$$

Increasing symmetry
Increasing melting point

Solubility

- Alkanes are soluble in organic solvents.
- Alkanes are insoluble in water.

Key: **bp** = boiling point; **mp** = melting point; **VDW** = van der Waals; **DD** = dipole–dipole; **HB** = hydrogen bonding; **MW** = molecular weight

4.9   Conformations of Acyclic Alkanes—Ethane

Let's now take a closer look at the three-dimensional structure of alkanes. The study of three-dimensional structure is called **stereochemistry.** In Chapter 4 we examine the effect of rotation around single bonds. In Chapter 5, we will learn about other aspects of stereochemistry.

Recall from Section 1.9A that **rotation occurs around carbon–carbon σ bonds.** Thus, the two CH_3 groups of ethane rotate, allowing the hydrogens on one to adopt different orientations relative to the hydrogens on the other. These arrangements are called **conformations.**

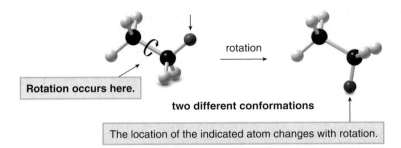

Rotation occurs here.

rotation

two different conformations

The location of the indicated atom changes with rotation.

◆ *Conformations* are different arrangements of atoms that are interconverted by rotation about single bonds. A particular conformation is called a *conformer.*

Names are given to two different arrangements.

◆ In the *eclipsed conformation,* the C–H bonds on one carbon are directly aligned with the C–H bonds on the adjacent carbon.
◆ In the *staggered conformation,* the C–H bonds on one carbon bisect the H–C–H bond angle on the adjacent carbon.

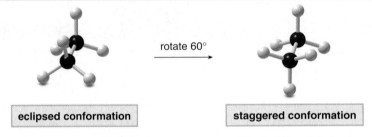

rotate 60°

eclipsed conformation

staggered conformation

The C–H bonds are all **aligned.**

The C–H bonds in front **bisect** the H–C–H bond angles in back.

Rotating the atoms on one carbon by 60° converts an eclipsed conformation into a staggered conformation, and vice versa. These conformations are often viewed **end-on**—that is, looking directly down the carbon–carbon bond. The angle that separates a bond on one atom from a bond on an adjacent atom is called a **dihedral angle.** For ethane in the staggered conformation, the dihedral angle for the C–H bonds is **60°.** For eclipsed ethane, it is **0°.**

End-on view: looking directly down the C–C bond

back carbon   0° dihedral angle   60° dihedral angle

rotate 60°

front carbon

eclipsed conformation

staggered conformation

End-on representations for conformations are commonly drawn using a convention called a **Newman projection.** A Newman projection is a graphic that shows the three groups bonded to each of the carbon atoms in a particular C–C bond, as well as the dihedral angle that separates them.

How To   Draw a Newman Projection

Step [1]   Look directly down the C–C bond (end-on), and draw a circle with a dot in the center to represent the carbons of the C–C bond.

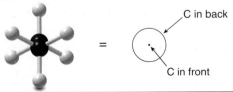

C in back

=

C in front

• The circle represents the back carbon and the dot represents the front carbon.

Step [2]   **Draw in the bonds.**

bonds in front

- Draw the bonds on the **front** C as three lines **meeting at the center** of the circle.
- Draw the bonds on the **back** C as three lines coming **out of the edge** of the circle.

bonds in back

Step [3]   **Add the atoms on each bond.**

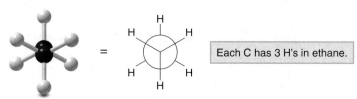

Each C has 3 H's in ethane.

Figure 4.6 illustrates the Newman projections for both the staggered and eclipsed **conformers** for ethane.

Follow this procedure for any C–C bond. With a Newman projection, **always consider *one* C–C bond only and draw the atoms bonded to the carbon atoms, *not* the carbon atoms in the bond itself.** Newman projections for the staggered and eclipsed conformers of propane are drawn in Figure 4.7.

PROBLEM 4.19   Draw the staggered and eclipsed conformations that result from rotation around the C–C bond in CH_3–CH_2Br.

Figure 4.6   Newman projections for the staggered and eclipsed conformations of ethane

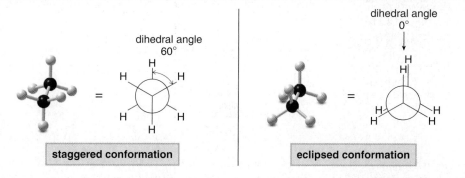

dihedral angle 60°

staggered conformation

dihedral angle 0°

eclipsed conformation

Figure 4.7
Newman projections for the staggered and eclipsed conformers of propane

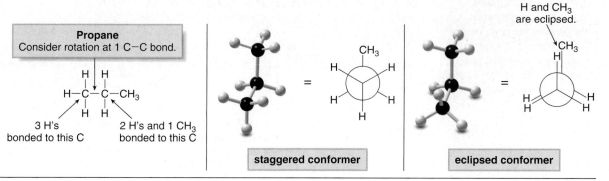

Propane
Consider rotation at 1 C–C bond.

3 H's bonded to this C

2 H's and 1 CH_3 bonded to this C

staggered conformer

H and CH_3 are eclipsed.

eclipsed conformer

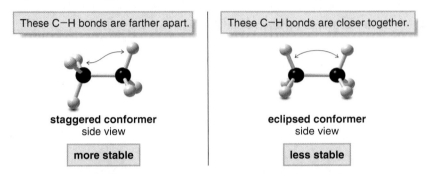

CH₃ bonded to the back C

CH₃ bonded to the front C

In a Newman projection it doesn't matter which C you pick to be in the front or the back. All of the Newman projections shown here represent the staggered conformation of propane.

Another explanation for the relative energies of the staggered and eclipsed conformations of ethane has recently been proposed (*Nature*, **2001**, *411*, 565–568.)

Strain results in an **increase in energy.** Torsional strain is the first of three types of strain discussed in this text. The other two are steric strain (Section 4.10) and angle strain (Section 4.11).

The staggered and eclipsed conformations of ethane interconvert at room temperature, but **each conformer is *not* equally stable.**

◆ **The staggered conformations are more stable (lower in energy) than the eclipsed conformations.**

Electron–electron repulsion between the bonds in the eclipsed conformation increases its energy compared to the staggered conformation, where the bonding electrons are farther apart.

These C–H bonds are farther apart.

These C–H bonds are closer together.

staggered conformer
side view

more stable

eclipsed conformer
side view

less stable

The difference in energy between the staggered and eclipsed conformers is ~3 kcal/mol, a small enough difference that the rotation is still very rapid at room temperature, and the conformers cannot be separated. Because three eclipsed C–H bonds increase the energy of a conformation by 3 kcal/mol, **each eclipsed C–H bond results in an increase in energy of 1 kcal/mol.** The energy difference between the staggered and eclipsed conformers is called **torsional energy.** Thus, eclipsing introduces **torsional strain** into a molecule.

◆ *Torsional strain* is an increase in energy caused by eclipsing interactions.

The graph in Figure 4.8 shows how the potential energy of ethane changes with dihedral angle as one CH₃ group rotates relative to the other. **The staggered conformation is the most stable arrangement, so it is at an *energy minimum*.** As the C–H bonds on one carbon are rotated relative to the C–H bonds on the other carbon, the energy increases as the C–H bonds get closer until a **maximum is reached after 60° rotation to the eclipsed conformation.** As rotation continues, the energy decreases until after 60° rotation, when the staggered conformer is reached once again.

Figure 4.8   Graph: Energy versus dihedral angle for ethane

At any given moment, all ethane molecules do not exist in the more stable staggered conformation; rather, a higher percentage of molecules is present in the more stable staggered conformation than any other possible arrangement.

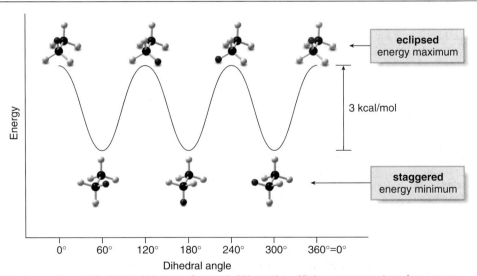

eclipsed
energy maximum

3 kcal/mol

staggered
energy minimum

0°   60°   120°   180°   240°   300°   360°=0°
Dihedral angle

• Note the position of the labeled H atom after each 60° rotation. All three staggered conformers are identical (except for the position of the label), and the same is true for all three eclipsed conformers.

Each H,H eclipsing interaction contributes 1 kcal/mol of destabilization to the eclipsed conformation.

◆ An energy minimum and maximum occur every 60° as the conformation changes from staggered to eclipsed. Conformations that are neither staggered nor eclipsed are intermediate in energy.

PROBLEM 4.20 Draw an energy versus rotation diagram similar to Figure 4.8 for rotation around a C−C bond in propane.

PROBLEM 4.21 The torsional energy in propane is ~3.5 kcal/mol. Because each H,H eclipsing interaction is worth 1 kcal/mol of destabilization, how much is one H,CH_3 eclipsing interaction worth in destabilization? (See Section 4.10 for an alternate way to arrive at this value.)

PROBLEM 4.22 Consider rotation around a C−C bond in neopentane, $(CH_3)_4C$.
a. Draw the most stable and least stable conformers that result from rotation around this bond using Newman projections.
b. Draw a graph of energy versus rotation similar to Figure 4.8 for rotation around this bond.

4.10 Conformations of Butane

Butane and higher molecular weight alkanes have several carbon–carbon bonds, all capable of rotation.

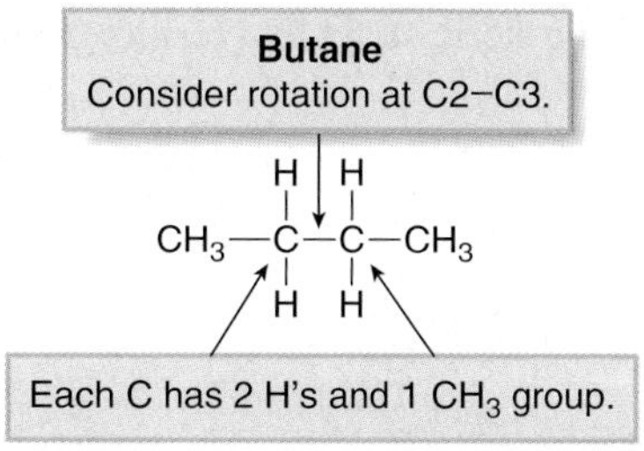

Butane
Consider rotation at C2−C3.

$$CH_3{-}C{-}C{-}CH_3$$

Each C has 2 H's and 1 CH_3 group.

To analyze the different conformations that result from rotation about the C2−C3 bond, begin arbitrarily with one—for example, the staggered conformation that places two CH_3 groups 180° from each other—then,

It takes six 60° rotations to return to the original conformation.

◆ Rotate one carbon atom in 60° increments either clockwise or counterclockwise, while keeping the other carbon fixed. Continue until you return to the original conformation.

Figure 4.9 illustrates the six possible conformers that result from this process.

Figure 4.9 Six different conformations of butane

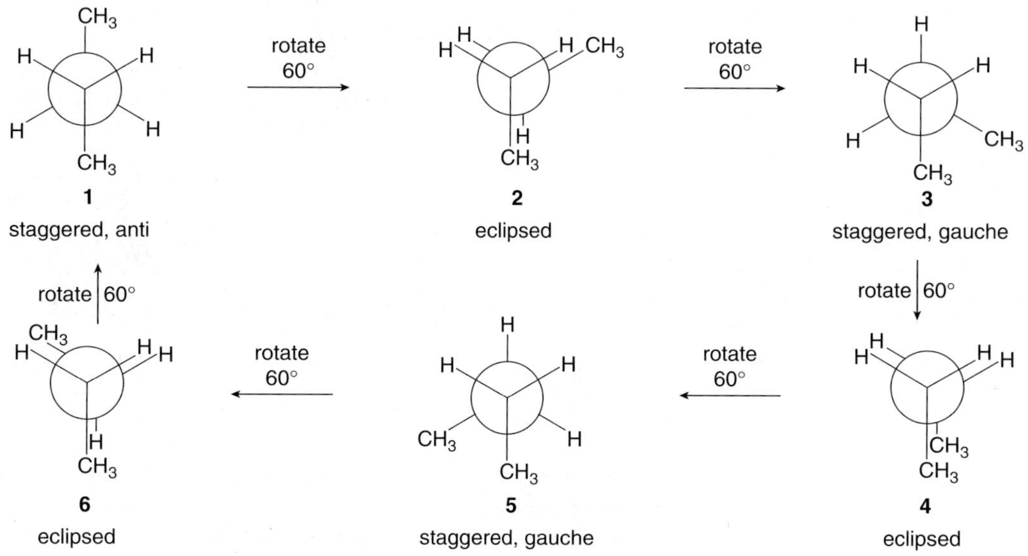

Although each 60° bond rotation converts a staggered conformer into an eclipsed conformer (or vice versa), neither all the staggered conformers nor all the eclipsed conformers are the same. For example, the dihedral angles between the methyl groups in staggered conformers **3** and **5** are both 60°, whereas it is 180° in staggered conformer **1**.

◆ A staggered conformation with two larger groups 180° from each other is called *anti*.
◆ A staggered conformation with two larger groups 60° from each other is called *gauche*.

Similarly, the methyl groups in conformers **2** and **6** both eclipse hydrogen atoms, whereas they eclipse each other in conformer **4**.

The staggered conformations (**1, 3,** and **5**) are lower in energy than the eclipsed conformations (**2, 4,** and **6**), but how do the energies of the individual staggered and eclipsed conformers compare to each other? The relative energies of the individual staggered conformers (or the individual eclipsed conformers) depend on their **steric strain.**

◆ *Steric* strain is an increase in energy resulting when atoms are forced too close to one another.

The methyl groups are further apart in the anti conformer (**1**) than in the gauche conformers (**3** and **5**), so amongst the staggered conformers, **1** is lower in energy (more stable) than **3** and **5**. In fact, the anti conformer is 0.9 kcal/mol lower in energy than either gauche conformer because of the steric strain that results from the proximity of the methyl groups in **3** and **5**.

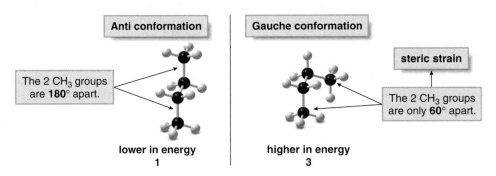

To graph energy versus dihedral angle, keep in mind two considerations:

◆ Gauche conformations are generally higher in energy than anti conformations because of steric strain.

Steric strain also affects the relative energies of eclipsed conformations. Conformer **4** is higher in energy than **2** or **6**, because the two larger CH_3 groups are forced close to each other, introducing considerable steric strain.

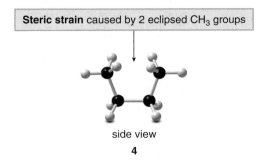

To graph energy versus dihedral angle, keep in mind two considerations:

◆ Staggered conformations are at energy minima and eclipsed conformations are at energy maxima.
◆ Unfavorable steric interactions increase energy.

Figure 4.10 Graph: Energy versus dihedral angle for butane

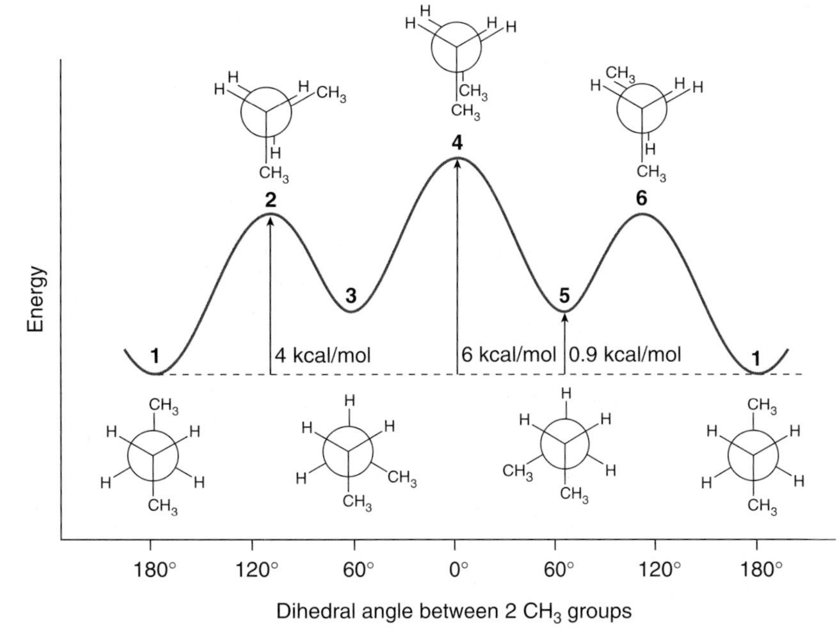

- Staggered conformers **1, 3,** and **5** are at energy minima.
- Anti conformer **1** is lower in energy than gauche conformers **3** and **5,** which possess steric strain.
- Eclipsed conformers **2, 4,** and **6** are at energy maxima.
- Eclipsed conformer **4,** which has additional steric strain due to two eclipsed CH_3 groups, is highest in energy.

For butane, this means that anti conformer **1** is lowest in energy, and conformer **4** with two eclipsed CH_3 groups is the highest in energy. The relative energy of other conformations is depicted in the energy versus rotation diagram for butane in Figure 4.10.

We can now use the values in Figure 4.10 to estimate the destabilization caused by other eclipsed groups. For example, conformer **4** is 6 kcal/mol less stable than the anti conformer **1.** Conformer **4** possesses two H,H eclipsing interactions, worth 1 kcal/mol each in destabilization (Section 4.9), and one CH_3,CH_3 eclipsing interaction. Thus the **CH_3,CH_3 interaction** is worth 6 − 2(1) = **4 kcal/mol** of destabilization.

Similarly, conformer **2** is 4 kcal/mol less stable than the anti conformer **1,** and possesses one H,H eclipsing interaction (worth 1 kcal/mol of destabilization), and two H,CH_3 interactions. Thus **each H,CH_3 interaction** is worth 1/2(4 − 1) = **1.5 kcal/mol** of destabilization. These values are summarized in Table 4.3.

TABLE 4.3   Summary: Torsional and Steric Strain Energies in Acyclic Alkanes

Type of interaction	Energy increase (kcal/mol)
H,H eclipsing	1
H,CH_3 eclipsing	1.5
CH_3,CH_3 eclipsing	4
gauche CH_3 groups	0.9

◆ The energy difference between the lowest and highest energy conformations is called the *barrier to rotation.*

We can use these same principles to determine conformations and relative energies for any acyclic alkane. Because the lowest energy conformation has all bonds staggered and all large groups anti, alkanes are often drawn in zigzag skeletal structures to indicate this.

A zigzag arrangement keeps all carbons **staggered** and **anti.**

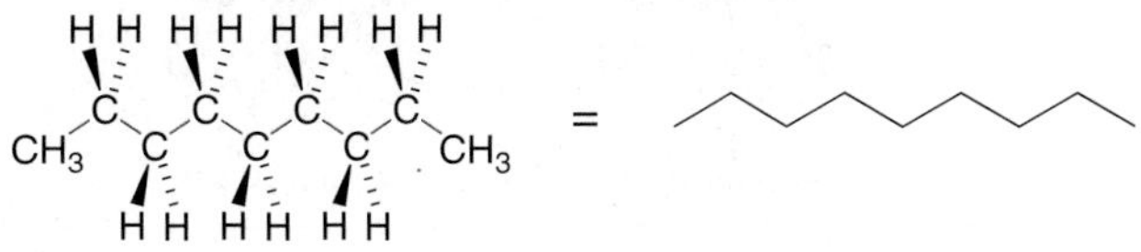

PROBLEM 4.23

rotation here

$$CH_3-\overset{\underset{|}{CH_3}}{\underset{|}{\overset{H}{C}}}-CH_2CH_3$$

a. Draw the three staggered and three eclipsed conformations that result from rotation around the designated bond using Newman projections.

b. Label the most stable and least stable conformation.

PROBLEM 4.24 Consider rotation around the carbon–carbon bond in 1,2-dichloroethane ($ClCH_2CH_2Cl$).

a. Using Newman projections, draw all of the staggered and eclipsed conformations that result from rotation around this bond.

b. Graph energy versus dihedral angle for rotation around this bond.

PROBLEM 4.25 Calculate the destabilization present in each eclipsed conformation.

a. 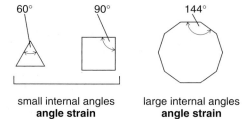 b.

4.11 An Introduction to Cycloalkanes

Besides torsional strain and steric strain, the conformations of cycloalkanes are also affected by **angle strain.**

> ◆ *Angle* strain is an increase in energy when bond angles deviate from the optimum tetrahedral angle of 109.5°.

Originally cycloalkanes were thought to be flat rings, with the bond angles between carbon atoms determined by the size of the ring. For example, a flat cyclopropane ring would have 60° internal bond angles, a flat cyclobutane ring would have 90° angles, and large flat rings would have very large angles. It was assumed that rings with bond angles so different from the tetrahedral bond angle would be very strained and highly reactive. This is called the Baeyer strain theory.

60°   90°   144°

small internal angles
angle strain   large internal angles
angle strain

It turns out, though, that **cycloalkanes with more than three C atoms in the ring are not flat molecules.** They are puckered to **reduce strain,** both angle strain and torsional strain. The three-dimensional structures of some simple cycloalkanes are shown in Figure 4.11.

Figure 4.11 Three-dimensional structure of some cycloalkanes

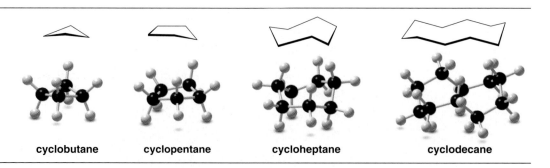

cyclobutane   cyclopentane   cycloheptane   cyclodecane

Many polycyclic hydrocarbons are of interest to chemists. For example, **dodecahedrane,** containing 12 five-membered rings bonded together, is one member of a family of three hydrocarbons that contain several rings of one size joined together. The two other members of this family are **tetrahedrane,** consisting of four three-membered rings, and **cubane,** consisting of six four-membered rings. These compounds are the simplest regular polyhedra whose structures resemble three of the highly symmetrical Platonic solids: the tetrahedron, the cube, and the dodecahedron.

Keep in mind the three different types of strain in organic molecules:

◆ **Torsional strain:** strain caused by eclipsing interactions.

◆ **Steric strain:** strain produced when atoms are forced too close to each other.

◆ **Angle strain:** strain produced when bond angles deviate from 109.5° (for sp^3 hybridized atoms).

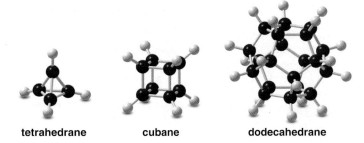

tetrahedrane   cubane   dodecahedrane

How stable are these compounds? Tetrahedrane (with internal 60° bond angles) is so strained that all attempts to prepare it have been thus far unsuccessful. Although cubane is also highly strained because of its 90° bond angles, it was first synthesized in 1964 and is a stable molecule at room temperature. Finally, dodecahedrane is very stable because it has bond angles very close to the tetrahedral bond angle (108° versus 109.5°). Its synthesis eluded chemists for years not because of its strain or inherent instability, but because of the enormous challenge of joining 12 five-membered rings together to form a sphere.

4.12 Cyclohexane

Let's now examine in detail the conformation of **cyclohexane,** the most common ring size in naturally occurring compounds.

4.12A   The Chair Conformation

A planar cyclohexane ring would experience angle strain, because the internal bond angle between the carbon atoms would be 120°, and torsional strain, because all of the hydrogens on adjacent carbon atoms would be eclipsed.

Visualizing the chair. If a cyclohexane conformer is tipped downward, we can more easily view it as a chair with a back, seat, and foot support.

If a cyclohexane ring were flat....

angle strain

120°

The internal bond angle is >109.5°.

torsional strain

All H's are aligned.

In reality, cyclohexane adopts a puckered conformation, called the **chair** form, which is more stable than any other possible conformation.

The carbon skeleton of chair cyclohexane

=

The chair conformation is so stable because it eliminates angle strain (**all C – C – C bond angles are 109.5°**) and torsional strain (all hydrogens on adjacent carbon atoms are **staggered,** not eclipsed).

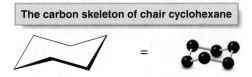

109.5°

All H's are **staggered.**

◆ In cyclohexane, three C atoms pucker up and three C atoms pucker down, alternating around the ring. These C atoms are called *up* C's and *down* C's.

Each carbon in cyclohexane has two different kinds of hydrogens.

◆ *Axial* hydrogens are located above and below the ring (along a perpendicular axis).
◆ *Equatorial* hydrogens are located in the plane of the ring (around the equator).

3 *up* C's and 3 *down* C's	Two kinds of H's
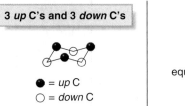 ● = *up* C ○ = *down* C	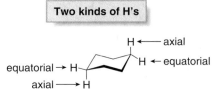

• **Axial** bonds are oriented **above** and **below**.
• **Equatorial** bonds are oriented around the **equator**.

Each cyclohexane carbon atom has one axial and one equatorial hydrogen.

A three-dimensional representation of the chair form is drawn in Figure 4.12. Before continuing, we must first learn how to draw the chair form of cyclohexane.

Figure 4.12 A three-dimensional model of the chair form of cyclohexane with all H atoms drawn

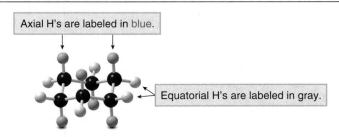

Axial H's are labeled in blue.

Equatorial H's are labeled in gray.

• Cyclohexane has **6 axial H's** and **6 equatorial H's**.

How To   Draw the Chair Form of Cyclohexane

Step [1]   Draw the carbon skeleton.

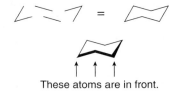

These atoms are in front.

• Draw three parts of the chair: **a wedge, a set of parallel lines,** and **another wedge.**
• Then, join them together.
• The bottom 3 C's come out of the page, and for this reason, bonds to them are often highlighted in bold.

Step [2]   Label the *up* C's and *down* C's on the ring.

● = *up* C   ○ = *down* C

• There are 3 *up* and 3 *down* C's, and they alternate around the ring.

Step [3]   Draw in the axial H atoms.

3 axial H's **above** the ring ⟶

3 axial H's **below** the ring ⟶

• On an *up* C the axial H is *up*.
• On a *down* C the axial H is *down*.

Step [4] **Draw in the equatorial H atoms.**

- The axial H is **down** on a down C, so the equatorial H must be up.
- The axial H is **up** on an up C, so the equatorial H must be down.

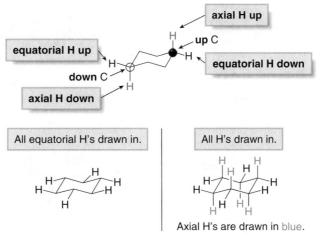

Axial H's are drawn in blue.

PROBLEM 4.26 Classify the ring carbons as *up* C's or *down* C's. Identify the bonds highlighted in bold as axial or equatorial.

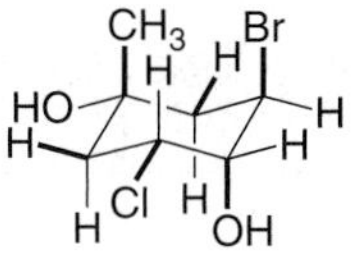

4.12B Ring-Flipping

Like acyclic alkanes, **cyclohexane does not remain in a single conformation.** The bonds twist and bend, resulting in new arrangements, but the movement is more restricted. One important conformational change involves **ring-flipping,** illustrated in the accompanying equation. Ring-flipping can be viewed as a two-step process.

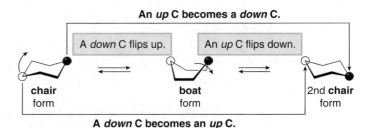

- **A *down* carbon flips up.** This forms a new conformer of cyclohexane called a **boat.** The boat form has two carbons oriented above a plane containing the other four carbons.

- The boat form can flip in two possible ways. The original carbon (labeled with an open circle) can flip down, re-forming the initial conformer; or the **second *up* carbon** (labeled with a solid circle) **can flip down. This forms a second chair conformer.**

Because of ring-flipping, the ***up* carbons become *down* carbons and the *down* carbons become *up* carbons.** Thus, cyclohexane exists as two different chair conformers of equal stability, which rapidly interconvert at room temperature.

The process of ring-flipping also affects the orientation of cyclohexane's hydrogen atoms.

- Axial and equatorial H atoms are interconverted during a ring flip. Axial H atoms become equatorial H atoms, and equatorial H atoms become axial H atoms (Figure 4.13).

Figure 4.13 Ring-flipping interconverts axial and equatorial hydrogens in cyclohexane

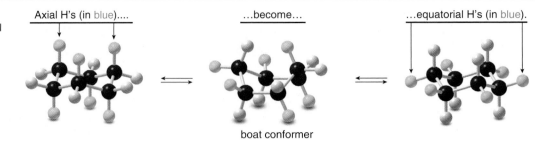

Axial H's (in blue)....   ...become...   ...equatorial H's (in blue).

boat conformer

Figure 4.14 Two views of the boat conformer of cyclohexane

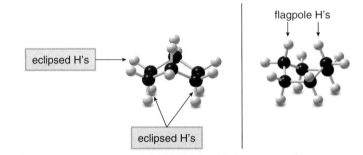

flagpole H's

eclipsed H's

eclipsed H's

The boat form of cyclohexane is less stable than the chair forms for two reasons.

- Eclipsing interactions between H's cause torsional strain.
- The proximity of the flagpole H's causes steric strain.

The chair forms of cyclohexane are 7 kcal/mol more stable than the boat forms. The boat conformer is destabilized by torsional strain because the hydrogens on the four carbon atoms in the plane are eclipsed. Additionally, there is steric strain because two hydrogens at either end of the boat—the **flagpole hydrogens**—are forced close to each other, as shown in Figure 4.14.

4.13   Substituted Cycloalkanes

What happens when one hydrogen on cyclohexane is replaced by a larger substituent? Is there a difference in the stability of the two cyclohexane conformers? To answer these questions, remember one rule.

◆ The equatorial position has more room than the axial position, so larger substituents are more stable in the equatorial position.

4.13A   Cyclohexane with One Substituent

There are two possible chair conformations of a monosubstituted cyclohexane, such as methyl-cyclohexane.

How To	Draw the Two Conformers for a Substituted Cyclohexane

Step [1]   **Draw one chair form and add the substituents.**

- Arbitrarily pick a ring carbon, classify it as an *up* or *down* carbon, and draw the bonds. Each C has one axial and one equatorial bond.
- Add the substituents, in this case H and CH_3, arbitrarily placing one axial and one equatorial. In this example, the CH_3 group is drawn equatorial.

- This forms one of the two possible chair conformers, labeled Conformer 1.

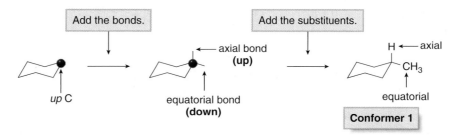

Step [2]   **Ring-flip the cyclohexane ring.**

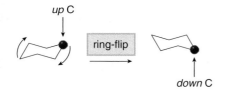

- Convert *up* C's to *down* C's and vice versa. The chosen *up* C now puckers down.

Step [3]   **Add the substituents to the second conformer.**

- Draw axial and equatorial bonds. On a *down* C the axial bond is *down*.
- Ring-flipping converts axial bonds to equatorial bonds, and vice versa. The equatorial methyl becomes axial.
- This forms the other possible chair conformer, labeled Conformer 2.

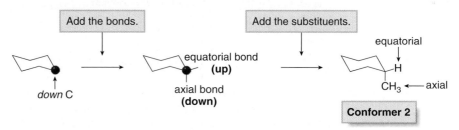

Note that although the CH$_3$ group flips from equatorial to axial, it starts on a down bond, and stays on a down bond. It never flips from below the ring to above the ring.

◆ A substituent always stays on the same side of the ring—either below or above—during the process of ring-flipping.

The two conformers of methylcyclohexane are different, so they are not equally stable. In fact, Conformer **1,** which places the larger methyl group in the roomier equatorial position, is considerably more stable than Conformer **2,** which places it axial.

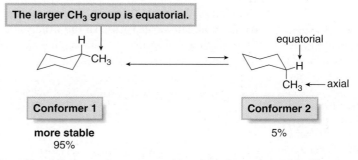

Why is a substituted cyclohexane ring more stable with a larger group in the equatorial position? Figure 4.15 shows that with an equatorial CH$_3$ group, steric interactions with nearby groups are minimized. An axial CH$_3$ group, however, is close to two other axial H atoms, creating two destabilizing steric interactions called **1,3-diaxial interactions.** Each unfavorable H,CH$_3$ interaction destabilizes the conformer by 0.9 kcal/mol, so Conformer 2 (on the right) is 1.8 kcal/mol less stable than Conformer 1 (on the left).

Each carbon atom has one *up* and one *down* bond. An *up* bond can be either axial or equatorial, depending on the carbon to which it is attached. On an *up* C, the axial bond is *up,* but on a *down* C, the equatorial bond is *up.*

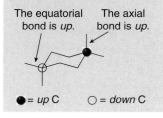

Figure 4.15 Three-dimensional representations for the two conformers of methylcyclohexane

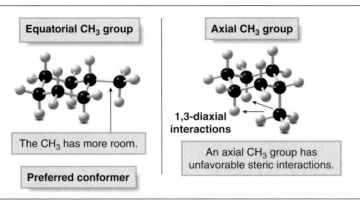

Equatorial CH$_3$ group	Axial CH$_3$ group

The CH$_3$ has more room.

Preferred conformer

1,3-diaxial interactions

An axial CH$_3$ group has unfavorable steric interactions.

Figure 4.16 The two conformers of *tert*-butylcyclohexane

axial *tert*-butyl group

equatorial *tert*-butyl group

very crowded

100%

highly destabilized

The large *tert*-butyl group anchors the cyclohexane ring in this conformation.

◆ **Larger axial substituents create unfavorable 1,3-diaxial interactions, destabilizing a cyclohexane conformer.**

The larger the substituent on the six-membered ring, the higher the percentage of the conformer containing the equatorial substituent at equilibrium. In fact, with a very large substituent like *tert*-butyl [(CH$_3$)$_3$C−], essentially none of the conformer containing an axial *tert*-butyl group is present at room temperature, so that **the ring is essentially anchored in a single conformer having an equatorial *tert*-butyl group.** This is illustrated in Figure 4.16.

PROBLEM 4.27   Draw a second chair conformer for each cyclohexane. Then decide which conformer is present in higher concentration at equilibrium.

a.   [structure with Br]   b.   [structure with Cl]   c.   [structure with CH$_2$CH$_3$]

4.13B   A Disubstituted Cycloalkane

Rotation around the C−C bonds in the ring of a cycloalkane is restricted, so **a group on one side of the ring can *never* rotate to the other side of the ring.** As a result, there are two different 1,2-dimethylcyclopentanes—one having two CH$_3$ groups on the **same side** of the ring and one having them on **opposite sides** of the ring.

Wedges indicate bonds in front of the plane of the ring and dashes indicate bonds behind. For a review of this convention, see Section 1.6B. In this text, dashes are drawn equal in length, as recommended in the latest IUPAC guidelines. If a ring carbon is bonded to a CH$_3$ group in **front** of the ring (on a wedge), it is *assumed* that the other atom bonded to this carbon is hydrogen, located **behind** the ring (on a dash).

A disubstituted cycloalkane: 1,2-dimethylcyclopentane

These two compounds cannot be interconverted.

A =   = **B**

CH$_3$ CH$_3$ CH$_3$ CH$_3$

2 CH$_3$'s above the ring   1 CH$_3$ above and 1 CH$_3$ below

cis isomer   **trans isomer**

two groups on the same side   two groups on opposite sides

A and **B** are **isomers,** because they are different compounds with the same molecular formula, but they represent the second major class of isomers called **stereoisomers.**

◆ *Stereoisomers* are isomers that differ *only* in the way the atoms are oriented in space.

The prefixes **cis** and **trans** are used to distinguish these stereoisomers.

◆ The cis isomer has two groups on the *same side* of the ring.
◆ The trans isomer has two groups on *opposite sides* of the ring.

Cis- and *trans-*1,2-dimethylcyclopentane can also be drawn as if the plane of the ring goes through the plane of the page. Each carbon in the ring then has one bond that points above the ring and one that points below.

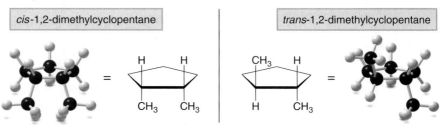

PROBLEM 4.28   Draw the structure for each compound using wedges and dashes.
a. *cis*-1,2-dimethylcyclopropane   b. *trans*-1-ethyl-2-methylcyclopentane

4.13C   A Disubstituted Cyclohexane

Each of the cis and trans isomers of a disubstituted cyclohexane, such as **1,4-dimethylcyclohexane,** has two possible chair conformers.

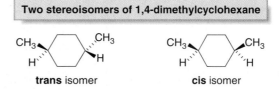

To draw both conformers for each stereoisomer, follow the procedure in Section 4.13A for a monosubstituted cyclohexane, keeping in mind that two substituents must now be added to the ring.

How To	Draw Two Conformers for a Disubstituted Cyclohexane

Step [1]   Draw one chair form and add the substituents.

- For *trans*-1,4-dimethylcyclohexane, arbitrarily pick two C's located 1,4- to each other, classify them as *up* or *down* C's, and draw in the substituents.
- The trans isomer must have one group *above* the ring (on an *up* bond) and one group *below* the ring (on a *down* bond). The substituents can be either axial or equatorial, as long as one is up and one is down. The easiest trans isomer to visualize has two axial CH₃ groups. This arrangement is said to be diaxial.
- This forms one of the two possible chair conformers, labeled Conformer 1.

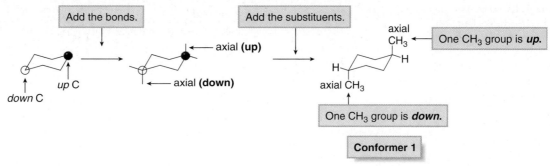

Step [2]   **Ring-flip the cyclohexane ring.**

- The *up* C flips down, and the *down* C flips up.

Step [3]   **Add the substituents to the second conformer.**

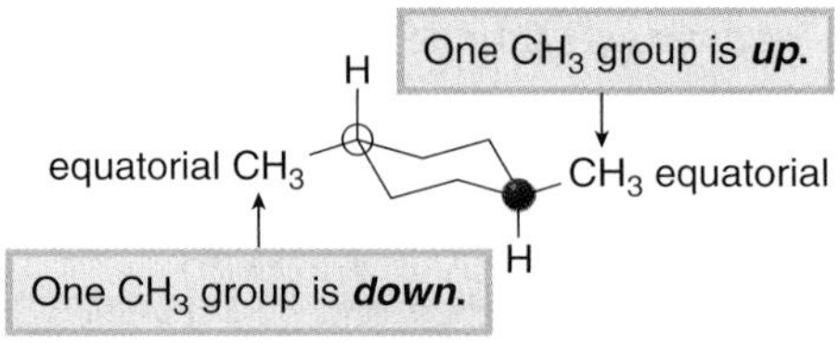

One CH$_3$ group is **up.**

equatorial CH$_3$ CH$_3$ equatorial

One CH$_3$ group is **down.**

Conformer 2

- Ring-flipping converts axial bonds to equatorial bonds, and vice versa. The diaxial CH$_3$ groups become diequatorial. This trans conformer is less obvious to visualize. It is still trans, because one CH$_3$ group is above the ring (on an *up* bond), and one is below (on a *down* bond).

Conformers **1** and **2** are not equally stable. **Because Conformer 2 has both larger CH$_3$ groups in the roomier equatorial position, it is lower in energy.**

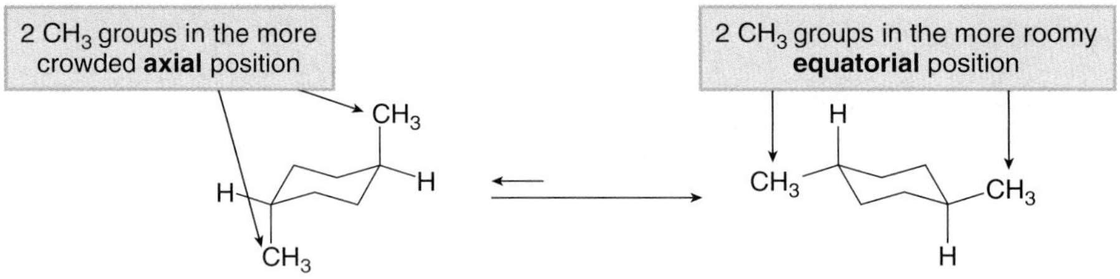

2 CH$_3$ groups in the more crowded **axial** position

2 CH$_3$ groups in the more roomy **equatorial** position

CH$_3$

H   H

H   CH$_3$ CH$_3$

CH$_3$ H

1
diaxial conformer

2
diequatorial conformer
more stable

The cis isomer of 1,4-dimethylcyclohexane also has two conformers, as shown in Figure 4.17. Because each conformer has one CH$_3$ group axial and one equatorial, they are **identical in energy.** At room temperature, therefore, the two conformers exist in a 50:50 mixture at equilibrium.

The relative stability of the two conformations of any disubstituted cyclohexane can be analyzed using this procedure.

- ◆ A cis isomer has two substituents on the same side, either both on *up* bonds or both on *down* bonds.
- ◆ A trans isomer has two substituents on opposite sides, one *up* and one *down*.
- ◆ Whether substituents are axial or equatorial depends on the relative location of the two substituents (on carbons 1,2-, 1,3-, or 1,4-).

Figure 4.17   The two conformers of *cis*-1,4-dimethylcyclohexane

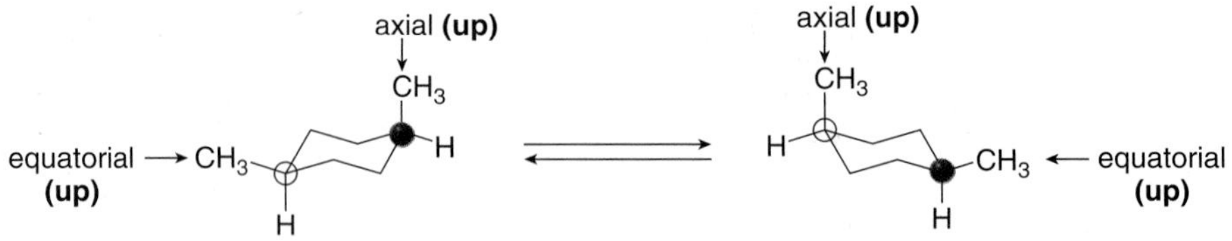

axial **(up)**   axial **(up)**

CH$_3$ CH$_3$

equatorial → CH$_3$ H   H   CH$_3$ ← equatorial
(up)   **(up)**

H   H

Conformer 1   **Conformer 2**
50%   50%

- A cis isomer has two groups on the same side of the ring, either both *up* or both *down*. In this example, Conformers **1** and **2** have two CH$_3$ groups drawn up.
- Both conformers have one CH$_3$ group axial and one equatorial, making them equally stable.

SAMPLE PROBLEM 4.4   Draw both chair conformers for *trans*-1,3-dimethylcyclohexane.

SOLUTION

Step [1]   Draw one chair form and add substituents.

Conformer 1

- Pick 2 C's 1,3- to each other.
- The trans isomer has two groups on **opposite sides.** In Conformer **1,** this means that one CH₃ is equatorial (on an *up* bond), and one group is axial (on a *down* bond).

Steps [2–3]   Ring-flip and add substituents.

Conformer 2

- The 2 *down* C's flip up.
- The axial CH₃ flips equatorial (still a *down* bond) and the equatorial CH₃ flips axial (still an *up* bond). Conformer **2** is trans because the two CH₃'s are still on opposite sides.
- **Conformers 1 and 2 are equally stable** because each has one CH₃ equatorial and one axial.

PROBLEM 4.29   Label each compound as cis or trans. Then draw the second chair conformer.

PROBLEM 4.30   Draw the two possible chair conformers for *cis*-1,3-dimethylcyclohexane. Which conformer, if either, is more stable?

PROBLEM 4.31   Consider 1,2-dimethylcyclohexane.
a. Draw structures for the cis and trans isomers using a hexagon for the six-membered ring.
b. Draw the two possible chair conformers for the cis isomer. Which conformer, if either, is more stable?
c. Draw the two possible chair conformers for the trans isomer. Which conformer, if either, is more stable?
d. Which isomer, cis or trans, is more stable and why?

4.14   Oxidation of Alkanes

Compounds that contain many C–H bonds and few C–Z bonds are said to be in a *reduced state,* whereas those that contain few C–H bonds and more C–Z bonds are in a *more oxidized state.* CH₄ has only C–H bonds and is thus highly reduced. CO₂ has only C–O bonds and is thus highly oxidized.

In Chapter 3 we learned that a functional group contains a heteroatom or π bond and constitutes **the reactive part of a molecule.** Alkanes are the only family of organic molecules that have no functional group, and therefore, **alkanes undergo few reactions.** In fact, alkanes are inert to reaction unless forcing conditions are used.

In Chapter 4, we consider only one reaction of alkanes—**combustion.** Combustion is an **oxidation–reduction** reaction. Let's begin this topic with a general discussion of oxidation and reduction.

4.14A   Oxidation and Reduction Reactions

- ◆ *Oxidation* is the *loss* of electrons.
- ◆ *Reduction* is the *gain* of electrons.

Oxidation and reduction are opposite processes. As in acid–base reactions, there are always two components in these reactions. **One component is oxidized and one is reduced.**

To determine if an organic compound undergoes oxidation or reduction, we concentrate on the carbon atoms of the starting material and product, and **compare the relative number of C–H and C–Z bonds,** where Z = an element *more electronegative* than carbon (usually O, N, or X). Oxidation and reduction are then defined in two complementary ways.

♦ *Oxidation* results in an *increase* in the number of C–Z bonds; *or*
♦ *Oxidation* results in a *decrease* in the number of C–H bonds.

> Because Z is more electronegative than C, replacing C–H bonds with C–Z bonds decreases the electron density around C. Loss of electron density = oxidation.

♦ *Reduction* results in a *decrease* in the number of C–Z bonds; *or*
♦ *Reduction* results in an *increase* in the number of C–H bonds.

Figure 4.18 illustrates the oxidation of CH_4 by replacing C–H bonds with C–O bonds (from left to right). The symbol **[O]** indicates oxidation. Because reduction is the reverse of oxidation, the molecules in Figure 4.18 are progressively reduced moving from right to left, from CO_2 to CH_4. The symbol **[H]** indicates reduction.

SAMPLE PROBLEM 4.5   Determine whether the organic compound is oxidized or reduced in each transformation.

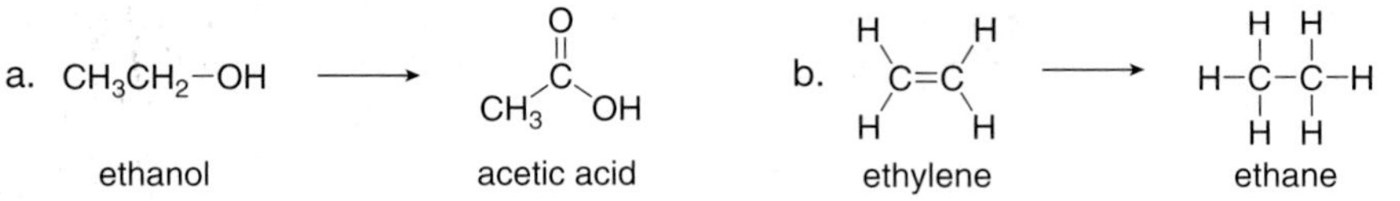

a. CH_3CH_2-OH ⟶ acetic acid

ethanol   acetic acid

b. ethylene ⟶ ethane

ethylene   ethane

SOLUTION

[a] The conversion of ethanol to acetic acid is an **oxidation** because the number of C–O bonds increases: CH_3CH_2OH has one C–O bond and CH_3COOH has three C–O bonds.

[b] The conversion of ethylene to ethane is a **reduction** because the number of C–H bonds increases: ethane has two more C–H bonds than ethylene.

PROBLEM 4.32   Classify each transformation as an oxidation, reduction, or neither.

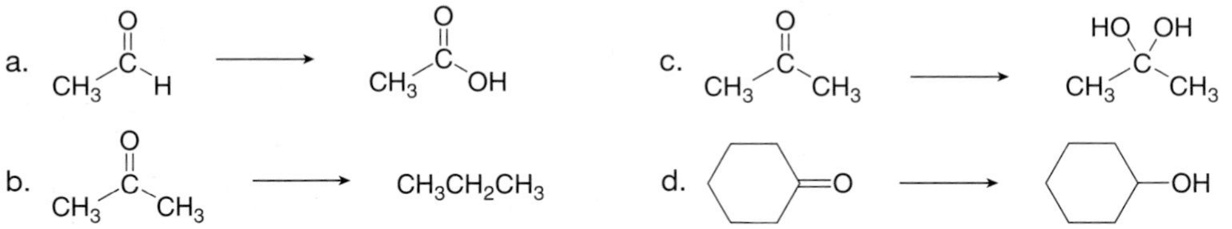

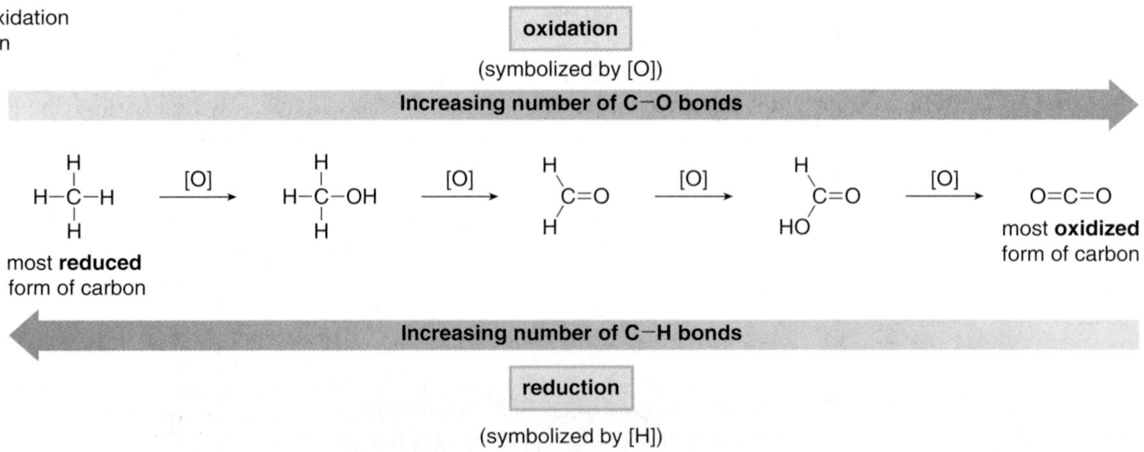

a.

b.

c.

d.

Figure 4.18 The oxidation and reduction of a carbon compound

4.14B   Combustion of Alkanes

When an organic compound is *oxidized* by a reagent, the reagent itself is *reduced*. Similarly, when an organic compound is *reduced* by a reagent, the reagent is *oxidized*. **Organic chemists identify a reaction as an oxidation or reduction by what happens to the** *organic* **component of the reaction.**

Alkanes undergo **combustion**—that is, **they burn in the presence of oxygen to form carbon dioxide and water.** This is a practical example of oxidation. Every $C-H$ and $C-C$ bond in the starting material is converted to a $C-O$ bond in the product. The reactions drawn show the combustion of two different alkanes. Note that the products, $CO_2 + H_2O$, are the same, regardless of the identity of the starting material. Combustion of alkanes in the form of natural gas, gasoline, or heating oil releases energy for heating homes, powering vehicles, and cooking food.

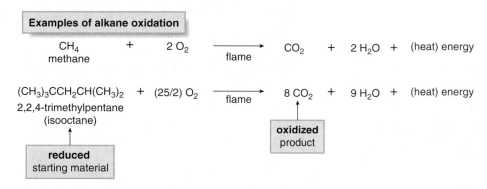

Combustion requires a spark or a flame to initiate the reaction. Gasoline, therefore, which is composed largely of alkanes, can be safely handled and stored in the air, but the presence of a spark or match causes immediate and violent combustion.

Driving an automobile 10,000 miles at 25 miles per gallon releases ~10,000 lb of CO_2 into the atmosphere.

The combustion of alkanes and other hydrocarbons obtained from fossil fuels adds a tremendous amount of CO_2 to the atmosphere each year. Quantitatively, data show more than a 10% increase in the atmospheric concentration of CO_2 in the last 40 years (from 315 parts per million in 1958 to 365 parts per million in 1998; Figure 4.19). Although the composition of the atmosphere has changed over the lifetime of the earth, this may be the first time that the actions of humankind have altered that composition significantly and so quickly.

An increased CO_2 concentration in the atmosphere may have long-range and far-reaching effects. CO_2 absorbs thermal energy that normally radiates from the earth's surface, and redirects it back to the surface. Higher levels of CO_2 may therefore contribute to an increase in the average temperature of the earth's atmosphere. This **global warming,** as it has been called, has many consequences—the melting of polar ice caps, the rise in sea level, and drastic global climate changes to name a few. How great a role CO_2 plays in this process is hotly debated.

Figure 4.19   The changing concentration of CO_2 in the atmosphere since 1957

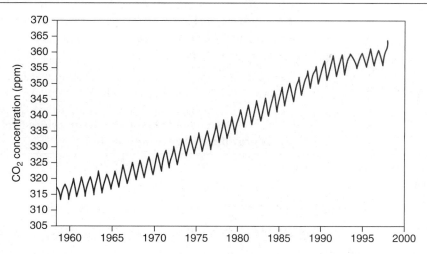

The increasing level of atmospheric CO_2 is clearly evident on the graph. Two data points are recorded each year. The sawtooth nature of the graph is due to seasonal variation of CO_2 level with the seasonal variation in photosynthesis. (Data recorded at Mauna Loa, Hawaii)

PROBLEM 4.33   Draw the products of each combustion reaction.

a. $CH_3CH_2CH_3$ + O_2 $\xrightarrow{\text{flame}}$ b. ⬡ + O_2 $\xrightarrow{\text{flame}}$

4.15 Lipids—Part 1

Lipids are biomolecules whose properties resemble those of alkanes and other hydrocarbons. They are unlike any other class of biomolecules, though, because they are defined by a **physical property,** not by the presence of a particular functional group.

> ◆ **Lipids are biomolecules that are soluble in organic solvents and insoluble in water.**

Lipids have varied sizes and shapes, and a diverse number of functional groups. Fat-soluble vitamins like vitamin A and the phospholipids that comprise cell membranes are two examples of lipids that were presented in Sections 3.6 and 3.8. Other examples are shown in Figure 4.20. One unifying feature accounts for their solubility.

Lipids are discussed in greater detail in Section 10.6, as examples of compounds that contain carbon–carbon double bonds.

> ◆ **Lipids are composed of many nonpolar C–H and C–C bonds, and have few polar functional groups.**

Waxes are lipids having two long alkyl chains joined by a single oxygen-containing functional group. Because of their many C–C and C–H bonds, **waxes are hydrophobic.** They form a protective coating on the feathers of birds to make them water repellent, and on leaves to prevent water evaporation. Bees secrete $CH_3(CH_2)_{14}COO(CH_2)_{29}CH_3$, a wax that forms the honeycomb in which they lay eggs.

$PGF_{2\alpha}$ belongs to a class of lipids called **prostaglandins.** Prostaglandins contain many C–C and C–H bonds and a single COOH group (a **carboxy group**). Prostaglandins possess a wide range of biological activities. They control inflammation, affect blood-platelet aggregation, and stimulate uterine contractions. Nonsteroidal anti-inflammatory drugs such as ibuprofen operate by blocking the synthesis of prostaglandins, as discussed in Sections 19.6 and 26.6.

More details concerning cholesterol's structure and properties are presented in Section 26.8.

Cholesterol is a member of the steroid family, a group of lipids having four rings joined together. Because it has just one polar OH group, cholesterol is insoluble in the aqueous medium of the blood. It is synthesized in the liver and transported to other cells bound to water-soluble organic molecules. Elevated cholesterol levels can lead to coronary artery disease.

Cholesterol is a vital component of the cell membrane. Its hydrophobic carbon chain is embedded in the interior of the lipid bilayer, and its hydrophilic hydroxy group is oriented toward the aqueous exterior (Figure 4.21). Because its tetracyclic carbon skeleton is quite rigid compared to the long floppy side chains of a phospholipid, cholesterol stiffens the cell membrane somewhat, giving it more strength.

Lipids have a high energy content, meaning that much energy is released on their metabolism. Because lipids are composed mainly of C–C and C–H bonds, they are oxidized with the release of energy, just like alkanes are. In fact, lipids are the most efficient biomolecules for the storage of energy. **The combustion of alkanes provides heat for our homes, and the metabolism of lipids provides energy for our bodies.**

Figure 4.20 Three representative lipid molecules

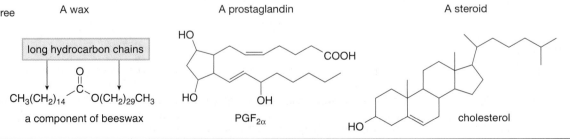

A wax — a component of beeswax
$CH_3(CH_2)_{14}$ C(=O) O(CH_2)_{29}CH_3
long hydrocarbon chains

A prostaglandin — $PGF_{2\alpha}$

A steroid — cholesterol

Figure 4.21   Cholesterol
embedded in a lipid bilayer of a
cell membrane

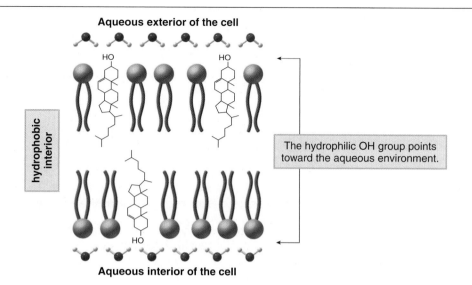

- The nonpolar hydrocarbon skeleton of cholesterol is embedded in the nonpolar interior of the cell
  membrane. Its rigid carbon skeleton stiffens the fluid lipid bilayer, giving it strength.
- Cholesterol's polar OH group is oriented toward the aqueous media inside and outside the cell.

PROBLEM 4.34   Which of the following compounds can be classified as lipids?

a. $CH_3(CH_2)_7CH=CH(CH_2)_7COOH$

  oleic acid

b.   HOOC

c. $CH_2-O-CO(CH_2)_{16}CH_3$
  $CH-O-CO(CH_2)_{16}CH_3$
  $CH_2-O-CO(CH_2)_{16}CH_3$

  tristearin

aspartame

PROBLEM 4.35   Explain why beeswax is insoluble in H_2O, slightly soluble in ethanol (CH_3CH_2OH), and soluble in
chloroform ($CHCl_3$).

4.16   Key Concepts—Alkanes

General Facts About Alkanes (4.1–4.3)

- Alkanes are composed of **tetrahedral, sp^3** hybridized C atoms.
- There are two types of alkanes: acyclic alkanes having molecular formula C_nH_{2n+2}, and cycloalkanes having molecular for-
  mula C_nH_{2n}.
- Alkanes have only **nonpolar C–C and C–H bonds** and no functional group, so they undergo few reactions.
- Alkanes are named with the suffix **-ane.**

Classifying C Atoms and H Atoms (4.1A)

- Carbon atoms are classified by the number of carbon atoms bonded to them; **a 1° carbon is bonded to one other carbon,**
  and so forth.
- Hydrogen atoms are classified by the type of carbon atom to which they are bonded; **a 1° H is bonded to a 1° carbon,** and
  so forth.

Conformations in Acyclic Alkanes (4.9, 4.10)

- Alkane conformations can be classified as **eclipsed, staggered, anti,** or **gauche** depending on the relative orientation of the groups on adjacent carbons.

eclipsed	staggered	anti	gauche
• dihedral angle = 0°	• dihedral angle = 60°	• dihedral angle of two CH₃ groups = 180°	• dihedral angle of two CH₃ groups = 60°

- A staggered conformation is **lower in energy** than an eclipsed conformation.
- An anti conformation is **lower in energy** than a gauche conformation.

Types of Strain

- **Torsional strain**—an increase in energy caused by eclipsing interactions (4.9).
- **Steric strain**—an increase in energy when atoms are forced too close to each other (4.10).
- **Angle strain**—an increase in energy when bond angles deviate from 109.5° (4.11).

Two Types of Isomers

[1] **Constitutional isomers**—isomers that differ in the way the atoms are connected to each other (4.1A).
[2] **Stereoisomers**—isomers that differ *only* in the way the atoms are oriented in space (4.13B).

constitutional isomers

stereoisomers

Conformations in Cyclohexane (4.12, 4.13)

- Cyclohexane exists as **two chair conformers** in rapid equilibrium at room temperature.
- Each carbon atom on a cyclohexane ring has **one axial** and **one equatorial hydrogen.** Ring-flipping converts axial H's to equatorial H's, and vice versa.

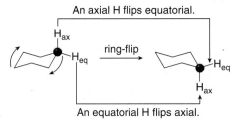

An axial H flips equatorial.

ring-flip

An equatorial H flips axial.

- In substituted cyclohexanes, groups larger than hydrogen are more stable in the **roomier equatorial position.**
- Disubstituted cyclohexanes with substituents on different atoms exist as two possible stereoisomers.
  - The **cis** isomer has two groups on the **same side** of the ring, either both up or both down.
  - The **trans** isomer has two groups on **opposite sides** of the ring, one up and one down.

Oxidation–Reduction Reactions (4.14)

- **Oxidation** results in an **increase in the number of C – Z bonds** or a **decrease in the number of C – H bonds.**
- **Reduction** results in a **decrease in the number of C – Z bonds** or an **increase in the number of C – H bonds.**

Problems

Classifying Carbons and Hydrogens

4.36 For each alkane: (a) classify each carbon atom as 1°, 2°, 3°, or 4°; (b) classify each hydrogen atom as 1°, 2°, or 3°.

(1)   (2)

4.37 Draw the structure of an alkane that:
  a. Contains only 1° and 4° carbons.
  b. Contains only 2° carbons.
  c. Contains only 1° and 2° hydrogens.
  d. Contains only 1° and 3° hydrogens.

Constitutional Isomers

4.38 Draw the structure of all compounds that fit the following descriptions.
  a. Five constitutional isomers having the molecular formula C_4H_8.
  b. Nine constitutional isomers having the molecular formula C_7H_{16}.
  c. Twelve constitutional isomers having the molecular formula C_6H_{12} and containing one ring.

IUPAC Nomenclature

4.39 Give the IUPAC name for each compound.
  a. $CH_3CH_2CHCH_2CHCH_2CH_2CH_3$
       | |
      CH_3 CH_2CH_3

  b. $CH_3CH_2CCH_2CH_2CHCHCH_2CH_2CH_3$
       CH_2CH_3 CH_3
      CH_2CH_3 CH_2CH_3

  c. $CH_3CH_2CH_2C(CH_3)_2C(CH_3)_2CH_2CH_3$

  d. $CH_3CH_2C(CH_2CH_3)_2CH(CH_3)CH(CH_2CH_2CH_3)_2$

  e. $(CH_3CH_2)_3CCH(CH_3)CH_2CH_2CH_3$

  f. $CH_3CH_2CH(CH_3)CH(CH_3)CH(CH_2CH_2CH_3)(CH_2)_3CH_3$

  g. $(CH_3CH_2CH_2)_4C$

  h.

  i.

  j.

  k.

  l. $\text{—}CH(CH_2CH_3)_2$

  m.

  n.

4.40 Give the structure and IUPAC name for each of the nine isomers having molecular formula C_9H_{20} that contain seven carbons in the longest chain and two methyl groups as substituents.

4.41 Draw the structure corresponding to each IUPAC name.
  a. 3-ethyl-2-methylhexane
  b. *sec*-butylcyclopentane
  c. 4-isopropyl-2,4,5-trimethylheptane
  d. cyclobutylcycloheptane
  e. 3-ethyl-1,1-dimethylcyclohexane
  f. 4-butyl-1,1-diethylcyclooctane
  g. 6-isopropyl-2,3-dimethylnonane
  h. 2,2,6,6,7-pentamethyloctane
  i. *cis*-1-ethyl-3-methylcyclopentane
  j. *trans*-1-*tert*-butyl-4-ethylcyclohexane

4.42 Each of the following IUPAC names is incorrect. Explain why it is incorrect and give the correct IUPAC name.
  a. 2,2-dimethyl-4-ethylheptane
  b. 5-ethyl-2-methylhexane
  c. 2-methyl-2-isopropylheptane
  d. 1,5-dimethylcyclohexane
  e. 1-ethyl-2,6-dimethylcycloheptane
  f. 5,5,6-trimethyloctane
  g. 3-butyl-2,2-dimethylhexane
  h. 1,3-dimethylbutane

Physical Properties

4.43 Rank each group of alkanes in order of increasing boiling point. Explain your choice of order.
  a. $CH_3CH_2CH_2CH_2CH_3$, $CH_3CH_2CH_2CH_3$, $CH_3CH_2CH_3$
  b. $CH_3CH_2CH_2CH(CH_3)_2$, $CH_3(CH_2)_4CH_3$, $(CH_3)_2CHCH(CH_3)_2$

4.44   The melting points and boiling points for two C_8H_{18} isomers are given below. Explain why $CH_3(CH_2)_6CH_3$ has a lower melting point but higher boiling point.

	mp (°C)	bp (°C)
$CH_3(CH_2)_6CH_3$	–57	126
$(CH_3)_3CC(CH_3)_3$	102	106

Conformation of Acyclic Alkanes

4.45   Which conformer in each pair is *higher* in energy? Calculate the energy difference between the two conformers using the values given in Table 4.3.

a. 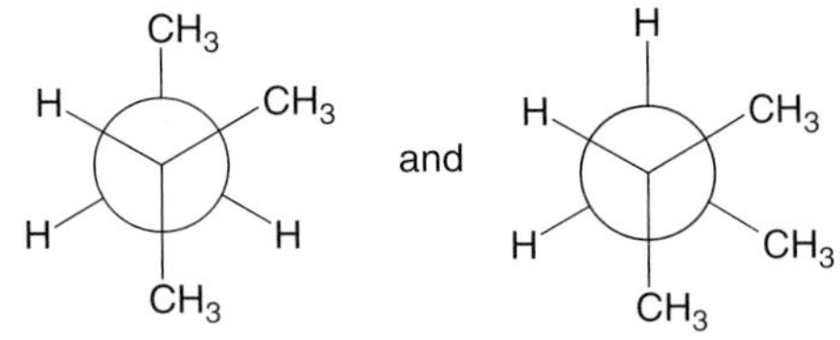   and   b. 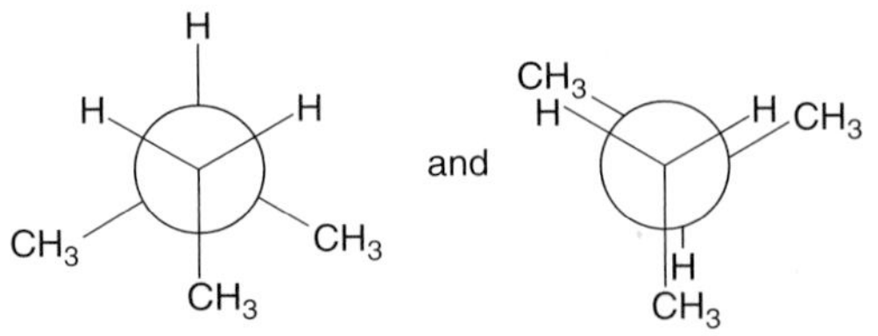   and

4.46   Considering rotation around the indicated bond in each compound, draw Newman projections for the most stable and least stable conformers.

a. $CH_3{-}CH_2CH_2CH_2CH_3$

b. $CH_3CH_2CH_2{-}CH_2CH_2CH_3$

4.47   (a) Using Newman projections, draw all staggered and eclipsed conformations that result from rotation around the indicated bond in each molecule; (b) draw a graph of energy versus dihedral angle for rotation around this bond.

(1) $CH_3CH_2{-}CH_2CH_2CH_3$ (2) $CH_3CH_2{-}CHCH_2CH_3$
                                                              $|CH_3$

4.48   Label the sites of torsional and steric strain in each conformation.

a. 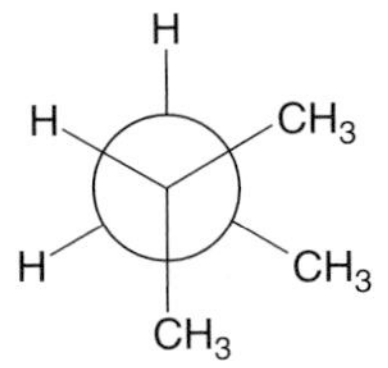   b. 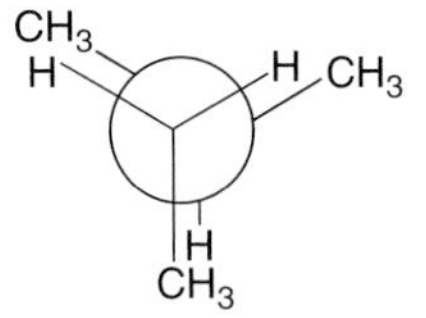   c.

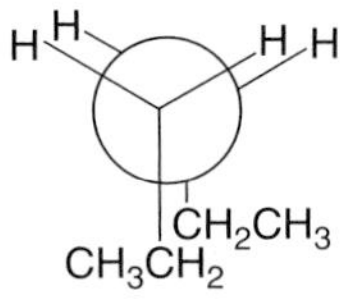

4.49   Calculate the barrier to rotation for each designated bond.

a. $CH_3{-}CH(CH_3)_2$

b. $CH_3{-}C(CH_3)_3$

4.50   The eclipsed conformation of CH_3CH_2Cl is 3.7 kcal/mol less stable than the staggered conformer. How much is the H,Cl eclipsing interaction worth in destabilization?

4.51   (a) Draw the anti and gauche conformers for ethylene glycol ($HOCH_2CH_2OH$). (b) Ethylene glycol is unusual in that the gauche conformer is more stable than the anti conformer. Offer an explanation.

Conformations and Stereoisomers in Cycloalkanes

4.52   For each compound drawn below:
   a.   Label each OH, Br, and CH_3 group as axial or equatorial.
   b.   Classify each conformer as cis or trans.
   c.   Translate each structure into a representation with a hexagon for the six-membered ring, and wedges and dashes for groups above and below the ring.
   d.   Draw the second possible chair conformation for each compound.

(1) 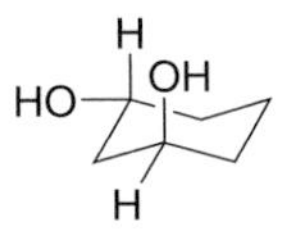   (2) 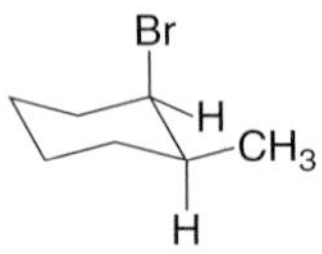   (3)

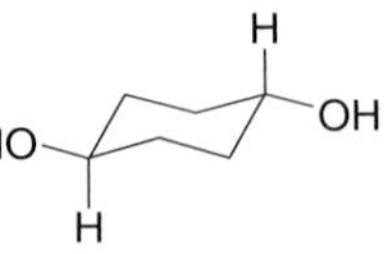

4.53 For each compound drawn below:
a. Draw representations for the cis and trans isomers using a hexagon for the six-membered ring, and wedges and dashes for substituents.
b. Draw the two possible chair conformers for the cis isomer. Which conformer, if either, is more stable?
c. Draw the two possible chair conformers for the trans isomer. Which conformer, if either, is more stable?
d. Which isomer, cis or trans, is more stable and why?

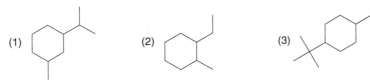

4.54 Which isomer in each pair of compounds is *lower* in energy?
a. *cis*- or *trans*-1,2-diethylcyclohexane
b. *cis*- or *trans*-1-ethyl-3-isopropylcyclohexane

4.55 Which compound in each pair is more stable?

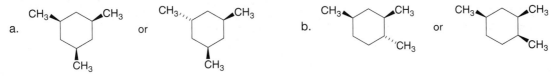

4.56 Glucose is a simple sugar with five substituents bonded to a six-membered ring.

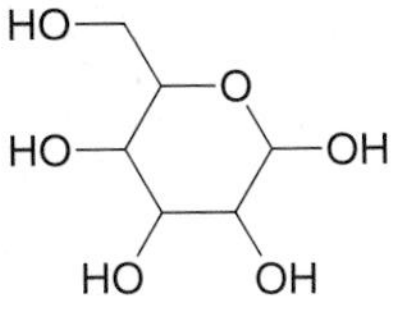

a. Using a chair representation, draw the most stable arrangement of these substituents on the six-membered ring.
b. Convert this representation into one that uses a hexagon with wedges and dashes.

Constitutional Isomers and Stereoisomers

4.57 Classify each pair of compounds as constitutional isomers, stereoisomers, identical molecules, or not isomers of each other.

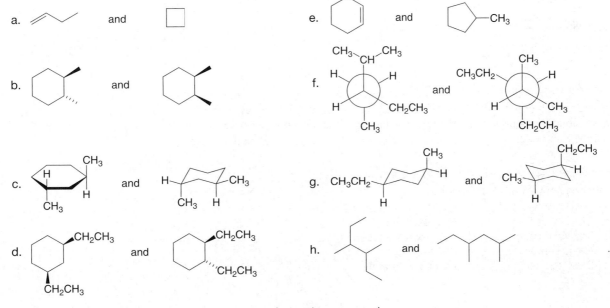

4.58 Draw a constitutional isomer and a stereoisomer for each compound.

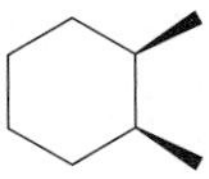

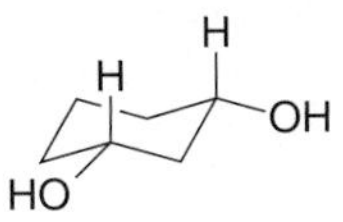

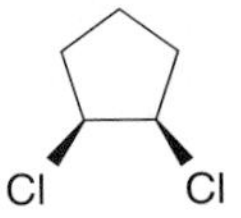

4.59 Draw the three constitutional isomers having molecular formula C_7H_{14} that contain a five-membered ring and two methyl groups as substituents. For each constitutional isomer that can have cis and trans isomers, draw the two stereoisomers.

Oxidation and Reduction

4.60 Classify each reaction as oxidation, reduction, or neither.

a. $CH_3CHO \longrightarrow CH_3CH_2OH$

d. $CH_2{=}CH_2 \longrightarrow H{-}C{\equiv}C{-}H$

b.

e.

c. $CH_2{=}CH_2 \longrightarrow HOCH_2CH_2OH$

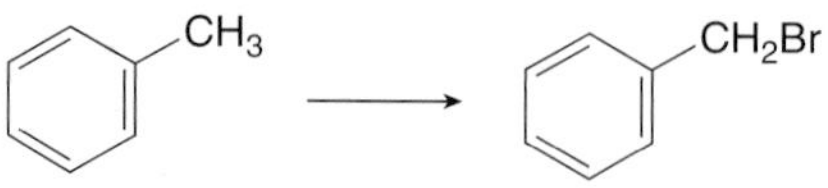

f. $CH_3CH_2OH \longrightarrow CH_2{=}CH_2$

4.61 Draw the products of combustion of each alkane.

a. $CH_3CH_2CH_2CH_2CH(CH_3)_2$

b.

4.62 Hydrocarbons like benzene are metabolized in the body to arene oxides, which rearrange to form phenols. This is an example of a general process in the body, in which an unwanted compound (benzene) is converted to a more water-soluble derivative called a *metabolite,* so that it can be excreted more readily from the body.

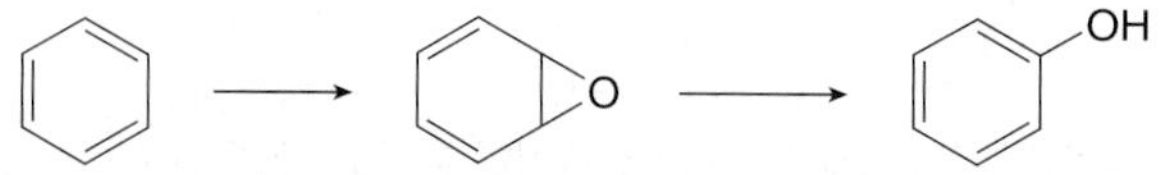

benzene   arene oxide   phenol

a. Classify each of these reactions as oxidation, reduction, or neither.
b. Explain why phenol is more water soluble than benzene. This means that phenol dissolves in urine, which is largely water, to a greater extent than benzene.

Lipids

4.63 Which of the following compounds are lipids?

a. 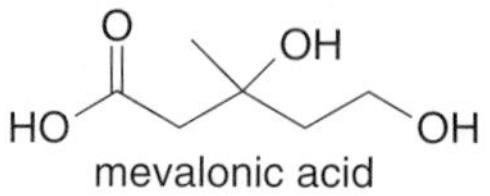
mevalonic acid

b. 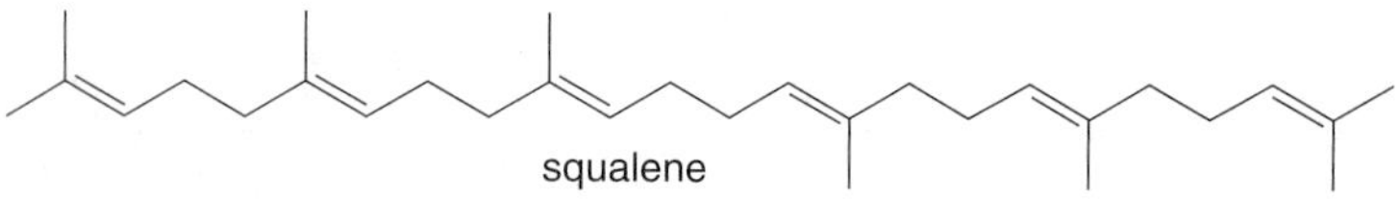
squalene

c. 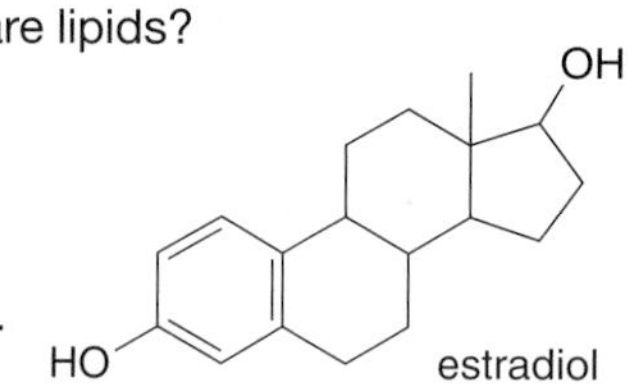
estradiol

d. 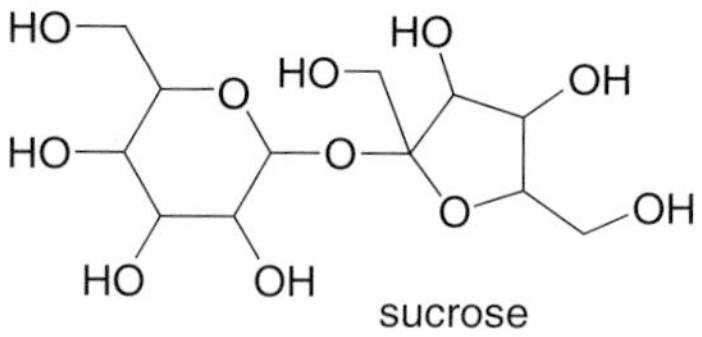
sucrose

4.64 Cholic acid, a compound called a **bile acid,** is converted to a **bile salt** in the body. Bile salts have properties similar to soaps, and they help transport lipids through aqueous solutions. Explain why this is so.

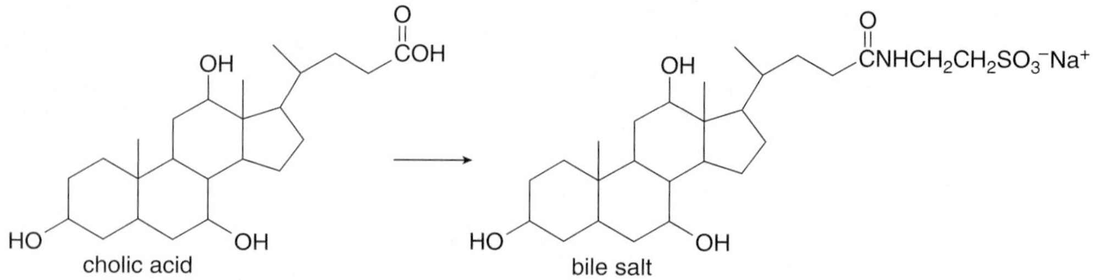

cholic acid
a bile acid

bile salt

Challenge Problems

4.65 Although penicillin G has two amide functional groups, one is much more reactive than the other. Which amide is more reactive and why?

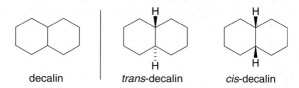

penicillin G

4.66 Which compound has the higher melting point: *cis*- or *trans*-1,4-dimethylcyclohexane? Explain your choice.

4.67 Haloethanes (CH_3CH_2X, X = Cl, Br, I) have similar barriers to rotation (3.2–3.7 kcal/mol) despite the fact that the size of the halogen increases, Cl → Br → I. Offer an explanation.

4.68 When two six-membered rings share a C–C bond, this bicyclic system is called a **decalin.** There are two possible arrangements: *trans*-decalin having two hydrogen atoms at the ring fusion on opposite sides of the rings, and *cis*-decalin having the two hydrogens at the ring fusion on the same side.

decalin   *trans*-decalin   *cis*-decalin

    a. Draw *trans*- and *cis*-decalin using the chair form for the cyclohexane rings.
    b. The trans isomer is more stable. Explain why.

CHAPTER

5

Stereochemistry

(S)-Naproxen is the active ingredient in the widely used pain relievers Naprosyn and Aleve. The three-dimensional orientation of two atoms at a single carbon in naproxen determines its therapeutic properties. Changing the position of these two atoms converts this anti-inflammatory agent into a liver toxin. In Chapter 5 we will learn more about stereochemistry and how small structural differences can have a large effect on the properties of a molecule.

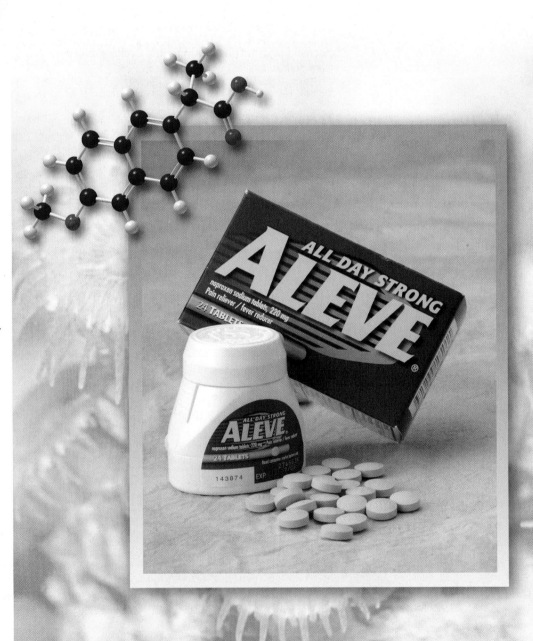

A re you left-handed or right-handed? If you're right-handed, you've probably spent little time thinking about your hand preference. If you're left-handed, though, you probably learned at an early age that many objects—like scissors and baseball gloves—"fit" for righties, but are "backwards" for lefties. **Hands, like many objects in the world around us, are mirror images that are *not* identical.**

In Chapter 5 we examine the "handedness" of molecules, and ask, "How important is the three-dimensional shape of a molecule?"

5.1 Starch and Cellulose

Recall from Chapter 4 that *stereochemistry* **is the three-dimensional structure of a molecule.** How important is stereochemistry? Two biomolecules—starch and cellulose—illustrate how apparently minute differences in structure can result in vastly different properties.

Starch and **cellulose** are two polymers that belong to the family of biomolecules called **carbohydrates** (Figure 5.1).

Starch is the main carbohydrate in the seeds and roots of plants. When we humans ingest wheat, rice, or potatoes, for example, we consume starch, which is then hydrolyzed to the simple sugar **glucose,** one of the compounds our bodies uses for energy. **Cellulose,** nature's most abundant organic material, gives rigidity to tree trunks and plant stems. Wood, cotton, and flax are composed largely of cellulose. Complete hydrolysis of cellulose also forms glucose, but unlike starch, humans cannot metabolize cellulose to glucose. In other words, we can digest starch but not cellulose.

Cellulose and starch are both composed of the same repeating unit—a six-membered ring containing an oxygen atom and three OH groups—joined by an oxygen atom. They differ in the position of the O atom joining the rings together.

In **cellulose,** the O occupies the **equatorial** position.

In **starch,** the O occupies the **axial** position.

repeating unit

♦ In cellulose, the O atom joins two rings using two equatorial bonds.
♦ In starch, the O atom joins two rings using one equatorial and one axial bond.

two equatorial bonds

one axial, one equatorial bond

Starch and cellulose are **isomers** because they are different compounds with the same molecular formula $(C_6H_{10}O_5)_n$. They are **stereoisomers** because only the three-dimensional arrangement of atoms is different.

How the six-membered rings are joined together has an enormous effect on the shape and properties of these carbohydrate molecules. Cellulose is composed of long chains held together by intermolecular hydrogen bonds, thus forming sheets that stack in an extensive three-dimensional network. The

Figure 5.1 Starch and cellulose—Two common carbohydrates

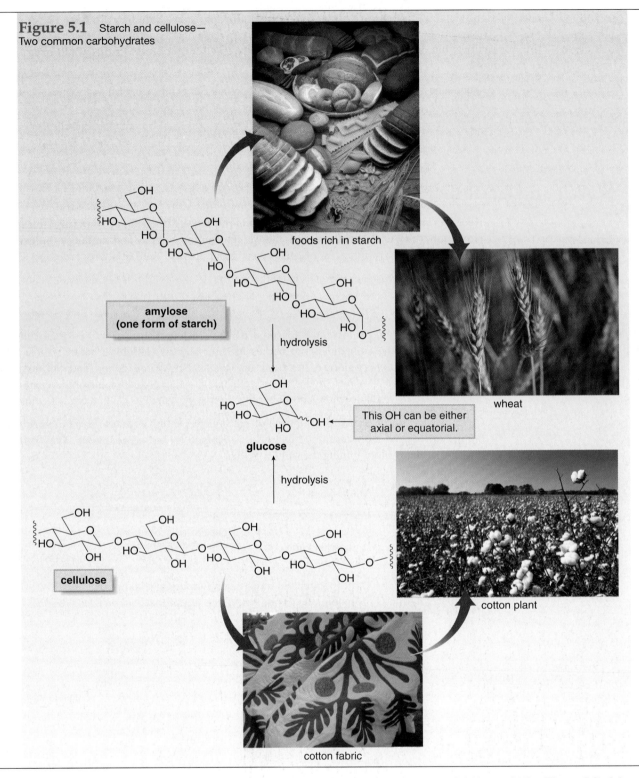

foods rich in starch

amylose (one form of starch)

hydrolysis

wheat

This OH can be either axial or equatorial.

glucose

hydrolysis

cellulose

cotton plant

cotton fabric

axial-equatorial ring junction in starch creates chains that fold into a helix (Figure 5.2). Moreover, the human digestive system contains the enzyme necessary to hydrolyze starch by cleaving its axial C–O bond, but not an enzyme to hydrolyze the equatorial C–O bond in cellulose.

Thus, an **apparently minor difference in the three-dimensional arrangement of atoms confers very different properties on starch and cellulose.**

PROBLEM 5.1   Cellulose is water insoluble, despite its many OH groups. Considering its three-dimensional structure, why do you think this is so?

Figure 5.2   Three-dimensional structure of cellulose and starch

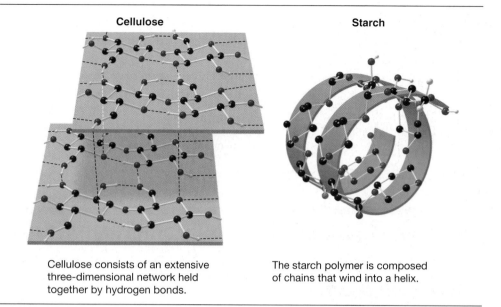

Cellulose

Starch

Cellulose consists of an extensive three-dimensional network held together by hydrogen bonds.

The starch polymer is composed of chains that wind into a helix.

5.2   The Two Major Classes of Isomers

Because an understanding of isomers is integral to the discussion of stereochemistry, let's begin with an overview of isomers.

◆ Isomers are different compounds with the same molecular formula.

There are two major classes of isomers: **constitutional isomers** and **stereoisomers.** *Constitutional (or structural) isomers* **differ in the way the atoms are connected to each other.** Constitutional isomers have:

- ◆ different IUPAC names;
- ◆ the same or different functional groups;
- ◆ different physical properties, so they are separable by physical techniques such as distillation; and
- ◆ different chemical properties. They behave differently or give different products in chemical reactions.

Stereoisomers **differ** *only* **in the way atoms are oriented in space.** Stereoisomers have identical IUPAC names (except for a prefix like cis or trans). Because they differ only in the three-dimensional arrangement of atoms, stereoisomers always have the same functional group(s).

A particular three-dimensional arrangement is called a *configuration.* Thus, stereoisomers differ in configuration. The cis and trans isomers in Section 4.13B and the biomolecules starch and cellulose in Section 5.1 are two examples of stereoisomers.

Figure 5.3 illustrates examples of both types of isomers. Most of Chapter 5 relates to the types and properties of stereoisomers.

PROBLEM 5.2   Classify each pair of compounds as constitutional isomers or stereoisomers.

a. CH$_3$CH$_2$CHCHCH$_3$ and   CH$_3$CHCH$_2$CHCH$_3$
       | | | |
       CH$_3$ CH$_3$ CH$_3$

c. [cube structure]   and   [cube structure]

b. [cyclobutanone]   and   [cyclopropane—OH]

d. [cube structure]   and   [cube structure]

Figure 5.3 A comparison of constitutional isomers and stereoisomers

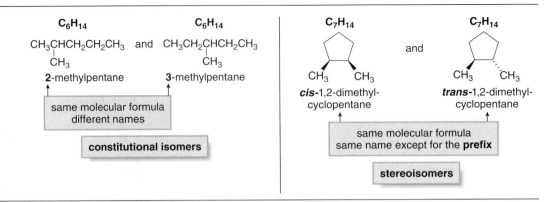

C_6H_{14} $CH_3CHCH_2CH_2CH_3$ and C_6H_{14} $CH_3CH_2CHCH_2CH_3$

$\underset{CH_3}{|}$ $\underset{CH_3}{|}$

2-methylpentane   **3**-methylpentane

same molecular formula
different names

constitutional isomers

C_7H_{14} and C_7H_{14}

CH_3 CH_3 CH_3 CH_3

***cis*-1,2-dimethyl-cyclopentane**   ***trans*-1,2-dimethyl-cyclopentane**

same molecular formula
same name except for the **prefix**

stereoisomers

5.3 Looking Glass Chemistry—Chiral and Achiral Molecules

The dominance of right-handedness over left-handedness occurs in all races and cultures. Despite this fact, even identical twins can exhibit differences in hand preference. Pictured are Zachary (left-handed) and Matthew (right-handed), identical twin sons of the author.

The adjective **chiral** comes from the Greek *cheir*, meaning *hand*.

Left and right hands are **chiral:** they are mirror images that do not superimpose on each other.

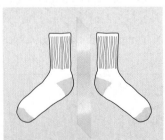

mirror
Socks are **achiral:** they are mirror images that are superimposable.

Everything has a mirror image. What's important in chemistry is **whether a molecule is *identical* to or *different* from its mirror image.**

Some molecules are like hands. **Left and right hands are mirror images of each other, but they are *not* identical.** If you try to mentally place one hand inside the other hand you can never superimpose either all the fingers, or the tops and palms. To *superimpose* an object on its mirror image means to align *all* parts of the object with its mirror image. With molecules, this means aligning all atoms and all bonds.

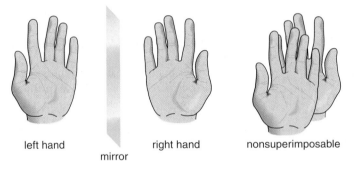

left hand   right hand   nonsuperimposable
mirror

◆ A molecule (or object) that is *not* superimposable on its mirror image is said to be *chiral*.

Other molecules are like socks. **Two socks from a pair are mirror images that *are* superimposable.** One sock can fit inside another, aligning toes and heels, and tops and bottoms. A sock and its mirror image are *identical*.

◆ A molecule (or object) that *is* superimposable on its mirror image is said to be *achiral*.

Let's determine whether three molecules—H_2O, CH_2BrCl and $CHBrClF$—are superimposable on their mirror images; that is, **are H_2O, CH_2BrCl and $CHBrClF$ chiral or achiral?**

To test chirality:

◆ Draw the molecule in three dimensions.

◆ Draw its mirror image.

◆ Try to align all bonds and atoms. To superimpose a molecule and its mirror image you can perform any rotation but **you cannot break bonds.**

Following this procedure, H_2O and CH_2BrCl are both **achiral** molecules because each molecule is superimposable on its mirror image.

Few beginning students of organic chemistry can readily visualize whether a compound and its mirror image are superimposable by looking at drawings on a two-dimensional page. Molecular models can help a great deal in this process.

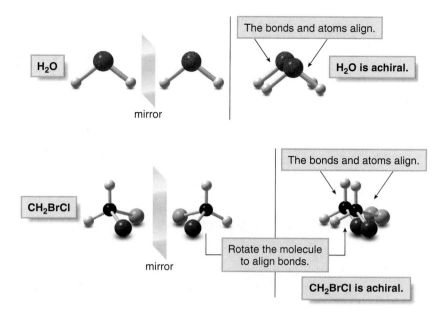

The bonds and atoms align.

H_2O

H_2O is achiral.

mirror

The bonds and atoms align.

CH_2BrCl

Rotate the molecule to align bonds.

CH_2BrCl is achiral.

mirror

With CHBrClF, the result is different. The molecule (labeled **A**) and its mirror image (labeled **B**) are not superimposable. No matter how you rotate **A** and **B**, all the atoms never align. **CHBrClF is thus a chiral molecule,** and **A** and **B** are different compounds.

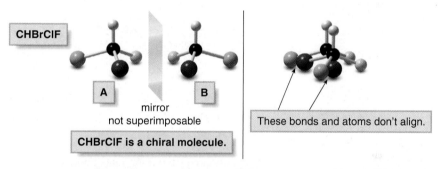

CHBrClF

A   B

mirror
not superimposable

These bonds and atoms don't align.

CHBrClF is a chiral molecule.

A and **B** are **stereoisomers** because they are isomers differing only in the three-dimensional arrangement of substituents. These stereoisomers are called **enantiomers.**

◆ *Enantiomers* are mirror images that are not superimposable.

Naming a carbon atom with four different groups is a topic that currently has no firm agreement among organic chemists. The IUPAC recommends the term *chirality center*, but the term has not gained wide acceptance among organic chemists since it was first suggested in 1996. Other terms in common use are chiral center, chiral carbon, asymmetric carbon, and stereogenic center, the term used in this text.

CHBrClF contains a carbon atom bonded to four different groups. **A carbon atom bonded to four different groups is commonly called a** *stereogenic center.* Most chiral molecules contain one or more stereogenic centers.

A stereogenic center is any site on a molecule at which the interchange of two groups forms a stereoisomer. A **carbon atom with four different groups is a** *tetrahedral* **stereogenic center,** because the interchange of two groups converts one enantiomer into another. We will learn about another type of stereogenic center in Section 8.2B.

We have now learned two related but different concepts, and it is necessary to distinguish between them.

◆ A molecule that is not superimposable on its mirror image is a *chiral molecule.*
◆ A carbon atom bonded to four different groups is a *stereogenic center.*

Molecules can contain zero, one, or more stereogenic centers.

◆ **With no stereogenic centers, a molecule generally is not chiral.** H_2O and CH_2BrCl have *no* stereogenic centers and are *achiral* molecules. (There are a few exceptions to this generalization, as we will learn in Section 17.5.)

◆ **With one tetrahedral stereogenic center, a molecule is *always* chiral.** CHBrClF is a *chiral* molecule containing *one* stereogenic center.

◆ **With two or more stereogenic centers, a molecule *may* or *may not* be chiral,** as we will learn in Section 5.8.

PROBLEM 5.3   Draw the mirror image of each compound. Label each molecule as chiral or achiral.

When trying to distinguish between chiral and achiral compounds, keep in mind the following:

◆ Achiral molecules contain a plane of symmetry but chiral molecules do not.

◆ A *plane of symmetry* is a mirror plane that cuts a molecule in half, so that one half of the molecule is a reflection of the other half.

The achiral molecule CH_2BrCl has a plane of symmetry, but the chiral molecule CHBrClF does not.

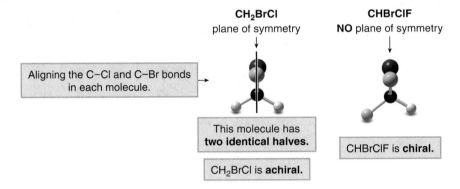

Figure 5.4 summarizes the main facts about chirality we have learned thus far.

PROBLEM 5.4   Draw in the plane of symmetry for each molecule.

a. 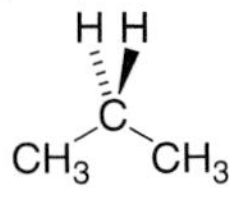 (structure with H H, C, CH₃, CH₃)
b. (structure with CH₃, H, CH₃, H, H, H)
c. (structure with CH₃, CH₃, H, C, C, H, Cl, Cl)

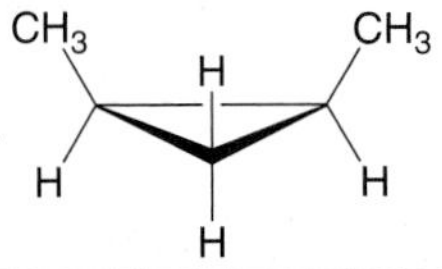

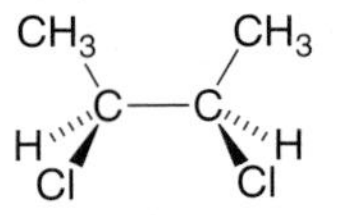

PROBLEM 5.5   A molecule is achiral if it has a plane of symmetry in *any* conformation. The given conformation of 2,3-dibromobutane does not have a plane of symmetry, but rotation around the C2–C3 bond forms a conformation that does have a plane of symmetry. Draw this conformation.

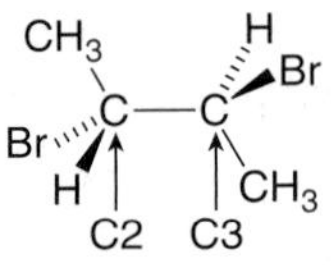

(structure with CH₃, H, Br, Br, C, C, H, CH₃, labeled C2 C3)

Figure 5.4   Summary:
The basic principles of chirality

- Everything has a mirror image. The fundamental question is whether a molecule and its mirror image are superimposable.
- If a molecule and its mirror image are *not* superimposable, the molecule and its mirror image are *chiral.*
- The terms *stereogenic center* and *chiral molecule* are related but distinct. In general, a chiral molecule must have one or more stereogenic centers.
- The presence of a *plane of symmetry* indicates that a molecule is achiral.

A species of snail, *Liguus virgineus,* possesses a chiral, right-handed helical shell.

Stereochemistry may seem esoteric, but chirality pervades our very existence. On a molecular level, many biomolecules fundamental to life are chiral. On a macroscopic level, many naturally occurring objects possess handedness. Examples include chiral helical seashells shaped like right-handed screws, and plants such as honeysuckle that wind in a chiral left-handed helix. The human body is chiral, and hands, feet, and ears are not superimposable.

5.4   Stereogenic Centers

A necessary skill in the study of stereochemistry is the ability to locate and draw tetrahedral stereogenic centers.

5.4A   Stereogenic Centers on Carbon Atoms That Are Not Part of a Ring

Recall from Section 5.3 that any carbon atom bonded to four different groups is a stereogenic center. To locate a stereogenic center, examine each *tetrahedral* carbon atom in a molecule, and look at the four **groups**—not the four *atoms*—bonded to it. CBrClFI has one stereogenic center because its central carbon atom is bonded to four different elements. 3-Bromohexane also has one stereogenic center because one carbon is bonded to H, Br, CH_2CH_3, and $CH_2CH_2CH_3$. We consider all atoms in a group as a *whole unit*, not just the atom directly bonded to the carbon in question.

Always omit from consideration all C atoms that can't be tetrahedral stereogenic centers. These include:

- CH_2 and CH_3 groups (more than one H bonded to C)
- any *sp* or sp^2 hybridized C (less than four groups around C)

SAMPLE PROBLEM 5.1   Locate the stereogenic center in each drug.

a.   albuterol (bronchodilator)

b.   brompheniramine (antihistamine)

SOLUTION

Omit all CH_2 and CH_3 groups and all doubly bonded (sp^2 hybridized) C atoms. In albuterol, one C has three CH_3 groups bonded to it, so it can be eliminated as well. This leaves one C in each molecule with four different groups bonded to it.

a.

b.

Heteroatoms surrounded by four different groups are also stereogenic centers. Stereogenic N atoms are discussed in Chapter 25.

PROBLEM 5.6

Label any stereogenic center in the given molecules. (Some compounds contain no stereogenic centers.)

a. $CH_3CH_2CH(Cl)CH_2CH_3$
b. $(CH_3)_3CH$
c. $CH_3CH(OH)CH=CH_2$

d. $CH_3CH_2CH_2OH$
e. $(CH_3)_2CHCH_2CH_2CH(CH_3)CH_2CH_3$
f. $CH_3CH_2CH(CH_3)CH_2CH_2CH_3$

Larger organic molecules can have two, three, or even hundreds of stereogenic centers. **Propoxyphene** and **ephedrine** each contain two stereogenic centers, and **fructose,** a simple carbohydrate, has three.

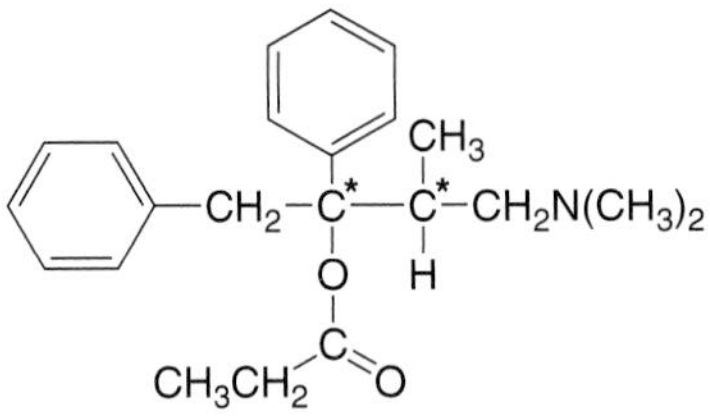

propoxyphene
Trade name: Darvon
(analgesic)

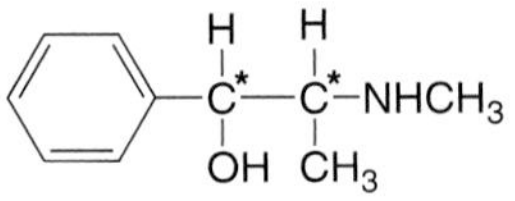

ephedrine
(bronchodilator, decongestant)

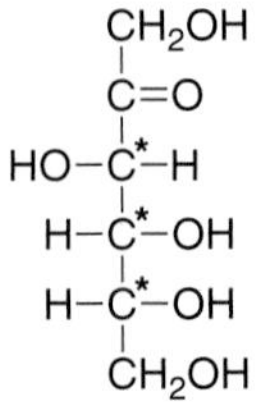

[* = **stereogenic center**]

fructose
(a simple sugar)

PROBLEM 5.7

Label the stereogenic centers in each molecule. Compounds may have one or more stereogenic centers.

a. $CH_3CH_2CH_2CH(OH)CH_3$
b. $(CH_3)_2CHCH_2CH(NH_2)COOH$ c. 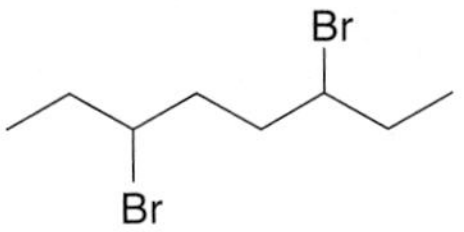   d.

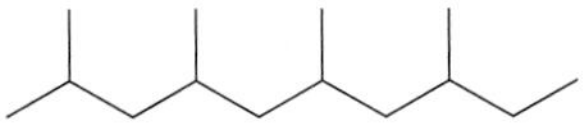

PROBLEM 5.8

Label the stereogenic centers in each biomolecule.

a.

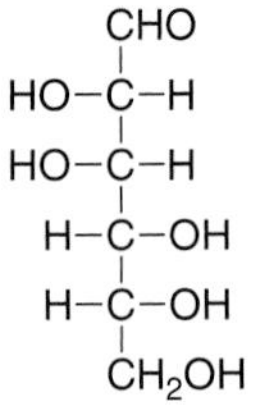

mannose
(a simple carbohydrate)

b.

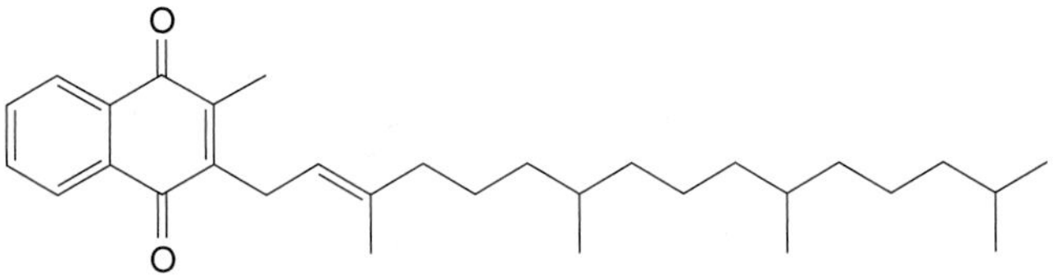

vitamin K₁

5.4B   Drawing a Pair of Enantiomers

◆ Any molecule with one tetrahedral stereogenic center is a chiral compound and exists as a pair of enantiomers.

2-Butanol, for example, has one stereogenic center. To draw both enantiomers, use the typical convention for depicting a tetrahedron: **place two bonds in the plane, one in front of the plane on a wedge, and one behind the plane on a dash.** Then, to form the first enantiomer **A,** arbitrarily place the four groups—H, OH, CH_3, and CH_2CH_3—on any bond to the stereogenic center.

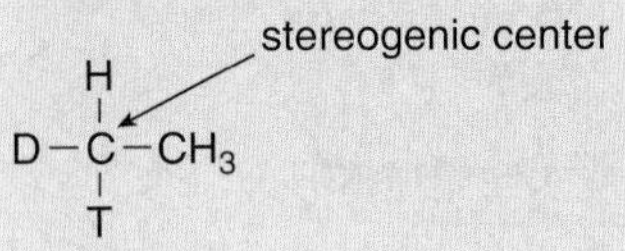

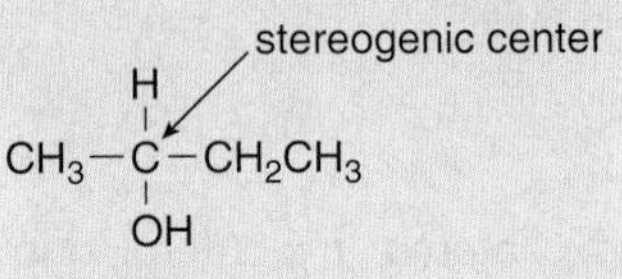

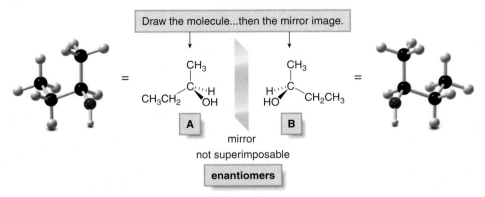

Draw the molecule...then the mirror image.

mirror

not superimposable

enantiomers

Figure 5.5 Three-dimensional representations for pairs of enantiomers

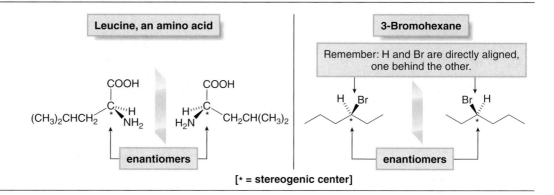

Then, draw a mirror plane and arrange the substituents in the mirror image so that they are a reflection of the groups in the first molecule, forming **B.** No matter how **A** and **B** are rotated, it is impossible to align all of their atoms. Because **A** and **B** are mirror images and not superimposable, **A** and **B** are a pair of **enantiomers.** Two other pairs of enantiomers are drawn in Figure 5.5.

PROBLEM 5.9   Locate the stereogenic center in each compound and draw both enantiomers.
a. $CH_3CH(Cl)CH_2CH_3$ b. $CH_3CH_2CH(OH)CH_2OH$

5.5   Stereogenic Centers in Cyclic Compounds

Stereogenic centers may also occur at carbon atoms that are part of a ring. To find stereogenic centers on ring carbons always draw the rings as flat polygons, and look for tetrahedral carbons that are bonded to four different groups, as usual. Each ring carbon is bonded to two other atoms in the ring, as well as two substituents attached to the ring. When these two substituents are *different,* we must compare the ring atoms equidistant from the atom in question.

Does methylcyclopentane have a stereogenic center? All of the carbon atoms are bonded to two or three hydrogen atoms except for C1, the ring carbon bonded to the methyl group. Next, compare the ring atoms and bonds on both sides equidistant from C1, and continue until a point of difference is reached, or until both sides meet, either at an atom or in the middle of a bond. In this case, there is no point of difference on either side, so C1 is bonded to identical alkyl groups that happen to be part of a ring. **C1 is therefore *not* a stereogenic center.**

In drawing a tetrahedron using solid lines, wedges, and dashes, always draw the two solid lines first; then draw the wedge and the dash on the *opposite side* of the solid lines.

If you draw the two solid lines down... then add the wedge and dash above.

If you draw the two solid lines on the left... then add the wedge and dash to the right.

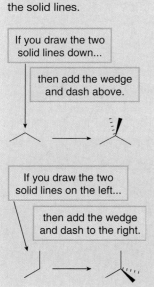

Is C1 a stereogenic center?

methylcyclopentane

two identical groups, equidistant from C1

NO, C1 is **not** a stereogenic center.

With 3-methylcyclohexene, the result is different. All carbon atoms are bonded to two or three hydrogen atoms or are sp^2 hybridized except for C3, the ring carbon bonded to the methyl group. In this case, the atoms equidistant from C3 are different, so C3 is bonded to *different* alkyl groups in the ring. **C3 is therefore bonded to four different groups, making it a stereogenic center.**

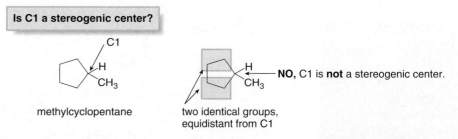

Is C3 a stereogenic center?

These 2 C's are different.

YES, C3 is a stereogenic center.

3-methylcyclohexene

Because 3-methylcyclohexene has one tetrahedral stereogenic center it is a chiral compound and exists as a pair of enantiomers. Substituents above and below the ring are drawn with wedges and dashes as usual.

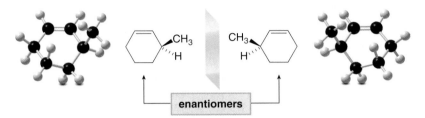

Two enantiomers are *different* compounds. To convert one enantiomer to another you must **switch the position of two atoms.** This amounts to breaking bonds.

Many biologically active compounds contain one or more stereogenic centers on ring carbons. For example, **thalidomide,** which contains one such stereogenic center, was used as a popular sedative and anti-nausea drug for pregnant women in Europe and Great Britain from 1959–1962.

Two enantiomers of thalidomide

stereogenic center   stereogenic center

anti-nausea drug   teratogen

Although it is a potent teratogen (a substance that causes fetal abnormalities), thalidomide exhibits several beneficial effects. For this reason it is now prescribed under strict control for the treatment of Hansen's disease (leprosy) and certain forms of cancer.

Unfortunately thalidomide was sold as a mixture of its two enantiomers, and each of these stereoisomers has a different biological activity. This is a property not uncommon in chiral drugs, as we will see in Section 5.13. Although one enantiomer had the desired therapeutic effect, the other enantiomer was responsible for thousands of serious birth defects in children born to women who took the drug during pregnancy.

Sucrose and **taxol** are two useful molecules with several stereogenic centers at ring carbons. Identify the stereogenic centers in these more complicated compounds in exactly the same way, looking at one carbon at a time. **Sucrose,** with nine stereogenic centers on two rings, is the carbohydrate used as table sugar. **Taxol,** with 11 stereogenic centers, is an anticancer agent active against ovarian, breast, and some lung tumors.

Initial studies with taxol were carried out with material isolated from the bark of the Pacific yew tree, but stripping the bark killed these magnificent trees. More recently, a precursor to taxol was isolated from the needles of the common English yew tree. Because this precursor can be converted to taxol in four steps with high yield, supplies are more available for treating cancer patients.

sucrose
(table sugar)

taxol
Trade name: Paclitaxel
(anticancer agent)

[* = stereogenic center]

PROBLEM 5.10   Label the stereogenic centers in each compound. A molecule may have zero, one, or more stereogenic centers.

a.   b.   c.   d.   e.   f.

PROBLEM 5.11   Label the eight stereogenic centers in the biomolecule cholesterol.

cholesterol

5.6   Labeling Stereogenic Centers with *R* or *S*

Naming enantiomers with the prefixes *R* or *S* is called the Cahn–Ingold–Prelog system after the three chemists who devised it.

Because enantiomers are two different compounds, we need to distinguish them by name. This is done by adding the prefix *R* or *S* to the IUPAC name of the enantiomer. To designate an enantiomer as *R* or *S*, first **assign a priority** (1, 2, 3, or 4) to each group bonded to the stereogenic center, and then use these priorities to label one enantiomer *R* and one *S*.

Rules Needed to Assign Priority

Rule 1   **Assign priorities (1, 2, 3, or 4) to the atoms directly bonded to the stereogenic center in order of decreasing atomic number. The atom of *highest* atomic number gets the *highest* priority (1).**

- In CHBrClF, priorities are assigned as follows: Br (1, highest) → Cl (2) → F (3)→ H (4, lowest). In many molecules the lowest priority group will be H.

$$4 \longrightarrow H$$
$$3 \longrightarrow F\text{--}\overset{\displaystyle |}{\underset{\displaystyle |}{C}}\text{--}Br \longleftarrow 1$$
$$2 \longrightarrow Cl$$

Rule 2   **If two atoms on a stereogenic center are the *same*, assign priority based on the atomic number of the atoms bonded to these atoms. *One* atom of higher atomic number determines a higher priority.**

- With 2-butanol, the O atom gets highest priority (1) and H gets lowest priority (4) using rule 1. To assign priority (either 2 or 3) to the two C atoms, look at what atoms (other than the stereogenic center) are bonded to each C.

Following rule 1:

4 (lowest atomic number)

↓ 2 or 3

2-butanol $CH_3\text{--}\overset{\displaystyle H}{\underset{\displaystyle OH}{C}}\text{--}CH_2CH_3$

2 or 3 ↑
1 (highest atomic number)

Adding rule 2:

$CH_3\text{--}\overset{\displaystyle H}{\underset{\displaystyle OH}{C}}\text{--}CH_2CH_3$

This C is bonded to 2 H's and 1 C. → $\text{--}\overset{\displaystyle H}{\underset{\displaystyle H}{C}}\text{--}CH_3$

higher priority group (2)

This C is bonded to 3 H's. → $H\text{--}\overset{\displaystyle H}{\underset{\displaystyle H}{C}}\text{--}$ lower priority group (3)

- The order of priority of groups in 2-butanol is: $-OH$ **(1)**, $-CH_2CH_3$ **(2)** $-CH_3$ **(3)**, and $-H$ **(4)**.

- If priority still cannot be assigned, continue along a chain until a point of difference is reached.

Rule 3   **If two isotopes are bonded to the stereogenic center, assign priorities in order of decreasing *mass* number.**

- In comparing the three isotopes of hydrogen, the order of priorities is:

	Mass number	Priority
T (tritium)	3 (1 proton + 2 neutrons)	1
D (deuterium)	2 (1 proton + 1 neutron)	2
H (hydrogen)	1 (1 proton)	3

Figure 5.6 Examples of assigning priorities to stereogenic centers

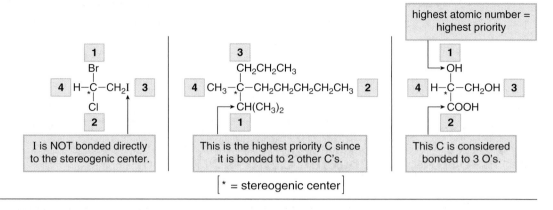

Rule 4   **To assign a priority to an atom that is part of a multiple bond, treat a multiply bonded atom as an equivalent number of singly bonded atoms.**

◆ For example, the C of a C=O is considered to be bonded to two O atoms.

bonded to a stereogenic center here

$$\text{C=O} \quad \text{equivalent to} \quad \text{C—O}$$

Consider this C bonded to **2 O's.**

◆ Other common multiple bonds are drawn below.

$$\text{—C=C—H} \quad \xrightarrow{\text{equivalent to}} \quad \text{—C—C—H} \qquad \text{—C≡C—H} \quad \xrightarrow{\text{equivalent to}} \quad \text{—C—C—H}$$

Each atom in the **double** bond is drawn **twice.**   Each atom in the **triple** bond is drawn **three** times.

Figure 5.6 gives examples of priorities assigned to stereogenic centers.

PROBLEM 5.12   Which group in each pair is assigned the *higher* priority?
a. $-CH_3$, $-CH_2CH_3$ c. $-H$, $-D$ e. $-CH_2CH_2Cl$, $-CH_2CH(CH_3)_2$
b. $-I$, $-Br$ d. $-CH_2Br$, $-CH_2CH_2Br$ f. $-CH_2OH$, $-CHO$

PROBLEM 5.13   Rank the following groups in order of *decreasing* priority.
a. $-COOH$, $-H$, $-NH_2$, $-OH$ c. $-CH_2CH_3$, $-CH_3$, $-H$, $-CH(CH_3)_2$
b. $-H$, $-CH_3$, $-Cl$, $-CH_2Cl$ d. $-CH=CH_2$, $-CH_3$, $-C≡CH$, $-H$

R is derived from the Latin word *rectus* meaning *right* and *S* is from the Latin word *sinister* meaning *left*.

Once we have learned how to assign priority to the four groups around a stereogenic center, we can use three steps to designate the center as either *R* or *S*.

How To   Assign *R* or *S* to a Stereogenic Center

Example   **Label each enantiomer as *R* or *S*.**

two enantiomers of 2-butanol

Step [1]   **Assign priorities from 1 to 4 to each group bonded to the stereogenic center.**

- The priorities for the four groups around the stereogenic center in 2-butanol were given in rule 2, on page 167.

−OH	−CH$_2$CH$_3$	−CH$_3$	−H
1	2	3	4
highest			lowest

Decreasing priority →

Step [2]   **Orient the molecule with the lowest priority group (4) *back* (on a *dash*), and visualize the relative positions of the remaining three groups (priorities 1, 2, and 3).**

- For each enantiomer of 2-butanol, look toward the lowest priority group, drawn behind the plane, down the C−H bond.

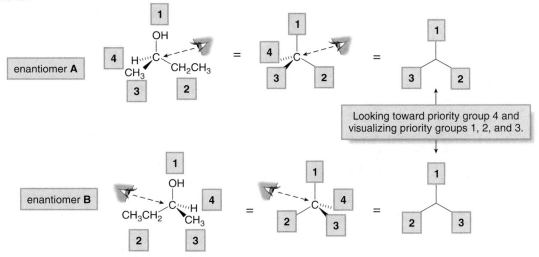

Looking toward priority group 4 and visualizing priority groups 1, 2, and 3.

Step [3]   **Trace a circle from priority group 1 → 2 → 3.**

- If tracing the circle goes in the **clockwise** direction—to the right from the noon position—the isomer is named **R.**
- If tracing the circle goes in the **counterclockwise** direction—to the left from the noon position—the isomer is named **S.**

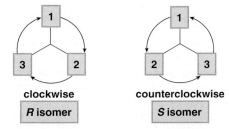

clockwise
R isomer

counterclockwise
S isomer

- The letters *R* or *S* precede the IUPAC name of the molecule. For the enantiomers of 2-butanol:

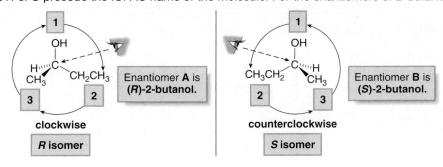

Enantiomer **A** is (**R**)-2-butanol.

Enantiomer **B** is (**S**)-2-butanol.

clockwise
R isomer

counterclockwise
S isomer

Figure 5.7 Examples: Orienting the lowest priority group in back

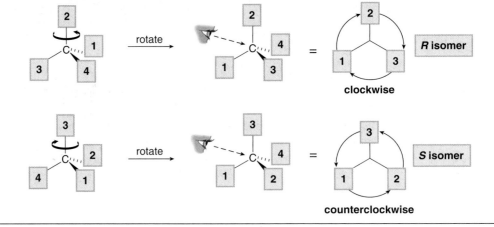

SAMPLE PROBLEM 5.2 Label the following compound as *R* or *S*.

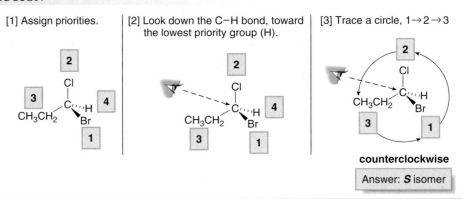

SOLUTION

[1] Assign priorities.

[2] Look down the C—H bond, toward the lowest priority group (H).

[3] Trace a circle, 1→2→3

counterclockwise

Answer: **S** isomer

How do you assign *R* or *S* to a molecule when the lowest priority group is not oriented toward the back, on a dashed line? You could rotate and flip the molecule until the lowest priority group is in the back, as shown in Figure 5.7, then follow the stepwise procedure for assigning the configuration. Or, if manipulating and visualizing molecules in three dimensions is difficult for you, try the procedure suggested in Sample Problem 5.3.

SAMPLE PROBLEM 5.3 Label the following compound as *R* or *S*.

$$(CH_3)_2CH-\overset{\displaystyle OH}{\underset{\displaystyle H}{C}}\text{''''}CH_2CH_3$$

SOLUTION

In this problem, the lowest priority group (H) is oriented in **front** of, not behind, the page. To assign *R* or *S* in this case:

- **Switch** the position of the lowest priority group (H) with the group located behind the page ($-CH_2CH_3$).
- Determine *R* or *S* in the usual manner.
- **Reverse the answer.** Because we switched the position of two groups on the stereogenic center to begin with, and there are only two possibilities, the answer is **opposite** to the correct answer.

[1] Assign priorities.

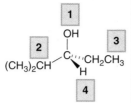

[2] Switch groups 4 and 3.

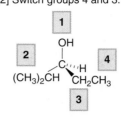

[3] Trace a circle, 1 → 2 → 3, and reverse the answer

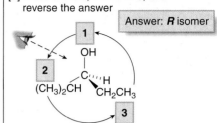

Answer: **R** isomer

counterclockwise
It looks like an *S* isomer, but we must reverse the answer because we switched groups 3 and 4, *S* → *R*.

PROBLEM 5.14   Label each compound as *R* or *S*.

a.   b.   c.   d.

PROBLEM 5.15   Draw the two enantiomers of each compound and label them as *R* or *S*.

a. $CH_3-\overset{\overset{\displaystyle H}{|}}{\underset{\underset{\displaystyle Cl}{|}}{C}}-CH_2CH_3$ b. $CH_3O-\overset{\overset{\displaystyle H}{|}}{\underset{\underset{\displaystyle CH_2CH_3}{|}}{C}}-CHO$

PROBLEM 5.16   Draw both enantiomers of fenfluramine, one component of the appetite suppressant Fen–Phen. The *S* enantiomer was sold independently under the name dexfenfluramine. Which enantiomer is dexfenfluramine? (Fen–Phen was withdrawn from the market in 1997, after it was shown to damage heart valves in some patients.)

fenfluramine

5.7 Diastereomers

We have now seen many examples of compounds containing one tetrahedral stereogenic center. The situation is more complex for compounds with two stereogenic centers, because more stereoisomers are possible. Moreover, a molecule with two stereogenic centers *may* or *may not* be chiral.

♦ For *n* stereogenic centers, the maximum number of stereoisomers is 2^n.

   ♦ When *n* = **1**, $2^1 = $ **2**. With one stereogenic center there are always two stereoisomers and they are **enantiomers.**
   ♦ When *n* = **2**, $2^2 = $ **4**. With two stereogenic centers, the maximum number of stereoisomers is four, although sometimes there are *fewer* than four.

PROBLEM 5.17   What is the maximum number of stereoisomers possible for a compound with three stereogenic centers? What is the maximum number of stereoisomers for a compound like cholesterol (Problem 5.11) with eight stereogenic centers?

Let's illustrate a stepwise procedure for finding all possible stereoisomers using 2,3-dibromopentane.

In testing to see if one compound is superimposable on another, rotate atoms and flip the entire molecule, but **do not break any bonds.**

$$CH_3-\overset{*}{\underset{|}{C}}-\overset{*}{\underset{|}{C}}-CH_2CH_3$$

2,3-dibromopentane
[* = stereogenic center]
maximum number of stereoisomers = **4**

Add substituents around stereogenic centers with the bonds **eclipsed,** for easier visualization.

eclipsed   rapidly   staggered
              interconvert

Don't forget, however, that the staggered arrangement is more stable.

How To   Find and Draw All Possible Stereoisomers for a Compound with Two Stereogenic Centers

Step [1]   **Draw one stereoisomer by arbitrarily arranging substituents around the stereogenic centers. Then draw its mirror image.**

Draw one stereoisomer of 2,3-dibromopentane...   ...then draw its mirror image.

A   **B**

- Arbitrarily add the H, Br, CH_3 and CH_2CH_3 groups to the stereogenic centers, forming **A**. Then draw the mirror image (**B**) so that substituents in **B** are a reflection of the substituents in **A**.
- Determine whether **A** and **B** are superimposable by flipping or rotating one molecule to see if all the atoms align.
- If you have drawn the compound and the mirror image in the described manner, you only have to do two things to see if the atoms align. Place **B** directly on top of **A** (either in your mind or use models); and, rotate **B** 180° and place it on top of **A** to see if the atoms align.

A and B are different compounds.

rotate

B   **A**

H and Br do not align.

180°

B

- In this case, the atoms of **A** and **B** do not align, making **A** and **B** nonsuperimposable mirror images—**enantiomers. A** and **B** are two of the four possible stereoisomers for 2,3-dibromopentane.

Step [2]   **Draw a third possible stereoisomer by switching the positions of any two groups on *one* stereogenic center only. Then draw its mirror image.**

- Switching the positions of H and Br (or any two groups) on one stereogenic center of either **A** or **B** forms a new stereoisomer (labeled **C** in this example), which is different from both **A** and **B.** We then draw the mirror image of **C**, labeled **D. C** and **D** are nonsuperimposable mirror images—**enantiomers.** We have now drawn four stereoisomers for 2,3-dibromopentane, the maximum number possible.

A   **C**   **D**

Switch H and Br on one stereogenic center.

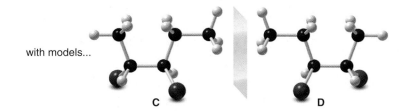

with models...

There are only two types of stereoisomers: *Enantiomers* **are stereoisomers that** *are* **mirror images.** *Diastereomers* **are stereoisomers that are** *not* **mirror images.**

There are four stereoisomers for 2,3-dibromopentane: enantiomers **A** and **B,** and enantiomers **C** and **D.** What is the relationship between two stereoisomers like **A** and **C?** **A** and **C** represent the second broad class of stereoisomers, called **diastereomers.** *Diastereomers* **are stereoisomers that are not mirror images of each other. A** and **B** are diastereomers of **C** and **D,** and vice versa. Figure 5.8 summarizes the relationships between the stereoisomers of 2,3-dibromopentane.

PROBLEM 5.18   Label the two stereogenic centers in each compound and draw all possible stereoisomers: (a) $CH_3CH_2CH(Cl)CH(OH)CH_2CH_3$; (b) $CH_3CH(Br)CH_2CH(Cl)CH_3$

5.8   Meso Compounds

Whereas 2,3-dibromopentane has two stereogenic centers and the maximum of four stereoisomers, **2,3-dibromobutane** has two stereogenic centers but fewer than the maximum number of stereoisomers.

$$CH_3-\overset{\overset{H}{|}}{\underset{\underset{Br}{|}}{C}}-\overset{\overset{H}{|}}{\underset{\underset{Br}{|}}{C}}-CH_3$$

2,3-dibromobutane
[* = stereogenic center]

With two stereogenic centers, the **maximum** number of stereoisomers = **4.**

To find and draw all the stereoisomers of 2,3-dibromobutane, we follow the same stepwise procedure outlined in Section 5.7. Arbitrarily add the H, Br, and CH₃ groups to the stereogenic centers, forming one stereoisomer **A,** and then draw its mirror image **B. A** and **B** are nonsuperimposable mirror images—**enantiomers.**

Figure 5.8   Summary: The four stereoisomers of 2,3-dibromopentane

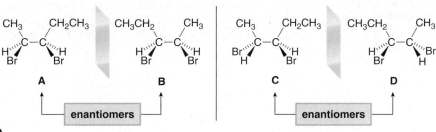

- Pairs of enantiomers: **A** and **B; C** and **D.**
- Pairs of diastereomers: **A** and **C; A** and **D; B** and **C; B** and **D.**

A and B are **diastereomers** of C and D.

To find the other two stereoisomers (if they exist), switch the position of two groups on *one* stereogenic center of *one* enantiomer only. In this case, switching the positions of H and Br on one stereogenic center of **A** forms **C**, which is different from both **A** and **B** and is thus a new stereoisomer.

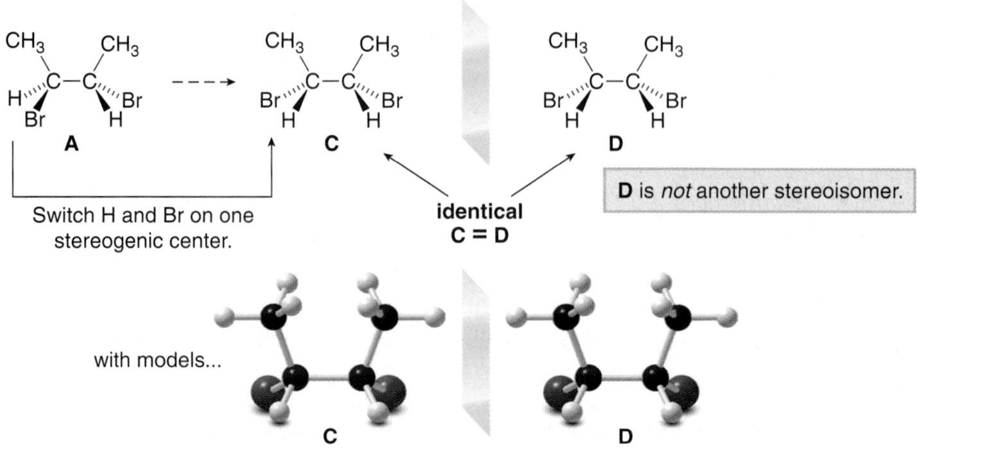

However, the mirror image of **C**, labeled **D**, is superimposable on **C**, so **C** and **D** are identical. Thus, **C** is achiral, even though it has two stereogenic centers. **C** is a **meso compound.**

◆ A *meso compound* is an achiral compound that contains tetrahedral stereogenic centers.

Like all achiral compounds, **C** contains a **plane of symmetry. All meso compounds have a plane of symmetry,** so that they possess two identical halves.

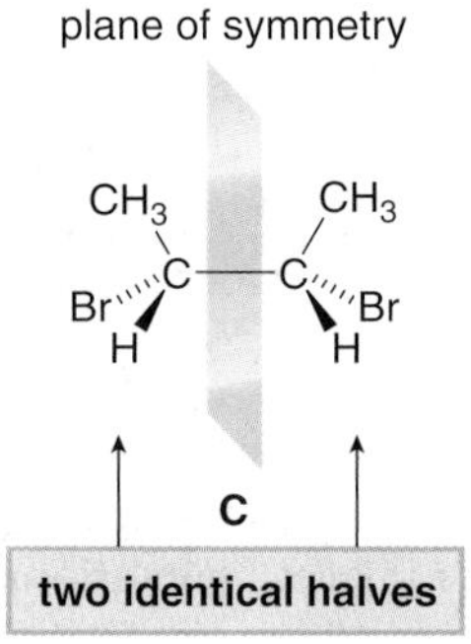

A meso compound must contain a plane of symmetry in one conformation and must have at least two stereogenic centers.

Because one stereoisomer of 2,3-dibromobutane is superimposable on its mirror image, there are only three stereoisomers and not four, as summarized in Figure 5.9.

PROBLEM 5.19   Draw all the possible stereoisomers for each compound and label pairs of enantiomers and diastereomers: (a) $CH_3CH(OH)CH(OH)CH_3$ (b) $CH_3CH(OH)CH(Cl)CH_3$

Figure 5.9   Summary: The three stereoisomers of 2,3-dibromobutane

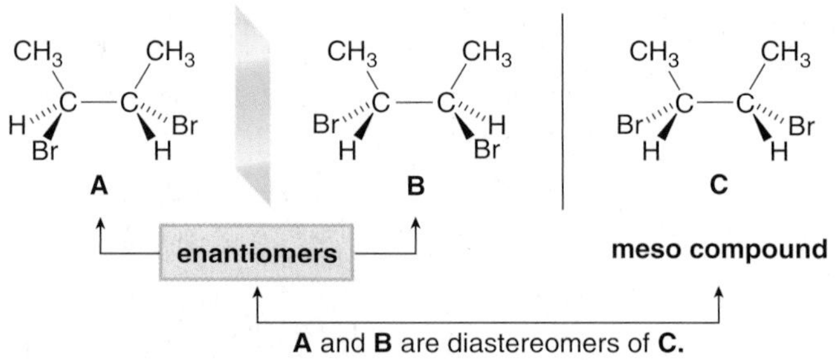

- Pair of enantiomers: **A** and **B**.
- Pairs of diastereomers: **A** and **C**; **B** and **C**.

PROBLEM 5.20   Draw the enantiomer and one diastereomer for each compound.

a.

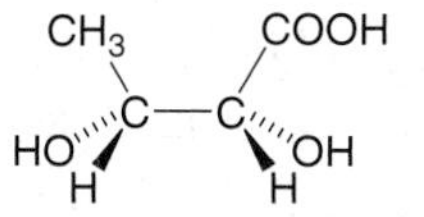

b.

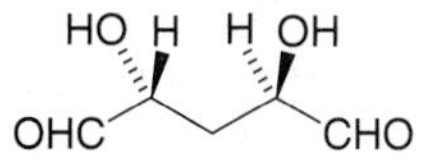

PROBLEM 5.21   Which compounds are meso compounds?

a.

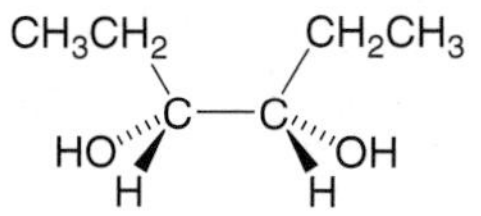

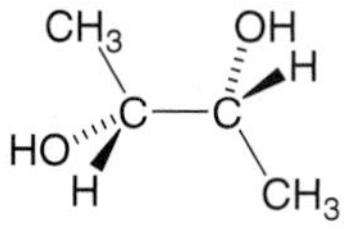

b.

c.

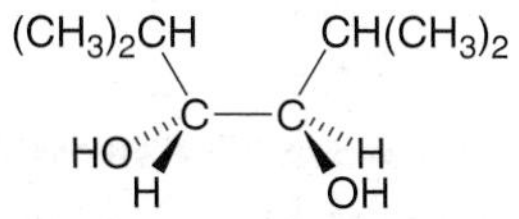

d.

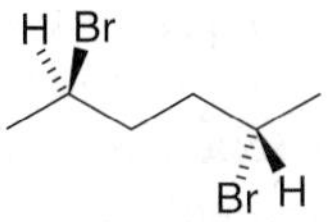

5.9   *R* and *S* Assignments in Compounds with Two or More Stereogenic Centers

When a compound has more than one stereogenic center, the *R* and *S* configuration must be assigned to each of them. In the stereoisomer of 2,3-dibromopentane drawn here, C2 has the *S* configuration and C3 has the *R*, so the complete name of the compound is (2*S*,3*R*)-2,3-dibromopentane.

S configuration ——→ C — C ←—— *R* configuration

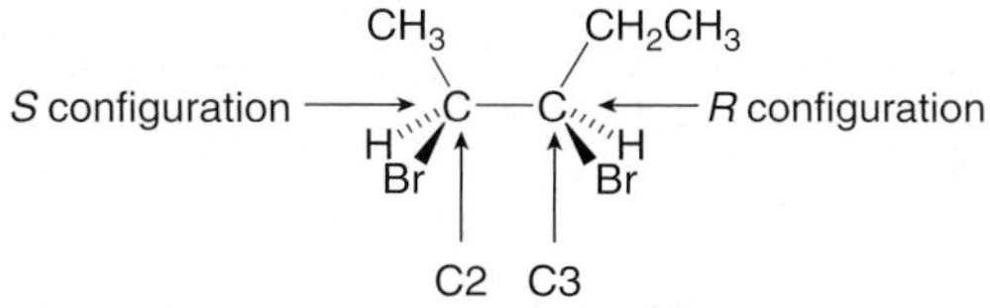

Complete name:
(2*S*,3*R*)-2,3-dibromopentane

C2   C3

one stereoisomer of 2,3-dibromopentane

R,*S* configurations can be used to determine whether two compounds are identical, enantiomers, or diastereomers.

◆ Identical compounds have the *same R,S* designations at every tetrahedral stereogenic center.
◆ Enantiomers have exactly *opposite R,S* designations.
◆ Diastereomers have the *same R,S* designation for at least one stereogenic center and the *opposite* for at least one of the other stereogenic centers.

For example, if a compound has two stereogenic centers, both with the *R* configuration, then its enantiomer is *S*,*S* and the diastereomers are either *R*,*S* or *S*,*R*.

PROBLEM 5.22   If the two stereogenic centers of a compound are *R*,*S* in configuration, what are the *R*,*S* assignments for its enantiomer and two diastereomers?

PROBLEM 5.23   Without drawing out the structures, label each pair of compounds as enantiomers or diastereomers.
a. (2*R*,3*S*)-2,3-hexanediol and (2*R*,3*R*)-2,3-hexanediol
b. (2*R*,3*R*)-2,3-hexanediol and (2*S*,3*S*)-2,3-hexanediol
c. (2*R*,3*S*,4*R*)-2,3,4-hexanetriol and (2*S*,3*R*,4*R*)-2,3,4-hexanetriol

PROBLEM 5.24   How is each compound related to **A?** Is it an enantiomer, diastereomer, or identical?

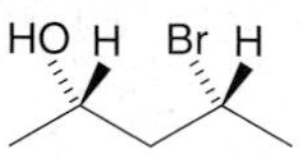

A

a.

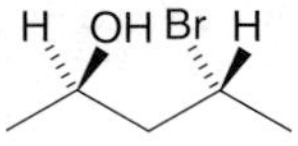

b.

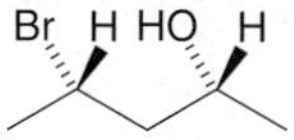

c.

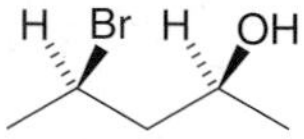

d.

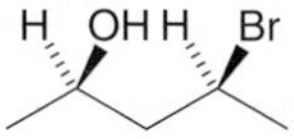

5.10 Disubstituted Cycloalkanes

Let us now turn our attention to disubstituted cycloalkanes, and draw all possible stereoisomers for **1,3-dibromocyclopentane.** Because 1,3-dibromocyclopentane has two stereogenic centers, it has a maximum of four stereoisomers.

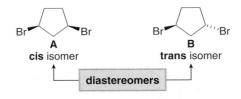

With two stereogenic centers, the **maximum number of stereoisomers = 4.**

1,3-dibromocyclopentane
[* = stereogenic center]

To draw all possible stereoisomers, remember that a disubstituted cycloalkane can have two substituents on the same side of the ring (**cis isomer,** labeled **A**) or on opposite sides of the ring (**trans isomer,** labeled **B**). These compounds are **stereoisomers but not mirror images of each other,** making them **diastereomers. A** and **B** are two of the four possible stereoisomers.

> **Remember:** In determining chirality in substituted cycloalkanes, always draw the rings as **flat polygons.** This is especially true for cyclohexane derivatives, where having two chair forms that interconvert can make analysis especially difficult.

To find the other two stereoisomers (if they exist), draw mirror images of each compound and determine whether the compound and its mirror image are superimposable.

> Like all meso compounds, *cis*-1,3-dibromocyclopentane contains a plane of symmetry.
>
> plane of symmetry
>
>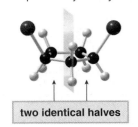
>
> two identical halves

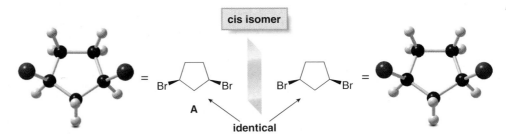

◆ The cis isomer is superimposable on its mirror image, making them *identical.* Thus, **A** is an **achiral meso compound.**

◆ The trans isomer **B** is *not* superimposable on its mirror image, labeled **C,** making **B** and **C** different compounds. Thus, **B** and **C** are **enantiomers.**

Because one stereoisomer of 1,3-dibromocyclopentane is superimposable on its mirror image, there are only three stereoisomers, not four. **A** is an achiral meso compound and **B** and **C** are a pair of chiral enantiomers. **A** and **B** are diastereomers, as are **A** and **C.**

PROBLEM 5.25 Which of the following cyclic molecules are *meso* compounds?

a.   b.   c.

PROBLEM 5.26   Draw all possible stereoisomers for each compound. Label pairs of enantiomers and diastereomers.

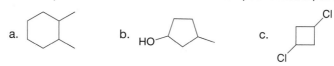

a.   b. HO   c.

5.11   Isomers—A Summary

Before moving on to other aspects of stereochemistry, take the time to review Figures 5.10 and 5.11. Keep in mind the following facts, and use Figure 5.10 to summarize the types of isomers.

◆ There are two major classes of isomers: constitutional isomers and stereoisomers.
◆ There are only two kinds of stereoisomers: enantiomers and diastereomers.

Then, to determine the relationship between two nonidentical molecules, refer to the flowchart in Figure 5.11.

Figure 5.10   Summary—
Types of isomers

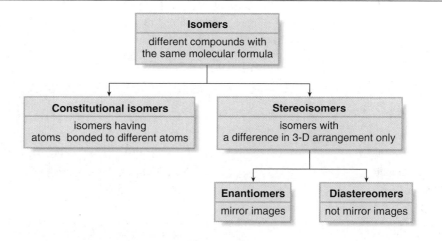

Figure 5.11   Determining
the relationship between two
nonidentical molecules

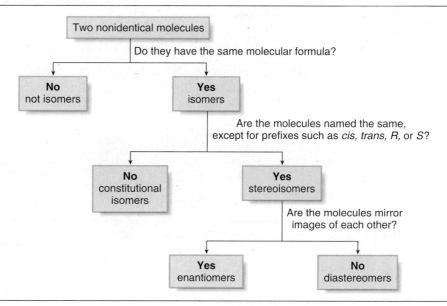

PROBLEM 5.27   State how each pair of compounds is related. Are they enantiomers, diastereomers, constitutional isomers, or identical?

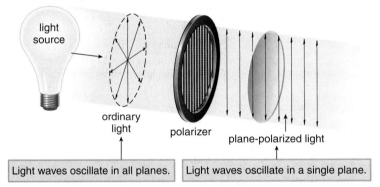

5.12   Physical Properties of Stereoisomers

Recall from Section 5.2 that constitutional isomers have different physical and chemical properties. How, then, do the physical and chemical properties of enantiomers compare?

> ◆ The chemical and physical properties of two enantiomers are *identical* except in their interaction with *chiral* substances.

5.12A   Optical Activity

Two enantiomers have identical physical properties—melting point, boiling point, solubility—except for how they interact with plane-polarized light.

What is plane-polarized light? Ordinary light consists of electromagnetic waves that oscillate in all planes perpendicular to the direction in which the light travels. Passing light through a polarizer allows light in only one plane to come through. This is **plane-polarized light** (or simply **polarized light**), and it has an electric vector that oscillates in a single plane.

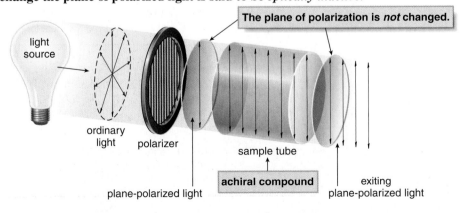

A **polarimeter** is an instrument that allows plane-polarized light to travel through a sample tube containing an organic compound. After the light exits the sample tube, an analyzer slit is rotated to determine the direction of the plane of the polarized light exiting the sample tube. There are two possible results.

With **achiral compounds,** the light exits the sample tube *unchanged*, and the plane of the polarized light is in the same position it was before entering the sample tube. **A compound that does not change the plane of polarized light is said to be** *optically inactive.*

With **chiral compounds,** the plane of the polarized light is rotated through an angle α. The angle α, measured in degrees (°), is called the **observed rotation. A compound that rotates the plane of polarized light is said to be** *optically active.*

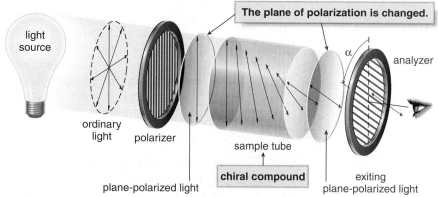

For example, the achiral compound CH_2BrCl is optically inactive, whereas a single enantiomer of $CHBrClF$, a chiral compound, is optically active.

The rotation of polarized light can be in the **clockwise** or **counterclockwise** direction.

◆ If the rotation is *clockwise* (to the right from the noon position), the compound is called *dextrorotatory.* The rotation is labeled *d* or (+).
◆ If the rotation is *counterclockwise* (to the left from noon), the compound is called *levorotatory.* The rotation is labeled *l* or (–).

No relationship exists between the *R* and *S* prefixes that designate configuration and the (+) and (–) designations indicating optical rotation. For example, the *S* enantiomer of lactic acid is dextrorotatory (+), whereas the *S* enantiomer of glyceraldehyde is levorotatory (–).

How does the rotation of two enantiomers compare?

◆ **Two enantiomers rotate plane-polarized light** *to an equal extent* **but in the** *opposite* **direction.**

Thus, if enantiomer **A** rotates polarized light +5°, then the same concentration of enantiomer **B** rotates it –5°.

5.12B   Racemic Mixtures

What is the observed rotation of an equal amount of two enantiomers? Because **two enantiomers rotate plane-polarized light to an equal extent but in opposite directions, the rotations cancel,** and no rotation is observed.

◆ **An equal amount of two enantiomers is called a** *racemic mixture* **or a** *racemate.* **A racemic mixture is optically inactive.**

Besides optical rotation, other physical properties of a racemate are not readily predicted. The melting point and boiling point of a racemic mixture are not necessarily the same as either pure enantiomer, and this fact is not easily explained. The physical properties of two enantiomers and their racemic mixture are summarized in Table 5.1.

(Left margin)

CHO
|
HOCH₂—C""H
OH

(S)-glyceraldehyde
[α] = –8.7°

S configuration—
levorotatory

COOH
|
CH₃—C""H
OH

(S)-lactic acid
[α] = +3.8°

S configuration—
dextrorotatory

TABLE 5.1   The Physical Properties of Enantiomers A and B Compared

Property	A alone	B alone	Racemic A + B
Melting point	identical to **B**	identical to **A**	may be different from **A** and **B**
Boiling point	identical to **B**	identical to **A**	may be different from **A** and **B**
Optical rotation	equal in magnitude but opposite in sign to **B**	equal in magnitude but opposite in sign to **A**	0°

PROBLEM 5.28   The amino acid (*S*)-alanine has the physical characteristics listed under the structure.

COOH
|
CH₃—C‴H
      NH₂

(*S*)-alanine
$[\alpha] = +8.5°$
mp = 297 °C

a. What is the melting point of (*R*)-alanine?
b. What is the melting point of a racemic mixture of (*R*)- and (*S*)-alanine?
c. What is the observed rotation of (*R*)-alanine, recorded under the same conditions as the reported rotation of (*S*)-alanine?
d. What is the observed rotation of a racemic mixture of (*R*)- and (*S*)-alanine?
e. Label each of the following as optically active or inactive: a solution of pure (*S*)-alanine; an equal mixture of (*R*)- and (*S*)-alanine; a solution that contains 75% (*S*)- and 25% (*R*)-alanine.

5.12C   Specific Rotation

The observed rotation depends on the number of chiral molecules that interact with polarized light. This in turn depends on the concentration of the sample and the length of the sample tube. To standardize optical rotation data, the quantity **specific rotation** ($[\alpha]$) is defined using a specific sample tube length (usually 1 dm), concentration, temperature (25 °C), and wavelength (589 nm, the D line emitted by a sodium lamp).

$$\text{specific rotation} = [\alpha] = \frac{\alpha}{l \times c}$$

α = observed rotation (°)
l = length of sample tube (dm)
c = concentration (g/mL)

$$\left[\begin{array}{l} \text{dm = decimeter} \\ \text{1 dm = 10 cm} \end{array} \right]$$

Specific rotations are physical constants just like melting points or boiling points, and are reported in chemical reference books for a wide variety of compounds.

PROBLEM 5.29   A natural product was isolated in the laboratory, and its observed rotation was +10° when measured in a 1 dm sample tube containing 1.0 g of compound in 10 mL of water. What is the specific rotation of this compound?

5.12D   Enantiomeric Excess

Sometimes in the laboratory we have neither a pure enantiomer nor a racemic mixture, but rather a mixture of two enantiomers in which one enantiomer is present in excess of the other. The **enantiomeric excess (*ee*),** also called the **optical purity,** tells how much more there is of one enantiomer.

◆ **Enantiomeric excess = *ee* = % of one enantiomer – % of the other enantiomer.**

Enantiomeric excess tells how much one enantiomer is present in excess of the racemic mixture. For example, if a mixture contains 75% of one enantiomer and 25% of the other, the enantiomeric excess is 75% – 25% = 50%. There is a 50% excess of one enantiomer over the racemic mixture.

PROBLEM 5.30   What is the *ee* for each of the following mixtures of enantiomers **A** and **B**?
a. 95% **A** and 5% **B**   b. 85% **A** and 15% **B**

Knowing the *ee* of a mixture makes it possible to calculate the amount of each enantiomer present, as shown in Sample Problem 5.4.

SAMPLE PROBLEM 5.4   If the enantiomeric excess is 95%, how much of each enantiomer is present?

SOLUTION

Label the two enantiomers **A** and **B** and assume that **A** is in excess. A 95% *ee* means that the solution contains an excess of 95% of **A**, and 5% of the racemic mixture of **A** and **B**. Because a racemic mixture is an equal amount of both enantiomers, it has 2.5% of **A** and 2.5% of **B**.

• Total amount of **A** = 95% + 2.5% = 97.5%
• Total amount of **B** = 2.5% (or 100% – 97.5%)

PROBLEM 5.31   For the given *ee* values, calculate the percentage of each enantiomer present.
a. 90% *ee*   b. 99% *ee*   c. 60% *ee*

The enantiomeric excess can also be calculated if two quantities are known—the specific rotation [α] of a mixture and the specific rotation [α] of a pure enantiomer.

$$ee = \frac{[\alpha]\ \text{mixture}}{[\alpha]\ \text{pure enantiomer}} \times 100\%$$

SAMPLE PROBLEM 5.5   Pure cholesterol has a specific rotation of –32°. A sample of cholesterol prepared in the lab had a specific rotation of –16°. What is the enantiomeric excess of this sample of cholesterol?

SOLUTION

Calculate the *ee* of the mixture using the given formula.

$$ee = \frac{[\alpha]\ \text{mixture}}{[\alpha]\ \text{pure enantiomer}} \times 100\% = \frac{-16°}{-32°} \times 100\% = \boxed{50\%\ ee}$$

PROBLEM 5.32   A pure compound exhibits a specific rotation of +25°.
a. Calculate the *ee* of a solution whose [α] is +15°.
b. If the *ee* of a solution of this same compound is 80%, what is [α] for this solution?

5.12E   The Physical Properties of Diastereomers

Diastereomers are not mirror images of each other, and as such, **their physical properties are different, including optical rotation.** Figure 5.12 compares the physical properties of the three stereoisomers of tartaric acid, consisting of a meso compound that is a diastereomer of a pair of enantiomers.

Whether the physical properties of a set of compounds are the same or different has practical applications in the lab. Physical properties characterize a compound's physical state, and two compounds can usually be separated only if their physical properties are different.

◆ Because two enantiomers have identical physical properties, they cannot be separated by common physical techniques like distillation.
◆ Diastereomers and constitutional isomers have different physical properties, and therefore they can be separated by common physical techniques.

Figure 5.12   The physical properties of the three stereoisomers of tartaric acid

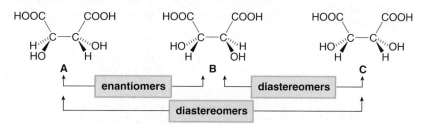

Property	A	B	C	A + B (1:1)
melting point (°C)	171	171	146	206
solubility (g/100 mL H_2O)	139	139	125	20.6
[α] (°)	+13	–13	0	0
R,S designation	R,R	S,S	R,S	—
d,l designation	d	l	none	d,l

• The physical properties of **A** and **B** differ from their diastereomer **C**.
• The physical properties of a racemic mixture of **A** and **B** (last column) also differ from either enantiomer and diastereomer **C**.
• **C** is an achiral meso compound, so it is optically inactive; [α] = 0°.

PROBLEM 5.33   Compare the physical properties of the three stereoisomers of 1,3-dimethylcyclopentane.

three stereoisomers of 1,3-dimethylcyclopentane

a. How do the boiling points of **A** and **B** compare? What about **A** and **C**?
b. Characterize a solution of each of the following as optically active or optically inactive: pure **A**; pure **B**; pure **C**; an equal mixture of **A** and **B**; an equal mixture of **A** and **C**.
c. A reaction forms a 1:1:1 mixture of **A, B,** and **C**. If this mixture is distilled, how many fractions would be obtained? Which fractions would be optically active and which would be optically inactive?

5.13 Chemical Properties of Enantiomers

When two enantiomers react with an achiral reagent they react at the same rate and give the same products. When they react with a chiral, non-racemic reagent, they react at a different rate and give different products.

◆ Two enantiomers have exactly the same chemical properties except for their reaction with chiral, non-racemic reagents.

For an everyday analogy, consider what happens when you are handed an achiral object like a pen and a chiral object like a right-handed glove. Your left and right hands are enantiomers, but they can both hold the achiral pen in the same way. With the glove, however, only your right hand can fit inside it, not your left.

We will examine specific reactions of chiral molecules with both chiral and achiral reagents later in this text. Here, we examine two more general applications.

5.13A Chiral Drugs

A living organism is a sea of chiral molecules. Many drugs are chiral, and often they must interact with a chiral receptor or a chiral enzyme to be effective. One enantiomer of a drug may effectively treat a disease whereas its mirror image may be ineffective. Alternatively, one enantiomer may trigger one biochemical response and its mirror image may elicit a totally different response.

Although (*R*)-ibuprofen shows no anti-inflammatory activity itself, it is slowly converted to the *S* enantiomer in vivo.

For example, the drugs ibuprofen and fluoxetine each contain one stereogenic center, and thus exist as a pair of enantiomers, only one of which exhibits biological activity. (*S*)-**Ibuprofen** is the active component of the anti-inflammatory agents Motrin and Advil, and (*R*)-**fluoxetine** is the active component in the antidepressant Prozac.

(S)-ibuprofen
anti-inflammatory agent

(R)-fluoxetine
antidepressant

The *S* enantiomer of **naproxen,** the molecule that introduced Chapter 5, is an active anti-inflammatory agent, but the *R* enantiomer is a harmful liver toxin. Changing the orientation of two substituents to form a mirror image can thus alter biological activity to produce an undesirable side effect in the other enantiomer.

(S)-naproxen
anti-inflammatory agent

(R)-naproxen
liver toxin

For more examples of two enantiomers that exhibit very different biochemical properties, see *Journal of Chemical Education*, **1996,** *73,* 481.

If a chiral drug could be sold as a single active enantiomer, it should be possible to use smaller doses with fewer side effects. Many chiral drugs continue to be sold as racemic mixtures, however, because it is more difficult and therefore more costly to obtain a single enantiomer. An enantiomer is not easily separated from a racemic mixture because the two enantiomers have the same physical properties. In Chapter 12 we will study reactions that can form a single active enantiomer, an important development in making chiral drugs more readily available.

Recent rulings by the Food and Drug Administration have encouraged the development of so-called *racemic switches,* the patenting and marketing of a single enantiomer that was originally sold as a racemic mixture. To obtain a new patent on a single enantiomer, however, a company must show evidence that it provides significant benefit over the racemate.

5.13B   Enantiomers and the Sense of Smell

Research suggests that the odor of a particular molecule is determined more by its shape than by the presence of a particular functional group. For example, hexachloroethane and cyclooctane have no obvious structural similarities, but they both have a camphor-like odor, a fact attributed to their similar spherical shape.

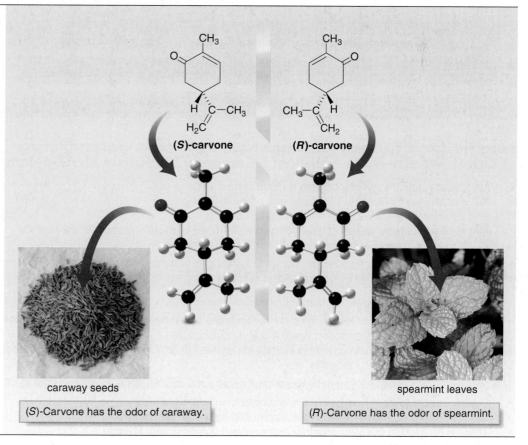

Because enantiomers interact with chiral smell receptors in the nose, some enantiomers have different odors. There are a few well-characterized examples of this phenomenon in nature. For example, (*S*)-carvone is responsible for the odor of caraway, whereas (*R*)-carvone is responsible for the odor of spearmint (Figure 5.13).

Figure 5.13   Two enantiomers can have different odors

(*S*)-Carvone has the odor of caraway.

(*R*)-Carvone has the odor of spearmint.

These examples demonstrate that understanding the three-dimensional structure of a molecule is very important in organic chemistry.

5.14 Key Concepts—Stereochemistry

Isomers are Different Compounds with the Same Molecular Formula (5.2, 5.11)

[1] **Constitutional isomers**—isomers that differ in the way the atoms are connected to each other. They have:
- different IUPAC names;
- the same or different functional groups; and
- different physical and chemical properties.

[2] **Stereoisomers**—isomers that differ only in the way atoms are oriented in space. They have the same functional group and the same IUPAC name except for prefixes such as *cis, trans, R,* and *S.*
- **Enantiomers**—stereoisomers that are nonsuperimposable mirror images of each other (5.4).
- **Diastereomers**—stereoisomers that are not mirror images of each other (5.7).

Some Basic Principles

- When a compound and its mirror image are **superimposable,** they are **identical achiral compounds.** An achiral compound has a plane of symmetry in one conformation (5.3).
- When a compound and its mirror image are **not superimposable,** they are **different chiral compounds** called **enantiomers.** A chiral compound has no plane of symmetry in any conformation (5.3).
- A **tetrahedral stereogenic center** is a carbon atom bonded to four different groups (5.4, 5.5).
- For *n* **stereogenic centers,** the maximum number of stereoisomers is 2^n (5.7).

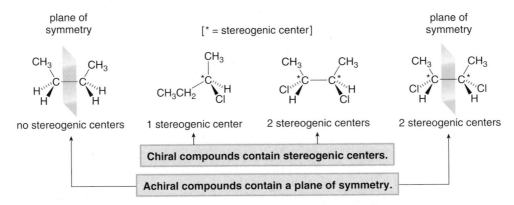

Optical Activity is the Ability of a Compound to Rotate Plane-polarized Light (5.12)

- An optically active solution contains a chiral compound.
- An optically inactive solution contains one of the following:
  - an achiral compound with no stereogenic centers
  - a meso compound—an achiral compound with two or more stereogenic centers
  - a racemic mixture—an equal amount of two enantiomers

The Prefixes *R* and *S* Compared with *d* and *l*

The prefixes *R* and *S* are labels used in nomenclature. Rules on assigning *R,S* are found in Section 5.6.

- An enantiomer has every stereogenic center opposite in configuration. If a compound with two stereogenic centers has the *R,R* configuration, its enantiomer has the *S,S* configuration.
- A diastereomer of this same compound has either the *R,S* or *S,R* configuration; one stereogenic center has the same configuration and one is opposite.

The prefixes *d* (or +) and *l* (or –) tell the direction a compound rotates plane-polarized light (5.12).

- Dextrorotatory (*d* or +) compounds rotate polarized light clockwise.
- Levorotatory (*l* or –) compounds rotate polarized light counterclockwise.
- There is no relation between whether a compound is *R* or *S* and whether it is *d* or *l.*

The Physical Properties of Isomers Compared (5.12)

Type of isomer	Physical properties
Constitutional isomers	Different
Enantiomers	Identical except for the direction polarized light is rotated
Diastereomers	Different
Racemic mixture	Possibly different from either enantiomer

Equations

- Specific rotation (5.12C):

$$\text{specific rotation} = [\alpha] = \frac{\alpha}{l \times c}$$

α = observed rotation (°)
l = length of sample tube (dm)
c = concentration (g/mL)

$$\left[\begin{array}{l} \text{dm = decimeter} \\ \text{1 dm = 10 cm} \end{array} \right]$$

- Enantiomeric excess (5.12D):

$$ee = \% \text{ of one enantiomer} - \% \text{ of the other enantiomer}$$

$$= \frac{[\alpha] \text{ mixture}}{[\alpha] \text{ pure enantiomer}} \times 100\%$$

Problems

Constitutional Isomers versus Stereoisomers

5.34 Label each pair of compounds as constitutional isomers, stereoisomers, or not isomers of each other.

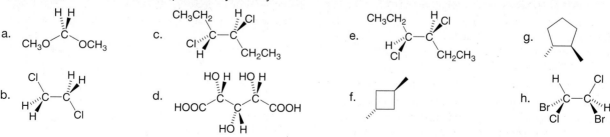

Mirror Images and Chirality

5.35 Draw the mirror image of each compound, and label the compound as chiral or achiral.

a.
b. cysteine (an amino acid)
c.
d.
e. threose (a simple sugar)

5.36 Determine if each compound is identical to or an enantiomer of **A**.

A

a.
b.
c.

5.37 Indicate the plane of symmetry for each molecule that contains one. Some molecules require rotation around a carbon–carbon bond to see the plane of symmetry.

a.
b.
c.
d.
e.
f.
g.
h.

Finding and Drawing Stereogenic Centers

5.38   Label the stereogenic center(s) in each compound. A molecule may have zero, one, or more stereogenic centers.

    a. $CH_3CH_2CH_2CH_2CH_2CH_3$

    b. $CH_3CH_2OCH(CH_3)CH_2CH_3$

    c. $(CH_3)_2CHCH(OH)CH(CH_3)_2$

    d. $(CH_3)_2CHCH_2CH(CH_3)CH_2CH(CH_3)CH(CH_3)CH_2CH_3$

    e. $CH_3-\overset{\overset{\displaystyle H}{|}}{\underset{\underset{\displaystyle D}{|}}{C}}-CH_2CH_3$

f.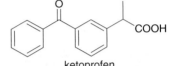

g.

h.

i.

j.

5.39   Draw the eight constitutional isomers having molecular formula $C_5H_{11}Cl$. Label any stereogenic centers.

5.40   Draw the structures of two constitutional isomers having molecular formula C_7H_{16} that contain a stereogenic center. For each constitutional isomer, draw all stereoisomers.

5.41   Draw both enantiomers for each biologically active compound.

    a.

amphetamine
(a powerful central nervous stimulant)

    b.

ketoprofen
(analgesic and anti-inflammatory agent)

5.42   Draw the structure for the lowest molecular weight alkane (having only C and H and no isotopes) that contains a stereogenic center.

Nomenclature

5.43   Which group in each pair is assigned the higher priority in *R,S* nomenclature?

    a. $-OH, -NH_2$

    b. $-CD_3, -CH_3$

    c. $-CH(CH_3)_2, -CH_2OH$

    d. $-CH_2Cl, -CH_2CH_2CH_2Br$

    e. $-CHO, -COOH$

    f. $-CH_2NH_2, -NHCH_3$

5.44   Rank the following groups in order of decreasing priority.

    a. $-F, -NH_2, -CH_3, -OH$

    b. $-CH_3, -CH_2CH_3, -CH_2CH_2CH_3, -(CH_2)_3CH_3$

    c. $-NH_2, -CH_2NH_2, -CH_3, -CH_2NHCH_3$

    d. $-COOH, -CH_2OH, -H, -CHO$

    e. $-Cl, -CH_3, -SH, -OH$

    f. $-C\equiv CH, -CH(CH_3)_2, -CH_2CH_3, -CH=CH_2$

5.45   Label each stereogenic center as *R* or *S*.

    a.

    b.

    c.

    d.

    e.

    f.

    g.

    h.

5.46   Draw the structure for each compound.

    a. (3*R*)-3-methylhexane

    b. (4*R*,5*S*)-4,5-diethyloctane

    c. (3*R*,5*S*,6*R*)-5-ethyl-3,6-dimethylnonane

    d. (3*S*,6*S*)-6-isopropyl-3-methyldecane

5.47   Draw the two enantiomers for the amino acid leucine, $HOOCCH(NH_2)CH_2CH(CH_3)_2$, and label each enantiomer as *R* or *S*. Only the *S* isomer exists in nature, and it has a bitter taste. Its enantiomer, however, is sweet.

5.48   Label the stereogenic center in each biologically active compound as *R* or *S*.

    a.

L-dopa
(used to treat
Parkinson's disease)

    b.

adrenalin
(hormone that increases heart rate,
dilates airways)

    c.

ketamine
(anesthetic)

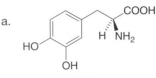

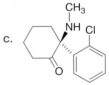

5.49 Methylphenidate (trade name: Ritalin) is prescribed for attention deficit hyperactivity disorder (ADHD). Ritalin is a mixture of R,R and S,S isomers, even though only the R,R isomer is active in treating ADHD. (The single R,R enantiomer, called dexmethylphenidate, is now sold under the trade name Focalin.) Draw the structure of the R,R and S,S isomers of methylphenidate.

methylphenidate

Compounds with More than One Stereogenic Center

5.50 Label the stereogenic centers in each drug.

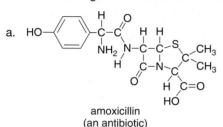

a. amoxicillin
(an antibiotic)

b. norethindrone
(oral contraceptive component)

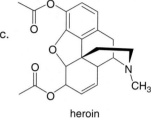

c. heroin
(an opiate)

5.51 What is the maximum number of stereoisomers possible for each compound?

  a. $CH_3CH(OH)CH(OH)CH_2CH_3$

  b. $CH_3CH_2CH_2CH(CH_3)_2$

  c.

5.52 Draw all possible stereoisomers for each compound. Label pairs of enantiomers and diastereomers. Label any meso compound.
  a. $CH_3CH(OH)CH(OH)CH_2CH_3$
  b. $CH_3CH(OH)CH_2CH_2CH(OH)CH_3$
  c. $CH_3CH(Cl)CH_2CH(Br)CH_3$
  d. $CH_3CH(Br)CH(Br)CH(Br)CH_3$

5.53 Threonine is a naturally occurring amino acid that contains two stereogenic centers.

threonine

  a. Label the two stereogenic centers in threonine.
  b. Draw all possible stereoisomers and assign the R,S configuration to each isomer.
  c. Only the $2S,3R$ isomer of threonine occurs in nature. (Numbering begins at the COOH group.) Which isomer in part (b) is naturally occurring?

5.54 Draw the enantiomer and a diastereomer for each compound.

a.   b.   c.   d.

5.55 Draw all possible stereoisomers for each cycloalkane. Label pairs of enantiomers and diastereomers. Label any meso compound.

a.   b.   c.

5.56 Draw all possible constitutional and stereoisomers for a compound of molecular formula C_6H_{12} having a cyclobutane ring and two methyl groups as substituents. Label each compound as chiral or achiral.

5.57 Explain why compound **A** has no enantiomer and why compound **B** has no diastereomer.

A

B

Comparing Compounds: Enantiomers, Diastereomers, and Constitutional Isomers

5.58 How is each compound related to the simple sugar D-erythrose? Is it an enantiomer, diastereomer, or identical?

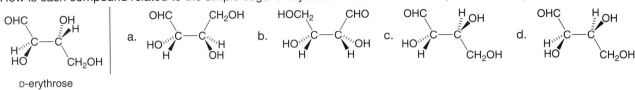

5.59 How are the compounds in each pair related to each other? Are they identical, enantiomers, diastereomers, constitutional isomers, or not isomers of each other?

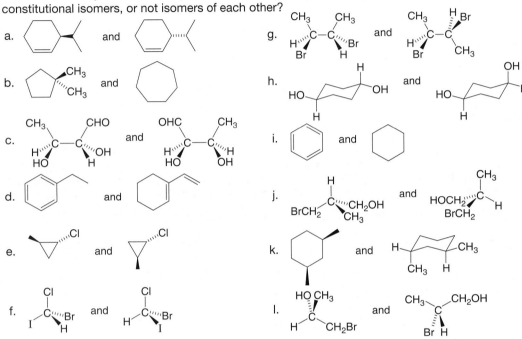

Physical Properties of Isomers

5.60 Drawn are four isomeric dimethylcyclopropanes.

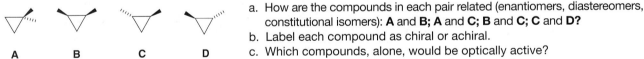

    A   **B**   **C**   **D**

  a. How are the compounds in each pair related (enantiomers, diastereomers, constitutional isomers): **A** and **B**; **A** and **C**; **B** and **C**; **C** and **D?**
  b. Label each compound as chiral or achiral.
  c. Which compounds, alone, would be optically active?
  d. Which compounds have a plane of symmetry?
  e. How do the boiling points of the compounds in each pair compare: **A** and **B**; **B** and **C**; **C** and **D?**
  f. Which of the compounds are meso compounds?
  g. Would an equal mixture of compounds **C** and **D** be optically active? What about an equal mixture of **B** and **C?**

5.61 The [α] of pure quinine, an antimalarial drug, is −165°.

  a. Calculate the *ee* of a solution with the following [α] values: −50°, −83°, and −120°.
  b. For each *ee*, calculate the percent of each enantiomer present.
  c. What is [α] for the enantiomer of quinine?
  d. If a solution contains 80% quinine and 20% of its enantiomer, what is the *ee* of the solution?
  e. What is [α] for the solution described in (d)?

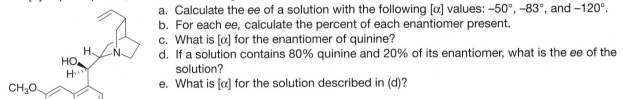

quinine
(antimalarial drug)

5.62 Amygdalin, a compound isolated from the pits of apricots, peaches, and wild cherries, is commonly known as *laetrile*. Although it has no known therapeutic value, amygdalin has been used as an unsanctioned anticancer drug both within and outside of the United States. One hydrolysis product formed from amygdalin is mandelic acid, used in treating common skin problems caused by photo-aging and acne.

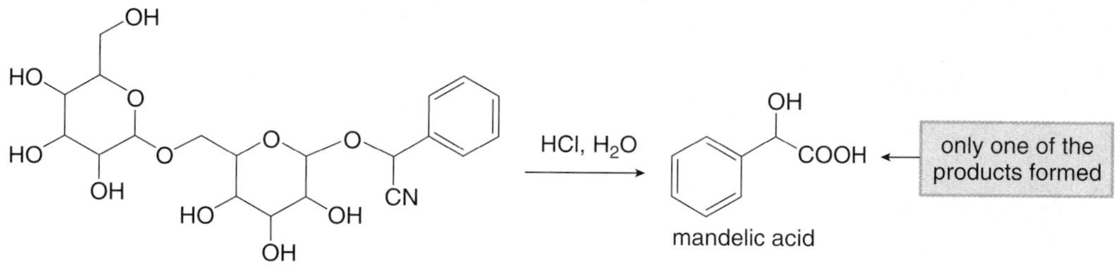

amygdalin
(laetrile)

mandelic acid

a. How many stereogenic centers are present in amygdalin? What is the maximum number of stereoisomers possible?
b. Draw both enantiomers of mandelic acid and label each stereogenic center as R or S.
c. Pure (R)-mandelic acid has a specific rotation of −154°. If a sample contains 60% of the R isomer and 40% of its enantiomer, what is [α] of this solution?
d. Calculate the ee of a solution of mandelic acid having [α] = +50°. What is the percentage of each enantiomer present?

Challenge Questions

5.63

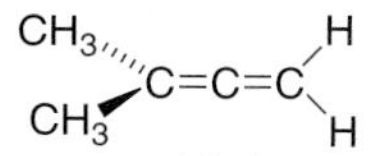

achiral
A

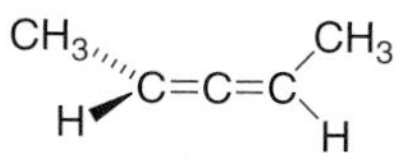

chiral
B

A limited number of chiral compounds having no stereogenic centers exist. For example, although **A** is achiral, constitutional isomer **B** is chiral. Make models and explain this observation. Compounds containing two double bonds that share a single carbon atom are called *allenes*.

5.64   a. Locate all tetrahedral stereogenic centers in palytoxin, the poisonous compound isolated from soft corals, first introduced in the Prologue.
b. Certain carbon–carbon double bonds can also be stereogenic centers. With reference to the definition in Section 5.3, explain how this can occur, and then, locate the seven additional stereogenic centers in palytoxin.
c. Considering all stereogenic centers, what is the maximum number of stereoisomers possible for palytoxin?

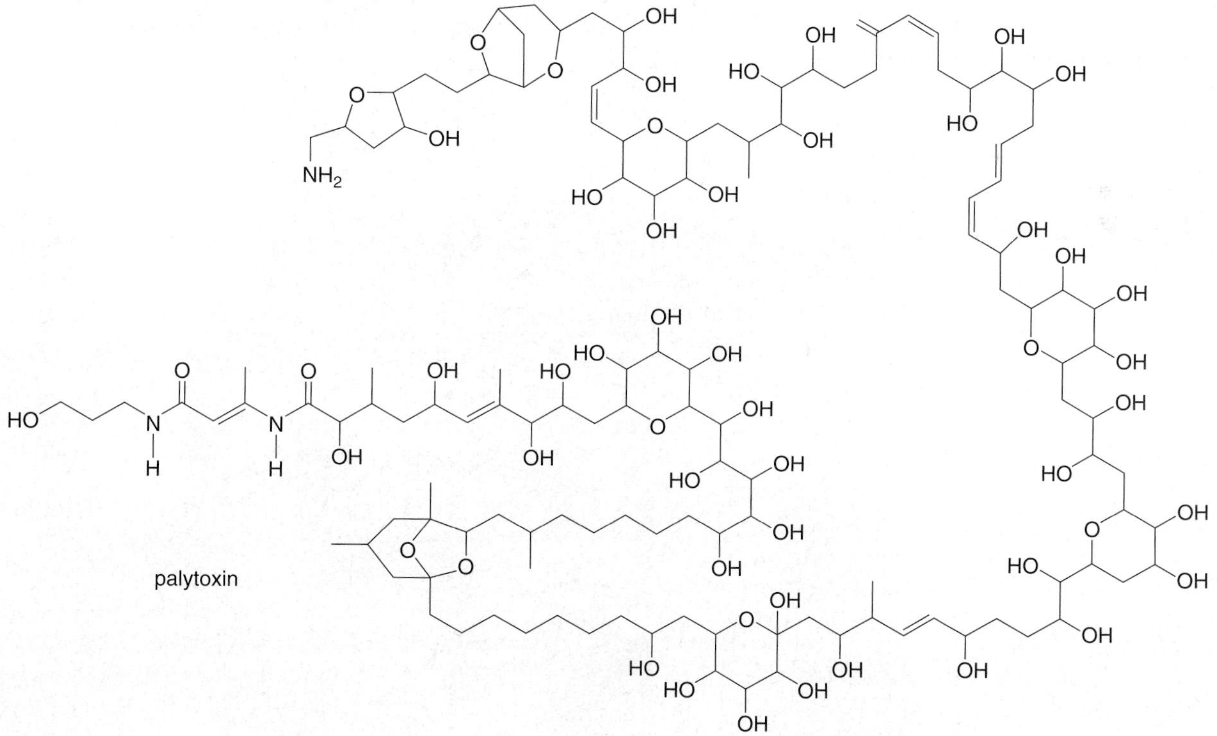

palytoxin

5.65

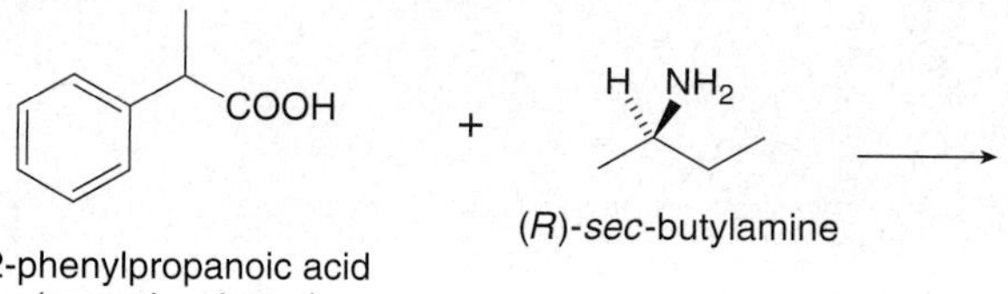

2-phenylpropanoic acid
(racemic mixture)

(R)-sec-butylamine

An acid–base reaction of (R)-sec-butylamine with a racemic mixture of 2-phenylpropanoic acid forms two products having different melting points and somewhat different solubilities. Draw the structure of these two products. Assign R and S to any stereogenic centers in the products. How are the two products related? Choose from enantiomers, diastereomers, constitutional isomers, or not isomers.

CHAPTER 6

Understanding Organic Reactions

Isooctane, a component of petroleum, and **glucose,** a simple sugar formed from starch during digestion, are very different organic molecules that share a common feature. On oxidation, both compounds release a great deal of energy. Isooctane is burned in gasoline to power automobiles, and glucose is metabolized in the body to provide energy for exercise. In Chapter 6, we learn about these energy changes that accompany chemical reactions.

Why do certain reactions occur when two compounds are mixed together whereas others do not? To answer this question we must learn how and why organic compounds react.

Reactions are at the heart of organic chemistry. An understanding of chemical processes has made possible the conversion of natural substances into new compounds with different, and sometimes superior properties. Aspirin, ibuprofen, nylon, and polyethylene are all products of chemical reactions between substances derived from petroleum.

Reactions are difficult to learn when each reaction is considered a unique and isolated event. *Avoid this tendency.* Virtually all chemical reactions are woven together by a few basic themes. After we learn the general principles, specific reactions then fit neatly into a general pattern.

In our study of organic reactions we will begin with the functional groups, looking for electron-rich and electron-deficient sites, and bonds that might be broken easily. These reactive sites give us a clue as to the general type of reaction a particular class of compound undergoes. Finally, we will learn about how a reaction occurs. Does it occur in one step or in a series of steps? Understanding the details of an organic reaction allows us to determine when it might be used in preparing interesting and useful organic compounds.

6.1 Writing Equations for Organic Reactions

Often the solvent and temperature of a reaction are omitted from chemical equations, to further focus attention on the main substances involved in the reaction.

Solvent. Although the solvent is often omitted in an equation for a chemical reaction, keep in mind that most organic reactions take place in a **liquid solvent.** Solvents solubilize key reaction components and serve as heat reservoirs to maintain a given temperature. Chapter 7 presents the two major types of organic solvents and how they affect substitution reactions.

Like other reactions, equations for organic reactions are usually drawn with a single reaction arrow ($\rightarrow$) between the starting material and product, but other conventions make these equations look different from those encountered in general chemistry.

The **reagent,** the chemical substance with which an organic compound reacts, is sometimes drawn on the left side of the equation with the other reactants. At other times, the reagent is drawn above the reaction arrow itself, to focus attention on the organic starting material by itself on the left side. The solvent and temperature of a reaction may be added above or below the arrow. **The symbols "*hv*" and "Δ" are used for reactions that require *light* or *heat,* respectively.** Figure 6.1 presents an organic reaction in different ways.

When two sequential reactions are carried out without drawing any intermediate compound, the steps are usually numbered above the reaction arrow. This convention signifies that the first step occurs *before* the second, and the reagents are added *in sequence,* not at the same time.

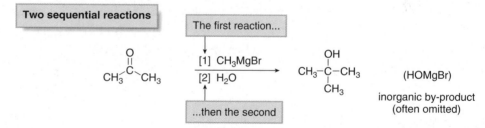

In this equation only the organic products are drawn on the right side of the arrow. Although the reagent CH_3MgBr contains both Mg and Br, these elements do not appear in the organic product,

Figure 6.1 Different ways of writing organic reactions

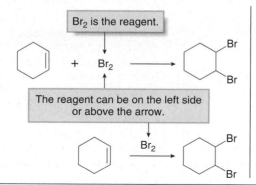

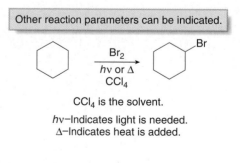

and they are often omitted on the product side of the equation. These elements have not disappeared. They are part of an inorganic by-product (HOMgBr in this case), and are often of little interest to an organic chemist.

6.2   Kinds of Organic Reactions

Like other compounds, organic molecules undergo acid–base and oxidation–reduction reactions, as discussed in Chapters 2 and 4. Organic molecules also undergo **substitution, elimination, and addition** reactions.

6.2A   Substitution Reactions

◆ *Substitution* is a reaction in which an atom or a group of atoms is *replaced* by another atom or group of atoms.

In a general substitution reaction, Y **replaces** Z on a carbon atom. **Substitution reactions involve σ bonds: one σ bond breaks and another forms at the same carbon atom.** The most common examples of substitution occur when Z is hydrogen or a heteroatom that is more electronegative than carbon.

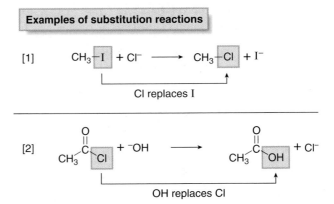

6.2B   Elimination Reactions

◆ *Elimination* is a reaction in which elements of the starting material are "lost" and a π bond is formed.

In an elimination reaction, two groups X and Y are removed from a starting material. **Two σ bonds are broken, and a π bond is formed between adjacent atoms.** The most common examples of elimination occur when X = H and Y is a heteroatom more electronegative than carbon.

Examples of elimination reactions

[1]
$$H-\underset{\underset{H}{|}}{\overset{\overset{H}{|}}{C}}-\underset{\underset{Br}{|}}{\overset{\overset{H}{|}}{C}}-H \quad + \quad {}^-OH \quad \longrightarrow \quad \underset{\underset{H}{}}{\overset{\overset{H}{}}{C}}=\underset{\underset{H}{}}{\overset{\overset{H}{}}{C}} \quad + \quad H_2O \quad + \quad Br^-$$

loss of HBr   π bond

[2]

$$\text{(cyclohexane with HO and H)} \quad \xrightarrow{H_2SO_4} \quad \text{(cyclohexene)} \quad + \quad H_2O$$

loss of H₂O   π bond

6.2C Addition Reactions

♦ *Addition* is a reaction in which elements are added to a starting material.

A general addition reaction

$$\overset{\diagdown}{\underset{\diagup}{C}}=\overset{\diagdown}{\underset{\diagup}{C} } \quad + \quad X-Y \quad \longrightarrow \quad -\overset{|}{\underset{X}{C}}-\overset{|}{\underset{Y}{C}}-$$

This π bond is broken.   Two σ bonds are formed.

In an addition reaction, new groups X and Y are added to a starting material. **A π bond is broken and two σ bonds are formed.**

Examples of addition reactions

[1]
$$\underset{\underset{H}{}}{\overset{\overset{H}{}}{C}}=\underset{\underset{H}{}}{\overset{\overset{H}{}}{C}} \quad + \quad H-Br \quad \longrightarrow \quad H-\underset{\underset{H}{|}}{\overset{\overset{H}{|}}{C}}-\underset{\underset{Br}{|}}{\overset{\overset{H}{|}}{C}}-H$$

This π bond is broken.   HBr is added.

[2]

$$\text{(cyclohexene)} \quad + \quad H_2O \quad \xrightarrow{H_2SO_4} \quad \text{(cyclohexane with H and OH)}$$

This π bond is broken.   H₂O is added.

A summary of the general types of organic reations is given in Appendix F.

Addition and elimination reactions are exactly opposite. A π bond is *formed* in elimination reactions, whereas a π bond is *broken* in addition reactions.

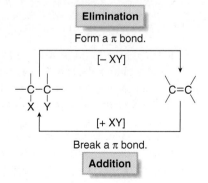

Elimination

Form a π bond.

[– XY]

$$-\overset{|}{\underset{X}{C}}-\overset{|}{\underset{Y}{C}}- \qquad \overset{\diagdown}{\underset{\diagup}{C}}=\overset{\diagdown}{\underset{\diagup}{C}}$$

[+ XY]

Break a π bond.

Addition

PROBLEM 6.1   Classify each transformation as substitution, elimination, or addition.

a.   (cyclohexanol with OH) ⟶ (cyclohexane with Br)

b.   (cyclohexene) ⟶ (cyclohexane)

c. $CH_3\overset{\overset{O}{\|}}{C}CH_3$ ⟶ $CH_3\overset{\overset{O}{\|}}{C}CH_2Cl$

d. $CH_3CH_2CH(OH)CH_3$ ⟶ $CH_3CH=CHCH_3$

6.3   Bond Breaking and Bond Making

Having now learned how to write and identify some common kinds of organic reactions, we can turn to a discussion of **reaction mechanism.**

◆ A *reaction mechanism* is a detailed description of how bonds are broken and formed as a starting material is converted to a product.

A reaction mechanism describes the relative order and rate of bond cleavage and formation. It explains all the known facts about a reaction and accounts for all products formed, and it is subject to modification or refinement as new details are discovered.

A reaction can occur either in one step or in a series of steps.

◆ **A one-step reaction is called a *concerted reaction.*** No matter how many bonds are broken or formed, a starting material is converted *directly* to a product.

Concerted reaction:
one step   A ⟶ B

◆ A **stepwise reaction** involves more than one step. A starting material is first converted to an unstable intermediate, called a **reactive intermediate,** which then goes on to form the product.

Stepwise reaction   A ⟶ reactive intermediate ⟶ B

6.3A   Bond Cleavage

Bonds are broken and formed in all chemical reactions. No matter how many steps there are in the reaction, however, there are only two ways to break (cleave) a bond: the electrons in the bond can be divided **equally** or **unequally** between the two atoms of the bond.

◆ Breaking a bond by **equally dividing the electrons** between the two atoms in the bond is called **homolysis** or **homolytic cleavage.**

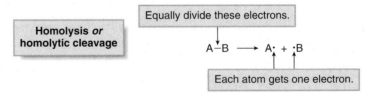

Homolysis *or* homolytic cleavage

Equally divide these electrons.

A—B ⟶ A· + ·B

Each atom gets one electron.

◆ Breaking a bond by **unequally dividing the electrons** between the two atoms in the bond is called **heterolysis** or **heterolytic cleavage.** Heterolysis of a bond between **A** and **B** can give either **A** or **B** the two electrons in the bond. When **A** and **B** have different electronegativities, the *electrons end up on the more electronegative atom.*

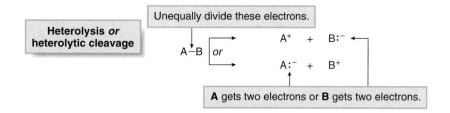

Homolysis and heterolysis require energy. Both processes generate reactive intermediates but the products are different in each case.

♦ Homolysis generates uncharged reactive intermediates with *unpaired* electrons.
♦ Heterolysis generates *charged* intermediates.

Each of these reactive intermediates has a very short lifetime, and each reacts quickly to form a stable organic product.

6.3B   Radicals, Carbocations, and Carbanions

The curved arrow notation first discussed in Section 1.5 works fine for heterolytic bond cleavage because it illustrates the movement of an **electron pair.** For homolytic cleavage, however, one electron moves to one atom in the bond and one electron moves to the other, so a different kind of curved arrow is needed.

♦ To illustrate the movement of a single electron, use a half-headed curved arrow, sometimes called a *fishhook.*

Full-headed curved arrows (⌒⇀) show the movement of an electron pair. **Half-headed curved arrows** (⌒⇁) show the movement of a single electron.

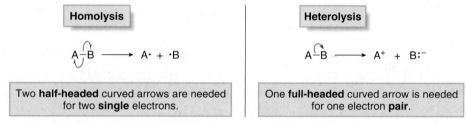

Figure 6.2 illustrates homolysis and two different heterolysis reactions for a carbon compound using curved arrows. Three different reactive intermediates are formed.

Homolysis of the C−Z bond generates two uncharged products with unpaired electrons.

Figure 6.2   Three reactive intermediates resulting from homolysis and heterolysis of a C−Z bond

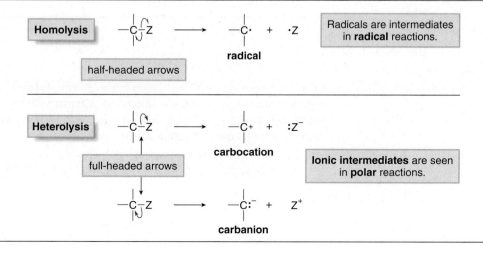

◆ A reactive intermediate with a single unpaired electron is called a *radical*.

Most radicals are highly unstable because they contain an atom that does not have an octet of electrons. Radicals typically have **no charge. They are intermediates in a group of reactions called *radical reactions,*** which are discussed in detail in Chapter 13.

Heterolysis of the C−Z bond can generate a **carbocation** or a **carbanion.**

◆ Giving two electrons to Z and none to carbon generates a positively charged carbon intermediate called a *carbocation.*
◆ Giving two electrons to C and none to Z generates a negatively charged carbon species called a *carbanion.*

Both carbocations and carbanions are unstable reactive intermediates. A carbocation contains a carbon atom surrounded by only six electrons, and a carbanion has a negative charge on carbon, which is not a very electronegative atom. **Carbocations (electrophiles)** and **carbanions (nucleophiles)** can be intermediates in *polar reactions*—**reactions in which a nucleophile reacts with an electrophile.**

Thus, homolysis and heterolysis generate radicals, carbocations, and carbanions, the three most important reactive intermediates in organic chemistry.

◆ Radicals and carbocations are electrophiles because they contain an electron-deficient carbon.
◆ Carbanions are nucleophiles because they contain a carbon with a lone pair.

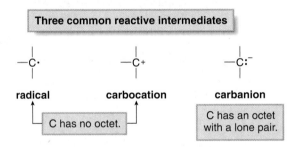

PROBLEM 6.2   By taking into account electronegativity differences, draw the products formed by heterolysis of the carbon–heteroatom bond in each molecule. Classify the organic reactive intermediate as a carbocation or a carbanion.

a. $CH_3-\overset{\displaystyle CH_3}{\underset{\displaystyle CH_3}{C}}-OH$ b. ⬡—Br   c. CH_3CH_2-Li

---

6.3C   Bond Formation

Like bond cleavage, bond formation occurs in two different ways. Two radicals can each donate **one electron** to form a two-electron bond. Alternatively, two ions with unlike charges can come together, with the negatively charged ion donating **both electrons** to form the resulting two-electron bond. **Bond formation always releases energy.**

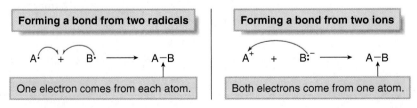

6.3D   All Kinds of Arrows

Table 6.1 summarizes the many kinds of arrows used in describing organic reactions. Curved arrows are especially important because they explicitly show what electrons are involved in a reaction, how these electrons move in forming and breaking bonds, and if a reaction proceeds via a radical or polar pathway.

A more complete summary of the arrows used in organic chemistry is given in Appendix A—Common Abbreviations, Arrows, and Symbols.

TABLE 6.1   A Summary of Arrow Types in Chemical Reactions

Arrow	Name	Use
$\longrightarrow$	Reaction arrow	Drawn between the starting materials and products in an equation (6.1)
$\rightleftharpoons$	Double reaction arrows (equilibrium arrows)	Drawn between the starting materials and products in an equilibrium equation (2.2)
$\longleftrightarrow$	Double-headed arrow	Drawn between resonance structures (1.5)
⤻	Full-headed curved arrow	Shows movement of an electron pair (1.5, 2.2)
⤻	Half-headed curved arrow (fishhook)	Shows movement of a single electron (6.3)

SAMPLE PROBLEM 6.1   Use full-headed or half-headed curved arrows to show the movement of electrons in each equation.

a.

b. $H-\overset{\displaystyle H}{\underset{\displaystyle H}{C}}-H \;+\; \cdot\ddot{\underset{..}{Cl}}: \;\longrightarrow\; H-\overset{\displaystyle H}{\underset{\displaystyle H}{C}}\cdot \;+\; H-\ddot{\underset{..}{Cl}}:$

SOLUTION

[a] In this reaction, the C—O bond is broken heterolytically. Because only one electron pair is involved, **one full-headed curved arrow** is needed. Notice that charge is balanced; that is, it is +1 on both sides.

$+\; H_2\ddot{O}:$ ⟵ The electron pair in the C—O bond ends up on O.

[b] This reaction involves radicals, so half-headed curved arrows are needed to show the movement of single electrons. One new two-electron bond is formed between H and Cl, and an unpaired electron is left on C. Because a total of three electrons are involved, **three half-headed curved arrows are needed.**

Two electrons form a bond.

An electron remains on C.

PROBLEM 6.3   Use curved arrows to show the movement of electrons in each equation.

a. $(CH_3)_3C-\overset{+}{N}\equiv N: \;\longrightarrow\; (CH_3)_3C^+ \;+\; :N\equiv N:$

b. $\cdot CH_3 \;+\; \cdot CH_3 \;\longrightarrow\; CH_3-CH_3$

c. $CH_3-\overset{\displaystyle CH_3}{\underset{\displaystyle CH_3}{C}}{}^+ \;+\; Br^- \;\longrightarrow\; CH_3-\overset{\displaystyle CH_3}{\underset{\displaystyle CH_3}{C}}-Br$

d. $HO-OH \;\longrightarrow\; 2HO\cdot$

6.4   Bond Dissociation Energy

Bond breaking can be quantified using the bond dissociation energy.

♦ The *bond dissociation energy* is the energy needed to homolytically cleave a covalent bond.

$$A-B \longrightarrow A\cdot \ + \ \cdot B \qquad \Delta H° = \text{bond dissociation energy}$$

↑
Homolysis requires energy.

The energy absorbed or released in any reaction, symbolized by $\Delta H°$, is called the **enthalpy change** or **heat of reaction.**

> The superscript (°) means that values are determined under standard conditions (pure compounds in their most stable state at 25 °C and 1 atm pressure).

♦ When $\Delta H°$ is positive (+), energy is absorbed and the reaction is *endothermic.*
♦ When $\Delta H°$ is negative (–), energy is released and the reaction is *exothermic.*

A bond dissociation energy is the $\Delta H°$ for a specific kind of reaction—the homolysis of a covalent bond to form two radicals. Because bond breaking requires energy, **bond dissociation energies are always *positive* numbers,** and homolysis is always **endothermic.** Conversely, **bond formation always *releases* energy,** so this reaction is always **exothermic.** The H–H bond requires +104 kcal/mol to cleave and releases –104 kcal/mol when formed.

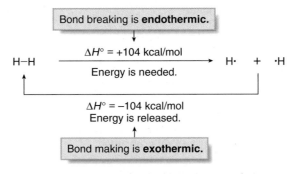

Bond breaking is **endothermic.**

$$H-H \xrightarrow{\quad \Delta H° = +104 \text{ kcal/mol} \quad} H\cdot \ + \ \cdot H$$
Energy is needed.

$\Delta H° = -104$ kcal/mol
Energy is released.

Bond making is **exothermic.**

> Additional bond dissociation energies for C–C multiple bonds and *sp*, *sp*^{2}, and *sp*^{3} hybridized C–H bonds are given in Table 1.3.

Table 6.2 contains a representative list of bond dissociation energies for many common bonds.

Comparing bond dissociation energies is equivalent to comparing **bond strength.**

> A table of bond dissociation energies also appears in Appendix C.

♦ The *stronger* the bond, the *higher* its bond dissociation energy.

For example, the H–H bond is stronger than the Cl–Cl bond because its bond dissociation energy is higher [Table 6.2: 104 kcal/mol (H$_2$) versus 58 kcal/mol (Cl$_2$)]. The data in Table 6.2 demonstrate that **bond dissociation energies *decrease* down a column of the periodic table as the valence electrons used in bonding are moved further from the nucleus.** Bond dissociation energies for a group of methyl–halogen bonds exemplify this trend.

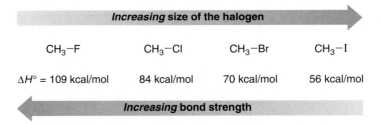

Increasing size of the halogen

| CH$_3$–F | CH$_3$–Cl | CH$_3$–Br | CH$_3$–I |

$\Delta H° = 109$ kcal/mol   84 kcal/mol   70 kcal/mol   56 kcal/mol

Increasing bond strength

Because bond length increases down a column of the periodic table, bond dissociation energies are a quantitative measure of the general phenomenon noted in Chapter 1—*shorter* bonds are *stronger* bonds.

TABLE 6.2 Bond Dissociation Energies for Some Common Bonds [A–B → A· + ·B]

Bond	$\Delta H°$ kcal/mol	(kJ/mol)	Bond	$\Delta H°$ kcal/mol	(kJ/mol)
H−Z bonds			**R−X bonds**		
H−F	136	(569)	CH_3−F	109	(456)
H−Cl	103	(431)	CH_3−Cl	84	(351)
H−Br	88	(368)	CH_3−Br	70	(293)
H−I	71	(297)	CH_3−I	56	(234)
H−OH	119	(498)	CH_3CH_2−F	107	(448)
			CH_3CH_2−Cl	81	(339)
Z−Z bonds			CH_3CH_2−Br	68	(285)
H−H	104	(435)	CH_3CH_2−I	53	(222)
F−F	38	(159)	$(CH_3)_2CH$−F	106	(444)
Cl−Cl	58	(242)	$(CH_3)_2CH$−Cl	80	(335)
Br−Br	46	(192)	$(CH_3)_2CH$−Br	68	(285)
I−I	36	(151)	$(CH_3)_2CH$−I	53	(222)
HO−OH	51	(213)	$(CH_3)_3C$−F	106	(444)
			$(CH_3)_3C$−Cl	79	(331)
R−H bonds			$(CH_3)_3C$−Br	65	(272)
CH_3−H	104	(435)	$(CH_3)_3C$−I	50	(209)
CH_3CH_2−H	98	(410)			
$CH_3CH_2CH_2$−H	98	(410)	**R−OH bonds**		
$(CH_3)_2CH$−H	95	(397)	CH_3−OH	91	(381)
$(CH_3)_3C$−H	91	(381)	CH_3CH_2−OH	91	(381)
$CH_2{=}CH$−H	104	(435)	$CH_3CH_2CH_2$−OH	91	(381)
$HC{\equiv}C$−H	125	(523)	$(CH_3)_2CH$−OH	91	(381)
$CH_2{=}CHCH_2$−H	87	(364)	$(CH_3)_3C$−OH	91	(381)
C_6H_5−H	110	(460)			
$C_6H_5CH_2$−H	85	(356)			
R−R bonds					
CH_3−CH_3	88	(368)			
CH_3−CH_2CH_3	85	(356)			
CH_3−$CH{=}CH_2$	92	(385)			
CH_3−$C{\equiv}CH$	117	(489)			

PROBLEM 6.4 Without looking at a table of bond dissociation energies, determine which bond in each pair has the higher bond dissociation energy.

a. H−Cl and H−Br   b. CH_3−OH and CH_3−SH   c. $(CH_3)_2C{=}O$ and $CH_3{-}OCH_3$
(↑)   ↑

Bond dissociation energies are also used to calculate the enthalpy change ($\Delta H°$) in a reaction in which several bonds are broken and formed. **$\Delta H°$ indicates the relative strength of bonds broken and formed in a reaction.**

◆ When $\Delta H°$ is positive, more energy is needed to break bonds than is released in forming bonds. The bonds broken in the starting material are *stronger* than the bonds formed in the product.

◆ When $\Delta H°$ is negative, more energy is released in forming bonds than is needed to break bonds. The bonds formed in the product are *stronger* than the bonds broken in the starting material.

To determine the overall $\Delta H°$ for a reaction:

[1] Beginning with a balanced equation, add the bond dissociation energies for all bonds broken in the starting materials. This (+) value represents the **energy needed** to break bonds.

[2] Add the bond dissociation energies for all bonds formed in the products. This (–) value represents the **energy released** in forming bonds.

[3] **The overall $\Delta H°$ is the sum in Step [1] _plus_ the sum in Step [2].**

$\Delta H°$ overall enthalpy change	=	sum of $\Delta H°$ of bonds broken	+	(–) sum of $\Delta H°$ of bonds formed

---

SAMPLE PROBLEM 6.2   Use the values in Table 6.2 to determine $\Delta H°$ for the following reaction.

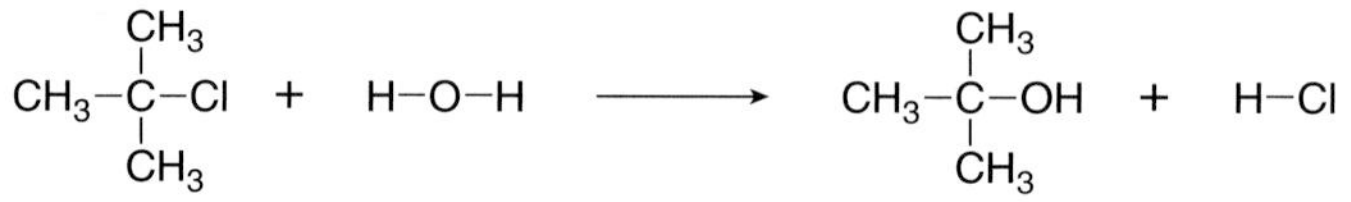

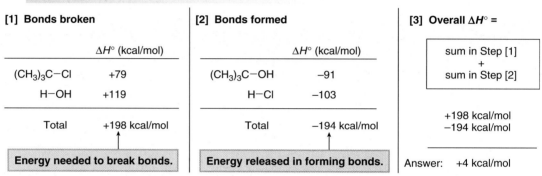

SOLUTION

[1] Bonds broken		**[2] Bonds formed**		**[3] Overall $\Delta H°$ =**

	$\Delta H°$ (kcal/mol)		$\Delta H°$ (kcal/mol)	
$(CH_3)_3C-Cl$	+79	$(CH_3)_3C-OH$	–91	sum in Step [1] + sum in Step [2]
$H-OH$	+119	$H-Cl$	–103	
Total	+198 kcal/mol	Total	–194 kcal/mol	+198 kcal/mol −194 kcal/mol

Energy needed to break bonds.	Energy released in forming bonds.	Answer: +4 kcal/mol

Because $\Delta H°$ is a positive value, this reaction is **endothermic** and energy is absorbed. **The bonds broken in the starting material are stronger than the bonds formed in the product.**

PROBLEM 6.5   Use the values in Table 6.2 to calculate $\Delta H°$ for each reaction. Classify each reaction as endothermic or exothermic.

a. $CH_3CH_2-Br + H_2O \longrightarrow CH_3CH_2-OH + HBr$ b. $CH_4 + Cl_2 \longrightarrow CH_3Cl + HCl$

The oxidation of both isooctane and glucose, the two molecules that introduced Chapter 6, forms CO_2 and H_2O.

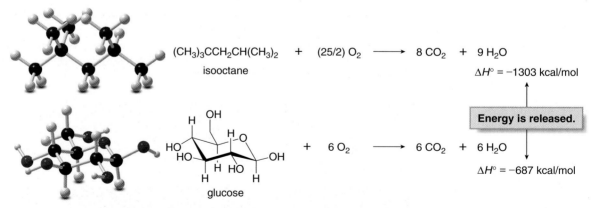

$\Delta H°$ is negative for both oxidations, so both reactions are exothermic. **Both isooctane and glucose release energy on oxidation because the bonds in the products are stronger than the bonds in the reactants.**

Bond dissociation energies have two important limitations. They present _overall_ energy changes only. They reveal nothing about the reaction mechanism or how fast a reaction proceeds. Moreover, bond dissociation energies are determined for reactions in the gas phase, whereas most organic reactions are carried out in a liquid solvent where solvation energy contributes to the overall enthalpy of a reaction. As such, bond dissociation energies are imperfect indicators of energy changes in a reaction. Despite these limitations, using bond dissociation energies to calculate $\Delta H°$ gives a useful approximation of the energy changes that occur when bonds are broken and formed in a reaction.

PROBLEM 6.6 Calculate $\Delta H°$ for each oxidation reaction. Each equation is balanced as written; remember to take into account the coefficients in determining the number of bonds broken or formed.
[$\Delta H°$ for O_2 = 119 kcal/mol; $\Delta H°$ for one $C=O$ in CO_2 = 128 kcal/mol]
a. $CH_4 + 2O_2 \longrightarrow CO_2 + 2H_2O$ b. $CH_3CH_3 + (7/2)O_2 \longrightarrow 2CO_2 + 3H_2O$

6.5 Thermodynamics

For a reaction to be practical, the equilibrium must favor the products, *and* the reaction rate must be fast enough to form them in a reasonable time. These two conditions depend on the **thermodynamics** and the **kinetics** of a reaction, respectively.

◆ *Thermodynamics* describes energy and equilibrium. How do the *energies* of the reactants and the products compare? What are the relative *amounts* of reactants and products at equilibrium?

Reaction kinetics are discussed in Section 6.9.

◆ *Kinetics* describes reaction rates. How *fast* are reactants converted to products?

6.5A Equilibrium Constant and Free Energy Changes

The **equilibrium constant, K_{eq},** is a mathematical expression that relates the amount of starting material and product at equilibrium. For example, when starting materials **A** and **B** react to form products **C** and **D,** the equilibrium constant is given by the following expression.

K_{eq} was first defined in Section 2.3 for acid–base reactions.

Reaction		Equilibrium constant
$A + B \rightleftharpoons C + D$	$K_{eq} =$	$\dfrac{[products]}{[starting\ materials]} = \dfrac{[C][D]}{[A][B]}$

The size of K_{eq} tells about the position of equilibrium; that is, it expresses whether the starting materials or products predominate once equilibrium has been reached.

◆ When $K_{eq} > 1$, equilibrium favors the products (**C** and **D**) and the equilibrium lies to the right as the equation is written.

◆ When $K_{eq} < 1$, equilibrium favors the starting materials (**A** and **B**) and the equilibrium lies to the left as the equation is written.

◆ For a reaction to be useful, the equilibrium must favor the products, and $K_{eq} > 1$.

What determines whether equilibrium favors the products in a given reaction? **The position of equilibrium is determined by the relative energies of the reactants and products.** The free energy of a molecule, also called its **Gibbs free energy,** is symbolized by $G°$. The **change in free energy** between reactants and products, symbolized by $\Delta G°$, determines whether the starting materials or products are favored at equilibrium.

◆ $\Delta G°$ is the overall energy difference between reactants and products.

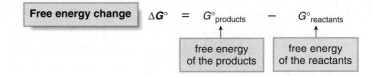

Free energy change $\Delta G° = G°_{products} - G°_{reactants}$

free energy of the products   free energy of the reactants

Figure 6.3 Summary of the relationship between $\Delta G°$ and K_{eq}

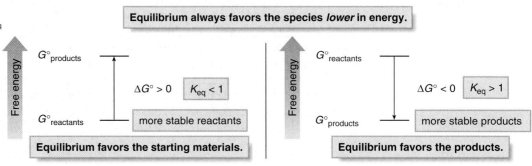

$\Delta G°$ is related to the equilibrium constant K_{eq} by the following equation:

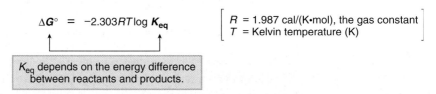

$$\Delta G° = -2.303RT \log K_{eq}$$

$\left[\begin{array}{l} R = 1.987 \text{ cal/(K·mol), the gas constant} \\ T = \text{Kelvin temperature (K)} \end{array}\right]$

K_{eq} depends on the energy difference between reactants and products.

Using this expression we can determine the relationship between the equilibrium constant and the free energy change between reactants and products.

$K_{eq} > 1$ when $\Delta G° < 0$, and equilibrium favors the products. $K_{eq} < 1$ when $\Delta G° > 0$, and equilibrium favors the starting materials.

◆ When $K_{eq} > 1$, log K_{eq} is positive, making $\Delta G°$ **negative,** and energy is released. Thus, **equilibrium favors the products when the energy of the products is** *lower* **than the energy of the reactants.**

◆ When $K_{eq} < 1$, log K_{eq} is negative, making $\Delta G°$ **positive,** and energy is absorbed. Thus, **equilibrium favors the reactants when the energy of the products is** *higher* **than the energy of the reactants.**

At 25 °C, $2.303RT = 1.4$ kcal/mol; thus, $\Delta G° = -1.4$ log K_{eq}.

Compounds that are lower in energy have increased stability. Thus, **equilibrium favors the products when they are more stable (lower in energy) than the starting materials of a reaction.** This is summarized in Figure 6.3.

The symbol ~ means *approximately.*

Because $\Delta G°$ depends on the logarithm of K_{eq}, **a small change in energy corresponds to a large difference in the relative amount of starting material and product at equilibrium.** Several values of $\Delta G°$ and K_{eq} are given in Table 6.3. For example, a difference in energy of only ~1 kcal/mol means that there is 10 times as much of the more stable species at equilibrium. A difference in energy of ~4 kcal/mol means that there is essentially only one compound, either starting material or product, at equilibrium.

TABLE 6.3   Representative Values for $\Delta G°$ and K_{eq} at 25 °C, for a Reaction A → B

$\Delta G°$ (kcal/mol)	K_{eq}	Relative amount of A and B at equilibrium	
+4.2	10^{-3}	**Essentially all A (99.9%)**	
+2.8	10^{-2}	100 times as much **A** as **B**	Increasing [product]
+1.4	10^{-1}	10 times as much **A** as **B**	
0	1	Equal amounts of **A** and **B**	
−1.4	10^{1}	10 times as much **B** as **A**	
−2.8	10^{2}	100 times as much **B** as **A**	
−4.2	10^{3}	**Essentially all B (99.9%)**	

A small difference in free energy means a large difference in the amount of **A** and **B** at equilibrium.

PROBLEM 6.7   a. Which K_{eq} corresponds to a negative value of $\Delta G°$, K_{eq} = 1000 or K_{eq} = .001?
       b. Which K_{eq} corresponds to a lower value of $\Delta G°$, K_{eq} = 10^{-2} or K_{eq} = 10^{-5}?

PROBLEM 6.8   Given each of the following values, is the starting material or product favored at equilibrium?
       a. K_{eq} = 5.5   b. $\Delta G°$ = 10.3 kcal

PROBLEM 6.9   Given each of the following values, is the starting material or product lower in energy?
       a. $\Delta G°$ = 2.0 kcal   b. K_{eq} = 10   c. $\Delta G°$ = –3.0 kcal   d. K_{eq} = 10^{-3}

6.5B   Energy Changes and Conformational Analysis

These equations can be used for any process with two states in equilibrium. As an example, monosubstituted cyclohexanes exist as two different chair conformers that rapidly interconvert at room temperature, with the conformer having the substituent in the roomier equatorial position favored (Section 4.13). Knowing the energy difference between the two conformers allows us to calculate the amount of each at equilibrium.

For example, the energy difference between the two chair conformers of phenylcyclohexane is –2.9 kcal/mol, as shown in the accompanying equation. Using the values in Table 6.3, this corresponds to an equilibrium constant of ~100, meaning that there is approximately 100 times more **B** (equatorial phenyl group) than **A** (axial phenyl group) at equilibrium.

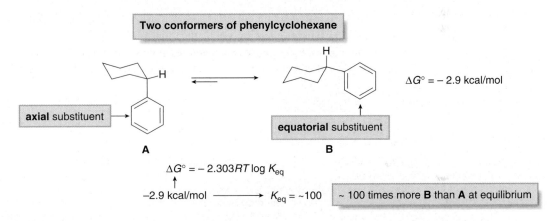

PROBLEM 6.10   The equilibrium constant for the conversion of the axial to the equatorial conformer of methoxycyclohexane is 2.7.

$$K_{eq} = 2.7$$

   a. Given these data, which conformer is present in the larger amount at equilibrium?
   b. Is $\Delta G°$ for this process positive or negative?
   c. From the values in Table 6.3, approximate the size of $\Delta G°$.

6.6   Enthalpy and Entropy

The **free energy change** ($\Delta G°$) depends on the **enthalpy change** ($\Delta H°$) and the **entropy change** ($\Delta S°$). $\Delta H°$ indicates relative bond strength, but what does $\Delta S°$ measure?

***Entropy** (S°)* **is a measure of the randomness in a system.** The more freedom of motion or the more disorder present, the higher the entropy. Gas molecules move more freely than liquid molecules and are higher in entropy. Cyclic molecules have more restricted bond rotation than similar acyclic molecules and are lower in entropy.

Entropy is a rather intangible concept that comes up again and again in chemistry courses. One way to remember the relation between entropy and disorder is to consider a handful of chopsticks. Dropped on the floor, they are arranged randomly (a state of high entropy). Placed end-to-end in a straight line, they are arranged intentionally (a state of low entropy). The more disordered, random arrangement is favored and easier to achieve energetically.

The *entropy change* (ΔS°) **is the change in the amount of disorder between reactants and products.** ΔS° is positive (+) when the products are more disordered than the reactants. ΔS° is negative (−) when the products are less disordered (more ordered) than the reactants.

◆ **Reactions resulting in an increase in entropy are favored.**

ΔG° is related to ΔH° and ΔS° by the following equation:

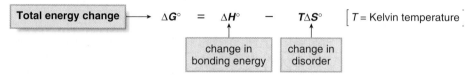

This equation tells us that the total energy change in a reaction is due to two factors: the change in the **bonding energy** and the change in **disorder.** The change in bonding energy can be calculated from bond dissociation energies (Section 6.4). Entropy changes, on the other hand, are more difficult to access, but they are important in the following two cases:

◆ When the number of molecules of starting material *differs* from the number of molecules of product in the balanced chemical equation.

◆ When an acyclic molecule is *cyclized* to a cyclic one, or a cyclic molecule is converted to an acyclic one.

For example, **when a single starting material forms two products,** as in the homolytic cleavage of a bond to form two radicals, **entropy increases** and favors formation of the products. In contrast, **entropy decreases when an acyclic compound forms a ring,** because a ring has fewer degrees of freedom. In this case, therefore, entropy does *not* favor formation of the product.

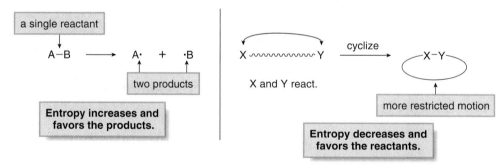

In most other reactions that are not carried out at high temperature, the entropy term (TΔS°) is small compared to the enthalpy term (ΔH°) and it can be neglected. Thus, **we will often approximate the overall free energy change of a reaction by the change in the bonding energy only.** Keep in mind that this is an approximation, but for many reactions it gives us a starting point from which to decide if the reaction is energetically favorable.

Recall from Section 6.4 that a reaction is endothermic when ΔH° is positive and exothermic when ΔH° is negative. A reaction is **endergonic when ΔG° is positive** and **exergonic when ΔG° is negative.** ΔG° is usually approximated by ΔH° in this text, so the terms endergonic and exergonic are rarely used.

$$\Delta G° \approx \Delta H°$$ • The total energy change is approximated by the change in bonding energy only.

According to this approximation:

◆ The product is favored in reactions in which ΔH° is a *negative* value; that is, the bonds in the product are *stronger* than the bonds in the starting material.

◆ The starting material is favored in a reaction in which ΔH° is a *positive* value; that is, the bonds in the starting material are *stronger* than the bonds in the product.

PROBLEM 6.11   Considering each of the following values, and neglecting entropy, tell whether the starting material or product is favored at equilibrium: (a) ΔH° = 20 kcal; (b) ΔH° = −10 kcal.

PROBLEM 6.12   For a reaction with $\Delta H° = 10$ kcal, decide which of the following statements is (are) true. Correct any false statement to make it true. (a) The reaction is exothermic; (b) $\Delta G°$ for the reaction is positive; (c) K_{eq} is greater than 1; (d) the bonds in the starting materials are stronger than the bonds in the product; and (e) the product is favored at equilibrium.

6.7   Energy Diagrams

An **energy diagram** is a schematic representation of the energy changes that take place as reactants are converted to products. An energy diagram indicates how readily a reaction proceeds, how many steps are involved, and how the energies of the reactants, products, and intermediates compare.

Consider, for example, a concerted reaction between molecule **A – B** with anion **C:⁻** to form products **A:⁻** and **B – C.** If the reaction occurs in a single step, the bond between **A** and **B** is broken *as* the bond between **B** and **C** is formed. Let's assume that the products are lower in energy than the reactants in this hypothetical reaction.

An energy diagram plots **energy on the y axis** versus the progress of reaction, often labeled the **reaction coordinate,** on the *x* axis. As the starting materials **A – B** and **C:⁻** approach one another, their electron clouds feel some repulsion, causing an increase in energy, until a maximum value is reached. This unstable energy maximum is called the **transition state.** In the transition state the bond between **A** and **B** is partially broken, and the bond between **B** and **C** is partially formed. Because it is at the top of an energy "hill," **a transition state can never be isolated.**

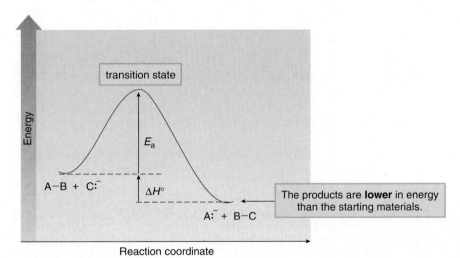

At the transition state, the bond between **A** and **B** can re-form to regenerate starting material, *or* the bond between **B** and **C** can form to generate product. As the bond forms between **B** and **C** the energy decreases until some stable energy minimum of the products is reached.

◆ The energy difference between the reactants and products is $\Delta H°$. Because the products are at lower energy than the reactants, this reaction is exothermic and energy is released.

◆ The energy difference between the transition state and the starting material is called the *energy of activation,* symbolized by E_a.

The *energy of activation* is the minimum amount of energy needed to break bonds in the reactants. It represents an **energy barrier** that must be overcome for a reaction to occur. The size of E_a tells us about the reaction rate.

> A slow reaction has a large E_a. A fast reaction has a low E_a.

◆ The larger the E_a, the greater the amount of energy that is needed to break bonds, and the slower the reaction rate.

How can we draw the structure of the unstable transition state? The structure of the transition state is somewhere in between the structures of the starting material and product. Any bond that is partially broken or formed is drawn with a dashed line. Any atom that gains or loses a charge contains a partial charge in the transition state. Transition states are drawn in brackets, with a superscript double dagger ($\ddagger$).

In the hypothetical reaction between **A–B** and **C:⁻** to form **A:⁻** and **B–C,** the bond between **A** and **B** is partially broken, and the bond between **B** and **C** is partially formed. Because **A** gains a negative charge and **C** loses a charge in the course of the reaction, each atom bears a partial negative charge in the transition state.

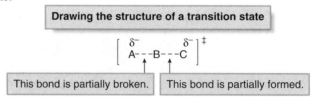

Drawing the structure of a transition state

$$\left[\overset{\delta^-}{A}\text{---}B\text{---}\overset{\delta^-}{C} \right]^{\ddagger}$$

This bond is partially broken.   This bond is partially formed.

Several energy diagrams are drawn in Figure 6.4. For any energy diagram:

Figure 6.4 Some representative energy diagrams

Example [1]

• Large E_a ⟶ slow reaction
• (+) $\Delta H°$ ⟶ endothermic reaction

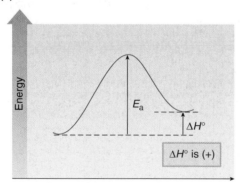

$\Delta H°$ is (+)

Reaction coordinate

Example [3]

• Low E_a ⟶ fast reaction
• (+) $\Delta H°$ ⟶ endothermic reaction

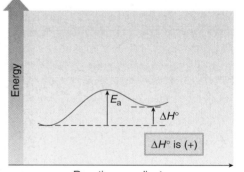

$\Delta H°$ is (+)

Reaction coordinate

Example [2]

• Large E_a ⟶ slow reaction
• (−) $\Delta H°$ ⟶ exothermic reaction

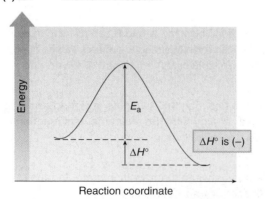

$\Delta H°$ is (−)

Reaction coordinate

Example [4]

• Low E_a ⟶ fast reaction
• (−) $\Delta H°$ ⟶ exothermic reaction

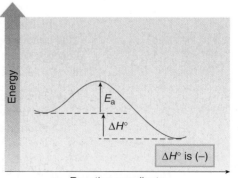

$\Delta H°$ is (−)

Reaction coordinate

Figure 6.5 Comparing $\Delta H°$ and E_a in two energy diagrams

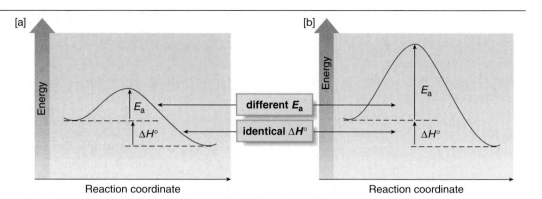

- Energy diagrams in [a] and [b] both depict exothermic reactions with the same negative value of $\Delta H°$.
- E_a in [a] is lower than E_a in [b], so reaction [a] is faster than reaction [b].

♦ E_a determines the height of the energy barrier.
♦ $\Delta H°$ determines the relative position of the reactants and products.

The two variables, E_a **and** $\Delta H°$, **are independent** of each other. Two reactions can have identical values for $\Delta H°$ but very different E_a values. In Figure 6.5, both reactions have the same negative $\Delta H°$ favoring the products, but the second reaction has a much higher E_a, so it proceeds more slowly.

PROBLEM 6.13 Draw an energy diagram for a reaction in which the products are higher in energy than the starting materials and E_a is large. Clearly label all of the following on the diagram: the axes, the starting materials, the products, the transition state, $\Delta H°$, and E_a.

PROBLEM 6.14 Draw the structure for the transition state in each reaction.

a. $CH_3-\overset{\underset{|}{CH_3}}{\underset{|}{CH_3}}{C}-\overset{+}{O}H_2 \longrightarrow CH_3-\overset{\underset{|}{CH_3}}{\underset{|}{CH_3}}{C}{}^+ + H_2O$ b. $CH_3O-H + {}^-OH \longrightarrow CH_3O^- + H_2O$

6.8 Energy Diagram for a Two-Step Reaction Mechanism

Although the hypothetical reaction in Section 6.7 is concerted, many reactions involve more than one step with formation of a reactive intermediate. Consider the same overall reaction, **A–B + C:⁻** to form products **A:⁻ + B–C,** but in this case begin with the assumption that the reaction occurs by a *stepwise* pathway—that is, bond breaking occurs *before* bond making. Once again, assume that the overall process is exothermic.

Same overall reaction   A–B + C:⁻ $\longrightarrow$ A:⁻ + B–C

This bond is broken...   *before*   ...this bond is formed.

One possible stepwise mechanism involves heterolysis of the **A–B** bond to form two ions **A:⁻** and **B⁺,** followed by reaction of **B⁺** with anion **C:⁻** to form product **B–C,** as outlined in the accompanying equations. Species **B⁺** is a reactive intermediate. It is formed as a product in Step [1], and then goes on to react with **C:⁻** in Step [2].

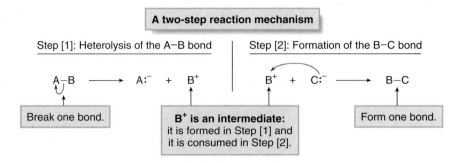

We must draw an energy diagram for each step, and then combine them in an energy diagram for the overall two-step mechanism. Each step has its own energy barrier, with a transition state at the energy maximum.

Step [1] is endothermic because energy is needed to cleave the **A–B** bond, making $\Delta H°$ a positive value and placing the products of Step [1] at higher energy than the starting materials. In the transition state, the **A–B** bond is partially broken.

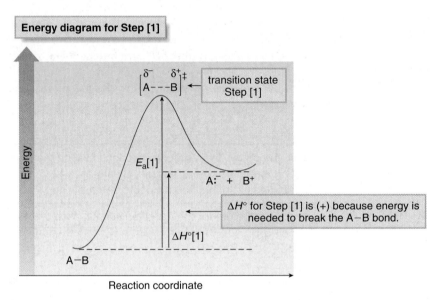

Step [2] is exothermic because energy is released in forming the **B–C** bond, making $\Delta H°$ a negative value and placing the products of Step [2] at lower energy than the starting materials of Step [2]. In the transition state, the **B–C** bond is partially formed.

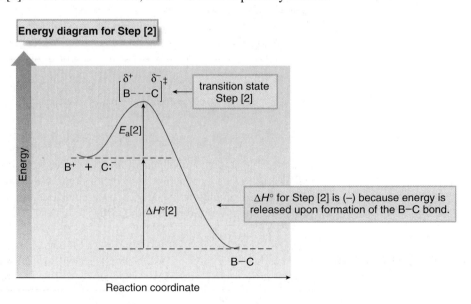

Figure 6.6 Complete energy diagram for the two-step conversion of $A-B + C{:}^- \rightarrow A{:}^- + B-C$

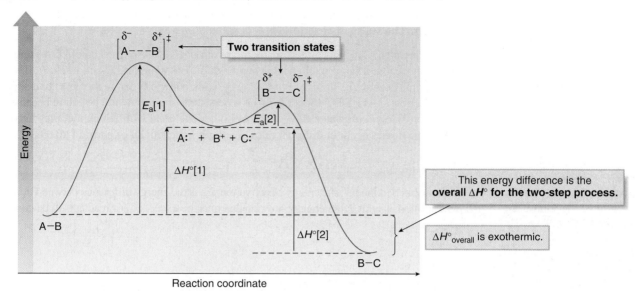

- **The transition states are located at energy maxima while the reactive intermediate B⁺ is located at an energy minimum.**
- Each step is characterized by its own value of $\Delta H°$ and E_a.
- The overall energy difference between starting material and products is labeled as $\Delta H°_{overall}$. In this example, the products of the two-step sequence are at lower energy than the starting materials.
- Since Step [1] has the higher energy transition state, it is the **rate-determining step.**

The overall process is shown in Figure 6.6 as a single energy diagram that combines both steps. Because the reaction has two steps, there are two transition states, each corresponding to an energy barrier. The transition states are separated by an energy minimum, at which the reactive intermediate **B⁺** is located. Because we made the assumption that the overall two-step process is exothermic, the overall energy difference between the reactants and products, labeled $\Delta H°_{overall}$, has a negative value, and the final products are at a lower energy than the starting materials.

The energy barrier for Step [1], labeled $E_a[1]$, is higher than the energy barrier for Step [2], labeled $E_a[2]$. This is because bond cleavage (Step [1]) is more difficult (requires more energy) than bond formation (Step [2]). A higher energy transition state for Step [1] makes it the slower step of the mechanism.

◆ **In a multistep mechanism, the step with the highest energy transition state is called the rate-determining step.**

In this reaction, the rate-determining step is Step [1].

PROBLEM 6.15 Consider the following energy diagram.

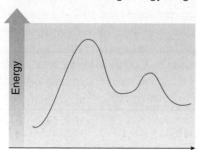

a. How many steps are involved in this reaction?
b. Label $\Delta H°$ and E_a for each step, and label $\Delta H°_{overall}$.
c. Label each transition state.
d. Which point on the graph corresponds to a reactive intermediate?
e. Which step is rate-determining?
f. Is the overall reaction endothermic or exothermic?

PROBLEM 6.16 Draw an energy diagram for a two-step reaction, $A \rightarrow B \rightarrow C$, where the relative energy of these compounds is $C < A < B$, and the conversion of $B \rightarrow C$ is rate-determining.

6.9 Kinetics

We now turn to a more detailed discussion of **reaction rate**—that is, how fast a particular reaction proceeds. **The study of reaction rates is called** *kinetics.*

The rate of chemical processes affects many facets of our lives. Aspirin is an effective anti-inflammatory agent because it rapidly inhibits the synthesis of prostaglandins (Section 19.6). Butter turns rancid with time because its lipids are only slowly oxidized by oxygen in the air to undesirable by-products (Section 13.11). DDT (Section 7.4) is a persistent environmental pollutant because it does not react appreciably with water, oxygen, or any other chemical with which it comes into contact. All of these processes occur at different rates, resulting in beneficial or harmful effects.

6.9A Energy of Activation

> **The temperature of a reaction.** A reaction in the laboratory is generally run at a temperature that allows it to proceed both within a reasonable time, and without undesired side reactions. Some reactions occur rapidly well below room temperature, and others need an external source of heat. In this text, the temperature of a reaction is often omitted under a reaction arrow, but the symbol Δ is used when high temperature is needed for the reaction to occur.

As we learned in Section 6.7, the energy of activation, E_a, is the energy difference between the reactants and the transition state. It is the **energy barrier** that must be exceeded for reactants to be converted to products.

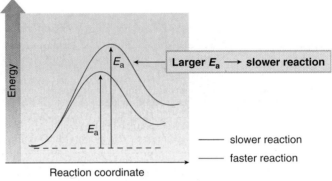

♦ The *larger the* E_a, the *slower* the reaction.

Concentration and temperature also affect reaction rate.

> ♦ The *higher* the concentration, the *faster* the rate. Increasing concentration increases the number of collisions between reacting molecules, which in turn increases the rate.

> **E_a, concentration, and temperature affect reaction rate.**

> ♦ The *higher* the temperature, the *faster* the rate. Increasing temperature increases the average kinetic energy of the reacting molecules. Because the kinetic energy of colliding molecules is used for bond cleavage, increasing the average kinetic energy increases the rate.

The E_a values of most organic reactions are 10–35 kcal. When $E_a < 20$ kcal, the reaction occurs readily at or below room temperature. When $E_a > 20$ kcal, higher temperatures are needed. As a rule of thumb, increasing the temperature by 10 °C doubles the reaction rate. Thus, reactions in the lab are often heated to increase their rates so they occur in a reasonable amount of time.

> Some reactions have a very favorable equilibrium constant ($K_{eq} \gg 1$), but the rate is very slow. The oxidation of alkanes like isooctane to form CO_2 and H_2O is an example of this phenomenon. Without a spark to initiate the reaction, isooctane does not react with O_2; and gasoline, which contains isooctane, can be safely handled in the air.

Keep in mind that certain **reaction quantities have** *no effect* **on reaction rate.**

> ♦ $\Delta G°$, $\Delta H°$, and K_{eq} do not determine the rate of a reaction. These quantities indicate the direction of equilibrium and the relative energy of reactants and products.

PROBLEM 6.17 Which value (if any) corresponds to a faster reaction: (a) $E_a = 10$ kcal or $E_a = 1$ kcal; (b) a reaction temperature of 0 °C or a reaction temperature of 25 °C; (c) $K_{eq} = 10$ or $K_{eq} = 100$; (d) $\Delta H° = -2$ kcal/mol or $\Delta H° = 2$ kcal/mol?

6.9B Rate Equations

The rate of a chemical reaction is determined by measuring the decrease in the concentration of the reactants over time, or the increase in the concentration of the products over time. A **rate law** (or **rate equation**) is an equation that shows the relationship between the rate of a reaction and the concentration of the reactants. A rate law is determined *experimentally,* and it depends on the mechanism of the reaction.

A rate law has two important terms: the **rate constant symbolized by *k*,** and the **concentration of the reactants.** Not all reactant concentrations may appear in the rate equation, as we shall soon see.

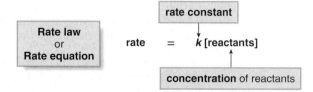

A rate constant *k* is a fundamental characteristic of a reaction. It is a complex mathematical term that takes into account the dependence of a reaction rate on temperature and the energy of activation.

> A rate constant *k* and the energy of activation E_a are inversely related. **A high E_a corresponds to a small *k*.**

♦ Fast reactions have large rate constants.
♦ Slow reactions have small rate constants.

What concentration terms appear in the rate equation? That depends on the mechanism. For the organic reactions we will encounter:

♦ A rate equation contains concentration terms for *all* reactants involved in a *one-step* mechanism.
♦ A rate equation contains concentration terms for *only* the reactants involved in the *rate-determining step* in a multistep reaction.

For example, in the one-step reaction of **A–B + C:⁻** to form **A:⁻ + B–C,** *both* reactants appear in the transition state of the only step of the mechanism. The **concentration of *both* reactants affects the reaction rate** and *both* terms appear in the rate equation. This type of reaction involving two reactants is said to be **bimolecular.**

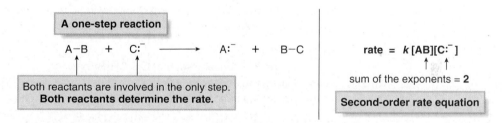

The *order* of a rate equation equals the sum of the exponents of the concentration terms in the rate equation. In the rate equation for the concerted reaction of **A–B + C:⁻,** there are two concentration terms, each with an exponent of one. Thus, the sum of the exponents is two and the **rate equation is *second order*** (the reaction follows second-order kinetics).

Because the rate of the reaction depends on the concentration of both reactants, doubling the concentration of *either* **A–B** or **C:⁻** doubles the rate of the reaction. Doubling the concentration of *both* **A–B** and **C:⁻** increases the reaction rate by a factor of *four.*

The situation is different in the stepwise conversion of **A–B + C:⁻** to form **A:⁻ + B–C.** The mechanism shown in Section 6.8 has two steps: a slow step (the **rate-determining** step) in which the **A–B** bond is broken, and a fast step in which the **B–C** bond is formed.

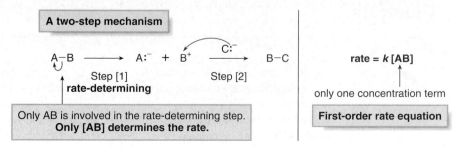

In a multistep mechanism, a reaction can occur no faster than its rate-determining step. **Only the concentrations of the reactants affecting the rate-determining step appear in the rate equation.** In this example, the rate depends on the concentration of **A−B** *only,* because only **A−B** appears in the rate-determining step. A reaction involving only one reactant is said to be **unimolecular.** Because there is only one concentration term (raised to the first power), the **rate equation is** *first order* (the reaction follows first-order kinetics).

Because the rate of the reaction depends on the concentration of only *one* reactant, doubling the concentration of **A−B** doubles the rate of the reaction, but **doubling the concentration of C:⁻ has** *no effect* **on the reaction rate.**

This might seem like a puzzling result. If **C:⁻** is involved in the reaction, why doesn't it affect the overall rate of the reaction? Not only can you change the concentration of **C:⁻** and not affect the rate, but you also can replace it by a different anion without affecting the rate. How can this be? **C:⁻ is not involved in the slow step of the reaction so neither its concentration nor its identity affect the reaction rate.**

The following analogy is useful. Let's say three students must make 20 peanut butter and jelly sandwiches for a class field trip. Student (**1**) spreads the peanut butter on the bread. Student (**2**) spreads on the jelly, and student (**3**) cuts the sandwiches in half. Suppose student (**2**) is very slow in spreading the jelly. It doesn't matter how fast students (**1**) and (**3**) are; they can't finish making sandwiches any faster than student (**2**) can add the jelly. Five more students can spread on the peanut butter, or an entirely different individual can replace student (**3**), and this doesn't speed up the process. How fast the sandwiches are made is determined entirely by the rate-determining step—that is, spreading the jelly.

Rate equations provide very important information about the mechanism of a reaction. Rate laws for new reactions with unknown mechanisms are determined by a set of experiments that measure how a reaction's rate changes with concentration. Then, a mechanism is suggested based on which reactants affect the rate.

PROBLEM 6.18   For each rate equation, what effect does the indicated concentration change have on the overall rate of the reaction?

(1) rate = $k[CH_3CH_2Br][^-OH]$
    a. tripling the concentration of CH_3CH_2Br only
    b. tripling the concentration of ^-OH only
    c. tripling the concentration of both CH_3CH_2Br and ^-OH

(2) rate = $k[(CH_3)_3COH]$
    a. doubling the concentration of $(CH_3)_3COH$
    b. increasing the concentration of $(CH_3)_3COH$ by a factor of 10

PROBLEM 6.19   Write a rate equation for each reaction, given the indicated mechanism.

a. CH_3CH_2-Br + ^-OH $\longrightarrow$ $CH_2=CH_2$ + H_2O + Br^-

b. $(CH_3)_3C-Br$ $\xrightarrow{\text{slow}}$ $(CH_3)_3C^+$ $\xrightarrow[\text{fast}]{^-OH}$ $(CH_3)_2C=CH_2$ + H_2O
                                          + Br^-

6.10   Catalysts

Some reactions do not occur in a reasonable time unless a **catalyst** is added.

◆ A *catalyst* is a substance that speeds up the rate of a reaction. A catalyst is recovered unchanged in a reaction, and it does not appear in the product.

Common catalysts in organic reactions are **acids** and **metals.** Two examples are shown in the accompanying equations.

Figure 6.7   The effect of a catalyst on a reaction

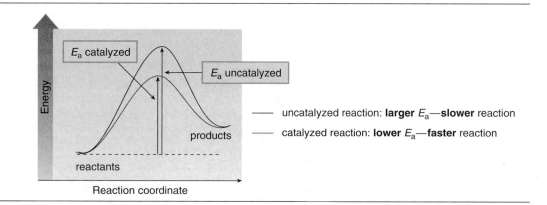

- The catalyst lowers the energy of activation, thus **increasing the rate of the catalyzed reaction.**
- The energy of the reactants and products is the same in both the uncatalyzed and catalyzed reactions, so the **position of equilibrium is unaffected.**

— uncatalyzed reaction: **larger** E_a—**slower** reaction
— catalyzed reaction: **lower** E_a—**faster** reaction

The reaction of acetic acid with ethanol to yield ethyl acetate and water occurs in the presence of an acid catalyst. The acid catalyst is written over or under the arrow to emphasize that it is not part of the starting materials or the products. The details of this reaction are discussed in Chapter 22.

The reaction of ethylene with hydrogen to form ethane occurs only in the presence of a metal catalyst such as palladium, platinum, or nickel. The metal provides a surface that binds both the ethylene and the hydrogen, and in doing so, facilitates the reaction. We return to this mechanism in Chapter 12.

Catalysts accelerate a reaction by lowering the energy of activation (Figure 6.7). They have no effect on the equilibrium constant, so they do not change the amount of reactant and product at equilibrium. Thus, catalysts affect how *quickly* equilibrium is achieved, but not the relative amounts of reactants and products at equilibrium. If a catalyst is somehow used up in one step of a reaction sequence, it must be regenerated in another step. Because only a small amount of a catalyst is needed relative to starting material, it is said to be present in a **catalytic amount.**

PROBLEM 6.20   Identify the catalyst in each equation.

a. $CH_2{=}CH_2 \xrightarrow[H_2SO_4]{H_2O} CH_3CH_2OH$ c.   (structure) $\xrightarrow[Pt]{H_2}$ (structure)

b. $CH_3Cl \xrightarrow[^-OH]{I^-} CH_3OH + Cl^-$

6.11   Enzymes

The catalysts that synthesize and break down biomolecules in living organisms are governed by the same principles as the acids and metals in organic reactions. The catalysts in living organisms, however, are usually protein molecules called **enzymes.**

◆ *Enzymes* are biochemical catalysts composed of amino acids held together in a very specific three-dimensional shape.

An enzyme contains a region called its **active site,** which binds an organic reactant, called a **substrate.** When bound, this unit is called the **enzyme–substrate complex** (Figure 6.8). Understanding the nature of the active sites of enzymes is often important in designing drugs to treat diseases. Once bound, the organic substrate undergoes a very specific reaction at an enhanced rate.

Figure 6.8 A schematic representation of an enzyme at work

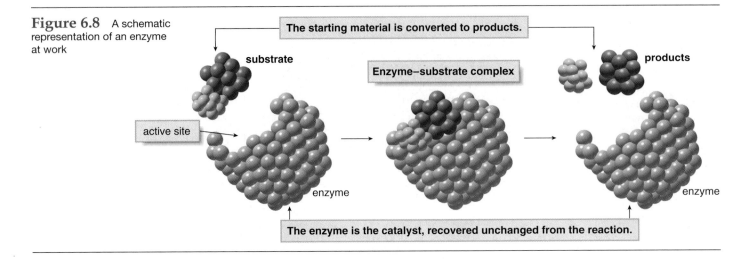

An enzyme speeds up a biological reaction in a variety of ways. It may hold reactants close in the proper conformation to facilitate reaction, or it may provide an acidic site needed for a particular transformation. Once the reaction is completed, the enzyme releases the substrate and it is then able to catalyze another reaction.

6.12 Key Concepts—Understanding Organic Reactions

Writing Equations for Organic Reactions (6.1)

- Use curved arrows to show the movement of electrons. Full-headed arrows are used for electron pairs and half-headed arrows are used for single electrons.
- Reagents can be drawn either on the left side of an equation or over the reaction arrow. Catalysts are drawn over or under the reaction arrow.

Types of Reactions (6.2)

[1] Substitution

$$-\overset{|}{\underset{|}{C}}-Z \ + \ Y \longrightarrow -\overset{|}{\underset{|}{C}}-Y \ + \ Z \qquad \left[Z = \text{H or a heteroatom} \right]$$

Y replaces Z

[2] Elimination

$$-\overset{|}{\underset{X}{C}}-\overset{|}{\underset{Y}{C}}- \ + \ \text{reagent} \longrightarrow \ \overset{\backslash}{\underset{/}{C}}=\overset{/}{\underset{\backslash}{C}} \ + \ X-Y$$

Two σ bonds are broken.   π bond

[3] Addition

$$\overset{\backslash}{\underset{/}{C}}=\overset{/}{\underset{\backslash}{C}} \ + \ X-Y \longrightarrow -\overset{|}{\underset{X}{C}}-\overset{|}{\underset{Y}{C}}-$$

This π bond is broken.   Two σ bonds are formed.

Important Trends

Values compared	Trend
Bond dissociation energy and bond strength	The **higher** the bond dissociation energy, the **stronger** the bond (6.4).
Energy and stability	The **higher** the energy, the **less stable** the species (6.5A).
E_a and reaction rate	The **larger** the energy of activation, the **slower** the reaction (6.9A).
E_a and rate constant	The **larger** the energy of activation, the **smaller** the rate constant (6.9B).

Reactive Intermediates (6.3)

- Breaking bonds generates reactive intermediates.
- Homolysis generates radicals with unpaired electrons.
- Heterolysis generates ions.

Reactive intermediate	General structure	Reactive feature	Reactivity
Radical	—C·	Unpaired electron	Electrophilic
Carbocation	—C$^+$	Positive charge; only six electrons around C	Electrophilic
Carbanion	—C:$^-$	Net negative charge; lone electron pair on C	Nucleophilic

Energy Diagrams (6.7, 6.8)

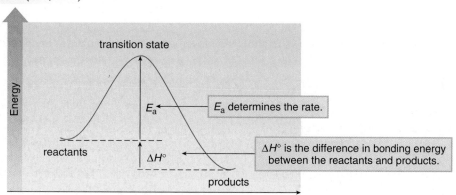

transition state

Energy

E_a

E_a determines the rate.

reactants

$\Delta H°$

$\Delta H°$ is the difference in bonding energy between the reactants and products.

products

Reaction coordinate

Conditions Favoring Product Formation (6.5, 6.6)

Variable	Value	Meaning
K_{eq}	$K_{eq} > 1$	More products than reactants are present at equilibrium.
$\Delta G°$	$\Delta G° < 0$	The free energy of the products is **lower** than the energy of the reactants.
$\Delta H°$	$\Delta H° < 0$	Bonds in the products are **stronger** than bonds in the reactants.
$\Delta S°$	$\Delta S° > 0$	The products are **more disordered** than the reactants.

Equations (6.5, 6.6)

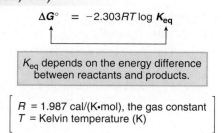

$$\Delta G° = -2.303 RT \log K_{eq}$$

K_{eq} depends on the energy difference between reactants and products.

$$\left[\begin{array}{l} R = 1.987 \text{ cal/(K·mol), the gas constant} \\ T = \text{Kelvin temperature (K)} \end{array} \right]$$

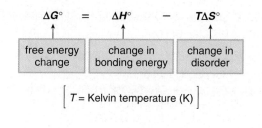

$$\Delta G° = \Delta H° - T\Delta S°$$

free energy change

change in bonding energy

change in disorder

$$\left[T = \text{Kelvin temperature (K)} \right]$$

Factors Affecting Reaction Rate (6.9)

Factor	Effect
Energy of activation	Larger E_a ---→ slower reaction
Concentration	Higher concentration ---→ faster reaction
Temperature	Higher temperature ---→ faster reaction

Problems

Types of Reactions

6.21 Classify each transformation as substitution, elimination, or addition.

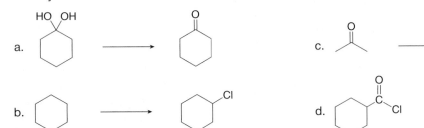

Bond Cleavage

6.22 Draw the products of homolysis or heterolysis of each indicated bond. Use electronegativity differences to decide on the location of charges in heterolysis reactions. Classify each carbon reactive intermediate as a radical, carbocation, or carbanion.

a. homolysis of CH_3–$\overset{\overset{H}{|}}{\underset{\underset{H}{|}}{C}}$–H

b. heterolysis of CH_3–O–H

c. heterolysis of CH_3–MgBr

Curved Arrows

6.23 Use full-headed or half-headed curved arrows to show the movement of electron pairs in each reaction.

a. (cyclohexyl)–Br ⟶ (cyclohexyl)$^+$ + Br$^-$

d. ⟶ (isopropyl with Br) + Br·, + Br$_2$

b. CH_3–$\overset{\overset{O^-}{|}}{\underset{\underset{Cl}{|}}{C}}$–$CH_3$ ⟶ $\underset{CH_3}{\overset{O}{\overset{||}{C}}}\underset{}{CH_3}$ + Cl$^-$

e. CH_3CH_2Br + ^-OH ⟶ CH_3CH_2OH + Br$^-$

c. ·CH_3 + ·Cl ⟶ CH_3–Cl

f. CH_3–$\overset{\overset{CH_3}{|}}{\underset{\underset{H}{|}}{\overset{+}{C}}}$–$\overset{\overset{H}{|}}{\underset{\underset{H}{|}}{C}}$–H + ^-OH ⟶ $\underset{CH_3}{\overset{CH_3}{}}C=C\underset{H}{\overset{H}{}}$ + H_2O

6.24 Draw the products of each reaction by following the curved arrows.

a. (cyclohexyl)–I + ^-OH ⟶

c. HO$^-$ H–$\overset{\overset{H}{|}}{\underset{\underset{H}{|}}{C}}$–$\overset{\overset{H}{|}}{\underset{\underset{Br}{|}}{C}}$–H ⟶

b. CH_3–$\overset{\overset{O^-}{|}}{\underset{\underset{OCH_2CH_3}{}}{C}}$–$CH_2CH_2CH_3$ ⟶

d. (phenyl)–$\overset{\overset{H}{|}}{\underset{\underset{H}{|}}{C}}$–H + ·Cl ⟶

Bond Dissociation Energy and Calculating $\Delta H°$

6.25 Rank each of the indicated bonds in order of increasing bond dissociation energy.

a. Cl–CCl_3, I–CCl_3, Br–CCl_3

b. N≡N, HN=NH, H_2N–NH_2

6.26 Calculate $\Delta H°$ for each reaction.

a. CH_3CH_3 + Br_2 ⟶ CH_3CH_2Br + HBr

b. HO· + CH_4 ⟶ ·CH_3 + H_2O

c. CH_3OH + HBr ⟶ CH_3Br + H_2O

d. Br· + CH_4 ⟶ H· + CH_3Br

6.27 Explain why the bond dissociation energy for the C–C σ bond in propane is lower than the bond dissociation energy for the C–C σ bond in propene, $CH_3CH=CH_2$.

CH_3–CH_2CH_3

$\Delta H° = 85$ kcal/mol

CH_3–$CH=CH_2$

$\Delta H° = 92$ kcal/mol

6.28 Homolysis of the indicated C–H bond in propene forms a resonance-stabilized free radical.

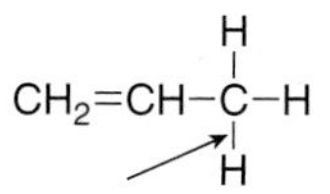

$$CH_2=CH-\overset{\overset{\displaystyle H}{|}}{\underset{\underset{\displaystyle H}{|}}{C}}-H$$

    a. Draw the two possible resonance structures for this radical.
    b. Use half-headed curved arrows to illustrate how one resonance structure can be converted to the other.
    c. Draw a structure for the resonance hybrid.

6.29 Because propane ($CH_3CH_2CH_3$) has both 1° and 2° carbon atoms, it has two different types of C–H bonds.
    a. Draw the carbon radical formed by homolysis of each of these C–H bonds.
    b. Use the values in Table 6.2 to determine which C–H bond is stronger.
    c. Explain how this information can also be used to determine the relative stability of the two radicals formed. Which radical formed from propane is more stable?

6.30 Use the bond dissociation energies in Table 1.3 (listed as bond strengths) to estimate the strength of the σ and π components of the double bond in ethylene.

Thermodynamics, $\Delta G°$, $\Delta H°$, $\Delta S°$, and K_{eq}

6.31 As we learned in Chapter 4, monosubstituted cyclohexanes exist as an equilibrium mixture of two conformers having either an axial or equatorial substituent.

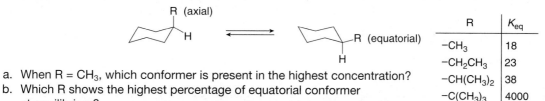

R	K_{eq}
$-CH_3$	18
$-CH_2CH_3$	23
$-CH(CH_3)_2$	38
$-C(CH_3)_3$	4000

    a. When R = CH_3, which conformer is present in the highest concentration?
    b. Which R shows the highest percentage of equatorial conformer at equilibrium?
    c. Which R shows the highest percentage of axial conformer at equilibrium?
    d. For which R is $\Delta G°$ most negative?
    e. How is the size of R related to the amount of axial and equatorial conformers at equilibrium?
    f. Challenge question: Explain why three monosubstituted cycloalkanes [R = $-CH_3$, $-CH_2CH_3$, $-CH(CH_3)_2$] have similar values of K_{eq}, but K_{eq} for *tert*-butylcyclohexane [R = $-C(CH_3)_3$] is much higher.

6.32 Given each value, determine whether the starting material or product is favored at equilibrium.
    a. $K_{eq} = 0.5$ d. $K_{eq} = 16$ g. $\Delta S° = 2$ cal/K•mol
    b. $\Delta G° = -25$ kcal/mol   e. $\Delta G° = 0.5$ kcal/mol   h. $\Delta S° = -2$ cal/K•mol
    c. $\Delta H° = 2.0$ kcal/mol   f. $\Delta H° = 100$ kcal/mol

6.33 a. Which value corresponds to a negative value of $\Delta G°$: $K_{eq} = 10^{-2}$ or $K_{eq} = 10^2$?
    b. In a unimolecular reaction with five times as much starting material as product at equilibrium, what is the value of K_{eq}? Is $\Delta G°$ positive or negative?
    c. Which value corresponds to a larger K_{eq}: $\Delta G° = -2$ kcal/mol or $\Delta G° = 5$ kcal/mol?

6.34 For which of the following reactions is $\Delta S°$ a positive value?

    a. 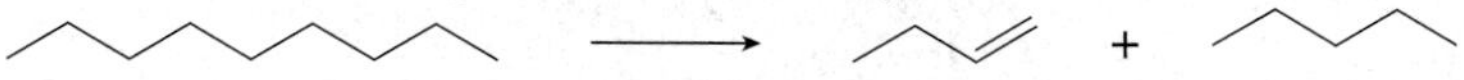 ![zigzag structure] ⟶ ![alkene] + ![alkane]

    b. $CH_3•$ + $CH_3•$ ⟶ CH_3CH_3

    c. $(CH_3)_2C(OH)_2$ ⟶ $(CH_3)_2C=O$ + H_2O

    d. CH_3COOCH_3 + H_2O ⟶ CH_3COOH + CH_3OH

Energy Diagrams and Transition States

6.35 Draw the transition state for each reaction.

    a. 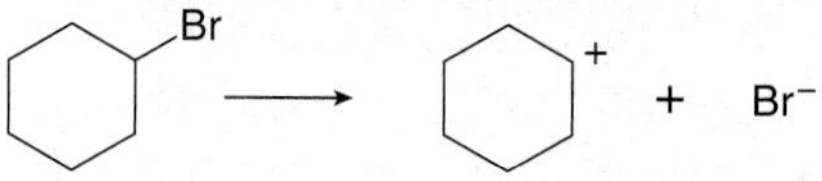 ![cyclohexyl bromide] ⟶ ![cyclohexyl cation]$^+$ + Br^- c. 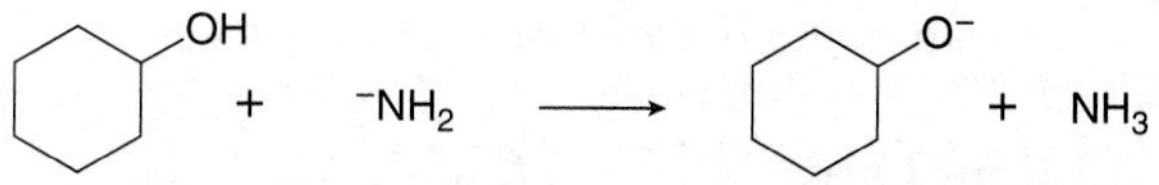 ![cyclohexanol] OH + $^-NH_2$ ⟶ ![cyclohexanolate] O^- + NH_3

    b. BF_3 + Cl^- ⟶ $F-\overset{\overset{\displaystyle F}{|}}{\underset{\underset{\displaystyle F}{|}}{B}}-Cl$ d. 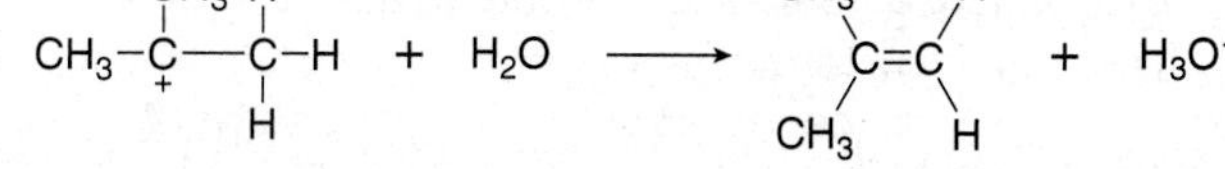 $CH_3-\overset{\overset{\displaystyle CH_3}{|}}{\underset{}{\overset{+}{C}}}-\overset{\overset{\displaystyle H}{|}}{\underset{\underset{\displaystyle H}{|}}{C}}-H$ + H_2O ⟶ $\overset{\displaystyle CH_3}{\underset{\displaystyle CH_3}{}}C=C\overset{\displaystyle H}{\underset{\displaystyle H}{}}$ + H_3O^+

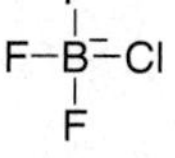

6.36 Draw an energy diagram for each reaction. Label the axes, the starting material, product, transition state, $\Delta H°$, and E_a.
    a. A concerted, exothermic reaction with a low energy of activation.
    b. A one-step endothermic reaction with a high energy of activation.
    c. A two-step reaction, **A → B → C**, in which the relative energy of the compounds is **A < C < B**, and the step **A → B** is rate-determining.
    d. A concerted reaction with $\Delta H° = -20$ kcal/mol and $E_a = 4$ kcal.

6.37 Consider the following reaction: $CH_4 + Cl\bullet \rightarrow \bullet CH_3 + HCl$.
   a. Use curved arrows to show the movement of electrons.
   b. Calculate $\Delta H°$ using the bond dissociation energies in Table 6.2.
   c. Draw an energy diagram assuming that $E_a = 4$ kcal.
   d. What is E_a for the reverse reaction ($\bullet CH_3 + HCl \rightarrow CH_4 + Cl\bullet$)?

6.38 Consider the following energy diagram for the conversion of **A → G.**

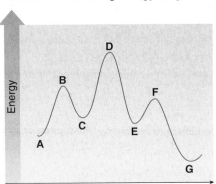

   a. Which points on the graph correspond to transition states?
   b. Which points on the graph correspond to reactive intermediates?
   c. How many steps are present in the reaction mechanism?

6.39 Consider the following two-step reaction:

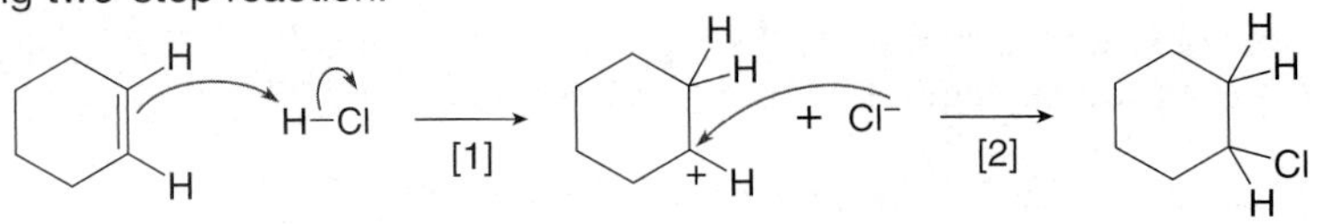

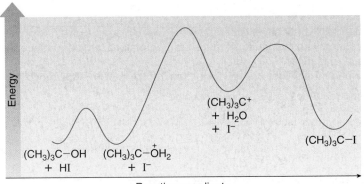

   a. How many bonds are broken and formed in Step [1]? Would you predict $\Delta H°$ of Step [1] to be positive or negative?
   b. How many bonds are broken and formed in Step [2]? Would you predict the $\Delta H°$ of Step [2] to be positive or negative?
   c. Which step is rate-determining?
   d. Draw the structure for the transition state in both steps of the mechanism.
   e. If $\Delta H°_{overall}$ is negative for this two-step reaction, draw an energy diagram illustrating all of the information in a–d.

6.40 Consider the following energy diagram for the overall reaction: $(CH_3)_3COH + HI \rightarrow (CH_3)_3CI + H_2O$.

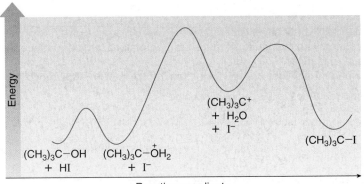

   a. How many steps are in the reaction mechanism?
   b. Label the E_a and $\Delta H°$ for each step, and the $\Delta H°_{overall}$ for the reaction.
   c. Draw the structure of the transition state for each step and indicate its location on the energy diagram.
   d. Which step is rate-determining? Why?

Kinetics and Rate Laws

6.41 Indicate which factors affect the rate of a reaction.
   a. $\Delta G°$ d. temperature   g. k
   b. $\Delta H°$ e. concentration   h. catalysts
   c. E_a f. K_{eq} i. $\Delta S°$

6.42 The following is a concerted, bimolecular reaction: $CH_3Br + NaCN \rightarrow CH_3CN + NaBr$.
   a. What is the rate equation for this reaction?
   b. What happens to the rate of the reaction if $[CH_3Br]$ is doubled?
   c. What happens to the rate of the reaction if $[NaCN]$ is halved?
   d. What happens to the rate of the reaction if $[CH_3Br]$ and $[NaCN]$ are both increased by a factor of five?

6.43 The conversion of acetyl chloride to methyl acetate occurs via the following two-step mechanism:

a. Write the rate equation for this reaction, assuming the first step is rate-determining.
b. If the concentration of $^-OCH_3$ were increased 10 times, what would happen to the rate of the reaction?
c. If the concentrations of both CH_3COCl and $^-OCH_3$ were increased 10 times, what would happen to the rate of the reaction?
d. Classify the conversion of acetyl chloride to methyl acetate as an addition, elimination, or substitution.

6.44 Label each statement as true or false. Correct any false statement to make it true.
a. Increasing temperature increases reaction rate.
b. If a reaction is fast, it has a large rate constant.
c. A fast reaction has a large negative ΔG° value.
d. When E_a is large, the rate constant k is also large.
e. Fast reactions have equilibrium constants > 1.
f. Increasing the concentration of a reactant always increases the rate of a reaction.

6.45 The conversion of $(CH_3)_3CI$ to $(CH_3)_2C=CH_2$ can occur by either a one-step or a two-step mechanism, as shown in Equations [1] and [2].

a. What rate equation would be observed for the mechanism in Equation [1]?
b. What rate equation would be observed for the mechanism in Equation [2]?
c. What is the order of each rate equation (i.e., first, second, and so forth)?
d. How can these rate equations be used to show which mechanism is the right one for this reaction?

Challenge Problems

6.46 Explain why $HC\equiv CH$ is more acidic than CH_3CH_3, even though the $C-H$ bond in $HC\equiv CH$ has a higher bond dissociation energy than the $C-H$ bond in CH_3CH_3.

6.47 Esterification is the reaction of a carboxylic acid (RCOOH) with an alcohol (R'OH) to form an ester (RCOOR') with loss of water. Equation [1] is an example of an *intermolecular* esterification reaction. Equation [2] is an example of an *intramolecular* esterification reaction; that is, the carboxylic acid and alcohol are contained in the same starting material, forming a cyclic ester as product. The equilibrium constants for both reactions are given. Explain why K_{eq} is different for these two apparently similar reactions.

6.48 Although K_{eq} of Equation [1] in Problem 6.47 does not greatly favor formation of the product, it is sometimes possible to use Le Châtelier's principle to increase the yield of ethyl acetate. Le Châtelier's principle states that if an equilibrium is disturbed, a system will react to counteract this disturbance. How can Le Châtelier's principle be used to drive the equilibrium to increase the yield of ethyl acetate? Another example of Le Châtelier's principle is given in Section 9.8.

6.49 ΔH° for the $O-O$ bond in H_2O_2 is lower than for many other bonds involving second-row elements ($\Delta H^\circ = 51$ kcal/mol). For example, ΔH° for the $C-C$ bond in CH_3CH_3 is 88 kcal/mol and ΔH° for the $C-O$ bond in CH_3OH is 91 kcal/mol. Offer an explanation.

CHAPTER 7

Alkyl Halides and Nucleophilic Substitution

Adrenaline (or **epinephrine**), a hormone secreted by the adrenal gland, increases blood pressure and heart rate, and dilates lung passages. Individuals often speak of the "rush of adrenaline" when undertaking a particularly strenuous or challenging activity. Adrenaline is made in the body by a simple organic reaction called **nucleophilic substitution.** In Chapter 7 you will learn about the mechanism of nucleophilic substitution and how adrenaline is synthesized in organisms.

T his is the first of three chapters dealing with an in-depth study of the organic reactions of compounds containing C−Z σ bonds, where Z is an element more electronegative than carbon. In Chapter 7 we learn about **alkyl halides** and one of their characteristic reactions, **nucleophilic substitution.** In Chapter 8, we look at **elimination,** a second general reaction of alkyl halides. We conclude this discussion in Chapter 9 by examining other molecules that also undergo nucleophilic substitution and elimination reactions.

7.1   Introduction to Alkyl Halides

Alkyl halides are organic molecules containing a halogen atom X bonded to an sp^3 hybridized carbon atom. Alkyl halides are classified as **primary (1°), secondary (2°),** or **tertiary (3°)** depending on the number of carbons bonded to the carbon with the halogen.

Alkyl halides have the general molecular formula $C_nH_{2n+1}X$, and are formally derived from an alkane by replacing a hydrogen atom with a halogen.

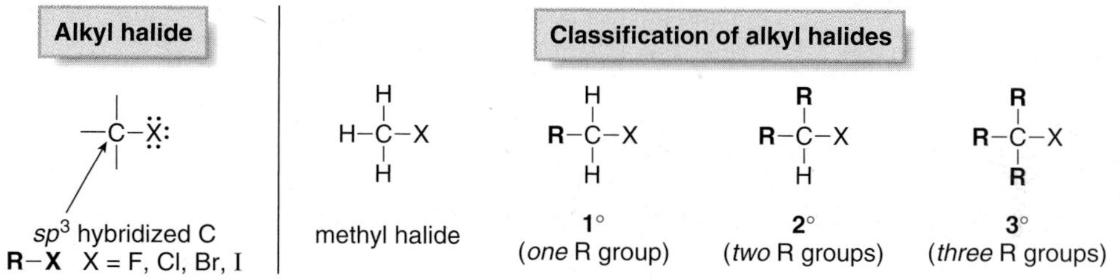

Alkyl halide		Classification of alkyl halides		

sp^3 hybridized C
R−X   X = F, Cl, Br, I

methyl halide

1°
(*one* R group)

2°
(*two* R groups)

3°
(*three* R groups)

Whether an alkyl halide is 1°, 2°, or 3° is the *most important factor* in determining the course of its chemical reactions. Figure 7.1 illustrates three examples.

Four types of organic halides having the halogen atom in close proximity to a π bond are illustrated in Figure 7.2. **Vinyl halides** have a halogen atom bonded to a carbon–carbon double bond, and **aryl halides** have a halogen atom bonded to a benzene ring. These two types of organic halides with X directly bonded to an sp^2 hybridized carbon atom do **not** undergo the reactions presented in Chapter 7, as discussed in Section 7.18.

Allylic halides and benzylic halides have halogen atoms bonded to sp^3 hybridized carbon atoms and **do** undergo the reactions described in Chapter 7. **Allylic halides** have X bonded to the carbon atom *adjacent* to a carbon–carbon double bond, and **benzylic halides** have X bonded to the carbon atom *adjacent* to a benzene ring. The synthesis of allylic and benzylic halides is discussed in Sections 13.10 and 18.13, respectively.

Figure 7.1   Examples of 1°, 2°, and 3° alkyl halides

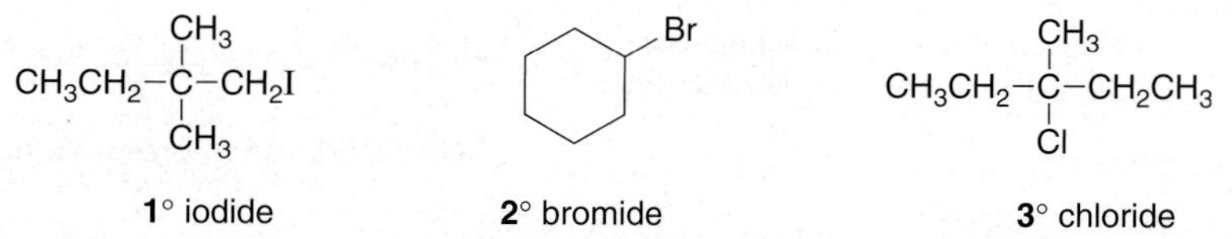

$$CH_3CH_2-\overset{\overset{\displaystyle CH_3}{|}}{\underset{\underset{\displaystyle CH_3}{|}}{C}}-CH_2I$$

1° iodide

(cyclohexane with Br)

2° bromide

$$CH_3CH_2-\overset{\overset{\displaystyle CH_3}{|}}{\underset{\underset{\displaystyle Cl}{|}}{C}}-CH_2CH_3$$

3° chloride

Figure 7.2   Four types of organic halides (RX) having X near a π bond

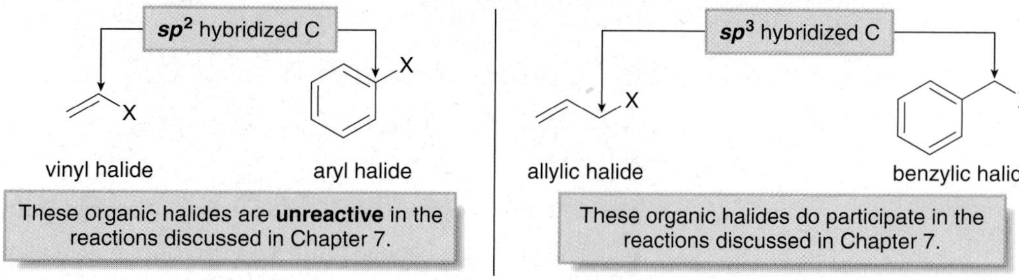

sp^2 hybridized C

sp^3 hybridized C

vinyl halide   aryl halide

allylic halide   benzylic halide

These organic halides are **unreactive** in the reactions discussed in Chapter 7.

These organic halides do participate in the reactions discussed in Chapter 7.

PROBLEM 7.1   Classify each alkyl halide as 1°, 2°, or 3°.

a. $CH_3CH_2CH_2CH_2CH_2-Br$ b.   c. $CH_3-\overset{\overset{\displaystyle CH_3}{|}}{\underset{\underset{\displaystyle CH_3}{|}}{C}}-\overset{\overset{\displaystyle Cl}{|}}{CHCH_3}$ d.

7.2   Nomenclature

The systematic (IUPAC) method for naming alkyl halides follows from the basic rules described in Chapter 4. The common method is also discussed in Section 7.2B, because many low molecular weight alkyl halides are often referred to by their common names.

7.2A   IUPAC System

An alkyl halide is named as an alkane with a halogen substituent—that is, as a **halo alkane.** To name a halogen substituent, change the **-ine** ending of the name of the halogen to the suffix **-o** (chlor*ine* → chlor*o*).

How To	Name an Alkyl Halide Using the IUPAC System

Example   Give the IUPAC name of the following alkyl halide:

$$\underset{\underset{}{}}{CH_3CH_2\overset{\overset{\displaystyle CH_3}{|}}{CH}CH_2CH_2\overset{\overset{\displaystyle Cl}{|}}{CH}CH_3}$$

Step [1]   **Find the parent carbon chain containing the halogen.**

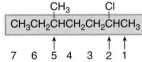

7 C's in the longest chain

7 C's ⟶ heptane

- Name the parent chain as an **alkane,** with the halogen as a substituent bonded to the longest chain.

Step [2]   **Apply all other rules of nomenclature.**

[a] **Number** the chain.

$$\underset{7\quad 6\quad 5\quad 4\quad 3\quad 2\quad 1}{CH_3CH_2\overset{\overset{\displaystyle CH_3}{|}}{CH}CH_2CH_2\overset{\overset{\displaystyle Cl}{|}}{CH}CH_3}$$

- Begin at the end nearest the first substituent, either alkyl or halogen.

[b] **Name** and **number** the substituents.

methyl at C5   **chloro at C2**

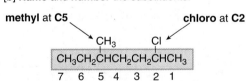

7   6   5   4   3   2   1

[c] **Alphabetize: c** for **chloro,** then **m** for **methyl.**

ANSWER: 2-chloro-5-methylheptane

7.2B   Common Names

Common names for alkyl halides are used only for simple alkyl halides. To assign a common name:

- Name all the carbon atoms of the molecule as a single **alkyl group.**
- Name the halogen bonded to the alkyl group. To name the halogen, change the **-ine** ending of the halogen name to the suffix **-ide;** for example, **brom*ine* → brom*ide.***

Figure 7.3 Examples: Nomenclature of alkyl halides

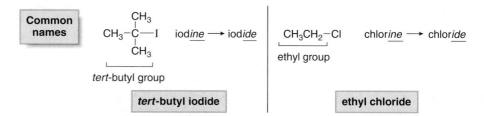

IUPAC: **1-chloro-2-methylpropane**
Common: isobutyl chloride

IUPAC: **1-ethyl-2-fluorocyclopentane**

• ethyl group at **C1**
• fluoro group at **C2**

earlier letter ⟶ lower number

[too complex to use a common name]

◆ Combine the names of the alkyl group and halide, separating the words with a space.

Common names

$CH_3-\underset{\underset{CH_3}{|}}{\overset{\overset{CH_3}{|}}{C}}-I$ iod*ine* ⟶ iod*ide*

tert-butyl group

tert-butyl iodide

CH_3CH_2-Cl chlor*ine* ⟶ chlor*ide*

ethyl group

ethyl chloride

Other examples of alkane nomenclature are given in Figure 7.3.

PROBLEM 7.2   Give the IUPAC name for each compound.

a. $(CH_3)_2CHCH(Cl)CH_2CH_3$ b.   c.   d.

PROBLEM 7.3   Give the structure corresponding to each name.
a. 3-chloro-2-methylhexane
b. 4-ethyl-5-iodo-2,2-dimethyloctane
c. *cis*-1,3-dichlorocyclopentane
d. 1,1,3-tribromocyclohexane
e. propyl chloride
f. *sec*-butyl bromide

7.3 Physical Properties

Alkyl halides are weakly polar molecules. They exhibit **dipole–dipole** interactions because of their polar C−X bond, but because the rest of the molecule contains only C−C and C−H bonds they are incapable of intermolecular hydrogen bonding. How this affects their physical properties is summarized in Table 7.1.

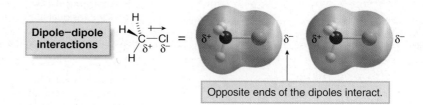

Opposite ends of the dipoles interact.

PROBLEM 7.4   Rank the compounds in each group in order of increasing boiling point.
a. $CH_3CH_2CH_2I$, $CH_3CH_2CH_2Cl$, $CH_3CH_2CH_2F$
b. $CH_3(CH_2)_4CH_3$, $CH_3(CH_2)_5Br$, $CH_3(CH_2)_5OH$

TABLE 7.1 Physical Properties of Alkyl Halides	
Property	**Observation**
Boiling point and melting point	• Alkyl halides have higher bp's and mp's than alkanes having the same number of C carbons. CH_3CH_3 and CH_3CH_2Br bp = −89 °C bp = 39 °C • Bp's and mp's increase as the size of R increases. CH_3CH_2Cl and $CH_3CH_2CH_2Cl$ ← **larger surface area— higher mp and bp** mp = −136 °C mp = −123 °C bp = 12 °C bp = 47 °C • Bp's and mp's increase as the size of X increases. CH_3CH_2Cl and CH_3CH_2Br ← **more polarizable halogen— higher mp and bp** mp = −136 °C mp = −119 °C bp = 12 °C bp = 39 °C
Solubility	• RX is soluble in organic solvents. • RX is insoluble in water.

7.4 Interesting Alkyl Halides

Many simple alkyl halides make excellent solvents because they are not flammable and dissolve a wide variety of organic compounds. Compounds in this category include $CHCl_3$ (chloroform or trichloromethane) and CCl_4 (carbon tetrachloride or tetrachloromethane). Large quantities of these solvents are produced industrially each year, but like many chlorinated organic compounds, both chloroform and carbon tetrachloride are toxic if inhaled or ingested. Other simple alkyl halides are shown in Figure 7.4.

Synthetic organic halides are also used in insulating materials, plastic wrap, and coatings. Two such compounds are **Teflon** and **polyvinyl chloride (PVC).**

Figure 7.4 Some simple alkyl halides

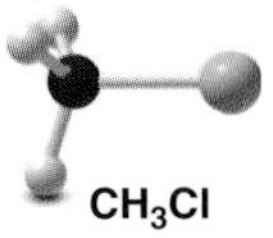

CH_3Cl

• **Chloromethane (CH₃Cl)** is produced by giant kelp and algae and also found in emissions from volcanoes such as Hawaii's Kilauea. Almost all of the atmospheric chloromethane results from these natural sources.

CH_2Cl_2

• **Dichloromethane (or methylene chloride, CH₂Cl₂)** is an important solvent, once used to decaffeinate coffee. Coffee is now decaffeinated by using supercritical CO_2 due to concerns over the possible ill effects of trace amounts of residual CH_2Cl_2 in the coffee. Subsequent studies on rats have shown, however, that no cancers occurred when animals ingested the equivalent of over 100,000 cups of decaffeinated coffee per day.

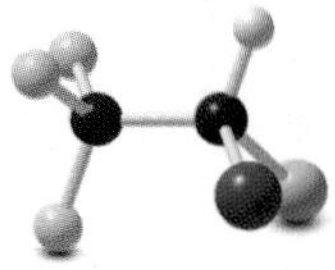

$CF_3CHClBr$

• **Halothane (CF₃CHClBr)** is a safe general anesthetic that has now replaced other organic anesthetics such as $CHCl_3$, which causes liver and kidney damage, and $CH_3CH_2OCH_2CH_3$ (diethyl ether), which is very flammable.

F F F F F F
‑C‑C‑C‑C‑C‑C‑
F F F F F F

H Cl H Cl H Cl
‑C‑C‑C‑C‑C‑C‑
H H H H H H

Teflon
(nonstick coating)

polyvinyl chloride (PVC)
(plastic used in films, pipes, and insulation)

Organic halides constitute a growing list of useful naturally occurring molecules, many produced by marine organisms. Some have irritating odors or an unpleasant taste and are synthesized by organisms for self-defense or feeding deterrents. Examples include $Br_2C{=}CHCHCl_2$ and $Br_2C{=}CHCHBr_2$, isolated from the red seaweed *Asparagopsis taxiformis,* known as *limu kohu* (supreme seaweed) in Hawaii. This seaweed has a strong and characteristic odor and flavor, in part probably because of these organic halides.

Although the beneficial effects of many organic halides are undisputed, certain synthetic chlorinated organics such as the **chlorofluorocarbons** and the pesticide **DDT** have caused lasting harm to the environment.

$CFCl_3$
CFC 11
Freon 11

Cl
DDT
H CCl₃

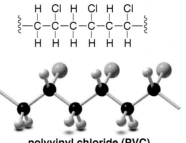

Chlorofluorocarbons (CFCs) have the general molecular structure CF_xCl_{4-x}. Trichlorofluoromethane [$CFCl_3$, CFC 11, or Freon 11 (trade name)] is an example of these easily vaporized compounds, having been extensively used as a refrigerant and an aerosol propellant. CFCs slowly rise to the stratosphere, where sunlight catalyzes their decomposition, a process that contributes to the destruction of the ozone layer, the thin layer of atmosphere that shields the earth's surface from harmful ultraviolet radiation as discussed in Section 13.9. Although it is now easy to second-guess the extensive use of CFCs, it is also easy to see why they were used so widely. **CFCs made refrigeration available to the general public.** Would you call your refrigerator a comfort or a necessity?

The story of the insecticide **DDT** (*d*ichloro*d*iphenyl*t*richloroethane) follows the same theme: DDT is an organic molecule with valuable short-term effects that has caused long-term problems. DDT kills insects that spread diseases such as malaria and typhus, and in controlling insect populations, DDT has saved millions of lives worldwide. DDT is a weakly polar and very stable organic compound, and so it (and compounds like it) persist in the environment for years. Because DDT is soluble in organic media, it accumulates in the fatty tissues of most animals. Most adults in the United States have low concentrations of DDT (or a degradation product of DDT) in their bodies. The long-term effect on humans is not known, but DDT has had a direct harmful effect on the eggshell formation of certain predator birds such as eagles and hawks.

PVC was used at one time for the plastic sheets used to store and display baseball cards and other archival papers. Residual acid from its preparation, however, seeped into the paper over time, eventually destroying it.

Asparagopsis taxiformis is an edible red seaweed that grows on the edges of reefs in areas of constant water motion. Almost 100 different organic halides have been isolated from this source.

Time Magazine, June 30, 1947.

Winston Churchill called DDT a "miraculous" chemical in 1945 for the many lives it saved during World War II. Fewer than 20 years later, as prolonged use of large quantities of this nonspecific pesticide led to harmful environmental effects, Rachel Carson called DDT the "elixir of death" in her book *Silent Spring.* DDT use was banned in the United States in 1973, but because of its effectiveness and low cost, it is still widely used to control insect populations in developing countries.

PROBLEM 7.5

Although nonpolar compounds tend to dissolve and remain in fatty tissues, polar substances are more water soluble, and more readily excreted into an environment where they may be degraded by other organisms. Explain why methoxychlor is more biodegradable than DDT.

H
CH_3O— —C— —OCH_3
CCl_3

methoxychlor

Figure 7.5 Electrostatic potential maps of four halomethanes (CH_3X)

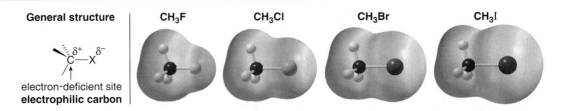

General structure

electron-deficient site
electrophilic carbon

- The polar C–X bond makes the carbon atom *electron deficient* in each CH_3X molecule.

7.5 The Polar Carbon–Halogen Bond

The properties of alkyl halides dictate their reactivity. The electrostatic potential maps of four simple alkyl halides in Figure 7.5 illustrate that the electronegative halogen X creates a polar C–X bond, making the carbon atom electron deficient. **The chemistry of alkyl halides is determined by this polar C–X bond.**

What kind of reactions do alkyl halides undergo? **The characteristic reactions of alkyl halides are substitution and elimination.** Because alkyl halides contain an electrophilic carbon, they react with electron-rich reagents—Lewis bases (nucleophiles) and Brønsted–Lowry bases.

◆ **Alkyl halides undergo substitution reactions with nucleophiles.**

$$R-X \;+\; :Nu^- \longrightarrow R-Nu \;+\; X:^-$$

nucleophile

substitution of X by Nu

In a substitution reaction of RX, **the halogen X is replaced by an electron-rich nucleophile :Nu⁻**. The C–X σ bond is broken and the C–Nu σ bond is formed.

◆ **Alkyl halides undergo elimination reactions with Brønsted–Lowry bases.**

new π bond
an alkene

base

elimination of HX

In an elimination reaction of RX, the **elements of HX are removed by a Brønsted–Lowry base :B.**

The remainder of Chapter 7 is devoted to a discussion of the substitution reactions of alkyl halides. Elimination reactions are discussed in Chapter 8.

7.6 General Features of Nucleophilic Substitution

Three components are necessary in any substitution reaction.

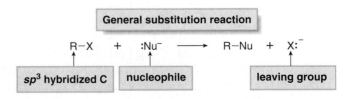

General substitution reaction

$$R-X \;+\; :Nu^- \longrightarrow R-Nu \;+\; X:^-$$

sp³ hybridized C   nucleophile   leaving group

[1] **R**—An alkyl group R containing an sp^3 hybridized carbon bonded to X.

[2] **X**—An atom X (or a group of atoms) called **a leaving group,** which is able to accept the electron density in the C–X bond. The most common leaving groups are halogen atoms ($-X$), but $-OH_2^+$ and $-N_2^+$ are also encountered.

[3] **:Nu⁻**—A **nucleophile.** Nucleophiles contain a lone pair or a π bond but not necessarily a negative charge.

Because these substitution reactions involve electron-rich nucleophiles, they are called *nucleophilic* **substitution reactions.** Examples are shown in Equations [1]–[3]. **Nucleophilic substitutions are Lewis acid–base reactions.** The nucleophile donates its electron pair, the alkyl halide (Lewis acid) accepts it, and the C–X bond is heterolytically cleaved. Curved arrow notation can be used to show the movement of electron pairs, as shown in Equation [3].

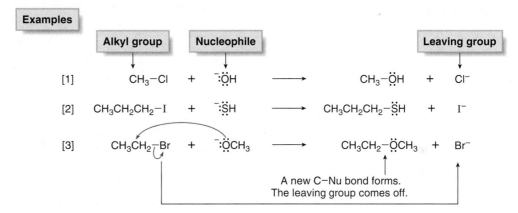

Negatively charged nucleophiles like ⁻OH and ⁻SH are used as **salts** with Li^+, Na^+, or K^+ counterions to balance charge. The identity of the cation is usually inconsequential, and therefore it is often omitted from the chemical equation.

$$CH_3CH_2\overset{\curvearrowleft}{C}H_2\!-\!Br \; + \; Na^+ \; {:}\ddot{O}H \longrightarrow CH_3CH_2CH_2\!-\!\ddot{O}H \; + \; Na^+Br^-$$

Na⁺ balances charge.

When a neutral nucleophile is used, the substitution product bears a positive charge. **Note that all atoms bonded to the nucleophile originally, stay bonded to it after substitution occurs.** All three CH_3 groups stay bonded to the N atom in the given example.

neutral nucleophile

$$CH_3CH_2\overset{\curvearrowleft}{C}H_2\!-\!Br \; + \; {:}N(CH_3)_3 \longrightarrow CH_3CH_2CH_2\!-\!\overset{+}{N}(CH_3)_3 \; + \; Br^-$$

All CH_3 groups remain in the product.

Furthermore, when the substitution product bears a positive charge and also contains a proton bonded to O or N, the initial substitution product readily loses a proton in a Brønsted–Lowry acid–base reaction, forming a neutral product.

$$CH_3CH_2\overset{\curvearrowleft}{C}H_2\!-\!Br \; + \; {:}NH_3 \longrightarrow CH_3CH_2CH_2\!-\!\overset{+}{\underset{H}{N}}\!-\!H \xrightarrow{\;:NH_3\;} CH_3CH_2CH_2\!-\!\overset{\cdot\cdot}{\underset{H}{N}}\!-\!H \; + \; NH_4^+$$

Step [1]:
nucleophilic substitution   + Br⁻   **Step [2]:**
proton transfer   The overall result:
a neutral product

All of these reactions are nucleophilic substitutions and have the same overall result—**replacement of the leaving group by the nucleophile,** regardless of the identity or charge of the nucleophile. To draw any nucleophilic substitution product:

◆ **Find the *sp*³ hybridized carbon** with the leaving group.

◆ **Identify the nucleophile,** the species with a lone pair or π bond.

◆ **Substitute the nucleophile for the leaving group** and assign charges (if necessary) to any atom that is involved in bond breaking or bond formation.

PROBLEM 7.6   Identify the nucleophile and leaving group and draw the products of each reaction.

a. [structure with Br] + ⁻OCH₂CH₃ ⟶   c. [structure with I] + N₃⁻ ⟶

b. [structure with Cl] + NaOH ⟶   d. [structure with Br] + NaCN ⟶

PROBLEM 7.7   Draw the product of nucleophilic substitution with each neutral nucleophile. When the initial substitution product can lose a proton to form a neutral product, draw that product as well.

a. [structure with Br] + :N(CH₂CH₃)₃ ⟶   c. [structure with Br] + CH₃ÖH ⟶

b. (CH₃)₃C—Cl + H₂Ö: ⟶

7.7 The Leaving Group

Nucleophilic substitution is a general reaction of organic compounds. Why, then, are alkyl halides the most common substrates, and halide anions the most common leaving groups? To answer this question, we must understand leaving group ability. **What makes a good leaving group?**

In a nucleophilic substitution reaction of R−X, the C−X bond is heterolytically cleaved, and the leaving group departs with the electron pair in that bond, forming X:⁻. **The more stable the leaving group X:⁻, the better able it is to accept an electron pair,** giving rise to the following generalization:

◆ In comparing two leaving groups, the better leaving group is the weaker base.

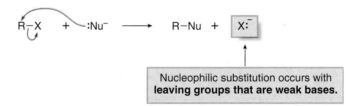

Nucleophilic substitution occurs with **leaving groups that are weak bases.**

For example, H₂O is a better leaving group than ⁻OH because H₂O is a weaker base. Moreover, the periodic trends in basicity listed in Section 2.9 can now be used to identify **periodic trends in leaving group ability:**

◆ Left-to-right across a row of the periodic table, basicity *decreases* so leaving group ability *increases.*

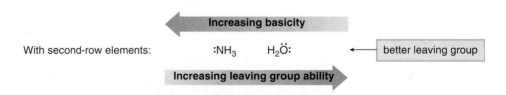

◆ Down a column of the periodic table, basicity *decreases* so leaving group ability *increases.*

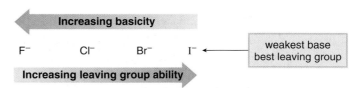

All good leaving groups are weak bases with strong conjugate acids having low pK_a values. Thus, all halide anions except F⁻ are good leaving groups because their conjugate acids (HCl, HBr, and HI) have low pK_a values. Tables 7.2 and 7.3 list good and poor leaving groups for nucleophilic substitution reactions, respectively. Nucleophilic substitution does not occur with any of the leaving groups in Table 7.3 because these leaving groups are strong bases.

TABLE 7.2 Good Leaving Groups for Nucleophilic Substitution

Starting material	Leaving group	Conjugate acid	pK_a
R—Cl	Cl⁻	HCl	−7
R—Br	Br⁻	HBr	−9
R—I	I⁻	HI	−10
R—OH₂⁺	H₂O	H₃O⁺	−1.7

These molecules undergo nucleophilic substitution.

good leaving groups

TABLE 7.3 Poor Leaving Groups for Nucleophilic Substitution

Starting material	Leaving group	Conjugate acid	pK_a
R—F	⁻F	HF	3.2
R—OH	⁻OH	H₂O	15.7
R—NH₂	⁻NH₂	NH₃	38
R—H	H⁻	H₂	35
R—R	R⁻	RH	50

These molecules do *not* undergo nucleophilic substitution.

poor leaving groups

PROBLEM 7.8 Which is the better leaving group in each pair?
a. Cl⁻, I⁻   b. NH₃, ⁻NH₂   c. H₂O, H₂S

PROBLEM 7.9 Which molecules contain good leaving groups?
a. $CH_3CH_2CH_2Br$ b. $CH_3CH_2CH_2OH$ c. $CH_3CH_2\overset{+}{O}H_2$ d. CH_3CH_3

Given a particular nucleophile and leaving group, how can we determine whether the equilibrium will favor products in a nucleophilic substitution? We can often correctly predict the direction of equilibrium by comparing the basicity of the nucleophile and the leaving group.

◆ **Equilibrium favors the products of nucleophilic substitution when the leaving group is a weaker base than the nucleophile.**

Sample Problem 7.1 illustrates how to apply this general rule.

SAMPLE PROBLEM 7.1   Will the following substitution reaction favor formation of the products?

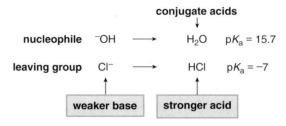

$$CH_3CH_2-Cl \ + \ :\overset{-}{O}H \ \longrightarrow \ CH_3CH_2-OH \ + \ Cl^-$$

SOLUTION

Compare the basicity of the nucleophile ($^-$OH) and the leaving group (Cl$^-$) by comparing the pK_a values of their conjugate acids. **The stronger the conjugate acid, the weaker the base, and the better the leaving group.**

conjugate acids

nucleophile	$^-$OH	$\longrightarrow$	H_2O	pK_a = 15.7
leaving group	Cl$^-$	$\longrightarrow$	HCl	pK_a = -7

weaker base   stronger acid

ANSWER

Because Cl$^-$, the leaving group, is a weaker base than $^-$OH, the nucleophile, the reaction favors the products.

PROBLEM 7.10   Does the equilibrium favor the reactants or products in each substitution reaction?

a. $CH_3CH_2-NH_2 \ + \ Br^- \ \longrightarrow \ CH_3CH_2-Br \ + \ ^-NH_2$

b.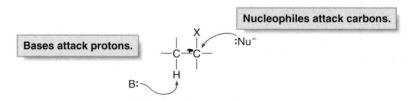

7.8   The Nucleophile

Nucleophiles and bases are structurally similar: both have a lone pair or a π bond. They differ in what they attack.

> We use the word *base* to mean *Brønsted–Lowry* base and the word *nucleophile* to mean a *Lewis base* that reacts with electrophiles *other than protons.*

♦ Bases attack protons. Nucleophiles attack other electron-deficient atoms (usually carbons).

Bases attack protons.

Nucleophiles attack carbons.

7.8A   Nucleophilicity Versus Basicity

> Although nucleophilicity and basicity are interrelated, they are fundamentally different. **Basicity** is a measure of how readily an atom donates its electron pair to a proton; it is characterized by an equilibrium constant K_a in an acid–base reaction, making it a **thermodynamic property. Nucleophilicity** is a measure of how readily an atom donates its electron pair to other atoms; it is characterized by the rate constant, *k*, of a nucleophilic substitution reaction, making it a **kinetic property.**

How is **nucleophilicity** (nucleophile strength) related to **basicity?** Although it is generally true that **a strong base is a strong nucleophile,** nucleophile size and steric factors can sometimes change this relationship.

Nucleophilicity parallels basicity in three instances:

[1]   **For two nucleophiles with the same nucleophilic atom, the stronger base is the stronger nucleophile.**

♦ The relative nucleophilicity of $^-$OH and CH_3COO^-, two oxygen nucleophiles, is determined by comparing the pK_a values of their conjugate acids (H_2O and CH_3COOH). CH_3COOH (pK_a = 4.8) is a stronger acid than H_2O (pK_a = 15.7), so $^-$OH is a stronger base and stronger nucleophile than CH_3COO^-.

[2]   **A negatively charged nucleophile is always stronger than its conjugate acid.**

♦ $^-$OH is a stronger base and stronger nucleophile than H_2O, its conjugate acid.

[3] Right-to-left across a row of the periodic table, nucleophilicity increases as basicity increases.

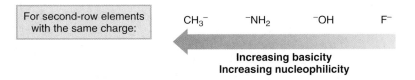

| For second-row elements with the same charge: | CH_3^- | $^-NH_2$ | ^-OH | F^- |

Increasing basicity
Increasing nucleophilicity

PROBLEM 7.11   Identify the stronger nucleophile in each pair.
a. NH_3, $^-NH_2$ b. CH_3^-, HO^- c. CH_3NH_2, CH_3OH d. CH_3COO^-, $CH_3CH_2O^-$

7.8B   Steric Effects and Nucleophilicity

All steric effects arise because two atoms cannot occupy the same space. In Chapter 4, for example, we learned that **steric strain** is an increase in energy when big groups (occupying a large volume) are forced close to each other.

Nucleophilicity does not parallel basicity when **steric hindrance** becomes important. *Steric hindrance* **is a decrease in reactivity resulting from the presence of bulky groups at the site of a reaction.**

For example, although pK_a tables indicate that *tert*-butoxide [$(CH_3)_3CO^-$], is a stronger base than ethoxide ($CH_3CH_2O^-$), **ethoxide is the *stronger* nucleophile.** The three CH_3 groups around the O atom of *tert*-butoxide create steric hindrance, making it more difficult for this big, bulky base to attack a tetravalent carbon atom.

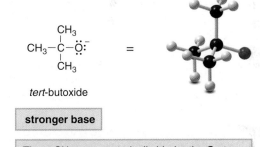

$= \quad CH_3CH_2-\ddot{\overset{..}{O}}{:}^-$

ethoxide

stronger nucleophile

$CH_3-\overset{\displaystyle CH_3}{\underset{\displaystyle CH_3}{\overset{|}{\underset{|}{C}}}}-\ddot{\overset{..}{O}}{:}^- \quad =$

tert-butoxide

stronger base

Three CH_3 groups sterically hinder the O atom, making it a **weaker nucleophile.**

Steric hindrance decreases nucleophilicity but *not* basicity. Because bases pull off small, easily accessible protons, they are unaffected by steric hindrance. Nucleophiles, on the other hand, must attack a crowded tetrahedral carbon, so bulky groups decrease reactivity.

Sterically hindered bases that are poor nucleophiles are called *nonnucleophilic bases*. Potassium *tert*-butoxide [$K^+{}^-OC(CH_3)_3$] is a strong, nonnucleophilic base.

7.8C   Comparing Nucleophiles of Different Size—Solvent Effects

Atoms vary greatly in size down a column of the periodic table, and in this case, **nucleophilicity depends on the solvent used in a substitution reaction.** Although solvent has thus far been ignored, most organic reactions take place in a liquid solvent that dissolves all reactants to some extent. Because substitution reactions involve polar starting materials, polar solvents are used to dissolve them. There are two main kinds of polar solvents—**polar *protic* solvents and polar *aprotic* solvents.**

Polar Protic Solvents

In addition to dipole–dipole interactions, **polar *protic* solvents are capable of intermolecular hydrogen bonding,** because they contain an $O-H$ or $N-H$ bond. The most common polar protic solvents are water and alcohols (ROH), as seen in the examples in Figure 7.6. **Polar protic solvents solvate *both* cations and anions well.**

◆ Cations are solvated by ion–dipole interactions.
◆ Anions are solvated by hydrogen bonding.

Figure 7.6 Examples of polar protic solvents

H_2O	CH_3OH	CH_3CH_2OH	$(CH_3)_3COH$	CH_3COOH
	methanol	ethanol	*tert*-butanol	acetic acid

For example, if the salt NaBr is used as a source of the nucleophile Br^- in H_2O, the Na^+ cations are solvated by ion–dipole interactions with H_2O molecules, and the Br^- anions are solvated by strong hydrogen bonding interactions.

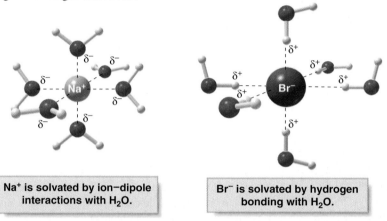

Na^+ is solvated by ion–dipole interactions with H_2O.

Br^- is solvated by hydrogen bonding with H_2O.

How do polar protic solvents affect nucleophilicity? **In polar protic solvents, nucleophilicity** *increases* **down a column of the periodic table as the size of the anion increases. This is** *opposite* **to basicity.** A small electronegative anion like F^- is very well solvated by hydrogen bonding, effectively shielding it from reaction. On the other hand, a large, less electronegative anion like I^- does not hold onto solvent molecules as tightly. The solvent does not "hide" a large nucleophile as well, and the nucleophile is much more able to donate its electron pairs in a reaction. Thus, **nucleophilicity increases down a column** even though basicity decreases, giving rise to the following trend in polar protic solvents:

I^- is a weak base but a strong nucleophile in polar protic solvents.

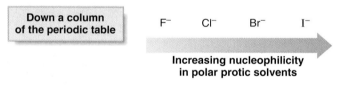

Down a column of the periodic table

F^- Cl^- Br^- I^-

Increasing nucleophilicity in polar protic solvents

Polar Aprotic Solvents

Polar *aprotic* **solvents** also exhibit dipole–dipole interactions, but they have no $O-H$ or $N-H$ bond so they are **incapable of hydrogen bonding.** Examples of polar aprotic solvents are shown in Figure 7.7. **Polar aprotic solvents solvate only cations well.**

- ◆ Cations are solvated by ion–dipole interactions.
- ◆ Anions are not well solvated because the solvent cannot hydrogen bond to them.

Figure 7.7 Examples of polar aprotic solvents

Abbreviations are often used in organic chemistry, instead of a compound's complete name. A list of common abbreviations is given in Appendix A.

$$CH_3 \overset{\overset{\displaystyle O}{\|}}{C} CH_3$$
acetone

$$CH_3-C{\equiv}N$$
acetonitrile

tetrahydrofuran
THF

$$CH_3 \overset{\overset{\displaystyle O}{\|}}{S} CH_3$$
dimethyl sulfoxide
DMSO

$$H \overset{\overset{\displaystyle O}{\|}}{C} N(CH_3)_2$$
dimethylformamide
DMF

$$(CH_3)_2N \overset{\overset{\displaystyle O}{\|}}{\underset{\underset{\displaystyle N(CH_3)_2}{|}}{P}} N(CH_3)_2$$
hexamethylphosphoramide
HMPA

When the salt NaBr is dissolved in acetone, $(CH_3)_2C=O$, the Na^+ cations are solvated by ion–dipole interactions with the acetone molecules, but, with no possibility for hydrogen bonding, the Br^- anions are not well solvated. Often these anions are called **naked anions** because they are not bound by tight interactions with solvent.

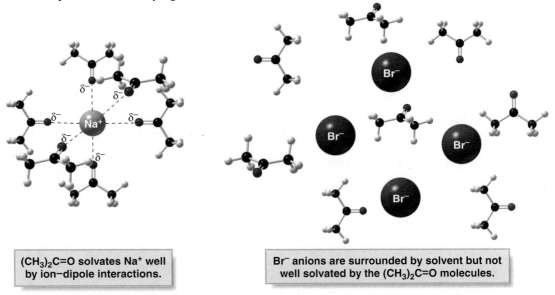

| $(CH_3)_2C=O$ solvates Na^+ well by ion–dipole interactions. | Br^- anions are surrounded by solvent but not well solvated by the $(CH_3)_2C=O$ molecules. |

How do polar aprotic solvents affect nucleophilicity? Because anions are not well solvated in polar aprotic solvents, there is no need to consider whether solvent molecules more effectively hide one anion than another. Nucleophilicity once again parallels basicity and **the stronger base is the stronger nucleophile.** Because basicity decreases with size down a column, nucleophilicity decreases as well:

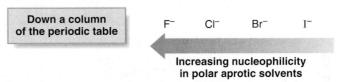

Down a column of the periodic table

F^- Cl^- Br^- I^-

Increasing nucleophilicity in polar aprotic solvents

PROBLEM 7.12   Classify each solvent as protic or aprotic.
a. $HOCH_2CH_2OH$ b. $CH_3CH_2OCH_2CH_3$ c. $CH_3COOCH_2CH_3$

PROBLEM 7.13   Identify the stronger nucleophile in each pair of anions.
a. Br^- and Cl^- in a polar protic solvent
b. HO^- and Cl^- in a polar aprotic solvent
c. HS^- and F^- in a polar protic solvent

7.8D Summary

This long discussion of nucleophilicity has brought together many new concepts, such as steric hindrance and solvent effects, both of which we will meet again in our study of organic chemistry. Keep in mind, however, the central relationship between nucleophilicity and basicity in comparing two nucleophiles.

◆ It is generally true that the *stronger* base is the *stronger* nucleophile.
◆ In polar protic solvents, however, nucleophilicity increases with increasing size of an anion (opposite to basicity).
◆ Steric hindrance decreases nucleophilicity without decreasing basicity, making $(CH_3)_3CO^-$ a stronger base but a weaker nucleophile than $CH_3CH_2O^-$.

Table 7.4. lists some common nucleophiles used in nucleophilic substitution reactions.

TABLE 7.4	Common Nucleophiles in Organic Chemistry				
	Negatively charged nucleophiles			**Neutral nucleophiles**	
Oxygen	^-OH	^-OR	CH_3COO^-	H_2O	ROH
Nitrogen	N_3^-			NH_3	RNH_2
Carbon	^-CN	$HC \equiv C^-$			
Halogen	Cl^-	Br^-	I^-		
Sulfur	HS^-	RS^-		H_2S	RSH

PROBLEM 7.14   Rank the nucleophiles in each group in order of increasing nucleophilicity.
a. ^-OH, $^-NH_2$, H_2O b. ^-OH, Br^-, F^- (polar aprotic solvent)   c. H_2O, ^-OH, CH_3COO^-

PROBLEM 7.15   What nucleophile is needed to carry out each reaction?
a. $CH_3CH_2CH_2-Br \longrightarrow CH_3CH_2CH_2-SH$

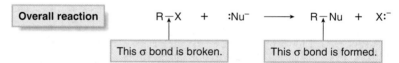

7.9 Possible Mechanisms for Nucleophilic Substitution

Now that you know something about the general features of nucleophilic substitution, you can begin to understand the mechanism.

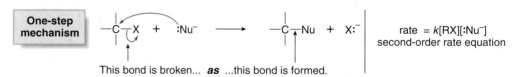

Nucleophilic substitution at an sp^3 hybridized carbon involves two σ bonds: the bond to the leaving group, which is broken, and the bond to the nucleophile, which is formed. To understand the mechanism of this reaction, though, we must know the timing of these two events; that is, **what is the order of bond breaking and bond making?** Do they happen at the same time, or does one event precede the other? There are three possibilities.

[1] **Bond breaking and bond making occur at the *same* time.**

One-step mechanism $-C-X + :Nu^- \longrightarrow -C-Nu + X:^-$ rate = $k[RX][:Nu^-]$
second-order rate equation

This bond is broken... ***as*** ...this bond is formed.

◆ If the $C-X$ bond is broken *as* the $C-Nu$ bond is formed, the mechanism has **one step**. As we learned in Section 6.9, the rate of such a bimolecular reaction depends on the concentration of both reactants; that is, the rate equation is **second order.**

[2] **Bond breaking occurs *before* bond making.**

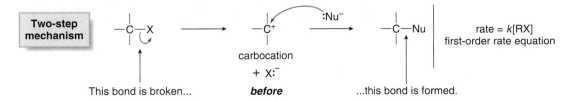

rate = $k[RX]$
first-order rate equation

◆ If the C−X bond is broken *first* and then the C−Nu bond is formed, the mechanism has **two steps** and a **carbocation** is formed as an intermediate. Because the first step is rate-determining, the rate depends on the concentration of RX only; that is, the rate equation is **first order.**

[3] **Bond making occurs *before* bond breaking.**

Ten electrons around C violates the octet rule.

Two-step mechanism

This bond is broken...   *after*   ...this bond is formed.

◆ If the C−Nu bond is formed *first* and then the C−X bond is broken, the mechanism has **two steps,** but this mechanism has an inherent problem. The intermediate generated in the first step has 10 electrons around carbon, violating the octet rule. Because two other mechanistic possibilities do not violate a fundamental rule, this last possibility can be disregarded.

The preceding discussion has generated two possible mechanisms for nucleophilic substitution: **a one-step mechanism in which bond breaking and bond making are simultaneous, and a two-step mechanism in which bond breaking comes before bond making.** In Section 7.10 we look at data for two specific nucleophilic substitution reactions and see if those data fit either of these proposed mechanisms.

7.10 Two Mechanisms for Nucleophilic Substitution

Rate equations for two different reactions give us insight into the possible mechanism for nucleophilic substitution.

Reaction of bromomethane (CH_3Br) with the nucleophile acetate (CH_3COO^-) affords the substitution product methyl acetate with loss of Br^- as the leaving group (Equation [1]). Kinetic data show that the reaction rate depends on the concentration of *both* reactants; that is, the rate equation is **second order.** This suggests a **bimolecular reaction with a one-step mechanism** in which the C−X bond is broken *as* the C−Nu bond is formed.

[1] $CH_3–Br$ +   →   + Br^- rate = $k[CH_3Br][CH_3COO^-]$
second-order kinetics

acetate   methyl acetate

Both reactants appear in the rate equation.

Equation [2] illustrates a similar nucleophilic substitution reaction with a different alkyl halide, $(CH_3)_3CBr$, which also leads to substitution of Br^- by CH_3COO^-. Kinetic data show that this reaction rate depends on the concentration of only *one* reactant, the alkyl halide; that is, the rate equation is **first order.** This suggests a **two-step mechanism in which the rate-determining step involves the alkyl halide only.**

[2] $(CH_3)_3C-Br$ + [acetate structure: $^-O-C(=O)-CH_3$] $\longrightarrow$ $(CH_3)_3C-O-C(=O)-CH_3$ + Br^- | rate = $k[(CH_3)_3CBr]$
first-order kinetics

acetate

Only **one** reactant appears in the rate equation.

The numbers **1** and **2** in the names S_N1 and S_N2 refer to the kinetic order of the reactions. For example, S_N**2** means that the kinetics are **second** order. The number 2 does *not* refer to the number of steps in the mechanism.

How can these two different results be explained? Although these two reactions have the same nucleophile and leaving group, **there must be two different mechanisms** because there are two different rate equations. These equations are specific examples of two well known mechanisms for nucleophilic substitution at an sp^3 hybridized carbon:

♦ The S_N**2 mechanism (substitution nucleophilic bimolecular),** illustrated by the reaction in Equation [1].

♦ The S_N**1 mechanism (substitution nucleophilic unimolecular),** illustrated by the reaction in Equation [2].

We will now examine the characteristics of the S_N2 and S_N1 mechanisms.

7.11   The S_N2 Mechanism

The reaction of CH_3Br with CH_3COO^- is an example of an **S_N2 reaction.** What are the general features of this mechanism?

An S_N2 reaction CH_3-Br + [acetate structure: $^-O-C(=O)-CH_3$] $\longrightarrow$ $CH_3-O-C(=O)-CH_3$ + Br^-

acetate

7.11A   Kinetics

An S_N2 reaction exhibits **second-order kinetics;** that is, the reaction is **bimolecular** and both the alkyl halide and the nucleophile appear in the rate equation.

♦ rate = $k[CH_3Br][CH_3COO^-]$

Changing the concentration of *either* reactant affects the rate. For example, doubling the concentration of *either* the nucleophile or the alkyl halide doubles the rate. Doubling the concentration of *both* reactants increases the rate by a factor of four.

PROBLEM 7.16   What happens to the rate of an S_N2 reaction under each of the following conditions?
a. [RX] is tripled, and [:Nu$^-$] stays the same.   c. [RX] is halved, and [:Nu$^-$] stays the same.
b. Both [RX] and [:Nu$^-$] are tripled.   d. [RX] is halved, and [:Nu$^-$] is doubled.

7.11B   A One-Step Mechanism

The most straightforward explanation for the observed second-order kinetics is a **concerted reaction**—bond breaking and bond making occur at the *same* time, as shown in Mechanism 7.1.

MECHANISM 7.1

The S_N2 Mechanism

One step   The C–Br bond breaks *as* the C–O bond forms.

CH_3-Br + $^-O-C(=O)-CH_3$ $\xrightarrow{\text{one step}}$ $CH_3-O-C(=O)-CH_3$ + Br^-

Figure 7.8   An energy
diagram for the S_N2 reaction:
$CH_3Br + CH_3COO^- \rightarrow CH_3COOCH_3 + Br^-$

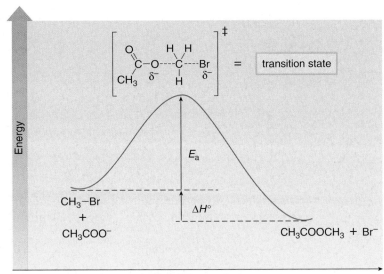

- In the transition state, the C–Br bond is partially broken, the C–O bond is partially formed, and both the attacking nucleophile and the departing leaving group bear a partial negative charge.

An energy diagram for the reaction of $CH_3Br + CH_3COO^-$ is shown in Figure 7.8. The reaction has one step, so there is one energy barrier between reactants and products. Because the equilibrium for this S_N2 reaction favors the products, they are drawn at lower energy than the starting materials.

PROBLEM 7.17   Draw the structure of the transition state in each of the following S_N2 reactions.

a. $CH_3CH_2CH_2-Cl$ + $^-OCH_3$ $\longrightarrow$ $CH_3CH_2CH_2-OCH_3$ + Cl^-

b.   ⌇⌇Br   + ^-SH $\longrightarrow$ ⌇⌇SH   + Br^-

PROBLEM 7.18   Draw an energy diagram for the reaction in Problem 7.17a. Label the axes, the starting material, the product, and the transition state. Assume the reaction is exothermic. Label ΔH° and E_a.

7.11C   Stereochemistry of the S_N2 Reaction

From what direction does the nucleophile approach the substrate in an S_N2 reaction? There are two possibilities.

◆ **Frontside attack:** The nucleophile approaches from the *same* side as the leaving group.
◆ **Backside attack:** The nucleophile approaches from the side *opposite* the leaving group.

The results of frontside and backside attack of a nucleophile are illustrated with $CH_3CH(D)Br$ as substrate and the general nucleophile $:Nu^-$. This substrate has the leaving group bonded to a stereogenic center, thus allowing us to see the structural difference that results when the nucleophile attacks from two different directions.

In **frontside attack,** the nucleophile approaches from the **same** side as the leaving group, forming **A.** In this example, the leaving group was drawn on the right, so the nucleophile attacks from the right, and all other groups remain in their original positions. Because the nucleophile and leaving group are in the same position relative to the other three groups on carbon, frontside attack results in **retention of configuration** around the stereogenic center.

| Frontside attack |

CH_3, H
    C—Br   +   :Nu$^-$ $\longrightarrow$
    D

CH_3, H
    C—Nu   +   Br$^-$
    D   **A**

Nu replaces Br on the **same** side.

In **backside attack,** the nucleophile approaches from the **opposite** side to the leaving group, forming **B.** In this example, the leaving group was drawn on the right, so the nucleophile attacks from the left. Because the nucleophile and leaving group are in the opposite position relative to the other three groups on carbon, backside attack results in **inversion of configuration** around the stereogenic center.

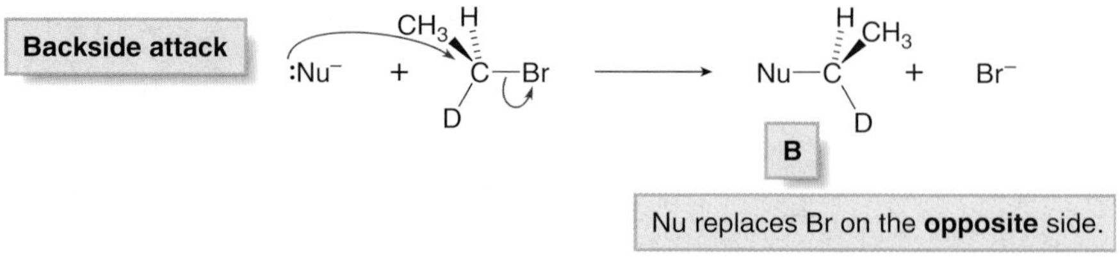

The products of frontside and backside attack are *different* compounds. **A** and **B** are stereoisomers that are nonsuperimposable—they are **enantiomers.**

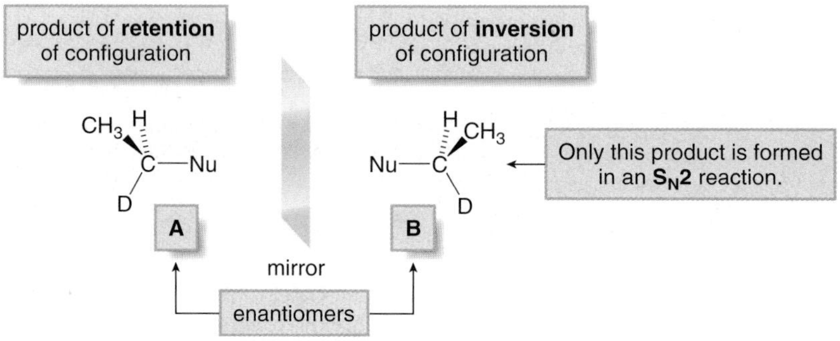

Which product is formed in an S_N2 reaction? When the stereochemistry of the product is determined, **only B, the product of backside attack, is formed.**

Inversion of configuration in an S_N2 reaction is often called **Walden inversion,** after Latvian chemist Dr. Paul Walden, who first observed this process in 1896.

◆ All S_N2 reactions proceed with backside attack of the nucleophile, resulting in *inversion* of configuration at a stereogenic center.

One explanation for backside attack is based on an electronic argument. Both the nucleophile and leaving group are electron rich and these like charges repel each other. Backside attack keeps these two groups as far away from each other as possible. In the transition state, the nucleophile and leaving group are 180° away from each other, and the other three groups around carbon occupy a plane, as illustrated in Figure 7.9.

Backside attack resulting in inversion of configuration occurs in all S_N2 reactions, but we can observe this change only when the leaving group is bonded to a stereogenic center.

Two additional examples of inversion of configuration in S_N2 reactions are given in Figure 7.10.

Figure 7.9 Stereochemistry of the S_N2 reaction

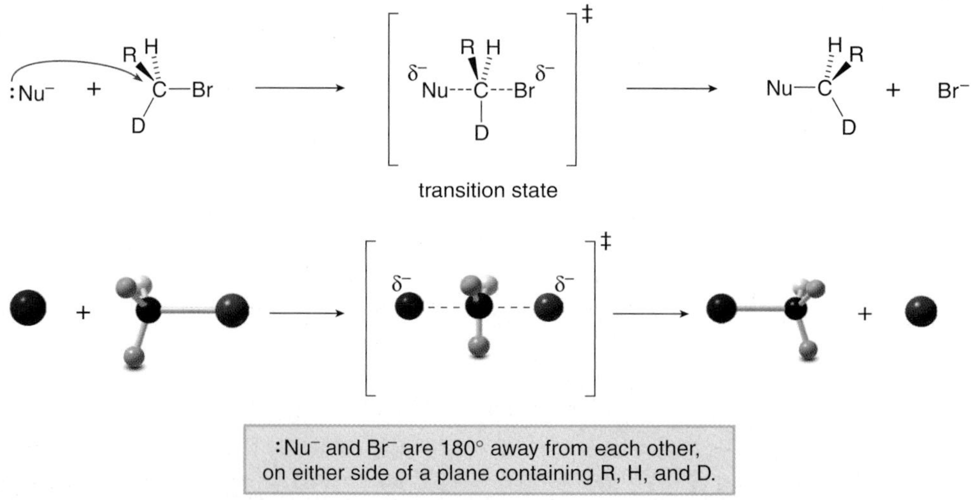

:Nu⁻ and Br⁻ are 180° away from each other, on either side of a plane containing R, H, and D.

Figure 7.10 · Two examples of inversion of configuration in the S$_N$2 reaction

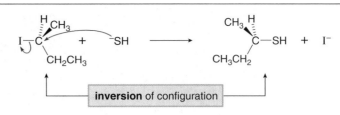

- The bond to the nucleophile in the product is always on the **opposite side** relative to the bond to the leaving group in the starting material.

SAMPLE PROBLEM 7.2   Draw the product (including stereochemistry) of the following S$_N$2 reaction.

$$CH_3 \cdots \bigcirc \cdots Br \quad + \quad {}^-CN \longrightarrow$$

SOLUTION

Br$^-$ is the leaving group and $^-$CN is the nucleophile. Because S$_N$2 reactions proceed with **inversion** of configuration and the leaving group is drawn above the ring (on a wedge), the nucleophile must come in from below.

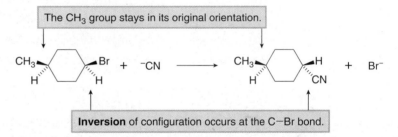

The CH$_3$ group stays in its original orientation.

Inversion of configuration occurs at the C—Br bond.

Note how the stereochemistry of the starting material and product compares. Backside attack converts the starting material, which has two groups **cis** to each other, to a product with two groups **trans** to each other because the nucleophile ($^-$CN) attacks from below the plane of the ring.

PROBLEM 7.19   Draw the product of each S$_N$2 reaction and indicate stereochemistry.

a.
$$\begin{array}{c} CH_3CH_2 \quad D \\ \diagdown \\ C-Br \\ \diagup \\ H \end{array} \quad + \quad {}^-OH \longrightarrow$$

b.
$$\begin{array}{c} H \quad Br \\ \diagdown \diagup \\ \diagup \diagdown \end{array} \quad + \quad {}^-OCH_2CH_3 \longrightarrow$$

c.
$$\bigcirc\!\!-\!\!I \quad + \quad {}^-CN \longrightarrow$$

7.11D   The Identity of the R Group

How does the rate of an S$_N$2 reaction change as the alkyl group in the substrate alkyl halide changes from CH$_3$ $\dashrightarrow$ 1° $\dashrightarrow$ 2° $\dashrightarrow$ 3°?

◆ As the number of R groups on the carbon with the leaving group *increases,* the rate of an S$_N$2 reaction *decreases.*

CH$_3$—X	RCH$_2$—X	R$_2$CH—X	R$_3$C—X
methyl	**1°**	**2°**	**3°**

⬅ **Increasing rate of an S$_N$2 reaction**

- ◆ Methyl and 1° alkyl halides undergo S_N2 reactions with ease.
- ◆ 2° Alkyl halides react more slowly.
- ◆ 3° Alkyl halides *do not* undergo S_N2 reactions.

This order of reactivity can be explained by steric effects. As small H atoms are replaced by larger alkyl groups, **steric hindrance caused by bulky R groups makes nucleophilic attack from the back side more difficult,** slowing the reaction rate. Figure 7.11 illustrates the effect of increasing steric hindrance around the carbon bearing the leaving group in a series of alkyl halides.

The effect of steric hindrance on the rate of an S_N2 reaction is reflected in the energy of the transition state, too. Let's compare the reaction of ⁻OH with two different alkyl halides, CH_3Br and $(CH_3)_2CHBr$. Steric hindrance around the 2° halide $(CH_3)_2CHBr$ is greater, making attack of the nucleophile more difficult, and *destabilizing* the transition state. This is shown in the energy diagrams for these two reactions in Figure 7.12.

- ◆ The *higher* the E_a, the *slower* the reaction rate. Thus, any factor that increases E_a (i.e., destabilizes the transition state), decreases the reaction rate.

Figure 7.11   Steric effects in the S_N2 reaction

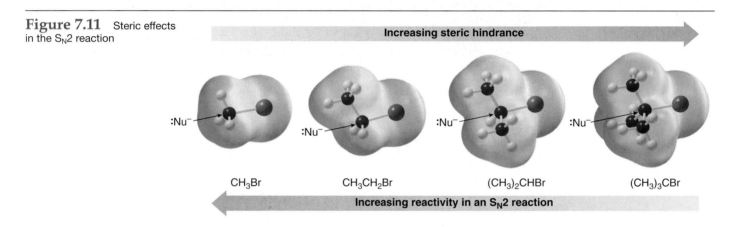

Increasing steric hindrance

:Nu⁻   :Nu⁻   :Nu⁻   :Nu⁻

CH_3Br CH_3CH_2Br $(CH_3)_2CHBr$ $(CH_3)_3CBr$

Increasing reactivity in an S_N2 reaction

Figure 7.12   Two energy diagrams depicting the effect of steric hindrance in S_N2 reactions

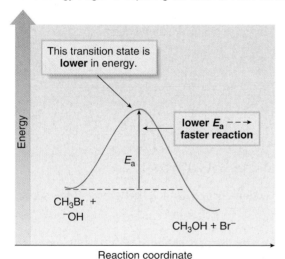

This transition state is **lower** in energy.

lower E_a - - →
faster reaction

Energy

E_a

CH_3Br +
⁻OH

$CH_3OH + Br^-$

Reaction coordinate

[a]: $CH_3Br + ^-OH \rightarrow CH_3OH + Br^-$

- • CH_3Br is an unhindered alkyl halide. The transition state in the S_N2 reaction is lower in energy, making E_a lower and increasing the reaction rate.

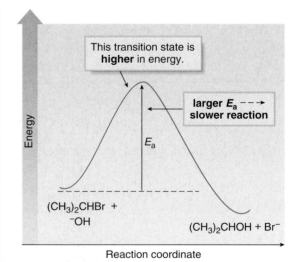

This transition state is **higher** in energy.

larger E_a - - →
slower reaction

Energy

E_a

$(CH_3)_2CHBr$ +
⁻OH

$(CH_3)_2CHOH + Br^-$

Reaction coordinate

[b]: $(CH_3)_2CHBr + ^-OH \rightarrow (CH_3)_2CHOH + Br^-$

- • $(CH_3)_2CHBr$ is a sterically hindered alkyl halide. The transition state in the S_N2 reaction is higher in energy, making E_a higher and decreasing the reaction rate.

Why does steric hindrance destabilize the transition state, making it higher in energy? The transition state of an S$_N$2 reaction consists of five groups around the central carbon atom—three bonds to either H or R groups and two partial bonds to the leaving group and the nucleophile. **Crowding around the central carbon atom increases as H atoms are successively replaced by R groups,** so the central carbon is much more sterically hindered in the transition state for (CH$_3$)$_2$CHBr than for CH$_3$Br. This increased crowding in the transition state makes it higher in energy (increases E_a), so the rate of the S$_N$2 reaction decreases.

less crowded transition state
lower in energy

faster S$_N$2 reaction

more crowded transition state
higher in energy

slower S$_N$2 reaction

◆ Increasing the number of R groups on the carbon with the leaving group *increases* *crowding* in the transition state, *decreasing* the rate of an S$_N$2 reaction.

◆ The S$_N$2 reaction is fastest with unhindered halides.

PROBLEM 7.20 Which compound in each pair undergoes a faster S$_N$2 reaction?

a. CH$_3$CH$_2$—Cl   or   CH$_3$—Cl

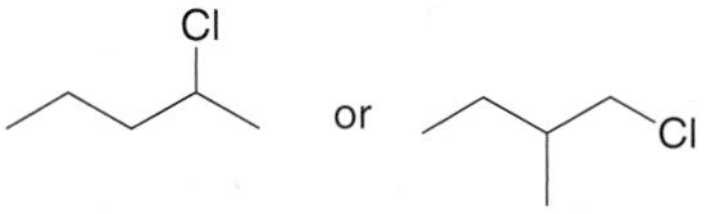

c.

b.

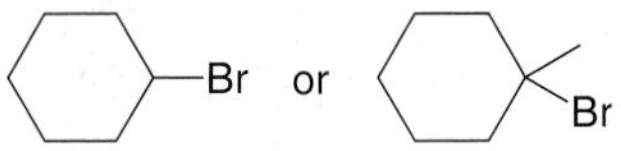

PROBLEM 7.21 Explain why (CH$_3$)$_3$CCH$_2$Br, a 1° alkyl halide, undergoes S$_N$2 reactions very slowly.

Table 7.5 summarizes what we have learned thus far about an S$_N$2 mechanism.

TABLE 7.5 Characteristics of the S$_N$2 Mechanism

Characteristic	Result
Kinetics	• Second-order kinetics; rate = k[RX][:Nu$^-$]
Mechanism	• One step
Stereochemistry	• Backside attack of the nucleophile • Inversion of configuration at a stereogenic center
Identity of R	• Unhindered halides react fastest. • Rate: CH$_3$X > RCH$_2$X > R$_2$CHX > R$_3$CX

7.12 Application: Useful S$_N$2 Reactions

Nucleophilic substitution by an S$_N$2 mechanism is common in the laboratory and in biological systems.

The S$_N$2 reaction is a key step in the laboratory synthesis of many drugs including **ethambutol** (trade name: Myambutol), used in the treatment of tuberculosis, and **fluoxetine** (trade name: Prozac), an antidepressant, as illustrated in Figure 7.13.

Figure 7.13 Nucleophilic substitution in the synthesis of two useful drugs

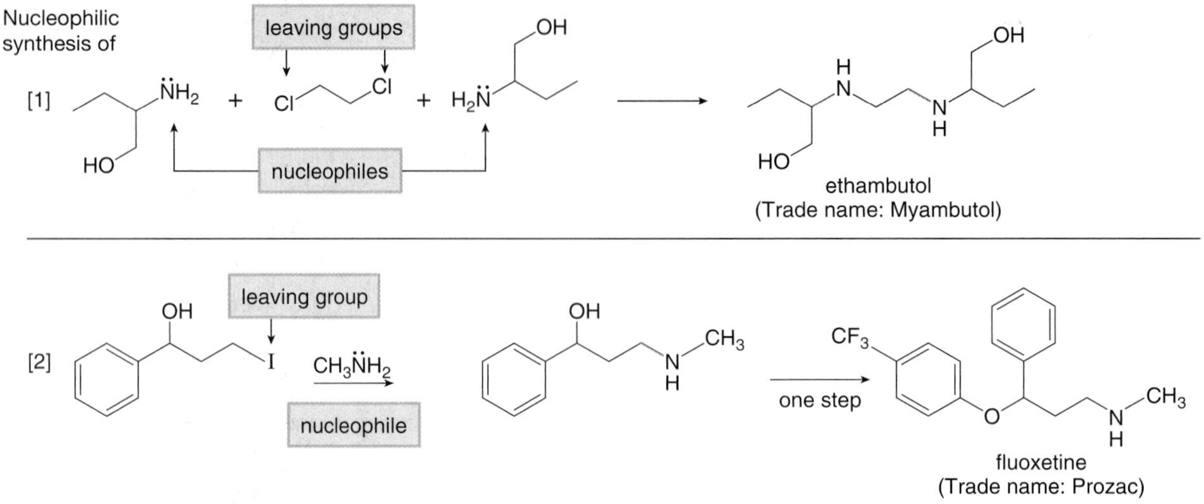

ethambutol
(Trade name: Myambutol)

[2]

fluoxetine
(Trade name: Prozac)

In both syntheses, the NH$_2$ group serves as a neutral nucleophile to displace halogen. The new bonds formed by nucleophilic substitution are drawn in red in the products.

Nucleophilic substitution reactions are important in biological systems as well. The most common reaction involves nucleophilic substitution at the CH$_3$ group in S-adenosylmethionine, or **SAM.** SAM is the cell's equivalent of CH$_3$I.

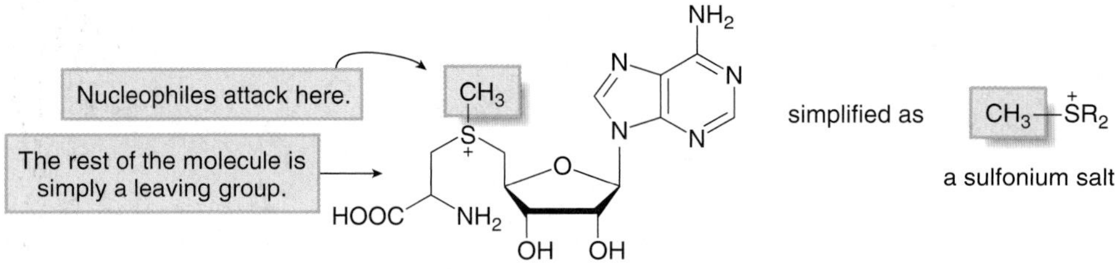

S-adenosylmethionine
SAM

simplified as

$CH_3-\overset{+}{S}R_2$

a sulfonium salt

The CH$_3$ group in SAM [abbreviated as (CH$_3$SR$_2$)$^+$] is part of a sulfonium salt, a positively charged sulfur species that contains a good leaving group. Nucleophilic attack at the CH$_3$ group of SAM displaces R$_2$S, a good neutral leaving group. This reaction is called **methylation,** because a CH$_3$ group is transferred from one compound (SAM) to another (:Nu$^-$).

Nucleophilic substitution   :Nu$^-$ + CH$_3-\overset{+}{S}R_2$ ⟶ CH$_3$–Nu + SR$_2$ ← leaving group

SAM   substitution product

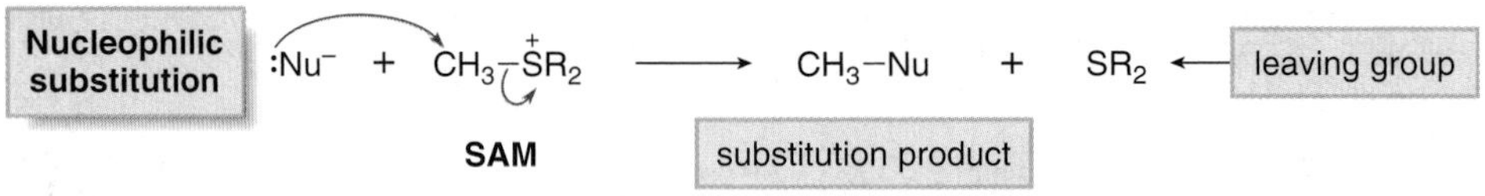

Adrenaline (epinephrine), the molecule that opened Chapter 7, is a hormone synthesized in the adrenal glands from noradrenaline (norepinephrine) by nucleophilic substitution using SAM (Figure 7.14). When an individual senses danger or is confronted by stress, the hypothalamus region of the brain signals the adrenal glands to synthesize and release adrenaline, which enters the bloodstream and then stimulates a response in many organs. Stored carbohydrates are metabolized in the liver to form glucose, which is further metabolized to provide an energy boost. Heart rate and blood pressure increase, and lung passages are dilated. These physiological changes prepare an individual for "fight or flight."

Cells use SAM, (CH$_3$SR$_2$)$^+$, instead of alkyl halides such as CH$_3$I in nucleophilic substitution reactions because alkyl halides are insoluble in the predominantly aqueous environment of the cell. In biological nucleophilic substitutions, therefore, the leaving group contains additional polar functional groups that make SAM, (CH$_3$SR$_2$)$^+$, water soluble.

Figure 7.14 Adrenaline synthesis from noradrenaline in response to stress

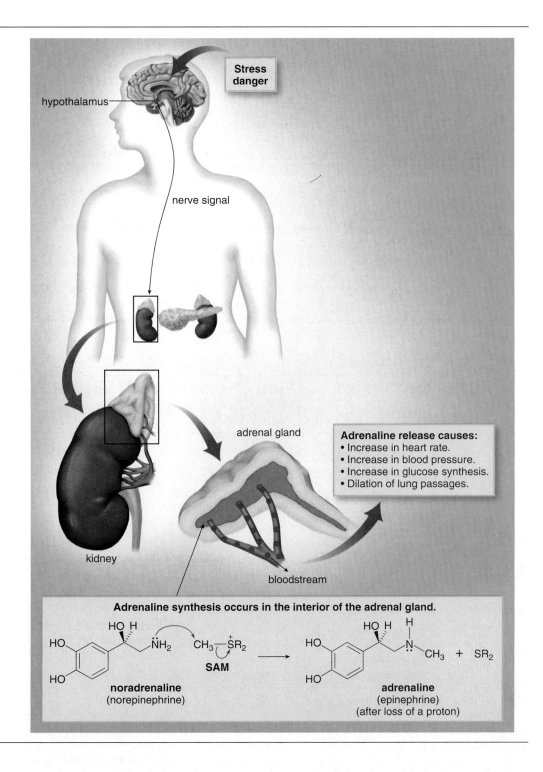

Adrenaline synthesis occurs in the interior of the adrenal gland.

PROBLEM 7.22   Nicotine, a toxic and addictive component of tobacco, is synthesized from **A** using SAM. Write out the reaction that converts **A** into nicotine.

7.13   The S_N1 Mechanism

The reaction of $(CH_3)_3CBr$ with CH_3COO^- is an example of the second mechanism for nucleophilic substitution, the **S_N1 mechanism.** What are the general features of this mechanism?

| An S_N1 reaction |

$(CH_3)_3C-Br$ + [acetate] $\longrightarrow$ $(CH_3)_3C-O...C...CH_3$ + Br^-

acetate

7.13A   Kinetics

The S_N1 reaction exhibits **first-order kinetics.**

♦ rate = $k[(CH_3)_3CBr]$

As we learned in Section 7.10, this suggests that the S_N1 mechanism involves **more than one step,** and that the slow step is **unimolecular,** involving *only* the alkyl halide. **The identity and concentration of the nucleophile have no effect on the reaction rate.** For example, doubling the concentration of $(CH_3)_3CBr$ doubles the rate, but doubling the concentration of the nucleophile has no effect.

PROBLEM 7.23   What happens to the rate of an S_N1 reaction under each of the following conditions?
a. [RX] is tripled, and [:Nu$^-$] stays the same.   c. [RX] is halved, and [:Nu$^-$] stays the same.
b. Both [RX] and [:Nu$^-$] are tripled.   d. [RX] is halved, and [:Nu$^-$] is doubled.

7.13B   A Two-Step Mechanism

The most straightforward explanation for the observed first-order kinetics is a **two-step mechanism** in which **bond breaking occurs** *before* **bond making,** as shown in Mechanism 7.2.

MECHANISM 7.2

The S_N1 Mechanism

Step [1] The C–Br bond is broken.

$CH_3-C(CH_3)(CH_3)-Br$ $\xrightarrow{slow}$ $CH_3-C^+(CH_3)-CH_3$ + Br^-

carbocation

♦ **Heterolysis of the C–Br bond** forms an intermediate **carbocation.** This step is rate-determining because it involves only bond cleavage.

Step [2] The C–O bond is formed.

$CH_3-C^+(CH_3)-CH_3$ [acetate] $\xrightarrow{fast}$ $(CH_3)_3C-O...C...CH_3$

acetate | new bond |

♦ **Nucleophilic attack of acetate** on the carbocation forms the new C–O bond in the product. This is a **Lewis acid–base reaction;** the nucleophile is the Lewis base and the carbocation is the Lewis acid. Step [2] is *faster* than Step [1] because no bonds are broken and one bond is formed.

The key features of the S_N1 mechanism are:

♦ The mechanism has two steps.
♦ Carbocations are formed as reactive intermediates.

An energy diagram for the reaction of $(CH_3)_3CBr + CH_3COO^-$ is shown in Figure 7.15. Each step has its own energy barrier, with a transition state at each energy maximum. Because Step [1]

Figure 7.15 An energy diagram for the S$_N$1 reaction: $(CH_3)_3CBr + CH_3COO^- \rightarrow$ $(CH_3)_3COCOCH_3 + Br^-$

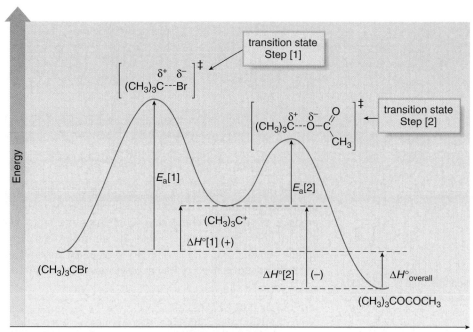

- Since the S$_N$1 mechanism has two steps, there are two energy barriers.
- Step [1] is rate-determining; $E_a[1] > E_a[2]$.
- In each step only one bond is broken or formed, so the transition state for each step has one partial bond.
- The reaction is drawn with $\Delta H°_{overall}$ as a negative value, since the products are lower in energy than the starting materials.

is rate-determining, its transition state is at higher energy. $\Delta H°$ for Step [1] has a positive value because only bond breaking occurs, whereas $\Delta H°$ of Step [2] has a negative value because only bond making occurs. The overall reaction is assumed to be exothermic, so the final product is drawn at lower energy than the initial starting material.

PROBLEM 7.24   Assume the following reaction has an S$_N$1 mechanism and draw the two steps.

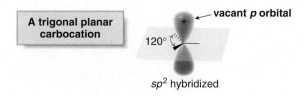

7.13C   Stereochemistry of the S$_N$1 Reaction

To understand the stereochemistry of the S$_N$1 reaction, we must examine the geometry of the carbocation intermediate.

A trigonal planar carbocation

vacant *p* orbital

120°

sp^2 hybridized

◆ A carbocation (with three groups around C) is *sp*^{2} hybridized and trigonal planar, and contains a vacant *p* orbital extending above and below the plane.

To illustrate the consequences of having a trigonal planar carbocation formed as a reactive intermediate, we examine the S$_N$1 reaction of a 3° alkyl halide **A** having the leaving group bonded to a stereogenic carbon.

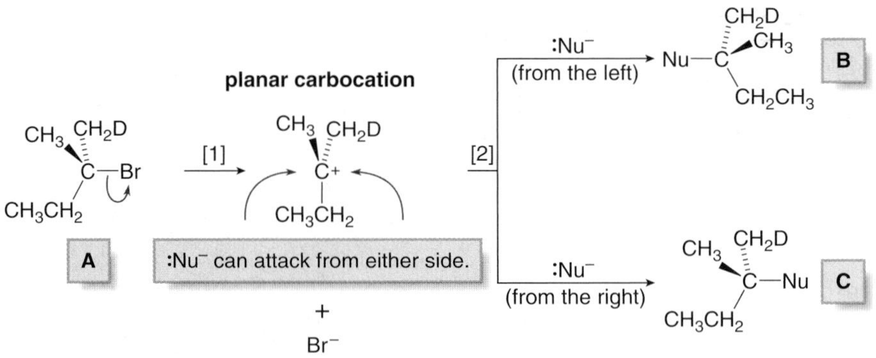

Loss of the leaving group in Step [1] generates a planar carbocation that is now achiral. Attack of the nucleophile in Step [2] can occur from either side to afford two products, **B** and **C.** These two products are *different* compounds containing one stereogenic center. **B** and **C** are stereoisomers that are not superimposable—they are **enantiomers.** Because there is no preference for nucleophilic attack from either direction, an equal amount of the two enantiomers is formed—a **racemic mixture.** We say that *racemization* has occurred.

Nucleophilic attack from both sides of a planar carbocation occurs in S_N1 reactions, but we see the result of this phenomenon only when the leaving group is bonded to a stereogenic center.

◆ *Racemization* is the formation of equal amounts of two enantiomeric products from a single starting material.
◆ S_N1 reactions proceed with *racemization* at a single stereogenic center.

Two additional examples of racemization in S_N1 reactions are given in Figure 7.16.

SAMPLE PROBLEM 7.3   Draw the products (including stereochemistry) of the following S_N1 reaction.

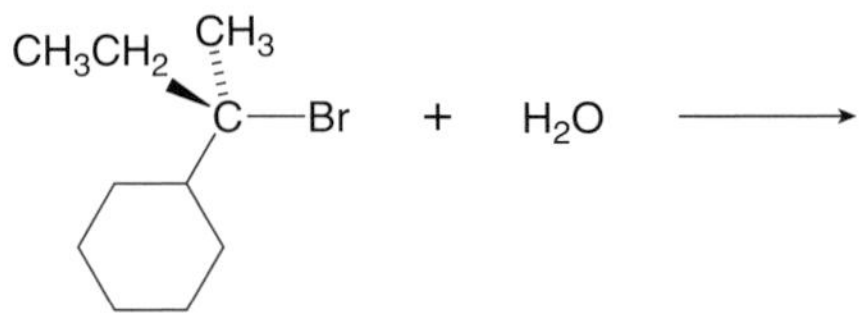

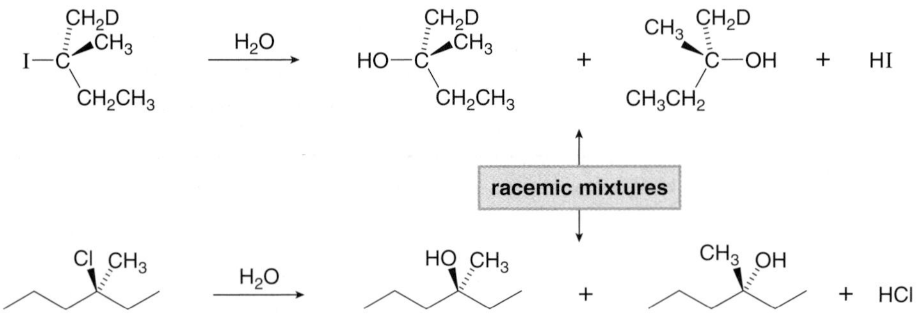

Figure 7.16   Two examples of racemization in the S_N1 reaction

- Nucleophilic substitution of each starting material by an S_N1 mechanism forms a **racemic mixture** of two products.
- With H_2O, a neutral nucleophile, the initial product of nucleophilic substitution (ROH_2^+) loses a proton to form the final neutral product, ROH (Section 7.6).

SOLUTION

Br$^-$ is the leaving group and H$_2$O is the nucleophile. Loss of the leaving group generates a trigonal planar carbocation, which can react with the nucleophile from either direction to form two products.

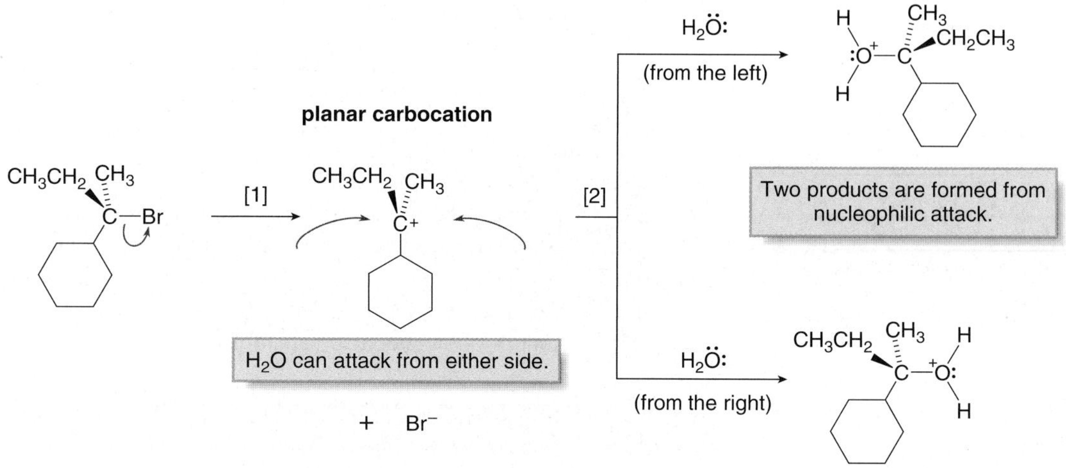

planar carbocation

H$_2$O can attack from either side.

Two products are formed from nucleophilic attack.

+   Br$^-$

In this example, the initial products of nucleophilic substitution bear a positive charge. They readily lose a proton to form neutral products. The overall process with a neutral nucleophile thus has **three steps:** the first two constitute the **two-step S$_N$1 mechanism** (loss of the leaving group and attack of the nucleophile), and the third is a **Brønsted–Lowry acid–base reaction** leading to a neutral organic product.

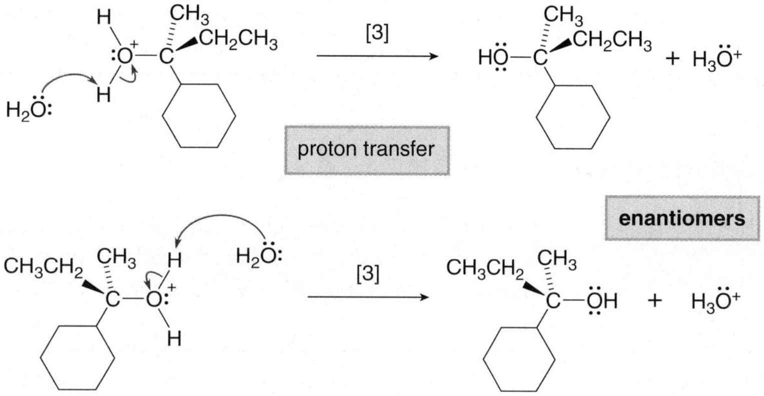

proton transfer

enantiomers

The two products in this reaction are nonsuperimposable mirror images—**enantiomers.** Because nucleophilic attack on the trigonal planar carbocation occurs with equal frequency from both directions, a **racemic mixture is formed.**

PROBLEM 7.25   Draw the products of each S$_N$1 reaction and indicate the stereochemistry of any stereogenic centers.

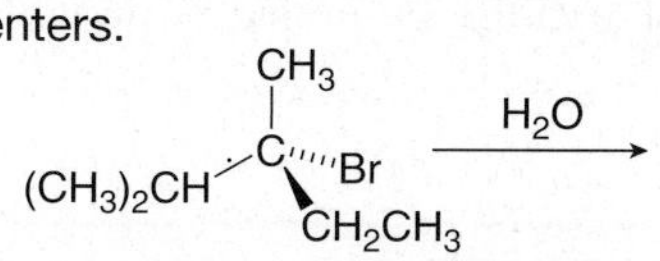

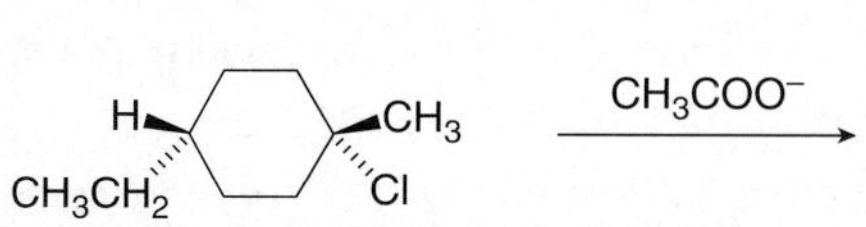

7.13D   The Identity of the R Group

How does the rate of an S$_N$1 reaction change as the alkyl group in the substrate alkyl halide changes from CH$_3$ --→ 1° --→ 2° --→ 3°?

◆ As the number of R groups on the carbon with the leaving group *increases,* the rate of an S$_N$1 reaction *increases.*

CH$_3$—X   RCH$_2$—X   R$_2$CH—X   R$_3$C—X

methyl   **1°**   **2°**   **3°**

Increasing rate of an S$_N$1 reaction →

◆ 3° Alkyl halides undergo S$_N$1 reactions rapidly.
◆ 2° Alkyl halides react more slowly.
◆ Methyl and 1° alkyl halides do *not* undergo S$_N$1 reactions.

Table 7.6 summarizes the characteristics of the S$_N$1 mechanism.

This trend is exactly opposite to that observed for the S$_N$2 mechanism. To explain this result, we must examine the rate-determining step, the formation of the carbocation, and learn about the effect of alkyl groups on **carbocation stability.**

TABLE 7.6   Characteristics of the S$_N$1 Mechanism

Characteristic	Result
Kinetics	• First-order kinetics; rate = k[RX]
Mechanism	• Two steps
Stereochemistry	• Trigonal planar carbocation intermediate • Racemization at a single stereogenic center
Identity of R	• More substituted halides react fastest. • Rate: R$_3$CX > R$_2$CHX > RCH$_2$X > CH$_3$X

7.14 Carbocation Stability

Carbocations are classified as **primary (1°), secondary (2°), or tertiary (3°)** by the number of R groups bonded to the charged carbon atom. As the number of R groups on the positively charged carbon atom increases, the stability of the carbocation **increases.**

$\overset{+}{C}H_3$ $R\overset{+}{C}H_2$ $R_2\overset{+}{C}H$ $R_3\overset{+}{C}$

methyl   **1°**   **2°**   **3°**

Increasing carbocation stability →

When we speak of carbocation stability, we really mean *relative* stability. Tertiary carbocations are too unstable to isolate, but they are more stable than secondary carbocations. We will examine the reason for this order of stability by invoking two different principles: **inductive effects** and **hyperconjugation.**

PROBLEM 7.26   Classify each carbocation as 1°, 2°, or 3°.

a. CH$_3\overset{+}{C}$HCH$_2$CH$_3$ b. 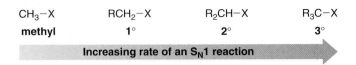   c. (CH$_3$)$_3$C$\overset{+}{C}$H$_2$ d.   e.

7.14A   Inductive Effects

Inductive effects are electronic effects that occur through σ bonds. They stabilize a species when electron density is dispersed over a larger volume. In Section 2.5B, for example, we learned that more electronegative atoms stabilize a negative charge by an electron-withdrawing inductive effect.

Electron donor groups (Z) stabilize a (+) charge; $Z \rightarrow Y^+$. Electron-withdrawing groups (W) stabilize a (–) charge; $W \leftarrow Y^-$.

To stabilize a positive charge, **electron-donating groups** are needed. **Alkyl groups are electron donor groups that stabilize a positive charge.** Because an alkyl group has several σ bonds, each containing electron density, it is more polarizable than a hydrogen atom, and more able to donate electron density. Thus, as R groups successively replace the H atoms in CH_3^+, **the positive charge is more dispersed on the electron donor R groups, and the carbocation is more stabilized.**

methyl   1°   2°   3°

Increasing number of electron-donating R groups
Increasing carbocation stability

Electrostatic potential maps for four different carbocations in Figure 7.17 illustrate how the positive charge on carbon becomes less concentrated (more dispersed) as the number of alkyl groups increases.

PROBLEM 7.27

Rank the following carbocations in order of increasing stability.

a. $(CH_3)_2\overset{+}{C}CH_2CH_3$ $(CH_3)_2CHCH_2\overset{+}{C}H_2$ $(CH_3)_2CH\overset{+}{C}HCH_3$

b.

PROBLEM 7.28

Which of the following 1° carbocations is more stable: $CH_3CH_2^+$ or $CCl_3CH_2^+$?

7.14B Hyperconjugation

A second explanation for the observed trend in carbocation stability is based on orbital overlap. A 3° carbocation is more stable than a 2°, 1°, or methyl carbocation because the positive charge is delocalized over more than one atom.

◆ Spreading out charge by the overlap of an empty *p* orbital with an adjacent σ bond is called *hyperconjugation.*

Figure 7.17 Electrostatic potential maps for different carbocations

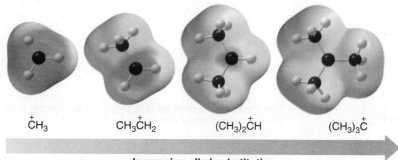

$\overset{+}{C}H_3$ $CH_3\overset{+}{C}H_2$ $(CH_3)_2\overset{+}{C}H$ $(CH_3)_3\overset{+}{C}$

Increasing alkyl substitution
Increasing dispersal of positive charge

• Dark blue areas in electrostatic potential plots indicate regions low in electron density. As alkyl substitution increases, the region of positive charge is less concentrated on carbon.

For example, CH_3^+ cannot be stabilized by hyperconjugation, but $(CH_3)_2CH^+$ can:

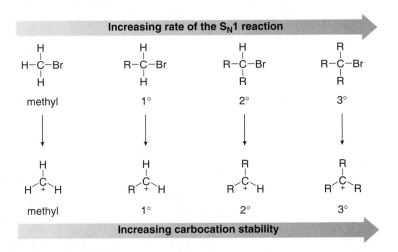

| This carbocation has no opportunity for orbital overlap with the vacant *p* orbital. | Overlap of the C−H σ bond with the adjacent vacant *p* orbital stabilizes the carbocation. |

Both carbocations contain an sp^2 hybridized carbon, so both are trigonal planar with a vacant *p* orbital extending above and below the plane. There are no adjacent C−H σ bonds with which the *p* orbital can overlap in CH_3^+, but there *are* adjacent C−H σ bonds in $(CH_3)_2CH^+$. This overlap (the **hyperconjugation**) delocalizes the positive charge on the carbocation, spreading it over a larger volume, and this stabilizes the carbocation.

The larger the number of alkyl groups on the adjacent carbons, the greater the possibility for hyperconjugation, and the larger the stabilization. Hyperconjugation thus provides an alternate way of explaining why **carbocations with a larger number of R groups are more stabilized.**

7.15 The Hammond Postulate

The rate of an S_N1 reaction depends on the rate of formation of the carbocation (the product of the rate-determining step) via heterolysis of the C−X bond.

♦ The rate of an S_N1 reaction *increases* as the number of R groups on the carbon with the leaving group *increases*.
♦ The stability of a carbocation *increases* as the number of R groups on the positively charged carbon *increases*.

Increasing rate of the S_N1 reaction

H	H	H	R
H−C−Br	R−C−Br	R−C−Br	R−C−Br
H	H	R	R
methyl	1°	2°	3°

H	H	R	R
C	C	C	C
H + H	R + H	R + H	R + R
methyl	1°	2°	3°

Increasing carbocation stability

♦ Thus, the rate of an S_N1 reaction *increases* as the stability of the carbocation *increases*.

$$-\overset{|}{\underset{|}{C}}-X \xrightarrow[\text{step}]{\text{rate-determining}} \overset{|}{\underset{+}{C}} \diagdown + \quad X^-$$

The reaction is faster with a more stable carbocation.

The rate of a reaction depends on the magnitude of E_a, and the stability of a product depends on $\Delta G°$. The **Hammond postulate**, first proposed in 1955, **relates rate to stability.**

7.15A   The General Features of the Hammond Postulate

The Hammond postulate provides a qualitative estimate of the energy of a transition state. Because the energy of the transition state determines the energy of activation and therefore the reaction rate, predicting the relative energy of two transition states allows us to determine the relative rates of two reactions.

According to the Hammond postulate, the transition state of a reaction resembles the structure of the species (reactant or product) to which it is closer in energy. A transition state is always higher in energy than both the reactants and products, so it will resemble the structure of either the reactant or product, whichever is higher in energy.

◆ In endothermic reactions, the transition state is closer in energy to the products.
◆ In exothermic reactions, the transition state is closer in energy to the reactants.

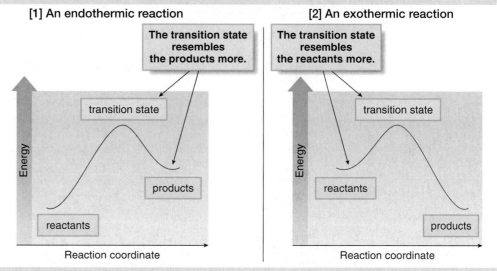

[1] An endothermic reaction   [2] An exothermic reaction

◆ Transition states in endothermic reactions resemble the products.
◆ Transition states in exothermic reactions resemble the reactants.

What happens to the reaction rate if the energy of the product is lowered? In an **endothermic reaction,** the transition state resembles the products more than the reactants, so anything that stabilizes the product stabilizes the transition state, too. **Lowering the energy of the transition state** *decreases* **the energy of activation (E_a), which** *increases* **the reaction rate.**

Suppose there are two possible products of an endothermic reaction, but one is more stable (lower in energy) than the other (Figure 7.18). According to the Hammond postulate, **the transition state to form the more stable product is lower in energy, so this reaction should occur faster.**

Figure 7.18   An endothermic reaction—How the energy of the transition state and products are related

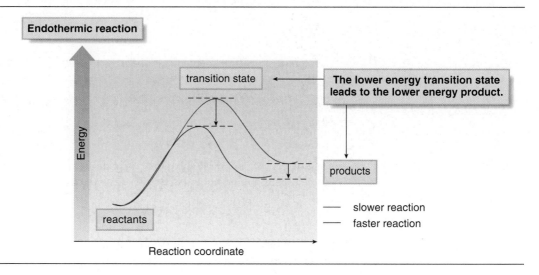

Figure 7.19 An exothermic reaction— How the energy of the transition state and products are related

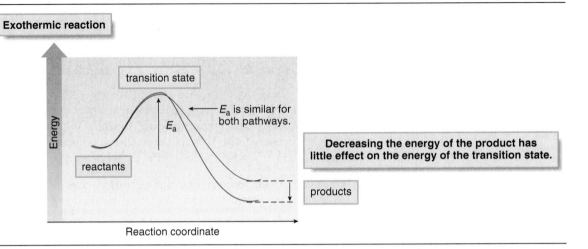

- ◆ Conclusion: In an endothermic reaction, the more stable product forms faster.

What happens to the reaction rate of an **exothermic reaction** if the energy of the product is lowered? The transition state resembles the reactants more than the products, so **lowering the energy of the products has little or no effect on the energy of the transition state.** If E_a is unaffected, then the reaction rate is unaffected, too, as shown in Figure 7.19.

- ◆ Conclusion: In an exothermic reaction, the more stable product may or may not form faster because E_a is similar for both products.

7.15B   The Hammond Postulate and the S_N1 Reaction

In the S_N1 reaction, the rate-determining step is the formation of the carbocation, an *endothermic* reaction. According to the Hammond postulate, the **stability of the carbocation determines the rate of its formation.**

For example, heterolysis of the C–Cl bond in CH_3Cl affords a highly unstable methyl carbocation, CH_3^+ (Equation [1]), whereas heterolysis of the C–Cl bond in $(CH_3)_3CCl$ affords a more stable 3° carbocation, $(CH_3)_3C^+$ (Equation [2]). The Hammond postulate states that Reaction [2] is much faster than Reaction [1], because the transition state to form the more stable 3° carbocation is lower in energy. Figure 7.20 depicts an energy diagram comparing these two endothermic reactions.

[1] slower reaction

$$H{-}\overset{\overset{\textstyle H}{|}}{\underset{\underset{\textstyle H}{|}}{C}}{-}Cl \longrightarrow \underset{H}{\overset{H}{C}}\overset{+}{}H \quad + \quad Cl^-$$

less stable carbocation

[2] faster reaction

$$CH_3{-}\overset{\overset{\textstyle CH_3}{|}}{\underset{\underset{\textstyle CH_3}{|}}{C}}{-}Cl \longrightarrow CH_3\overset{+}{\underset{}{C}}CH_3 \quad + \quad Cl^-$$

3°   3°

more stable carbocation

In conclusion, the Hammond postulate estimates the relative energy of transition states, and thus it can be used to predict the relative rates of two reactions.

Figure 7.20 Energy diagram for carbocation formation in two different S_N1 reactions

- Since CH_3^+ is less stable than $(CH_3)_3C^+$, $E_a[1] > E_a[2]$, and Reaction [1] is slower.

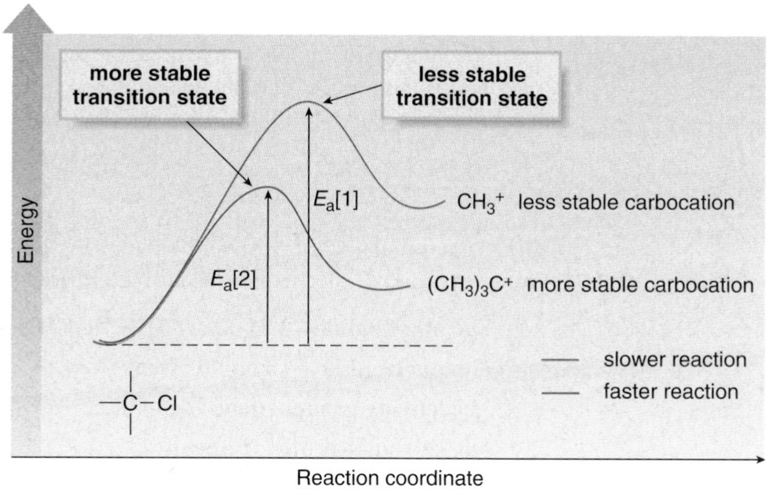

PROBLEM 7.29 Which alkyl halide in each pair reacts faster in an S_N1 reaction?

a. $(CH_3)_3CBr$ and $(CH_3)_3CCH_2Br$

b. 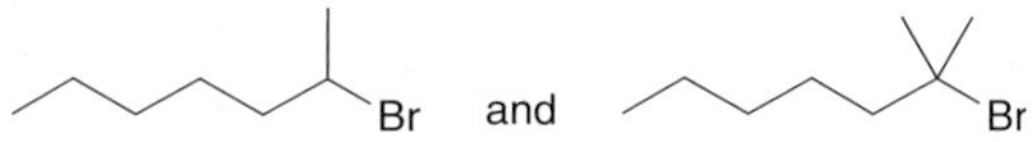

c. 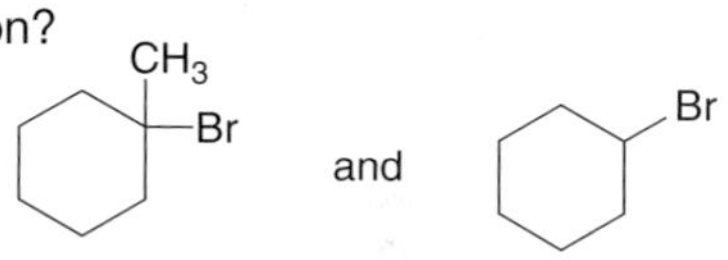

7.16 Application: S_N1 Reactions, Nitrosamines, and Cancer

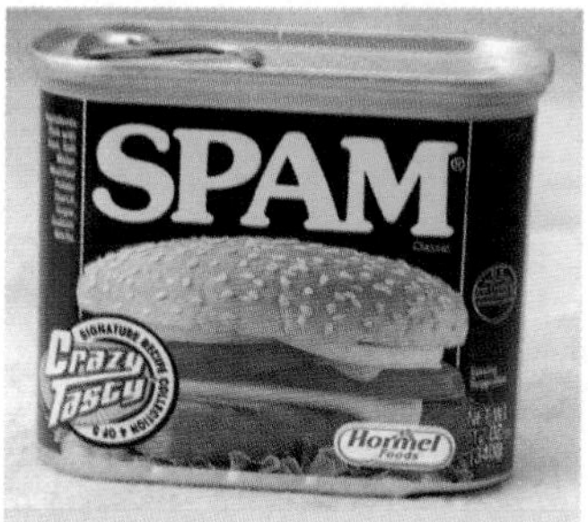

Spam, a widely consumed canned meat in Alaska, Hawaii, and other parts of the United States, contains sodium nitrite.

Examples of two common nitrosamines:

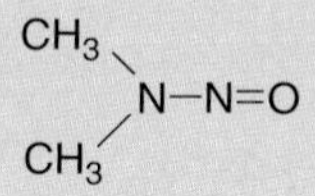

N-nitrosodimethylamine

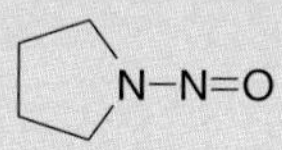

N-nitrosopyrrolidine

S_N1 reactions are thought to play a role in how **nitrosamines,** compounds having the general structure $\mathbf{R_2NN{=}O}$, act as toxins and carcinogens. Nitrosamines are present in many foods, especially cured meats and smoked fish, and they are also found in tobacco smoke, alcoholic beverages, and cosmetics. Nitrosamines cause many forms of cancer.

Nitrosamines are formed when amines that occur naturally in food react with sodium nitrite, $NaNO_2$, a preservative added to meats such as ham, bacon, and hot dogs to inhibit the growth of *Clostridium botulinum,* a bacterium responsible for a lethal form of food poisoning. Nitrosamines are also formed in vivo in the gastrointestinal tract when bacteria in the body convert nitrates (NO_3^-) into nitrites (NO_2^-), which then react with amines.

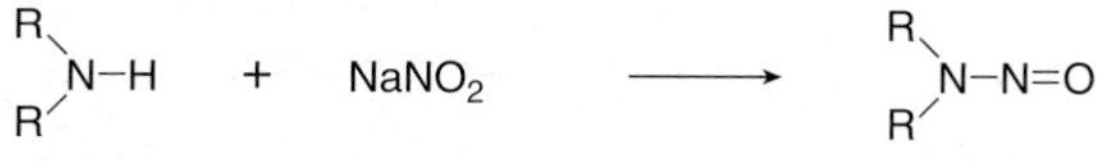

In the presence of acid or heat, nitrosamines are converted to **diazonium ions,** which contain a very good leaving group, N_2. With certain R groups, these diazonium compounds form carbocations, which then react with biological nucleophiles (such as DNA or an enzyme) in the cell. If this nucleophilic substitution reaction occurs at a crucial site in a biomolecule, it can disrupt normal cell function leading to cancer or cell death. This two-step process—loss of N_2 as a leaving group and reaction with a nucleophile—is an $\mathbf{S_N1}$ **reaction.**

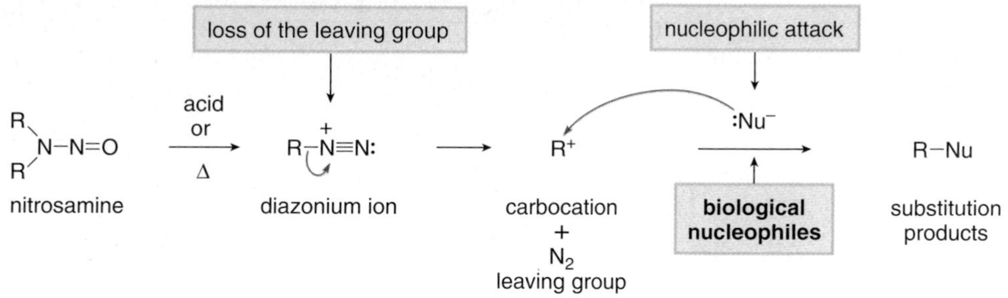

The preparation and reactions of diazonium ions (RN$_2^+$) are discussed in Chapter 25.

The use of sodium nitrite as a preservative is a classic example of the often delicate balance between risk and benefit. On the one hand, there is an enormous benefit in reducing the prevalence of fatal toxins in meats by the addition of sodium nitrite. On the other, there is the added risk that sodium nitrite may increase the level of nitrosamines in certain foods. Nitrites are still used as food additives, but the allowable level of nitrites in cured meats has been reduced.

7.17   When Is the Mechanism S$_N$1 or S$_N$2?

Given a particular starting material and nucleophile, how do we know whether a reaction occurs by the S$_N$1 or S$_N$2 mechanism? Four factors are examined:

◆ **The alkyl halide—CH$_3$X, RCH$_2$X, R$_2$CHX, or R$_3$CX**
◆ **The nucleophile—strong or weak**
◆ **The leaving group—good or poor**
◆ **The solvent—protic or aprotic**

7.17A   The Alkyl Halide—The Most Important Factor

The most important factor in determining whether a reaction follows the S$_N$1 or S$_N$2 mechanism is the *identity of the alkyl halide*.

◆ Increasing alkyl substitution favors S$_N$1.
◆ Decreasing alkyl substitution favors S$_N$2.

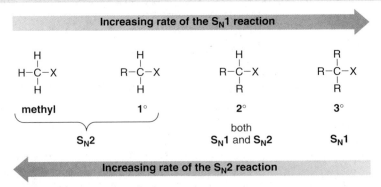

◆ Methyl and 1° halides (CH$_3$X and RCH$_2$X) undergo S$_N$2 reactions only.
◆ 3° Alkyl halides (R$_3$CX) undergo S$_N$1 reactions only.
◆ 2° Alkyl halides (R$_2$CHX) undergo both S$_N$1 and S$_N$2 reactions. Other factors determine the mechanism.

Examples are given in Figure 7.21.

PROBLEM 7.30   What is the likely mechanism of nucleophilic substitution for each alkyl halide?

a. CH$_3$—C(CH$_3$)(CH$_3$)—C(H)(CH$_3$)—Br

b. [structure with Br]

c. [cyclopentyl]—Br

d. [structure with Br]

Figure 7.21 Examples: The identity of RX and the mechanism of nucleophilic substitution

1° halide — S$_N$2

2° halide — Both S$_N$1 and S$_N$2 are possible.

3° halide — S$_N$1

7.17B   The Nucleophile

How does the strength of the nucleophile affect an S$_N$1 or S$_N$2 mechanism? The rate of the S$_N$1 reaction is unaffected by the identity of the nucleophile because the nucleophile does not appear in the rate equation (rate = k[RX]). The identity of the nucleophile *is* important for the S$_N$2 reaction, however, because the nucleophile does appear in the rate equation for this mechanism (rate = k[RX][:Nu$^-$]).

> ◆ Strong nucleophiles present in high concentration favor S$_N$2 reactions.
> ◆ Weak nucleophiles favor S$_N$1 reactions by decreasing the rate of any competing S$_N$2 reaction.

The most common nucleophiles in S$_N$2 reactions bear a net negative charge. The most common nucleophiles in S$_N$1 reactions are weak nucleophiles such as H$_2$O and ROH. The identity of the nucleophile is especially important in determining the mechanism and therefore the stereochemistry of nucleophilic substitution when 2° alkyl halides are starting materials.

Let's compare the substitution products formed when the 2° alkyl halide **A** (*cis*-1-bromo-4-methylcyclohexane) is treated with either the strong nucleophile $^-$OH or the weak nucleophile H$_2$O. Because a 2° alkyl halide can react by either mechanism, the strength of the nucleophile determines which mechanism takes place.

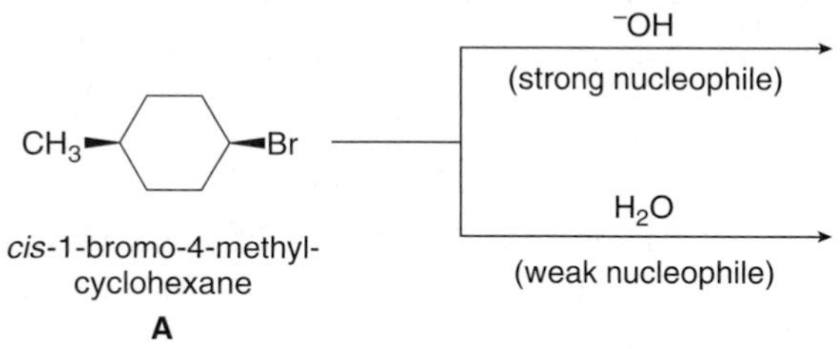

The **strong nucleophile $^-$OH favors an S$_N$2 reaction,** which occurs with **backside** attack of the nucleophile, resulting in **inversion of configuration.** Because the leaving group Br$^-$ is above the plane of the ring, the nucleophile attacks from below, and a single product **B** is formed.

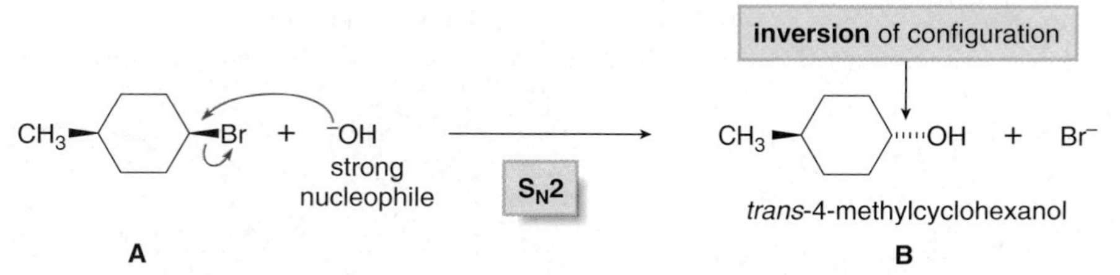

The **weak nucleophile H$_2$O favors an S$_N$1 reaction,** which occurs by way of an intermediate carbocation. Loss of the leaving group in **A** forms the carbocation, which undergoes nucleophilic attack from both above and below the plane of the ring to afford two products, **C** and **D.** Loss of a proton by proton transfer forms the final products, **B** and **E. B** and **E** are diastereomers of each other (**B** is a trans isomer and **E** is a cis isomer).

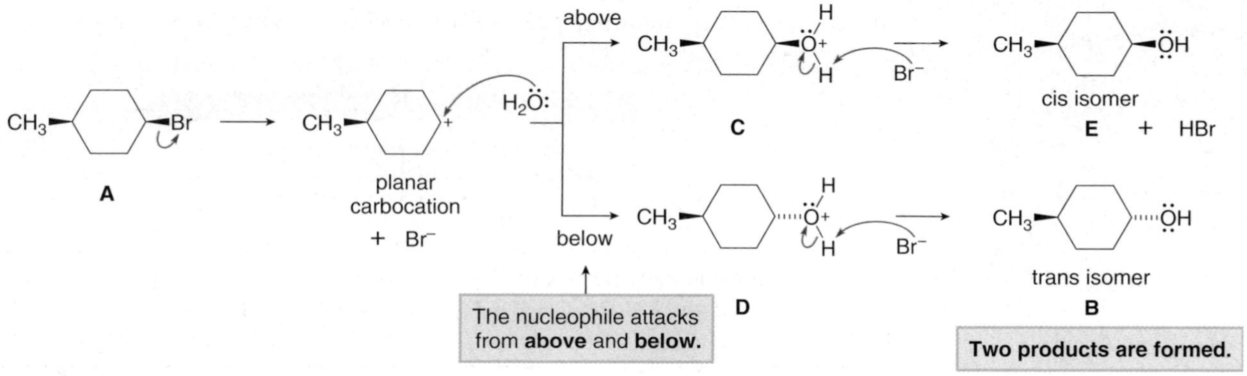

Thus, the mechanism of nucleophilic substitution determines the stereochemistry of the products formed.

PROBLEM 7.31   What is the likely mechanism (S_N1 or S_N2) for each reaction?

a.   [structure with Cl] $\xrightarrow{CH_3OH}$ [structure with OCH$_3$] $+$ HCl

b.   [cyclohexylmethyl Br] $\xrightarrow{^-SH}$ [cyclohexylmethyl SH] $+$ Br$^-$

c.   [cyclohexyl I] $\xrightarrow{CH_3CH_2O^-}$ [cyclohexyl OCH$_2$CH$_3$] $+$ I$^-$

d.   [structure with Br] $\xrightarrow{CH_3OH}$ [structure with OCH$_3$] $+$ HBr

---

PROBLEM 7.32   Draw the products (including stereochemistry) for each reaction.

a.   [structure with H Br] $+$ H$_2$O $\longrightarrow$ b.   [structure with Cl, H D] $+$ $^-$:C≡C–H $\longrightarrow$

7.17C   The Leaving Group

How does the identity of the leaving group affect an S_N1 or S_N2 reaction? Unlike nucleophilicity, the identity of the leaving group does *not* favor one mechanism over the other.

◆ **A better leaving group increases the rate of both S_N1 and S_N2 reactions.**

Because the bond to the leaving group is partially broken in the transition state of the only step of the S_N2 mechanism and the slow step of the S_N1 mechanism, **a better leaving group increases the rate of both reactions.** The better the leaving group, the more willing it is to accept the electron pair in the C–X bond, and the faster the reaction.

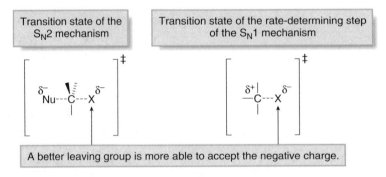

Transition state of the S_N2 mechanism

Transition state of the rate-determining step of the S_N1 mechanism

A better leaving group is more able to accept the negative charge.

For alkyl halides, the following order of reactivity is observed for the S_N1 and the S_N2 mechanisms:

R–F   R–Cl   R–Br   R–I

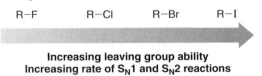

Increasing leaving group ability
Increasing rate of S_N1 and S_N2 reactions

PROBLEM 7.33   Which compound in each pair reacts faster in nucleophilic substitution?
a. $CH_3CH_2CH_2Cl$ and $CH_3CH_2CH_2I$ c. $(CH_3)_3COH$ and $(CH_3)_3COH_2^+$
b. $(CH_3)_3CBr$ and $(CH_3)_3CI$ d. $CH_3CH_2CH_2OH$ and $CH_3CH_2CH_2OCOCH_3$

7.17D   The Solvent

See Section 7.8C to review the differences between polar protic solvents and polar aprotic solvents.

Polar protic solvents and polar aprotic solvents affect the rates of S$_N$1 and S$_N$2 reactions differently.

> ◆ Polar *protic* solvents are especially good for S$_N$1 reactions.
> ◆ Polar *aprotic* solvents are especially good for S$_N$2 reactions.

Summary of solvent effects:
◆ Polar protic solvents favor S$_N$1 reactions because the ionic intermediates are stabilized by solvation.
◆ Polar aprotic solvents favor S$_N$2 reactions because nucleophiles are not well solvated, and therefore are more nucleophilic.

Polar protic solvents like H_2O and ROH solvate both cations and anions well, and this characteristic is important for the S$_N$1 mechanism, in which two ions (a carbocation and a leaving group) are formed by heterolysis of the $C-X$ bond. The carbocation is solvated by ion–dipole interactions with the polar solvent, and the leaving group is solvated by hydrogen bonding, in much the same way that Na^+ and Br^- are solvated in Section 7.8C. These interactions stabilize the reactive intermediate. In fact, a polar protic solvent is generally needed for an S$_N$1 reaction. S$_N$1 reactions do not occur in the gas phase, because there is no solvent to stabilize the intermediate ions.

Polar aprotic solvents exhibit dipole–dipole interactions but not hydrogen bonding, and as a result, they do not solvate anions well. This has a pronounced effect on the nucleophilicity of anionic nucleophiles. Because these nucleophiles are not "hidden" by strong interactions with the solvent, they are **more nucleophilic.** Because stronger nucleophiles favor S$_N$2 reactions, **polar aprotic solvents are especially good for S$_N$2 reactions.**

PROBLEM 7.34   Which solvents favor S$_N$1 reactions and which favor S$_N$2 reactions?
a. CH_3CH_2OH b. CH_3CN c. CH_3COOH d. $CH_3CH_2OCH_2CH_3$

PROBLEM 7.35   For each reaction, use the identity of the alkyl halide and nucleophile to determine which substitution mechanism occurs. Then determine which solvent affords the faster reaction.

a. $(CH_3)_3CBr$ + H_2O $\xrightarrow[\substack{\text{or} \\ (CH_3)_2C=O}]{H_2O}$ $(CH_3)_3COH$ + HBr

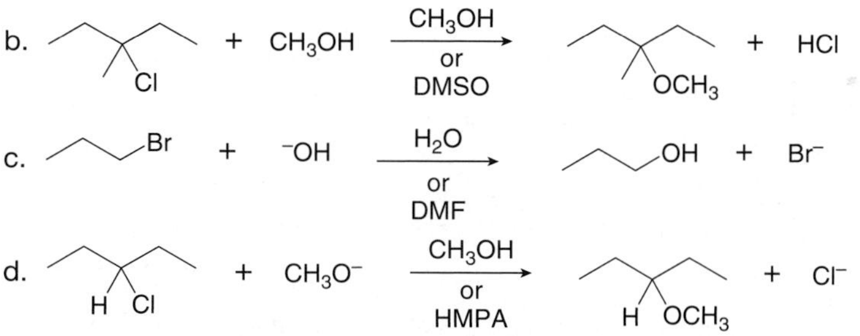

b. + CH_3OH $\xrightarrow[\substack{\text{or} \\ \text{DMSO}}]{CH_3OH}$ + HCl

c. + ^-OH $\xrightarrow[\substack{\text{or} \\ \text{DMF}}]{H_2O}$ + Br^-

d. + CH_3O^- $\xrightarrow[\substack{\text{or} \\ \text{HMPA}}]{CH_3OH}$ + Cl^-

---

7.17E   Summary of Factors that Determine Whether the S$_N$1 or S$_N$2 Mechanism Occurs

Table 7.7 summarizes the factors that determine whether a reaction occurs by the S$_N$1 or S$_N$2 mechanism. Sample Problems 7.4 and 7.5 illustrate how these factors are used to determine the mechanism of a given reaction.

SAMPLE PROBLEM 7.4   Determine the mechanism of nucleophilic substitution for each reaction and draw the products.

a. $CH_3CH_2CH_2-Br$ + $^-:C\equiv CH$ $\longrightarrow$ b. 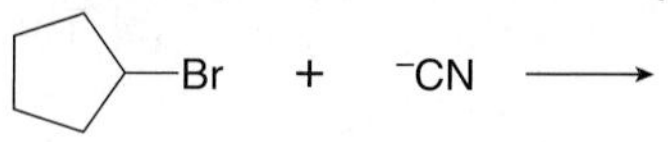 —Br + ^-CN $\longrightarrow$

SOLUTION

[a] The alkyl halide is 1° so it must react by an S$_N$2 mechanism with the nucleophile $^-:C\equiv CH$.

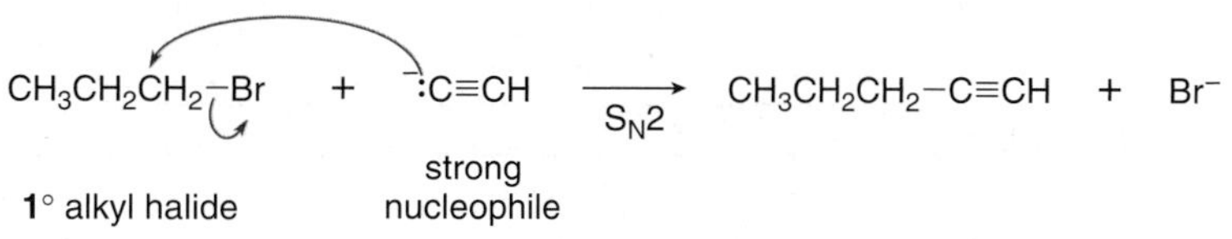

$$CH_3CH_2CH_2-Br \;+\; {}^-:C\equiv CH \xrightarrow{S_N2} CH_3CH_2CH_2-C\equiv CH \;+\; Br^-$$

1° alkyl halide   strong nucleophile

TABLE 7.7   Summary of Factors that Determine the S$_N$1 or S$_N$2 Mechanism

Alkyl halide	Mechanism	Other factors
CH$_3$X RCH$_2$X (1°)	**S$_N$2**	Favored by • **strong nucleophiles** (usually a net negative charge) • polar **aprotic** solvents
R$_3$CX (3°)	**S$_N$1**	Favored by • **weak nucleophiles** (usually neutral) • polar **protic** solvents
R$_2$CHX (2°)	**S$_N$1 or S$_N$2**	The mechanism depends on the conditions. • **Strong nucleophiles favor the S$_N$2 mechanism over the S$_N$1 mechanism.** For example, RO$^-$ is a stronger nucleophile than ROH, so RO$^-$ favors the S$_N$2 reaction and ROH favors the S$_N$1 reaction. • **Protic solvents favor the S$_N$1 mechanism and aprotic solvents favor the S$_N$2 mechanism.** For example, H$_2$O and CH$_3$OH are polar protic solvents that favor the S$_N$1 mechanism, whereas acetone [(CH$_3$)$_2$C=O] and DMSO [(CH$_3$)$_2$S=O] are polar aprotic solvents that favor the S$_N$2 mechanism.

[b]   The alkyl halide is 2° so it can react by either the S$_N$1 or S$_N$2 mechanism. The strong nucleophile ($^-$CN) favors the S$_N$2 mechanism.

2° alkyl halide   strong nucleophile

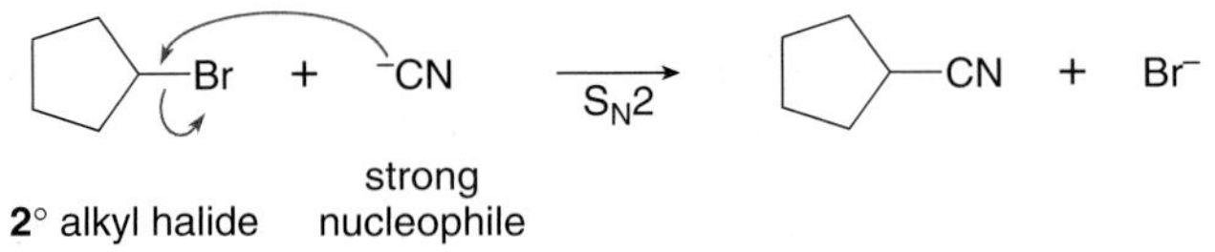

SAMPLE PROBLEM 7.5   Determine the mechanism of nucleophilic substitution for each reaction and draw the products, including stereochemistry.

SOLUTION

[a]   The 2° alkyl halide can react by either the S$_N$1 or S$_N$2 mechanism. The strong nucleophile ($^-$OCH$_3$) favors the S$_N$2 mechanism, as does the polar aprotic solvent (DMSO). S$_N$2 reactions proceed with inversion of configuration.

[b]   The alkyl halide is 3° so it reacts by an S$_N$1 mechanism with the weak nucleophile CH$_3$OH. S$_N$1 reactions proceed with racemization at a single stereogenic center, so two products are formed.

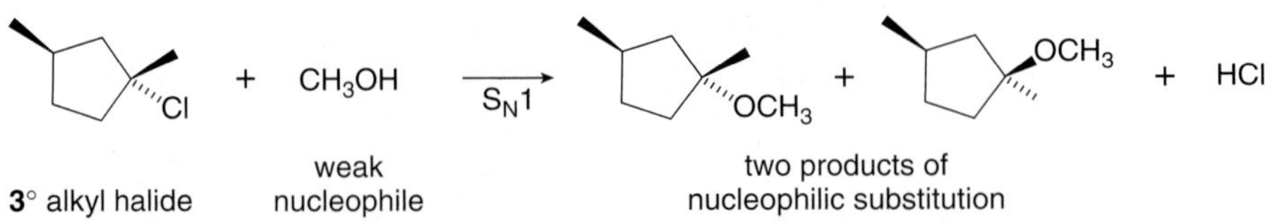

3° alkyl halide   weak nucleophile   two products of nucleophilic substitution

PROBLEM 7.36   Determine the mechanism and draw the products of each reaction. Include the stereochemistry at all stereogenic centers.

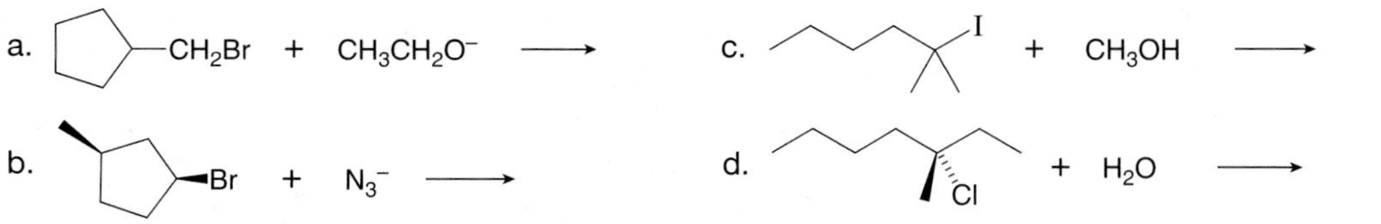

a. [cyclopentyl]—CH₂Br + CH₃CH₂O⁻ ⟶

b. [cyclopentane with methyl wedge and Br] + N₃⁻ ⟶

c. [branched chain with I] + CH₃OH ⟶

d. [branched chain with Cl] + H₂O ⟶

7.18 Vinyl Halides and Aryl Halides

S_N1 and S_N2 reactions occur only at sp^3 hybridized carbon atoms. Now that we have learned about the mechanisms for nucleophilic substitution we can understand why **vinyl halides** and **aryl halides,** which have a halogen atom bonded to an sp^2 hybridized C, do not undergo nucleophilic substitution by either the S_N1 or S_N2 mechanism. The discussion here centers on vinyl halides, but similar arguments hold for aryl halides as well.

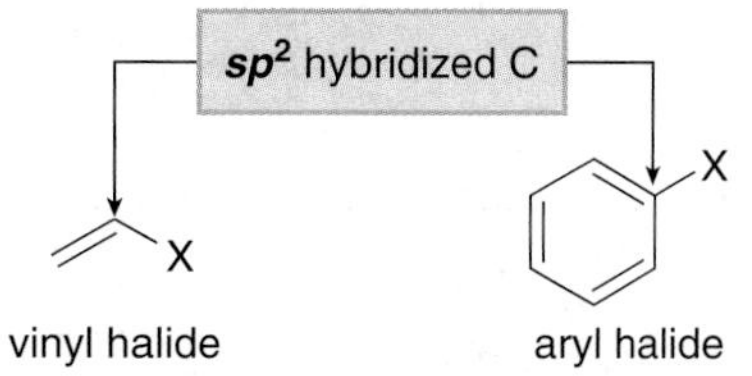

$$\boxed{sp^2 \text{ hybridized C}}$$

vinyl halide   aryl halide

Vinyl halides do not undergo an S_N2 reaction in part for electronic reasons, because the electron-rich π bond prevents backside attack by the electron-rich nucleophile. The nucleophile would have to approach the leaving group in the plane of the double bond (Figure 7.22a). Steric factors also make approach in the plane of the double bond difficult.

Vinyl halides do not undergo S_N1 reactions because heterolysis of the C–X bond would form a **highly unstable vinyl carbocation.** Because this carbocation has only two groups around the positively charged carbon, it is sp hybridized (Figure 7.22b). These carbocations are even less stable than 1° carbocations, so the S_N1 reaction does not take place.

PROBLEM 7.37   Rank the following carbocations in order of increasing stability.

a. CH₃CH₂CH₂CH₂CH=C⁺H   b. CH₃CH₂CH₂CH₂C⁺HCH₃   c. CH₃CH₂CH₂CH₂CH₂C⁺H₂

Figure 7.22 Vinyl halides and nucleophilic substitution mechanisms

[a] Vinyl halides and the S_N2 mechanism:
Backside attack of the nucleophile is not possible.

:Nu⁻

Backside attack would force the nucleophile to approach in the plane of the double bond.

[b] Vinyl halides and the S_N1 mechanism:
Heterolysis of the C–X bond forms a very unstable carbocation, making the rate-determining step very slow.

sp hybridized

H\C=C\Br (with H's) ⟶ H\C=C⁺–H + Br⁻

a vinyl carbocation
highly unstable

7.19   Organic Synthesis

Thus far we have concentrated on the starting material in nucleophilic substitution—the alkyl halide—and have not paid much attention to the product formed. Nucleophilic substitution reactions, and in particular S_N2 reactions, introduce a wide variety of different functional groups in molecules, depending on the nucleophile. For example, when ^-OH, ^-OR, and ^-CN are used as nucleophiles, the products are alcohols (ROH), ethers (ROR), and nitriles (RCN), respectively. Table 7.8 lists some functional groups readily introduced using nucleophilic substitution.

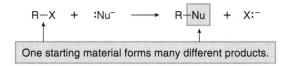

One starting material forms many different products.

TABLE 7.8 Molecules Synthesized from R–X by the S_N2 Reaction			
	Nucleophile (:Nu⁻)	**Product**	**Name**
Oxygen compounds	^-OH	R–OH	alcohol
	$^-OR'$	R–OR'	ether
	⁻O–C(=O)–R'	R–O–C(=O)–R'	ester
Carbon compounds	^-CN	R–CN	nitrile
	:C≡C–H	R–C≡C–H	alkyne
Nitrogen compounds	N_3^-	R–N_3	azide
	:NH_3	R–NH_2	amine
Sulfur compounds	^-SH	R–SH	thiol
	$^-SR'$	R–SR'	sulfide

the products of nucleophilic substitution

By thinking of **nucleophilic substitution as a reaction that *makes* a particular kind of organic compound,** we begin to think about *synthesis.*

◆ Organic synthesis is the systematic preparation of a compound from a readily available starting material by one or many steps.

7.19A   Background on Organic Synthesis

Chemists synthesize molecules for many reasons. Sometimes a **natural product,** a compound isolated from natural sources, has useful medicinal properties, but is produced by an organism in only minute quantities. Synthetic chemists then prepare this molecule from simpler starting materials so that it can be made available to a large number of people. **Taxol** (Section 5.5), the complex anticancer compound isolated in small amounts from the bark of the Pacific yew tree, is one such natural product. It is now synthesized commercially from a compound isolated from the leaves and needles of the European yew.

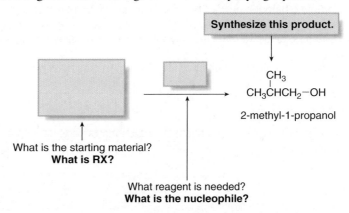

taxol

aspirin

Sometimes, chemists prepare molecules that do not occur in nature (although they may be similar to those in nature), because these molecules have superior properties to their naturally occurring relatives. **Aspirin, or acetylsalicylic acid** (Section 2.7), is a well known example. Acetylsalicylic acid is prepared from phenol, a product of the petroleum industry, by a two-step procedure (Figure 7.23). Aspirin has become one of the most popular and widely used drugs in the world because it has excellent analgesic and anti-inflammatory properties, *and* it is cheap and readily available.

> Phenol, the starting material for the aspirin synthesis, is a petroleum product, like most of the starting materials used in large quantities in industrial syntheses. A shortage of petroleum reserves thus affects the availability not only of fuels for transportation, but also of raw materials needed for most chemical synthesis.

7.19B   Nucleophilic Substitution and Organic Synthesis

To carry out synthesis we must think backwards. We examine a compound and ask: **What starting material and reagent are needed to make it?** If we are using nucleophilic substitution, we must determine what alkyl halide and what nucleophile can be used to form a specific product. This is the simplest type of synthesis because it involves only one step. In Chapter 11 we will learn about multistep syntheses.

Suppose, for example, that we are asked to prepare $(CH_3)_2CHCH_2OH$ (2-methyl-1-propanol) from an alkyl halide and any required reagents. To accomplish this synthesis, we must "fill in the boxes" for the starting material and reagent in the accompanying equation.

Synthesize this product.

$$CH_3CHCH_2-OH$$
with CH_3 above

2-methyl-1-propanol

What is the starting material?
What is RX?

What reagent is needed?
What is the nucleophile?

Figure 7.23   Synthesis of aspirin

- Aspirin is synthesized by a two-step procedure from simple, cheap starting materials.

phenol   [1] Na   [2] CO_2 [3] H_2O → (with OH and $COOH$) $(CH_3CO)_2O$ / acid → aspirin

To determine the two components needed for the synthesis, remember that the carbon atoms come from the organic starting material, in this case a 1° alkyl halide [(CH$_3$)$_2$CHCH$_2$Br]. The functional group comes from the nucleophile, ⁻OH in this case. With these two components, we can "fill in the boxes" to complete the synthesis.

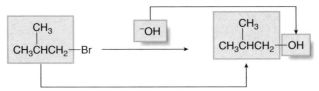

After any synthesis is proposed, check to see if it is reasonable, given what we know about reactions. Will the reaction written give a high yield of product? The synthesis of (CH$_3$)$_2$CHCH$_2$OH is reasonable, because the starting material is a 1° alkyl halide and the nucleophile (⁻OH) is strong, and both facts contribute to a successful S$_N$2 reaction.

PROBLEM 7.38   What alkyl halide and nucleophile are needed to prepare each compound?

a. [structure] CN   b. (CH$_3$)$_3$CCH$_2$CH$_2$SH   c. [structure] OH   d. CH$_3$CH$_2$—C≡C—H

PROBLEM 7.39   The ether, CH$_3$OCH$_2$CH$_3$, can be prepared by two different nucleophilic substitution reactions, one using CH$_3$O⁻ as nucleophile and the other using CH$_3$CH$_2$O⁻ as nucleophile. Draw both routes.

7.20   Key Concepts—Alkyl Halides and Nucleophilic Substitution

General Facts about Alkyl Halides

- Alkyl halides contain a halogen atom X bonded to an sp^3 hybridized carbon (7.1).
- Alkyl halides are named as halo alkanes, with the halogen as a substituent (7.2).
- Alkyl halides have a polar C−X bond, so they exhibit dipole–dipole interactions but are incapable of intermolecular hydrogen bonding (7.3).
- The polar C−X bond containing an electrophilic carbon makes alkyl halides reactive towards nucleophiles and bases (7.5).

The Central Theme (7.6)

- Nucleophilic substitution is one of the two main reactions of alkyl halides. A nucleophile replaces a leaving group on an sp^3 hybridized carbon.

- One σ bond is broken and one σ bond is formed.
- There are two possible mechanisms: S$_N$1 and S$_N$2.

S_N1 and S_N2 Mechanisms Compared

	S_N2 mechanism	S_N1 mechanism
[1] Mechanism	• One step (7.11B)	• Two steps (7.13B)
[2] Alkyl halide	• Order of reactivity: $CH_3X > RCH_2X >$ $R_2CHX > R_3CX$ (7.11D)	• Order of reactivity: $R_3CX > R_2CHX > RCH_2X >$ CH_3X (7.13D)
[3] Rate equation	• Rate = $k[RX][:Nu^-]$	• Rate = $k[RX]$
	• Second-order kinetics (7.11A)	• First-order kinetics (7.13A)
[4] Stereochemistry	• Backside attack of the nucleophile (7.11C)	• Trigonal planar carbocation intermediate (7.13C)
	• Inversion of configuration at a stereogenic center	• Racemization at a single stereogenic center
[5] Nucleophile	• Favored by stronger nucleophiles (7.17B)	• Favored by weaker nucleophiles (7.17B)
[6] Leaving group	• Better leaving group $--\rightarrow$ faster reaction (7.17C)	• Better leaving group $--\rightarrow$ faster reaction (7.17C)
[7] Solvent	• Favored by polar aprotic solvents (7.17D)	• Favored by polar protic solvents (7.17D)

Important Trends

- The best leaving group is the weakest base. Leaving group ability increases left-to-right across a row and down a column of the periodic table (7.7).
- Nucleophilicity decreases left-to-right across a row of the periodic table (7.8A).
- Nucleophilicity decreases down a column of the periodic table in polar aprotic solvents (7.8C).
- Nucleophilicity increases down a column of the periodic table in polar protic solvents (7.8C).
- The stability of a carbocation increases as the number of R groups bonded to the positively charged carbon increases (7.14).

Important Principles

Principle

- Electron-donating groups (such as R groups) stabilize a positive charge (7.14A).

- Steric hindrance decreases nucleophilicity but not basicity (7.8B).

- Hammond postulate: In an endothermic reaction, the more stable product is formed faster. In an exothermic reaction, this is not necessarily true (7.15).

- Planar, sp^2 hybridized atoms react with reagents from both sides of the plane (7.13C).

Example

- 3° Carbocations (R_3C^+) are more stable than 2° carbocations (R_2CH^+), which are more stable than 1° carbocations (RCH_2^+).

- $(CH_3)_3CO^-$ is a stronger base but a weaker nucleophile than $CH_3CH_2O^-$.

- S_N1 reactions are faster when more stable (more substituted) carbocations are formed, because the rate-determining step is endothermic.

- A trigonal planar carbocation reacts with nucleophiles from both sides of the plane.

Problems

Nomenclature

7.40 Give the IUPAC name for each compound.

a. $CH_3-\underset{\underset{CH_3}{|}}{\overset{\overset{CH_3}{|}}{C}}-CH_2CH_2F$

c. $(CH_3)_3CCH_2Br$

e.

g. $(CH_3)_3CCH_2CH(Cl)CH_2Cl$

b.

d.

f.

h.

(Also, label this compound as *R* or *S*.)

7.41 Give the structure corresponding to each name.
- a. isopropyl bromide
- b. 3-bromo-4-ethylheptane
- c. 1,1-dichloro-2-methylcyclohexane
- d. *trans*-1-chloro-3-iodocyclobutane
- e. 1-bromo-4-ethyl-3-fluorooctane

7.42 Classify each alkyl halide in Problem 7.40 as 1°, 2°, or 3°. When a compound has more than one halogen, assign each separately.

7.43 Draw the eight constitutional isomers having the molecular formula $C_5H_{11}Cl$.
- a. Give the IUPAC name for each compound (ignoring R and S designations).
- b. Label any stereogenic centers.
- c. For each constitutional isomer that contains a stereogenic center, draw all possible stereoisomers, and label each stereogenic center as R or S.

Physical Properties

7.44 Which compound in each pair has the higher boiling point?
- a. $(CH_3)_3CBr$ and $CH_3CH_2CH_2CH_2Br$
- b. 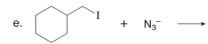 and
- c. and

General Nucleophilic Substitution, Leaving Groups, and Nucleophiles

7.45 Draw the substitution product that results when $CH_3CH_2CH_2CH_2Br$ reacts with each nucleophile.
- a. ⁻OH
- b. ⁻SH
- c. ⁻CN
- d. ⁻OCH(CH₃)₂
- e. ⁻C≡CH
- f. H_2O
- g. NH_3
- h. NaI
- i. NaN_3

7.46 Draw the products of each nucleophilic substitution reaction.

a. [structure with Cl] + [CH₃C(=O)O⁻] ⟶

b. [CH₃CH₂CH₂CH₂CH₂I] + NaCN ⟶

c. [cyclohexane with I] + H_2O ⟶

d. [cyclopentane with Cl] + CH_3CH_2OH ⟶

e. [structure with I] + N_3^- ⟶

f. [cyclohexane with Br] + $NaOCH_3$ ⟶

g. $(CH_3)_3CBr$ + CH_3COOH ⟶

h. [structure with Cl] + CH_3SCH_3 ⟶

7.47 Which of the following molecules contain a good leaving group?

a. [cyclohexane with OH]
b. $CH_3CH_2CH_2CH_2Cl$
c. [cyclohexane]
d. [cyclohexane with $\overset{+}{O}H_2$]
e. $CH_3CH_2NH_2$
f. $CH_3CH_2CH_2I$

7.48 Rank the species in each group in order of increasing leaving group ability.
- a. ⁻OH, F⁻, ⁻NH₂
- b. H_2O, ⁻NH₂, ⁻OH
- c. Br⁻, Cl⁻, I⁻
- d. NH_3, H_2S, H_2O

7.49 Which of the following nucleophilic substitution reactions will take place?

a. [cyclohexane with NH₂] + I⁻ ⟶ [cyclohexane with I] + ⁻NH₂

b. CH_3CH_2I + CH_3O^- ⟶ $CH_3CH_2OCH_3$ + I⁻

c. [structure with OH] + F⁻ ⟶ [structure with F] + ⁻OH

d. [structure with CN] + I⁻ ⟶ [structure with I] + ⁻CN

7.50 What nucleophile is needed to carry out each reaction?

a. $CH_3CH_2CH_2CH_2Br \longrightarrow CH_3CH_2CH_2CH_2OH$

b.

c.

d. $CH_3CH_2Cl \longrightarrow CH_3CH_2OCH_2CH_3$

e.

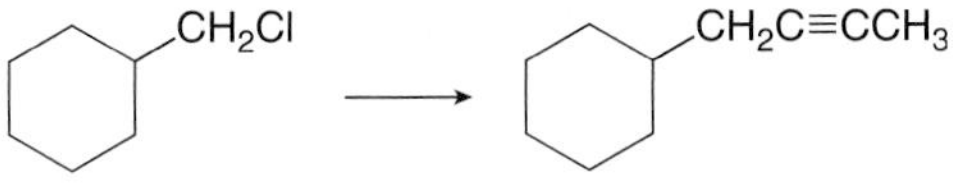

f. $CH_3CH_2Br \longrightarrow CH_3CH_2\overset{+}{N}(CH_3)_3 \ Br^-$

7.51 Rank the species in each group in order of increasing nucleophilicity.

a. CH_3^-, ^-OH, $^-NH_2$

b. H_2O, ^-OH, ^-SH in CH_3OH

c. $CH_3CH_2S^-$, $CH_3CH_2O^-$, CH_3COO^- in CH_3OH

d. CH_3NH_2, CH_3SH, CH_3OH in acetone

e. ^-OH, F^-, Cl^- in acetone

f. HS^-, F^-, Cl^- in CH_3OH

7.52 We have learned that nucleophilicity depends on the solvent when nucleophiles of very different size are compared. For example, in polar protic solvents nucleophilicity follows the trend $F^- < Cl^- < Br^- < I^-$, but the reverse is true in polar aprotic solvents. Predict the trend in nucleophilicity for these halide anions in the gas phase, with no solvent.

7.53 Classify each solvent as protic or aprotic.

a. $(CH_3)_2CHOH$

b. CH_3NO_2

c. CH_2Cl_2

d. NH_3

e. $N(CH_3)_3$

f. $HCONH_2$

The S_N2 Reaction

7.54 Consider the following S_N2 reaction:

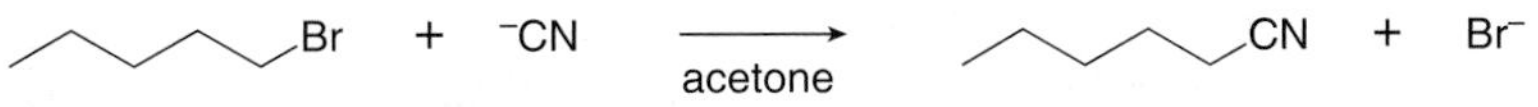

a. Draw a mechanism using curved arrows.

b. Draw an energy diagram. Label the axes, the reactants, products, E_a, and $\Delta H°$. Assume that the reaction is exothermic.

c. Draw the structure of the transition state.

d. What is the rate equation?

e. What happens to the reaction rate in each of the following instances? [1] The leaving group is changed from Br^- to I^-; [2] The solvent is changed from acetone to CH_3CH_2OH; [3] The alkyl halide is changed from $CH_3(CH_2)_4Br$ to $CH_3CH_2CH_2CH(Br)CH_3$; [4] The concentration of ^-CN is increased by a factor of five; and [5] The concentrations of both the alkyl halide and ^-CN are increased by a factor of five.

7.55 Rank the alkyl halides in each group in order of increasing S_N2 reactivity.

a.

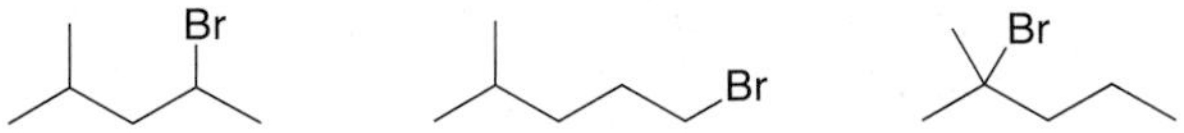

b.

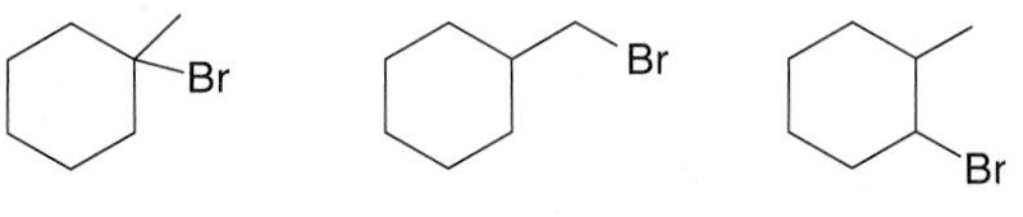

c.

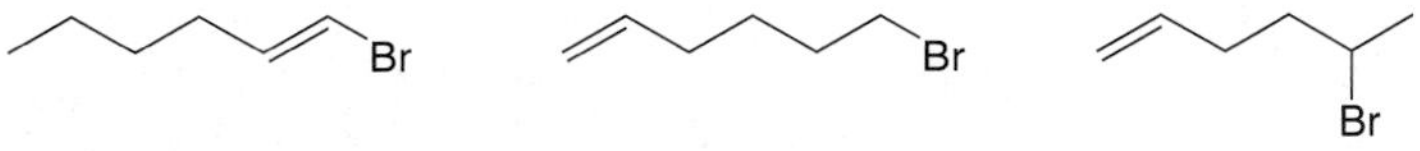

7.56 Which S_N2 reaction in each pair is faster?

a. $CH_3CH_2Br + {}^-OH \longrightarrow$

$CH_3CH_2Cl + {}^-OH \longrightarrow$

b.

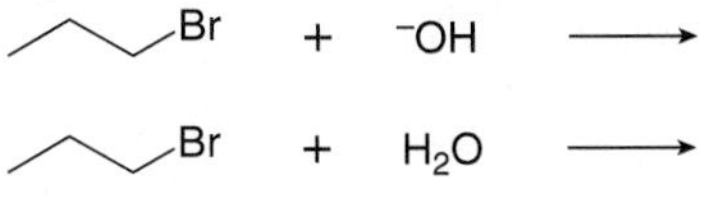

c.

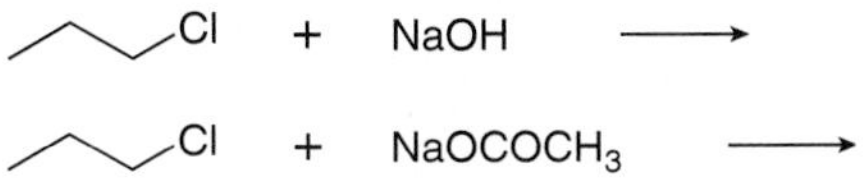

d.

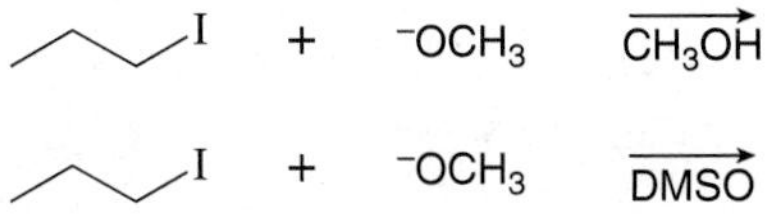

e.

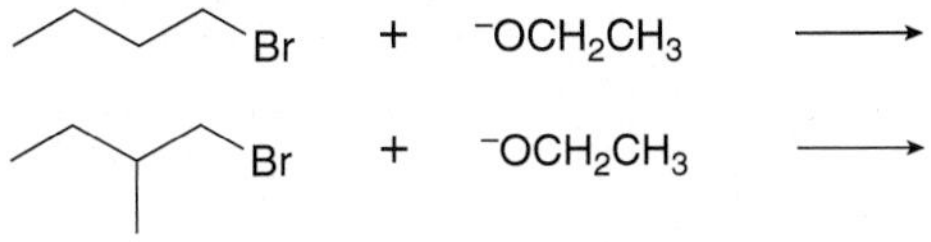

7.57 Draw the products of each S$_N$2 reaction and indicate the stereochemistry where appropriate.

a. 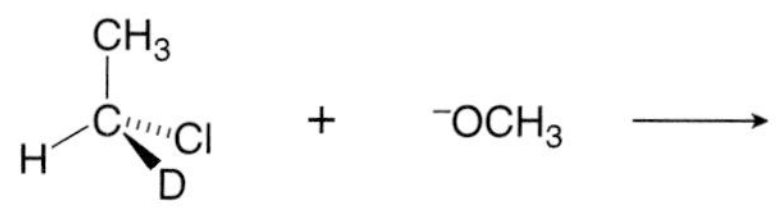 + $^-$OCH$_3$ ⟶

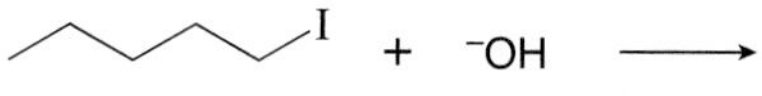

c. 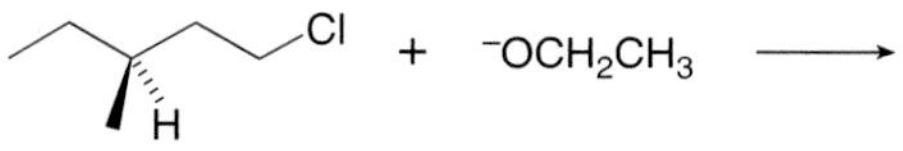 + $^-$OCH$_2$CH$_3$ ⟶

b. + $^-$OH ⟶

d. 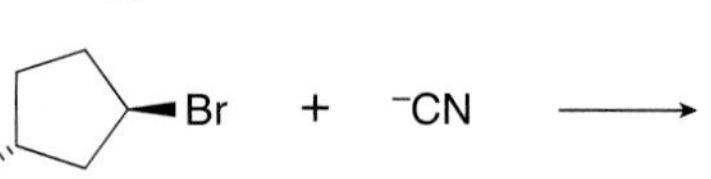 + $^-$CN ⟶

Carbocations

7.58 Classify each carbocation as 1°, 2°, or 3°.

a. CH$_3$CH$_2$$\overset{+}{C}HCH_2CH_3$ b.   c. (CH$_3$)$_2$CHCH$_2$$\overset{+}{C}H_2$ d. 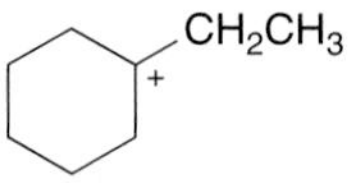   e. 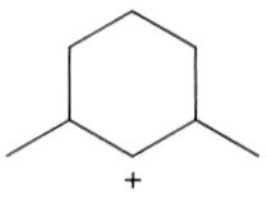   f.

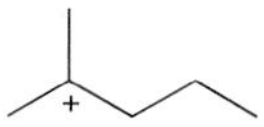

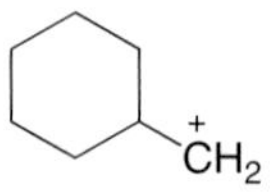

7.59 Rank the carbocations in each group in order of increasing stability.

a.

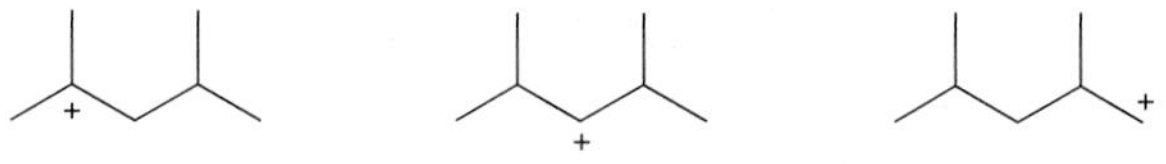

b.

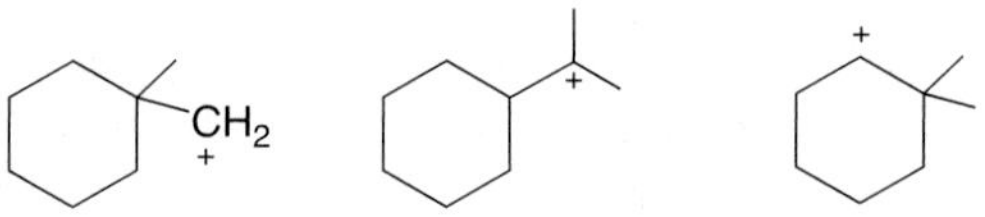

The S$_N$1 Reaction

7.60 Consider the following S$_N$1 reaction.

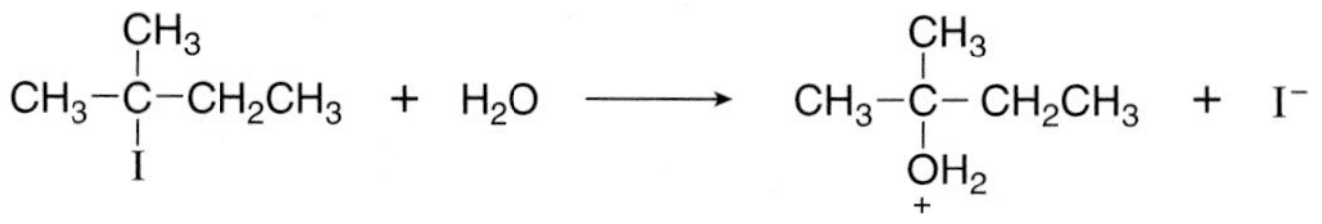

a. Draw a mechanism for this reaction using curved arrows.
b. Draw an energy diagram. Label the axes, starting material, product, E_a, and $\Delta H°$. Assume that the starting material and product are equal in energy.
c. Draw the structure of any transition states.
d. What is the rate equation for this reaction?
e. What happens to the reaction rate in each of the following instances? [1] The leaving group is changed from I$^-$ to Cl$^-$; [2] The solvent is changed from H$_2$O to DMF; [3] The alkyl halide is changed from (CH$_3$)$_2$C(I)CH$_2$CH$_3$ to (CH$_3$)$_2$CHCH(I)CH$_3$; [4] The concentration of H$_2$O is increased by a factor of five; and [5] The concentrations of both the alkyl halide and H$_2$O are increased by a factor of five.

7.61 Rank the alkyl halides in each group in order of increasing S$_N$1 reactivity.

a.

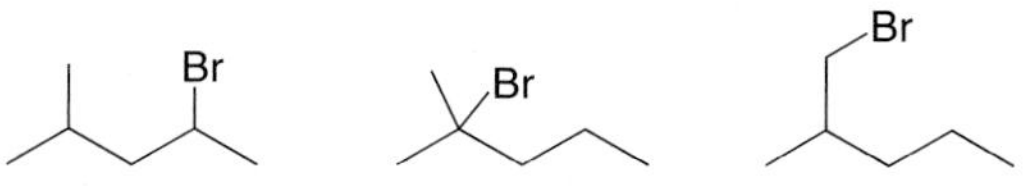

c.

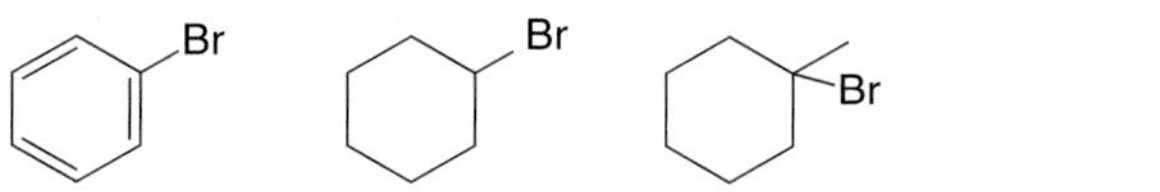

b.

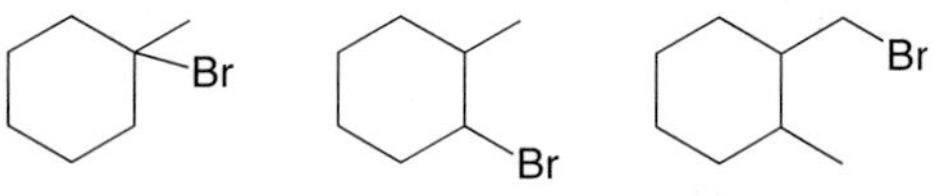

7.62 Which S$_N$1 reaction in each pair is faster?

a. (CH$_3$)$_3$CCl + H$_2$O ⟶

(CH$_3$)$_3$CI + H$_2$O ⟶

c. 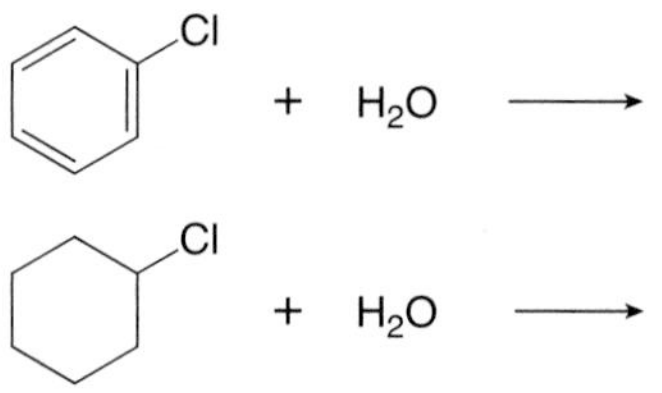 + H$_2$O ⟶

+ H$_2$O ⟶

b. 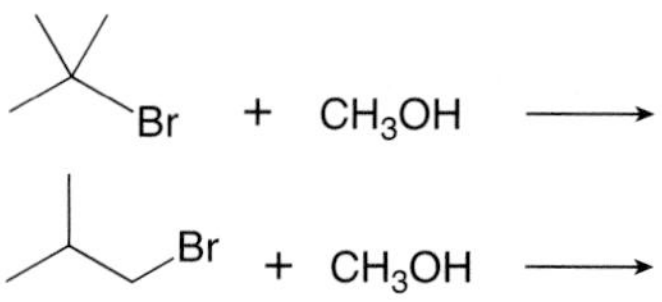 + CH$_3$OH ⟶

+ CH$_3$OH ⟶

d. 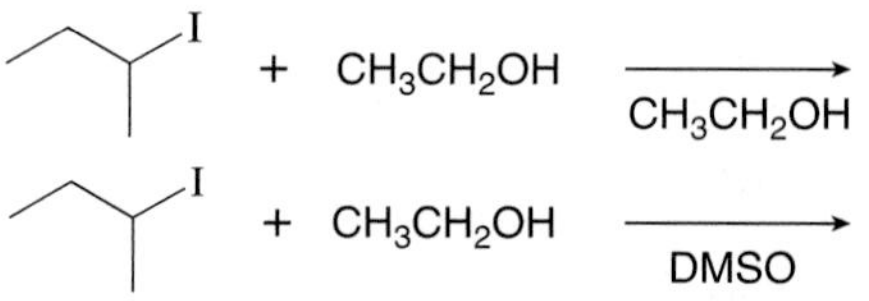 + CH$_3$CH$_2$OH $\xrightarrow{\text{CH}_3\text{CH}_2\text{OH}}$

+ CH$_3$CH$_2$OH $\xrightarrow{\text{DMSO}}$

7.63 Draw the products of each S$_N$1 reaction and indicate the stereochemistry when necessary.

a. 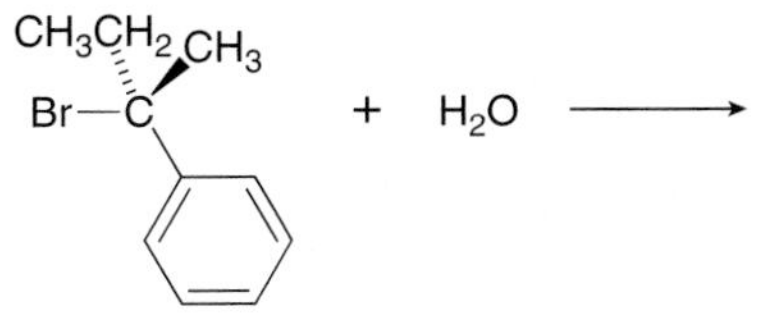   Br—C(CH$_3$CH$_2$)(CH$_3$)(C$_6$H$_5$)   +   H$_2$O   ⟶

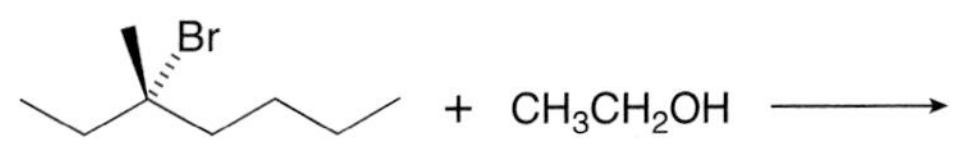

b. 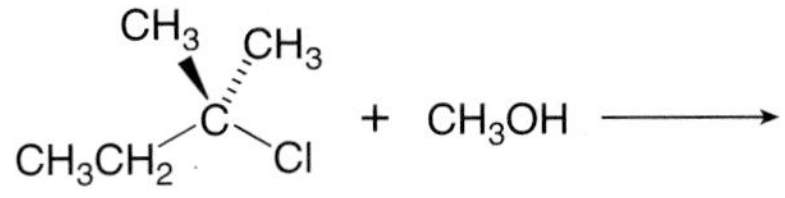   CH$_3$CH$_2$—C(CH$_3$)(CH$_3$)—Cl   +   CH$_3$OH   ⟶

c.   (structure with Br)   +   CH$_3$CH$_2$OH   ⟶

d. 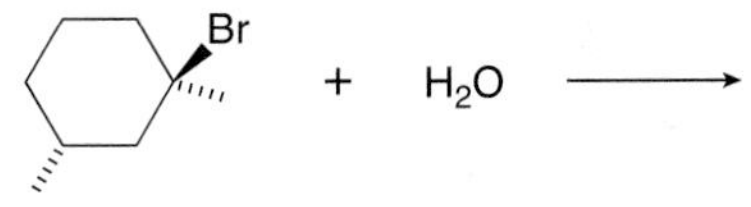   (cyclohexane with Br)   +   H$_2$O   ⟶

7.64 Draw a stepwise mechanism for the following reaction.

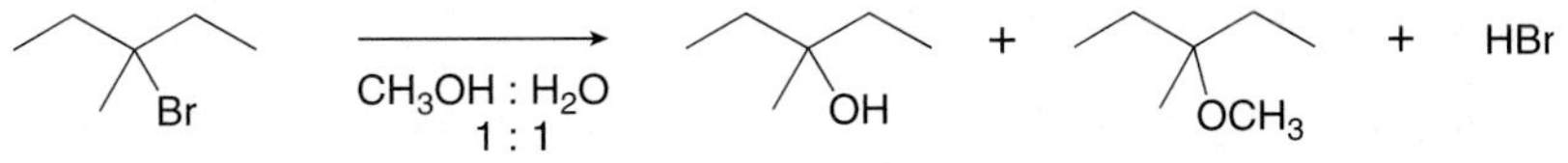

(structure)—Br $\xrightarrow{\text{CH}_3\text{OH : H}_2\text{O} \atop 1:1}$ (structure)—OH   +   (structure)—OCH$_3$ +   HBr

7.65 H$_2$O is a more polar medium than CH$_3$CH$_2$OH because the nonpolar CH$_3$CH$_2$ group of CH$_3$CH$_2$OH decreases the number of polar interactions in a given volume. Explain what happens to the rate of an S$_N$1 reaction as the solvent is changed from 100% H$_2$O to 50% H$_2$O/50% CH$_3$CH$_2$OH.

S$_N$1 and S$_N$2 Reactions

7.66 Draw a stepwise, detailed mechanism for each reaction. Use curved arrows to show the movement of electrons.

a. 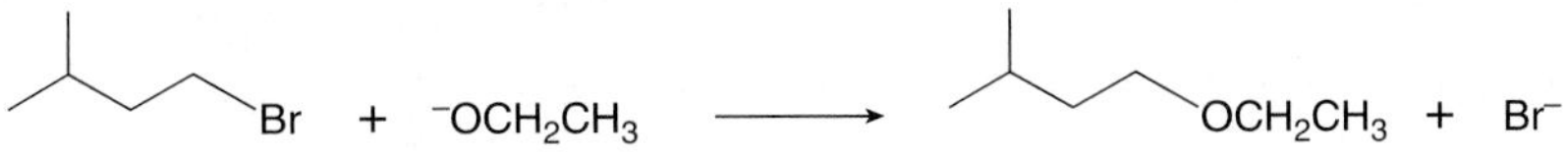   (structure)—Br   +   ⁻OCH$_2$CH$_3$ ⟶   (structure)—OCH$_2$CH$_3$ +   Br⁻

b.    (cyclohexane)—Br   +   CH$_3$CH$_2$OH   ⟶   (cyclohexane)—OCH$_2$CH$_3$ +   HBr

c.   CH$_3$CH$_2$CH$_2$CH$_2$NH$_2$ +   CH$_3$I $\xrightarrow[\text{(excess)}]{\text{K}_2\text{CO}_3}$ CH$_3$CH$_2$CH$_2$CH$_2$N(CH$_3$)$_3$⁺   +   I⁻

7.67 Determine the mechanism of nucleophilic substitution of each reaction and draw the products, including stereochemistry.

a. 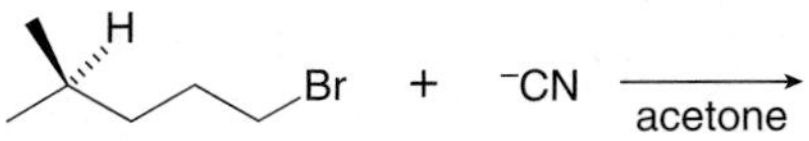   (structure)—Br   +   ⁻CN $\xrightarrow{\text{acetone}}$

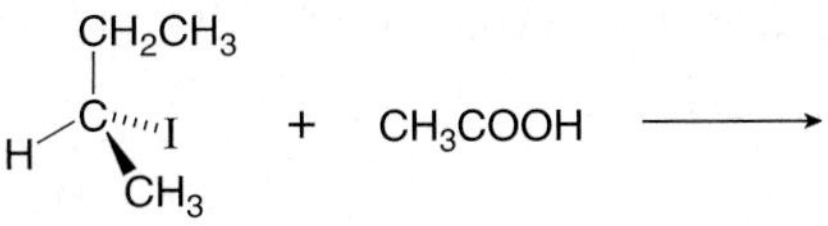

b. 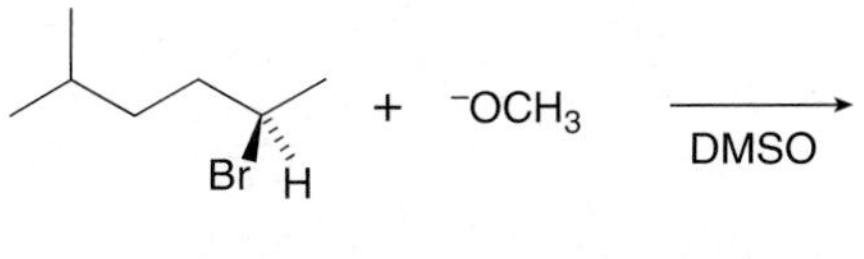   (structure)   +   ⁻OCH$_3$ $\xrightarrow{\text{DMSO}}$

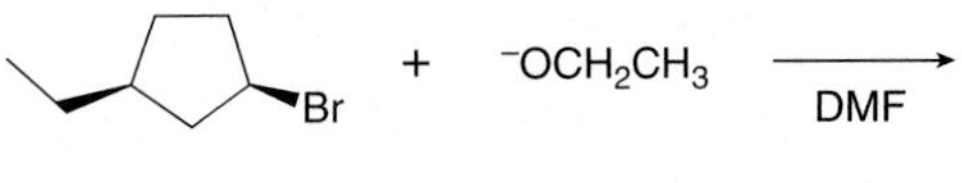

c. 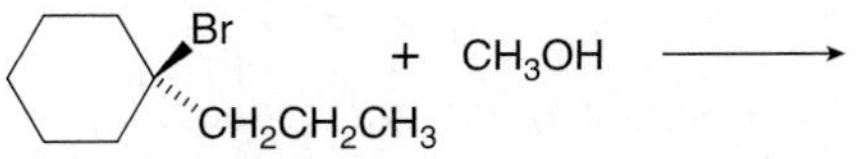   (cyclohexane structure)   +   CH$_3$OH   ⟶

d.   (structure)   +   CH$_3$COOH   ⟶

e.   (cyclopentane)—Br   +   ⁻OCH$_2$CH$_3$ $\xrightarrow{\text{DMF}}$

f. 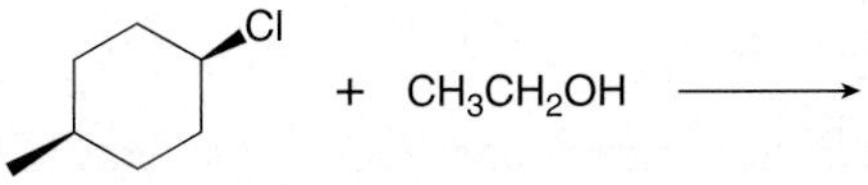   (cyclohexane)—Cl   +   CH$_3$CH$_2$OH   ⟶

7.68 Draw the products of each nucleophilic substitution reaction.

a. 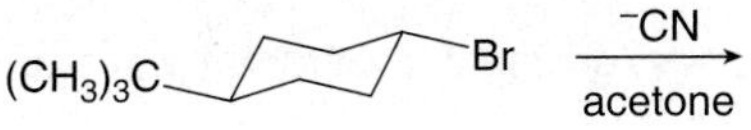   (CH$_3$)$_3$C—(cyclohexane)—Br $\xrightarrow[\text{acetone}]{⁻\text{CN}}$

b. 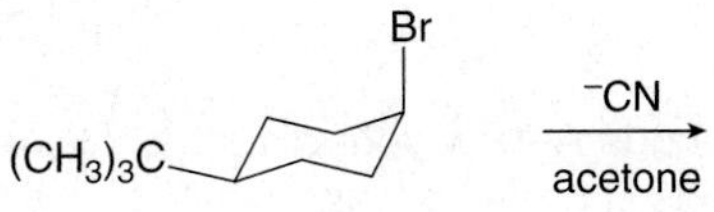   (CH$_3$)$_3$C—(cyclohexane)—Br $\xrightarrow[\text{acetone}]{⁻\text{CN}}$

7.69 When a single compound contains both a nucleophile and a leaving group, an **intramolecular** reaction may occur. With this in mind, draw the product of each reaction.

a. 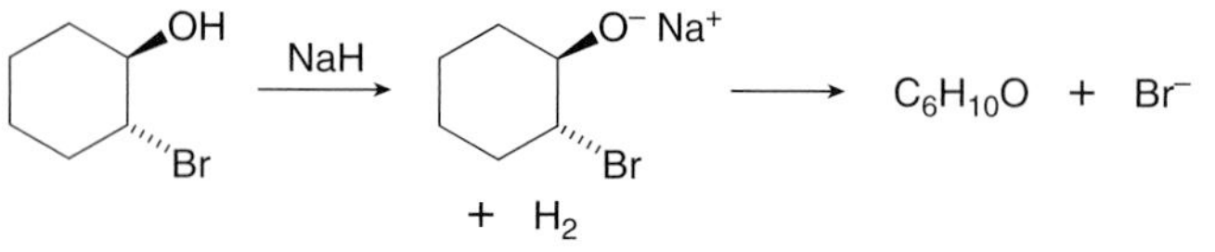

b. 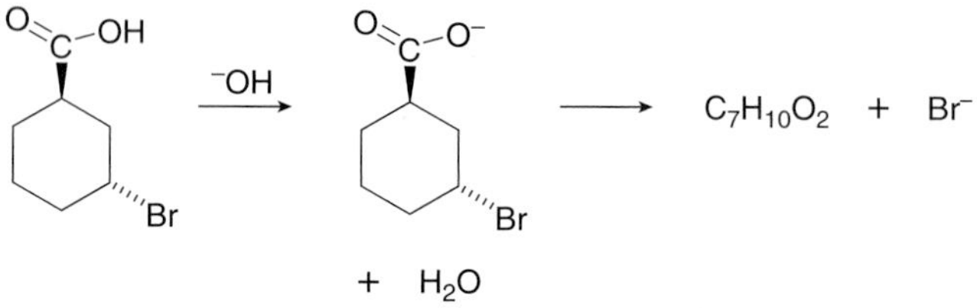

7.70 When CH_3I is reacted with $NaOCH_3$ and $NaSCH_3$, both CH_3OCH_3 and CH_3SCH_3 form, but the amount of each depends on the solvent. Predict which is the major product when CH_3OH is the solvent and when $(CH_3)_2S{=}O$ is the solvent. Explain why this happens.

7.71 Explain each of the following statements.

a. Hexane is not a common solvent for either S_N1 or S_N2 reactions.
b. $(CH_3)_3CO^-$ is a stronger base than $CH_3CH_2O^-$.
c. $(CH_3)_3CBr$ is more reactive than $(CH_3)_2C(CF_3)Br$ in S_N1 reactions.
d. $(CH_3)_3CBr$ reacts at the same rate with F^- and H_2O in substitution reactions even though F^- has a net negative charge.
e. When optically active (*R*)-2-bromobutane is added to a solution of NaBr in acetone, the solution gradually loses optical activity until it becomes optically inactive.

7.72 Simple organic halides like CH_3Cl and CH_2Cl_2 are slowly decomposed in the environment by reaction with H_2O. For example, the half-life (the time it takes for the initial concentration of a compound to be halved) for the conversion of CH_3Cl to CH_3OH by nucleophilic substitution is 339 days. Explain why this reaction is slow, and why CH_2Cl_2 reacts even more slowly with H_2O.

Synthesis

7.73 Fill in the appropriate reagent or starting material in each of the following reactions.

a. 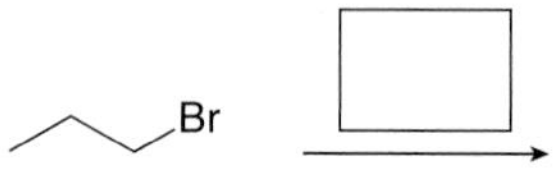

d. 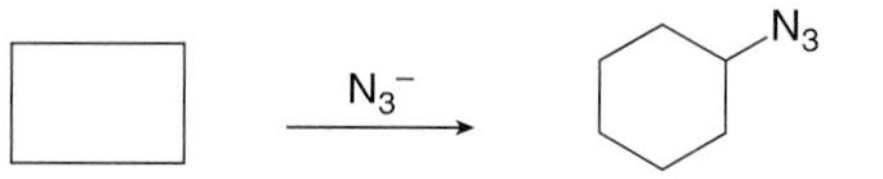

b. 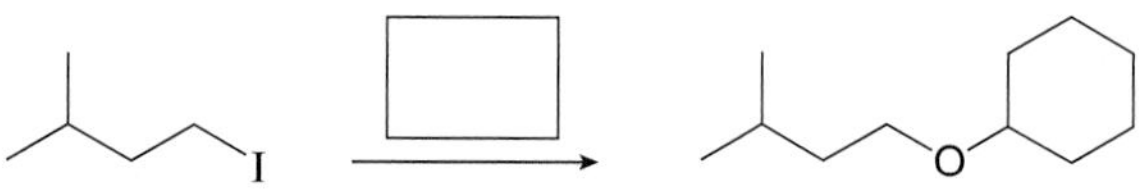

e. 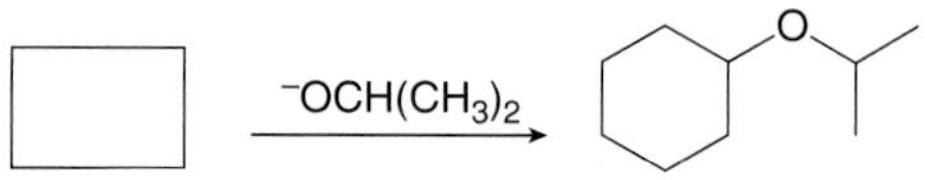

c. 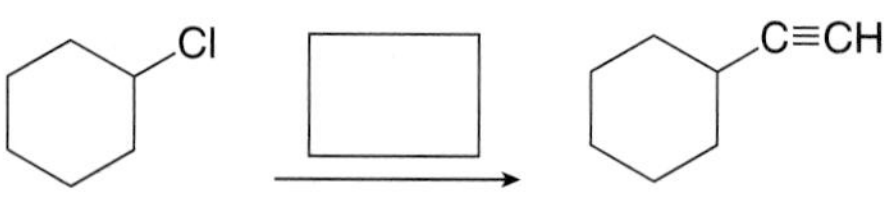

f. 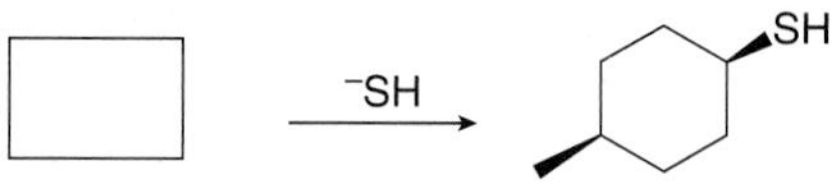

7.74 Devise a synthesis of each compound from an alkyl halide using any other organic or inorganic reagents.

a. 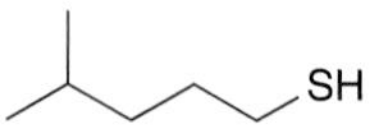

b. 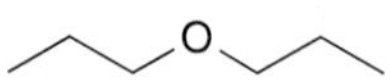

c. CH_3CH_2CN
d. 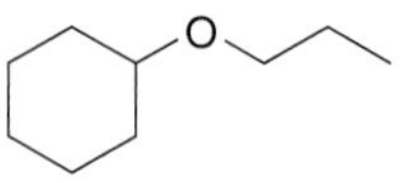

e. $CH_3CH_2OCOCH_3$

7.75 Suppose you have compounds **A–D** at your disposal. Using these compounds, devise two different ways to make **E**. Which one of these methods is preferred, and why?

CH_3I

A

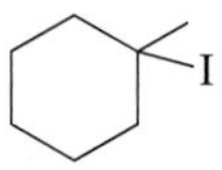

B

$NaOCH_3$

C

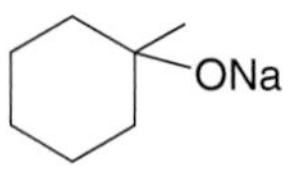

D

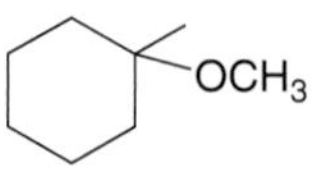

E

Challenge Questions

7.76 Explain why alkyl halides **A** and **B** do not undergo nucleophilic substitution by either an S_N1 or S_N2 mechanism.

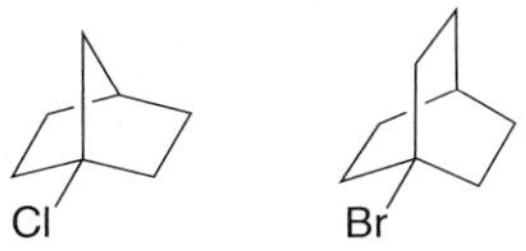

A   **B**

7.77 Explain why quinuclidine is a much more reactive nucleophile than triethylamine, even though both compounds have N atoms bonded to three R groups.

quinuclidine 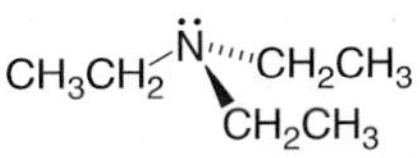   triethylamine

7.78 Draw a stepwise mechanism for the following reaction sequence.

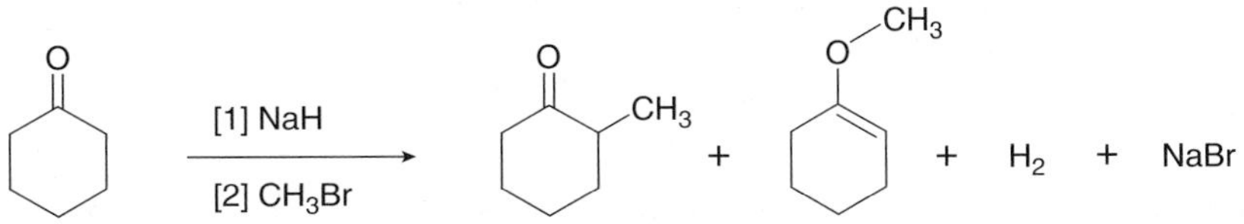

major product   minor product

7.79 Explain why compound **A** reacts readily with LiI in DMF to afford products of nucleophilic substitution, but isomer **B** does not react under similar reaction conditions.

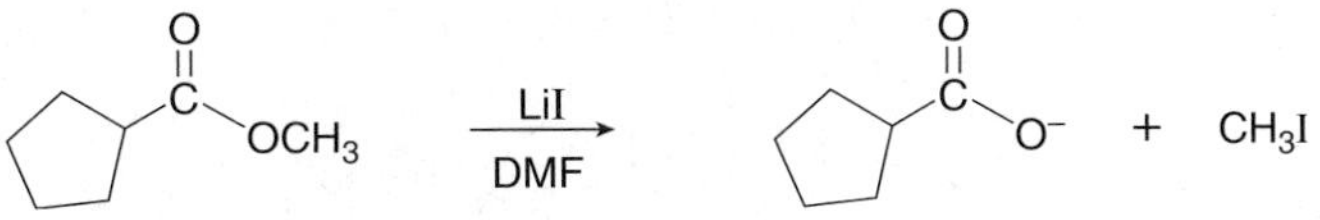

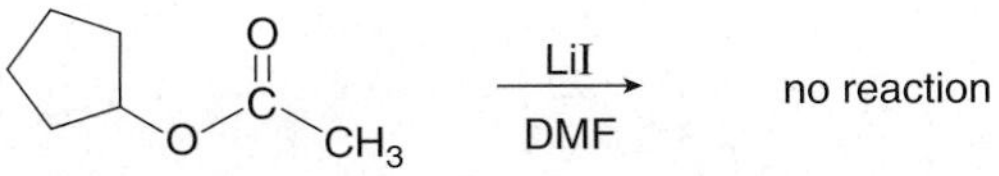

7.80 Explain why the alkyl halide $CH_3CH_2OCH_2Cl$ reacts rapidly with CH_3CH_2OH under S_N1 conditions to afford $CH_3CH_2OCH_2OCH_2CH_3$.

Alkyl Halides
and Elimination Reactions

Ethylene, the simplest hydrocarbon containing a carbon–carbon double bond, is a hormone that regulates plant growth and fruit ripening. As a result, bananas and tomatoes are often picked green, and then sprayed with ethylene when ripening is desired. Ethylene is also an important industrial starting material for the synthesis of millions of tons of the plastic polyethylene each year. Ethylene and other alkenes are prepared by **elimination reactions.** In Chapter 8 we learn about the details of elimination, the second general reaction of organic compounds.

Elimination reactions introduce π bonds into organic compounds, so they can be used to synthesize **alkenes** and **alkynes**—hydrocarbons that contain one and two π bonds, respectively. Like nucleophilic substitution, elimination reactions can occur by two different pathways, depending on the conditions. By the end of Chapter 8, therefore, you will have learned four different organic mechanisms, two for nucleophilic substitution (S_N1 and S_N2) and two for elimination (E1 and E2).

The biggest challenge with this material is learning how to sort out two different reactions that follow four different mechanisms. **Will a particular alkyl halide undergo substitution or elimination with a given reagent, and by which of the four possible mechanisms?** To answer this question, we conclude Chapter 8 with a summary that allows you to predict which reaction and mechanism are likely for a given substrate.

8.1 General Features of Elimination

All **elimination reactions** involve loss of elements from the starting material to form a new π bond in the product.

◆ Alkyl halides undergo elimination reactions with Brønsted–Lowry bases. The elements of HX are lost and an alkene is formed.

General elimination reaction

$$\text{—C—C—} + \text{:B} \longrightarrow \text{C=C} + \text{H—B}^+ + \text{X:}^-$$

base   new π bond
an alkene

elimination of HX

Equations [1] and [2] illustrate examples of elimination reactions. In both reactions a base removes the elements of an acid, HBr or HCl, from the organic starting material.

Examples	Base	Alkene	By-products
[1] $CH_3CH_2-\overset{H}{\underset{H}{C}}-\overset{H}{\underset{Br}{C}}-H$	$K^+\ {}^-OC(CH_3)_3$ [–HBr]	CH_3CH_2 $C=C$ H H H	$HOC(CH_3)_3$ $+ K^+ Br^-$
[2]	$Na^+\ {}^-OCH_2CH_3$ [–HCl]		$HOCH_2CH_3$ $+ Na^+ Cl^-$

Removal of the elements of HX, called **dehydrohalogenation,** is one of the most common methods to introduce a π bond and prepare an alkene. Dehydrohalogenation is an example of **β elimination,** because it involves loss of elements from two adjacent atoms: the **α carbon** bonded to the leaving group X, and the **β carbon** adjacent to it. Three curved arrows illustrate how four bonds are broken or formed in the process.

$$\text{B:} \quad \overset{\beta}{\text{—C—}}\overset{\alpha}{\text{C—}} \longrightarrow \overset{\beta}{\text{C}}=\overset{\alpha}{\text{C}} + \text{H—B}^+ + \text{X:}^-$$

new π bond   leaving group

◆ The base (B:) removes a proton on the β carbon, thus forming H−B⁺.

◆ The electron pair in the β C−H bond forms the new π bond between the α and β carbons.

◆ The electron pair in the C−X bond ends up on halogen, forming the leaving group :X⁻.

The most common bases used in elimination reactions are negatively charged oxygen compounds such as ⁻OH and its alkyl derivatives, ⁻OR, called **alkoxides,** listed in Table 8.1. Potassium *tert*-butoxide, K⁺ ⁻OC(CH₃)₃, a bulky nonnucleophilic base, is especially useful (Section 7.8B).

TABLE 8.1	Common Bases Used in Dehydrohalogenation
Na⁺ ⁻OH	sodium hydroxide
K⁺ ⁻OH	potassium hydroxide
Na⁺ ⁻OCH₃	sodium methoxide
Na⁺ ⁻OCH₂CH₃	sodium ethoxide
K⁺ ⁻OC(CH₃)₃	potassium *tert*-butoxide

To draw any product of dehydrohalogenation:

◆ **Find the α carbon—the sp^3 hybridized carbon bonded to the leaving group.**

◆ **Identify all β carbons with H atoms.**

◆ **Remove the elements of H and X from the α and β carbons and form a π bond.**

For example, 2-bromo-2-methylpropane has three β carbons (three CH₃ groups), but because all three are *identical*, only *one* alkene is formed upon elimination of HBr. In contrast, 2-bromobutane has two *different* β carbons (labeled β₁ and β₂), so elimination affords *two* constitutional isomers by loss of HBr across either the α and β₁ carbons, or the α and β₂ carbons. We learn about which product predominates and why in Section 8.5.

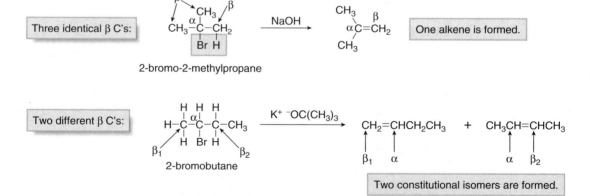

An elimination reaction is the first step in the slow degradation of the **pesticide DDT** (Section 7.4) to simpler compounds. Elimination of HCl from DDT forms the degradation product **DDE** (**d**ichloro**d**iphenyl**d**ichloro**e**thene). This stable alkene is found in minute concentration in the fatty tissues of most adults in the United States.

PROBLEM 8.1   Label the α and β carbons in each alkyl halide, and draw all possible elimination products in each reaction.

a. $CH_3CH_2CH_2CH_2CH_2-Cl$ $\xrightarrow{K^+ \ ^-OC(CH_3)_3}$

c. $\xrightarrow{K^+ \ ^-OH}$

b. $\xrightarrow{Na^+ \ ^-OCH_2CH_3}$

d. $\xrightarrow{K^+ \ ^-OH}$

8.2   Alkenes—The Products of Elimination Reactions

Because elimination reactions of alkyl halides form alkenes, let's review earlier material on alkene structure and learn some additional facts as well.

8.2A   Bonding in a Carbon–Carbon Double Bond

Recall from Section 1.9B that alkenes are hydrocarbons containing a carbon–carbon double bond. Each carbon of the double bond is sp^2 hybridized and trigonal planar, and all bond angles are 120°.

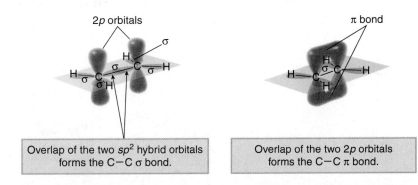

The double bond of an alkene consists of a σ bond and a π bond.

2p orbitals   π bond

Overlap of the two sp^2 hybrid orbitals forms the C—C σ bond.

Overlap of the two 2p orbitals forms the C—C π bond.

◆ The σ bond, formed by end-on overlap of the two sp^2 hybrid orbitals, lies in the plane of the molecule.
◆ The π bond, formed by side-by-side overlap of two 2p orbitals, lies perpendicular to the plane of the molecule. The π bond is formed during elimination.

Alkenes are classified according to the number of carbon atoms bonded to the carbons of the double bond. A **monosubstituted alkene** has one carbon atom bonded to the carbons of the double bond. A **disubstituted alkene** has two carbon atoms bonded to the carbons of the double bond, and so forth.

monosubstituted
(*one* R group)

disubstituted
(*two* R groups)

trisubstituted
(*three* R groups)

tetrasubstituted
(*four* R groups)

Figure 8.1 Classifying alkenes by the number of R groups bonded to the double bond

CH3CH2
H
C=C
H
H
monosubstituted
H
H
disubstituted
H
CH3
trisubstituted

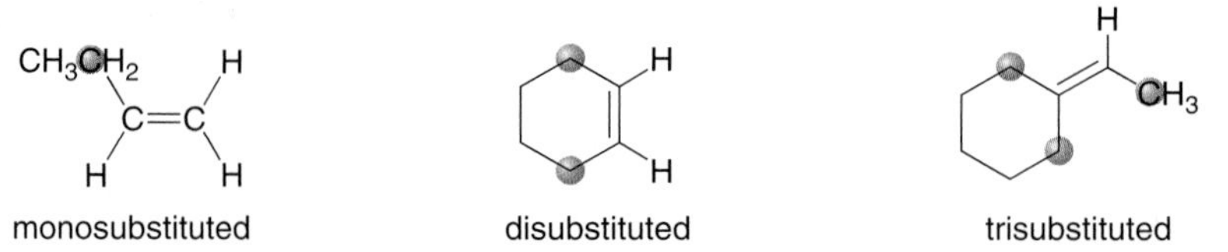

- Carbon atoms bonded to the double bond are screened in red.

Figure 8.1 shows several alkenes and how they are classified. You must be able to classify alkenes in this way to determine the major and minor products of elimination reactions, when a mixture of alkenes is formed.

PROBLEM 8.2 Classify each alkene by the number of carbon substituents bonded to the double bond.

a. $CH_3CH{=}CHCH_2CH_3$

b.

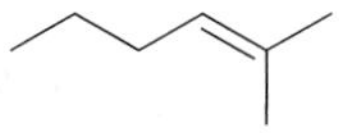

c.
CH3
CH3

d.
=CH2

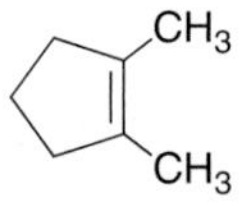

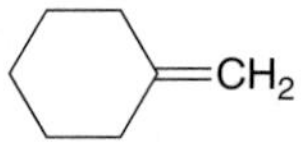

8.2B Restricted Rotation

Figure 8.2 shows that there is free rotation about the carbon–carbon single bonds of butane, but *not* around the carbon–carbon double bond of 2-butene.

Because of restricted rotation, two stereoisomers of 2-butene are possible.

- ◆ The cis isomer has two groups on the *same side* of the double bond.
- ◆ The trans isomer has two groups on *opposite sides* of the double bond.

Figure 8.2 Rotation around C–C and C=C compared

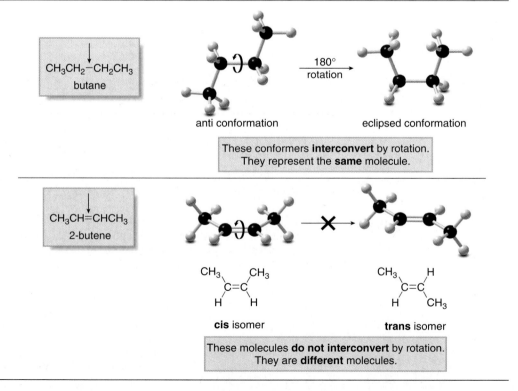

CH3CH2–CH2CH3
butane
180°
rotation
anti conformation
eclipsed conformation
These conformers interconvert by rotation.
They represent the same molecule.
CH3CH=CHCH3
2-butene
CH3
CH3
C=C
H
H
cis isomer
CH3
H
C=C
H
CH3
trans isomer
These molecules do not interconvert by rotation.
They are different molecules.

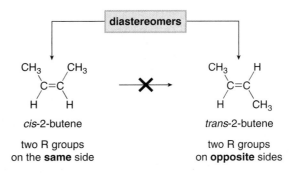

The concept of cis and trans isomers was first introduced for disubstituted cycloalkanes in Chapter 4. In both cases, a ring or a double bond restricts motion, preventing the rotation of a group from one side of the ring or double bond to the other.

cis-2-Butene and *trans*-2-butene are stereoisomers, but not mirror images of each other, so they are **diastereomers.**

The cis and trans isomers of 2-butene are a specific example of a general type of stereoisomer occurring at carbon–carbon double bonds. **Whenever the two groups on *each* end of a carbon–carbon double bond are *different from each other*, two diastereomers are possible.**

Stereoisomers on a C=C are possible when:

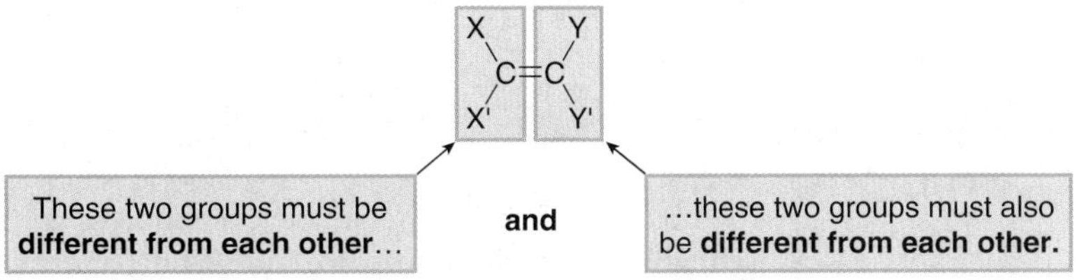

These two groups must be **different from each other**…   **and**   …these two groups must also be **different from each other.**

PROBLEM 8.3   For which alkenes are cis and trans isomers possible?

a. 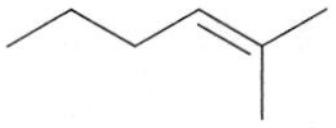 [structure]

b. $CH_3CH_2CH=CHCH_3$

c. $CH_2=CHCH_2CH_2CH_3$

d. 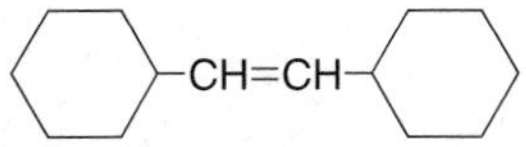 [structure]

PROBLEM 8.4   Label each pair of alkenes as constitutional isomers, stereoisomers, or identical.

a. 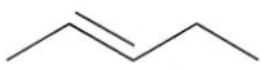 [structure] and [structure]

b. 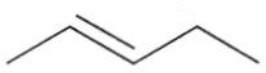 [structure] and [structure]

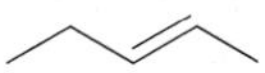

c. 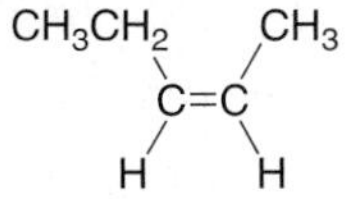 [structure] and [structure]

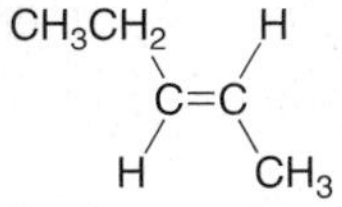

d. 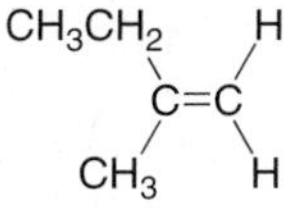 [structure] and [structure]

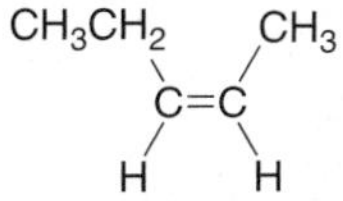

8.2C   Stability of Alkenes

Some alkenes are more stable than others. For example, **trans alkenes are more stable than cis alkenes** because the groups bonded to the double bond carbons are further apart, reducing steric interactions.

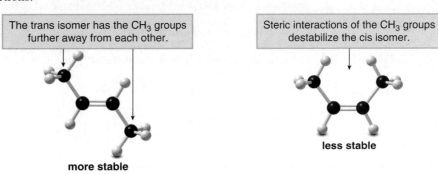

The trans isomer has the CH_3 groups further away from each other.

Steric interactions of the CH_3 groups destabilize the cis isomer.

more stable

less stable

The stability of an alkene increases, moreover, as the **number of R groups bonded to the double bond carbons increases.**

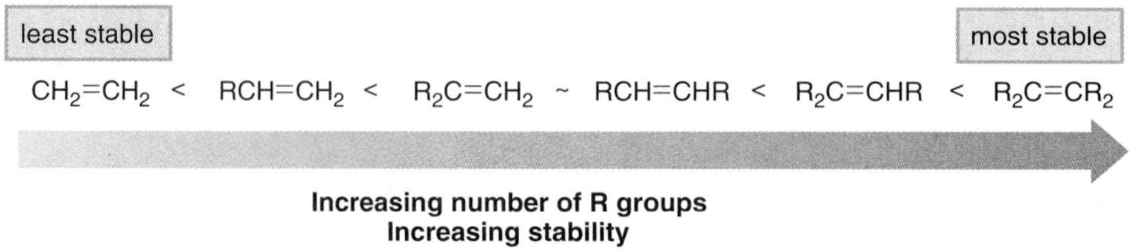

| least stable | | | | | | | | most stable |

$$CH_2{=}CH_2 \;<\; RCH{=}CH_2 \;<\; R_2C{=}CH_2 \;\sim\; RCH{=}CHR \;<\; R_2C{=}CHR \;<\; R_2C{=}CR_2$$

Increasing number of R groups
Increasing stability

R groups increase the stability of an alkene because R groups are sp^3 hybridized (able to donate electron density), whereas the carbon atoms of the double bond are sp^2 hybridized (able to accept electron density). Recall from Sections 1.10B and 2.5D that the percent s-character of a hybrid orbital increases from 25% to 33% in going from sp^3 to sp^2.

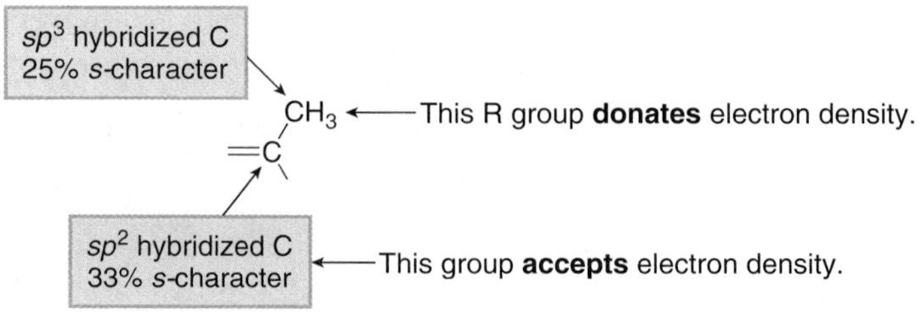

sp^3 hybridized C
25% s-character

CH₃ ◄——— This R group **donates** electron density.

sp^2 hybridized C
33% s-character ◄——— This group **accepts** electron density.

The higher the percent s-character, the more readily an atom accepts electron density. Thus, sp^2 **hybridized carbon atoms are more able to *accept* electron density and sp^3 hybridized carbon atoms are more able to *donate* electron density.**

◆ As a result, increasing the number of electron donating R groups on a carbon atom able to accept electron density makes the alkene more stable.

Thus, *trans*-2-butene (a disubstituted alkene) is more stable than *cis*-2-butene (another disubstituted alkene), but both are more stable than 1-butene (a monosubstituted alkene).

| One electron donor R group | Two electron donor R groups | |

1-butene   *cis*-2-butene   *trans*-2-butene
                                           fewer steric interactions

Increasing stability

In summary:

◆ Trans alkenes are more stable than cis alkenes because they have fewer steric interactions.
◆ Increasing alkyl substitution stabilizes an alkene by an electron donating inductive effect.

PROBLEM 8.5   Which alkene in each pair is more stable?

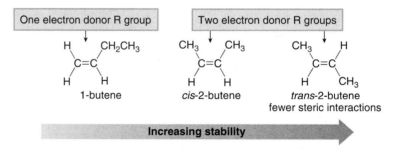

a.  ⟋⟍⟍   and   ⟍⟍⟍

c.    (ring with CH₃)   and   (ring with CH₃)

b. 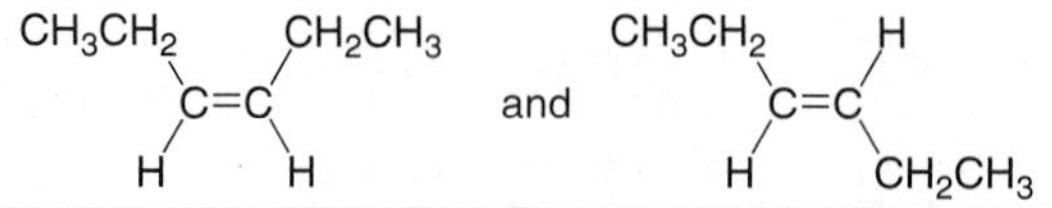
$$\underset{H}{\overset{CH_3CH_2}{}}C{=}C\underset{H}{\overset{CH_2CH_3}{}} \quad \text{and} \quad \underset{H}{\overset{CH_3CH_2}{}}C{=}C\underset{CH_2CH_3}{\overset{H}{}}$$

8.3   The Mechanisms of Elimination

What is the mechanism for elimination? What is the order of bond breaking and bond making? Is the reaction a one-step process or does it occur in many steps?

There are two mechanisms for elimination—E2 and E1—just as there are two mechanisms for nucleophilic substitution—S_N2 and S_N1.

- The **E2 mechanism (bimolecular elimination)**
- The **E1 mechanism (unimolecular elimination)**

The E2 and E1 mechanisms differ in the timing of bond cleavage and bond formation, analogous to the S_N2 and S_N1 mechanisms. In fact, E2 and S_N2 reactions have some features in common, as do E1 and S_N1 reactions.

8.4   The E2 Mechanism

The most common mechanism for dehydrohalogenation is the E2 mechanism. For example, $(CH_3)_3CBr$ reacts with ^-OH to form $(CH_3)_2C=CH_2$ via an E2 mechanism.

$$\boxed{\text{An E2 reaction}} \quad CH_3-\underset{\underset{Br}{|}}{\overset{\overset{CH_3}{|}}{C}}-CH_3 \quad \xrightarrow{^-OH} \quad \underset{CH_3}{\overset{CH_3}{\diagdown}}C=CH_2 \quad + \quad H_2O \quad + \quad Br^-$$

8.4A   Kinetics

An E2 reaction exhibits **second-order kinetics;** that is, the reaction is **bimolecular** and both the alkyl halide and the base appear in the rate equation.

- $rate = k[(CH_3)_3CBr][^-OH]$

8.4B   A One-Step Mechanism

The most straightforward explanation for the second-order kinetics is a **concerted reaction: all bonds are broken and formed in a single step,** as shown in Mechanism 8.1.

MECHANISM 8.1

The E2 Mechanism

$$CH_3-\underset{\underset{\underset{\beta}{}}{\overset{|}{Br}}}{\overset{\overset{\alpha}{\overset{CH_3\ H}{|\ |}}}{C}}-CH_2 \quad \xrightarrow{^-OH} \quad \underset{CH_3}{\overset{CH_3}{\diagdown}}C=CH_2 \quad + \quad H_2O \quad + \quad Br^-$$

new π bond

- The base ^-OH removes a proton from the β carbon, forming H_2O (a by-product).
- The electron pair in the β C–H bond forms the new π bond.
- The leaving group Br^- comes off with the electron pair in the C–Br bond.

An energy diagram for the reaction of $(CH_3)_3CBr$ with ^-OH is shown in Figure 8.3. The reaction has one step, so there is one energy barrier between reactants and products. Two bonds are broken (C–H and C–Br) and two bonds are formed (H–OH and the π bond) in a single step, so the transition state contains four partial bonds, with the negative charge distributed over the base and the leaving group. Entropy favors the products of an E2 reaction because two molecules of starting material form three molecules of product.

Figure 8.3 An energy diagram for an E2 reaction: $(CH_3)_3CBr + {}^-OH \rightarrow (CH_3)_2C{=}CH_2 + H_2O + Br^-$

- In the transition state, the C−H and C−Br bonds are partially broken, the O−H and π bonds are partially formed, and both the base and the departing leaving group bear a partial negative charge.

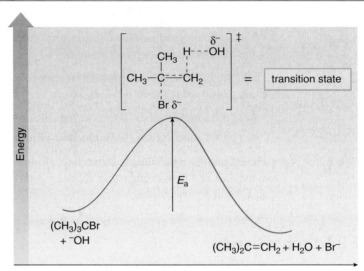

PROBLEM 8.6 Use curved arrows to show the movement of electrons in the following E2 mechanism. Draw the structure of the transition state.

$$(CH_3CH_2)_3CBr \xrightarrow{{}^-OCH_2CH_3} (CH_3CH_2)_2C{=}CHCH_3 + HOCH_2CH_3 + Br^-$$

PROBLEM 8.7 Why is $(CH_3)_3CCH_2Br$ inert to an E2 elimination?

There are close parallels between the E2 and S_N2 mechanisms in how the identity of the base, the leaving group, and the solvent affect the rate.

The Base

◆ The base appears in the rate equation, so the rate of the E2 reaction increases as the strength of the base increases.

E2 reactions are generally run with strong, negatively charged bases like ^-OH and ^-OR. Two strong, sterically hindered nitrogen bases, called **DBN** and **DBU,** are also sometimes used. An example of an E2 reaction with DBN is shown in Figure 8.4.

The IUPAC names for **DBN** and **DBU** are rarely used because the names are complex. **DBN** stands for 1,5-diazabicyclo[4.3.0]non-5-ene, and **DBU** stands for 1,8-diazabicyclo[5.4.0]-undec-7-ene.

Two useful bases for E2 reactions

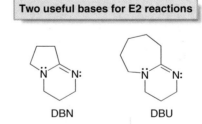

DBN   DBU

The Leaving Group

◆ Because the bond to the leaving group is partially broken in the transition state, the better the leaving group the faster the E2 reaction.

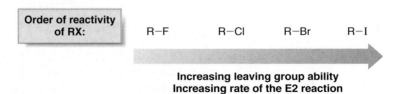

Order of reactivity of RX: R−F   R−Cl   R−Br   R−I

Increasing leaving group ability
Increasing rate of the E2 reaction

Figure 8.4 An E2 elimination with DBN used as the base

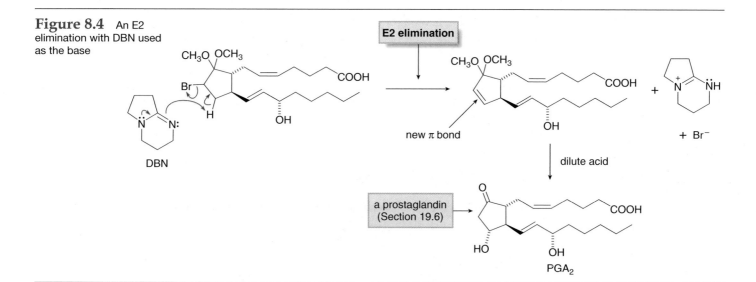

E2 elimination

new π bond

dilute acid

a prostaglandin (Section 19.6)

PGA₂

The Solvent

◆ **Polar aprotic solvents increase the rate of E2 reactions.**

Because **polar aprotic solvents** like $(CH_3)_2C=O$ do not solvate anions well, a negatively charged base is not "hidden" by strong interactions with the solvent (Section 7.17D), and the base is stronger. **A stronger base increases the reaction rate.**

PROBLEM 8.8   Which E2 reaction in each pair is faster?

a. CH_3CH_2-Br + $^-OH \longrightarrow$

    CH_3CH_2-Br + $^-OC(CH_3)_3 \longrightarrow$

b. CH_3CH_2-Br + $^-OC(CH_3)_3 \longrightarrow$

    CH_3CH_2-Cl + $^-OC(CH_3)_3 \longrightarrow$

8.4C   The Identity of the Alkyl Halide

The S_N2 and E2 mechanisms differ in how the R group affects the reaction rate.

◆ **As the number of R groups on the carbon with the leaving group increases, the rate of the E2 reaction increases.**

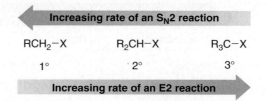

Increasing rate of an S_N2 reaction

RCH_2-X R_2CH-X R_3C-X

1°   2°   3°

Increasing rate of an E2 reaction

This trend is exactly *opposite* to the reactivity of alkyl halides in S_N2 reactions, where increasing alkyl substitution decreases the rate of reaction (Section 7.11D).

Why does increasing alkyl substitution increase the rate of an E2 reaction? In the transition state, the double bond is partially formed, so increasing the stability of the double bond with alkyl substituents stabilizes the transition state (that is, it lowers E_a, which increases the rate of the reaction according to the Hammond postulate).

Transition state for an E2 reaction with ^-OH as base

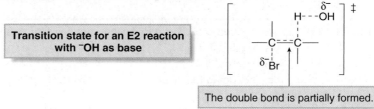

The double bond is partially formed.

◆ Increasing the number of R groups on the carbon with the leaving group forms more highly substituted, more stable alkenes in E2 reactions.

For example, the E2 reaction of a 1° alkyl halide (1-bromobutane) forms a monosubstituted alkene, whereas the E2 reaction of a 3° alkyl halide (2-bromo-2-methylpropane) forms a disubstituted alkene. The disubstituted alkene is more stable, so the 3° alkyl halide reacts faster than the 1° alkyl halide.

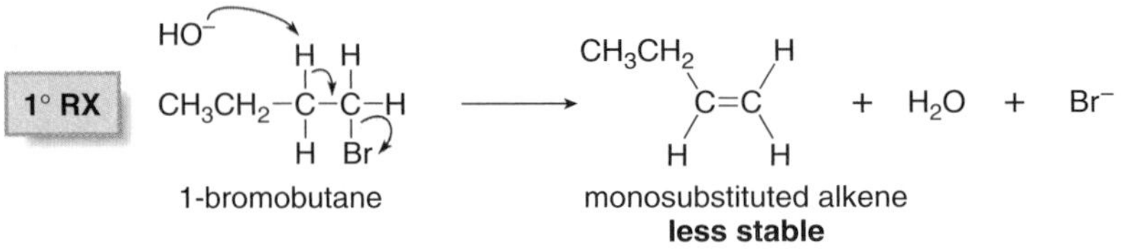

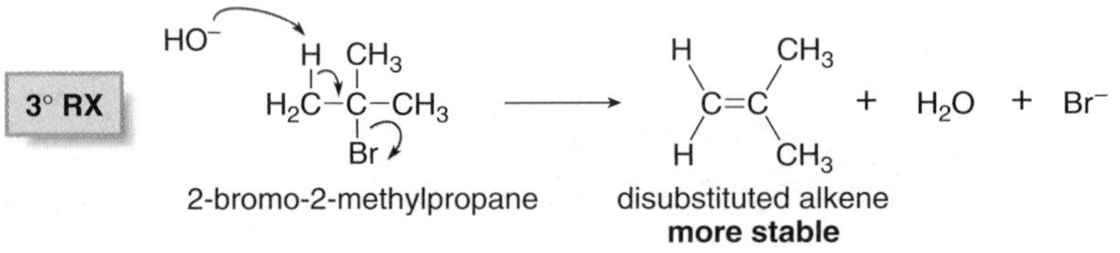

1° RX

1-bromobutane → monosubstituted alkene **less stable** + H$_2$O + Br$^-$

3° RX

2-bromo-2-methylpropane → disubstituted alkene **more stable** + H$_2$O + Br$^-$

PROBLEM 8.9   Rank the alkyl halides in each group in order of increasing reactivity in an E2 reaction.

a. (CH$_3$)$_2$C(Br)CH$_2$CH$_2$CH$_3$ (CH$_3$)$_2$CHCH$_2$CH$_2$CH$_2$Br   (CH$_3$)$_2$CHCH$_2$CH(Br)CH$_3$

b.

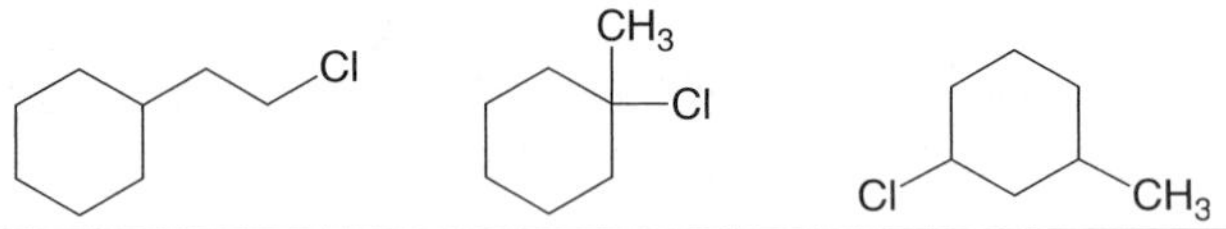

Table 8.2 summarizes the characteristics of the E2 mechanism, and Figure 8.5 illustrates two examples of E2 reactions used in the synthesis of **quinine** and **estradiol,** two naturally occurring compounds.

TABLE 8.2   Characteristics of the E2 Mechanism

Characteristic	Result
Kinetics	• Second order
Mechanism	• One step
Identity of R	• More substituted halides react fastest • Rate: R$_3$CX > R$_2$CHX > RCH$_2$X
Base	• Favored by strong bases
Leaving group	• Better leaving group ---→ faster reaction
Solvent	• Favored by polar aprotic solvents

PROBLEM 8.10   How does each of the following changes affect the rate of an E2 reaction?

a. tripling [RX]

b. halving [B:]

c. changing the solvent from CH$_3$OH to DMSO

d. changing the leaving group from I$^-$ to Br$^-$

e. changing the base from $^-$OH to H$_2$O

f. changing the alkyl halide from CH$_3$CH$_2$Br to (CH$_3$)$_2$CHBr

Figure 8.5 Two examples of the E2 reaction used in organic synthesis

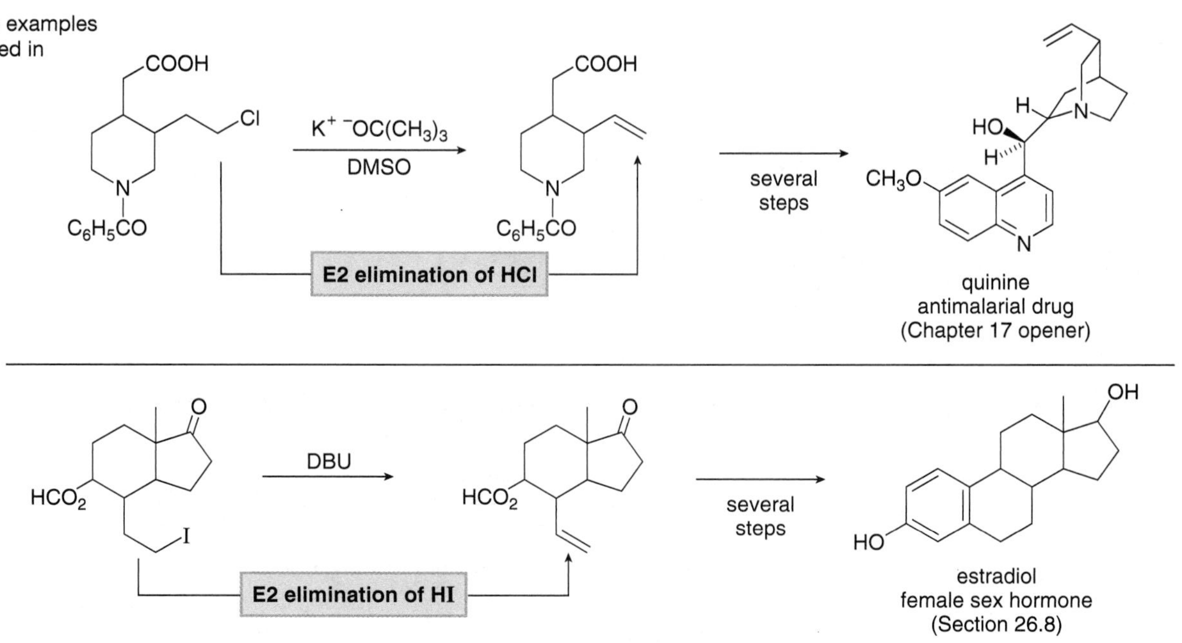

quinine
antimalarial drug
(Chapter 17 opener)

estradiol
female sex hormone
(Section 26.8)

8.5   The Zaitsev Rule

Recall from Section 8.1 that a mixture of alkenes can form from the dehydrohalogenation of alkyl halides having two or more different β carbon atoms. When this occurs, one of the products usually predominates. The **major product is the more stable product—the one with the more substituted double bond.** For example, elimination of the elements of H and I from 1-iodo-1-methylcyclohexane yields two constitutional isomers: the trisubstituted alkene **A** (the major product) and the disubstituted alkene **B** (the minor product).

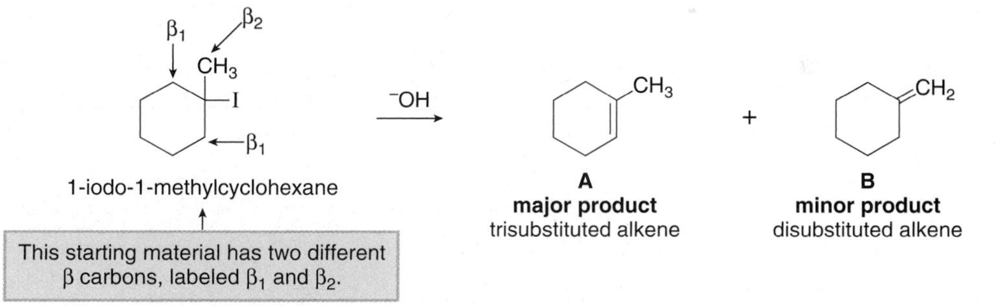

1-iodo-1-methylcyclohexane

This starting material has two different β carbons, labeled $β_1$ and $β_2$.

A
major product
trisubstituted alkene

B
minor product
disubstituted alkene

This phenomenon is called the **Zaitsev rule** (also called the **Saytzeff rule,** depending on the translation) for the Russian chemist who first noted this trend.

◆ **The Zaitsev rule: The major product in β elimination has the more substituted double bond.**

A reaction is *regioselective* when it yields predominantly or exclusively one constitutional isomer when more than one is possible. The E2 reaction is **regioselective** because the more substituted alkene predominates.

The Zaitsev rule results because the double bond is partially formed in the transition state for the E2 reaction. Thus, increasing the stability of the double bond by adding R groups lowers the energy of the transition state, which increases the reaction rate. For example, E2 elimination of HBr from 2-bromo-2-methylbutane yields alkenes **C** and **D. D,** having the more substituted double bond, is the major product, because the transition state leading to its formation is lower in energy.

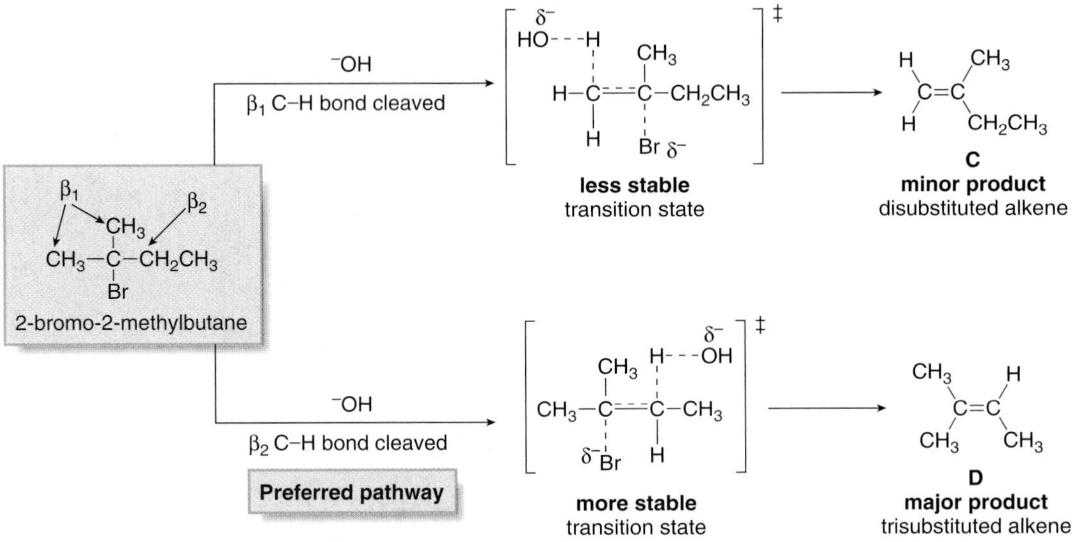

When a mixture of stereoisomers is possible from dehydrohalogenation, the **major product is the more stable stereoisomer.** For example, dehydrohalogenation of alkyl halide **X** forms a mixture of trans and cis alkenes, **Y** and **Z**. The trans alkene **Y** is the major product because it is most stable.

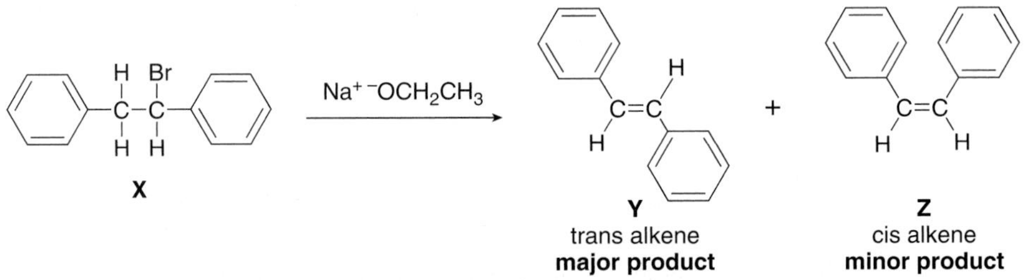

A reaction is *stereoselective* when it forms predominantly or exclusively one stereoisomer when two or more are possible. The E2 reaction is stereoselective because one stereoisomer is formed preferentially.

SAMPLE PROBLEM 8.1   Predict the major product in the following E2 reaction.

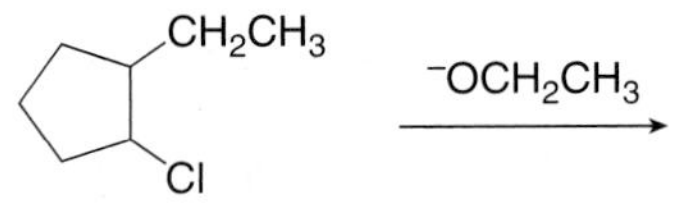

SOLUTION

The alkyl halide has two different β C atoms (labeled β_1 and β_2), so two different alkenes are possible: one formed by removal of HCl across the α and β_1 carbons, and one formed by removal of HCl across the α and β_2 carbons. Using the Zaitsev rule, the major product should be **A,** because it has the more substituted double bond.

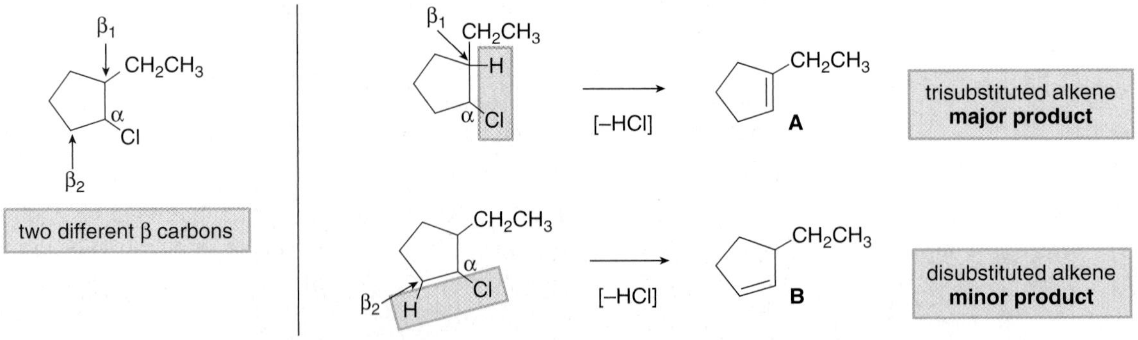

PROBLEM 8.11   What alkenes are formed from each alkyl halide by an E2 reaction? Use the Zaitsev rule to predict the major product.

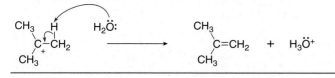

8.6   The E1 Mechanism

The dehydrohalogenation of $(CH_3)_3CI$ with H_2O to form $(CH_3)_2C=CH_2$ can be used to illustrate the second general mechanism of elimination, the **E1 mechanism.**

| An E1 reaction |

$$CH_3-\underset{\underset{I}{|}}{\overset{\overset{CH_3}{|}}{C}}-CH_3 \xrightarrow{H_2O} \underset{\underset{CH_3}{|}}{\overset{\overset{CH_3}{|}}{C}}=CH_2 \;+\; H_3O^+ \;+\; I^-$$

8.6A   Kinetics

An E1 reaction exhibits first-order kinetics.

◆ rate $= k[(CH_3)_3CI]$

Like the S_N1 mechanism, the kinetics suggest that the reaction mechanism involves more than one step, and that the slow step is **unimolecular,** involving *only* the alkyl halide.

8.6B   A Two-Step Mechanism

The most straightforward explanation for the observed first-order kinetics is a **two-step reaction: the bond to the leaving group breaks first *before* the π bond is formed,** as shown in Mechanism 8.2.

MECHANISM 8.2

The E1 Mechanism

Step [1] The C–I bond is broken.

$$CH_3-\underset{\underset{CH_3}{|}}{\overset{\overset{CH_3}{|}}{C}}-I \xrightarrow{\text{slow}} \underset{\underset{CH_3}{|}}{\overset{\overset{CH_3}{|}}{C}}{}^+\!-CH_3 \;+\; I^-$$

carbocation

◆ **Heterolysis of the C–I bond** forms an intermediate **carbocation.** This is the same first step as the S_N1 mechanism. It is responsible for the first-order kinetics because it is rate-determining.

Step [2] A C–H bond is cleaved and the π bond is formed.

$$\underset{\underset{CH_3}{|}}{\overset{\overset{CH_3}{|}}{C}}{}^+\!-CH_2-H \quad H_2\ddot{O}: \longrightarrow \underset{\underset{CH_3}{|}}{\overset{\overset{CH_3}{|}}{C}}=CH_2 \;+\; H_3\ddot{O}^+$$

◆ **A base** (such as H_2O or I^-) **removes a proton from a carbon adjacent to the carbocation** (a β carbon). The electron pair in the C–H bond is used to form the new π bond.

The E1 and E2 mechanisms both involve the same number of bonds broken and formed. **The only difference is the timing.**

◆ In an E1 reaction, the leaving group comes off *before* the β proton is removed, and the reaction occurs in two steps.
◆ In an E2 reaction, the leaving group comes off *as* the β proton is removed, and the reaction occurs in one step.

Figure 8.6 Energy diagram for an E1 reaction: $(CH_3)_3CI + H_2O \rightarrow$ $CH_2=C(CH_3)_2 + H_3O^+ + I^-$

- Since the E1 mechanism has two steps, there are two energy barriers.
- Step [1] is rate-determining; $E_a[1] > E_a[2]$.

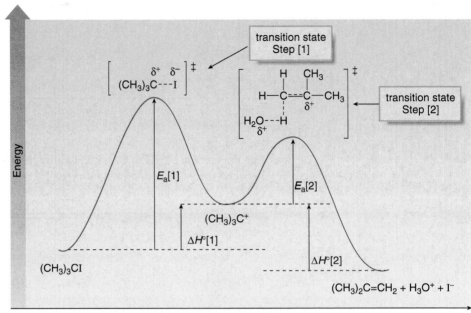

Reaction coordinate

An energy diagram for the reaction of $(CH_3)_3CI + H_2O$ is shown in Figure 8.6. Each step has its own energy barrier, with a transition state at each energy maximum. Because its transition state is higher in energy, Step [1] is rate-determining. $\Delta H°$ for Step [1] is positive because only bond breaking occurs, whereas $\Delta H°$ of Step [2] is negative because two bonds are formed and only one is broken.

PROBLEM 8.12   Draw an E1 mechanism for the following reaction. Draw the structure of the transition state for each step.

$$(CH_3)_2C(Cl)CH_2CH_3 + CH_3OH \longrightarrow (CH_3)_2C=CHCH_3 + CH_3\overset{+}{O}H_2 + Cl^-$$

8.6C   Other Characteristics of E1 Reactions

Three other features of E1 reactions are worthy of note.

[1] The rate of an E1 reaction increases as the number of R groups on the carbon with the leaving group increases.

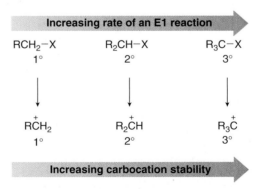

Increasing alkyl substitution has the same effect on the rate of *both* an E1 and E2 reaction; increasing rate of the E1 and E2 reactions: RCH_2X (1°) < R_2CHX (2°) < R_3CX (3°).

Like an S_N1 reaction, more substituted alkyl halides yield more substituted (and more stable) carbocations in the rate-determining step. Increasing the stability of a carbocation, in turn, decreases E_a for the slow step, which increases the rate of the E1 reaction according to the Hammond postulate.

[2] Because the base does not appear in the rate equation, weak bases favor E1 reactions.

The strength of the base usually determines whether a reaction follows the E1 or E2 mechanism.

◆ Strong bases like ⁻OH and ⁻OR favor E2 reactions, whereas weaker bases like H_2O and ROH favor E1 reactions.

[3] E1 reactions are regioselective, favoring formation of the more substituted, more stable alkene.

The Zaitsev rule applies to E1 reactions, too. For example, E1 elimination of HBr from 1-bromo-1-methylcyclopentane yields alkenes **A** and **B. A,** having the more substituted double bond, is the major product.

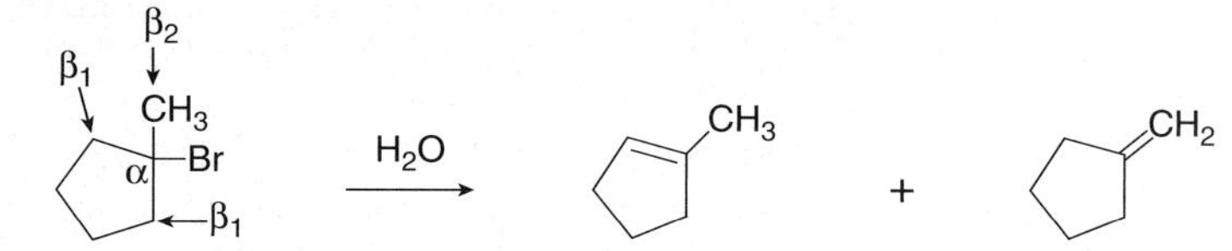

1-bromo-1-methyl-
cyclopentane

H_2O →

A
trisubstituted alkene
major product

+

B
disubstituted alkene
minor product

two different β carbons
labeled β₁ and β₂

PROBLEM 8.13   What alkenes are formed from each alkyl halide by an E1 reaction? Use the Zaitsev rule to predict the major product.

a. $CH_3CH_2-\underset{\underset{Cl}{|}}{\overset{\overset{CH_3}{|}}{C}}-CH_2CH_3$

b.

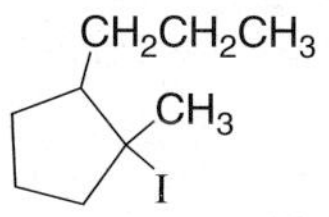

Table 8.3 summarizes the characteristics of E1 reactions.

TABLE 8.3   Characteristics of the E1 Mechanism

Characteristic	Result
Kinetics	• First order
Mechanism	• Two steps
Identity of R	• More substituted halides react fastest • Rate: $R_3CX > R_2CHX > RCH_2X$
Base	• Favored by weaker bases such as H_2O and ROH
Leaving group	• A better leaving group makes the reaction faster because the bond to the leaving group is partially broken in the rate-determining step.
Solvent	• Polar protic solvents that solvate the ionic intermediates are needed.

PROBLEM 8.14   How does each of the following changes affect the rate of an E1 reaction?
a. doubling [RX]
b. doubling [B:]
c. changing the halide from $(CH_3)_3CBr$ to $CH_3CH_2CH_2Br$
d. changing the leaving group from Cl⁻ to Br⁻
e. changing the solvent from DMSO to CH_3OH

8.7   S_N1 and E1 Reactions

S_N1 and E1 reactions have exactly the same first step—formation of a carbocation. They differ in what happens to the carbocation.

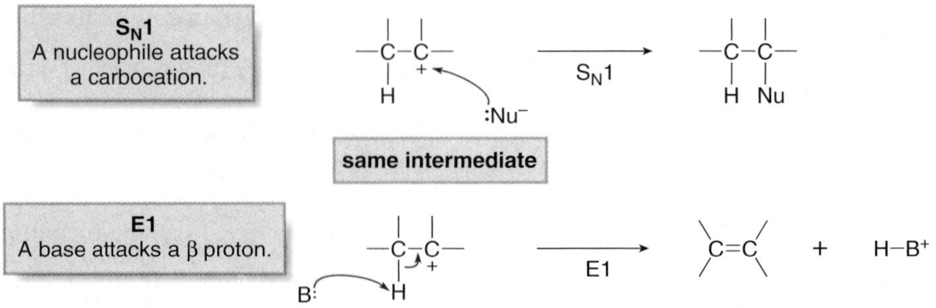

- In an S_N1 reaction, a nucleophile attacks the carbocation, forming a substitution product.
- In an E1 reaction, a base removes a proton, forming a new π bond.

The same conditions that favor substitution by an S_N1 mechanism also favor elimination by an E1 mechanism: a 3° alkyl halide as substrate, a weak nucleophile or base as reagent, and a polar protic solvent. As a result, both reactions usually occur in the same reaction mixture to afford a mixture of products, as illustrated in Sample Problem 8.2.

SAMPLE PROBLEM 8.2   Draw the S_N1 and E1 products formed in the reaction of $(CH_3)_3CBr$ with H_2O.

SOLUTION

The first step in both reactions is heterolysis of the C−Br bond to form a **carbocation.**

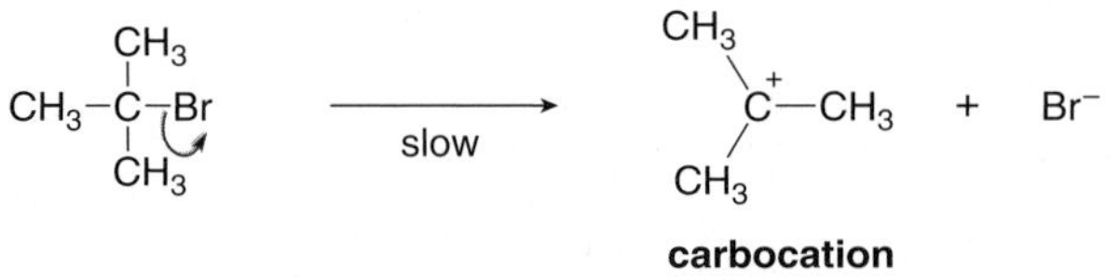

Reaction of the carbocation with H_2O as a nucleophile affords the substitution product (Reaction [1]). Alternatively, H_2O acts as a base to remove a proton, affording the elimination product (Reaction [2]). **Two products are formed.**

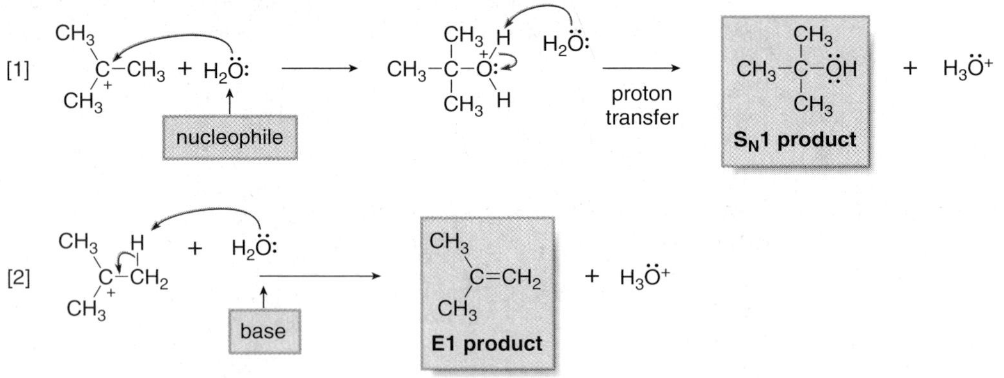

Because E1 reactions often occur with a competing S_N1 reaction, **E1 reactions of alkyl halides are *much less useful* than E2 reactions.**

PROBLEM 8.15   Draw both the S_N1 and E1 products of each reaction.

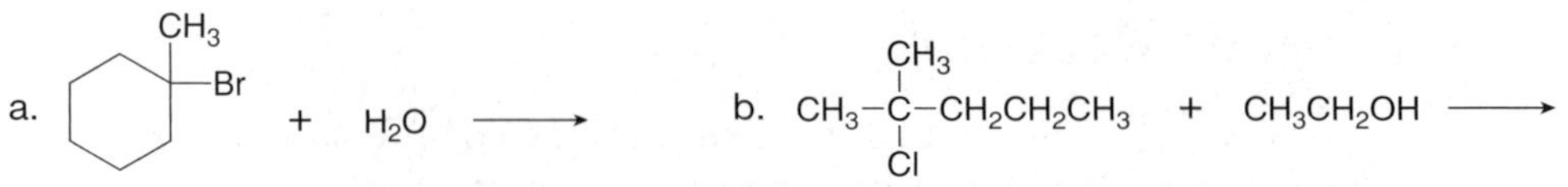

8.8   Stereochemistry of the E2 Reaction

Although the E2 reaction does not produce products with tetrahedral stereogenic centers, its transition state consists of four atoms that react at the same time, and they react only if they possess a particular stereochemical arrangement.

8.8A   General Stereochemical Features

The transition state of an E2 reaction consists of **four atoms** from the alkyl halide—one hydrogen atom, two carbon atoms, and the leaving group (X)—**all aligned in a plane.** There are two ways for the C−H and C−X bonds to be coplanar.

H and X are on the **same** side. | H and X are on **opposite** sides.

syn periplanar | **anti periplanar**

> The dihedral angle for the C−H and C−X bonds equals 0° for the syn periplanar arrangement and 180° for the anti periplanar arrangement.

◆ The H and X atoms can be oriented on the same side of the molecule. This geometry is called *syn periplanar.*
◆ The H and X atoms can be oriented on opposite sides of the molecule. This geometry is called *anti periplanar.*

All evidence suggests that **E2 elimination occurs most often in the anti periplanar geometry.** This arrangement allows the molecule to react in the lower energy *staggered* conformation. It also allows two electron-rich species, the incoming base and the departing leaving group, to be further away from each other, as illustrated in Figure 8.7.

Anti periplanar geometry is the preferred arrangement for any alkyl halide undergoing E2 elimination, regardless of whether it is cyclic or acyclic. This stereochemical requirement has important consequences for compounds containing six-membered rings.

PROBLEM 8.16   Draw the anti periplanar geometry for the E2 reaction of $(CH_3)_2CHCH_2Br$ with base. Then draw the product that results after elimination of HBr.

PROBLEM 8.17   Given that an E2 reaction proceeds with anti periplanar stereochemistry, draw the products of each elimination. The alkyl halides in (a) and (b) are diastereomers of each other. How are the products of these two reactions related?

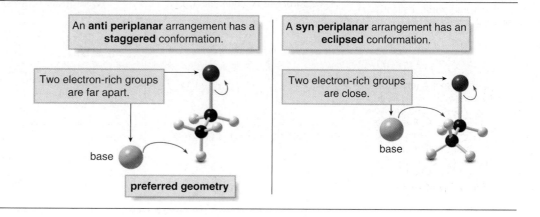

Figure 8.7 Two possible geometries for the E2 reaction

An **anti periplanar** arrangement has a **staggered** conformation. | A **syn periplanar** arrangement has an **eclipsed** conformation.

Two electron-rich groups are far apart. | Two electron-rich groups are close.

base | base

preferred geometry

8.8B   Anti Periplanar Geometry and Halocyclohexanes

Recall from Section 4.13 that cyclohexane exists as two chair conformers that rapidly intercon-vert, and that substituted cyclohexanes are more stable with substituents in the roomier equato-rial position. Thus, chlorocyclohexane exists as two chair conformers, but **A** is preferred because the Cl group is equatorial.

For E2 elimination, **the C–Cl bond must be anti periplanar to a C–H bond on a β carbon,** and this occurs only when the H and Cl atoms are both in the **axial** position. This requirement for **trans diaxial geometry** means that E2 elimination must occur from the less stable conformer **B,** as shown in Figure 8.8.

Sometimes this rigid stereochemical requirement affects the regioselectivity of the E2 reaction of substituted cyclohexanes. Dehydrohalogenation of *cis*- and *trans*-1-chloro-2-methylcyclohexane via an E2 mechanism illustrates this phenomenon.

cis-1-chloro-2-methyl-cyclohexane   *trans*-1-chloro-2-methyl-cyclohexane

The **cis isomer** exists as two conformers (**A** and **B**), each of which has one group axial and one group equatorial. E2 reaction must occur from conformer **B,** which contains an axial Cl atom.

Because conformer **B** has two different axial β H atoms, labeled H_a and H_b, E2 reaction occurs in two different directions to afford two alkenes. **The major product contains the more stable trisubstituted double bond, as predicted by the Zaitsev rule.**

Figure 8.8   The trans diaxial geometry for the E2 elimination in chlorocyclohexane

- In conformer **A** (**equatorial** Cl group), a β C–H bond and a C–Cl bond are never anti periplanar; therefore, no E2 elimination can occur.
- In conformer **B** (**axial** Cl group), a β C–H bond and a C–Cl bond are **trans diaxial;** therefore, E2 elimination occurs.

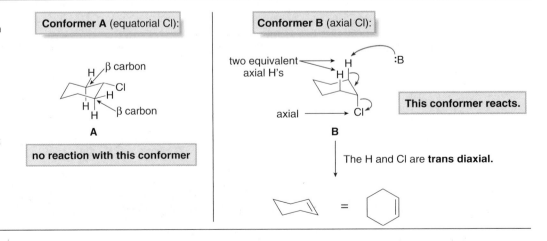

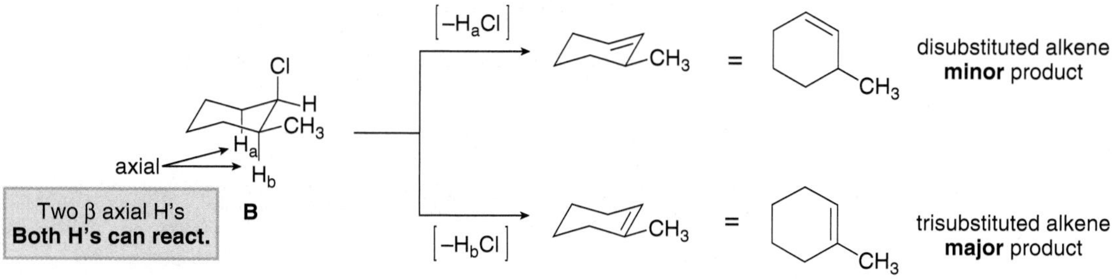

disubstituted alkene
minor product

$[-H_aCl]$

$[-H_bCl]$

trisubstituted alkene
major product

Two β axial H's
Both H's can react.

axial

B

The **trans isomer** exists as two conformers, **C**, having two equatorial substituents, and **D**, having two axial substituents. E2 reaction must occur from conformer **D**, which contains an axial Cl atom.

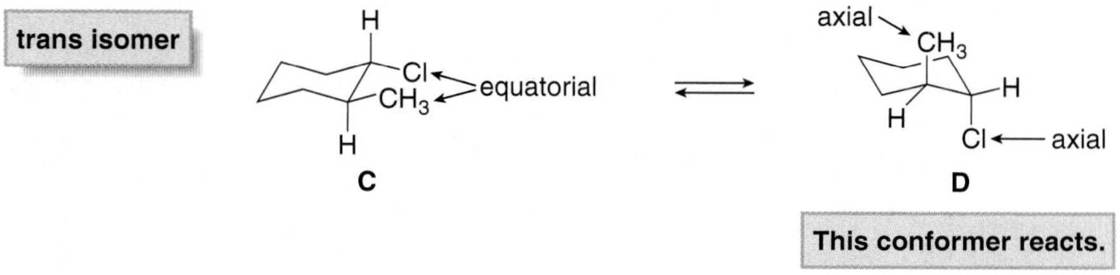

trans isomer

equatorial

axial

C   **D**

This conformer reacts.

Because conformer **D** has **only one axial β H**, E2 reaction occurs in only one direction to afford a **single product**, having the disubstituted double bond. This is *not* predicted by the Zaitsev rule. E2 reaction requires H and Cl to be trans and diaxial, and with the trans isomer, this is possible only when the less stable alkene is formed as product.

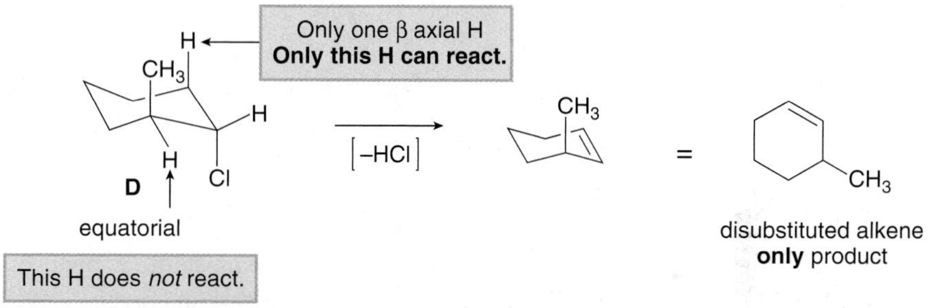

Only one β axial H
Only this H can react.

D

equatorial

$[-HCl]$

disubstituted alkene
only product

This H does *not* react.

◆ In conclusion, with substituted cyclohexanes, E2 elimination must occur with a trans diaxial arrangement of H and X, and as a result of this requirement, the more substituted alkene is not necessarily the major product.

PROBLEM 8.18   Draw the major E2 elimination products from each of the following alkyl halides.

a. 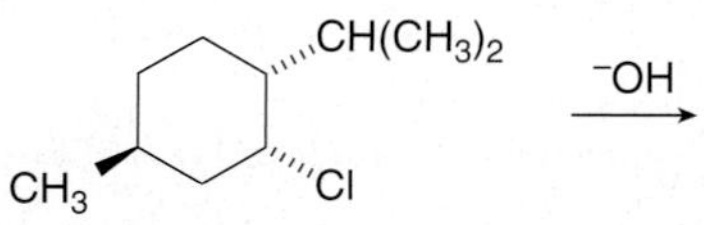 ^-OH b. 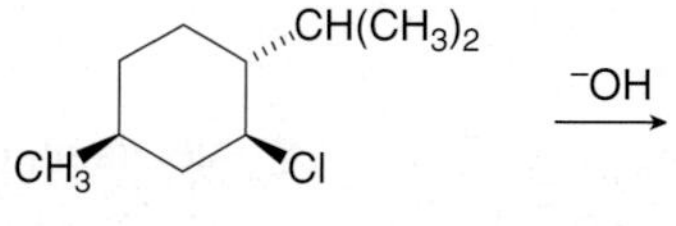 ^-OH

PROBLEM 8.19   Explain why *cis*-1-chloro-2-methylcyclohexane undergoes E2 elimination much faster than its trans isomer.

8.9 When Is the Mechanism E1 or E2?

Given a particular starting material and base, how do we know whether a reaction occurs by the E1 or E2 mechanism?

Because the rate of *both* the E1 and E2 reactions increases as the number of R groups on the carbon with the leaving group increases, you cannot use the identity of the alkyl halide to decide which elimination mechanism occurs. This makes determining the mechanisms for substitution and elimination very different processes.

◆ **The strength of the base is the most important factor in determining the mechanism for elimination. Strong bases favor the E2 mechanism. Weak bases favor the E1 mechanism.**

Table 8.4 compares the E1 and E2 mechanisms.

TABLE 8.4	A Comparison of the E1 and E2 Mechanisms
E2 mechanism	• Much more common and useful.
	• Favored by strong, negatively charged bases, especially ^-OH and ^-OR.
	• The reaction occurs with 1°, 2°, and 3° alkyl halides. Order of reactivity: $R_3CX > R_2CHX > RCH_2X$.
E1 mechanism	• Much less useful because a mixture of S_N1 and E1 products usually results.
	• Favored by weaker, neutral bases, such as H_2O and ROH.
	• This mechanism does not occur with 1° RX because they form highly unstable 1° carbocations.

PROBLEM 8.20   Which mechanism, E1 or E2, will occur in each reaction?

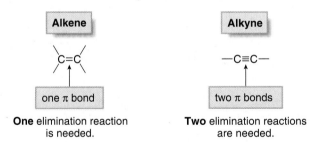

a. $CH_3-\overset{\overset{\displaystyle CH_3}{|}}{\underset{\underset{\displaystyle Cl}{|}}{C}}-CH_3$ + $^-OCH_3$ ⟶

b. [cyclohexane with I] + H_2O ⟶

c. [cyclohexane with $C(CH_3)_3$ and Cl] + CH_3OH ⟶

d. CH_3CH_2Br + $^-OC(CH_3)_3$ ⟶

8.10 E2 Reactions and Alkyne Synthesis

Recall from Section 1.9C that the carbon–carbon triple bond of alkynes consists of one σ and two π bonds.

A single elimination reaction produces the π bond of an alkene. Two consecutive elimination reactions produce the two π bonds of an alkyne.

Alkene	**Alkyne**
$\overset{\diagdown}{}C=C\overset{\diagup}{}$	$-C\equiv C-$
one π bond	two π bonds
One elimination reaction is needed.	**Two** elimination reactions are needed.

◆ **Alkynes are prepared by two successive dehydrohalogenation reactions.**

Two elimination reactions are needed to remove two moles of HX from a **dihalide** as substrate. Two different starting materials can be used.

$$R-\overset{\overset{\displaystyle H}{|}}{\underset{\underset{\displaystyle X}{\boxed{}}}{C}}-\overset{\overset{\displaystyle H}{|}}{\underset{\underset{\displaystyle X}{\boxed{}}}{C}}-R \qquad R-\overset{\overset{\displaystyle H}{|}}{\underset{\underset{\displaystyle H}{|}}{C}}-\overset{\overset{\displaystyle \boxed{X}}{}}{\underset{\underset{\displaystyle \boxed{X}}{}}{C}}-R$$

vicinal dihalide   geminal dihalide

◆ A **vicinal dihalide** has two X atoms on *adjacent* carbon atoms.

◆ A **geminal dihalide** has two X atoms on the *same* carbon atom.

The word *geminal* comes from the Latin *geminus*, meaning *twin*.

Equations [1] and [2] illustrate how two moles of HX can be removed from these dihalides with base. Two equivalents of strong base are used and each step follows an E2 mechanism.

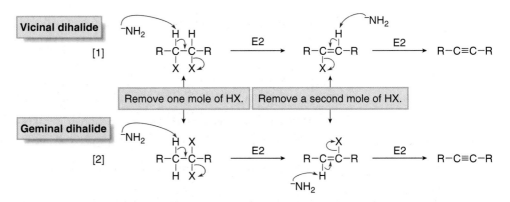

Stronger bases are needed to synthesize alkynes by dehydrohalogenation than are needed to synthesize alkenes. The typical base is **amide (⁻NH₂),** used as the sodium salt NaNH₂ (sodium amide). KOC(CH₃)₃ can also be used with DMSO as solvent. Because DMSO is a polar aprotic solvent, the anionic base is not well solvated, thus **increasing its basicity** and making it strong enough to remove two equivalents of HX. Examples are given in Figure 8.9.

Why is a stronger base needed in this dehydrohalogenation? The transition state for the second elimination reaction includes partial cleavage of a C–H bond. In this case, however, the carbon atom is sp^2 hybridized and sp^2 **hybridized C–H bonds are stronger than** sp^3 **hybridized C–H bonds.** As a result, a stronger base is needed to cleave this bond.

The relative strength of C–H bonds depends on the hybridization of the carbon atom: $sp > sp^2 > sp^3$. For more information, review Section 1.10B.

Starting material for Step [2]	Transition state for Step [2]
sp^2 hybridized	δ^-
	H---NH₂ partially broken
R–C=C–R	R–C≡C–R
X	X δ^-

PROBLEM 8.21   Draw the alkynes formed in each reaction.

a. (cyclohexyl)–C(Cl)(H)–C(Cl)(H)–CH₂CH₃ $\xrightarrow{^-NH_2}$

b. CH₃CH₂CH₂CHCl₂ $\xrightarrow[\text{DMSO}]{\text{KOC(CH}_3)_3}$

c. CH₃–C(Br)(Br)–CH₂CH₃ $\xrightarrow{^-NH_2}$

d. (phenyl)CH(Br)–CH(Br)(phenyl) $\xrightarrow{^-NH_2}$

Figure 8.9   Example of dehydrohalogenation of dihalides to afford alkynes

(phenyl)–C(H)(Cl)–C(H)(Cl)–(phenyl) $\xrightarrow[\text{-2 HCl}]{\text{Na}^+ \ ^-NH_2 \ (2 \text{ equiv})}$ (phenyl)–C≡C–(phenyl)

two new π bonds

CH₃–C(CH₃)–C(H)(Br)–C(H)(Br)–H $\xrightarrow[\text{-2 HBr}]{\text{K}^+ \ ^-OC(CH_3)_3 \ (2 \text{ equiv}) \\ \text{DMSO}}$ CH₃–C(CH₃)(CH₃)–C≡C–H

8.11   When Is the Reaction S_N1, S_N2, E1, or E2?

We have now considered two different kinds of reactions (substitution and elimination) and four different mechanisms (S_N1, S_N2, E1, and E2) that begin with one class of compounds (alkyl halides). How do we know if a given alkyl halide will undergo substitution or elimination with a given base or nucleophile, and by what mechanism?

Unfortunately, there is no easy answer, and often mixtures of products result. Two generalizations help to determine whether substitution or elimination occurs.

> **[1] Good nucleophiles that are weak bases favor substitution over elimination.**

Certain anions always give products of substitution because they are good nucleophiles but weak bases. These include: **I^-, Br^-, HS^-,** and **CH_3COO^-.**

$$CH_3CH_2{-}Br \; + \; I^- \xrightarrow[\text{CH}_3\text{OH}]{} \; CH_3CH_2{-}I \; + \; Br^-$$

good nucleophile
weak base ⟶ **substitution**

> **[2] Bulky, nonnucleophilic bases favor elimination over substitution.**

$KOC(CH_3)_3$, DBU, and **DBN** are too sterically hindered to attack a tetravalent carbon, but are able to remove a small proton, favoring elimination over substitution.

$$H{-}\overset{\displaystyle H}{\underset{\displaystyle H}{C}}{-}CH_2{-}Br \longrightarrow CH_2{=}CH_2 \; + \; (CH_3)_3COH \; + \; KBr$$

$K^+ \; {}^-OC(CH_3)_3$
strong, nonnucleophilic base ⟶ **elimination**

Most often, however, we will have to rely on other criteria to predict the outcome of these reactions. To determine the product of a reaction with an alkyl halide:

[1] Classify the alkyl halide as 1°, 2°, or 3°.
[2] Classify the base or nucleophile as strong, weak, or bulky.

Predicting the substitution and elimination products of a reaction can then be organized by the type of alkyl halide, as shown in Figure 8.10.

Figure 8.10   **Determining whether an alkyl halide reacts by an S_N1, S_N2, E1, or E2 mechanism**

[1] 3° Alkyl halides (R_3CX react by all mechanisms except S_N2.)

- With strong bases
- **Elimination occurs by an E2 mechanism.**
- Rationale: A strong base or nucleophile favors an S_N2 or E2 mechanism, but 3° halides are too sterically hindered to undergo an S_N2 reaction, so only E2 elimination occurs.
- Example:

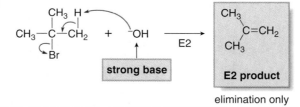

elimination only

- With weak nucleophiles
  or bases

- **A mixture of S$_N$1 and E1 products results.**
- Rationale: A weak base or nucleophile favors S$_N$1 and E1 mechanisms, and both occur.
- Example:

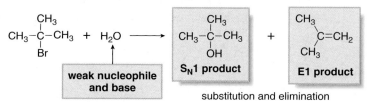

---

[2] 1° Alkyl halides (RCH$_2$X react by S$_N$2 and E2 mechanisms.)

- With strong nucleophiles

- **Substitution occurs by an S$_N$2 mechanism.**
- Rationale: A strong base or nucleophile favors S$_N$2 or E2, but 1° halides are the least reactive halide type in elimination; therefore, only an S$_N$2 reaction occurs.
- Example:

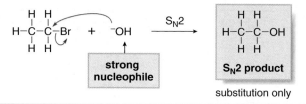

- With strong, sterically
  hindered bases

- **Elimination occurs by an E2 mechanism.**
- Rationale: A strong, sterically hindered base cannot act as a nucleophile, so elimination occurs and the mechanism is E2.
- Example:

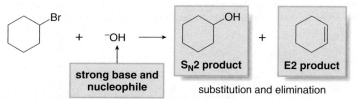

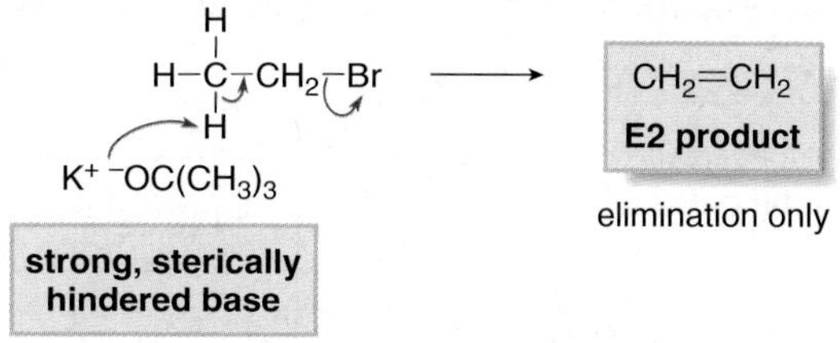

[3] 2° Alkyl halides (R$_2$CHX react by all mechanisms.)

- With strong bases and
  nucleophiles

- **A mixture of S$_N$2 and E2 products results.**
- Rationale: A strong base that is also a strong nucleophile gives a mixture of S$_N$2 and E2 products.
- Example:

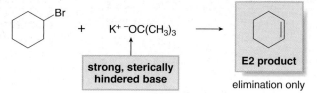

- With strong, sterically
  hindered bases

- **Elimination occurs by an E2 mechanism.**
- Rationale: A strong, sterically hindered base cannot act as a nucleophile, so elimination occurs and the mechanism is E2.
- Example:

Br + K$^+$ $^-$OC(CH$_3$)$_3$ ⟶

**strong, sterically
hindered base**

E2 product

elimination only

(continued on next page)

Figure 8.10 Continued

- With weak nucleophiles or bases

- **A mixture of SN1 and E1 products results.**
- Rationale: A weak base or nucleophile favors S_N1 and E1 mechanisms, and both occur.
- Example:

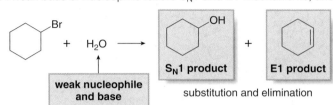

S_N1 product   E1 product

weak nucleophile and base   substitution and elimination

Sample Problems 8.3–8.5 illustrate how to apply the information in Figure 8.10 to specific alkyl halides.

SAMPLE PROBLEM 8.3   Draw the products of the following reaction.

$$C(CH_3)_3 \quad -Br \quad + \quad H_2O \longrightarrow$$

SOLUTION

[1] Classify the halide as 1°, 2°, or 3° and the reagent as a strong or weak base (and nucleophile) to determine the mechanism. In this case, the alkyl halide is 3° and the reagent (H_2O) is a weak base and nucleophile, so products of both **S_N1** and **E1** mechanisms are formed.

[2] To draw the products of substitution and elimination:

S_N1 product	E1 product
Substitute the nucleophile (H_2O) for the leaving group (Br^-), and draw the neutral product after loss of a proton.	Remove the elements of H and Br from the α and β carbons. There are two identical β C atoms with H atoms so only one elimination product is possible.

C(CH₃)₃ −Br + H₂O ⟶ C(CH₃)₃ −OH

leaving group   nucleophile   **S_N1 product**

β C(CH₃)₃ α −Br + H₂O ⟶ α C(CH₃)₃ β
←β   base   **E1 product**

SAMPLE PROBLEM 8.4   Draw the products of the following reaction.

$$-Br \quad + \quad CH_3O^- \xrightarrow{\quad CH_3OH \quad}$$

SOLUTION

[1] Classify the halide as 1°, 2°, or 3° and the reagent as a strong or weak base (and nucleophile) to determine the mechanism. In this case, the alkyl halide is 2° and the reagent (CH_3O^-) is a strong base and nucleophile, so products of both **S_N2** and **E2** mechanisms are formed.

[2] To draw the products of substitution and elimination:

S_N2 product	E2 product
Substitute the nucleophile (CH_3O^-) for the leaving group (Br^-).	Remove the elements of H and Br from the α and β carbons. There are two identical β C atoms with H atoms so only one elimination product is possible.

−Br + CH₃O⁻ ⟶ −OCH₃

nucleophile   **S_N2 product**

β
α −Br
β   α
CH₃O⁻   H   **E2 product**
β

SAMPLE PROBLEM 8.5   Draw the products of the following reaction, and include the mechanism showing how each product is formed.

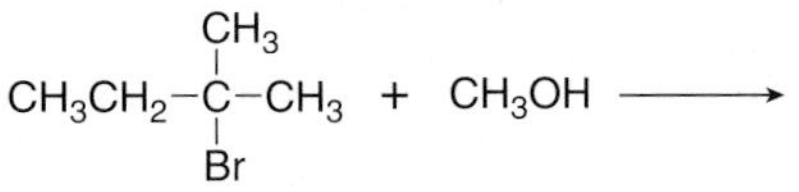

$$CH_3CH_2-\underset{\underset{Br}{|}}{\overset{\overset{CH_3}{|}}{C}}-CH_3 \ + \ CH_3OH \longrightarrow$$

SOLUTION

[1] Classify the halide as 1°, 2°, or 3°, and the reagent as a strong or weak base (and nucleophile) to determine the mechanism. In this case, the alkyl halide is 3° and the reagent (CH_3OH) is a weak base and nucleophile, so products of both **S$_N$1** and **E1** mechanisms are formed.

[2] Draw the steps of the mechanisms to give the products. Both mechanisms begin with the same first step: loss of the leaving group to form a carbocation.

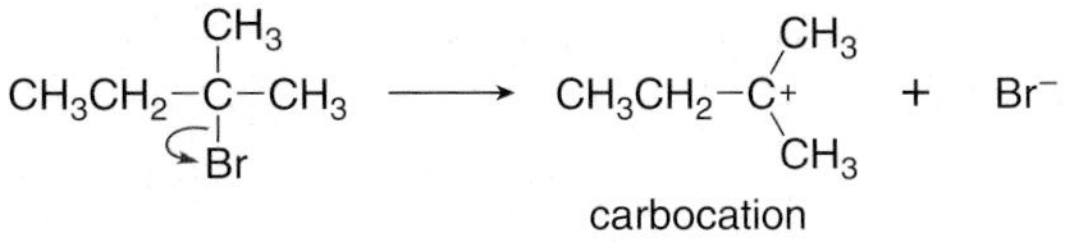

$$CH_3CH_2-\underset{\underset{Br}{|}}{\overset{\overset{CH_3}{|}}{C}}-CH_3 \longrightarrow CH_3CH_2-\underset{\underset{CH_3}{|}}{\overset{\overset{CH_3}{|}}{C}}{}^+ \ + \ Br^-$$

carbocation

- **For S$_N$1:** The carbocation reacts with a nucleophile. Nucleophilic attack of CH_3OH on the carbocation generates a positively charged intermediate that loses a proton to afford the neutral S$_N$1 product.

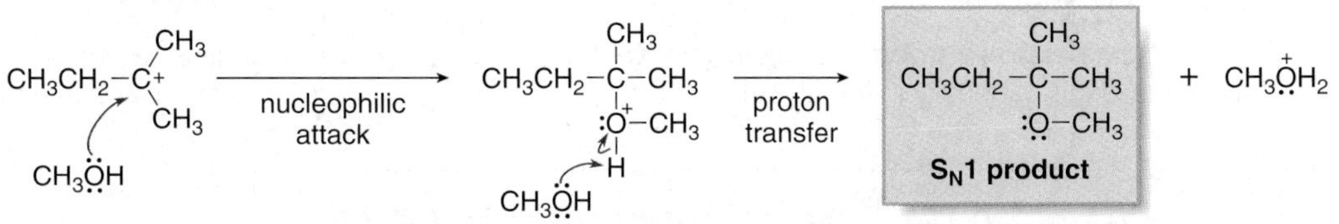

- **For E1:** The carbocation reacts with a base. Two different products of elimination can form because the carbocation has two different β carbons.

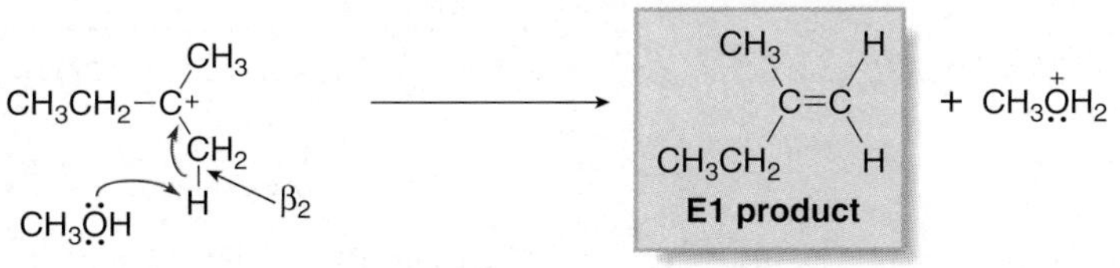

In this problem, three products are formed: one from an S$_N$1 reaction and two from E1 reactions.

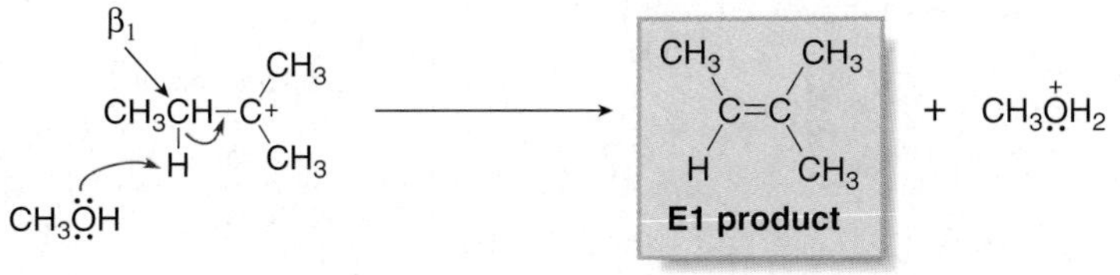

PROBLEM 8.22   Draw the products in each reaction.

a. 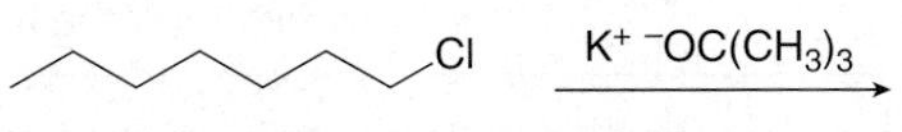 $\xrightarrow{K^+ \ {}^-OC(CH_3)_3}$

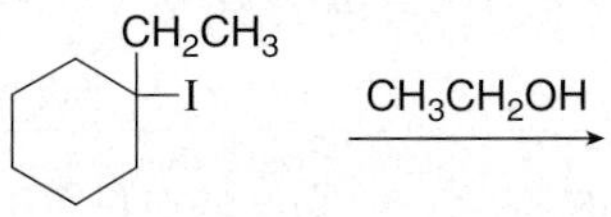

b. 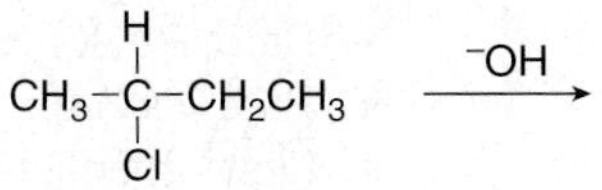 $CH_3-\underset{\underset{Cl}{|}}{\overset{\overset{H}{|}}{C}}-CH_2CH_3 \xrightarrow{\ {}^-OH\ }$

c. $\xrightarrow{CH_3CH_2OH}$

d. 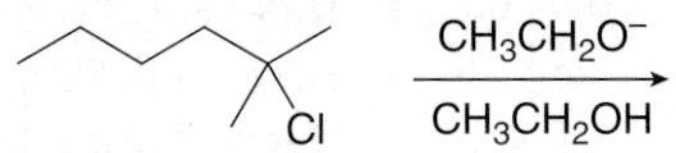 $\xrightarrow[CH_3CH_2OH]{CH_3CH_2O^-}$

PROBLEM 8.23   Draw a stepwise mechanism for the following reaction.

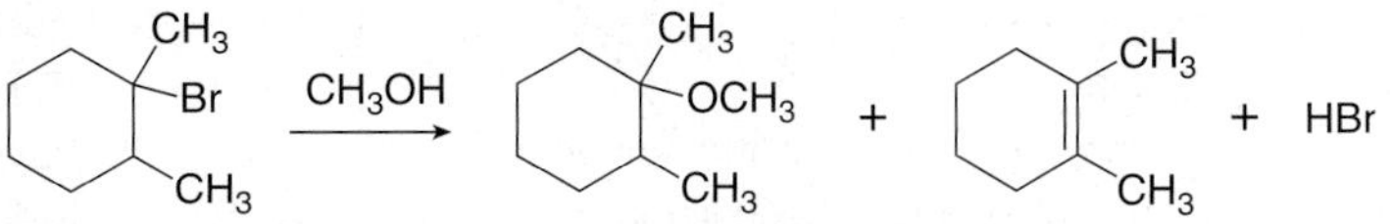

8.12   Key Concepts—Alkyl Halides and Elimination Reactions

A Comparison Between Nucleophilic Substitution and β Elimination

Nucleophilic substitution—A nucleophile attacks a carbon atom (7.6).

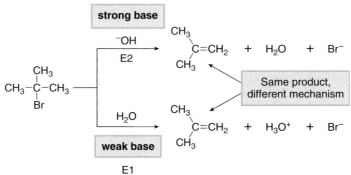

β **Elimination**—A base attacks a proton (8.1).

Similarities	Differences
• In both reactions RX acts as an electrophile, reacting with an electron-rich reagent.	• In substitution, a nucleophile attacks a single carbon atom.
• Both reactions require a **good leaving group X:⁻** that can accept the electron density in the C−X bond.	• In elimination, a Brønsted–Lowry base removes a proton to form a π bond, and two carbons are involved in the reaction.

The Importance of the Base in E2 and E1 Reactions (8.9)

The strength of the base determines the mechanism of elimination.
* Strong bases favor E2 reactions.
* Weak bases favor E1 reactions.

E1 and E2 Mechanisms Compared

	E2 mechanism	E1 mechanism
Mechanism	• One step (8.4B)	• Two steps (8.6B)
Alkyl halide	• Rate: $R_3CX > R_2CHX > RCH_2X$ (8.4C)	• Rate: $R_3CX > R_2CHX > RCH_2X$ (8.6C)
Rate equation	• Rate = $k[RX][B:]$	• Rate = $k[RX]$
	• Second-order kinetics (8.4A)	• First-order kinetics (8.6A)
Stereochemistry	• Anti periplanar arrangement of H and X (8.8)	• Trigonal planar carbocation intermediate (8.6B)
Base	• Favored by strong bases (8.4B)	• Favored by weak bases (8.6C)
Leaving group	• Better leaving group --→ faster reaction (8.4B)	• Better leaving group --→ faster reaction (Table 8.3)
Solvents	• Favored by polar aprotic solvents (8.4B)	• Favored by polar protic solvents (Table 8.3)
Product	• More substituted alkene favored (Zaitsev rule, 8.5)	• More substituted alkene favored (Zaitsev rule, 8.6C)

Summary Chart on the Four Mechanisms: S_N1, S_N2, E1, or E2

Alkyl halide type	Conditions	Mechanism
1° RCH_2X	strong nucleophile $\dashrightarrow$	S_N2
	strong bulky base $\dashrightarrow$	E2
2° R_2CHX	strong base and nucleophile $\dashrightarrow$	S_N2 + E2
	strong bulky base $\dashrightarrow$	E2
	weak base and nucleophile $\dashrightarrow$	S_N1 + E1
3° R_3CX	weak base and nucleophile $\dashrightarrow$	S_N1 + E1
	strong base $\dashrightarrow$	E2

Problems

General Elimination

8.24 Draw all possible constitutional isomers formed by dehydrohalogenation of each alkyl halide.

a. $CH_3CH_2CH_2CH_2CH_2CH_2Br$ b. 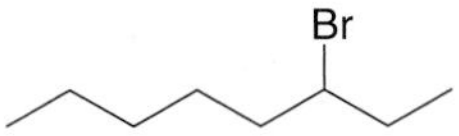   c. 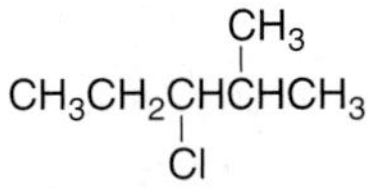 $CH_3CH_2CHCHCH_3$ d.

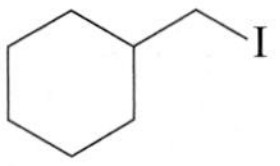

8.25 What alkyl halide forms each of the following alkenes as the *only* product in an elimination reaction?

a. CH_2=$CHCH_2CH_2CH_3$ b. $(CH_3)_2CHCH$=CH_2 c. 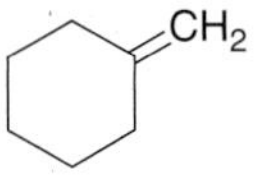   d. 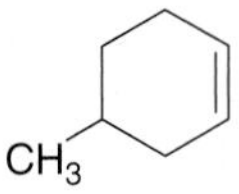   e.

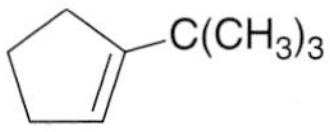

Alkenes

8.26 Farnesene is an alkene found in the waxy coating of apple skins. Which double bonds in farnesene can have cis and trans configurations?

$$(CH_3)_2C=CHCH_2CH_2C(CH_3)=CHCH_2CH_2CH(CH_3)CH=CH_2$$

farnesene

8.27 Label each pair of alkenes as constitutional isomers, stereoisomers, or identical.

a. 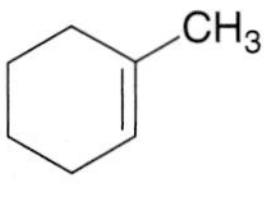   and

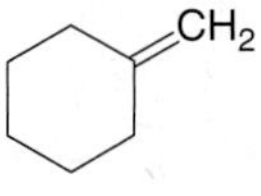

b.

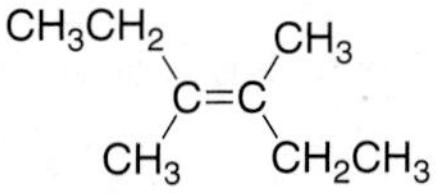

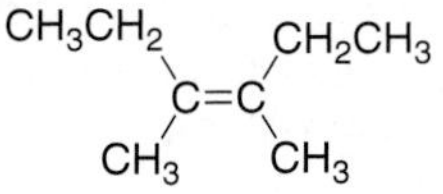

c. 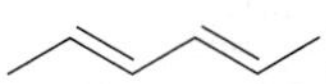   and

d. 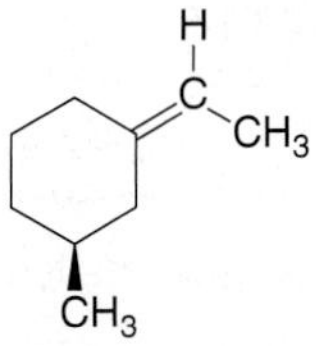   and

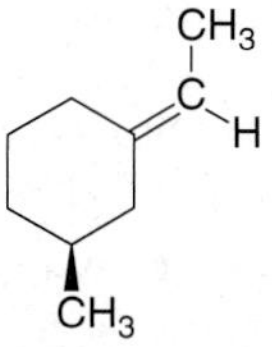

8.28 Draw all isomers of molecular formula C_2H_2BrCl. Label pairs of diastereomers and constitutional isomers.

8.29 PGF$_{2\alpha}$ is a prostaglandin, a group of compounds that are responsible for inflammation (Section 19.6). (a) How many tetrahedral stereogenic centers does PGF$_{2\alpha}$ contain? (b) How many double bonds can exist as cis and trans isomers? (c) Considering both double bonds and tetrahedral stereogenic centers, what is the maximum number of stereoisomers that can exist for PGF$_{2\alpha}$?

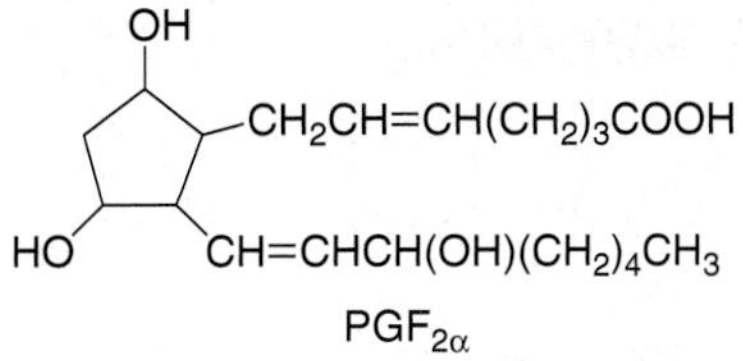

OH

$CH_2CH=CH(CH_2)_3COOH$

HO $CH=CHCH(OH)(CH_2)_4CH_3$

PGF$_{2\alpha}$

8.30 Rank the alkenes in each group in order of increasing stability.

a. $CH_2=CHCH_2CH_2CH_3$

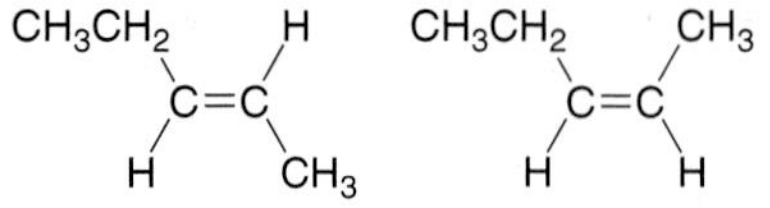

b. $CH_2=C(CH_3)CH_2CH_3$ $CH_2=CHCH(CH_3)_2$ $(CH_3)_2C=CHCH_3$

8.31 $\Delta H°$ values obtained for a series of similar reactions are one set of experimental data used to determine the relative stability of alkenes. Explain how the following data suggest that *cis*-2-butene is more stable than 1-butene (Section 12.3A).

$CH_2=CHCH_2CH_3 + H_2 \longrightarrow CH_3CH_2CH_2CH_3$ $\Delta H° = -30.3$ kcal/mol
1-butene

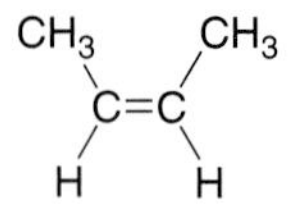

$+ H_2 \longrightarrow CH_3CH_2CH_2CH_3$ $\Delta H° = -28.6$ kcal/mol

cis-2-butene

E2 Reaction

8.32 Draw all constitutional isomers formed in each E2 reaction and predict the major product using the Zaitsev rule.

a. 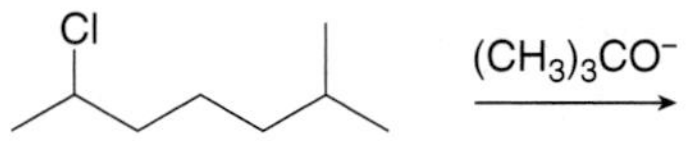 $(CH_3)_3CO^-$

c. 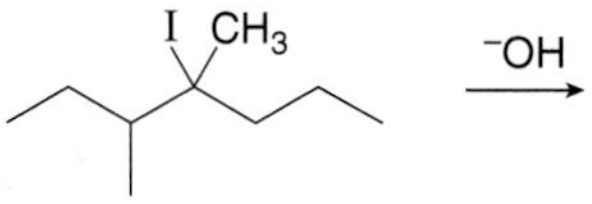 ^-OH

e. 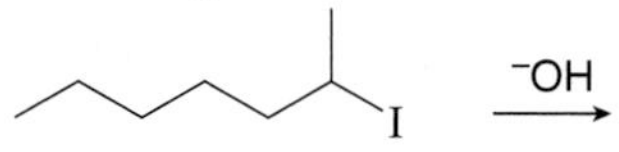 ^-OH

b. 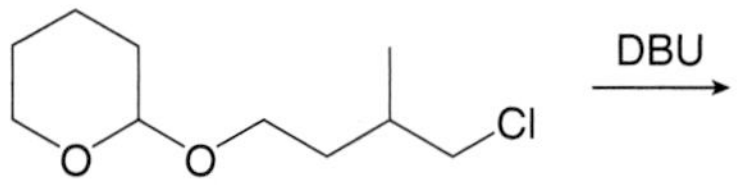   DBU

d. 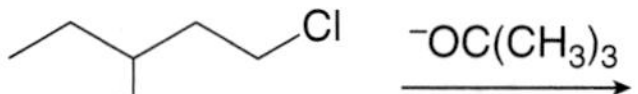 $^-OC(CH_3)_3$

f. 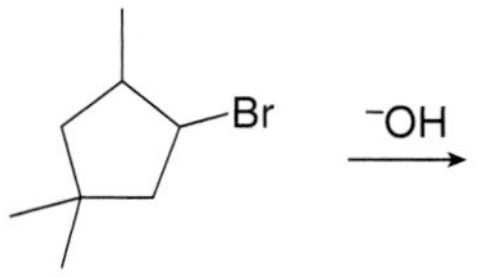 ^-OH

8.33 What two different alkyl halides yield $(CH_3)_2C=CH_2$ as the only product of dehydrohalogenation?

8.34 Explain why $(CH_3)_2CHCH(Br)CH_2CH_3$ reacts faster than $(CH_3)_2CHCH_2CH(Br)CH_3$ in an E2 reaction, even though both alkyl halides are 2°.

8.35 Consider the following E2 reaction.

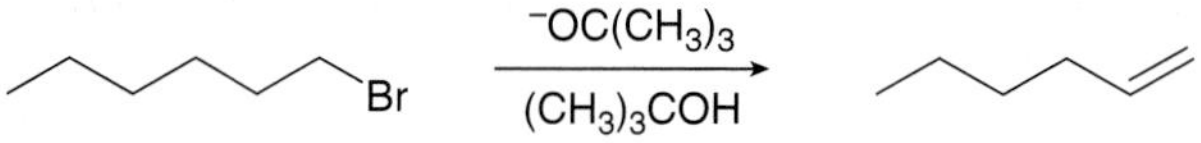

$\dfrac{^-OC(CH_3)_3}{(CH_3)_3COH}$

a. Draw the by-products of the reaction and use curved arrows to show the movement of electrons.
b. What happens to the reaction rate with each of the following changes? [1] The solvent is changed to DMF; [2] The concentration of $^-OC(CH_3)_3$ is decreased; [3] The base is changed to ^-OH; [4] The halide is changed to $CH_3CH_2CH_2CH_2CH(Br)CH_3$; and [5] The leaving group is changed to I^-.

8.36 Sometimes sterically hindered bases give the less substituted alkene as the major product of an E2 reaction. Explain the following result.

	21%	79%
$CH_3CH_2O^-$	21%	79%
$(CH_3)_3CO^-$	73%	27%

8.37 What is the major stereoisomer formed in each reaction?

a. 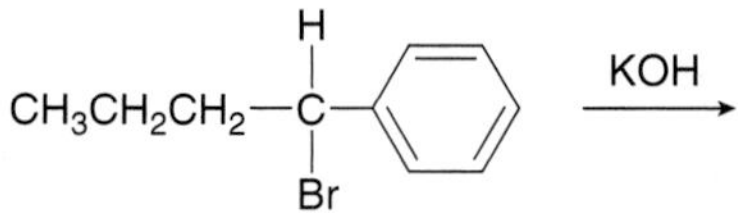 $CH_3CH_2CH_2-\overset{H}{\underset{Br}{C}}-$ KOH

b. 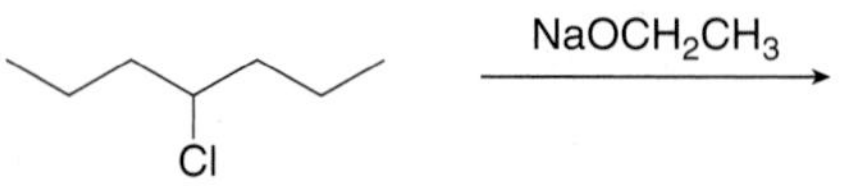   NaOCH$_2$CH$_3$

E1 Reaction

8.38 What alkene is the major product formed from each alkyl halide in an E1 reaction?

a.

b.

c.

E1 and E2

8.39 Draw all constitutional isomers formed in each elimination reaction. Label the mechanism as E2 or E1.

a. $\xrightarrow{\text{}^-\text{OCH}_3}$

c. $\xrightarrow{\text{}^-\text{OC(CH}_3)_3}$

e. 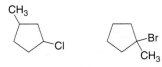 $\xrightarrow{\text{}^-\text{OH}}$

b. $\xrightarrow{\text{CH}_3\text{OH}}$

d. $\xrightarrow{\text{H}_2\text{O}}$

f. $\xrightarrow{\text{}^-\text{OH}}$

8.40 Rank the alkyl halides in each group in order of increasing E2 reactivity. Then do the same for E1 reactivity.

a.

b.

8.41 Which elimination reaction in each pair is faster?

a. $\xrightarrow{\text{}^-\text{OH}}$

b. $\xrightarrow{\text{H}_2\text{O}}$

c. $(\text{CH}_3)_3\text{CCl} \xrightarrow[\text{H}_2\text{O}]{\text{}^-\text{OH}}$

$\xrightarrow{\text{}^-\text{OH}}$

$\xrightarrow{\text{H}_2\text{O}}$

$(\text{CH}_3)_3\text{CCl} \xrightarrow[\text{DMSO}]{\text{}^-\text{OH}}$

Stereochemistry and the E2 Reaction

8.42 Taking into account anti periplanar geometry, predict the major E2 product formed from each starting material.

a.

b.

c.

d.

8.43 a. Draw three-dimensional representations for all stereoisomers of 2-chloro-3-methylpentane, and label pairs of enantiomers.

b. Considering dehydrohalogenation across C2 and C3 only, draw the E2 product that results from each of these alkyl halides. How many different products have you drawn?

c. How are these products related to each other?

Alkynes

8.44 Draw the products of each reaction.

a. $\xrightarrow[\text{(2 equiv)}]{\text{}^-\text{NH}_2}$

c. $\text{CH}_3-\overset{\overset{\text{Cl}}{|}}{\underset{\underset{\text{Cl}}{|}}{\text{C}}}-\text{CH}_2\text{CH}_3 \xrightarrow[\text{(excess)}]{\text{}^-\text{NH}_2}$

b. $\text{CH}_3\text{CH}_2-\overset{\overset{\text{CH}_3}{|}}{\underset{\underset{\text{CH}_3}{|}}{\text{C}}}-\text{CHCH}_2\text{Br} \xrightarrow[\text{(2 equiv)}]{\text{}^-\text{NH}_2}$ with Br on the CHCH₂Br

d. $\xrightarrow[\text{(2 equiv)}]{\text{}^-\text{NH}_2}$

8.45 Draw the structure of a dihalide that could be used to prepare each alkyne. There may be more than one possible dihalide.

a. $CH_3C\equiv CCH_3$

b. $CH_3-\overset{\overset{\displaystyle CH_3}{|}}{\underset{\underset{\displaystyle CH_3}{|}}{C}}-C\equiv CH$

c. 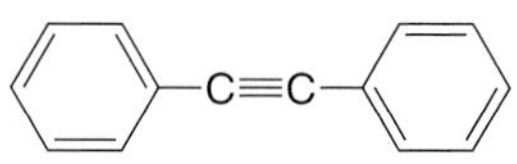 [benzene]—C≡C—[benzene]

8.46 Under certain reaction conditions, 2,3-dibromobutane reacts with two equivalents of base to give three products, each of which contains two new π bonds. Product **A** has two *sp* hybridized carbon atoms, product **B** has one *sp* hybridized carbon atom, and product **C** has none. What are the structures of **A**, **B**, and **C**?

S_N1, S_N2, E1, and E2 Mechanisms

8.47 Draw the organic products formed in each reaction.

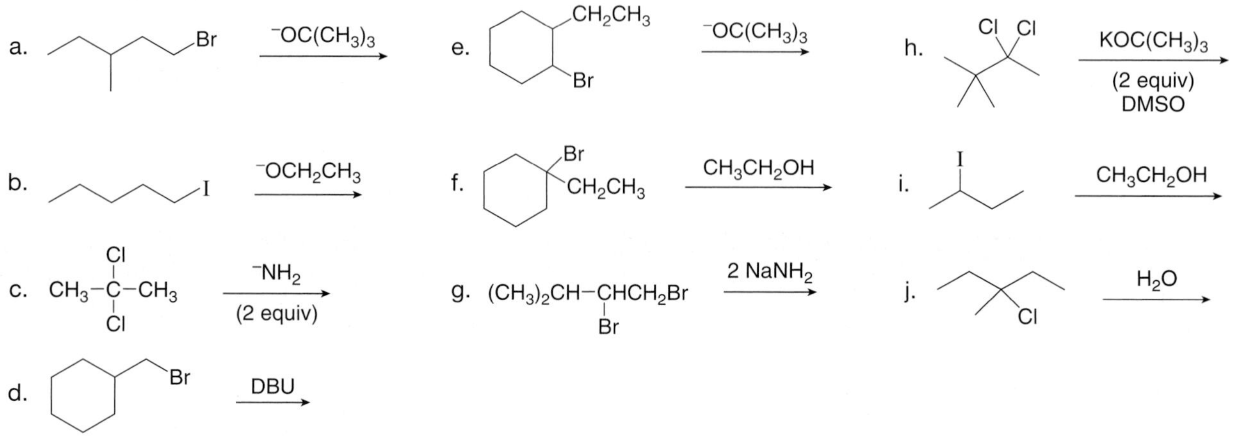

a. [structure] $\xrightarrow{\;^-OC(CH_3)_3\;}$

b. [structure] $\xrightarrow{\;^-OCH_2CH_3\;}$

c. $CH_3-\overset{\overset{\displaystyle Cl}{|}}{\underset{\underset{\displaystyle Cl}{|}}{C}}-CH_3 \xrightarrow[\text{(2 equiv)}]{\;^-NH_2\;}$

d. [structure] $\xrightarrow{\;DBU\;}$

e. [structure with CH₂CH₃ and Br] $\xrightarrow{\;^-OC(CH_3)_3\;}$

f. [structure with Br and CH₂CH₃] $\xrightarrow{\;CH_3CH_2OH\;}$

g. $(CH_3)_2CH-\overset{\overset{\displaystyle }{}}{\underset{\underset{\displaystyle Br}{|}}{C}}HCH_2Br \xrightarrow{\;2\,NaNH_2\;}$

h. [structure with Cl Cl] $\xrightarrow[\substack{\text{(2 equiv)}\\\text{DMSO}}]{\;KOC(CH_3)_3\;}$

i. [structure with I] $\xrightarrow{\;CH_3CH_2OH\;}$

j. [structure with Cl] $\xrightarrow{\;H_2O\;}$

8.48 Draw all products, including stereoisomers, in each reaction.

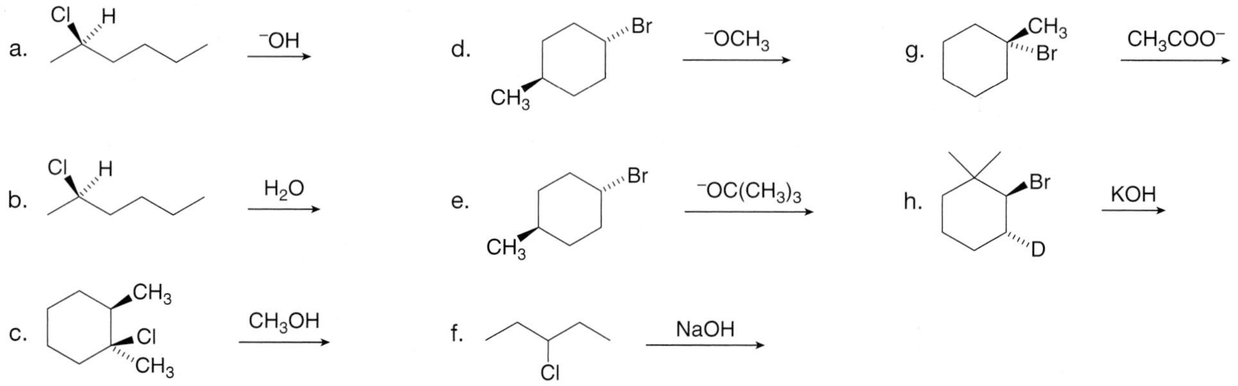

a. [structure with Cl, H] $\xrightarrow{\;^-OH\;}$

b. [structure with Cl, H] $\xrightarrow{\;H_2O\;}$

c. [structure with CH₃, Cl, CH₃] $\xrightarrow{\;CH_3OH\;}$

d. [cyclohexane with Br and CH₃] $\xrightarrow{\;^-OCH_3\;}$

e. [cyclohexane with Br and CH₃] $\xrightarrow{\;^-OC(CH_3)_3\;}$

f. [structure with Cl] $\xrightarrow{\;NaOH\;}$

g. [cyclohexane with CH₃ and Br] $\xrightarrow{\;CH_3COO^-\;}$

h. [cyclohexane with Br and D] $\xrightarrow{\;KOH\;}$

8.49 The following reactions do not afford the major product that is given. Explain why this is so, and draw the structure of the major product actually formed.

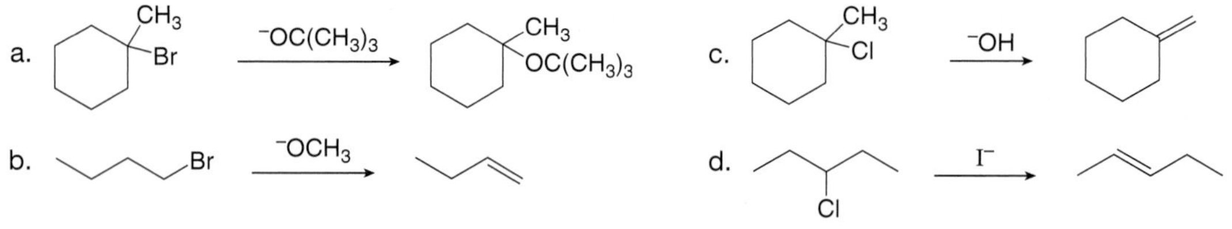

a. [cyclohexane with CH₃, Br] $\xrightarrow{\;^-OC(CH_3)_3\;}$ [cyclohexane with CH₃, OC(CH₃)₃]

b. [structure with Br] $\xrightarrow{\;^-OCH_3\;}$ [alkene structure]

c. [cyclohexane with CH₃, Cl] $\xrightarrow{\;^-OH\;}$ [methylenecyclohexane]

d. [structure with Cl] $\xrightarrow{\;I^-\;}$ [alkene structure]

8.50 Draw a stepwise, detailed mechanism for each reaction.

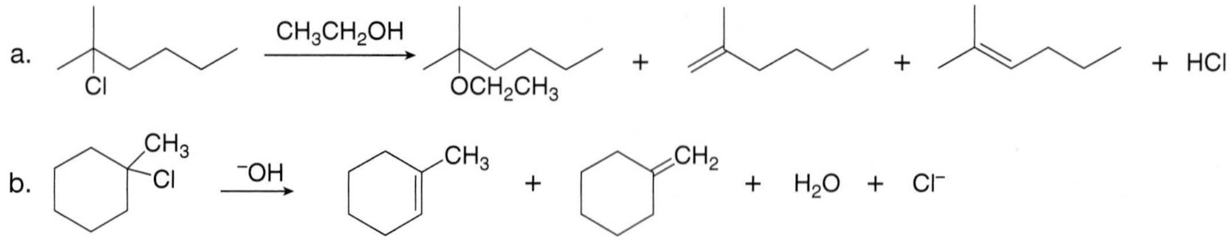

a. [structure with Cl] $\xrightarrow{\;CH_3CH_2OH\;}$ [structure with OCH₂CH₃] + [alkene] + [alkene] + HCl

b. [cyclohexane with CH₃, Cl] $\xrightarrow{\;^-OH\;}$ [cyclohexene with CH₃] + [cyclohexane with CH₂] + H_2O + Cl^-

8.51 Explain why 2-bromopropane gives different amounts of substitution and elimination products when the nucleophile/base is changed from acetate (CH_3COO^-) to ethoxide ($CH_3CH_2O^-$).

$$(CH_3)_2CHBr \ + \ CH_3COO^- \ \longrightarrow \ (CH_3)_2CHOCOCH_3$$
$$\text{only}$$

$$(CH_3)_2CHBr \ + \ CH_3CH_2O^- \ \longrightarrow \ (CH_3)_2CHOCH_2CH_3 \ + \ CH_3CH=CH_2$$
$$\text{20\%} \qquad\qquad \text{80\%}$$

Challenge Problems

8.52 Although there are nine stereoisomers of 1,2,3,4,5,6-hexachlorocyclohexane, one stereoisomer reacts 7000 times more slowly than any of the others in an E2 elimination. Draw the structure of this isomer and explain why this is so.

8.53 Explain the regiochemistry of the following reactions.

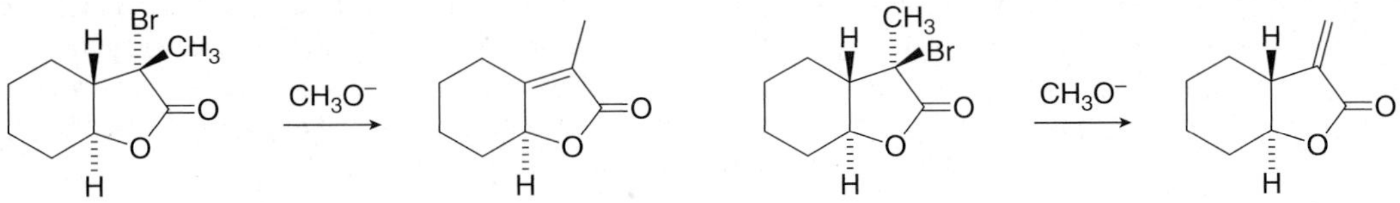

8.54 Draw a stepwise mechanism for the following reaction. Use the principles of β elimination and nucleophilic substitution to convert the five-membered ring to a six-membered ring.

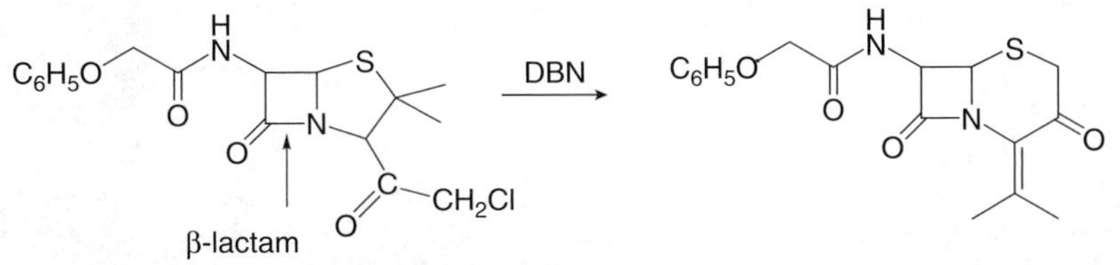

The four-membered ring in the starting material and product is called a β-lactam. This functional group confers biological activity on penicillin and many related antibiotics, as is discussed in Chapter 22.

Alcohols, Ethers, and Epoxides

Brevetoxin B is a complex neurotoxin produced by *Karenia brevis,* a single-celled organism that proliferates during "red tides," vast algae blooms that turn the ocean red, brown, or green. Red tides are responsible for massive fish kills worldwide, and for poisoning humans who eat seafood containing high levels of such neurotoxins. Brevetoxin is a polyether, a molecule containing many ether oxygen atoms bonded to two alkyl groups. In Chapter 9, we learn about the properties of ethers and related functional groups.

In Chapter 9, we take the principles learned in Chapters 7 and 8 about leaving groups, nucleophiles, and bases, and apply them to **alcohols, ethers,** and **epoxides,** three new functional groups that contain polar $C-O$ bonds. In the process, you will discover that all of the reactions in Chapter 9 follow one of the four general mechanisms introduced in Chapters 7 and 8—S_N1, S_N2, E1, or E2—so there are **no new mechanisms to learn.**

Although alcohols, ethers, and epoxides share many characteristics, each functional group has its own distinct reactivity, making each unique and different from the alkyl halides studied in Chapters 7 and 8. Appreciate the similarities but pay attention to the differences.

9.1 Introduction

Alcohols, ethers, and **epoxides** are three functional groups that contain carbon–oxygen σ bonds.

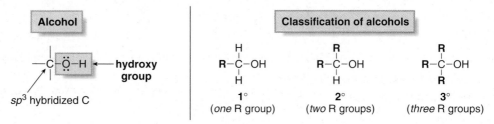

alcohol   **ether**   **epoxide**

Alcohols **contain a hydroxy group (OH** group) bonded to an sp^3 hybridized carbon atom. Alcohols are classified as **primary (1°), secondary (2°),** or **tertiary (3°)** based on the number of carbon atoms bonded to the carbon with the OH group.

Alcohol	Classification of alcohols		

sp³ hybridized C → hydroxy group

1° (*one* R group)   2° (*two* R groups)   3° (*three* R groups)

Compounds having a hydroxy group on an sp^2 hybridized carbon atom—**enols** and **phenols**—undergo different reactions than alcohols and are discussed in Chapters 11 and 19, respectively. **Enols** have an OH group on a carbon of a $C-C$ double bond. **Phenols** have an OH group on a benzene ring.

sp² hybridized C

enol   phenol

Ethers **have two alkyl groups bonded to an oxygen atom.** An ether is **symmetrical** if the two alkyl groups are the same, and **unsymmetrical** if they are different. Both alcohols and ethers are organic derivatives of H_2O, formed by replacing one or both of the hydrogens on the oxygen atom by R groups, respectively.

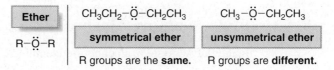

Ether	$CH_3CH_2-\ddot{O}-CH_2CH_3$	$CH_3-\ddot{O}-CH_2CH_3$
$R-\ddot{O}-R$	**symmetrical ether**	**unsymmetrical ether**
	R groups are the **same.**	R groups are **different.**

Epoxides **are ethers having the oxygen atom in a three-membered ring.** Epoxides are also called **oxiranes.**

epoxide or **oxirane**

An epoxide is a special type of ether.

Figure 9.1 Electrostatic potential maps for a simple alcohol, ether, and epoxide

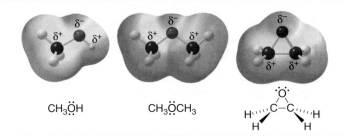

- Electron-rich regions are shown by the red around the O atoms.

$CH_3\ddot{O}H$ $CH_3\ddot{O}CH_3$

PROBLEM 9.1   Draw all constitutional isomers having molecular formula $C_4H_{10}O$. Classify each compound as a 1°, 2°, or 3° alcohol, or a symmetrical or unsymmetrical ether.

9.2 Structure and Bonding

Alcohols, ethers, and epoxides each contain an oxygen atom surrounded by two atoms and two nonbonded electron pairs, making the O atom **tetrahedral** and sp^3 hybridized. Because only two of the four groups around O are atoms, alcohols and ethers have a **bent** shape like H_2O.

sp^3 hybridized

sp^3 hybridized

$CH_3\ddot{O}H$ $CH_3\ddot{O}CH_3$
109°   111°

The bond angle around the O atom in an alcohol or ether is similar to the tetrahedral bond angle of **109.5°**. In contrast, the C–O–C bond angle of an epoxide must be **60°**, a considerable deviation from the tetrahedral bond angle. For this reason, **epoxides have angle strain,** making them much more reactive than other ethers.

60°

a strained, three-membered ring

Because oxygen is much more electronegative than carbon or hydrogen, the C–O and O–H bonds are all polar, with the O atom electron rich and the C and H atoms electron poor. The electrostatic potential maps in Figure 9.1 show these polar bonds for all three functional groups.

9.3 Nomenclature

To name an alcohol, ether, or epoxide using the IUPAC system, we must learn how to name the functional group either as a substituent or by using a suffix added to the parent name.

9.3A Naming Alcohols

In the IUPAC system, alcohols are identified by the suffix **-ol.**

How To   Name an Alcohol Using the IUPAC System

Example   **Give the IUPAC name of the following alcohol:**

$$CH_3CHCH_2CHCH_2CH_3$$
with substituents CH_3 and OH

Step [1]   **Find the longest carbon chain containing the OH group.**

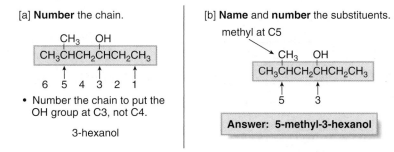

6 C's in the longest chain

6 C's --→ **hexan*e*** --→ **hexan*ol***

• Change the **-*e*** ending of the parent alkane to the suffix **-*ol.***

Step [2]   **Number the carbon chain to give the OH group the lower number, and apply all other rules of nomenclature.**

[a] **Number** the chain.

• Number the chain to put the OH group at C3, not C4.

3-hexanol

[b] **Name** and **number** the substituents.

methyl at C5

Answer: 5-methyl-3-hexanol

The 1979 IUPAC recommendations for alcohol nomenclature place the number indicating the location of the OH group before the name of the *parent;* thus, $CH_3CH_2CH_2CH_2OH$ is named 1-butanol. The 1993 IUPAC recommendations place this number immediately preceding the *suffix* -ol; thus, $CH_3CH_2CH_2CH_2OH$ is named butan-1-ol. The first of these conventions is more widely used, so we follow it in this text.

When an OH group is bonded to a ring, the **ring is numbered beginning with the OH group.** Because the functional group is always at C1, the *1* is usually omitted from the name. The ring is then numbered in a clockwise or counterclockwise fashion to give the next substituent the lower number. Representative examples are given in Figure 9.2.

Common names are often used for simple alcohols. To assign a common name:

◆ Name all the carbon atoms of the molecule as a single **alkyl group.**

◆ Add the word **alcohol,** separating the words with a space.

$CH_3-\overset{\overset{\displaystyle H}{|}}{\underset{\underset{\displaystyle CH_3}{|}}{C}}-OH$ ← alcohol →   **isopropyl alcohol**

isopropyl group

a common name

Compounds with two hydroxy groups are called **diols** (using the IUPAC system) or **glycols.** Compounds with three hydroxy groups are called **triols,** and so forth.

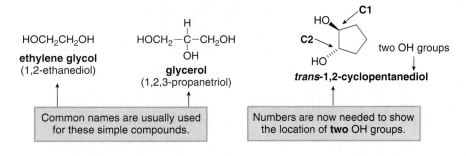

$HOCH_2CH_2OH$

ethylene glycol
(1,2-ethanediol)

$HOCH_2-\overset{\overset{\displaystyle H}{|}}{\underset{\underset{\displaystyle OH}{|}}{C}}-CH_2OH$

glycerol
(1,2,3-propanetriol)

trans-1,2-cyclopentanediol

two OH groups

Common names are usually used for these simple compounds.

Numbers are now needed to show the location of **two** OH groups.

Figure 9.2   Examples: Naming cyclic alcohols

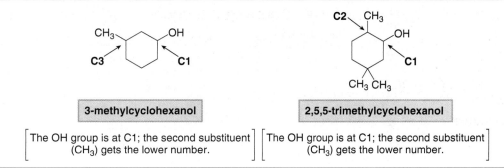

3-methylcyclohexanol

2,5,5-trimethylcyclohexanol

The OH group is at C1; the second substituent (CH_3) gets the lower number.

The OH group is at C1; the second substituent (CH_3) gets the lower number.

PROBLEM 9.2 Give the IUPAC name for each compound.

a.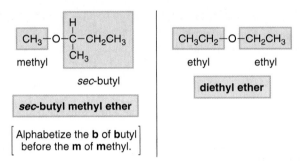

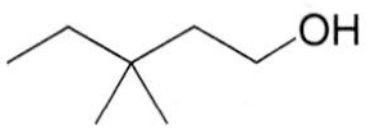

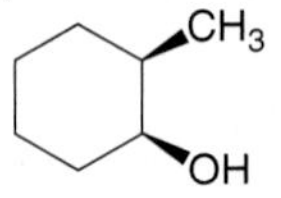

b. $(CH_3CH_2)_2CHCH(OH)CH_2CH_3$

c.

d.

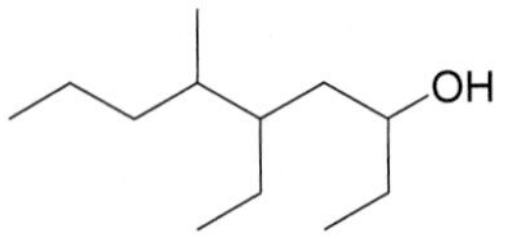

PROBLEM 9.3 Give the structure corresponding to each name.
a. 7,7-dimethyl-4-octanol
b. 5-methyl-4-propyl-3-heptanol
c. 2-*tert*-butyl-3-methylcyclohexanol
d. *trans*-1,2-cyclohexanediol

9.3B Naming Ethers

Simple ethers are usually assigned common names. To do so, **name both alkyl groups** bonded to the oxygen, arrange these names alphabetically, and add the word **ether.** For symmetrical ethers, name the alkyl group and add the prefix **di-.**

CH₃—O—C(H)(CH₃)—CH₂CH₃

methyl   sec-butyl

sec-butyl methyl ether

⎡ Alphabetize the **b** of **butyl** before the **m** of **methyl**. ⎤

CH₃CH₂—O—CH₂CH₃

ethyl   ethyl

diethyl ether

More complex ethers are named using the IUPAC system. One alkyl group is named as a hydrocarbon chain, and the other is named as part of a substituent bonded to that chain.

◆ Name the simpler alkyl group + O atom as an **alkoxy** substituent by changing the **-yl** ending of the alkyl group to **-oxy.**

Common alkoxy groups

CH₃O—   CH₃CH₂O—   CH₃—C(CH₃)₂—O—

methoxy   ethoxy   *tert*-butoxy

◆ Name the remaining alkyl group as an alkane, with the alkoxy group as a substituent bonded to this chain.

SAMPLE PROBLEM 9.1 Give the IUPAC name for the following ether.

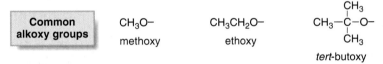

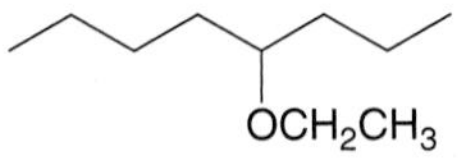

OCH₂CH₃

SOLUTION

[1] Name the longer chain as an alkane and the shorter chain as an alkoxy group.

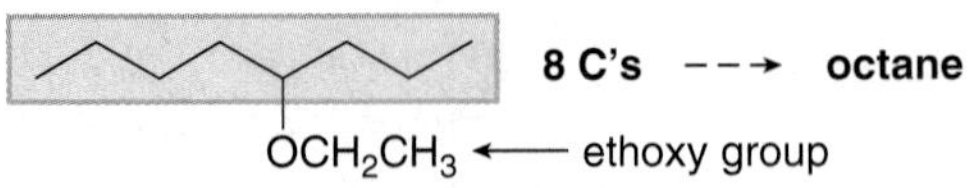

8 C's -- → octane

OCH₂CH₃ ←— ethoxy group

[2] Apply the other nomenclature rules to complete the name.

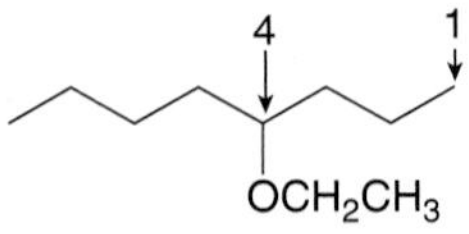

4   1

OCH₂CH₃

Answer: 4-ethoxyoctane

PROBLEM 9.4 Name each of the following ethers.

a. $CH_3-O-CH_2CH_2CH_2CH_3$

b.

c. $CH_3CH_2CH_2-O-CH_2CH_2CH_3$

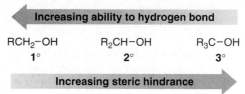

tetrahydrofuran
THF

> Any cyclic compound containing a heteroatom is called a *heterocycle.*

Cyclic ethers have an O atom in a ring. A common cyclic ether is **tetrahydrofuran (THF),** a somewhat polar aprotic solvent used in nucleophilic substitution (Section 7.8C) and many other organic reactions.

9.3C Naming Epoxides

Epoxides bonded to a chain of carbon atoms are named as derivatives of **oxirane,** the simplest epoxide having two carbons and one oxygen atom in a ring. When an epoxide is bonded to a ring, the O atom is considered a substituent, called an **epoxy** group, bonded to a cycloalkane, and two numbers are needed to designate the atoms to which the O atoms are bonded.

oxirane   **1,1-dimethyloxirane**   **1,2-epoxycyclohexane**

Epoxides are also named as **alkene oxides,** because they are often prepared by adding an O atom to an alkene (Chapter 12). To name an epoxide this way, mentally replace the epoxide oxygen by a double bond, name the alkene (Section 10.3), and then add the word *oxide.* For example, the common name for oxirane is ethylene oxide, because it is an epoxide derived from the alkene ethylene.

$CH_2=CH_2$ - - - →
ethylene

ethylene oxide (common name)
oxirane   (IUPAC name)

PROBLEM 9.5 Name each of the following epoxides.

a.   b.

9.4 Physical Properties

Alcohols, ethers, and epoxides exhibit dipole–dipole interactions because they have a bent structure with two polar bonds. **Alcohols are also capable of intermolecular hydrogen bonding,** because they possess a hydrogen atom on an oxygen, making alcohols much more polar than ethers and epoxides.

hydrogen bond

Steric factors affect the extent of hydrogen bonding. Although all alcohols can hydrogen bond, increasing the number of R groups around the carbon atom bearing the OH group decreases the extent of hydrogen bonding. Thus, 3° alcohols are least able to hydrogen bond, whereas 1° alcohols are most able to.

◄ Increasing ability to hydrogen bond

RCH_2-OH R_2CH-OH R_3C-OH
1°   **2°**   **3°**

Increasing steric hindrance ►

How these factors affect the physical properties of alcohols, ethers, and epoxides is summarized in Table 9.1.

TABLE 9.1   Physical Properties of Alcohols, Ethers, and Epoxides

Property	Observation
Boiling point (bp) and melting point (mp)	• For compounds of comparable molecular weight, the **stronger** the intermolecular forces, the **higher** the bp or mp. $CH_3CH_2CH_2CH_3$ — VDW — bp 0 °C $CH_3OCH_2CH_3$ — VDW, DD — bp 11 °C $CH_3CH_2CH_2OH$ — VDW, DD, HB — bp 97 °C Increasing boiling point → • Bp's **increase** as the extent of hydrogen bonding **increases**. $(CH_3)_3C-OH$ 3° bp 83 °C $CH_3CH_2\overset{\text{OH}}{\underset{}{C}}HCH_3$ 2° bp 98 °C $CH_3CH_2CH_2CH_2-OH$ 1° bp 118 °C Increasing ability to hydrogen bond Increasing boiling point →
Solubility	• Alcohols, ethers, and epoxides having ≤ 5 C's are H_2O soluble because they each have an oxygen atom capable of hydrogen bonding to H_2O (Section 3.5C). • Alcohols, ethers, and epoxides having > 5 C's are H_2O insoluble because the nonpolar alkyl portion is too large to dissolve in H_2O. • Alcohols, ethers, and epoxides of any size are soluble in organic solvents.

Key: VDW = van der Waals forces; DD = dipole–dipole; HB = hydrogen bonding

---

PROBLEM 9.6   Rank the following compounds in order of increasing boiling point.

a.

b.

PROBLEM 9.7   Explain why dimethyl ether $(CH_3)_2O$ and ethanol (CH_3CH_2OH) are both water soluble, but the boiling point of ethanol (78 °C) is much higher than the boiling point of dimethyl ether (–24 °C).

9.5   Interesting Alcohols, Ethers, and Epoxides

A large number of alcohols, ethers, and epoxides have interesting and useful properties.

9.5A   Interesting Alcohols

Ethanol (Figure 9.3), formed by the fermentation of the carbohydrates in grains, grapes, and potatoes, is the alcohol present in alcoholic beverages. It is perhaps the first organic compound synthesized by humans, because alcohol production has been known for at least 4000 years. Ethanol depresses the central nervous system, increases the production of stomach acid, and dilates blood vessels, producing a flushed appearance.

Ethanol is also a common laboratory solvent, which is sometimes made unfit to ingest by adding small amounts of benzene or methanol (both of which are toxic). Ethanol is used as a gasoline

Figure 9.3 Ethanol— The alcohol in alcoholic beverages

- Ethanol is the alcohol in red wine, obtained by the fermentation of grapes.

Figure 9.4 Some simple alcohols

CH₃OH

- **Methanol (CH₃OH)** is also called wood alcohol, because it can be obtained by heating wood at high temperatures in the absence of air. Methanol is extremely toxic because of the oxidation products formed when it is metabolized in the liver (Section 12.13). Ingestion of as little as 15 mL causes blindness, and 100 mL causes death.

(CH₃)₂CHOH

- **2-Propanol [(CH₃)₂CHOH]** is the major component of rubbing alcohol. When rubbed on the skin it evaporates readily, producing a pleasant cooling sensation. Because it has weak antibacterial properties, 2-propanol is used to clean skin before minor surgery and to sterilize medical instruments.

HOCH₂CH₂OH

- **Ethylene glycol (HOCH₂CH₂OH)** is the major component of antifreeze. It is readily prepared from ethylene oxide by reactions discussed in Section 9.15. It is sweet tasting but toxic.

additive because it readily combusts with the release of energy. **Gasohol** contains 10% ethanol. Other simple alcohols are listed in Figure 9.4.

Starch and **cellulose,** two polymeric carbohydrates with many OH groups, were first discussed in Section 5.1. Starch is used for short-term energy storage, and cellulose is the major structural compound in plant and bacterial cell walls. The many hydroxy groups allow for substantial hydrogen bonding interactions either between atoms in the same molecule or within the aqueous environment of the cell.

Other complex naturally occurring alcohols include palytoxin, toxic component of a soft coral (Prologue); cholesterol, a major building block of cell membranes (Sections 3.5C and 4.15); and taxol, an anticancer drug (Section 5.5).

cellulose

[OH groups are labeled in red.]

amylose (one form of starch)

diethyl ether

This painting by Robert Hinckley depicts a public demonstration of the use of diethyl ether as an anesthetic at the Massachusetts General Hospital in Boston in the 1840s.

Recall from Section 3.8B that crown ethers are named as *x*-crown-*y,* where *x* is the total number of atoms in the ring and *y* is the number of O atoms.

Karenia brevis is the single-celled organism that produces brevetoxin B.

9.5B   Interesting Ethers

The discovery that **diethyl ether** ($CH_3CH_2OCH_2CH_3$) is a general anesthetic revolutionized surgery in the nineteenth century. For years, a heated controversy existed over who first discovered diethyl ether's anesthetic properties and recognized the enormous benefit in its use. Early experiments were performed by a dentist, Dr. William Morton, resulting in a public demonstration of diethyl ether as an anesthetic in Boston in 1846. In fact, Dr. Crawford Long, a Georgia physician, had been using diethyl ether in surgery and obstetrics for several years, but had not presented his findings to a broader audience.

Diethyl ether is an imperfect anesthetic, but considering the alternatives in the nineteenth century, it was a miracle drug. It is safe, easy to administer, and causes little patient mortality, but it is highly flammable and causes nausea in many patients. For these reasons, it has largely been replaced by halothane (Figure 7.4), which is non-flammable and causes little patient discomfort.

Recall from Section 3.8B that some cyclic **polyethers**—compounds with two or more ether linkages—contain cavities that can complex specific-sized cations. For example, 18-crown-6 binds K^+, whereas 12-crown-4 binds Li^+.

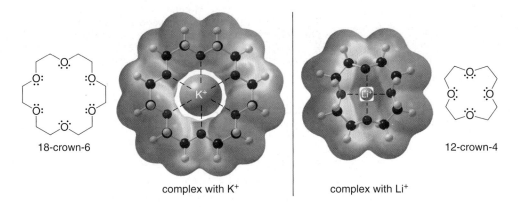

18-crown-6   complex with K^+ complex with Li^+ 12-crown-4

- ◆ A crown ether–cation complex is called a *host–guest* complex. The crown ether is the *host* and the cation is the *guest.*
- ◆ The ability of a host molecule to bind specific guests is called *molecular recognition.*

Brevetoxin B, the complex neurotoxin that opened Chapter 9, is a naturally occurring polyether that interferes with Na^+ ion transport across cell membranes.

brevetoxin B
[Ether linkages are labeled in red.]

The ability of crown ethers to complex cations can be exploited in nucleophilic substitution reactions, as shown in Figure 9.5. Nucleophilic substitution reactions are usually run in polar solvents to dissolve both the polar organic substrate and the ionic nucleophile. With a crown ether, though, the reaction can be run in a nonpolar solvent under conditions that enhance nucleophilicity.

When 18-crown-6 is added to the reaction of CH_3CH_2Br with KCN, for example, the crown ether forms a tight complex with K^+ that has nonpolar C−H bonds on the outside, making the complex soluble in nonpolar solvents like benzene (C_6H_6) or hexane. When the crown ether/K^+ complex dissolves in the nonpolar solvent, it carries the ⁻CN along with it to maintain electrical

Figure 9.5
The use of crown ethers in nucleophilic substitution reactions

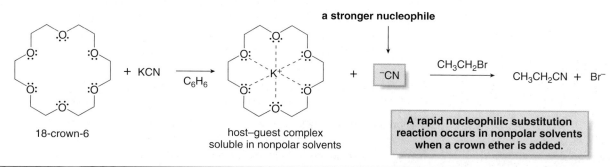

KCN is insoluble in nonpolar solvents alone, but with 18-crown-6:

18-crown-6

host–guest complex
soluble in nonpolar solvents

a stronger nucleophile

A rapid nucleophilic substitution reaction occurs in nonpolar solvents when a crown ether is added.

neutrality. The result is a solution of tightly complexed cation and relatively unsolvated anion (nucleophile). The anion, therefore, is extremely nucleophilic because it is not hidden from the substrate by solvent molecules.

In many cases, crown ether assisted nucleophilic substitution reactions in nonpolar solvents are *faster* than similar reactions in polar aprotic solvents, even though polar aprotic solvents enhance nucleophilicity by solvating anions poorly.

PROBLEM 9.8   Which mechanism is favored by the use of crown ethers in nonpolar solvents, S_N1 or S_N2?

PROBLEM 9.9   What reaction conditions could be used to carry out the following nucleophilic substitution?

$$CH_3CH_2CH_2CH_2-Br \longrightarrow CH_3CH_2CH_2CH_2-F$$

9.5C   Interesting Epoxides

Naturally occurring compounds that contain epoxides include **periplanone B,** the sex pheromone of the female American cockroach, and **epothilone B,** a novel anticancer drug. The epothilones represent a new group of anticancer drugs first isolated in 1987 from soil bacteria in southern Africa. Because they are active against many taxol-resistant tumors and are readily produced by the fermentation of soil bacteria, they hold great promise for cancer treatment in the future.

periplanone B
cockroach sex pheromone

epothilone B
anticancer drug

PROBLEM 9.10   To be an active chemotherapeutic agent, a drug must be somewhat water soluble. Predict the solubility properties of epothilone B in water and organic solvents.

9.6   Preparation of Alcohols, Ethers, and Epoxides

Alcohols and ethers are both common products of nucleophilic substitution. They are synthesized from alkyl halides by S_N2 reactions using strong nucleophiles. As in all S_N2 reactions, highest yields of products are obtained with unhindered methyl and 1° alkyl halides.

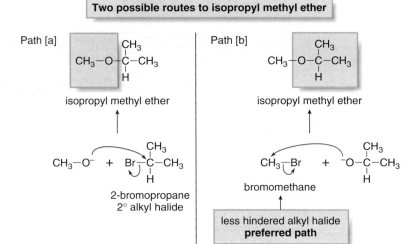

nucleophile   product
            ↓   ↓

CH_3CH_2—Br + ⁻OH $\xrightarrow{S_N2}$ CH_3CH_2—OH   alcohol

$CH_3CH_2CH_2$—Cl + ⁻OCH_3 $\xrightarrow{S_N2}$ $CH_3CH_2CH_2$—OCH_3 unsymmetrical ether

CH_3CH_2—Br + ⁻OCH_2CH_3 $\xrightarrow{S_N2}$ CH_3CH_2—OCH_2CH_3 symmetrical ether

The preparation of ethers by this method is called the **Williamson ether synthesis,** and, although it was first reported in the 1800s, it is still the most general method to prepare an ether. Unsymmetrical ethers can be synthesized in two different ways, but often one path is preferred.

For example, isopropyl methyl ether can be prepared from CH_3O^- and 2-bromopropane (Path [a]), or from $(CH_3)_2CHO^-$ and bromomethane (Path [b]). Because the mechanism is S_N2, **the preferred path uses the less sterically hindered halide, CH_3Br—Path [b].**

Two possible routes to isopropyl methyl ether

Path [a]
CH_3
CH_3—O—C—CH_3
|
H

isopropyl methyl ether

CH_3
CH_3—O⁻ + Br—C—CH_3
|
H
2-bromopropane
2° alkyl halide

Path [b]
CH_3
CH_3—O—C—CH_3
|
H

isopropyl methyl ether

CH_3
CH_3—Br + ⁻O—C—CH_3
|
H
bromomethane

less hindered alkyl halide
preferred path

PROBLEM 9.11   Draw the organic product of each reaction and classify the product as an alcohol, symmetrical ether, or unsymmetrical ether.

a. $CH_3CH_2CH_2CH_2$—Br + ⁻OH ⟶

c. [cyclohexyl]—CH_2CH_2I + ⁻$OCH(CH_3)_2$ ⟶

b. [hexyl chain]—Cl + ⁻OCH_3 ⟶

d. [branched chain]—Br + ⁻OCH_2CH_3 ⟶

PROBLEM 9.12   Draw two different routes to each ether and state which route, if any, is preferred.

a. CH_3—O—[cyclohexyl]

b. CH_3CH_2—O—C—CH_3 (with CH_3 up and H down)

A **hydroxide** nucleophile is needed to synthesize an alcohol, and salts such as **NaOH** and **KOH** are inexpensive and commercially available. An **alkoxide** salt is needed to make an ether. Simple alkoxides such as sodium methoxide ($NaOCH_3$) can be purchased, but others are prepared from alcohols by a Brønsted–Lowry acid–base reaction. For example, **sodium ethoxide ($NaOCH_2CH_3$)** is prepared by treating ethanol with NaH.

NaH is an especially good base for forming an alkoxide, because the by-product of the reaction, H_2, is a gas that just bubbles out of the reaction mixture.

An acid–base reaction

CH_3CH_2O—H + Na^+H^- ⟶ $CH_3CH_2O^-\ Na^+$ + H_2
ethanol   base   sodium ethoxide

an alkoxide nucleophile

SAMPLE PROBLEM 9.2 Draw the product of the following two-step reaction sequence.

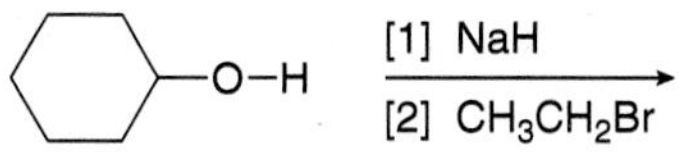

SOLUTION

[1] The base removes a proton from the OH group, forming an alkoxide.	[2] The alkoxide acts as a nucleophile in an S_N2 reaction, forming an ether.

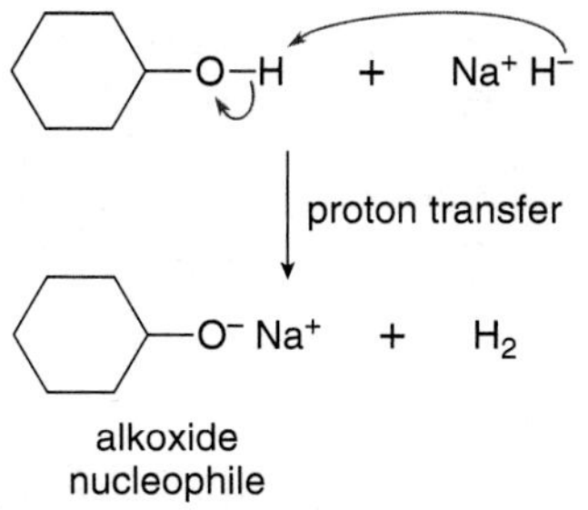

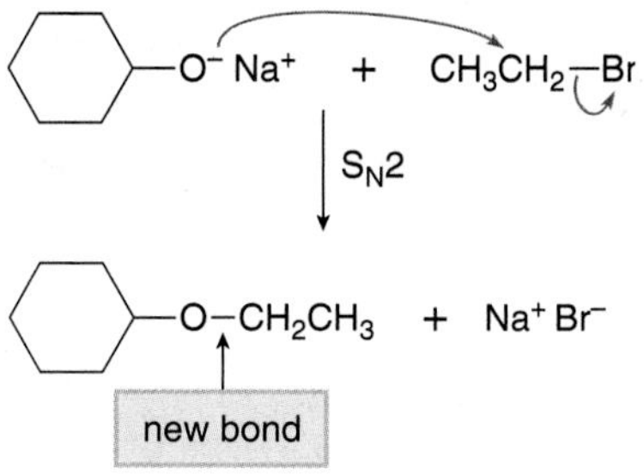

proton transfer

alkoxide
nucleophile

S_N2

new bond

- This two-step sequence converts an alcohol to an ether.

When an organic compound contains both a hydroxy group and a halogen atom on adjacent carbon atoms, an *intramolecular* version of this reaction forms an epoxide. The starting material for this two-step sequence, a **halohydrin**, is prepared from an alkene, as we will learn in Chapter 10.

Epoxide synthesis—A two-step procedure

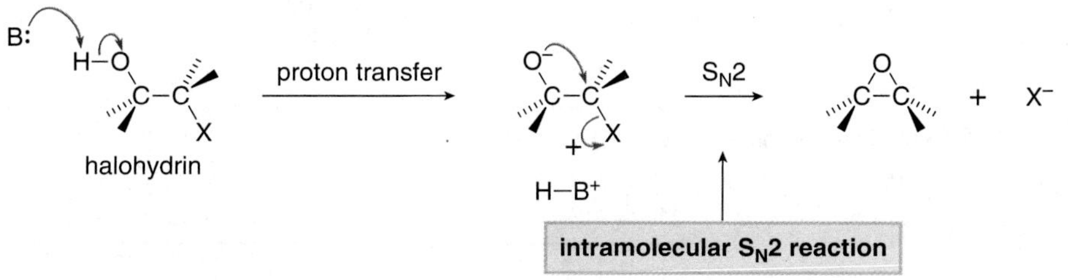

halohydrin

proton transfer

$H-B^+$

S_N2

$+ X^-$

intramolecular S_N2 reaction

PROBLEM 9.13 Draw the products of each reaction.

a. $CH_3CH_2CH_2-OH + NaH \longrightarrow$

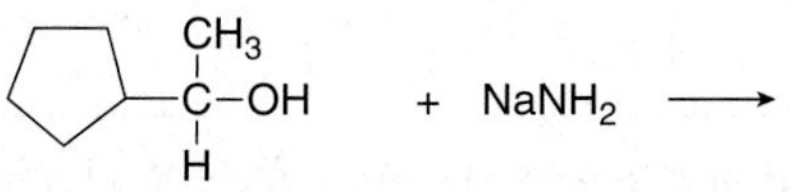

c. 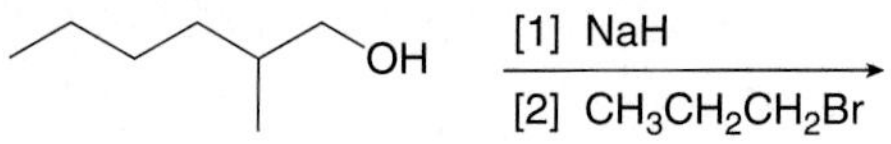 $\dfrac{[1]\ NaH}{[2]\ CH_3CH_2CH_2Br}$

b. $+ NaNH_2 \longrightarrow$

d. 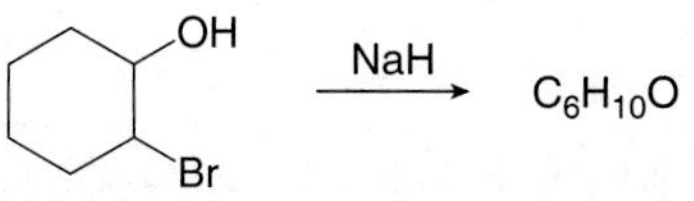 $\xrightarrow{NaH}$ $C_6H_{10}O$

9.7 General Features—Reactions of Alcohols, Ethers, and Epoxides

We begin our discussion of the chemical reactions of alcohols, ethers, and epoxides, with a look at the general reactive features of each functional group.

9.7A Alcohols

Unlike many families of molecules, the reactions of alcohols do *not* fit neatly into a single reaction class. In Chapter 9, we discuss only the substitution and β elimination reactions of alcohols. Alcohols are also key starting materials in oxidation reactions (Chapter 12), and their polar O−H bond makes them more acidic than many other organic compounds, a feature we will explore in Chapter 19.

Alcohols are similar to alkyl halides in that both contain an electronegative element bonded to an sp^3 hybridized carbon atom. Alkyl halides contain a good leaving group (X^-), however, whereas alcohols do not. Nucleophilic substitution with ROH as starting material would displace $^-$OH, a **strong base and therefore a poor leaving group.**

$$R\!-\!X \quad + \quad :Nu^- \quad \longrightarrow \quad R\!-\!Nu \quad + \quad \boxed{X^-} \longleftarrow \textbf{good} \text{ leaving group}$$

$$R\!-\!OH \quad + \quad :Nu^- \quad \overset{\times}{\longrightarrow} \quad R\!-\!Nu \quad + \quad \boxed{^-OH} \longleftarrow \textbf{poor} \text{ leaving group}$$

For an alcohol to undergo a nucleophilic substitution or elimination reaction, the **OH group must be converted into a better leaving group.** This can be done by reaction with acid. Treatment of an alcohol with a strong acid like HCl or H_2SO_4 protonates the O atom via an acid–base reaction. This transforms the $^-$OH leaving group into **H_2O, a weak base and therefore a good leaving group.**

$$R\!-\!\overset{..}{\underset{..}{O}}H \quad + \quad H\!-\!Cl \quad \rightleftharpoons \quad R\!-\!\overset{+}{\underset{..}{O}}H_2 \quad + \quad Cl^-$$

Strong acid

weak base
good leaving group

If the OH group of an alcohol is made into a good leaving group, alcohols *can* undergo β elimination and nucleophilic substitution, as described in Sections 9.8–9.12.

β elimination	
$- H_2O$	Alkenes are formed by β elimination.
	(Sections 9.8–9.10)
nucleophilic substitution	Alkyl halides are formed by nucleophilic substitution.
	(Sections 9.11–9.12)

> Because the pK_a of $(ROH_2)^+$ is ~–2, protonation of an alcohol occurs only with very strong acids—namely, those having a $pK_a \le -2$.

PROBLEM 9.14   Which of the following acids are strong enough to protonate an alcohol?
a. HF ($pK_a = 3$)   c. $C_6H_5SO_3H$ ($pK_a = -7$)
b. $HClO_4$ ($pK_a = -10$)   d. CH_3COOH ($pK_a = 5$)

9.7B   Ethers and Epoxides

Like alcohols, **ethers do not contain a good leaving group,** which means that nucleophilic substitution and β elimination do not occur directly. Ethers undergo fewer useful reactions than alcohols.

$$R\!-\!\overset{..}{\underset{..}{O}}R \longleftarrow \textbf{poor leaving group}$$

Epoxides don't have a good leaving group either, but they have one characteristic that neither alcohols nor ethers have: **the "leaving group" is contained in a strained three-membered ring.** Nucleophilic attack opens the three-membered ring and relieves angle strain, making nucleophilic attack a favorable process that occurs even with the poor leaving group. Specific examples are presented in Section 9.15.

9.8   Dehydration of Alcohols to Alkenes

The dehydrohalogenation of alkyl halides, discussed in Chapter 8, is one way to introduce a π bond into a molecule. Another way is to eliminate water from an alcohol in a **dehydration** reaction. **Dehydration, like dehydrohalogenation, is a β elimination reaction in which the elements of OH and H are removed from the α and β carbon atoms, respectively.**

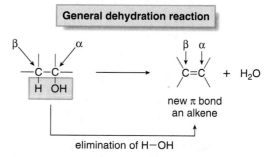

Dehydration is typically carried out using H_2SO_4 and other strong acids, or phosphorus oxychloride ($POCl_3$) in the presence of an amine base. We consider dehydration in acid first, followed by dehydration with $POCl_3$ in Section 9.10.

9.8A General Features of Dehydration in Acid

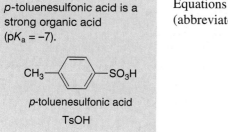

Recall from Section 2.6 that *p*-toluenesulfonic acid is a strong organic acid ($pK_a = -7$).

p-toluenesulfonic acid

TsOH

Alcohols undergo dehydration in the presence of strong acid to afford alkenes, as illustrated in Equations [1] and [2]. Typical acids used for this conversion are H_2SO_4 or *p*-toluenesulfonic acid (abbreviated as TsOH).

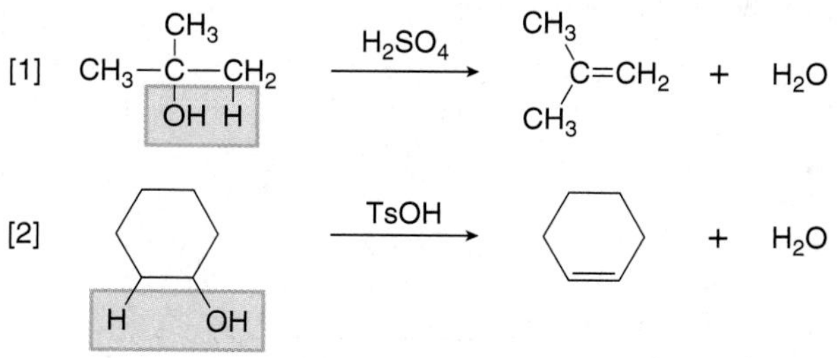

More substituted alcohols dehydrate more readily, giving rise to the following order of reactivity:

$$RCH_2-OH \qquad R_2CH-OH \qquad R_3C-OH$$
$$1° \qquad\qquad 2° \qquad\qquad 3°$$

Increasing rate of dehydration

When an alcohol has two or three different β carbons, dehydration is regioselective and follows the Zaitsev rule. **The more substituted alkene is the major product when a mixture of constitutional isomers is possible.** For example, elimination of H and OH from 2-methyl-2-butanol yields two constitutional isomers: the trisubstituted alkene **A** as major product and the disubstituted alkene **B** as minor product.

PROBLEM 9.15 Draw the products of each dehydration reaction and label the major product when a mixture results.

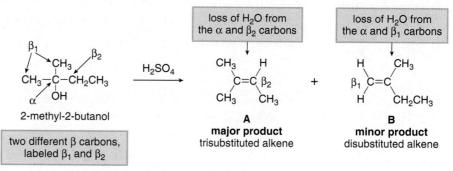

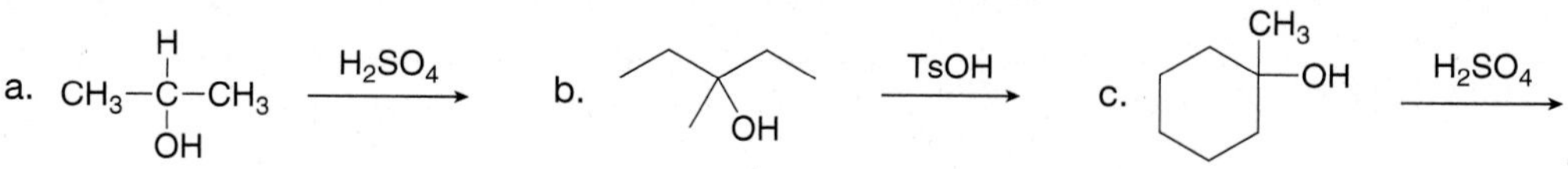

PROBLEM 9.16   Rank the alcohols in each group in order of increasing reactivity when dehydrated with H_2SO_4.

a. $(CH_3)_2C(OH)CH_2CH_2CH_3$ $(CH_3)_2CHCH_2CH_2CH_2OH$ $(CH_3)_2CHCH_2CH(OH)CH_3$

b.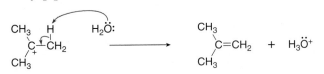

9.8B   The E1 Mechanism for the Dehydration of 2° and 3° Alcohols

The mechanism of dehydration depends on the structure of the alcohol: **2° and 3° alcohols react by an E1 mechanism, whereas 1° alcohols react by an E2 mechanism.** Regardless of the type of alcohol, however, strong acid is *always* needed to protonate the O atom to form a good leaving group.

The E1 dehydration of 2° and 3° alcohols is illustrated with $(CH_3)_3COH$ (a 3° alcohol) as starting material to form $(CH_3)_2C=CH_2$ as product (Mechanism 9.1). The mechanism consists of **three steps.**

MECHANISM 9.1

Dehydration of 2° and 3° ROH—An E1 Mechanism

Step [1]   The O atom is protonated.

$$CH_3-\underset{\overset{|}{:\ddot{O}H}}{\overset{\overset{CH_3}{|}}{C}}-CH_3 \quad \xrightarrow{\text{proton transfer}} \quad CH_3-\underset{\overset{|}{\underset{+}{:\ddot{O}H_2}}}{\overset{\overset{CH_3}{|}}{C}}-CH_3 \quad + \quad HSO_4^-$$

with $H-OSO_3H$

good leaving group

◆ **Protonation of the oxygen atom** of the alcohol converts a poor leaving group (^-OH) into a good leaving group (H_2O).

Step [2]   The C–O bond is broken.

$$CH_3-\underset{\overset{|}{\underset{+}{:OH_2}}}{\overset{\overset{CH_3}{|}}{C}}-CH_3 \quad \xrightarrow{\text{slow}} \quad \underset{\overset{|}{CH_3}}{\overset{\overset{CH_3}{|}}{\underset{+}{C}}}-CH_3 \quad + \quad H_2\ddot{O}:$$

carbocation   good leaving group

◆ **Heterolysis of the C–O bond forms a carbocation.** This step is rate-determining because it involves only bond cleavage.

Step [3]   A C–H bond is cleaved and the π bond is formed.

$$\underset{\overset{|}{CH_3}}{\overset{\overset{CH_3}{|}}{\underset{+}{C}}}-CH_2 \quad H_2\ddot{O}: \quad \longrightarrow \quad \underset{\overset{|}{CH_3}}{\overset{\overset{CH_3}{|}}{C}}=CH_2 \quad + \quad H_3\ddot{O}^+$$

◆ **A base (such as HSO_4^- or H_2O) removes a proton** from a carbon adjacent to the carbocation (a β carbon). The electron pair in the C–H bond is used to form the new π bond.

Thus, **dehydration of 2° and 3° alcohols occurs via an E1 mechanism with an added first step.**

◆ Step [1] protonates the OH group to make a good leaving group.
◆ Steps [2] and [3] are the two steps of an E1 mechanism: loss of a leaving group (H_2O in this case) to form a carbocation, followed by removal of a β proton to form a π bond.
◆ The acid used to protonate the alcohol in Step [1] is regenerated upon removal of the proton in Step [3], so dehydration is **acid-catalyzed.**

The E1 dehydration of 2° and 3° alcohols with acid gives clean elimination products without by-products formed from an S_N1 reaction. This makes the E1 dehydration of alcohols much more synthetically useful than the E1 dehydrohalogenation of alkyl halides (Section 8.7). Clean elimination takes place because the reaction mixture contains no good nucleophile to react with the intermediate carbocation, so no competing S_N1 reaction occurs.

PROBLEM 9.17   Draw the structure of each transition state in the three-step mechanism for the reaction, $(CH_3)_3COH + H_2SO_4 \rightarrow (CH_3)_2C=CH_2 + H_2O$.

9.8C   The E2 Mechanism for the Dehydration of 1° Alcohols

Because 1° carbocations are highly unstable, the dehydration of 1° alcohols cannot occur by an E1 mechanism involving a carbocation intermediate. With 1° alcohols, therefore, **dehydration follows an E2 mechanism.** This two-step process for the conversion of $CH_3CH_2CH_2OH$ (a 1° alcohol) to $CH_3CH=CH_2$ with H_2SO_4 as acid catalyst is shown in Mechanism 9.2.

MECHANISM 9.2

Dehydration of a 1° ROH—An E2 Mechanism

Step [1]   The O atom is protonated.

$$CH_3-\overset{\overset{\displaystyle H}{|}}{\underset{\underset{\displaystyle H}{|}}{C}}-CH_2 \quad \xrightarrow{\text{proton transfer}} \quad CH_3-\overset{\overset{\displaystyle H}{|}}{\underset{\underset{\displaystyle H}{|}}{C}}-CH_2 \quad + \quad HSO_4^-$$

good leaving group

◆ **Protonation of the oxygen atom** of the alcohol converts a poor leaving group (^-OH) into a good leaving group (H_2O).

Step [2]   The C–H and C–O bonds are broken and the π bond is formed.

$$CH_3-\overset{\overset{\displaystyle H}{|}}{\underset{\underset{\displaystyle \beta}{|}}{C}}-CH_2 \quad \xrightarrow{\quad} \quad CH_3CH=CH_2 \quad + \quad H_2\overset{\displaystyle ..}{O}: \quad + \quad H_2SO_4$$

good leaving group

◆ **Two bonds are broken and two bonds are formed in a single step:** the base (HSO_4^- or H_2O) removes a proton from the β carbon; the electron pair in the β C–H bond forms the new π bond; the leaving group (H_2O) comes off with the electron pair in the C–O bond.

The dehydration of a 1° alcohol begins with the protonation of the OH group to form a good leaving group, just as in the dehydration of a 2° or 3° alcohol. With 1° alcohols, however, loss of the leaving group and removal of a β proton occur at the same time, so that no highly unstable 1° carbocation is generated.

PROBLEM 9.18   Draw the structure of each transition state in the two-step mechanism for the reaction, $CH_3CH_2CH_2OH + H_2SO_4 \rightarrow CH_3CH=CH_2 + H_2O$.

9.8D   Le Châtelier's Principle

Although entropy favors product formation in dehydration (one molecule of reactant forms two molecules of products), enthalpy does not, because the two σ bonds broken in the reactant are stronger than the σ and π bonds formed in the products. For example, $\Delta H°$ for the dehydration of CH_3CH_2OH to $CH_2=CH_2$ is +6 kcal/mol (Figure 9.6).

Figure 9.6   The dehydration of CH_3CH_2OH to $CH_2=CH_2$—An endothermic reaction

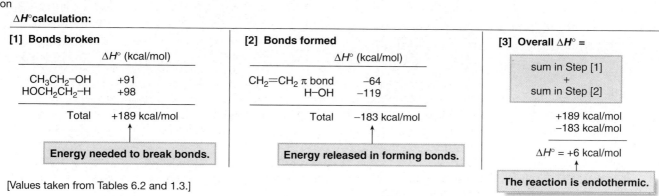

Overall reaction: $CH_3CH_2OH \xrightarrow{H_2SO_4} CH_2=CH_2 + H_2O$

$\Delta H°$calculation:

[1] Bonds broken

	$\Delta H°$ (kcal/mol)
CH_3CH_2–OH	+91
$HOCH_2CH_2$–H	+98
Total	+189 kcal/mol

Energy needed to break bonds.

[2] Bonds formed

	$\Delta H°$ (kcal/mol)
$CH_2=CH_2$ π bond	–64
H–OH	–119
Total	–183 kcal/mol

Energy released in forming bonds.

[3] Overall $\Delta H°$ =

sum in Step [1]
+
sum in Step [2]

+189 kcal/mol
–183 kcal/mol

$\Delta H°$ = +6 kcal/mol

The reaction is endothermic.

[Values taken from Tables 6.2 and 1.3.]

According to **Le Châtelier's principle, a system at equilibrium will react to counteract any disturbance to the equilibrium.** Thus, removing a product from a reaction mixture as it is formed drives the equilibrium to the right, forming more product.

Le Châtelier's principle can be used to favor products in dehydration reactions because the alkene product has a lower boiling point than the alcohol reactant. Thus, the alkene can be distilled from the reaction mixture as it is formed, leaving the alcohol and acid to react further, forming more product.

9.9   Carbocation Rearrangements

Sometimes "unexpected" products are formed in dehydration; that is, the carbon skeletons of the starting material and product might be different, or the double bond might be in an unexpected location. For example, the dehydration of 3,3-dimethyl-2-butanol yields two alkenes, whose carbon skeletons do not match the carbon framework of the starting material.

| **The carbon skeletons of the reactant and products are different.** |

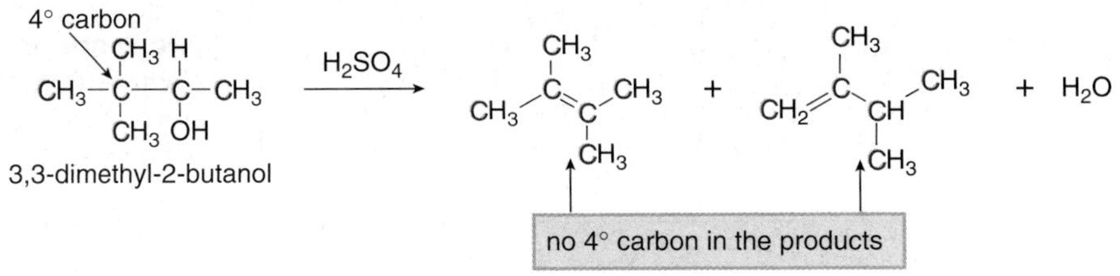

This phenomenon sometimes occurs when carbocations are reactive intermediates. **A less stable carbocation can rearrange to a more stable carbocation by shift of a hydrogen atom or an alkyl group.** These rearrangements are called **1,2-shifts,** because they involve migration of an alkyl group or hydrogen atom from one carbon to an adjacent carbon atom. The migrating group moves with the two electrons that bonded it to the carbon skeleton.

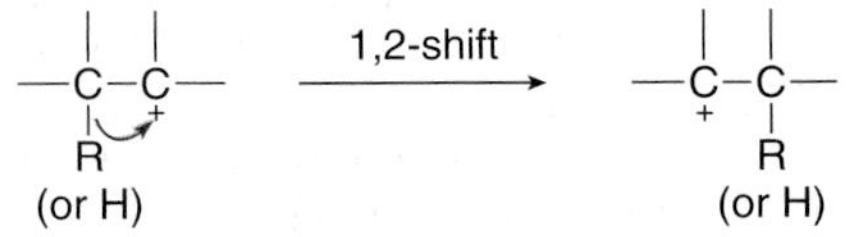

Because the migrating group in a 1,2-shift moves with two bonding electrons, the carbon it leaves behind now has only three bonds (six electrons), giving it a net positive (+) charge.

◆ Movement of a hydrogen atom is called a 1,2-hydride shift.
◆ Movement of an alkyl group is called a 1,2-alkyl shift.

The dehydration of 3,3-dimethyl-2-butanol illustrates the rearrangement of a 2° to a 3° carbocation by a **1,2-methyl shift,** as shown in Mechanism 9.3. The carbocation rearrangement occurs in Step [3] of the four-step mechanism.

MECHANISM 9.3

A 1,2-Methyl Shift—Carbocation Rearrangement During Dehydration

Steps [1] and [2] Formation of a 2° carbocation

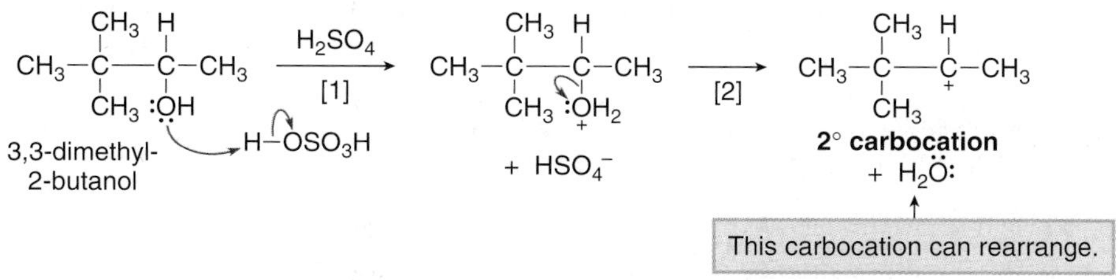

◆ **Protonation of the oxygen atom** on the alcohol in Step [1] forms a good leaving group (H_2O), and loss of H_2O in Step [2] forms a **2° carbocation.**

This carbocation can rearrange.

Step [3] Rearrangement of the carbocation by a 1,2-CH_3 shift

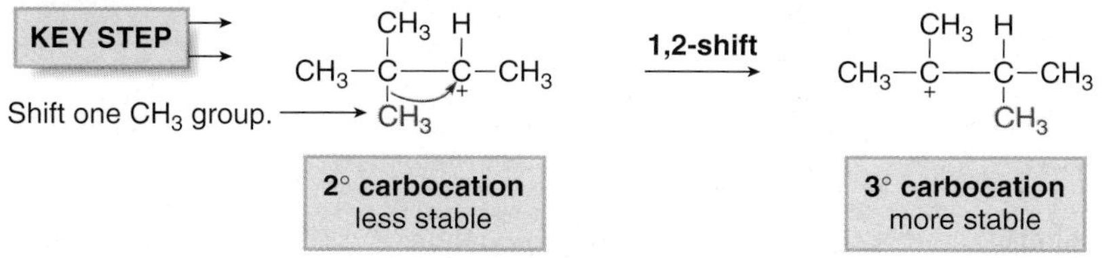

KEY STEP

Shift one CH_3 group.

1,2-shift

2° carbocation
less stable

3° carbocation
more stable

◆ **1,2-Shift** of a CH_3 group from one carbon to the adjacent carbon converts the 2° carbocation to a more stable 3° carbocation.

Step [4] Loss of a proton to form the π bond

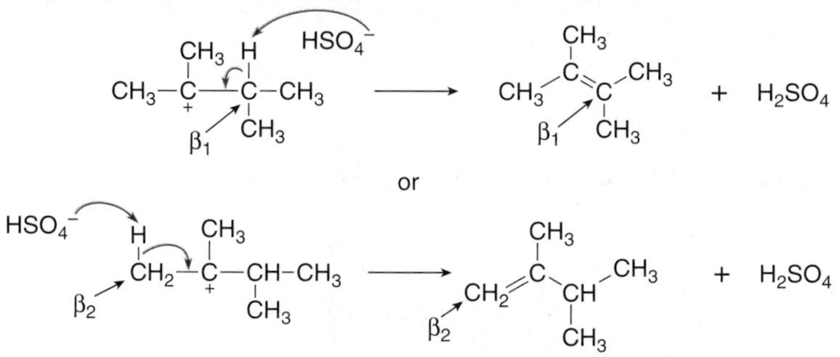

or

◆ **Loss of a proton** from a β carbon (labeled β_1 and β_2) forms two different alkenes.

Steps [1], [2], and [4] in the mechanism for the dehydration of 3,3-dimethyl-2-butanol are exactly the same steps previously seen in dehydration: protonation, loss of H_2O, and loss of a proton. Only Step [3], rearrangement of the less stable 2° carbocation to the more stable 3° carbocation, is new.

◆ All 1,2-shifts convert a less stable carbocation to a more stable carbocation.

For example, 2° carbocation **A** rearranges to the more stable 3° carbocation by a 1,2-hydride shift, whereas carbocation **B** does not rearrange because it is 3° to begin with.

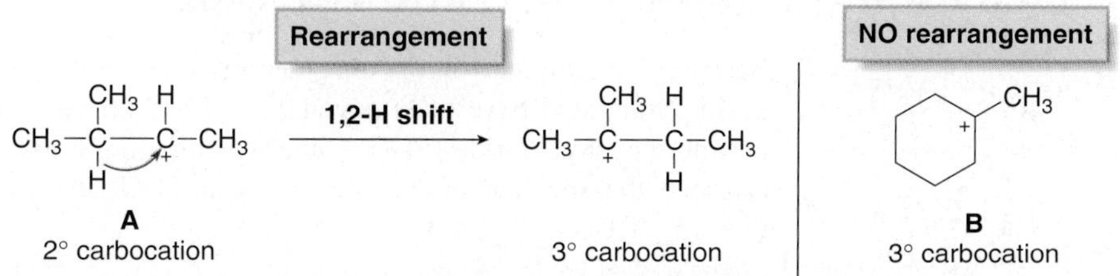

Rearrangement

1,2-H shift

A
2° carbocation

A → 3° carbocation

NO rearrangement

B
3° carbocation

Sample Problem 9.3 illustrates a dehydration reaction that occurs with a **1,2-hydride** shift.

SAMPLE PROBLEM 9.3 Show how the dehydration of alcohol **X** forms alkene **Y** using a 1,2-hydride shift.

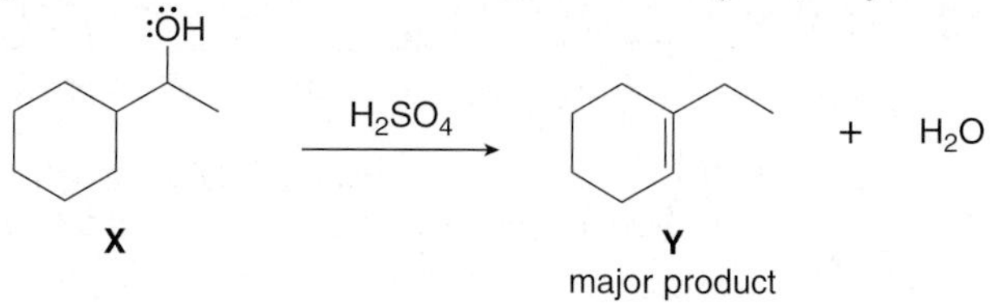

X → H_2SO_4 → **Y** major product + H_2O

SOLUTION

Steps [1] and [2]   Protonation of **X** and loss of H_2O form a 2° carbocation.

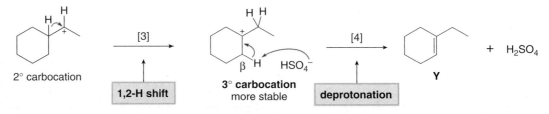

Steps [3] and [4]   Rearrangement of the 2° carbocation by a 1,2-hydride shift forms a more stable 3° carbocation. Loss of a proton from a β carbon forms alkene **Y**.

PROBLEM 9.19   What other alkene is also formed along with **Y** in Sample Problem 9.3? What alkenes would form from **X** if no carbocation rearrangement occurred?

Rearrangements are not unique to dehydration reactions. **Rearrangements can occur whenever a carbocation is formed as reactive intermediate,** meaning any S_N1 or E1 reaction. In fact, the presence of rearranged products often indicates the presence of a carbocation intermediate.

PROBLEM 9.20   Show how a 1,2-shift forms a more stable carbocation from each intermediate.

a. $(CH_3)_2\overset{+}{C}HCHCH_2CH_3$ b.   c.

PROBLEM 9.21   Explain how the reaction of $(CH_3)_2CHCH(Cl)CH_3$ with H_2O yields two substitution products, $(CH_3)_2CHCH(OH)CH_3$ and $(CH_3)_2C(OH)CH_2CH_3$.

9.10   Dehydration Using POCl₃ and Pyridine

Because some organic compounds decompose in the presence of strong acid, other methods that avoid strong acid have been developed to convert alcohols to alkenes. A common method uses phosphorus oxychloride ($POCl_3$) and pyridine (an amine base) in place of H_2SO_4 or TsOH. For example, the treatment of cyclohexanol with $POCl_3$ and pyridine forms cyclohexene in good yield.

$POCl_3$ serves much the same role as strong acid does in acid-catalyzed dehydration. **It converts a poor leaving group (⁻OH) into a good leaving group.** Dehydration then proceeds by an **E2 mechanism,** as shown in Mechanism 9.4. Pyridine is the base that removes a β proton during elimination.

MECHANISM 9.4

Dehydration Using POCl$_3$ + Pyridine—An E2 Mechanism

Steps [1] and [2] Conversion of OH to a good leaving group

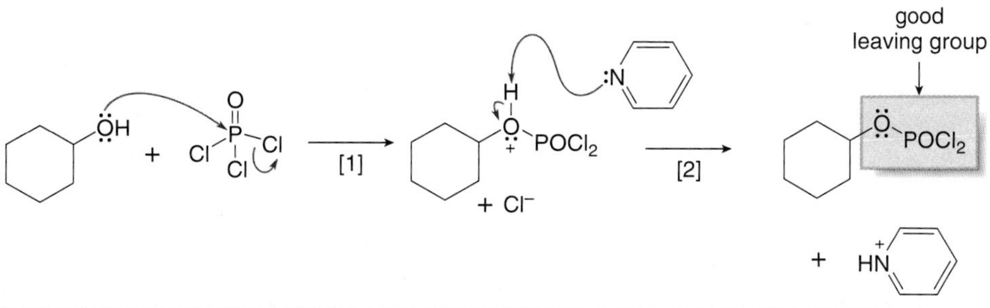

◆ A two-step process converts an OH group into OPOCl$_2$, **a good leaving group:** reaction of the OH group with POCl$_3$ followed by removal of a proton.

Step [3] The C−H and C−O bonds are broken and the π bond is formed.

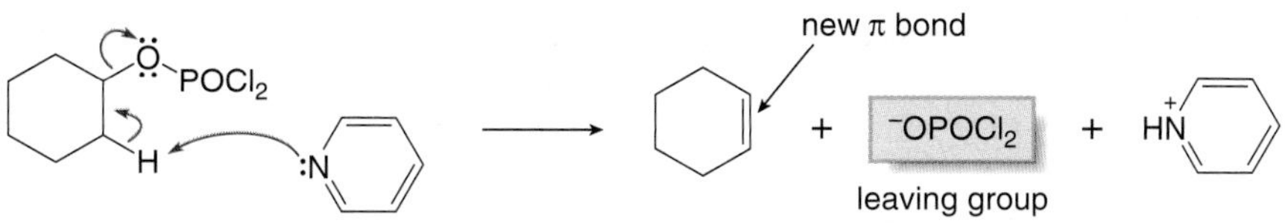

◆ **Two bonds are broken and two bonds are formed in a single step:** the base (pyridine) removes a proton from the β carbon; the electron pair in the β C−H bond forms the new π bond; the leaving group (⁻OPOCl$_2$) comes off with the electron pair from the C−O bond.

◆ Steps [1] and [2] of the mechanism convert the OH group to a good leaving group.

◆ In Step [3], the C−H and C−O bonds are broken and the π bond is formed.

◆ No rearrangements occur during dehydration with POCl$_3$, suggesting that carbocations are not formed as intermediates in this reaction.

We have now learned about two different reagents for alcohol dehydration—strong acid (H$_2$SO$_4$ or TsOH) and POCl$_3$ + pyridine. The best dehydration method for a given alcohol is often hard to know ahead of time, and this is why organic chemists develop more than one method for a given type of transformation. Two examples of dehydration reactions used in the synthesis of natural products are given in Figure 9.7.

Figure 9.7   Dehydration reactions in the synthesis of two natural products

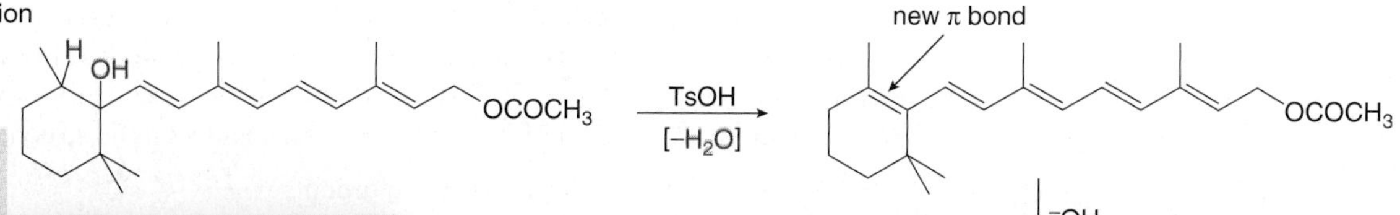

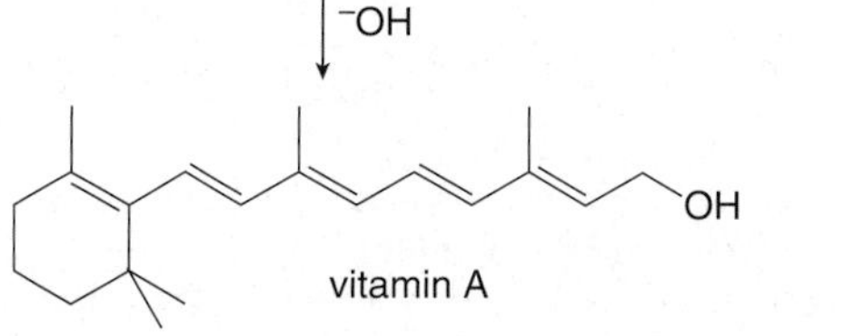

Patchouli alcohol, a major component of the essential oil of the patchouli plant native to Malaysia, has been used in perfumery because of its exotic fragrance. In the 1800s shawls imported from India were often packed with patchouli leaves to ward off insects, thus permeating the clothing with the distinctive odor.

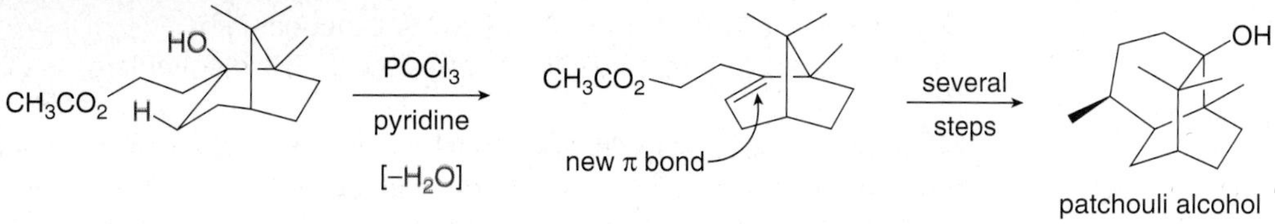

9.11 Conversion of Alcohols to Alkyl Halides with HX

Alcohols undergo nucleophilic substitution reactions only if the OH group is converted into a better leaving group before nucleophilic attack. Thus, substitution does *not* occur when an alcohol is treated with X^- because ^-OH is a poor leaving group (Reaction [1]), but substitution *does* occur on treatment of an alcohol with HX because H_2O is now the leaving group (Reaction [2]).

The reaction of alcohols with HX (X = Cl, Br, I) is a general method to prepare 1°, 2°, and 3° alkyl halides.

More substituted alcohols usually react more rapidly with HX:

$$RCH_2{-}OH \quad\quad R_2CH{-}OH \quad\quad R_3C{-}OH$$
$$1° \quad\quad\quad\quad 2° \quad\quad\quad\quad 3°$$

Increasing rate of reaction with HX →

PROBLEM 9.22 Draw the products of each reaction.

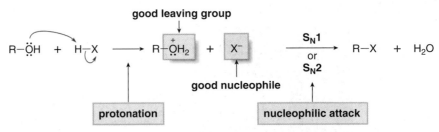

9.11A Two Mechanisms for the Reaction of ROH with HX

How does the reaction of ROH with HX occur? Acid–base reactions are very fast, so the strong acid HX protonates the OH group of the alcohol, forming a **good leaving group** (H_2O) and a **good nucleophile** (the conjugate base, X^-). Both components are needed for nucleophilic substitution. The mechanism of substitution of X^- for H_2O then depends on the structure of the R group.

Whenever there is an oxygen-containing reactant and a strong acid, the first step in the mechanism is always protonation of the oxygen atom.

good leaving group

$$R{-}\ddot{O}H \; + \; H{-}X \longrightarrow R{-}\overset{+}{O}H_2 \; + \; X^- \xrightarrow[\text{or}\; S_N2]{S_N1} R{-}X \; + \; H_2O$$

protonation   **good nucleophile**   **nucleophilic attack**

♦ Methyl and 1° ROH form RX by an S_N2 mechanism.
♦ Secondary (2°) and 3° ROH form RX by an S_N1 mechanism.

The reaction of CH_3CH_2OH with HBr illustrates the S_N2 mechanism of a 1° alcohol (Mechanism 9.5). Nucleophilic attack on the protonated alcohol occurs in one step: **the bond to the nucleophile X^- is formed *as* the bond to the leaving group (H_2O) is broken.**

MECHANISM 9.5

Reaction of a 1° ROH with HX—An S$_N$2 Mechanism

Step [1]   The O atom is protonated.

$$CH_3CH_2-\overset{..}{\underset{..}{O}}H \ + \ H-Br \longrightarrow CH_3CH_2-\overset{+}{\underset{|}{O}}H_2 \ + \ Br^-$$

good
leaving group

◆ **Protonation of the OH group** forms a good leaving group (H_2O).

Step [2]   The C–O bond is broken as the C–Br bond is formed.

$$CH_3CH_2-\overset{+}{\underset{..}{O}}H_2 \ + \ \boxed{Br^-} \longrightarrow CH_3CH_2-Br \ + \ H_2\overset{..}{\underset{..}{O}}:$$

good
nucleophile

◆ **Nucleophilic attack of Br$^-$ and loss of the leaving group** occur in a single step.

The reaction of $(CH_3)_3COH$ with HBr illustrates the S$_N$1 mechanism of a 3° alcohol (Mechanism 9.6). Nucleophilic attack on the protonated alcohol occurs in two steps: **the bond to the leaving group (H_2O) is broken *before* the bond to the nucleophile X$^-$ is formed.**

MECHANISM 9.6

Reaction of 2° and 3° ROH with HX—An S$_N$1 Mechanism

Step [1]   The O atom is protonated.

$$CH_3-\overset{\overset{\displaystyle CH_3}{|}}{\underset{\underset{\displaystyle CH_3}{|}}{C}}-\overset{..}{\underset{..}{O}}H \ + \ H-Br \longrightarrow CH_3-\overset{\overset{\displaystyle CH_3}{|}}{\underset{\underset{\displaystyle CH_3}{|}}{C}}-\overset{+}{\underset{..}{O}}H_2 \ + \ Br^-$$

good
leaving group

◆ **Protonation of the OH group** forms a good leaving group (H_2O).

Steps [2] and [3]   The C–O bond is broken, and then the C–Br bond is formed.

$$CH_3-\overset{\overset{\displaystyle CH_3}{|}}{\underset{\underset{\displaystyle CH_3}{|}}{C}}-\overset{+}{\underset{..}{O}}H_2 \ \xrightarrow{[2]} \ CH_3-\overset{+}{\underset{\underset{\displaystyle CH_3}{}}{C}}-CH_3 \ + \ \boxed{Br^-} \ \xrightarrow{[3]} \ CH_3-\overset{\overset{\displaystyle CH_3}{|}}{\underset{\underset{\displaystyle CH_3}{|}}{C}}-Br$$

carbocation   good
nucleophile

$+ \ H_2\overset{..}{\underset{..}{O}}:$

| loss of the leaving group |

| nucleophilic attack |

◆ Loss of the leaving group in Step [2] forms a **carbocation,** which reacts with the nucleophile (Br$^-$) in Step [3] to form the substitution product.

Both mechanisms begin with the same first step—protonation of the O atom to form a good leaving group—and both mechanisms give an alkyl halide (RX) as product. The mechanisms differ only in the timing of bond breaking and bond making.

The reactivity of hydrogen halides increases with increasing acidity:

$$H-Cl \qquad\qquad H-Br \qquad\qquad H-I$$

Increasing reactivity toward ROH

Because Cl$^-$ is a poorer nucleophile than Br$^-$ or I$^-$, the reaction of 1° alcohols with HCl occurs only when an additional Lewis acid catalyst, usually **ZnCl$_2$**, is added. ZnCl$_2$ complexes with the O atom of the alcohol in a Lewis acid–base reaction, making an especially good leaving group and facilitating the S$_N$2 reaction.

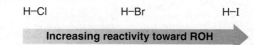

| **1° Alcohols** | $RCH_2-\overset{..}{\underset{..}{O}}H$ + ZnCl$_2$ | $\longrightarrow$ | $RCH_2-\overset{\overset{\displaystyle -ZnCl_2}{+}}{\underset{}{O}}-H$ | $\xrightarrow{S_N2}$ | RCH_2-Cl + ZnCl$_2$(OH)$^-$ |

Lewis base   Lewis acid   Cl$^-$ leaving group

Knowing the mechanism allows us to predict the stereochemistry of the products when reaction occurs at a stereogenic center.

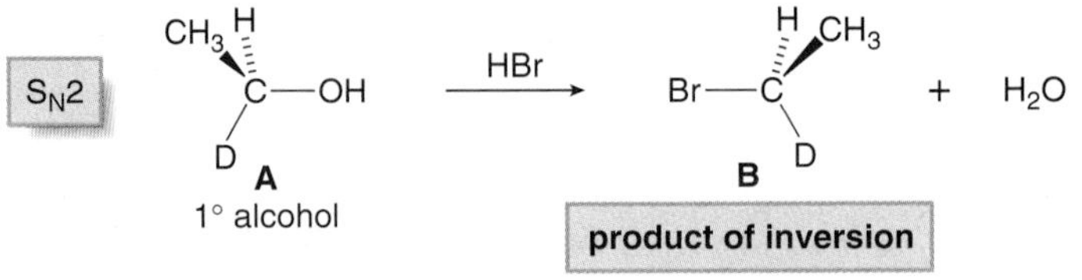

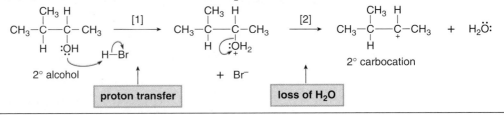

S_N2

Br—C

A
1° alcohol

B
product of inversion

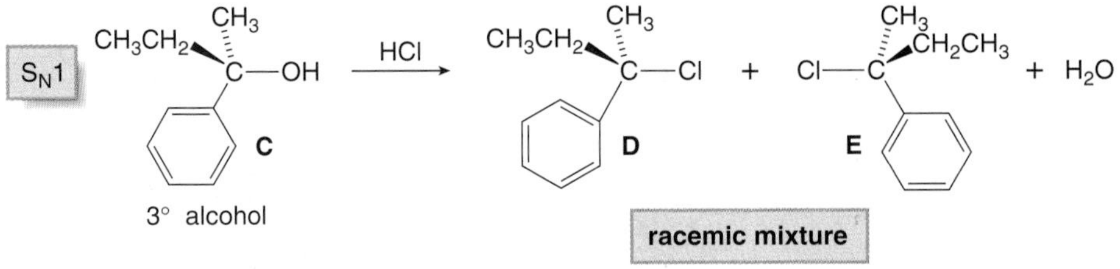

+ H₂O

S_N1

C
3° alcohol

D   E
racemic mixture

+ H₂O

◆ The 1° alcohol **A** reacts with HBr via an S_N2 mechanism to yield the alkyl bromide **B** with **inversion** of stereochemistry at the stereogenic center.

◆ The 3° alcohol **C** reacts with HCl via an S_N1 mechanism to yield a **racemic mixture** of alkyl chlorides **D** and **E,** because a trigonal planar carbocation intermediate is formed.

PROBLEM 9.23   Draw the products of each reaction, indicating the stereochemistry around any stereogenic centers.

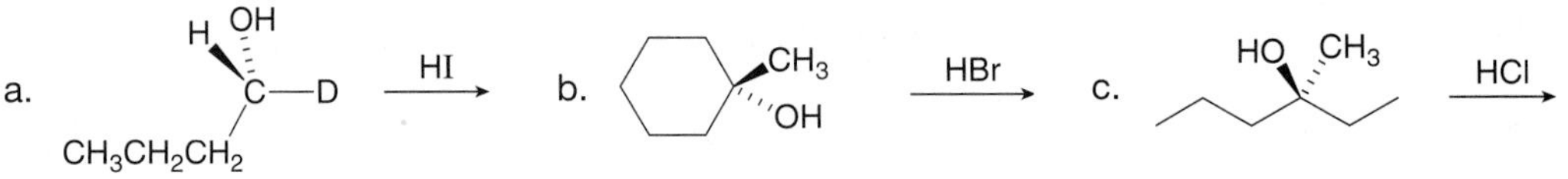

a.   [structure] →HI→   b.   [structure] →HBr→   c.   [structure] →HCl→

9.11B   Carbocation Rearrangement in the S_N1 Reaction

Because carbocations are formed in the S_N1 reaction of 2° and 3° alcohols with HX, **carbocation rearrangements are possible,** as illustrated in Sample Problem 9.4.

SAMPLE PROBLEM 9.4   Draw a stepwise mechanism for the following reaction.

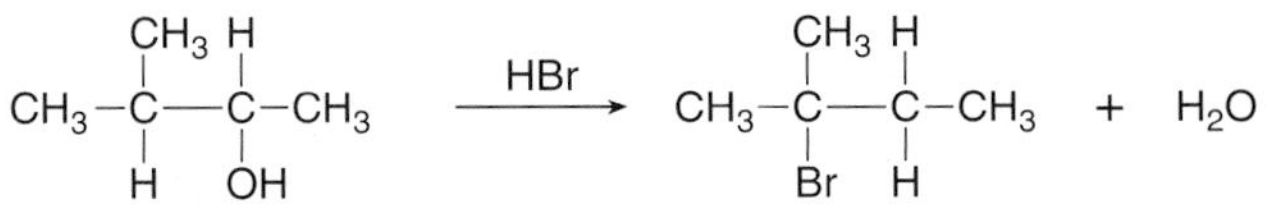

CH₃—C—C—CH₃ →HBr→ CH₃—C—C—CH₃ + H₂O

SOLUTION

A 2° alcohol reacts with HBr by an S_N1 mechanism. Because substitution converts a 2° alcohol to a 3° alkyl halide in this example, a carbocation rearrangement must occur.

Steps [1] and [2]   Protonation of the O atom and then loss of H₂O form a 2° carbocation.

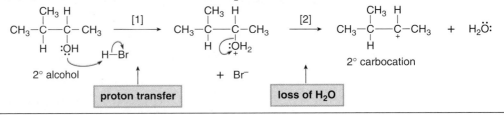

CH₃—C—C—CH₃   [1]   CH₃—C—C—CH₃   [2]   CH₃—C—C—CH₃   +   H₂Ö:

H :ÖH
2° alcohol
H—Br

+ Br⁻

proton transfer

:ÖH₂

loss of H₂O

2° carbocation

Steps [3] and [4]   **Rearrangement of the 2° carbocation by a 1,2-hydride shift forms a more stable 3° carbocation.** Nucleophilic attack forms the substitution product.

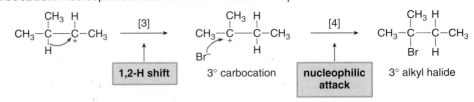

CH₃—C—C—CH₃   [3]   CH₃—C—C—CH₃   [4]   CH₃—C—C—CH₃

1,2-H shift   3° carbocation   Br⁻   **nucleophilic attack**   3° alkyl halide

PROBLEM 9.24   What product is formed when each alcohol is treated with HCl?

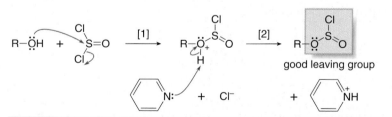

a.   b.   c.

9.12   Conversion of Alcohols to Alkyl Halides with SOCl$_2$ and PBr$_3$

Primary and secondary alcohols can be converted to alkyl halides using SOCl$_2$ and PBr$_3$.

> ◆ **SOCl$_2$ (thionyl chloride) converts alcohols into alkyl chlorides.**
> ◆ **PBr$_3$ (phosphorus tribromide) converts alcohols into alkyl bromides.**

Both reagents convert ¯OH into a good leaving group *in situ*—that is, directly in the reaction mixture—as well as provide the **nucleophile,** either Cl¯ or Br¯, to displace the leaving group.

9.12A   Reaction of ROH with SOCl$_2$

The treatment of a 1° or 2° alcohol with thionyl chloride, SOCl$_2$, and pyridine forms an alkyl chloride, with SO$_2$ and HCl as by-products.

General reaction	R—OH + SOCl$_2$ $\xrightarrow{\text{pyridine}}$ R—Cl + SO$_2$ + HCl

Examples   CH$_3$CH$_2$—OH + SOCl$_2$ $\xrightarrow{\text{pyridine}}$ CH$_3$CH$_2$—Cl

⬡—OH + SOCl$_2$ $\xrightarrow{\text{pyridine}}$ ⬡—Cl

> 1° and 2° RCl are formed.

The mechanism for this reaction consists of two parts: **conversion of the OH group into a better leaving group, and nucleophilic attack by Cl¯ via an S$_N$2 reaction,** as shown in Mechanism 9.7.

MECHANISM 9.7

Reaction of ROH with SOCl$_2$ + Pyridine—An S$_N$2 Mechanism

Steps [1] and [2]   **The OH group is converted into a good leaving group.**

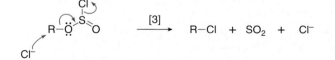

good leaving group

◆ Reaction of the alcohol with SOCl$_2$ forms an intermediate that loses a proton by reaction with pyridine in Step [2]. This two-step process converts the OH group into OSOCl, a **good leaving group,** and also generates the **nucleophile (Cl¯)** needed for Step [3].

Step [3]   **The C–O bond is broken *as* the C–Cl bond is formed.**

R—O—S(=O)—Cl $\xrightarrow{[3]}$ R—Cl + SO$_2$ + Cl¯

◆ **Nucleophilic attack of Cl¯ and loss of the leaving group (SO$_2$ + Cl¯) occur in a single step.**

PROBLEM 9.25   If the reaction of an alcohol with SOCl$_2$ and pyridine follows an S$_N$2 mechanism, what is the stereochemistry of the alkyl chloride formed from (*R*)-2-butanol?

9.12B   Reaction of ROH with PBr$_3$

In a similar fashion, the treatment of a 1° or 2° alcohol with phosphorus tribromide, PBr$_3$, forms an alkyl bromide.

| General reaction | R—OH | + PBr₃ | ⟶ | R—Br | + HOPBr₂ |

General reaction   R⊟OH + PBr₃ ⟶ R⊟Br + HOPBr₂

Examples   CH₃CH₂—OH + PBr₃ ⟶ CH₃CH₂—Br

> **1° and 2° RBr are formed.**

⬡—OH + PBr₃ ⟶ ⬡—Br

The mechanism for this reaction also consists of two parts: **conversion of the OH group into a better leaving group, and nucleophilic attack by Br⁻ via an S_N2 reaction,** as shown in Mechanism 9.8.

MECHANISM 9.8

Reaction of ROH with PBr₃—An S_N2 Mechanism

Step [1] The OH group is converted into a good leaving group.

R—ÖH + [Br—P—Br with Br] ⟶ R—Ö⁺—P—Br (with H, Br, Br) + Br⁻

good leaving group   good nucleophile

◆ Reaction of the alcohol with PBr₃ converts the OH group into a **better leaving group,** and also generates the **nucleophile (Br⁻)** needed for Step [2].

Step [2] The C–O bond is broken *as* the C–Br bond is formed.

R—Ö⁺—P—Br (with Br, H) ⟶ R—Br + HOPBr₂

Br⁻

◆ Nucleophilic attack of Br⁻ and loss of the leaving group **(HOPBr₂)** occur in a single step.

Table 9.2 summarizes the methods for converting an alcohol to an alkyl halide presented in Sections 9.11 and 9.12.

TABLE 9.2 Summary of Methods for ROH → RX		
Overall reaction	**Reagent**	**Comment**
ROH → RCl	HCl	• Useful for all ROH • An S_N1 mechanism for 2° and 3° ROH; an S_N2 mechanism for CH₃OH and 1° ROH
	SOCl₂	• Best for CH₃OH, and 1° and 2° ROH • An S_N2 mechanism
ROH → RBr	HBr	• Useful for all ROH • An S_N1 mechanism for 2° and 3° ROH; an S_N2 mechanism for CH₃OH and 1° ROH
	PBr₃	• Best for CH₃OH, and 1° and 2° ROH • An S_N2 mechanism
ROH → RI	HI	• Useful for all ROH • An S_N1 mechanism for 2° and 3° ROH; an S_N2 mechanism for CH₃OH and 1° ROH

PROBLEM 9.26   If the reaction of an alcohol with PBr₃ follows an S_N2 mechanism, what is the stereochemistry of the alkyl bromide formed from (R)-2-butanol?

PROBLEM 9.27   Draw the organic products formed in each reaction, and indicate the stereochemistry of products that contain stereogenic centers.

a. $\ce{\diagonal{}—OH}$ $\underset{\text{pyridine}}{\xrightarrow{\text{SOCl}_2}}$ b. (structure) $\xrightarrow{\text{HI}}$ c. (structure)$-\text{OH}$ $\xrightarrow{\text{PBr}_3}$

9.12C   The Importance of Making RX from ROH

We have now learned two methods to prepare an alkyl chloride and two methods to prepare an alkyl bromide from an alcohol. If there is one good way to carry out a reaction, why search for more? A particular reagent might work well for one starting material, but not so well for another. Thus, organic chemists try to devise several different ways to perform the same overall reaction. For now, though, concentrate on one or two of the most general methods, so you can better understand the underlying concepts.

Why are there so many ways to convert an alcohol to an alkyl halide? Alkyl halides are versatile starting materials in organic synthesis, as shown in Sample Problem 9.5.

SAMPLE PROBLEM 9.5   Convert 1-propanol to butanenitrile **(A)**.

$$\underset{\substack{\text{1-propanol}}}{\ce{CH3CH2CH2-OH}} \xrightarrow{?} \underset{\substack{\text{butanenitrile} \\ \textbf{A}}}{\ce{CH3CH2CH2-CN}}$$

SOLUTION

Direct conversion of 1-propanol to **A** using $^-$CN as a nucleophile is not possible because $^-$OH is a poor leaving group. However, conversion of the OH group to a Br atom forms a good leaving group, which can then readily undergo an S_N2 reaction with $^-$CN to yield **A. This two-step sequence is our first example of a multistep synthesis.**

Direct substitution is NOT possible.	$\ce{CH3CH2CH2-OH}$ + $^-$CN $\xrightarrow{\large\times}$ $\ce{CH3CH2CH2-CN}$ + $^-$OH

poor leaving group

Two steps are needed for substitution.	$\ce{CH3CH2CH2-OH}$ $\xrightarrow{\text{PBr}_3}$ $\ce{CH3CH2CH2-Br}$ $\underset{S_N2}{\xrightarrow{^-\text{CN}}}$ $\ce{CH3CH2CH2-CN}$ + $\ce{Br-}$

good leaving group

The overall result: $^-$CN replaces $^-$OH.

PROBLEM 9.28   Draw two steps to convert $(CH_3)_2CHOH$ into each of the following compounds: $(CH_3)_2CHN_3$ and $(CH_3)_2CHOCH_2CH_3$?

9.13   Tosylate—Another Good Leaving Group

We have now learned two methods to convert the OH group of an alcohol to a better leaving group: treatment with strong acids (Section 9.8), and conversion to an alkyl halide (Sections 9.11–9.12). Alcohols can also be converted to **alkyl tosylates.**

Recall from Section 1.4B that a third-row element like sulfur can have 10 or 12 electrons around it in a valid Lewis structure.

An alkyl tosylate

$$\ce{R-OH} \longrightarrow \ce{R-O-\underset{O}{\overset{O}{S}}-C6H4-CH3}$$

poor leaving group

tosylate good leaving group

An alkyl tosylate is often called simply a **tosylate.**

An **alkyl tosylate** is composed of two parts: the **alkyl group R,** derived from an alcohol; and the **tosylate** (short for ***p*-toluenesulfonate**), which is a good leaving group. A tosyl group, $CH_3C_6H_4SO_2-$, is abbreviated as **Ts,** so an alkyl tosylate becomes **ROTs.**

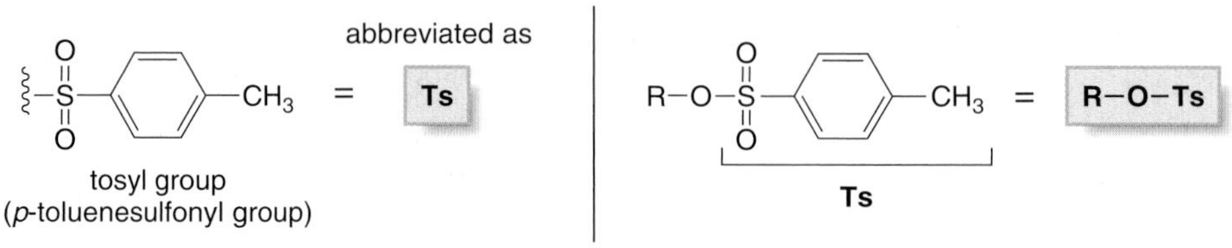

tosyl group
(*p*-toluenesulfonyl group)

abbreviated as

9.13A   Conversion of Alcohols to Tosylates

> A tosylate (TsO⁻) is similar to I⁻ in leaving group ability.

Alcohols are converted to tosylates by treatment with *p*-toluenesulfonyl chloride (TsCl) in the presence of pyridine. This overall process converts a poor leaving group (⁻OH) into a good one (⁻OTs). A tosylate is a good leaving group because its conjugate acid, *p*-toluenesulfonic acid ($CH_3C_6H_4SO_3H$, TsOH) is a strong acid ($pK_a = -7$, Section 2.6).

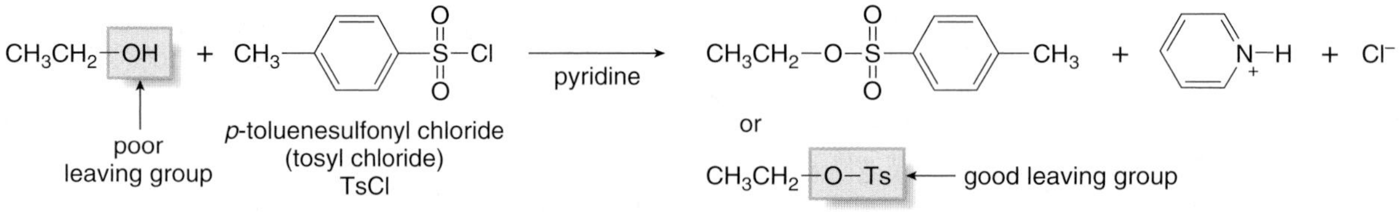

p-toluenesulfonyl chloride
(tosyl chloride)
TsCl

poor
leaving group

good leaving group

(*S*)-2-Butanol is converted to its tosylate with **retention of configuration** at the stereogenic center. Thus, the C–O bond of the alcohol must *not* be broken when the tosylate is formed.

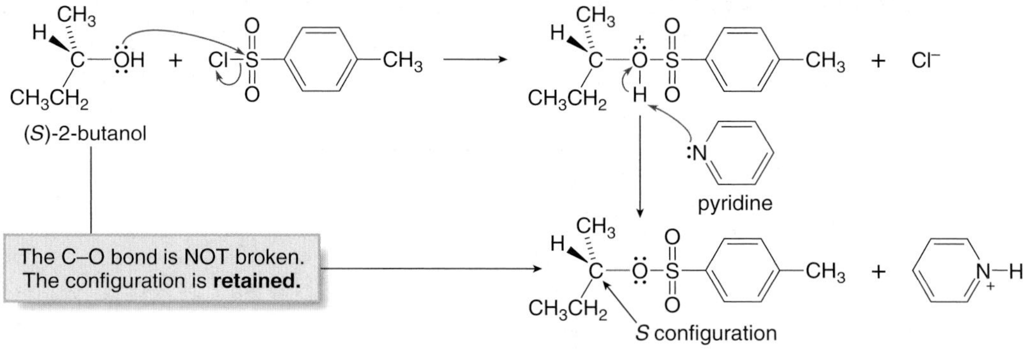

(*S*)-2-butanol

pyridine

> The C–O bond is NOT broken. The configuration is **retained.**

S configuration

PROBLEM 9.29   Draw the products of each reaction, and indicate the stereochemistry at any stereogenic centers.

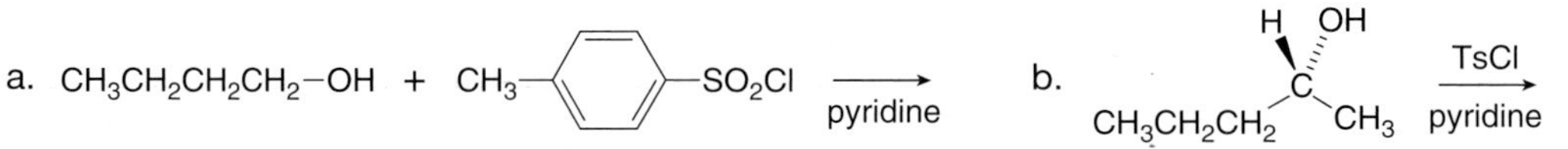

9.13B   Reactions of Tosylates

Because alkyl tosylates have good leaving groups, **they undergo both nucleophilic substitution and β elimination,** exactly as alkyl halides do. Generally, alkyl tosylates are treated with strong nucleophiles and bases, so that the mechanism of substitution is **S$_N$2** and the mechanism of elimination is **E2.**

For example, ethyl tosylate, which has the leaving group on a 1° carbon, reacts with $NaOCH_3$ to yield ethyl methyl ether, the product of nucleophilic substitution by an S$_N$2 mechanism. It reacts with $KOC(CH_3)_3$, a strong sterically hindered base, to yield ethylene by an E2 mechanism.

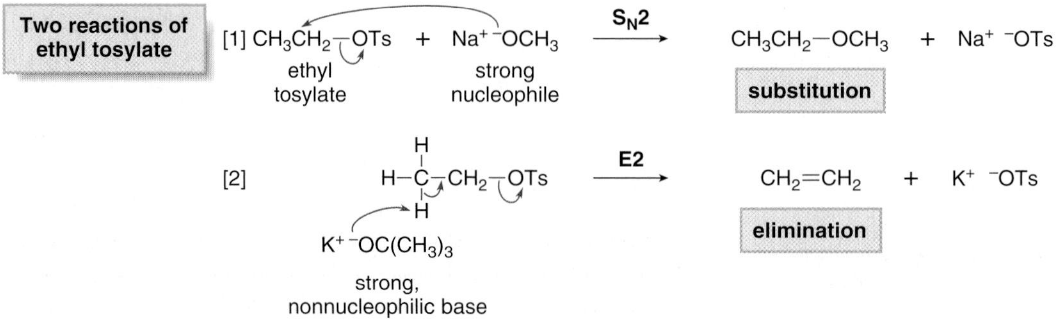

Two reactions of ethyl tosylate

[1]   ethyl tosylate   +   strong nucleophile   →   **substitution**

[2]   →   **elimination**

strong, nonnucleophilic base

Because substitution occurs via an S_N2 mechanism, **inversion of configuration** results when the leaving group is bonded to a stereogenic center.

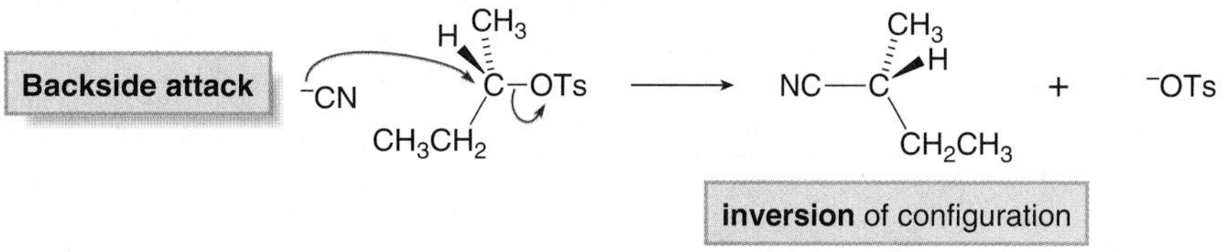

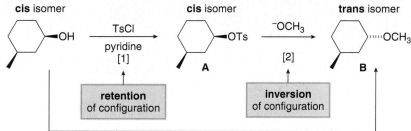

SAMPLE PROBLEM 9.6   Draw the product of the following reaction, including stereochemistry.

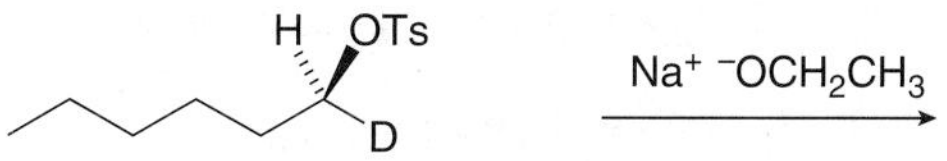

SOLUTION

The 1° alkyl tosylate and the strong, unhindered nucleophile both favor substitution by an S_N2 mechanism, which proceeds by backside attack, resulting in **inversion** of configuration at the stereogenic center.

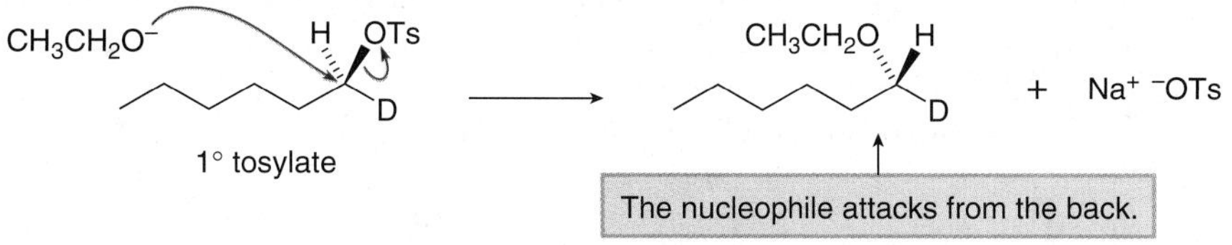

The nucleophile attacks from the back.

PROBLEM 9.30   Draw the products of each reaction, and include the stereochemistry at any stereogenic centers in the products.

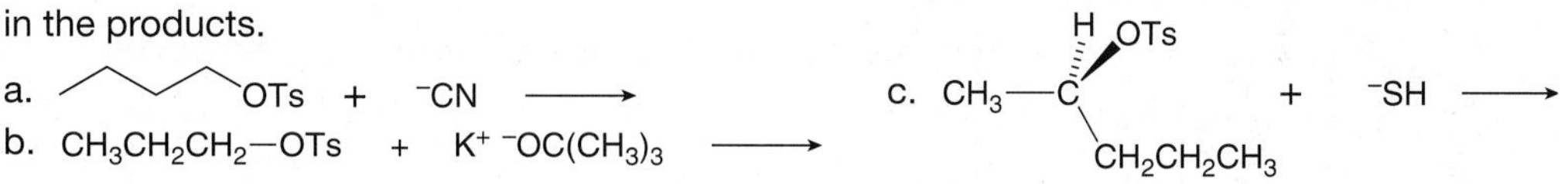

a.   ~~~OTs + ⁻CN ⟶

b. $CH_3CH_2CH_2$—OTs + K^+ ⁻$OC(CH_3)_3$ ⟶

c. CH_3—C(H,OTs)($CH_2CH_2CH_3$) + ⁻SH ⟶

9.13C   The Two-Step Conversion of an Alcohol to a Substitution Product

We now have another **two-step method to convert an alcohol to a substitution product:** reaction of an alcohol with TsCl and pyridine to form a tosylate (Step [1]), followed by nucleophilic attack on the tosylate (Step [2]).

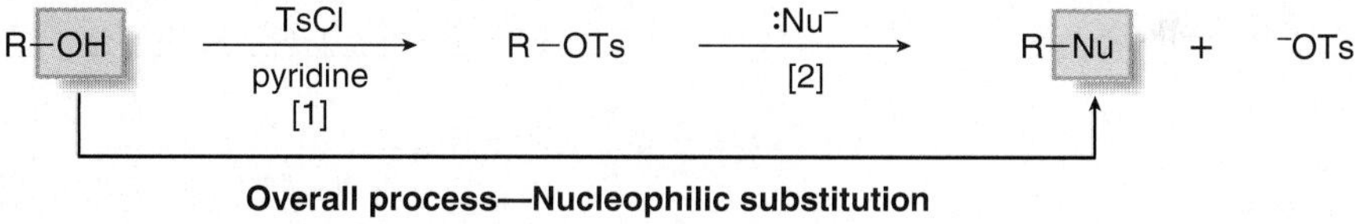

$$R{-}OH \xrightarrow[\substack{\text{pyridine}\\[1]}]{\text{TsCl}} R{-}OTs \xrightarrow[\substack{[2]}]{:Nu^-} R{-}Nu + {}^-OTs$$

Overall process—Nucleophilic substitution

Let's look at the stereochemistry of this two-step process.

◆ Step [1], formation of the tosylate, proceeds with **retention** of configuration at a stereogenic center because the C–O bond remains intact.

◆ Step [2] is an S_N2 reaction, so it proceeds with **inversion of configuration** because the nucleophile attacks from the back side.

◆ Overall there is a **net inversion of configuration** at a stereogenic center.

For example, the treatment of *cis*-3-methylcyclohexanol with *p*-toluenesulfonyl chloride and pyridine forms a cis tosylate **A,** which undergoes backside attack by the nucleophile ⁻OCH_3 to yield the trans ether **B.**

cis isomer   **cis** isomer   **trans** isomer

retention of configuration   inversion of configuration

Overall result—Inversion

PROBLEM 9.31 Draw the products formed when (S)-2-butanol is treated with TsCl and pyridine, followed by NaOH. Label the stereogenic center in each compound as R or S. What is the stereochemical relationship between the starting alcohol and the final product?

9.13D A Summary of Substitution and Elimination Reactions of Alcohols

The reactions of alcohols in Sections 9.8–9.13C share two similarities:

♦ The OH group is converted into a better leaving group by treatment with acid or another reagent.
♦ The resulting product undergoes either elimination or substitution, depending on the reaction conditions.

Figure 9.8 summarizes these reactions with cyclohexanol as starting material.

PROBLEM 9.32 Draw the product formed when $(CH_3)_2CHOH$ is treated with each reagent.
a. $SOCl_2$, pyridine   c. H_2SO_4 e. PBr_3, then NaCN
b. TsCl, pyridine   d. HBr   f. $POCl_3$, pyridine

9.14 Reaction of Ethers with Strong Acid

Because ethers are so unreactive, diethyl ether and tetrahydrofuran (THF) are often used as solvents for organic reactions.

Recall from Section 9.7B that ethers have a poor leaving group, so they cannot undergo nucleophilic substitution or β elimination reactions directly. Instead, they must first be converted to a good leaving group by reaction with strong acids. Only **HBr** and **HI** can be used, though, because they are strong acids that are also sources of good nucleophiles (Br⁻ and I⁻, respectively). **When ethers react with HBr or HI, both C–O bonds are cleaved and two alkyl halides are formed as products.**

General reaction R–O–R' + H–X ⟶ R–X + R'–X + H_2O
[X = Br or I]
Two C–O bonds are broken. Two new C–X bonds are formed.

Examples CH_3–O–CH_2CH_3 + HBr ⟶ CH_3–Br + CH_3CH_2–Br + H_2O

CH_3–C(–CH_3)(CH_3)–O–CH_3 + HI ⟶ CH_3–C(–CH_3)(CH_3)–I + CH_3–I + H_2O

HBr or HI serves as a strong acid that both protonates the O atom of the ether and is the source of a good nucleophile (Br⁻ or I⁻). Because both C–O bonds in the ether are broken, **two successive nucleophilic substitution reactions occur.**

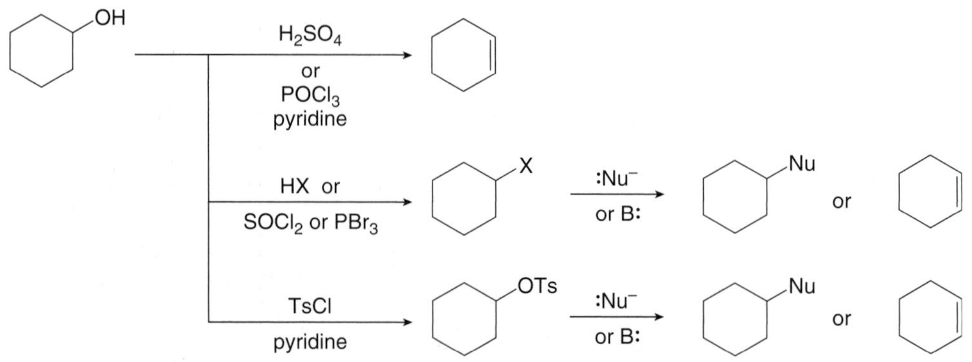

Figure 9.8 Summary: Nucleophilic substitution and β elimination reactions of alcohols

- ◆ The mechanism of ether cleavage is S$_N$1 or S$_N$2, depending on the identity of R.
- ◆ With 2° or 3° alkyl groups bonded to the ether oxygen, the C−O bond is cleaved by an S$_N$1 mechanism involving a carbocation; with methyl or 1° R groups, the C−O bond is cleaved by an S$_N$2 mechanism.

For example, cleavage of (CH$_3$)$_3$COCH$_3$ with HI occurs at two bonds, as shown in Mechanism 9.9. The 3° alkyl group undergoes nucleophilic substitution by an S$_N$1 mechanism, resulting in the cleavage of one C−O bond. The methyl group undergoes nucleophilic substitution by an S$_N$2 mechanism, resulting in the cleavage of the second C−O bond.

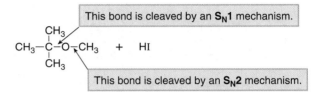

MECHANISM 9.9

Mechanism of Ether Cleavage in Strong Acid—(CH$_3$)$_3$COCH$_3$ + HI → (CH$_3$)$_3$CI + CH$_3$I + H$_2$O

Part [1] Cleavage of the 3° C−O bond by an S$_N$1 mechanism

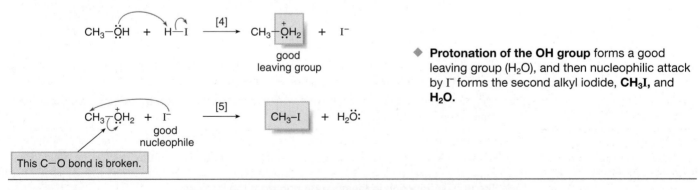

- ◆ **Protonation** of the O atom forms a good leaving group in Step [1]. Cleavage of the C−O bond then occurs in two steps: the bond to the leaving group is broken to form a **carbocation,** and then the bond to the nucleophile (I$^-$) is formed. This generates one of the alkyl iodides, **(CH$_3$)$_3$CI.**

Part [2] Cleavage of the CH$_3$−O bond by an S$_N$2 mechanism

- ◆ **Protonation of the OH group** forms a good leaving group (H$_2$O), and then nucleophilic attack by I$^-$ forms the second alkyl iodide, **CH$_3$I,** and **H$_2$O.**

The mechanism illustrates the central role of HX in the reaction:

- ◆ HX protonates the ether oxygen, thus making a good leaving group.
- ◆ HX provides a source of X$^-$ for nucleophilic attack.

PROBLEM 9.33 What alkyl halides are formed when each ether is treated with HBr?

a. CH$_3$CH$_2$−O−CH$_2$CH$_3$ b. (CH$_3$)$_2$CH−O−CH$_2$CH$_3$ c. ⬡−O−CH$_3$

9.15   Reaction of Epoxides

Although epoxides do not contain a good leaving group, they contain a strained three-membered ring with two polar bonds. **Nucleophilic attack opens the strained three-membered ring,** making it a favorable process even with the poor leaving group.

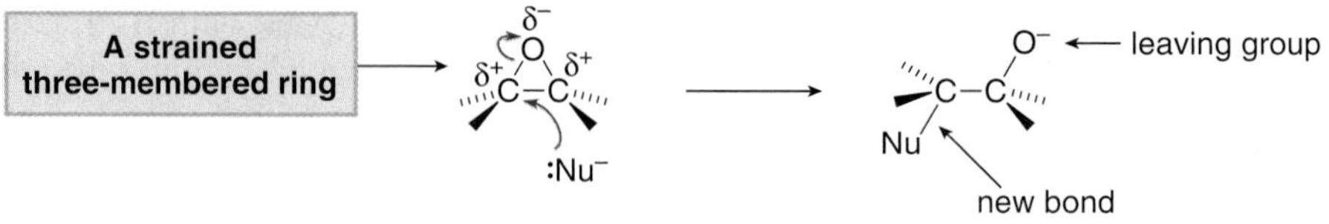

This reaction occurs readily with strong nucleophiles, and with acids like HZ, where Z is a nucleophilic atom.

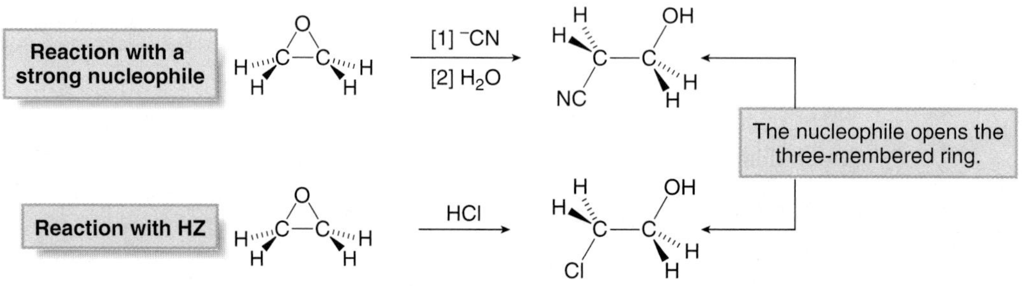

PROBLEM 9.34   Explain why cyclopropane, which has a strained three-membered ring like an epoxide, does *not* react readily with nucleophiles.

9.15A   Opening of Epoxide Rings with Strong Nucleophiles

Virtually all strong nucleophiles open an epoxide ring by a two-step reaction sequence:

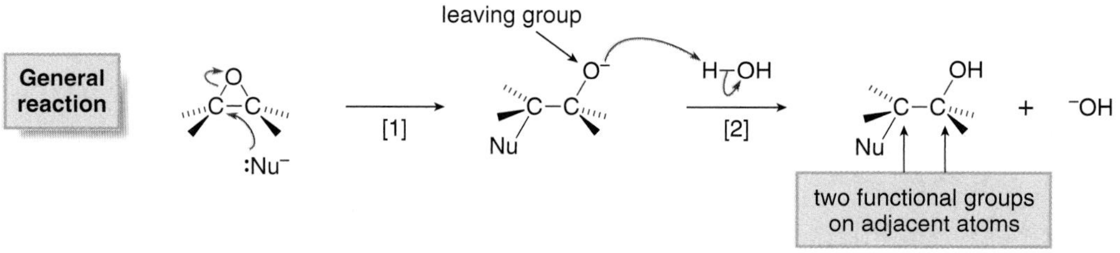

◆ **Step [1]:** The nucleophile attacks an electron-deficient carbon of the epoxide, thus cleaving a C–O bond and relieving the strain of the three-membered ring.

◆ **Step [2]:** Protonation of the alkoxide with water generates a neutral product with two functional groups on adjacent atoms.

Common nucleophiles that open epoxide rings include $^-$OH, $^-$OR, $^-$CN, $^-$SR, and NH_3. With these strong nucleophiles, the reaction occurs via an **S_N2 mechanism,** resulting in two consequences:

◆ **The nucleophile opens the epoxide ring from the back side.**

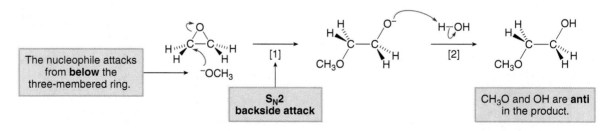

Other examples of the nucleophilic opening of epoxide rings are presented in Sections 12.6 and 20.14.

◆ In an unsymmetrical epoxide, the nucleophile attacks at the less substituted carbon atom.

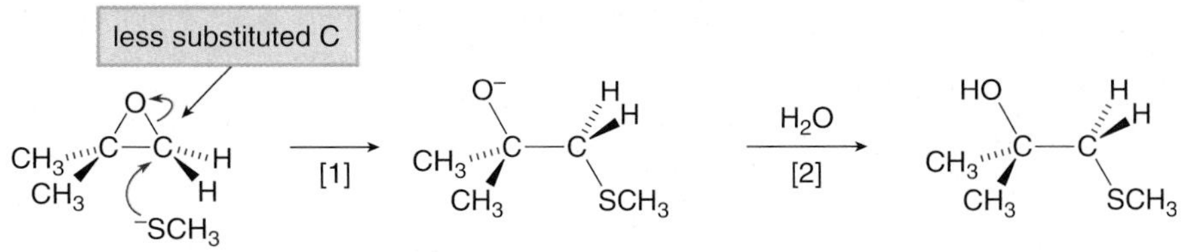

PROBLEM 9.35   Draw the product of each reaction, indicating the stereochemistry at any stereogenic centers.

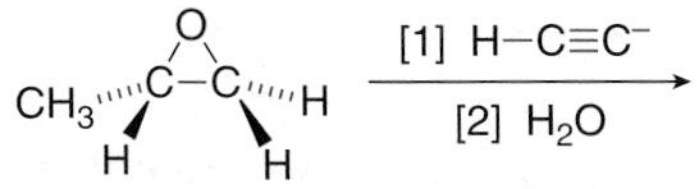

1,2-Epoxycyclohexane is a symmetrical epoxide that is achiral because it possesses a plane of symmetry. It reacts with $^-OCH_3$, however, to yield two *trans*-1,2-disubstituted cyclohexanes, **A** and **B**, which are **enantiomers;** each has two stereogenic centers.

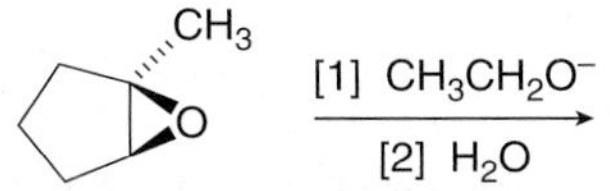

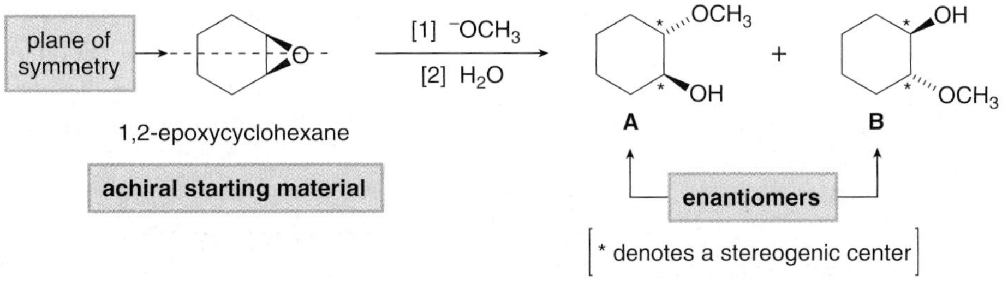

In this case, **nucleophilic attack of $^-OCH_3$ occurs from the back side at either C–O bond, because both ends are equally substituted.** Because attack at either side occurs with equal probability, an equal amount of the two enantiomers is formed—**a racemic mixture.** This is a specific example of a general rule concerning the stereochemistry of products obtained from an achiral reactant.

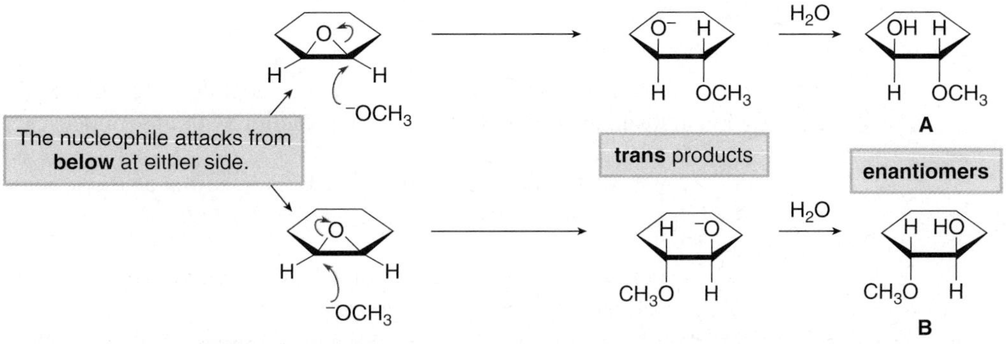

◆ Whenever an achiral reactant yields a product with stereogenic centers, the product must be achiral (meso) or racemic.

This general rule can be restated in terms of optical activity. Recall from Section 5.12 that achiral compounds and racemic mixtures are optically inactive.

◆ Optically inactive starting materials give optically inactive products.

PROBLEM 9.36   The cis and trans isomers of 2,3-dimethyloxirane both react with ^-OH to give 2,3-butanediol. One stereoisomer gives a single achiral product, and one gives two chiral enantiomers. Which epoxide gives one product and which gives two?

cis-2,3-dimethyloxirane

one enantiomer of
trans-2,3-dimethyloxirane

9.15B Reaction with Acids HZ

Acids **HZ** that contain a nucleophile **Z** also open epoxide rings by a two-step reaction sequence:

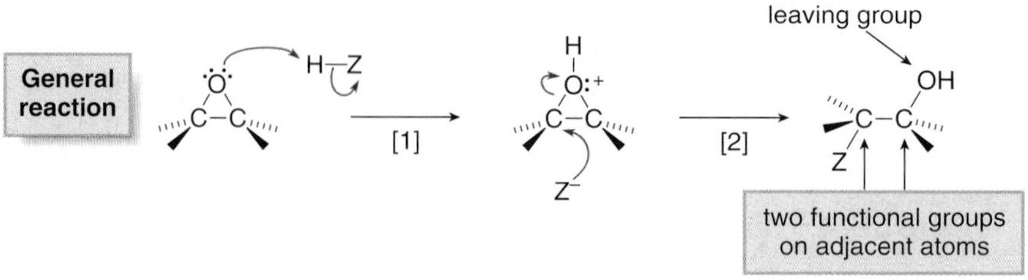

- ◆ **Step [1]:** Protonation of the epoxide oxygen with HZ makes the epoxide oxygen into a good leaving group (OH). It also provides a source of a good nucleophile (Z⁻) to open the epoxide ring.
- ◆ **Step [2]:** The nucleophile Z⁻ then opens the protonated epoxide ring by **backside** attack.

These two steps—**protonation followed by nucleophilic attack**—are the exact reverse of the opening of epoxide rings with strong nucleophiles, where nucleophilic attack precedes protonation.

HCl, HBr, and **HI** all open an epoxide ring in this manner. **H_2O** and **ROH** can, too, but acid must also be added. Regardless of the reaction, the product has an OH group from the epoxide on one carbon and a new functional group Z from the nucleophile on the adjacent carbon. With epoxides fused to rings, *trans*-**1,2-disubstituted cycloalkanes** are formed.

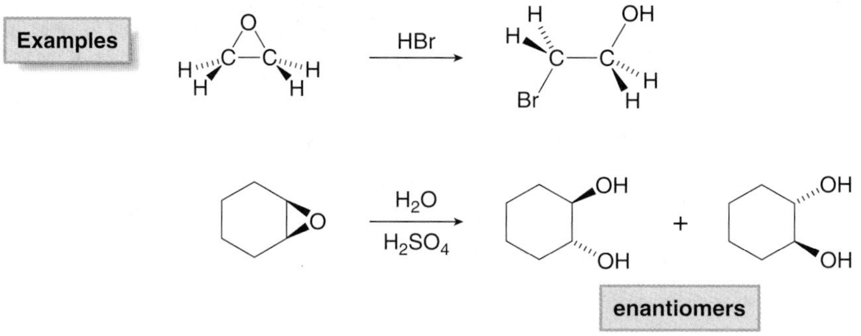

Although backside attack of the nucleophile suggests that this reaction follows an S_N2 mechanism, the regioselectivity of the reaction with unsymmetrical epoxides does not.

◆ With unsymmetrical epoxides, nucleophilic attack occurs at the *more* substituted carbon atom.

For example, the treatment of 1,1-dimethyloxirane with HCl results in nucleophilic attack at the carbon with two methyl groups.

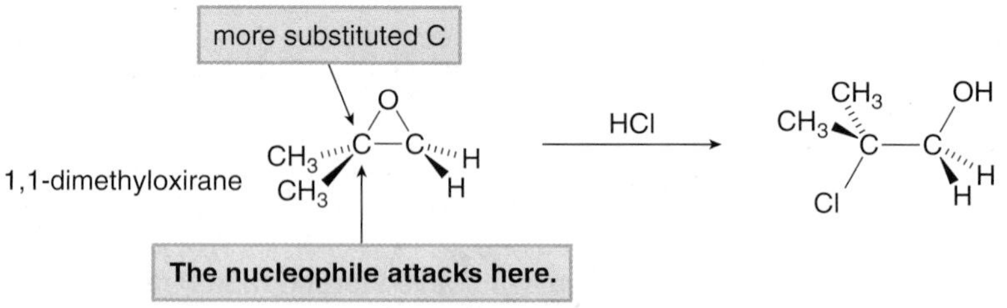

Backside attack of the nucleophile suggests an S_N2 mechanism, but attack at the more substituted carbon suggests an S_N1 mechanism. To explain these results, the **mechanism of nucleophilic attack is thought to be somewhere in between S_N1 and S_N2.**

Figure 9.9 illustrates two possible pathways for the reaction of 1,1-dimethyloxirane with HCl. Backside attack of Cl⁻ at the more substituted carbon proceeds via transition state **A,** whereas backside attack of Cl⁻ at the less substituted carbon proceeds via transition state **B. Transition state A has a partial positive charge on a more substituted carbon, making it more stable.** Thus, the preferred reaction path takes place by way of the lower energy transition state **A.**

Figure 9.9 Opening of an unsymmetrical epoxide ring with HCl

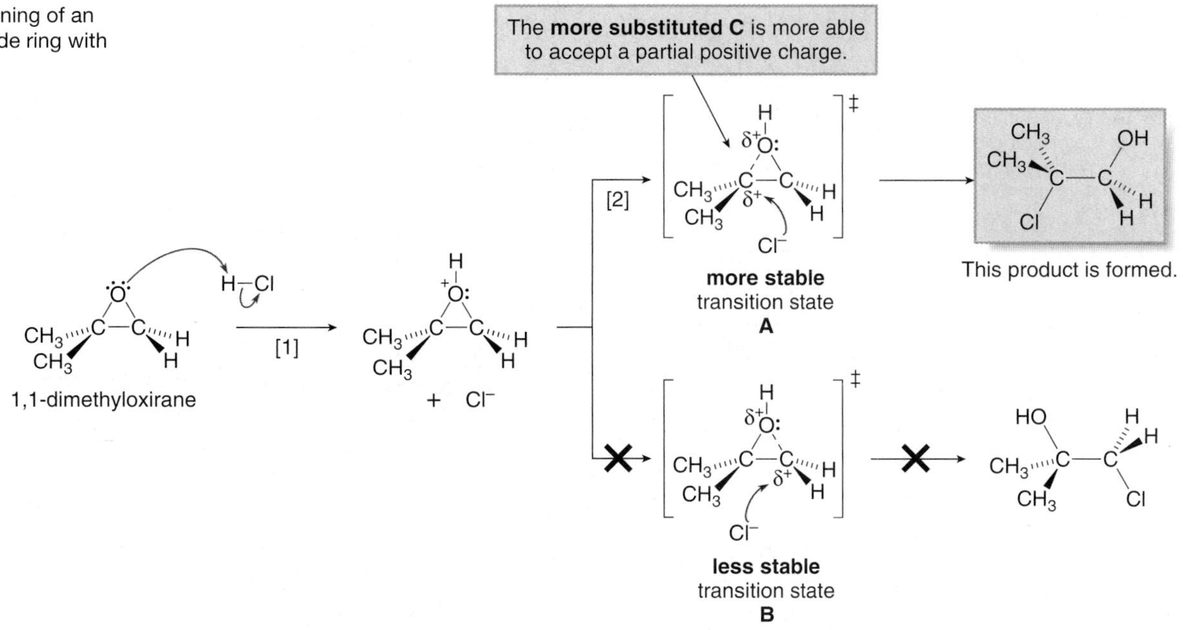

The **more substituted C** is more able to accept a partial positive charge.

more stable transition state **A**

This product is formed.

less stable transition state **B**

1,1-dimethyloxirane

- Transition state **A** is lower in energy because the partial positive charge (δ^+) is located on the more substituted carbon. In this case, therefore, nucleophilic attack occurs from the back side (an S_N2 characteristic) at the more substituted carbon (an S_N1 characteristic).

Opening of an epoxide ring with either a strong nucleophile :Nu⁻ or an acid HZ is **regioselective,** because one constitutional isomer is the major or exclusive product. The **site selectivity of these two reactions, however, is** *exactly the opposite.*

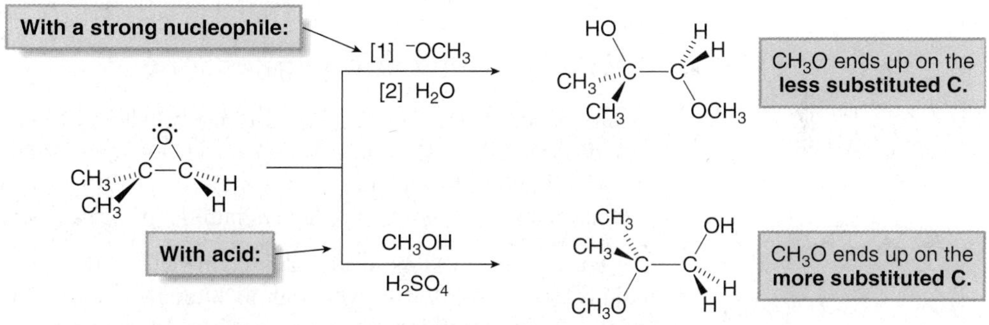

With a strong nucleophile:

[1] ⁻OCH₃
[2] H₂O

CH₃O ends up on the **less substituted C.**

With acid:

CH₃OH
H₂SO₄

CH₃O ends up on the **more substituted C.**

- With a strong nucleophile, :Nu⁻ attacks at the less substituted carbon.
- With an acid HZ, the nucleophile attacks at the more substituted carbon.

The reaction of epoxide rings with nucleophiles is important for the synthesis of many biologically active compounds, including **albuterol** and **salmeterol,** two bronchodilators used in the treatment of asthma (Figure 9.10).

PROBLEM 9.37   Draw the product of each reaction.

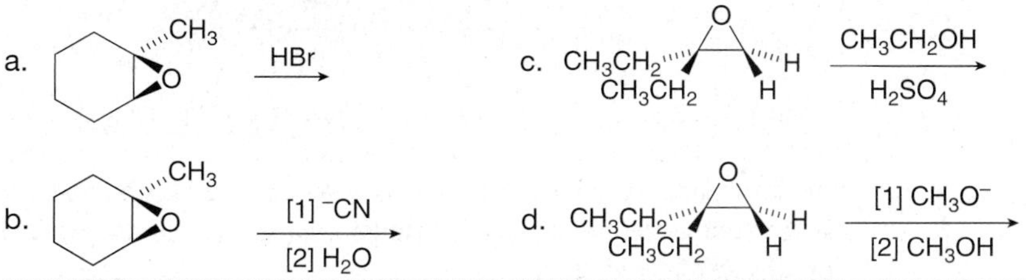

a.   HBr

b.   [1] ⁻CN
           [2] H₂O

c.   CH₃CH₂ · · · CH₃CH₂ · · · H   CH₃CH₂OH / H₂SO₄

d.   CH₃CH₂ · · · CH₃CH₂ · · · H   [1] CH₃O⁻ / [2] CH₃OH

Figure 9.10 The synthesis of two bronchodilators uses opening of an epoxide ring.

- A key step in each synthesis is the opening of an epoxide ring with a nitrogen nucleophile to form a new C−N bond.

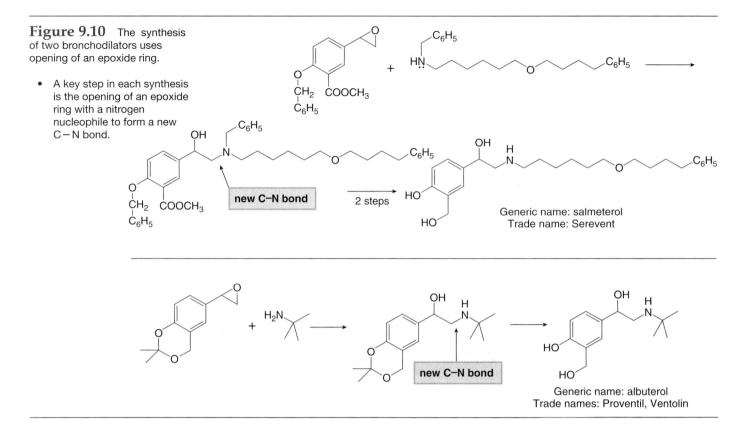

9.16   Application: Epoxides, Leukotrienes, and Asthma

The opening of epoxide rings with nucleophiles is a key step in some important biological processes.

9.16A   Asthma and Leukotrienes

Asthma is an obstructive lung disease that affects millions of Americans. Because it involves episodic constriction of small airways, bronchodilators such as albuterol (Figure 9.10) are used to treat symptoms by widening airways. Because asthma is also characterized by chronic inflammation, inhaled steroids that reduce inflammation are also commonly used.

The **leukotrienes** are molecules that contribute to the asthmatic response. A typical example, **leukotriene C$_4$,** is shown. Although its biological activity was first observed in the 1930s, the chemical structure of leukotriene C$_4$ was not determined until 1979. Structure determination and chemical synthesis were difficult because leukotrienes are highly unstable and extremely potent, and are therefore present in tissues in exceedingly small amounts.

Leukotrienes were first synthesized in 1980 in the laboratory of Professor E. J. Corey, the 1990 recipient of the Nobel Prize in Chemistry.

9.16B   Leukotriene Synthesis and Asthma Drugs

Leukotrienes are synthesized in cells by the oxidation of **arachidonic acid** to 5-HPETE, which is then converted to an epoxide, **leukotriene A$_4$.** Opening of the epoxide ring with a sulfur nucleophile **RSH** yields leukotriene C$_4$.

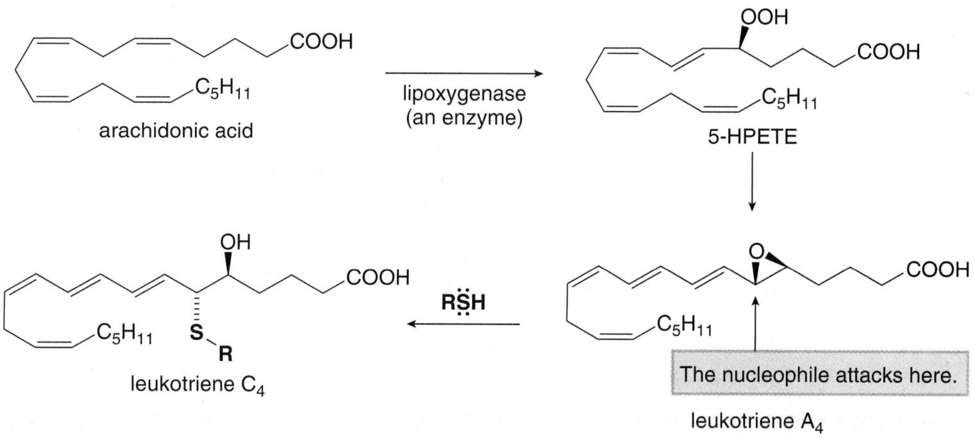

New asthma drugs act by blocking the synthesis of leukotriene C$_4$ from arachidonic acid. For example, **zileuton** (trade name: Zyflo) inhibits the enzyme (called a lipoxygenase) needed for the first step of this process. By blocking the synthesis of leukotriene C$_4$, a compound responsible for the disease, zileuton treats the **cause of asthma,** not just its symptoms.

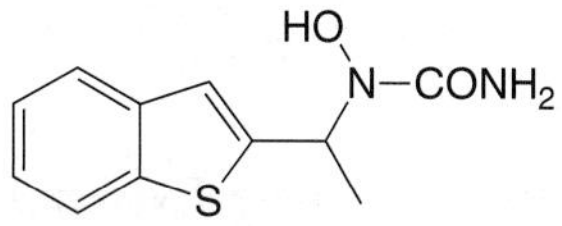

Generic name: zileuton
Trade name: Zyflo
anti-asthma drug

9.17   Application: Benzo[*a*]pyrene, Epoxides, and Cancer

Benzo[*a*]pyrene is a widespread environmental pollutant, produced during the combustion of all types of organic material—gasoline, fuel oil, wood, garbage, and cigarettes. It is a **polycyclic aromatic hydrocarbon (PAH),** a class of compounds that are discussed further in Chapter 17.

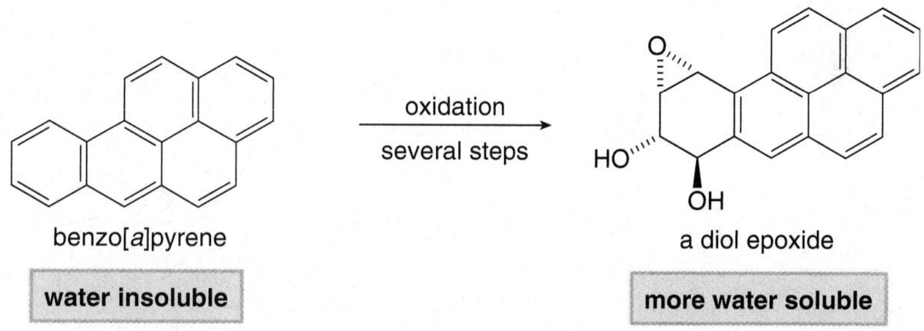

benzo[*a*]pyrene

| water insoluble |

a diol epoxide

| more water soluble |

After this nonpolar and water insoluble hydrocarbon is inhaled or ingested, it is oxidized in the liver to a diol epoxide. Oxidation is a common fate of foreign substances that are not useful nutrients for the body. The oxidation product has three oxygen-containing functional groups, making it much more water soluble, and more readily excreted in urine. It is also a potent carcinogen. The strained three-membered ring of the epoxide reacts readily with biological nucleophiles (such as DNA or an enzyme), leading to ring-opened products that often disrupt normal cell function, causing cancer or cell death.

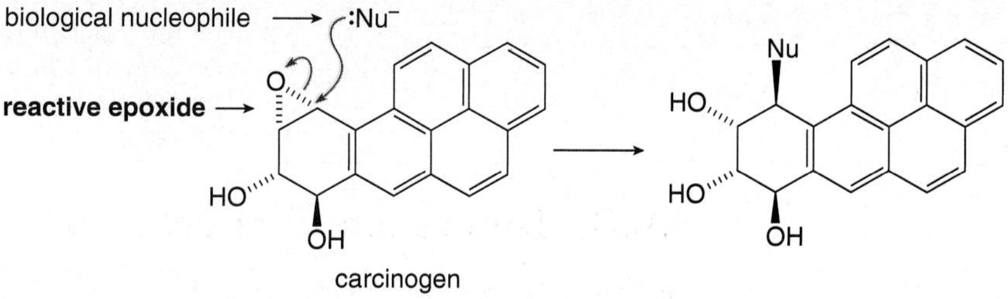

carcinogen

These examples illustrate the central role of the nucleophilic opening of epoxide rings in two well-defined cellular processes.

9.18 Key Concepts—Alcohols, Ethers, and Epoxides

General Facts about ROH, ROR, and Epoxides

- All three compounds contain an O atom that is sp^3 hybridized and tetrahedral (9.2).
- All three compounds have polar C–O bonds, but only alcohols have an O–H bond for intermolecular hydrogen bonding (9.4).
- Alcohols and ethers do not contain a good leaving group. Nucleophilic substitution can occur only after the OH (or OR) group is converted to a better leaving group (9.7A).
- Epoxides have a leaving group located in a strained three-membered ring, making them reactive to strong nucleophiles and acids HZ that contain a nucleophilic atom Z (9.15).

A New Reaction of Carbocations (9.9)

- Less stable carbocations rearrange to more stable carbocations by the shift of a hydrogen atom or an alkyl group.

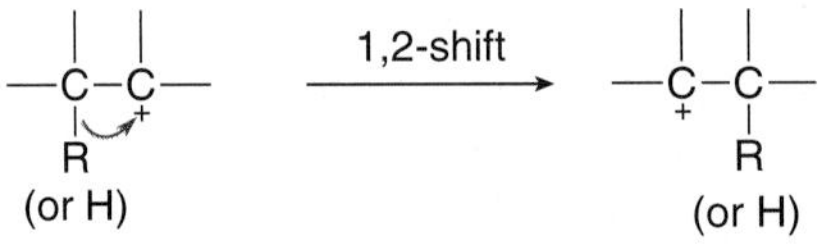

- Besides rearranging, carbocations also react with a nucleophile (7.13) and a base (8.6).

Preparation of Alcohols, Ethers, and Epoxides (9.6)

[1] Preparation of alcohols

$$R{-}X \; + \; {}^{-}OH \longrightarrow R{-}OH \; + \; X^{-}$$

- The mechanism is S_N2.
- The reaction works best for CH_3X and 1° RX.

[2] Preparation of alkoxides—A Brønsted–Lowry acid–base reaction

$$R{-}O{-}H \; + \; Na^{+}H^{-} \longrightarrow R{-}O^{-} \; Na^{+} \; + \; H_2$$

alkoxide

[3] Preparation of ethers (Williamson ether synthesis)

$$R{-}X \; + \; {}^{-}OR' \longrightarrow R{-}OR' \; + \; X^{-}$$

- The mechanism is S_N2.
- The reaction works best for CH_3X and 1° RX.

[4] Preparation of epoxides—Intramolecular S_N2 reaction

halohydrin

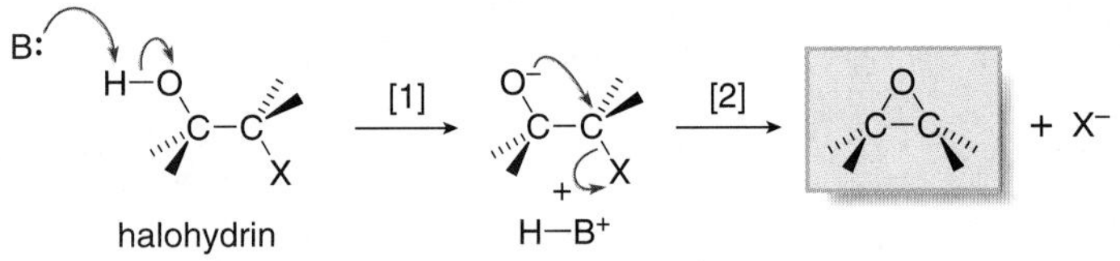

- A two-step reaction sequence:
  - [1] The removal of a proton with base forms an alkoxide.
  - [2] An intramolecular $\mathbf{S_N2}$ reaction forms the epoxide.

Reactions of Alcohols

[1] Dehydration to form alkenes

[a] Using strong acid (9.8, 9.9)

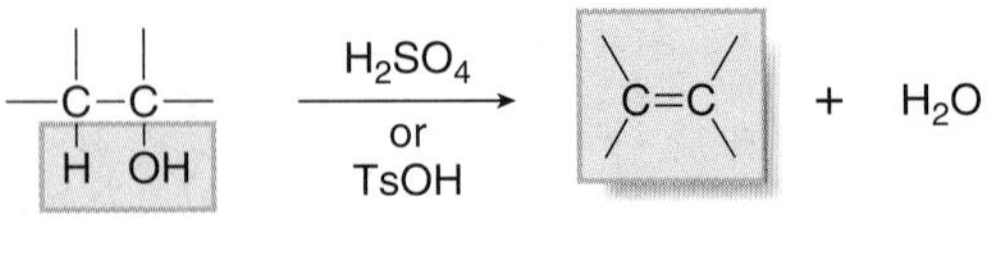

- Order of reactivity: $R_3COH > R_2CHOH > RCH_2OH$.
- The mechanism for 2° and 3° ROH is E1—carbocations are intermediates and rearrangements occur.
- The mechanism for 1° ROH is E2.
- The Zaitsev rule is followed.

[b] Using $POCl_3$ and pyridine (9.10)

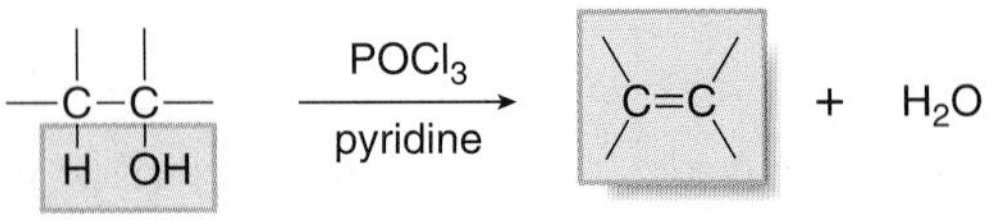

- The mechanism is E2.
- No carbocation rearrangements occur.

[2] Reaction with HX to form RX (9.11)

$$R-OH \ + \ H-X \ \longrightarrow \ \boxed{R-X} \ + \ H_2O$$

- Order of reactivity: $R_3COH > R_2CHOH > RCH_2OH$.
- The mechanism for 2° and 3° ROH is S_N1—carbocations are intermediates and rearrangements occur.
- The mechanism for CH_3OH and 1° ROH is S_N2.

[3] Reaction with other reagents to form RX (9.12)

$$R-OH \ + \ SOCl_2 \ \xrightarrow{\text{pyridine}} \ \boxed{R-Cl}$$

$$R-OH \ + \ PBr_3 \ \longrightarrow \ \boxed{R-Br}$$

- Reactions occur with CH_3OH and 1° and 2° ROH.
- The reactions follow an S_N2 mechanism.

[4] Reaction with tosyl chloride to form tosylates (9.13A)

$$R-OH + Cl-\overset{\displaystyle O}{\underset{\displaystyle O}{S}}-\!\!\!\!\!\!\!\!\!\!\bigcirc\!\!\!\!\!\!-CH_3 \ \xrightarrow{\text{pyridine}} \ R-O-\overset{\displaystyle O}{\underset{\displaystyle O}{S}}-\!\!\!\!\!\!\!\!\!\!\bigcirc\!\!\!\!\!\!-CH_3$$

$$\boxed{R-OTs}$$

- The C−O bond is not broken so the configuration at a stereogenic center is retained.

Reactions of Tosylates

Tosylates undergo either substitution or elimination, depending on the reagent (9.13B).

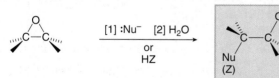

- Substitution is carried out with a strong $:Nu^-$ so the mechanism is S_N2.

- Elimination is carried out with a strong base so the mechanism is E2.

Reactions of Ethers

Only one reaction is useful: cleavage with strong acids (9.14).

$$R-O-R' \ + \ H-X \ \longrightarrow \ \boxed{R-X} \ + \ \boxed{R'-X} \ + \ H_2O$$
$$[X = Br \text{ or } I]$$

- With 2° and 3° R groups, the mechanism is S_N1.
- With 1° R groups the mechanism is S_N2.

Reactions of Epoxides

Epoxide rings are opened with nucleophiles $:Nu^-$ and acids HZ (9.15).

$$\xrightarrow[\text{or} \ \text{HZ}]{[1] \ :Nu^- \ \ [2] \ H_2O}$$

- The reaction occurs with backside attack, resulting in trans or anti products.
- With $:Nu^-$, the mechanism is S_N2, and nucleophilic attack occurs at the less substituted C.
- With HZ, the mechanism is between S_N1 and S_N2, and attack of Z^- occurs at the more substituted C.

Problems

Nomenclature

9.38 Give the IUPAC name for each alcohol.

a. $(CH_3)_2CHCH_2CH_2CH_2OH$

d. HO——◯——OH

g.

b. $(CH_3)_2CHCH_2CH(CH_2CH_3)CH(OH)CH_2CH_3$

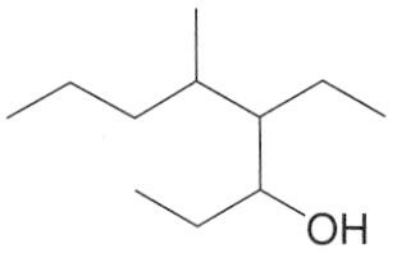

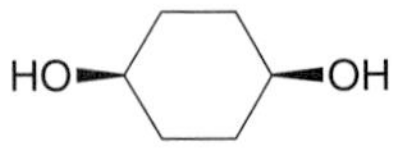

e.

c.

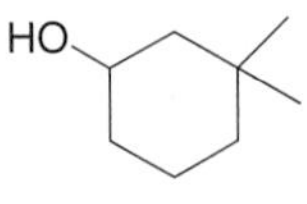

f.

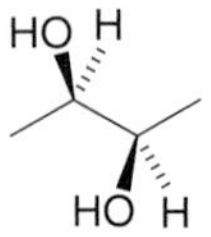

[Also label the stereogenic centers as R or S.]

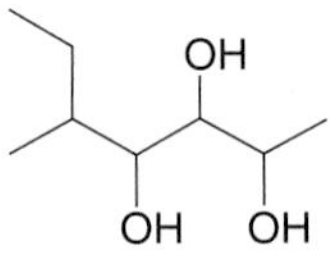

h.

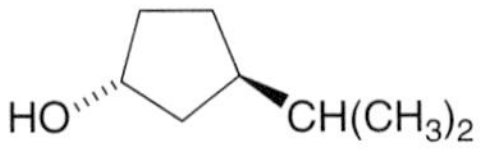

9.39 Name each ether and epoxide.

a.

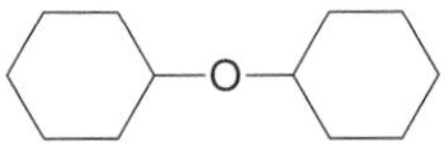

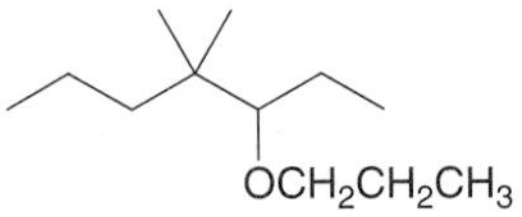

c.

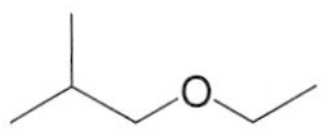

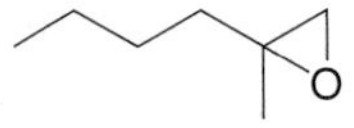

e.

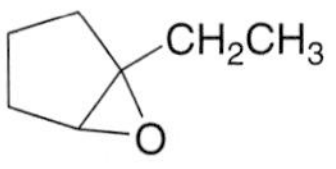

b.

d.

f. 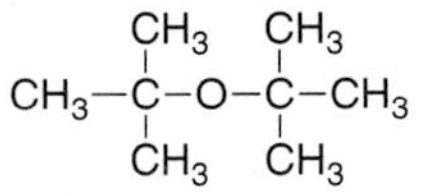 $CH_3-\underset{\underset{CH_3}{|}}{\overset{\overset{CH_3}{|}}{C}}-O-\underset{\underset{CH_3}{|}}{\overset{\overset{CH_3}{|}}{C}}-CH_3$

9.40 Give the structure corresponding to each name.

a. 4-ethyl-3-heptanol
b. *trans*-2-methylcyclohexanol
c. 2,3,3-trimethyl-2-butanol
d. 6-*sec*-butyl-7,7-diethyl-4-decanol

e. 3-chloro-1,2-propanediol
f. diisobutyl ether
g. 1,2-epoxy-1,3,3-trimethylcyclohexane
h. 1-ethoxy-3-ethylheptane

9.41 Draw the eight constitutional isomers with molecular formula $C_5H_{12}O$ that contain an OH group. Give the IUPAC name for each compound.

Physical Properties

9.42 Rank each group of compounds in order of:

a. increasing boiling point: $CH_3CH_2CH_2OH$, $(CH_3)_2CHOH$, $CH_3CH_2OCH_3$
b. increasing water solubility: $CH_3(CH_2)_5OH$, $HO(CH_2)_6OH$, $CH_3(CH_2)_4CH_3$

Alcohols

9.43 Draw the organic product(s) formed when $CH_3CH_2CH_2OH$ is treated with each reagent.

a. H_2SO_4
b. NaH
c. $HCl + ZnCl_2$

d. HBr
e. $SOCl_2$, pyridine
f. PBr_3

g. TsCl, pyridine
h. [1] NaH; [2] CH_3CH_2Br
i. [1] TsCl, pyridine; [2] NaSH

9.44 Draw the organic product(s) formed when 1-methylcyclohexanol is treated with each reagent. In some cases no reaction occurs.

a. NaH
b. NaCl

c. HBr
d. HCl

e. H_2SO_4
f. $NaHCO_3$

g. [1] NaH; [2] CH_3CH_2Br
h. $POCl_3$, pyridine

9.45 What alkenes are formed when each alcohol is dehydrated with TsOH? Label the major product when a mixture results.

a. 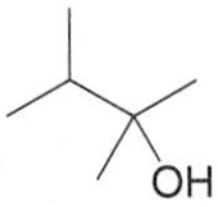
b. 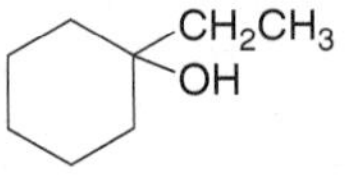
c. 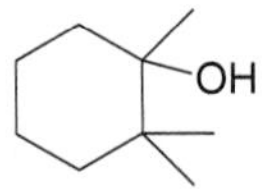
d. $CH_3CH_2CH_2CH_2OH$
e.

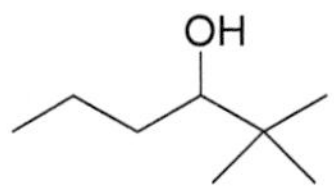

9.46 What three alkenes are formed when $CH_3CH_2CH_2CH(OH)CH_3$ is treated with H_2SO_4? Label the major product.

9.47 Draw the products formed when $CH_3CH_2CH_2CH_2OTs$ is treated with each reagent.

a. CH_3SH
b. $NaOCH_2CH_3$
c. NaOH
d. $KOC(CH_3)_3$

9.48 Draw the products of each reaction and indicate stereochemistry around stereogenic centers.

a. 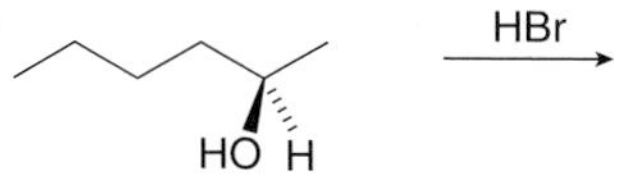 [structure] $\xrightarrow{\text{HBr}}$

c. 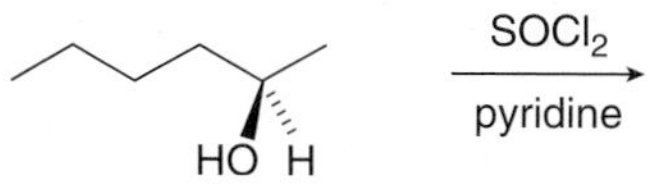 [structure] $\xrightarrow[\text{pyridine}]{\text{SOCl}_2}$

b. 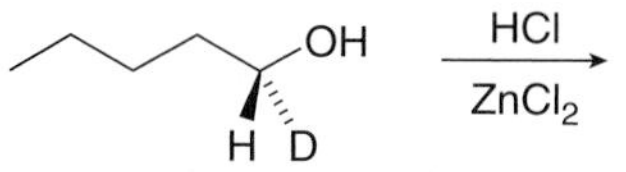 [structure] $\xrightarrow[\text{ZnCl}_2]{\text{HCl}}$

d. 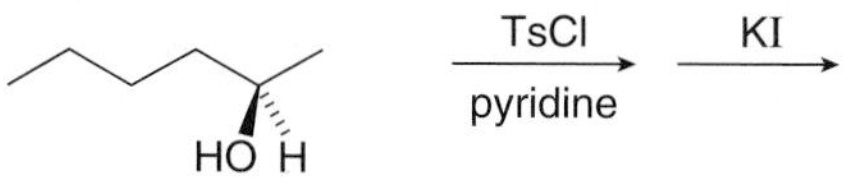 [structure] $\xrightarrow[\text{pyridine}]{\text{TsCl}}$ $\xrightarrow{\text{KI}}$

9.49 (a) Identify compounds **A–F** in the following reactions. (b) How are compounds **B** and **D** related? (c) How are compounds **B** and **F** related?

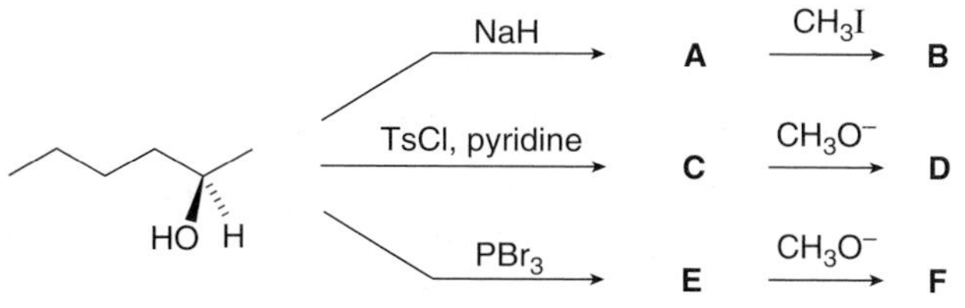

[structure] $\xrightarrow{\text{NaH}}$ **A** $\xrightarrow{\text{CH}_3\text{I}}$ **B**

$\xrightarrow{\text{TsCl, pyridine}}$ **C** $\xrightarrow{\text{CH}_3\text{O}^-}$ **D**

$\xrightarrow{\text{PBr}_3}$ **E** $\xrightarrow{\text{CH}_3\text{O}^-}$ **F**

9.50 Draw a stepwise mechanism for each reaction.

a. 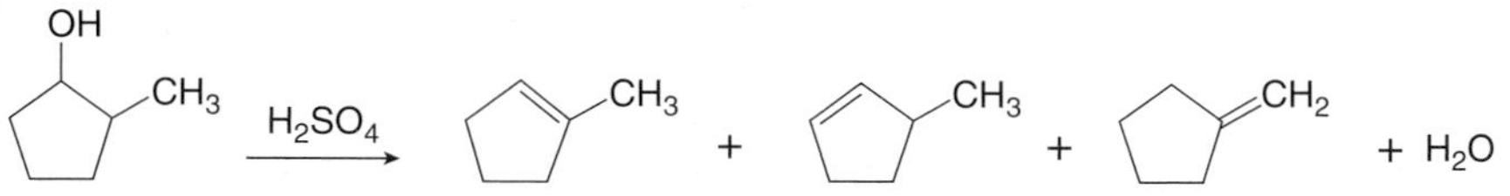 [structure] $\xrightarrow{\text{H}_2\text{SO}_4}$ [structure] + [structure] + [structure] + H_2O

b. 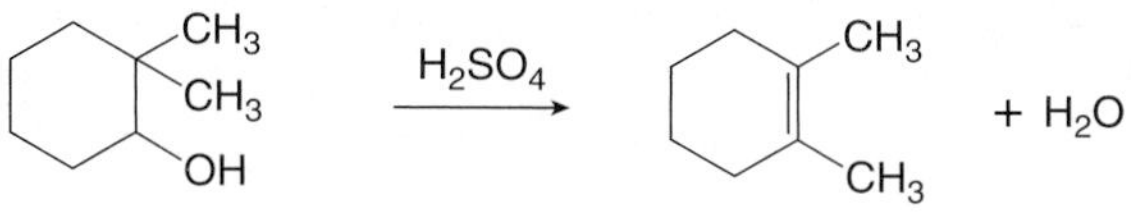 [structure] $\xrightarrow{\text{H}_2\text{SO}_4}$ [structure] + H_2O

9.51 Sometimes carbocation rearrangements can change the size of a ring. Draw a stepwise, detailed mechanism for the following reaction.

[structure] $\xrightarrow{\text{H}_2\text{SO}_4}$ [structure] + H_2O

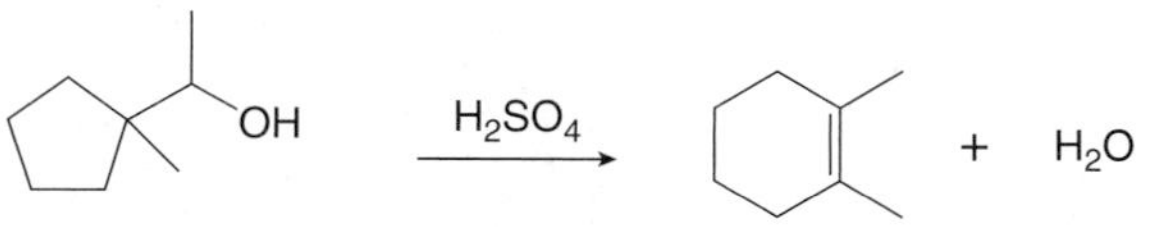

9.52 Indicate the stereochemistry of the alkyl halide formed when (S)-3-hexanol is treated with (a) HBr; (b) PBr$_3$; (c) HCl; (d) SOCl$_2$ and pyridine.

9.53 Explain the following observation. When 3-methyl-2-butanol is treated with HBr, a single alkyl bromide is isolated, resulting from a 1,2-shift. When 2-methyl-1-propanol is treated with HBr, no rearrangement occurs to form an alkyl bromide.

9.54 To convert a 1° alcohol into a 1° alkyl chloride with HCl, a Lewis acid such as ZnCl$_2$ must be added to the reaction mixture. Explain why it is possible to omit the Lewis acid if a polar aprotic solvent such as HMPA, [(CH$_3$)$_2$N]$_3$P=O, is used.

9.55 When CH$_3$CH$_2$CH$_2$CH$_2$OH is treated with H$_2$SO$_4$ + NaBr, CH$_3$CH$_2$CH$_2$CH$_2$Br is the major product, and CH$_3$CH$_2$CH=CH$_2$ and CH$_3$CH$_2$CH$_2$CH$_2$OCH$_2$CH$_2$CH$_2$CH$_3$ are isolated as minor products. Draw a mechanism that accounts for the formation of each of these products.

9.56 Identify the product of the following two-step reaction sequence used in the synthesis of leukotriene C$_4$ (Section 9.16).

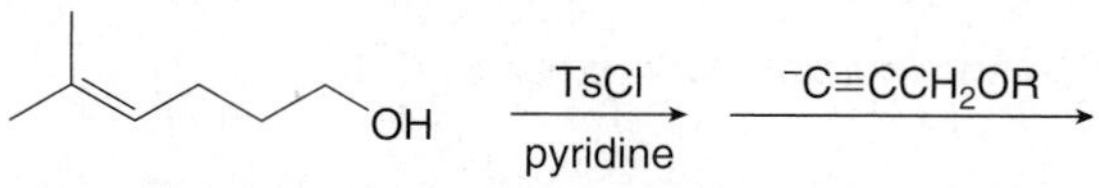

[structure] $\xrightarrow[\text{pyridine}]{\text{TsCl}}$ $\xrightarrow{^-\text{C}\equiv\text{CCH}_2\text{OR}}$

9.57 Alkyl tosylates undergo E2 elimination when treated with strong base, in exactly the same manner as alkyl halides. Thus, when the two groups removed (H and OTs) are bonded to a cyclohexane ring, they must adopt a trans diaxial arrangement. Keeping this in mind, draw the alkenes formed when each tosylate is treated with a strong base. Label the major product when a mixture results.

a. 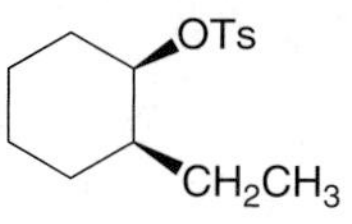 [structure OTs / CH$_2$CH$_3$]

b. 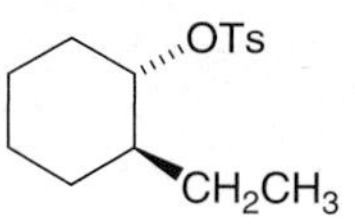 [structure OTs / CH$_2$CH$_3$]

Ethers

9.58 Draw two different routes to each of the following ethers using a Williamson ether synthesis. Indicate the preferred route (if there is one).

a. 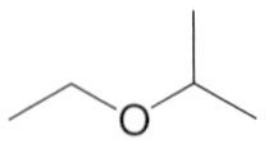   b. 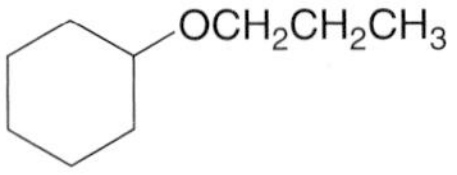 $OCH_2CH_2CH_3$ c. $CH_3CH_2OCH_2CH_2CH_3$

9.59 Explain why it is not possible to prepare the following ether using a Williamson ether synthesis.

$OC(CH_3)_3$

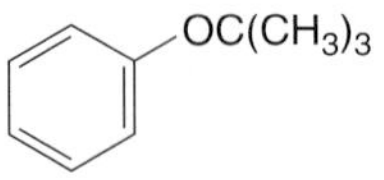

9.60 Draw the products formed when each ether is treated with HBr.

a. $(CH_3)_3COCH_2CH_2CH_3$ b. 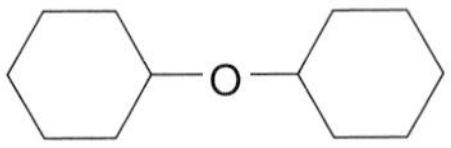   c. 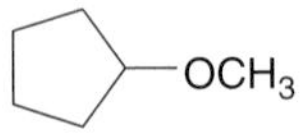 OCH_3

9.61 Draw a stepwise mechanism for each reaction.

a. 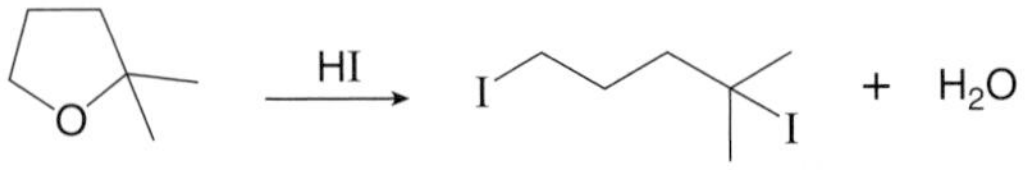 $\xrightarrow{\text{HI}}$ I $\diagup\diagup\diagup$ I $+ H_2O$ b. 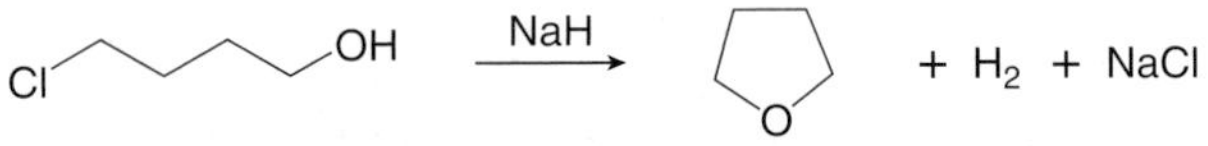 Cl $\diagup\diagup\diagup$ OH $\xrightarrow{\text{NaH}}$ $+ H_2 + NaCl$

9.62 Alcohols can be converted to methyl ethers using dimethyl sulfate in the following two-step sequence. Draw a mechanism for this sequence and explain why dimethyl sulfate is a very reactive methylating agent—that is, it readily transfers a methyl group from itself to another compound.

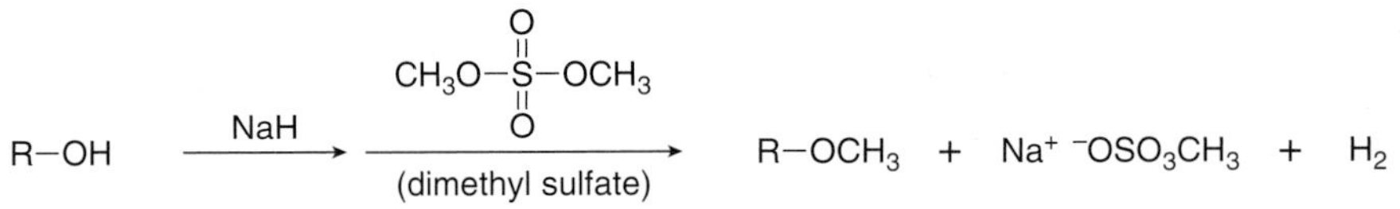

$$CH_3O-\overset{\overset{\displaystyle O}{\|}}{\underset{\underset{\displaystyle O}{\|}}{S}}-OCH_3$$

$$R-OH \xrightarrow[\text{(dimethyl sulfate)}]{\text{NaH}} R-OCH_3 + Na^+ \ ^-OSO_3CH_3 + H_2$$

Epoxides

9.63 Draw the products formed when ethylene oxide is treated with each reagent.
a. HBr
b. H_2O (H_2SO_4)
c. [1] $CH_3CH_2O^-$; [2] H_2O
d. [1] $HC\equiv C^-$; [2] H_2O
e. [1] ^-OH; [2] H_2O
f. [1] CH_3S^-; [2] H_2O

9.64 Draw the products of each reaction.

a. 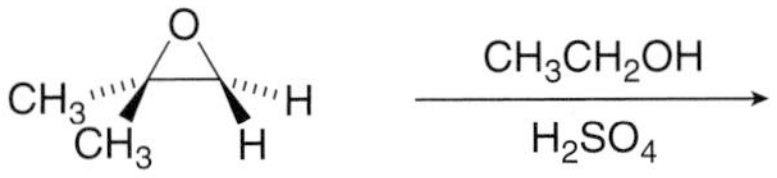 CH_3 H $\xrightarrow[H_2SO_4]{CH_3CH_2OH}$ c. 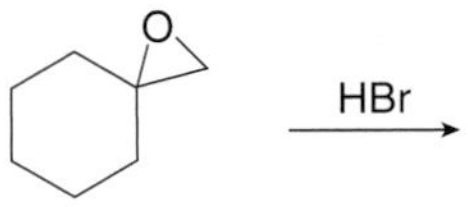 $\xrightarrow{\text{HBr}}$

b. $\xrightarrow[\text{[2] } H_2O]{\text{[1] } CH_3CH_2O^- \ Na^+}$ d. 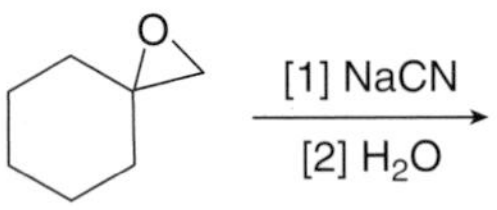 $\xrightarrow[\text{[2] } H_2O]{\text{[1] NaCN}}$

9.65 Draw the product of each reaction and indicate the stereochemistry.

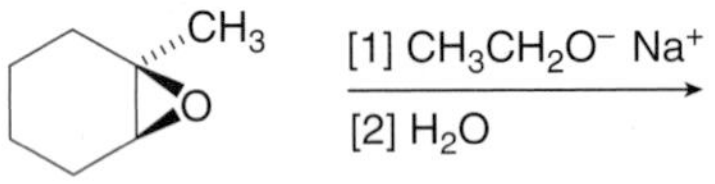

a. 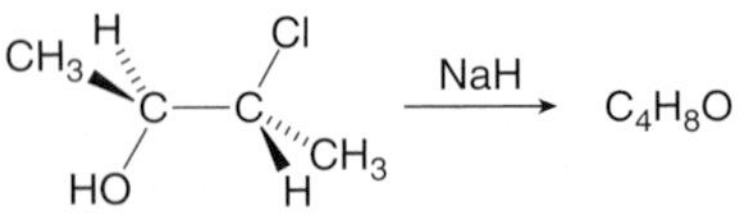 $\xrightarrow{\text{NaH}}$ C_4H_8O b. 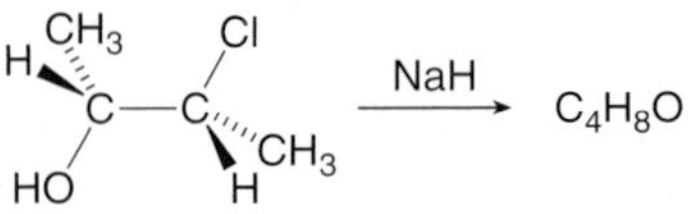 $\xrightarrow{\text{NaH}}$ C_4H_8O

General Problems

9.66 Draw the products of each reaction, and indicate the stereochemistry where appropriate.

a.  OTs $\xrightarrow{\text{KOC(CH}_3)_3}$

f. $\xrightarrow[\text{pyridine}]{\text{TsCl}}$ $\xrightarrow{\text{CH}_3\text{CO}_2^-}$

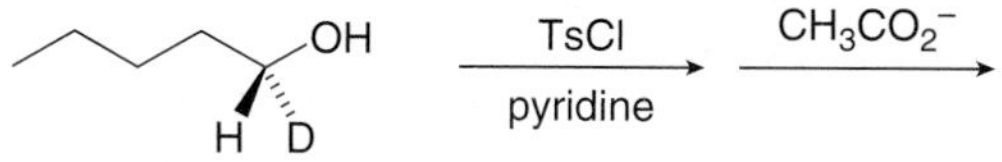

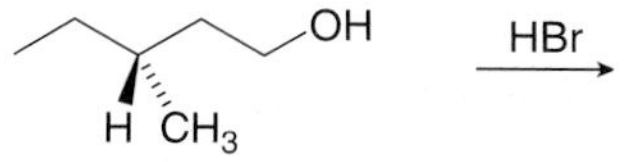

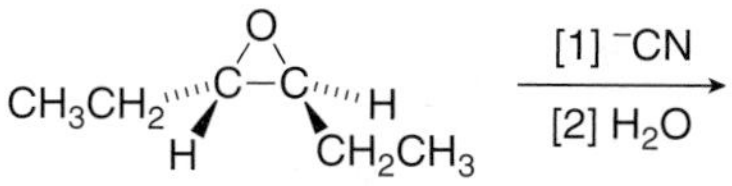

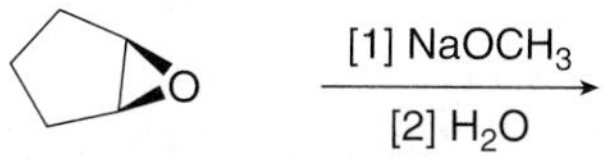

b. $\xrightarrow{\text{HBr}}$

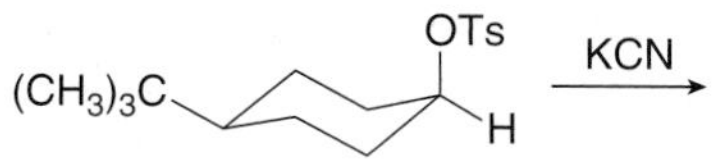

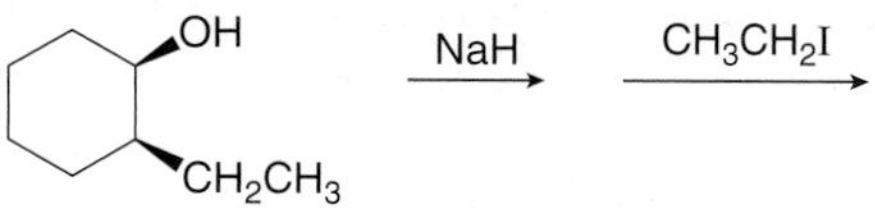

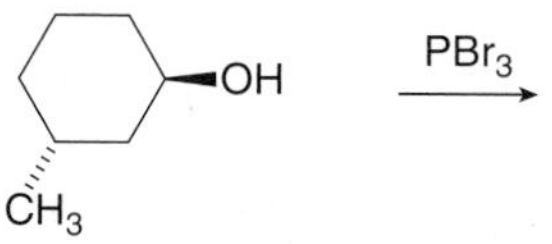

g. $\xrightarrow{\text{HBr}}$

c. CH_3CH_2 $\xrightarrow[\text{[2] H}_2\text{O}]{\text{[1]}^-\text{CN}}$

h. $\xrightarrow[\text{[2] H}_2\text{O}]{\text{[1] NaOCH}_3}$

d. $(\text{CH}_3)_3\text{C}$ $\xrightarrow{\text{KCN}}$

i. $\xrightarrow{\text{NaH}}$ $\xrightarrow{\text{CH}_3\text{CH}_2\text{I}}$

e. $\xrightarrow{\text{PBr}_3}$

j. $\text{CH}_3\text{CH}_2-\overset{\overset{\text{CH}_3}{|}}{\underset{\underset{\text{CH}_3}{|}}{\text{C}}}-\text{O}-\text{CH}_3$ $\xrightarrow{\text{HI}}$

9.67 Prepare each compound from $\text{CH}_3\text{CH}_2\text{CH}_2\text{CH}_2\text{OH}$. In some cases, more than one step is needed.
a. $\text{CH}_3\text{CH}_2\text{CH}_2\text{CH}_2\text{Br}$
b. $\text{CH}_3\text{CH}_2\text{CH}_2\text{CH}_2\text{Cl}$
c. $\text{CH}_3\text{CH}_2\text{CH}_2\text{CH}_2\text{OCH}_2\text{CH}_3$
d. $\text{CH}_3\text{CH}_2\text{CH}_2\text{CH}_2\text{N}_3$

9.68 Prepare each compound from cyclopentanol. In some cases, more than one step is needed.

a. 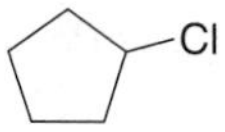 Cl   b.    c. 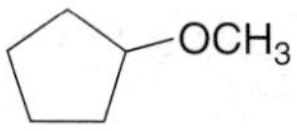 OCH$_3$ d. 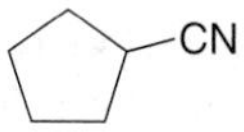 CN

9.69 Identify the reagents (a–h) needed to carry out each reaction.

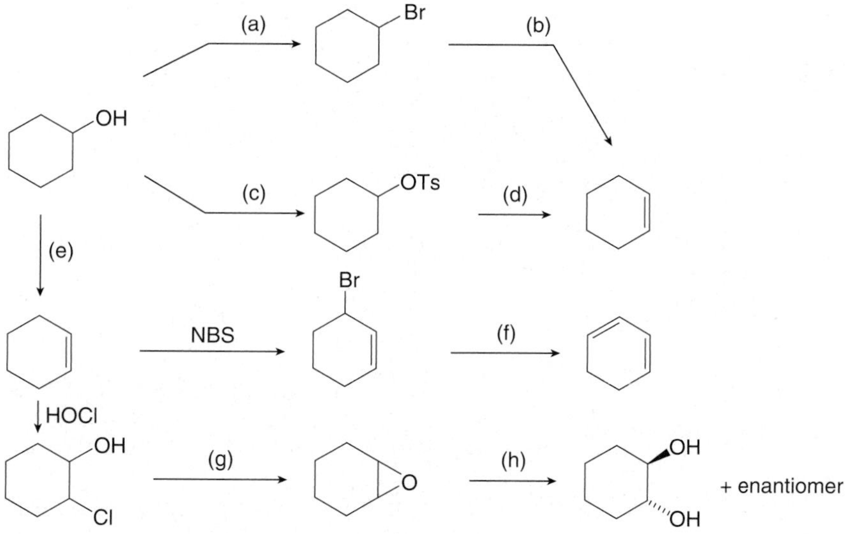

9.70 In Section 9.12, we learned that an alcohol is converted to an alkyl halide using PBr₃ because PBr₃ makes OH a better leaving group and provides the nucleophile Br⁻ for substitution. Mitsunobu and co-workers devised a more general variation of this reaction that converts ROH into a variety of substitution products R–Nu, using nucleophiles having the general structure H–Nu: or :Nu⁻.

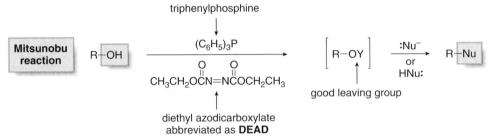

Draw the products of each Mitsunobu reaction.

a.

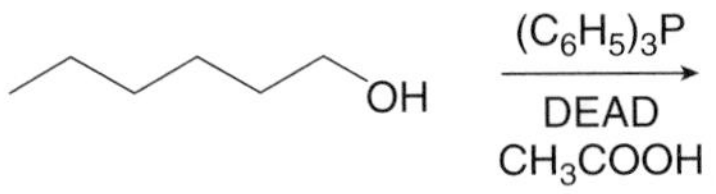

$$\xrightarrow[\substack{DEAD \\ CH_3COOH}]{(C_6H_5)_3P}$$

b.

$$\xrightarrow[\substack{DEAD \\ CH_3SH}]{(C_6H_5)_3P}$$

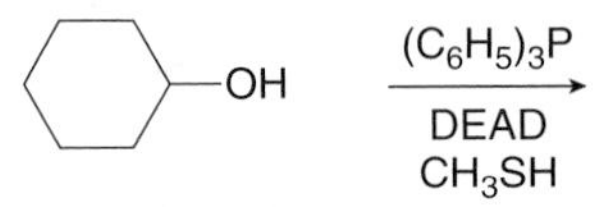

Challenge Questions

9.71 Drawn below are three isomeric halohydrins. One reacts rapidly to form an epoxide, one is intermediate in reactivity, and one does not react at all. Identify which halohydrin corresponds to each kind of reactivity and explain why.

a. 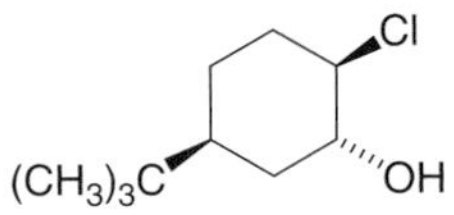   b. 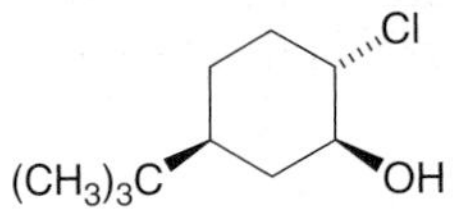   c.

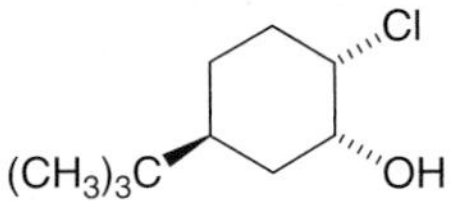

9.72 1,2-Diols are converted to carbonyl compounds when treated with strong acids, in a reaction called the *pinacol rearrangement.* Draw a stepwise mechanism for this reaction.

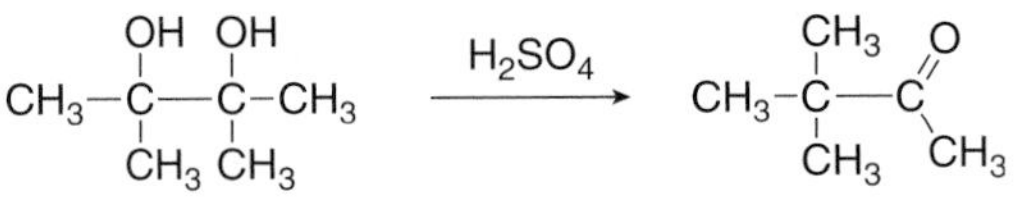

$$\underset{\text{pinacol}}{CH_3-\underset{\underset{CH_3}{|}}{\overset{\overset{OH}{|}}{C}}-\underset{\underset{CH_3}{|}}{\overset{\overset{OH}{|}}{C}}-CH_3} \quad \xrightarrow{H_2SO_4} \quad \underset{\text{pinacolone}}{CH_3-\underset{\underset{CH_3}{|}}{\overset{\overset{CH_3}{|}}{C}}-\overset{\overset{O}{||}}{C}-CH_3}$$

9.73 Draw a stepwise mechanism for the following reaction.

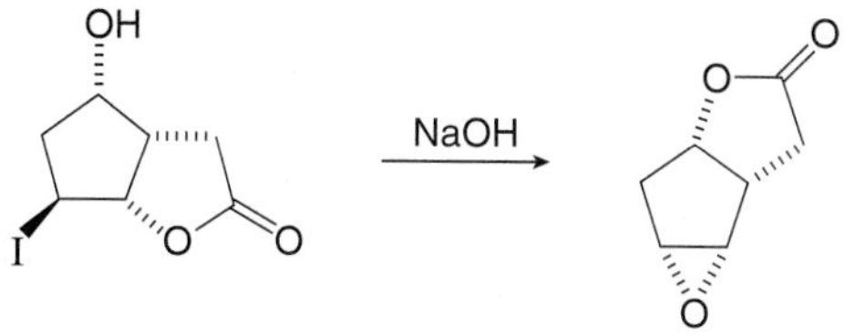

9.74 Epoxide rings bonded to six-membered rings are opened in a **trans diaxial** fashion; that is, the nucleophile attacks from the axial direction to give a leaving group that is also axial. Because of this, only one product is formed in each of the following reactions, even though both ends of the epoxide are equally substituted with alkyl groups. Draw the product of each reaction, using chair forms to indicate the stereochemistry of substituents on the cyclohexane ring.

a. 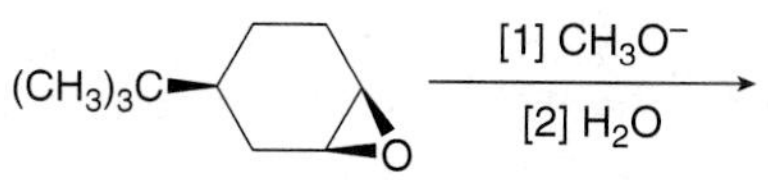 (CH$_3$)$_3$C —

      [1] CH$_3$O$^-$
      [2] H$_2$O

b. 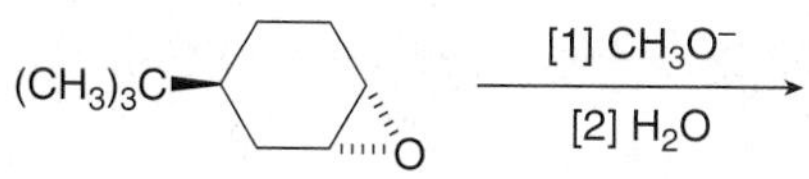 (CH$_3$)$_3$C —

      [1] CH$_3$O$^-$
      [2] H$_2$O

Alkenes

Stearic acid and **oleic acid** are fatty acids, compounds that contain a carboxy (COOH) group attached to the end of a long carbon chain. Stearic acid is a **saturated fatty acid** because each carbon atom in its long chain has the maximum number of bonds to hydrogen. Oleic acid is an **unsaturated fatty acid** because its carbon chain contains one (cis) double bond. The presence of a double bond greatly affects the chemical and physical properties of these fatty acids. In Chapter 10 we will learn about alkenes, organic compounds that contain carbon–carbon double bonds.

I n Chapters 10 and 11 we turn our attention to **alkenes** and **alkynes,** compounds that contain one and two π bonds, respectively. Because π bonds are easily broken, alkenes and alkynes undergo **addition,** the third general type of organic reaction. These multiple bonds also make carbon atoms electron rich, so alkenes and alkynes react with a wide variety of electrophilic reagents in addition reactions.

In Chapter 10 we review the properties and synthesis of alkenes first, and then concentrate on reactions. **Every new reaction in Chapter 10 is an *addition reaction.*** The most challenging part is learning the reagents, mechanism, and stereochemistry that characterize each individual reaction.

10.1   Introduction

Alkenes are also called **olefins.**

Alkenes are compounds that contain a carbon–carbon double bond. **Terminal alkenes** have the double bond at the end of the carbon chain, whereas **internal alkenes** have at least one carbon atom bonded to each end of the double bond. **Cycloalkenes** contain a double bond in a ring.

Alkene

double bond   terminal alkene   internal alkene   cycloalkene

The double bond of an alkene consists of one σ bond and one π bond. Each carbon is sp^2 hybridized and trigonal planar, and all bond angles are approximately 120° (Section 8.2A).

π bond

σ bond

ethylene sp^2 hybridized

Bond dissociation energies of the C–C bonds in ethane (a σ bond only) and ethylene (one σ and one π bond) can be used to estimate the strength of the π component of the double bond. If we assume that the σ bond in ethylene is similar in strength to the σ bond in ethane (88 kcal/mol), then the π bond is worth 64 kcal/mol.

$$CH_2=CH_2 \qquad CH_3-CH_3$$

152 kcal/mol   −   88 kcal/mol   =   64 kcal/mol
(σ + π bond)   (σ bond)

π bond only

◆ The π bond is much weaker than the σ bond of a C–C double bond, making it much more easily broken. As a result, alkenes undergo many reactions that alkanes do not.

Other features of the carbon–carbon double bond, which were presented in Chapter 8, are summarized in Table 10.1.

Cycloalkenes having fewer than eight carbon atoms have a cis geometry. A trans cycloalkene must have a carbon chain long enough to connect the ends of the double bond without introducing too much strain. *trans*-Cyclooctene is the smallest, isolable trans cycloalkene, but it is considerably less stable than *cis*-cyclooctene, making it one of the few alkenes having a higher energy trans isomer.

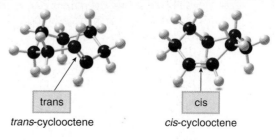

trans   cis

trans-cyclooctene   *cis*-cyclooctene

TABLE 10.1	**Properties of the Carbon–Carbon Double Bond**
Restricted rotation	• **The rotation around the C–C double bond is restricted.** Rotation can only occur if the π bond breaks and then re-forms, a process that is unfavorable (Section 8.2B).
Stereoisomerism	• Whenever the two groups on each end of a C=C are different from each other, two diastereomers are possible. *Cis-* and *trans*-2-butene (drawn at the bottom of Table 10.1) are diastereomers (Section 8.2B).
Stability	• **Trans** alkenes are generally more stable than **cis** alkenes. • **The stability of an alkene increases as the number of R groups on the C=C increases** (Section 8.2C).

| 1-butene | *cis*-2-butene | *trans*-2-butene |

Increasing stability

PROBLEM 10.1   Draw the six alkenes of molecular formula C_5H_{10}. Label one pair of diastereomers.

10.2 Calculating Degrees of Unsaturation

An acyclic alkene has the general molecular formula C_nH_{2n}, giving it two fewer hydrogens than an acyclic alkane with the same number of carbons.

◆ Alkenes are *unsaturated hydrocarbons* because they have fewer than the maximum number of hydrogen atoms per carbon.

In Chapter 12 we will learn how to use the hydrogenation of π bonds to determine how many degrees of unsaturation result from π bonds and how many result from rings.

Cycloalkanes also have the general molecular formula C_nH_{2n}. Thus, **each π bond or ring removes two hydrogen atoms from a molecule, and this introduces one** *degree of unsaturation.* The number of degrees of unsaturation for a given molecular formula can be calculated by comparing the actual number of H atoms in a compound and the maximum number of H atoms possible. Remember that for *n* carbons, the maximum number of H atoms is $2n + 2$ (Section 4.1). This procedure gives the total number of rings and π bonds in a molecule.

SAMPLE PROBLEM 10.1   Calculate the number of degrees of unsaturation in a compound of molecular formula C_4H_6, and propose possible structures.

SOLUTION

[1] Calculate the maximum number of H's possible.

- For *n* carbons, the maximum number of H's is $2n + 2$; in this example, $2n + 2 = 2(4) + 2 = 10$.

[2] Subtract the actual number of H's from the maximum number and divide by two.

- 10 H's (maximum) – 6 H's (actual) = 4 H's fewer than the maximum number.

$$\frac{4 \text{ H's fewer than the maximum}}{2 \text{ H's removed for each degree of unsaturation}} =$$

Two degrees of unsaturation

A compound with two degrees of unsaturation has:

| two rings | *or* | two π bonds | *or* | one ring and one π bond |

Possible structures for C_4H_6:

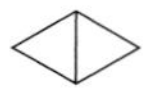

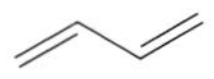

$$H-C \equiv C-CH_2CH_3$$

This procedure can be extended to compounds that contain heteroatoms such as oxygen and halogen, as illustrated in Sample Problem 10.2.

SAMPLE PROBLEM 10.2 Calculate the number of degrees of unsaturation for each molecular formula: [a] C_5H_8O; [b] $C_6H_{11}Cl$. Propose two possible structures for each compound.

SOLUTION

[a] **You can ignore the presence of an O atom** when calculating degrees of unsaturation; that is, use only the given number of C's and H's for the calculation (C_5H_8).

[1] For 5 C's, the maximum number of H's = $2n + 2 = 2(5) + 2 = 12$.
[2] Because the compound contains only 8 H's, it has $12 - 8 = 4$ H's fewer than the maximum number.
[3] Each degree of unsaturation removes 2 H's, so the answer in [2] must be divided by 2.
     Answer: two degrees of unsaturation

---

[b] **A compound with a halogen atom is equivalent to a hydrocarbon having one more H;** that is, $C_6H_{11}Cl$ is equivalent to C_6H_{12} when calculating degrees of unsaturation.

[1] For 6 C's, the maximum number of H's = $2n + 2 = 2(6) + 2 = 14$.
[2] Because the compound contains only 12 H's, it has $14 - 12 = 2$ H's fewer than the maximum number.
[3] Each degree of unsaturation removes 2 H's, so the answer in [2] must be divided by 2.
     Answer: one degree of unsaturation

Two possible structures for [a]:   Two possible structures for [b]:

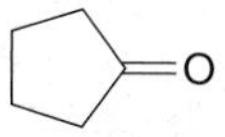

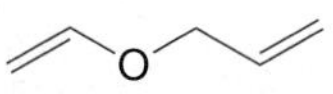

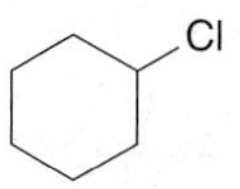

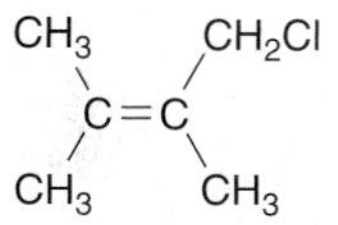

PROBLEM 10.2 How many degrees of unsaturation are present in each compound?
a. C_2H_2 b. C_6H_6 c. C_8H_{18} d. C_7H_8O e. $C_7H_{11}Br$

PROBLEM 10.3 How many rings and π bonds does a compound with molecular formula $C_{10}H_{14}$ possess? List all possibilities.

10.3 Nomenclature

◆ In the IUPAC system, an alkene is identified by the suffix *-ene*.

10.3A General IUPAC Rules

How To Name an Alkene

Example Give the IUPAC name of the following alkene:

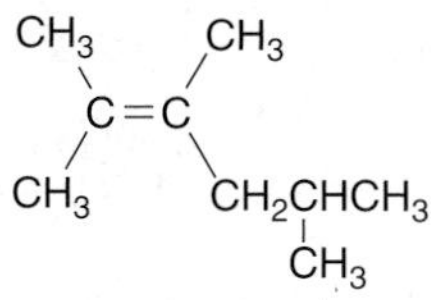

Step [1] **Find the longest chain that contains *both* carbon atoms of the double bond.**

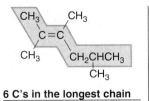

6 C's in the longest chain

hex*ane* - - - → hex*ene*

• Change the *-ane* ending of the parent alkane to *-ene*.

Step [2] **Number the carbon chain to give the double bond the lower number, and apply all other rules of nomenclature.**

[a] **Number** the chain, and name using the *first number* assigned to the C=C.

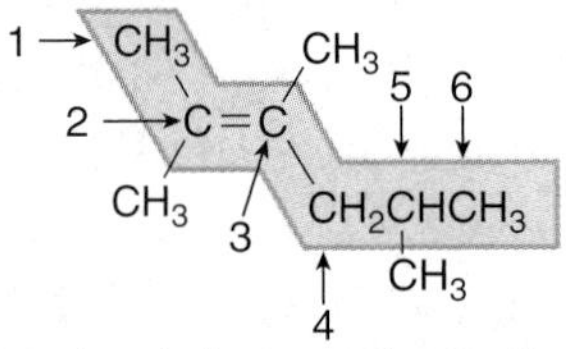

• Number the chain to put the C=C at C2, not C4.

2-hexene

[b] **Name** and **number** the substituents.

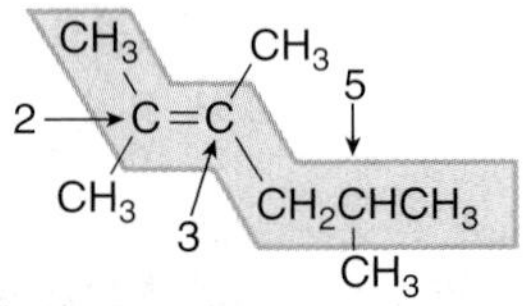

three methyl groups at C2, C3, and C5

Answer: 2,3,5-trimethyl-2-hexene

Compounds with two double bonds are called **dienes,** those with three are called **trienes,** and so forth. Always choose the longest chain that contains *both* atoms of the double bond. In Figure 10.1, the alkene is named as a derivative of heptene because the seven-carbon chain contains both atoms of the double bond, but the eight-carbon chain does not.

In naming cycloalkenes, the double bond is located between C1 and C2, and the 1 is usually omitted in the name. The ring is numbered clockwise or counterclockwise to give the first substituent the lower number. Representative examples are given in Figure 10.2.

Figure 10.1 Naming an alkene in which the longest carbon chain does not contain both atoms of the double bond

CH₂CH₃
CH₂=C
CH₂CH₂CH₂CH₂CH₃

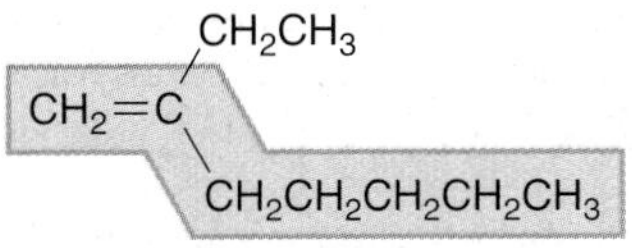

7 C's - - - → **heptene**

Both C's of the C=C are contained in this long chain.

Correct: 2-ethyl-1-heptene

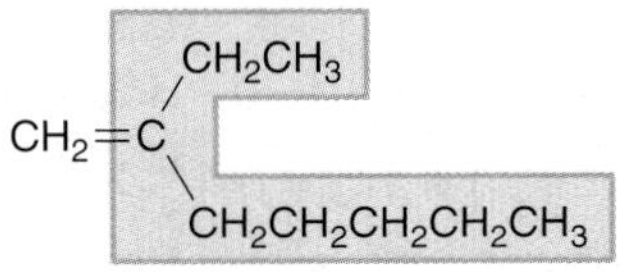

CH₂CH₃
CH₂=C
CH₂CH₂CH₂CH₂CH₃

8 C's

Both C's of the C=C are NOT contained in this long chain.

Incorrect

Figure 10.2 Examples of cycloalkene nomenclature

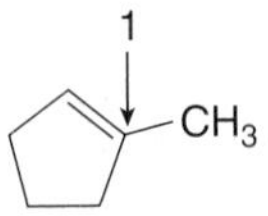

1-methylcyclopentene

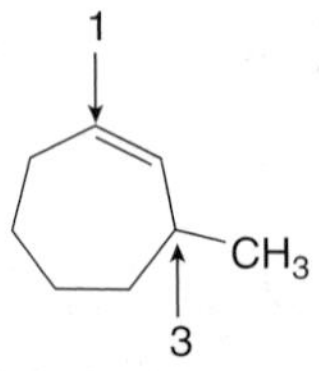

3-methylcycloheptene

Number clockwise beginning at the C=C and place the CH₃ at C3.

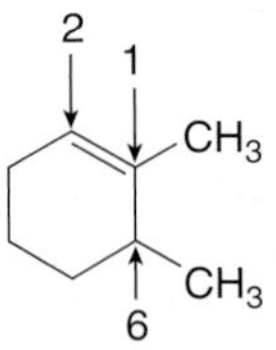

1,6-dimethylcyclohexene

Number counterclockwise beginning at the C=C and place the first CH₃ at C1.

PROBLEM 10.4 Give the IUPAC name for each alkene.

a. $CH_2=CHCH(CH_3)CH_2CH_3$ c.   d.   e.

b. $(CH_3CH_2)_2C=CHCH_2CH_2CH_3$

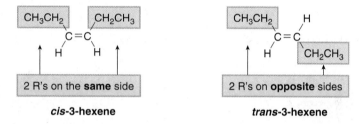

10.3B   Naming Stereoisomers

A prefix is needed to distinguish two alkenes when diastereomers are possible.

Using cis and trans as Prefixes

An alkene having one alkyl group bonded to each carbon atom can be named using the prefixes **cis** and **trans** to designate the relative location of the two alkyl groups. For example, *cis*-3-hexene has two ethyl groups on the **same side** of the double bond, whereas *trans*-3-hexene has two ethyl groups on **opposite sides** of the double bond.

CH_3CH_2 CH_2CH_3 CH_3CH_2 H

C=C   C=C

H H H CH_2CH_3

2 R's on the **same** side   2 R's on **opposite** sides

***cis*-3-hexene**   ***trans*-3-hexene**

Using the Prefixes E and Z

Although the prefixes cis and trans can be used to distinguish diastereomers when two alkyl groups are bonded to the C=C, they cannot be used when there are three or four alkyl groups bonded to the C=C.

CH_3 CH_3 CH_3 CH_2CH_3

C=C   C=C

H CH_2CH_3 H CH_3

A   **B**

3-methyl-2-pentene

> *E* stands for the German word *entgegen* meaning *opposite*. *Z* stands for the German word *zusammen*, meaning *together*.

For example, alkenes **A** and **B** are two *different* compounds that are both called 3-methyl-2-pentene. In **A** the two CH_3 groups are cis, whereas in **B** the CH_3 and CH_2CH_3 groups are cis. The **E,Z system of nomenclature** has been devised to unambiguously name these kinds of alkenes.

How To   Assign the Prefixes E and Z to an Alkene

Step [1]   **Assign priorities to the two substituents on each end of the C=C by using the priority rules for R,S nomenclature.**

- Divide the double bond in half, and assign the numbers **1** and **2** to indicate the relative priority of the two groups on each end—the higher priority group is labeled **1,** and the lower priority group is labeled **2.**

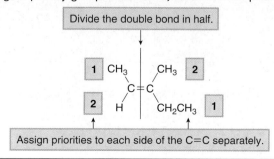

Step [2]   Assign *E* or *Z* based on the location of the two highest priority groups (1).

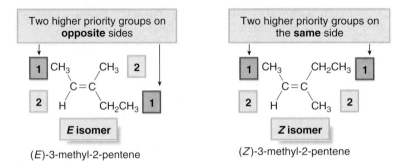

(*E*)-3-methyl-2-pentene   (*Z*)-3-methyl-2-pentene

- The **E** isomer has the two higher priority groups on the **opposite sides.**
- The **Z** isomer has the two higher priority groups on the **same side.**

PROBLEM 10.5   Label each alkene as *E* or *Z*.

Review the rules on assigning priority in Section 5.6.

a.

b.

c.

Tamoxifen binds to estrogen receptors, slowing the growth of some breast tumors.

tamoxifen

PROBLEM 10.6   Give the IUPAC name for each alkene.

a.

b.

PROBLEM 10.7   Draw the structure corresponding to each IUPAC name.
a. (*Z*)-4-ethyl-3-heptene   b. (*E*)-3,5,6-trimethyl-2-octene   c. (*Z*)-2-bromo-1-iodo-1-hexene

---

10.3C   Common Names

Using E,Z nomenclature, a cis isomer has the Z configuration and a trans isomer has the E configuration.

The simplest alkene, **CH₂=CH₂,** named in the IUPAC system as **ethene,** is often called **ethylene,** its common name. The common names for three **alkyl groups** derived from alkenes are also used. Two examples of naming organic molecules using these common names are shown in Figure 10.3.

$CH_2=$

methylene group

vinyl group

allyl group

Figure 10.3 Naming alkenes with common substituent names

methylenecyclohexane

1-vinylcyclohexene

10.4 Physical Properties

Most alkenes exhibit only weak van der Waals interactions, so their physical properties are similar to alkanes of comparable molecular weight.

> ◆ Alkenes have low melting points and boiling points.
> ◆ Melting points and boiling points increase as the number of carbons increases because of increased surface area.
> ◆ Alkenes are soluble in organic solvents and insoluble in water.

Cis and trans alkenes often have somewhat different physical properties. For example, *cis*-2-butene has a higher boiling point (4 °C) than *trans*-2-butene (1 °C). This difference arises because the C–C single bond between an alkyl group and one of the double-bond carbons of an alkene is slightly polar. **The *sp*³ hybridized alkyl carbon donates electron density to the *sp*² hybridized alkenyl carbon.**

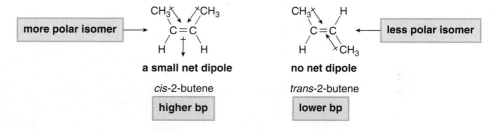

The bond dipole places a partial negative charge on the alkenyl carbon (*sp*²) relative to the alkyl carbon (*sp*³) because an *sp*² hybridized orbital has greater *s*-character (33%) than an *sp*³ hybridized orbital (25%). **In a cis isomer, the two C_{sp^3}–C_{sp^2} bond dipoles reinforce each other, yielding a small net molecular dipole. In a trans isomer, the two bond dipoles cancel.**

more polar isomer →

$$CH_3 \quad CH_3$$
$$C=C$$
$$H \quad H$$

a small net dipole

cis-2-butene

higher bp

← less polar isomer

$$CH_3 \quad H$$
$$C=C$$
$$H \quad CH_3$$

no net dipole

trans-2-butene

lower bp

Related arguments involving C_{sp^3}–C_{sp^2} bonds were used in Section 8.2C to explain why the stability of an alkene increases with increasing alkyl substitution.

> ◆ A cis alkene is more polar than a trans alkene, giving it a slightly higher boiling point and making it more soluble in polar solvents.

PROBLEM 10.8   Rank the following isomers in order of increasing boiling point.

$$CH_3 \quad CH_3$$
$$C=C$$
$$CH_3 \quad CH_3$$

$$CH_3CH_2 \quad CH_2CH_3$$
$$C=C$$
$$H \quad H$$

$$CH_3CH_2 \quad H$$
$$C=C$$
$$H \quad CH_2CH_3$$

10.5 Interesting Alkenes

The alkenes in Figure 10.5 all originate from the same five-carbon starting material in nature, as we will learn in Chapter 26.

Ethylene is prepared from petroleum by a process called **cracking**. Ethylene is the most widely produced organic chemical, serving as the starting material not only for the polymer **polyethylene,** a widely used plastic, but also for many other useful organic compounds, as shown in Figure 10.4.

Numerous organic compounds containing carbon–carbon double bonds have been isolated from natural sources (Figure 10.5).

Figure 10.4 Ethylene, an industrial starting material for many useful products

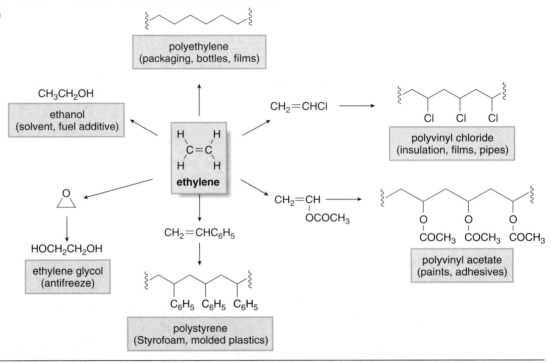

Figure 10.5 Five naturally occurring alkenes

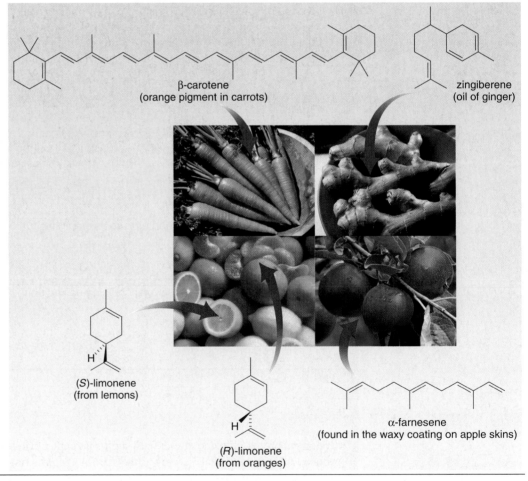

10.6 Lipids—Part 2

Understanding the geometry of C−C double bonds provides an insight into the properties of **triacylglycerols,** the most abundant lipids. Triacylglycerols contain three ester groups, each having a long carbon chain (abbreviated as R, R', and R'') bonded to a carbonyl group (C=O).

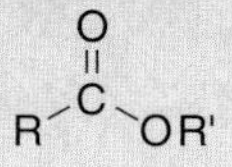

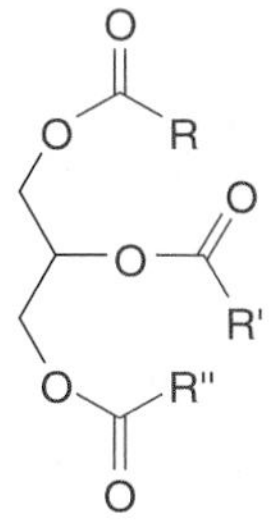

R groups have 11–19 C's.

[Three ester groups are labeled in red.]

triacylglycerol

> Lipids are water insoluble biomolecules composed largely of nonpolar C−C and C−H bonds (Section 4.15).

> General structure of an ester:
>

10.6A   Fatty Acids

Triacylglycerols are hydrolyzed to glycerol (a triol), and three **fatty acids** of general structure RCOOH. Naturally occurring fatty acids contain 12–20 carbon atoms, with a carboxy group (COOH) at one end.

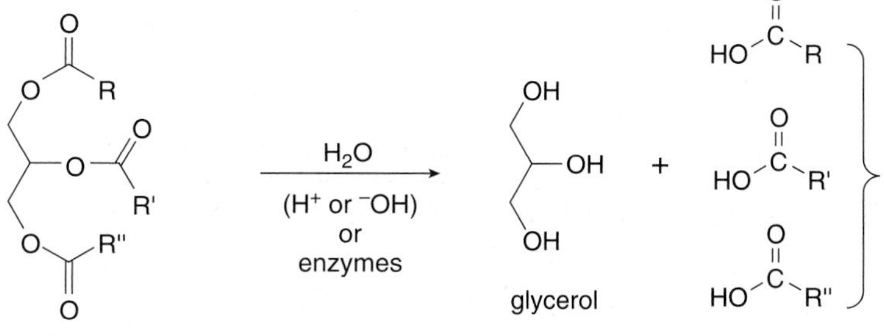

H_2O
(H$^+$ or $^-$OH)
or
enzymes

These fatty acids have **12–20 C's.**

triacylglycerol   glycerol   fatty acids

> Linoleic and linolenic acids are essential fatty acids, meaning they cannot be synthesized in the body and must therefore be obtained in the diet. A common source of these essential fatty acids is whole milk. Babies fed a diet of nonfat milk in their early months do not thrive because they do not obtain enough of these essential fatty acids.

◆ Saturated fatty acids have no double bonds in their long hydrocarbon chains, and unsaturated fatty acids have one or more double bonds in their hydrocarbon chains.
◆ All double bonds in naturally occurring fatty acids have the *Z* configuration.

Table 10.2 lists the structure and melting point of four fatty acids containing 18 carbon atoms. Stearic acid is one of the two most common saturated fatty acids, and oleic and linoleic acids are the most common unsaturated ones. The data show the effect of *Z* double bonds on the melting point of fatty acids.

◆ As the number of double bonds in the fatty acid *increases*, the melting point *decreases*.

TABLE 10.2   The Effect of Double Bonds on the Melting Point of Fatty Acids

Name	Structure	Mp (°C)
Stearic acid (0 C=C)	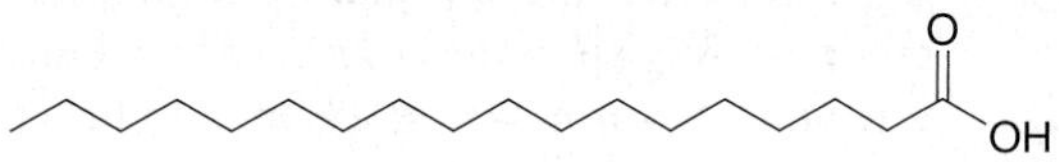	69
Oleic acid (1 C=C)	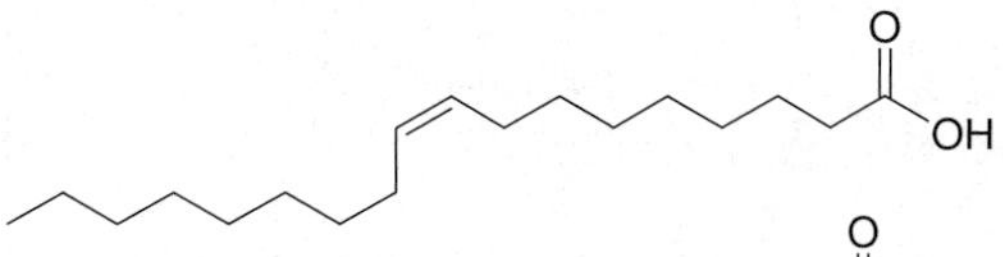	4
Linoleic acid (2 C=C)	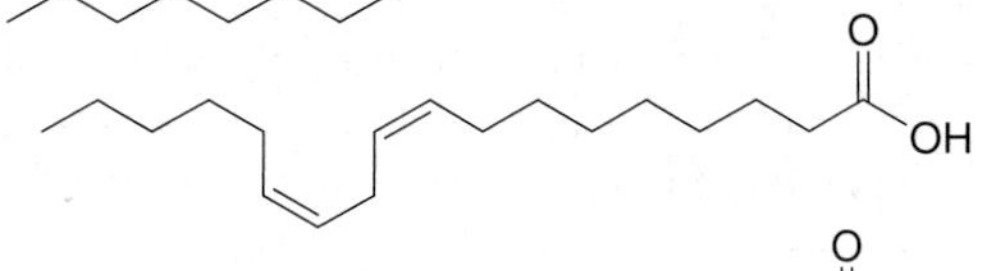	−5
Linolenic acid (3 C=C)	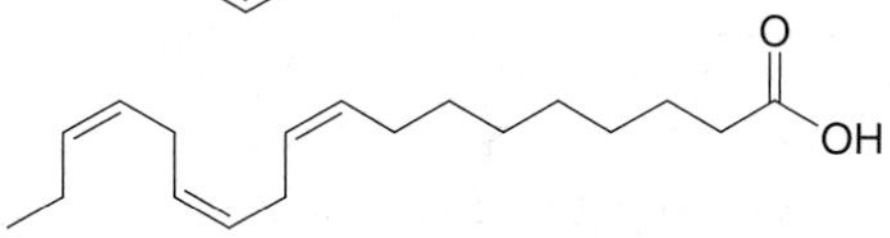	−11

Increasing number of double bonds

Figure 10.6 Three-dimensional structure of four C_{18} fatty acids

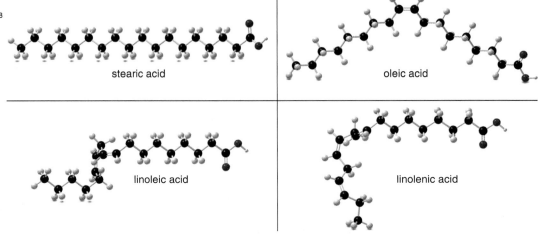

stearic acid

oleic acid

linoleic acid

linolenic acid

The three-dimensional structures of the fatty acids in Figure 10.6 illustrate how *Z* double bonds introduce kinks in the long hydrocarbon chain, decreasing the ability of the fatty acid to pack well in a crystalline lattice. **The larger the number of *Z* double bonds, the more kinks in the hydrocarbon chain, and the lower the melting point.**

PROBLEM 10.9   Explain why arachidonic acid has a lower melting point (–50 °C) than linolenic acid, even though it contains two more carbon atoms.

arachidonic acid

10.6B   Fats and Oils

Fats and **oils** are triacylglycerols with different physical properties.

- ◆ Fats have higher melting points—they are *solids* at room temperature.
- ◆ Oils have lower melting points—they are *liquids* at room temperature.

The identity of the three fatty acids in the triacylglycerol determines whether it is a fat or an oil. **Increasing the number of double bonds in the fatty acid side chains decreases the melting point of the triacylglycerol.**

- ◆ Fats are derived from fatty acids having few double bonds.
- ◆ Oils are derived from fatty acids having a larger number of double bonds.

Saturated fats are typically obtained from animal sources, whereas unsaturated oils are common in vegetable sources. Thus, butter and lard are high in saturated triacylglycerols, and olive oil and safflower oil are high in unsaturated triacylglycerols. An exception to this generalization is coconut oil, which is largely composed of saturated alkyl side chains.

Considerable evidence suggests that an elevated cholesterol level is linked to increased risk of heart disease. Saturated fats stimulate cholesterol synthesis in the liver, thus increasing the cholesterol concentration in the blood.

10.7   Preparation of Alkenes

Recall from Chapters 8 and 9 that alkenes can be prepared from alkyl halides and alcohols via elimination reactions. For example, **dehydrohalogenation of alkyl halides with strong base yields alkenes via an E2 mechanism** (Sections 8.4 and 8.5).

- ◆ Typical bases include ⁻OH and ⁻OR [especially ⁻OC(CH₃)₃], and nonnucleophilic bases such as DBU and DBN.
- ◆ Alkyl tosylates can also be used as starting materials under similar reaction conditions (Section 9.13).

Examples

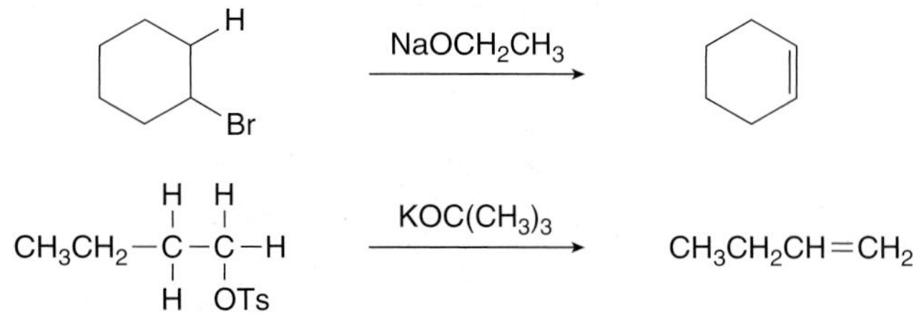

The acid-catalyzed dehydration of alcohols with H_2SO_4 or TsOH yields alkenes, too (Sections 9.8 and 9.9). The reaction occurs via an E1 mechanism for 2° and 3° alcohols, and an E2 mechanism for 1° alcohols. E1 reactions involve carbocation intermediates, so rearrangements are possible. **Dehydration can also be carried out with $POCl_3$ and pyridine** by an E2 mechanism (Section 9.10).

Examples

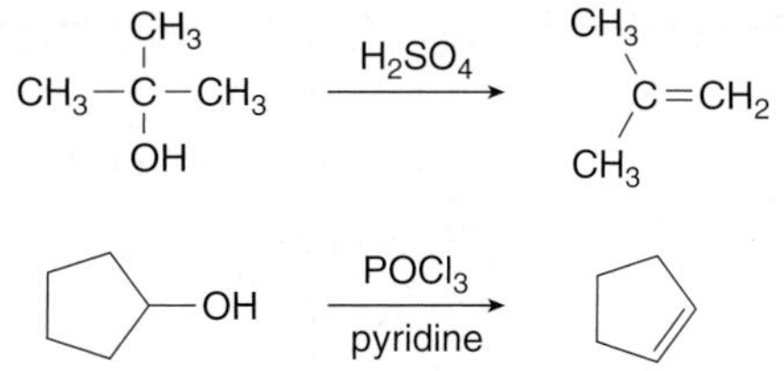

These elimination reactions are stereoselective and regioselective, so the most stable alkene is usually formed as the major product.

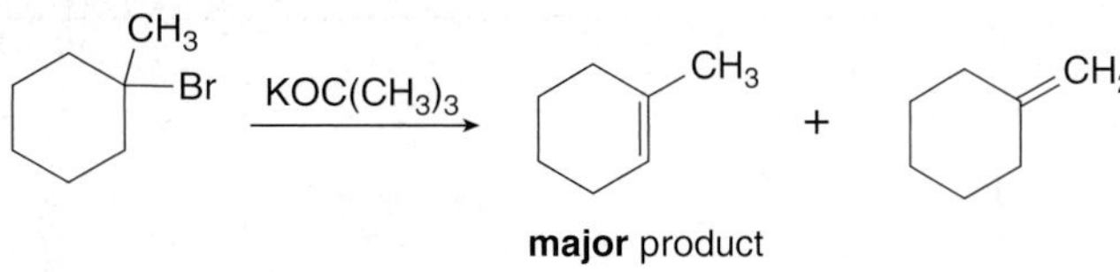

major product

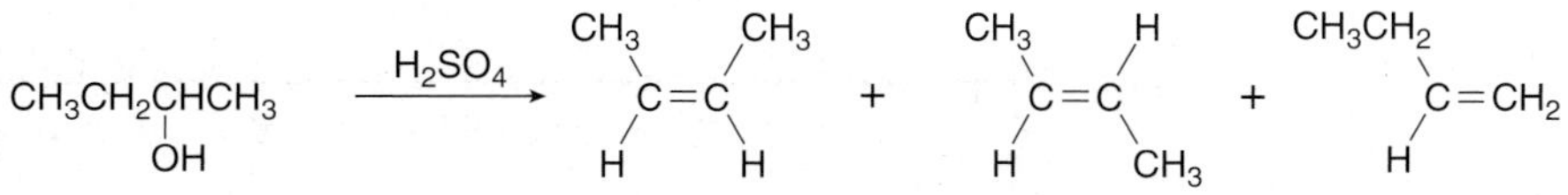

major product
trans disubstituted alkene

PROBLEM 10.10   Draw the products of each elimination reaction.

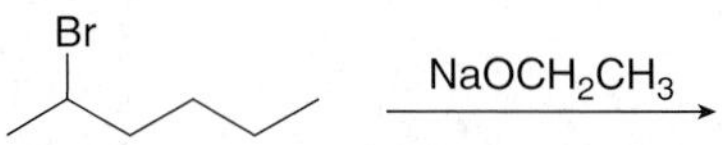

10.8   Introduction to Addition Reactions

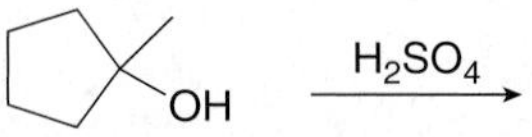

Because the C–C π bond of an alkene is much weaker than a C–C σ bond, the characteristic reaction of alkenes is **addition: the π bond is broken and two new σ bonds are formed.**

A general addition reaction

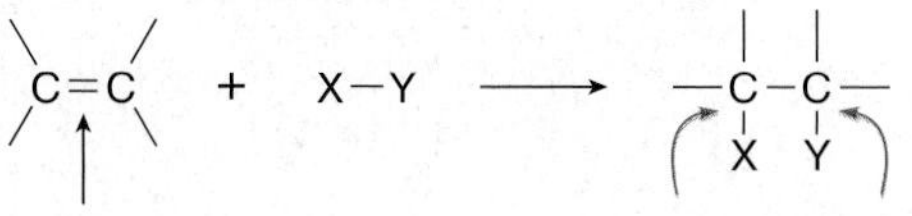

This π bond is broken.   Two σ bonds are formed.

Alkenes are electron rich, as seen in the electrostatic potential plot in Figure 10.7. The electron density of the π bond is concentrated above and below the plane of the molecule, making the π bond more exposed than the σ bond.

What kinds of reagents add to the weak, electron-rich π bond of alkenes? There are many of them, and that can make alkene chemistry challenging. To help you organize this information, keep in mind the following:

Figure 10.7 Electrostatic
potential plot of ethylene

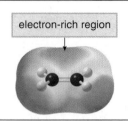

electron-rich region

- The red electron-rich region of the π bond is located
  above and below the plane of the molecule. Because
  the plane of the alkene depicted in this electrostatic
  potential plot is tipped, only the red region above the
  molecule is visible.

The addition reactions of
alkenes are discussed in
Sections 10.9–10.16 and in
Chapter 12 (Oxidation and
Reduction).

◆ **Every reaction of alkenes involves addition: the π bond is always broken.**
◆ **Because alkenes are electron rich, simple alkenes do *not* react with nucleophiles or
bases, reagents that are themselves electron rich. Alkenes react with *electrophiles*.**

The stereochemistry of addition is often important in delineating a reaction's mechanism.
Because the carbon atoms of a double bond are both trigonal planar, the elements of X and Y can
be added to them from the **same side** or from **opposite sides.**

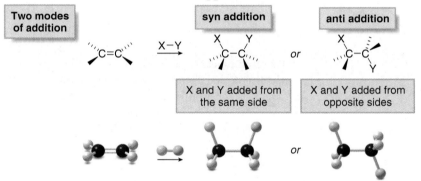

◆ *Syn addition* takes place when both X and Y are added from the *same* side.
◆ *Anti addition* takes place when X and Y are added from *opposite* sides.

Five reactions of alkenes are discussed in Chapter 10 and each is illustrated in Figure 10.8, using
cyclohexene as the starting material.

Figure 10.8 Five addition
reactions of cyclohexene

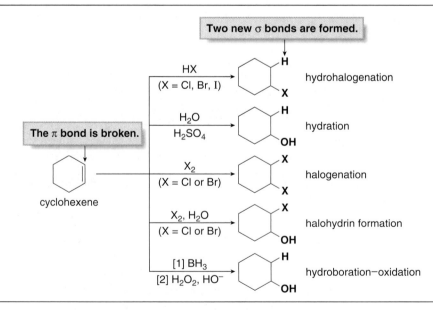

10.9   Hydrohalogenation—Electrophilic Addition of HX

Hydrohalogenation is the addition of hydrogen halides HX (X = Cl, Br, and I) to alkenes to form
alkyl halides.

Hydrohalogenation—General reaction

$$\ce{C=C} + \overset{\delta^+ \;\; \delta^-}{\ce{H-X}} \longrightarrow \ce{-C-C-}$$

(X = Cl, Br, I)   H X ← HX is added.

alkyl halide

This π bond is broken.

Hydrohalogenation of an alkene to form an alkyl halide is the reverse of the dehydrohalogenation of an alkyl halide to form an alkene, a reaction discussed in detail in Sections 8.4 and 8.5.

Two bonds are broken in this reaction—the weak π bond of the alkene and the HX bond—and two new σ bonds are formed—one to H and one to X. Because X is more electronegative than H, the H–X bond is polarized, with a partial positive charge on H. Because the electrophilic (H) end of HX is attracted to the electron-rich double bond, these reactions are called **electrophilic additions.**

To draw the products of an addition reaction:

 ◆ **Locate the C–C double bond.**

 ◆ **Identify the σ bond of the reagent that breaks—namely, the H–X bond in hydro-halogenation.**

 ◆ **Break the π bond of the alkene and the σ bond of the reagent, and form two new σ bonds to the C atoms of the double bond.**

Examples

$$\underset{\underset{H}{|}}{\overset{\overset{H}{|}}{C}}=\underset{\underset{H}{|}}{\overset{\overset{H}{|}}{C}} + \text{HBr} \longrightarrow \ce{H-C-C-H}$$

H Br

$$\text{(cyclohexene)} + \text{HCl} \longrightarrow \text{(cyclohexane with H and Cl)}$$

H   Cl

PROBLEM 10.11   Draw the products of each reaction.

a. (cyclopentene) + HI ⟶

b. (1,2-dimethylcyclohexene) + HCl ⟶   with CH₃ groups

Addition reactions are exothermic because the two σ bonds formed in the product are stronger than the σ and π bonds broken in the reactants. For example, $\Delta H°$ for the addition of HBr to ethylene is –14 kcal/mol, as illustrated in Figure 10.9.

Figure 10.9 The addition of HBr to $CH_2=CH_2$—An exothermic reaction

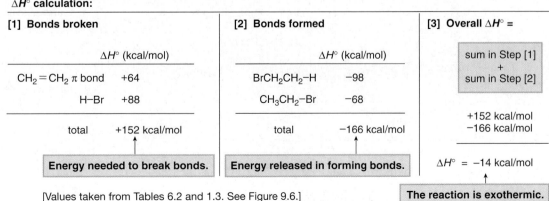

Overall reaction:

$$\underset{\underset{H}{|}}{\overset{\overset{H}{|}}{C}}=\underset{\underset{H}{|}}{\overset{\overset{H}{|}}{C}} + \text{HBr} \longrightarrow \ce{H-C-C-H}$$

H Br

$\Delta H°$ calculation:

[1] Bonds broken		[2] Bonds formed		[3] Overall $\Delta H°$ =
	$\Delta H°$ (kcal/mol)		$\Delta H°$ (kcal/mol)	sum in Step [1] + sum in Step [2]
$CH_2=CH_2$ π bond	+64	$BrCH_2CH_2$–H	–98	
H–Br	+88	CH_3CH_2–Br	–68	
total	+152 kcal/mol	total	–166 kcal/mol	+152 kcal/mol −166 kcal/mol
Energy needed to break bonds.		Energy released in forming bonds.		$\Delta H°$ = −14 kcal/mol

[Values taken from Tables 6.2 and 1.3. See Figure 9.6.]

The reaction is exothermic.

The mechanism of electrophilic addition of HX consists of **two steps:** addition of H⁺ to form a carbocation, followed by nucleophilic attack of X⁻. The mechanism is illustrated for the reaction of *cis*-2-butene with HBr in Mechanism 10.1.

MECHANISM 10.1

Electrophilic Addition of HX to an Alkene

Step [1] Addition of the electrophile (H⁺) to the π bond

◆ The π bond attacks the H atom of HBr, thus forming a new C–H bond while breaking the H–Br bond. Because the remaining carbon atom of the original double bond is left with only six electrons, a **carbocation** intermediate is formed. This step is **rate-determining** because two bonds are broken but only one bond is formed.

Step [2] Nucleophilic attack of Br⁻

◆ **Nucleophilic attack of Br⁻** on the carbocation forms the new C–Br bond.

The mechanism of electrophilic addition consists of two successive Lewis acid–base reactions. In Step [1], the **alkene is the Lewis base** that donates an electron pair to **H–Br, the Lewis acid,** while in Step [2], **Br⁻ is the Lewis base** that donates an electron pair to the **carbocation, the Lewis acid.**

An energy diagram for the reaction of $CH_3CH=CHCH_3$ with HBr is given in Figure 10.10. Each step has its own energy barrier with a transition state at each energy maximum. Because Step [1] has a higher energy transition state, it is rate-determining. $\Delta H°$ for Step [1] is positive because more bonds are broken than formed, whereas $\Delta H°$ for Step [2] is negative because only bond making occurs.

Figure 10.10
Energy diagram for electrophilic addition:
$CH_3CH=CHCH_3$
+ HBr →
$CH_3CH_2CH(Br)CH_3$

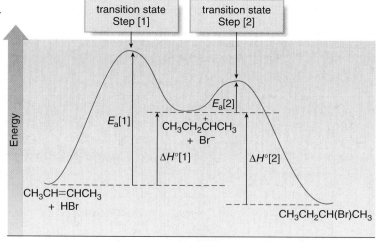

• The mechanism has two steps, so there are two energy barriers.
• Step [1] is rate-determining.

PROBLEM 10.12   Draw a stepwise mechanism for the following reaction.

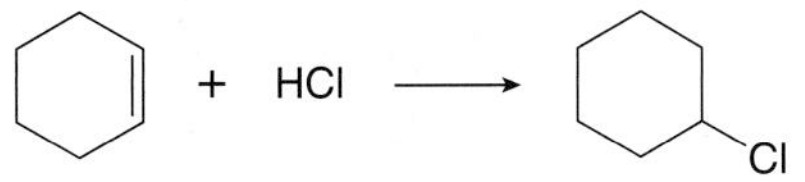

10.10   Markovnikov's Rule

With an unsymmetrical alkene, HX can add to the double bond to give two constitutional isomers.

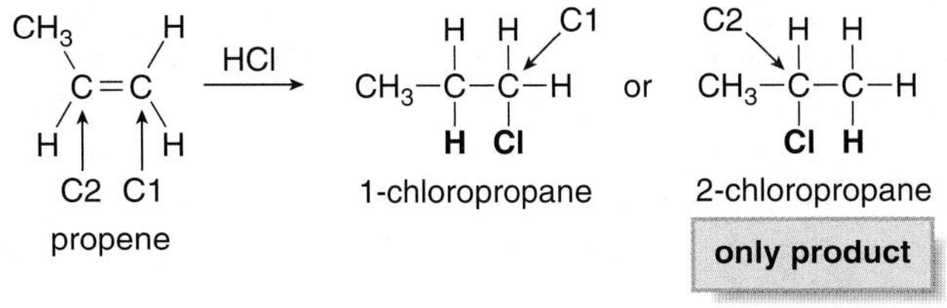

For example, HCl addition to propene could in theory form 1-chloropropane by addition of H and Cl to C2 and C1, respectively, and 2-chloropropane by addition of H and Cl to C1 and C2, respectively. In fact, **electrophilic addition forms *only* 2-chloropropane.** This is a specific example of a general trend called **Markovnikov's rule,** named for the Russian chemist who first determined the regioselectivity of electrophilic addition of HX.

◆ In the addition of HX to an unsymmetrical alkene, the H atom bonds to the less substituted carbon atom—that is, the carbon that has more H atoms to begin with.

The basis of Markovnikov's rule is the formation of a carbocation in the rate-determining step of the mechanism. With propene, there are two possible paths for this first step, depending on which carbon atom of the double bond forms the new bond to hydrogen.

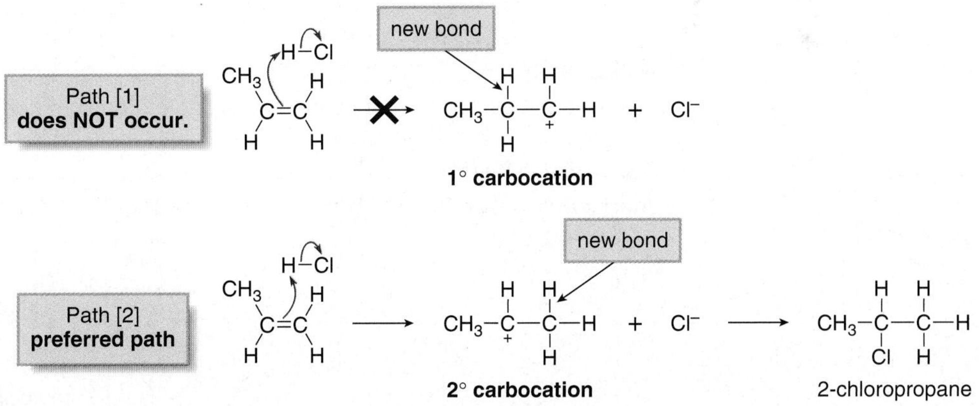

The Hammond postulate was first introduced in Section 7.15 to explain the relative rate of S$_N$1 reactions with 1°, 2°, and 3° RX.

Path [1] forms a highly unstable 1° carbocation, whereas Path [2] forms a **more stable 2° carbocation.** According to the Hammond postulate, Path [2] is preferred because formation of the carbocation is an endothermic process, so **the more stable intermediate is formed faster.** Path [2] is much faster than Path [1] because **the transition state to form the more stable 2° carbocation is lower in energy,** as depicted in the energy diagram in Figure 10.11.

◆ In the addition of HX to an unsymmetrical alkene, the H atom is added to the less substituted carbon to form the more stable, more substituted carbocation.

Similar results are seen in any electrophilic addition involving an intermediate carbocation: **the more stable, more substituted carbocation is formed by addition of the electrophile to the less substituted carbon.**

Figure 10.11   Electrophilic addition and the Hammond postulate

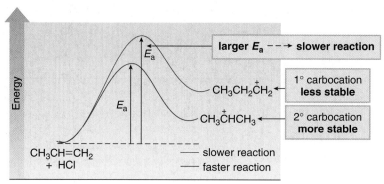

- The E_a for formation of the more stable 2° carbocation is lower than the E_a for formation of the 1° carbocation. The 2° carbocation is formed faster.

PROBLEM 10.13   Draw the products formed when each alkene is treated with HCl.

a. [structure: cyclohexene with CH₃]   b. [structure: $(CH_3)_2C=CH_2$]   c. [structure: alkene]

PROBLEM 10.14   Use the Hammond postulate to explain why $(CH_3)_2C=CH_2$ reacts faster than $CH_3CH=CH_2$ in electrophilic addition of HX.

Because carbocations are formed as intermediates in hydrohalogenation, carbocation rearrangements can occur, as illustrated in Sample Problem 10.3.

SAMPLE PROBLEM 10.3   Draw a stepwise mechanism for the following reaction.

$$CH_3-\underset{\underset{CH_3}{|}}{\overset{\overset{CH_3}{|}}{C}}-CH=CH_2 \xrightarrow{HBr} CH_3-\underset{\underset{Br}{|}}{\overset{\overset{CH_3}{|}}{C}}-\underset{\underset{CH_3}{|}}{\overset{\overset{H}{|}}{C}}-CH_3$$

SOLUTION

Because the carbon skeletons of the starting material and product are different—the alkene reactant has a 4° carbon and the product alkyl halide does not—a carbocation rearrangement must have occurred.

Step [1]   **Markovnikov addition of HBr** adds H⁺ to the less substituted end of the double bond, forming a 2° carbocation.

$$CH_3-\underset{\underset{CH_3}{|}}{\overset{\overset{CH_3}{|}}{\underset{4°}{C}}}-CH=CH_2 \xrightarrow{[1]} CH_3-\underset{\underset{CH_3}{|}}{\overset{\overset{CH_3}{|}}{C}}-\overset{+}{\underset{\underset{H}{}}{C}}-CH_2 \quad + \quad Br^-$$

H—Br

2° carbocation   new bond

Steps [2] and [3]:   **Rearrangement of the 2° carbocation by a 1,2-methyl shift forms a more stable 3° carbocation.** Nucleophilic attack of Br⁻ forms the product, a 3° alkyl halide.

$$CH_3-\underset{\underset{CH_3}{|}}{\overset{\overset{CH_3}{|}}{C}}-\overset{+}{C}-CH_3 \xrightarrow{[2]} CH_3-\overset{+}{C}-\underset{\underset{CH_3}{|}}{\overset{\overset{H}{|}}{C}}-CH_3 \xrightarrow{[3]} CH_3-\underset{\underset{Br}{|}}{\overset{\overset{CH_3}{|}}{C}}-\underset{\underset{CH_3}{|}}{\overset{\overset{H}{|}}{C}}-CH_3$$

1,2-CH₃ shift   Br⁻   3° carbocation   nucleophilic attack   3° alkyl halide

---

PROBLEM 10.15   Treatment of 3-methylcyclohexene with HCl yields two products, 1-chloro-3-methylcyclohexane and 1-chloro-1-methylcyclohexane. Draw a mechanism to explain this result.

10.11 Stereochemistry of Electrophilic Addition of HX

To understand the stereochemistry of electrophilic addition, recall two stereochemical principles learned in Chapters 7 and 9.

> ◆ Trigonal planar atoms react with reagents from two directions with equal probability (Section 7.13C).
> ◆ Achiral starting materials yield achiral or racemic products (Section 9.15).

Many hydrohalogenation reactions begin with an **achiral reactant** and form an **achiral product.** For example, the addition of HBr to cyclohexene, an achiral alkene, forms bromocyclohexane, an achiral alkyl halide.

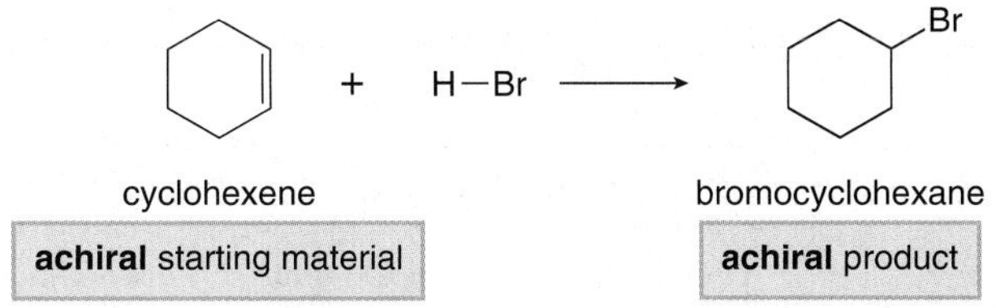

Because addition converts sp^2 hybridized carbons to sp^3 hybridized carbons, however, sometimes new stereogenic centers are formed from hydrohalogenation. For example, Markovnikov addition of HCl to 2-ethyl-1-pentene, an achiral alkene, forms one constitutional isomer, 3-chloro-3-methylhexane. Because this product now has a stereogenic center at one of the newly formed sp^3 hybridized carbons, **an equal amount of two enantiomers—a racemic mixture—**must form.

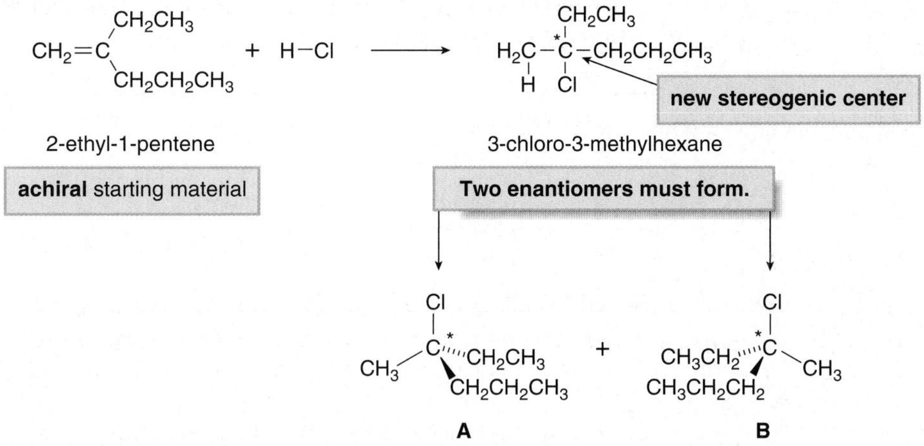

The mechanism of hydrohalogenation illustrates why two enantiomers are formed. Initial addition of the electrophile H^+ (from HCl) occurs from **either side of the planar double bond** to form a carbocation. Both modes of addition (from above and below) generate the same **achiral carbocation.** Either representation of this carbocation can then be used to draw the second step of the mechanism.

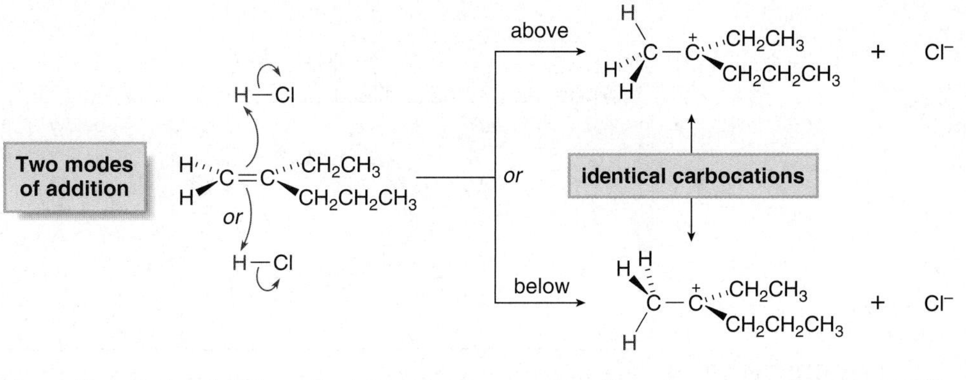

Nucleophilic attack of Cl^- on the trigonal planar carbocation also occurs from two different directions, forming two products, **A** and **B,** having a new stereogenic center. **A** and **B** are not superimposable, so they are **enantiomers.** Because attack from either direction occurs with equal probability, a **racemic mixture** of **A** and **B** is formed.

above —

$=$ **A** | **syn addition** |

H and Cl are added from the **same side.**

or — **enantiomers**

below —

$=$ **B** | **anti addition** |

H and Cl are added from **opposite sides.**

Because hydrohalogenation begins with a **planar** double bond and forms a **planar** carbocation, addition of H and Cl occurs in two different ways. The elements of H and Cl can both be added from the same side of the double bond—that is, **syn addition**—or they can be added from opposite sides—that is, **anti addition.** *Both* modes of addition occur in this two-step reaction mechanism.

◆ **Hydrohalogenation occurs with syn and anti addition of HX.**

The terms **cis** and **trans** refer to the arrangement of groups in a particular compound, usually an alkene or a disubstituted cycloalkane. The terms **syn** and **anti** describe the stereochemistry of a process—for example, how two groups are added to a double bond.

Finally, addition of HCl to 1,2-dimethylcyclohexene forms **two** new stereogenic centers. Initial addition of H⁺ (from HCl) forms two enantiomeric carbocations that react with the Cl⁻ nucleophile from two different directions, forming **four stereoisomers, A–D—two pairs of enantiomers** (Figure 10.12).

1,2-dimethyl-
cyclohexene

HCl →

| Two new stereogenic centers are formed. |

[* denotes a stereogenic center]

Figure 10.12   Reaction of 1,2-dimethylcyclohexene with HCl

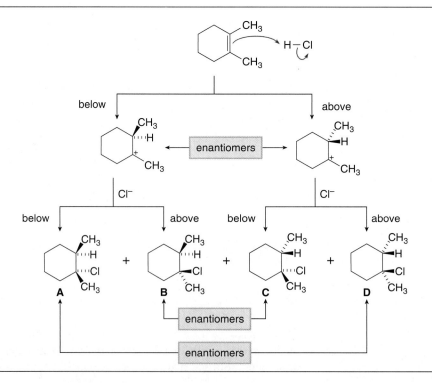

Four stereoisomers are formed:
• Compounds **A** and **D** are enantiomers formed in equal amounts.
• Compounds **B** and **C** are enantiomers formed in equal amounts.

Table 10.3 summarizes the characteristics of electrophilic addition of HX to alkenes.

TABLE 10.3	Summary: Electrophilic Addition of HX to Alkenes
	Observation
Mechanism	• The mechanism involves two steps. • The rate-determining step forms a carbocation. • Rearrangements can occur.
Regioselectivity	• Markovnikov's rule is followed. In unsymmetrical alkenes, H bonds to the less substituted C to form the more stable carbocation.
Stereochemistry	• Syn and anti addition occur.

PROBLEM 10.16 Draw the products, including stereochemistry, of each reaction.

a. 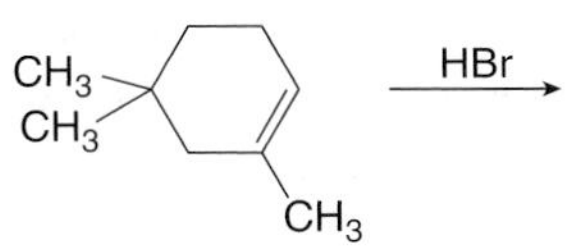 (structure with CH_3, CH_3, CH_3) $\xrightarrow{\text{HBr}}$ b. 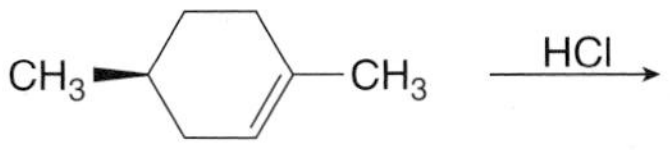 CH_3—(cyclohexene)—CH_3 $\xrightarrow{\text{HCl}}$

PROBLEM 10.17 Which compounds (**A–D**) in Figure 10.12 are formed by syn addition of HCl and which are formed by anti addition?

10.12 Hydration—Electrophilic Addition of Water

Hydration is the addition of water to an alkene to form an alcohol. H_2O itself is too weak an acid to protonate an alkene, but with added H_2SO_4, H_3O^+ is formed and addition readily occurs.

Hydration— General reaction

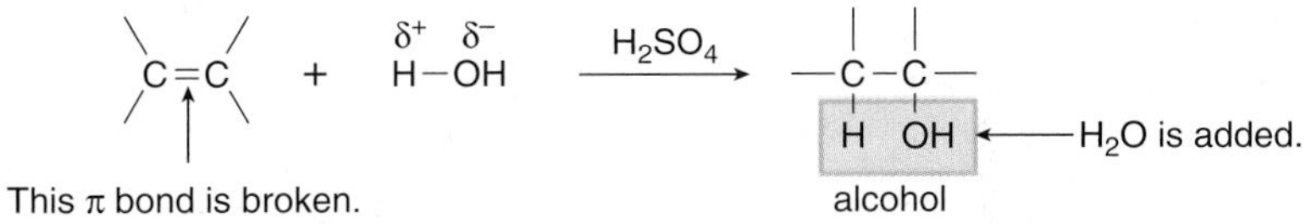

$$\underset{\text{This } \pi \text{ bond is broken.}}{\overset{}{\text{C}=\text{C}}} + \overset{\delta^+ \ \delta^-}{\text{H}-\text{OH}} \xrightarrow{H_2SO_4} \underset{\underset{\text{alcohol}}{\text{H} \ \text{OH}}}{-\text{C}-\text{C}-} \longleftarrow H_2O \text{ is added.}$$

Examples

$$CH_3CH_2CH=CH_2 + H_2O \xrightarrow{H_2SO_4} CH_3CH_2-CH-CH_2 \quad (\text{HO} \ \ \text{H})$$

$$\text{(cyclohexene with } CH_3 \text{ and H)} + H_2O \xrightarrow{H_2SO_4} \text{(cyclohexane with } CH_3, \text{OH, H)}$$

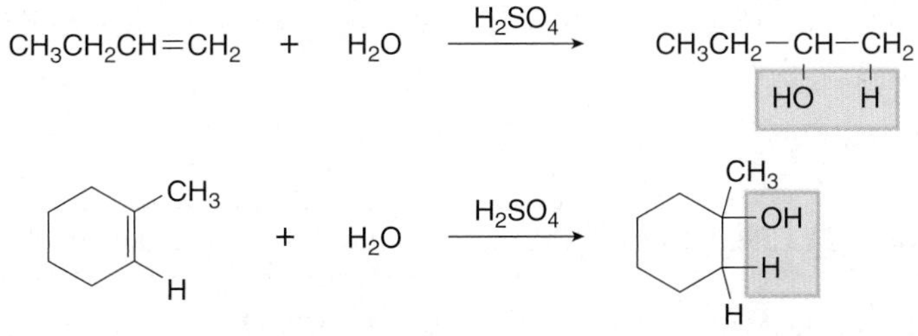

Hydration is simply another example of **electrophilic addition.** The first two steps of the mechanism are similar to those of electrophilic addition of HX—that is, addition of H^+ (from H_3O^+) to generate a carbocation, followed by nucleophilic attack of H_2O. Mechanism 10.2 illustrates the addition of H_2O to cyclohexene to form cyclohexanol.

There are three consequences to the formation of carbocation intermediates:

Hydration of an alkene to form an alcohol is the reverse of the dehydration of an alcohol to form an alkene, a reaction discussed in detail in Section 9.8.

◆ In unsymmetrical alkenes, H adds to the less substituted carbon to form the more stable carbocation; that is, Markovnikov's rule holds.
◆ Addition of H and OH occurs in both a syn and anti fashion.
◆ Carbocation rearrangements can occur.

Alcohols add to alkenes, forming ethers, using the same mechanism. Addition of CH_3OH to 2-methylpropene, for example, forms *tert*-butyl methyl ether (**MTBE**), a high octane fuel additive described in Section 3.5C.

MECHANISM 10.2

Electrophilic Addition of H₂O to an Alkene—Hydration

Step [1] Addition of the electrophile (H⁺) to the π bond

cyclohexene → carbocation

- The π bond attacks H_3O^+, thus forming a new C–H bond while breaking the H–O bond. Because the remaining carbon atom of the original double bond is left with only six electrons, a **carbocation** intermediate is formed. This step is **rate-determining** because two bonds are broken but only one bond is formed.

Step [2] Nucleophilic attack of H₂O

- **Nucleophilic attack of H₂O** on the carbocation forms the new C–O bond.

Step [3] Loss of a proton

cyclohexanol

- Removal of a proton with a base (H₂O) forms a neutral alcohol. Because the acid used in Step [1] is regenerated in Step [3], hydration is **acid-catalyzed.**

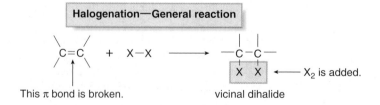

CH_3
$C=CH_2$ + CH_3O-H $\xrightarrow{H_2SO_4}$ CH_3-C-CH_2 ← **an ether**
CH_3 methanol

tert-butyl methyl ether
MTBE

PROBLEM 10.18   What two alkenes give rise to each alcohol as the major product by acid-catalyzed hydration?

a. $CH_3-\underset{\underset{OH}{|}}{\overset{\overset{CH_3}{|}}{C}}-CH_2CH_2CH_3$ b. (cyclohexane with CH₃ and OH)   c. (structure with OH)

PROBLEM 10.19   What stereoisomers are formed when 1-pentene is treated with H₂O and H₂SO₄?

10.13 Halogenation—Addition of Halogen

Halogenation is the addition of halogen X₂ (X = Cl or Br) to an alkene, forming a **vicinal dihalide.**

Halogenation—General reaction

$C=C$ + X–X ⟶ –C–C–
                              X X   ← X₂ is added.

This π bond is broken.   vicinal dihalide

Halogenation is synthetically useful only with Cl_2 and Br_2, because the addition of I_2 is often too slow and the addition of F_2 is too explosive. The dichlorides and dibromides formed in this reaction serve as starting materials for the synthesis of alkynes, as we learned in Section 8.10.

Bromination is a simple chemical test for the presence of π bonds in unknown compounds. When bromine, a fuming red liquid, is added to an alkene, the bromine adds to the double bond and the red color disappears. The disappearance of the red color is therefore a positive test for π bonds.

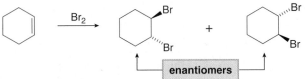

| Examples |

Halogens add to π bonds because halogens are **polarizable.** The electron-rich double bond induces a dipole in an approaching halogen molecule, making one halogen atom electron deficient and the other electron rich. **The electrophilic halogen atom is then attracted to the nucleophilic double bond,** making addition possible.

induced dipole

The electron-rich double bond induces a dipole.

The electrophilic end of X₂ is now able to react with the nucleophilic C=C.

Two facts demonstrate that halogenation follows a different mechanism from that of hydrohalogenation or hydration. First, no rearrangements occur, and second, only anti addition of X₂ is observed. For example, treatment of cyclohexene with Br₂ yields two **trans** enantiomers formed by **anti addition.**

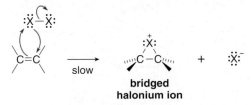

| enantiomers |

These facts suggest that **carbocations are *not* intermediates in halogenation.** Unstable carbocations rearrange, and both syn and anti addition is possible with carbocation intermediates. The accepted mechanism for halogenation comprises **two steps,** but it does *not* proceed with formation of a carbocation, as shown in Mechanism 10.3.

MECHANISM 10.3

Addition of X₂ to an Alkene—Halogenation

Step [1] Addition of the electrophile (X⁺) to the π bond

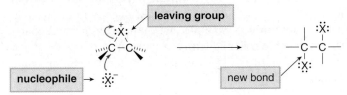

bridged
halonium ion

- Four bonds are broken or formed in this step: the electron pair in the π bond and a lone pair on a halogen atom are used to form two new C−X bonds. The X−X bond is also cleaved heterolytically, forming X⁻. This step is rate-determining.
- The three-membered ring containing a positively charged halogen atom is called a **bridged halonium ion.** This strained three-membered ring is highly unstable, making it amenable to opening of the ring in the second step.

Step [2] Nucleophilic attack of X⁻

leaving group

nucleophile

new bond

- **Nucleophilic attack of X⁻** opens the ring of the halonium ion, forming a new C−X bond and relieving the strain in the three-membered ring.

Bridged halonium ions resemble carbocations in that they are short-lived intermediates that react readily with nucleophiles. Carbocations are inherently unstable because only six electrons surround carbon, whereas **halonium ions are unstable because they contain a strained three-membered ring** with a positively charged halogen atom.

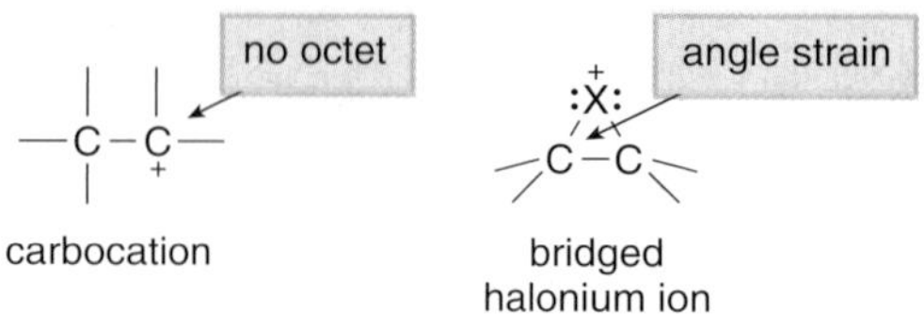

no octet

carbocation

angle strain

bridged
halonium ion

PROBLEM 10.20   Draw the halonium ion formed when Br_2 reacts with $(CH_3)_2C=CH_2$.

PROBLEM 10.21   Draw the transition state for each step in the general mechanism for the halogenation of an alkene.

10.14   Stereochemistry of Halogenation

How does the proposed mechanism invoking a bridged halonium ion intermediate explain the observed **trans products of halogenation**? For example, chlorination of cyclopentene affords both enantiomers of *trans*-1,2-dichlorocyclopentane, with *no* cis products.

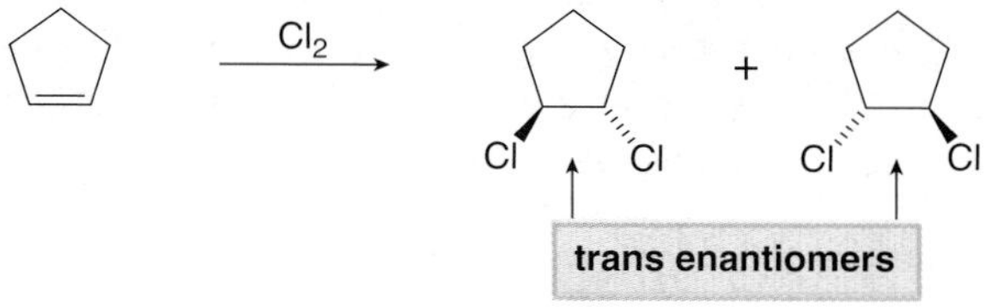

trans enantiomers

Initial addition of the electrophile Cl^+ (from Cl_2) occurs from either side of the planar double bond to form the bridged chloronium ion. In this example, both modes of addition (from above and below) generate the same **achiral** intermediate, so either representation can be used to draw the second step.

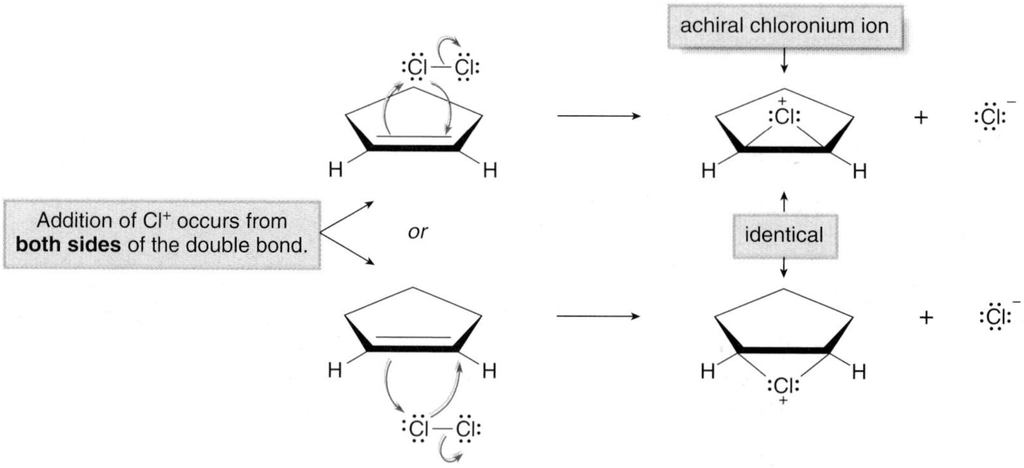

Addition of Cl^+ occurs from **both sides** of the double bond.

or

achiral chloronium ion

identical

In the second step, **nucleophilic attack of Cl^- must occur from the back side**—that is, from the side of the five-membered ring opposite to the side having the bridged chloronium ion. Because the nucleophile attacks from below in this example and the leaving group departs from above, the two Cl atoms in the product are oriented **trans** to each other. Backside attack occurs with equal probability at either carbon of the three-membered ring to yield an equal amount of two enantiomers—a **racemic mixture.**

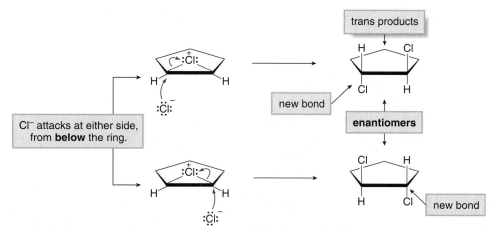

The ring-opening of bridged halonium ion intermediates resembles the opening of epoxide rings with nucleophiles discussed in Section 9.15.

In summary, the mechanism for halogenation of alkenes occurs in two steps:

◆ Addition of X⁺ forms an unstable bridged halonium ion in the rate-determining step.
◆ Nucleophilic attack of X⁻ occurs from the back side to form trans products. The overall result is anti addition of X_2 across the double bond.

PROBLEM 10.22   Draw the products of each reaction, including stereochemistry.

a. [] $\xrightarrow{Br_2}$ b. [cyclohexene] $\xrightarrow{Cl_2}$

Because halogenation occurs exclusively in an anti fashion, cis and trans alkenes yield different stereoisomers. Halogenation of alkenes is a **stereospecific reaction.**

◆ A reaction is stereospecific when each of two specific stereoisomers of a starting material yields a particular stereoisomer of a product.

cis-2-Butene yields two enantiomers, whereas *trans*-2-butene yields a single achiral meso compound, as shown in Figure 10.13.

Figure 10.13   Halogenation of *cis*- and *trans*-2-butene

CH₃ C=C CH₃ / H H   *cis*-2-butene $\xrightarrow{Br_2}$ Br CH₃ / CH₃ C-C H / H Br   +   CH₃ Br / H C-C CH₃ / Br H

enantiomers

achiral alkenes

CH₃ C=C H / H CH₃   *trans*-2-butene $\xrightarrow{Br_2}$ Br H / CH₃ C-C CH₃ / H Br   +   CH₃ Br / H C-C H / Br CH₃

an achiral **meso** compound

To draw the products of halogenation:
• Add Br_2 in an **anti** fashion across the double bond, leaving all other groups in their original orientations. Draw the products such that a given Br atom is above the plane in one product and below the plane in the other product.
• Sometimes this reaction produces two stereoisomers, as in the case of *cis*-2-butene, which forms an equal amount of two enantiomers. Sometimes it produces a single compound, as in the case of *trans*-2-butene, where a meso compound is formed.

PROBLEM 10.23 Draw all stereoisomers formed in each reaction.

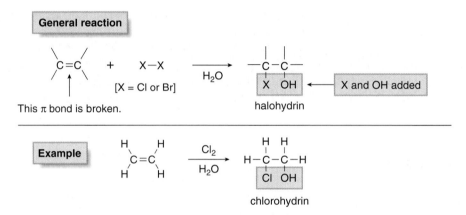

PROBLEM 10.24 Draw a stepwise mechanism for the conversion of *trans*-2-butene to the meso dibromide in Figure 10.13.

10.15 Halohydrin Formation

Treatment of an alkene with a halogen X_2 and H_2O forms a **halohydrin** by addition of the elements of **X** and **OH** to the double bond.

General reaction

$$\text{C=C} + \text{X—X} \xrightarrow{\text{H}_2\text{O}} -\overset{|}{\underset{X}{C}}-\overset{|}{\underset{OH}{C}}-$$

This π bond is broken.   [X = Cl or Br]   X and OH added → halohydrin

Example

$$\text{H}_2\text{C=CH}_2 \xrightarrow[\text{H}_2\text{O}]{\text{Cl}_2} \text{H—}\overset{|}{\underset{Cl}{C}}\text{H—}\overset{|}{\underset{OH}{C}}\text{H—H}$$

chlorohydrin

The mechanism for halohydrin formation is similar to the mechanism for halogenation: addition of the electrophile X^+ (from X_2) to form a **bridged halonium ion,** followed by nucleophilic attack by H_2O from the back side on the three-membered ring (Mechanism 10.4). Even though X^- is formed in Step [1] of the mechanism, its concentration is small compared to H_2O (often the solvent), so H_2O and *not* X^- is the nucleophile.

MECHANISM 10.4

The Mechanism of Halohydrin Formation

Step [1] Addition of the electrophile (X^+) to the π bond

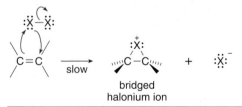

◆ Four bonds are broken or formed in this step: the electron pair in the π bond and a lone pair on a halogen atom are used to form two new C–X bonds in the bridged halonium ion. The X–X bond is also cleaved heterolytically, forming X^-. This step is rate-determining.

Steps [2] and [3] Nucleophilic attack of H_2O and loss of a proton

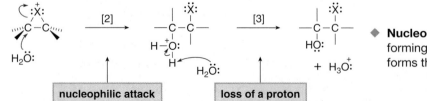

◆ **Nucleophilic attack of H_2O** opens the halonium ion ring, forming a new C–X bond. Subsequent loss of a proton forms the neutral halohydrin.

Although the combination of Br_2 and H_2O effectively forms bromohydrins from alkenes, other reagents can also be used. Bromohydrins are also formed with **N-bromosuccinimide** (abbreviated as **NBS**) in **aqueous DMSO** [$(CH_3)_2S=O$]. In H_2O, NBS decomposes to form Br_2, which then goes on to form a bromohydrin by the same reaction mechanism.

Recall from Section 7.8C that DMSO (dimethyl sulfoxide) is a polar aprotic solvent.

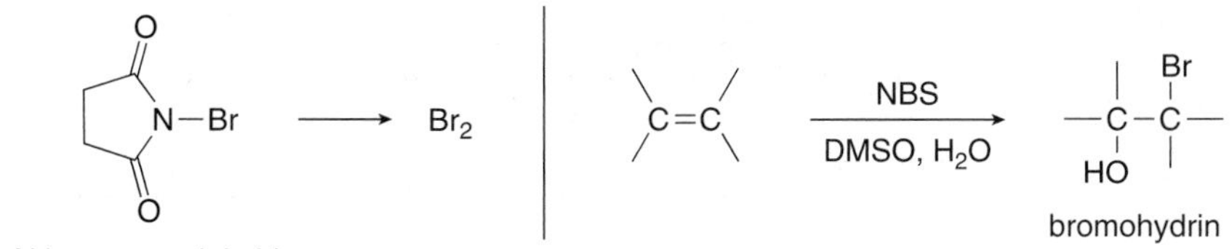

N-bromosuccinimide
NBS

10.15A   Stereochemistry and Regioselectivity of Halohydrin Formation

Because the bridged halonium ion ring is opened by backside attack of H_2O, addition of X and OH occurs in an **anti** fashion and **trans** products are formed.

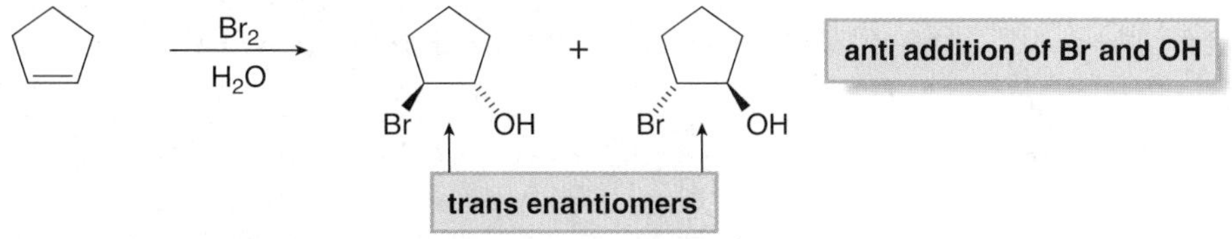

anti addition of Br and OH

trans enantiomers

SAMPLE PROBLEM 10.4   Draw the product of the following reaction, including stereochemistry.

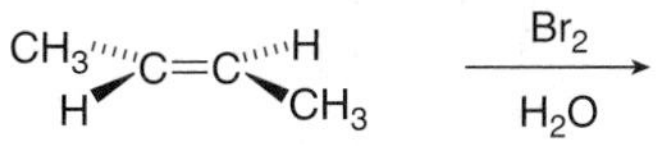

trans-2-butene

SOLUTION

The reagent ($Br_2 + H_2O$) adds the elements of Br and OH to a double bond in an **anti** fashion—that is, from **opposite** sides. To draw two products of anti addition: add Br from above and OH from below in one product; then add Br from below and OH from above in the other product. In this example, the two products are nonsuperimposable mirror images—**enantiomers.**

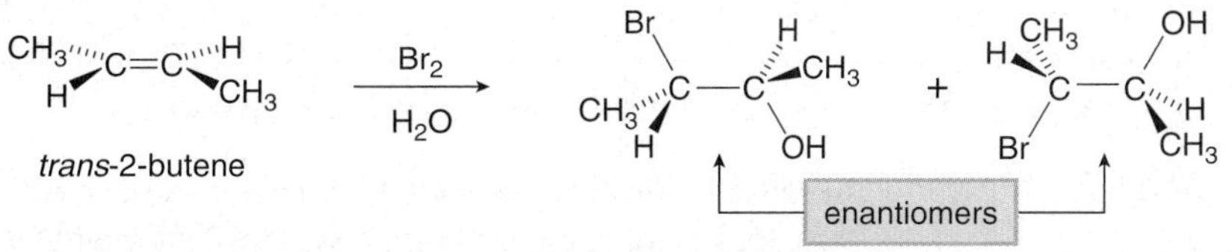

trans-2-butene

enantiomers

With unsymmetrical alkenes, two constitutional isomers are possible from addition of X and OH, but only one is formed. **The preferred product has the electrophile X^+ bonded to the less substituted carbon atom**—that is, the carbon that has more H atoms to begin with in the reacting alkene. Thus, the **nucleophile (H_2O) bonds to the more substituted carbon.**

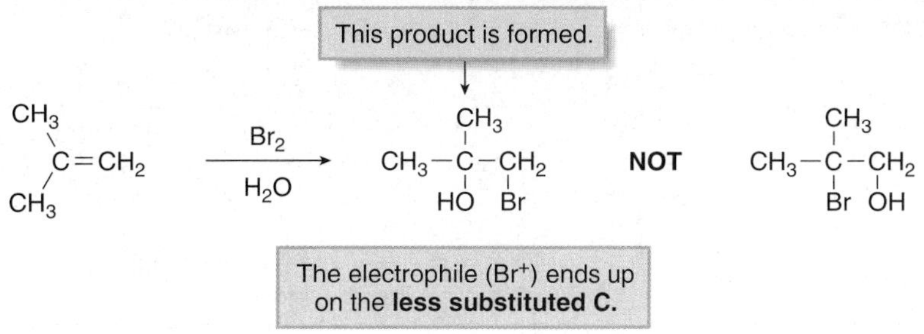

This product is formed.

The electrophile (Br^+) ends up on the **less substituted C.**

This result is reminiscent of the opening of epoxide rings with acids HZ (Z = a nucleophile), which we encountered in Section 9.15B. As in the opening of an epoxide ring, **nucleophilic attack occurs at the more substituted carbon end of the bridged halonium ion** because that carbon is better able to accommodate a partial positive charge in the transition state.

Halohydrin formation in an unsymmetrical alkene

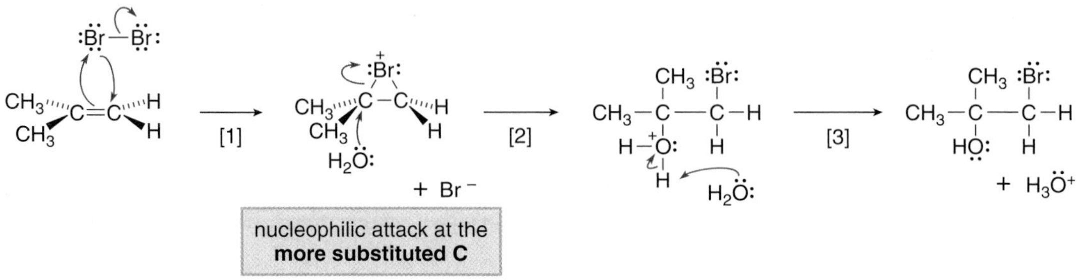

Table 10.4 summarizes the characteristics of halohydrin formation.

TABLE 10.4   Summary: Conversion of Alkenes to Halohydrins

	Observation
Mechanism	• The mechanism involves two steps. • The rate-determining step forms a bridged halonium ion. • No rearrangements can occur.
Regioselectivity	• Markovnikov's rule is followed. X$^+$ bonds to the less substituted carbon.
Stereochemistry	• Anti addition occurs.

PROBLEM 10.25   Draw the products of each reaction and indicate their stereochemistry.

PROBLEM 10.26   The elements of Br and Cl are added to a double bond with Br$_2$ + NaCl. Draw the product formed when an unsymmetrical alkene such as 2-methyl-1-propene is used as the starting material.

10.15B   Halohydrins: Useful Compounds in Organic Synthesis

Because halohydrins are easily converted to epoxides by intramolecular S$_N$2 reaction (Section 9.6), they have been used in the synthesis of many naturally occurring compounds. Key steps in the synthesis of estrone, a female sex hormone, are illustrated in Figure 10.14.

Figure 10.14 The synthesis of estrone from a chlorohydrin

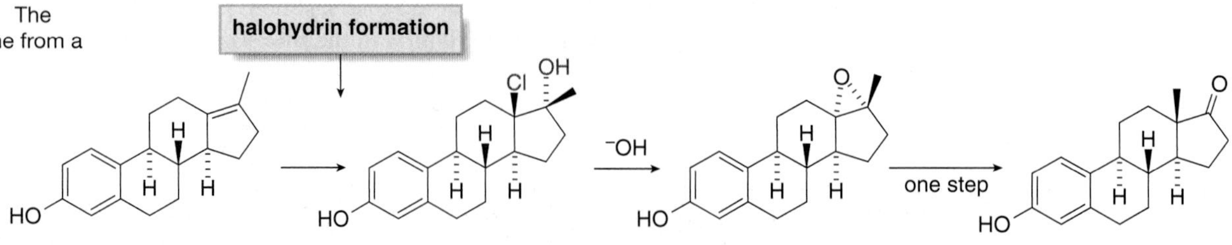

• Chlorohydrin **B**, prepared from alkene **A** by addition of Cl and OH, is converted to epoxide **C** with base. **C** is converted to estrone in one step.

10.16   Hydroboration–Oxidation

Hydroboration–oxidation is a two-step reaction sequence that converts an alkene to an alcohol.

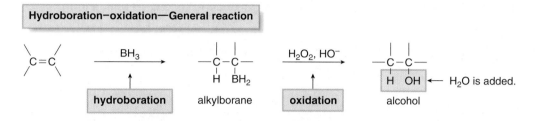

Hydroboration–oxidation—General reaction

- ◆ *Hydroboration* is the addition of borane (BH_3) to an alkene, forming an alkylborane.
- ◆ *Oxidation* converts the C–B bond of the alkylborane to a C–O bond.

Hydroboration–oxidation results in **addition of H_2O** to an alkene.

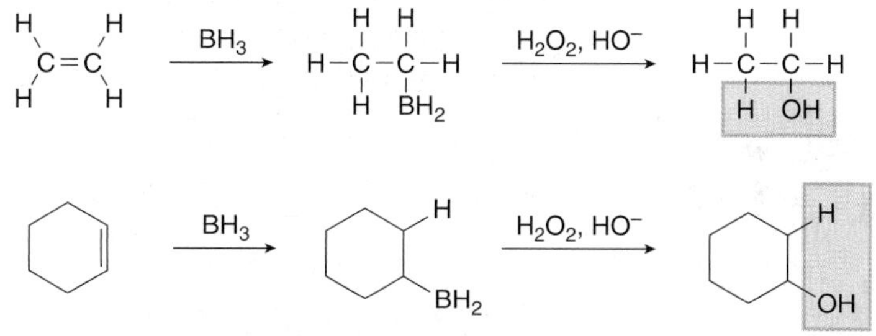

Borane (BH_3) is a reactive gas that exists mostly as the dimer, diborane (B_2H_6). Borane is a strong **Lewis acid** that reacts readily with Lewis bases. For ease in handling in the laboratory, it is commonly used as a complex with tetrahydrofuran (THF).

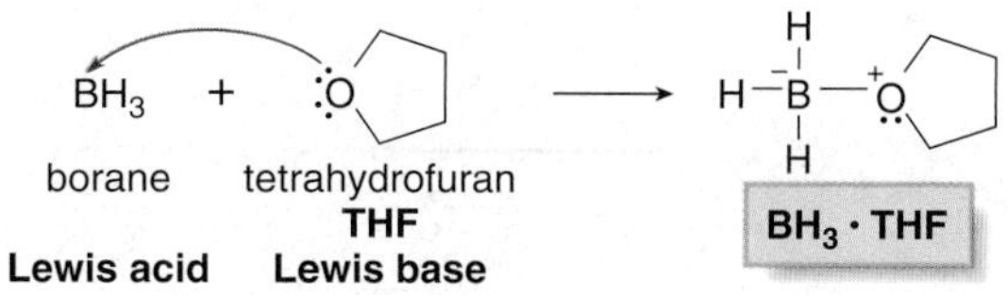

PROBLEM 10.27   Borane is sold for laboratory use as a complex with many other Lewis bases. Draw the Lewis acid–base complex that forms between BH_3 and each compound.
a. $(CH_3)_2S$ b. $(CH_3CH_2)_3N$ c. $(CH_3CH_2CH_2CH_2)_3P$

10.16A   Hydroboration

The first step in hydroboration–oxidation is **addition of the elements of H and BH_2 to the π bond** of the alkene, forming an intermediate alkylborane.

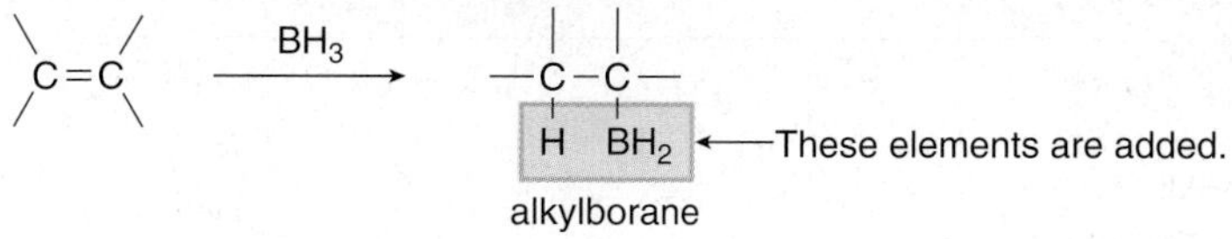

Because syn addition to the double bond occurs and no carbocation rearrangements are observed, carbocations are *not* formed during hydroboration, as shown in Mechanism 10.5. The proposed mechanism involves a **concerted addition of H and BH_2 from the same side of the planar double bond:** the π bond and H–BH_2 bond are broken as two new σ bonds are formed. Because four atoms are involved, the transition state is said to be **four-centered.**

MECHANISM 10.5

The Mechanism of Hydroboration

One step The π bond and H−BH₂ bonds break *as* the C−H and C−B bonds form.

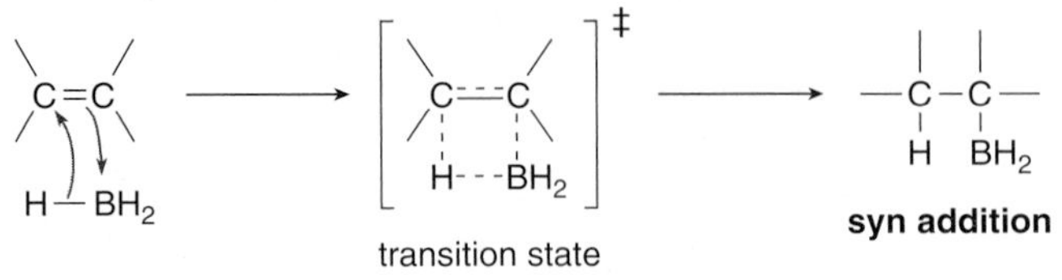

$$C=C \longrightarrow \left[\begin{array}{c} C \cdots C \\ H \cdots BH_2 \end{array} \right]^{\ddagger} \longrightarrow \underset{H \quad BH_2}{-C-C-}$$

transition state

syn addition

Because the alkylborane formed by reaction with one equivalent of alkene still has two B−H bonds, it can react with two more equivalents of alkene to form a trialkylborane. This is illustrated in Figure 10.15 for the reaction of $CH_2=CH_2$ with BH_3.

Because only one B−H bond is needed for hydroboration, commercially available dialkylboranes having the general structure **R₂BH** are sometimes used instead of BH_3. A common example is 9-borabicyclo[3.3.1]nonane **(9-BBN)**. 9-BBN undergoes hydroboration in the same manner as BH_3.

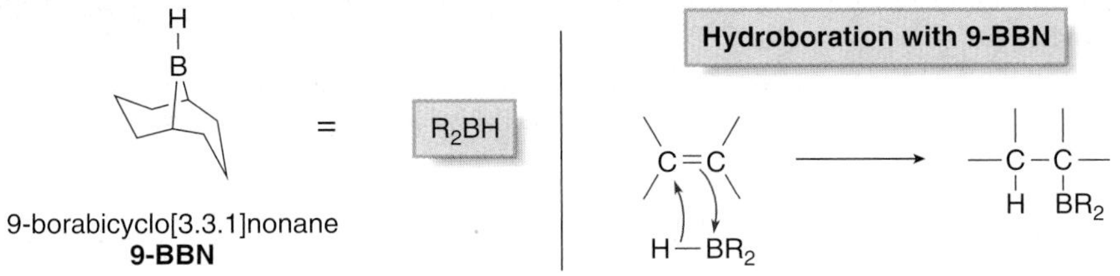

9-borabicyclo[3.3.1]nonane
9-BBN

= R₂BH

Hydroboration with 9-BBN

$$C=C \xrightarrow{\quad} \underset{H \quad BR_2}{-C-C-}$$

Hydroboration is regioselective. **With unsymmetrical alkenes, the boron atom bonds to the less substituted carbon atom.** For example, addition of BH_3 to propene forms an alkylborane with the B bonded to the terminal carbon atom.

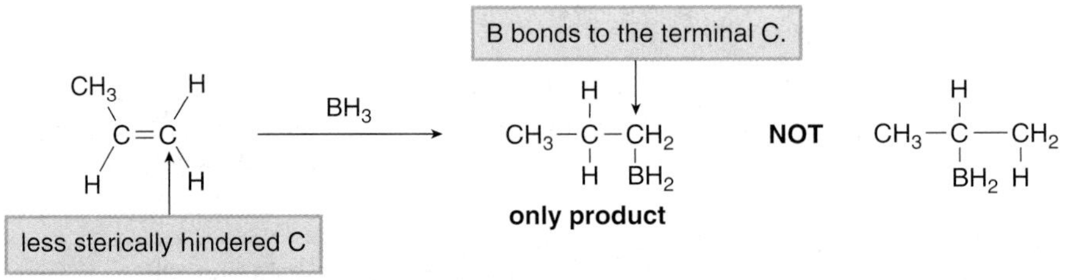

B bonds to the terminal C.

$$CH_3-\underset{H}{\overset{H}{C}}-\underset{BH_2}{CH_2}$$

only product

NOT

$$CH_3-\underset{BH_2}{\overset{H}{C}}-\underset{H}{CH_2}$$

less sterically hindered C

Steric factors explain this regioselectivity. The larger boron atom bonds to the less sterically hindered, more accessible carbon atom.

> Because H is more electronegative than B, the B−H bond is polarized to give boron a partial positive charge ($H^{\delta-}-B^{\delta+}$), making BH₂ the electrophile in hydroboration.

Electronic factors are also used to explain this regioselectivity. If bond breaking and bond making are not completely symmetrical, boron bears a partial negative charge in the transition state and carbon bears a partial positive charge. Because alkyl groups stabilize a positive charge, the more stable transition state has the partial positive charge on the more substituted carbon, as illustrated in Figure 10.16.

Figure 10.15

Conversion of BH_3 to a trialkylborane with three equivalents of $CH_2=CH_2$

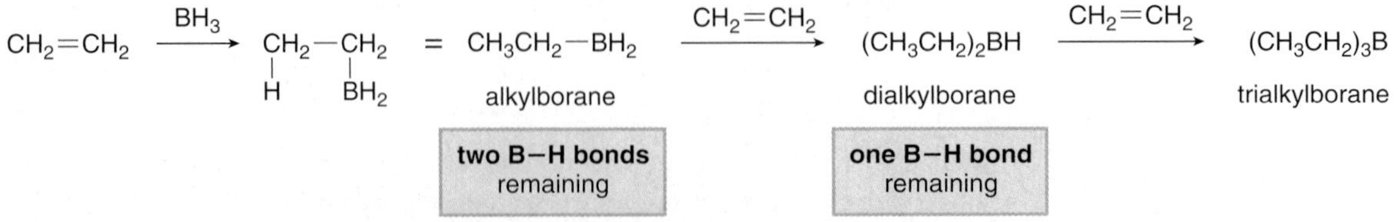

$$CH_2=CH_2 \xrightarrow{BH_3} \underset{H \quad BH_2}{CH_2-CH_2} = CH_3CH_2-BH_2 \xrightarrow{CH_2=CH_2} (CH_3CH_2)_2BH \xrightarrow{CH_2=CH_2} (CH_3CH_2)_3B$$

alkylborane

two B−H bonds remaining

dialkylborane

one B−H bond remaining

trialkylborane

• We often draw hydroboration as if addition stopped after one equivalent of alkene reacts with BH_3. Instead, all three B−H bonds actually react with three equivalents of an alkene to form a trialkylborane. The term organoborane is used for any compound with a carbon–boron bond.

Figure 10.16
Hydroboration of an
unsymmetrical alkene

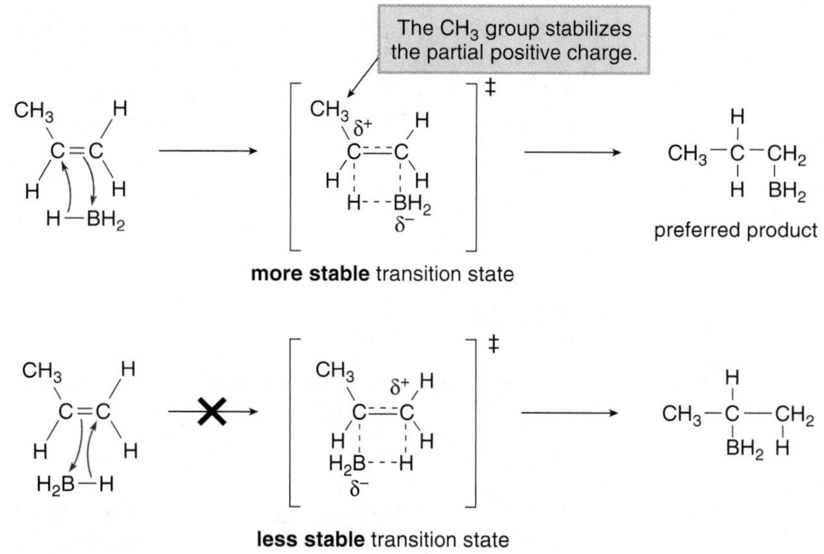

The CH$_3$ group stabilizes
the partial positive charge.

more stable transition state

preferred product

less stable transition state

◆ In hydroboration, the boron atom bonds to the less substituted carbon.

PROBLEM 10.28   What alkylborane is formed from hydroboration of each alkene?

a. 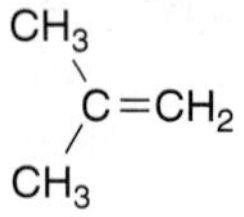   b. 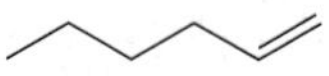   c.

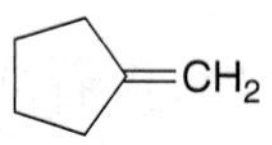

10.16B   Oxidation of the Alkylborane

Because alkylboranes react rapidly with water and spontaneously burn when exposed to the air, they are oxidized, without isolation, with basic hydrogen peroxide (H_2O_2, HO^-). **Oxidation replaces the C–B bond with a C–O bond, forming a new OH group with retention of configuration**; that is, the **OH group replaces the BH$_2$ group in the same position** relative to the other three groups on carbon.

retention of configuration

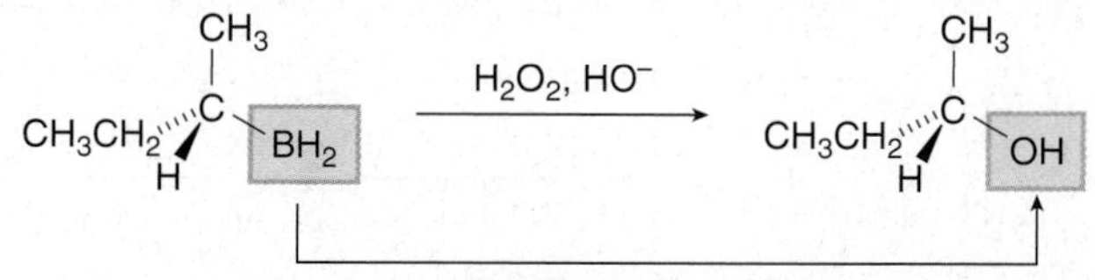

Thus, to draw the product of a hydroboration–oxidation reaction, keep in mind two stereochemical facts:

◆ Hydroboration occurs with syn addition.
◆ Oxidation occurs with retention of configuration.

The overall result of this two-step sequence is **syn addition of the elements of H and OH** to a double bond, as illustrated in Sample Problem 10.5. The OH group bonds to the less substituted carbon.

SAMPLE PROBLEM 10.5   Draw the product of the following reaction sequence, including stereochemistry.

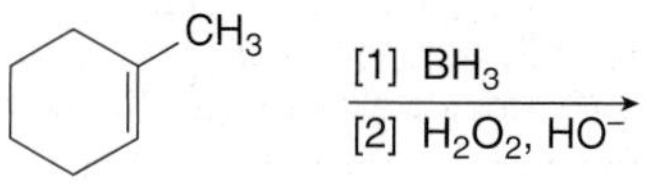

[1] BH$_3$
[2] H_2O_2, HO^-

SOLUTION

In Step [1], **syn addition of BH₃ to the unsymmetrical alkene adds the BH₂ group to the less substituted carbon from above and below the planar double bond.** Two enantiomeric alkylboranes are formed. In Step [2], oxidation replaces the BH₂ group with OH in each enantiomer with **retention of configuration** to yield two alcohols that are also enantiomers.

Hydroboration–oxidation results in the **addition of H and OH in a syn fashion** across the double bond. The achiral alkene is converted to an equal mixture of two enantiomers—that is, a **racemic mixture of alcohols.**

PROBLEM 10.29   Draw the products formed when each alkene is treated with BH₃ followed by H₂O₂, HO⁻. Include the stereochemistry at all stereogenic centers.

a. $CH_3CH_2CH=CH_2$ b. (cyclopentene)—CH_2CH_3 c. CH_3—(cyclohexene)—CH_3

Table 10.5 summarizes the features of hydroboration–oxidation.

With increased resistance of the malarial parasite to chloroquine, a common antimalarial drug, the search for new medicines to treat malaria is active.

chloroquine

TABLE 10.5   Summary: Hydroboration–Oxidation of Alkenes

	Observation
Mechanism	• The addition of H and BH₂ occurs in one step. • No rearrangements can occur.
Regioselectivity	• The OH group bonds to the less substituted carbon atom.
Stereochemistry	• Syn addition occurs. • OH replaces BH₂ with retention of configuration.

Hydroboration–oxidation is a very common method for adding H₂O across a double bond. One example is shown in the synthesis of **artemisinin** (or **qinghaosu**), the active component of **qinghao,** a Chinese herbal remedy used for the treatment of malaria (Figure 10.17).

Figure 10.17   An example of hydroboration–oxidation in synthesis

[1] BH₃
[2] H₂O₂, HO⁻

several steps

Hydroboration–oxidation takes place here.

A

artemisinin
(antimalarial drug)

• The carbon atoms of artemisinin that come from alcohol **A** are indicated in **red.**

10.16C   A Comparison of Hydration Methods

Hydration (H_2O, H^+) and hydroboration–oxidation (BH_3 followed by H_2O_2, HO^-) both add the elements of H_2O across a double bond. Despite their similarities, these reactions often form different constitutional isomers, as shown in Sample Problem 10.6.

SAMPLE PROBLEM 10.6   Draw the product formed in each reaction.

$$CH_3CH_2CH_2CH_2CH=CH_2$$
$$\xrightarrow[\text{H}_2\text{SO}_4]{\text{H}_2\text{O}}$$
$$[1]\ BH_3$$
$$[2]\ H_2O_2,\ HO^-$$

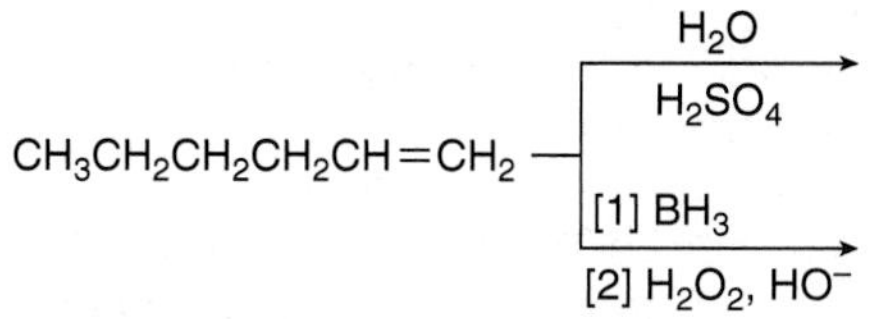

SOLUTION

With H_2O + H_2SO_4, electrophilic addition of H and OH places the **H atom on the less substituted carbon** of the alkene to yield **a 2° alcohol**. In contrast, addition of BH_3 gives an alkylborane with the **BH_2 group on the less substituted terminal carbon** of the alkene. Oxidation replaces BH_2 by OH to yield a **1° alcohol**.

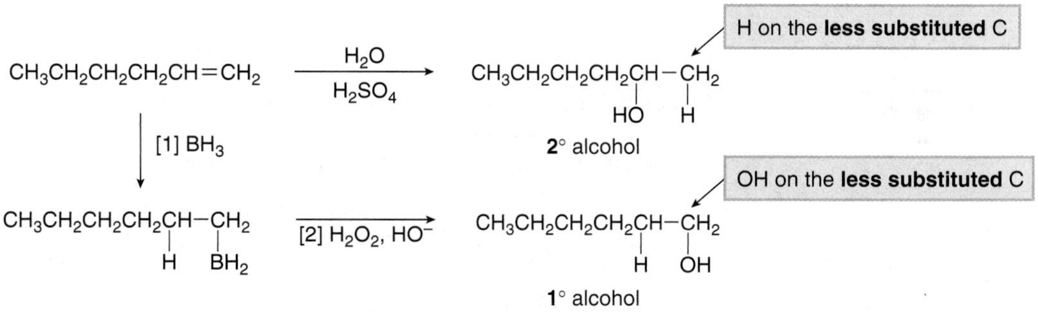

H on the **less substituted** C

$$CH_3CH_2CH_2CH_2CH=CH_2 \xrightarrow[\text{H}_2\text{SO}_4]{\text{H}_2\text{O}} CH_3CH_2CH_2CH_2CH-CH_2$$
HO   H
2° alcohol

$$[1]\ BH_3 \Big\downarrow$$

$$CH_3CH_2CH_2CH_2CH-CH_2 \xrightarrow{[2]\ H_2O_2,\ HO^-} CH_3CH_2CH_2CH_2CH-CH_2$$
H BH_2 H   OH

OH on the **less substituted** C

1° alcohol

PROBLEM 10.30   Draw the constitutional isomer formed in each reaction.

a. 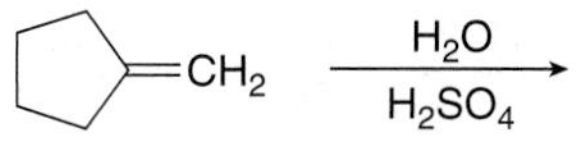
$$\xrightarrow[\text{H}_2\text{SO}_4]{\text{H}_2\text{O}}$$

c. 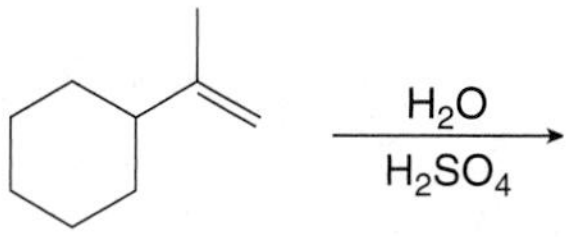
$$\xrightarrow[\text{H}_2\text{SO}_4]{\text{H}_2\text{O}}$$

b. 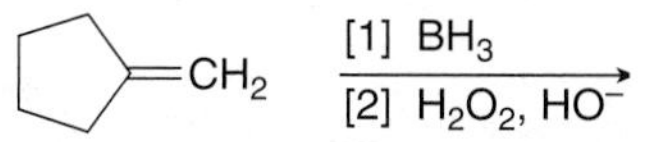
$$\xrightarrow[[2]\ H_2O_2,\ HO^-]{[1]\ BH_3}$$

d. 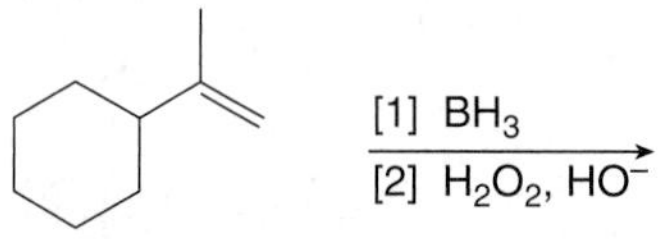
$$\xrightarrow[[2]\ H_2O_2,\ HO^-]{[1]\ BH_3}$$

10.17   Keeping Track of Reactions

Chapters 7–10 have introduced three basic kinds of organic reactions: **nucleophilic substitution, β elimination,** and **addition.** In the process, many specific reagents have been discussed and the stereochemistry that results from many different mechanisms has been examined. **How can we keep track of all the reactions?**

To make the process easier, **remember that most organic molecules undergo only one or two different kinds of reactions.** For example:

◆ Alkyl halides undergo substitution and elimination because they have good leaving groups.
◆ Alcohols also undergo substitution and elimination, but can only do so when OH is made into a good leaving group.
◆ Alkenes undergo addition because they have easily broken π bonds.

You must still learn many reaction details, and in truth, there is no one method to learn them. *You must practice these reactions over and over again, not by merely looking at them, but by writing them.* Some students do this by making a list of specific reactions for each functional group, and then rewriting them with different starting materials. Others make flash cards: index cards that have the starting material and reagent on one side and the product on the other. Whatever method you choose, **the details must become second nature,** much like the answers to simple addition problems, such as, what is the sum of 2 + 2?

Learning reactions is really a two-step process.

◆ First, learn the basic type of reaction for a functional group. This provides an overall organization to the reactions.
◆ Then, learn the specific reagents for each reaction. It helps to classify the reagent according to its properties. Is it an acid or a base? Is it a nucleophile or an electrophile? Is it an oxidizing agent or a reducing agent?

Sample Problem 10.7 illustrates this process.

SAMPLE PROBLEM 10.7   Draw the product of each reaction.

a. $(CH_3CH_2)_2CHCH_2$—Br $\xrightarrow{KOC(CH_3)_3}$

b. $(CH_3CH_2)_2CHCH{=}CH_2$ $\xrightarrow[H_2O]{Br_2}$

SOLUTION

In each problem, identify the functional group to determine the general reaction type—substitution, elimination, or addition. Then, determine if the reagent is an electrophile, nucleophile, acid, base, and so forth.

[a] The reactant is a **1° alkyl halide,** which can undergo substitution and elimination. The reagent [KOC(CH₃)₃] is a **strong nonnucleophilic base,** favoring elimination by an E2 mechanism (Figure 8.10).

elimination

CH₃CH₂
          C=CH₂
CH₃CH₂
E2 product

[b] The reactant is an **alkene,** which undergoes addition reactions to its π bond. The reagent ($Br_2 + H_2O$) serves as the source of the **electrophile Br⁺,** resulting in **addition** of Br and OH to the double bond (Section 10.15).

This π bond is broken.

$(CH_3CH_2)_2CHCH{=}CH_2$

$Br_2 + H_2O$

$(CH_3CH_2)_2CHCH{-}CH_2$
                  HO   Br

addition product

PROBLEM 10.31   Draw the products of each reaction using the two-part strategy from Sample Problem 10.7.

a. [cyclohexane =CH₂] $\xrightarrow{HBr}$

b. [structure with Cl] $\xrightarrow{NaOCH_3}$

c. [cyclohexane CH₂CH₂CH₃ / OH] $\xrightarrow{H_2SO_4}$

10.18 Alkenes in Organic Synthesis

Alkenes are a central functional group in organic chemistry. **Alkenes are easily prepared by elimination reactions** such as dehydrohalogenation and dehydration. **Because their π bond is easily broken, they undergo many addition reactions** to prepare a variety of useful compounds.

Suppose, for example, that we must synthesize 1,2-dibromocyclohexane from cyclohexanol, a cheap and readily available starting material. Because there is no way to accomplish this transformation in one step, this synthesis must have at least two steps.

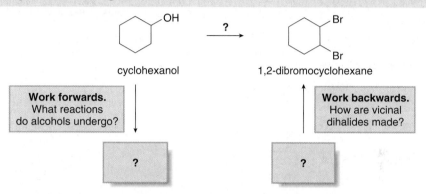

cyclohexanol

starting material

1,2-dibromocyclohexane

product

To solve this problem we must:

◆ Work backwards from the product by asking: What type of reactions introduce the functional groups in the product?

◆ Work forwards from the starting material by asking: What type of reactions does the starting material undergo?

cyclohexanol

1,2-dibromocyclohexane

Work forwards.
What reactions
do alcohols undergo?

Work backwards.
How are vicinal
dihalides made?

?

?

In Chapter 11 we will learn about retrosynthetic analysis in more detail.

Working backwards from the product to determine the starting material from which it is made is called *retrosynthetic analysis*.

We know reactions that answer each of these questions.

Working backwards:

[1] 1,2-Dibromocyclohexane, a vicinal dibromide, can be prepared by the addition of Br_2 to **cyclohexene.**

cyclohexene $\xrightarrow{Br_2}$ 1,2-dibromocyclohexane

Working forwards:

[2] Cyclohexanol can undergo acid-catalyzed dehydration to form **cyclohexene.**

cyclohexanol $\xrightarrow[H_2SO_4]{H_2O}$ cyclohexene

*A **reactive intermediate** is an unstable intermediate like a carbocation, which is formed during the conversion of a stable starting material to a stable product. A **synthetic intermediate** is a stable compound that is the product of one step and the starting material of another in a multistep synthesis.*

Cyclohexene is called a **synthetic intermediate,** or simply an **intermediate,** because it is the **product of one step and the starting material of another.** We now have a two-step sequence to convert cyclohexanol to 1,2-dibromocyclohexane, and the synthesis is complete. Take note of the central role of the alkene in this synthesis.

A two-step synthesis

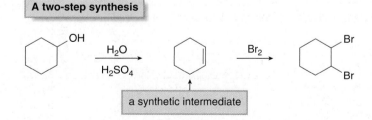

a synthetic intermediate

PROBLEM 10.32   Devise a synthesis of each compound from the indicated starting material.

a.

starting material $\xrightarrow{?}$ product

b.

starting material $\xrightarrow{?}$ product

10.19   Key Concepts—Alkenes

General Facts About Alkenes

- Alkenes contain a carbon–carbon double bond consisting of a stronger σ bond and a weaker π bond. Each carbon is sp^2 hybridized and trigonal planar (10.1).
- Alkenes are named using the suffix -*ene* (10.3).
- Alkenes with different groups on each end of the double bond exist as a pair of diastereomers, identified by the prefixes *E* and *Z* (10.3B).
- Alkenes have weak intermolecular forces, giving them low mp's and bp's, and making them water insoluble. A cis alkene is more polar than a trans alkene, giving it a slightly higher boiling point (10.4).
- Because a π bond is electron rich and much weaker than a σ bond, alkenes undergo addition reactions with electrophiles (10.8).

Stereochemistry of Alkene Addition Reactions (10.8)

A reagent XY adds to a double bond in one of three different ways:

- **Syn addition**—X and Y add from the same side.

- Syn addition occurs in **hydroboration.**

- **Anti addition**—X and Y add from opposite sides.

- Anti addition occurs in **halogenation** and **halohydrin formation.**

- **Both syn and anti addition** occur when carbocations are intermediates.

- Syn and anti addition occur in **hydrohalogenation** and **hydration.**

Addition Reactions of Alkenes

[1] Hydrohalogenation—Addition of HX (X = Cl, Br, I) (10.9–10.11)

$$RCH=CH_2 \ + \ H-X \longrightarrow \underset{\text{alkyl halide}}{R-\underset{X}{\underset{|}{C}}H-\underset{H}{\underset{|}{C}}H_2}$$

- The mechanism has two steps.
- Carbocations are formed as intermediates.
- Carbocation rearrangements are possible.
- Markovnikov's rule is followed. H bonds to the less substituted C to form the more stable carbocation.
- Syn and anti addition occur.

[2] Hydration and related reactions (Addition of H_2O or ROH) (10.12)

$$RCH=CH_2 \ + \ H-OH \xrightarrow{\ H_2SO_4\ } \underset{\text{alcohol}}{R-\underset{OH}{\underset{|}{C}}H-\underset{H}{\underset{|}{C}}H_2}$$

For both reactions:
- The mechanism has three steps.
- Carbocations are formed as intermediates.
- Carbocation rearrangements are possible.
- Markovnikov's rule is followed. H bonds to the less substituted C to form the more stable carbocation.
- Syn and anti addition occur.

$$RCH=CH_2 \ + \ H-OR \xrightarrow{\ H_2SO_4\ } \underset{\text{ether}}{R-\underset{OR}{\underset{|}{C}}H-\underset{H}{\underset{|}{C}}H_2}$$

[3] Halogenation (Addition of X_2; X = Cl or Br) (10.13–10.14)

$$RCH=CH_2 \ + \ X-X \ \longrightarrow \ R-\underset{\underset{X}{|}}{C}H-\underset{\underset{X}{|}}{C}H_2$$

vicinal dihalide

- The mechanism has two steps.
- Bridged halonium ions are formed as intermediates.
- No rearrangements can occur.
- Anti addition occurs.

[4] Halohydrin formation (Addition of OH and X; X = Cl, Br) (10.15)

$$RCH=CH_2 \ + \ X-X \ \xrightarrow{\ H_2O\ } \ R-\underset{\underset{OH}{|}}{C}H-\underset{\underset{X}{|}}{C}H_2$$

halohydrin

- The mechanism has three steps.
- Bridged halonium ions are formed as intermediates.
- No rearrangements can occur.
- X bonds to the less substituted C.
- Anti addition occurs.
- NBS in DMSO and H_2O adds Br and OH in the same fashion.

[5] Hydroboration–oxidation (Addition of H_2O) (10.16)

$$RCH=CH_2 \ \xrightarrow[\text{[2] } H_2O_2,\ HO^-]{\text{[1] } BH_3 \text{ or } 9\text{-BBN}} \ R-\underset{\underset{H}{|}}{C}H-\underset{\underset{OH}{|}}{C}H_2$$

alcohol

- Hydroboration has a one-step mechanism.
- No rearrangements can occur.
- OH bonds to the less substituted C.
- Syn addition of H_2O results.

Problems

Nomenclature

10.33 Give the IUPAC name for each alkene.

a. $CH_2=CHCH_2CH(CH_3)CH_2CH_3$

b.

c.

d.

$$\underset{H}{\overset{CH_3}{\diagdown}} C = C \underset{CH_2CH(CH_3)_2}{\overset{CH_3}{\diagup}}$$

e.

f.

10.34 Give the structure corresponding to each name.
- a. (*E*)-4-ethyl-3-heptene
- b. 3,3-dimethylcyclopentene
- c. *cis*-4-octene
- d. 4-vinylcyclopentene
- e. (*Z*)-3-isopropyl-2-heptene
- f. *cis*-3,4-dimethylcyclopentene
- g. *trans*-2-heptene
- h. 1-isopropyl-4-propylcyclohexene

10.35 Each of the following names is incorrect. Explain why it is incorrect and give the correct IUPAC name.
- a. 2-butyl-3-methyl-1-pentene
- b. (*Z*)-2-methyl-2-hexene
- c. (*E*)-1-isopropyl-1-butene
- d. 5-methylcyclohexene
- e. 4-isobutyl-2-methylcyclohexene
- f. 1-*sec*-butyl-2-cyclopentene

10.36

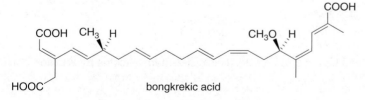

bongkrekic acid

Bongkrekic acid is a toxic compound produced by *Pseudomonas cocovenenans,* and isolated from a mold that grows on bongkrek, a fermented Indonesian coconut dish. (a) Label each double bond in bongkrekic acid as *E* or *Z*. (b) Label each tetrahedral stereogenic center as *R* or *S*. (c) How many stereoisomers are possible for bongkrekic acid?

10.37 Draw all stereoisomers having molecular formula C_6H_{12} that contain one double bond *and* a five-carbon chain with a one-carbon branch. Name each compound, including its *E,Z* or *R,S* designation when necessary.

Degrees of Unsaturation

10.38 Calculate the number of degrees of unsaturation for each molecular formula.
- a. C_3H_4
- b. C_6H_8
- c. $C_{40}H_{56}$
- d. C_8H_8O
- e. $C_{10}H_{16}O_2$
- f. C_8H_9Br
- g. C_8H_9ClO
- h. C_7H_9Br

10.39 Give an example of a compound with molecular formula C_6H_{10} that satisfies each criterion.
a. a compound with two π bonds
b. a compound with one ring and one π bond
c. a compound with two rings
d. a compound with one triple bond

Lipids

10.40 Although naturally occurring unsaturated fatty acids have the *Z* configuration, elaidic acid, a C_{18} fatty acid having an *E* double bond, is present in processed foods such as margarine and cooking oils. Predict how the melting point of elaidic acid compares with the melting points of stearic and oleic acids (Table 10.2).

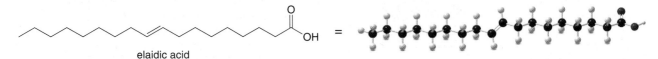

elaidic acid

10.41 (a) Draw two possible triacylglycerols formed from one molecule of stearic acid and two molecules of oleic acid. (b) One of these molecules contains a tetrahedral stereogenic center. Draw both enantiomers, and label the stereogenic center as *R* or *S*.

Energy Diagram and ΔH° Calculations

10.42 Draw an energy diagram for the two-step mechanism for the addition of Br_2 to $CH_2=CH_2$ to form 1,2-dibromoethane. Draw the structure of the transition state for each step.

10.43 By using the bond dissociation energies in Tables 1.3 and 6.2, calculate ΔH° for the addition of HCl and HI to ethylene to form chloroethane and iodoethane, respectively. Assuming entropy changes for both reactions are similar, which reaction has the largest K_{eq}?

Reactions of Alkenes

10.44 Draw the products formed when cyclohexene is treated with each of the following reagents.
a. HBr
b. HI
c. H_2O, H_2SO_4
d. CH_3CH_2OH, H_2SO_4
e. Cl_2
f. Br_2, H_2O
g. NBS (aqueous DMSO)
h. [1] BH_3; [2] H_2O_2, HO^-
i. [1] 9-BBN; [2] H_2O_2, HO^-

10.45 Repeat Problem 10.44 with $(CH_3)_2C=CH_2$ as the starting material.

10.46 Draw the product formed when 1-butene is treated with each reagent: (a) Br_2 in CCl_4; (b) Br_2 in H_2O; (c) Br_2 in CH_3OH.

10.47 What alkene can be used to prepare each alkyl halide or dihalide as the exclusive or major product of an addition reaction?

a. [cyclopentane with Br]
b. [cyclohexane with Cl, Cl]
c. [cyclohexane-C(CH₃)₂ with Br]
d. [chain with Br, Br]
e. [cyclopentane with CH₃, Cl]
f. $(CH_3CH_2)_3CBr$

10.48 Draw all constitutional isomers formed in each reaction.

a. $(CH_3CH_2)_2C=CHCH_2CH_3$ $\xrightarrow{HCl}$

b. $(CH_3CH_2)_2C=CH_2$ $\xrightarrow[H_2SO_4]{H_2O}$

c. $(CH_3)_2C=CHCH_3$ $\xrightarrow[2) H_2O_2, HO^-]{1) BH_3}$

d. [cycloheptene] $\xrightarrow{Cl_2}$

e. [cyclopentane=CH_2] $\xrightarrow[H_2O]{Br_2}$

f. [cyclohexane=CH_2] $\xrightarrow[2) H_2O_2, HO^-]{1) 9-BBN}$

g. [alkene] $\xrightarrow[DMSO, H_2O]{NBS}$

h. [alkene] $\xrightarrow{Br_2}$

10.49 What three alkenes (excluding stereoisomers) can be used to prepare 3-chloro-3-methylhexane by addition of HCl?

10.50 Draw all stereoisomers formed in each reaction.

a. 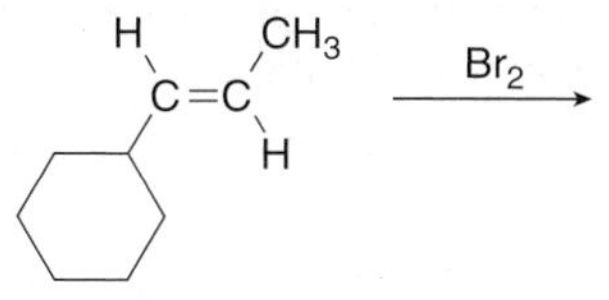 $\xrightarrow{Br_2}$

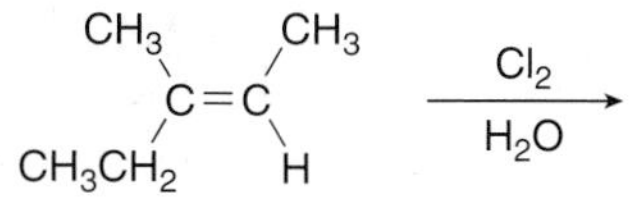

b. 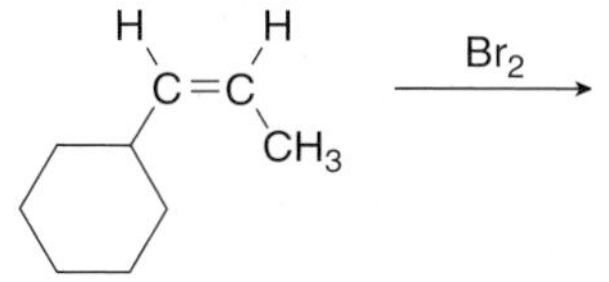 $\xrightarrow{Br_2}$

c. $\xrightarrow[H_2O]{Cl_2}$

d. 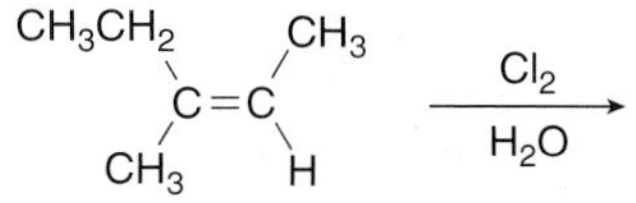 $\xrightarrow[H_2O]{Cl_2}$

10.51 Draw the products of each reaction, including stereoisomers.

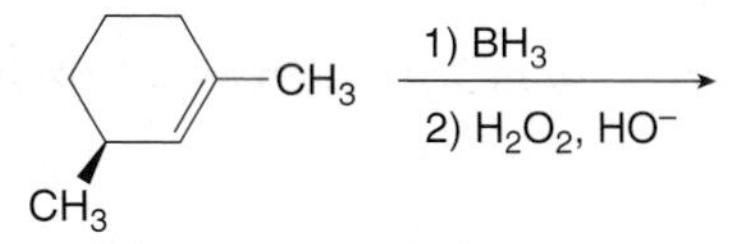

a. 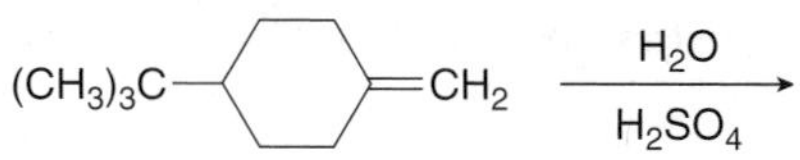 $(CH_3)_3C$—$=CH_2$ $\xrightarrow[H_2SO_4]{H_2O}$

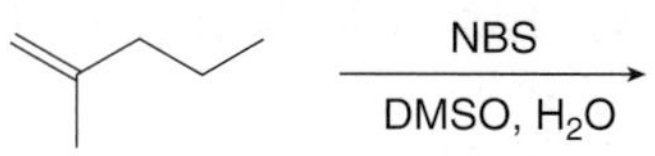

b.  $\xrightarrow{HI}$

c. $\xrightarrow{Cl_2}$

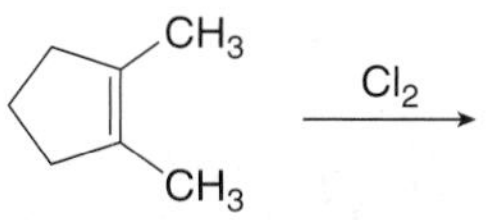

d. $\xrightarrow[2)\ H_2O_2,\ HO^-]{1)\ BH_3}$

e. 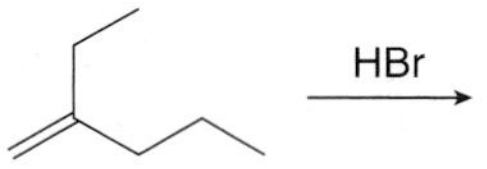 $\xrightarrow{HBr}$

f. $\xrightarrow[H_2O]{Cl_2}$

g. $\xrightarrow[DMSO,\ H_2O]{NBS}$

h. $CH_3CH{=}CHCH_2CH_3$ $\xrightarrow[H_2SO_4]{H_2O}$

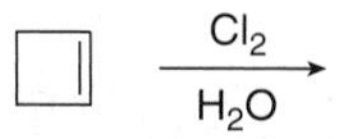

10.52 Explain why $CH_2{=}CH_2$ reacts rapidly with Br_2 at room temperature, but $Cl_2C{=}CCl_2$ does not.

10.53 Several reactions add two electronegative atoms to a double bond via a halonium ion intermediate. With this in mind, draw the products formed in each reaction.

a. $CH_3CH{=}CHCH_3 + CH_3SCl \longrightarrow$

b. $(CH_3)_2C{=}CH_2 + ICl \longrightarrow$

10.54 Using cis- and trans-3-hexene, demonstrate that the addition of HCl is not a stereospecific reaction. Draw the structure of the stereoisomers formed from each alkene.

Mechanisms

10.55 Draw a stepwise mechanism for each reaction.

a. 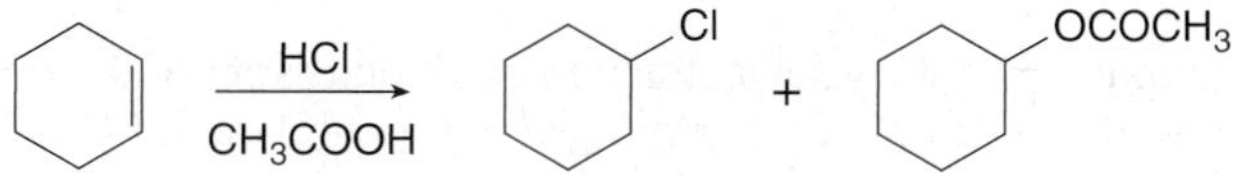 $\xrightarrow[CH_3COOH]{HCl}$

b.

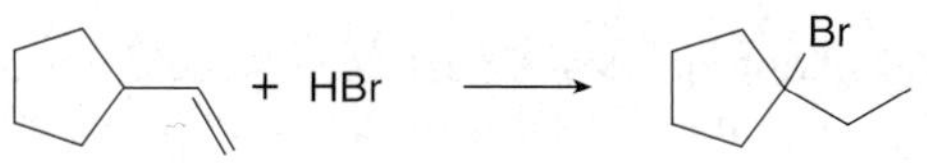

10.56 Draw a stepwise mechanism for each reaction.

a. 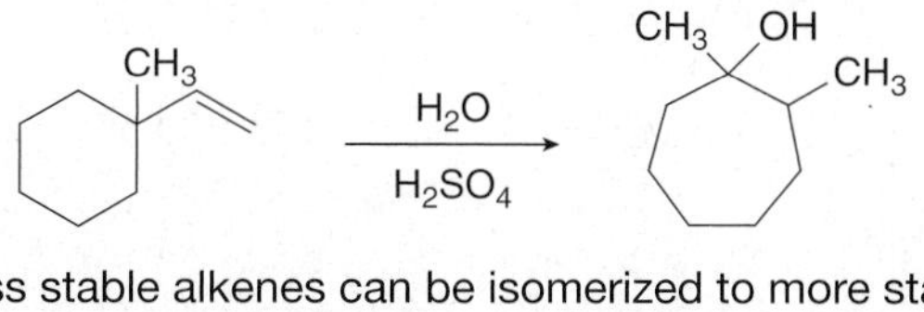 $\xrightarrow[H_2SO_4]{H_2O}$

b. 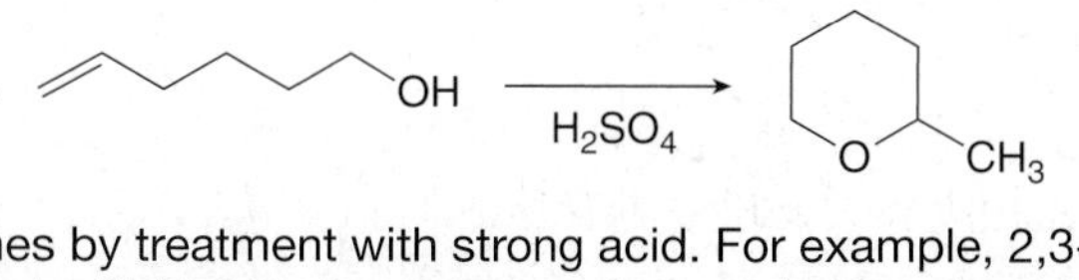 $\xrightarrow{H_2SO_4}$

10.57 Less stable alkenes can be isomerized to more stable alkenes by treatment with strong acid. For example, 2,3-dimethyl-1-butene is converted to 2,3-dimethyl-2-butene when treated with H_2SO_4. Draw a stepwise mechanism for this isomerization process.

10.58 When 1,3-butadiene ($CH_2{=}CH{-}CH{=}CH_2$) is treated with HBr, two constitutional isomers are formed, $CH_3CHBrCH{=}CH_2$ and $BrCH_2CH{=}CHCH_3$. Draw a stepwise mechanism that accounts for the formation of both products.

10.59 Explain why the addition of HBr to alkenes **A** and **C** is regioselective, forming addition products **B** and **D,** respectively.

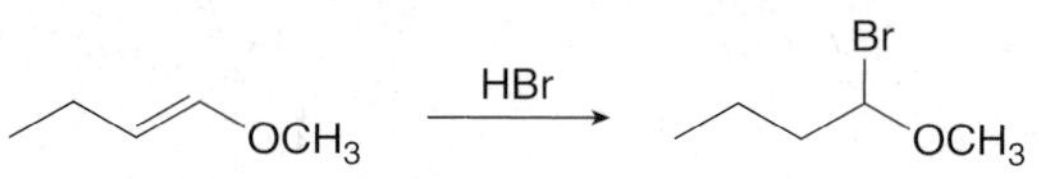

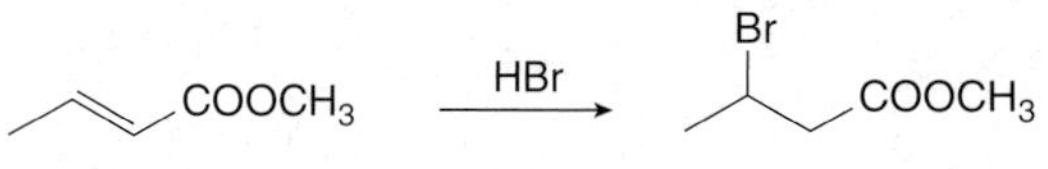

10.60 Bromoetherification, the addition of the elements of Br and OR to a double bond, is a common method for constructing rings containing oxygen atoms. This reaction has been used in the synthesis of the polyether antibiotic monensin (Problem 21.38). Draw a stepwise mechanism for the following intramolecular bromoetherification reaction.

Bromoetherification

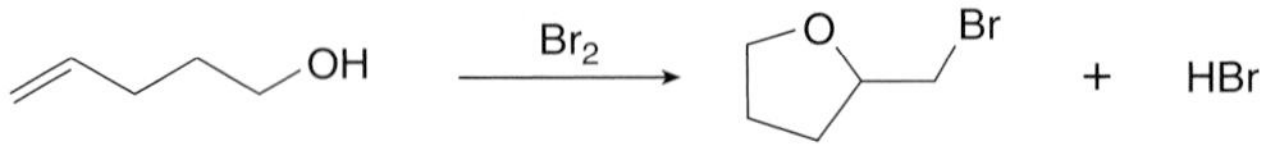

Synthesis

10.61 Devise a synthesis of each product from the given starting material. More than one step is required.

a.

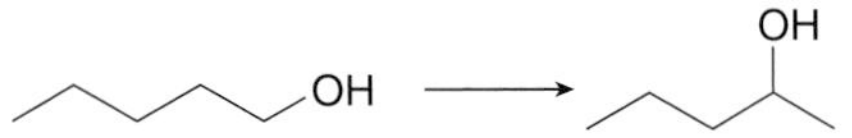

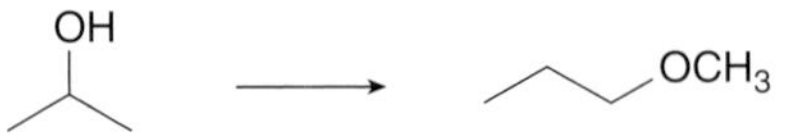

b. $CH_3-\overset{Br}{\underset{H}{C}}-CH_3 \longrightarrow CH_3-\overset{Br}{\underset{H}{C}}-CH_2Br$

c.

d. $(CH_3)_2CHCH_2I \longrightarrow (CH_3)_3CCl$

10.62 Devise a synthesis of each product from the given starting material. More than one step is required.

a.

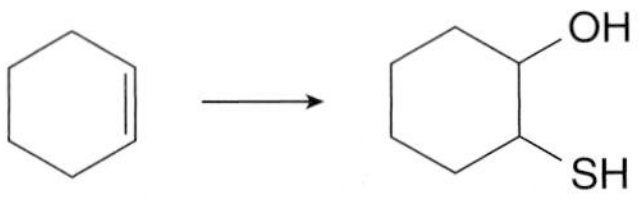

b. $CH_3CH=CH_2 \longrightarrow CH_3C\equiv CH$

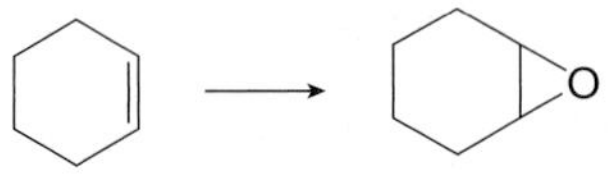

c.

d.

e.

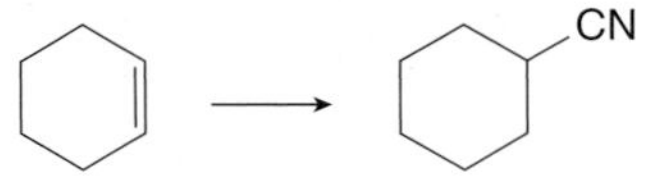

Challenge Questions

10.63 Alkene **A** can be isomerized to isocomene, a natural product isolated from goldenrod, by treatment with TsOH. Draw a stepwise mechanism for this conversion. (Hint: Look for a carbocation rearrangement.)

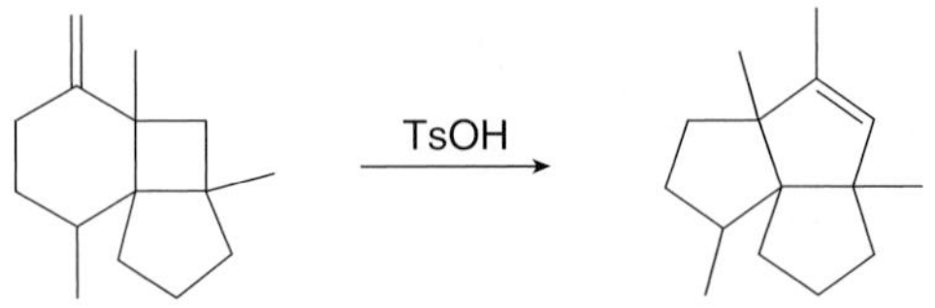

A   isocomene

10.64 Lactones, cyclic esters such as compound **A,** are prepared by halolactonization, an addition reaction to an alkene. For example, iodolactonization of **B** forms lactone **C,** a key intermediate in the synthesis of prostaglandin PGF$_{2\alpha}$ (Section 4.15). Draw a stepwise mechanism for this addition reaction.

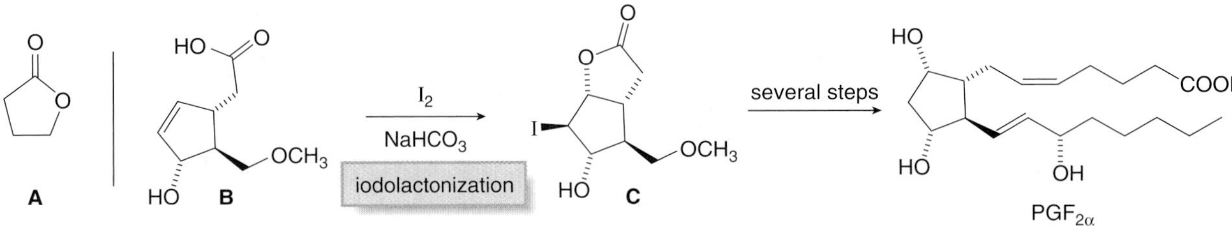

10.65 Like other electrophiles, carbocations add to alkenes to form new carbocations, which can then undergo substitution or elimination, depending on the reaction conditions.

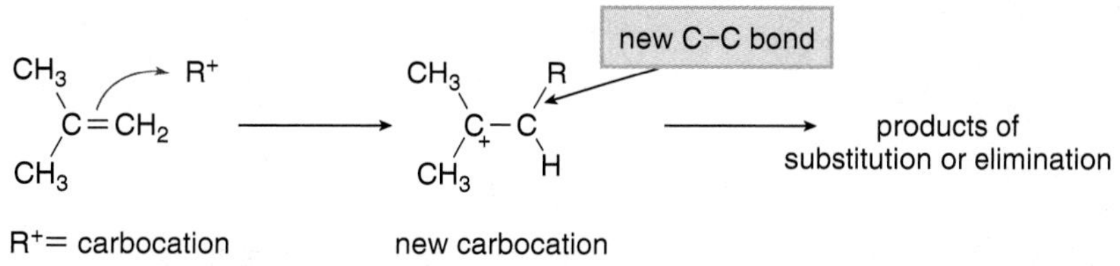

With this in mind, draw a stepwise mechanism for the following reaction.

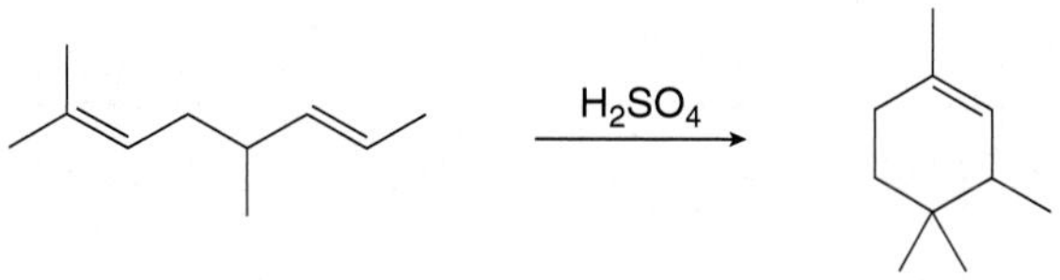

Alkynes

Ethynylestradiol is a synthetic compound whose structure closely resembles the carbon skeleton of female estrogen hormones. Because it is more potent than its naturally occurring analogues, it is a component of several widely used oral contraceptives. Ethynylestradiol and related compounds with similar biological activity contain a carbon–carbon triple bond. In Chapter 11 we will learn about alkynes, hydrocarbons that contain triple bonds.

$\mathbf{I}$n Chapter 11 we continue our focus on organic molecules with electron-rich functional groups by examining *alkynes,* **compounds that contain a carbon–carbon triple bond.** Like alkenes, **alkynes are nucleophiles with easily broken π bonds,** and as such, they undergo **addition** reactions with electrophilic reagents.

Alkynes also undergo a reaction that has no analogy in alkene chemistry. Because a C–H bond of an alkyne is more acidic than a C–H bond in an alkene or an alkane, alkynes are readily deprotonated with strong base. The resulting nucleophiles react with electrophiles to form new carbon–carbon σ bonds, so that complex molecules can be prepared from simple starting materials. The study of alkynes thus affords an opportunity to learn a great deal about organic synthesis.

11.1   Introduction

Alkynes contain a carbon–carbon triple bond. A **terminal alkyne** has the triple bond at the end of the carbon chain, so that a hydrogen atom is directly bonded to a carbon atom of the triple bond. An **internal alkyne** has a carbon atom bonded to each carbon atom of the triple bond.

Alkyne

—C≡C—
↑
triple bond

$CH_3CH_2CH_2$—C≡C—H

terminal alkyne

$CH_3CH_2CH_2$—C≡C—CH_2CH_3

internal alkyne

An alkyne has the general molecular formula C_nH_{2n-2}, giving it four fewer hydrogens than the maximum number possible. Because every degree of unsaturation removes two hydrogens, a **triple bond introduces two degrees of unsaturation.**

PROBLEM 11.1   Draw structures for the three alkynes having molecular formula C_5H_8 and classify each as an internal or terminal alkyne.

Each carbon of a triple bond is *sp* hybridized and **linear,** and all bond angles are **180°** (Section 1.9C).

180°

H—C≡C—H   =
acetylene

sp hybridized

The triple bond of an alkyne consists of one σ bond and two π bonds.

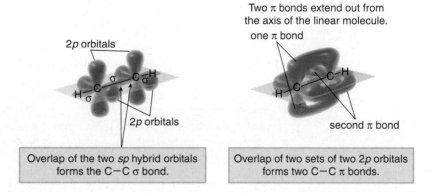

2*p* orbitals

2*p* orbitals

Two π bonds extend out from the axis of the linear molecule.

one π bond

second π bond

Overlap of the two *sp* hybrid orbitals forms the C–C σ bond.	Overlap of two sets of two 2*p* orbitals forms two C–C π bonds.

◆ The σ bond is formed by end-on overlap of the two *sp* hybrid orbitals.
◆ Each π bond is formed by side-by-side overlap of two 2*p* orbitals.

Bond dissociation energies of the C–C bonds in ethylene (one σ and one π bond) and acetylene (one σ and two π bonds) can be used to estimate the strength of the second π bond of the triple bond. If we assume that the σ bond and first π bond in acetylene are similar in strength to the σ and π bonds in ethylene (88 and 64 kcal/mol, respectively), then the second π bond is worth 48 kcal/mol.

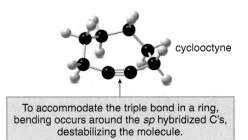

$$HC\equiv CH \qquad CH_2{=}CH_2$$

| 200 kcal/mol | − | 152 kcal/mol | = | 48 kcal/mol |
| (σ + two π bonds) | | (σ + π bond) | | |

second π bond

> ◆ Both π bonds of a C–C triple bond are weaker than a C–C σ bond, making them much more easily broken. As a result, alkynes undergo many addition reactions.
> ◆ Alkynes are more polarizable than alkenes because the electrons in their π bonds are more loosely held.

Like trans cycloalkenes, cycloalkynes with small rings are unstable. The carbon chain must be long enough to connect the two ends of the triple bond without introducing too much strain. Cyclooctyne is the smallest isolated cycloalkyne, though it decomposes upon standing at room temperature after a short time.

cyclooctyne

To accommodate the triple bond in a ring, bending occurs around the *sp* hybridized C's, destabilizing the molecule.

PROBLEM 11.2 What orbitals are used to form each labeled bond in the following molecule?

$$\underset{(a)}{\underset{CH_3-C}{\overset{\displaystyle \underset{H}{\overset{(b)\quad(c)}{C-C\equiv C-CH_3}}}{}}}$$

PROBLEM 11.3 Would you predict an internal or terminal alkyne to be more stable? Why?

11.2 Nomenclature

Alkynes are named in the same way that alkenes were named in Section 10.3.

> ◆ In the IUPAC system, change the *-ane* ending of the parent alkane to the suffix *-yne.*
> ◆ Choose the longest carbon chain that contains both atoms of the triple bond and number the chain to give the triple bond the lower number.
> ◆ Compounds with two triple bonds are called *diynes,* those with three are called *triynes,* and so forth.

SAMPLE PROBLEM 11.1 Give the IUPAC name for the following alkyne.

$$CH_3CH_2-C\equiv C-CH_2-\underset{\underset{CH_3}{|}}{\overset{\overset{CH_2CH_3}{|}}{C}}-CH_3$$

SOLUTION

[1] **Find the longest chain that contains both carbons of the triple bond.**	[2] **Number the long chain; then name and number the substituents.**

$CH_3CH_2-C{\equiv}C-CH_2-\overset{\displaystyle CH_2CH_3}{\underset{\displaystyle CH_3}{\overset{|}{\underset{|}{C}}}}-CH_3$

8 C's in the longest chain

oct*ane* – – – ➞ oct*yne*

$CH_3CH_2-\underset{1}{C}{\equiv}C-\underset{3}{CH_2}-\underset{4}{\overset{\displaystyle CH_2CH_3}{\underset{\displaystyle CH_3}{\overset{|}{\underset{6}{C}}}}}-CH_3$

two methyl groups at C6

Answer: 6,6-dimethyl-3-octyne

The simplest alkyne, $HC{\equiv}CH$, named in the IUPAC system as **ethyne,** is more often called **acetylene,** its common name. The two-carbon alkyl group derived from acetylene is called an **ethynyl** group.

$H-C{\equiv}C-$
ethynyl group

$\langle hexagon \rangle-C{\equiv}C-H$
ethynylcyclohexane

PROBLEM 11.4 Draw the seven isomeric alkynes having molecular formula C_6H_{10}, and give the IUPAC name for each compound. Consider constitutional isomers only.

PROBLEM 11.5 Give the IUPAC name for each alkyne.
a. $H-C{\equiv}C-CH_2C(CH_2CH_2CH_3)_3$ b. $CH_3C{\equiv}CC(CH_3)ClCH_2CH_3$

PROBLEM 11.6 Give the structure corresponding to each of the following names.
a. *trans*-2-ethynylcyclopentanol   b. 4-*tert*-butyl-5-decyne   c. 3-methylcyclononyne

11.3 Physical Properties

The physical properties of alkynes resemble those of hydrocarbons having a similar shape and molecular weight.

- ◆ Alkynes have low melting points (mp) and boiling points (bp).
- ◆ Melting point and boiling point increase as the number of carbons increases.
- ◆ Alkynes are soluble in organic solvents and insoluble in water.

PROBLEM 11.7 Explain why an alkyne often has a slightly higher boiling point than an alkene of similar molecular weight. For example, the bp of 1-pentyne is 39 °C, and the bp of 1-pentene is 30 °C.

11.4 Interesting Alkynes

Acetylene, $HC{\equiv}CH$, is a colorless gas with an ethereal odor that burns in oxygen to form CO_2 and H_2O. Because the combustion of acetylene releases more energy per mole of product formed than other hydrocarbons, it burns with a very hot flame, making it an excellent fuel for welding torches.

Ethynylestradiol, the molecule that opened Chapter 11, and **norethindrone** are two components of oral contraceptives that contain a carbon–carbon triple bond. Both molecules are synthetic analogues of the naturally occurring female hormones estradiol and progesterone, but are more potent so they can be administered in lower doses. Most oral contraceptives contain two of these synthetic hormones. They act by artificially elevating hormone levels in a woman, thereby preventing ovulation.

Synthetic hormones in oral contraceptives

ethynylestradiol

norethindrone

Naturally occurring female sex hormones

estradiol

progesterone

RU 486, also called **mifepristone,** is another synthetic hormone that blocks the effect of progesterone, and because of this, prevents implantation of a fertilized egg. It has been used in France since 1988 as a "morning after" pill to induce abortions, and is now available in the United States for this purpose as well. Like ethynylestradiol and norethindrone, RU 486 contains four rings joined together in the central portion of its carbon skeleton, with an alkynyl appendage bonded to the five-membered ring.

RU 486
mifepristone

> Recall from general chemistry that an **allotrope** is one of two or more structural forms of an element. For carbon, the two most common allotropes are diamond and graphite.

Histrionicotoxin is a diyne isolated in small quantities from the skin of *Dendrobates histrionicus,* a colorful South American frog (Figure 11.1). This toxin, secreted by the frog as a natural defense mechanism, was used as a poison on arrow tips by the Choco tribe of South America.

Unusual polyynes containing several carbon–carbon triple bonds joined in a ring or long chain have been prepared. Because these polyynes consist entirely of carbon, they are esoteric allotropes of carbon. Cyclo[18]carbon, a highly reactive compound, contains an 18-membered ring with alternating single and triple bonds. Long polymers of triple bonds containing up to 500 *sp* hybridized carbons have also been prepared.

Figure 11.1
Histrionicotoxin

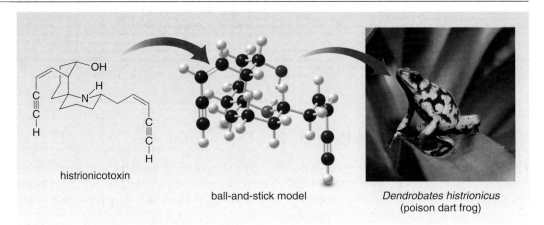

histrionicotoxin

ball-and-stick model

Dendrobates histrionicus
(poison dart frog)

- Histrionicotoxin is a defensive toxin that protects *Dendrobates histrionicus* from potential predators. These small "poison dart" frogs inhabit the moist humid floor of tropical rainforests, and are commonly found in western Ecuador and Colombia. Histrionicotoxin acts by interfering with nerve transmission in mammals, resulting in prolonged muscle contraction.

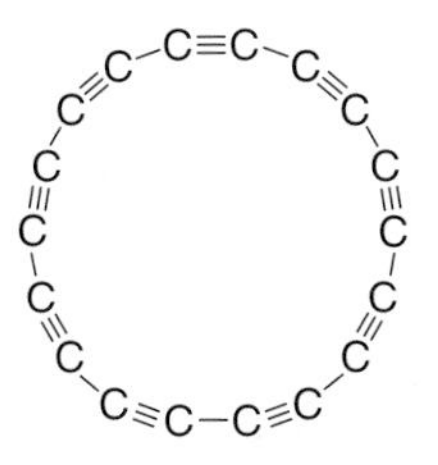

cyclo[18]carbon

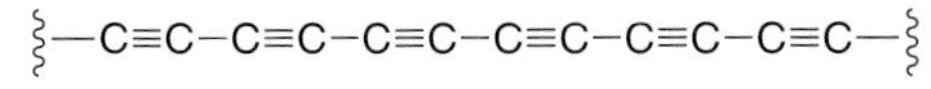

a polyyne with alternating single and triple bonds in a long carbon chain

11.5 Preparation of Alkynes

Alkynes are prepared by elimination reactions, as discussed in Section 8.10. **A strong base removes two equivalents of HX from a vicinal or geminal dihalide to yield an alkyne** by two successive E2 eliminations.

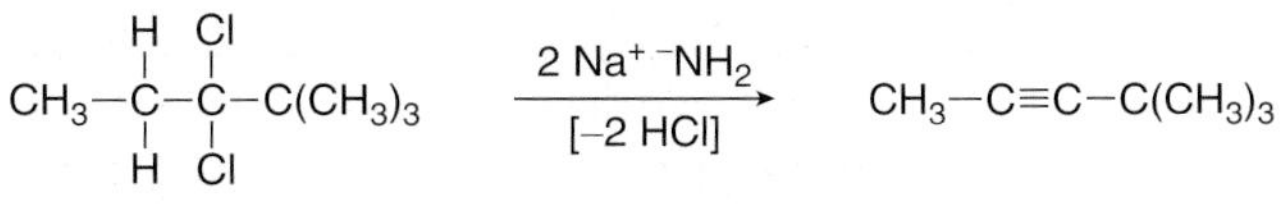

geminal dichloride

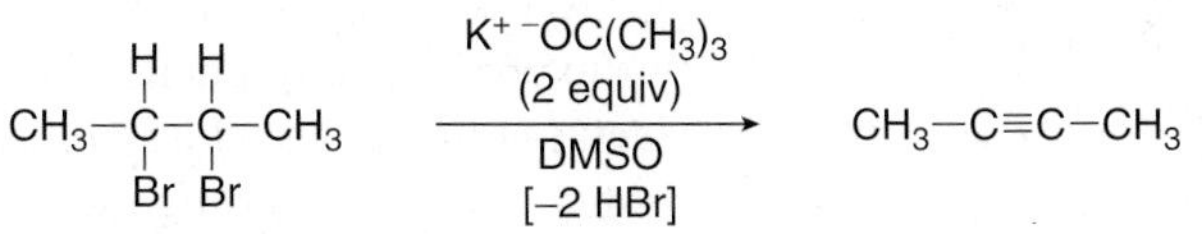

vicinal dibromide

Because vicinal dihalides are synthesized by adding halogens to alkenes, an alkene can be converted to an alkyne by the two-step process illustrated in Sample Problem 11.2.

SAMPLE PROBLEM 11.2 Convert alkene **A** into alkyne **B** by a stepwise method.

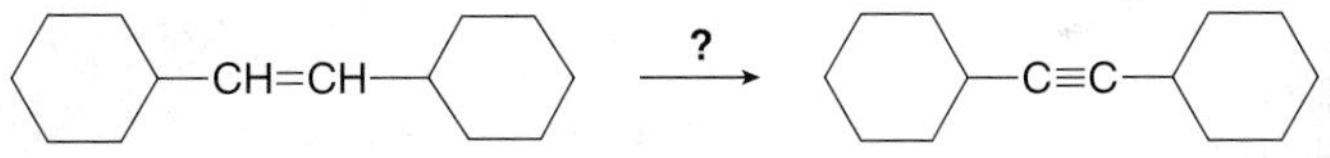

A **B**

SOLUTION

A two-step method is needed:

- **Addition of X_2** forms a vicinal dihalide.
- **Elimination** of two equivalents of HX forms two π bonds.

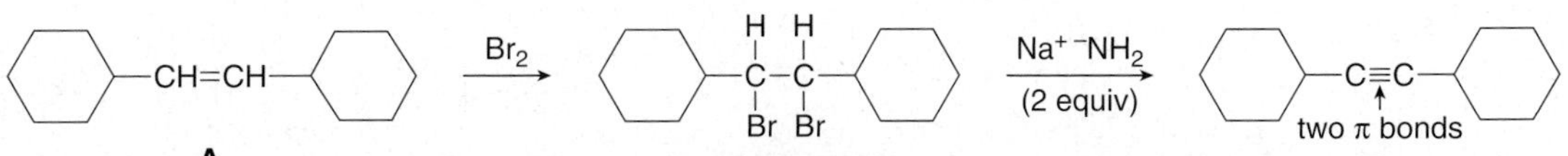

A vicinal dibromide **B**

- **This two-step process introduces one degree of unsaturation:** an alkene with one π bond is converted to an alkyne with two π bonds.

PROBLEM 11.8 Convert each compound to 1-hexyne, $HC{\equiv}CCH_2CH_2CH_2CH_3$.

a. $Br_2CH(CH_2)_4CH_3$ b. $CH_2{=}CCl(CH_2)_3CH_3$ c. $CH_2{=}CH(CH_2)_3CH_3$

PROBLEM 11.9 Treatment of 2,2-dibromobutane with two equivalents of strong base affords 1-butyne and 2-butyne, as well as a small amount of 1,2-butadiene. Draw a mechanism showing how each compound is formed. Which alkyne should be the major product?

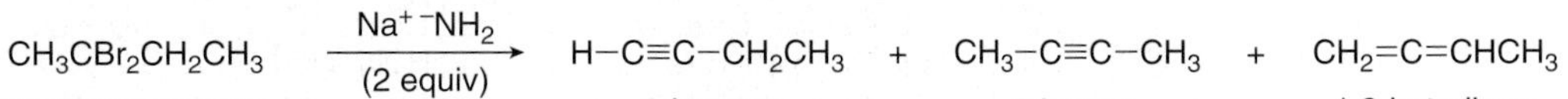

2,2-dibromobutane 1-butyne 2-butyne 1,2-butadiene

11.6   Introduction to Alkyne Reactions

All reactions of alkynes occur because they contain **easily broken π bonds** or, in the case of terminal alkynes, an **acidic, *sp* hybridized C–H bond.**

11.6A   Addition Reactions

Like alkenes, **alkynes undergo addition reactions because they contain weak π bonds.** Two sequential reactions take place: addition of one equivalent of reagent forms an alkene, which then adds a second equivalent of reagent to yield a product having **four new bonds.**

Alkynes are electron rich, as shown in the electrostatic potential map of acetylene in Figure 11.2. The two π bonds form a cylinder of electron density between the two *sp* hybridized carbon atoms, and this exposed electron density makes a triple bond nucleophilic. As a result, **alkynes react with electrophiles.** Four addition reactions are discussed in Chapter 11 and illustrated in Figure 11.3 with 1-butyne as starting material.

> The oxidation and reduction of alkynes, reactions that also involve addition, are discussed in Chapter 12.

Figure 11.2 Electrostatic potential map of acetylene

- The red electron-rich region is located between the two carbon atoms, forming a cylinder of electron density.

Figure 11.3 Four addition reactions of 1-butyne

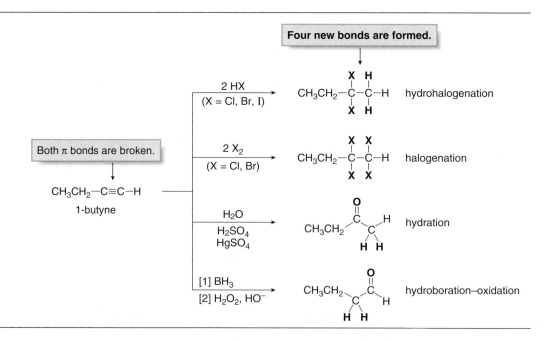

Recall from Section 2.5D that the acidity of a C–H bond increases as the percent *s*-character of C increases. Thus, *sp* hybridized C–H bonds (having a C atom with 50% *s*-character) are more acidic than *sp*^{2} and *sp*^{3} hybridized C–H bonds (having C atoms with 33% and 25% *s*-character, respectively).

11.6B Terminal Alkynes—Reaction as an Acid

Because *sp* **hybridized C–H bonds are more acidic than** *sp*^{2} **and** *sp*^{3} **hybridized C–H bonds,** terminal alkynes are readily deprotonated with strong base in a Brønsted–Lowry acid–base reaction. The resulting anion is called an **acetylide anion.**

$$R-C\equiv C-H \;+\; :B \;\rightleftharpoons\; R-C\equiv C:^- \;+\; H-B^+$$

terminal alkyne
p$K_a \approx 25$

acetylide anion

What bases can be used for this reaction? Because an acid–base equilibrium favors the weaker acid and base, only **bases having conjugate acids with pK_a values *higher* than the terminal alkyne—that is, pK_a values > 25—are strong enough** to form a significant concentration of acetylide anion. As shown in Table 11.1, $^-NH_2$ and H^- are strong enough to deprotonate a terminal alkyne, but ^-OH and ^-OR are not.

TABLE 11.1   A Comparison of Bases for Alkyne Deprotonation

	Base	pK_a of the conjugate acid
These bases are **strong** enough to deprotonate an alkyne.	$^-NH_2$	38
	H^-	35
These bases are **not** strong enough to deprotonate an alkyne.	^-OH	15.7
	^-OR	15.5–18

Why is this reaction useful? The acetylide anions formed by deprotonating terminal alkynes are **strong nucleophiles** that can react with a variety of electrophiles, as shown in Section 11.11.

$$R-C\equiv C:^- \;+\; E^+ \;\longrightarrow\; R-C\equiv C-E$$

nucleophile   electrophile

new bond

PROBLEM 11.10   Which bases can deprotonate acetylene? The pK_a values of the conjugate acids are given in parentheses.
a. CO_3^{2-} (pK_a = 10.2)   b. $CH_2=CH^-$ (pK_a = 44)   c. $(CH_3)_3CO^-$ (pK_a = 18)

11.7 Addition of Hydrogen Halides

Alkynes undergo **hydrohalogenation, the addition of hydrogen halides, HX** (X = Cl, Br, I). Two equivalents of HX are usually used: addition of one mole forms a **vinyl halide,** which then reacts with a second mole of HX to form a **geminal dihalide.**

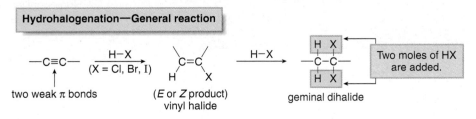

Addition of HX to an alkyne is another example of **electrophilic addition,** because the electrophilic (H) end of the reagent is attracted to the electron-rich triple bond.

♦ With two equivalents of HX, both H atoms bond to the *same* carbon.
♦ With a terminal alkyne, both H atoms bond to the *terminal* carbon; that is, the hydrohalogenation of alkynes follows Markovnikov's rule.

Examples   [1]   CH$_3$–C≡C–CH$_3$ $\xrightarrow{\text{H–Cl}}$ [first equivalent]

$$\underset{\substack{H \quad\quad Cl}}{\overset{\substack{CH_3 \quad CH_3}}{C=C}}$$

[Product can be *E* or *Z*.]

$\xrightarrow{\text{H–Cl}}$ [second equivalent]

CH$_3$–C–C–CH$_3$ with H Cl / H Cl

[2]   H–C≡C–CH$_2$CH$_3$ $\xrightarrow{\text{H–Br}}$

$$\underset{\substack{H \quad\quad Br}}{\overset{\substack{H \quad CH_2CH_3}}{C=C}}$$

$\xrightarrow{\text{H–Br}}$ H–C–C–CH$_2$CH$_3$ with H Br / H Br

Both H's end up on the terminal C.

♦ With only one equivalent of HX, the reaction stops with formation of the vinyl halide.

H–C≡C–CH$_3$ $\xrightarrow[\text{(1 equiv)}]{\text{H–Cl}}$ $\underset{\substack{H \quad\quad Cl}}{\overset{\substack{H \quad CH_3}}{C=C}}$

a vinyl chloride
(2-chloropropene)

PROBLEM 11.11   Draw the organic products formed in each reaction.

a.   CH$_3$CH$_2$CH$_2$CH$_2$–C≡C–H $\xrightarrow{\text{2 HI}}$ b.   CH$_3$–C≡C–CH$_2$CH$_3$ $\xrightarrow{\text{2 HBr}}$

One currently accepted mechanism for the addition of two equivalents of HX to an alkyne involves **two steps for each addition of HX:** addition of H$^+$ (from HX) to form a carbocation, followed by nucleophilic attack of X$^-$. Mechanism 11.1 illustrates the addition of HBr to 1-butyne to yield 2,2-dibromobutane. Each two-step mechanism is similar to the two-step addition of HBr to *cis*-2-butene discussed in Section 10.9.

MECHANISM 11.1

Electrophilic Addition of HX to an Alkyne

Part [1] Addition of HBr to form a vinyl halide

H–C≡C–CH$_2$CH$_3$ $\xrightarrow{[1]}$ $\underset{\substack{H}}{\overset{\substack{H}}{C}}=\overset{+}{C}$–CH$_2CH_3$ $\xrightarrow{[2]}$ $\underset{\substack{H \quad\quad Br}}{\overset{\substack{H \quad CH_2CH_3}}{C=C}}$

1-butyne

H–Br

vinyl carbocation

Br$^-$

a vinyl bromide

♦ The π bond attacks the H atom of HBr to form a new C–H bond, generating a **vinyl carbocation.** Addition follows Markovnikov's rule: H$^+$ adds to the less substituted carbon atom to form the **more substituted, more stable carbocation.** Nucleophilic attack of Br$^-$ then forms a vinyl bromide; one mole of HBr has now been added.

Part [2] Addition of HBr to form a geminal dihalide

H–Br

$\underset{\substack{H \quad\quad Br}}{\overset{\substack{H \quad CH_2CH_3}}{C=C}}$ $\xrightarrow{[3]}$ H–C–$\overset{+}{C}$ with H / CH$_2$CH$_3$ / H Br $\xrightarrow{[4]}$ H–C–C–CH$_2$CH$_3$ with H Br / H Br

Br$^-$

carbocation

2,2-dibromobutane

♦ **The second addition of HBr occurs in the same two-step manner.** Addition of H$^+$ to the π bond of the vinyl bromide generates a carbocation. Nucleophilic attack of Br$^-$ then forms a geminal dibromide (2,2-dibromobutane), and two moles of HBr have now been added.

Because of the instability of a vinyl carbocation, other mechanisms for HX addition have been proposed that avoid formation of a discrete carbocation. It is likely that more than one mechanism occurs, depending in part on the identity of the alkyne substrate.

The formation of both carbocations (in Steps [1] and [3]) deserves additional scrutiny. **The vinyl carbocation formed in Step [1] is *sp* hybridized and therefore less stable than a 2° *sp²* hybridized carbocation** (Section 7.18). This makes electrophilic addition of HX to an alkyne *slower* than electrophilic addition of HX to an alkene, even though alkynes are more polarizable and have more loosely held π electrons than alkenes.

sp hybridized

This unstable carbocation slows the first step, making **an alkyne less reactive than an alkene** towards HX.

vinyl carbocation

In Step [3] two carbocations are possible but only one is formed. Markovnikov addition in Step [3] places the H on the terminal carbon (C1) to form the more substituted carbocation **A,** rather than the less substituted carbocation **B.** Because the more stable carbocation is formed faster—another example of the Hammond postulate—carbocation **A** must be more stable than carbocation **B.**

new bond

new bond

NOT

more stable carbocation

Why is carbocation **A,** having a positive charge on a carbon that also has a Br atom, more stable? Shouldn't the electronegative Br atom withdraw electron density from the positive charge, and thus destabilize it? It turns out that **A is stabilized by resonance** but **B** is not. Two resonance structures can be drawn for carbocation **A,** but only one Lewis structure can be drawn for carbocation **B.**

The positive charge is delocalized.

two resonance structures for **A**

hybrid

◆ **Resonance stabilizes a molecule by delocalizing charge and electron density.**
◆ **Thus, halogens stabilize an adjacent positive charge by resonance.**

Markovnikov's rule applies to the addition of HX to vinyl halides because **addition of H⁺ forms a resonance-stabilized carbocation.** As a result, addition of each equivalent of HX to a triple bond forms the more stable carbocation, so that both H atoms bond to the less substituted C.

PROBLEM 11.12   Draw an additional resonance structure for each cation.

a.   b. $CH_3—\overset{..}{O}—\overset{+}{C}H_2$

11.8   Addition of Halogen

Halogens, X_2 (X = Cl or Br), add to alkynes in much the same way they add to alkenes (Section 10.13). Addition of one mole of X_2 forms a **trans dihalide,** which can then react with a second mole of X_2 to yield a **tetrahalide.**

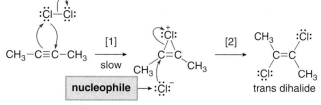

Each addition of X_2 involves a two-step process with a **bridged halonium ion** intermediate, reminiscent of the addition of X_2 to alkenes (Section 10.13). A trans dihalide is formed after addition of one equivalent of X_2 because the intermediate **halonium ion ring is opened upon backside attack of the nucleophile.** Mechanism 11.2 illustrates the addition of two equivalents of Cl_2 to $CH_3C{\equiv}CCH_3$ to form $CH_3CCl_2CCl_2CH_3$.

MECHANISM 11.2

Addition of X_2 to an Alkyne—Halogenation

Part [1] Addition of X_2 to form a trans dihalide

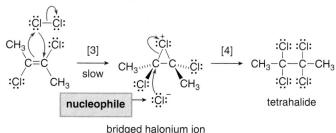

◆ Two bonds are broken and two are formed in Step [1] to generate a **bridged halonium ion.** This strained three-membered ring is highly unstable, making it amenable to opening of the ring in the second step.

◆ Nucleophilic attack by Cl^- from the back side forms the trans dihalide in Step [2].

Part [2] Addition of X_2 to form a tetrahalide

◆ **Electrophilic addition** of Cl^+ in Step [3] forms the bridged halonium ion ring, which is opened with Cl^- to form the tetrahalide in Step [4].

PROBLEM 11.13   Draw the products formed when 3-hexyne is treated with each reagent: (a) Br_2 (2 equiv); (b) Cl_2 (1 equiv).

PROBLEM 11.14   Explain the following result. Although alkenes are generally more reactive than alkynes towards electrophiles, the reaction of Cl_2 with 2-butyne can be stopped after one equivalent of Cl_2 has been added.

11.9 Addition of Water

Although the addition of H_2O to an alkyne resembles the acid-catalyzed addition of H_2O to an alkene in some ways, an important difference exists. In the presence of strong acid or Hg^{2+} catalyst, the **elements of H_2O add to the triple bond,** but the initial addition product, an **enol,** is unstable and rearranges to a product containing a **carbonyl group**—that is, a **C=O.** A carbonyl compound having two alkyl groups bonded to the C=O carbon is called a **ketone.**

Hydration—General reaction

R−C≡C−R $\xrightarrow[\substack{H_2SO_4 \\ HgSO_4}]{H_2O}$ $\left[\begin{array}{c} \underset{H}{\overset{R}{}}C=C\underset{OH}{\overset{R}{}} \end{array} \right]$ ⇌ $\underset{H\ \ H}{\overset{O}{R-\overset{||}{C}-C-R}}$ ← **carbonyl** group

less stable **enol**   **ketone**

H₂O has been added.

Internal alkynes undergo hydration with concentrated acid, whereas terminal alkynes require the presence of an additional Hg^{2+} catalyst—usually $HgSO_4$—to yield methyl ketones by **Markovnikov addition of H$_2$O.**

Examples

Because an enol contains both a C=C and a hydroxy group, the name **enol** comes from alk**ene** + alcoh**ol**.

$CH_3-C\equiv C-CH_3$ $\xrightarrow[H_2SO_4]{H_2O}$ $\left[\begin{array}{c} \underset{H}{\overset{CH_3}{}}C=C\underset{OH}{\overset{CH_3}{}} \end{array} \right]$ ⇌ $\underset{H\ \ H}{\overset{O}{CH_3-\overset{||}{C}-C-CH_3}}$

enol

HgSO₄ is often used in the hydration of internal alkynes as well, because hydration can be carried out under milder reaction conditions.

$H-C\equiv C-CH_3$ $\xrightarrow[\substack{H_2SO_4 \\ HgSO_4}]{H_2O}$ $\left[\begin{array}{c} \underset{H}{\overset{H}{}}C=C\underset{OH}{\overset{CH_3}{}} \end{array} \right]$ ⇌ $\underset{H\ \ H}{\overset{O}{H-\overset{||}{C}-C-CH_3}}$

enol   methyl ketone

ketones

Markovnikov addition of H$_2$O

H adds to the terminal C.

Let's first examine the conversion of a general enol **A** to the carbonyl compound **B**. **A** and **B** are called **tautomers: A** is the *enol form* and **B** is the *keto form* of the tautomer.

◆ *Tautomers* are constitutional isomers that differ in the location of a double bond and a hydrogen atom. Two tautomers are in equilibrium with each other.

enol   **ketone**

$\underset{\underset{A}{\text{enol form}}}{\overset{}{}C=C}$ ⇌ $\underset{\underset{B}{\text{keto form}}}{\overset{}{}-\overset{|}{C}-\overset{|}{C}}$

Tautomers differ in the position of a double bond and a hydrogen atom. In Chapter 23 an in-depth discussion of keto–enol tautomers is presented.

◆ An enol tautomer has an O−H group bonded to a C=C.
◆ A keto tautomer has a C=O and an additional C−H bond.

Equilibrium favors the keto form largely because a C=O is much stronger than a C=C. Tautomerization, the process of converting one tautomer into another, is catalyzed by both acid and base. Under the strongly acidic conditions of hydration, tautomerization of the enol to the keto form occurs rapidly by a two-step process: **protonation,** followed by **deprotonation** as shown in Mechanism 11.3

MECHANISM 11.3

Tautomerization in Acid

Step [1] Protonation of the enol double bond

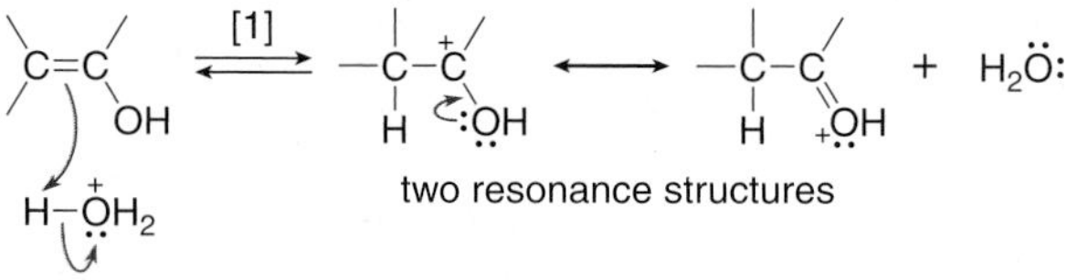

two resonance structures

◆ **Protonation** of the enol C=C with acid (H_3O^+) adds H^+ to form a **resonance-stabilized carbocation.**

Step [2] Deprotonation of the OH group

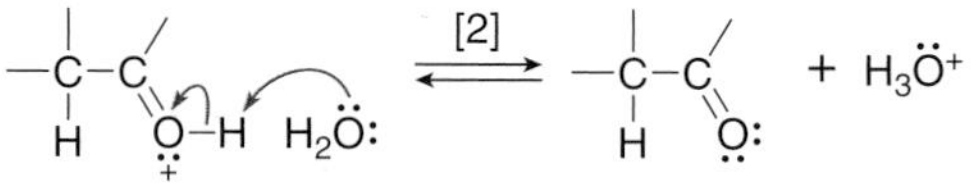

◆ **Loss of a proton** forms the carbonyl group. This step can be drawn with either resonance structure as starting material. Because the acid used in Step [1] is re-formed in Step [2], tautomerization is **acid catalyzed.**

Hydration of an internal alkyne with strong acid forms an enol by a mechanism similar to that of the acid-catalyzed hydration of an alkene (Section 10.12). Mechanism 11.4 illustrates the hydration of 2-butyne with H_2O and H_2SO_4. Once formed, the enol then tautomerizes to the more stable keto form by protonation followed by deprotonation.

MECHANISM 11.4

Hydration of an Alkyne

Step [1] Addition of the electrophile (H^+) to a π bond

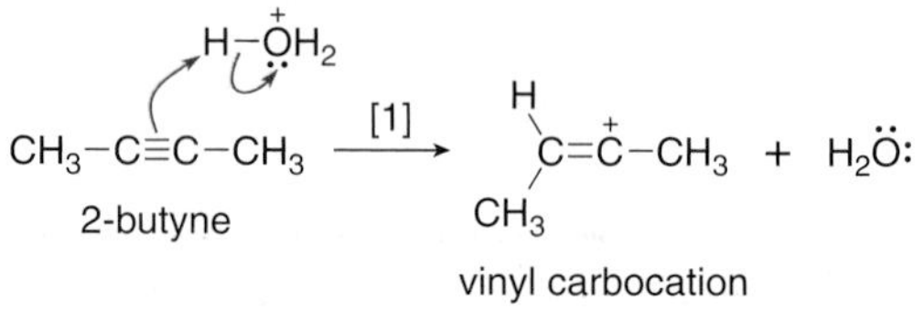

2-butyne

vinyl carbocation

◆ Addition of H^+ (from H_3O^+) forms an *sp* hybridized **vinyl carbocation.**

Steps [2] and [3] Nucleophilic attack of H_2O and loss of a proton

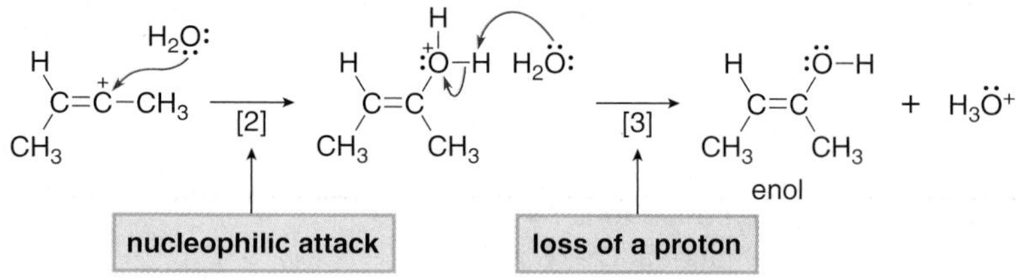

| nucleophilic attack | loss of a proton |

enol

◆ **Nucleophilic attack of H_2O** on the carbocation followed by loss of a proton forms the enol.

Steps [4] and [5] Tautomerization

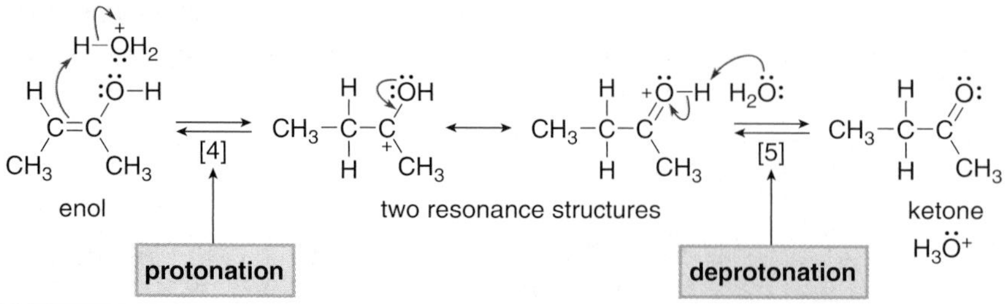

enol

two resonance structures

ketone

| protonation | deprotonation |

◆ **Tautomerization of the enol to the keto form** occurs by protonation of the double bond to form a carbocation. Loss of a proton from this **resonance-stabilized carbocation** generates the more stable keto form.

SAMPLE PROBLEM 11.3   Draw the enol intermediate and the ketone product formed in the following reaction.

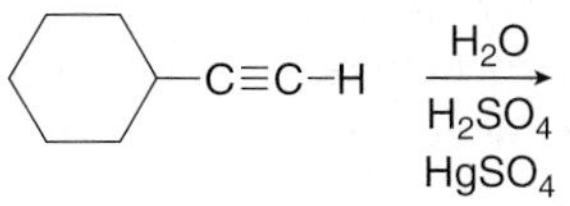

$$\text{cyclohexyl}-C\equiv C-H \xrightarrow[\substack{H_2SO_4 \\ HgSO_4}]{H_2O}$$

SOLUTION

First, form the enol by adding H_2O to the triple bond with the **H bonded to the less-substituted terminal carbon.**

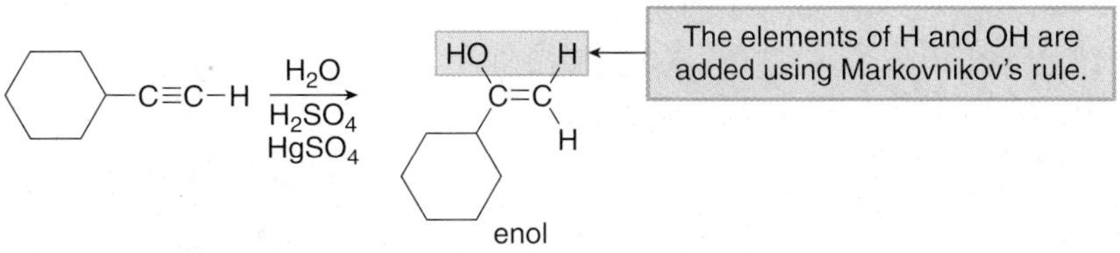

$$\text{cyclohexyl}-C\equiv C-H \xrightarrow[\substack{H_2SO_4 \\ HgSO_4}]{H_2O}$$

| The elements of H and OH are added using Markovnikov's rule. |

enol

To convert the enol to the keto tautomer, add a proton to the $C=C$ and remove a proton from the OH group. In tautomerization, the $C-OH$ bond is converted to a $C=O$, and a new $C-H$ bond is formed on the other enol carbon.

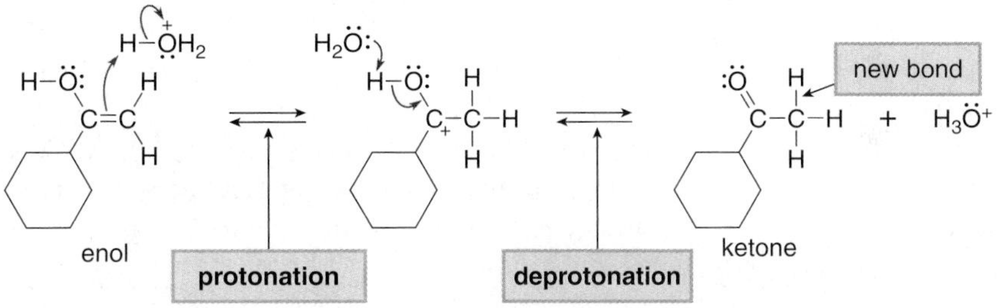

enol   **protonation**   **deprotonation**   ketone $+ \ H_3\ddot{O}^+$

| new bond |

• **The overall result is the addition of H_2O to a triple bond to form a ketone.**

PROBLEM 11.15   Draw the keto tautomer of each enol.

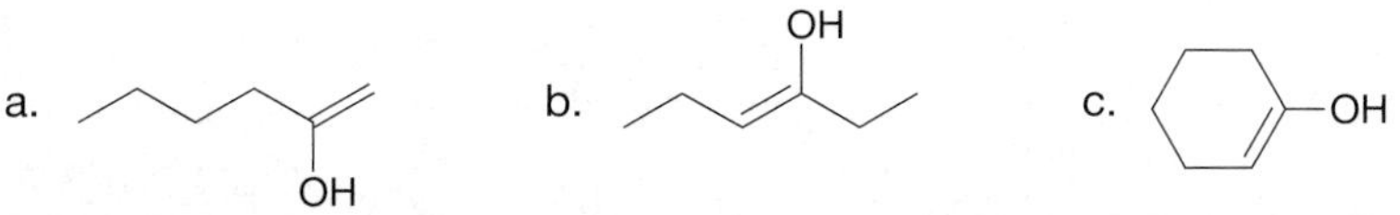

a.   b.   c.

PROBLEM 11.16   What two enols are formed when 2-pentyne is treated with H_2O, H_2SO_4, and $HgSO_4$? Draw the ketones formed from these enols after tautomerization.

PROBLEM 11.17   What two different alkynes yield 2-butanone from hydration with H_2O, H_2SO_4, and $HgSO_4$?

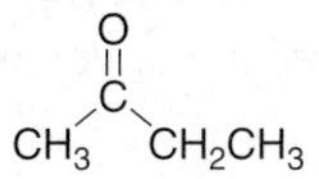

$$\underset{\text{2-butanone}}{CH_3\overset{\overset{\displaystyle O}{\|}}{C}CH_2CH_3}$$

11.10   Hydroboration–Oxidation

Hydroboration–oxidation is a two-step reaction sequence that converts an alkyne to a carbonyl compound.

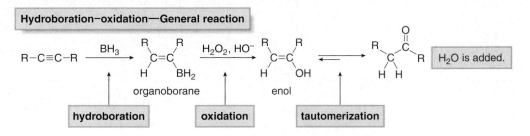

| Hydroboration–oxidation—General reaction |

$$R-C\equiv C-R \xrightarrow{BH_3} \underset{\text{organoborane}}{\overset{R}{\underset{H}{C}}=\overset{R}{\underset{BH_2}{C}}} \xrightarrow{H_2O_2,\ HO^-} \underset{\text{enol}}{\overset{R}{\underset{H}{C}}=\overset{R}{\underset{OH}{C}}} \rightleftharpoons \underset{H\ H}{\overset{R}{C}\overset{\overset{\displaystyle O}{\|}}{C}R}$$

| H_2O is added. |

hydroboration   **oxidation**   **tautomerization**

◆ Addition of borane forms an organoborane.
◆ Oxidation with basic H_2O_2 forms an enol.
◆ Tautomerization of the enol forms a carbonyl compound.
◆ The overall result is addition of H_2O to a triple bond.

Hydroboration–oxidation of an *internal* alkyne forms a ketone. **Hydroboration of a *terminal* alkyne adds BH_2 to the less substituted, terminal carbon.** After oxidation to the enol, tautomerization yields an **aldehyde,** a carbonyl compound having a hydrogen atom bonded to the carbonyl carbon.

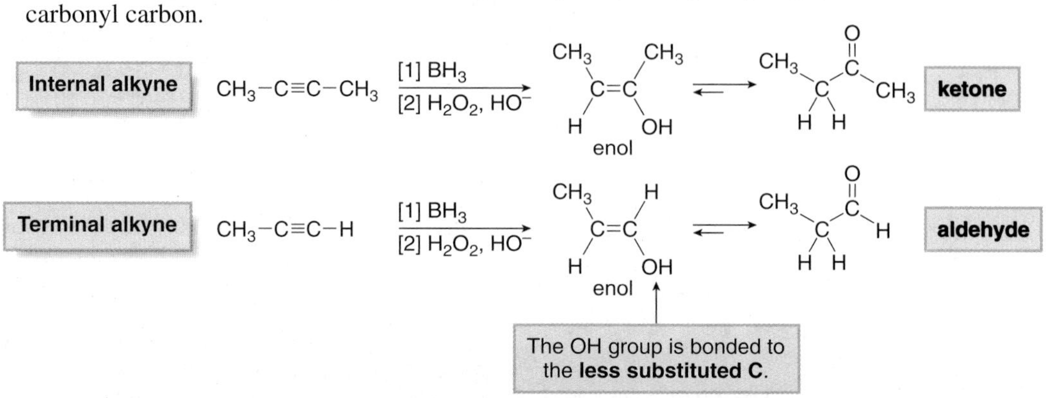

Hydration (H_2O, H_2SO_4, and $HgSO_4$) and **hydroboration–oxidation** (BH_3 followed by H_2O_2, HO^-) both **add the elements of H_2O across a triple bond.** Sample Problem 11.4 shows that different constitutional isomers are formed from terminal alkynes in these two reactions despite their similarities.

SAMPLE PROBLEM 11.4   Draw the product formed in each reaction.

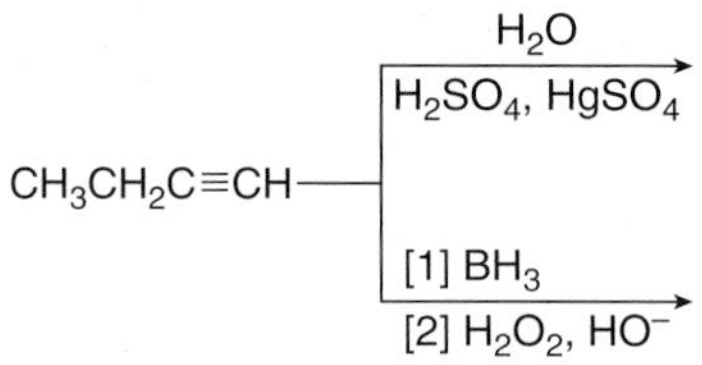

SOLUTION

With H_2O + H_2SO_4 + $HgSO_4$, electrophilic addition of H and OH places the **H atom on the less substituted carbon** of the alkyne to form a **ketone** after tautomerization. In contrast, addition of BH_3 places the **BH_2 group on the less substituted terminal carbon** of the alkyne. Oxidation and tautomerization yield an **aldehyde.**

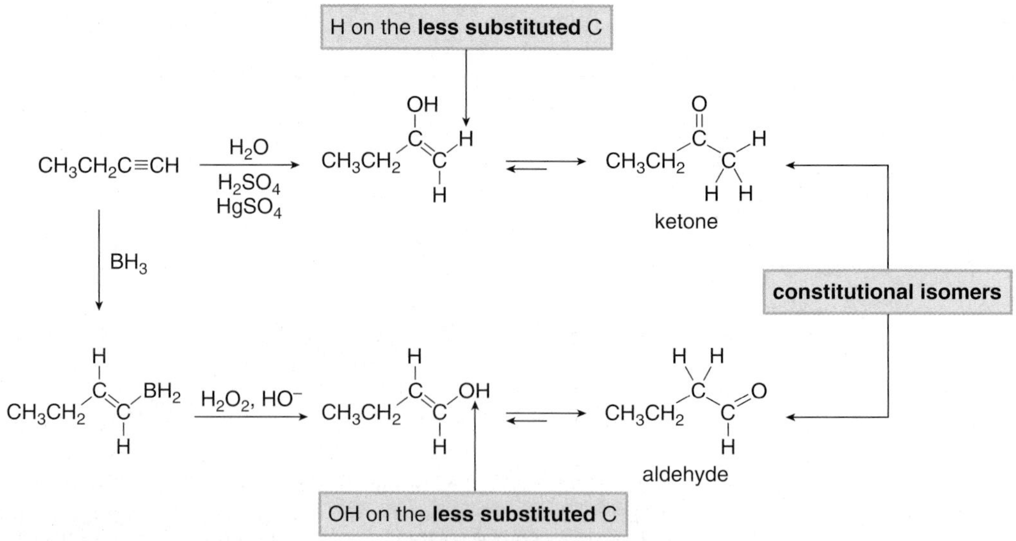

♦ Addition of H_2O using H_2O, H_2SO_4, and $HgSO_4$ forms methyl ketones from terminal alkynes.
♦ Addition of H_2O using BH_3, then H_2O_2, HO^- forms aldehydes from terminal alkynes.

PROBLEM 11.18   Draw the products of each reaction.

a. $(CH_3)_2CHCH_2-C{\equiv}C-H$ $\xrightarrow[H_2SO_4,\ HgSO_4]{H_2O}$

b. $(CH_3)_2CHCH_2-C{\equiv}C-H$ $\xrightarrow[[2]\ H_2O_2,\ HO^-]{[1]\ BH_3}$

c. ⬡$-C{\equiv}CH$ $\xrightarrow[H_2SO_4,\ HgSO_4]{H_2O}$

d. ⬡$-C{\equiv}CH$ $\xrightarrow[[2]\ H_2O_2,\ HO^-]{[1]\ BH_3}$

11.11   Reaction of Acetylide Anions

Terminal alkynes are readily converted to acetylide anions with strong base. These anions are strong nucleophiles, capable of reacting with electrophiles such as alkyl halides and epoxides.

nucleophile	electrophile

$$R-C{\equiv}C-H \ +\ \ :B \longrightarrow R-C{\equiv}C:^- \ +\ E^+ \longrightarrow R-C{\equiv}C-E$$

terminal alkyne
p$K_a \approx 25$

acetylide anion

new bond

common bases (:B): **NaNH₂, NaH**

11.11A   Reaction of Acetylide Anions with Alkyl Halides

Acetylide anions react with unhindered alkyl halides to yield products of nucleophilic substitution.

$$R-X \ +\ \ ^-:C{\equiv}C-R' \xrightarrow{S_N2} R-C{\equiv}C-R' \ +\ X^-$$

nucleophile

new bond

Because acetylide anions are strong nucleophiles, the mechanism of nucleophilic substitution is **S_N2,** and thus the **reaction is fastest with CH_3X and 1° alkyl halides.** Terminal alkynes (Reaction [1]) or internal alkynes (Reaction [2]) can be prepared depending on the identity of the acetylide anion.

nucleophile			leaving group

[1] $CH_3-Cl \ +\ \ ^-:C{\equiv}C-H \xrightarrow{S_N2} CH_3-C{\equiv}C-H \ +\ Cl^-$

new C–C bond

[2] $CH_3CH_2-Br \ +\ \ ^-:C{\equiv}C-CH_3 \xrightarrow{S_N2} CH_3CH_2-C{\equiv}C-CH_3 \ +\ Br^-$

2 C's   3 C's   5 C's

♦ **Nucleophilic substitution with acetylide anions forms new carbon–carbon bonds.**

Because organic compounds consist of a carbon framework, reactions that form carbon–carbon bonds are especially useful. In Reaction [2], for example, nucleophilic attack of a three-carbon acetylide anion on a two-carbon alkyl halide yields a five-carbon alkyne as product.

Although nucleophilic substitution with acetylide anions is a very valuable carbon–carbon bond-forming reaction, it has the same limitations as any S_N2 reaction. **Steric hindrance around the leaving group causes 2° and 3° alkyl halides to undergo elimination by an E2 mechanism,** as shown with 2-bromo-2-methylpropane. Thus, nucleophilic substitution with acetylide anions leads to carbon–carbon bond formation only with CH_3X and 1° alkyl halides.

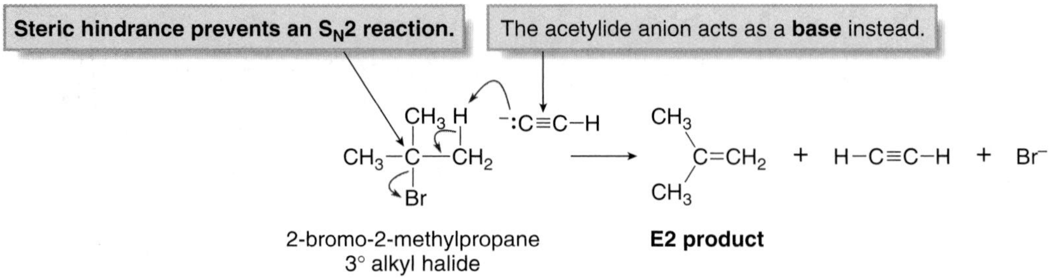

Steric hindrance prevents an S_N2 reaction.	The acetylide anion acts as a **base** instead.

2-bromo-2-methylpropane
3° alkyl halide   **E2 product**

SAMPLE PROBLEM 11.5

Draw the organic products formed in each reaction.

a. $CH_3CH_2CH_2CH_2-Cl$ + $^-$:C≡C–H ⟶

b. 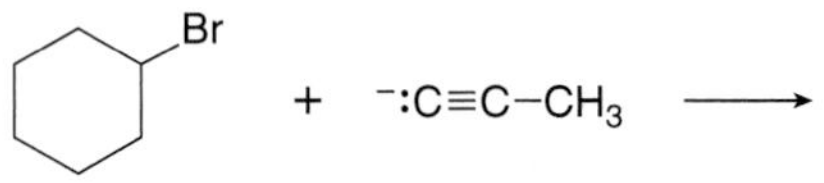 + $^-$:C≡C–CH_3 ⟶

SOLUTION

[a] Because the alkyl halide is **1°** and the acetylide anion is a strong nucleophile, substitution occurs by an **S_N2** mechanism, resulting in a new C–C bond.

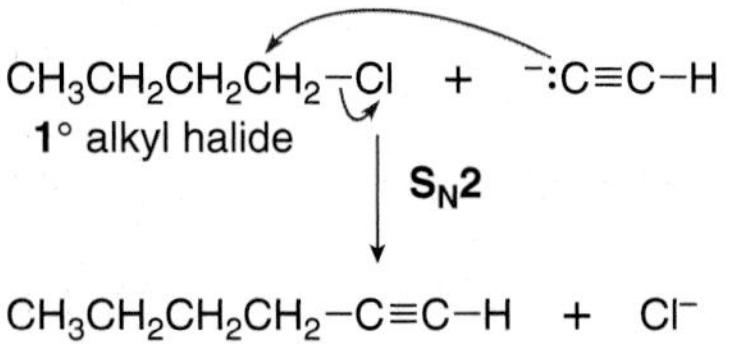

$CH_3CH_2CH_2CH_2-Cl$ + $^-$:C≡C–H
1° alkyl halide

S_N2

$CH_3CH_2CH_2CH_2-C≡C–H$ + Cl^-

[b] Because the alkyl halide is **2°,** elimination occurs by an **E2** mechanism. A new C–C bond is *not* formed in this case.

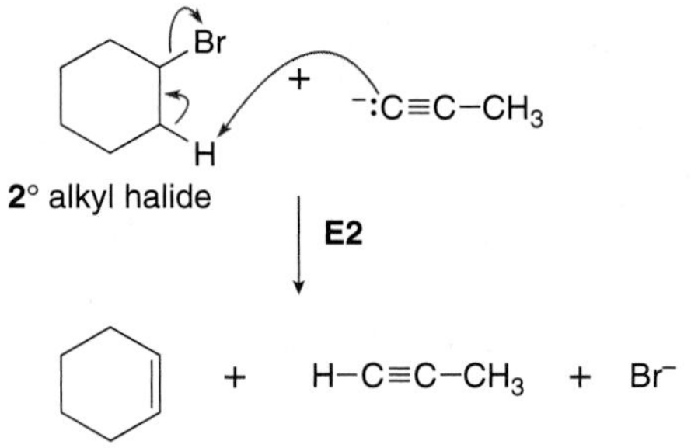

2° alkyl halide

E2

+ H–C≡C–CH_3 + Br^-

PROBLEM 11.19

Draw the organic products formed in each reaction.

a. H–C≡C–H $\xrightarrow[\text{[2] }(CH_3)_2CHCH_2Cl]{\text{[1] NaH}}$

b.

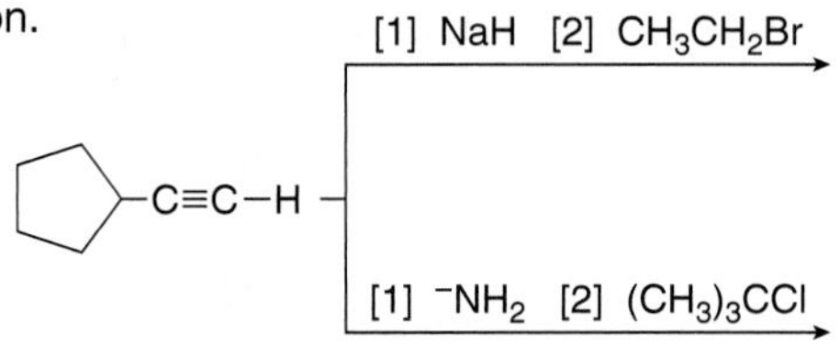

[1] NaH [2] CH_3CH_2Br

[1] $^-NH_2$ [2] $(CH_3)_3CCl$

PROBLEM 11.20

What acetylide anion and alkyl halide can be used to prepare each alkyne? Indicate all possibilities when more than one route will work.
a. $(CH_3)_2CHCH_2C≡CH$ b. $CH_3C≡CCH_2CH_2CH_2CH_3$ c. $(CH_3)_3CC≡CCH_2CH_3$

Because acetylene has two *sp* hybridized C–H bonds, two sequential reactions can occur to form **two new carbon–carbon bonds,** as shown in Sample Problem 11.6.

SAMPLE PROBLEM 11.6

Identify the terminal alkyne **A** and the internal alkyne **B** in the following reaction sequence.

$$H–C≡C–H \xrightarrow[\text{[2] }CH_3Br]{\text{[1] }^-NH_2} \textbf{A} \xrightarrow[\text{[2] }CH_3CH_2Cl]{\text{[1] }^-NH_2} \textbf{B}$$

SOLUTION

In each step, the base $^-NH_2$ removes an *sp* hybridized proton, and the resulting acetylide anion reacts as a nucleophile with an alkyl halide to yield an S_N2 product. The first two-step reaction sequence forms the **terminal alkyne A** by nucleophilic attack of the acetylide anion on CH_3Br.

first new C—C bond

$$H-C\equiv C-H \quad + \quad {}^-NH_2 \quad\longrightarrow\quad H-C\equiv C:^- \quad + \quad CH_3-Br \quad\longrightarrow\quad H-C\equiv C-CH_3 \quad + \quad Br^-$$

acetylide anion   terminal alkyne

+ NH₃   **A**

The second two-step reaction sequence forms the **internal alkyne B** by nucleophilic attack of the acetylide anion on CH_3CH_2Cl.

second new C—C bond

$$^-NH_2 \quad + \quad H-C\equiv C-CH_3 \quad\longrightarrow\quad {}^-:C\equiv C-CH_3 \quad\longrightarrow\quad CH_3CH_2-C\equiv C-CH_3 \quad + \quad Cl^-$$

acetylide anion   internal alkyne

$$CH_3CH_2-Cl \quad + \quad NH_3$$ **B**

The soft coral *Capnella imbricata*

Sample Problem 11.6 illustrates how a five-carbon product can be prepared from three smaller molecules by forming two new carbon–carbon bonds.

$$CH_3CH_2-Cl \qquad H-C\equiv C-H \qquad CH_3-Br$$

The internal alkyne is prepared from three simpler reactants. →→ $$CH_3CH_2-C\equiv C-CH_3$$

new C—C bonds

Carbon–carbon bond formation with acetylide anions is a valuable reaction used in the synthesis of numerous natural products. Two examples include **capnellene,** isolated from the soft coral *Capnella imbricata,* and **niphatoxin B,** isolated from a red sea sponge, as shown in Figure 11.4.

PROBLEM 11.21   Show how $HC\equiv CH$, CH_3CH_2Br, and $(CH_3)_2CHCH_2CH_2Br$ can be used to prepare $CH_3CH_2C\equiv CCH_2CH_2CH(CH_3)_2$. Show all reagents, and use curved arrows to show movement of electron pairs.

PROBLEM 11.22   Explain why 2,2,5,5-tetramethyl-3-hexyne can't be made using acetylide anions.

Figure 11.4   Use of acetylide anion reactions in the synthesis of two marine natural products

[1] $^-:C\equiv CH$
[2] H_2O

new C—C bond

—C≡CH   several steps   capnellene

RO———Br

[1] $HC\equiv C$——OH   + base
[2] H_2O

RO————C≡C——OH   several steps   niphatoxin B

new C—C bond

11.11B   Reaction of Acetylide Anions with Epoxides

Acetylide anions are strong nucleophiles that open epoxide rings by an S_N2 mechanism. This reaction also results in the formation of a **new carbon–carbon bond.** Backside attack occurs at the less substituted end of the epoxide.

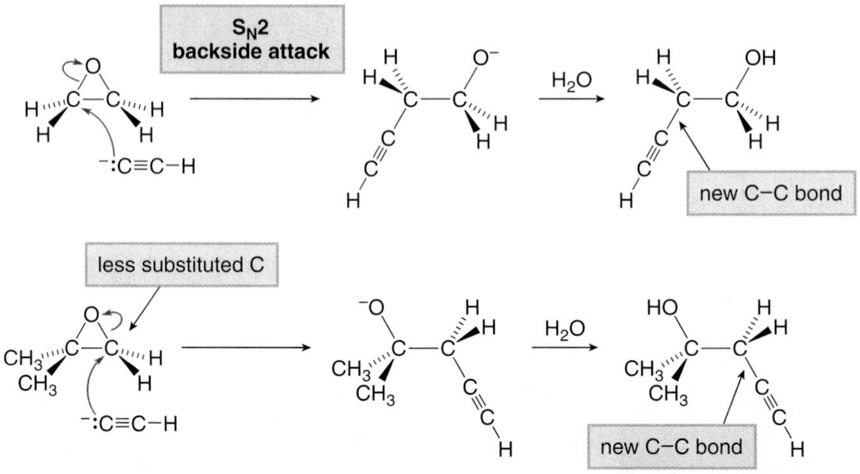

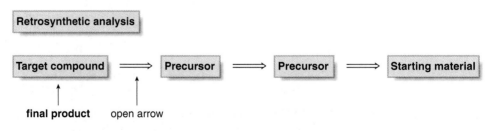

> Opening of epoxide rings with strong nucleophiles was first discussed in Section 9.15A.

PROBLEM 11.23   Draw the products of each reaction.

a. [1] $^-:C\equiv C-H$ [2] H_2O

b. [1] $^-:C\equiv C-H$ [2] H_2O

11.12   Synthesis

The reactions of acetylide anions give us an opportunity to examine organic synthesis more systematically. Performing a multistep synthesis can be difficult. Not only must you know the reactions for a particular functional group, but you must also put these reactions in a logical order, a process that takes much practice to master.

11.12A   General Terminology and Conventions

To plan a synthesis of more than one step, we use the process of **retrosynthetic analysis**—that is, working backwards from the desired product to determine the starting materials from which it is made (Section 10.18). To write a synthesis working backwards from the product to the starting material, an **open arrow** ($\Rightarrow$) is used to indicate that the product is drawn on the left and the starting material on the right.

> Carefully read the directions for each synthesis problem. Sometimes a starting material is specified, whereas at other times you must begin with a compound that meets a particular criterion; for example, you may be asked to synthesize a compound from alcohols having five or fewer carbon atoms. These limitations are meant to give you some direction in planning a multistep synthesis.

The product of a synthesis is often called the **target compound.** Using retrosynthetic analysis, we must determine what compound can be converted to the target compound by a single reaction. That is, **what is the immediate precursor of the target compound?** After an appropriate precursor is identified, this process is continued until we reach a specified starting material. Sometimes multiple retrosynthetic pathways are examined before a particular route is decided upon.

Retrosynthetic analysis

| Target compound | $\Rightarrow$ | Precursor | $\Rightarrow$ | Precursor | $\Rightarrow$ | Starting material |

final product   open arrow

In designing a synthesis, reactions are often divided into two categories:

♦ Those that form new carbon–carbon bonds.
♦ Those that convert one functional group into another—that is, functional group interconversions, abbreviated as FGIs.

Appendix H lists the carbon–carbon bond forming reactions encountered in this text.

Carbon–carbon bond forming reactions are central to organic synthesis because simpler and less valuable starting materials can be converted to more complex products. Keep in mind that whenever the product of a synthesis has more carbon–carbon bonds than the starting material, the synthesis must contain at least one of these reactions.

How To Develop a Retrosynthetic Analysis

Step [1] **Compare the carbon skeletons of the starting material and product.**
- If the product has more carbon–carbon σ bonds than the starting material, the synthesis must form one or more C–C bonds. If not, only functional group interconversion occurs.
- Match the carbons in the starting material with those in the product, to see where new C–C bonds must be added or where functional groups must be changed.

Step [2] **Concentrate on the functional groups in the starting material and product and ask:**
- What methods introduce the functional groups in the product?
- What kind of reactions does the starting material undergo?

Step [3] **Work backwards from the product and forwards from the starting material.**
- Ask: What is the immediate precursor of the product?
- Compare each precursor to the starting material to determine if there is a one-step reaction that converts one to the other. Continue this process until the starting material is reached.
- Always generate simpler precursors when working backwards.
- Use fewer steps when multiple routes are possible.
- Keep in mind that you may need to evaluate several different precursors for a given compound.

Step [4] **Check the synthesis by writing it in the synthetic direction.**
- To check a retrosynthetic analysis, write out the steps beginning with the starting material, indicating all necessary reagents.

11.12B Examples of Multistep Synthesis

Retrosynthetic analysis with acetylide anions is illustrated in Sample Problems 11.7 and 11.8.

SAMPLE PROBLEM 11.7 Devise a synthesis of $HC \equiv CCH_2CH_2CH_3$ from $HC \equiv CH$ and any other organic or inorganic reagents.

RETROSYNTHETIC ANALYSIS

The two C's in the starting material match up with the two *sp* hybridized C's in the product, so a three-carbon unit must be added.

Thinking backwards . . .
[1] Form a new C–C bond using an acetylide anion and a 1° alkyl halide.
[2] Prepare the acetylide anion from acetylene by treatment with base.

SYNTHESIS

Deprotonation of $HC \equiv CH$ with NaH forms the acetylide anion, which undergoes S_N2 reaction with an alkyl halide to form the target compound, a five-carbon alkyne.

A two-step process:

$$H-C≡C-H \xrightarrow{Na^+H^-} H-C≡C:^- + Cl-CH_2CH_2CH_3 \longrightarrow H-C≡C-CH_2CH_2CH_3 + Cl^-$$
$$+ H_2$$

acid–base reaction S_N2 reaction   target compound

SAMPLE PROBLEM 11.8   Devise a synthesis of the following compound from starting materials having two carbons or fewer.

$$CH_3CH_2 \overset{O}{\underset{}{\overset{||}{C}}} CH_3 \implies \text{compounds having} \leq 2 \text{ C's}$$

RETROSYNTHETIC ANALYSIS

A carbon–carbon bond-forming reaction must be used to convert the two-carbon starting materials to the four-carbon product.

Target compound   **Starting material**

$$CH_3CH_2 \overset{O}{\underset{}{\overset{||}{C}}} CH_3 \overset{[1]}{\Longrightarrow} CH_3CH_2-C≡C-H \overset{[2]}{\Longrightarrow} {}^-:C≡C-H \overset{[3]}{\Longrightarrow} H-C≡C-H$$
$$+$$
$$CH_3CH_2X$$

new C–C bond

Thinking backwards . . .
[1] Form the carbonyl group by hydration of a triple bond.
[2] Form a new C–C bond using an acetylide anion and a 1° alkyl halide.
[3] Prepare the acetylide anion from acetylene by treatment with base.

SYNTHESIS

Three steps are needed to complete the synthesis. Treatment of HC≡CH with NaH forms the acetylide anion, which undergoes an S_N2 reaction with an alkyl halide to form a four-carbon terminal alkyne. Hydration of the alkyne with H_2O, H_2SO_4, and $HgSO_4$ yields the target compound.

$$H-C≡C-H \xrightarrow{Na^+H^-} H-C≡C:^- + Cl-CH_2CH_3 \longrightarrow H-C≡C-CH_2CH_3 \xrightarrow[H_2SO_4]{\overset{H_2O}{}} CH_3 \overset{O}{\underset{}{\overset{||}{C}}} CH_2CH_3$$
$$+ H_2 \qquad\qquad\qquad + Cl^- \quad HgSO_4$$

acid–base reaction S_N2 reaction   hydration   target compound

---

These examples illustrate the synthesis of organic compounds by multistep routes. In Chapter 12 we will learn other useful reactions that expand our capability to do synthesis.

PROBLEM 11.24   Use retrosynthetic analysis to show how 3-hexyne can be prepared from acetylene and any other organic and inorganic compounds. Then draw the synthesis in the synthetic direction, showing all needed reagents.

PROBLEM 11.25   Devise a synthesis of $CH_3CH_2CH_2CHO$ from two-carbon starting materials.

11.13   Key Concepts—Alkynes

General Facts About Alkynes

- Alkynes contain a carbon–carbon triple bond consisting of a strong σ bond and two weak π bonds. Each carbon is *sp* hybridized and linear (11.1).
- Alkynes are named using the suffix **-yne** (11.2).
- Alkynes have weak intermolecular forces, giving them low mp's and low bp's, and making them water insoluble (11.3).
- Because its weaker π bonds make an alkyne electron rich, alkynes undergo addition reactions with electrophiles (11.6).

Addition Reactions of Alkynes

[1] Hydrohalogenation—Addition of HX (X = Cl, Br, I) (11.7)

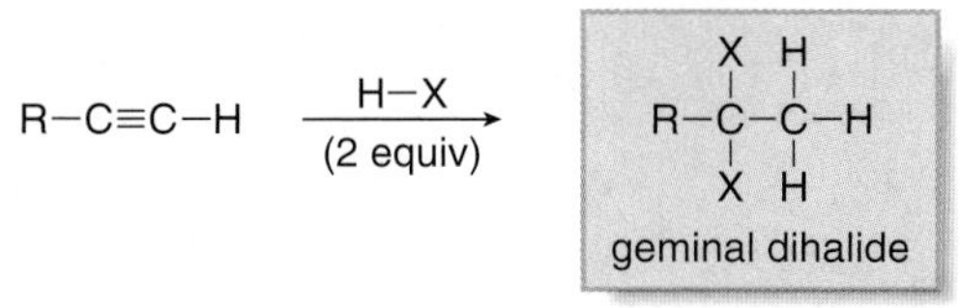

geminal dihalide

- Markovnikov's rule is followed. H bonds to the less substituted C to form the more stable carbocation.

[2] Halogenation—Addition of X_2 (X = Cl or Br) (11.8)

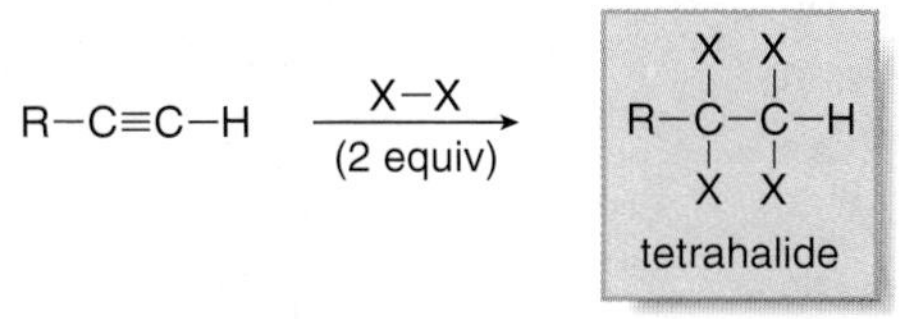

tetrahalide

- Bridged halonium ions are formed as intermediates.
- Anti addition of X_2 occurs.

[3] Hydration—Addition of H_2O (11.9)

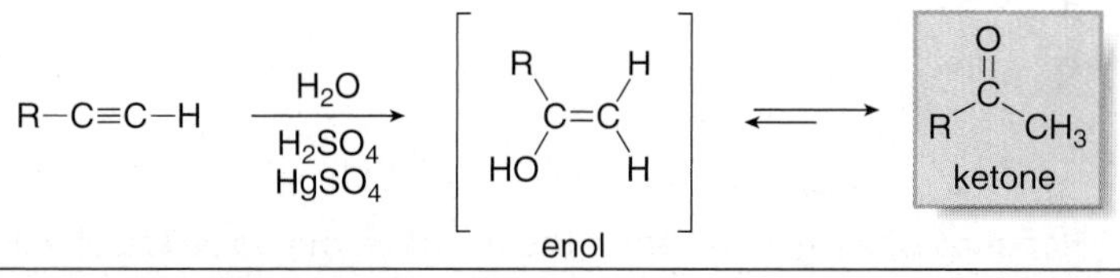

enol

ketone

- Markovnikov's rule is followed. H bonds to the less substituted C to form the more stable carbocation.
- An unstable enol is first formed, which rearranges to a carbonyl group.

[4] Hydroboration–oxidation—Addition of H_2O (11.10)

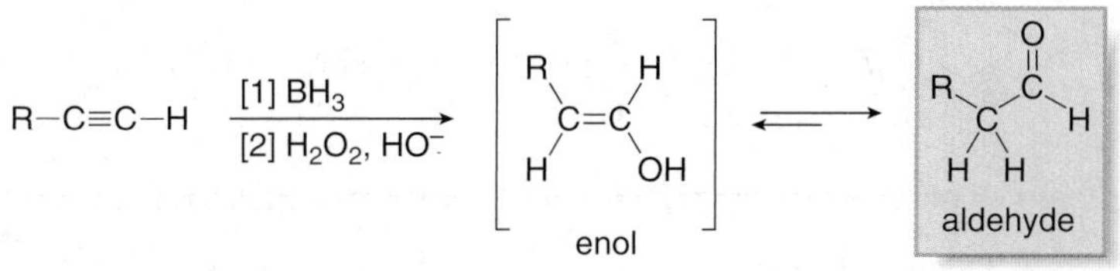

enol

aldehyde

- The unstable enol, first formed after oxidation, rearranges to a carbonyl group.

Reactions Involving Acetylide Anions

[1] Formation of acetylide anions from terminal alkynes (11.6B)

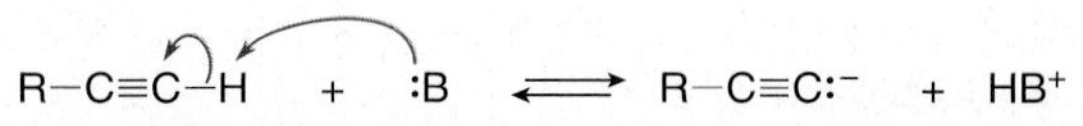

$R-C\equiv C-H$ + :B $\rightleftharpoons$ $R-C\equiv C:^-$ + HB$^+$

- Typical bases used for the reaction are $NaNH_2$ and NaH.

[2] Reaction of acetylide anions with alkyl halides (11.11A)

$H-C\equiv C:^-$ + $R-X$ $\longrightarrow$ $H-C\equiv C-R$ + X^-

- The reaction follows an S_N2 mechanism.
- The reaction works best with CH_3X and RCH_2X.

[3] Reaction of acetylide anions with epoxides (11.11B)

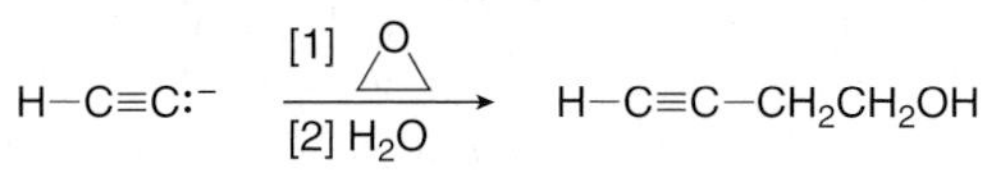

$H-C\equiv C:^-$ $\xrightarrow[\text{[2] } H_2O]{\text{[1]}}$ $H-C\equiv C-CH_2CH_2OH$

- The reaction follows an S_N2 mechanism.
- Opening of the ring occurs from the back side at the less substituted end of the epoxide.

Problems

Nomenclature

11.26 Give the IUPAC name for each alkyne.

a. CH₃CH₂CH(CH₃)C≡CCH₂CH₃

c. (CH₃CH₂)₂CHC≡CCH(CH₂CH₃)CH(CH₃)CH₂CH₃

e.

b. (CH₃)₂CHC≡CCH(CH₃)₂

d. HC≡C–CH(CH₂CH₃)CH₂CH₂CH₃

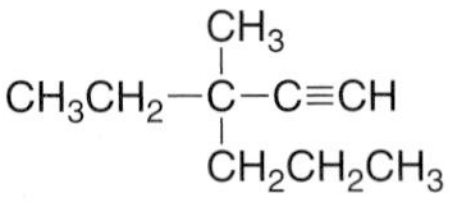

f. CH₃CH₂C≡CCH₂C≡CCH₃

11.27 Give the structure corresponding to each name.
a. 5,6-dimethyl-2-heptyne
b. 5-*tert*-butyl-6,6-dimethyl-3-nonyne
c. (S)-4-chloro-2-pentyne
d. *cis*-1-ethynyl-2-methylcyclopentane

Tautomers

11.28 Which of the following pairs of compounds represent keto–enol tautomers?

a.

c. 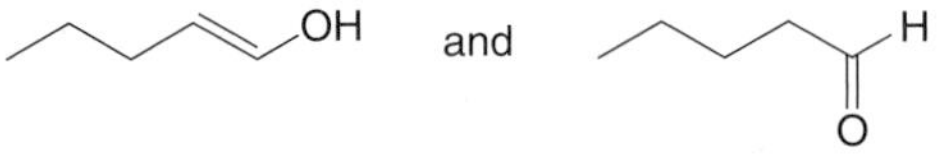 ⌇⌇⌇–OH and ⌇⌇⌇–H (aldehyde)

b. (cyclohexanone) and (cyclohexenol)

d.  (ketone) and (OH alkene)

11.29 Draw the enol form of each keto tautomer.

a. 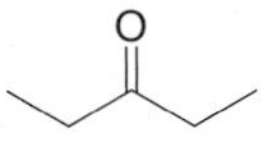 (ketone)   b. CH₃CH₂CHO   c. 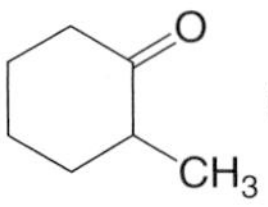 (2-methylcyclohexanone) (two different enols)

11.30 Draw the keto form of each enol tautomer.

a. (enol with OH)   b. 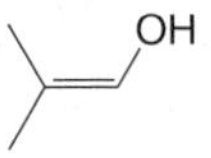 (enol with OH)   c. 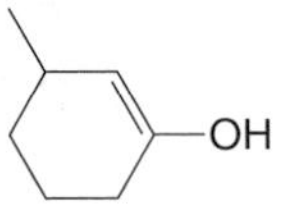 (cyclohexenol with methyl) –OH

11.31 Draw a stepwise mechanism for the conversion of cyclopentanone (**A**) to its enol tautomer (**B**) in the presence of acid.

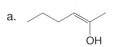

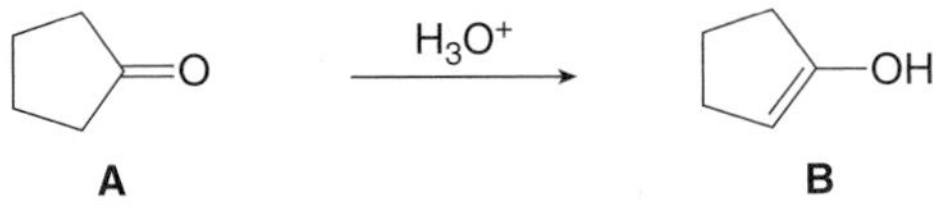

A   H₃O⁺   B

Reactions

11.32 Draw the products of each acid–base reaction. Indicate whether equilibrium favors the starting materials or the products.

a. HC≡C⁻ + CH₃OH ⟶

c. HC≡CH + NaBr ⟶

b. CH₃C≡CH + CH₃⁻ ⟶

d. CH₃CH₂C≡C⁻ + CH₃COOH ⟶

11.33 Draw the products formed when 1-hexyne is treated with each reagent.
a. HCl (2 equiv)
b. HBr (2 equiv)
c. Cl₂ (2 equiv)
d. H₂O + H₂SO₄ + HgSO₄
e. [1] BH₃; [2] H₂O₂, HO⁻
f. NaH
g. [1] ⁻NH₂; [2] CH₃CH₂Br
h. [1] ⁻NH₂; [2] (cyclopropane)

11.34 Draw the products formed when 3-hexyne is treated with each reagent.
a. HBr (2 equiv)
b. Br₂ (2 equiv)
c. H₂O, H₂SO₄
d. [1] BH₃; [2] H₂O₂, HO⁻

11.35 What reagents are needed to carry out each reaction?

a. 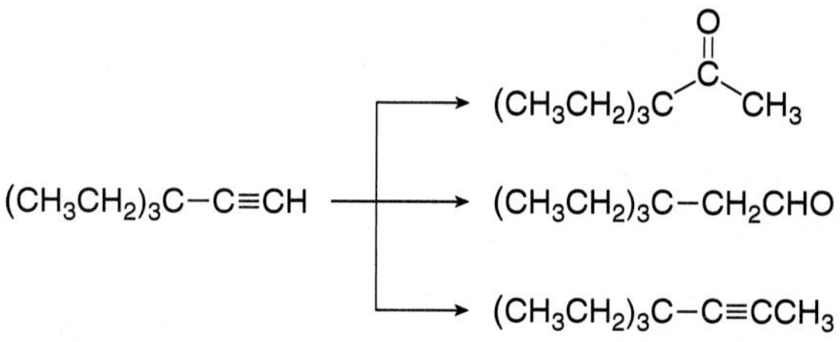 $(CH_3CH_2)_3C-C\equiv CH$

→ $(CH_3CH_2)_3C\overset{\displaystyle O}{\overset{\|}{C}}CH_3$

→ $(CH_3CH_2)_3C-CH_2CHO$

→ $(CH_3CH_2)_3C-C\equiv CCH_3$

b. $CH_3CH_2CH_2CH(CH_3)CH_2CHBr_2 \longrightarrow CH_3CH_2CH_2CH(CH_3)C\equiv CH$

c. $CH_3CH_2CH=CHCH_2CH_3 \xrightarrow[\text{(two steps)}]{} CH_3CH_2C\equiv CCH_2CH_3$

11.36 What alkyne gives each of the following ketones as the only product after hydration with H_2O, H_2SO_4, and $HgSO_4$?

a. 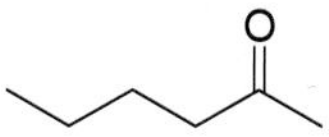 [structure]   b. 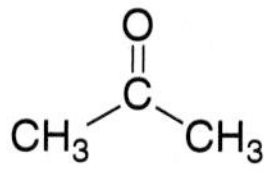 [structure $CH_3\overset{O}{\overset{\|}{C}}CH_3$]   c. [structure]   d. 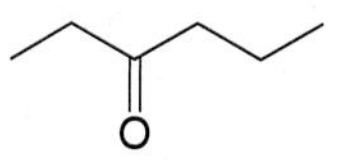 [structure]

11.37 What alkyne gives each of the following compounds as the only product after hydroboration–oxidation?

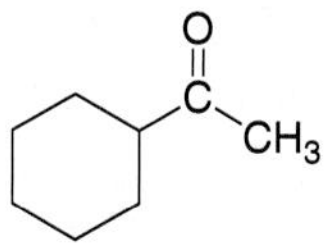

a. 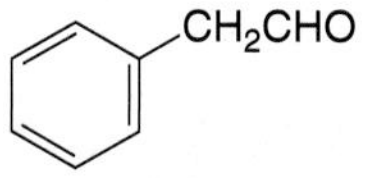 [structure with CH_2CHO]   b. 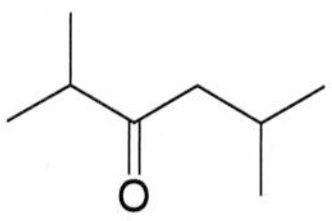 [structure]

11.38 What is the structure of the only alkyne that gives the same aldehyde upon treatment with (1) H_2O, H_2SO_4, $HgSO_4$, and (2) BH_3 followed by H_2O_2, HO^-?

11.39 Draw the organic products formed in each reaction.

a. 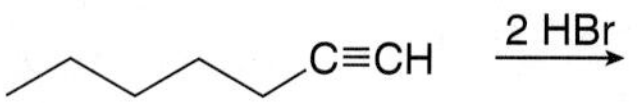 [structure] $C\equiv CH \xrightarrow{2\ HBr}$

g. [structure] $-C\equiv C-CH_3 \xrightarrow[H_2SO_4]{H_2O}$

b. $(CH_3)_3CC\equiv CH \xrightarrow{2\ Cl_2}$

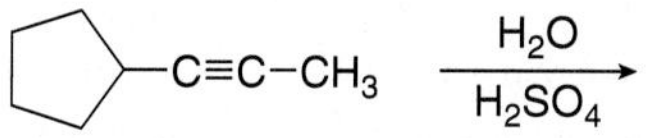

h. $CH_3CH_2C\equiv C^- + CH_3CH_2CH_2OTs \longrightarrow$

c. [structure] $-CH=CH-$ [structure] $\xrightarrow[\substack{[2]\ NaNH_2 \\ (2\ \text{equiv})}]{[1]\ Cl_2}$

i. 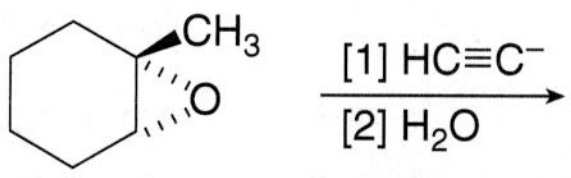 [structure with CH_3, epoxide O] $\xrightarrow[\text{[2] } H_2O]{[1]\ HC\equiv C^-}$

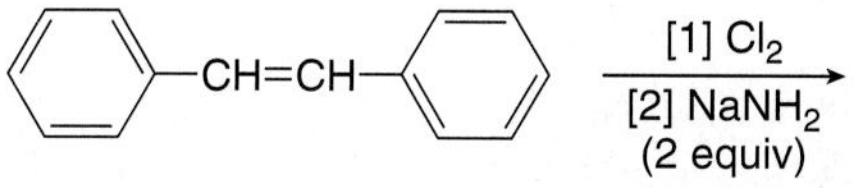

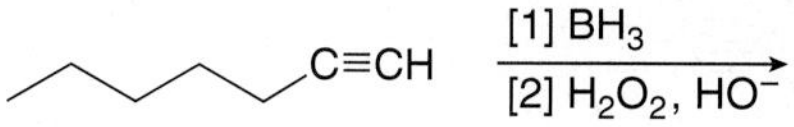

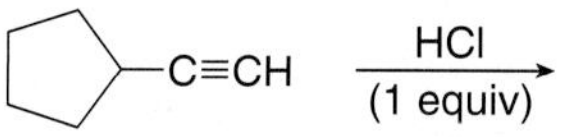

d. [structure] $C\equiv CH \xrightarrow[\text{[2] } H_2O_2,\ HO^-]{[1]\ BH_3}$

j. $CH_3C\equiv CH \xrightarrow[\text{[2]}]{[1]\ NaH}$ [structure with Br]

e. [structure] $-C\equiv CH \xrightarrow[\text{(1 equiv)}]{HCl}$

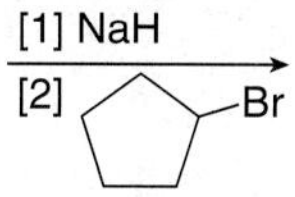

k. $CH_3CH_2C\equiv CH \xrightarrow[\text{[2]}]{[1]\ NaNH_2}$ [structure with CH_2I]

f. $HC\equiv C^- + D_2O \longrightarrow$

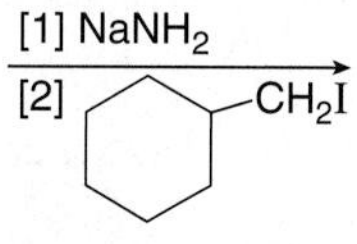

l. 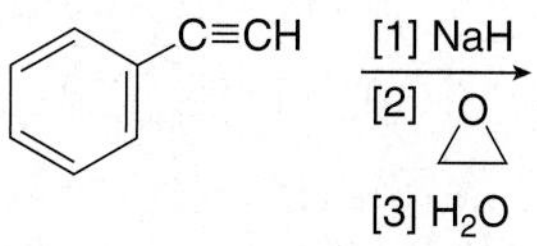 [structure] $C\equiv CH \xrightarrow[\substack{[2]\ \text{[epoxide]} \\ [3]\ H_2O}]{[1]\ NaH}$

11.40 Draw the structure of compounds **A–E** in the following reaction scheme.

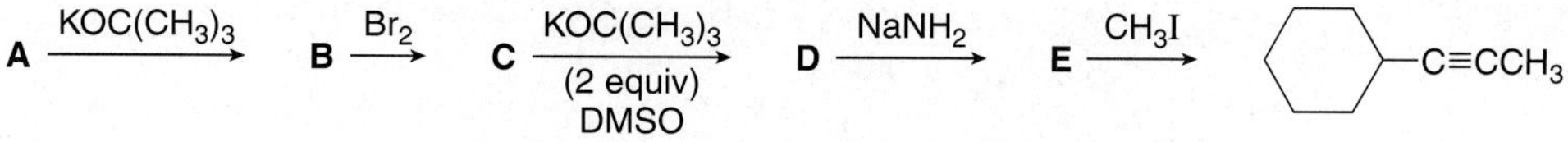

$A \xrightarrow{KOC(CH_3)_3} B \xrightarrow{Br_2} C \xrightarrow[\substack{(2\ \text{equiv}) \\ DMSO}]{KOC(CH_3)_3} D \xrightarrow{NaNH_2} E \xrightarrow{CH_3I}$ [structure] $-C\equiv CCH_3$

11.41 When alkyne **A** is treated with $NaNH_2$ followed by CH_3I, a product having molecular formula $C_6H_{10}O$ is formed, but it is *not* compound **B**. What is the structure of the product and why is it formed?

$H-C\equiv C-CH_2CH_2CH_2OH \xrightarrow[\substack{1)\ NaNH_2 \\ 2)\ CH_3I}]{\times} CH_3-C\equiv C-CH_2CH_2CH_2OH$

$\quad\quad\quad\quad\quad$ **A** $\quad\quad\quad\quad\quad\quad\quad\quad\quad\quad\quad\quad$ **B**

11.42 Draw the products formed in each reaction and indicate stereochemistry.

a. 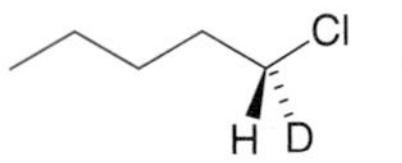 [structure: pentyl chain with Cl, H, D stereocenter] →(HC≡C⁻)

b. 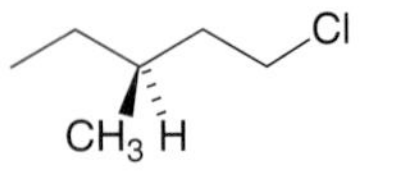 [structure with CH₃, H stereocenter, Cl] →(HC≡C⁻)

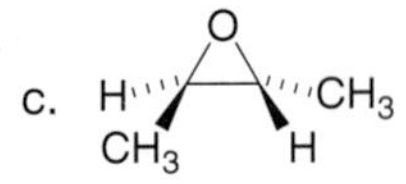

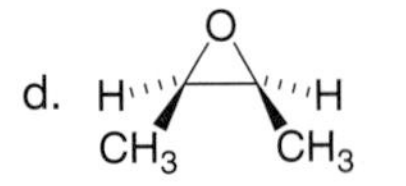

c. [epoxide: H, CH₃, CH₃, H] →([1] HC≡C⁻ / [2] H₂O)

d. [epoxide: H, H, CH₃, CH₃] →([1] HC≡C⁻ / [2] H₂O)

11.43 Identify the lettered compounds in the following reaction schemes. Each reaction sequence was used in the synthesis of a natural product.

a. 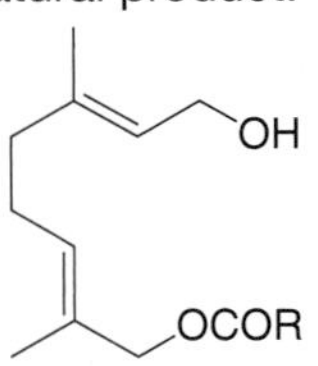 [structure with OH and OCOR groups] →(PBr₃) **A** →(HC≡CCH₂OR' / CH₃⁻ Li⁺ **B**) **C**

b. 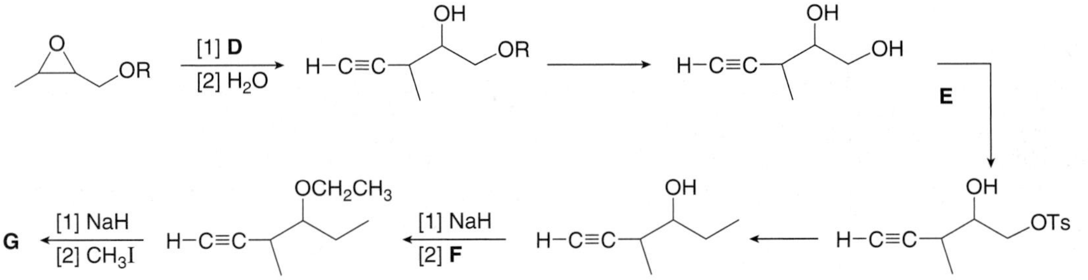 [epoxide-OR] →([1] **D** / [2] H₂O) H–C≡C [structure with OH and OR] → H–C≡C [structure with OH and OH] **E**

→ H–C≡C [structure with OH and OTs] → H–C≡C [structure with OH] →([1] NaH / [2] **F**) H–C≡C [structure with OCH₂CH₃] →([1] NaH / [2] CH₃I) **G**

Mechanisms

11.44 Draw a diagram illustrating the orbitals in the vinyl cation drawn below. Show how each carbon atom is hybridized, and in what orbital the positive charge resides. Explain why this vinyl cation is less stable than $(CH_3)_2CH^+$.

$$CH_2=\overset{+}{C}-CH_3$$
vinyl cation

11.45 Draw a stepwise mechanism for the following reaction and explain why a mixture of E and Z isomers is formed.

$$CH_3C≡CCH_3 \xrightarrow{HCl}$$ [structure: CH₃ CH₃ C=C H Cl] + [structure: CH₃ Cl C=C H CH₃]

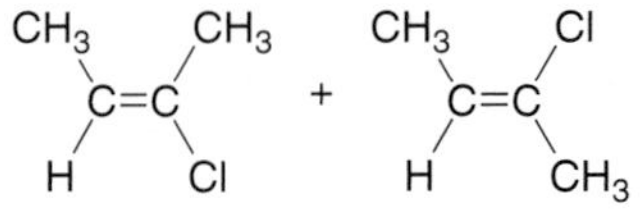

11.46 Draw a stepwise mechanism for the following reaction.

[alkyne aldehyde structure] →([1] CH₃CH₂⁻ Li⁺ / [2] CH₂=O / [3] H₂O) [alkyne alcohol structure with OH]

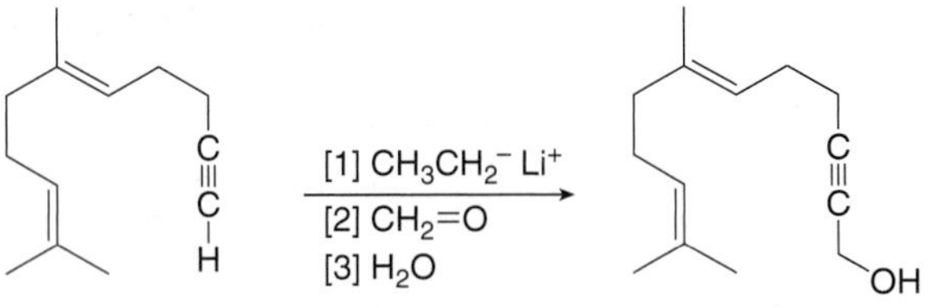

11.47 Draw a stepwise mechanism for each reaction.

a. 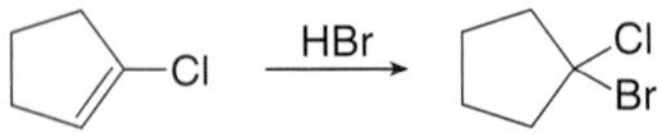 [cyclopentene with Cl] →(HBr) [cyclopentane with Cl and Br]

b. $CH_3-\underset{\underset{H}{|}}{\overset{\overset{OH}{|}}{C}}-C≡CH \xrightarrow[H_2SO_4]{H_2O} CH_3-CH=CH-CHO$

Synthesis

11.48   What acetylide anion and alkyl halide are needed to synthesize each alkyne?

a. $HC\equiv CCH_2CH_2CH(CH_3)_2$

b. $CH_3-C\equiv C-\overset{\overset{\displaystyle CH_3}{|}}{\underset{\underset{\displaystyle CH_3}{|}}{C}}-CH_2CH_3$

c. 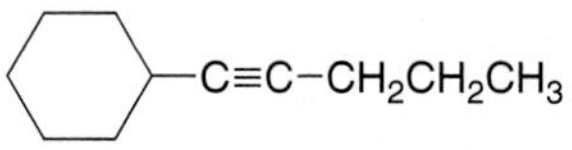   [cyclohexyl]$-C\equiv C-CH_2CH_2CH_3$

11.49   Devise a synthesis of 1-hexyne from each starting material.

a. 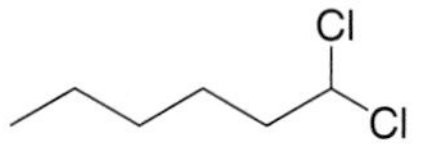   [structure with CCl and Cl]

b. 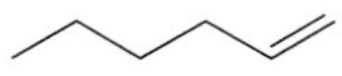   [alkene structure]

c. 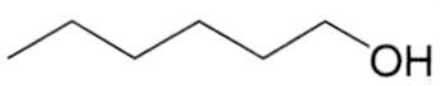   [structure ending in OH]

11.50   Synthesize each compound from acetylene. You may use any other organic or inorganic reagents.

a. $(CH_3)_2CHCH_2C\equiv CH$

b. $CH_3CH_2CH_2C\equiv CCH_2CH_2CH_3$

c. $CH_3CH_2CH_2CH_2CHO$

d. $CH_3CH_2CH_2\overset{\overset{\displaystyle O}{||}}{C}CH_3$

e. $CH_3CH_2CH_2CCl_2CH_3$

f. $CH_3CH_2CH_2\overset{\overset{\displaystyle O}{||}}{C}CH_2CH_2CH_2CH_3$

11.51   Synthesize 2-hexyne from $CH_3CH_2CH_2OH$ as the only organic starting material.

Challenge Questions

11.52   Conversion of an enol to a ketone also occurs in the presence of base. Draw a stepwise mechanism for the following tautomerization.

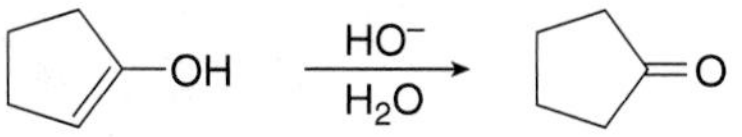

[cyclopentene-OH] $\xrightarrow[\text{H}_2\text{O}]{\text{HO}^-}$ [cyclopentanone =O]

11.53   Draw a stepwise mechanism for the following reaction.

$CH_3-C\equiv C-H$ $\xrightarrow[\text{H}_2\text{O}]{\text{Br}_2}$ $CH_3\overset{\overset{\displaystyle O}{||}}{C}CH_2Br$

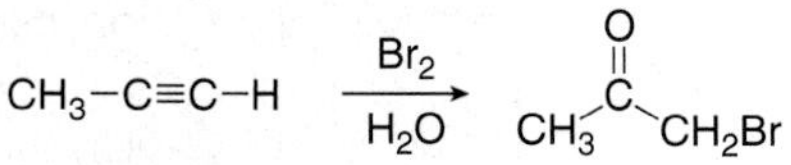

11.54   Why is compound **X** formed in the following reaction, instead of its constitutional isomer **Y**?

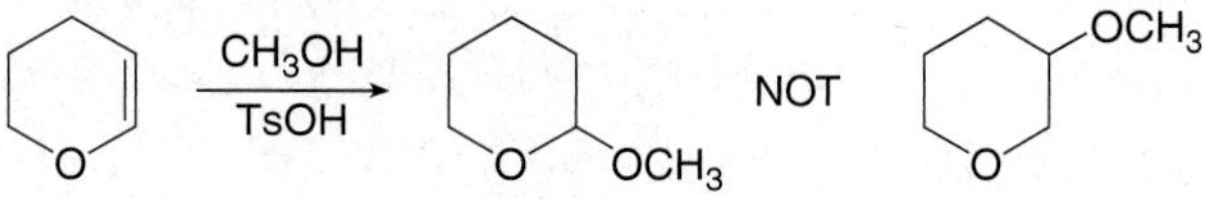

[dihydropyran] $\xrightarrow[\text{TsOH}]{\text{CH}_3\text{OH}}$ [tetrahydropyran with OCH$_3$] **X**   NOT   [tetrahydropyran with OCH$_3$] **Y**

11.55   Draw a stepwise mechanism for the following intramolecular reaction.

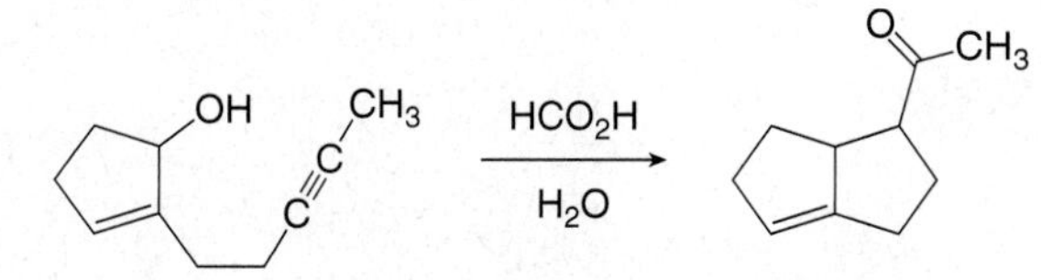

[cyclopentene with OH and CH$_3$ alkyne chain] $\xrightarrow[\text{H}_2\text{O}]{\text{HCO}_2\text{H}}$ [bicyclic product with C(=O)CH$_3$]

Oxidation and Reduction

Disparlure, the sex pheromone of the female gypsy moth, has been used to control the spread of the gypsy moth caterpillar, a pest that has periodically devastated forests in the northeastern United States by defoliating many shade and fruit-bearing trees. The active pheromone is placed in a trap containing a poison or sticky substance, and the male moth is lured to the trap by the pheromone. Such a species-specific method presents a new way of controlling an insect population that avoids the widespread use of harmful, nonspecific pesticides. Disparlure is synthesized by oxidation of an alkene using chemistry presented in Chapter 12.

In Chapter 12, we discuss the oxidation and reduction of **alkenes** and **alkynes,** as well as compounds with **polar C – X σ bonds**—alcohols, alkyl halides, and epoxides. Although there will be many different reagents and mechanisms, discussing these reactions as a group allows us to more easily compare and contrast them.

The word *mechanism* will often be used loosely here. In contrast to the S_N1 reaction of alkyl halides or the electrophilic addition reactions of alkenes, the details of some of the mechanisms presented in Chapter 12 are known with less certainty. For example, although the identity of a particular intermediate might be confirmed by experiment, other details of the mechanism are suggested by the structure or stereochemistry of the final product.

Oxidation and reduction reactions are very versatile, and knowing them allows us to design many more complex organic syntheses.

12.1 Introduction

Recall from Section 4.14 that the way to determine whether an organic compound has been oxidized or reduced is to compare the **relative number of C – H and C – Z bonds** (Z = an element *more electronegative* than carbon) in the starting material and product.

◆ *Oxidation* results in an *increase* in the number of C – Z bonds (usually C – O bonds) *or* a *decrease* in the number of C – H bonds.

◆ *Reduction* results in a *decrease* in the number of C – Z bonds (usually C – O bonds) *or* an *increase* in the number of C – H bonds.

> Two components are always present in an oxidation or reduction reaction—**one component is oxidized and one is reduced.** When an organic compound is *oxidized* by a reagent, the reagent itself must be *reduced*. Similarly, when an organic compound is *reduced* by a reagent, the reagent becomes *oxidized*.

Thus, an organic compound such as CH_4 can be oxidized by replacing C – H bonds with C – O bonds, as shown in Figure 12.1. Reduction is the opposite of oxidation, so Figure 12.1 also shows how a compound can be reduced by replacing C – O bonds with C – H bonds. The symbols **[O]** and **[H]** indicate oxidation and reduction, respectively.

Sometimes two carbon atoms are involved in a single oxidation or reduction reaction, and the net change in the number of C – H or C – Z bonds at *both* atoms must be taken into account. The conversion of an **alkyne to an alkene** and an **alkene to an alkane** are examples of reduction, because each process adds two new C – H bonds to the starting material, as shown in Figure 12.2.

PROBLEM 12.1 Classify each reaction as oxidation, reduction, or neither.

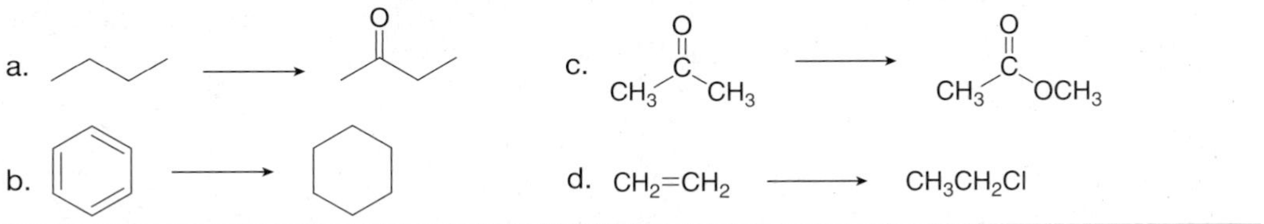

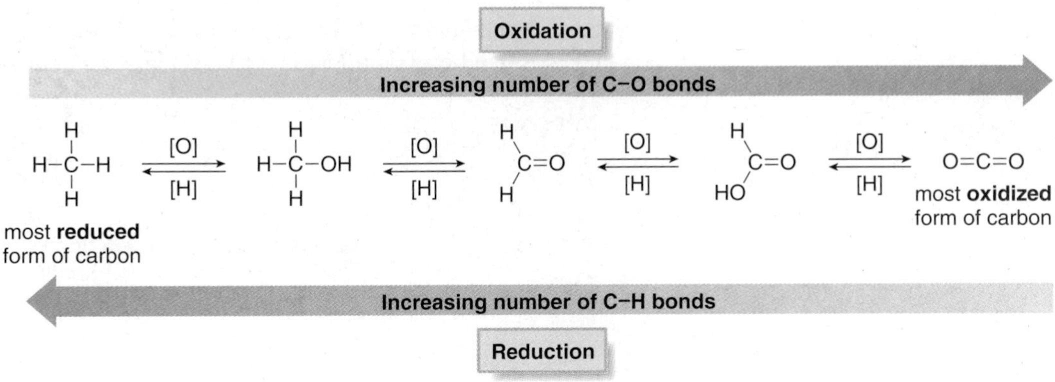

a.

b.

c.

d. $CH_2{=}CH_2 \longrightarrow CH_3CH_2Cl$

Figure 12.1 A general scheme for the oxidation and reduction of a carbon compound

Figure 12.2 Oxidation and reduction of hydrocarbons

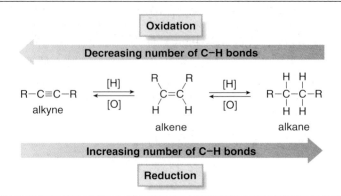

12.2 Reducing Agents

All reducing agents provide the equivalent of two hydrogen atoms, but **there are three types of reductions,** differing in how H_2 is added. The simplest reducing agent is molecular H_2. Reductions of this sort are carried out in the presence of a metal catalyst that acts as a surface on which reaction occurs.

The second way to deliver H_2 in a reduction is to add two protons and two electrons to a substrate—that is, $H_2 = 2H^+ + 2e^-$. Reducing agents of this sort use alkali metals as a source of electrons and liquid ammonia (NH_3) as a source of protons. Reductions with **Na in NH_3** are called **dissolving metal reductions.**

$$2\,Na \longrightarrow 2\,Na^+ \;+\; \boxed{2\,e^-}$$

$$2\,NH_3 \longrightarrow 2\,{}^-NH_2 \;+\; \boxed{2\,H^+}$$

an equivalent of H_2 for reduction

The third way to deliver the equivalent of two hydrogen atoms is to add **hydride (H^-)** and a **proton (H^+).** The most common hydride reducing agents contain a hydrogen atom bonded to boron or aluminum. Simple examples include **sodium borohydride ($NaBH_4$)** and **lithium aluminum hydride ($LiAlH_4$).** These reagents deliver H^- to a substrate, and then a proton is added from H_2O or an alcohol.

> Lithium aluminum hydride is often abbreviated as LAH.

$NaBH_4$

$$Na^+ \quad H-\overset{\overset{\displaystyle H}{|}}{\underset{\underset{\displaystyle H}{|}}{B}}{}^{-}-H$$

sodium borohydride

$LiAlH_4$

$$Li^+ \quad H-\overset{\overset{\displaystyle H}{|}}{\underset{\underset{\displaystyle H}{|}}{Al}}{}^{-}-H$$

lithium aluminum hydride
LAH

◆ Metal hydride reagents act as a source of H^- because they contain polar metal–hydrogen bonds that place a partial negative charge on hydrogen.

$$\overset{\longleftrightarrow}{M-H} \underset{\delta^+ \ \ \delta^-}{} \;=\; \boxed{H^-}$$

a polar metal–hydrogen bond

$M = B$ or Al

12.3   Reduction of Alkenes

Reduction of an alkene forms an alkane by addition of H₂. Two bonds are broken—the **weak π bond** of the alkene and the H₂ σ bond—and two new C–H σ bonds are formed.

Hydrogenation—General reaction

$$\text{C=C} \quad + \quad \text{H–H} \quad \xrightarrow[\text{catalyst}]{\text{metal}} \quad \text{—C–C—}$$

weak π bond   H H ← H₂ is added.
                                                                        alkane

The addition of H₂ occurs only in the presence of a **metal catalyst,** and thus, the reaction is called **catalytic hydrogenation.** The catalyst consists of a metal—usually Pd, Pt, or Ni—adsorbed onto a finely divided inert solid, such as charcoal. For example, the catalyst 10% Pd on carbon is composed of 10% Pd and 90% carbon, by weight. H₂ adds in a **syn** fashion, as shown in Equation [2].

Hydrogenation catalysts are insoluble in common solvents, thus creating a **heterogeneous** reaction mixture. This insolubility has a practical advantage. These catalysts contain expensive metals, but they can be filtered away from the other reactants after the reaction is complete, and then reused.

Examples

[1] $\text{H}_2\text{C=CH}_2 \xrightarrow[\text{Pd-C}]{\text{H}_2} \text{H–C–C–H}$

H H ← H₂ is added.

[2]   (cyclohexene with two CH₃ groups) $\xrightarrow[\text{Pd-C}]{\text{H}_2}$ ← **syn** addition of H₂

PROBLEM 12.2   What alkane is formed when each alkene is treated with H₂ and a Pd catalyst?

a. $\underset{\text{CH}_3}{\overset{\text{CH}_3}{\text{C=C}}}\overset{\text{CH}_2\text{CH(CH}_3)_2}{\underset{\text{H}}{}}$

b. (alkene structure)

c. (methylcyclohexene structure)

12.3A   Hydrogenation and Alkene Stability

Hydrogenation reactions are **exothermic** because the bonds in the product are stronger than the bonds in the starting materials, making them similar to other alkene addition reactions. The $\Delta H°$ for hydrogenation, called the **heat of hydrogenation,** can be used as a measure of the relative stability of two different alkenes that are hydrogenated to the same alkane.

Recall from Chapter 8 that **trans alkenes are generally more stable than cis alkenes.**

For example, both *cis-* and *trans-*2-butene are hydrogenated to butane, and the heat of hydrogenation for the trans isomer is less than that for the cis isomer. **Because less energy is released in converting the trans alkene to butane, it must be lower in energy (more stable) to begin with.** The relative energies of the butene isomers are illustrated in Figure 12.3.

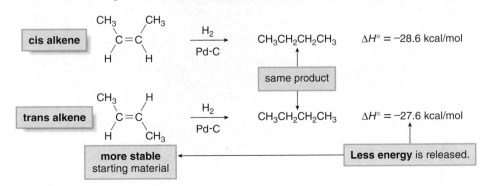

cis alkene $\underset{\text{H}}{\overset{\text{CH}_3}{\text{C=C}}}\underset{\text{H}}{\overset{\text{CH}_3}{}}$ $\xrightarrow[\text{Pd-C}]{\text{H}_2}$ $\text{CH}_3\text{CH}_2\text{CH}_2\text{CH}_3$ $\Delta H° = -28.6$ kcal/mol

same product

trans alkene $\underset{\text{H}}{\overset{\text{CH}_3}{\text{C=C}}}\underset{\text{CH}_3}{\overset{\text{H}}{}}$ $\xrightarrow[\text{Pd-C}]{\text{H}_2}$ $\text{CH}_3\text{CH}_2\text{CH}_2\text{CH}_3$ $\Delta H° = -27.6$ kcal/mol

more stable starting material ← **Less energy** is released.

Figure 12.3   Relative energies of *cis*- and *trans*-2-butene

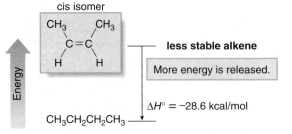

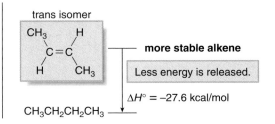

◆ **When hydrogenation of two alkenes gives the same alkane, the more stable alkene has the *smaller* heat of hydrogenation.**

PROBLEM 12.3   Which alkene in each pair has the larger heat of hydrogenation?

PROBLEM 12.4   Explain why heats of hydrogenation cannot be used to determine the relative stability of 2-methyl-2-pentene and 3-methyl-1-pentene.

12.3B   The Mechanism of Catalytic Hydrogenation

In the generally accepted mechanism for catalytic hydrogenation, the surface of the metal catalyst binds both H_2 and the alkene, and H_2 is transferred to the π bond in a rapid but stepwise process (Mechanism 12.1).

MECHANISM 12.1

Addition of H_2 to an Alkene—Hydrogenation

Steps [1] and [2] Complexation of H_2 and the alkene to the catalyst

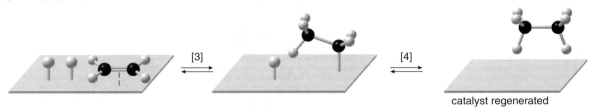

◆ **H_2 adsorbs to the catalyst surface** with partial or complete cleavage of the H−H bond.
◆ The π bond of the alkene complexes with the metal.

Steps [3] and [4] Sequential addition of the elements of H_2

◆ **Two H atoms are transferred sequentially** to the π bond in Steps [3] and [4], forming the alkane.
◆ Because the product alkane no longer has a π bond with which to complex to the metal, it is released from the catalyst surface.

The mechanism explains two facts about hydrogenation:

> ◆ Rapid, sequential addition of H_2 occurs from the side of the alkene complexed to the metal surface, resulting in **syn** addition.
>
> ◆ Less crowded double bonds complex more readily to the catalyst surface, resulting in faster reaction.

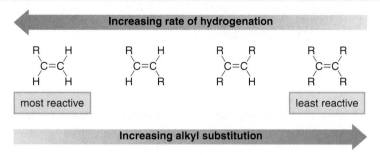

PROBLEM 12.5   What product is formed when limonene is treated with one equivalent of H_2 and a Pd catalyst?

limonene

PROBLEM 12.6   Given that syn addition of H_2 occurs from both sides of a trigonal planar double bond, draw all stereoisomers formed when each alkene is treated with H_2.

a.   b.

12.3C   Hydrogenation Data and Degrees of Unsaturation

Recall from Section 10.2 that the **number of degrees of unsaturation gives the *total* number of rings and π bonds in a molecule.** Because H_2 adds to π bonds but does not add to the C–C σ bonds of rings, hydrogenation allows us to determine how many degrees of unsaturation are due to π bonds and how many are due to rings. This is done by comparing the number of degrees of unsaturation before and after a molecule is treated with H_2, as illustrated in Sample Problem 12.1.

SAMPLE PROBLEM 12.1   How many rings and π bonds are contained in a compound of molecular formula C_8H_{12} that is hydrogenated to a compound of molecular formula C_8H_{14}?

SOLUTION

[1] Determine the number of degrees of unsaturation in the compounds before and after hydrogenation.

Before H_2 addition — C_8H_{12}
- The maximum number of H's possible for n C's is $2n + 2$; in this example, $2n + 2 = 2(8) + 2 = 18$.
- 18 H's (maximum) – 12 H's (actual) = 6 H's fewer than the maximum number.

$$\frac{6 \text{ H's fewer than the maximum}}{2 \text{ H's removed for each degree of unsaturation}} =$$

> **three degrees of unsaturation**

After H_2 addition — C_8H_{14}
- The maximum number of H's possible for n C's is $2n + 2$; in this example, $2n + 2 = 2(8) + 2 = 18$.
- 18 H's (maximum) – 14 H's (actual) = 4 H's fewer than the maximum number.

$$\frac{4 \text{ H's fewer than the maximum}}{2 \text{ H's removed for each degree of unsaturation}} =$$

> **two degrees of unsaturation**

[2] Assign the number of degrees of unsaturation to rings or π bonds as follows:

- The number of degrees of unsaturation that remain in the product after H_2 addition = the **number of rings** in the starting material.

- The number of degrees of unsaturation that react with H_2 = the **number of π bonds.**

In this example, **two** degrees of unsaturation remain after hydrogenation so the starting material has **two** rings. Thus:

Before H₂ addition:		After H₂ addition:

three degrees of unsaturation   —   **two** degrees of unsaturation   =   one degree of unsaturation that reacted with H_2

three rings or π bonds in C_8H_{12}   =   **two rings**   +   **one π bond**   ANSWER

PROBLEM 12.7   Complete the missing information for compounds **A, B,** and **C,** each subjected to hydrogenation. The number of rings and π bonds refers to the reactant (**A, B,** or **C**) prior to hydrogenation.

Compound	Molecular formula before hydrogenation	Molecular formula after hydrogenation	Number of rings	Number of π bonds
A	$C_{10}H_{12}$	$C_{10}H_{16}$	?	?
B	?	C_4H_{10}	0	1
C	C_6H_8	?	1	?

12.3D Hydrogenation of Other Double Bonds

Compounds that contain a carbonyl group also react with H_2 and a metal catalyst. For example, **aldehydes and ketones are reduced to 1° and 2° alcohols,** respectively. We return to this reaction in Chapter 20.

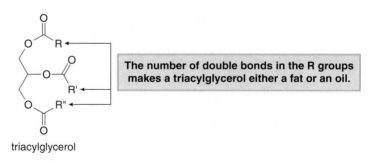

aldehyde   1° alcohol   ketone   2° alcohol

12.4   Application: Hydrogenation of Oils

Peanut butter is a common consumer product that contains partially hydrogenated vegetable oil.

Many processed foods, such as peanut butter, margarine, and some brands of crackers, contain *partially hydrogenated* vegetable oils. These oils are produced by hydrogenating the long hydrocarbon chains of triacylglycerols.

In Section 10.6 we learned that **fats and oils are triacylglycerols that differ in the number of degrees of unsaturation** in their long alkyl side chains.

The number of double bonds in the R groups makes a triacylglycerol either a fat or an oil.

triacylglycerol

- ◆ Fats—usually animal in origin—are solids with triacylglycerols having few degrees of unsaturation.
- ◆ Oils—usually vegetable in origin—are liquids with triacylglycerols having a larger number of degrees of unsaturation.

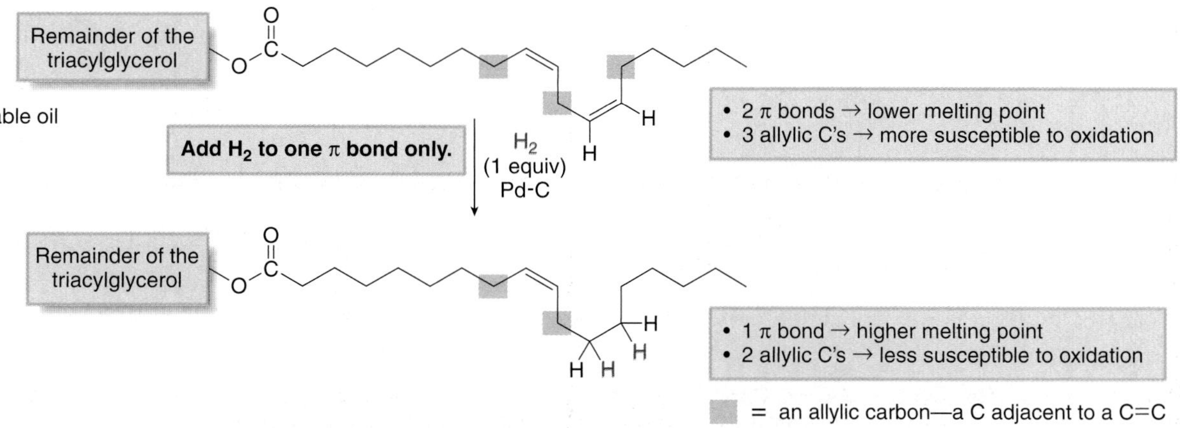

Figure 12.4
Partial hydrogenation of the double bonds in a vegetable oil

- Decreasing the number of degrees of unsaturation **increases** the melting point.
- When an oil is *partially* hydrogenated, some double bonds react with H_2, whereas some double bonds remain in the product.
- Partial hydrogenation **decreases** the number of allylic sites, making a triacylglycerol **less** susceptible to oxidation, thereby increasing its shelf life.

Margarine produced by partial hydrogenation of vegetable oil has no color and no taste. Several ingredients are then added to make margarine more closely resemble butter: orange β-carotene (Section 10.5) is often added for color, salt for flavor, and 3-hydroxy-2-butanone [$CH_3COCH(OH)CH_3$] or 2,3-butanedione ($CH_3COCOCH_3$) to mimic the flavor of butter. Vitamins A and D are also sometimes added, since these vitamins are present in butter.

When an unsaturated vegetable oil is treated with hydrogen, some (or all) of the π bonds add H_2, decreasing the number of degrees of unsaturation (Figure 12.4). This increases the melting point of the oil. For example, margarine is prepared by partially hydrogenating vegetable oil to give a product having a semi-solid consistency that more closely resembles butter. This process is sometimes called ***hardening.***

If unsaturated oils are healthier than saturated fats, why does the food industry hydrogenate oils? There are two reasons—aesthetics and shelf life. Consumers prefer the semi-solid consistency of margarine to a liquid oil. Imagine pouring vegetable oil on a piece of toast or pancakes.

Furthermore, unsaturated oils are more susceptible than saturated fats to oxidation at the **allylic carbon atoms**—the carbons adjacent to the double bond carbons—a process discussed in Chapter 13. Oxidation makes the oil rancid and inedible. Hydrogenating the double bonds reduces the number of allylic carbons (also illustrated in Figure 12.4), thus reducing the likelihood of oxidation and increasing the shelf life of the food product. This process reflects a delicate balance between providing consumers with healthier food products, while maximizing shelf life to prevent spoilage.

PROBLEM 12.8 Draw the products formed when triacylglycerol **A** is treated with each reagent, forming compounds **B** and **C**. Rank **A, B,** and **C** in order of increasing melting point.

$CH_2OCO(CH_2)_{16}CH_3$
$CHOCO(CH_2)_6(CH_2CH=CH)_2(CH_2)_4CH_3$
$CH_2OCO(CH_2)_{16}CH_3$
A

a. H_2 (excess), Pd-C (Compound **B**)
b. H_2 (1 equiv), Pd-C (Compound **C**)

12.5 Reduction of Alkynes

Reduction of an alkyne adds H_2 to one or both of the π bonds. There are three different ways by which the elements of H_2 can be added to a triple bond.

◆ Adding two equivalents of H_2 forms an alkane.

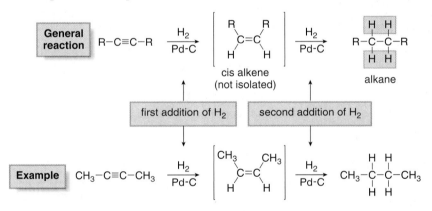

$$R-C\equiv C-R \xrightarrow[\text{(2 equiv)}]{H_2} R-\underset{\underset{H}{|}}{\overset{\overset{H}{|}}{C}}-\underset{\underset{H}{|}}{\overset{\overset{H}{|}}{C}}-R$$

alkane

◆ **Adding one equivalent of H₂ in a syn fashion forms a cis alkene.**

$$R-C\equiv C-R \xrightarrow[\text{(1 equiv)}]{H_2} \underset{H}{\overset{R}{\diagdown}}C=C\underset{H}{\overset{R}{\diagup}}$$ **syn** addition

cis alkene

◆ **Adding one equivalent of H₂ in an anti fashion forms a trans alkene.**

$$R-C\equiv C-R \xrightarrow[\text{(1 equiv)}]{H_2} \underset{H}{\overset{R}{\diagdown}}C=C\underset{R}{\overset{H}{\diagup}}$$ **anti** addition

trans alkene

12.5A   Reduction of an Alkyne to an Alkane

When an alkyne is treated with two or more equivalents of H₂ and a Pd catalyst, reduction of *both* π bonds occurs. **Syn addition** of one equivalent of H₂ forms a cis alkene, which adds a second equivalent of H₂ to form an **alkane. Four new C–H bonds are formed.** By using a Pd-C catalyst, it is not possible to stop the reaction after addition of only one equivalent of H₂.

General reaction $R-C\equiv C-R \xrightarrow{\underset{\text{Pd-C}}{H_2}} \left[\underset{H}{\overset{R}{\diagdown}}C=C\underset{H}{\overset{R}{\diagup}} \right] \xrightarrow{\underset{\text{Pd-C}}{H_2}} R-\overset{\overset{H}{|}}{C}-\overset{\overset{H}{|}}{C}-R$

cis alkene (not isolated)   alkane

first addition of H₂   second addition of H₂

Example $CH_3-C\equiv C-CH_3 \xrightarrow{\underset{\text{Pd-C}}{H_2}} \left[\underset{H}{\overset{CH_3}{\diagdown}}C=C\underset{H}{\overset{CH_3}{\diagup}} \right] \xrightarrow{\underset{\text{Pd-C}}{H_2}} CH_3-\overset{\overset{H}{|}}{\underset{\underset{H}{|}}{C}}-\overset{\overset{H}{|}}{\underset{\underset{H}{|}}{C}}-CH_3$

PROBLEM 12.9   What two different alkynes are reduced to pentane?

12.5B   Reduction of an Alkyne to a Cis Alkene

Palladium metal is too active a catalyst to allow the hydrogenation of an alkyne to stop after one equivalent of H₂. To prepare a cis alkene from an alkyne and H₂, a less active Pd catalyst is used—Pd adsorbed onto CaCO₃ with added lead(II) acetate and quinoline. This catalyst is called the **Lindlar catalyst** after the chemist who first prepared it. Compared to Pd metal, the **Lindlar catalyst is deactivated or "poisoned."**

Pd on CaCO₃
+ Pb(OCOCH₃)₂ + quinoline

Lindlar catalyst

quinoline

With the Lindlar catalyst, one equivalent of H₂ adds to an alkyne, and the cis alkene product is unreactive to further reduction.

Reduction of an alkyne to a cis alkene is a **stereo-selective reaction,** because only one stereoisomer is formed.

General reaction $R-C\equiv C-R \xrightarrow[\text{Lindlar catalyst}]{H_2}$ cis alkene

Example $CH_3-C\equiv C-CH_3 \xrightarrow[\text{Lindlar catalyst}]{H_2}$

PROBLEM 12.10   What is the structure of *cis*-jasmone, a natural product isolated from jasmine flowers, formed by treatment of alkyne **A** with H₂ in the presence of the Lindlar catalyst?

$$\text{A} \quad \xrightarrow[\text{Lindlar catalyst}]{H_2} \quad \begin{array}{c}\textit{cis}\text{-jasmone}\\ \text{(perfume component}\\ \text{isolated from jasmine flowers)}\end{array}$$

PROBLEM 12.11   Draw the organic products formed in each hydrogenation.

a. $CH_2=CHCH_2CH_2-C\equiv C-CH_3 \xrightarrow[\text{Pd-C}]{H_2 \text{ (excess)}}$ b. $CH_2=CHCH_2CH_2-C\equiv C-CH_3 \xrightarrow[\substack{\text{Lindlar}\\ \text{catalyst}}]{H_2 \text{ (excess)}}$

12.5C   Reduction of an Alkyne to a Trans Alkene

NH₃ has a boiling point of −33 °C, making it a gas at room temperature. To carry out a Na, NH₃ reduction, NH₃ gas is condensed into a flask kept at −78 °C by a cooling bath of solid CO₂ in acetone. When Na is added to the liquid NH₃, a brilliant blue solution is formed.

Although catalytic hydrogenation is a convenient method for preparing cis alkenes from alkynes, it cannot be used to prepare trans alkenes. With a dissolving metal reduction (such as Na in NH₃), however, the elements of H₂ are added in an **anti** fashion to the triple bond, thus forming a **trans alkene.** For example, 2-butyne reacts with Na in NH₃ to form *trans*-2-butene.

General reaction $R-C\equiv C-R \xrightarrow[\text{NH}_3]{\text{Na}}$ **trans** alkene

Example $CH_3-C\equiv C-CH_3 \xrightarrow[\text{NH}_3]{\text{Na}}$
2-butyne
trans-2-butene

The **mechanism** for the dissolving metal reduction using Na in NH₃ features sequential addition of electrons and protons to the triple bond. Half-headed arrows denoting the movement of a single electron must be used in two steps when Na donates a *single* electron. The mechanism can be divided conceptually into two parts, each of which consists of two steps: **addition of an electron followed by protonation of the resulting negative charge,** as shown in Mechanism 12.2.

The trans alkene is formed during dissolving metal reduction because the vinyl carbanion formed in Step [3] is more stable; it has the larger R groups further away from each other to avoid steric interactions. Protonation of this anion leads to the more stable trans product.

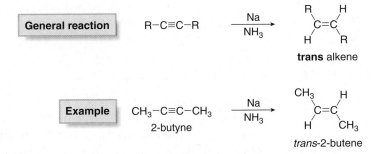

The larger R groups are further away from each other.

more stable vinyl carbanion

↓

trans alkene

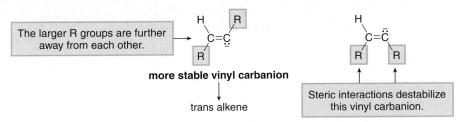

Steric interactions destabilize this vinyl carbanion.

MECHANISM 12.2

Dissolving Metal Reduction of an Alkyne to a Trans Alkene

Steps [1] and [2]   Addition of one electron and one proton to form a radical

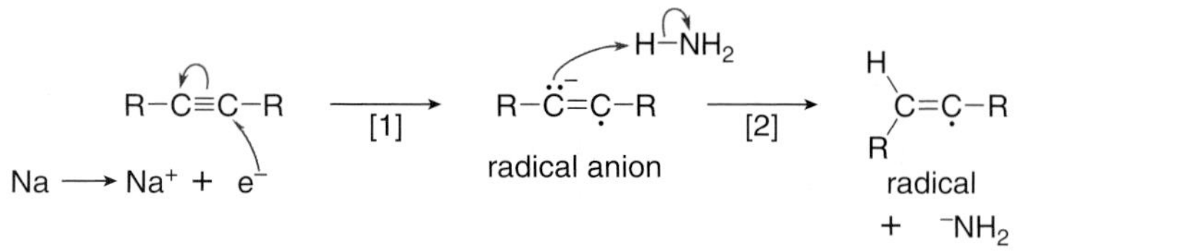

R–C≡C–R —[1]→ R–Ċ=Ċ–R —[2]→ (H)(R)C=C(Ċ)–R

Na ⟶ Na⁺ + e⁻

radical anion

radical
+ ⁻NH₂

◆ Addition of an electron to the triple bond in Step [1] forms a **radical anion,** a species containing *both* a negative charge and an unpaired electron.
◆ Protonation of the anion with the solvent NH₃ in Step [2] yields a **radical.** The net effect of Steps [1] and [2] is to add one hydrogen atom (H•) to the triple bond.

Steps [3] and [4]   Addition of one electron and one proton to form the trans alkene

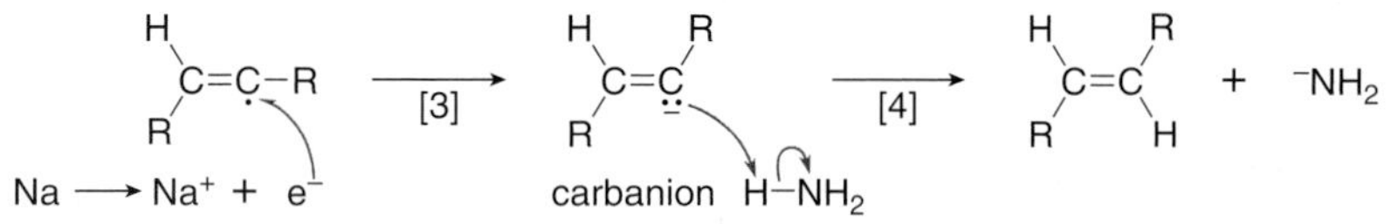

(H)(R)C=C(•)–R —[3]→ carbanion —[4]→ (H)(R)C=C(R)(H) + ⁻NH₂

Na ⟶ Na⁺ + e⁻

◆ Addition of a second electron to the radical in Step [3] forms a **carbanion.**
◆ Protonation of the carbanion in Step [4] forms the trans alkene. These last two steps add the second hydrogen atom (H•) to the triple bond.

Dissolving metal reduction of a triple bond with Na in NH₃ is a **stereoselective reaction** because it forms a trans product exclusively.

◆ **Dissolving metal reductions always form the more stable trans product preferentially.**

The three methods to reduce a triple bond are summarized in Figure 12.5 using 3-hexyne as starting material.

PROBLEM 12.12   What product is formed when $CH_3OCH_2CH_2C{\equiv}CCH_2CH(CH_3)_2$ is treated with each reagent: (a) H_2 (excess), Pd-C; (b) H_2 (1 equiv), Lindlar catalyst; (c) H_2 (excess), Lindlar catalyst; (d) Na, NH_3?

PROBLEM 12.13   Deuterium is introduced into a molecule by using reducing agents that contain D atoms instead of H atoms. Draw the products formed when 2-hexyne is treated with each reagent: (a) D_2, Pd; (b) D_2, Lindlar catalyst; (c) Na, ND_3.

12.6   The Reduction of Polar C–X σ Bonds

Compounds containing polar $C-X$ σ bonds that react with strong nucleophiles are reduced with metal hydride reagents, most commonly lithium aluminum hydride. Two functional groups possessing both of these characteristics are **alkyl halides** and **epoxides.** Alkyl halides are reduced to alkanes with loss of X^- as the leaving group. Epoxide rings are opened to form alcohols.

Figure 12.5   Summary: Three methods to reduce a triple bond

$CH_3CH_2-C{\equiv}C-CH_2CH_3$
3-hexyne

$\xrightarrow[\text{Pd-C}]{2\ H_2}$ $CH_3CH_2-\overset{H}{\underset{H}{C}}-\overset{H}{\underset{H}{C}}-CH_2CH_3$
hexane

$\xrightarrow[\substack{\text{Lindlar}\\\text{catalyst}}]{H_2}$ $(CH_3CH_2)(H)C{=}C(H)(CH_2CH_3)$
cis-3-hexene

$\xrightarrow[NH_3]{Na}$ $(CH_3CH_2)(H)C{=}C(H)(CH_2CH_3)$
trans-3-hexene

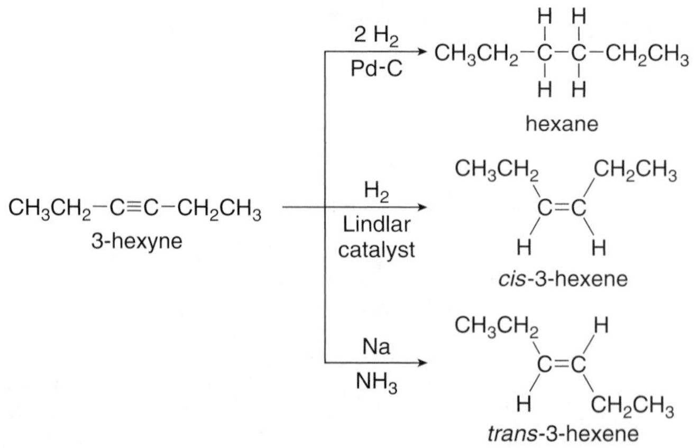

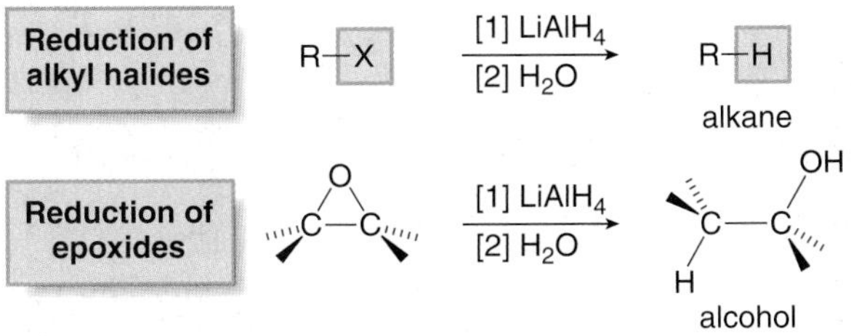

Reduction of alkyl halides	R—X	$\xrightarrow[\text{[2] } H_2O]{\text{[1] LiAlH}_4}$	R—H alkane

Reduction of epoxides	(epoxide)	$\xrightarrow[\text{[2] } H_2O]{\text{[1] LiAlH}_4}$	(alcohol)

Reduction of these C—X σ bonds is another example of nucleophilic substitution, in which LiAlH$_4$ serves as a source of a hydride nucleophile (H⁻). Because H⁻ is a strong nucleophile, the reaction follows an **S$_N$2 mechanism,** illustrated for the one-step reduction of an alkyl halide in Mechanism 12.3.

MECHANISM 12.3

Reduction of RX with LiAlH$_4$

One step   The nucleophile H⁻ substitutes for X⁻ in a single step.

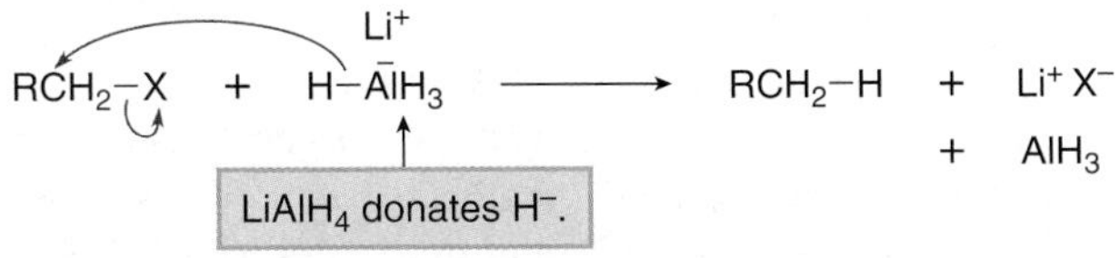

$$RCH_2{-}X \;+\; H{-}\bar{A}lH_3 \quad\xrightarrow{\quad Li^+\quad}\quad RCH_2{-}H \;+\; Li^+ X^-$$
$$+ \;\; AlH_3$$

LiAlH$_4$ donates H⁻.

Because the reaction follows an S$_N$2 mechanism:

- Unhindered CH$_3$X and 1° alkyl halides are more easily reduced than more substituted 2° and 3° halides.
- In unsymmetrical epoxides, nucleophilic attack of H⁻ (from LiAlH$_4$) occurs at the less substituted carbon atom.

Examples are shown in Figure 12.6.

PROBLEM 12.14   Draw the products of each reaction.

a. 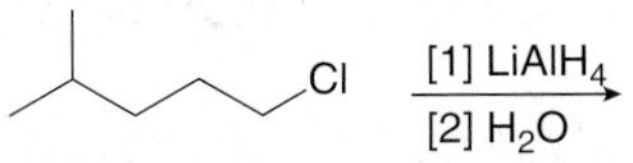 (structure) —Cl $\xrightarrow[\text{[2] } H_2O]{\text{[1] LiAlH}_4}$

b. 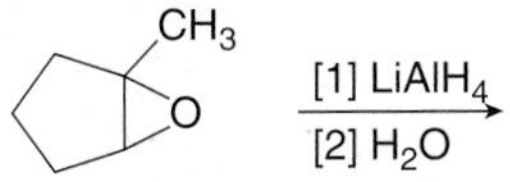 (structure with CH$_3$, O) $\xrightarrow[\text{[2] } H_2O]{\text{[1] LiAlH}_4}$

12.7   Oxidizing Agents

Oxidizing agents fall into two main categories:

- **Reagents that contain an oxygen–oxygen bond**
- **Reagents that contain metal–oxygen bonds**

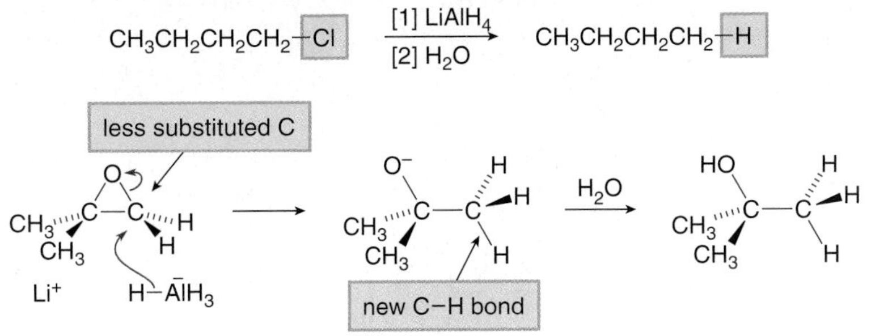

Figure 12.6   Examples of reduction of C—X σ bonds with LiAlH$_4$

$$CH_3CH_2CH_2CH_2{-}Cl \quad\xrightarrow[\text{[2] } H_2O]{\text{[1] LiAlH}_4}\quad CH_3CH_2CH_2CH_2{-}H$$

less substituted C

(epoxide structure with CH$_3$, CH$_3$, H; Li⁺, H—$\bar{A}$lH$_3$) → (intermediate with O⁻, new C—H bond) $\xrightarrow{H_2O}$ (HO, CH$_3$, H product)

Figure 12.7 Common peroxyacids

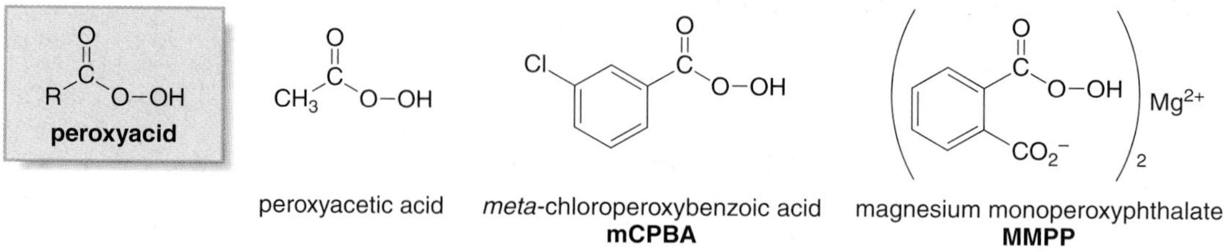

peroxyacetic acid   *meta*-chloroperoxybenzoic acid
mCPBA

magnesium monoperoxyphthalate
MMPP

Oxidizing agents containing an O—O bond include O_2, O_3 (ozone), H_2O_2 (hydrogen peroxide), $(CH_3)_3COOH$ (*tert*-butyl hydroperoxide), and peroxyacids. **Peroxyacids** (or peracids), a group of reagents having the general structure RCO_3H, have one more O atom than carboxylic acids (RCO_2H). Some peroxyacids are commercially available whereas others are prepared and used without isolation. Examples are shown in Figure 12.7. All of these reagents contain a weak O—O bond that is cleaved during oxidation.

The most common oxidizing agents with metal–oxygen bonds contain either chromium in the +6 oxidation state (six Cr—O bonds) or manganese in the +7 oxidation state (seven Mn—O bonds). Common Cr^{6+} reagents include chromium(VI) oxide (CrO_3) and sodium or potassium dichromate ($Na_2Cr_2O_7$ and $K_2Cr_2O_7$). **These reagents are strong oxidants** used in the presence of strong aqueous acid such as H_2SO_4, H_2O. **Pyridinium chlorochromate (PCC)**, a Cr^{6+} reagent that is soluble in halogenated organic solvents, can be used without strong acid present. This makes it a **more selective Cr^{6+} oxidant,** as described in Section 12.12.

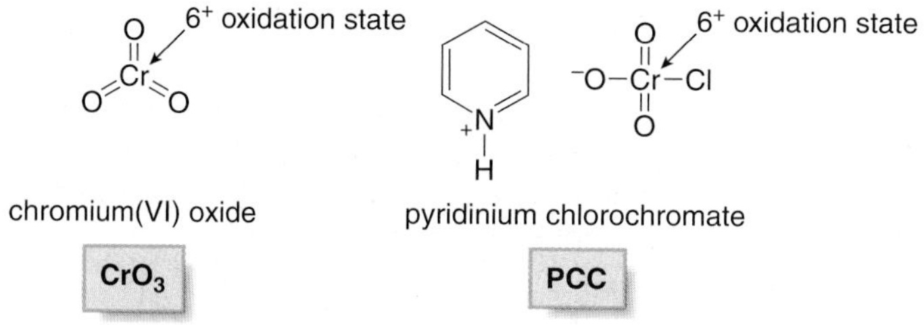

chromium(VI) oxide

CrO₃

pyridinium chlorochromate

PCC

The most common Mn^{7+} reagent is **$KMnO_4$ (potassium permanganate)**, a strong, water-soluble oxidant. Other oxidizing agents that contain metals include **OsO_4** (osmium tetroxide) and **Ag_2O** [silver(I) oxide].

In the remainder of Chapter 12, the oxidation of alkenes, alkynes, and alcohols—three functional groups already introduced in this text—is presented (Figure 12.8). Addition reactions to alkenes and alkynes that increase the number of C—O bonds are described in Sections 12.8–12.11. Oxidation of alcohols to carbonyl compounds appears in Section 12.12.

Figure 12.8 Oxidation reactions of alkenes, alkynes, and alcohols

epoxidation (Sections 12.8, 12.14)

dihydroxylation (Section 12.9)

oxidative cleavage (Section 12.10)

oxidative cleavage (Section 12.11)

(Section 12.12)

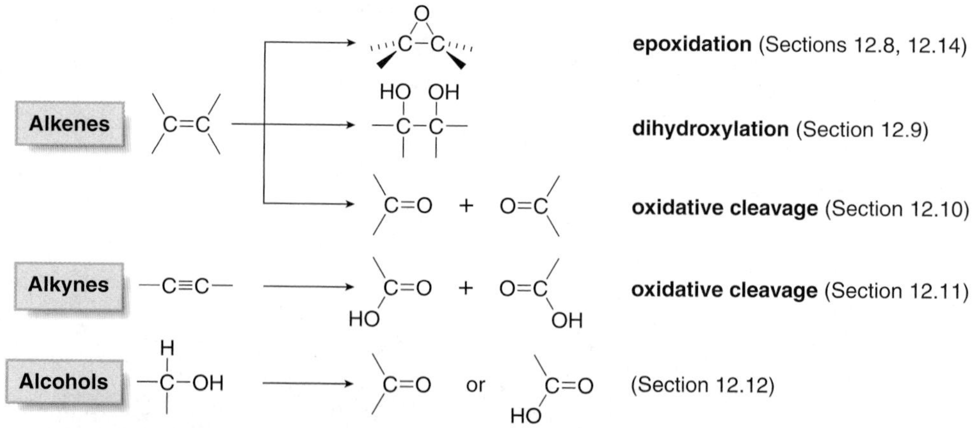

12.8 Epoxidation

Epoxidation is the addition of a single oxygen atom to an alkene to form an **epoxide.**

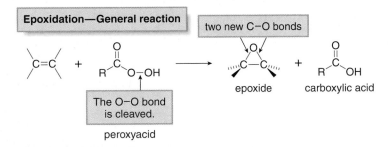

Epoxidation—General reaction

two new C–O bonds

The O–O bond is cleaved.

peroxyacid

epoxide   carboxylic acid

The weak π bond of the alkene is broken and two new C–O σ bonds are formed. Epoxidation is typically carried out with a peroxyacid, resulting in cleavage of the weak O–O bond of the reagent.

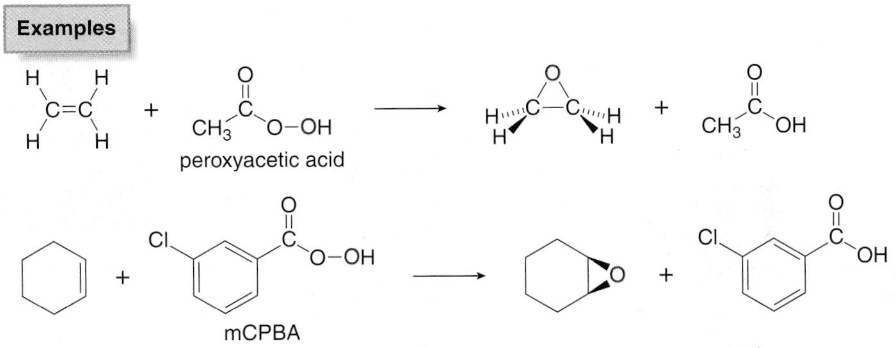

Examples

peroxyacetic acid

mCPBA

Epoxidation occurs via the concerted addition of one oxygen atom of the peroxyacid to the π bond as shown in Mechanism 12.4. Epoxidation resembles the formation of the bridged halonium ion in Section 10.13, in that two bonds in a three-membered ring are formed in one step.

MECHANISM 12.4

Epoxidation of an Alkene with a Peroxyacid

One step All bonds are broken or formed in a single step.

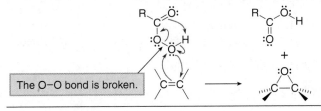

The O–O bond is broken.

◆ Two C–O bonds are formed to one O atom with one electron pair from the peroxyacid and one from the π bond.
◆ The weak O–O bond is broken.

PROBLEM 12.15   What epoxide is formed when each alkene is treated with mCPBA?

a. $(CH_3)_2C=CH_2$ b. $(CH_3)_2C=C(CH_3)_2$ c. 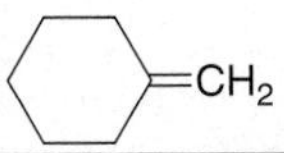 ⬡=CH₂

12.8A The Stereochemistry of Epoxidation

Epoxidation occurs via **syn addition** of an O atom from either side of the planar double bond, so that both C–O bonds are formed on the same side. The relative position of substituents in the alkene reactant is **retained** in the epoxide product.

◆ A cis alkene gives an epoxide with cis substituents. A trans alkene gives an epoxide with trans substituents.

Epoxidation is a **stereospecific** reaction because cis and trans alkenes yield different stereo-isomers as products, as illustrated in Sample Problem 12.2.

SAMPLE PROBLEM 12.2   Draw the stereoisomers formed when *cis*- and *trans*-2-butene are epoxidized with mCPBA.

SOLUTION

To draw each product of epoxidation, add an O atom from either side of the alkene, and keep all substituents in their *original* orientations. The **cis** methyl groups in *cis*-2-butene become **cis** substituents in the epoxide. Addition of an O atom from either side of the trigonal planar alkene leads to the same compound—an **achiral meso compound that contains two stereogenic centers.**

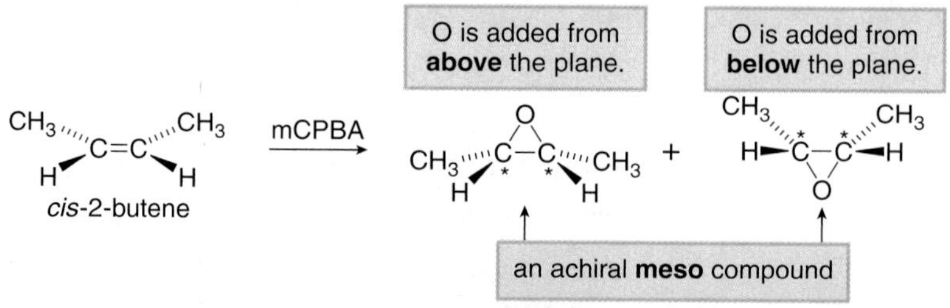

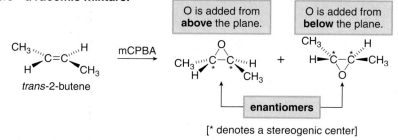

[* denotes a stereogenic center]

The **trans** methyl groups in *trans*-2-butene, become **trans** substituents in the epoxide. Addition of an O atom from either side of the trigonal planar alkene yields an equal mixture of two enantiomers—a **racemic mixture.**

Epoxidation of *cis*- and *trans*-2-butene illustrates the general rule about the stereochemistry of reactions: **an achiral starting material gives achiral or racemic products.**

[* denotes a stereogenic center]

PROBLEM 12.16   Draw all stereoisomers formed when each alkene is treated with mCPBA.

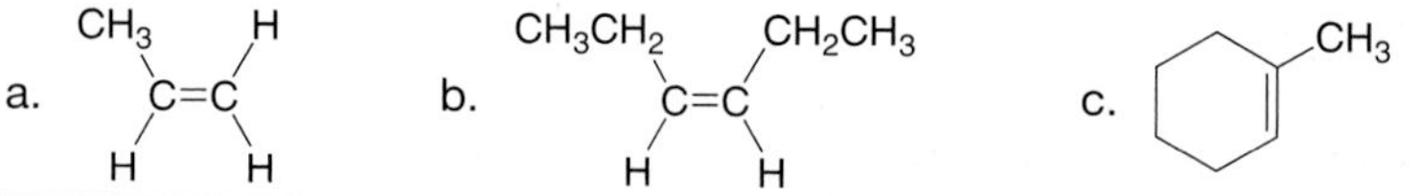

a., b., c.

12.8B   The Synthesis of Disparlure

Disparlure, the sex pheromone of the female gypsy moth and the epoxide that opened Chapter 12, is synthesized by a stepwise reaction sequence that uses an epoxidation reaction as the final step.

Retrosynthetic analysis of disparlure illustrates three key operations:

In 1869, the gypsy moth was introduced into New England in an attempt to develop a silk industry. Some moths escaped into the wild and the population flourished. Mature gypsy moth caterpillars eat an average of one square foot of leaf surface per day, defoliating shade trees and entire forests. Many trees die after a single defoliation.

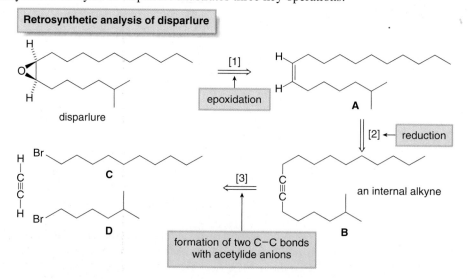

◆ **Step [1]** The cis epoxide in disparlure is prepared from a cis alkene **A** by epoxidation.

◆ **Step [2]** **A** is prepared from an internal alkyne **B** by reduction.

◆ **Step [3]** **B** is prepared from acetylene and two 1° alkyl halides (**C** and **D**) by using S_N2 reactions with acetylide anions.

Figure 12.9 illustrates the synthesis of disparlure beginning with acetylene.

The synthesis of disparlure is conceptually divided into three parts:

◆ **Part [1]** Acetylene is converted to an internal alkyne **B** by forming two C–C bonds. Each bond is formed by treating an alkyne with base ($NaNH_2$) to form an acetylide anion, which reacts with an alkyl halide (**C** or **D**) in an S_N2 reaction (Section 11.11).

◆ **Part [2]** The internal alkyne **B** is reduced to a cis alkene **A** by syn addition of H_2 using the Lindlar catalyst (Section 12.5B).

◆ **Part [3]** The cis alkene **A** is epoxidized to disparlure using a peroxyacid such as mCPBA.

Epoxidation of the cis alkene **A** from two different sides of the double bond affords two cis epoxides in the last step—a racemic mixture of two enantiomers. Thus, half of the product is the desired pheromone disparlure, but the other half is its biologically inactive enantiomer. Separating the desired from the undesired enantiomer is difficult and expensive, because both compounds have identical physical properties. A reaction that affords a chiral epoxide from an achiral precursor without forming a racemic mixture is discussed in Section 12.14.

Figure 12.9 The synthesis of disparlure

Part [1] Formation of two C–C bonds using acetylide anions

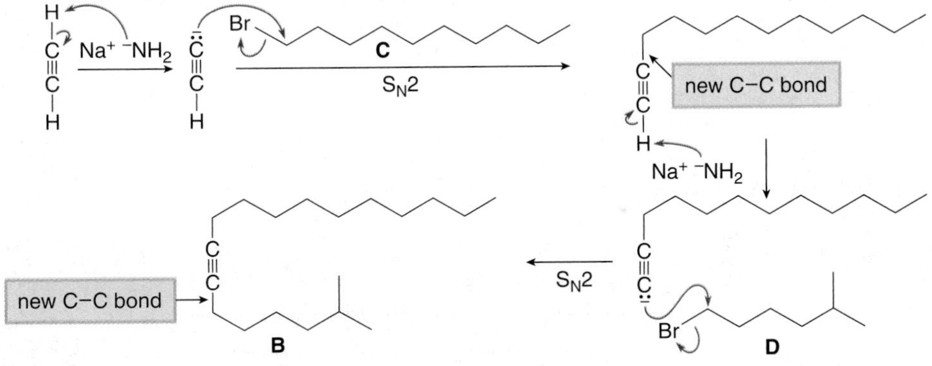

Part [2] Reduction of alkyne **B** to form cis alkene **A**

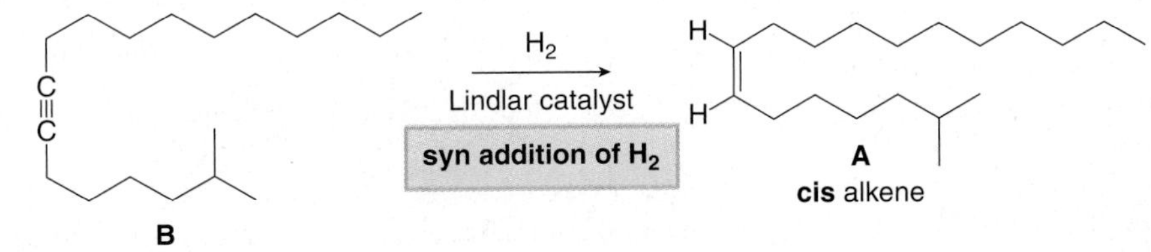

Part [3] Epoxidation of **A** to form disparlure

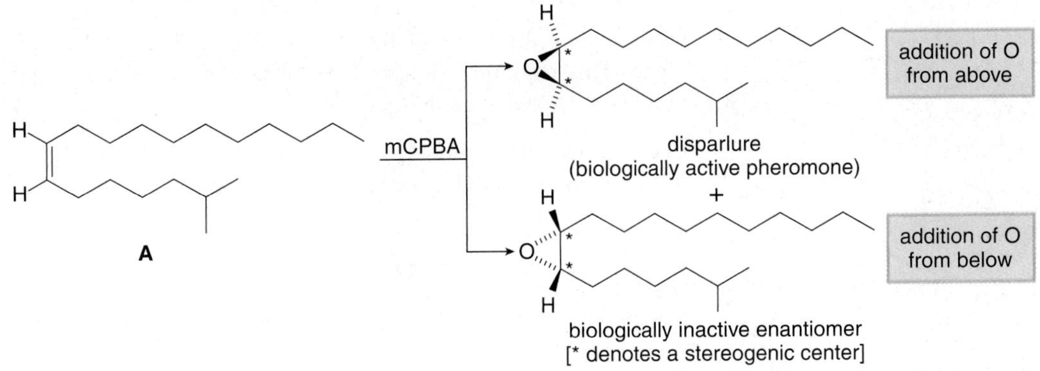

12.9   Dihydroxylation

Dihydroxylation is the addition of two hydroxy groups to a double bond, forming a **1,2-diol** or **glycol.** Depending on the reagent, the two new OH groups can be added to the opposite sides (**anti** addition) or the same side (**syn** addition) of the double bond.

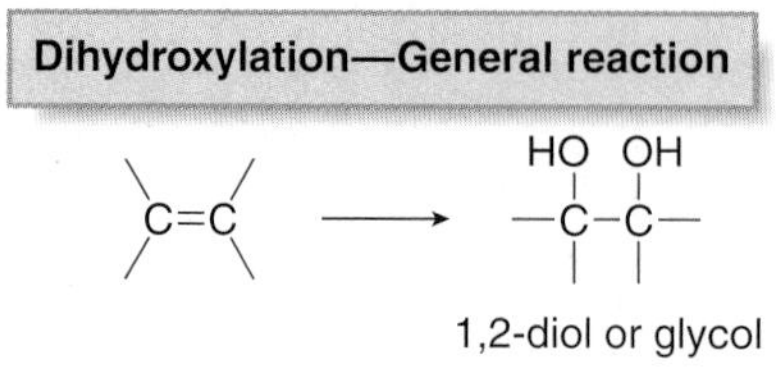

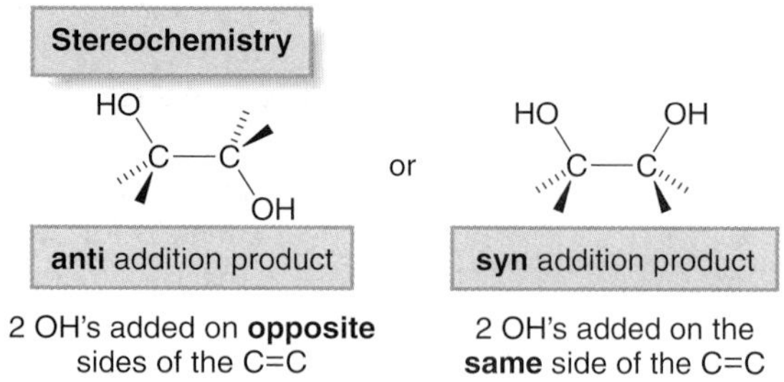

12.9A   Anti Dihydroxylation

Anti dihydroxylation is achieved in two steps—epoxidation followed by opening of the ring with ⁻OH or H_2O. Cyclohexene, for example, is converted to a racemic mixture of two *trans*-1,2-cyclohexanediols by anti addition of two OH groups.

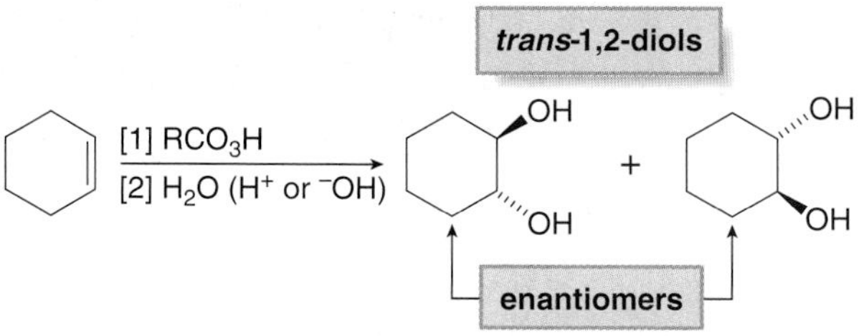

The stereochemistry of the products can be understood by examining the stereochemistry of each step.

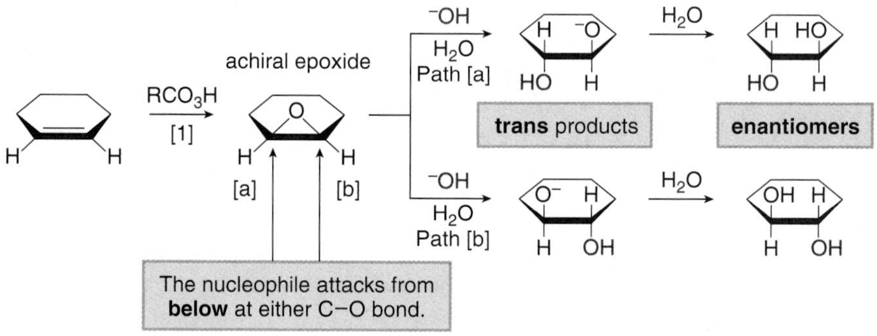

Epoxidation of cyclohexene adds an O atom from either above or below the plane of the double bond to form a single **achiral epoxide,** so only one representation is shown. Opening of the epoxide ring then occurs with **backside attack at either C–O bond.** Because the epoxide is drawn above the plane of the six-membered ring, nucleophilic attack occurs from **below** the plane. This reaction is a specific example of the opening of epoxide rings with strong nucleophiles, first presented in Section 9.15.

Because one OH group of the 1,2-diol comes from the epoxide and one OH group comes from the nucleophile (⁻OH), the overall result is **anti addition of two OH groups** to an alkene.

PROBLEM 12.17   Draw the products formed when both *cis*- and *trans*-2-butene are treated with a peroxyacid followed by ⁻OH (in H_2O). Explain how these reactions illustrate that anti dihydroxylation is stereospecific.

12.9B Syn Dihydroxylation

Syn dihydroxylation results when an alkene is treated with either **KMnO₄** or **OsO₄.**

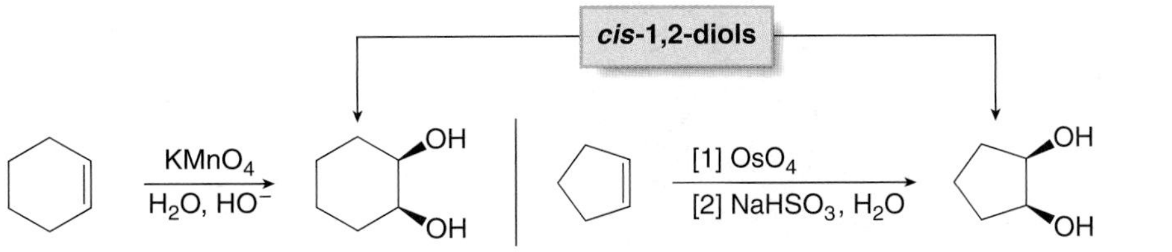

cis-1,2-cyclohexanediol   *cis*-1,2-cyclopentanediol

Each reagent adds two oxygen atoms to the same side of the double bond—that is, in a **syn** fashion—to yield a cyclic intermediate. Hydrolysis of the cyclic intermediate cleaves the metal–oxygen bonds, forming the *cis*-1,2-diol. With OsO₄, sodium bisulfite (NaHSO₃) is also added in the hydrolysis step.

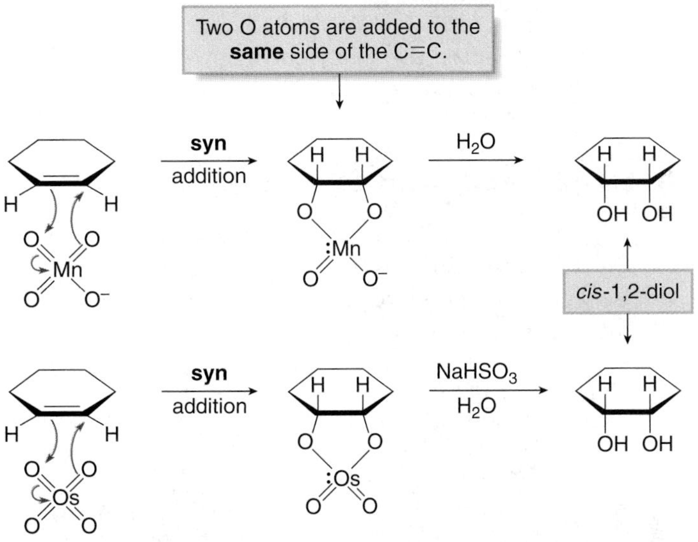

Although KMnO₄ is inexpensive and readily available, its use is limited by its insolubility in organic solvents. To prevent further oxidation of the product 1,2-diol, the reaction mixture must be kept basic with added ⁻OH.

Although OsO₄ is a more selective oxidant than KMnO₄ and is soluble in organic solvents, it is toxic and expensive. To overcome these limitations, dihydroxylation can be carried out by using a catalytic amount of OsO₄, if the oxidant **N-methylmorpholine N-oxide (NMO)** is also added.

N-methylmorpholine *N*-oxide

NMO

In the catalytic process, dihydroxylation of the double bond converts the Os^{8+} oxidant into an Os^{6+} product, which is then re-oxidized by NMO to Os^{8+}. This Os^{8+} reagent can then be used for dihydroxylation once again, and the catalytic cycle continues.

> NMO is an **amine oxide.** It is not possible to draw a Lewis structure of an amine oxide having only neutral atoms.
>
>
> amine oxide

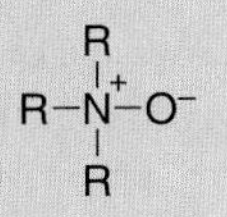

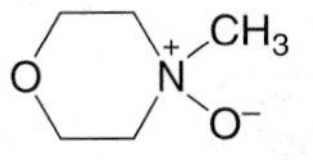

Dihydroxylation with Os^{8+} + NMO

$$\text{C=C} + \text{Os}^{8+} \text{ oxidant} \longrightarrow \text{C-C} + \text{Os}^{6+} \text{ product}$$

catalyst

NMO oxidizes the **Os^{6+} product** back to **Os^{8+}** to begin the cycle again.

PROBLEM 12.18   Draw the products formed when both *cis-* and *trans-*2-butene are treated with OsO$_4$, followed by hydrolysis with NaHSO$_3$ + H$_2$O. Explain how these reactions illustrate that syn dihydroxylation is stereospecific.

PROBLEM 12.19   Explain why treatment of ethylene with either RCO$_3$H followed by H$_2$O ($^-$OH) or KMnO$_4$ + H$_2$O + $^-$OH gives the same 1,2-diol.

12.10   Oxidative Cleavage of Alkenes

Oxidative cleavage of an alkene breaks both the σ and π bonds of the double bond to form two carbonyl groups. Depending on the number of R groups bonded to the double bond, oxidative cleavage yields either **ketones** or **aldehydes**.

Oxidative cleavage—General reaction

$$\underset{R}{\overset{R}{\text{C}}}=\underset{H}{\overset{R}{\text{C}}} \xrightarrow{[O]} \underset{R}{\overset{R}{\text{C}}}=O + O=\underset{H}{\overset{R}{\text{C}}}$$

The σ and π bonds are broken.   ketone   aldehyde

One method of oxidative cleavage relies on a two-step procedure using **ozone (O$_3$) as the oxidant** in the first step. Cleavage with ozone is called **ozonolysis**.

Examples

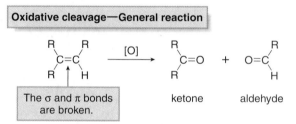

Addition of ozone to the π bond of the alkene forms an unstable intermediate called a **molozonide**, which then rearranges to an **ozonide** by a stepwise process. The unstable ozonide is then reduced without isolation to afford carbonyl compounds. **Zn (in H$_2$O) or dimethyl sulfide (CH$_3$SCH$_3$)** are two common reagents used to convert the ozonide to carbonyl compounds.

The pungent odor around a heavily used photocopy machine is O$_3$ produced from O$_2$ during the process. O$_3$ at ground level is an unwanted atmospheric pollutant. In the stratosphere, however, it protects us from harmful ultraviolet radiation, as discussed in Chapter 13.

The key intermediates in ozonolysis

$$\text{C=C} \xrightarrow{\text{addition}} \text{molozonide} \longrightarrow \text{ozonide} \xrightarrow[\text{CH}_3\text{SCH}_3]{\overset{\text{Zn, H}_2\text{O}}{\text{or}}} \text{C=O:} + :\text{O=C}$$

molozonide   ozonide   by-product: Zn(OH)$_2$ or (CH$_3$)$_2$S=O

To draw the product of any oxidative cleavage:

◆ **Locate all π bonds in the molecule.**

◆ **Replace each C=C by two C=O bonds.**

SAMPLE PROBLEM 12.3   Draw the products when each alkene is treated with O_3 followed by CH_3SCH_3.

a. 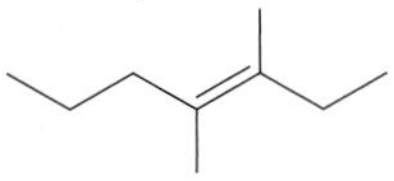 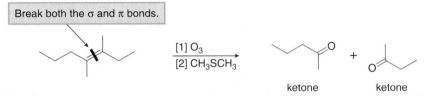   b.

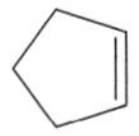

SOLUTION

[a] Cleave the double bond and replace it with two carbonyl groups.

Break both the σ and π bonds.

$$\xrightarrow[\text{[2] } CH_3SCH_3]{\text{[1] } O_3}$$

ketone   +   ketone

[b] For a cycloalkene, oxidative cleavage results in a **single molecule with two carbonyl groups—a dicarbonyl compound.**

Break both the σ and π bonds.

$$\xrightarrow[\text{[2] } CH_3SCH_3]{\text{[1] } O_3}$$

two aldehyde groups

dicarbonyl compound

PROBLEM 12.20   Draw the products formed when each alkene is treated with O_3 followed by Zn, H_2O.

a. $(CH_3)_2C=CHCH_2CH_2CH_2CH_3$ b. 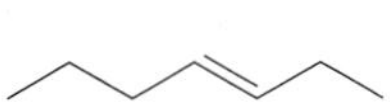   c.

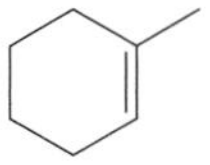

Ozonolysis of dienes (and other polyenes) results in oxidative cleavage of all C=C bonds. The number of carbonyl groups formed in the products is *twice* the number of double bonds in the starting material.

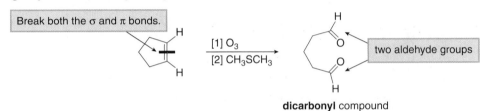

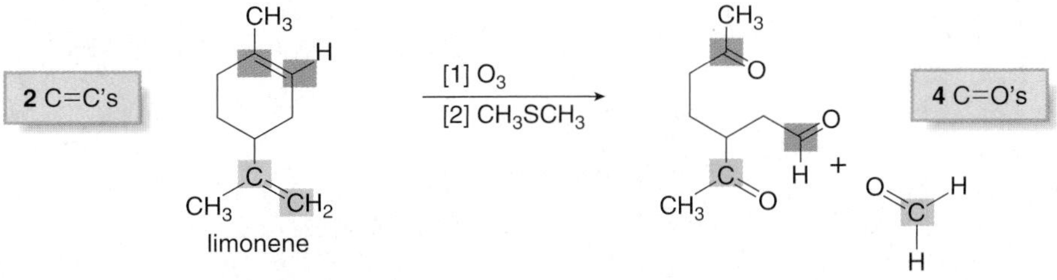

2 C=C's

$$\xrightarrow[\text{[2] } CH_3SCH_3]{\text{[1] } O_3}$$

4 C=O's

limonene

Oxidative cleavage is a valuable tool for structure determination of unknown compounds. The ability to determine what alkene gives rise to a particular set of oxidative cleavage products is thus a useful skill, illustrated in Sample Problem 12.4.

SAMPLE PROBLEM 12.4   What alkene forms the following products after reaction with O_3 followed by CH_3SCH_3?

+   H—C=O

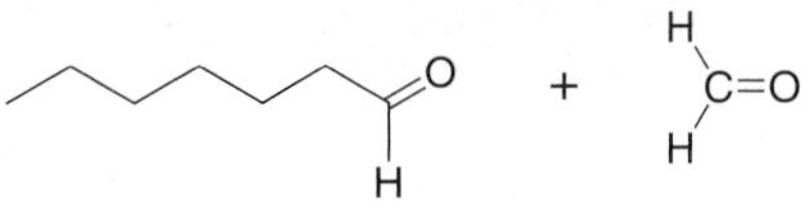

SOLUTION

To draw the starting material, ignore the O atoms in the carbonyl groups and join the carbonyl carbons together by a C=C.

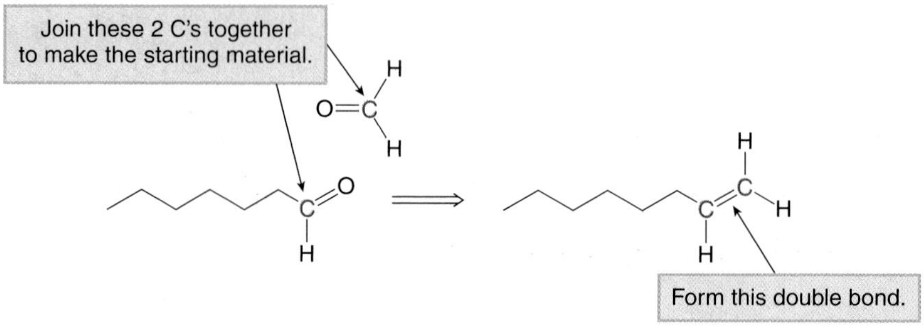

Join these 2 C's together
to make the starting material.

Form this double bond.

PROBLEM 12.21   What alkene yields each set of oxidative cleavage products?

a. $(CH_3)_2C=O$ + $(CH_3CH_2)_2C=O$

b. 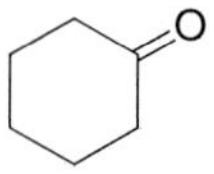 [cyclohexanone] + CH_3CHO

c. $\underset{CH_3}{\overset{CH_3}{C}}=O$ only

Oxidative cleavage reactions occur in cells as well. Recent research has shown that ozone gener-ated in vivo oxidatively cleaves the double bond in cholesterol to form a dicarbonyl product. This oxidation product is thought to contribute to arterial deposits, which are observed in coronary artery disease.

HO — cholesterol

This σ and π bond are broken.

oxidative cleavage

dicarbonyl product
(possible contributor to
coronary artery disease)

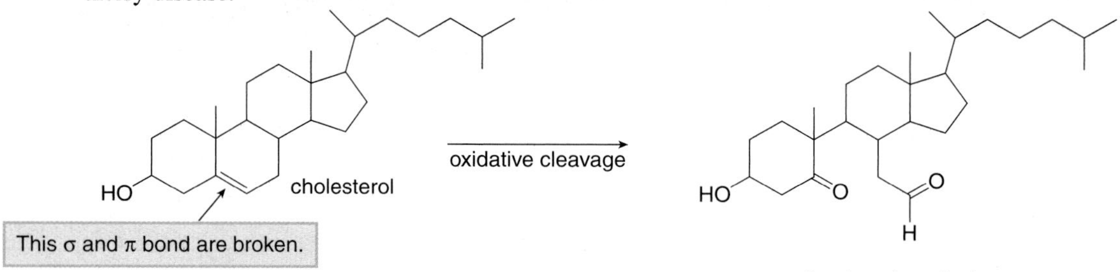

12.11   Oxidative Cleavage of Alkynes

Alkynes also undergo oxidative cleavage of the σ bond and both π bonds of the triple bond. Internal alkynes are oxidized to **carboxylic acids (RCOOH),** whereas terminal alkynes afford carboxylic acids and CO_2 from the *sp* hybridized C–H bond.

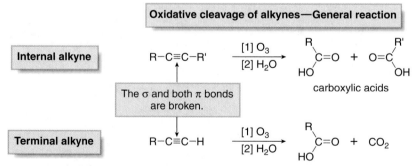

Oxidative cleavage of alkynes—General reaction

Internal alkyne $R-C\equiv C-R'$ $\xrightarrow[\text{[2] } H_2O]{\text{[1] } O_3}$ $\underset{HO}{\overset{R}{C}}=O$ + $O=\underset{OH}{\overset{R'}{C}}$

The σ and both π bonds are broken.

carboxylic acids

Terminal alkyne $R-C\equiv C-H$ $\xrightarrow[\text{[2] } H_2O]{\text{[1] } O_3}$ $\underset{HO}{\overset{R}{C}}=O$ + CO_2

Oxidative cleavage is commonly carried out with O_3, followed by cleavage of the intermediate ozonide with H_2O.

Examples $CH_3-C\equiv C-CH_2CH_3$ $\xrightarrow[\text{[2] } H_2O]{\text{[1] } O_3}$ $\underset{HO}{\overset{CH_3}{C}}=O$ + $O=\underset{OH}{\overset{CH_2CH_3}{C}}$

[cyclohexyl] $-C\equiv C-H$ $\xrightarrow[\text{[2] } H_2O]{\text{[1] } O_3}$ [cyclohexyl carboxylic acid] + CO_2

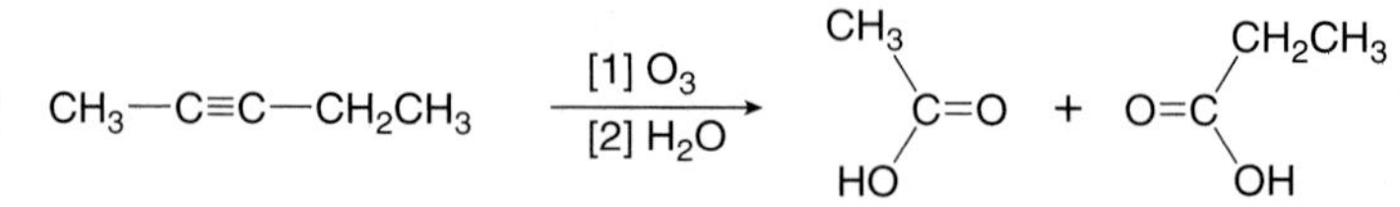

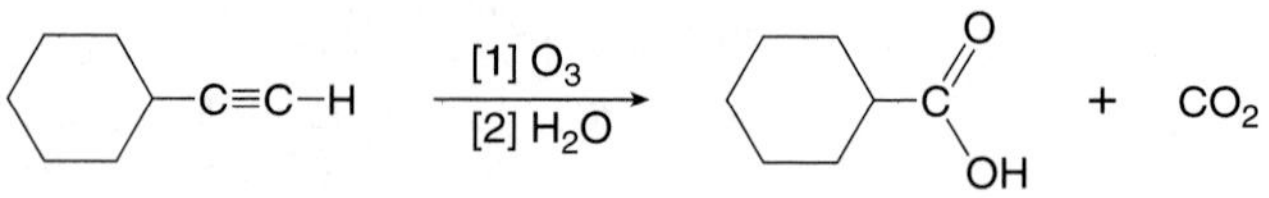

PROBLEM 12.22 Draw the products formed when each alkyne is treated with O_3 followed by H_2O.

a. $CH_3CH_2-C{\equiv}C-CH_2CH_2CH_3$
b. 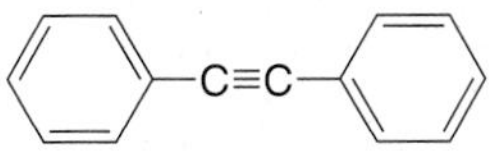 (phenyl)$-C{\equiv}C-$(phenyl)
c. $HC{\equiv}C-CH_2CH_2-C{\equiv}C-CH_3$

12.12 Oxidation of Alcohols

Alcohols are oxidized to a variety of carbonyl compounds, depending on the type of alcohol and reagent. Oxidation occurs by replacing the $C-H$ bonds *on the carbon bearing the OH group* by $C-O$ bonds.

- **1° Alcohols** are oxidized to either **aldehydes** or **carboxylic acids** by replacing either one or two $C-H$ bonds by $C-O$ bonds.

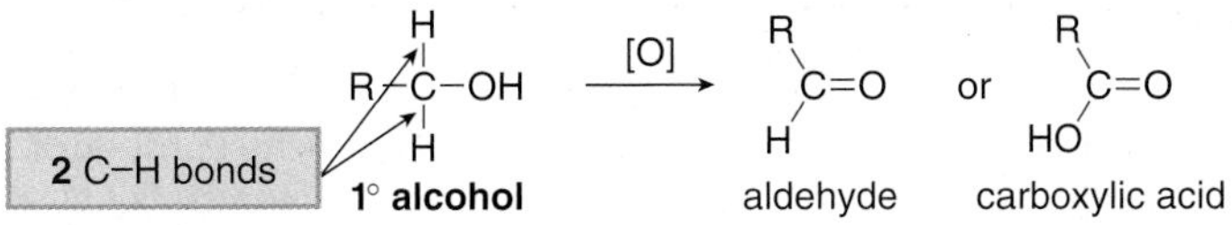

- **2° Alcohols** are oxidized to **ketones** by replacing the one $C-H$ bond by a $C-O$ bond.

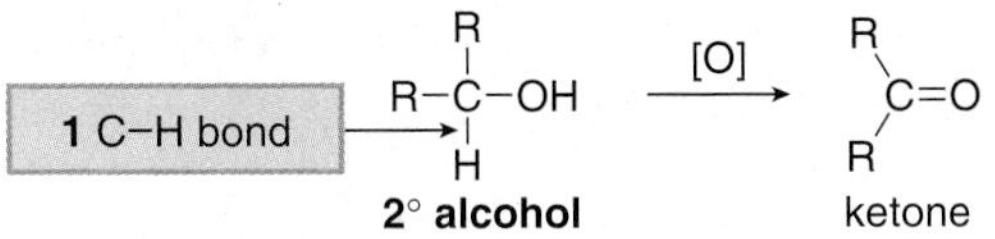

- **3° Alcohols have no H atoms on the carbon with the OH group,** so they are not easily oxidized.

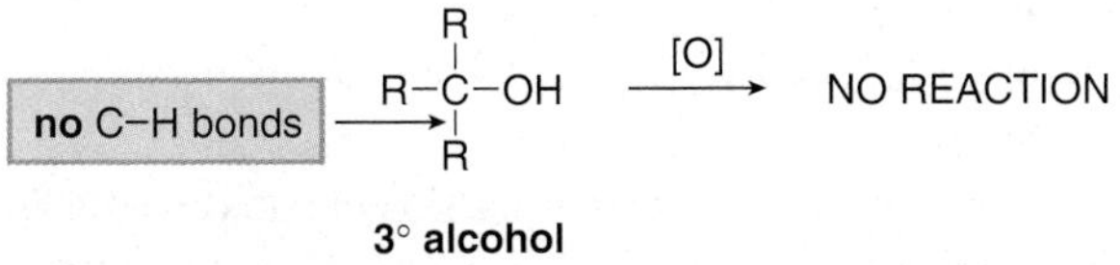

The oxidation of alcohols to carbonyl compounds is typically carried out with Cr^{6+} oxidants, which are reduced to Cr^{3+} products.

- **CrO_3, $Na_2Cr_2O_7$, and $K_2Cr_2O_7$** are **strong, nonselective oxidants** used in aqueous acid ($H_2SO_4 + H_2O$).
- **PCC** (Section 12.7) is soluble in CH_2Cl_2 (dichloromethane), and can be used without strong acid present, making it a **more selective, milder oxidant.**

12.12A Oxidation of 2° Alcohols

Any of the Cr^{6+} oxidants effectively oxidize 2° alcohols to ketones.

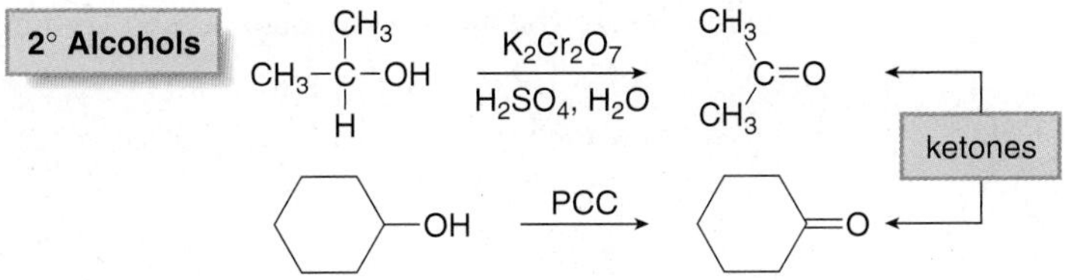

The mechanism for alcohol oxidation has two key parts: **formation of a chromate ester and loss of a proton.** Mechanism 12.5 is drawn for the oxidation of a general 2° alcohol with CrO_3.

MECHANISM 12.5

Oxidation of an Alcohol with CrO_3

Steps [1] and [2] Formation of the chromate ester

- Nucleophilic attack of the alcohol on the electrophilic metal followed by proton transfer forms a **chromate ester.** The C–H bond in the starting material (the 2° alcohol) is still present in the chromate ester, so there is no net oxidation in Steps [1] and [2].

Step [3] Removal of a proton to form the carbonyl group

- In Step [3], a base (H_2O or a molecule of the starting alcohol) removes a proton, with the electron pair in the C–H bond forming the new π bond of the C=O. **Oxidation at carbon occurs in this step** because the number of C–H bonds decreases and the number of C–O bonds increases.

These three steps convert the Cr^{6+} oxidant to a Cr^{4+} product, which is then further reduced to a Cr^{3+} product by a series of steps.

12.12B Oxidation of 1° Alcohols

1° Alcohols are oxidized to either aldehydes or carboxylic acids, depending on the reagent.

- 1° Alcohols are oxidized to aldehydes (RCHO) under mild reaction conditions—using PCC in CH_2Cl_2.
- 1° Alcohols are oxidized to carboxylic acids (RCOOH) under harsher reaction conditions: $Na_2Cr_2O_7$, $K_2Cr_2O_7$, or CrO_3 in the presence of H_2O and H_2SO_4.

The mechanism for the oxidation of 1° alcohols to aldehydes parallels the oxidation of 2° alcohols to ketones detailed in Section 12.12A. Oxidation of a 1° alcohol to a carboxylic acid requires three operations: **oxidation first to the aldehyde, reaction with water,** and then further **oxidation to the carboxylic acid,** as shown in Mechanism 12.6.

MECHANISM 12.6

Oxidation of a 1° Alcohol to a Carboxylic Acid

Part [1]   Oxidation of a 1° alcohol to an aldehyde

$$R-\underset{\underset{H}{|}}{\overset{\overset{H}{|}}{C}}-OH \quad + \quad Cr^{6+} \quad \xrightarrow[\text{[Mechanism 12.5]}]{\textbf{3 steps}} \quad \underset{H}{\overset{R}{}}C=O \quad + \quad Cr^{4+}$$

1° alcohol   aldehyde

♦ Oxidation of a 1° alcohol to an aldehyde occurs by the three-step mechanism detailed in Mechanism 12.5.

Part [2]   Addition of H_2O to form a hydrate

$$\underset{H}{\overset{R}{}}C=O \quad \xrightarrow[\text{H}_2\text{SO}_4]{\text{H}_2\text{O}} \quad R-\underset{\underset{OH}{|}}{\overset{\overset{H}{|}}{C}}-OH$$

aldehyde   hydrate

> The elements of H and OH have been added.

♦ The aldehyde reacts with H_2O to form a **hydrate,** a compound with two OH groups on the same carbon atom. Hydrates are discussed in greater detail in Section 21.13.

Part [3]   Oxidation of the hydrate to a carboxylic acid

$$R-\underset{\underset{OH}{|}}{\overset{\overset{H}{|}}{C}}-OH \quad \xrightarrow[\textbf{2 steps}]{Cr^{6+}} \quad H_2\ddot{O}: \quad R-\underset{\underset{OH}{|}}{\overset{\overset{H}{|}}{C}}-\ddot{O}-\underset{\underset{O}{\|}}{Cr}-OH \quad \longrightarrow \quad \underset{HO}{\overset{R}{}}C=O \quad + \quad Cr^{4+} \quad + \quad H_3\ddot{O}^+$$

hydrate   chromate ester   carboxylic acid

♦ The C–H bond of the hydrate is then oxidized with the Cr^{6+} reagent, following Mechanism 12.5. Because the hydrate contains two OH groups, the product of oxidation is a carboxylic acid.

Cr^{6+} oxidations are characterized by a color change, as the **red-orange Cr^{6+} reagent** is reduced to **green Cr^{3+}.** The first devices used to measure blood alcohol content in individuals suspected of "driving under the influence" made use of this color change. Oxidation of CH_3CH_2OH, the 1° alcohol in alcoholic beverages, with orange $K_2Cr_2O_7$ forms CH_3COOH and green Cr^{3+}.

$$CH_3CH_2-OH \quad + \quad \boxed{\begin{array}{c}\textbf{K}_2\textbf{Cr}_2\textbf{O}_7 \\ \textbf{red-orange}\end{array}} \quad \longrightarrow \quad \underset{CH_3}{\overset{\overset{O}{\|}}{}}\underset{}{C}\underset{}{-OH} \quad + \quad \boxed{\begin{array}{c}\textbf{Cr}^{3+} \\ \textbf{green}\end{array}}$$

ethanol
"alcohol"   acetic acid

Blood alcohol level can be determined by having an individual blow into a tube containing $K_2Cr_2O_7$, H_2SO_4, and an inert solid. The alcohol in the exhaled breath is oxidized by the Cr^{6+} reagent, which turns green in the tube (Figure 12.10). The higher the concentration of CH_3CH_2OH in the breath, the more Cr^{6+} is reduced, and the farther the green Cr^{3+} color extends down the length of the sample tube. This value is then correlated with blood alcohol content to determine if an individual has surpassed the legal blood alcohol limit.

Figure 12.10   Blood alcohol screening

• The oxidation of CH_3CH_2OH with $K_2Cr_2O_7$ to form CH_3COOH and Cr^{3+} was the first available method for the routine testing of alcohol concentration in exhaled air. Some consumer products for alcohol screening are still based on this technology.

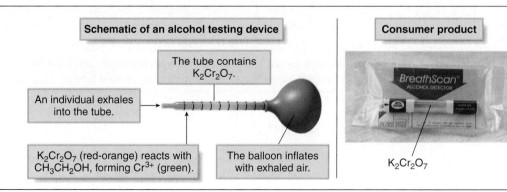

Schematic of an alcohol testing device

The tube contains $K_2Cr_2O_7$.

An individual exhales into the tube.

$K_2Cr_2O_7$ (red-orange) reacts with CH_3CH_2OH, forming Cr^{3+} (green).

The balloon inflates with exhaled air.

Consumer product

BreathScan ALCOHOL DETECTOR

$K_2Cr_2O_7$

PROBLEM 12.23   Draw the organic products in each of the following reactions.

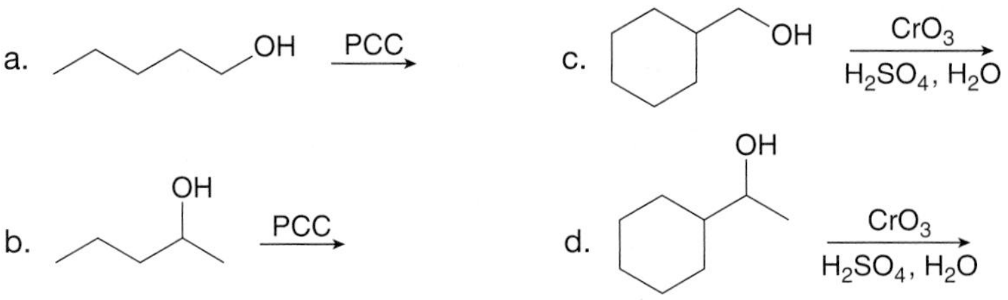

a. $\text{CH}_3(\text{CH}_2)_3\text{CH}_2$—OH $\xrightarrow{\text{PCC}}$

c. cyclohexyl-CH₂—OH $\xrightarrow[\text{H}_2\text{SO}_4,\ \text{H}_2\text{O}]{\text{CrO}_3}$

b. alcohol with OH $\xrightarrow{\text{PCC}}$

d. cyclohexyl-CH(OH)CH₃ $\xrightarrow[\text{H}_2\text{SO}_4,\ \text{H}_2\text{O}]{\text{CrO}_3}$

12.13 Application: The Oxidation of Ethanol

Many reactions in biological systems involve oxidation or reduction. Instead of using Cr^{6+} reagents for oxidation, cells use two organic compounds—a high molecular weight **enzyme** and a simpler **coenzyme** that serves as the oxidizing agent.

For example, when $\text{CH}_3\text{CH}_2\text{OH}$ (ethanol) is ingested, it is oxidized in the liver first to CH_3CHO (acetaldehyde), and then to CH_3COO^- (the acetate anion, the conjugate base of acetic acid). Acetate is the starting material for the synthesis of fatty acids and cholesterol. Both oxidations are catalyzed by the enzyme **alcohol dehydrogenase.**

$$\underset{\text{ethanol}}{\text{CH}_3\text{CH}_2\text{—OH}} \xrightarrow[\text{dehydrogenase}]{\text{[O]}\quad\text{alcohol}} \underset{\text{acetaldehyde}}{\underset{\text{CH}_3}{\overset{\overset{\text{O}}{\|}}{\text{C}}}\text{H}} \xrightarrow[\text{dehydrogenase}]{\text{[O]}\quad\text{alcohol}} \underset{\text{acetate anion}}{\underset{\text{CH}_3}{\overset{\overset{\text{O}}{\|}}{\text{C}}}\text{O}^-}$$

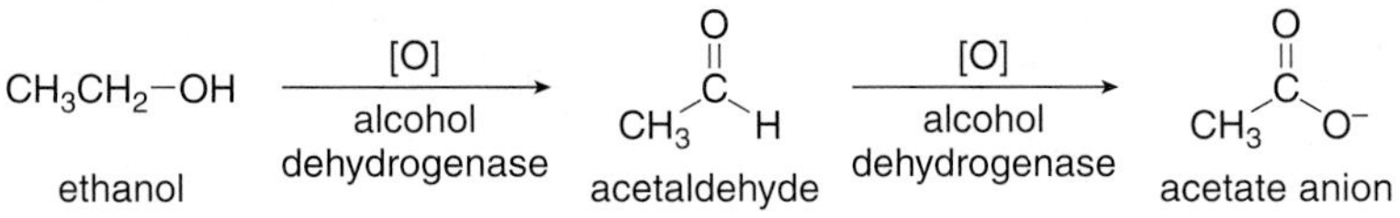

If more ethanol is ingested than can be metabolized in a given time, the concentration of acetaldehyde builds up. This toxic compound is responsible for the feelings associated with a hangover.

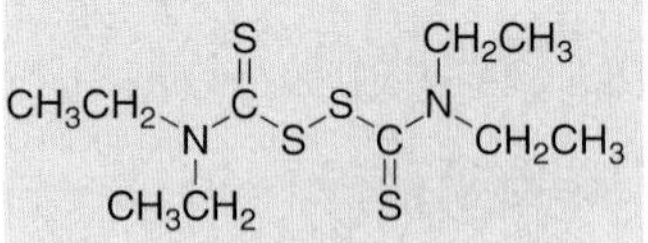

Antabuse, a drug given to alcoholics to prevent them from consuming alcoholic beverages, acts by interfering with the normal oxidation of ethanol. Antabuse inhibits the oxidation of acetaldehyde to the acetate anion. Because the first step in ethanol metabolism occurs but the second does not, the concentration of acetaldehyde rises, causing an individual to become violently ill.

antabuse

Like ethanol, methanol is oxidized by the same enzyme to give an aldehyde and an acid: formaldehyde and formic acid. These oxidation products are extremely toxic because they cannot be used by the body. As a result, the pH of the blood decreases, and blindness and death can follow.

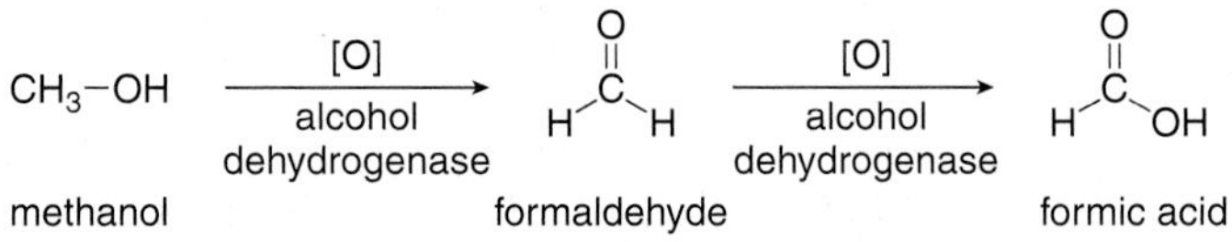

$$\underset{\text{methanol}}{\text{CH}_3\text{—OH}} \xrightarrow[\text{dehydrogenase}]{\text{[O]}\quad\text{alcohol}} \underset{\text{formaldehyde}}{\underset{\text{H}}{\overset{\overset{\text{O}}{\|}}{\text{C}}}\text{H}} \xrightarrow[\text{dehydrogenase}]{\text{[O]}\quad\text{alcohol}} \underset{\text{formic acid}}{\underset{\text{H}}{\overset{\overset{\text{O}}{\|}}{\text{C}}}\text{OH}}$$

Because alcohol dehydrogenase has a higher affinity for ethanol than methanol, methanol poisoning is treated by giving ethanol to the afflicted individual. With both methanol and ethanol in the patient's system, alcohol dehydrogenase reacts more readily with ethanol, allowing the methanol to be excreted unchanged without the formation of methanol's toxic oxidation products.

PROBLEM 12.24   Ethylene glycol, $\text{HOCH}_2\text{CH}_2\text{OH}$, is an extremely toxic diol because its oxidation products are also toxic. Draw the oxidation products formed during the metabolism of ethylene glycol.

12.14 Sharpless Epoxidation

In all of the reactions discussed so far, an **achiral starting material has reacted with an achiral reagent to give either an achiral product or a racemic mixture of two enantiomers.** If you are trying to make a chiral product, this means that only half of the product mixture is the desired enantiomer and the other half is the undesired one. The synthesis of disparlure, outlined in Figure 12.9, exemplifies this dilemma.

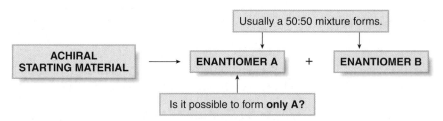

K. Barry Sharpless, currently at the Scripps Research Institute, reasoned that using a chiral reagent might make it possible to favor the formation of one enantiomer over the other.

◆ An *enantioselective* reaction affords predominantly or exclusively one enantiomer.
◆ A reaction that converts an achiral starting material into predominantly one enantiomer is also called an *asymmetric reaction*.

The Sharpless asymmetric epoxidation is an enantioselective reaction that oxidizes alkenes to epoxides. Only the double bonds of **allylic alcohols**—that is, alcohols having a hydroxy group on the carbon adjacent to a C=C—are oxidized in this reaction.

> K. Barry Sharpless shared the 2001 Nobel Prize in Chemistry for his work on chiral oxidation reactions.

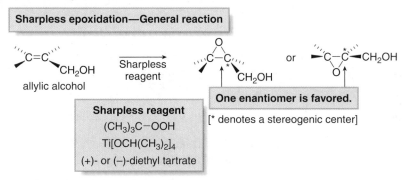

The **Sharpless reagent** consists of three components: *tert*-butyl hydroperoxide, **(CH₃)₃COOH;** a titanium catalyst—usually titanium(IV) isopropoxide, **Ti[OCH(CH₃)₂]₄;** and **diethyl tartrate (DET).** There are two different chiral diethyl tartrate isomers, labeled as (+)-DET or (–)-DET to indicate the direction in which they rotate polarized light.

The identity of the DET isomer determines which enantiomer is the major product obtained in the epoxidation of an allylic alcohol with the Sharpless reagent.

> (+)-DET is prepared from (+)-(R,R)-tartaric acid [HO₂CCH(OH)CH(OH)CO₂H], a naturally occurring carboxylic acid found in grapes and sold as a by-product of the wine industry.

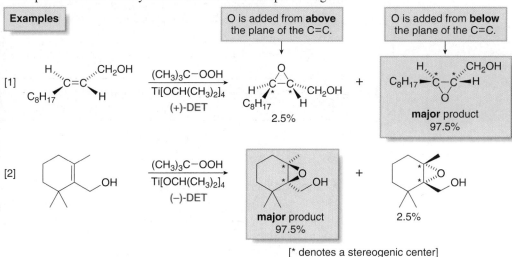

[* denotes a stereogenic center]

Enantiomeric excess = ***ee*** = % of one enantiomer – % of the other enantiomer.

The degree of enantioselectivity of a reaction is measured by its enantiomeric excess ***(ee)*** (Section 5.12D). Reactions [1] and [2] are highly enantioselective because each has an enantiomeric excess of 95% (97.5% of the major enantiomer – 2.5% of the minor enantiomer).

To determine which enantiomer is formed for a given isomer of DET, draw the allylic alcohol in a plane, with the **OH group in the bottom right corner;** then:

- **Epoxidation with (–)-DET adds an oxygen atom from above the plane.**
- **Epoxidation with (+)-DET adds an oxygen atom from below the plane.**

(–)-DET—The O is added from above.
(+)-DET—The O is added from below.
major product
major product

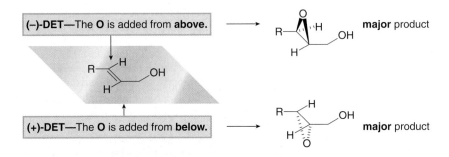

SAMPLE PROBLEM 12.5 Predict the major product in each epoxidation.

a.

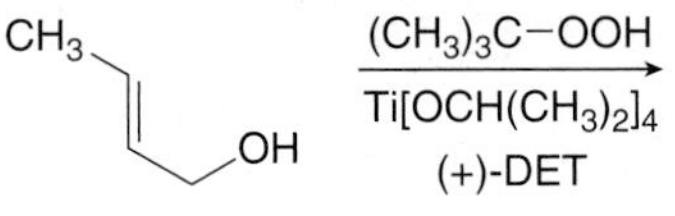

$(CH_3)_3C{-}OOH$
$Ti[OCH(CH_3)_2]_4$
(+)-DET

b.

$(CH_3)_3C{-}OOH$
$Ti[OCH(CH_3)_2]_4$
(–)-DET

SOLUTION

To draw an epoxidation product:

- Draw the allylic alcohol with the OH group on the **bottom right corner of the alkene.** Re-draw the alkene if necessary.
- **(+)-DET** adds the O atom from **below,** and **(–)-DET adds** the O atom from **above.**

[a] The OH group is drawn on the bottom right bond of the double bond and **(+)-DET** is used, so the O atom is added from **below.**

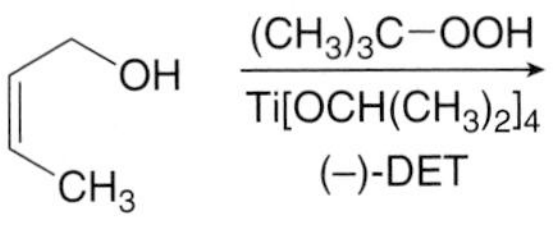

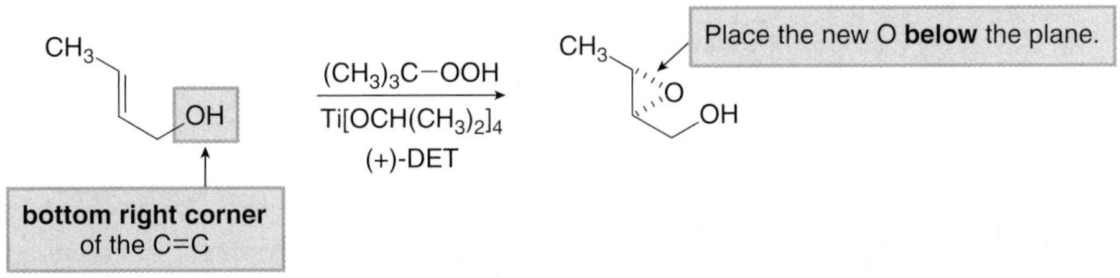

bottom right corner of the C=C
$(CH_3)_3C{-}OOH$
$Ti[OCH(CH_3)_2]_4$
(+)-DET
Place the new O below the plane.

[b] The allylic alcohol must be re-drawn with the OH group in the bottom right corner. Because **(–)-DET** is used, the O atom is then added from **above.**

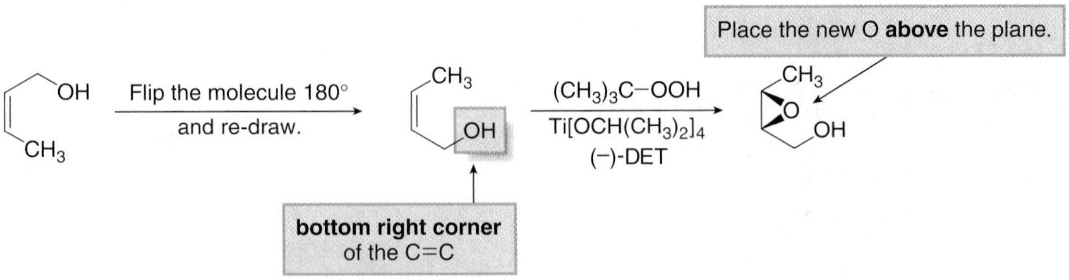

Flip the molecule 180° and re-draw.
bottom right corner of the C=C
$(CH_3)_3C{-}OOH$
$Ti[OCH(CH_3)_2]_4$
(–)-DET
Place the new O above the plane.

PROBLEM 12.25 Draw the products of each Sharpless epoxidation.

a.

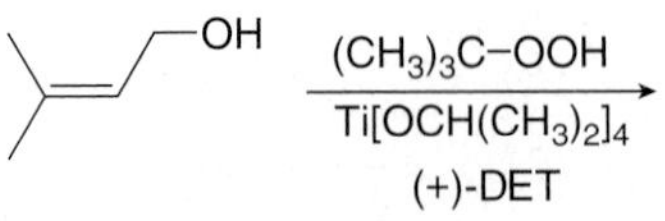

$(CH_3)_3C{-}OOH$
$Ti[OCH(CH_3)_2]_4$
(+)-DET

b.

$(CH_3)_3C{-}OOH$
$Ti[OCH(CH_3)_2]_4$
(–)-DET

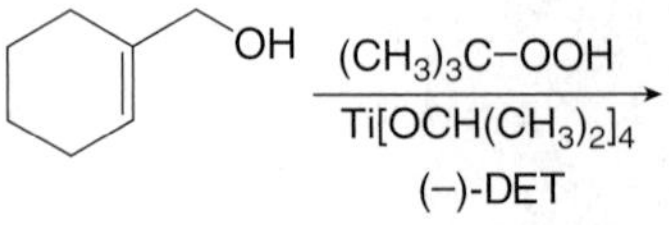

Figure 12.11 The synthesis of chiral insect pheromones using asymmetric epoxidation

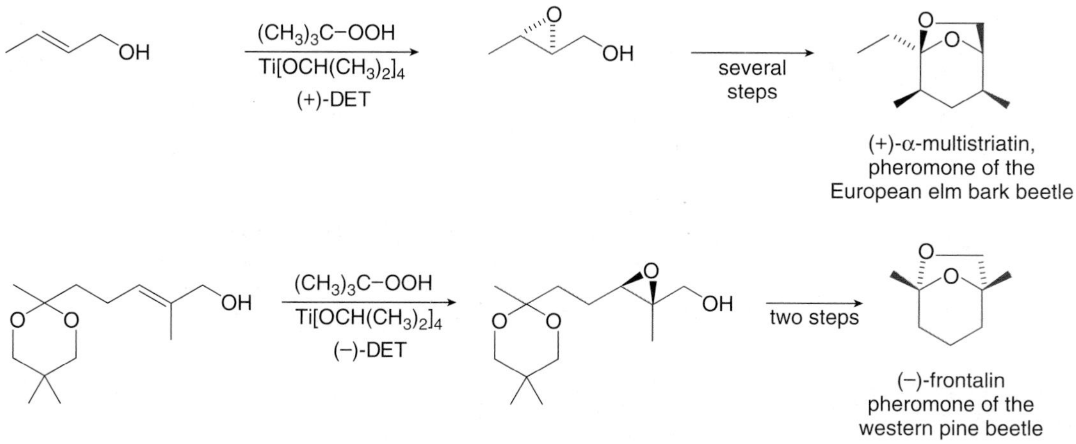

(+)-α-multistriatin, pheromone of the European elm bark beetle

- The bonds in the products that originate from the epoxide intermediate are indicated in red.

(–)-frontalin pheromone of the western pine beetle

PROBLEM 12.26   Explain why only one C=C of geraniol is epoxidized with the Sharpless reagent.

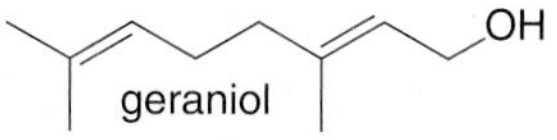

geraniol

The Sharpless epoxidation has been used to synthesize many chiral natural products, including two insect pheromones—(+)-α-multistriatin and (–)-frontalin, as shown in Figure 12.11.

12.15   Key Concepts—Oxidation and Reduction

Summary: Terms that Describe Reaction Selectivity

- A **regioselective reaction** forms predominantly or exclusively one constitutional isomer (Section 8.5).
- A **stereoselective reaction** forms predominantly or exclusively one stereoisomer (Section 8.5).
- An **enantioselective reaction** forms predominantly or exclusively one enantiomer (Section 12.14).

Definitions of Oxidation and Reduction (12.1)

Oxidation reactions result in:
- an increase in the number of C–Z bonds, *or*
- a decrease in the number of C–H bonds

Reduction reactions result in:
- a decrease in the number of C–Z bonds, *or*
- an increase in the number of C–H bonds

Reduction Reactions

[1] Reduction of alkenes—Catalytic hydrogenation (12.3)

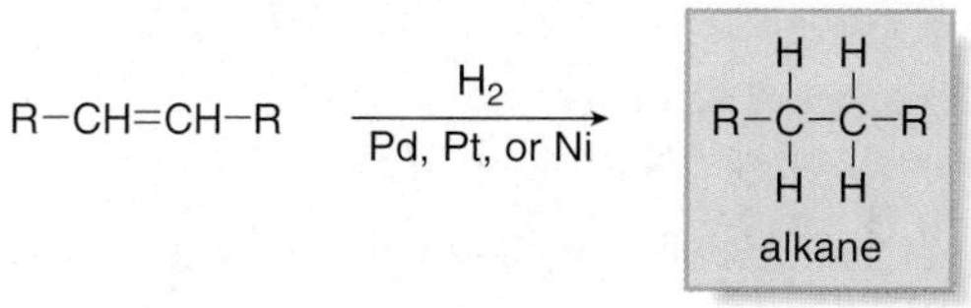

- **Syn addition** of H_2 occurs.
- Increasing alkyl substitution on the C=C decreases the rate of reaction.

[2] Reduction of alkynes

$$R-C\equiv C-R \xrightarrow[\text{Pd-C}]{2\ H_2}$$

alkane

- Two equivalents of H_2 are added and four new C–H bonds are formed (12.5A).

$$R-C\equiv C-R \xrightarrow[\substack{\text{Lindlar}\\\text{catalyst}}]{H_2}$$

cis alkene

- **Syn addition** of H_2 occurs, forming a **cis** alkene (12.5B).
- The Lindlar catalyst is deactivated; reaction stops after one equivalent of H_2 has added.

$$R-C\equiv C-R \xrightarrow[\text{NH}_3]{\text{Na}}$$

trans alkene

- **Anti addition** of H_2 occurs, forming a **trans** alkene (12.5C).

[3] Reduction of alkyl halides (12.6)

$$R-X \xrightarrow[\text{[2] H}_2\text{O}]{\text{[1] LiAlH}_4}$$

R–H
alkane

- The reaction follows an S_N2 mechanism.
- CH_3X and RCH_2X react faster than a more substituted RX.

[4] Reduction of epoxides (12.6)

$$\xrightarrow[\text{[2] H}_2\text{O}]{\text{[1] LiAlH}_4}$$

alcohol

- The reaction follows an S_N2 mechanism.
- In unsymmetrical epoxides, H^- (from $LiAlH_4$) attacks at the less substituted carbon.

Oxidation Reactions

[1] Oxidation of alkenes

[a] Epoxidation (12.8)

$$\text{C=C} + \text{RCO}_3\text{H} \longrightarrow$$

epoxide

- The mechanism has **one step.**
- **Syn addition** of an O atom occurs.
- The reaction is stereospecific.

[b] Anti dihydroxylation (12.9A)

$$\text{C=C} \xrightarrow[\text{[2] H}_2\text{O (H}^+ \text{ or HO}^-\text{)}]{\text{[1] RCO}_3\text{H}}$$

1,2-diol

- Opening of an epoxide ring intermediate with $^-$OH or H_2O forms a 1,2-diol with two OH groups added in an **anti** fashion.

[c] Syn dihydroxylation (12.9B)

$$\text{C=C} \xrightarrow[\substack{\text{or}\\\text{[1] OsO}_4\text{, NMO; [2] NaHSO}_3\\\text{or}\\\text{KMnO}_4\text{, H}_2\text{O, HO}^-}]{\text{[1] OsO}_4\text{; [2] NaHSO}_3}$$

1,2-diol

- Each reagent adds two new C–O bonds to the C=C in a **syn** fashion.

[d] Oxidative cleavage (12.10)

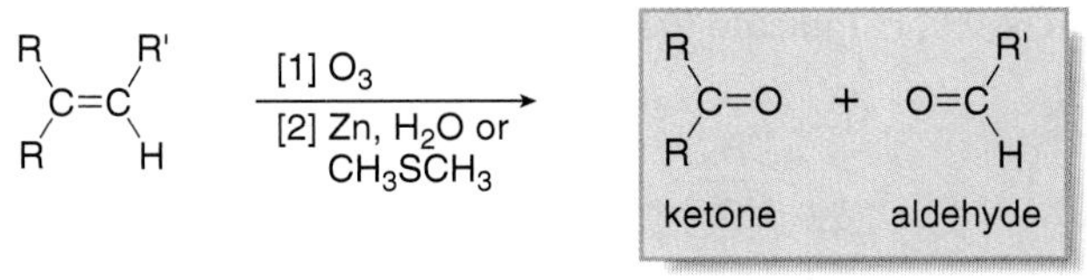

R₂C=CR'H   [1] O₃ / [2] Zn, H₂O or CH₃SCH₃ →   R₂C=O (ketone) + O=CR'H (aldehyde)

- Both the σ and π bond of the alkene are cleaved to form two carbonyl groups.

---

[2] Oxidative cleavage of alkynes (12.11)

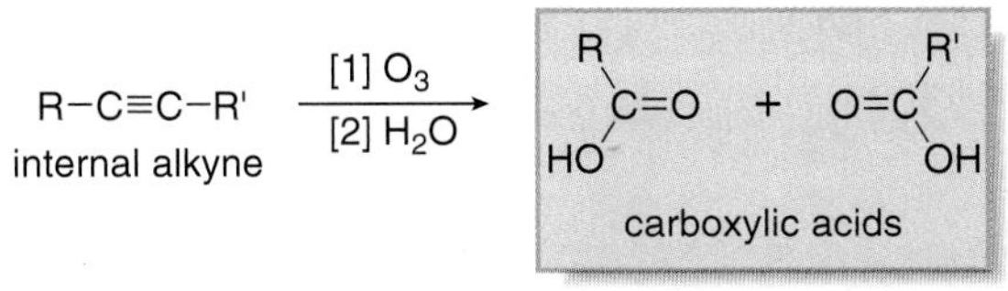

R−C≡C−R' (internal alkyne)   [1] O₃ / [2] H₂O →   R(HO)C=O + O=C(R')OH   carboxylic acids

- The σ bond and both π bonds of the alkyne are cleaved.

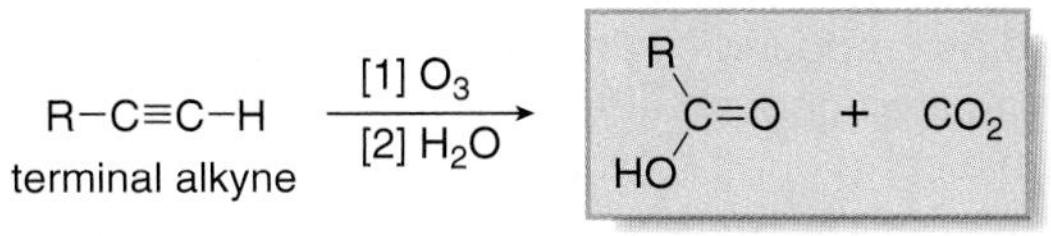

R−C≡C−H (terminal alkyne)   [1] O₃ / [2] H₂O →   R(HO)C=O + CO₂

---

[3] Oxidation of alcohols (12.12)

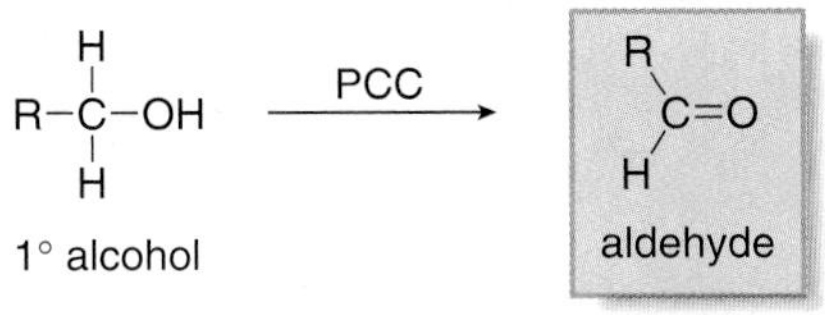

R−CH₂−OH (1° alcohol)   PCC →   R(H)C=O   aldehyde

- Oxidation of a 1° alcohol with PCC stops at the aldehyde stage. Only one C−H bond is replaced by a C−O bond.

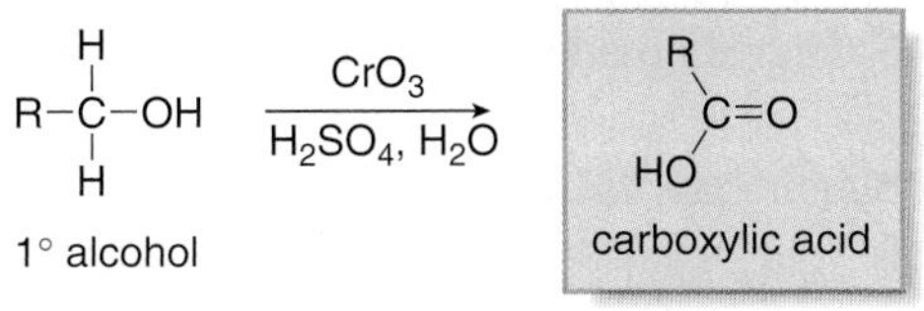

R−CH₂−OH (1° alcohol)   CrO₃ / H₂SO₄, H₂O →   R(HO)C=O   carboxylic acid

- Oxidation of a 1° alcohol under harsher reaction conditions— CrO₃ (or Na₂Cr₂O₇ or K₂Cr₂O₇) + H₂O + H₂SO₄—leads to a RCOOH. Two C−H bonds are replaced by two C−O bonds.

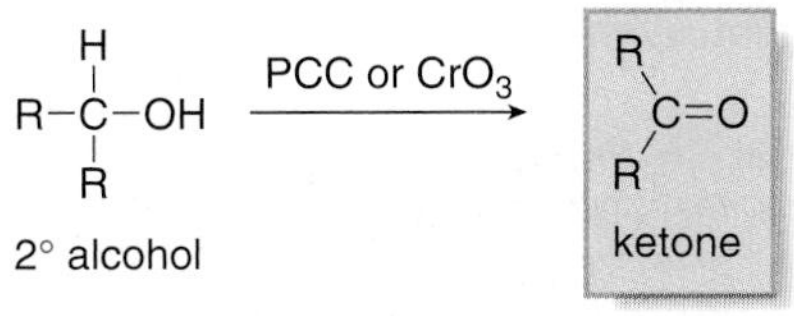

R−CH(R)−OH (2° alcohol)   PCC or CrO₃ →   R(R)C=O   ketone

- Because a 2° alcohol has only one C−H bond on the carbon bearing the OH group, all Cr⁶⁺ reagents—PCC, CrO₃, Na₂Cr₂O₇, or K₂Cr₂O₇—oxidize a 2° alcohol to a ketone.

---

[4] Asymmetric epoxidation of allylic alcohols (12.14)

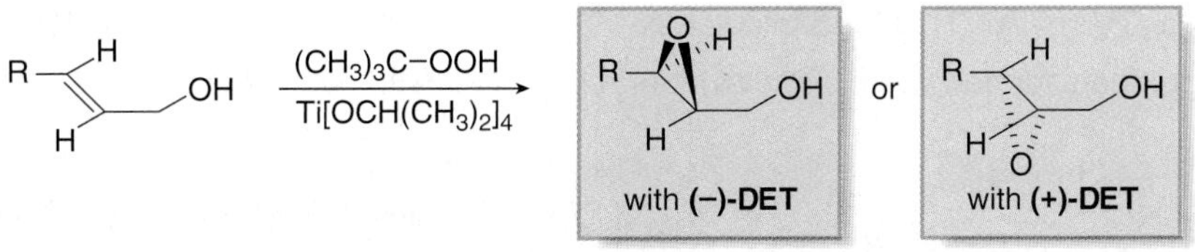

R−CH=CH−CH₂OH   (CH₃)₃C−OOH / Ti[OCH(CH₃)₂]₄ →   epoxide with (−)-DET   or   epoxide with (+)-DET

Problems

Classifying Reactions as Oxidation or Reduction

12.27 Label each reaction as oxidation, reduction, or neither.

a. 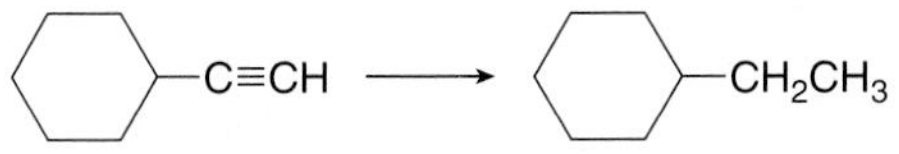 cyclohexyl−C≡CH ⟶ cyclohexyl−CH₂CH₃

b. CH₃CH₂CH₂CH₂OH ⟶ CH₃CH₂CH₂COOH

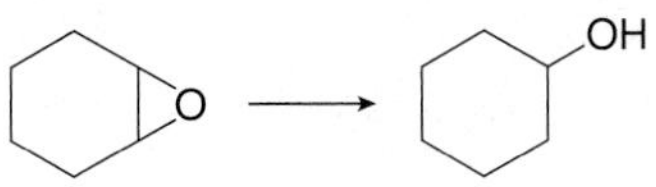

c. CH₃CH₂Br ⟶ CH₂=CH₂

d.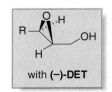

e. CH₂=CH₂ ⟶ ClCH₂CH₂Cl

f.  HO−C₆H₄−OH ⟶ O=C₆H₄=O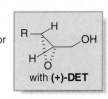

Hydrogenation

12.28 Draw the organic products formed in each hydrogenation reaction. Indicate the three-dimensional structure of all stereoisomers formed.

a. $\xrightarrow[\text{Pd-C}]{\text{H}_2}$

b. $\xrightarrow[\text{Pd-C}]{\text{H}_2}$

c. $\xrightarrow[\text{Pd-C}]{\text{H}_2}$

d. $\xrightarrow[\text{Pd-C}]{\text{H}_2}$

12.29 Match each alkene to its heat of hydrogenation.
Alkenes: 3-methyl-1-butene, 2-methyl-1-butene, 2-methyl-2-butene
$\Delta H°$ (hydrogenation) kcal/mol: –28.5, –30.3, –26.9

12.30 How many rings and π bonds are contained in compounds **A–C?** Draw one possible structure for each compound.
a. Compound **A** has molecular formula C_5H_8 and is hydrogenated to a compound having molecular formula C_5H_{10}.
b. Compound **B** has molecular formula $C_{10}H_{16}$ and is hydrogenated to a compound having molecular formula $C_{10}H_{18}$.
c. Compound **C** has molecular formula C_8H_8 and is hydrogenated to a compound having molecular formula C_8H_{16}.

12.31 For alkenes **A, B,** and **C:** (a) Rank **A, B,** and **C** in order of increasing heat of hydrogenation; (b) rank **A, B,** and **C** in order of increasing rate of reaction with H_2, Pd-C; (c) draw the products formed when each alkene is treated with ozone, followed by Zn, H_2O.

A   **B**   **C**

12.32 Draw the structure of all alkenes (excluding stereoisomers) that are hydrogenated to 2-methylpentane.

12.33 A chiral compound **X** having the molecular formula C_6H_{12} is converted to 3-methylpentane with H_2, Pd-C. Draw all possible structures for **X.**

12.34 Explain why the hydrogenation of alkenes is a kinetically slow, but thermodynamically favorable process.

Reactions—General

12.35 Draw the organic products formed when cyclopentene is treated with each reagent. With some reagents, no reaction occurs.

a. H_2 + Pd-C
b. H_2 + Lindlar catalyst
c. Na, NH_3
d. CH_3CO_3H

e. [1] CH_3CO_3H; [2] H_2O, HO^-
f. [1] OsO_4 + NMO; [2] $NaHSO_3$
g. $KMnO_4$, H_2O, HO^-
h. [1] $LiAlH_4$; [2] H_2O

i. [1] O_3; [2] CH_3SCH_3
j. $(CH_3)_3COOH$, $Ti[OCH(CH_3)_2]_4$, (–)-DET
k. mCPBA
l. Product in (k); then [1] $LiAlH_4$; [2] H_2O

12.36 Draw the organic products formed when 4-octyne is treated with each reagent.
a. H_2 (excess) + Pd-C   b. H_2 + Lindlar catalyst   c. Na, NH_3 d. [1] O_3; [2] H_2O

12.37 Draw the organic products formed when allylic alcohol **A** is treated with each reagent.

A

a. H_2 + Pd-C
b. mCPBA
c. PCC
d. CrO_3, H_2SO_4, H_2O

e. $(CH_3)_3COOH$, $Ti[OCH(CH_3)_2]_4$, (+)-DET
f. $(CH_3)_3COOH$, $Ti[OCH(CH_3)_2]_4$, (–)-DET
g. [1] PBr_3; [2] $LiAlH_4$; [3] H_2O

12.38 Draw the products formed when allylic alcohol **B** is treated with each reagent. Indicate the stereochemistry of any stereoisomers formed.

B

a. H_2 + Pd-C
b. $Na_2Cr_2O_7$, H_2SO_4, H_2O
c. PCC
d. CF_3CO_3H

e. [1] OsO_4; [2] $NaHSO_3$
f. [1] HCO_3H; [2] H_2O, HO^-
g. $(CH_3)_3COOH$, $Ti[OCH(CH_3)_2]_4$, (+)-DET
h. $KMnO_4$, H_2O, HO^-

12.39 Draw the organic products formed in each reaction.

a. 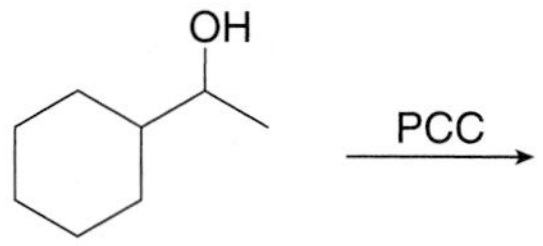
OH
PCC

c. $CH_3CH_2CH_2CH_2OH \xrightarrow{PCC}$

b. 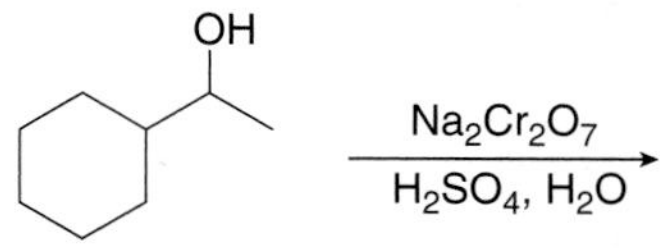
OH
$Na_2Cr_2O_7$
H_2SO_4, H_2O

d. 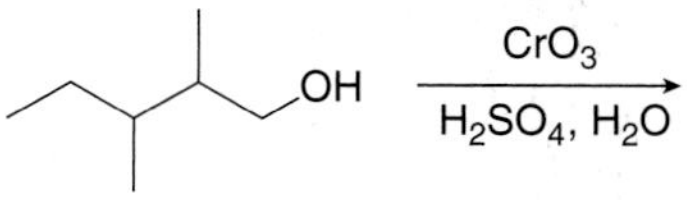
OH
CrO_3
H_2SO_4, H_2O

12.40 Draw the organic products formed in each reaction.

a. 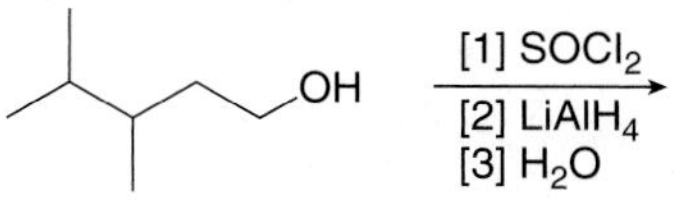
OH
[1] $SOCl_2$
[2] $LiAlH_4$
[3] H_2O

c. 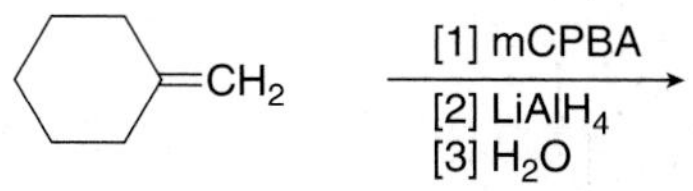
$=CH_2$
[1] mCPBA
[2] $LiAlH_4$
[3] H_2O

b. 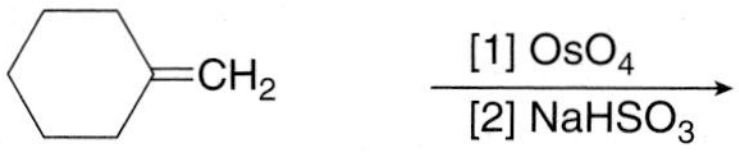
$=CH_2$
[1] OsO_4
[2] $NaHSO_3$

d. 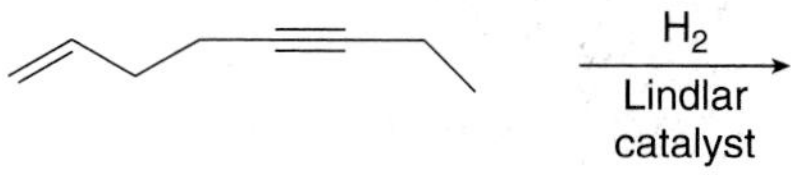
H_2
Lindlar
catalyst

12.41 Identify the reagents needed to carry out each transformation.

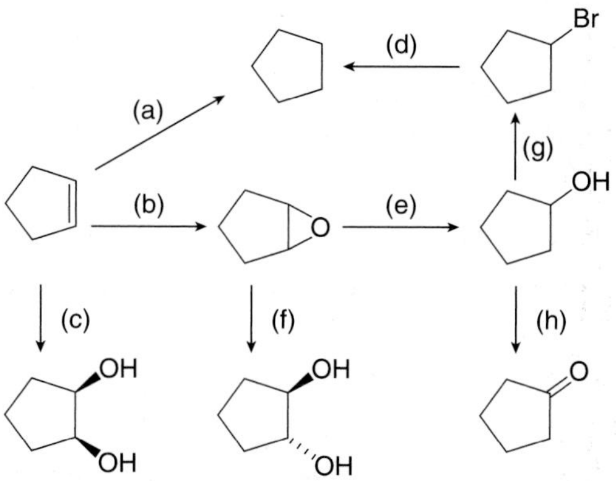
Br
(d)
(a)
(g)
OH
(b)
O
(e)
(c)
(f)
(h)
OH
OH
OH
OH
O

(+ enantiomer)

12.42 Identify compounds **A, B,** and **C.**

a. Compound **A** has molecular formula C_8H_{12} and reacts with two equivalents of H_2. **A** gives $HCOCH_2CH_2CHO$ as the only product of oxidative cleavage with O_3 followed by CH_3SCH_3.

b. Compound **B** has molecular formula C_6H_{10} and gives $(CH_3)_2CHCH_2CH_2CH_3$ when treated with excess H_2 in the presence of Pd. **B** reacts with $NaNH_2$ and CH_3I to form compound **C** (molecular formula C_7H_{12}).

Oxidative Cleavage

12.43 Draw the products formed in each oxidative cleavage.

a. $(CH_3CH_2)_2C{=}CHCH_2CH_3 \xrightarrow[\text{[2] } CH_3SCH_3]{\text{[1] } O_3}$

c. 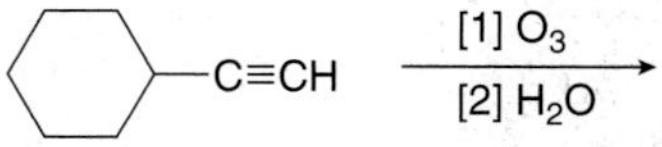
C≡CH
[1] O_3
[2] H_2O

b. 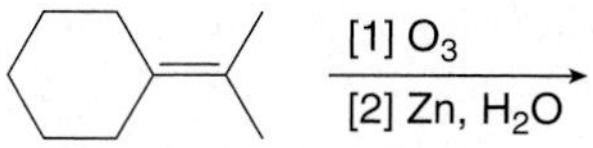
[1] O_3
[2] Zn, H_2O

d. 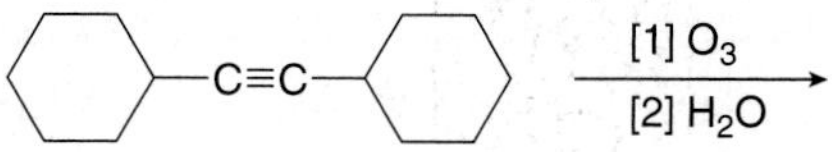
C≡C
[1] O_3
[2] H_2O

12.44 What alkene yields each set of products after treatment with O_3 followed by CH_3SCH_3?

a. $(CH_3)_2C{=}O$ and $CH_2{=}O$

c. $CH_3CH_2CH_2CHO$ only

b. 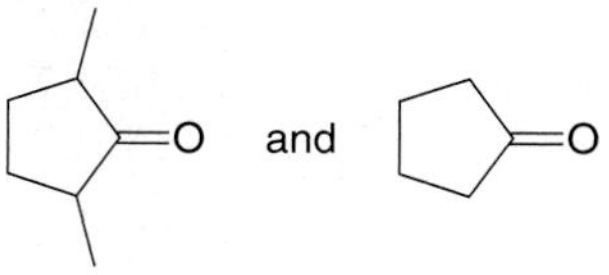
O
and
O

d. 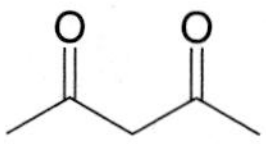 and two equivalents of $CH_2{=}O$

12.45 Identify the starting material in each reaction.

a. $C_{10}H_{18}$ $\xrightarrow[\text{[2] CH}_3\text{SCH}_3]{\text{[1] O}_3}$

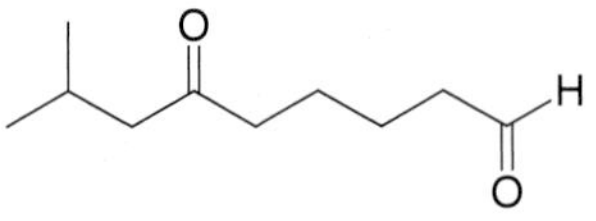

b. $C_{10}H_{16}$ $\xrightarrow[\text{[2] CH}_3\text{SCH}_3]{\text{[1] O}_3}$

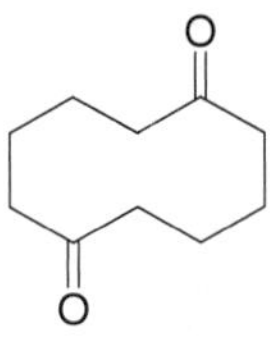

12.46 What alkyne gives each set of products after treatment with O_3 followed by H_2O?

a. $CH_3CH_2CH_2CH_2COOH$ and CO_2

b. CH_3CH_2COOH and $CH_3CH_2CH_2COOH$

c. 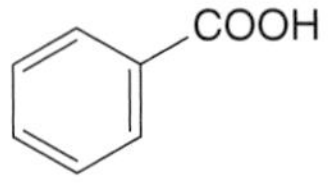 and CH_3COOH

12.47 Draw the products formed when each naturally occurring compound is treated with O_3 followed by Zn, H_2O.

a.

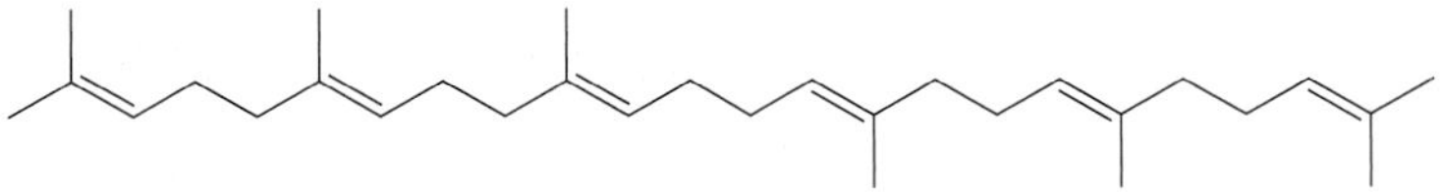

squalene

b.

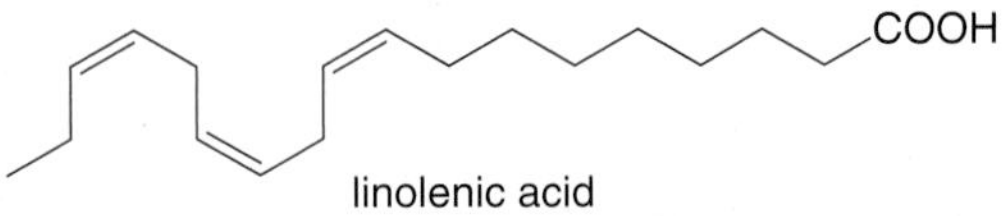

linolenic acid

c.

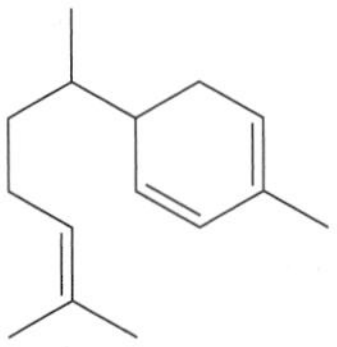

zingiberene

12.48 Oximene and myrcene, two hydrocarbons isolated from alfalfa that have the molecular formula $C_{10}H_{16}$, both yield 2,6-dimethyloctane when treated with H_2 and a Pd catalyst. Ozonolysis of oximene forms $(CH_3)_2C=O$, $CH_2=O$, $CH_2(CHO)_2$, and CH_3COCHO. Ozonolysis of myrcene yields $(CH_3)_2C=O$, $CH_2=O$ (two equiv), and $HCOCH_2CH_2COCHO$. Identify the structures of oximene and myrcene.

Sharpless Asymmetric Epoxidation

12.49 Draw the product of each asymmetric epoxidation reaction.

a. 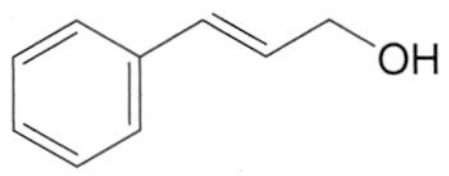 $\xrightarrow[\text{(−)-DET}]{\substack{(CH_3)_3COOH \\ Ti[OC(CH_3)_2]_4}}$

b. $\xrightarrow[\text{(+)-DET}]{\substack{(CH_3)_3COOH \\ Ti[OC(CH_3)_2]_4}}$

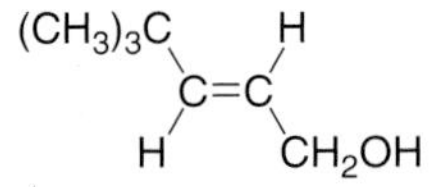

12.50 Epoxidation of the following allylic alcohol using the Sharpless reagent with (−)-DET gives two epoxy alcohols in a ratio of 87:13.

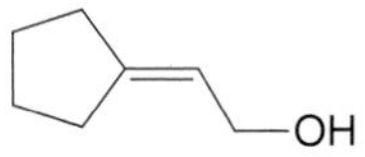

a. Assign structures to the major and minor product.
b. What is the enantiomeric excess in this reaction?

12.51 What allylic alcohol and DET isomer are needed to make each chiral epoxide using a Sharpless asymmetric epoxidation reaction?

a.

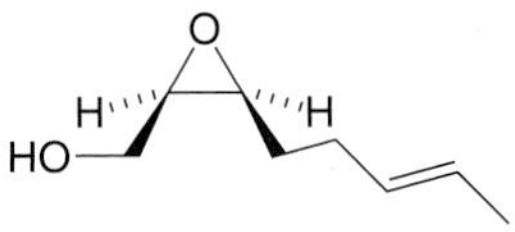

b.

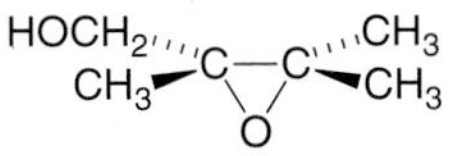

c.

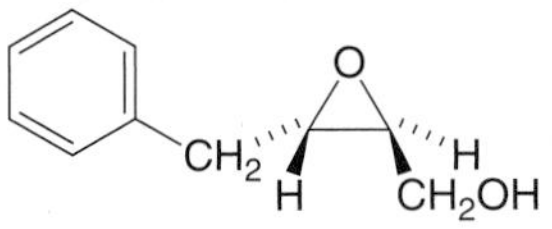

Synthesis

12.52 Devise a synthesis of each hydrocarbon from acetylene, and any other needed reagents.

a. $CH_3CH_2CH=CH_2$

b.

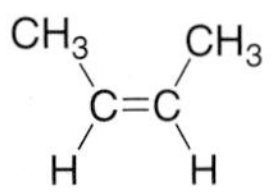

c.

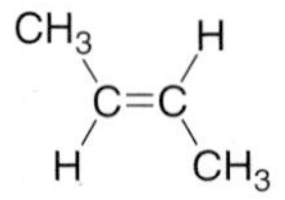

d. $(CH_3)_2CHCH_2CH_2CH_2CH_2CH(CH_3)_2$

12.53 Devise a synthesis of muscalure, the sex pheromone of the common housefly, from acetylene and any other required reagents.

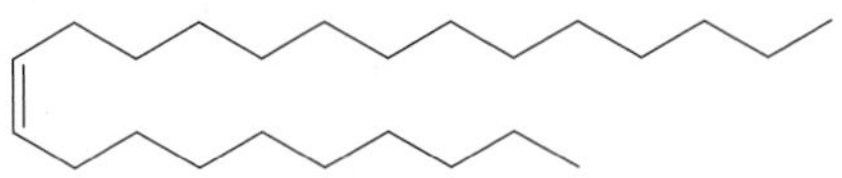

muscalure

12.54 Devise a synthesis of each compound from acetylene and any other required reagents.

a.

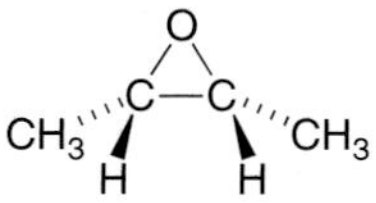

b.

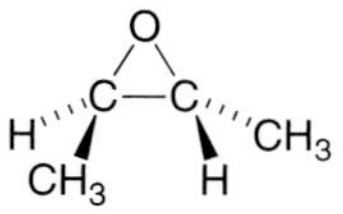

(+ enantiomer)

c.

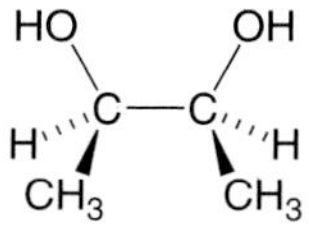

d.

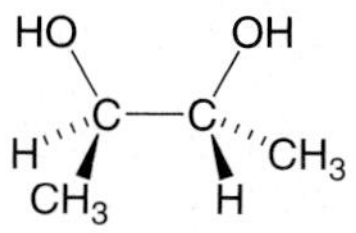

(+ enantiomer)

12.55 Give two methods to synthesize the following epoxide from an alkene.

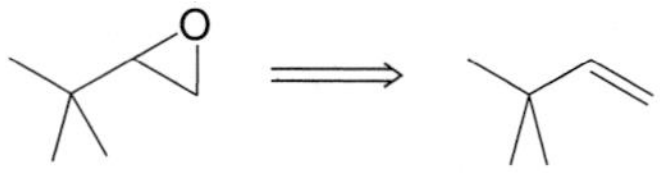

12.56 Devise a synthesis of each compound from the indicated starting material and any other required reagents.

a. $CH_3CH_2CH=CH_2 \longrightarrow CH_3CH_2CH_2COOH$

c.

b. $CH_3CH_2CH_2CH_2OH \longrightarrow$

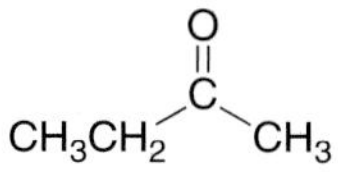

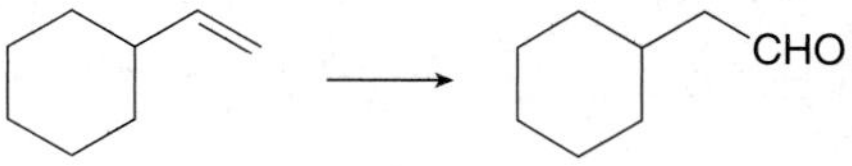

d.

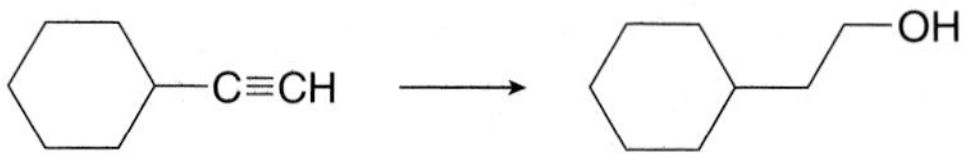

12.57 Identify the lettered reagents in the synthesis of optically active disparlure.

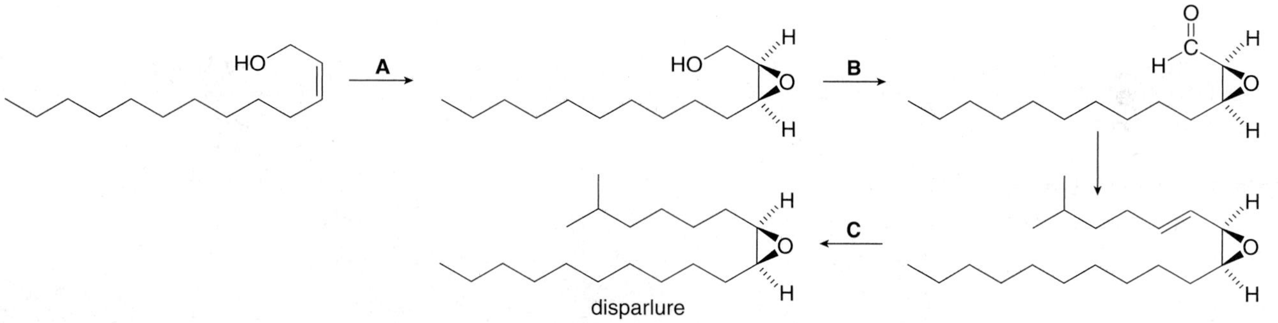

disparlure

Challenge Questions

12.58 Write a stepwise mechanism analogous to that of the dissolving metal reaction that reduces alkynes to trans alkenes for the following reaction:

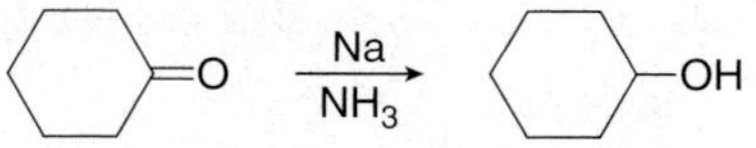

12.59 In the Cr^{6+} oxidation of cyclohexanols, it is generally true that sterically hindered axial alcohols react faster than equatorial alcohols. Which of the following alcohols should be oxidized more rapidly?

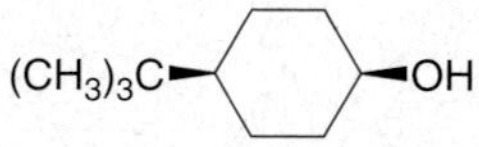

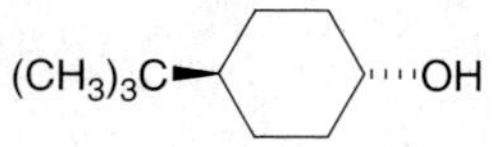

$(CH_3)_3C$ —OH $(CH_3)_3C$ ···OH

12.60 Draw a stepwise mechanism for the following reaction.

mCPBA

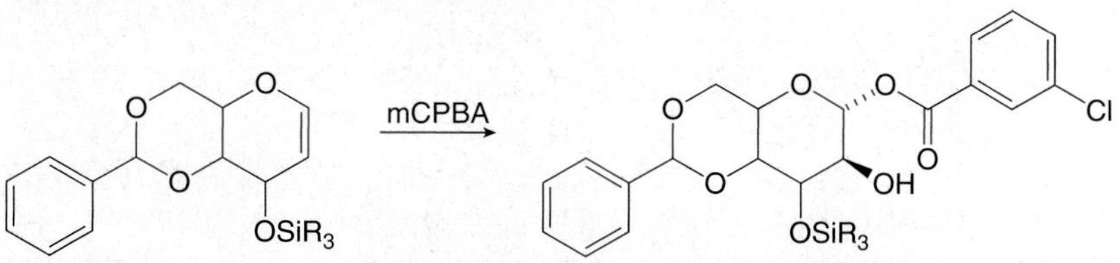

R = alkyl group   R = alkyl group

Radical Reactions

Vitamin E, or **α-tocopherol,** is a natural antioxidant found in fish oil, peanut oil, wheat germ, and leafy greens. Although the molecular details of its function remain obscure, it is thought that vitamin E traps radicals, thus preventing the unwanted oxidation of unsaturated fatty acid residues in cell membranes. In this way, vitamin E helps retard the aging process. In Chapter 13 we learn about radical reactions and the role of vitamin E and other antioxidants in inhibiting radical processes.

A small but significant group of reactions involves the homolysis of nonpolar bonds to form highly reactive **radical inter-mediates.** Although they are unlike other organic reactions, radical transformations are important in many biological and industrial processes. The gases O_2 and NO (nitric oxide) are both radicals. This means that many oxidation reactions with O_2 involve radical intermediates, and biological processes mediated by NO such as blood clotting and neurotransmission may involve radicals. Many useful industrial products such as Styrofoam and polyethylene are prepared by radical processes.

In Chapter 13 we examine the cleavage of nonpolar bonds by radical reactions.

13.1 Introduction

Radicals were first discussed in Section 6.3.

◆ A *radical* is a reactive intermediate with a single unpaired electron, formed by homolysis of a covalent bond.

radicals

A radical contains an atom that does not have an octet of electrons, making it reactive and unstable. Radical processes involve single electrons, so half-headed arrows are used to show the movement of electrons. **One half-headed arrow is used for each electron.**

Carbon radicals are classified as **primary (1°), secondary (2°),** or **tertiary (3°)** by the number of R groups bonded to the carbon with the unpaired electron. A carbon radical is sp^2 hybridized and **trigonal planar,** like sp^2 hybridized carbocations. The unhybridized p orbital contains the unpaired electron and extends above and below the trigonal planar carbon.

Classification of carbon radicals	The trigonal planar geometry of a carbon radical
$R\overset{\cdot}{C}H_2$ $R_2\overset{\cdot}{C}H$ $R_3\overset{\cdot}{C}$	120°
1° 2° 3°	The p orbital contains a single electron.
	sp^2 hybridized

Bond dissociation energies for the cleavage of C—H bonds are used as a measure of radical stability. For example, two different radicals can be formed by cleavage of the C—H bonds in $CH_3CH_2CH_3$.

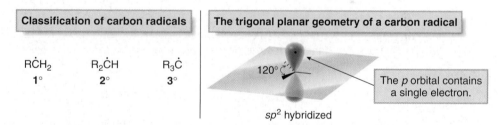

Cleavage of the stronger 1° C—H bond to form the 1° radical ($CH_3CH_2CH_2\cdot$) requires more energy than cleavage of the weaker 2° C—H bond to form the 2° radical [$(CH_3)_2CH\cdot$]—98 versus 95 kcal/mol. This makes the 2° radical more stable, because less energy is required for its formation, as illustrated in Figure 13.1. Thus, **cleavage of the weaker bond forms the more stable radical.** This is a specific example of a general trend.

Figure 13.1 The relative stability of 1° and 2° carbon radicals

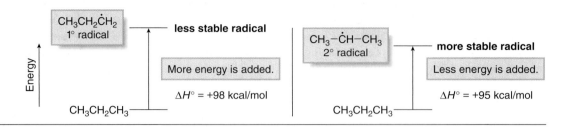

◆ The stability of a radical increases as the number of alkyl groups bonded to the radical carbon increases.

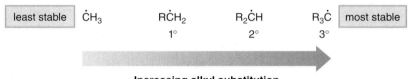

Increasing alkyl substitution
Increasing radical stability

> The **lower** the bond dissociation energy for a C−H bond, the **more stable** the resulting carbon radical.

Thus, a 3° radical is more stable than a 2° radical, and a 2° radical is more stable than a 1° radical. Increasing alkyl substitution increases radical stability in the same way it increases carbocation stability. Alkyl groups are more polarizable than hydrogen atoms, so they can more easily donate electron density to the electron-deficient carbon radical, thus increasing stability.

Unlike carbocations, however, less stable radicals do *not* rearrange to more stable radicals. This difference can be used to distinguish between reactions involving radical intermediates and those involving carbocations.

PROBLEM 13.1   Classify each radical as 1°, 2°, or 3°.

a. CH₃CH₂−ĊHCH₂CH₃   b. [structure]   c. [structure]   d. [structure]

PROBLEM 13.2   Rank each group of radicals in order of increasing stability.

a. (CH₃)₂CHĊH₂   CH₃CH₂ĊHCH₃   (CH₃)₃C·

b. [structures]

13.2   General Features of Radical Reactions

Radicals are formed from covalent bonds by adding energy in the form of **heat (Δ)** or **light (hv).** Some radical reactions are carried out in the presence of a **radical initiator,** a compound that contains an especially weak bond that serves as a source of radicals. **Peroxides,** compounds with the general structure **RO−OR,** are the most commonly used radical initiators. Heating a peroxide readily causes homolysis of the weak O−O bond, forming two RO· radicals.

13.2A   Two Common Reactions of Radicals

Radicals undergo two main types of reactions: **they react with σ bonds,** and **they add to π bonds,** in both cases achieving an octet of electrons.

[1] Reaction of a Radical X· with a C−H Bond

A radical X· abstracts a hydrogen atom from a C−H σ bond to form H−X and a carbon radical. One electron from the C−H bond is used to form the new H−X bond, and the other electron in the C−H bond remains on carbon. The result is that the original radical X· is now surrounded by an octet of electrons, and a new radical is formed.

> • One electron comes from the radical.
> • One electron comes from the C−H bond.

$$-\overset{|}{\underset{|}{C}}-H \;+\; \cdot\ddot{X}: \longrightarrow -\overset{|}{\underset{|}{C}}\cdot \;+\; H-\ddot{X}:$$

new radical

This radical reaction is typically seen with the nonpolar C−H bonds of **alkanes,** which cannot react with polar or ionic electrophiles and nucleophiles.

[2] Reaction of a Radical X· with a C=C

A radical X· also adds to the π bond of a carbon–carbon double bond. One electron from the double bond is used to form a new C−X bond, and the other electron remains on the other carbon originally part of the double bond.

> • One electron comes from the radical.
> • One electron comes from the π bond.

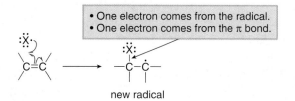

new radical

> **Whenever a radical reacts with a stable single or double bond, a new radical is formed in the products.**

Although the electron-rich double bond of an **alkene** reacts with electrophiles by ionic addition mechanisms, it also reacts with radicals because these reactive intermediates are also electron deficient.

13.2B   Two Radicals Reacting with Each Other

A radical, once formed, rapidly reacts with whatever is available. Usually that means a stable σ or π bond. Occasionally, however, two radicals come into contact with each other, and they react to form a σ bond.

$$:\ddot{X}\cdot \;+\; \cdot\ddot{X}: \longrightarrow :\ddot{X}-\ddot{X}:$$

> One electron comes from each radical.

The reaction of a radical with oxygen, a diradical in its ground state electronic configuration, is another example of two radicals reacting with each other. In this case, the reaction of O_2 with X· forms a new radical, thus preventing X· from reacting with an organic substrate.

> **O_2 is a radical inhibitor.**

$$\cdot\ddot{O}-\ddot{O}\cdot \;+\; \cdot\ddot{X}: \longrightarrow \cdot\ddot{O}-\ddot{O}-X$$

a diradical

Compounds that prevent radical reactions from occurring are called *radical inhibitors* **or** *radical scavengers.* Besides O_2, vitamin E and related compounds, discussed in Section 13.12, are radical scavengers, too. The fact that these compounds inhibit a reaction often suggests that the reaction occurs via radical intermediates.

PROBLEM 13.3   Draw the products formed when a chlorine atom (Cl·) reacts with each species.

a. CH_3-CH_3 b. $CH_2{=}CH_2$ c. :$\ddot{Cl}\cdot$

13.3 Halogenation of Alkanes

In the presence of light or heat, alkanes react with halogens to form alkyl halides. Halogenation is a **radical substitution reaction,** because a halogen atom X replaces a hydrogen via a mechanism that involves radical intermediates.

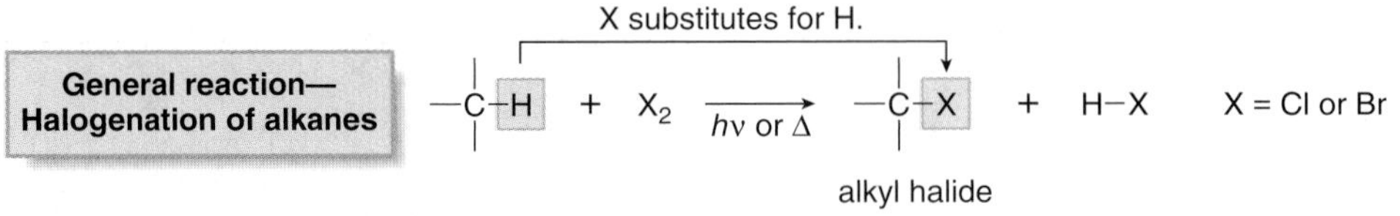

X substitutes for H.

General reaction—
Halogenation of alkanes

—C—H + X₂ $\xrightarrow{h\nu \text{ or } \Delta}$ —C—X + H—X   X = Cl or Br

alkyl halide

Halogenation of alkanes is only useful with Cl_2 and Br_2. Reaction with F_2 is too violent and reaction with I_2 is too slow to be useful. With an alkane that has more than one type of hydrogen atom, a mixture of alkyl halides may result (Reaction [3]).

Examples

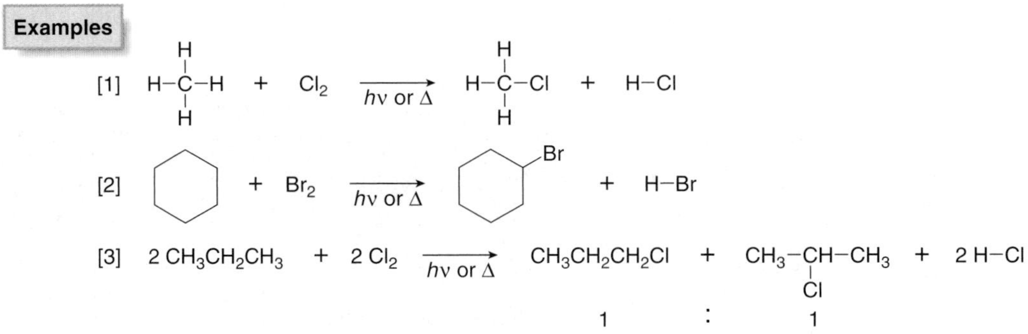

In these examples of halogenation, a halogen has replaced a single hydrogen atom on the alkane. Can the other hydrogen atoms be replaced, too? Figure 13.2 shows that when CH_4 is treated with excess Cl_2, all four hydrogen atoms can be successively replaced by Cl to form CCl_4. **Monohalogenation**—the substitution of a single H by X—can be achieved experimentally by adding halogen X_2 to an excess of alkane.

> When asked to draw the products of halogenation of an alkane, **draw the products of monohalogenation only,** unless specifically directed to do otherwise.

SAMPLE PROBLEM 13.1   Draw all the constitutional isomers formed by monohalogenation of $(CH_3)_2CHCH_2CH_3$ with Cl_2 and $h\nu$.

SOLUTION

Substitute Cl for H on every carbon, and then check to see if any products are identical. The starting material has five C atoms, but replacement of one H atom on two C atoms gives the same product. Thus, **$(CH_3)_2CHCH_2CH_3$ affords four monochloro substitution products.**

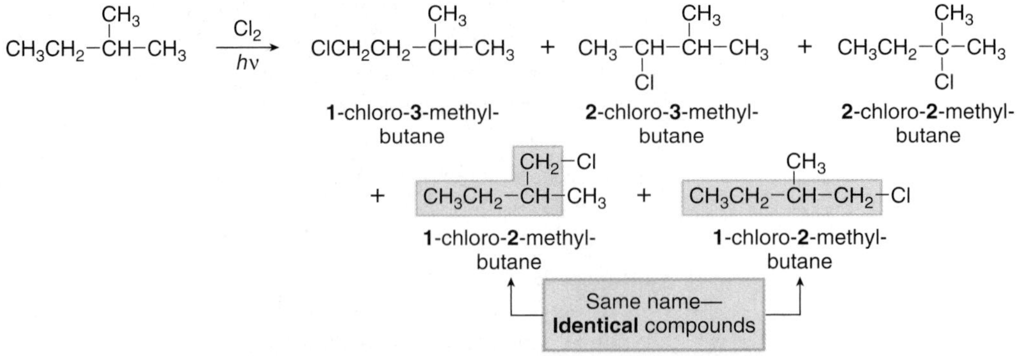

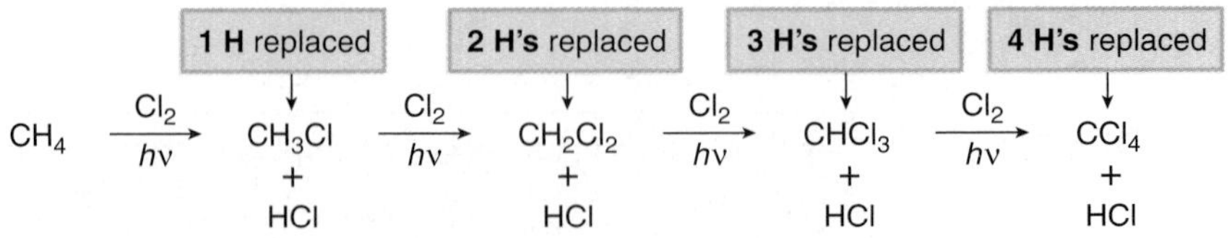

Figure 13.2 Complete halogenation of CH_4 using excess Cl_2

PROBLEM 13.4 Draw all constitutional isomers formed by monochlorination of each alkane.

a. 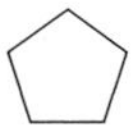 [pentagon]   b. $CH_3CH_2CH_2CH_2CH_2CH_3$ c. $(CH_3)_3CH$

PROBLEM 13.5 What alkane of molecular formula C_5H_{12} gives a single product of monohalogenation when heated with Cl_2?

13.4 The Mechanism of Halogenation

Unlike nucleophilic substitution, which proceeds by two different mechanisms depending on the starting material and reagent, all halogenation reactions of alkanes—regardless of the halogen and alkane used—proceed by the *same* mechanism. Three facts about halogenation suggest that the mechanism involves radical, not ionic, intermediates.

Fact	Explanation
[1] Light, heat, or added peroxide is necessary for the reaction.	• Light or heat provides the energy needed for homolytic bond cleavage to form radicals. Breaking the weak O–O bond of peroxides initiates radical reactions as well.
[2] O_2 inhibits the reaction.	• The diradical O_2 removes radicals from a reaction mixture, thus preventing reaction.
[3] No rearrangements are observed.	• Radicals do not rearrange.

13.4A The Steps of Radical Halogenation

The chlorination of ethane illustrates the **three distinct parts of radical halogenation** (Mechanism 13.1):

Overall reaction CH_3CH_3 + Cl_2 $\xrightarrow{h\nu \text{ or } \Delta}$ CH_3CH_2Cl + HCl

◆ *Initiation:* Two radicals are formed by homolysis of a σ bond and this begins the reaction.
◆ *Propagation:* A radical reacts with another reactant to form a new σ bond and another radical.
◆ *Termination:* Two radicals combine to form a stable bond. Removing radicals from the reaction mixture without generating any new radicals stops the reaction.

MECHANISM 13.1

Radical Halogenation of Alkanes

Initiation

Step [1] Bond cleavage forms two radicals.

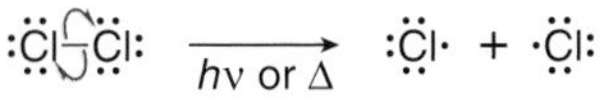

:Cl̈–Cl̈: ──────▶ :Cl̈· + ·Cl̈:
          hν or Δ

◆ The reaction begins with homolysis of the weakest bond in the starting materials using energy from light or heat.

◆ Thus, the Cl–Cl bond ($\Delta H° = 58$ kcal/mol), which is weaker than either the C–C or C–H bond in ethane ($\Delta H° = 88$ and 98 kcal/mol, respectively), is broken to form two chlorine radicals.

Propagation

Steps [2] and [3] One radical reacts and a new radical is formed.

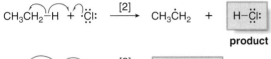

Repeat Steps [2], [3], [2], [3], again and again.

◆ The Cl· radicals are highly reactive (they lack an octet of electrons), so they abstract a hydrogen atom from ethane (Step [2]). This forms H–Cl and leaves one unpaired electron on carbon, generating the ethyl radical ($CH_3CH_2·$).

◆ $CH_3CH_2·$ is highly reactive, so it can abstract a chlorine atom from Cl_2 (Step [3]), forming CH_3CH_2Cl and a new chlorine radical (Cl·).

◆ The Cl· radical formed in Step [3] is a reactant in Step [2], so Steps [2] and [3] can occur repeatedly without an additional initiation reaction (Step [1]).

◆ In each propagation step, one radical is consumed and one radical is formed. The two products—CH_3CH_2Cl and HCl—are formed during propagation.

Termination

Step [4] Two radicals react to form a σ bond.

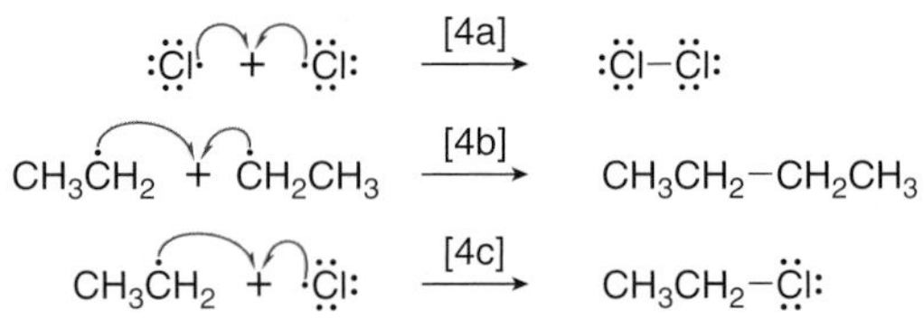

:Cl̈· + ·Cl̈: ──[4a]──▶ :Cl̈–Cl̈:

CH₃ĊH₂ + ĊH₂CH₃ ──[4b]──▶ CH₃CH₂–CH₂CH₃

CH₃ĊH₂ + ·Cl̈: ──[4c]──▶ CH₃CH₂–Cl̈:

◆ To terminate the chain, two radicals react with each other in one of three ways (Steps [4a, b, and c]). Because these reactions remove reactive radicals and form stable bonds, they prevent further propagation via Steps [2] and [3].

Although initiation generates the Cl· radicals needed to begin the reaction, the propagation steps ([2] and [3]) form the two reaction products—**CH₃CH₂Cl** and **HCl.** Once the process has begun, propagation occurs over and over without the need for Step [1] to occur. **A mechanism such as radical halogenation that involves two or more repeating steps is called a *chain mechanism.*** Each propagation step involves a reactive radical abstracting an atom from a stable bond to form a new bond and another radical that continues the chain.

Usually a radical reacts with a stable bond to propagate the chain, but occasionally two radicals combine, and this reaction terminates the chain. Depending on the reaction and the reaction conditions, some radical chain mechanisms can repeat thousands of times before termination occurs.

Termination Step [4a] forms Cl₂, a reactant, whereas Step [4c] forms CH₃CH₂Cl, one of the reaction products. Termination Step [4b] forms CH₃CH₂–CH₂CH₃, which is neither a reactant nor a desired product. The formation of a small quantity of CH₃CH₂–CH₂CH₃, however, is evidence that ethyl radicals are formed in the reaction.

The most important steps of radical halogenation are those that lead to product formation— the propagation steps—so subsequent discussion of this reaction concentrates on these steps only.

PROBLEM 13.6   Using Mechanism 13.1 as a guide, write the mechanism for the reaction of CH₄ with Br₂ to form CH₃Br and HBr. Classify each step as initiation, propagation, or termination.

PROBLEM 13.7   Write a stepwise mechanism that shows how a very small amount of CH₃CH₂Cl could form during the chlorination of CH₄.

Figure 13.3 Energy changes in the propagation steps during the chlorination of ethane

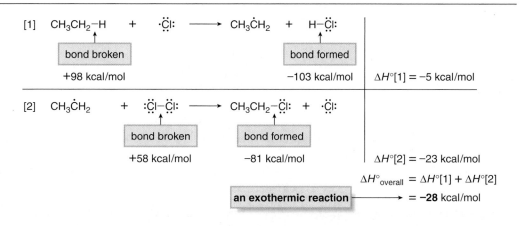

Figure 13.4 Energy diagram for the propagation steps in the chlorination of ethane

- Because radical halogenation consists of two propagation steps, the energy diagram has two energy barriers.
- The first step is rate-determining because its transition state is at higher energy.
- The reaction is exothermic because $\Delta H^{\circ}_{overall}$ is negative.

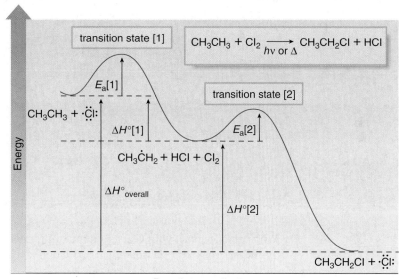

Reaction coordinate

13.4B   Energy Changes During the Chlorination of Ethane

Figure 13.3 shows how bond dissociation energies (Section 6.4) can be used to calculate ΔH° for the two propagation steps in the chlorination of ethane. Because the overall ΔH° is negative, the reaction is **exothermic.** Moreover, because the transition state for the first propagation step is higher in energy than the transition state for the second propagation step, the **first step is rate-determining.** Both of these facts are illustrated in the energy diagram in Figure 13.4.

PROBLEM 13.8 Draw the structure of the transition state for each propagation step in the reaction of CH_3CH_3 with Cl_2 (Figure 13.3).

PROBLEM 13.9 Calculate ΔH° for the two propagation steps in the reaction of CH_4 with Br_2 to form CH_3Br and HBr (Problem 13.6).

PROBLEM 13.10 Calculate ΔH° for the rate-determining step of the reaction of CH_4 with I_2. Explain why this result illustrates that this reaction is extremely slow.

13.5   Chlorination of Other Alkanes

Recall from Section 13.3 that the chlorination of $CH_3CH_2CH_3$ affords a 1:1 mixture of $CH_3CH_2CH_2Cl$ (formed by removal of a 1° hydrogen) and $(CH_3)_2CHCl$ (formed by removal of a 2° hydrogen).

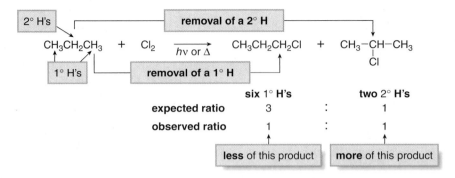

$CH_3CH_2CH_3$ has six 1° hydrogen atoms and only two 2° hydrogens, so the expected product ratio of $CH_3CH_2CH_2Cl$ to $(CH_3)_2CHCl$ (assuming all hydrogens are *equally* reactive) is 3:1. Because the observed ratio is 1:1, however, the 2° C–H bonds must be more reactive; that is, **it must be easier to homolytically cleave a 2° C–H bond than a 1° C–H bond.** Recall from Section 13.2 that 2° C–H bonds are weaker than 1° C–H bonds. Thus,

♦ **The weaker the C–H bond, the more readily the hydrogen atom is removed in radical halogenation.**

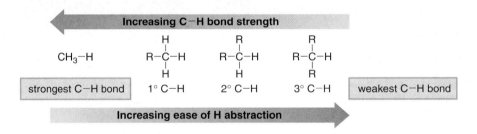

When alkanes react with Cl_2, a mixture of products results, with more product formed by cleavage of the weaker C–H bond than you would expect on statistical grounds.

PROBLEM 13.11   Which C–H bond in each compound is most readily broken during radical halogenation?

a.   b.   c. $CH_3CH_2CH_2CH_3$

13.6   Chlorination versus Bromination

Although alkanes undergo radical substitution reactions with both Cl_2 and Br_2, chlorination and bromination exhibit two important differences:

♦ Chlorination is *faster* than bromination.
♦ Although chlorination is *unselective*, yielding a mixture of products, bromination is often *selective*, yielding one major product.

For example, propane reacts rapidly with Cl_2 to form a 1:1 mixture of 1° and 2° alkyl chlorides. On the other hand, propane reacts with Br_2 much more slowly and forms 99% $(CH_3)_2CHBr$.

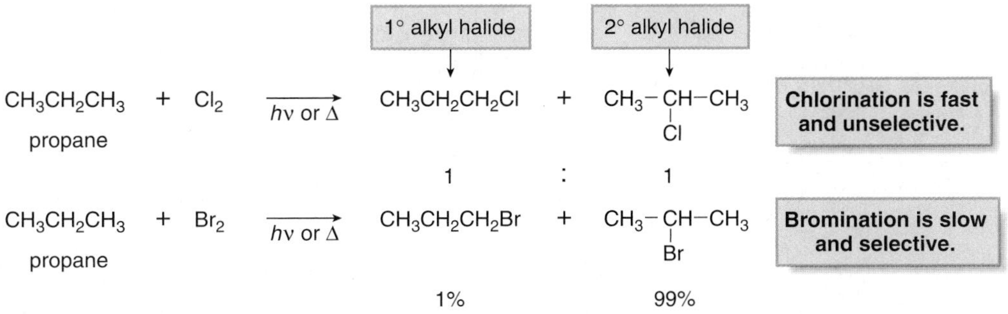

	1° alkyl halide		2° alkyl halide	

$CH_3CH_2CH_3$ + Cl_2 $\xrightarrow{h\nu \text{ or } \Delta}$ $CH_3CH_2CH_2Cl$ + $CH_3-\overset{\displaystyle |}{\underset{\displaystyle Cl}{CH}}-CH_3$ **Chlorination is fast and unselective.**

propane

1 : 1

$CH_3CH_2CH_3$ + Br_2 $\xrightarrow{h\nu \text{ or } \Delta}$ $CH_3CH_2CH_2Br$ + $CH_3-\overset{\displaystyle |}{\underset{\displaystyle Br}{CH}}-CH_3$ **Bromination is slow and selective.**

propane

1% 99%

This is a specific example of the **reactivity–selectivity principle: less reactive reagents are more selective. In bromination, the major (and sometimes exclusive) product results from cleavage of the *weakest* C–H bond.**

SAMPLE PROBLEM 13.2   Draw the major product formed when 3-ethylpentane is heated with Br_2.

SOLUTION

Keep in mind: **the more substituted the carbon atom, the weaker the C–H bond.** The major bromination product in 3-ethylpentane is formed by cleavage of the sole **3° C–H bond,** its weakest C–H bond.

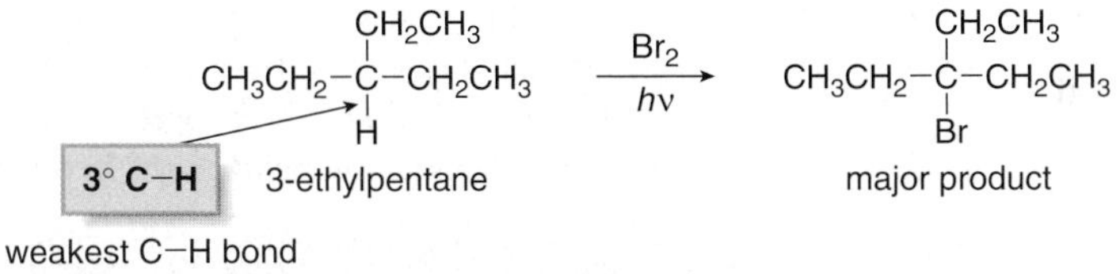

PROBLEM 13.12   Draw the structure of the single alkyl halide formed when methylcyclohexane is heated with Br_2.

To explain the difference between chlorination and bromination, we return to the Hammond postulate (Section 7.15) to estimate the relative energy of the transition states of the rate-determining steps of these reactions. The **rate-determining step is the abstraction of a hydrogen atom by the halogen radical,** so we must compare these steps for bromination and chlorination. Keep in mind:

◆ Transition states in endothermic reactions resemble the products. The more stable product is formed faster.
◆ Transition states in exothermic reactions resemble the starting materials. The relative stability of the products does not greatly affect the relative energy of the transition states, so a mixture of products often results.

Bromination: $CH_3CH_2CH_3$ + Br_2

A bromine radical can abstract either a 1° or a 2° hydrogen from propane, generating either a 1° radical or a 2° radical. Calculating $\Delta H°$ using bond dissociation energies reveals that both reactions are **endothermic,** but **it takes less energy to form the more stable 2° radical.**

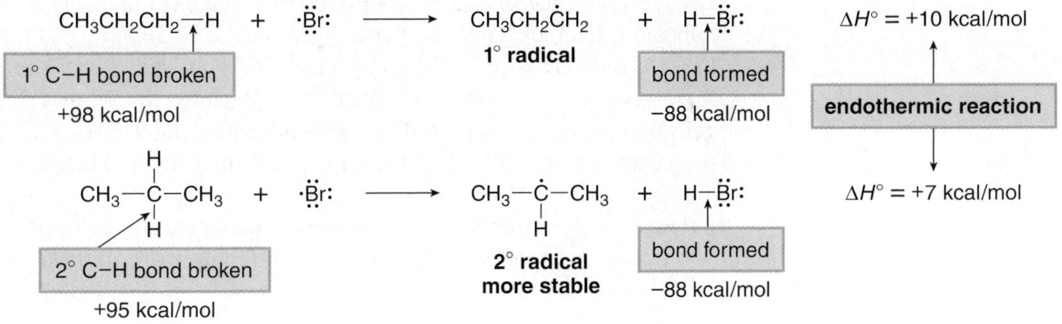

Figure 13.5 Energy diagram for the endothermic reaction: $CH_3CH_2CH_3 + Br\cdot \rightarrow$ $CH_3CH_2CH_2\cdot$ or $(CH_3)_2CH\cdot + HBr$

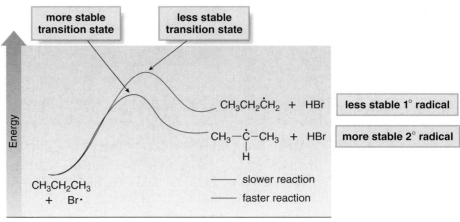

• The transition state to form the less stable 1° radical ($CH_3CH_2CH_2\cdot$) is higher in energy than the transition state to form the more stable 2° radical [$(CH_3)_2CH\cdot$]. Thus, **the 2° radical is formed faster.**

According to the Hammond postulate, the transition state of an endothermic reaction resembles the products, so the energy of activation to form the more stable 2° radical is lower and it is formed faster, as shown in the energy diagram in Figure 13.5. Because the 2° radical [$(CH_3)_2CH\cdot$] is converted to 2-bromopropane [$(CH_3)_2CHBr$] in the second propagation step, this **2° alkyl halide is the major product of bromination.**

◆ Conclusion: Because the rate-determining step in bromination is endothermic, the more stable radical is formed faster, and often a single radical halogenation product predominates.

Chlorination: $CH_3CH_2CH_3 + Cl_2$

A chlorine radical can also abstract either a 1° or a 2° hydrogen from propane, generating either a 1° radical or a 2° radical. Calculating $\Delta H°$ using bond dissociation energies reveals that both reactions are **exothermic.**

$$CH_3CH_2CH_2{-}H + \cdot\ddot{C}\ddot{l}: \longrightarrow CH_3CH_2\dot{C}H_2 + H{-}\ddot{C}\ddot{l}: \qquad \Delta H° = -5\ \text{kcal/mol}$$

1° C–H bond broken		**1° radical**	bond formed	**exothermic reaction**

+98 kcal/mol   −103 kcal/mol

$$\underset{\underset{\text{H}}{|}}{\overset{\overset{\text{H}}{|}}{CH_3{-}C{-}CH_3}} + \cdot\ddot{C}\ddot{l}: \longrightarrow CH_3{-}\dot{C}{-}CH_3 + H{-}\ddot{C}\ddot{l}: \qquad \Delta H° = -8\ \text{kcal/mol}$$

2° C–H bond broken		**2° radical**	bond formed

+95 kcal/mol   −103 kcal/mol

Because chlorination has an exothermic rate-determining step, the transition state to form both radicals resembles the same starting material, $CH_3CH_2CH_3$. As a result, the relative stability of the two radicals is much less important and both radicals are formed. An energy diagram for these processes is drawn in Figure 13.6. Because the 1° and 2° radicals are converted to 1-chloropropane ($CH_3CH_2CH_2Cl$) and 2-chloropropane [$(CH_3)_2CHCl$], respectively, in the second propagation step, **both alkyl halides are formed in chlorination.**

◆ Conclusion: Because the rate-determining step in chlorination is exothermic, the transition state resembles the starting material, both radicals are formed, and a mixture of products results.

Figure 13.6 Energy diagram for the exothermic reaction: $CH_3CH_2CH_3 + Cl\cdot \rightarrow$ $CH_3CH_2CH_2\cdot$ or $(CH_3)_2CH\cdot + HCl$

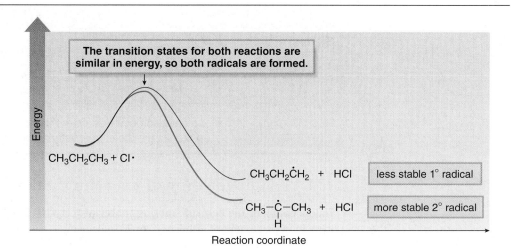

PROBLEM 13.13   Why is the reaction of methylcyclohexane with Cl_2 not a useful method to prepare 1-chloro-1-methylcyclohexane? What other constitutional isomers are formed in the reaction mixture?

PROBLEM 13.14   Reaction of $(CH_3)_3CH$ with Cl_2 forms two products: $(CH_3)_2CHCH_2Cl$ (63%) and $(CH_3)_3CCl$ (37%). Why is the major product formed by cleavage of the stronger 1° C−H bond?

13.7   Halogenation as a Tool in Organic Synthesis

Halogenation is a useful tool because it adds a functional group to a previously unfunctionalized molecule, making an **alkyl halide.** These alkyl halides can then be converted to alkenes by elimination, and to alcohols and ethers by nucleophilic substitution.

SAMPLE PROBLEM 13.3   Show how cyclohexane can be converted to cyclohexene by a stepwise sequence.

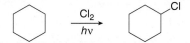

cyclohexane   cyclohexene

SOLUTION

There is no one-step method to convert an alkane to an alkene. A two-step method is needed:

[1]   **Radical halogenation** produces an alkyl halide.

[2]   **Elimination of HCl** with a strong base produces cyclohexene.

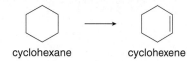

PROBLEM 13.15   Synthesize each compound from $(CH_3)_3CH$.
a. $(CH_3)_3CBr$ b. $(CH_3)_2C=CH_2$ c. $(CH_3)_3COH$ d. $(CH_3)_2C(Cl)CH_2Cl$

13.8   The Stereochemistry of Halogenation Reactions

The stereochemistry of a reaction product depends on whether the reaction occurs at a stereogenic center or at another atom, and whether a new stereogenic center is formed. The rules predicting the stereochemistry of reaction products are summarized in Table 13.1.

TABLE 13.1	Rules for Predicting the Stereochemistry of Reaction Products
Starting material	**Result**
Achiral	• An achiral starting material always gives either an achiral or a racemic product.
Chiral	• If a reaction does not occur at a stereogenic center, the configuration at a stereogenic center is retained in the product.
	• If a reaction occurs at a stereogenic center, we must know the mechanism to predict the stereochemistry of the product.

13.8A   Halogenation of an Achiral Starting Material

Halogenation of the **achiral starting material CH₃CH₂CH₂CH₃**, forms two constitutional isomers by replacement of either a 1° or 2° hydrogen.

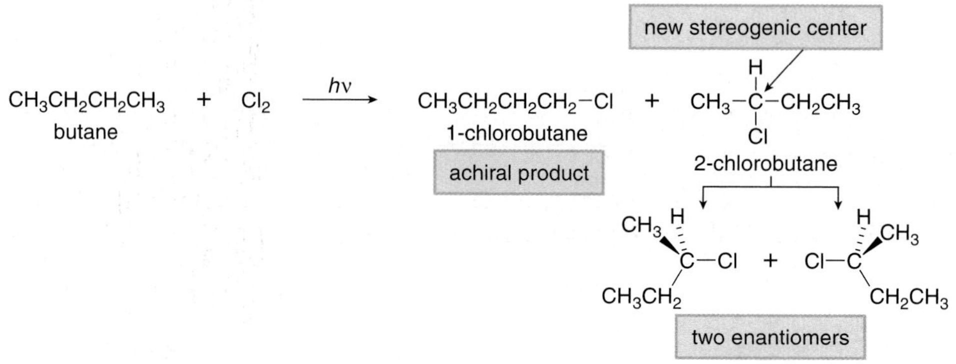

◆ 1-Chlorobutane (CH₃CH₂CH₂CH₂Cl) has no stereogenic centers and thus it is an **achiral** compound.

◆ 2-Chlorobutane [CH₃CH(Cl)CH₂CH₃] has a new stereogenic center, and so an **equal amount of two enantiomers** must form—**a racemic mixture.**

A racemic mixture results when a new stereogenic center is formed because the first propagation step generates a **planar, sp^2 hybridized radical.** Cl₂ then reacts with the planar radical from either side to form an equal amount of two enantiomers.

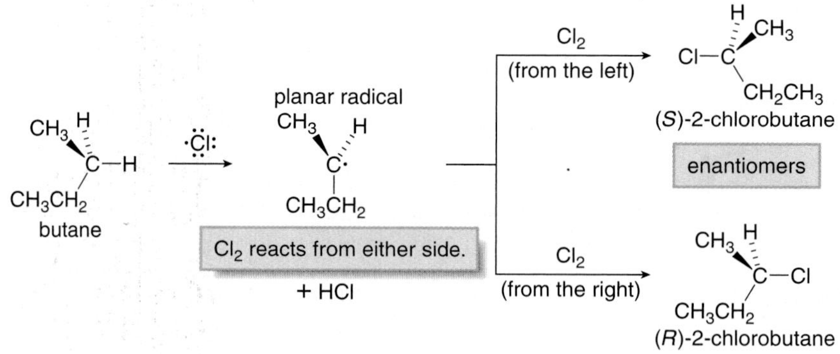

Thus, the achiral starting material butane forms an achiral product (1-chlorobutane) and a racemic mixture of two enantiomers [(R)- and (S)-2-chlorobutane].

13.8B   Halogenation of a Chiral Starting Material

Let's now examine chlorination of the chiral starting material (R)-2-bromobutane at C2 and C3.

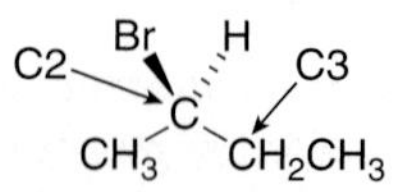

(R)-2-bromobutane

Chlorination at C2 occurs at the stereogenic center. Abstraction of a hydrogen atom at C2 forms a trigonal planar sp^2 hybridized radical that is now achiral. This achiral radical then reacts with Cl_2 from either side to form a new stereogenic center, resulting in an **equal amount of two enantiomers**—a **racemic mixture.**

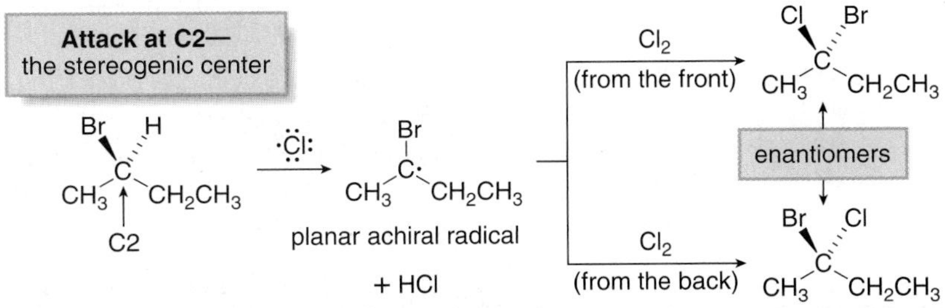

♦ **Radical halogenation reactions at a stereogenic center occur with racemization.**

Chlorination at C3 does *not* occur at the stereogenic center, but it forms a new stereogenic center. Because no bond is broken to the stereogenic center at C2, **its configuration is retained** during the reaction. Abstraction of a hydrogen atom at C3 forms a trigonal planar sp^2 hybridized radical that still contains this stereogenic center. Reaction of the radical with Cl_2 from either side forms a new stereogenic center, so the products have two stereogenic centers: the configuration at C2 is the same in both compounds, but the configuration at C3 is different, making them **diastereomers.**

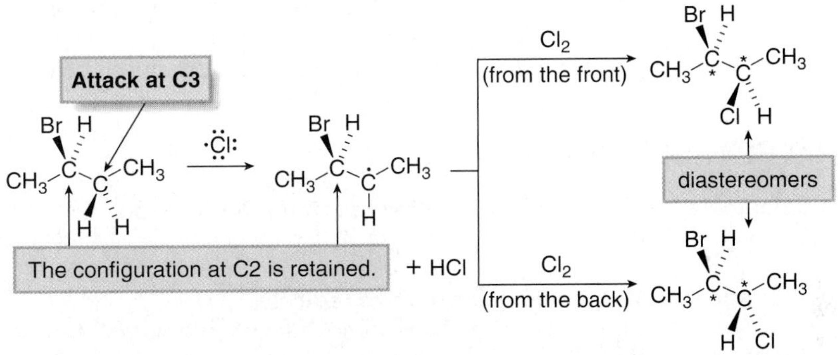

[* denotes a stereogenic center]

Thus, four isomers are formed by chlorination of (*R*)-2-bromobutane at C2 and C3. Attack at the stereogenic center (C2) gives a product with one stereogenic center, resulting in a mixture of enantiomers. Attack at C3 forms a new stereogenic center, giving a mixture of diastereomers.

PROBLEM 13.16   What products are formed from monochlorination of (*R*)-2-bromobutane at C1 and C4? Assign *R* and *S* designations to each stereogenic center.

PROBLEM 13.17   Draw the monochlorination products formed when each compound is heated with Cl_2. Include the stereochemistry at any stereogenic centers.

a. $CH_3CH_2CH_2CH_2CH_3$ b. 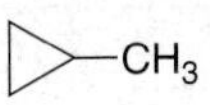 ▷—CH_3 c. $(CH_3CH_2)_3CH$ d.

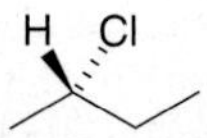

(Consider attack at C2 and C3 only.)

13.9   Application: The Ozone Layer and CFCs

Ozone is formed in the upper atmosphere by reaction of oxygen molecules with oxygen atoms. Ozone is also decomposed with sunlight back to these same two species. The overall result of these reactions is to convert high-energy ultraviolet light into heat.

The synthesis and decomposition of O_3 in the upper atmosphere

$$O_2 + \cdot \ddot{O} \cdot \longrightarrow O_3 + \text{heat}$$
ozone

$$O_3 \xrightarrow{h\nu} O_2 + \cdot \ddot{O} \cdot$$
ozone

Ozone is vital to life; it acts like a shield, protecting the earth's surface from destructive ultraviolet radiation. A decrease in ozone concentration in this protective layer would have some immediate consequences, including an increase in the incidence of skin cancer and eye cataracts. Other long-term effects include a reduced immune response, interference with photosynthesis in plants, and harmful effects on the growth of plankton, the mainstay of the ocean food chain.

Current research suggests that **chlorofluorocarbons (CFCs)** are responsible for destroying ozone in the upper atmosphere. **CFCs** are simple halogen-containing organic compounds manufactured under the trade name Freons.

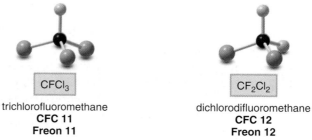

trichlorofluoromethane	dichlorodifluoromethane
CFC 11	**CFC 12**
Freon 11	**Freon 12**

CFCs are inert, odorless, and nontoxic, and they have been used as refrigerants, solvents, and aerosol propellants. Because CFCs are volatile and water insoluble, they readily escape into the upper atmosphere, where they are decomposed by high-energy sunlight to form radicals that destroy ozone by the radical chain mechanism shown in Mechanism 13.2 (Figure 13.7).

MECHANISM 13.2

A Radical Chain Mechanism Leading to Ozone Destruction

Initiation

Step [1] Bond cleavage in a CFC molecule forms two radicals.

$$CFCl_3 \xrightarrow{h\nu} \cdot CFCl_2 + \boxed{\cdot \ddot{Cl} :}$$

◆ The reaction begins with homolysis of a C−Cl bond in $CFCl_3$. The Cl· radical formed in this step initiates the radical process.

Propagation

Steps [2] and [3] One radical reacts and a new radical is formed in each step.

$$\cdot \ddot{Cl} : + O_3 \xrightarrow{[2]} : \ddot{Cl} - \ddot{O} \cdot + O_2$$

◆ Reaction of Cl· with O_3 forms chlorine monoxide (ClO·) in Step [2], which then reacts with oxygen atoms to form O_2 and Cl·.

$$: \ddot{Cl} - \ddot{O} \cdot + \cdot \ddot{O} \cdot \xrightarrow{[3]} \cdot \ddot{Cl} : + O_2$$

◆ Because the Cl· radical formed in Step [3] is a reactant in Step [2], Steps [2] and [3] can occur repeatedly to continue the chain.

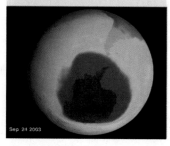

The overall result is that O_3 is consumed as a reactant and O_2 molecules are formed. In this way, a small amount of CFC can destroy a large amount of O_3. These findings led to a ban on the use of CFCs in aerosol propellants in the United States in 1978 and to the phasing out of their use in refrigeration systems.

Newer alternatives to CFCs are **hydrochlorofluorocarbons (HCFCs)** and **hydrofluorocarbons (HFCs)** such as CH_2FCF_3. These compounds have many properties in common with CFCs, but they are decomposed by HO· before they reach the stratosphere and therefore they do not take part in the radical reactions resulting in O_3 destruction.

Ozone (Dobson Units)

100   200   300   400   500

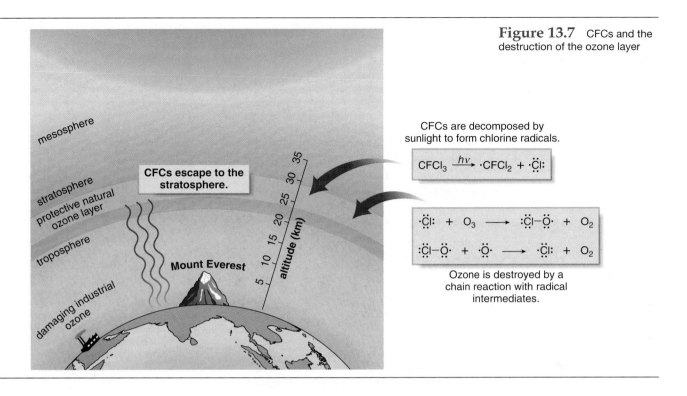

Figure 13.7 CFCs and the destruction of the ozone layer

CFCs are decomposed by sunlight to form chlorine radicals.

$$CFCl_3 \xrightarrow{h\nu} \cdot CFCl_2 + \cdot \ddot{C}l\colon$$

$$\cdot \ddot{C}l\colon + O_3 \longrightarrow \colon \ddot{C}l - \ddot{O}\cdot + O_2$$

$$\colon \ddot{C}l - \ddot{O}\cdot + \cdot \ddot{O}\cdot \longrightarrow \cdot \ddot{C}l\colon + O_2$$

Ozone is destroyed by a chain reaction with radical intermediates.

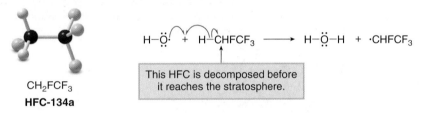

$$H - \ddot{O}\cdot \; + \; H - CHFCF_3 \longrightarrow H - \ddot{O} - H \; + \; \cdot CHFCF_3$$

This HFC is decomposed before it reaches the stratosphere.

CH₂FCF₃
HFC-134a

PROBLEM 13.18   Nitric oxide, NO·, is another radical also thought to cause ozone destruction by a similar mechanism. One source of NO· in the stratosphere is supersonic aircraft whose jet engines convert small amounts of N_2 and O_2 to NO·. Write the propagation steps for the reaction of O_3 with NO·.

13.10   Radical Halogenation at an Allylic Carbon

Now let's examine radical halogenation at an *allylic carbon*—**the carbon adjacent to a double bond.** Homolysis of the allylic C−H bond of propene generates the **allyl radical,** which has an unpaired electron on the carbon adjacent to the double bond.

$$CH_2{=}CH{-}CH_2{-}H \longrightarrow CH_2{=}CH{-}\dot{C}H_2 \; + \; \cdot H \quad \Delta H^\circ = +87 \text{ kcal/mol}$$

allyl radical

allylic C−H bond

The bond dissociation energy for this process (87 kcal/mol) is even less than that for a 3° C−H bond (91 kcal/mol). Because the weaker the C−H bond, the more stable the resulting radical, an **allyl radical is more stable than a 3° radical,** and the following order of radical stability results:

least stable	$\dot{C}H_3$	$R\dot{C}H_2$	$R_2\dot{C}H$	$R_3\dot{C}$	$CH_2{=}CH{-}\dot{C}H_2$	most stable
		1°	2°	3°	allyl radical	

Increasing radical stability

The position of the atoms and the σ bonds stays the same in drawing resonance structures. Resonance structures differ in the location of only π bonds and nonbonded electrons.

The allyl radical is more stable than other radicals because two resonance structures can be drawn for it.

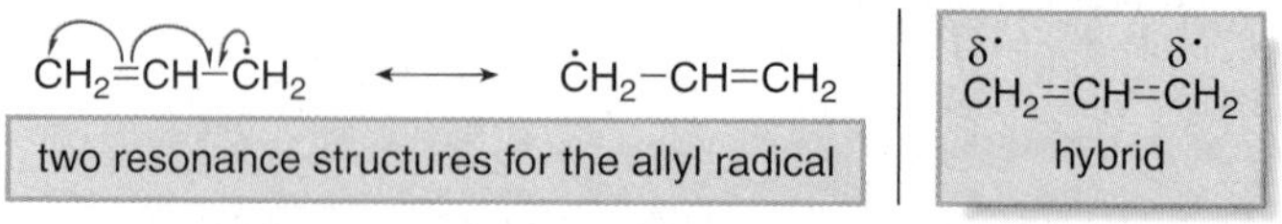

$$CH_2{=}CH{-}\overset{\cdot}{C}H_2 \longleftrightarrow \overset{\cdot}{C}H_2{-}CH{=}CH_2 \quad \Big| \quad \overset{\delta^{\cdot}}{CH_2}{=}CH{=}\overset{\delta^{\cdot}}{CH_2}$$

| two resonance structures for the allyl radical | hybrid |

◆ The "true" structure of the allyl radical is a hybrid of the two resonance structures. In the hybrid, the π bond and the unpaired electron are delocalized.
◆ Delocalizing electron density lowers the energy of the hybrid, thus stabilizing the allyl radical.

PROBLEM 13.19   Draw a second resonance structure for each radical. Then draw the hybrid.

a. $CH_3CH{=}CH{-}\overset{\cdot}{C}H_2$ b.

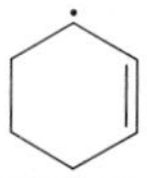

13.10A   Selective Bromination at Allylic C–H Bonds

Because allylic C–H bonds are weaker than other sp^3 hybridized C–H bonds, the **allylic carbon can be selectively halogenated** by using *N*-bromosuccinimide (**NBS,** Section 10.15) in the presence of light or peroxides. Under these conditions only the allylic C–H bond in cyclohexene reacts to form an allylic halide.

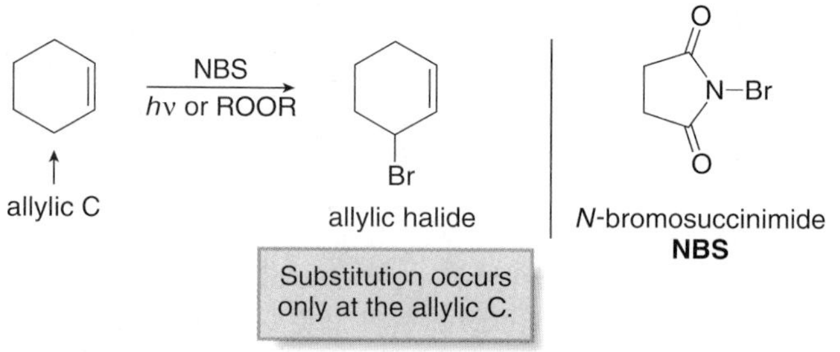

allylic C

allylic halide

N-bromosuccinimide
NBS

Substitution occurs only at the allylic C.

NBS contains a weak N–Br bond that is homolytically cleaved with light to generate a bromine radical, initiating an allylic halogenation reaction. Propagation then consists of the usual two steps of radical halogenation as shown in Mechanism 13.3.

MECHANISM 13.3

Allylic Bromination with NBS

Initiation

Step [1] Cleavage of the N − Br bond forms two radicals.

◆ The reaction begins with homolysis of the weak N − Br bond in NBS using light energy. This generates a Br· radical that begins the radical halogenation process.

Propagation

Steps [2] and [3] One radical reacts and a new radical is formed in each step.

◆ The Br· radical abstracts an allylic hydrogen atom to afford an allylic radical in Step [2]. (Only one Lewis structure of the allylic radical is drawn.)

◆ The allylic radical reacts with Br_2 in the second propagation step to form the product of allylic halogenation. Because the Br· radical formed in Step [3] is also a reactant in Step [2], Steps [2] and [3] repeatedly occur without the need for Step [1].

Besides acting as a source of Br· to initiate the reaction, NBS generates a low concentration of Br_2 needed in the second chain propagation step (Step [3] of the mechanism). The HBr formed in Step [2] reacts with NBS to form Br_2, which is then used for halogenation in Step [3] of the mechanism.

NBS   succinimide

used in Step [3] of allylic bromination

A **low concentration of Br_2** (from NBS) **favors allylic substitution** (over addition) in part because bromine is needed for only *one* step of the mechanism. When Br_2 adds to a double bond, a low Br_2 concentration would first form a low concentration of bridged bromonium ion (Section 10.13), which must then react with more bromine (in the form of Br⁻) in a second step to form a dibromide. **If concentrations of both intermediates— bromonium ion and Br⁻— are low, the overall rate of addition is very slow.**

Thus, an alkene with allylic C − H bonds undergoes two different reactions depending on the reaction conditions.

Br_2

Addition via ionic intermediates

vicinal dibromide

NBS
hv or ROOR

Substitution via radical intermediates

Br
allylic bromide

◆ Treatment of cyclohexene with Br_2 (in an organic solvent like CCl_4) leads to **addition** via **ionic intermediates** (Section 10.13).

◆ Treatment of cyclohexene with NBS (+ *hv* or ROOR) leads to **allylic substitution,** via **radical intermediates.**

PROBLEM 13.20   Draw the products of each reaction.

a. [cyclopentene structure] $\xrightarrow[h\nu]{NBS}$ b. $CH_2=CH-CH_3$ $\xrightarrow[h\nu]{NBS}$ c. $CH_2=CH-CH_3$ $\xrightarrow{Br_2}$

13.10B   Product Mixtures in Allylic Halogenation

Halogenation at an allylic carbon often results in a mixture of products. For example, bromination of 1-butene under radical conditions forms a mixture of 3-bromo-1-butene and 1-bromo-2-butene.

$$CH_2=CHCH_2CH_3 \xrightarrow[h\nu \text{ or ROOR}]{NBS} \underset{\underset{Br}{|}}{CH_2=CHCHCH_3} + BrCH_2CH=CHCH_3$$

1-butene   3-bromo-1-butene   1-bromo-2-butene

A mixture is obtained because the reaction proceeds by way of a **resonance-stabilized radical.** Abstraction of an allylic hydrogen from 1-butene with a Br· radical (from NBS) forms an allylic radical for which **two different Lewis structures** can be drawn.

two nonidentical resonance structures

$$CH_2=CHCHCH_3 \longrightarrow CH_2=CH\dot{C}HCH_3 \longleftrightarrow \dot{C}H_2-CH=CHCH_3 + H-\ddot{Br}:$$

$\delta^{\cdot} \qquad \delta^{\cdot}$
$CH_2\!=\!CH\!=\!CHCH_3$
hybrid

$\downarrow Br_2$ $\downarrow Br_2$

$$\underset{\underset{Br}{|}}{CH_2=CHCHCH_3} + BrCH_2CH=CHCH_3 + \cdot\ddot{Br}:$$

3-bromo-1-butene   1-bromo-2-butene

As a result, two different C atoms have partial radical character, so that Br_2 reacts at two different sites and two allylic halides are formed.

◆ Whenever two different resonance structures can be drawn for an allylic radical, two different allylic halides are formed by radical substitution.

SAMPLE PROBLEM 13.4   Draw the products formed when **A** is treated with NBS + $h\nu$.

[cyclohexane ring]$=CH_2$ $\xrightarrow[h\nu]{NBS}$
A

SOLUTION

Hydrogen abstraction at the allylic C forms a resonance-stabilized radical (with two different resonance structures) that reacts with Br_2 to form two constitutional isomers as products.

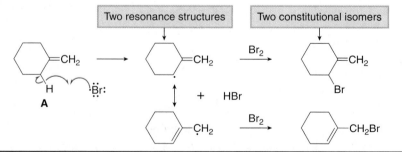

Two resonance structures   Two constitutional isomers

PROBLEM 13.21   Draw all constitutional isomers formed when each alkene is treated with NBS + $h\nu$.

a. $CH_3CH=CHCH_3$ b. [cyclohexene structure with $-CH_3$ and CH_3]   c. $CH_2=C(CH_2CH_3)_2$

PROBLEM 13.22   Draw the structure of the four allylic halides formed when 3-methylcyclohexene undergoes allylic halogenation with NBS + $h\nu$.

13.11   Application: Oxidation of Unsaturated Lipids

Oils—triacylglycerols having one or more sites of unsaturation in their long carbon chains—are susceptible to oxidation at their allylic carbon atoms. Oxidation occurs by way of a radical chain mechanism, as shown in Figure 13.8.

◆ Step [1] Oxygen in the air abstracts an allylic hydrogen atom to form an allylic radical because the allylic C—H bond is weaker than the other C—H bonds.

◆ Step [2] The allylic radical reacts with another molecule of O_2 to form a peroxy radical.

◆ Step [3] The peroxy radical abstracts an allylic hydrogen from another lipid molecule to form a hydroperoxide and another allylic radical that continues the chain. Steps [2] and [3] can repeat again and again until some other radical terminates the chain.

The hydroperoxides formed by this process are unstable and decompose to other oxidation products, many of which have a disagreeable odor and taste. **This process turns an oil rancid. Unsaturated lipids are more easily oxidized than saturated ones** because they contain weak allylic C—H bonds that are readily cleaved in Step [1] of this reaction, forming resonance-stabilized allylic radicals. Because saturated fats have no double bonds and thus no weak allylic C—H bonds, they are much less susceptible to air oxidation, resulting in increased shelf life of products containing them.

Figure 13.8 The oxidation of unsaturated lipids with O_2

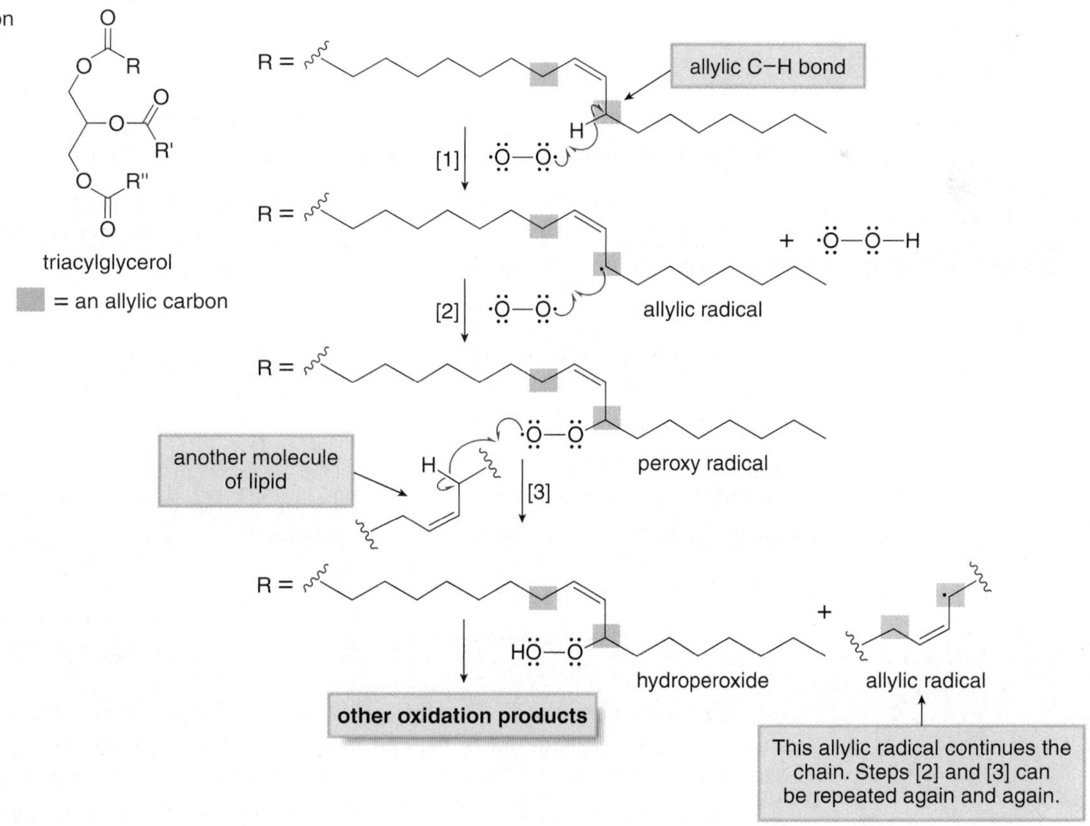

• Oxidation is shown at one allylic carbon only. Reaction at the other labeled allylic carbon is also possible.

PROBLEM 13.23   Draw a second resonance structure for the allylic radical formed as a product of Step [1] in Figure 13.8. What hydroperoxide is formed using this Lewis structure?

PROBLEM 13.24   Which C−H bond is most readily cleaved in linoleic acid? Draw all possible resonance structures for the resulting radical. Draw all the hydroperoxides formed by reaction of this resonance-stabilized radical with O_2.

linoleic acid

13.12   Application: Antioxidants

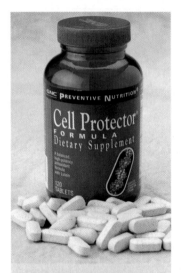

The purported health benefits of antioxidants have made them a popular component in anti-aging formulations.

An *antioxidant* is a compound that stops an oxidation reaction from occurring.

♦ Naturally occurring antioxidants such as **vitamin E** (the chapter-opening molecule) prevent radical reactions that can cause cell damage.

♦ Synthetic antioxidants such as **BHT**—**b**utylated **h**ydroxy **t**oluene—are added to packaged and prepared foods to prevent oxidation and spoilage.

vitamin E

BHT
(**b**utylated **h**ydroxy **t**oluene)

Vitamin E and BHT are radical inhibitors, so they terminate radical chain mechanisms by reacting with radicals. How do they trap radicals? Both vitamin E and BHT use a hydroxy group bonded to a benzene ring—a general structure called a **phenol**.

Radicals (R·) abstract a hydrogen atom from the OH group of an antioxidant, forming a new resonance-stabilized radical. **This new radical does not participate in chain propagation,** but rather terminates the chain and halts the oxidation process. All phenols (including vitamin E and BHT) inhibit oxidation by this radical process.

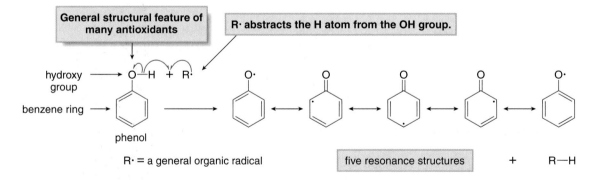

The many nonpolar C−C and C−H bonds of vitamin E make it fat soluble, and thus it dissolves in the nonpolar interior of the cell membrane, where it is thought to inhibit the oxidation of the unsaturated fatty acid residues in the phospholipids. Oxidative damage to lipids in cells via radical mechanisms is thought to play an important role in the aging process. For this reason, many anti-aging formulas with antioxidants like vitamin E are now popular consumer products.

PROBLEM 13.25   Draw all resonance structures for the radical that results from hydrogen atom abstraction from BHT.

13.13 Radical Addition Reactions to Double Bonds

We now turn our attention to the second common reaction of radicals, addition to double bonds. Because an alkene contains an electron-rich, easily broken π bond, it reacts with an electron-deficient radical.

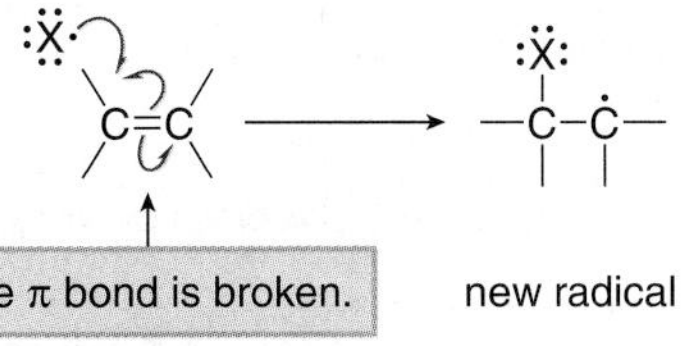

Radicals react with alkenes via a radical chain mechanism that consists of initiation, propagation, and termination steps analogous to those discussed previously for radical substitution.

13.13A Addition of HBr

HBr adds to alkenes to form alkyl bromides in the presence of light, heat, or peroxides.

Radical addition of HBr—General reaction

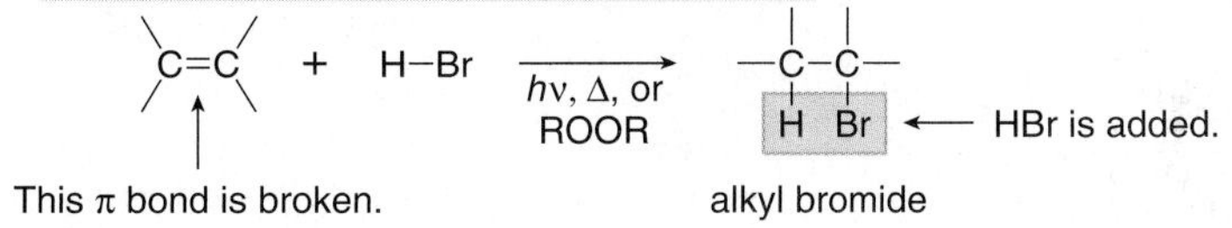

The regioselectivity of addition to an unsymmetrical alkene is different from the addition of HBr without added light, heat, or peroxides.

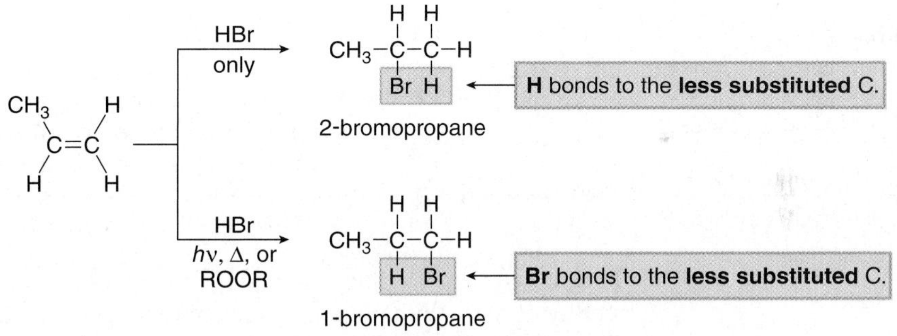

- HBr addition to propene *without* added light, heat, or peroxides gives 2-bromopropane: the **H atom is added to the less substituted carbon.** This reaction occurs via **carbocation** intermediates (Section 10.10).
- HBr addition to propene *with* added light, heat, or peroxides gives 1-bromopropane: the **Br atom is added to the less substituted carbon.** This reaction occurs via **radical** intermediates.

PROBLEM 13.26 Draw the products formed in each reaction.

a. $CH_2{=}CHCH_2CH_2CH_2CH_3 \xrightarrow{HBr}$

b. 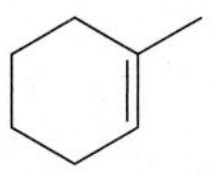 $\xrightarrow[ROOR]{HBr}$

c. $(CH_3)_2C{=}CHCH_3 \xrightarrow{HBr}$

d. $CH_3CH{=}CHCH_2CH_2CH_3 \xrightarrow[ROOR]{HBr}$

13.13B The Mechanism of the Radical Addition of HBr to an Alkene

In the presence of added light, heat, or peroxides, HBr addition to an alkene forms radical intermediates, and like other radical reactions, proceeds by a mechanism with three distinct parts:

initiation, propagation, and termination. Mechanism 13.4 is written for the reaction of $CH_3CH=CH_2$ with HBr and ROOR to form $CH_3CH_2CH_2Br$.

MECHANISM 13.4

Radical Addition of HBr to an Alkene

Initiation

Steps [1] and [2] Abstraction of H from HBr occurs by a two-step process.

◆ With ROOR to initiate the reaction, two steps are needed to form Br·.

◆ Homolysis of the weak O–O bond of the peroxide forms RO·, which abstracts a hydrogen atom from HBr to form Br·.

Propagation

Steps [3] and [4] The π bond is broken and the C–H and C–Br σ bonds are formed.

◆ Chain propagation occurs in two steps, and in each step one radical is consumed and another is formed.

◆ The first step of propagation forms the C–Br bond when the Br· radical adds to the terminal carbon, leading to a 2° carbon radical.

◆ The 2° radical abstracts a H atom from HBr, forming the new C–H bond and completing the addition reaction. Because a new Br· radical is also formed in this step, Steps [3] and [4] occur repeatedly.

Repeat Steps [3], [4], [3], [4], and so forth.

Termination

Step [5] Two radicals react to form a bond.

◆ To terminate the chain, two radicals (for example two Br· radicals) react with each other to form a stable bond, preventing further propagation via Steps [3] and [4].

The first propagation step (Step [3] of the mechanism, the addition of Br· to the double bond) is worthy of note. With propene there are two possible paths for this step, depending on which carbon atom of the double bond forms the new bond to bromine. Path [A] forms a less stable 1° radical whereas Path [B] forms a more stable 2° radical. **The more stable 2° radical forms faster, so Path [B] is preferred.**

Path [A]:
Does NOT occur

less stable
1° radical

Path [B]:
Preferred path

more stable
2° radical

The mechanism also illustrates why the regioselectivity of HBr addition is different depending on the reaction conditions. In both reactions, H and Br add to the double bond, but the *order* of addition depends on the mechanism.

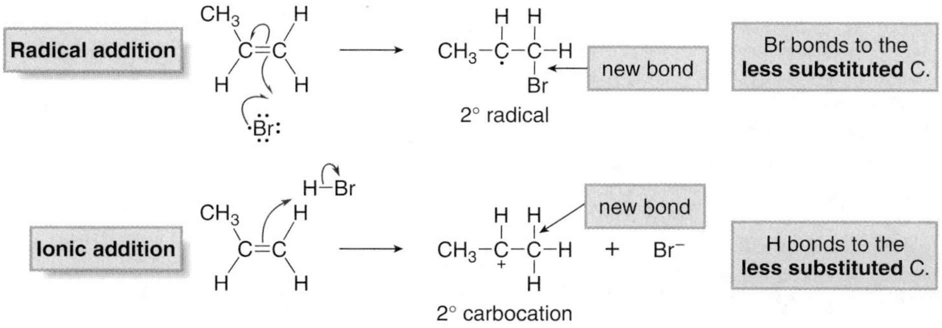

- ◆ In radical addition (HBr with added light, heat, or ROOR), *Br· adds first* to generate the more stable radical.
- ◆ In ionic addition (HBr alone), *H⁺ adds first* to generate the more stable carbocation.

PROBLEM 13.27   When HBr adds to $(CH_3)_2C=CH_2$ under radical conditions, two radicals are possible products in the first step of chain propagation. Draw the structure of both radicals and indicate which one is formed. Then draw the preferred product from HBr addition under radical conditions.

13.13C   Energy Changes in the Radical Addition of HBr

The energy changes during propagation in the radical addition of HBr to $CH_2=CH_2$ can be calculated from bond dissociation energies, as shown in Figure 13.9.

Both propagation steps for the addition of HBr are exothermic, so propagation is exothermic (energetically favorable) overall. For the addition of HCl or HI, however, one of the chain-propagating steps is quite endothermic, and thus too difficult to be part of a repeating chain mechanism. Thus, **HBr adds to alkenes under radical conditions, but HCl and HI do not.**

PROBLEM 13.28   Draw an energy diagram for the two propagation steps in the radical addition of HBr to propene. Draw the transition state for each step.

PROBLEM 13.29   Write out the two propagation steps for the addition of HCl to propene and calculate $\Delta H°$ for each step. Which step prohibits chain propagation from repeatedly occurring?

Figure 13.9 Energy changes during the propagation steps: $CH_2=CH_2 + HBr \rightarrow CH_3CH_2Br$

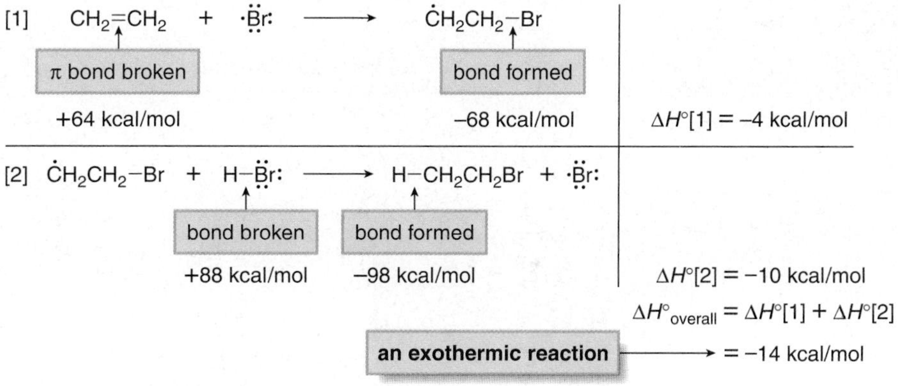

13.14   Polymers and Polymerization

HDPE (high-density polyethylene) and **LDPE** (low-density polyethylene) are two common types of polyethylene prepared under different reaction conditions and having different physical properties. HDPE is opaque and rigid, and is used in milk containers and water jugs. LDPE is less opaque and more flexible, and is used in plastic bags and electrical insulation. Products containing HDPE and LDPE (and other plastics) are often labeled with a symbol indicating recycling ease: the lower the number, the easier to recycle.

High-density polyethylene—HDPE

Low-density polyethylene—LDPE

Polymers—large molecules made up of repeating units of smaller molecules called *monomers*—include such biologically important compounds as proteins and carbohydrates. They also include such industrially important plastics as polyethylene, poly(vinyl chloride) (PVC), and polystyrene.

13.14A   Synthetic Polymers

Many synthetic polymers—that is, those synthesized in the lab—are among the most widely used organic compounds in modern society. Although some synthetic polymers resemble natural substances, many have different and unusual properties that make them more useful than naturally occurring materials. Soft drink bottles, plastic bags, food wrap, compact discs, teflon, and Styrofoam are all made of synthetic polymers.

♦ *Polymerization* is the joining together of monomers to make polymers.

For example, joining **ethylene monomers** together forms the polymer **polyethylene,** a plastic used in milk containers and sandwich bags.

$$CH_2{=}CH_2 \quad + \quad CH_2{=}CH_2 \quad + \quad CH_2{=}CH_2$$

$$\downarrow \text{ polymerization}$$

$$\{-CH_2CH_2 + CH_2CH_2 + CH_2CH_2 -\} \quad = \quad$$

three monomer units joined together

Many ethylene derivatives having the general structure **CH₂=CHZ** are also used as monomers for polymerization. The identity of Z affects the physical properties of the resulting polymer, making some polymers more suitable for one consumer product (e.g., plastic bags or food wrap) than another (e.g., soft drink bottles or compact discs). Polymerization of CH₂=CHZ usually affords polymers with the Z groups on every other carbon atom in the chain. Table 13.2 lists some common monomers and polymers prepared industrially.

$$CH_2{=}CHZ \quad + \quad CH_2{=}CHZ \quad + \quad CH_2{=}CHZ$$

$$\downarrow \text{ polymerization}$$

$$\{-CH_2CH + CH_2CH + CH_2CH -\}$$
$$\quad\; | \qquad\quad\; | \qquad\quad\; |$$
$$\quad\; Z \qquad\quad\; Z \qquad\quad\; Z$$

three monomer units joined together

TABLE 13.2 Common Industrial Monomers and Polymers

Monomer $\longrightarrow$	Polymer	Consumer product

CH_2=CHCl
vinyl chloride

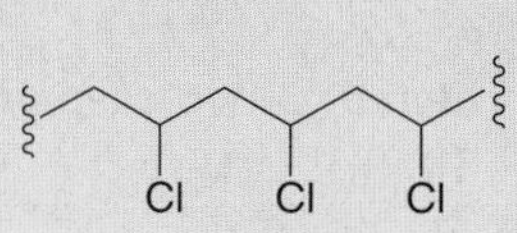

**poly(vinyl chloride)
PVC**

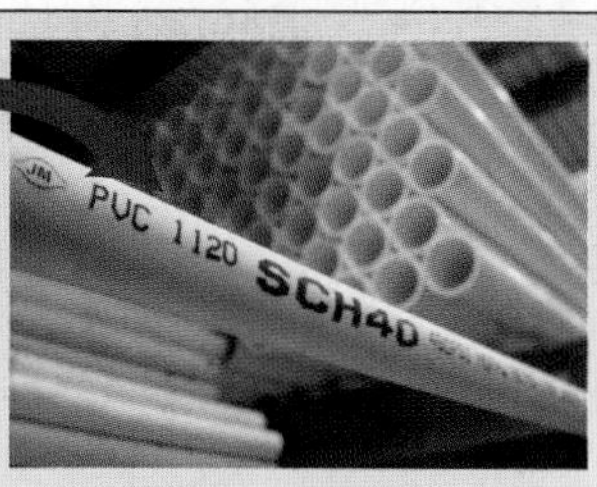

PVC pipes

CH_2=CHCH$_3$
propene

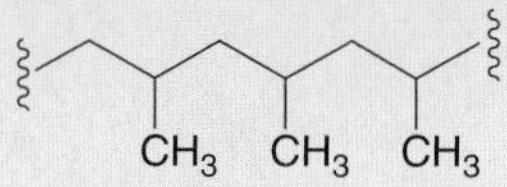

polypropylene

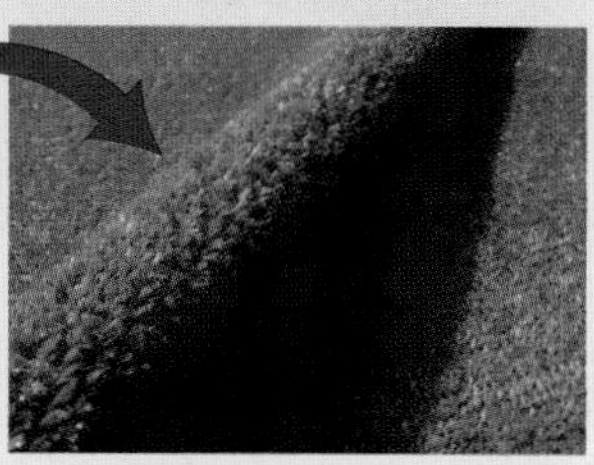

polypropylene carpeting

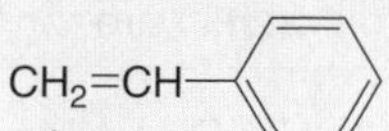

CH_2=CH—
styrene

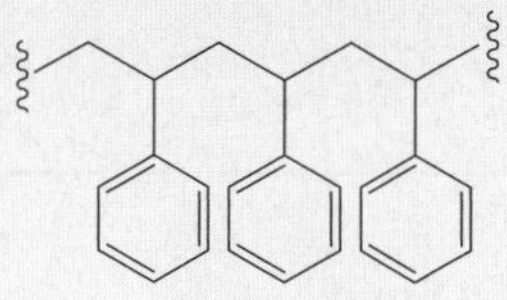

polystyrene

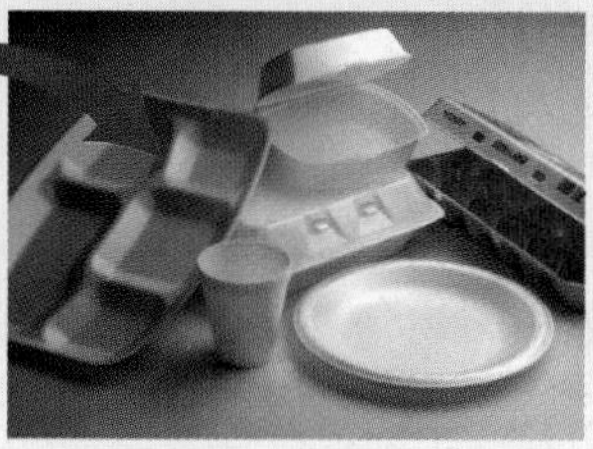

Styrofoam products

PROBLEM 13.30   (a) What polymer is formed on polymerization of CH_2=$C(CH_3)_2$? (b) What monomer is used to form poly(vinyl acetate), a polymer used in paints and adhesives?

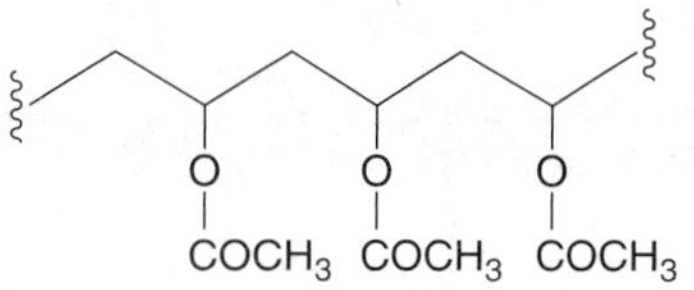

poly(vinyl acetate)

O   O   O

COCH$_3$ COCH$_3$ COCH$_3$

13.14B   Radical Polymerization

All polymers described in Section 13.14A are prepared by polymerization of alkene monomers by **adding a radical to a π bond.** The mechanism resembles the radical addition of HBr to an alkene, except that a **carbon radical rather than a bromine atom is added to the double bond.** Mechanism 13.5 is written with the general monomer CH_2=CHZ, and again has three parts: initiation, propagation, and termination.

MECHANISM 13.5

Radical Polymerization of CH$_2$=CHZ

Initiation

Steps [1] and [2]   A carbon radical is formed by a two-step process.

RÖ⌢ÖR $\xrightarrow{[1]}$ 2 RÖ·   +   CH$_2$=C$\overset{Z}{\underset{H}{\vert}}$ $\xrightarrow{[2]}$ RÖCH$_2$–C·$\overset{Z}{\underset{H}{\vert}}$

                                                                carbon radical

- Chain initiation begins with homolysis of the weak O–O bond of the peroxide to form RO·, which then adds to a molecule of monomer to form a carbon radical.

Propagation

Step [3]   The polymer chain grows.

RÖCH$_2$–C·$\overset{Z}{\underset{H}{\vert}}$ +   CH$_2$=C$\overset{Z}{\underset{H}{\vert}}$ $\xrightarrow{[3]}$ RÖCH$_2$–C$\overset{Z}{\underset{H}{\vert}}$–CH$_2$–C·$\overset{Z}{\underset{H}{\vert}}$

Repeat Step [3] over and over.

| new C–C bond |

- Chain propagation consists of a single step that joins monomer units together.
- In Step [3], the carbon radical formed during initiation adds to another alkene molecule to form a new C–C bond and another carbon radical. Addition always forms the more substituted carbon radical—that is, the **unpaired electron is always located on the carbon atom having the Z substituent.**
- This carbon radical reacts with more monomer, so that Step [3] occurs repeatedly, and the polymer chain grows. Each time a carbon radical adds to a double bond, a **new C–C bond** and a new carbon radical are formed.

Termination

Step [4]   Two radicals combine to form a bond.

∿∿CH$_2$–C·$\overset{Z}{\underset{H}{\vert}}$ ·C–CH$_2$∿∿$\overset{Z}{\underset{H}{\vert}}$ $\xrightarrow{[4]}$ ∿∿CH$_2$–C$\overset{Z}{\underset{H}{\vert}}$–C$\overset{Z}{\underset{H}{\vert}}$–CH$_2$∿∿

- To terminate the chain, two radicals combine to form a stable bond, thus ending the polymerization process.

In radical polymerization, the more substituted radical always adds to the less substituted end of the monomer, a process called **head-to-tail polymerization.**

> The **more substituted radical** adds to the **less substituted end** of the double bond.

RÖCH$_2$–C·$\overset{Z}{\underset{H}{\vert}}$ +   CH$_2$=C$\overset{Z}{\underset{H}{\vert}}$ $\longrightarrow$ RÖCH$_2$–C$\overset{Z}{\underset{H}{\vert}}$–CH$_2$–C·$\overset{Z}{\underset{H}{\vert}}$

> The new radical is always located on the C bonded to Z.

PROBLEM 13.31   Draw the steps of the mechanism that converts vinyl chloride (CH$_2$=CHCl) into poly(vinyl chloride).

13.15   Key Concepts—Radical Reactions

General Features of Radicals

- A radical is a reactive intermediate with a single unpaired electron (13.1).
- A carbon radical is sp^2 hybridized and trigonal planar (13.1).
- The stability of a radical increases as the number of C atoms bonded to the radical carbon increases (13.1).
- Allylic radicals are stabilized by resonance, making them more stable than 3° radicals (13.10).

Radical Reactions

[1] Halogenation of alkanes (13.4)

- The reaction follows a radical chain mechanism.
- The weaker the C–H bond, the more readily the hydrogen is replaced by X.
- Chlorination is faster and less selective than bromination (13.6).
- Radical substitution at a stereogenic center results in racemization (13.8).

[2] Allylic halogenation (13.10)

$$CH_2=CH-CH_3 \xrightarrow[\text{h}\nu \text{ or ROOR}]{\text{NBS}} \boxed{\begin{array}{c} CH_2=CHCH_2Br \\ \text{allylic halide} \end{array}}$$

- The reaction follows a radical chain mechanism.

[3] Radical addition of HBr to an alkene (13.13)

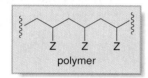

- A radical addition mechanism is followed.
- Br bonds to the less substituted carbon atom to form the more substituted, more stable radical.

[4] Radical polymerization of alkenes (13.14)

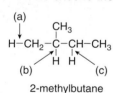

- A radical addition mechanism is followed.

Problems

Radicals and Bond Strength

13.32 With reference to the indicated C–H bonds in 2-methylbutane:

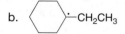

2-methylbutane

a. Rank the C–H bonds in order of increasing bond strength.
b. Draw the radical resulting from cleavage of each C–H bond, and classify it as 1°, 2°, or 3°.
c. Rank the radicals in order of increasing stability.
d. Rank the C–H bonds in order of increasing ease of H abstraction in a radical halogenation reaction.

13.33 Rank each group of radicals in order of increasing stability.

a. $(CH_3)_2\dot{C}CH_2CH(CH_3)_2$ $(CH_3)_2CH\dot{C}HCH(CH_3)_2$ $(CH_3)_2CHCH_2CH(CH_3)\dot{C}H_2$

b.

13.34 Why is a benzylic C–H bond unusually weak?

benzylic C–H bond

Halogenation of Alkanes

13.35 Rank the indicated hydrogen atoms in order of increasing ease of abstraction in a radical halogenation reaction.

$$\begin{array}{c} H_b\longrightarrow H \quad H\longleftarrow H_d \\ CH_2=CHCHCHC(CH_3)CH_2-H\longleftarrow H_c \\ H_a\longrightarrow H \end{array}$$

13.36   Draw all constitutional isomers formed by monochlorination of each alkane with Cl_2 and $h\nu$.

a. ☐   b. $(CH_3)_3CCH_2CH_2CH_2CH_3$ c.   d. ⬠—CH_3

13.37   What is the major monobromination product formed by heating each alkane with Br_2?

a. 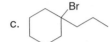   b. $(CH_3)_3CCH_2CH(CH_3)_2$ c. ⬠—CH_3 d. $(CH_3)_3CCH_2CH_3$

13.38   Five isomeric alkanes (**A–E**) having the molecular formula C_6H_{14} are each treated with Cl_2 + $h\nu$ to give alkyl halides having molecular formula $C_6H_{13}Cl$. **A** yields five constitutional isomers. **B** yields four constitutional isomers. **C** yields two constitutional isomers. **D** yields three constitutional isomers, two of which possess stereogenic centers. **E** yields three constitutional isomers, only one of which possesses a stereogenic center. Identify the structures of **A–E**.

13.39   What alkane is needed to make each alkyl halide by radical halogenation?

a. ⬠—Cl b.   c.   d. $(CH_3)_3CCH_2Cl$

13.40   Which alkyl halides can be prepared in good yield by radical halogenation of an alkane?

a. ⬠—Cl b. (structure with Br)   c. (structure with Cl)   d. (structure with Br)

13.41   Explain why chlorination of cyclohexane with two equivalents of Cl_2 in the presence of light is a poor method to prepare 1,2-dichlorocyclohexane.

Resonance

13.42   Draw resonance structures for each radical.

a. (structure with $\dot{C}H_2$)   b. (cyclohexadienyl radical)

Allylic Halogenation

13.43   Draw the products formed when each alkene is treated with NBS + $h\nu$.

a. ☐   b. $CH_3CH_2CH{=}CHCH_2CH_3$ c. $(CH_3)_2C{=}CHCH_3$ d. (structure)

13.44   Which compounds can be prepared in good yield by allylic halogenation of an alkene?

a.   b. $CH_3CH_2CH{=}CHCH_2Br$ c. (cyclohexene with Br)

Reactions

13.45   Draw the organic products formed in each reaction.

a. (structure) $\xrightarrow[h\nu]{Cl_2}$ d. (cyclopentane=CH_2) $\xrightarrow[h\nu]{NBS}$ g. (cyclopentane=CH_2) $\xrightarrow[ROOR]{HBr}$

b. $\xrightarrow[\Delta]{Br_2}$ e. (structure) $\xrightarrow{HBr}$ h. (structure) $\xrightarrow{Br_2}$

c. (structure) $\xrightarrow[h\nu]{Br_2}$ f. (structure) $\xrightarrow[ROOR]{HBr}$ i. (structure) $\xrightarrow[h\nu]{NBS}$

13.46 Identify the reagents needed to carry out each transformation.

a.

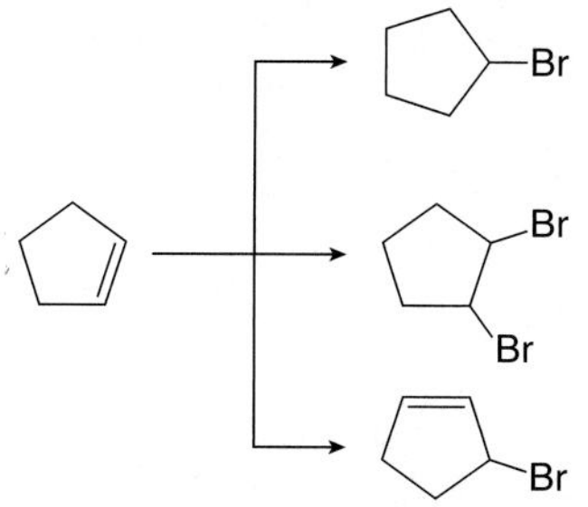

b.

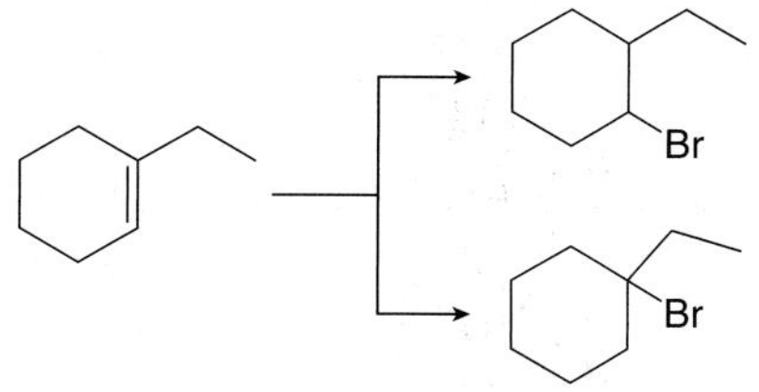

Stereochemistry and Reactions

13.47 Draw the products formed in each reaction and include the stereochemistry around any stereogenic centers.

a.

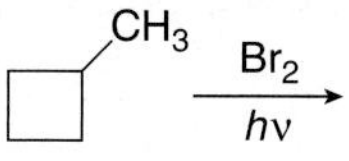

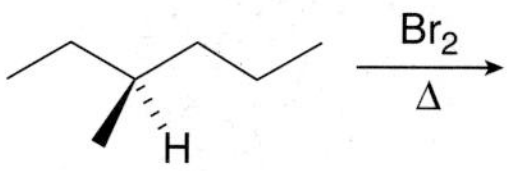

b.

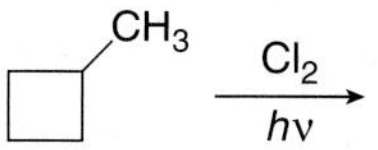

c.

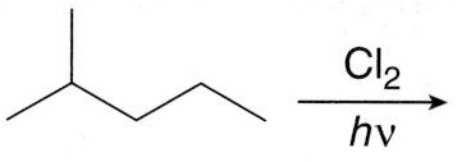

d.

e.

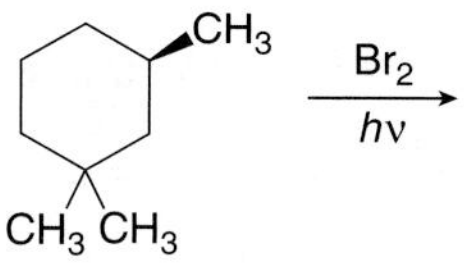

f.

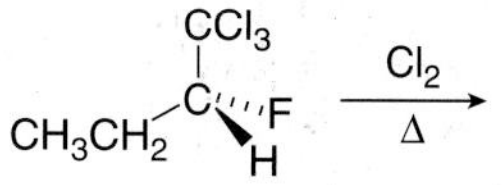

13.48 (a) Draw all stereoisomers of molecular formula $C_5H_{10}Cl_2$ formed when (R)-2-chloropentane is heated with Cl_2. (b) Assuming that products having different physical properties can be separated into fractions by some physical method (such as fractional distillation), how many different fractions would be obtained? (c) Which of these fractions would be optically active?

13.49 Draw the six products (including stereoisomers) formed when **A** is treated with NBS + hv.

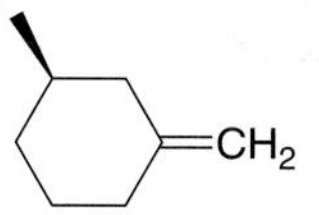

A

Mechanisms

13.50 Consider the following bromination: $(CH_3)_3CH + Br_2 \xrightarrow{\Delta} (CH_3)_3CBr + HBr$.
   a. Calculate $\Delta H°$ for this reaction by using the bond dissociation energies in Table 6.2.
   b. Draw out a stepwise mechanism for the reaction, including the initiation, propagation, and termination steps.
   c. Calculate $\Delta H°$ for each propagation step.
   d. Draw an energy diagram for the propagation steps.
   e. Draw the structure of the transition state of each propagation step.

13.51 Draw a stepwise mechanism for the following reaction.

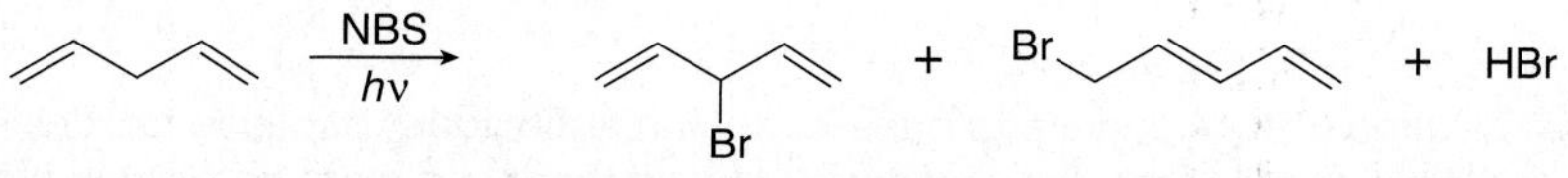

13.52 Although CH_4 reacts with Cl_2 to form CH_3Cl and HCl, the corresponding reaction of CH_4 with I_2 does not occur at an appreciable rate, even though the I–I bond is much weaker than the Cl–Cl bond. Explain why this is so.

13.53 An alternative mechanism for the propagation steps in the radical chlorination of CH_4 is drawn below. Calculate $\Delta H°$ for these steps and explain why this pathway is unlikely.

$$CH_4 + :\overset{..}{\underset{..}{Cl}}\cdot \longrightarrow CH_3Cl + H\cdot$$

$$H\cdot + Cl_2 \longrightarrow HCl + :\overset{..}{\underset{..}{Cl}}\cdot$$

13.54 When 3,3-dimethyl-1-butene is treated with HBr alone, the major product is 2-bromo-2,3-dimethylbutane. When the same alkene is treated with HBr and peroxide, the sole product is 1-bromo-3,3-dimethylbutane. Explain these results by referring to the mechanisms.

Synthesis

13.55 Devise a synthesis of each compound from cyclopentane and any other required organic or inorganic reagents.

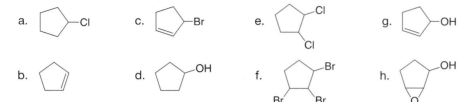

13.56 Devise a synthesis of each target compound from the indicated starting material.

a. $CH_3C{\equiv}CH \Longrightarrow CH_3CH_2CH_3$

c. ⌁⌁Br $\Longrightarrow HC{\equiv}CH$

b. ⌁Br $\Longrightarrow$ (isopropyl bromide)

d. (1-methylcyclohexanol) $\Longrightarrow$ (methylcyclohexane)

Radical Oxidation Reactions

13.57 As described in Section 9.16, the leukotrienes, important components in the asthmatic response, are synthesized from arachidonic acid via the hydroperoxide 5-HPETE. Write a stepwise mechanism for the conversion of arachidonic acid to 5-HPETE with O_2.

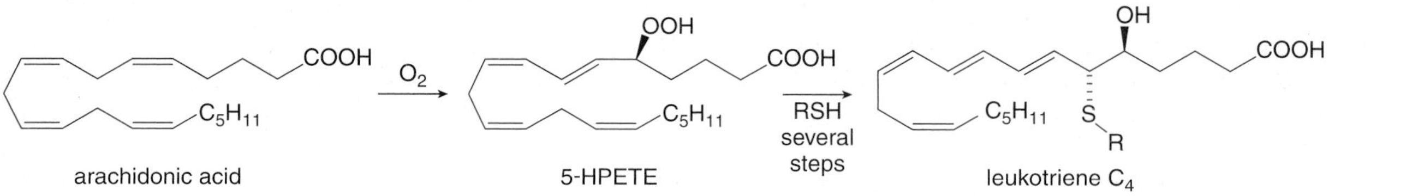

arachidonic acid O_2 5-HPETE   RSH several steps   leukotriene C_4

13.58 Ethers are oxidized with O_2 to form hydroperoxides that decompose violently when heated. Draw a stepwise mechanism for this reaction.

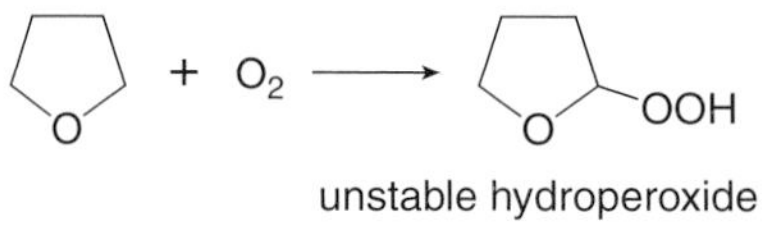

(THF) $+ O_2 \longrightarrow$ (THF-OOH)

unstable hydroperoxide

Antioxidants

13.59 Draw all resonance structures of the radical resulting from abstraction of a hydrogen atom from the antioxidant BHA (**b**utylated **h**ydroxy **a**nisole)

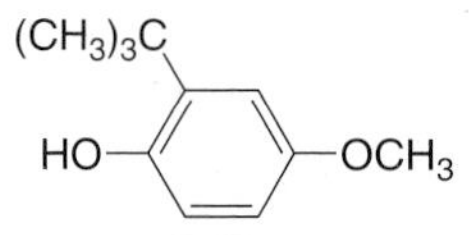

$(CH_3)_3C$

$HO-$ ⬡ $-OCH_3$

BHA

13.60 In cells, vitamin C exists largely as its conjugate base **X**. **X** is an antioxidant because radicals formed in oxidation processes abstract the indicated H atom, forming a new radical that halts oxidation. Draw the structure of the radical formed by H abstraction, and explain why this H atom is most easily removed.

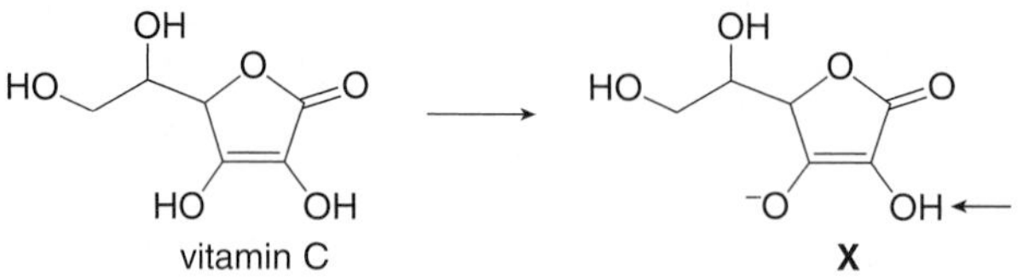

vitamin C → **X**

Polymers and Polymerization

13.61 What monomer is needed to form each polymer?

a.

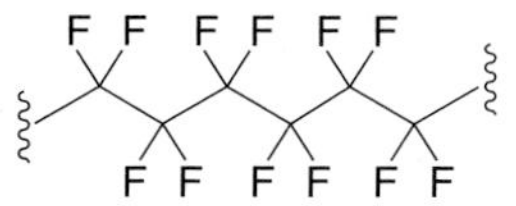

Teflon
(nonstick surface coating)

b.

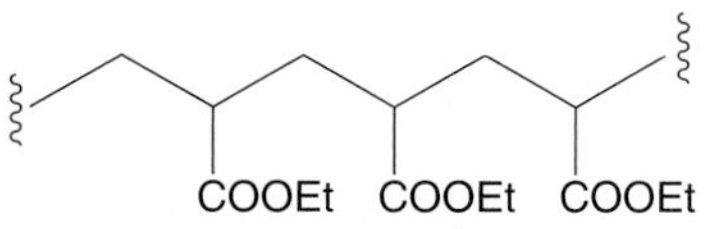

COOEt COOEt COOEt   Et = CH_2CH_3

poly(ethyl acrylate)
(used in latex paints)

13.62 Draw the structure of poly(methyl methacrylate), a polymer used in Lucite and Plexiglas, and formed from radical polymerization of methyl methacrylate [$CH_2=C(CH_3)COOCH_3$].

13.63 Draw a stepwise mechanism for the following polymerization reaction.

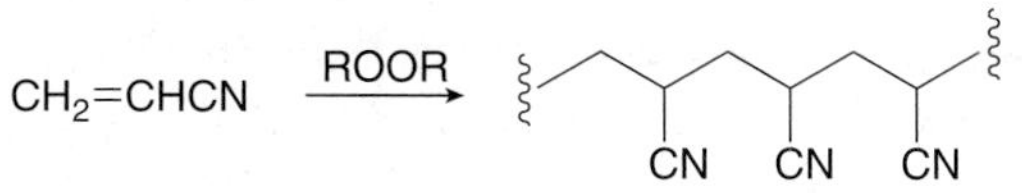

$$CH_2=CHCN \xrightarrow{\text{ROOR}}$$

CN   CN   CN

Challenge Questions

13.64 Draw a stepwise mechanism for the following addition reaction to an alkene.

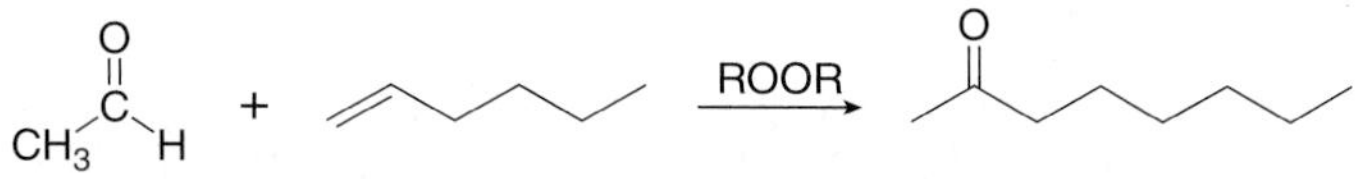

13.65 In the presence of a radical initiator (Z·), tributyltin hydride (**R_3SnH, R = $CH_3CH_2CH_2CH_2$**) reduces alkyl halides to alkanes: R'X + R_3SnH → R'H + R_3SnX. The mechanism consists of a radical chain process with an intermediate tin radical:

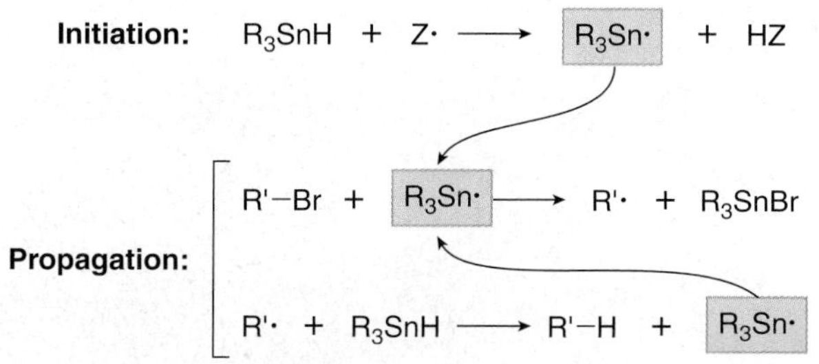

Initiation: R_3SnH + Z· ⟶ R_3Sn· + HZ

Propagation:

R'−Br + R_3Sn· ⟶ R'· + R_3SnBr

R'· + R_3SnH ⟶ R'−H + R_3Sn·

This reaction has been employed in many radical cyclization reactions. Draw a stepwise mechanism for the following reaction.

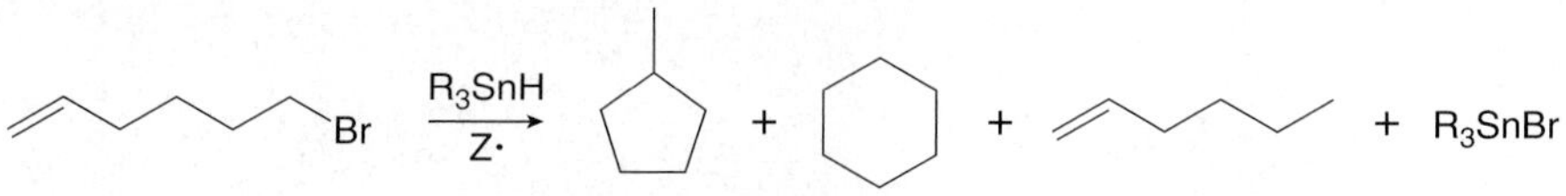

Mass Spectrometry and Infrared Spectroscopy

Tetrahydrocannabinol (THC), first isolated from Indian hemp, is the primary active constituent of marijuana. Although recreational use of cannabis is illegal in the United States, the FDA has approved THC in capsule form as an anti-nausea agent for chemotherapy patients and as an appetite stimulant for AIDS-related anorexia. Like other controlled substances, THC can be detected in minute amounts by modern instrumental methods. In Chapter 14, we learn about mass spectrometry and infrared spectroscopy, two techniques used for characterizing organic compounds.

Whether a compound is prepared in the laboratory or isolated from a natural source, a chemist must determine its identity. Fifty years ago, determining the structure of an organic compound involved a series of time-consuming operations: measuring physical properties (melting point, boiling point, solubility, and density), identifying the functional groups using a series of chemical tests, and converting an unknown compound into another compound whose physical and chemical properties were then characterized as well.

Although still a challenging task, structure determination has been greatly simplified by modern instrumental methods. These techniques have both decreased the time needed for compound characterization, and increased the complexity of compounds whose structures can be completely determined.

In Chapter 14 we examine **mass spectrometry (MS),** which is used to determine the molecular weight and molecular formula of a compound, and **infrared (IR) spectroscopy,** a tool used to identify a compound's functional groups. Chapter 15 is devoted to **nuclear magnetic resonance (NMR) spectroscopy,** which is used to identify the carbon–hydrogen framework in a compound, making it the most powerful spectroscopic tool for organic structure analysis. Each of these methods relies on the interaction of an energy source with a molecule to produce a change that is recorded in a spectrum.

◆

14.1   Mass Spectrometry

Mass spectrometry **is a technique used for measuring the molecular weight and determining the molecular formula of an organic molecule.**

14.1A   General Features

In a mass spectrometer, a molecule is vaporized and ionized, usually by bombardment with a beam of high-energy electrons, as shown in Figure 14.1. The energy of these electrons is typically about 1600 kcal, or 70 electron volts (eV). Because it takes ~100 kcal of energy to cleave a typical σ bond, 1600 kcal is an enormous amount of energy to come into contact with a molecule. This electron beam ionizes a molecule by causing it to eject an electron.

Figure 14.1   Schematic of a mass spectrometer

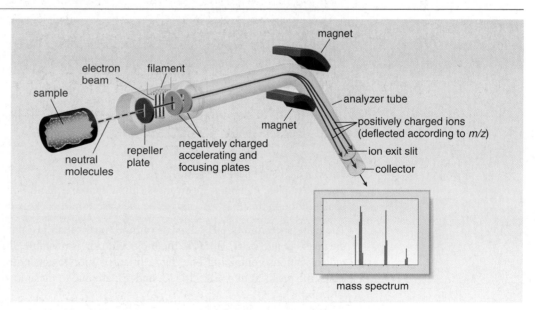

In a mass spectrometer, a sample is vaporized and bombarded by a beam of electrons to form an unstable radical cation, which then decomposes to smaller fragments. The positively charged ions are accelerated toward a negatively charged plate, and then passed through a curved analyzer tube in a magnetic field, where they are deflected by different amounts depending on their ratio of mass to charge. A mass spectrum plots the intensity of each ion versus its *m/z* ratio.

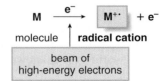

The species formed is a **radical cation,** symbolized **M⁺•.** It is a radical because it has an unpaired electron and it is a cation because it has one fewer electron than it started with.

◆ The radical cation **M⁺•** is called the *molecular ion* or the *parent ion.*

A single electron has a negligible mass, so the **mass of M⁺• represents the molecular weight** of M. For organic chemists, this is the most useful information available from mass spectrometry.

Because the molecular ion **M⁺•** is inherently unstable, it decomposes. Single bonds break to form *fragments,* **radicals and cations having a lower molecular weight than the molecular ion.** A mass spectrometer analyzes the masses of cations only. The cations are accelerated in an electric field and deflected in a curved path in a magnetic field, thus sorting the molecular ion and its fragments by their mass-to-charge (m/z) ratio. Because z is almost always +1, m/z actually measures the mass (m) of the individual ions.

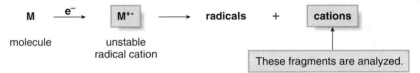

◆ A *mass spectrum* plots the amount of each cation (its relative abundance) versus its mass.

A mass spectrometer analyzes the masses of *individual* molecules, not the weighted average mass of a group of molecules, so the whole-number masses of the most common individual isotopes must be used to calculate the mass of the molecular ion. Thus, the mass of the molecular ion for CH_4 should be 16. As a result, the mass spectrum of CH_4 shows a line for the molecular ion—the parent peak or **M** peak—at $m/z = 16$.

The whole-number mass of CH_4 is (1C × 12 amu) + (4H × 1 amu) = 16 amu; amu = atomic mass units.

Mass spectrum of CH₄

m/z	relative abundance
17	1.2
16	100.0 ← base peak
15	85.9
14	16.1
13	8.1
12	2.8

The tallest peak in a mass spectrum is called the **base peak.** For CH_4, the base peak is also the M peak, although this may *not* always be the case for all organic compounds.

The mass spectrum of CH_4 consists of more peaks than just the M peak. What is responsible for the peaks at $m/z < 16$? Because the molecular ion is unstable, it fragments into other cations and radical cations containing one, two, three, or four fewer hydrogen atoms than methane itself. Thus, the peaks at $m/z = 15, 14, 13,$ and 12, are due to these lower molecular weight fragments.

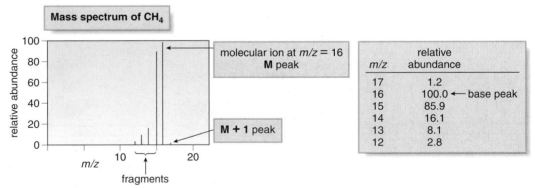

Figure 14.2 Mass spectrum of hexane ($CH_3CH_2CH_2CH_2CH_2CH_3$)

- The molecular ion for hexane (molecular formula C_6H_{14}) is at $m/z = 86$.
- The base peak (relative abundance = 100) occurs at $m/z = 57$.
- A small M + 1 peak occurs at $m/z = 87$.

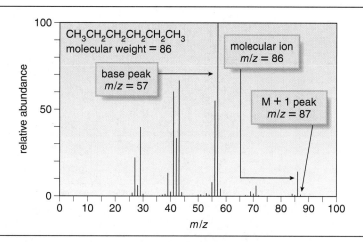

What is responsible for the small peak at $m/z = 17$ in the mass spectrum of CH_4? Although most carbon atoms have an atomic mass of 12, 1.1% of them have an additional neutron in the nucleus, giving them an atomic mass of 13. When one of these carbon-13 isotopes forms methane, it gives a molecular ion peak at $m/z = 17$ in the mass spectrum. This peak is called the **M + 1** peak.

These key features—the molecular ion, the base peak, and the M + 1 peak—are illustrated in the mass spectrum of hexane in Figure 14.2.

14.1B Analyzing Unknowns Using the Molecular Ion

Because the **mass of the molecular ion equals the molecular weight of a compound,** a mass spectrum can be used to distinguish between compounds that have similar physical properties but different molecular weights, as illustrated in Sample Problem 14.1.

SAMPLE PROBLEM 14.1 Pentane, 1-pentene, and 1-pentyne are low boiling hydrocarbons that have different molecular ions in their mass spectra. Match each hydrocarbon to its mass spectrum.

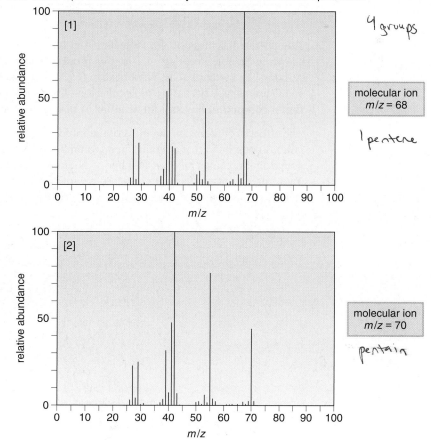

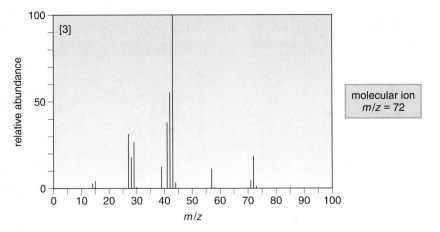

SOLUTION

To solve this problem, first determine the molecular formula and molecular weight of each compound. Then, because the molecular weight of the compound equals the mass of the molecular ion, match the molecular weight to m/z for the molecular ion:

Compound	Molecular formula	Molecular weight = m/z of molecular ion	Spectrum
pentane $CH_3CH_2CH_2CH_2CH_3$	C_5H_{12}	72	[3]
1-pentene $CH_2=CHCH_2CH_2CH_3$	C_5H_{10}	70	[2]
1-pentyne $HC\equiv CCH_2CH_2CH_3$	C_5H_8	68	[1]

PROBLEM 14.1 What is the mass of the molecular ion formed from compounds having each molecular formula: (a) C_3H_6O; (b) $C_{10}H_{20}$; (c) $C_8H_8O_2$; (d) methamphetamine ($C_{10}H_{15}N$)?

PROBLEM 14.2 An unknown compound **X** gave a molecular ion of 88 in its mass spectrum. Which structures are possible for **X**: (a) $CH_3CH_2OCH_2CH_2CH_3$; (b) $CH_3COCH_2CH_2CH_3$; (c) $CH_3CO_2CH_2CH_3$?

How to use the mass of the molecular ion to propose molecular formulas for an unknown is shown in Sample Problem 14.2. In this process, keep in mind the following useful fact. Hydrocarbons like methane (CH_4) and hexane (C_6H_{14}), as well as compounds that contain only C, H, and O atoms, always have a molecular ion with an *even* mass. **An odd molecular ion indicates that a compound has an odd number of nitrogen atoms.**

The effect of N atoms on the mass of the molecular ion in a mass spectrum is called the **nitrogen rule:** A compound that contains an *odd* number of N atoms gives an *odd* molecular ion. A compound that contains an *even* number of N atoms (including *zero*) gives an *even* molecular ion. Two "street" drugs that mimic the effects of heroin illustrate this principle: **3-methylfentanyl** (two N atoms, even molecular weight) and **MPPP** (one N atom, odd molecular weight).

3-methylfentanyl
$C_{23}H_{30}N_2O$
molecular weight = 350

MPPP
(1-methyl-4-phenyl-4-propionoxypiperidine)
$C_{15}H_{21}NO_2$
molecular weight = 247

SAMPLE PROBLEM 14.2   Propose possible molecular formulas for a compound with a molecular ion at $m/z = 86$.

SOLUTION

Because the molecular ion has an **even** mass, the compound likely contains C, H, and possibly O atoms. Begin by determining the molecular formula for a hydrocarbon having a molecular ion at 86. Then, because the mass of an O atom is 16 (the mass of CH_4), replace CH_4 by O to give a molecular formula containing one O atom. Repeat this last step to give possible molecular formulas for compounds with two or more O atoms.

For a molecular ion at $m/z = 86$:

Possible hydrocarbons:	**Possible compounds with C, H, and O:**
• Divide 86 by 12 (mass of 1 C atom). This gives the maximum number of C's possible.	• Substitute 1 O for CH_4. (This can't be done for C_7H_2.)
$\dfrac{86}{12}$ = 7 C's maximum $\longrightarrow$ $\boxed{C_7H_2}$ (remainder = 2)	$C_6H_{14} \xrightarrow[+\,1\,O]{-\,CH_4} \boxed{C_5H_{10}O}$
• Replace one C by 12 H's for another possible molecular formula.	• Repeat the process.
$C_7H_2 \xrightarrow[+\,12\,H's]{-\,1\,C} \boxed{C_6H_{14}}$	$C_5H_{10}O \xrightarrow[+\,1\,O]{-\,CH_4} \boxed{C_4H_6O_2}$

PROBLEM 14.3   Propose two molecular formulas for each of the following molecular ions: (a) 72; (b) 100; (c) 73.

14.2   Alkyl Halides and the M + 2 Peak

Most of the elements found in organic compounds, such as carbon, hydrogen, oxygen, nitrogen, sulfur, phosphorus, fluorine, and iodine, have one major isotope. **Chlorine** and **bromine,** on the other hand, have two, giving characteristic patterns to the mass spectra of their compounds.

Chlorine has two common isotopes, ^{35}Cl and ^{37}Cl, which occur naturally in a 3:1 ratio. Thus, **there are two peaks in a 3:1 ratio for the molecular ion of an alkyl chloride.** The larger peak—the **M** peak—corresponds to the compound containing ^{35}Cl, and the smaller peak—the **M + 2** peak—corresponds to the compound containing ^{37}Cl.

◆   When the molecular ion consists of two peaks (M and M + 2) in a 3:1 ratio, a Cl atom is present.

SAMPLE PROBLEM 14.3   What molecular ions will be present in a mass spectrum of 2-chloropropane, $(CH_3)_2CHCl$?

SOLUTION

Calculate the molecular weight using each of the common isotopes of Cl.

Molecular formula	Mass of molecular ion (m/z)
$C_3H_7{}^{35}Cl$	78 (M peak)
$C_3H_7{}^{37}Cl$	80 (M + 2 peak)

ANSWER

There should be two peaks in a ratio of 3:1, at $m/z = 78$ and 80, as illustrated in the mass spectrum of 2-chloropropane in Figure 14.3.

Bromine has two common isotopes, ^{79}Br and ^{81}Br, which occur naturally in a 1:1 ratio. Thus, **there are two peaks in a 1:1 ratio for the molecular ion of an alkyl bromide.** In the mass spectrum of 2-bromopropane (Figure 14.4), for example, there is an M peak at $m/z = 122$ and an M + 2 peak at $m/z = 124$.

Figure 14.3 Mass spectrum of 2-chloropropane [(CH₃)₂CHCl]

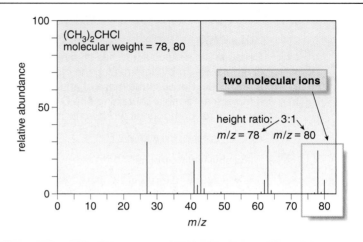

Figure 14.4 Mass spectrum of 2-bromopropane [(CH₃)₂CHBr]

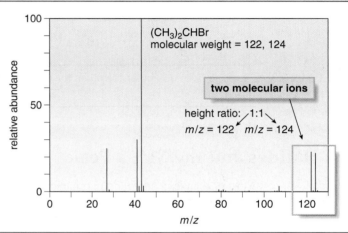

◆ When the molecular ion consists of two peaks (M and M + 2) in a 1:1 ratio, a Br atom is present in the molecule.

PROBLEM 14.4   What molecular ions would you expect for compounds having each of the following molecular formulas: (a) C_4H_9Cl; (b) C_3H_7F; (c) $C_6H_{11}Br$; (d) $C_4H_{11}N$?

14.3   Other Types of Mass Spectrometry

Although using the molecular ion to determine the molecular weight of an organic compound is indeed valuable, recent advances have greatly expanded the information obtained from mass spectrometry.

14.3A   High-Resolution Mass Spectrometry

TABLE 14.1   Exact Masses of Some Common Isotopes

Isotope	Mass
^{12}C	12.0000
^{1}H	1.00783
^{16}O	15.9949
^{14}N	14.0031

The mass spectra described thus far have been low-resolution spectra; that is, they report m/z values to the nearest whole number. As a result, the mass of a given molecular ion can correspond to many different molecular formulas, as shown in Sample Problem 14.2.

High-resolution mass spectrometers measure m/z ratios to four (or more) decimal places. This is valuable because except for carbon-12, whose mass is defined as 12.0000, the masses of all other nuclei are very close to—but not exactly—whole numbers. Table 14.1 lists the exact mass values of a few common nuclei. Using these values it is possible to determine the single molecular formula that gives rise to a molecular ion.

For example, a compound having a molecular ion at $m/z = 60$ using a low-resolution mass spectrometer could have the following molecular formulas:

Formula	Exact mass
C_3H_8O	60.0575
$C_2H_4O_2$	60.0211
$C_2H_8N_2$	60.0688

If the molecular ion had an exact mass of 60.0578, the compound's molecular formula is C_3H_8O, because its mass is closest to the observed value.

PROBLEM 14.5 The low-resolution mass spectrum of an unknown analgesic **X** had a molecular ion of 151. Possible molecular formulas include $C_7H_5NO_3$, $C_8H_9NO_2$, and $C_{10}H_{17}N$. High-resolution mass spectrometry gave an exact mass of 151.0640. What is the molecular formula of **X**?

14.3B   Gas Chromatography–Mass Spectrometry (GC–MS)

Two analytical tools—**gas chromatography (GC)** and **mass spectrometry (MS)**—can be combined into a single instrument **(GC–MS)** to analyze mixtures of compounds (Figure 14.5a). The gas chromatograph separates the mixture, and then the mass spectrometer records a spectrum of the individual components.

Figure 14.5 Compound analysis using GC–MS

[a] Schematic of a GC–MS instrument

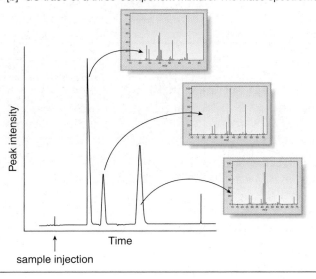

[b] GC trace of a three-component mixture. The mass spectrometer gives a spectrum for each component.

A gas chromatograph consists of a thin capillary column containing a viscous, high-boiling liquid, all housed in an oven. When a sample is injected into the GC, it is vaporized and swept by an inert gas through the column. The components of the mixture travel through the column at different rates, often separated by boiling point, with lower boiling compounds exiting the column before higher boiling compounds. Each compound then enters the mass spectrometer, where it is ionized to form its molecular ion and lower molecular weight fragments. The GC–MS records a gas chromatogram for the mixture, which plots the amount of each component versus its **retention time**—that is, the time required to travel through the column. Each component of a mixture is characterized by its retention time in the gas chromatogram and its molecular ion in the mass spectrum (Figure 14.5b).

GC–MS is widely used for characterizing mixtures containing environmental pollutants. It is also used to analyze urine and hair samples for the presence of illegal drugs or banned substances thought to improve athletic performance.

To analyze a urine sample for THC (the chapter-opening molecule), the principal psychoactive component of marijuana, the organic compounds are extracted from urine, purified, concentrated, and injected into the GC–MS. THC appears as a GC peak with a characteristic retention time (for a given set of experimental parameters), and gives a molecular ion at 314, its molecular weight, as shown in Figure 14.6.

PROBLEM 14.6   Benzene, toluene, and *p*-xylene **(BTX)** are often added to gasoline to boost octane ratings. What would be observed if a mixture of these three compounds were subjected to GC–MS analysis? How many peaks would be present in the gas chromatogram? What would be the relative order of the peaks? What molecular ions would be observed in the mass spectra?

benzene   toluene   *p*-xylene

14.3C   Mass Spectra of High Molecular Weight Biomolecules

Dr. John Fenn, professor emeritus of chemical engineering of Yale University, shared the 2002 Nobel Prize in Chemistry for his development of ESI mass spectrometry.

Until the 1980s mass spectra were limited to molecules that could be readily vaporized with heat under vacuum, and thus had molecular weights of < 800. In the last 25 years, new methods have been developed to generate gas phase ions of large molecules, allowing mass spectra to be recorded for large biomolecules such as proteins and carbohydrates. **Electrospray ionization (ESI),** for example, forms ions by creating a fine spray of charged droplets in an electric field. Evaporation of the charged droplets forms gaseous ions that are then analyzed by their *m/z* ratio. ESI and related techniques have extended mass spectrometry into the analysis of nonvolatile compounds with molecular weights greater than 100,000 daltons (atomic mass units).

Figure 14.6   Mass spectrum of tetrahydrocannabinol (THC)

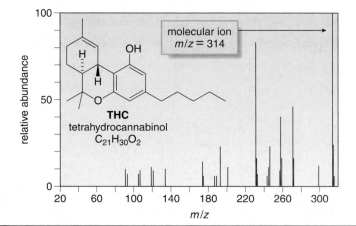

molecular ion
m/z = 314

THC
tetrahydrocannabinol
$C_{21}H_{30}O_2$

14.4   Electromagnetic Radiation

Infrared (IR) spectroscopy and **nuclear magnetic resonance (NMR)** spectroscopy (Chapter 15) both use a form of electromagnetic radiation as their energy source. To understand IR and NMR, therefore, you need to understand some of the properties of **electromagnetic radiation**—radiant energy having dual properties of both waves and particles.

The particles of electromagnetic radiation are called **photons,** each having a discrete amount of energy called a **quantum.** Because electromagnetic radiation also has wave properties, it can be characterized by its **wavelength** and **frequency.**

Length units used to report wavelength include:

Unit	Length
meter (m)	1 m
centimeter (cm)	10^{-2} m
micrometer (μm)	10^{-6} m
nanometer (nm)	10^{-9} m
Angstrom (Å)	10^{-10} m

◆ Wavelength (λ) is the distance from one point on a wave (e.g., the peak or trough) to the same point on the adjacent wave. A variety of different length units are used for λ, depending on the type of radiation.
◆ Frequency (ν) is the number of waves passing a point per unit time. It is reported in cycles per second (s^{-1}), which is also called hertz (Hz).

You come into contact with many different kinds of electromagnetic radiation in your daily life. For example, you use visible light to see the words on this page, you may cook with microwaves, and you should use sunscreen to protect your skin from the harmful effects of ultraviolet radiation.

The different forms of electromagnetic radiation make up the **electromagnetic spectrum.** The spectrum is arbitrarily divided into different regions, as shown in Figure 14.7. All electromagnetic radiation travels at the speed of light (c), 3.0×10^8 m/s.

The speed of electromagnetic radiation (c) is directly proportional to its wavelength and frequency:

$$c = \lambda\nu$$

The speed of light (c) is a constant, so wavelength and frequency are *inversely* related:

◆ $\lambda = c/\nu$: Wavelength increases as frequency decreases.
◆ $\nu = c/\lambda$: Frequency increases as wavelength decreases.

The energy (E) of a photon is directly proportional to its frequency:

$$E = h\nu \qquad h = \text{Planck's constant } (1.58 \times 10^{-34} \text{ cal·s})$$

Figure 14.7 The electromagnetic spectrum

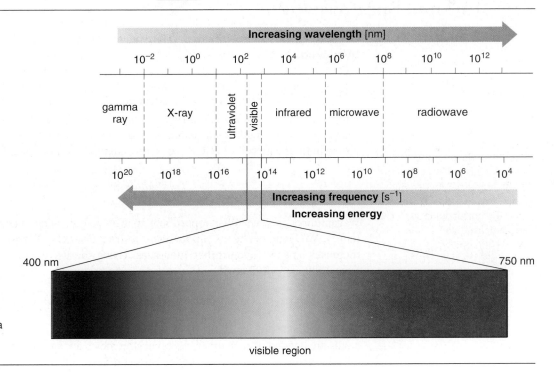

• Visible light occupies only a small region of the electromagnetic spectrum.

Frequency and wavelength are *inversely* proportional ($\nu = c/\lambda$), however, so energy and wavelength are *inversely* proportional:

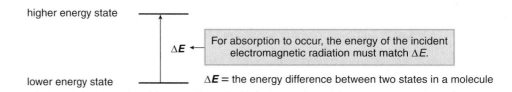

$$E = h\nu = \frac{hc}{\lambda}$$

- E increases as ν increases.
- E decreases as λ increases.

When electromagnetic radiation strikes a molecule, some wavelengths—but not all—are absorbed. Only some wavelengths are absorbed because molecules have discrete energy levels. The energies of their electronic, vibrational, and nuclear spin states are *quantized,* not *continuous.*

◆ **For absorption to occur, the energy of the photon must match the difference between two energy states in a molecule.**

higher energy state ————

ΔE ◄— | For absorption to occur, the energy of the incident electromagnetic radiation must match ΔE.

lower energy state ———— ΔE = the energy difference between two states in a molecule

◆ **The *larger* the energy difference between two states, the *higher* the energy of radiation needed for absorption, the *higher* the frequency, and the *shorter* the wavelength.**

PROBLEM 14.7 Which of the following has the higher frequency: (a) light having a wavelength of 10^2 or 10^4 nm; (b) light having a wavelength of 100 nm or 100 μm; (c) red light or blue light?

PROBLEM 14.8 Which of the following has the higher energy: (a) light having a ν of 10^4 Hz or 10^8 Hz; (b) light having a λ of 10 nm or 1000 nm; (c) red light or blue light?

PROBLEM 14.9 The difference in energy between two electronic states is ~100 kcal/mol, whereas the difference in the energy between two vibrational states is ~5 kcal/mol. Which transition requires the higher ν of radiation?

14.5 Infrared Spectroscopy

Organic chemists use infrared (IR) spectroscopy to identify the functional groups in a compound.

14.5A Background

Infrared radiation (λ = 2.5–25 μm) is the energy source in infrared spectroscopy. These are somewhat longer wavelengths than visible light, so they are lower in frequency and lower in energy than visible light. Frequencies in IR spectroscopy are reported using a unit called the **wavenumber** ($\tilde{\nu}$):

$$\tilde{\nu} = \frac{1}{\lambda}$$

Using the wavenumber scale results in IR frequencies in a numerical range that is easier to report than the corresponding frequencies given in hertz (4000–400 cm^{-1} compared to 1.2×10^{14}–1.2×10^{15} Hz).

Wavenumber is *inversely* proportional to wavelength and reported in reciprocal centimeters **(cm^{-1}).** Wavenumber ($\tilde{\nu}$) is *proportional* to frequency (ν). **Frequency (and therefore energy) increases as the wavenumber increases.** Using the wavenumber scale, IR absorptions occur from **4000 cm^{-1}–400 cm^{-1}.**

◆ **Absorption of IR light causes changes in the vibrational motions of a molecule.**

Covalent bonds are not static. They are more like springs with weights on each end. When two atoms are bonded to each other, the bond stretches back and forth. When three or more atoms are joined together, bonds can also bend. These bond stretching and bending vibrations represent the different vibrational modes available to a molecule.

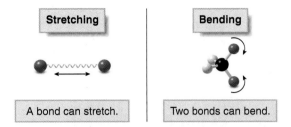

These vibrations are quantized, so they occur only at specific frequencies, which correspond to the frequency of IR light. When the frequency of IR light matches the frequency of a particular vibrational mode, the IR light is absorbed, causing the amplitude of the particular bond stretch or bond bend to increase.

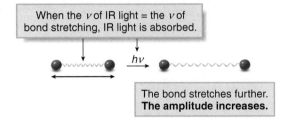

◆ **Different kinds of bonds vibrate at different frequencies, so they absorb different frequencies of IR light.**
◆ **IR spectroscopy distinguishes between the different kinds of bonds in a molecule, so it is possible to determine the functional groups present.**

PROBLEM 14.10   Which of the following has higher energy: (a) IR light of 3000 cm^{-1} or 1500 cm^{-1} in wavenumber; (b) IR light having a wavelength of 10 μm or 20 μm?

14.5B   Characteristics of an IR Spectrum

In an IR spectrometer, light passes through a sample. Frequencies that match vibrational frequencies are absorbed, and the remaining light is transmitted to a detector. A spectrum plots the amount of transmitted light versus its wavenumber. The IR spectrum of 1-propanol, $CH_3CH_2CH_2OH$, illustrates several important features of IR spectroscopy.

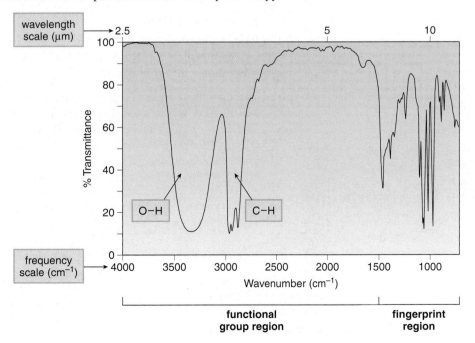

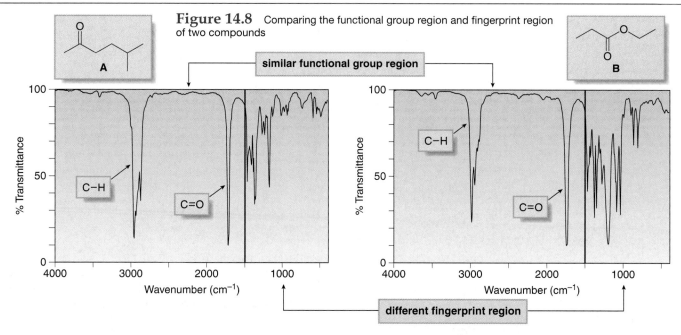

Figure 14.8 Comparing the functional group region and fingerprint region of two compounds

- **A** and **B** show peaks in the same regions for their C=O group and sp^3 hybridized C–H bonds.
- **A** and **B** are different compounds so their fingerprint regions are quite different.

- An IR spectrum has broad lines.

- The absorption peaks go *down* on a page. The *y* axis measures **percent transmittance:** 100% transmittance means that all the light shone on a sample is transmitted and none is absorbed; 0% transmittance means that none of the light shone on a sample is transmitted and all is absorbed. Most absorptions lie between these two extremes.

- **Each peak corresponds to a particular kind of bond, and each bond type (such as O–H and C–H) occurs at a characteristic frequency.**

- IR spectra have both a wavelength and a wavenumber scale on the *x* axis. Wavelengths are recorded in μm (2.5–25). Wavenumber, frequency, and energy *decrease* from left to right. Where a peak occurs is reported in reciprocal centimeters (cm^{-1}).

Conceptually, the IR spectrum is divided into two regions:

- The **functional group region** occurs at ≥ **1500 cm⁻¹.** Common functional groups give one or two peaks in this region, at a characteristic frequency.

- The **fingerprint region** occurs at **< 1500 cm⁻¹.** This region often contains a complex set of peaks and is unique for every compound.

We will analyze only the functional group region (above 1500 cm⁻¹) because it provides the most unambiguous structural information. Figure 14.8 shows, for example, that the IR spectra of 5-methyl-2-hexanone (**A**) and ethyl propanoate (**B**), two compounds that contain both a carbonyl group (C=O) and several sp^3 hybridized C–H bonds, look similar in their functional group regions, but very different in their fingerprint regions.

14.6 IR Absorptions

14.6A Where Particular Bonds Absorb in the IR

Where a particular bond absorbs in the IR depends on **bond strength** and **atom mass.**

- Bond strength: stronger bonds vibrate at higher frequency, so they absorb at higher $\tilde{\nu}$.
- Atom mass: bonds with lighter atoms vibrate at higher frequency, so they absorb at higher $\tilde{\nu}$.

Figure 14.9 Hooke's law: How the frequency of bond vibration depends on atom mass and bond strength

The frequency of bond vibration can be derived from Hooke's law, which describes the motion of a vibrating spring:

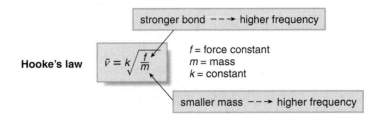

stronger bond $--\!\rightarrow$ higher frequency

Hooke's law $\tilde{\nu} = k\sqrt{\dfrac{f}{m}}$

f = force constant
m = mass
k = constant

smaller mass $--\!\rightarrow$ higher frequency

- The force constant (f) is the strength of the bond (or spring). The larger the value of f, the stronger the bond, and the higher the $\tilde{\nu}$ of vibration.
- The mass (m) is the mass of atoms (or weights). The smaller the value of m, the higher the $\tilde{\nu}$ of vibration.

Thinking of bonds as springs with weights on each end illustrates these trends. The strength of the spring is analogous to bond strength, and the mass of the weights is analogous to atomic mass. For two springs with the same weights on each end, the **stronger spring vibrates at a higher frequency.** For two springs of the same strength, **springs with lighter weights vibrate at higher frequency** than those with heavier weights. Hooke's law, as shown in Figure 14.9, describes the relationship of frequency to mass and bond strength.

As a result, **bonds absorb in four predictable regions in an IR spectrum.** These four regions, and the bonds that absorb there, are summarized in Figure 14.10. Remembering the information in this figure will help you analyze the spectra of unknown compounds. To help you remember it, keep in mind the following two points:

- ◆ Absorptions for bonds to hydrogen always occur on the left side of the spectrum (the high wavenumber region). H has so little mass that H−Z bonds (where Z = C, O, and N) vibrate at high frequencies.
- ◆ Bond strength decreases in going from C≡C → C=C → C−C, so the frequency of vibration decreases—that is, the absorptions for these bonds move further to the right side of the spectrum.

The functional group region consists of absorptions for single bonds to hydrogen (all H−Z bonds), as well as absorptions for all multiple bonds. Most absorptions in the functional group region are due to bond stretching (rather than bond bending). The fingerprint region consists of absorptions due to all other single bonds (except H−Z bonds), making it often a complex region that is very difficult to analyze.

Besides learning the general regions of the IR spectrum, it is also important to learn the specific absorption values for common bonds. Table 14.2 lists the most important IR absorptions in the

Figure 14.10 Summary: The four regions of the IR spectrum

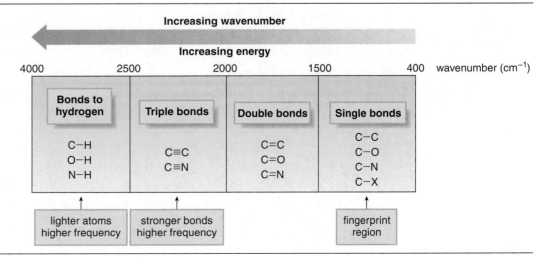

functional group region. Other details of IR absorptions will be presented in later chapters when new functional groups are introduced. Appendix D contains a detailed list of the characteristic IR absorption frequencies for common bonds.

Even subtle differences that affect bond strength affect the frequency of an IR absorption. Recall from Section 1.10 that the strength of a C−H bond increases as the percent s-character of the hybrid orbital on the carbon increases; thus:

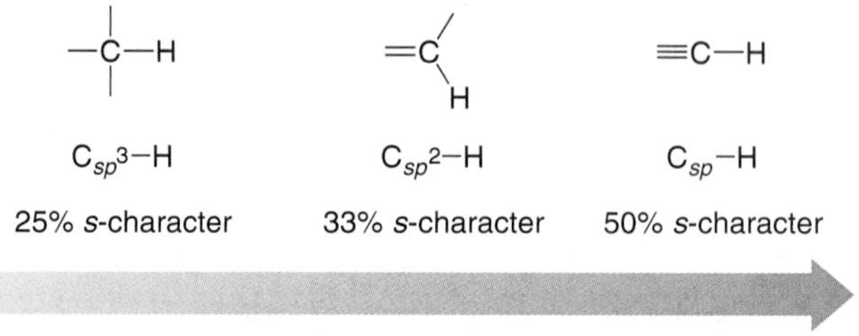

$$-\overset{|}{\underset{|}{C}}-H \qquad =C\overset{/}{\underset{H}{}} \qquad \equiv C-H$$

$$C_{sp^3}-H \qquad\qquad C_{sp^2}-H \qquad\qquad C_{sp}-H$$

25% s-character   33% s-character   50% s-character

Increasing percent s-character
Increasing $\tilde{\nu}$

◆ **The higher the percent s-character, the stronger the bond and the higher the wavenumber of absorption.**

PROBLEM 14.11   Which bond in each pair absorbs at higher wavenumber?

a. $CH_3-C\equiv C-CH_2CH_3$ or $CH_2=C(CH_3)_2$ c. CH_3-H or CH_3-D

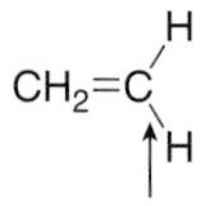

b. CH_3CH_2-H or $CH_2=C\overset{H}{\underset{H}{}}$

TABLE 14.2	Important IR Absorptions	
Bond type	**Approximate $\tilde{\nu}$ (cm⁻¹)**	**Intensity**
O−H	3600–3200	strong, broad
N−H	3500–3200	medium
C−H	~3000	
• $C_{sp^3}-H$	3000–2850	strong
• $C_{sp^2}-H$	3150–3000	medium
• $C_{sp}-H$	3300	medium
C≡C	2250	medium
C≡N	2250	medium
C=O	1800–1650 (often ~1700)	strong
C=C	1650	medium
 ⬡	1600, 1500	medium

Finally, almost all bonds in a molecule give rise to an absorption peak in an IR spectrum, but a few do not. **For a bond to absorb in the IR, there must be a change in dipole moment during the vibration.** Thus, symmetrical, nonpolar bonds do not absorb in the IR. The carbon–carbon triple bond of 2-butyne, for example, does not have an IR stretching absorption at 2250 cm⁻¹ because the C≡C bond is nonpolar and there is no change in dipole moment when the bond stretches along its axis. This type of vibration is said to be **IR inactive**.

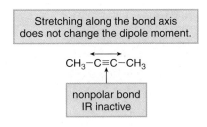

Stretching along the bond axis does not change the dipole moment.

$$CH_3-C\equiv C-CH_3$$

nonpolar bond
IR inactive

14.6B   IR Absorptions in Hydrocarbons

The IR spectra of hexane, 1-hexene, and 1-hexyne illustrate the important differences that characterize the IR spectra of hydrocarbons above 1500 cm^{-1}. Although all three compounds contain $C-C$ bonds and sp^3 hybridized $C-H$ bonds, the absorption peaks due to $C=C$ and $C\equiv C$ readily distinguish the alkene and alkyne.

Note, too, that the $C-H$ absorptions in alkanes, alkenes, and alkynes have a characteristic appearance and position. The sp^3 hybridized $C-H$ bonds are often seen as a broad, strong absorption at < 3000 cm^{-1}, whereas sp^2 and sp hybridized $C-H$ bonds absorb at somewhat higher frequency.

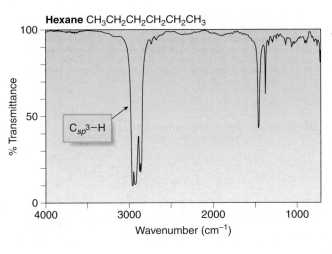

◆ The **alkane** CH$_3$CH$_2$CH$_2$CH$_2$CH$_2$CH$_3$ has only $C-C$ single bonds and sp^3 hybridized C atoms. Therefore, it has only one major absorption above 1500 cm^{-1}, its $C_{sp^3}-H$ absorption at 3000–2850 cm^{-1}.

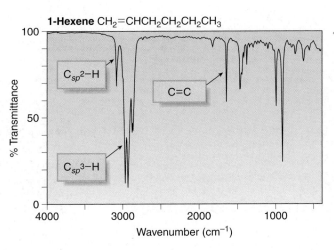

◆ The **alkene** CH$_2$=CHCH$_2$CH$_2$CH$_2$CH$_3$ has a $C=C$ and $C_{sp^2}-H$, in addition to its sp^3 hybridized C atoms. Therefore, there are three major absorptions above 1500 cm^{-1}:

◆ $C_{sp^2}-H$ at 3150–3000 cm^{-1}
◆ $C_{sp^3}-H$ at 3000–2850 cm^{-1}
◆ $C=C$ at 1650 cm^{-1}

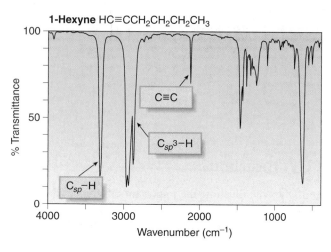

1-Hexyne HC≡CCH₂CH₂CH₂CH₃

- The **alkyne** HC≡CCH₂CH₂CH₂CH₃ has a C≡C and C$_{sp}$−H, in addition to its sp^3 hybridized C atoms. Therefore, there are three major absorptions:
  - C$_{sp}$−H at 3300 cm⁻¹
  - C$_{sp^3}$−H at 3000–2850 cm⁻¹
  - C≡C at 2250 cm⁻¹

PROBLEM 14.12 How do the IR spectra of the isomers cyclopentane and 1-pentene differ?

The exact location of the C=O absorption depends on the particular functional group, whether the carbonyl carbon is part of a ring, and whether there are nearby double bonds. These details are discussed in Chapters 21 and 22.

14.6C IR Absorptions in Oxygen-Containing Compounds

The most important IR absorptions for oxygen-containing compounds occur at **3600–3200 cm⁻¹ for an OH group,** and at approximately **1700 cm⁻¹ for a C=O,** as illustrated in the IR spectra of an **alcohol** (2-butanol), a **ketone** (2-butanone), and an **ether** (diethyl ether). The peak at ~3000 cm⁻¹ in each spectrum is due to C$_{sp^3}$−H bonds.

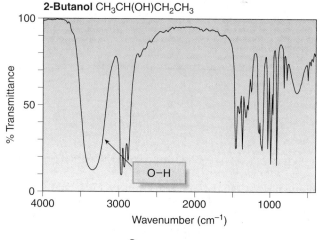

2-Butanol CH₃CH(OH)CH₂CH₃

- The **OH** group in the alcohol CH₃CH(OH)CH₂CH₃ shows a strong absorption at 3600–3200 cm⁻¹.

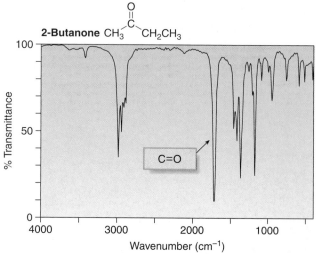

2-Butanone CH₃ —C(=O)— CH₂CH₃

- The **C=O** group in the ketone CH₃COCH₂CH₃ shows a strong absorption at ~1700 cm⁻¹.

Diethyl ether CH₃CH₂OCH₂CH₃

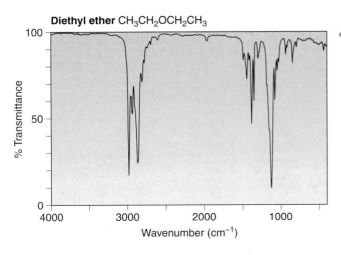

♦ (CH₃CH₂)₂O, has neither an OH group nor a C=O, so its only absorption above 1500 cm⁻¹ occurs at ~3000 cm⁻¹, due to sp^3 hybridized C–H bonds. Compounds that contain an oxygen atom but do not show an OH or C=O absorption are **ethers.**

SAMPLE PROBLEM 14.4   How can the two isomers having molecular formula C₂H₆O be distinguished by IR spectroscopy?

SOLUTION

First, draw the structures of the compounds and then locate the functional groups. One compound is an alcohol and one is an ether.

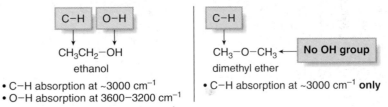

| C–H | O–H |
| C–H |

CH₃CH₂—OH
ethanol
• C–H absorption at ~3000 cm⁻¹
• O–H absorption at 3600–3200 cm⁻¹

CH₃—O—CH₃   **No OH group**
dimethyl ether
• C–H absorption at ~3000 cm⁻¹ **only**

Although both compounds have sp^3 hybridized C–H bonds, ethanol has an OH group that gives a strong absorption at 3600–3200 cm⁻¹, and dimethyl ether does not. This feature distinguishes the two isomers.

PROBLEM 14.13   How do the three isomers of molecular formula C₃H₆O (**A, B,** and **C**) differ in their IR spectra?

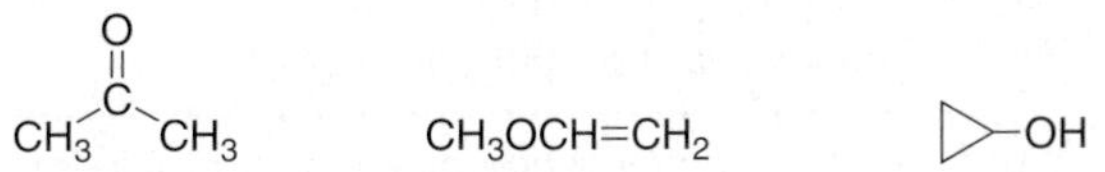

$$\underset{\textbf{A}}{\overset{\displaystyle O}{\underset{CH_3}{\overset{\|}{C}}}CH_3} \qquad \underset{\textbf{B}}{CH_3OCH=CH_2} \qquad \underset{\textbf{C}}{\triangleright\!\!-OH}$$

14.6D   IR Absorptions in Nitrogen-Containing Compounds

Additional details on the IR spectra of amines, amides, and nitriles are given in Chapters 22 and 25.

Common functional groups that contain nitrogen atoms are also distinguishable by their IR absorptions above 1500 cm⁻¹, as illustrated by the IR spectra of an **amine** (octylamine), an **amide** (propanamide), and a **nitrile** (octanenitrile).

Octylamine CH₃CH₂CH₂CH₂CH₂CH₂CH₂CH₂NH₂

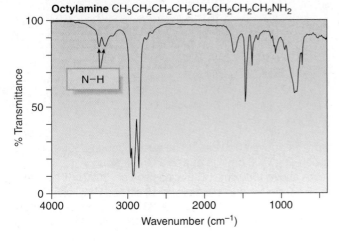

♦ The **N–H** bonds in the amine CH₃(CH₂)₇NH₂ give rise to two weak absorptions at 3300 and 3400 cm⁻¹.

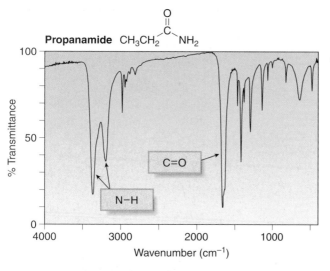

Propanamide $CH_3CH_2\overset{\overset{\displaystyle O}{\displaystyle \|}}{C}NH_2$

◆ The **amide** $CH_3CH_2CONH_2$ exhibits absorptions above 1500 cm^{-1} for both its **N–H** and **C=O** groups:
  ◆ N–H (two peaks) at 3200 and 3400 cm^{-1}
  ◆ C=O at 1660 cm^{-1}

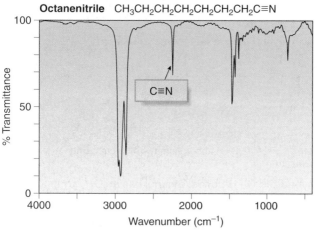

Octanenitrile $CH_3CH_2CH_2CH_2CH_2CH_2CH_2C\equiv N$

◆ The **C≡N** group of the **nitrile** $CH_3(CH_2)_6CN$ absorbs in the triple bond region at ~2250 cm^{-1}.

Sample Problem 14.5 shows how the region above 1500 cm^{-1} in an IR spectrum can be used for functional group identification.

SAMPLE PROBLEM 14.5 What functional groups are responsible for the absorptions above 1500 cm^{-1} in compounds **A** and **B**?

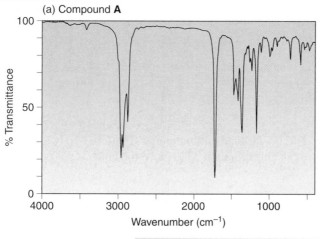

(a) Compound **A**

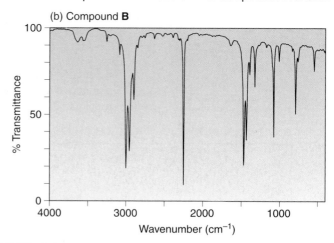

(b) Compound **B**

SOLUTION

[a] Compound **A** has two major absorptions above 1500 cm^{-1}: The absorption at ~3000 cm^{-1} is due to C–H bonds and the absorption at ~1700 cm^{-1} is due to a C=O group.

[b] Compound **B** has two major absorptions above 1500 cm^{-1}: The absorption at ~3000 cm^{-1} is due to C–H bonds and the absorption at ~2250 cm^{-1} is due to a triple bond, either a C≡C or a C≡N. Because there is no absorption due to an *sp* hybridized C–H bond at 3300 cm^{-1}, this IR spectrum can *not* be due to a terminal alkyne (HC≡CR) but may still be due to an internal alkyne.

PROBLEM 14.14   What functional groups are responsible for the absorptions above 1500 cm^{-1} in the IR spectra for compounds **A** and **B**?

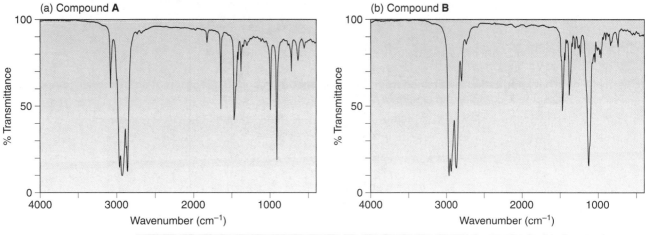

(a) Compound **A**

(b) Compound **B**

PROBLEM 14.15   What are the major IR absorptions in the functional group region for each compound?

a.   b. ⬡—OH   c.   d. ⬡=O

14.7   IR and Structure Determination

New instruments for determining blood alcohol concentration use IR spectroscopy for analyzing the C–H absorption of CH$_3$CH$_2$OH in exhaled air (compare Figure 12.10 for a description of an earlier method).

Since its introduction, IR spectroscopy has proven to be a valuable tool for determining the functional groups in organic molecules.

In the 1940s, IR spectroscopy played a key role in elucidating the structure of the antibiotic penicillin G. **β-Lactams,** four-membered rings that contain an amide unit, have a carbonyl group that absorbs at a much higher frequency (~1760 cm^{-1}) than the frequency observed for most other carbonyl groups. Because penicillin G exhibited an IR absorption at this frequency, **A** became the leading candidate for the structure of penicillin rather than **B**, a possibility originally considered more likely. Structure **A** was later confirmed by X-ray analysis.

Correct structure

β-lactam

A
penicillin G

β-lactam

Incorrect structure, ruled out using IR spectroscopy

B

IR spectroscopy is often used to determine the outcome of a chemical reaction. For example, oxidation of the hydroxy group in **C** to form the carbonyl group in periplanone B (Section 9.5C), is accompanied by the disappearance of the OH absorption (3600–3200 cm^{-1}) and the appearance of a carbonyl absorption near 1700 cm^{-1} in the IR spectrum of the product.

The absorption at 3600–3200 cm^{-1} disappears.

The absorption at ~1700 cm^{-1} appears.

C   Cr^{6+} oxidant →

periplanone B
sex pheromone of the
female American cockroach

The combination of IR and mass spectral data provides key information on the structure of an unknown compound. The mass spectrum reveals the molecular weight of the unknown (and the molecular formula if an exact mass is available), and the IR spectrum helps to identify the important functional groups.

How To   Use MS and IR for Structure Determination

Example   **What information is obtained from the mass spectrum and IR spectrum of an unknown compound X? Assume X contains the elements C, H, and O.**

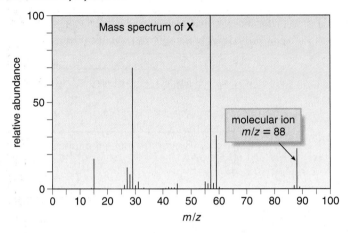

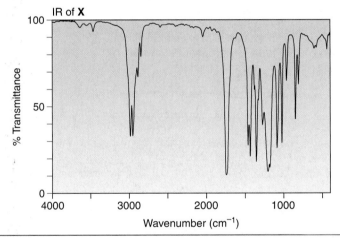

Step [1]   **Use the molecular ion to determine possible molecular formulas. Use an exact mass (when available) to determine a molecular formula.**

- Use the procedure outlined in Sample Problem 14.2 to calculate possible molecular formulas. For a molecular ion at $m/z = 88$:

$$\frac{88}{12} = 7 \text{ C's} \xrightarrow{} C_7H_4 \xrightarrow[+1\ O]{-CH_4} C_6O \xrightarrow[+12\ H's]{-1\ C} \boxed{C_5H_{12}O} \xrightarrow[+1\ O]{-CH_4} \boxed{C_4H_8O_2} \xrightarrow[+1\ O]{-CH_4} \boxed{C_3H_4O_3}$$

maximum (remainder = 4)

three possible formulas

- Discounting C_7H_4 (a hydrocarbon) and C_6O (because it contains no H's) gives three possible formulas for **X**.
- If high-resolution mass spectral data are available, the molecular formula can be determined directly. If the molecular ion had an exact mass of 88.0580, the molecular formula of **X** is **$C_4H_8O_2$** (exact mass = 88.0524) rather than $C_5H_{12}O$ (exact mass = 88.0888) or C_3H_4O (exact mass = 88.0160).

Step [2]   **Calculate the number of degrees of unsaturation (Section 10.2).**

- For a compound of molecular formula $C_4H_8O_2$, the maximum number of H's = $2n + 2 = 2(4) + 2 = 10$.
- Because the compound contains only 8 H's, it has $10 - 8 = 2$ H's fewer than the maximum number.
- Because each degree of unsaturation removes 2 H's, **X** has one degree of unsaturation. **X has one ring or one π bond.**

Step [3]   **Determine what functional group is present from the IR spectrum.**

- The two major absorptions in the IR spectrum above 1500 cm^{-1} are due to sp^3 hybridized C—H bonds (~3000–2850 cm^{-1}) and a C=O group (1740 cm^{-1}). Thus, the one degree of unsaturation in **X** is due to the presence of the **C=O.**

Mass spectrometry and IR spectroscopy give valuable but limited information on the identity of an unknown. Although the mass spectral and IR data reveal that **X** has a molecular formula of $C_4H_8O_2$ and contains a carbonyl group, more data are needed to determine its complete structure. In Chapter 15, we will learn how other spectroscopic data can be used for that purpose.

PROBLEM 14.16   Which of the following possible structures for **X** can be excluded on the basis of its IR spectrum: (a) $CH_3COOCH_2CH_3$; (b) $HOCH_2CH_2CH_2CHO$; (c) $CH_3CH_2COOCH_3$; (d) $CH_3CH_2CH_2COOH$?

PROBLEM 14.17   Propose structures consistent with each set of data: (a) a hydrocarbon with a molecular ion at $m/z = 68$ and IR absorptions at 3310, 3000–2850, and 2120 cm^{-1}; (b) a compound containing C, H, and O with a molecular ion at $m/z = 60$ and IR absorptions at 3600–3200 and 3000–2850 cm^{-1}.

14.8   Key Concepts—Mass Spectrometry and Infrared Spectroscopy

Mass Spectrometry (MS; 14.1–14.3)

- Mass spectrometry measures the molecular weight of a compound (14.1A).
- The mass of the molecular ion (**M**) = the molecular weight of a compound. Except for isotope peaks at M + 1 and M + 2, the molecular ion has the highest mass in a mass spectrum (14.1A).
- The base peak is the tallest peak in a mass spectrum (14.1A).
- A compound with an odd number of N atoms gives an odd molecular ion. A compound with an even number of N atoms (including zero) gives an even molecular ion (14.1B).
- Organic monochlorides show two peaks for the molecular ion (M and M + 2) in a 3:1 ratio (14.2).
- Organic monobromides show two peaks for the molecular ion (M and M + 2) in a 1:1 ratio (14.2).
- High-resolution mass spectrometry gives the molecular formula of a compound (14.3A).

Electromagnetic Radiation (14.4)

- The wavelength and frequency of electromagnetic radiation are *inversely* related by the following equations: $\lambda = c/v$ or $v = c/\lambda$ (14.4).
- The energy of a photon is proportional to its frequency; the higher the frequency the higher the energy: $E = hv$ (14.4).

Infrared Spectroscopy (IR; 14.5 and 14.6)

- Infrared spectroscopy identifies functional groups.
- IR absorptions are reported in wavenumbers, $\tilde{v} = 1/\lambda$.
- The functional group region from **4000–1500 cm^{-1}** is the most useful region of an IR spectrum.
- C—H, O—H, and N—H bonds absorb at high frequency, ≥ 2500 cm^{-1}.
- As bond strength increases, the $\tilde{v}$ of absorption increases; thus, triple bonds absorb at higher $\tilde{v}$ than double bonds.

C=C   C≡C
~ 1650 cm^{-1} ~ 2250 cm^{-1}

Increasing bond strength
Increasing $\tilde{v}$

- The higher the percent *s*-character, the stronger the bond, and the higher the $\tilde{v}$ of an IR absorption.

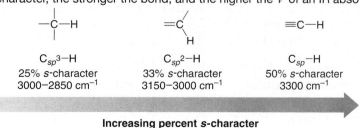

C_{sp^3}—H C_{sp^2}—H C_{sp}—H
25% *s*-character   33% *s*-character   50% *s*-character
3000–2850 cm^{-1} 3150–3000 cm^{-1} 3300 cm^{-1}

Increasing percent *s*-character
Increasing $\tilde{v}$

Problems

Mass Spectrometry

14.18 What molecular ion is expected for each compound?

a.  b.  c. 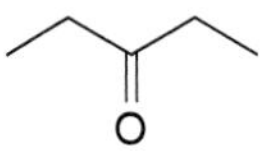 d. 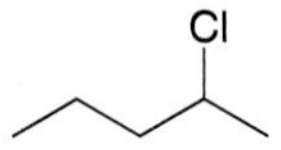 e. $(CH_3)_3CCH(Br)CH(CH_3)_2$

14.19 Which compound gives a molecular ion at $m/z = 122$?

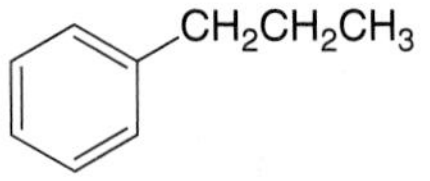

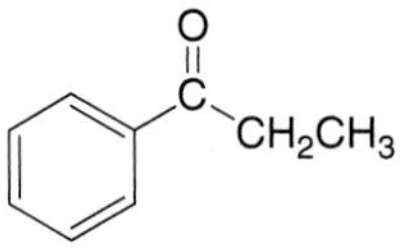

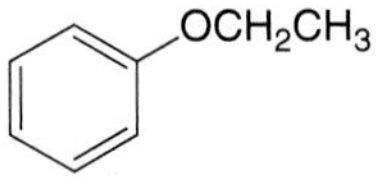

$CH_2CH_2CH_3$ CH_2CH_3 OCH_2CH_3

14.20 Propose two molecular formulas for each molecular ion: (a) 102; (b) 98; (c) 119; (d) 74.

14.21 Propose four possible structures for a hydrocarbon with a molecular ion at $m/z = 112$.

14.22 Match each structure to its mass spectrum.

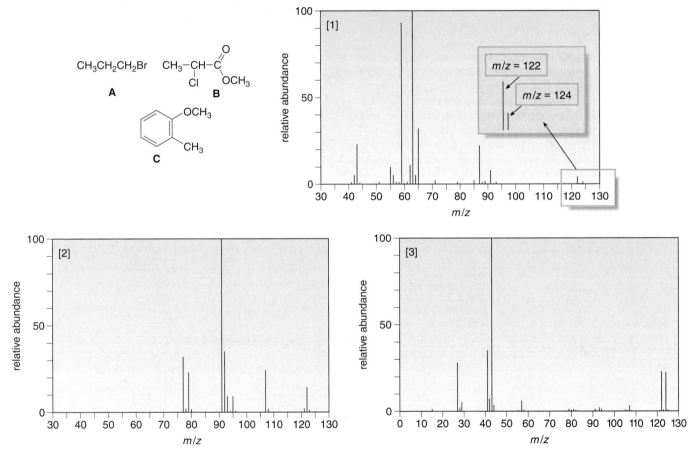

$CH_3CH_2CH_2Br$ CH_3-CH-C
                        Cl OCH_3
A   **B**

OCH_3
CH_3
C

14.23 Propose two possible structures for a hydrocarbon having an exact mass of 96.0939 that forms ethylcyclopentane upon hydrogenation with H_2 and Pd-C.

14.24 Propose a structure consistent with each set of data.
   a. A compound that contains a benzene ring and has a molecular ion at $m/z = 107$
   b. A hydrocarbon that contains only sp^3 hybridized carbons and a molecular ion at $m/z = 84$
   c. A compound that contains a carbonyl group and gives a molecular ion at $m/z = 114$
   d. A compound that contains C, H, N, and O and has an exact mass for the molecular ion at 101.0841

14.25 A low-resolution mass spectrum of the neurotransmitter dopamine gave a molecular ion at $m/z = 153$. Two possible molecular formulas for this molecular ion are $C_8H_{11}NO_2$ and $C_7H_{11}N_3O$. A high-resolution mass spectrum provided an exact mass at 153.0680. Which of the possible molecular formulas is the correct one?

14.26 Explain why compounds containing an odd number of nitrogen atoms have an odd molecular ion in their mass spectra.

14.27 Can the exact mass obtained in a high-resolution mass spectrum distinguish between two isomers such as $CH_2=CHCH_2CH_2CH_2CH_3$ and $(CH_3)_2C=CHCH_2CH_3$?

Infrared Spectroscopy

14.28 Which of the indicated bonds absorbs at higher $\tilde{v}$ in an IR spectrum?

a. $(CH_3)_2C=O$ or $(CH_3)_2CH-OH$ b. $(CH_3)_2C=NCH_3$ or $(CH_3)_2CH-NHCH_3$ c. (ring with =C–H) or (ring with C–H)

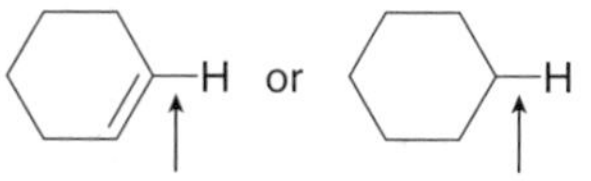

14.29 What major IR absorptions are present above 1500 cm^{-1} for each compound?

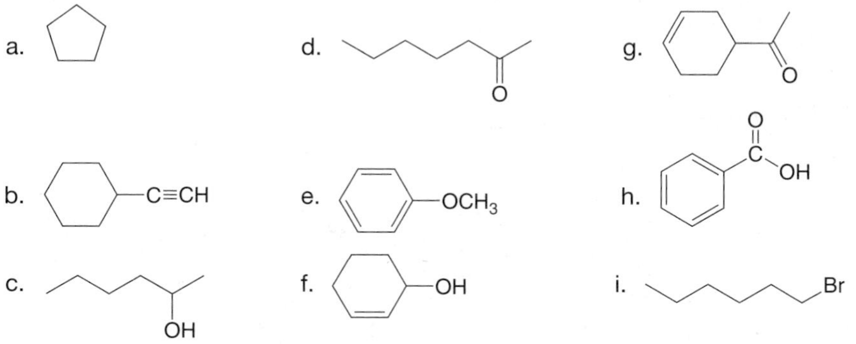

a. (cyclopentane)

b. (cyclohexyl–C≡CH)

c. (CH$_3$CH$_2$CH$_2$CH(OH)CH$_3$ structure with OH)

d. (CH$_3$CH$_2$CH$_2$CH$_2$CH$_2$C(=O)CH$_3$ ketone)

e. (benzene–OCH$_3$)

f. (cyclohexene–OH)

g. (cyclohexene with acetyl, C=O)

h. (benzene–C(=O)OH)

i. (CH$_3$CH$_2$CH$_2$CH$_2$CH$_2$CH$_2$Br)

14.30 How would each of the following pairs of compounds differ in their IR spectra?

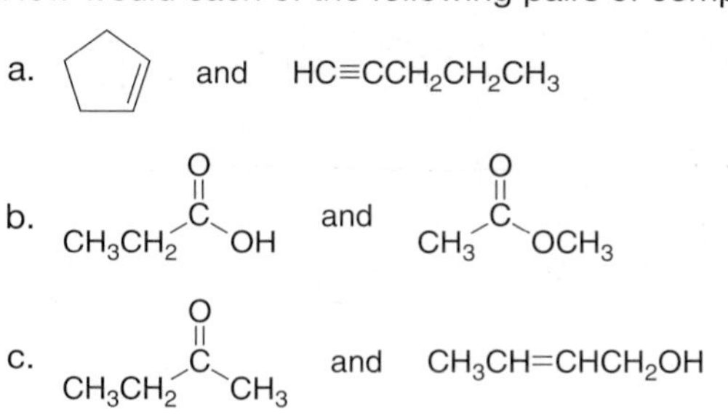

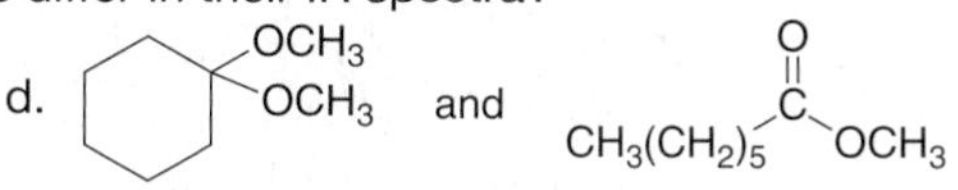

a. (cyclopentene) and $HC\equiv CCH_2CH_2CH_3$

b. $CH_3CH_2C(=O)OH$ and $CH_3C(=O)OCH_3$

c. $CH_3CH_2C(=O)CH_3$ and $CH_3CH=CHCH_2OH$

d. (cyclohexane with two OCH$_3$) and $CH_3(CH_2)_5C(=O)OCH_3$

e. $CH_3C\equiv CCH_3$ and $CH_3CH_2C\equiv CH$

f. $HC\equiv CCH_2N(CH_2CH_3)_2$ and $CH_3(CH_2)_5C\equiv N$

14.31 Tell how IR spectroscopy could be used to determine when each reaction is complete.

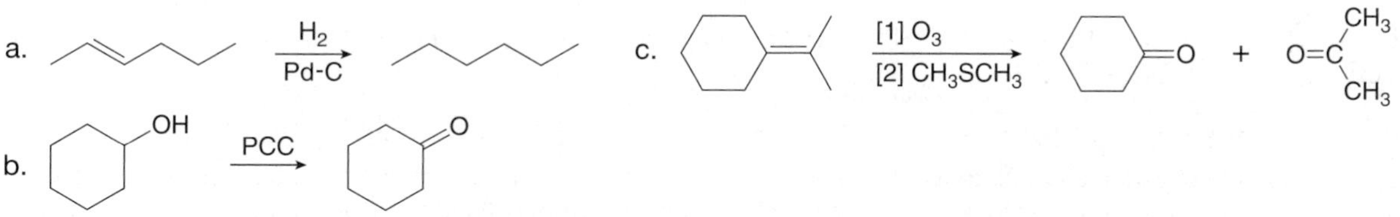

a. (CH$_3$CH$_2$CH=CHCH$_2$CH$_3$) $\xrightarrow[\text{Pd-C}]{\text{H}_2}$ (CH$_3$CH$_2$CH$_2$CH$_2$CH$_2$CH$_3$)

b. (cyclohexanol, OH) $\xrightarrow{\text{PCC}}$ (cyclohexanone, =O)

c. (cyclohexane=C(CH$_3$)$_2$) $\xrightarrow[\text{[2] CH}_3\text{SCH}_3]{\text{[1] O}_3}$ (cyclohexanone =O) + $O=C(CH_3)CH_3$

14.32 Match each compound to its IR spectrum.

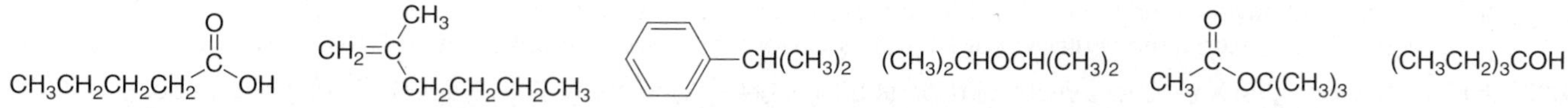

$CH_3CH_2CH_2CH_2C(=O)OH$ $CH_2=C(CH_3)CH_2CH_2CH_2CH_3$ (benzene–CH(CH$_3$)$_2$) $(CH_3)_2CHOCH(CH_3)_2$ $CH_3C(=O)OC(CH_3)_3$ $(CH_3CH_2)_3COH$

A   **B**   **C**   **D**   **E**   **F**

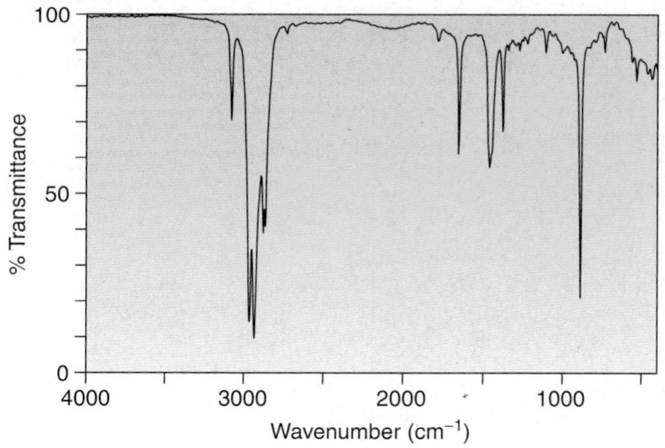

Spectrum [1]   Spectrum [2]

% Transmittance vs Wavenumber (cm^{-1})

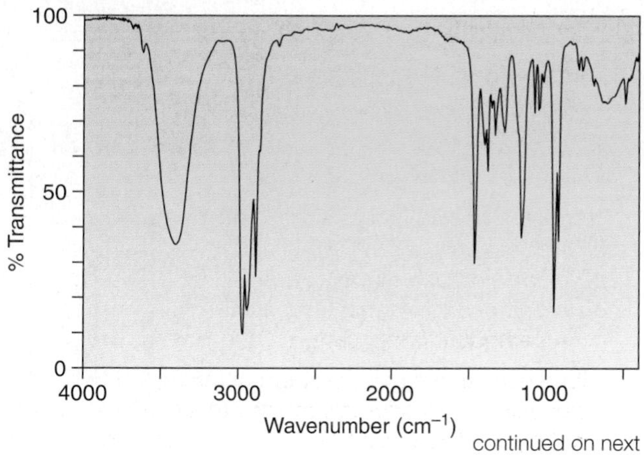

continued on next page

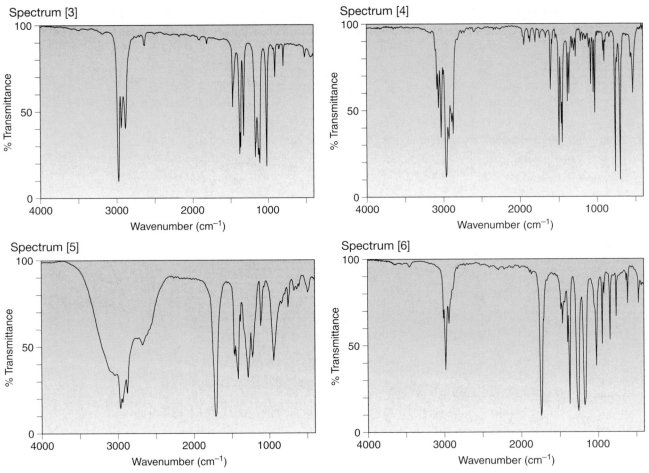

Spectrum [3]

Spectrum [4]

Spectrum [5]

Spectrum [6]

14.33 Draw the structures of the seven compounds of molecular formula C_3H_6O. For each compound tell what prominent IR absorptions it exhibits above 1500 cm^{-1}.

14.34 Rank the carbon–halogen bonds in the following compounds in order of increasing frequency of absorption in an IR spectrum: CH_3-Cl, CH_3-Br, CH_3-I.

Combined Spectroscopy Problems

14.35 Propose possible structures consistent with each set of data. Assume each compound has an sp^3 hybridized $C-H$ absorption in its IR spectrum, and that other major IR absorptions above 1500 cm^{-1} are listed.

   a. A compound having a molecular ion at 72 and an absorption in its IR spectrum at 1725 cm^{-1}.

   b. A compound having a molecular ion at 55 and an absorption in its IR spectrum at ~2250 cm^{-1}.

   c. A compound having a molecular ion of 74 and an absorption in its IR spectrum at 3600–3200 cm^{-1}.

14.36 A chiral hydrocarbon **X** exhibits a molecular ion at 82 in its mass spectrum. The IR spectrum of **X** shows peaks at 3300, 3000–2850, and 2250 cm^{-1}. Propose a structure for **X.**

14.37 A chiral compound **Y** has a strong absorption at 2970–2840 cm^{-1} in its IR spectrum and gives the following mass spectrum. Propose a structure for **Y.**

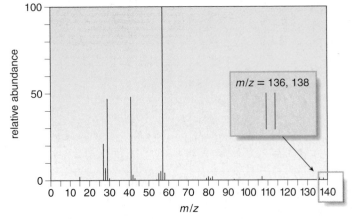

$m/z = 136, 138$

14.38 Reaction of *tert*-butyl pentyl ether [$CH_3CH_2CH_2CH_2CH_2OC(CH_3)_3$] with HBr forms 1-bromopentane ($CH_3CH_2CH_2CH_2CH_2Br$) and compound **Z**. **Z** has a molecular ion in its mass spectrum at 56 and gives peaks in its IR spectrum at 3150–3000, 3000–2850, and 1650 cm^{-1}. Propose a structure for **Z** and draw a stepwise mechanism that accounts for its formation.

14.39 Reaction of $BrCH_2CH_2CH_2CH_2NH_2$ with NaH forms compound **W,** which gives the IR and mass spectra shown below. Propose a structure for **W** and draw a stepwise mechanism that accounts for its formation.

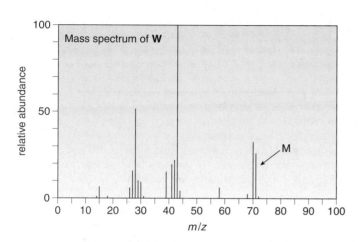

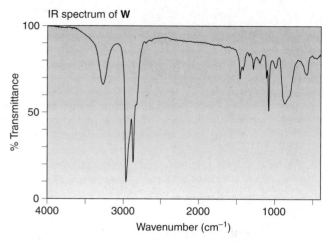

Challenge Problems

14.40 Explain why a carbonyl absorption shifts to lower frequency in an α,β-unsaturated carbonyl compound—a compound having a carbonyl group bonded directly to a carbon–carbon double bond. For example, the carbonyl absorption occurs at 1720 cm^{-1} for cyclohexanone, and at 1685 cm^{-1} for 2-cyclohexenone.

cyclohexanone

2-cyclohexenone
an α,β-unsaturated
carbonyl compound

14.41 What isotope pattern would be observed for the molecular ion of 4,4'-dichlorobiphenyl, a polychlorinated biphenyl (PCB) that is a widespread environmental pollutant (Section 3.5)?

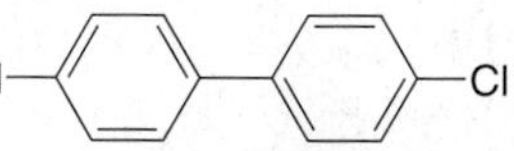

4,4'-dichlorobiphenyl
(a common PCB)

14.42 Oxidation of citronellol, a constituent of rose and geranium oils, with PCC in the presence of added $NaOCOCH_3$ forms compound **A**. **A** has a molecular ion in its mass spectrum at 154 and a strong peak in its IR spectrum at 1730 cm^{-1}, in addition to C–H stretching absorptions. Without added $NaOCOCH_3$, oxidation of citronellol with PCC yields isopulegone, which is then converted to **B** with aqueous base. **B** has a molecular ion at 152, and a peak in its IR spectrum at 1680 cm^{-1} in addition to C–H stretching absorptions.

$$A \xleftarrow[\text{NaOCOCH}_3]{\text{PCC}} \text{citronellol} \xrightarrow{\text{PCC}} \text{isopulegone} \xrightarrow[\text{H}_2\text{O}]{^-\text{OH}} B$$

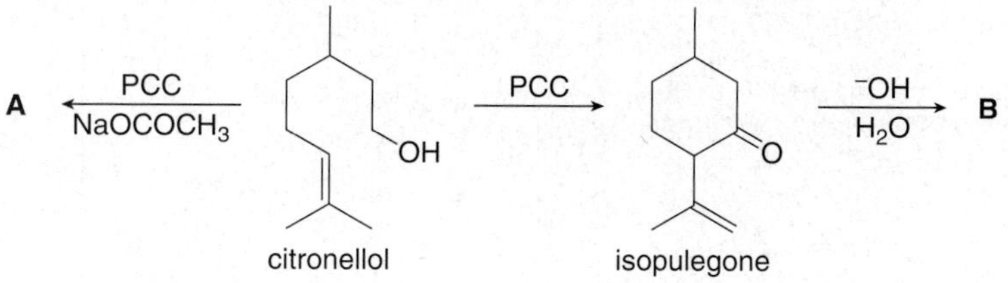

a. Identify the structures of **A** and **B**.
b. Draw a mechanism for the conversion of citronellol to isopulegone.
c. Draw a mechanism for the conversion of isopulegone to **B**.

Nuclear Magnetic Resonance Spectroscopy

Melatonin, a hormone synthesized by the pineal gland, is thought to induce sleep. Melatonin levels in the body rise as less light falls upon the eye, and drop quickly at dawn. For this reason, melatonin has become a popular dietary supplement for travelers suffering from jetlag and individuals with mild sleep disorders. Modern spectroscopic techniques have been used to characterize the structure of melatonin. In Chapter 15, we learn how nuclear magnetic resonance spectroscopy plays a key role in organic structure determination.

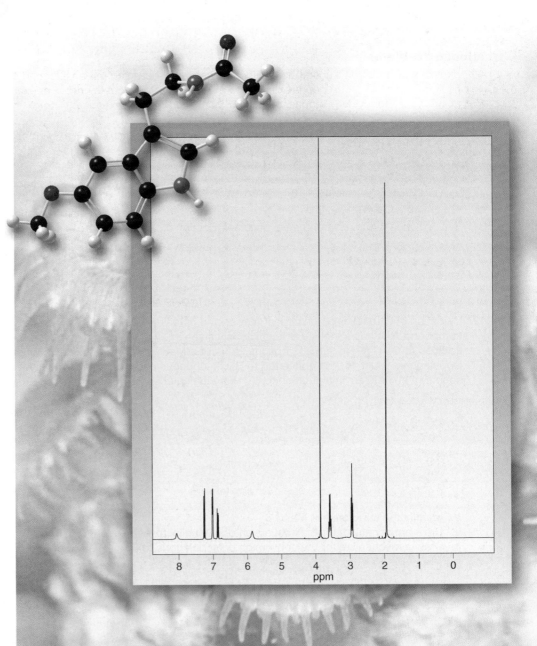

In Chapter 15 we continue our study of organic structure determination by learning about **nuclear magnetic resonance (NMR)** spectroscopy. NMR spectroscopy is the most powerful tool for characterizing organic molecules, because it can be used to identify the carbon–hydrogen framework in a compound.

15.1   An Introduction to NMR Spectroscopy

Two common types of NMR spectroscopy are used to characterize organic structure:

◆ **^{1}H NMR** (proton NMR) is used to determine the number and type of hydrogen atoms in the molecule; and

◆ **^{13}C NMR** (carbon NMR) is used to determine the type of carbon atoms in the molecule.

Before you can learn how to use NMR spectroscopy to determine the structure of a compound, you need to understand a bit about the physics behind it. Keep in mind, though, that NMR stems from the same basic principle as all other forms of spectroscopy. Energy interacts with a molecule, and absorptions occur only when the incident energy matches the energy difference between two states.

15.1A   The Basis of NMR Spectroscopy

The source of energy in NMR is radio waves. Radiation in the radiofrequency region of the electromagnetic spectrum (so-called **RF** radiation) has very long wavelengths, so its corresponding frequency and energy are both low. **When these low-energy radio waves interact with a molecule, they can change the nuclear spins of some elements, including ^{1}H and ^{13}C.**

When a charged particle such as a proton spins on its axis, it creates a magnetic field. For the purpose of this discussion, therefore, a nucleus is a tiny bar magnet, symbolized by ⚡. Normally these nuclear magnets are randomly oriented in space, but in the presence of an external magnetic field, $B_\circ$, they are oriented with or against this applied field. More nuclei are oriented *with* the applied field because this arrangement is lower in energy, but the **energy difference between these two states is very small** (< 0.1 cal).

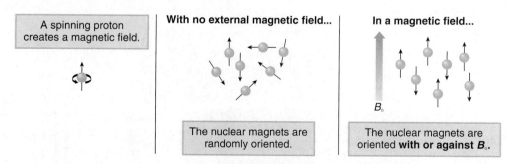

In a magnetic field, there are now two different energy states for a proton:

◆ A *lower energy state* with the nucleus aligned in the *same direction* as $B_\circ$.

◆ A *higher energy state* with the nucleus aligned *opposed* to $B_\circ$.

When an external energy source ($h\nu$) that matches the energy difference (ΔE) between these two states is applied, energy is absorbed, causing the **nucleus to "spin flip" from one orientation to another.** The energy difference between these two nuclear spin states corresponds to the low-frequency radiation in the RF region of the electromagnetic spectrum.

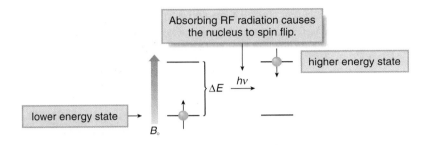

◆ A nucleus is in *resonance* when it absorbs RF radiation and "spin flips" to a higher energy state.

Thus, two variables characterize NMR:

An older unit used for magnetic field strength is gauss (G). 1 T = 10^4 G

◆ **An applied magnetic field, $B_\circ$.** Magnetic field strength is measured in tesla (T).
◆ **The frequency ν of radiation used for resonance,** measured in hertz (Hz) or megahertz (MHz); (1 MHz = 10^6 Hz)

The frequency needed for resonance and the applied magnetic field strength are proportionally related:

◆ The stronger the magnetic field, the larger the energy difference between the two nuclear spin states, and the higher the ν needed for resonance.

NMR spectrometers are referred to as 300 MHz instruments, 500 MHz instruments, and so forth, depending on the frequency of RF radiation used for resonance.

Early NMR spectrometers used a magnetic field strength of ~1.4 T, which required RF radiation of 60 MHz for resonance. Modern NMR spectrometers use stronger magnets, thus requiring higher frequencies of RF radiation for resonance. For example, a magnetic field strength of 7.05 T requires a frequency of 300 MHz. These spectrometers use very powerful magnetic fields to create a small, but measurable energy difference between the two possible spin states. A schematic of an NMR spectrometer is shown in Figure 15.1.

If all protons absorbed at the same frequency in a given magnetic field, the spectra of all compounds would consist of a single absorption, rendering NMR useless for structure determination. Fortunately, however, this is not the case.

Figure 15.1 Schematic of an NMR spectrometer

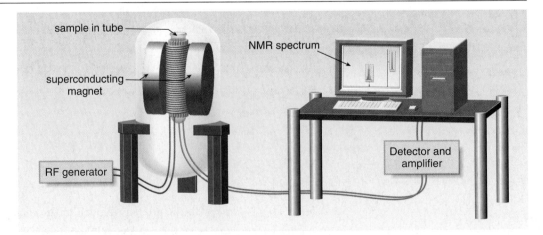

An NMR spectrometer. The sample is dissolved in a solvent, usually $CDCl_3$ (deuterochloroform), and placed in a magnetic field. A radiofrequency generator then irradiates the sample with a short pulse of radiation, causing resonance. When the nuclei fall back to their lower energy state, the detector measures the energy released, and a spectrum is recorded. The superconducting magnets in modern NMR spectrometers have coils that are cooled in liquid helium and conduct electricity with essentially no resistance.

♦ All protons do *not* absorb at the same frequency. Protons in different environments absorb at slightly different frequencies, and so they are distinguishable by NMR.

The frequency at which a particular proton absorbs is determined by its electronic environment, as discussed in Section 15.3. Because electrons are moving charged particles, they create a magnetic field opposed to the applied field $B_\circ$, and the size of the magnetic field generated by the electrons around a proton determines where it absorbs. Modern NMR spectrometers use a constant magnetic field strength $B_\circ$, and then a narrow range of frequencies is applied to achieve the resonance of all protons.

Only nuclei that contain odd mass numbers (such as ^1H, ^{13}C, ^{19}F, and ^{31}P) or odd atomic numbers (such as ^2H and ^{14}N) give rise to NMR signals. Because both ^1H and ^{13}C, the less abundant isotope of carbon, are NMR active, NMR allows us to map the carbon and hydrogen framework of an organic molecule.

tert-Butyl methyl ether (MTBE) is the high-octane gasoline additive that has contaminated the water supply in some areas (Section 3.5).

15.1B A ^{1}H NMR Spectrum

An NMR spectrum plots the **intensity of a peak** against its **chemical shift** measured in **parts per million (ppm)**. The common scale of chemical shifts is called the **δ (delta) scale**. The proton NMR spectrum of *tert*-butyl methyl ether [$CH_3OC(CH_3)_3$] illustrates several important features:

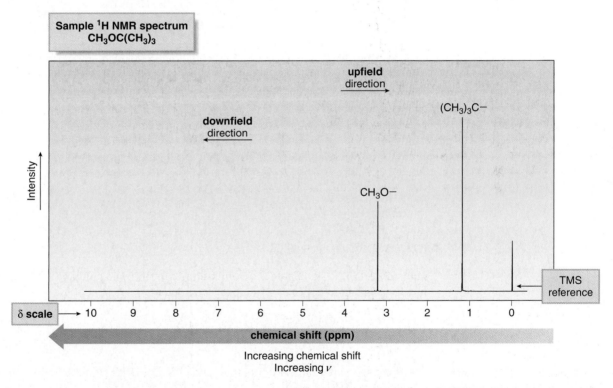

♦ NMR absorptions generally appear as sharp peaks. The ^{1}H NMR spectrum of $CH_3OC(CH_3)_3$ consists of two peaks: a tall peak at 1.2 ppm due to the $(CH_3)_3C-$ group, and a smaller peak at 3.2 ppm due to the CH_3O- group.

♦ **Increasing chemical shift is plotted from *right to left*.** Most protons absorb somewhere from 0–10 ppm.

♦ The terms **upfield** and **downfield** describe the relative location of peaks. Upfield means to the *right*. The $(CH_3)_3C-$ peak is upfield from the CH_3O- peak. Downfield means to the *left*. The CH_3O- peak is downfield from the $(CH_3)_3C-$ peak.

$(CH_3)_4Si$
tetramethylsilane
TMS

NMR absorptions are measured relative to the position of a reference peak at 0 ppm on the δ scale due to **tetramethylsilane (TMS)**. **TMS** is a volatile and inert compound that gives a single peak upfield from other typical NMR absorptions.

Although chemical shifts are measured relative to the TMS peak at 0 ppm, this reference is often not plotted on a spectrum.

The **chemical shift** on the x axis gives the position of an NMR signal, measured in ppm, according to the following equation:

$$\begin{array}{c} \text{chemical shift} \\ \text{(in ppm on the } \delta \text{ scale)} \end{array} = \frac{\text{observed chemical shift (in Hz) downfield from TMS}}{\nu \text{ of the NMR spectrometer (in MHz)}}$$

The *positive* direction of the δ scale is *downfield* from TMS. A very small number of absorptions occur upfield from the TMS peak, which is defined as the negative direction of the δ scale. (See Problem 15.59.)

A chemical shift gives absorptions as a fraction of the NMR operating frequency, making it independent of the spectrometer used to record a spectrum. Because the frequency of the radiation required for resonance is proportional to the strength of the applied magnetic field, $B_\circ$, reporting NMR absorptions in frequency is meaningless unless the value of $B_\circ$ is also reported. By reporting the absorption as a fraction of the NMR operating frequency, though, we get units—ppm—that are independent of the spectrometer.

SAMPLE PROBLEM 15.1   Calculate the chemical shift of an absorption that occurs at 1500 Hz downfield from TMS using a 300 MHz NMR spectrometer.

SOLUTION

Use the equation that defines the chemical shift in ppm:

$$\text{chemical shift} = \frac{1500 \text{ Hz downfield from TMS}}{300 \text{ MHz operating frequency}} = \boxed{5 \text{ ppm}}$$

PROBLEM 15.1   Calculate the chemical shift in ppm for each absorption: (a) an absorption 60 Hz downfield from TMS using a 60 MHz NMR spectrometer; (b) an absorption 1600 Hz downfield from TMS using a 200 MHz NMR spectrometer.

Four different features of a ^{1}H NMR spectrum provide information about a compound's structure:

[1] **Number of signals** (Section 15.2)
[2] **Position of signals** (Sections 15.3 and 15.4)
[3] **Intensity of signals** (Section 15.5)
[4] **Spin–spin splitting of signals** (Sections 15.6–15.8)

15.2 ^{1}H NMR: Number of Signals

How many NMR signals does a compound exhibit? The number of NMR signals *equals* the number of different types of protons in a compound.

15.2A   General Principles

Any CH$_3$ group is different from any CH$_2$ group, which is different from any CH group in a molecule. Two CH$_3$ groups may be identical (as in CH$_3$OCH$_3$) or different (as in CH$_3$OCH$_2$CH$_3$), depending on what each CH$_3$ group is bonded to.

♦ Protons in different environments give different NMR signals. Equivalent protons give the same NMR signal.

In many compounds, deciding whether two protons are in identical or different environments is intuitive.

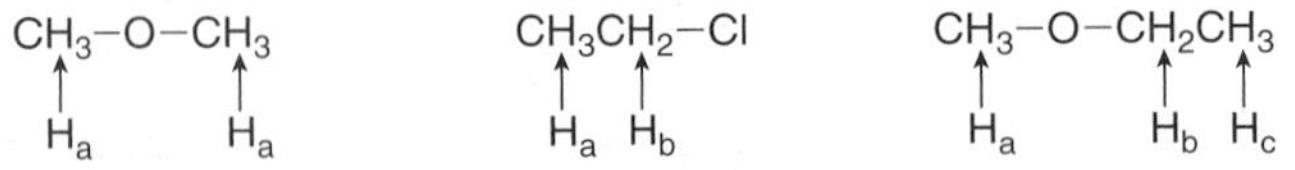

$$CH_3-O-CH_3 \qquad CH_3CH_2-Cl \qquad CH_3-O-CH_2CH_3$$
$$\uparrow \quad\quad \uparrow \qquad\qquad \uparrow \quad \uparrow \qquad\qquad \uparrow \quad\quad \uparrow \quad \uparrow$$
$$H_a \quad\quad H_a \qquad\qquad H_a \quad H_b \qquad\quad H_a \quad\quad H_b \quad H_c$$

| All equivalent H's | 2 types of H's | 3 types of H's |
| 1 NMR signal | 2 NMR signals | 3 NMR signals |

♦ **CH$_3$OCH$_3$:** Each CH$_3$ group is bonded to the same group ($-$OCH$_3$), making both CH$_3$ groups equivalent.

tert-Butyl methyl ether [CH₃OC(CH₃)₃] (Section 15.1) exhibits two NMR signals because it contains two different kinds of protons: one CH₃ group is bonded to $-OC(CH_3)_3$, whereas the other three CH₃ groups are each bonded to the same group, $[-C(CH_3)_2]OCH_3$.

◆ **CH₃CH₂Cl:** The protons of the CH₃ group are different from those of the CH₂ group.

◆ **CH₃OCH₂CH₃:** The protons of the CH₂ group are different from those in each CH₃ group. The two CH₃ groups are also different from each other; one CH₃ group is bonded to $-OCH_2CH_3$ and the other is bonded to $-CH_2OCH_3$.

In some cases, it is less obvious by inspection if two protons are equivalent or different. To rigorously determine whether two protons are in identical environments (and therefore give rise to one NMR signal), replace each H atom in question by another atom Z (for example, Z = Cl). If substitution by Z yields the same compound or enantiomers, the two protons are equivalent, as shown in Sample Problem 15.2.

SAMPLE PROBLEM 15.2   How many different kinds of H atoms does CH₃CH₂CH₂CH₂CH₃ contain?

SOLUTION

In comparing two H atoms, replace each H by Z (for example, Z = Cl), and examine the substitution products that result. The two CH₃ groups are identical because substitution of one H by Cl gives CH₃CH₂CH₂CH₂CH₂Cl (1-chloropentane). There are two different types of CH₂ groups, because substitution of one H by Cl gives two different products:

CH₃CH₂CH₂CH₂CH₃   CH₃CHCH₂CH₂CH₃   CH₃CH₂CHCH₂CH₃
                         | |
                         Cl   Cl

[different H's]   2-chloropentane   3-chloropentane

[different products]

Hᵦ   Hᵦ
↓   ↓
CH₃CH₂CH₂CH₂CH₃
↑   ↑   ↑
Hₐ   Hᵪ   Hₐ

Thus, CH₃CH₂CH₂CH₂CH₃ has **three** different types of protons and gives **three** NMR signals.

Figure 15.2 gives the number of NMR signals exhibited by four additional molecules. All protons—not just protons bonded to carbon atoms—give rise to NMR signals. Ethanol (CH₃CH₂OH), for example, gives three NMR signals, one of which is due to its OH proton.

PROBLEM 15.2   How many NMR signals does each of the following compounds show?
a. CH₃CH₃   c. CH₃CH₂CH₂CH₃   e. CH₃CH₂CO₂CH₂CH₃   g. CH₃CH₂OCH₂CH₃
b. CH₃CH₂CH₃   d. (CH₃)₂CHCH(CH₃)₂   f. CH₃OCH₂CH(CH₃)₂   h. CH₃CH₂CH₂OH

PROBLEM 15.3   How many different types of protons does CH₃CH₂CH₂CH₂CH₂CH₂CH₂CH₂Cl contain?

15.2B Determining Equivalent Protons in Alkenes and Cycloalkanes

To determine equivalent protons in cycloalkanes and alkenes that have restricted bond rotation, always draw in all bonds to hydrogen.

Draw   H   H   Cl   **NOT**   ▷—Cl   **Draw**   Cl   H   **NOT** ClCH=CH₂
            H   H   C=C
              H   H   H

Figure 15.2 The number of NMR signals of some representative organic compounds

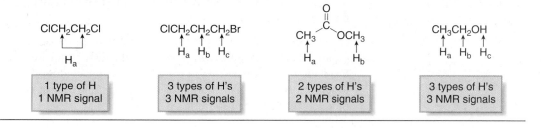

ClCH₂CH₂Cl
   Hₐ

[1 type of H]
[1 NMR signal]

ClCH₂CH₂CH₂Br
   Hₐ Hᵦ Hᵪ

[3 types of H's]
[3 NMR signals]

    O
    ‖
CH₃ C OCH₃
    Hₐ   Hᵦ

[2 types of H's]
[2 NMR signals]

CH₃CH₂OH
Hₐ Hᵦ Hᵪ

[3 types of H's]
[3 NMR signals]

Then, in comparing two H atoms on a ring or double bond, **two protons are equivalent only if they are cis (or trans) to the same groups,** as illustrated with 1,1-dichloroethylene, 1-bromo-1-chloroethylene, and chloroethylene.

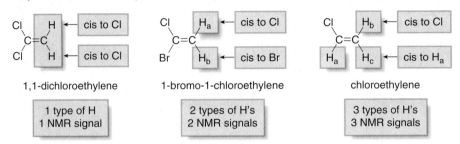

◆ **1,1-Dichloroethylene:** The two H atoms on the C=C are both cis to a Cl atom. Thus, all H atoms are equivalent.

◆ **1-Bromo-1-chloroethylene:** H_a is cis to a Cl atom and H_b is cis to a Br atom. Thus, H_a and H_b are different, giving rise to two NMR signals.

◆ **Chloroethylene:** H_a is bonded to the carbon with the Cl atom, making it different from H_b and H_c. Of the remaining two H atoms, H_b is cis to a Cl atom and H_c is cis to a H atom, making them different. All three H atoms in this compound are different.

Proton equivalency in cycloalkanes can be determined similarly.

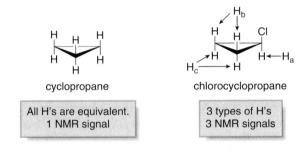

◆ **Cyclopropane:** All H atoms are equivalent so there is only one NMR signal.

◆ **Chlorocyclopropane:** There are now three kinds of H atoms: H_a is bonded to a carbon bonded to a Cl; both H_b protons are cis to the Cl whereas both H_c protons are cis to another H.

PROBLEM 15.4   Draw the three isomeric alkenes of molecular formula $C_2H_2Br_2$ and indicate how many different types of protons each has.

PROBLEM 15.5   How many NMR signals does each dimethylcyclopropane show?

a. [structure with CH₃ and CH₃]   b. [structure with CH₃ and CH₃]   c. [structure with CH₃ and CH₃]

15.2C   Enantiotopic and Diastereotopic Protons

Let's look more closely at the protons of a single sp^3 hybridized CH_2 group to determine whether these two protons are always equivalent to *each other*. Two examples illustrate different outcomes.

CH_3CH_2Br has two different types of protons—those of the CH_3 group and those of the CH_2 group—meaning that the two H atoms of the CH_2 group are equivalent to each other. To confirm this fact, we replace each H of the CH_2 group by an atom Z and examine the products of substitution. In this case, substitution of each H by Z creates a new stereogenic center, forming two products that are **enantiomers.**

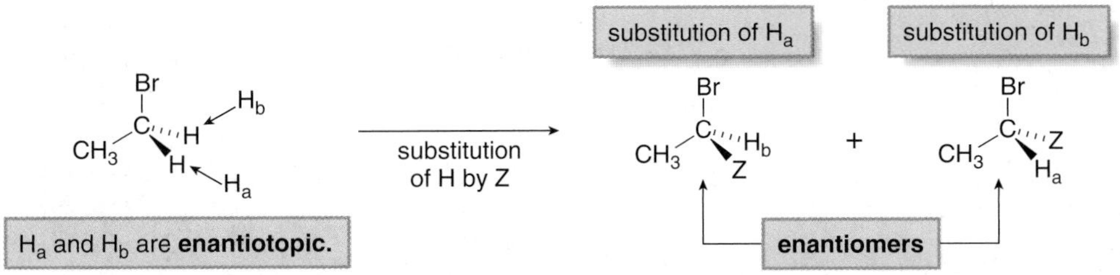

substitution of H$_a$ substitution of H$_b$

H$_a$ and H$_b$ are **enantiotopic**.

enantiomers

◆ **When substitution of two H atoms by Z forms enantiomers, the two H atoms are equivalent and give a single NMR signal. These two H atoms are called *enantiotopic* protons.**

In contrast, the two H atoms of the CH$_2$ group in (*R*)-2-chlorobutane, which contains one stereogenic center, are *not* equivalent to each other. Substitution of each H by Z forms two **diastereomers,** and thus, these two H atoms give *different* NMR signals.

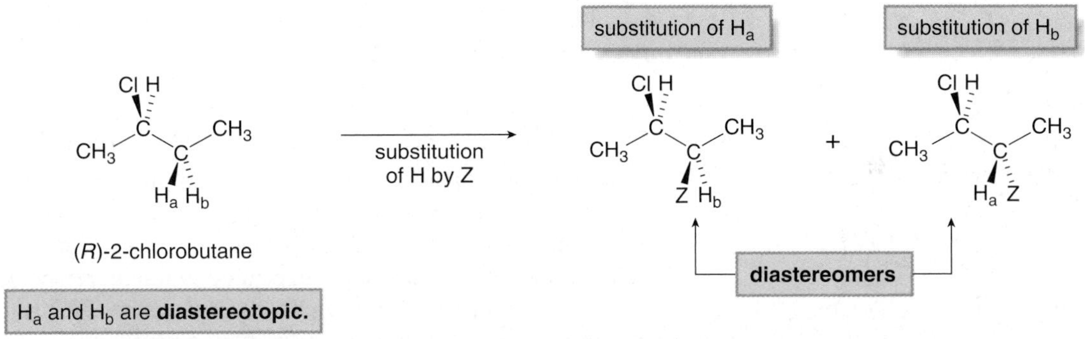

substitution of H$_a$ substitution of H$_b$

(*R*)-2-chlorobutane

H$_a$ and H$_b$ are **diastereotopic**.

diastereomers

◆ **When substitution of two H atoms by Z forms diastereomers, the two H atoms are not equivalent, and give two NMR signals. These two H atoms are called *diastereotopic* protons.**

PROBLEM 15.6   How many different types of protons does 2-chlorobutane contain?

PROBLEM 15.7   Label the indicated protons in each CH$_2$ group as enantiotopic, diastereotopic, or neither.

a. CH$_3$CH$_2$CH$_2$CH$_2$CH$_2$CH$_3$ b. CH$_3$CH$_2$CH$_2$CH$_2$CH$_3$ c. CH$_3$CH(OH)CH$_2$CH$_2$CH$_3$

PROBLEM 15.8   How many NMR signals would you expect for each compound: (a) ClCH$_2$CH(CH$_3$)OCH$_3$; (b) CH$_3$CH(Br)CH$_2$CH$_2$CH$_3$?

15.3   ¹H NMR: Position of Signals

In the NMR spectrum of *tert*-butyl methyl ether in Section 15.1B, why does the CH$_3$O– group absorb downfield from the –C(CH$_3$)$_3$ group?

◆ **Where a particular proton absorbs depends on its electronic environment.**

15.3A   Shielding and Deshielding Effects

To understand how the electronic environment around a nucleus affects its chemical shift, recall that in a magnetic field, an electron creates a small magnetic field that opposes the applied magnetic field, B_o. **Electrons are said to *shield* the nucleus from B_o.**

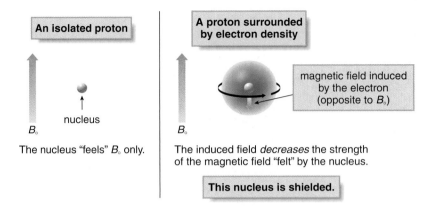

An isolated proton	A proton surrounded by electron density
nucleus	magnetic field induced by the electron (opposite to $B_\circ$)
$B_\circ$	$B_\circ$
The nucleus "feels" $B_\circ$ only.	The induced field *decreases* the strength of the magnetic field "felt" by the nucleus.
	This nucleus is shielded.

In the vicinity of the nucleus, therefore, the magnetic field generated by the circulating electron *decreases* the external magnetic field that the proton "feels." Because the proton experiences a lower magnetic field strength, it needs a lower frequency to achieve resonance. Lower frequency is to the right in an NMR spectrum, toward lower chemical shift, so **shielding shifts an absorption *upfield*,** as shown in Figure 15.3a.

What happens if the electron density around a nucleus is *decreased,* instead? For example, how do the chemical shifts of the protons in CH_4 and CH_3Cl compare?

The less shielded the nucleus becomes, the more of the applied magnetic field ($B_\circ$) it feels. This *deshielded* nucleus experiences a higher magnetic field strength, so it needs a higher frequency to achieve resonance. Higher frequency is to the *left* in an NMR spectrum, toward higher chemical shift, so **deshielding shifts an absorption downfield,** as shown in Figure 15.3b for CH_3Cl versus CH_4. The electronegative Cl atom withdraws electron density from the carbon and hydrogen atoms in CH_3Cl, thus deshielding them relative to those in CH_4.

◆ **Protons near electronegative atoms are deshielded, so they absorb downfield.**

Remember the trend:
Decreased electron density *deshields* a nucleus and an absorption moves *downfield*.

Figure 15.4 summarizes the effects of shielding and deshielding.

These electron density arguments explain the relative position of NMR signals in many compounds.

CH_3CH_2Cl

H_a H_b

◆ The H_b protons are **deshielded** because they are closer to the electronegative Cl atom, so they absorb **downfield** from H_a.

Figure 15.3   How chemical shift is affected by electron density around a nucleus

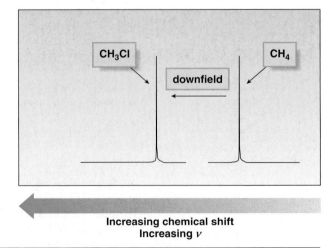

[a] Shielding effects
- An electron shields the nucleus.
- The absorption shifts *upfield*.

[b] Deshielding effects
- Decreased electron density deshields a nucleus.
- The absorption shifts *downfield*.

Figure 15.4 Shielding and deshielding effects

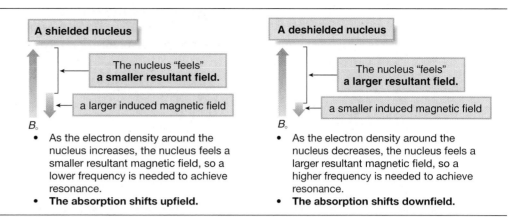

- As the electron density around the nucleus increases, the nucleus feels a smaller resultant magnetic field, so a lower frequency is needed to achieve resonance.
- **The absorption shifts upfield.**

- As the electron density around the nucleus decreases, the nucleus feels a larger resultant magnetic field, so a higher frequency is needed to achieve resonance.
- **The absorption shifts downfield.**

$BrCH_2CH_2F$
H_a H_b

◆ Because F is more electronegative than Br, the H_b protons are more **deshielded** than the H_a protons and absorb further **downfield.**

$ClCH_2CHCl_2$
H_a H_b

◆ The larger number of electronegative Cl atoms (two versus one) **deshield** H_b more than H_a, so it absorbs **downfield** from H_a.

SAMPLE PROBLEM 15.3 Which of the underlined protons in each pair absorbs further downfield: (a) $CH_3CH_2C\underline{H}_3$ or $CH_3OC\underline{H}_3$; (b) $CH_3OC\underline{H}_3$ or $CH_3SC\underline{H}_3$?

SOLUTION

[a] The CH_3 group in CH_3OCH_3 is deshielded by the electronegative O atom. **Deshielding shifts the absorption downfield.**

[b] Because oxygen is more electronegative than sulfur, the CH_3 group in CH_3OCH_3 is more **deshielded** and absorbs **downfield.**

PROBLEM 15.9 For each compound, which of the underlined protons absorbs further downfield:
(a) $FC\underline{H}_2CH_2C\underline{H}_2Cl$; (b) $CH_3CH_2CH_2C\underline{H}_2OCH_3$; (c) $C\underline{H}_3OC(C\underline{H}_3)_3$?

15.3B Chemical Shift Values

Not only is the *relative* position of NMR absorptions predictable, but it is also possible to predict the approximate chemical shift value for a given type of proton.

◆ Protons in a given environment absorb in a predictable region in an NMR spectrum.

A more detailed list of characteristic chemical shift values is found in Appendix E.

Table 15.1 lists the typical chemical shift values for the most common bonds encountered in organic molecules.

Table 15.1 illustrates that absorptions for a given type of C–H bond occur in a narrow range of chemical shift values, usually 1–2 ppm. For example, all sp^3 hybridized C–H bonds in alkanes and cycloalkanes absorb between 0.9 and 2.0 ppm. By contrast, absorptions due to N–H and O–H protons can occur over a broader range. For example, the OH proton of an alcohol is found anywhere in the 1–5 ppm range. The position of these absorptions is affected by the extent of hydrogen bonding, making it more variable.

The chemical shift of a particular type of C–H bond is also affected by the number of R groups bonded to the carbon atom.

$RCH_2{-}H$	$R_2CH{-}H$	$R_3C{-}H$
~ 0.9 ppm	~ 1.3 ppm	~ 1.7 ppm

Increasing alkyl substitution
Increasing chemical shift

TABLE 15.1   Characteristic Chemical Shifts of Common Types of Protons

Type of proton	Chemical shift (ppm)	Type of proton	Chemical shift (ppm)
sp^3 —C—H	0.9–2	sp^2 C=C—H	4.5–6
• RCH_3	~0.9	benzene ring —H	6.5–8
• R_2CH_2	~1.3		
• R_3CH	~1.7		
Z=C—C—H, Z = C, O, N	1.5–2.5	R—C(=O)—H	9–10
—C≡C—H	~2.5	R—C(=O)—OH	10–12
sp^3 —C—H with Z, Z = N, O, X	2.5–4	RO—H or R—N—H	1–5

◆ The chemical shift of a C–H bond increases with increasing alkyl substitution.

PROBLEM 15.10   For each compound, first label each different type of proton and then rank the protons in order of increasing chemical shift.

a. $ClCH_2CH_2CH_2Br$ b. $CH_3OCH_2OC(CH_3)_3$ c. 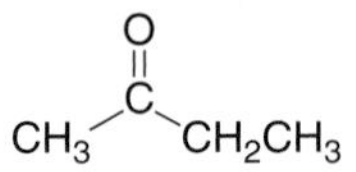 CH_3—C(=O)—CH_2CH_3

15.4   The Chemical Shift of Protons on sp^2 and sp Hybridized Carbons

The chemical shift of protons bonded to benzene rings, C–C double bonds, and C–C triple bonds merits additional comment.

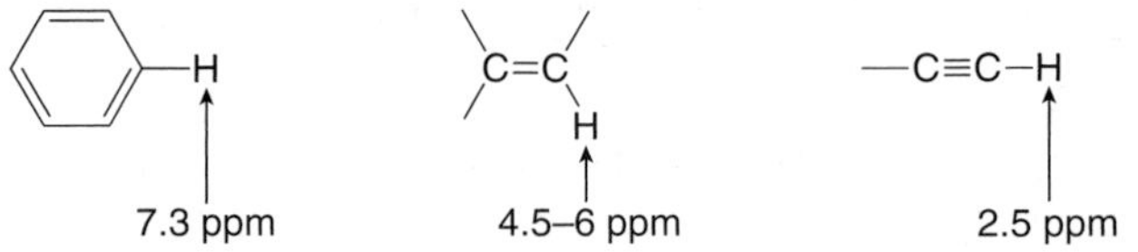

benzene—H   C=C(—H)   —C≡C—H
7.3 ppm   4.5–6 ppm   2.5 ppm

Each of these functional groups contains π bonds with **loosely held π electrons.** When placed in a magnetic field, these π electrons move in a circular path, inducing a new magnetic field. How this induced magnetic field affects the chemical shift of a proton depends on the direction of the induced field *in the vicinity of the absorbing proton.*

Protons on Benzene Rings

In a magnetic field, the six π electrons in **benzene** circulate around the ring, creating a ring current. The magnetic field induced by these moving electrons reinforces the applied magnetic field in the vicinity of the protons. The protons thus feel a stronger magnetic field and a higher frequency is needed for resonance, so the **protons are deshielded and the absorption is** *downfield.*

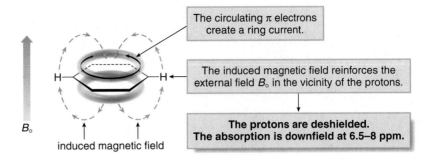

The circulating π electrons create a ring current.

The induced magnetic field reinforces the external field B_o in the vicinity of the protons.

The protons are deshielded. The absorption is downfield at 6.5–8 ppm.

B_o

induced magnetic field

Protons on Carbon–Carbon Double Bonds

A similar phenomenon occurs with protons on carbon–carbon double bonds. In a magnetic field, the loosely held π electrons create a magnetic field that reinforces the applied field in the vicinity of the protons. Because the protons now feel a stronger magnetic field, they require a higher frequency for resonance. **The protons are deshielded and the absorption is *downfield*.**

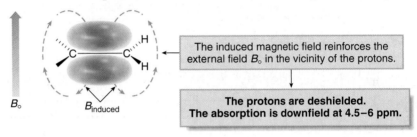

The induced magnetic field reinforces the external field B_o in the vicinity of the protons.

The protons are deshielded. The absorption is downfield at 4.5–6 ppm.

B_o $B_{induced}$

Protons on Carbon–Carbon Triple Bonds

In a magnetic field, the π electrons of a carbon–carbon triple bond are induced to circulate, but in this case the induced magnetic field *opposes* the applied magnetic field (B_o). The proton thus feels a weaker magnetic field, so a lower frequency is needed for resonance. **The nucleus is shielded and the absorption is *upfield*.**

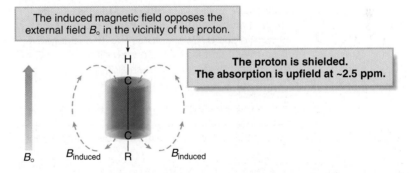

The induced magnetic field opposes the external field B_o in the vicinity of the proton.

The proton is shielded. The absorption is upfield at ~2.5 ppm.

B_o $B_{induced}$ R $B_{induced}$

Table 15.2 summarizes the shielding and deshielding effects due to circulating π electrons.

TABLE 15.2	Effect of π Electrons on Chemical Shift Values	
Proton type	**Effect**	**Chemical shift (ppm)**
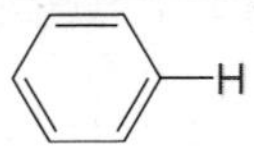 ⬡—H	highly deshielded	6.5–8
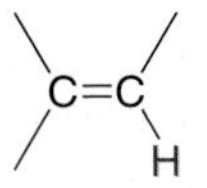 C=C, H	deshielded	4.5–6
—C≡C—H	shielded	~2.5

Figure 15.5   Regions in the
^{1}H NMR spectrum

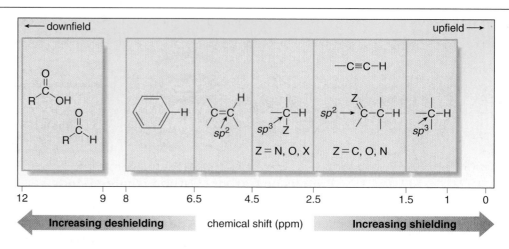

- Shielded protons absorb at lower chemical shift (to the right).
- Deshielded protons absorb at higher chemical shift (to the left).

To remember the chemical shifts of some common bond types, it is helpful to think of a ^{1}H NMR spectrum as being divided into six different regions (Figure 15.5).

SAMPLE PROBLEM 15.4   Rank H_a, H_b, and H_c in order of increasing chemical shift.

$$
\begin{array}{c}
H_c \rightarrow H \\
| \\
C=CH_2 \\
| \\
CH_3CH_2O \\
\uparrow \quad \uparrow \\
H_a \quad H_b
\end{array}
$$

SOLUTION

The H_a protons are bonded to an sp^3 hybridized carbon, so they are shielded and absorb upfield compared to H_b and H_c. Because the H_b protons are deshielded by the electronegative oxygen atom on the C to which they are bonded, they absorb downfield from H_a. The H_c proton is deshielded by two factors. The electronegative O atom withdraws electron density from H_c. Moreover, because H_c is bonded directly to a C=C, the magnetic field induced by the π electrons causes further deshielding.

ANSWER

In order of increasing chemical shift, $H_a < H_b < H_c$.

PROBLEM 15.11   Rank each group of protons in order of increasing chemical shift.

a. $CH_3-C\equiv C-H$ $CH_3CH=CH_2$ $CH_3CH_2CH_3$ b.
         $\uparrow$ $\uparrow$ $\uparrow$
         H_a H_b H_c

$$
\begin{array}{c}
O \\
\| \\
CH_3 \overset{C}{} OCH_2CH_3 \\
\uparrow \qquad \uparrow \quad \uparrow \\
H_a \qquad H_b \ H_c
\end{array}
$$

15.5 ^{1}H NMR: Intensity of Signals

The relative intensity of ^{1}H NMR signals also provides information about a compound's structure.

◆ **The area under an NMR signal is proportional to the number of absorbing protons.**

For example, in the ^{1}H NMR spectrum of $CH_3OC(CH_3)_3$, the ratio of the area under the downfield peak (due to the CH_3O- group) to the upfield peak [due to the $-C(CH_3)_3$ group] is 1:3. An NMR spectrometer automatically integrates the area under the peaks, and prints out a stepped curve (an **integral**) on the spectrum. **The height of each step is proportional to the area under the peak, which is in turn proportional to the number of absorbing protons.**

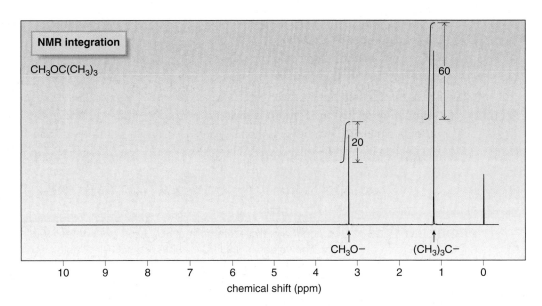

Integrals can be manually measured, but modern NMR spectrometers automatically calculate and plot the value of each integral in arbitrary units. If the heights of two integrals are 20 units and 60 units, the ratio of absorbing protons is 20:60, or 1:3, or 2:6, or 3:9, and so forth. This tells the *ratio,* not the absolute number of protons. Integration ratios are approximate, and often values must be rounded to the nearest whole number.

PROBLEM 15.12   Which compounds give an NMR spectrum with two signals in a ratio of 2:3?
a. CH_3CH_2Cl b. $CH_3CH_2CH_3$ c. $CH_3CH_2OCH_2CH_3$ d. $CH_3OCH_2CH_2OCH_3$

Knowing the molecular formula of a compound and integration values from its ^1H NMR spectrum gives the *actual number* of protons that give rise to a particular signal.

How To   Use a Molecular Formula and an Integrated ^1H NMR Spectrum to Determine the Number of Protons Giving Rise to an NMR Signal

Example   A compound of molecular formula $C_9H_{10}O_2$ gives the following integrated ^1H NMR spectrum. How many protons give rise to each signal?

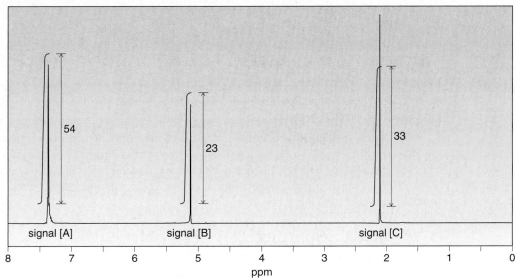

Step [1]   **Determine the number of integration units per proton by dividing the total number of integration units by the total number of protons.**

- Total number of integration units: $54 + 23 + 33 = 110$ units
- Total number of protons = 10
- Divide: 110 units/10 protons = **11 units per proton**

Step [2]   **Determine the number of protons giving rise to each signal.**

- To determine the number of H atoms giving rise to each signal, divide each integration value by the answer of Step [1] and round to the nearest whole number.

Signal [A]:   Signal [B]:   Signal [C]:

Answer: $\dfrac{54}{11} = 4.9 \approx$ | 5 H | $\dfrac{23}{11} = 2.1 \approx$ | 2 H | $\dfrac{33}{11} = $ | 3 H |

PROBLEM 15.13   A compound of molecular formula $C_8H_{14}O_2$ gives three NMR signals having the indicated integration values: signal [A] 14 units, signal [B] 12 units, and signal [C] 44 units. How many protons give rise to each signal?

15.6 ^{1}H NMR: Spin–Spin Splitting

The ^{1}H NMR spectra you have seen up to this point have been limited to one or more single absorptions called **singlets.** In the ^{1}H NMR spectrum of $BrCH_2CHBr_2$, however, the two signals for the two different kinds of protons are each split into more than one peak. The splitting patterns, the result of **spin–spin splitting,** can be used to determine how many protons reside on the carbon atoms near the absorbing proton.

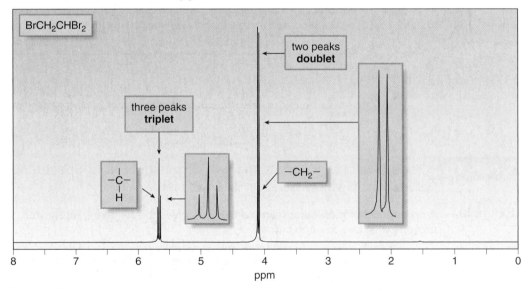

To understand spin–spin splitting, we must distinguish between the **absorbing protons** that give rise to an NMR signal, and the **adjacent protons** that cause the signal to split. **The number of adjacent protons determines the observed splitting pattern.**

- The CH_2 signal appears as **two peaks,** called a **doublet.** The relative area under the peaks of a doublet is 1:1.
- The CH signal appears as **three peaks,** called a **triplet.** The relative area under the peaks of a triplet is 1:2:1.

Spin–spin splitting occurs only between nonequivalent protons on the same carbon or adjacent carbons. To illustrate how spin–spin splitting arises, we'll examine nonequivalent protons on adjacent carbons, the more common example. Spin–spin splitting arises because protons are little magnets that can be aligned with or against an applied magnetic field, and this affects the magnetic field that a nearby proton feels.

15.6A   Splitting: How a Doublet Arises

First, let's examine how the doublet due to the CH_2 group in $BrCH_2CHBr_2$ arises. The CH_2 group contains the absorbing protons and the CH group contains the adjacent proton that causes the splitting.

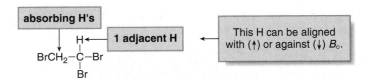

When placed in an applied magnetic field ($B_\circ$), the adjacent proton ($CHBr_2$) can be aligned with ($\uparrow$) or against ($\downarrow$) $B_\circ$. As a result, the absorbing protons (CH_2Br) feel two slightly different magnetic fields—one slightly larger than $B_\circ$ and one slightly smaller than $B_\circ$. Because the absorbing protons feel two different magnetic fields, they absorb at two different frequencies in the NMR spectrum, thus splitting a single absorption into a doublet.

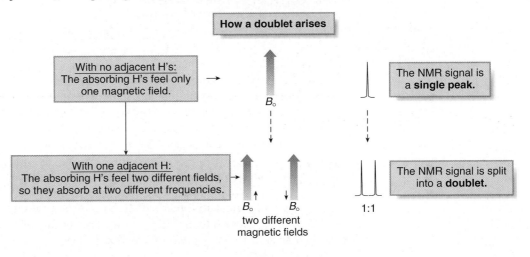

Keep in mind the difference between an NMR **signal** and an NMR **peak**. An NMR signal is the entire absorption due to a particular kind of proton. NMR peaks are contained within a signal. A doublet constitutes one signal that is split into two peaks.

◆ **One adjacent proton splits an NMR signal into a doublet.**

The two peaks of a doublet are approximately equal in area. The area under both peaks—the entire NMR signal—is due to both protons of the CH_2 group of $BrCH_2CHBr_2$.

The frequency difference (measured in Hz) between the two peaks of the doublet is called the **coupling constant**, denoted by J. Coupling constants are usually in the range of 0–18 Hz, and are independent of the strength of the applied magnetic field $B_\circ$.

coupling constant, *J*, in Hz

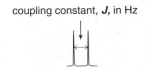

15.6B   Splitting: How a Triplet Arises

Now let's examine how the triplet due to the CH group in $BrCH_2CHBr_2$ arises. The CH group contains the absorbing proton and the CH_2 group contains the adjacent protons (H_a and H_b) that cause the splitting.

When placed in an applied magnetic field ($B_\circ$), the adjacent protons H_a and H_b can each be aligned with ($\uparrow$) or against ($\downarrow$) $B_\circ$. As a result, the absorbing proton feels three slightly different magnetic fields—one slightly larger than $B_\circ$, one slightly smaller than $B_\circ$, and one the same strength as $B_\circ$.

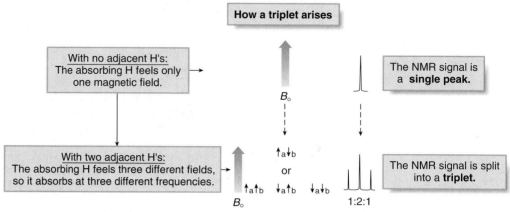

three different magnetic fields

Because the absorbing proton feels three different magnetic fields, it absorbs at three different frequencies in the NMR spectrum, thus splitting a single absorption into a triplet. Because there are two different ways to align one proton with $B_\circ$ and one proton against $B_\circ$—that is, $\uparrow_a\downarrow_b$ and $\downarrow_a\uparrow_b$—the middle peak of the triplet is twice as intense as the two outer peaks, making the ratio of the areas under the three peaks 1:2:1.

◆ **Two adjacent protons split an NMR signal into a triplet.**

When two protons split each other's NMR signals they are said to be *coupled.* In $BrCH_2CHBr_2$, the CH proton is coupled to the CH_2 protons. The spacing between peaks in a split NMR signal, measured by the *J* value, is *equal* for coupled protons.

15.6C   Splitting: The Rules and Examples

Three general rules describe the splitting patterns commonly seen in the ^{1}H NMR spectra of organic compounds.

Rule [1]	Equivalent protons don't split each other's signals.

Rule [2]	A set of *n* nonequivalent protons splits the signal of a nearby proton into *n* + 1 peaks.

◆ In $BrCH_2CHBr_2$, for example, *one* adjacent CH proton splits an NMR signal into *two* peaks (a doublet), and *two* adjacent CH_2 protons split an NMR signal into *three* peaks (a triplet). Names for split NMR signals containing two to seven peaks are given in Table 15.3. An NMR signal having more than seven peaks is called a **multiplet.**

◆ The inside peaks of a split NMR signal are always most intense, with the area under the peaks decreasing from the inner to the outer peaks in a given splitting pattern.

Rule [3]	Splitting is observed for nonequivalent protons on the same carbon or adjacent carbons.

If H$_a$ and H$_b$ are not equivalent, splitting is observed when:

$$
\begin{array}{ccc}
\text{H}_a & \quad & \text{H}_a \\
| & & \\
-\text{C}- & \quad & =\text{C} \\
| & & \\
\text{H}_b & \quad & \text{H}_b
\end{array}
\qquad\qquad
\begin{array}{c}
|\quad\; | \\
-\text{C}-\text{C}- \\
|\quad\; | \\
\text{H}_a\;\; \text{H}_b
\end{array}
$$

H$_a$ and H$_b$ are on the **same** carbon.   H$_a$ and H$_b$ are on **adjacent** carbons.

The splitting of an NMR signal reveals the number of nearby nonequivalent protons. It tells nothing about the absorbing proton itself.

Splitting is not generally observed between protons separated by more than three σ bonds. Although H$_a$ and H$_b$ are not equivalent to each other in ethyl methyl ether and 2-butanone, H$_a$ and H$_b$ are separated by four σ bonds and so they are too far away to split each other's NMR signals.

$$
\begin{array}{cc}
& \overset{\displaystyle O}{\underset{\displaystyle \parallel}{}} \\
\overset{\sigma}{CH_2}\overset{\sigma}{\underset{\sigma}{C}}\overset{\sigma}{CHCH_3} \\
\overset{\sigma|}{\underset{H_a}{}} \qquad \overset{|\sigma}{\underset{H_b}{}}
\end{array}
$$

2-butanone
H$_a$ and H$_b$ are separated by four σ bonds.

no splitting between H$_a$ and H$_b$

$$
\overset{\sigma}{CH_2}\overset{\sigma}{\underset{}{O}}\overset{\sigma}{CHCH_3} \\
\overset{\sigma|}{\underset{H_a}{}} \qquad \overset{|\sigma}{\underset{H_b}{}}
$$

ethyl methyl ether
H$_a$ and H$_b$ are separated by four σ bonds.

no splitting between H$_a$ and H$_b$

TABLE 15.3	Names for a Given Number of Peaks in an NMR Signal		
Number of peaks	**Name**	**Number of peaks**	**Name**
1	singlet	5	quintet
2	doublet	6	sextet
3	triplet	7	septet
4	quartet	> 7	multiplet

Table 15.4 illustrates common splitting patterns observed for adjacent nonequivalent protons.

TABLE 15.4	Common Splitting Patterns Observed in ^{1}H NMR	
	Example / **Pattern**	**Analysis (H$_a$ and H$_b$ are not equivalent.)**
[1]		• H$_a$: one adjacent H$_b$ proton ----→ two peaks ----→ a **doublet** • H$_b$: one adjacent H$_a$ proton ----→ two peaks ----→ a **doublet**
[2]		• H$_a$: two adjacent H$_b$ protons ----→ three peaks ----→ a **triplet** • H$_b$: one adjacent H$_a$ proton ----→ two peaks ----→ a **doublet**
[3]		• H$_a$: two adjacent H$_b$ protons ----→ three peaks ----→ a **triplet** • H$_b$: two adjacent H$_a$ protons ----→ three peaks ----→ a **triplet**
[4]		• H$_a$: three adjacent H$_b$ protons ----→ four peaks ----→ a **quartet;** the relative area under the peaks of a quartet is 1:3:3:1. • H$_b$: two adjacent H$_a$ protons ----→ three peaks ----→ a **triplet**
[5]		• H$_a$: three adjacent H$_b$ protons ----→ four peaks ----→ a **quartet** • H$_b$: one adjacent H$_a$ proton ----→ two peaks ----→ a **doublet**

Predicting splitting is always a two-step process:

◆ **Determine if two protons are equivalent or different.** Only nonequivalent protons split each other.

◆ **Determine if two nonequivalent protons are close enough to split each other's signals.** Splitting is observed only for nonequivalent protons on the *same* carbon or *adjacent* carbons.

Several examples of spin–spin splitting in specific compounds illustrate the result of this two-step strategy.

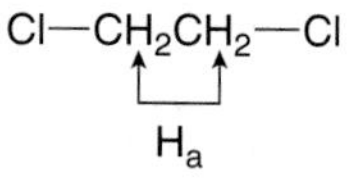

$Cl—CH_2CH_2—Cl$
H_a

◆ All protons are equivalent (H_a), so there is no splitting and the NMR signal is one singlet.

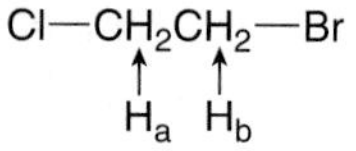

$Cl—CH_2CH_2—Br$
H_a H_b

◆ There are two NMR signals. H_a and H_b are nonequivalent protons bonded to adjacent C atoms, so they are close enough to split each other's NMR signals. The H_a signal is split into a triplet by the two H_b protons. The H_b signal is split into a triplet by the two H_a protons.

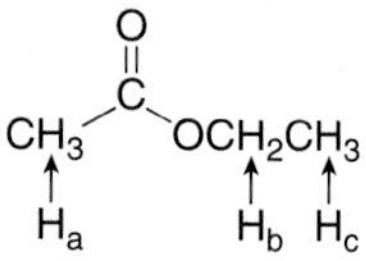

CH_3 OCH_2CH_3
H_a H_b H_c

◆ There are three NMR signals. H_a has no adjacent nonequivalent protons so its signal is a singlet. The H_b signal is split into a quartet by the three H_c protons. The H_c signal is split into a triplet by the two H_b protons.

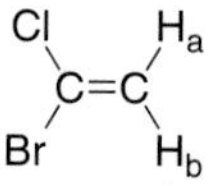

Cl H_a
$C=C$
Br H_b

◆ There are two NMR signals. H_a and H_b are nonequivalent protons on the same carbon, so they are close enough to split each other's NMR signals. The H_a signal is split into a doublet by H_b. The H_b signal is split into a doublet by H_a.

PROBLEM 15.14   Into how many peaks will each indicated proton be split?

a. 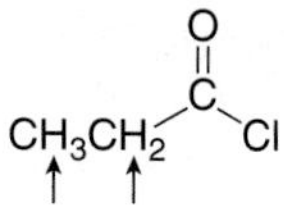 $CH_3CH_2—C(=O)—Cl$

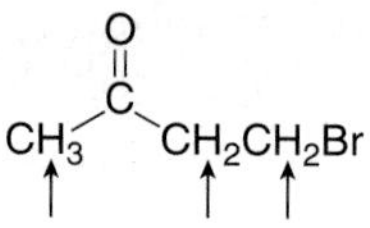

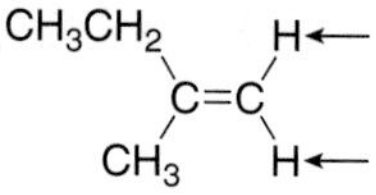

b. 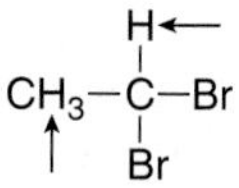 $CH_3—C(H)(Br)—Br$

c. $CH_3—C(=O)—CH_2CH_2Br$

d. 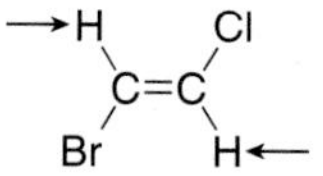 $H—C(=O)—C(Cl)=C(Br)—H$ (configuration as drawn)

e. $CH_3CH_2, CH_3—C=C—H, H$

PROBLEM 15.15   For each compound give the number of NMR signals, and then determine how many peaks are present for each NMR signal.

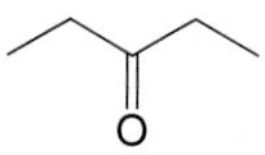

a. $Cl_2CHCHCl_2$

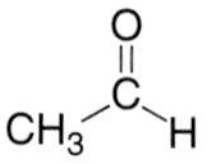

b. $Cl_2CHCHBr_2$

c. (structure: pentan-3-one)

d. 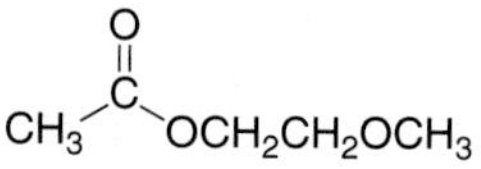 $CH_3—C(=O)—OCH_2CH_2OCH_3$

e. $CH_3—C(=O)—H$

PROBLEM 15.16   Sketch the NMR spectrum of CH_3CH_2Cl, giving the approximate location of each NMR signal.

15.7   More Complex Examples of Splitting

Up to now you have studied examples of spin–spin splitting where the absorbing proton has nearby protons on *one* adjacent carbon only. What happens when the absorbing proton has nonequivalent protons on *two* adjacent carbons? Two different outcomes are possible, depending on whether the adjacent nonequivalent protons are *equivalent to* or *different from* each other.

For example, 2-bromopropane [$(CH_3)_2CHBr$] has two types of protons—H_a and H_b—so it exhibits two NMR signals, as shown in Figure 15.6.

◆ The H_a protons have only one adjacent nonequivalent proton (H_b) so they are split into two peaks, a **doublet.**

◆ H_b has three H_a protons on each side. Because the six H_a protons are equivalent to each other, the ***n* + 1** rule can be used to determine splitting: 6 + 1 = 7 peaks, a **septet.**

This is a specific example of a general rule:

◆ Whenever two (or three) sets of adjacent protons are equivalent to each other, use the *n* + 1 rule to determine the splitting pattern.

Figure 15.6   The ^{1}H NMR spectrum of 2-bromopropane, [$(CH_3)_2CHBr$]

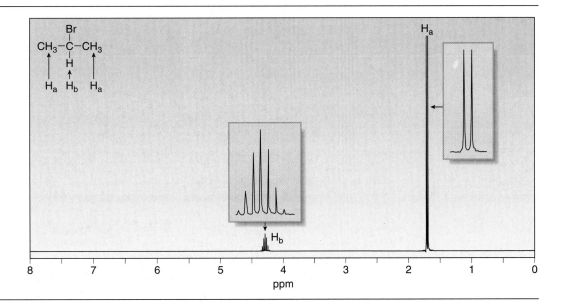

The ^{1}H NMR spectrum of 1-bromopropane ($CH_3CH_2CH_2Br$) illustrates a different result.

$CH_3CH_2CH_2{-}Br$
   $\uparrow$ $\uparrow$ $\uparrow$
   H_a H_b H_c

$CH_3CH_2CH_2Br$ has three different types of protons—H_a, H_b, and H_c—so it exhibits three NMR signals (Figure 15.7). H_a and H_c are each triplets because they are adjacent to two H_b protons. H_b has protons on both adjacent carbons, but because H_a and H_c are *not equivalent to each other,* we cannot merely add them together and use the $n + 1$ rule.

Instead, to determine the splitting of H_b, we must consider the effect of the H_a protons and the H_c protons *separately.* The three H_a protons split the H_b signal into four peaks, and the two H_c protons split each of these four peaks into three peaks—that is, the NMR signal due to H_b consists of $4 \times 3 = $ **12 peaks.** Figure 15.8 shows a splitting diagram illustrating how these 12 peaks arise. Often, an NMR signal with so many lines has several overlapping peaks, as is the case with the multiplet for H_b in Figure 15.7.

Figure 15.7   The ^{1}H NMR spectrum of 1-bromopropane, $CH_3CH_2CH_2Br$

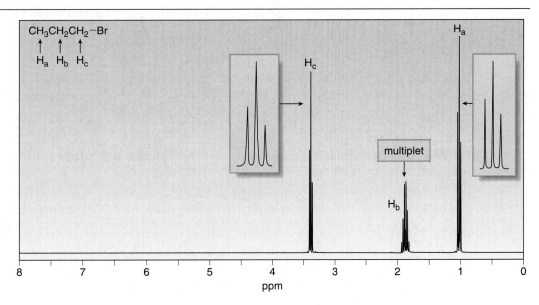

- H_a and H_c are both triplets.
- H_b is split into 12 peaks, labeled as a multiplet. Fewer peaks are seen because some peaks overlap.

Figure 15.8 A splitting diagram for the H_b protons in 1-bromopropane

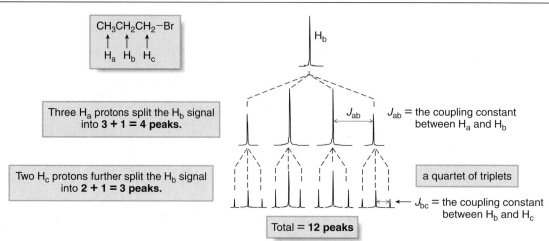

- The H_b signal is split into 12 peaks, a quartet of triplets. The number of peaks actually seen for the signal depends on the relative size of the coupling constants, J_{ab} and J_{bc}. When $J_{ab} \gg J_{bc}$, as drawn in this diagram, all 12 lines of the pattern are visible. When J_{ab} and J_{bc} are similar in magnitude, peaks overlap and fewer lines are observed.

◆ **When two sets of adjacent protons are *different from each other* (n protons on one adjacent carbon and m protons on the other), the number of peaks in an NMR signal = $(n + 1)(m + 1)$.**

SAMPLE PROBLEM 15.5   How many peaks are present in the NMR signal of each indicated proton?

a. $ClCH_2CH_2CH_2Cl$ b. $ClCH_2CH_2CH_2Br$

SOLUTION

[a] $ClCH_2CH_2CH_2Cl$
        H_a H_b H_a

- H_b has two H_a protons on each adjacent C. Because the four H_a protons are equivalent to each other, the $n + 1$ rule can be used to determine splitting: $4 + 1 = $ **5 peaks,** a quintet.

[b] $ClCH_2CH_2CH_2Br$
        H_a H_b H_c

- H_b has two H_a protons on one adjacent C and two H_c protons on the other. Because H_a and H_c are not equivalent to each other, the number of peaks due to $H_b = (n + 1)(m + 1) = (2 + 1)(2 + 1) = $ **9 peaks.**

PROBLEM 15.17   How many peaks are present in the NMR signal of each indicated proton?

a. $(CH_3)_2CHCO_2CH_3$ b. $CH_3CH_2CH_2CH_2CH_3$ c.
$$\begin{array}{cc} Cl & CH_2Br \\ C=C \\ \rightarrow H & H \leftarrow \end{array}$$
d.
$$\begin{array}{cc} H & H \\ C=C \\ Br & H \end{array}$$ (all H atoms)

PROBLEM 15.18   Identify compounds **A** and **B,** two isomers of molecular formula $C_3H_5Cl_3$, from the given ^1H NMR data.

Compound **A:** singlet at 2.23 and singlet at 4.04 ppm
Compound **B:** doublet at 1.69, multiplet at 4.34, and doublet at 5.85 ppm

PROBLEM 15.19   Describe the ^1H NMR spectrum of each compound. State how many NMR signals are present, the splitting pattern for each signal, and the approximate chemical shift.

a. $CH_3OCH_2CH_3$

c. $CH_3OCH_2CH_2CH_2OCH_3$

b.
$$\begin{array}{c} O \\ \parallel \\ CH_3CH_2 \quad C \quad OCH(CH_3)_2 \end{array}$$

d.
$$\begin{array}{cc} CH_3CH_2 & CH_2CH_3 \\ C=C \\ H & H \end{array}$$

15.8 Spin–Spin Splitting in Alkenes

Protons on carbon–carbon double bonds often give characteristic splitting patterns. A disubstituted double bond can have two **geminal protons** (on the same carbon atom), two **cis protons,** or two **trans protons.** When these protons are different, each proton splits the NMR signal of the other, so that each proton appears as a doublet. **The magnitude of the coupling constant J for these doublets depends on the arrangement of hydrogen atoms.**

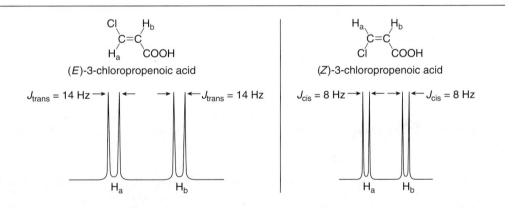

geminal H's   cis H's   trans H's

$J_{geminal}$ < J_{cis} < J_{trans}

0–3 Hz   5–10 Hz   11–18 Hz

characteristic coupling constants for three types of disubstituted alkenes

Thus, the E and Z isomers of 3-chloropropenoic acid both exhibit two doublets for the two alkenyl protons, but the coupling constant is larger when the protons are trans compared to when the protons are cis, as shown in Figure 15.9.

When a double bond is monosubstituted, there are three nonequivalent protons, and the pattern is more complicated because all three protons are coupled to each other.

For example, vinyl acetate (CH_2=CHOCOCH₃) has four different types of protons, three of which are bonded to the double bond. Besides the singlet for the CH_3 group, each proton on the double bond is coupled to two other different protons on the double bond, giving the spectrum in Figure 15.10.

> Vinyl acetate is polymerized to poly(vinyl acetate) (Problem 13.30), a polymer used in paints, glues, and adhesives.

Figure 15.9 ¹H NMR spectra for the alkenyl protons of (E)- and (Z)-3-chloropropenoic acid

- Although both (E)- and (Z)-3-chloropropenoic acid show two doublets in their ¹H NMR spectra for their alkenyl protons, $J_{trans} > J_{cis}$.

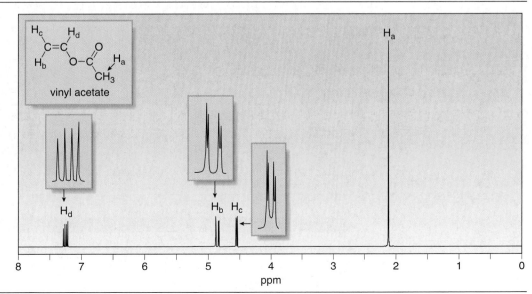

Figure 15.10 The ¹H NMR spectrum of vinyl acetate (CH_2=CHOCOCH₃)

Figure 15.11 Splitting diagram for the alkenyl protons in vinyl acetate (CH_2=CHOCOCH$_3$)

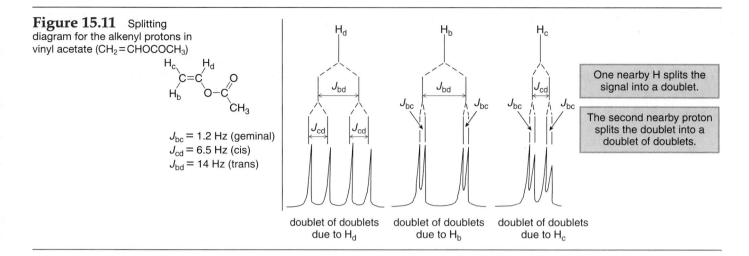

J_{bc} = 1.2 Hz (geminal)
J_{cd} = 6.5 Hz (cis)
J_{bd} = 14 Hz (trans)

One nearby H splits the signal into a doublet.

The second nearby proton splits the doublet into a doublet of doublets.

doublet of doublets due to H$_d$ doublet of doublets due to H$_b$ doublet of doublets due to H$_c$

◆ H$_b$ has two nearby nonequivalent protons that split its signal, the geminal proton H$_c$ and the trans proton H$_d$. H$_d$ splits the H$_b$ signal into a doublet, and the H$_c$ proton splits the doublet into two doublets. This pattern of four peaks is called a **doublet of doublets.**

◆ H$_c$ has two nearby nonequivalent protons that split its signal, the geminal proton H$_b$ and the cis proton H$_d$. H$_d$ splits the H$_c$ signal into a doublet, and the H$_b$ proton splits the doublet into two doublets, forming another **doublet of doublets.**

◆ H$_d$ has two nearby nonequivalent protons that split its signal, the trans proton H$_b$ and the cis proton H$_c$. H$_b$ splits the H$_d$ signal into a doublet, and the H$_c$ proton splits the doublet into two doublets, forming another **doublet of doublets.**

Splitting diagrams for the three alkenyl protons in vinyl acetate are drawn in Figure 15.11. Note that each pattern is different in appearance because the magnitude of the coupling constants forming them is different.

PROBLEM 15.20   Label the signals due to H$_a$, H$_b$, and H$_c$ in the ^{1}H NMR spectrum of acrylonitrile (CH_2=CHCN). Draw a splitting diagram for the absorption due to the H$_a$ proton.

J_{ab} = 11.8 Hz
J_{bc} = 0.9 Hz
J_{ac} = 18 Hz

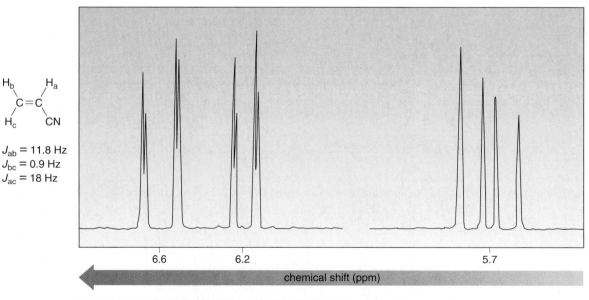

6.6   6.2   5.7

chemical shift (ppm)

Figure 15.12   The ^{1}H NMR spectrum of ethanol (CH$_3$CH$_2$OH)

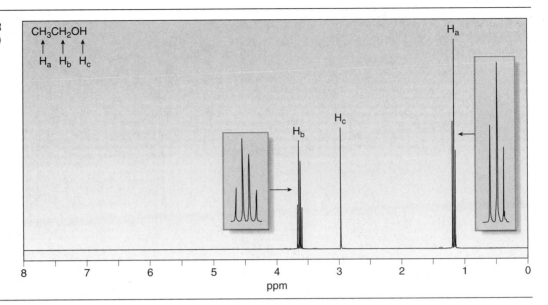

15.9   Other Facts About ^{1}H NMR Spectroscopy

15.9A   OH Protons

◆ Under usual conditions, an OH proton does not split the NMR signal of adjacent protons.
◆ The signal due to an OH proton is not split by adjacent protons.

Ethanol (CH$_3$CH$_2$OH), for example, has three different types of protons, so there are three signals in its ^{1}H NMR spectrum, as shown in Figure 15.12.

◆ The H$_a$ signal is split by the two H$_b$ protons into three peaks, a **triplet.**
◆ The H$_b$ signal is split only by the three H$_a$ protons into four peaks, a **quartet.** The adjacent OH proton does *not* split the signal due to H$_b$.
◆ H$_c$ is a **singlet** because OH protons are *not* split by adjacent protons.

Why is a proton bonded to an oxygen atom a singlet in a ^{1}H NMR spectrum? Protons on electronegative elements rapidly **exchange** between molecules in the presence of trace amounts of acid or base. It is as if the CH$_2$ group in ethanol never "feels" the presence of the OH proton, because the OH proton is rapidly moving from one molecule to another. We therefore see a peak due to the OH proton, but it is a single peak with no splitting. This phenomenon usually occurs with NH and OH protons.

PROBLEM 15.21   How many signals are present in the ^{1}H NMR spectrum for each molecule? What splitting is observed in each signal? (a) (CH$_3$)$_3$CCH$_2$OH; (b) CH$_3$CH$_2$CH$_2$OH

15.9B   Cyclohexane Conformers

How does the rotation around carbon–carbon σ bonds and the ring flip of cyclohexane rings affect an NMR spectrum? Because these processes are rapid at room temperature, an NMR spectrum records an **average** of all conformers that interconvert.

Thus, even though each cyclohexane carbon has two different types of hydrogens—one axial and one equatorial—the two chair forms of cyclohexane rapidly interconvert them, and an **NMR spectrum shows a single signal for the average environment** that it "sees."

Figure 15.13 The 6.5–8 ppm region of the ^{1}H NMR spectrum of three benzene derivatives

- The appearance of the signals in the 6.5–8 ppm region of the ^{1}H NMR spectrum depends on the identity of Z in C_6H_5Z.

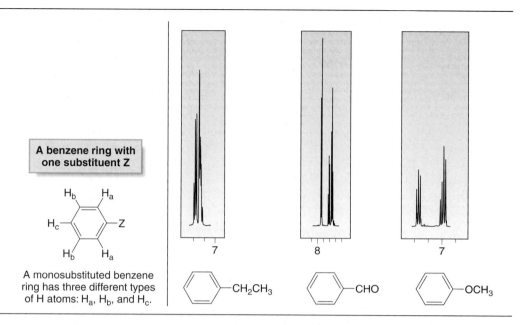

A benzene ring with one substituent Z

A monosubstituted benzene ring has three different types of H atoms: H_a, H_b, and H_c.

axial → H_a equatorial

Axial and equatorial H's rapidly interconvert. NMR sees an average environment and shows one signal.

15.9C Protons on Benzene Rings

Benzene has six equivalent, deshielded protons and exhibits a single peak in its ^{1}H NMR spectrum at 7.27 ppm. Monosubstituted benzene derivatives—that is, benzene rings with one H atom replaced by another substituent Z—contain five deshielded protons that are no longer all equivalent to each other, and the appearance of these signals is highly variable. Depending on the identity of Z, this region of a ^{1}H NMR spectrum (6.5–8 ppm) can have a singlet or multiplets, as shown in Figure 15.13. We will not analyze the splitting patterns observed for the ring protons of monosubstituted benzenes.

We will learn more about the spectroscopic absorptions of benzene derivatives in Chapter 17.

PROBLEM 15.22 What protons in alcohol **A** give rise to each signal in its ^{1}H NMR spectrum? Explain all splitting patterns observed for absorptions between 0–7 ppm.

A

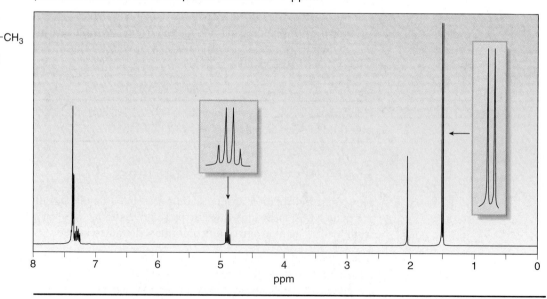

15.10   Using ^{1}H NMR to Identify an Unknown

Once we know a compound's molecular formula from its mass spectral data and the identity of its functional group from its IR spectrum, we can then use its ^{1}H NMR spectrum to determine its structure. A suggested procedure is illustrated for compound **X,** whose molecular formula ($C_4H_8O_2$) and functional group (C=O) were determined in Section 14.7.

How To   Use ^{1}H NMR Data to Determine a Structure

Example   Using its ^{1}H NMR spectrum, determine the structure of an unknown compound X that has molecular formula $C_4H_8O_2$ and contains a C=O absorption in its IR spectrum.

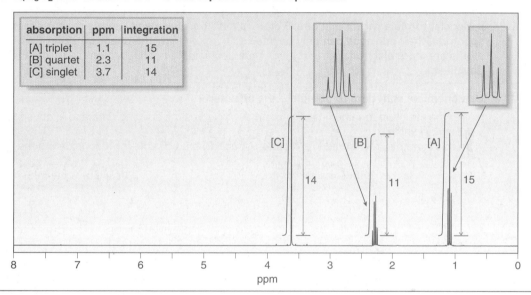

absorption	ppm	integration
[A] triplet	1.1	15
[B] quartet	2.3	11
[C] singlet	3.7	14

Step [1]   **Determine the number of different kinds of protons.**

- The number of NMR signals equals the number of different types of protons.
- This molecule has three NMR signals ([A], [B], and [C]) and therefore **three** types of protons (H_a, H_b, and H_c).

Step [2]   **Use the integration data to determine the number of H atoms giving rise to each signal (Section 15.5).**

- Total number of integration units: 14 + 11 + 15 = 40 units
- Total number of protons = 8
- Divide: 40 units/8 protons = **5 units per proton**
- Then, divide each integration value by this answer (5 units per proton) and round to the nearest whole number.

$$\frac{15}{5} = \boxed{3\ H_a\ \text{protons}} \qquad \frac{11}{5} = 2.2 \approx \boxed{2\ H_b\ \text{protons}} \qquad \frac{14}{5} = 2.8 \approx \boxed{3\ H_c\ \text{protons}}$$

signal [A]   signal [B]   signal [C]

Three equivalent H's usually means a CH$_3$ group.	Two equivalent H's usually means a CH$_2$ group.	Three equivalent H's usually means a CH$_3$ group.

Step [3]   **Use individual splitting patterns to determine what C atoms are bonded to each other.**

- Start with the singlets. Signal [C] is due to a CH$_3$ group with no adjacent nonequivalent H atoms. Possible structures include:

$$CH_3O- \quad \text{or} \quad \overset{\overset{\textstyle O}{\|}}{\underset{CH_3}{C}} \quad \text{or} \quad CH_3-\overset{|}{\underset{|}{C}}-$$

- Because signal [A] is a **triplet,** there must be **2 H's** (CH_2 group) on the adjacent carbon.
- Because signal [B] is a **quartet,** there must be **3 H's** (CH_3 group) on the adjacent carbon.
- This information suggests that **X** has an **ethyl** group – – –→ CH_3CH_2-

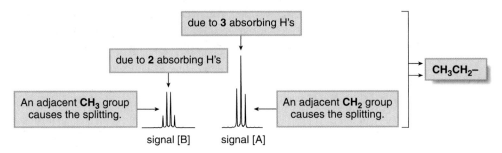

To summarize, **X** contains CH_3-, CH_3CH_2-, and $C=O$ (from the IR). Comparing these atoms with the molecular formula shows that one O atom is missing. Because O atoms do not absorb in a ^1H NMR spectrum, their presence can only be inferred by examining the chemical shift of protons near them. O atoms are more electronegative than C, thus deshielding nearby protons, and shifting their absorption downfield.

Step [4]   **Use chemical shift data to complete the structure.**

- Put the structure together in a manner that preserves the splitting data and is consistent with the reported chemical shifts.
- In this example, two isomeric structures (**A** and **B**) are possible for **X** considering the splitting data only:

| Structural pieces | | | Possible structures | |

CH_3-
 ↑
 H_c

CH_3CH_2-
 ↑   ↑
 H_a H_b

O
‖
C

$-O-$

– – –→

O
‖
$CH_3CH_2\!-\!C\!-\!OCH_3$
 ↑   ↑   ↑
 H_a H_b H_c

A

or

O
‖
$CH_3\!-\!C\!-\!OCH_2CH_3$
 ↑   ↑   ↑
 H_c H_b H_a

B

- Chemical shift information distinguishes the two possibilities. The electronegative O atom deshields adjacent H's, shifting them downfield between 3 and 4 ppm. If **A** is the correct structure, the singlet due to the CH_3 group (H_c) should occur downfield, whereas if **B** is the correct structure, the quartet due to the CH_2 group (H_b) should occur downfield.
- Because the NMR of **X** has a singlet (not a quartet) at 3.7, **A is the correct structure.**

PROBLEM 15.23   Propose a structure for a compound of molecular formula $C_7H_{14}O_2$ with an IR absorption at 1740 cm^{-1} and the following NMR data:

Absorption	ppm	Integration value
singlet	1.2	26
triplet	1.3	10
quartet	4.1	6

PROBLEM 15.24   Propose a structure for a compound of molecular formula C_3H_8O with an IR absorption at 3600–3200 cm^{-1} and the following NMR spectrum:

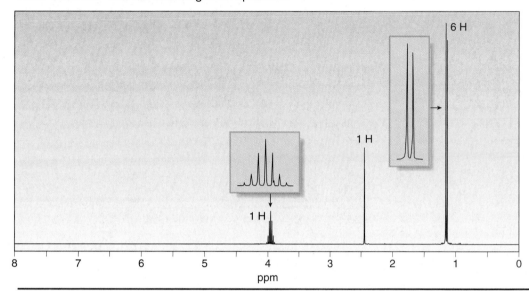

PROBLEM 15.25   The ^{1}H NMR spectrum of melatonin, the chapter-opening molecule, is more complex than other examples we have encountered, but the chemical shift and splitting patterns observed for several peaks can be explained by what we have learned about ^{1}H NMR thus far. (a) Which protons in melatonin give rise to signals [A]–[D]? (b) Explain the splitting pattern observed in signal [C].

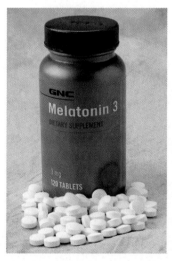

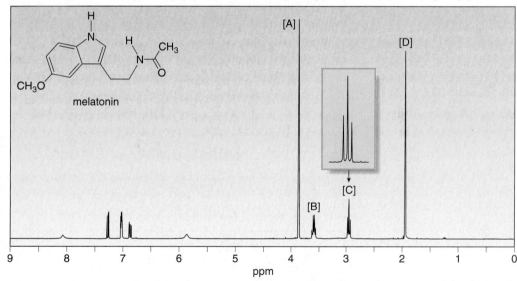

15.11 ^{13}C NMR Spectroscopy

Although less commonly used than ^{1}H NMR, ^{13}C NMR spectroscopy is also an important tool for organic structure analysis. The physical basis for ^{13}C NMR is the same as for ^{1}H NMR. When placed in a magnetic field, $B_\circ$, ^{13}C nuclei can align themselves with or against $B_\circ$. More nuclei are aligned with $B_\circ$ because this arrangement is lower in energy, but these nuclei can be made to spin flip against the applied field by applying RF radiation of the appropriate frequency.

^{13}C NMR spectra, like ^{1}H NMR spectra, plot peak intensity versus chemical shift, using TMS as the reference peak at 0 ppm. ^{13}C occurs in only 1.1% natural abundance, however, so ^{13}C NMR signals are much weaker than ^{1}H NMR signals. To overcome this limitation, modern spectrometers irradiate samples with many pulses of RF radiation and use complex mathematical tools to increase signal sensitivity and decrease background noise. The spectrum of acetic acid (CH$_3$COOH) illustrates the general features of a ^{13}C NMR spectrum.

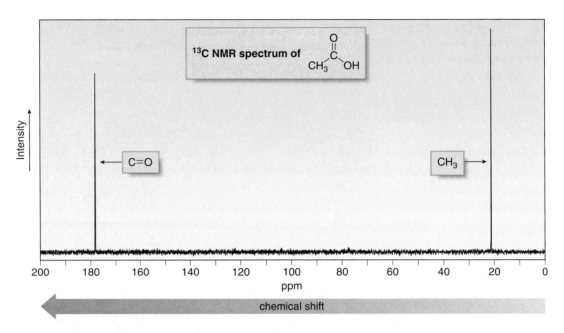

^{13}C NMR spectra are easier to analyze than ^1H spectra because signals are not split. **Each type of carbon atom appears as a single peak.**

Why aren't ^{13}C signals split by nearby carbon atoms? Recall from Section 15.6 that splitting occurs when two NMR active nuclei—like two protons—are close to each other. Because of the low natural abundance of ^{13}C nuclei (1.1%), the chance of two ^{13}C nuclei being bonded to each other is very small (0.01%), and so no carbon–carbon splitting is observed.

A ^{13}C NMR signal can also be split by nearby protons. This $^1H-^{13}C$ splitting is usually eliminated from a spectrum, however, by using an instrumental technique that decouples the proton–carbon interactions, so that every peak in a ^{13}C NMR spectrum is a singlet.

Two features of ^{13}C NMR spectra provide the most structural information: the **number of signals** observed and the **chemical shifts** of those signals.

15.11A ^{13}C NMR: Number of Signals

◆ The number of signals in a ^{13}C spectrum gives the number of different types of carbon atoms in a molecule.

Carbon atoms in the same environment give the same NMR signal, whereas carbons in different environments give different NMR signals. The ^{13}C NMR spectrum of CH_3COOH has two signals because there are two different types of carbon atoms—the C of the CH_3 group and the C of the carbonyl (C=O).

◆ Because ^{13}C NMR signals are not split, the number of signals equals the number of lines in the ^{13}C NMR spectrum.

Thus, the ^{13}C NMR spectra of dimethyl ether, chloroethane, and methyl acetate exhibit one, two, and three lines, respectively, because these compounds contain one, two, and three different types of carbon atoms.

$$C_a \qquad\qquad C_a$$
$$CH_3-O-CH_3$$
dimethyl ether

1 ^{13}C NMR signal

Both C's are equivalent.

$$C_a \qquad C_b$$
$$CH_3-CH_2-Cl$$
chloroethane

2 ^{13}C NMR signals

$$C_a \quad O \quad C_b$$
$$CH_3 \quad\quad OCH_3 \quad C_c$$
methyl acetate

3 ^{13}C NMR signals

In contrast to the proton NMR situation, peak intensity is not proportional to the number of absorbing carbons, so ^{13}C NMR signals are not integrated.

SAMPLE PROBLEM 15.6   How many lines are observed in the ^{13}C NMR spectrum of each compound?

a. $CH_3CH_2CH_2CH_2CH_3$

b.

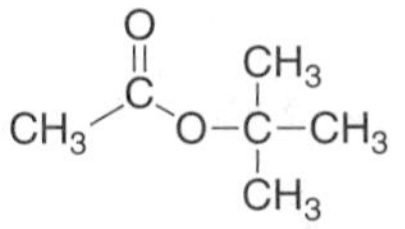

c.

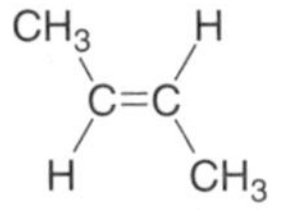

SOLUTION

The number of different types of carbons equals the number of lines in a ^{13}C NMR spectrum.

a.

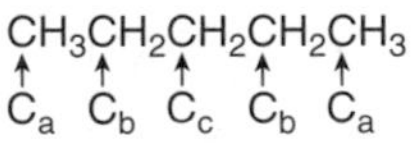

3 types of C's
3 ^{13}C NMR signals

b.

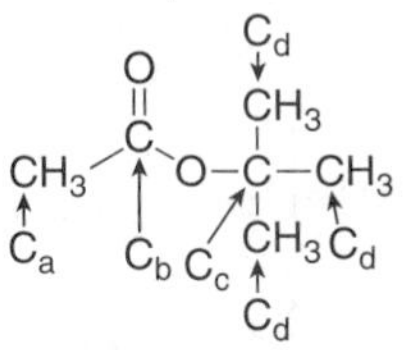

4 types of C's
4 ^{13}C NMR signals

c.

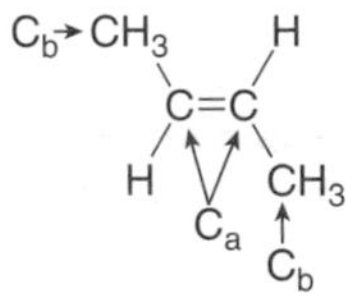

2 types of C's
2 ^{13}C NMR signals

PROBLEM 15.26   How many lines are observed in the ^{13}C NMR spectrum of each compound?

a. $CH_3CH_2CH_2CH_3$

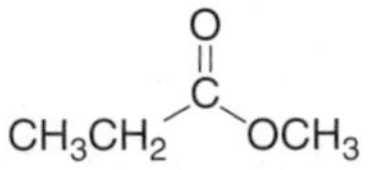

c. $CH_3CH_2CH_2-O-CH_2CH_2CH_3$

b.

d.

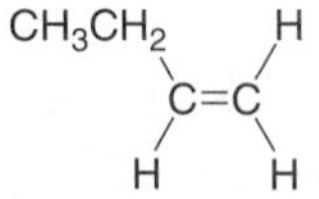

PROBLEM 15.27   Which compound(s) will give only one peak in both its ^{1}H and ^{13}C NMR spectra?

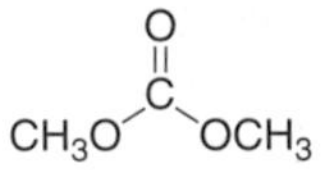

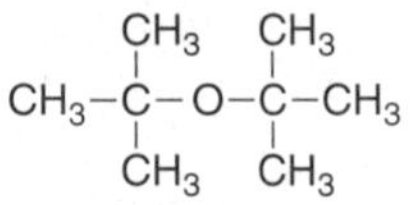

15.11B ^{13}C NMR: Position of Signals

In contrast to the small range of chemical shifts in ^{1}H NMR (0–10 ppm usually), ^{13}C NMR absorptions occur over a much broader range, 0–220 ppm. The chemical shifts of carbon atoms in ^{13}C NMR depend on the same effects as the chemical shifts of protons in ^{1}H NMR:

♦ The sp^3 hybridized C atoms of alkyl groups are shielded and absorb upfield.

♦ Electronegative elements like halogen, nitrogen, and oxygen shift absorptions downfield.

♦ The sp^2 hybridized C atoms of alkenes and benzene rings absorb downfield.

♦ Carbonyl carbons are highly deshielded, and absorb further downfield than other carbon types.

Table 15.5 lists common ^{13}C chemical shift values. The ^{13}C NMR spectra of 1-propanol ($CH_3CH_2CH_2OH$) and methyl acetate ($CH_3CO_2CH_3$) in Figure 15.14 illustrate these principles.

TABLE 15.5   Common ^{13}C Chemical Shift Values

Type of carbon	Chemical shift (ppm)	Type of carbon	Chemical shift (ppm)
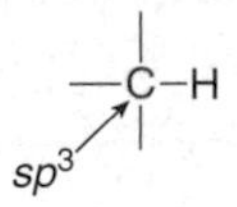 sp^3	5–45	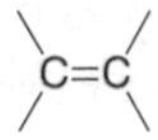	100–140
—C≡C—	65–100	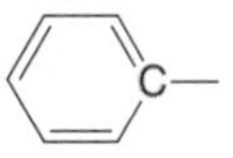	120–150
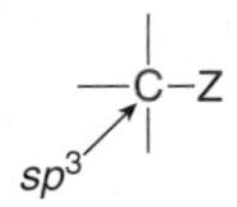 sp^3 Z = N, O, X	30–80	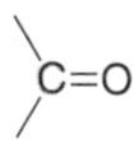	160–210

Figure 15.14
Representative ^{13}C NMR spectra

[a] 1-Propanol

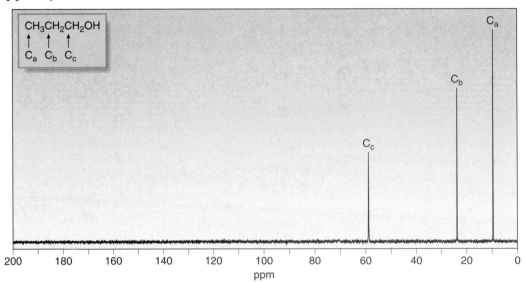

- The three types of C's in 1-propanol—identified as C_a, C_b, and C_c—give rise to three ^{13}C NMR signals.
- Deshielding increases with increasing proximity to the electronegative O atom, and the absorption shifts downfield; thus, in order of increasing chemical shift: $C_a < C_b < C_c$.

[b] Methyl acetate

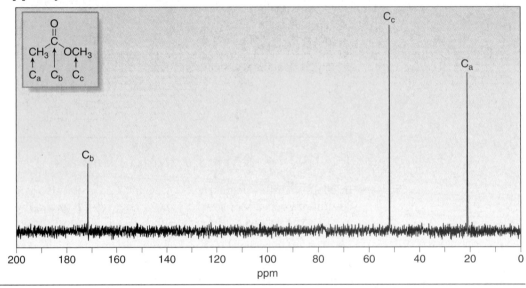

- The three types of C's in methyl acetate—identified as C_a, C_b, and C_c—give rise to three ^{13}C NMR signals.
- The carbonyl carbon (C_b) is highly deshielded so it absorbs furthest downfield.
- C_a, an sp^3 hybridized C that is not bonded to an O atom, is the most shielded, and so it absorbs furthest upfield.
- Thus, in order of increasing chemical shift: $C_a < C_c < C_b$.

PROBLEM 15.28   Which of the indicated carbon atoms in each molecule absorbs further downfield?

a. $CH_3CH_2OCH_2CH_3$

b. $BrCH_2CHBr_2$

c. $H-C(=O)-OCH_3$

d. $CH_3CH=CH_2$

PROBLEM 15.29   Identify the carbon atoms that give rise to each NMR signal.

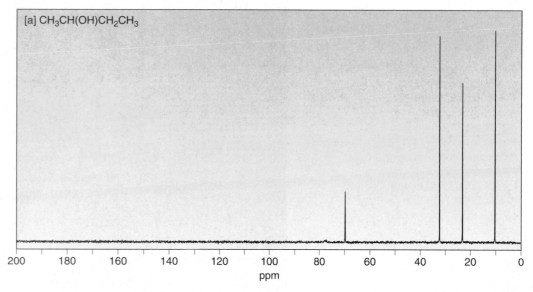

[a] CH₃CH(OH)CH₂CH₃

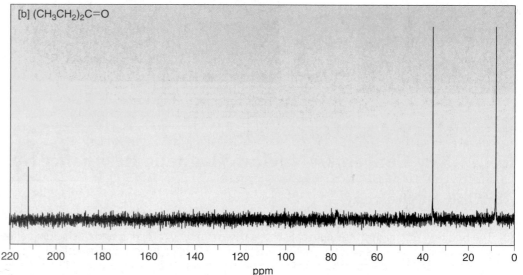

[b] (CH₃CH₂)₂C=O

PROBLEM 15.30   A compound of molecular formula $C_4H_8O_2$ shows no IR peaks at 3600–3200 or 1700 cm⁻¹. It exhibits one singlet in its ¹H NMR spectrum at 3.69 ppm, and one line in its ¹³C NMR spectrum at 67 ppm. What is the structure of this unknown?

15.12   Magnetic Resonance Imaging (MRI)

Magnetic resonance imaging (MRI)—NMR spectroscopy in medicine—is a powerful diagnostic technique. The "sample" is the patient, who is placed in a large cavity in a magnetic field, and then irradiated with RF energy. Because RF energy has very low frequency and low energy, the method is safer than X-rays or computed tomography (CT) scans that employ high-frequency, high-energy radiation that is known to damage living cells.

Living tissue contains protons in different concentrations and environments. When irradiated with RF energy, these protons are excited to a higher energy spin state, and then fall back to the lower energy spin state. These data are analyzed by a computer that generates a plot that delineates tissues of different proton density (Figure 15.15). MRIs can be recorded in any plane. Moreover, because the calcium present in bones is not NMR active, an MRI instrument can "see through" bones such as the skull and visualize the soft tissue underneath.

Figure 15.15   An MRI image of the neck

An MRI instrument is especially useful for visualizing soft tissue. In 2002, 60 million MRI procedures were performed. The 2003 Nobel Prize in Physiology or Medicine was awarded to chemist Paul C. Lauterbur and physicist Sir Peter Mansfield for their contributions in developing magnetic resonance imaging.

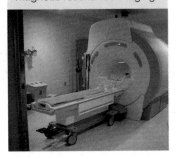

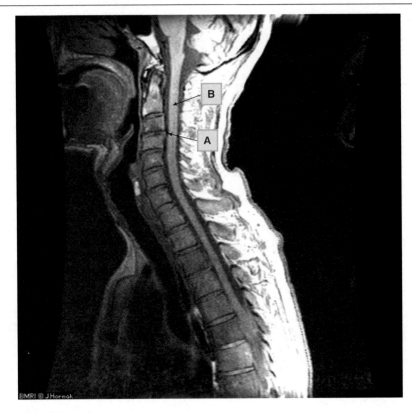

A: Spinal cord compression from a herniated disc
B: Spinal cord (would not be visualized with conventional X-rays)

15.13   Key Concepts—Nuclear Magnetic Resonance Spectroscopy

^{1}H NMR Spectroscopy

[1] The **number of signals** equals the number of different types of protons (15.2).
[2] The **position of a signal** (its chemical shift) is determined by shielding and deshielding effects.
  - Shielding shifts an absorption upfield; deshielding shifts an absorption downfield.
  - Electronegative atoms withdraw electron density, deshield a nucleus, and shift an absorption downfield (15.3).

—C–H	This proton is shielded. Its absorption is upfield, 0.9–2 ppm.

—C–H with X below	This proton is deshielded. Its absorption is further downfield, 2.5–4 ppm.

  - Loosely held π electrons can either shield or deshield a nucleus. Protons on benzene rings and double bonds are deshielded and absorb downfield, whereas protons on triple bonds are shielded and absorb upfield (15.4).

deshielded H
downfield absorption

—C≡C–H

shielded H
upfield absorption

[3] The **area under an NMR signal** is proportional to the number of absorbing protons (15.5).
[4] **Spin–spin splitting** tells about nearby nonequivalent protons (15.6–15.8).
  - Equivalent protons do not split each other's signals.
  - A set of n nonequivalent protons on the same carbon or adjacent carbons splits an NMR signal into $n + 1$ peaks.
  - OH and NH protons do not cause splitting (15.9).
  - When an absorbing proton has two sets of nearby nonequivalent protons that are equivalent to each other, use the $n + 1$ rule to determine splitting.
  - When an absorbing proton has two sets of nearby nonequivalent protons that are not equivalent to each other, the number of peaks in the NMR signal = $(n + 1)(m + 1)$.

^{13}C NMR Spectroscopy (15.11)

[1] **The number of signals equals the number of different types of carbon atoms.** All signals are **single peaks.**
[2] The relative position of ^{13}C signals is determined by shielding and deshielding effects.
   - Carbons that are sp^3 hybridized are shielded and absorb upfield.
   - Electronegative elements (N, O, and halogen) shift absorptions downfield.
   - The carbons of alkenes and benzene rings absorb downfield.
   - Carbonyl carbons are highly deshielded, and absorb further downfield than other carbon types.

Problems

^{1}H NMR Spectroscopy—Determining Equivalent Protons

15.31 How many different types of protons are present in each compound?

a. $(CH_3)_3CH$

b. $(CH_3)_3CC(CH_3)_3$

c. $CH_3CH_2OCH_2CH_2CH_2CH_2CH_3$

d. $CH_3CH{=}CH_2$

e.

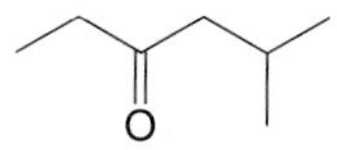

f.

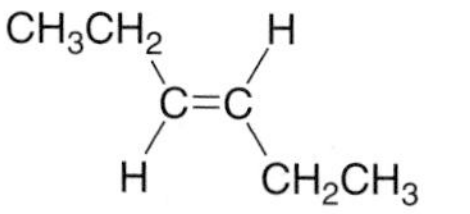

g. $CH_3CH_2CH_2OCH_2CH_2CH_3$

h.

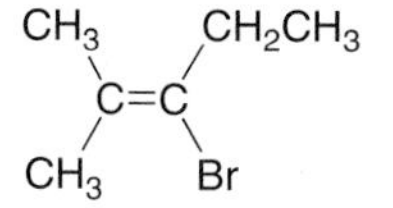

i. $CH_3CH(OH)CH_2CH_3$

j.

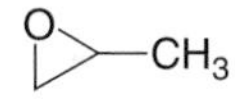

15.32 Draw the structures for the four isomeric dihalides with molecular formula $C_3H_6Cl_2$ and label each different type of proton.

15.33 How many ^{1}H NMR signals does each compound give?

a. 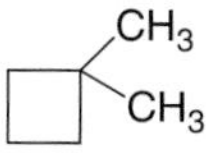 b. 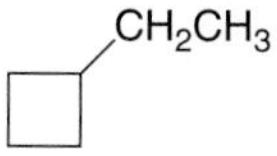 c. 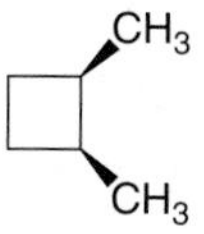 d.

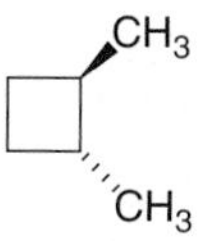

15.34 How many ^{1}H NMR signals does each natural product exhibit?

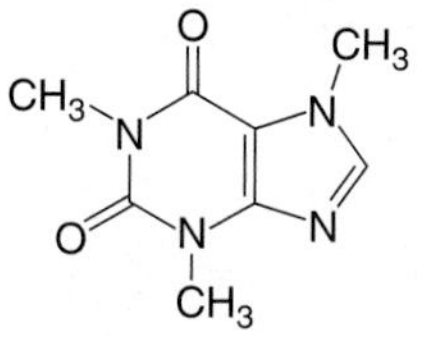

caffeine
(from coffee beans and tea leaves)

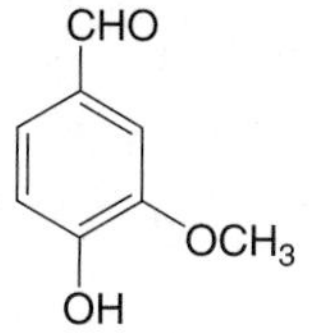

vanillin
(from the vanilla bean)

^{1}H NMR—Chemical Shift and Integration

15.35 Using a 300 MHz NMR instrument:
   a. How many Hz downfield from TMS is a peak at 2.5 ppm?
   b. If a signal comes at 1200 Hz downfield from TMS, at what ppm does it occur?
   c. If two peaks are separated by 2 ppm, how many Hz does this correspond to?

15.36 Which of the indicated protons in each pair absorbs further downfield?

a. $CH_3CH_2CH_2CH_2CH_3$ or $CH_3CH_2CH_2OCH_3$

b. $CH_3CH_2CH_2I$ or $CH_3CH_2CH_2F$

c. $CH_3OCH_2CH_3$ or

d. $CH_3CH_2CHBr_2$ or $CH_3CH_2CH_2Br$

15.37 A compound of molecular formula C_6H_{10} gives three signals in its ^{1}H NMR spectrum with the following integration units: 13, 33, 73 units. How many protons are responsible for each signal?

^{1}H NMR—Splitting

15.38 Which compounds give one singlet in the ^{1}H NMR spectrum?

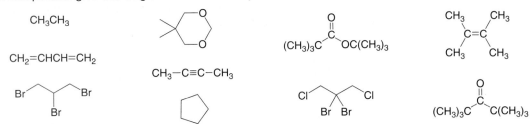

CH_3CH_3

$CH_2=CHCH=CH_2$

$CH_3-C\equiv C-CH_3$

$(CH_3)_3C\,\overset{\overset{\displaystyle O}{\|}}{C}\,OC(CH_3)_3$

15.39 For the five isomeric alkanes of molecular formula C_6H_{14}, label each type of proton and indicate how many peaks each will exhibit in its ^{1}H NMR signal.

15.40 Into how many peaks will the signal for each of the indicated protons be split?

a. $CH_3CH(OCH_3)_2$

b. $CH_3OCH_2CH_2\,\overset{\overset{\displaystyle O}{\|}}{C}\,OCH_3$

c. (benzene ring)—CH_2CH_3

d. $CH_3OCH_2CHCl_2$

e. $(CH_3)_2CH\,\overset{\overset{\displaystyle O}{\|}}{C}\,OCH_2CH_3$

f. $HOCH_2CH_2CH_2OH$

g. $CH_3CH_2CH_2CH_2OH$

h. $CH_3CH_2CH_2\,\overset{\overset{\displaystyle O}{\|}}{C}\,OH$

i. $CH_3CH_2\,\overset{\overset{\displaystyle O}{\|}}{C}\,H$

j. (alkene structure)

k. (alkene structure)

l. (alkene structure)

Using ^{1}H NMR, IR, and MS to Determine a Structure

15.41 Propose a structure consistent with each set of spectral data:

a. $C_4H_8Br_2$: IR peak at 3000–2850 cm^{-1}; NMR (ppm):
   1.87 (singlet, 6 H)
   3.86 (singlet, 2 H)

b. $C_9H_{18}O$: IR peak at 1710 cm^{-1}; NMR (ppm):
   1.2 (singlet)

c. $C_2H_4Cl_2$: IR peak at 3000–2850 cm^{-1}; NMR (ppm):
   2.1 (doublet)
   5.9 (quartet)

d. $C_3H_6Br_2$: IR peak at 3000–2850 cm^{-1}; NMR (ppm):
   2.4 (quintet)
   3.5 (triplet)

e. $C_5H_{10}O_2$: IR peak at 1740 cm^{-1}; NMR (ppm):
   1.15 (triplet, 3 H)   2.30 (quartet, 2 H)
   1.25 (triplet, 3 H)   4.72 (quartet, 2 H)

f. $C_6H_{14}O$: IR peak at 3600–3200 cm^{-1}; NMR (ppm):
   0.8 (triplet, 6 H)   1.5 (quartet, 4 H)
   1.0 (singlet, 3 H)   1.6 (singlet, 1 H)

g. $C_6H_{14}O$: IR peak at 3000–2850 cm^{-1}; NMR (ppm):
   1.10 (doublet, 30 units)
   3.60 (septet, 5 units)

h. C_3H_6O: IR peak at 1730 cm^{-1}; NMR (ppm):
   1.11 (triplet)
   2.46 (multiplet)
   9.79 (triplet)

15.42 Identify the structures of isomers **A** and **B** (molecular formula $C_9H_{10}O$).

Compound A	Compound B
IR absorption at 1742 cm^{-1}	IR absorption at 1688 cm^{-1}
NMR data:	NMR data:

Absorption	ppm	Absorption	ppm
singlet	2.15 (3 H)	triplet	1.22 (3 H)
singlet	3.70 (2 H)	quartet	2.98 (2 H)
broad singlet	7.20 (5 H)	multiplet	7.28–7.95 (5 H)

15.43 How would each pair of compounds be different in their ^{1}H NMR spectra?

a. (benzene ring)—OCH_3 and (benzene ring)—CH_2OH

b. (benzene ring)—CH_2CH_3 and (benzene ring)—OCH_2CH_3

15.44 Identify the structures of isomers **C** and **D** (molecular formula $C_4H_8O_2$).

Compound C: IR absorption at 1743 cm^{-1}

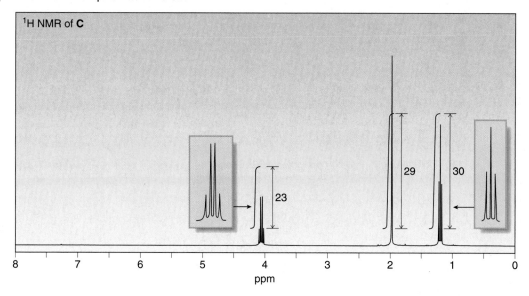

Compound D: IR absorption at 1730 cm^{-1}

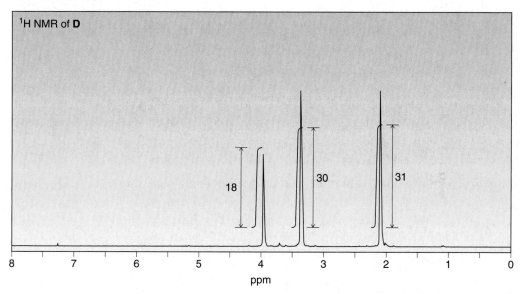

15.45 Propose a structure consistent with each set of data.

   a. $C_9H_{10}O_2$:   IR absorption at 1718 cm^{-1}

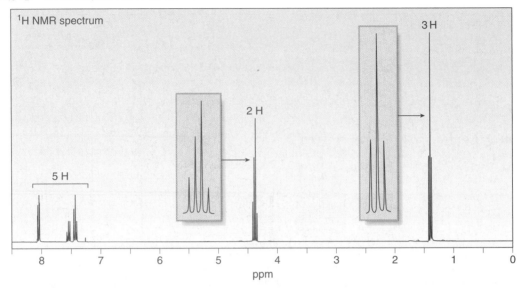

   b. C_9H_{12}:   IR absorption at 2850–3150 cm^1

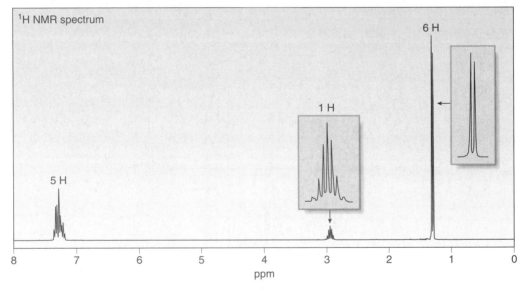

15.46 Propose a structure consistent with each set of data.

   a. Compound **A** has a molecular ion at 72 and gives a peak in its IR spectrum at 1710 cm^{-1}. ^{1}H NMR data (ppm):

   1.0 (triplet, 3 H)
   2.1 (singlet, 3 H)
   2.4 (quartet, 2 H)

   b. Compound **B** has a molecular ion at 88 and gives a peak in its IR spectrum at 3600–3200 cm^{-1}. ^{1}H NMR data (ppm):

   0.9 (triplet, 3 H)
   1.2 (singlet, 6 H)
   1.5 (quartet, 2 H)
   1.6 (singlet, 1 H)

15.47 A solution of acetone, [(CH$_3$)$_2$C=O], in ethanol (CH$_3$CH$_2$OH) in the presence of a trace of acid was allowed to stand for several days, and a new compound of molecular formula C$_7$H$_{16}$O$_2$ was formed. The IR spectrum showed only one major peak in the functional group region around 3000 cm^{-1}, and the ^{1}H NMR spectrum is given here. What is the structure of the product?

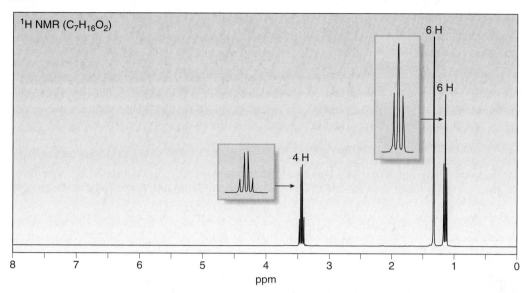

^{1}H NMR (C$_7$H$_{16}$O$_2$)

15.48 Treatment of 2-butanone (CH$_3$COCH$_2$CH$_3$) with strong base followed by CH$_3$I forms a compound **W,** which gives a molecular ion in its mass spectrum at 86. The IR (>1500 cm^{-1} only) and ^{1}H NMR spectrum of **W** are given below. What is the structure of **W**?

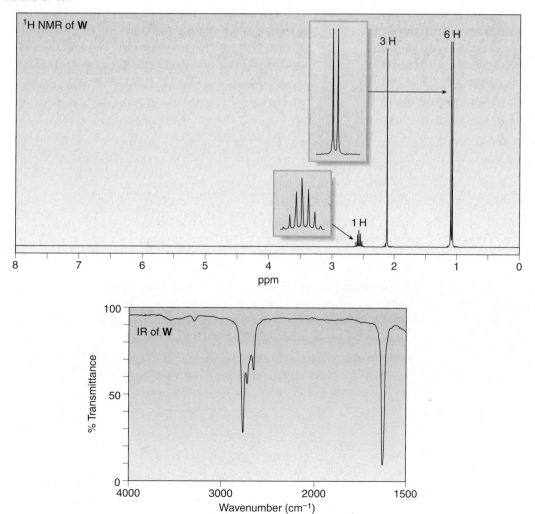

^{1}H NMR of **W**

IR of **W**

15.49 Low molecular weight esters (RCO_2R) often have characteristic odors. Using its molecular formula and ^1H NMR spectral data, identify each ester.
   a. Compound **A**, the odor of banana: $C_7H_{14}O_2$; ^1H NMR: 0.93 (doublet, 6 H), 1.52 (multiplet, 2 H), 1.69 (multiplet, 1 H), 2.04 (singlet, 3 H), and 4.10 (triplet, 2 H) ppm
   b. Compound **B**, the odor of rum: $C_7H_{14}O_2$; ^1H NMR: 0.94 (doublet, 6 H), 1.15 (triplet, 3 H), 1.91 (multiplet, 1 H), 2.33 (quartet, 2 H), and 3.86 (doublet, 2 H) ppm

15.50 When 2-bromo-3,3-dimethylbutane is treated with $K^+ \, ^-OC(CH_3)_3$, a single product **A** having molecular formula C_6H_{12} is formed. When 3,3-dimethyl-2-butanol is treated with H_2SO_4, the major product **B** has the same molecular formula. Given the following ^1H NMR data, what are the structures of **A** and **B**? Explain in detail the splitting patterns observed for the three split signals in **A**.
   ^1H NMR of **A**: 1.01 (singlet, 9 H), 4.82 (doublet of doublets, 1 H, J = 10, 1.7 Hz), 4.93 (doublet of doublets, 1 H, J = 18, 1.7 Hz), and 5.83 (doublet of doublets, 1 H, J = 18, 10 Hz) ppm
   ^1H NMR of **B**: 1.60 (singlet) ppm

15.51 In a Baeyer–Villiger reaction, ketones ($R_2C=O$) are converted to esters (RCO_2R) by using peroxy acids. With an unsymmetrical ketone, two possible esters can be formed, as shown for 3,3-dimethyl-2-butanone as starting material. How could you use spectroscopic techniques—^1H NMR, IR, and MS—to determine which ester (**A** or **B**) is formed?

^{13}C NMR

15.52 Draw the four constitutional isomers having molecular formula C_4H_9Br and indicate how many different kinds of carbon atoms each has.

15.53 Which compounds in Problem 15.38 give one signal in their ^{13}C NMR spectra?

15.54 How many ^{13}C NMR signals does each compound exhibit?

   a. $HC(CH_3)_3$
   b.
   c. $CH_3OCH(CH_3)_2$
   d.
   e.
   f.
   g.
   h.
   i.

15.55 Rank the indicated carbon atoms in each compound in order of increasing chemical shift.

15.56 Identify the carbon atoms that give rise to the signals in the ^{13}C NMR spectrum of each compound.
   a. $CH_3CH_2CH_2CH_2OH$; ^{13}C NMR: 14, 19, 35, and 62 ppm
   b. $(CH_3)_2CHCHO$; ^{13}C NMR: 16, 41, and 205 ppm
   c. $CH_2=CHCH(OH)CH_3$; ^{13}C NMR: 23, 69, 113, and 143 ppm

15.57 Propose a structure consistent with each set of data.

a. A compound **X** (molecular formula $C_6H_{12}O_2$) gives a strong peak in its IR spectrum at 1740 cm^{-1}. The ^{1}H NMR spectrum of **X** shows only two singlets, including one at 3.5 ppm. The ^{13}C NMR spectrum is given below. Propose a structure for **X**.

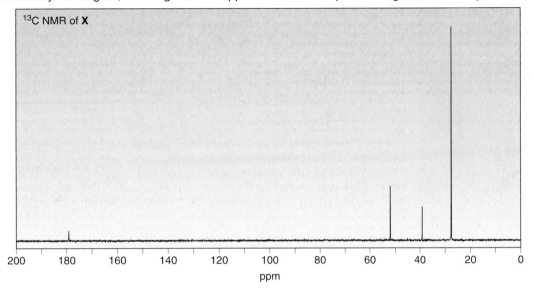

^{13}C NMR of **X**

b. A compound **Y** (molecular formula C_6H_{10}) gives four lines in its ^{13}C NMR spectrum (27, 30, 67, and 93 ppm), and the IR spectrum given here. Propose a structure for **Y**.

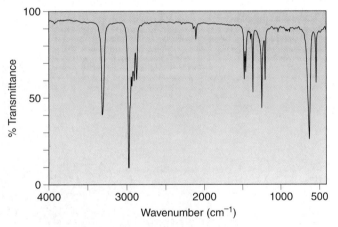

% Transmittance

Wavenumber (cm^{-1})

Challenge Problems

15.58 The ^{1}H NMR spectrum of *N,N*-dimethylformamide shows three singlets at 2.9, 3.0, and 8.0 ppm. Explain why the two CH$_3$ groups are not equivalent to each other, thus giving rise to two NMR signals.

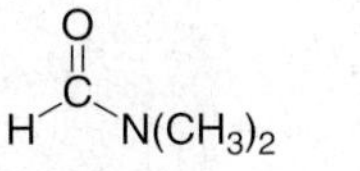

N,N-dimethylformamide

15.59 18-Annulene shows two signals in its ^{1}H NMR spectrum, one at 8.9 (12 H) and one at –1.8 (6 H) ppm. Using a similar argument to that offered for the chemical shift of benzene protons, explain why both shielded and deshielded values are observed for 18-annulene.

18-annulene

15.60 Explain why the ^{13}C NMR spectrum of 3-methyl-2-butanol shows five signals.

CHAPTER 16

Conjugation, Resonance, and Dienes

Lycopene is a red pigment found in tomatoes, watermelon, papaya, guava, and pink grape-fruit. An antioxidant like vitamin E, lycopene contains many conjugated double bonds—double bonds separated by only one single bond—that allow π electron density to delocalize and give the molecule added stability. In Chapter 16 we learn about such conjugated unsaturated systems.

Chapter 16 is the first of three chapters that discusses the chemistry of conjugated molecules—molecules with *p* orbitals on three or more adjacent atoms. Chapter 16 focuses mainly on acyclic conjugated compounds, whereas Chapters 17 and 18 discuss the chemistry of benzene and related compounds that have a *p* orbital on every atom in a ring.

Much of Chapter 16 is devoted to the properties and reactions of 1,3-dienes. To understand these compounds, however, we must first learn about the consequences of having *p* orbitals on three or more adjacent atoms. Because the ability to draw resonance structures is also central to mastering this material, the key aspects of resonance theory are presented in detail.

16.1 Conjugation

The word *conjugation* is derived from the Latin *conjugatus*, meaning "to join."

***Conjugation* occurs whenever *p* orbitals are located on three or more adjacent atoms.** Two common conjugated systems are 1,3-dienes and allylic carbocations.

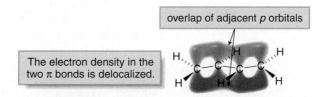

1,3-diene   allylic carbocation

16.1A 1,3-Dienes

1,3-Dienes such as 1,3-butadiene contain two carbon–carbon double bonds joined by a single σ bond. Each carbon atom of a 1,3-diene is bonded to three other atoms and has no nonbonded electron pairs, so each carbon atom is *sp²* hybridized and has one *p* orbital containing an electron. **The four *p* orbitals on adjacent atoms make a 1,3-diene a conjugated system.**

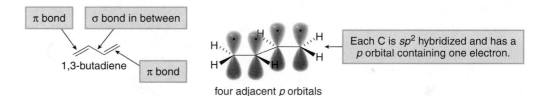

four adjacent *p* orbitals

What is special about conjugation? Having three or more *p* orbitals on adjacent atoms allows *p* orbitals to overlap and electrons to delocalize.

overlap of adjacent *p* orbitals

The electron density in the two π bonds is delocalized.

◆ **When *p* orbitals overlap, the electron density in each of the π bonds is spread out over a larger volume, thus lowering the energy of the molecule and making it more stable.**

Conjugation makes 1,3-butadiene inherently different from 1,4-pentadiene, a compound having two double bonds separated by more than one σ bond. The π bonds in 1,4-pentadiene are too far apart to be conjugated.

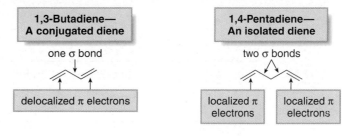

1,3-Butadiene—A conjugated diene	1,4-Pentadiene—An isolated diene
one σ bond	two σ bonds
delocalized π electrons	localized π electrons / localized π electrons

Figure 16.1   Electrostatic
potential plots for a conjugated
and an isolated diene

**1,3-Butadiene—
A conjugated diene**

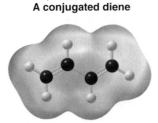

The red electron-rich region is
spread over four adjacent atoms.

**1,4-Pentadiene—
An isolated diene**

The red electron-rich regions are localized in
the π bonds on the two ends of the molecule.

1,4-Pentadiene is an **isolated diene.** The electron density in each π bond of an isolated diene is
localized between two carbon atoms. In 1,3-butadiene, however, the electron density of both π
bonds is *delocalized* over the four atoms of the diene. Electrostatic potential maps in Figure 16.1
clearly indicate the difference between these localized and delocalized π bonds.

PROBLEM 16.1   Classify each diene as isolated or conjugated.

a.   b.   c.   d.

16.1B   Allylic Carbocations

The **allyl carbocation** is another example of a conjugated system. The three carbon atoms of the
allyl carbocation—the positively charged carbon atom and the two that form the double bond—
are sp^2 hybridized with a p orbital. The p orbitals for the double bond carbons each contain an
electron, whereas the p orbital for the carbocation is empty.

$CH_2=CH-\overset{+}{C}H_2$
allyl carbocation

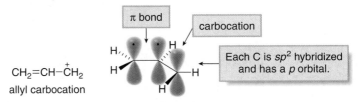

◆ **Three p orbitals on three adjacent atoms, even if one of the p orbitals is empty, make
the allyl carbocation conjugated.**

Conjugation stabilizes the allyl carbocation because overlap of three adjacent p orbitals delocalizes the electron density of the π bond over three atoms.

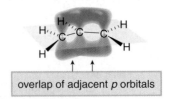

PROBLEM 16.2   Which of the following species are conjugated?

a. $CH_2=CH-CH=CH-CH=CH_2$ b.   c.   d.   e.

16.2   Resonance and Allylic Carbocations

The word *resonance* is used in two different contexts. In NMR spectroscopy, a nucleus is *in resonance* when it absorbs energy, promoting it to a higher energy state. In drawing molecules, there is *resonance* when two different Lewis structures can be drawn for the same arrangement of atoms.

Recall from Section 1.5 that resonance structures are two or more different Lewis structures for the same arrangement of atoms. Being able to draw correct resonance structures is crucial to understanding conjugation and the reactions of conjugated dienes.

◆ Two resonance structures differ in the placement of π bonds and nonbonded electrons. The placement of atoms and σ bonds stays the same.

We have already drawn resonance structures for the acetate anion (Section 2.5C) and the allyl radical (Section 13.10). The **conjugated allyl carbocation** is another example of a species for which two resonance structures can be drawn. Drawing resonance structures for the allyl carbocation is a way to use Lewis structures to illustrate how conjugation delocalizes electrons.

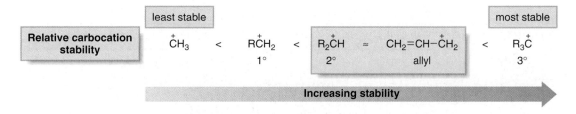

The true structure of the allyl carbocation is a hybrid of the two resonance structures. In the hybrid, the π bond is delocalized over all three atoms. As a result, the positive charge is also delocalized over the two terminal carbons. Delocalizing electron density lowers the energy of the hybrid, thus stabilizing the allyl carbocation and making it more stable than a normal 1° carbocation. Experimental data show that its stability is comparable to a more highly substituted 2° carbocation.

Relative carbocation stability

least stable

$\overset{+}{C}H_3$ < $R\overset{+}{C}H_2$ < $R_2\overset{+}{C}H$ ≈ $CH_2=CH-\overset{+}{C}H_2$ < $R_3\overset{+}{C}$

1°   2°   allyl   3°

most stable

Increasing stability

The electrostatic potential maps in Figure 16.2 compare the resonance-stabilized allyl carbocation with $CH_3CH_2CH_2^+$, a localized 1° carbocation. The electron-deficient region—the site of the positive charge—is concentrated on a single carbon atom in the 1° carbocation $CH_3CH_2CH_2^+$. In the allyl carbocation, however, the electron-poor region is spread out on both terminal carbons.

Figure 16.2   Electrostatic potential maps for a localized and a delocalized carbocation

$CH_3CH_2CH_2^+$
a localized carbocation

$CH_2=CHCH_2^+$ ⟷ $^+CH_2CH=CH_2$
a delocalized carbocation

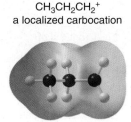

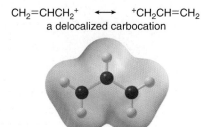

The electron-deficient region (in blue) of a **1° carbocation** is concentrated on a single carbon atom.

The electron-deficient region (in blue-green) of the **allyl carbocation** is distributed over both terminal carbons.

Figure 16.3 ^{13}C chemical shifts for a localized and a resonance-stabilized carbocation

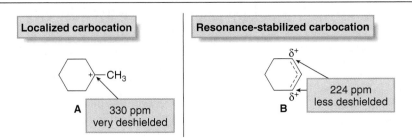

• The absorption shifts *upfield* as the amount of positive charge *decreases*.

PROBLEM 16.3   Draw a second resonance structure for each carbocation. Then draw the hybrid.

a.   b.   c.

How "real" is the delocalization of charge and electron density in the allyl carbocation? Recall from the discussion of NMR spectroscopy in Chapter 15 that an NMR absorption shifts downfield (to higher chemical shift) as the electron density around the nucleus decreases (is deshielded). Thus, a positively charged carbocation, such as **A** in Figure 16.3, is highly deshielded, so its ^{13}C NMR absorption is far downfield at 330 ppm. A resonance-stabilized carbocation, however, such as **B** in Figure 16.3, has less positive charge concentrated on any given carbon atom (because the charge is delocalized), so its ^{13}C NMR absorption is further upfield at 224 ppm.

PROBLEM 16.4   How many ^{13}C NMR signals are predicted for carbocations **A** and **B** in Figure 16.3?

16.3   Common Examples of Resonance

When are resonance structures drawn for a molecule or reactive intermediate? Because resonance involves delocalizing π bonds and nonbonded electrons, one or both of these structural features must be present to draw additional resonance forms. There are four common bonding patterns for which more than one Lewis structure can be drawn.

Type [1] The Three Atom "Allyl" System, X=Y–Z*

◆ **For any group of three atoms having a double bond X=Y and an atom Z that contains a *p* orbital with zero, one, or two electrons, two resonance structures are possible:**

$$X=Y-Z \longleftrightarrow X-Y=Z$$

0, 1, or 2 electrons

The asterisk [*] corresponds to a charge, a radical, or a lone pair

$* = +, -, \cdot, \text{ or } \cdot\cdot$

This is called **allyl** type resonance because it can be drawn for allylic carbocations, allylic carbanions, and allylic radicals.

X, Y, and Z may all be carbon atoms, as in the case of an allylic carbocation (resonance structures **A** and **B**), or they may be heteroatoms, as in the case of the acetate anion (resonance structures **C** and **D**). The atom Z bonded to the multiple bond can be charged (a net positive or negative charge) or neutral (having zero, one, or two nonbonded electrons). **The two resonance** structures differ in the location of the double bond, and either the charge, the radical, or the lone pair, generalized by [*].

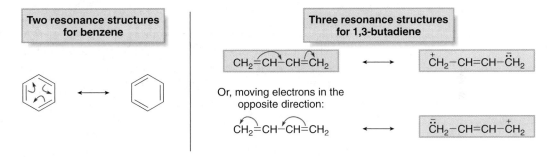

| Allylic carbocation | Acetate anion |

A   B   C   D

Type [2] Conjugated Double Bonds

Cyclic, completely conjugated rings like benzene have two resonance structures, drawn by moving the electrons in a cyclic manner around the ring. Three resonance structures can be drawn for other conjugated dienes, two of which involve charge separation.

Two resonance structures for benzene

Three resonance structures for 1,3-butadiene

$$CH_2=CH-CH=CH_2 \longleftrightarrow \overset{+}{C}H_2-CH=CH-\overset{..}{C}H_2$$

Or, moving electrons in the opposite direction:

$$CH_2=CH-CH=CH_2 \longleftrightarrow \overset{..}{C}H_2-CH=CH-\overset{+}{C}H_2$$

Type [3] Cations Having a Positive Charge Adjacent to a Lone Pair

◆ When a lone pair and a positive charge are located on adjacent atoms, two resonance structures can be drawn.

General case

$$\overset{..}{X}-\overset{+}{Y} \longleftrightarrow \overset{+}{X}=Y$$

Specific example

$$CH_3-\overset{..}{O}-\overset{+}{C}H_2 \longleftrightarrow CH_3-\overset{+}{\underset{..}{O}}=CH_2$$

The overall charge is the same in both resonance structures. Based on formal charge, a neutral X in one structure must bear a (+) charge in the other.

Type [4] Double Bonds Having One Atom More Electronegative Than the Other

◆ For a double bond X=Y in which the electronegativity of Y > X, a second resonance structure can be drawn by moving the π electrons onto Y.

General case

$$X=Y \longleftrightarrow \overset{+}{X}-\overset{..}{Y}:$$

Electronegativity of Y > X.

Specific example

$$\underset{CH_3}{\overset{CH_3}{C}}=\overset{..}{O}: \longleftrightarrow \underset{CH_3}{\overset{CH_3}{\overset{+}{C}}}-\overset{..}{\underset{..}{O}}:^-$$

| Charge separation results. |

Sample Problem 16.1 illustrates how to apply these different types of resonance to actual molecules.

SAMPLE PROBLEM 16.1   Draw two more resonance structures for each species.

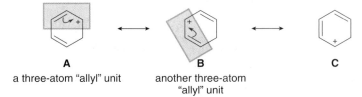

SOLUTION

Mentally breaking a molecule into two- or three-atom units can make it easier to draw additional resonance structures.

[a] Think of the top three atoms of the six-membered ring in **A** as an "allyl" unit. Moving the π bond forms a new "allyl" unit in **B,** and moving the π bond in **B** generates a third resonance structure **C.** No new valid resonance structures are generated by moving electrons in **C.**

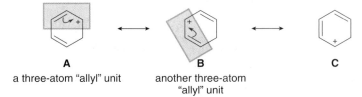

A	**B**	**C**
a three-atom "allyl" unit	another three-atom "allyl" unit	

[b] Compound **D** contains a carbonyl group, so moving the electron pair in the double bond to the more electronegative oxygen atom separates the charge and generates structure **E. E** now has a three-atom "allyl" unit, so the remaining π bond can be moved to form structure **F.**

D	**E**	**F**
	a three-atom "allyl" unit	

PROBLEM 16.5   Draw additional resonance structures for each ion.

a. $CH_2=CH-\ddot{C}H-CH=CH_2$ b. 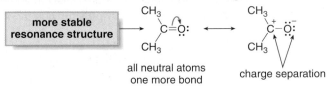   c. $CH_3-\overset{+}{C}H-\ddot{\underset{\cdot\cdot}{C}l}:$

16.4   The Resonance Hybrid

Although the resonance hybrid is some combination of all of its valid resonance structures, the **hybrid more closely resembles the most stable resonance structure.** Recall from Section 1.5C that the most stable resonance structure is called the **major contributor** to the hybrid, and the less stable resonance structures are called the **minor contributors.**

Use the following three rules to evaluate the relative stabilities of two or more valid resonance structures.

Rule [1]   **Resonance structures with more bonds and fewer charges are more stable.**

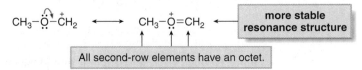

Rule [2]   **Resonance structures in which every atom has an octet are more stable.**

$$CH_3-\overset{\frown}{\ddot{O}}-\overset{+}{C}H_2 \quad\longleftrightarrow\quad CH_3-\overset{+}{\ddot{O}}=CH_2$$

All second-row elements have an octet.

| Rule [3] | Resonance structures that place a negative charge on a more electronegative atom are more stable. |

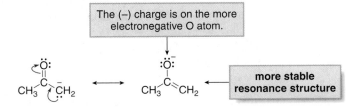

The (−) charge is on the more electronegative O atom.

more stable resonance structure

Sample Problem 16.2 illustrates how to determine the relative energy of contributing resonance structures and the hybrid.

SAMPLE PROBLEM 16.2

Draw a second resonance structure for carbocation **A,** as well as the hybrid of both resonance structures. Then use Rules [1]–[3] to rank the relative energy of both resonance structures and the hybrid.

A

SOLUTION

Because **A** contains a positive charge and a lone pair on adjacent atoms, a second resonance structure **B** can be drawn. Because **B** has more bonds and all second-row atoms have octets, **B** is more stable than **A,** making it the major contributor to the hybrid **C.** Because the hybrid is more stable than either resonance contributor, the order of stability is:

| **A** | **B** | **C** |
| minor contributor | major contributor | hybrid |

Increasing stability

PROBLEM 16.6

Draw a second resonance structure and the hybrid for each species, and then rank the two resonance structures and the hybrid in order of increasing stability.

a. $(CH_3)_2\overset{+}{C}-\ddot{N}H_2$ b.

$$CH_3 \quad \overset{:\ddot{O}:}{\underset{}{\overset{\|}{C}}} \quad \overset{-}{\ddot{N}}H$$

16.5   Electron Delocalization, Hybridization, and Geometry

To delocalize nonbonded electrons or electrons in π bonds, there must be *p* orbitals that can overlap. This may mean that the hybridization of an atom is different than would have been predicted using the rules first outlined in Chapter 1.

For example, there are two Lewis structures (**A** and **B**) for the resonance-stabilized anion $(CH_3COCH_2)^-$.

A   **B**

| The C is surrounded by four groups—three atoms and one nonbonded electron pair. **Is it sp^3 hybridized?** | Here the C is surrounded by **three** groups—three atoms and no nonbonded electron pairs. **Is it sp^2 hybridized?** |

Based on structure **A,** the indicated carbon is sp^3 hybridized, with the lone pair of electrons in an sp^3 hybrid orbital. Based on structure **B,** though, it is sp^2 hybridized with the unhybridized *p* orbital forming the π portion of the double bond.

Delocalizing electrons stabilizes a molecule. The electron pair on the carbon atom adjacent to the C=O can only be delocalized, though, if it has a *p* orbital that can overlap with two other *p* orbitals on two adjacent atoms. Thus, the terminal carbon atom is sp^2 hybridized with trigonal planar geometry. **Three adjacent *p* orbitals make the anion conjugated.**

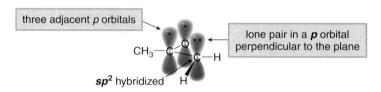

◆ In any system X=Y−Z:, Z is sp^2 hybridized, and the nonbonded electron pair occupies a *p* orbital to make the system conjugated.

SAMPLE PROBLEM 16.3 Determine the hybridization around the indicated carbon atom in the following anion.

$$CH_2=CH-\overset{\displaystyle \cdot\cdot}{C}H-CH_3$$
$$\uparrow$$

SOLUTION

Because this is an example of an allyl-type system (X=Y−Z*), a second resonance structure can be drawn that "moves" the lone pair and the π bond. To delocalize the lone pair and make the system conjugated, the labeled carbon atom must be sp^2 hybridized with the lone pair occupying a *p* orbital.

Two resonance structures	$CH_2=CH-\overset{\cdot\cdot}{C}H-CH_3$ ⟷ $\overset{\cdot\cdot}{C}H_2-CH=CH-CH_3$

The indicated C atom must be sp^2 hybridized, with the lone pair in a *p* orbital.

PROBLEM 16.7 Determine the hybridization of the indicated atom in each species.

a. (cyclopentanone structure with indicated atom) =O

b. $CH_3-\overset{\displaystyle :\overset{\cdot\cdot}{O}:}{\underset{\displaystyle :\overset{\cdot\cdot}{O}:^-}{C}}$ ←

c. (benzene ring with $\overset{\cdot\cdot}{C}H_2^-$)

16.6 Conjugated Dienes

*Compounds with many π bonds are called **polyenes**.*

In the remainder of Chapter 16 we examine **conjugated dienes,** compounds having two double bonds joined by one σ bond. Conjugated dienes are also called **1,3-dienes.** 1,3-Butadiene ($CH_2=CH-CH=CH_2$) is the simplest conjugated diene.

Three stereoisomers are possible for 1,3-dienes with alkyl groups bonded to each end carbon of the diene (RCH=CH−CH=CHR).

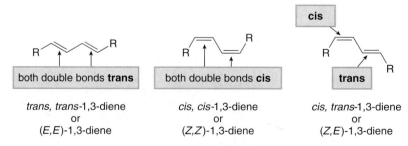

trans, trans-1,3-diene or (*E,E*)-1,3-diene	*cis, cis*-1,3-diene or (*Z,Z*)-1,3-diene	*cis, trans*-1,3-diene or (*Z,E*)-1,3-diene

Two possible conformations result from rotation around the C−C bond that joins the two double bonds.

♦ The **s-cis** conformer has two double bonds on the **same side** of the single bond.

♦ The **s-trans** conformer has two double bonds on **opposite sides** of the single bond.

Keep in mind that stereoisomers are discrete molecules, whereas conformers interconvert. Three structures drawn for 2,4-hexadiene illustrate the differences between stereoisomers and conformers in a 1,3-diene:

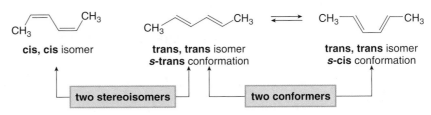

PROBLEM 16.8   Draw the structure consistent with each description.
a. (2*E*,4*E*)-2,4-octadiene in the *s*-trans conformation
b. (3*E*,5*Z*)-3,5-nonadiene in the *s*-cis conformation
c. (3*Z*,5*Z*)-4,5-dimethyl-3,5-decadiene. Draw both the *s*-cis and *s*-trans conformers.

16.7   Interesting Dienes and Polyenes

Isoprene and **lycopene** are two naturally occurring compounds with conjugated double bonds.

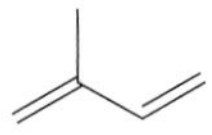

isoprene
(2-methyl-1,3-butadiene)

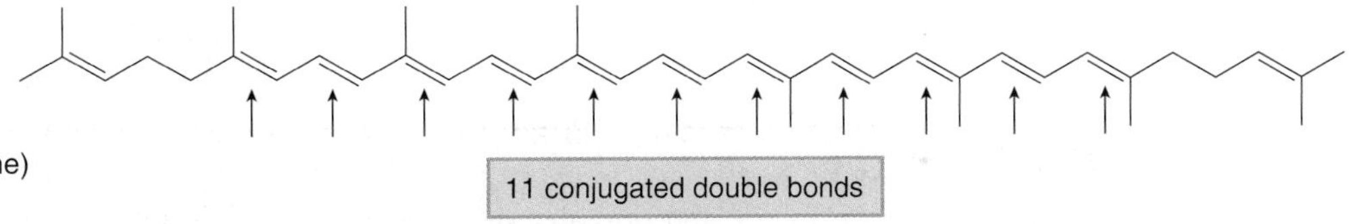

11 conjugated double bonds

lycopene

The Blue Ridge Mountains

Isoprene, the common name for 2-methyl-1,3-butadiene, is given off by plants as the temperature rises, a process thought to increase a plant's tolerance for heat stress. Isoprene is a component of the blue haze seen above forested hillsides, such as Virginia's Blue Ridge Mountains.

Unlike most of the organic compounds encountered up to this point, **lycopene,** the chapter-opening molecule, is colored. The 11 conjugated double bonds of lycopene cause its red color, a phenomenon discussed in Section 16.15.

Many biologically active compounds contain conjugated double bonds along with other functional groups (Figure 16.4). These include **vitamin D₃,** which is needed for forming and maintaining healthy bones; the cholesterol-lowering medicine **simvastatin** (trade name Zocor); and **linearmycin B,** a polyene antibiotic that exhibits both antifungal and antibacterial properties.

16.8   The Carbon–Carbon σ Bond Length in 1,3-Butadiene

Four features distinguish conjugated dienes from isolated dienes.

[1] The C–C single bond joining the two double bonds is unusually short.
[2] Conjugated dienes are more stable than similar isolated dienes.
[3] Some reactions of conjugated dienes are different than reactions of isolated double bonds.
[4] Conjugated dienes absorb longer wavelengths of ultraviolet light.

The bond lengths of the carbon–carbon double bonds in 1,3-butadiene are similar to an isolated double bond, but the central carbon–carbon single bond is shorter than the C–C bond in ethane.

Figure 16.4 Biologically active organic compounds that contain conjugated double bonds

vitamin D₃

simvastatin (Zocor)

linearmycin B

The observed bond distances can be explained by looking at hybridization. Each carbon atom in 1,3-butadiene is sp^2 hybridized, so the central C−C single bond is formed by the overlap of two sp^2 hybridized orbitals, rather than the sp^3 hybridized orbitals used to form the C−C bond in CH_3CH_3.

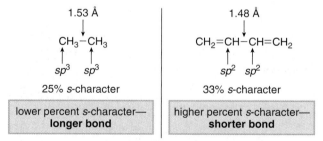

1.53 Å
CH_3-CH_3
sp^3 sp^3
25% *s*-character
lower percent *s*-character— **longer bond**

1.48 Å
$CH_2=CH-CH=CH_2$
sp^2 sp^2
33% *s*-character
higher percent *s*-character— **shorter bond**

Recall from Section 1.10B that increasing percent *s*-character decreases bond length.

◆ Based on hybridization, a $C_{sp^2}-C_{sp^2}$ bond should be shorter than a $C_{sp^3}-C_{sp^3}$ bond because it is formed from orbitals having a higher percent *s*-character.

A resonance argument can also be used to explain the shorter C−C σ bond length of 1,3-butadiene. 1,3-Butadiene can be represented by three resonance structures:

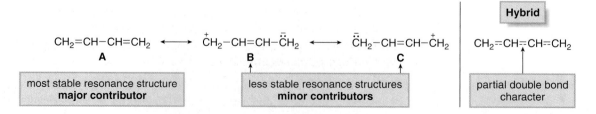

				Hybrid
$CH_2=CH-CH=CH_2$	⟷	$\overset{+}{C}H_2-CH=CH-\overset{..}{\overset{-}{C}}H_2$	⟷	$\overset{..}{\overset{-}{C}}H_2-CH=CH-\overset{+}{C}H_2$
A		**B**		**C**
most stable resonance structure **major contributor**		less stable resonance structures **minor contributors**		partial double bond character

Structures **B** and **C** have charge separation and fewer bonds than **A**, making them less stable resonance structures and only minor contributors to the resonance hybrid. **B** and **C** both contain a double bond between the central carbon atoms, however, so the hybrid must have a partial double bond there. This makes the central C–C bond shorter than a C–C single bond in an alkane.

◆ Based on resonance, the central C–C bond in 1,3-butadiene is shorter because it has partial double bond character.

Finally, 1,3-butadiene is a conjugated molecule with four overlapping *p* orbitals on adjacent atoms. As a result, the π electrons are not localized between the carbon atoms of the double bonds, but rather delocalized over four atoms. This places more electron density between the central two carbon atoms of 1,3-butadiene than would normally be present. This *shortens* the bond. Drawing resonance structures illustrates this delocalization.

The overlap of adjacent *p* orbitals increases the electron density in the C–C σ bond.

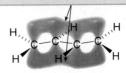

PROBLEM 16.9   Using hybridization, predict how the bond length of the C–C σ bond in HC≡C–C≡CH should compare with the C–C σ bonds in CH_3CH_3 and CH_2=CH–CH=CH_2.

PROBLEM 16.10   Rank the following bonds in order of increasing bond length.

$$CH_3CH_2\text{—}CH_3 \qquad CH_2\text{=CH—}CH_3 \qquad CH_2\text{=CH—CH=}CH_2$$

16.9 Stability of Conjugated Dienes

In Section 12.3 we learned that hydrogen adds to alkenes to form alkanes, and that the heat released in this reaction, the **heat of hydrogenation,** can be used as a measure of alkene stability.

Recall: $\overset{\diagdown}{\underset{\diagup}{C}}=\overset{\diagup}{\underset{\diagdown}{C}}$ + H_2 $\xrightarrow{\text{Pd-C}}$ $-\overset{|}{\underset{H}{C}}-\overset{|}{\underset{H}{C}}-$ $\Delta H° =$ **heat of hydrogenation**

The relative stability of conjugated and isolated dienes can also be determined by comparing their heats of hydrogenation.

◆ When hydrogenation gives the same alkane from two dienes, the more stable diene has the smaller heat of hydrogenation.

For example, both 1,4-pentadiene (an isolated diene) and (3*E*)-1,3-pentadiene (a conjugated diene) are hydrogenated to pentane with two equivalents of H_2. Because less energy is released in converting the conjugated diene to pentane, it must be lower in energy (more stable) to begin with. The relative energies of these isomeric pentadienes are illustrated in Figure 16.5.

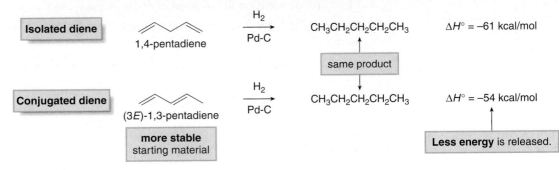

Figure 16.5
Relative energies
of an isolated and
conjugated diene

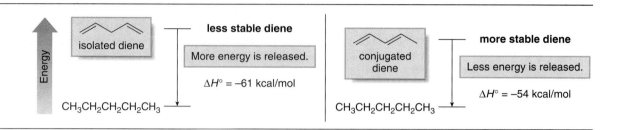

◆ **A conjugated diene has a smaller heat of hydrogenation and is more stable than a similar isolated diene.**

Why is a conjugated diene more stable than an isolated diene? Because a conjugated diene has overlapping p orbitals on four adjacent atoms, its π electrons are not localized between the carbon atoms of the double bonds, but rather delocalized over four atoms. This delocalization, which adds stability to the diene, cannot occur in an isolated diene. Drawing resonance structures illustrates this delocalization.

No resonance structures can be drawn for 1,4-pentadiene, but three can be drawn for ($3E$)-1,3-pentadiene (or any other conjugated diene). Thus, the hybrid illustrates that the two adjacent π bonds are delocalized in a conjugated diene, making it lower in energy and therefore more stable than an isolated diene.

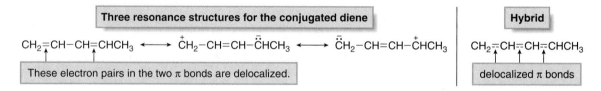

PROBLEM 16.11   Which diene in each pair has the higher heat of hydrogenation?

a. [structure] or [structure]   b. [structure] or [structure]

PROBLEM 16.12   Rank the following compounds in order of increasing stability.

[structures]

16.10   Electrophilic Addition: 1,2- Versus 1,4-Addition

Recall from Chapters 10 and 11 that the characteristic reaction of compounds with π bonds is **addition**. The π bonds in conjugated dienes undergo addition reactions, too, but they differ in two ways from the addition reactions to isolated double bonds.

◆ **Electrophilic addition in conjugated dienes gives a mixture of products.**

◆ **Conjugated dienes undergo a unique addition reaction not seen in alkenes or isolated dienes.**

We learned in Chapter 10 that HX adds to the π bond of alkenes to form alkyl halides.

Recall: $\overset{\diagdown}{\diagup}C=C\overset{\diagup}{\diagdown}$ + $\overset{\delta^+ \ \ \delta^-}{H-X}$ ⟶ $-\overset{|}{\underset{H}{C}}-\overset{|}{\underset{X}{C}}-$

(X = Cl, Br, I)   alkyl halide

This π bond is broken.

With an **isolated diene,** electrophilic addition of one equivalent of HBr yields *one* product and Markovnikov's rule is followed. The H atom bonds to the less substituted carbon—that is, the carbon atom of the double bond that had more H atoms to begin with.

$$\boxed{\text{Isolated diene}} \quad CH_2{=}CH{-}CH_2{-}CH{=}CH_2 \quad \xrightarrow[\text{(1 equiv)}]{\text{HBr}} \quad \underset{\underset{H}{|}\quad\underset{Br}{|}}{CH_2{-}CH{-}CH_2{-}CH{=}CH_2}$$

$$\boxed{\text{H bonds to the less substituted C.}}$$

With a **conjugated diene,** electrophilic addition of one equivalent of HBr affords *two* products.

$$\boxed{\text{Conjugated diene}} \quad CH_2{=}CH{-}CH{=}CH_2 \quad \xrightarrow[\text{(1 equiv)}]{\text{HBr}} \quad \overset{C1\quad C2}{\underset{\underset{H}{|}\;\underset{Br}{|}}{CH_2{-}CH{-}CH{=}CH_2}} \;+\; \overset{C1\qquad\qquad C4}{\underset{\underset{H}{|}\qquad\quad\underset{Br}{|}}{CH_2{-}CH{=}CH{-}CH_2}}$$

$$\text{1,2-product} \qquad\qquad \text{1,4-product}$$

◆ The **1,2-addition product** results from Markovnikov addition of HBr across two adjacent carbon atoms (C1 and C2) of the diene.

◆ The **1,4-addition product** results from addition of HBr to the two end carbons (C1 and C4) of the diene. 1,4-Addition is also called **conjugate addition.**

The mechanism of electrophilic addition of HX involves **two steps:** addition of H$^+$ (from HX) to form a resonance-stabilized carbocation, followed by nucleophilic attack of X$^-$ at either electrophilic end of the carbocation to form two products. Mechanism 16.1 illustrates the reaction of 1,3-butadiene with HBr.

MECHANISM 16.1

Electrophilic Addition of HBr to a 1,3-Diene—1,2- and 1,4-Addition

Step [1]   Addition of the electrophile (H$^+$) to the π bond

$$CH_2{=}CH{-}CH{=}CH_2 \quad \xrightarrow{\text{slow}} \quad \underset{\underset{H}{|}}{CH_2{-}\overset{+}{CH}{-}CH{=}CH_2} \;\longleftrightarrow\; \underset{\underset{H}{|}}{CH_2{-}CH{=}CH{-}\overset{+}{CH_2}}$$
$$H{-}Br \qquad\qquad\qquad +\;\; Br^-$$
$$\text{allylic carbocation}$$

◆ Addition of H$^+$ (from HBr) forms a new C−H bond, as the H−Br bond is broken. By Markovnikov's rule, H$^+$ adds to the less substituted carbon to form a 2° allylic carbocation, which is resonance stabilized. This is the slow step of the mechanism because two bonds are broken and only one is formed.

Step [2]   Nucleophilic attack of Br$^-$

$$\underset{\underset{H}{|}}{CH_2{-}\overset{+}{CH}{-}CH{=}CH_2} \quad\longrightarrow\quad \underset{\boxed{H}\quad\boxed{Br}}{CH_2{-}CH{-}CH{=}CH_2} \quad \boxed{\begin{array}{c}\textbf{1,2-addition}\\\textbf{product}\end{array}}$$
$$Br^-$$

$$\underset{\underset{H}{|}}{CH_2{-}CH{=}CH{-}\overset{+}{CH_2}} \quad\longrightarrow\quad \underset{\boxed{H}\qquad\quad\boxed{Br}}{CH_2{-}CH{=}CH{-}CH_2} \quad \boxed{\begin{array}{c}\textbf{1,4-addition}\\\textbf{product}\end{array}}$$
$$Br^-$$

◆ **Nucleophilic attack of Br$^-$** can occur at either site of the resonance-stabilized carbocation that bears a (+) charge, forming either the 1,2-addition product or 1,4-addition product.

Like the electrophilic addition of HX to an alkene, the addition of HBr to a conjugated diene forms the more stable carbocation in Step [1], the rate-determining step. In this case, however, the carbocation is both 2° and **allylic,** and thus two Lewis structures can be drawn for it. In the

The ends of the 1,3-diene are called C1 and C4 arbitrarily, without regard to IUPAC numbering.

second step, nucleophilic attack of Br⁻ can then occur at two different electrophilic sites, forming two different products.

◆ **Addition of HX to a conjugated diene forms 1,2- and 1,4-products because of the resonance-stabilized allylic carbocation intermediate.**

SAMPLE PROBLEM 16.4   Draw the products of the following reaction.

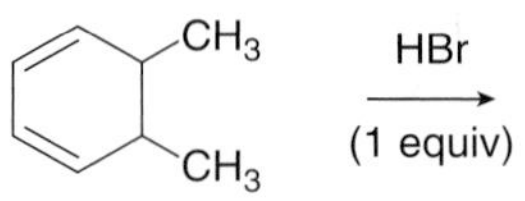

SOLUTION

Write the steps of the mechanism to determine the structure of the products. Addition of H⁺ forms the more stable 2° allylic carbocation, for which two resonance structures can be drawn. Nucleophilic attack of Br⁻ at either end of the allylic carbocation gives two constitutional isomers, formed by 1,2-addition and 1,4-addition to the diene.

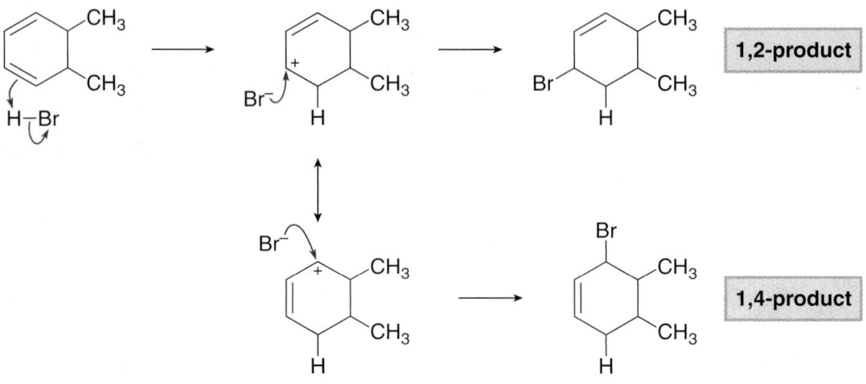

PROBLEM 16.13   Draw the products formed when each diene is treated with one equivalent of HCl.

a. CH₃CH=CH−CH=CHCH₃

b.

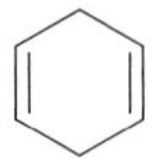

c.

d.

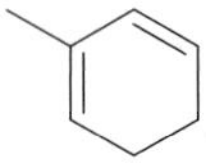

PROBLEM 16.14   Draw a stepwise mechanism for the following reaction.

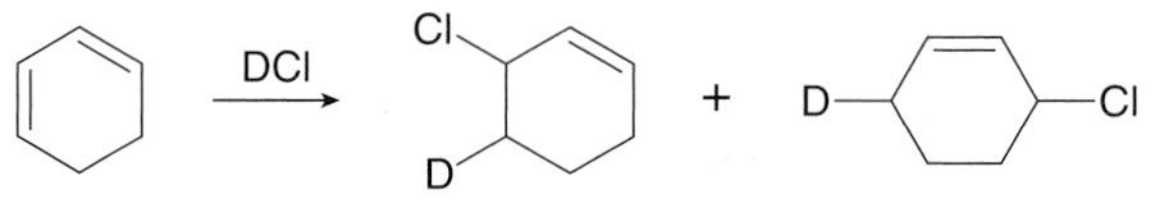

16.11   Kinetic Versus Thermodynamic Products

The amount of 1,2- and 1,4-addition products formed in the electrophilic addition reactions of conjugated dienes depends greatly on the reaction conditions.

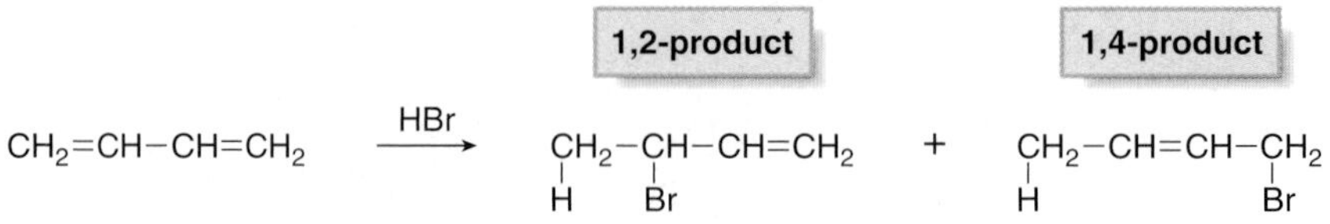

	1,2-product	1,4-product
low temperature (−80 °C)	80%	20%
high temperature (40 °C)	20%	80%

◆ **At low temperature the major product is formed by 1,2-addition.**

◆ **At higher temperature the major product is formed by 1,4-addition.**

Moreover, when a mixture containing predominately the 1,2-product is heated, the 1,4-addition product becomes the major product at equilibrium.

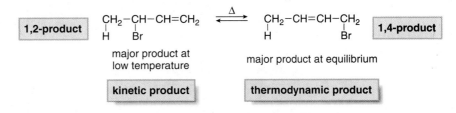

In many of the reactions we have learned thus far, the more stable product is formed faster—that is, the kinetic and thermodynamic products are the same. The electrophilic addition of HBr to 1,3-butadiene is different, in that **the more stable product is formed more slowly**—that is, the kinetic and thermodynamic products are *different*. Why is the more stable product formed more slowly?

To answer this question, recall that the rate of a reaction is determined by its energy of activation (E_a), whereas the amount of product present at equilibrium is determined by its stability (Figure 16.6). When a single starting material **A** forms two different products (**B** and **C**) by two exothermic pathways, the relative height of the energy barriers determines how fast **B** and **C** are formed, whereas the relative energies of **B** and **C** determine the amount of each at equilibrium. In an exothermic reaction, the relative energies of **B** and **C** do not determine the relative energies of activation to form **B** and **C**.

Why, in the addition of HBr to 1,3-butadiene, is the 1,2-product formed faster, but the 1,4-product more stable? The 1,4-product (1-bromo-2-butene) is more stable because it has two alkyl groups bonded to the carbon–carbon double bond, whereas the 1,2-product (3-bromo-1-butene) has only one.

Figure 16.6 How kinetic and thermodynamic products form in a reaction: **A → B + C**

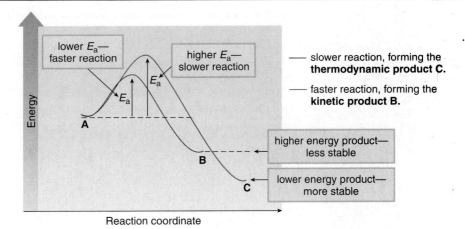

- The conversion of **A → B** is a faster reaction because the energy of activation leading to **B** is lower. **B** is the kinetic product.
- Because **C** is lower in energy, **C** is the thermodynamic product.

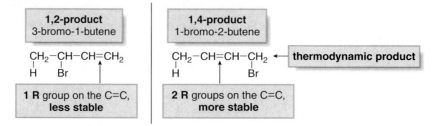

◆ More substituted alkenes are more stable, so 1-bromo-2-butene is the thermodynamic product.

The 1,2-product is the kinetic product because of a **proximity effect.** When H⁺ (from HBr) adds to the double bond, Br⁻ is closer to the adjacent carbon (C2) than it is to C4. Even though the resonance-stabilized carbocation bears a partial positive charge on both C2 and C4, attack at C2 is faster simply because Br⁻ is closer to this carbon.

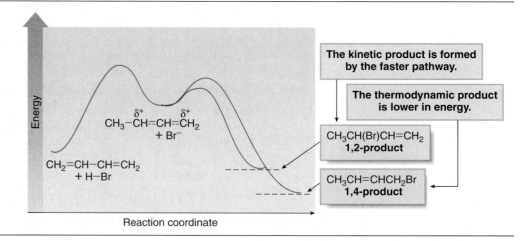

A **proximity effect** occurs because one species is close to another.

◆ The 1,2-product forms faster because of the proximity of Br⁻ to C2.

The overall two-step mechanism for addition of HBr to 1,3-butadiene, forming a 1,2-addition product and 1,4-addition product, is illustrated with the energy diagram in Figure 16.7.

Why is the ratio of products temperature dependent?

◆ At low temperature, the energy of activation is the more important factor. Because most molecules do not have enough kinetic energy to overcome the higher energy barrier at lower temperature, they react by the faster pathway, forming the kinetic product.

◆ At higher temperature, most molecules have enough kinetic energy to reach either transition state. The two products are in equilibrium with each other, and the more stable compound—which is lower in energy—becomes the major product.

Figure 16.7 Energy diagram for the two-step mechanism: $CH_2=CH-CH=CH_2$ + HBr → $CH_3CH(Br)CH=CH_2$ + $CH_3CH=CHCH_2Br$

PROBLEM 16.15   Label each product in the following reaction as a 1,2-product or 1,4-product, and decide which is the kinetic product and which is the thermodynamic product.

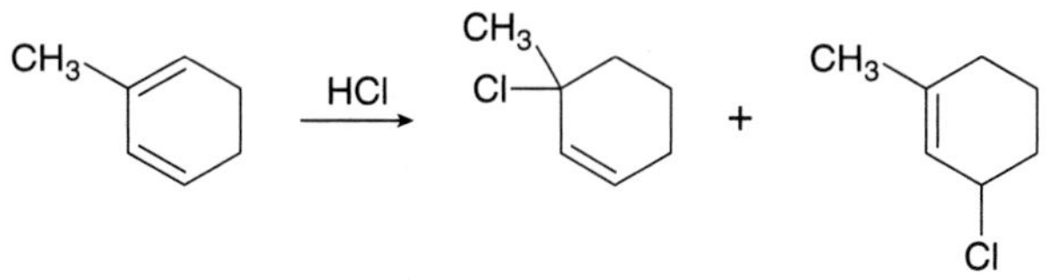

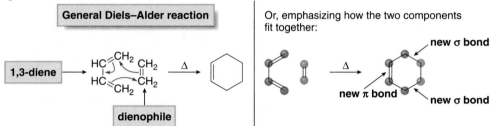

16.12   The Diels–Alder Reaction

Diels and Alder shared the 1950 Nobel Prize in Chemistry for unraveling the intricate details of this remarkable reaction.

The arrows may be drawn in a clockwise or counter-clockwise direction to show the flow of electrons in a Diels–Alder reaction.

The **Diels–Alder reaction,** named for German chemists Otto Diels and Kurt Alder, is an addition reaction between a **1,3-diene** and an alkene called a **dienophile,** to form a new six-membered ring.

Three curved arrows are needed to show the cyclic movement of electron pairs because three π bonds break and two σ bonds and one π bond form. Because each new σ bond is ~20 kcal/mol stronger than a π bond that is broken, a typical Diels–Alder reaction releases ~40 kcal/mol of energy. The following equations illustrate three examples of the Diels–Alder reaction:

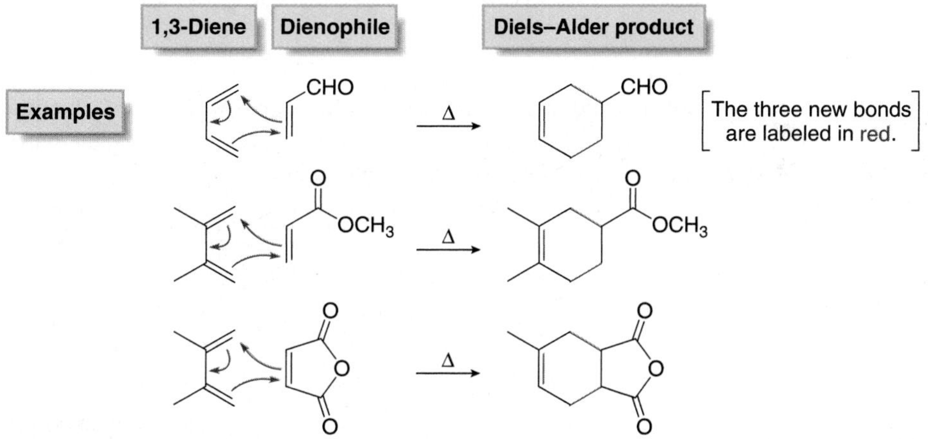

All Diels–Alder reactions have the following features in common:

[1]   **They are initiated by heat; that is, the Diels–Alder reaction is a *thermal* reaction.**
[2]   **They form new six-membered rings.**
[3]   **Three π bonds break, and two new C – C σ bonds and one new C – C π bond form.**
[4]   **They are concerted; that is, all old bonds are broken and all new bonds are formed in a single step.**

The Diels–Alder reaction forms new carbon–carbon bonds, so it can be used to synthesize larger, more complex molecules from smaller ones. For example, Figure 16.8 illustrates a Diels–Alder reaction used in the synthesis of tetrodotoxin, a toxin isolated from many different types of puffer fish.

Diels–Alder reactions may seem complicated at first, but they are really less complicated than many of the reactions you have already learned, especially those with multistep mechanisms and carbocation intermediates. The key is to learn how to arrange the starting materials to more easily visualize the structure of the product.

Figure 16.8 Synthesis of a natural product using the Diels–Alder reaction

Bonds formed from the Diels–Alder reaction are shown in red.

Diels–Alder reaction

many steps

tetrodotoxin   Japanese puffer fish

- Tetrodotoxin, a complex natural product containing several six-membered rings joined together, is a poison isolated from the ovaries and liver of the puffer fish, so named because the fish inflates itself into a ball when alarmed. Eating fish tainted with trace amounts of this potent toxin results in weakness, paralysis, and eventually death. One step in the synthesis of tetrodotoxin involves forming a six-membered ring by a Diels–Alder reaction.

How To Draw the Product of a Diels–Alder Reaction

Example Draw the product of the following Diels–Alder reaction:

$$\xrightarrow{\Delta}$$

Step [1] Arrange the 1,3-diene and the dienophile next to each other, with the diene drawn in the *s-cis* conformation.

- This step is key: **Rotate the diene** so that it is drawn in the *s-cis* conformation, and place the end C's of the diene close to the double bond of the dienophile.

Place these atoms near each other.

rotate

dienophile

1,3-diene
s-trans

s-cis

Place these atoms near each other.

Step [2] Cleave the three π bonds and use arrows to show where the new bonds will be formed.

diene dienophile   **Diels–Alder** product

PROBLEM 16.16 Draw the product formed when each diene and dienophile react in a Diels–Alder reaction.

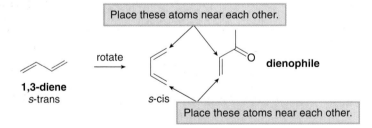

a.   + COOH   b.   +   CH₃ COOCH₃   c.   +

16.13   Specific Rules Governing the Diels–Alder Reaction

Several rules govern the course of the Diels–Alder reaction.

16.13A   Diene Reactivity

Rule [1]   The diene can react only when it adopts the *s-cis* conformation.

Both ends of the conjugated diene must be close to the π bond of the dienophile for reaction to occur. Thus, an acyclic diene in the *s-trans* conformation must rotate about the central C–C σ bond to form the *s-cis* conformation before reaction can take place.

This rotation is prevented in cyclic dienes. As a result:

◆ When the two double bonds are constrained in the ***s-cis*** conformer, the diene is unusually *reactive*.

◆ When the two double bonds are constrained in the ***s-trans*** conformer, the diene is *unreactive*.

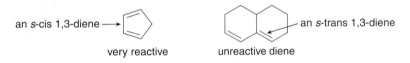

PROBLEM 16.17   Rank the following dienes in order of increasing reactivity in a Diels–Alder reaction.

PROBLEM 16.18   Why is (*Z,Z*)-2,4-hexadiene considerably less reactive than (*E,E*)-2,4-hexadiene in a Diels–Alder reaction?

16.13B   Dienophile Reactivity

Rule [2]   Electron-withdrawing substituents in the dienophile increase the reaction rate.

In a Diels–Alder reaction, the conjugated diene acts as a nucleophile and the dienophile acts as an electrophile. As a result, electron-withdrawing groups make the dienophile more electrophilic (and, thus, more reactive) by withdrawing electron density from the carbon–carbon double bond. If Z is an electron-withdrawing group, then the reactivity of the dienophile increases as follows:

$$CH_2=CH_2 \qquad CH_2=CHZ \qquad ZCH=CHZ$$

Increasing reactivity →

A carbonyl group is an effective electron-withdrawing group because the carbonyl carbon bears a partial positive charge (δ^+), which withdraws electron density from the carbon–carbon double bond of the dienophile. Common dienophiles that contain a carbonyl group are shown in Figure 16.9.

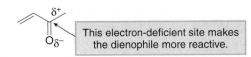

This electron-deficient site makes the dienophile more reactive.

Figure 16.9 Common dienophiles in the Diels–Alder reaction

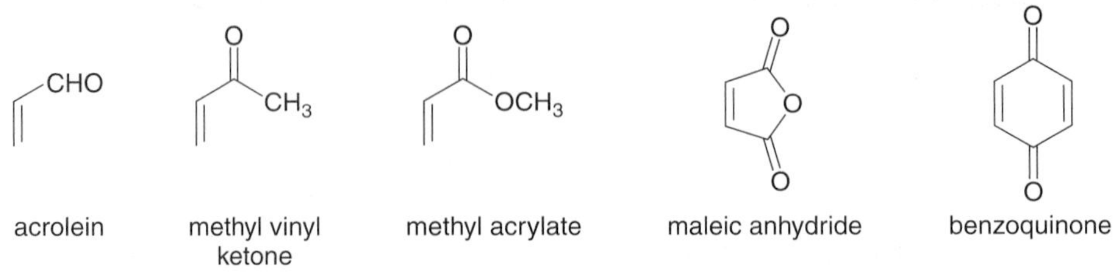

acrolein   methyl vinyl ketone   methyl acrylate   maleic anhydride   benzoquinone

PROBLEM 16.19   Rank the following dienophiles in order of increasing reactivity.

$CH_2=CHCOOH$ $CH_2=CH_2$

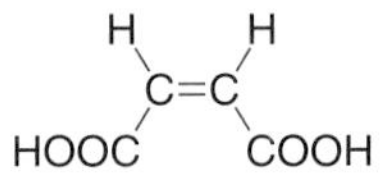

16.13C Stereospecificity

Rule [3]	The stereochemistry of the dienophile is retained in the product.

- ◆ **A cis dienophile forms a cis-substituted cyclohexene.**
- ◆ **A trans dienophile forms a trans-substituted cyclohexene.**

The two **cis** COOH groups of maleic acid become two **cis** substituents in a Diels–Alder adduct. The COOH groups can be drawn both above or both below the plane to afford a single achiral **meso** compound. The **trans dienophile** fumaric acid yields two enantiomers with **trans COOH** groups.

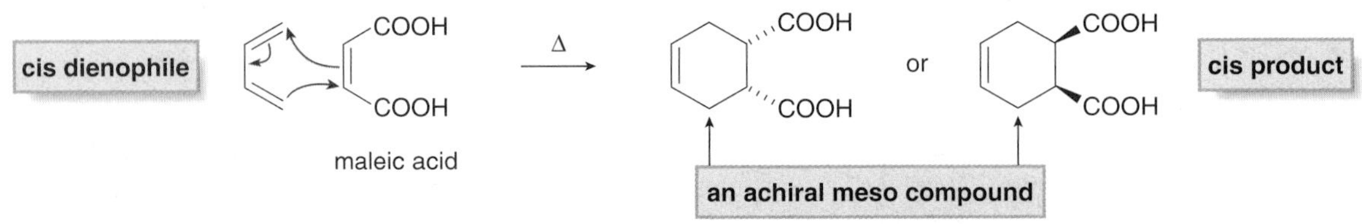

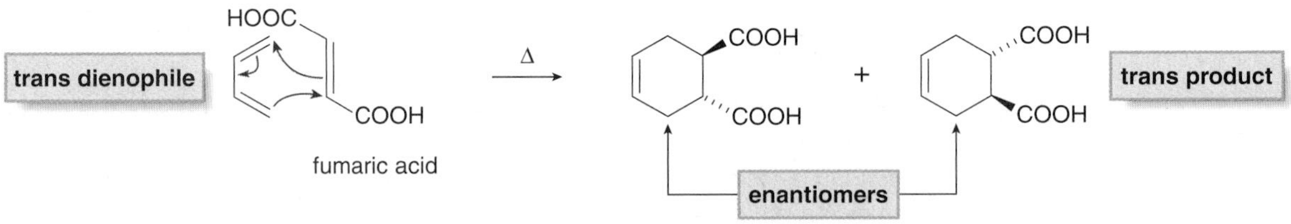

A **cyclic dienophile** forms a **bicyclic product.** A bicyclic system in which the two rings share a common C—C bond is called a **fused ring system.** The two H atoms at the ring fusion must be cis, because they were cis in the starting dienophile. A bicyclic system of this sort is said to be **cis-fused.**

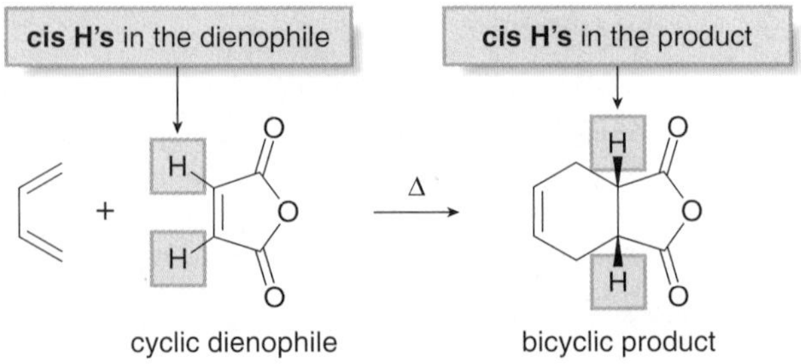

cyclic dienophile   bicyclic product

PROBLEM 16.20   Draw the products of each Diels–Alder reaction, and indicate the stereochemistry.

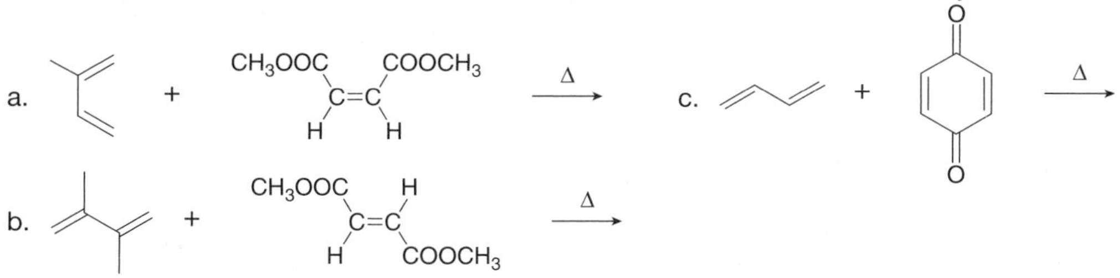

16.13D   The Rule of Endo Addition

Rule [4]	When endo and exo products are possible, the endo product is preferred.

To understand the rule of endo addition, we must first examine Diels–Alder products that result from cyclic 1,3-dienes. When cyclopentadiene reacts with a dienophile such as ethylene, a new six-membered ring forms, and above the ring there is a one atom "bridge." This carbon atom originated as the sp^3 hybridized carbon of the diene that was not involved in the reaction.

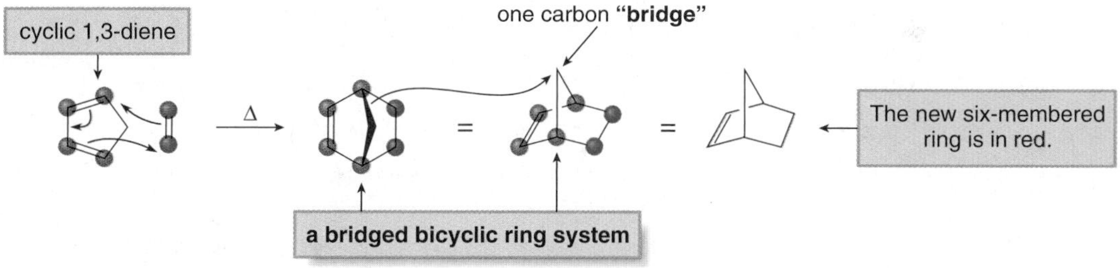

The product of the Diels–Alder reaction of a cyclic 1,3-diene is bicyclic, but the carbon atoms shared by both rings are *non-adjacent*. Thus, this bicyclic product differs from the fused ring system obtained when the dienophile is cyclic.

◆ **A bicyclic ring system in which the two rings share non-adjacent carbon atoms is called a *bridged* ring system.**

Fused and bridged bicyclic ring systems are compared in Figure 16.10.

When cyclopentadiene reacts with a substituted alkene as the dienophile (CH_2=CHZ), the substituent Z can be oriented in one of two ways in the product. The terms **endo** and **exo** are used to indicate the position of Z.

Figure 16.10 Fused and bridged bicyclic ring systems compared

A fused bicyclic system	**A bridged bicyclic system**
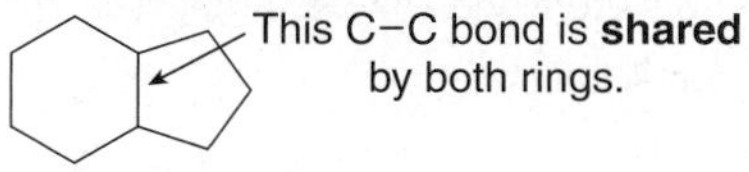 This C–C bond is **shared** by both rings.	These C's are **shared** by two rings.
• One bond is shared by two rings. • The shared C's are adjacent.	• Two non-adjacent atoms are shared by both rings.

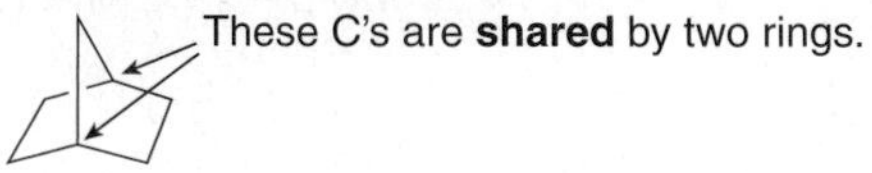

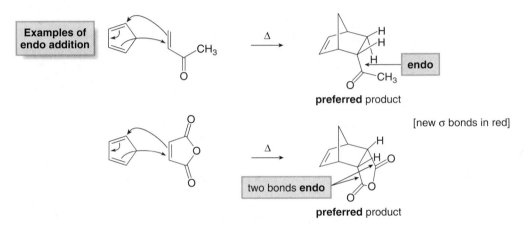

To help you distinguish endo and exo, remember that *endo* is *under* the newly formed six-membered ring.

◆ A substituent on one bridge is *endo* if it is closer to the *longer* bridge that joins the two carbons common to both rings.
◆ A substituent is *exo* if it is closer to the *shorter* bridge that joins the carbons together.

In a Diels–Alder reaction, the **endo** product is preferred, as shown in two examples.

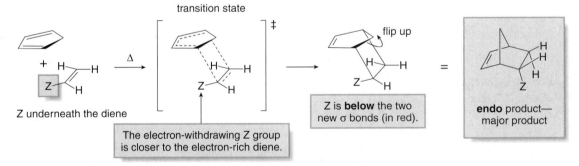

The Diels–Alder reaction is concerted, and the reaction occurs with the diene and the dienophile arranged one above the other, as shown in Figure 16.11, not side-by-side. In theory, the substituent

---

Figure 16.11
How endo and exo products are formed in the Diels–Alder reaction

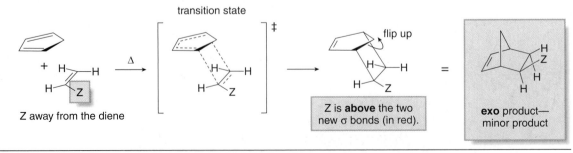

Pathway [1] With Z oriented under the diene, the endo product is formed.

Pathway [2] With Z oriented away from the diene, the exo product is formed.

Z can be oriented either directly under the diene to form the endo product (Pathway [1] in Figure 16.11) or away from the diene to form the exo product (Pathway [2] in Figure 16.11). In practice, though, the **endo product is the major product.** The transition state leading to the endo product allows more interaction between the electron-rich diene and the electron-withdrawing substituent Z on the dienophile, an energetically favorable arrangement.

PROBLEM 16.21   Draw the product of each Diels–Alder reaction.

16.14   Other Facts About the Diels–Alder Reaction

16.14A   Retrosynthetic Analysis of a Diels–Alder Product

The Diels–Alder reaction is used widely in organic synthesis, so you must be able to look at a compound and determine what conjugated diene and what dienophile were used to make it. To draw the starting materials from a given Diels–Alder adduct:

◆ **Locate the six-membered ring that contains the C=C.**

◆ **Draw three arrows around the cyclohexene ring, beginning with the π bond. Each arrow moves two electrons to the adjacent bond, cleaving one π bond and two σ bonds, and forming three π bonds.**

◆ **Retain the stereochemistry of substituents on the C=C of the dienophile. Cis substituents on the six-membered ring give a cis dienophile.**

This stepwise retrosynthetic analysis gives the 1,3-diene and dienophile needed for any Diels–Alder reaction, as shown in the two examples in Figure 16.12.

PROBLEM 16.22   What diene and dienophile are needed to prepare each product?

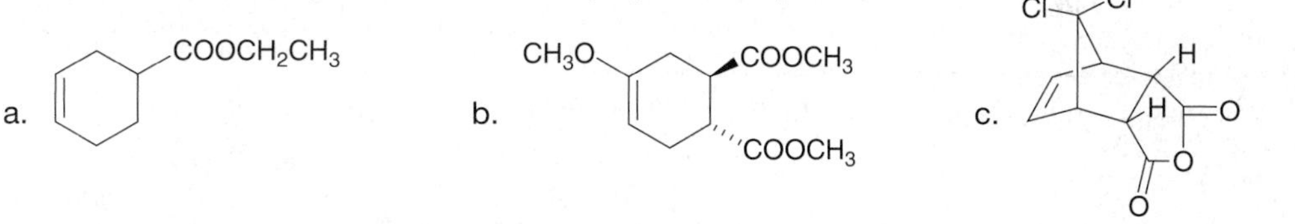

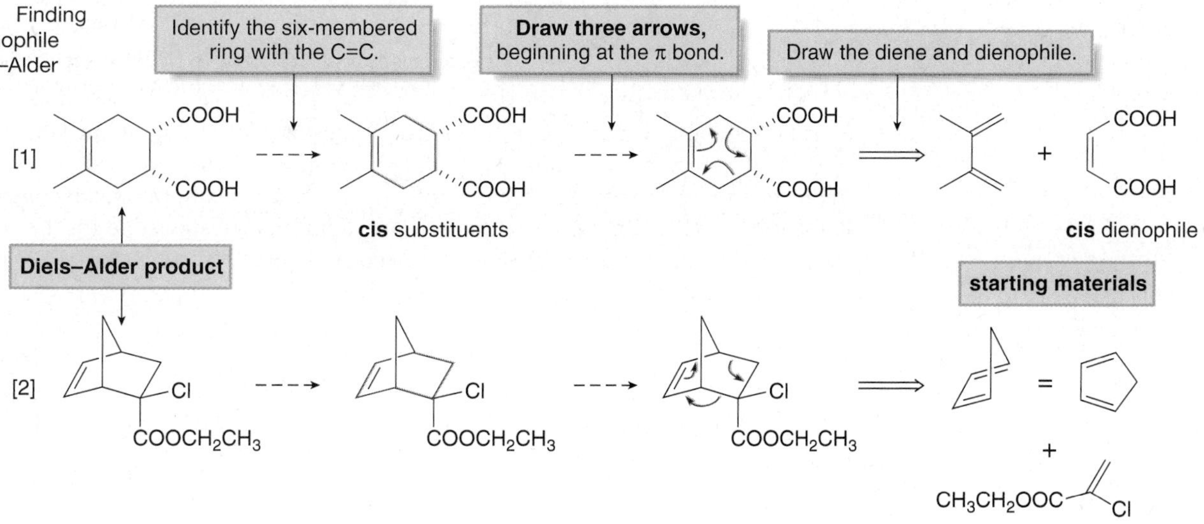

Figure 16.12   Finding the diene and dienophile needed for a Diels–Alder reaction

16.14B   Retro Diels–Alder Reaction

A reactive diene like 1,3-cyclopentadiene readily undergoes a Diels–Alder reaction with *itself;* that is, **1,3-cyclopentadiene dimerizes because one molecule acts as the diene and another acts as the dienophile.**

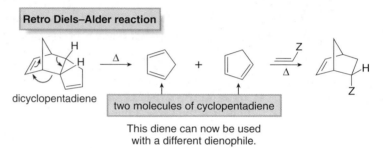

diene   dienophile   dicyclopentadiene   =

This ring is **endo**.

dimer

The formation of dicyclopentadiene is so rapid that it takes only a few hours at room temperature for cyclopentadiene to completely dimerize. How, then, can cyclopentadiene be used in a Diels–Alder reaction if it really exists as a dimer?

When heated, dicyclopentadiene undergoes a **retro Diels–Alder reaction,** and two molecules of cyclopentadiene are re-formed. If cyclopentadiene is immediately treated with a different dienophile, it reacts to form a new Diels–Alder adduct with this dienophile.

Retro Diels–Alder reaction

dicyclopentadiene

two molecules of cyclopentadiene

This diene can now be used
with a different dienophile.

16.14C   Application: Diels–Alder Reaction in the Synthesis of Steroids

Steroids **are tetracyclic lipids containing three six-membered rings and one five-membered ring.** The four rings are designated as **A, B, C,** and **D.**

Recall from Section 4.15 that lipids are water-insoluble biomolecules that have diverse structures.

The steroid skeleton

C   D

A   B

three-dimensional view
from above

carbon skeleton
viewed from the side

[Note the chair conformations
of the three cyclohexane rings.]

Steroids exhibit a wide range of biological properties, depending on the substitution pattern of functional groups on the rings. They include **cholesterol** (used to stiffen cell membranes and implicated in cardiovascular disease), **estrone** (a female sex hormone responsible for the regulation of the menstrual cycle), and **cortisone** (a hormone responsible for the control of inflammation and the regulation of carbohydrate metabolism).

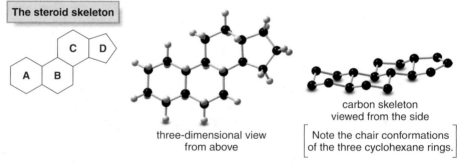

cholesterol   estrone   cortisone

Diels–Alder reactions have been used widely in the laboratory syntheses of steroids. The key Diels–Alder reactions used to prepare the C ring of estrone and the B ring of cortisone are as follows:

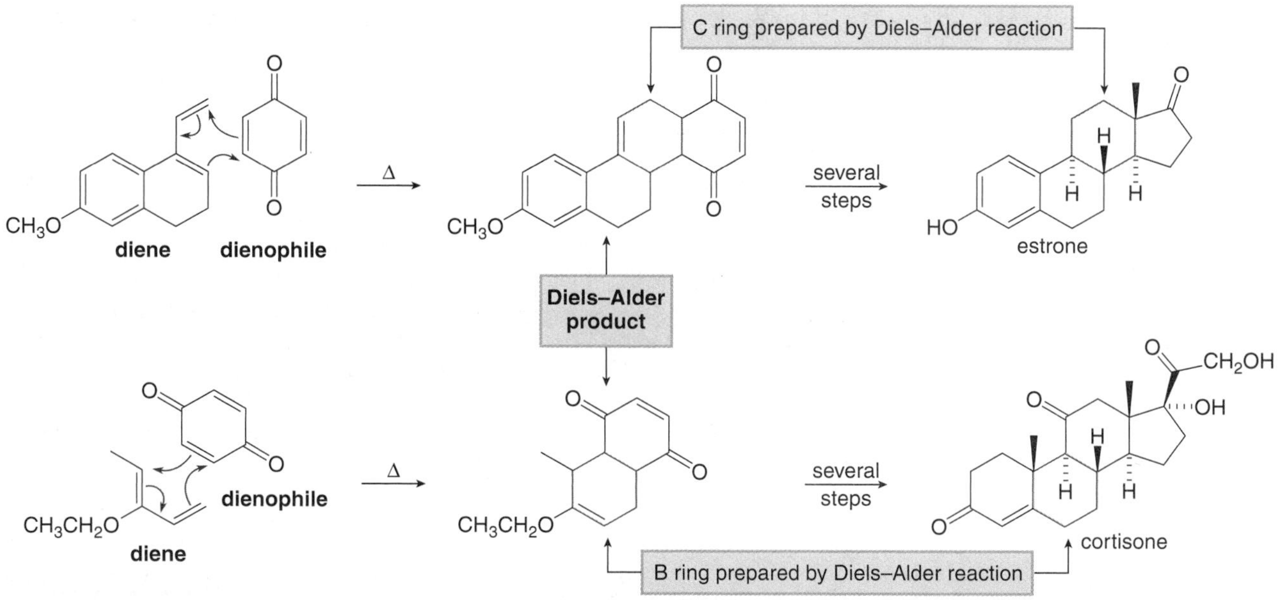

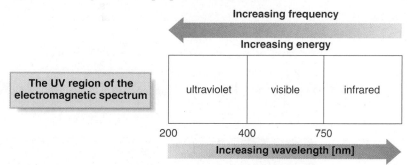

PROBLEM 16.23 Draw the product (**A**) of the following Diels–Alder reaction. **A** was a key intermediate in the synthesis of the addicting pain reliever morphine, isolated from the opium poppy.

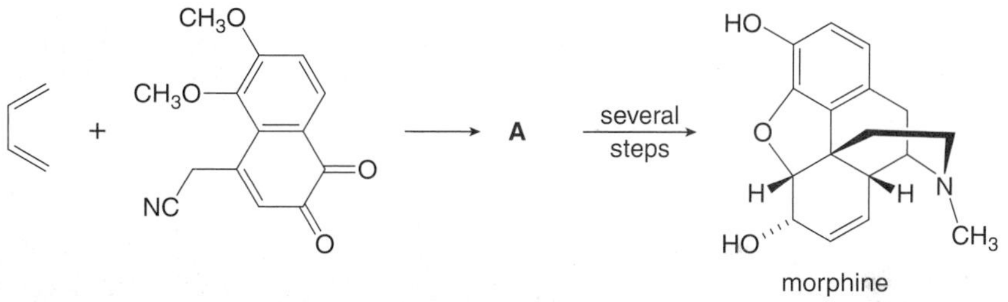

Opium poppy

16.15 Conjugated Dienes and Ultraviolet Light

Recall from Chapter 14 that the absorption of infrared energy can promote a molecule from a lower vibrational state to a higher one. In a similar fashion, the absorption of ultraviolet (UV) light can promote an electron from a lower electronic state to a higher one. Ultraviolet light has a slightly shorter wavelength (and, thus, higher frequency) than visible light. The most useful region of UV light for this purpose is **200–400 nm.**

Increasing frequency

Increasing energy

The UV region of the electromagnetic spectrum	ultraviolet	visible	infrared

200   400   750

Increasing wavelength [nm]

16.15A General Principles

When electrons in a lower energy state (the **ground state**) absorb light having the appropriate energy, an electron is promoted to a higher electronic state (the **excited state**).

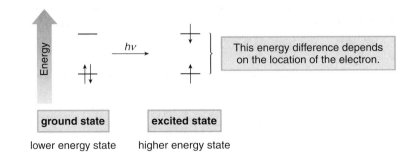

The energy difference between the two states depends on the location of the electron. The promotion of electrons in σ bonds and unconjugated π bonds requires light having a wavelength of < 200 nm; that is, it has a shorter wavelength and higher energy than light in the UV region of the electromagnetic spectrum. With conjugated dienes, however, the energy difference between the ground and excited states decreases, so longer wavelengths of light can be used to promote electrons. The wavelength of UV light absorbed by a compound is often referred to as its λ_{max}. 1,3-Butadiene, for example, absorbs UV light at $\lambda_{max} = 217$ nm and 1,3-cyclohexadiene has a λ_{max} of 256 nm.

$\lambda_{max} = 217$ nm $\lambda_{max} = 256$ nm

◆ **Conjugated dienes and polyenes absorb light in the UV region of the electromagnetic spectrum (200–400 nm).**

As the number of conjugated π bonds increases, the energy difference between the ground and excited state decreases, shifting the absorption to longer wavelengths.

$\lambda_{max} = 217$ nm $\lambda_{max} = 268$ nm $\lambda_{max} = 364$ nm

Increasing conjugation
Increasing λ_{max}

With molecules having eight or more conjugated π bonds, the absorption shifts from the UV to the visible region and the compound takes on the color of those wavelengths of visible light it does *not* absorb. For example, lycopene absorbs visible light at $\lambda_{max} = 470$ nm, in the blue-green region of the visible spectrum. Because it does not absorb light in the red region, lycopene appears bright red (Figure 16.13).

Figure 16.13 Why lycopene appears red

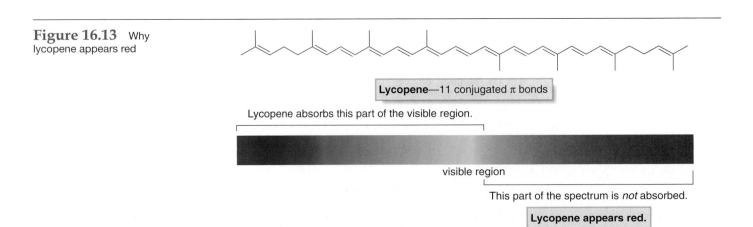

Lycopene—11 conjugated π bonds

Lycopene absorbs this part of the visible region.

visible region

This part of the spectrum is *not* absorbed.

Lycopene appears red.

PROBLEM 16.24   Which compound in each pair absorbs UV light at longer wavelength?

a. 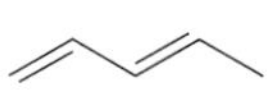   and    b. 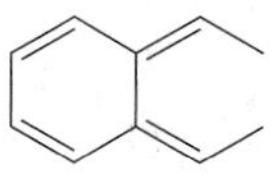   and

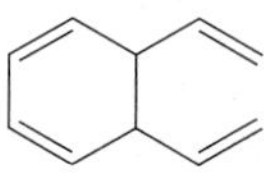

16.15B   Sunscreens

Ultraviolet radiation from the sun is high enough in energy to cleave bonds, forming radicals that can prematurely age skin and cause skin cancers. Fortunately, much of the highest energy UV light is filtered out by the ozone layer, so that only UV light having wavelengths > 290 nm reaches the skin's surface. Much of this UV light is absorbed by **melanin,** the highly conjugated colored pigment in the skin that serves as the body's natural protection against the harmful effects of UV radiation.

Prolonged exposure to the sun can allow more UV radiation to reach your skin than melanin can absorb. A commercial sunscreen can offer added protection, however, because it contains conjugated compounds that absorb UV light, thus shielding your skin (for a time) from the harmful effects of UV radiation. Two sunscreens that have been used for this purpose are *para*-aminobenzoic acid (PABA) and padimate O.

Commercial sunscreens are given an **SPF** rating (sun protection factor), according to the amount of sunscreen present. The higher the number, the greater the protection.

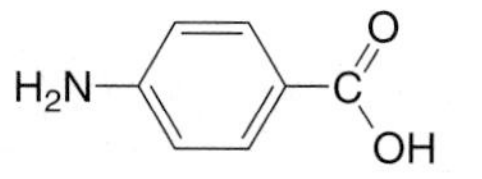

para-aminobenzoic acid
(PABA)

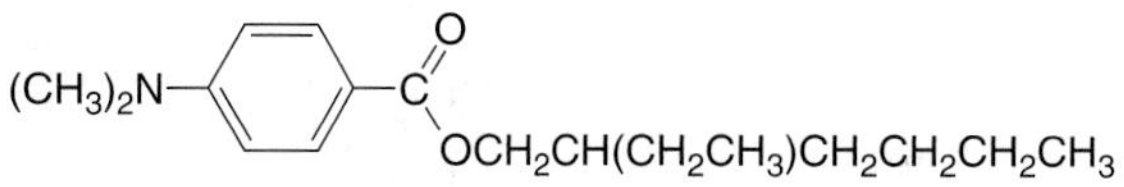

padimate O

PROBLEM 16.25   Which of the following compounds might be an ingredient in a commercial sunscreen? Explain why or why not.

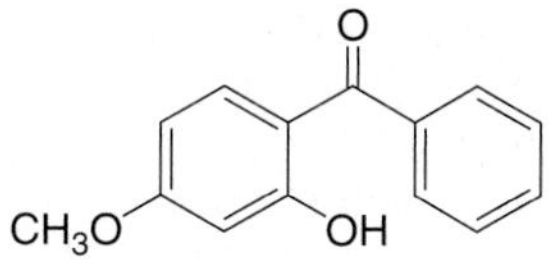

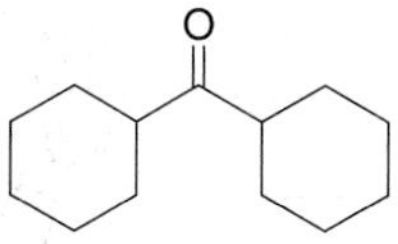

16.16   Key Concepts—Conjugation, Resonance, and Dienes

Conjugation and Delocalization of Electron Density

- Having p orbitals on three or more adjacent atoms allows electron density to delocalize, thus adding stability (16.1).
- An allyl carbocation (CH_2=$CHCH_2^+$) is more stable than a 1° carbocation because of p orbital overlap (16.2).
- In any system X=Y–Z:, Z is sp^2 hybridized to allow the lone pair to occupy a p orbital, making the system conjugated (16.5).

Four Common Examples of Resonance (16.3)

[1] The three-atom "allyl" system: X=Y–$\underset{*}{Z}$ ⟷ $\underset{*}{X}$–Y=Z $* = +, -, \cdot, \text{ or } \cdot\cdot$

[2] Conjugated double bonds:

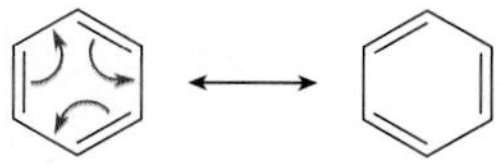

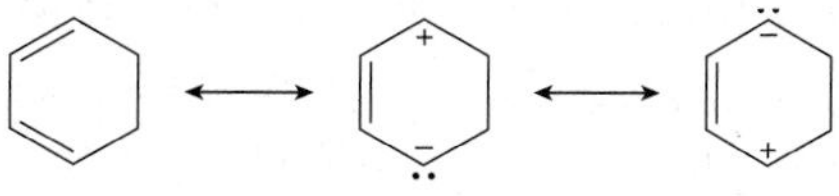

[3] Cations having a positive charge adjacent to a lone pair: $\overset{\cdot\cdot}{X}$–$\overset{+}{Y}$ ⟷ $\overset{+}{X}$=Y

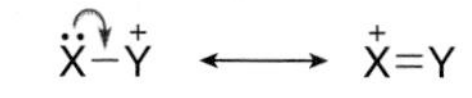

[4] Double bonds involving one atom more electronegative than the other: X=Y ⟷ $\overset{+}{X}$–$\overset{-}{Y}$: [electronegativity of Y > X]

Rules on Evaluating the Relative "Stability" of Resonance Structures (16.4)

[1] Structures with more bonds and fewer charges are more stable.
[2] Structures in which every atom has an octet are more stable.
[3] Structures that place a negative charge on a more electronegative atom are more stable.

The Unusual Properties of Conjugated Dienes

[1] The C–C σ bond joining the two double bonds is unusually short (16.8).
[2] Conjugated dienes are more stable than the corresponding isolated dienes. $\Delta H°$ of hydrogenation is smaller for a conjugated diene than for an isolated diene converted to the same product (16.9).
[3] The reactions are unusual:
   • Electrophilic addition affords products of 1,2-addition and 1,4-addition (16.10, 16.11).
   • Conjugated dienes undergo the Diels–Alder reaction, a reaction that does not occur with isolated dienes (16.12–16.14).
[4] Conjugated dienes absorb UV light in the 200–400 nm region. As the number of conjugated π bonds increases, the absorption shifts to longer wavelength (16.15).

Reactions of Conjugated Dienes

[1] Electrophilic addition of HX (X = halogen) (16.10–16.11)

$$CH_2=CH-CH=CH_2 \xrightarrow[\text{(1 equiv)}]{\text{HX}}$$

CH₂–CH–CH=CH₂
  | |
  H   X
1,2-product
kinetic product
	CH₂–CH=CH–CH₂
	 | |
	 H X
	1,4-product
	thermodynamic product

• The mechanism has two steps.
• Markovnikov's rule is followed. Addition of H⁺ forms the more stable allylic carbocation.
• The 1,2-product is the kinetic product. When H⁺ adds to the double bond, X⁻ adds to the end of the allylic carbocation to which it is closer (C2 not C4). The kinetic product is formed faster at low temperature.
• The thermodynamic product has the more substituted, more stable double bond. The thermodynamic product predominates at equilibrium. With 1,3-butadiene, the thermodynamic product is the 1,4-product.

[2] Diels–Alder reaction (16.12–16.14)

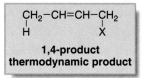

1,3-diene dienophile

[The three new bonds are labeled in red.]

• The reaction forms two σ and one π bond in a six-membered ring.
• The reaction is initiated by heat.
• The mechanism is concerted: All bonds are broken and formed in a single step.
• The diene must react in the s-cis conformation (16.13A).
• Electron-withdrawing groups in the dienophile increase the reaction rate (16.13B).
• The stereochemistry of the dienophile is retained in the product (16.13C).
• Endo products are preferred (16.13D).

Problems

Conjugation

16.26 Which of the following systems are conjugated?

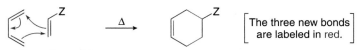

CH₂=CHCN

Resonance

16.27 Draw all reasonable resonance structures for each species.

a. $(CH_3)_2\overset{+}{C}CH=CH_2$

c. 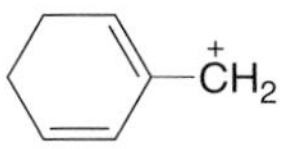 [structure: benzene ring with $\overset{+}{C}H_2$ substituent]

e. $CH_3O\overset{..}{C}H=CH\overset{+}{C}H_2$

g. 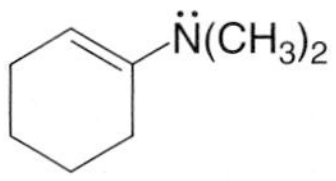 [structure: cyclohexene ring with $\overset{..}{N}(CH_3)_2$ substituent]

b. 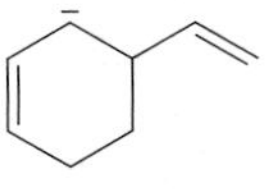 [structure: cyclohexene ring with vinyl group and negative charge]

d. 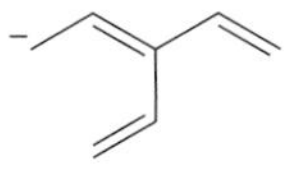 [structure: diene with negative charge]

f. 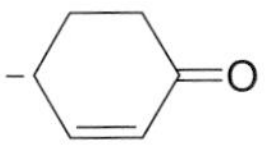 [structure: cyclohexadienone with negative charge]

h. 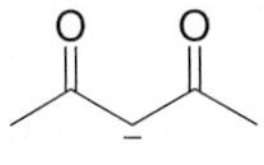 [structure: 2,4-pentanedione anion]

16.28 Draw four more resonance structures for each species.

a. 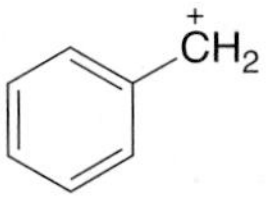 [structure: benzene ring with $\overset{+}{C}H_2$]

b. 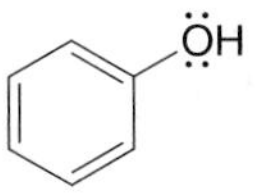 [structure: benzene ring with $\overset{..}{O}H$]

16.29 Considering acetic acid:

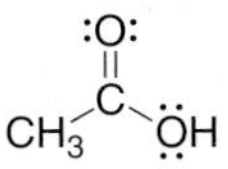

[structure: acetic acid, $CH_3-C(=\overset{..}{O}:)-\overset{..}{O}H$]
acetic acid

a. Draw two more resonance structures.
b. Rank all three resonance structures in order of increasing stability.
c. Draw a structure for the resonance hybrid.

16.30 Explain why the cyclopentadienide anion **A** gives only one signal in its ^{13}C NMR spectrum.

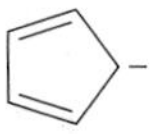

[structure: cyclopentadienide anion] $^- = \mathbf{A}$

16.31 Determine the hybridization at the carbon atom indicated in each species.

a. 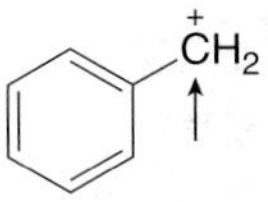 [structure: benzene ring with $\overset{+}{C}H_2$, arrow pointing up]

b. 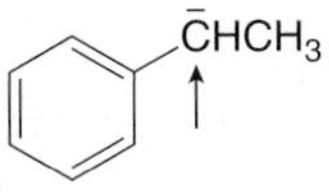 [structure: benzene ring with $\overset{..}{C}HCH_3$, arrow pointing up]

c. [structure: benzene ring with $CH_2\overset{..}{C}H_2$, arrow pointing up]

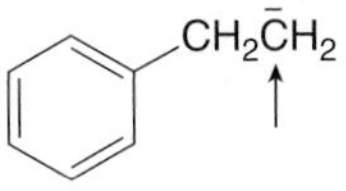

16.32 Explain each statement using resonance theory.

a. Both $C-O$ bond lengths are equal in the acetate anion.

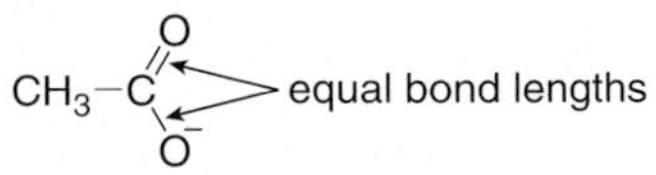

[structure: CH_3-C with two O's] equal bond lengths
acetate

b. The indicated $C-H$ bond in propene is more acidic than the indicated $C-H$ bond in propane.

more acidic   less acidic

$CH_2=CHCH_2-H$ $CH_3CH_2CH_2-H$
propene   propane

c. 3-Chloro-1-propene is more reactive than 1-chloropropane under S_N1 conditions.

$CH_2=CHCH_2-Cl$ $CH_3CH_2CH_2-Cl$
3-chloro-1-propene   1-chloropropane
more reactive   less reactive

d. The bond dissociation energy for the $C-C$ bond in ethane is much higher than the bond dissociation energy for the indicated $C-C$ bond in 1-butene.

+88 kcal/mol   +72 kcal/mol

CH_3-CH_3 $CH_3-CH_2CH=CH_2$
ethane   1-butene

16.33 Draw a stepwise mechanism for the following reaction.

$CH_3CH=CHCH_2OH \xrightarrow{HBr} CH_3CH=CHCH_2Br + CH_3\underset{Br}{CH}CH=CH_2 + H_2O$

Nomenclature and Stereoisomers in Conjugated Dienes

16.34 Draw the structure of each compound.
a. (3Z)-1,3-pentadiene in the s-trans conformation
b. (2E,4Z)-1-bromo-3-methyl-2,4-hexadiene
c. (2E,4E,6E)-2,4,6-octatriene
d. (E,E)-3-methyl-2,4-hexadiene in the s-cis conformation

16.35 Draw all possible stereoisomers of 2,4-heptadiene and label each double bond as *E* or *Z*.

16.36 Label each pair of compounds as stereoisomers or conformers.

a.  and 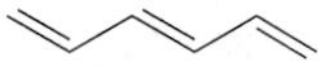   c.  and

b. 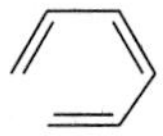 and

16.37 Which compound in each pair has the higher heat of hydrogenation?

a. 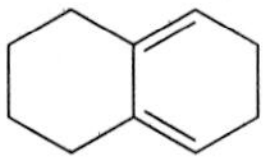 or 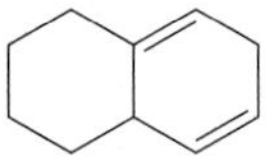   b. 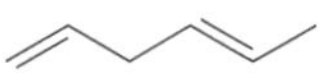 or

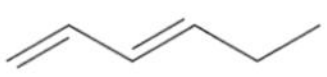

Electrophilic Addition

16.38 Draw the products of each reaction.

a. 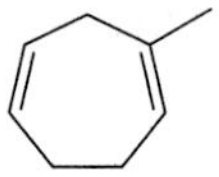 $\xrightarrow[\text{(1 equiv)}]{\text{HBr}}$ c.  $\xrightarrow[\text{(1 equiv)}]{\text{DCl}}$

b. 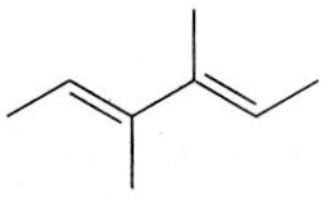 $\xrightarrow[\text{(1 equiv)}]{\text{HCl}}$ d. 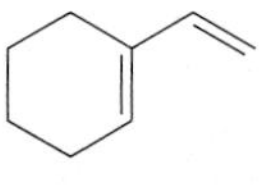 $\xrightarrow[\text{(1 equiv)}]{\text{HBr}}$

16.39 Treatment of alkenes **A** and **B** with HBr gives the same alkyl halide **C**. Draw a mechanism for each reaction, including all reasonable resonance structures for any intermediate.

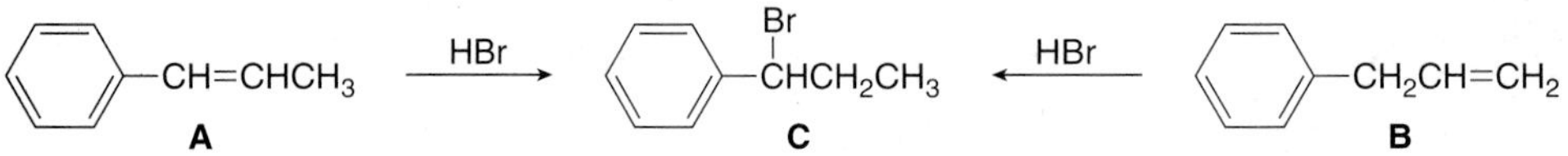

16.40 Draw a stepwise mechanism for the following reaction.

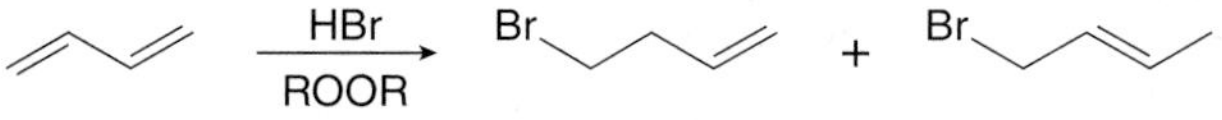

16.41 Addition of HCl to alkene **X** forms two alkyl halides **Y** and **Z**.

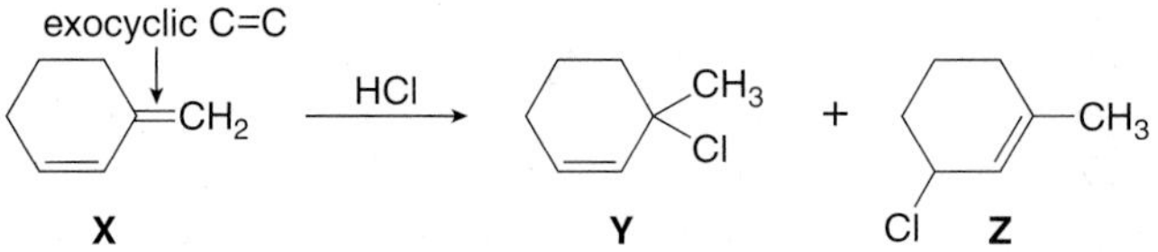

a. Label **Y** and **Z** as a 1,2-addition product or a 1,4-addition product.
b. Label **Y** and **Z** as the kinetic or thermodynamic product and explain why.
c. Explain why addition of HCl occurs at the indicated C=C (called an exocyclic double bond), rather than the other C=C (called an endocyclic double bond).

16.42 Explain, with reference to the mechanism, why addition of one equivalent of HCl to diene **A** forms only two products of electrophilic addition, even though four constitutional isomers are possible.

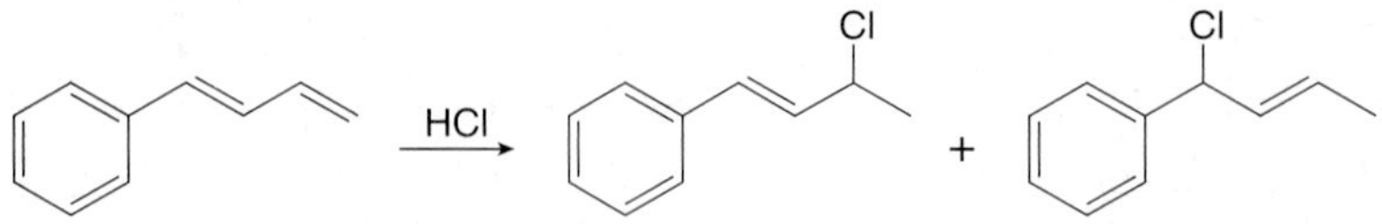

Diels–Alder Reaction

16.43 Explain why methyl vinyl ether (CH_2=$CHOCH_3$) is not a reactive dienophile in the Diels–Alder reaction.

16.44 Draw the products of the following Diels–Alder reactions. Indicate stereochemistry where appropriate.

a. [structure] + [structure] $\xrightarrow{\Delta}$

b. [structure] + [structure with COOCH$_3$, Cl] $\xrightarrow{\Delta}$

c. [structure] + [structure with COOCH$_3$, Cl] $\xrightarrow{\Delta}$

d. [structure] + [structure] $\xrightarrow{\Delta}$

e. [structure with CN, CN] + [structure] $\xrightarrow{\Delta}$

f. [structure] + [structure] $\xrightarrow{\Delta}$

16.45 What diene and dienophile are needed to prepare each Diels–Alder product?

a. [structure with CH$_3$, COOCH$_3$]

b. [structure with COOCH$_3$, CH$_3$]

c. [structure with H, COOCH$_3$]

d. [structure]

e. [structure with Cl, O—C(=O)CH$_3$]

f. [structure]

16.46 Give two different ways to prepare the following compound by the Diels–Alder reaction. Explain which method is preferred.

[structure]

16.47 Compounds containing triple bonds are also Diels–Alder dienophiles. With this in mind, draw the products of each reaction.

a. [structure] + HC≡C—COOCH$_3$ $\xrightarrow{\Delta}$

b. [structure] + CH$_3$O$_2$C—C≡C—CO$_2$CH$_3$ $\xrightarrow{\Delta}$

16.48 What two constitutional isomers are formed in the following Diels–Alder reaction?

[structure] + [structure] $\xrightarrow{\Delta}$

16.49 What is the structure of the product formed when **A** is heated in the presence of maleic acid? Explain why only one product is formed even though **A** has four double bonds.

[structure] **A** + HOOC—[structure]—COOH maleic acid $\xrightarrow{\Delta}$

16.50 Draw the structure of the product formed (molecular formula C$_{14}$H$_{16}$O$_2$) when benzoquinone is heated with excess 1,3-butadiene.

[structure] excess + [structure] benzoquinone $\xrightarrow{\Delta}$ C$_{14}$H$_{16}$O$_2$

16.51 What diene and dienophile are needed to prepare **X,** an intermediate in the synthesis of the antibiotic fumagillin?

[structure with CHO, Br, COOCH$_3$] **X** $\xrightarrow{\text{several steps}}$ [structure of fumagillin with HOOC, OCH$_3$] fumagillin

16.52 The following reactions have been used to synthesize dieldrin and aldrin (named for Diels and Alder), two pesticides having a similar story to DDT (Section 7.4). Identify the lettered compounds in this reaction scheme.

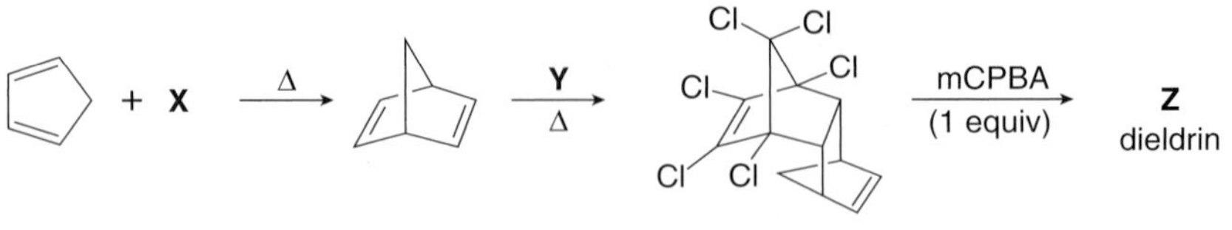

aldrin

16.53 Devise a stepwise synthesis of each compound from dicyclopentadiene using a Diels–Alder reaction as one step.

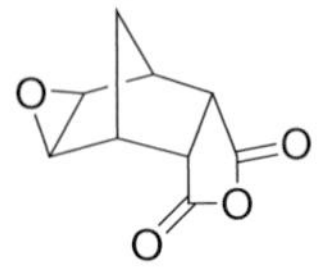

16.54 Intramolecular Diels–Alder reactions are possible when a substrate contains both a 1,3-diene and a dienophile, as shown in the following general reaction.

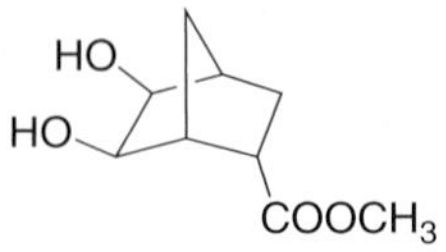

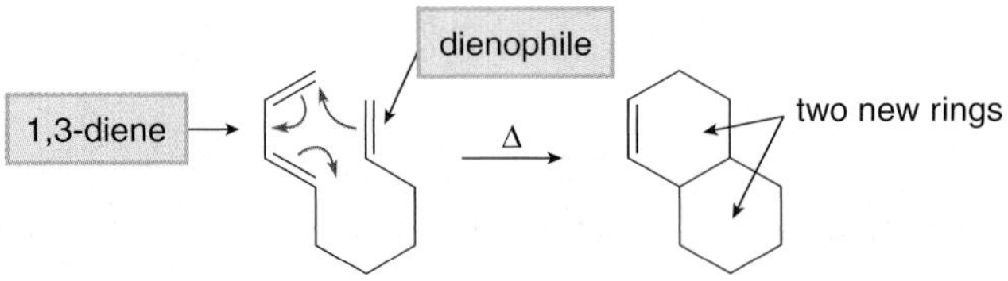

With this in mind, draw the product of each intramolecular Diels–Alder reaction.

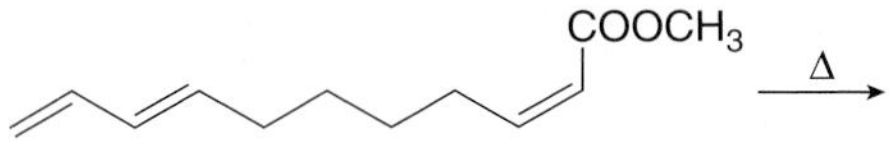

General Reactions of Dienes

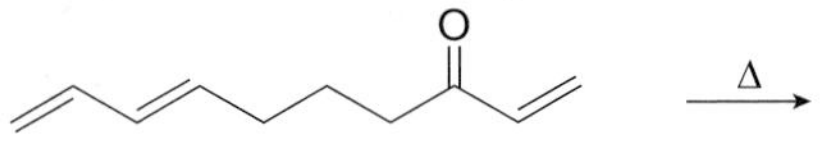

16.55 Draw the products of each reaction and indicate stereochemistry where appropriate.

a. $(CH_3)_2C=CHCH_2CH_2CH=CH_2$ $\xrightarrow[\text{(1 equiv)}]{\text{HCl}}$

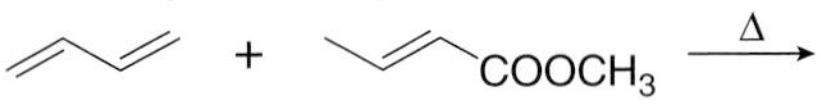

b. 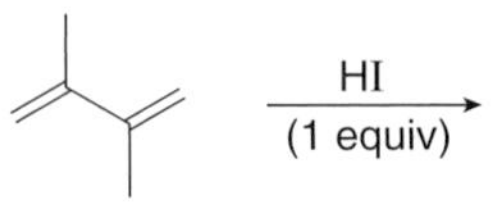 $\xrightarrow[\text{(1 equiv)}]{\text{HI}}$

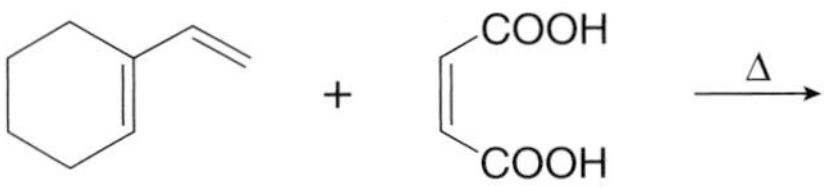

c. 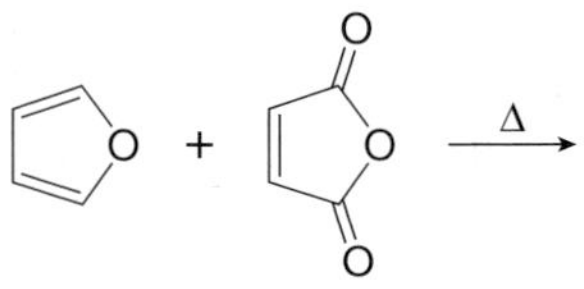 $\xrightarrow{\Delta}$

d. $\xrightarrow{\Delta}$

e. $\xrightarrow{\Delta}$

f. 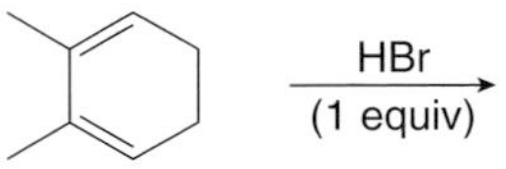 $\xrightarrow[\text{(1 equiv)}]{\text{HBr}}$

UV Absorption

16.56 Rank the following compounds in the order of increasing λ_{max}.

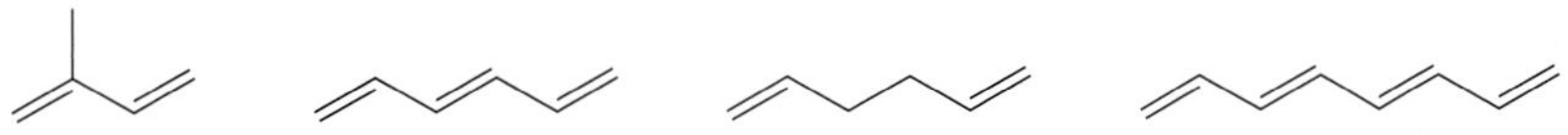

Challenge Problems

16.57 One step in the synthesis of occidentalol, a natural product isolated from the eastern white cedar tree, involved the following reaction. Identify the structure of **A** and show how **A** is converted to **B**.

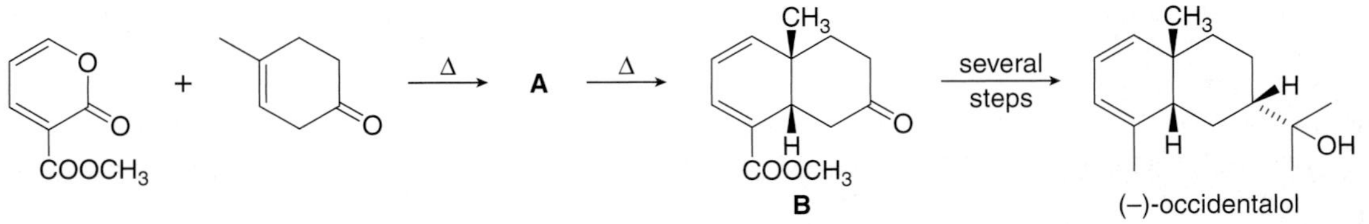

16.58 One step in the synthesis of dodecahedrane (Chapter 4 opening molecule) involved reaction of the tetraene **A** with dimethylacetylene dicarboxylate (**B**) to afford two compounds having molecular formula $C_{16}H_{16}O_4$. This reaction has been called a domino Diels–Alder reaction. Identify the two products formed.

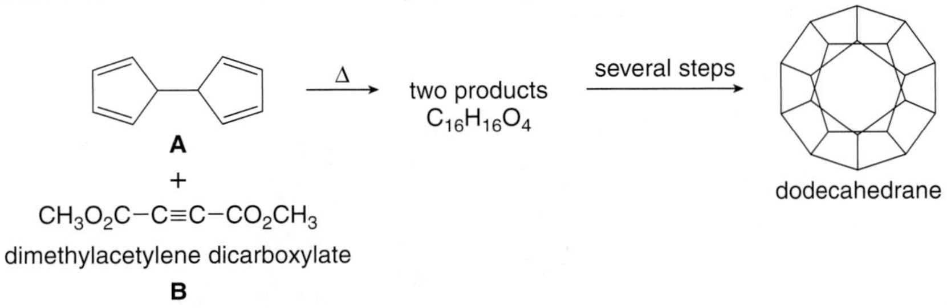

Benzene and Aromatic Compounds

The elegant synthesis of **quinine** in 1944 is considered by many to be the beginning of modern-day organic synthesis. Quinine, a natural product isolated from the bark of the Cinchona tree native to the Andes Mountains, is a powerful antipyretic—that is, it reduces fever—and so, for centuries, it was the only effective treatment for malaria. Its bitter taste gives tonic water its characteristic flavor. Quinine and many other natural products contain a benzene ring. In Chapter 17, we will learn about the special properties of benzene and related aromatic compounds.

The hydrocarbons we have examined thus far—including the alkanes, alkenes, and alkynes, as well as the conjugated dienes and polyenes of Chapter 16—have been aliphatic hydrocarbons. In Chapter 17, we continue our study of conjugated systems with **aromatic hydrocarbons.**

We begin with **benzene** and then examine other cyclic, planar, and conjugated ring systems to learn the modern definition of what it means to be aromatic. Then, in Chapter 18, we will learn about the reactions of aromatic compounds, highly unsaturated hydrocarbons that do not undergo addition reactions like other unsaturated compounds. An explanation of this behavior relies on an understanding of the structure of aromatic compounds presented in Chapter 17.

17.1   Background

For 6 C's, the maximum number of H's = $2n + 2 = 2(6) + 2 = 14$. Because benzene contains only 6 H's, it has $14 - 6 = 8$ H's fewer than the maximum number. This corresponds to 8 H's/2 H's for each degree of unsaturation = **four degrees of unsaturation in benzene.**

Benzene (C_6H_6) is the simplest aromatic hydrocarbon (or arene). Since its isolation by Michael Faraday from the oily residue remaining in the illuminating gas lines in London in 1825, it has been recognized as an unusual compound. Based on the calculation introduced in Section 10.2, **benzene has four degrees of unsaturation, making it a highly unsaturated hydrocarbon.** But, whereas unsaturated hydrocarbons such as alkenes, alkynes, and dienes readily undergo addition reactions, benzene does not. For example, bromine adds to ethylene to form a dibromide, but benzene is inert under similar conditions.

Benzene **does** react with bromine, but only in the presence of $FeBr_3$ (a Lewis acid), and the reaction is a **substitution,** *not* an addition.

Thus, any structure proposed for benzene must account for its high degree of unsaturation and its lack of reactivity towards electrophilic addition.

In the last half of the nineteenth century August Kekulé proposed structures that were close to the modern description of benzene. In the Kekulé model, benzene was thought to be a rapidly equilibrating mixture of two compounds, each containing a six-membered ring with three alternating π bonds. These structures are now called **Kekulé structures.** In the Kekulé description, the bond between any two carbon atoms is sometimes a single bond and sometimes a double bond.

Although benzene is still drawn as a six-membered ring with three alternating π bonds, in reality **there is no equilibrium between two different kinds of benzene molecules.** Instead, current descriptions of benzene are based on resonance and electron delocalization due to orbital overlap, as described in Section 17.2.

In the nineteenth century, many other compounds having properties similar to those of benzene were isolated from natural sources. Because these compounds possessed strong and characteristic odors, they were called *aromatic* compounds. It is their chemical properties, though, not their odor that make these compounds special.

◆ **Aromatic compounds resemble benzene—they are unsaturated compounds that do not undergo the addition reactions characteristic of alkenes.**

17.2 The Structure of Benzene

Any structure for benzene must account for the following:

◆ **It contains a six-membered ring and three additional degrees of unsaturation.**

◆ **It is planar.**

◆ **All C−C bond lengths are equal.**

Although the Kekulé structures satisfy the first two criteria, they break down with the third, because having three alternating π bonds means that benzene should have three short double bonds alternating with three longer single bonds.

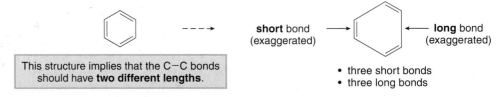

This structure implies that the C−C bonds should have **two different lengths**.

short bond (exaggerated) ⟶ ⟵ **long** bond (exaggerated)

• three short bonds
• three long bonds

Resonance

Benzene is conjugated, so we must use resonance and orbitals to describe its structure. The resonance description of benzene consists of two equivalent Lewis structures, each with three double bonds that alternate with three single bonds.

The hybrid

The electrons in the π bonds are **delocalized** around the ring.

The resonance description of benzene matches the Kekulé description with one important exception. **The two Kekulé representations are *not* in equilibrium with each other.** Instead, the true structure of benzene is a resonance **hybrid** of the two Lewis structures, with the dashed lines of the hybrid indicating the position of the π bonds.

We will use one of the two Lewis structures and not the hybrid in drawing benzene, because it is easier to keep track of the electron pairs in the π bonds (the π electrons).

◆ **Because each π bond has two electrons, benzene has six π electrons.**

The resonance hybrid of benzene explains why all C−C bond lengths are the same. Each C−C bond is single in one resonance structure and double in the other, so the actual bond length (1.39 Å) is intermediate in length between a carbon–carbon single bond (1.53 Å) and a carbon–carbon double bond (1.34 Å).

> Some texts draw benzene as a hexagon with an inner circle to emphasize that the six π electrons are spread over the ring.
>
> The circle represents the **six π electrons**, distributed over the six atoms of the ring.

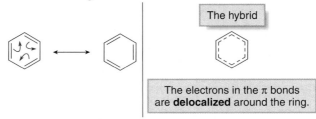

CH_3-CH_3
↑
1.53 Å

$CH_2=CH_2$
↑
1.34 Å

1.39 Å

The C−C bonds in benzene are equal and intermediate in length.

Hybridization and Orbitals

Each carbon atom in a benzene ring is surrounded by three atoms and no lone pairs of electrons, making it *sp²* **hybridized and trigonal planar with all bond angles 120°**. Each carbon also has a *p* orbital with one electron that extends above and below the plane of the molecule.

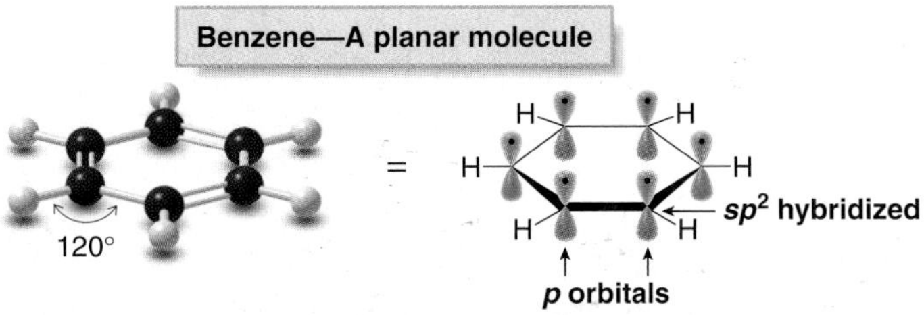

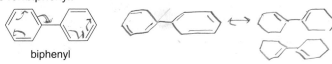

The six adjacent *p* orbitals overlap, delocalizing the six electrons over the six atoms of the ring and making benzene a conjugated molecule. Because each *p* orbital has two lobes, one above and one below the plane of the benzene ring, the overlap of the *p* orbitals creates two "doughnuts" of electron density, as shown in Figure 17.1[a]. The electrostatic potential plot shown in Figure 17.1[b] also shows that the electron-rich region is concentrated above and below the plane of the molecule, where the six π electrons are located.

◆ Benzene's six π electrons make it electron rich and so it readily reacts with electrophiles.

PROBLEM 17.1   Draw all possible resonance structures for biphenyl.

biphenyl

PROBLEM 17.2   What orbitals are used to form the bonds indicated in each molecule?

a.

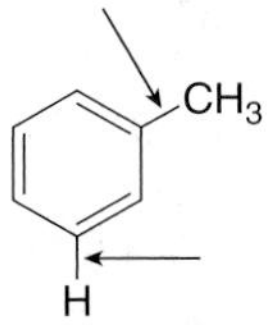

b.

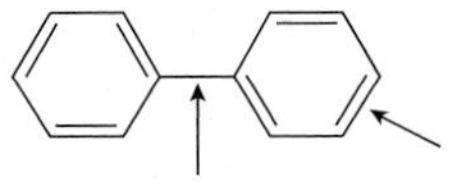

17.3   Nomenclature of Benzene Derivatives

Many organic molecules contain a benzene ring with one or more substituents, so we must learn how to name them. Many common names are recognized by the IUPAC system, however, so this complicates the nomenclature of benzene derivatives somewhat.

17.3A   Monosubstituted Benzenes

To name a benzene ring with one substituent, **name the substituent and add the word *benzene.*** Carbon substituents are named as alkyl groups.

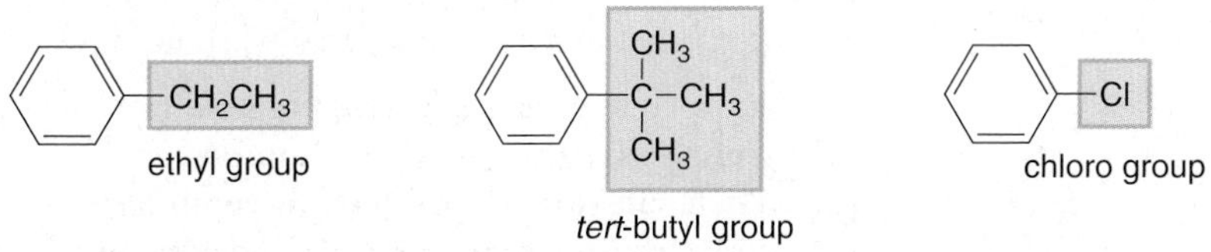

—CH₂CH₃	CH₃ | —C—CH₃ | CH₃	—Cl
ethyl group	*tert*-butyl group	chloro group
ethylbenzene	*tert*-butylbenzene	chlorobenzene

Figure 17.1   Two views of the electron density in a benzene ring

[a] View of the *p* orbital overlap

- Overlap of six adjacent *p* orbitals creates two rings of electron density, one above and one below the plane of the benzene ring.

[b] Electrostatic potential plot

- The electron-rich region (in red) is concentrated above and below the ring carbons, where the six π electrons are located. (The electron-rich region below the plane is hidden from view.)

Many monosubstituted benzenes, such as those with methyl (CH_3-), hydroxy ($-OH$), and amino ($-NH_2$) groups, have common names that you must learn, too.

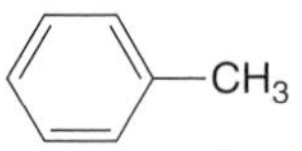

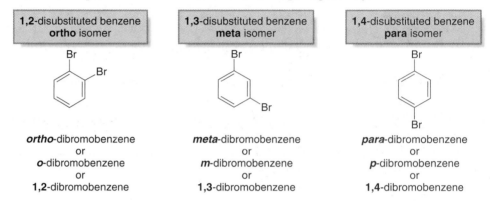

toluene
(methylbenzene)

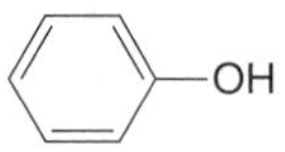

phenol
(hydroxybenzene)

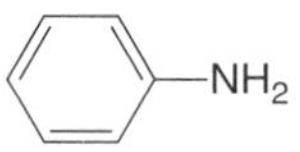

aniline
(aminobenzene)

17.3B   Disubstituted Benzenes

There are three different ways that two groups can be attached to a benzene ring, so a prefix—**ortho, meta,** or **para**—can be used to designate the relative position of the two substituents. Ortho, meta, and para are also abbreviated as *o, m,* and *p,* respectively.

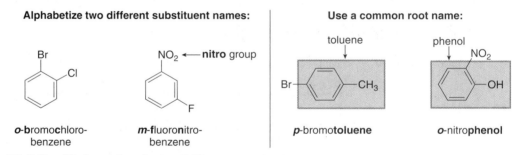

1,2-disubstituted benzene **ortho** isomer	**1,3**-disubstituted benzene **meta** isomer	**1,4**-disubstituted benzene **para** isomer
***ortho*-dibromobenzene** or ***o*-dibromobenzene** or **1,2-dibromobenzene**	***meta*-dibromobenzene** or ***m*-dibromobenzene** or **1,3-dibromobenzene**	***para*-dibromobenzene** or ***p*-dibromobenzene** or **1,4-dibromobenzene**

If the two groups on the benzene ring are different, **alphabetize the names of the substituents** preceding the word benzene. If one of the substituents is part of a **common root,** name the **molecule as a derivative of that monosubstituted benzene.**

Alphabetize two different substituent names: **Use a common root name:**

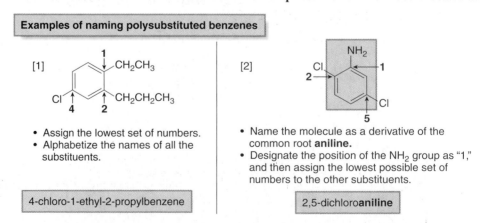

o-**b**romochloro-
benzene   *m*-**f**luoronitro-
benzene   *p*-bromo**toluene**   *o*-nitro**phenol**

17.3C   Polysubstituted Benzenes

For three or more substituents on a benzene ring:

[1]   Number to give the lowest possible numbers around the ring.

[2]   Alphabetize the substituent names.

[3]   When substituents are part of common roots, name the molecule as a derivative of that monosubstituted benzene. The substituent that comprises the common root is located at C1.

Examples of naming polysubstituted benzenes

[1]

- Assign the lowest set of numbers.
- Alphabetize the names of all the substituents.

[2]

- Name the molecule as a derivative of the common root **aniline.**
- Designate the position of the NH_2 group as "1," and then assign the lowest possible set of numbers to the other substituents.

4-chloro-1-ethyl-2-propylbenzene   **2,5-dichloroaniline**

17.3D   Naming Aromatic Rings as Substituents

A benzene substituent (C_6H_5-) is called a **phenyl group,** and it can be abbreviated in a structure as **Ph–.**

phenyl group
C_6H_5-

abbreviated as   Ph–

◆ A phenyl group (C_6H_5-) is formed by removing one hydrogen from benzene (C_6H_6).

Benzene, therefore, can be represented as PhH, and phenol would be PhOH.

benzene   = C_6H_5-H
PhH

phenol   = C_6H_5-OH
PhOH

The **benzyl** group, another common substituent that contains a benzene ring, differs from a phenyl group.

an extra CH_2 group

benzyl group
$C_6H_5CH_2-$

phenyl group
C_6H_5-

Finally, substituents derived from other substituted aromatic rings are collectively called **aryl groups.**

Examples of aryl groups

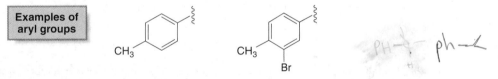

PH-   ph-

PROBLEM 17.3   Give the IUPAC name for each compound.

a. $PhCH(CH_3)_2$ b.   (structure with CH_2CH_3 and I)   c.   (structure with OH)   d.   (structure with CH_3, Br, Cl)

PROBLEM 17.4   Draw the structure corresponding to each name:
a. isobutylbenzene
b. o-dichlorobenzene
c. cis-1,2-diphenylcyclohexane
d. m-bromoaniline
e. 4-chloro-1,2-diethylbenzene

17.4   Spectroscopic Properties

The important IR and NMR absorptions of aromatic compounds are summarized in Table 17.1.

The absorption at 6.5–8.0 ppm in the ^1H NMR spectrum is particularly characteristic of compounds containing benzene rings. **All aromatic compounds have highly deshielded protons due to the ring current effect of the circulating π electrons,** as discussed in Section 15.4. Observing whether a new compound absorbs in this region of a ^1H NMR spectrum is one piece of data used to determine if it is aromatic.

^{13}C NMR spectroscopy is used to determine the substitution patterns in disubstituted benzenes, because each line in a spectrum corresponds to a different kind of carbon atom. For example, o-, m-, and p-dibromobenzene each exhibit a different number of lines in its ^{13}C NMR spectrum, as shown in Figure 17.2.

Figure 17.2 ^{13}C NMR absorptions of the three isomeric dibromobenzenes

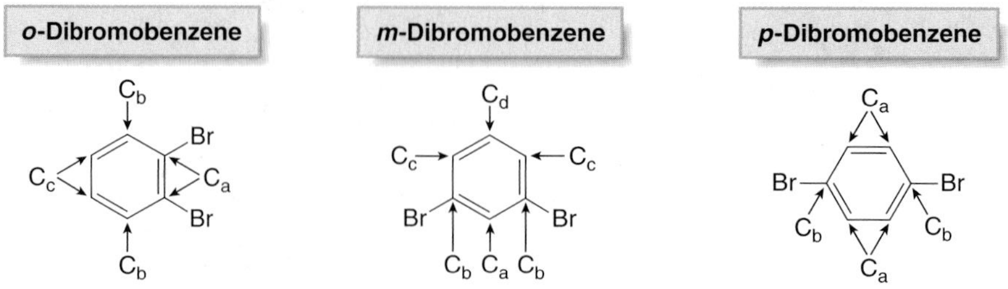

o-Dibromobenzene	**m-Dibromobenzene**	**p-Dibromobenzene**
3 types of C's	**4** types of C's	**2** types of C's
3 ^{13}C NMR signals	**4** ^{13}C NMR signals	**2** ^{13}C NMR signals

- The number of signals (lines) in the ^{13}C NMR spectrum of a disubstituted benzene with two identical groups indicates whether they are ortho, meta, or para to each other.

TABLE 17.1   Characteristic Spectroscopic Absorptions of Benzene Derivatives

IR absorptions	$C_{sp2}-H$	3150–3000 cm^{-1}
	C=C (arene)	1600, 1500 cm^{-1}
^{1}H NMR absorptions	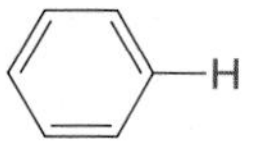 (aryl H)	6.5–8 ppm (highly deshielded protons)
	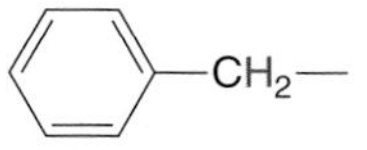 (benzylic H)	1.5–2.5 ppm (somewhat deshielded $C_{sp3}-H$)
^{13}C NMR absorption	C_{sp2} of arenes	120–150 ppm

PROBLEM 17.5   What is the structure of a compound of molecular formula $C_{10}H_{14}O_2$ that shows a strong IR absorption at 3150–2850 cm^{-1} and gives the following ^{1}H NMR absorptions: 1.4 (triplet, 6 H), 4.0 (quartet, 4 H), and 6.8 (singlet, 4 H) ppm?

PROBLEM 17.6   How many ^{13}C NMR signals does each compound exhibit?

a. 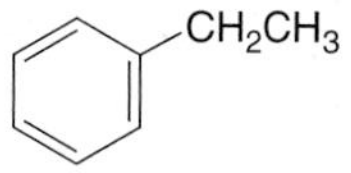 (structure with CH$_2$CH$_3$)   b. 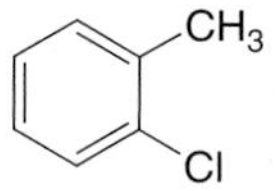 (structure with CH$_3$ and Cl)   c. 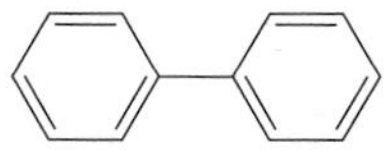 (biphenyl structure)

17.5   Interesting Aromatic Compounds

BTX contains **b**enzene, **t**oluene, and **x**ylene (the common name for dimethylbenzene).

Benzene and **toluene,** the simplest aromatic hydrocarbons obtained from petroleum refining, are useful starting materials for synthetic polymers. They are two components of the **BTX** mixture added to gasoline to boost octane ratings.

The components of the gasoline additive BTX

benzene 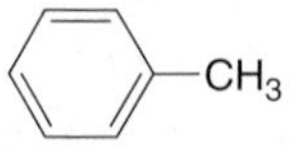   toluene    p-xylene 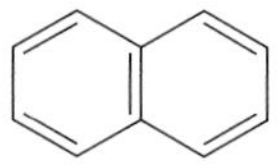   naphthalene (used in mothballs)

Compounds containing two or more benzene rings that share carbon–carbon bonds are called **polycyclic aromatic hydrocarbons (PAHs).** Naphthalene, the simplest PAH, is the active ingredient in mothballs.

Benzo[a]pyrene, a more complicated PAH shown in Figure 17.3, is formed by the incomplete combustion of organic materials. It is found in cigarette smoke, automobile exhaust, and the fumes from charcoal grills. When ingested or inhaled, benzo[a]pyrene and other similar PAHs are oxidized to carcinogenic products, as discussed in Section 9.17.

Helicene and **twistoflex** are two synthetic PAHs whose unusual shapes are shown in Figure 17.4. Helicene consists of six benzene rings. Because the rings at both ends are not bonded to each other, all of the rings twist slightly, creating a rigid helical shape that prevents the hydrogen atoms on both ends from crashing into each other. Similarly, to reduce steric hindrance between the hydrogen atoms on nearby benzene rings, twistoflex is also nonplanar.

Both helicene and twistoflex are chiral molecules—that is, they are not superimposable on their mirror images, even though neither of them contains a stereogenic center. It's their shape that makes them chiral, not the presence of carbon atoms bonded to four different groups. Each ring system is twisted into a shape that lacks a mirror plane, and each structure is rigid, thus creating the chirality.

Many widely used drugs contain a benzene ring. Six examples are shown in Figure 17.5.

17.6 Benzene's Unusual Stability

Considering benzene as the hybrid of two resonance structures adequately explains its equal C−C bond lengths, but does not account for its unusual stability and lack of reactivity towards addition.

Heats of hydrogenation, which were used in Section 16.9 to show that conjugated dienes are more stable than isolated dienes, can also be used to estimate the stability of benzene. Equations [1]–[3] compare the heats of hydrogenation of cyclohexene, 1,3-cyclohexadiene, and benzene, all of which give cyclohexane when treated with excess hydrogen in the presence of a metal catalyst.

Figure 17.3
Benzo[a]pyrene, a common PAH

- Benzo[a]pyrene, produced by the incomplete oxidation of organic compounds in tobacco, is found in cigarette smoke.

benzo[a]pyrene
(a polycyclic aromatic hydrocarbon)

tobacco plant

Figure 17.4   Helicene and twistoflex—Two synthetic polycyclic aromatic hydrocarbons

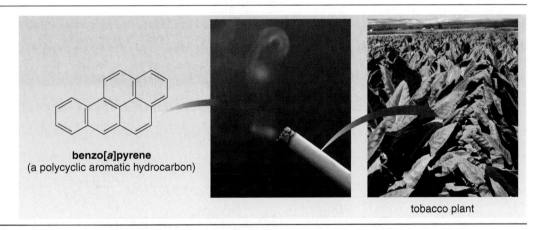

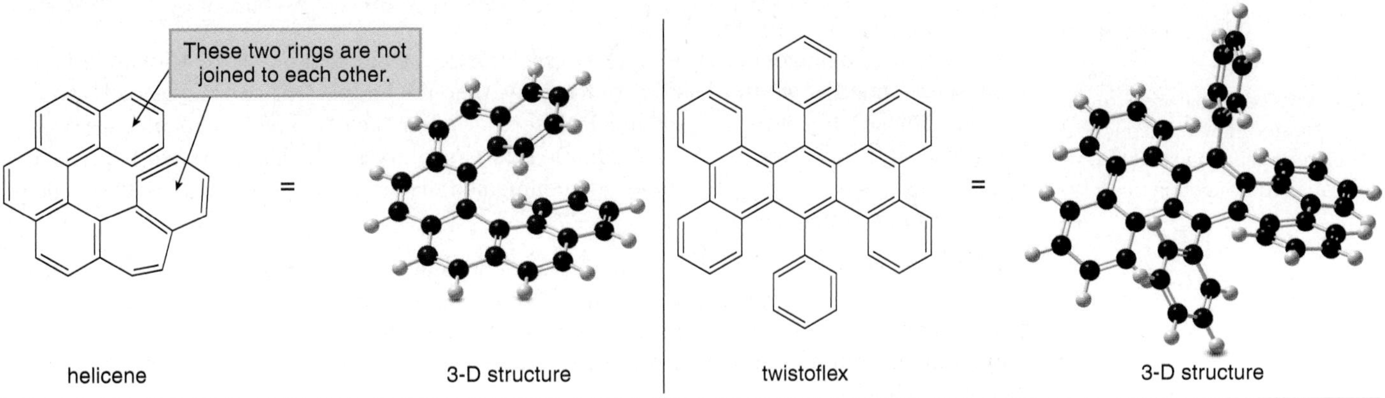

These two rings are not joined to each other.

helicene   3-D structure   twistoflex   3-D structure

Figure 17.5 Selected drugs that contain a benzene ring

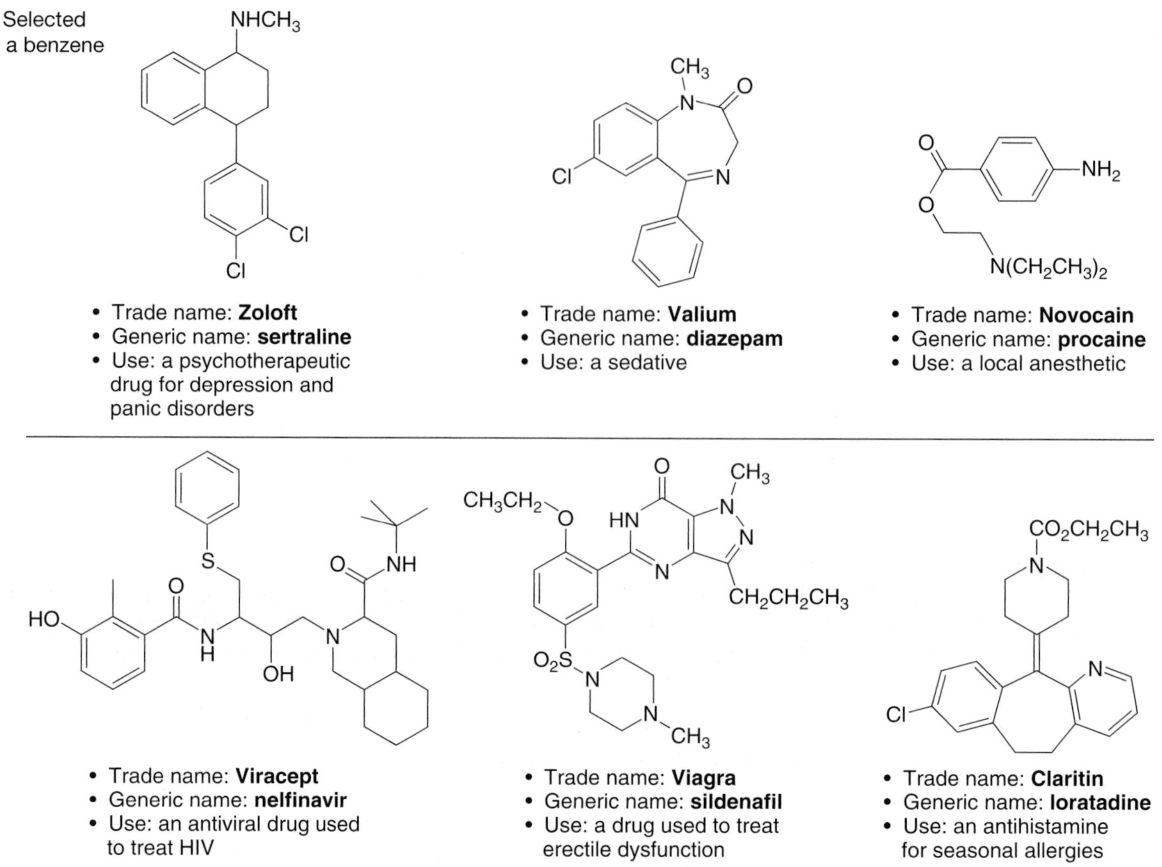

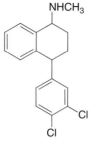

- Trade name: **Zoloft**
- Generic name: **sertraline**
- Use: a psychotherapeutic drug for depression and panic disorders

- Trade name: **Valium**
- Generic name: **diazepam**
- Use: a sedative

- Trade name: **Novocain**
- Generic name: **procaine**
- Use: a local anesthetic

- Trade name: **Viracept**
- Generic name: **nelfinavir**
- Use: an antiviral drug used to treat HIV

- Trade name: **Viagra**
- Generic name: **sildenafil**
- Use: a drug used to treat erectile dysfunction

- Trade name: **Claritin**
- Generic name: **loratadine**
- Use: an antihistamine for seasonal allergies

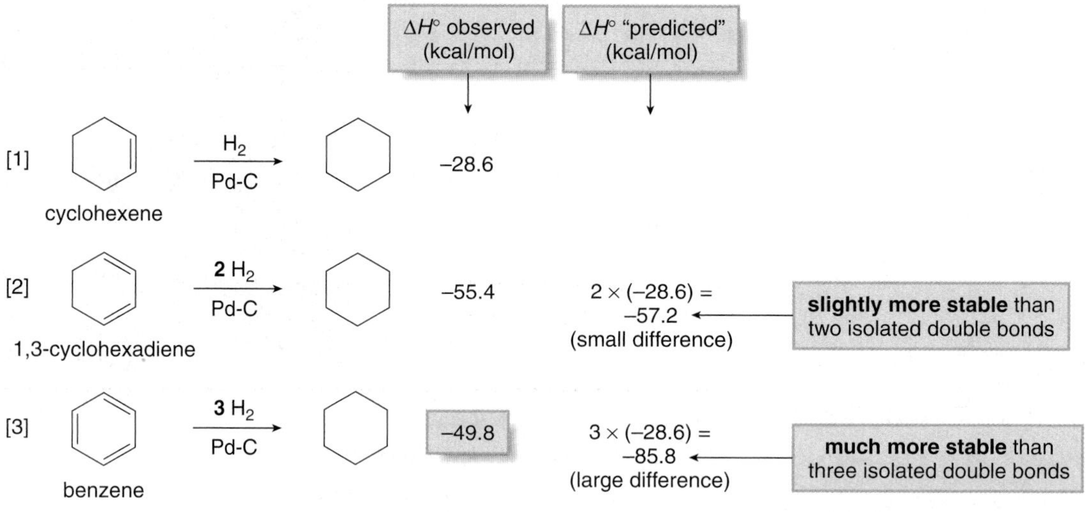

				$\Delta H°$ observed (kcal/mol)	$\Delta H°$ "predicted" (kcal/mol)	
[1]	cyclohexene	$\xrightarrow[\text{Pd-C}]{H_2}$		−28.6		
[2]	1,3-cyclohexadiene	$\xrightarrow[\text{Pd-C}]{2\ H_2}$		−55.4	$2 \times (-28.6) = -57.2$ (small difference)	**slightly more stable** than two isolated double bonds
[3]	benzene	$\xrightarrow[\text{Pd-C}]{3\ H_2}$		−49.8	$3 \times (-28.6) = -85.8$ (large difference)	**much more stable** than three isolated double bonds

The relative stability of conjugated dienes versus isolated dienes was first discussed in Section 16.9.

The addition of one mole of H_2 to cyclohexene releases −28.6 kcal of energy (Equation [1]). If each double bond is worth −28.6 kcal of energy, then the addition of two moles of H_2 to 1,3-cyclohexadiene (Equation [2]) should release 2×-28.6 kcal = −57.2 kcal of energy. The observed value, however, is −55.4 kcal/mol. This is slightly lower than expected because 1,3-cyclohexadiene is a conjugated diene and conjugated dienes are more stable than two isolated carbon–carbon double bonds.

The hydrogenations of cyclohexene and 1,3-cyclohexadiene occur readily at room temperature, but benzene can be hydrogenated only under forcing conditions, and even then the reaction is extremely slow. If each double bond is worth −28.6 kcal of energy, then the addition of three moles of H_2 to benzene should release 3×-28.6 kcal = −85.8 kcal of energy. In fact, the observed heat of hydrogenation is only −49.8 kcal/mol, which is 36 kcal/mol lower than predicted and

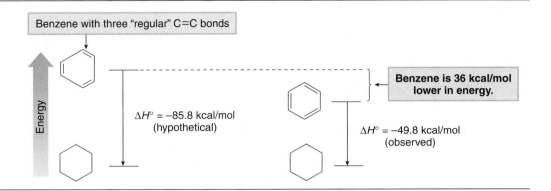

Figure 17.6 A comparison between the observed and hypothetical heats of hydrogenation for benzene

even lower than the observed value for 1,3-cyclohexadiene. Figure 17.6 compares the hypothetical and observed heats of hydrogenation for benzene.

The huge difference between the hypothetical and observed heats of hydrogenation for benzene cannot be explained solely on the basis of resonance and conjugation.

◆ The low heat of hydrogenation of benzene means that benzene is especially stable— even more so than the conjugated compounds introduced in Chapter 16. This unusual stability is characteristic of aromatic compounds.

Benzene's unusual behavior in chemical reactions is not limited to hydrogenation. As mentioned in Section 17.1, **benzene does not undergo addition reactions typical of other highly unsaturated compounds, including conjugated dienes.** Benzene does not react with Br_2 to yield an addition product. Instead, in the presence of a Lewis acid, bromine *substitutes* for a hydrogen atom, thus yielding a product that retains the benzene ring.

Addition does *not* occur.   ❌   An addition product would no longer contain a benzene ring.

Substitution occurs.   A substitution product still contains a benzene ring.

This behavior is characteristic of aromatic compounds. The structural features that distinguish aromatic compounds from the rest are discussed in Section 17.7.

PROBLEM 17.7   Compounds **A** and **B** are both hydrogenated to methylcyclohexane. Which compound has the higher heat of hydrogenation? Which compound is more stable?

A   **B**

17.7   The Criteria for Aromaticity—Hückel's Rule

Four structural criteria must be satisfied for a compound to be aromatic:

◆ A molecule must be cyclic, planar, completely conjugated, and contain a particular number of π electrons.

[1]   A molecule must be cyclic.

◆ To be aromatic, each *p* orbital must overlap with *p* orbitals on adjacent atoms.

The *p* orbitals on all six carbons of benzene continuously overlap, so benzene is aromatic. 1,3,5-Hexatriene has six *p* orbitals, too, but the two on the terminal carbons cannot overlap with each other, so **1,3,5-hexatriene is not aromatic.**

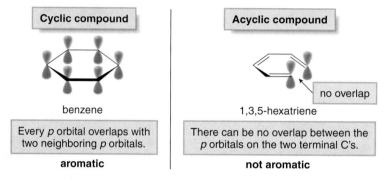

Cyclic compound	Acyclic compound
benzene	1,3,5-hexatriene
Every *p* orbital overlaps with two neighboring *p* orbitals.	There can be no overlap between the *p* orbitals on the two terminal C's.
aromatic	**not aromatic**

[2]   A molecule must be planar.

◆ All adjacent *p* orbitals must be aligned so that the π electron density can be delocalized.

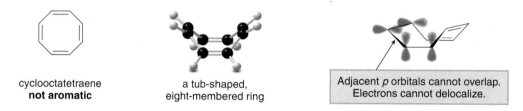

cyclooctatetraene
not aromatic

a tub-shaped,
eight-membered ring

Adjacent *p* orbitals cannot overlap.
Electrons cannot delocalize.

For example, cyclooctatetraene resembles benzene in that it is a cyclic molecule with alternating double and single bonds. Cyclooctatetraene is tub shaped, however, **not planar,** so overlap between adjacent π bonds is impossible. **Cyclooctatetraene, therefore, is *not* aromatic,** so it undergoes addition reactions like those of other alkenes.

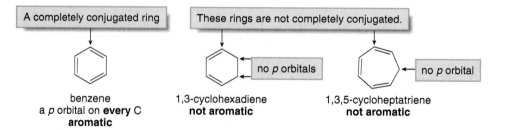

cyclooctatetraene   **addition** product

[3]   A molecule must be completely conjugated.

◆ Aromatic compounds must have a *p* orbital on every atom.

A completely conjugated ring	These rings are not completely conjugated.	
benzene	1,3-cyclohexadiene	1,3,5-cycloheptatriene
a *p* orbital on **every** C	**not aromatic**	**not aromatic**
aromatic	no *p* orbitals	no *p* orbital

Both 1,3-cyclohexadiene and 1,3,5-cycloheptatriene contain at least one carbon atom that does not have a *p* orbital, and so they are not completely conjugated and therefore ***not* aromatic.**

[4] **A molecule must satisfy Hückel's rule, and contain a particular number of π electrons.**

Some compounds satisfy the first three criteria for aromaticity, but still they show none of the stability typical of aromatic compounds. For example, **cyclobutadiene** is so highly reactive that it can only be prepared at extremely low temperatures.

cyclobutadiene ← a planar, cyclic, completely conjugated molecule that is *not* aromatic

It turns out that in addition to being cyclic, planar, and completely conjugated, a compound needs a particular number of π electrons to be aromatic. Erich Hückel first recognized in 1931 that the following criterion, expressed in two parts and now known as **Hückel's rule,** had to be satisfied, as well:

♦ An aromatic compound must contain [4n + 2] π electrons (*n* = 0, 1, 2, and so forth).
♦ Cyclic, planar, and completely conjugated compounds that contain 4n π electrons are especially unstable, and are said to be *antiaromatic.*

> Hückel's rule refers to the number of π electrons, *not* the number of atoms in a particular ring.

Thus, compounds that contain 2, 6, 10, 14, 18, and so forth π electrons are aromatic, as shown in Table 17.2. **Benzene is aromatic and especially stable because it contains 6 π electrons. Cyclobutadiene is antiaromatic and especially unstable because it contains 4 π electrons.**

TABLE 17.2 The Number of π Electrons that Satisfy Hückel's Rule

n	4n + 2
0	2
1	6
2	10
3	14
4, etc.	18

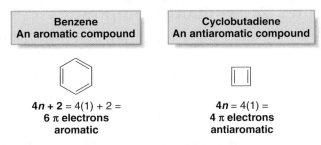

Benzene
An aromatic compound

$4n + 2 = 4(1) + 2 =$
6 π electrons
aromatic

Cyclobutadiene
An antiaromatic compound

$4n = 4(1) =$
4 π electrons
antiaromatic

Considering aromaticity, all compounds can be classified in one of three ways:

[1] Aromatic • **A cyclic, planar, completely conjugated compound with 4n + 2 π electrons.**

[2] Antiaromatic • **A cyclic, planar, completely conjugated compound with 4n π electrons.**

[3] Not aromatic (or nonaromatic) • **A compound that lacks one (or more) of the following requirements for aromaticity: being cyclic, planar, and completely conjugated.**

Note, too, the relationship between each compound type and a similar open-chained molecule having the same number of π electrons.

♦ An aromatic compound is *more* stable than a similar acyclic compound having the same number of π electrons. Benzene is more stable than 1,3,5-hexatriene.

♦ An antiaromatic compound is *less* stable than an acyclic compound having the same number of π electrons. Cyclobutadiene is less stable than 1,3-butadiene.

♦ A compound that is not aromatic is *similar* in stability to an acyclic compound having the same number of π electrons. 1,3-Cyclohexadiene is similar in stability to *cis,cis*-2,4-hexadiene, so it is not aromatic.

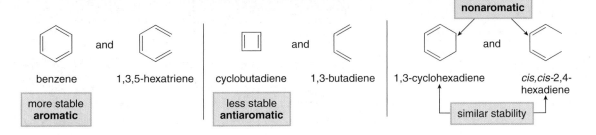

benzene   and   1,3,5-hexatriene
more stable
aromatic

cyclobutadiene   and   1,3-butadiene
less stable
antiaromatic

nonaromatic

1,3-cyclohexadiene   and   *cis,cis*-2,4-hexadiene
similar stability

^{1}H NMR spectroscopy readily indicates whether a compound is aromatic. The protons on sp^2 hybridized carbons in aromatic hydrocarbons are highly deshielded and absorb at 6.5–8 ppm, whereas hydrocarbons that are not aromatic absorb at 4.5–6 ppm, typical of protons bonded to the C=C of an alkene. Thus, benzene absorbs at 7.3 ppm, whereas cyclooctatetraene, which is not aromatic, absorbs further upfield, at 5.8 ppm for the protons on its sp^2 hybridized carbons.

Many compounds in addition to benzene are aromatic. Several examples are presented in Section 17.8.

PROBLEM 17.8   Estimate where the protons bonded to sp^2 hybridized carbons will absorb in the ^{1}H NMR spectrum of each compound.

a.   b.   c.

17.8   Examples of Aromatic Compounds

In Section 17.8 we look at many different types of aromatic compounds.

17.8A   Aromatic Compounds with a Single Ring

Benzene is the most common aromatic compound having a single ring. **Completely conjugated rings larger than benzene are also aromatic if they are planar and have $4n + 2$ π electrons.**

> ◆ Hydrocarbons containing a single ring with alternating double and single bonds are called *annulenes*.

To name an annulene, indicate the number of atoms in the ring in brackets and add the word *annulene*. Thus, benzene is [6]-annulene. Both **[14]-annulene** and **[18]-annulene** are cyclic, planar, completely conjugated molecules that follow Hückel's rule, and so they are aromatic.

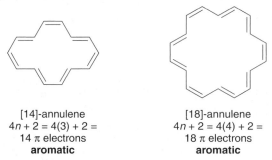

[14]-annulene
$4n + 2 = 4(3) + 2 =$
14 π electrons
aromatic

[18]-annulene
$4n + 2 = 4(4) + 2 =$
18 π electrons
aromatic

[10]-Annulene has 10 π electrons, which satisfies Hückel's rule, but a planar molecule would place the two H atoms inside the ring too close to each other, so the ring puckers to relieve this strain. Because **[10]-annulene is not planar,** the 10 π electrons can't delocalize over the entire ring and it is **not aromatic.**

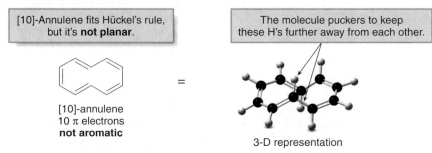

[10]-Annulene fits Hückel's rule, but it's **not planar**.

The molecule puckers to keep these H's further away from each other.

[10]-annulene
10 π electrons
not aromatic

3-D representation

PROBLEM 17.9 Would [16]-, [20]- or [22]-annulene be aromatic if each ring is planar?

PROBLEM 17.10 Explain why an annulene cannot have an odd number of carbon atoms in the ring.

17.8B   Aromatic Compounds with More than One Ring

Hückel's rule for determining aromaticity can be applied only to monocyclic systems, but many aromatic compounds containing several benzene rings joined together are also known. Two or more six-membered rings with alternating double and single bonds can be fused together to form **polycyclic aromatic hydrocarbons (PAHs).** Joining two benzene rings together forms **naphthalene.** There are two different ways to join three rings together, forming **anthracene** and **phenanthrene,** and many more complex hydrocarbons are known.

naphthalene
10 π electrons

anthracene
14 π electrons

phenanthrene
14 π electrons

As the number of fused benzene rings increases, the number of resonance structures increases as well. Although two resonance structures can be drawn for benzene, naphthalene is a hybrid of three resonance structures.

Three resonance structures
for naphthalene

PROBLEM 17.11 Draw the four resonance structures for anthracene.

17.8C   Aromatic Heterocycles

> Recall from Section 9.3 that a **heterocycle** is a ring that contains at least one heteroatom.

Heterocycles containing oxygen, nitrogen, or sulfur—atoms that also have at least one lone pair of electrons—can also be aromatic. With heteroatoms, we must always determine whether the lone pair is localized on the heteroatom or part of the delocalized π system. Two examples, **pyridine** and **pyrrole,** illustrate these different possibilities.

Pyridine

Pyridine is a heterocycle containing a six-membered ring with three π bonds and one nitrogen atom. Like benzene, two resonance structures can be drawn.

two resonance structures for pyridine
6 π electrons

Pyridine is cyclic, planar, and completely conjugated, because the three single and double bonds alternate around the ring. **Pyridine has six π electrons, two from each π bond, thus satisfying Hückel's rule and making pyridine aromatic.** The nitrogen atom of pyridine also has a non-bonded electron pair, which is localized on the N atom, so it is not part of the delocalized π electron system of the aromatic ring.

How is the nitrogen atom of the pyridine ring hybridized? The N atom is surrounded by three groups (two atoms and a lone electron pair), making it sp^2 **hybridized,** and leaving one unhybridized p orbital with one electron that overlaps with adjacent p orbitals. The lone pair on N resides in an sp^2 hybrid orbital that is perpendicular to the delocalized π electrons.

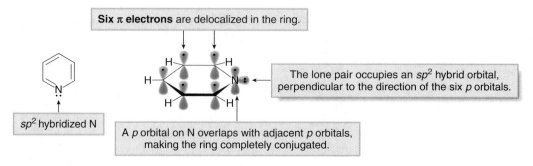

Six π **electrons** are delocalized in the ring.

The lone pair occupies an sp^2 hybrid orbital, perpendicular to the direction of the six p orbitals.

sp^2 hybridized N

A p orbital on N overlaps with adjacent p orbitals, making the ring completely conjugated.

Pyrrole

Pyrrole contains a five-membered ring with two π bonds and one nitrogen atom. The N atom also has a lone pair of electrons.

pyrrole

Pyrrole is cyclic and planar, with a total of four π electrons from the two π bonds. Is the non-bonded electron pair localized on N or part of a delocalized π electron system? The lone pair on N is *adjacent* to a double bond. Recall the following general rule from Section 16.5:

◆ In any system X=Y–Z:, Z is sp^2 hybridized and the lone pair occupies a p orbital to make the system conjugated.

If the lone pair on the N atom occupies a p orbital:

◆ **Pyrrole has a p orbital on every adjacent atom, so it is completely conjugated.**

◆ **Pyrrole has six π electrons—four from the π bonds and two from the lone pair.**

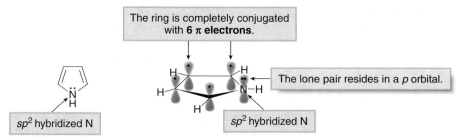

The ring is completely conjugated with **6 π electrons**.

The lone pair resides in a p orbital.

sp^2 hybridized N

sp^2 hybridized N

Because pyrrole is cyclic, planar, completely conjugated, and has $4n + 2$ π electrons, **pyrrole is aromatic. The number of electrons—not the size of the ring—determines whether a compound is aromatic.**

Electrostatic potential maps, shown in Figure 17.7 for pyridine and pyrrole, confirm that the **nonbonded electron pair in pyridine is localized on N,** whereas the **lone pair in pyrrole is part of the delocalized π system.**

Figure 17.7 Electrostatic potential maps of pyridine and pyrrole

pyridine

pyrrole

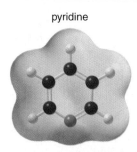

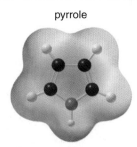

• In pyridine, the nonbonded electron pair is localized on the N atom in an sp^2 hybridized orbital, as shown by the region of high electron density (in red) on N.

• In pyrrole, the nonbonded electron pair is in a p orbital and is delocalized over the ring, so the entire ring is electron rich (red).

Histamine

Histamine, a biologically active amine formed in many tissues, has an aromatic heterocycle with two N atoms, one of which is similar to the N atom of pyridine and one of which is similar to the N atom of pyrrole.

histamine

Histamine has a five-membered ring with two π bonds and two nitrogen atoms, each of which contains a lone pair of electrons. The heterocycle has four π electrons from the two double bonds. The lone pair on N1 also occupies a *p* orbital, making the heterocycle completely conjugated, and giving it a total of six π electrons. The lone pair on N1 is thus delocalized over the five-membered ring and the heterocycle is aromatic. The lone pair on N2 occupies an *sp*^{2} hybrid orbital perpendicular to the delocalized π electrons.

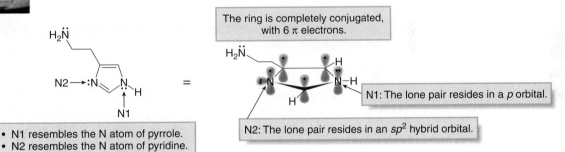

The ring is completely conjugated, with 6 π electrons.

N1: The lone pair resides in a *p* orbital.

N2: The lone pair resides in an *sp*^{2} hybrid orbital.

• N1 resembles the N atom of pyrrole.
• N2 resembles the N atom of pyridine.

Histamine produces a wide range of physiological effects in the body. Excess histamine is responsible for the runny nose and watery eyes symptomatic of hay fever. It also stimulates the overproduction of stomach acid, and contributes to the formation of hives. These effects result from the interaction of histamine with two different cellular receptors. We will learn more about antihistamines and antiulcer drugs, compounds that block the effects of histamine, in Section 25.6.

PROBLEM 17.12   Which heterocycles are aromatic?

a.   b.   c.

PROBLEM 17.13   (a) How is each N atom in quinine, the chapter-opening molecule, hybridized? (b) In what type of orbital does the lone pair on each N reside?

CH$_3$O

HO

H

H

quinine
(antimalarial drug)

17.8D   Charged Aromatic Compounds

Both negatively and positively charged ions can also be aromatic if they possess all the necessary elements.

Cyclopentadienyl Anion

The **cyclopentadienyl anion** is a cyclic and planar anion with two double bonds and a non-bonded electron pair. In this way it resembles pyrrole. The two π bonds contribute four electrons and the lone pair contributes two more, for a total of six. By Hückel's rule, having **six π electrons confers aromaticity.** Like the N atom in pyrrole, the **negatively charged carbon atom must be** sp^2 **hybridized,** and the **nonbonded electron pair must occupy a p orbital** for the ring to be completely conjugated.

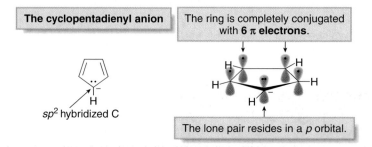

The cyclopentadienyl anion

The ring is completely conjugated with **6 π electrons.**

sp^2 hybridized C

The lone pair resides in a p orbital.

◆ The cyclopentadienyl anion is aromatic because it is cyclic, planar, completely conjugated, and has six π electrons.

We can draw **five equivalent resonance structures for the cyclopentadienyl anion,** delocalizing the negative charge over every carbon atom of the ring.

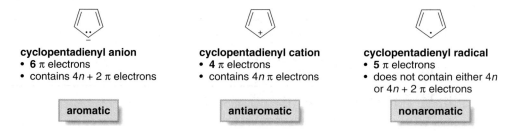

five equivalent resonance structures for the cyclopentadienyl anion

Although five resonance structures can also be drawn for both the **cyclopentadienyl cation** and **radical,** only the cyclopentadienyl anion has six π electrons, a number that satisfies Hückel's rule. The cyclopentadienyl cation has four π electrons, making it antiaromatic and especially unstable. The cyclopentadienyl radical has five π electrons, so it is neither aromatic nor antiaromatic. Having the "right" number of electrons is necessary for a species to be unusually stable by virtue of aromaticity.

cyclopentadienyl anion
• 6 π electrons
• contains $4n + 2$ π electrons

aromatic

cyclopentadienyl cation
• 4 π electrons
• contains $4n$ π electrons

antiaromatic

cyclopentadienyl radical
• 5 π electrons
• does not contain either $4n$ or $4n + 2$ π electrons

nonaromatic

The cyclopentadienyl anion is readily formed from cyclopentadiene by a Brønsted–Lowry acid–base reaction.

no p orbital

:B

cyclopentadiene
not aromatic
pK_a = 15

cyclopentadienyl anion
aromatic
a stabilized conjugate base

+ H–B$^+$

Cyclopentadiene itself is not aromatic because it is not fully conjugated. The cyclopentadienyl anion, however, is both aromatic and resonance stabilized, so it is a very stable base. As such, it makes cyclopentadiene more acidic than other hydrocarbons. In fact, the pK_a of cyclopentadiene is 15, much lower (more acidic) than the pK_a of any C–H bond discussed thus far.

◆ Cyclopentadiene is more acidic than many hydrocarbons because its conjugate base is aromatic.

PROBLEM 17.14   Draw five resonance structures for the cyclopentadienyl cation.

PROBLEM 17.15   Draw the product formed when 1,3,5-cycloheptatriene (pK_a = 39) is treated with a strong base. Why is its pK_a so much higher than the pK_a of cyclopentadiene?

1,3,5-cycloheptatriene
pK_a = 39

PROBLEM 17.16   Rank the following compounds in order of increasing acidity.

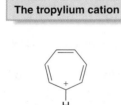

Tropylium Cation

The cyclopentadienyl anion and the tropylium cation both illustrate an important principle: The **number of π electrons determines aromaticity,** not the number of atoms in a ring or the number of p orbitals that overlap. The cyclopentadienyl anion and tropylium cation are aromatic because they each have six π electrons.

The **tropylium cation** is a planar carbocation with three double bonds and a positive charge contained in a seven-membered ring. This carbocation is completely conjugated, because the positively charged carbon is sp^2 hybridized and has a vacant p orbital that overlaps with the six p orbitals from the carbons of the three double bonds. **Because the tropylium cation has three π bonds and no other nonbonded electron pairs, it contains six π electrons,** thereby satisfying Hückel's rule.

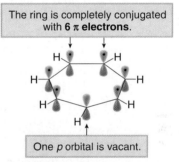

◆ The tropylium cation is aromatic because it is cyclic, planar, completely conjugated, and has six π electrons delocalized over the seven atoms of the ring.

PROBLEM 17.17   Draw the seven resonance structures for the tropylium cation.

PROBLEM 17.18   Assuming the rings are planar, which ions are aromatic?

a.   b.   c.   d.

PROBLEM 17.19   Compound **A** exhibits a peak in its ^{1}H NMR spectrum at 7.6 ppm, indicating that it is aromatic. How are the carbon atoms of the triple bonds hybridized? In what type of orbitals are the π electrons of the triple bonds contained? How many π electrons are delocalized around the ring in **A**?

A =   H ◄——— absorbs at 7.6 ppm

17.9   What Is the Basis of Hückel's Rule?

Why does the number of π electrons determine whether a compound is aromatic? Cyclobutadiene is cyclic, planar, and completely conjugated, just like benzene, but why is benzene aromatic and cyclobutadiene antiaromatic?

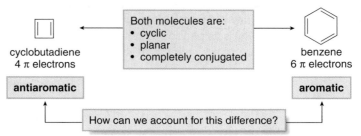

cyclobutadiene
4 π electrons

Both molecules are:
• cyclic
• planar
• completely conjugated

benzene
6 π electrons

antiaromatic

aromatic

How can we account for this difference?

A complete explanation is beyond the scope of an introductory organic chemistry text, but nevertheless, you can better understand the basis of aromaticity by learning more about orbitals and bonding.

17.9A   Bonding and Antibonding Orbitals

So far we have used the following basic concepts to describe how bonds are formed:

◆ Hydrogen uses its 1*s* orbital to form σ bonds with other elements.
◆ Second-row elements use hybrid orbitals (*sp*, *sp*^{2}, or *sp*^{3}) to form σ bonds.
◆ Second-row elements use *p* orbitals to form π bonds.

This description of bonding is called **valence bond theory.** In valence bond theory, a covalent bond is formed by the overlap of two atomic orbitals, and the electron pair in the resulting bond is shared by both atoms. Thus, a carbon–carbon double bond consists of a σ bond, formed by overlap of two *sp*^{2} hybrid orbitals, each containing one electron, and a π bond, formed by overlap of two *p* orbitals, each containing one electron.

This description of bonding works well for most of the organic molecules we have encountered thus far. Unfortunately, it is inadequate for describing systems with many adjacent *p* orbitals that overlap, as there are in aromatic compounds. To more fully explain the bonding in these systems, we must utilize **molecular orbital (MO) theory.**

MO theory describes bonds as the mathematical combination of atomic orbitals that form a new set of orbitals called **molecular orbitals (MOs).** A molecular orbital occupies a region of space *in a molecule* where electrons are likely to be found. When forming molecular orbitals from atomic orbitals, keep in mind:

◆ A set of *n* atomic orbitals forms *n* molecular orbitals.

If *two* atomic orbitals combine, *two* molecular orbitals are formed. This is fundamentally different than valence bond theory. Because aromaticity is based on *p* orbital overlap, what does MO theory predict will happen when two *p* (atomic) orbitals combine?

The two lobes of each *p* orbital are opposite in phase, with a node of electron density at the nucleus. When two *p* orbitals combine, two molecular orbitals should form. The two *p* orbitals can add together constructively—that is, with like phases interacting—or destructively—that is, with opposite phases interacting.

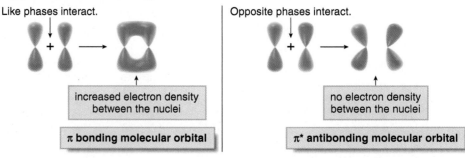

Like phases interact.

increased electron density
between the nuclei

π bonding molecular orbital

Opposite phases interact.

no electron density
between the nuclei

π* antibonding molecular orbital

Figure 17.8 Combination of two *p* orbitals to form π and π* molecular orbitals

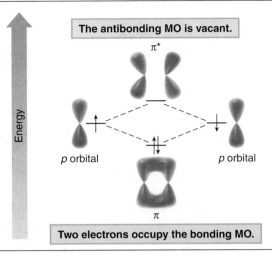

The antibonding MO is vacant.

π*

p orbital   *p* orbital

π

Two electrons occupy the bonding MO.

- Two atomic *p* orbitals combine to form two molecular orbitals. The bonding π MO is lower in energy than the two *p* orbitals from which it was formed, and the antibonding π* MO is higher in energy than the two *p* orbitals from which it was formed.
- Two electrons fill the lower energy bonding MO first.

◆ When two *p* orbitals of similar phase overlap side-by-side, a π bonding molecular orbital results.

◆ When two *p* orbitals of opposite phase overlap side-by-side, a π* antibonding molecular orbital results.

A π bonding MO is lower in energy than the two atomic *p* orbitals from which it is formed because a stable bonding interaction results, holding nuclei together when orbitals of similar phase combine. Similarly, a π* antibonding MO is higher in energy because a destabilizing node results, which pushes nuclei apart when orbitals of opposite phase combine.

If two atomic *p* orbitals each have one electron and then combine to form MOs, the two electrons will occupy the lower energy π bonding MO, as shown in Figure 17.8.

17.9B   Molecular Orbitals Formed When More than Two *p* Orbitals Combine

The molecular orbital description of benzene is much more complex than the two MOs formed in Figure 17.8. Because each of the six carbon atoms of benzene has a *p* orbital, six atomic *p* orbitals combine to form six π molecular orbitals, as shown in Figure 17.9. A description of the exact appearance and energies of these six MOs requires more sophisticated mathematics and understanding of MO theory than is presented in this text. Nevertheless, note that the six MOs are labeled ψ_1–ψ_6, with ψ_1 being the lowest in energy and ψ_6 the highest.

The most important features of the six benzene MOs are as follows:

◆ **The larger the number of bonding interactions, the lower in energy the MO.** The lowest energy molecular orbital (ψ_1) has all bonding interactions between the *p* orbitals.

◆ **The larger the number of nodes, the higher in energy the MO.** The highest energy MO (ψ_6*) has all nodes between the *p* orbitals.

◆ Three MOs are lower in energy than the starting *p* orbitals, making them bonding MOs (ψ_1, ψ_2, ψ_3), whereas three MOs are higher in energy than the starting *p* orbitals, making them antibonding MOs (ψ_4*, ψ_5*, ψ_6*).

◆ The two pairs of MOs (ψ_2 and ψ_3; ψ_4* and ψ_5*) with the same energy are called **degenerate orbitals.**

◆ **The highest energy orbital that contains electrons is called the *highest occupied molecular orbital* (HOMO).** For benzene, the degenerate orbitals ψ_2 and ψ_3 are the HOMOs.

◆ **The lowest energy orbital that does *not* contain electrons is called the *lowest unoccupied molecular orbital* (LUMO).** For benzene, the degenerate orbitals ψ_4* and ψ_5* are the LUMOs.

Figure 17.9 The six
molecular orbitals of benzene

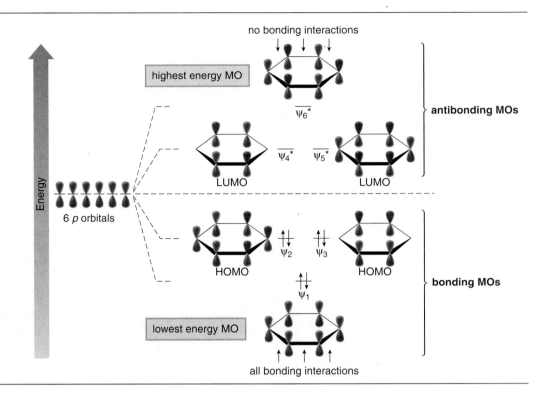

To fill the MOs, the six electrons are added, two to an orbital, beginning with the lowest energy orbital. As a result, the six electrons completely fill the bonding MOs, leaving the antibonding MOs empty. This is what gives benzene and other aromatic compounds their special stability and this is why six π electrons satisfies Hückel's $4n + 2$ rule.

◆ All bonding MOs (and HOMOs) are completely filled in aromatic compounds. No π
electrons occupy antibonding MOs.

17.10 The Inscribed Polygon Method for Predicting Aromaticity

An inscribed polygon is also called a **Frost circle**.

To predict whether a compound has π electrons completely filling bonding MOs, we must know how many bonding molecular orbitals and how many π electrons it has. It is possible to predict the relative energies of cyclic, completely conjugated compounds, without sophisticated math (or knowing what the resulting MOs look like) by using the **inscribed polygon method.**

| *How To* | Use the Inscribed Polygon Method to Determine the Relative Energies of MOs for Cyclic, Completely Conjugated Compounds |

Example Plot the relative energies of the MOs of benzene

Step [1] Draw the polygon in question inside a circle with its vertices touching the circle and one of the vertices pointing down. Mark the points at which the polygon intersects the circle.

• Inscribe a hexagon inside a circle for benzene. The six vertices of the hexagon form six points of intersection, corresponding to the six MOs of benzene. The pattern—a single MO having the lowest energy, two degenerate pairs of MOs, and a single highest energy MO—matches that found in Figure 17.9.

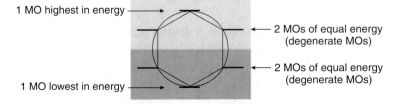

Step [2]   **Draw a line horizontally through the center of the circle and label MOs as bonding, nonbonding, or antibonding.**

- **MOs below this line are bonding,** and lower in energy than the *p* orbitals from which they were formed. Benzene has three bonding MOs.
- **MOs at this line are nonbonding,** and equal in energy to the *p* orbitals from which they were formed. Benzene has no nonbonding MOs.
- **MOs above this line are antibonding,** and higher in energy than the *p* orbitals from which they were formed. Benzene has three antibonding MOs.

Step [3]   **Add the electrons, beginning with the lowest energy MO.**

- **All the bonding MOs (and the HOMOs) are completely filled in aromatic compounds. No π electrons occupy antibonding MOs.**
- Benzene is aromatic because it has six π electrons that completely fill the bonding MOs.

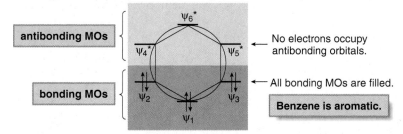

This method works for all monocyclic, completely conjugated hydrocarbons regardless of ring size. Figure 17.10 illustrates MOs for completely conjugated five- and seven-membered rings using this method. The total number of MOs always equals the number of vertices of the polygon. Because both systems have three bonding MOs, each needs six π electrons to fully occupy them, making the cyclopentadienyl anion and the tropylium cation aromatic, as we learned in Section 17.8D.

The inscribed polygon method is consistent with Hückel's 4*n* + 2 rule; that is, there is always one lowest energy bonding MO that can hold two π electrons and the other bonding MOs come in degenerate pairs that can hold a total of four π electrons. For the compound to be aromatic, these MOs must be completely filled with electrons, so the "magic numbers" for aromaticity fit Hückel's 4*n* + 2 rule (Figure 17.11).

Figure 17.10   Using the inscribed polygon method for five- and seven-membered rings

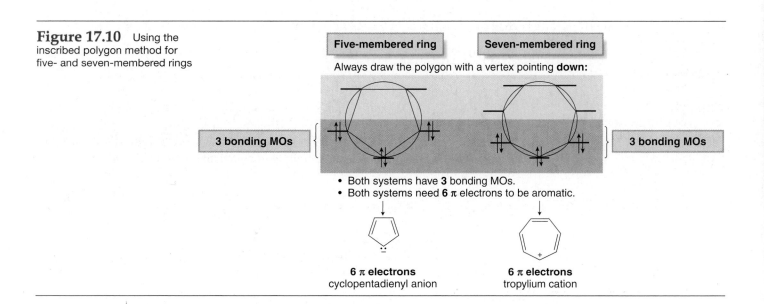

Figure 17.11 MO patterns for cyclic, completely conjugated systems

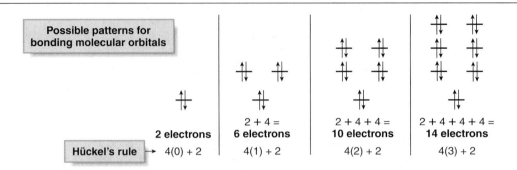

SAMPLE PROBLEM 17.1

Use the inscribed polygon method to show why cyclobutadiene is not aromatic.

cyclobutadiene
4 π electrons

SOLUTION

Cyclobutadiene has four MOs (formed from its four *p* orbitals), to which its four π electrons must be added.

Step [1]

Inscribe a square with a vertex down and mark its four points of intersection with the circle.

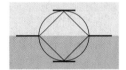

- The four points of intersection correspond to the four MOs of cyclobutadiene.

Steps [2] and [3]

Draw a line through the center of the circle, label the MOs, and add the electrons.

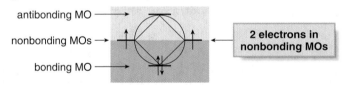

- Cyclobutadiene has four MOs—one bonding, two nonbonding, and one antibonding.
- Adding cyclobutadiene's four π electrons to these orbitals places two in the lowest energy bonding MO and one each in the two nonbonding MOs.
- Separating electrons in two degenerate MOs keeps like charges further away from each other.

Conclusion: Cyclobutadiene is not aromatic because its HOMOs, two degenerate nonbonding MOs, are not completely filled.

The procedure followed in Sample Problem 17.1 also illustrates why cyclobutadiene is antiaromatic. Having the two unpaired electrons in nonbonding MOs suggests that cyclobutadiene should be a highly unstable diradical. In fact, antiaromatic compounds resemble cyclobutadiene because their HOMOs contain two unpaired electrons, making them especially unstable.

PROBLEM 17.20

Use the inscribed polygon method to show why the following cation is aromatic:

PROBLEM 17.21

Use the inscribed polygon method to show why the cyclopentadienyl cation and radical are not aromatic.

17.11 Buckminsterfullerene—Is It Aromatic?

The two most common elemental forms of carbon are diamond and graphite. Diamond, one of the hardest substances known, is used for industrial cutting tools, whereas graphite, a slippery black substance, is used as a lubricant. Their physical characteristics are so different because their molecular structures are very different.

The structure of diamond consists of a continuous tetrahedral network of sp^3 hybridized carbon atoms, thus creating an infinite array of chair cyclohexane rings (without the hydrogen atoms). The structure of graphite, on the other hand, consists of parallel sheets of sp^2 hybridized carbon atoms, thus creating an infinite array of benzene rings. The parallel sheets are then held together by weak intermolecular interactions.

Three sheets of graphite, viewed edge-on

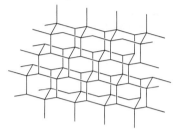

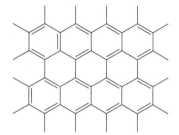

diamond
an "infinite" array of six-membered rings, covalently bonded in three dimensions

graphite
an "infinite" array of benzene rings, covalently bonded in two dimensions

Graphite exists in planar sheets of benzene rings, held together by weak intermolecular forces.

Buckminsterfullerene (or buckyball) was discovered by Smalley, Curl, and Kroto, who shared the 1996 Nobel Prize in Chemistry for their work. Its unusual name stems from its shape, which resembles the geodesic dome invented by R. Buckminster Fuller. The pattern of five- and six-membered rings also resembles the pattern of rings on a soccer ball.

Buckminsterfullerene (C_{60}) is a third elemental form of carbon. Its structure consists of 20 hexagons and 12 pentagons of sp^2 hybridized carbon atoms joined in a spherical arrangement. It is completely conjugated because each carbon atom has a p orbital with an electron in it.

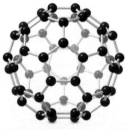

buckminsterfullerene, C_{60}

20 hexagons + 12 pentagons of carbon atoms joined together

The 60 C's of buckminsterfullerene are drawn. Each C also contains a p orbital with one electron, which is not drawn.

Is C_{60} aromatic? Although it is completely conjugated, it is not planar. Because of its curvature, it is not as stable as benzene. In fact, it undergoes addition reactions with electrophiles in much the same way as ordinary alkenes. Benzene, on the other hand, undergoes substitution reactions with electrophiles, which preserves the unusually stable benzene ring intact. These reactions are the subject of Chapter 18.

PROBLEM 17.22   How many ^{13}C NMR signals does C_{60} exhibit?

Diamond and graphite are two elemental forms of carbon.

17.12 Key Concepts—Benzene and Aromatic Compounds

Comparing Aromatic, Antiaromatic, and Nonaromatic Compounds (17.7)

- **Aromatic compound**
  - A cyclic, planar, completely conjugated compound that contains $4n + 2$ π electrons ($n = 0, 1, 2, 3$, and so forth).
  - An aromatic compound is more stable than a similar acyclic compound having the same number of π electrons.

- **Antiaromatic compound**
  - A cyclic, planar, completely conjugated compound that contains $4n$ π electrons ($n = 0, 1, 2, 3$, and so forth).
  - An antiaromatic compound is less stable than a similar acyclic compound having the same number of π electrons.

- **Nonaromatic compound**
  - A compound that is either not cyclic or not planar or not completely conjugated.

Properties of Aromatic Compounds

- Every atom in the ring has a *p* orbital to delocalize electron density (17.2).
- They are unusually stable. $\Delta H°$ for hydrogenation is much less than expected, given the number of degrees of unsaturation (17.6).
- They do not undergo the usual addition reactions of alkenes (17.6).
- ¹H NMR spectra show highly deshielded protons because of ring currents that reinforce the applied magnetic field (17.4).
- All bonding MOs and HOMOs are completely filled and no electrons occupy antibonding orbitals (17.9).

Examples of Aromatic Compounds with Six π Electrons (17.8)

benzene   pyridine   pyrrole   cyclopentadienyl anion   tropylium cation

Examples of Compounds that Are Not Aromatic (17.8)

not cyclic   not planar   not completely conjugated

Problems

Benzene

17.23 Early structural studies on benzene had to explain the following experimental evidence. When benzene was treated with Br_2 (plus a Lewis acid), a single substitution product of molecular formula C_6H_5Br was formed. When this product was treated with another equivalent of Br_2, three different compounds of molecular formula $C_6H_4Br_2$ were formed.

   a. Explain why a single Kekulé structure is consistent with the first result, but does not explain the second result.

   b. Then explain why a resonance description of benzene is consistent with the results of both reactions.

Nomenclature

17.24 Give the IUPAC name for each compound.

a.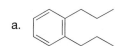

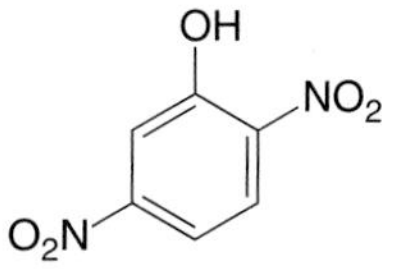

b.

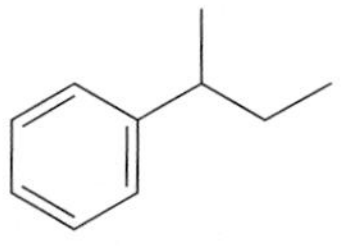

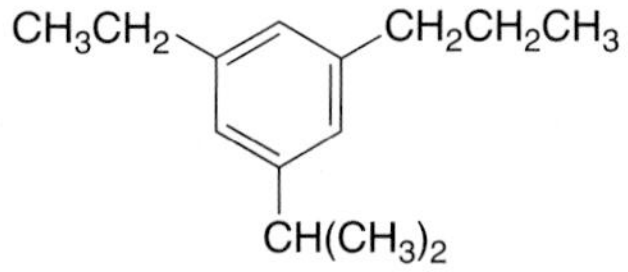

c.

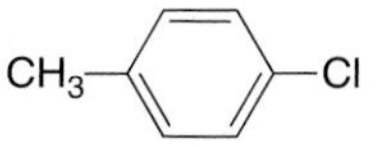

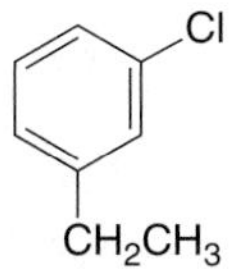

d. CH₃— —Cl

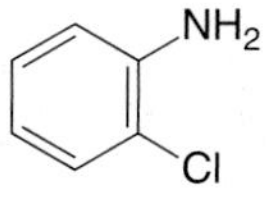

e.

f.

g.

h.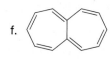

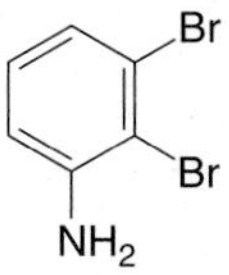

i.

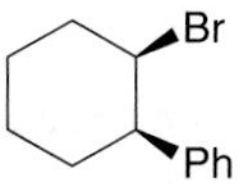

17.25 Draw a structure corresponding to each name.

a. *p*-dichlorobenzene
b. *m*-chlorophenol
c. *p*-iodoaniline
d. *o*-bromonitrobenzene

e. 2,6-dimethoxytoluene
f. 2-phenyl-1-butene
g. 2-phenyl-2-propen-1-ol
h. *trans*-1-benzyl-3-phenylcyclopentane

17.26 Draw and name all the isomeric trichlorobenzenes (molecular formula C₆H₃Cl₃).

Aromaticity

17.27 How many π electrons are contained in each molecule?

a.

b.

c.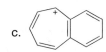

17.28 Which of the following compounds are aromatic? For any compound that is not aromatic, state why this is so.

a.

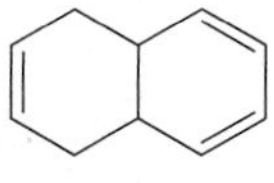

b.

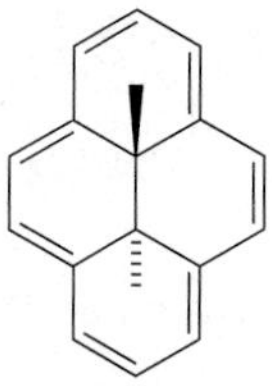

c.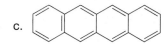

d.

e.

f.

17.29 Which of the following heterocycles are aromatic?

a.

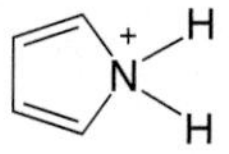

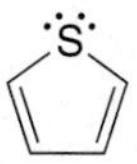

b.

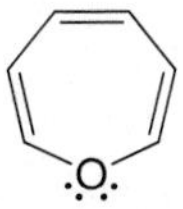

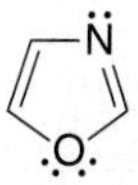

c.

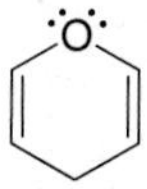

d.

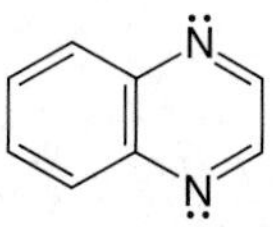

e.

f.

g.

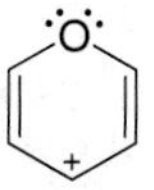

h.

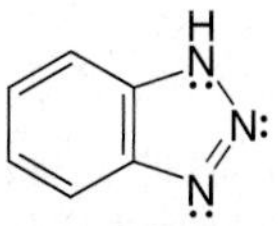

17.30 Label each compound as aromatic, antiaromatic, or not aromatic. Assume all completely conjugated compounds have planar rings.

a.

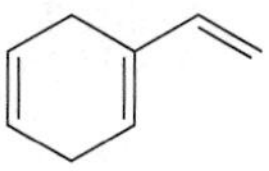

b.

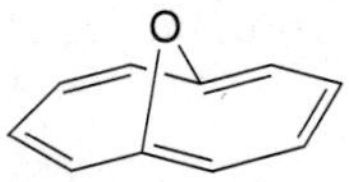

c.

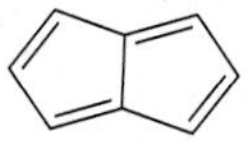

d.

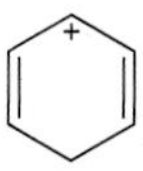

e.

17.31 Hydrocarbons **A** and **B** both possess a significant dipole, even though each is composed only of C−C and C−H bonds. Explain why the dipole arises in each compound. Use resonance structures to illustrate the direction of the dipole. Which ring is more electron rich in each compound?

azulene

A   **B**

17.32 Rank the indicated C−C bonds in order of increasing bond length, and explain why you chose this order.

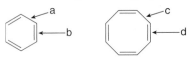

17.33 The purine heterocycle occurs commonly in the structure of DNA.

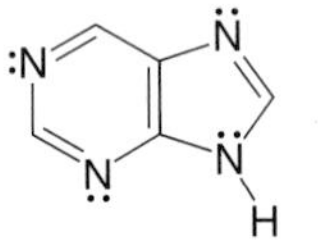

purine

    a. How is each N atom hybridized?
    b. In what type of orbital does each lone pair on a N atom reside?
    c. How many π electrons does purine contain?
    d. Why is purine aromatic?

17.34

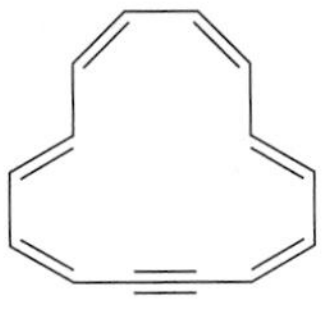

C

    a. How many π electrons does **C** contain?
    b. How many π electrons are delocalized in the ring?
    c. Explain why **C** is aromatic.

17.35 Explain the observed rate of reactivity of the following 2° alkyl halides in an S_N1 reaction.

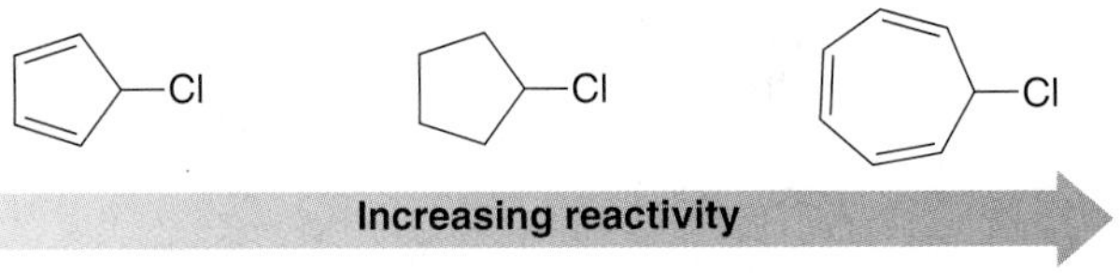

Increasing reactivity

17.36 Draw a stepwise mechanism for the following reaction.

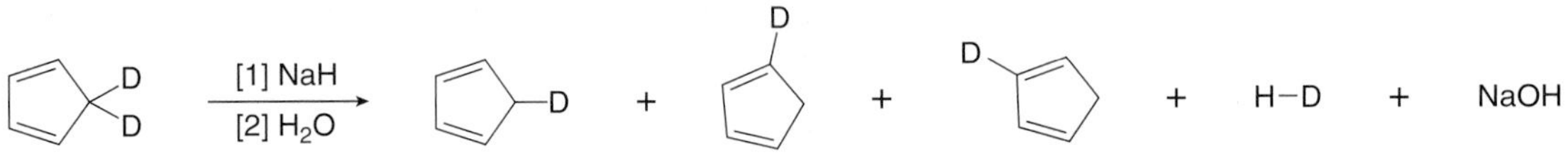

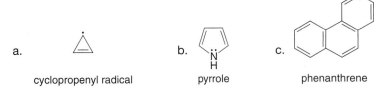

17.37 Explain why α-pyrone reacts with Br_2 to yield a substitution product (like benzene does), rather than an addition product to one of its C=C bonds.

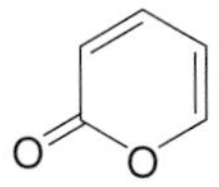

α-pyrone

Resonance

17.38 Draw additional resonance structures for each species.

    a.   b.   c.

    cyclopropenyl radical   pyrrole   phenanthrene

17.39 The carbon–carbon bond lengths in naphthalene are not equal. Use a resonance argument to explain why bond (a) is shorter than bond (b).

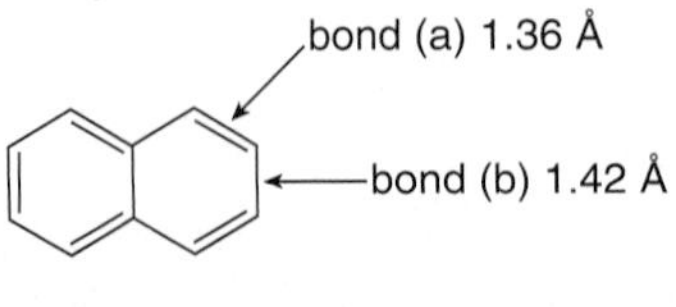

bond (a) 1.36 Å

bond (b) 1.42 Å

Acidity

17.40 Which compound in each pair is the stronger acid?

a. and [cyclopentadiene structure]   b. [cyclopentene structure] and

17.41 Treatment of indene with NaNH₂ forms its conjugate base in a Brønsted–Lowry acid–base reaction. Draw all reasonable resonance structures for indene's conjugate base, and explain why the pK_a of indene is lower than the pK_a of most hydrocarbons.

indene
pK_a = 20

+ NaNH₂ ⟶ [indenyl anion structure] + NH₃

Na⁺

17.42 Explain why **A** (the conjugate acid of pyrrole) is much more acidic than **B** (the conjugate acid of pyridine).

pK_a = 0.4   pK_a = 5.2

A   **B**

Inscribed Polygon Method

17.43 Use the inscribed polygon method to show the pattern of molecular orbitals in cyclooctatetraene.

[cyclooctatetraene structure] $\xrightarrow{2\ K}$ [dianion structure] (one resonance structure) + 2 K⁺

cyclooctatetraene   dianion of
                           cyclooctatetraene

a. Label the MOs as bonding, antibonding, or nonbonding.
b. Indicate the arrangement of electrons in these orbitals for cyclooctatetraene, and explain why cyclooctatetraene is not aromatic.
c. Treatment of cyclooctatetraene with potassium forms a dianion. How many π electrons does this dianion contain?
d. How are the π electrons in this dianion arranged in the molecular orbitals?
e. Classify the dianion of cyclooctatetraene as aromatic, antiaromatic, or not aromatic, and explain why this is so.

17.44 Use the inscribed polygon method to show the pattern of molecular orbitals in 1,3,5,7-cyclononatetraene and use it to label its cation, radical, and anion as aromatic, antiaromatic, or not aromatic.

cyclononatetraenyl   cyclononatetraenyl   cyclononatetraenyl
cation   radical   anion

Spectroscopy

17.45 How many ¹³C NMR signals does each compound exhibit?

a. [toluene/methylbenzene structure]   b. [1-ethyl-2-methylbenzene structure with CH₃ and CH₂CH₃]   c. [naphthalene structure]   d. [anthracene structure]

17.46 Which of the diethylbenzene isomers (ortho, meta, or para) corresponds to each set of ¹³C NMR spectral data?

[A] ¹³C NMR (ppm)	[B] ¹³C NMR (ppm)	[C] ¹³C NMR (ppm)
16	15	16
29	26	29
125	126	128
127.5	128	141
128.4	142	
144		

17.47 Propose a structure consistent with each set of data.

a. $C_{10}H_{14}$: IR absorptions at 3150–2850, 1600, and 1500 cm^{-1}.

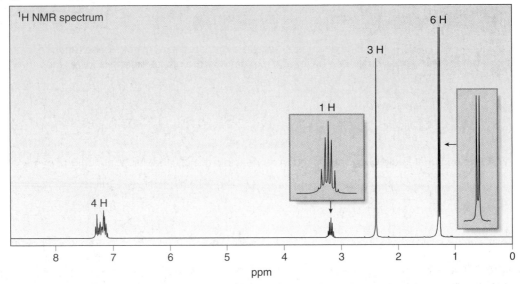

b. C_9H_{12}: ^{13}C NMR signals at 21, 127, and 138 ppm.

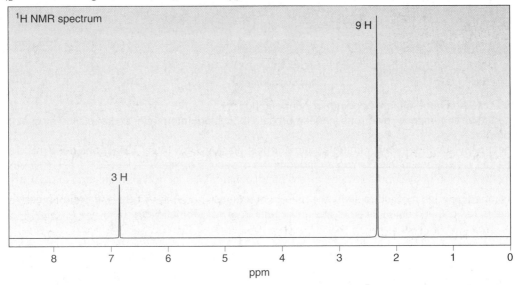

c. C_8H_{10}: IR absorptions at 3108–2875, 1606, and 1496 cm^{-1}.

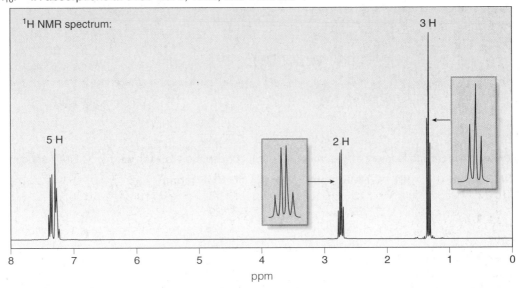

17.48 Propose a structure consistent with each set of data.

    a. Compound **A:**

        Molecular formula: $C_8H_{10}O$

        IR absorption at 3150–2850 cm^{-1}

        ^1H NMR data: 1.4 (triplet, 3 H), 3.95 (quartet, 2 H), and 6.8–7.3 (multiplet, 5 H) ppm

    b. Compound **B:**

        Molecular formula: $C_9H_{10}O_2$

        IR absorption at 1669 cm^{-1}

        ^1H NMR data: 2.5 (singlet, 3 H), 3.8 (singlet, 3 H), 6.9 (doublet, 2 H) and 7.9 (doublet, 2 H) ppm

17.49 [7]-Paracyclophane is an unusual aromatic compound with a bridge connecting two para carbons. Explain why the labeled protons absorb in different regions of the ^1H NMR spectrum, even though both are bonded to sp^3 hybridized C atoms.

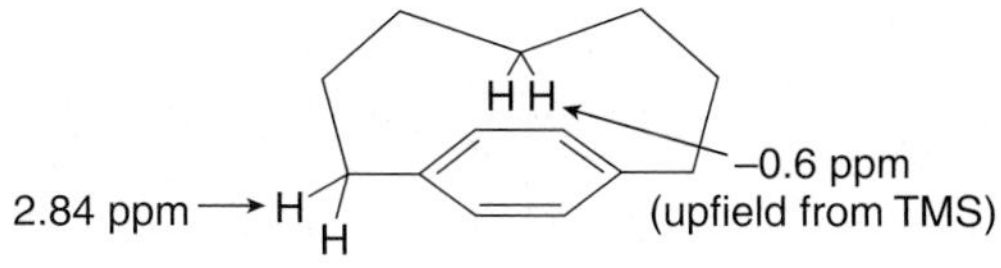

[7]-paracyclophane

Challenge Problems

17.50 Explain why compound **A** is much more stable than compound **B**.

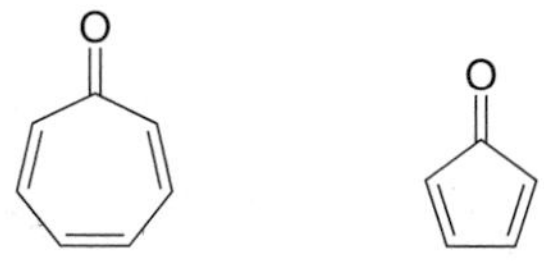

      A   **B**

17.51 Rank the N atoms in histamine in order of increasing basicity.

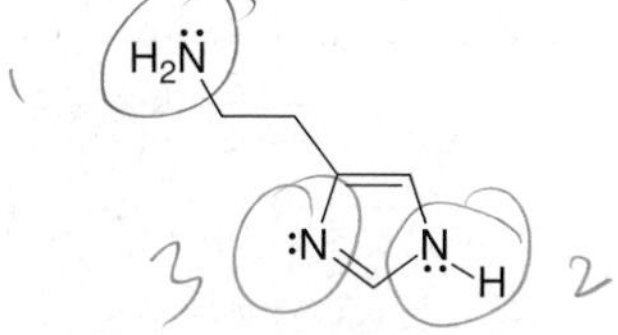

histamine

17.52 Explain why triphenylene resembles benzene in that it does not undergo addition reactions with Br_2, but phenanthrene reacts with Br_2 to yield the addition product drawn. (Hint: Draw resonance structures for both triphenylene and phenanthrene, and use them to determine how delocalized each π bond is.)

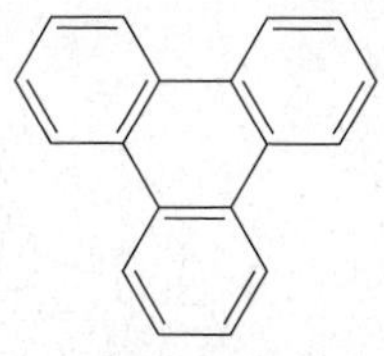

    triphenylene 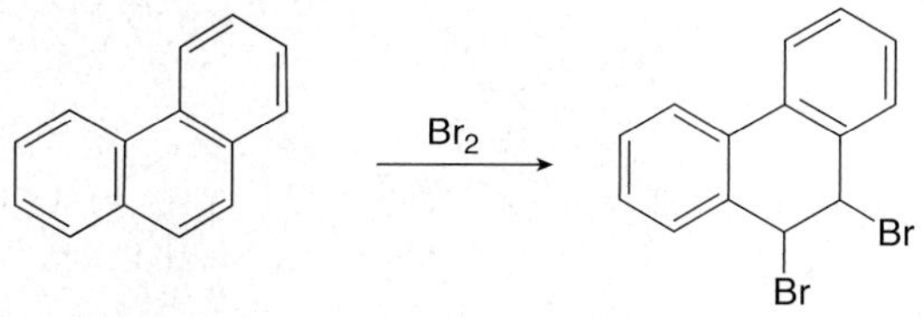   phenanthrene

17.53 Although benzene itself absorbs at 7.3 ppm in its ^1H NMR spectrum, the protons on substituted benzenes absorb either upfield or downfield from this value, depending on the substituent. Explain the observed values for the para disubstituted benzene derivatives **X** and **Y.**

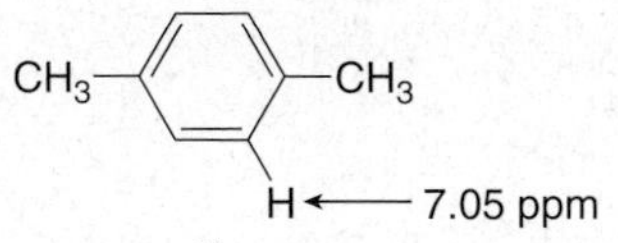

            X 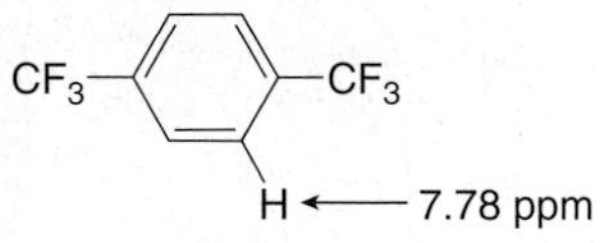   **Y**

Then explain why *p*-difluorobenzene shows a single peak in its ^1H NMR spectrum at 7.00 ppm.

Electrophilic Aromatic Substitution

LSD, commonly referred to as "acid," is a powerful hallucinogen prepared from lysergic acid, the principal organic compound derived from one of the ergot fungi. Immortalized in the 1967 Beatles' song, "*L*ucy in the *S*ky with *D*iamonds," LSD produces sensory illusions, making it difficult for the user to distinguish between reality and fantasy. Given its potent biological properties, LSD has been the target of several different laboratory syntheses. A key step in one of them involves carbon–carbon bond formation using electrophilic aromatic substitution, the most common reaction of aromatic compounds and the subject of Chapter 18.

hapter 18 discusses the chemical reactions of benzene and other aromatic compounds. Although aromatic rings are unusually stable, making benzene unreactive in most of the reactions discussed so far, benzene acts as a nucleophile with certain electrophiles, yielding substitution products with an intact aromatic ring.

We begin with the basic features and mechanism of electrophilic aromatic substitution (Sections 18.1–18.5), the basic reaction of benzene. Next, we discuss the electrophilic aromatic substitution of substituted benzenes (Sections 18.6–18.12), and conclude with other useful reactions of benzene derivatives (Sections 18.13–18.14). The ability to interconvert resonance structures and evaluate their relative stabilities is crucial to understanding this material.

18.1   Electrophilic Aromatic Substitution

Based on its structure and properties, what kinds of reactions should benzene undergo? Are any of its bonds particularly weak? Does it have electron-rich or electron-deficient atoms?

♦ Benzene has six π electrons delocalized in six *p* orbitals that overlap above and below the plane of the ring. These loosely held π electrons make the benzene ring electron rich, and so it reacts with electrophiles.
♦ Because benzene's six π electrons satisfy Hückel's rule, benzene is especially stable. Reactions that keep the aromatic ring intact are therefore favored.

As a result, **the characteristic reaction of benzene is** *electrophilic aromatic substitution—a* **hydrogen atom is replaced by an electrophile.**

Benzene does *not* undergo addition reactions like other unsaturated hydrocarbons, because addition would yield a product that is not aromatic. Substitution of a hydrogen, on the other hand, keeps the aromatic ring intact.

Five specific examples of electrophilic aromatic substitution are shown in Figure 18.1. The basic mechanism, discussed in Section 18.2, is the same in all five cases. The reactions differ only in the identity of the electrophile, E^+.

PROBLEM 18.1   Why is benzene less reactive towards electrophiles than an alkene, even though it has more π electrons than an alkene (six versus two)?

18.2   The General Mechanism

No matter what electrophile is used, all electrophilic aromatic substitution reactions occur via a **two-step mechanism:** addition of the electrophile E^+ to form a resonance-stabilized carbocation, followed by deprotonation with base, as shown in Mechanism 18.1.

MECHANISM 18.1

General Mechanism—Electrophilic Aromatic Substitution

Step [1] Addition of the electrophile (E⁺) to form a carbocation

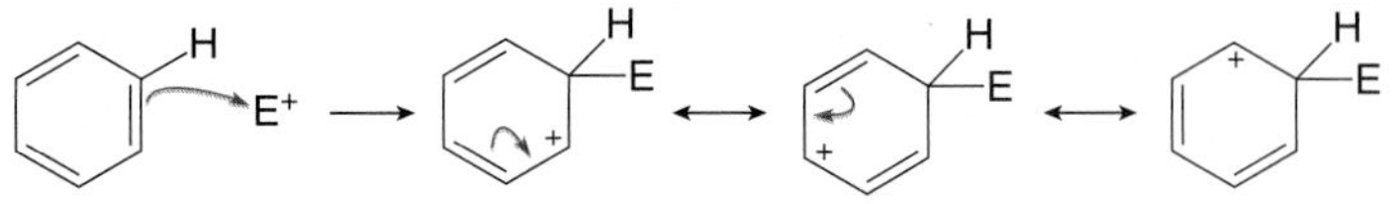

resonance-stabilized carbocation

◆ Addition of the electrophile (E⁺) forms a new C–E bond using two π electrons from the benzene ring, and generating a carbocation. This carbocation intermediate is not aromatic, but it is resonance stabilized—**three resonance structures can be drawn.**

◆ Step [1] is rate-determining because the aromaticity of the benzene ring is lost.

Step [2] Loss of a proton to re-form the aromatic ring

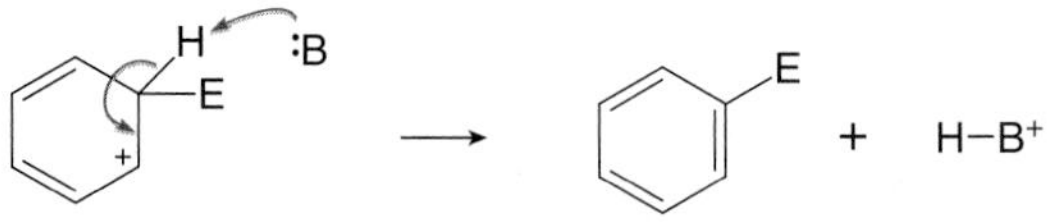

◆ In Step [2], a base (B:) removes the proton from the carbon bearing the electrophile, thus re-forming the aromatic ring. This step is fast because the aromaticity of the benzene ring is restored.

◆ Any of the three resonance structures of the carbocation intermediate can be used to draw the product. The choice of resonance structure affects how curved arrows are drawn, but not the identity of the product.

The first step in electrophilic aromatic substitution forms a carbocation, for which three resonance structures can be drawn. To help keep track of the location of the positive charge:

◆ Always draw in the H atom on the carbon bonded to E. This serves as a reminder that it is the only sp^3 hybridized carbon in the carbocation intermediate.

◆ Notice that the positive charge in a given resonance structure is always located ortho or para to the new C–E bond. In the hybrid, therefore, the charge is delocalized over three atoms of the ring.

Always draw in the H atom at the site of electrophilic attack.			The hybrid
(+) ortho to E	**(+) para to E**	**(+) ortho to E**	

This two-step mechanism for electrophilic aromatic substitution applies to all of the electrophiles in Figure 18.1. **The net result of addition of an electrophile (E⁺) followed by elimination of a proton (H⁺) is substitution of E for H.**

The energy changes in electrophilic aromatic substitution are shown in Figure 18.2. The mechanism consists of two steps, so the energy diagram has two energy barriers. Because the first step is rate-determining, its transition state is higher in energy.

PROBLEM 18.2   Draw two more resonance structures for each cation.

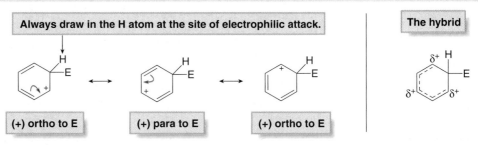

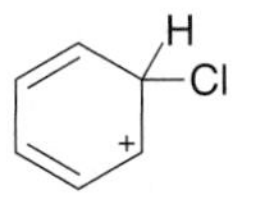

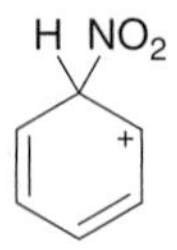

PROBLEM 18.3   In Step [2] of Mechanism 18.1, loss of a proton to form the substitution product was drawn using one resonance structure only. Use curved arrows to show how the other two resonance structures can be converted to the substitution product (PhE) by removal of a proton with :B.

Figure 18.1 Five examples of electrophilic aromatic substitution

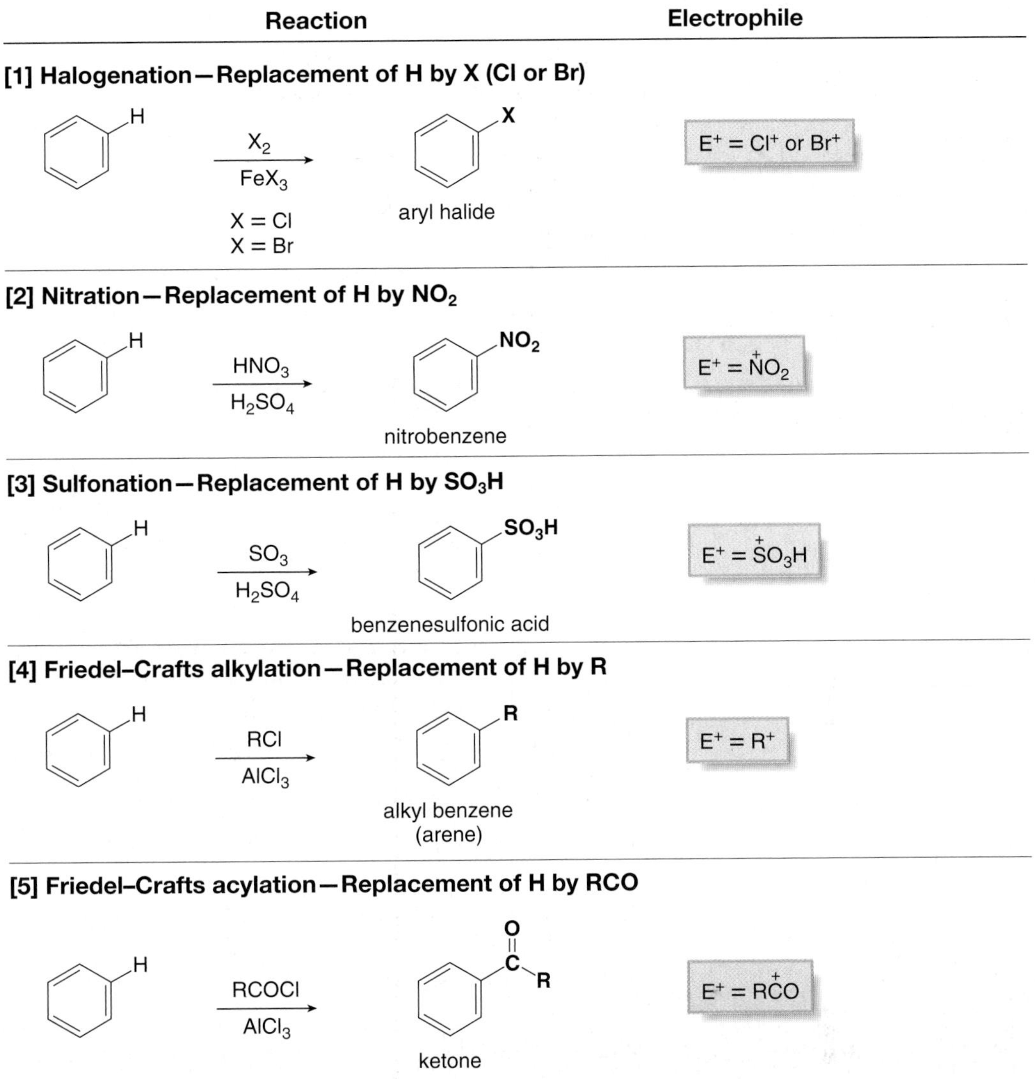

Reaction	Electrophile

[1] Halogenation—Replacement of H by X (Cl or Br)

$X = Cl$
$X = Br$

aryl halide

$E^+ = Cl^+$ or Br^+

[2] Nitration—Replacement of H by NO_2

HNO$_3$ / H$_2$SO$_4$

nitrobenzene

$E^+ = \overset{+}{N}O_2$

[3] Sulfonation—Replacement of H by SO_3H

SO$_3$ / H$_2$SO$_4$

benzenesulfonic acid

$E^+ = \overset{+}{S}O_3H$

[4] Friedel–Crafts alkylation—Replacement of H by R

RCl / AlCl$_3$

alkyl benzene (arene)

$E^+ = R^+$

Friedel–Crafts alkylation and acylation, named for Charles Friedel and James Crafts who discovered the reactions in the nineteenth century, form new carbon–carbon bonds.

[5] Friedel–Crafts acylation—Replacement of H by RCO

RCOCl / AlCl$_3$

ketone

$E^+ = R\overset{+}{C}O$

Figure 18.2 Energy diagram for electrophilic aromatic substitution: PhH + E$^+$ → PhE + H$^+$

- The mechanism has two steps so there are two energy barriers.
- Step [1] is rate-determining; its transition state is at higher energy.

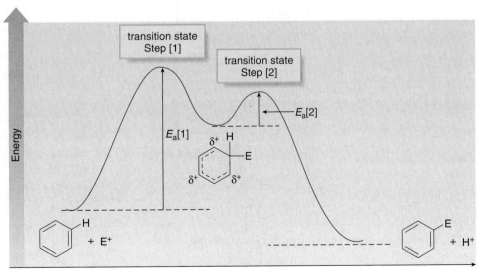

transition state Step [1]

transition state Step [2]

$E_a[2]$

$E_a[1]$

Energy

Reaction coordinate

18.3   Halogenation

The general mechanism outlined in Mechanism 18.1 can now be applied to each of the five specific examples of electrophilic aromatic substitution shown in Figure 18.1. For each mechanism we must learn how to generate a specific electrophile. This step is *different* with each electrophile. Then, the electrophile reacts with benzene by the two-step process of Mechanism 18.1. These two steps are the *same* for all five reactions.

In **halogenation,** benzene reacts with Cl_2 or Br_2 in the presence of a Lewis acid catalyst, such as $FeCl_3$ or $FeBr_3$, to give the **aryl halides** chlorobenzene or bromobenzene, respectively. Analogous reactions with I_2 and F_2 are not synthetically useful because I_2 is too unreactive and F_2 reacts too violently.

Chlorination

$$\xrightarrow[\text{FeCl}_3]{\text{Cl}_2}$$

chlorobenzene

Bromination

$$\xrightarrow[\text{FeBr}_3]{\text{Br}_2}$$

bromobenzene

In bromination (Mechanism 18.2), the Lewis acid $FeBr_3$ reacts with Br_2 to form a Lewis acid–base complex that weakens and polarizes the $Br-Br$ bond, making it more electrophilic. This reaction is Step [1] of the mechanism for the bromination of benzene. The remaining two steps follow directly from the general mechanism for electrophilic aromatic substitution: addition of the electrophile (Br^+ in this case) forms a resonance-stabilized carbocation, and loss of a proton regenerates the aromatic ring.

Chlorination proceeds by a similar mechanism. Reactions that introduce a halogen substituent on a benzene ring are widely used, and many halogenated aromatic compounds with a range of biological activity have been synthesized, as shown in Figure 18.3.

MECHANISM 18.2

Bromination of Benzene

Step [1]   Generation of the electrophile

$$:\ddot{Br}-\ddot{Br}: \; + \; FeBr_3 \longrightarrow \; :\ddot{Br}-\overset{+}{\underset{}{\ddot{Br}}}-\overset{-}{Fe}Br_3$$

Lewis base   Lewis acid   electrophile
(serves as a source of Br^+)

♦ Lewis acid–base reaction of Br_2 with $FeBr_3$ forms a species with a weakened and polarized $Br-Br$ bond. This adduct serves as a source of Br^+ in the next step.

Step [2]   Addition of the electrophile to form a carbocation

resonance-stabilized carbocation
+ $FeBr_4^-$

♦ Addition of the electrophile forms a new $C-Br$ bond and generates a carbocation. This carbocation intermediate is resonance stabilized—**three resonance structures can be drawn.**

♦ The $FeBr_4^-$ also formed in this reaction is the base used in Step [3].

Step [3]   Loss of a proton to re-form the aromatic ring

$$+ \; HBr \; + \; FeBr_3$$

The catalyst is regenerated.

♦ $FeBr_4^-$ removes the proton from the carbon bearing the Br, thus re-forming the aromatic ring.

♦ $FeBr_3$, a catalyst, is also regenerated for another reaction cycle.

Figure 18.3 Examples of biologically active aryl chlorides

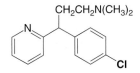

Generic name: **bupropion**
Trade names: **Wellbutrin, Zyban**
antidepressant,
also used to reduce nicotine cravings

chlorpheniramine
antihistamine

Herbicides were used extensively during the Vietnam War to defoliate dense jungle areas. The concentration of certain herbicide by-products in the soil remains high today.

2,4-D
2,4-dichlorophenoxy-
acetic acid
herbicide

2,4,5-T
2,4,5-trichlorophenoxy-
acetic acid
herbicide

the active components in **Agent Orange**,
a defoliant used in the Vietnam War

PROBLEM 18.4   Draw a detailed mechanism for the chlorination of benzene using Cl_2 and $FeCl_3$.

18.4   Nitration and Sulfonation

Chloramphenicol, an aromatic nitro compound, is an antibiotic that became available in 1949. Today it is often replaced by newer drugs, but it remains an important antibiotic for treating certain infections.

Nitration and **sulfonation** of benzene introduce two different functional groups on an aromatic ring. Nitration is an especially useful reaction because a nitro group can then be reduced to an NH_2 group, a common benzene substituent, in a reaction discussed in Section 18.14.

Nitration

$$\xrightarrow[H_2SO_4]{HNO_3}$$

nitrobenzene

$$\xrightarrow{\text{Section } 18.14}$$

aniline

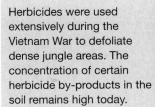

chloramphenicol

Sulfonation

$$\xrightarrow[H_2SO_4]{SO_3}$$

benzenesulfonic acid

Generation of the electrophile in both nitration and sulfonation requires strong acid. In **nitration,** the electrophile is $^+NO_2$ (the **nitronium ion**), formed by protonation of HNO_3 followed by loss of water.

Formation of the Nitronium Ion ($^+NO_2$) for Nitration

$$H-\overset{..}{\underset{..}{O}}-NO_2 \; + \; H-OSO_3H \;\longrightarrow\; H-\overset{+}{\underset{H}{O}}-NO_2 \;\longrightarrow\; H_2\overset{..}{O}: \; + \; \overset{+}{NO_2} \; = \; \overset{..}{\underset{..}{O}}=\overset{+}{N}=\overset{..}{\underset{..}{O}}$$

$$+ \; HSO_4^-$$

electrophile   Lewis structure

In **sulfonation,** protonation of sulfur trioxide, SO_3, forms a positively charged sulfur species ($^+SO_3H$) that acts as an electrophile.

Formation of the Electrophile $^+SO_3H$ for Sulfonation

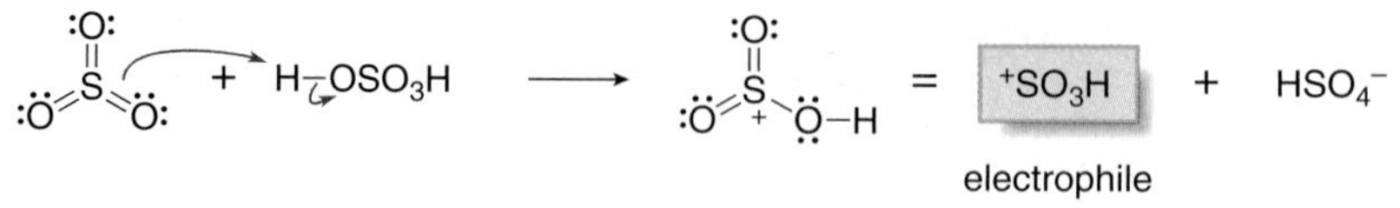

These steps illustrate how to generate the electrophile E^+ for nitration and sulfonation, the process that begins any mechanism for electrophilic aromatic substitution. To complete either of these mechanisms, you must replace the electrophile E^+ by either $^+NO_2$ or $^+SO_3H$ in the general mechanism (Mechanism 18.1). Thus, **the two-step sequence that replaces H by E is the same regardless of E^+.** This is shown in Sample Problem 18.1 using the reaction of benzene with the nitronium ion.

SAMPLE PROBLEM 18.1 Draw a stepwise mechanism for the nitration of a benzene ring.

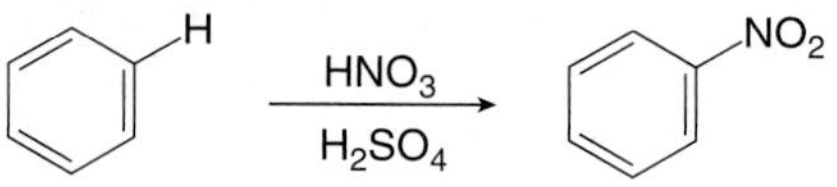

nitrobenzene

SOLUTION

We must first generate the electrophile and then write the two-step mechanism for electrophilic aromatic substitution using it.

Generation of the electrophile $^+NO_2$:

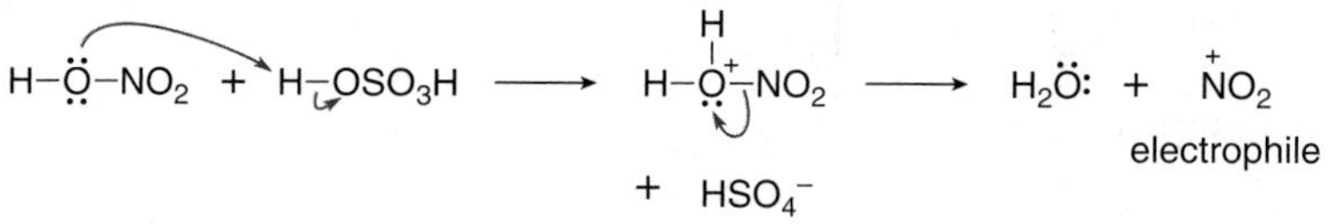

Two-step mechanism for substitution:

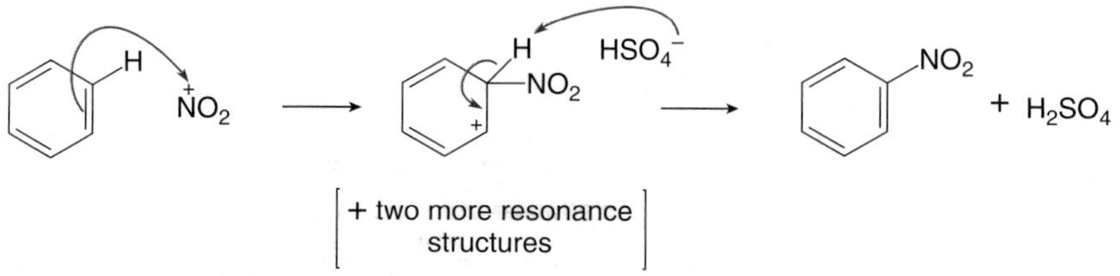

[+ two more resonance structures]

Any species with a lone pair of electrons can be used to remove the proton in the last step. In this case, the mechanism is drawn with HSO_4^-, formed when $^+NO_2$ is generated as the electrophile.

PROBLEM 18.5 Draw a stepwise mechanism for the sulfonation of benzene with SO_3 and H_2SO_4.

18.5 Friedel–Crafts Alkylation and Friedel–Crafts Acylation

Friedel–Crafts alkylation and **Friedel–Crafts acylation** form new carbon–carbon bonds.

18.5A General Features

In **Friedel–Crafts alkylation,** treatment of benzene with an alkyl halide and a Lewis acid ($AlCl_3$) forms an alkyl benzene. This reaction is an **alkylation** because it results in transfer of an alkyl group from one atom to another (from Cl to benzene).

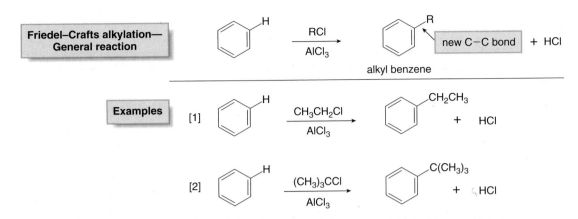

In **Friedel–Crafts acylation,** a benzene ring is treated with an **acid chloride** (RCOCl) and AlCl$_3$ to form a ketone. Because the new group bonded to the benzene ring is called an **acyl group,** the transfer of an acyl group from one atom to another is an **acylation.**

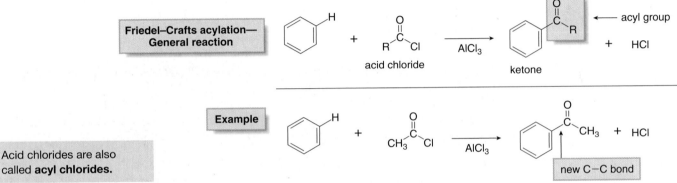

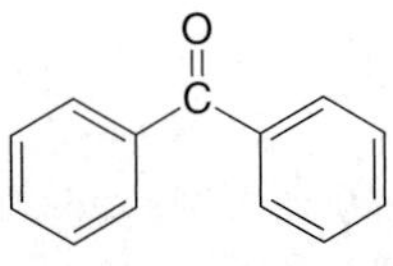

Acid chlorides are also called **acyl chlorides.**

PROBLEM 18.6 What product is formed when benzene is treated with each organic halide in the presence of AlCl$_3$?

a. (CH$_3$)$_2$CHCl   b. 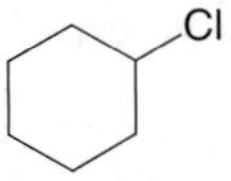 [structure: cyclohexane with Cl]   c. 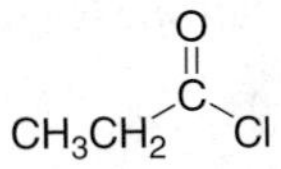 [structure: CH$_3$CH$_2$C(=O)Cl]

PROBLEM 18.7 What acid chloride would be needed to prepare each of the following ketones from benzene using a Friedel–Crafts acylation?

a. 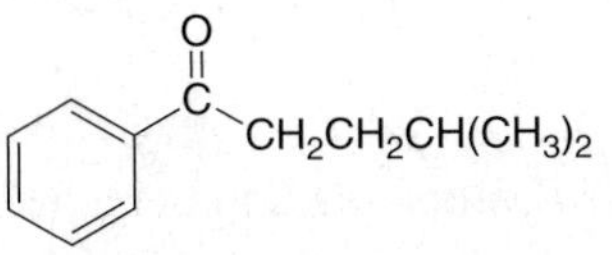 [structure: benzene–C(=O)–CH$_2$CH$_2$CH(CH$_3$)$_2$]   b. [structure: benzophenone]   c. [structure: cyclopentyl phenyl ketone]

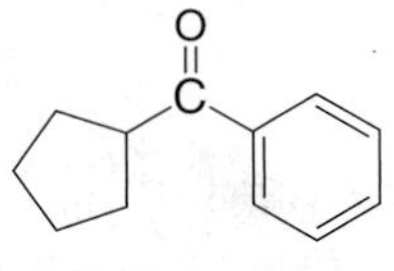

18.5B Mechanism

The mechanisms of alkylation and acylation proceed in a manner analogous to those for halogenation, nitration, and sulfonation. The unique feature in each reaction is how the electrophile is generated.

In **Friedel–Crafts alkylation,** the Lewis acid AlCl$_3$ reacts with the alkyl chloride to form a Lewis acid–base complex, illustrated with CH$_3$CH$_2$Cl and (CH$_3$)$_3$CCl as alkyl chlorides. The identity of the alkyl chloride determines the exact course of the reaction.

Formation of the Electrophile in Friedel–Crafts Alkylation—Two Possibilities

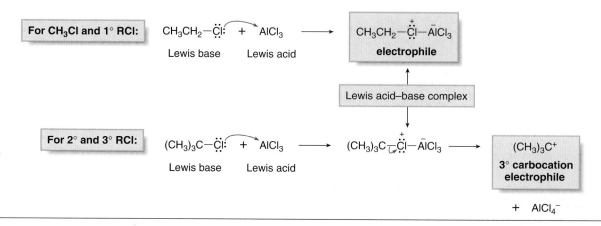

◆ For CH_3Cl and 1° RCl, the Lewis acid–base complex itself serves as the electrophile for electrophilic aromatic substitution.

◆ With 2° and 3° RCl, the Lewis acid–base complex reacts further to give a 2° or 3° carbocation, which serves as electrophile. Carbocation formation occurs only with 2° and 3° alkyl chlorides, because they afford more stable carbocations.

In either case, the electrophile goes on to react with benzene in the two-step mechanism characteristic of electrophilic aromatic substitution, illustrated in Mechanism 18.3 using the 3° carbocation, $(CH_3)_3C^+$.

MECHANISM 18.3

Friedel–Crafts Alkylation Using a 3° Carbocation

◆ Addition of the electrophile (a 3° carbocation) forms a new carbon–carbon bond in Step [1].

◆ $AlCl_4^-$ removes a proton on the carbon bearing the new substituent, thus re-forming the aromatic ring in Step [2].

In **Friedel–Crafts acylation,** the Lewis acid $AlCl_3$ ionizes the carbon–halogen bond of the acid chloride, thus forming a positively charged carbon electrophile called an **acylium ion,** which is resonance stabilized. The positively charged carbon atom of the acylium ion then goes on to react with benzene in the two-step mechanism of electrophilic aromatic substitution.

Formation of the Electrophile in Friedel–Crafts Acylation

This C serves as the electrophilic site.

R–C(=O:)–Cl + AlCl₃ ⟶ R–C⁺=Ö: ⟷ R–C≡O⁺: + AlCl₄⁻

Lewis acid

a resonance-stabilized **acylium ion**

electrophile

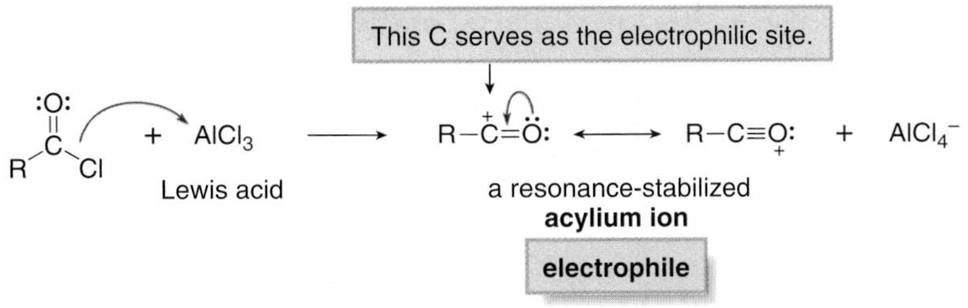

To complete the mechanism for acylation, insert the electrophile into the general mechanism and draw the last two steps, as illustrated in Sample Problem 18.2.

SAMPLE PROBLEM 18.2   Draw a stepwise mechanism for the following Friedel–Crafts acylation.

benzene + CH₃–C(=O)–Cl —AlCl₃→ benzene–C(=O)–CH₃ + HCl

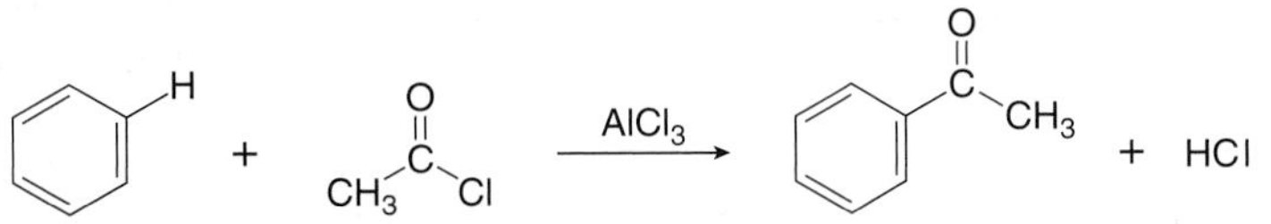

SOLUTION

We must first generate the acylium ion, and then write the two-step mechanism for electrophilic aromatic substitution using it for the electrophile.

Generation of the electrophile (CH₃CO)⁺:

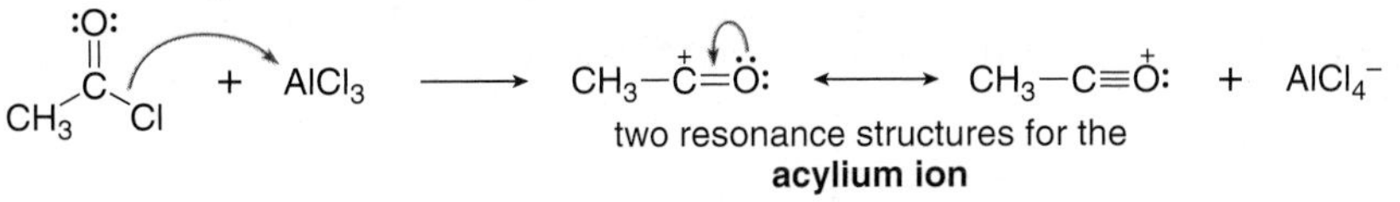

CH₃–C(=O:)–Cl + AlCl₃ ⟶ CH₃–C⁺=Ö: ⟷ CH₃–C≡O⁺: + AlCl₄⁻

two resonance structures for the **acylium ion**

Two-step mechanism for substitution:

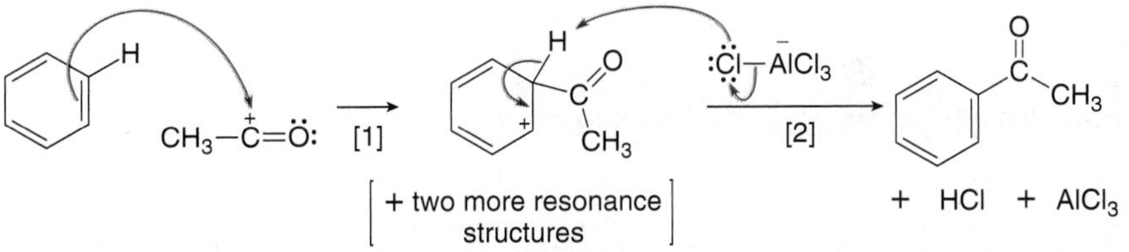

benzene + CH₃–C⁺=Ö: [1] ⟶ arenium ion intermediate [+ two more resonance structures] —[2]→ benzene–C(=O)–CH₃ + HCl + AlCl₃

PROBLEM 18.8   Draw a stepwise mechanism for the Friedel–Crafts alkylation of benzene with CH₃CH₂Cl and AlCl₃.

18.5C   Other Facts About Friedel–Crafts Alkylation

Three additional facts about Friedel–Crafts alkylations must be kept in mind.

[1]   Vinyl halides and aryl halides do *not* react in Friedel–Crafts alkylation.

Most Friedel–Crafts reactions involve carbocation electrophiles. Because the carbocations derived from vinyl halides and aryl halides are highly unstable and do not readily form, these organic halides do *not* undergo Friedel–Crafts alkylation.

Unreactive halides in the Friedel–Crafts alkylation

CH₂=CHCl

vinyl halide

aryl halide (chlorobenzene)

PROBLEM 18.9   Which halides are unreactive in a Friedel–Crafts alkylation reaction?

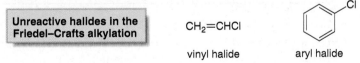

a. 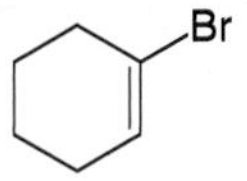   b. 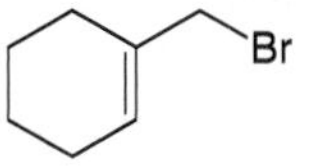   c. 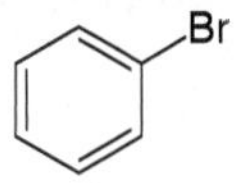   d.

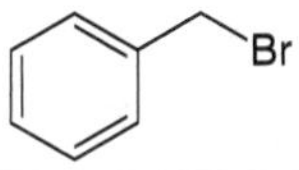

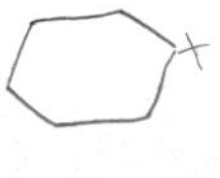

| [2] | **Rearrangements can occur.** |

The Friedel–Crafts reaction can yield products having rearranged carbon skeletons when 1° and 2° alkyl halides are used as starting materials, as shown in Equations [1] and [2]. In both reactions, the carbon atom bonded to the halogen in the starting material (labeled in red) is not bonded to the benzene ring in the product, thus indicating that a rearrangement has occurred.

> Recall from Section 9.9 that a 1,2-shift converts a less stable carbocation to a more stable carbocation by shift of a hydrogen atom or an alkyl group.

> Rearrangements do not occur in Friedel–Crafts acylation because the acylium ion is resonance stabilized.

The result in Equation [1] is explained by a carbocation rearrangement involving a 1,2-hydride shift: **the less stable 2° carbocation (formed from the 2° halide) rearranges to a more stable 3° carbocation,** as illustrated in Mechanism 18.4.

MECHANISM 18.4

Friedel–Crafts Alkylation Involving Carbocation Rearrangement

Steps [1] and [2] Formation of a 2° carbocation

◆ Reaction of the alkyl chloride with $AlCl_3$ forms a complex that decomposes in Step [2] to form a **2° carbocation.**

Step [3] Carbocation rearrangement

◆ **1,2-Hydride shift** converts the less stable 2° carbocation to a more stable 3° carbocation.

Steps [4] and [5] Addition of the carbocation and loss of a proton

◆ Friedel–Crafts alkylation occurs by the usual two-step process: **addition of the carbocation** followed by **loss of a proton** to form the alkylated product.

Rearrangements can occur even when no free carbocation is formed initially. For example, the 1° alkyl chloride in Equation [2] forms a complex with $AlCl_3$, which does *not* decompose to an unstable 1° carbocation. Instead, a **1,2-hydride shift** forms a 2° carbocation, which then serves as the electrophile in the two-step mechanism for electrophilic aromatic substitution.

A Rearrangement Reaction Beginning with a 1° Alkyl Chloride

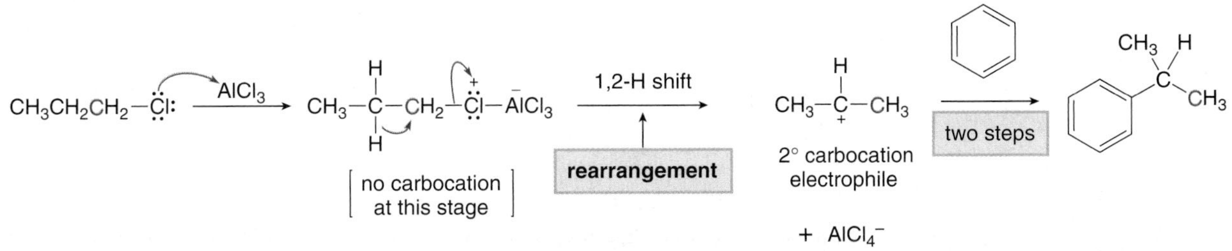

$$CH_3CH_2CH_2-\overset{..}{\underset{..}{Cl}}: \xrightarrow{\ AlCl_3\ } CH_3-\overset{\overset{H}{|}}{\underset{\underset{H}{|}}{C}}-CH_2-\overset{+}{\underset{..}{Cl}}-\bar{A}lCl_3$$

[no carbocation at this stage]

$$\xrightarrow[\text{rearrangement}]{\text{1,2-H shift}} CH_3-\overset{\overset{H}{|}}{\underset{+}{C}}-CH_3 \xrightarrow{\text{two steps}}$$

2° carbocation electrophile

$$+\ AlCl_4^-$$

PROBLEM 18.10   Draw a stepwise mechanism for the following reaction.

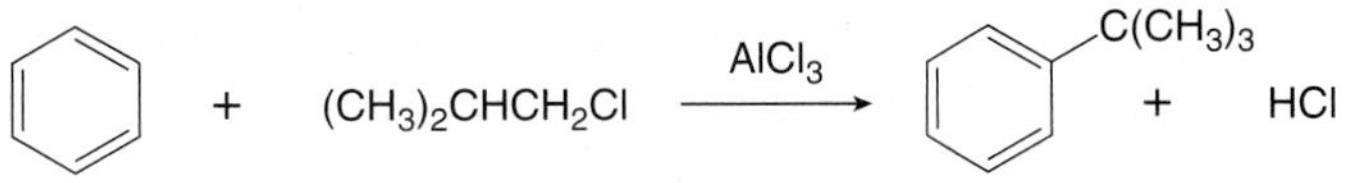

$$\bigcirc\ +\ (CH_3)_2CHCH_2Cl \xrightarrow{\ AlCl_3\ } \bigcirc\text{--}C(CH_3)_3\ +\ HCl$$

[3]   **Other functional groups that form carbocations can also be used as starting materials.**

Although Friedel–Crafts alkylation works well with alkyl halides, any compound that readily forms a carbocation can be used instead. The two most common alternatives are alkenes and alcohols, both of which afford carbocations in the presence of strong acid.

◆ Protonation of an alkene forms a carbocation, which can then serve as an electrophile in a Friedel–Crafts alkylation.

◆ Protonation of an alcohol, followed by loss of water, likewise forms a carbocation.

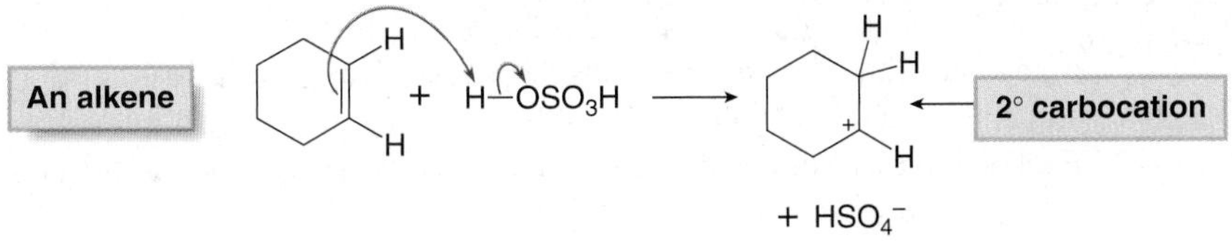

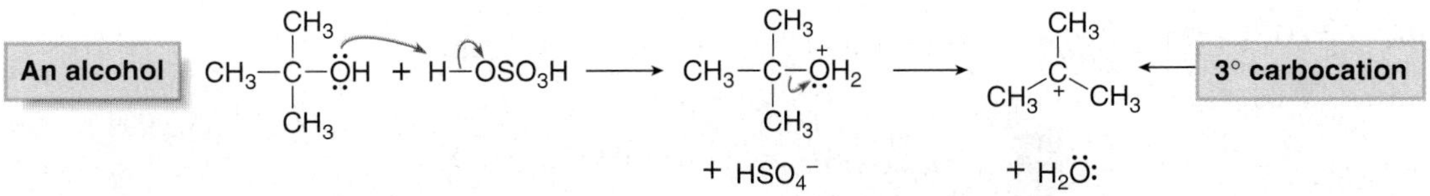

An alkene

$$\bigcirc\text{=}\overset{H}{\underset{H}{}}\ +\ H-OSO_3H \longrightarrow \bigcirc\overset{H}{\underset{\overset{+}{H}}{}}_H \longleftarrow \text{2° carbocation}$$

$$+\ HSO_4^-$$

An alcohol

$$CH_3-\overset{\overset{CH_3}{|}}{\underset{\underset{CH_3}{|}}{C}}-\overset{..}{O}H\ +\ H-OSO_3H \longrightarrow CH_3-\overset{\overset{CH_3}{|}}{\underset{\underset{CH_3}{|}}{C}}-\overset{+}{O}H_2 \longrightarrow \overset{CH_3}{\underset{CH_3}{}}\overset{|}{\underset{}{C^+}}CH_3 \longleftarrow \text{3° carbocation}$$

$$+\ HSO_4^-\qquad\qquad\qquad +\ H_2\overset{..}{O}:$$

Each carbocation can then go on to react with benzene to form a product of electrophilic aromatic substitution. For example:

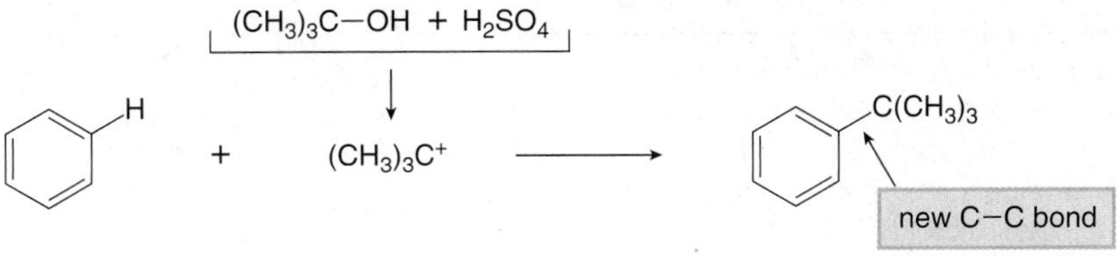

$$\underset{(CH_3)_3C-OH\ +\ H_2SO_4}{\Big\downarrow}$$

$$\bigcirc\text{--}H\ +\ (CH_3)_3C^+ \longrightarrow \bigcirc\text{--}C(CH_3)_3$$

new C–C bond

PROBLEM 18.11 Draw the product of each reaction.

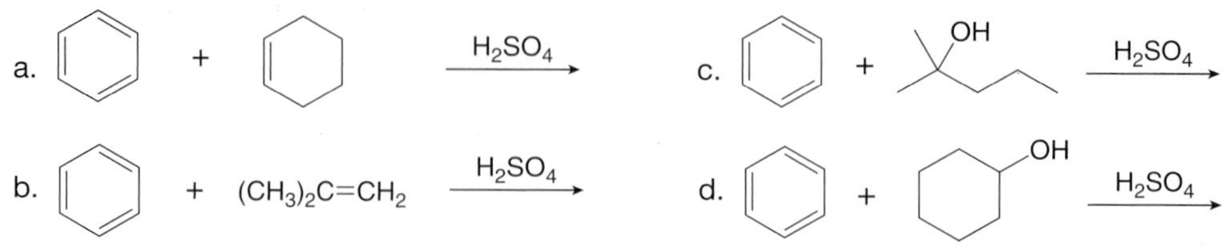

a. ⬡ + ⬡ $\xrightarrow{H_2SO_4}$ c. ⬡ + (structure with OH) $\xrightarrow{H_2SO_4}$

b. ⬡ + $(CH_3)_2C{=}CH_2$ $\xrightarrow{H_2SO_4}$ d. ⬡ + (cyclohexanol with OH) $\xrightarrow{H_2SO_4}$

18.5D Intramolecular Friedel–Crafts Reactions

All of the Friedel–Crafts reactions discussed thus far have resulted from intermolecular reaction of a benzene ring with an electrophile. Starting materials that contain both units are capable of **intramolecular reaction,** and this forms a new ring. For example, treatment of compound **A,** which contains both a benzene ring and an acid chloride, with AlCl$_3$, forms α-tetralone by an intramolecular Friedel–Crafts acylation reaction.

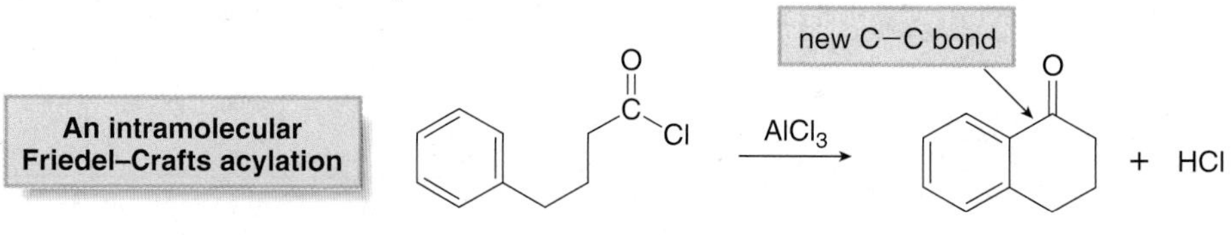

An intramolecular Friedel–Crafts acylation

new C–C bond

$\xrightarrow{AlCl_3}$ + HCl

A   α-tetralone

Such an intramolecular Friedel–Crafts acylation was a key step in the synthesis of LSD, the molecule that introduced Chapter 18, as shown in Figure 18.4.

PROBLEM 18.12 Draw a stepwise mechanism for the following intramolecular reaction.

$\xrightarrow{AlCl_3}$ + HCl

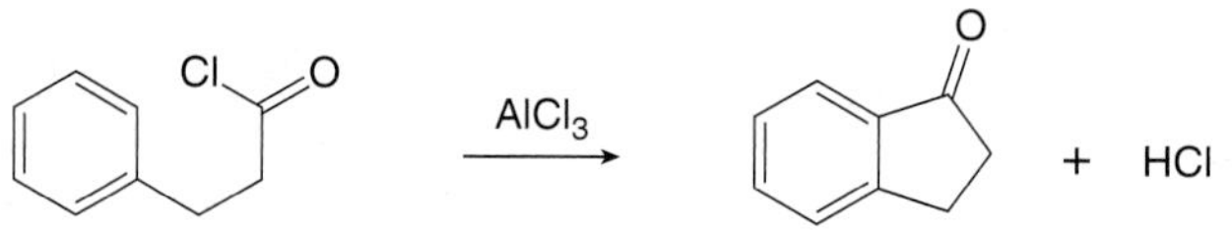

18.6 Substituted Benzenes

Many substituted benzene rings undergo electrophilic aromatic substitution. Common substituents include halogens, OH, NH$_2$, alkyl, and many functional groups that contain a carbonyl. Each substituent either increases or decreases the electron density in the benzene ring, and this affects the course of electrophilic aromatic substitution, as we will learn in Section 18.7.

Figure 18.4 Intramolecular Friedel–Crafts acylation in the synthesis of LSD

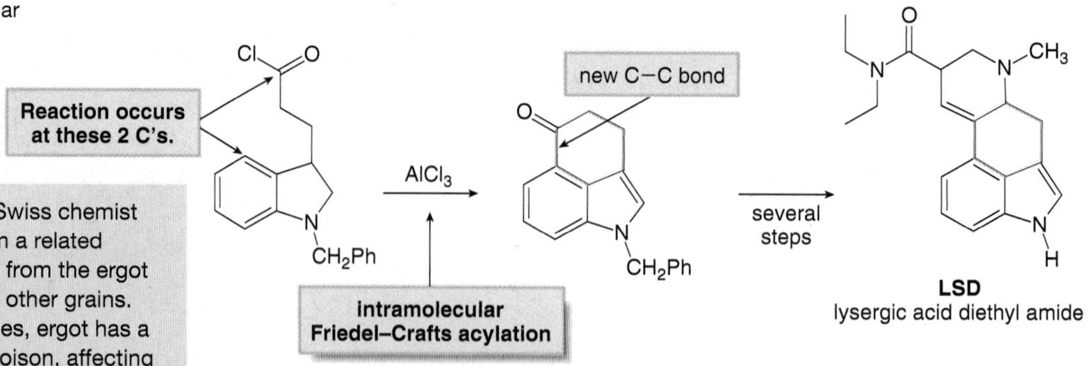

Reaction occurs at these 2 C's.

$\xrightarrow{AlCl_3}$

new C–C bond

$\xrightarrow{\text{several steps}}$

LSD
lysergic acid diethyl amide

intramolecular Friedel–Crafts acylation

LSD was first prepared by Swiss chemist Albert Hoffman in 1938 from a related organic compound isolated from the ergot fungus that attacks rye and other grains. Known since the Middle Ages, ergot has a long history as a dreaded poison, affecting individuals who become ill from eating ergot-contaminated bread. The hallucinogenic effects of LSD were first discovered when Hoffman himself ingested small amounts of the drug.

- Intramolecular Friedel–Crafts acylation formed a product containing a new six-membered ring (in red), which was converted to LSD in several steps.

♦ Donation of electron density to the ring makes benzene more electron rich.
♦ Withdrawal of electron density from the ring makes benzene less electron rich.

What makes a substituent on a benzene ring electron donating or electron withdrawing? The answer is **inductive effects** or **resonance effects,** both of which can add or remove electron density.

Inductive Effects

Inductive effects stem from the **electronegativity** of the atoms in the substituent and the **polarizability** of the substituent group.

♦ Atoms more electronegative than carbon—including N, O, and X—pull electron density away from carbon and thus exhibit an electron-withdrawing inductive effect.
♦ Polarizable alkyl groups donate electron density, and thus exhibit an electron-donating inductive effect.

Inductive and resonance effects were first discussed in Sections 2.5B and 2.5C, respectively.

Considering inductive effects *only,* an NH_2 group withdraws electron density and CH_3 donates electron density.

Electron-withdrawing inductive effect	Electron-donating inductive effect

• N is **more electronegative** than C.
• N inductively withdraws electron density.

• Alkyl groups are **polarizable,** making them electron-donating groups.

PROBLEM 18.13 Which substituents have an electron-withdrawing and which have an electron-donating inductive effect: (a) $CH_3CH_2CH_2CH_2-$; (b) $Br-$; (c) CH_3CH_2O- ?

Resonance Effects

Resonance effects can either donate or withdraw electron density, depending on whether they place a positive or negative charge on the benzene ring. **Resonance effects are only observed with substituents containing lone pairs or π bonds.**

♦ A resonance effect is electron donating when resonance structures place a negative charge on carbons of the benzene ring.
♦ A resonance effect is electron withdrawing when resonance structures place a positive charge on carbons of the benzene ring.

An electron-donating resonance effect is observed whenever an atom Z having a lone pair of electrons is directly bonded to a benzene ring (general structure—C_6H_5-Z:). Resonance delocalizes the lone pair on the atoms of the ring, making the ring more electron rich. Common examples of Z include N, O, and halogen.

For example, five resonance structures can be drawn for aniline ($C_6H_5NH_2$). Because three of them place a negative charge on a carbon atom of the benzene ring, an **NH_2 group donates electron density to a benzene ring by a resonance effect.**

aniline

Three resonance structures place a (−) charge on atoms in the ring.

In contrast, **an electron-withdrawing resonance effect is observed in substituted benzenes having the general structure $C_6H_5-Y=Z$,** where Z is more electronegative than Y. Resonance delocalizes a positive charge on the atoms of the ring in this case, making the ring less electron rich than benzene.

For example, seven resonance structures can be drawn for benzaldehyde (C_6H_5CHO). Because three of them place a positive charge on a carbon atom of the benzene ring, a CHO group withdraws electron density from a benzene ring by a resonance effect.

benzaldehyde

Three resonance structures place a (+) charge on atoms in the ring.

PROBLEM 18.14   Draw all resonance structures for each compound and use the resonance structures to determine if the substituent has an electron-donating or electron-withdrawing resonance effect.

a.

b.

Considering Both Inductive and Resonance Effects

To predict whether a substituted benzene is more or less electron rich than benzene itself, we must consider the **net balance of *both* the inductive and the resonance effects.** Alkyl groups, for instance, donate electrons by an inductive effect, but they have no resonance effect because they lack nonbonded electron pairs or π bonds. Thus, any **alkyl-substituted benzene is more electron rich than benzene itself.**

- R donates electrons by an inductive effect.
- R has no resonance effect.

Alkyl benzenes are more electron rich than benzene.

When electronegative atoms, such as N, O, or halogen, are bonded to the benzene ring, they inductively withdraw electron density from the ring. All of these groups also have a nonbonded pair of electrons, so they donate electron density to the ring by resonance. **Whether the overall effect adds or subtracts electron density from the ring depends on the net balance of these two opposing effects.**

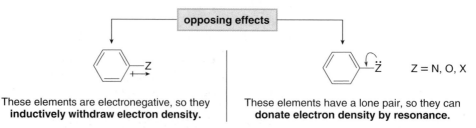

opposing effects

Z = N, O, X

These elements are electronegative, so they **inductively withdraw electron density.**

These elements have a lone pair, so they can **donate electron density by resonance.**

- ◆ When a neutral O or N atom is directly bonded to a benzene ring, the resonance effect dominates and the net effect is electron donation.
- ◆ When a halogen X is bonded to a benzene ring, the inductive effect dominates and the net effect is electron withdrawal.

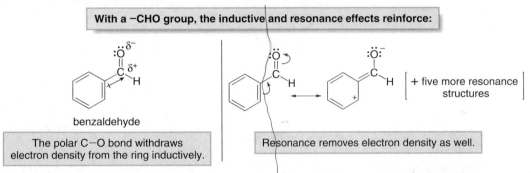

Electron-donating groups

The electron-donating resonance effect predominates.

Electron-withdrawing groups

The electron-withdrawing inductive effect predominates.

Finally, the inductive and resonance effects in compounds having the general structure $C_6H_5-Y=Z$ (with Z more electronegative than Y) are **both electron withdrawing;** in other words, the two effects *reinforce* each other. This is true for benzaldehyde (C_6H_5CHO) and all other compounds that contain a carbonyl group directly bonded to the benzene ring.

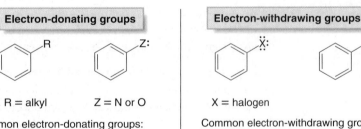

With a −CHO group, the inductive and resonance effects reinforce:

benzaldehyde

The polar C−O bond withdraws electron density from the ring inductively.

+ five more resonance structures

Resonance removes electron density as well.

Thus, on balance, an **NH_2 group is electron donating,** so the benzene ring of aniline ($C_6H_5NH_2$) has more electron density than benzene. An **aldehyde group (CHO), on the other hand, is electron withdrawing,** so the benzene ring of benzaldehyde (C_6H_5CHO) has less electron density than benzene. These effects are illustrated in the electrostatic potential maps in Figure 18.5.

These compounds represent examples of the general structural features in electron-donating and electron-withdrawing substituents:

Electron-donating groups

R = alkyl   Z = N or O

Common electron-donating groups:
• Alkyl groups
• Groups with an N or O atom bonded to the benzene ring; N or O must have a lone pair.

Electron-withdrawing groups

X = halogen

Common electron-withdrawing groups:
• Halogens
• Groups with an atom Y bearing a positive charge (δ^+ or +) bonded to the benzene ring.

Figure 18.5 The effect of substituents on the electron density in substituted benzenes

• The −NH_2 group donates electron density, making the benzene ring more electron rich (redder), whereas the −CHO group withdraws electron density, making the benzene ring less electron rich (greener).

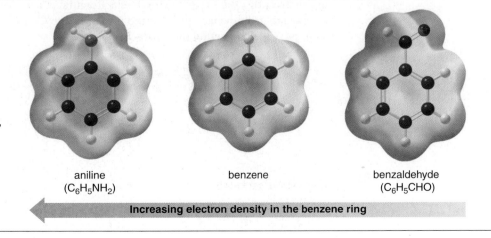

aniline ($C_6H_5NH_2$)

benzene

benzaldehyde (C_6H_5CHO)

Increasing electron density in the benzene ring

The net effect of electron donation and withdrawal on the reactions of substituted aromatics is discussed in Sections 18.7–18.9.

SAMPLE PROBLEM 18.3   Classify each substituent as electron donating or electron withdrawing.

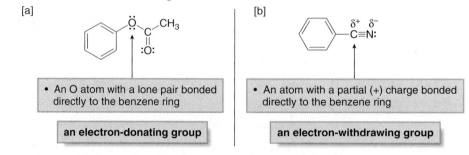

SOLUTION

Always look at the atom directly bonded to the benzene ring to determine electron-donating or electron-withdrawing effects. An O or N atom with a lone pair of electrons makes a substituent electron donating. A halogen or an atom with a partial positive charge makes a substituent electron withdrawing.

[a]

- An O atom with a lone pair bonded directly to the benzene ring

an electron-donating group

[b]

δ^+ δ^-

- An atom with a partial (+) charge bonded directly to the benzene ring

an electron-withdrawing group

PROBLEM 18.15   Classify each substituent as electron donating or electron withdrawing.

a. OCH$_3$ b. I   c. C(CH$_3$)$_3$

18.7   Electrophilic Aromatic Substitution of Substituted Benzenes

Electrophilic aromatic substitution is a general reaction of *all* aromatic compounds, including polycyclic aromatic hydrocarbons, heterocycles, and substituted benzene derivatives. A substituent affects two aspects of electrophilic aromatic substitution:

◆ **The rate of reaction:** A substituted benzene reacts faster or slower than benzene itself.

◆ **The orientation:** The new group is located either ortho, meta, or para to the existing substituent. The identity of the first substituent determines the position of the second substituent.

Toluene ($C_6H_5CH_3$) and nitrobenzene ($C_6H_5NO_2$) illustrate two possible outcomes.

[1] Toluene

Toluene reacts **faster** than benzene in all substitution reactions. Thus, its **electron-donating CH$_3$ group activates the benzene ring** to electrophilic attack. Although three products are possible, compounds with the new group ortho or para to the CH$_3$ group predominate. The CH$_3$ group is therefore called an **ortho, para director.**

CH$_3$ → (Br$_2$ / FeBr$_3$) → CH$_3$ Br (ortho) + CH$_3$ Br (meta) + CH$_3$ Br (para)

ortho **40%**

meta trace

para **60%**

[2] Nitrobenzene

Nitrobenzene reacts **more slowly** than benzene in all substitution reactions. Thus, its **electron-withdrawing NO$_2$ group deactivates the benzene ring** to electrophilic attack. Although three products are possible, the compound with the new group meta to the NO_2 group predominates. The NO_2 group is called a **meta director.**

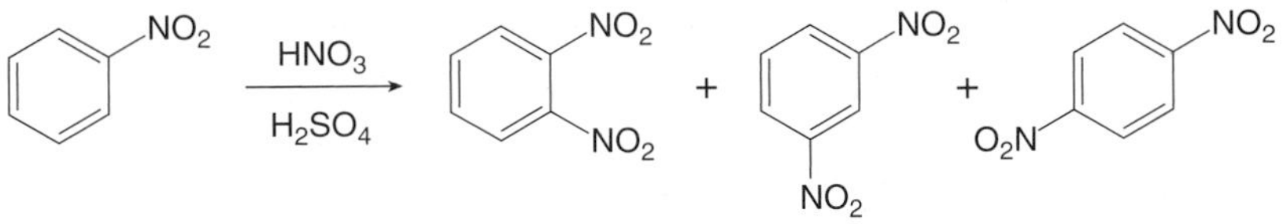

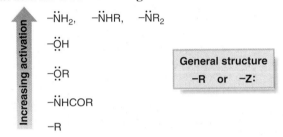

ortho	meta	para
7%	93%	trace

Substituents either activate or deactivate a benzene ring towards electrophiles, and direct selective substitution at specific sites on the ring. **All substituents can be divided into three general types.**

---

[1]   ortho, para directors and activators

◆ Substituents that *activate* a benzene ring and direct substitution ortho and para.

$-\ddot{N}H_2$, $-\ddot{N}HR$, $-\ddot{N}R_2$

$-\ddot{O}H$

$-\ddot{O}R$

$-\ddot{N}HCOR$

$-R$

(Increasing activation ↑)

> **General structure**
> $-R$ or $-Z:$

---

[2]   ortho, para deactivators

◆ Substituents that *deactivate* a benzene ring and direct substitution ortho and para.

$-\ddot{\underset{\cdot\cdot}{F}}:$ $-\ddot{\underset{\cdot\cdot}{Cl}}:$ $-\ddot{\underset{\cdot\cdot}{Br}}:$ $-\ddot{\underset{\cdot\cdot}{I}}:$

---

[3]   meta directors

◆ Substituents that direct substitution meta.

◆ All meta directors *deactivate* the ring.

$-CHO$

$-COR$

$-COOR$

$-COOH$

$-CN$

$-SO_3H$

$-NO_2$

$-\overset{+}{N}R_3$

(Increasing deactivation ↓)

> **General structure**
> $-Y$ $(\delta^+$ or $+)$

To learn these lists: **Keep in mind that the halogens are in a class by themselves.** Then learn the general structures for each type of substituent.

◆ All ortho, para directors are R groups or have a nonbonded electron pair on the atom bonded to the benzene ring.

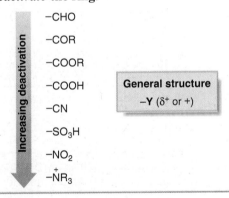

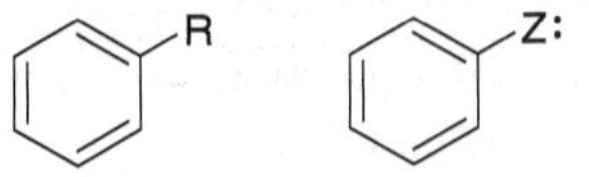

Z = N or O - - → The ring is **activated.**
Z = halogen - → The ring is **deactivated.**

◆ All meta directors have a full or partial positive charge on the atom bonded to the benzene ring.

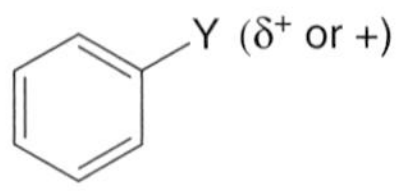

Sample Problem 18.4 shows how this information can be used to predict the products of electrophilic aromatic substitution reactions.

SAMPLE PROBLEM 18.4

Draw the products of each reaction and state whether each reaction is faster or slower than a similar reaction with benzene.

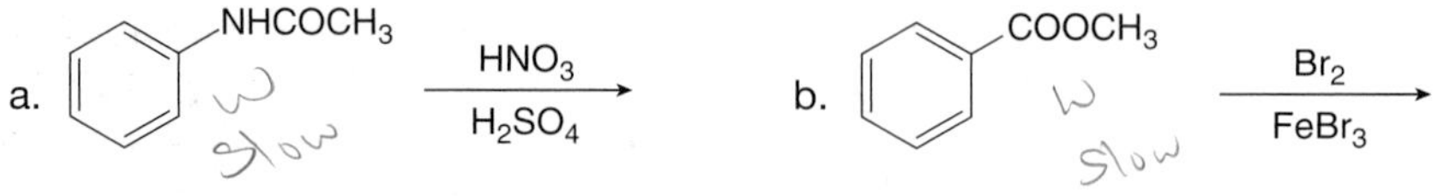

SOLUTION

To draw the products:
- Draw the Lewis structure for the substituent to see if it has a lone pair or partial positive charge on the atom bonded to the benzene ring.
- Classify the substituent—ortho, para activating, ortho, para deactivating, or meta deactivating—and draw the products.

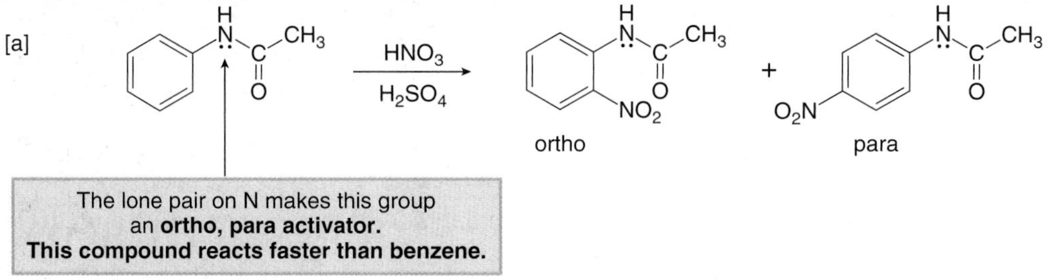

[a]

ortho   para

> The lone pair on N makes this group an **ortho, para activator.**
> **This compound reacts faster than benzene.**

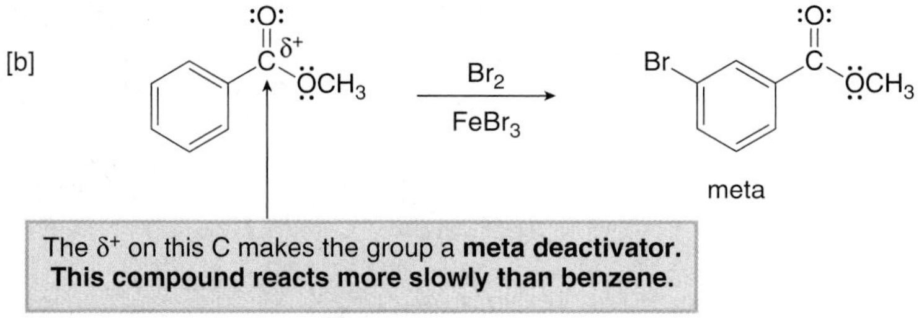

[b]

meta

> The δ+ on this C makes the group a **meta deactivator.**
> **This compound reacts more slowly than benzene.**

PROBLEM 18.16

Draw the products of each reaction.

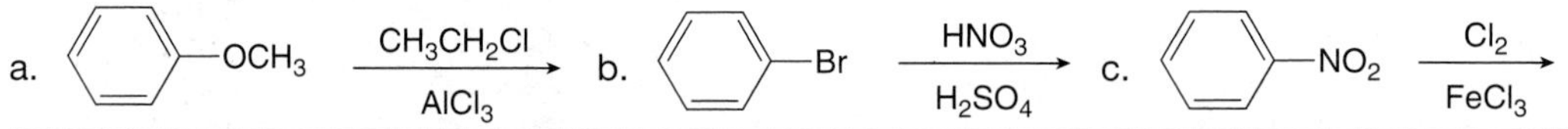

PROBLEM 18.17

Draw the products formed when each compound is treated with HNO_3 and H_2SO_4. State whether the reaction occurs faster or slower than a similar reaction with benzene.

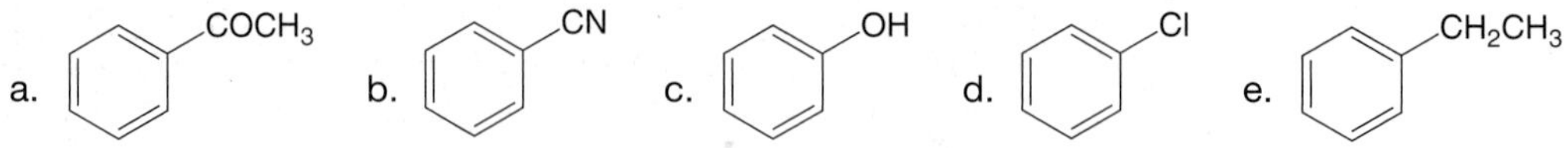

18.8 Why Substituents Activate or Deactivate a Benzene Ring

◆ **Why do substituents activate or deactivate a benzene ring?**

◆ **Why are particular orientation effects observed?** Why are some groups ortho, para directors and some groups meta directors?

To understand why some substituents make a benzene ring react faster than benzene itself (activators), whereas others make it react slower (deactivators), we must evaluate the rate-determining step (the first step) of the mechanism. Recall from Section 18.2 that the first step in electrophilic aromatic substitution is the addition of an electrophile (E^+) to form a resonance-stabilized carbocation. The Hammond postulate (Section 7.15) makes it possible to predict the relative rate of the reaction by looking at the stability of the carbocation intermediate.

♦ **The more stable the carbocation, the lower in energy the transition state that forms it, and the faster the reaction.**

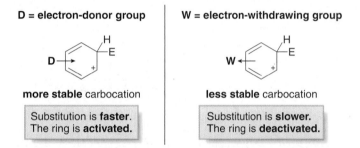

[+ two resonance structures]

Stabilizing the carbocation makes the reaction faster.

The principles of inductive effects and resonance effects, first introduced in Section 18.6, can now be used to predict carbocation stability.

♦ **Electron-donating groups stabilize the carbocation, making the reaction faster.**
♦ **Electron-withdrawing groups destabilize the carbocation, making the reaction slower.**

D = electron-donor group   **W = electron-withdrawing group**

more stable carbocation   **less stable** carbocation

Substitution is **faster**.   Substitution is **slower**.
The ring is **activated**.   The ring is **deactivated**.

♦ **In other words, electron-donating groups activate a benzene ring and electron-withdrawing groups deactivate a benzene ring towards electrophilic attack.**

The energy diagrams in Figure 18.6 illustrate the effect of electron-donating and electron-withdrawing groups on the energy of the transition state of the rate-determining step in electrophilic aromatic substitution.

Figure 18.6
Energy diagrams comparing the rate of electrophilic aromatic substitution of substituted benzenes

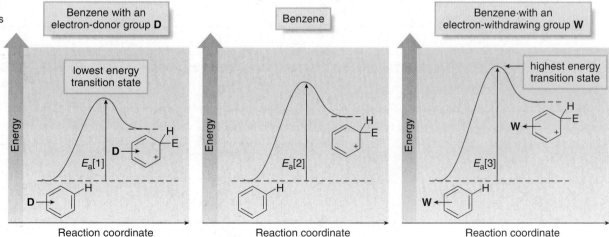

- Electron-donor groups **D** stabilize the carbocation intermediate, lower the energy of the transition state, and increase the rate of reaction.
- Electron-withdrawing groups **W** destabilize the carbocation intermediate, raise the energy of the transition state, and decrease the rate of reaction.

We learned in Section 18.6 which groups are electron donating and electron withdrawing. As a result, we know which groups increase or decrease the rate of reaction of substituted benzenes with electrophiles.

◆ **All activators are either R groups or they have an N or O atom with a lone pair directly bonded to the benzene ring. These are the electron-donor groups of Section 18.6.**

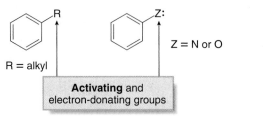

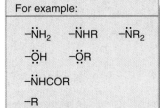

◆ **All deactivators are either halogens or they have an atom with a partial or full positive charge bonded directly to the benzene ring. These are the electron-withdrawing groups of Section 18.6.**

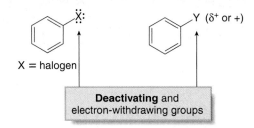

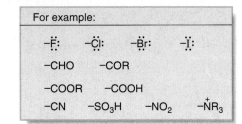

PROBLEM 18.18 Label each compound as more or less reactive than benzene in electrophilic aromatic substitution.

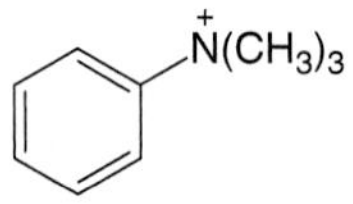

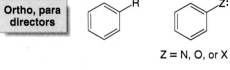

PROBLEM 18.19 Rank the compounds in each group in order of increasing reactivity in electrophilic aromatic substitution.

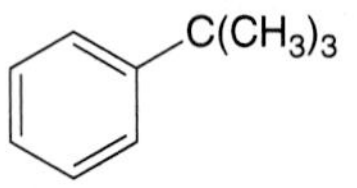

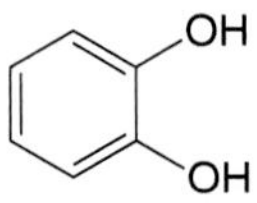

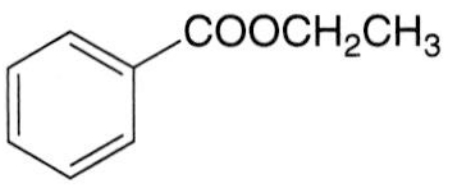

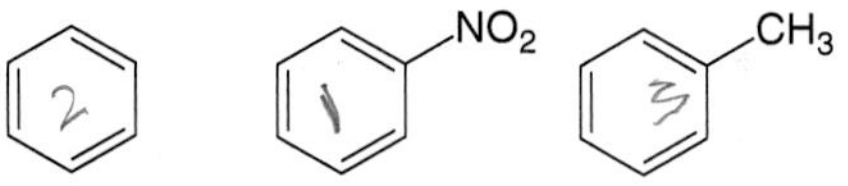

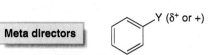

18.9 Orientation Effects in Substituted Benzenes

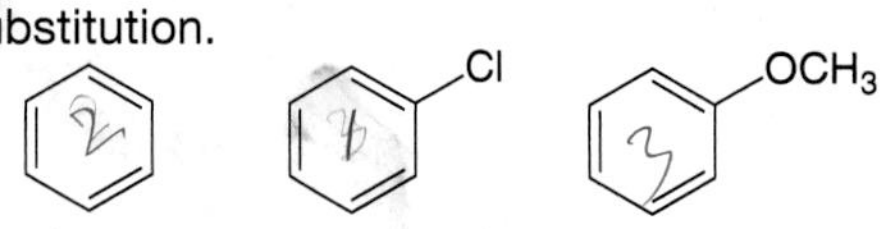

To understand why particular orientation effects arise, you must keep in mind the general structures for ortho, para directors and for meta directors already given in Section 18.7. There are two general types of ortho, para directors and one general type of meta director:

Ortho, para directors

Z = N, O, or X

• All ortho, para directors are R groups or have a nonbonded electron pair on the atom bonded to the benzene ring.

Meta directors

Y (δ⁺ or +)

• All meta directors have a full or partial positive charge on the atom bonded to the benzene ring.

To evaluate the directing effects of a given substituent, we can follow a stepwise procedure.

How To	Determine the Directing Effects of a Particular Substituent

Step [1]   **Draw all resonance structures for the carbocation formed from attack of an electrophile E$^+$ at the ortho, meta, and para positions of a substituted benzene (C$_6$H$_5$ – A).**

original substituent ⟶ A

- ortho
- meta
- para

- There are at least three resonance structures for each site of reaction.
- Each resonance structure places a positive charge ortho or para to the new C – E bond.

Step [2]   **Evaluate the stability of the intermediate resonance structures. The electrophile attacks at those positions that give the most stable carbocation.**

Sections 18.9A–C show how this two-step procedure can be used to determine the directing effects of the CH$_3$ group in toluene, the NH$_2$ group in aniline, and the NO$_2$ group in nitrobenzene, respectively.

18.9A   The CH$_3$ Group—An ortho, para Director

To determine why a **CH$_3$ group directs electrophilic aromatic substitution to the ortho and para positions,** first draw all resonance structures that result from electrophilic attack at the ortho, meta, and para positions to the CH$_3$ group.

ortho attack

CH$_3$ stabilizes the (+) charge

preferred product

meta attack

para attack

CH$_3$ stabilizes the (+) charge

preferred product

> Always draw in the H atom at the site of electrophilic attack. This will help you keep track of where the charges go.

Note that the positive charge in all resonance structures is always ortho or para to the new C – E bond. It is *not* necessarily ortho or para to the CH$_3$ group.

To evaluate the stability of the resonance structures, determine whether any are especially stable or unstable. In this example, **attack ortho or para to CH$_3$ generates a resonance structure that places a positive charge on a carbon atom with the CH$_3$ group.** The electron-donating CH$_3$ group *stabilizes* the adjacent positive charge. In contrast, attack meta to the CH$_3$ group does *not* generate any resonance structure stabilized by electron donation. Other alkyl groups are ortho, para directors for the same reason.

> ◆ **Conclusion: The CH$_3$ group directs electrophilic attack ortho and para to itself because an electron-donating inductive effect stabilizes the carbocation intermediate.**

18.9B The NH_2 Group—An ortho, para Director

To determine why an **amino group (NH_2) directs electrophilic aromatic substitution to the ortho and para positions,** follow the same procedure.

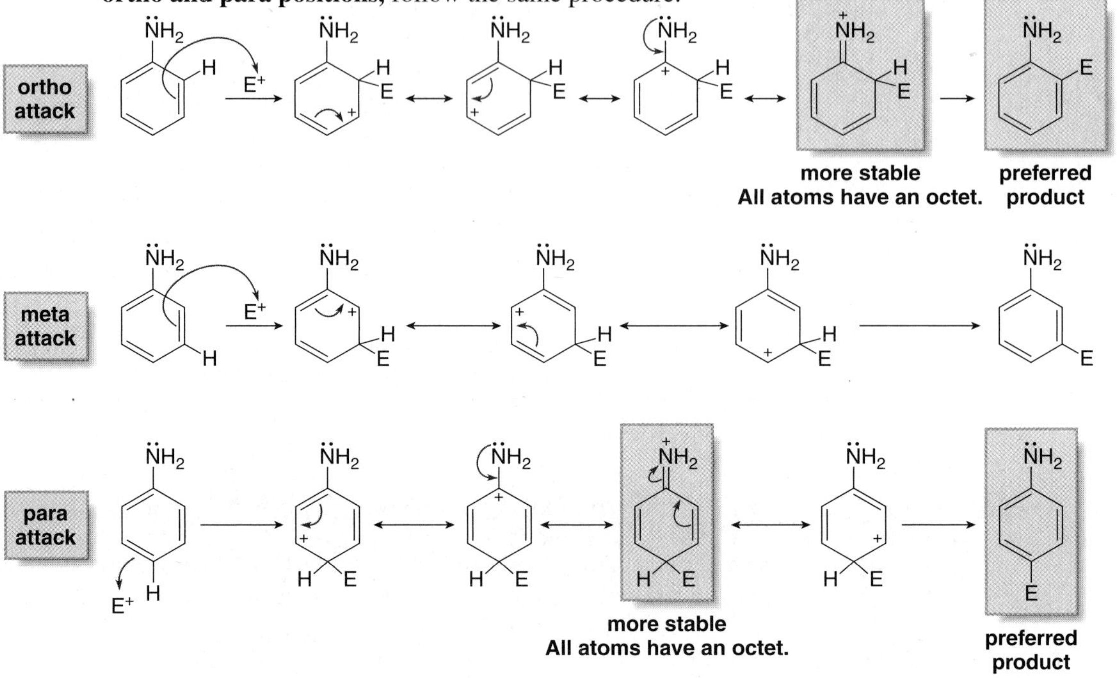
ortho attack
more stable
All atoms have an octet.
preferred product
meta attack
para attack
more stable
All atoms have an octet.
preferred product

Attack at the meta position generates the usual three resonance structures. Because of the lone pair on the N atom, attack at the ortho and para positions generates a fourth resonance structure, which is stabilized because **every atom has an octet of electrons. This additional resonance structure can be drawn for all substituents that have an N, O, or halogen atom directly bonded to the benzene ring.**

- Conclusion: The NH_2 group directs electrophilic attack ortho and para to itself because the carbocation intermediate has additional resonance stabilization.

18.9C The NO_2 Group—A meta Director

To determine why a **nitro group (NO_2) directs electrophilic aromatic substitution to the meta position,** follow the same procedure.

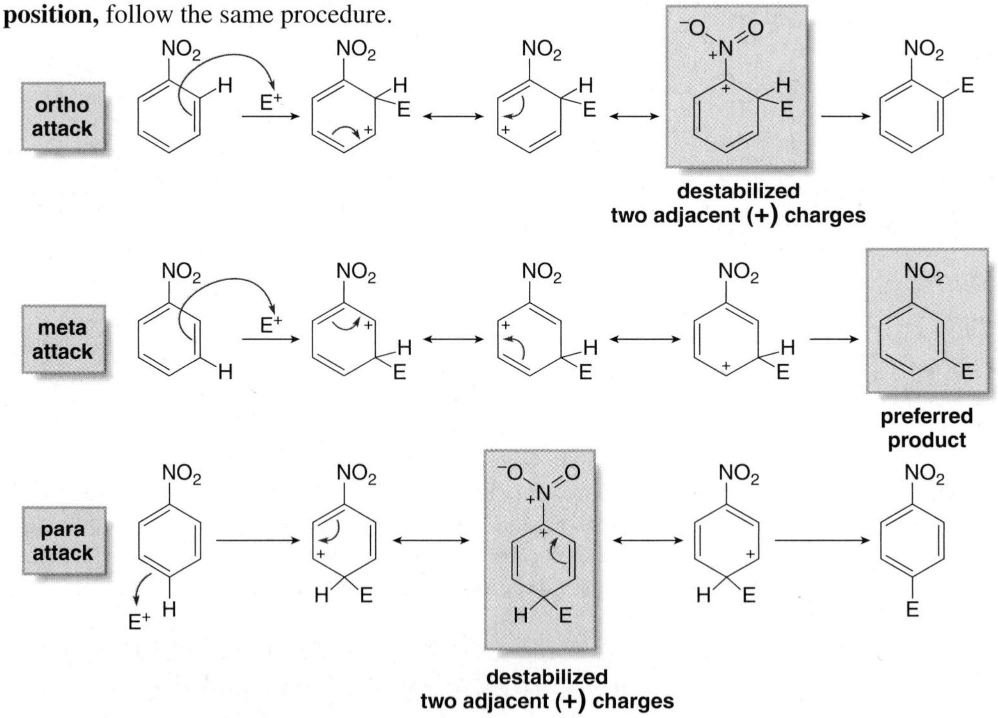
ortho attack
destabilized
two adjacent (+) charges
meta attack
preferred product
para attack
destabilized
two adjacent (+) charges

Attack at each position generates three resonance structures. One resonance structure resulting from attack at the ortho and para positions is especially destabilized, because it contains a positive charge on two adjacent atoms. Attack at the meta position does not generate any particularly unstable resonance structures.

> ◆ **Conclusion:** With the NO_2 group (and all meta directors), meta attack occurs because attack at the ortho or para position gives a destabilized carbocation intermediate.

PROBLEM 18.20 Draw all resonance structures for the carbocation formed by ortho attack of the electrophile $^+NO_2$ on each starting material. Label any resonance structures that are especially stable or unstable.

a. $C(CH_3)_3$ b. OH c. CHO

PROBLEM 18.21 Use the procedure illustrated in Sections 18.9A–C to show why chlorine is an ortho, para director.

Figure 18.7 summarizes the reactivity and directing effects of the common substituents on benzene rings. Do not memorize this list. Instead, follow the general procedure outlined in Sections 18.9A–C to predict particular substituent effects.

18.10   Limitations on Electrophilic Substitution Reactions with Substituted Benzenes

Although electrophilic aromatic substitution works well with most substituted benzenes, halogenation and the Friedel–Crafts reactions have some additional limitations that must be kept in mind.

Figure 18.7 The reactivity and directing effects of common substituted benzenes

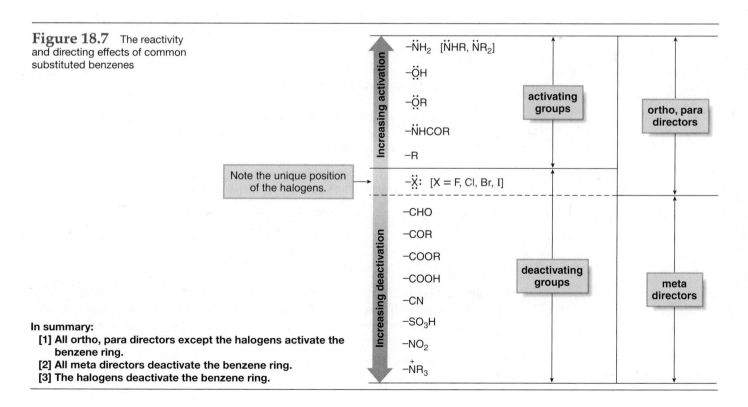

In summary:
[1] All ortho, para directors except the halogens activate the benzene ring.
[2] All meta directors deactivate the benzene ring.
[3] The halogens deactivate the benzene ring.

18.10A   Halogenation of Activated Benzenes

Considering all electrophilic aromatic substitution reactions, halogenation occurs the most readily. As a result, benzene rings activated by strong electron donating groups—OH, NH$_2$, and their derivatives (OR, NHR, and NR$_2$)—undergo **polyhalogenation** when treated with X$_2$ and FeX$_3$. For example, aniline (C$_6$H$_5$NH$_2$) and phenol (C$_6$H$_5$OH) both give a tribromo derivative when treated with Br$_2$ and FeBr$_3$. Substitution occurs at all hydrogen atoms ortho and para to the NH$_2$ and OH groups.

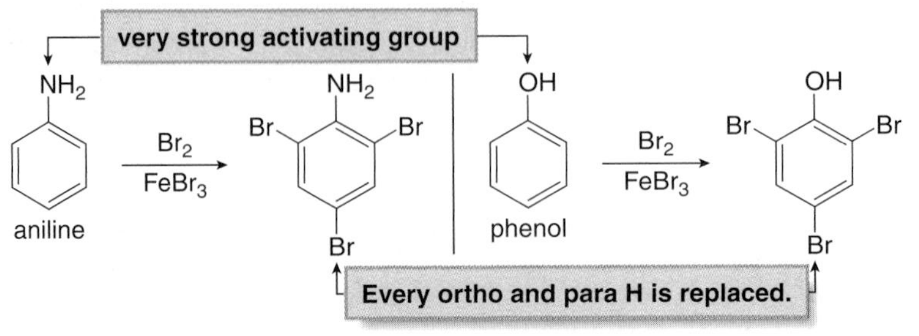

Monosubstitution of H by Br occurs with Br$_2$ *alone* without added catalyst to form a mixture of ortho and para products.

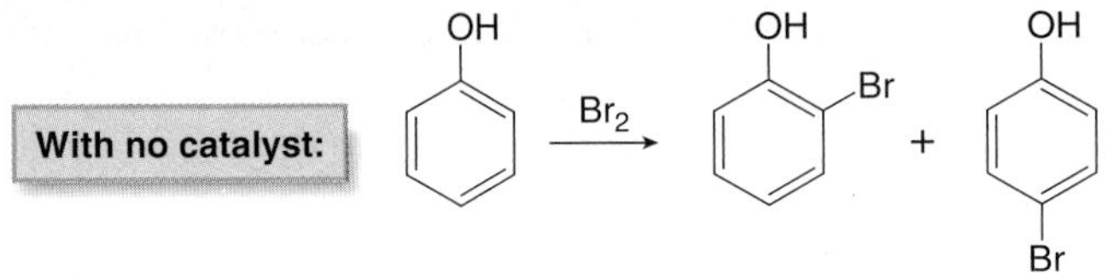

PROBLEM 18.22   Draw the products of each reaction.

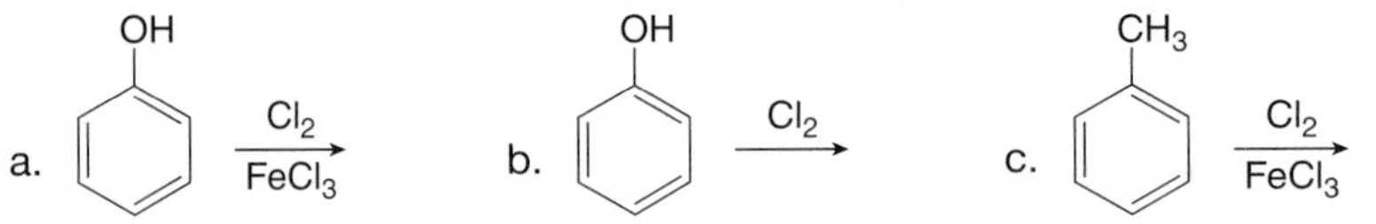

18.10B   Limitations in Friedel–Crafts Reactions

Friedel–Crafts reactions are the most difficult electrophilic aromatic substitution reactions to carry out in the laboratory. For example, they do not occur when the benzene ring is substituted with NO$_2$ (a strong deactivator) or with NH$_2$, NHR, or NR$_2$ (strong activators).

A benzene ring deactivated by a strong electron-withdrawing group—that is, any of the meta directors—is not electron rich enough to undergo Friedel–Crafts reactions.

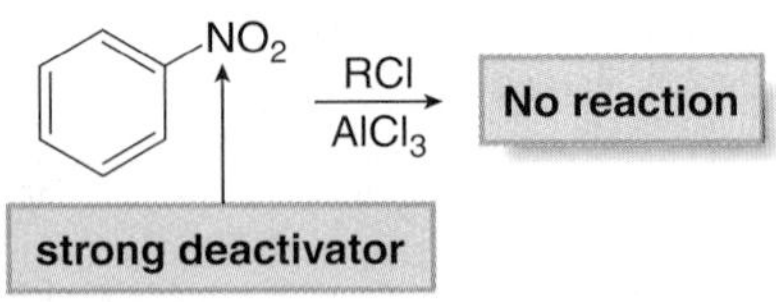

Friedel–Crafts reactions also do not occur with NH$_2$ groups, which are strong activating groups. NH$_2$ groups are strong Lewis bases (due to the nonbonded electron pair on N), so they react with AlCl$_3$, the Lewis acid needed for alkylation or acylation. The resulting product contains a positive charge adjacent to the benzene ring, so the ring is now strongly deactivated and therefore unreactive in Friedel–Crafts reactions.

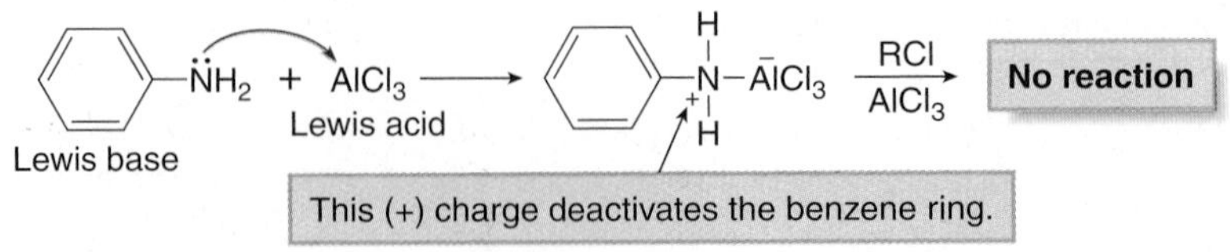

PROBLEM 18.23   Which of the following compounds undergo Friedel–Crafts alkylation with CH_3Cl and $AlCl_3$? Draw the products formed when a reaction occurs.

a. [benzene ring]—SO_3H b. [benzene ring]—Cl   c. [benzene ring]—$N(CH_3)_2$ d. [benzene ring]—$NHCOCH_3$

Another limitation of the Friedel–Crafts alkylation arises because of **polyalkylation.** Treatment of benzene with an alkyl halide and $AlCl_3$ places an electron-donor R group on the ring. Because R groups activate a ring, the alkylated product (C_6H_5R) is now *more reactive* than benzene itself towards further substitution, and it reacts again with RCl to give products of polyalkylation.

> To minimize polyalkylation a large excess of benzene is used relative to the amount of alkyl halide.

[reaction scheme: benzene $\xrightarrow[\text{AlCl}_3]{\text{RCl}}$ benzene—R (an electron-donor group) $\xrightarrow[\text{AlCl}_3]{\text{RCl}}$ ortho product with R groups + para product R—[ring]—R (major products)]

Polysubstitution does not occur with Friedel–Crafts acylation, because the product now has an electron-withdrawing group that deactivates the ring towards another electrophilic substitution.

[reaction scheme: benzene + R—C(=O)—Cl $\xrightarrow{\text{AlCl}_3}$ phenyl ketone C(=O)R (a deactivating group)]

18.11   Disubstituted Benzenes

What happens in electrophilic aromatic substitution when a disubstituted benzene ring is used as starting material? **To predict the products, look at the directing effects of both substituents and then determine the net result,** using the following three guidelines.

Rule [1]   **When the directing effects of two groups reinforce, the new substituent is located on the position directed by both groups.**

For example, the CH_3 group in *p*-nitrotoluene is an ortho, para director and the NO_2 group is a meta director. These two effects reinforce each other so that one product is formed on treatment with Br_2 and $FeBr_3$. Notice that the position para to the CH_3 group is "blocked" by a nitro group so no substitution can occur on that carbon.

[reaction scheme: *p*-nitrotoluene with CH₃ (ortho, para director) and NO₂ (meta director) $\xrightarrow[\text{FeBr}_3]{\text{Br}_2}$ product with CH₃, Br, NO₂]

The new group is ortho to the CH_3 group and meta to the NO_2 group.

p-nitrotoluene

Rule [2]   **If the directing effects of two groups oppose each other, the more powerful activator "wins out."**

In compound **A,** the $NHCOCH_3$ group activates its two ortho positions, and the CH_3 group activates its two ortho positions to reaction with electrophiles. Because the $NHCOCH_3$ is a stronger activator, substitution occurs ortho to it.

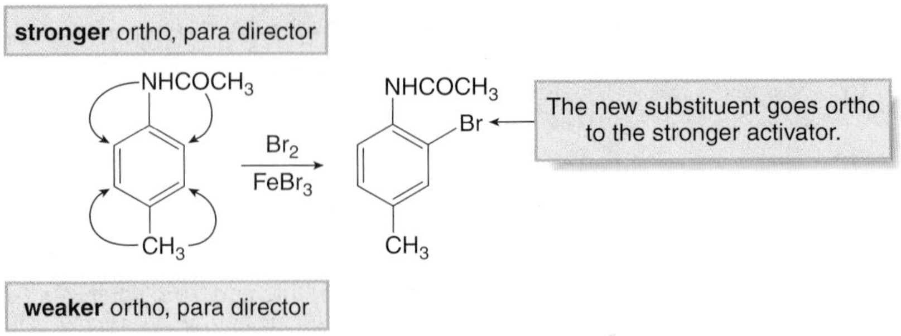

stronger ortho, para director

Br₂ / FeBr₃

The new substituent goes ortho to the stronger activator.

weaker ortho, para director

A

Rule [3]   **No substitution occurs between two meta substituents because of crowding.**

For example, no substitution occurs at the carbon atom between the two CH_3 groups in *m*-xylene, even though two CH_3 groups activate that position.

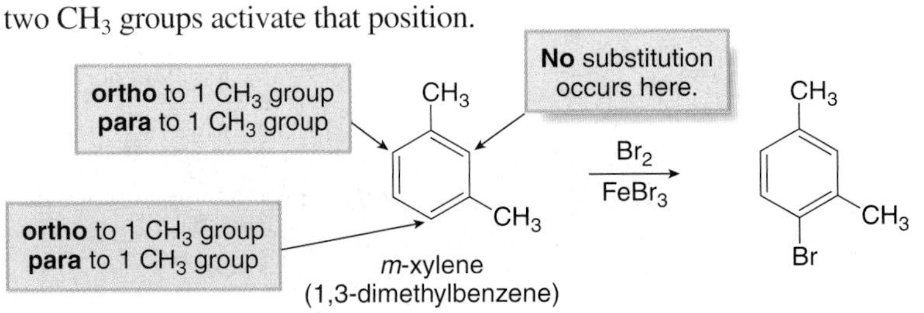

ortho to 1 CH₃ group
para to 1 CH₃ group

No substitution occurs here.

ortho to 1 CH₃ group
para to 1 CH₃ group

Br₂ / FeBr₃

m-xylene
(1,3-dimethylbenzene)

SAMPLE PROBLEM 18.5   Draw the products formed from nitration of each compound.

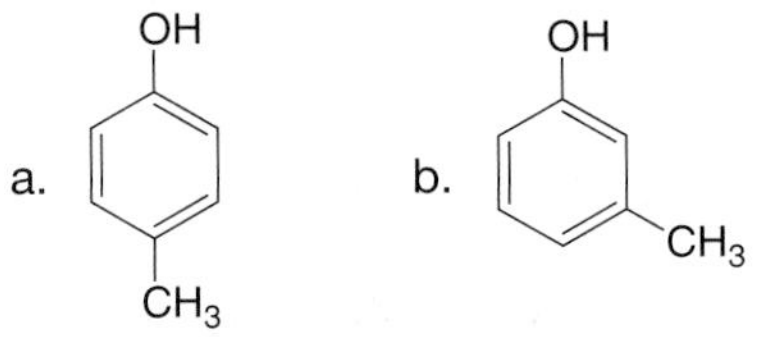

a.   b.

SOLUTION

[a] Both the OH and CH_3 groups are ortho, para directors. Because the OH group is a stronger activator, substitution occurs ortho to it.

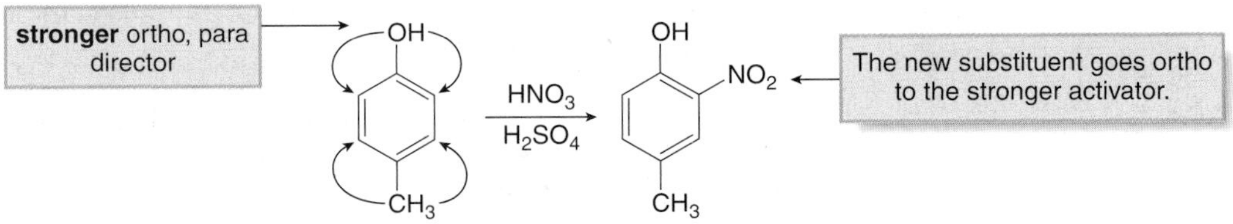

stronger ortho, para director

HNO₃ / H₂SO₄

The new substituent goes ortho to the stronger activator.

[b] Both the OH and CH_3 groups are ortho, para directors whose directing effects reinforce each other in this case. No substitution occurs between the two meta substituents, however, so two products result.

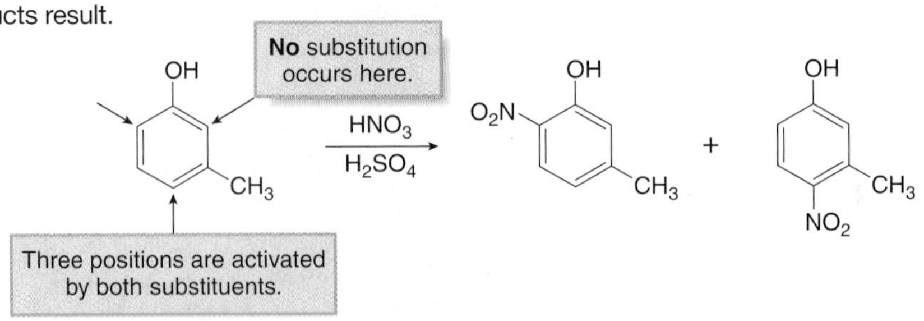

No substitution occurs here.

HNO₃ / H₂SO₄

Three positions are activated by both substituents.

PROBLEM 18.24   Draw the products formed when each compound is treated with HNO₃ and H₂SO₄.

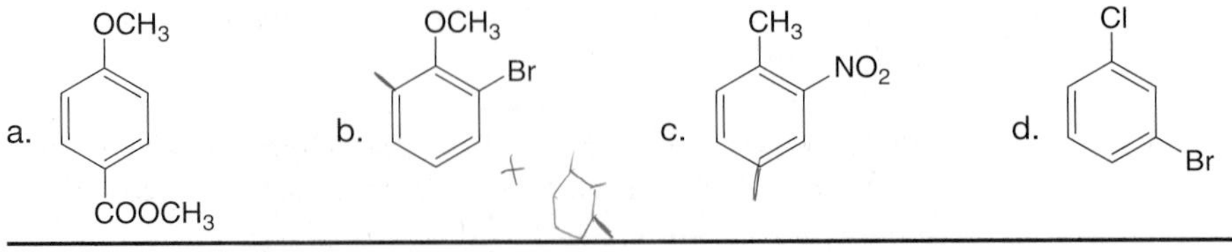

a.   b.   c.   d.

18.12 Synthesis of Benzene Derivatives

To synthesize benzene derivatives with more than one substituent, we must always take into account the directing effects of each substituent. In a disubstituted benzene, for example, **the directing effects indicate which substituent must be added to the ring first.**

For example, the Br group in *p*-bromonitrobenzene is an ortho, para director and the NO_2 group is a meta director. Because the two substituents are para to each other, the ortho, para director must be introduced *first* when synthesizing this compound from benzene.

p-bromonitrobenzene

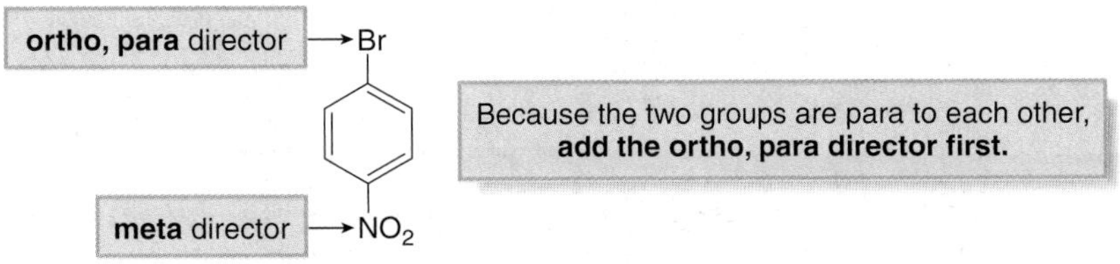

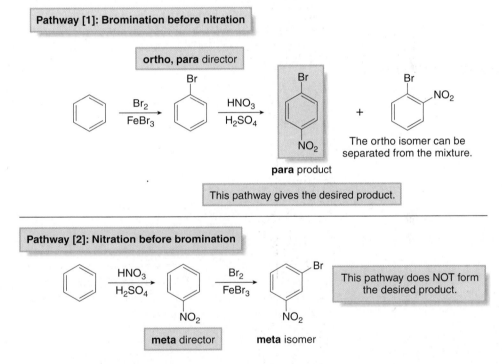

Thus, Pathway [1], in which bromination precedes nitration, yields the desired para product, whereas Pathway [2], in which nitration precedes bromination, yields the undesired meta isomer.

Pathway [1]: Bromination before nitration

Pathway [2]: Nitration before bromination

Pathway [1] yields both the desired para product as well as the undesired ortho isomer. Because these compounds are constitutional isomers, they are separable. Obtaining such a mixture of ortho and para isomers is often unavoidable.

SAMPLE PROBLEM 18.6 Devise a synthesis of *o*-nitrotoluene from benzene.

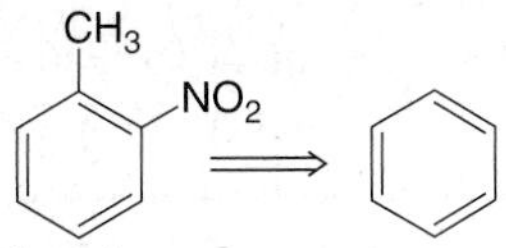

o-nitrotoluene

SOLUTION

The CH_3 group in *o*-nitrotoluene is an ortho, para director and the NO_2 group is a meta director. Because the two substituents are ortho to each other, the **ortho, para director must be introduced first.** The synthesis thus involves two steps: Friedel–Crafts alkylation followed by nitration.

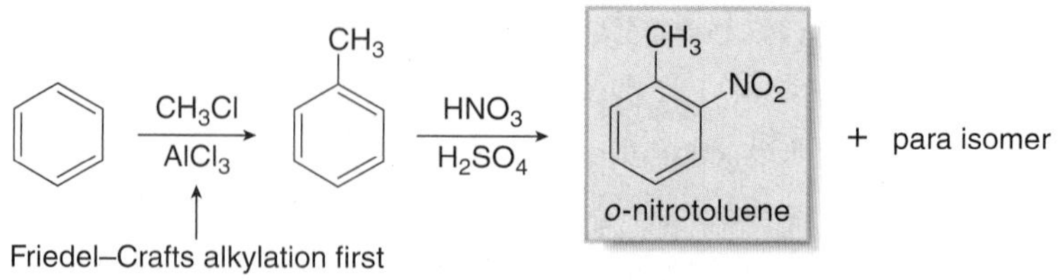

o-nitrotoluene + para isomer

Friedel–Crafts alkylation first

PROBLEM 18.25 Devise a synthesis of each compound from the indicated starting material.

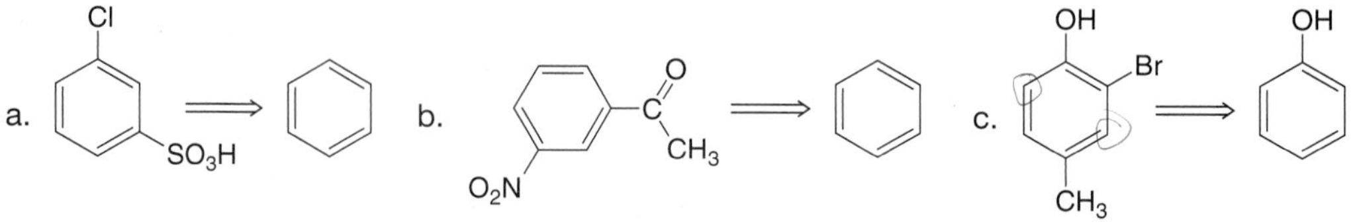

18.13 Halogenation of Alkyl Benzenes

Radical halogenation of alkanes was discussed in Chapter 13. The mechanism of radical halogenation at an allylic carbon was given in Section 13.10.

We finish Chapter 18 by learning some additional reactions of substituted benzenes that greatly expand the ability to synthesize benzene derivatives. These reactions do not involve the benzene ring itself, so they are not further examples of electrophilic aromatic substitution. In Section 18.13 we return to radical halogenation, and in Section 18.14 we examine useful oxidation and reduction reactions.

Benzylic C–H bonds are weaker than most other sp^3 hybridized C–H bonds, because homolysis forms a resonance-stabilized benzylic radical.

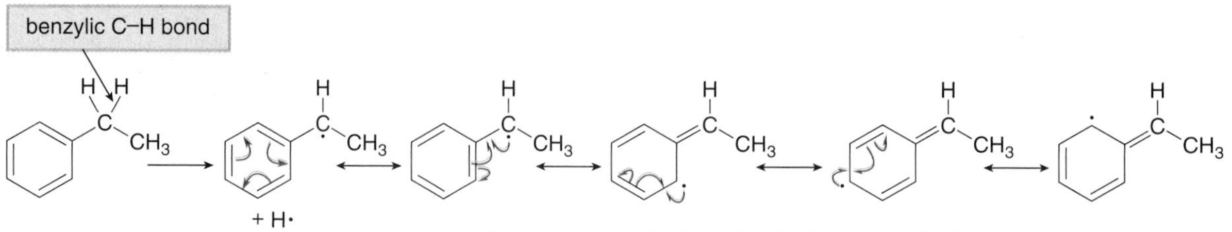

benzylic C–H bond

+ H·

five resonance structures for the benzylic radical

The bond dissociation energy for a benzylic C–H bond (85 kcal/mol) is even less than the bond dissociation energy for a 3° C–H bond (91 kcal/mol).

As a result, an alkyl benzene undergoes selective bromination at the weak benzylic C–H bond under radical conditions to form a **benzylic halide.** For example, radical bromination of ethylbenzene using either Br_2 (in the presence of light or heat) or *N*-bromosuccinimide (NBS, in the presence of light or peroxides) forms a benzylic bromide as the sole product.

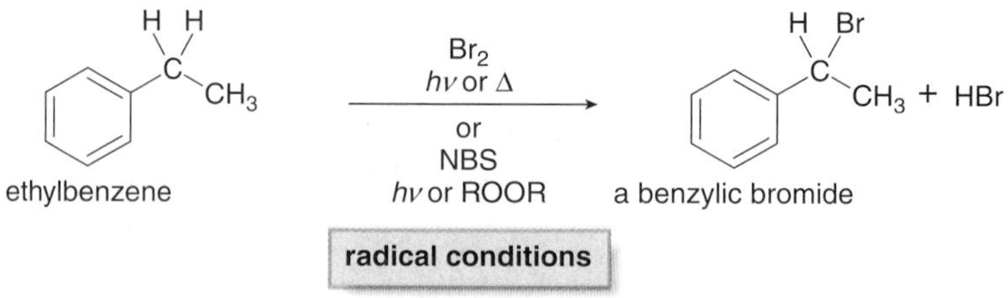

ethylbenzene

Br_2
hv or Δ
or
NBS
hv or ROOR

a benzylic bromide

radical conditions

The mechanism for halogenation at the benzylic position resembles other radical halogenation reactions, and so it involves initiation, propagation, and termination. Mechanism 18.5 illustrates the radical bromination of ethylbenzene using Br_2 (*hv* or Δ).

MECHANISM 18.5

Benzylic Bromination

Initiation
Step [1]   Bond cleavage forms two radicals.

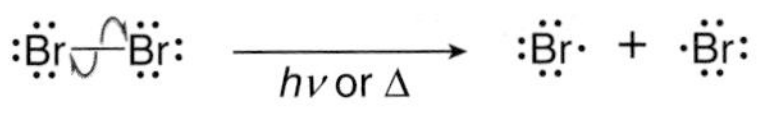

:Br̈–Br̈: →(*hv* or Δ)→ :Br̈· + ·Br̈:

◆ The reaction begins with homolysis of the Br–Br bond using energy from light or heat to form two Br· radicals.

Propagation
Steps [2] and [3]   One radical reacts and a new radical is formed.

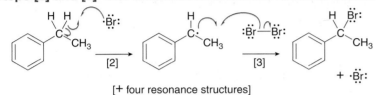

[+ four resonance structures]
+ HBr
Repeat Steps [2], [3], [2], [3], again and again.

◆ Abstraction of a benzylic hydrogen by a Br· radical forms the resonance-stabilized benzylic radical in Step [2], which reacts with Br_2 in Step [3] to form the bromination product.

◆ Because the Br· radical formed in Step [3] is a reactant in Step [2], Steps [2] and [3] can occur repeatedly without additional initiation.

Termination
Step [4]   Two radicals react to form a bond.

:Br̈· + ·Br̈: → :Br̈–Br̈:

◆ To terminate the reaction, two radicals, for example two Br· radicals, react with each other to form a stable bond.

PROBLEM 18.26   Explain why $C_6H_5CH_2CH_2Br$ is not formed during the radical bromination of $C_6H_5CH_2CH_3$.

Thus, an alkyl benzene undergoes two different reactions with Br_2, depending on the reaction conditions.

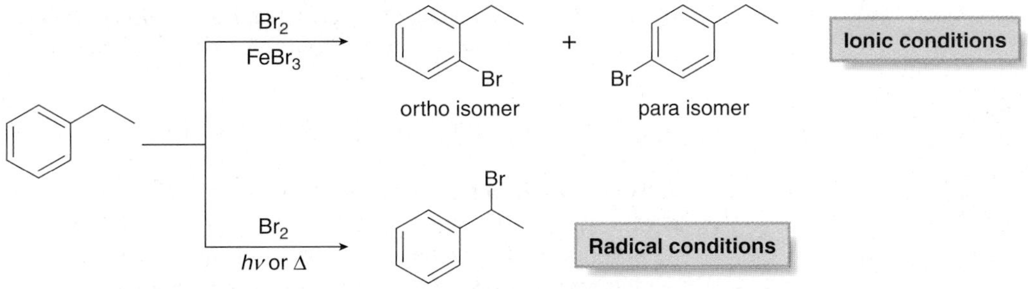

With Br_2 and $FeBr_3$ → ortho isomer + para isomer (**Ionic conditions**)

With Br_2, *hv* or Δ → (**Radical conditions**)

◆ With Br_2 and $FeBr_3$ (**ionic conditions**), electrophilic aromatic substitution occurs, resulting in replacement of H by Br on the aromatic ring to form ortho and para isomers.

◆ With Br_2 and light or heat (**radical conditions**), substitution of H by Br occurs at the *benzylic* carbon of the alkyl group.

PROBLEM 18.27   Draw the products formed when isopropylbenzene $[C_6H_5CH(CH_3)_2]$ is treated with each reagent: (a) Br_2, $FeBr_3$; (b) Br_2, *hv*; (c) Cl_2, $FeCl_3$.

The radical bromination of alkyl benzenes is a useful reaction because the resulting benzylic halide can serve as starting material for a variety of substitution and elimination reactions, thus making it possible to form many new substituted benzenes. Sample Problem 18.7 illustrates one possibility.

SAMPLE PROBLEM 18.7   Design a synthesis of styrene from ethylbenzene.

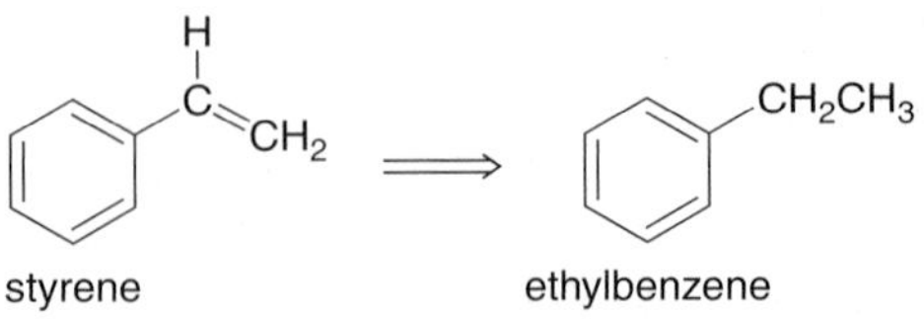

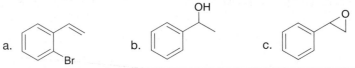

styrene   ethylbenzene

SOLUTION

The double bond can be introduced by a two-step reaction sequence: bromination at the benzylic position under radical conditions, followed by elimination of HBr with strong base to form the π bond.

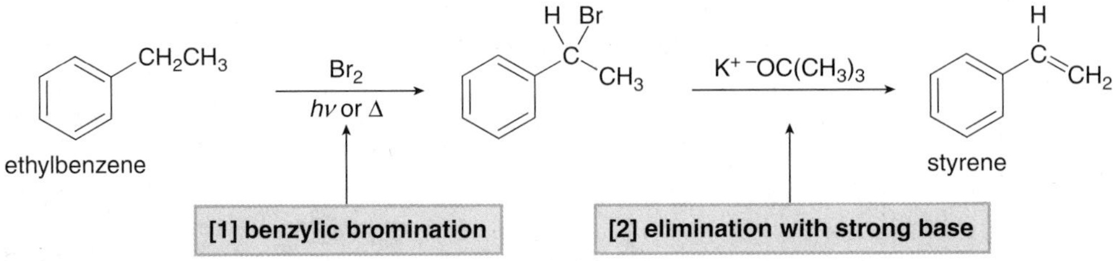

ethylbenzene
Br_2
hv or Δ

K^+ $^-OC(CH_3)_3$

styrene

[1] benzylic bromination   **[2] elimination with strong base**

PROBLEM 18.28   How could you use ethylbenzene to prepare each compound? More than one step is required.

a.   b.   c.

PROBLEM 18.29   What reagents are needed to carry out each reaction: $C_6H_6 \rightarrow C_6H_5CH_3 \rightarrow C_6H_5CH_2Br \rightarrow$ $C_6H_5CH_2CN$?

18.14   Oxidation and Reduction of Substituted Benzenes

Oxidation and reduction reactions are valuable tools for preparing many other benzene derivatives. Because the mechanisms are complex and do not have general applicability, reagents and reactions are presented only, without reference to the detailed mechanism.

18.14A   Oxidation of Alkyl Benzenes

Arenes containing at least one benzylic C−H bond are oxidized with $KMnO_4$ to benzoic acid, a carboxylic acid with the carboxy group (COOH) bonded directly to the benzene ring. With some alkyl benzenes, this also results in the cleavage of carbon–carbon bonds, so the product has fewer carbon atoms than the starting material.

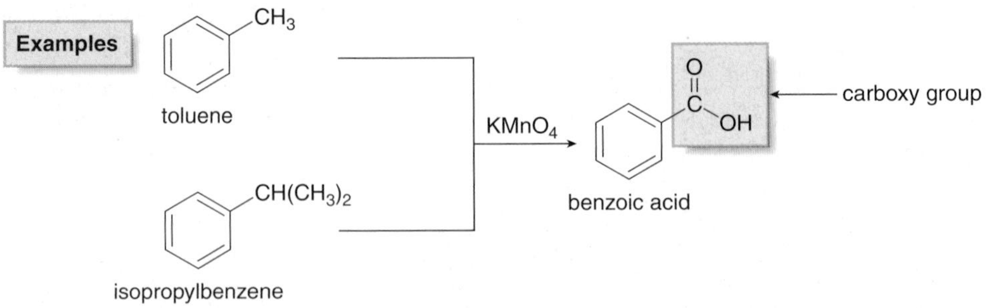

Examples

toluene

isopropylbenzene

$KMnO_4$

carboxy group

benzoic acid

Substrates with more than one alkyl group are oxidized to dicarboxylic acids. Compounds without a benzylic C−H bond are inert to oxidation.

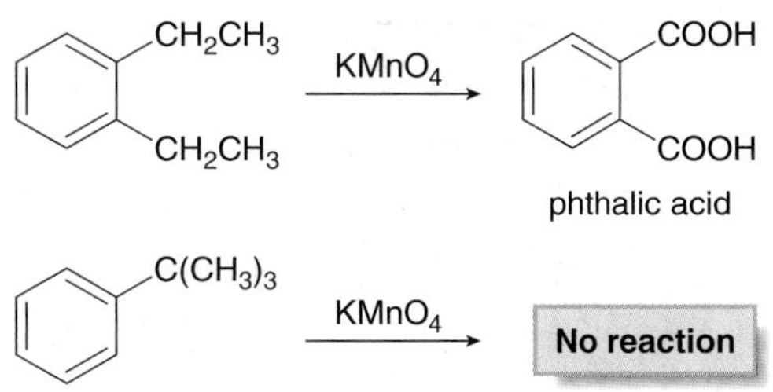

phthalic acid

18.14B   Reduction of Aryl Ketones to Alkyl Benzenes

Ketones formed as products in Friedel–Crafts acylation can be reduced to alkyl benzenes by two different methods.

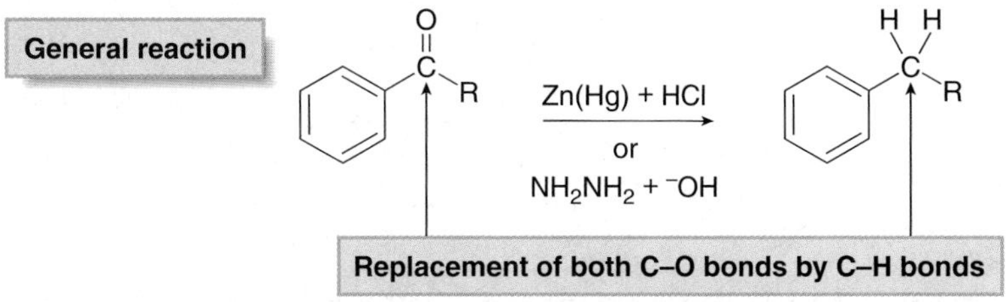

Replacement of both C–O bonds by C–H bonds

◆ The **Clemmensen reduction** uses zinc and mercury in the presence of strong acid.
◆ The **Wolff–Kishner reduction** uses hydrazine (NH_2NH_2) and strong base (KOH).

Because both $C-O$ bonds in the starting material are converted to $C-H$ bonds in the product, the reduction is difficult and the reaction conditions must be harsh.

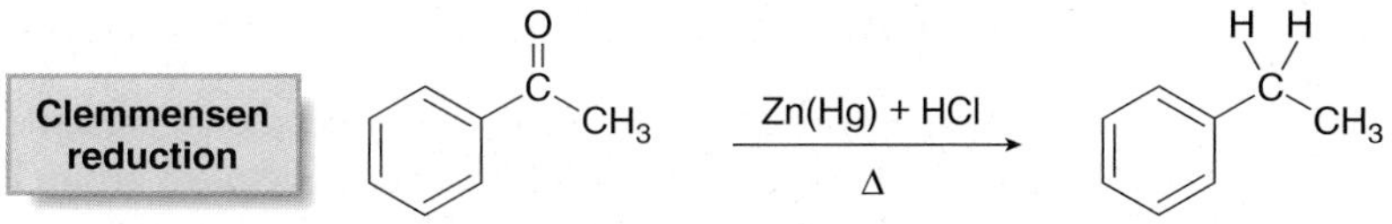

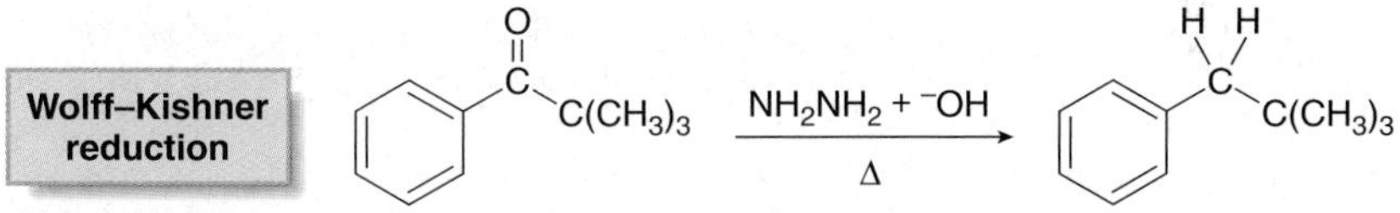

We now know two different ways to introduce an alkyl group on a benzene ring (Figure 18.8):

◆ **A one-step method using Friedel–Crafts alkylation**
◆ **A two-step method using Friedel–Crafts acylation to form a ketone, followed by reduction**

Although the two-step method seems more roundabout, it must be used to synthesize certain alkyl benzenes that cannot be prepared by the one-step Friedel–Crafts alkylation because of rearrangements.

Figure 18.8   Two methods to prepare an alkyl benzene

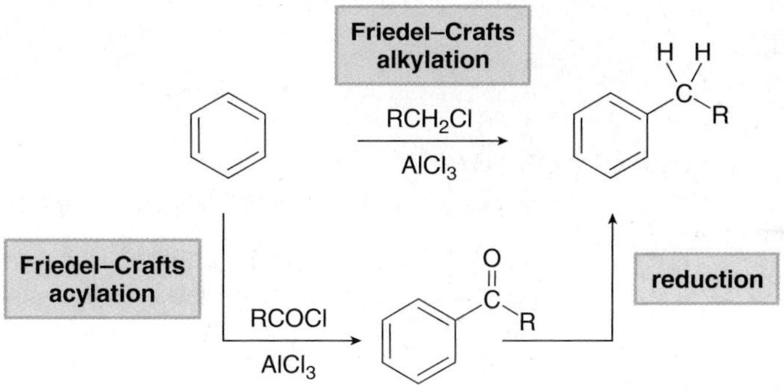

Recall from Section 18.5C that propylbenzene cannot be prepared by a Friedel–Crafts alkylation. Instead, when benzene is treated with 1-chloropropane and AlCl$_3$, isopropylbenzene is formed by a rearrangement reaction. Propylbenzene can be made, however, by a two-step procedure using Friedel–Crafts acylation followed by reduction.

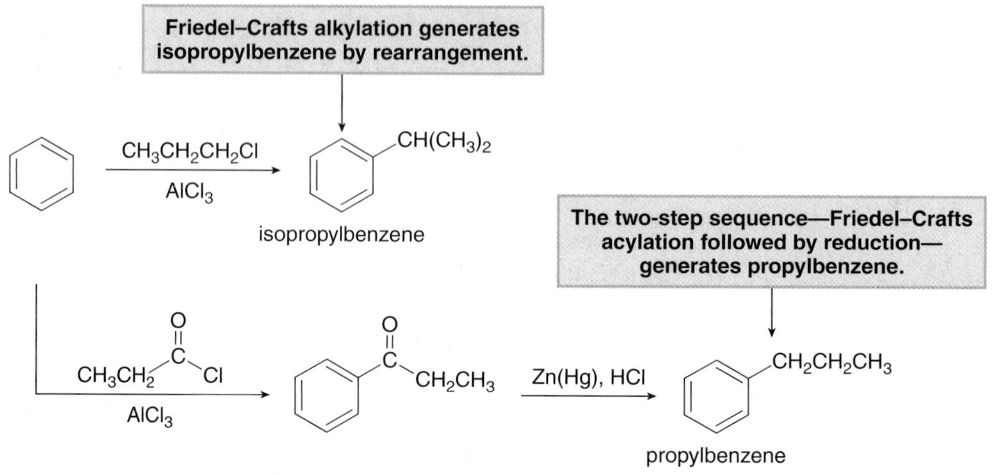

PROBLEM 18.30   Write out the two-step sequence that converts benzene to each compound:
(a) C$_6$H$_5$CH$_2$CH$_2$CH$_2$CH$_2$CH$_3$; (b) C$_6$H$_5$CH$_2$C(CH$_3$)$_3$.

PROBLEM 18.31   Explain why C$_6$H$_5$C(CH$_3$)$_3$ can be made by Friedel–Crafts alkylation, but cannot be made by Friedel–Crafts acylation followed by reduction.

18.14C Reduction of Nitro Groups

A nitro group (NO$_2$) is easily introduced on a benzene ring by nitration with strong acid (Section 18.4). This process is useful because the nitro group is readily reduced to an amino group (NH$_2$) under a variety of conditions. The most common methods use H$_2$ and a catalyst, or a metal (such as Fe or Sn) and a strong acid like HCl.

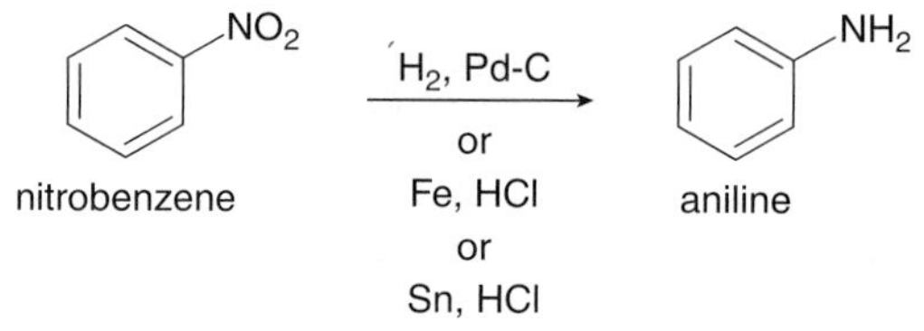

Sample Problem 18.8 illustrates the utility of this process in a short synthesis.

SAMPLE PROBLEM 18.8   Design a synthesis of *m*-bromoaniline from benzene.

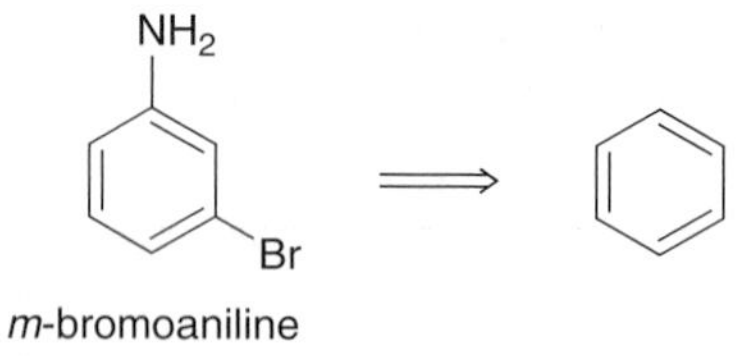

SOLUTION

To devise a retrosynthetic plan, keep in mind:

• The NH$_2$ group cannot be introduced directly on the ring by electrophilic aromatic substitution. It must be added by a two-step process: nitration followed by reduction.
• Both the Br and NH$_2$ groups are ortho, para directors, but they are located meta to each other on the ring. However, an NO$_2$ group (from which an NH$_2$ group is made) *is* a meta director, and we can use this fact to our advantage.

RETROSYNTHETIC ANALYSIS

Working backwards gives the following **three-step retrosynthetic analysis:**

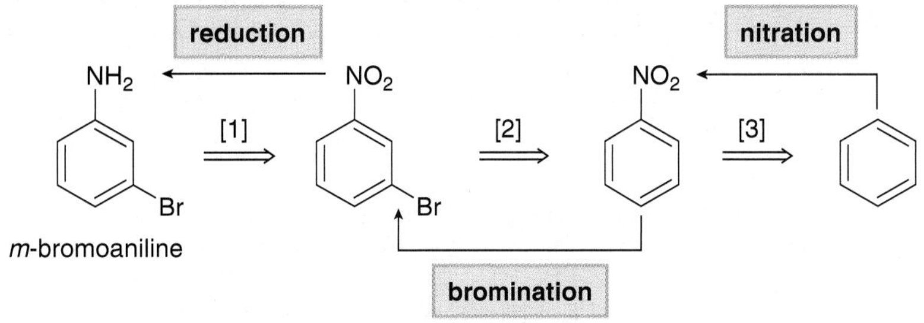

- [1] Form the NH_2 group by reduction of NO_2.
- [2] Introduce the Br group meta to the NO_2 group by halogenation.
- [3] Add the NO_2 group by nitration.

SYNTHESIS

The synthesis then involves three steps, and the order is crucial for success. Halogenation (Step [2] of the synthesis) must occur *before* reduction (Step [3]) in order to form the meta substitution product.

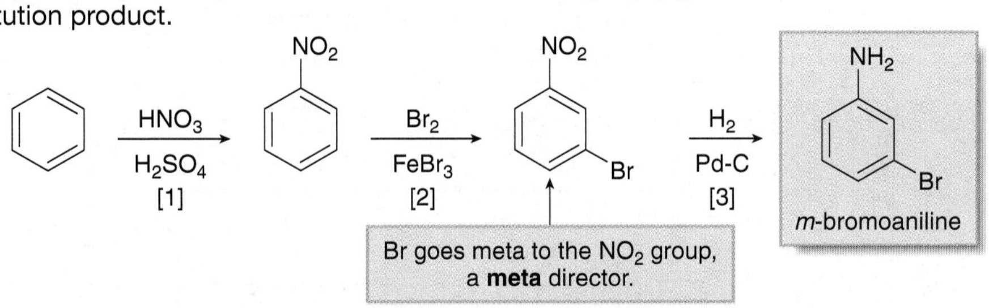

Br goes meta to the NO_2 group, a **meta** director.

PROBLEM 18.32 | Synthesize each compound from benzene.

a. 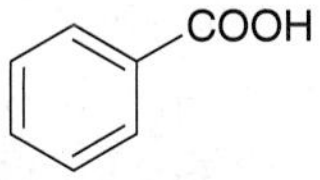   b. 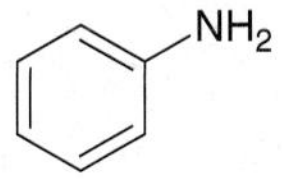   c.

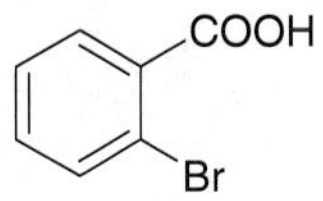

18.15 Multistep Synthesis

The reactions learned in Chapter 18 make it possible to synthesize a wide variety of substituted benzenes, as shown in Sample Problems 18.9–18.11.

SAMPLE PROBLEM 18.9 | Synthesize *p*-nitrobenzoic acid from benzene.

p-nitrobenzoic acid

SOLUTION

Both groups on the ring (NO_2 and COOH) are meta directors. To place these two groups para to each other, remember that the COOH group is prepared by oxidizing an alkyl group, which is an ortho, para director.

RETROSYNTHETIC ANALYSIS

p-nitrobenzoic acid

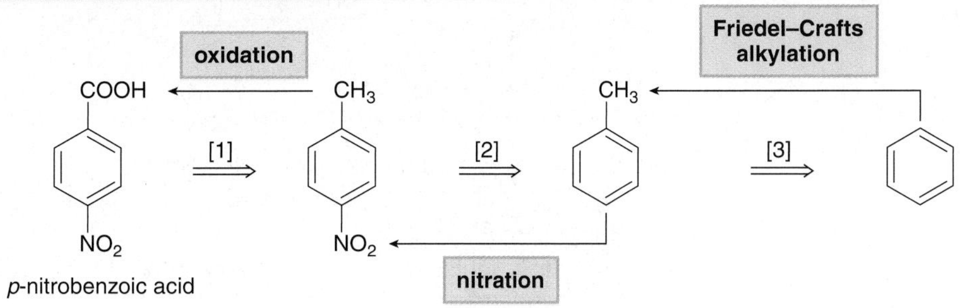

Working backwards:
- [1] Form the COOH group by oxidation of an alkyl group.
- [2] Introduce the NO_2 group para to the CH_3 group (an ortho, para director) by nitration.
- [3] Add the CH_3 group by Friedel–Crafts alkylation.

SYNTHESIS

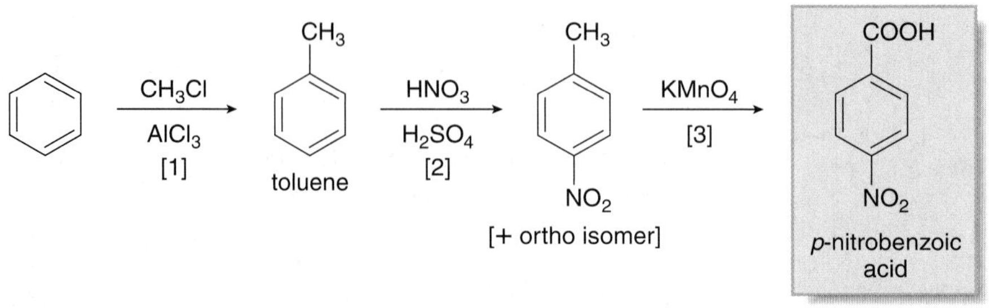

- Friedel–Crafts alkylation with CH_3Cl and $AlCl_3$ forms toluene in Step [1]. Because CH_3 is an ortho, para director, nitration yields the desired para product, which can be separated from its ortho isomer (Step [2]).
- Oxidation with $KMnO_4$ converts the CH_3 group into a COOH group, giving the desired product in Step [3].

SAMPLE PROBLEM 18.10   Synthesize *p*-chlorostyrene from benzene.

p-chlorostyrene

SOLUTION

Both groups on the ring are ortho, para directors located para to each other. To introduce the double bond in the side chain, we must follow the two-step sequence in Sample Problem 18.7.

RETROSYNTHETIC ANALYSIS

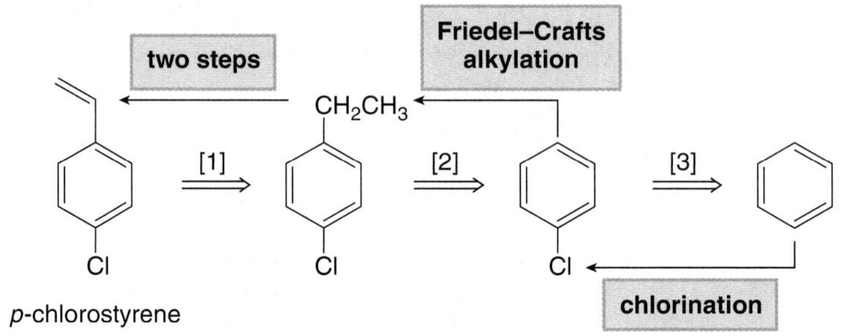

p-chlorostyrene

Working backwards:
- [1] Form the double bond by two steps: benzylic halogenation followed by elimination.
- [2] Introduce the CH_3CH_2 group by Friedel–Crafts alkylation.
- [3] Add the Cl atom by chlorination.

SYNTHESIS

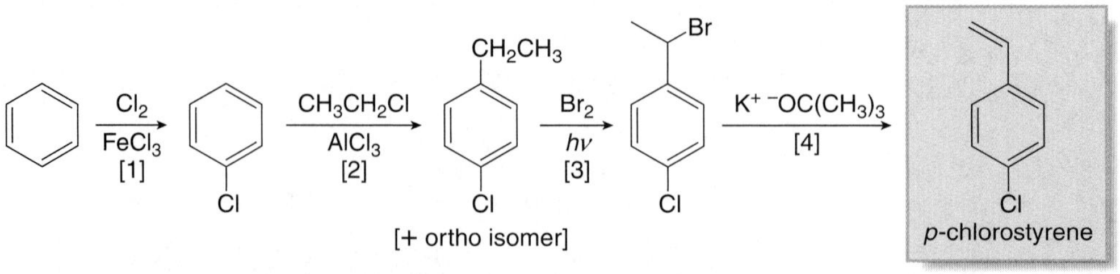

- Chlorination in Step [1] followed by Friedel–Crafts alkylation in Step [2] forms the desired para product, which can be separated from its ortho isomer.
- Benzylic bromination followed by elimination with strong base [$KOC(CH_3)_3$] (Steps [3] and [4]) forms the double bond of the target compound, *p*-chlorostyrene.

SAMPLE PROBLEM 18.11   Synthesize the trisubstituted benzene **A** from benzene.

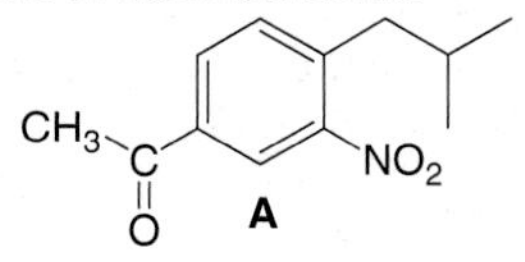

SOLUTION

Two groups (CH_3CO and NO_2) in **A** are meta directors located meta to each other, and the third substituent, an alkyl group, is an ortho, para director.

RETROSYNTHETIC ANALYSIS

With three groups on the benzene ring, begin by determining the possible disubstituted benzenes that are immediate precursors of the target compound, and then eliminate any that cannot be converted to the desired product. For example, three different disubstituted benzenes (**B–D**) can theoretically be precursors to **A.** However, conversion of compounds **B** or **D** to **A** would require a Friedel–Crafts reaction on a deactivated benzene ring, a reaction that does not occur. Thus, only **C** is a feasible precursor of **A.**

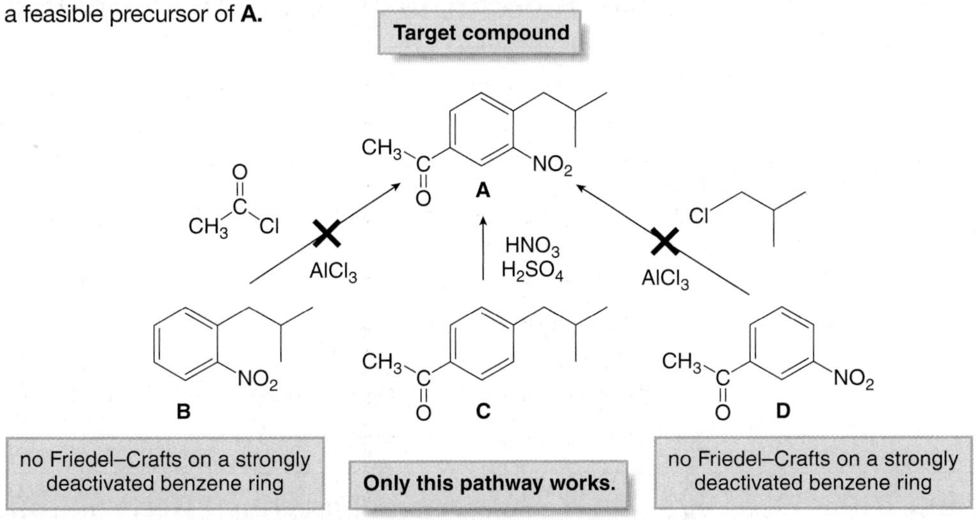

To complete the retrosynthetic analysis, prepare **C** from benzene:

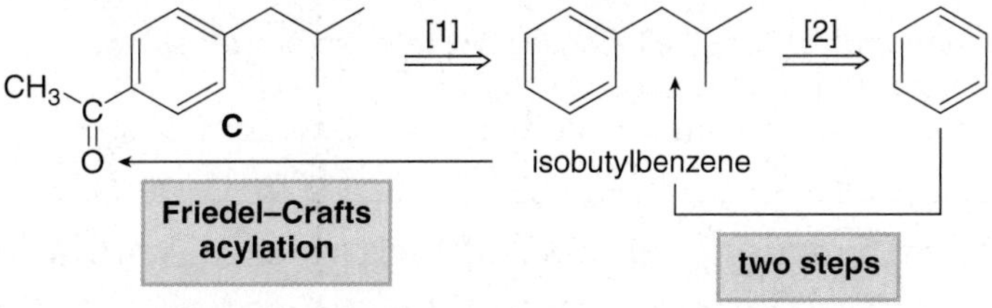

- [1] Add the ketone by Friedel–Crafts acylation.
- [2] Add the alkyl group by the two-step process—Friedel–Crafts acylation followed by reduction. It is not possible to prepare isobutylbenzene by a one-step Friedel–Crafts alkylation because of a rearrangement reaction (Problem 18.10).

SYNTHESIS

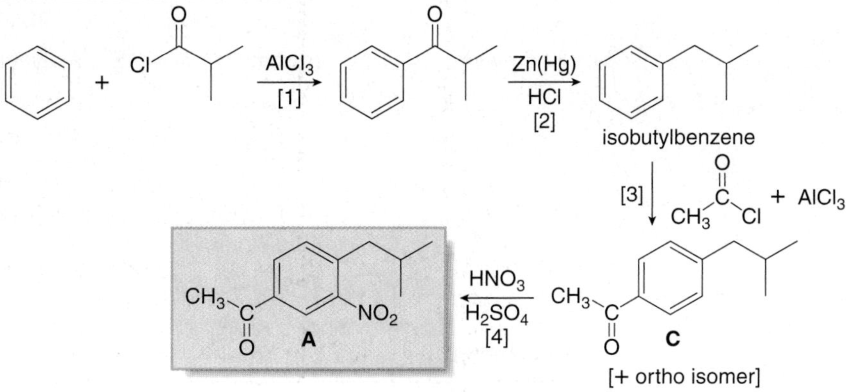

- Friedel–Crafts acylation followed by reduction with Zn(Hg), HCl yields isobutylbenzene (Steps [1]–[2]).
- Friedel–Crafts acylation gives the para product **C,** which can be separated from its ortho isomer (Step [3]).
- Nitration in Step [4] introduces the NO_2 group ortho to the alkyl group (an ortho, para director) and meta to the CH_3CO group (a meta director).

PROBLEM 18.33   Synthesize each compound from benzene.

a.

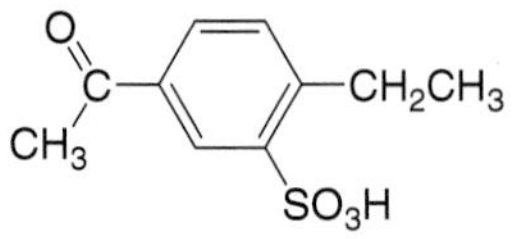

b.

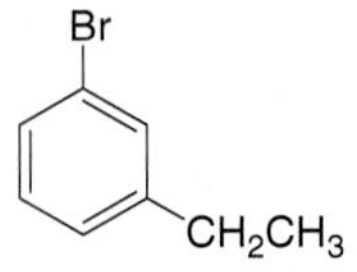

18.16   Key Concepts—Electrophilic Aromatic Substitution

Mechanism of Electrophilic Aromatic Substitution (18.2)

- Electrophilic aromatic substitution follows a two-step mechanism. Reaction of the aromatic ring with an electrophile forms a carbocation, and loss of a proton regenerates the aromatic ring.
- The first step is rate-determining.
- The intermediate carbocation is stabilized by resonance; a minimum of three resonance structures can be drawn. The positive charge is always located ortho or para to the new C–E bond.

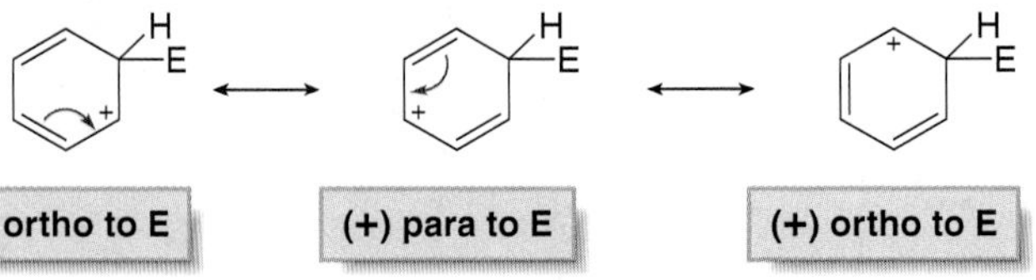

| (+) ortho to E | (+) para to E | (+) ortho to E |

Three Rules Describing the Reactivity and Directing Effects of Common Substituents (18.7–18.9)

[1] All ortho, para directors except the halogens activate the benzene ring.
[2] All meta directors deactivate the benzene ring.
[3] The halogens deactivate the benzene ring and direct ortho, para.

Summary of Substituent Effects in Electrophilic Aromatic Substitution (18.6–18.9)

Substituent	Inductive effect	Resonance effect	Reactivity	Directing effect
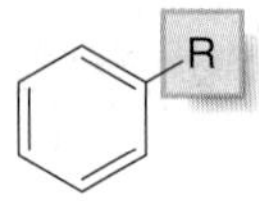 [1] R = alkyl	donating	none	activating	ortho, para
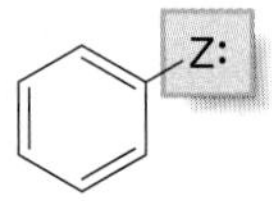 [2] Z = N or O	withdrawing	donating	activating	ortho, para
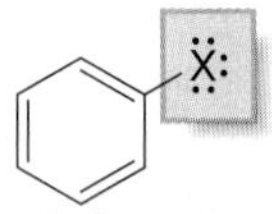 [3] X = halogen	withdrawing	donating	deactivating	ortho, para
[4] Y (δ^+ or +)	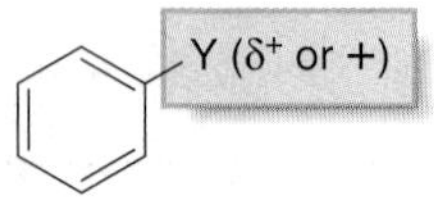 withdrawing	withdrawing	deactivating	meta

Five Examples of Electrophilic Aromatic Substitution

[1] Halogenation—Replacement of H by Cl or Br (18.3)

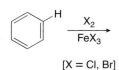

H
X_2
FeX_3
[X = Cl, Br]
Cl
aryl chloride
or
Br
aryl bromide

- Polyhalogenation occurs on benzene rings substituted by OH and NH_2 (and related substituents) (18.10A).

[2] Nitration—Replacement of H by NO_2 (18.4)

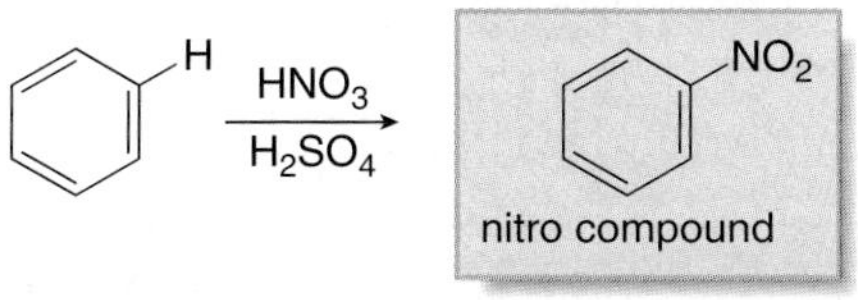

H
HNO_3
H_2SO_4
NO_2
nitro compound

[3] Sulfonation—Replacement of H by SO_3H (18.4)

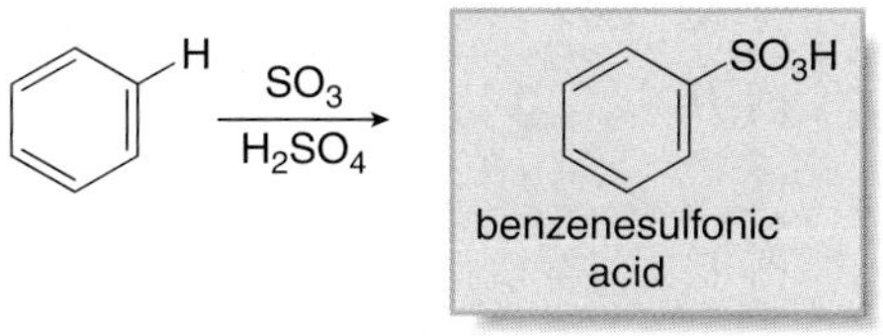

H
SO_3
H_2SO_4
SO_3H
benzenesulfonic acid

[4] Friedel–Crafts alkylation—Replacement of H by R (18.5)

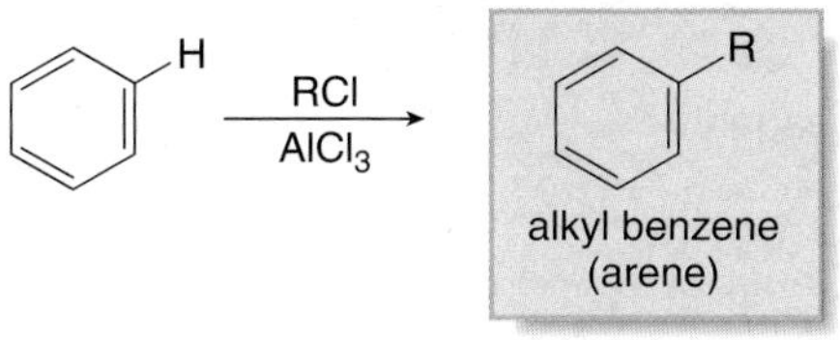

H
RCl
$AlCl_3$
R
alkyl benzene (arene)

- Rearrangements can occur.
- Vinyl halides and aryl halides are unreactive.
- The reaction does not occur on benzene rings substituted by meta deactivating groups or NH_2 groups (18.10B).
- Polyalkylation can occur.

Variations:

[1] with alcohols

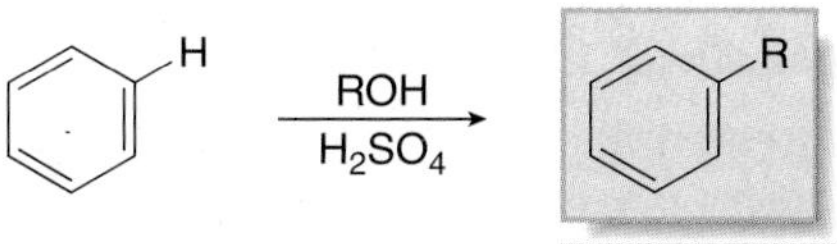

H
ROH
H_2SO_4
R

[2] with alkenes

H
$CH_2{=}CHR$
H_2SO_4
R
CH_3

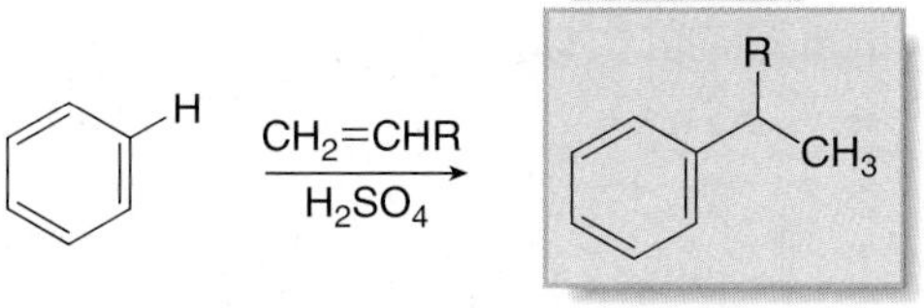

[5] Friedel–Crafts acylation—Replacement of H by RCO (18.5)

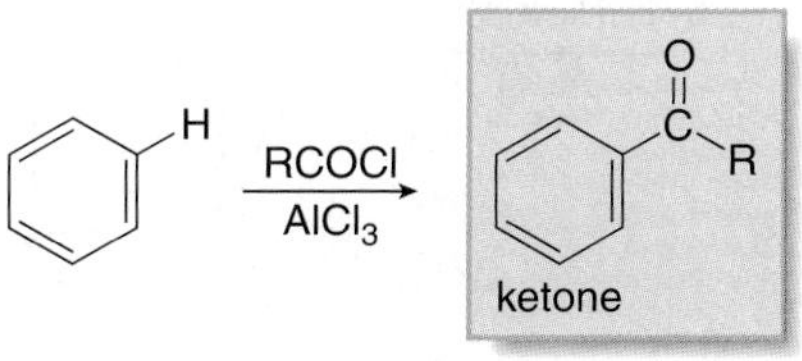

H
RCOCl
$AlCl_3$
O
C
R
ketone

- The reaction does not occur on benzene rings substituted by meta deactivating groups or NH_2 groups (18.10B).

Other Reactions of Benzene Derivatives

[1] Benzylic halogenation (18.13)

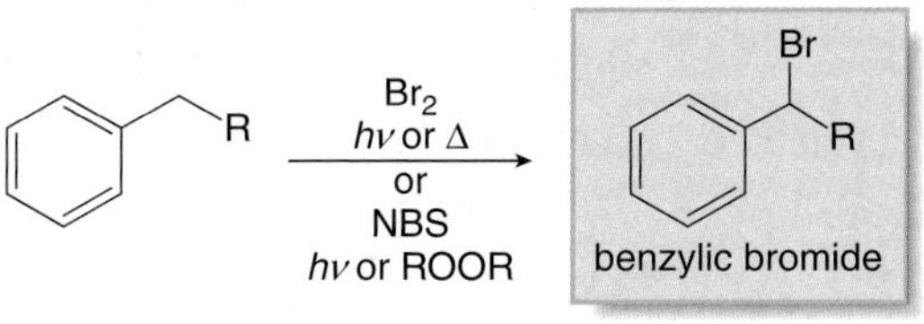

R
Br_2
$h\nu$ or Δ
or
NBS
$h\nu$ or ROOR
Br
R
benzylic bromide

[2] Oxidation of alkyl benzenes (18.14A)

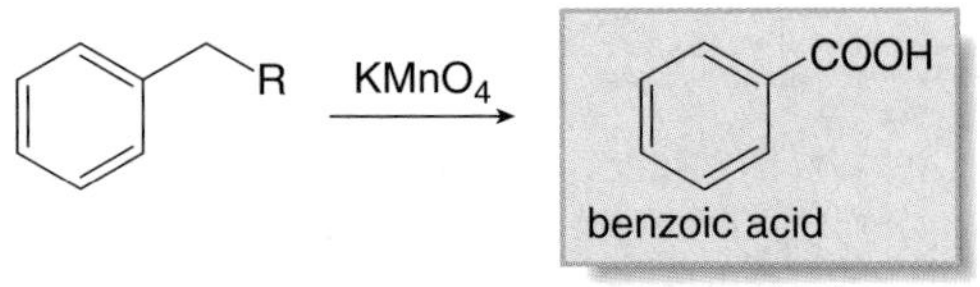

C₆H₅CH₂R + KMnO₄ → benzoic acid (C₆H₅COOH)

- A benzylic C–H bond is needed for reaction.

[3] Reduction of ketones to alkyl benzenes (18.14B)

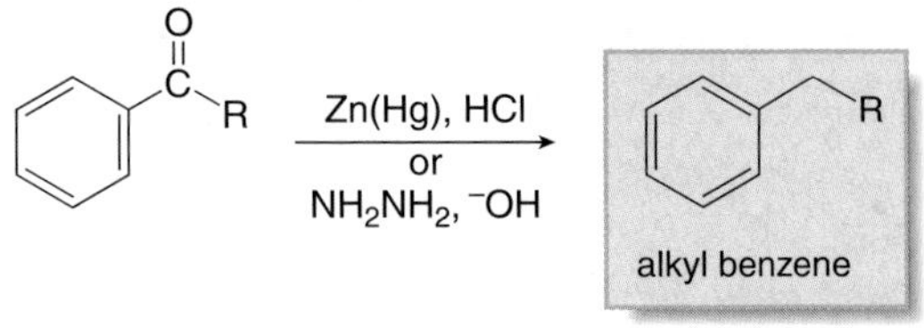

C₆H₅COR + Zn(Hg), HCl or NH₂NH₂, ⁻OH → alkyl benzene

[4] Reduction of nitro groups to amino groups (18.14C)

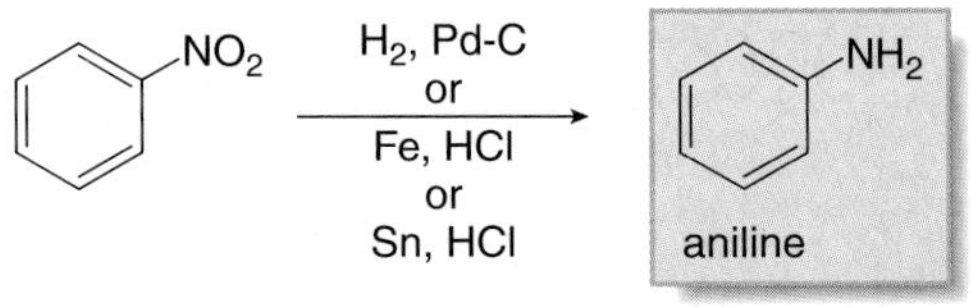

C₆H₅NO₂ + H₂, Pd-C or Fe, HCl or Sn, HCl → aniline (C₆H₅NH₂)

Problems

Reactions

18.34 Draw the products formed when phenol (C₆H₅OH) is treated with each reagent.
  a. HNO₃, H₂SO₄
  b. SO₃, H₂SO₄
  c. CH₃CH₂Cl, AlCl₃
  d. (CH₃CH₂)₂CHCOCl, AlCl₃
  e. Br₂, FeBr₃
  f. Br₂

  g. Cl₂, FeCl₃
  h. product in (a), then Sn, HCl
  i. product in (d), then Zn(Hg), HCl
  j. product in (d), then NH₂NH₂, ⁻OH
  k. product in (c), then Br₂, *hv*
  l. product in (c), then KMnO₄

18.35 Draw the products formed when benzonitrile (C₆H₅CN) is treated with each reagent.
  a. Br₂, FeBr₃   b. HNO₃, H₂SO₄   c. SO₃, H₂SO₄   d. CH₃CH₂CH₂Cl, AlCl₃   e. CH₃COCl, AlCl₃

18.36 Draw the products formed when each compound is treated with CH₃CH₂COCl, AlCl₃.

  a. 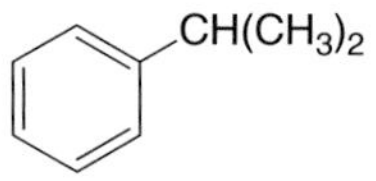 C₆H₅CH(CH₃)₂
  b. 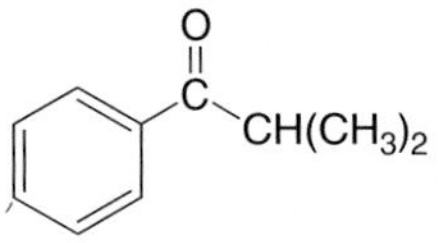 C₆H₅C(O)CH(CH₃)₂
  c. 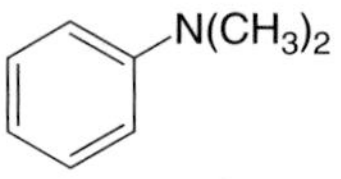 C₆H₅N(CH₃)₂
  d. 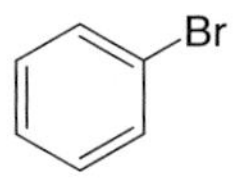 C₆H₅Br
  e. 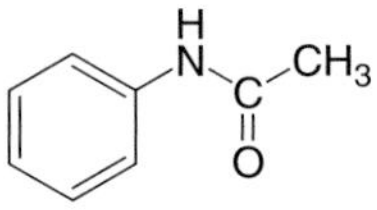 C₆H₅NHC(O)CH₃

18.37 Draw the products of each reaction.

  a. 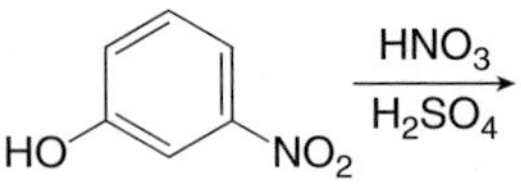 (3-hydroxy, nitro benzene) + HNO₃ / H₂SO₄ →

  b. 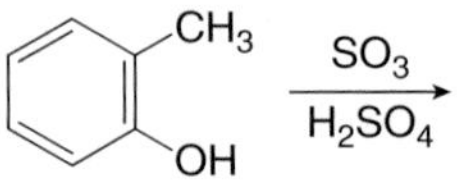 (2-methylphenol) + SO₃ / H₂SO₄ →

  c.  Cl-C₆H₄-OCOCH₃ + CH₃CH₂Cl / AlCl₃ →

  d. 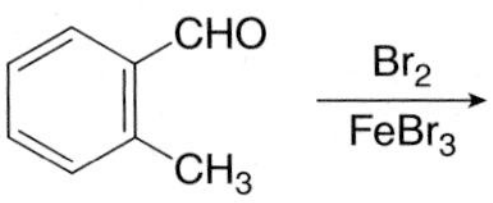 (2-methyl benzaldehyde, CHO) + Br₂ / FeBr₃ →

  e. 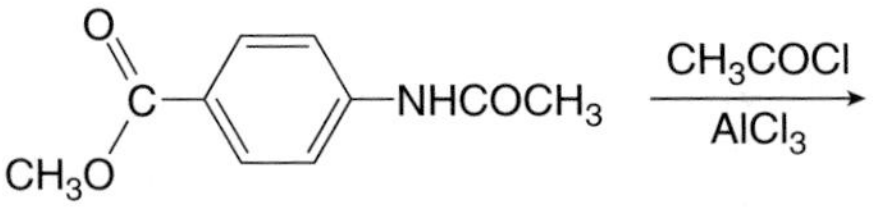 CH₃O-C(O)-C₆H₄-NHCOCH₃ + CH₃COCl / AlCl₃ →

  f. 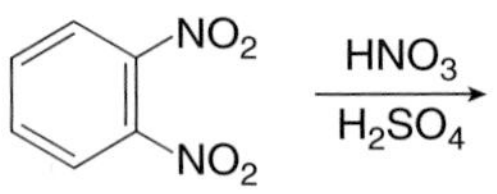 (1,2-dinitrobenzene) + HNO₃ / H₂SO₄ →

  g. 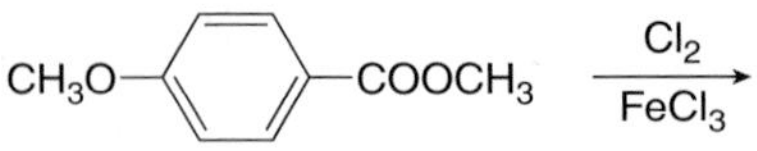 CH₃O-C₆H₄-COOCH₃ + Cl₂ / FeCl₃ →

  h. 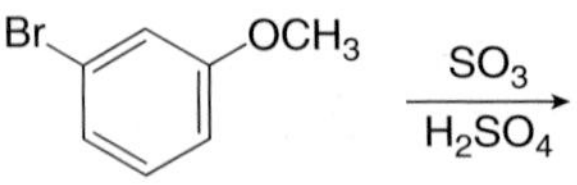 Br-C₆H₄-OCH₃ + SO₃ / H₂SO₄ →

18.38 What products are formed when benzene is treated with each alkyl chloride and AlCl₃?

  a. 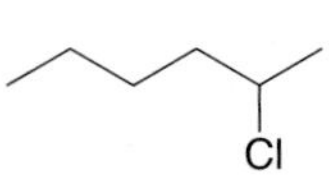 CH₃CH₂CH₂CH(Cl)CH₃
  b. 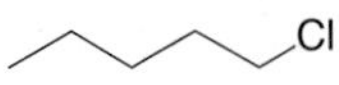 CH₃CH₂CH₂CH₂CH₂Cl
  c. 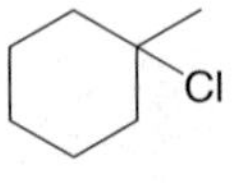 1-chloro-1-methylcyclohexane
  d. 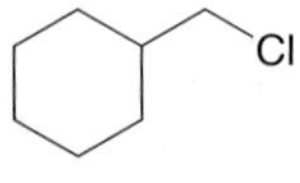 cyclohexylmethyl chloride

18.39 Write out two different routes to ketone **A** using a Friedel–Crafts acylation.

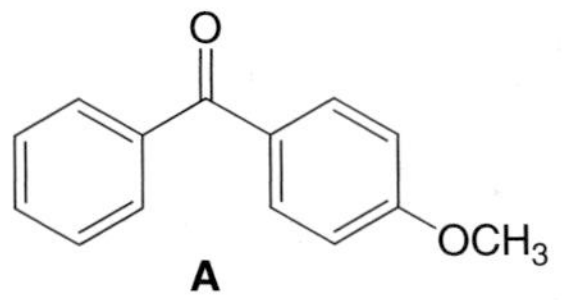

A

18.40 Draw the products of each reaction.

a. 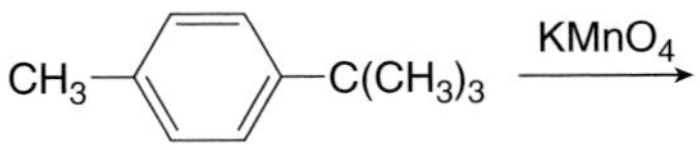 CH$_3$—[benzene]—C(CH$_3$)$_3$ $\xrightarrow{\text{KMnO}_4}$

d. 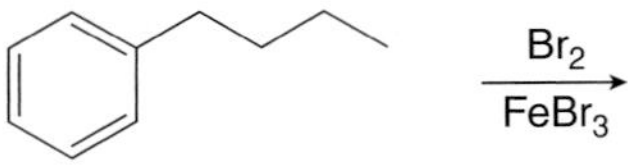 [benzene with butyl chain] $\xrightarrow{\text{Br}_2}{\text{FeBr}_3}$

b. 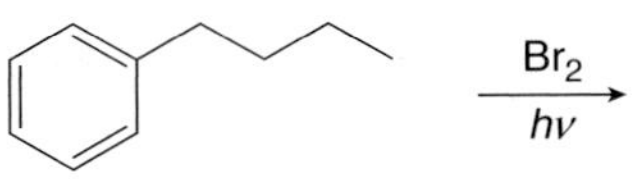 [benzene with butyl chain] $\xrightarrow{\text{Br}_2}{hv}$

e. 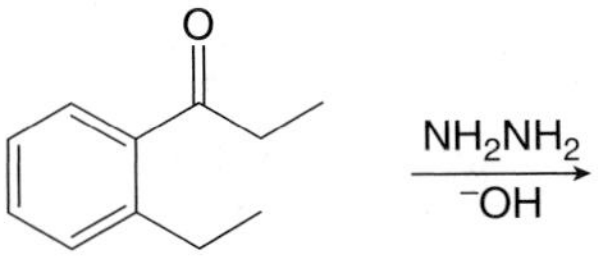 [ortho-ethyl benzene with propanoyl group] $\xrightarrow{\text{NH}_2\text{NH}_2}{^-\text{OH}}$

c. CH$_3$—[benzene]—C(=O)CH$_2$CH$_2$CH$_3$ $\xrightarrow{\text{Zn(Hg), HCl}}$

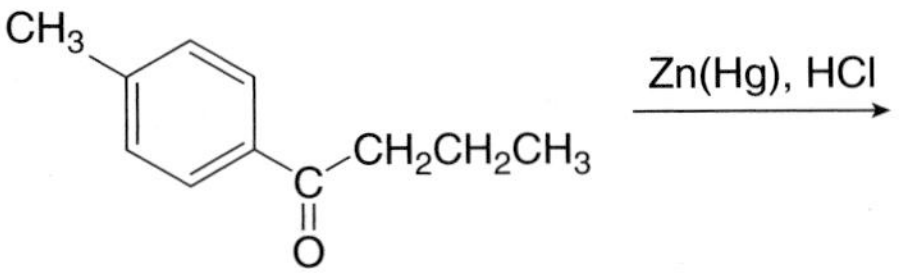

f. 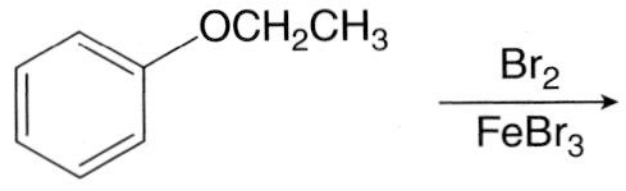 [benzene]—OCH$_2$CH$_3$ $\xrightarrow{\text{Br}_2}{\text{FeBr}_3}$

18.41 You have learned two ways to make an alkyl benzene: Friedel–Crafts alkylation, and Friedel–Crafts acylation followed by reduction. Although some alkyl benzenes can be prepared by both methods, it is often true that only one method can be used to prepare a given alkyl benzene. Which method(s) can be used to prepare each of the following compounds from benzene? Show the steps that would be used.

a. 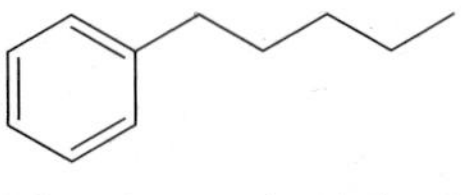 [benzene with pentyl chain]

b. 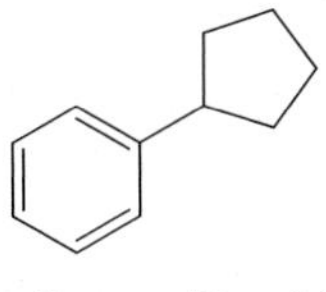 [benzene with cyclopentyl]

c. 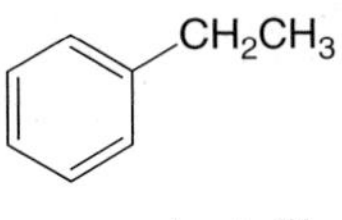 [benzene]—CH$_2$CH$_3$

18.42 Explain why each of the following reactions will not form the given product. Then, design a synthesis of **A** from benzene and **B** from phenol (C$_6$H$_5$OH).

a. 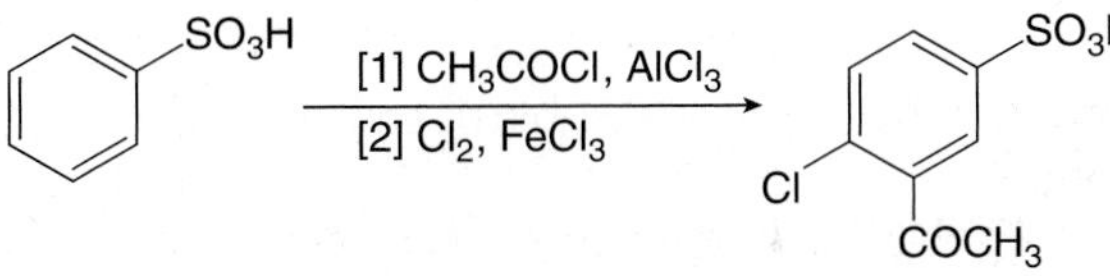 [benzene]—SO$_3$H $\xrightarrow{\text{[1] CH}_3\text{COCl, AlCl}_3}{\text{[2] Cl}_2\text{, FeCl}_3}$ [benzene with SO$_3$H, Cl, COCH$_3$] = **A**

b. 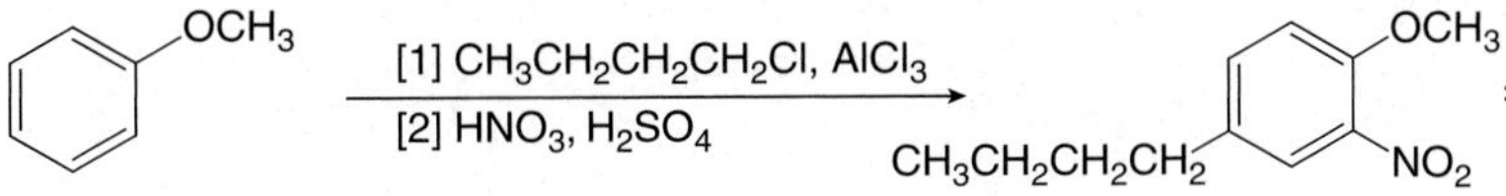 [benzene]—OCH$_3$ $\xrightarrow{\text{[1] CH}_3\text{CH}_2\text{CH}_2\text{CH}_2\text{Cl, AlCl}_3}{\text{[2] HNO}_3\text{, H}_2\text{SO}_4}$ CH$_3$CH$_2$CH$_2$CH$_2$—[benzene with OCH$_3$, NO$_2$] = **B**

Substituent Effects

18.43 Rank the compounds in each group in order of increasing reactivity in electrophilic aromatic substitution.
   a. C$_6$H$_5$NO$_2$, C$_6$H$_6$, C$_6$H$_5$OH
   b. C$_6$H$_6$, C$_6$H$_5$Cl, C$_6$H$_5$CHO
   c. C$_6$H$_6$, C$_6$H$_5$NO$_2$, C$_6$H$_5$NH$_2$
   d. C$_6$H$_6$, C$_6$H$_5$CH$_2$Cl, C$_6$H$_5$CHCl$_2$
   e. C$_6$H$_5$CH$_3$, C$_6$H$_5$NH$_2$, C$_6$H$_5$CH$_2$NH$_2$

18.44 Draw all resonance structures for each compound, and explain why a particular substituent has an electron-donating or electron-withdrawing resonance effect: (a) C$_6$H$_5$NO$_2$; (b) C$_6$H$_5$F.

18.45 For each of the following substituted benzenes: [1] C$_6$H$_5$Br; [2] C$_6$H$_5$CN; [3] C$_6$H$_5$OCOCH$_3$:
   a. Does the substituent donate or withdraw electron density by an inductive effect?
   b. Does the substituent donate or withdraw electron density by a resonance effect?
   c. On balance, does the substituent make a benzene ring more or less electron rich than benzene itself?
   d. Does the substituent activate or deactivate the benzene ring in electrophilic aromatic substitution?

18.46 Which benzene ring in each compound is more reactive in electrophilic aromatic substitution?

a. 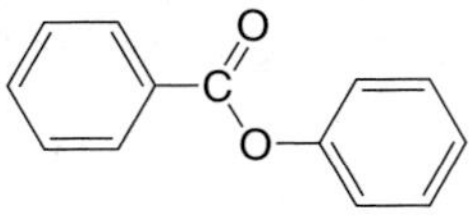 [benzene–C(=O)–O–benzene]

b. 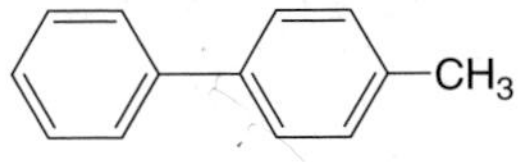 [biphenyl with CH$_3$]

18.47 Explain in detail using resonance structures:
   a. Why is an OH group an ortho, para director that activates a benzene ring?
   b. Why is a COOCH$_3$ group a meta director that deactivates a benzene ring?

18.48 Explain in detail, using resonance structures, why a phenyl group (C$_6$H$_5$–) is an ortho, para director. Would you expect a phenyl group to activate or deactivate a benzene ring in electrophilic aromatic substitution?

18.49 Which one of the xylene isomers reacts faster in electrophilic aromatic substitution and why?

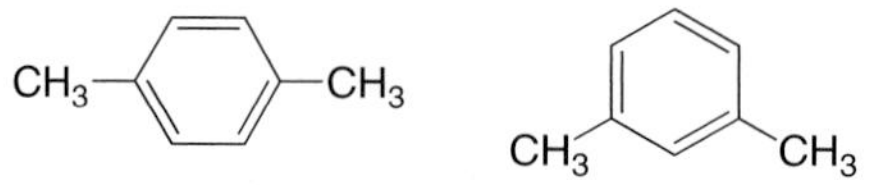

p-xylene   m-xylene

18.50 Explain why the meta product is formed in the following reaction despite the fact that – N(CH$_3$)$_2$ is usually an ortho, para director.

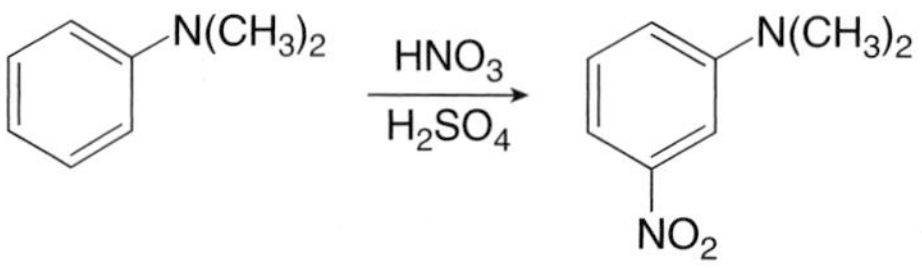

Mechanisms

18.51 Draw a stepwise mechanism for each reaction.

a.

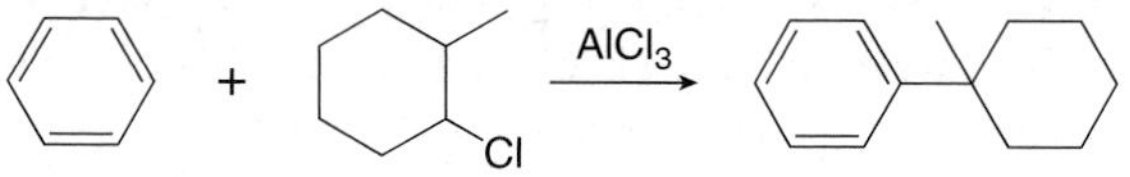

b.

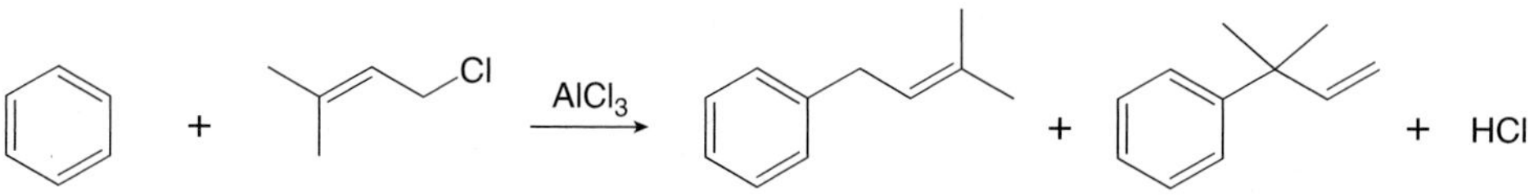

c.

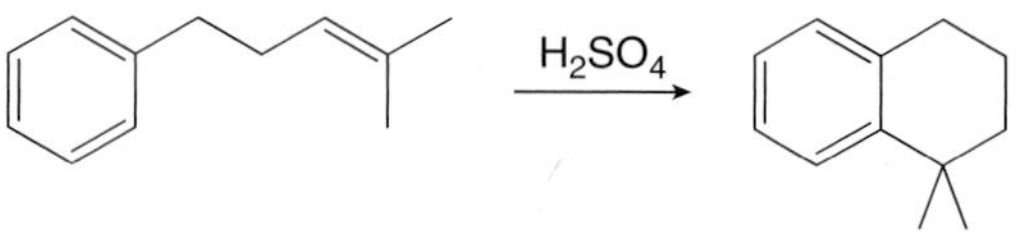

18.52 Friedel–Crafts alkylation of benzene with (R)-2-chlorobutane and AlCl$_3$ affords sec-butylbenzene.
   a. How many stereogenic centers are present in the product?
   b. Would you expect the product to exhibit optical activity? Explain, with reference to the mechanism.

18.53 Although two products (**A** and **B**) are possible when naphthalene undergoes electrophilic aromatic substitution, only **A** is formed. Draw resonance structures for the intermediate carbocation to explain why this is observed.

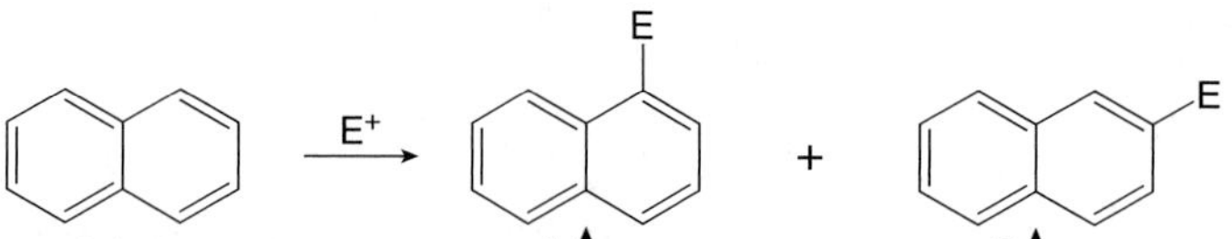

naphthalene   **A** ↑   **B** ↑
          This product is formed.   This product is not formed.

18.54 Draw a stepwise mechanism for the following reaction, which is used to prepare the pesticide DDT.

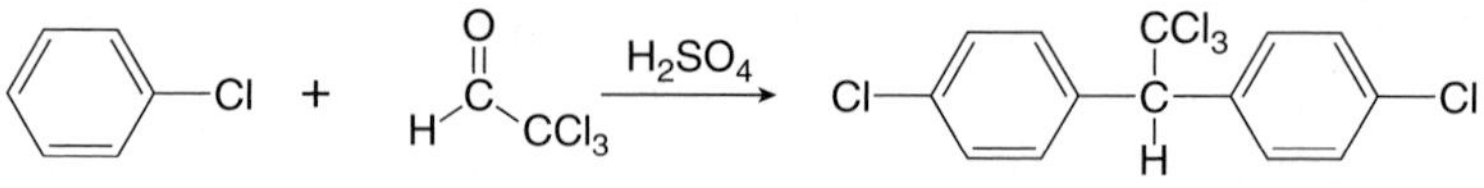

18.55 Benzyl bromide (C$_6$H$_5$CH$_2$Br) reacts rapidly with CH$_3$OH to afford benzyl methyl ether (C$_6$H$_5$CH$_2$OCH$_3$). Draw a stepwise mechanism for the reaction, and explain why this 1° alkyl halide reacts rapidly with a weak nucleophile under conditions that favor an S$_N$1 mechanism.

18.56 Explain why HBr addition to C$_6$H$_5$CH=CHCH$_3$ forms only one alkyl halide, C$_6$H$_5$CH(Br)CH$_2$CH$_3$.

Synthesis

18.57 Synthesize each compound from benzene and any other organic or inorganic reagents.

a. isopropylbenzene
b. butylbenzene
c. *o*-butylchlorobenzene
d. *m*-bromonitrobenzene
e. *o*-bromonitrobenzene

f. (structure: benzene ring with COOH and SO$_3$H in ortho positions)

g. (structure: benzene ring with COOH and NO$_2$ in meta positions)

h. Br—(benzene ring)—COOH (para)

i. (structure: benzene ring with NO$_2$, CH$_2$CH$_2$CH$_3$, and Br substituents)

18.58 How would you convert benzene into each compound: (a) C$_6$H$_5$C(CH$_3$)=CH$_2$; (b) C$_6$H$_5$C≡CH?

18.59 Synthesize each compound from benzene and any other organic or inorganic reagents.

a. (benzene ring with NH$_2$ and Br para)

b. (benzene ring with COOH, two Br in positions 2,6 and Br at position 4)

c. (benzene ring with CH$_3$, Br, and NO$_2$)

d. (benzene ring with NO$_2$, isopropyl, and C(=O)CH$_3$)

e. HOOC—(benzene ring)—NH$_2$ (PABA) sunscreen component

18.60 Synthesize each compound from toluene (C$_6$H$_5$CH$_3$) and any other organic or inorganic reagents.

a. C$_6$H$_5$CH$_2$Br
b. C$_6$H$_5$CH$_2$OH
c. C$_6$H$_5$CH$_2$OC(CH$_3$)$_3$
d. C$_6$H$_5$CHO

e. (benzene ring with CHO and Br meta)

f. (benzene ring with COOH and Cl ortho)

g. Cl—(benzene ring)—COOH (para)

h. CH$_3$—(benzene ring)—(CH$_2$)$_5$CH$_3$ (para)

i. CH$_3$C(=O)—(benzene ring)—COOH (para)

j. (benzene ring with COOH, HO$_3$S, and NO$_2$)

k. HOOC—(benzene ring)—NO$_2$ with O$_2$N substituent

18.61 Carboxylic acid **X** is an intermediate in the multistep synthesis of proparacaine, a local anesthetic. Devise a synthesis of **X** from phenol and any needed organic or inorganic reagents.

(structure of **X**: benzene ring with O$_2$N, propoxy O group, and COOH) → several steps → (structure of proparacaine: benzene ring with H$_2$N, propoxy O group, and ester C(=O)O-CH$_2$CH$_2$-N(CH$_2$CH$_3$)$_2$)

X   proparacaine

Spectroscopy

18.62 Compound **X** (molecular formula $C_{10}H_{12}O$) was treated with NH_2NH_2, ^-OH to yield compound **Y** (molecular formula $C_{10}H_{14}$). Based on the ^1H NMR spectra of **X** and **Y** given below, what are the structures of **X** and **Y**?

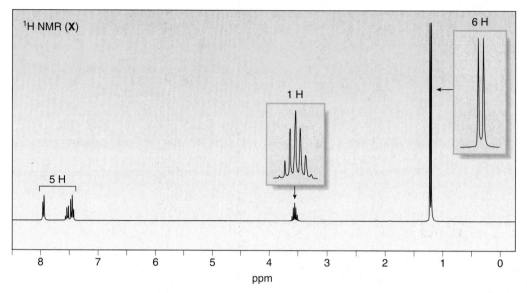

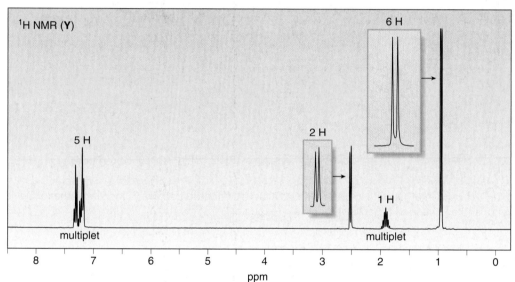

18.63 Reaction of *p*-cresol with two equivalents of 2-methyl-1-propene affords BHT, a preservative with molecular formula $C_{15}H_{24}O$. BHT gives the following ^1H NMR spectral data: 1.4 (singlet, 18 H), 2.27 (singlet, 3 H), 5.0 (singlet, 1 H), and 7.0 (singlet, 2 H) ppm. What is the structure of BHT? Draw a stepwise mechanism illustrating how it is formed.

$$CH_3\text{-}\bigcirc\text{-}OH \quad + \quad \underset{CH_3}{\overset{CH_3}{C}}=CH_2 \quad \xrightarrow{H_2SO_4} \quad BHT\ (C_{15}H_{24}O)$$

p-cresol   2-methyl-1-propene
(2 equiv)

18.64 Compound **Z** (molecular formula C_9H_9ClO) can be converted to the antidepressant bupropion (Figure 18.3) by a series of reactions. **Z** shows a strong peak in its IR spectrum at 1683 cm^{-1}. The ^1H NMR spectrum of **Z** shows peaks at 1.2 (triplet, 3 H), 2.9 (quartet, 2 H), and 7.2–8.0 (multiplet, 4 H) ppm. Propose a structure for **Z**.

$$Z \quad \xrightarrow[\text{steps}]{\text{several}} \quad \text{bupropion}$$

bupropion

Challenge Problems

18.65 The ^{1}H NMR spectrum of phenol (C_6H_5OH) shows three absorptions in the aromatic region: 6.70 (2 ortho H's), 7.14 (2 meta H's), and 6.80 (1 para H) ppm. Explain why the ortho and para absorptions occur at lower chemical shift than the meta absorption.

18.66 Heterocycles also undergo electrophilic aromatic substitution. Explain why furan undergoes substitution with electrophiles mainly at the 2-position.

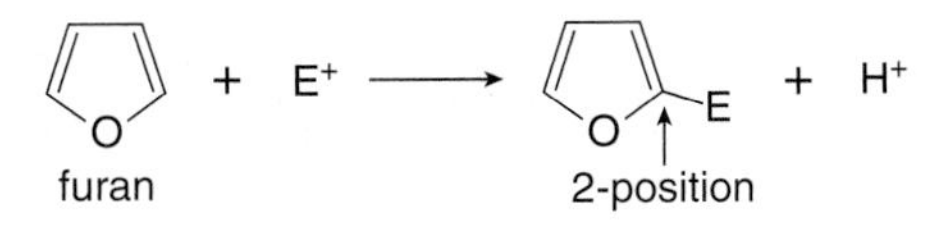

18.67 Draw a stepwise mechanism for the following intramolecular reaction, which is used in the synthesis of the female sex hormone estrone.

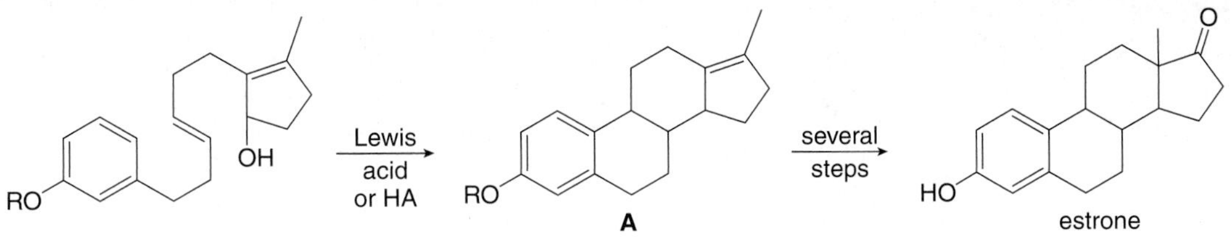

CHAPTER 19

Carboxylic Acids and the Acidity of the O–H Bond

PGF$_{2\alpha}$ is one of a group of biologically active fatty acids called prostaglandins. Prostaglandins are involved in the regulation of a variety of physiological phenomena, including inflammation, blood clotting, and the induction of labor. Aspirin and many other nonsteroidal anti-inflammatory drugs (NSAIDs) act by blocking the biosynthesis of prostaglandins in the cell. In Chapter 19 we learn about the properties of prostaglandins and other carboxylic acids.

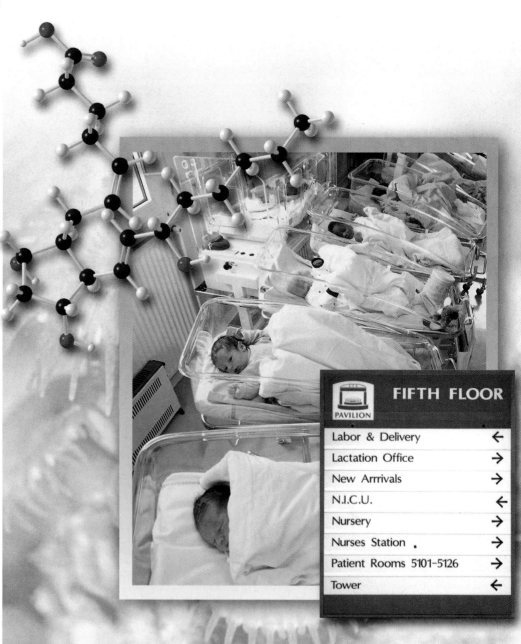

FIFTH FLOOR

PAVILION		
Labor & Delivery	←
Lactation Office	→
New Arrrivals	→
N.I.C.U.	←
Nursery	→
Nurses Station .	→
Patient Rooms 5101–5126	→
Tower	←

hapter 19 serves as a transition between the preceding discussion of resonance and aromaticity, and the subsequent treatment of carbonyl chemistry. We pause to study the chemistry of the OH group by examining **carboxylic acids (RCOOH),** and to a lesser extent, **phenols (PhOH)** and **alcohols (ROH).**

In Chapter 19 we concentrate on the acidity of carboxylic acids, and revisit some of the factors that determine acidity, a topic first discussed in Chapter 2. Then, in Chapters 20 and 22 we will learn other reactions of carboxylic acids that occur at the carbonyl group.

19.1   Structure and Bonding

Carboxylic acids **are organic compounds containing a carboxy group (COOH).** Although the structure of a carboxylic acid is often abbreviated as **RCOOH** or **RCO$_2$H,** keep in mind that the central carbon atom of the functional group is doubly bonded to one oxygen atom and singly bonded to another.

The word **carboxy** (for a COOH group) is derived from **carbo**nyl (C=O) + hydro**xy** (OH).

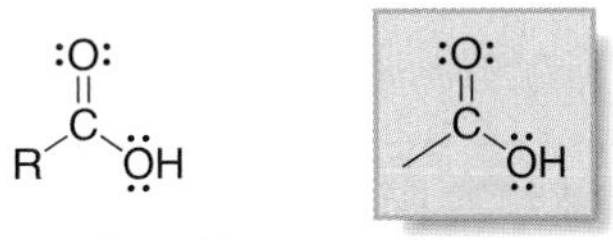

carboxylic acid   carboxy group

The carbon atom of a carboxy group is surrounded by three groups, making it *sp^2* **hybridized** and **trigonal planar,** with bond angles of approximately 120°. The C=O of a carboxylic acid is shorter than its C—O.

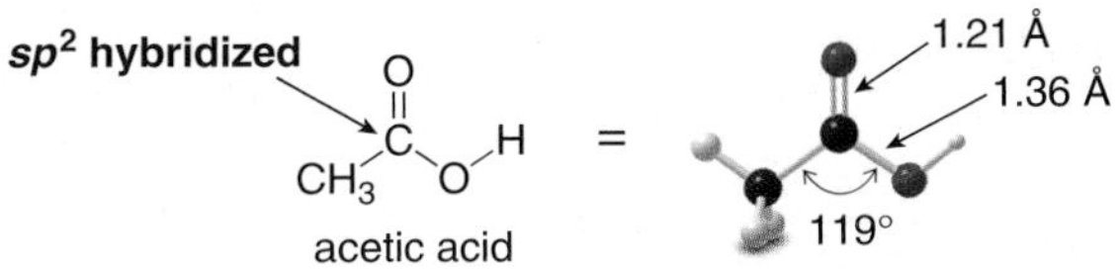

sp^2 hybridized

acetic acid

1.21 Å
1.36 Å

The C=O is shorter than the C—O.

119°

The C—O single bond of a carboxylic acid is shorter than the C—O single bond of an alcohol. This can be explained by looking at the hybridization of the respective carbon atoms. In the alcohol, the carbon is *sp^3* hybridized, whereas in the carboxylic acid the carbon is *sp^2* hybridized. As a result, the higher percentage *s*-character in the *sp^2* hybrid orbital shortens the C—O bond in the carboxylic acid.

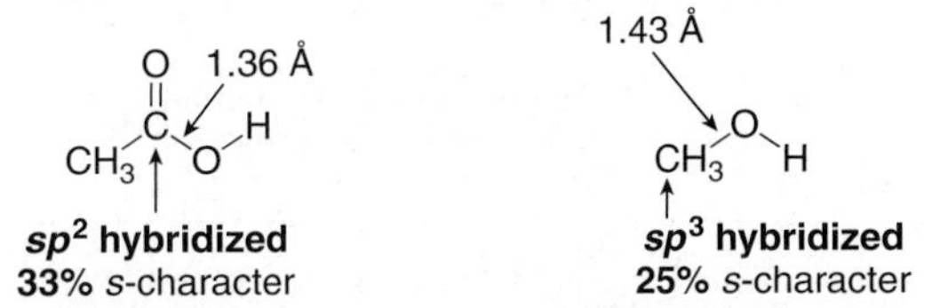

sp^2 **hybridized**
33% *s*-character

sp^3 **hybridized**
25% *s*-character

higher percentage *s*-character shorter bond

lower percentage *s*-character longer bond

Because oxygen is more electronegative than either carbon or hydrogen, **the C—O and O—H bonds are polar.** The electrostatic potential plot of acetic acid in Figure 19.1 shows that the carbon and hydrogen atoms are electron poor and the oxygen atoms are electron rich.

Figure 19.1   Electrostatic potential plot of acetic acid (CH$_3$COOH)

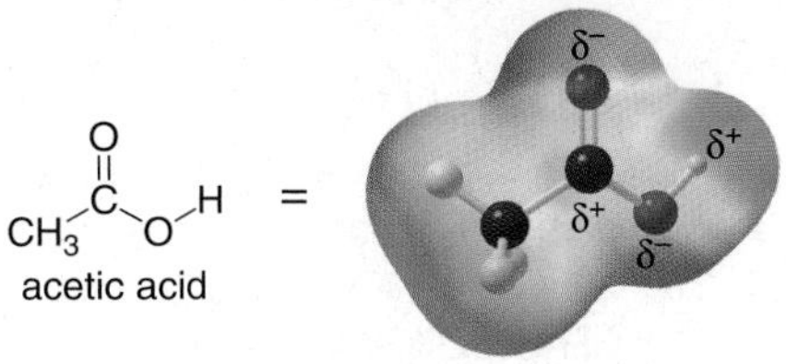

acetic acid

Acetic acid contains two electron-rich oxygen atoms (in red). Its carbonyl carbon and hydroxy hydrogen are both electron deficient.

19.2 Nomenclature

Both IUPAC and common names are used for carboxylic acids.

19.2A IUPAC System

In IUPAC nomenclature, carboxylic acids are identified by a suffix added to the parent name of the longest chain, and two different endings are used depending on whether the carboxy group is bonded to a chain or ring.

To name a carboxylic acid using the IUPAC system:

[1] If the COOH is bonded to a chain of carbons, find the longest chain containing the COOH group, and change the **-e** ending of the parent alkane to the suffix **-oic acid.** If the COOH group is bonded to a ring, name the ring and add the words **carboxylic acid.**

[2] Number the carbon chain or ring to put the **COOH group at C1,** but omit this number from the name. Apply all of the other usual rules of nomenclature.

SAMPLE PROBLEM 19.1 Give the IUPAC name of each compound.

[a]
$$CH_3-\underset{\underset{CH_3}{|}}{\overset{\overset{CH_3}{|}}{CH}}-CH-CH_2CH_2COOH$$

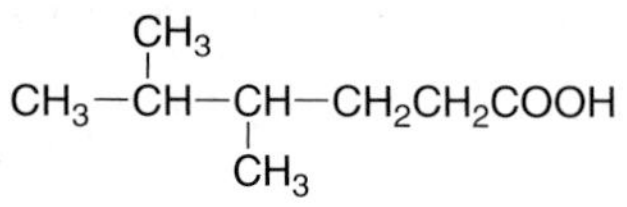

[b]

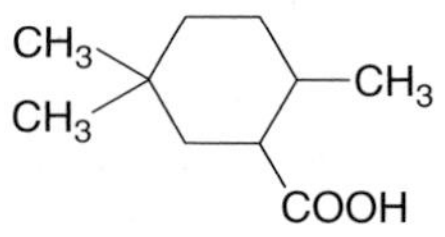

SOLUTION

[a] [1] Find and name the longest chain containing COOH:

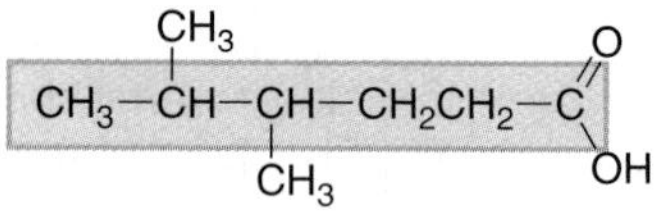

hexane ⟶ hexanoic acid
(6 C's)

> The COOH contributes one C to the longest chain.

[2] Number and name the substituents:

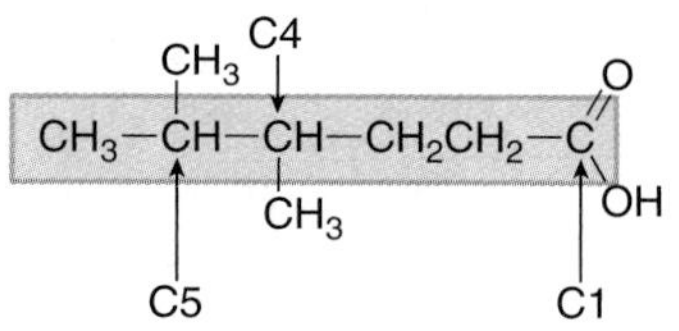

two methyl substituents on C4 and C5

Answer: 4,5-dimethylhexanoic acid

---

[b] [1] Find and name the ring bonded to COOH.

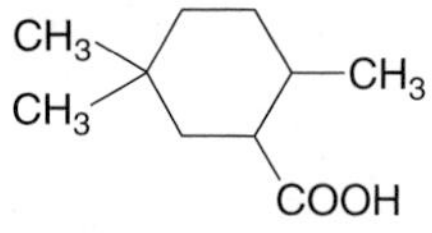

cyclohexane + carboxylic acid
(6 C's)

[2] Number and name the substituents:

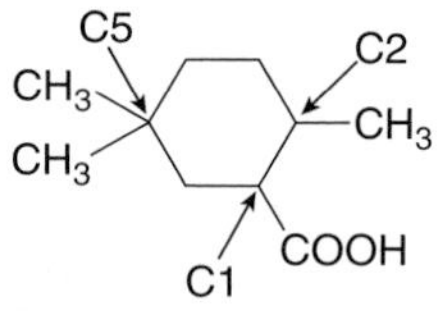

> Number to put COOH at C1 and give the second substituent (CH₃) the lower number (C2).

Answer: 2,5,5-trimethylcyclohexanecarboxylic acid

PROBLEM 19.1 Give the IUPAC name for each compound.

a. $CH_3CH_2CH_2C(CH_3)_2CH_2COOH$

b. $CH_3CH(Cl)CH_2CH_2COOH$

c. $(CH_3CH_2)_2CHCH_2CH(CH_2CH_3)COOH$

d.

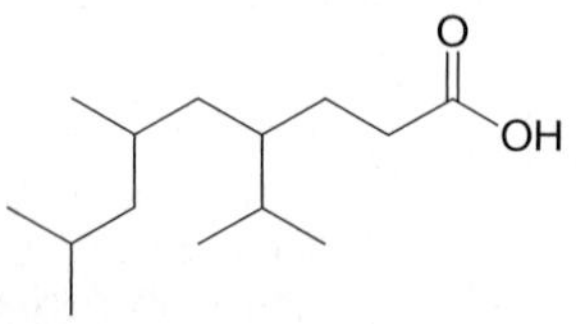

PROBLEM 19.2   Give the structure corresponding to each IUPAC name.

a. 2-bromobutanoic acid
b. 2,3-dimethylpentanoic acid
c. 3,3,4-trimethylheptanoic acid
d. 2-*sec*-butyl-4,4-diethylnonanoic acid
e. 3,4-diethylcyclohexanecarboxylic acid
f. 1-isopropylcyclobutanecarboxylic acid

19.2B   Common Names

Most simple carboxylic acids have common names that are more widely used than their IUPAC names.

◆ A common name is formed by using a common parent name followed by the suffix *-ic acid.*

Table 19.1 lists common parent names for some simple carboxylic acids. These parent names are used in the nomenclature of many other compounds with carbonyl groups (Chapters 21 and 22).

TABLE 19.1   Common Names for Some Simple Carboxylic Acids

Number of C atoms	Structure	Parent name	Common name
1	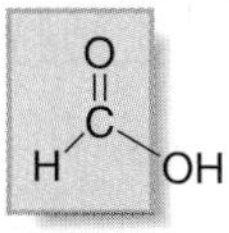	**form-**	formic acid
2	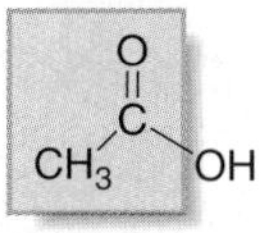	**acet-**	acetic acid
3	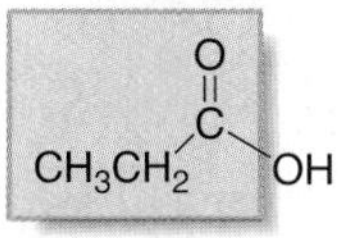	**propion-**	propionic acid
4	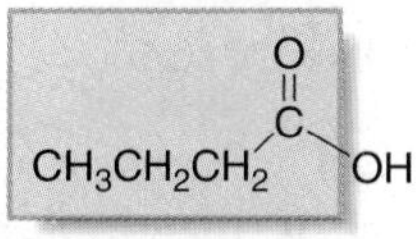	**butyr-**	butyric acid
5	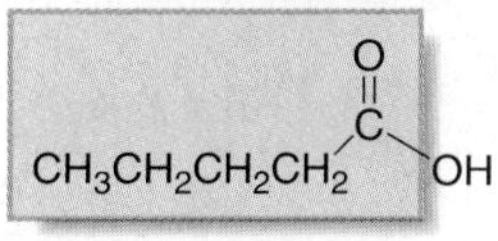	**valer-**	valeric acid
6	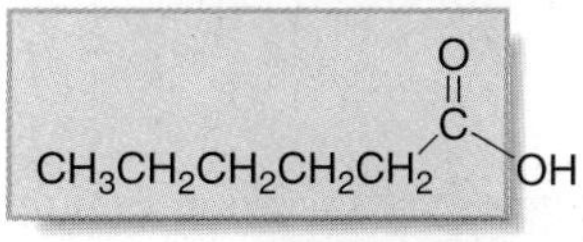	**capro-**	caproic acid
	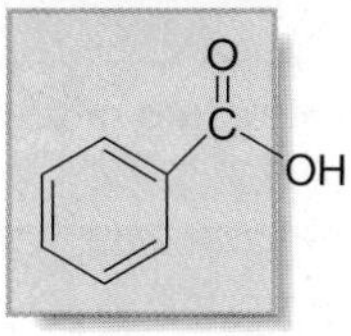	**benzo-**	benzoic acid

Greek letters are used to designate the location of substituents in common names.

◆ The carbon adjacent to the COOH is called the α carbon.
◆ The carbon bonded to the α carbon is the β carbon, followed by the γ (gamma) carbon, the δ (delta) carbon, and so forth down the chain. The last carbon in the chain is sometimes called the Ω (omega) carbon.

The α carbon in the common system is numbered C2 in the IUPAC system.

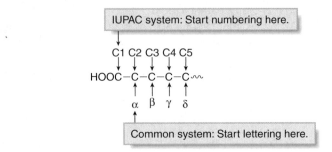

PROBLEM 19.3   Draw the structure corresponding to each common name:
a. α-methoxyvaleric acid   c. α,β-dimethylcaproic acid
b. β-phenylpropionic acid   d. α-chloro-β-methylbutyric acid

PROBLEM 19.4   Give an IUPAC and common name for each carboxylic acid.

a. [structure: chain with Cl and Br substituents, COOH]   b. [structure: benzene ring with ethyl, COOH, and I]

19.2C   Other Nomenclature Facts

Many compounds containing two carboxy groups are also known. In the IUPAC system, **diacids** are named using the suffix *-dioic acid.* The three simplest diacids are most often identified by their common names, as shown.

[structures of diacids]

oxalic acid   malonic acid   succinic acid

Metal salts of carboxylate anions are formed from carboxylic acids in many reactions in Chapter 19. To name the **metal salt of a carboxylate anion,** put three parts together:

name of the metal cation   +   parent   +   suffix

common
or   ⟶   *-ate*   (for a common parent name)
or
IUPAC   ⟶   *-oate*   (for an IUPAC parent name)

Two examples are shown in Figure 19.2.

PROBLEM 19.5   Draw the structure of sodium benzoate, a common preservative. (You can check your answer in Section 19.5.)

Figure 19.2   Naming the metal salts of carboxylate anions

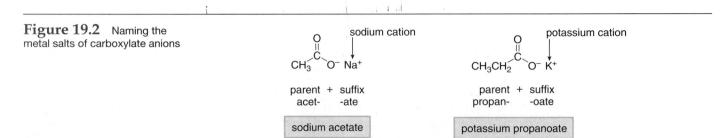

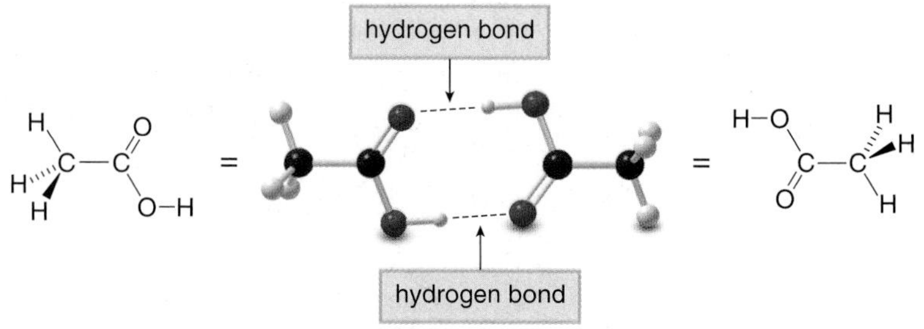

Figure 19.3 Two molecules of acetic acid (CH_3COOH) held together by two hydrogen bonds

19.3 Physical Properties

Carboxylic acids exhibit **dipole–dipole** interactions because they have polar C–O and O–H bonds. They also exhibit intermolecular **hydrogen bonding** because they possess a hydrogen atom bonded to an electronegative oxygen atom. Carboxylic acids often exist as **dimers,** held together by *two* intermolecular hydrogen bonds. The carbonyl oxygen atom of one molecule hydrogen bonds to the hydrogen atom of another molecule (Figure 19.3), and thus they are the **most polar** organic compounds we have seen so far.

How these intermolecular forces affect the physical properties of carboxylic acids is summarized in Table 19.2.

TABLE 19.2 Physical Properties of Carboxylic Acids

Property	Observation
Boiling point and melting point	• Carboxylic acids have higher boiling points and melting points than other compounds of comparable molecular weight. $CH_3CH_2CH_2CH_3$ VDW, MW = 58, bp 0 °C CH_3CH_2CHO VDW, DD, MW = 58, bp 48 °C $CH_3CH_2CH_2OH$ VDW, DD, HB, MW = 60, bp 97 °C CH_3COOH VDW, DD, two HB, MW = 60, bp 118 °C **Increasing strength of intermolecular forces** **Increasing boiling point**
Solubility	• Carboxylic acids are soluble in organic solvents regardless of size. • Carboxylic acids having ≤ 5 C's are water soluble because they can hydrogen bond with H_2O (Section 3.5C). • Carboxylic acids having > 5 C's are water insoluble because the nonpolar alkyl portion is too large to dissolve in the polar H_2O solvent. These "fatty" acids dissolve in a nonpolar fat-like environment but do not dissolve in water.

Key: VDW = van der Waals, DD = dipole–dipole, HB = hydrogen bonding, MW = molecular weight

PROBLEM 19.6 Rank the following compounds in order of increasing boiling point. Which compound is the most water soluble? Which compound is the least water soluble?

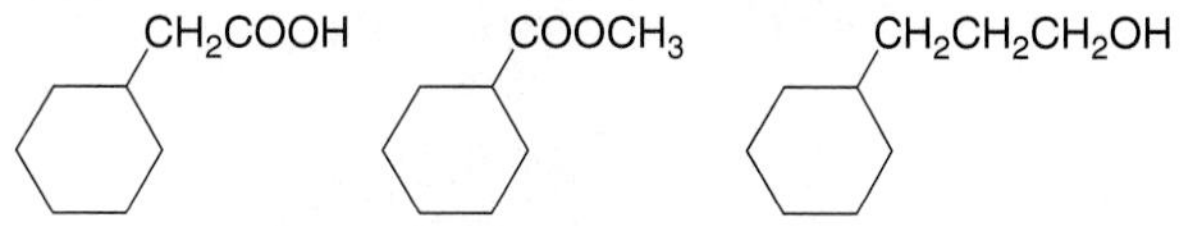

Figure 19.4   The IR spectrum of butanoic acid, $CH_3CH_2CH_2COOH$

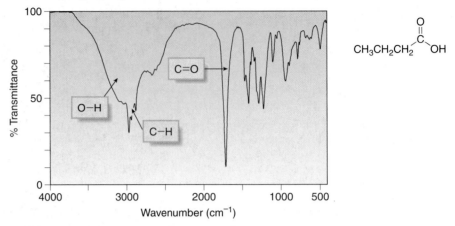

- A strong C=O absorption occurs at 1712 cm^{-1}.
- The broad O–H absorption (2500–3500 cm^{-1}) nearly obscures the C–H peak at ~3000 cm^{-1}.

19.4   Spectroscopic Properties

Carboxylic acids have very characteristic IR and NMR absorptions. In the IR, carboxylic acids show two strong absorptions.

◆ The **C=O group absorbs at about 1710 cm^{-1}**, in the usual region for a carbonyl.

◆ The **O–H absorption** occurs from **2500–3500 cm^{-1}**. This very broad absorption sometimes obscures the C–H peak at 3000 cm^{-1}.

The IR spectrum of butanoic acid in Figure 19.4 illustrates these characteristic peaks.

Carboxylic acids have two noteworthy ^{1}H NMR absorptions and one noteworthy ^{13}C NMR absorption.

◆ **The highly deshielded OH proton absorbs in the ^{1}H NMR spectrum somewhere between 10 and 12 ppm,** further *downfield* than all other absorptions of common organic compounds. Like the OH signal of an alcohol, the exact location depends on the degree of hydrogen bonding and the concentration of the sample.

◆ The protons on the α carbon to the carboxy group are somewhat deshielded, absorbing at 2–2.5 ppm.

◆ In the ^{13}C NMR spectrum, the carbonyl absorption is highly deshielded, appearing at 170–210 ppm.

Figure 19.5 illustrates the ^{1}H and ^{13}C NMR spectra of propanoic acid.

PROBLEM 19.7   Explain how you could use IR spectroscopy to distinguish among the following three compounds.

$$CH_3CH_2CH_2CH_2\overset{\overset{\displaystyle O}{\|}}{C}{-}OH \qquad CH_3CH_2CH_2CH_2\overset{\overset{\displaystyle O}{\|}}{C}{-}OCH_3$$

PROBLEM 19.8   Identify the structure of a compound of molecular formula $C_4H_8O_2$ that gives the following ^{1}H NMR data: 0.95 (triplet, 3 H), 1.65 (multiplet, 2 H), 2.30 (triplet, 2 H), and 11.8 (singlet, 1 H) ppm.

Figure 19.5   The ^1H and ^{13}C NMR spectra of propanoic acid

Sometimes the OH absorption of a carboxylic acid is very broad, so that it is almost buried in the baseline of the ^1H NMR spectrum, making it difficult to see.

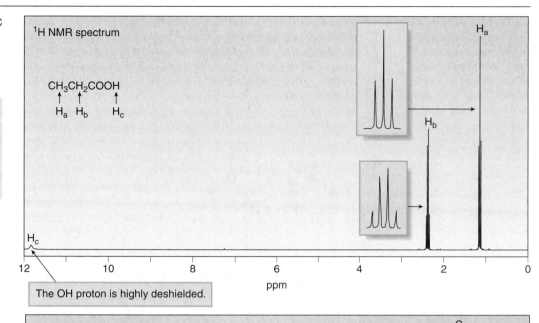

^1H NMR spectrum

CH_3CH_2COOH
H_a H_b H_c

H_a
H_b
H_c

The OH proton is highly deshielded.

ppm

- **^1H NMR spectrum:** There are three signals due to three different kinds of H atoms. The H_a and H_b signals are split into a triplet and quartet, respectively, but the H_c signal is a singlet.
- **^{13}C NMR spectrum:** There are three signals due to three different kinds of carbon atoms.

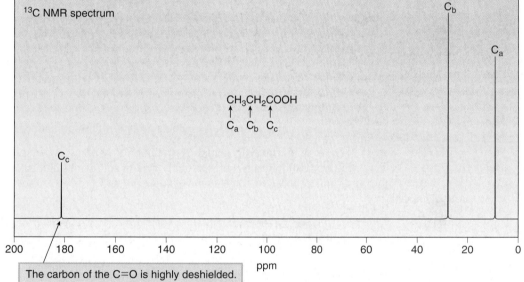

^{13}C NMR spectrum

C_b
C_a
C_c

CH_3CH_2COOH
C_a C_b C_c

The carbon of the C=O is highly deshielded.

ppm

19.5   Interesting Carboxylic Acids

Pure acetic acid is often called *glacial* acetic acid, because it freezes just below room temperature (mp = 17 °C), forming white crystals reminiscent of the ice in a glacier.

Several simple carboxylic acids have characteristic odors and flavors.

HCOOH

- **Formic acid,** a carboxylic acid with an acrid odor and a biting taste, is responsible for the sting of some types of ants. The name is derived from the Latin word *formica,* meaning "ant."

CH₃COOH

- **Acetic acid** is the sour-tasting component of vinegar. The name comes from the Latin *acetum,* meaning "vinegar." The air oxidation of ethanol to acetic acid is the process that makes "bad" wine taste sour. Acetic acid is an industrial starting material for polymers used in paints and adhesives.

CH₃CH₂CH₂COOH

- **Butanoic acid** is an oxidation product that contributes to the disagreeable smell of body odor. Its common name, butyric acid, is derived from the Latin word *butyrum,* meaning "butter," because butyric acid gives rancid butter its peculiar odor and taste.

Several carboxylic acids have been discussed in earlier chapters—for example, naproxen (Section 5.13), leukotrienes (Section 9.16), and fatty acids (Section 10.6).

CH₃CH₂CH₂CH₂CH₂COOH

- Caproic acid, the common name for **hexanoic acid,** has the foul odor associated with dirty socks and locker rooms. Its name is derived from the Latin word *caper,* meaning "goat."

Although oxalic acid is toxic, you would have to eat about nine pounds of spinach at one time to ingest a fatal dose.

Oxalic acid and **lactic acid** are simple carboxylic acids quite prevalent in nature.

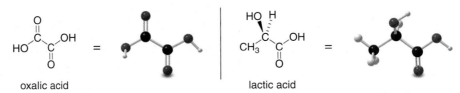

oxalic acid   lactic acid

Oxalic acid occurs naturally in spinach and rhubarb. **Lactic acid** gives sour milk its distinctive taste. Lactic acid is also formed during the metabolism of glucose to CO_2 and H_2O (Section 6.4). During vigorous exercise, lactic acid forms faster than it can be oxidized, resulting in the aching feeling of tired muscles. As lactic acid is further metabolized, this sensation disappears.

Salts of carboxylic acids are commonly used as preservatives. **Sodium benzoate,** a fungal growth inhibitor, is a preservative used in soft drinks and baked goods.

Soaps, the sodium salts of fatty acids, were discussed in Section 3.7.

sodium benzoate
(a preservative)

19.6 Aspirin, Arachidonic Acid, and Prostaglandins

The word *aspirin* is derived from the prefix *a-* for *acetyl* + *spir* from the Latin name *spirea* for the meadowsweet plant.

Recall from Chapter 2 that **aspirin (acetylsalicylic acid)** is a synthetic carboxylic acid, similar in structure to **salicin,** a naturally occurring compound isolated from willow bark, and **salicylic acid,** found in meadowsweet.

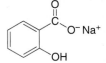

aspirin
(acetylsalicylic acid)

salicin
(isolated from
willow bark)

salicylic acid
(isolated from
meadowsweet)

sodium salicylate
(sweet carboxylate salt)

Aspirin is the most widely used pain reliever and anti-inflammatory agent in the world, yet its mechanism of action remained unknown until the 1970s. John Vane, Bengt Samuelsson, and Sune Bergstrom shared the 1982 Nobel Prize in Physiology or Medicine for unraveling the details of its mechanism.

Both salicylic acid and sodium salicylate (its sodium salt) were widely used analgesics in the nineteenth century, but both had undesirable side effects. Salicylic acid irritated the mucous membranes of the mouth and stomach, and sodium salicylate was too sweet for most patients. Aspirin, a synthetic compound, was first sold in 1899 after Felix Hoffman, a German chemist at Bayer Company, developed a feasible commercial synthesis. Hoffman's work was motivated by personal reasons; his father suffered from rheumatoid arthritis and was unable to tolerate the sweet taste of sodium salicylate.

How does aspirin relieve pain and reduce inflammation? Aspirin blocks the synthesis of **prostaglandins,** 20-carbon fatty acids with a five-membered ring that are responsible for pain, inflammation, and a wide variety of other biological functions. **$PGF_{2\alpha}$,** the chapter-opening molecule, contains the typical carbon skeleton of a prostaglandin.

$PGF_{2\alpha}$
a prostaglandin

Prostaglandins are not stored in cells. Rather they are synthesized from arachidonic acid, a polyunsaturated fatty acid having four cis double bonds. Aspirin acts by blocking the synthesis of prostaglandins from arachidonic acid. Aspirin inactivates cyclooxygenase, an enzyme that converts arachidonic acid to PGG_2, an unstable precursor of $PGF_{2\alpha}$ and other prostaglandins. **Aspirin lessens pain and decreases inflammation because it prevents the synthesis of prostaglandins, the compounds responsible for both of these physiological responses.**

Unlike hormones, which are transported in the bloodstream to their sites of action, prostaglandins act where they are synthesized.

| Aspirin acts here. |

arachidonic acid → cyclo-oxygenase → PGG₂ unstable intermediate → $PGF_{2\alpha}$ and other prostaglandins

PROBLEM 19.9   How many tetrahedral stereogenic centers does $PGF_{2\alpha}$ contain? Draw its enantiomer. How many of its double bonds can exhibit cis-trans isomerism? Considering both its double bonds and its tetrahedral stereogenic centers, how many stereoisomers are possible for $PGF_{2\alpha}$?

19.7   Preparation of Carboxylic Acids

Our discussion of the reactions involving carboxylic acids begins with a brief list of reactions that synthesize them. This list serves as a reminder of where you have seen this functional group before. In these reactions, the carboxy group is formed in the *product,* and many different functional groups serve as starting materials. Reactions that produce a particular functional group are called **preparations.**

In the remainder of Chapter 19 (and Chapters 20 and 22) we discuss reactions in which a carboxylic acid is a *starting material* that may be converted to a variety of different products. Keep in mind that **reactions of a particular functional group follow a common theme.** For example, alkenes undergo addition reactions. As a result, these reactions are easier to learn than the list of preparations, in which vastly different functional groups undergo a wide variety of reactions to form the same kind of product.

Where have we encountered carboxylic acids as reaction products before? The carbonyl carbon is highly oxidized, because it has three C–O bonds, so **carboxylic acids are typically prepared by oxidation reactions.** Three oxidation methods are summarized below. Two other useful methods to prepare carboxylic acids are presented in Chapter 20.

[1]   By oxidation of 1° alcohols (Section 12.12B)

1° Alcohols are converted to carboxylic acids with $Na_2Cr_2O_7$, $K_2Cr_2O_7$, or CrO_3 in the presence of H_2O and H_2SO_4.

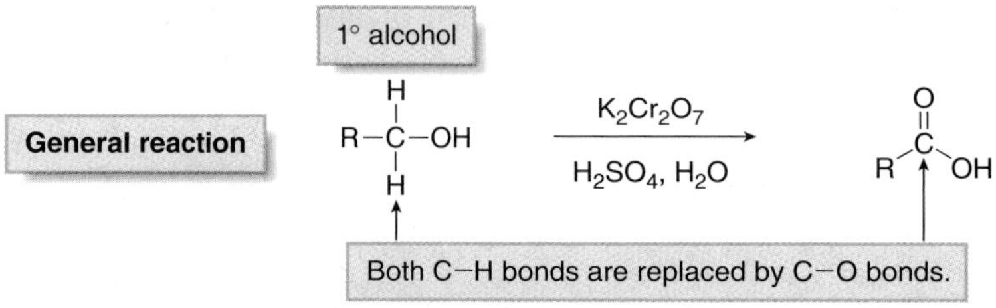

[2]   By oxidation of alkyl benzenes (Section 18.14A)

Alkyl benzenes having at least one benzylic C–H bond are oxidized with $KMnO_4$ to benzoic acid.

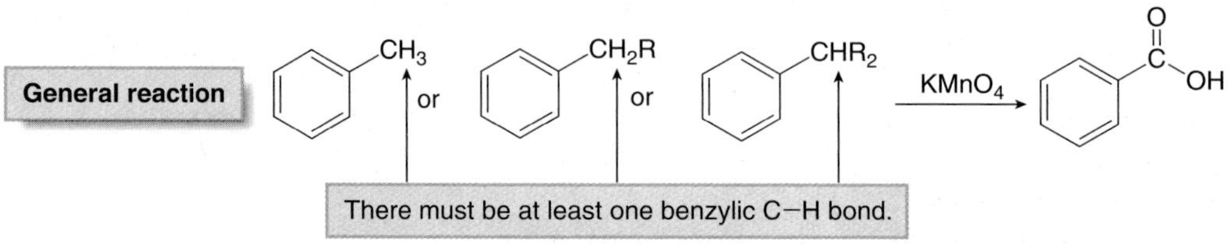

Benzoic acid is *always* the product regardless of the alkyl benzene used as starting material.

[3]   By oxidative cleavage of alkynes (Section 12.11)

Both internal and terminal alkynes are oxidatively cleaved with ozone to give carboxylic acids.

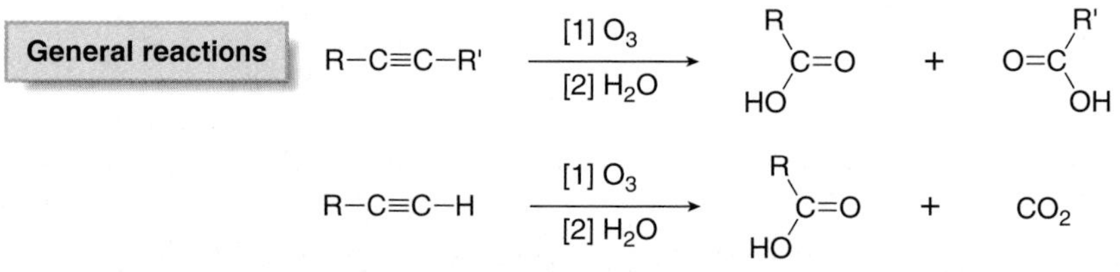

With internal alkynes two carboxylic acids are formed as products. With terminal alkynes, the *sp* hybridized C–H bond is converted to CO_2.

PROBLEM 19.10   What alcohol can be oxidized to each carboxylic acid?

a. 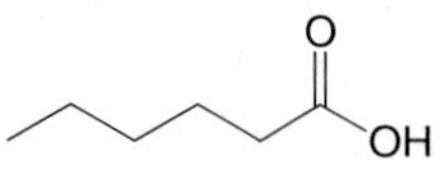   b. $(CH_3)_2CHCOOH$ c.

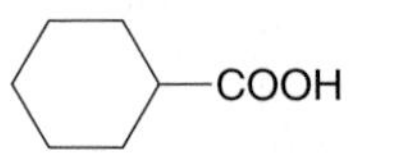

PROBLEM 19.11   Identify **A**, **B**, and **C** in the following reactions.

a. 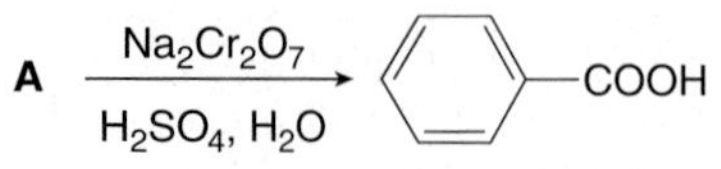 **A** $\xrightarrow{\begin{array}{c} Na_2Cr_2O_7 \\ \hline H_2SO_4, H_2O \end{array}}$ c. 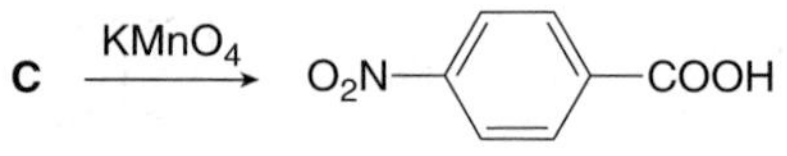 **C** $\xrightarrow{KMnO_4}$

b. **B** $\xrightarrow{\begin{array}{c} [1]\ O_3 \\ \hline [2]\ H_2O \end{array}}$ $\underset{\text{(2 equiv)}}{CH_3COOH}$

19.8   Reactions of Carboxylic Acids—General Features

The polar C–O and O–H bonds, nonbonded electron pairs on oxygen, and the π bond give a carboxylic acid many reactive sites, complicating its chemistry somewhat. By far, **the most important reactive feature of a carboxylic acid is its polar O–H bond, which is readily cleaved with base.**

◆ Carboxylic acids react as Brønsted–Lowry acids—that is, as proton donors.

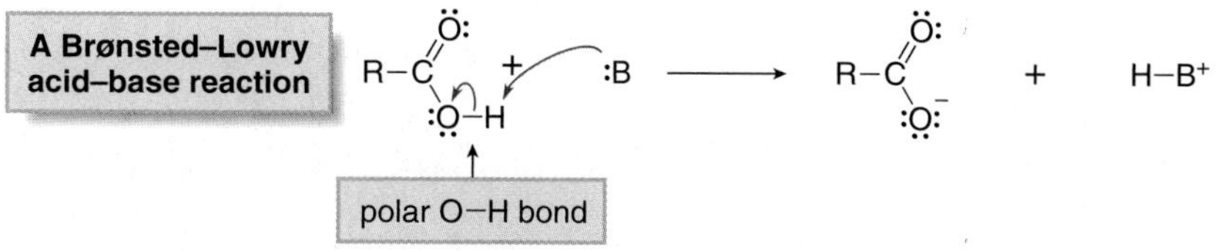

Much of the rest of Chapter 19 is devoted to the acidity of carboxylic acids, as well as some related acid–base reactions. Two other structural features are less important in the reactions of carboxylic acids, but they play a role in the reactions of Chapters 20 and 22.

The nonbonded electron pairs on oxygen create electron-rich sites that can be protonated by strong acids (H–A). Protonation occurs at the carbonyl oxygen because the resulting conjugate acid is resonance stabilized (Possibility [1]). The product of protonation of the OH group (Possibility [2]) cannot be resonance stabilized. As a result, **carboxylic acids are weakly basic—they react with strong acids by protonation of the carbonyl oxygen.** This reaction plays an important role in several mechanisms in Chapter 22.

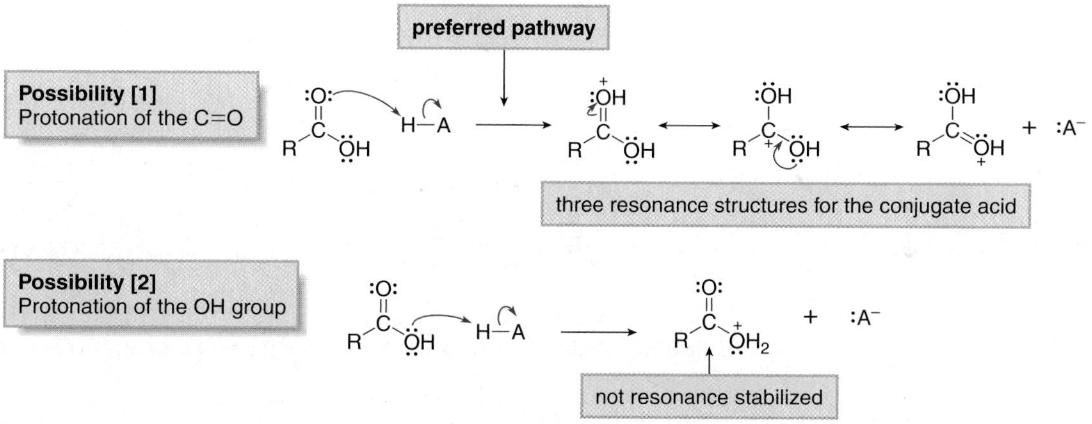

Finally, the polar C–O bonds make the carboxy carbon electrophilic, so carboxylic acids react with nucleophiles. Nucleophilic attack occurs at an sp^2 hybridized carbon atom, so it results in the cleavage of the π bond, as well. This reaction is also discussed in Chapter 22.

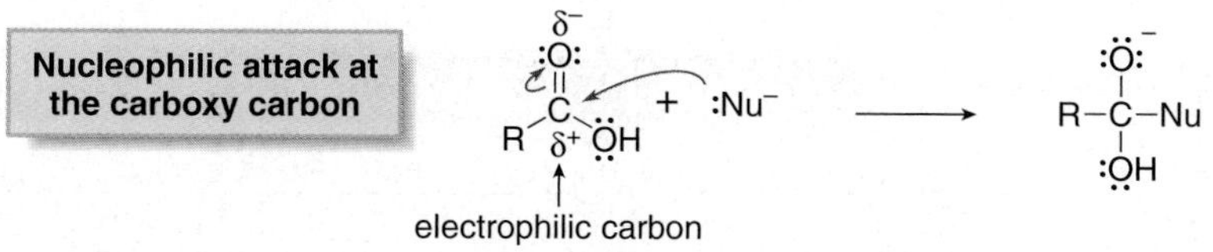

PROBLEM 19.12   Carboxylic acids react with all of the following reagents except one. Which reagent does not react with RCOOH and why?

a. NaOH   b. HCl   c. LiAlH$_4$ d. CrO$_3$ e. NaOCH$_3$

19.9 Carboxylic Acids—Strong Organic Brønsted–Lowry Acids

Carboxylic acids are strong organic acids, and as such, readily react with Brønsted–Lowry bases to form carboxylate anions.

> Recall from Section 2.3 that the lower the pK_a, the stronger the acid.

General acid–base reaction

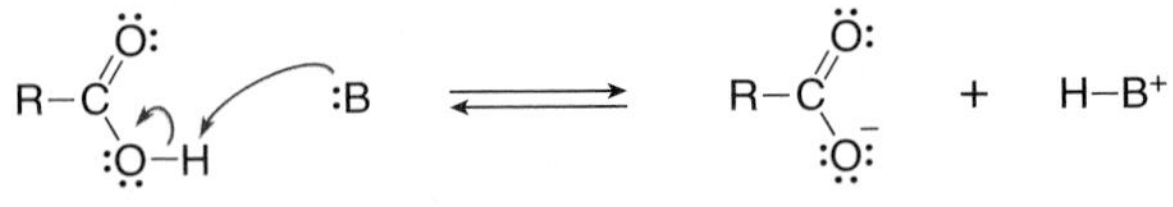

carboxylate anion

What bases are used to deprotonate a carboxylic acid? As we learned in Section 2.3, equilibrium favors the products of an acid–base reaction when the weaker base and acid are formed. Because a weaker acid has a higher pK_a, the following general rule results:

◆ An acid can be deprotonated by a base that has a conjugate acid with a higher pK_a.

Because the pK_a values of many carboxylic acids are ~5, bases that have conjugate acids with pK_a values higher than 5 are strong enough to deprotonate them. Thus, acetic acid ($pK_a = 4.8$) and benzoic acid ($pK_a = 4.2$) can be deprotonated with NaOH and NaHCO₃, as shown in the following equations.

Examples

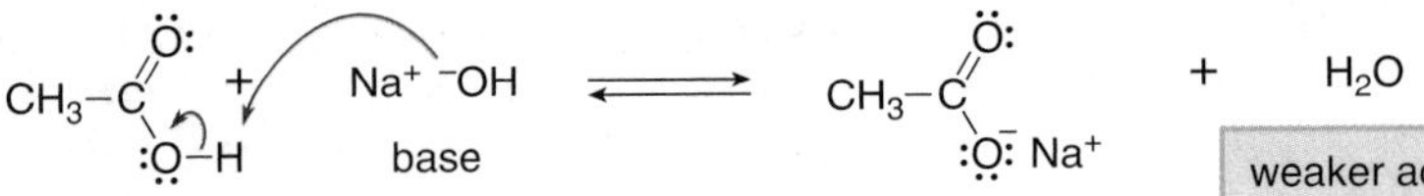

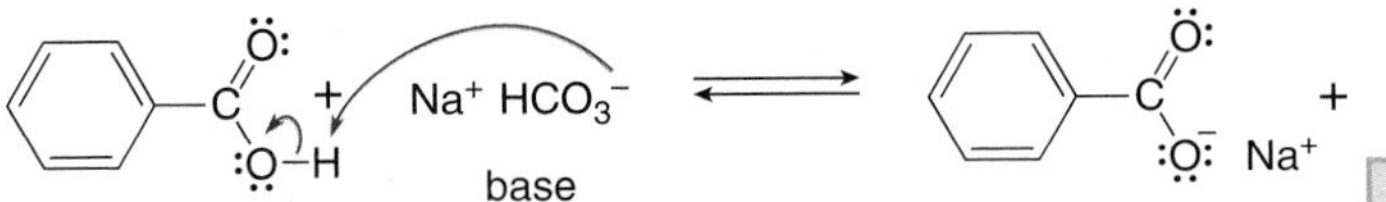

Table 19.3 lists common bases that can be used to deprotonate carboxylic acids. It is noteworthy that even a weak base like NaHCO₃ is strong enough to remove a proton from RCOOH.

Why are carboxylic acids such strong organic acids? Remember that a strong acid has a weak, stabilized conjugate base. **Deprotonation of a carboxylic acid forms a resonance-stabilized conjugate base—a carboxylate anion.** For example, two equivalent resonance structures can be drawn for acetate (the conjugate base of acetic acid), both of which place a negative charge on an electronegative O atom. In the resonance hybrid, therefore, the negative charge is delocalized over two oxygen atoms.

TABLE 19.3   Common Bases Used to Deprotonate Carboxylic Acids

Base	Conjugate acid (pK_a)
$Na^+ \ HCO_3^-$	H_2CO_3 (6.4)
NH_3	NH_4^+ (9.4)
Na_2CO_3	HCO_3^- (10.2)
$Na^+ \ {}^-OCH_3$	CH_3OH (15.5)
$Na^+ \ {}^-OH$	H_2O (15.7)
$Na^+ \ {}^-OCH_2CH_3$	CH_3CH_2OH (16)
$Na^+ \ H^-$	H_2 (35)

Increasing basicity (↓)

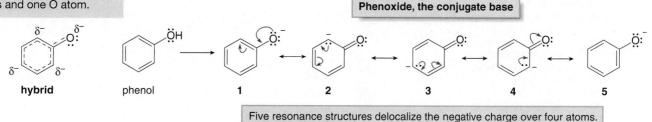

acetic acid   **two resonance structures for acetate,**   **hybrid**
the conjugate base

The (−) is delocalized over 2 O's.

How resonance affects acidity was first discussed in Section 2.5C.

Experimental data support this resonance description of acetate. The acetate anion has two C−O bonds of equal length (1.27 Å) and intermediate between the length of a C−O single bond (1.36 Å) and C=O (1.21 Å).

CH_3-C 1.27 Å   **Acetate has two equivalent C−O bonds.**

acetate hybrid

Resonance stabilization accounts for why carboxylic acids are more acidic than other compounds with O−H bonds—namely, alcohols and phenols. For example, the pK_a values of ethanol (CH_3CH_2OH) and phenol (C_6H_5OH) are 16 and 10, respectively, both higher than the pK_a of acetic acid (4.8).

CH_3CH_2OH

ethanol
$pK_a = 16$

phenol
$pK_a = 10$

acetic acid
$pK_a = 4.8$

Increasing acidity

To understand the relative acidity of ethanol, phenol, and acetic acid, we must compare the stability of their conjugate bases and use the following rule:

◆ **Anything that stabilizes a conjugate base A:⁻ makes the starting acid H−A more acidic.**

Ethoxide, the conjugate base of ethanol, bears a negative charge on an oxygen atom, but there are no additional factors to further stabilize the anion. Because ethoxide is less stable than acetate, **ethanol is a weaker acid than acetic acid.**

$CH_3CH_2\ddot{O}H$ ⟶ $CH_3CH_2\ddot{O}:^-$ no additional resonance stabilization

ethanol   ethoxide

The resonance hybrid of phenoxide illustrates that its negative charge is dispersed over four atoms—three C atoms and one O atom.

Like acetate, **phenoxide** ($C_6H_5O^-$, the conjugate base of phenol) is also resonance stabilized. In the case of phenoxide, however, there are five resonance structures that disperse the negative charge over a total of four different atoms (three different carbons and the oxygen).

Phenoxide, the conjugate base

hybrid   phenol   1   2   3   4   5

Five resonance structures delocalize the negative charge over four atoms.

Phenoxide is more stable than ethoxide, but less stable than acetate, because acetate has two electronegative oxygen atoms upon which to delocalize the negative charge, whereas phenoxide has only one. Additionally, phenoxide resonance structures **2–4** have the negative charge on a carbon, a less electronegative element than oxygen. As a result, structures **2–4** are less stable than structures **1** and **5,** which have the negative charge on oxygen.

Figure 19.6 The relative energies of the five resonance structures for phenoxide and its hybrid

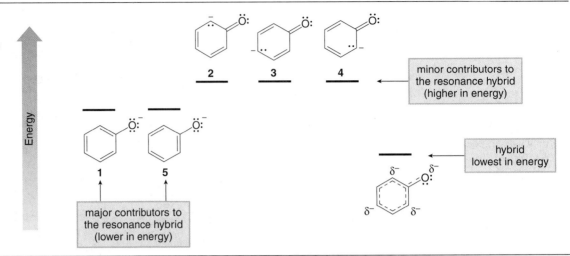

Moreover, resonance structures **1** and **5** have intact aromatic rings, whereas structures **2–4** do not. This, too, makes structures **2–4** less stable than **1** and **5**. Figure 19.6 summarizes this information about phenoxide by displaying the approximate relative energies of its five resonance structures and its hybrid.

As a result, resonance stabilization of the conjugate base is important in determining acidity, but **the absolute number of resonance structures alone is not what's important.** We must evaluate their relative contributions to predict the relative stability of the conjugate bases.

Keep in mind that although carboxylic acids are strong organic acids, they are still much weaker than strong inorganic acids like HCl and H_2SO_4, which have pK_a values < 0.

◆ Because of their O–H bond, RCOOH, ROH, and C_6H_5OH are more acidic than most organic hydrocarbons.

◆ A carboxylic acid is a stronger acid than an alcohol or phenol because its conjugate base is most effectively resonance stabilized.

The relationship between acidity and stability of the conjugate base is summarized for acetic acid, phenol, and ethanol in Figure 19.7.

Figure 19.7 Summary: The relationship between acidity and conjugate base stability for acetic acid, phenol, and ethanol

• **Acetate is the most stable conjugate base** because it has two equivalent resonance structures, both of which place a negative charge on an O atom.
• **Phenoxide** has only one O atom to accept the negative charge. The two resonance structures that contain an intact aromatic ring and place a negative charge on an O atom are major contributors to the hybrid. Resonance stabilizes phenoxide but not as much as resonance stabilizes acetate.
• **Ethoxide is the least stable conjugate base** because it has no additional resonance stabilization.

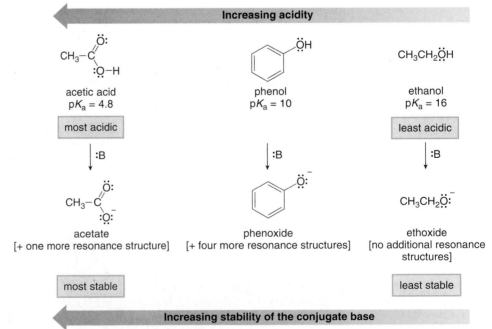

Because alcohols and phenols are weaker acids than carboxylic acids, stronger bases are needed to deprotonate them. To deprotonate C_6H_5OH (pK_a = 10), a base whose conjugate acid has a pK_a > 10 is needed. Thus, of the bases listed in Table 19.3, $NaOCH_3$, $NaOH$, $NaOCH_2CH_3$, and NaH are strong enough. To deprotonate CH_3CH_2OH (pK_a = 16), only NaH is strong enough.

PROBLEM 19.13   Draw the products of each acid–base reaction.

a. [cyclohexane]—COOH $\xrightarrow{\text{NaOH}}$

b. CH_3—[benzene]—OH $\xrightarrow{\text{NaOCH}_3}$

c. $CH_3-\overset{\overset{\displaystyle CH_3}{|}}{\underset{\underset{\displaystyle CH_3}{|}}{C}}-OH \xrightarrow{\text{NaH}}$

d. [benzene]—COOH $\xrightarrow{\text{NaHCO}_3}$

PROBLEM 19.14   Given the pK_a values in Appendix B, which of the following bases are strong enough to deprotonate CH_3COOH: (a) F^-; (b) $(CH_3)_3CO^-$; (c) CH_3^-; (d) $^-NH_2$; (e) Cl^-?

PROBLEM 19.15   Explain why phenol (C_6H_5OH) is a stronger acid than aniline ($C_6H_5NH_2$), even though both compounds give resonance-stabilized anions on treatment with base.

19.10   Inductive Effects in Aliphatic Carboxylic Acids

The pK_a of a carboxylic acid is affected by nearby groups that inductively donate or withdraw electron density.

> ◆ Electron-withdrawing groups stabilize a conjugate base, making a carboxylic acid more acidic.
> ◆ Electron-donating groups destabilize the conjugate base, making a carboxylic acid less acidic.

The relative acidity of CH_3COOH, $ClCH_2COOH$, and $(CH_3)_3CCOOH$ illustrates these principles in the following equations.

We first learned about inductive effects and acidity in Section 2.5B.

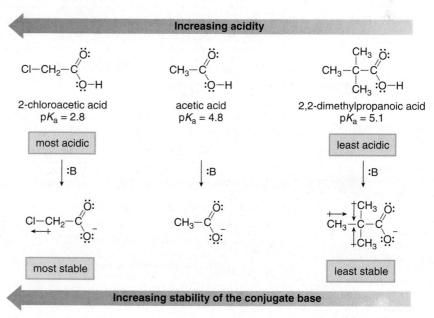

◆ $ClCH_2COOH$ is more acidic (pK_a = 2.8) than CH_3COOH (pK_a = 4.8) because its conjugate base is stabilized by the electron-withdrawing inductive effect of the electronegative Cl.

◆ $(CH_3)_3CCOOH$ is less acidic (pK_a = 5.1) than CH_3COOH because the three polarizable CH_3 groups donate electron density and destabilize the conjugate base.

The number, electronegativity, and location of substituents also affect acidity.

◆ **The larger the number of electronegative substituents, the stronger the acid.**

$$ClCH_2-COOH \qquad Cl_2CH-COOH \qquad Cl_3C-COOH$$
$$pK_a = 2.8 \qquad\qquad pK_a = 1.3 \qquad\qquad pK_a = 0.9$$

Increasing acidity
Increasing number of electronegative Cl atoms

◆ **The more electronegative the substituent, the stronger the acid.**

$$ClCH_2-COOH \qquad\qquad FCH_2-COOH$$
$$pK_a = 2.8 \qquad\qquad\qquad pK_a = 2.6$$

F is more electronegative than Cl.

stronger acid

◆ **The closer the electron-withdrawing group to the COOH, the stronger the acid.**

$$ClCH_2CH_2CH_2COOH \qquad CH_3\overset{\underset{\displaystyle |}{Cl}}{C}HCH_2COOH \qquad CH_3CH_2\overset{\underset{\displaystyle |}{Cl}}{C}HCOOH$$

4-chlorobutanoic acid   3-chlorobutanoic acid   2-chlorobutanoic acid
$pK_a = 4.5$ $pK_a = 4.1$ $pK_a = 2.9$

Increasing acidity
Increasing proximity of Cl to COOH

PROBLEM 19.16 Match each of the following pK_a values (3.2, 4.9, and 0.2) to the appropriate carboxylic acid: (a) CH_3CH_2COOH; (b) CF_3COOH; (c) ICH_2COOH.

PROBLEM 19.17 Explain why HCOOH (formic acid) has a lower pK_a than acetic acid (3.8 versus 4.8).

PROBLEM 19.18 Rank the compounds in each group in order of increasing acidity.
a. CH_3COOH, $HSCH_2COOH$, $HOCH_2COOH$ b. ICH_2COOH, $I_2CHCOOH$, ICH_2CH_2COOH

19.11   Substituted Benzoic Acids

Recall from Chapter 18 that substituents on a benzene ring either donate or withdraw electron density, depending on the balance of their inductive and resonance effects. These same effects also determine the acidity of substituted benzoic acids. There are two rules to keep in mind.

[1]   Electron-donor groups destabilize a conjugate base, making an acid less acidic.

An electron-donor group destabilizes a conjugate base by donating electron density onto a negatively charged carboxylate anion. A benzoic acid substituted by an electron-donor group has a higher pK_a than benzoic acid ($pK_a = 4.2$).

D = Electron-donor group

This acid is less acidic than benzoic acid.

$pK_a > 4.2$

D destabilizes the carboxylate anion.

| **[2]** | **Electron-withdrawing groups stabilize a conjugate base, making an acid more acidic.** |

An electron-withdrawing group stabilizes a conjugate base by removing electron density from the negatively charged carboxylate anion. A benzoic acid substituted by an electron-withdrawing group has a lower pK_a than benzoic acid ($pK_a = 4.2$).

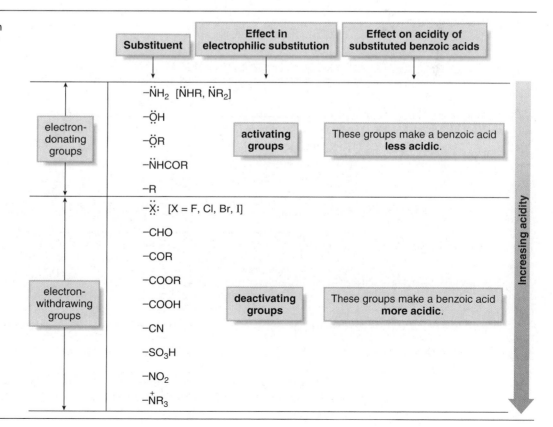

How do we know which groups are electron-donating or electron-withdrawing on a benzene ring? We already learned the characteristics of electron-donating and electron-withdrawing groups in Chapter 18, and how they affect the rate of electrophilic aromatic substitution. These principles can now be extended to substituted benzoic acids.

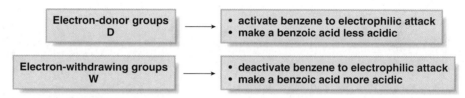

Figure 19.8 illustrates how common electron-donating and electron-withdrawing groups affect both the rate of reaction of a benzene ring towards electrophiles and the acidity of substituted benzoic acids.

Figure 19.8   How common substituents affect the reactivity of a benzene ring towards electrophiles and the acidity of substituted benzoic acids

- **Groups that donate electron density activate a benzene ring towards electrophilic attack and make a benzoic acid *less* acidic.** Common electron-donating groups are R groups, or groups that have an N or O atom (with a lone pair) bonded to the benzene ring.
- **Groups that withdraw electron density deactivate a benzene ring towards electrophilic attack, and make a benzoic acid *more* acidic.** Common electron-withdrawing groups are the halogens, or groups with an atom Y (with a full or partial positive charge) bonded to the benzene ring.

SAMPLE PROBLEM 19.2 Rank the following three carboxylic acids in order of increasing acidity.

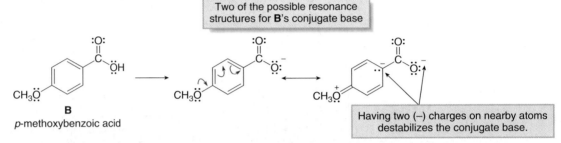

A	**B**	**C**
benzoic acid	p-methoxybenzoic acid	p-nitrobenzoic acid

SOLUTION

p-Methoxybenzoic acid (B): The CH_3O group is an electron-donor group because its electron-donating resonance effect is stronger than its electron-withdrawing inductive effect (Section 18.6). This destabilizes the conjugate base by donating electron density to the negatively charged carboxylate anion, making **B** less acidic than benzoic acid **A.**

> Two of the possible resonance structures for **B**'s conjugate base

B
p-methoxybenzoic acid

> Having two (–) charges on nearby atoms destabilizes the conjugate base.

p-Nitrobenzoic acid (C): The NO_2 group is an electron-withdrawing group because of both inductive effects and resonance (Section 18.6). This stabilizes the conjugate base by removing electron density from the negatively charged carboxylate anion, making **C** more acidic than benzoic acid **A.**

> Two of the possible resonance structures for **C**'s conjugate base

C
p-nitrobenzoic acid

> Having unlike charges on nearby atoms stabilizes the conjugate base.

ANSWER

By this analysis, the order of acidity is **B < A < C.**

PROBLEM 19.19 Rank the compounds in each group in order of increasing acidity.

a. [structures] —COOH   Cl—[ring]—COOH CH_3—[ring]—COOH

b. CH_3—[ring]—COOH   [acetyl]—[ring]—COOH CH_3O—[ring]—COOH

PROBLEM 19.20 Substituted phenols show similar substituent effects. Explain the following observed pK_a values.

[ring]—OH O_2N—[ring]—OH

$pK_a = 10$ $pK_a = 7.2$

19.12 Extraction

An organic chemist in the laboratory must separate and purify mixtures of compounds. One particularly useful technique is **extraction,** which uses solubility differences and acid–base principles to separate and purify compounds.

Extraction has long been and remains the first step in isolating a natural product from its source.

Two solvents are used in extraction: water or an aqueous solution such as 10% $NaHCO_3$ or 10% NaOH; and an organic solvent such as dichloromethane (CH_2Cl_2), diethyl ether, or hexane. **Compounds are separated by their solubility differences in an aqueous and organic solvent.**

A piece of glassware called a **separatory funnel,** depicted in Figure 19.9, is used for the extraction. When two insoluble liquids are added to the separatory funnel, two layers form, with the less dense liquid on top and the more dense liquid on the bottom.

Suppose a mixture of benzoic acid (C_6H_5COOH) and NaCl is added to a separatory funnel containing H_2O and CH_2Cl_2. The benzoic acid would dissolve in the organic layer and the NaCl would dissolve in the water layer. Separating the organic and aqueous layers and placing them in different flasks separates the benzoic acid and NaCl from each other.

How could we separate a mixture of benzoic acid and cyclohexanol? Both compounds are organic, and as a result, both are soluble in an organic solvent such as CH_2Cl_2 and insoluble in water. If a mixture of benzoic acid and cyclohexanol were added to a separatory funnel with CH_2Cl_2 and water, both would dissolve in the CH_2Cl_2 layer, and the two compounds would *not* be separated from each other. Is it possible to use extraction to separate two compounds of this sort that have similar solubility properties?

Recall from Tables 9.1 and 19.2 that alcohols and carboxylic acids having more than five carbons are water insoluble.

COOH

Both compounds have similar solubility properties.

OH

benzoic acid

- insoluble in water
- soluble in CH_2Cl_2

cyclohexanol

- insoluble in water
- soluble in CH_2Cl_2

If a carboxylic acid is one of the compounds, the answer is *yes,* because we can use acid–base chemistry to change its solubility properties.

When benzoic acid (a strong organic acid) is treated with aqueous NaOH, benzoic acid is deprotonated, forming sodium benzoate. Because sodium benzoate is ionic, it is soluble in water, but insoluble in organic solvents.

benzoic acid
$pK_a = 4.2$

- insoluble in water
- soluble in CH_2Cl_2

Na^+ ^-OH
base

sodium benzoate

- soluble in water
- insoluble in CH_2Cl_2

$+$ H_2O
$pK_a = 15.7$

The solubility properties of the conjugate base are *different* from those of the starting acid.

Figure 19.9 Using a separatory funnel for extraction

- When two insoluble liquids are added to a separatory funnel, two layers are visible, and the less dense liquid forms the upper layer.
- To separate the layers, the lower layer can be drained from the bottom of the separatory funnel by opening the stopcock. The top layer can then be poured out the top neck of the funnel.

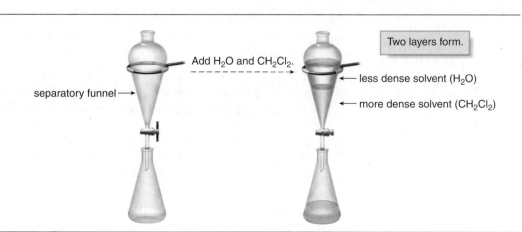

separatory funnel →

Add H_2O and CH_2Cl_2.

Two layers form.

← less dense solvent (H_2O)

← more dense solvent (CH_2Cl_2)

Figure 19.10
Separation of
benzoic acid and
cyclohexanol by
an extraction
procedure

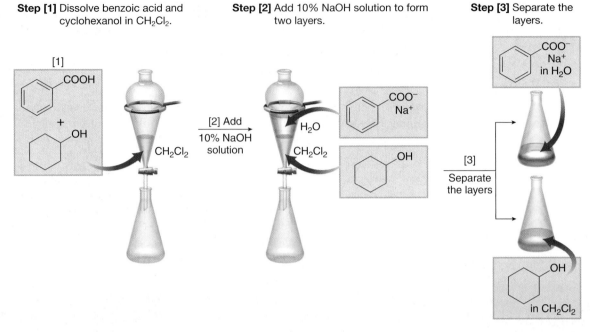

Step [1] Dissolve benzoic acid and cyclohexanol in CH_2Cl_2.

Step [2] Add 10% NaOH solution to form two layers.

Step [3] Separate the layers.

- Both compounds dissolve in the organic solvent CH_2Cl_2.

- Adding 10% aqueous NaOH solution forms two layers. When the two layers are mixed, the NaOH deprotonates C_6H_5COOH to form $C_6H_5COO^-Na^+$, which dissolves in the aqueous layer.
- The cyclohexanol remains in the CH_2Cl_2 layer.

- Draining the lower layer out the bottom stopcock separates the two layers, and the separation process is complete.
- Cyclohexanol (dissolved in CH_2Cl_2) is in one flask. The sodium salt of benzoic acid, $C_6H_5COO^-Na^+$ (dissolved in water) is in another flask.

A similar acid–base reaction does *not* occur when cyclohexanol is treated with NaOH because organic alcohols are much weaker organic acids, so they can only be deprotonated by a *very strong base* such as NaH. NaOH is not strong enough to form significant amounts of the sodium alkoxide.

cyclohexanol
$pK_a \sim 17$

+ $Na^+ \ ^-OH$ ⇌   base

$pK_a = 15.7$

+ H_2O

Since equilibrium favors the starting materials, little alkoxide is formed.

This difference in acid–base chemistry can be used to separate benzoic acid and cyclohexanol by the stepwise extraction procedure illustrated in Figure 19.10. This extraction scheme relies on two basic principles:

◆ Extraction can separate only compounds having different solubility properties. One compound must dissolve in the aqueous layer and one must dissolve in the organic layer.

◆ A carboxylic acid can be separated from other organic compounds by converting it to a water-soluble carboxylate anion by an acid–base reaction.

Thus, the water-soluble salt, $C_6H_5COO^-Na^+$ (derived from C_6H_5COOH by an acid–base reaction) can be separated from water-insoluble cyclohexanol by an extraction procedure.

PROBLEM 19.21   Which of the following pairs of compounds can be separated from each other by an extraction procedure?
a. $CH_3(CH_2)_6COOH$ and $CH_3CH_2CH_2CH_2CH=CH_2$
b. $CH_3CH_2CH_2CH_2CH=CH_2$ and $(CH_3CH_2CH_2)_2O$
c. $CH_3(CH_2)_6COOH$ and $NaCl$
d. $NaCl$ and KCl

19.13   Sulfonic Acids

Recall from Section 9.13 that $CH_3C_6H_4SO_2-$ is called a **tosyl group,** abbreviated by the letters **Ts.** For this reason, *p*-toluenesulfonic acid (also called tosic acid) is abbreviated as **TsOH.**

Although much less common than carboxylic acids, **sulfonic acids** constitute a useful group of organic acids. Sulfonic acids have the general structure **RSO_3H.** The most widely used sulfonic acid, **p-toluenesulfonic acid,** was first discussed in Section 2.6.

General structure	Example

R—S—O—H

sulfonic acid

CH_3—〈 〉—S—O—H   =   **TsOH**

p-toluenesulfonic acid

Sulfonic acids are very strong acids (pK_a values ≈ -7) because their conjugate bases are resonance stabilized, and all the resonance structures delocalize a negative charge on oxygen. The conjugate base of a sulfonic acid is called a **sulfonate anion.**

R—S—O—H   +   :B   ⟶   R—S—O: ⟷   R—S=O: ⟷   R—S=O   +   H—B⁺

strong acid
$pK_a \approx -7$

Three resonance structures—
All have a negative charge on oxygen.

Because sulfonate anions are such weak bases, they make **good leaving groups** in nucleophilic substitution reactions, as we learned in Section 9.13.

PROBLEM 19.22   Two other commonly used sulfonic acids are methanesulfonic acid (CH_3SO_3H) and trifluoromethanesulfonic acid (CF_3SO_3H). Which has the weaker conjugate base? Which conjugate base is the better leaving group? Which of these acids has the higher pK_a?

19.14   Amino Acids

Amino acids, one of four kinds of small biomolecules that have important biological functions in the cell (Section 3.3), also undergo proton transfer reactions.

19.14A   Introduction

Chapter 28 discusses the synthesis of amino acids and their conversion to proteins.

Amino acids contain two functional groups—an amino group (NH_2) and a carboxy group ($COOH$). In most naturally occurring amino acids, the amino group is bonded to the α carbon, and so they are called **α-amino acids.** Amino acids are the building blocks of proteins, biomolecules that comprise muscle, hair, fingernails, and many other biological tissues.

A complete list of the 20 naturally occurring amino acids is found in Figure 28.2.

amino group → H_2N—C—H   COOH ← carboxy group

R
α carbon

α-amino acid

Humans can synthesize only 10 of the 20 amino acids needed for protein synthesis. The remaining 10, called **essential amino acids,** must be obtained from the diet and consumed on a regular, almost daily basis. Diets that include animal products readily supply all the needed amino acids. Because no one plant source has sufficient amounts of all the essential amino acids, vegetarian diets must be carefully balanced. Grains—wheat, rice, and corn—are low in lysine, and legumes—beans, peas, and peanuts—are low in methionine, but a combination of these foods provides all the needed amino acids. Thus, a diet of corn tortillas and beans, or rice and tofu, provides all essential amino acids. A peanut butter sandwich on wheat bread does the same.

The 20 amino acids that occur naturally in proteins differ in the identity of the R group bonded to the α carbon. **The simplest amino acid, called glycine, has R = H.** When the R group is any other substituent, **the α carbon is a stereogenic center,** and there are two possible enantiomers.

Amino acids exist in nature as only one of these enantiomers. Except when the R group is CH_2SH, the stereogenic center on the α carbon has the *S* configuration. An older system of nomenclature names the **naturally occurring enantiomer of an amino acid as the L isomer, and its unnatural enantiomer the D isomer.**

The R group of an amino acid can be H, alkyl, aryl, or an alkyl chain containing an N, O, or S atom. Representative examples are listed in Table 19.4. All amino acids have common names, which are abbreviated by a three-letter or one-letter designation. For example, glycine is often written as the three-letter abbreviation **gly,** or the one-letter abbreviation **G.** These abbreviations are also given in Table 19.4.

TABLE 19.4   Representative Amino Acids

General structure:
$$H_2N-\underset{\underset{R}{|}}{\overset{\overset{COOH}{|}}{C}}-H$$

R group	Name	Three-letter abbreviation	One-letter abbreviation
H	glycine	gly	G
CH_3	alanine	ala	A
$CH_2C_6H_5$	phenylalanine	phe	F
CH_2OH	serine	ser	S
CH_2SH	cysteine	cys	C
$CH_2CH_2SCH_3$	methionine	met	M
CH_2CH_2COOH	glutamic acid	glu	E
$(CH_2)_4NH_2$	lysine	lys	K

PROBLEM 19.23   Draw both enantiomers of each amino acid and label them as *R* or *S:* (a) phenylalanine; (b) methionine.

19.14B   Acid–Base Properties

An amino acid is both an acid and a base.

◆ The NH_2 group has a nonbonded electron pair, making it a base.
◆ The COOH group has an acidic proton, making it an acid.

Amino acids are never uncharged neutral compounds. They exist as salts, so they have very high melting points and are very soluble in water.

◆ Proton transfer from the acidic carboxy group to the basic amino group forms a salt called a *zwitterion*, which contains both a positive and a negative charge.

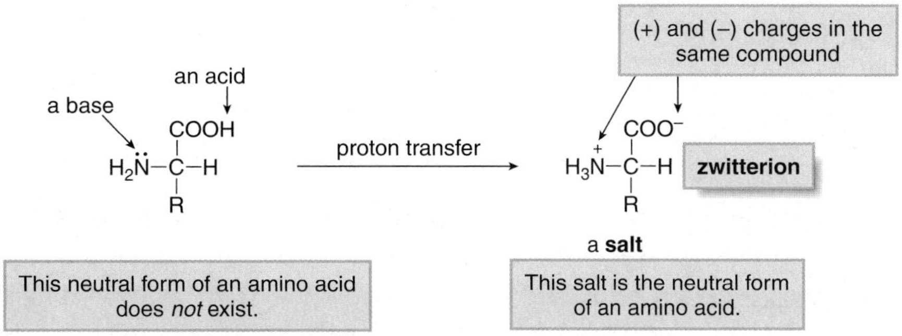

This neutral form of an amino acid does *not* exist.

(+) and (−) charges in the same compound

This salt is the neutral form of an amino acid.

In actuality, an amino acid can exist in three different forms, depending on the pH of the aqueous solution in which it is dissolved.

When the pH of a solution is ~7, alanine ($R = CH_3$) exists in its zwitterionic form (**A**), having no net charge. In this form the carboxy group bears a negative charge—it is a **carboxylate anion**—and the amino group bears a net positive charge (an **ammonium cation**).

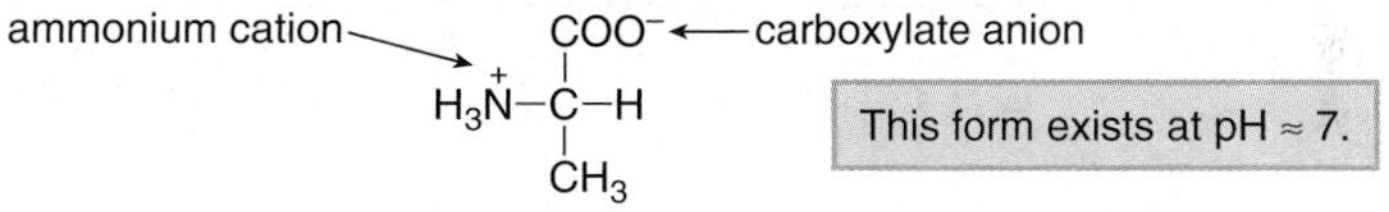

This form exists at pH ≈ 7.

When strong acid is added to lower the pH (≤ 2), the carboxylate anion is protonated and the **amino acid has a net positive charge** (form **B**).

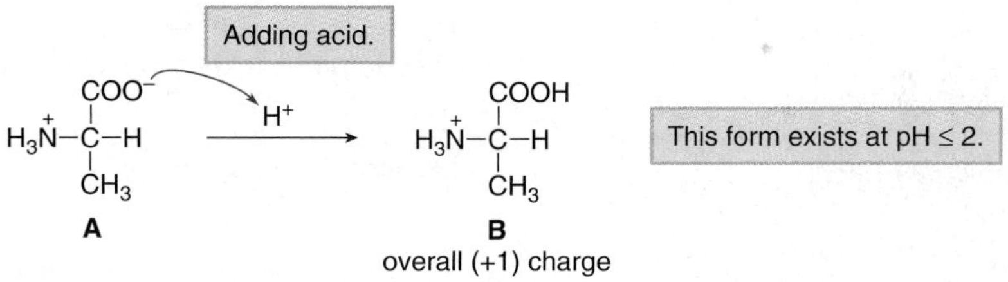

Adding acid.

This form exists at pH ≤ 2.

overall (+1) charge

When strong base is added to **A** to raise the pH (≥ 10), the ammonium cation is deprotonated and the **amino acid has a net negative charge** (form **C**).

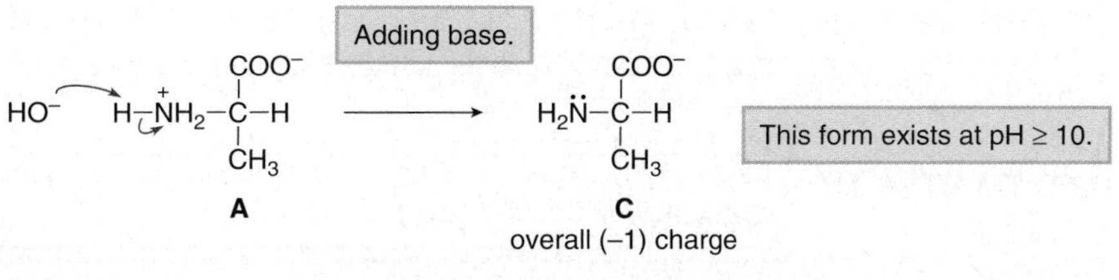

Adding base.

This form exists at pH ≥ 10.

overall (−1) charge

Thus, **alanine exists in one of three different forms depending on the pH of the solution in which it is dissolved.** If the pH of a solution is gradually increased from 2 to 10, the following process occurs.

◆ At low pH alanine has a net (+) charge (form B).
◆ As the pH is increased to ~7, the carboxy group is deprotonated, and the amino acid exists as a zwitterion with no overall charge (form A).
◆ At high pH, the ammonium cation is deprotonated, and the amino acid has a net (−) charge (form C).

Figure 19.11 Summary of the acid–base reactions of alanine

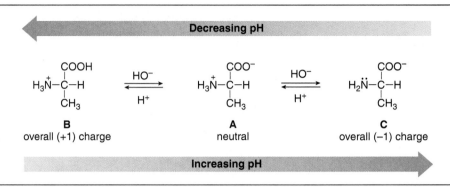

These reactions are summarized in Figure 19.11.

PROBLEM 19.24   Explain why amino acids, unlike most other organic compounds, are insoluble in organic solvents like diethyl ether.

PROBLEM 19.25   Draw the positively charged, neutral, and negatively charged forms for the amino acid glycine. Which species predominates at pH 11? Which species predominates at pH 1?

19.14C   Isoelectric Point

Because a protonated amino acid has at least two different protons that can be removed, a pK_a value is reported for each of these protons. For example, the pK_a of the carboxy proton of alanine is 2.35 and the pK_a of the ammonium proton is 9.87. Table 28.1 lists these values for all 20 amino acids.

◆ The pH at which the amino acid exists primarily in its neutral form is called its *isoelectric point,* abbreviated as p*I.*

More information on the isoelectric point can be found in Section 28.1 and Problem 28.31.

Generally, the isoelectric point is the average of both pK_a values of an amino acid:

$$\text{Isoelectric point} = \text{p}I = \frac{pK_a\,(\text{COOH}) + pK_a\,(\text{NH}_3^+)}{2}$$

$$\text{For alanine:}\quad \text{p}I = \frac{2.35 + 9.87}{2} = \boxed{\begin{array}{c}6.12\\ \text{p}I\text{ (alanine)}\end{array}}$$

PROBLEM 19.26   The pK_a values for the carboxy and ammonium protons of phenylalanine are 2.58 and 9.24, respectively. What is the isoelectric point of phenylalanine? Draw the structure of phenylalanine at its isoelectric point.

PROBLEM 19.27   Explain why the pK_a of the COOH group of glycine is much lower than the pK_a of the COOH of acetic acid (2.35 compared to 4.8).

19.15   Key Concepts—Carboxylic Acids and the Acidity of the O–H Bond

General Facts

- Carboxylic acids contain a carboxy group (COOH). The central carbon is sp^2 hybridized and trigonal planar (19.1).
- Carboxylic acids are identified by the suffixes -*oic acid, carboxylic acid,* or -*ic acid* (19.2).
- Carboxylic acids are polar compounds that exhibit hydrogen bonding interactions (19.3).

Summary of Spectroscopic Absorptions (19.4)

IR absorptions	C=O	~1710 cm^{-1}
	O–H	3500–2500 cm^{-1} (very broad and strong)
^{1}H NMR absorptions	O–H	10–12 ppm (highly deshielded proton)
	C–H α to COOH	2–2.5 ppm (somewhat deshielded C_{sp^3}–H)
^{13}C NMR absorption	C=O	170–210 ppm (highly deshielded carbon)

General Acid–Base Reaction of Carboxylic Acids (19.9)

$pK_a \approx 5$ carboxylate anion

- Carboxylic acids are especially acidic because carboxylate anions are resonance stabilized.
- For equilibrium to favor the products, the base must have a conjugate acid with a $pK_a > 5$. Common bases are listed in Table 19.3.

Factors that Affect Acidity

Resonance effects.

A carboxylic acid is more acidic than an alcohol or phenol because its conjugate base is more effectively stabilized by resonance (19.9).

ROH   R–C(=O)OH

pK_a = 16–18 pK_a = 10 $pK_a \approx 5$

Increasing acidity

Inductive effects.

Acidity increases with the presence of electron-withdrawing groups (like the electronegative halogens) and decreases with the presence of electron-donating groups (like polarizable alkyl groups) (19.10).

Substituted benzoic acids.

- Electron-donor groups **(D)** make a substituted benzoic acid less acidic than benzoic acid.
- Electron-withdrawing groups **(W)** make a substituted benzoic acid more acidic than benzoic acid.

D → COOH	COOH	W ← COOH
less acidic **higher pK_a** pK_a > 4.2	pK_a = 4.2	**more acidic** **lower pK_a** pK_a < 4.2

Increasing acidity

Other Facts

- Extraction is a useful technique for separating compounds having different solubility properties. Carboxylic acids can be separated from other organic compounds by extraction, because aqueous base converts a carboxylic acid into a water-soluble carboxylate anion (19.12).
- A sulfonic acid (RSO$_3$H) is a strong acid because it forms a weak, resonance-stabilized conjugate base on deprotonation (19.13).
- Amino acids have an amino group on the α carbon to the carboxy group [RCH(NH$_2$)COOH]. Amino acids exist as zwitterions at pH ≈ 7. Adding acid forms a species with a net (+1) charge [RCH(NH$_3$)COOH]$^+$. Adding base forms a species with a net (–1) charge [RCH(NH$_2$)COO]$^-$ (19.14).

Problems

Nomenclature

19.28 Give the IUPAC name for each compound.

a. $(CH_3)_2CHCH_2CH_2CO_2H$
b. $BrCH_2COOH$

e.

h. CH_3CH_2—⬡—COOH

c.

f.

i.

d. $CH_3CH_2CH_2COO^-Li^+$

g.

j.

19.29 Draw the structure corresponding to each name.

a. 3,3-dimethylpentanoic acid
b. 4-chloro-3-phenylheptanoic acid
c. (R)-2-chloropropanoic acid
d. β,β-dichloropropionic acid

e. m-hydroxybenzoic acid
f. o-chlorobenzoic acid
g. potassium acetate
h. sodium α-bromobutyrate

19.30 Draw the structures and give the IUPAC names for the carboxylic acids having molecular formula $C_5H_{10}O_2$. Then give the IUPAC names for the sodium salts that result from treatment of each carboxylic acid with NaOH.

Physical Properties

19.31 Rank the compounds in each group in order of increasing boiling point.

a. $CH_3CH_2CH_2CH_2COOH$, $(CH_3CH_2CH_2)_2O$, $CH_3(CH_2)_5OH$
b. $CH_3COCH_2CH(CH_3)_2$, $(CH_3)_2CHCH_2COOH$, $(CH_3)_2CHCH_2CH(OH)CH_3$

Preparation of Carboxylic Acids

19.32 Draw the organic products formed in each reaction.

a. $\xrightarrow[\text{H}_2\text{SO}_4, \text{H}_2\text{O}]{\text{CrO}_3}$

c. ⬡—C≡C–H $\xrightarrow[\text{[2] H}_2\text{O}]{\text{[1] O}_3}$

b. $(CH_3)_2CH$—⬡—CH_3 $\xrightarrow{\text{KMnO}_4}$

d. $CH_3(CH_2)_6CH_2OH$ $\xrightarrow[\text{H}_2\text{SO}_4, \text{H}_2\text{O}]{\text{Na}_2\text{Cr}_2\text{O}_7}$

19.33 Identify the lettered compounds in each reaction sequence.

a. $=CH_2$ $\xrightarrow[\text{[2] H}_2\text{O}_2, \text{HO}^-]{\text{[1] BH}_3}$ **A** $\xrightarrow[\text{H}_2\text{SO}_4, \text{H}_2\text{O}]{\text{CrO}_3}$ **B**

c. ⬡ $\xrightarrow[\text{AlCl}_3]{(CH_3)_2CHCl}$ **G** $\xrightarrow{\text{KMnO}_4}$ **H**

b. $HC≡CH$ $\xrightarrow[\text{[2] CH}_3\text{I}]{\text{[1] NaNH}_2}$ **C** $\xrightarrow[\text{[2] CH}_3\text{CH}_2\text{I}]{\text{[1] NaNH}_2}$ **D** $\xrightarrow[\text{[2] H}_2\text{O}]{\text{[1] O}_3}$ **E** + **F**

Acid–Base Reactions; General Questions on Acidity

19.34 Using the pK_a table in Appendix B, determine whether each of the following bases is strong enough to deprotonate the three compounds listed below. Bases: [1] $^-$OH; [2] $CH_3CH_2^-$; [3] $^-NH_2$; [4] NH_3; [5] $HC≡C^-$.

a. CH_3—⬡—COOH

$pK_a = 4.3$

b. Cl—⬡—OH

$pK_a = 9.4$

c. $(CH_3)_3COH$

$pK_a = 18$

19.35 Draw the products of each acid–base reaction, and using the pK_a table in Appendix B, determine if equilibrium favors the reactants or products.

a. [benzene ring]—COOH + KOC(CH$_3$)$_3$ ⇌

b. [chain]—OH + NH$_3$ ⇌

c. [benzene ring]—OH + NaNH$_2$ ⇌

d. [benzene ring with CH$_3$]—COOH + CH$_3$Li ⇌

e. (CH$_3$)$_2$CHCH$_2$OH + NaH ⇌

f. CH$_3$—[benzene ring]—OH + Na$_2$CO$_3$ ⇌

19.36 Which compound in each pair has the lower pK_a? Which compound in each pair has the stronger conjugate base?

a. [benzene ring with COOH]   and   [cyclohexane ring with CH$_2$OH]

b. ClCH$_2$COOH   and   FCH$_2$COOH

c. CH$_3$—[benzene ring]—COOH   and   Cl—[benzene ring]—COOH

d. NCCH$_2$COOH   and   CH$_3$COOH

19.37 Rank the compounds in each group in order of increasing acidity.

a. [CH$_3$CHClCH$_2$COOH]   [CH$_3$CH$_2$CH$_2$COOH]   [CH$_3$CHBrCH$_2$COOH]

c. [benzene—COOH]   [CH$_3$—benzene—COOH]   [CF$_3$—benzene—COOH]

b. [CH$_3$—benzene—OH]   [O$_2$N—benzene—OH]   [Cl—benzene—OH]

d. [Br—benzene—OH]   [O$_2$N—benzene—OH]   [O$_2$N—benzene—OH with NO$_2$]

19.38 Rank the compounds in each group in order of increasing basicity.

a. BrCH$_2$COO$^-$ (CH$_3$)$_3$CCOO$^-$ BrCH$_2$CH$_2$COO$^-$

b. C$_6$H$_5$NH$^-$ C$_6$H$_5$O$^-$ C$_6$H$_5$CH$_2^-$

c. [cyclohexane—O$^-$]   [benzene—O$^-$]   [O$_2$N—benzene—O$^-$]

19.39 Match the pK_a values to the appropriate structure. pK_a values: 0.28, 1.24, 2.66, 2.86, and 3.12. Compounds: (a) FCH$_2$COOH; (b) CF$_3$COOH; (c) F$_2$CHCOOH; (d) ICH$_2$COOH; (e) BrCH$_2$COOH.

19.40 Although all nitro substituted phenols have lower pK_a values than phenol, the pK_a of the para isomer is lower than the pK_a of the meta isomer (7.2 versus 8.3). Explain this observation.

19.41 Explain why carboxylic acids **A** and **B** are both more acidic than CH$_3$CH$_2$COOH.

[CH$_3$C(=O)CH$_2$COOH]   [CH$_2$=CHCH$_2$COOH]
     A   **B**

19.42 Explain the following result. Acetic acid (CH$_3$COOH), labeled at its OH oxygen with the uncommon ^{18}O isotope, was treated with aqueous base, and then the solution was acidified. Two products having the ^{18}O label at different locations were formed.

[CH$_3$C(=O)—*OH] →[1] NaOH →[2] H$_3$O$^+$ [CH$_3$C(=*O)OH] + [CH$_3$C(=O)—*OH]

| labeled O atom | | The label is now in two different locations. |

19.43 Draw all resonance structures of the conjugate bases formed by removal of each labeled proton (H$_a$, H$_b$, and H$_c$) in 1,3-cyclohexanedione. Rank these protons in order of increasing acidity and explain the order you chose.

[structure of 1,3-cyclohexanedione with H$_a$, H$_b$, H$_c$ labeled]

1,3-cyclohexanedione

19.44 Explain why sulfonic acids (RSO$_3$H) are stronger organic acids than carboxylic acids (RCOOH).

19.45 Identify **X** in the following equation, and explain how hexanoic acid is formed by this stepwise reaction sequence.

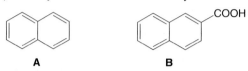

$$CH_3COOH \xrightarrow[\substack{\text{strong base} \\ \text{(2 equiv)}}]{} \quad X \quad \xrightarrow[\substack{[2] \ H_3O^+}]{[1] \ CH_3CH_2CH_2CH_2Br} \quad CH_3CH_2CH_2CH_2CH_2COOH$$
$$\text{hexanoic acid}$$

19.46 The pK_a of acetamide is 16. Draw the structure for its conjugate base and explain why acetamide is less acidic than CH_3COOH.

$$\underset{\text{acetamide}}{CH_3 - \overset{\overset{\displaystyle O}{\|}}{C} - NH_2}$$

Extraction

19.47 Write out the steps needed to separate hydrocarbon **A** and carboxylic acid **B** by using an extraction procedure.

A   COOH

                                 B

19.48 Because phenol (C_6H_5OH) is less acidic than a carboxylic acid, it can be deprotonated by NaOH but not by the weaker base $NaHCO_3$. Using this information, write out an extraction sequence that can be used to separate C_6H_5OH from cyclohexanol. Show what compound is present in each layer at each stage of the process, and if it is present in its neutral or ionic form.

19.49 Can octane and 1-octanol be separated using an aqueous extraction procedure? Explain why or why not.

Spectroscopy

19.50 Identify each compound from its spectral data.

   a. Molecular formula: $C_3H_5ClO_2$
                IR: $3500–2500 \ cm^{-1}$, $1714 \ cm^{-1}$
       ^1H NMR data: 2.87 (triplet, 2 H), 3.76 (triplet, 2 H), and 11.8 (singlet, 1 H) ppm

   b. Molecular formula: $C_8H_8O_3$
                IR: $3500–2500 \ cm^{-1}$, $1688 \ cm^{-1}$
       ^1H NMR data: 3.8 (singlet, 3 H), 7.0 (doublet, 2 H), 7.9 (doublet, 2 H), and 12.7 (singlet, 1 H) ppm

   c. Molecular formula: $C_8H_8O_3$
                IR: $3500–2500 \ cm^{-1}$, $1710 \ cm^{-1}$
       ^1H NMR data: 4.7 (singlet, 2 H), 6.9–7.3 (multiplet, 5 H), and 11.3 (singlet, 1 H) ppm

19.51 Use the ^1H NMR and IR spectra given below to identify the structures of two isomers (**A** and **B**) having molecular formula $C_4H_8O_2$.

Compound **A**:

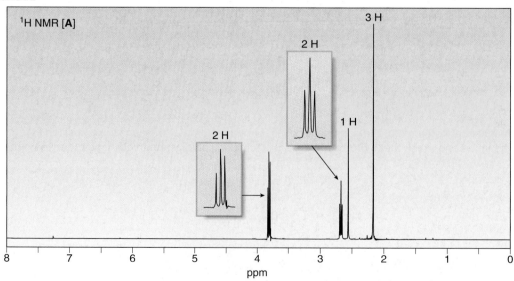

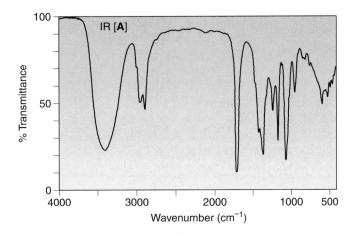

Compound B:

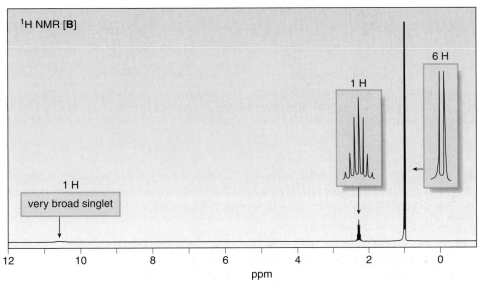

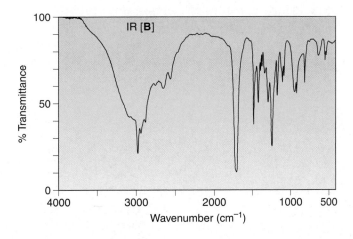

19.52 What is the structure of a carboxylic acid (molecular formula $C_6H_{12}O_2$) that gives three singlets in its ^1H NMR spectrum at 1.1, 2.2, and 11.9 ppm?

19.53 A monomer needed to synthesize polyethylene terephthalate (PET), a polymer used to make plastic sheeting and soft drink bottles (Section 22.16), shows a strong absorption in its IR spectrum at 1692 cm^{-1} and two singlets in its ^1H NMR spectrum at 8.2 and 10.0 ppm. What is the structure of this monomer (molecular formula $C_8H_6O_4$)?

19.54 Match the ^{13}C NMR data to the appropriate structure.
Spectrum [1]: peaks at 14, 22, 27, 34, 181 ppm
Spectrum [2]: peaks at 27, 39, 186 ppm
Spectrum [3]: peaks at 22, 26, 43, 180 ppm

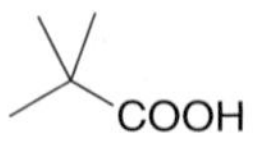

A    B 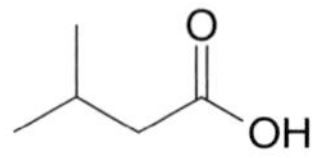   C

Amino Acids

19.55 Threonine is a naturally occurring amino acid that has two stereogenic centers.

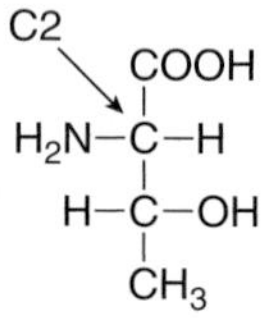

threonine

a. Draw the four possible stereoisomers using wedges and dashes.
b. The naturally occurring amino acid has the 2S,3R configuration at its two stereogenic centers. Which structure does this correspond to?

19.56 Proline is an unusual amino acid because its N atom on the α carbon is part of a five-membered ring.

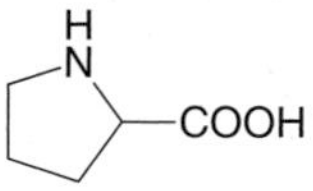

proline

a. Draw both enantiomers of proline.
b. Draw proline in its zwitterionic form.

19.57 For each amino acid [$RCH(NH_2)COOH$], draw its neutral, positively charged, and negatively charged forms. Which form predominates at pH = 1, 7, and 11? What is the structure of each amino acid at its isoelectric point?
a. methionine (R = $CH_2CH_2SCH_3$)   b. serine (R = CH_2OH)

19.58 Calculate the isoelectric point for each amino acid.
a. cysteine: pK_a (COOH) = 2.05; pK_a (α-NH_3^+) = 10.25   b. methionine: pK_a (COOH) = 2.28; pK_a (α-NH_3^+) = 9.21

19.59 Amino acids can be prepared from α-halo carboxylic acids [$RCH(X)COOH$] by reaction with excess NH_3. Why is excess NH_3 needed for this reaction?

Challenge Questions

19.60 Explain why using one or two equivalents of NaH results in different products in the following reactions.

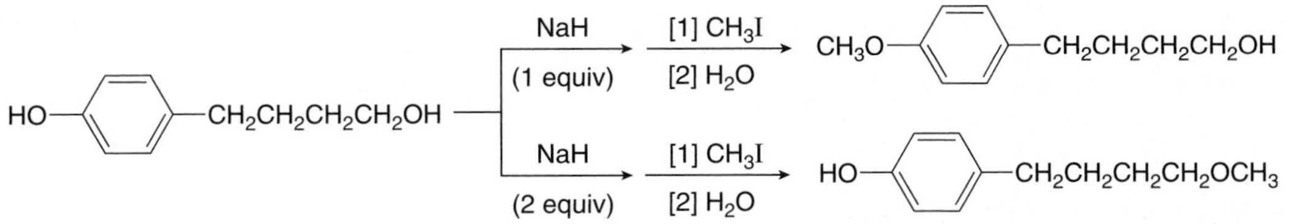

HO—⟨benzene⟩—CH$_2$CH$_2$CH$_2$CH$_2$OH

$\xrightarrow[\text{(1 equiv)}]{\text{NaH}}$ $\xrightarrow[\text{[2] H}_2\text{O}]{\text{[1] CH}_3\text{I}}$ CH$_3$O—⟨benzene⟩—CH$_2$CH$_2$CH$_2$CH$_2$OH

$\xrightarrow[\text{(2 equiv)}]{\text{NaH}}$ $\xrightarrow[\text{[2] H}_2\text{O}]{\text{[1] CH}_3\text{I}}$ HO—⟨benzene⟩—CH$_2$CH$_2$CH$_2$CH$_2$OCH$_3$

19.61 Although *p*-hydroxybenzoic acid is less acidic than benzoic acid, *o*-hydroxybenzoic acid is slightly more acidic than benzoic acid. Explain this result.

HO—⟨benzene⟩—COOH

p-hydroxybenzoic acid

⟨benzene with OH⟩—COOH

o-hydroxybenzoic acid

19.62 Adrenaline, the hormone that introduced Chapter 7, contains four hydrogen atoms (H$_a$–H$_d$) bonded to heteroatoms. Rank these hydrogens in order of increasing acidity.

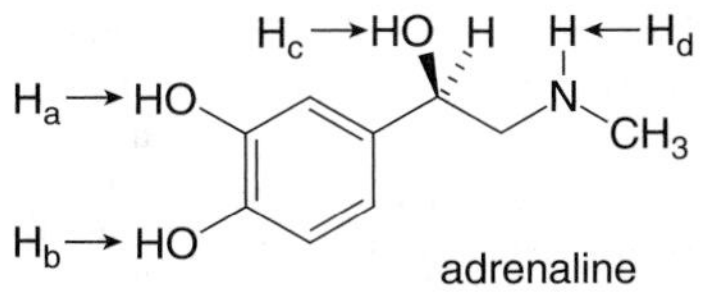

H$_c$ →HO H H H←H$_d$
H$_a$ → HO ⟨benzene⟩ ... N—CH$_3$
H$_b$ → HO

adrenaline

Introduction to Carbonyl Chemistry; Organometallic Reagents; Oxidation and Reduction

Juvenile hormones are a group of structurally related molecules that regulate the complex life cycle of an insect. In particular, they maintain the juvenile stage until an insect is ready for adulthood. This property has been exploited to control mosquitoes and other pests infecting both livestock and crops. Application of synthetic juvenile hormones to the egg or larva of an insect prevents maturation. With no sexually mature adults to propagate the next generation, the insect population is reduced. Juvenile hormones are synthesized by reactions that form new carbon–carbon bonds, thus allowing complex organic molecules to be prepared from simple starting materials. In Chapter 20 we will learn about these useful organic reactions.

© Bill Scifres

Chapters 20 through 24 of this text discuss carbonyl compounds—aldehydes, ketones, acid halides, esters, amides, and carboxylic acids. **The carbonyl group is perhaps the most important functional group in organic chemistry,** because its electron-deficient carbon and easily broken π bond make it susceptible to a wide variety of useful reactions.

We begin by examining the similarities and differences between two broad classes of carbonyl compounds. We will then spend the remainder of Chapter 20 on reactions that are especially important in organic synthesis. Chapters 21 and 22 present specific reactions that occur at the carbonyl carbon, and Chapters 23 and 24 concentrate on reactions occurring at the α carbon to the carbonyl group.

Although Chapter 20 is "jam-packed" with reactions, most of them follow one of two general pathways, so they can be classified in a well-organized fashion, provided you remember a few basic principles. Keep in mind the following fundamental themes about reactions:

♦ Nucleophiles attack electrophiles.

♦ π Bonds are easily broken.

♦ Bonds to good leaving groups are easily cleaved heterolytically.

20.1 Introduction

Two broad classes of compounds contain a *carbonyl group:*

carbonyl group

[1] Compounds that have only carbon and hydrogen atoms bonded to the carbonyl group

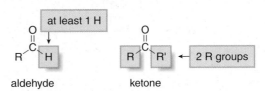

aldehyde   ketone

♦ An **aldehyde** has at least one H atom bonded to the carbonyl group.

♦ A **ketone** has two alkyl or aryl groups bonded to the carbonyl group.

[2] Compounds that contain an electronegative atom bonded to the carbonyl group

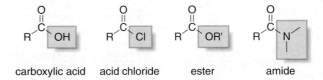

carboxylic acid   acid chloride   ester   amide

These include **carboxylic acids, acid chlorides, esters,** and **amides,** as well as other similar compounds discussed in Chapter 22. Each of these compounds contains an electronegative atom (Cl, O, or N) capable of acting as a **leaving group.** Acid chlorides, esters, and amides are often called **carboxylic acid** *derivatives,* because they can be synthesized from carboxylic acids (Chapter 22).

♦ The presence or absence of a leaving group on the carbonyl carbon determines the type of reactions these compounds undergo (Section 20.2).

Figure 20.1 Electrostatic potential map of formaldehyde, $CH_2=O$

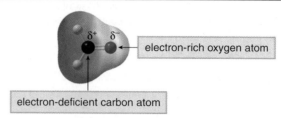

- An electrostatic potential map shows the electron-deficient carbon and the electron-rich oxygen atom of the carbonyl group.

The carbonyl carbon atom is sp^2 hybridized and trigonal planar, and all bond angles are ~120°. The double bond of a carbonyl group consists of one σ bond and one π bond. The π bond is formed by the overlap of two p orbitals, and extends above and below the plane. In these features the carbonyl group resembles the trigonal planar, sp^2 hybridized carbons of a C–C double bond.

In one important way, though, a C=O and C=C are very different. **The electronegative oxygen atom in the carbonyl group means that the bond is polarized, making the carbonyl carbon electron deficient.** Using a resonance description, the carbonyl group is represented by two resonance structures, with a charge-separated resonance structure a minor contributor to the hybrid. An electrostatic potential plot for formaldehyde, the simplest aldehyde, is shown in Figure 20.1. It clearly indicates the polarized carbonyl group.

PROBLEM 20.1

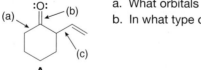

a. What orbitals are used to form the indicated bonds in **A**?

b. In what type of orbitals do the lone pairs on O reside?

20.2 General Reactions of Carbonyl Compounds

With what types of reagents should a carbonyl group react? The electronegative oxygen makes the carbonyl carbon electrophilic, and because it is trigonal planar, a carbonyl carbon is uncrowded. Moreover, a carbonyl group has an easily broken π bond.

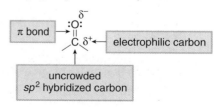

As a result, **carbonyl compounds react with nucleophiles.** The outcome of nucleophilic attack, however, depends on the identity of the carbonyl starting material.

◆ Aldehydes and ketones undergo nucleophilic addition.

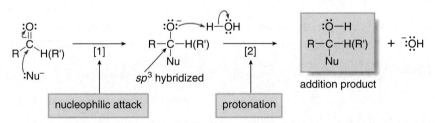

◆ Carbonyl compounds that contain leaving groups undergo nucleophilic substitution.

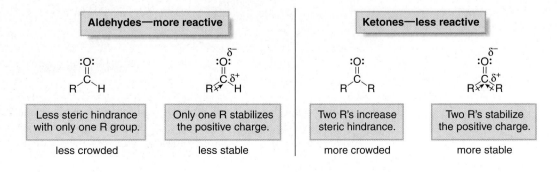

Let's examine each of these general reactions individually.

20.2A   Nucleophilic Addition to Aldehydes and Ketones

Aldehydes and ketones react with nucleophiles to form addition products by a two-step process: **nucleophilic attack** followed by **protonation.**

Nucleophilic Addition—A Two-Step Process

◆ In Step [1], **the nucleophile (:Nu⁻) attacks the electrophilic carbonyl.** As the new bond to the nucleophile forms, the π bond is broken, moving an electron pair out on the oxygen atom. This forms an sp^3 hybridized intermediate.

◆ In Step [2], **protonation of the negatively charged oxygen atom by H₂O** (or another proton source) forms the addition product.

More examples of nucleophilic addition to aldehydes and ketones are discussed in Chapter 21.

The net result is that the π bond is broken, two new σ bonds are formed, and the elements of H and Nu are added across the π bond. Nucleophilic addition with two different nucleophiles—**hydride (H:⁻)** and **carbanions (R:⁻)**—is discussed in Chapter 20.

Aldehydes are more reactive than ketones towards nucleophilic attack for both steric and electronic reasons.

◆ The two R groups bonded to the ketone carbonyl group make it more crowded, so nucleophilic attack is more difficult.
◆ The two electron-donor R groups stabilize the partial charge on the carbonyl carbon of a ketone, making it more stable and less reactive.

20.2B Nucleophilic Substitution of RCOZ (Z = Leaving Group)

Carbonyl compounds with leaving groups react with nucleophiles to form substitution products by a two-step process: **nucleophilic attack,** followed by **loss of the leaving group.**

Nucleophilic Substitution—A Two-Step Process

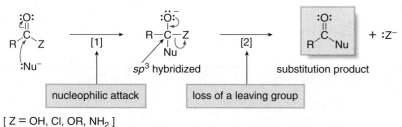

[Z = OH, Cl, OR, NH₂]

◆ In Step [1], **the nucleophile (:Nu⁻) attacks the electrophilic carbonyl,** forming an sp^3 hybridized intermediate. This step is identical to nucleophilic addition.

◆ Step [2] is different. Because the intermediate contains an electronegative atom Z, **Z can act as a leaving group.** To do so, an electron pair on O re-forms the π bond, and Z leaves with the electron pair in the C–Z bond.

The net result is that Nu replaces Z—a nucleophilic substitution reaction. This reaction is often called **nucleophilic *acyl* substitution** to distinguish it from the nucleophilic substitution reactions at sp^3 hybridized carbons discussed in Chapter 7. Nucleophilic substitution with two different nucleophiles—**hydride (H:⁻)** and **carbanions (R:⁻)**—is discussed in Chapter 20. Other nucleophiles are examined in Chapter 22.

Carboxylic acid derivatives differ greatly in their reactivity towards nucleophiles. The order in which they react parallels the leaving group ability of the group Z bonded to the carbonyl carbon.

Recall from Section 7.7 that the weaker the base, the better the leaving group.

◆ The better the leaving group Z, the more reactive RCOZ is in nucleophilic acyl substitution.

Thus, the following trends result:

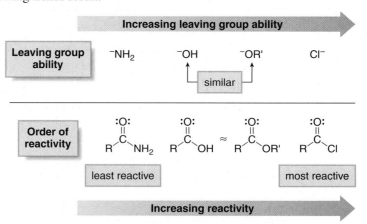

◆ Acid chlorides (RCOCl), which have the best leaving group (Cl⁻), are the most reactive carboxylic acid derivatives, and amides (RCONH₂), which have the worst leaving group (⁻NH₂), are the least reactive.

◆ Carboxylic acids (RCOOH) and esters (RCOOR'), which have leaving groups of similar basicity (⁻OH and ⁻OR'), fall in the middle.

Nucleophilic addition and nucleophilic acyl substitution involve the *same* first step—**nucleophilic attack on the electrophilic carbonyl group** to form a tetrahedral intermediate. The difference between them is what then happens to this intermediate. Aldehydes and ketones cannot undergo substitution because they have no leaving group bonded to the newly formed sp^3 hybridized carbon. Nucleophilic substitution with an aldehyde, for example, would form $H:^-$, an extremely strong base and therefore a very poor (and highly unlikely) leaving group.

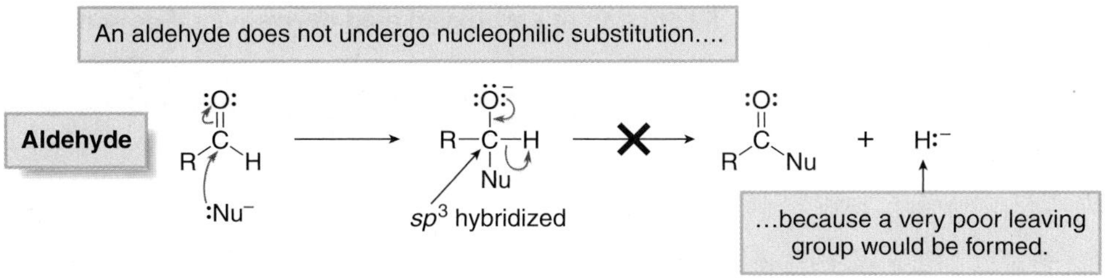

An aldehyde does not undergo nucleophilic substitution....

Aldehyde

sp^3 hybridized

...because a very poor leaving group would be formed.

PROBLEM 20.2 Which compounds undergo nucleophilic addition and which undergo substitution?
a. $(CH_3)_2C=O$ b. $CH_3CH_2CH_2COCl$ c. CH_3COOCH_3 d. C_6H_5CHO

PROBLEM 20.3 Which compound in each pair is more reactive towards nucleophilic attack?
a. $CH_3CH_2CH_2CHO$ and $CH_3CH_2CH_2COCH_3$ c. CH_3CH_2COCl and CH_3COOCH_3
b. $CH_3CH_2COCH_3$ and $CH_3CH(CH_3)COCH_2CH_3$ d. CH_3COOCH_3 and $CH_3CONHCH_3$

To show how these general principles of nucleophilic substitution and addition apply to carbonyl compounds, we are going to discuss oxidation and reduction reactions, and reactions with organometallic reagents—compounds that contain carbon–metal bonds. We begin with reduction to build on what you learned previously in Chapter 12.

20.3 A Preview of Oxidation and Reduction

Recall the definitions of oxidation and reduction presented in Section 12.1:

◆ Oxidation results in an increase in the number of C−Z bonds (usually C−O bonds) or a decrease in the number of C−H bonds.
◆ Reduction results in a decrease in the number of C−Z bonds (usually C−O bonds) or an increase in the number of C−H bonds.

Carbonyl compounds are either reactants or products in many of these reactions, as illustrated in the accompanying diagram. For example, because aldehydes fall in the middle of this scheme, they can be both oxidized and reduced. Carboxylic acids and their derivatives (RCOZ), on the other hand, are already highly oxidized, so their only useful reaction is reduction.

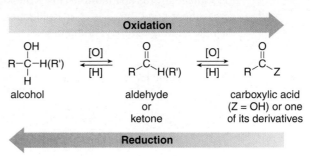

The three most useful oxidation and reduction reactions of carbonyl starting materials can be summarized as follows:

[1]   Reduction of aldehydes and ketones to alcohols (Sections 20.4–20.6)

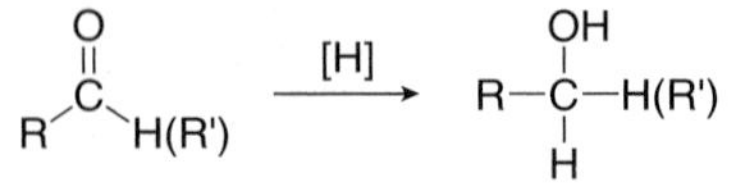

aldehyde or ketone   1° or 2° alcohol

Aldehydes and ketones are reduced to 1° and 2° alcohols, respectively.

[2]   Reduction of carboxylic acid derivatives (Section 20.7)

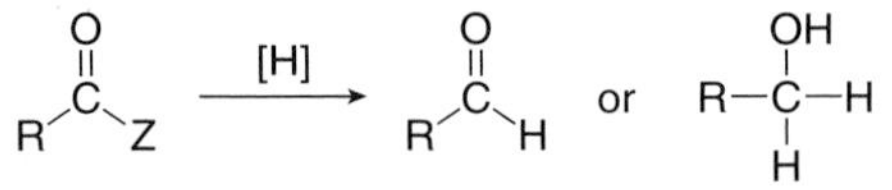

aldehyde   1° alcohol

The reduction of carboxylic acids and their derivatives gives a variety of products, depending on the identity of Z and the nature of the reducing agent. The usual products are aldehydes or 1° alcohols.

[3]   Oxidation of aldehydes to carboxylic acids (Section 20.8)

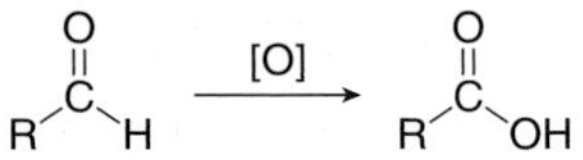

aldehyde   carboxylic acid

The most useful oxidation reaction of carbonyl compounds is the oxidation of aldehydes to carboxylic acids.

We begin with reduction, because the mechanisms of reduction reactions follow directly from the general mechanisms for nucleophilic addition and substitution.

20.4   Reduction of Aldehydes and Ketones

LiAlH$_4$ and NaBH$_4$ serve as a source of H:$^-$, but there are no free H:$^-$ ions present in reactions with these reagents.

The most useful reagents for reducing aldehydes and ketones are the metal hydride reagents (Section 12.2). The two most common metal hydride reagents are **sodium borohydride (NaBH$_4$)** and **lithium aluminum hydride (LiAlH$_4$)**. These reagents contain a polar metal–hydrogen bond that serves as a source of the nucleophile hydride, **H:$^-$**. LiAlH$_4$ is a stronger reducing agent than NaBH$_4$, because the Al–H bond is more polar than the B–H bond.

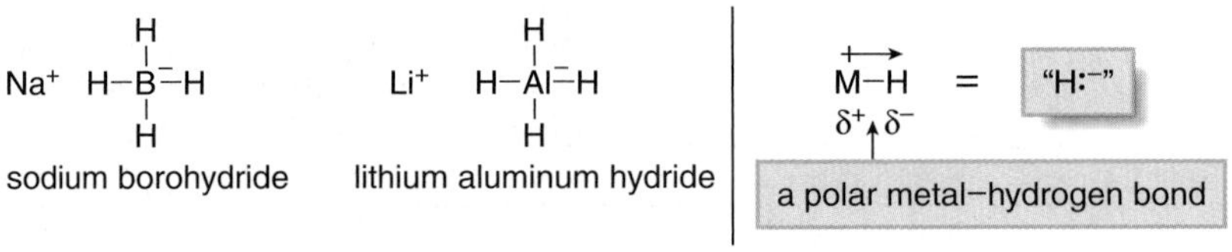

sodium borohydride   lithium aluminum hydride   a polar metal–hydrogen bond

20.4A   Reduction with Metal Hydride Reagents

Treating an aldehyde or a ketone with NaBH$_4$ or LiAlH$_4$, followed by water or some other proton source, affords an **alcohol.** This is an addition reaction because the elements of H$_2$ are added across the π bond, but it is also a reduction because the product alcohol has fewer C–O bonds than the starting carbonyl compound.

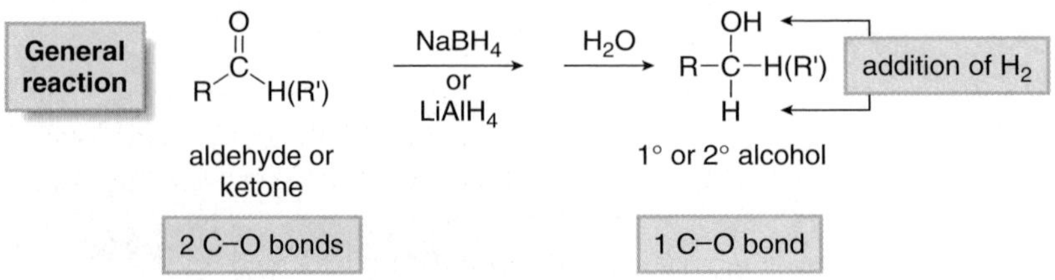

aldehyde or ketone   1° or 2° alcohol
2 C–O bonds   1 C–O bond

LiAlH$_4$ reductions must be carried out under anhydrous conditions, because water reacts violently with the reagent. Water is added to the reaction mixture (to serve as a proton source) *after* the reduction with LiAlH$_4$ is complete.

The product of this reduction reaction is a **1° alcohol** when the starting carbonyl compound is an aldehyde, and a **2° alcohol** when it is a ketone.

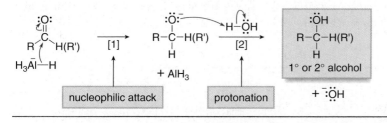

NaBH$_4$ selectively reduces aldehydes and ketones in the presence of most other functional groups. Reductions with NaBH$_4$ are typically carried out in CH$_3$OH as solvent. LiAlH$_4$ reduces aldehydes and ketones and many other functional groups as well (Sections 12.6 and 20.7).

PROBLEM 20.4   What alcohol is formed when each compound is treated with NaBH$_4$ in CH$_3$OH?

a. CH$_3$CH$_2$CH$_2$C(=O)H   b.   c.

PROBLEM 20.5   What aldehyde or ketone is needed to prepare each alcohol by metal hydride reduction?

a.   b.   c.

PROBLEM 20.6   Why can't 1-methylcyclohexanol be prepared from a carbonyl compound by reduction?

20.4B   The Mechanism of Hydride Reduction

Hydride reduction of aldehydes and ketones occurs via the general mechanism of nucleophilic addition—that is, **nucleophilic attack** followed by **protonation.** Mechanism 20.1 is shown using LiAlH$_4$, but an analogous mechanism can be written for NaBH$_4$.

MECHANISM 20.1

LiAlH$_4$ Reduction of RCHO and R$_2$C=O

◆ In Step [1], **the nucleophile (AlH$_4^-$) donates H:$^-$ to the carbonyl group,** cleaving the π bond, and moving an electron pair onto oxygen. This forms a new C–H bond.

◆ In Step [2], **the alkoxide is protonated by H$_2$O (or CH$_3$OH)** to form the alcohol reduction product. This acid–base reaction forms a new O–H bond.

◆ The net result of adding H:$^-$ (from NaBH$_4$ or LiAlH$_4$) and H$^+$ (from H$_2$O) is the addition of the elements of H$_2$ to the carbonyl π bond.

PROBLEM 20.7   What product is formed when cyclohexanone is reduced with NaBD$_4$ in CH$_3$OH?

20.4C   Catalytic Hydrogenation of Aldehydes and Ketones

Catalytic hydrogenation also reduces aldehydes and ketones to 1° and 2° alcohols, respectively, using H$_2$ and Pd-C (or another metal catalyst). H$_2$ adds to the C=O in much the same way that it adds to the C=C of an alkene (Section 12.3). The metal catalyst (Pd-C) provides a surface that binds the carbonyl starting material and H$_2$, and two H atoms are sequentially transferred with cleavage of the π bond.

Examples

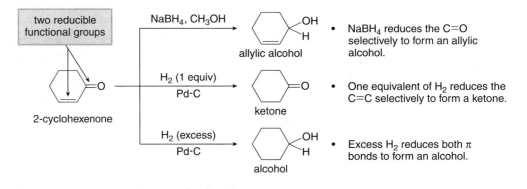

When a compound contains both a carbonyl group and a carbon–carbon double bond, selective reduction of one functional group can be achieved by proper choice of reagent.

- ◆ A C=C is reduced faster than a C=O with H_2 (Pd-C).
- ◆ A C=O is readily reduced with $NaBH_4$ and $LiAlH_4$, but a C=C is inert.

Thus, 2-cyclohexenone, a compound that contains both a carbon–carbon double bond and a carbonyl group, can be reduced to three different compounds—an allylic alcohol, a carbonyl compound, or an alcohol—depending on the reagent.

two reducible functional groups		
2-cyclohexenone	**NaBH₄, CH₃OH** → allylic alcohol	• NaBH₄ reduces the C=O selectively to form an allylic alcohol.
	H₂ (1 equiv) Pd-C → ketone	• One equivalent of H₂ reduces the C=C selectively to form a ketone.
	H₂ (excess) Pd-C → alcohol	• Excess H₂ reduces both π bonds to form an alcohol.

PROBLEM 20.8   Draw the products formed when $CH_3COCH_2CH_2CH=CH_2$ is treated with each reagent: (a) $LiAlH_4$, then H_2O; (b) $NaBH_4$ in CH_3OH; (c) H_2 (1 equiv), Pd-C; (d) H_2 (excess), Pd-C; (e) $NaBH_4$ (excess) in CH_3OH.

20.4D   Examples in Synthesis

The reduction of aldehydes and ketones is a common reaction used in the synthesis of many useful natural products. Two examples are shown in Figure 20.2.

Figure 20.2 $NaBH_4$ reductions used in organic synthesis

The male musk deer, a small antlerless deer found in the mountain regions of China and Tibet, has long been hunted for its musk, a strongly scented liquid used in early medicine and later in perfumery.

- **Muscone** is the major compound in musk, one of the oldest known ingredients in perfumes. Musk was originally isolated from the male musk deer, but it can now be prepared synthetically in the laboratory in a variety of ways.

(Figure shows: A → [NaBH₄, CH₃OH] → B → [3 steps] → ibuprofen (anti-inflammatory agent in Motrin and Advil); and a cyclic ketone → [NaBH₄, CH₃OH] → alcohol → [4 steps] → muscone, odor of musk (perfume component))

20.5   The Stereochemistry of Carbonyl Reduction

The stereochemistry of carbonyl reduction follows the same principles we have previously learned.

Recall from Section 9.15 that an achiral starting material gives a racemic mixture when a new stereogenic center is formed.

Reduction converts a **planar sp^2 hybridized carbonyl carbon to a tetrahedral sp^3 hybridized carbon.** What happens when a new stereogenic center is formed in this process? With an achiral reagent like $NaBH_4$ or $LiAlH_4$, a racemic product is obtained. For example, $NaBH_4$ in CH_3OH solution reduces 2-butanone, an achiral ketone, to 2-butanol, an alcohol that contains a new stereogenic center. Both enantiomers of 2-butanol are formed in equal amounts.

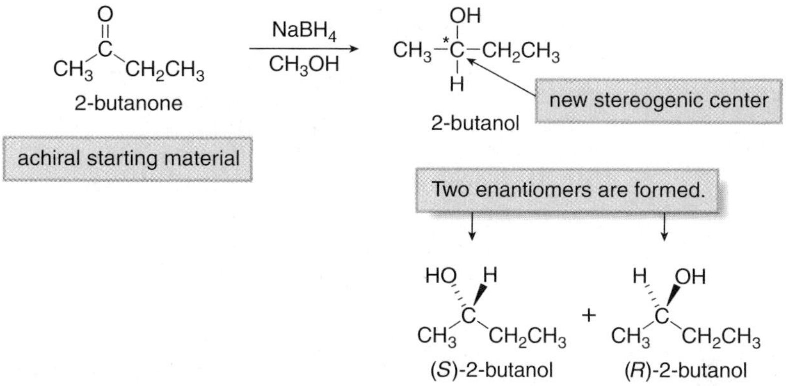

Why is a racemic mixture formed? Because the carbonyl carbon is sp^2 hybridized and planar, hydride can approach the double bond with equal probability from both sides of the plane, forming two alkoxides, which are **enantiomers** of each other. Protonation of the alkoxides gives an equal amount of two alcohols, which are also **enantiomers.**

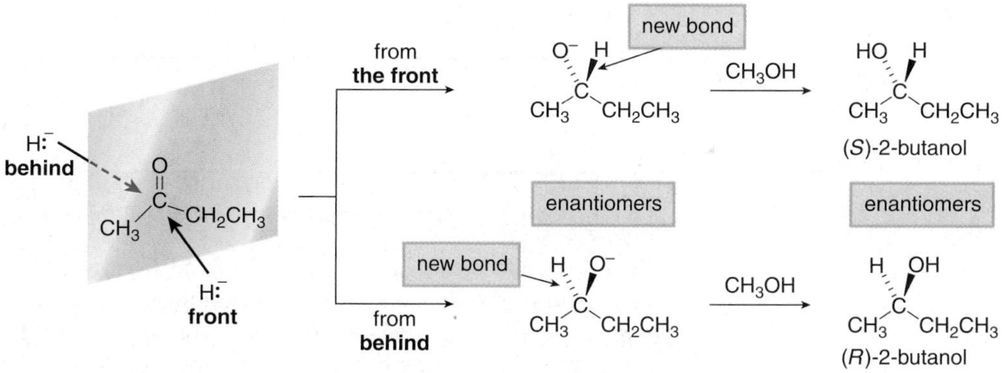

◆ Conclusion: Hydride reduction of an achiral ketone with $LiAlH_4$ or $NaBH_4$ gives a racemic mixture of two alcohols when a new stereogenic center is formed.

PROBLEM 20.9   Draw the products formed (including stereoisomers) when each compound is reduced with $NaBH_4$ in CH_3OH.

a. 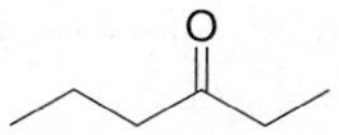   b.    c. 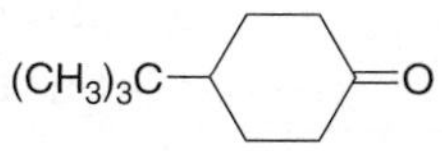 $(CH_3)_3C$

20.6   Enantioselective Carbonyl Reductions

20.6A   CBS Reagents

One enantiomer can be formed selectively from the reduction of a carbonyl group, provided a **chiral reducing agent** is used. This strategy is identical to that employed in the Sharpless asymmetric epoxidation reaction (Section 12.14). A reduction that forms one enantiomer predominantly or exclusively is an **enantioselective** or **asymmetric reduction.**

Many different chiral reducing agents have now been prepared for this purpose. One such reagent, formed by reacting borane (**BH₃**) with a heterocycle called an **oxazaborolidine,** has one stereogenic center (and thus two enantiomers).

Two enantiomers of the chiral CBS reducing agent

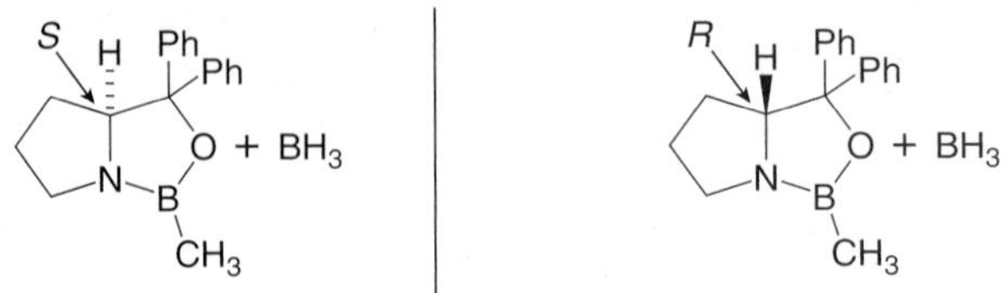

(S)-2-methyl-**CBS**-oxazaborolidine

(S)-CBS reagent

(R)-2-methyl-**CBS**-oxazaborolidine

(R)-CBS reagent

These reagents are called the **(S)-CBS reagent** and the **(R)-CBS reagent.** CBS refers to *Corey, Bakshi,* and *Shibata,* the chemists who developed these versatile reagents. One B−H bond of BH₃ serves as the source of hydride in this reduction. CBS reagents predictably give one enantiomer as the major product of ketone reduction, as illustrated with acetophenone as the starting material.

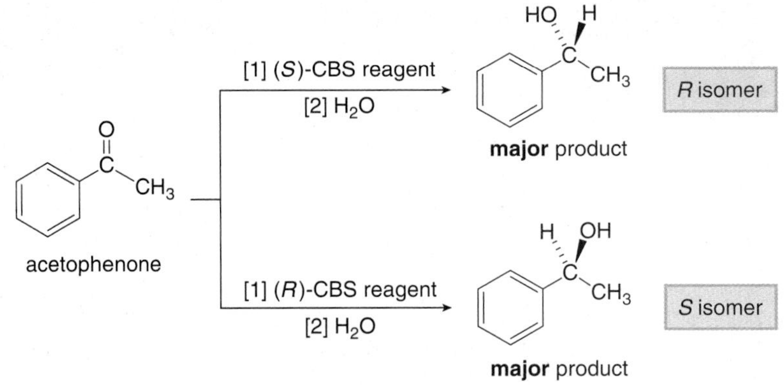

♦ The (S)-CBS reagent generally gives the R alcohol as major product.
♦ The (R)-CBS reagent generally gives the S alcohol as major product.

These reagents are highly enantioselective. Treatment of propiophenone with the (S)-CBS reagent forms the **R** alcohol in 97% enantiomeric excess (ee). Enantioselective reduction using these reagents has provided the key step in the synthesis of several widely used drugs, including salmeterol, a long-acting bronchodilator shown in Figure 20.3.

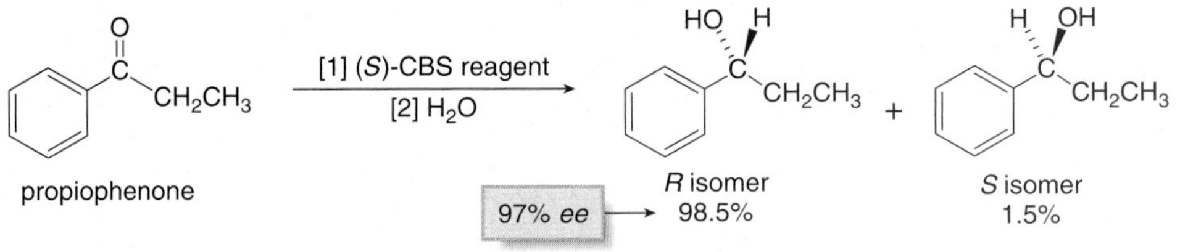

This new technology provides access to single enantiomers of biologically active compounds, often previously available only as a racemic mixture.

Figure 20.3
Enantioselective reduction—A key step in the synthesis of salmeterol

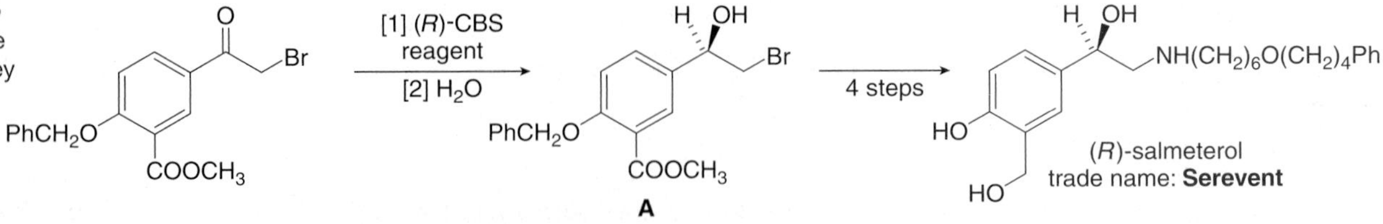

- (R)-Salmeterol is a long-acting bronchodilator used for the treatment of asthma.
- In this example, the (R)-CBS reagent adds the new H atom from behind, the same result observed with acetophenone and propiophenone. In this case, however, alcohol **A** has the R configuration using the rules for assigning priority in Chapter 5.

PROBLEM 20.10   What reagent is needed to reduce **A** to **B,** an intermediate in the synthesis of the antidepressant (*R*)-fluoxetine (trade name: Prozac)?

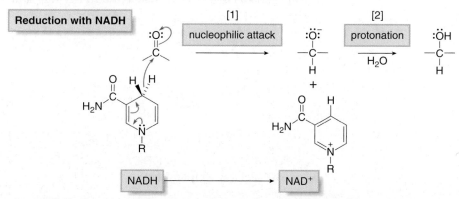

A   **B**   (*R*)-fluoxetine
                                                             Trade name: **Prozac**

20.6B   Enantioselective Biological Reduction

Although laboratory reduction reactions often do not proceed with 100% enantioselectivity, biological reductions that occur in cells *always* proceed with complete selectivity, forming a single enantiomer. NaBH$_4$ or chiral boranes are not the reducing agents for these processes. In cells, the reducing agent is **NADH.**

the **reactive** part

NADH
(abbreviated structure)

NADH
nicotinamide adenine dinucleotide
(reduced form)

NADH is a **coenzyme,** an organic molecule that can function only in the presence of an enzyme. The active site of the enzyme binds both the carbonyl substrate and NADH, keeping them in close proximity. **NADH then donates H:⁻** in much the same way as a metal hydride reagent; that is, reduction consists of nucleophilic attack followed by protonation.

 ◆ In Step [1], **NADH donates H:⁻ to the carbonyl group** to form an alkoxide. In the process, NADH is converted to NAD⁺.

 ◆ In Step [2], the alkoxide is protonated by the aqueous medium.

Reduction with NADH

[1] nucleophilic attack

[2] protonation
H$_2$O

NADH ———————→ NAD⁺

This reaction is completely enantioselective. For example, reduction of pyruvic acid with NADH catalyzed by lactate dehydrogenase affords a single enantiomer of lactic acid with the *S* configuration. NADH reduces a variety of different carbonyl compounds in biological systems. The configuration of the product (*R* or *S*) depends on the enzyme used to catalyze the process.

pyruvic acid

NADH
(H⁺ source)

lactate
dehydrogenase

(*S*)-lactic acid
only product

not formed

[*] denotes a new stereogenic center

Niacin can be obtained from foods such as soybeans, which contain it naturally, and from breakfast cereals, which are fortified with it.

NAD$^+$, the oxidized form of NADH, is a biological oxidizing agent capable of oxidizing alcohols to carbonyl compounds (it forms NADH in the process). NAD$^+$ is synthesized from the vitamin niacin, which can be obtained from soybeans among other dietary sources. Breakfast cereals are fortified with niacin to help people consume their recommended daily allowance of this B vitamin.

niacin
vitamin B$_3$

NAD$^+$

20.7   Reduction of Carboxylic Acids and Their Derivatives

The reduction of carboxylic acids and their derivatives (RCOZ) is complicated because the products obtained depend on the identity of both the leaving group (Z) and the reducing agent. Metal hydride reagents are the most useful reducing reagents. **Lithium aluminum hydride is a strong reducing agent that reacts with *all* carboxylic acid derivatives.** Two other related but milder reducing agents are also used.

[1] **Diisobutylaluminum hydride, [(CH$_3$)$_2$CHCH$_2$]$_2$AlH,** abbreviated as **DIBAL-H,** has two bulky isobutyl groups, which make this reagent less reactive than LiAlH$_4$. The single H atom is donated as H:$^-$ in hydride reductions.

[2] **Lithium tri-*tert*-butoxyaluminum hydride, LiAlH[OC(CH$_3$)$_3$]$_3$,** has three electronegative oxygen atoms bonded to aluminum, which make this reagent less nucleophilic than LiAlH$_4$.

LiAlH$_4$ is a strong reducing agent. DIBAL-H and LiAlH[OC(CH$_3$)$_3$]$_3$ are milder, more selective reducing agents.

Al–H   =   **[(CH$_3$)$_2$CHCH$_2$]$_2$AlH**

diisobutylaluminum hydride
DIBAL-H

Li$^+$ H–Al=OC(CH$_3$)$_3$, with OC(CH$_3$)$_3$ groups   =   **LiAlH[OC(CH$_3$)$_3$]$_3$**

lithium tri-*tert*-butoxyaluminum hydride

20.7A   Reduction of Acid Chlorides and Esters

Acid chlorides and esters can be reduced to either aldehydes or 1° alcohols, depending on the reagent.

Acid chloride

R–C(=O)–Cl

or

Ester

R–C(=O)–OR'

[H] →

R–C(=O)–H   or   RCH$_2$OH

aldehyde   1° alcohol

♦ LiAlH$_4$ converts RCOCl and RCOOR' to 1° alcohols.
♦ A milder reducing agent (DIBAL-H or LiAlH[OC(CH$_3$)$_3$]$_3$) converts RCOCl or RCOOR' to RCHO at low temperatures.

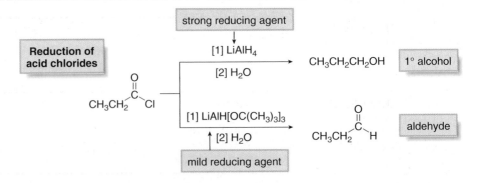

strong reducing agent

Reduction of acid chlorides

CH$_3$CH$_2$–C(=O)–Cl

[1] LiAlH$_4$
[2] H$_2$O
→ CH$_3$CH$_2$CH$_2$OH   1° alcohol

[1] LiAlH[OC(CH$_3$)$_3$]$_3$
[2] H$_2$O
→ CH$_3$CH$_2$–C(=O)–H   aldehyde

mild reducing agent

In the reduction of an acid chloride, Cl⁻ comes off as the leaving group.

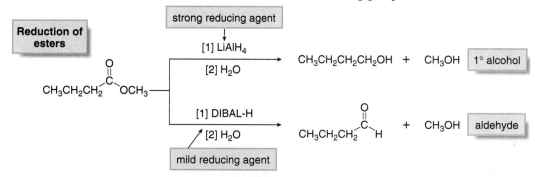

In the reduction of the ester, CH_3O^- comes off as the leaving group, which is then protonated by H_2O to form CH_3OH.

Mechanism 20.2 illustrates why two different products are possible. It can be conceptually divided into two parts: **nucleophilic substitution** to form an aldehyde, followed by **nucleophilic addition** to the aldehyde to form a 1° alcohol. A general mechanism is drawn using $LiAlH_4$ as reducing agent.

MECHANISM 20.2

Reduction of RCOCl and RCOOR' with a Metal Hydride Reagent

Part [1] Nucleophilic substitution forms an aldehyde.

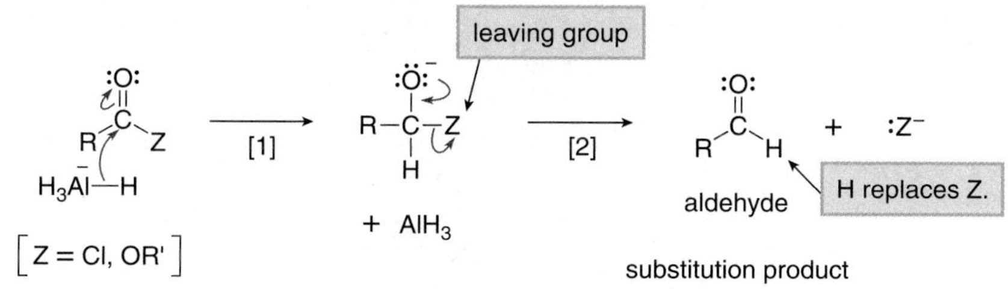

◆ **Nucleophilic attack of H:⁻** (from $LiAlH_4$) in Step [1] forms a tetrahedral intermediate with a leaving group Z.

◆ In Step [2], the π bond is re-formed and **Z⁻ comes off.** The overall result of the addition of H:⁻ and elimination of Z⁻ is the substitution of H for Z.

Part [2] Nucleophilic addition forms a 1° alcohol.

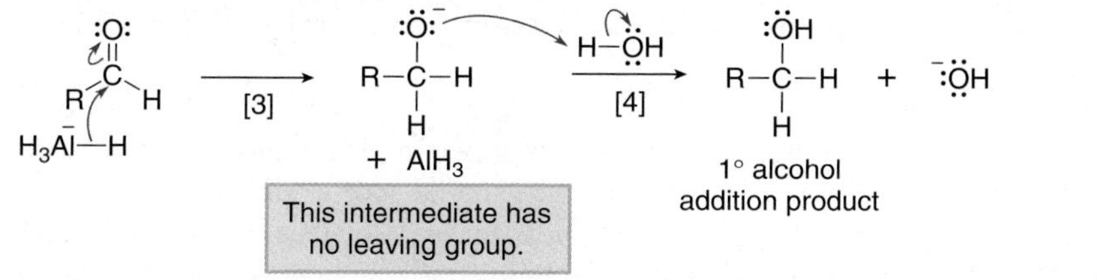

◆ **Nucleophilic attack of H:⁻** (from $LiAlH_4$) in Step [3] forms an alkoxide.

◆ **Protonation of the alkoxide** by H_2O in Step [4] forms the alcohol reduction product. The overall result of Steps [3] and [4] is addition of H_2.

With milder reducing agents such as DIBAL-H and $LiAlH[OC(CH_3)_3]_3$, the reaction stops after reaction with one equivalent of H:⁻ and the aldehyde is formed as product. With a stronger reducing agent like $LiAlH_4$, two equivalents of H:⁻ are added and a 1° alcohol is formed.

PROBLEM 20.11   Draw a stepwise mechanism for the following reaction.

$$CH_3COCl \xrightarrow[\text{[2] } H_2O]{\text{[1] } LiAlH_4} CH_3CH_2OH$$

PROBLEM 20.12   Draw the structure of both an acid chloride and an ester that can be used to prepare each compound by reduction.

a. 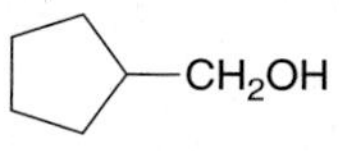 (cyclopentyl)—CH_2OH b. 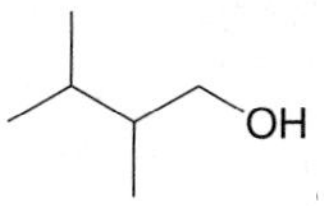 (isopropyl-branched chain)—OH

Figure 20.4 The DIBAL-H reduction of an ester to an aldehyde in the synthesis of the marine neurotoxin ciguatoxin CTX3C

Only the ester is reduced.

[1] DIBAL-H
[2] H₂O

several steps

ciguatoxin CTX3C

- One step in a lengthy synthesis of ciguatoxin CTX3C involved selective reduction of an ester to an aldehyde using DIBAL-H.

Thousands of people contract ciguatera seafood poisoning each year from ingesting tropical reef fish containing ciguatoxin. Even very low concentrations of ciguatoxin CTX3C cause gastrointestinal and neurological problems, leading to paralysis and sometimes death.

Selective reductions are routinely used in the synthesis of highly complex natural products such as **ciguatoxin CTX3C,** a potent neurotoxin found in more than 400 species of warm-water fish. Interest in providing a practical supply of ciguatoxin CTX3C for biological studies led to its laboratory synthesis in 2001. One reaction in the synthesis involved the reduction of an ester to an aldehyde using DIBAL-H, as shown in Figure 20.4.

20.7B Reduction of Carboxylic Acids and Amides

Carboxylic acids are reduced to 1° alcohols with LiAlH₄. LiAlH₄ is too strong a reducing agent to stop the reaction at the aldehyde stage, but milder reagents are not strong enough to initiate the reaction in the first place, so this is the only useful reduction reaction of carboxylic acids.

Reduction of a RCOOH— General reaction

$$R-\overset{O}{\underset{}{C}}-OH \xrightarrow[\text{[2] H}_2\text{O}]{\text{[1] LiAlH}_4} RCH_2OH$$
1° alcohol

Example

$$CH_3CH_2CH_2-\overset{O}{\underset{}{C}}-OH \xrightarrow[\text{[2] H}_2\text{O}]{\text{[1] LiAlH}_4} CH_3CH_2CH_2CH_2OH$$

Two C–O bonds are replaced by C–H bonds.

Unlike the LiAlH$_4$ reduction of all other carboxylic acid derivatives, which affords 1° alcohols, the **LiAlH$_4$ reduction of amides forms amines.**

Reduction of an amide—General reaction

$$R-\overset{\overset{\displaystyle O}{\|}}{C}-\overset{\overset{\displaystyle H(R')}{|}}{\underset{\underset{\displaystyle H(R')}{|}}{\ddot{N}}} \xrightarrow[\text{[2] H}_2\text{O}]{\text{[1] LiAlH}_4} RCH_2-\overset{\overset{\displaystyle H(R')}{}}{\underset{\underset{\displaystyle H(R')}{|}}{\ddot{N}}}-H(R')$$ amine

Both C–O bonds are replaced by C–H bonds.

Both C–O bonds are reduced to C–H bonds by LiAlH$_4$, and any H atom or R group bonded to the amide nitrogen atom remains bonded to it in the product. Because $^-$NH$_2$ (or $^-$NHR or $^-$NR$_2$) is a poorer leaving group than Cl$^-$ or $^-$OR, $^-$NH$_2$ is never lost during reduction, and therefore it forms an amine in the final product.

Examples

$$CH_3CH_2-\overset{\overset{\displaystyle O}{\|}}{C}-NH_2 \xrightarrow[\text{[2] H}_2\text{O}]{\text{[1] LiAlH}_4} CH_3CH_2CH_2-NH_2$$

(cyclohexyl)$-\overset{\overset{\displaystyle O}{\|}}{C}-$NHCH$_3$ $\xrightarrow[\text{[2] H}_2\text{O}]{\text{[1] LiAlH}_4}$ (cyclohexyl)$-$CH$_2$NHCH$_3$

Imines and related compounds are discussed in Chapter 21.

The mechanism, illustrated in Mechanism 20.3 with RCONH$_2$ as starting material, is somewhat different than the previous reductions of carboxylic acid derivatives. Amide reduction proceeds with formation of an intermediate *imine,* **a compound containing a C–N double bond,** which is then further reduced to an amine.

MECHANISM 20.3

Reduction of an Amide to an Amine with LiAlH$_4$

Part [1]   Reduction of an amide to an imine

$$R-\overset{\overset{\displaystyle :O:}{\|}}{C}-\overset{|}{\underset{|}{\overset{|}{N}}}-H \xrightarrow[\text{[1]}]{H-\bar{A}lH_3} \overset{:\ddot{O}:^-}{\underset{+ H_2}{R-\overset{}{C}-\ddot{N}H}} \xrightarrow[\underset{H_3\bar{A}l-H}{\text{[2]}}]{} \overset{:\ddot{O}:^{\diagdown AlH_3}}{R-\overset{}{C}-\ddot{N}H} \xrightarrow[\text{[3]}]{} \overset{:\ddot{O}:^{\diagdown AlH_3}}{R-\overset{|}{\underset{H}{C}}-\ddot{N}H} \xrightarrow[+ AlH_3]{\text{[4]}} \underset{H}{\overset{R}{\diagdown}}C=NH$$

imine

+ $^-$OAlH$_3$

◆ In Part [1], the amide is converted to an **imine** by a series of steps: proton transfer, nucleophilic attack of H:$^-$, and loss of AlH$_3$O$^-$ (a leaving group).

◆ An imine is similar to a carbonyl compound. Its polarized C=N makes it susceptible to nucleophilic attack.

Part [2]   Reduction of an imine to an amine

$$H_3\bar{A}l-H \quad \underset{H}{\overset{R}{\diagdown}}C=NH \xrightarrow[\text{[5]}]{} \underset{\underset{+ AlH_3}{|}}{R-\overset{|}{\underset{H}{C}}-\ddot{N}H} \xrightarrow[\text{[6]}]{H-\ddot{O}H} \underset{H}{R-\overset{|}{\underset{|}{C}}-\ddot{N}H_2} + :\ddot{O}H$$

amine

◆ **Nucleophilic attack of H:$^-$** (from LiAlH$_4$) in Step [5], followed by protonation of the intermediate nitrogen anion, forms the **amine.** The overall result of Steps [5] and [6] is addition of H$_2$ to the intermediate imine.

PROBLEM 20.13   Draw the products formed from LiAlH$_4$ reduction of each compound.

a. (structure: 2,2-dimethylbutanoic acid, $\overset{\overset{\displaystyle O}{\|}}{C}$—OH)
b. (structure: hexanamide, $\overset{\overset{\displaystyle O}{\|}}{C}$—NH$_2$)
c. (structure: cyclohexyl amide, $\overset{\overset{\displaystyle O}{\|}}{C}$—N(CH$_3$)$_2$)
d. (structure: cyclic lactam with NH)

PROBLEM 20.14   What amide will form each of the following amines on treatment with LiAlH$_4$?

a. (structure: benzyl CH$_2$NH$_2$)
b. (structure: cyclohexylmethyl—N(CH$_2$CH$_3$)$_2$)

20.7C   A Summary of the Reagents for Reduction

The many available metal hydride reagents reduce a wide variety of functional groups. Keep in mind that $LiAlH_4$ is such a strong reducing agent that it *nonselectively* reduces most polar functional groups. All other metal hydride reagents are milder, and each has its particular reactions that best utilize its reduced reactivity. The reagents and their uses are summarized in Table 20.1.

TABLE 20.1   A Summary of Metal Hydride Reducing Agents

	Reagent	Starting material	→	Product
strong reagent	$LiAlH_4$	RCHO	→	RCH_2OH
		R_2CO	→	R_2CHOH
		RCOOH	→	RCH_2OH
		RCOOR'	→	RCH_2OH
		RCOCl	→	RCH_2OH
		$RCONH_2$	→	RCH_2NH_2
milder reagents	$NaBH_4$	RCHO	→	RCH_2OH
		R_2CO	→	R_2CHOH
	$LiAlH[OC(CH_3)_3]_3$	RCOCl	→	RCHO
	DIBAL-H	RCOOR'	→	RCHO

PROBLEM 20.15   What product is formed when each compound is treated with either $LiAlH_4$ (followed by H_2O), or $NaBH_4$ in CH_3OH?

a. [structure: ketone with $COOCH_3$]   b. CH_3O [structure with two C=O, ending in OH]   c. CH_3O [cyclohexanone with methoxy substituent]

20.8   Oxidation of Aldehydes

Aldehydes give a positive Tollens test; that is, they react with Ag^+ to form RCOOH and Ag. Other functional groups give a negative Tollens test, because no silver mirror forms. A silver mirror forms when an aldehyde reacts with Tollens reagent.

The most common oxidation reaction of carbonyl compounds is the oxidation of **aldehydes to carboxylic acids.** A variety of oxidizing agents can be used, including CrO_3, $Na_2Cr_2O_7$, $K_2Cr_2O_7$, and $KMnO_4$. Aldehydes are also oxidized selectively in the presence of other functional groups using **silver(I) oxide in aqueous ammonium hydroxide.** This is called **Tollens reagent.** Because ketones have no H on the carbonyl carbon, they do not undergo this oxidation reaction.

Oxidation of RCHO

$CH_3CH_2CH_2$ [C=O, H] $\xrightarrow[H_2SO_4, H_2O]{CrO_3}$ $CH_3CH_2CH_2$ [C=O, OH] $+ Cr^{3+}$

[cyclohexane ring with CHO, HO substituent] $\xrightarrow{Ag_2O, NH_4OH}$ [cyclohexane ring with COOH, HO substituent] $+ Ag \leftarrow$ a silver mirror

Only the aldehyde is oxidized.

Oxidation with Tollens reagent provides a distinct color change, because the Ag^+ reagent is reduced to silver metal (Ag), which precipitates out of solution. When the reaction is carried out in a glass flask, a silver mirror is formed on its walls.

PROBLEM 20.16   What product is formed when each compound is treated with either Ag_2O, NH_4OH or $Na_2Cr_2O_7$, H_2SO_4, H_2O: (a) $C_6H_5CH_2OH$; (b) $CH_3CH(OH)CH_2CH_2CH_2CHO$?

PROBLEM 20.17   Classify each reaction as oxidation or reduction. What reagent is needed to carry out each reaction?

a. $CH_3CH_2CHO \longrightarrow CH_3CH_2CH_2OH$

b.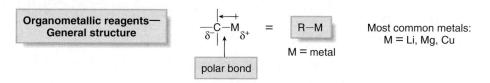

c. $CH_3CH_2CHO \longrightarrow CH_3CH_2COOH$

d.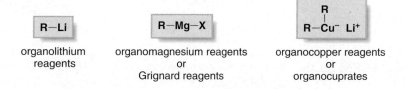

PROBLEM 20.18   Draw the product formed when compound **B** is treated with each reagent.

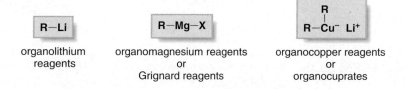

a. $NaBH_4$, CH_3OH
b. [1] $LiAlH_4$; [2] H_2O
c. PCC
d. Ag_2O, NH_4OH
e. CrO_3, H_2SO_4, H_2O

20.9   Organometallic Reagents

We will now discuss the reactions of carbonyl compounds with organometallic reagents, another class of nucleophiles.

◆ *Organometallic reagents* contain a carbon atom bonded to a metal.

Organometallic reagents—General structure

$$-\overset{|}{\underset{|}{C}}\!\!-\!\!M \quad = \quad \boxed{R-M}$$

M = metal

polar bond

Most common metals:
M = Li, Mg, Cu

Lithium, magnesium, and copper are the most commonly used metals in organometallic reagents, but others (such as Sn, Si, Tl, Al, Ti, and Hg) are known. General structures of the three common organometallic reagents are shown. R can be alkyl, aryl, allyl, benzyl, sp^2 hybridized, and with M = Li or Mg, sp hybridized. Because metals are *more electropositive* (less electronegative) than carbon, they donate electron density towards carbon, so that **carbon bears a partial negative charge.**

$$\boxed{R-Li}$$

organolithium reagents

$$\boxed{R-Mg-X}$$

organomagnesium reagents
or
Grignard reagents

$$\boxed{\overset{R}{\underset{}{\overset{|}{R-Cu^-}}}\ Li^+}$$

organocopper reagents
or
organocuprates

◆ **The more polar the carbon–metal bond, the more reactive the organometallic reagent.**

Because both Li and Mg are very electropositive metals, **organolithium (RLi)** and **organomagnesium reagents (RMgX)** contain very polar carbon–metal bonds and are therefore very reactive reagents. Organomagnesium reagents are called **Grignard reagents,** after Victor Grignard, who received the Nobel Prize in Chemistry in 1912 for his work with them.

Organocopper reagents (R$_2$CuLi), also called **organocuprates,** have a less polar carbon–metal bond and are therefore less reactive. Although organocuprates contain two alkyl groups bonded to copper, only one R group is utilized in a reaction.

Regardless of the metal, organometallic reagents are useful synthetically because they react as if they were free carbanions; that is, carbon bears a partial *negative* charge, so the **reagents react as bases and nucleophiles.**

Electronegativity values for carbon and the common metals in R−M reagents are C (2.5), Li (1.0), Mg (1.3), and Cu (1.8).

$$\underset{|}{\overset{|}{-}C-M} \quad \text{reacts like} \quad \underset{|}{\overset{|}{-}C:^{-}} \quad M^{+}$$

a base and a nucleophile

carbanion

20.9A   Preparation of Organometallic Reagents

Organolithium and Grignard reagents are typically prepared by reaction of an alkyl halide with the corresponding metal, as shown in the accompanying equations.

Organolithium reagents   **Grignard reagents**

General reaction $R-X + 2\,Li \longrightarrow R-Li + LiX$ $R-X + Mg \xrightarrow{(CH_3CH_2)_2O} R-Mg-X$

Example $CH_3-Br + 2\,Li \longrightarrow CH_3-Li + LiBr$ $CH_3-Br + Mg \xrightarrow{(CH_3CH_2)_2O} CH_3-Mg-Br$

methyllithium   methylmagnesium bromide

With lithium, the halogen and metal exchange to form the organolithium reagent. With magnesium, the metal inserts in the carbon–halogen bond, forming the Grignard reagent. Grignard reagents are usually prepared in diethyl ether ($CH_3CH_2OCH_2CH_3$) as solvent. It is thought that two ether oxygen atoms complex with the magnesium atom, stabilizing the reagent.

$$CH_3CH_2 \overset{\cdot\cdot}{O} CH_2CH_3$$
$$\downarrow$$
$$R-Mg-X$$
$$\uparrow$$
$$CH_3CH_2 \overset{\cdot\cdot}{O} CH_2CH_3$$

Two molecules of diethyl ether complex with the Mg atom of the Grignard reagent.

Organocuprates are prepared from organolithium reagents by reaction with a Cu^{+} salt, often CuI.

Organocopper reagents

General reaction $2\,R-Li + CuI \longrightarrow \underset{|}{\overset{R}{R}}-Cu^{-}\,Li^{+} + LiI$

Example $2\,CH_3-Li + CuI \longrightarrow \underset{|}{\overset{CH_3}{CH_3}}-Cu^{-}\,Li^{+} + LiI$

lithium dimethyl cuprate

PROBLEM 20.19   Write the step(s) needed to convert CH_3CH_2Br to each reagent: (a) CH_3CH_2Li; (b) CH_3CH_2MgBr; (c) $(CH_3CH_2)_2CuLi$.

20.9B   Acetylide Anions

The **acetylide anions** discussed in Chapter 11 are another example of organometallic compounds. These reagents are prepared by an acid–base reaction of an alkyne with a base such as $NaNH_2$ or NaH. We can think of these compounds as **organosodium** reagents. Because sodium is even more electropositive (less electronegative) than lithium, the C—Na bond of these organosodium compounds is best described as **ionic**, rather than polar covalent.

an ionic carbon–sodium bond

$$R-C\equiv C-H + Na^{+}\,{}^{-}NH_2 \rightleftharpoons R-C\equiv C:^{-}\,Na^{+} + NH_3$$

acetylide anion
an organosodium compound

An acid–base reaction can also be used to prepare *sp* hybridized organolithium compounds. Treatment of a terminal alkyne with CH_3Li affords a lithium acetylide. Equilibrium favors the products because the *sp* hybridized C−H bond of the terminal alkyne is more acidic than the sp^3 hybridized conjugate acid, CH_4, that is formed.

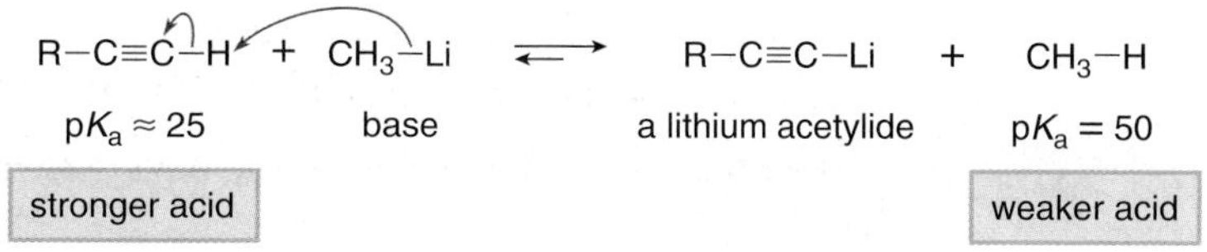

$$R-C\equiv C-H \ + \ CH_3-Li \ \rightleftharpoons \ R-C\equiv C-Li \ + \ CH_3-H$$

$pK_a \approx 25$	base	a lithium acetylide	$pK_a = 50$
stronger acid			weaker acid

20.9C   Reaction as a Base

◆ **Organometallic reagents are strong bases that readily abstract a proton from water to form hydrocarbons.**

The electron pair in the carbon–metal bond is used to form a new bond to the proton. Equilibrium favors the products of this acid–base reaction because H_2O is a much stronger acid than the alkane product.

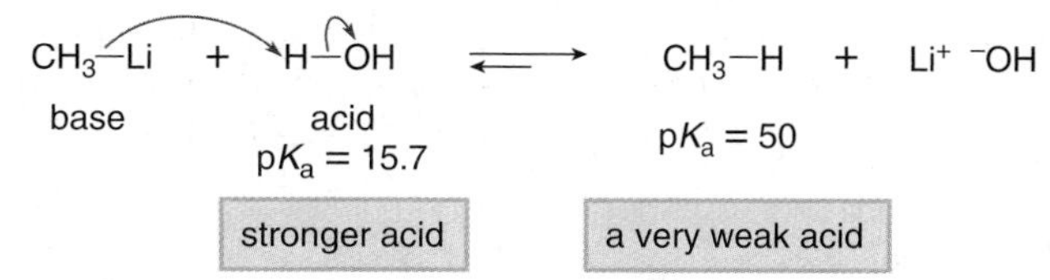

$$CH_3-Li \ + \ H-OH \ \rightleftharpoons \ CH_3-H \ + \ Li^+ \ ^-OH$$

base	acid	$pK_a = 50$
	$pK_a = 15.7$	
	stronger acid	a very weak acid

Similar reactions occur for the same reason with the O−H proton in alcohols and carboxylic acids, and the N−H protons of amines.

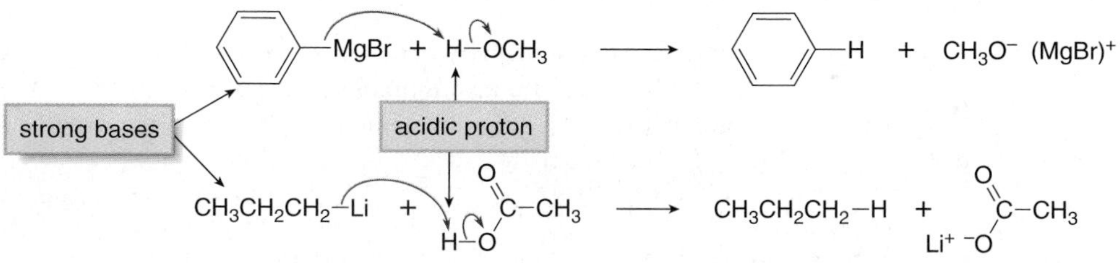

Because organolithium and Grignard reagents are themselves prepared from alkyl halides, a two-step method converts an alkyl halide into an alkane (or another hydrocarbon).

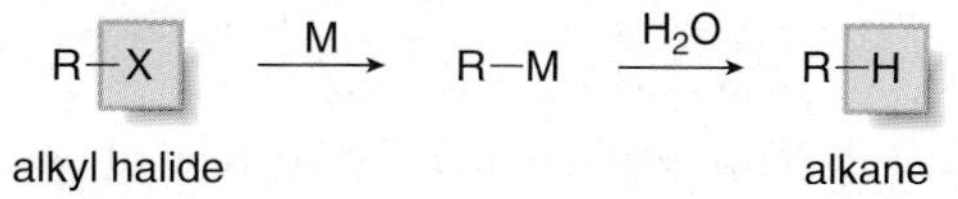

$$R-X \ \xrightarrow{M} \ R-M \ \xrightarrow{H_2O} \ R-H$$

alkyl halide   alkane

PROBLEM 20.20 Draw the product formed when each organometallic reagent is treated with H_2O.

a. 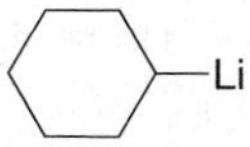 (cyclohexyl)−Li   b. $(CH_3)_3CMgBr$ c. 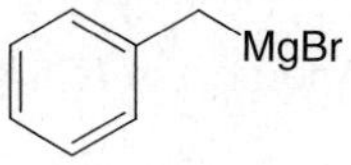 (benzyl)CH_2MgBr   d. $CH_3CH_2C\equiv C-Li$

20.9D   Reaction as a Nucleophile

Organometallic reagents are also strong nucleophiles that react with electrophilic carbon atoms to form new carbon–carbon bonds. These reactions are very valuable in forming the carbon skeletons of complex organic molecules. The following reactions of organometallic reagents are examined in Sections 20.10, 20.13, and 20.14:

[1]   Reaction of R–M with aldehydes and ketones to afford alcohols (Section 20.10)

Aldehydes and ketones are converted to 1°, 2°, or 3° alcohols with R″Li or R″MgX.

[2]   Reaction of R–M with carboxylic acid derivatives (Section 20.13)

Acid chlorides and esters can be converted to ketones or 3° alcohols with organometallic reagents. The identity of the product depends on the identity of R–M and the leaving group Z.

[3]   Reaction of R–M with other electrophilic functional groups (Section 20.14)

Organometallic reagents also react with CO_2 to form carboxylic acids and with epoxides to form alcohols.

20.10   Reaction of Organometallic Reagents with Aldehydes and Ketones

Treatment of an aldehyde or ketone with either an organolithium or Grignard reagent followed by water forms an alcohol with a new carbon–carbon bond. This reaction is an **addition reaction** because the elements of R″ and H are added across the π bond.

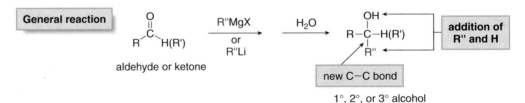

20.10A   General Features

This reaction follows the general mechanism for nucleophilic addition (Section 20.2A)—that is, **nucleophilic attack** by a carbanion followed by **protonation.** Mechanism 20.4 is shown using R″MgX, but the same steps occur with organolithium reagents and acetylide anions.

MECHANISM 20.4

Nucleophilic Addition of R″MgX to RCHO and $R_2C=O$

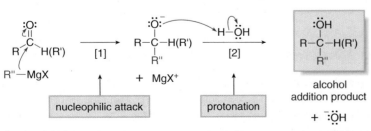

- In Step [1], the **nucleophile (R″)⁻ attacks the carbonyl carbon** and the π bond cleaves, forming an alkoxide. **This step forms a new carbon–carbon bond.**

- In Step [2], **protonation of the alkoxide by H_2O** forms the alcohol addition product. This acid–base reaction forms a new O–H bond.

- The overall result is addition of (R″)⁻ (from R″MgX) and H⁺ (from H_2O) to the carbonyl group.

This reaction is used to prepare 1°, 2°, and 3° alcohols, depending on the number of alkyl groups bonded to the carbonyl carbon of the aldehyde or ketone.

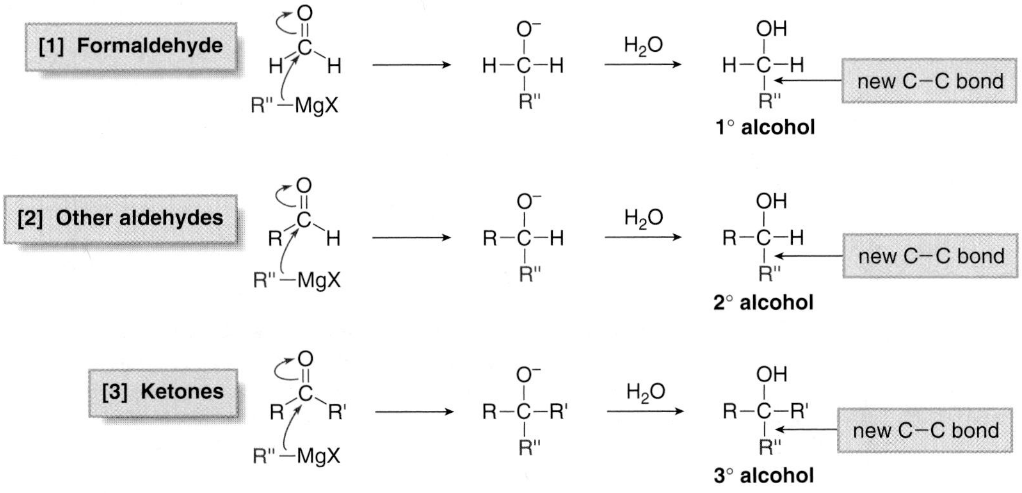

[1] **Formaldehyde**

[2] **Other aldehydes**

[3] **Ketones**

---

[1] Addition of R"MgX to formaldehyde ($CH_2=O$) forms a 1° alcohol.
[2] Addition of R"MgX to all other aldehydes forms a 2° alcohol.
[3] Addition of R"MgX to ketones forms a 3° alcohol.

Each reaction results in addition of one new alkyl group to the carbonyl carbon, and forms one new carbon–carbon bond. The reaction is general for all organolithium and Grignard reagents, and works for acetylide anions as well, as illustrated in Equations [1]–[3].

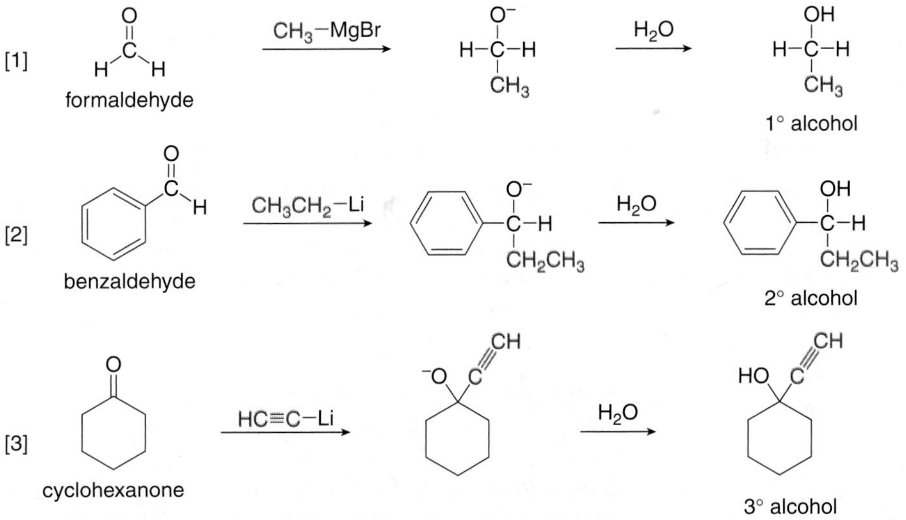

[1]   formaldehyde   →   1° alcohol

[2]   benzaldehyde   →   2° alcohol

[3]   cyclohexanone   →   3° alcohol

Because organometallic reagents are strong bases that rapidly react with H_2O (Section 20.9C), the addition of the new alkyl group must be carried out under anhydrous conditions to prevent traces of water from reacting with the reagent, thus reducing the yield of the desired alcohol. Water is added after the addition to protonate the alkoxide.

PROBLEM 20.21   Draw the product formed when each carbonyl compound is treated with C_6H_5MgBr, followed by protonation with H_2O.

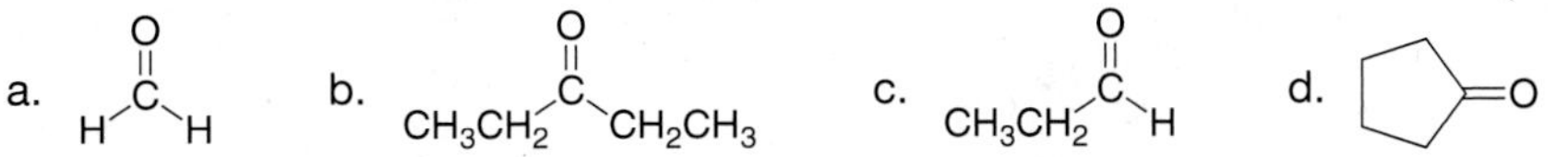

a.   b.   c.   d.

PROBLEM 20.22   Draw the product of each reaction.

a. [structure] $\xrightarrow[\text{[2] H}_2\text{O}]{\text{[1] CH}_3\text{CH}_2\text{CH}_2\text{Li}}$

c. [structure] $\xrightarrow[\text{[2] H}_2\text{O}]{\text{[1] C}_6\text{H}_5\text{Li}}$

b. [structure] $\xrightarrow[\text{[2] H}_2\text{O}]{\text{[1]} \quad \text{—Li}}$

d. [structure] $\xrightarrow[\text{[2] H}_2\text{O}]{\text{[1] CH}_2\text{=O}}$

20.10B   Stereochemistry

Like reduction, addition of organometallic reagents converts an sp^2 hybridized carbonyl carbon to a tetrahedral sp^3 hybridized carbon. Addition of R—M always occurs from both sides of the trigonal planar carbonyl group. **When a new stereogenic center is formed from an achiral starting material, an equal mixture of enantiomers results,** as shown in Sample Problem 20.1.

SAMPLE PROBLEM 20.1   Draw all stereoisomers formed in the following reaction.

$$\text{CH}_3\text{CH}_2 \overset{\overset{\text{O}}{\|}}{\text{C}} \text{CH}_2\text{CH}_2\text{CH}_3 \xrightarrow[\text{[2] H}_2\text{O}]{\text{[1] CH}_3\text{MgBr}}$$

SOLUTION

The Grignard reagent adds from both sides of the trigonal planar carbonyl group, forming two alkoxides, each containing a new stereogenic center. Protonation with water yields **an equal amount of two enantiomers—a racemic mixture.**

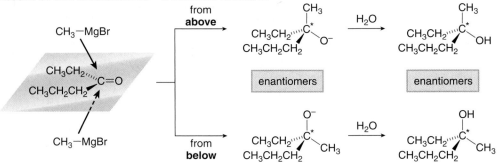

[* denotes a stereogenic center]

PROBLEM 20.23   Draw the products (including stereochemistry) of the following reactions.

a. [structure] $\xrightarrow[\text{[2] H}_2\text{O}]{\text{[1] CH}_3\text{CH}_2\text{MgBr}}$

b. CH_3-[structure]$=\text{O}$ $\xrightarrow[\text{[2] H}_2\text{O}]{\text{[1] CH}_3\text{CH}_2\text{Li}}$

20.10C   Applications in Synthesis

Many syntheses of useful compounds utilize the nucleophilic addition of a Grignard or organolithium reagent to form carbon–carbon bonds. For example, a key step in the synthesis of ethynylestradiol (Section 11.4), an oral contraceptive component, is the addition of lithium acetylide to a ketone, as shown in Figure 20.5.

Figure. 20.5   The synthesis of ethynylestradiol

Nucleophilic addition occurs here.

[structure] $\xrightarrow[\text{[2] H}_2\text{O}]{\text{[1] HC}\equiv\text{C}-\text{Li}}$ [structure] $\xrightarrow{\text{one step}}$ [structure]

ethynylestradiol

Figure 20.6 C$_{18}$ juvenile hormone

- Addition of CH$_3$MgCl to ketone **A** gives an alkoxide, **B**, which is protonated with H$_2$O to form 3° alcohol **C**. Although the ester group (–COOCH$_3$) can also react with the Grignard reagent (Section 20.13), it is less reactive than the ketone carbonyl. Thus, with control of reaction conditions, nucleophilic addition occurs selectively at the ketone.
- Treatment of halohydrin **C** with K$_2$CO$_3$ forms the C$_{18}$ juvenile hormone in one step. Conversion of a halohydrin to an epoxide was discussed in Section 9.6.

The synthesis of C$_{18}$ juvenile hormone, the molecule that opened Chapter 20, is another example. The last steps of the synthesis are outlined in Figure 20.6.

Although juvenile hormone itself is too unstable in light and too expensive to synthesize for use in controlling insect populations, related compounds, called **juvenile hormone** *mimics,* have been used effectively. The best known example is called **methoprene,** sold under such trade names as Altocid, Precor, and Diacon. Methoprene is used in cattle salt blocks to control horn-flies, in stored tobacco to control pests, and on dogs and cats to control fleas.

methoprene
juvenile hormone mimic

20.11   Retrosynthetic Analysis of Grignard Products

To use the Grignard addition in synthesis, you must be able to determine what carbonyl and Grignard components are needed to prepare a given compound—that is, **you must work backwards, in the retrosynthetic direction.** This involves a two-step process:

Step [1] Find the carbon bonded to the OH group in the product.
Step [2] Break the molecule into two components: One alkyl group bonded to the carbon with the OH group comes from the organometallic reagent. The rest of the molecule comes from the carbonyl component.

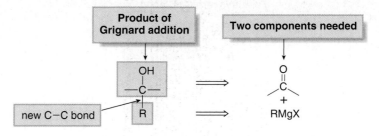

For example, to synthesize 3-pentanol [(CH$_3$CH$_2$)$_2$CHOH] by a Grignard reaction, you must first locate the carbon bonded to the OH group, then break the molecule into two components at this carbon. Thus, retrosynthetic analysis shows that one of the ethyl groups on this carbon comes from a Grignard reagent (CH_3CH_2MgX), and the rest of the molecule comes from the carbonyl component, a three-carbon aldehyde.

Retrosynthetic analysis for preparing 3-pentanol

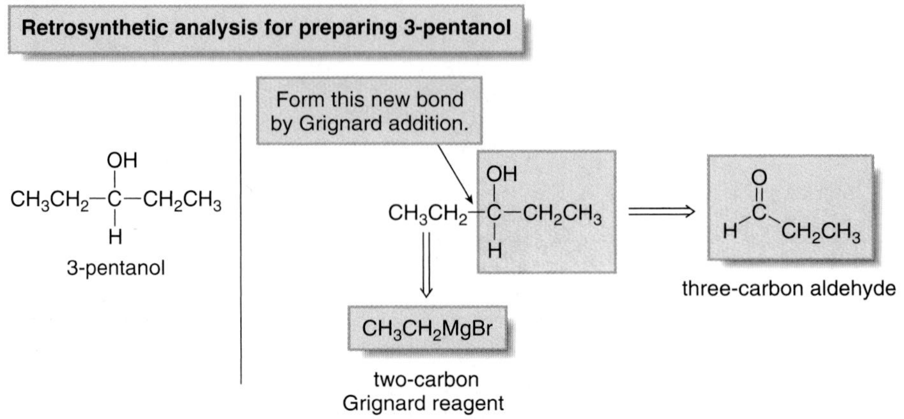

Then, writing the reaction in the synthetic direction—that is, from starting material to product—shows whether the synthesis is feasible and the analysis is correct. In this example, a three-carbon aldehyde reacts with CH_3CH_2MgBr to form an alkoxide, which can then be protonated by H_2O to form 3-pentanol, the desired alcohol.

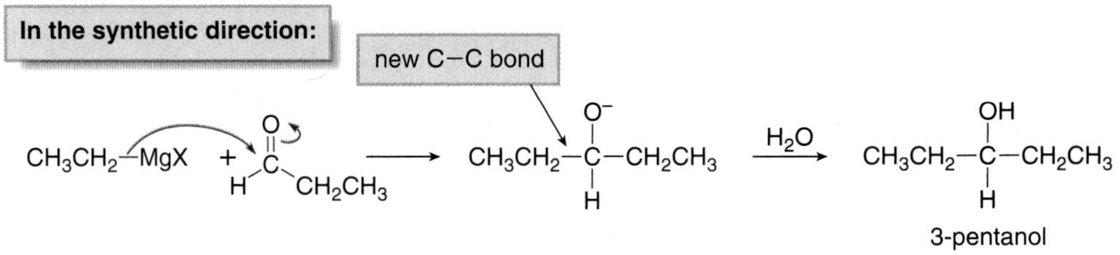

In the synthetic direction:

There is often more than one way to synthesize a 2° alcohol by Grignard addition, as shown in Sample Problem 20.2.

SAMPLE PROBLEM 20.2   Show two different methods to synthesize 2-butanol using a Grignard reaction.

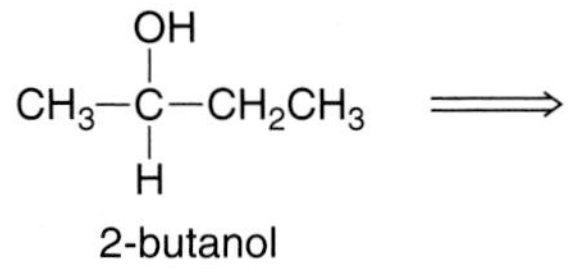

2-butanol

SOLUTION

Because 2-butanol has two different alkyl groups bonded to the carbon bearing the OH group, there are two different ways to form a new carbon–carbon bond by Grignard addition.

Possibility [1] Use CH_3MgX and a three-carbon aldehyde.

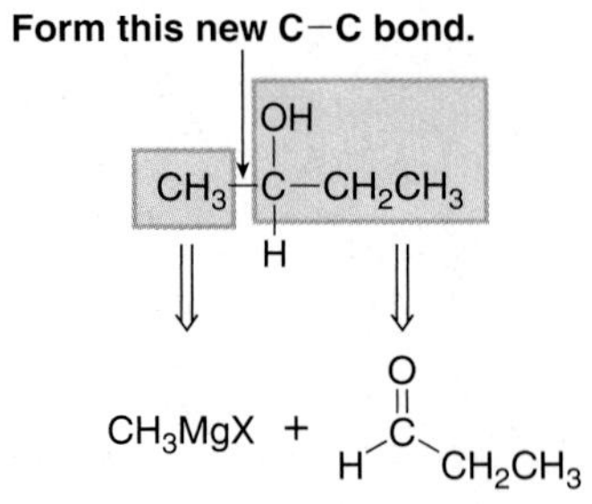

Possibility [2] Use CH_3CH_2MgX and a two-carbon aldehyde.

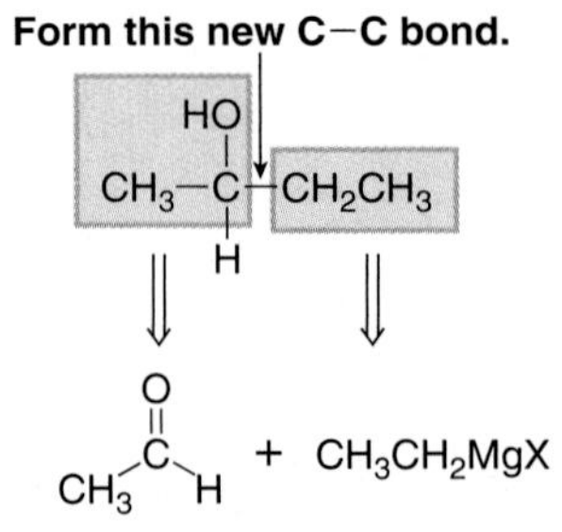

Both methods give the desired product, 2-butanol, as can be seen by writing the reaction from starting material to product.

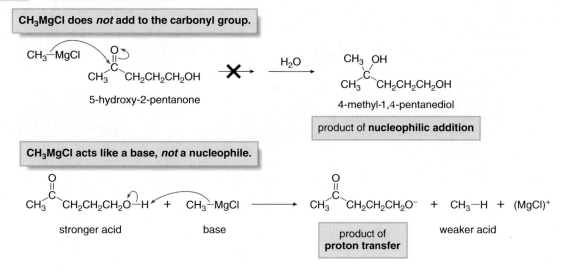

PROBLEM 20.24   What Grignard reagent and carbonyl compound are needed to prepare each alcohol?

a. $CH_3-\underset{\underset{H}{|}}{\overset{\overset{OH}{|}}{C}}-CH_3$

b. (cyclohexane)$-CH_2OH$

c. (cyclohexane)$-\underset{\underset{H}{|}}{\overset{\overset{OH}{|}}{C}}-CH_2CH_3$
(two methods)

d. (chain with OH)
(two methods)

PROBLEM 20.25   3° Alcohols having three different R groups on the carbon bonded to the OH group can be prepared by three different Grignard reactions. Give three different ways to synthesize each alcohol.

a. (structure with OH)

b. $CH_3-\underset{\underset{(phenyl)}{|}}{\overset{\overset{OH}{|}}{C}}-(cyclohexane)$

20.12   Protecting Groups

Although the addition of organometallic reagents to carbonyls is a very versatile reaction, it cannot be used with molecules that contain both a carbonyl group and N–H or O–H bonds.

> ◆ **Carbonyl compounds that also contain N–H or O–H bonds undergo an acid–base reaction with organometallic reagents, not nucleophilic addition.**

Rapid acid–base reactions occur between organometallic reagents and all of the following functional groups: ROH, RCOOH, RNH₂, R₂NH, RCONH₂, RCONHR, and RSH.

Suppose, for example, that you wanted to add methylmagnesium chloride (CH₃MgCl) to the carbonyl group of 5-hydroxy-2-pentanone to form a diol. Nucleophilic addition will *not* occur with this substrate. Instead, **because Grignard reagents are strong bases and proton transfer reactions are fast, CH₃MgCl removes the O–H proton before nucleophilic addition takes place.** The stronger acid and base react to form the weaker conjugate acid and conjugate base, as we learned in Section 20.9C.

CH₃MgCl does *not* add to the carbonyl group.

CH_3-MgCl $\underset{CH_3}{\overset{O}{\underset{}{C}}}CH_2CH_2CH_2OH$ ✗ $\xrightarrow{H_2O}$ $\underset{CH_3}{\overset{CH_3\ OH}{C}}CH_2CH_2CH_2OH$

5-hydroxy-2-pentanone

4-methyl-1,4-pentanediol

product of **nucleophilic addition**

CH₃MgCl acts like a base, *not* a nucleophile.

$\underset{CH_3}{\overset{O}{\underset{}{C}}}CH_2CH_2CH_2O-H$ + CH_3-MgCl $\longrightarrow$ $\underset{CH_3}{\overset{O}{\underset{}{C}}}CH_2CH_2CH_2O^-$ + CH_3-H + $(MgCl)^+$

stronger acid   base   product of **proton transfer**   weaker acid

Solving this problem requires a three-step strategy:

| Step [1] | Convert the OH group into another functional group that does not interfere with the desired reaction. **This new blocking group is called a protecting group, and the reaction that creates it is called *protection*.** |

| Step [2] | Carry out the desired reaction. |

| Step [3] | Remove the protecting group. This reaction is called *deprotection*. |

Application of the general strategy to the Grignard addition of CH_3MgCl to 5-hydroxy-2-pentanone is illustrated in Figure 20.7.

A common OH protecting group is a **silyl ether**. A silyl ether has a new O–Si bond in place of the O–H bond of the alcohol. The most widely used silyl ether protecting group is the ***tert*-butyldimethylsilyl ether.**

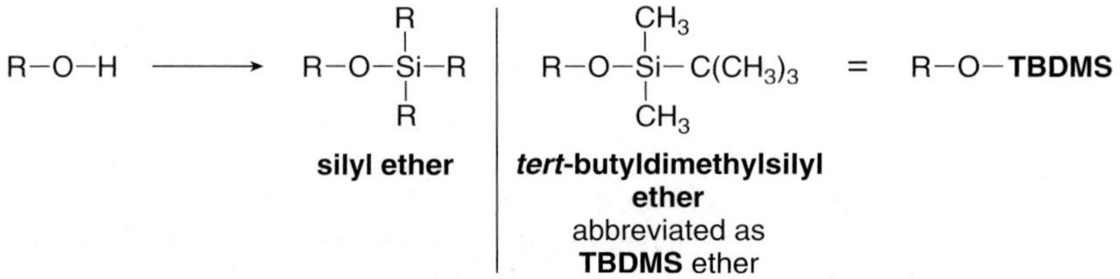

$$R-O-H \longrightarrow R-O-\underset{\underset{R}{|}}{\overset{\overset{R}{|}}{Si}}-R \quad \Big| \quad R-O-\underset{\underset{CH_3}{|}}{\overset{\overset{CH_3}{|}}{Si}}-C(CH_3)_3 \quad = \quad R-O-\textbf{TBDMS}$$

silyl ether   ***tert*-butyldimethylsilyl ether** abbreviated as **TBDMS** ether

tert-Butyldimethylsilyl ethers are prepared from alcohols by reaction with *tert*-butyldimethylsilyl chloride and an amine base, usually imidazole.

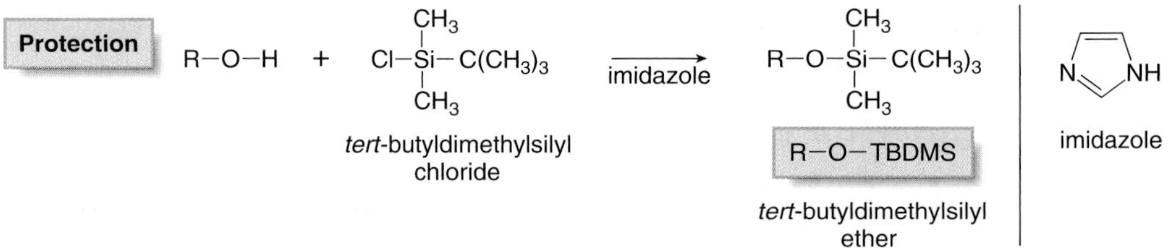

| **Protection** | $R-O-H$ + $Cl-\underset{\underset{CH_3}{|}}{\overset{\overset{CH_3}{|}}{Si}}-C(CH_3)_3$ $\xrightarrow{\text{imidazole}}$ $R-O-\underset{\underset{CH_3}{|}}{\overset{\overset{CH_3}{|}}{Si}}-C(CH_3)_3$ |

tert-butyldimethylsilyl chloride

$\boxed{R-O-TBDMS}$

tert-butyldimethylsilyl ether

imidazole

The silyl ether is typically removed with a fluoride salt, usually **tetrabutylammonium fluoride** $(CH_3CH_2CH_2CH_2)_4N^+ F^-$.

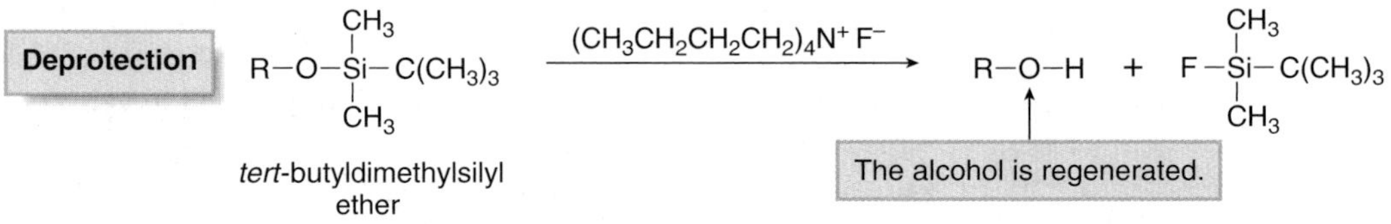

| **Deprotection** | $R-O-\underset{\underset{CH_3}{|}}{\overset{\overset{CH_3}{|}}{Si}}-C(CH_3)_3$ $\xrightarrow{(CH_3CH_2CH_2CH_2)_4N^+ F^-}$ $R-O-H$ + $F-\underset{\underset{CH_3}{|}}{\overset{\overset{CH_3}{|}}{Si}}-C(CH_3)_3$ |

tert-butyldimethylsilyl ether

The alcohol is regenerated.

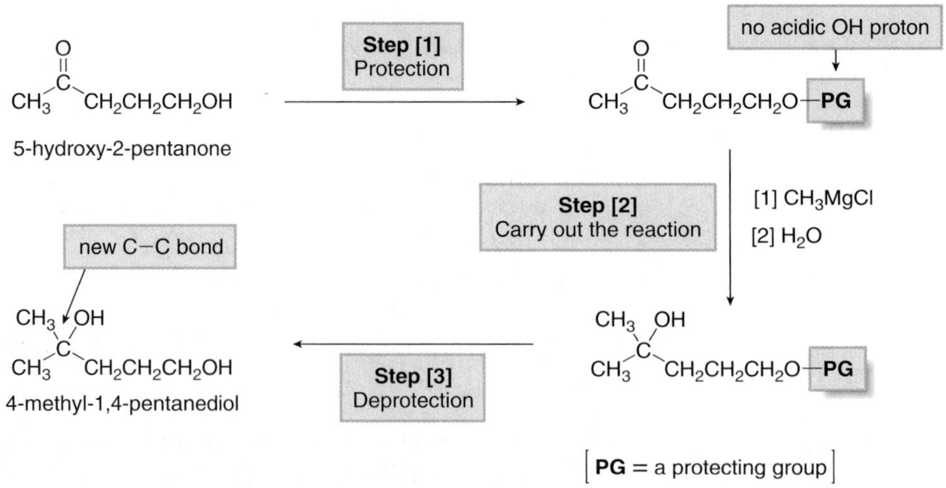

Figure 20.7 General strategy for using a protecting group

- In Step [1], the OH proton in 5-hydroxy-2-pentanone is replaced with a protecting group, written as **PG.** Because the product of Step [1] no longer has an OH proton, it can now undergo nucleophilic addition.
- In Step [2], CH_3MgCl adds to the carbonyl group to yield a 3° alcohol after protonation with water.
- Removal of the protecting group in Step [3] forms the desired product, 4-methyl-1,4-pentanediol.

5-hydroxy-2-pentanone

Step [1] Protection

no acidic OH proton

Step [2] Carry out the reaction

[1] CH_3MgCl
[2] H_2O

Step [3] Deprotection

new C–C bond

4-methyl-1,4-pentanediol

$\big[$ **PG** = a protecting group $\big]$

The use of a *tert*-butyldimethylsilyl ether as a protecting group makes possible the synthesis of 4-methyl-1,4-pentanediol by a three-step sequence.

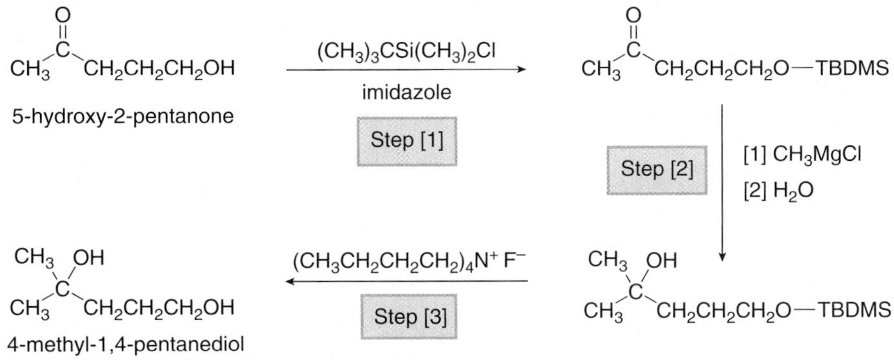

- ◆ **Step [1] Protect the OH group** as a *tert*-butyldimethylsilyl ether by reaction with *tert*-butyldimethylsilyl chloride and imidazole.
- ◆ **Step [2] Carry out nucleophilic addition** by using CH_3MgCl, followed by protonation.
- ◆ **Step [3] Remove the protecting group** with tetrabutylammonium fluoride to form the desired addition product.

Protecting groups block interfering functional groups, and in this way, a wider variety of reactions can take place with a particular substrate. For more on protecting groups, see the discussion of acetals in Section 21.15.

PROBLEM 20.26   Using protecting groups, show how the following product can be made from the given starting material.

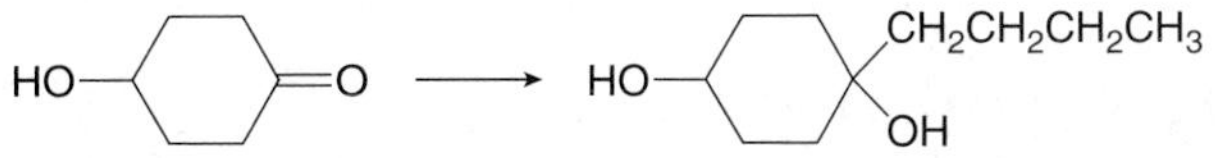

20.13 Reaction of Organometallic Reagents with Carboxylic Acid Derivatives

Organometallic reagents react with carboxylic acid derivatives (RCOZ) to form two different products, depending on the identity of both the leaving group Z and the reagent R−M. The most useful reactions are carried out with esters and acid chlorides, forming either **ketones** or **3° alcohols**.

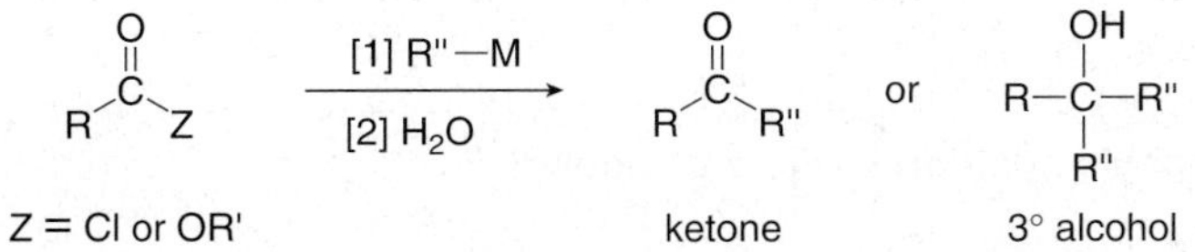

◆ Throughout this discussion, keep in mind that RLi and RMgX are very reactive reagents, whereas R_2CuLi is much less reactive. This reactivity difference makes selective reactions possible.

20.13A Reaction of RLi and RMgX with Esters and Acid Chlorides

Both esters and acid chlorides form 3° alcohols when treated with two equivalents of either Grignard or organolithium reagents. Two new carbon–carbon bonds are formed in the product.

General reaction

$$
\underset{R}{\overset{O}{\underset{\parallel}{C}}}\!-\!Z \quad \xrightarrow[\text{(2 equiv)}]{R''\text{Li or }R''\text{MgX}} \quad \xrightarrow{H_2O} \quad \underset{R''}{\overset{OH}{\underset{|}{R\!-\!C\!-\!R''}}}
$$

Z = Cl or OR'   → new C–C bonds

3° alcohol

Examples

$$
\underset{CH_3}{\overset{O}{\underset{\parallel}{C}}}\!-\!Cl \quad \xrightarrow[\text{(2 equiv)}]{CH_3CH_2MgCl} \quad \xrightarrow{H_2O} \quad CH_3\!-\!\underset{CH_2CH_3}{\overset{OH}{\underset{|}{C}}}\!-\!CH_2CH_3
$$

new bond / new bond

$$
\text{cyclohexyl}\!-\!\underset{}{\overset{O}{\underset{\parallel}{C}}}\!-\!OCH_2CH_3 \quad \xrightarrow[\text{(2 equiv)}]{CH_3MgI} \quad \xrightarrow{H_2O} \quad \text{cyclohexyl}\!-\!\underset{CH_3}{\overset{OH}{\underset{|}{C}}}\!-\!CH_3
$$

new C–C bonds

PROBLEM 20.27 Draw the product formed when each compound is treated with $CH_3CH_2CH_2CH_2MgBr$ followed by H_2O.

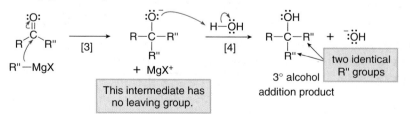

a.   b.   c.

The mechanism for this addition reaction resembles the mechanism for the metal hydride reduction of acid chlorides and esters discussed in Section 20.7A. The mechanism is conceptually divided into two parts: **nucleophilic substitution** to form a ketone, followed by **nucleophilic addition** to form a 3° alcohol, as shown in Mechanism 20.5.

MECHANISM 20.5

Reaction of R''MgX or R''Li with RCOCl and RCOOR'

Part [1] Nucleophilic substitution forms a ketone.

$$
\underset{R''-MgX}{\overset{O}{\underset{\parallel}{R-C-Z}}} \quad \xrightarrow{[1]} \quad \underset{R''}{\overset{:\ddot{O}:^-}{\underset{|}{R-C-Z}}} \quad \xrightarrow{[2]} \quad \underset{R}{\overset{:O:}{\underset{\parallel}{C-R''}}} + :Z^-
$$

leaving group

$\left[Z = Cl, OR' \right]$ + MgX⁺   ketone / substitution product   R'' replaces Z.

♦ **Nucleophilic attack of (R'')⁻** (from R''MgX) in Step [1] forms a tetrahedral intermediate with a leaving group Z.

♦ In Step [2], the π bond is re-formed and **Z⁻ comes off.** The overall result of the addition of (R'')⁻ and elimination of Z⁻ is the substitution of R'' for Z.

♦ Because the product of Part [1] is a ketone, it can react with a second equivalent of R''MgX to form an alcohol by nucleophilic addition in Part [2].

Part [2] Nucleophilic addition forms a 3° alcohol.

$$
\underset{R''-MgX}{\overset{O}{\underset{\parallel}{R-C-R''}}} \quad \xrightarrow{[3]} \quad \underset{R''}{\overset{:\ddot{O}:^-}{\underset{|}{R-C-R''}}} \quad \xrightarrow[H-\ddot{O}H]{[4]} \quad \underset{R''}{\overset{:\ddot{O}H}{\underset{|}{R-C-R''}}} + {}^-:\ddot{O}H
$$

+ MgX⁺

This intermediate has no leaving group.   3° alcohol / addition product   two identical R'' groups

♦ **Nucleophilic attack of (R'')⁻** (from R''MgX) in Step [3] forms an alkoxide.

♦ **Protonation of the alkoxide** by H_2O in Step [4] forms a 3° alcohol.

Reactive organolithium and Grignard reagents always afford 3° alcohols when they react with esters and acid chlorides. As soon as the ketone forms by addition of one equivalent of reagent to RCOZ (Part [1] of the mechanism), it reacts with a second equivalent of reagent to form the 3° alcohol.

This reaction is more limited than the Grignard addition to aldehydes and ketones, because only 3° alcohols having two identical alkyl groups can be prepared. Nonetheless, it is still a valuable reaction because it forms two new carbon–carbon bonds.

SAMPLE PROBLEM 20.3   What ester and Grignard reagent are needed to prepare the following alcohol?

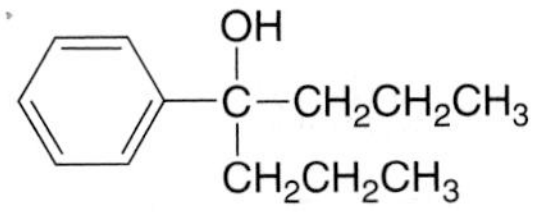

SOLUTION

A 3° alcohol formed from an ester and Grignard reagent must have **two identical R groups,** and these R groups come from RMgX. The remainder of the molecule comes from the ester.

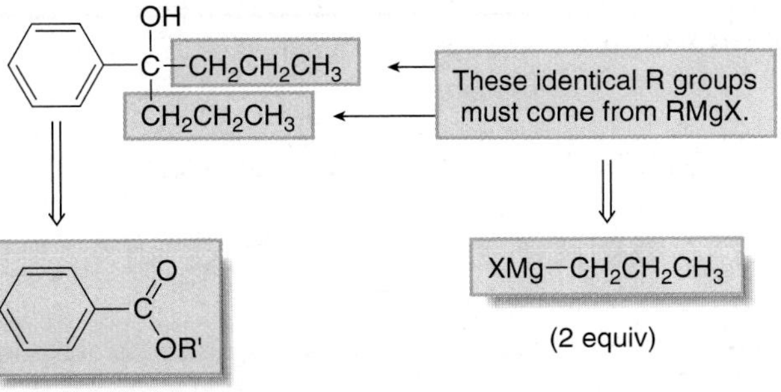

All other C's come from the ester.
R' = any alkyl group

Checking in the synthetic direction:

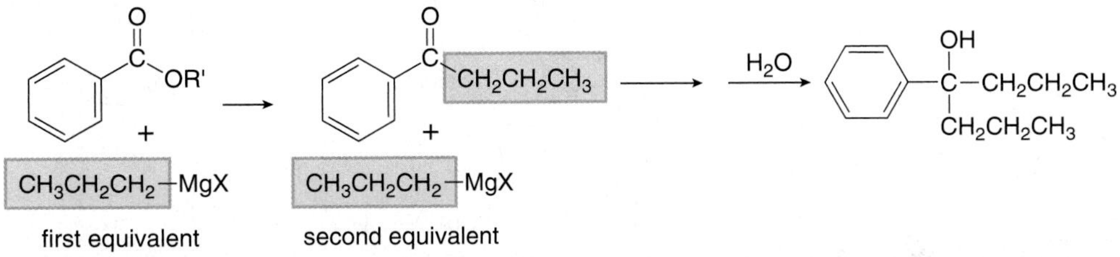

PROBLEM 20.28   What ester and Grignard reagent are needed to prepare each alcohol?

a. 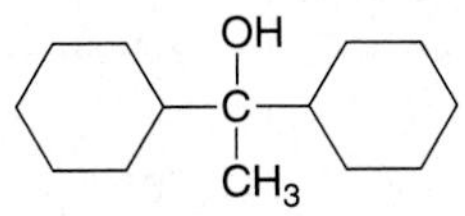   b. $(CH_3CH_2CH_2)_3COH$ c. 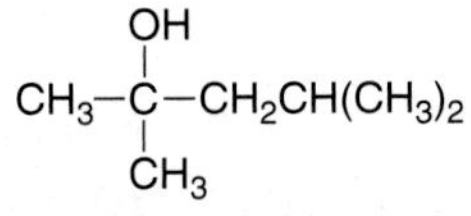 $CH_3-\overset{\overset{\displaystyle OH}{|}}{\underset{\underset{\displaystyle CH_3}{|}}{C}}-CH_2CH(CH_3)_2$

20.13B   Reaction of R₂CuLi with Acid Chlorides

To form a ketone from a carboxylic acid derivative, a less reactive organometallic reagent— namely an **organocuprate**—is needed. **Acid chlorides, which have the best leaving group (Cl⁻) of the carboxylic acid derivatives, react with R'₂CuLi, to give a ketone as product.** Esters, which contain a poorer leaving group (⁻OR), do not react with R'₂CuLi.

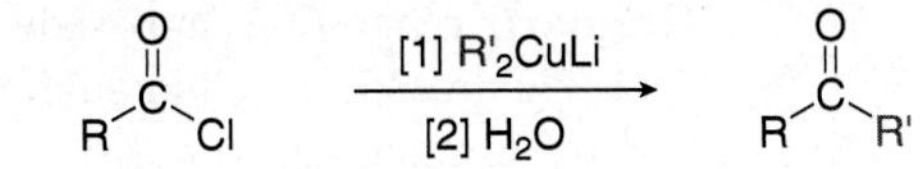

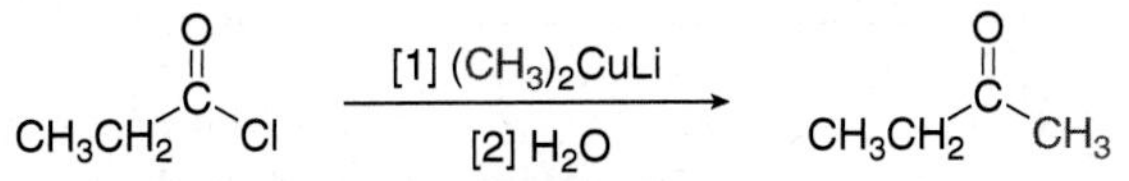

This reaction results in nucleophilic substitution of an alkyl group R' for the leaving group Cl, forming one new carbon–carbon bond.

PROBLEM 20.29   What organocuprate reagent is needed to convert CH_3CH_2COCl to each ketone?

a.

$$CH_3CH_2 \overset{\overset{\displaystyle O}{\|}}{C} CH_3$$

b.

c.

$$\overset{\overset{\displaystyle O}{\|}}{} Ph$$

PROBLEM 20.30   What reagent is needed to convert $(CH_3)_2CHCH_2COCl$ into each compound?

a. $(CH_3)_2CHCH_2CHO$ b. (HO structure)   c. (O structure)   d. $(CH_3)_2CHCH_2CH_2OH$

A ketone with two different R groups bonded to the carbonyl carbon can be made by two different methods, as illustrated in Sample Problem 20.4.

SAMPLE PROBLEM 20.4   Show two different ways to prepare 2-pentanone from an acid chloride and an organocuprate reagent.

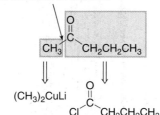

$$CH_3 \overset{\overset{\displaystyle O}{\|}}{C} CH_2CH_2CH_3$$

2-pentanone

SOLUTION

In each case, one alkyl group comes from the organocuprate and one comes from the acid chloride.

Possibility [1] Use $(CH_3)_2CuLi$ and a four-carbon acid chloride.

Form this new C–C bond.

$$CH_3 \overset{\overset{\displaystyle O}{\|}}{C} CH_2CH_2CH_3$$

$(CH_3)_2CuLi$ $$Cl \overset{\overset{\displaystyle O}{\|}}{C} CH_2CH_2CH_3$$

Possibility [2] Use $(CH_3CH_2CH_2)_2CuLi$ and a two-carbon acid chloride.

Form this new C–C bond.

$$CH_3 \overset{\overset{\displaystyle O}{\|}}{C} CH_2CH_2CH_3$$

$$CH_3 \overset{\overset{\displaystyle O}{\|}}{C} Cl$$ $(CH_3CH_2CH_2)_2CuLi$

PROBLEM 20.31   Draw two different ways to prepare each ketone from an acid chloride and an organocuprate reagent: (a) $C_6H_5COCH_3$; (b) $(CH_3)_2CHCOC(CH_3)_3$.

20.14   Reaction of Organometallic Reagents with Other Compounds

Because organometallic reagents are strong nucleophiles, they react with many other electrophiles in addition to carbonyl groups. Because these reactions always lead to the formation of new carbon–carbon bonds, they are also valuable in organic synthesis. In Section 20.14, we examine the reactions of organometallic reagents with **carbon dioxide** and **epoxides.**

20.14A   Reaction of Grignard Reagents with Carbon Dioxide

Grignard reagents react with CO_2 to give carboxylic acids after protonation with aqueous acid. This reaction, called **carboxylation,** forms a carboxylic acid with one more carbon atom than the Grignard reagent from which it is prepared.

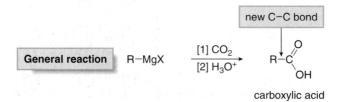

carboxylic acid

Because Grignard reagents are made from alkyl halides, an alkyl halide can be converted to a carboxylic acid having one more carbon atom by a two-step reaction sequence: **formation of a Grignard reagent,** followed by **reaction with CO_2.**

Example

$$\text{(chlorobenzene)} \xrightarrow{\text{Mg}} \text{(phenyl–MgCl)} \xrightarrow[\text{[2] } H_3O^+]{\text{[1] } CO_2} \text{(benzoic acid)}$$

new C–C bond

The mechanism resembles earlier reactions of nucleophilic Grignard reagents with carbonyl groups, as shown in Mechanism 20.6.

MECHANISM 20.6

Carboxylation—Reaction of RMgX with CO_2

$$R\!\!-\!\!MgX + \underset{\delta^-}{\overset{\delta^-}{\underset{:O:}{\overset{:O:}{C}}}}{}^{\delta^+} \xrightarrow{\text{[1]}} R\!\!-\!\!\overset{:O:}{\underset{:O:^- \cdots H-OH_2}{C}} \xrightarrow{\text{[2]}} R\!\!-\!\!\overset{:O:}{\underset{:OH}{C}} + H_2\ddot{O}:$$

$+ \ MgX^+$

♦ In Step [1], the nucleophilic Grignard reagent attacks the electrophilic carbon atom of CO_2, cleaving a π bond and forming a new carbon–carbon bond.

♦ The carboxylate anion is protonated with aqueous acid in Step [2] to form the carboxylic acid.

PROBLEM 20.32   What carboxylic acid is formed from each alkyl halide on treatment with [1] Mg; [2] CO_2; [3] H_3O^+?

a. (cyclohexyl bromide, Br)

b. (1-chlorobutane)

c. $CH_3O{-}\!\!\!\langle\!\!\!\rangle\!\!\!-CH_2Br$

20.14B   Reaction of Organometallic Reagents with Epoxides

Like other strong nucleophiles, **organometallic reagents—RLi, RMgX, and R_2CuLi—open epoxide rings to form alcohols.**

General reaction

$$\underset{\text{epoxide}}{\overset{O}{C-C}} \xrightarrow[\text{[2] } H_2O]{\text{[1] RLi, RMgX, or } R_2CuLi} \underset{\text{alcohol}}{\overset{OH}{\underset{R}{C-C}}}$$

Example

$$\underset{H\ \ \ \ \ H}{\overset{O}{\underset{H}{C-C}}}\!\!H \xrightarrow[\text{[2] } H_2O]{\text{[1] } \langle\!\!\!\rangle\!\!-MgBr} (\text{product}) = \langle\!\!\!\rangle\!\!-CH_2CH_2OH$$

The opening of epoxide rings with negatively charged nucleophiles was discussed in Section 9.15A.

The reaction follows the same two-step process as the opening of epoxide rings with other negatively charged nucleophiles—that is, **nucleophilic attack from the back side of the epoxide ring, followed by protonation of the resulting alkoxide.** In unsymmetrical epoxides, nucleophilic attack occurs at the less substituted carbon atom.

less substituted C

$$\underset{CH_3CH_2-Li}{\overset{O}{\underset{H}{\overset{H}{C}-\underset{CH_3}{\overset{CH_3}{C}}}}} \xrightarrow{\text{[1]}} \underset{CH_3CH_2}{\overset{O^-}{\underset{CH_3}{\overset{H}{C}-\overset{CH_3}{C}}}} \xrightarrow{\overset{H-OH}{\text{[2]}}} \underset{CH_3CH_2}{\overset{OH}{\underset{CH_3}{\overset{H}{C}-\overset{CH_3}{C}}}} + \ ^-OH$$

S_N2 backside attack

protonation

PROBLEM 20.33   Draw a stepwise mechanism for the following epoxide ring opening.

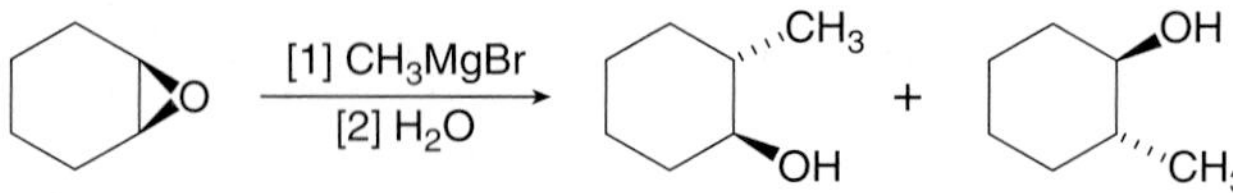

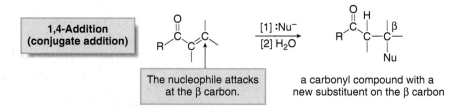

20.15   α,β-Unsaturated Carbonyl Compounds

α,β-Unsaturated carbonyl compounds are conjugated molecules containing a carbonyl group and a carbon–carbon double bond, separated by a single σ bond.

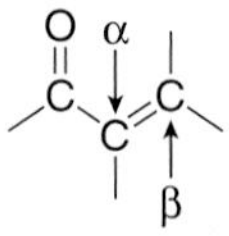

α,β-unsaturated carbonyl compound

Both functional groups of α,β-unsaturated carbonyl compounds have π bonds, but individually, they react with very different kinds of reagents. Carbon–carbon double bonds react with electrophiles (Chapter 10) and carbonyl groups react with nucleophiles (Section 20.2). What happens, then, when these two functional groups having opposite reactivity are in close proximity?

Because the two π bonds are conjugated, the electron density in an α,β-unsaturated carbonyl compound is delocalized over four atoms. Three resonance structures show that the carbonyl carbon and the β carbon bear a partial positive charge. This means that **α,β-unsaturated carbonyl compounds can react with nucleophiles at two different sites.**

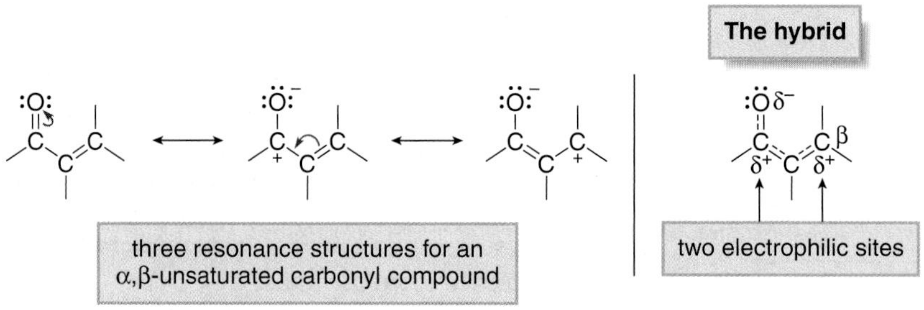

The hybrid

three resonance structures for an
α,β-unsaturated carbonyl compound

two electrophilic sites

◆ Addition of a nucleophile to the carbonyl carbon, called **1,2-addition,** adds the elements of H and Nu across the C=O, forming an allylic alcohol.

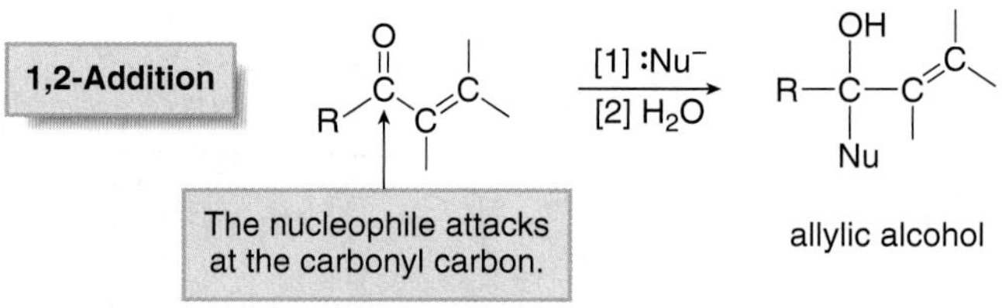

1,2-Addition

The nucleophile attacks
at the carbonyl carbon.

allylic alcohol

◆ Addition of a nucleophile to the β carbon, called **1,4-addition** or **conjugate addition,** forms a carbonyl compound.

**1,4-Addition
(conjugate addition)**

The nucleophile attacks
at the β carbon.

a carbonyl compound with a
new substituent on the β carbon

Both 1,2- and 1,4-addition result in nucleophilic addition of the elements of H and Nu.

20.15A   The Mechanisms for 1,2-Addition and 1,4-Addition

The steps for the mechanism of 1,2-addition are exactly the same as those for the nucleophilic addition to an aldehyde or ketone—that is, **nucleophilic attack,** followed by **protonation** (Section 20.2A), as shown in Mechanism 20.7.

MECHANISM 20.7

1,2-Addition to an α,β-Unsaturated Carbonyl Compound

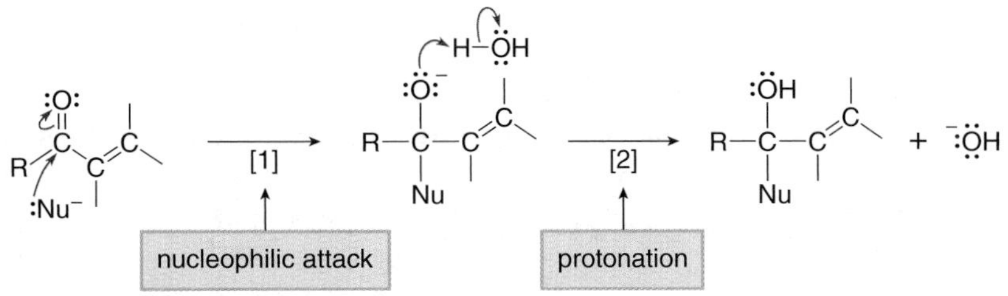

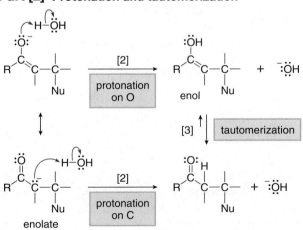

The mechanism for 1,4-addition also begins with nucleophilic attack, and then protonation and tautomerization add the elements of H and Nu to the α and β carbons of the carbonyl compound, as shown in Mechanism 20.8.

MECHANISM 20.8

1,4-Addition to an α,β-Unsaturated Carbonyl Compound

Part [1]   Nucleophilic attack at the β carbon

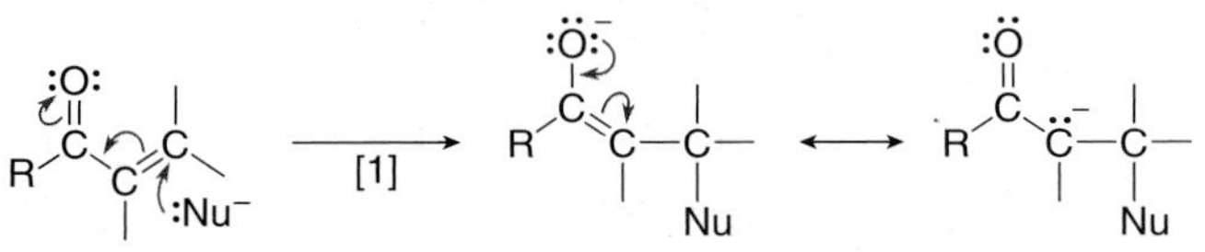

a resonance-stabilized
enolate anion

◆ In Part [1], nucleophilic attack at the β carbon forms a resonance-stabilized anion called an **enolate.** Either resonance structure can be used to continue the mechanism in Part [2].

Part [2]   Protonation and tautomerization

enol

enolate

◆ Protonation on the carbon end of the enolate forms the 1,4-addition product directly.

◆ Protonation of the oxygen end of the enolate forms an **enol.** Recall from Section 11.9 that enols are unstable and tautomerize (by a two-step process) to carbonyl compounds. Tautomerization forms the same 1,4-addition product that results from protonation on carbon.

20.15B   Reaction of α,β-Unsaturated Carbonyl Compounds with Organometallic Reagents

The identity of the metal in an organometallic reagent determines whether it reacts with an α,β-unsaturated aldehyde or ketone by 1,2-addition or 1,4-addition.

Why is conjugate addition also called 1,4-addition? If the atoms of the enol are numbered beginning with the O atom, then the elements of H and Nu add to atoms "1" and "4," respectively.

The enol has H and Nu added to atoms 1 and 4.

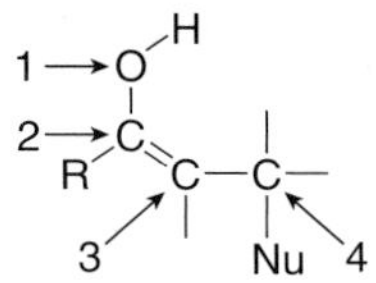

- **Organolithium and Grignard reagents form 1,2-addition products.**

1,2-Addition— General reaction

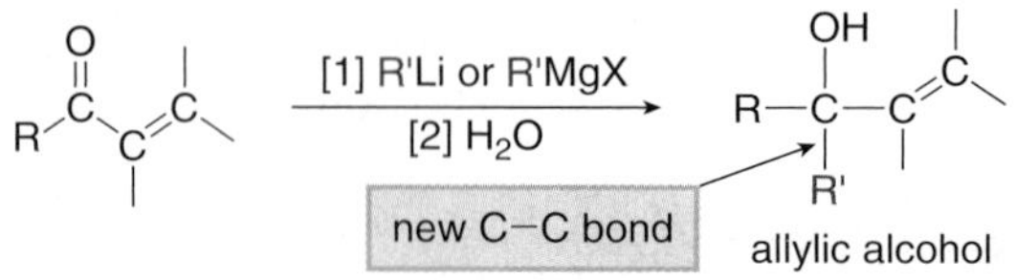

Example

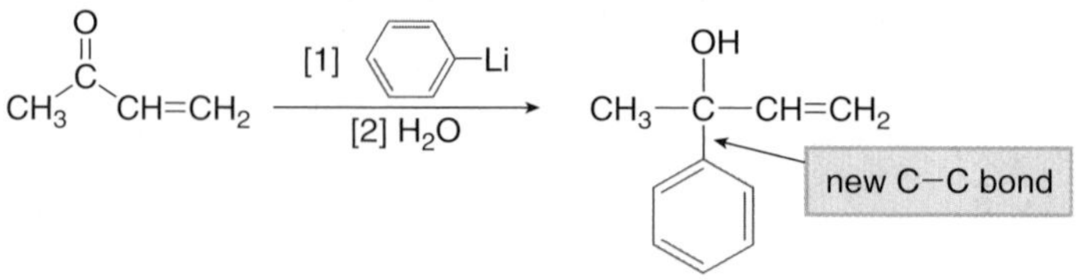

- **Organocuprate reagents form 1,4-addition products.**

1,4-Addition— General reaction

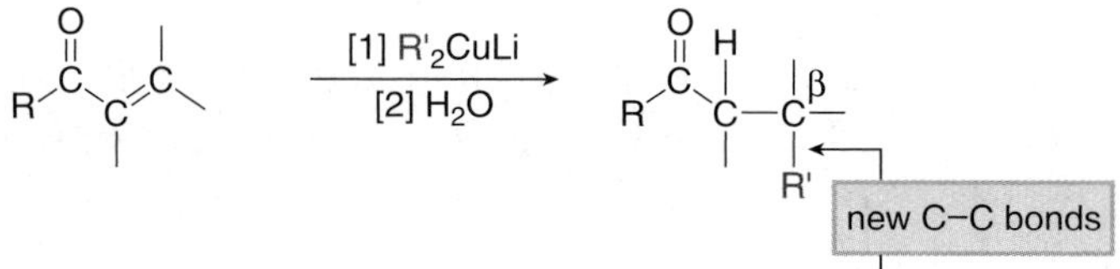

Example

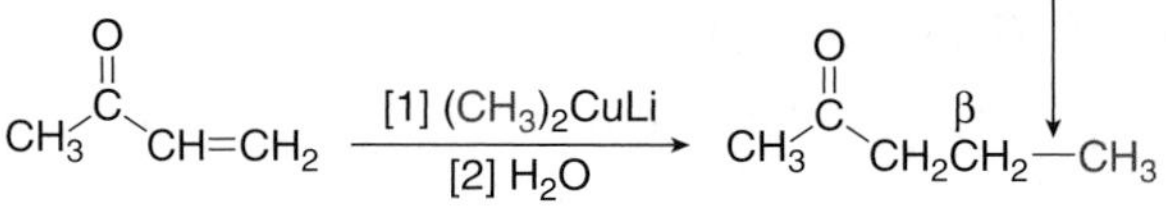

SAMPLE PROBLEM 20.5 Draw the products of each reaction.

a. 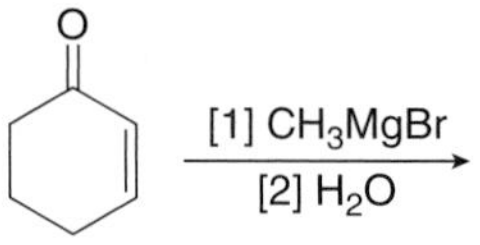

b. 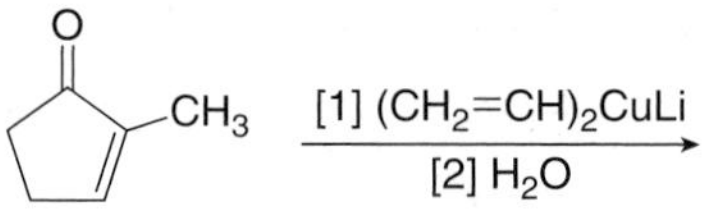

SOLUTION

The characteristic reaction of α,β-unsaturated carbonyl compounds is nucleophilic addition. The reagent determines the mode of addition (1,2- or 1,4-).

[a] **Grignard reagents undergo 1,2-addition.** CH_3MgBr adds a new CH_3 group at the carbonyl carbon.

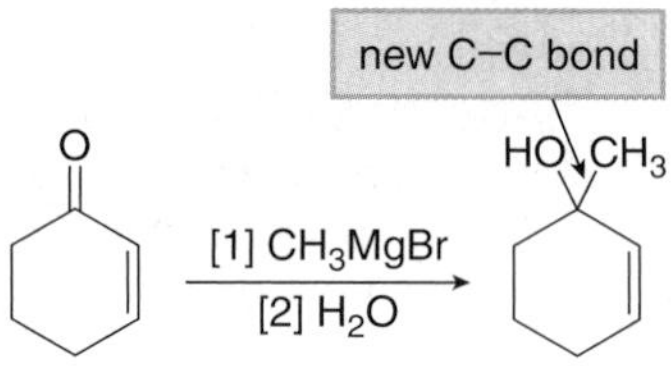

[b] **Organocuprate reagents undergo 1,4-addition.** The cuprate reagent adds a new vinyl group ($CH_2{=}CH$) at the β carbon.

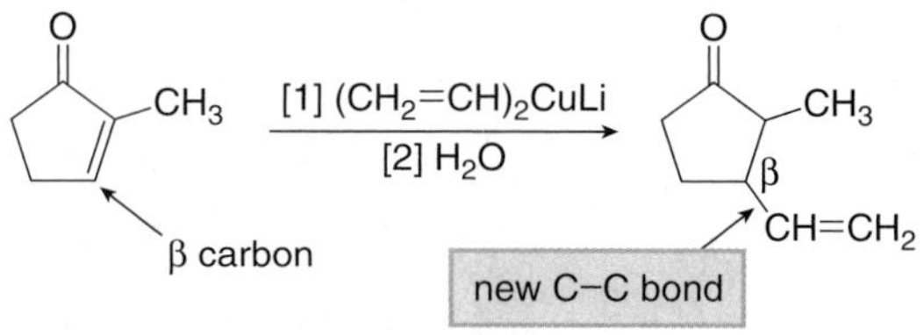

PROBLEM 20.34 Draw the product when each compound is treated with either $(CH_3)_2CuLi$, followed by H_2O, or HC≡CLi, followed by H_2O.

a. 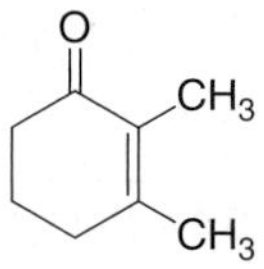

b. 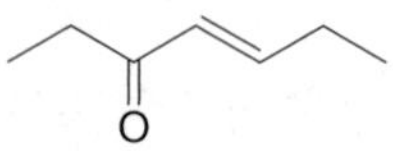

20.16   Summary—The Reactions of Organometallic Reagents

We have now seen many different reactions of organometallic reagents with a variety of functional groups, and you may have some difficulty keeping them all straight. Rather than memorizing them all, keep in mind the following three concepts:

[1]   **Organometallic reagents (R–M) attack electrophilic carbon atoms, especially the carbonyl carbon.**

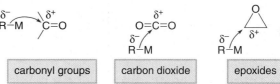

| carbonyl groups | carbon dioxide | epoxides |

[2]   **After an organometallic reagent adds to a carbonyl group, the fate of the intermediate depends on the presence or absence of a leaving group.**

- Without a leaving group, the characteristic reaction is *nucleophilic addition.*
- With a leaving group, it is *nucleophilic substitution.*

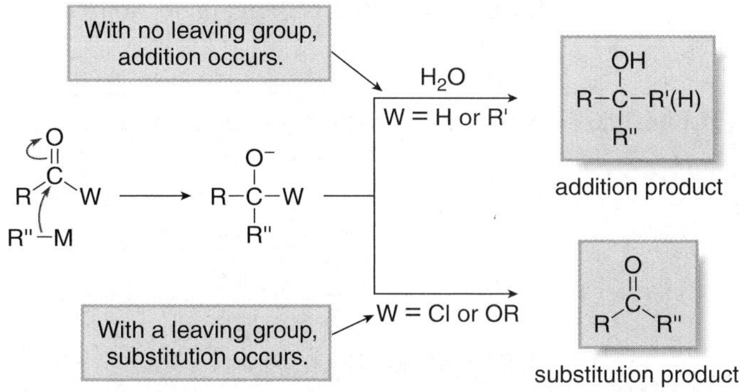

With no leaving group, addition occurs.

$$R-\overset{\underset{\|}{O}}{C}-W \quad \longrightarrow \quad R-\overset{\underset{R''}{|}}{\underset{|}{C}}-W$$

R''–M

With a leaving group, substitution occurs.

W = H or R'   →   addition product

$$\text{OH} \atop R-C-R'(H) \atop R''$$

W = Cl or OR   →   substitution product

[3]   **The polarity of the R–M bond determines the reactivity of the reagents.**

- RLi and RMgX are very reactive reagents.
- R_2CuLi is much less reactive.

20.17   Synthesis

The reactions learned in Chapter 20 have proven extremely useful in organic synthesis. Oxidation and reduction reactions interconvert two functional groups that differ in oxidation state. Organometallic reagents form new carbon–carbon bonds.

Synthesis is perhaps the most difficult aspect of organic chemistry. It requires you to remember both the new reactions you've just learned, and the ones you've encountered in previous chapters. In a successful synthesis, you must also put these reactions in a logical order. Don't be discouraged. Learn the basic reactions and then practice them over and over again with synthesis problems.

In Sample Problems 20.6–20.8 that follow, keep in mind that the products formed by the reactions of Chapter 20 can themselves be transformed into many other functional groups. For example, 2-hexanol, the product of Grignard addition of butylmagnesium chloride to acetaldehyde, can be transformed into a variety of other compounds, as shown in Figure 20.8.

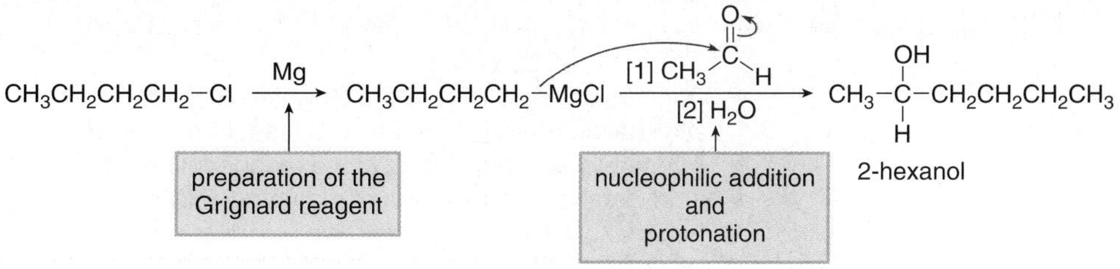

$$CH_3CH_2CH_2CH_2-Cl \xrightarrow{\text{Mg}} CH_3CH_2CH_2CH_2-MgCl \xrightarrow[\text{[2] } H_2O]{\text{[1] } CH_3\overset{O}{\underset{}{C}}H} CH_3-\overset{\underset{H}{|}}{\underset{|}{C}}-CH_2CH_2CH_2CH_3$$

preparation of the Grignard reagent

nucleophilic addition and protonation

2-hexanol

Figure 20.8 Conversion of
2-hexanol into other compounds

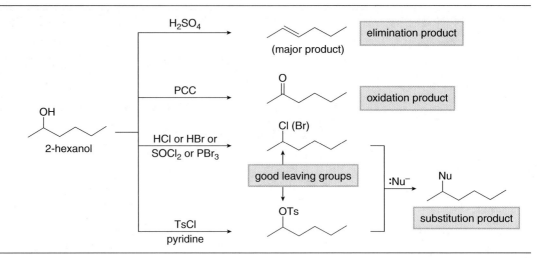

Before proceeding with Sample Problems 20.6–20.8, you should review the stepwise strategy for designing a synthesis found in Section 11.12.

SAMPLE PROBLEM 20.6   Synthesize 1-methylcyclohexene from cyclohexanone and any organic alcohol.

1-methylcyclohexene ⟹ cyclohexanone

RETROSYNTHETIC ANALYSIS

[1]   [2]

+

CH₃MgX ⟹ CH₃OH
     [3]

Thinking backwards:
- [1] Form the double bond by dehydration of an alcohol.
- [2] Make the 3° alcohol by Grignard addition of CH₃MgX.
- [3] Prepare the Grignard reagent stepwise from an alcohol.

SYNTHESIS

Four steps are needed:

$$CH_3OH \xrightarrow[\substack{\text{or} \\ PBr_3 \\ [1]}]{HBr} CH_3Br \xrightarrow[[2]]{Mg} CH_3MgBr \xrightarrow[[3]]{} \xrightarrow[[3]]{H_2O} \xrightarrow[[4]]{H_2SO_4} +$$

major product
trisubstituted alkene

- Conversion of CH₃OH to the Grignard reagent CH₃MgBr requires two steps: formation of an alkyl halide (Step [1]), followed by reaction with Mg (Step [2]).
- Addition of CH₃MgBr to cyclohexanone followed by protonation forms an alcohol in Step [3].
- Acid-catalyzed elimination of water in Step [4] forms a mixture of alkenes, with the desired trisubstituted alkene as the major product.

SAMPLE PROBLEM 20.7   Synthesize 2,4-dimethyl-3-hexanone from four-carbon alcohols.

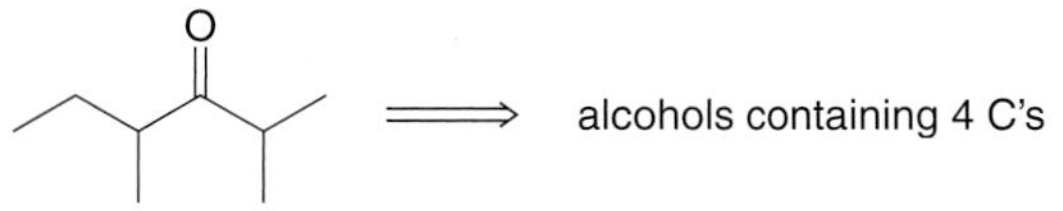

2,4-dimethyl-3-hexanone $\Longrightarrow$ alcohols containing 4 C's

RETROSYNTHETIC ANALYSIS

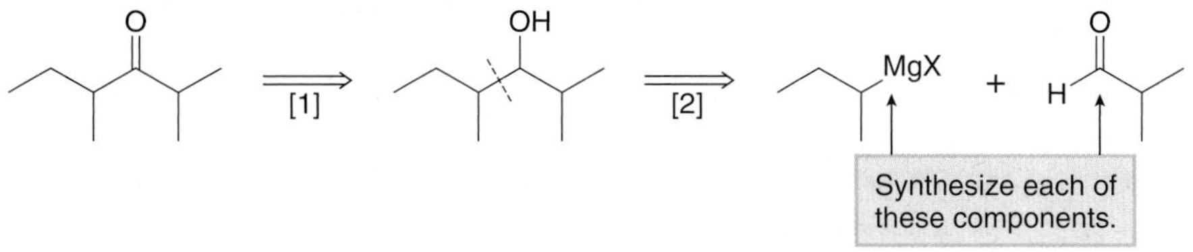

Synthesize each of these components.

Thinking backwards:

- [1] Form the ketone by oxidation of a 2° alcohol.
- [2] Make the 2° alcohol by Grignard addition to an aldehyde. Both of these compounds have 4 C's, and each must be synthesized from an alcohol.

SYNTHESIS

First, make both components needed for the Grignard reaction.

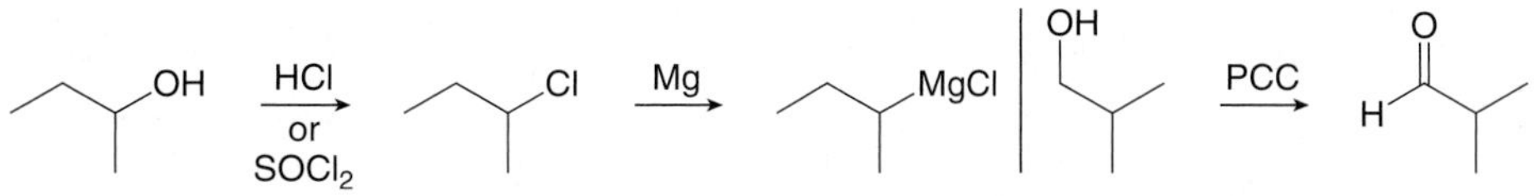

Then, complete the synthesis with Grignard addition, followed by oxidation of the alcohol to the ketone.

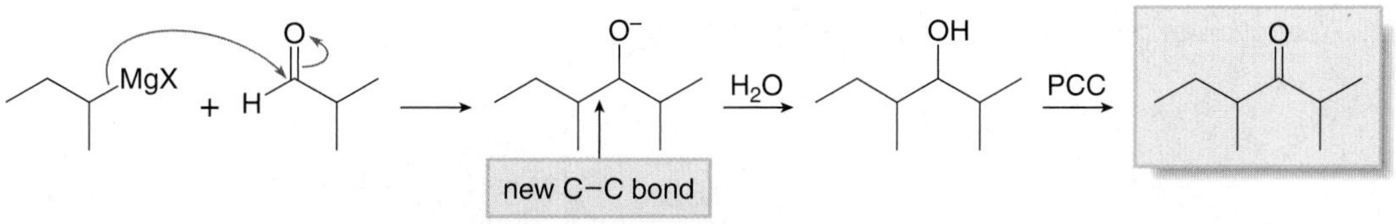

new C–C bond

SAMPLE PROBLEM 20.8   Synthesize isopropylcyclopentane from alcohols having ≤ 5 C's.

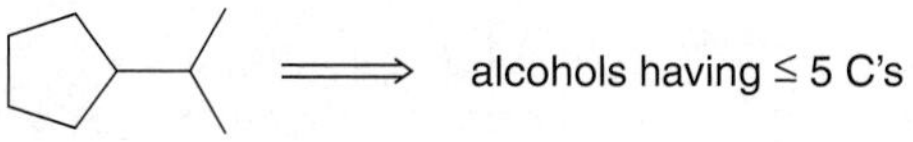

isopropylcyclopentane $\Longrightarrow$ alcohols having ≤ 5 C's

RETROSYNTHETIC ANALYSIS

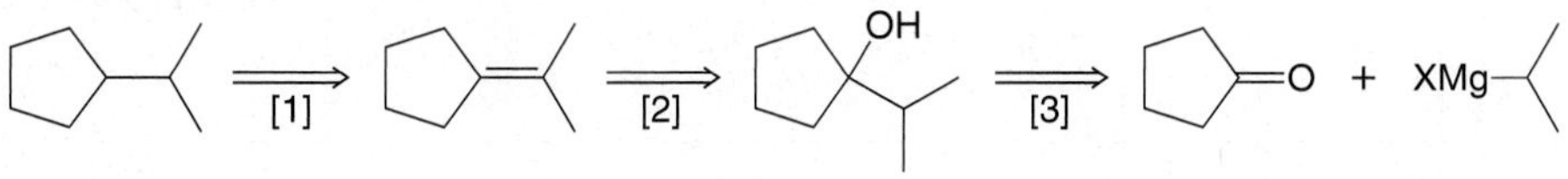

Thinking backwards:

- [1] Form the alkane by hydrogenation of an alkene.
- [2] Introduce the double bond by dehydration of an alcohol.
- [3] Form the 3° alcohol by Grignard addition to a ketone. Both components of the Grignard reaction must then be synthesized.

SYNTHESIS

First make both components needed for the Grignard reaction.

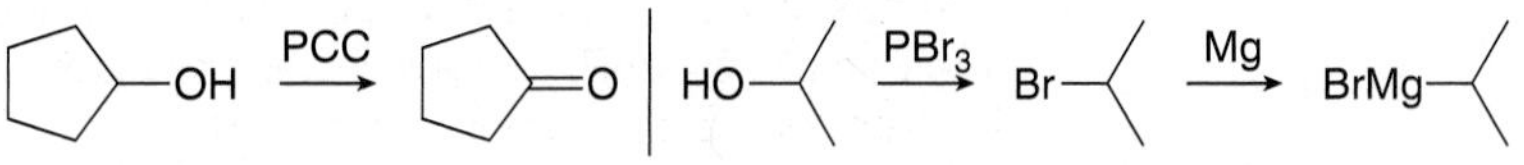

Complete the synthesis with Grignard addition, dehydration, and hydrogenation.

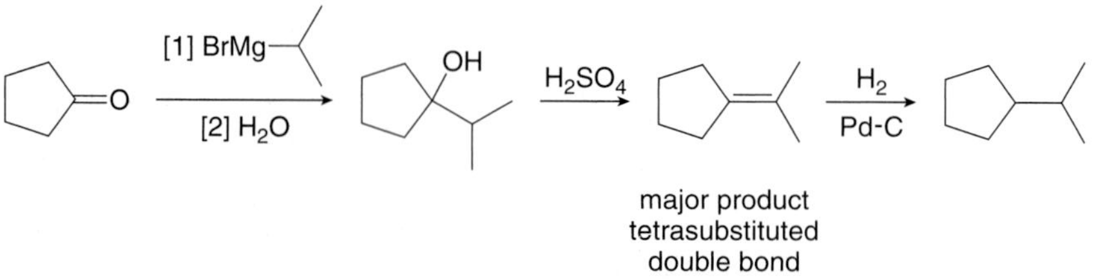

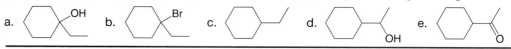

major product
tetrasubstituted
double bond

PROBLEM 20.35   Synthesize each compound from cyclohexanol, ethanol, and any other inorganic reagents.

a.   b.   c.   d.   e.

20.18   Key Concepts—Introduction to Carbonyl Chemistry; Organometallic Reagents; Oxidation and Reduction

Reduction Reactions

[1]   Reduction of aldehydes and ketones to 1° and 2° alcohols (20.4)

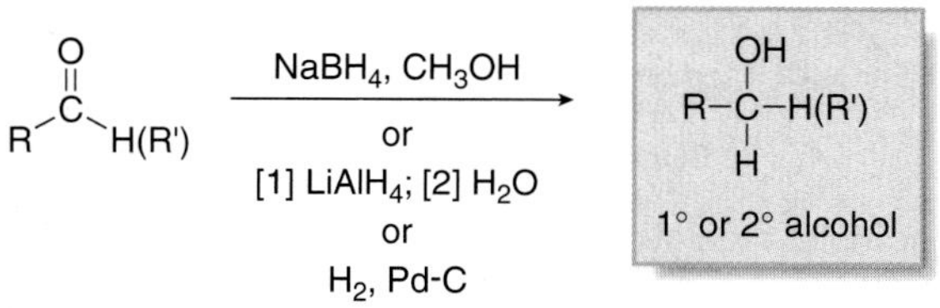

$$R-\overset{O}{\underset{}{C}}-H(R') \xrightarrow[\substack{\text{or} \\ [1]\ LiAlH_4;\ [2]\ H_2O \\ \text{or} \\ H_2,\ Pd\text{-}C}]{NaBH_4,\ CH_3OH}$$

OH
|
R–C–H(R')
|
H
1° or 2° alcohol

[2]   Reduction of α,β-unsaturated aldehydes and ketones (20.4C)

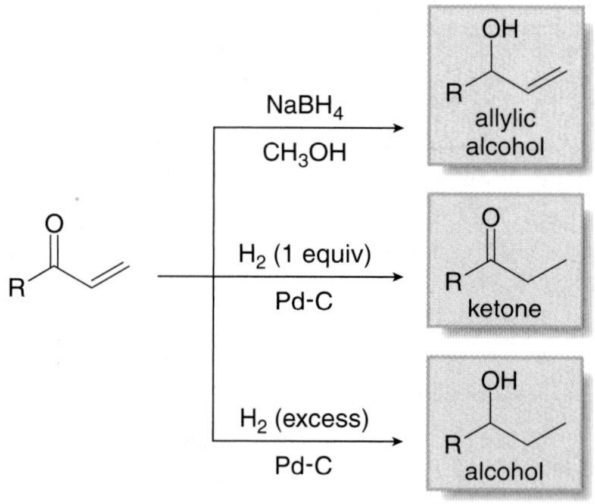

NaBH₄
CH₃OH
→ OH / R / allylic alcohol   • reduction of the C=O only

H₂ (1 equiv)
Pd-C
→ O / R / ketone   • reduction of the C=C only

H₂ (excess)
Pd-C
→ OH / R / alcohol   • reduction of both π bonds

[3]   Enantioselective ketone reduction (20.6)

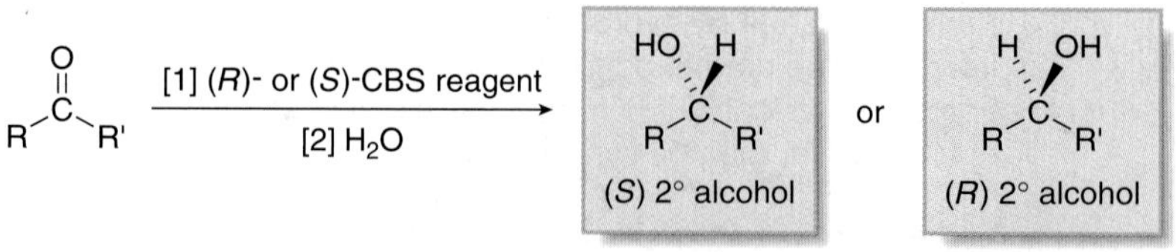

$$R-\overset{O}{\underset{R'}{C}} \xrightarrow[\substack{[2]\ H_2O}]{[1]\ (R)\text{-}\ or\ (S)\text{-}CBS\ reagent}}$$

HO   H
\ /
C
/ \
R   R'
(S) 2° alcohol

or

H   OH
\ /
C
/ \
R   R'
(R) 2° alcohol

• A single enantiomer is formed.

[4] Reduction of acid chlorides (20.7A)

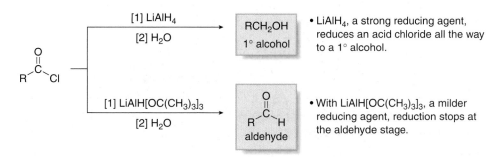

- LiAlH$_4$, a strong reducing agent, reduces an acid chloride all the way to a 1° alcohol.

- With LiAlH[OC(CH$_3$)$_3$]$_3$, a milder reducing agent, reduction stops at the aldehyde stage.

[5] Reduction of esters (20.7A)

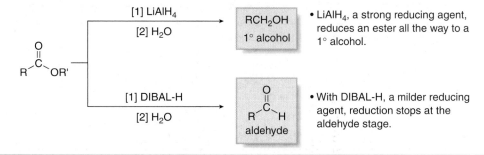

- LiAlH$_4$, a strong reducing agent, reduces an ester all the way to a 1° alcohol.

- With DIBAL-H, a milder reducing agent, reduction stops at the aldehyde stage.

[6] Reduction of carboxylic acids to 1° alcohols (20.7B)

[7] Reduction of amides to amines (20.7B)

Oxidation Reactions

Oxidation of aldehydes to carboxylic acids (20.8)

Preparation of Organometallic Reagents (20.9)

[1] Organolithium reagents: R—X + 2 Li ⟶ R—Li + LiX

[2] Grignard reagents: R—X + Mg ⟶(CH$_3$CH$_2$)$_2$O R—Mg—X

[3] Organocuprate reagents: R—X + 2 Li ⟶ R—Li + LiX

2 R—Li + CuI ⟶ R—Cu⁻ Li⁺ + LiI

[4] Lithium and sodium acetylides:

$$R-C\equiv C-H \xrightarrow{\text{Na}^+ \;^-\text{NH}_2} \boxed{R-C\equiv C^- \; \text{Na}^+} + \text{NH}_3$$

a sodium acetylide

$$R-C\equiv C-H \xrightarrow{\text{R}'-\text{Li}} \boxed{R-C\equiv C-\text{Li}} + \text{R}'-\text{H}$$

a lithium acetylide

Reactions with Organometallic Reagents

[1] Reaction as a base (20.9C)

$$R-M + H-\overset{..}{\underset{..}{O}}-R \longrightarrow \boxed{R-H} + M^+ \; ^-\overset{..}{\underset{..}{O}}-R$$

- RM = RLi, RMgX, R$_2$CuLi
- This acid–base reaction occurs with H$_2$O, ROH, RNH$_2$, R$_2$NH, RSH, RCOOH, RCONH$_2$, and RCONHR.

[2] Reaction with aldehydes and ketones to form 1°, 2°, and 3° alcohols (20.10)

$$\underset{R}{\overset{\overset{\displaystyle O}{\parallel}}{C}}_{\!\!\!\!\!\!\!\!\!\!\!\!\! H(R')} \xrightarrow[\text{[2] H}_2\text{O}]{\text{[1] R''MgX or R''Li}} \boxed{\underset{R''}{\overset{\overset{\displaystyle OH}{|}}{R-C-H(R')}}}$$

1°, 2°, or 3° alcohol

[3] Reaction with esters to form 3° alcohols (20.13A)

$$\underset{R}{\overset{\overset{\displaystyle O}{\parallel}}{C}}_{\!\!\!\!\!\!\!\!\!\!\!\!\! OR'} \xrightarrow[\text{[2] H}_2\text{O}]{\substack{\text{[1] R''Li or R''MgX} \\ \text{(2 equiv)}}} \boxed{\underset{R''}{\overset{\overset{\displaystyle OH}{|}}{R-C-R''}}}$$

3° alcohol

[4] Reaction with acid chlorides (20.13B)

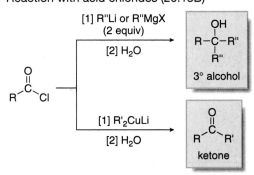

- More reactive organometallic reagents—R''Li and R''MgX—add two equivalents of R'' to an acid chloride to form a 3° alcohol with two identical R'' groups.

- Less reactive organometallic reagents—R'$_2$CuLi—add only one equivalent of R' to an acid chloride to form a ketone.

[5] Reaction with carbon dioxide—Carboxylation (20.14A)

$$R-MgX \xrightarrow[\text{[2] H}_3\text{O}^+]{\text{[1] CO}_2} \boxed{\underset{OH}{\overset{\overset{\displaystyle O}{\parallel}}{R-C}}}$$

carboxylic acid

[6] Reaction with epoxides (20.14B)

$$\underset{C-C}{\overset{\displaystyle O}{\triangle}} \xrightarrow[\text{[2] H}_2\text{O}]{\text{[1] RLi, RMgX, or R}_2\text{CuLi}} \boxed{\underset{R}{C}-\underset{OH}{C}}$$

alcohol

[7] Reaction with α,β-unsaturated aldehydes and ketones (20.15B)

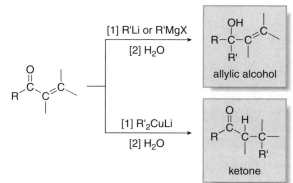

- More reactive organometallic reagents—R'Li and R'MgX—react with α,β-unsaturated carbonyls by 1,2-addition.

- Less reactive organometallic reagents—R'₂CuLi—react with α,β-unsaturated carbonyls by 1,4-addition.

Protecting Groups (20.12)

[1] Protecting an alcohol as a *tert*-butyldimethylsilyl ether

$$R-O-H \ + \ \underset{[Cl-TBDMS]}{\overset{CH_3}{\underset{CH_3}{Cl-Si-C(CH_3)_3}}} \ \xrightarrow{\quad\underset{N\quad NH}{}\quad} \ \underset{\substack{[R-O-TBDMS] \\ \textit{tert}\text{-butyldimethylsilyl ether}}}{\overset{CH_3}{\underset{CH_3}{R-O-Si-C(CH_3)_3}}}$$

[2] Deprotecting a *tert*-butyldimethylsilyl ether to re-form an alcohol

$$\underset{[R-O-TBDMS]}{\overset{CH_3}{\underset{CH_3}{R-O-Si-C(CH_3)_3}}} \ \xrightarrow{(CH_3CH_2CH_2CH_2)_4N^+ \ F^-} \ R-O-H \ + \ \underset{[F-TBDMS]}{\overset{CH_3}{\underset{CH_3}{F-Si-C(CH_3)_3}}}$$

Problems

Reactions and Reagents

20.36 Draw the product formed when pentanal ($CH_3CH_2CH_2CH_2CHO$) is treated with each reagent. With some reagents, no reaction occurs.

a. $NaBH_4$, CH_3OH
b. [1] $LiAlH_4$; [2] H_2O
c. H_2, Pd-C
d. PCC
e. $Na_2Cr_2O_7$, H_2SO_4, H_2O
f. Ag_2O, NH_4OH

g. [1] CH_3MgBr; [2] H_2O
h. [1] C_6H_5Li; [2] H_2O
i. [1] $(CH_3)_2CuLi$; [2] H_2O
j. [1] $HC\equiv CNa$; [2] H_2O
k. [1] $CH_3C\equiv CLi$; [2] H_2O
l. The product in (a), then TBDMS–Cl, imidazole

20.37 Repeat Problem 20.36 using 2-pentanone ($CH_3COCH_2CH_2CH_3$) as the starting material.

20.38 Draw the product formed when 1-bromobutane is treated with each reagent.

a. Li (2 equiv)
b. Mg in $(CH_3CH_2)_2O$ solvent
c. Li (2 equiv), then CuI (0.5 equiv)

d. The answer in (a), then H_2O
e. The answer in (b), then D_2O
f. The answer in (a), then $CH_3C\equiv CH$

20.39 Draw the product formed when $CH_3CH_2CH_2MgBr$ is treated with each compound.

a. $CH_2=O$, then H_2O

b. ⬠=O, then H_2O

c. CH_3CH_2COCl, then H_2O

d. $CH_3CH_2COOCH_3$, then H_2O

e. H_2O

f. CH_3CH_2OH

g. CH_3COOH

h. $HC\equiv CH$

i. CO_2, then H_3O^+

j. triangle(O)–CH_3, then H_2O, with CH_3

k. D_2O

l. cyclohexenone=O, then H_2O

20.40 Draw the product formed when $(CH_3CH_2CH_2CH_2)_2CuLi$ is treated with each compound. In some cases, no reaction occurs.

a.   (benzoyl chloride, Ph–C(=O)–Cl)

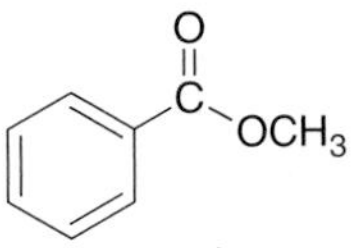

b. methyl benzoate, Ph–C(=O)–OCH₃

c. 2-methylcyclohex-2-enone, then H₂O

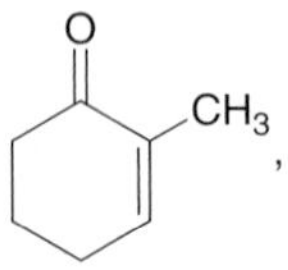

d. 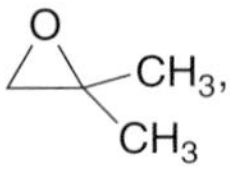 epoxide with two CH₃ groups –CH₃, then H₂O

20.41 The stereochemistry of the products of reduction depends on the reagent used, as you learned in Sections 20.5 and 20.6. With this in mind, how would you convert 3,3-dimethyl-2-butanone $[CH_3COC(CH_3)_3]$ to: (a) racemic 3,3-dimethyl-2-butanol $[CH_3CH(OH)C(CH_3)_3]$; (b) only (R)-3,3-dimethyl-2-butanol; (c) only (S)-3,3-dimethyl-2-butanol?

20.42 Draw the product formed when the α,β-unsaturated ketone **A** is treated with each reagent.

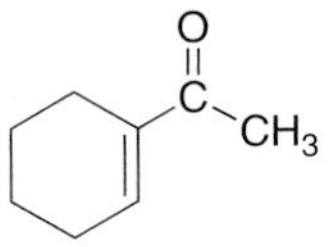

(structure of **A**: 1-cyclohexenyl methyl ketone)

a. $NaBH_4$, CH_3OH
b. H_2 (1 equiv), Pd-C
c. H_2 (excess), Pd-C
d. [1] CH_3Li; [2] H_2O
e. [1] CH_3CH_2MgBr; [2] H_2O
f. [1] $(CH_2=CH)_2CuLi$; [2] H_2O

20.43 What reagent is needed to convert $(CH_3)_2CHCH_2CH_2COCl$ to each compound?

a. $(CH_3)_2CHCH_2CH_2CHO$
b. $(CH_3)_2CHCH_2CH_2COCH=CH_2$
c. $(CH_3)_2CHCH_2CH_2C(OH)(C_6H_5)_2$
d. $(CH_3)_2CHCH_2CH_2CH_2OH$

20.44 What reagent is needed to convert $CH_3CH_2COOCH_2CH_2CH_3$ to each compound?

a. $CH_3CH_2CH_2OH$
b. $CH_3CH_2C(OH)(CH_2CH_2CH_3)_2$
c. CH_3CH_2CHO

20.45 As discussed in Sections 12.12 and 20.8, some oxidizing agents selectively oxidize a particular functional group, whereas others oxidize many different functional groups. Draw the product formed when $HOCH_2CH_2CH_2CH_2CHO$ is treated with each reagent: (a) CrO_3, H_2SO_4, H_2O; (b) PCC; (c) Ag_2O, NH_4OH; (d) $Na_2Cr_2O_7$, H_2SO_4, H_2O.

20.46 Draw the products of each reduction reaction.

a. 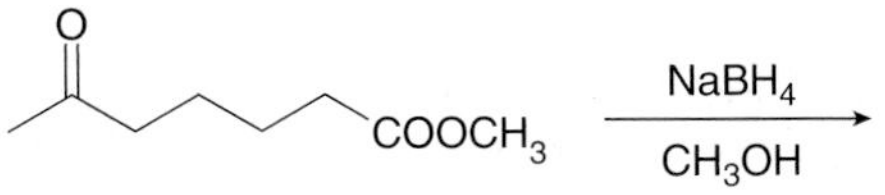 (methyl ketoester) $\xrightarrow[\text{CH}_3\text{OH}]{\text{NaBH}_4}$

b. 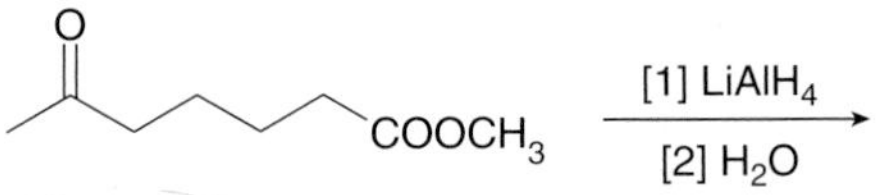 (methyl ketoester) $\xrightarrow[\text{[2] H}_2\text{O}]{\text{[1] LiAlH}_4}$

c. 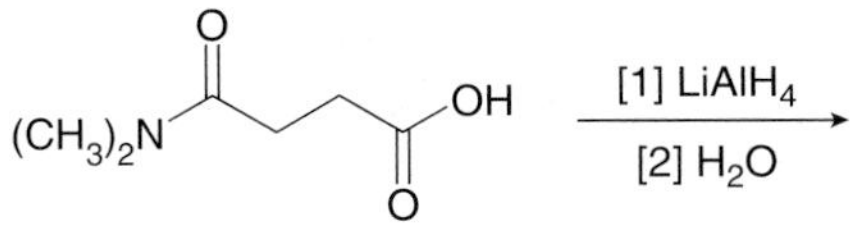 $(CH_3)_2N$-(amide-acid) $\xrightarrow[\text{[2] H}_2\text{O}]{\text{[1] LiAlH}_4}$

d. 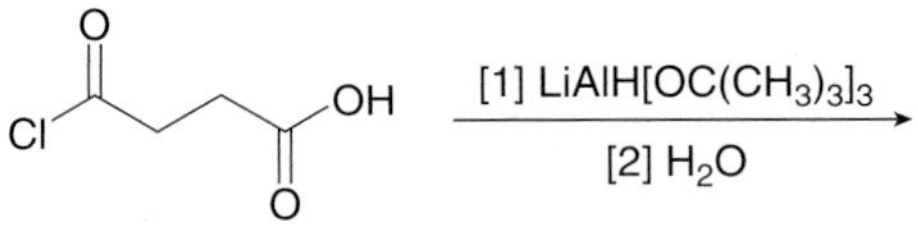 Cl-(acid chloride-acid) $\xrightarrow[\text{[2] H}_2\text{O}]{\text{[1] LiAlH[OC(CH}_3)_3]_3}$

20.47 Draw the products of the following reactions with organometallic reagents.

a. 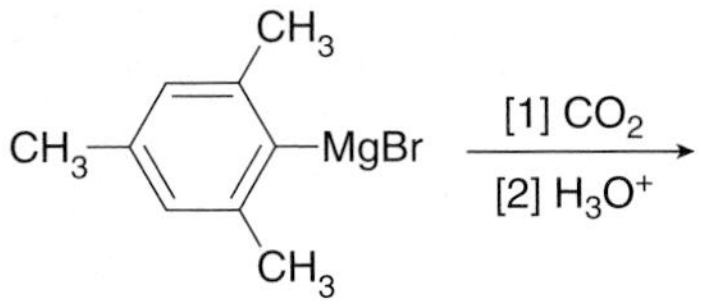 (trimethylphenyl)-MgBr $\xrightarrow[\text{[2] H}_3\text{O}^+]{\text{[1] CO}_2}$

b. (2-tetralone) $\xrightarrow[\text{[2] H}_2\text{O}]{\text{[1] CH}_3\text{CH}_2\text{MgBr}}$

c. $\sim\sim$CHO $\xrightarrow[\text{[2] H}_2\text{O}]{\text{[1] C}_6\text{H}_5\text{Li}}$

d. (benzoyl chloride)-COCl $\xrightarrow[\text{[2] H}_2\text{O}]{\substack{\text{[1] C}_6\text{H}_5\text{MgBr} \\ \text{(excess)}}}$

e. (ethyl benzoate)-COOCH₂CH₃ $\xrightarrow[\text{[2] H}_2\text{O}]{\substack{\text{[1] CH}_3\text{MgCl} \\ \text{(excess)}}}$

f. $\sim\sim$MgBr $\xrightarrow[\text{[2] H}_2\text{O}]{\text{[1] CH}_2=\text{O}}$

g. (unsaturated ketone) $\xrightarrow[\text{[2] H}_2\text{O}]{\text{[1] (CH}_3)_2\text{CuLi}}$

h. (alkylidene cyclohexane ketone) $\xrightarrow[\text{[2] H}_2\text{O}]{\text{[1] CH}_3\text{MgBr}}$

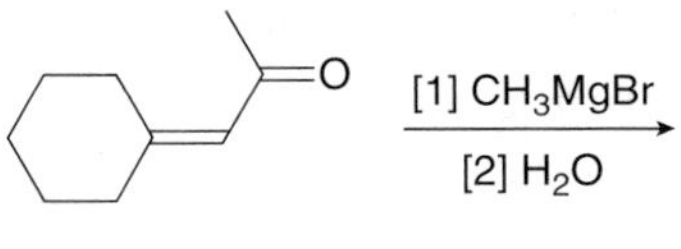

i. 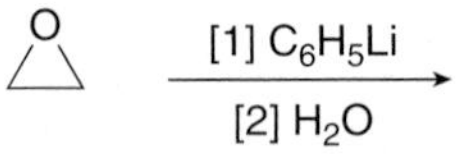 (cyclopropane oxide/epoxide) $\xrightarrow[\text{[2] H}_2\text{O}]{\text{[1] C}_6\text{H}_5\text{Li}}$

j. 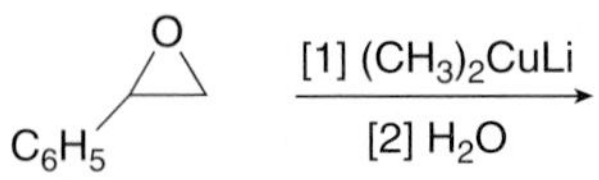 (phenyl epoxide, C_6H_5) $\xrightarrow[\text{[2] H}_2\text{O}]{\text{[1] (CH}_3)_2\text{CuLi}}$

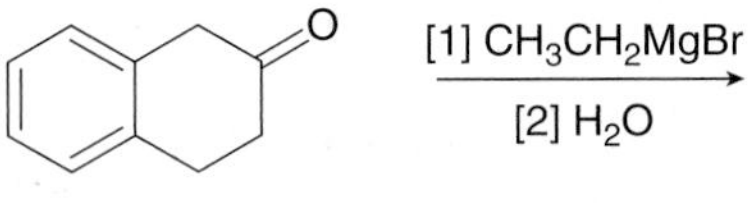

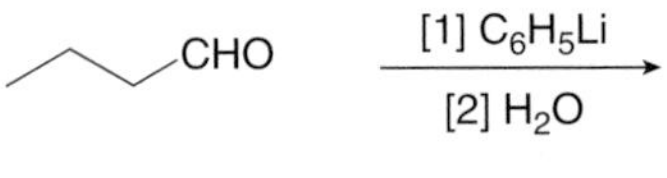

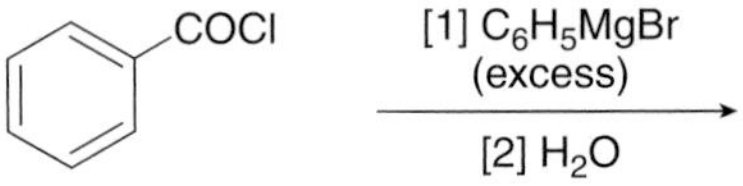

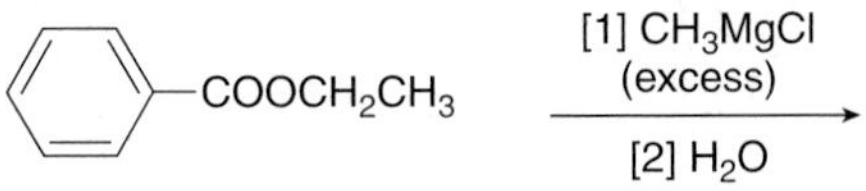

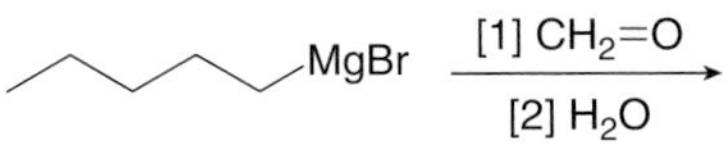

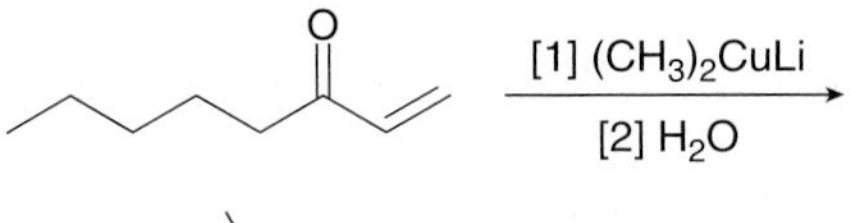

20.48 Draw all stereoisomers formed in each reaction.

a. 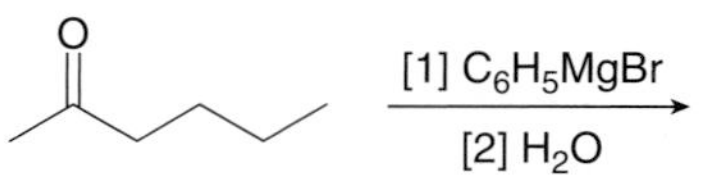 $\xrightarrow[\text{[2] H}_2\text{O}]{\text{[1] C}_6\text{H}_5\text{MgBr}}$

b. $(CH_3)_3C$— $\xrightarrow[\text{[2] H}_2\text{O}]{\text{[1] CH}_3\text{Li}}$

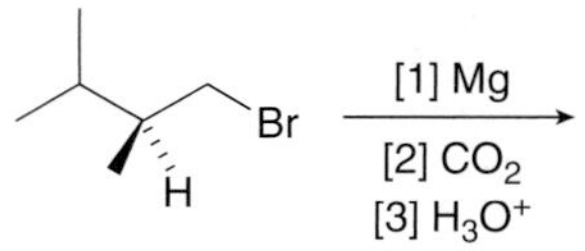

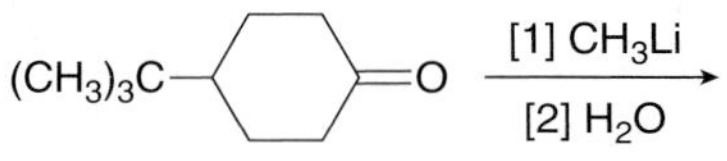

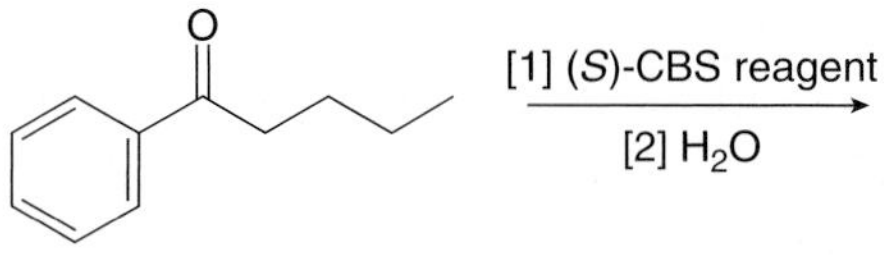

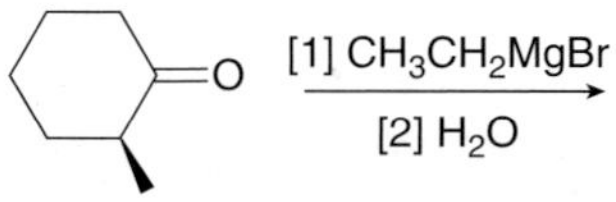

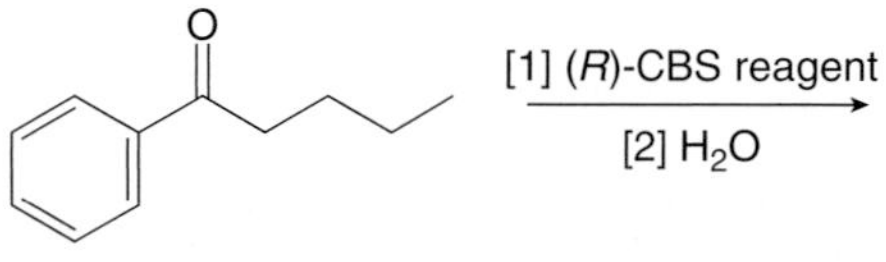

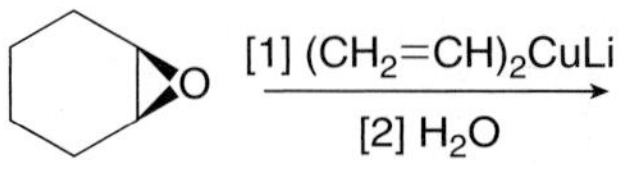

c. $\xrightarrow[\text{[2] H}_2\text{O}]{\text{[1] CH}_3\text{CH}_2\text{MgBr}}$

d. $\xrightarrow[\text{[2] H}_2\text{O}]{\text{[1] (CH}_2\text{=CH)}_2\text{CuLi}}$

e. $\xrightarrow[\substack{\text{[2] CO}_2 \\ \text{[3] H}_3\text{O}^+}]{\text{[1] Mg}}$

f. $\xrightarrow[\text{[2] H}_2\text{O}]{\text{[1] (}S\text{)-CBS reagent}}$

g. $\xrightarrow[\text{[2] H}_2\text{O}]{\text{[1] (}R\text{)-CBS reagent}}$

h. $\xrightarrow[\text{[2] H}_2\text{O}]{\text{[1] LiAlH}_4}$

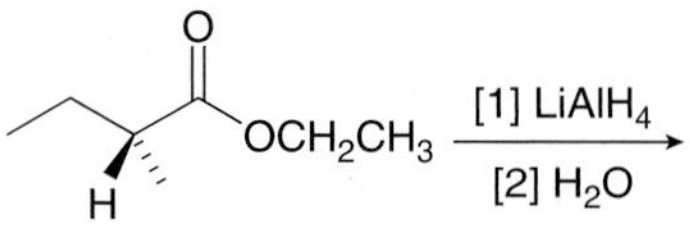

20.49 A student tried to carry out the following reaction sequence, but none of diol **A** was formed. Explain what was wrong with this plan, and design a successful stepwise synthesis of **A**.

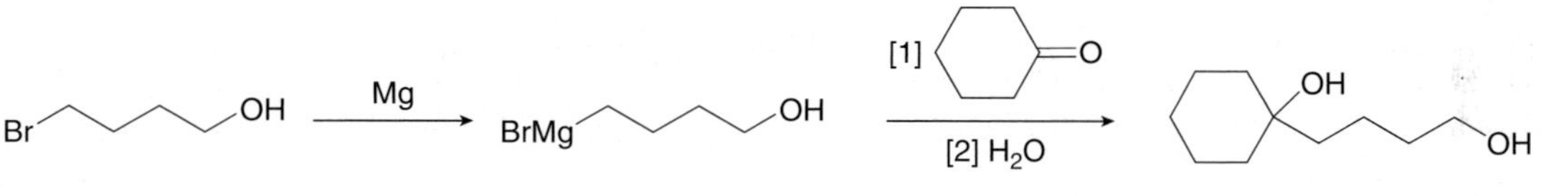

20.50 Identify the lettered compounds in the following reaction scheme. Compounds **F, G,** and **K** are isomers of molecular formula $C_{13}H_{18}O$. How could ^{1}H NMR spectroscopy distinguish these three compounds from each other?

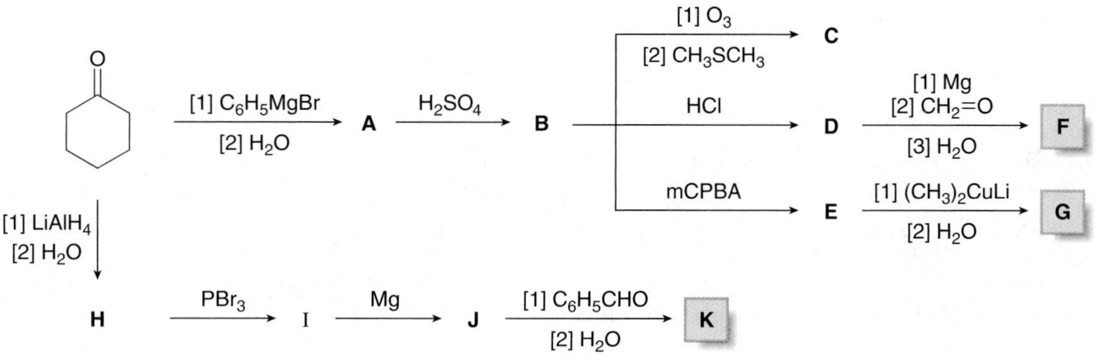

20.51 Fill in the lettered products (**A–G**) in the synthesis of the three biologically active compounds drawn below.

a.

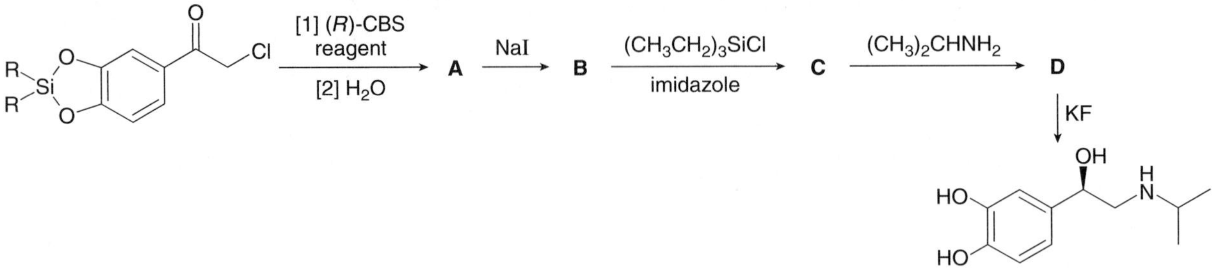

[1] (*R*)-CBS reagent → **A** → NaI → **B** → $(CH_3CH_2)_3SiCl$ / imidazole → **C** → $(CH_3)_2CHNH_2$ → **D**
[2] H_2O

KF

(*R*)-isoproterenol
trade name: Isuprel

b.

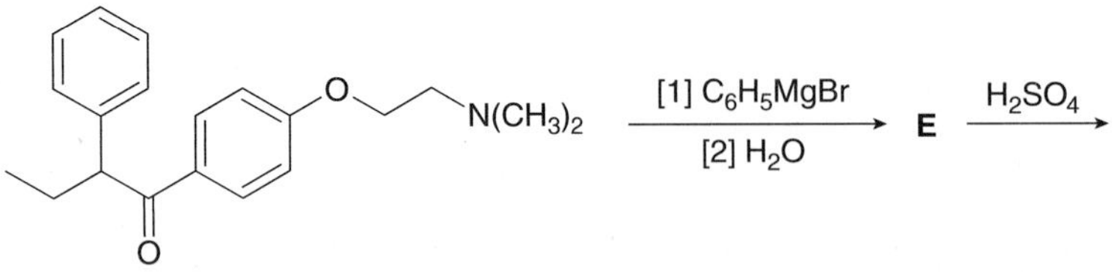

[1] C_6H_5MgBr → **E** → H_2SO_4 → **F**   (*Z* and *E* isomers)
[2] H_2O

The *Z* isomer is tamoxifen.
(used in treatment of breast cancer)

c.

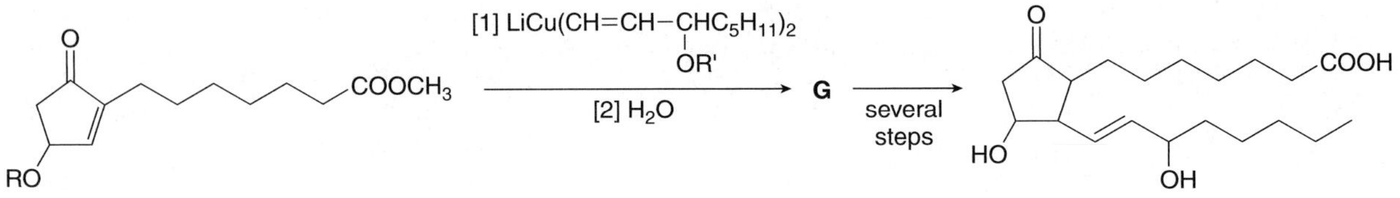

[1] LiCu(CH=CH–CHC$_5$H$_{11}$)$_2$
                    OR'
                                          → **G** → several steps
[2] H_2O

PGE$_1$
(a prostaglandin)

Mechanism

20.52 Draw a stepwise mechanism for each reaction.

a. 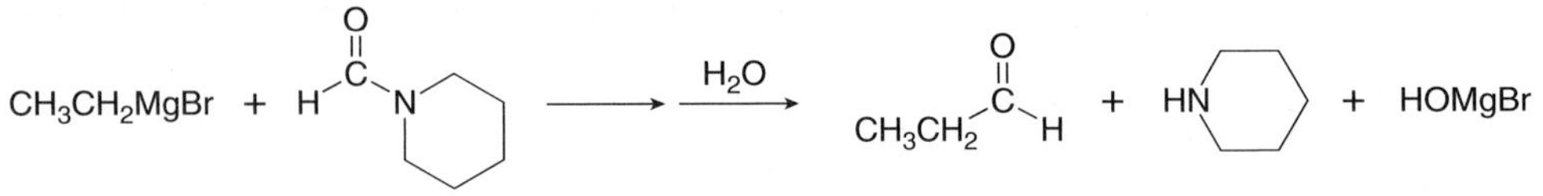 CH_3CH_2MgBr +

$\xrightarrow{H_2O}$ $CH_3CH_2\overset{O}{\underset{}{C}}H$ + HN + HOMgBr

b. 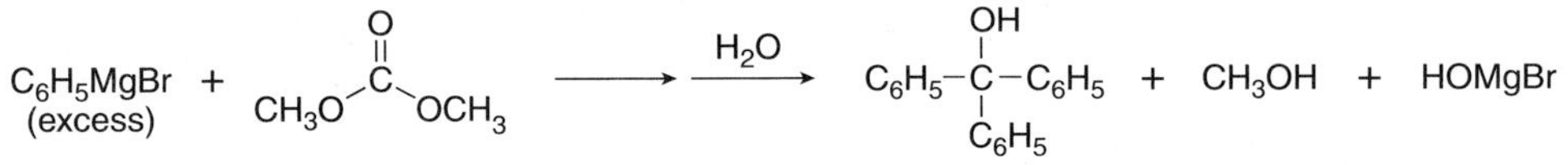 C_6H_5MgBr +
    (excess)

$\xrightarrow{H_2O}$ $C_6H_5\overset{OH}{\underset{C_6H_5}{C}}C_6H_5$ + CH_3OH + HOMgBr

20.53 Draw a stepwise mechanism for the following reaction.

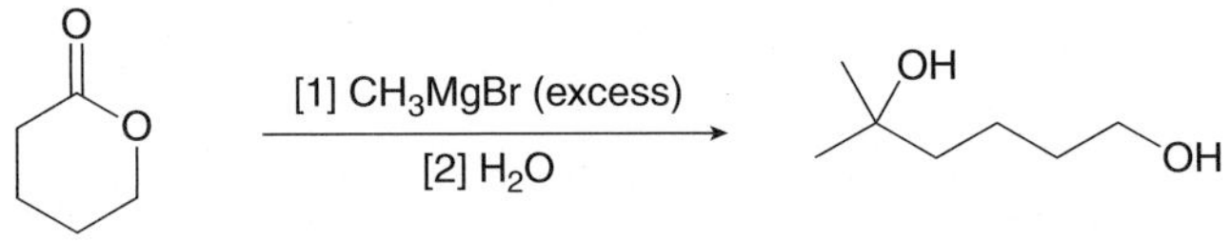

[1] CH_3MgBr (excess)
[2] H_2O

Synthesis

20.54 What Grignard reagent and aldehyde (or ketone) are needed to prepare each of the following alcohols? Show all possible routes.

a.

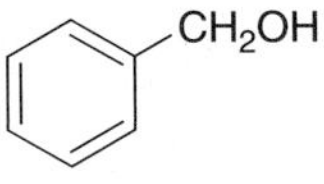

b.

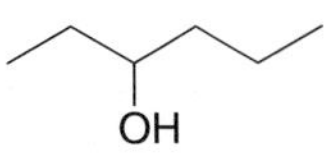

c. $(C_6H_5)_3COH$

d.

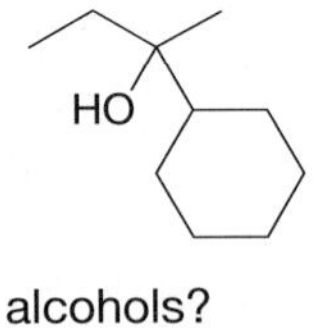

e.

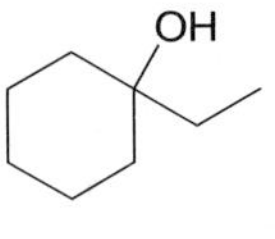

20.55 What ester and Grignard reagent are needed to synthesize each of the following alcohols?

a.

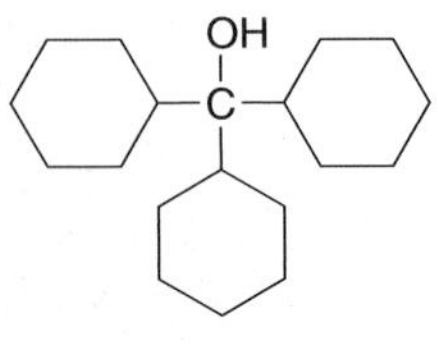

b. $CH_3-\overset{OH}{\underset{CH_3}{C}}-CH_2CH_2CH(CH_3)_2$

c. $(CH_3CH_2CH_2CH_2)_2C(OH)CH_3$

20.56 What organolithium reagent and carbonyl compound can be used to prepare each of the following compounds? You may use aldehydes, ketones, or esters as carbonyl starting materials.

a. 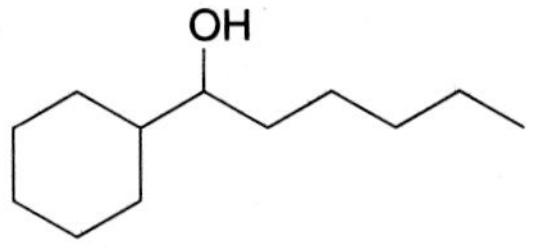
(two ways)

b. 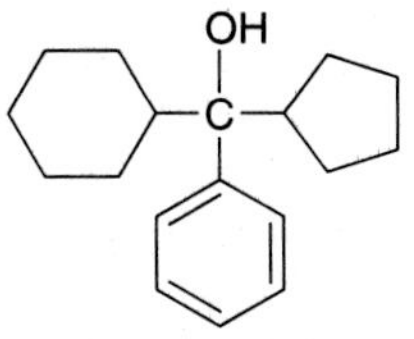
(three ways)

c. 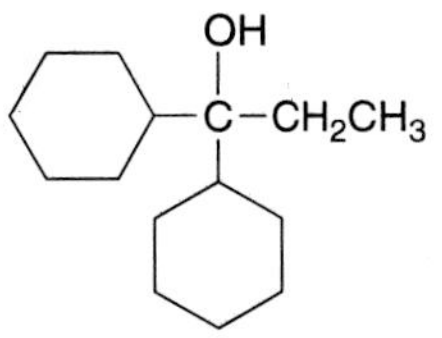
(three ways)

20.57 Ethynylestradiol can be made directly from estrone if two equivalents of lithium acetylide (Li−C≡CH) are used in the reaction.
   a. Explain why two equivalents of Li−C≡CH are needed.
   b. Devise a stepwise synthesis of ethynylestradiol from estrone that uses only one equivalent of Li−C≡CH.

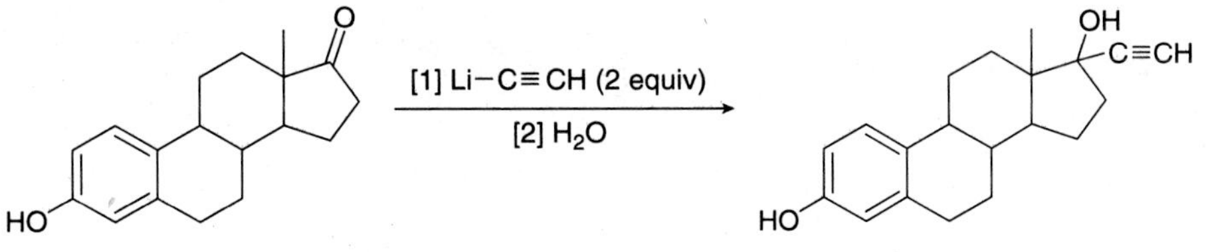

[1] Li−C≡CH (2 equiv)
[2] H$_2$O

estrone   ethynylestradiol

20.58 Synthesize each compound from cyclohexanol using any other organic or inorganic reagents.

a.

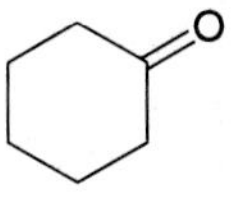

b.

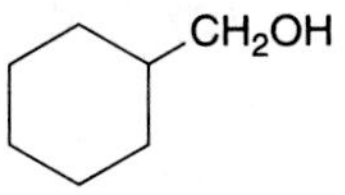

c.

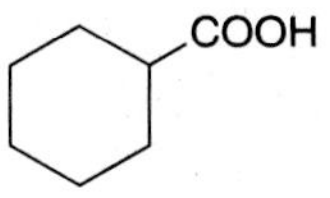

d.

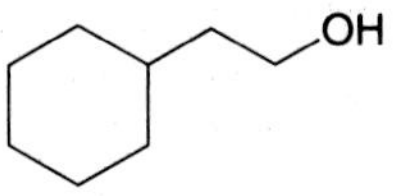

e.

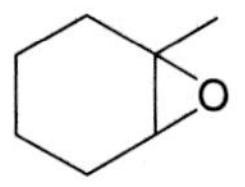

f.

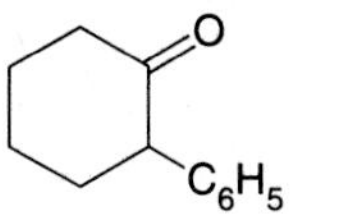

g.

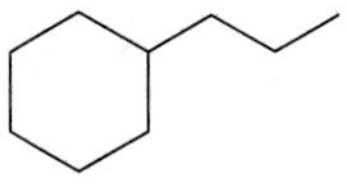

h.

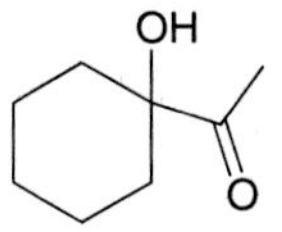

20.59 Convert 2-propanol [(CH$_3$)$_2$CHOH] into each compound. You may use any other organic or inorganic compounds.
   a. (CH$_3$)$_2$CHCl
   b. (CH$_3$)$_2$C=O
   c. (CH$_3$)$_2$CHCH(OH)CH$_3$
   d. (CH$_3$)$_2$CHCH$_2$CH$_2$OH
   e. (CH$_3$)$_2$CHCOOH
   f. (CH$_3$)$_2$CHCHO
   g. (CH$_3$)$_2$CHD

h.

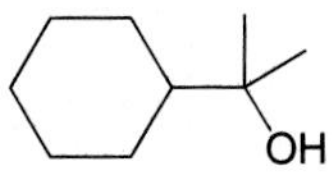

i.

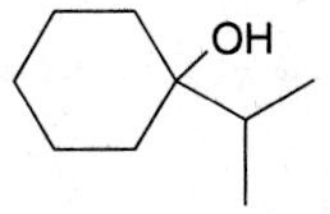

j.

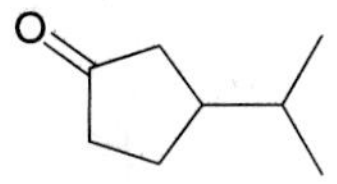

20.60 Convert benzene into each compound. You may also use any inorganic reagents and organic alcohols having three carbons or fewer. One step of the synthesis must use a Grignard reagent.

a.

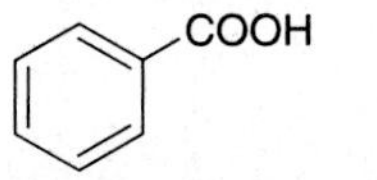

b.

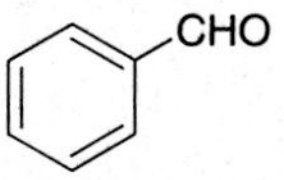

c.

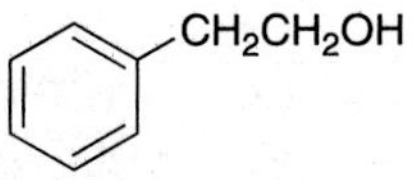

d.

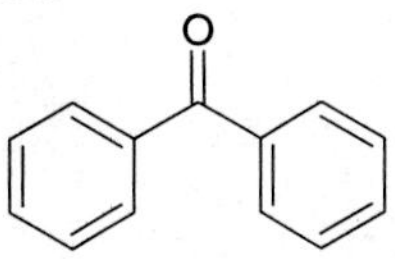

e.

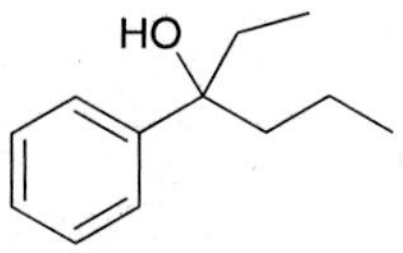

20.61 Design a synthesis of each compound from alcohols having four carbons or fewer as the only organic starting materials. You may use any other inorganic reagents you choose.

a.

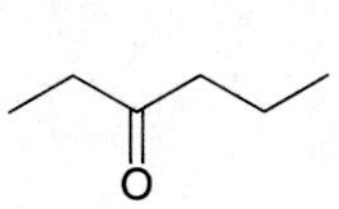

b.

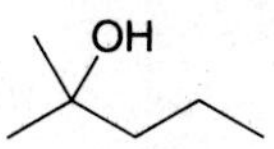

c.

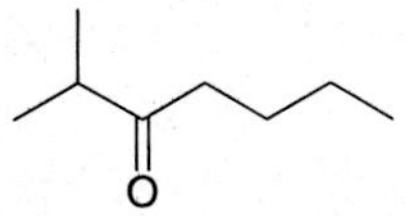

d.

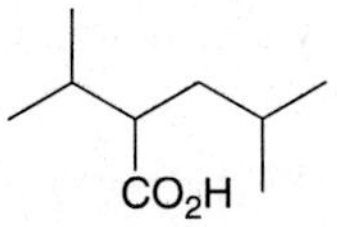

20.62 Design a synthesis of each compound from the given starting material. You may use any other inorganic reagents you choose.

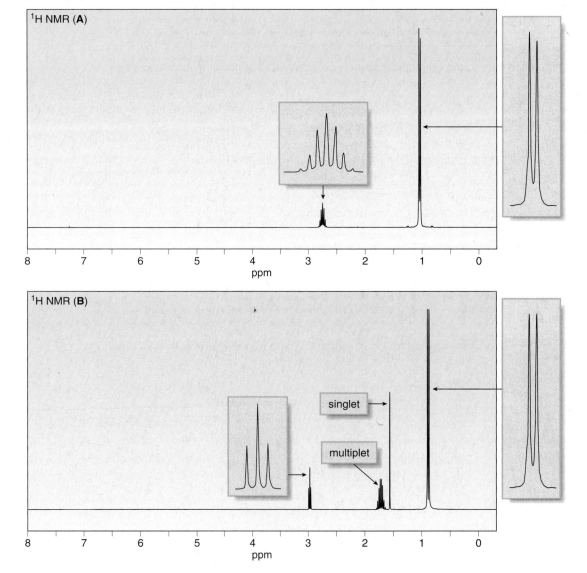

a.    ⟹   =O + any organic halides

b. 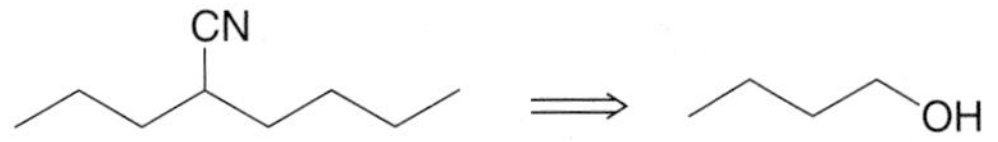   CN ⟹ ⟶OH   (the only organic starting material)

c. 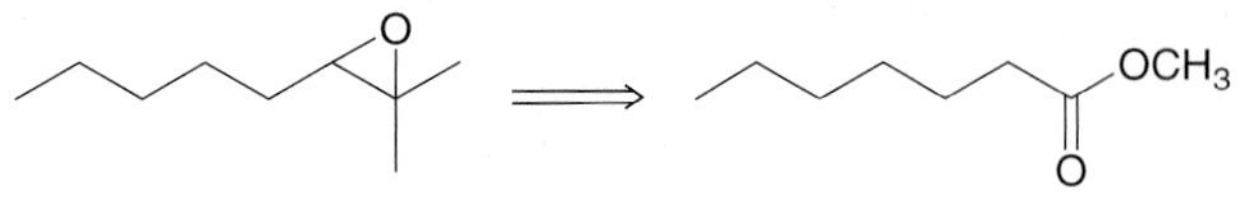   ⟹

Spectroscopy

20.63 An unknown compound **A** (molecular formula $C_7H_{14}O$) was treated with $NaBH_4$ in CH_3OH to form compound **B** (molecular formula $C_7H_{16}O$). Compound **A** has a strong absorption in its IR spectrum at 1716 cm^{-1}. Compound **B** has a strong absorption in its IR spectrum at 3600–3200 cm^{-1}. The ^1H NMR spectra of **A** and **B** are given. What are the structures of **A** and **B**?

^1H NMR (**A**)

^1H NMR (**B**)

singlet

multiplet

20.64 Treatment of compound **C** (molecular formula C_4H_8O) with C_6H_5MgBr, followed by H_2O, affords compound **D** (molecular formula $C_{10}H_{14}O$). Compound **D** has a strong peak in its IR spectrum at 3600–3200 cm^{-1}. The ^1H NMR spectral data of **C** and **D** are given. What are the structures of **C** and **D**?

Compound **C** peaks at 1.3 (singlet, 6 H) and 2.4 (singlet, 2 H) ppm

Compound **D** peaks at 1.2 (singlet, 6 H), 1.6 (singlet, 1 H), 2.7 (singlet, 2 H), and 7.2 (multiplet, 5 H) ppm

20.65 Treatment of compound **E** (molecular formula $C_4H_8O_2$) with excess CH_3CH_2MgBr yields compound **F** (molecular formula $C_6H_{14}O$) after protonation with H_2O. **E** shows a strong absorption in its IR spectrum at 1743 cm^{-1}. **F** shows a strong IR absorption at 3600–3200 cm^{-1}. The ^1H NMR spectral data of **E** and **F** are given. What are the structures of **E** and **F**?

Compound **E** peaks at 1.2 (triplet, 3 H), 2.0 (singlet, 3 H), and 4.1 (quartet, 2 H) ppm

Compound **F** peaks at 0.9 (triplet, 6 H), 1.1 (singlet, 3 H), 1.5 (quartet, 4 H), and 1.55 (singlet, 1 H) ppm

Challenge Problems

20.66 Design a synthesis of (*R*)-salmeterol (Figure 20.3) from the following starting materials.

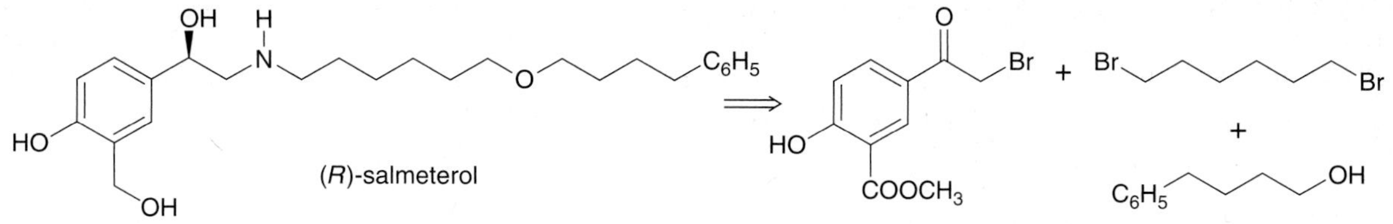

20.67 Lithium tri-*sec*-butylborohydride, also known as L-selectride, is a metal hydride reagent that contains three *sec*-butyl groups bonded to boron. When this reagent is used to reduce cyclic ketones, one stereoisomer often predominates as product. Explain why the reduction of 4-*tert*-butylcyclohexanone with L-selectride forms the cis alcohol as the major product.

$LiBH[CH(CH_3)CH_2CH_3]_3$

lithium tri-*sec*-butylborohydride
L-selectride

4-*tert*-butylcyclohexanone

cis-4-*tert*-butylcyclohexanol

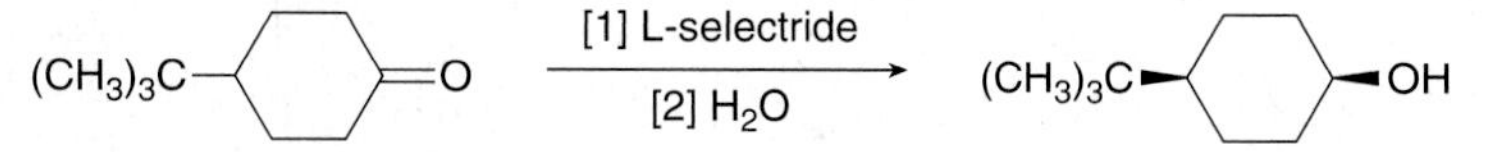

Aldehydes and Ketones—Nucleophilic Addition

11-*cis*-Retinal is a light-sensitive conjugated aldehyde that plays a key role in the complex chemistry of vision for all vertebrates, arthropods, and mollusks. In the human retina, 11-*cis*-retinal is bonded to the protein opsin, forming rhodopsin or visual purple. When light is absorbed by rhodopsin, the 11-cis double bond is isomerized to its more stable trans isomer and rhodopsin changes shape, generating a nerve impulse, thereby converting light into electrical signals that are processed into images by the brain. In Chapter 21, we learn about these and other reactions of aldehydes and ketones.

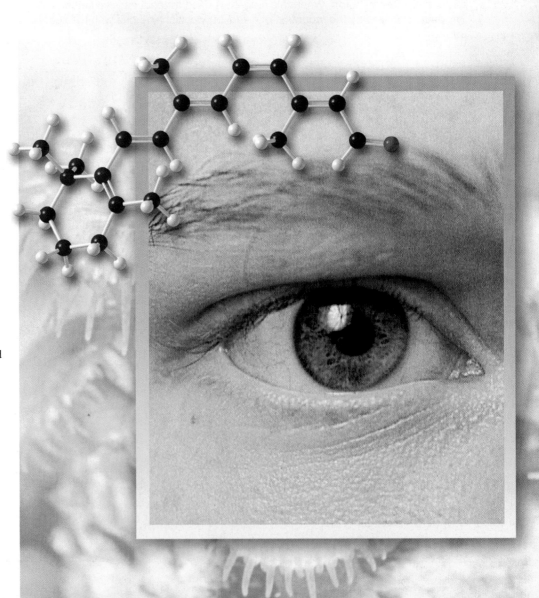

In Chapter 21 we continue the study of carbonyl compounds with a detailed look at **aldehydes** and **ketones.** We will first learn about the nomenclature, physical properties, and spectroscopic absorptions that characterize aldehydes and ketones. The remainder of Chapter 21 is devoted to **nucleophilic addition** reactions. Although we have already learned two examples of this reaction in Chapter 20, nucleophilic addition to aldehydes and ketones is a general reaction that occurs with many nucleophiles, forming a wide variety of products.

Every new reaction in Chapter 21 involves nucleophilic addition, so the challenge lies in learning the specific reagents and mechanisms that characterize each reaction.

21.1 Introduction

An aldehyde is often written as **RCHO.** Remember that the **H atom is bonded to the carbon atom,** not the oxygen. Likewise, a ketone is written as **RCOR,** or if both alkyl groups are the same, **R₂CO.** Each structure must contain a C=O for every atom to have an octet.

As we learned in Chapter 20, **aldehydes and ketones contain a carbonyl group.** An aldehyde contains at least one H atom bonded to the carbonyl carbon, whereas a ketone has two alkyl or aryl groups bonded to it.

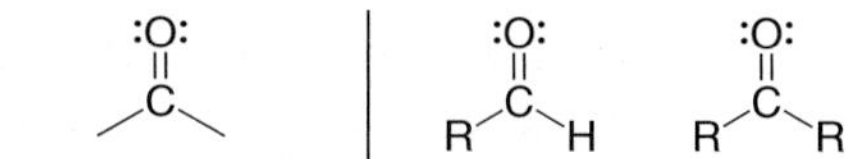

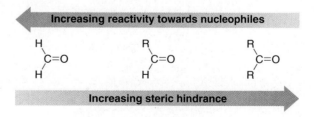

carbonyl group | **aldehyde** | **ketone**

Two structural features determine the chemistry and properties of aldehydes and ketones.

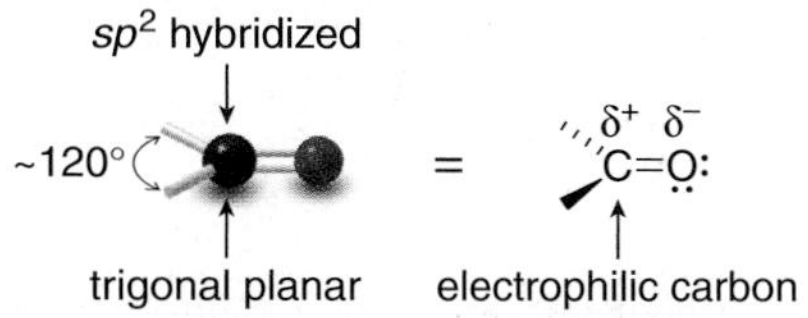

sp^2 hybridized

~120° trigonal planar   electrophilic carbon

◆ The carbonyl group is sp^2 hybridized and trigonal planar, making it relatively uncrowded.
◆ The electronegative oxygen atom polarizes the carbonyl group, making the carbonyl carbon electrophilic.

As a result, **aldehydes and ketones react with nucleophiles.** The relative reactivity of the carbonyl group is determined by the number of R groups bonded to it. **As the number of R groups around the carbonyl carbon increases, the reactivity of the carbonyl compound decreases,** resulting in the following order of reactivity:

Increasing the number of alkyl groups on the carbonyl carbon decreases reactivity for both steric and electronic reasons, as discussed in Section 20.2B.

Increasing reactivity towards nucleophiles

H—C=O   R—C=O   R—C=O
H   H   R

Increasing steric hindrance

PROBLEM 21.1 Rank the compounds in each group in order of increasing reactivity towards nucleophilic attack.

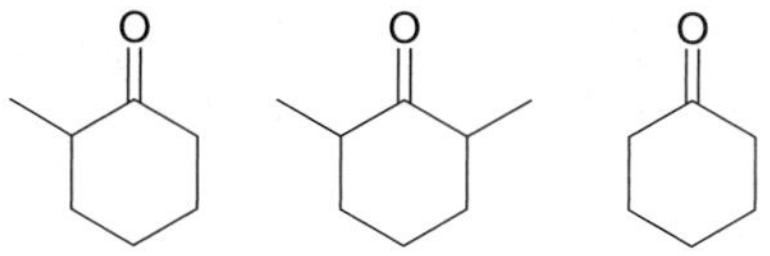

a. CH₃CH=O   CH₂=O   (CH₃)₂C=O   b.

PROBLEM 21.2   Explain why benzaldehyde is less reactive than cyclohexanecarbaldehyde towards nucleophilic attack.

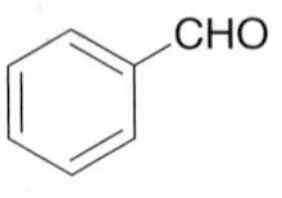

benzaldehyde 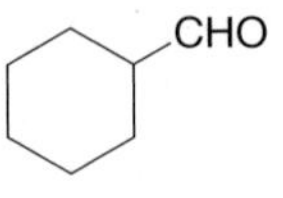   cyclohexanecarbaldehyde

21.2   Nomenclature

Both IUPAC and common names are used for aldehydes and ketones.

21.2A   Naming Aldehydes in the IUPAC System

In IUPAC nomenclature, aldehydes are identified by a suffix added to the parent name of the longest chain. Two different suffixes are used, depending on whether the CHO group is bonded to a chain or a ring.

To name an aldehyde using the IUPAC system:

[1] If the CHO is bonded to a chain of carbons, find the longest chain containing the CHO group, and change the *-e* ending of the parent alkane to the suffix *-al.* If the CHO group is bonded to a ring, name the ring and add the suffix *-carbaldehyde.*

[2] Number the chain or ring to put the CHO group at C1, but omit this number from the name. Apply all of the other usual rules of nomenclature.

SAMPLE PROBLEM 21.1   Give the IUPAC name for each compound.

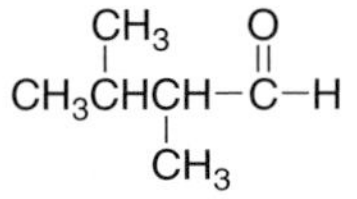

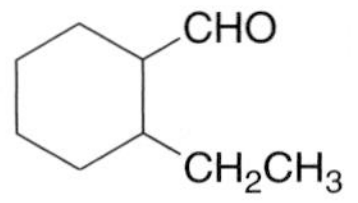

SOLUTION

a.   [1] Find and name the longest chain containing the CHO.

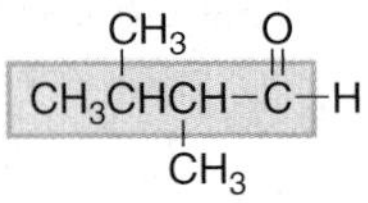

butane ⟶ butan*al*
(4 C's)

[2] Number and name substituents:

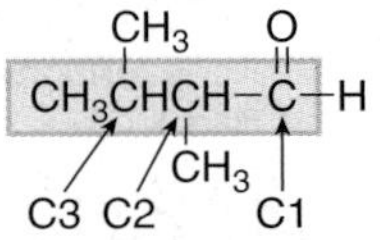

Answer: 2,3-dimethylbutanal

b.   [1] Find and name the ring bonded to the CHO group:

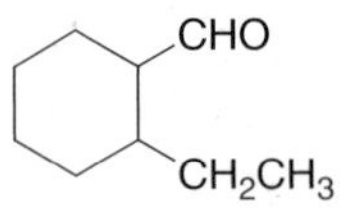

cyclohexane + carbaldehyde
(6 C's)

[2] Number and name substituents:

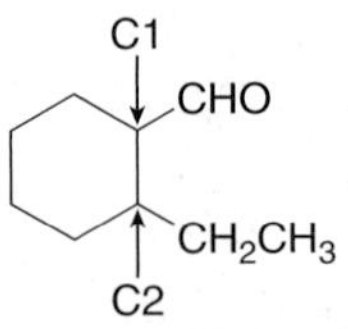

**Answer:
2-ethylcyclohexanecarbaldehyde**

PROBLEM 21.3   Give the IUPAC name for each aldehyde.

a. $(CH_3)_3CC(CH_3)_2CH_2CHO$ b. 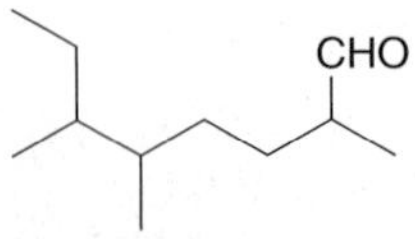   c.

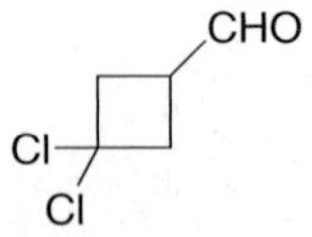

PROBLEM 21.4   Give the structure corresponding to each IUPAC name.

a. 2-isobutyl-3-isopropylhexanal

c. 1-methylcyclopropanecarbaldehyde

b. *trans*-3-methylcyclopentanecarbaldehyde

d. 3,6-diethylnonanal

21.2B   Common Names for Aldehydes

Like carboxylic acids, many simple aldehydes have common names that are widely used.

◆ **A common name for an aldehyde is formed by taking the common parent name and adding the suffix -aldehyde.**

The common parent names are similar to those used for carboxylic acids, listed in Table 19.1. The common names **formaldehyde, acetaldehyde,** and **benzaldehyde** are virtually always used instead of their IUPAC names.

formaldehyde
(methanal)   acetaldehyde
(ethanal)   benzaldehyde
(benzenecarbaldehyde)

(IUPAC names are in parentheses.)

Greek letters are used to designate the location of substituents in common names. **The carbon adjacent to the CHO group is the α carbon,** and so forth down the chain.

Start lettering here.

$$H-\overset{\overset{\displaystyle O}{\|}}{C}-C-C-C-C\sim$$

α   β   γ   δ

Figure 21.1 gives the common and IUPAC names for three aldehydes.

Figure 21.1   Three examples of aldehyde nomenclature

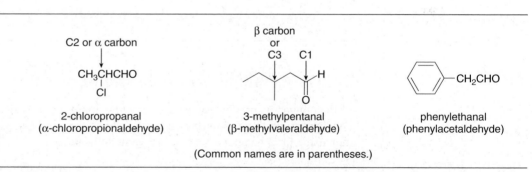

C2 or α carbon

$$CH_3\underset{\underset{\displaystyle Cl}{|}}{CH}CHO$$

2-chloropropanal
(α-chloropropionaldehyde)

β carbon
or
C3   C1

3-methylpentanal
(β-methylvaleraldehyde)

—CH$_2$CHO

phenylethanal
(phenylacetaldehyde)

(Common names are in parentheses.)

21.2C   Naming Ketones in the IUPAC System

◆ **In the IUPAC system all ketones are identified by the suffix -one.**

To name an acyclic ketone using IUPAC rules:

[1]   Find the longest chain containing the carbonyl group, and change the **-e** ending of the parent alkane to the suffix **-one.**

[2]   Number the carbon chain to give the carbonyl carbon the lower number. Apply all of the other usual rules of nomenclature.

With cyclic ketones, numbering always begins at the carbonyl carbon, but the "*1*" is usually omitted from the name. The ring is then numbered clockwise or counterclockwise to give the first substituent the lower number.

SAMPLE PROBLEM 21.2   Give IUPAC names for each ketone.

a. 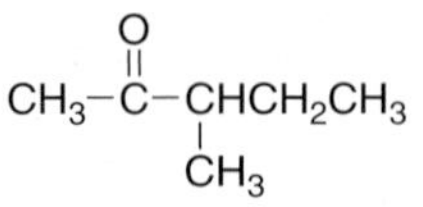 CH₃–C(=O)–CHCH₂CH₃
                  |
                  CH₃

b.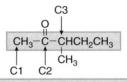

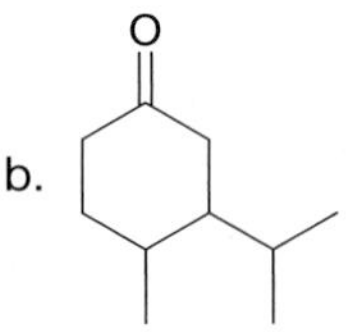

SOLUTION

a. [1] Find and name the longest chain containing the carbonyl group:

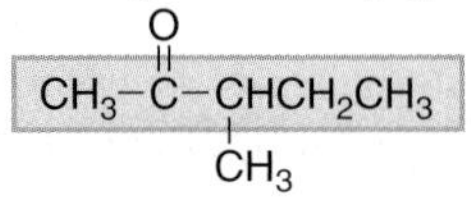

$$CH_3-C(=O)-CHCH_2CH_3 \;|\; CH_3$$

pentane $\longrightarrow$ pentanone
(5 C's)

[2] Number and name substituents:

C3

$$CH_3-C(=O)-CHCH_2CH_3 \;|\; CH_3$$

C1   C2

Answer: 3-methyl-2-pentanone

---

b. [1] Name the ring:

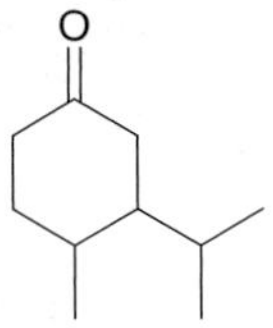

cyclohexane $\longrightarrow$ cyclohexanone
(6 C's)

[2] Number and name substituents:

O C1

C4

C3

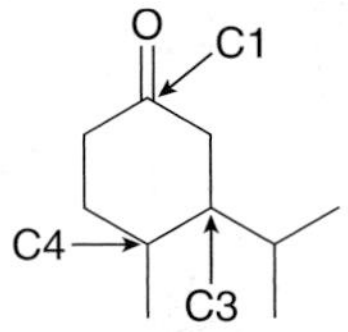

**Answer:
3-isopropyl-4-methylcyclohexanone**

PROBLEM 21.5   Give the IUPAC name for each ketone.

a. 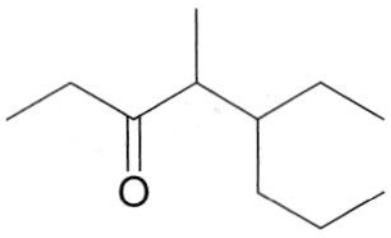 [structure]

b. 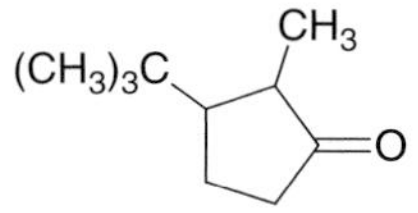 $(CH_3)_3C$ — cyclopentanone with CH₃

c. $(CH_3)_3CCOC(CH_3)_3$

21.2D   Common Names for Ketones

Most common names for ketones are formed by **naming both alkyl groups** on the carbonyl carbon, **arranging them alphabetically,** and adding the word **ketone.** Using this method, the common name for 2-butanone becomes ethyl methyl ketone.

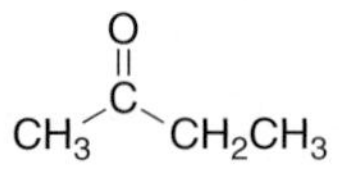

$$CH_3-C(=O)-CH_2CH_3$$

IUPAC name: **2-butanone**

methyl group   ethyl group

$$CH_3-C(=O)-CH_2CH_3$$

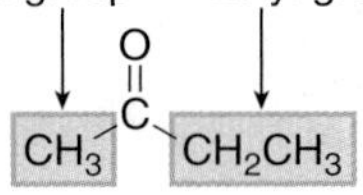

Common name: **ethyl methyl ketone**

Three widely used common names for some simple ketones do not follow this convention:

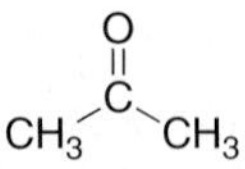

$$CH_3-C(=O)-CH_3$$

acetone

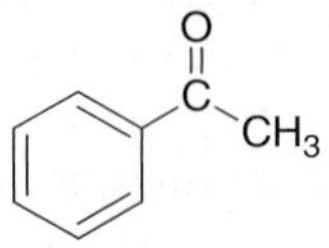

acetophenone

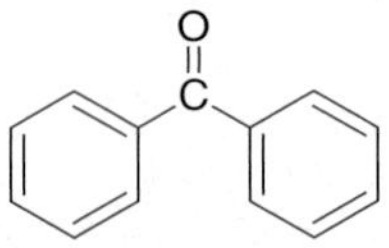

benzophenone

Figure 21.2 gives acceptable names for two ketones.

Figure 21.2 Two examples
of ketone nomenclature

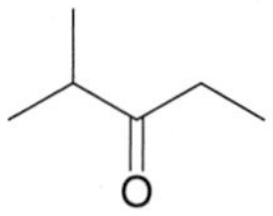

IUPAC name: 2-methyl-3-pentanone
Common name: ethyl isopropyl ketone

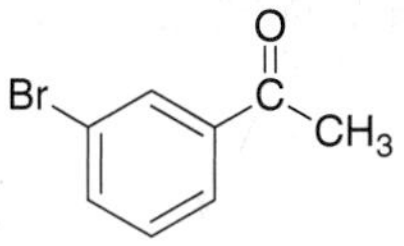

m-bromoacetophenone
or
3-bromoacetophenone

21.2E Naming Acyl Groups

> Do not confuse a **benzyl**
> group with a **benzoyl** group.

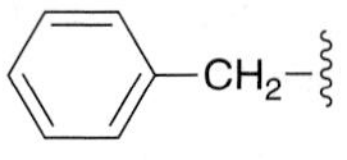

Sometimes acyl groups (RCO−) must be named as substituents. To name an acyl group, take
either the IUPAC or common parent name and add the suffix **-yl** or **-oyl.** The three most common
acyl groups are drawn below.

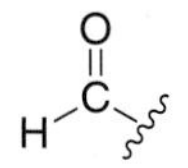

benz*yl* group

formyl group

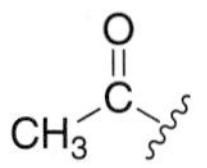

acetyl group

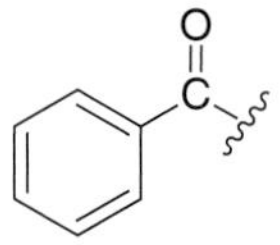

benzoyl group

PROBLEM 21.6 Give the structure corresponding to each name: (a) *sec*-butyl ethyl ketone; (b) methyl vinyl
ketone; (c) *p*-ethylacetophenone; (d) 2-benzyl-3-benzoylcyclopentanone.

21.3 Physical Properties

Aldehydes and ketones exhibit dipole–dipole interactions because of their polar carbonyl
group. Because they have no O−H bond, two molecules of RCHO or RCOR are incapable of
intermolecular hydrogen bonding, making them less polar than alcohols and carboxylic acids.
How these intermolecular forces affect the physical properties of aldehydes and ketones is summarized in Table 21.1.

TABLE 21.1 Physical Properties of Aldehydes and Ketones

Property	Observation
Boiling point and melting point	• For compounds of comparable molecular weight, bp's and mp's follow the usual trend: The stronger the intermolecular forces, the higher the bp or mp.

$CH_3CH_2CH_2CH_2CH_3$

VDW
MW = 72
bp 36 °C

$CH_3CH_2CH_2CHO$
VDW, DD   MW = 72
bp 76 °C

$CH_3CH_2COCH_3$
VDW, DD   MW = 72
bp 80 °C

$CH_3CH_2CH_2CH_2OH$
VDW, DD, HB
MW = 74
bp 118 °C

Increasing strength of intermolecular forces
Increasing boiling point

Solubility	• RCHO and RCOR are soluble in organic solvents regardless of size.
	• RCHO and RCOR having ≤ 5 C's are H_2O soluble because they can hydrogen bond with H_2O (Section 3.5C).
	• RCHO and RCOR having > 5 C's are H_2O insoluble because the nonpolar alkyl portion is too large to dissolve in the polar H_2O solvent.

Key: VDW = van der Waals, DD = dipole–dipole, HB = hydrogen bonding, MW = molecular weight

PROBLEM 21.7   The boiling point of 2-butanone (80 °C) is significantly higher than the boiling point of diethyl ether (35 °C), even though both compounds exhibit dipole–dipole interactions and have comparable molecular weights. Offer an explanation.

21.4 Spectroscopic Properties

The presence of the carbonyl group in aldehydes and ketones gives them characteristic absorptions in their IR and NMR spectra.

21.4A IR Spectra

Aldehydes and ketones exhibit the following characteristic IR absorptions:

♦ Like all carbonyl compounds, **aldehydes and ketones give a strong peak at ~1700 cm^{-1} due to the C=O.**

♦ **The sp^2 hybridized C–H bond of an aldehyde shows one or two peaks at ~2700–2830 cm^{-1}.**

The IR spectrum of propanal in Figure 21.3 illustrates these characteristic peaks.

The exact position of the carbonyl absorption often provides additional information about a compound. For example, most aldehydes have a C=O peak around **1730 cm^{-1}**, whereas for ketones, it is typically around **1715 cm^{-1}**. Two other structural features—**ring size** (for cyclic ketones) and **conjugation**—affect the location of the carbonyl absorption in a predictable manner.

[1] The carbonyl absorption of cyclic ketones shifts to higher wavenumber as the size of the ring decreases and the ring strain increases.

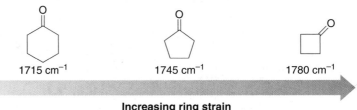

| 1715 cm^{-1} | 1745 cm^{-1} | 1780 cm^{-1} |

Increasing ring strain
Increasing wavenumber of the C=O absorption

Figure 21.3 The IR spectrum of propanal, CH$_3$CH$_2$CHO

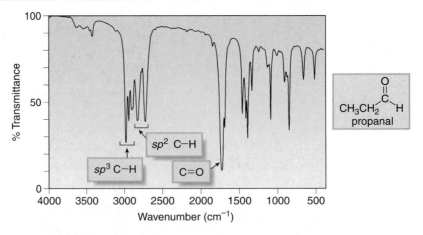

• A strong C=O occurs at 1739 cm^{-1}.
• The sp^2 C–H of the CHO appears as two peaks at 2813 and 2716 cm^{-1}.

[2] **Conjugation of the carbonyl group with a C=C or a benzene ring shifts the absorption to lower wavenumber by ~30 cm^{-1}.**

The effect of conjugation on the frequency of the C=O absorption is explained by **resonance.** An α,β-unsaturated carbonyl compound has three resonance structures, two of which place a single bond between the carbon and oxygen atoms of the carbonyl group. Thus, the π bond of the carbonyl group is delocalized, giving the conjugated carbonyl group some single bond character, and making it somewhat **weaker** than an unconjugated C=O. **Weaker bonds absorb at lower frequency (lower wavenumber) in an IR spectrum.**

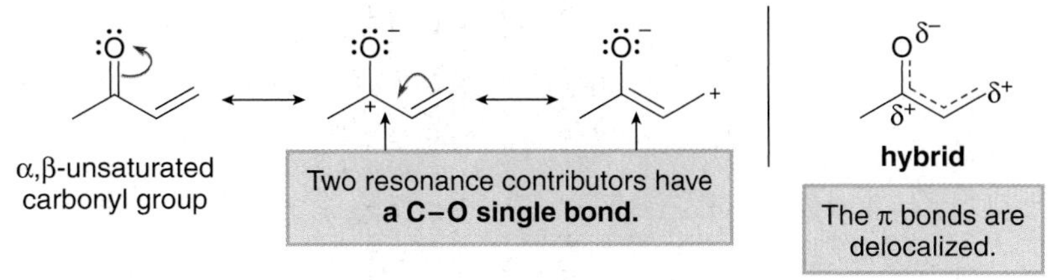

α,β-unsaturated carbonyl group

Two resonance contributors have **a C–O single bond.**

hybrid

The π bonds are delocalized.

Figure 21.4 illustrates the effects of conjugation on the location of the carbonyl absorption in some representative compounds.

PROBLEM 21.8   Which carbonyl group in each pair absorbs at a higher frequency?

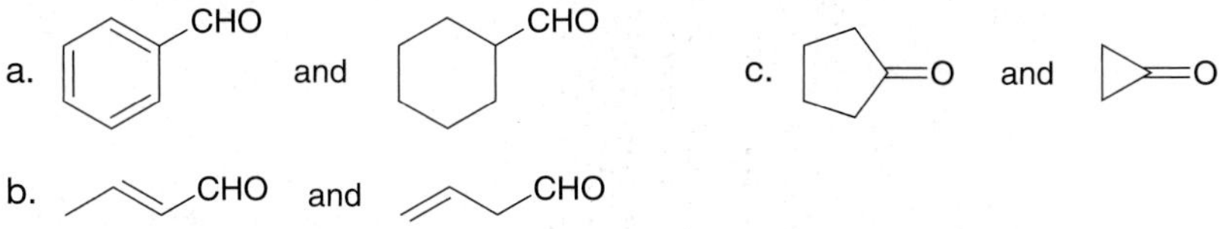

21.4B NMR Spectra

Aldehydes and ketones exhibit the following characteristic ^{1}H and ^{13}C NMR absorptions:

◆ **The sp^2 hybridized C–H proton of an aldehyde is highly deshielded and absorbs far downfield at 9–10 ppm.** Splitting occurs with protons on the α carbon, but the coupling constant is often very small (J = 1–3 Hz).

◆ **Protons on the α carbon to the carbonyl group absorb at 2–2.5 ppm.** Methyl ketones, for example, give a characteristic singlet at ~2.1 ppm.

◆ In a ^{13}C NMR spectrum, **the carbonyl carbon is highly deshielded, appearing in the 190–215 ppm region.**

The ^{1}H and ^{13}C NMR spectra of propanal are illustrated in Figure 21.5.

PROBLEM 21.9   Three isomeric carbonyl compounds (**A–C**) have molecular formula C_4H_8O. **A** shows three peaks in its ^{13}C NMR spectrum. **B** has a peak at 9.8 ppm in its ^{1}H NMR spectrum. **C** has a singlet at 2.1 ppm in its ^{1}H NMR spectrum. What are the structures of **A, B,** and **C**?

Figure 21.4 The effect of conjugation on the carbonyl absorption in an IR spectrum

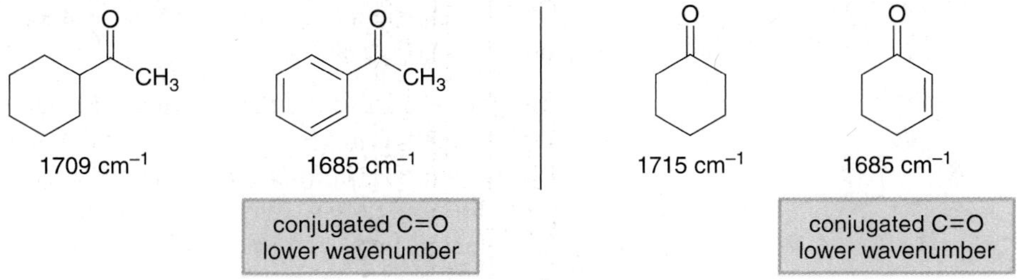

1709 cm^{-1} 1685 cm^{-1} 1715 cm^{-1} 1685 cm^{-1}

conjugated C=O lower wavenumber

conjugated C=O lower wavenumber

Figure 21.5 The ^{1}H and ^{13}C NMR spectra of propanal, CH_3CH_2CHO

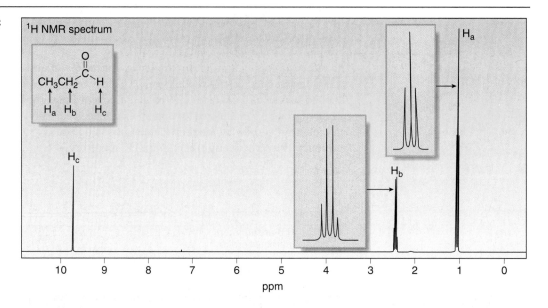

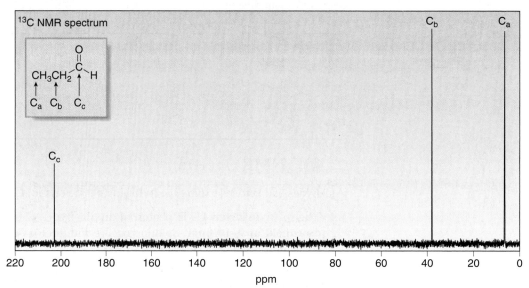

- **^{1}H NMR:** There are three signals due to the three different kinds of hydrogens, labeled H_a, H_b, and H_c. The deshielded CHO proton occurs downfield at 9.8 ppm.
- **^{13}C NMR:** There are three signals due to the three different kinds of carbons, labeled C_a, C_b, and C_c. The deshielded carbonyl carbon absorbs downfield at 203 ppm.

21.5   Interesting Aldehydes and Ketones

formaldehyde
[CH_2=O]

acetone
[$(CH_3)_2$C=O]

Because it is a starting material for the synthesis of many resins and plastics, billions of pounds of **formaldehyde** are produced annually in the United States by the oxidation of methanol (CH_3OH). Formaldehyde is also sold as a 37% aqueous solution called **formalin,** which has been used as a disinfectant, antiseptic, and preservative for biological specimens. Formaldehyde, a product of the incomplete combustion of coal and other fossil fuels, is partly responsible for the irritation caused by smoggy air.

Acetone is an industrial solvent and a starting material in the synthesis of some organic polymers. Acetone is produced in vivo during the breakdown of fatty acids. In diabetes, a common endocrine disease in which normal metabolic processes are altered because of the inadequate secretion of insulin, individuals often have unusually high levels of acetone in their bloodstreams. The characteristic odor of acetone can be detected on the breath of diabetic patients when their disease is poorly controlled.

Figure 21.6   Some naturally occurring aldehydes and ketones with strong odors

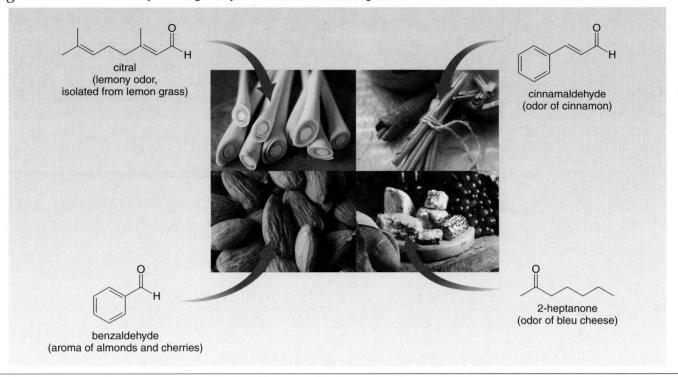

citral
(lemony odor,
isolated from lemon grass)

cinnamaldehyde
(odor of cinnamon)

benzaldehyde
(aroma of almonds and cherries)

2-heptanone
(odor of bleu cheese)

cortisone
(naturally occurring)

prednisone
(synthetic)

Many aldehydes and ketones with characteristic odors occur in nature, as shown in Figure 21.6.

Many steroid hormones contain a carbonyl along with other functional groups. **Cortisone** and **prednisone** are two anti-inflammatory steroids with closely related structures. Cortisone is secreted by the body's adrenal gland, whereas prednisone is a synthetic analogue used in the treatment of inflammatory diseases such as arthritis and asthma.

21.6   Preparation of Aldehydes and Ketones

Aldehydes and ketones can be prepared by a variety of methods. Because these reactions are needed for many multistep syntheses, Section 21.6 briefly summarizes earlier reactions that synthesize an aldehyde or ketone.

21.6A   Common Methods to Synthesize Aldehydes

Aldehydes are prepared from 1° alcohols, esters, acid chlorides, and alkynes.

- **By oxidation of 1° alcohols with PCC**

$$RCH_2-OH \xrightarrow{PCC} \underset{R}{\overset{O}{\underset{}{\parallel}}}CH$$

1° alcohol

(Section 12.12B)

- **By reduction of esters and acid chlorides**

$$\underset{R}{\overset{O}{\underset{}{\parallel}}}C-OR' \xrightarrow[\text{[2] } H_2O]{\text{[1] DIBAL-H}} \underset{R}{\overset{O}{\underset{}{\parallel}}}CH$$

ester

(Section 20.7A)

$$\underset{R}{\overset{O}{\underset{}{\parallel}}}C-Cl \xrightarrow[\text{[2] } H_2O]{\text{[1] LiAlH[OC(CH_3)_3]_3}} \underset{R}{\overset{O}{\underset{}{\parallel}}}CH$$

acid chloride

- **By hydration of an alkyne using hydroboration–oxidation**

$$R-C\equiv C-H \xrightarrow[\text{[2] } H_2O_2, \ ^-OH]{\text{[1] } BH_3} \underset{RCH_2}{\overset{O}{\underset{}{\parallel}}}CH$$

alkyne

(Section 11.10)

21.6B Common Methods to Synthesize Ketones

Ketones are prepared from 2° alcohols, acid chlorides, and alkynes.

- **By oxidation of 2° alcohols with Cr⁶⁺ reagents**

$$\underset{H}{\overset{OH}{\underset{}{\mid}}}R-C-R' \xrightarrow[\substack{K_2Cr_2O_7 \text{ or} \\ PCC}]{\substack{CrO_3 \text{ or} \\ Na_2Cr_2O_7 \text{ or}}} \underset{R}{\overset{O}{\underset{}{\parallel}}}C-R'$$

2° alcohol

(Section 12.12A)

- **By reaction of acid chlorides with organocuprates**

$$\underset{R}{\overset{O}{\underset{}{\parallel}}}C-Cl \xrightarrow[\text{[2] } H_2O]{\text{[1] } R'_2CuLi} \underset{R}{\overset{O}{\underset{}{\parallel}}}C-R'$$

acid chloride

(Section 20.13)

- **By Friedel–Crafts acylation**

$$\bigcirc + \underset{R}{\overset{O}{\underset{}{\parallel}}}C-Cl \xrightarrow{AlCl_3} \bigcirc-\underset{}{\overset{O}{\underset{}{\parallel}}}C-R$$

acid chloride

(Section 18.5)

- **By hydration of an alkyne**

$$R-C\equiv C-H \xrightarrow[\substack{H_2SO_4 \\ HgSO_4}]{H_2O} \underset{R}{\overset{O}{\underset{}{\parallel}}}C-CH_3$$

alkyne

(Section 11.9)

Aldehydes and ketones are also both obtained as products of the oxidative cleavage of alkenes (Section 12.10).

$$\underset{R}{\overset{R}{\underset{}{}}}C=C\underset{R'}{\overset{H}{\underset{}{}}} \xrightarrow[\substack{\text{or} \\ CH_3SCH_3}]{\underset{Zn, H_2O}{O_3}} \underset{R}{\overset{R}{\underset{}{}}}C=O + O=C\underset{R'}{\overset{H}{\underset{}{}}}$$

alkene   ketone   aldehyde

PROBLEM 21.10   What reagents are needed to convert each compound into butanal ($CH_3CH_2CH_2CHO$):
(a) $CH_3CH_2CH_2COOCH_3$; (b) $CH_3CH_2CH_2CH_2OH$; (c) $HC\equiv CCH_2CH_3$;
(d) $CH_3CH_2CH_2CH=CHCH_2CH_2CH_3$?

PROBLEM 21.11   What reagents are needed to convert each compound into acetophenone ($C_6H_5COCH_3$):
(a) benzene; (b) C_6H_5COCl; (c) $C_6H_5C\equiv CH$?

21.7   Reactions of Aldehydes and Ketones—General Considerations

Let's begin our discussion of carbonyl reactions by looking at the two general kinds of reactions that aldehydes and ketones undergo.

[1] Reaction at the carbonyl carbon

Recall from Chapter 20 that the uncrowded, electrophilic carbonyl carbon makes aldehydes and ketones susceptible to **nucleophilic addition** reactions.

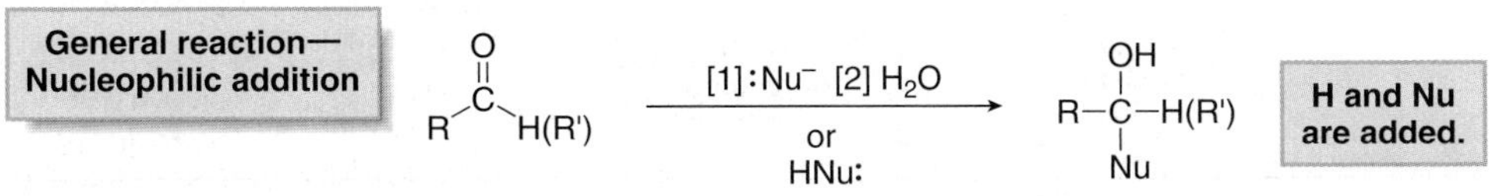

The elements of H and Nu are added to the carbonyl group. In Chapter 20 you learned about this reaction with hydride (H:⁻) and carbanions (R:⁻) as nucleophiles. In Chapter 21, we will discuss similar reactions with other nucleophiles.

[2] Reaction at the α carbon

A second general reaction of aldehydes and ketones involves reaction at the **α carbon.** A C−H bond on the α carbon to a carbonyl group is more acidic than many other C−H bonds, because reaction with base forms a resonance-stabilized enolate anion.

◆ Enolates are nucleophiles, and so they react with electrophiles to form new bonds on the α carbon.

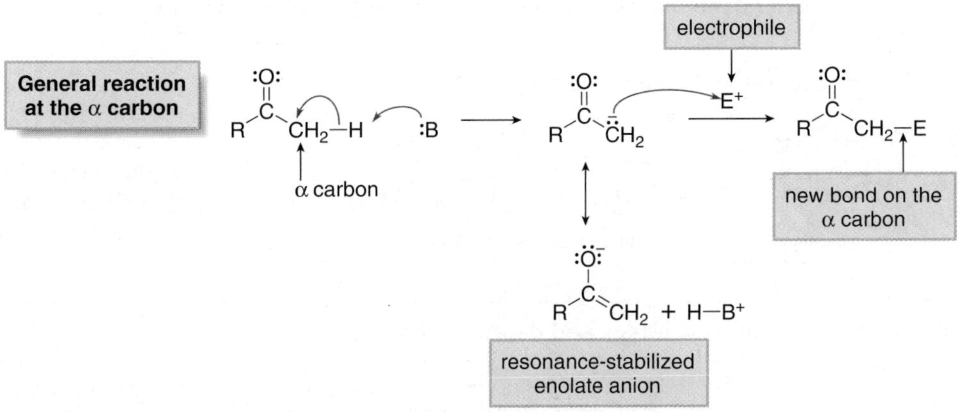

Chapters 23 and 24 are devoted to reactions at the α carbon to a carbonyl group.

◆ Aldehydes and ketones react with nucleophiles at the carbonyl carbon.
◆ Aldehydes and ketones form enolates that react with electrophiles at the α carbon.

21.7A   The General Mechanism of Nucleophilic Addition

Two general mechanisms are usually drawn for nucleophilic addition, depending on the nucleophile (negatively charged versus neutral) and the presence or absence of an acid catalyst. With negatively charged nucleophiles, nucleophilic addition follows the two-step process first discussed in Chapter 20—**nucleophilic attack** followed by **protonation,** as shown in Mechanism 21.1.

MECHANISM 21.1

General Mechanism—Nucleophilic Addition

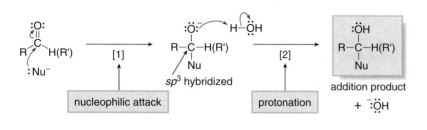

◆ In Step [1], **the nucleophile attacks the carbonyl group,** cleaving the π bond and moving an electron pair onto oxygen. This forms an sp^3 hybridized intermediate with a new C–Nu bond.

◆ In Step [2], **protonation of the negatively charged O atom by H₂O** forms the addition product.

In this mechanism **nucleophilic attack *precedes* protonation.** This process occurs with strong neutral or negatively charged nucleophiles.

With some neutral nucleophiles, however, nucleophilic addition does not occur unless an acid catalyst is added. The mechanism for this reaction consists of three steps (not two), but the same product results because H and Nu add across the carbonyl π bond. In this mechanism, **protonation *precedes* nucleophilic attack.** Mechanism 21.2 is shown with the neutral nucleophile H–Nu: and a general acid H–A.

MECHANISM 21.2

General Mechanism—Acid-Catalyzed Nucleophilic Addition

Step [1] Protonation of the carbonyl group

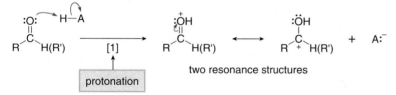

◆ Protonation of the carbonyl oxygen forms a resonance-stabilized cation that bears a full positive charge.

Steps [2]–[3] Nucleophilic attack and deprotonation

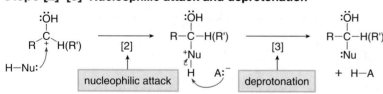

◆ In Step [2], the nucleophile attacks, and then deprotonation forms the neutral addition product in Step [3].

◆ The overall result is the addition of H and Nu to the carbonyl group.

The effect of protonation is to convert a neutral carbonyl group to one having a net positive charge. **This protonated carbonyl group is much more electrophilic,** and much more susceptible to attack by a nucleophile. This step is unnecessary with strong nucleophiles like hydride (H:⁻) that were used in Chapter 20. With weaker nucleophiles, however, nucleophilic attack does not occur unless the carbonyl group is first protonated.

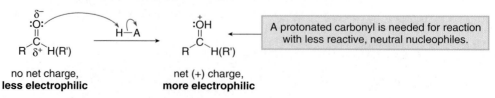

no net charge,
less electrophilic   net (+) charge,
more electrophilic

A protonated carbonyl is needed for reaction with less reactive, neutral nucleophiles.

This step is a specific example of a general phenomenon.

◆ Any reaction involving a carbonyl group and a strong acid begins with the same first step—protonation of the carbonyl oxygen.

21.7B   The Nucleophile

What nucleophiles add to carbonyl groups? This cannot be predicted solely on the trends in nucleophilicity learned in Chapter 7. Only *some* of the nucleophiles that react well in nucleophilic substitution at sp^3 hybridized carbons give reasonable yields of nucleophilic addition products.

Cl⁻, Br⁻, and I⁻ are good nucleophiles in substitution reactions at sp^3 hybridized carbons, but they are ineffective nucleophiles in addition. Addition of Cl⁻ to a carbonyl group, for example, would cleave the C−O π bond, forming an alkoxide. Because Cl⁻ is a much weaker base than the alkoxide formed, equilibrium favors the starting materials (the weaker base, Cl⁻), *not* the addition product.

The situation is further complicated because some of the initial nucleophilic addition adducts are unstable and undergo elimination to form a stable product. For example, amines (RNH_2) add to carbonyl groups in the presence of mild acid to form unstable **carbinolamines,** which readily lose water to form **imines.** This addition–elimination sequence replaces a C=O by a C=N. The details of this process are discussed in Section 21.11.

Figure 21.7 lists nucleophiles that add to a carbonyl group, as well as the products obtained from nucleophilic addition using cyclohexanone as a representative ketone. These reactions are discussed in the remaining sections of Chapter 21. In cases in which the initial addition adduct is unstable, it is enclosed within parentheses, followed by the final product.

PROBLEM 21.12   Why does equilibrium favor the product when H⁻ adds to a carbonyl group?

21.8   Nucleophilic Addition of H⁻ and R⁻—A Review

We begin our study of nucleophilic additions to aldehydes and ketones by briefly reviewing nucleophilic addition of hydride and carbanions, two reactions examined in Sections 20.4 and 20.10, respectively.

Treatment of an aldehyde or ketone with either $NaBH_4$ or $LiAlH_4$ followed by protonation forms a 1° or 2° alcohol. $NaBH_4$ and $LiAlH_4$ serve as a source of **hydride, H:⁻—the nucleophile**—and the reaction results in addition of the elements of H_2 across the C−O π bond. Addition of H_2 reduces the carbonyl group to an alcohol.

Figure 21.7 Specific examples of nucleophilic addition

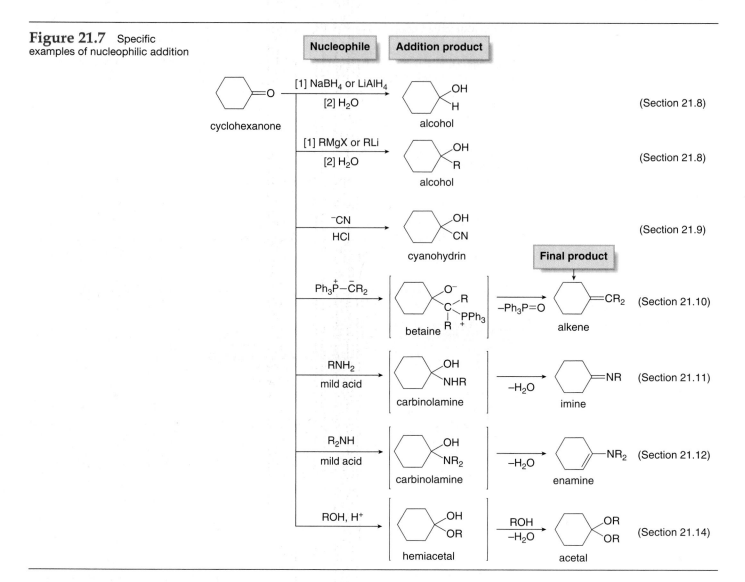

Hydride reduction of aldehydes and ketones occurs via the two-step mechanism of nucleophilic addition—that is, **nucleophilic attack of H:⁻ followed by protonation**—shown previously in Section 20.4B.

Treatment of an aldehyde or ketone with either an organolithium (R″Li) or Grignard reagent (R″MgX) followed by water forms a 1°, 2°, or 3° alcohol containing a new carbon–carbon bond. R″Li and R″MgX serve as a source of a **carbanion (R″)⁻—the nucleophile**—and the reaction results in addition of the elements of R″ and H across the C—O π bond.

The nucleophilic addition of carbanions to aldehydes and ketones occurs via the two-step mechanism of nucleophilic addition—that is, **nucleophilic attack of (R″)⁻ followed by protonation**—shown previously in Section 20.10.

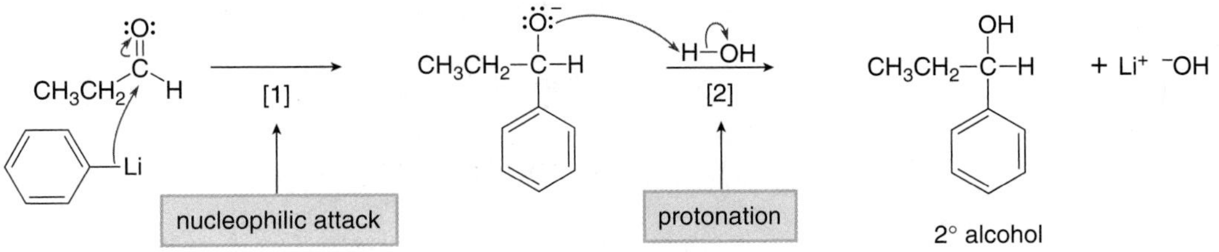

The stereochemistry of hydride reduction and Grignard addition was discussed previously in Sections 20.5 and 20.10B, respectively.

In both reactions, the nucleophile—either hydride or a carbanion—attacks the trigonal planar sp^2 hybridized carbonyl from both sides, so that when a new stereogenic center is formed, a mixture of stereoisomers results, as shown in Sample Problem 21.3.

SAMPLE PROBLEM 21.3   Draw the products (including the stereochemistry) formed in the following reaction.

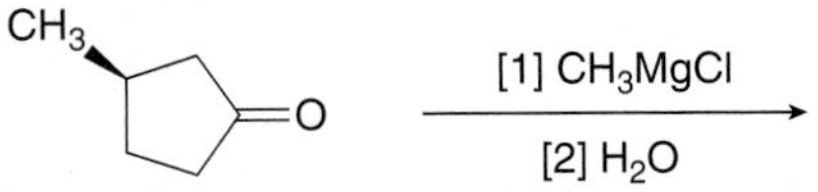

(R)-3-methylcyclopentanone

SOLUTION

The Grignard reagent adds CH_3^- from both sides of the trigonal planar carbonyl group, yielding a mixture of 3° alcohols after protonation with water. In this example, the starting ketone and both alcohol products are chiral. The two products, which contain two stereogenic centers, are stereoisomers but not mirror images—that is, they are **diastereomers.**

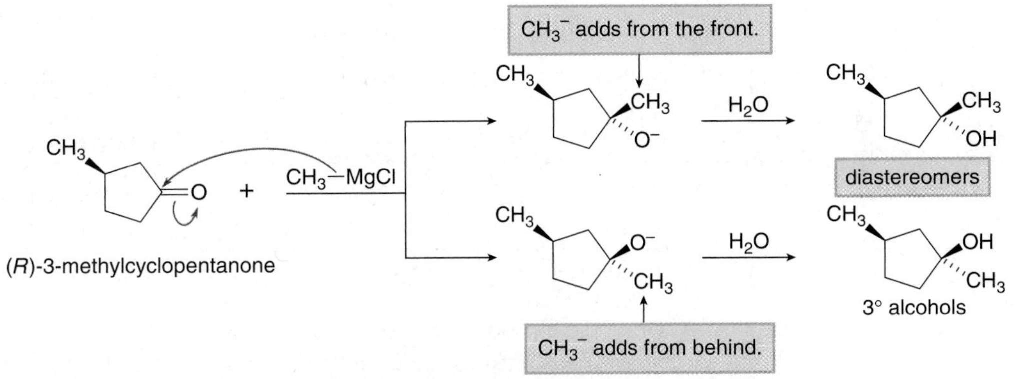

PROBLEM 21.13   Draw the products of each reaction. Include all stereoisomers formed.

a.

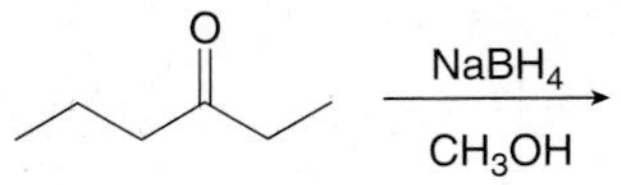

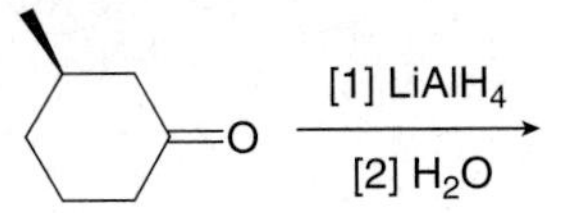

b. 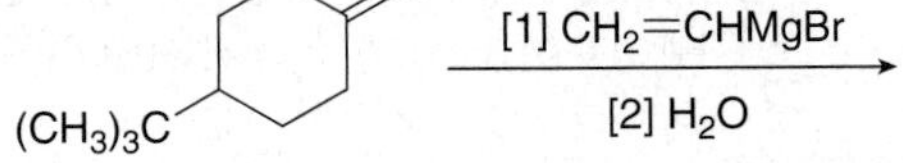 $(CH_3)_3C$

c.

PROBLEM 21.14   Alcohols containing deuterium atoms can be synthesized using reagents that contain deuterium atoms. Draw the products formed when cyclohexanone is treated with each reagent.

a. [1] $LiAlD_4$; [2] H_2O c. [1] $LiAlD_4$; [2] D_2O e. [1] CH_3MgCl; [2] D_2O
b. [1] $LiAlH_4$; [2] D_2O d. [1] CD_3MgCl; [2] H_2O

21.9 Nucleophilic Addition of ⁻CN

Treatment of an aldehyde or ketone with NaCN and a strong acid such as HCl adds the elements of HCN across the carbon–oxygen π bond, forming a **cyanohydrin.**

Nucleophilic addition of HCN— General reaction	

$$\underset{\substack{\text{``HCN''}}}{R-\overset{\displaystyle O}{\overset{\|}{C}}-H(R')} \xrightarrow[\text{HCl}]{\text{NaCN}} \underset{\text{cyanohydrin}}{R-\overset{\displaystyle OH}{\underset{\displaystyle CN}{C}}-H(R')}$$

This reaction adds one carbon to the aldehyde or ketone, forming a new carbon–carbon bond.

Example	

$$CH_3-\overset{\displaystyle O}{\overset{\|}{C}}-H \xrightarrow[\text{HCl}]{\text{NaCN}} CH_3-\overset{\displaystyle OH}{\underset{\displaystyle CN}{C}}-H \quad \leftarrow \boxed{\text{new C–C bond}}$$

acetaldehyde
cyanohydrin

21.9A The Mechanism

The mechanism of cyanohydrin formation involves the usual two steps of nucleophilic addition: **nucleophilic attack followed by protonation** as shown in Mechanism 21.3.

MECHANISM 21.3

Nucleophilic Addition of ⁻CN—Cyanohydrin Formation

$$\underset{\substack{\\ :C\equiv N:}}{R-\overset{\displaystyle :O:}{\overset{\|}{C}}-H(R')} \xrightarrow{[1]} \underset{\substack{\text{nucleophilic attack}}}{} R-\overset{\displaystyle :\overset{..}{O}:^-}{\underset{\displaystyle C\equiv N:}{C}}-H(R') \quad H-CN \xrightarrow{[2]} \underset{\substack{\text{addition product}\\ +\ ^-CN}}{} R-\overset{\displaystyle :\overset{..}{O}H}{\underset{\displaystyle C\equiv N:}{C}}-H(R')$$

♦ In Step [1], **nucleophilic attack of ⁻CN** forms a new carbon–carbon bond with cleavage of the C–O π bond.

♦ In Step [2], **protonation of the negatively charged O atom by HCN** forms the addition product. The hydrogen cyanide (HCN) used in this step is formed by the acid–base reaction of cyanide (⁻CN) with the strong acid, HCl.

This reaction does not occur with HCN alone. The cyanide anion makes addition possible because it is a strong nucleophile that attacks the carbonyl group.

Cyanohydrins can be reconverted to carbonyl compounds by treatment with base. This process is just the reverse of the addition of HCN: **deprotonation followed by elimination of ⁻CN.**

$$\underset{\substack{CN}}{R-\overset{\displaystyle :\overset{..}{O}-H}{C}-H(R')} \quad ^-OH \xrightarrow[\substack{\text{deprotonation}\\ +\ H_2O}]{[1]} \underset{\substack{\\ CN}}{R-\overset{\displaystyle :\overset{..}{O}:^-}{C}-H(R')} \xrightarrow[\text{loss of ⁻CN}]{[2]} \underset{\substack{\\ }}{R-\overset{\displaystyle :O:}{\overset{\|}{C}}-H(R')} + \ ^-CN$$

The cyano group (CN) of a cyanohydrin is readily hydrolyzed to a carboxy group (COOH) by heating with aqueous acid or base. Hydrolysis replaces the three C–N bonds by three C–O bonds.

Note the difference between two similar terms.
Hydration results in *adding water* to a compound.
Hydrolysis results in *cleaving bonds* with water.

Hydrolysis of a cyano group	

$$\underset{\substack{C\equiv N}}{R-\overset{\displaystyle OH}{C}-R'} \xrightarrow[\substack{(H^+ \text{ or } ^-OH)\\ \Delta}]{H_2O} \underset{\substack{COOH}}{R-\overset{\displaystyle OH}{C}-R'}$$

PROBLEM 21.15 Draw the products of each reaction.

a. [benzaldehyde structure] CHO → NaCN / HCl →

b. [cyclopentane with OH and CN] → H₃O⁺, Δ →

21.9B Application: Naturally Occurring Cyanohydrin Derivatives

Although the cyanohydrin is an uncommon functional group, **linamarin** and **amygdalin** are two naturally occurring cyanohydrin derivatives. Both contain a carbon atom bonded to both an oxygen atom and a cyano group, analogous to a cyanohydrin.

[structure of linamarin]
linamarin
(found in cassava root)

[structure of amygdalin]
amygdalin
(commonly called laetrile)

Linamarin is isolated from cassava, a woody shrub grown as a root crop in the humid tropical regions of South America and Africa. **Amygdalin** is present in the seeds and pits of apricots, peaches, and wild cherries. Amygdalin, also known as **laetrile,** was once touted as an anticancer drug, and is still available in some countries for this purpose, although its effectiveness is unproven.

Both linamarin and amygdalin are toxic compounds because they are metabolized to cyanohydrins, which are hydrolyzed to carbonyl compounds and toxic HCN gas. This second step is merely the reconversion of a cyanohydrin to a carbonyl compound, a process that occurs with base in reactions run in the laboratory (Section 21.9A). If cassava root is processed with care, linamarin is enzymatically metabolized by this reaction sequence and the toxic HCN is released before the root is ingested, making it safe to eat.

> Cassava is a widely grown root crop, first introduced to Africa by Portuguese traders from Brazil in the sixteenth century. The peeled root is eaten after boiling or roasting, and some varieties yield a type of flour. Toxic cyanohydrin derivatives present in the plant are removed during processing. If the root is eaten without processing, illness and even death can result from high levels of HCN. HCN is a cellular poison with a characteristic almond odor.

The breakdown of linamarin to HCN

[structure of linamarin] → enzyme → [acetone cyanohydrin structure] → enzyme → [acetone structure] + HCN

linamarin

↑ cyanohydrin derivative

acetone cyanohydrin

toxic by-product

PROBLEM 21.16 What cyanohydrin and carbonyl compound are formed when amygdalin is metabolized in a similar manner?

21.10 The Wittig Reaction

The additions of H⁻, R⁻, and ⁻CN all involve the same two steps—nucleophilic attack followed by protonation. Other examples of nucleophilic addition in Chapter 21 are somewhat different. Although they still involve attack of a nucleophile, the initial addition adduct is converted to another product by a series of reactions.

The first reaction in this category is the **Wittig reaction,** named for German chemist Georg Wittig, who was awarded the Nobel Prize in Chemistry in 1979 for its discovery. The Wittig reaction uses a carbon nucleophile, the **Wittig reagent,** to form **alkenes.** When a carbonyl compound is treated with a Wittig reagent, the carbonyl oxygen atom is replaced by the negatively charged alkyl group bonded to the phosphorus—that is, **the C=O is converted to a C=C.**

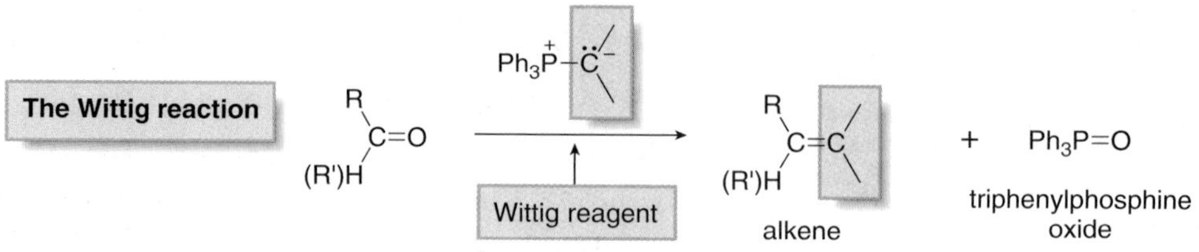

A Wittig reaction forms two new carbon–carbon bonds—one new σ bond and one new π bond—as well as a phosphorus by-product, Ph$_3$P=O (triphenylphosphine oxide).

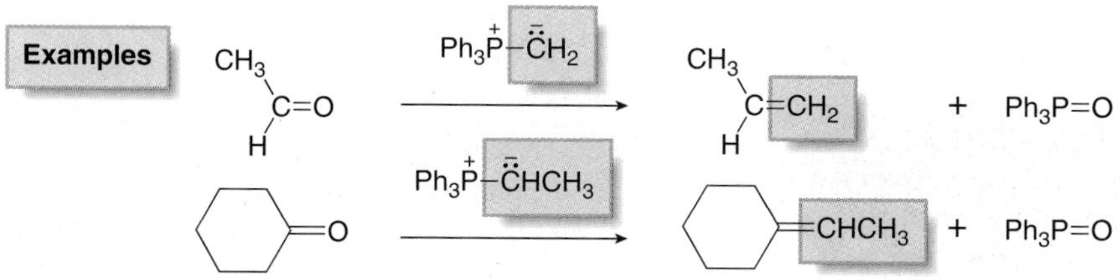

21.10A   The Wittig Reagent

A **Wittig reagent** is an **organophosphorus reagent**—a reagent that contains a carbon–phosphorus bond. A typical Wittig reagent has a phosphorus atom bonded to three phenyl groups, plus another alkyl group that bears a negative charge.

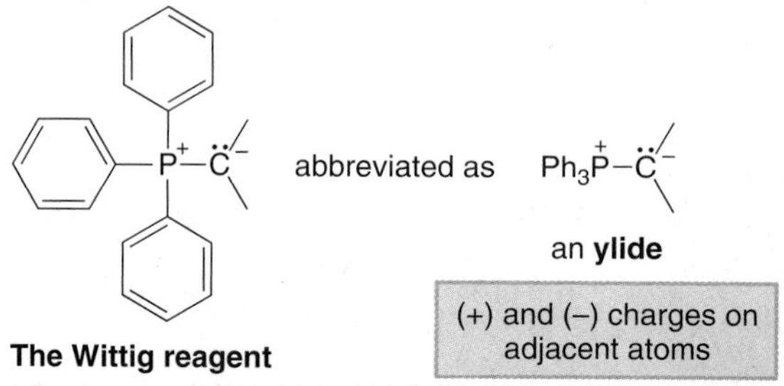

> Phosphorus ylides are also called **phosphoranes.**

A Wittig reagent is an *ylide,* **a species that contains two oppositely charged atoms bonded to each other, and both atoms have octets.** In a Wittig reagent, a negatively charged carbon atom is bonded to a positively charged phosphorus atom.

Because phosphorus is a third-row element, it can be surrounded by more than eight electrons. As a result, a second resonance structure can be drawn that places a double bond between carbon and phosphorus. Regardless of which resonance structure is drawn, a **Wittig reagent has no net charge.** In one resonance structure, though, the carbon atom bears a net negative charge, so it is nucleophilic.

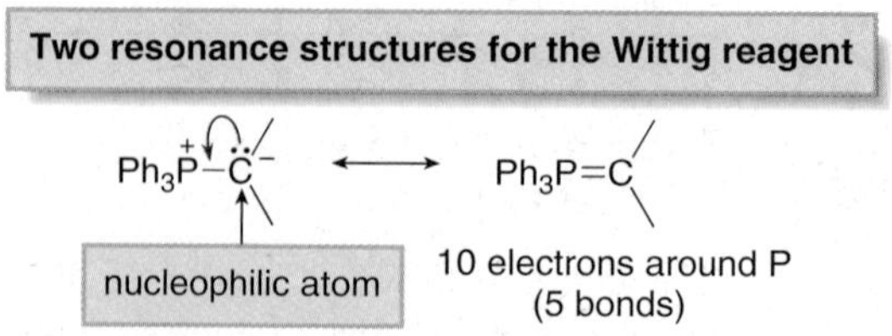

Wittig reagents are synthesized by a two-step procedure.

Step [1]   **S$_N$2 reaction of triphenylphosphine with an alkyl halide forms a phosphonium salt.**

$$Ph_3P: \quad + \quad RCH_2{-}X \quad \xrightarrow{S_N2} \quad Ph_3\overset{+}{P}{-}CH_2R \quad + \quad X^-$$

triphenylphosphine
nucleophile

phosphonium salt

Because phosphorus is located below nitrogen in the periodic table, a neutral phosphorus atom with three bonds also has a lone pair of electrons.

Triphenylphosphine (Ph$_3$P:), which contains a lone pair of electrons on P, is the nucleophile. Because the reaction follows an S$_N$2 mechanism, it works best with unhindered CH$_3$X and 1° alkyl halides (RCH$_2$X). Secondary alkyl halides (R$_2$CHX) can also be used, although yields are often lower.

Step [2]   **Deprotonation of the phosphonium salt with strong base forms the ylide.**

strong base

$$Ph_3\overset{+}{P}{-}\overset{H}{\underset{X^-}{\overset{|}{C}}}HR \quad :B \longrightarrow \quad Ph_3\overset{+}{P}{-}\overset{..}{C}HR \quad + \quad H{-}B^+$$

phosphonium salt

ylide

Typical strong base:

$$CH_3CH_2CH_2CH_2{-}Li$$

Bu$-$Li

Section 20.9C discussed the reaction of organometallic reagents as strong bases.

Because removal of a proton from a carbon bonded to phosphorus generates a resonance-stabilized carbanion (the ylide), this proton is somewhat more acidic than other protons on an alkyl group in the phosphonium salt. Very strong bases are still needed, though, to favor the products of this acid–base reaction. Common bases used for this reaction are the organolithium reagents such as butyllithium, CH$_3$CH$_2$CH$_2$CH$_2$Li, abbreviated as BuLi.

To synthesize the Wittig reagent, Ph$_3$P=CH$_2$, use these two steps:

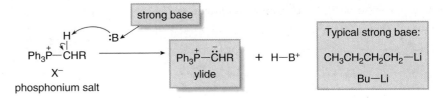

methyltriphenyl-phosphonium bromide

two resonance structures for the ylide

butane

- **Step [1]** Form the phosphonium salt by S$_N$2 reaction of Ph$_3$P: and CH$_3$Br.
- **Step [2]** Form the ylide by removal of a proton using BuLi as a strong base.

PROBLEM 21.17   Draw the products of the following Wittig reactions.

a. (CH$_3$)$_2$C=O + Ph$_3$P=CH$_2$ $\longrightarrow$

b. ⬠=O + Ph$_3$P=CHCH$_2$CH$_2$CH$_2$CH$_3$ $\longrightarrow$

PROBLEM 21.18   Outline a synthesis of each Wittig reagent from Ph$_3$P and an alkyl halide.
a. Ph$_3$P=CHCH$_3$ b. Ph$_3$P=C(CH$_3$)$_2$ c. Ph$_3$P=CHC$_6$H$_5$

PROBLEM 21.19   Why is Ph$_3$P and not (CH$_3$CH$_2$)$_3$P used to generate the phosphonium salt?

21.10B   Mechanism of the Wittig Reaction

Wittig reactions occur by an addition–elimination sequence that involves three steps. Like other nucleophiles, the Wittig reagent attacks an electrophilic carbonyl carbon, but then the initial addition adduct undergoes elimination to form an alkene. Mechanism 21.4 is drawn using Ph$_3$P=CH$_2$.

MECHANISM 21.4

The Wittig Reaction

Steps [1]–[2] Nucleophilic addition and cyclization form a four-membered ring.

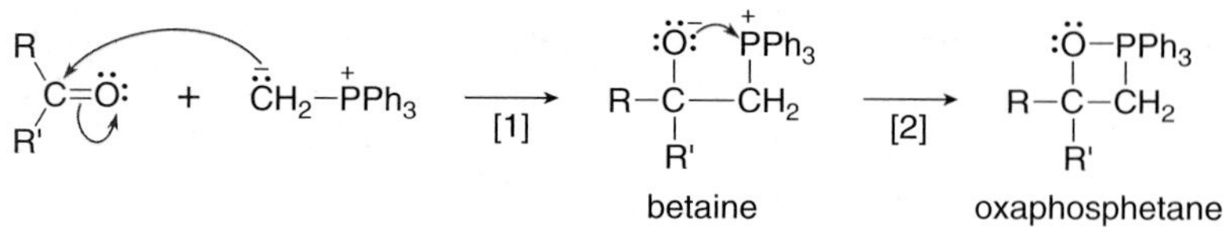

R$\cdot$C=O: + $\ddot{C}H_2$–$\overset{+}{P}Ph_3$ → R–C–CH$_2$ (betaine) → R–C–CH$_2$ (oxaphosphetane)

- In Step [1], nucleophilic attack of the negatively charged carbon atom of the ylide forms a new carbon–carbon σ bond, with cleavage of the C–O π bond. The product of this step is called a **betaine** (pronounced baita-ene).

- In Step [2], cyclization generates an **oxaphosphetane,** a four-membered ring containing a strong P–O bond.

Step [3] Elimination of Ph₃P=O forms the alkene.

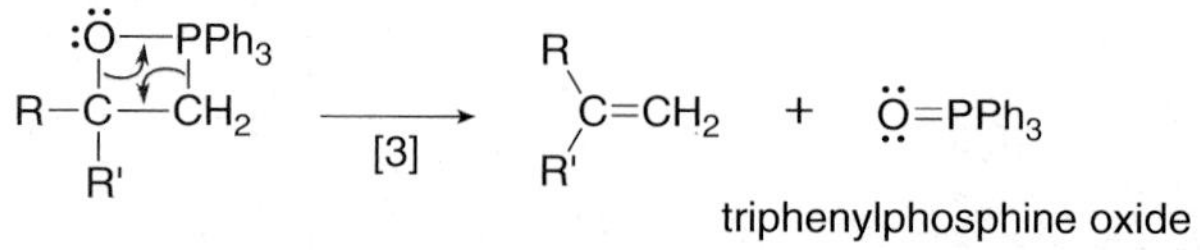

R–C–CH$_2$ → $\underset{R'}{\overset{R}{C}}$=CH$_2$ + $\ddot{O}$=PPh$_3$ (triphenylphosphine oxide)

- In Step [3], **Ph₃P=O (triphenylphosphine oxide) is eliminated,** forming two new π bonds. The formation of the very strong P–O double bond provides the driving force for the Wittig reaction.

One limitation of the Wittig reaction is that a mixture of alkene stereoisomers sometimes forms. For example, reaction of propanal (CH_3CH_2CHO) with a Wittig reagent forms the mixture of *E* and *Z* isomers shown.

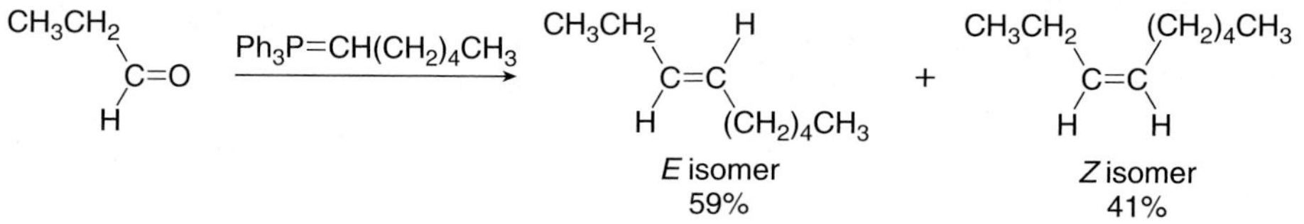

$\underset{H}{\overset{CH_3CH_2}{C}}$=O $\xrightarrow{Ph_3P=CH(CH_2)_4CH_3}$ $\underset{H}{\overset{CH_3CH_2}{C}}$=C$\underset{(CH_2)_4CH_3}{\overset{H}{}}$ + $\underset{H}{\overset{CH_3CH_2}{C}}$=C$\underset{H}{\overset{(CH_2)_4CH_3}{}}$

E isomer   *Z* isomer
59%   41%

Because the Wittig reaction forms two carbon–carbon bonds in a single reaction, it has been used to synthesize many natural products, including β-carotene, shown in Figure 21.8.

PROBLEM 21.20 Draw the products (including stereoisomers) formed when benzaldehyde (C_6H_5CHO) is treated with each Wittig reagent: (a) $Ph_3P=CHCH_2CH_3$; (b) $Ph_3P=CHC_6H_5$; (c) $Ph_3P=CHCOOCH_3$.

21.10C Retrosynthetic Analysis

To use the Wittig reaction in synthesis, you must be able to determine what carbonyl compound and Wittig reagent are needed to prepare a given compound—that is, **you must work backwards, in the retrosynthetic direction.** There can be two different Wittig routes to a given alkene, but one is often preferred on steric grounds.

Figure 21.8 A Wittig reaction used to synthesize β-carotene

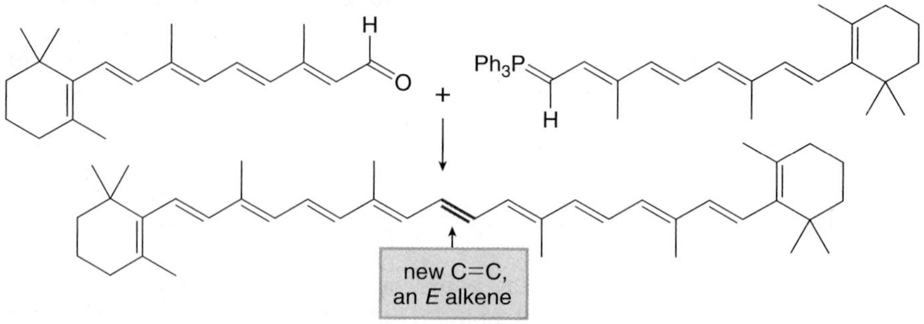

new C=C, an *E* alkene

β-carotene
orange pigment found in carrots
(vitamin A precursor)

- The more stable *E* alkene is the major product in this Wittig reaction.

How To   Determine the Starting Materials for a Wittig Reaction Using Retrosynthetic Analysis

Example   **What starting materials are needed to synthesize alkene A by a Wittig reaction?**

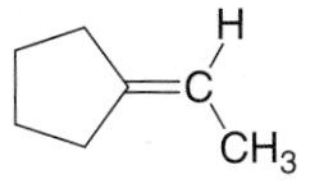

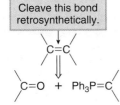

A

Step [1]   **Cleave the carbon–carbon double bond into two components.**

Cleave this bond retrosynthetically.

$$\text{C=C} \Longrightarrow \text{C=O} + \text{Ph}_3\text{P=C}$$

- Part of the molecule becomes the carbonyl component and the other part becomes the Wittig reagent.

There are usually two routes to a given alkene using a Wittig reaction:

Possibility [1]	**Possibility [2]**

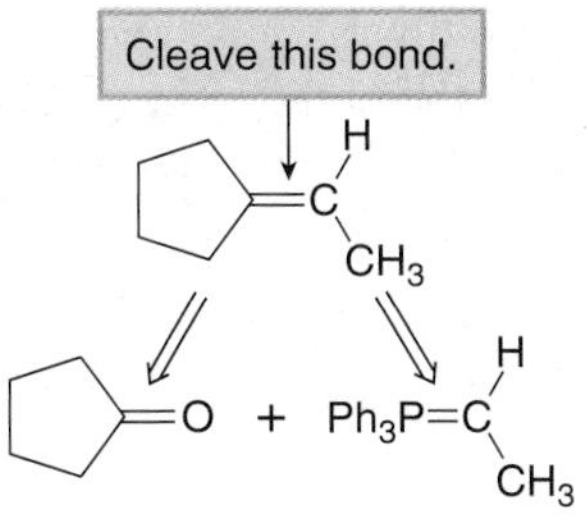

Cleave this bond.

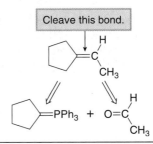

Step [2]   **Compare the Wittig reagents. The preferred pathway uses a Wittig reagent derived from an unhindered alkyl halide—CH_3X or RCH_2X.**

Determine what alkyl halide is needed to prepare each Wittig reagent:

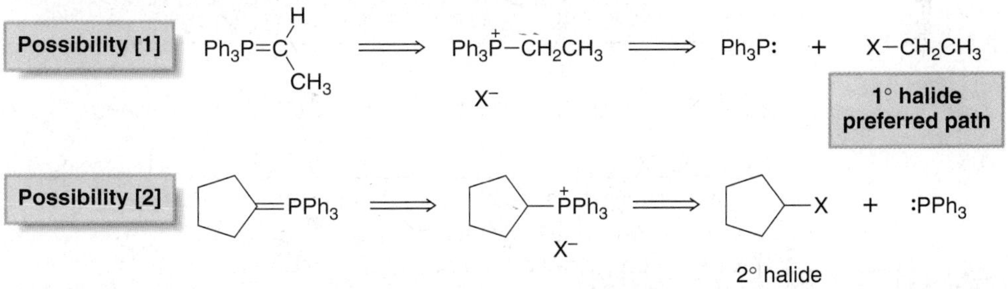

Possibility [1] $\text{Ph}_3\text{P=C} \overset{H}{\underset{CH_3}{}} \Longrightarrow \text{Ph}_3\overset{+}{P}-CH_2CH_3 \quad X^- \Longrightarrow \text{Ph}_3\text{P:} + X-CH_2CH_3$

1° halide preferred path

Possibility [2] $\bigcirc = PPh_3 \Longrightarrow \bigcirc - \overset{+}{P}Ph_3 \quad X^- \Longrightarrow \bigcirc - X + :PPh_3$

2° halide

Because the synthesis of the Wittig reagent begins with an S_N2 reaction, **the preferred pathway begins with an unhindered methyl halide or 1° alkyl halide.** In this example, retrosynthetic analysis of both Wittig reagents indicates that only one of them ($\text{Ph}_3\text{P=CHCH}_3$) can be synthesized from a 1° alkyl halide, making Possibility [1] the preferred pathway.

PROBLEM 21.21   What starting materials are needed to prepare each alkene by a Wittig reaction? When there are two possible routes, indicate which route, if any, is preferred: (a) $(CH_3)_2C=CHCH_2CH_3$; (b) $CH_3CH_2CH=CHCH_2CH_3$.

21.10D   Comparing Methods of Alkene Synthesis

An advantage in using the Wittig reaction over other elimination methods to synthesize alkenes is that **you always know the location of the double bond.** Whereas other methods of alkene synthesis often give a mixture of constitutional isomers, the Wittig reaction always gives a single constitutional isomer.

For example, two methods can be used to convert cyclohexanone into alkene **B** (methylenecyclohexane): a two-step method consisting of Grignard addition followed by dehydration, or a one-step Wittig reaction.

cyclohexanone   **B**

In a two-step method, treatment of cyclohexanone with CH_3MgBr forms a 3° alcohol after protonation. Dehydration of the alcohol with H_2SO_4 forms a mixture of alkenes, in which the desired disubstituted alkene is the minor product.

> Recall from Section 9.8 that the major product formed in acid-catalyzed dehydration of an alcohol is the more substituted alkene.

cyclohexanone   3° alcohol   trisubstituted C=C   **B**
                                                          major product   disubstituted C=C
                                                                               minor product

By contrast, reaction of cyclohexanone with $Ph_3P=CH_2$ affords the desired alkene as the only product. The newly formed double bond always joins the carbonyl carbon with the negatively charged carbon of the Wittig reagent. In other words, **the position of the double bond is always unambiguous in the Wittig reaction.** This makes the Wittig reaction an especially attractive method for use in preparing many alkenes.

cyclohexanone   **B**

PROBLEM 21.22   Show two methods to synthesize each alkene: a one-step method using a Wittig reagent, and a two-step method that forms a carbon–carbon bond with an organometallic reagent in one of the steps.

a.   b.

21.11   Addition of 1° Amines

We now move on to the reaction of aldehydes and ketones with nitrogen and oxygen heteroatoms. **Amines, for example, are organic nitrogen compounds that contain a nonbonded electron pair on the N atom.** Amines are classified as 1°, 2°, or 3° by the number of alkyl groups bonded to the *nitrogen* atom.

1° amine   **2° amine**   **3° amine**
(1 R group on N)   (2 R groups on N)   (3 R groups on N)

Both 1° and 2° amines react with aldehydes and ketones. We begin by examining the reaction of aldehydes and ketones with 1° amines.

21.11A Formation of Imines

Treatment of an aldehyde or ketone with a 1° amine affords an **imine** (also called a **Schiff base**). Nucleophilic attack of the 1° amine on the carbonyl group forms an unstable **carbinolamine,** which loses water to form an imine. The overall reaction results in **replacement of C=O by C=NR.**

Imine formation— General reaction

$$\underset{\substack{\text{R' = H or alkyl}}}{\overset{R}{\underset{R'}{\diagdown}}C{=}\ddot{O}} \quad \xrightarrow[\text{mild acid}]{R''\ddot{N}H_2} \quad \left[\underset{\substack{R'}}{\overset{:\ddot{O}H}{R{-}\underset{|}{\overset{|}{C}}{-}\ddot{N}HR''}} \right] \quad \xrightarrow{-H_2O} \quad \underset{\substack{R' \quad R''}}{\overset{R}{\diagdown}}C{=}\ddot{N}$$

carbinolamine   imine

Because the N atom of an imine is surrounded by three groups (two atoms and a lone pair), it is sp^2 hybridized, making the C−N−R″ bond angle ~120° (*not* 180°). Imine formation is fastest when the reaction medium is weakly acidic.

Examples

The mechanism of imine formation (Mechanism 21.5) can be divided into two distinct parts: **nucleophilic addition of the 1° amine, followed by elimination of H₂O.** Each step involves a reversible equilibrium, so that the reaction is driven to completion by removing H₂O.

MECHANISM 21.5

Imine Formation from an Aldehyde or Ketone

Part [1] Nucleophilic addition forms a carbinolamine.

◆ **Nucleophilic attack** of the amine followed by proton transfer forms the unstable carbinolamine (Steps [1]–[2]). These steps result in the addition of H and NHR″ to the carbonyl group.

Part [2] Elimination of H₂O forms an imine.

◆ Elimination of H₂O forms the imine in three steps. Protonation of the OH group in Step [3] forms a good leaving group, leading to **loss of water** in Step [4], giving a resonance-stabilized **iminium ion.** Loss of a proton forms the imine in Step [5].

◆ Except for Steps [1] (nucleophilic addition) and [4] (H₂O elimination), all other steps in the mechanism are acid–base reactions—that is, moving a proton from one atom to another.

Imine formation is most rapid at pH 4–5. Mild acid is needed for protonation of the hydroxy group in Step [3] to form a **good leaving group.** Under strongly acidic conditions, the reaction rate decreases because the amine nucleophile is protonated. With no free electron pair, it is no longer a nucleophile, and so nucleophilic addition cannot occur.

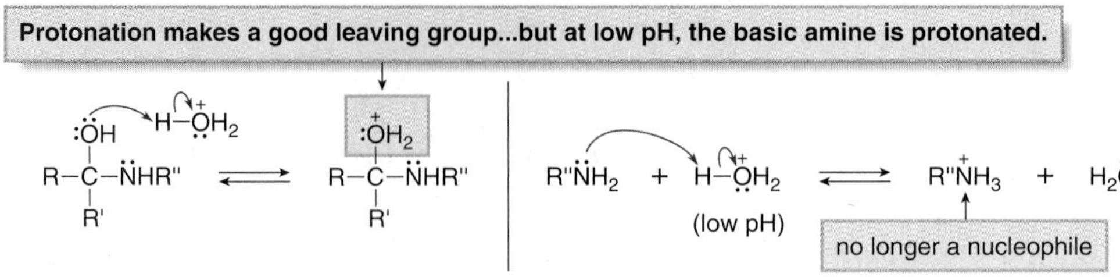

PROBLEM 21.23   Draw the product formed when $CH_3CH_2CH_2CH_2NH_2$ reacts with each carbonyl compound.

a. 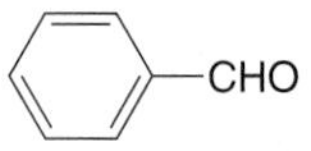 [benzene ring]—CHO   b. 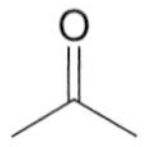 [acetone structure with O]   c. 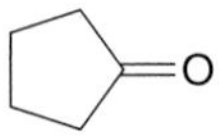 [cyclopentanone]=O

PROBLEM 21.24   What 1° amine and carbonyl compound are needed to prepare each imine?

a. 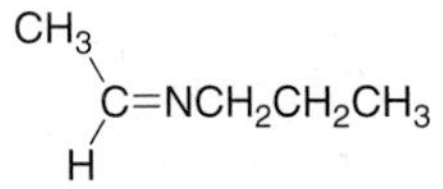 $\underset{H}{\overset{CH_3}{C}}=NCH_2CH_2CH_3$ b. 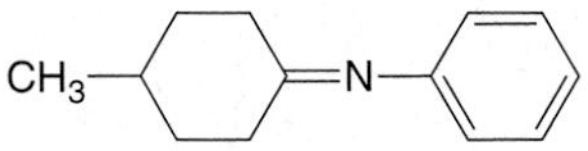 CH_3—[cyclohexane ring]=N—[benzene ring]

21.11B   Application: Retinal, Rhodopsin, and the Chemistry of Vision

Many imines play vital roles in biological systems. A key molecule in the chemistry of vision is the highly conjugated imine **rhodopsin,** which is synthesized in the rod cells of the eye from **11-*cis*-retinal** (the molecule introducing Chapter 21) and a 1° amine in the protein **opsin.**

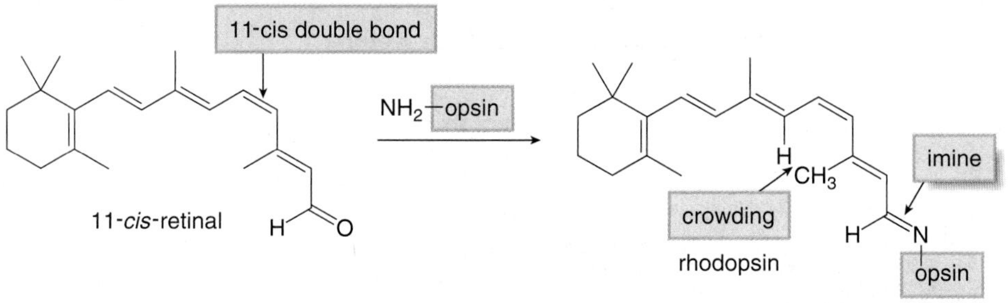

The complex process of vision centers around this imine derived from retinal (Figure 21.9). The 11-cis double bond in rhodopsin creates crowding in the rather rigid side chain. When light strikes the rod cells of the retina, it is absorbed by the conjugated double bonds of rhodopsin, and the 11-cis double bond is isomerized to the 11-trans arrangement. This isomerization is accompanied by a drastic change in shape in the protein, altering the concentration of Ca^{2+} ions moving across the cell membrane, and sending a nerve impulse to the brain, which is then processed into a visual image.

The central role of rhodopsin in the visual process was delineated by Nobel Laureate George Wald of Harvard University.

PROBLEM 21.25   Draw a stepwise mechanism for the formation of rhodopsin from 11-*cis*-retinal and an NH_2 group in opsin.

21.12   Addition of 2° Amines

A 2° amine reacts with an aldehyde or ketone to give an **enamine.** *Enamines* **have a nitrogen atom bonded to a double bond** (alk*ene* + am*ine* = en*amine*).

Figure 21.9   The key
reaction in the chemistry of vision

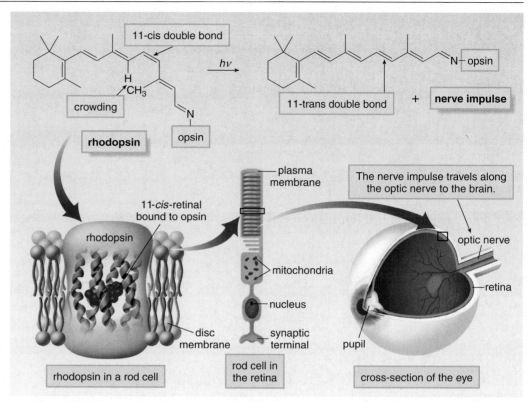

- Rhodopsin is a light-sensitive compound located in the membrane of the rod cells in the retina of the eye. Rhodopsin contains the protein opsin bonded to 11-*cis*-retinal via an imine linkage. When light strikes this molecule, the crowded 11-cis double bond isomerizes to the 11-trans isomer, and a nerve impulse is transmitted to the brain by the optic nerve.

Like imines, enamines are also formed by the addition of a nitrogen nucleophile to a carbonyl group followed by elimination of water. In this case, however, elimination occurs across two adjacent *carbon* atoms to form a new carbon–carbon π bond.

The mechanism for enamine formation (Mechanism 21.6) is identical to the mechanism for imine formation except for the last step, involving formation of the π bond. The mechanism can be divided into two distinct parts: **nucleophilic addition of the 2° amine, followed by elimination of H₂O.** Each step involves a reversible equilibrium once again, so that the reaction is driven to completion by removing H₂O.

MECHANISM 21.6

Enamine Formation from an Aldehyde or Ketone

Part [1] Nucleophilic addition forms a carbinolamine.

◆ **Nucleophilic attack** of the amine followed by proton transfer forms the unstable carbinolamine (Steps [1]–[2]).

Part [2] Elimination of H₂O forms an enamine.

◆ Protonation of the OH group in Step [3] forms a good leaving group, leading to **loss of water** in Step [4], giving a resonance-stabilized **iminium ion**.

◆ Removal of a proton from the adjacent C−H bond forms the enamine in Step [5].

The mechanisms illustrate why **the reaction of 1° amines with carbonyl compounds forms imines, but the reaction with 2° amines forms** *enamines*. In Figure 21.10, the last step of both mechanisms is compared using cyclohexanone as starting material. The position of the double bond depends on which proton is removed in the last step. Removal of an N−H proton forms a C=N, whereas removal of a C−H proton forms a C=C.

Because imines and enamines are formed by a set of reversible reactions, both can be converted back to carbonyl compounds by hydrolysis with mild acid. The mechanism of these reactions is exactly the *reverse* of the mechanism written for the formation of imines and enamines. In the hydrolysis of enamines, the carbonyl carbon in the product comes from the sp^2 hybridized carbon bonded to the N atom in the starting material.

Figure 21.10 The formation of imines and enamines compared

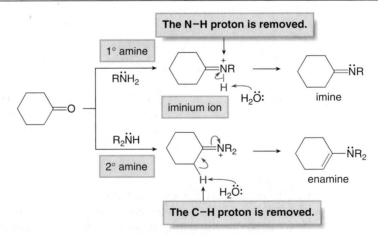

- With a **1° amine,** the intermediate iminium ion still has a proton on the N atom that may be removed to form a C=N.
- With a **2° amine,** the intermediate iminium ion has no proton on the N atom. A proton must be removed from an adjacent C−H bond, and this forms a C=C.

◆ **Hydrolysis of imines and enamines forms aldehydes and ketones.**

Imine hydrolysis

$$\underset{CH_3}{\overset{CH_3}{C}}=NCH_2CH_2CH_2CH_3 \quad \xrightarrow{H_3O^+} \quad \underset{CH_3}{\overset{CH_3}{C}}=O \quad + \quad H_2NCH_2CH_2CH_2CH_3$$

Enamine hydrolysis

CH₂CH₂CH₃ / N—CH₃ $\xrightarrow{H_3O^+}$ =O   +   HN(CH₂CH₂CH₃)(CH₃)

PROBLEM 21.26   What two enamines are formed when 2-methylcyclohexanone is treated with $(CH_3)_2NH$?

PROBLEM 21.27   Explain why benzaldehyde cannot form an enamine with a 2° amine.

PROBLEM 21.28   What carbonyl compound and amine are formed by hydrolysis of each compound?

a.   (benzene ring)—CH=N—(cyclopentane ring)

b.   (benzene ring)—CH₂—N(CH₃)—(cyclopentene ring)

21.13   Addition of H₂O—Hydration

Treatment of a carbonyl compound with H_2O in the presence of an acid or base catalyst **adds the elements of H and OH across the carbon–oxygen π bond,** forming a *gem*-diol or hydrate.

Nucleophilic addition of H_2O— General reaction

$$\underset{R \quad R'}{\overset{O}{\underset{\|}{C}}} \quad \xrightarrow[H^+ \text{ or } ^-OH]{H_2O} \quad R{-}\overset{OH}{\underset{OH}{\overset{|}{\underset{|}{C}}}}{-}R' \quad \boxed{\text{addition of } H_2O}$$

R' = H or alkyl   *gem*-diol (hydrate)

Hydration of a carbonyl group gives a good yield of *gem*-diol only with an unhindered aldehyde like formaldehyde, and with aldehydes containing nearby electron-withdrawing groups.

Examples

$$\underset{H \quad H}{\overset{O}{\underset{\|}{C}}} \quad \xrightarrow{H_2O} \quad H{-}\overset{OH}{\underset{OH}{\overset{|}{\underset{|}{C}}}}{-}H$$

formaldehyde   formaldehyde hydrate

$$\underset{Cl_3C \quad H}{\overset{O}{\underset{\|}{C}}} \quad \xrightarrow{H_2O} \quad Cl_3C{-}\overset{OH}{\underset{OH}{\overset{|}{\underset{|}{C}}}}{-}H$$

chloral   chloral hydrate

21.13A   The Thermodynamics of Hydrate Formation

Whether addition of H_2O to a carbonyl group affords a good yield of the *gem*-diol depends on the relative energies of the starting material and the product. With less stable carbonyl starting materials, equilibrium favors the hydrate product, whereas with more stable carbonyl starting materials, equilibrium favors the carbonyl starting material. Because **alkyl groups stabilize a carbonyl group** (Section 20.2B):

◆ **Increasing the number of alkyl groups on the carbonyl carbon decreases the amount of hydrate at equilibrium.**

This can be illustrated by comparing the amount of hydrate formed from formaldehyde, acetaldehyde, and acetone.

Chloral hydrate, a sedative sometimes administered to calm a patient prior to a surgical procedure, has also been used for less reputable purposes. Adding it to an alcoholic beverage makes a so-called knock-out drink, causing an individual who drinks it to pass out. Because it is addictive and care must be taken in its administration, it is a controlled substance.

Increasing number of R groups

Increasing stability of the carbonyl compound →

$$
\begin{array}{ccc}
\underset{\substack{\text{H} \quad \text{H} \\ \text{formaldehyde}}}{\overset{\text{O}}{\underset{\|}{\text{C}}}} &
\underset{\substack{\text{CH}_3 \quad \text{H} \\ \text{acetaldehyde}}}{\overset{\text{O}}{\underset{\|}{\text{C}}}} &
\underset{\substack{\text{CH}_3 \quad \text{CH}_3 \\ \text{acetone}}}{\overset{\text{O}}{\underset{\|}{\text{C}}}}
\end{array}
$$

$\updownarrow$ H_2O $\updownarrow$ H_2O $\updownarrow$ H_2O

$$
\begin{array}{ccc}
\underset{\substack{\text{OH} \\ | \\ \text{OH} \\ \text{99.9\% product}}}{\text{H}-\text{C}-\text{H}} &
\underset{\substack{\text{OH} \\ | \\ \text{OH} \\ \text{58\% product}}}{\text{CH}_3-\text{C}-\text{H}} &
\underset{\substack{\text{OH} \\ | \\ \text{OH} \\ \text{0.2\% product}}}{\text{CH}_3-\text{C}-\text{CH}_3}
\end{array}
$$

← **Increasing amount of hydrate present at equilibrium**

Formaldehyde, the least stable carbonyl compound, forms the largest percentage of hydrate. On the other hand, acetone and other ketones, which have two electron-donor R groups, form < 1% of the hydrate at equilibrium. Other electronic factors come into play as well.

◆ **Electron-donating groups near the carbonyl carbon stabilize the carbonyl group, decreasing the amount of the hydrate at equilibrium.**
◆ **Electron-withdrawing groups near the carbonyl carbon destabilize the carbonyl group, increasing the amount of hydrate at equilibrium.**

This explains why chloral (trichloroacetaldehyde) forms a large amount of hydrate at equilibrium. Three electron-withdrawing Cl atoms place a partial positive charge on the α carbon to the carbonyl, destabilizing the carbonyl group, and therefore increasing the amount of hydrate at equilibrium.

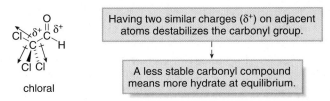

Having two similar charges (δ⁺) on adjacent atoms destabilizes the carbonyl group.
↓
A less stable carbonyl compound means more hydrate at equilibrium.

PROBLEM 21.29   Which compound in each pair forms the higher percentage of *gem*-diol at equilibrium?

a. $CH_3CH_2CH_2CHO$ and $CH_3CH_2COCH_3$ c.

b. CH_3CF_2CHO and CH_3CH_2CHO

21.13B The Kinetics of Hydrate Formation

Although H_2O itself adds slowly to a carbonyl group, both acid and base catalyze the addition. In base, the nucleophile is ⁻OH, and the mechanism follows the usual two steps for nucleophilic addition: **nucleophilic attack followed by protonation,** as shown in Mechanism 21.7.

MECHANISM 21.7

Base-Catalyzed Addition of H₂O to a Carbonyl Group

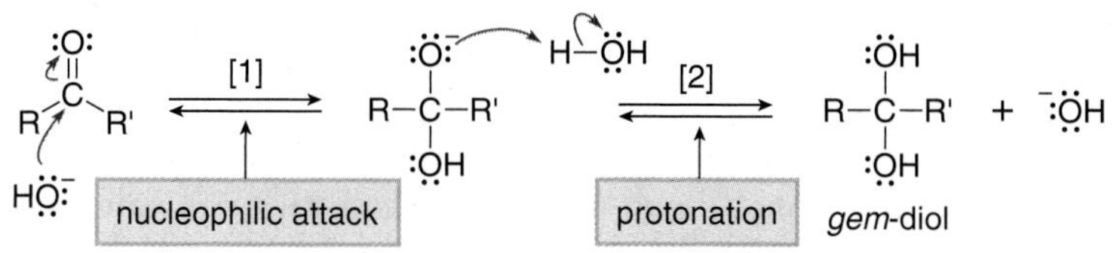

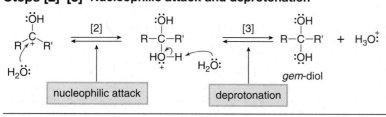

♦ In Step [1], **the nucleophile (⁻OH) attacks the carbonyl group,** cleaving the π bond, and moving an electron pair onto oxygen.

♦ In Step [2], **protonation of the negatively charged O atom by H₂O** forms the *gem*-diol.

The acid-catalyzed addition follows the general mechanism presented in Section 21.7A. For a poorer nucleophile like H₂O to attack a carbonyl group, the **carbonyl must be protonated by acid first; thus, protonation *precedes* nucleophilic attack.** The overall mechanism has three steps, as shown in Mechanism 21.8.

MECHANISM 21.8

Acid-Catalyzed Addition of H₂O to a Carbonyl Group

Step [1]   Protonation of the carbonyl group

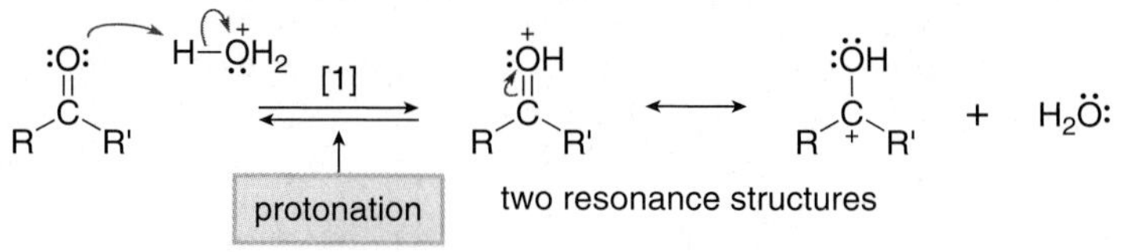

♦ Protonation of the carbonyl oxygen forms a resonance-stabilized cation that bears a full positive charge.

Steps [2]–[3]   Nucleophilic attack and deprotonation

♦ In Step [2], the nucleophile (H₂O) attacks, and then deprotonation forms the neutral addition product in Step [3].

♦ The overall result is the addition of H and OH to the carbonyl group and regeneration of the acid catalyst.

Acid and base increase the rate of reaction for different reasons.

♦ **Base converts H₂O into ⁻OH, a *stronger nucleophile*.**

♦ **Acid protonates the carbonyl group, making it *more electrophilic* towards nucleophilic attack.**

These catalysts increase the rate of the reaction, but they do not affect the equilibrium constant. Starting materials that give a low yield of *gem*-diol do so whether or not a catalyst is present. Because these reactions are reversible, the conversion of *gem*-diols to aldehydes and ketones is also catalyzed by acid and base, and the steps of the mechanism are reversed.

PROBLEM 21.30   Draw a stepwise mechanism for the following reaction.

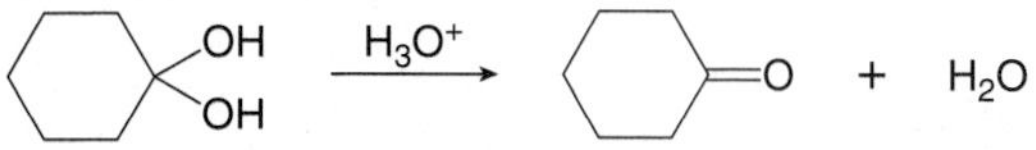

21.14 Addition of Alcohols—Acetal Formation

Aldehydes and ketones react with *two* equivalents of alcohol to form acetals. In an acetal, the carbonyl carbon from the aldehyde or ketone is now singly bonded to two OR" (alkoxy) groups.

Acetal formation— General reaction

$$
\underset{R' = H \text{ or alkyl}}{\overset{O}{\underset{R}{\overset{\|}{C}}}R'} \quad + \quad \underset{(2\text{ equiv})}{R''OH} \quad \underset{\longleftarrow}{\overset{H^+}{\longrightarrow}} \quad \underset{acetal}{\overset{R''O\quad OR''}{\underset{R}{\overset{|}{C}}R'}} \quad + \quad H_2O
$$

This reaction differs from other additions we have seen thus far, because **two equivalents of alcohol are added to the carbonyl group,** and two new C—O σ bonds are formed. Acetal formation is catalyzed by acids, commonly *p*-toluenesulfonic acid (TsOH).

Example

two new σ bonds

$$
\underset{CH_3CH_2}{\overset{O}{\overset{\|}{C}}}H \quad + \quad \underset{(2\text{ equiv})}{CH_3OH} \quad \underset{\longleftarrow}{\overset{TsOH}{\longrightarrow}} \quad \underset{CH_3CH_2\quad \quad H}{\overset{CH_3O\quad OCH_3}{\underset{acetal}{\overset{|}{C}}}} \quad + \quad H_2O
$$

Acetals are *not* ethers, even though both functional groups contain a C—O σ bond. Having two C—O σ bonds on the *same* carbon atom makes an acetal very different from an ether.

$$
\underset{acetal}{\overset{RO\quad OR}{\underset{R}{\overset{|}{C}}R}} \quad \neq \quad \underset{ether}{R-O-R}
$$

When a diol such as ethylene glycol is used in place of two equivalents of ROH, a cyclic acetal is formed. Both oxygen atoms in the cyclic acetal come from the diol.

$$
\text{cyclohexanone} \quad + \quad HOCH_2CH_2OH \quad \underset{\longleftarrow}{\overset{TsOH}{\longrightarrow}} \quad \text{a cyclic acetal} \quad + \quad H_2O
$$

ethylene glycol

a cyclic acetal

Driving an equilibrium to the right by removing one of the products is an application of Le Châtelier's principle. To review Le Châtelier's principle, see Section 9.8.

Like *gem*-diol formation, the synthesis of acetals is reversible, and often the equilibrium favors reactants, not products. In acetal synthesis, however, water is formed as a by-product, so the equilibrium can be driven to the right by removing the water as it is formed. This can be done in a variety of ways in the laboratory. A drying agent can be added that reacts with the water, or more commonly, the water can be distilled from the reaction mixture as it is formed by using a Dean–Stark trap, as pictured in Figure 21.11.

Figure 21.11 A Dean–Stark trap for removing water

A Dean–Stark trap is an apparatus used for removing water from a reaction mixture. To use a Dean–Stark trap to convert a carbonyl compound to an acetal:

The carbonyl compound, an alcohol, and an acid are dissolved in benzene. As the mixture is heated, the carbonyl compound is converted to the acetal with water as a by-product. Benzene and water co-distill from the reaction mixture. When the hot vapors reach the cold condenser, they condense, forming a liquid that then collects in the glass tube below. Water, the more dense liquid, forms the lower layer, so that as it collects, it can be drained through the stopcock into a flask. In this way, water can be removed from a reaction mixture, driving the equilibrium.

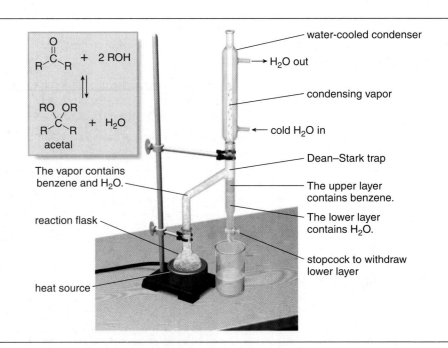

The vapor contains benzene and H₂O.

reaction flask

heat source

water-cooled condenser

H₂O out

condensing vapor

cold H₂O in

Dean–Stark trap

The upper layer contains benzene.

The lower layer contains H₂O.

stopcock to withdraw lower layer

PROBLEM 21.31   Draw the products of each reaction.

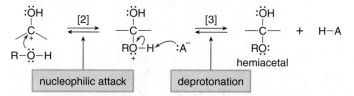

21.14A   The Mechanism

The mechanism for acetal formation can be divided into two parts: **the addition of one equivalent of alcohol** to form a **hemiacetal,** followed by the **conversion of the hemiacetal** to the **acetal.** A hemiacetal has a carbon atom bonded to one OH group and one OR group.

Part [1] | **Part [2]**
1st equivalent | 2nd equivalent

hemiacetal   acetal

Removing H$_2$O drives the equilibrium.

Like *gem*-diols, hemiacetals are often higher in energy than their carbonyl starting materials, making the direction of equilibrium unfavorable for hemiacetal formation. The elimination of H$_2$O, which can be removed from the reaction mixture to drive the equilibrium to favor product, occurs during the conversion of the hemiacetal to the acetal. This explains why two equivalents of ROH react with a carbonyl compound, forming the acetal as product.

The mechanism is written in two parts (Mechanisms 21.9 and 21.10) with a general acid H–A.

MECHANISM 21.9

Acetal Formation—Part [1] Formation of a Hemiacetal

Step [1]   Protonation of the carbonyl group

resonance-stabilized cation

◆ **Protonation** of the carbonyl oxygen forms a resonance-stabilized cation that bears a full positive charge.

Steps [2]–[3]   Nucleophilic attack and deprotonation

hemiacetal

nucleophilic attack   deprotonation

◆ In Step [2], the **nucleophile (ROH) attacks,** and then deprotonation forms the neutral addition product in Step [3].

◆ **The overall result is the addition of H and OR to the carbonyl group.**

Because formation of an acetal involves a carbonyl group and an acid catalyst, the first step of the mechanism is **protonation of the carbonyl oxygen.**

Formation of the acetal requires a second nucleophilic attack of ROH. For this to occur, however, **H$_2$O must first come off as a leaving group,** as shown in Mechanism 21.10.

MECHANISM 21.10

Acetal Formation—Part [2] Formation of the Acetal

Steps [4]–[5] Elimination of H₂O

◆ Protonation of the OH group in the hemiacetal in Step [4] forms a **good leaving group** (H₂O). Loss of H₂O in Step [5] forms a resonance-stabilized cation.

Steps [6]–[7] Nucleophilic attack and deprotonation

◆ Nucleophilic attack on the cation in Step [6] followed by loss of a proton forms the acetal.

◆ **The overall result of Steps [4]–[7] is the addition of a second OR group to the carbonyl group.**

Although this mechanism is lengthy—there are seven steps altogether—there are only three different kinds of reactions: **addition of a nucleophile, elimination of a leaving group, and proton transfer.** Steps [2] and [6] involve nucleophilic attack and Step [5] eliminates H₂O. The other four steps in the mechanism shuffle protons from one oxygen atom to another, to make a better leaving group or a more electrophilic carbonyl group.

PROBLEM 21.32   Label each compound as an acetal, a hemiacetal, or an ether.

a.   b.   c.   d.

PROBLEM 21.33   Draw a stepwise mechanism for the following reaction.

21.14B Hydrolysis of Acetals

Conversion of an aldehyde or ketone to an acetal is a **reversible reaction,** so **an acetal can be hydrolyzed to an aldehyde or ketone by treatment with aqueous acid.** Because this reaction is also an equilibrium process, it is driven to the right by using a large excess of water for hydrolysis.

Acetal hydrolysis— General reaction

acetal
R′ = H or alkyl

large excess

R″OH
(2 equiv)

Example

ethylene glycol

The mechanism for this reaction is the reverse of acetal synthesis, as illustrated in Sample Problem 21.4.

SAMPLE PROBLEM 21.4   Draw a stepwise mechanism for the following reaction.

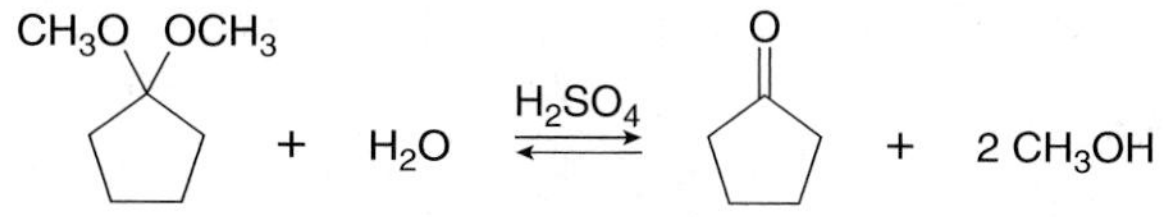

SOLUTION

The mechanism is the reverse of acetal formation and involves two parts—conversion of the acetal to a hemiacetal, followed by conversion of the hemiacetal to the carbonyl compound.

Part [1]   **Conversion of the acetal to a hemiacetal**

To convert this acetal to a hemiacetal, one molecule of CH_3OH must be eliminated and one molecule of H_2O must be added.

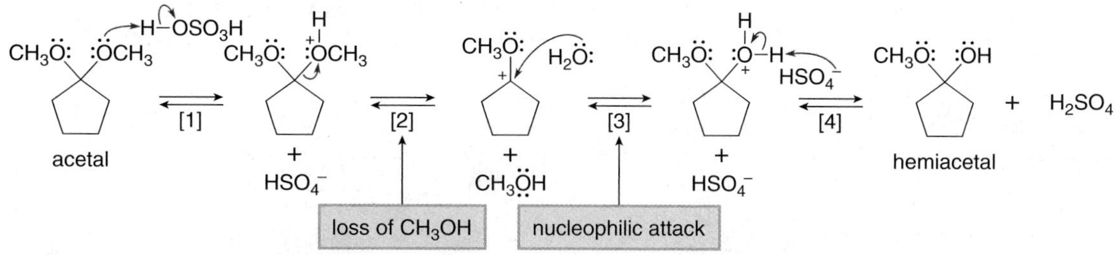

Part [2]   **Conversion of the hemiacetal to the carbonyl compound**

To convert the hemiacetal to a carbonyl compound, one molecule of CH_3OH must be eliminated and the C—O π bond must be formed.

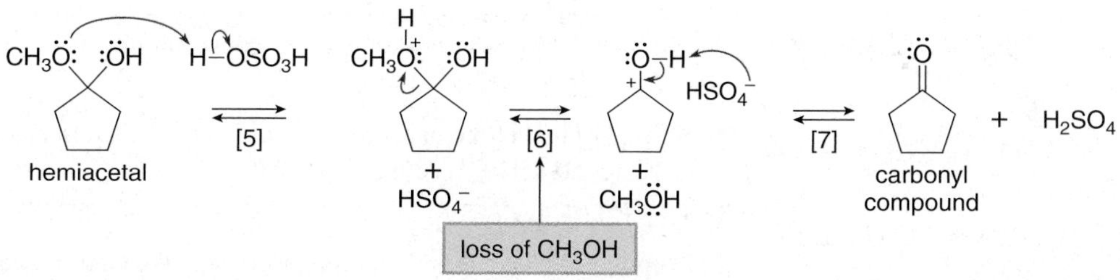

Steps [2] and [6] involve loss of the leaving group (CH_3OH), and Step [3] involves nucleophilic attack of H_2O. The other four steps in the mechanism shuffle protons from one oxygen atom to another.

Acetal hydrolysis requires a strong acid to make a good leaving group (ROH). In Sample Problem 21.4, H_2SO_4 converts CH_3O^- into CH_3OH, a weak base and neutral leaving group. Acetal hydrolysis does not occur in base.

PROBLEM 21.34   Draw the products of each reaction.

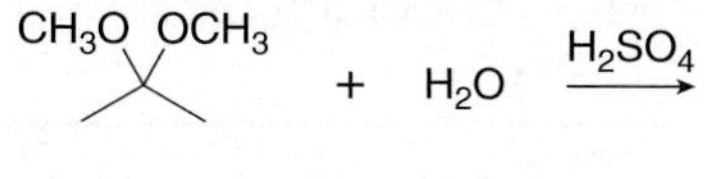

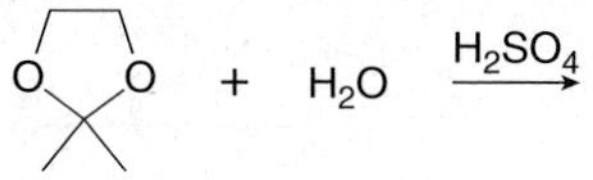

21.15 Acetals as Protecting Groups

Just as the *tert*-butyldimethylsilyl ethers are used as protecting groups for alcohols (Section 20.12), **acetals are valuable protecting groups for aldehydes and ketones.**

Suppose a starting material **A** contains both a ketone and an ester, and it is necessary to selectively reduce the ester to an alcohol (6-hydroxy-2-hexanone), leaving the ketone untouched. Such a selective reduction is *not* possible in one step. Because ketones are more readily reduced, methyl 5-hydroxyhexanoate is formed instead.

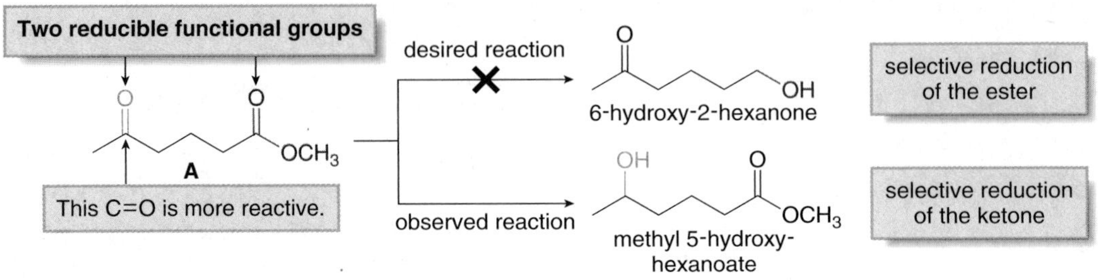

To solve this problem we can use a protecting group to block the more reactive ketone carbonyl group. The overall process requires three steps.

[1] Protect the interfering functional group—the ketone carbonyl.
[2] Carry out the desired reaction—reduction.
[3] Remove the protecting group.

The following three-step sequence using a cyclic acetal leads to the desired product.

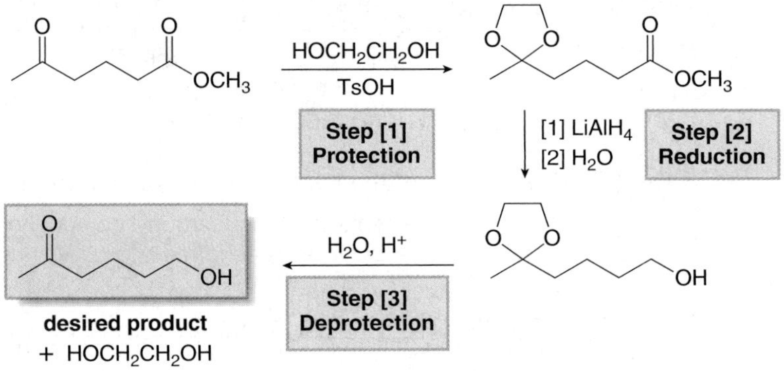

◆ **Step [1]** The ketone carbonyl is protected as a cyclic acetal by reaction of the starting material with $HOCH_2CH_2OH$ and TsOH.
◆ **Step [2]** Reduction of the ester is then carried out with $LiAlH_4$, followed by treatment with H_2O.
◆ **Step [3]** The acetal is then converted back to a ketone carbonyl group with aqueous acid.

Acetals are widely used protecting groups for aldehydes and ketones because they are easy to add and easy to remove, and they are stable to a wide variety of reaction conditions. Acetals do not react with base, oxidizing agents, reducing agents, or nucleophiles. Good protecting groups must survive a variety of reaction conditions that take place at other sites in a molecule, but they must also be selectively removed under mild conditions when needed.

PROBLEM 21.35   How would you use a protecting group to carry out the following transformation?

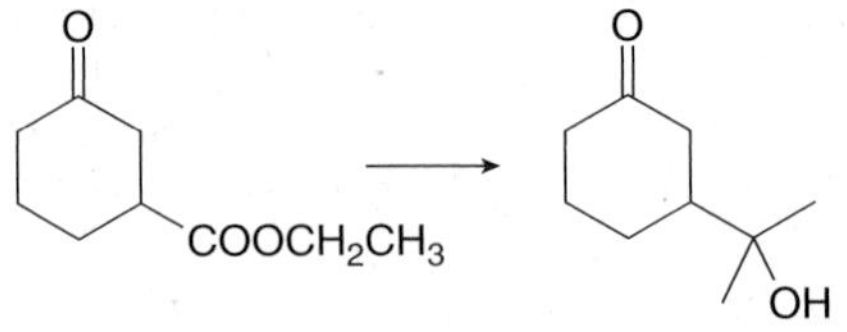

21.16 Cyclic Hemiacetals

Cyclic hemiacetals are also called **lactols**.

Although acyclic hemiacetals are generally unstable and therefore not present in appreciable amounts at equilibrium, **cyclic hemiacetals containing five- and six-membered rings are stable compounds** that are readily isolated.

A hemiacetal— General structure	Cyclic hemiacetals

One C is bonded to:
• an OH group
• an OR group

Each indicated C is bonded to:
• an OH group
• an OR group that is part of a ring

21.16A   Forming Cyclic Hemiacetals

All hemiacetals are formed by nucleophilic addition of a hydroxy group to a carbonyl group. In the same way, cyclic hemiacetals are formed by **intramolecular cyclization of hydroxy aldehydes.**

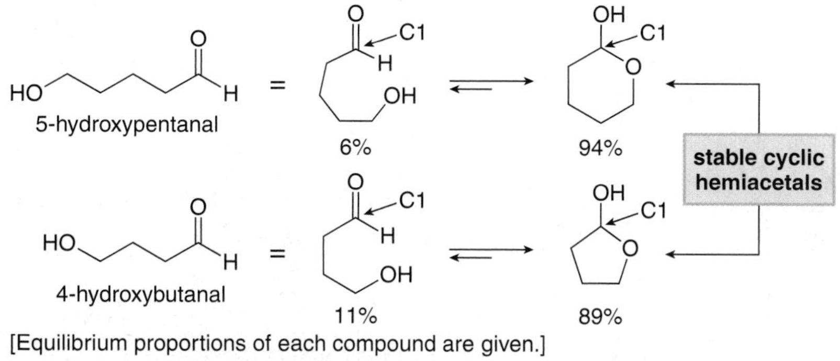

5-hydroxypentanal   6%   94%   **stable cyclic hemiacetals**

4-hydroxybutanal   11%   89%

[Equilibrium proportions of each compound are given.]

Such intramolecular reactions to form five- and six-membered rings are faster than the corresponding intermolecular reactions. The two reacting functional groups, in this case OH and C=O, are held in close proximity, increasing the probability of reaction.

PROBLEM 21.36   What lactol (cyclic hemiacetal) is formed from intramolecular cyclization of each hydroxy aldehyde?

a. 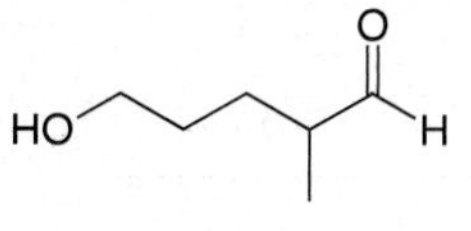   b.

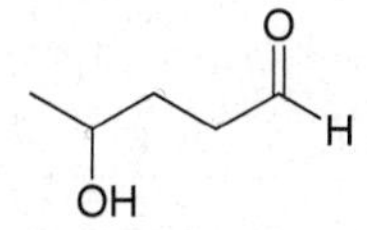

Hemiacetal formation is catalyzed by both acid and base. The acid-catalyzed mechanism is identical to Mechanism 21.9, except that the reaction occurs in an **intramolecular** fashion, as shown for the acid-catalyzed cyclization of 5-hydroxypentanal to form a six-membered cyclic hemiacetal in Mechanism 21.11.

MECHANISM 21.11

Acid-Catalyzed Cyclic Hemiacetal Formation

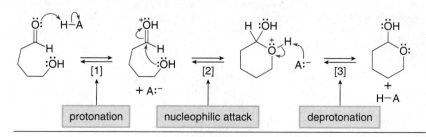

◆ Protonation of the carbonyl oxygen in Step [1] followed by **intramolecular nucleophilic attack** in Step [2] forms the six-membered ring.

◆ Deprotonation in Step [3] forms the neutral cyclic hemiacetal.

Intramolecular cyclization of a hydroxy aldehyde forms a **hemiacetal with a new stereogenic center, so that an equal amount of two enantiomers** results.

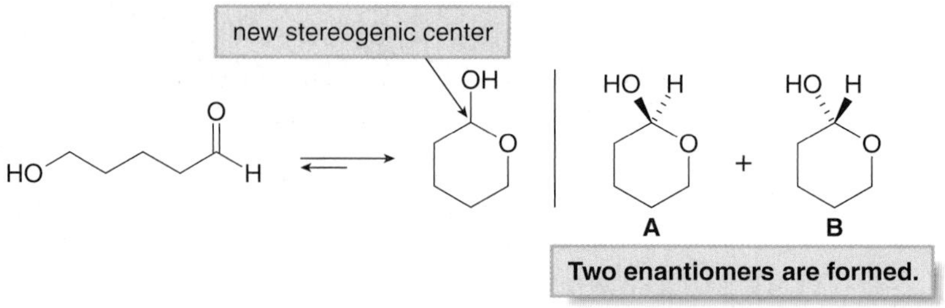

Two enantiomers are formed.

Redrawing the starting material and products in a three-dimensional representation results in the following:

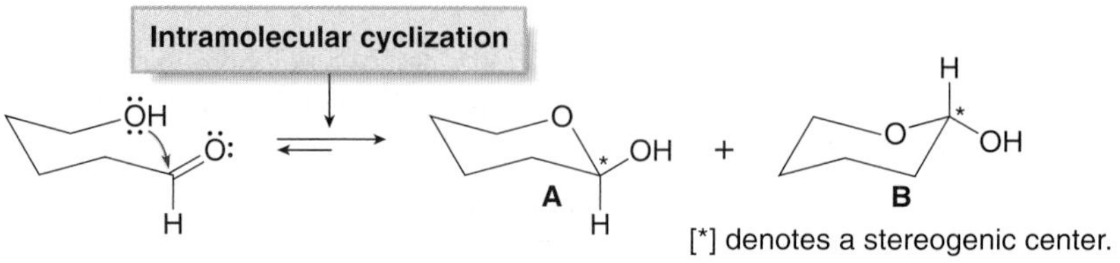

Intramolecular cyclization

[*] denotes a stereogenic center.

21.16B   The Conversion of Hemiacetals to Acetals

Cyclic hemiacetals can be converted to acetals by treatment with an alcohol and acid. This reaction converts the OH group that is part of the hemiacetal to an OR group.

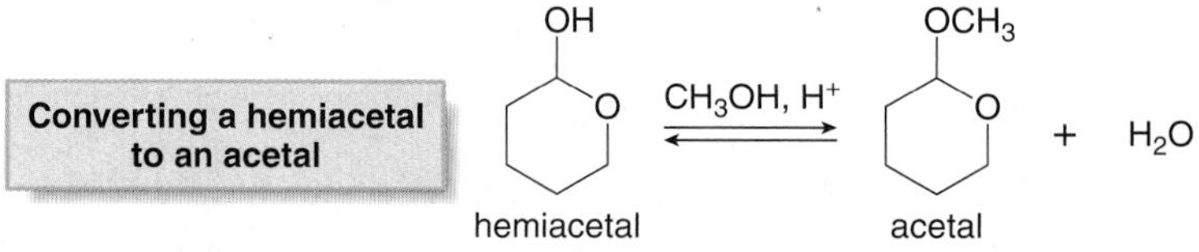

Converting a hemiacetal to an acetal

CH_3OH, H^+

hemiacetal   acetal $+$ H_2O

Mechanism 21.12 for this reaction is identical to Mechanism 21.10, which illustrates the conversion of an acyclic hemiacetal to an acetal.

MECHANISM 21.12

A Cyclic Acetal from a Cyclic Hemiacetal

Steps [1]–[2]   Protonation and loss of the leaving group

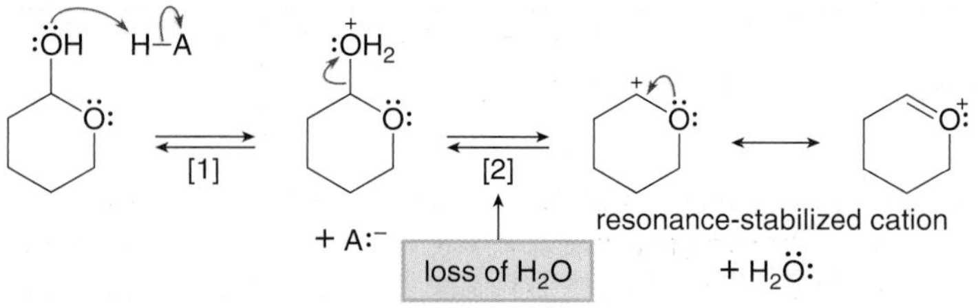

resonance-stabilized cation

loss of H_2O

$+ H_2\ddot{O}:$

◆ Protonation of the OH group followed by **loss of H_2O** forms a resonance-stabilized cation (Steps [1] and [2]).

Steps [3]–[4]   Nucleophilic attack and deprotonation

This O atom comes from CH_3OH.

nucleophilic attack

◆ **Nucleophilic attack of CH_3OH** followed by deprotonation forms the acetal (Steps [3] and [4]). The mechanism illustrates that the O atom in the OCH_3 group comes from CH_3OH.

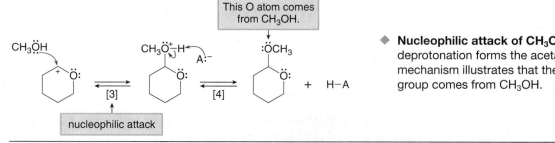

The overall result of this reaction is the **replacement of the hemiacetal OH group by an OCH$_3$ group.** This substitution reaction readily occurs because the carbocation formed in Step [2] is stabilized by resonance. This fact makes the OH group of a hemiacetal different from the hydroxy group in other alcohols.

Thus, when a compound that contains both an alcohol OH group and a hemiacetal OH group is treated with an alcohol and acid, only the hemiacetal OH group reacts to form an acetal. The alcohol OH group does *not* react.

> The conversion of cyclic hemiacetals to acetals is an important reaction in carbohydrate chemistry, as discussed in Chapter 27.

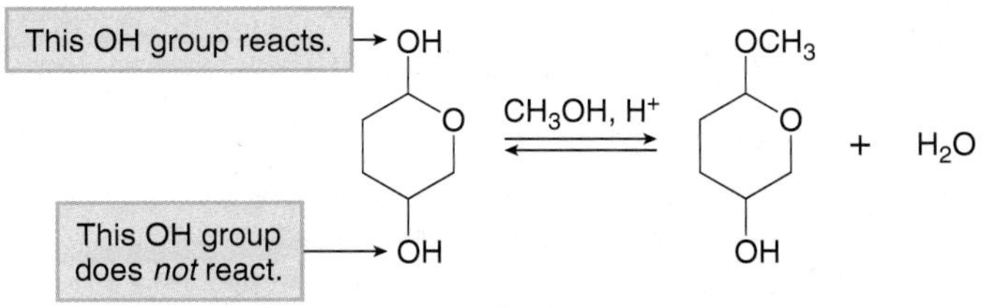

PROBLEM 21.37 Draw the products of each reaction.

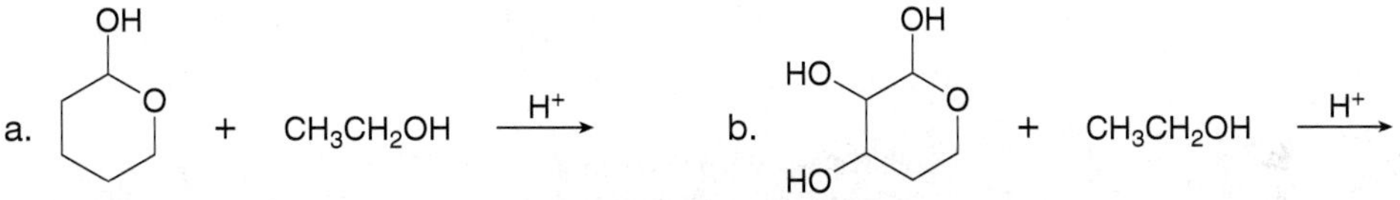

PROBLEM 21.38 Two naturally occurring compounds that contain stable cyclic hemiacetals and acetals are monensin and paeoniflorin. Monensin, a polyether antibiotic produced by *Streptomyces cinamonensis,* is used as an additive in cattle feed. Paeoniflorin, a sedative, anticoagulant, and anti-inflammatory agent, is obtained from the root of the white peony, an herb known in China for its medicinal properties since the time of Confucius. Label each acetal and hemiacetal in both compounds.

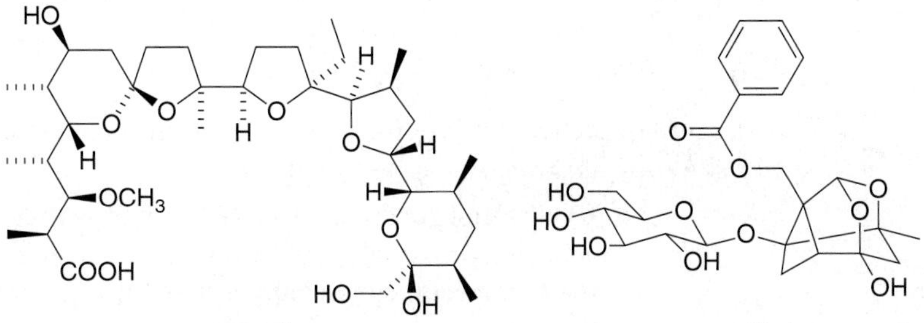

monensin   paeoniflorin

21.17   An Introduction to Carbohydrates

> Glucose is the carbohydrate that is transported in the blood to be metabolized by individual cells. The hormone insulin regulates the level of glucose in the blood. Diabetes is a common disease that results from a deficiency of insulin, resulting in increased glucose levels in the blood and other metabolic abnormalities. Insulin injections control glucose levels.

Carbohydrates, commonly referred to as sugars and starches, are polyhydroxy aldehydes and ketones, or compounds that can be hydrolyzed to them. Along with proteins, fatty acids, and nucleotides, they form one of the four main groups of biomolecules responsible for the structure and function of all living cells.

Many carbohydrates contain cyclic acetals or hemiacetals. Examples include **glucose,** the most common simple sugar, and **lactose,** the principal carbohydrate in milk.

3-D structure

OH

HO—
  HO—   O
      HO   OH

β-D-glucose
(one form of glucose)

hemiacetal

OH   acetal
HO—
HO   O
      HO   O   OH
         HO   HO   3-D structure
            H

lactose

HO
     OH   hemiacetal

Hemiacetals in sugars are formed in the same way that other hemiacetals are formed—that is, by **cyclization of hydroxy aldehydes.** Thus, the hemiacetal of glucose is formed by cyclization of an acyclic *poly*hydroxy aldehyde (**A**), as shown in the accompanying equation. This process illustrates two important features.

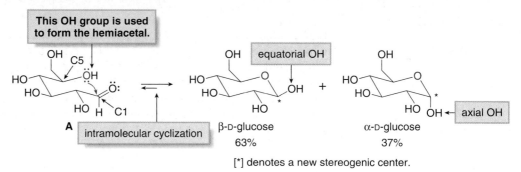

This OH group is used to form the hemiacetal.

OH   C5
HO—   ÖH
  HO—   Ö:
      HO H C1

A   intramolecular cyclization

OH   equatorial OH
HO—   O
  HO—   OH
        HO   *

β-D-glucose
63%

+

OH
HO—   O
  HO—   *
       HO   OH ← axial OH

α-D-glucose
37%

[*] denotes a new stereogenic center.

◆ When the OH group on C5 is the nucleophile, **cyclization yields a six-membered ring,** and this ring size is preferred.

◆ **Cyclization forms a new stereogenic center,** exactly analogous to the cyclization of the simpler hydroxy aldehyde (5-hydroxypentanal) in Section 21.16A. **The new OH group of the hemiacetal can occupy either the equatorial or axial position.**

For glucose, this results in two cyclic forms, called **β-D-glucose** (having an equatorial OH group) and **α-D-glucose** (having an axial OH group). Because β-D-glucose has the new OH group in the more roomy equatorial position, this cyclic form of glucose is the major product. At equilibrium, only a trace of the acyclic hydroxy aldehyde **A** is present.

Many more details on this process and other aspects of carbohydrate chemistry are presented in Chapter 27.

PROBLEM 21.39   How many stereogenic centers are present in β-D-glucose? What type of isomer are α- and β-D-glucose? How are **A** and β-D-glucose related—stereoisomers, constitutional isomers, or not isomers of each other?

PROBLEM 21.40

OH
HO
        O
HO
     HO
        OH

α-D-galactose

a. Label the hemiacetal carbon in α-D-galactose.
b. Draw the structure of β-D-galactose.
c. Draw the structure of the polyhydroxy aldehyde that cyclizes to α- and β-D-galactose.
d. From what you learned in Section 21.16B, what product(s) is (are) formed when α-D-galactose is treated with CH_3OH and an acid catalyst?

21.18   Key Concepts—Aldehydes and Ketones—Nucleophilic Addition

General Facts

- Aldehydes and ketones contain a carbonyl group bonded only to H atoms or R groups. The carbonyl carbon is sp^2 hybridized and trigonal planar (21.1).
- Aldehydes are identified by the suffix -*al*, whereas ketones are identified by the suffix -*one* (21.2).
- Aldehydes and ketones are polar compounds that exhibit dipole–dipole interactions (21.3).

Summary of Spectroscopic Absorptions of RCHO and R$_2$CO (21.4)

IR absorptions	C=O	~1715 cm^{-1} for ketones (increasing frequency with decreasing ring size)
		~1730 cm^{-1} for aldehydes
		• For both RCHO and R$_2$CO, the frequency decreases with conjugation.
	C_{sp^2}–H of CHO	~2700–2830 cm^{-1} (one or two peaks)
^{1}H NMR absorptions	CHO	9–10 ppm (highly deshielded proton)
	C–H α to C=O	2–2.5 ppm (somewhat deshielded C_{sp^3}–H)
^{13}C NMR absorption	C=O	190–215 ppm

Nucleophilic Addition Reactions

[1] Addition of hydride (H$^-$) (21.8)

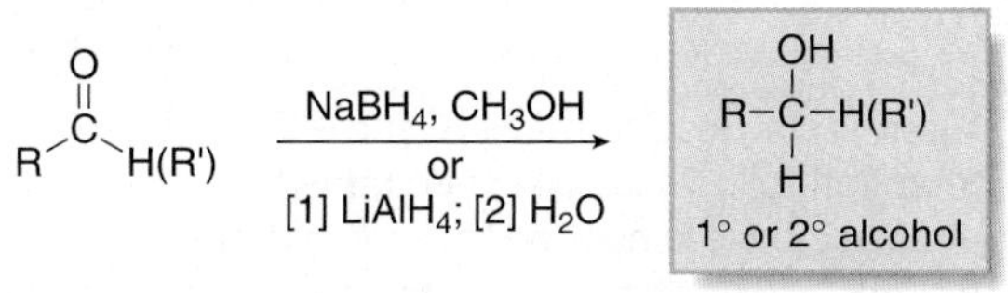

- The mechanism has two steps.
- H:$^-$ adds to the planar C=O from both sides.

[2] Addition of organometallic reagents (R$^-$) (21.8)

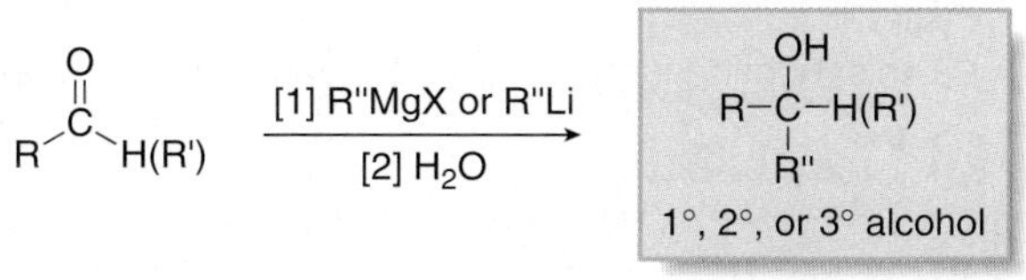

- The mechanism has two steps.
- (R″)$^-$ adds to the planar C=O from both sides.

[3] Addition of cyanide ($^-$CN) (21.9)

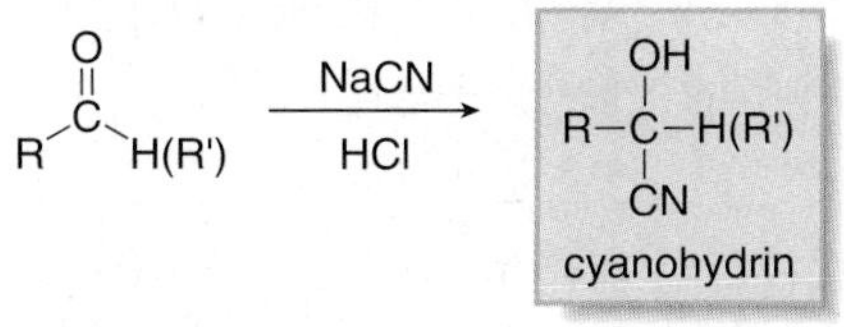

- The mechanism has two steps.
- $^-$CN adds to the planar C=O from both sides.

[4] Wittig reaction (21.10)

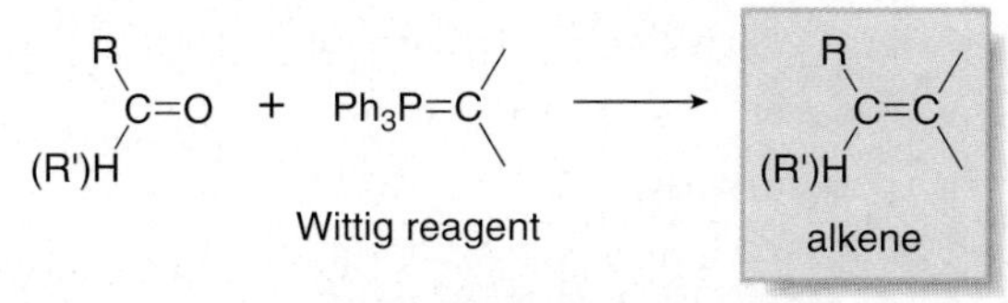

- The reaction forms a new C–C σ bond and a new C–C π bond.
- Ph$_3$P=O is formed as by-product.

[5] Addition of 1° amines (21.11)

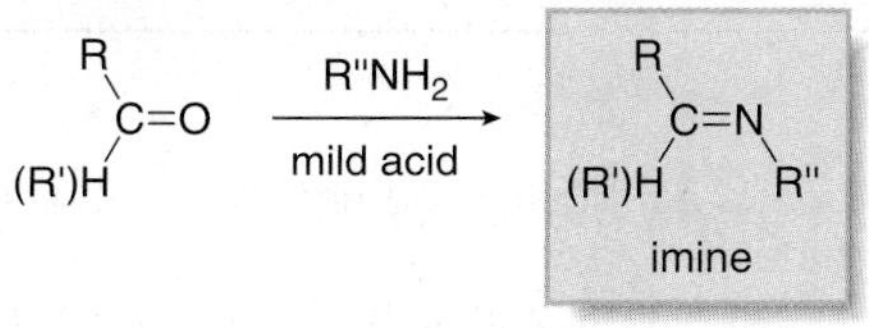

- The reaction is fastest at pH 4–5.
- The intermediate carbinolamine is unstable, and loses H$_2$O to form the C=N.

[6] Addition of 2° amines (21.12)

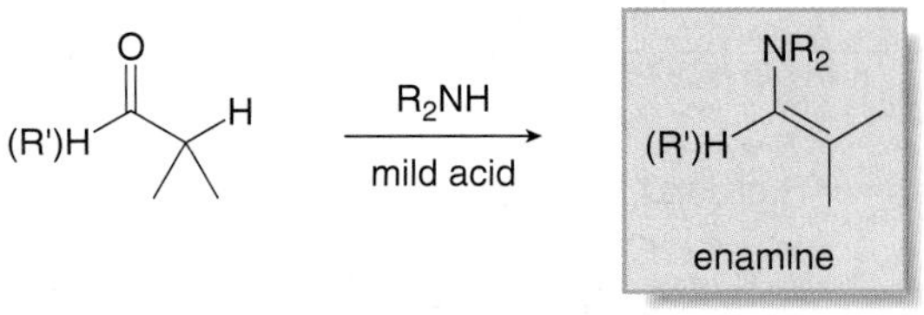

- The reaction is fastest at pH 4–5.
- The intermediate carbinolamine is unstable, and loses H_2O to form the $C=C$.

[7] Addition of H_2O—Hydration (21.13)

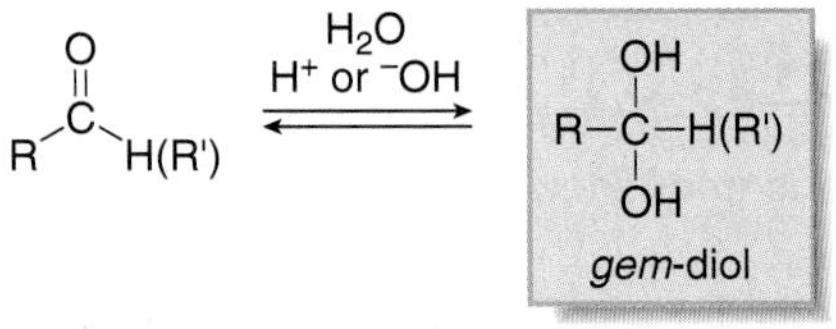

- The reaction is reversible. Equilibrium favors the product only with less stable carbonyl compounds (e.g., H_2CO and Cl_3CCHO).
- The reaction is catalyzed by either H^+ or ^-OH.

[8] Addition of alcohols (21.14)

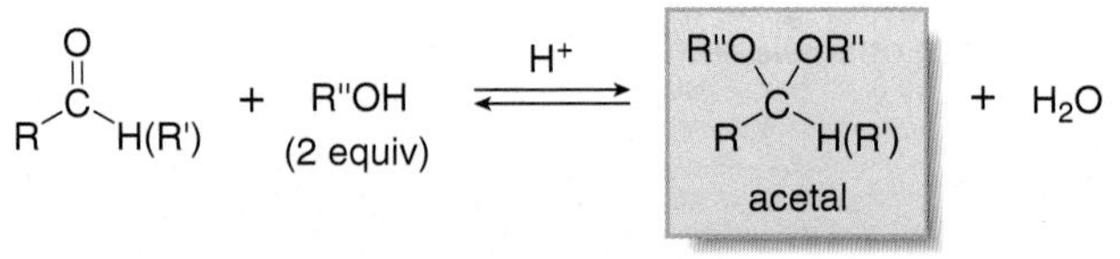

- The reaction is reversible.
- The reaction is catalyzed by acid.
- Removal of H_2O drives the equilibrium to favor the products.

Other Reactions

[1] Synthesis of Wittig reagents (21.10A)

$$RCH_2X \xrightarrow[\text{[2] Bu—Li}]{\text{[1] Ph}_3\text{P:}} Ph_3P=CHR$$

- Step [1] is best with CH_3X and RCH_2X because the reaction follows an S_N2 mechanism.
- A strong base is needed for proton removal in Step [2].

[2] Conversion of cyanohydrins to aldehydes and ketones (21.9)

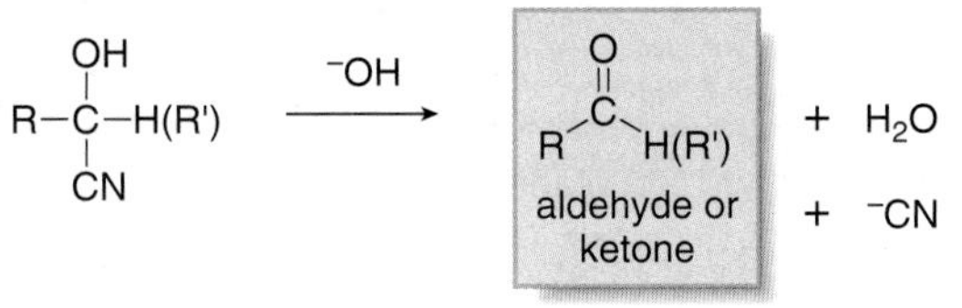

- This reaction is the reverse of cyanohydrin formation.

[3] Hydrolysis of nitriles (21.9)

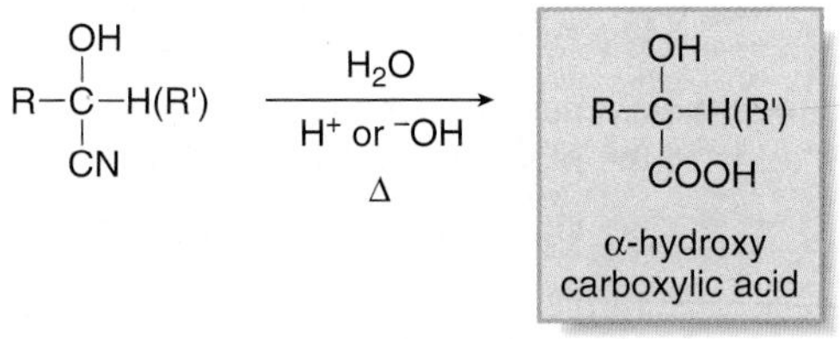

[4] Hydrolysis of imines and enamines (21.12)

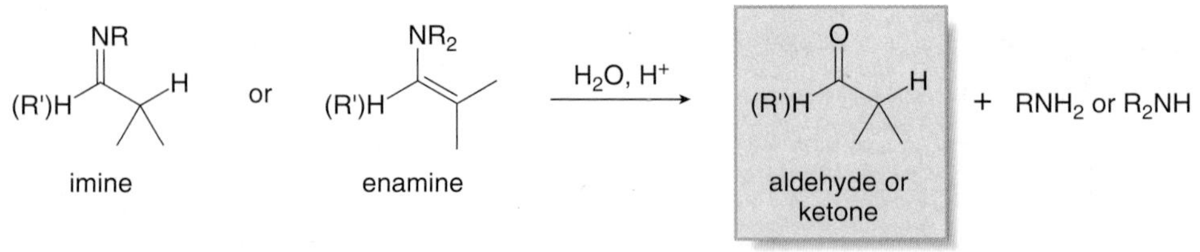

[5] Hydrolysis of acetals (21.14)

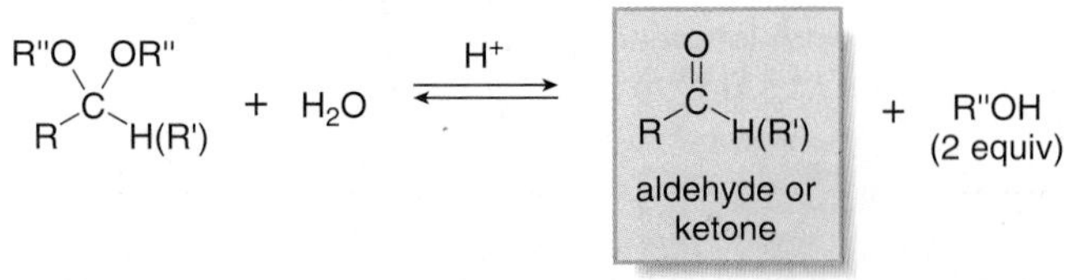

- The reaction is acid catalyzed and is the reverse of acetal synthesis.
- A large excess of H_2O drives the equilibrium to favor the products.

Problems

Nomenclature

21.41 Give the IUPAC name for each compound.

a. $(CH_3)_3CCH_2CHO$

b. 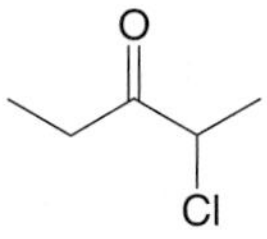 [structure: pentan-3-one with Cl substituent]

c. 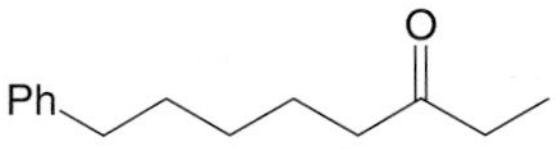 Ph [chain structure with ketone]

d. 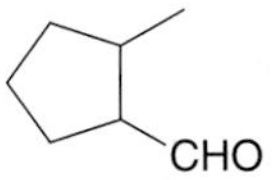 [cyclopentane with methyl and CHO substituents]

e. 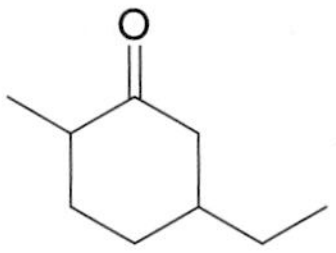 [cyclohexanone with methyl and ethyl substituents]

f. 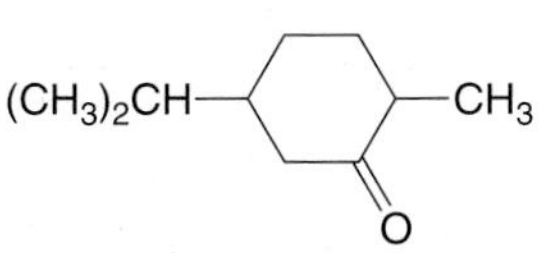 $(CH_3)_2CH$—[cyclohexanone with CH_3]

g. 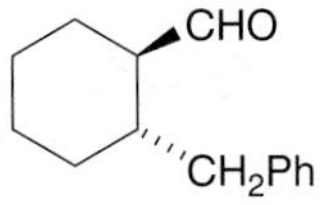 [cyclohexane with CHO and CH_2Ph substituents]

h. 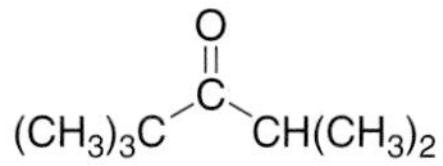 $(CH_3)_3C$—C(=O)—$CH(CH_3)_2$

i. 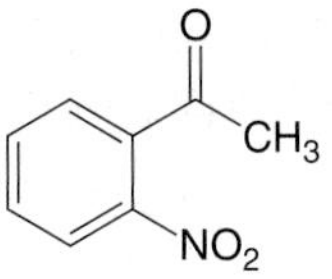 [benzene ring with C(=O)CH_3 and NO_2 substituents]

j. 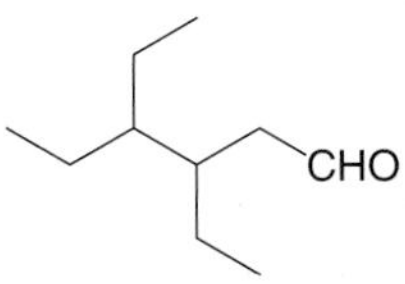 [branched chain with ethyl groups and CHO]

21.42 Give the structure corresponding to each name.
   a. 2-methyl-3-phenylbutanal
   b. dipropyl ketone
   c. 3,3-dimethylcyclohexanecarbaldehyde
   d. α-methoxypropionaldehyde
   e. 3-benzoylcyclopentanone
   f. 2-formylcyclopentanone
   g. (R)-3-methyl-2-heptanone
   h. m-acetylbenzaldehyde

21.43 Ignoring stereoisomers, draw the six ketones and eight aldehydes having molecular formula $C_6H_{12}O$, and give the IUPAC name for each compound.

Reactions

21.44 Draw the product formed when phenylacetaldehyde ($C_6H_5CH_2CHO$) is treated with each reagent.
   a. $NaBH_4$, CH_3OH
   b. [1] $LiAlH_4$; [2] H_2O
   c. [1] CH_3MgBr; [2] H_2O
   d. NaCN, HCl
   e. $Ph_3P=CHCH_3$
   f. $(CH_3)_2CHNH_2$, mild acid
   g. $(CH_3CH_2)_2NH$, mild acid
   h. CH_3CH_2OH (excess), H^+
   i. 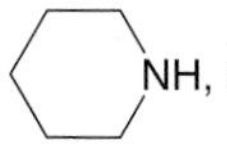 [piperidine] NH, mild acid
   j. $HOCH_2CH_2OH$, H^+

21.45 Answer Problem 21.44 using 2-butanone ($CH_3COCH_2CH_3$) as starting material.

21.46 Draw the products formed in each Wittig reaction. Draw all stereoisomers formed when a mixture of products results.

a. 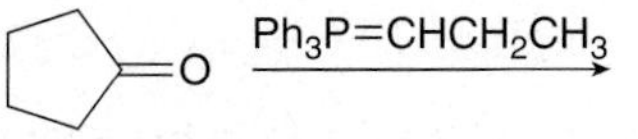 [cyclopentanone] =O $\xrightarrow{Ph_3P=CHCH_2CH_3}$

b. 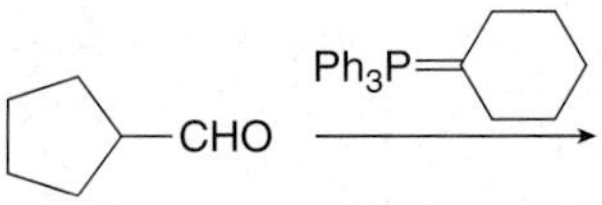 [cyclopentane]—CHO $\xrightarrow{Ph_3P=[cyclohexylidene]}$

c. 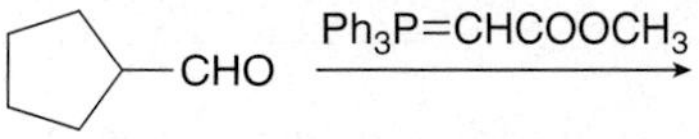 [cyclopentane]—CHO $\xrightarrow{Ph_3P=CHCOOCH_3}$

d. 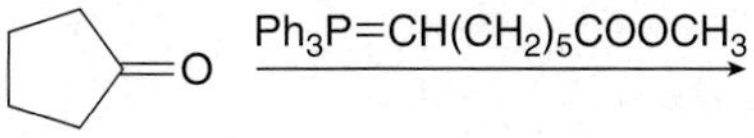 [cyclopentanone]=O $\xrightarrow{Ph_3P=CH(CH_2)_5COOCH_3}$

21.47 Draw the products formed in each reaction sequence.

a. CH_3CH_2Cl $\xrightarrow[\text{[2] BuLi}]{\text{[1] Ph}_3\text{P}}$ [3] $(CH_3)_2C=O$

b. $C_6H_5CH_2Br$ $\xrightarrow[\text{[2] BuLi}]{\text{[1] Ph}_3\text{P}}$ [3] $C_6H_5CH_2CH_2CHO$

c. 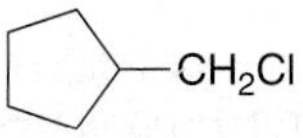 [cyclopentane]—CH_2Cl $\xrightarrow[\text{[2] BuLi}]{\text{[1] Ph}_3\text{P}}$ [3] $CH_3CH_2CH_2CHO$

21.48 What alkyl halide is needed to prepare each Wittig reagent?
   a. $Ph_3P=CHCH_2CH_2CH_3$
   b. $Ph_3P=C(CH_2CH_2CH_3)_2$
   c. $Ph_3P=CHCH=CH_2$

21.49 Fill in the lettered reagents (**A–G**) in the following reaction scheme.

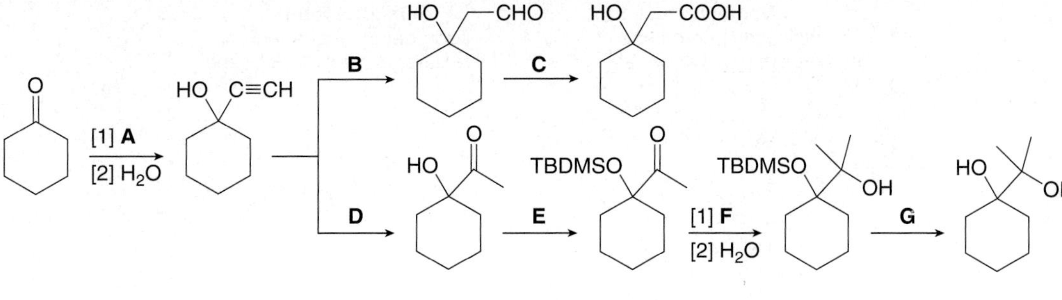

[reaction scheme: cyclohexanone with [1] A, [2] H_2O → 1-ethynylcyclohexanol (HO, C≡CH); branching to **B** → HO—CH₂—CHO cyclohexane → **C** → HO—CH₂—COOH cyclohexane; and **D** → HO—C(=O) cyclohexane → **E** → TBDMSO—C(=O) cyclohexane → [1] **F**, [2] H_2O → TBDMSO—C(OH) cyclohexane → **G** → HO—C(OH) cyclohexane]

21.50 Draw the products of each reaction.

a. 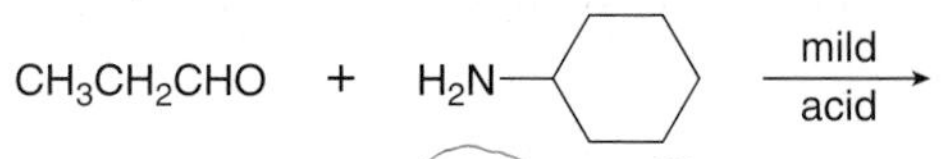 CH₃CH₂CHO + H₂N—⬡ →(mild acid)

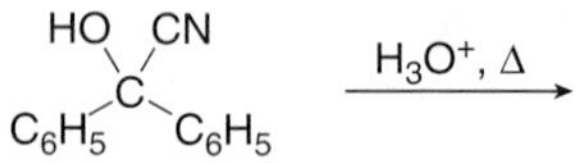

b. 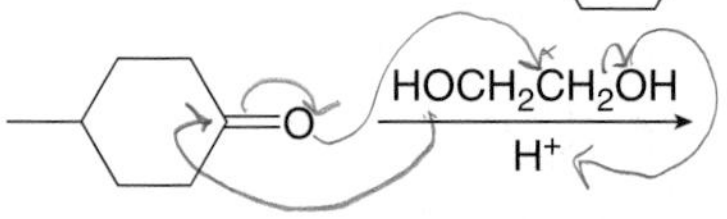 (cyclohexanone) + HOCH₂CH₂OH / H⁺

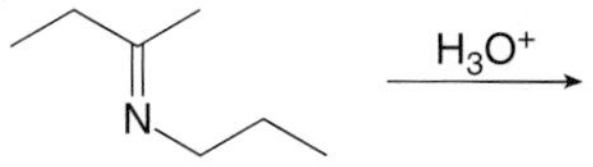

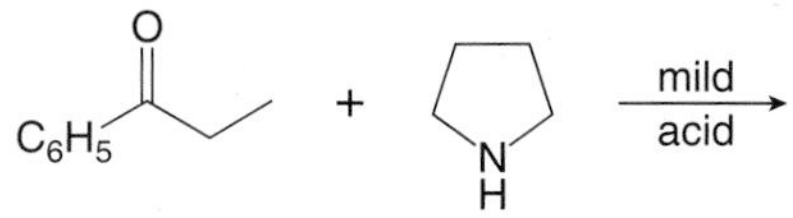

c. (imine) →(H₃O⁺)

d. C₆H₅—C(=O)—CH₂CH₃ + (pyrrolidine) →(mild acid)

e. HO—C(CN)(C₆H₅)(C₆H₅) →(H₃O⁺, Δ)

f. 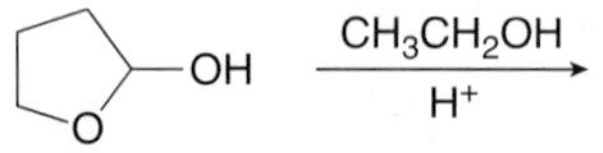 (tetrahydrofuran-2-ol) →(CH₃CH₂OH / H⁺)

g. 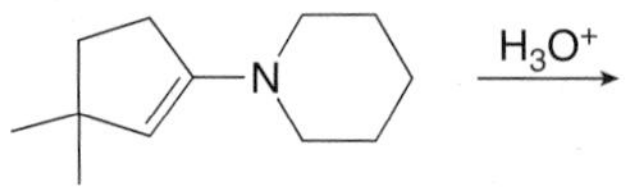 (cyclopentene with piperidine) →(H₃O⁺)

h. CH₃O—⬡(OCH₃)(OCH₃) →(H₃O⁺)

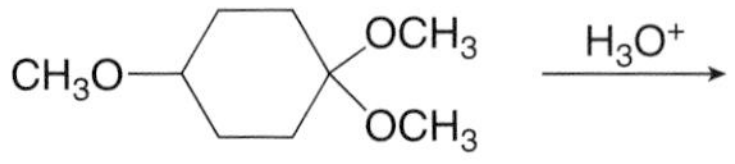

21.51 What carbonyl compound and alcohol are formed by hydrolysis of each acetal?

a. 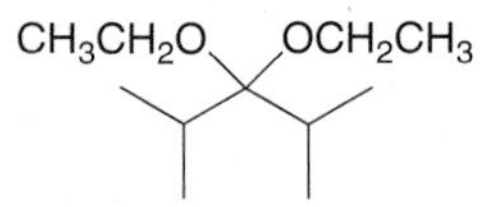 CH₃CH₂O OCH₂CH₃ (acetal)

b. 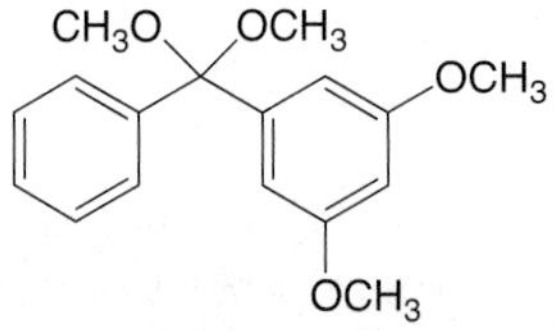 CH₃O OCH₃ / OCH₃ aromatic acetal with OCH₃ groups

c. 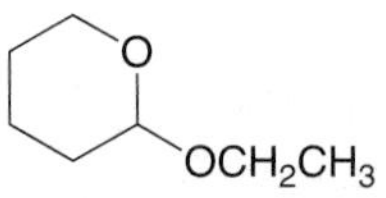 (tetrahydropyran) OCH₂CH₃

21.52 Identify the lettered intermediates in the following reaction sequence. When a mixture of ortho and para products results in electrophilic aromatic substitution, consider the para product only. The ¹H NMR spectrum of **G** shows two singlets at 2.6 and 8.18 ppm.

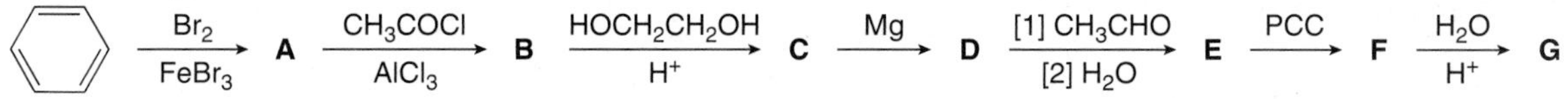

⬡ →(Br₂ / FeBr₃) **A** →(CH₃COCl / AlCl₃) **B** →(HOCH₂CH₂OH / H⁺) **C** →(Mg) **D** →([1] CH₃CHO [2] H₂O) **E** →(PCC) **F** →(H₂O / H⁺) **G**

21.53 Draw all stereoisomers formed in each reaction.

a. 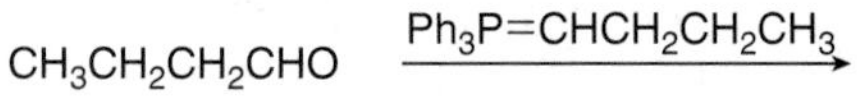 CH₃CH₂CH₂CHO →(Ph₃P=CHCH₂CH₂CH₃)

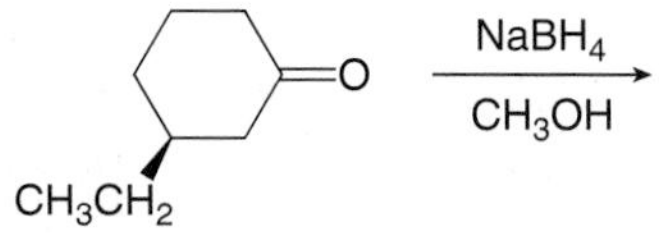

b. 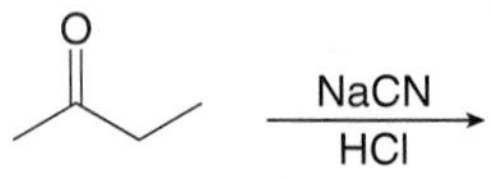 (ketone) →(NaCN / HCl)

c. (cyclohexanone with CH₃CH₂ substituent) →(NaBH₄ / CH₃OH)

d. 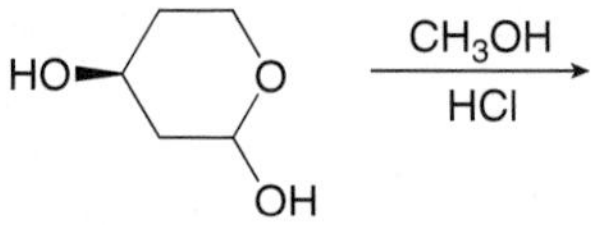 HO—(tetrahydropyran)—OH →(CH₃OH / HCl)

21.54 Hydroxy aldehydes **A** and **B** readily cyclize to form hemiacetals. Draw the stereoisomers formed in this reaction from both **A** and **B**. Explain why this process gives an optically inactive product mixture from **A**, and an optically active product mixture from **B**.

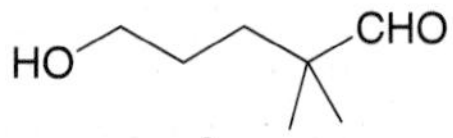

HO—(chain)—CHO
A

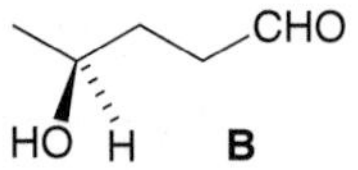

HO H—(chain)—CHO
B

21.55 Frontalin and multistriatin are two insect pheromones that contain a cyclic acetal. What hydrolysis products are formed when each compound is treated with aqueous acid?

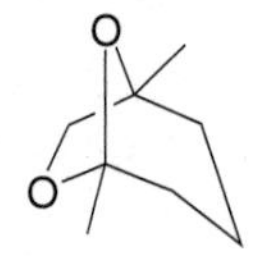

frontalin
communication pheromone of
the western pine bark beetle

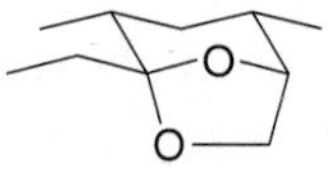

multistriatin
aggregating pheromone of
the European elm bark beetle

Properties of Aldehydes and Ketones

21.56 Rank the carbonyl compounds in each group in order of increasing stability. Which compound in each group forms the highest percentage of hydrate at equilibrium?

a.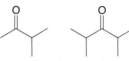

b.

21.57 Rank the compounds in each group in order of increasing reactivity in nucleophilic addition.

a.

b.

21.58 Explain why a *gem*-diol is the major species present at equilibrium when cyclopropanone is dissolved in H_2O.

Synthesis

21.59 What Wittig reagent and carbonyl compound are needed to prepare each alkene? When two routes are possible, indicate which route, if any, is preferred.

a. $(CH_3CH_2)_2C=CHCH_2CH_2CH_3$

b.

c.

d.

21.60 What amine and carbonyl compound are needed to prepare each product?

a.

b.

c.

d.

21.61 What carbonyl compound and alcohol (or diol) are needed to synthesize each acetal?

a. $(CH_3CH_2)_2C(OCH_2CH_3)_2$

b.

c.

d.

21.62 What reagents are needed to convert each compound to benzaldehyde (C_6H_5CHO)? More than one step may be required.
a. $C_6H_5CH_2OH$
b. C_6H_5COCl
c. $C_6H_5COOCH_3$
d. C_6H_5COOH
e. $C_6H_5CH_3$
f. $C_6H_5CH=CH_2$
g. $C_6H_5CH=NCH_2CH_2CH_3$
h. $C_6H_5CH(OCH_2CH_3)_2$

21.63 What reagents are needed to convert each compound to 2-butanone ($CH_3COCH_2CH_3$)?

a.

b.

c. CH_3COCl

d. $CH_3CH_2C\equiv CH$

e. $CH_3C\equiv CCH_3$

21.64 Show two different methods to carry out each transformation: a one-step method using a Wittig reagent, and a two-step method using a Grignard reagent. Which route, if any, is preferred for each compound?

a.

b.

21.65 Devise a synthesis of each compound from the given starting material. You may use any other organic or inorganic reagents you choose.

a.

b.

c. $CH_3-\overset{\overset{\displaystyle OCH_2CH_3}{|}}{\underset{\underset{\displaystyle CH_2CH_3}{|}}{C}}-OCH_2CH_3 \implies CH_3CH_2OH$ (the only source of carbon atoms)

d.

e. CH_3O—⟨⟩—$CH=CH$—⟨⟩—$C(CH_3)_3$ and

Protecting Groups

21.66 Design a stepwise synthesis to convert cyclopentanone and 4-bromobutanal to hydroxy aldehyde **A.**

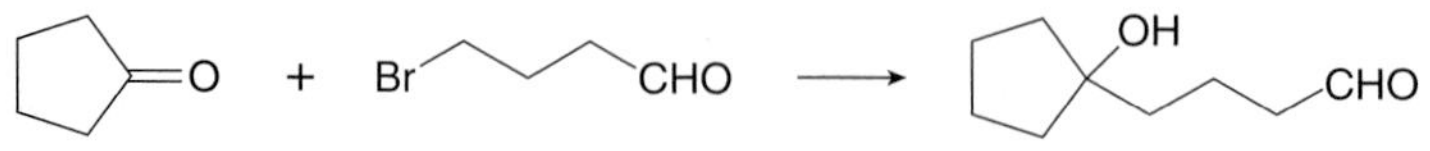

cyclopentanone   4-bromobutanal   **A**

21.67 The ketone carbonyl in **B** is more reactive towards nucleophilic attack than its α,β-unsaturated carbonyl. (a) Explain this difference in reactivity. (b) Design a synthesis of both **C** and **D** from **B**.

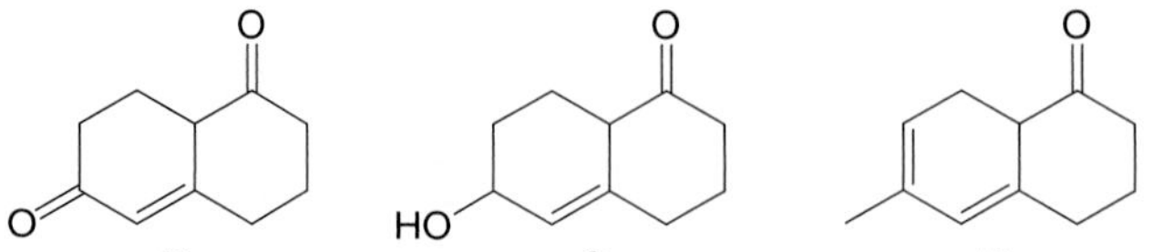

B   **C**   **D**

Mechanism

21.68 Draw a stepwise mechanism for the following reaction.

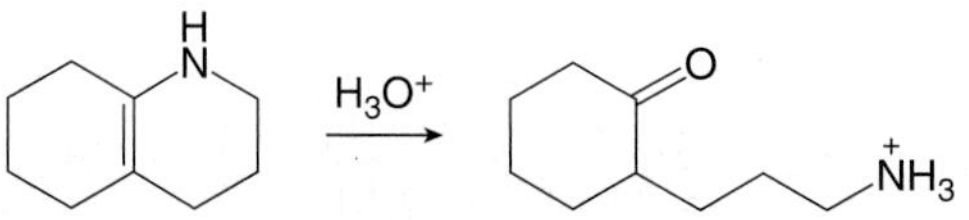

21.69 When acetone is dissolved in aqueous acid containing isotopically labeled H_2O ($H_2{}^{18}O$), the carbonyl group becomes labeled with ^{18}O. Draw a mechanism that explains this observation.

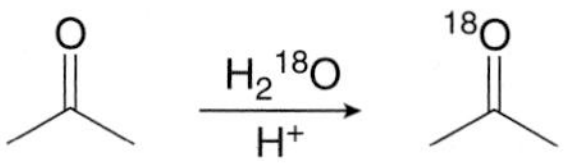

21.70 Draw a stepwise mechanism for each hydrolysis.

a. 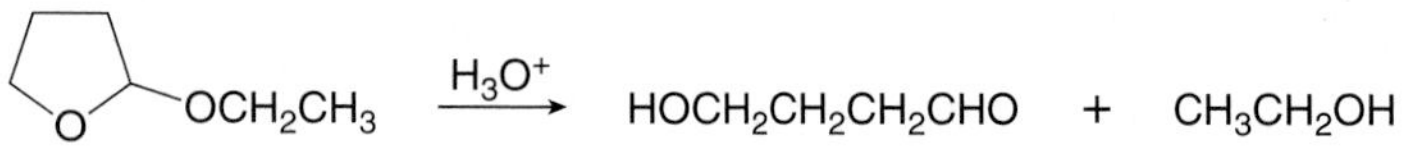 $\xrightarrow{H_3O^+}$ $HOCH_2CH_2CH_2CHO$ + CH_3CH_2OH

b.

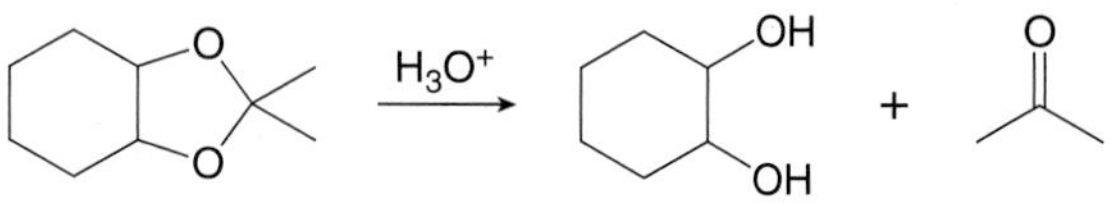

21.71 Another way to synthesize cyclic acetals uses enol ethers (not carbonyl compounds) as starting materials. Draw a stepwise mechanism for the following synthesis of an acetal from an enol ether and ethylene glycol.

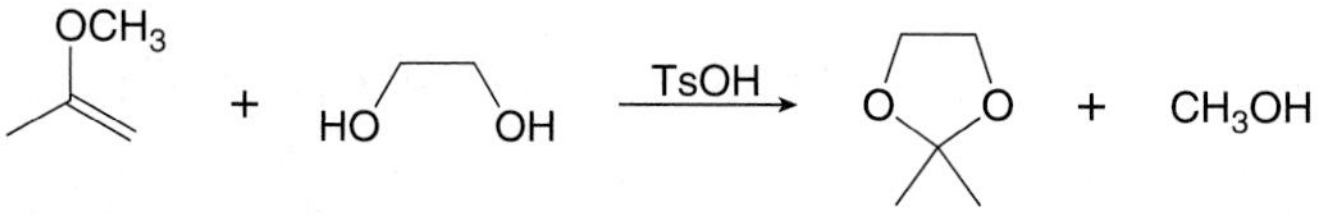

enol ether   acetal

21.72 A two-step method to reduce a C=O to a CH_2 group involves the conversion of a carbonyl group to a tosylhydrazone, followed by reduction with $NaBH_4$. Draw a mechanism for the conversion of cyclohexanone to its tosylhydrazone, which is illustrated in the following equation.

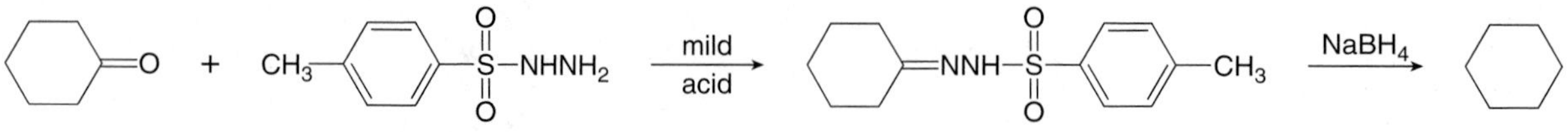

tosylhydrazine   tosylhydrazone

21.73 Sulfur ylides, like Wittig reagents, are useful intermediates in organic synthesis. Sulfur ylides are formed by the treatment of sulfonium salts with butyllithium. They react with carbonyl compounds to form epoxides. Draw the mechanism for formation of epoxide **X** from cyclohexanone using a sulfur ylide.

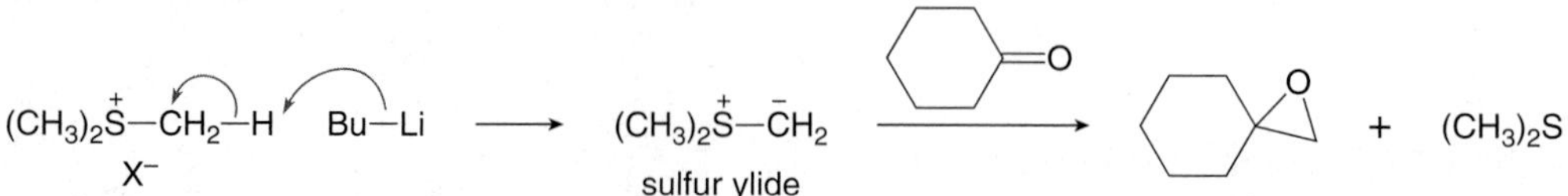

sulfonium salt   sulfur ylide   **X**

+ Bu—H + LiX

21.74 Explain how NaBH$_4$ in CH$_3$OH can reduce hemiacetal **A** to 1,4-butanediol (HOCH$_2$CH$_2$CH$_2$CH$_2$OH).

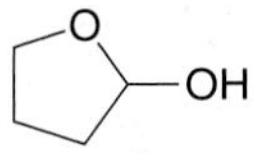

A

Spectroscopy

21.75 How would the compounds in each pair differ in their IR spectra?

a. 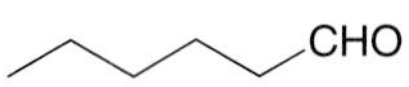   ⟍⟍⟍⟍CHO

and

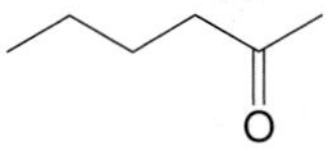

b. 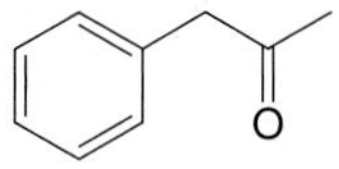   and

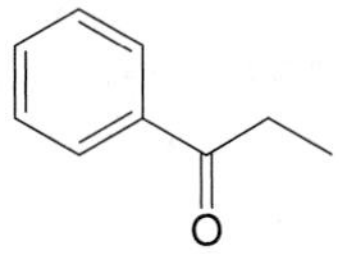

c. 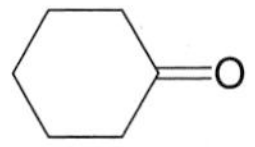   and

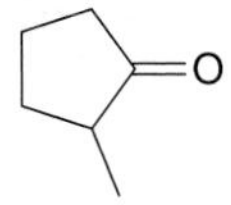

21.76 Use the ^{1}H NMR and IR data to determine the structure of each compound.

Compound **A**	Molecular formula:	C$_5$H$_{10}$O
	IR absorptions at	1728, 2791, and 2700 cm^{-1}
	^{1}H NMR data:	1.08 (singlet, 9 H) and 9.48 (singlet, 1 H) ppm
Compound **B**	Molecular formula:	C$_5$H$_{10}$O
	IR absorption at	1718 cm^{-1}
	^{1}H NMR data:	1.10 (doublet, 6 H), 2.14 (singlet, 3 H), and 2.58 (septet, 1 H) ppm
Compound **C**	Molecular formula:	C$_{10}$H$_{12}$O
	IR absorption at	1686 cm^{-1}
	^{1}H NMR data:	1.21 (triplet, 3 H), 2.39 (singlet, 3 H), 2.95 (quartet, 2 H), 7.24 (doublet, 2 H), and 7.85 (doublet, 2 H) ppm
Compound **D**	Molecular formula:	C$_{10}$H$_{12}$O
	IR absorption at	1719 cm^{-1}
	^{1}H NMR data:	1.02 (triplet, 3 H), 2.45 (quartet, 2 H), 3.67 (singlet, 2 H), and 7.06–7.48 (multiplet, 5 H) ppm

21.77 Compounds **A** and **B** have molecular formula C$_9$H$_{10}$O. Identify their structures from the ^{1}H NMR and IR spectra given.

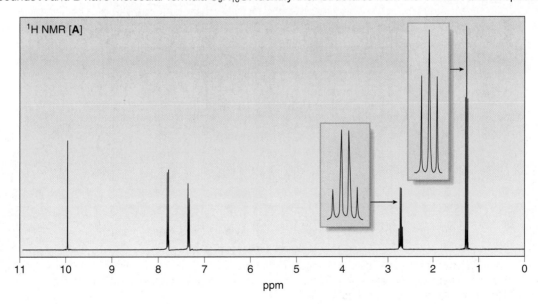

Problem continued on next page.

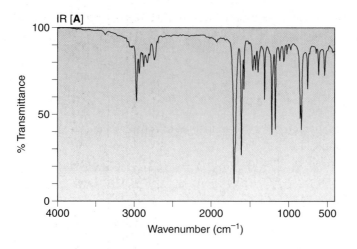

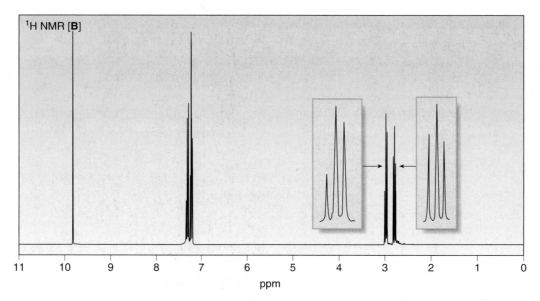

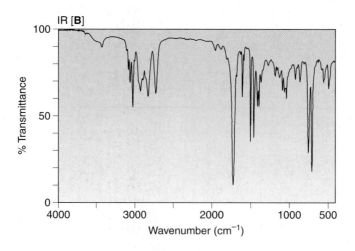

Carbohydrates

21.78 Draw the structure of the acyclic polyhydroxy aldehyde that cyclizes to each hemiacetal.

a.

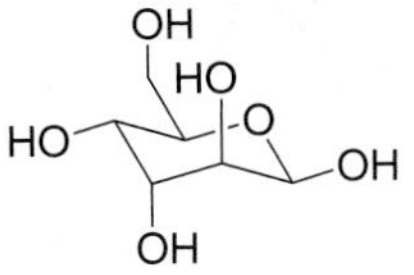

b.

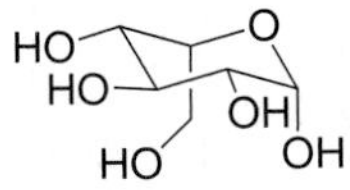

21.79 β-D-Glucose, a hemiacetal, can be converted to an acetal on treatment with CH_3OH in the presence of acid. Draw a stepwise mechanism for this reaction.

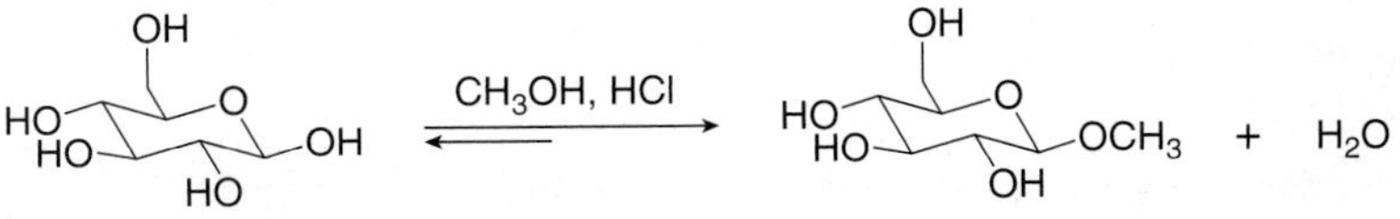

β-D-glucose   acetal

Challenge Problems

21.80 One variation of the Wittig reaction converts aldehydes into allylic alcohols having trisubstituted double bonds. This reaction sequence is outlined below, and **A, B,** and **C** are intermediates (not stable products) formed along the way. Assign structures to **A, B,** and **C.**

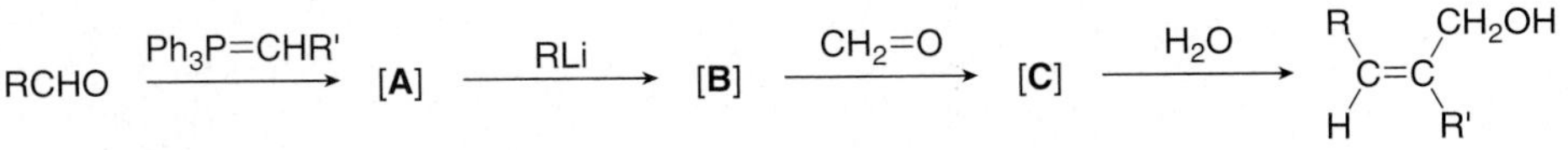

21.81 Draw a stepwise mechanism for the following reaction.

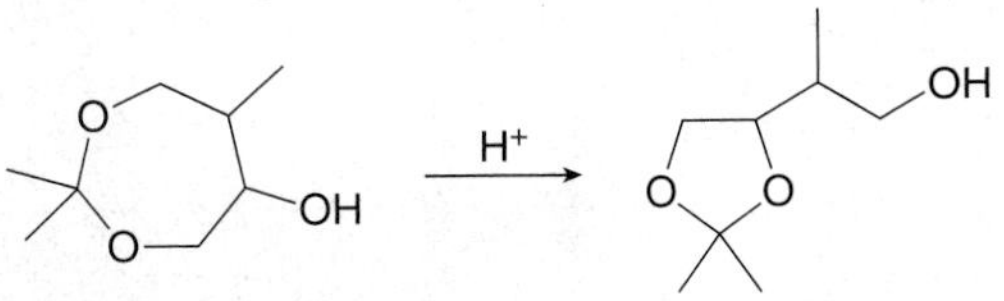

21.82 Identify the lettered intermediates in the following reaction sequence. Work Problem 21.73 before attempting this problem.

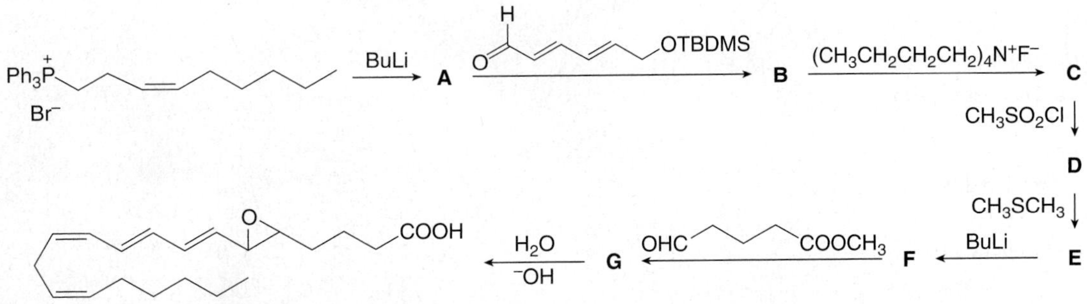

leukotriene
LTA$_4$
(Section 9.16B)

Carboxylic Acids and Their Derivatives— Nucleophilic Acyl Substitution

The serendipitous discovery of **penicillin** by Scottish bacteriologist Sir Alexander Fleming in 1928 is considered one of the single most important events in the history of medicine. **Penicillin G** and related compounds are members of the β-lactam family of antibiotics. Although each β-lactam is active against certain types of bacteria, these drugs all function by interfering with bacterial cell wall synthesis. They contain a strained four-membered amide ring that is responsible for this biological activity. In Chapter 22 we learn about the chemistry of amides and other derivatives of carboxylic acids.

hapter 22 continues the study of carbonyl compounds with a detailed look at **nucleophilic acyl substitution,** a key reaction of carboxylic acids and their derivatives. Substitution at sp^2 hybridized carbon atoms was introduced in Chapter 20 with reactions involving carbon and hydrogen nucleophiles. In Chapter 22, we learn that nucleophilic acyl substitution is a general reaction that occurs with a variety of heteroatomic nucleophiles. This reaction allows the conversion of one carboxylic acid derivative into another. *Every* **reaction in Chapter 22 that begins with a carbonyl compound involves nucleophilic substitution.** Chapter 22 also discusses the properties and chemical reactions of **nitriles,** compounds that contain a carbon–nitrogen triple bond. Nitriles are in the same carbon oxidation state as carboxylic acids, and they undergo reactions that form related products.

22.1   Introduction

Chapter 22 focuses on carbonyl compounds that contain an **acyl group bonded to an electronegative atom.** These include the **carboxylic acids,** as well as carboxylic acid derivatives that can be prepared from them: **acid chlorides, anhydrides, esters, and amides.**

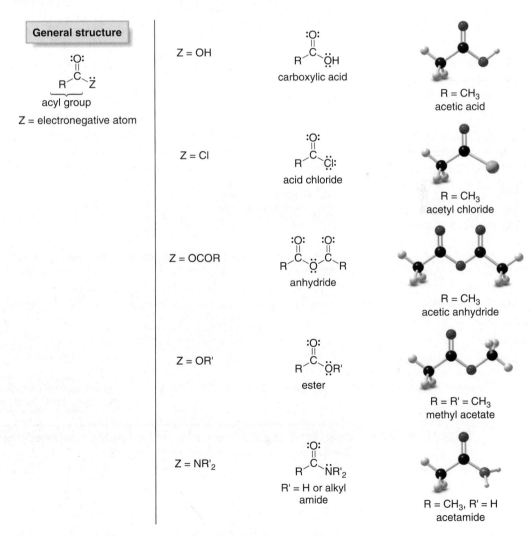

Anhydrides contain two carbonyl groups joined by a single oxygen atom. **Symmetrical anhydrides** have two identical alkyl groups bonded to the carbonyl carbons, and **mixed anhydrides** have two different alkyl groups. **Cyclic anhydrides** are also known.

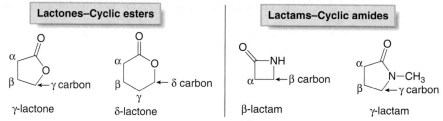

symmetrical anhydride   mixed anhydride   cyclic anhydride

Amides are classified as **1°, 2°,** or **3°** depending on the number of carbon atoms directly bonded to the *nitrogen* atom.

1° amide
1 C—N bond

2° amide
2 C—N bonds

3° amide
3 C—N bonds

Cyclic esters and amides are called **lactones** and **lactams,** respectively. The ring size of the heterocycle is indicated by a Greek letter. An amide in a four-membered ring is called a **β-lactam,** because the β carbon to the carbonyl is bonded to the heteroatom. An ester in a five-membered ring is called a **γ-lactone.**

Lactones–Cyclic esters

γ-lactone   δ-lactone

Lactams–Cyclic amides

β-lactam   γ-lactam

Nucleophilic acyl substitution was first discussed in Chapter 20 with R⁻ and H⁻ as the nucleophiles. This substitution reaction is general for a variety of nucleophiles, making it possible to form many different substitution products, as discussed in Sections 22.8–22.13.

All of these compounds contain an acyl group bonded to an electronegative atom Z that can serve as a **leaving group.** As a result, these compounds undergo **nucleophilic acyl substitution.** Recall from Chapters 20 and 21 that aldehydes and ketones do *not* undergo nucleophilic substitution because they have no leaving group on the carbonyl carbon.

Nucleophilic substitution— General reaction

Z = OH, Cl, OCOR, OR', NR'₂

Nu replaces Z.

Nitriles **are compounds that contain a cyano group, C≡N, bonded to an alkyl group.** Nitriles have no carbonyl group, so they are structurally distinct from carboxylic acids and their derivatives. The carbon atom of the cyano group, however, has the same oxidation state as the carbonyl carbon of carboxylic acid derivatives, so there are certain parallels in their chemistry.

General structure—Nitriles

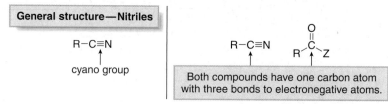

R—C≡N
cyano group

Both compounds have one carbon atom with three bonds to electronegative atoms.

PROBLEM 22.1   Dolastatin is an anticancer compound isolated from an Indian seahare *Dolabella auricularia.* Classify each of its four amides as 1°, 2°, or 3°.

dolastatin

22.2 Structure and Bonding

The two most important features of any carbonyl group, regardless of the other groups bonded to it, are the following:

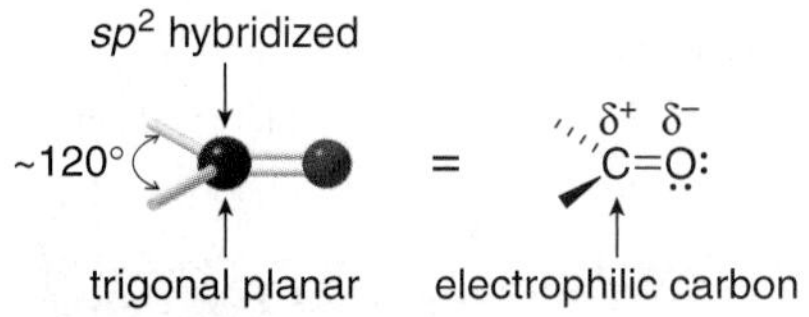

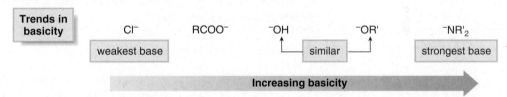

- ◆ The carbonyl carbon is sp^2 hybridized and trigonal planar, making it relatively uncrowded.
- ◆ The electronegative oxygen atom polarizes the carbonyl group, making the carbonyl carbon electrophilic.

Because carboxylic acid derivatives (RCOZ) all contain an atom Z with a nonbonded electron pair, three resonance structures can be drawn for RCOZ, compared to just two for aldehydes and ketones (Section 20.1). These three resonance structures stabilize RCOZ by delocalizing electron density. In fact, the more resonance structures **2** and **3** contribute to the resonance hybrid, the more stable RCOZ is.

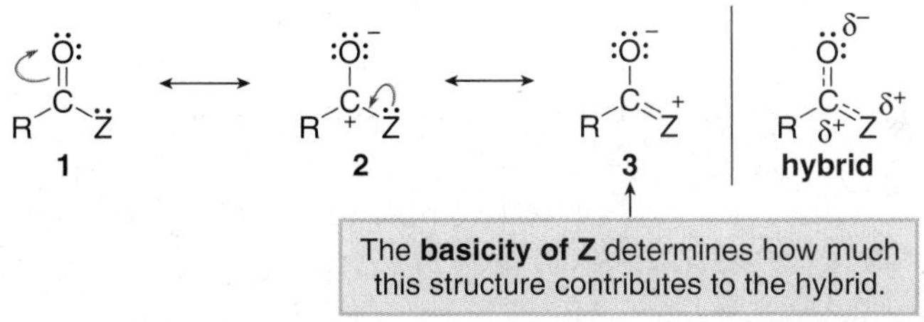

The **basicity of Z** determines how much this structure contributes to the hybrid.

- ◆ The more basic Z is, the more it donates its electron pair, and the more resonance structure 3 contributes to the hybrid.

To determine the relative basicity of the leaving group Z, we compare the pK_a values of the conjugate acids HZ, given in Table 22.1. The following order of basicity results:

Trends in basicity

| Cl^- | $RCOO^-$ | ^-OH | $^-OR'$ | $^-NR'_2$ |

weakest base   similar   strongest base

Increasing basicity →

TABLE 22.1 pK_a Values of the Conjugate Acids (HZ) for Common Z Groups of Acyl Compounds (RCOZ)

Structure	Leaving group (Z^-)	Conjugate Acid (HZ)	pK_a
RCOCl acid chloride	Cl^-	HCl	−7
RCOOCOR anhydride	$RCOO^-$	RCOOH	3–5
RCOOH carboxylic acid	^-OH	H_2O	15.7
RCOOR' ester	$^-OR'$	R'OH	15.5–18
RCONR'₂ amide	$^-NR'_2$	R'_2NH	38–40

Increasing basicity of Z ↓   Increasing acidity of HZ ↑

Figure 22.1 Electrostatic potential plots of three carboxylic acid derivatives

The electrostatic potential plots of an acid chloride, ester, and amide all show an electron-deficient carbonyl carbon (blue-green region), but contain some subtle differences in electron density at the carbonyl group.

- The carbonyl carbon of the acid chloride is the most electron deficient (darkest blue), because it is the least stabilized by electron donation from the Cl atom.
- As the basicity of the substituent increases (Cl → OR → NH_2), more electron density is donated to the carbonyl group, and the carbonyl oxygen becomes more electron rich. Thus, the amide oxygen is the most electron rich (reddest), because NH_2 is the most basic substituent bonded to the carbonyl group.

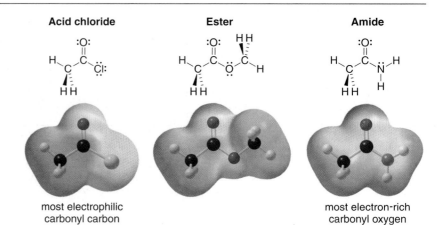

Acid chloride   **Ester**   **Amide**

most electrophilic carbonyl carbon

most electron-rich carbonyl oxygen

Because the basicity of Z determines the relative stability of the carboxylic acid derivatives, the following **order of stability** results:

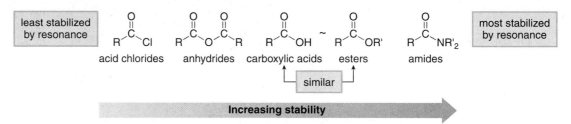

least stabilized by resonance

acid chlorides   anhydrides   carboxylic acids   esters   amides

most stabilized by resonance

similar

Increasing stability

Thus, an acid chloride is the least stable carboxylic acid derivative because Cl^- is the weakest base. An amide is the most stable carboxylic acid derivative because $^-NR'_2$ is the strongest base.

◆ **In summary: As the basicity of Z increases, the stability of RCOZ increases because of added resonance stabilization.**

The electrostatic potential plots of a simple acid chloride, ester, and amide in Figure 22.1 show subtle differences suggested by this resonance description.

PROBLEM 22.2   Draw the three possible resonance structures for an acid bromide, RCOBr. Then, using the pK_a values in Appendix B, decide if RCOBr is more or less stabilized by resonance than a carboxylic acid (RCOOH).

PROBLEM 22.3   How do the following experimental results support the resonance description of the relative stability of acid chlorides compared to amides? The C–Cl bond lengths in CH_3Cl and CH_3COCl are identical (1.78 Å), but the C–N bond in $HCONH_2$ is shorter than the C–N bond in CH_3NH_2 (1.35 Å versus 1.47 Å).

The structure and bonding in nitriles is very different from the carboxylic acid derivatives, and resembles the carbon–carbon triple bond of alkynes.

$$CH_3-C\equiv N: \quad = \qquad = $$

sp hybridized

180°

$\delta^+ \quad \delta^-$

Nucleophiles attack here.

♦ The carbon atom of the C≡N group is *sp* hybridized, making it linear with a bond angle of 180°.
♦ The triple bond consists of one σ and two π bonds.

Like the carboxylic acid derivatives, **nitriles contain an electrophilic carbon atom,** making them susceptible to nucleophilic attack.

PROBLEM 22.4   How is the N atom of CH_3CN hybridized? In what type of orbital does its nonbonded electron pair reside?

22.3   Nomenclature

The names of carboxylic acid derivatives are formed from the names of the parent carboxylic acids discussed in Section 19.2. Keep in mind that the common names **formic acid, acetic acid,** and **benzoic acid** are virtually always used for the parent acid, so these common parent names are used for their derivatives as well.

22.3A   Naming an Acid Chloride—RCOCl

Acid chlorides are named by naming the acyl group and adding the word *chloride.* Two different methods are used.

[1]   For acyclic acid chlorides: Change the suffix *-ic acid* of the parent carboxylic acid to the suffix *-yl chloride;* or
[2]   When the −COCl group is bonded to a ring: Change the suffix *-carboxylic acid* to *-carbonyl chloride.*

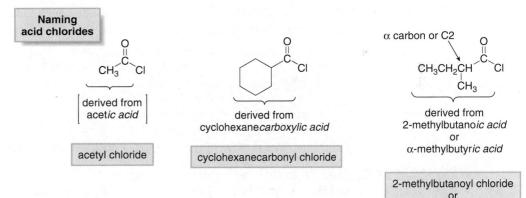

22.3B   Naming an Anhydride

The word anhydride means *without water.* Removing one molecule of water from two molecules of carboxylic acid forms an anhydride.

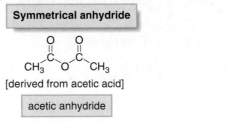

Symmetrical anhydrides are named by changing the *acid* ending of the parent carboxylic acid to the word *anhydride.* **Mixed anhydrides,** which are derived from two different carboxylic acids, are named by alphabetizing the names for both acids and replacing the word *acid* by the word *anhydride.*

22.3C Naming an Ester—RCOOR'

An ester has two alkyl groups, each of which must be named: an **acyl group** (**RCO–**) and an **alkyl group** (designated as **R'**) bonded to an oxygen atom.

[1] **Name R' as an alkyl group.** This becomes the *first* part of the name.
[2] **Name the acyl group** by changing the *-ic acid* ending of the parent carboxylic acid to the suffix *-ate.*

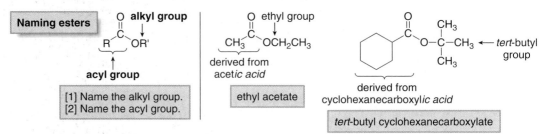

22.3D Naming an Amide

All 1° amides are named by replacing the *-ic acid, -oic acid,* or *-ylic acid* ending with the suffix **amide.**

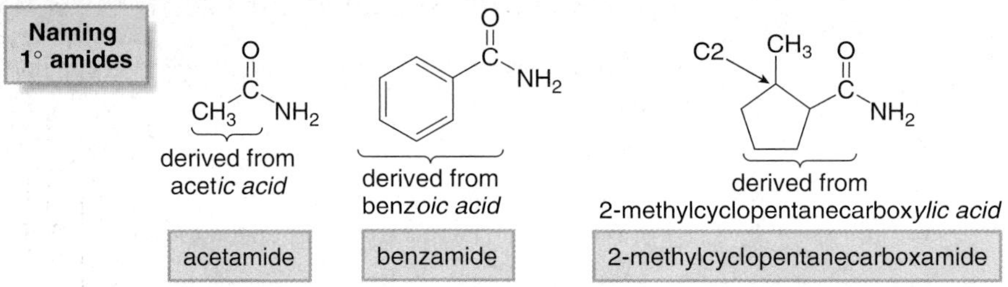

Naming 1° amides

CH₃—C(=O)—NH₂
derived from
acet*ic acid*

acetamide

benzene-C(=O)—NH₂
derived from
benz*oic acid*

benzamide

C2 → CH₃ cyclopentane-C(=O)—NH₂
derived from
2-methylcyclopentanecarbox*ylic acid*

2-methylcyclopentanecarboxamide

Naming 2° and 3° amides requires two parts: the **one or two alkyl groups** bonded to the N atom, and the **acyl group.** To name a 2° or 3° amide:

[1] **Name the alkyl group (or groups)** bonded to the N atom of the amide. Use the prefix "*N-*" preceding the name of **each** alkyl group to show that it is bonded to a nitrogen atom. This becomes the *first* part of the name.
[2] For 3° amides, use the prefix **di-** if the two alkyl groups on N are the same. If the two alkyl groups are different, **alphabetize** their names. One "*N-*" is needed for each alkyl group, even if both R groups are identical.
[3] Name the acyl group by replacing the *-ic acid, -oic acid,* or *-ylic acid* ending by the suffix **amide.**

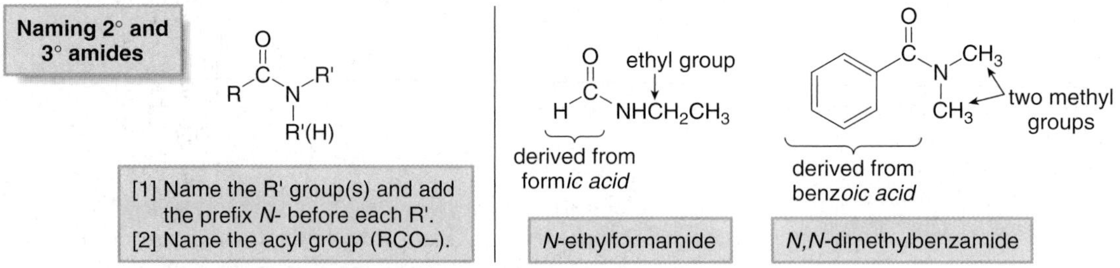

Naming 2° and 3° amides

R—C(=O)—N—R' R'(H)

[1] Name the R' group(s) and add the prefix *N-* before each R'.
[2] Name the acyl group (RCO–).

H—C(=O)—NHCH₂CH₃ ← ethyl group
derived from
form*ic acid*

N-ethylformamide

benzene-C(=O)—N(CH₃)CH₃ ← two methyl groups
derived from
benz*oic acid*

N,N-dimethylbenzamide

22.3E Naming a Nitrile

In contrast to the carboxylic acid derivatives, **nitriles are named as alkane derivatives.** To name a nitrile using IUPAC rules:

◆ Find the longest chain that contains the CN and add the word *nitrile* to the name of the parent alkane. Number the chain to put CN at C1, but omit this number from the name.

Common names for nitriles are derived from the names of the carboxylic acid having the same number of carbon atoms by replacing the *-ic acid* ending of the carboxylic acid by the suffix *-onitrile.*

When CN is named as a substituent it is called a *cyano* group.

In naming a nitrile, the CN carbon is one carbon atom of the longest chain. **CH₃CH₂CN** is propanenitrile, *not* ethanenitrile.

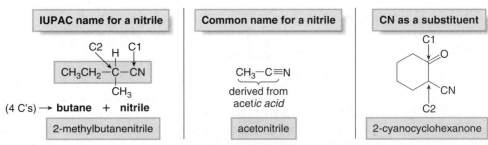

IUPAC name for a nitrile	Common name for a nitrile	CN as a substituent
C2 H C1 CH₃CH₂—C—CN CH₃ (4 C's) → **butane** + **nitrile** **2-methylbutanenitrile**	CH₃—C≡N derived from acet*ic acid* **acetonitrile**	C1 CN C2 **2-cyanocyclohexanone**

Table 22.2 summarizes the most important points about the nomenclature of carboxylic acid derivatives.

TABLE 22.2   Summary: Nomenclature of Carboxylic Acid Derivatives and Nitriles

Compound	Name ending	Example	Name
acid chloride	**-yl chloride** or **-carbonyl chloride**	C₆H₅—C(=O)—Cl	benzoyl chloride
anhydride	**anhydride**	C₆H₅—C(=O)—O—C(=O)—C₆H₅	benzoic anhydride
ester	**-ate**	C₆H₅—C(=O)—OCH₂CH₃	ethyl benzoate
amide	**-amide**	C₆H₅—C(=O)—NHCH₃	*N*-methylbenzamide
nitrile	**-nitrile** or **-onitrile**	C₆H₅—C≡N	benzonitrile

SAMPLE PROBLEM 22.1   Give the IUPAC name for each compound.

a.
$$\text{CH}_3\text{CH}_2\overset{\overset{\displaystyle CH_3}{|}}{C}H\text{CH}_2\overset{\overset{\displaystyle CH_3}{|}}{C}H\text{COCl}$$

b.
$$\text{CH}_3\text{CH}_2\text{CH}_2\overset{\overset{\displaystyle CH_3}{|}}{C}H-\overset{\overset{\displaystyle O}{\parallel}}{C}\text{OCH(CH}_3)_2$$

SOLUTION

[a] The functional group is an acid chloride bonded to a chain of atoms, so the name ends in **-yl chloride.**

[1] Find and name the longest chain containing the COCl:

CH₃ CH₃
 CH₃CH₂CHCH₂CHCOCl

hexano*ic acid* ⟶ **hexano*yl chloride***
 (6 C's)

[2] Number and name the substituents:

CH₃ CH₃
 CH₃CH₂CHCH₂CHCOCl
 C4 C2 C1

Answer: 2,4-dimethylhexanoyl chloride

[b] The functional group is an ester, so the name ends in **-ate.**

[1] Find and name the longest chain containing the carbonyl group:

CH₃ O
 CH₃CH₂CH₂CH—C
 OCH(CH₃)₂

pentano*ic acid* ⟶ **pentano*ate***
 (5 C's)

[2] Number and name the substituents:

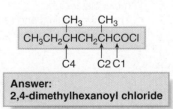

CH₃ O
 CH₃CH₂CH₂CH—C ← C1
 C2 OCHCH₃
 CH₃ <br> isopropyl group

Answer: isopropyl 2-methylpentanoate

The name of the alkyl group on the O atom goes **first** in the name.

PROBLEM 22.5   Give an IUPAC or common name for each compound.

a. $(CH_3CH_2)_2CHCOCl$
b. $C_6H_5COOCH_3$
c. $CH_3CH_2CON(CH_3)CH_2CH_3$

d. 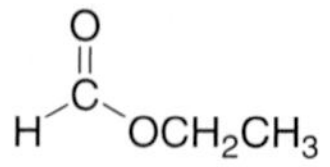 (structure: H–C(=O)–OCH₂CH₃)

e. 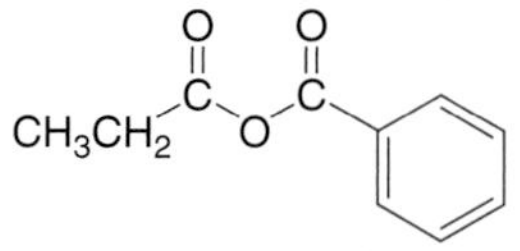 (structure: CH₃CH₂–C(=O)–O–C(=O)–C₆H₅)

f. 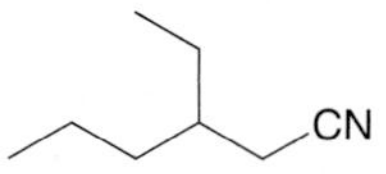 (structure: branched chain with CN group)

PROBLEM 22.6   Draw the structure corresponding to each name.
a. 5-methylheptanoyl chloride
b. isopropyl propanoate
c. acetic formic anhydride
d. *N*-isobutyl-*N*-methylbutanamide

e. 3-methylpentanenitrile
f. *o*-cyanobenzoic acid
g. *sec*-butyl 2-methylhexanoate
h. *N*-ethylhexanamide

22.4   Physical Properties

Because all carbonyl compounds have a polar carbonyl group, they exhibit **dipole–dipole interactions.** Nitriles also have dipole–dipole interactions because they have a polar $C\equiv N$ group. Because they contain one or two N–H bonds, 1° and 2° amides are capable of intermolecular hydrogen bonding. The N–H bond of one amide intermolecularly hydrogen bonds to the C=O of another amide, as shown using two acetamide molecules (CH_3CONH_2) in Figure 22.2.

How these factors affect the physical properties of carboxylic acid derivatives is summarized in Table 22.3.

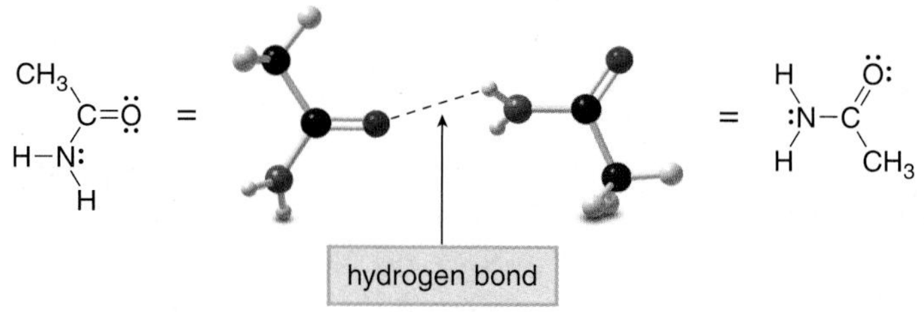

Figure 22.2   Intermolecular hydrogen bonding between two CH_3CONH_2 molecules

hydrogen bond

TABLE 22.3   Physical Properties of Carboxylic Acid Derivatives

Property	Observation
Boiling point and melting point	• Primary (1°) and 2° amides have *higher* boiling points and melting points than compounds of comparable molecular weight. • The boiling points and melting points of other carboxylic acid derivatives are similar to those of other polar compounds of comparable size and shape. 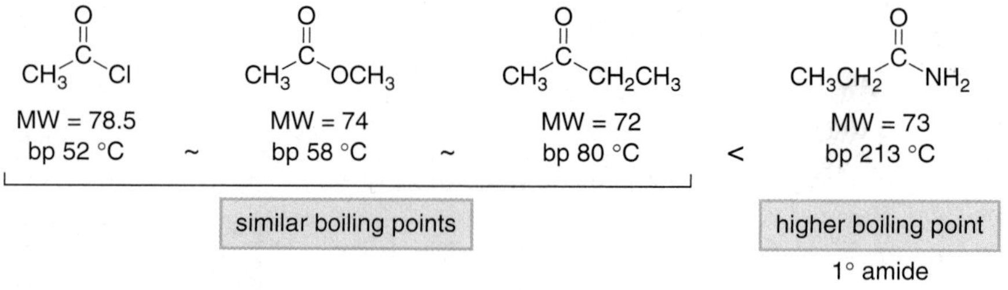 CH_3–C(=O)–Cl MW = 78.5 bp 52 °C ~ CH_3–C(=O)–OCH_3 MW = 74 bp 58 °C ~ CH_3–C(=O)–CH_2CH_3 MW = 72 bp 80 °C < CH_3CH_2–C(=O)–NH_2 MW = 73 bp 213 °C similar boiling points higher boiling point 1° amide
Solubility	• Carboxylic acid derivatives are soluble in organic solvents regardless of size. • Most carboxylic acid derivatives having ≤ 5 C's are H_2O soluble because they can hydrogen bond with H_2O (Section 3.5C). • Carboxylic acid derivatives having > 5 C's are H_2O insoluble because the nonpolar alkyl portion is too large to dissolve in the polar H_2O solvent.

Key: MW = molecular weight

PROBLEM 22.7   Explain why the boiling point of CH_3CONH_2 (221 °C) is significantly higher than the boiling point of $CH_3CON(CH_3)_2$ (166 °C), even though the latter compound has a higher molecular weight and more surface area.

22.5   Spectroscopic Properties

22.5A   IR Spectra

The most prominent IR absorptions for carboxylic acid derivatives and nitriles are as follows:

[1]   Like all carbonyl compounds, carboxylic acid derivatives have a **strong C=O absorption between 1600 and 1850 cm^{-1}.**
[2]   **Primary (1°) and 2° amides** have two additional absorptions due to the N–H bonds:
  • one or two N–H stretching peaks at **3200–3400 cm^{-1}.**
  • an N–H bending absorption at **~1640 cm^{-1}.**
[3]   **Nitriles** have an absorption at **2250 cm^{-1} for the C≡N.**

The exact location of the carbonyl absorption varies with the identity of Z in the carbonyl compound RCOZ. As detailed in Section 22.2, as the basicity of Z increases, resonance stabilization of RCOZ increases, resulting in the following trend:

◆ As the carbonyl π bond becomes more delocalized, the C=O absorption shifts to lower frequency.

Thus, the carbonyl group of an acid chloride and anhydride, which are least stabilized by resonance, absorb at higher frequency than the carbonyl group of an amide, which is more stabilized by resonance. Table 22.4 lists specific values for the carbonyl absorptions of the carboxylic acid derivatives.

Conjugation and ring size affect the location of these carbonyl absorptions.

◆ Conjugation shifts a carbonyl absorption to lower frequencies.
◆ For cyclic carboxylic acid derivatives, decreasing ring size shifts a carbonyl absorption to higher frequencies.

The effects of conjugation and ring size on the location of a carbonyl absorption were first discussed in Section 21.4A.

TABLE 22.4   IR Absorptions for the Carbonyl Group of Carboxylic Acid Derivatives

Compound type	Structure (RCOZ)	Carbonyl absorption ($\tilde{\nu}$)
acid chloride	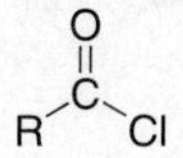 $\underset{R}{\overset{O}{\parallel}}\!\!\!C\!\!-\!Cl$	~1800
anhydride	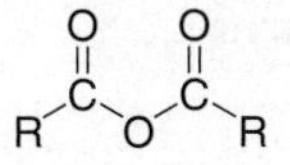 $R\!-\!\underset{\;}{\overset{O}{\parallel}}C\!-\!O\!-\!\underset{\;}{\overset{O}{\parallel}}C\!-\!R$	1820 and 1760 (2 peaks)
ester	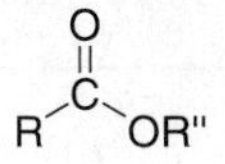 $\underset{R}{\overset{O}{\parallel}}\!\!\!C\!\!-\!OR''$	1735–1745
amide	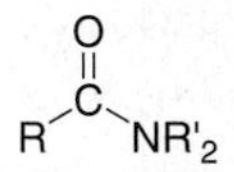 $\underset{R}{\overset{O}{\parallel}}\!\!\!C\!\!-\!NR'_2$ R' = H or alkyl	1630–1680

Increasing basicity of Z (downward) · Increasing $\tilde{\nu}$ of absorption (upward)

PROBLEM 22.8   How would the compounds in each pair differ in their IR spectra?

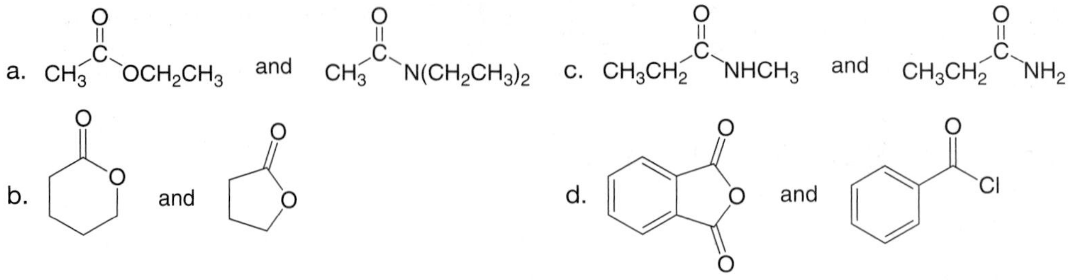

a. CH_3—C(=O)—OCH_2CH_3 and CH_3—C(=O)—$N(CH_2CH_3)_2$ c. CH_3CH_2—C(=O)—$NHCH_3$ and CH_3CH_2—C(=O)—NH_2

b.   and   d.   and

22.5B   NMR Spectra

Carboxylic acid derivatives have two characteristic ^{1}H NMR absorptions.

[1] **Protons on the α carbon to the carbonyl absorb at 2–2.5 ppm.**
[2] **The N–H protons of 1° and 2° amides absorb at 7.5–8.5 ppm.**

In their ^{13}C NMR spectra, carboxylic acid derivatives give a highly deshielded peak at 160–180 ppm due to the carbonyl carbon. This is somewhat upfield from the carbonyl absorption of aldehydes and ketones, which occurs at 190–215 ppm.

Nitriles give a peak at 115–120 ppm in their ^{13}C NMR spectrum due to the *sp* hybridized carbon. This is further downfield than the signal due to the *sp* hybridized carbon of an alkyne, which occurs at 65–100 ppm.

PROBLEM 22.9   How would α-methoxyacetone ($CH_3COCH_2OCH_3$) and methyl propanoate ($CH_3CH_2COOCH_3$) differ in their IR, ^{1}H NMR, and ^{13}C NMR spectra?

22.6   Interesting Esters and Amides

22.6A   Esters

Many low molecular weight esters have pleasant and very characteristic odors.

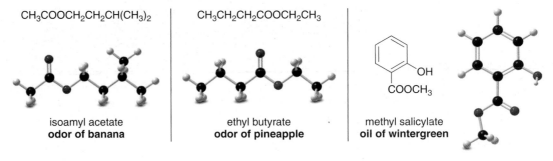

$CH_3COOCH_2CH_2CH(CH_3)_2$

isoamyl acetate
odor of banana

$CH_3CH_2CH_2COOCH_2CH_3$

ethyl butyrate
odor of pineapple

methyl salicylate
oil of wintergreen

Several lactones have important biological activities.

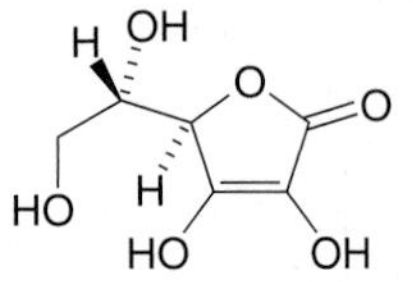

vitamin C

◆ **Vitamin C** (or **ascorbic acid**) is a water-soluble vitamin containing a five-membered lactone that we first discussed in Section 3.6B. Although vitamin C is synthesized in plants, humans do not have the necessary enzymes to make it, and so they must obtain it from their diet.

Peptides and proteins are
discussed in detail in
Chapter 28.

Ginkgo biloba has existed
on earth for over 280 million
years. The fossil record
indicates that it has
undergone little significant
evolutionary change for
eons. Extracts of the ginkgo
tree have long been used in
China, India, and Japan in
medicine and cooking.

Capsaicin (Chapter 1),
melatonin (Chapter 15), and
aspartame (Chapter 28) are
three chapter-opening
molecules that also contain
an amide.

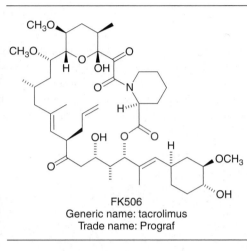

ginkgolide B
[three ester units labeled in red]

FK506
Generic name: tacrolimus
Trade name: Prograf

◆ **Ginkgolide B** is one of the active components of ginkgo extract, which has been used
for centuries in traditional Chinese medicine.
Ginkgo extracts are obtained from the ginkgo
tree, an ancient tree native to the Orient, now
introduced in Europe and the Americas as
well.

◆ **FK506** is an immunosuppressant known by
the trade name Prograf. One functional group
in its complex structure is a cyclic ester in a
21-membered ring. A cyclic ester contained
in such a large ring is called a **macrocyclic
lactone** or a **macrolide**. FK506 is used to
suppress rejection after organ transplants.

22.6B   Amides

An important group of naturally occurring amides consists of **proteins, polymers of amino
acids joined together by amide linkages.** Proteins differ in the length of the polymer chain, as
well as in the identity of the R groups bonded to it. The word *protein* is usually reserved for high
molecular weight polymers composed of 40 or more amino acid units, while the designation *peptide* is given to polymers of lower molecular weight.

| Portion of a protein molecule | | The R groups come from amino acids joined together to form the protein. |

[Amide bonds are shown in red.]

Proteins and peptides have diverse functions in the cell. They form the structural components of
muscle, connective tissue, hair, and nails. They catalyze reactions and transport ions and molecules across cell membranes. **Met-enkephalin,** for example, a peptide with four amide bonds
found predominately in nerve tissue cells, relieves pain and acts as an opiate by producing
morphine-like effects.

met-enkephalin
[The four amide bonds are shown in red.]

3-D structure

Several useful drugs are amides. For example, **Gleevec** (generic name: imatinib mesylate), an
amide sold as a salt with methanesulfonic acid (CH_3SO_3H), is an anticancer drug approved in
2001 for the treatment of chronic myeloid leukemia. Gleevec represents a new approach to

cancer chemotherapy, which targets a single molecule to disable the molecular mechanism responsible for a specific type of cancer.

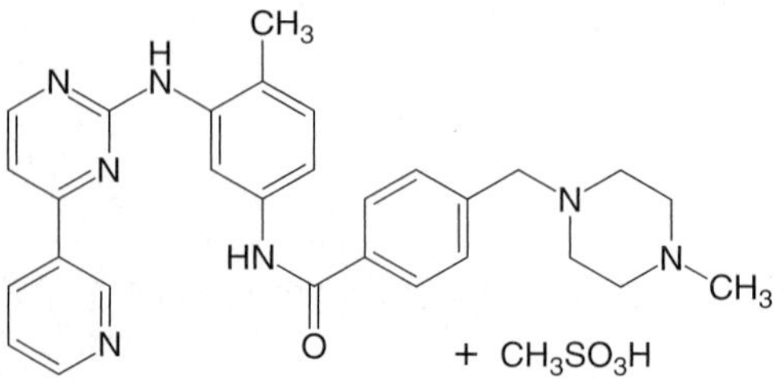

Trade name: Gleevec
Generic name: imatinib mesylate

Penicillins are a group of structurally related antibiotics, known since the pioneering work of Sir Alexander Fleming led to the discovery of penicillin G in the 1920s (chapter opener). All penicillins contain a strained β-lactam fused to a five-membered ring, as well as a second amide located α to the β-lactam carbonyl group. Particular penicillins differ in the identity of the R group in the amide side chain.

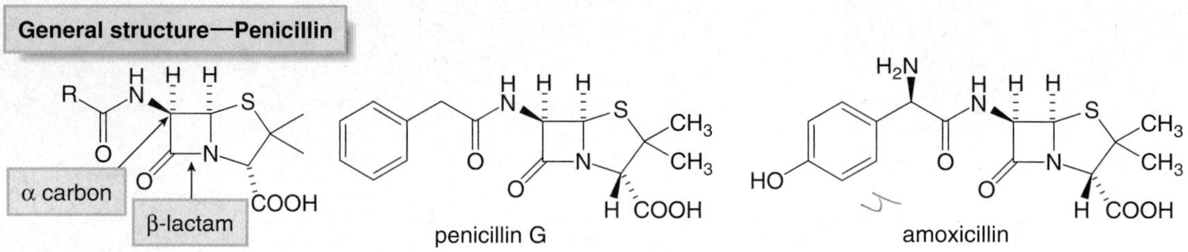

General structure—Penicillin

α carbon

β-lactam

penicillin G

amoxicillin

Cephalosporins represent a second group of β-lactam antibiotics that contain a four-membered ring fused to a six-membered ring. Cephalosporins are generally active against a broader range of bacteria than penicillins.

General structure—Cephalosporin

β-lactam

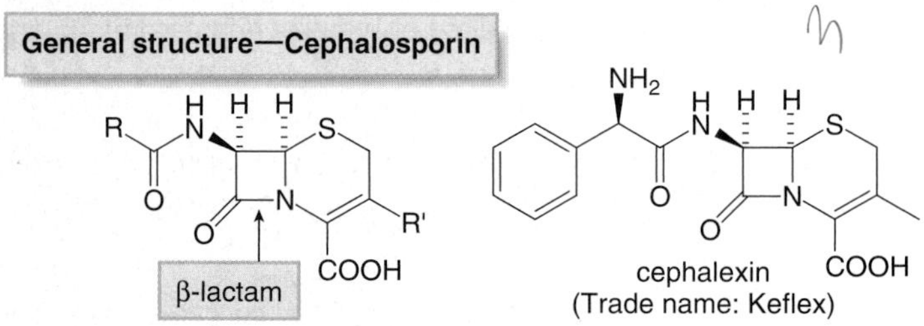

cephalexin
(Trade name: Keflex)

Section 22.14 describes how these β-lactam antibiotics function as antibacterial agents.

PROBLEM 22.10   For both amoxicillin and cephalexin: (a) How many stereogenic centers does each compound contain? (b) What is the maximum number of stereoisomers possible? (c) Draw the enantiomer of each compound.

22.7 Introduction to Nucleophilic Acyl Substitution

The characteristic reaction of carboxylic acid derivatives is *nucleophilic acyl substitution*. This is a general reaction that occurs with both negatively charged nucleophiles (Nu:⁻) and neutral nucleophiles (HNu:).

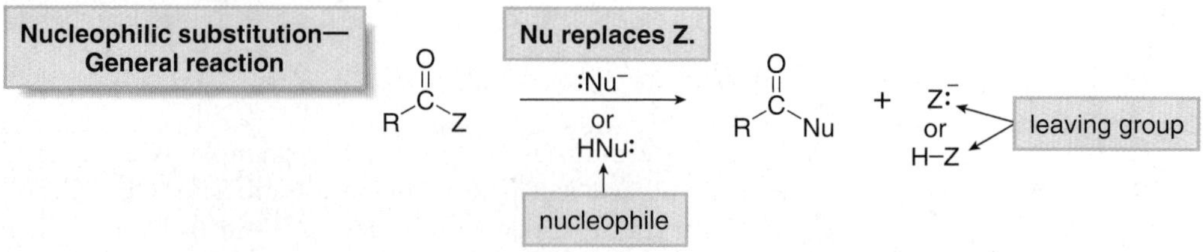

Nucleophilic substitution—
General reaction

Nu replaces Z.

leaving group

nucleophile

◆ Carboxylic acid derivatives (RCOZ) react with nucleophiles because they contain an electrophilic, unhindered carbonyl carbon.
◆ Substitution occurs, *not* addition, because carboxylic acid derivatives (RCOZ) have a leaving group Z on the carbonyl carbon.

The mechanism for nucleophilic acyl substitution was first presented in Section 20.2.

22.7A   The Mechanism

The general mechanism for nucleophilic acyl substitution is a two-step process: **nucleophilic attack** followed by **loss of the leaving group,** as shown in Mechanism 22.1.

MECHANISM 22.1

General Mechanism—Nucleophilic Acyl Substitution

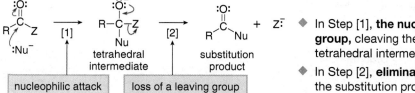

◆ In Step [1], **the nucleophile attacks the carbonyl group,** cleaving the π bond, and forming a tetrahedral intermediate with a new C—Nu bond.

◆ In Step [2], **elimination of the leaving group** forms the substitution product.

[Z = Cl, OCOR, OH, OR', NR'$_2$]

The overall result of addition of a nucleophile and elimination of a leaving group is substitution of the nucleophile for the leaving group. Recall from Chapter 20 that nucleophilic substitution occurs with carbanions (R$^-$) and hydride (H$^-$) as nucleophiles. A variety of oxygen and nitrogen nucleophiles also participate in this reaction.

Oxygen nucleophiles				Nitrogen nucleophiles		
$^-$ÖH	H$_2$Ö:	RÖH	R—C(=Ö)Ö:	N̈H$_3$	RN̈H$_2$	R$_2$N̈H

Nucleophilic acyl substitution using heteroatomic nucleophiles results in the conversion of one carboxylic acid derivative into another, as shown in two examples.

Examples	nucleophile	product	by-product
CH$_3$—C(=O)—Cl	NH$_3$	CH$_3$—C(=O)—NH$_2$ (1° amide)	+ HCl
C$_6$H$_5$—C(=O)—OH	CH$_3$OH, H$^+$	C$_6$H$_5$—C(=O)—OCH$_3$ (ester)	+ H$_2$O

Each reaction results in the replacement of the leaving group by the nucleophile, regardless of the identity of or charge on the nucleophile. To draw any nucleophilic acyl substitution product:

◆ **Find the *sp*^{2} hybridized carbon with the leaving group.**

◆ **Identify the nucleophile.**

◆ **Substitute the nucleophile for the leaving group.** With a neutral nucleophile a proton must be lost to obtain a neutral substitution product.

PROBLEM 22.11   Draw the products of each reaction.

a. CH₃COCl —CH₃OH→   b. CH₃COOCH₂CH₃ —NH₃→

22.7B Relative Reactivity of Carboxylic Acids and Their Derivatives

Recall that the **best leaving group is the weakest base.** The relative basicity of the common leaving groups, Z, is given in Table 22.1.

As discussed in Section 20.2B, carboxylic acids and their derivatives differ greatly in reactivity toward nucleophiles. The order of reactivity parallels the leaving group ability of the group Z.

◆ **The better the leaving group, the more reactive RCOZ is in nucleophilic acyl substitution.**

Thus, the following trends result:

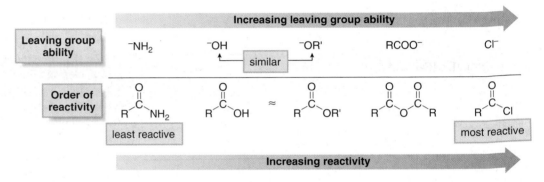

Based on this order of reactivity, **more reactive acyl compounds (acid chlorides and anhydrides) can be converted to less reactive ones (carboxylic acids, esters, and amides). The reverse is not usually true.**

To see why this is so, recall that nucleophilic addition to a carbonyl group forms a tetrahedral intermediate with two possible leaving groups, Z⁻ or :Nu⁻. The group that is subsequently eliminated is the *better* of the two leaving groups. For a reaction to form a substitution product, therefore, Z⁻ must be the better leaving group, making the starting material RCOZ a more reactive acyl compound.

To evaluate whether a nucleophilic substitution reaction will occur, **compare the leaving group ability of the incoming nucleophile and the departing leaving group,** as shown in Sample Problem 22.2.

SAMPLE PROBLEM 22.2   Determine whether each nucleophilic acyl substitution is likely to occur.

a. CH₃COCl —?→ CH₃COOCH₂CH₃   b. C₆H₅CONH₂ —?→ C₆H₅COOCOC₆H₅

SOLUTION

[a] Conversion of CH_3COCl to $CH_3COOCH_2CH_3$ requires the substitution of Cl^- by $^-OCH_2CH_3$. Because Cl^- is a weaker base and therefore a better leaving group than $^-OCH_2CH_3$, **this reaction occurs.**

[b] Conversion of $C_6H_5CONH_2$ to $C_6H_5COOCOC_6H_5$ requires the substitution of $^-NH_2$ by $^-OCOC_6H_5$. Because $^-NH_2$ is a stronger base and therefore a poorer leaving group than $^-OCOC_6H_5$, **this reaction does *not* occur.**

Learn the order of reactivity of carboxylic acid derivatives. Keeping this in mind allows you to organize a very large number of reactions.

To summarize:

◆ Nucleophilic substitution occurs when the leaving group Z⁻ is a weaker base and therefore better leaving group than the attacking nucleophile :Nu⁻.

◆ More reactive acyl compounds can be converted to less reactive acyl compounds by nucleophilic substitution.

PROBLEM 22.12 Rank the compounds in each group in order of increasing reactivity in nucleophilic acyl substitution.
a. $C_6H_5COOCH_3$, C_6H_5COCl, $C_6H_5CONH_2$
b. CH_3CH_2COOH, $(CH_3CH_2CO)_2O$, $CH_3CH_2CONHCH_3$

PROBLEM 22.13 Without reading ahead in Chapter 22, state whether it should be possible to carry out each of the following nucleophilic substitution reactions.

a. $CH_3COCl \longrightarrow CH_3COOH$ c. $CH_3COOCH_3 \longrightarrow CH_3COCl$

b. $CH_3CONHCH_3 \longrightarrow CH_3COOCH_3$

PROBLEM 22.14 Explain why trichloroacetic anhydride [$(Cl_3CCO)_2O$] is more reactive than acetic anhydride [$(CH_3CO)_2O$] in nucleophilic acyl substitution reactions.

22.7C   A Preview of Specific Reactions

Sections 22.8–22.14 are devoted to specific examples of nucleophilic acyl substitution using heteroatoms as nucleophiles. There are a great many reactions, and it is easy to confuse them unless you learn the general order of reactivity of carboxylic acid derivatives. **Keep in mind that every reaction that begins with an acyl starting material involves nucleophilic substitution.**

In this text, all of the nucleophilic substitution reactions are grouped according to the carboxylic acid derivative used as a starting material. We begin with the reactions of acid chlorides, the most reactive acyl compounds, then proceed to less and less reactive carboxylic acid derivatives, ending with amides. Acid chlorides undergo many reactions, because they have the best leaving group of all acyl compounds, whereas amides undergo only one reaction, which must be carried out under harsh reaction conditions, because amides have a poor leaving group.

In general, we will examine nucleophilic acyl substitution with four different nucleophiles, as shown in the following equations.

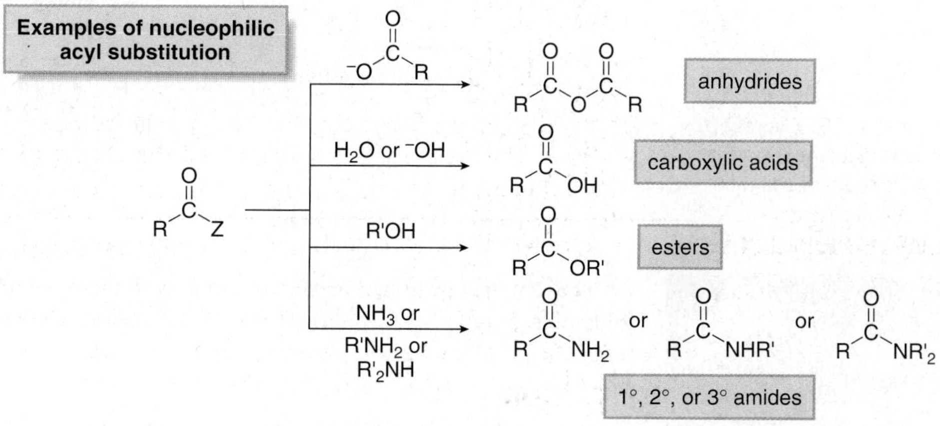

These reactions are used to make anhydrides, carboxylic acids, esters, and amides, but not acid chlorides, from other acyl compounds. Acid chlorides are the most reactive acyl compounds (they have the best leaving group), so they are not easily formed as a product of nucleophilic substitution reactions. They can only be prepared from carboxylic acids using special reagents, as discussed in Section 22.10A.

22.8   Reactions of Acid Chlorides

The reaction of acid chlorides with water is rapid. Exposure of an acid chloride to moist air on a humid day leads to some hydrolysis, giving the acid chloride a very acrid odor, due to the HCl formed as a by-product.

Acid chlorides readily react with nucleophiles to form nucleophilic substitution products, with HCl usually formed as a reaction by-product. A weak base like pyridine is added to the reaction mixture to remove this strong acid, forming an ammonium salt.

Acid chlorides react with oxygen nucleophiles to form anhydrides, carboxylic acids, and esters.

Acid chlorides also react with ammonia and 1° and 2° amines to form 1°, 2°, and 3° amides, respectively. Two equivalents of NH_3 or amine are used. One equivalent acts as a nucleophile to replace Cl and form the substitution product, while the second equivalent reacts as a base with the HCl by-product to form an ammonium salt.

Insect repellents containing DEET have become particularly popular because of the recent spread of many insect-borne diseases such as West Nile virus and Lyme disease. DEET does not kill insects—it repels them. It is thought that DEET somehow confuses insects so that they can no longer sense the warm moist air that surrounds a human body.

As an example, reaction of an acid chloride with diethylamine forms the 3° amide *N,N*-diethyl-*m*-toluamide, popularly known as **DEET**. DEET, the active ingredient in the most widely used insect repellents, is effective against mosquitoes, fleas, and ticks.

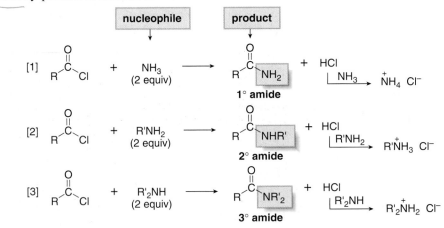

PROBLEM 22.15   Draw the products formed when benzoyl chloride (C_6H_5COCl) is treated with each nucleophile: (a) H_2O, pyridine; (b) CH_3COO^-; (c) NH_3 (excess); (d) $(CH_3)_2NH$ (excess).

With a carboxylate nucleophile the mechanism follows the general, two-step mechanism discussed in Section 22.7: **nucleophilic attack followed by loss of the leaving group,** as shown in Mechanism 22.2.

MECHANISM 22.2

Conversion of Acid Chlorides to Anhydrides

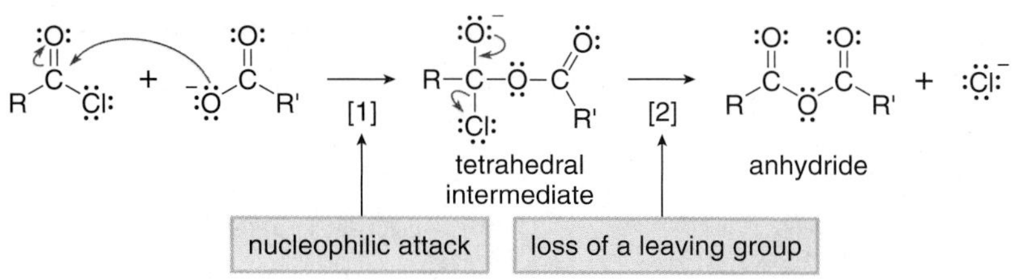

♦ In Step [1], **nucleophilic addition of R'COO⁻** forms a tetrahedral intermediate.

♦ In Step [2], **elimination of the leaving group (Cl⁻)** forms the substitution product, an **anhydride.**

Nucleophilic substitution with the neutral nucleophiles (H_2O, $R'OH$, NH_3, and so forth) requires an additional step for proton transfer. For example, the reaction of an acid chloride with H_2O as nucleophile converts an acid chloride to a carboxylic acid in three steps (Mechanism 22.3).

MECHANISM 22.3

Conversion of Acid Chlorides to Carboxylic Acids

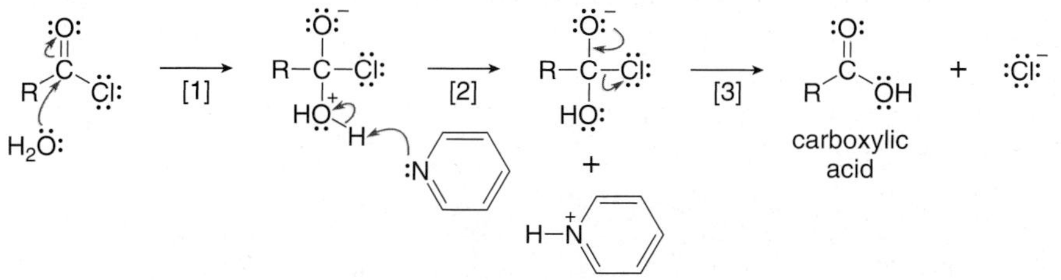

♦ In Step [1], **nucleophilic attack by H_2O** forms a tetrahedral intermediate.

♦ Removal of a proton followed by **elimination of the leaving group, Cl⁻** (Steps [2]–[3]), forms the substitution product, a **carboxylic acid.**

The exact same three-step process can be written for any neutral nucleophile that reacts with acid chlorides.

PROBLEM 22.16   Draw a stepwise mechanism for the following reaction.

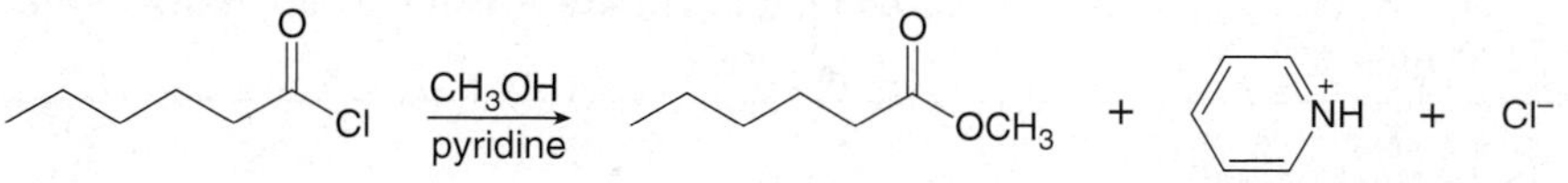

22.9 Reactions of Anhydrides

Although somewhat less reactive than acid chlorides, anhydrides nonetheless readily react with most nucleophiles to form substitution products. Nucleophilic substitution reactions of anhydrides are no different than the reactions of other carboxylic acid derivatives, even though anhydrides contain two carbonyl groups. **Nucleophilic attack occurs at one carbonyl group, while the second carbonyl becomes part of the leaving group.**

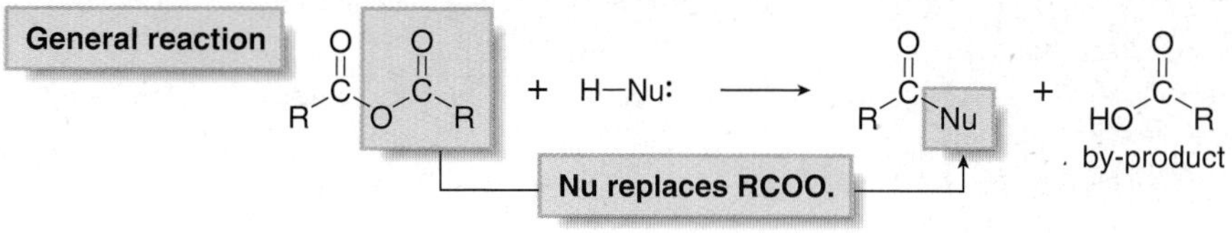

Nucleophilic substitution occurs only when the leaving group is a weaker base and therefore a better leaving group than the attacking nucleophile.

Anhydrides can't be used to make acid chlorides, because RCOO⁻ is a stronger base and therefore a poorer leaving group than Cl⁻. Anhydrides can be used to make all other acyl derivatives, however. Reaction with water and alcohols yields **carboxylic acids** and **esters,** respectively. Reaction with two equivalents of NH_3 or amines forms **1°, 2°,** and **3° amides.** A molecule of carboxylic acid (or a carboxylate salt) is always formed as a by-product.

Replacing NH_3 with a 1° or 2° amine forms a 2° or 3° amide, respectively.

PROBLEM 22.17   Draw the products formed when benzoic anhydride [$(C_6H_5CO)_2O$] is treated with each nucleophile: (a) H_2O, pyridine; (b) CH_3OH; (c) NH_3 (excess); (d) $(CH_3)_2NH$ (excess).

The conversion of an anhydride to an amide illustrates the mechanism of nucleophilic acyl substitution with an anhydride as starting material (Mechanism 22.4). Besides the usual steps of **nucleophilic addition** and **elimination of the leaving group,** an additional proton transfer is needed.

MECHANISM 22.4

Conversion of an Anhydride to an Amide

◆ In Step [1], **nucleophilic attack by NH_3** forms a tetrahedral intermediate.
◆ Removal of a proton followed by **elimination of the leaving group, RCOO⁻** (Steps [2]–[3]), forms the substitution product, a **1° amide.**

Acetaminophen reduces pain and fever, but it is not anti-inflammatory, so it is ineffective in treating conditions like arthritis, which have a significant inflammatory component. Although it was first used as early as 1893, acetaminophen did not receive formal approval from the FDA until 1950. In large doses, acetaminophen causes liver damage, so dosage recommendations must be carefully followed.

Anhydrides react with alcohols and amines with ease, so they are often used in the laboratory to prepare esters and amides. For example, acetic anhydride is used to prepare two analgesics, **acetylsalicylic acid** (aspirin) and **acetaminophen** (the active ingredient in Tylenol).

acetylsalicylic acid (aspirin)

acetaminophen (active ingredient in Tylenol)

The psychological effects of the opium poppy have been known for probably 6000 years, and opium has been widely used as a recreational drug and pain-killing remedy for centuries. The analgesic and narcotic effects of opium are largely due to morphine. Heroin is two to three times more potent than morphine, probably because it is somewhat less polar and therefore more lipid-soluble.

These are called **acetylation** reactions because they result in the transfer of an acetyl group, CH_3CO-, from one heteroatom to another.

Heroin is prepared by the acetylation of morphine, an analgesic compound isolated from the opium poppy. Both OH groups of morphine are readily acetylated with acetic anhydride to form the diester present in heroin.

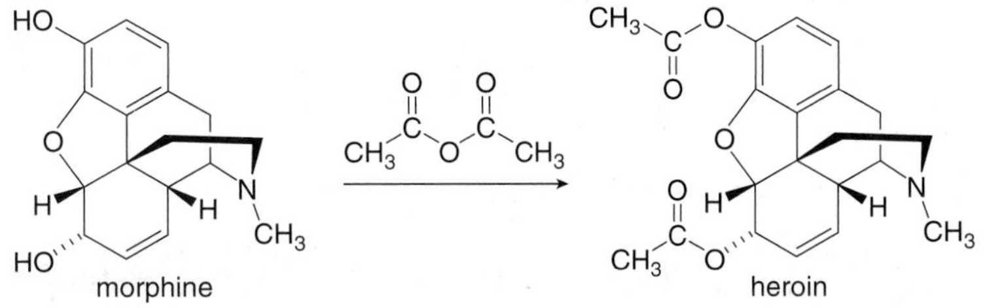

PROBLEM 22.18   Draw a stepwise mechanism for the conversion of morphine to heroin with acetic anhydride.

22.10   Reactions of Carboxylic Acids

Carboxylic acids are strong organic acids. Because acid–base reactions proceed rapidly, any nucleophile that is also a strong base will react with a carboxylic acid by removing a proton *first,* before any nucleophilic substitution reaction can take place.

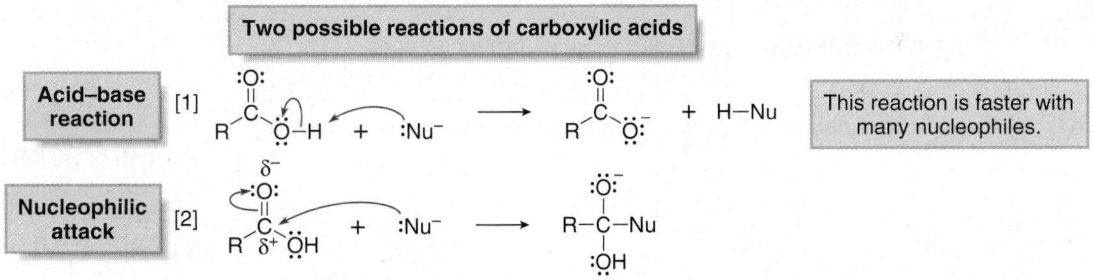

An acid–base reaction (Reaction [1]) occurs with ^-OH, NH_3, and amines, all common nucleophiles used in nucleophilic acyl substitution reactions. Nonetheless, carboxylic acids can be converted to a variety of other acyl derivatives using special reagents, with acid catalysis, or sometimes, by using rather forcing reaction conditions. These reactions are summarized in Figure 22.3 and detailed in Sections 22.10A–22.10D.

Figure 22.3   Nucleophilic acyl substitution reactions of carboxylic acids

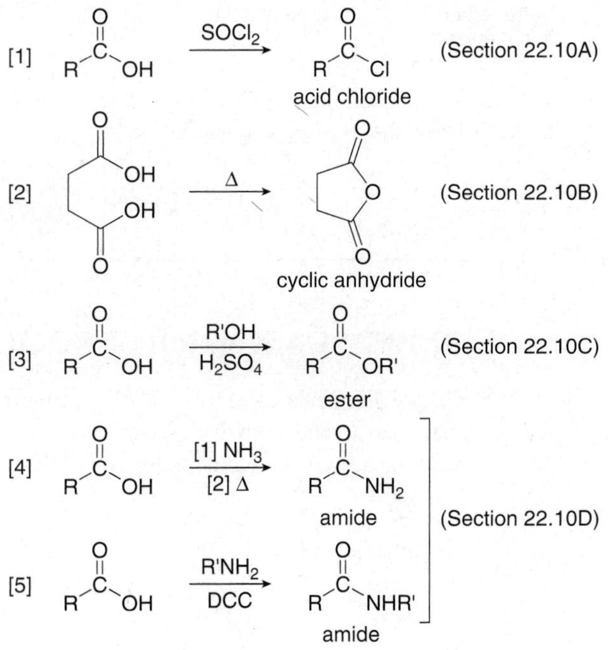

22.10A   Conversion of RCOOH to RCOCl

Carboxylic acids can't be converted to acid chlorides by using Cl⁻ as a nucleophile, because the attacking nucleophile Cl⁻ is a weaker base than the departing leaving group, ⁻OH. But carboxylic acids *can* be converted to acid chlorides using thionyl chloride, **SOCl₂,** a reagent that was introduced in Section 9.12 to convert alcohols to alkyl chlorides.

General reaction $$\underset{R}{\overset{O}{\underset{\;}{C}}}\text{OH} \xrightarrow{\;SOCl_2\;} \underset{R}{\overset{O}{\underset{\;}{C}}}\text{Cl} \;+\; SO_2 \;+\; HCl$$

Example

This reaction converts a less reactive acyl derivative (a carboxylic acid) into a more reactive one (an acid chloride). This is possible because thionyl chloride converts the OH group of the acid into a better leaving group, and because it provides the nucleophile (Cl⁻) to displace the leaving group. The steps in the process are illustrated in Mechanism 22.5.

MECHANISM 22.5

Conversion of Carboxylic Acids to Acid Chlorides

Steps [1] and [2]   Conversion of the OH group into a good leaving group

◆ Reaction of the OH group with SOCl₂ forms an intermediate that loses a proton in Step [2]. This two-step process converts the OH group into OSOCl, a **good leaving group.**

Steps [3] and [4]   Substitution of the leaving group by Cl

◆ **Nucleophilic attack by Cl⁻ and loss of the leaving group** (SO₂ + Cl⁻) forms the acid chloride.

PROBLEM 22.19   Draw the products of each reaction.

a. $$\underset{CH_3CH_2}{\overset{O}{\underset{\;}{C}}}\text{OH} \xrightarrow{\;SOCl_2\;}$$

b. $$\xrightarrow[\text{[2] }(CH_3CH_2)_2NH\text{ (excess)}]{\text{[1] }SOCl_2}$$

22.10B   Conversion of RCOOH to (RCO)₂O

Carboxylic acids cannot be readily converted to anhydrides, but dicarboxylic acids can be converted to cyclic anhydrides by heating to high temperatures. This is a **dehydration** reaction because a water molecule is lost from the diacid.

22.10C   Conversion of RCOOH to RCOOR'

Treatment of a carboxylic acid with an alcohol in the presence of an acid catalyst forms an ester. This reaction is called a **Fischer esterification.**

Fischer esterification—General reaction

$$\underset{R}{\overset{O}{\underset{\|}{C}}}\!\!-\!\!OH \;+\; R'OH \;\overset{H_2SO_4}{\rightleftharpoons}\; \underset{R}{\overset{O}{\underset{\|}{C}}}\!\!-\!\!\boxed{OR'} \;+\; H_2O$$

This reaction is an equilibrium. According to Le Châtelier's principle, it is driven to the right by using excess alcohol or by removing the water as it is formed.

Examples

$$CH_3\!\!-\!\!\overset{O}{\underset{\|}{C}}\!\!-\!\!OH \;+\; CH_3CH_2OH \;\overset{H_2SO_4}{\rightleftharpoons}\; CH_3\!\!-\!\!\overset{O}{\underset{\|}{C}}\!\!-\!\!OCH_2CH_3 \;+\; H_2O$$

ethyl acetate

$$C_6H_5\!\!-\!\!\overset{O}{\underset{\|}{C}}\!\!-\!\!OH \;+\; CH_3OH \;\overset{H_2SO_4}{\rightleftharpoons}\; C_6H_5\!\!-\!\!\overset{O}{\underset{\|}{C}}\!\!-\!\!OCH_3 \;+\; H_2O$$

methyl benzoate

The mechanism for the Fischer esterification involves the usual two steps of nucleophilic acyl substitution—that is, **addition of a nucleophile followed by elimination of a leaving group.** Because the reaction is acid catalyzed, however, there are additional protonation and deprotonation steps. As always, though, the first step of any mechanism with an oxygen-containing starting material and an acid is to **protonate an oxygen atom** as shown with a general acid H−A in Mechanism 22.6.

MECHANISM 22.6

Fischer Esterification—Acid-Catalyzed Conversion of Carboxylic Acids to Esters

Part [1]   Addition of the nucleophile R'OH

$$\underset{R}{\overset{:O:}{\underset{\|}{C}}}\!\!-\!\!\overset{..}{O}H \;\;\;\;H\!\!-\!\!A \;\;\overset{[1]}{\rightleftharpoons}\;\; \underset{R}{\overset{+}{\underset{\|}{C}}}\overset{:OH}{\underset{..}{}}\!\!-\!\!\overset{..}{O}H \;\;\overset{[2]}{\rightleftharpoons}\;\; R\!\!-\!\!\overset{:\overset{..}{O}H}{\underset{\underset{R'\overset{+}{O}}{\overset{|}{\underset{H}{}}}}{\overset{|}{C}}}\!\!-\!\!\overset{..}{O}H \;\;\overset{[3]}{\rightleftharpoons}\;\; R\!\!-\!\!\overset{:\overset{..}{O}H}{\underset{R'\overset{..}{O}:}{\overset{|}{C}}}\!\!-\!\!\overset{..}{O}H$$

R'ÖH + :A⁻   :A⁻   + H−A

nucleophilic addition

- **Protonation** in Step [1] makes the carbonyl group more electrophilic.
- **Nucleophilic addition of R'OH** forms a tetrahedral intermediate, and loss of a proton forms the neutral addition product (Steps [2]–[3]).

Part [2]   Elimination of the leaving group H₂O

$$R\!\!-\!\!\overset{:\overset{..}{O}H}{\underset{R'\overset{..}{O}:}{\overset{|}{C}}}\!\!-\!\!\overset{..}{O}H \;\;\;\;H\!\!-\!\!A \;\;\overset{[4]}{\rightleftharpoons}\;\; R\!\!-\!\!\overset{:\overset{..}{O}H}{\underset{R'\overset{..}{O}:}{\overset{|}{C}}}\!\!-\!\!\overset{+}{O}H_2 \;\;\overset{[5]}{\rightleftharpoons}\;\; \underset{R}{\overset{\overset{+}{O}\!\!-\!\!H}{\underset{\|}{C}}}\!\!-\!\!OR' \;\;\overset{[6]}{\rightleftharpoons}\;\; \underset{R}{\overset{:O:}{\underset{\|}{C}}}\!\!-\!\!OR'$$

+ :A⁻   + H₂Ö: ester

:A⁻   + H−A

loss of H₂O

- Protonation of an OH group in Step [4] forms a good leaving group that is **eliminated in Step [5]**.
- Loss of a proton in Step [6] forms the ester.

Esterification of a carboxylic acid occurs in the presence of acid but *not* in the presence of base. Base removes a proton from the carboxylic acid, forming a carboxylate anion, which does not react with an electron-rich nucleophile.

Both species are electron rich.

$$R\!\!-\!\!\overset{:O:}{\underset{\|}{C}}\!\!-\!\!\overset{..}{O}\!\!-\!\!H \;+\; :\overset{..}{O}H \;\longrightarrow\; R\!\!-\!\!\overset{:O:}{\underset{\|}{C}}\!\!-\!\!\overset{..}{O}:^- \;\;\overset{R'\overset{..}{O}H}{\underset{\times}{\longrightarrow}}\;\; R\!\!-\!\!\overset{:O:}{\underset{\|}{C}}\!\!-\!\!\overset{..}{O}R'$$

+ H₂Ö:

⁻OH acts as a base, **not** a nucleophile.

Ethyl acetate is a common organic solvent with a characteristic odor. It is used in nail polish remover and model airplane glue.

Intramolecular esterification of γ- and δ-hydroxy carboxylic acids forms five- and six-membered lactones.

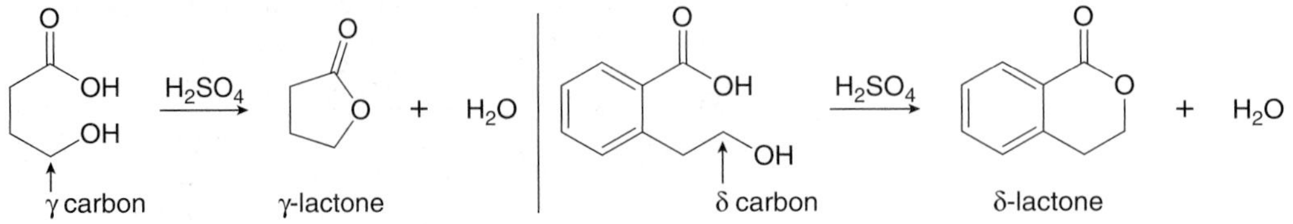

PROBLEM 22.20 Draw the products of each reaction.

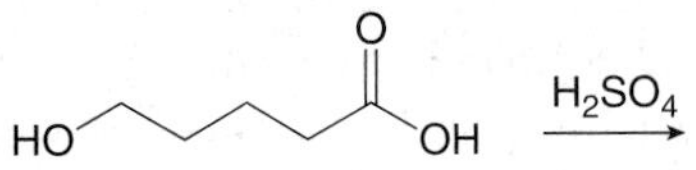

PROBLEM 22.21 Draw the products formed when benzoic acid (C_6H_5COOH) is treated with CH_3OH having its O atom labeled with ^{18}O ($CH_3^{18}OH$). Indicate where the labeled oxygen atom resides in the products.

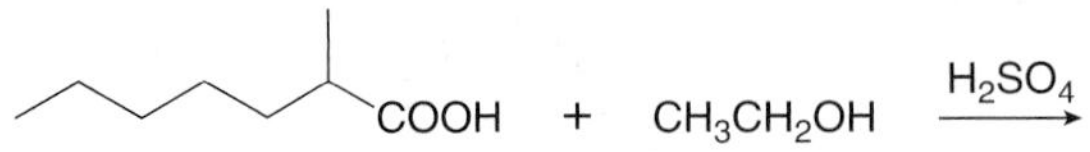

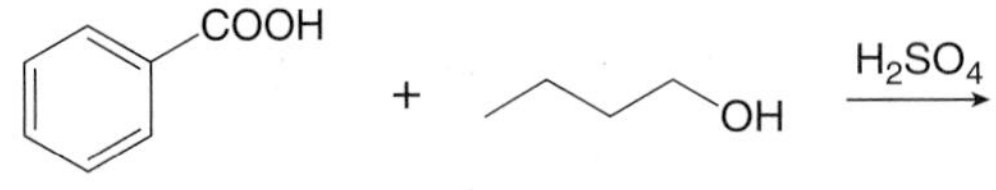

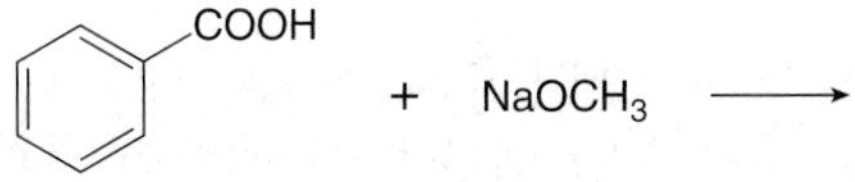

PROBLEM 22.22 Draw a stepwise mechanism for the following reaction.

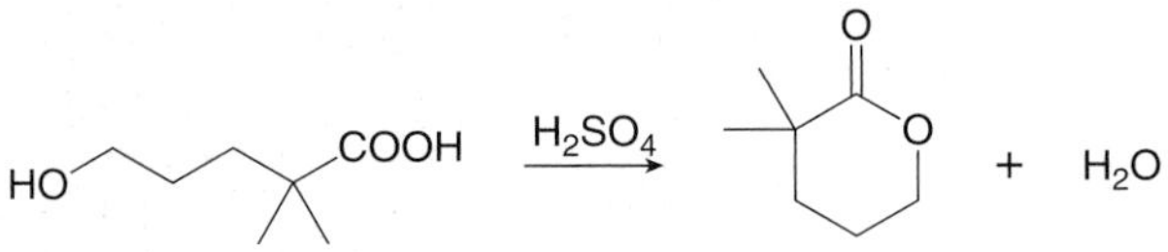

22.10D Conversion of RCOOH to RCONR'$_2$

The direct conversion of a carboxylic acid to an amide with NH_3 or an amine is very difficult, even though a more reactive acyl compound is being transformed into a less reactive one. The problem is that carboxylic acids are strong organic acids and NH_3 and amines are bases, so they undergo an **acid–base reaction to form an ammonium salt** before any nucleophilic substitution occurs.

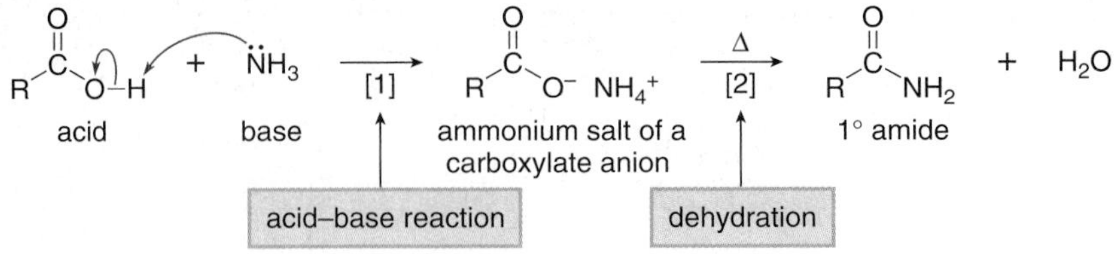

Heating at high temperature (>100 °C) dehydrates the resulting ammonium salt of the carboxylate anion to form an amide, though the yield can be low.

Therefore, the overall conversion of RCOOH to RCONH$_2$ requires two steps:

[1] **Acid–base reaction of RCOOH with NH$_3$ to form an ammonium salt**
[2] **Dehydration at high temperature (>100 °C)**

Amides are much more easily prepared from acid chlorides and anhydrides, as discussed in Sections 22.8 and 22.9.

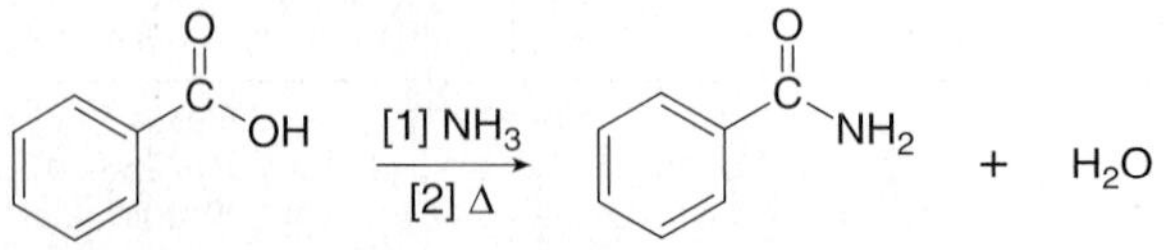

A carboxylic acid and an amine readily react to form an amide in the presence of an additional reagent, **dicyclohexylcarbodiimide (DCC),** which is converted to the by-product dicyclohexylurea in the course of the reaction.

Amide formation using DCC

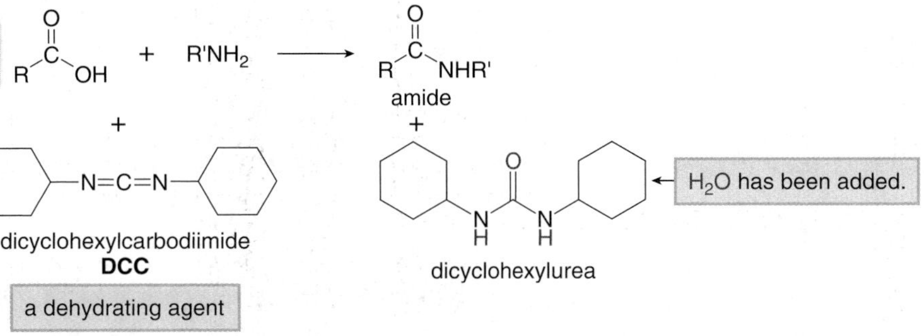

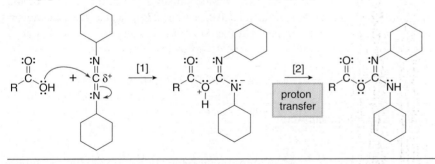

DCC is a dehydrating agent. The dicyclohexylurea by-product is formed by adding the elements of H_2O to DCC. DCC promotes amide formation by converting the carboxy OH group into a better leaving group.

Example

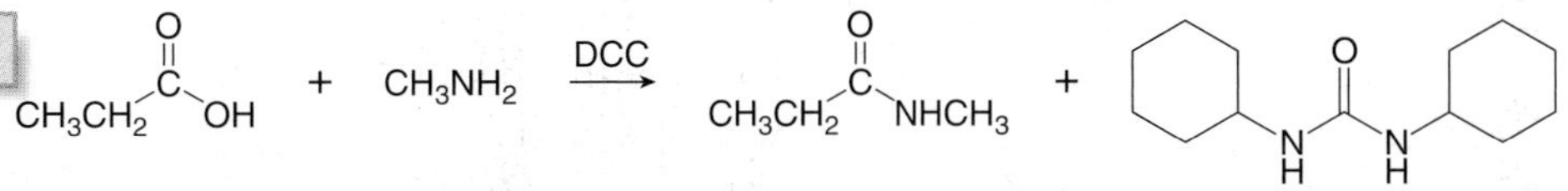

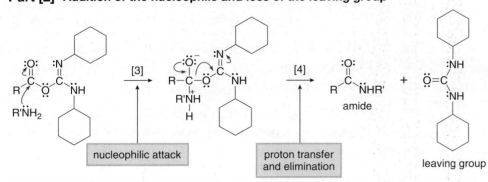

The mechanism consists of two parts: [1] conversion of the OH group into a better leaving group, followed by [2] **addition of the nucleophile and loss of the leaving group** to form the product of nucleophilic acyl substitution (Mechanism 22.7).

MECHANISM 22.7

Conversion of Carboxylic Acids to Amides with DCC

Part [1]   Conversion of OH into a better leaving group

- The first part of the mechanism consists of two steps that convert the carboxy OH group into a **better leaving group.**
- This process *activates* the carboxy group towards nucleophilic attack in Part [2].

Part [2]   Addition of the nucleophile and loss of the leaving group

- In Steps [3] and [4], **nucleophilic attack of the amine** on the activated carboxy group, followed by **elimination of dicyclohexylurea** as the leaving group, forms the amide.

The reaction of an acid and an amine with DCC is often used in the laboratory to form the amide bond in peptides, as is discussed in Chapter 28.

PROBLEM 22.23   What product is formed when acetic acid is treated with each reagent: (a) CH_3NH_2; (b) CH_3NH_2, then heat; (c) CH_3NH_2 + DCC?

22.11   Reactions of Esters

Esters can be converted into carboxylic acids and amides.

◆ **Esters are hydrolyzed with water in the presence of either acid or base to form carboxylic acids or carboxylate anions.**

Ester hydrolysis—General reaction

$$R\text{-}C(=O)\text{-}OR' \xrightarrow[(H^+ \text{ or } ^-OH)]{H_2O} R\text{-}C(=O)\text{-}OH \quad or \quad R\text{-}C(=O)\text{-}O^- \quad + \quad R'OH$$

carboxylic acid (in acid)   carboxylate anion (in base)

◆ **Esters react with NH_3 and amines to form 1°, 2°, or 3° amides.**

Reaction with nitrogen nucleophiles

$$R\text{-}C(=O)\text{-}OR' \xrightarrow{\quad \backslash N / \quad} R\text{-}C(=O)\text{-}NH_2 \quad or \quad R\text{-}C(=O)\text{-}NHR'' \quad or \quad R\text{-}C(=O)\text{-}NR''_2 \quad + \quad R'OH$$

(with NH_3)   (with $R''NH_2$)   (with R''_2NH)
1° amide   2° amide   3° amide

22.11A   Ester Hydrolysis in Aqueous Acid

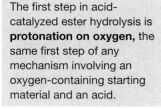
The first step in acid-catalyzed ester hydrolysis is **protonation on oxygen,** the same first step of any mechanism involving an oxygen-containing starting material and an acid.

The hydrolysis of esters in aqueous acid is a reversible equilibrium reaction that is driven to the right by using a large excess of water.

$$CH_3\text{-}C(=O)\text{-}OCH_2CH_3 \quad + \quad H_2O \xrightleftharpoons{H_2SO_4} CH_3\text{-}C(=O)\text{-}OH \quad + \quad CH_3CH_2OH$$

The mechanism of ester hydrolysis in acid (shown in Mechanism 22.8) is the reverse of the mechanism of ester synthesis from carboxylic acids (Mechanism 22.6). Thus, the mechanism consists of the addition of the nucleophile and the elimination of the leaving group, the two steps common to all nucleophilic acyl substitutions, as well as several proton transfers, because the reaction is acid-catalyzed.

MECHANISM 22.8

Acid-Catalyzed Hydrolysis of an Ester to a Carboxylic Acid

Part [1] Addition of the nucleophile H_2O

nucleophilic addition

◆ **Protonation** in Step [1] makes the carbonyl group more electrophilic.

◆ **Nucleophilic addition of H_2O** forms a tetrahedral intermediate, and loss of a proton forms the neutral addition product (Steps [2]–[3]).

Part [2] Elimination of the leaving group R'OH

loss of R'OH

◆ Protonation of the OR' group in Step [4] forms a **good leaving group (R'OH) that is eliminated** in Step [5].

◆ Loss of a proton in Step [6] forms the carboxylic acid.

PROBLEM 22.24   In Mechanism 22.8, only one Lewis structure is drawn for each intermediate. Draw all other resonance structures for the resonance-stabilized intermediates.

PROBLEM 22.25   Draw a stepwise mechanism for the following reaction.

$$C_6H_5COOCH_3 \quad + \quad H_2O \quad \xrightarrow{H_2SO_4} \quad C_6H_5COOH \quad + \quad CH_3OH$$

22.11B   Ester Hydrolysis in Aqueous Base

Esters are hydrolyzed in aqueous base to form carboxylate anions. Basic hydrolysis of an ester is called **saponification.**

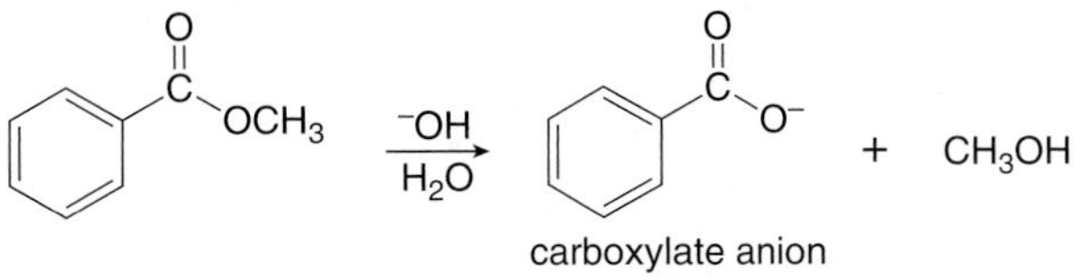

The word **saponification** comes from the Latin *sapo* meaning **soap.** Soap is prepared by hydrolyzing esters in fats with aqueous base, as explained in Section 22.12B.

carboxylate anion

The mechanism for this reaction has the usual two steps of the general mechanism for nucleophilic acyl substitution presented in Section 22.7—**addition of the nucleophile** followed by **loss of a leaving group**—plus an additional step involving proton transfer (Mechanism 22.9).

MECHANISM 22.9

Base-Promoted Hydrolysis of an Ester to a Carboxylic Acid

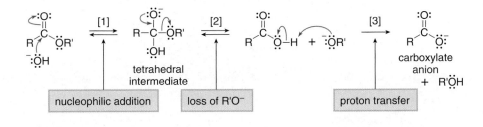

◆ Steps [1] and [2] result **in addition of the nucleophile, ⁻OH, followed by elimination of the leaving group, ⁻OR'.** These two steps, which form the carboxylic acid, are reversible, because the stability of the reactants and products is comparable.

◆ Next, the carboxylic acid is a strong organic acid and the leaving group (⁻OR') is a strong base, so an **acid–base reaction** occurs in Step [3] to form the carboxylate anion.

The carboxylate anion is resonance stabilized, and this drives the equilibrium in its favor. Once the reaction is complete and the carboxylate anion is formed, it can be protonated with strong acid to form the neutral carboxylic acid.

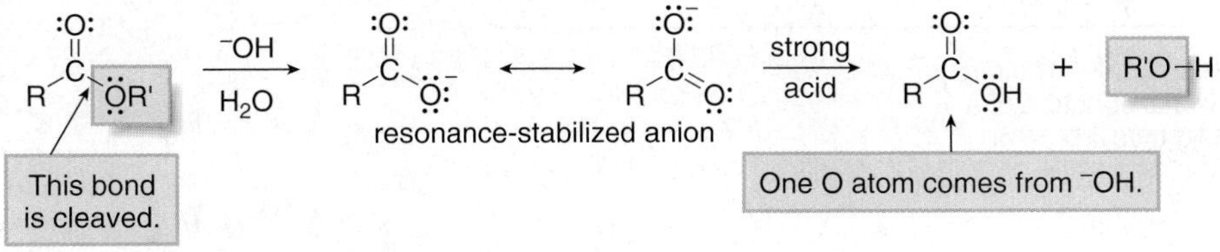

Hydrolysis is base promoted, *not* **base catalyzed,** because the base (⁻OH) is the nucleophile that adds to the ester and forms part of the product. It participates in the reaction and is not regenerated later.

Where do the oxygen atoms in the product come from? The C—OR' bond in the ester is cleaved, so the OR' group becomes the alcohol by-product (R'OH) and one of the oxygens in the carboxylate anion product comes from ⁻OH (the nucleophile).

PROBLEM 22.26   Draw the products formed when each ester is treated with ⁻OH, H_2O, followed by treatment with strong acid.

a.

$$CH_3-\overset{\overset{\displaystyle O}{\|}}{C}-OCH_2CH_3$$

b.

COOCH(CH₃)₂

OCH₃

c.

$$\overset{\overset{\displaystyle O}{\|}}{C}-OCH_3$$

PROBLEM 22.27   When each isotopically-labeled starting material is hydrolyzed with aqueous base, where does the label end up in the products?

a.

$$CH_3CH_2-\overset{\overset{\displaystyle O}{\|}}{C}-{}^{18}OCH_3$$

b.

$$CH_3CH_2-\overset{\overset{\displaystyle {}^{18}O}{\|}}{C}-OCH_3$$

22.12   Application: Lipid Hydrolysis

22.12A   Olestra—A Synthetic Fat

The most prevalent naturally occurring esters are the **triacylglycerols,** which were first discussed in Section 10.6. **Triacylglycerols are the lipids that comprise animal fats and vegetable oils.**

◆ Each triacylglycerol is a triester, containing three long hydrocarbon side chains.
◆ Unsaturated triacylglycerols have one or more double bonds in their long hydrocarbon chains, whereas saturated triacylglycerols have none.

R groups have 11–19 C's.
[Three ester groups are labeled in red.]

triacylglycerol
the most common type of lipid

Figure 22.4 contains a ball-and-stick model of a saturated fat.

Animals store energy in the form of triacylglycerols, kept in a layer of fat cells below the surface of the skin. This fat serves to insulate the organism, as well as provide energy for its metabolic needs for long periods. The first step in the metabolism of a triacylglycerol is hydrolysis of the ester bonds to form glycerol and three fatty acids. **This reaction is simply ester hydrolysis.** In cells, this reaction is carried out with enzymes called **lipases.**

Figure 22.4   The three-dimensional structure of a saturated triacylglycerol

• This triacylglycerol has no double bonds in the three R groups (each with 11 C's) bonded to the ester carbonyl, making it a saturated fat.

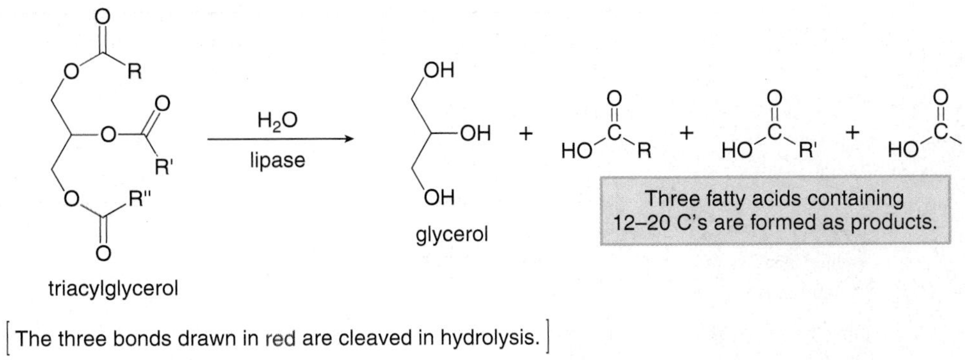

The three bonds drawn in red are cleaved in hydrolysis.

The fatty acids produced on hydrolysis are then oxidized in a stepwise fashion, ultimately yielding CO_2 and H_2O, as well as a great deal of energy. Oxidation of fatty acids yields twice as much energy per gram as oxidation of an equivalent weight of carbohydrate.

Diets high in fat content lead to a large amount of stored fat, ultimately causing an individual to be overweight. One recent attempt to reduce calories in common snack foods has been to substitute "fake fats" such as **olestra** (trade name: **Olean**) for triacylglycerols.

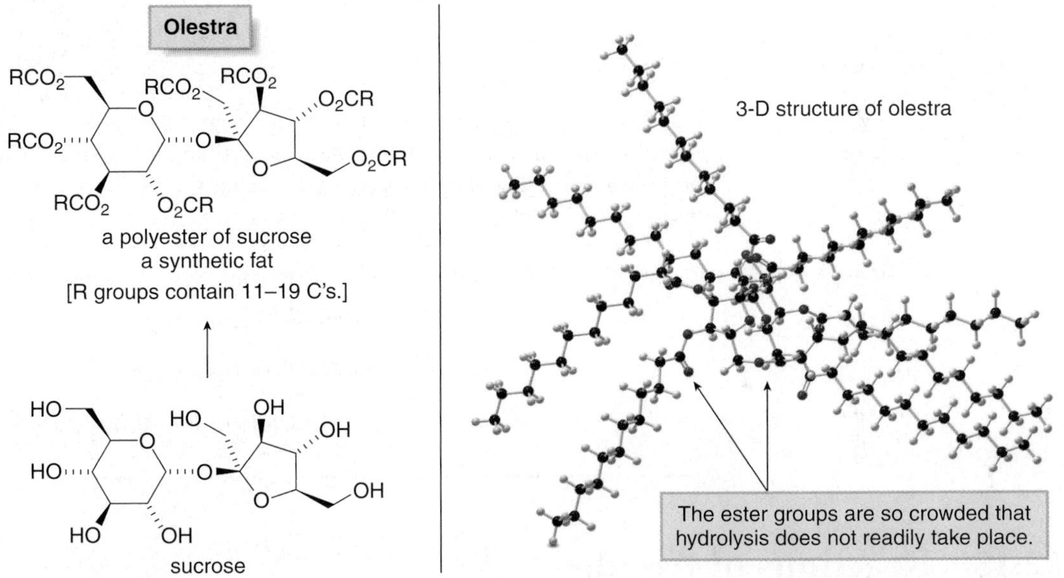

Olestra is a polyester formed from long-chain fatty acids and sucrose, the sweet-tasting carbohydrate in table sugar. Naturally occurring triacylglycerols are also polyesters formed from long-chain fatty acids, but olestra has so many ester units clustered together in close proximity that they are too hindered to be hydrolyzed. As a result, olestra is not metabolized. Instead, it passes through the body unchanged, providing no calories to the consumer.

Thus, olestra's many C–C and C–H bonds make it similar in solubility to naturally occurring triacylglycerols, but its three-dimensional structure makes it inert to hydrolysis because of steric hindrance.

PROBLEM 22.28   How would you synthesize olestra from sucrose?

22.12B   The Synthesis of Soap

Soap has been previously discussed in Section 3.7.

Soap is prepared by the basic hydrolysis or saponification of a triacylglycerol. Heating an animal fat or vegetable oil with aqueous base hydrolyzes the three esters to form glycerol and sodium salts of three fatty acids. These carboxylate salts are **soaps,** which clean away dirt

All soaps are salts of fatty acids. The main difference between soaps is the addition of other ingredients that do not alter their cleaning properties: dyes for color, scents for a pleasing odor, and oils for lubrication. Soaps that float are aerated so that they are less dense than water.

because of their two structurally different regions. The nonpolar tail dissolves grease and oil and the polar head makes it soluble in water (Figure 3.7). Most triacylglycerols have two or three different R groups in their hydrocarbon chains, so soaps are usually mixtures of two or three different carboxylate salts.

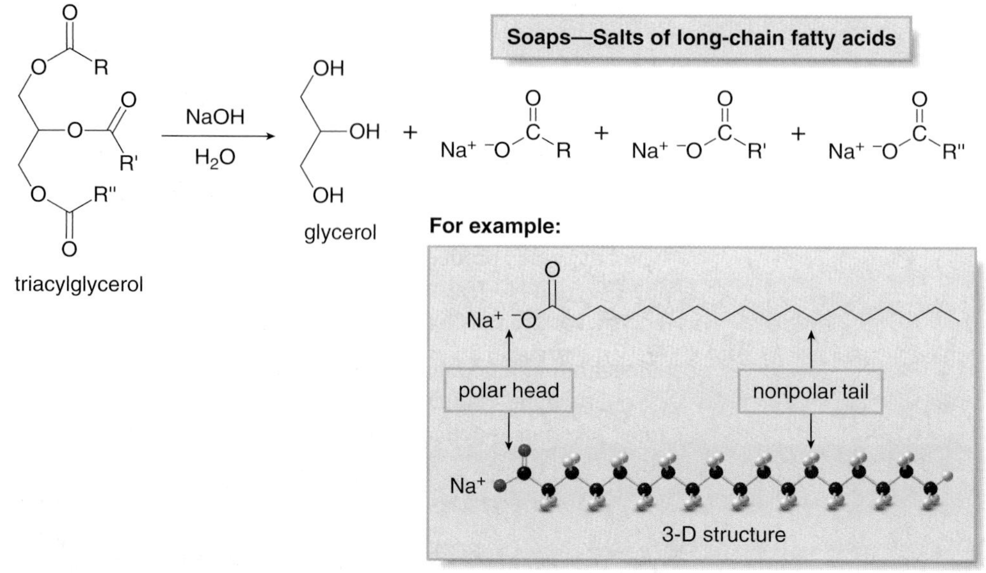

Soaps—Salts of long-chain fatty acids

triacylglycerol

glycerol

For example:

polar head

nonpolar tail

3-D structure

Soaps are typically made from lard (from hogs), tallow (from cattle or sheep), coconut oil, or palm oil. All soaps work in the same way, but have somewhat different properties depending on the lipid source. The length of the carbon chain in the fatty acids and the number of degrees of unsaturation affect the properties of the soap to some extent.

PROBLEM 22.29   What is the composition of the soap prepared by hydrolysis of this triacylglycerol?

$$CH_2OCO(CH_2)_{15}CH_3$$
$$CHOCO(CH_2)_{15}CH_3$$
$$CH_2OCO(CH_2)_7CH=CH(CH_2)_7CH_3$$
$$cis$$

22.13   Reactions of Amides

Because amides have the poorest leaving group of all the carboxylic acid derivatives, they are the least reactive. In fact, they have only one useful reaction. Under strenuous reaction conditions, **amides are hydrolyzed in acid or base to form carboxylic acids or carboxylate anions.**

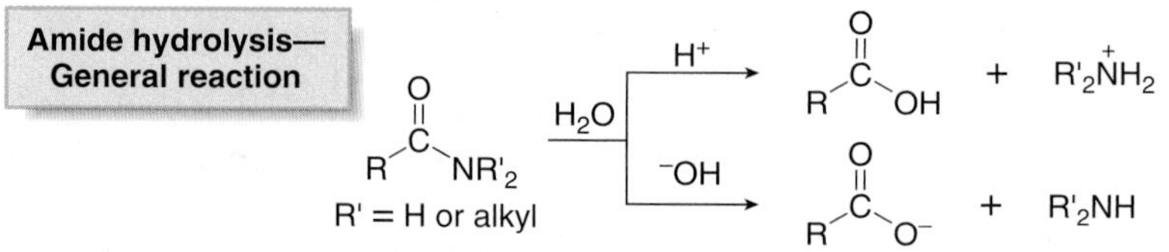

Amide hydrolysis— General reaction

In acid, the amine by-product is protonated as an ammonium ion, whereas in base, a neutral amine is formed.

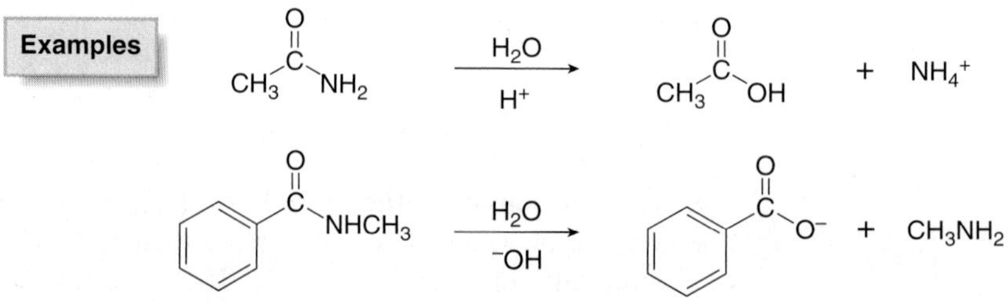

Examples

The relative lack of reactivity of the amide bond is notable in proteins, which are polymers of amino acids connected by amide linkages (Section 22.6B). Proteins are stable in aqueous solution in the absence of acid or base, so they can perform their various functions in the aqueous cellular environment without breaking down. The hydrolysis of the amide bonds in proteins requires a variety of specific enzymes.

The mechanism of amide hydrolysis in acid is exactly the same as the mechanism of ester hydrolysis in acid (Section 22.11A) except that the leaving group is different.

The mechanism of amide hydrolysis in base has the usual two steps of the general mechanism for nucleophilic acyl substitution—**addition of the nucleophile** followed by **loss of a leaving group**—plus an additional proton transfer. The initially formed carboxylic acid reacts further under basic conditions to form the resonance-stabilized carboxylate anion, and this drives the reaction to completion. Mechanism 22.10 is written for a 1° amide.

MECHANISM 22.10

Amide Hydrolysis in Base

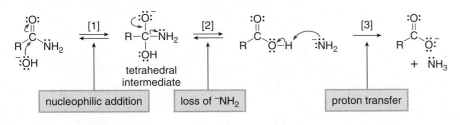

- Steps [1] and [2] result **in addition of the nucleophile, ⁻OH,** followed by **elimination of the leaving group, ⁻NH₂.**

- Because the carboxylic acid is a strong organic acid and the leaving group (⁻NH₂) is a strong base, an **acid–base reaction** occurs in Step [3] to form the carboxylate anion.

For amide hydrolysis to occur, the tetrahedral intermediate must lose ⁻NH₂, a stronger base and therefore poorer leaving group than ⁻OH (Step [2]). This means that loss of ⁻NH₂ does not often happen. Instead, ⁻OH is lost as the leaving group most of the time, and the starting material is regenerated. But, when ⁻NH₂ is occasionally eliminated, the carboxylic acid product is converted to a lower energy carboxylate anion in Step [3], and this drives the equilibrium to favor its formation.

PROBLEM 22.30   Draw a stepwise mechanism for the following reaction.

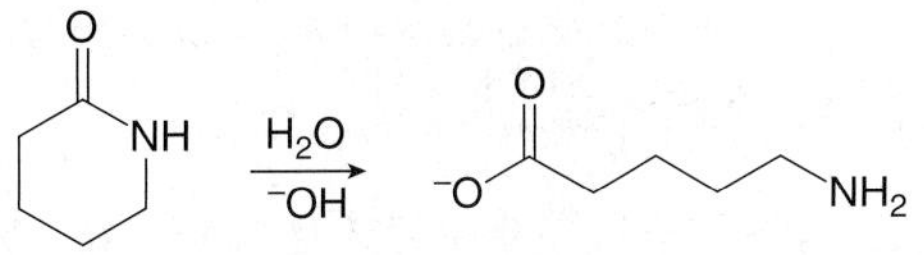

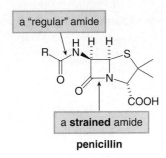

22.14   Application: The Mechanism of Action of β-Lactam Antibiotics

Penicillin and related β-lactams kill bacteria by a nucleophilic acyl substitution reaction. All penicillins have an unreactive amide side chain and a very reactive amide that is part of a β-lactam. The β-lactam is more reactive than other amides because it is part of a strained, four-membered ring that is readily opened with nucleophiles.

a "regular" amide

a **strained** amide

penicillin

Unlike mammalian cells, bacterial cells are surrounded by a fairly rigid cell wall, which allows the bacterium to live in many different environments. This protective cell wall is composed of carbohydrates linked together by peptide chains containing amide linkages, formed using the enzyme **transpeptidase.**

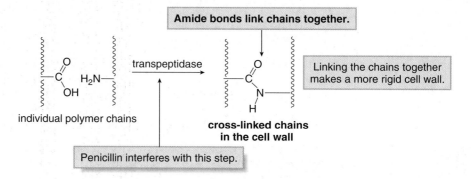

Penicillin interferes with the synthesis of the bacterial cell wall. Penicillin reacts with transpeptidase by a nucleophilic substitution reaction, resulting in cleavage of the β-lactam ring. The opened ring of the penicillin molecule remains covalently bonded to the enzyme, thus deactivating the enzyme, halting cell wall construction, and killing the bacterium.

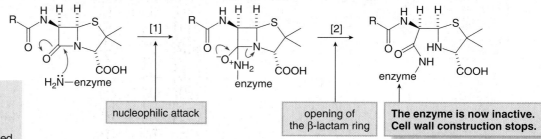

Penicillin has no effect on mammalian cells because they are surrounded by a flexible membrane composed of a lipid bilayer (Chapter 3) and not a cell wall.

Thus, penicillin and other β-lactam antibiotics are biologically active precisely because they undergo a nucleophilic acyl substitution reaction with an important bacterial enzyme.

PROBLEM 22.31   Some penicillins cannot be administered orally because their β-lactam is rapidly hydrolyzed by the acidic environment of the stomach. What product is formed in the following hydrolysis reaction?

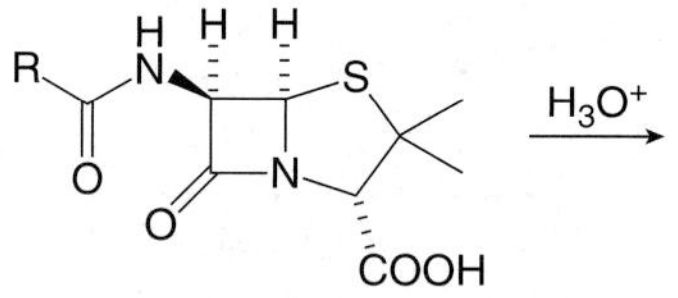

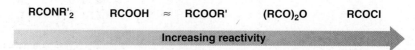

22.15 Summary of Nucleophilic Acyl Substitution Reactions

To help you organize and remember all of the nucleophilic acyl substitution reactions that can occur at a carbonyl carbon, keep in mind the following two principles:

> ◆ The better the leaving group, the more reactive the carboxylic acid derivative.
> ◆ More reactive acyl compounds can always be converted to less reactive ones. The reverse is not usually true.

This results in the following order of reactivity:

$RCONR'_2$ $RCOOH$ ≈ $RCOOR'$ $(RCO)_2O$ $RCOCl$

Increasing reactivity →

Table 22.5 summarizes the specific nucleophilic acyl substitution reactions. Use it as a quick reference to remind you which products can be formed from a given starting material.

| **TABLE 22.5** | **Summary of the Nucleophilic Substitution Reactions of Carboxylic Acids and Their Derivatives** | | | | |

	RCOCl	**(RCO)₂O**	**RCOOH**	**RCOOR'**	**RCONR'₂**
[1] RCOCl →	–	✓	✓	✓	✓
[2] (RCO)₂O →	✗	–	✓	✓	✓
[3] RCOOH →	✓	✓	–	✓	✓
[4] RCOOR' →	✗	✗	✓	–	✓
[5] RCONR'₂ →	✗	✗	✓	✗	–

Table key: ✓ = A reaction occurs.
✗ = No reaction occurs.

22.16   Natural and Synthetic Fibers

All natural and synthetic fibers are high molecular weight polymers. Natural fibers are obtained from either plant or animal sources, and this determines the fundamental nature of their chemical structure. Fibers like **wool and silk obtained from animals are proteins,** and so they are formed from amino acids joined together by many amide linkages. **Cotton and linen,** on the other hand, are derived from plants and so they are **carbohydrates having the general structure of cellulose,** formed from glucose monomers. General structures for these polymers are shown in Figure 22.5.

An important practical application of organic chemistry has been the synthesis of synthetic fibers, many of which have properties that are different from and sometimes superior to their naturally occurring counterparts. The two most common classes of synthetic polymers are based on polyamides and polyesters.

22.16A   Nylon—A Polyamide

The search for a synthetic fiber with properties similar to silk led to the discovery of **nylon** (Section 3.4B), a **polyamide.** There are several different kinds of nylon, but the most well known is called nylon 6,6.

Figure 22.5   The general structure of the common natural fibers

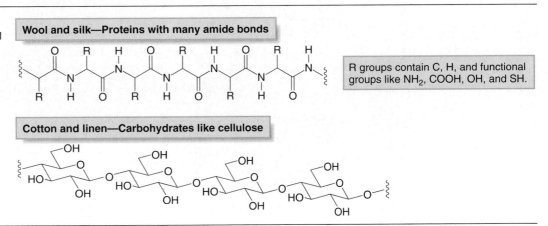

Wool and silk—Proteins with many amide bonds

R groups contain C, H, and functional groups like NH₂, COOH, OH, and SH.

Cotton and linen—Carbohydrates like cellulose

DuPont built the first commercial nylon plant in 1938. Although it was first used by the military to make parachutes, nylon quickly replaced silk in many common clothing articles after World War II.

Nylon 6,6

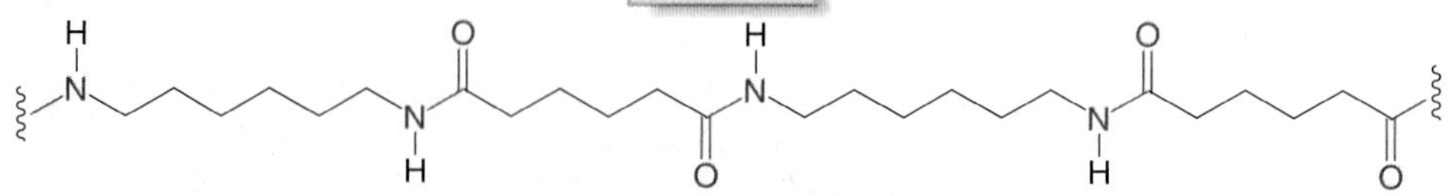

[The amide bonds are labeled in red.]

Nylon 6,6 can be synthesized from two six-carbon monomers (hence its name)—adipoyl chloride **(ClOCCH$_2$CH$_2$CH$_2$CH$_2$COCl)** and hexamethylenediamine **(H$_2$NCH$_2$CH$_2$CH$_2$CH$_2$CH$_2$CH$_2$NH$_2$).** This diacid chloride and diamine react together to form new amide bonds, yielding the polymer. Nylon is called a **condensation polymer** because a small molecule, in this case HCl, is eliminated during its synthesis.

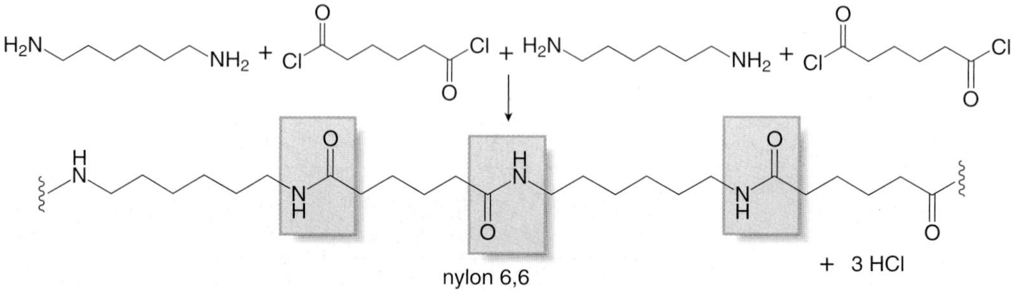

nylon 6,6

+ 3 HCl

Three new amide bonds are shown.

PROBLEM 22.32   What two monomers are needed to prepare nylon 6,10?

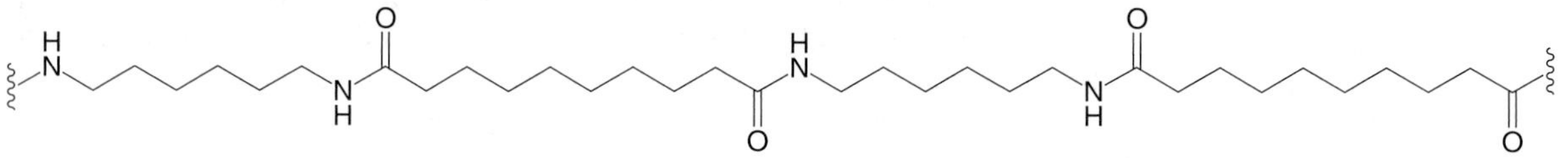

nylon 6,10

22.16B   Polyesters

Polyesters constitute a second major class of condensation polymers. The most common polyester is polyethylene terephthalate (**PET**), which is sold under a variety of trade names (Dacron, Terylene, and Mylar) depending on its use.

Polyethylene terephthalate
PET
(Dacron, Terylene, and Mylar)

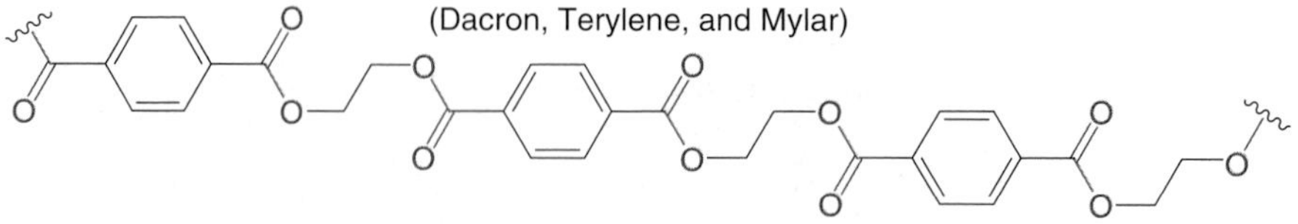

Ester bonds (in red) join the carbon skeleton together.

One method of synthesizing a polyester is by acid-catalyzed esterification of a diacid with a diol (Fischer esterification).

Synthesis of PET

terephthalic acid   ethylene glycol

two monomers needed
for PET synthesis

acid catalyst

+ 3 H₂O

Three new ester bonds are shown.

Because these polymers are easily and cheaply prepared and form strong and chemically stable materials, they have been used in clothing, films, tires, and many other products.

PROBLEM 22.33

Two polyesters have received recent attention because a starting material needed for their preparation can be obtained from carbohydrates, rather than petroleum, making them much more "environmentally friendly." What is the structure of each polymer?
a. PLA (polylactic acid), formed by polymerization of lactic acid, $CH_3CH(OH)COOH$.
b. PTT (polytrimethylene terephthalate), formed by polymerization of 1,3-propanediol $(HOCH_2CH_2CH_2OH)$ and terephthalic acid.

Spandex is lighter in weight than many other elastic polymers, and it does not break down when exposed to perspiration and detergents.

22.16C   Molecular Structure and Macroscopic Properties

The macroscopic properties of polymers, like all molecules, depend on their structure at the molecular level. **Spandex,** for example, was first used in ladies' corsets, girdles, and support hose, but is now routinely used in both men's and women's active wear. Spandex is strong and lends "support" to the wearer, but it also stretches. On the molecular level, it has rigid regions that are joined together by soft, flexible segments. The flexible regions allow the polymer to expand and then recover its original shape. The rigid regions strengthen the polymer.

flexible portion

rigid segment

spandex
(Trade name: Lycra)

22.17   Biological Acylation Reactions

Nucleophilic acyl substitution is a common reaction in biological systems. These acylation reactions are called **acyl transfer reactions** because they result in the transfer of an acyl group from one atom to another (from Z to Nu in this case).

Acyl transfer reaction

:Nu⁻   +   :Z⁻

The acyl group is transferred from Z to Nu.

In cells, such acylations occur with the sulfur analogue of an ester, called a **thioester,** having the general structure **RCOSR'.** The most common thioester is called **acetyl coenzyme A,** often referred to merely as **acetyl CoA.**

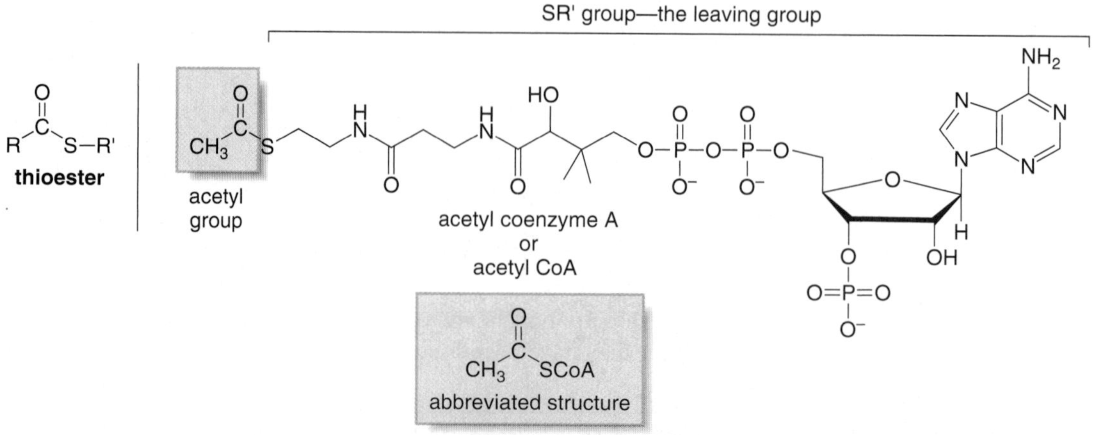

♦ A thioester (RCOSR') has a good leaving group (⁻SR) so, like other acyl compounds, it undergoes substitution reactions with other nucleophiles.

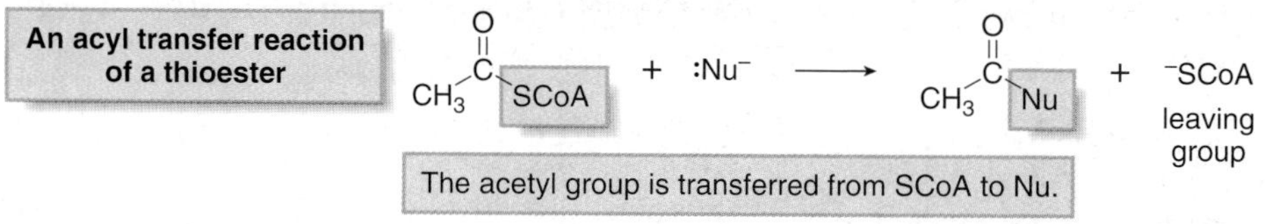

For example, acetyl CoA undergoes enzyme-catalyzed nucleophilic acyl substitution with choline, forming acetylcholine, a charged compound that transmits nerve impulses between nerve cells.

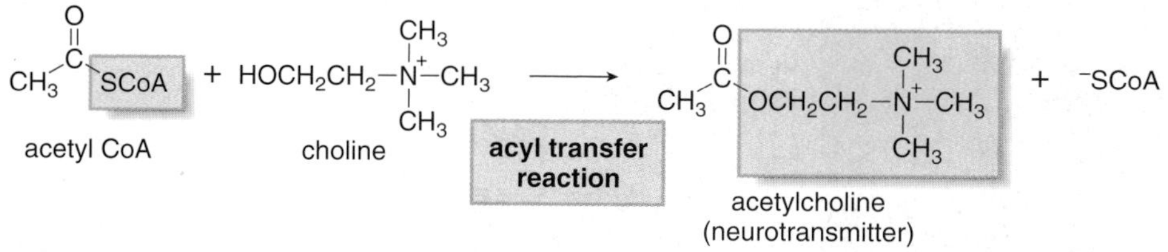

Many other acyl transfer reactions are important cellular processes. Thioesters of fatty acids react with cholesterol, forming **cholesteryl esters** in an enzyme-catalyzed reaction (Figure 22.6). These esters are the principal form in which cholesterol is stored and transported in the body. Because cholesterol is a lipid, insoluble in the aqueous environment of the blood, it travels through the bloodstream in particles that also contain proteins and phospholipids. These particles are classified by their density.

♦ **LDL particles** (low density lipoproteins) transport cholesterol from the liver to the tissues.
♦ **HDL particles** (high density lipoproteins) transport cholesterol from the tissues back to the liver, where it is metabolized or converted to other steroids.

Atherosclerosis is a disease that results from the build up of fatty deposits on the walls of arteries, forming deposits called **plaque.** They are composed largely of the cholesterol (esterified as an ester) of LDL particles. LDL is often referred to as "bad cholesterol" for this reason. In contrast, HDL particles are called "good cholesterol" because they reduce the amount of cholesterol in the bloodstream by transporting it back to the liver.

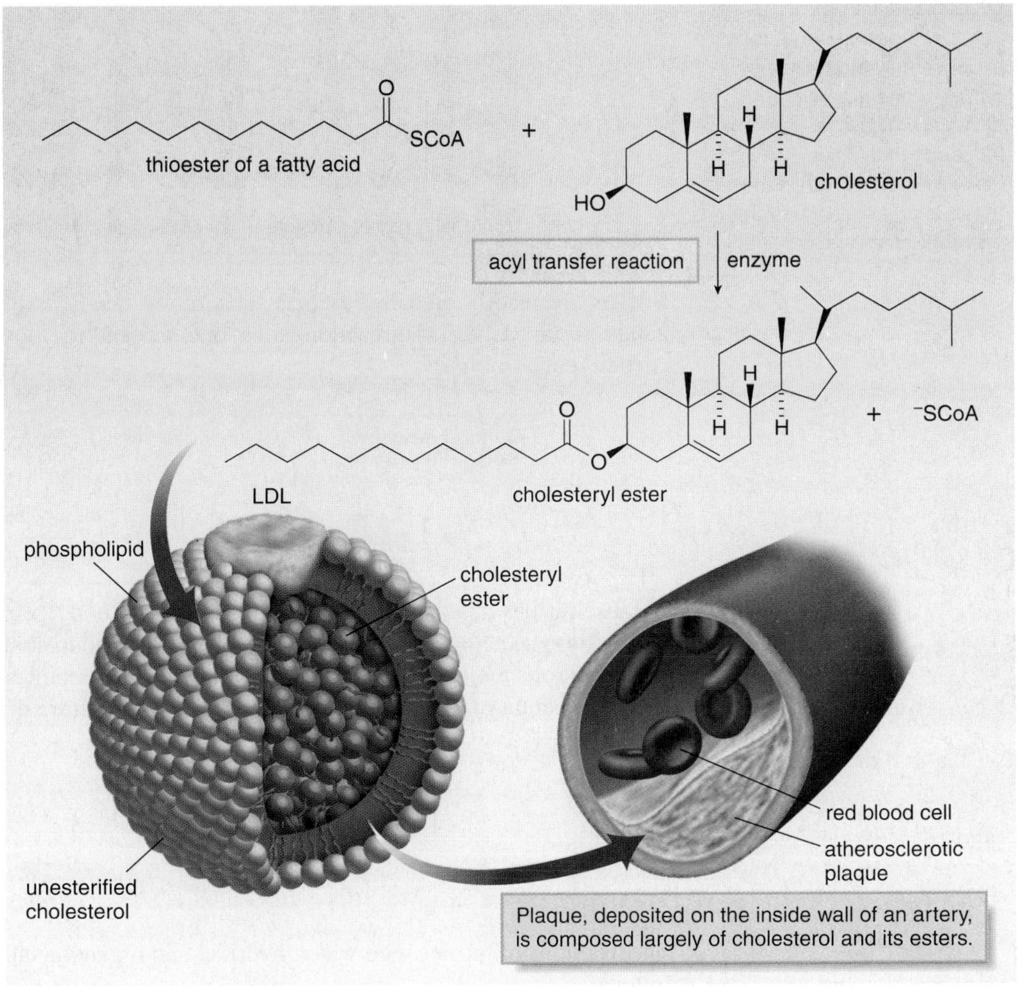

Figure 22.6 Cholesteryl esters and LDL particles

thioester of a fatty acid + cholesterol

acyl transfer reaction | enzyme

cholesteryl ester + ⁻SCoA

LDL

phospholipid

cholesteryl ester

unesterified cholesterol

red blood cell

atherosclerotic plaque

Plaque, deposited on the inside wall of an artery, is composed largely of cholesterol and its esters.

PROBLEM 22.34 Draw the products of each reaction.

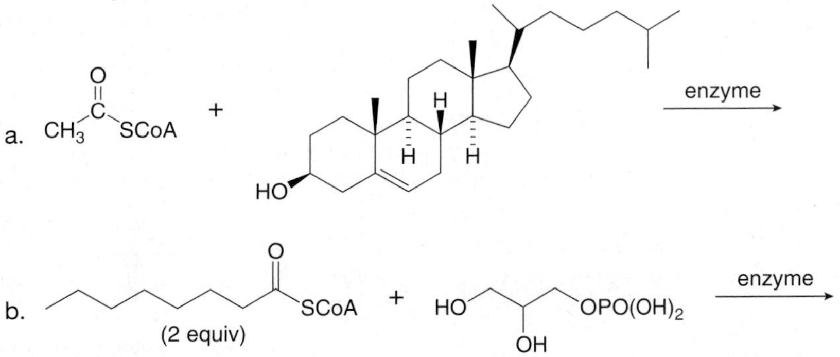

a. CH_3—C(=O)—SCoA + [steroid with HO] $\xrightarrow{\text{enzyme}}$

b. [fatty acyl]—SCoA (2 equiv) + HO—CH$_2$—CH(OH)—CH$_2$—OPO(OH)$_2$ $\xrightarrow{\text{enzyme}}$

22.18 Nitriles

We end Chapter 22 with the chemistry of **nitriles (RC≡N).** Nitriles have a carbon atom in the same oxidation state as in the acyl compounds that are the principal focus of Chapter 22. Moreover, the chemical reactions of nitriles illustrate some of the concepts first discussed earlier in Chapter 22 and in Chapters 20 and 21.

In addition to the cyanohydrins discussed in Section 21.9, two useful biologically active nitriles are **letrozole** and **anastrozole,** new drugs that reduce the recurrence of breast cancer in women whose tumors are estrogen positive.

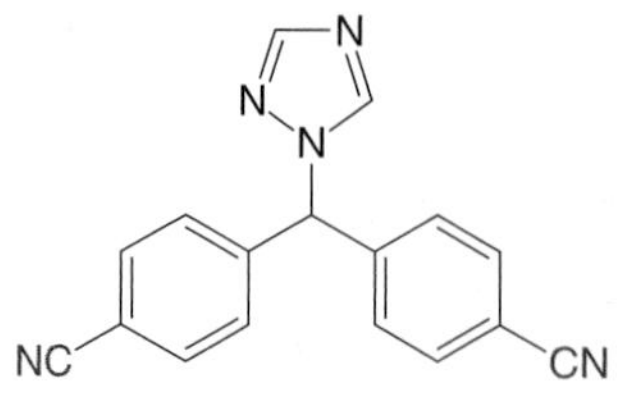

Generic name: letrozole
Trade name: Femara

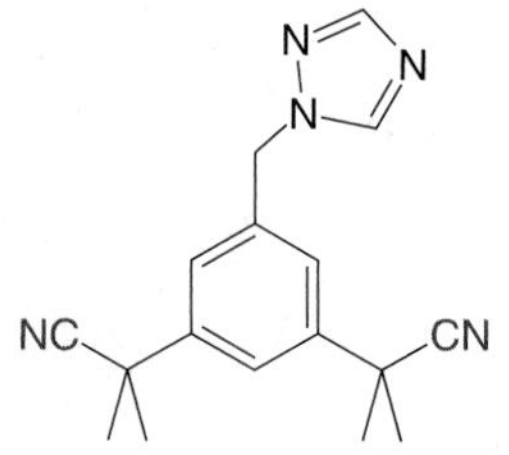

Generic name: anastrozole
Trade name: Arimidex

Nitriles are readily prepared by **S_N2** substitution reactions of unhindered methyl and 1° alkyl halides with ^-CN. This reaction adds one carbon to the alkyl halide and **forms a new carbon–carbon bond.**

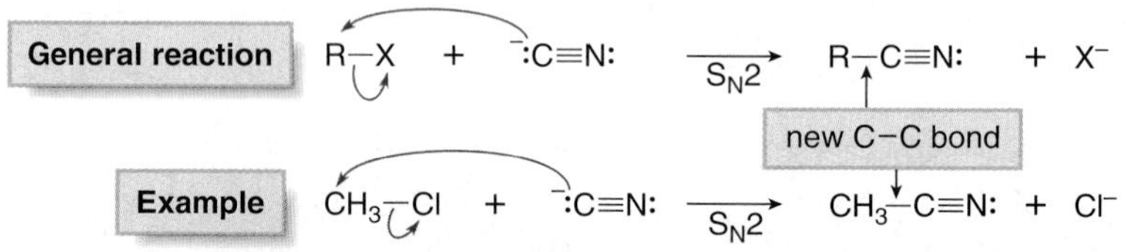

General reaction $R-X$ + $:C\equiv N:$ $\xrightarrow{S_N2}$ $R-C\equiv N:$ + X^-

new C–C bond

Example CH_3-Cl + $:C\equiv N:$ $\xrightarrow{S_N2}$ $CH_3-C\equiv N:$ + Cl^-

Because nitriles have no leaving group, they do not undergo nucleophilic substitution reactions like carboxylic acid derivatives. Because the cyano group contains an electrophilic carbon atom that is part of a multiple bond, a nitrile reacts with nucleophiles by a **nucleophilic addition reaction.** The nature of the nucleophile determines the structure of the product.

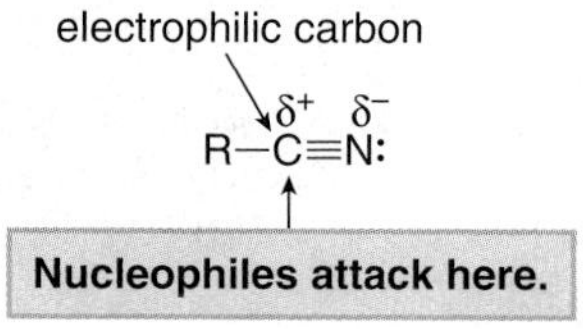

electrophilic carbon

$\overset{\delta^+}{R-C}\equiv\overset{\delta^-}{N:}$

Nucleophiles attack here.

The reactions of nitriles with water, hydride, and organometallic reagents as nucleophiles are as follows:

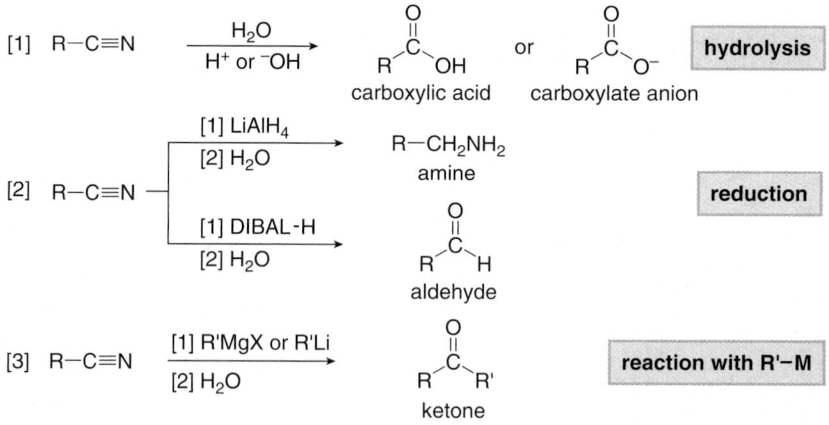

[1] $R-C\equiv N$ $\xrightarrow[\text{H}^+\text{ or }^-\text{OH}]{\text{H}_2\text{O}}$

$\underset{\text{carboxylic acid}}{R-\overset{\overset{\textstyle O}{\|}}{C}-OH}$ or $\underset{\text{carboxylate anion}}{R-\overset{\overset{\textstyle O}{\|}}{C}-O^-}$ **hydrolysis**

[2] $R-C\equiv N$

$\xrightarrow[\text{[2] H}_2\text{O}]{\text{[1] LiAlH}_4}$ $\underset{\text{amine}}{R-CH_2NH_2}$

$\xrightarrow[\text{[2] H}_2\text{O}]{\text{[1] DIBAL-H}}$ $\underset{\text{aldehyde}}{R-\overset{\overset{\textstyle O}{\|}}{C}-H}$ **reduction**

[3] $R-C\equiv N$ $\xrightarrow[\text{[2] H}_2\text{O}]{\text{[1] R'MgX or R'Li}}$ $\underset{\text{ketone}}{R-\overset{\overset{\textstyle O}{\|}}{C}-R'}$ **reaction with R'–M**

22.18A   Hydrolysis of Nitriles

Nitriles are hydrolyzed with water in the presence of acid or base to yield **carboxylic acids** or **carboxylate anions.** In this reaction, the three C–N bonds are replaced by three C–O bonds.

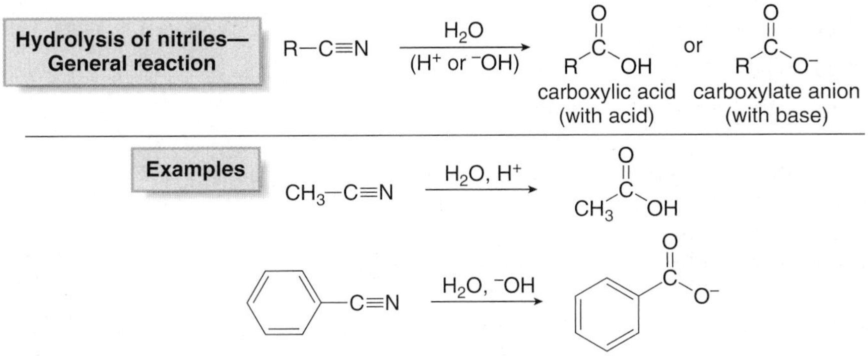

Hydrolysis of nitriles—General reaction

$$R-C\equiv N \xrightarrow[\text{(H}^+\text{ or }^-\text{OH)}]{\text{H}_2\text{O}} \underset{\substack{\text{carboxylic acid}\\\text{(with acid)}}}{R-C(=O)-OH} \quad or \quad \underset{\substack{\text{carboxylate anion}\\\text{(with base)}}}{R-C(=O)-O^-}$$

Examples

$$CH_3-C\equiv N \xrightarrow{\text{H}_2\text{O, H}^+} CH_3-C(=O)-OH$$

$$\text{Ph}-C\equiv N \xrightarrow{\text{H}_2\text{O, }^-\text{OH}} \text{Ph}-C(=O)-O^-$$

The mechanism of this reaction involves the formation of an amide tautomer. Two tautomers can be drawn for any carbonyl compound, and those for a 1° amide are as follows:

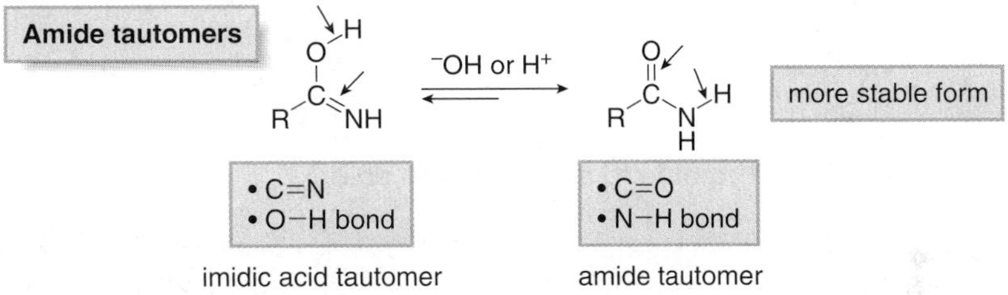

Amide tautomers

$$\underset{\substack{\bullet\, C=N\\ \bullet\, O-H \text{ bond}\\ \text{imidic acid tautomer}}}{R-C(=N-H)(O-H)} \xrightleftharpoons[^-\text{OH or H}^+]{} \underset{\substack{\bullet\, C=O\\ \bullet\, N-H \text{ bond}\\ \text{amide tautomer}}}{R-C(=O)(NH_2)} \quad \boxed{\text{more stable form}}$$

Recall from Chapter 11 that tautomers are constitutional isomers that differ in the location of a double bond and a proton.

◆ The amide form is the more stable tautomer, having a C=O and an N–H bond.
◆ The imidic acid tautomer is the less stable form, having a C=N and an O–H bond.

The imidic acid and amide tautomers are interconverted by treating with acid or base, analogous to the keto–enol tautomers of other carbonyl compounds. In fact, the two amide tautomers are exactly the same as keto–enol tautomers except that a nitrogen atom replaces a carbon atom bonded to the carbonyl group.

Recall from Chapter 11 that the keto and enol tautomers of a carbonyl compound are in equilibrium, but the keto form is lower in energy, so it is highly favored in most cases.

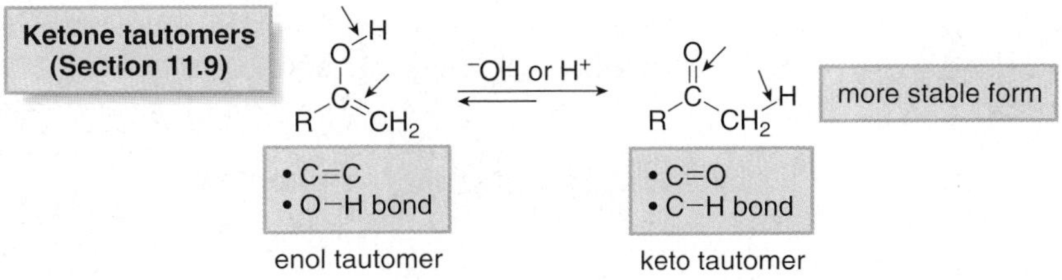

Ketone tautomers (Section 11.9)

$$\underset{\substack{\bullet\, C=C\\ \bullet\, O-H \text{ bond}\\ \text{enol tautomer}}}{R-C(=CH_2)(O-H)} \xrightleftharpoons[^-\text{OH or H}^+]{} \underset{\substack{\bullet\, C=O\\ \bullet\, C-H \text{ bond}\\ \text{keto tautomer}}}{R-C(=O)(CH_3)} \quad \boxed{\text{more stable form}}$$

The mechanism of nitrile hydrolysis in both acid and base consists of three parts: [1] **nucleophilic addition** of H_2O or $^-$OH to form the imidic acid tautomer; [2] **tautomerization** to form the amide, and [3] **hydrolysis of the amide** to form RCOOH or RCOO$^-$. The mechanism is shown for the basic hydrolysis of RCN to RCOO$^-$ (Mechanism 22.11).

PROBLEM 22.35 Draw the products of each reaction.

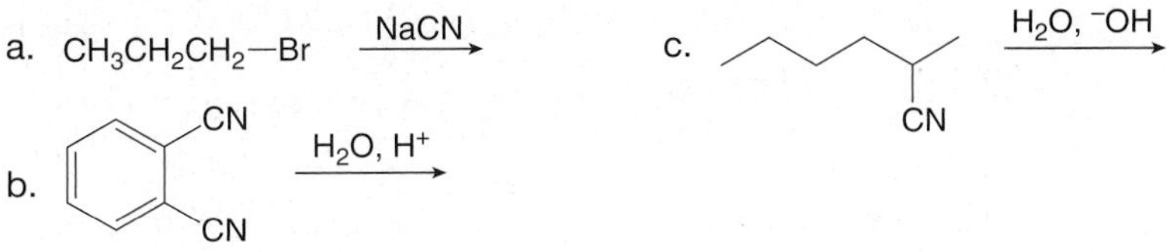

a. $CH_3CH_2CH_2-Br \xrightarrow{\text{NaCN}}$

b. (benzene ring with two CN groups ortho) $\xrightarrow{\text{H}_2\text{O, H}^+}$

c. (structure with CN) $\xrightarrow{\text{H}_2\text{O, }^-\text{OH}}$

PROBLEM 22.36 Draw a tautomer of each compound.

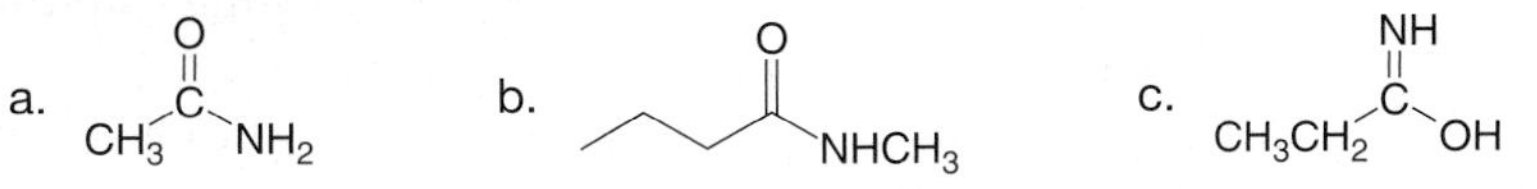

a. $CH_3-C(=O)-NH_2$

b. (structure with $C(=O)-NHCH_3$)

c. $CH_3CH_2-C(=NH)-OH$

MECHANISM 22.11

Hydrolysis of a Nitrile in Base

Part [1] Addition of the nucleophile (⁻OH) to form an imidic acid

◆ **Nucleophilic attack of ⁻OH** followed by protonation forms an imidic acid.

Part [2] Tautomerization of the imidic acid to an amide

◆ Tautomerization occurs by a two-step sequence—**deprotonation followed by protonation.**

Part [3] Hydrolysis of the 1° amide to a carboxylate anion

◆ Conversion of the amide to the carboxylate anion occurs by the multistep sequence detailed in Section 22.13.

22.18B   Reduction of Nitriles

Nitriles are reduced with metal hydride reagents to form either 1° amines or aldehydes, depending on the reducing agent.

◆ Treatment of a nitrile with **LiAlH₄** followed by H_2O adds two equivalents of H_2 across the triple bond, forming a **1° amine.**

General reaction $R-C\equiv N$ $\xrightarrow[\text{[2] } H_2O]{\text{[1] LiAlH}_4}$ $R-CH_2NH_2$ ← **addition of two equivalents of H_2**
1° amine

Example $(CH_3)_3C-C\equiv N$ $\xrightarrow[\text{[2] } H_2O]{\text{[1] LiAlH}_4}$ $(CH_3)_3C-CH_2NH_2$

◆ Treatment of a nitrile with a milder reducing agent such as DIBAL-H followed by H_2O forms an **aldehyde.**

General reaction $R-C\equiv N$ $\xrightarrow[\text{[2] } H_2O]{\text{[1] DIBAL-H}}$ (aldehyde)   ← **addition of one equivalent of H_2**

Example   (phenyl)$-C\equiv N$ $\xrightarrow[\text{[2] } H_2O]{\text{[1] DIBAL-H}}$ (benzaldehyde)

The mechanism of both reactions involves **nucleophilic addition of hydride (H⁻) to the polarized C−N triple bond.** With LiAlH₄, two equivalents of hydride are sequentially added to yield a dianion (Steps [1]–[2]), which is then protonated with H_2O to form the amine in Step [3], as shown in Mechanism 22.12.

MECHANISM 22.12

Reduction of a Nitrile with LiAlH₄

With **DIBAL-H,** nucleophilic addition of one equivalent of hydride forms an anion (Step [1]), which is protonated with water to generate an **imine,** as shown in Mechanism 22.13. As described in Sections 21.11–21.12, imines are hydrolyzed in water to form aldehydes.

MECHANISM 22.13

Reduction of a Nitrile with DIBAL-H

PROBLEM 22.37 Draw the product of each reaction.

a. CH₃CH₂—Br $\xrightarrow[\substack{[2]\ LiAlH_4 \\ [3]\ H_2O}]{[1]\ NaCN}$

b. CH₃CH₂CH₂—CN $\xrightarrow[{[2]\ H_2O}]{[1]\ DiBAL-H}$

22.18C   Addition of Grignard and Organolithium Reagents to Nitriles

Both Grignard and organolithium reagents react with nitriles to form ketones with a new carbon–carbon bond.

The reaction occurs by nucleophilic addition of the organometallic reagent to the polarized C−N triple bond to form an anion (Step [1]), which is protonated with water to form an **imine.** Water then hydrolyzes the imine, replacing the C=N by C=O as described in Sections 21.11–21.12. The final product is a ketone with a new carbon–carbon bond (Mechanism 22.14).

MECHANISM 22.14

Addition of Grignard and Organolithium Reagents (R–M) to Nitriles

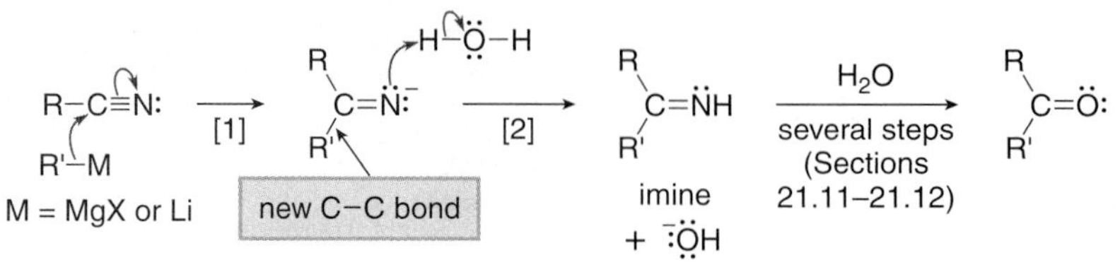

R—C≡N:
R'—M
M = MgX or Li

[1] →

R
C=N:⁻
R'
new C–C bond

[2] →

R
C=ṄH
R'
imine
+ :ÖH⁻

H₂O
several steps
(Sections
21.11–21.12)

R
C=Ö:
R'

PROBLEM 22.38   Draw the products of each reaction.

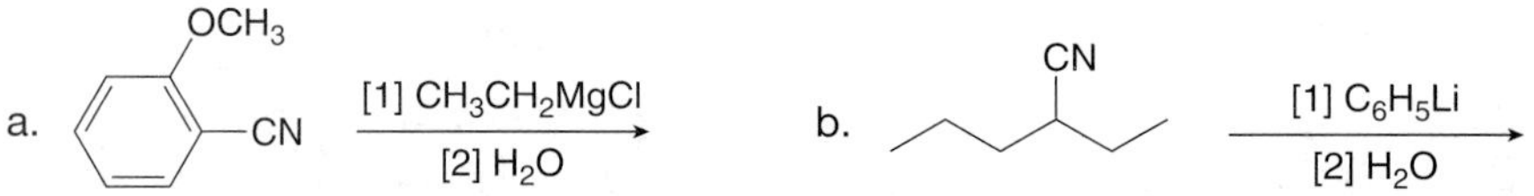

a.

OCH₃
CN

[1] CH₃CH₂MgCl
→
[2] H₂O

b.

CN

[1] C₆H₅Li
→
[2] H₂O

PROBLEM 22.39   What reagents are needed to convert phenylacetonitrile (C₆H₅CH₂CN) to each compound:
(a) C₆H₅CH₂COCH₃; (b) C₆H₅CH₂COC(CH₃)₃; (c) C₆H₅CH₂CHO; (d) C₆H₅CH₂COOH?

PROBLEM 22.40   Outline two different ways that 2-butanone can be prepared from a nitrile and a Grignard reagent.

22.19   Key Concepts—Carboxylic Acids and Their Derivatives— Nucleophilic Acyl Substitution

Summary of Spectroscopic Absorptions of RCOZ (22.5)

IR absorptions
- All RCOZ compounds have a $C=O$ absorption in the region 1600–1850 cm⁻¹.
  - RCOCl: 1800 cm⁻¹
  - (RCO)₂O: 1820 and 1760 cm⁻¹ (two peaks)
  - RCOOR': 1735–1745 cm⁻¹
  - RCONR'₂: 1630–1680 cm⁻¹
- Additional amide absorptions occur at 3200–3400 cm⁻¹ (N–H stretch) and 1640 cm⁻¹ (N–H bending).
- Decreasing the ring size of a cyclic lactone, lactam, or anhydride increases the frequency of the $C=O$ absorption.
- Conjugation shifts the $C=O$ to lower wavenumber.

¹H NMR absorptions
- C–H α to the $C=O$ absorbs at 2–2.5 ppm.
- N–H of an amide absorbs at 7.5–8.5 ppm.

¹³C NMR absorption
- $C=O$ absorbs at 160–180 ppm.

Summary of Spectroscopic Absorptions of RCN (22.5)

IR absorption
- $C≡N$ absorbs at ~2250 cm⁻¹.

¹³C NMR absorption
- $C≡N$ absorbs at 115–120 ppm.

Summary: The Relationship between the Basicity of Z⁻ and the Properties of RCOZ

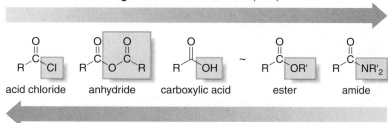

- **Increasing basicity of the leaving group** (22.2)
- **Increasing resonance stabilization** (22.2)

acid chloride   anhydride   carboxylic acid   ester   amide

- **Increasing leaving group ability** (22.7B)
- **Increasing reactivity** (22.7B)
- **Increasing frequency of the C=O absorption in the IR** (22.5)

General Features of Nucleophilic Acyl Substitution

- The characteristic reaction of compounds having the general structure RCOZ is nucleophilic acyl substitution (22.1).
- The mechanism consists of two steps (22.7A):
  [1] Addition of a nucleophile to form a tetrahedral intermediate
  [2] Elimination of a leaving group
- More reactive acyl compounds can be used to prepare less reactive acyl compounds. The reverse is not necessarily true (22.7B).

Nucleophilic Acyl Substitution Reactions

[1] Reactions that produce acid chlorides (RCOCl)

From RCOOH (22.10A):

$$\underset{R}{\overset{O}{\|}}C\text{—OH} \; + \; SOCl_2 \; \longrightarrow \; \underset{R}{\overset{O}{\|}}C\text{—Cl} \; + \; SO_2 \; + \; HCl$$

[2] Reactions that produce anhydrides [(RCO)₂O]

[a] From RCOCl (22.8):

$$R\text{COCl} \; + \; {}^{-}O\text{—COR'} \; \longrightarrow \; R\text{CO—O—COR'} \; + \; Cl^{-}$$

[b] From dicarboxylic acids (22.10B):

$$\xrightarrow{\Delta} \quad \text{cyclic anhydride} \; + \; H_2O$$

cyclic anhydride

[3] Reactions that produce carboxylic acids (RCOOH)

[a] From RCOCl (22.8):

$$R\text{COCl} \; + \; H_2O \; \xrightarrow{\text{pyridine}} \; R\text{COOH} \; + \; \text{pyridinium} \; Cl^{-}$$

[b] From (RCO)₂O (22.9):

$$R\text{CO—O—COR} \; + \; H_2O \; \longrightarrow \; 2 \; R\text{COOH}$$

[c] From RCOOR' (22.11):

$$R\text{COOR'} \; + \; H_2O \; \xrightarrow{(H^+ \text{ or } {}^{-}OH)} \; \underset{\text{(with acid)}}{R\text{COOH}} \quad \text{or} \quad \underset{\text{(with base)}}{R\text{COO}^{-}} \; + \; R'\text{OH}$$

[d] From RCONR'$_2$ (R' = H or alkyl, 22.13):

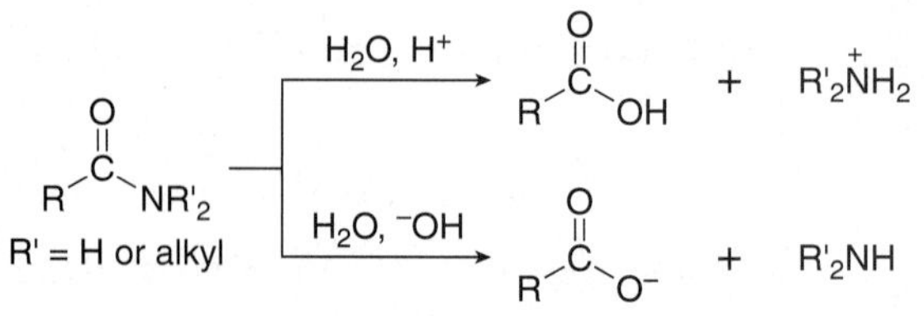

[4] Reactions that produce esters (RCOOR')

[a] From RCOCl (22.8):

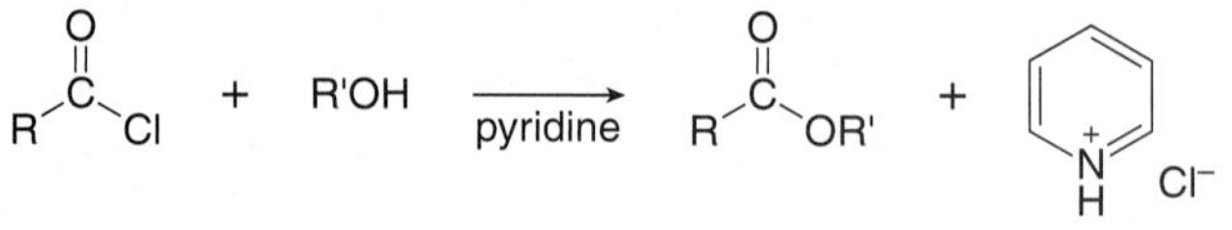

[b] From (RCO)$_2$O (22.9):

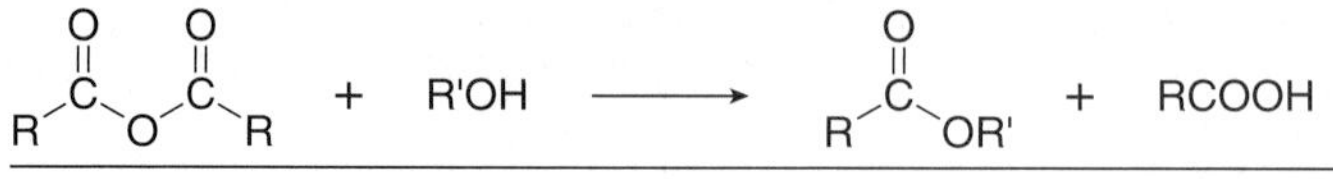

[c] From RCOOH (22.10C):

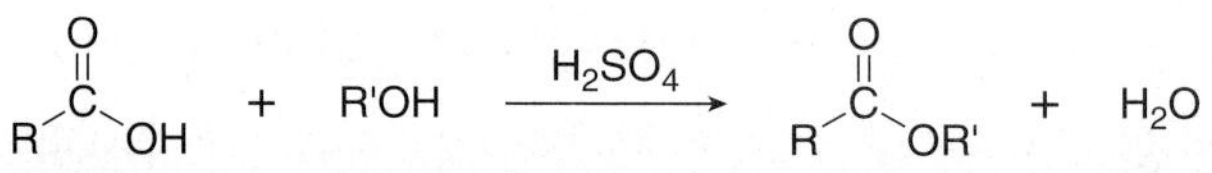

[5] Reactions that produce amides (RCONH$_2$) [The reactions are written with NH$_3$ as nucleophile to form RCONH$_2$. Similar reactions occur with R'NH$_2$ to form RCONHR', and with R'$_2$NH to form RCONR'$_2$.]

[a] From RCOCl (22.8):

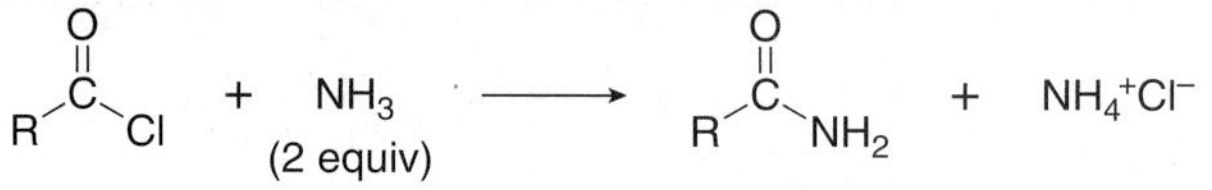

[b] From (RCO)$_2$O (22.9):

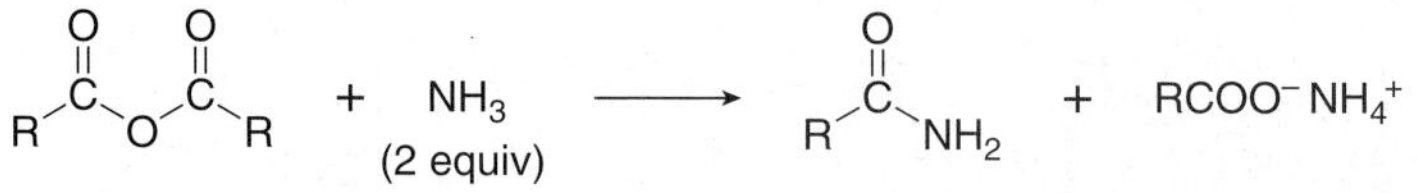

[c] From RCOOH (22.10D):

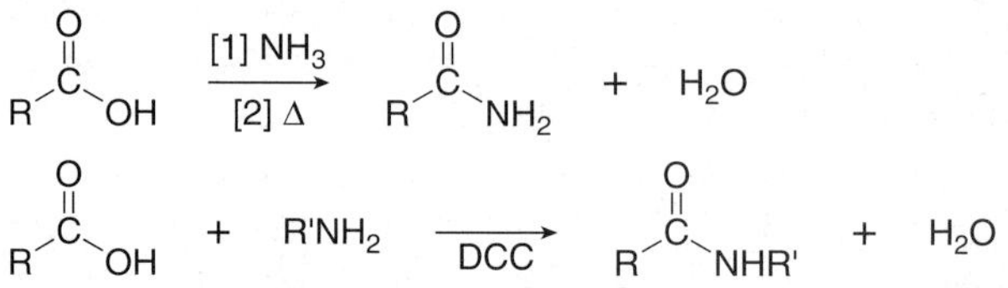

[d] From RCOOR' (22.11):

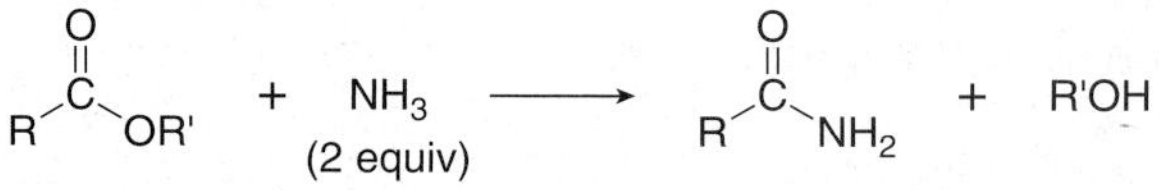

Nitrile Synthesis (22.18)

Nitriles are prepared by S$_N$2 substitution using unhindered alkyl halides as starting materials.

$$R-X \ + \ ^-CN \ \xrightarrow{\ S_N2\ } \ R-C\equiv N \ + \ X^-$$
R = CH$_3$, 1°

Reactions of Nitriles

[1] Hydrolysis (22.18A)

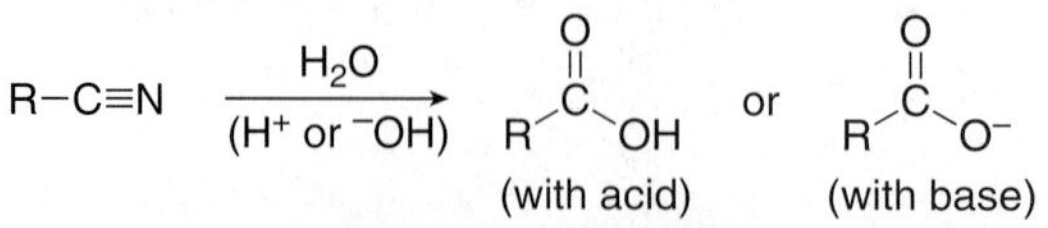

[2] Reduction (22.18B)

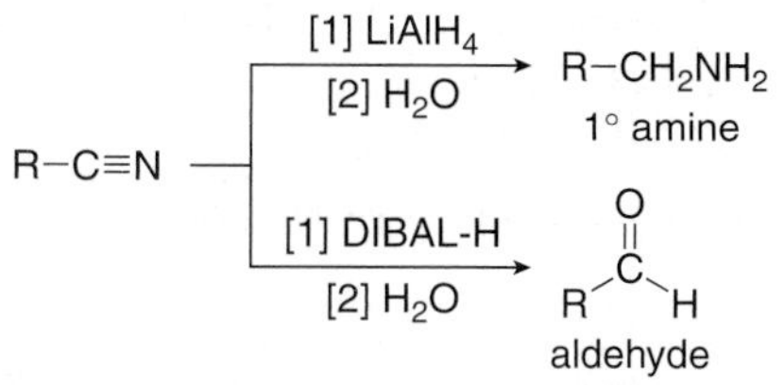

[3] Reaction with organometallic reagents (22.18C)

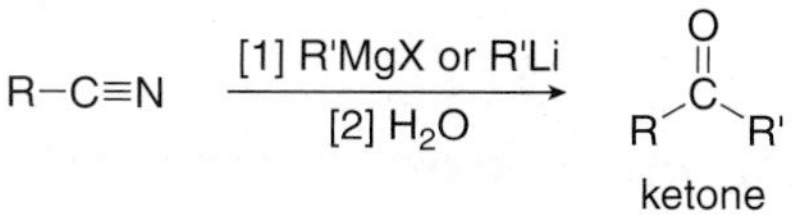

Problems

Nomenclature

22.41 Give the IUPAC or common name for each compound.

a. $(CH_3)_3CCOCl$

b.

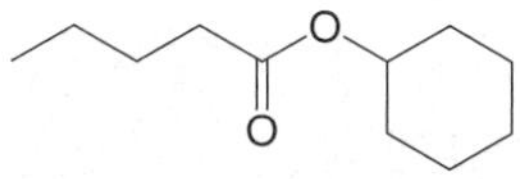

c. $(CH_3)_3\overset{\text{O}}{C}COOCH_2CH(CH_3)_2$

d.

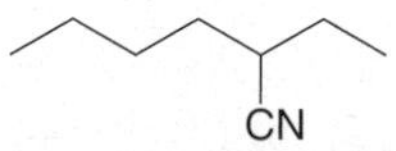

e.

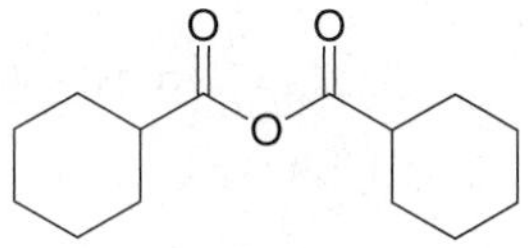

f. $C_6H_5CH_2COOC_6H_5$

g. 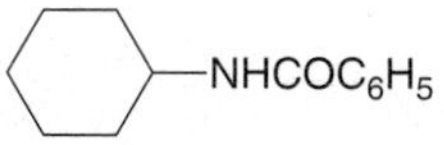 —NHCOC$_6$H$_5$

h.

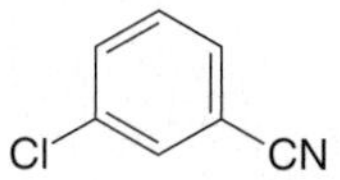

i. $C_6H_5CH_2CH_2COCl$

j.

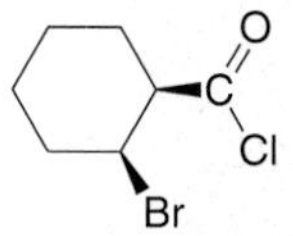

k. 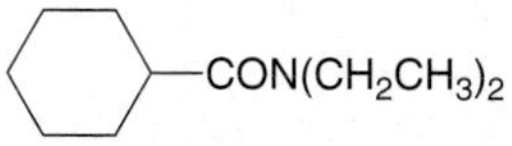 —CON(CH$_2$CH$_3$)$_2$

l.

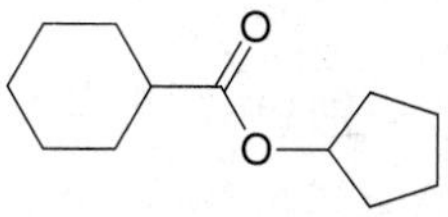

22.42 Give the structure corresponding to each name.
a. propanoic anhydride
b. α-chlorobutyryl chloride
c. cyclohexyl propanoate
d. cyclohexanecarboxamide
e. isopropyl formate
f. N-cyclopentylpentanamide
g. 4-methylheptanenitrile
h. vinyl acetate
i. benzoic propanoic anhydride
j. 3-methylhexanoyl chloride
k. octyl butanoate
l. N,N-dibenzylformamide

Properties of Carboxylic Acid Derivatives

22.43 Rank the compounds in each group in order of increasing reactivity in nucleophilic acyl substitution.
a. $CH_3CH_2CH_2CONH_2$, $CH_3CH_2CH_2COCl$, $CH_3CH_2CH_2COOCH_2CH_2CH_3$
b. $(CH_3CH_2CO)_2O$, $(CF_3CO)_2O$, $CH_3CH_2CO_2CH_2CH_2CH_3$
c. CH_3COOH, CH_3COSH, CH_3COCl

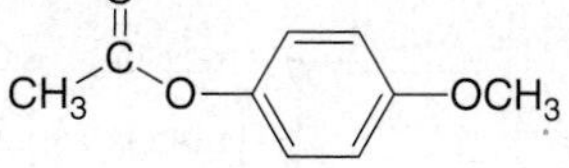

22.44 Explain why ester **A** is more reactive than ester **B** in nucleophilic acyl substitution.

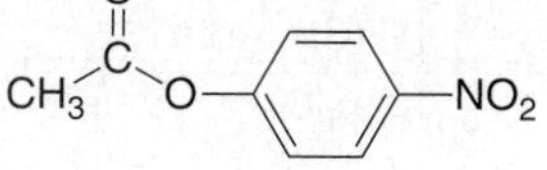

22.45 Explain why imidazolides are much more reactive than other amides in nucleophilic acyl substitution.

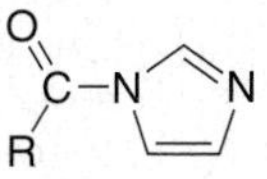

imidazolide

22.46 Explain why CH_3CONH_2 is a stronger acid and a weaker base than $CH_3CH_2NH_2$.

Reactions

22.47 Draw the product formed when pentanoyl chloride ($CH_3CH_2CH_2CH_2COCl$) is treated with each reagent.

a. H_2O, pyridine   c. CH_3COO^- e. $(CH_3CH_2)_2NH$ (excess)

b. CH_3CH_2OH, pyridine   d. NH_3 (excess)   f. $C_6H_5NH_2$ (excess)

22.48 Draw the product formed when pentanoic anhydride [$(CH_3CH_2CH_2CH_2CO)_2O$] is treated with each reagent. With some reagents, no reaction occurs.

a. $SOCl_2$ c. CH_3OH e. $(CH_3CH_2)_2NH$ (excess)

b. H_2O d. $NaCl$ f. $CH_3CH_2NH_2$ (excess)

22.49 Draw the product formed when phenylacetic acid ($C_6H_5CH_2COOH$) is treated with each reagent. With some reagents, no reaction occurs.

a. $NaHCO_3$ e. NH_3 (1 equiv)   i. [1] $NaOH$; [2] CH_3COCl

b. $NaOH$ f. NH_3, Δ j. CH_3NH_2, DCC

c. $SOCl_2$ g. CH_3OH, H_2SO_4 k. [1] $SOCl_2$; [2] $CH_3CH_2CH_2NH_2$ (excess)

d. $NaCl$ h. CH_3OH, ^-OH l. [1] $SOCl_2$; [2] $(CH_3)_2CHOH$

22.50 Draw the product formed when ethyl butanoate ($CH_3CH_2CH_2COOCH_2CH_3$) is treated with each reagent. With some reagents, no reaction occurs.

a. $SOCl_2$ b. H_3O^+ c. H_2O, ^-OH d. NH_3 e. $CH_3CH_2NH_2$

22.51 Draw the products formed when phenylacetamide ($C_6H_5CH_2CONH_2$) is treated with each reagent.

a. H_3O^+ b. H_2O, ^-OH

22.52 Draw the products formed when phenylacetonitrile ($C_6H_5CH_2CN$) is treated with each reagent.

a. H_3O^+ c. [1] CH_3MgBr; [2] H_2O e. [1] DIBAL-H; [2] H_2O

b. H_2O, ^-OH d. [1] CH_3CH_2Li; [2] H_2O f. [1] $LiAlH_4$; [2] H_2O

22.53 Draw the organic products formed in each reaction.

a. 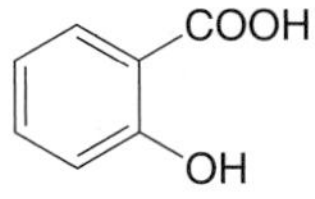 salicylic acid structure with $SOCl_2$ / pyridine

g. $CH_3CH_2CH_2CH_2Br$ [1] NaCN / [2] H_2O, ^-OH

b. C_6H_5COCl + pyrrolidine (excess) $\longrightarrow$

h. $C_6H_5CH_2COOH$ [1] $SOCl_2$ / [2] $CH_3CH_2CH_2CH_2NH_2$ / [3] $LiAlH_4$ / [4] H_2O

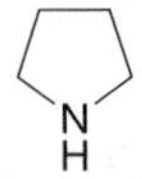

c. C_6H_5CN [1] $CH_3CH_2CH_2MgBr$ / [2] H_2O

i. $C_6H_5CH_2CH_2CH_2CN$ H_3O^+

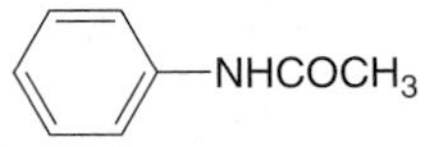

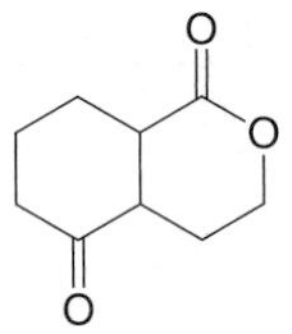

d. $(CH_3)_2CHCOOH$ + $CH_3CH_2CHOH(CH_3)$ H_2SO_4

j.  HOOC–CH=CH–COOH Δ

e. C_6H_5–$NHCOCH_3$ H_2O / ^-OH

k. $(CH_3CO)_2O$ + cyclohexyl–NH_2 (excess) $\longrightarrow$

f. bicyclic lactone structure H_3O^+

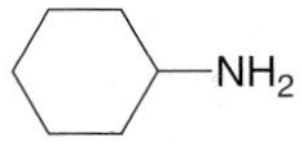

l. $C_6H_5CH_2CH_2COOCH_2CH_3$ H_2O / ^-OH

22.54 Identify compounds **A–M** in the following reaction sequence.

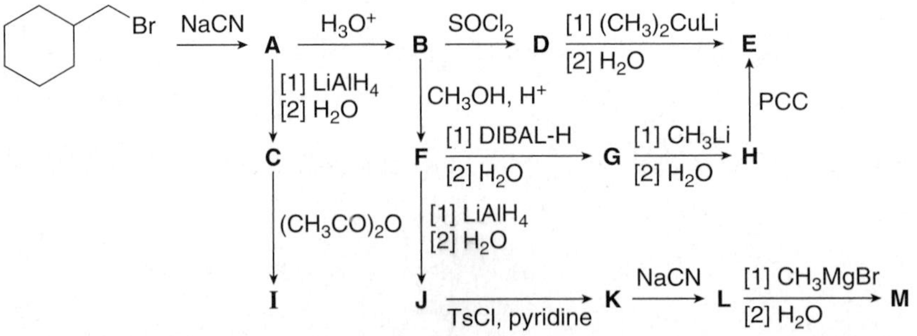

22.55 Draw the products of each reaction and indicate the stereochemistry at any stereogenic centers.

a. (cyclohexane with OH and CH₃) $\xrightarrow[\text{pyridine}]{CH_3COCl}$

b. (structure with Br, H, D) $\xrightarrow{NaCN}$

c. (structure with H, D, CH₃, COOH) $\xrightarrow[\text{H}^+]{CH_3CH_2OH}$

d. CH_3—C(=O)—Cl + (structure with CH₃, C₆H₅, H, NH₂) (2 equiv) $\longrightarrow$

22.56 The reactions of phosgene ($Cl_2C=O$), a toxic gas that was used as a chemical weapon in World War I, resemble those of acid chlorides.
   a. Draw the product formed when phosgene reacts with excess CH_3NH_2.
   b. What compound is formed when phosgene is treated with ethylene glycol ($HOCH_2CH_2OH$)? The product exhibits one singlet in its ^1H NMR spectrum at 4.54 ppm and two peaks in its IR spectrum at 1803 and 1776 cm^{-1}.
   c. Explain why phosgene increases the acidity of the lungs when it is inhaled.

22.57 Draw the products formed by acidic hydrolysis of **aspartame,** the artificial sweetener used in Equal and many diet beverages. One of the products of this hydrolysis reaction is the amino acid phenylalanine. Infants afflicted with phenylketonuria cannot metabolize this amino acid, so it accumulates, causing mental retardation. When the affliction is identified early, a diet limiting the consumption of phenylalanine (and compounds like aspartame that are converted to it) can make a normal life possible.

aspartame
artificial sweetener

22.58 Treatment of $CH_3CH(Br)CH_2COOH$ with NaOH forms a compound **A** of molecular formula $C_4H_6O_2$ that exhibits the following IR absorptions: 3500–2500, 1703, and 1656 cm^{-1}. Treatment of $BrCH_2CH_2CH_2COOH$ with NaOH forms an isomer **B** that shows a peak in its IR spectrum at 1770 cm^{-1}. Identify **A** and **B** and explain why different products are formed.

Mechanism

22.59 Draw a stepwise mechanism for each reaction.

a. (benzoyl chloride) + NH_2NH_2 $\xrightarrow{\text{pyridine}}$ (benzoyl hydrazide, NHNH₂) + (pyridinium, $\overset{+}{N}H$ Cl^-)

b. (lactone) $\xrightarrow[^-OH]{H_2O}$ (HO...C(=O)O⁻ structure)

c. (benzene ring with COOH and CH₂OH) $\xrightarrow{H_2SO_4}$ (phthalide-type structure) + H_2O

22.60 When acetic acid (CH_3COOH) is treated with a trace of acid in water labeled with ^{18}O, the label gradually appears in both oxygen atoms of the carboxylic acid. Draw a mechanism that explains this phenomenon.

CH_3—C(=O)—OH + $H_2^{18}O$ $\xrightarrow{H^+}$ CH_3—C(=O)—^{18}OH + CH_3—C(=^{18}O)—OH

22.61 Transesterification converts one ester into another by reaction with an alcohol in the presence of an acid catalyst. Draw a stepwise mechanism for the following transesterification.

(methyl benzoate, OCH_3) + $CH_3CH_2CH_2CH_2OH$ $\underset{}{\overset{H_2SO_4}{\rightleftarrows}}$ (butyl benzoate, $OCH_2CH_2CH_2CH_3$) + CH_3OH

22.62 Early research on the mechanism of ester hydrolysis in aqueous base considered the following one-step S_N2 mechanism as a possibility.

Using the chiral ester **X** as a starting material, draw the carboxylate anion and alcohol formed (including stereochemistry) from hydrolysis of **X** via the accepted mechanism (having a tetrahedral intermediate) and the one-step S_N2 alternative. Given that only one alcohol, (R)-2-butanol, is formed in this reaction, what does this indicate about the mechanism?

22.63 Ester **A** and amide **B** are isomers of each other. **A** is converted to **B** in the presence of base, and **B** is converted to **A** in the presence of acid. Draw mechanisms for both conversions and explain why this phenomenon occurs.

22.64 Treatment of the amino alcohol **X** with diethyl carbonate forms the heterocycle **Y**. Draw a stepwise mechanism for this process.

22.65 Although alkyl chlorides (RCH_2Cl) and acid chlorides (RCOCl) both undergo nucleophilic substitution reactions, acid chlorides are much more reactive. Suggest reasons for this difference in reactivity.

Synthesis

22.66 What carboxylic acid and alcohol are needed to prepare each ester by Fischer esterification?

a. $(CH_3)_3CCO_2CH_2CH_3$ b. [structure]   c. [structure]   d. [structure]

22.67 Devise a synthesis of each compound using 1-bromobutane ($CH_3CH_2CH_2CH_2Br$) as the only organic starting material. You may use any other inorganic reagents.

22.68 Convert 1-bromohexane ($CH_3CH_2CH_2CH_2CH_2CH_2Br$) into each compound. More than one step may be required. You may use any other organic or inorganic reagents.

a. $CH_3CH_2CH_2CH_2CH_2CH_2CN$
b. $CH_3CH_2CH_2CH_2CH_2CH_2COOH$
c. $CH_3CH_2CH_2CH_2CH_2CH_2COCl$
d. $CH_3CH_2CH_2CH_2CH_2CH_2CO_2CH_2CH_3$
e. $CH_3CH_2CH_2CH_2CH_2CH_2COCH_3$
f. $CH_3CH_2CH_2CH_2CH_2CH_2CHO$
g. $CH_3CH_2CH_2CH_2CH_2CH_2CH_2NH_2$
h. $CH_3CH_2CH_2CH_2CH_2CH_2CH_2NHCOCH_3$

22.69 Two methods convert an alkyl halide into a carboxylic acid having one more carbon atom.

[1]   R—X   + $^-$CN $\longrightarrow$ R—CN $\xrightarrow{H_3O^+}$ R—COOH   (Section 22.18)
                                              + X$^-$

new C—C bond

[2]   R—X   +   Mg $\longrightarrow$ R—MgX $\xrightarrow[\text{[2] } H_3O^+]{\text{[1] } CO_2}$ R—COOH   (Section 20.14)

Depending on the structure of the alkyl halide, one or both of these methods may be employed. For each alkyl halide, write out a stepwise sequence that converts it to a carboxylic acid with one more carbon atom. If both methods work, draw both routes. If one method cannot be used, state why it can't.

a. CH_3Cl
b. (benzene ring with Br)
c. $(CH_3)_3CCl$
d. $HOCH_2CH_2CH_2CH_2Br$

22.70 Design a synthesis of dibutyl phthalate (a plasticizer added to many polymers to keep them from becoming brittle), from benzene and any other organic or inorganic reagents.

dibutyl phthalate

22.71 Devise a synthesis of each analgesic compound from phenol (C_6H_5OH) and any other organic or inorganic reagents.

a. salicylamide
b. acetaminophen
c. *p*-acetophenetidin

22.72 Devise a synthesis of each ester from benzene, organic alcohols, and any other needed inorganic reagents.

a. $CH_2COOCH_2CH_3$
ethyl phenylacetate
(odor of honey)

b. methyl anthranilate
(odor of grape)

c. benzyl acetate
(odor of peach)

22.73 Both monomers needed for the synthesis of nylon 6,6 can be prepared from 1,4-dichlorobutane. Write out the steps illustrating these syntheses.

22.74 Devise a synthesis of each labeled compound using $H_2{}^{18}O$ and $CH_3{}^{13}CH_2OH$ as the only sources of labeled starting materials. You may use any other unlabeled organic compounds and inorganic reagents.

a. $CH_3\overset{O}{\overset{\|}{C}}O{}^{13}CH_2CH_3$
b. $CH_3{}^{13}\overset{O}{\overset{\|}{C}}OCH_2CH_3$
c. $CH_3\overset{O}{\overset{\|}{C}}{}^{18}OCH_2CH_3$
d. $CH_3{}^{13}\overset{{}^{18}O}{\overset{\|}{C}}OCH_2CH_3$

Polymers

22.75 What polyester or polyamide can be prepared from each pair of monomers?

a. HO—(cyclohexane)—OH and $HOOC$—$COOH$
b. $ClOC$—(benzene)—$COCl$ and H_2N—NH_2

22.76 What two monomers are needed to prepare each polymer?

a.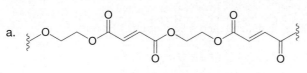

c.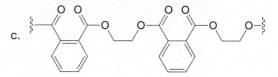

b.

22.77 Although Lexan, a polymer used in bicycle helmets, is similar to a polyester, the carbonyl carbon is bonded to two other oxygen atoms, not just one, and so it is called a **polycarbonate**. Draw a stepwise mechanism for the formation of Lexan from its two monomers.

diphenyl carbonate   bisphenol A

Lexan

Spectroscopy

22.78 How can IR spectroscopy be used to distinguish between each pair of isomers?

22.79 Rank the compounds in each group in order of increasing frequency of the C=O absorption in their IR spectra.

a. $CH_3CH_2COOCH_2CH_3$ $C_6H_5COOCH_2CH_3$ b. CH_3COCl CH_3CONH_2 CH_3COOCH_3

22.80 Identify the structures of each compound from the given data.

a. Molecular formula $C_6H_{12}O_2$
   IR absorption: 1738 cm^{-1}
   ^{1}H NMR: 1.12 (triplet, 3 H), 1.23 (doublet, 6 H), 2.28 (quartet, 2 H), and 5.00 (septet, 1 H) ppm

b. Molecular formula $C_6H_{12}O_2$
   IR absorption: 1746 cm^{-1}
   ^{1}H NMR: 0.94 (doublet, 6 H), 1.93 (multiplet, 1 H), 2.05 (singlet, 3 H), and 3.85 (doublet, 2 H) ppm

c. Molecular formula C_4H_7N
   IR absorption: 2250 cm^{-1}
   ^{1}H NMR: 1.08 (triplet, 3 H), 1.70 (multiplet, 2 H), and 2.34 (triplet, 2 H) ppm

d. Molecular formula C_8H_9NO
   IR absorptions: 3328 and 1639 cm^{-1}
   ^{1}H NMR: 2.95 (singlet, 3 H), 6.95 (singlet, 1 H), and 7.3–7.7 (multiplet, 5 H) ppm

e. Molecular formula C_4H_7ClO
   IR absorption: 1802 cm^{-1}
   ^{1}H NMR: 0.95 (triplet, 3 H), 1.07 (multiplet, 2 H), and 2.90 (triplet, 2 H) ppm

f. Molecular formula $C_5H_{10}O_2$
   IR absorption: 1750 cm^{-1}
   ^{1}H NMR: 1.20 (doublet, 6 H), 2.00 (singlet, 3 H), and 4.95 (septet, 1 H) ppm

g. Molecular formula $C_{10}H_{12}O_2$
   IR absorption: 1740 cm^{-1}
   ^{1}H NMR: 1.2 (triplet, 3 H), 2.4 (quartet, 2 H), 5.1 (singlet, 2 H), and 7.1–7.5 (multiplet, 5 H) ppm

h. Molecular formula $C_8H_{14}O_3$
   IR absorptions: 1810 and 1770 cm^{-1}
   ^{1}H NMR: 1.25 (doublet, 12 H) and 2.65 (septet, 2 H) ppm

22.81   Identify the structures of **A** and **B,** isomers of molecular formula $C_{10}H_{12}O_2$, from their IR data and ^1H NMR spectra.

a. IR absorption for **A** at 1718 cm^{-1}

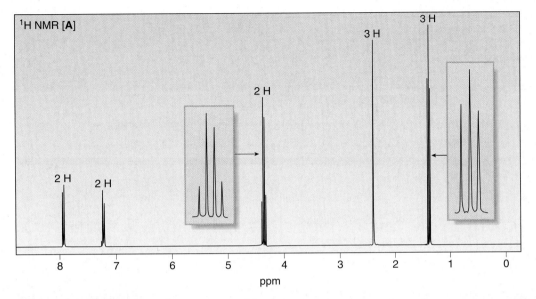

b. IR absorption for **B** at 1740 cm^{-1}

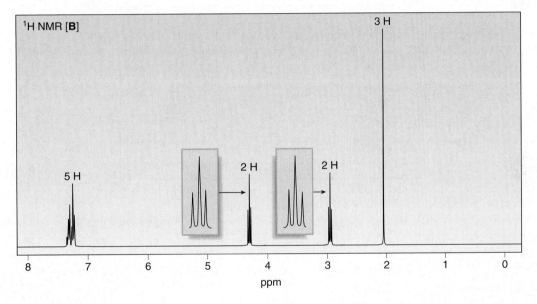

22.82 Phenacetin is an analgesic compound having molecular formula $C_{10}H_{13}NO_2$. Once a common component in over-the-counter pain relievers such as APC (**a**spirin, **p**henacetin, **c**affeine), phenacetin is no longer used because of its liver toxicity. Deduce the structure of phenacetin from its ^1H NMR and IR spectra.

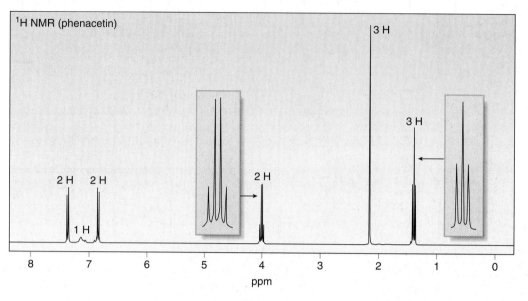

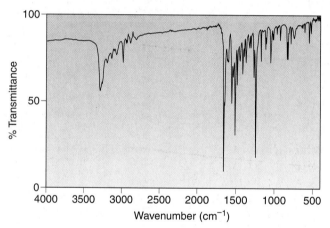

Challenge Problems

22.83 The ^1H NMR spectrum of *N,N*-dimethylformamide [$HCON(CH_3)_2$, DMF] shows three signals at 2.93, 3.03, and 8.0 ppm. What protons of DMF give rise to each signal? Explain why three signals are observed.

22.84 Compelling evidence for the existence of a tetrahedral intermediate in nucleophilic acyl substitution was obtained in a series of elegant experiments carried out by Myron Bender in 1951. The key experiment was the reaction of aqueous ^-OH with ethyl benzoate ($C_6H_5COOCH_2CH_3$) labeled at the carbonyl oxygen with ^{18}O. Bender did not allow the hydrolysis to go to completion, and then examined the presence of label in the *recovered starting material.* He found that some of the recovered ethyl benzoate no longer contained a label at the carbonyl oxygen. With reference to the accepted mechanism of nucleophilic acyl substitution, explain how this provides evidence for a tetrahedral intermediate.

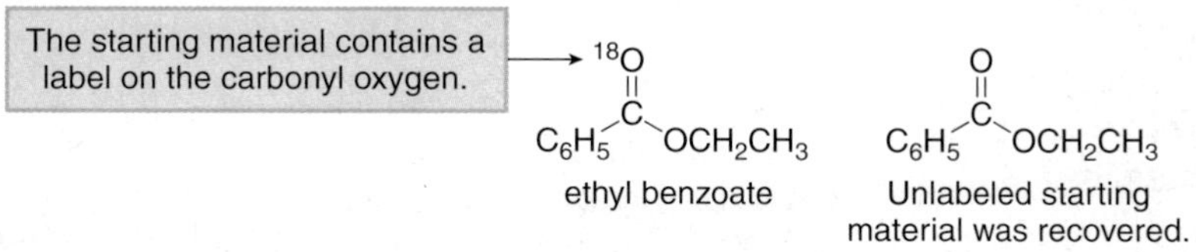

22.85 Draw a stepwise mechanism for the following reactions, two steps in R. B. Woodward's classic synthesis of reserpine in 1958. Reserpine, which is isolated from the extracts of the Indian snakeroot *Rauwolfia serpentina Benth,* has been used to manage mild hypertension associated with anxiety.

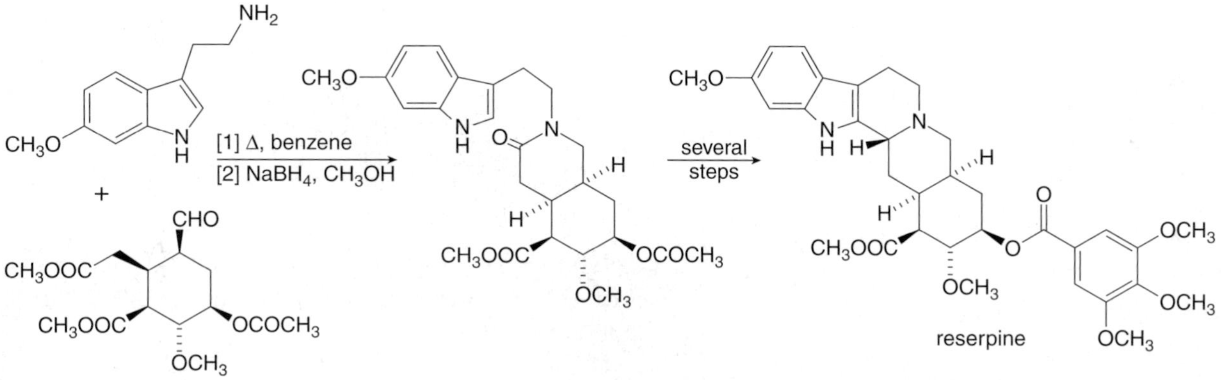

Substitution Reactions of Carbonyl Compounds at the α Carbon

Tamoxifen is a potent anticancer drug used widely in the treatment of breast cancer. Tamoxifen binds to estrogen receptors, and in this way inhibits the growth of breast cancers that are estrogen dependent. One method to synthesize tamoxifen forms a new carbon–carbon bond on the α carbon to a carbonyl group using an intermediate enolate. In Chapter 23 we will learn about these and other carbon–carbon bond–forming reactions that occur on the α carbon.

BREAST CANCER

FUND THE FIGHT. FIND A CURE.

1998

USA   First-Class

Chapters 23 and 24 focus on reactions that occur at the α carbon to a carbonyl group. These reactions are different from the reactions of Chapters 20–22, all of which involved nucleophilic attack at the electrophilic carbonyl carbon. In reactions at the α carbon, the carbonyl compound serves as a *nucleophile* that reacts with a carbon or halogen electrophile to form a new bond to the α carbon.

Chapter 23 concentrates on **substitution reactions at the α carbon,** whereas Chapter 24 concentrates on reactions between two carbonyl compounds, one of which serves as the nucleophile and one of which is the electrophile. Many of the reactions in Chapter 23 form new carbon–carbon bonds, thus adding to your repertoire of reactions that can be used to synthesize more complex organic molecules from simple precursors. As you will see, the reactions introduced in Chapter 23 have been used to prepare a wide variety of interesting and useful compounds.

23.1   Introduction

Up to now, the discussion of carbonyl compounds has centered on their reactions with nucleophiles at the electrophilic carbonyl carbon. **Two general reactions are observed,** depending on the structure of the carbonyl starting material.

◆ *Nucleophilic addition* occurs when there is no electronegative atom Z on the carbonyl carbon (as with aldehydes and ketones).

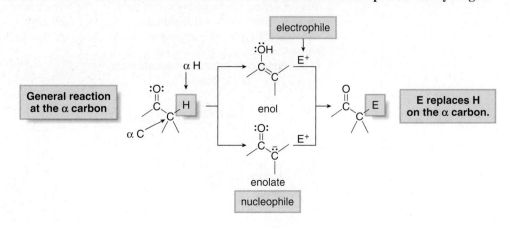

◆ *Nucleophilic acyl substitution* occurs when there is an electronegative atom Z on the carbonyl carbon (as with carboxylic acids and their derivatives).

Reactions can also occur at the α carbon to the carbonyl group. These reactions proceed by way of **enols** or **enolates,** two electron-rich intermediates that react with electrophiles, forming a new bond on the α carbon. This reaction results in the **substitution of the electrophile E for hydrogen.**

Hydrogen atoms on the α carbon are called α **hydrogens.**

23.2   Enols

Recall from Chapter 11 that the **enol and keto forms are tautomers of the carbonyl group that differ in the position of the double bond and a proton.** These constitutional isomers are in equilibrium with each other.

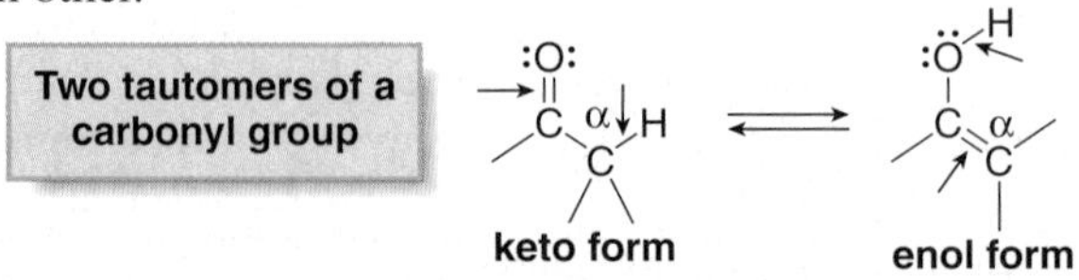

Two tautomers of a carbonyl group

keto form   enol form

- ◆ A keto tautomer has a C=O and an additional C–H bond.
- ◆ An enol tautomer has an O–H group bonded to a C=C.

Equilibrium favors the keto form for most carbonyl compounds largely because a C=O is much stronger than a C=C. For simple carbonyl compounds, < 1% of the enol is present at equilibrium. With unsymmetrical ketones, moreover, two different enols are possible, yet they still total < 1%.

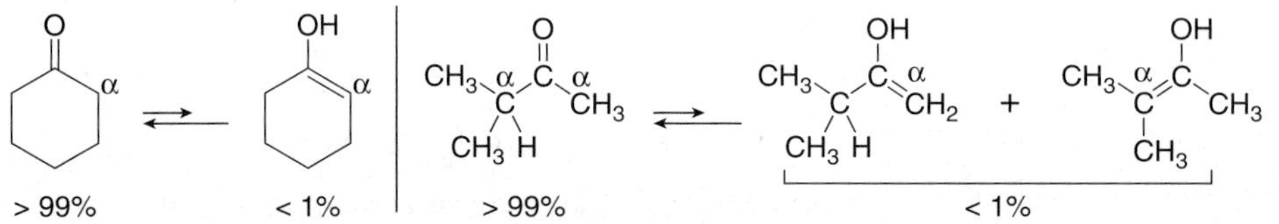

> 99%   < 1%   > 99%   < 1%

With compounds containing two carbonyl groups separated by a single carbon (called β-dicarbonyl compounds or 1,3-dicarbonyl compounds), however, the concentration of the enol form exceeds the concentration of the keto form.

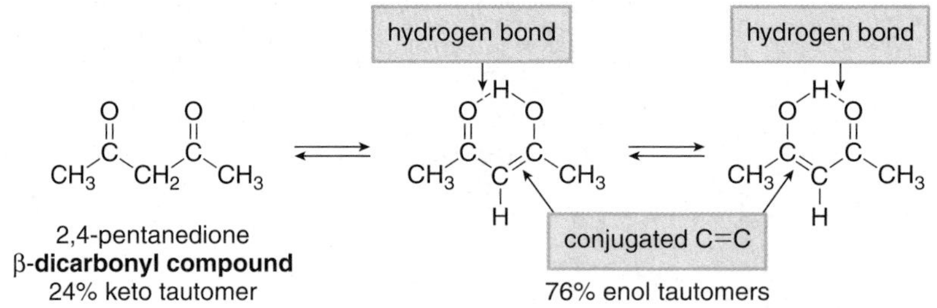

2,4-pentanedione
β-dicarbonyl compound
24% keto tautomer

76% enol tautomers

Two factors stabilize the enol of β-dicarbonyl compounds: **conjugation** and **intramolecular hydrogen bonding.** The C=C of the enol is conjugated with the carbonyl group, allowing delocalization of the electron density in the π bonds. Moreover, the OH of the enol can hydrogen bond to the oxygen of the nearby carbonyl group. Such intramolecular hydrogen bonds are especially stabilizing when they form a six-membered ring, as in this case.

SAMPLE PROBLEM 23.1   Convert each compound to its enol or keto tautomer.

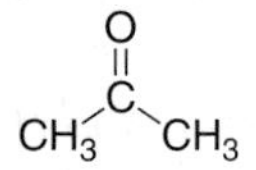

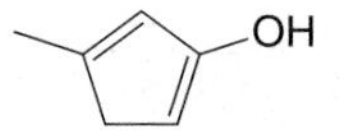

SOLUTION

[a] To convert a carbonyl compound to its enol tautomer, draw a double bond between the carbonyl carbon and the α carbon, and change the C=O to C–OH. In this case, both α carbons are identical, so only one enol is possible.

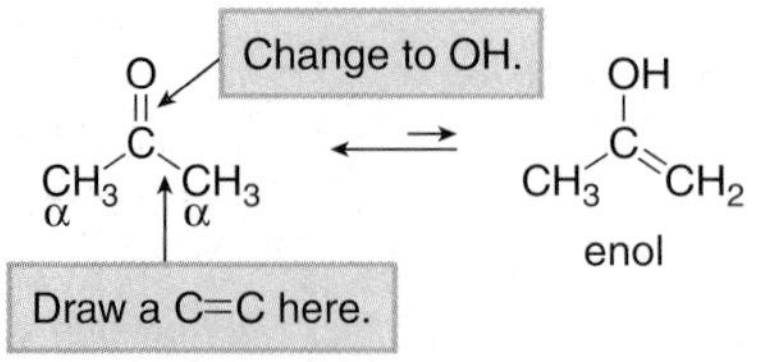

[b] To convert an enol to its keto tautomer, change the C–OH to C=O and add a proton to the other end of the C=C.

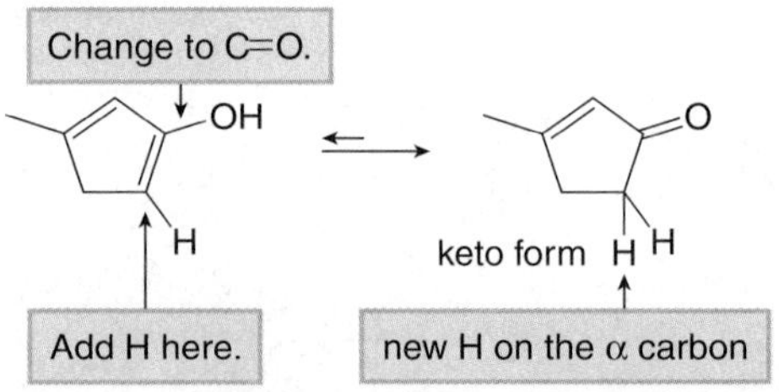

PROBLEM 23.1   Draw the enol or keto tautomer(s) of each compound.

a.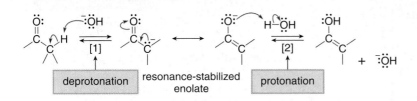

c. C₆H₅... (structure)

e. $C_6H_5CH_2CH_2CO_2CH_2CH_3$

b. $C_6H_5CH_2CHO$

d. (structure with HO)

f. (structure)   [Draw mono enol tautomers only.]

PROBLEM 23.2   Ignoring stereoisomers, draw the two possible enols for 2-butanone ($CH_3COCH_2CH_3$), and predict which one is more stable.

PROBLEM 23.3   Vitamin C is a compound in which the enol tautomer is *more* stable than its keto form. Vitamin C is actually an ene*diol,* because it contains two OH groups on the C=C. Draw the two possible keto tautomers of this enediol.

(structure of vitamin C)

vitamin C

23.2A   The Mechanism of Tautomerization

Tautomerization, the process of converting one tautomer into another, is catalyzed by both acid and base. Tautomerization always requires two steps (**protonation** and **deprotonation**), but the order of these steps depends on whether the reaction takes place in acid or base. In Mechanisms 23.1 and 23.2 for tautomerization, the keto form is converted to the enol form. All of the steps are reversible, though, so they equally apply to the conversion of the enol form to the keto form.

MECHANISM 23.1

Tautomerization in Acid

(mechanism scheme)

protonation   resonance-stabilized cation   deprotonation

- With acid, **protonation *precedes* deprotonation.**
- **Protonation** of the carbonyl oxygen forms a resonance-stabilized cation in Step [1], and deprotonation in Step [2] forms the enol. The net result of these two steps is the movement of a double bond and a proton.

MECHANISM 23.2

Tautomerization in Base

(mechanism scheme)

deprotonation   resonance-stabilized enolate   protonation

- With base, **deprotonation *precedes* protonation.**
- **Removal of a proton from the α carbon** forms a resonance-stabilized enolate in Step [1].
- **Protonation** of the enolate with H_2O forms the enol in Step [2].

PROBLEM 23.4   Draw a stepwise mechanism for the following reaction.

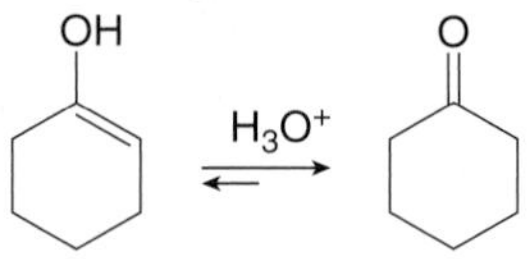

OH $\xrightarrow[\;]{\;H_3O^+\;}$ O

23.2B   How Enols React

Like other compounds with carbon–carbon double bonds, **enols are electron rich, so they react as nucleophiles.** Enols are even more electron rich than alkenes, though, because the OH group has a powerful electron-donating resonance effect. For example, a second resonance structure can be drawn for the enol that places a negative charge on one of the carbon atoms. As a result, this carbon atom is especially nucleophilic, and it can react with an electrophile E^+ to form a new bond to carbon. Loss of a proton then forms a neutral product.

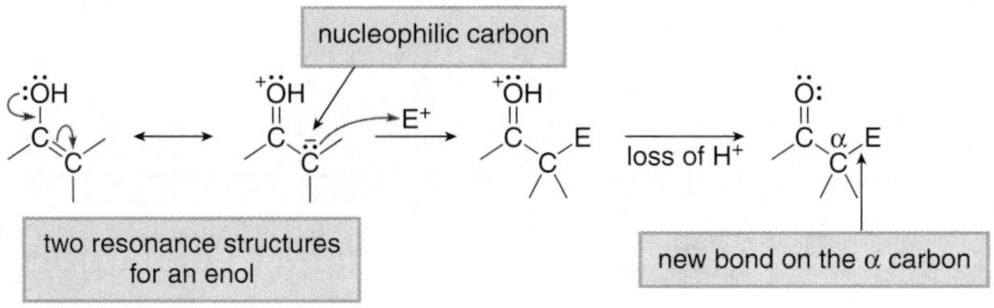

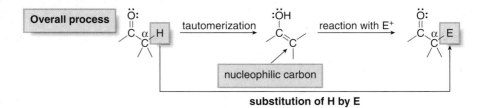

♦ Reaction of an enol with an electrophile E^+ forms a new C−E bond on the α carbon. The net result is substitution of H by E on the α carbon.

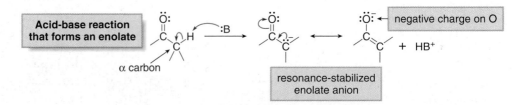

substitution of H by E

PROBLEM 23.5   When phenylacetaldehyde ($C_6H_5CH_2CHO$) is dissolved in D_2O with added DCl, the hydrogen atoms α to the carbonyl are gradually replaced by deuterium atoms. Write a mechanism for this process that involves enols as intermediates.

23.3   Enolates

Enolates are formed when a base removes a proton on the α carbon to a carbonyl group. A C−H bond on the α carbon is more acidic than many other sp^3 hybridized C−H bonds, because the resulting enolate is resonance stabilized. Moreover, one of the resonance structures is especially stable because it places a negative charge on an electronegative oxygen atom.

Acid-base reaction that forms an enolate

α carbon

resonance-stabilized enolate anion

negative charge on O

$+ \; HB^+$

Forming enolates from carbonyl compounds was first discussed in Section 21.7.

Enolates are always formed by removal of a proton on the **α carbon.**

Examples

propanal

cyclohexanone

The pK_a of the α hydrogen in an aldehyde or ketone is ~**20**. As shown in Table 23.1, this makes it considerably more acidic than the C−H bonds in CH_3CH_3 and $CH_3CH=CH_2$. Although C−H bonds α to a carbonyl are more acidic than many other C−H bonds, they are still less acidic than O−H bonds that always place the negative charge of the conjugate base on an electronegative oxygen atom (c.f. CH_3CH_2OH and CH_3COOH in Table 23.1).

The electrostatic potential plots in Figure 23.1 compare the electron density of the acetone enolate, which is resonance stabilized and delocalized, with that of $(CH_3)_2CHO^-$, an alkoxide that is not resonance stabilized.

TABLE 23.1   A Comparison of pK_a Values

	Compound	pK_a	Conjugate base	Structural features of the conjugate base
Increasing acidity Increasing stability of the conjugate base	CH_3CH_3	50	$CH_3\overset{..}{C}H_2$	• The conjugate base has a (−) charge on C, but is not resonance stabilized.
	$CH_2=CHCH_3$	43	$CH_2=CH-\overset{..}{C}H_2 \longleftrightarrow \overset{..}{C}H_2-CH=CH_2$	• The conjugate base has a (−) charge on C, and is resonance stabilized.
	$(CH_3)_2C=O$	19.2		**• The conjugate base has two resonance structures, one of which has a (−) charge on O.**
	CH_3CH_2OH	16	$CH_3CH_2\overset{..}{\underset{..}{O}}{:}^-$	• The conjugate base has a (−) charge on O, but is not resonance stabilized.
	CH_3CO_2H	4.8		• The conjugate base has two resonance structures, both of which have a (−) charge on O.

◆ **Resonance stabilization of the conjugate base increases acidity.**
   • $CH_2=CHCH_3$ is more acidic than $CH_3CH_2CH_3$.
   • CH_3COOH is more acidic than CH_3CH_2OH.

◆ **Placing a negative charge on O in the conjugate base increases acidity.**
   • CH_3CH_2OH is more acidic than $CH_3CH_2CH_3$.
   • CH_3COCH_3 is more acidic than $CH_2=CHCH_3$.
   • CH_3COOH (with two O atoms) is more acidic than CH_3COCH_3.

Figure 23.1 Electron density in an enolate and an alkoxide

• The acetone enolate is resonance stabilized. The negative charge is delocalized on the oxygen atom (pale red) and the carbon atom (pale green).
• The alkoxide anion is not resonance stabilized. The negative charge is concentrated on the oxygen atom only (deep red).

acetone enolate

The negative charge is delocalized on C and O.

The negative charge is concentrated on O.

an alkoxide anion

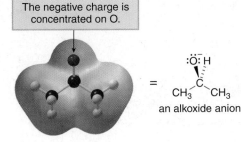

PROBLEM 23.6   Rank the indicated protons in order of increasing acidity.

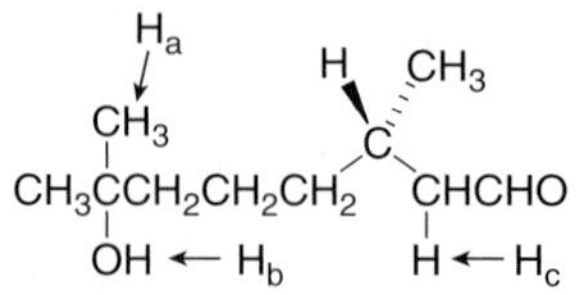

(S)-hydroxydihydrocitronellal
fragrance of lily of the valley

23.3A   Examples of Enolates and Related Anions

In addition to enolates from aldehydes and ketones, **enolates from esters and 3° amides can be formed as well,** although the α hydrogen is somewhat less acidic. **Nitriles** also have acidic protons on the carbon atom adjacent to the cyano group, because the negative charge of the conjugate base is stabilized by delocalization onto an electronegative nitrogen atom.

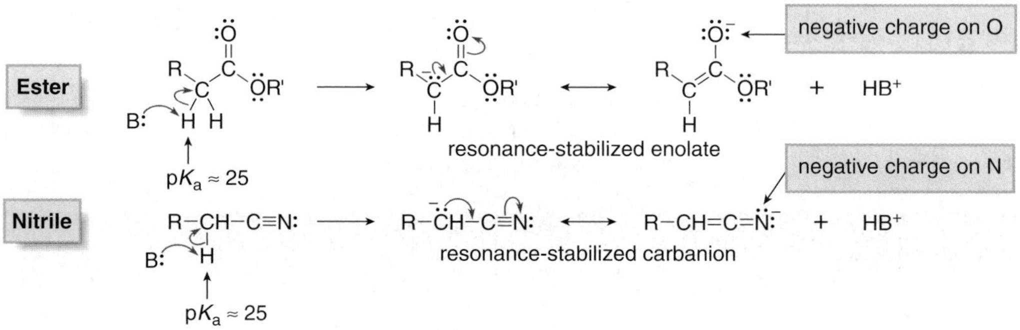

The protons on the carbon between the two carbonyl groups of a β-dicarbonyl compound are especially acidic because resonance delocalizes the negative charge on two different oxygen atoms. Table 23.2 lists pK_a values for β-dicarbonyl compounds as well as other carbonyl compounds and nitriles.

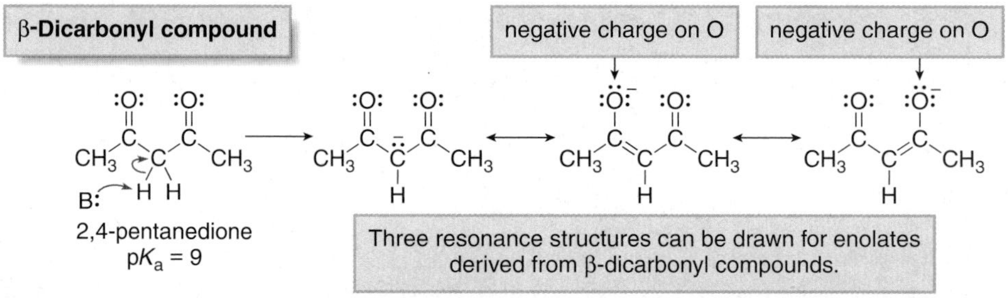

TABLE 23.2 pK_a Values for Some Carbonyl Compounds and Nitriles

Compound type	Example	pK_a	Compound type	Example	pK_a
[1] Amide	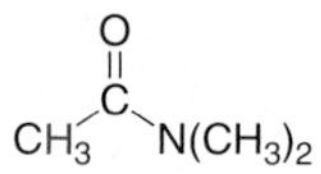 $CH_3CON(CH_3)_2$	30	[6] 1,3-Diester	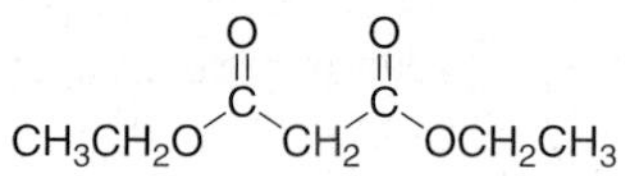 $CH_3CH_2O\text{-CO-}CH_2\text{-CO-}OCH_2CH_3$	13.3
[2] Nitrile	$CH_3-C\equiv N$	25	[7] 1,3-Dinitrile	$N\equiv C-CH_2-C\equiv N$	11
[3] Ester	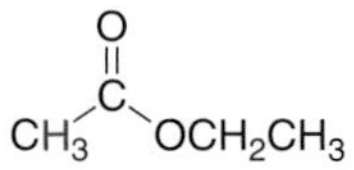 $CH_3CO\text{-}OCH_2CH_3$	25	[8] β-Keto ester	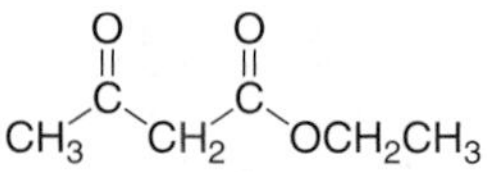 $CH_3\text{-CO-}CH_2\text{-CO-}OCH_2CH_3$	10.7
[4] Ketone	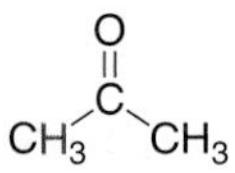 $CH_3\text{-CO-}CH_3$	19.2	[9] β-Diketone	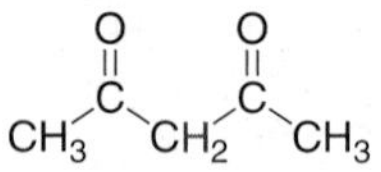 $CH_3\text{-CO-}CH_2\text{-CO-}CH_3$	8.9
[5] Aldehyde	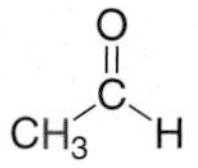 $CH_3\text{-CO-}H$	17			

PROBLEM 23.7   Draw additional resonance structures for each anion.

a. $^-$CH(COOCH$_2$CH$_3$)$_2$

b.   CH$_3$—C(=O)—$\overset{-}{C}$H—C(=O)—OCH$_2$CH$_3$

c.   CH$_3$—C(=O)—$\overset{-}{C}$HCN

PROBLEM 23.8   Which C–H bonds in the following molecules are acidic because the resulting conjugate base is resonance stabilized?

a.   CH$_3$—C(=O)—CH$_2$CH$_2$CH$_3$

b.   CH$_3$CH$_2$CH$_2$—CN

c.   CH$_3$CH$_2$CH$_2$—C(=O)—OCH$_2$CH$_3$

d.   (cyclohexane-1,3-dione with CH$_3$ substituent)

PROBLEM 23.9   Explain why H$_a$ is more acidic than H$_b$ in the following compound.

(phenyl)—CH$_2$—C(=O)—CH$_3$
        H$_a$ H$_b$

23.3B   The Base

The formation of an enolate is an acid–base equilibrium, so the **stronger the base, the more enolate that forms.**

We can predict the extent of an acid–base reaction by comparing the pK_a of the starting acid (the carbonyl compound in this case) with the pK_a of the conjugate acid formed. **The equilibrium favors the side with the weaker acid (the acid with the higher pK_a value).** The pK_a of many carbonyl compounds is ~20, so a significant amount of enolate will form only if the pK_a of the conjugate acid is > 20.

The common bases used to form enolates are hydroxide ($^-$OH), various alkoxides ($^-$OR), hydride (H$^-$), and dialkylamides ($^-$NR$_2$). How much enolate is formed using each of these bases is indicated in Table 23.3.

We have now used the term *amide* in two different ways—first as a functional group (e.g., the carboxylic acid derivative RCONH$_2$) and now as a base (e.g., $^-$NH$_2$, which can be purchased as a sodium or lithium salt, NaNH$_2$ or LiNH$_2$, respectively). In Chapter 23 we will use dialkylamides, $^-$NR$_2$, in which the two H atoms of $^-$NH$_2$ have been replaced by R groups.

TABLE 23.3   Enolate Formation with Various Bases:
RCOCH$_3$ (pK_a ≈ 20) + B: → RCOCH$_2^-$ + HB$^+$

Base (B:)	Conjugate acid (HB$^+$)	pK_a of HB$^+$	% Enolate
[1] Na$^+$ $^-$OH	H$_2$O	15.7	< 1%
[2] Na$^+$ $^-$OCH$_2$CH$_3$	CH$_3$CH$_2$OH	16	< 1%
[3] K$^+$ $^-$OC(CH$_3$)$_3$	(CH$_3$)$_3$COH	18	1–10% (depending on the carbonyl compound)
[4] Na$^+$ H$^-$	H$_2$	35	100%
[5] Li$^+$ $^-$N[CH(CH$_3$)$_2$]$_2$	HN[CH(CH$_3$)$_2$]$_2$	40	100%

Recall that **THF** stands for tetrahydrofuran, a polar aprotic solvent.

tetrahydrofuran
THF

Enolate formation with LDA is typically carried out at –78 °C, a convenient temperature to maintain in the laboratory because it is the temperature at which dry ice (solid CO$_2$) sublimes. A low-temperature cooling bath can be made by adding dry ice to acetone until the acetone cools to –78 °C. Immersing a reaction flask in this cooling bath keeps its contents at a constant low temperature.

When the pK_a of the conjugate acid is < 20, as it is for ⁻OH and all ⁻OR (entries 1–3), only a small amount of enolate is formed at equilibrium. These bases are more useful in forming enolates when more acidic 1,3-dicarbonyl compounds are used as starting materials. They are also used when both the enolate and the carbonyl starting material are involved in the reaction, as is the case for reactions described in Chapter 24.

To form an enolate in essentially 100% yield, a much stronger base such as lithium diisopropylamide, **Li⁺ ⁻N[CH(CH₃)₂]₂**, abbreviated as **LDA**, is used (entry 5). **LDA is a strong nonnucleophilic base.** Like the other nonnucleophilic bases (Sections 7.8B and 8.1), its bulky isopropyl groups make the nitrogen atom too hindered to serve as a nucleophile. It is still able, though, to remove a proton in an acid–base reaction.

3-D representation

lithium diisopropylamide

LDA

The N atom is too crowded to be a nucleophile.

LDA quickly deprotonates essentially all of the carbonyl starting material, even at –78 °C, to form the enolate product. THF is the typical solvent for these reactions.

pK_a = 20   +   LDA   ⇌ THF –78 °C   +   HN[CH(CH₃)₂]₂
diisopropylamine
pK_a = 40

Equilibrium greatly favors the products.
Essentially all of the ketone is converted to enolate.

LDA can be prepared by deprotonating diisopropylamine with an organolithium reagent such as butyllithium, and then used immediately in a reaction.

Preparation of LDA

CH₃CH₂CH₂CH₂—Li + H—N[CH(CH₃)₂]₂ ⟶ CH₃CH₂CH₂CH₂—H + Li⁺ ⁻N[CH(CH₃)₂]₂
diisopropylamine
LDA

PROBLEM 23.10   Draw the product formed when each starting material is treated with LDA in THF solution at –78 °C.

a.   b.   c.   CH₃—C(=O)—OCH₂CH₃   d.   CN

23.3C   General Reactions of Enolates

Enolates are nucleophiles, and as such they react with many electrophiles. Because an enolate is resonance stabilized, however, it has two reactive sites—the carbon and oxygen atoms that bear the negative charge. **A nucleophile with two reactive sites is called an** *ambident nucleophile.* In theory, each of these atoms could react with an electrophile to form two different products, one with a new bond to carbon, and one with a new bond to oxygen.

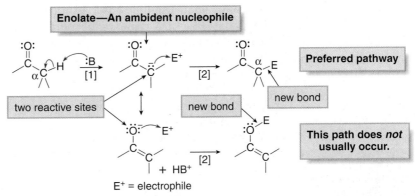

Enolate—An ambident nucleophile

Preferred pathway

two reactive sites

new bond

new bond

This path does *not* usually occur.

E⁺ = electrophile

<table>
<tr><td>Because enolates usually react at carbon instead of oxygen, the resonance structure that places the negative charge on oxygen will often be omitted in multistep mechanisms.</td><td>

An enolate usually reacts at the carbon end, however, because this site is more nucleophilic. Thus, **enolates generally react with electrophiles on the α carbon** so that many reactions in Chapter 23 follow a two-step path:

</td></tr>
</table>

[1] Reaction of a carbonyl compound with base forms an enolate.
[2] Reaction of the enolate with an electrophile forms a new bond on the α carbon.

23.4   Enolates of Unsymmetrical Carbonyl Compounds

What happens when an unsymmetrical carbonyl compound like 2-methylcyclohexanone is treated with base? **Two enolates are possible,** one formed by removal of a 2° hydrogen, and one formed by removal of a 3° hydrogen.

Path [1]
removal of a 2° H

less substituted enolate

kinetic enolate

This enolate is formed **faster.**

2° H

3° H

2-methylcyclohexanone

Path [2]
removal of a 3° H

more substituted enolate

more substituted C=C

thermodynamic enolate

This enolate is **more stable.**

Path [1] occurs *faster* than Path [2] because it results in removal of the less hindered 2° hydrogen, forming an enolate on the less substituted α carbon. Path [2] results in removal of a 3° hydrogen, forming the *more stable* enolate with the more substituted double bond. This enolate predominates at equilibrium.

◆ The kinetic enolate is formed faster because it is the less substituted enolate.
◆ The thermodynamic enolate is lower in energy because it is the more substituted enolate.

It is possible to regioselectively form one or the other enolate by the proper use of reaction conditions, because the base, solvent, and reaction temperature all affect the identity of the enolate formed.

Kinetic Enolates

The kinetic enolate forms faster, so mild reaction conditions favor it over slower processes with higher energies of activation. It is the less stable enolate, so it must not be allowed to equilibrate to the more stable thermodynamic enolate. **The kinetic enolate is favored by:**

[1] **A strong nonnucleophilic base.** A strong base assures that the enolate is formed rapidly. A **bulky base like LDA removes the more accessible proton on the less substituted carbon** much faster than a more hindered proton.

[2] **Polar aprotic solvent.** The solvent must be polar to dissolve the polar starting materials and intermediates. It must be aprotic so that it does not protonate any enolate that is formed. **THF** is both polar and aprotic.

[3] **Low temperature.** The temperature must be low (**−78 °C**) to prevent the kinetic enolate from equilibrating to the thermodynamic enolate.

> **A kinetic enolate is formed with a strong, nonnucleophilic base (LDA) in a polar aprotic solvent (THF) at low temperature (−78 °C).**

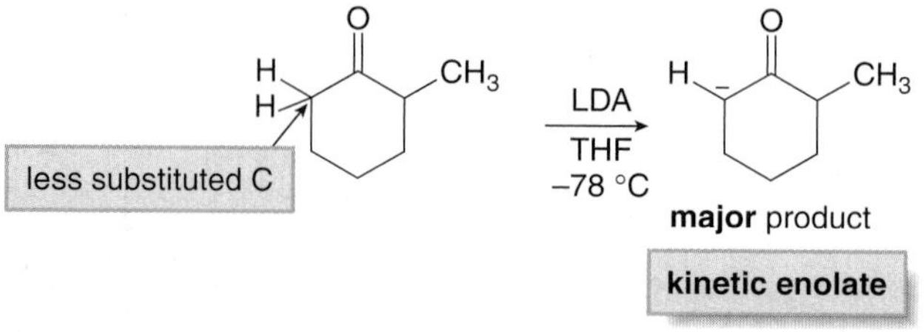

Thermodynamic Enolates

A thermodynamic enolate is favored by equilibrating conditions. This is often achieved using a **strong base in a protic solvent.** A strong base yields both enolates, but in a protic solvent, enolates can also be protonated to re-form the carbonyl starting material. At equilibrium, the lower energy intermediate always wins out, so that **the more stable, more substituted enolate is present in higher concentration.** Thus, the **thermodynamic enolate is favored by:**

[1] **A strong base. Na$^+$ $^-$OCH$_2$CH$_3$, K$^+$ $^-$OC(CH$_3$)$_3$** or other alkoxides are common.

[2] **Protic solvent. CH$_3$CH$_2$OH** or other alcohols.

[3] **Room temperature (25 °C).**

> **A thermodynamic enolate is formed with a strong base (RO$^-$) in a polar protic solvent (ROH) at room temperature.**

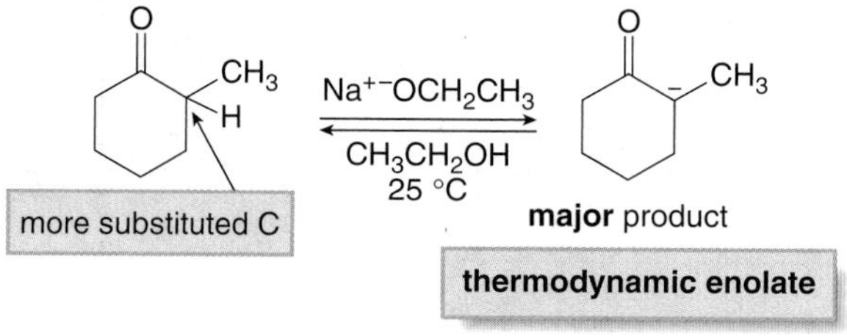

SAMPLE PROBLEM 23.2   What is the major enolate formed in each reaction?

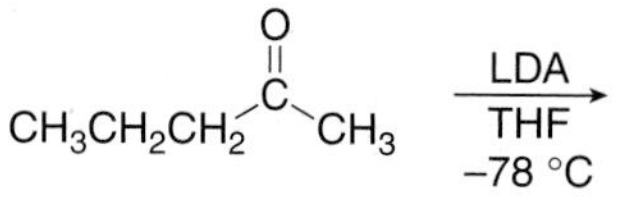

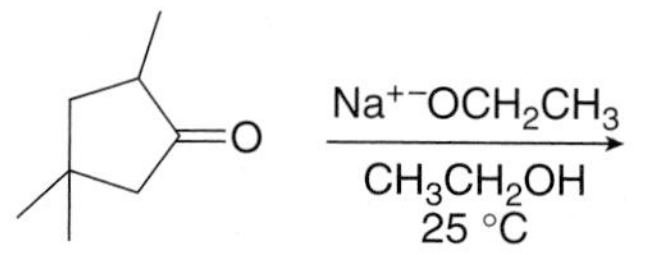

SOLUTION

[a] LDA is a strong, nonnucleophilic base that removes a proton on the less substituted α carbon to form the kinetic enolate.

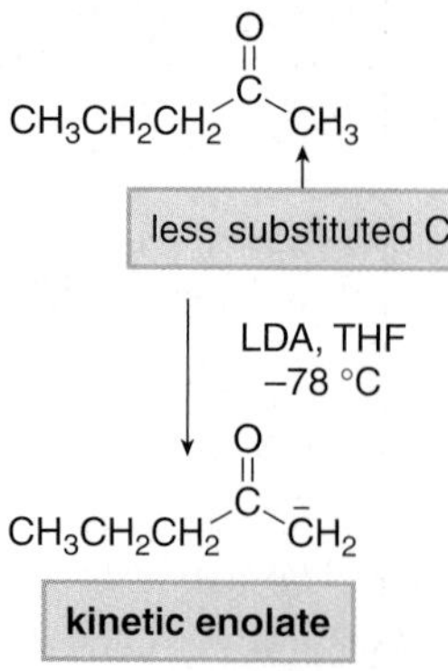

[b] NaOCH$_2$CH$_3$ (a strong base) and CH$_3$CH$_2$OH (a protic solvent) favor removal of a proton from the more substituted α carbon to form the thermodynamic enolate.

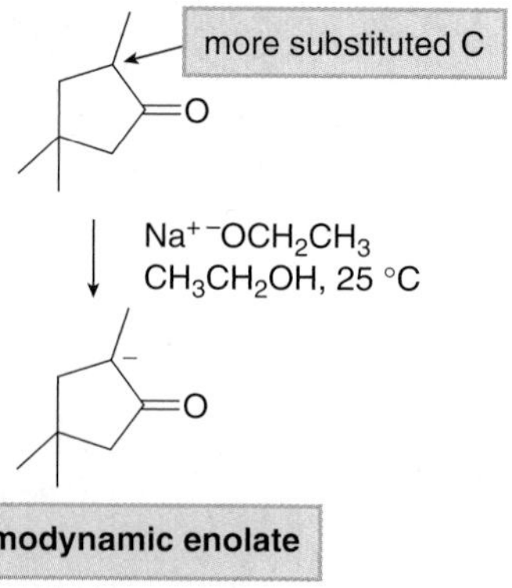

PROBLEM 23.11 What enolate is formed when each ketone is treated with LDA in THF solution? What enolate is formed when these same ketones are treated with NaOCH₃ in CH₃OH solution?

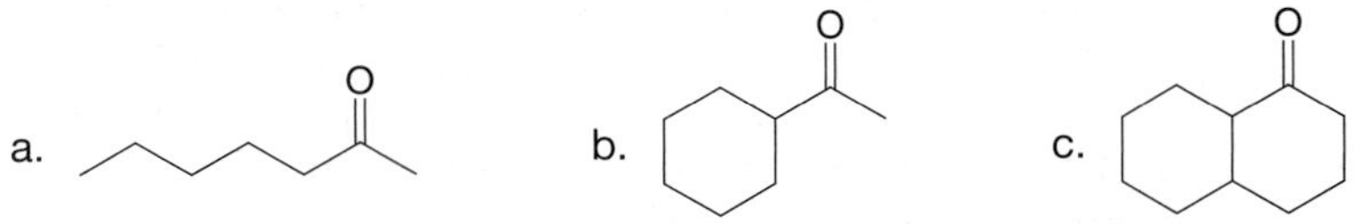

a.   b.   c.

23.5   Racemization at the α Carbon

Recall from Section 16.5 that an enolate can be stabilized by the delocalization of electron density only if it possesses the proper geometry and hybridization.

◆ The electron pair on the carbon adjacent to the C=O must occupy a *p* orbital that overlaps with the two other *p* orbitals of the C=O, making an enolate conjugated.

◆ Thus, all three atoms of the enolate are *sp²* hybridized and trigonal planar.

These bonding features are shown in the acetone enolate in Figure 23.2.

When the α carbon to the carbonyl is a stereogenic center, treatment with aqueous base leads to **racemization** by a two-step process: **deprotonation to form an enolate and protonation to re-form the carbonyl compound.** For example, chiral ketone **A** reacts with aqueous ⁻OH to form an achiral enolate having an *sp²* hybridized α carbon. Because the enolate is planar, it can be protonated with H₂O with equal probability from both directions, yielding a racemic mixture of two ketones.

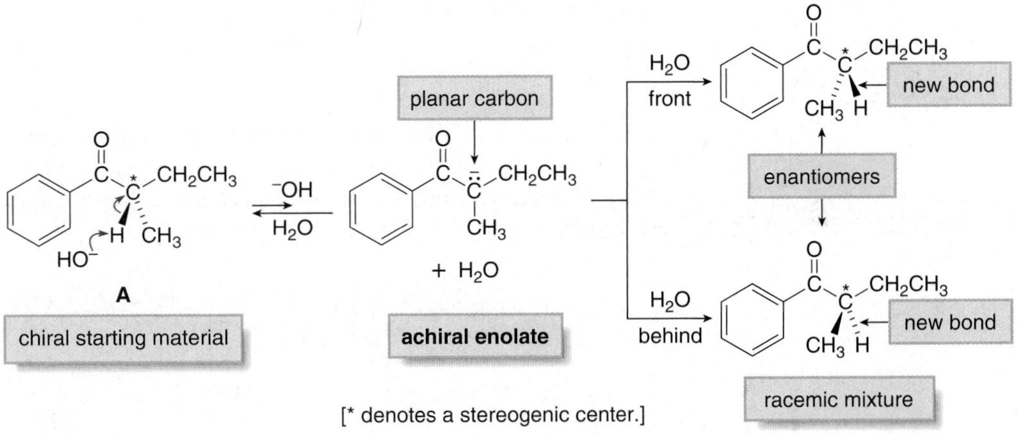

[* denotes a stereogenic center.]

Figure 23.2 The hybridization and geometry of the acetone enolate (CH₃COCH₂)⁻

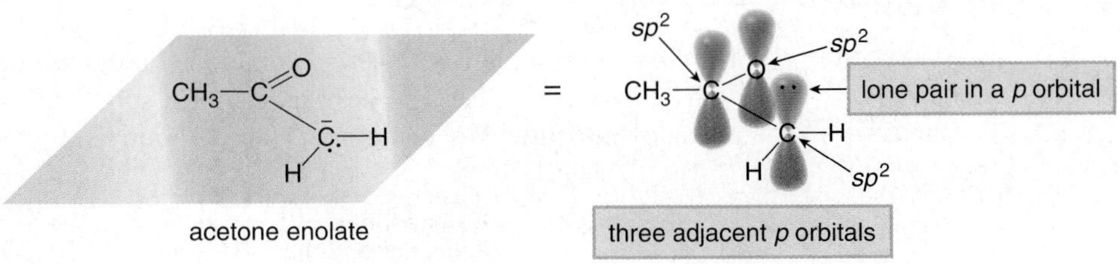

- The O atom and both C's of the enolate are *sp²* hybridized and lie in a plane.
- Each atom has a *p* orbital extending above and below the plane; these orbitals overlap to delocalize electron density.

PROBLEM 23.12   Explain each observation: (a) When (R)-2-methylcyclohexanone is treated with NaOH in H_2O, the optically active solution gradually loses optical activity. (b) When (R)-3-methylcyclohexanone is treated with NaOH in H_2O, the solution remains optically active.

PROBLEM 23.13   Treatment of propiophenone ($C_6H_5COCH_2CH_3$) with NaOD in D_2O forms two products having molecular formula C_9H_9OD. What are their structures and how are they formed?

23.6   A Preview of Reactions at the α Carbon

Having learned about the synthesis and properties of enolates, we can now turn our attention to their reactions. Like enols, **enolates are nucleophiles,** but because they are negatively charged, enolates are much more nucleophilic than neutral enols. Consequently, they undergo a wider variety of reactions.

Two general types of reactions of enolates—**substitutions** and **reactions with other carbonyl compounds**—will be discussed in the remainder of Chapter 23 and in Chapter 24. Both reactions form new bonds to the carbon α to the carbonyl.

◆ **Enolates react with electrophiles to afford substitution products.** Two different kinds of substitution reactions are examined: **halogenation** with X_2 and **alkylation** with alkyl halides RX. These reactions are detailed in Sections 23.7–23.10.

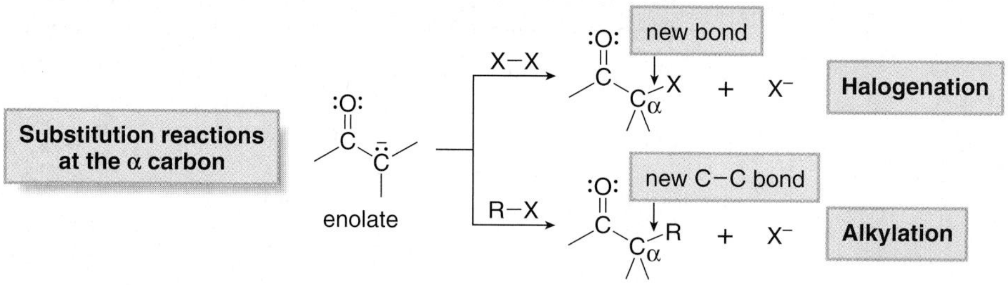

◆ **Enolates react with other carbonyl groups at the electrophilic carbonyl carbon.** These reactions are more complicated because the initial addition adduct goes on to form different products depending on the structure of the carbonyl group. These reactions form the subject of Chapter 24.

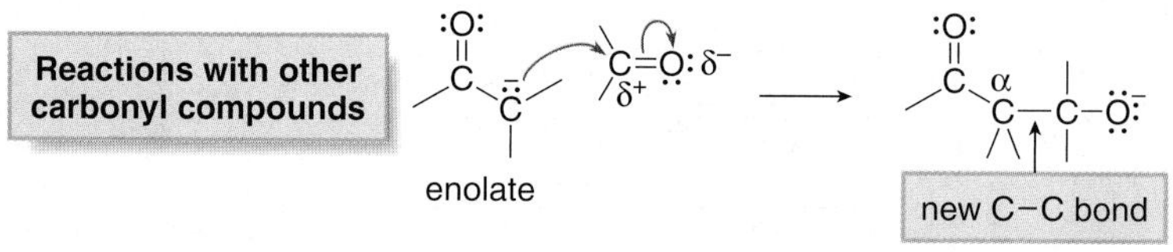

23.7   Halogenation at the α Carbon

The first substitution reaction we examine is **halogenation.** Treatment of a ketone or aldehyde with halogen and either acid or base results in **substitution of X for H on the α carbon,** forming an **α-halo aldehyde or ketone.** Halogenation readily occurs with Cl_2, Br_2, and I_2.

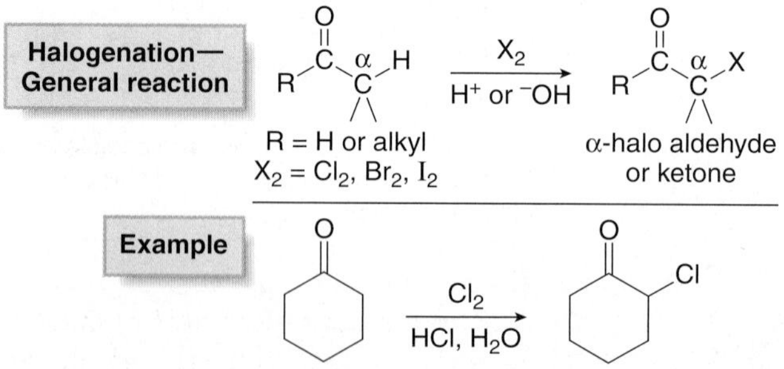

The mechanisms of halogenation in acid and base are somewhat different.

♦ Reactions done in acid generally involve *enol* intermediates.
♦ Reactions done in base generally involve *enolate* intermediates.

23.7A   Halogenation in Acid

Halogenation is often carried out by treating a carbonyl compound with a halogen in acetic acid. In this way, acetic acid is both the solvent and the acid catalyst for the reaction.

$$CH_3\text{-}C(\text{=}O)\text{-}CH_3 \xrightarrow[CH_3COOH]{Br_2} CH_3\text{-}C(\text{=}O)\text{-}CH_2Br + HBr$$

substitution of one H by Br

The mechanism of acid-catalyzed halogenation consists of two parts: **tautomerization** of the carbonyl compound to the enol form, and **reaction of the enol with halogen.** Mechanism 23.3 illustrates the reaction of $(CH_3)_2C{=}O$ with Br_2 in CH_3COOH.

MECHANISM 23.3

Acid-Catalyzed Halogenation at the α Carbon

Part [1] Tautomerization to the enol

In Part [1], the ketone is converted to its enol tautomer by the usual two-step process: **protonation** of the carbonyl oxygen, followed by **deprotonation** of the α carbon atom.

Part [2] Reaction of the enol with halogen

In Part [2], addition of the halogen to the enol followed by deprotonation forms the neutral substitution product (Steps [3]–[4]). **The overall process results in substitution of H by Br on the α carbon.**

α-bromoacetone

These α-halo carbonyl compounds can be converted to **α,β-unsaturated carbonyl compounds** by heating in the presence of a base such as pyridine. This reaction is an example of the E2 elimination discussed in Chapter 8. Thus, a two-step method can convert a carbonyl compound **A** into an α,β-unsaturated carbonyl compound **B**.

A new π bond is formed in two steps.

halogenation   elimination

α,β-Unsaturated carbonyl compounds undergo a variety of 1,2- and 1,4-addition reactions as discussed in Section 20.15.

[1] **Bromination at the α carbon** is accomplished with Br_2 in CH_3COOH.
[2] **E2 elimination** of Br and H from the α and β carbons occurs by heating in the presence of pyridine. This yields the α,β-unsaturated ketone **B** as product.

PROBLEM 23.14   Draw the products of each reaction.

a. [cyclopentanone structure] $\xrightarrow[\text{H}_2\text{O, HCl}]{\text{Cl}_2}$ b. [propanal structure] CHO $\xrightarrow[\text{CH}_3\text{CO}_2\text{H}]{\text{Br}_2}$ c. [α-tetralone structure] $\xrightarrow[\text{[2] pyridine, }\Delta]{\text{[1] Br}_2\text{, CH}_3\text{CO}_2\text{H}}$

23.7B   Halogenation in Base

Halogenation in base is much less useful, because it is difficult to stop the reaction after addition of just one halogen atom to the α carbon. For example, treatment of propiophenone with Br_2 and aqueous ⁻OH yields a dibromo ketone.

Reactions of carbonyl compounds with base invariably involve enolates because the α hydrogens of the carbonyl compound are easily removed.

[structure of propiophenone] $\xrightarrow[^-\text{OH}]{\text{Br}_2}$ [structure of dibromo ketone]

propiophenone

Both α H's are replaced by Br.

The mechanism for introduction of each Br atom involves the same two steps: **deprotonation with base followed by reaction with Br_2** to form a new C–Br bond, as shown in Mechanism 23.4.

MECHANISM 23.4

Halogenation at the α Carbon in Base

[mechanism structures: propiophenone → enolate → monosubstitution product]

propiophenone   enolate   One α H is replaced by Br.

◆ Treatment of the ketone with base forms a nucleophilic enolate in Step [1], which reacts with Br_2 (the electrophile) to form the monosubstitution product—that is, one α H is replaced by one Br on the α carbon.

Although all ketones with α hydrogens react with base and I_2, only **methyl** ketones form CHI_3 (iodoform), a pale yellow solid that precipitates from the reaction mixture. This reaction is the basis of the **iodoform test,** once a common chemical method to detect methyl ketones. Methyl ketones give a positive iodoform test (appearance of a yellow solid), whereas other ketones give a negative iodoform test (no change in the reaction mixture).

Only a small amount of the enolate forms at equilibrium using ⁻OH as base, but the enolate is such a strong nucleophile that it readily reacts with Br_2, thus driving the equilibrium to the right. Then, the same two steps introduce the second Br atom on the α carbon: **deprotonation** followed by **nucleophilic attack.**

[mechanism structures: α-bromopropiophenone → enolate → disubstitution product]

α-bromopropiophenone

The electronegative Br stabilizes the negative charge.

disubstitution product

It is difficult to stop this reaction after the addition of one Br atom because the electron-withdrawing inductive effect of Br stabilizes the second enolate. As a result, the α H of α-bromopropiophenone is more acidic than the α H atoms of propiophenone, making it easier to remove with base.

Halogenation of a methyl ketone with excess halogen, called the **haloform reaction,** results in cleavage of a carbon–carbon σ bond and formation of two products, a carboxylate anion and CHX_3 (commonly called **haloform**).

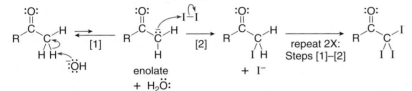

The haloform reaction— General reaction

This C–C bond is cleaved.

Example

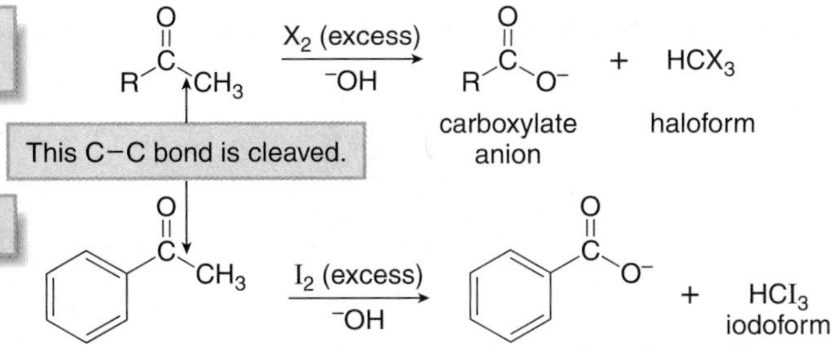

In the haloform reaction, the three H atoms of the CH₃ group are successively replaced by X to form an intermediate that is oxidatively cleaved with base. Mechanism 23.5 is written with I₂ as halogen, forming CHI₃ (iodoform) as product.

MECHANISM 23.5

The Haloform Reaction

Part [1] The conversion of CH₃ to CI₃

◆ In Part [1], the three hydrogen atoms of the CH₃ group are replaced with iodine via the two-step mechanism previously discussed—that is, base deprotonates the methyl group, forming a nucleophilic enolate that reacts with iodine to form the substitution product.

◆ Steps [1] and [2] are then repeated twice more to form the triiodo substitution product.

Part [2] Oxidative cleavage with ⁻OH

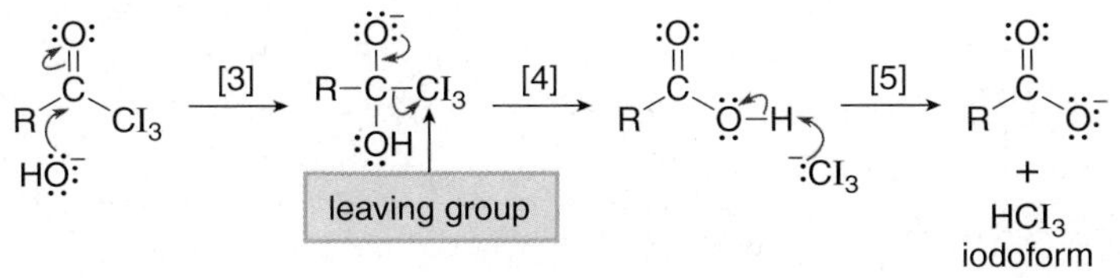

◆ In Step [3], ⁻OH adds to the carbonyl group in a typical **nucleophilic addition** reaction of a ketone, but the three I atoms give this ketone a good leaving group, ⁻CI₃.

◆ **Elimination of ⁻CI₃ in Step [4] results in cleavage of a carbon–carbon bond,** and then in the last step, proton transfer forms the carboxylate anion and iodoform.

Steps [3] and [4] result in a **nucleophilic *substitution*** reaction of a **ketone.** Because ketones normally undergo **nucleophilic *addition,*** this two-step sequence makes the haloform reaction unique. Substitution occurs because the three electronegative halogen atoms make CX₃ (CI₃ in the example) a good leaving group.

Figure 23.3 summarizes the three possible outcomes of halogenation at the α carbon, depending on the chosen reaction conditions.

Figure 23.3   Summary: Halogenation reactions at the α carbon to a carbonyl group

[1] Halogenation in acid

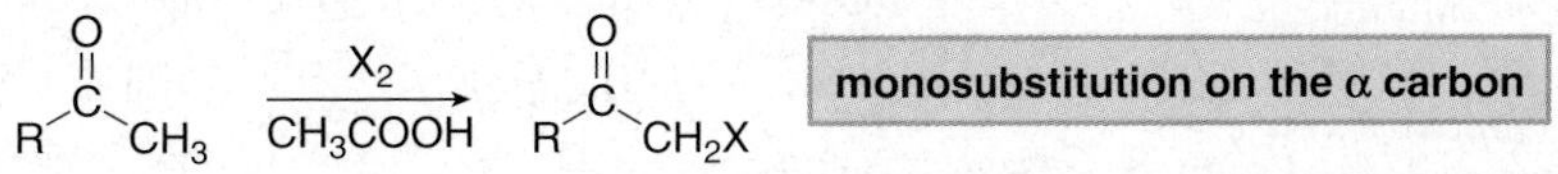

monosubstitution on the α carbon

[2] General halogenation in base

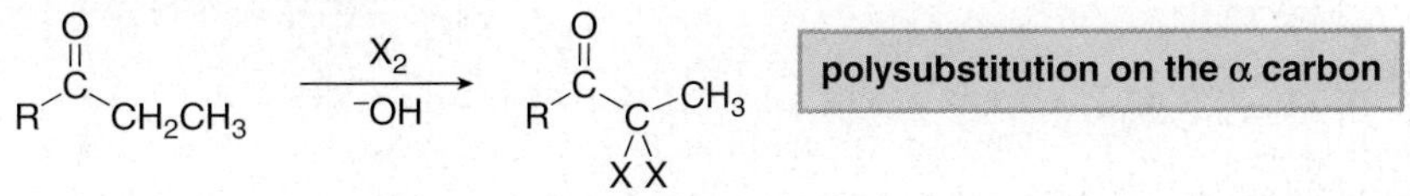

polysubstitution on the α carbon

[3] Halogenation of *methyl* ketones with excess X₂ and base

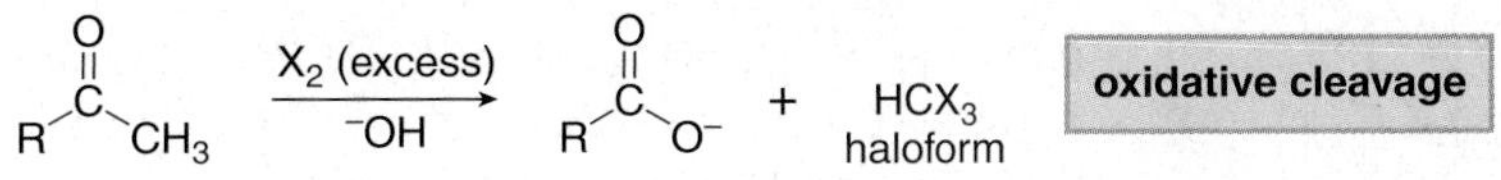

oxidative cleavage

PROBLEM 23.15 Draw the products of each reaction. Assume excess halogen is present.

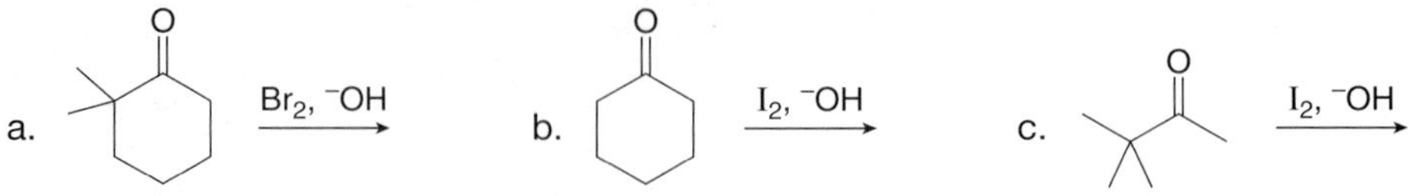

a. $\xrightarrow{Br_2, \ ^-OH}$

b. $\xrightarrow{I_2, \ ^-OH}$

c. $\xrightarrow{I_2, \ ^-OH}$

23.8 Direct Enolate Alkylation

Treatment of an aldehyde or ketone with base and an alkyl halide (RX) results in **alkylation—the substitution of R for H on the α carbon atom.** Alkylation forms a new carbon–carbon bond on the α carbon.

Alkylation at the α carbon— General reaction

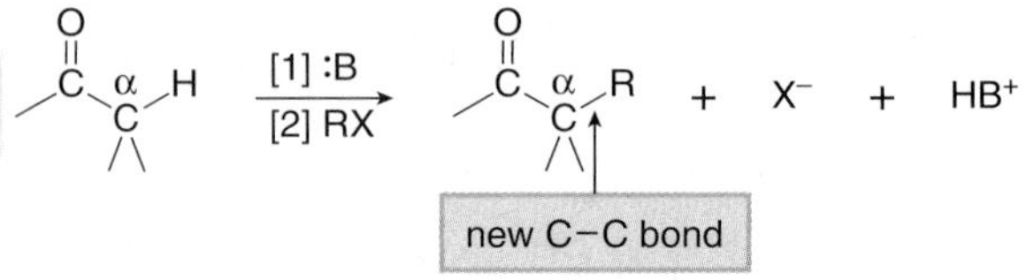

23.8A General Features

We will begin with the most direct method of alkylation, and then (in Sections 23.9 and 23.10) examine two older, multistep methods that are still used today. Direct alkylation is carried out by a two-step process:

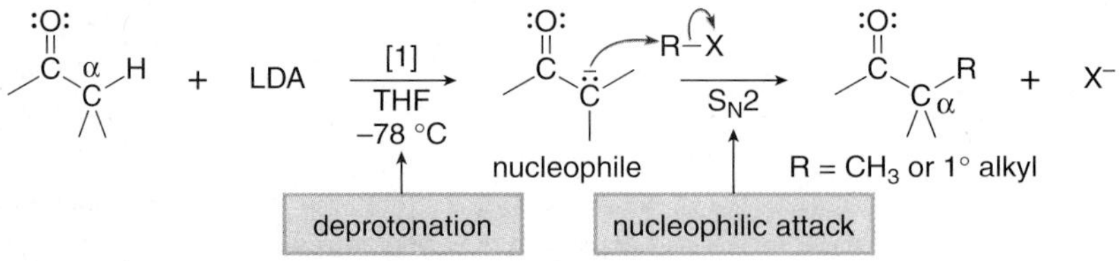

[1] **Deprotonation:** Base removes a proton from the α carbon to generate an enolate. The reaction works best with a strong nonnucleophilic base like LDA in THF solution at low temperature (−78 °C).

[2] **Nucleophilic attack:** The nucleophilic enolate attacks the alkyl halide, displacing the halide (a good leaving group) and forming the alkylation product by an S_N2 reaction.

Because Step [2] is an S_N2 reaction, it works well only with unhindered methyl and 1° alkyl halides. Hindered alkyl halides and those with halogens bonded to sp^2 hybridized carbons do not undergo substitution.

R_3CX, CH_2=CHX, and C_6H_5X do not undergo alkylation reactions with enolates, because they are unreactive in S_N2 reactions.

Examples

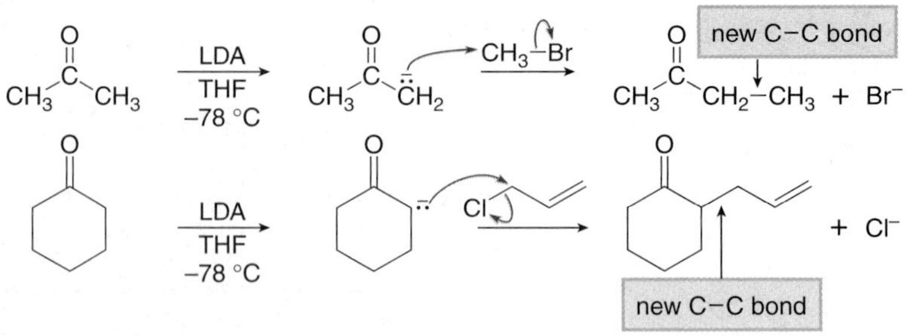

Ester enolates and carbanions derived from nitriles are also alkylated under these conditions.

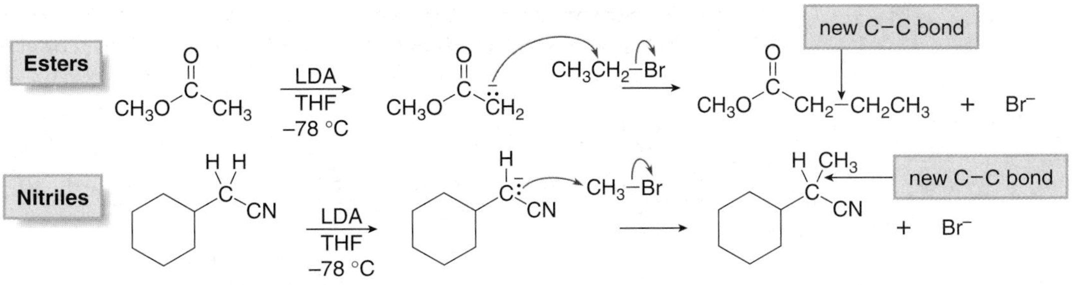

Esters

Nitriles

PROBLEM 23.16   What product is formed when each compound is treated first with LDA in THF solution at low temperature, followed by CH_3CH_2I?

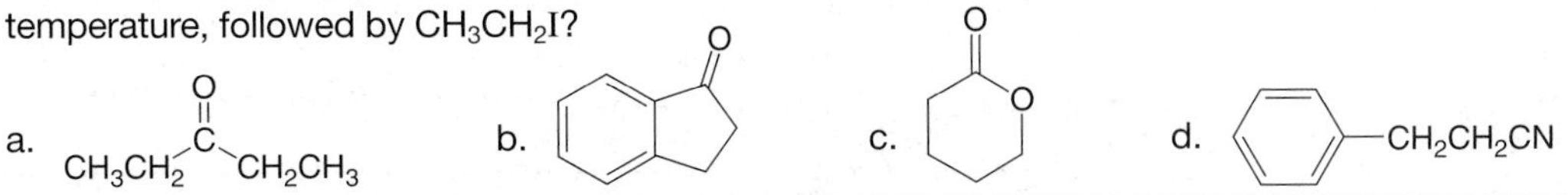

a.

b.

c.

d.

The stereochemistry of enolate alkylation follows the general rule governing the stereochemistry of reactions: **an achiral starting material yields an achiral or racemic product.** For example, when cyclohexanone (an achiral starting material) is converted to 2-ethylcyclohexanone by treatment with base and CH_3CH_2I, a new stereogenic center is introduced, and both enantiomers of the product are formed in equal amounts—that is, a **racemic mixture.**

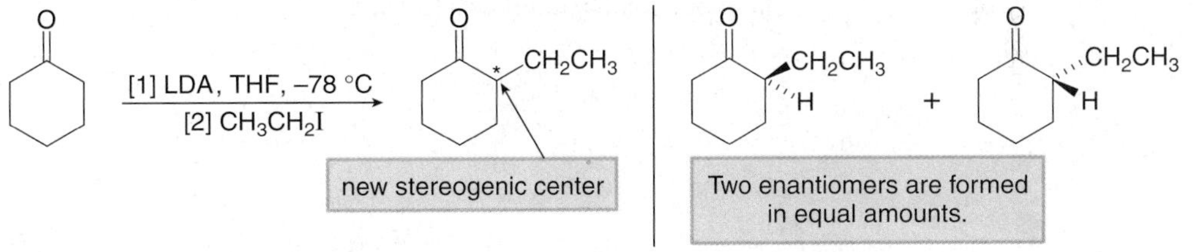

PROBLEM 23.17   Draw the products obtained (including stereochemistry) when each ketone is treated with LDA, followed by CH_3I.

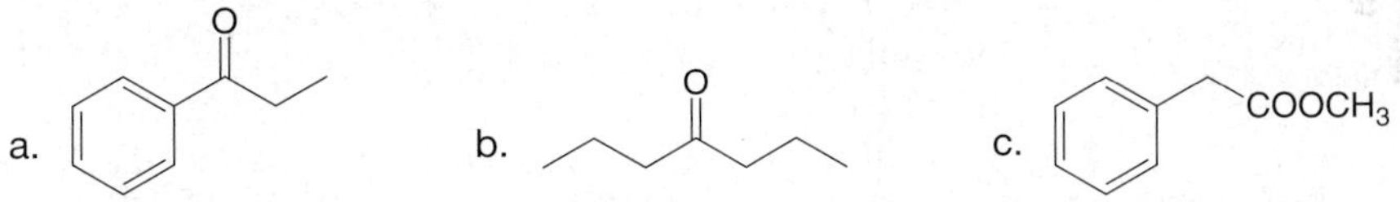

a.

b.

c.

23.8B   Alkylation of Unsymmetrical Ketones

An unsymmetrical ketone can be regioselectively alkylated to yield one major product. The strategy depends on the use of the appropriate base, solvent, and temperature to form the kinetic or thermodynamic enolate (Section 23.4), which is then treated with an alkyl halide to form the alkylation product.

For example, 2-methylcyclohexanone can be converted to either 2,6-dimethylcyclohexanone (**A**) or 2,2-dimethylcyclohexanone (**B**) by proper choice of reaction conditions.

◆ **Treatment of 2-methylcyclohexanone with LDA in THF solution at −78 °C gives the less substituted kinetic enolate,** which then reacts with CH_3I to form **A.**

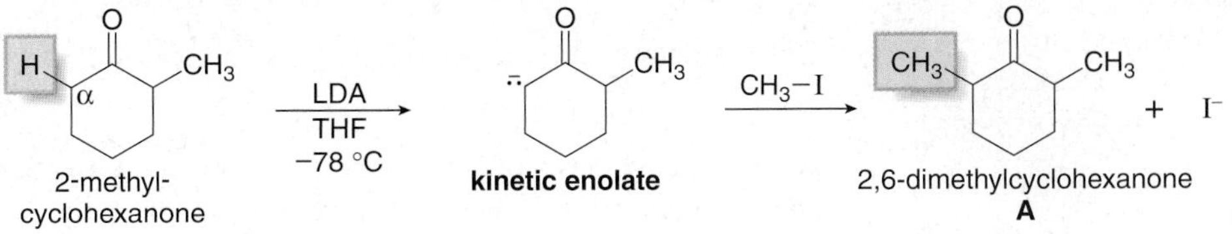

◆ **Treatment of 2-methylcyclohexanone with NaOCH₂CH₃ in CH₃CH₂OH solution at room temperature forms the more substituted thermodynamic enolate,** which then reacts with CH₃I to form **B**.

2-methyl-cyclohexanone → [NaOCH₂CH₃ / CH₃CH₂OH / 25 °C] → **thermodynamic enolate** → [CH₃—I] → 2,2-dimethylcyclohexanone **B** + I⁻

PROBLEM 23.18   How can 2-pentanone be converted into each compound?

a.   b.   c.   d.

23.8C   Application of Enolate Alkylation: Tamoxifen Synthesis

More of the details of this synthesis can be found in Problem 23.51 at the end of the chapter.

Tamoxifen, the chapter-opening molecule, is a potent anticancer drug that has been used to treat certain forms of breast cancer for many years. One step in the synthesis of tamoxifen involves the treatment of ketone **A** with NaH as base to form an enolate. Alkylation of this enolate with CH₃CH₂I forms **B** in high yield. **B** is converted to tamoxifen in several steps, some of which are reactions you have already learned.

A → [NaH] → **enolate** → [CH₃CH₂I] → **B** → [several steps] → tamoxifen

Only the *Z* isomer of the C=C provides beneficial effects.

Tamoxifen has been commercially available since the 1970s, sold under the brand name of Nolvadex.

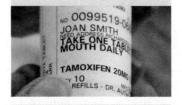

23.8D   Application of Enolate Alkylation: Synthesis of β-Vetivone

Vetiver is a perennial grass with long narrow leaves and an extensive root system. It yields a complex mixture of organic compounds that have been used in traditional medicine, pest control, and fragrance, particularly in India, Sri Lanka, and Thailand.

β-Vetivone is a major component of vetiver oil, an essential oil that is obtained from the roots of *Vetiveria zizanoides,* a tall perennial grass found in tropical and subtropical regions of the world. β-Vetivone contains a *spiro* **ring system—that is, two rings that share a single carbon atom.** Because the shared carbon atom is *sp*³ hybridized, the two rings are approximately perpendicular to each other, a fact that is not obvious from its two-dimensional line drawing.

β-vetivone   =   shared carbon

An elegant synthesis of β-vetivone was carried out in the laboratory of Professor Gilbert Stork and co-workers at Columbia University in the early 1970s. The key step in this synthesis

Figure 23.4 Enolate alkylation in the synthesis of β-vetivone

- Steps [1] and [2] form the first carbon–carbon bond. Treatment of **A** with LDA forms an enolate that reacts with one of the 1° alkyl halides in **B** to form **D**.
- Steps [3] and [4] form the second carbon–carbon bond. Because **D** contains both an acidic α hydrogen and a leaving group, an intramolecular reaction can form the five-membered ring. Treatment of **D** with LDA followed by intramolecular cyclization forms **C**.

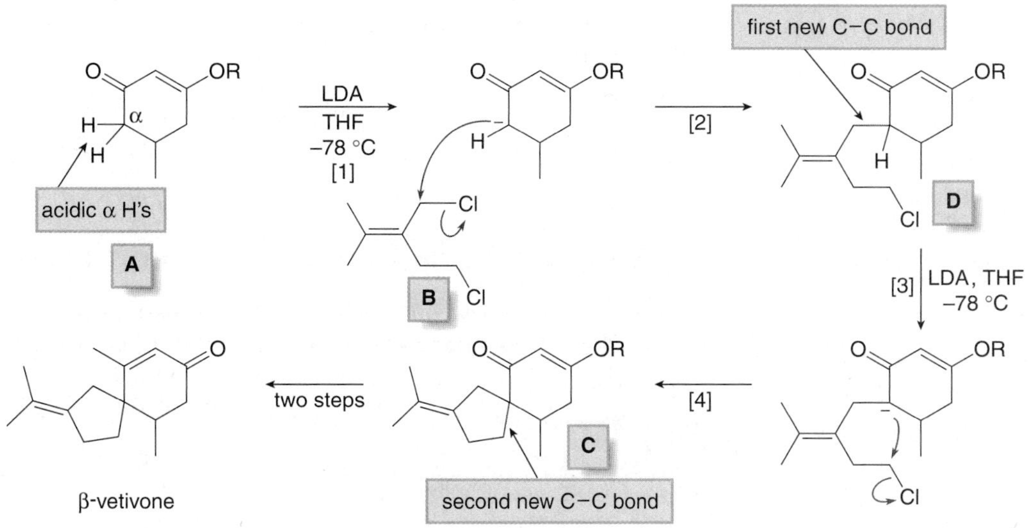

involves two sequential enolate alkylations of a carbonyl compound **A** with a dihalide **B** that forms two new carbon–carbon bonds and the five-membered ring of **C**. **C** is converted to β-vetivone in a single step. Figure 23.4 illustrates the intermediates involved in the process of enolate alkylation.

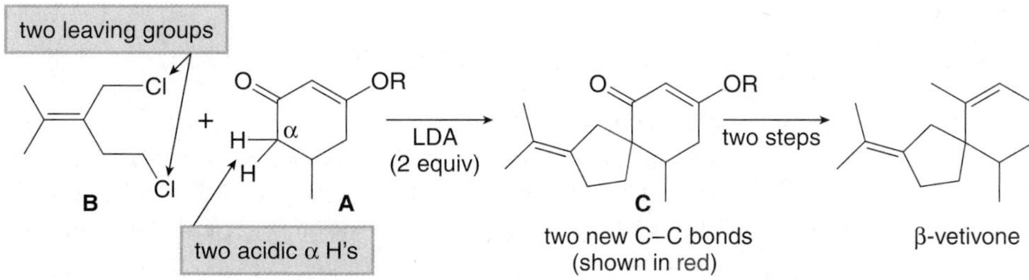

PROBLEM 23.19   Identify **A, B,** and **C,** intermediates in the synthesis of the five-membered ring called an α-methylene-γ-butyrolactone. This heterocyclic ring system is present in some antitumor agents.

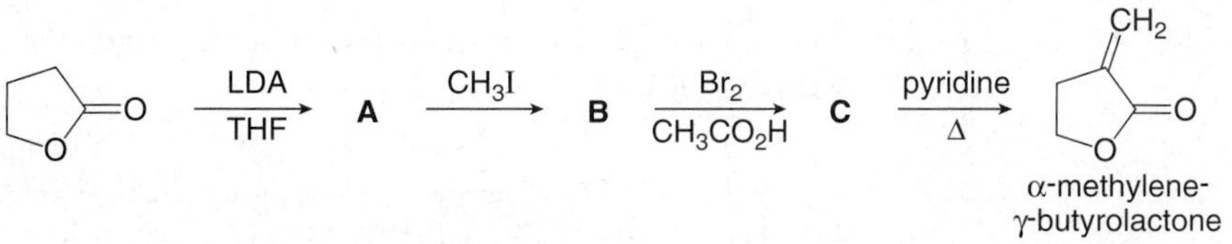

23.9 Malonic Ester Synthesis

Besides the direct method of enolate alkylation discussed in Section 23.8, a new alkyl group can also be introduced on the α carbon using the malonic ester synthesis and the acetoacetic ester synthesis.

◆ **The malonic ester synthesis prepares α-substituted carboxylic acids** having two general structures:

$$R-CH_2COOH \qquad R-CHCOOH$$
$$\underset{R'}{|}$$

- The **acetoacetic ester synthesis prepares α-substituted ketones** having two general structures:

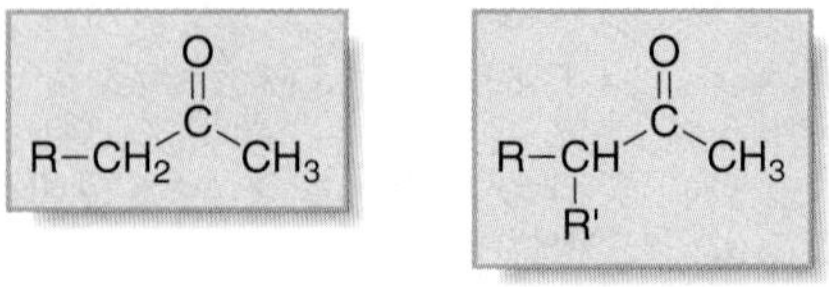

23.9A Background for the Malonic Ester Synthesis

The malonic ester synthesis is a stepwise method for converting diethyl malonate into a carboxylic acid having one or two alkyl groups on the α carbon. To simplify the structures, the CH_3CH_2 groups of the esters are abbreviated as Et.

Malonic ester synthesis

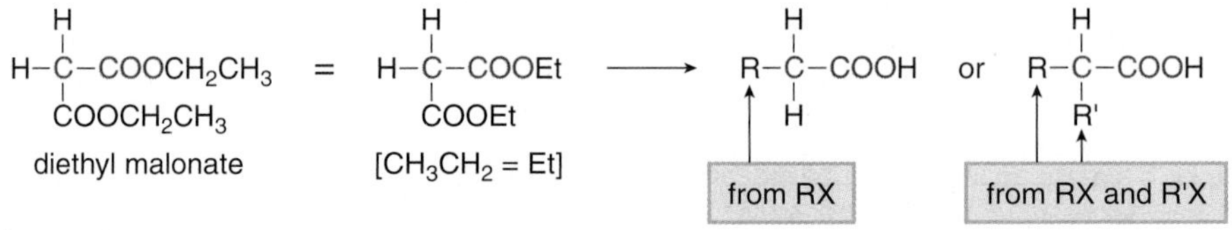

Before writing out the steps in the malonic ester synthesis, recall from Section 22.11 that esters are hydrolyzed by aqueous acid. Thus, heating diethyl malonate with acid and water hydrolyzes both esters to carboxy groups, forming a β-diacid (1,3-diacid).

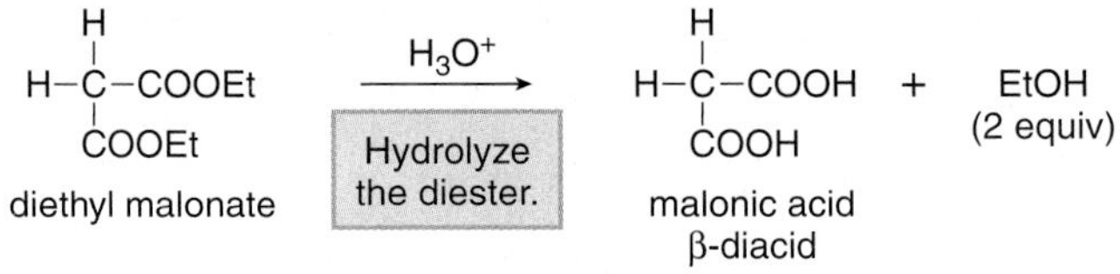

β-Diacids are unstable to heat. They **decarboxylate (lose CO_2)**, resulting in cleavage of a carbon–carbon bond and formation of a carboxylic acid. Decarboxylation is not a general reaction of all carboxylic acids. It occurs with β-diacids, however, because CO_2 can be eliminated through a cyclic, six-atom transition state. This forms an enol of a carboxylic acid, which in turn tautomerizes to the more stable keto form.

Decarboxylation of a β-diacid

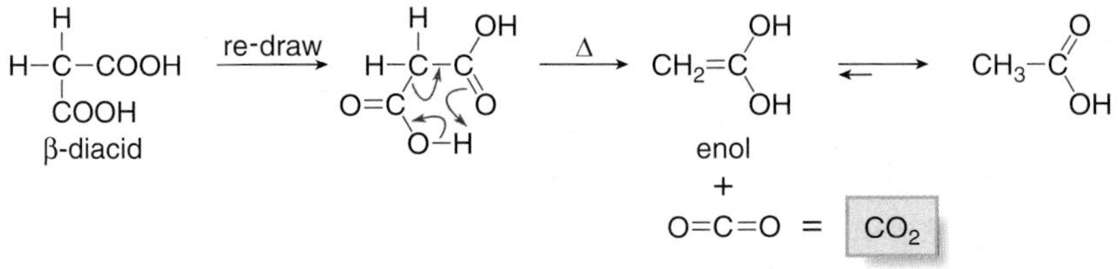

The net result of decarboxylation is cleavage of a carbon–carbon bond on the α carbon, with loss of CO_2.

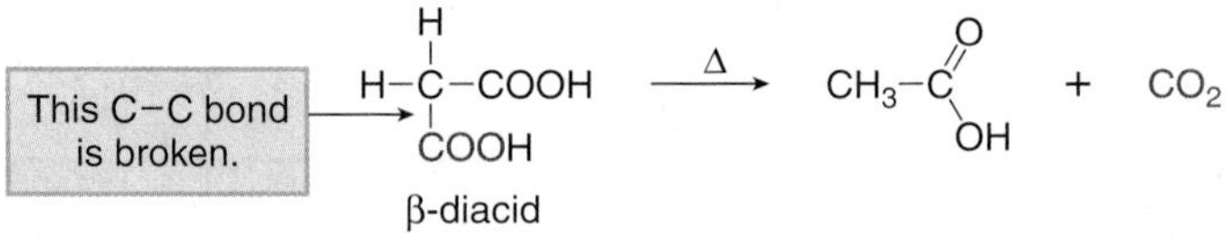

Decarboxylation occurs readily whenever a carboxy group (COOH) is bonded to the α carbon of another carbonyl group. For example, β-keto acids also readily lose CO_2 on heating to form ketones.

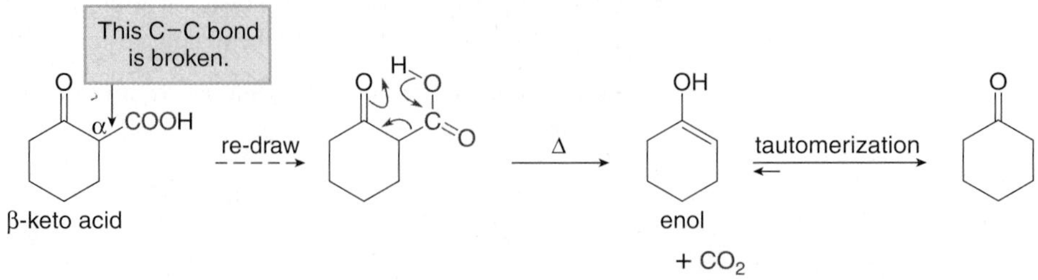

PROBLEM 23.20   Which of the following compounds will readily lose CO_2 when heated?

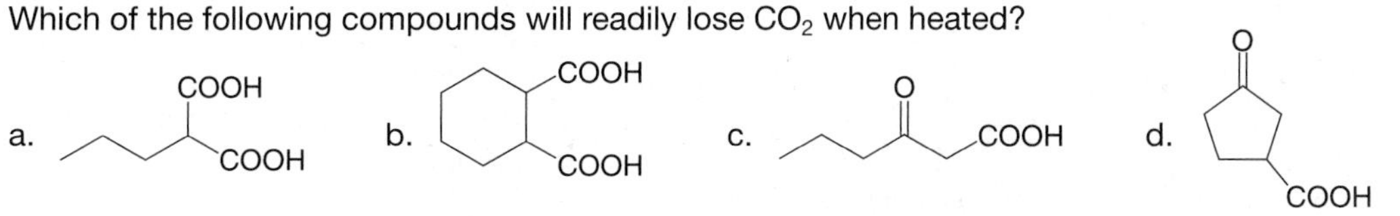

a. [structure: pentane chain with COOH and COOH groups]   b. [cyclohexane ring with COOH and COOH groups]   c. [structure with O and COOH]   d. [cyclopentanone ring with COOH]

23.9B   Steps in the Malonic Ester Synthesis

The malonic ester synthesis converts diethyl malonate to an α-substituted carboxylic acid in three steps.

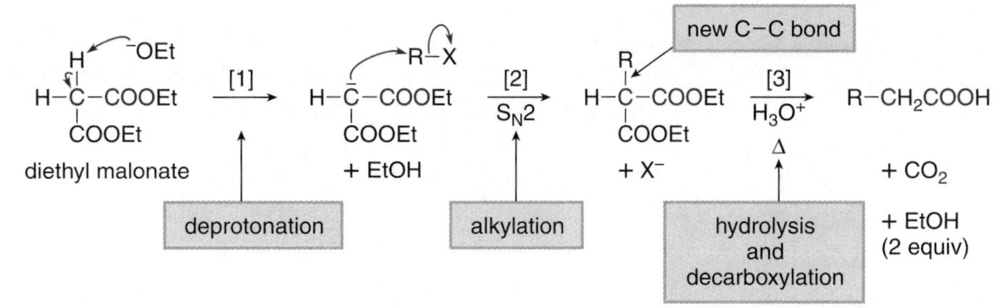

[1] **Deprotonation.** Treatment of diethyl malonate with ⁻OEt removes the acidic α proton between the two carbonyl groups. Recall from Section 23.3A that these protons are more acidic than other α protons because the enolate is stabilized by three resonance structures, instead of the usual two. Thus, ⁻OEt, rather than the stronger base LDA, can be used for this reaction.

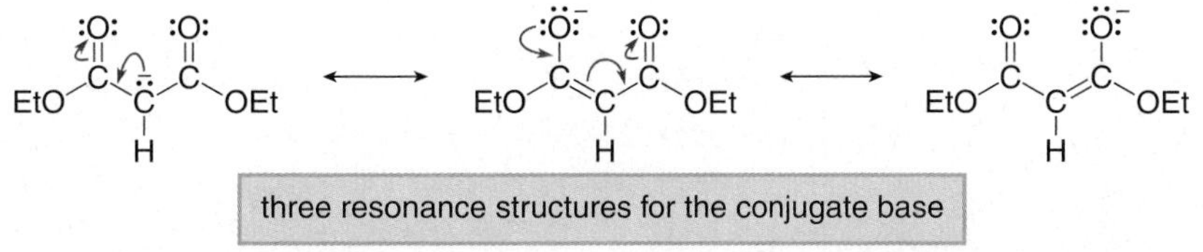

three resonance structures for the conjugate base

[2] **Alkylation.** The nucleophilic enolate reacts with an alkyl halide in an **S$_N$2** reaction to form a substitution product. Because the mechanism is S$_N$2, R must be CH_3 or a 1° alkyl group.

[3] **Hydrolysis and decarboxylation.** Heating the diester with aqueous acid hydrolyzes the diester to a β-diacid, which loses CO_2 to form a carboxylic acid.

The synthesis of 2-butanoic acid ($CH_3CH_2CH_2COOH$) from diethyl malonate illustrates the basic process:

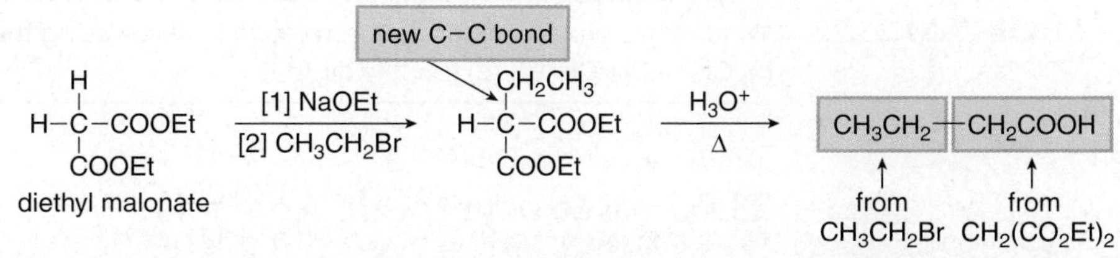

If the first two steps of the reaction sequence are repeated *prior* to hydrolysis and decarboxylation, then a carboxylic acid having *two new alkyl groups* on the α carbon can be synthesized. This is illustrated in the synthesis of 2-benzylbutanoic acid [$CH_3CH_2CH(CH_2C_6H_5)COOH$] from diethyl malonate:

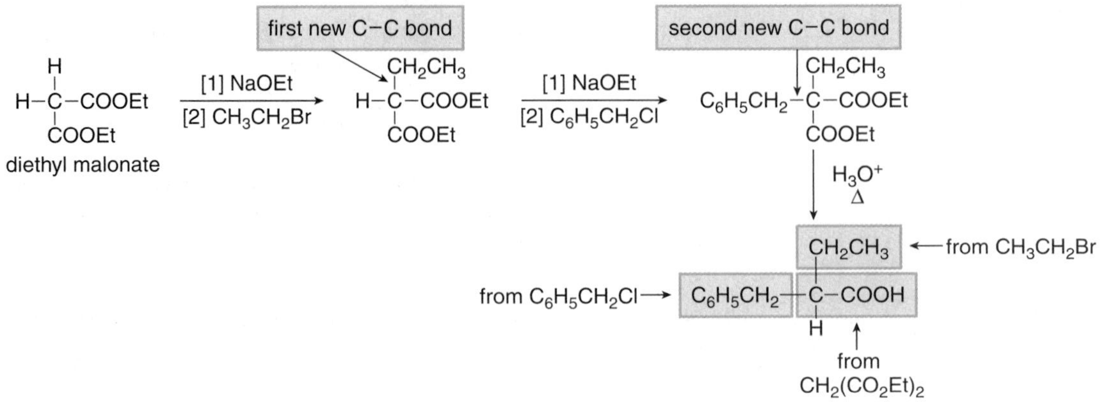

An intramolecular malonic ester synthesis can be used to form rings having three to six atoms, provided the appropriate dihalide is used as starting material. For example, cyclopentanecarboxylic acid can be prepared from diethyl malonate and 1,4-dibromobutane (BrCH$_2$CH$_2$CH$_2$CH$_2$Br) by the following sequence of reactions:

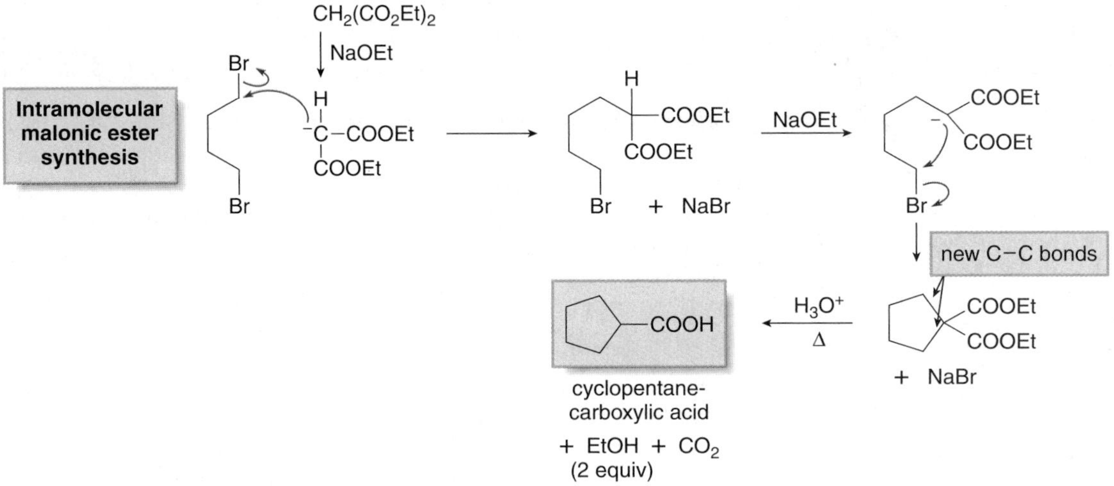

PROBLEM 23.21   Draw the products of each reaction.

a. 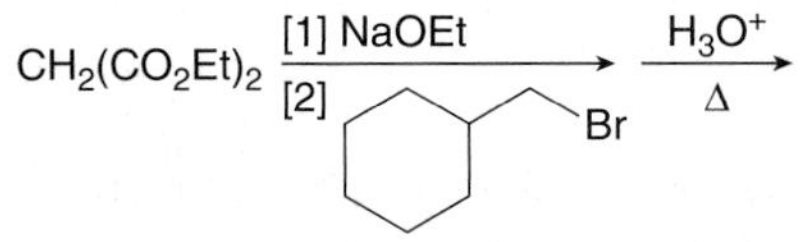   CH$_2$(CO$_2$Et)$_2$ $\xrightarrow{\text{[1] NaOEt}}$ $\xrightarrow[\Delta]{\text{H}_3\text{O}^+}$

b.   CH$_2$(CO$_2$Et)$_2$ $\xrightarrow{\text{[1] NaOEt}}$ $\xrightarrow{\text{[1] NaOEt}}$ $\xrightarrow[\Delta]{\text{H}_3\text{O}^+}$

PROBLEM 23.22   What cyclic product is formed from each dihalide using the malonic ester synthesis:
(a) ClCH$_2$CH$_2$CH$_2$Cl; (b) (BrCH$_2$CH$_2$)$_2$O?

23.9C   Retrosynthetic Analysis

To use the malonic ester synthesis you must be able to determine what starting materials are needed to prepare a given compound—that is, you must **work backwards in the retrosynthetic direction.** This involves a two-step process:

[1] Locate the α carbon to the COOH group, and identify all alkyl groups bonded to the α carbon.

[2] Break the molecule into two (or three) components: Each alkyl group bonded to the α carbon comes from an alkyl halide. The remainder of the molecule comes from CH$_2$(COOEt)$_2$.

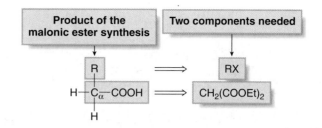

SAMPLE PROBLEM 23.3   What starting materials are needed to prepare 2-methylhexanoic acid [CH₃CH₂CH₂CH₂CH(CH₃)COOH] using a malonic ester synthesis?

SOLUTION

The target molecule has two different alkyl groups bonded to the α carbon, so three components are needed for the synthesis:

$$CH_3CH_2CH_2CH_2 \underset{H}{\overset{CH_3}{-C_\alpha-COOH}} \implies CH_2(CO_2Et)_2 \quad\implies CH_3I$$

$$CH_3CH_2CH_2CH_2Br$$

Writing the synthesis in the synthetic direction:

$$\underset{COOEt}{\overset{H}{H-C-COOEt}} \xrightarrow[\text{[2] CH}_3\text{I}]{\text{[1] NaOEt}} \underset{COOEt}{\overset{CH_3}{H-C-COOEt}} \xrightarrow[\text{[2] CH}_3(\text{CH}_2)_3\text{Br}]{\text{[1] NaOEt}} \underset{COOEt}{\overset{CH_3}{CH_3(CH_2)_3-C-COOEt}} \xrightarrow[\Delta]{H_3O^+} \underset{H}{\overset{CH_3}{CH_3(CH_2)_3-C-COOH}}$$

PROBLEM 23.23   What alkyl halides are needed to prepare each carboxylic acid by the malonic ester synthesis?

a. (CH₃)₂CHCH₂CH₂CH₂CH₂CH₂COOH   b. [structure with COOH]   c. (CH₃CH₂CH₂)₂CHCOOH

PROBLEM 23.24   Explain why each of the following carboxylic acids cannot be prepared by a malonic ester synthesis: (a) (CH₃)₃CCH₂COOH; (b) C₆H₅CH₂COOH; (c) (CH₃)₃CCOOH.

23.10   Acetoacetic Ester Synthesis

The acetoacetic ester synthesis is a stepwise method for converting ethyl acetoacetate into a ketone having one or two alkyl groups on the α carbon.

Acetoacetic ester synthesis

$$\underset{\text{ethyl acetoacetate}}{\underset{COOEt}{CH_3\overset{O}{\overset{\|}{C}}CH_2}} \longrightarrow \underset{\text{from RX}}{CH_3\overset{O}{\overset{\|}{C}}CH_2-R} \quad \text{or} \quad \underset{\text{from RX and R'X}}{\underset{R'}{CH_3\overset{O}{\overset{\|}{C}}CH-R}}$$

23.10A   Steps in the Acetoacetic Ester Synthesis

$$\underset{\alpha}{R\underset{\beta}{\overset{O}{\overset{\|}{C}}}\overset{O}{\overset{\|}{C}}OR'}$$

General structure of
a β-keto ester

The steps in the acetoacetic ester synthesis are exactly the same as those in the malonic ester synthesis. Because the starting material, CH₃COCH₂COOEt, is a β-keto ester, the final product is a **ketone,** not a carboxylic acid.

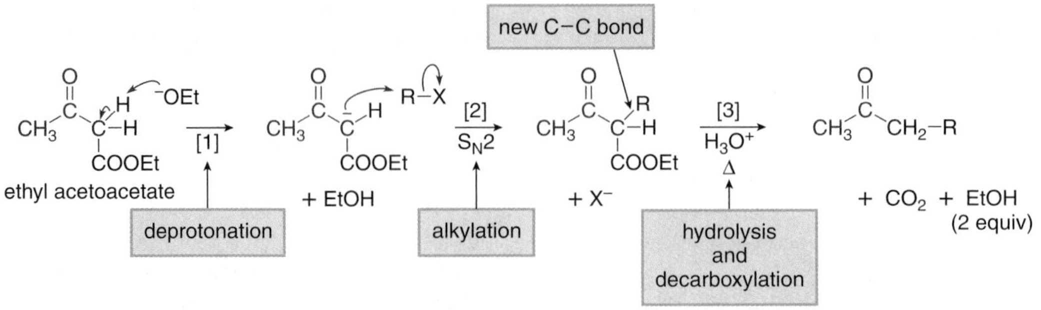

[1] **Deprotonation.** Treatment of ethyl acetoacetate with ⁻OEt removes the acidic proton between the two carbonyl groups.

[2] **Alkylation.** The nucleophilic enolate reacts with an alkyl halide (RX) in an S_N2 reaction to form a substitution product. Because the mechanism is S_N2, R must be CH_3 or a 1° alkyl group.

[3] **Hydrolysis and decarboxylation.** Heating the β-keto ester with aqueous acid hydrolyzes the ester to a β-keto acid, which loses CO_2 to form a ketone.

If the first two steps of the reaction sequence are repeated *prior* to hydrolysis and decarboxylation, then a ketone having *two new alkyl groups* on the α carbon can be synthesized.

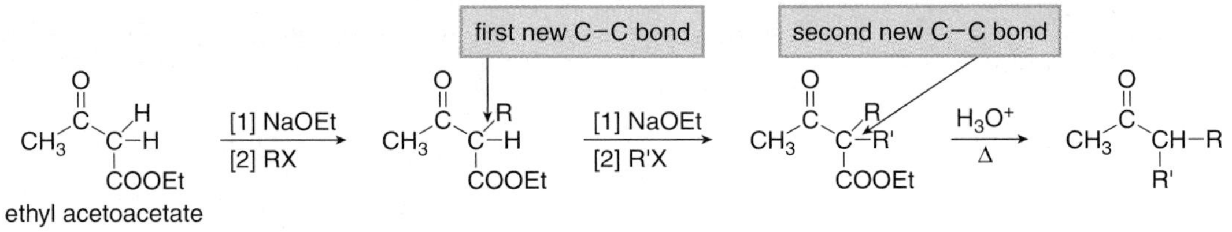

PROBLEM 23.25   What ketones are prepared by the following reactions?

a.

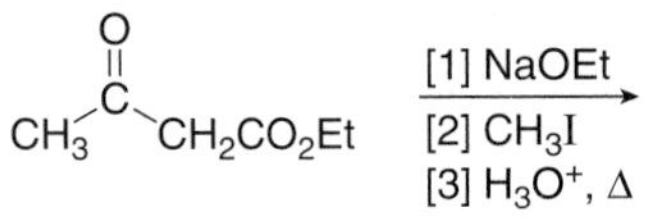

[1] NaOEt
[2] CH₃I
[3] H₃O⁺, Δ

b.

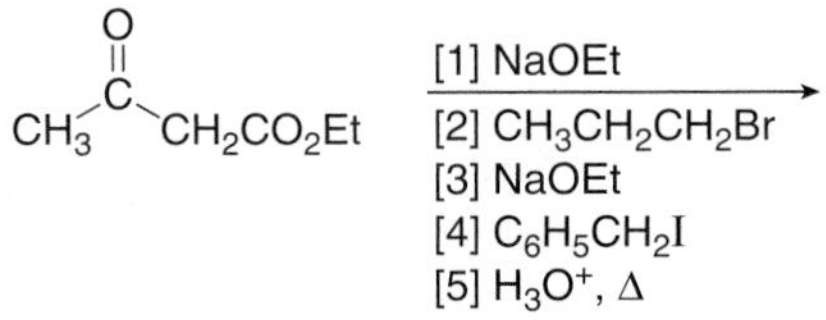

[1] NaOEt
[2] CH₃CH₂CH₂Br
[3] NaOEt
[4] C₆H₅CH₂I
[5] H₃O⁺, Δ

To determine what starting materials are needed to prepare a given ketone using the acetoacetic ester synthesis, you must again work in the **retrosynthetic** direction. This involves a two-step process:

[1] Identify the alkyl groups bonded to the α carbon to the carbonyl group.
[2] Break the molecule into two (or three) components: Each alkyl group bonded to the α carbon comes from an alkyl halide. The remainder of the molecule comes from CH_3COCH_2COOEt.

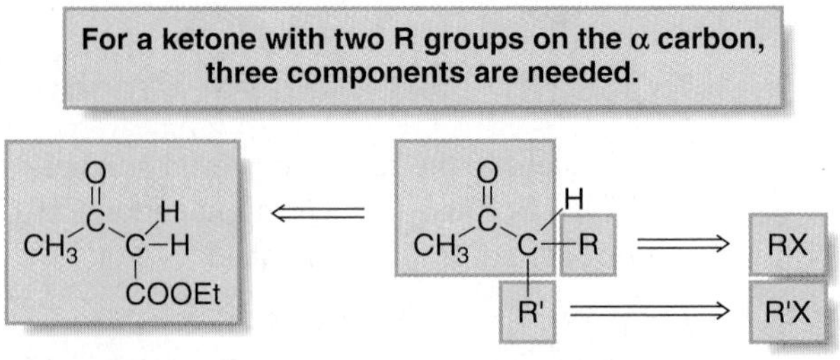

For a ketone with two R groups on the α carbon, three components are needed.

SAMPLE PROBLEM 23.4   What starting materials are needed to synthesize 2-heptanone using the acetoacetic ester synthesis?

$$CH_3\overset{\overset{\displaystyle O}{\|}}{C}CH_2CH_2CH_2CH_2CH_3$$

2-heptanone

SOLUTION

2-Heptanone has only one alkyl group bonded to the α carbon, so only one alkyl halide is needed in the acetoacetic ester synthesis.

$$CH_3\overset{\overset{\displaystyle O}{\|}}{C}\overset{\overset{\displaystyle H}{|}}{\underset{\underset{\displaystyle COOEt}{|}}{C}}H \iff CH_3\overset{\overset{\displaystyle O}{\|}}{C}\overset{\alpha}{CH_2}\!\!-\!\!CH_2CH_2CH_2CH_3 \implies CH_3CH_2CH_2CH_2Br$$

Writing the acetoacetic ester synthesis in the synthetic direction:

$$CH_3\overset{\overset{\displaystyle O}{\|}}{C}\overset{\overset{\displaystyle H}{|}}{\underset{\underset{\displaystyle COOEt}{|}}{C}}H \xrightarrow[\text{[2] } CH_3CH_2CH_2CH_2Br]{\text{[1] NaOEt}} CH_3\overset{\overset{\displaystyle O}{\|}}{C}\overset{\overset{\displaystyle H}{|}}{\underset{\underset{\displaystyle COOEt}{|}}{C}}\!\!-\!\!CH_2CH_2CH_2CH_3 \xrightarrow[\Delta]{H_3O^+} CH_3\overset{\overset{\displaystyle O}{\|}}{C}CH_2\!-\!CH_2CH_2CH_2CH_3$$

ethyl acetoacetate   2-heptanone

PROBLEM 23.26   What alkyl halides are needed to prepare each ketone using the acetoacetic ester synthesis?

a. $CH_3\overset{\overset{\displaystyle O}{\|}}{C}CH_2CH_2CH_3$ b. $CH_3\overset{\overset{\displaystyle O}{\|}}{C}CH(CH_2CH_3)_2$ c.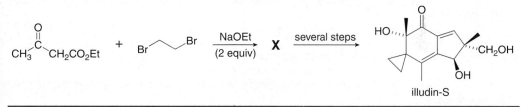

PROBLEM 23.27   Treatment of ethyl acetoacetate with NaOEt (2 equiv) and BrCH₂CH₂Br forms compound **X**. This reaction is the first step in the synthesis of illudin-S, an antitumor substance isolated from the jack-o'-lantern, a poisonous, saffron-colored mushroom. What is the structure of **X**?

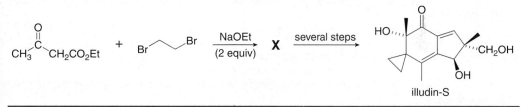

The jack-o-lantern, source of the antitumor agent illudin-S.

The acetoacetic ester synthesis and direct enolate alkylation are two different methods that prepare similar α-substituted ketones. 2-Butanone, for example, can be synthesized from acetone by direct enolate alkylation with CH₃I (Method [1]), or by alkylation of ethyl acetoacetate followed by hydrolysis and decarboxylation (Method [2]).

Method [1]
Direct enolate alkylation

$$CH_3\overset{\overset{\displaystyle O}{\|}}{C}CH_3 \xrightarrow[\text{[2] } CH_3I]{\text{[1] LDA}} CH_3\overset{\overset{\displaystyle O}{\|}}{C}CH_2\!-\!CH_3$$

acetone   2-butanone

The same product is formed by two different routes.

Method [2]
Acetoacetic ester synthesis

$$CH_3\overset{\overset{\displaystyle O}{\|}}{C}\overset{\overset{\displaystyle H}{|}}{\underset{\underset{\displaystyle COOEt}{|}}{C}}H \xrightarrow[\text{[2] } CH_3I]{\text{[1] NaOEt}} CH_3\overset{\overset{\displaystyle O}{\|}}{C}\overset{\overset{\displaystyle H}{|}}{\underset{\underset{\displaystyle COOEt}{|}}{C}}\!-\!CH_3 \xrightarrow[\Delta]{H_3O^+} CH_3\overset{\overset{\displaystyle O}{\|}}{C}CH_2\!-\!CH_3$$

ethyl acetoacetate   2-butanone

Why would you ever make 2-butanone from ethyl acetoacetate when you could make it in fewer steps from acetone? There are many factors to consider. First of all, synthetic organic chemists like to have a variety of methods to accomplish a single kind of reaction. Sometimes subtle changes in the structure of a starting material make one reaction work better than another.

In the chemical industry, moreover, cost is an important issue. Any reaction needed to make a large quantity of a useful drug or other consumer product must use cheap starting materials. Direct enolate alkylation usually requires a very strong base like LDA to be successful, whereas the acetoacetic ester synthesis utilizes NaOEt. NaOEt can be prepared from cheaper starting materials, and this makes the acetoacetic ester synthesis an attractive method, even though it involves more steps.

Thus, each method has its own advantages and disadvantages, depending on the starting material, the availability of reagents, the cost, and the occurrence of side reactions.

PROBLEM 23.28   Nabumetone is a pain reliever and anti-inflammatory agent sold under the brand name of Relafen.

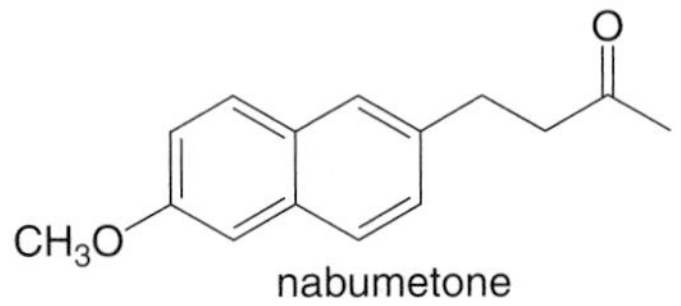

nabumetone

a. Write out a synthesis of nabumetone from ethyl acetoacetate.
b. What ketone and alkyl halide are needed to synthesize nabumetone by direct enolate alkylation?

23.11 Key Concepts—Substitution Reactions of Carbonyl Compounds at the α Carbon

Kinetic Versus Thermodynamic Enolates (23.4)

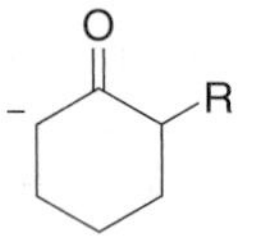

Kinetic enolate
- The less substituted enolate
- Favored by strong base, polar aprotic solvent, low temperature: LDA, THF, −78 °C

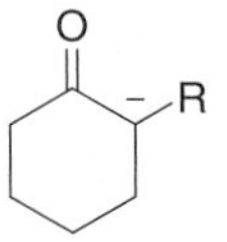

kinetic enolate

Thermodynamic enolate
- The more substituted enolate
- Favored by strong base, protic solvent, higher temperature: $NaOCH_2CH_3$, CH_3CH_2OH, room temperature

thermodynamic enolate

Halogenation at the α Carbon

[1] Halogenation in acid (23.7A)

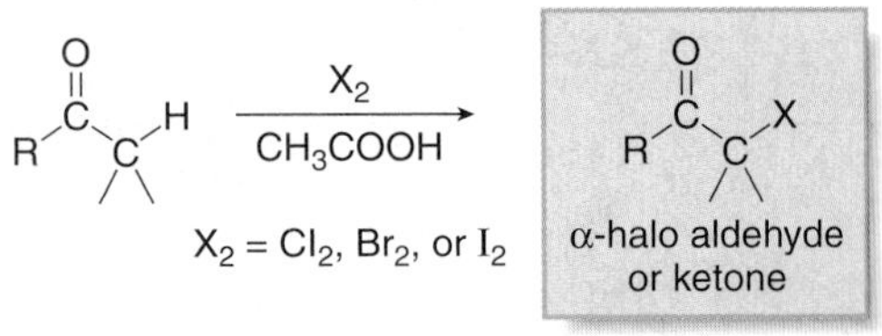

$X_2 = Cl_2$, Br_2, or I_2

α-halo aldehyde or ketone

- The reaction occurs via enol intermediates.
- Monosubstitution of X for H occurs on the α carbon.

[2] Halogenation in base (23.7B)

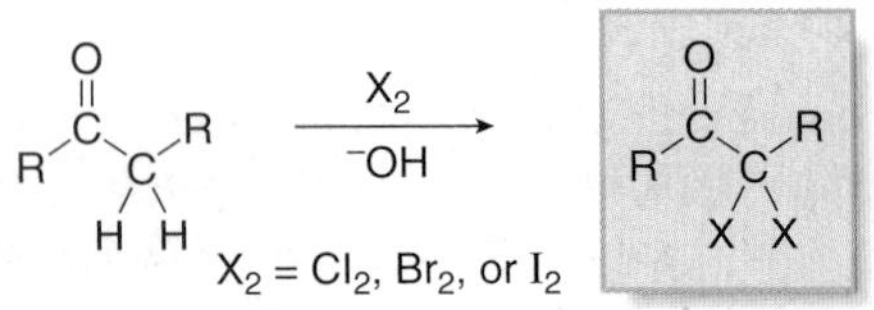

$X_2 = Cl_2$, Br_2, or I_2

- The reaction occurs via enolate intermediates.
- Polysubstitution of X for H occurs on the α carbon.

[3] Halogenation of *methyl* ketones in base—The haloform reaction (23.7B)

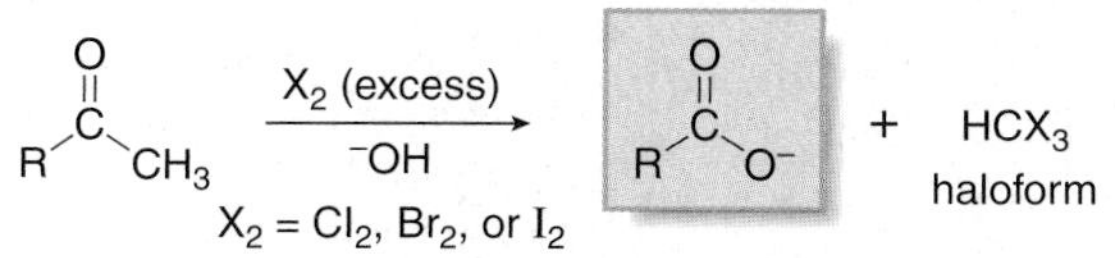

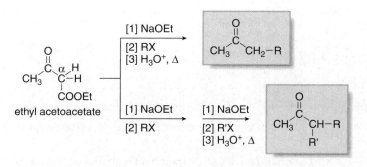

- The reaction occurs with methyl ketones and results in cleavage of a carbon–carbon σ bond.

Alkylation Reactions at the α Carbon

[1] Direct alkylation at the α carbon (23.8)

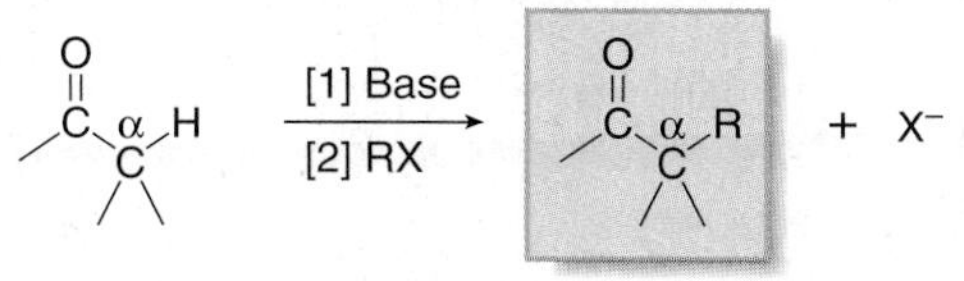

- The reaction forms a new C–C bond to the α carbon.
- LDA is a common base used to form an intermediate enolate.
- The alkylation in Step [2] follows an S$_N$2 mechanism.

[2] Malonic ester synthesis (23.9)

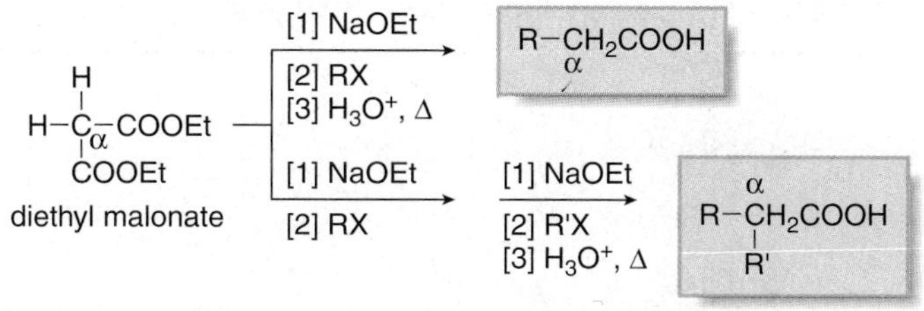

diethyl malonate

- The reaction is used to prepare α-substituted carboxylic acids with one or two alkyl groups on the α carbon.
- The alkylation in Step [2] follows an S$_N$2 mechanism.

[3] Acetoacetic ester synthesis (23.10)

ethyl acetoacetate

- The reaction is used to prepare α-substituted ketones with one or two alkyl groups on the α carbon.
- The alkylation in Step [2] follows an S$_N$2 mechanism.

Problems

Enols, Enolates, and Acidic Protons

23.29 Draw enol tautomer(s) for each compound.

a. $CH_3CH_2CH_2CHO$

b.

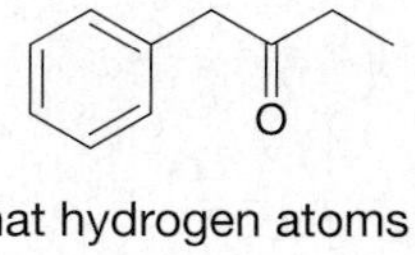

c. $CH_3CH_2CH_2CH_2CO_2CH_2CH_3$

d. 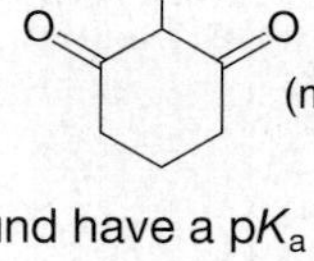 (mono enol form)

e. $CH_3CH_2CHCOOH$ with CH_3

f.

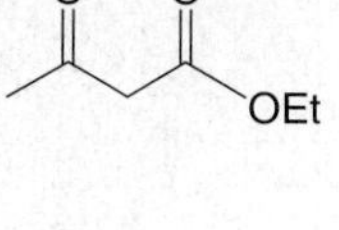

23.30 What hydrogen atoms in each compound have a p$K_a \le 25$?

a. $CH_3CH_2CH_2CO_2CH(CH_3)_2$

b.

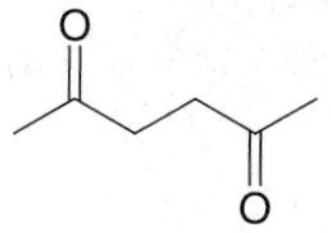

c.

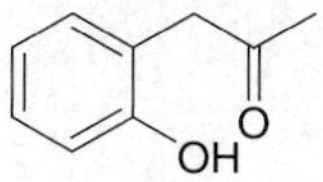

d. 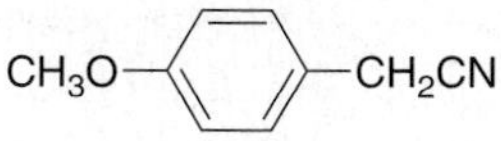 CH_3O—〈 〉—CH_2CN

e. 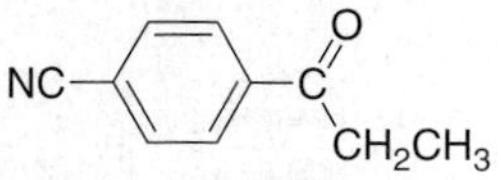 NC—〈 〉—C(=O)CH_2CH_3

f.

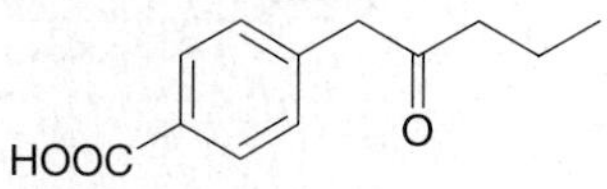

23.31 Rank the labeled protons in each compound in order of increasing acidity.

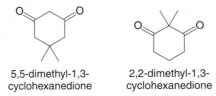

a. CH_3CH_2-C ⟨O, OH⟩ with H_a H_b labeled and H_c

b. (diketone with phenyl, CH_3, labeled H_a, H_b, H_c)

c. (cyclohexanone with COOH, H_c, $H \leftarrow H_b$, $H \leftarrow H_a$)

d. HO— ... labeled H_a, H_b, H_c

23.32 What is the major enolate (or carbanion) formed when each compound is treated with LDA?

a. (branched ketone)

b. (ester, OCH$_3$)

c. (cyclopentyl ketone)

d. (phenyl CN)

e. (decalone)

f. (methyl cyclohexanedione)

23.33 How could IR spectroscopy be used to detect the presence of enol tautomers?

23.34 Explain why 5,5-dimethyl-1,3-cyclohexanedione exists predominantly in its enol form, but 2,2-dimethyl-1,3-cyclohexanedione does not.

5,5-dimethyl-1,3-cyclohexanedione

2,2-dimethyl-1,3-cyclohexanedione

23.35 Explain why an optically active solution of (R)-α-methylbutyrophenone loses its optical activity when either dilute acid or base is added to the solution.

(R)-α-methylbutyrophenone

23.36 Explain why the α protons of an ester are less acidic than the α protons of a ketone by ~5 pK_a units.

23.37 Explain why reactions that use LDA as base must be carried out under anhydrous conditions; that is, all traces of H_2O must be rigorously excluded.

23.38 Although it is possible to alkylate methyl acetate (CH_3COOCH_3) on the α carbon by treatment with one equivalent of LDA and CH_3I to form $CH_3CH_2COOCH_3$, it is not possible to carry out this same reaction with propanoic acid (CH_3CH_2COOH) to form $CH_3CH(CH_3)COOH$. Explain why this is so.

Halogenation

23.39 Acid-catalyzed bromination of 2-pentanone ($CH_3COCH_2CH_2CH_3$) forms two products: $BrCH_2COCH_2CH_2CH_3$ (**A**) and $CH_3COCH(Br)CH_2CH_3$ (**B**). Explain why the major product is **B,** with the Br atom on the more substituted side of the carbonyl group.

23.40 Draw a stepwise mechanism for each reaction.

a. (cyclohexenone) $\xrightarrow[CH_3CO_2H]{Br_2}$ (bromo cyclohexenone) + HBr

b. (cyclohexyl methyl ketone) $\xrightarrow[^-OH]{I_2 \text{ (excess)}}$ (cyclohexyl carboxylate) + CHI$_3$

Malonic Ester Synthesis

23.41 What alkyl halides are needed to prepare each carboxylic acid using the malonic ester synthesis?

    a. $CH_3OCH_2CH_2COOH$ b. 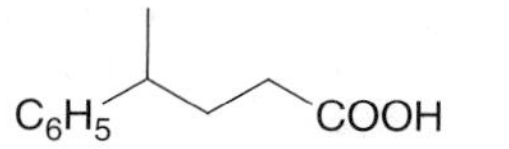 C_6H_5 ⟋⟍COOH   c. ⟋⟍COOH

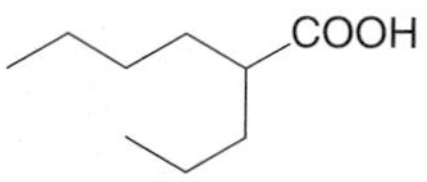

23.42 Use the malonic ester synthesis to prepare each carboxylic acid.

    a. $CH_3CH_2CH_2CH_2CH_2CH_2COOH$ b. 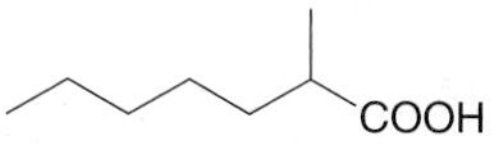 ⟋⟍COOH   c. 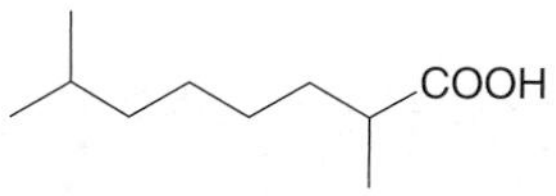 ⟋⟍COOH

23.43 Synthesize each compound from diethyl malonate. You may use any other organic or inorganic reagents.

    a. 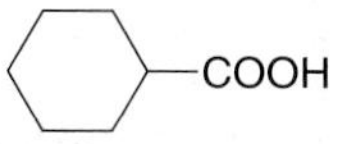 ⬡—COOH   b. 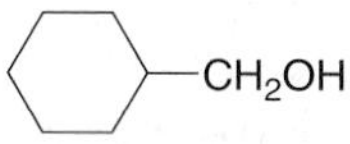 ⬡—CH_2OH c. 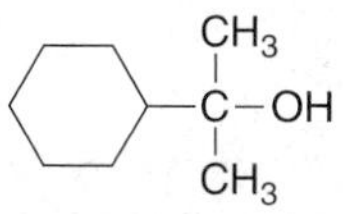 ⬡—C(CH₃)₂—OH   d. 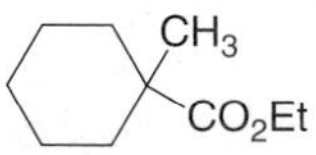 ⬡—CH₃, CO₂Et

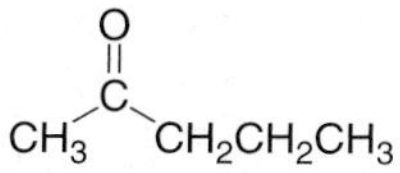

23.44 The enolate derived from diethyl malonate reacts with a variety of electrophiles (not just alkyl halides) to form new carbon–carbon bonds. With this in mind, draw the products formed when $Na^+ \ ^-CH(COOEt)_2$ reacts with each electrophile, followed by treatment with H_2O.

    a. 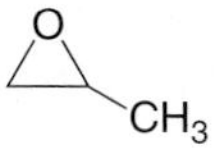 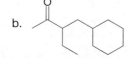   b. $CH_2{=}O$ c. 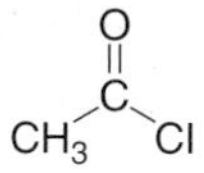 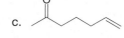   d.

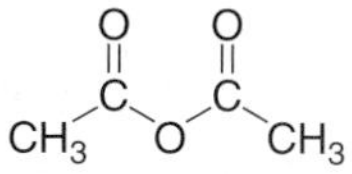

Acetoacetic Ester Synthesis

23.45 What alkyl halides are needed to prepare each ketone using the acetoacetic ester synthesis?

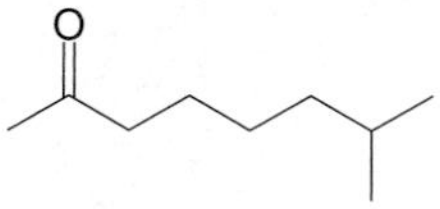

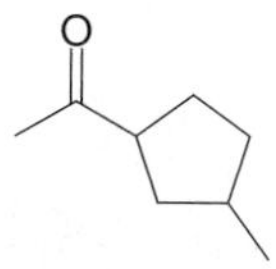

23.46 Synthesize each compound from ethyl acetoacetate. You may use any other organic or inorganic reagents.

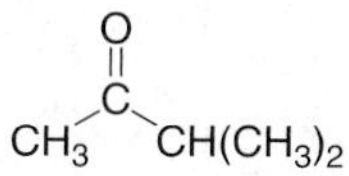

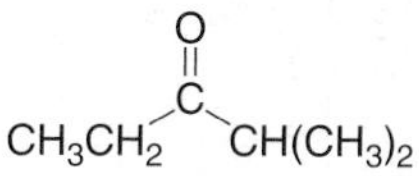

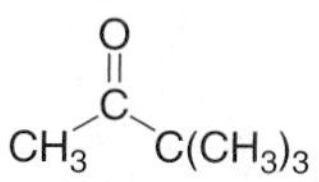

Reactions

23.47 Draw the organic products formed in each reaction.

    a. 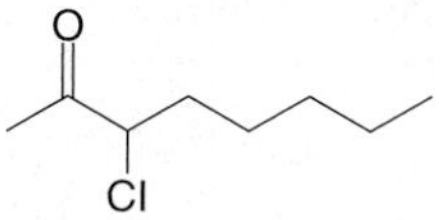 → pyridine / Δ

    e. → [1] Br₂, CH₃CO₂H   [2] pyridine, Δ

    b. 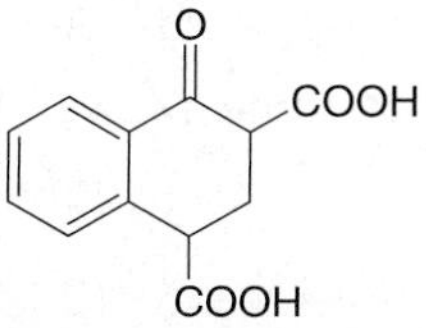 → Δ

    f. → I₂ (excess) / ⁻OH

    c. $CH_3CH_2CH_2CO_2Et$ → [1] LDA / [2] CH_3CH_2I

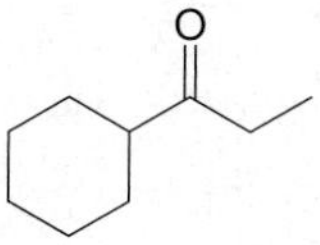

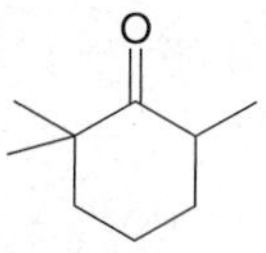

    g. 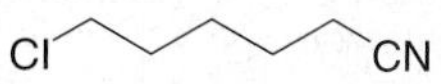 Cl⟋⟍CN → NaH → C_6H_9N

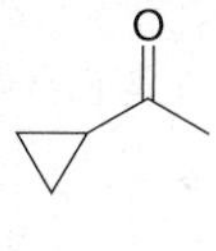

    d. → [1] LDA / [2] CH_3CH_2I

    h. → Br₂ (excess) / ⁻OH

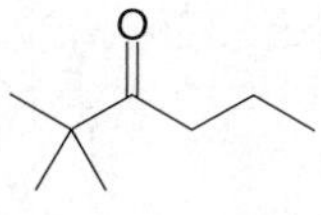

23.48 Draw the products formed (including stereoisomers) in each reaction.

    a. 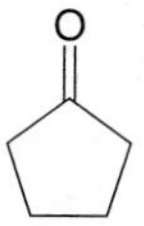 → [1] LDA / [2] ⟋⟍Cl

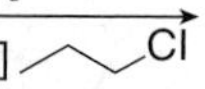

    b. 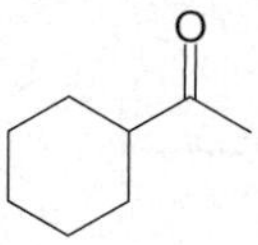 → [1] LDA / [2] CH₃—C(H)(D)—I

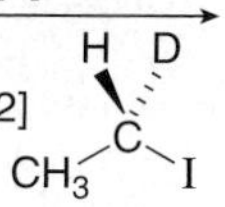

    c. → [1] LDA / [2] CH_3I

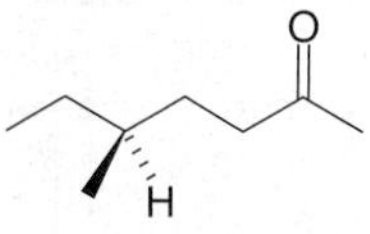

23.49 What reaction conditions—base, solvent, and temperature—are needed to convert ketone **A** to either **B** or **C** by an intramolecular alkylation reaction?

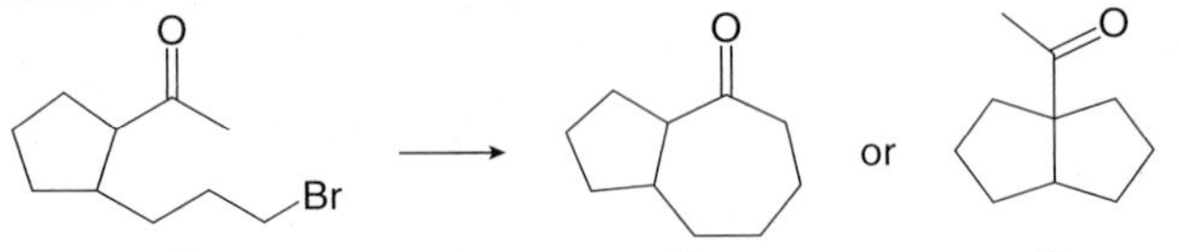

A	**B**	**C**

23.50 Explain why each of the following reactions will **not** proceed as written.

a.

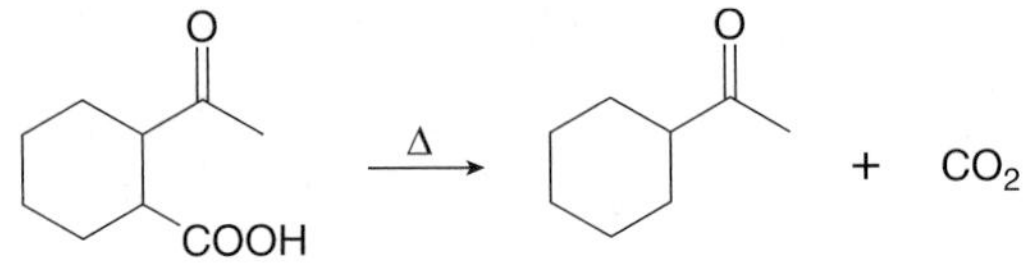

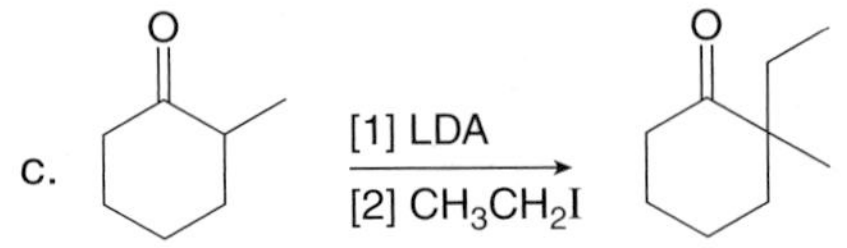

b. $CH_2(CO_2Et)_2$ $\xrightarrow[\text{[2] (CH}_3\text{CH}_2)_3\text{CBr}]{\text{[1] NaOEt}}$ $(CH_3CH_2)_3CCH(CO_2Et)_2$

c.

23.51 Identify the lettered intermediates in the following synthesis of tamoxifen.

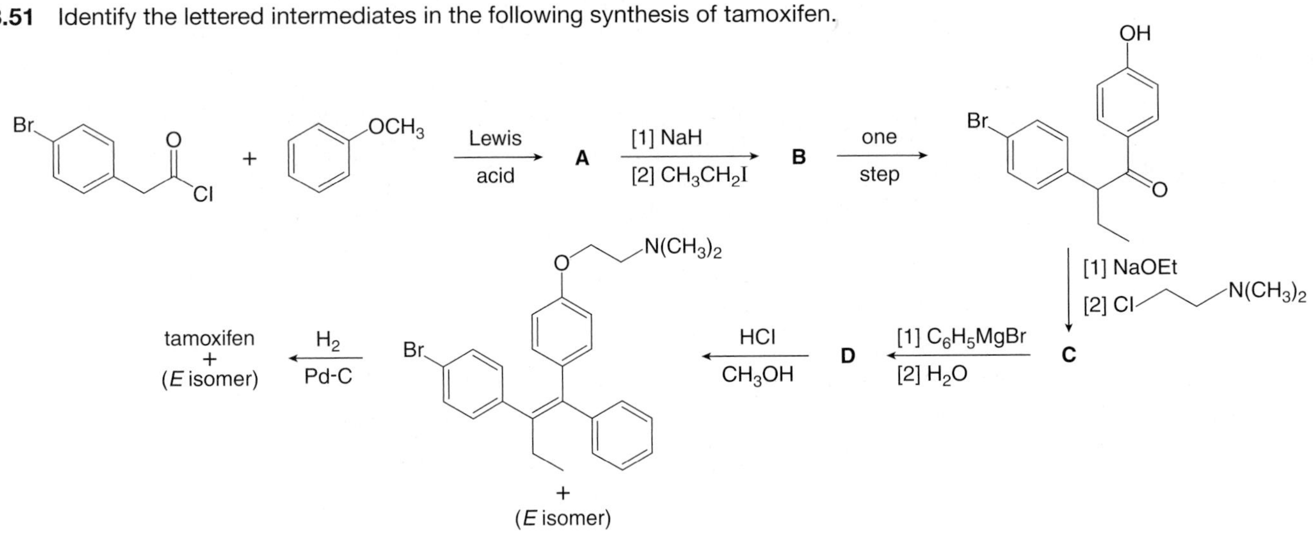

Mechanism

23.52 Draw a stepwise mechanism for the following reaction.

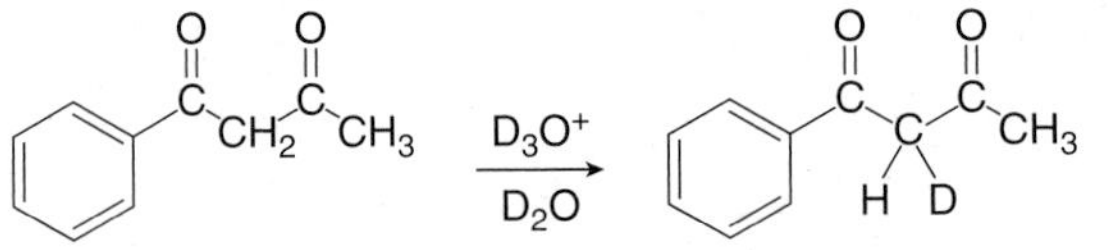

23.53 3-Methylenecyclohexanone is readily isomerized in dilute base to form 3-methyl-2-cyclohexenone. Draw a stepwise mechanism for this process.

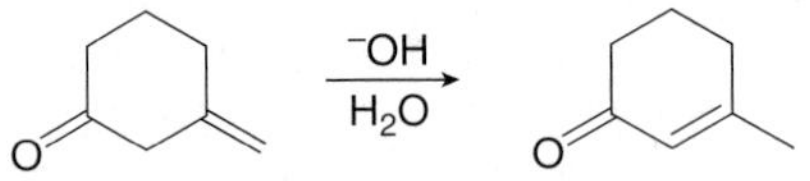

3-methylene-cyclohexanone   3-methyl-2-cyclohexenone

23.54 Draw a stepwise mechanism showing how two alkylation products are formed in the following reaction.

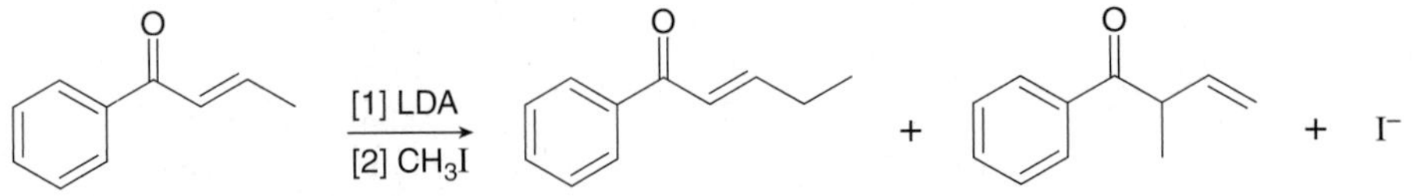

23.55 The cis ketone **A** is isomerized to the trans ketone **B** with aqueous NaOH. A similar isomerization reaction does not occur with the cis ketone **C**. Explain this difference in reactivity.

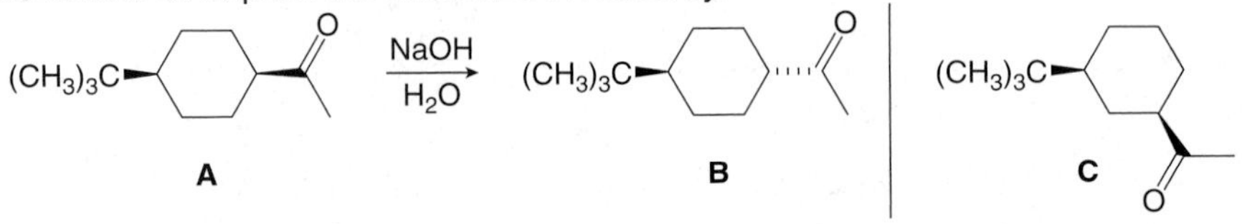

23.56 Draw stepwise mechanisms illustrating how each product is formed.

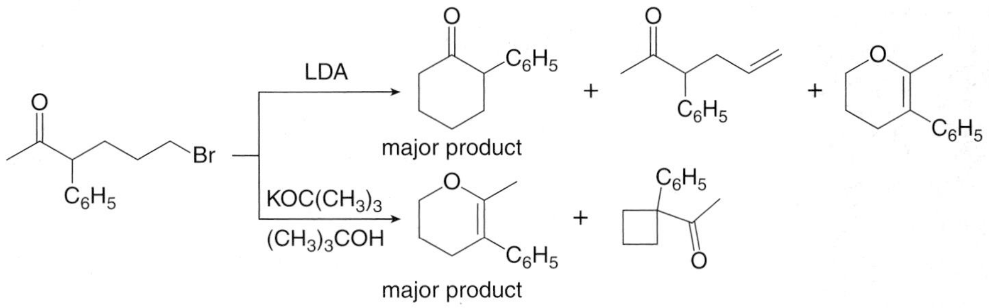

Synthesis

23.57 Convert acetophenone ($C_6H_5COCH_3$) into each of the following compounds. You may use any other organic or inorganic reagents. More than one step may be required.

a. 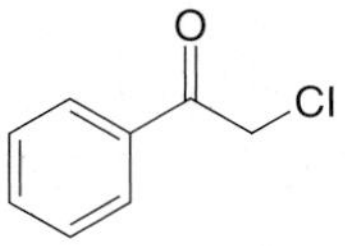   b. 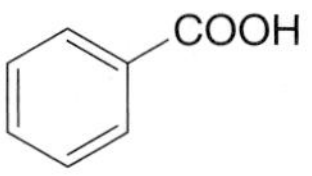   c. 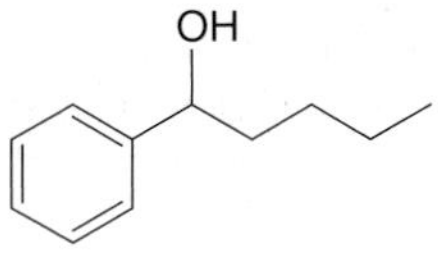   d.

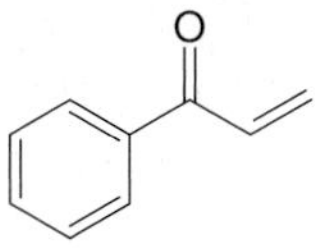

23.58 Synthesize each compound from cyclohexanone. You may use any other inorganic reagents.

a. 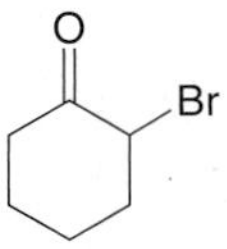   c. 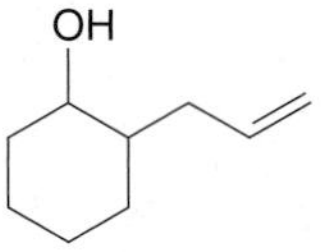   e. 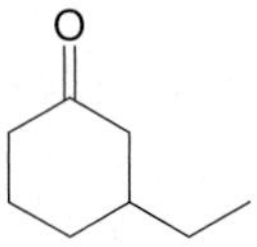   g.

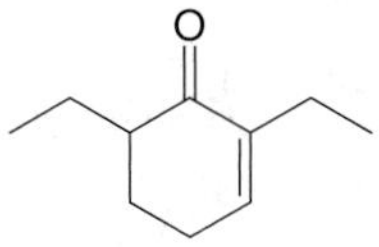

b.   d.   f.   h.

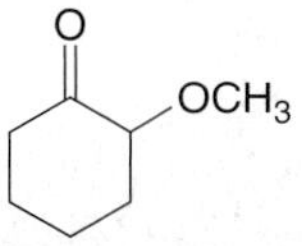

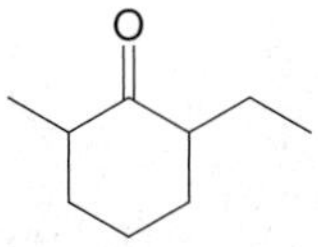

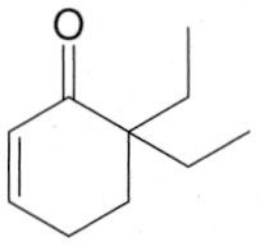

23.59 Synthesize each product from ethyl acetoacetate ($CH_3COCH_2CO_2Et$) and the given starting material.

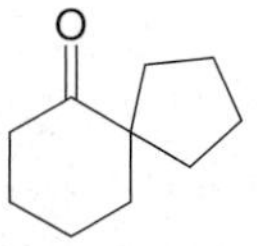

a. 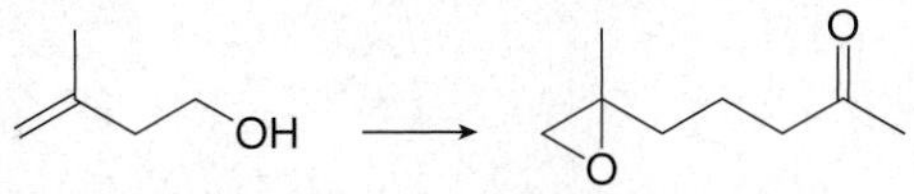   b. 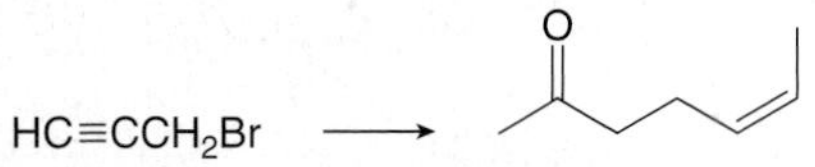   HC≡CCH₂Br ⟶

23.60 Treatment of acetone with LDA followed by $BrCH_2CH_2CH_2CH_2OH$ did not form the desired alkylation product, 7-hydroxy-2-heptanone. What products are formed instead? Devise a multistep method to convert these two starting materials to the desired hydroxy ketone.

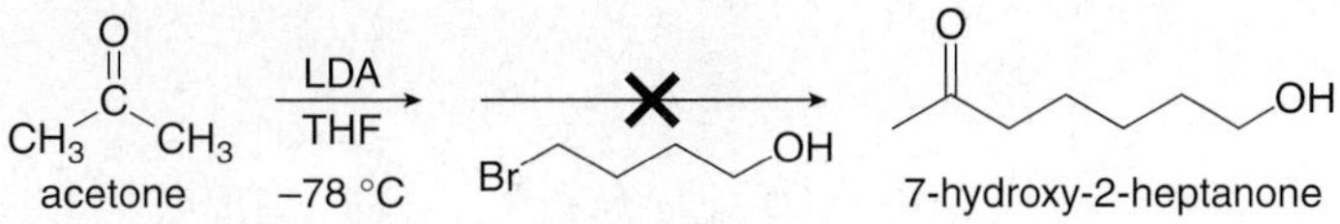

Challenge Problems

23.61 Explain why H_a is much less acidic than H_b. Then draw a mechanism for the following reaction.

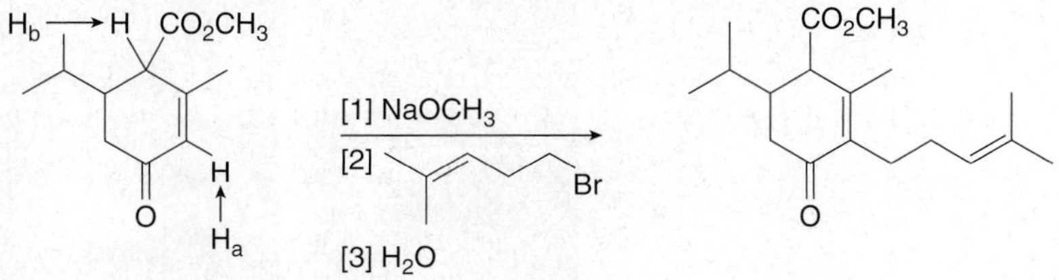

23.62 The last step in the synthesis of β-vetivone involves treatment of **C** with CH_3Li to form an intermediate **X**, which forms β-vetivone with aqueous acid. Identify the structure of **X** and draw a mechanism for converting **X** to β-vetivone.

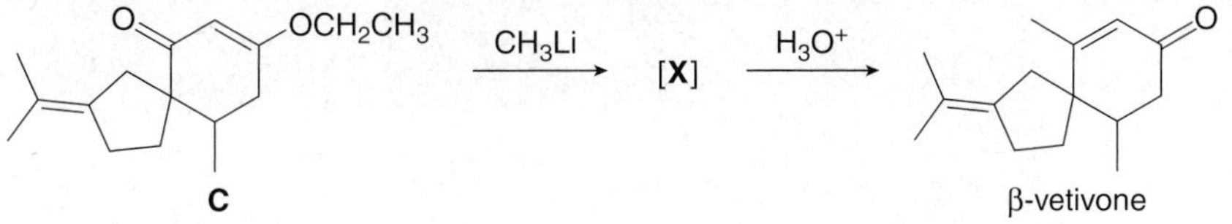

Carbonyl Condensation Reactions

Gibberellic acid is a member of a group of plant growth hormones called the gibberellins, which induce cell division and elongation, and thus they make plants tall and leaves large. Gibberellins also promote the germination of seeds and tubers, and induce flowering. One laboratory synthesis of gibberellic acid uses two intramolecular aldol reactions to construct two carbon–carbon bonds of the complex carbon skeleton. In Chapter 24, we learn about aldol reactions and other methods that form new carbon–carbon bonds at the α carbon of a carbonyl group.

Grown without added gibberellic acid

Treated with gibberellic acid

In Chapter 24, we examine **carbonyl condensations**—that is, condensations between two carbonyl compounds—a second type of reaction that occurs at the α carbon of a carbonyl group. Much of what is presented in Chapter 24 applies principles you have already learned. Many of the reactions may look more complicated than those in previous chapters, but they are fundamentally the same. Nucleophiles attack electrophilic carbonyl groups to form the products of nucleophilic addition or substitution, depending on the structure of the carbonyl starting material.

Every reaction in Chapter 24 forms a new carbon–carbon bond at the α carbon to a carbonyl group, so these reactions are extremely useful in the synthesis of complex natural products.

◆

24.1   The Aldol Reaction

Chapter 24 concentrates on the second general reaction of enolates—**reaction with other carbonyl compounds.** In these reactions, one carbonyl component serves as the nucleophile and one serves as the electrophile, and a new carbon–carbon bond is formed.

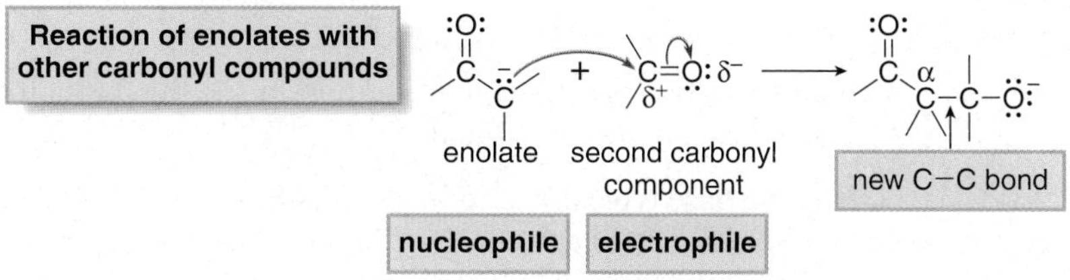

The presence or absence of a leaving group on the electrophilic carbonyl carbon determines the structure of the product. Even though they appear somewhat more complicated, these reactions are often reminiscent of the nucleophilic addition and nucleophilic acyl substitution reactions of Chapters 21 and 22. Four types of reactions are examined:

◆ **Aldol reaction** (Sections 24.1–24.4)

◆ **Claisen reaction** (Sections 24.5–24.7)

◆ **Michael reaction** (Section 24.8)

◆ **Robinson annulation** (Section 24.9)

24.1A   General Features of the Aldol Reaction

In the **aldol reaction,** two molecules of an aldehyde or ketone react with each other in the presence of base to form a **β-hydroxy carbonyl compound.** For example, treatment of acetaldehyde with aqueous ⁻OH forms 3-hydroxybutanal, a **β-hydroxy aldehyde.**

Many aldol products contain an *ald*ehyde and an alco*hol*; hence the name *aldol.*

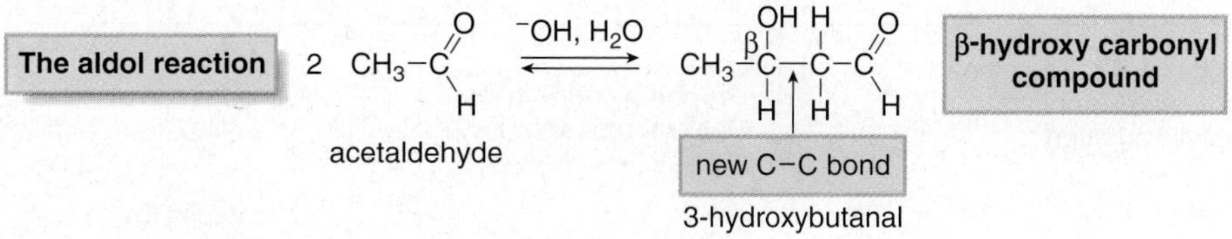

The mechanism of the aldol reaction has **three steps,** as shown in Mechanism 24.1. Carbon–carbon bond formation occurs in Step [2], when the nucleophilic enolate reacts with the electrophilic carbonyl carbon.

MECHANISM 24.1

The Aldol Reaction

Step [1] Formation of a nucleophilic enolate

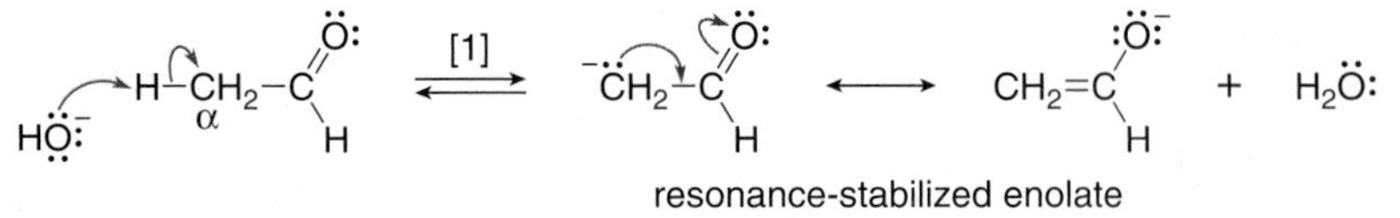

resonance-stabilized enolate

♦ In Step [1], the base removes a proton from the α carbon to form a **resonance-stabilized enolate.**

Steps [2]–[3] Nucleophilic addition and protonation

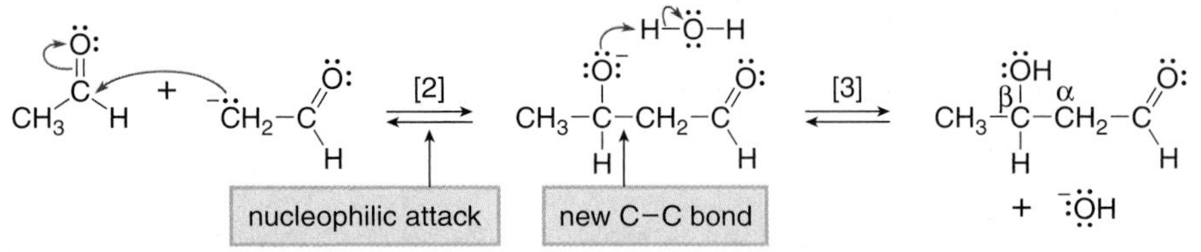

nucleophilic attack   new C–C bond

♦ In Step [2], the nucleophilic enolate attacks the electrophilic carbonyl carbon of another molecule of aldehyde, thus forming a new carbon–carbon bond. **This joins the α carbon of one aldehyde to the carbonyl carbon of a second aldehyde.**

♦ Protonation of the alkoxide in Step [3] forms the **β-hydroxy aldehyde.**

The aldol reaction is a reversible equilibrium, so the position of the equilibrium depends on the base and the carbonyl compound.

⁻OH is the base typically used in an aldol reaction. Recall from Section 23.3B that only a small amount of enolate forms with ⁻OH. In this case, that's appropriate because the starting aldehyde is needed to react with the enolate in the second step of the mechanism.

Aldol reactions can be carried out with either aldehydes or ketones. With aldehydes, the equilibrium usually favors the products, but with ketones the equilibrium favors the starting materials. There are ways of driving this equilibrium to the right, however, so we will write aldol products whether the substrate is an aldehyde or a ketone.

♦ The characteristic reaction of aldehydes and ketones is nucleophilic addition (Section 21.7). An aldol reaction is a nucleophilic addition in which an enolate is the nucleophile. See the comparison in Figure 24.1.

A **second example of an aldol** reaction is shown with propanal as starting material. The two molecules of the aldehyde that participate in the aldol reaction react in opposite ways.

♦ One molecule of propanal becomes an enolate—an electron-rich *nucleophile.*
♦ One molecule of propanal serves as the *electrophile* because its carbonyl carbon is electron deficient.

Figure 24.1 The aldol reaction—An example of nucleophilic addition

Nucleophilic addition—General reaction

nucleophile

The enolate is the nucleophile.

Aldol reaction—An example

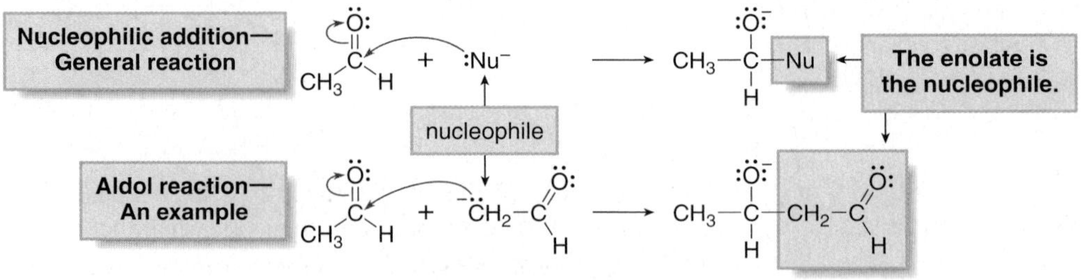

• Aldehydes and ketones react by nucleophilic addition. In an aldol reaction, an enolate is the nucleophile that adds to the carbonyl group.

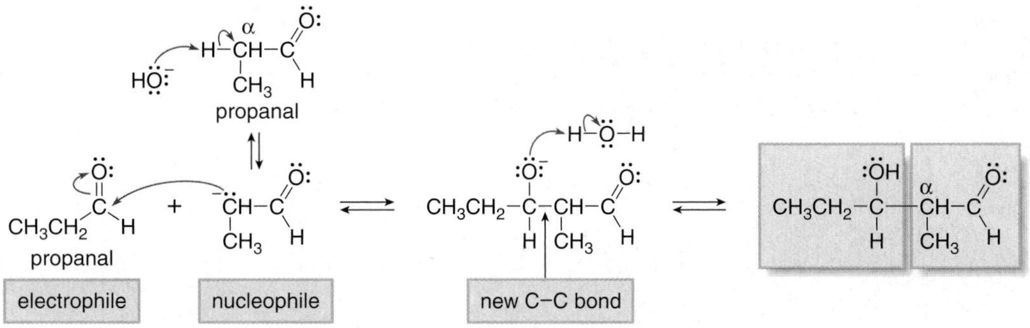

These two examples illustrate the general features of the aldol reaction. **The α carbon of one carbonyl component becomes bonded to the carbonyl carbon of the other component.**

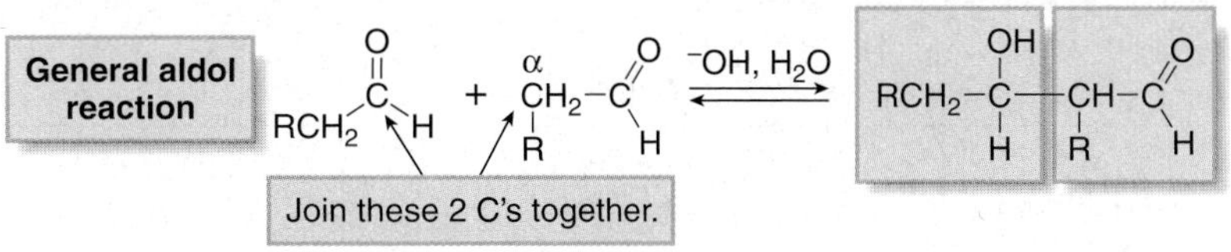

PROBLEM 24.1 Draw the aldol product formed from each compound.

a. 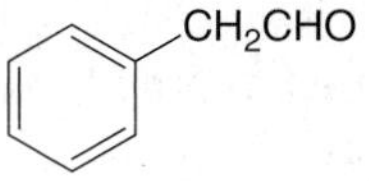

b. $(CH_3)_3CCH_2CHO$

c. 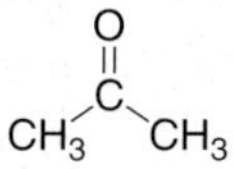

d. 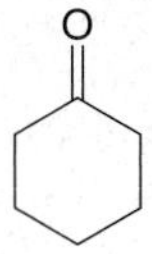

24.1B Dehydration of the Aldol Product

The β-hydroxy carbonyl compounds formed in the aldol reaction dehydrate more readily than other alcohols. In fact, under the basic reaction conditions, the initial aldol product is often not isolated. Instead, **it loses the elements of H_2O from the α and β carbons to form an α,β-unsaturated carbonyl compound.**

The acid-catalyzed dehydration of alcohols was discussed in Section 9.8.

All alcohols—including β-hydroxy carbonyl compounds—dehydrate in the presence of acid. Only β-hydroxy carbonyl compounds dehydrate in the presence of base.

An aldol reaction is often called an **aldol condensation,** because the β-hydroxy carbonyl compound that is initially formed loses H_2O by dehydration. **A *condensation reaction* is one in which a small molecule, in this case H_2O, is eliminated during a reaction.**

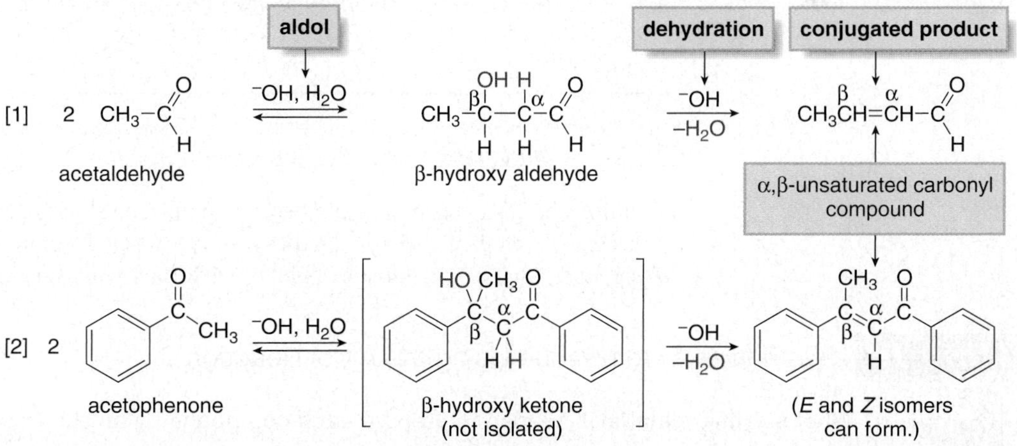

It may or may not be possible to isolate the β-hydroxy carbonyl compound under the conditions of the aldol reaction. When the α,β-unsaturated carbonyl compound is further conjugated with a carbon–carbon double bond or a benzene ring, as in the case of Reaction [2], the β-hydroxy carbonyl compound cannot be isolated.

The mechanism of dehydration consists of two steps: **deprotonation followed by loss of $^{-}$OH,** as shown in Mechanism 24.2.

MECHANISM 24.2

Dehydration of β-Hydroxy Carbonyl Compounds

◆ In Step [1], base removes a proton from the α carbon, thus forming a resonance-stabilized enolate.

◆ In Step [2], the electron pair of the enolate forms the π bond as ⁻OH is eliminated.

resonance-stabilized enolate

Like E1 elimination, E1cB requires two steps. Unlike E1, though, the intermediate in E1cB is a carbanion, not a carbocation. E1cB stands for **Elimination, unimolecular, conjugate base.**

This elimination mechanism, called the **E1cB mechanism,** differs from the two more general mechanisms of elimination, E1 and E2, which were discussed in Chapter 8. The E1cB mechanism involves two steps, and proceeds by way of an **anionic** intermediate.

Regular alcohols dehydrate only in the presence of acid but not base, because hydroxide is a poor leaving group. When the hydroxy group is β to a carbonyl group, however, loss of H and OH from the α and β carbons forms a **conjugated double bond,** and the stability of the conjugated system makes up for having such a poor leaving group.

Dehydration of the initial β-hydroxy carbonyl compound drives the equilibrium of an aldol reaction to the right, thus favoring product formation. Once the conjugated α,β-unsaturated carbonyl compound forms, it is *not* re-converted to the β-hydroxy carbonyl compound.

PROBLEM 24.2   What unsaturated carbonyl compound is formed by dehydration of each β-hydroxy carbonyl compound?

a.

b.

c.

24.1C   Retrosynthetic Analysis

To utilize the aldol reaction in synthesis, you must be able to determine which aldehyde or ketone is needed to prepare a particular β-hydroxy carbonyl compound or α,β-unsaturated carbonyl compound—that is, you must be able to **work backwards, in the retrosynthetic direction.**

How To   Synthesize a Compound Using the Aldol Reaction

Example   What starting material is needed to prepare each compound by an aldol reaction?

a.

b.

Step [1]   Locate the α and β carbons of the carbonyl group.

• When a carbonyl group has two different α carbons, choose the side that contains the OH group (in a β-hydroxy carbonyl compound) or is part of the C=C (in an α,β-unsaturated carbonyl compound).

Step [2]   **Break the molecule into two components between the α and β carbons.**

- The α carbon and all remaining atoms bonded to it belong to one carbonyl component. The β carbon and all remaining atoms bonded to it belong to the other carbonyl component. Both components are identical in all aldols we have thus far examined.

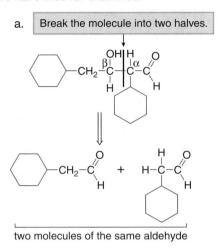

a. Break the molecule into two halves.

two molecules of the same aldehyde

b. Break the molecule into two halves.

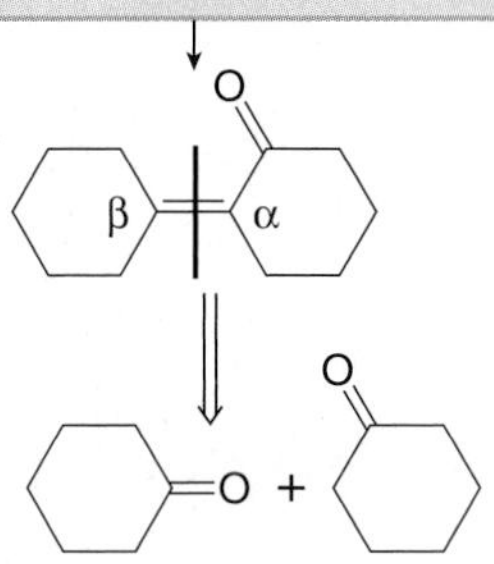

two molecules of cyclohexanone

PROBLEM 24.3   What aldehyde or ketone is needed to prepare each compound by an aldol reaction?

a.

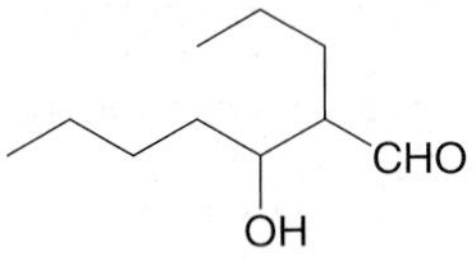

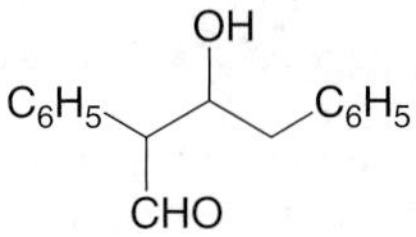

b.

c. C_6H_5 ... C_6H_5

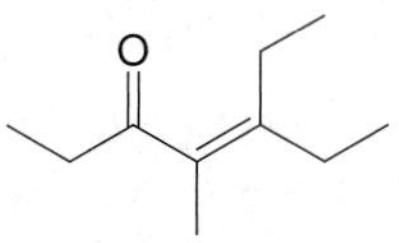

d.

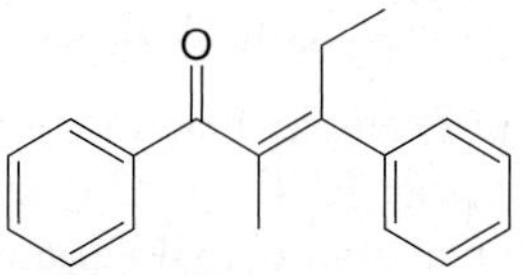

24.2   Crossed Aldol Reactions

In all of the aldol reactions discussed so far, the electrophilic carbonyl and the nucleophilic enolate have originated from the *same* aldehyde or ketone. Sometimes, though, it is possible to carry out an aldol reaction between two *different* carbonyl compounds.

◆ An aldol reaction between two different carbonyl compounds is called a *crossed aldol* or *mixed aldol reaction.*

24.2A   A Crossed Aldol Reaction with Two Different Aldehydes, Both Having α H Atoms

When two different aldehydes, both having α H atoms, are combined in an aldol reaction, four different β-hydroxy carbonyl compounds are formed. Four products form, not one, because both aldehydes can lose an acidic α hydrogen atom and form an enolate in the presence of base. Both enolates can then react with both carbonyl compounds, as shown for acetaldehyde and propanal in the following reaction scheme.

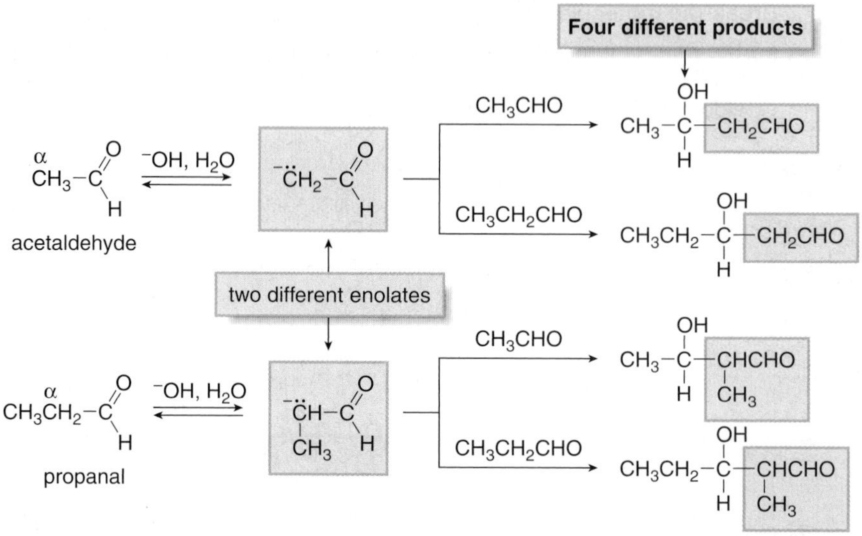

◆ Conclusion: When two different aldehydes have α hydrogens, a crossed aldol reaction is *not* synthetically useful.

24.2B   Synthetically Useful Crossed Aldol Reactions

Crossed aldols are synthetically useful in two different situations.

[1]   A crossed aldol when only one carbonyl component has α H atoms.

When one carbonyl compound has no α hydrogens, a crossed aldol reaction often leads to one product. Two common carbonyl compounds with no α hydrogens used for this purpose are **formaldehyde ($CH_2=O$) and benzaldehyde (C_6H_5CHO).**

For example, reaction of C_6H_5CHO (as the electrophile) with either acetaldehyde (CH_3CHO) or acetone [$(CH_3)_2C=O$] in the presence of base forms a single α,β-unsaturated carbonyl compound after dehydration.

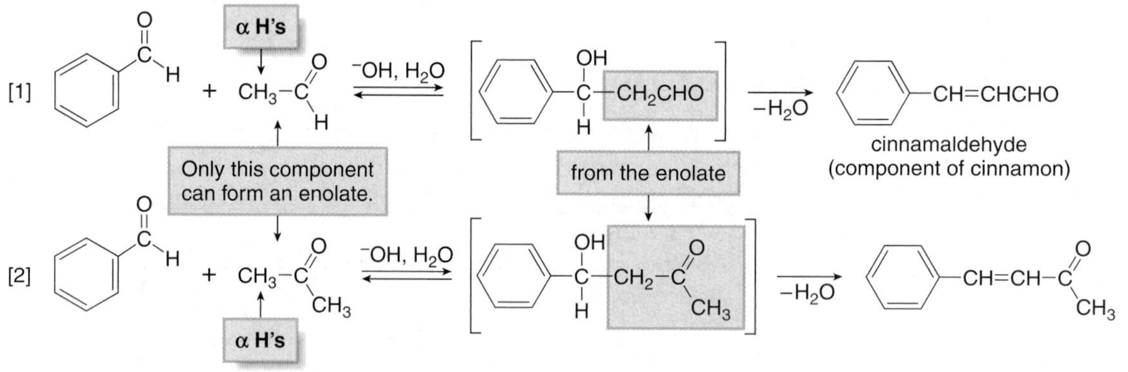

The yield of a single crossed aldol product is increased further if the electrophilic carbonyl component is relatively unhindered (as is the case with most aldehydes), and if it is used in excess.

PROBLEM 24.4   Draw the mechanism for the crossed aldol reaction between C_6H_5CHO and CH_3CHO to form $C_6H_5CH=CHCHO$ (Reaction [1]).

PROBLEM 24.5   Draw the products formed in each crossed aldol reaction.

a. $CH_3CH_2CH_2CHO$ and $CH_2=O$ c. C_6H_5CHO and

b. $C_6H_5COCH_3$ and $CH_2=O$

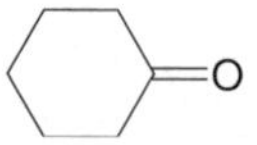

PROBLEM 24.6   A published synthesis of the analgesic nabumetone uses a crossed aldol reaction to form **X**. What is the structure of **X**? **X** is converted to nabumetone in one step by hydrogenation with H_2 and Pd-C. (See Problem 23.28 for another way to make nabumetone.)

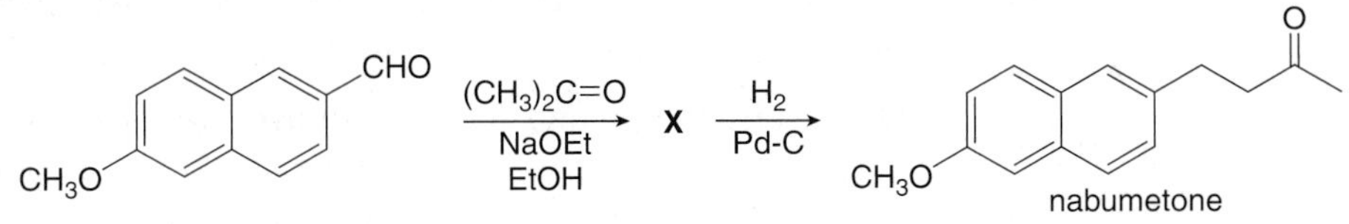

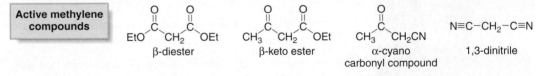

[2]   A crossed aldol when one carbonyl component has especially acidic α H atoms.

Another useful crossed aldol reaction takes place between an aldehyde or ketone and a β-dicarbonyl (or similar) compound.

General reaction

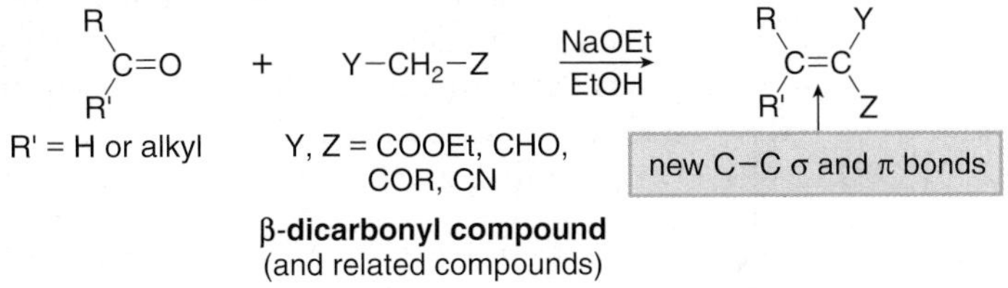

$$
\begin{array}{c} R \\ \diagdown \\ C{=}O \\ \diagup \\ R' \end{array}
\quad + \quad Y{-}CH_2{-}Z \quad
\xrightarrow[\text{EtOH}]{\text{NaOEt}}
\quad
\begin{array}{c} R \qquad Y \\ \diagdown \quad \diagup \\ C{=}C \\ \diagup \quad \diagdown \\ R' \qquad Z \end{array}
$$

R' = H or alkyl   Y, Z = COOEt, CHO, COR, CN

β-dicarbonyl compound
(and related compounds)

new C–C σ and π bonds

Example

benzaldehyde   + $CH_2(COOEt)_2$ $\xrightarrow[\text{EtOH}]{\text{NaOEt}}$

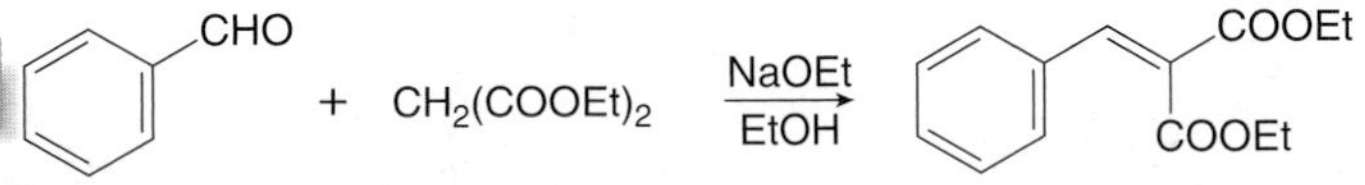

benzaldehyde   diethyl malonate

As we learned in Section 23.3, the α hydrogens between two carbonyl groups are especially acidic, and so they are more readily removed than other α H atoms. As a result, **the β-dicarbonyl compound always becomes the enolate component of the aldol reaction.** Figure 24.2 shows the steps for the crossed aldol reaction between diethyl malonate and benzaldehyde. In this type of crossed aldol reaction, the initial β-hydroxy carbonyl compound *always* loses water to form the highly conjugated product.

β-Dicarbonyl compounds are sometimes called **active methylene compounds** because they are more reactive towards base than other carbonyl compounds. **1,3-Dinitriles** and **α-cyano carbonyl compounds** are also active methylene compounds.

Active methylene compounds

$$\underset{\text{β-diester}}{\overset{\displaystyle O \qquad O}{EtO\overset{\|}{C}CH_2\overset{\|}{C}OEt}} \qquad \underset{\text{β-keto ester}}{\overset{\displaystyle O \qquad O}{CH_3\overset{\|}{C}CH_2\overset{\|}{C}OEt}} \qquad \underset{\substack{\text{α-cyano}\\\text{carbonyl compound}}}{\overset{\displaystyle O}{CH_3\overset{\|}{C}CH_2CN}} \qquad \underset{\text{1,3-dinitrile}}{N{\equiv}C{-}CH_2{-}C{\equiv}N}$$

PROBLEM 24.7   Draw the products formed in the crossed aldol reaction of phenylacetaldehyde ($C_6H_5CH_2CHO$) with each compound: (a) $CH_2(COOEt)_2$; (b) $CH_2(COCH_3)_2$; (c) CH_3COCH_2CN.

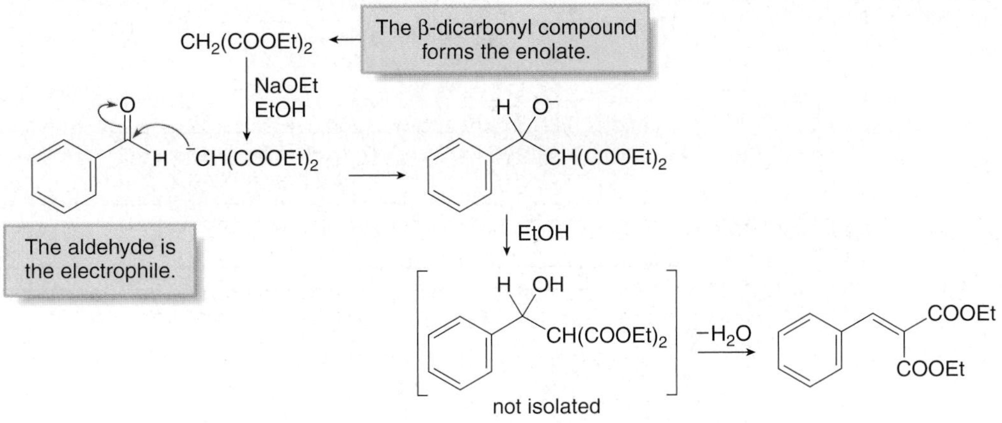

Figure 24.2   Crossed aldol reaction between benzaldehyde and $CH_2(COOEt)_2$

$CH_2(COOEt)_2$

The β-dicarbonyl compound forms the enolate.

NaOEt
EtOH

The aldehyde is the electrophile.

$CH(COOEt)_2$

EtOH

not isolated

$-H_2O$

24.3   Directed Aldol Reactions

A **directed aldol reaction** is a variation of the crossed aldol reaction that clearly defines which carbonyl compound becomes the nucleophilic enolate and which reacts at the electrophilic carbonyl carbon. The strategy of a directed aldol reaction is as follows:

[1] Prepare the enolate of one carbonyl component with LDA.
[2] Add the second carbonyl compound (the electrophile) to this enolate.

Because the steps are done sequentially and a strong nonnucleophilic base is used to form the enolate of one carbonyl component only, a variety of carbonyl substrates can be used in the reaction. Both carbonyl components can have α hydrogens because only one enolate is prepared with LDA. Also, when an unsymmetrical ketone is used, LDA selectively forms the **less substituted, kinetic enolate.**

Sample Problem 24.1 illustrates the steps of a directed aldol reaction between a ketone and an aldehyde, both of which have α hydrogens.

SAMPLE PROBLEM 24.1   Draw the product of the following directed aldol reaction.

(structure of 2-methylcyclohexanone)

[1] LDA, THF
[2] CH₃CHO
[3] H₂O

2-methylcyclohexanone

> Periplanone B is an extremely active compound produced in small amounts by the female American cockroach. Its structure was determined using 200 μg of periplanone B from more than 75,000 female cockroaches. This structure was confirmed by synthesis in the laboratory in 1979.

SOLUTION

2-Methylcyclohexanone forms an enolate on the less substituted carbon, which then reacts with the electrophile, CH₃CHO.

(reaction scheme showing enolate formation with LDA/THF, nucleophilic addition, and H₂O workup)

less substituted
kinetic enolate

new C–C bond

Figure 24.3 illustrates how a directed aldol reaction was used in the synthesis of **periplanone B,** the sex pheromone of the female American cockroach.

To determine the needed carbonyl components for a directed aldol, follow the same strategy used for a regular aldol reaction in Section 24.1C, as shown in Sample Problem 24.2.

Figure 24.3
A directed aldol
reaction in the
synthesis of
periplanone B

(reaction scheme for synthesis of periplanone B)

LDA
THF

[1]
[2] H₂O

several
steps

deprotonation   nucleophilic addition of the enolate
to the aldehyde carbonyl

periplanone B
sex pheromone of the female
American cockroach

SAMPLE PROBLEM 24.2   What starting materials are needed to prepare each compound using a directed aldol reaction?

a. (cyclopentanone with CH₂OH substituent structure)

b. (cyclohexylidene ketone structure)

SOLUTION

Identify the α and β carbons to the carbonyl group and **break the molecule into two components between these carbons.**

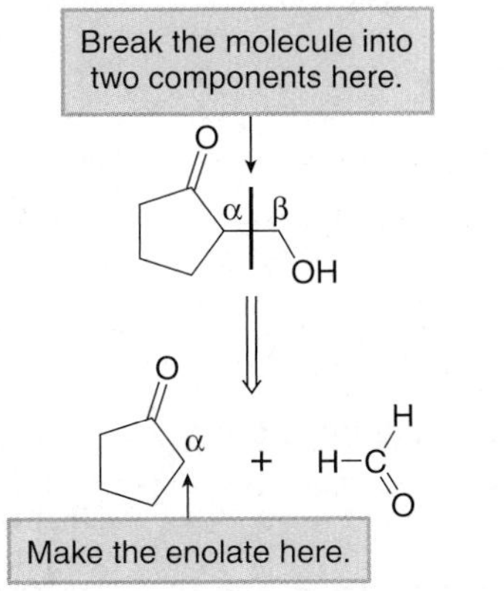

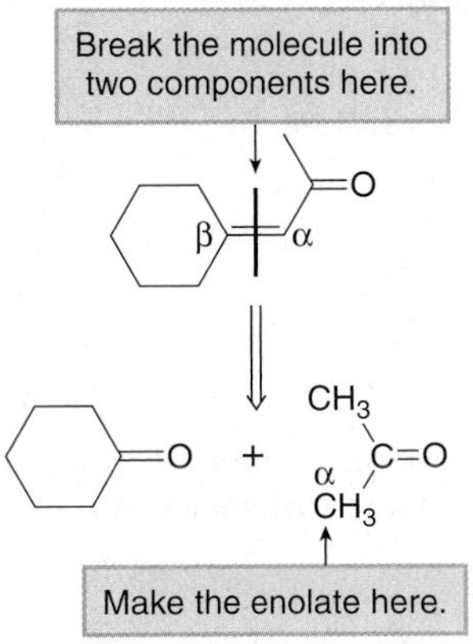

PROBLEM 24.8 What carbonyl starting materials are needed to prepare each compound using a directed aldol reaction?

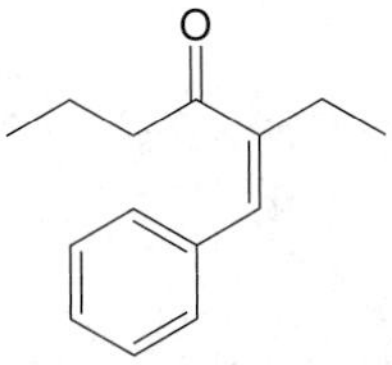

24.4 Intramolecular Aldol Reactions

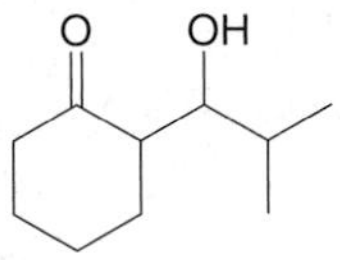

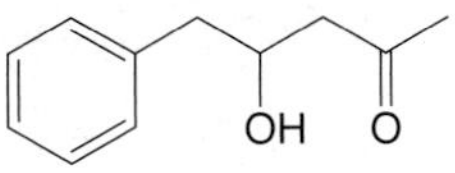

Aldol reactions with dicarbonyl compounds can be used to make five- and six-membered rings. The enolate formed from one carbonyl group is the nucleophile, and the carbonyl carbon of the other carbonyl group is the electrophile. For example, treatment of 2,5-hexanedione with base forms a five-membered ring.

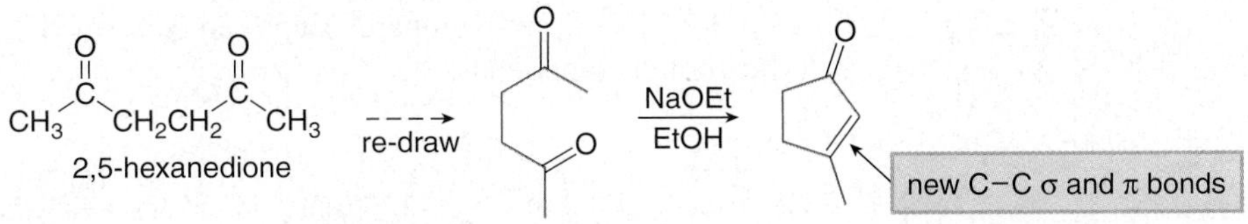

> 2,5-Hexanedione is called a **1,4-dicarbonyl compound** to emphasize the relative positions of its carbonyl groups. 1,4-Dicarbonyl compounds are starting materials for synthesizing five-membered rings.

The steps in this process, shown in Mechanism 24.3, are no different from the general mechanisms of the aldol reaction and dehydration described previously in Section 24.1.

When 2,5-hexanedione is treated with base in Step [1], two different enolates are possible—enolates **A** and **B,** formed by removal of H_a and H_b, respectively. Although enolate **A** goes on to form the five-membered ring, intramolecular cyclization using enolate **B** would lead to a strained three-membered ring.

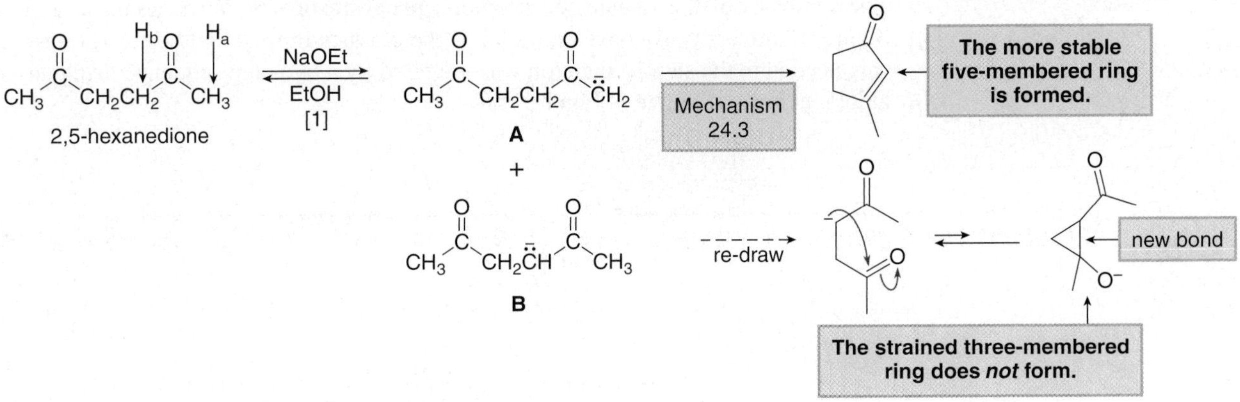

MECHANISM 24.3

The Intramolecular Aldol Reaction

Step [1] Formation of an enolate

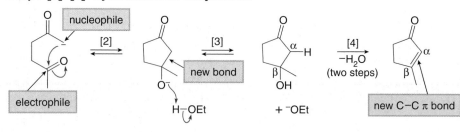

2,5-hexanedione

◆ Deprotonation of the CH$_3$ group with base forms a nucleophilic enolate, which is re-drawn to more clearly show the intramolecular reaction in Step [2].

Steps [2]–[4] Cyclization and dehydration

◆ In Step [2], the nucleophilic enolate attacks the electrophilic carbonyl carbon in the same molecule, forming a new carbon–carbon σ bond. **This generates the five-membered ring.**

◆ Protonation of the alkoxide in Step [3] and loss of H$_2$O by the two steps outlined in Mechanism 24.2 forms a new C–C π bond.

◆ The overall result is formation of an α,β-unsaturated carbonyl compound in the new five-membered ring.

Because the three-membered ring is much higher in energy than the enolate starting material, equilibrium greatly favors the starting materials and the **three-membered ring does not form.** Under the reaction conditions, enolate **B** is re-protonated to form 2,5-hexanedione, because all steps except dehydration are equilibria. **Thus, equilibrium favors formation of the more stable five-membered ring over the much less stable three-membered ring.**

In a similar fashion, six-membered rings can be formed from the intramolecular aldol reaction of 1,5-dicarbonyl compounds.

2,6-heptanedione — re-draw → a 1,5-dicarbonyl compound — NaOEt, EtOH → new C–C σ and π bonds

PROBLEM 24.9   Draw a stepwise mechanism for the conversion of 2,6-heptanedione to 3-methyl-2-cyclohexenone with NaOEt, EtOH.

The synthesis of the female sex hormone **progesterone** by W. S. Johnson and co-workers at Stanford University is considered one of the classics in total synthesis. The last six-membered ring needed in the steroid skeleton was prepared by a two-step sequence using an intramolecular aldol reaction, as shown in Figure 24.4.

The synthesis of progesterone

two steps → progesterone

The five-membered ring must be converted to a six-membered ring.

Figure 24.4   The synthesis of progesterone using an intramolecular aldol reaction

Ozone oxidatively cleaves the C=C.

1,5-dicarbonyl compound

Intramolecular aldol reaction forms the six-membered ring.

progesterone

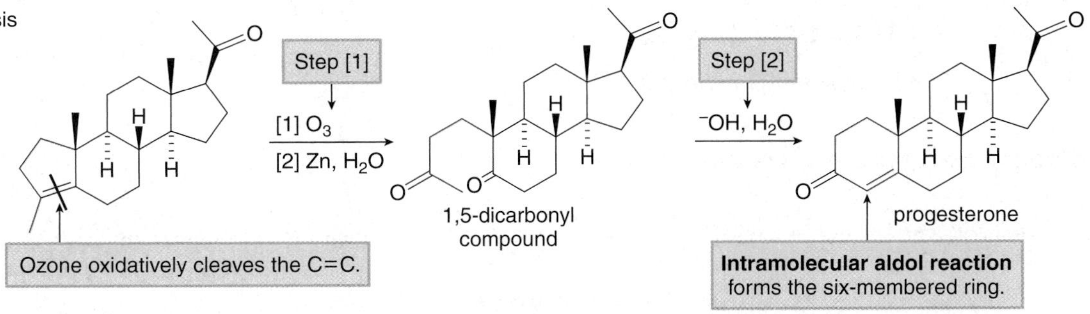

- **Step [1]:** Oxidative cleavage of the alkene with O_3, followed by Zn, H_2O (Section 12.10), gives the 1,5-dicarbonyl compound.
- **Step [2]:** Intramolecular aldol reaction of the 1,5-dicarbonyl compound with dilute ⁻OH in H_2O solution forms progesterone.
- This two-step reaction sequence converts a five-membered ring into a six-membered ring. Reactions that synthesize larger rings from smaller ones are called **ring expansion reactions.**

PROBLEM 24.10   What cyclic product is formed when the following 1,5-dicarbonyl compound is treated with aqueous ⁻OH?

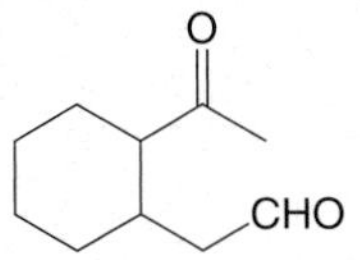

PROBLEM 24.11   In theory, the intramolecular aldol reaction of 6-oxoheptanal could yield the three compounds shown. It turns out, though, that 1-acetylcyclopentene is by far the major product. Why are the other two compounds formed in only minor amounts?

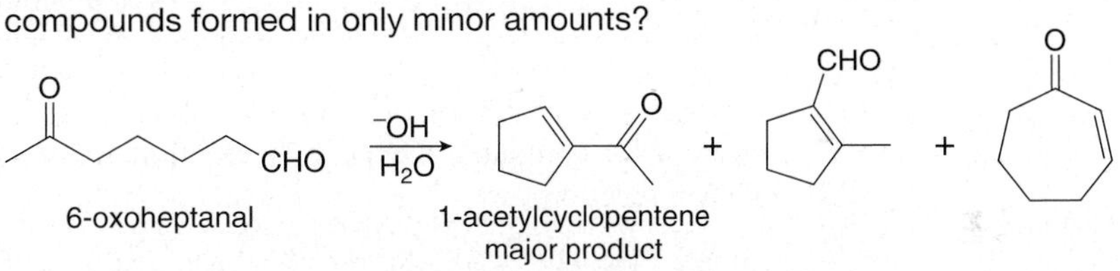

6-oxoheptanal

1-acetylcyclopentene major product

24.5   The Claisen Reaction

The **Claisen reaction** is the second general reaction of enolates with other carbonyl compounds. In the Claisen reaction, two molecules of an ester react with each other in the presence of an alkoxide base to form a **β-keto ester.** For example, treatment of ethyl acetate with NaOEt forms ethyl acetoacetate after protonation with aqueous acid.

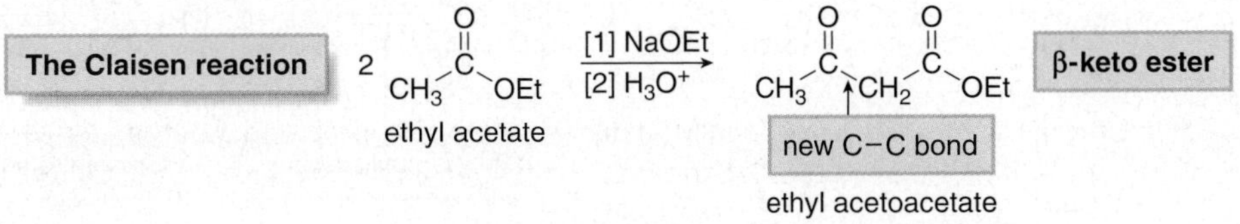

The Claisen reaction

ethyl acetate

β-keto ester

new C–C bond

ethyl acetoacetate

Unlike the aldol reaction, which is base-catalyzed, a full equivalent of base is needed to deprotonate the β-keto ester formed in Step [3] of the Claisen reaction.

The mechanism for the Claisen reaction (Mechanism 24.4) resembles the mechanism of an aldol reaction in that it involves nucleophilic addition of an enolate to an electrophilic carbonyl group. Because esters have a leaving group on the carbonyl carbon, however, loss of a leaving group occurs to form the product of **substitution,** *not* **addition.**

◆ **Keep in mind: The characteristic reaction of esters is nucleophilic substitution. A Claisen reaction is a nucleophilic substitution in which an enolate is the nucleophile.**

MECHANISM 24.4

The Claisen Reaction

Step [1] Formation of a nucleophilic enolate

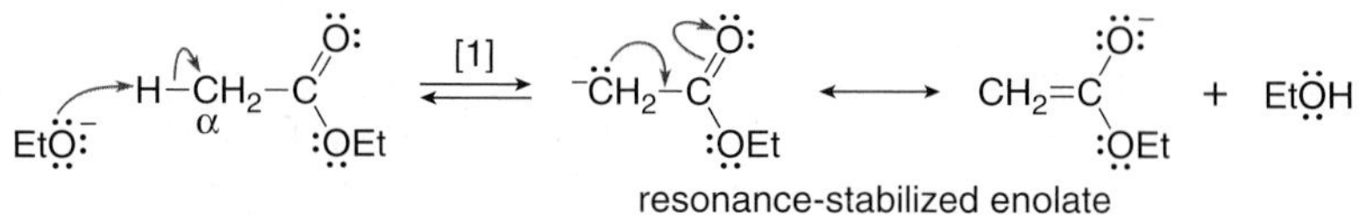

◆ In Step [1], the base removes a proton from the α carbon to form a **resonance-stabilized enolate.**

Steps [2]–[3] Nucleophilic addition and loss of the leaving group

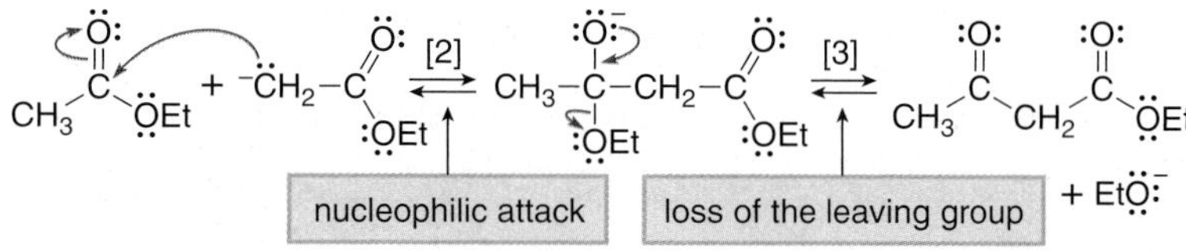

◆ In Step [2], the nucleophilic enolate attacks the electrophilic carbonyl carbon of another molecule of ester, forming a new carbon–carbon bond. **This joins the α carbon of one ester to the carbonyl carbon of a second ester.**

◆ Elimination of the leaving group, EtO⁻, forms a β-keto ester in Step [3]. Steps [1]–[3] are reversible equilibria.

Steps [4]–[5] Deprotonation and protonation

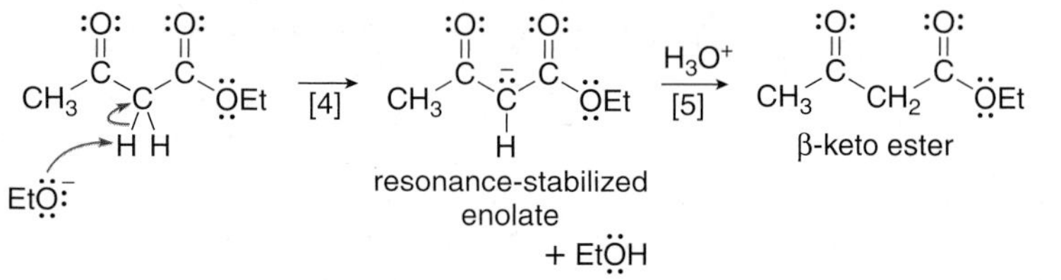

◆ Because the β-keto ester formed in Step [3] has especially acidic protons between its two carbonyl groups, a proton is removed under the basic reaction conditions to form an enolate (Step [4]). **The formation of this resonance-stabilized enolate drives the equilibrium in the Claisen reaction.**

◆ Protonation of this enolate with strong acid re-forms the neutral β-keto ester to complete the reaction.

Figure 24.5 compares the general reaction for nucleophilic substitution of an ester with the Claisen reaction.

Sample Problem 24.3 reinforces the basic features of the Claisen reaction.

Figure 24.5   The Claisen reaction—An example of nucleophilic substitution

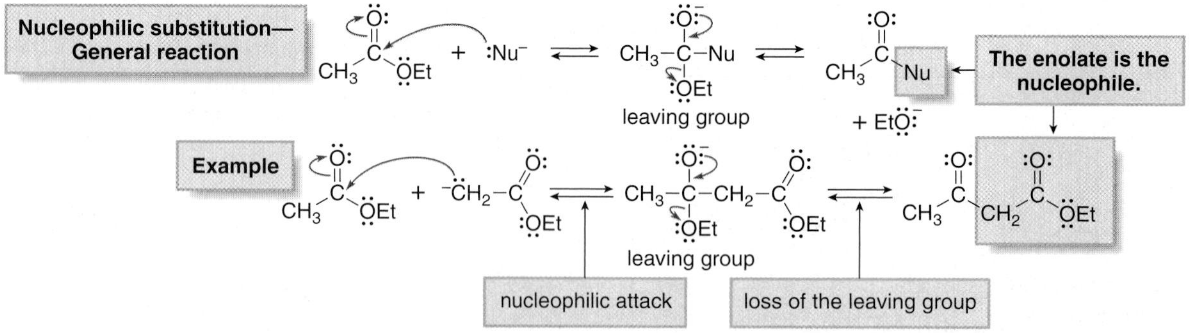

• Esters react by nucleophilic substitution. In a Claisen reaction, an enolate is the nucleophile that adds to the carbonyl group.

SAMPLE PROBLEM 24.3   Draw the product of the following Claisen reaction.

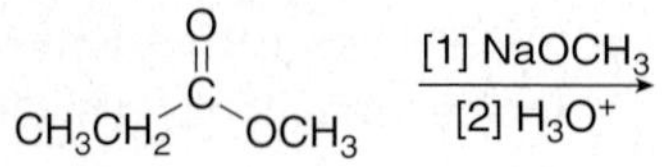

SOLUTION

To draw the product of any Claisen reaction, form a new carbon–carbon bond between the α carbon of one ester and the carbonyl carbon of another ester, with elimination of the leaving group ($^-$OCH$_3$ in this case).

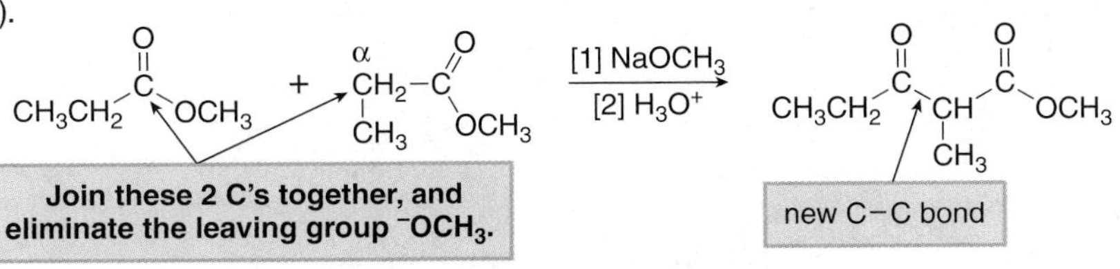

Join these 2 C's together, and eliminate the leaving group $^-$OCH$_3$.

new C–C bond

Next, write out the steps of the reaction to verify this product.

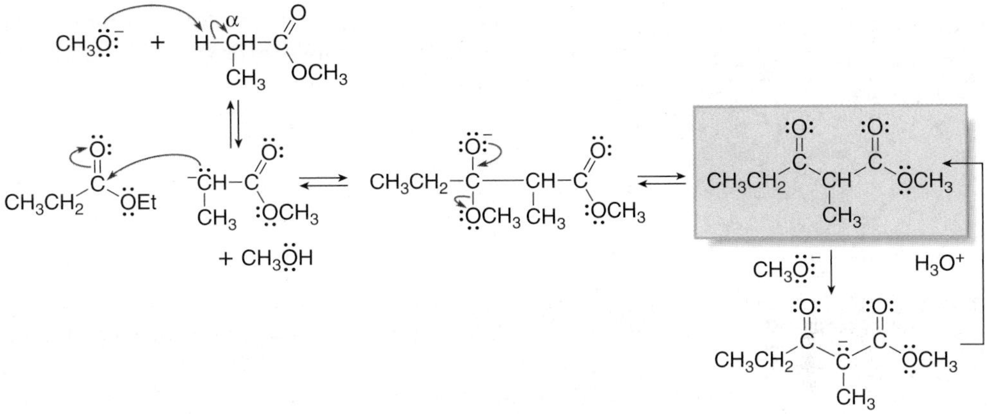

PROBLEM 24.12   What β-keto ester is formed when each ester is used in a Claisen reaction?

a.

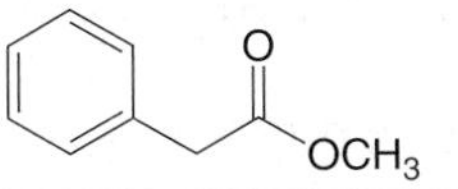

b.

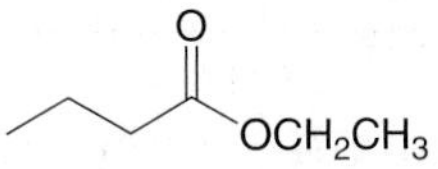

PROBLEM 24.13   What ester is needed to form each β-keto ester by a Claisen reaction?

a.

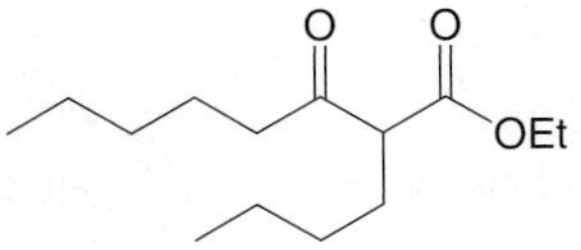

b.

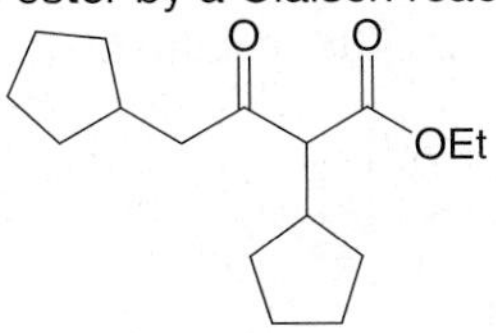

24.6 The Crossed Claisen and Related Reactions

Like the aldol reaction, it is sometimes possible to carry out a Claisen reaction with two different carbonyl components as starting materials.

◆ A Claisen reaction between two different carbonyl compounds is called a *crossed Claisen reaction.*

24.6A Two Useful Crossed Claisen Reactions

A crossed Claisen reaction is synthetically useful in two different instances.

[1] Between two different esters when only one has α hydrogens.

When one ester has no α hydrogens, a crossed Claisen reaction often leads to one product. Common esters with no α H atoms include ethyl formate (HCO$_2$Et) and ethyl benzoate (C$_6$H$_5$CO$_2$Et). For example, the reaction of ethyl benzoate (as the electrophile) with ethyl acetate (which forms the enolate) in the presence of base forms predominately one β-keto ester.

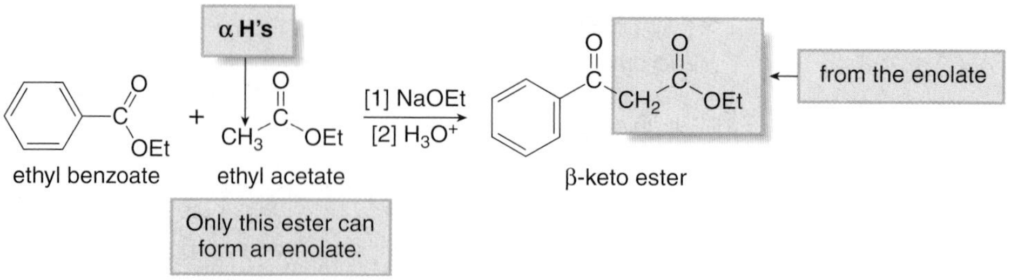

ethyl benzoate + ethyl acetate → β-keto ester

α H's

Only this ester can form an enolate.

from the enolate

[2] Between a ketone and an ester.

The reaction of a ketone and an ester in the presence of base also forms the product of a crossed Claisen reaction. The enolate is always formed from the ketone component, and the reaction works best when the ester has no α hydrogens. The product of this crossed Claisen reaction is a **β-dicarbonyl compound,** but *not* a β-keto ester.

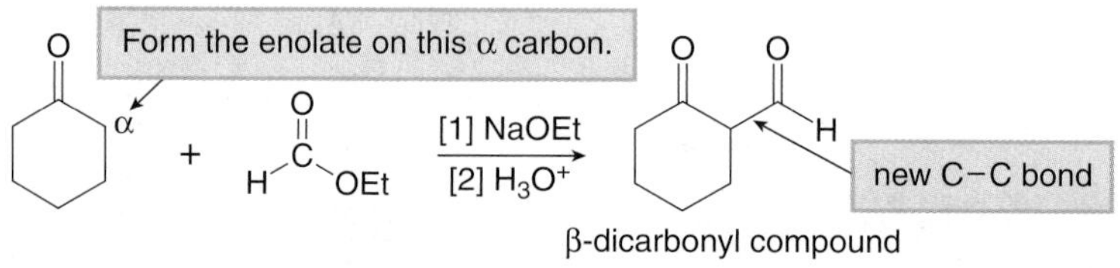

Form the enolate on this α carbon.

new C–C bond

β-dicarbonyl compound

PROBLEM 24.14   What crossed Claisen product is formed from each pair of compounds?

a. CH_3CH_2COOEt and HCO_2Et
b. $CH_3(CH_2)_5CO_2Et$ and HCO_2Et
c. $(CH_3)_2C=O$ and CH_3CO_2Et

d.

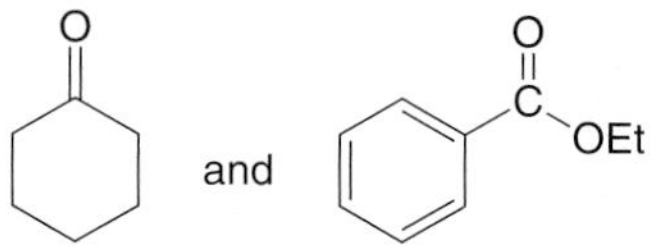

24.6B   Other Useful Variations of the Crossed Claisen Reaction

β-Dicarbonyl compounds are also prepared by reacting an enolate with **ethyl chloroformate** and **diethyl carbonate.**

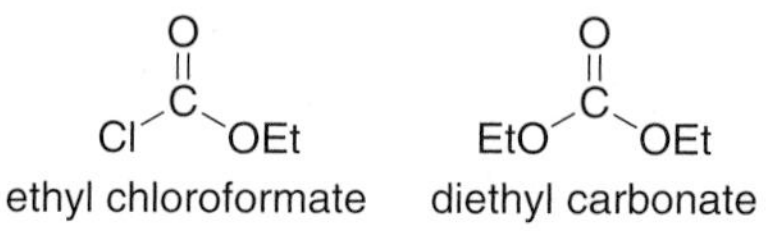

ethyl chloroformate   diethyl carbonate

These reactions resemble a Claisen reaction because they involve the same three steps:

[1] Formation of an enolate
[2] Nucleophilic addition to a carbonyl group
[3] Elimination of a leaving group

For example, reaction of an ester enolate with diethyl carbonate yields a β-diester (Reaction [1]), whereas reaction of a ketone enolate with ethyl chloroformate forms a β-keto ester (Reaction [2]).

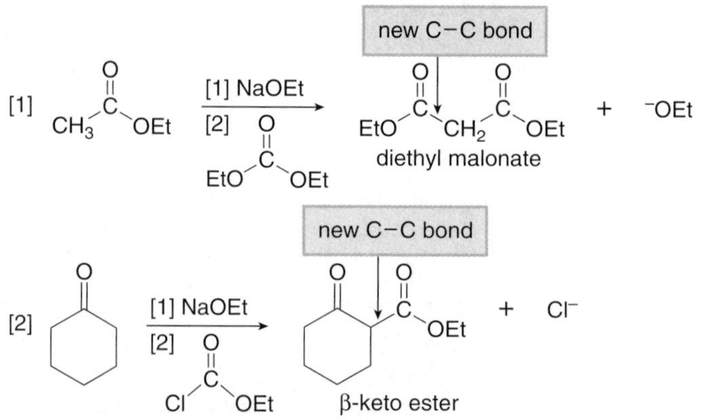

new C–C bond

diethyl malonate

new C–C bond

β-keto ester

Reaction [2] is noteworthy because it provides easy access to **β-keto esters,** which are useful starting materials in the acetoacetic ester synthesis (Section 23.10). In this reaction, Cl⁻ is eliminated rather than ⁻OEt in Step [3], because Cl⁻ is a better leaving group, as shown in the following steps.

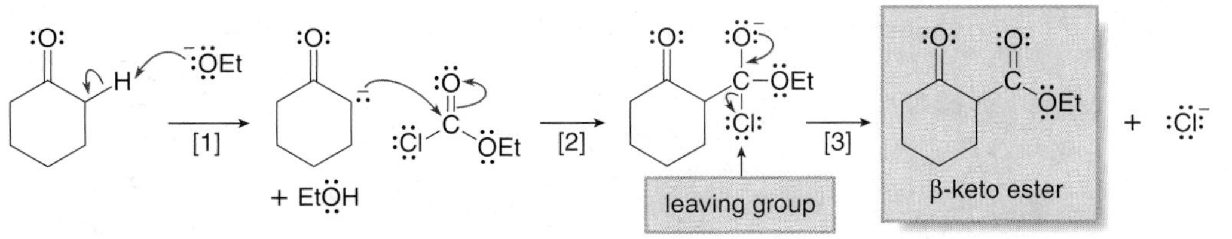

PROBLEM 24.15   Draw the products of each reaction.

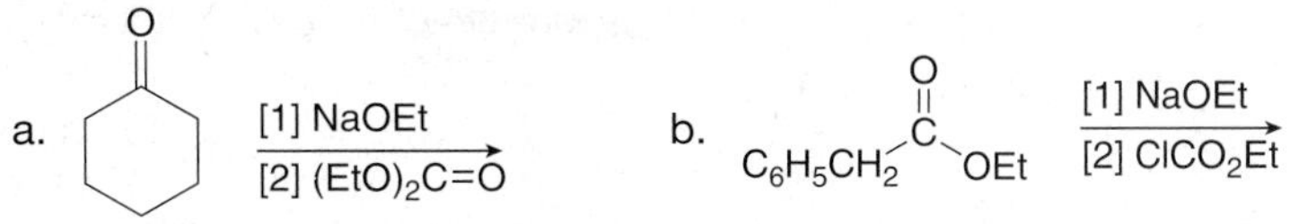

a.   [1] NaOEt
     [2] (EtO)₂C=O

b.   C₆H₅CH₂   [1] NaOEt
               [2] ClCO₂Et

PROBLEM 24.16   What product would you expect in the following reaction sequence?

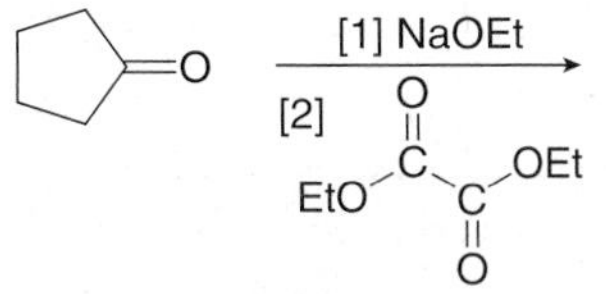

[1] NaOEt

[2]

24.7   The Dieckmann Reaction

Intramolecular Claisen reactions of diesters form five- and six-membered rings. The enolate of one ester is the nucleophile, and the carbonyl carbon of the other is the electrophile. An intramolecular Claisen reaction is called a **Dieckmann reaction.** Two types of diesters give good yields of cyclic products.

◆ **1,6-Diesters yield five-membered rings by the Dieckmann reaction.**

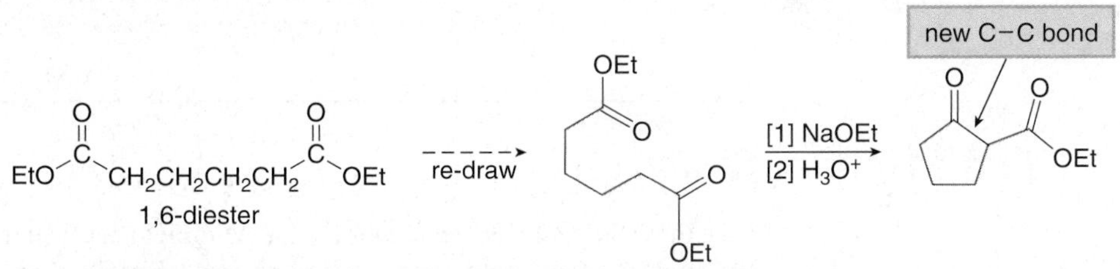

◆ **1,7-Diesters yield six-membered rings by the Dieckmann reaction.**

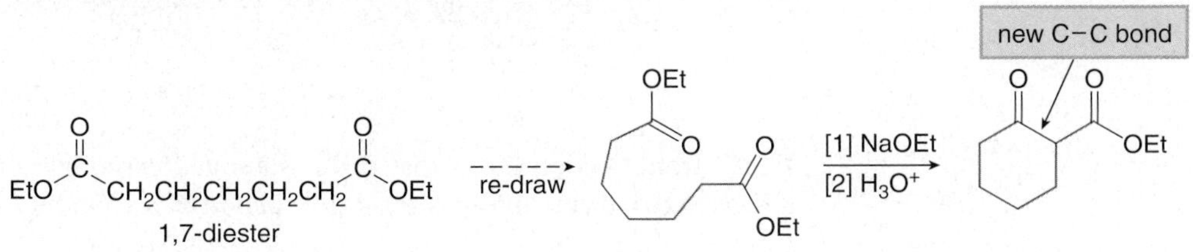

The mechanism of the Dieckmann reaction is exactly the same as the mechanism of an intermolecular Claisen reaction. It is illustrated in Mechanism 24.5 for the formation of a six-membered ring.

MECHANISM 24.5

The Dieckmann Reaction

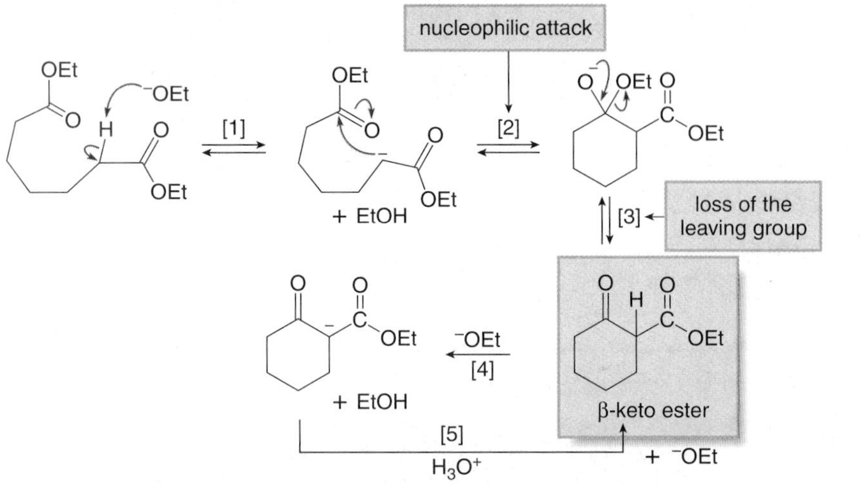

◆ In Step [1], the base removes a proton to form an enolate, which attacks the carbonyl group of the second ester in Step [2], thus forming a new carbon–carbon bond.

◆ In Step [3], elimination of ⁻OEt forms the β-keto ester.

◆ To complete the reaction, the proton between the two carbonyl groups is removed with base, and then protonation of the enolate re-forms the β-keto ester (Steps [4] and [5]).

PROBLEM 24.17 One synthesis of prostaglandin PGA$_2$ involved a Dieckmann reaction as a key step. What is the structure of **X** in the reactions drawn below?

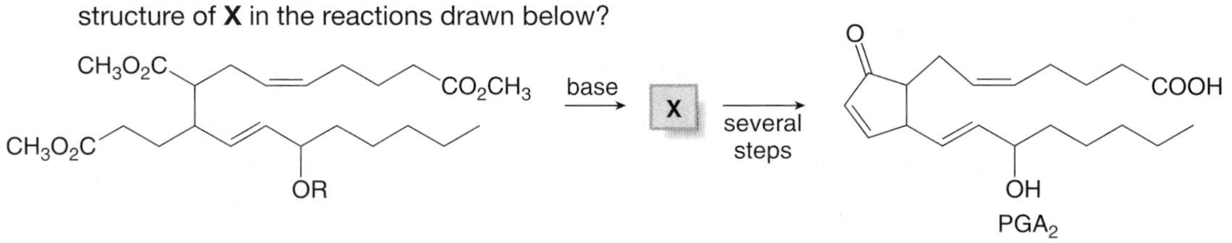

PROBLEM 24.18 What two β-keto esters are formed in the Dieckmann reaction of the following diester?

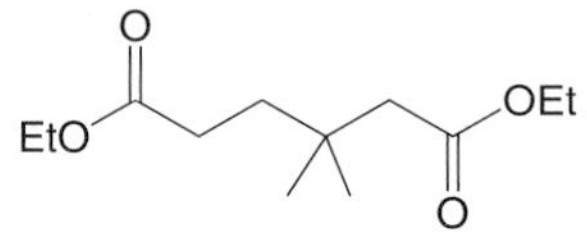

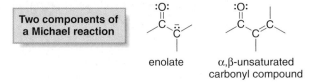

24.8 The Michael Reaction

Like the aldol and Claisen reactions, the **Michael reaction involves two carbonyl components—the enolate of one carbonyl compound and an α,β-unsaturated carbonyl compound.**

Two components of a Michael reaction

enolate   α,β-unsaturated carbonyl compound

Recall from Section 20.15 that α,β-unsaturated carbonyl compounds are resonance stabilized and have **two electrophilic sites—the carbonyl carbon and the β carbon.**

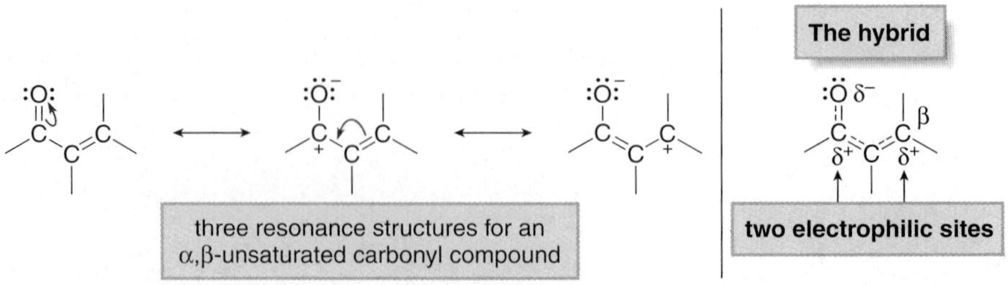

The hybrid

three resonance structures for an α,β-unsaturated carbonyl compound

two electrophilic sites

◆ The Michael reaction involves the conjugate addition (1,4-addition) of a resonance-stabilized enolate to the β carbon of an α,β-unsaturated carbonyl system.

All conjugate additions add the **elements of H and Nu across the α and β carbons.** In the Michael reaction, the **nucleophile is an enolate.** Enolates of active methylene compounds are particularly common. The α,β-unsaturated carbonyl component is often called a **Michael acceptor.**

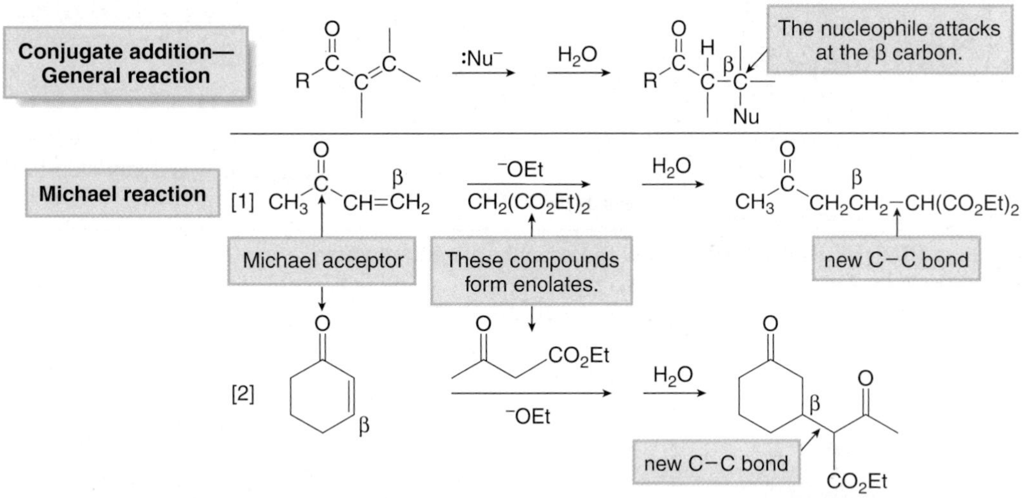

Conjugate addition— General reaction

Michael reaction

[1]

[2]

The Michael reaction always forms a new carbon–carbon bond on the β carbon of the Michael acceptor. Reaction [2] is used to illustrate the mechanism of the Michael reaction in Mechanism 24.6. The key step is nucleophilic addition of the enolate to the β carbon of the Michael acceptor in Step [2].

MECHANISM 24.6

The Michael Reaction

Step [1] Enolate formation

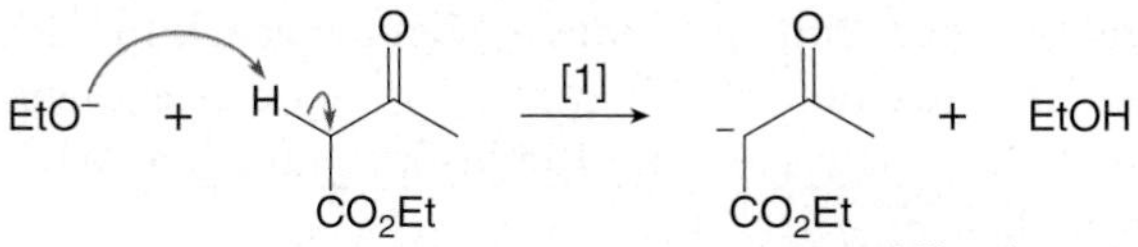

◆ Base removes the acidic proton between the two carbonyl groups, **forming the enolate** in Step [1].

Steps [2]–[3] Nucleophilic attack at the β carbon and protonation

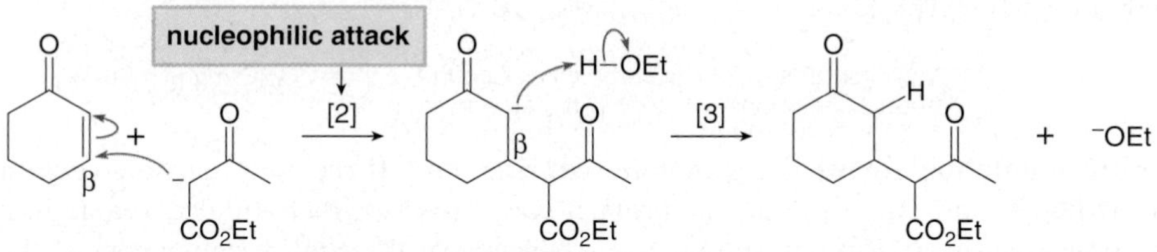

◆ **The nucleophilic enolate adds to the β carbon** of the α,β-unsaturated carbonyl compound, forming a new carbon–carbon bond and a resonance-stabilized enolate.

◆ Protonation of the enolate forms the 1,4-addition product in Step [3].

When the product of a Michael reaction is also a β-keto ester, it can be hydrolyzed and decarboxylated by heating in aqueous acid, as discussed in Section 23.9. This forms a **1,5-dicarbonyl compound.**

1,5-Dicarbonyl compounds are starting materials for intramolecular aldol reactions, as described in Section 24.4.

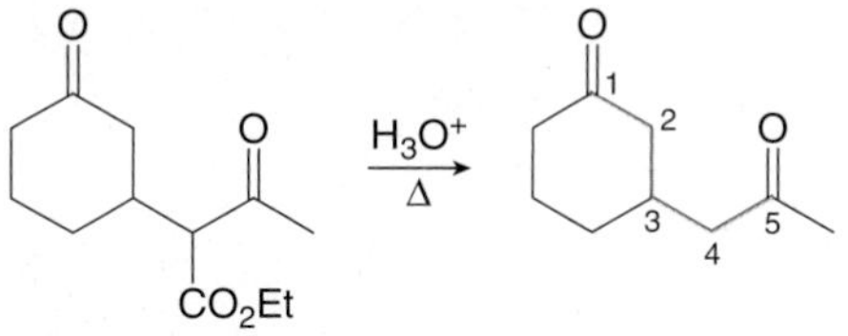

Michael reaction product

1,5-dicarbonyl compound

Figure 24.6 Using a Michael reaction in the synthesis of the steroid estrone

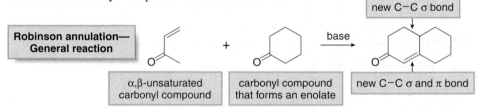

Figure 24.6 shows a Michael reaction that was a key step in the synthesis of **estrone,** a female sex hormone.

PROBLEM 24.19 Draw the product of each reaction.

a. CH_2=$CHCO_2Et$ + [structure: CH_3–C(=O)–CH_2CO_2Et] $\xrightarrow{[1]\ NaOEt,\ [2]\ H_2O}$

b. [bicyclic enone structure] + $CH_2(CO_2Et)_2$ $\xrightarrow{[1]\ NaOEt,\ [2]\ H_2O}$

c. [methylenecyclohexanone structure] + [structure: CH_3–C(=O)–CH_2CN] $\xrightarrow{[1]\ NaOEt,\ [2]\ H_2O}$

PROBLEM 24.20 What starting materials are needed to prepare each compound by the Michael reaction?

a. [diketone structure with CO_2Et]   b. [cyclopentanone with ketone side chain]

24.9   The Robinson Annulation

The word **annulation** comes from the Greek word *annulus* for ring. The Robinson annulation is named for English chemist Sir Robert Robinson, who was awarded the 1947 Nobel Prize in Chemistry.

The Robinson annulation is a ring-forming reaction that combines a Michael reaction with an intramolecular aldol reaction. Like the other reactions in Chapter 24, it involves enolates and it forms carbon–carbon bonds. The two starting materials for a Robinson annulation are an α,β-unsaturated carbonyl compound and an enolate.

Robinson annulation— General reaction

[methyl vinyl ketone] + [cyclohexanone] $\xrightarrow{base}$ [bicyclic enone product with new C–C σ bond and new C–C σ and π bond]

α,β-unsaturated carbonyl compound | carbonyl compound that forms an enolate | new C–C σ and π bond

The Robinson annulation forms a six-membered ring and three new carbon–carbon bonds—two σ bonds and one π bond. The product contains an α,β-unsaturated ketone in a cyclohexane ring—that is, a **2-cyclohexenone** ring. To generate the enolate component of the Robinson annulation, ⁻OH in H_2O and ⁻OEt in EtOH are typically used.

Examples

[1] [structure] + [cyclohexanone] $\xrightarrow{\text{⁻OH}, H_2O}$ [bicyclic product]

[2] [methyl vinyl ketone] + [2-methyl-1,3-cyclohexanedione] $\xrightarrow{\text{⁻OH}, H_2O}$ [bicyclic product]

methyl vinyl ketone   2-methyl-1,3-cyclohexanedione   [New C–C bonds are shown in red.]

The mechanism of the Robinson annulation consists of two parts: a **Michael addition** to the α,β-unsaturated carbonyl compound to form a 1,5-dicarbonyl compound, followed by an **intramolecular aldol reaction** to form the six-membered ring. The mechanism is written out in two parts (Mechanisms 24.7 and 24.8) for Reaction [2] between methyl vinyl ketone and 2-methyl-1,3-cyclohexanedione.

Part [1] illustrates the three-step mechanism for the Michael addition that forms the first carbon–carbon σ bond, generating the 1,5-dicarbonyl compound. The first step always involves removal of the most acidic proton to form an enolate.

MECHANISM 24.7

The Robinson Annulation—Part [1] Michael Addition to Form a 1,5-Dicarbonyl Compound

Step [1] Enolate formation

♦ Base removes the most acidic proton—that is, the proton between the two carbonyl groups—forming the enolate in Step [1].

Steps [2]–[3] Nucleophilic attack at the β carbon and protonation

♦ **Conjugate addition of the enolate to the β carbon** of the α,β-unsaturated carbonyl compound forms a new carbon–carbon bond and a resonance-stabilized enolate.

♦ Protonation of the enolate forms the 1,5-dicarbonyl compound.

In Part [2] of the mechanism, an intramolecular aldol reaction followed by dehydration forms the six-membered ring.

MECHANISM 24.8

The Robinson Annulation—Part [2] Intramolecular Aldol Reaction to Form a 2-Cyclohexenone

Steps [4]–[6] Intramolecular aldol reaction to form a β-hydroxy ketone

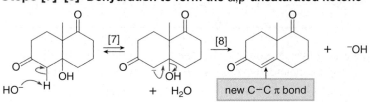

♦ The **intramolecular aldol reaction** consists of three steps: [4] **enolate formation,** [5] **nucleophilic attack,** and [6] **protonation.** This forms another carbon–carbon σ bond and a β-hydroxy carbonyl compound (compare Section 24.4).

Steps [7]–[8] Dehydration to form the α,β-unsaturated ketone

♦ Dehydration consists of two steps: **deprotonation** and **loss of ⁻OH** (Section 24.1B). This reaction forms the new π bond in the α,β-unsaturated ketone.

All of the parts of this mechanism have been discussed in previous sections of Chapter 24. However, the end result of the Robinson annulation—the formation of a 2-cyclohexenone ring—is new.

To draw the product of Robinson annulation without writing out the mechanism each time, place the α carbon of the compound that becomes the enolate next to the β carbon of the α,β-unsaturated carbonyl compound. Then, join the appropriate carbons together as shown. If you follow this method of drawing the starting materials, the double bond in the product always ends up in the same position in the six-membered ring.

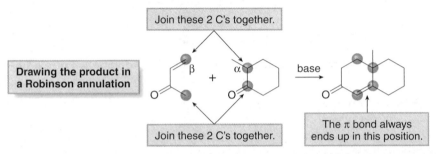

SAMPLE PROBLEM 24.4 Draw the Robinson annulation product formed from the following starting materials.

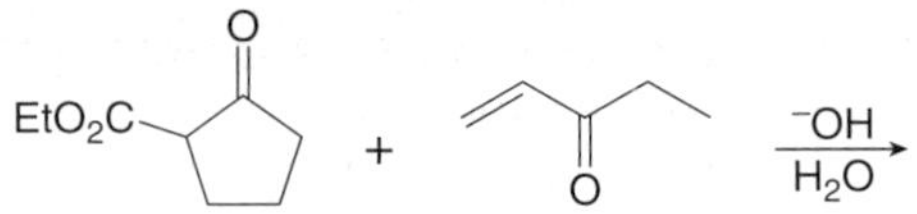

SOLUTION

Arrange the starting materials to put the reactive atoms next to each other. For example:

- Place the α,β-unsaturated carbonyl compound *to the left* of the carbonyl compound.
- Determine which α carbon will become the enolate. The most acidic H is always removed with base first, which in this case is the H on the α carbon between the two carbonyl groups. This α carbon is drawn adjacent to the β carbon of the α,β-unsaturated carbonyl compound.

Then draw the bonds to form the new six-membered ring.

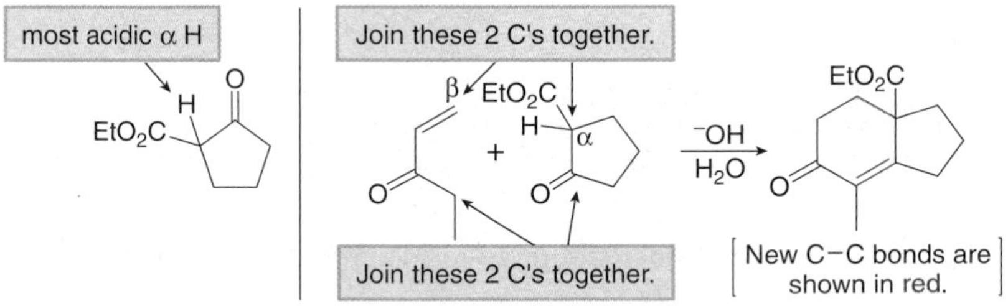

PROBLEM 24.21 Draw the products of each Robinson annulation.

a.

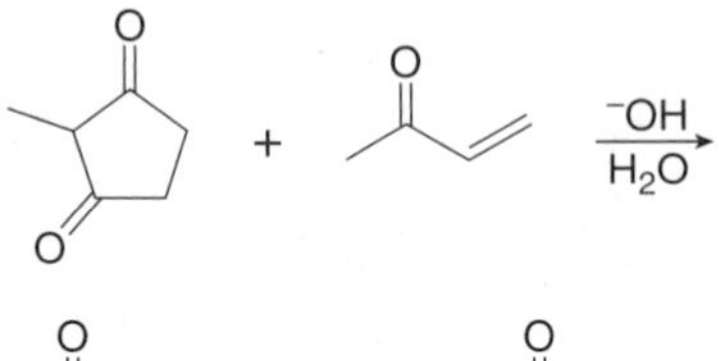

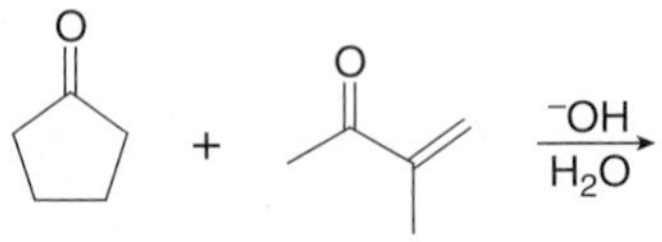

b.

c.

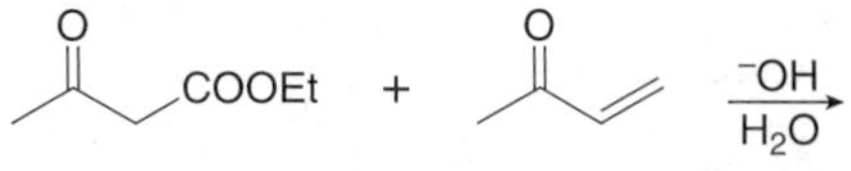

d.

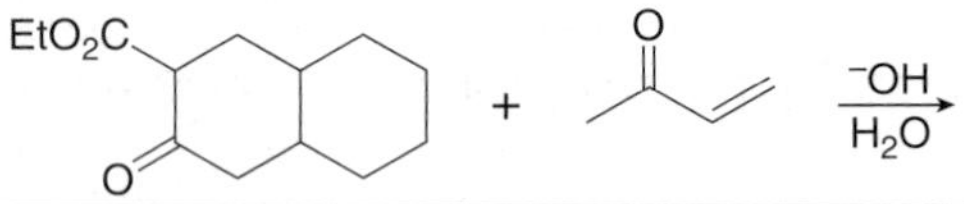

To use the Robinson annulation in synthesis, you must be able to determine what starting materials are needed to prepare a given compound, by working in the retrosynthetic direction.

How To   Synthesize a Compound Using the Robinson Annulation

Example   **What starting materials are needed to synthesize the following compound using a Robinson annulation?**

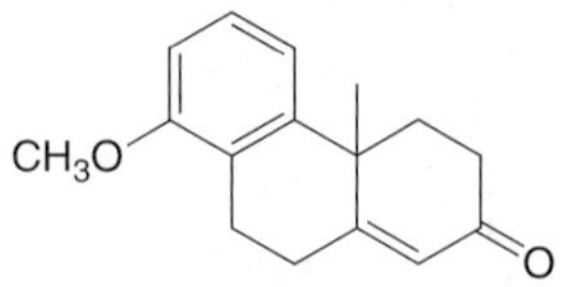

Step [1]   **Locate the 2-cyclohexenone ring and re-draw the target molecule if necessary.**

- To most easily determine the starting materials, always arrange the α,β-unsaturated carbonyl system in the same location. The target compound may have to be flipped or rotated, and you must be careful not to move any bonds to the wrong location during this process.

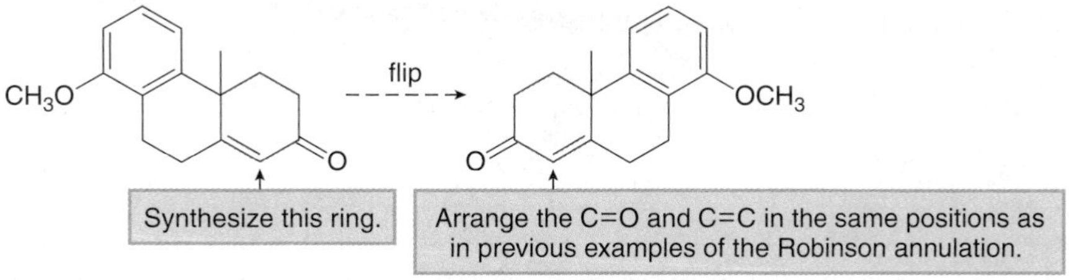

flip

| Synthesize this ring. | Arrange the C=O and C=C in the same positions as in previous examples of the Robinson annulation. |

Step [2]   **Break the 2-cyclohexenone ring into two components.**

- Break the C=C. One half becomes the carbonyl group of the enolate component.
- Break the bond between the β carbon and the carbon to which it is bonded.

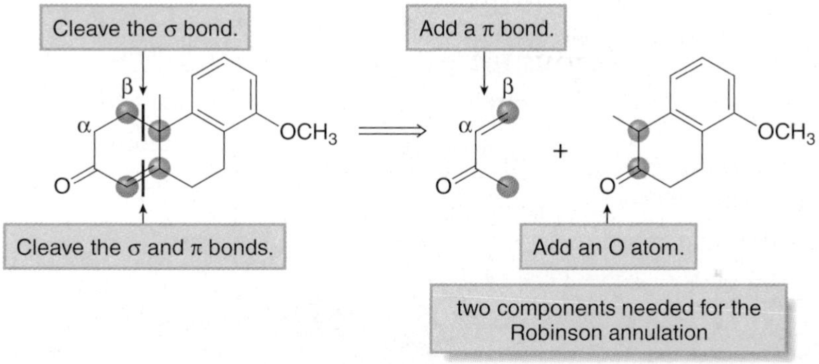

| Cleave the σ bond. | Add a π bond. |

| Cleave the σ and π bonds. | Add an O atom. |

| two components needed for the Robinson annulation |

PROBLEM 24.22   What starting materials are needed to synthesize each compound by a Robinson annulation?

a. 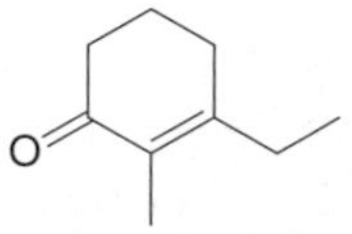   b. 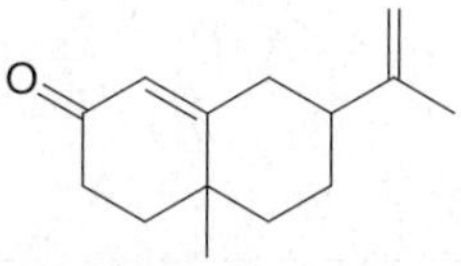   c.

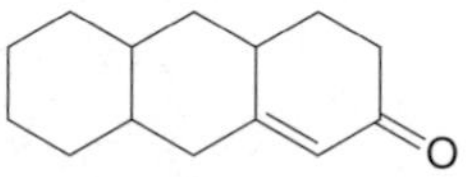

24.10   Key Concepts—Carbonyl Condensation Reactions

The Four Major Carbonyl Condensation Reactions

Reaction type	Reaction [New C−C bonds are shown in red.]		
[1] Aldol reaction (24.1)			
	aldehyde (or ketone)	β-hydroxy carbonyl compound	(E and Z) α,β-unsaturated carbonyl compound

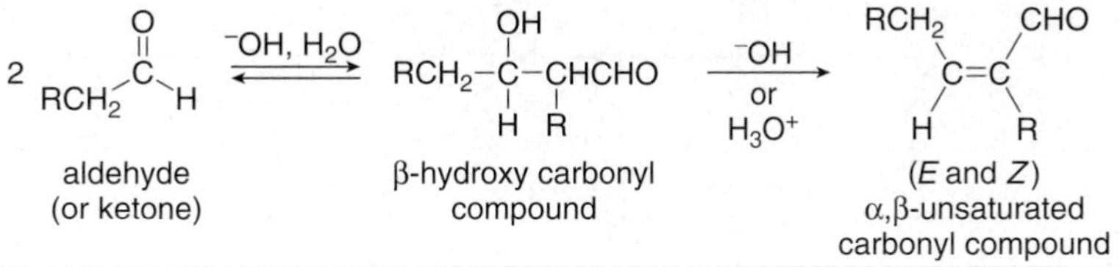

[2] Claisen reaction (24.5)

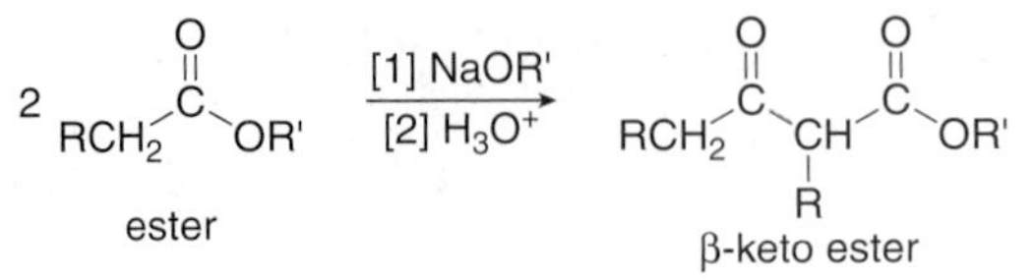

$$2 \; \underset{\text{ester}}{RCH_2\overset{\overset{\displaystyle O}{\|}}{C}OR'} \quad \xrightarrow[\text{[2] } H_3O^+]{\text{[1] NaOR'}} \quad \underset{\underset{\underset{R}{|}}{\beta\text{-keto ester}}}{RCH_2\overset{\overset{\displaystyle O}{\|}}{C}-CH-\overset{\overset{\displaystyle O}{\|}}{C}OR'}$$

[3] Michael reaction (24.8)

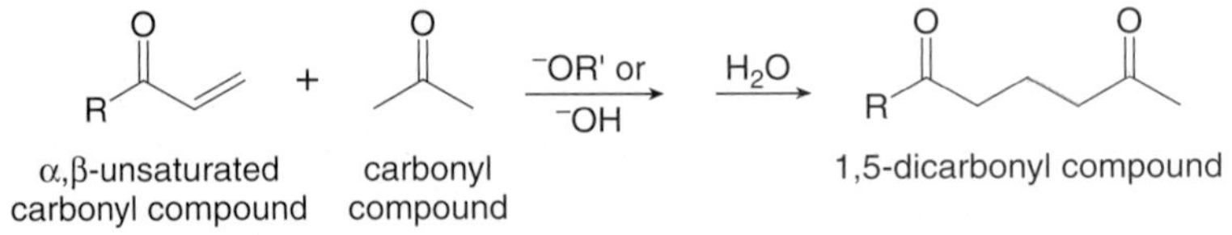

$$\underset{\substack{\alpha,\beta\text{-unsaturated} \\ \text{carbonyl compound}}}{\overset{\overset{\displaystyle O}{\|}}{R-C}=} \; + \; \underset{\substack{\text{carbonyl} \\ \text{compound}}}{\overset{\overset{\displaystyle O}{\|}}{C}} \quad \xrightarrow[\text{}^-\text{OH}]{\text{}^-\text{OR' or}} \; \xrightarrow{H_2O} \quad \underset{\text{1,5-dicarbonyl compound}}{R\overset{\overset{\displaystyle O}{\|}}{C}}$$

[4] Robinson annulation (24.9)

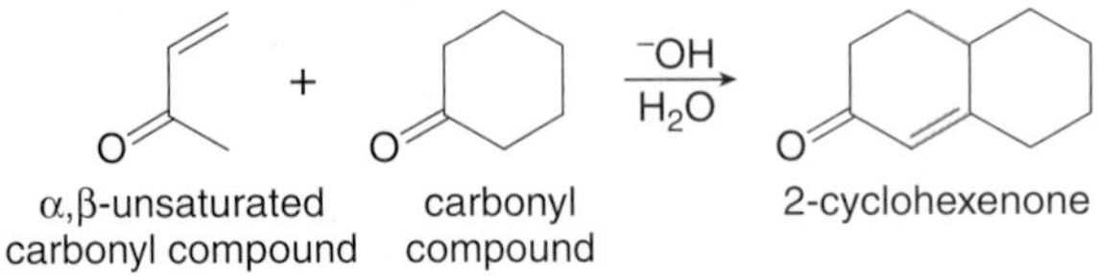

$$+ \quad \xrightarrow[H_2O]{^-\text{OH}} \quad$$

α,β-unsaturated   carbonyl   2-cyclohexenone
carbonyl compound   compound

Useful Variations

[New C−C bonds are shown in red.]

[1] Directed aldol reaction (24.3)

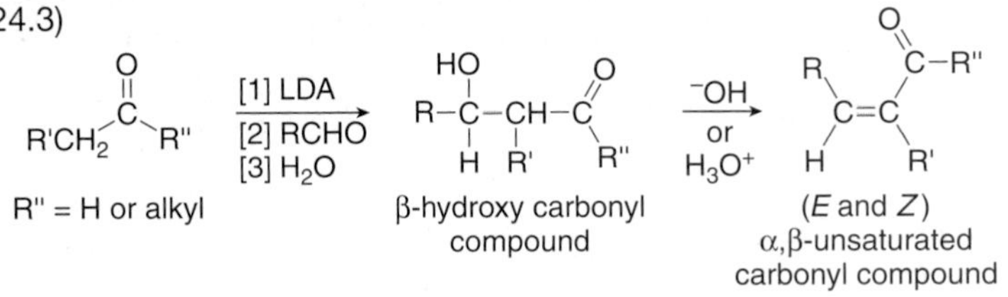

$$\underset{\substack{R'CH_2 \quad R'' \\ R''= H \text{ or alkyl}}}{\overset{\overset{\displaystyle O}{\|}}{C}} \quad \xrightarrow[\substack{\text{[2] RCHO} \\ \text{[3] } H_2O}]{\text{[1] LDA}} \quad \underset{\substack{H \quad R' \\ \beta\text{-hydroxy carbonyl} \\ \text{compound}}}{\overset{HO}{R-C-CH-\overset{\overset{\displaystyle O}{\|}}{C}R''}} \quad \xrightarrow[H_3O^+]{\overset{\text{}^-\text{OH}}{\text{or}}} \quad \underset{\substack{H \quad R' \\ (E \text{ and } Z) \\ \alpha,\beta\text{-unsaturated} \\ \text{carbonyl compound}}}{\overset{\overset{\overset{\displaystyle O}{\|}}{C-R''}}{\underset{}{C=C}}}$$

---

[2] Intramolecular aldol reaction (24.4)

   [a] With 1,4-dicarbonyl compounds:

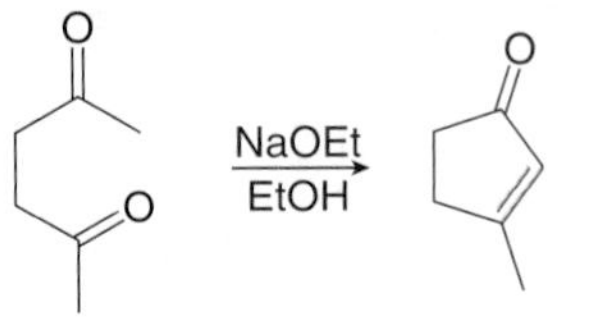

$$\xrightarrow[\text{EtOH}]{\text{NaOEt}}$$

   [b] With 1,5-dicarbonyl compounds:

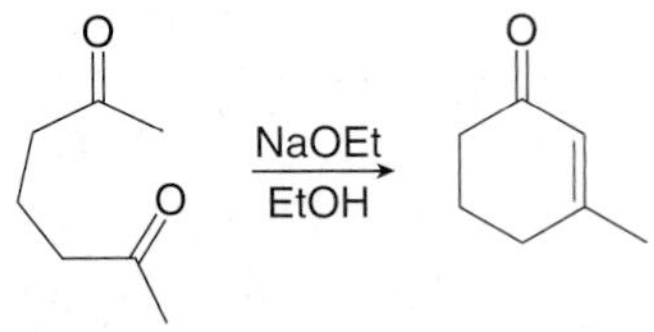

$$\xrightarrow[\text{EtOH}]{\text{NaOEt}}$$

---

[3] Dieckmann reaction (24.7)

   [a] With 1,6-diesters:

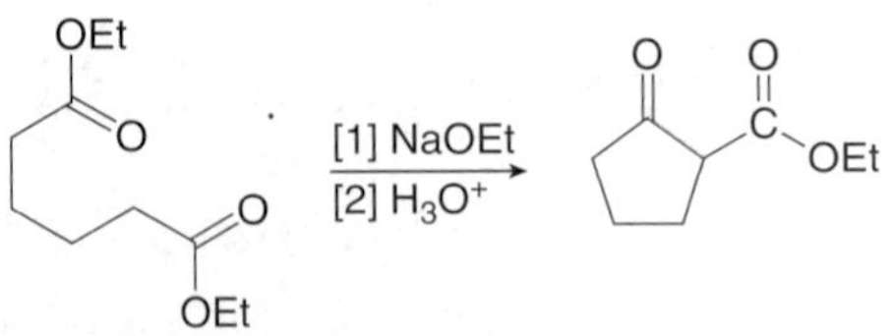

$$\xrightarrow[\text{[2] } H_3O^+]{\text{[1] NaOEt}}$$

   [b] With 1,7-diesters:

$$\xrightarrow[\text{[2] } H_3O^+]{\text{[1] NaOEt}}$$

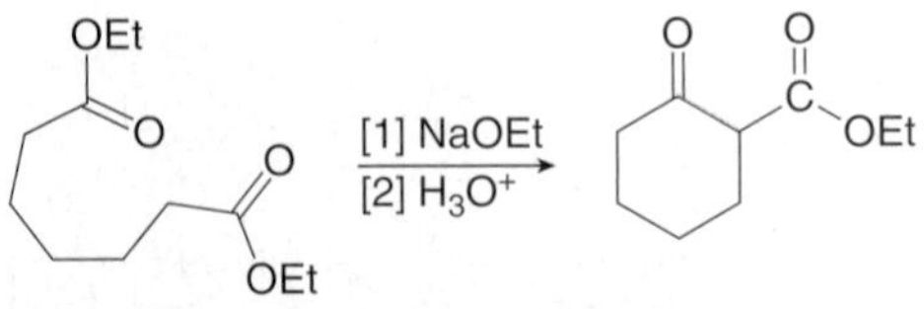

Problems

The Aldol Reaction

24.23 Draw the product formed from an aldol reaction with the given starting material(s) using ⁻OH, H_2O.

    a. $(CH_3)_2CHCHO$ only
    b. $(CH_3)_2CHCHO + CH_2=O$
    c. $C_6H_5CHO + CH_3CH_2CH_2CHO$
    d. $(CH_3CH_2)_2C=O$ only
    e. $(CH_3CH_2)_2C=O + CH_2=O$

    f. (cyclopentanone) $+ C_6H_5CHO$

24.24 What four β-hydroxy aldehydes are formed by a crossed aldol reaction of $CH_3CH_2CH_2CHO$ and $C_6H_5CH_2CHO$?

24.25 Draw the product formed in each directed aldol reaction.

    a. CH_3COCH_3 — [1] LDA / [2] $CH_3CH_2CH_2CHO$ / [3] H_2O

    b. $CH_3CH_2CO_2Et$ — [1] LDA / [2] (tetrahydropyranyl ether with CHO) / [3] H_2O

24.26 Draw the product formed when each dicarbonyl compound undergoes an intramolecular aldol reaction followed by dehydration.

    a. (keto aldehyde) ... CHO
    b. (diketone)
    c. OHC ... CHO

24.27 What starting materials are needed to synthesize each compound using an aldol or similar reaction?

    a. (keto alcohol OH)
    b. (C_6H_5 enone)
    c. (methylenecyclopentanone)
    d. (cyclohexenone with C_6H_5, C_6H_5)
    e. (CH_3-substituted aryl $-CH=CHCN$)

24.28 What dicarbonyl compound is needed to prepare each compound by an intramolecular aldol reaction?

    a. (cyclohexenone)
    b. (methyl cyclopentenone)
    c. (bicyclic enone)
    d. (bicyclic ketone with HO)

The Claisen and Dieckmann Reactions

24.29 Draw the Claisen product formed from each ester.

    a. $C_6H_5CH_2CH_2CH_2CO_2Et$
    b. $(CH_3)_2CHCH_2CH_2CH_2CO_2Et$
    c. CH_3O-(aryl)$-CH_2COOEt$

24.30 What four compounds are formed from the crossed Claisen reaction of $CH_3CH_2CH_2CH_2CO_2Et$ and $CH_3CH_2CO_2Et$?

24.31 Draw the product formed from a Claisen reaction with the given ester(s) using ⁻OEt, EtOH.

    a. $CH_3CH_2CH_2CO_2Et$ only
    b. $CH_3CH_2CH_2CO_2Et + C_6H_5CO_2Et$
    c. $CH_3CH_2CH_2CO_2Et + (CH_3)_2C=O$
    d. $EtO_2CC(CH_3)_2CH_2CH_2CH_2CO_2Et$
    e. $C_6H_5COCH_2CH_3 + C_6H_5CO_2Et$
    f. $CH_3CH_2CO_2Et + (EtO)_2C=O$

    g. (butyrolactone) $+ HCO_2Et$
    h. (cyclopentanone) $+ Cl-CO-OEt$

24.32 What starting materials are needed to synthesize each compound by a crossed Claisen reaction?

    a. CH_3O-(aryl ketone with CO_2Et)
    b. (diketone with C_6H_5)
    c. (keto ... CHO)
    d. $C_6H_5CH(COOEt)_2$

24.33 The 1,3-diketone shown below can be prepared by two different Claisen reactions—namely, one that forms bond (a) and one that forms bond (b). What starting materials are needed for each of these reactions?

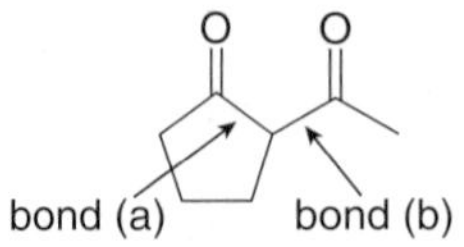

bond (a)   bond (b)

24.34 The Dieckmann reaction of $EtO_2C(CH_2)_4CH(CH_3)CO_2Et$ yields only one of the two possible cyclic β-keto esters. Draw structures for both possible products and explain why only one is formed.

Michael Reaction

24.35 Draw the product formed from a Michael reaction with the given starting materials using ⁻OEt, EtOH.

a. 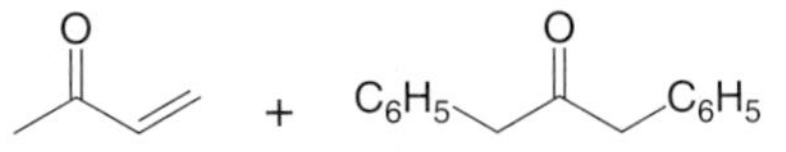 + C_6H_5 C_6H_5

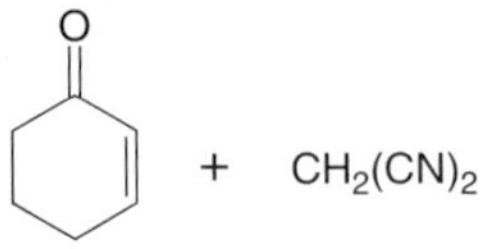

c. + $CH_2(CN)_2$

b. 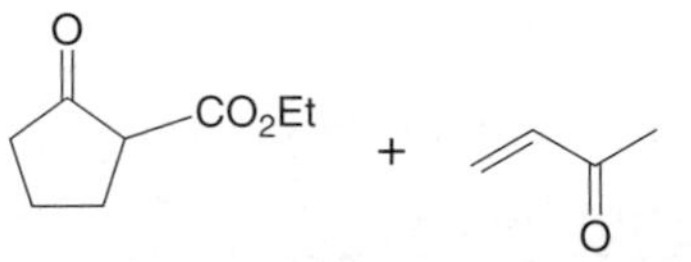 +

d. 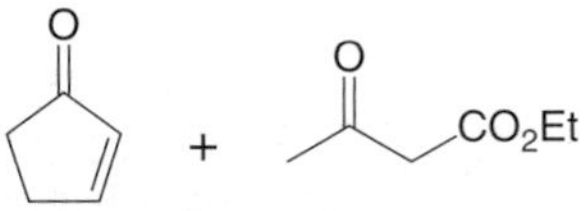 + CO_2Et

24.36 What starting materials are needed to prepare each compound using a Michael reaction?

a.

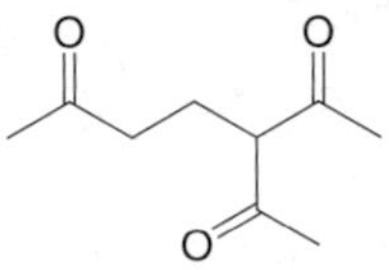

b.

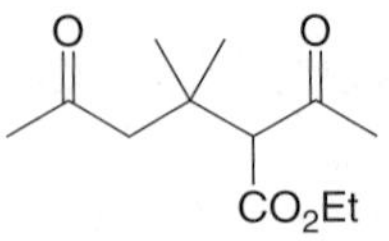

c.

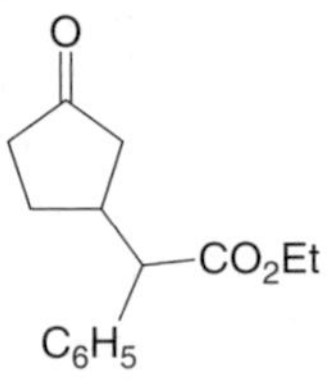

d.

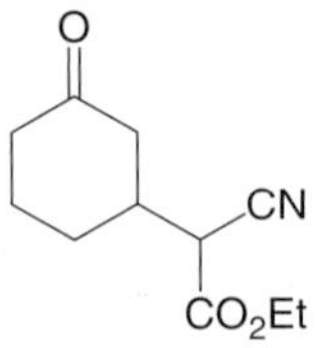

24.37 In Section 23.8D, the synthesis of β-vetivone using an intramolecular alkylation reaction was described. In another synthesis, ketone **A** is converted to β-vetivone by a two-step process: Michael reaction, followed by intramolecular aldol reaction. What Michael acceptor is needed for the conjugate addition?

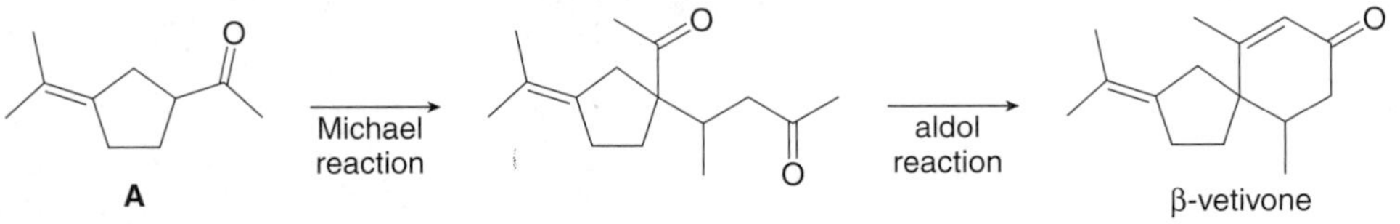

A   Michael reaction   aldol reaction   β-vetivone

Robinson Annulation

24.38 Draw the product of each Robinson annulation from the given starting materials using ⁻OH in H_2O solution.

a.

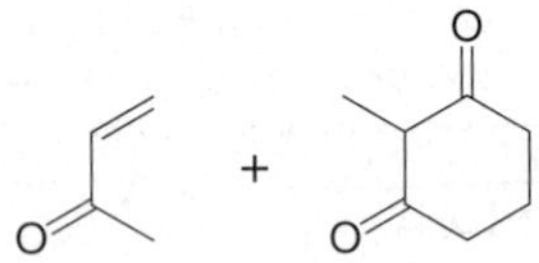

c.

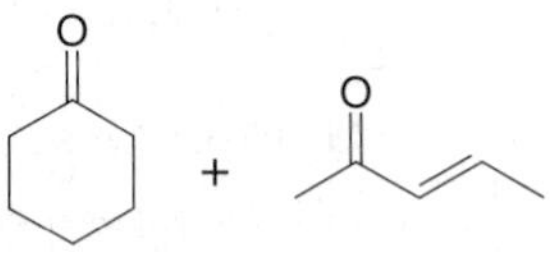

b.

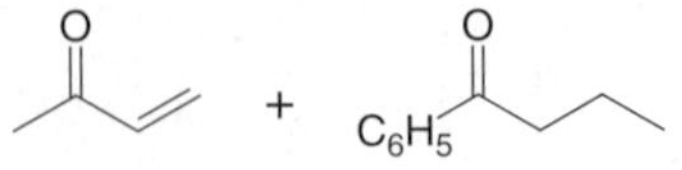

d.

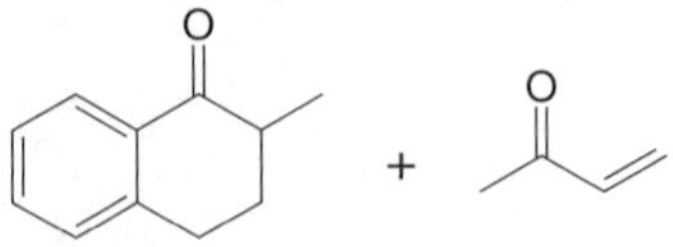

24.39 What starting materials are needed to synthesize each compound using a Robinson annulation?

a.

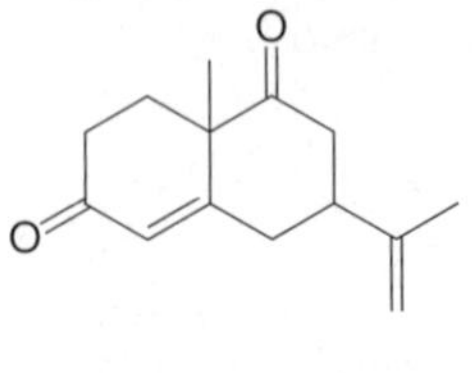

b.

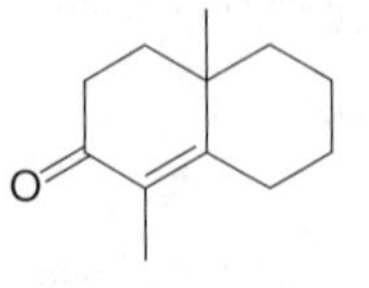

c.

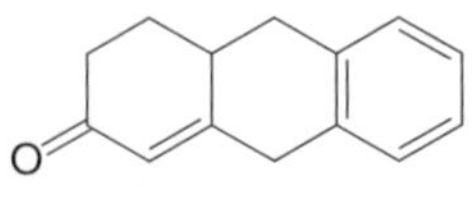

d.

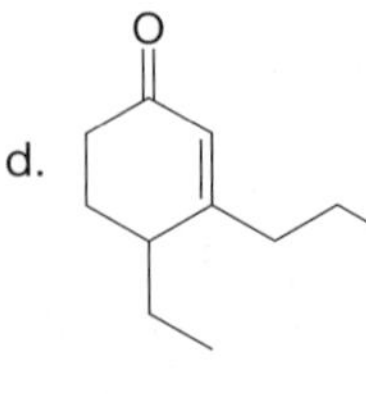

Reactions

24.40 Draw the organic products formed when butanal ($CH_3CH_2CH_2CHO$) is treated with each reagent.

a. ⁻OH, H_2O
b. ⁻OH, CH_2=O, H_2O
c. [1] LDA; [2] CH_3CHO; [3] H_2O
d. $CH_2(CO_2Et)_2$, NaOEt, EtOH
e. [1] CH_3Li; [2] H_2O

f. $NaBH_4$, CH_3OH
g. H_2, Pd-C
h. $HOCH_2CH_2OH$, TsOH
i. CH_3NH_2, mild acid
j. $(CH_3)_2NH$, mild acid

k. CrO_3, H_2SO_4
l. Br_2, CH_3COOH
m. Ph_3P=CH_2
n. NaCN, HCl
o. [1] LDA; [2] CH_3I

24.41 Draw the organic products formed in each reaction.

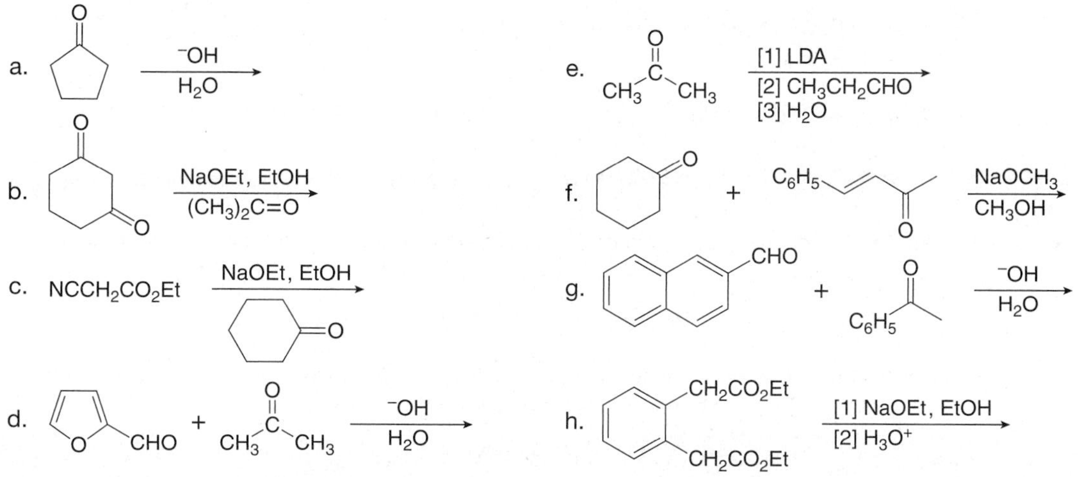

a.

b.

c. $NCCH_2CO_2Et$

d.

e.

f.

g.

h.

24.42 Fill in the lettered reagents needed for each reaction.

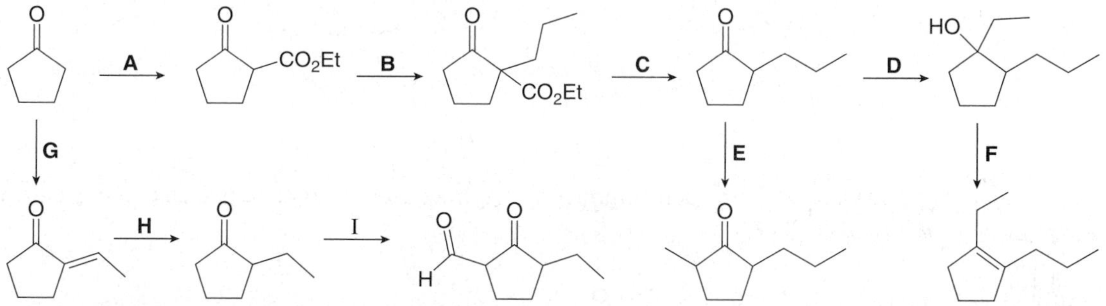

24.43 The 1979 synthesis of gibberellic acid (chapter opener) by Corey and Smith included these two steps. Identify products **A** and **B** in this reaction sequence.

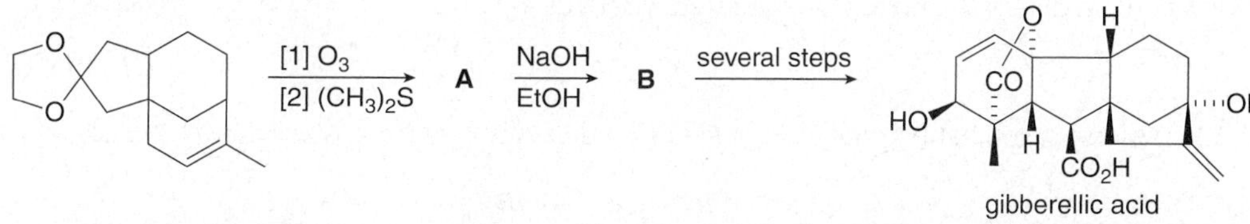

gibberellic acid

24.44 An intramolecular aldol reaction of the 1,4-dicarbonyl compound **A** forms *cis*-jasmone, a perfume component isolated from jasmine flowers. Why is *cis*-jasmone formed as the major product and not its constitutional isomer **B**?

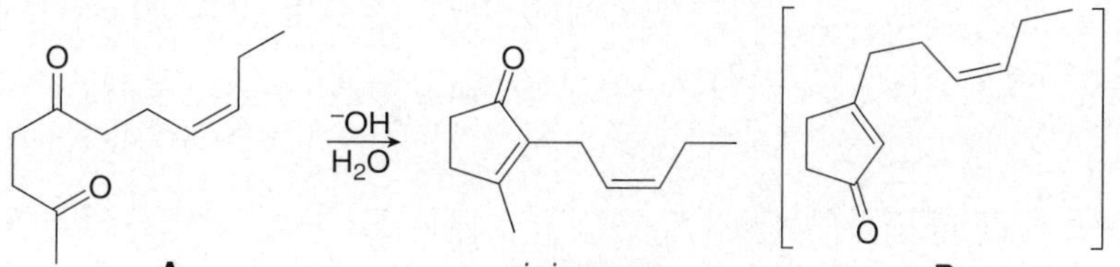

A   *cis*-jasmone   **B**
not formed

Mechanisms

24.45 When acetaldehyde (CH_3CHO) is treated with three equivalents of formaldehyde (CH_2=O) in the presence of aqueous Na_2CO_3, $(HOCH_2)_3CCHO$ is formed as product. Draw a stepwise mechanism for this process.

24.46 Draw a stepwise mechanism for each cyclization reaction.

a. 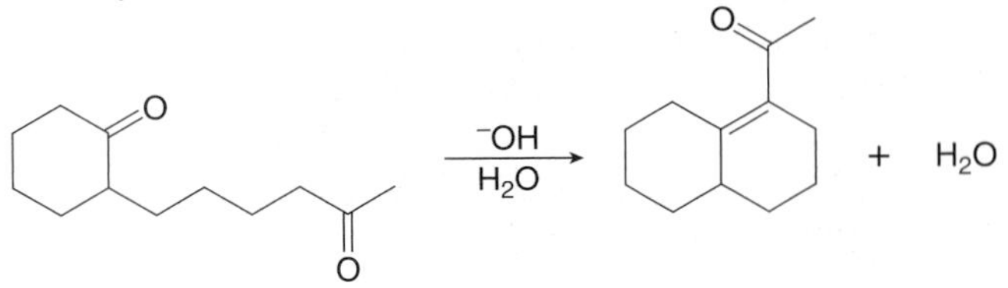 $\xrightarrow[\text{H}_2\text{O}]{\text{-OH}}$ + H$_2$O

b. 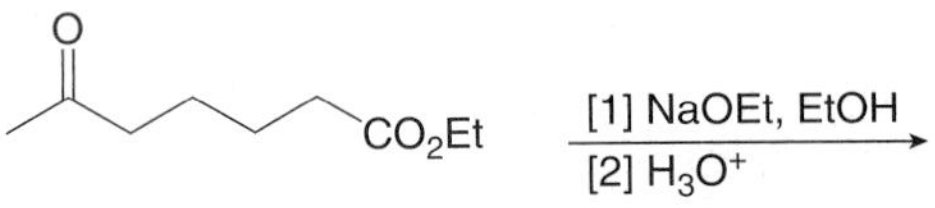 $\xrightarrow[\text{[2] H}_3\text{O}^+]{\text{[1] NaOEt, EtOH}}$

c. 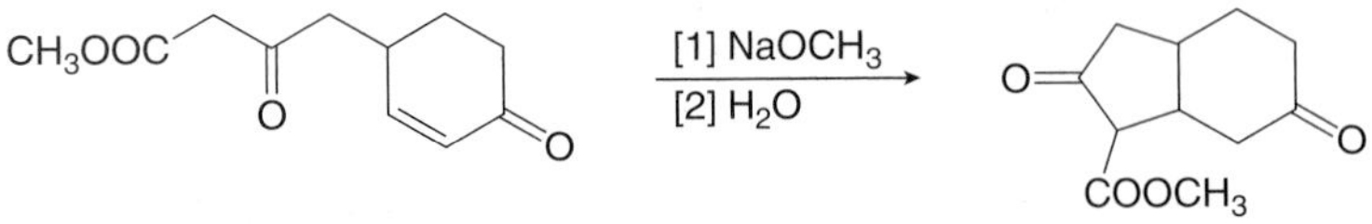 $\xrightarrow[\text{[2] H}_2\text{O}]{\text{[1] NaOCH}_3}$

24.47 Draw a stepwise mechanism for the following variation of the aldol reaction, often called a nitro aldol reaction.

+ CH$_3$NO$_2$ $\xrightarrow[\text{H}_2\text{O}]{\text{-OH}}$ C$_6$H$_5$CH=CHNO$_2$

24.48 Draw a stepwise mechanism for the following Robinson annulation. This reaction was a key step in a synthesis of the steroid cortisone by R. B. Woodward and co-workers at Harvard University in 1951.

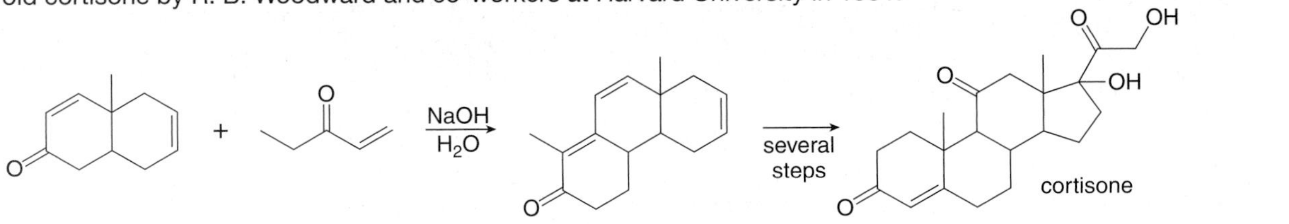

+ $\xrightarrow[\text{H}_2\text{O}]{\text{NaOH}}$ $\xrightarrow{\text{several steps}}$ cortisone

Synthesis

24.49 Convert acetophenone (C$_6$H$_5$COCH$_3$) into each compound. In some cases, more than one step is required. You may use any other organic or inorganic reagents.

a. 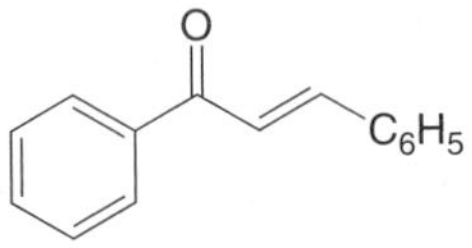   b. 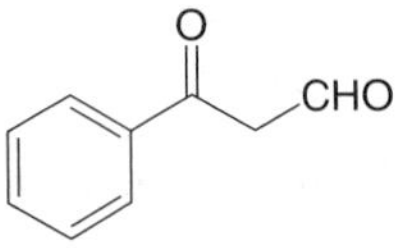   c. 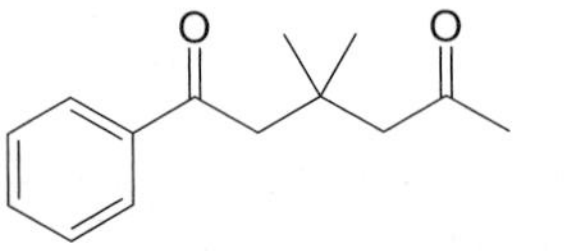   d.

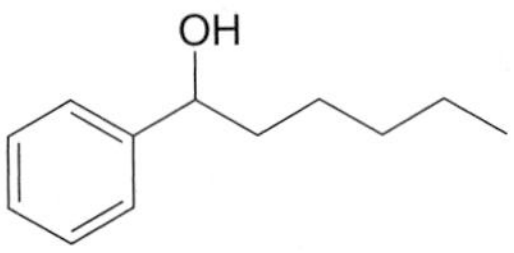

24.50 How would you convert alkene **A** into α,β-unsaturated aldehyde **B**?

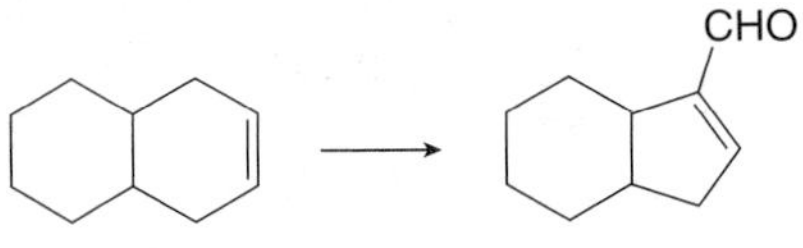

A   B

24.51 Synthesize each compound from cyclohexanone and any other organic or inorganic reagents. More than one step may be required.

a. 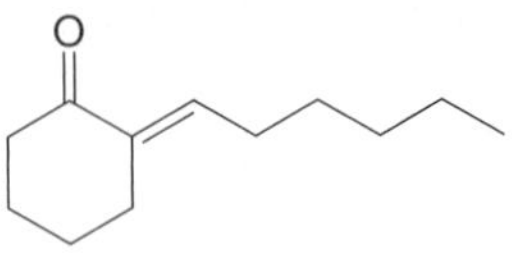   c. 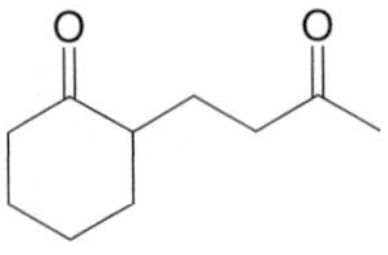   e. 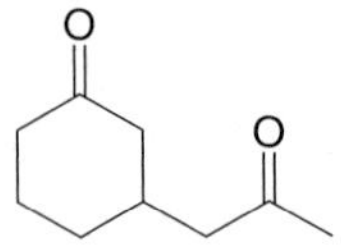   g.

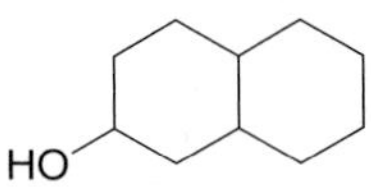

b. 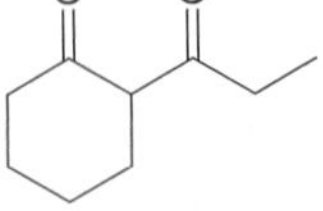   d. 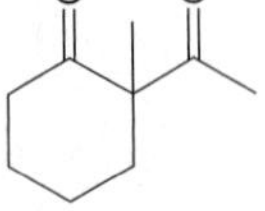   f. 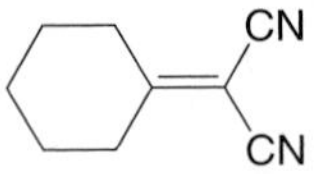   h.

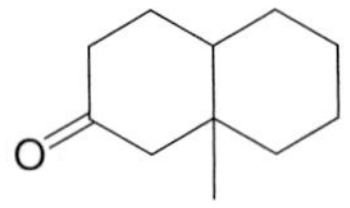

24.52 Synthesize each compound from cyclopentanone.

a.   b.

24.53 A retrosynthetic analysis for ketone **A** from acetone [(CH$_3$)$_2$C=O] and diethyl malonate [CH$_2$(CO$_2$Et)$_2$] is given in the following scheme. Following this plan, design a synthesis of **A**. Write out all steps and indicate all needed reagents.

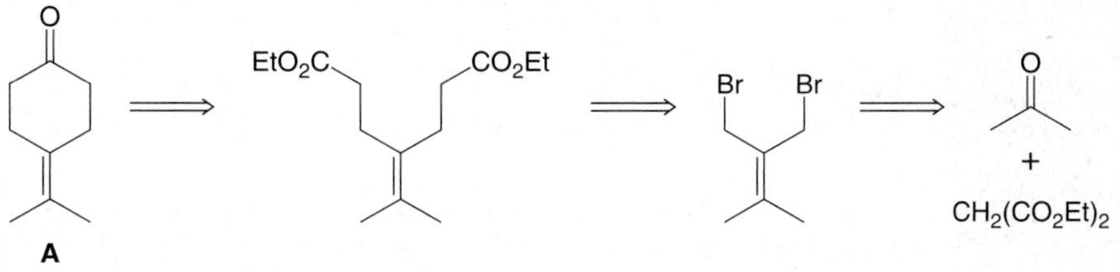

A

Challenge Problems

24.54 All Robinson annulations in Section 24.9 form a six-membered ring in which four carbons come from the α,β-unsaturated carbonyl compound and two carbons come from the enolate. A less common type of Robinson annulation forms a six-membered ring in which three carbons come from the α,β-unsaturated carbonyl compound and three carbons come from the enolate. Draw a stepwise mechanism for this type of Robinson annulation, shown in the following equation:

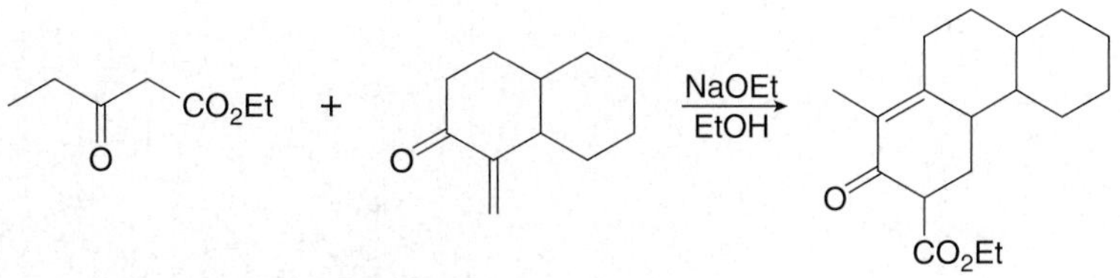

$$\xrightarrow[\text{EtOH}]{\text{NaOEt}}$$

24.55 Isophorone is formed from three molecules of acetone [(CH$_3$)$_2$C=O] in the presence of base. Draw a mechanism for this process.

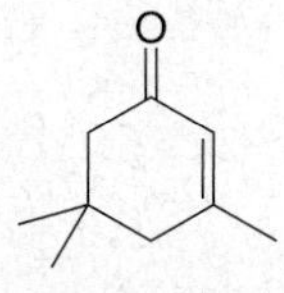

isophorone

24.56 Draw a stepwise mechanism for the following reaction. [*Hint:* Two Michael reactions are needed.]

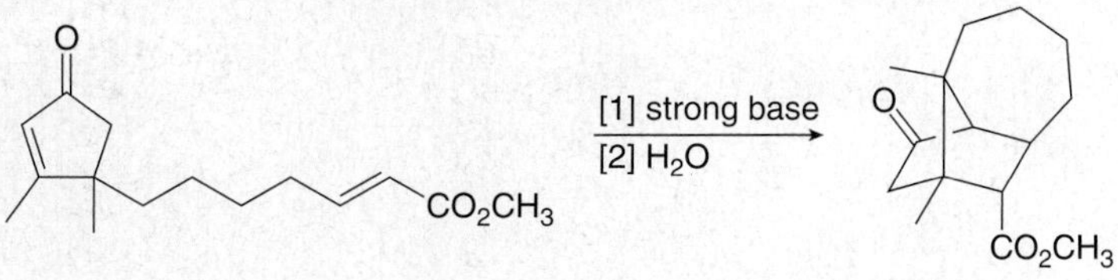

$$\xrightarrow[\text{[2] H}_2\text{O}]{\text{[1] strong base}}$$

Amines

The synthesis of the purple dye **mauveine** in 1856 marked the beginning of the chemical industry. Until that time all dyes were obtained from plant or animal sources, making them scarce and expensive, and thus available only to the wealthy. William Henry Perkin, an 18-year-old student with a makeshift home laboratory, changed this when he serendipitously prepared a colored synthetic dye during his failed attempt to synthesize the antimalarial drug quinine. Perkin realized the potential economic value of his discovery, and within a few years built the first chemical factory in West London, to produce the dye on a large scale. Mauveine is a mixture of two compounds, both of which contain several nitrogen atoms. In Chapter 25 we learn about the chemistry of amines, one type of organic compound that contains nitrogen atoms.

W e now leave the chemistry of carbonyl compounds to concentrate on **amines,** organic compounds that contain an sp^3 hybridized nitrogen atom. Amines are organic derivatives of ammonia (NH_3), formed by replacing one or more hydrogen atoms by alkyl or aryl groups. **Amines are stronger bases and better nucleophiles than other neutral organic compounds,** so much of Chapter 25 focuses on these properties.

Like that of alcohols, the chemistry of amines does not always fit neatly into one reaction class, and this can make learning the reactions of amines challenging. Many interesting natural products and widely used drugs are amines, so you also need to know how to introduce this functional group into organic molecules.

25.1 Introduction

Classifying amines as 1°, 2°, or 3° is reminiscent of classifying amides in Chapter 22, but is *different* from classifying other atoms and functional groups as 1°, 2°, and 3°. Compare, for example, a 2° amine and a 2° alcohol. A 2° amine (R_2NH) has *two* C–N bonds. A 2° alcohol (R_2CHOH), on the other hand, has only *one* C–O bond, but *two* C–C bonds on the carbon bonded to oxygen.

Amines **are organic nitrogen compounds,** formed by replacing one or more hydrogen atoms of ammonia (NH_3) with alkyl groups. As discussed in Section 21.11, amines are classified as 1°, 2°, or 3° by the number of alkyl groups bonded to the *nitrogen* atom.

$$R-\overset{\cdot\cdot}{N}-H$$
$$\underset{H}{\overset{|}{}}$$
1° amine
(1 R group on N)

$$R-\overset{\cdot\cdot}{N}-H$$
$$\underset{R}{\overset{|}{}}$$
2° amine
(2 R groups on N)

$$R-\overset{\cdot\cdot}{N}-R$$
$$\underset{R}{\overset{|}{}}$$
3° amine
(3 R groups on N)

Like ammonia, **the amine nitrogen atom has a nonbonded electron pair,** making it both a base and a nucleophile. As a result, amines react with electrophiles to form **quaternary ammonium salts**—compounds with four bonds to nitrogen.

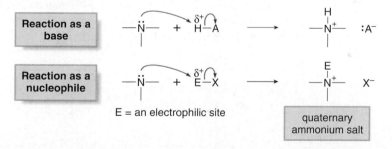

E = an electrophilic site

quaternary ammonium salt

◆ The chemistry of amines is dominated by the nonbonded electron pair on the nitrogen atom.

PROBLEM 25.1 Classify each amine in the following compounds as 1°, 2°, or 3°.

a. H_2N ⟶ N–H ⟶ N–H ⟶ NH_2
spermine
(isolated from semen)

b. CH_3CH_2O—C(=O)—(ring, C_6H_5)—N—CH_3
meperidine
(a narcotic)
Trade name: Demerol

c. (ring with methylenedioxy) ⟶ $NHCH_3$
MDMA
3,4-methylenedioxy-
methamphetamine
(an illegal stimulant)
"Ecstasy"

25.2   Structure and Bonding

An amine nitrogen atom is surrounded by three atoms and one nonbonded electron pair, making the N atom sp^3 hybridized and trigonal pyramidal, with bond angles of approximately 109.5°.

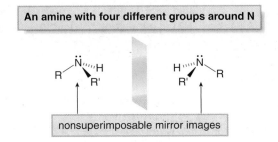

Because nitrogen is much more electronegative than carbon or hydrogen, **the C–N and N–H bonds are all polar,** with the N atom electron rich and the C and H atoms electron poor. The electrostatic potential maps in Figure 25.1 show the polar C–N and N–H bonds in CH_3NH_2 (methylamine) and $(CH_3)_3N$ (trimethylamine).

An amine nitrogen atom bonded to an electron pair and three different alkyl groups is technically a stereogenic center, so two nonsuperimposable trigonal pyramids can be drawn.

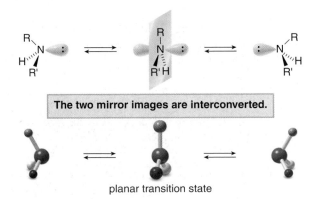

This does not mean, however, that such an amine exists as two different enantiomers, because one is rapidly converted to the other at room temperature. The amine flips inside out, passing through a trigonal planar (achiral) transition state. **Because the two enantiomers interconvert, we can ignore the chirality of the amine nitrogen.**

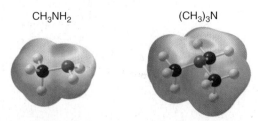

Figure 25.1   Electrostatic potential plots of CH_3NH_2 and $(CH_3)_3N$

CH_3NH_2 $(CH_3)_3N$

• Both amines clearly show the electron-rich region (in red) at the N atom.

In contrast, the chirality of a quaternary ammonium salt with four different groups cannot be ignored. Because there is no nonbonded electron pair on the nitrogen atom, interconversion cannot occur, and the N atom is just like a carbon atom with four different groups around it.

| Two enantiomers of a quaternary ammonium salt |

◆ **The N atom of a quaternary ammonium salt is a stereogenic center when N is surrounded by four different groups.**

PROBLEM 25.2   Label the stereogenic centers in each molecule.

Atropine, which blocks the action of acetylcholine, a neurotransmitter (Section 22.17), has a number of clinical uses. It increases heart rate and dilates the pupils, but in high doses can cause confusion and eventually death.

a. $CH_3 \overset{+}{N}-CH_2CH_2-\overset{+}{N}CH_2CH_3$

b.   dobutamine
(heart stimulant used in stress tests to measure cardiac fitness)

c.   atropine
(isolated from *Atropa belladona,* the poisonous nightshade plant)

PROBLEM 25.3   The $C-N$ bond length in CH_3NH_2 is 1.47 Å, whereas the $C-N$ bond length in $C_6H_5NH_2$ is only 1.40 Å. Why is the $C-N$ bond in the aromatic amine shorter?

25.3   Nomenclature

25.3A   Primary Amines

Primary amines are named using either systematic or common names.

◆ To assign the **systematic name,** find the longest continuous carbon chain bonded to the amine nitrogen, and change the *-e* ending of the parent alkane to the suffix *-amine.* Then use the usual rules of nomenclature to number the chain and name the substituents.

◆ To assign a **common name,** name the alkyl group bonded to the nitrogen atom and add the word *amine,* forming a single word.

| Examples | CH_3NH_2 |

Systematic name: **methanamine**
Common name: **methylamine**

Systematic name: **cyclohexanamine**
Common name: **cyclohexylamine**

25.3B   Secondary and Tertiary Amines

Secondary and tertiary amines having identical alkyl groups are named by using the prefix **di-** or **tri-** with the name of the primary amine.

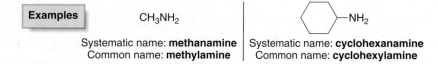

triethylamine   diisopropylamine

Secondary and tertiary amines having more than one kind of alkyl group are named as *N*-**substituted primary amines,** using the following procedure.

How To   Name 2° and 3° Amines with Different Alkyl Groups

Example   Name the following 2° amine: $(CH_3)_2CHNHCH_3$.

Step [1]   Designate the longest alkyl chain (or largest ring) bonded to the N atom as the parent amine and assign a common or systematic name.

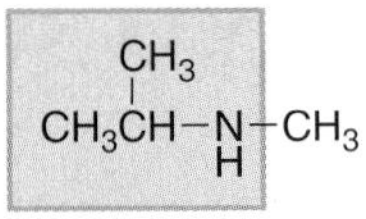

$$CH_3$$
$$|$$
$$CH_3CH-N-CH_3$$
$$\overset{|}{H}$$

3 C's in the longest chain

→   isopropylamine (common name)
or
2-propanamine (systematic name)

Step [2]   Name the other groups on the N atom as alkyl groups, alphabetize the names, and put the prefix *N-* before the name.

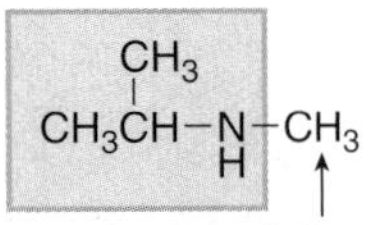

$$CH_3$$
$$|$$
$$CH_3CH-N-CH_3$$
$$\overset{|}{H}$$
↑
one methyl substituent

Answer: *N*-methylisopropylamine (common name)
or
***N*-methyl-2-propanamine (systematic name)**

SAMPLE PROBLEM 25.1   Name each amine.

a. 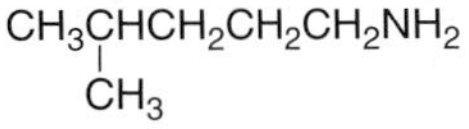 $CH_3CHCH_2CH_2CH_2NH_2$
         $|$
         CH_3

b. 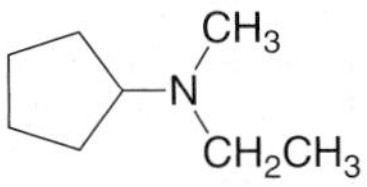   [cyclopentane ring]—N
                   CH_3
                   CH_2CH_3

SOLUTION

a.   [1]   A 1° amine: Find and name the longest chain containing the amine nitrogen.

$$CH_3CHCH_2CH_2CH_2NH_2$$
$$|$$
$$CH_3$$

pentane → **pentan*amine***
(5 C's)

[2]   Number and name the substituents:

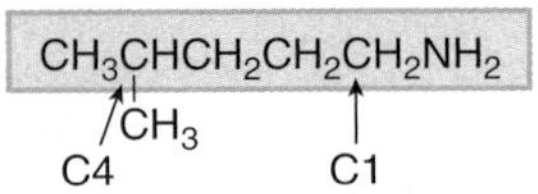

$$CH_3CHCH_2CH_2CH_2NH_2$$
$$\nearrow$$
$$CH_3$$
$$C4 \qquad C1$$

You must use a number to show the location of the NH_2 group.

Answer: 4-methyl-1-pentanamine

---

b.   For a 3° amine, one alkyl group on N is the principal R group and the others are substituents.

[1]   Name the ring bonded to the N:

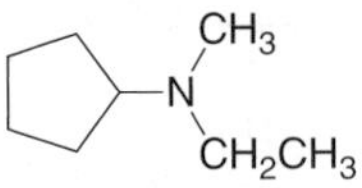

[cyclopentane ring]—N
                   CH_3
                   CH_2CH_3

cyclopentanamine
or
cyclopentylamine

[2]   Name the substituents:

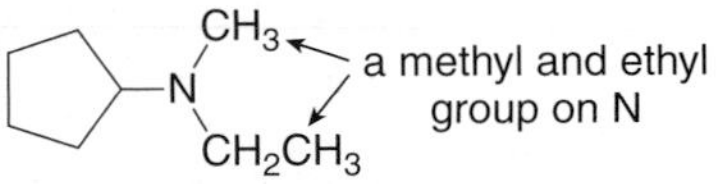

[cyclopentane ring]—N
                   CH_3 ← a methyl and ethyl group on N
                   CH_2CH_3

Two N's are needed, one for each alkyl group.

Answer: *N*-ethyl-*N*-methylcyclopentanamine
or
***N*-ethyl-*N*-methylcyclopentylamine**

PROBLEM 25.4   Name each amine.

a. $CH_3CH_2CH(NH_2)CH_3$

b. $(CH_3CH_2CH_2CH_2)_2NH$

c. 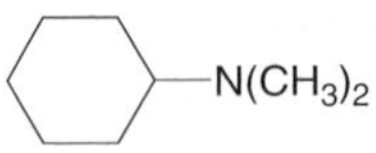   [cyclohexane ring]—$N(CH_3)_2$

d. 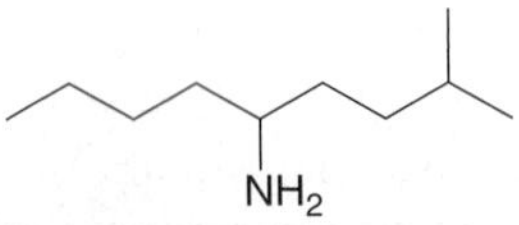   [chain structure with NH_2]

e. 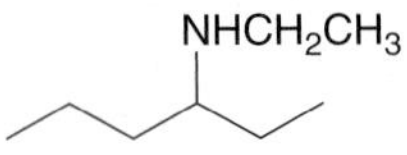   [chain structure with $NHCH_2CH_3$]

f. 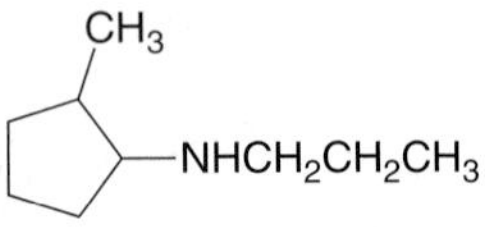   [cyclopentane ring with CH_3]—$NHCH_2CH_2CH_3$

25.3C   Aromatic Amines

Aromatic amines are named as derivatives of aniline.

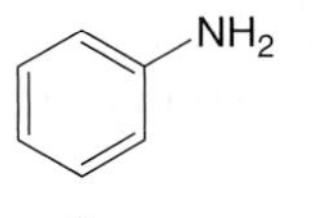

aniline 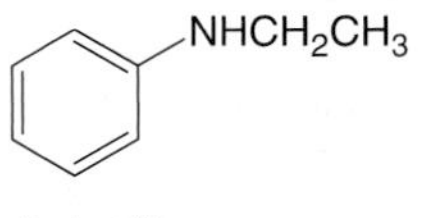   *N*-ethylaniline 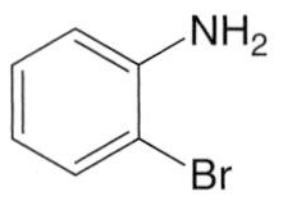   *o*-bromoaniline

---

PROBLEM 25.5   Draw a structure corresponding to each name: (a) *N*-methylaniline; (b) *m*-ethylaniline; (c) 3,5-diethylaniline; (d) *N,N*-diethylaniline.

25.3D   Miscellaneous Nomenclature Facts

An **NH₂** group named as a substituent is called an **amino group.**

There are many different **nitrogen heterocycles,** and each ring type is named differently depending on the number of N atoms in the ring, the ring size, and whether it is aromatic or not. The structures and names of four common nitrogen heterocycles are shown. In numbering these heterocycles, the **N atom is always placed at the "1" position.**

pyridine   pyrrole   piperidine   pyrrolidine

PROBLEM 25.6   Draw a structure corresponding to each name.

a. 2,4-dimethyl-3-hexanamine
b. *N*-methylpentylamine
c. *N*-isopropyl-*p*-nitroaniline
d. *N*-methylpiperidine

e. *N,N*-dimethylethylamine
f. 2-aminocyclohexanone
g. 1-propylcyclohexylamine
h. *p*-butyl-*N*-ethylaniline

25.4   Physical Properties

Amines exhibit dipole–dipole interactions because of the polar C−N and N−H bonds. **Primary and secondary amines are also capable of intermolecular hydrogen bonding,** because they contain N−H bonds. Because nitrogen is less electronegative than oxygen, however, intermolecular hydrogen bonds between N and H are *weaker* than those between O and H. How these factors affect the physical properties of amines is summarized in Table 25.1.

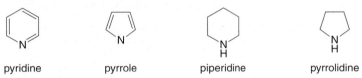

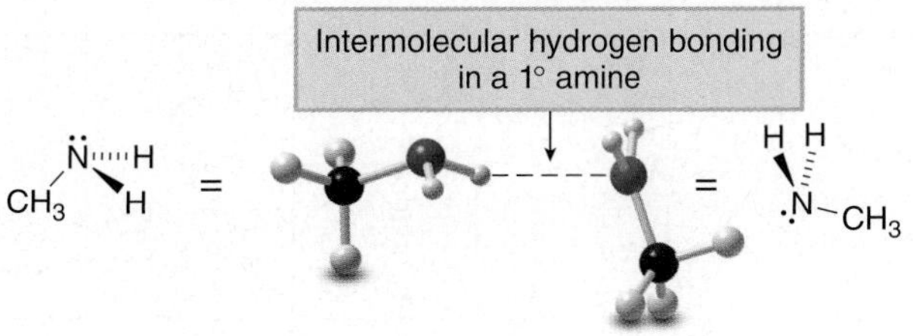

Intermolecular hydrogen bonding in a 1° amine

PROBLEM 25.7   Arrange each group of compounds in order of increasing boiling point.

a.

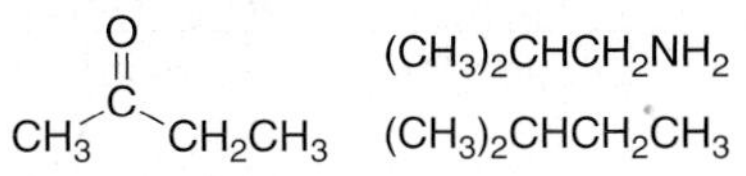

b.

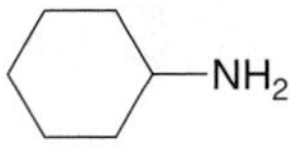

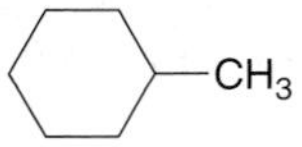

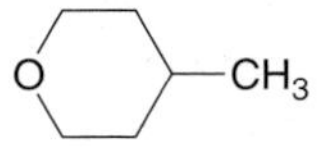

TABLE 25.1   Physical Properties of Amines

Property	Observation
Boiling point and melting point	• Primary (1°) and 2° amines have higher bp's than similar compounds (like ethers) incapable of hydrogen bonding, but lower bp's than alcohols that have stronger intermolecular hydrogen bonds.

$CH_3CH_2OCH_2CH_3$ $CH_3CH_2CH_2CH_2NH_2$ $CH_3CH_2CH_2CH_2OH$

MW = 74   MW = 73   MW = 74
bp 38 °C   bp 78 °C   bp 117 °C

Increasing intermolecular forces
Increasing boiling point

• Tertiary (3°) amines have lower boiling points than 1° and 2° amines of comparable molecular weight, because they have no N–H bonds and are incapable of hydrogen bonding.

| 3° amine | $CH_3CH_2N(CH_3)_2$ | $CH_3CH_2\text{–}\underset{H}{N}\text{–}CH_2CH_3$ ← | 2° amine higher bp |

MW = 73   MW = 73
bp 38 °C   bp 56 °C
no N–H bond   **N–H bond**

Solubility	• Amines are soluble in organic solvents regardless of size.
	• All amines having ≤ 5 C's are H_2O soluble because they can hydrogen bond with H_2O (Section 3.5C).
	• Amines having > 5 C's are H_2O insoluble because the nonpolar alkyl portion is too large to dissolve in the polar H_2O solvent.

MW = molecular weight

25.5   Spectroscopic Properties

Amines exhibit characteristic features in their mass spectra, IR spectra, and ^1H and ^{13}C NMR spectra.

25.5A   Mass Spectra

Amines differ from compounds that contain only C, H, and O atoms, which always have a molecular ion with an *even* mass in their mass spectra.

> The general molecular formula for an amine with one N atom is $C_nH_{2n+3}N$.

◆ **Amines with an odd number of N atoms give an odd molecular ion in their mass spectra.**

This is apparent in the mass spectrum of butylamine, which is shown in Figure 25.2.

Figure 25.2   Mass spectrum of butylamine

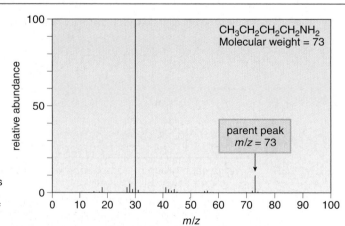

$CH_3CH_2CH_2CH_2NH_2$
Molecular weight = 73

parent peak
m/z = 73

• The molecular ion for $CH_3CH_2CH_2CH_2NH_2$ occurs at m/z = 73. This odd mass for a molecular ion is characteristic of an amine with an odd number of N atoms.

Figure 25.3   The single bond region of the IR spectra for a 1°, 2°, and 3° amine

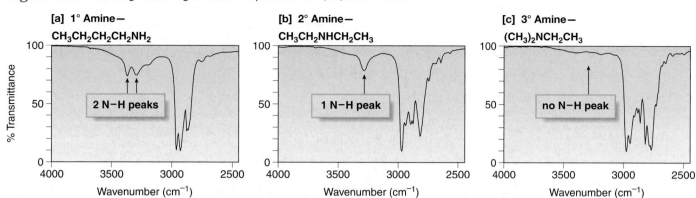

25.5B   IR Spectra

Amines with N–H bonds show characteristic absorptions in their IR spectra.

◆ 1° Amines show *two* N–H absorptions at 3300–3500 cm^{-1}.
◆ 2° Amines show *one* N–H absorption at 3300–3500 cm^{-1}.

Because 3° amines have no N–H bonds, they do *not* absorb in this region in their IR spectra. The single bond region (> 2500 cm^{-1}) of the IR spectra for 1°, 2°, and 3° amines illustrates these features in Figure 25.3.

PROBLEM 25.8   Only one amine shows a molecular ion in its mass spectrum at *m/z* = 59 and has one peak in its IR spectrum at ~3300 cm^{-1}. What is its structure?

PROBLEM 25.9   Draw the structures of the eight isomeric amines that have a molecular ion in the mass spectrum at *m/z* = 87 and show two peaks in their IR spectra at 3300–3500 cm^{-1}.

25.5C   NMR Spectra

Amines exhibit the following characteristic ^{1}H NMR and ^{13}C NMR absorptions.

◆ **The NH signal appears between 0.5 and 5.0 ppm.** The exact location depends on the degree of hydrogen bonding and the concentration of the sample.

◆ The protons on the carbon bonded to the amine nitrogen are deshielded and typically absorb at **2.3–3.0 ppm.**

◆ In the ^{13}C NMR spectrum, the carbon bonded to the N atom is deshielded and typically absorbs at **30–50 ppm.**

Like the OH absorption of an alcohol, the **NH absorption is not split by adjacent protons, nor does it cause splitting of adjacent C–H absorptions in a ^{1}H NMR spectrum.** The NH peak of an amine is sometimes somewhat broader than other peaks in the spectrum. The ^{1}H NMR spectrum of *N*-methylaniline is shown in Figure 25.4.

PROBLEM 25.10   What is the structure of an unknown compound with molecular formula $C_6H_{15}N$ that gives the following ^{1}H NMR absorptions: 0.9 (singlet, 1 H), 1.10 (triplet, 3 H), 1.15 (singlet, 9 H), and 2.6 (quartet, 2 H) ppm?

25.6   Interesting and Useful Amines

A great many simple and complex amines occur in nature, and others with biological activity have been synthesized in the lab.

Figure 25.4 The ^{1}H NMR spectrum of *N*-methylaniline

- The CH$_3$ group appears as a singlet at 2.7 ppm because there is no splitting by the adjacent NH proton.
- The NH proton appears as a broad singlet at 3.6 ppm.
- The five H atoms of the aromatic ring appear as a complex pattern at 6.6–7.2 ppm.

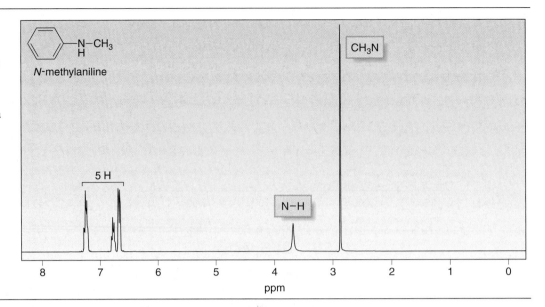

25.6A   Simple Amines and Alkaloids

Many low molecular weight amines have *very* foul odors. **Trimethylamine** [(CH$_3$)$_3$N], formed when enzymes break down certain fish proteins, has the characteristic odor of rotting fish. **Putrescine** (NH$_2$CH$_2$CH$_2$CH$_2$CH$_2$NH$_2$) and **cadaverine** (NH$_2$CH$_2$CH$_2$CH$_2$CH$_2$CH$_2$NH$_2$) are both poisonous diamines with putrid odors. They, too, are present in rotting fish, and are partly responsible for the odors of semen, urine, and bad breath.

The word **alkaloid** is derived from the word *alkali,* because aqueous solutions of alkaloids are slightly basic.

Naturally occurring amines derived from plant sources are called **alkaloids.** Alkaloids previously encountered in the text include **quinine** (Chapter 17 introductory molecule), **morphine** (Section 22.9), and **cocaine** (Problem 3.35). Three other common alkaloids are **caffeine, nicotine,** and **coniine,** illustrated in Figure 25.5.

25.6B   Histamine and Antihistamines

Histamine, a rather simple triamine first discussed in Section 17.8, is responsible for a wide variety of physiological effects. Histamine is a vasodilator (it dilates capillaries), so it is released at the site of an injury or infection to increase blood flow. It is also responsible for the symptoms of allergies, including a runny nose and watery eyes. In the stomach, histamine stimulates the secretion of acid.

histamine

Understanding the central role of histamine in these biochemical processes has helped chemists design drugs to counteract some of its undesirable effects.

brompheniramine
antihistamine in several
over-the-counter allergy remedies

cimetidine
(Tagamet)
antiulcer drug

Antihistamines bind to the same active site of the enzyme that binds histamine in the cell, but they evoke a different response. An antihistamine like **brompheniramine,** for example, inhibits vasodilation, so it is used to treat the symptoms of the common cold and allergies. **Cimetidine** (trade name: Tagamet) is a histamine mimic that blocks the secretion of hydrochloric acid in the stomach, so it is used to treat individuals with ulcers.

Figure 25.5   Three common alkaloids—Caffeine, nicotine, and coniine

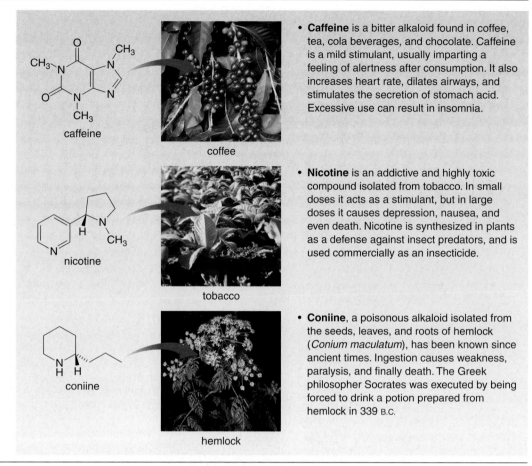

- **Caffeine** is a bitter alkaloid found in coffee, tea, cola beverages, and chocolate. Caffeine is a mild stimulant, usually imparting a feeling of alertness after consumption. It also increases heart rate, dilates airways, and stimulates the secretion of stomach acid. Excessive use can result in insomnia.

coffee

- **Nicotine** is an addictive and highly toxic compound isolated from tobacco. In small doses it acts as a stimulant, but in large doses it causes depression, nausea, and even death. Nicotine is synthesized in plants as a defense against insect predators, and is used commercially as an insecticide.

tobacco

- **Coniine**, a poisonous alkaloid isolated from the seeds, leaves, and roots of hemlock (*Conium maculatum*), has been known since ancient times. Ingestion causes weakness, paralysis, and finally death. The Greek philosopher Socrates was executed by being forced to drink a potion prepared from hemlock in 339 B.C.

hemlock

caffeine

nicotine

coniine

25.6C   Derivatives of 2-Phenylethylamine

A large number of physiologically active compounds are derived from **2-phenylethylamine, C₆H₅CH₂CH₂NH₂.** Some of these compounds are synthesized in cells and needed to maintain healthy mental function. Others are isolated from plant sources or are synthesized in the laboratory and have a profound effect on the brain because they interfere with normal neurochemistry. These compounds include **adrenaline, noradrenaline, methamphetamine,** and **mescaline.** Each contains a benzene ring bonded to a two-carbon unit with a nitrogen atom (shown in red).

adrenaline
(epinephrine)

a hormone secreted in response to stress
(Chapter 7, introductory molecule)

noradrenaline
(norepinephrine)

a neurotransmitter that increases heart rate
and dilates air passages

methamphetamine

an addictive stimulant sold as
speed, meth, or crystal meth

mescaline

a hallucinogen isolated from peyote, a cactus native
to the southwestern United States and Mexico

Another example, **dopamine,** is a neurotransmitter, a chemical messenger released by one nerve cell (neuron), which then binds to a receptor in a neighboring target cell (Figure 25.6). Dopamine

Figure 25.6   Dopamine—A neurotransmitter

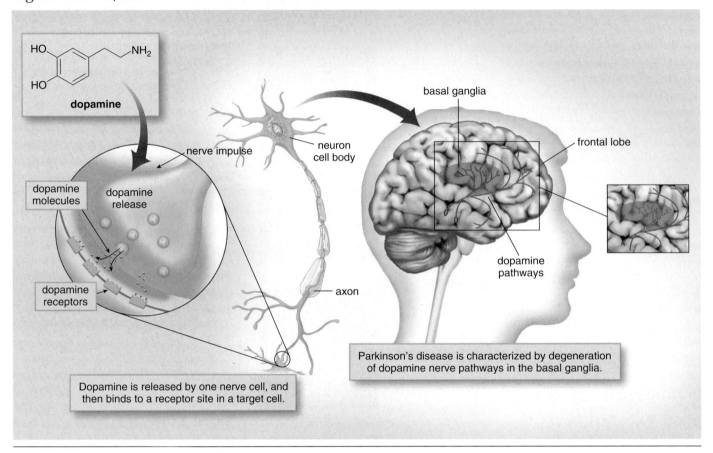

Dopamine is released by one nerve cell, and then binds to a receptor site in a target cell.

Parkinson's disease is characterized by degeneration of dopamine nerve pathways in the basal ganglia.

Cocaine, amphetamines, and several other addicting drugs increase the level of dopamine in the brain, which results in a pleasurable "high." With time, the brain adapts to increased dopamine levels, so more drug is required for the same sensation.

affects brain processes that control movement and emotions, so proper dopamine levels are necessary to maintain an individual's mental and physical health. For example, when dopamine-producing neurons die, the level of dopamine drops, resulting in the loss of motor control symptomatic of Parkinson's disease.

Understanding the neurochemistry of these compounds has led to the synthesis and availability of several useful drugs. **Fentanyl** is a common narcotic pain reliever used in surgical procedures, and **sumatriptan** (trade name: Imitrex) is used to relieve pain and light sensitivity in patients who suffer from migraine headaches.

fentanyl
a narcotic pain reliever

sumatriptan
Trade name: Imitrex

PROBLEM 25.11 LSD (a hallucinogen) and codeine (a narcotic) are structurally more complex derivatives of 2-phenylethylamine. Identify the atoms of 2-phenylethylamine in each of the following compounds.

a. 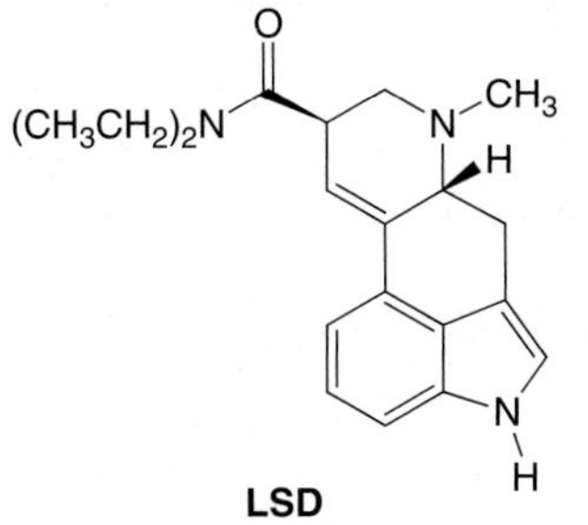 (CH$_3$CH$_2$)$_2$N

LSD
lysergic acid diethyl amide

b.

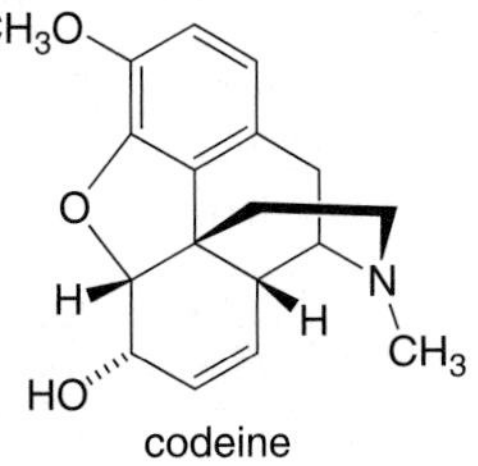

codeine

25.7 Preparation of Amines

In the preparations of a given functional group, many different starting materials form a common product (amines, in this case).

Three types of reactions are used to prepare an amine:

[1] **Nucleophilic substitution** using nitrogen nucleophiles
[2] **Reduction** of other nitrogen-containing functional groups
[3] **Reductive amination** of aldehydes and ketones

25.7A Nucleophilic Substitution Routes to Amines

Nucleophilic substitution is the key step in two different methods for synthesizing amines: direct nucleophilic substitution and the Gabriel synthesis of 1° amines.

Direct Nucleophilic Substitution

Conceptually, the simplest method to synthesize an amine is by **S$_N$2 reaction of an alkyl halide with NH$_3$ or an amine.** The method requires two steps:

[1] **Nucleophilic attack** of the nitrogen nucleophile forms an ammonium salt.
[2] **Removal of a proton** on N forms the amine.

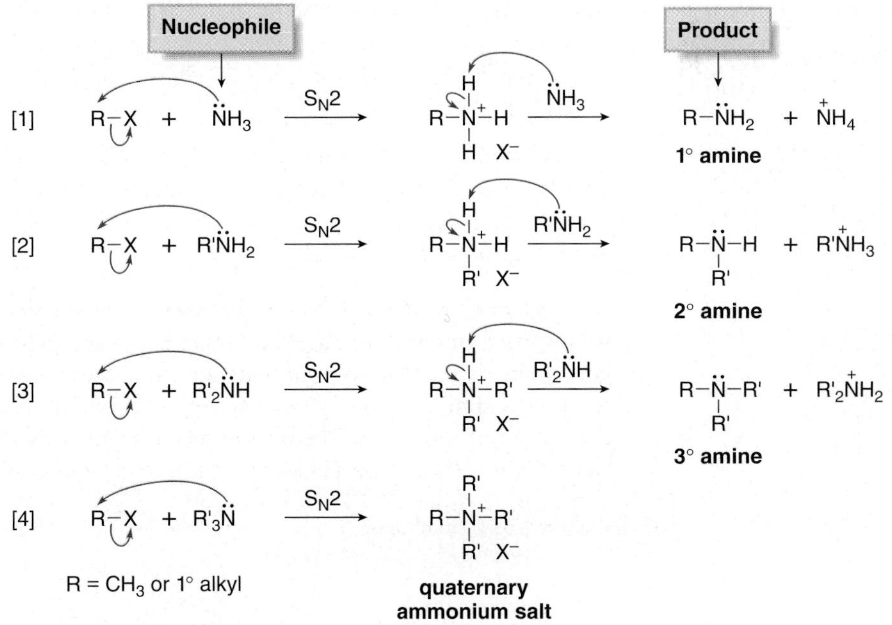

The identity of the nitrogen nucleophile determines the type of amine or ammonium salt formed as product. One new carbon–nitrogen bond is formed in each reaction. Because the reaction follows an S$_N$2 mechanism, the alkyl halide must be unhindered—that is, CH$_3$X or RCH$_2$X.

Although this process seems straightforward, polyalkylation of the nitrogen nucleophile limits its usefulness. **Any amine formed by nucleophilic substitution still has a nonbonded electron**

pair, making it a nucleophile as well. It will react with remaining alkyl halide to form a more substituted amine. Because of this, a mixture of 1°, 2°, and 3° amines often results. Only the final quaternary ammonium salt, with four alkyl groups on N, cannot react further, and so the reaction stops.

As a result, this reaction is most useful for preparing 1° amines by using a very large excess of NH_3 (a relatively inexpensive starting material) and for preparing quaternary ammonium salts by alkylating any nitrogen nucleophile with one or more equivalents of alkyl halide.

Useful S_N2 substitutions	

Useful S_N2 substitutions

$CH_3CH_2CH_2-Br$ + $\ddot{N}H_3$ (excess) $\longrightarrow$ $CH_3CH_2CH_2-\ddot{N}H_2$ (1° amine) + NH_4^+ Br^-

$CH_3\!\!-\!Br$ + $CH_3-\overset{CH_3}{\underset{\ddot{}}{\ddot{N}}}\!\!-\!\!\bigcirc$ $\longrightarrow$ $CH_3-\overset{CH_3}{\underset{CH_3}{\overset{+}{N}}}\!\!-\!\!\bigcirc$ + Br^-

quaternary ammonium salt

PROBLEM 25.12   Draw the product of each reaction.

a. ⌇⌇⌇—Cl + NH_3 (excess) $\longrightarrow$

b. ⬡—$CH_2CH_2NH_2$ + CH_3CH_2Br (excess) $\longrightarrow$

The Gabriel Synthesis of 1° Amines

To avoid polyalkylation, a nitrogen nucleophile is used that can react only in a single nucleophilic substitution reaction—that is, form a product that does not contain a nucleophilic nitrogen atom capable of reacting further.

The **Gabriel synthesis** consists of two steps and uses a nucleophile derived from phthalimide to synthesize 1° amines via nucleophilic substitution. The Gabriel synthesis begins with **phthalimide,** one of a group of compounds called **imides.** The **N−H bond of an imide is especially acidic** because the resulting anion is resonance stabilized by the two flanking carbonyl groups.

phthalimide
$pK_a = 10$

resonance-stabilized anion

An acid–base reaction forms a nucleophilic anion that can react with an unhindered alkyl halide—that is, CH_3X or RCH_2X—in an S_N2 reaction to form a substitution product. This alkylated imide is then hydrolyzed with aqueous base to give a 1° amine and a dicarboxylate. This reaction is similar to the hydrolysis of amides to afford carboxylate anions and amines, as discussed in Section 22.13. The overall result of this two-step sequence is **nucleophilic substitution of X by NH₂,** so the Gabriel synthesis can be used to prepare 1° amines only.

Steps in the Gabriel synthesis

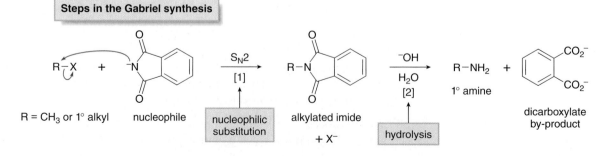

R = CH₃ or 1° alkyl   nucleophile

nucleophilic substitution   alkylated imide   + X⁻

hydrolysis

1° amine

dicarboxylate by-product

◆ The Gabriel synthesis converts an alkyl halide into a 1° amine by a two-step process: nucleophilic substitution followed by hydrolysis.

Example

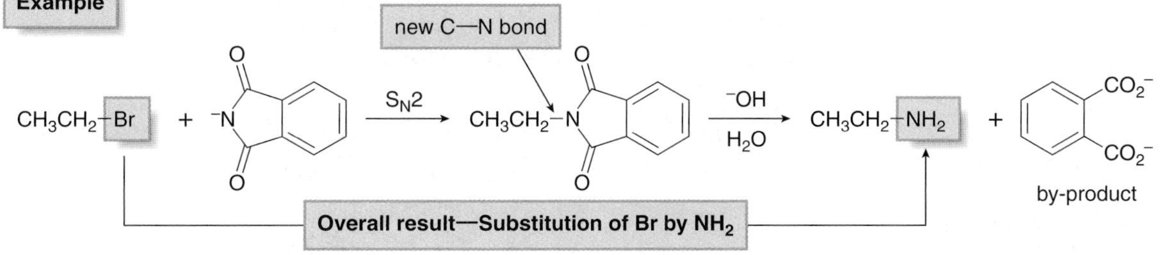

Overall result—Substitution of Br by NH$_2$

PROBLEM 25.13   Identify compounds **A, B,** and **C** in the given reaction sequence.

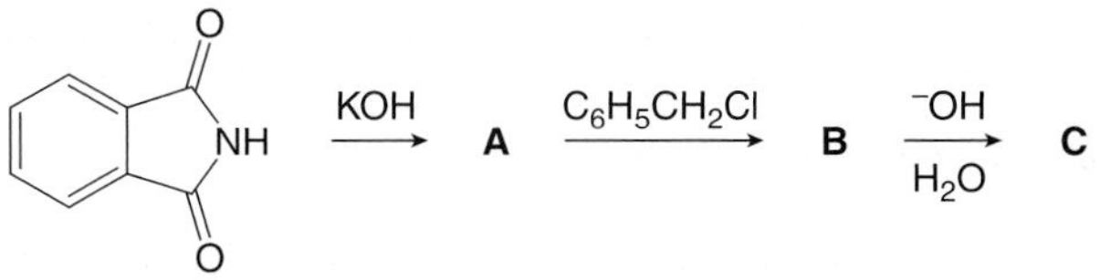

PROBLEM 25.14   What alkyl halide is needed to prepare each 1° amine by a Gabriel synthesis?

a. 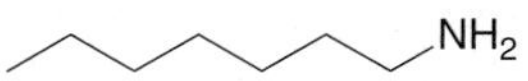   b. $(CH_3)_2CHCH_2CH_2NH_2$ c.

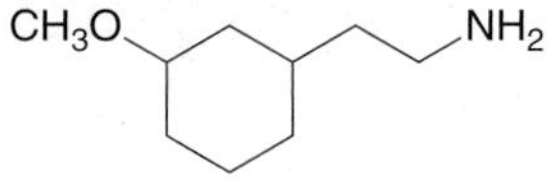

25.7B Reduction of Other Functional Groups that Contain Nitrogen

Amines can be prepared by reduction of nitro compounds, nitriles, and amides. Because the details of these reactions have been discussed previously, they are presented here in summary form only.

[1]   From nitro compounds (Section 18.14C)

Nitro groups are reduced to 1° amines using a variety of reducing agents.

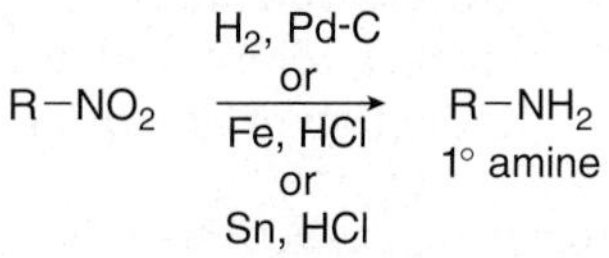

$$R-NO_2 \xrightarrow[\substack{\text{or} \\ \text{Fe, HCl} \\ \text{or} \\ \text{Sn, HCl}}]{H_2,\ Pd\text{-}C} \underset{\text{1° amine}}{R-NH_2}$$

[2]   From nitriles (Section 22.18B)

Nitriles are reduced to 1° amines with LiAlH$_4$.

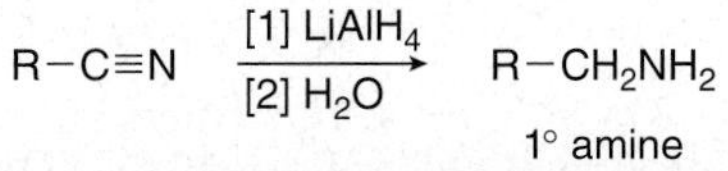

$$R-C\equiv N \xrightarrow[\text{[2] } H_2O]{\text{[1] LiAlH}_4} \underset{\text{1° amine}}{R-CH_2NH_2}$$

Because a cyano group is readily introduced by S_N2 substitution of alkyl halides with ^-CN, this provides **a two-step method to convert an alkyl halide to a 1° amine with one more carbon atom.** The conversion of CH_3Br to $CH_3CH_2NH_2$ illustrates this two-step sequence.

Example $CH_3-Br \xrightarrow[S_N2]{NaCN} CH_3-C\equiv N \xrightarrow[\text{[2] } H_2O]{\text{[1] LiAlH}_4} \underset{\text{1° amine}}{CH_3-CH_2NH_2}$

new C—C bond

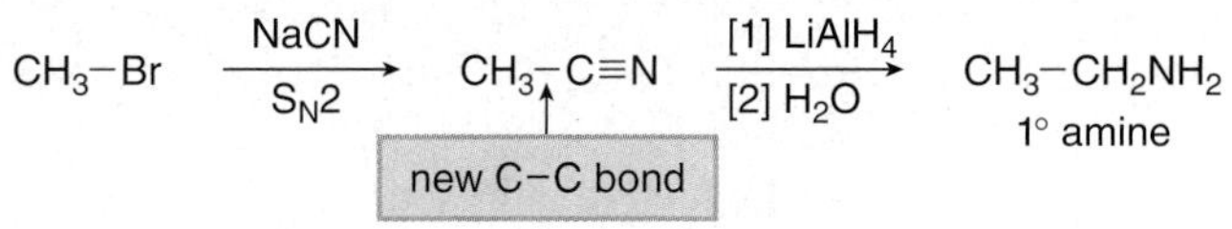

[3]   From amides (Section 20.7B)

1°, 2°, and 3° amides are reduced to 1°, 2°, and 3° amines, respectively, by using LiAlH$_4$.

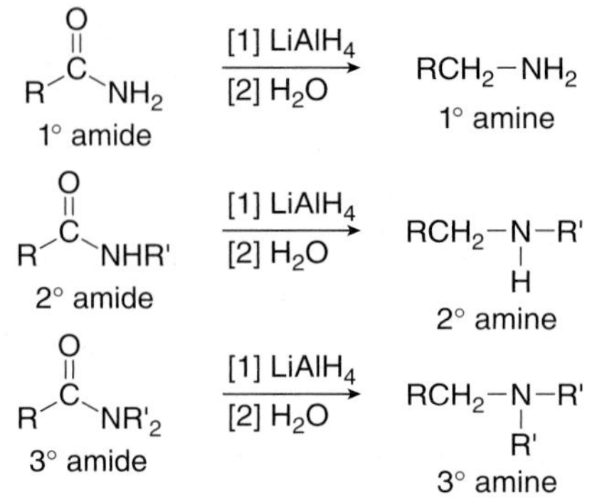

$$\underset{\text{1° amide}}{R-\overset{\displaystyle O}{\overset{\|}{C}}-NH_2} \quad \xrightarrow[\text{[2] H}_2\text{O}]{\text{[1] LiAlH}_4} \quad \underset{\text{1° amine}}{RCH_2-NH_2}$$

$$\underset{\text{2° amide}}{R-\overset{\displaystyle O}{\overset{\|}{C}}-NHR'} \quad \xrightarrow[\text{[2] H}_2\text{O}]{\text{[1] LiAlH}_4} \quad \underset{\text{2° amine}}{RCH_2-\underset{\overset{|}{H}}{N}-R'}$$

$$\underset{\text{3° amide}}{R-\overset{\displaystyle O}{\overset{\|}{C}}-NR'_2} \quad \xrightarrow[\text{[2] H}_2\text{O}]{\text{[1] LiAlH}_4} \quad \underset{\text{3° amine}}{RCH_2-\underset{\overset{|}{R'}}{N}-R'}$$

PROBLEM 25.15   What nitro compound, nitrile, and amide are reduced to each compound?

a. $CH_3\underset{\overset{|}{CH_3}}{CH}CH_2NH_2$ b. 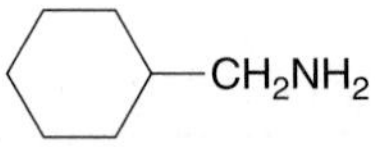 ⬡—CH$_2$NH$_2$ c. 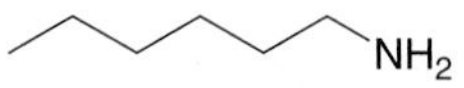 ⟋⟍⟋⟍⟋NH$_2$

PROBLEM 25.16   What amine is formed by reduction of each amide?

a. 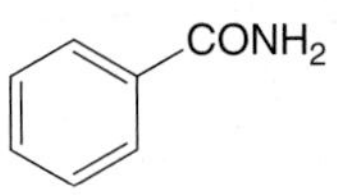 ⬡—CONH$_2$ b. 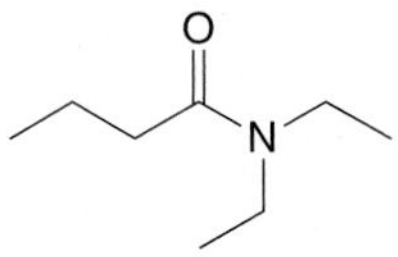 (structure)   c. 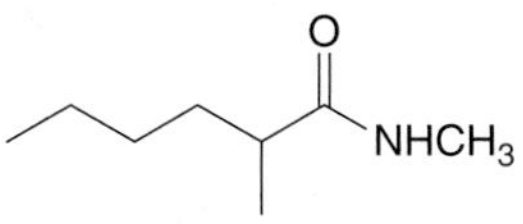 (structure with NHCH$_3$)

PROBLEM 25.17   Explain why isopropylamine [(CH$_3$)$_2$CHNH$_2$] can be prepared by reduction of a nitro compound, but cannot be prepared by reduction of a nitrile, even though it is a 1° amine.

25.7C   Reductive Amination of Aldehydes and Ketones

Reductive amination is a two-step method that converts aldehydes and ketones into 1°, 2°, and 3° amines. Let's first examine this method using NH$_3$ to prepare 1° amines. There are two distinct parts in reductive amination:

[1] **Nucleophilic attack of NH$_3$ on the carbonyl group forms an imine** (Section 21.11A), which is not isolated; then,

[2] **Reduction of the imine forms an amine** (Section 20.7B).

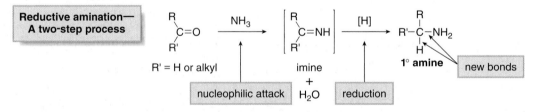

◆ Reductive amination replaces a C=O by a C–H and C–N bond.

The most effective reducing agent for this reaction is sodium cyanoborohydride (NaBH$_3$CN). This hydride reagent is a derivative of sodium borohydride (NaBH$_4$), formed by replacing one H atom by CN.

NaBH$_3$CN
sodium cyanoborohydride

Reductive amination combines two reactions we have already learned in a different way. Two examples are shown. The second reaction is noteworthy because the product is **amphetamine,** a potent central nervous system stimulant.

Examples

$$CH_3-C(=O)-CH_3 \xrightarrow[\text{NaBH}_3\text{CN}]{\text{NH}_3} CH_3-\underset{\underset{H}{|}}{C}(CH_3)-NH_2$$

new bonds

amphetamine
a powerful stimulant

new bonds

With a 1° or 2° amine as starting material, reductive amination is used to prepare 2° and 3° amines, respectively. Note the result: Reductive amination uses an aldehyde or ketone to replace one H atom on a nitrogen atom by an alkyl group, making a more substituted amine.

[1] $$\underset{R'}{\overset{R}{C}}=O \;+\; R''NH_2 \longrightarrow \left[\underset{R'}{\overset{R}{C}}=NR''\right] \xrightarrow{\text{NaBH}_3\text{CN}} R'-\underset{\underset{H}{|}}{\overset{\overset{R}{|}}{C}}-\underset{\underset{H}{|}}{N}-R''$$

1° amine   imine   2° amine

[2] $$\underset{R'}{\overset{R}{C}}=O \;+\; R''_2NH \longrightarrow \left[\underset{R'}{\overset{R}{C}}=\overset{+}{N}R''_2\right] \xrightarrow{\text{NaBH}_3\text{CN}} R'-\underset{\underset{H}{|}}{\overset{\overset{R}{|}}{C}}-\underset{\underset{R''}{|}}{N}-R''$$

R' = H or alkyl   2° amine   iminium ion   3° amine

The synthesis of methamphetamine (Section 25.6C) by reductive amination is illustrated in Figure 25.7.

PROBLEM 25.18   Draw the product of each reaction.

a.   [benzene ring]—CHO $\xrightarrow[\text{NaBH}_3\text{CN}]{\text{CH}_3\text{NH}_2}$

c.   [cyclohexanone] $\xrightarrow[\text{NaBH}_3\text{CN}]{(\text{CH}_3\text{CH}_2)_2\text{NH}}$

b.   [cyclohexyl–CH₂–C(=O)–CH₃] $\xrightarrow[\text{NaBH}_3\text{CN}]{\text{NH}_3}$

To use reductive amination in synthesis, you must be able to determine what aldehyde or ketone and nitrogen compound are needed to prepare a given amine—that is, you must work backwards in the retrosynthetic direction. Keep in mind the following two points:

◆ One alkyl group on N comes from the carbonyl compound.
◆ The remainder of the molecule comes from NH₃ or an amine.

Product of reductive amination	**Two components needed**

—N—   ⟹   —N— amine or NH₃
                         |
                         H

—C—H   ⟹   C=O
  | aldehyde or ketone

Figure 25.7   Synthesis of methamphetamine
by reductive amination

• In reductive amination, one of the H atoms bonded to N is replaced by an alkyl group. As a result, a 1° amine is converted to a 2° amine and a 2° amine is converted to a 3° amine. In this reaction, CH₃NH₂ (a 1° amine) is converted to methamphetamine (a 2° amine).

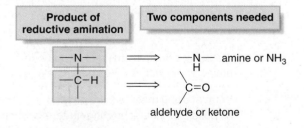

1° amine

$\xrightarrow[\text{NaBH}_3\text{CN}]{\text{CH}_3\text{NH}_2}$

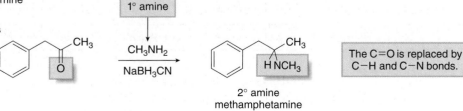

2° amine
methamphetamine

The C=O is replaced by C—H and C—N bonds.

For example, 2-phenylethylamine is a 1° amine, so it has only one alkyl group bonded to N. This alkyl group must come from the carbonyl compound, and the rest of the molecule then comes from the nitrogen component. **For a 1° amine, the nitrogen component must be NH$_3$.**

Retrosynthetic analysis for preparing 2-phenylethylamine:

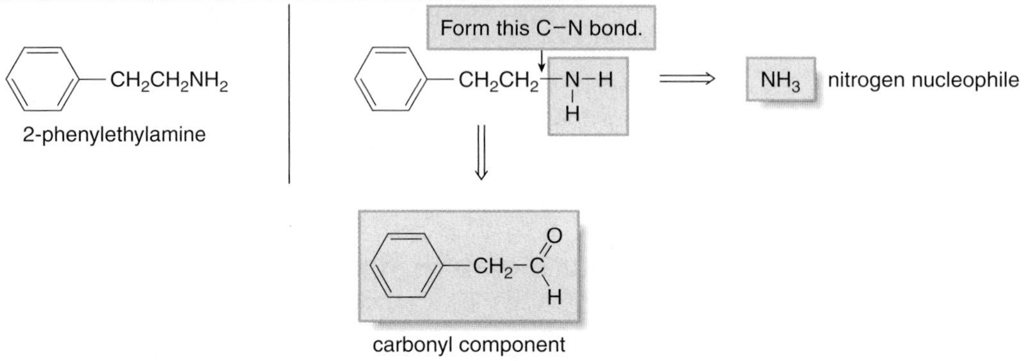

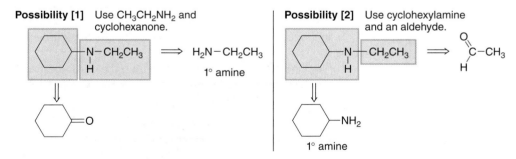

There is usually more than one way to use reductive amination to synthesize 2° and 3° amines, as shown in Sample Problem 25.2 for a 2° amine.

SAMPLE PROBLEM 25.2   What aldehyde or ketone and nitrogen component are needed to synthesize *N*-ethylcyclohexylamine by a reductive amination reaction?

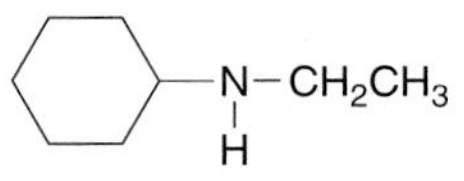

N-ethylcyclohexylamine

SOLUTION

Because *N*-ethylcyclohexylamine has two different alkyl groups bonded to the N atom, either R group can come from the carbonyl component and there are two different ways to form a C−N bond by reductive amination.

Possibility [1] Use CH$_3$CH$_2$NH$_2$ and cyclohexanone.

Possibility [2] Use cyclohexylamine and an aldehyde.

Because reductive amination adds one R group to a nitrogen atom, both routes to form the 2° amine begin with a 1° amine.

PROBLEM 25.19   What starting materials are needed to prepare each compound using reductive amination? Give all possible pairs of starting materials when more than one possibility exists.

a. 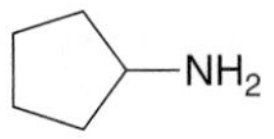 ⬠—NH$_2$

b. CH$_3$CH$_2$—N—CH$_2$CH$_3$
                    |
                    CH$_3$

c. H—N—CH(CH$_3$)$_2$
        |
        CH$_3$

PROBLEM 25.20   Explain why *tert*-butylamine cannot be made by reductive amination.

25.8   Reactions of Amines—General Features

◆ The chemistry of amines is dominated by the lone electron pair on nitrogen.

Only three elements in the second row of the periodic table have nonbonded electron pairs in neutral organic compounds: nitrogen, oxygen, and fluorine. Because basicity and nucleophilicity decrease across the row, **nitrogen is the most basic and most nucleophilic** of these elements.

$$-\ddot{N}- \qquad -\ddot{O}- \qquad -\ddot{F}:$$

Increasing basicity and nucleophilicity

◆ Amines are stronger bases and nucleophiles than other neutral organic compounds.

| Reaction as a base | $-\ddot{N}-$ + H—A $\longrightarrow$ $-\overset{+}{N}\overset{H}{-}$:A$^-$ |
| Reaction as a nucleophile | $-\ddot{N}-$ + E—X $\longrightarrow$ $-\overset{+}{N}\overset{E}{-}$ X$^-$ |

E = an electrophilic site

◆ Amines react as bases with compounds that contain acidic protons.
◆ Amines react as nucleophiles with compounds that contain electrophilic carbons.

25.9   Amines as Bases

Amines react as bases with a variety of organic and inorganic acids.

A Brønsted–Lowry acid–base reaction	R—$\ddot{N}$H$_2$ + H—A $\rightleftharpoons$ R—$\overset{+}{N}$H$_3$ + :A$^-$
	base acid conjugate acid
	p$K_a \approx 10$–11

To favor the products, the **pK_a of HA must be < 10.**

What acids can be used to protonate an amine? Equilibrium favors the products of an acid–base reaction when the weaker acid and base are formed. Because the pK$_a$ of many protonated amines is 10–11, the pK$_a$ of the starting acid must be **less than 10** for equilibrium to favor the products. Amines are thus readily protonated by strong inorganic acids like HCl and H$_2$SO$_4$, and by carboxylic acids as well.

Examples

$$CH_3CH_2-\ddot{N}H_2 + H-Cl \rightleftharpoons CH_3CH_2-\overset{+}{N}H_3 + Cl^-$$
$$pK_a = -7 \qquad\qquad pK_a = 10.8$$

$$(CH_3CH_2)_3N: + \;\; H-O\overset{\displaystyle O}{\underset{\displaystyle}{\overset{\|}{C}}}CH_3 \rightleftharpoons (CH_3CH_2)_3\overset{+}{N}H + \;\;{}^-O\overset{\displaystyle O}{\underset{\displaystyle}{\overset{\|}{C}}}CH_3$$
$$pK_a = 4.8 \qquad\qquad pK_a = 11.0$$

Equilibrium favors the products.

PROBLEM 25.21   Draw the products of each acid–base reaction. Indicate whether equilibrium favors the reactants or products.

a. CH$_3$CH$_2$CH$_2$CH$_2$—NH$_2$ + HCl $\rightleftharpoons$

b. C$_6$H$_5$COOH + (CH$_3$)$_2$NH $\rightleftharpoons$

c. + H$_2$O $\rightleftharpoons$

The principles used in an extraction procedure were detailed in Section 19.12.

Because amines are protonated by aqueous acid, they can be separated from other organic compounds by extraction using a separatory funnel. **Extraction separates compounds based on solubility differences.** When an amine is protonated by aqueous acid, its solubility properties change.

For example, when cyclohexylamine is treated with aqueous HCl, it is protonated, forming an ammonium salt. Because the ammonium salt is ionic, it is soluble in water, but insoluble in organic solvents. A similar acid–base reaction does not occur with other organic compounds like alcohols, which are much less basic.

cyclohexylamine

- insoluble in H_2O
- soluble in CH_2Cl_2

cyclohexylammonium chloride

- soluble in H_2O
- insoluble in CH_2Cl_2

This difference in acid–base chemistry can be used to separate cyclohexylamine and cyclohexanol by the stepwise extraction procedure illustrated in Figure 25.8.

◆ **An amine can be separated from other organic compounds by converting it to a water-soluble ammonium salt by an acid–base reaction.**

Thus, the water-soluble salt $C_6H_{11}NH_3^+Cl^-$ (obtained by protonation of $C_6H_{11}NH_2$), can be separated from water-insoluble cyclohexanol by an aqueous extraction procedure.

Figure 25.8   Separation of cyclohexylamine and cyclohexanol by an extraction procedure

Step [1] Dissolve cyclohexylamine and cyclohexanol in CH_2Cl_2.

Step [2] Add 10% HCl solution to form two layers.

Step [3] Separate the layers.

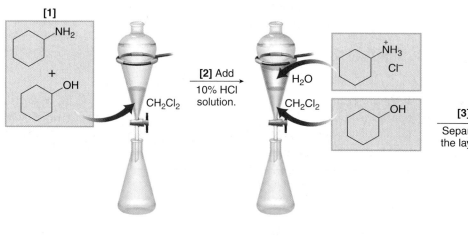

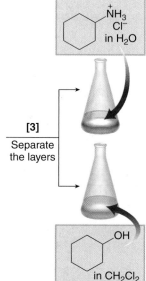

- Both compounds dissolve in the organic solvent CH_2Cl_2.

- Adding 10% aqueous HCl solution forms two layers. When the two layers are mixed, the HCl protonates the amine (RNH_2) to form $RNH_3^+Cl^-$, which dissolves in the aqueous layer.
- The cyclohexanol remains in the CH_2Cl_2 layer.

- Draining the lower layer out the bottom stopcock separates the two layers, and the separation process is complete.
- Cyclohexanol (dissolved in CH_2Cl_2) is in one flask. The ammonium salt, $RNH_3^+Cl^-$ (dissolved in water), is in another flask.

PROBLEM 25.22   Write out steps to show how each of the following pairs of compounds can be separated by an extraction procedure.

a.  ⬡—NH₂   and   ⬡—CH₃   b. $(CH_3CH_2CH_2CH_2)_3N$ and $(CH_3CH_2CH_2CH_2)_2O$

25.10 Relative Basicity of Amines and Other Compounds

The relative acidity of different compounds can be compared using their pK_a values. The relative *basicity* of different compounds (such as amines) can be compared using the pK_a values of their *conjugate acids*.

◆ **The weaker the conjugate acid, the higher its pK_a and the stronger the base.**

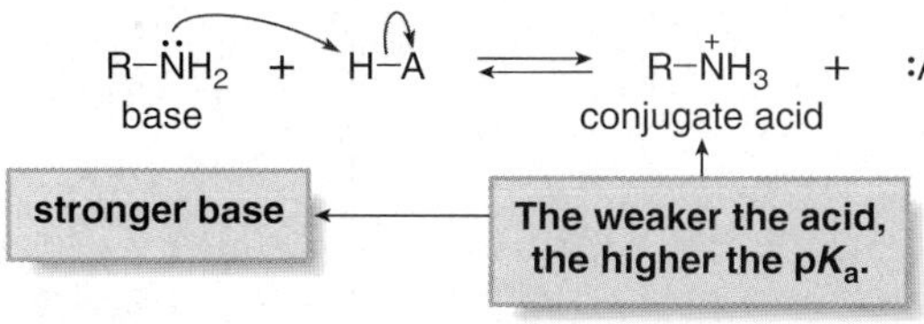

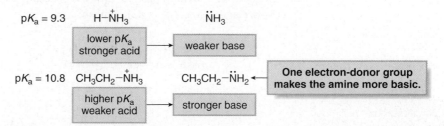

PROBLEM 25.23   Which amine in each pair is the stronger base? The pK_a values given are for the conjugate acid of each amine.

a. CH_3NH_2 ($pK_a = 10.7$) and $CH_3CH_2NH_2$ ($pK_a = 10.8$)
b. $(CH_3CH_2)_3N$ ($pK_a = 11.0$) and $(CH_3)_3N$ ($pK_a = 9.8$)

To compare the basicity of two compounds, keep in mind the following:

◆ **Any factor that increases the electron density on the N atom increases an amine's basicity.**
◆ **Any factor that decreases the electron density on N decreases an amine's basicity.**

25.10A   Comparing an Amine and NH₃

Because alkyl groups are electron donating, they increase the electron density on nitrogen, which makes an amine like $CH_3CH_2NH_2$ more basic than NH_3. In fact, the pK_a of $CH_3CH_2NH_3^+$ is higher than the pK_a of NH_4^+, so **$CH_3CH_2NH_2$ is a stronger base than NH_3.**

$pK_a = 9.3$ $H-\overset{+}{N}H_3$ $\overset{..}{N}H_3$

lower pK_a
stronger acid → weaker base

$pK_a = 10.8$ $CH_3CH_2-\overset{+}{N}H_3$ $CH_3CH_2-\overset{..}{N}H_2$ ← **One electron-donor group makes the amine more basic.**

higher pK_a
weaker acid → stronger base

The relative basicity of 1°, 2°, and 3° amines depends on additional factors, and will not be considered in this text.

◆ **Primary (1°), 2°, and 3° alkylamines are more basic than NH₃ because of the electron-donating inductive effect of the R groups.**

PROBLEM 25.24   Which compound in each pair is more basic: (a) $(CH_3)_2NH$ and NH_3; (b) $CH_3CH_2NH_2$ and $ClCH_2CH_2NH_2$?

25.10B   Comparing an Alkylamine and an Arylamine

To compare an alkylamine ($CH_3CH_2NH_2$) and an arylamine ($C_6H_5NH_2$, aniline), we must look at the availability of the nonbonded electron pair on N. With $CH_3CH_2NH_2$, the electron pair is localized on the N atom. With an arylamine, however, the electron pair is now delocalized on the benzene ring. This *decreases* the electron density on N, and makes $C_6H_5NH_2$ less basic than $CH_3CH_2NH_2$.

$$CH_3CH_2-\ddot{N}H_2 \longleftarrow \boxed{\textbf{The electron pair is localized on the N atom.}}$$

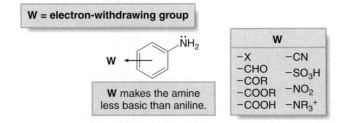

Once again, pK_a values support this reasoning. Because the pK_a of $CH_3CH_2NH_3^+$ is higher than the pK_a of $C_6H_5NH_3^+$, **$CH_3CH_2NH_2$ is a stronger base than $C_6H_5NH_2$.**

$$C_6H_5-\overset{+}{N}H_3 \qquad C_6H_5-\ddot{N}H_2 \qquad | \qquad CH_3CH_2-\overset{+}{N}H_3 \qquad CH_3CH_2-\ddot{N}H_2$$

| lower pK_a stronger acid | → | weaker base | | higher pK_a weaker acid | → | stronger base |

pK_a = 4.6   pK_a = 10.8

◆ Arylamines are less basic than alkylamines because the electron pair on N is delocalized.

Substituted anilines are more or less basic than aniline depending on the nature of the substituent.

◆ Electron-donor groups add electron density to the benzene ring, making the arylamine more basic than aniline.

$$\boxed{\textbf{D = electron-donor group}}$$

D
$-NH_2$
$-OH$
$-OR$
$-NHCOR$
$-R$

D makes the amine more basic than aniline.

◆ Electron-withdrawing groups remove electron density from the benzene ring, making the arylamine less basic than aniline.

$$\boxed{\textbf{W = electron-withdrawing group}}$$

W	
$-X$	$-CN$
$-CHO$	$-SO_3H$
$-COR$	$-NO_2$
$-COOR$	$-NR_3^+$
$-COOH$	

W makes the amine less basic than aniline.

The effect of electron-donating and electron-withdrawing groups on the acidity of substituted benzoic acids was discussed in Section 19.11.

Whether a substituent donates or withdraws electron density depends on the balance of its inductive and resonance effects (Section 18.6 and Figure 18.7).

SAMPLE PROBLEM 25.3 Rank the following compounds in order of increasing basicity.

aniline   p-nitroaniline   p-methylaniline
                                             (p-toluidine)

SOLUTION

p-Nitroaniline: NO_2 is an electron-withdrawing group, making the amine **less basic** than aniline.

The lone pair on N is delocalized on the O atom, decreasing the basicity of the amine.

p-Methylaniline: CH_3 has an electron-donating inductive effect, making the amine **more basic** than aniline.

CH_3 inductively donates electron density, increasing the basicity of the amine.

ANSWER

p-nitroaniline   aniline   p-methylaniline
                                             (p-toluidine)

Increasing basicity

The electrostatic potential plots in Figure 25.9 demonstrate that the electron density of the nitrogen atoms in these anilines increases in the order shown.

Figure 25.9 Electrostatic potential plots of substituted anilines

p-Nitroaniline   **Aniline**   **p-Methylaniline**
                                             (p-toluidine)

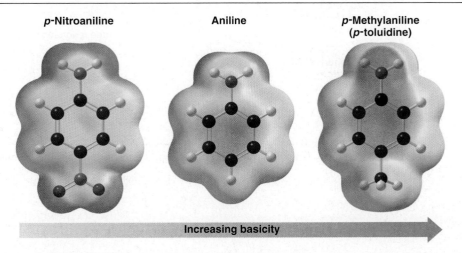

Increasing basicity

The NH_2 group gets more electron rich as the para substituent changes from $NO_2 \rightarrow H \rightarrow CH_3$. This is indicated by the color change around NH_2 (from green to yellow to red) in the electrostatic potential plot.

PROBLEM 25.25   Rank the compounds in each group in order of increasing basicity.

a.

b.

25.10C   Comparing an Alkylamine and an Amide

To compare the basicity of an alkylamine (RNH_2) and an amide ($RCONH_2$), we must once again compare the availability of the nonbonded electron pair on nitrogen. With RNH_2, the electron pair is localized on the N atom. With an amide, however, the electron pair is delocalized on the carbonyl oxygen by resonance. This *decreases* the electron density on N, making **an amide much less basic than an alkylamine.**

The electron pair on N is delocalized on O by resonance.

◆ **Amides are much less basic than amines because the electron pair on N is delocalized.**

In fact, amides are not much more basic than any carbonyl compound. When an amide is treated with acid, **protonation occurs at the carbonyl oxygen, *not* the nitrogen,** because the resulting cation is resonance stabilized. The product of protonation of the NH_2 group cannot be resonance stabilized. Thus, protonation on oxygen is the preferred pathway.

Preferred pathway

Protonation of the O atom

three resonance structures for the conjugate acid

Protonation of the N atom

not resonance stabilized

PROBLEM 25.26   Rank the following compounds in order of increasing basicity.

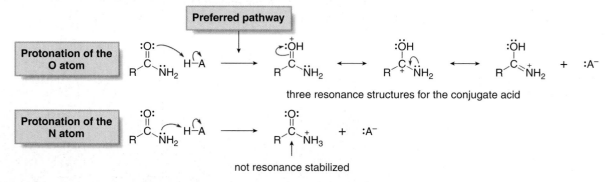

25.10D   Heterocyclic Aromatic Amines

To determine the relative basicity of nitrogen heterocycles that are also aromatic, you must know whether the nitrogen lone pair is part of the aromatic π system.

For example, pyridine and pyrrole are both aromatic, but the nonbonded electron pair on the N atom in these compounds is located in different orbitals. Recall from Section 17.8C that the **lone pair of electrons in pyridine occupies an sp^2 hybridized orbital,** perpendicular to the plane of the molecule, so it is *not* part of the aromatic system, whereas that of pyrrole resides in a *p* orbital, making it part of the aromatic system. **The lone pair on pyrrole, therefore, is delocalized on all of the atoms of the five-membered ring,** making pyrrole a much weaker base than pyridine.

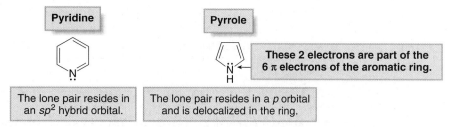

As a result, the pK_a of the conjugate acid of pyrrole is much less than that for the conjugate acid of pyridine.

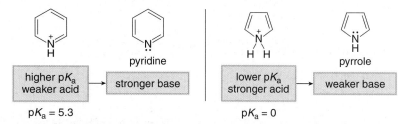

◆ Pyrrole is much less basic than pyridine because its lone pair of electrons is part of the aromatic π system.

25.10E   Hybridization Effects

The effect of hybridization on the acidity of an H–A bond was first discussed in Section 2.5D.

The hybridization of the orbital that contains an amine's lone pair also affects its basicity. This is illustrated by comparing the basicity of **piperidine** and **pyridine,** two nitrogen heterocycles. The lone pair in piperidine resides in an sp^3 hybrid orbital that has 25% s-character. The lone pair in pyridine resides in an sp^2 hybrid orbital that has 33% s-character.

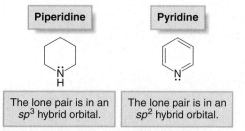

◆ The higher the percent s-character of the orbital containing the lone pair, the more tightly the lone pair is held, and the weaker the base.

Pyridine is a weaker base than piperidine because its nonbonded pair of electrons resides in an sp^2 hybrid orbital. Although pyridine is an aromatic amine, its lone pair is *not* part of the delocalized π system, so its basicity is determined by the hybridization of its N atom. As a result, the pK_a value for the conjugate acid of pyridine is much lower than that for the conjugate acid of piperidine, making pyridine the weaker base.

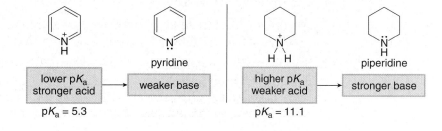

PROBLEM 25.27   Which nitrogen atom in each compound is more basic?

a.

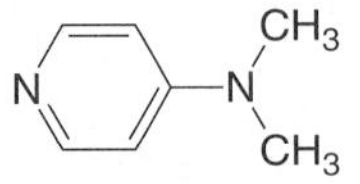

DMAP
4-(*N*,*N*-dimethylamino)pyridine

b.

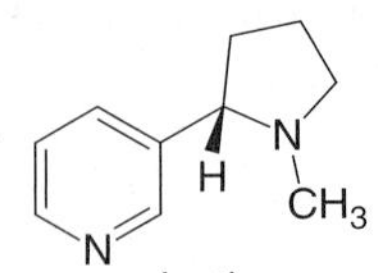

nicotine

25.10F   Summary

Acid–base chemistry is central to many processes in organic chemistry, so it has been a constant theme throughout the text. Tables 25.2 and 25.3 organize and summarize the acid–base principles discussed in Section 25.10. The principles in these tables can be used to determine the most basic site in a molecule that has more than one nitrogen atom, as shown in Sample Problem 25.4.

SAMPLE PROBLEM 25.4   Which N atom in LSD is the strongest base?

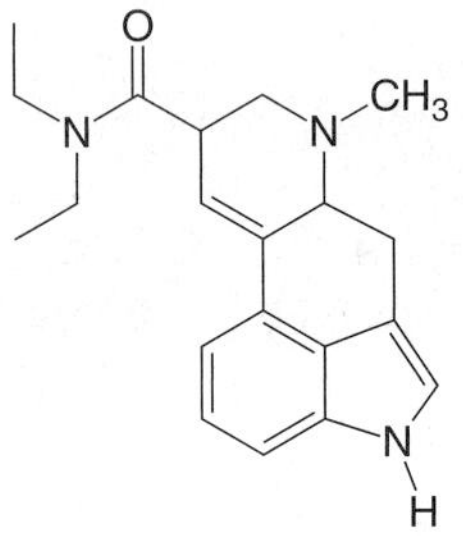

LSD
lysergic acid diethyl amide
a hallucinogen
(Chapter 18 opening molecule)

SOLUTION

LSD has a complex structure, but we need look only at the nitrogen atoms to determine their relative basicities.

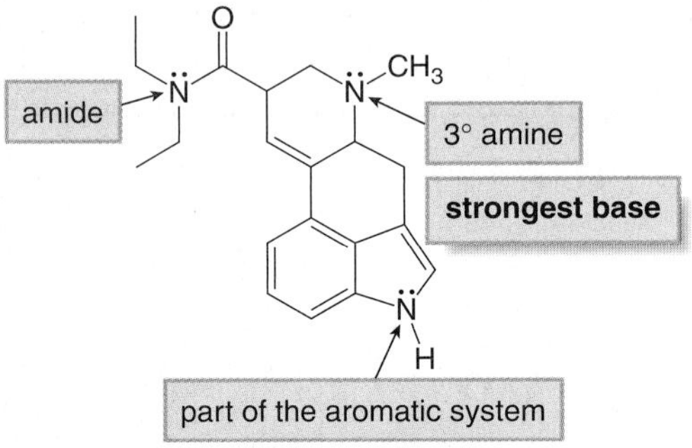

amide   3° amine

strongest base

part of the aromatic system

LSD has three nitrogen atoms. One is part of an amide, which makes it a very weak base. One is a 3° alkylamine, and these are usually the strongest amine bases (see Table 25.3). The third is part of a nitrogen heterocycle. The lone pair on the N atom of the nitrogen heterocycle is part of an aromatic π system containing 10 π electrons. (Recall from Section 17.7 that 10 π electrons are aromatic by Hückel's rule.) As a result, the lone pair on the nitrogen atom in the heterocycle is delocalized so the nitrogen atom is a very weak base.

PROBLEM 25.28   Which N atom in each drug is more basic?

a.

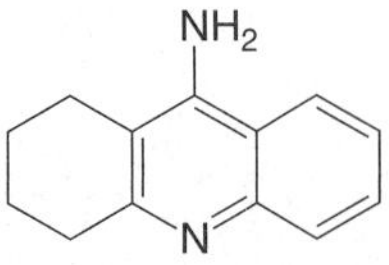

tacrine
drug used to treat
Alzheimer's disease

b.

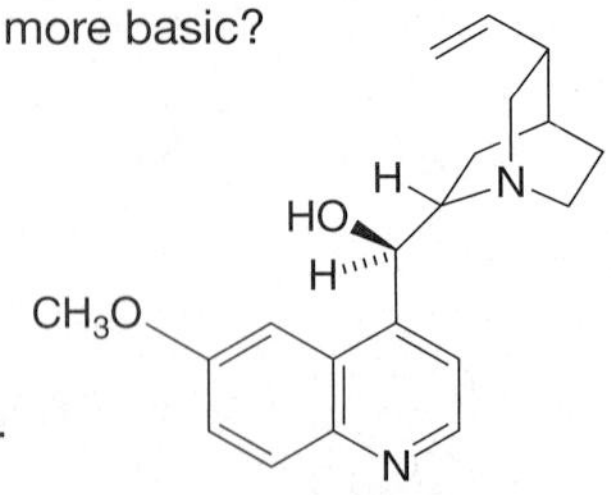

quinine
antimalarial drug

c.

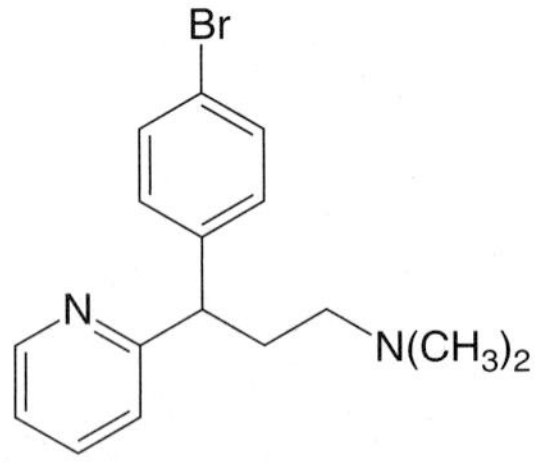

brompheniramine
antihistamine

TABLE 25.2 Factors that Determine Amine Basicity

Factor	Example
[1] **Inductive effects:** Electron-donating groups bonded to N increase basicity.	• RNH_2, R_2NH, and R_3N are more basic than NH_3.
[2] **Resonance effects:** Delocalizing the lone pair on N decreases basicity.	• Arylamines ($C_6H_5NH_2$) are less basic than alkylamines (RNH_2). • Amides ($RCONH_2$) are much less basic than amines (RNH_2).
[3] **Aromaticity:** Having the lone pair on N as part of the aromatic π system decreases basicity.	• Pyrrole is less basic than pyridine. 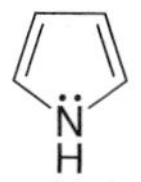 less basic 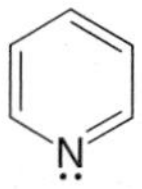 more basic
[4] **Hybridization effects:** Increasing the percent s-character in the orbital with the lone pair decreases basicity.	• Pyridine is less basic than piperidine. 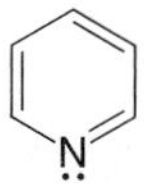 less basic 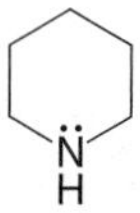 more basic

TABLE 25.3 Table of pK_a Values of Some Representative Organic Nitrogen Compounds

	Compound	pK_a of the conjugate acid	Comment
Ammonia	NH_3	9.3	
Alkylamines	 ⬡NH	11.1	
	$(CH_3CH_2)_2NH$	11.1	Alkylamines have pK_a values of ~10–11.
	$(CH_3CH_2)_3N$	11.0	
	$CH_3CH_2NH_2$	10.8	
Arylamines	p-$CH_3OC_6H_4NH_2$	5.3	
	p-$CH_3C_6H_4NH_2$	5.1	The pK_a decreases as the electron density of the benzene ring decreases.
	$C_6H_5NH_2$	4.6	
	p-$NO_2C_6H_4NH_2$	1.0	
Heterocyclic aromatic amines	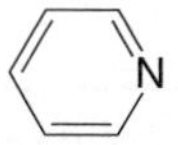 ⬡N	5.3	The pK_a depends on whether the lone pair on N is localized or delocalized.
	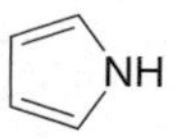 NH	0	
Amides	$RCONH_2$	−1	

25.11   Amines as Nucleophiles

Amines react as nucleophiles with electrophilic carbon atoms. The details of these reactions have been described in Chapters 21 and 22, so they are only summarized here to emphasize the similar role that the amine nitrogen plays.

◆ **Amines attack carbonyl groups to form products of nucleophilic addition or substitution.**

The nature of the product depends on the carbonyl electrophile. These reactions are limited to 1° and 2° amines, because only these compounds yield neutral organic products.

[1] **Reaction of 1° and 2° amines with aldehydes and ketones (Sections 21.11–21.12)**

Aldehydes and ketones react with 1° amines to form **imines** and with 2° amines to form **enamines.** Both reactions involve nucleophilic addition of the amine to the carbonyl group to form a carbinolamine, which then loses water to form the final product.

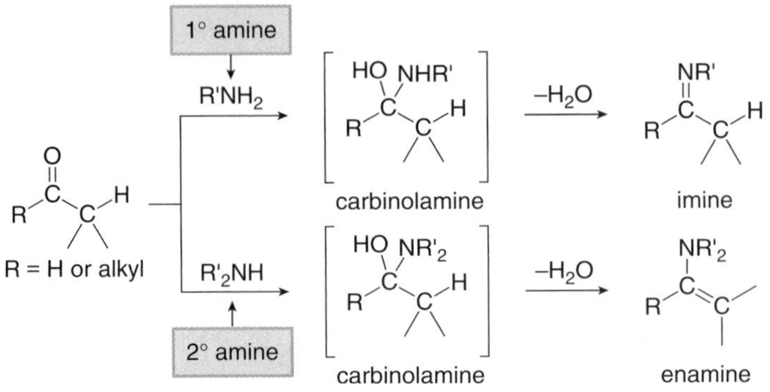

[2] **Reaction of NH₃ and 1° and 2° amines with acid chlorides and anhydrides (Sections 22.8–22.9)**

Acid chlorides and anhydrides react with NH_3, 1° amines, and 2° amines to form **1°, 2°, and 3° amides,** respectively. Both reactions involve attack of the nitrogen nucleophile on the carbonyl group followed by elimination of a leaving group (Cl⁻ or RCOO⁻). The overall result of this reaction is substitution of the leaving group by the nitrogen nucleophile.

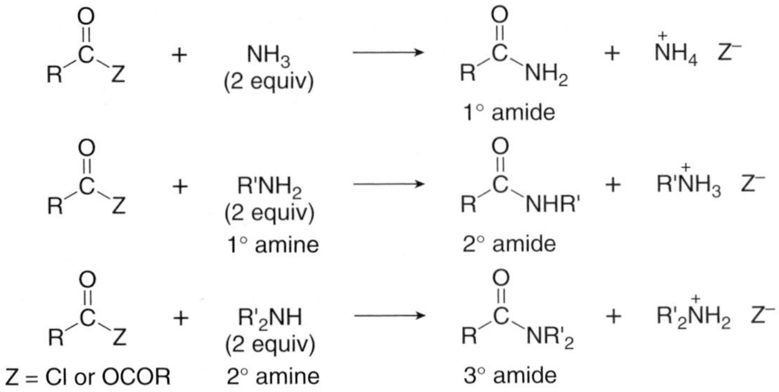

PROBLEM 25.29 Draw the products formed when each carbonyl compound reacts with the following amines: [1] $CH_3CH_2CH_2NH_2$; [2] $(CH_3CH_2)_2NH$.

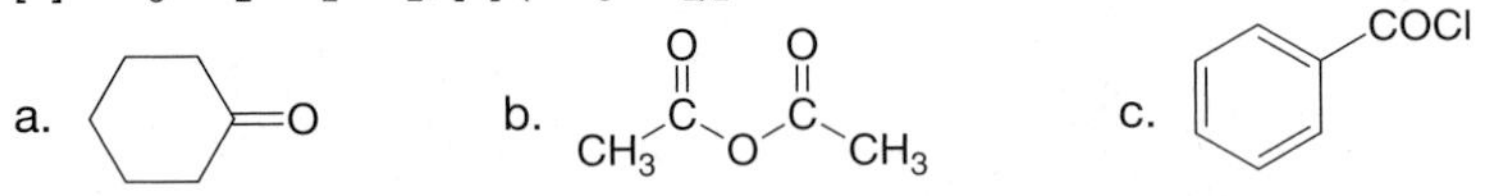

The conversion of amines to amides is useful in the synthesis of substituted anilines. For example, aniline itself does not undergo Friedel–Crafts reactions (Section 18.10B). Instead, its basic lone pair on N reacts with the Lewis acid ($AlCl_3$) to form a deactivated complex that does not undergo further reaction.

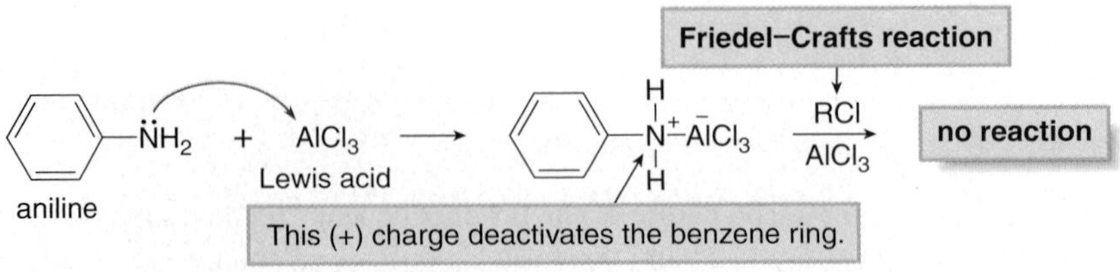

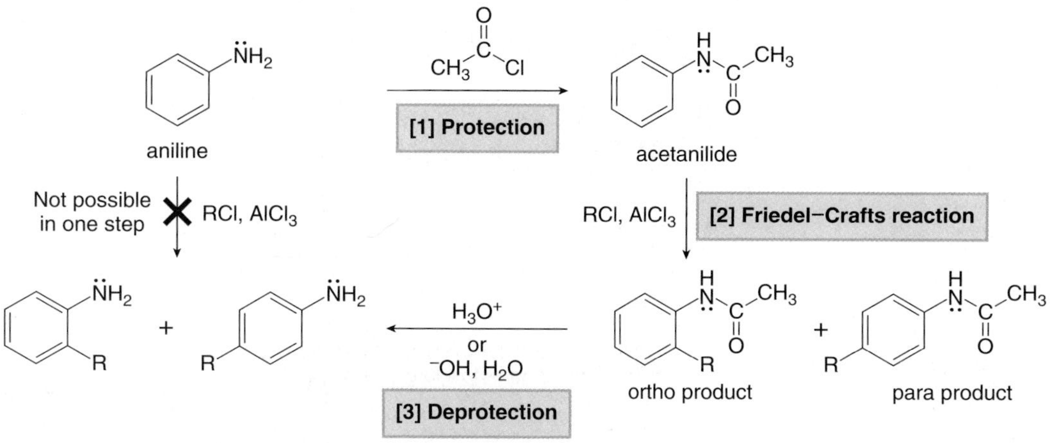

Figure 25.10   An amide as
a protecting group for an amine

A three-step sequence uses an amide as a protecting group.
[1] Treatment of aniline with acetyl chloride (CH_3COCl) forms an amide (acetanilide).
[2] Acetanilide, having a much less basic N atom compared to aniline, undergoes electrophilic aromatic
      substitution under Friedel–Crafts conditions, forming a mixture of ortho and para products.
[3] Hydrolysis of the amide forms the Friedel–Crafts substitution products.

The N atom of an amide, however, is much less basic than the N atom of an amine, so it does not undergo a similar Lewis acid–base reaction with $AlCl_3$. A three-step reaction sequence involving an intermediate amide can thus be used to form the products of the Friedel–Crafts reaction.

[1] **Convert the amine (aniline) into an amide (acetanilide).**
[2] **Carry out the Friedel–Crafts reaction.**
[3] **Hydrolyze the amide** to generate the free amino group.

This three-step procedure is illustrated in Figure 25.10. In this way, **the amide serves as a protecting group for the NH_2 group,** in much the same way that *tert*-butyldimethylsilyl ethers and acetals are used to protect alcohols and carbonyls, respectively (Sections 20.12 and 21.15).

PROBLEM 25.30   Devise a synthesis of each compound from aniline ($C_6H_5NH_2$).

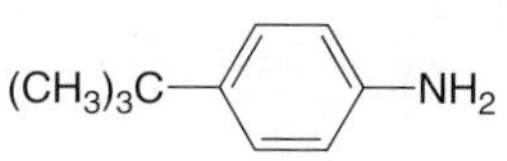

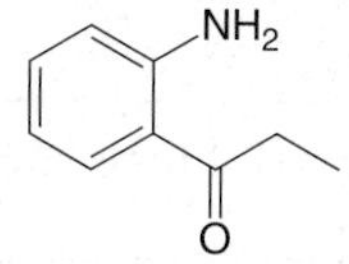

25.12   Hofmann Elimination

Amines, like alcohols, contain a poor leaving group. To undergo a β elimination reaction, for example, a 1° amine would need to lose the elements of NH_3 across two adjacent atoms. The leaving group, $^-NH_2$, is such a strong base, however, that this reaction does not occur.

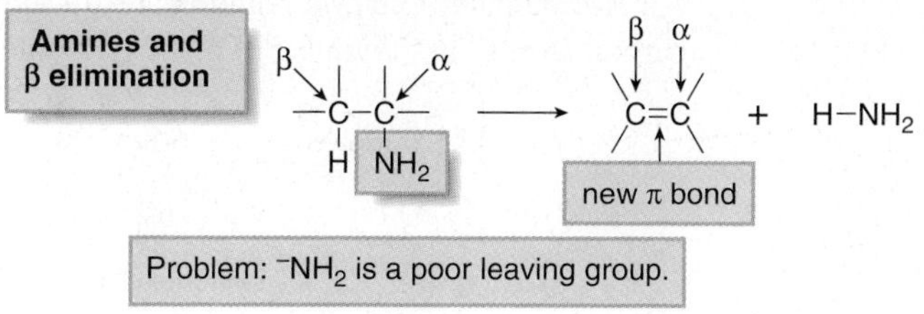

The only way around this obstacle is to convert $^-NH_2$ into a better leaving group. The most common method to accomplish this is called a **Hofmann elimination,** which converts an amine into a quaternary ammonium salt prior to β elimination.

25.12A   Details of the Hofmann Elimination

The **Hofmann elimination** converts an amine into an alkene.

Hofmann elimination— Overall reaction

$$\overset{\beta}{\underset{H}{\overset{|}{C}}}-\overset{\alpha}{\underset{NH_2}{\overset{|}{C}}} \quad \xrightarrow[\substack{[2]\ Ag_2O \\ [3]\ \Delta}]{[1]\ CH_3I\ (excess)} \quad \overset{\beta}{\underset{}{C}}=\overset{\alpha}{\underset{}{C}} \quad + \quad H_2O \quad + \quad N(CH_3)_3 \quad + \quad AgI$$

by-products

loss of H–NH₂

The Hofmann elimination consists of three steps, as shown for the conversion of propylamine to propene.

The steps in the Hofmann elimination

$$\underset{\substack{| \\ H}}{\overset{\beta}{CH_3}}-\underset{\substack{| \\ NH_2}}{\overset{}{CH}}-\overset{\alpha}{CH_2} \xrightarrow[\substack{(excess) \\ [1]}]{CH_3I} \underset{\substack{| \\ H}}{\overset{\beta}{CH_3}}-\underset{\substack{| \\ N(CH_3)_3 \\ + \\ I^-}}{\overset{}{CH}}-\overset{\alpha}{CH_2} \xrightarrow[\substack{[2]}]{Ag_2O} \underset{\substack{| \\ H}}{\overset{\beta}{CH_3}}-\underset{\substack{| \\ N(CH_3)_3 \\ + \\ + \ ^-OH \\ + \ AgI}}{\overset{}{CH}}-\overset{\alpha}{CH_2} \xrightarrow[\substack{[3]}]{\Delta} \overset{\beta}{CH_3}-\overset{\alpha}{CH}=CH_2$$

propylamine

+ H₂O

+ N(CH₃)₃

leaving group

quaternary ammonium salts

- In Step [1], the amine reacts as a nucleophile in an **S$_N$2** reaction with excess CH$_3$I to form a quaternary ammonium salt. **The N(CH$_3$)$_3$ group thus formed is a much better leaving group than $^-$NH$_2$.**

- Step [2] converts one ammonium salt into another one with a different anion. The silver(I) oxide, Ag$_2$O, replaces the I$^-$ anion with $^-$OH, a strong base.

- When the ammonium salt is heated in Step [3], $^-$**OH removes a proton from the β carbon atom,** forming the new π bond of the alkene. The mechanism of elimination is **E2,** so:

- All bonds are broken and formed in a single step.
- Elimination occurs through an anti periplanar geometry—that is, H and N(CH$_3$)$_3$ are oriented on opposite sides of the molecule.

The general E2 mechanism for the Hofmann elimination is shown in Mechanism 25.1.

MECHANISM 25.1

The E2 Mechanism for the Hofmann Elimination

$$HO^- \curvearrowright \overset{H}{\underset{N(CH_3)_3}{\overset{|}{C}}}-\overset{}{C} \xrightarrow{\Delta} \overset{}{C}=\overset{}{C} \quad + \quad H_2O \quad + \quad N(CH_3)_3$$

anti periplanar arrangement of H and N(CH₃)₃

leaving group

All Hofmann elimination reactions result in the formation of a new π bond between the α and β carbon atoms, as shown for cyclohexylamine and 2-phenylethylamine.

Examples

cyclohexylamine

$$\xrightarrow[\substack{[2]\ Ag_2O \\ [3]\ \Delta}]{[1]\ CH_3I\ (excess)}$$

$$\underset{2\text{-phenylethylamine}}{\overset{\beta\quad\alpha}{CH_2CH_2NH_2}} \xrightarrow[\substack{[2]\ Ag_2O \\ [3]\ \Delta}]{[1]\ CH_3I\ (excess)} \overset{\beta\quad\alpha}{CH=CH_2}$$

To help remember the reagents needed for the steps of the Hofmann elimination, keep in mind what happens in each step.

- ◆ **Step [1]** makes a good leaving group by forming a quaternary ammonium salt.
- ◆ **Step [2]** provides the strong base, $^-$OH, needed for elimination.
- ◆ **Step [3]** is the E2 elimination that forms the new π bond.

PROBLEM 25.31   Draw an energy diagram for the following reaction. Label the axes, the starting materials, E_a, and $\Delta H°$. Assume the reaction is exothermic. Draw a structure for the likely transition state.

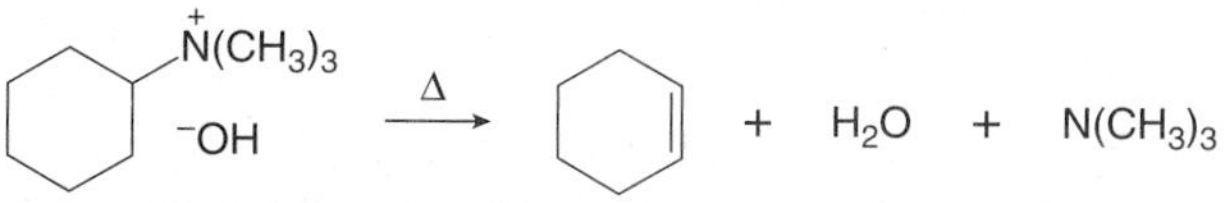

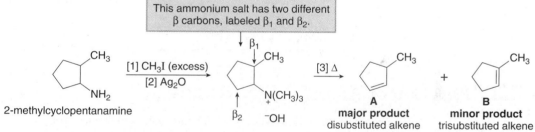

PROBLEM 25.32   Draw the product formed by treating each compound with excess CH_3I, followed by Ag_2O, and then heat.

a. $CH_3CH_2CH_2CH_2-NH_2$ b. $(CH_3)_2CHNH_2$ c. 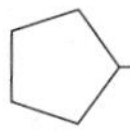 (cyclopentyl)$-NH_2$

25.12B   Regioselectivity of the Hofmann Elimination

There is one major difference between a Hofmann elimination and other E2 eliminations.

> ◆ When constitutional isomers are possible, the major alkene has the *less* substituted double bond in a Hofmann elimination.

For example, Hofmann elimination of the elements of H and $N(CH_3)_3$ from 2-methylcyclopentanamine yields two constitutional isomers: the disubstituted alkene **A** (the major product) and the trisubstituted alkene **B** (the minor product).

This ammonium salt has two different β carbons, labeled $β_1$ and $β_2$.

2-methylcyclopentanamine   **[1]** CH_3I (excess) **[2]** Ag_2O **[3]** Δ   **A** major product disubstituted alkene   +   **B** minor product trisubstituted alkene

This regioselectivity distinguishes a Hofmann elimination from other E2 eliminations, which form the more substituted double bond by the Zaitsev rule (Section 8.5). This result is sometimes explained by the size of the leaving group, $N(CH_3)_3$. **In a Hofmann elimination, the base removes a proton from the less substituted, more accessible β carbon atom, because of the bulky leaving group on the nearby α carbon.**

SAMPLE PROBLEM 25.5   Draw the major product formed from Hofmann elimination of the following amine.

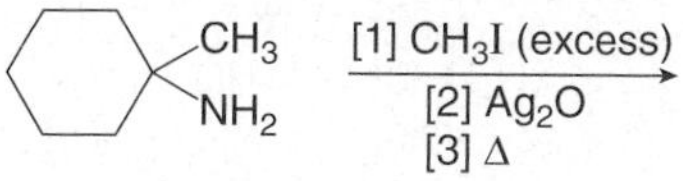

(1-methylcyclohexanamine)   **[1]** CH_3I (excess) **[2]** Ag_2O **[3]** Δ

SOLUTION

The amine has three β carbons but two of them are identical, so two alkenes are possible. Draw elimination products by forming alkenes having a C=C between the α and β carbons. The major product has the **less substituted double bond**—that is, the alkene with the C=C between the α and $β_1$ carbons in this example.

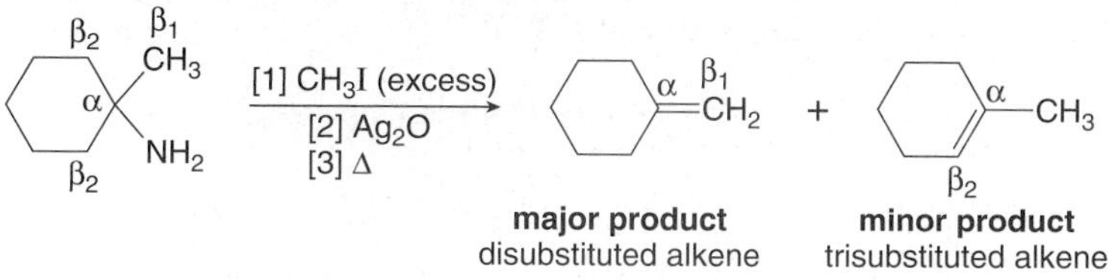

$β_2$ $β_1$ **[1]** CH_3I (excess) **[2]** Ag_2O **[3]** Δ   α $β_1$ =CH_2 major product disubstituted alkene   +   α $-CH_3$ $β_2$ minor product trisubstituted alkene

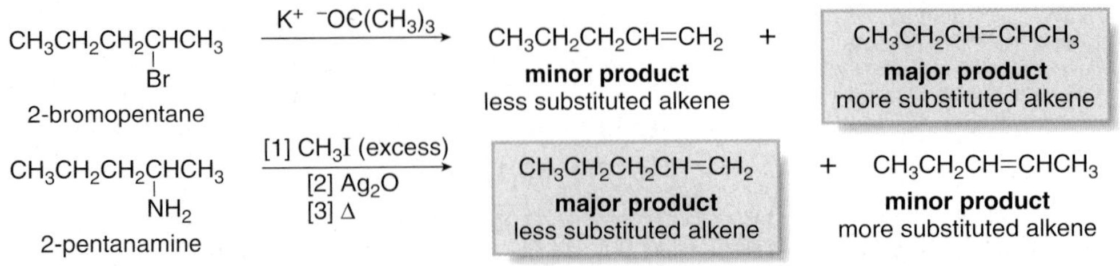

Figure 25.11 A comparison of E2 elimination reactions using alkyl halides and amines

$$CH_3CH_2CH_2\underset{\underset{Br}{|}}{C}HCH_3 \xrightarrow{K^+\ {}^-OC(CH_3)_3} CH_3CH_2CH_2CH=CH_2 + \boxed{CH_3CH_2CH=CHCH_3}$$

2-bromopentane

$CH_3CH_2CH_2CH=CH_2$
minor product
less substituted alkene

$\boxed{CH_3CH_2CH=CHCH_3}$
major product
more substituted alkene

$$CH_3CH_2CH_2\underset{\underset{NH_2}{|}}{C}HCH_3 \xrightarrow[\substack{[2]\ Ag_2O \\ [3]\ \Delta}]{[1]\ CH_3I\ (excess)} \boxed{CH_3CH_2CH_2CH=CH_2} + CH_3CH_2CH=CHCH_3$$

2-pentanamine

$\boxed{CH_3CH_2CH_2CH=CH_2}$
major product
less substituted alkene

$CH_3CH_2CH=CHCH_3$
minor product
more substituted alkene

Figure 25.11 contrasts the products formed by E2 elimination reactions using an alkyl halide and an amine as starting materials. Treatment of the alkyl halide (2-bromopentane) with base forms the more substituted alkene as the major product, following the **Zaitsev rule.** In contrast, the three-step Hofmann sequence of an amine (2-pentanamine) forms the less substituted alkene as major product.

PROBLEM 25.33 Draw the major product formed by treating each amine with excess CH_3I, followed by Ag_2O, and then heat.

a. 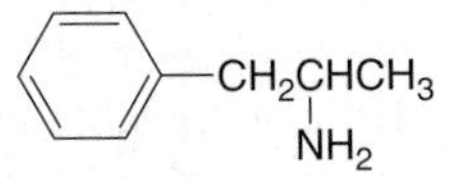   b. 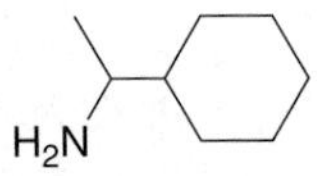   c.

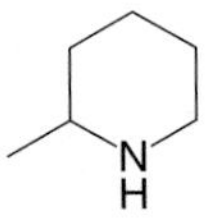

PROBLEM 25.34 Draw the major product formed in each reaction.

a. 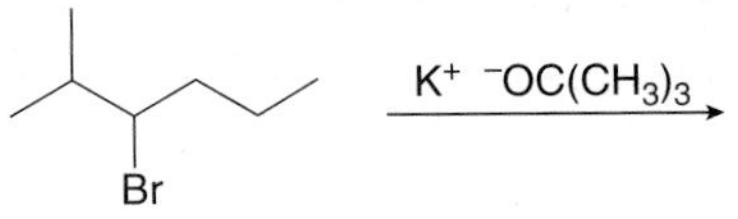 $\xrightarrow{K^+\ {}^-OC(CH_3)_3}$

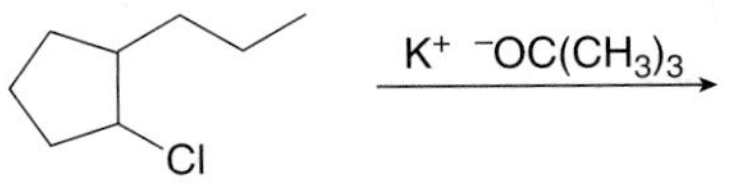

b. 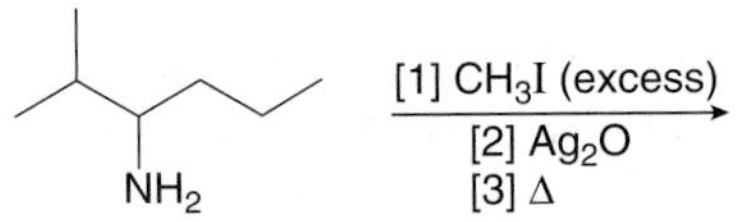 $\xrightarrow[\substack{[2]\ Ag_2O \\ [3]\ \Delta}]{[1]\ CH_3I\ (excess)}$

c. $\xrightarrow{K^+\ {}^-OC(CH_3)_3}$

d. 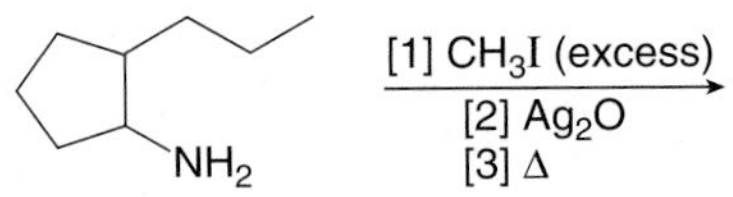 $\xrightarrow[\substack{[2]\ Ag_2O \\ [3]\ \Delta}]{[1]\ CH_3I\ (excess)}$

25.13   Reaction of Amines with Nitrous Acid

Nitrous acid, HNO_2, is a weak, unstable acid formed from $NaNO_2$ and a strong acid like HCl.

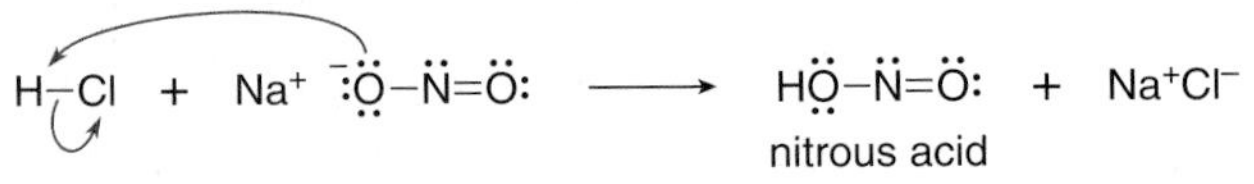

$$H{-}Cl + Na^+\ {}^-\ddot{O}{-}\ddot{N}{=}\ddot{O}: \longrightarrow H\ddot{O}{-}\ddot{N}{=}\ddot{O}: + Na^+Cl^-$$

nitrous acid

In the presence of acid, nitrous acid decomposes to $^+$**NO,** the **nitrosonium ion.** This electrophile then goes on to react with the nucleophilic nitrogen atom of amines to form **diazonium salts** $(RN_2{}^+Cl^-)$ from 1° amines and **N-nitrosamines** $(R_2NN{=}O)$ from 2° amines.

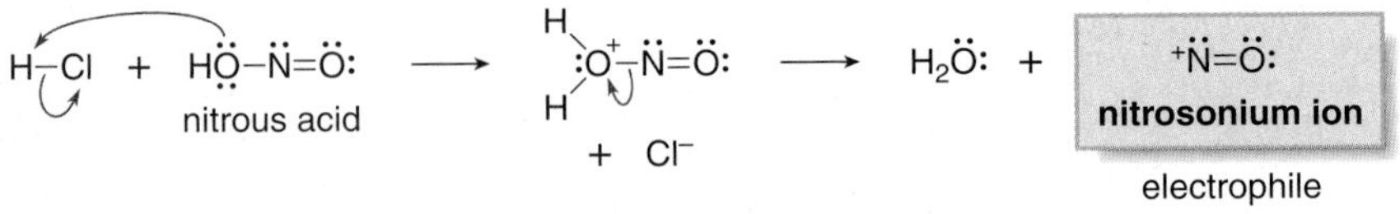

$$H{-}Cl + H\ddot{O}{-}\ddot{N}{=}\ddot{O}: \longrightarrow \underset{H}{\overset{H}{:\!\ddot{O}^+{-}\ddot{N}{=}\ddot{O}:}} \longrightarrow H_2\ddot{O}: + \boxed{{}^+\ddot{N}{=}\ddot{O}:}$$

nitrous acid

$+ \ Cl^-$

$\boxed{{}^+\ddot{N}{=}\ddot{O}:}$
nitrosonium ion

electrophile

25.13A   Reaction of $^+$NO with 1° Amines

Nitrous acid reacts with 1° alkylamines and arylamines to form **diazonium salts.** This reaction is called **diazotization.**

Preparation of diazonium salts

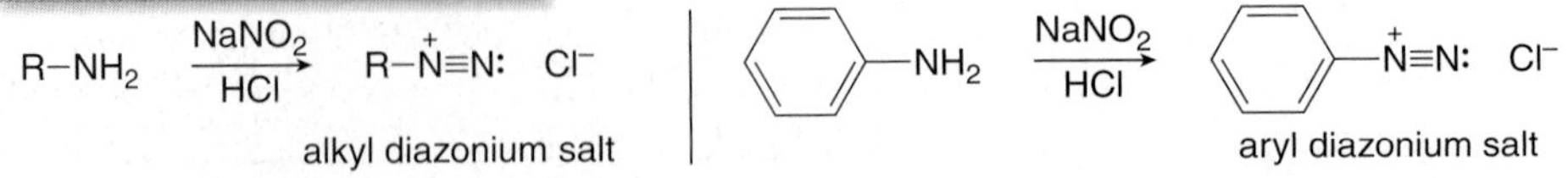

$$R{-}NH_2 \xrightarrow[HCl]{NaNO_2} R{-}\overset{+}{N}{\equiv}N: \ \ Cl^-$$

alkyl diazonium salt

$\xrightarrow[HCl]{NaNO_2}$

aryl diazonium salt

The mechanism for this reaction consists of many steps. It begins with nucleophilic attack of the amine on the nitrosonium ion, and it can conceptually be divided into two parts: formation of an *N*-nitrosamine, followed by loss of H_2O, as shown in Mechanism 25.2.

MECHANISM 25.2

Formation of a Diazonium Salt from a 1° Amine

Part [1] Formation of an *N*-nitrosamine

$$R-\overset{..}{N}H_2 \;+\; {}^{+}\overset{..}{N}=\overset{..}{O}: \;\xrightarrow{[1]}\; R-\overset{H}{\underset{H}{\overset{Cl}{\overset{|}{\underset{|}{N^+}}}}}-\overset{..}{N}=\overset{..}{O}: \;\xrightarrow{[2]}\; R-\overset{..}{\underset{H}{N}}-\overset{..}{N}=\overset{..}{O}: \;+\; HCl$$

(from NaNO₂ + HCl)   *N*-nitrosamine

◆ In Part [1], the amine is converted to an **N-nitrosamine** by nucleophilic attack of the amino group on ⁺NO, followed by loss of a proton.

Part [2] Loss of H_2O to form the diazonium salt

$$R-\overset{..}{\underset{H}{N}}-\overset{..}{N}=\overset{..}{O}: \;+\; H-Cl \;\xrightarrow{[3]}\; R-\overset{..}{\underset{\underset{H}{|}}{N}}-\overset{..}{N}=\overset{+}{\underset{}{O}}-H \;\xrightarrow{[4]}\; R-\overset{..}{N}=\overset{..}{N}-\overset{..}{O}H$$

Cl⁻   H–Cl

N-nitrosamine

◆ In Part [2], three proton transfer reactions lead to loss of H_2O in Step [6] and formation of the diazonium ion.

$$H_2\overset{..}{O}: \;+\; \boxed{R-\overset{+}{N}\equiv N:} \;\xleftarrow{[6]}\; R-\overset{..}{N}=\overset{+}{N}-\overset{+}{O}H_2 \;+\; Cl⁻$$

diazonium salt

Alkyl diazonium salts are generally not useful compounds. They readily decompose below room temperature to form carbocations with loss of N_2, a very good leaving group. These carbocations usually form a complex mixture of substitution, elimination, and rearrangement products.

$$CH_3-\overset{\overset{CH_3}{|}}{\underset{\underset{CH_3}{|}}{C}}-NH_2 \;\xrightarrow[\text{HCl}]{NaNO_2}\; \left[CH_3-\overset{\overset{CH_3}{|}}{\underset{\underset{CH_3}{|}}{C}}-N\equiv N: \;\;Cl⁻ \right] \;\longrightarrow\; \overset{CH_3}{\underset{CH_3 \quad CH_3}{\overset{|}{C^+}}} \;\longrightarrow\;$$

1° alkylamine   unstable diazonium salt   **carbocation**
+
N_2
good leaving group

products of substitution, elimination, and (in some cases) rearrangement

Care must be exercised in handling diazonium salts, because they can explode if allowed to dry.

On the other hand, **aryl diazonium salts are very useful synthetic intermediates.** Although they are rarely isolated and are generally unstable above 0 °C, they are useful starting materials in two general kinds of reactions described in Section 25.14.

25.13B   Reaction of ⁺NO with 2° Amines

Secondary alkylamines and arylamines react with nitrous acid to form *N*-nitrosamines.

$$R-\overset{..}{\underset{R}{N}}-H \;\xrightarrow[\text{HCl}]{NaNO_2}\; R-\overset{..}{\underset{R}{N}}-\overset{..}{N}=\overset{..}{O}:$$

2° amine   *N*-nitrosamine

As mentioned in Section 7.16, many *N*-nitrosamines are potent carcinogens found in some food and tobacco smoke. Nitrosamines in food are formed in the same way they are formed in the laboratory: **reaction of a 2° amine with the nitrosonium ion,** formed from nitrous acid (HNO_2). Mechanism 25.3 is shown for the conversion of dimethylamine [$(CH_3)_2NH$] to *N*-nitrosodimethylamine [$(CH_3)_2NN=O$].

MECHANISM 25.3

Formation of an *N*-Nitrosamine from a 2° Amine

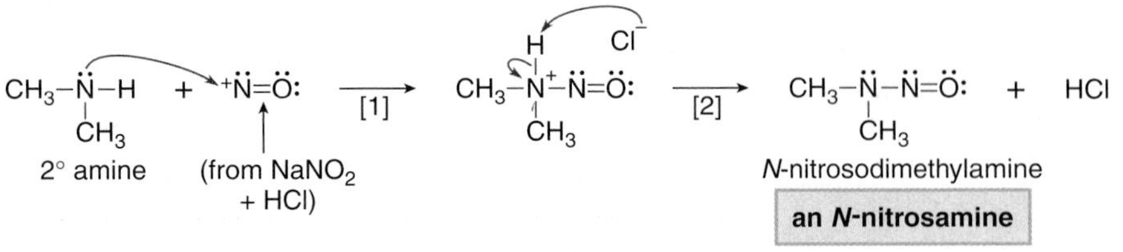

CH₃–N̈–H + ⁺N̈=Ö: —[1]→ CH₃–N̈⁺–N̈=Ö: —[2]→ CH₃–N̈–N̈=Ö: + HCl

2° amine   (from NaNO₂ + HCl)

N-nitrosodimethylamine

an *N*-nitrosamine

◆ The amine is converted to an ***N*-nitrosamine** by nucleophilic attack of the amino group on ⁺NO, followed by loss of a proton.

PROBLEM 25.35   Draw the product formed when each compound is treated with NaNO₂ and HCl.

a. 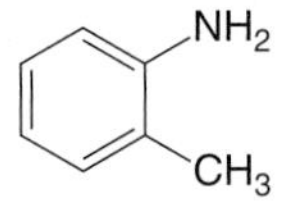 (benzene ring with NH₂ and CH₃ substituents)   b. CH₃CH₂–N(H)–CH₃   c. 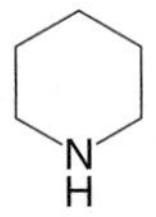 (piperidine, N–H)   d. 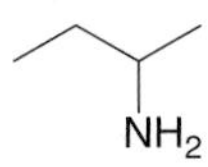 (branched alkyl chain with NH₂)

25.14   Substitution Reactions of Aryl Diazonium Salts

Aryl diazonium salts undergo two general reactions:

◆ **Substitution** of N_2 by an atom or a group of atoms **Z.**

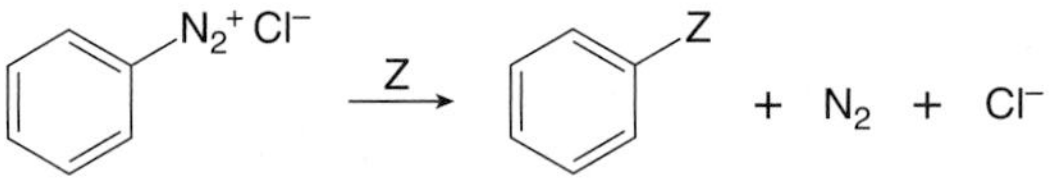

(benzene)–N_2^+ Cl⁻ —Z→ (benzene)–Z + N_2 + Cl⁻

◆ **Coupling** of a diazonium salt with another benzene derivative to form an **azo compound,** a compound containing a nitrogen–nitrogen double bond.

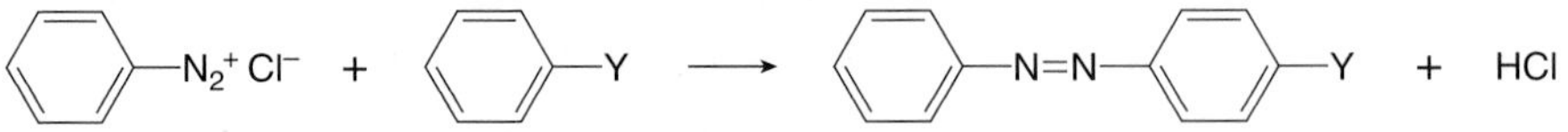

(benzene)–N_2^+ Cl⁻ + (benzene)–Y ⟶ (benzene)–N=N–(benzene)–Y + HCl

azo compound

Y = NH_2, NHR, NR_2, OH (a strong electron-donor group)

25.14A   Specific Substitution Reactions

Aryl diazonium salts react with a variety of reagents to form products in which Z (an atom or group of atoms) replaces N_2, a very good leaving group. The mechanism of these reactions varies with the identity of Z, so we will concentrate on the products of the reactions, not the mechanisms.

| General substitution reaction | (benzene)–N_2^+ Cl⁻ —Z→ (benzene)–Z + N_2 + Cl⁻ |

Z replaces N_2 good leaving group

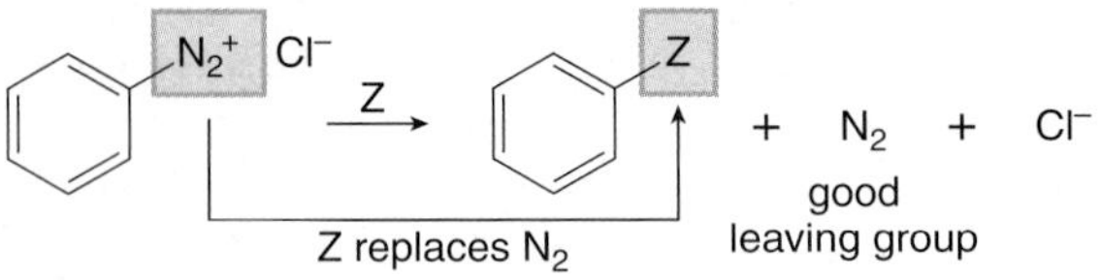

[1]   Substitution by OH—Synthesis of phenols

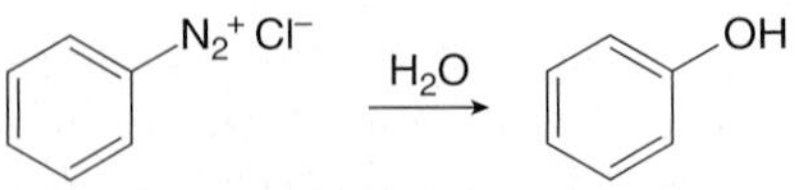

(benzene)–N_2^+ Cl⁻ —H_2O→ (benzene)–OH

phenol

A diazonium salt reacts with H_2O to form a **phenol.**

[2]   Substitution by Cl or Br—Synthesis of aryl chlorides and bromides

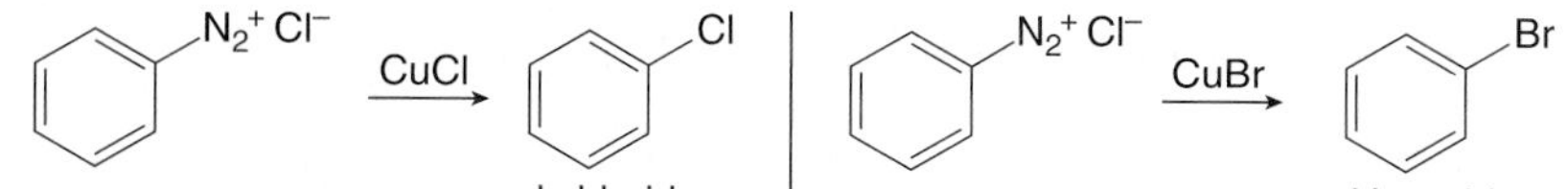

aryl chloride   aryl bromide

A diazonium salt reacts with copper(I) chloride or copper(I) bromide to form an **aryl chloride** or **aryl bromide,** respectively. This is called the **Sandmeyer reaction.** It provides an alternative to direct chlorination and bromination of an aromatic ring using Cl_2 or Br_2 and a Lewis acid catalyst.

[3]   Substitution by F—Synthesis of aryl fluorides

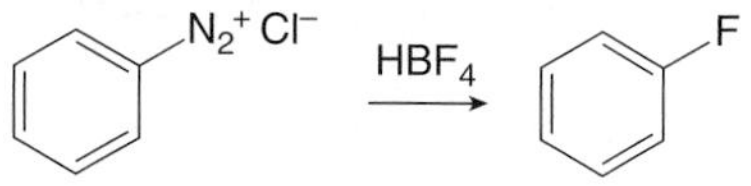

aryl fluoride

A diazonium salt reacts with fluoroboric acid (HBF_4) to form an **aryl fluoride.** This is a useful reaction because aryl fluorides cannot be produced by direct fluorination with F_2 and a Lewis acid catalyst, as F_2 reacts too violently (Section 18.3).

[4]   Substitution by I—Synthesis of aryl iodides

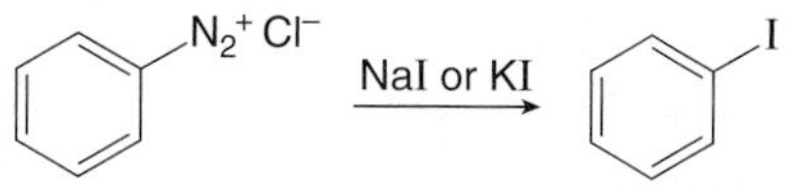

aryl iodide

A diazonium salt reacts with sodium or potassium iodide to form an aryl iodide. This, too, is a useful reaction because aryl iodides cannot be produced by direct iodination with I_2 and a Lewis acid catalyst, as I_2 reacts too slowly (Section 18.3).

[5]   Substitution by CN—Synthesis of benzonitriles

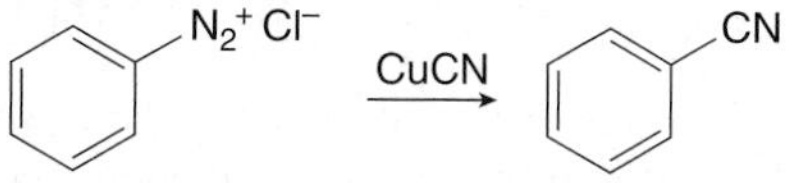

benzonitrile

A diazonium salt reacts with copper(I) cyanide to form a benzonitrile. Because a cyano group can be hydrolyzed to a carboxylic acid, reduced to an amine or aldehyde, or converted to a ketone with organometallic reagents, this reaction provides easy access to a wide variety of benzene derivatives using chemistry described in Section 22.18.

[6]   Substitution by H—Synthesis of benzene

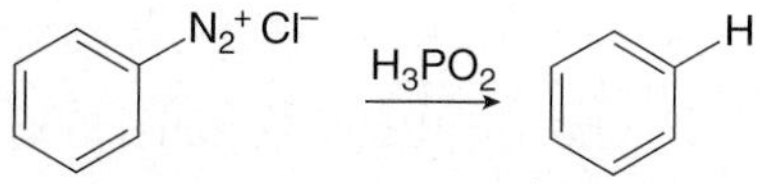

benzene

A diazonium salt reacts with hypophosphorus acid (H_3PO_2) to form benzene. This reaction has limited utility because it reduces the functionality of the benzene ring by replacing N_2 with a hydrogen atom. Nonetheless, this reaction *is* useful in synthesizing compounds that have substitution patterns that are not available by other means.

For example, it is not possible to synthesize 1,3,5-tribromobenzene from benzene by direct bromination. Because Br is an ortho, para director, bromination with Br_2 and $FeBr_3$ will not add Br substituents meta to each other on the ring.

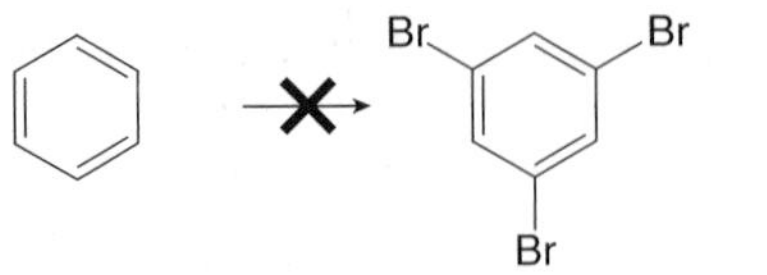

The Br atoms are ortho, para directors located meta to each other.

1,3,5-tribromobenzene

It is possible, however, to add three Br atoms meta to each other when aniline is the starting material. Because an NH_2 group is a very powerful ortho, para director, three Br atoms are introduced in a single step on halogenation (Section 18.10A). Then, the NH_2 group can be removed by diazotization and reaction with H_3PO_2.

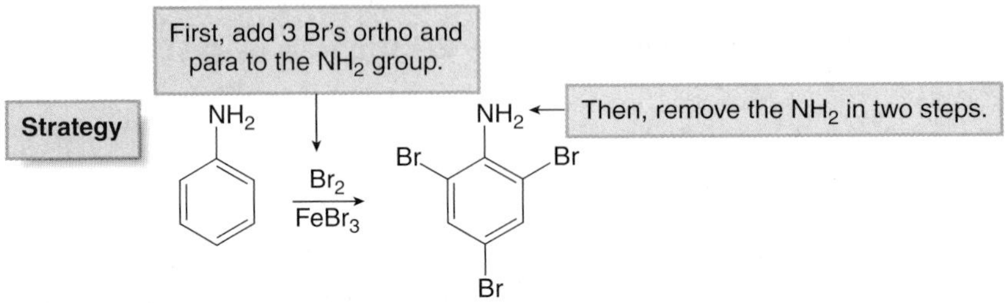

First, add 3 Br's ortho and para to the NH_2 group.

Strategy

Then, remove the NH_2 in two steps.

The complete synthesis of 1,3,5-tribromobenzene from benzene is outlined in Figure 25.12.

Figure 25.12
The synthesis of 1,3,5-tribromobenzene from benzene

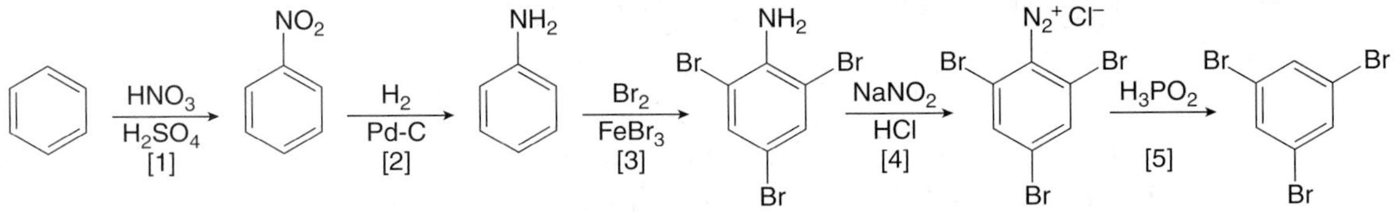

- Nitration followed by reduction forms aniline ($C_6H_5NH_2$) from benzene (Steps [1] and [2]).
- Bromination of aniline yields the tribromo derivative in Step [3].
- The NH_2 group is removed by a two-step process: diazotization with $NaNO_2$ and HCl (Step [4]), followed by substitution of the diazonium ion by H with H_3PO_2.

25.14B Using Diazonium Salts in Synthesis

Diazonium salts provide easy access to many different benzene derivatives. Keep in mind the following four-step sequence, because it will be used to synthesize many substituted benzenes.

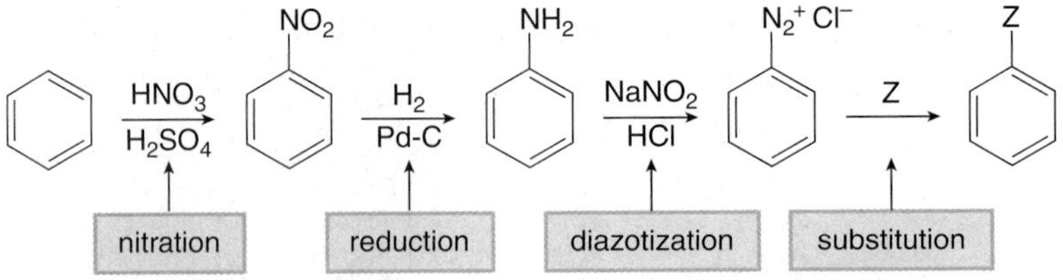

Sample Problems 25.6 and 25.7 apply these principles to two different multistep syntheses.

SAMPLE PROBLEM 25.6   Synthesize *m*-chlorophenol from benzene.

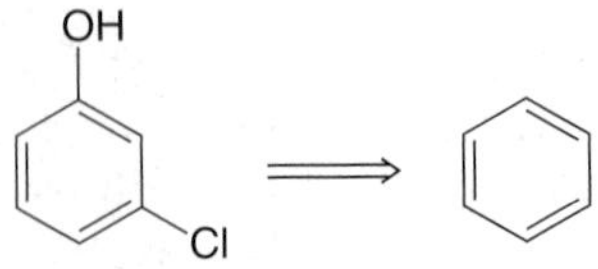

SOLUTION
Both OH and Cl are ortho, para directors, but they are located meta to each other. The OH group must be formed from a diazonium salt, which can be made from an NO_2 group by a stepwise method.

RETROSYNTHETIC ANALYSIS

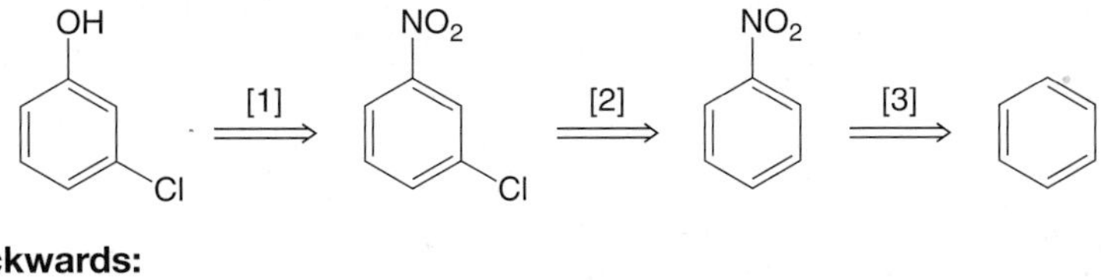

Working backwards:

- [1] Form the OH group from NO_2 by a three-step procedure using a diazonium salt.
- [2] Introduce Cl meta to NO_2 by halogenation.
- [3] Add the NO_2 group by nitration.

SYNTHESIS

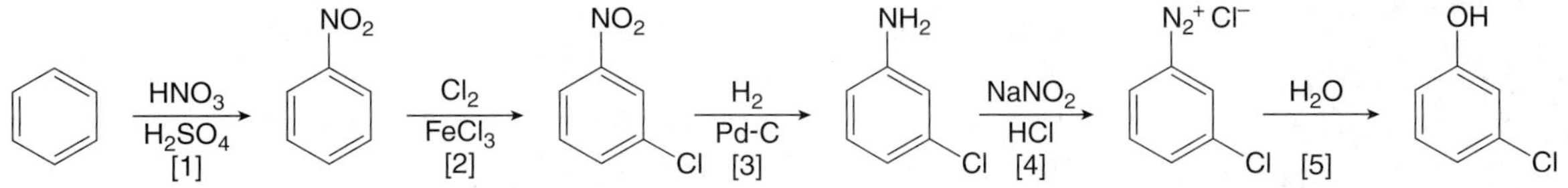

- Nitration followed by chlorination meta to the NO_2 group forms the meta disubstituted benzene (Steps [1]–[2]).
- Reduction of the nitro group followed by diazotization forms the diazonium salt in Step [4], which is then converted to the desired phenol by treatment with H_2O (Step [5]).

SAMPLE PROBLEM 25.7   Synthesize p-bromobenzaldehyde from benzene.

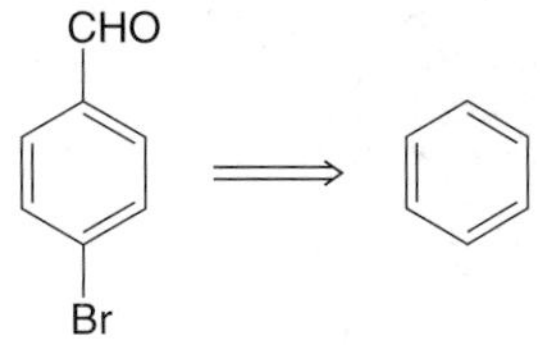

SOLUTION

Because the two groups are located para to each other and Br is an ortho, para director, Br should be added to the ring first. To add the CHO group, recall that it can be formed from CN by reduction.

RETROSYNTHETIC ANALYSIS

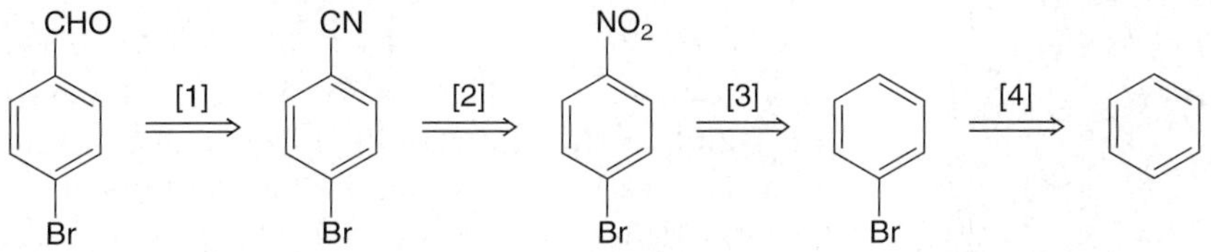

Working backwards:

- [1] Form the CHO group by reduction of CN.
- [2] Prepare the CN group from an NO_2 group by a three-step sequence using a diazonium salt.
- [3] Introduce the NO_2 group by nitration, para to the Br atom.
- [4] Introduce Br by bromination with Br_2 and $FeBr_3$.

SYNTHESIS

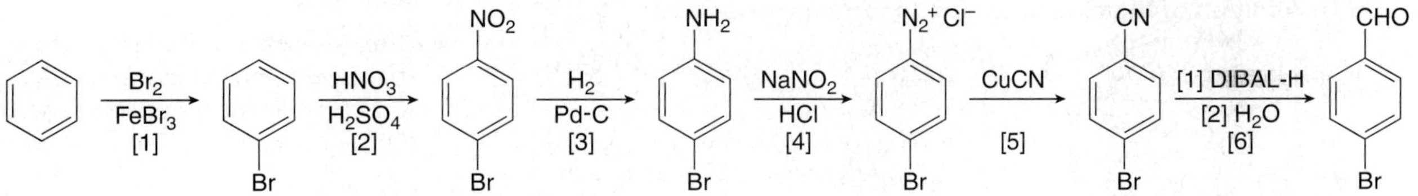

- Bromination followed by nitration forms a disubstituted benzene with two para substituents (Steps [1]–[2]).
- Reduction of the NO_2 group followed by diazotization forms the diazonium salt in Step [4], which is converted to a nitrile by reaction with CuCN (Step [5]).
- Reduction of the CN group with DIBAL-H (a mild reducing agent) forms the CHO group and completes the synthesis.

PROBLEM 25.36   Draw the product formed in each reaction.

a. [benzene ring]—NH_2 $\xrightarrow[\text{[2] CuBr}]{\text{[1] NaNO}_2, \text{ HCl}}$

c. CH_3O—[benzene ring]—NH_2 $\xrightarrow[\text{[2] HBF}_4]{\text{[1] NaNO}_2, \text{ HCl}}$

b. [benzene ring with O_2N]—NH_2 $\xrightarrow[\text{[2] H}_2\text{O}]{\text{[1] NaNO}_2, \text{ HCl}}$

d. [benzene ring with Cl]—N_2^+ Cl^- $\xrightarrow[\text{[3] H}_2\text{O}]{\text{[1] CuCN} \quad \text{[2] LiAlH}_4}$

PROBLEM 25.37   Devise a synthesis of each compound from benzene.

a. [benzene ring with F]

b. [benzene ring with HO and OH]

c. [benzene ring with CH_3 and I]

d. [benzene ring with Cl, Cl, Cl]

25.15   Coupling Reactions of Aryl Diazonium Salts

The second general reaction of diazonium salts is **coupling.** When a diazonium salt is treated with an aromatic compound that contains a strong electron-donor group, the two rings join together to form an **azo compound,** a compound with a nitrogen–nitrogen double bond.

Azo coupling—General reaction

[benzene]—N_2^+ Cl^- + [benzene]—Y $\longrightarrow$ [benzene]—N=N—[benzene]—Y + HCl

Y = NH_2, NHR, NR_2, OH
(a strong electron-donor group)

azo compound

> Synthetic dyes are described in more detail in Section 25.16.

Azo compounds are highly conjugated, rendering them colored (Section 16.15). Many of these compounds, such as the azo compound "butter yellow," are synthetic dyes. Butter yellow was once used to color margarine.

Example   [benzene]—N_2^+ Cl^- + [benzene]—$N(CH_3)_2$ $\longrightarrow$ [benzene]—N=N—[benzene]—$N(CH_3)_2$

a yellow azo dye
"butter yellow"

This reaction is another example of **electrophilic aromatic substitution,** with the **diazonium salt acting as the electrophile.** Like all electrophilic substitutions (Section 18.2), the mechanism has two steps: addition of the electrophile (the diazonium ion) to form a resonance-stabilized carbocation, followed by deprotonation, as shown in Mechanism 25.4.

MECHANISM 25.4

Azo Coupling

Step [1]   Addition of the diazonium ion to form a carbocation

[benzene]—$\overset{+}{N}$=N: + H—[benzene]—$\ddot{Y}$ $\longrightarrow$ [benzene]—$\ddot{N}$=$\ddot{N}$—[benzene ring with +]—$\ddot{Y}$

(+ three additional resonance structures)
resonance-stabilized carbocation

♦ **Step [1]** The electrophilic diazonium ion reacts with the electron-rich benzene ring to form a resonance-stabilized carbocation. (Only one resonance structure is drawn.)

Step [2]   Loss of a proton to re-form the aromatic ring

[benzene]—$\ddot{N}$=$\ddot{N}$—[benzene ring with +]—$\ddot{Y}$ ... $Cl^- \rightarrow H$ $\longrightarrow$ [benzene]—$\ddot{N}$=$\ddot{N}$—[benzene]—$\ddot{Y}$ + HCl

♦ **Step [2]** Loss of a proton regenerates the aromatic ring.

Because a diazonium salt is only weakly electrophilic, the reaction occurs only when the benzene ring has **a strong electron-donor group Y, where Y = NH₂, NHR, NR₂, or OH.** Although these groups activate both the ortho and para positions, para substitution occurs unless the para position already has another substituent present.

To determine what starting materials are needed to synthesize a particular azo compound, always divide the molecule into two components: **one has a benzene ring with a diazonium ion, and one has a benzene ring with a very strong electron-donor group.**

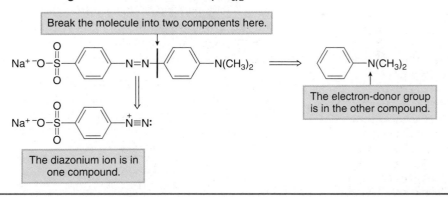

SAMPLE PROBLEM 25.8   What starting materials are needed to synthesize the following azo compound?

methyl orange
an orange dye

SOLUTION
Both benzene rings in methyl orange have a substituent, but only one group, $N(CH_3)_2$, is a strong electron donor. In determining the two starting materials, the **diazonium ion must be bonded to the ring that is *not* bonded to $N(CH_3)_2$.**

PROBLEM 25.38   Draw all possible resonance structures for the resonance-stabilized cation formed as an intermediate in azo coupling (the product of Step [1], Mechanism 25.4).

PROBLEM 25.39   Draw the product formed when $C_6H_5N_2{}^+Cl^-$ reacts with each compound.

a. (structure with NH₂)   b. HO–   c. HO––OH

PROBLEM 25.40   What starting materials are needed to synthesize each azo compound?

a. H_2N––N=N– (ring with O_2N)   b. HO––N=N– (ring with Cl and CH_3)

25.16 Application: Perkin's Mauveine and Synthetic Dyes

Azo compounds have two important applications: as dyes and as sulfa drugs, the first synthetic antibiotics (Section 25.17).

25.16A Natural and Synthetic Dyes

Until the synthesis of mauveine by William Henry Perkin in 1856, all dyes were naturally occurring, obtained from plants, animals, or minerals. Three natural dyes known for centuries are **indigo, tyrian purple,** and **alizarin.**

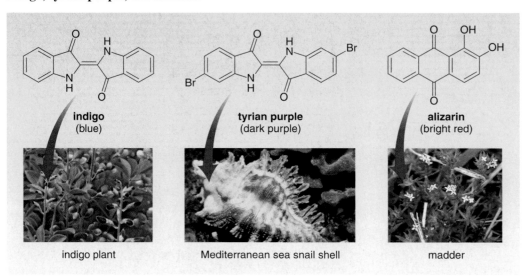

indigo
(blue)

tyrian purple
(dark purple)

alizarin
(bright red)

indigo plant

Mediterranean sea snail shell

madder

> Perkin's synthesis of mauveine is just one example of an important accidental scientific discovery. He was trying to synthesize the antimalarial drug quinine (the introductory molecule in Chapter 17) by oxidizing aniline. Instead, his reaction yielded a black tar from which he isolated a purple substance in 5% yield. Perkin initially named his dye tyrian purple, after the natural dye isolated from Mediterranean snails. It was later named **mauve** by the French, and it is that name that remains with it today.

The blue dye **indigo,** derived from the plant *Indigofera tinctoria,* has been used in India for thousands of years. Traders introduced it to the Mediterranean area and then to Europe. **Tyrian purple,** a natural dark purple dye obtained from the mucous gland of a Mediterranean snail of the genus *Murex,* was a symbol of royalty before the collapse of the Roman empire. **Alizarin,** a bright red dye obtained from madder root (*Rubia tinctorum*), a plant native to India and northeastern Asia, has been found in cloth entombed with Egyptian mummies.

Because all three of these dyes were derived from natural sources, they were difficult to obtain, making them expensive and available only to the privileged. As mentioned in the chapter opener, this all changed when Perkin synthesized mauveine. Mauveine is a mixture of two compounds that differ only in the presence of one methyl group on one of the aromatic rings.

Two components of Perkin's mauveine

major component

minor component

> Mauveine had been used as a dye for almost 150 years when the generally accepted structure was proven incorrect. Revised structures were reported in 1994.

Perkin's realization of the potential industrial importance of his dye was perhaps even more significant than its discovery. He patented the dye and went on to build a factory to commercially produce it on a large scale. This event began the surge of research in organic chemistry, not just in the synthesis of dyes, but in the production of perfumes, anesthetics, inks, and drugs as well. Perkin was a wealthy man when he retired at the age of 36 to devote the rest of his life to basic chemical research. The most prestigious award given by the American Chemical Society is named the Perkin Medal in his honor.

Many common synthetic dyes, such as alizarine yellow R, para red, and Congo red, are **azo compounds,** prepared by the diazonium coupling reaction described in Section 25.15.

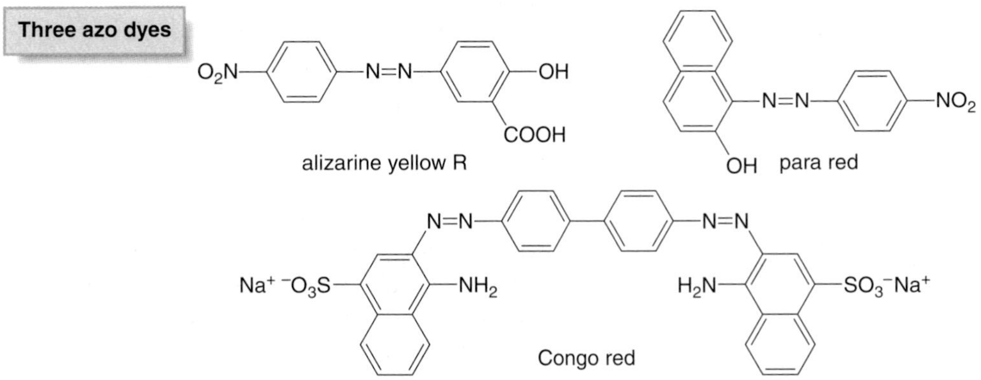

Three azo dyes

alizarine yellow R   para red

Congo red

Although natural and synthetic dyes are quite varied in structure, **all of them are colored because they are highly conjugated.** A molecule with eight or more π bonds in conjugation absorbs light in the visible region of the electromagnetic spectrum (Section 16.15A), taking on the color from the visible spectrum that it does *not* absorb.

PROBLEM 25.41   (a) What two components are needed to prepare para red by azo coupling? (b) What two components are needed to prepare alizarine yellow R?

25.16B   How Dyes Bind to Fabric

To be classified as a dye, a compound must be colored *and* it must bind to fabric. There are many ways for this binding to occur. Compounds that bind to fabric by some type of attractive forces are called **direct dyes.** These attractive forces may involve electrostatic interactions, van der Waals forces, hydrogen bonding, or sometimes, even covalent bonding. The type of interaction depends on the structure of the dye and the fiber. Thus, a compound that is good for dyeing wool or silk, both polyamides, may be poor for dyeing cotton, a carbohydrate (Figure 22.5).

Wool and silk contain charged functional groups, such as NH_3^+ and COO^-. Because of this, they bind to ionic dyes by electrostatic interactions. For example, positively charged NH_3^+ groups bonded to the protein backbone are electrostatically attracted to anionic groups in a dye like methyl orange.

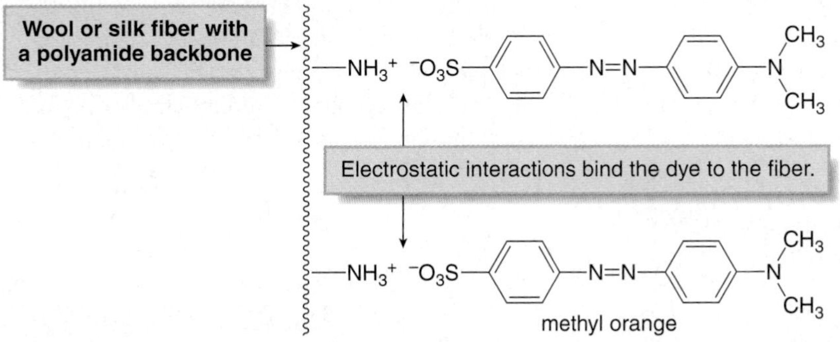

Wool or silk fiber with a polyamide backbone

Electrostatic interactions bind the dye to the fiber.

methyl orange

Cotton, on the other hand, binds dyes by hydrogen bonding interactions with its many OH groups. Thus, Congo red is bound to the cellulose backbone by hydrogen bonds.

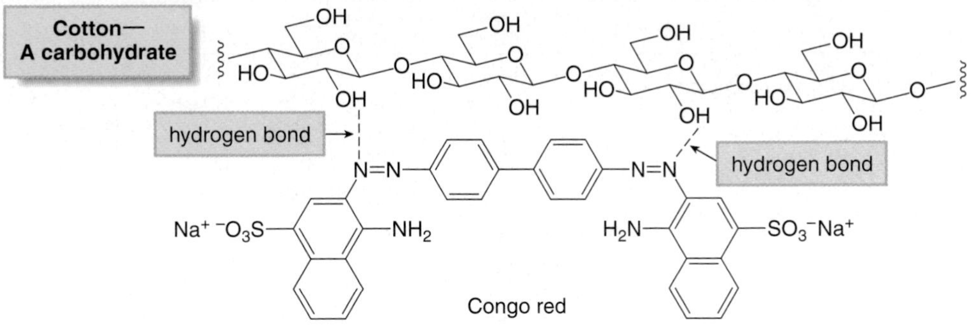

Cotton— A carbohydrate

hydrogen bond   hydrogen bond

Congo red

PROBLEM 25.42 Explain why Dacron, a polyester first discussed in Section 22.16B, does not bind well with an anionic dye such as methyl orange.

25.17 Application: Sulfa Drugs

Although they may seem quite unrelated, the synthesis of colored dyes led to the development of the first synthetic antibiotics. Much of the early effort in this field was done by the German chemist Paul Ehrlich, who worked with synthetic dyes and used them to stain tissues. This led him on a search for dyes that were lethal to bacteria without affecting other tissue cells, hoping that these dyes could treat bacterial infections. For many years this effort was unsuccessful.

Then, in 1935, Gerhard Domagk, a German physician working for a dye manufacturer, first used a synthetic dye as a drug to kill bacteria. His daughter had contracted a streptococcal infection, and as she neared death, he gave her **prontosil,** an azo dye that inhibited the growth of certain bacteria in mice. His daughter recovered, and the modern era of synthetic antibiotics was initiated. For his pioneering work, Domagk was awarded the Nobel Prize in Physiology or Medicine in 1939.

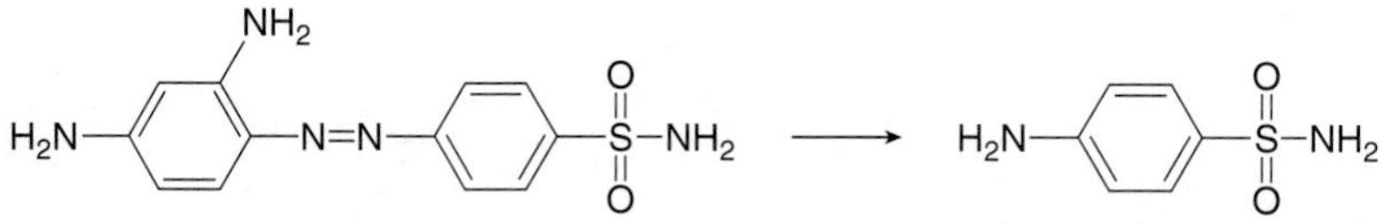

prontosil

sulfanilamide
active antibacterial agent

Prontosil and other sulfur-containing antibiotics are collectively called **sulfa drugs.** Prontosil is not the active agent itself. In cells, it is metabolized to **sulfanilamide,** the active drug. To understand how sulfanilamide functions as an antibacterial agent we must examine **folic acid,** which microorganisms synthesize from *p*-aminobenzoic acid.

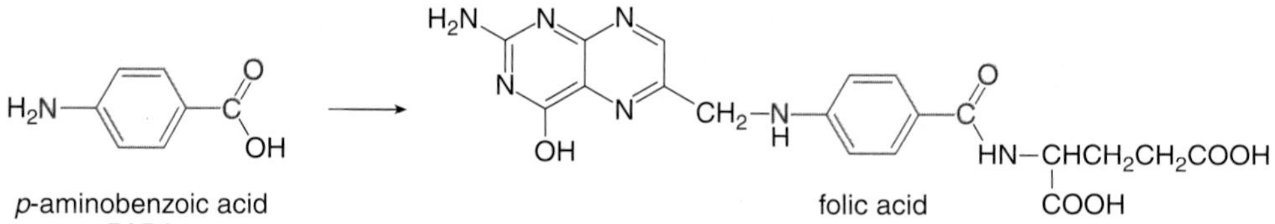

p-aminobenzoic acid
PABA

folic acid

Sulfanilamide and *p*-aminobenzoic acid are similar in size and shape and have related functional groups. Thus, when sulfanilamide is administered, bacteria attempt to use it in place of *p*-aminobenzoic acid to synthesize folic acid. Derailing folic acid synthesis means that the bacteria cannot grow and reproduce. Sulfanilamide only affects bacterial cells, though, because humans do not synthesize folic acid, and must obtain it from their diets.

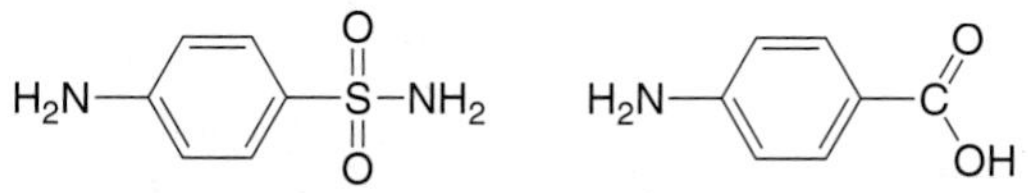

sulfanilamide

p-aminobenzoic acid

These compounds are similar in size and shape.

Many other compounds of similar structure have been prepared and are still widely used as antibiotics. The structures of two other sulfa drugs are shown in Figure 25.13.

Figure 25.13 Two common sulfa drugs

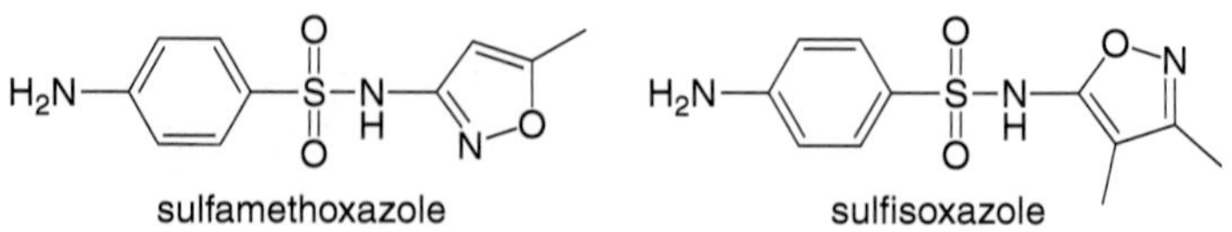

sulfamethoxazole

sulfisoxazole

- Sulfamethoxazole is the sulfa drug in Bactrim, and sulfisoxazole is sold as Gantrisin. Both drugs are commonly used in the treatment of ear and urinary tract infections.

25.18   Key Concepts—Amines

General Facts

- Amines are organic nitrogen compounds having the general structure RNH_2, R_2NH, or R_3N, with a lone pair of electrons on N (25.2).
- Amines are named using the suffix -amine (25.3).
- All amines have polar $C-N$ bonds. Primary (1°) and 2° amines have polar $N-H$ bonds and are capable of intermolecular hydrogen bonding (25.4).
- The lone pair on N makes amines strong organic bases and nucleophiles (25.8).

Summary of Spectroscopic Absorptions (25.5)

Mass spectra	Molecular ion	Amines with an odd number of N atoms give an odd molecular ion.
IR absorptions	$N-H$	3300–3500 cm^{-1} (two peaks for RNH_2, one peak for R_2NH)
^{1}H NMR absorptions	NH	0.5–5 ppm (no splitting with adjacent protons)
	$CH-N$	2.3–3.0 ppm (deshielded $C_{sp^3}-H$)
^{13}C NMR absorption	$C-N$	30–50 ppm

Comparing the Basicity of Amines and Other Compounds (25.10)

- Alkylamines (RNH_2, R_2NH, and R_3N) are more basic than NH_3 because of the electron-donating R groups (25.10A).
- Alkylamines (RNH_2) are more basic than aryl amines ($C_6H_5NH_2$), which have a delocalized lone pair from the N atom (25.10B).
- Arylamines with electron-donor groups are more basic than arylamines with electron-withdrawing groups (25.10B).
- Alkylamines (RNH_2) are more basic than amides ($RCONH_2$), which have a delocalized lone pair from the N atom (25.10C).
- Aromatic heterocycles with a localized electron pair on N are more basic than those with a delocalized lone pair from the N atom (25.10D).
- Alkylamines with a lone pair in an sp^3 hybrid orbital are more basic than those with a lone pair in an sp^2 hybrid orbital (25.10E).

Preparation of Amines (25.7)

[1]   Direct nucleophilic substitution with NH_3 and amines (25.7A)

- The mechanism is S_N2.
- The reaction works best for CH_3X or RCH_2X.
- The reaction works best to prepare 1° amines and quaternary ammonium salts.

[2]   Gabriel synthesis (25.7A)

- The mechanism is S_N2.
- The reaction works best for CH_3X or RCH_2X.
- Only 1° amines can be prepared.

[3]   Reduction methods (25.7B)

   [a] From nitro compounds

   [b] From nitriles

   [c] From amides

[4] Reductive amination (25.7C)

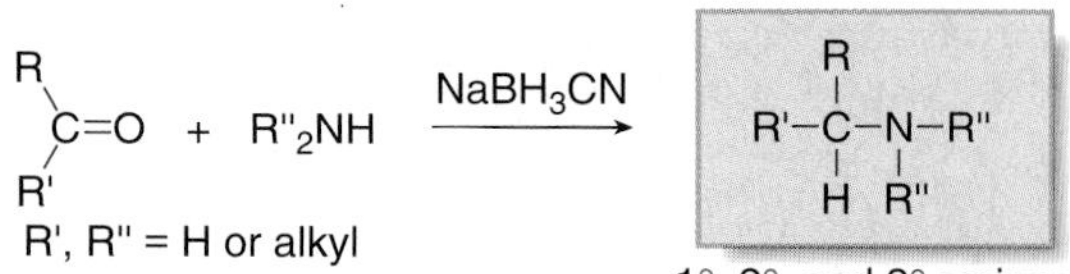

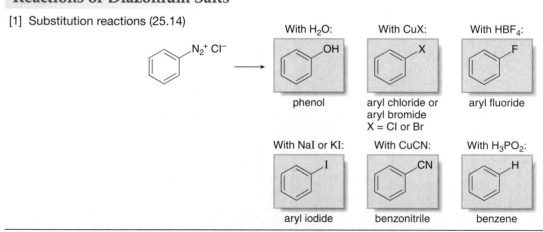

$$R-C=O \ + \ R''_2NH \xrightarrow{NaBH_3CN} \begin{array}{c} R \\ R'-C-N-R'' \\ H \ \ R'' \end{array}$$

R', R'' = H or alkyl

1°, 2°, and 3° amines

- Reductive amination adds one alkyl group (from an aldehyde or ketone) to a nitrogen nucleophile.
- Primary (1°), 2°, and 3° amines can be prepared.

Reactions of Amines

[1] Reaction as a base (25.9)

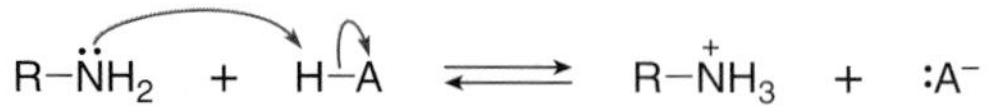

$$R-\ddot{N}H_2 \ + \ H-A \ \rightleftharpoons \ R-\overset{+}{N}H_3 \ + \ :A^-$$

[2] Nucleophilic addition to aldehydes and ketones (25.11)

With 1° amines:

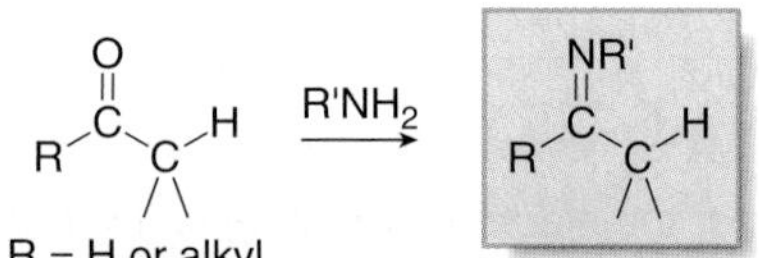

$$\xrightarrow{R'NH_2}$$

imine

With 2° amines:

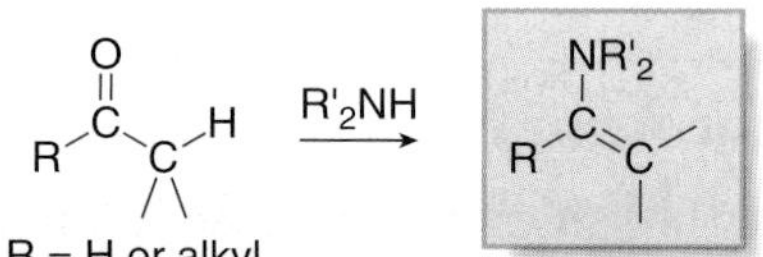

$$\xrightarrow{R'_2NH}$$

enamine

R = H or alkyl

[3] Nucleophilic substitution with acid chlorides and anhydrides (25.11)

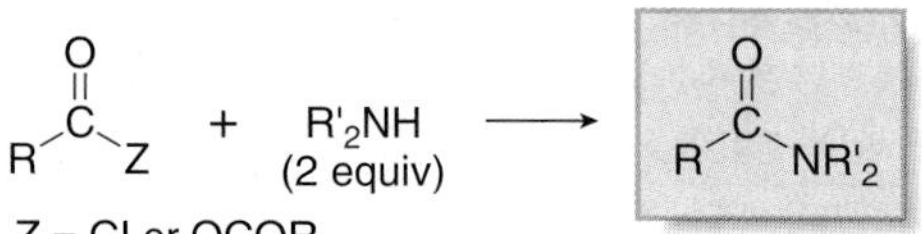

$$R-\overset{O}{\underset{Z}{C}} \ + \ R'_2NH \longrightarrow R-\overset{O}{\underset{NR'_2}{C}}$$

(2 equiv)

Z = Cl or OCOR
R' = H or alkyl

1°, 2°, and 3° amides

[4] Hofmann elimination (25.12)

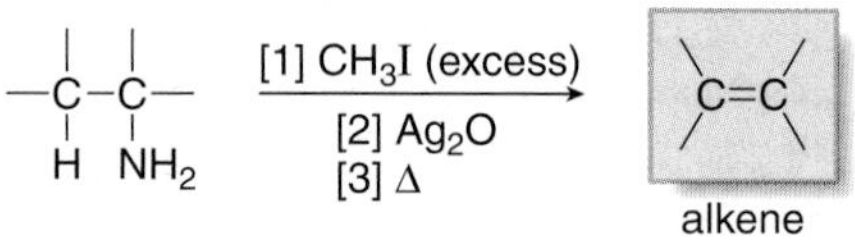

$$-\overset{|}{\underset{H}{C}}-\overset{|}{\underset{NH_2}{C}}- \xrightarrow[\substack{[2] \ Ag_2O \\ [3] \ \Delta}]{[1] \ CH_3I \ (excess)} C=C$$

alkene

- The less substituted alkene is the major product.

[5] Reaction with nitrous acid (25.13)

With 1° amines:

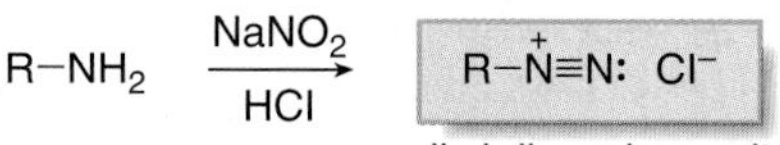

$$R-NH_2 \xrightarrow[HCl]{NaNO_2} R-\overset{+}{N}\equiv N: \ Cl^-$$

alkyl diazonium salt

With 2° amines:

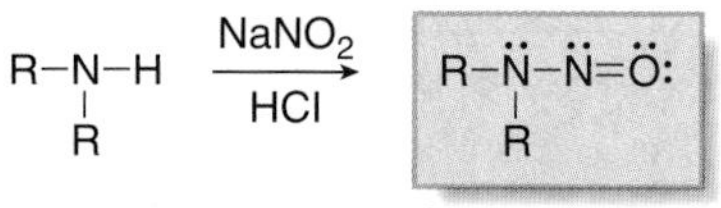

$$R-\underset{R}{N}-H \xrightarrow[HCl]{NaNO_2} R-\ddot{N}-\ddot{N}=\ddot{O}:$$

N-nitrosamine

Reactions of Diazonium Salts

[1] Substitution reactions (25.14)

N₂⁺ Cl⁻

With H₂O:

phenol

With CuX:

aryl chloride or aryl bromide
X = Cl or Br

With HBF₄:

aryl fluoride

With NaI or KI:

aryl iodide

With CuCN:

benzonitrile

With H₃PO₂:

benzene

[2] Coupling to form azo compounds (25.15)

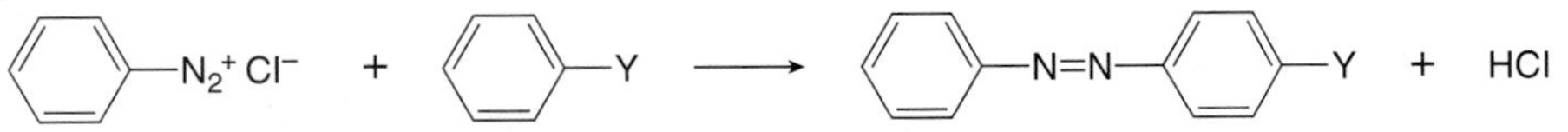

C₆H₅—N₂⁺ Cl⁻   +   C₆H₅—Y   ⟶   C₆H₅—N=N—C₆H₄—Y   +   HCl

Y = NH₂, NHR, NR₂, OH
(a strong electron-
donor group)

azo compound

Problems

Nomenclature

25.43 Give a systematic or common name for each compound.

a. CH₃NHCH₂CH₂CH₂CH₃

b. 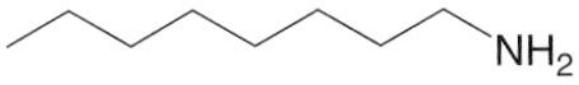 [structure: straight chain with NH₂]

c. 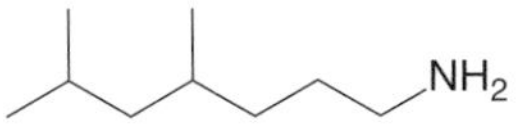 [structure with NH₂]

d. 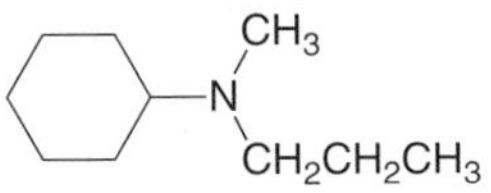 [cyclohexyl-N(CH₃)(CH₂CH₂CH₃)]

e. (CH₃CH₂CH₂)₃N

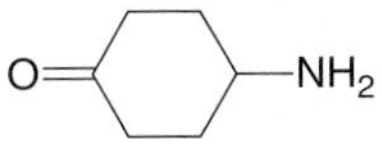

f. (C₆H₅)₂NH

g. 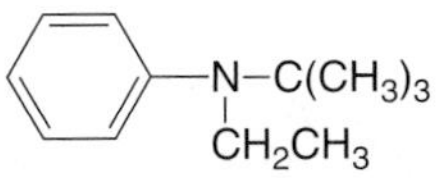 [phenyl—N—C(CH₃)₃ with CH₂CH₃]

h. O=[cyclohexane]—NH₂

i. [pyrrolidine N–H with CH₂CH₃]

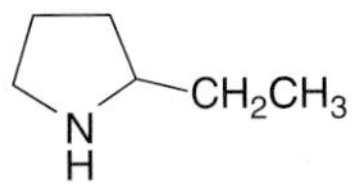

j. CH₃CH₂CH₂CH(NH₂)CH(CH₃)₂

k. 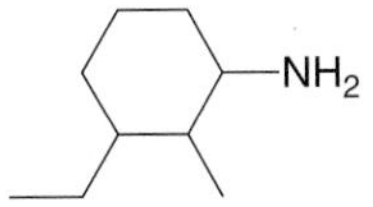 [cyclohexane with NH₂, ethyl, methyl]

l. 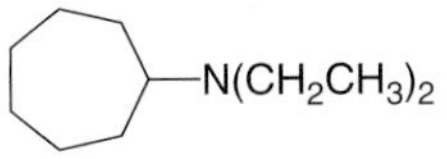 [cycloheptyl—N(CH₂CH₃)₂]

25.44 Draw the structure that corresponds to each name.

a. cyclobutylamine
b. N-isobutylcyclopentylamine
c. tri-*tert*-butylamine
d. N,N-diethylaniline
e. N-methylpyrrole
f. N-methylcyclopentylamine
g. *cis*-2-aminocyclohexanol
h. 3-methyl-2-hexanamine
i. 2-*sec*-butylpiperidine
j. (S)-2-heptanamine

Chiral Compounds

25.45 How many stereogenic centers are present in each compound? Draw all possible stereoisomers.

a. 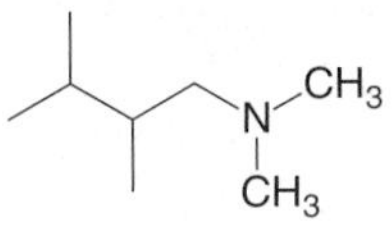 [structure with N(CH₃)₂]

b. CH₃CH₂CHCH₂CH₂CH₂—N⁺(CH₂CH₃)(CH₃)—CH₂CH₂CH₂CH₃ with CH₂CH₃ and CH₃ substituents   Cl⁻

Basicity

25.46 Which compound in each pair is the stronger base?

a. (CH₃CH₂)₂NH   or   [pyridine]

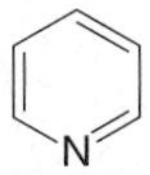

b. C₆H₅NHCH₃   or   C₆H₅CH₂NH₂

c. HCON(CH₃)₂   or   (CH₃)₃N

d. (CH₃CH₂)₂NH   or   (ClCH₂CH₂)₂NH

25.47 Rank the compounds in each group in order of increasing basicity.

a. NH₃   [C₆H₅—NH₂]   [cyclohexyl—NH₂]

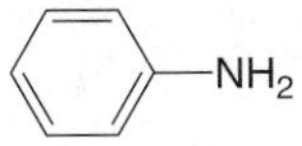

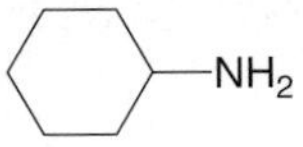

b. 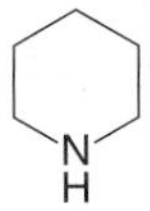 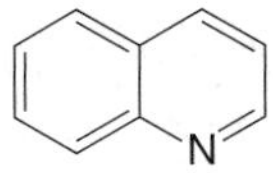 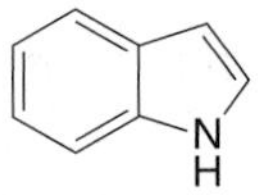 [piperidine]   [quinoline]   [indole]

c. 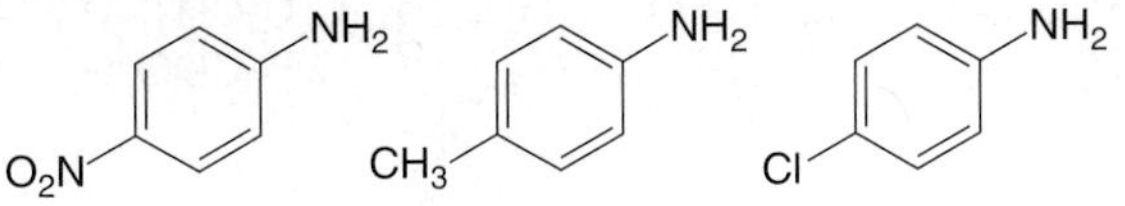 [O₂N—C₆H₄—NH₂]   [CH₃—C₆H₄—NH₂]   [Cl—C₆H₄—NH₂]

d. C₆H₅NH₂   (C₆H₅)₂NH   [cyclohexyl—NH₂]

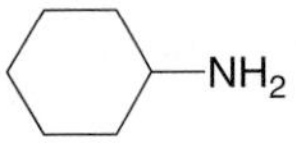

25.48 Which nitrogen atom in each compound is more basic?

a. 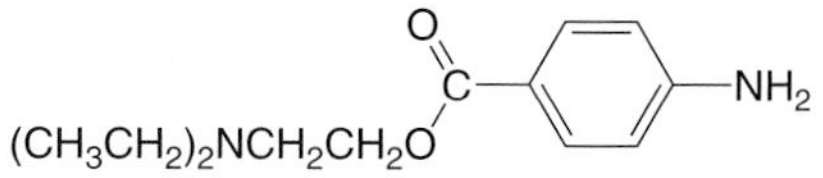
(CH₃CH₂)₂NCH₂CH₂O— [structure with C=O and benzene ring and NH₂]

novocaine
(local anesthetic)

b. 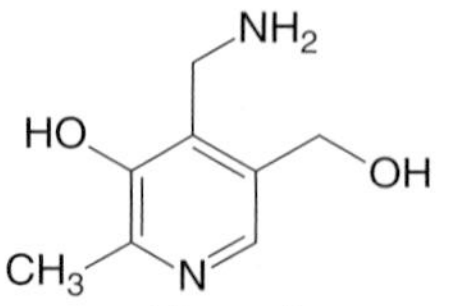
pyridoxamine

25.49 Rank the nitrogen atoms in each compound in order of increasing basicity.

a. 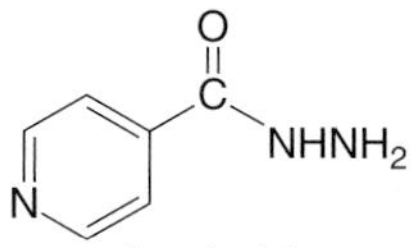
isoniazid
(a drug used to treat
tuberculosis)

b. 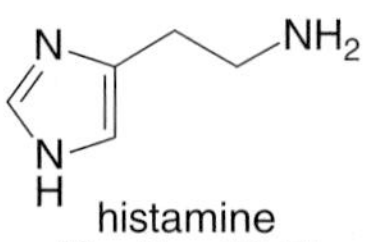
histamine
(Section 25.6)

25.50 Explain why *m*-nitroaniline is a stronger base than *p*-nitroaniline.

25.51 Explain the observed difference in the pK_a values of the conjugate acids of amines **A** and **B**.

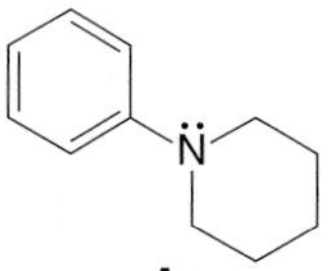

A
pK_a = 5.2

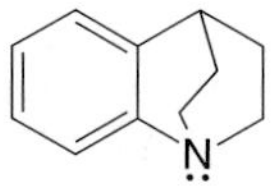

B
pK_a = 7.29

Preparation of Amines

25.52 How would you prepare 3-phenyl-1-propanamine (C₆H₅CH₂CH₂CH₂NH₂) from each compound?
a. C₆H₅CH₂CH₂CH₂Br   c. C₆H₅CH₂CH₂CH₂NO₂   e. C₆H₅CH₂CH₂CHO
b. C₆H₅CH₂CH₂Br   d. C₆H₅CH₂CH₂CONH₂

25.53 What amide is needed to prepare each amine by reduction?

a. (CH₃CH₂)₂NH   b.  [structure] NH₂   c. 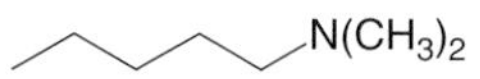 [structure] N(CH₃)₂   d. 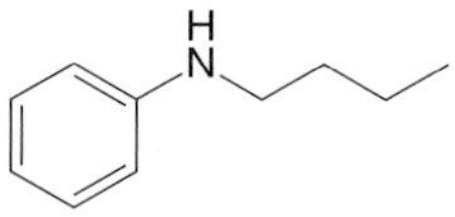 [structure]

25.54 What carbonyl and nitrogen compounds are needed to make each compound by reductive amination? When more than one set of starting materials is possible, give all possible methods.

a. 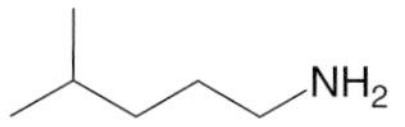 [structure] NH₂   b. 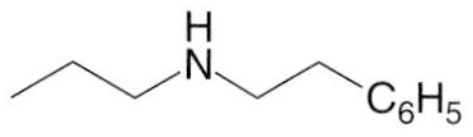 [structure] C₆H₅   c. (CH₃CH₂CH₂)₂N(CH₂)₂CH(CH₃)₂   d. 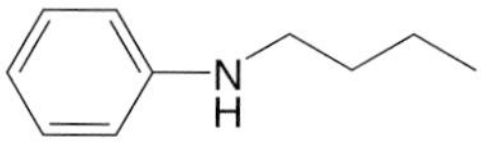 [structure]

25.55 Draw the product of each reductive amination reaction.

a. 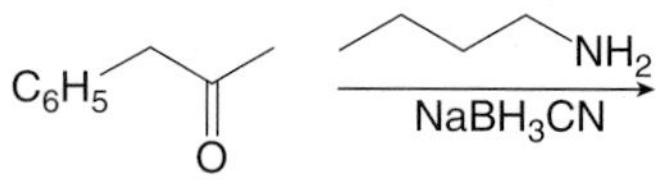 C₆H₅ [structure with O] + [structure] NH₂ → (NaBH₃CN)

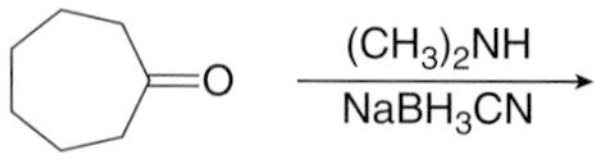

c. 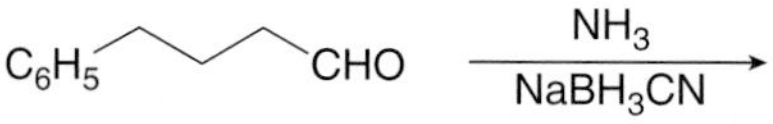 C₆H₅ [structure] CHO → (NaBH₃CN / NH₃)

b. [cycloheptanone structure] =O + (CH₃)₂NH → (NaBH₃CN)

d. 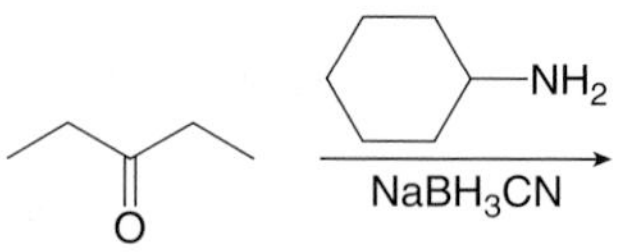 [structure] =O + [cyclohexyl] NH₂ → (NaBH₃CN)

25.56 How would you prepare benzylamine (C₆H₅CH₂NH₂) from each compound? In some cases, more than one step is required.
a. C₆H₅CH₂Br   c. C₆H₅CONH₂   e. C₆H₅CH₃   g. C₆H₅NH₂
b. C₆H₅CN   d. C₆H₅CHO   f. C₆H₅COOH   h. benzene

Extraction

25.57 How would you separate toluene (C₆H₅CH₃), benzoic acid (C₆H₅COOH), and aniline (C₆H₅NH₂) by an extraction procedure?

Reactions

25.58 What products are formed when *N*-ethylaniline ($C_6H_5NHCH_2CH_3$) is treated with each reagent?

a. HCl
b. CH_3COOH
c. $(CH_3)_2C=O$
d. CH_2O, $NaBH_3CN$

e. CH_3I (excess)
f. CH_3I (excess), followed by Ag_2O and Δ
g. CH_3CH_2COCl

h. The product in (g), then HNO_3, H_2SO_4
i. The product in (g), then [1] $LiAlH_4$; [2] H_2O
j. The product in (h), then H_2, Pd-C

25.59 Draw the products formed when *p*-methylaniline (*p*-$CH_3C_6H_4NH_2$) is treated with each reagent.

a. HCl
b. CH_3COCl
c. $(CH_3CO)_2O$
d. excess CH_3I

e. $(CH_3)_2C=O$
f. CH_3COCl, $AlCl_3$
g. CH_3COOH

h. $NaNO_2$, HCl
i. Step (b), then CH_3COCl, $AlCl_3$
j. CH_3CHO, $NaBH_3CN$

25.60 How would you convert $CH_3CH_2CH_2CH_2NH_2$ into each compound?

a. $CH_3CH_2CH_2CH_2NHCOC_6H_5$
b. $CH_3CH_2CH_2CH_2N=C(CH_2CH_3)_2$
c. $CH_3CH_2CH=CH_2$

d. $CH_3CH_2CH_2CH_2NHCH_2C_6H_5$
e. $CH_3CH_2CH_2CH_2NHCH_2CH_3$
f. $[CH_3CH_2CH_2CH_2N(CH_3)_3]^+I^-$

25.61 Draw the products formed when each amine is treated with [1] CH_3I (excess); [2] Ag_2O; [3] Δ. Indicate the major product when a mixture results.

a. $CH_3(CH_2)_6NH_2$
b. 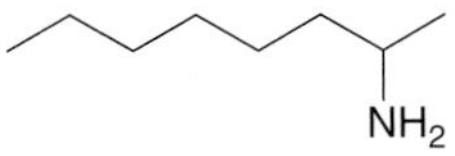 [structure: long chain with NH_2]
c. 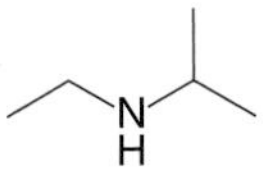 [structure: N-isopropyl secondary amine]
d. 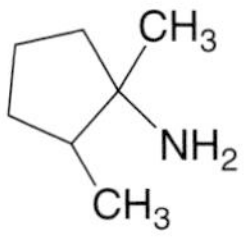 [structure: cyclopentane with CH_3, NH_2, CH_3 substituents]
e. 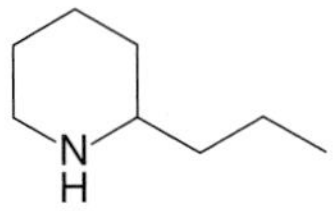 [structure: piperidine with propyl substituent]

25.62 Fill in the lettered reagents needed in each reaction.

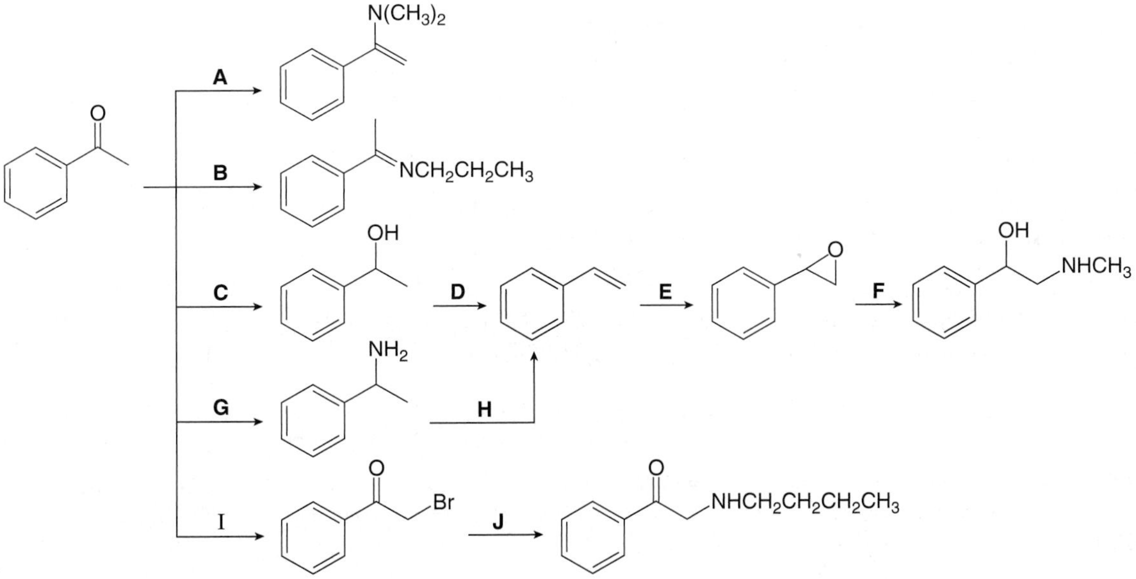

25.63 Draw the organic products formed in each reaction.

a. [cyclohexyl]-CH$_2$CH$_2$Cl $\xrightarrow[\text{excess}]{\text{NH}_3}$

f. C$_6$H$_5$CH$_2$CH$_2$NH$_2$ + (C$_6$H$_5$CO)$_2$O

b. [isobutyl]–Cl + [phthalimide N$^-$] $\longrightarrow$ $\xrightarrow[\text{H}_2\text{O}]{^-\text{OH}}$

g. [pyrrolidine]NH $\xrightarrow[\text{HCl}]{\text{NaNO}_2}$

c. Br–[benzene]–NO$_2$ $\xrightarrow[\text{HCl}]{\text{Sn}}$

h. [pyrrolidine]NH + C$_6$H$_5$CHO $\xrightarrow{\text{NaBH}_3\text{CN}}$

d. [CH$_3$CH$_2$CH(CH$_3$)]–CN $\xrightarrow[\text{[2] H}_2\text{O}]{\text{[1] LiAlH}_4}$

i. [piperidine] N–H + [cyclohexanone]=O $\longrightarrow$

e. [cyclohexyl]–CONHCH$_2$CH$_3$ $\xrightarrow[\text{[2] H}_2\text{O}]{\text{[1] LiAlH}_4}$

j. CH$_3$CH$_2$CH$_2$–N(H)–CH(CH$_3$)$_2$ $\xrightarrow[\text{[3] }\Delta]{\substack{\text{[1] CH}_3\text{I (excess)} \\ \text{[2] Ag}_2\text{O}}}$

25.64 What is the stereochemistry of the alkene formed from the Hofmann elimination of each quaternary ammonium salt?

a. [structure: CH$_3$, C$_6$H$_5$, C$_6$H$_5$, H, N(CH$_3$)$_3$] $^-$OH

b. [structure: CH$_3$, C$_6$H$_5$, H, C$_6$H$_5$, N(CH$_3$)$_3$] $^-$OH

25.65 Draw the product formed when **A** is treated with each reagent.

[structure: benzene ring with N$_2^+$ Cl$^-$ and Cl, labeled **A**]

a. H$_2$O
b. H$_3$PO$_2$
c. CuCl
d. CuBr
e. CuCN
f. HBF$_4$
g. NaI
h. C$_6$H$_5$NH$_2$
i. C$_6$H$_5$OH
j. KI

25.66 Explain why so much meta product is formed when aniline is nitrated with HNO$_3$ and H$_2$SO$_4$. For this reason, nitration of aniline is **not** a useful reaction to prepare either o- or p-nitroaniline.

[structure: aniline NH$_2$] $\xrightarrow[\text{H}_2\text{SO}_4]{\text{HNO}_3}$ [O$_2$N–benzene–NH$_2$, para] + [benzene–NH$_2$ with NO$_2$, meta] + [benzene–NH$_2$ with NO$_2$, ortho]

51% para   47% meta   2% ortho

25.67 A chiral amine **A** having the R configuration undergoes Hofmann elimination to form an alkene **B** as the major product. **B** is oxidatively cleaved with ozone, followed by CH$_3$SCH$_3$, to form CH$_2$=O and CH$_3$CH$_2$CH$_2$CHO. What are the structures of **A** and **B**?

Mechanism

25.68 Propose a reason why aryl diazonium salts are more stable than alkyl diazonium salts.

25.69 Draw a stepwise mechanism for each reaction.

a. Br[CH$_2$CH$_2$CH$_2$CH$_2$]Br + CH$_3$CH$_2$NH$_2$ $\xrightarrow{\text{NaOH}}$ [pyrrolidine with N–CH$_2$CH$_3$] + H$_2$O + NaBr

b. [cyclohexanone with CH$_2$CH$_2$NH$_2$ substituent] $\xrightarrow[\text{CH}_3\text{OH}]{\text{NaBH}_4}$ [bicyclic amine with N–H] + H$_2$O

25.70 Alkyl diazonium salts are unstable even at low temperature. They decompose to form carbocations, which go on to form products of substitution, elimination, and (sometimes) rearrangement. Keeping this in mind, draw a stepwise mechanism that forms all of the following products.

$$\xrightarrow[\text{HCl, H}_2\text{O}]{\text{NaNO}_2}$$

Synthesis

25.71 Devise a synthesis of each compound from benzene. You may use any other organic or inorganic reagents.

a. (compound with NH₂)

b. (compound with Br and NHCOCH₃)

c. I—◯—CH₃

d. (compound with NC and Br)

e. (compound with OH and Br)

f. HO—◯—COOH

g. (compound with I and C(=O)CH₃)

h. ◯—N=N—◯—OH

25.72 Devise a synthesis of each compound from aniline (C₆H₅NH₂) as starting material.

a. (compound with CONHCH₃)

b. (compound with CH₃ and Br)

c. (compound with COOCH₂CH₃)

d. Br—◯(Br)—CH₂OH (with two Br)

e. ◯—N=N—◯ (with HO and CH₃)

25.73 Synthesize each compound from benzene. Use a diazonium salt as one of the synthetic intermediates.

a. (compound with COOH and Cl)

b. (compound with OH and two Cl)

c. CH₃—◯—CH₂NH₂

d. CH₃CH₂—◯—COOH

e. HO—◯—CH₃ (with Br)

f. (compound with two Cl)—N=N—◯—NH₂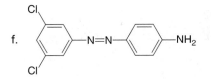

25.74 Devise a synthesis of each biologically active compound from benzene.

a.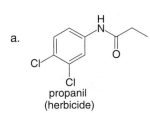
propanil
(herbicide)

b. (acetaminophen structure)
acetaminophen
(analgesic)

c. (pseudoephedrine structure with OH and NHCH₃)
pseudoephedrine
(nasal decongestant)

Spectroscopy

25.75 Compound **A** (molecular formula C₈H₇N) is converted to compound **B** (molecular formula C₈H₁₁N) with LiAlH₄ followed by H₂O. The ¹H NMR and IR spectral data for **A** and **B** are given below. What are the structures of **A** and **B**?

Compound **A:**
IR absorption at: 2230 cm⁻¹
¹H NMR peaks at: 2.4 (singlet, 3 H)
                              7.2 (2 H)
                              7.5 (2 H) ppm

Compound **B:**
IR absorption at: 3370, 3290 cm⁻¹
¹H NMR peaks at: 1.4 (singlet, 2 H)
                              2.3 (singlet, 3 H)
                              3.8 (singlet, 2 H)
                              7.0–7.3 (4 H) ppm

25.76 Three isomeric compounds, **A, B,** and **C,** all have molecular formula $C_8H_{11}N$. The ^1H NMR and IR spectral data of **A, B,** and **C** are given below. What are their structures?

Compound **A:** IR peak at 3400 cm^{-1}

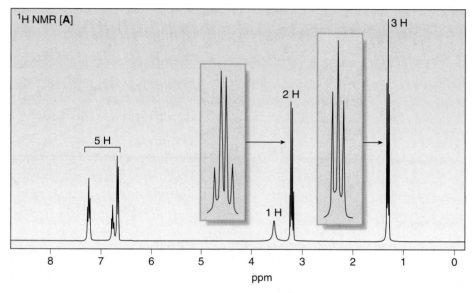

Compound **B:** IR peak at 3310 cm^{-1}

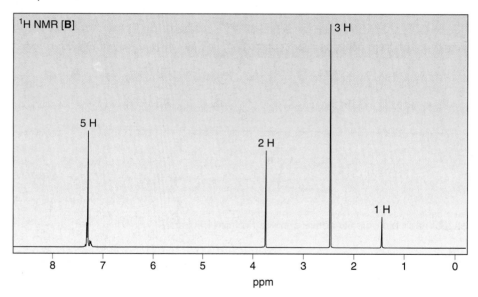

Compound **C:** IR peaks at 3430 and 3350 cm^{-1}

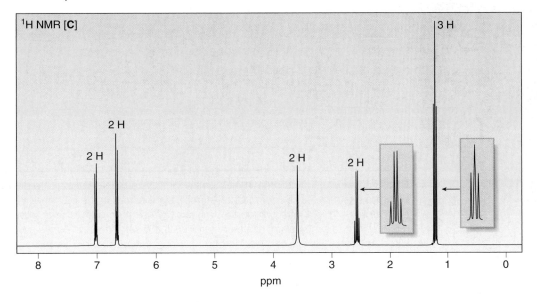

Challenge Problems

25.77 The pK_a of the conjugate acid of guanidine is 13.6, making it one of the strongest neutral organic bases. Offer an explanation.

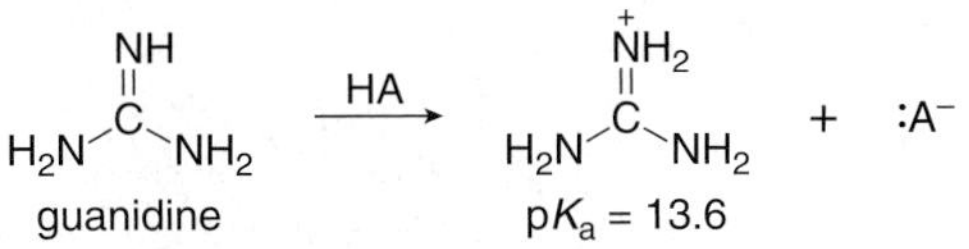

25.78 Draw the product (**Y**) of the following reaction sequence. **Y** was an intermediate in the remarkable synthesis of cyclooctatetraene by Wilstatter in 1911.

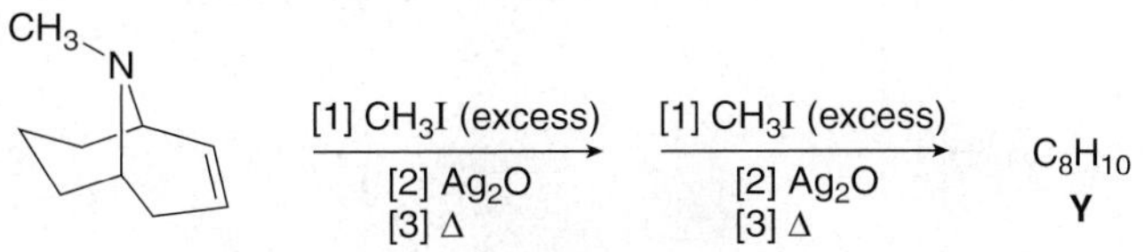

25.79 Explain why only one product is formed in the following Hofmann elimination reaction.

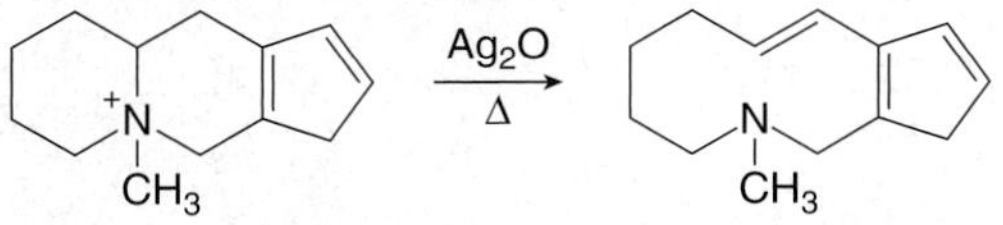

25.80 Devise a synthesis of the bronchodilator albuterol from the given starting material.

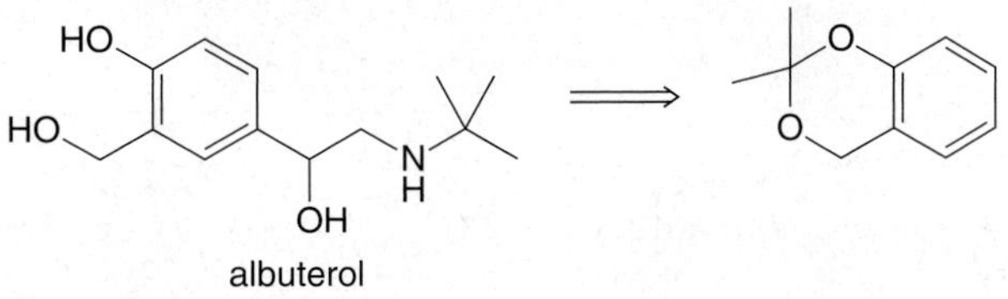

albuterol

Lipids

Cholesterol is the most prominent member of the steroid family, a group of organic lipids that contains a tetracyclic structure. Cholesterol is synthesized in the liver and is found in almost all body tissues. It is a vital component for healthy cell membranes and serves as the starting material for the synthesis of all other steroids. But, as the general public now knows well, elevated cholesterol levels can lead to coronary artery disease. For this reason, consumer products are now labeled with their cholesterol content. In Chapter 26, we will learn about the properties of cholesterol and other lipids.

Chapters 26, 27, and 28 discuss *biomolecules,* **organic compounds found in biological systems.** You have already learned many facts about these compounds in previous chapters while you studied other organic compounds having similar properties. In Chapter 10 (Alkenes), for example, you learned that the presence of double bonds determines whether a fatty acid is part of a fat or an oil. In Chapter 19 (Carboxylic Acids and the Acidity of the O−H Bond), you learned that amino acids are the building blocks of proteins.

Chapter 26 focuses on lipids, and Chapters 27 and 28 discuss carbohydrates and amino acids and proteins, respectively. These compounds are all organic molecules, so many of the same principles and chemical reactions that you have already studied will be examined once again. But, as you will see, each class of compound has its own unique features that we must learn as well.

26.1   Introduction

The word lipid comes from the Greek word *lipos* for *fat.*

◆ *Lipids* are biomolecules that are soluble in organic solvents.

Lipids are unique among organic molecules because their identity is defined on the basis of a *physical property* and not by the presence of a particular functional group. Because of this, lipids come in a wide variety of structures and they have many different functions in the cell. Three examples are given in Figure 26.1.

The large number of **carbon–carbon and carbon–hydrogen σ bonds in lipids makes them very soluble in organic solvents and insoluble in water.** Monosaccharides (from which carbohydrates are formed) and amino acids (from which proteins are formed), on the other hand, are very polar, so they tend to be water soluble. Because lipids share many properties with hydrocarbons, several features of lipid structure and properties have already been discussed. Table 26.1 summarizes sections of the text where aspects of lipid chemistry were covered previously.

TABLE 26.1 Summary of Lipid Chemistry Discussed Prior to Chapter 26			
Topic	**Section**	**Topic**	**Section**
• Vitamin A	3.6	• Lipid oxidation	13.11
• Soap	3.7	• Vitamin E	13.12
• Phospholipids, the cell membrane	3.8	• Steroid synthesis	16.14
• Lipids Part 1	4.15	• Prostaglandins	19.6
• Leukotrienes	9.16	• Lipid hydrolysis	22.12A
• Fats and oils	10.6	• Soap	22.12B
• Oral contraceptives	11.4	• Cholesteryl esters	22.17
• Hydrogenation of oils	12.4	• Steroid synthesis	24.8

Figure 26.1   Three examples of lipids

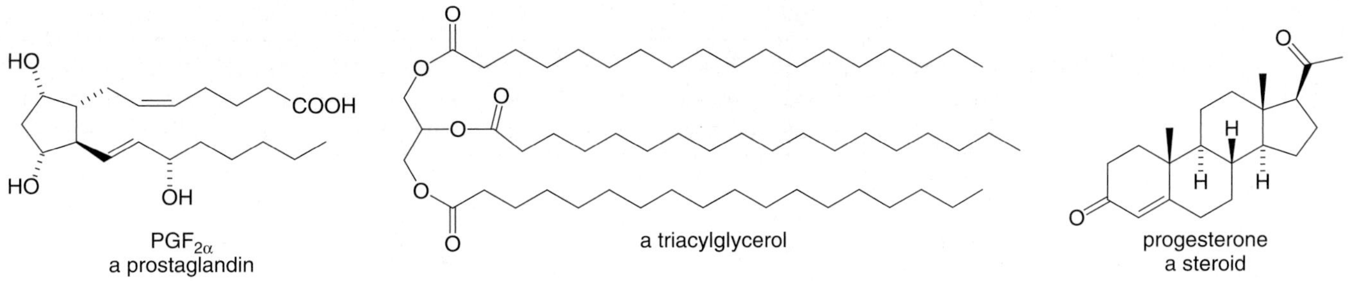

PGF$_{2\alpha}$
a prostaglandin

a triacylglycerol

progesterone
a steroid

All lipids have many C−C and C−H bonds, but there is no one functional group common to all lipids.

Lipids can be categorized as hydrolyzable or nonhydrolyzable.

[1] *Hydrolyzable lipids* **can be cleaved into smaller molecules by hydrolysis with water.** Most hydrolyzable lipids contain an ester unit. We will examine three subgroups: waxes, triacylglycerols, and phospholipids.

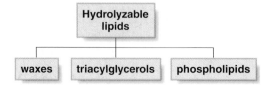

[2] *Nonhydrolyzable lipids* **cannot be cleaved into smaller units by aqueous hydrolysis.** Nonhydrolyzable lipids tend to be more varied in structure. We will examine four different types: fat-soluble vitamins, eicosanoids, terpenes, and steroids.

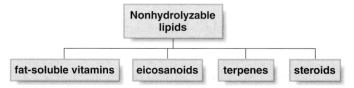

26.2   Waxes

Water beads up on the surface of a leaf because of the leaf's waxy coating.

Waxes are the simplest hydrolyzable lipids. **Waxes are esters (RCOOR') formed from a high molecular weight alcohol (R'OH) and a fatty acid (RCOOH).**

Because of their long hydrocarbon chains, **waxes are very hydrophobic.** They form a protective coating on the feathers of birds to make them water repellent, and on leaves to prevent water evaporation. **Lanolin,** a wax composed of a complex mixture of high molecular weight esters, coats the wool fibers of sheep. **Spermaceti wax,** isolated from the heads of sperm whales, is largely $CH_3(CH_2)_{14}COO(CH_2)_{15}CH_3$. The three-dimensional structure of this compound shows how small the ester group is compared to the long hydrocarbon chains.

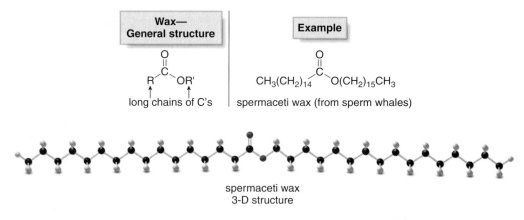

spermaceti wax
3-D structure

PROBLEM 26.1   Carnauba wax, a wax that coats the leaves of the Brazilian palm tree, is used for hard, high-gloss finishes for floors, boats, and automobiles. One component of carnauba wax is formed from an unbranched 32-carbon carboxylic acid and a straight chain 34-carbon alcohol. Draw its structure.

26.3   Triacylglycerols

Triacylglycerols, or triglycerides, are the most abundant lipids, and for this reason we have already discussed many of their properties in earlier sections of this text.

◆ Triacylglycerols are triesters that produce glycerol and three molecules of fatty acid upon hydrolysis.

triacylglycerol
the most common type of lipid

Three fatty acids containing 12–20 C's are formed as products.

Line structures of stearic, oleic, linoleic, and linolenic acids can be found in Table 10.2. Ball-and-stick models of these fatty acids are shown in Figure 10.6.

Simple triacylglycerols are composed of three identical fatty acid side chains, whereas **mixed triacylglycerols** have two or three different fatty acids. Table 26.2 lists the most common fatty acids used to form triacylglycerols.

TABLE 26.2 The Most Common Fatty Acids in Triacylglycerols

Number of C atoms	Number of C=C bonds	Structure	Name	Mp (°C)
		Saturated fatty acids		
12	0	$CH_3(CH_2)_{10}COOH$	lauric acid	44
14	0	$CH_3(CH_2)_{12}COOH$	myristic acid	58
16	0	$CH_3(CH_2)_{14}COOH$	palmitic acid	63
18	0	$CH_3(CH_2)_{16}COOH$	stearic acid	70
20	0	$CH_3(CH_2)_{18}COOH$	arachidic acid	77
		Unsaturated fatty acids		
16	1	$CH_3(CH_2)_5CH=CH(CH_2)_7COOH$	palmitoleic acid	1
18	1	$CH_3(CH_2)_7CH=CH(CH_2)_7COOH$	oleic acid	16
18	2	$CH_3(CH_2)_4(CH=CHCH_2)_2(CH_2)_6COOH$	linoleic acid	−5
18	3	$CH_3CH_2(CH=CHCH_2)_3(CH_2)_6COOH$	linolenic acid	−11
20	4	$CH_3(CH_2)_4(CH=CHCH_2)_4(CH_2)_2COOH$	arachidonic acid	−49

The most common saturated fatty acids are palmitic and stearic acid. The most common unsaturated fatty acid is oleic acid.

Linoleic and linolenic acids are called **essential fatty acids** because we cannot synthesize them and must acquire them in our diets.

More details on fatty acid structure and properties appear in Section 10.6.

What are the characteristics of these fatty acids?

◆ All fatty acid chains are unbranched, but they may be saturated or unsaturated.
◆ Naturally occurring fatty acids have an even number of carbon atoms.
◆ All double bonds in naturally occurring fatty acids have the *Z* configuration.
◆ The melting point of a fatty acid depends on the degree of unsaturation.

Fats and oils are triesters of glycerol and these fatty acids.

◆ Fats have higher melting points, making them solids at room temperature.
◆ Oils have lower melting points, making them liquids at room temperature.

This melting point difference correlates with the number of degrees of unsaturation present in the fatty acid side chains. **As the number of double bonds *increases*, the melting point *decreases*, as it does for the constituent fatty acids as well.**

Figure 26.2 Three-dimensional structures of a saturated and unsaturated triacylglycerol

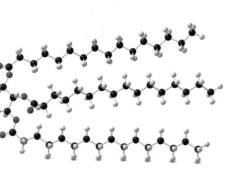

| A saturated triacylglycerol | An unsaturated triacylglycerol |

- Three saturated side chains lie parallel to each other, making a compact lipid.

- One *Z* double bond in a fatty acid side chain produces a twist so the lipid is no longer so compact.

Three-dimensional structures of a saturated and unsaturated triacylglycerol are shown in Figure 26.2. With no double bonds, the three side chains of the saturated lipid lie parallel to each other, making it possible for this compound to pack relatively efficiently in a crystalline lattice, thus leading to a high melting point. In the unsaturated lipid, however, a single *Z* double bond places a kink in the side chain, making it more difficult to pack efficiently in the solid state, thus leading to a lower melting point.

Solid fats have a relatively high percentage of saturated fatty acids and are generally of animal origin. Liquid oils have a higher percentage of unsaturated fatty acids and are generally of vegetable origin. Table 26.3 lists the fatty acid composition of some common fats and oils.

Unlike other vegetable oils, oils from palm and coconut trees are very high in saturated fats. Considerable evidence currently suggests that diets high in saturated fats lead to a greater risk of heart disease. For this reason, the demand for coconut and palm oils has decreased considerably in recent years, and many coconut plantations previously farmed in the South Pacific are no longer in commercial operation.

TABLE 26.3   Fatty Acid Composition of Some Fats and Oils

Source	% Saturated fatty acids	% Oleic acid	% Linoleic acid
beef	49–62	37–43	2–3
milk	37	33	3
coconut	86	7	—
corn	11–16	19–49	34–62
olive	11	84	4
palm	43	40	8
safflower	9	13	78
soybean	15	20	52

Data from *Merck Index,* 10th ed. Rahway, NJ: Merck and Co.; and Wilson, et al., 1967, *Principles of Nutrition,* 2nd ed. New York: Wiley.

Fish oils, such as cod liver and herring oils, are very rich in polyunsaturated triacylglycerols. These triacylglycerols pack so poorly that they have very low melting points; thus, they remain liquids even in the cold water inhabited by these fish.

The hydrolysis, hydrogenation, and oxidation of triacylglycerols—reactions originally discussed in Chapters 12, 13 and 22—are summarized here for your reference.

[1]   Hydrolysis of triacylglycerols (Section 22.12A)

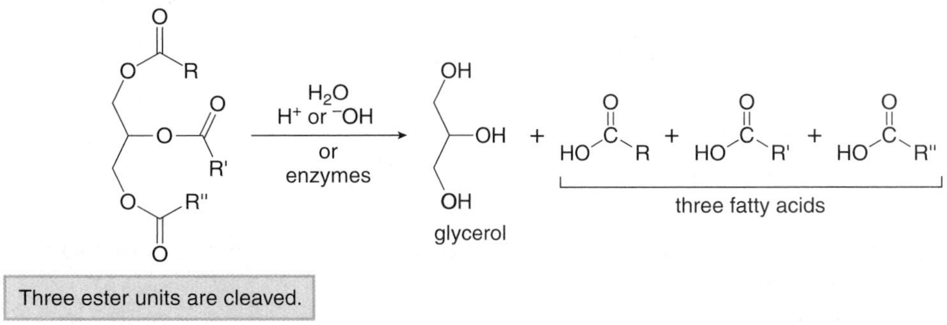

Three ester units are cleaved.

three fatty acids

glycerol

Hydrolysis of a triacylglycerol with water in the presence of either acid, base, or an enzyme yields glycerol and three fatty acids. This cleavage reaction follows the same mechanism as any other ester hydrolysis (Section 22.11). This reaction is the first step in triacylglycerol metabolism.

[2]   Hydrogenation of unsaturated fatty acids (Section 12.4)

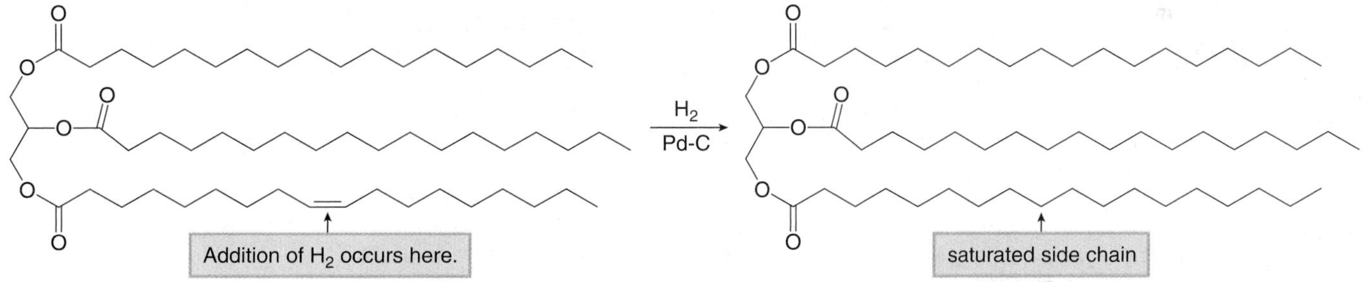

Addition of H_2 occurs here.

saturated side chain

The double bonds of an unsaturated fatty acid can be hydrogenated by using H_2 in the presence of a transition metal catalyst. Hydrogenation converts a liquid oil to a solid fat. This process, sometimes called **hardening,** is used to prepare margarine from vegetable oils.

[3]   Oxidation of unsaturated fatty acids (Section 13.11)

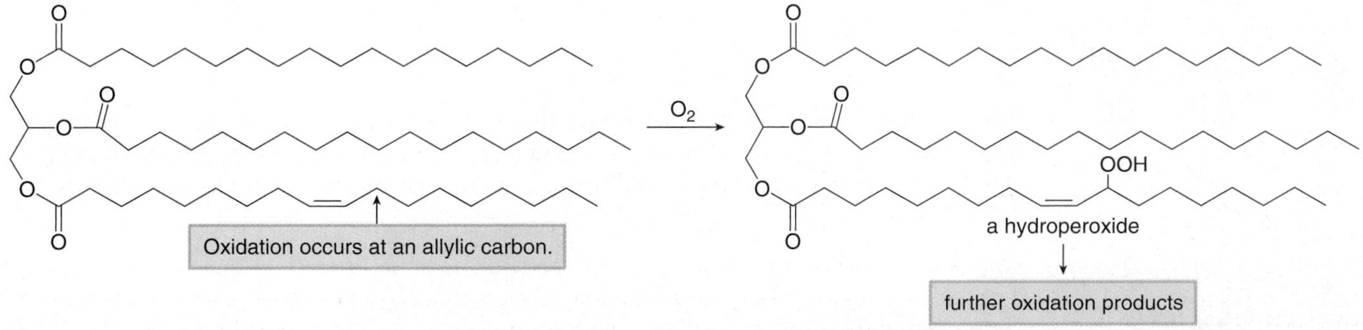

Oxidation occurs at an allylic carbon.

a hydroperoxide

further oxidation products

Allylic $C-H$ bonds are weaker than other $C-H$ bonds and are thus susceptible to oxidation with molecular oxygen by a radical process. The hydroperoxide formed by this process is unstable, and it undergoes further oxidation to products that often have a disagreeable odor. This oxidation process turns an oil rancid.

In the cell, the principal function of triacylglycerols is energy storage. Complete metabolism of a triacylglycerol yields CO_2 and H_2O, and a great deal of energy. This overall reaction is reminiscent of the combustion of alkanes in fossil fuels, a process that also yields CO_2 and H_2O and provides energy to heat homes and power automobiles (Section 4.14B). Fundamentally both processes convert $C-C$ and $C-H$ bonds to $C-O$ bonds, a highly exothermic reaction.

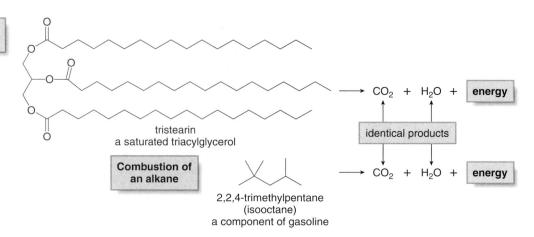

tristearin
a saturated triacylglycerol

Metabolism of a lipid

Combustion of an alkane

2,2,4-trimethylpentane
(isooctane)
a component of gasoline

identical products

$\longrightarrow$ CO_2 + H_2O + **energy**

$\longrightarrow$ CO_2 + H_2O + **energy**

> The average body fat content of men and women is ~20% and ~25%, respectively. (For elite athletes, however, the averages are more like <10% for men and <15% for women.) This stored fat can fill the body's energy needs for two or three months.

Carbohydrates provide an energy boost, but only for the short term, such as during strenuous exercise. Our long-term energy needs are met by triacylglycerols, because they store ~9 kcal/g, whereas carbohydrates and proteins store only ~4 kcal/g.

PROBLEM 26.2　How would you expect the melting point of eicosapentaenoic acid [$CH_3CH_2(CH=CHCH_2)_5(CH_2)_2COOH$] to compare with the melting points of the fatty acids listed in Table 26.2?

PROBLEM 26.3　Draw the products formed when triacylglycerol **A** is treated with each reagent.

a. H_2O, H^+
b. H_2 (excess), Pd-C → **B**
c. H_2 (1 equiv), Pd-C → **C**

A

PROBLEM 26.4　Rank compounds **A, B,** and **C** in Problem 26.3 in order of increasing melting point.

26.4　Phospholipids

Phospholipids are hydrolyzable lipids that contain a phosphorus atom. There are two common types of phospholipids: **phosphoacylglycerols** and **sphingomyelins.** Both classes are found almost exclusively in the cell membranes of plants and animals, as discussed in Section 3.8.

Phospholipids are organic derivatives of phosphoric acid, formed by replacing two of the H atoms by R groups. This type of functional group is called a **phosphodiester,** or a **phosphoric acid diester.** These compounds are phosphorus analogues of carboxylic esters. In cells, the remaining OH group on phosphorus loses its proton, giving the phosphodiester a net negative charge.

> The phosphorus atom in a phosphodiester shares 10 electrons. Recall, though, that third-row elements (such as P and S) can be surrounded by more than eight electrons.

phosphoric acid
H_3PO_4

phosphodiester

This form exists in cells.

26.4A   Phosphoacylglycerols

Phosphoacylglycerols (or phosphoglycerides) are the second most abundant type of lipid. They form the principal lipid component of most cell membranes. Their structure resembles the triacylglycerols of the preceding section with one important difference. In phosphoacylglycerols, only two of the hydroxy groups of glycerol are esterified with fatty acids. The third OH group is part of a phosphodiester, which is also bonded to another low molecular weight alcohol.

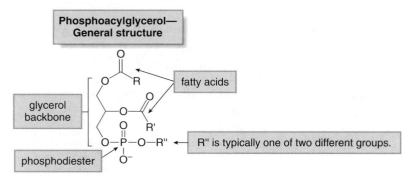

There are two prominent types of phosphoacylglycerols. They differ in the identity of the R" group in the phosphodiester.

◆ When R" = $CH_2CH_2NH_3^+$, the compound is called a **phosphatidylethanolamine** or **cephalin**.

◆ When R" = $CH_2CH_2N(CH_3)_3^+$, the compound is called a **phosphatidylcholine,** or **lecithin**.

phosphatidylethanolamine
or
cephalin

phosphatidylcholine
or
lecithin

The middle carbon of the glycerol backbone of all of these compounds is a stereogenic center, usually with the *R* configuration.

The phosphorus side chain of a phosphoacylglycerol makes it different from a triacylglycerol. The two fatty acid side chains form two nonpolar "tails" that lie parallel to each other, while the phosphodiester end of the molecule is a charged or polar "head." A three-dimensional structure of a phosphoacylglycerol is shown in Figure 26.3.

As discussed in Section 3.8, when these phospholipids are mixed with water, they assemble in an arrangement called a **lipid bilayer.** The ionic heads of the phospholipid are oriented on the outside and the nonpolar tails on the inside. The identity of the fatty acids in the phospholipid determines the rigidity of this bilayer. When the fatty acids are saturated, they pack well in the interior of the lipid bilayer, and the membrane is quite rigid. When there are many unsaturated fatty acids, the nonpolar tails cannot pack as well and the bilayer is more fluid. Thus, important characteristics of this lipid bilayer are determined by the three-dimensional structure of the molecules that comprise it.

Cell membranes are composed of these lipid bilayers (see Figure 3.8). Proteins and cholesterol are embedded in the membranes as well, but the phospholipid bilayer forms the main fabric of the insoluble barrier that protects the cell.

PROBLEM 26.5   Draw the structure of a lecithin containing oleic acid and palmitic acid as the fatty acid side chains.

PROBLEM 26.6   Phosphoacylglycerols should remind you of soaps (Section 3.7). In what ways are these compounds similar?

Figure 26.3 Three-dimensional structure of a phosphoacylglycerol

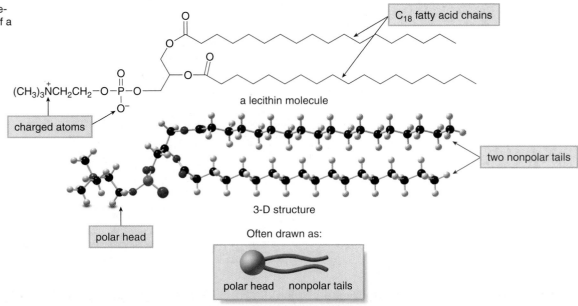

- A phosphoacylglycerol has two distinct regions: two nonpolar tails due to the long-chain fatty acids, and a very polar head from the charged phosphodiester.

26.4B   Sphingomyelins

Sphingomyelins, the second major class of phospholipids, are derivatives of the amino alcohol **sphingosine,** in much the same way that triacylglycerols and phosphoacylglycerols are derivatives of glycerol. Other notable features of a sphingomyelin include:

◆ **A phosphodiester at C1.**

◆ **An amide formed with a fatty acid at C2.**

Examples of sphingomyelins

The phosphodiester group is located at the terminal carbon.

Like phosphoacylglycerols, **sphingomyelins are also a component of the lipid bilayer of cell membranes.** The coating that surrounds and insulates nerve cells, the **myelin sheath,** is particularly rich in sphingomyelins, and is vital for proper nerve function. Deterioration of the myelin sheath as seen in multiple sclerosis leads to disabling neurological problems.

Figure 26.4 compares the structural features of the most common hydrolyzable lipids: a triacylglycerol, a phosphoacylglycerol, and a sphingomyelin.

PROBLEM 26.7   Why are phospholipids, but not triacylglycerols, found in cell membranes?

26.5   Fat-soluble Vitamins

Vitamins are organic compounds required in small quantities for normal metabolism (Section 3.6). Because our cells cannot synthesize these compounds, they must be obtained in the

Figure 26.4   A comparison of a triacylglycerol, a phosphoacylglycerol, and a sphingomyelin

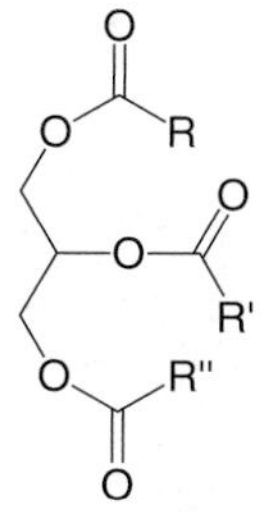

- A triacylglycerol has three nonpolar side chains.
- The three OH groups of glycerol are esterified with three fatty acids.

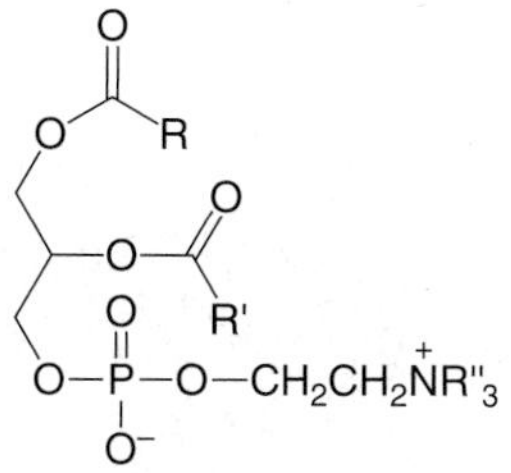

R" = H, CH$_3$

- A phosphoacylglycerol has two nonpolar side chain tails and one ionic head.
- Two OH groups of glycerol are esterified with fatty acids.
- A phosphodiester is located on a terminal carbon.

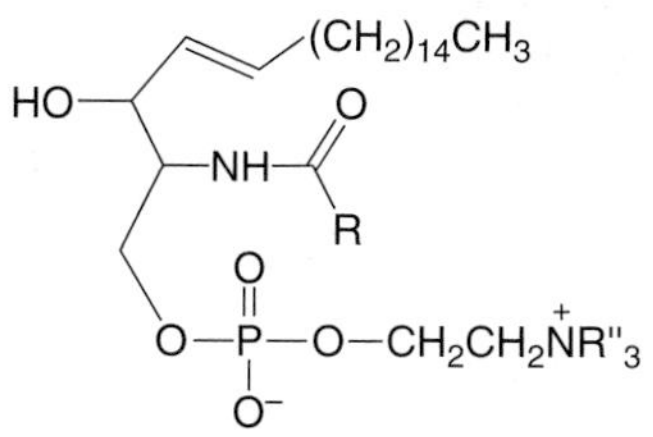

R" = H, CH$_3$

- A sphingomyelin has two nonpolar side chain tails and one ionic head.
- A sphingomyelin is formed from sphingosine, not glycerol. One of the nonpolar tails is an amide.
- A phosphodiester is located on a terminal carbon.

diet. Vitamins can be categorized as fat soluble or water soluble. **The fat-soluble vitamins are lipids.**

The four fat-soluble vitamins—**A, D, E,** and **K**—are found in fruits and vegetables, fish, liver, and dairy products. Although fat-soluble vitamins must be obtained from the diet, they do not have to be ingested every day. Excess vitamins are stored in fat cells, and then used when needed. Figure 26.5 shows the structure of these vitamins and summarizes their functions.

Electrostatic potential plots of vitamins A and E (Figure 26.6) show that the electron density is virtually uniform in these compounds. The large regions of nonpolar C—C and C—H bonds tend

Figure 26.5   The fat-soluble vitamins

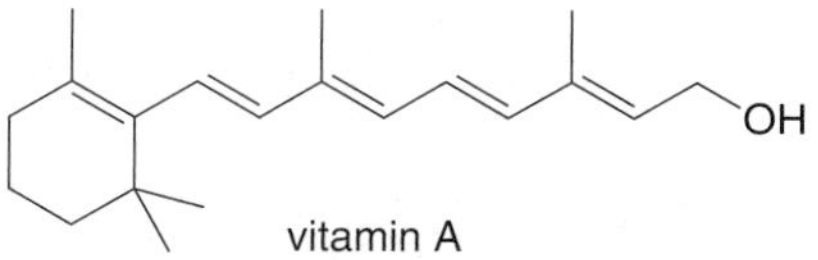

vitamin A

- **Vitamin A** (retinol, Section 3.6) is obtained from fish liver oils and dairy products, and is synthesized from β-carotene, the orange pigment in carrots.
- In the body, vitamin A is converted to 11-*cis*-retinal, the light-sensitive compound responsible for vision in all vertebrates (Section 21.11B). It is also needed for healthy mucous membranes.
- A deficiency of vitamin A causes night blindness, as well as dry eyes and skin.

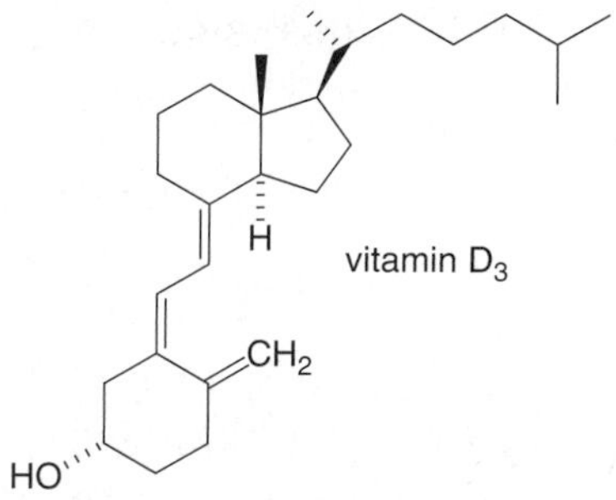

vitamin D$_3$

- **Vitamin D$_3$** is the most abundant of the D vitamins. Strictly speaking, it is not a vitamin because it can be synthesized in the body from cholesterol. Nevertheless, it is classified as such, and many foods (particularly milk) are fortified with vitamin D$_3$ so that we get enough of this vital nutrient.
- Vitamin D helps regulate both calcium and phosphorus metabolism.
- A deficiency of vitamin D causes rickets, a bone disease characterized by knock-knees, spinal curvature, and other deformities.

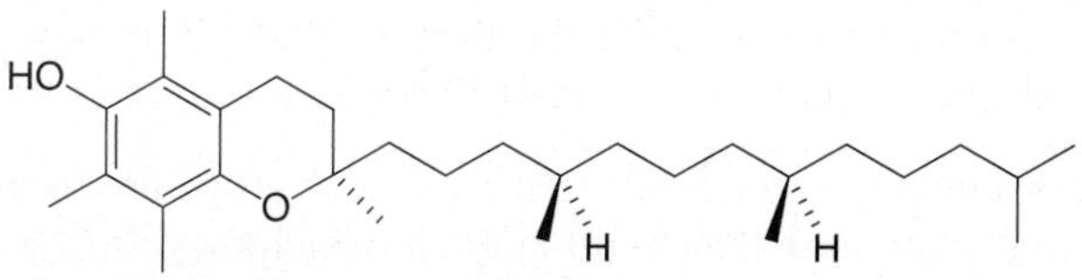

vitamin E
(α-tocopherol)

- The term **vitamin E** refers to a group of structurally similar compounds, the most potent being α-tocopherol (Section 13.12).
- Vitamin E is an antioxidant, so it protects unsaturated side chains in fatty acids from oxidation.
- A deficiency of vitamin E causes numerous neurologic problems.

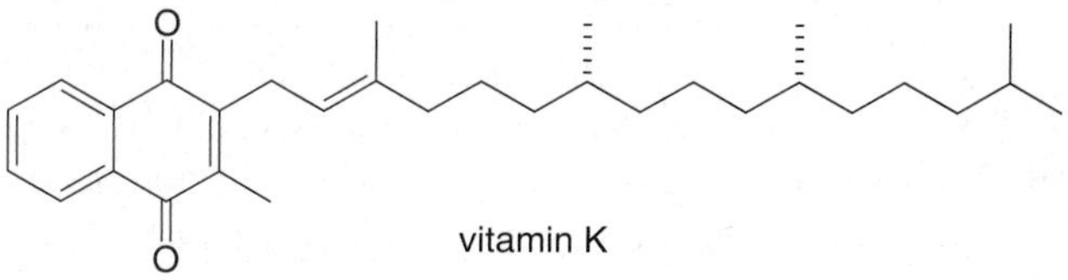

vitamin K

- **Vitamin K** (phylloquinone) regulates the synthesis of prothrombin and other proteins needed for blood to clot.
- A deficiency of vitamin K leads to excessive and sometimes fatal bleeding because of inadequate blood clotting.

Figure 26.6 Electrostatic potential plots of vitamins A and E

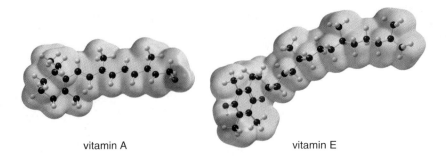

vitamin A   vitamin E

- The electron density is distributed fairly evenly among the carbon atoms of these vitamins due to their many nonpolar C−C and C−H bonds.

to obscure small dipoles that occur in the one or two polar bonds, making these vitamins nonpolar and hydrophobic.

PROBLEM 26.8 Explain why regularly ingesting a large excess of a fat-soluble vitamin can lead to severe health problems, whereas ingesting a large excess of a water-soluble vitamin often causes no major health problems.

26.6 Eicosanoids

The word *eicosanoid* is derived from the Greek word *eikosi,* meaning **20.**

The **eicosanoids** are a group of biologically active compounds containing 20 carbon atoms derived from arachidonic acid. The **prostaglandins** (Section 19.6) and the **leukotrienes** (Section 9.16) are two types of eicosanoids. Two others are the **thromboxanes** and **prostacyclins.**

$PGF_{2\alpha}$
a prostaglandin

LTC_4
a leukotriene

TA_4
a thromboxane

PGI_2
a prostacyclin

All eicosanoids are very potent compounds present in low concentration in cells. They are **local mediators,** meaning that they perform their function in the environment in which they are synthesized. This distinguishes them from **hormones,** which are first synthesized and then transported in the bloodstream to their site of action. Eicosanoids are not stored in cells; rather, they are synthesized from arachidonic acid in response to an external stimulus.

The synthesis of prostaglandins, thromboxanes, and prostacyclins begins with the oxidation of arachidonic acid with O_2 by a **cyclooxygenase** enzyme, which forms an unstable cyclic intermediate, PGG_2. PGG_2 is then converted via different pathways to these three classes of compounds. Leukotrienes are formed by a different pathway, using an enzyme called a **lipoxygenase.** These four paths for arachidonic acid are summarized in Figure 26.7.

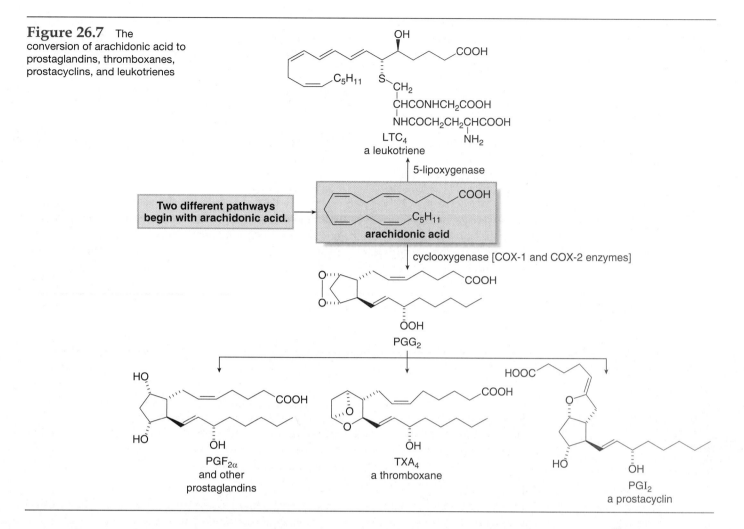

Figure 26.7 The conversion of arachidonic acid to prostaglandins, thromboxanes, prostacyclins, and leukotrienes

Other details of the biosynthesis of leukotrienes and prostaglandins were given in Sections 9.16 and 19.6, respectively.

Each eicosanoid is associated with specific types of biological activity (Table 26.4). In some cases, the effects oppose one another. For example, thromboxanes are vasoconstrictors that trigger blood platelet aggregation, whereas prostacyclins are vasodilators that inhibit platelet aggregation. The levels of these two eicosanoids must be in the right balance for cells to function properly.

TABLE 26.4 **Biological Activity of the Eicosanoids**

Eicosanoid	Effect
Prostaglandins	• Lower blood pressure • Inhibit blood platelet aggregation • Control inflammation • Lower gastric secretions • Stimulate uterine contractions • Relax smooth muscles of the uterus
Thromboxanes	• Constrict blood vessels • Trigger blood platelet aggregation
Prostacyclins	• Dilate blood vessels • Inhibit blood platelet aggregation
Leukotrienes	• Constrict smooth muscle, especially in the lungs

Because of their wide range of biological functions, prostaglandins and their analogues have found several clinical uses. For example, **dinoprostone,** the generic name for **PGE₂,** is administered to relax the smooth muscles of the uterus when labor is induced, and to terminate pregnancies in the early stages.

PGE₂
(dinoprostone)

Because prostaglandins themselves are unstable in the body, often having half-lives of only minutes, more stable analogues have been developed that retain their important biological activity longer. For example, **misoprostol,** an analogue of PGE₁, is sold as a mixture of stereoisomers. Misoprostol is administered to prevent gastric ulcers in patients who are at high risk of developing them.

PGE₁

misoprostol
(sold as a mixture of stereoisomers)

Studying the biosynthesis of eicosanoids has led to other discoveries as well. For example, aspirin and other non-steroidal anti-inflammatory drugs (**NSAIDs**) inactivate the cyclooxygenase enzyme needed for prostaglandin synthesis. In this way, NSAIDs block the synthesis of the prostaglandins that cause inflammation (Section 19.6).

More recently, it has been discovered that two *different* cyclooxygenase enzymes, called **COX-1** and **COX-2,** are responsible for prostaglandin synthesis. COX-1 is involved with the usual production of prostaglandins, but COX-2 is responsible for the synthesis of additional prostaglandins in inflammatory diseases like arthritis. **NSAIDs like aspirin and ibuprofen inactivate both the COX-1 and COX-2 enzymes.** This activity also results in an increase in gastric secretions, making an individual more susceptible to ulcer formation.

New anti-inflammatory drugs that block only the COX-2 enzyme have recently been developed. These drugs do *not* cause an increase in gastric secretions. One such drug is **valdecoxib** (trade name: **Bextra**).

Generic name: valdecoxib
Trade name: Bextra

A COX-3 enzyme was also reported in 2002. Its activity is inhibited by acetaminophen, the active ingredient in the pain reliever Tylenol.

The discovery of drugs such as Bextra illustrates how basic research in organic chemistry can lead to important practical applications. Elucidating the structure and biosynthesis of prostaglandins began as a project in basic research. It has now resulted in a number of applications that benefit many individuals with various illnesses.

PROBLEM 26.9 How are the two isomers of misoprostol related?

26.7   Terpenes

bayberry plant
(source of myrcene)

peppermint plant
(source of menthol)

Terpenes **are lipids composed of repeating five-carbon units called isoprene units. An iso-prene unit** has five carbons: four in a row, with a one-carbon branch on a middle carbon.

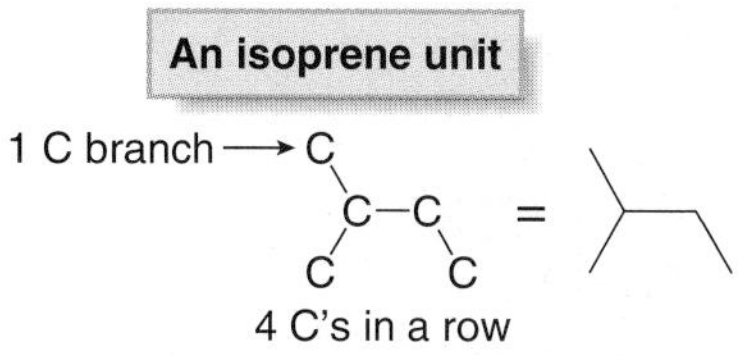

An isoprene unit

1 C branch → C
C—C
C   C
4 C's in a row

Terpenes have a wide variety of structures. They can be acyclic or have one or more rings. They may have only carbon and hydrogen atoms, or they may have heteroatoms as well. The most common heteroatom in terpenes is oxygen. Many **essential oils,** a group of compounds isolated from plant sources by distillation, are terpenes. Examples include myrcene and menthol.

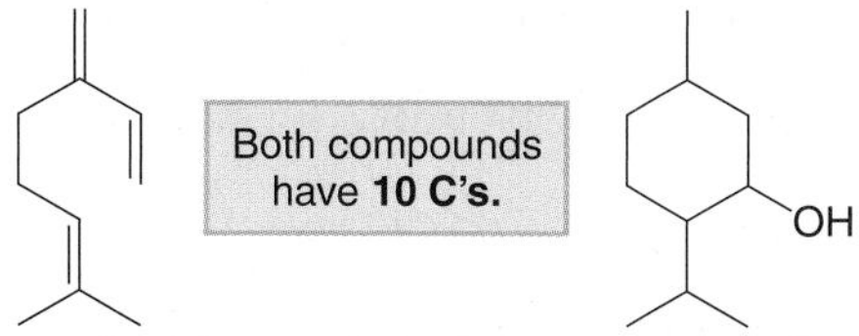

Both compounds
have **10 C's.**

myrcene
(isolated from bayberry oil)

menthol
(isolated from peppermint oil)

26.7A   Locating Isoprene Units in Terpenes

How do we identify the isoprene units in these molecules? Start at one end of the molecule near a branch point. Then **look for a four-carbon chain with a one-carbon branch.** This forms one isoprene unit. Continue along the chain or around the ring until all the carbons are part of an isoprene unit. Keep in mind the following:

♦ An isoprene unit may be composed of C–C σ bonds only, or there may be π bonds at any position.
♦ Isoprene units are always connected by one or more carbon–carbon bonds.
♦ Each carbon atom is part of one isoprene unit only.
♦ Every isoprene unit has five carbon atoms. Heteroatoms may be present but their presence is ignored in locating isoprene units.

Myrcene and menthol, for example, each have 10 carbon atoms, so they are composed of two isoprene units.

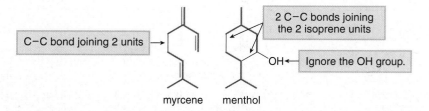

C–C bond joining 2 units

2 C–C bonds joining
the 2 isoprene units

OH ← Ignore the OH group.

myrcene   menthol

Terpenes are classified by the number of isoprene units they contain. A *monoterpene* **contains 10 carbons** and has two isoprene units, a *sesquiterpene* **contains 15 carbons** and has three iso-prene units, and so forth. The different terpene classes are summarized in Table 26.5.

TABLE 26.5   Classes of Terpenes

Name	Number of C atoms	Number of isoprene units
Monoterpene	10	2
Sesquiterpene	15	3
Diterpene	20	4
Sesterterpene	25	5
Triterpene	30	6
Tetraterpene	40	8

Several examples, with the isoprene units labeled in red, are given in Figure 26.8.

PROBLEM 26.10   Locate the isoprene units in each compound.

a.

geraniol
(roses and geraniums)

c.

grandisol
(sex pheromone of the
male boll weevil)

b.

vitamin A

d.

camphor

PROBLEM 26.11   Manoalide, a sesterterpene isolated from the Pacific marine sponge *Luffariella veriabilis* by Scheuer and co-workers at the University of Hawai'i at Mānoa, has anti-inflammatory, analgesic, and antifungal properties. Find the isoprene units in manoalide.

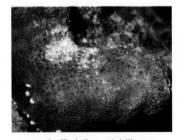

Luffariella veriabilis
(source of manoalide)

manoalide

Figure 26.8   Examples of some common terpenes

citral
(lemon grass)

farnesol
(lily of the valley)

zingiberene
(ginger)

cedrol
(cedar)

squalene
(shark oil)

α-phellandrene
(eucalyptus)

- Isoprene units are labeled in red, with C−C bonds (in black) joining two units.
- The source of each terpene is given in parentheses.

26.7B The Biosynthesis of Terpenes

Terpene biosynthesis is an excellent example of how syntheses in nature occur with high efficiency. There are two ways this is accomplished.

[1] The same reaction is used over and over again to prepare progressively more complex compounds.

[2] Key intermediates along the way serve as the starting materials for a wide variety of other compounds.

All terpenes are synthesized from dimethylallyl pyrophosphate and isopentenyl pyrophosphate. Both of these five-carbon compounds are synthesized, in turn, in a multistep process from three molecules of acetyl CoA (Section 22.17).

The starting materials for terpene biosynthesis

$$3\ CH_3\overset{\overset{\displaystyle O}{\|}}{C}\text{—SCoA}$$

acetyl CoA

dimethylallyl pyrophosphate   and   isopentenyl pyrophosphate

Pyrophosphate, abbreviated as **OPP,** is often used as a leaving group in biological systems. **It is a good leaving group because it is a weak, resonance-stabilized base.**

pyrophosphate
a good leaving group

R—OPP

$$R\text{—OPP} \longrightarrow R\text{—Nu} + \ ^-OPP =$$

good leaving group

:Nu$^-$

The overall strategy of terpene biosynthesis from dimethylallyl pyrophosphate and isopentenyl pyrophosphate is summarized in Figure 26.9.

Figure 26.9 An outline of terpene biosynthesis

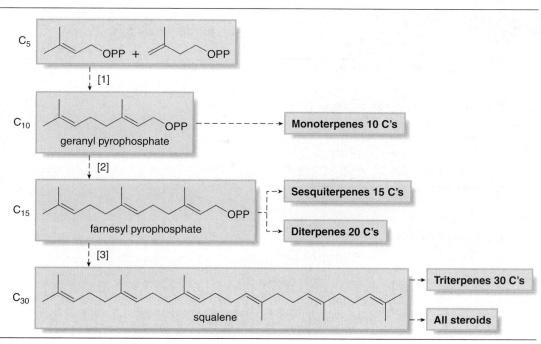

There are three basic parts:

[1] The two C_5 pyrophosphates are converted to **geranyl pyrophosphate, a C_{10} monoterpene.** Geranyl pyrophosphate is the starting material for all other monoterpenes.

[2] Geranyl pyrophosphate is converted to **farnesyl pyrophosphate, a C_{15} sesquiterpene,** by addition of a five-carbon unit. Farnesyl pyrophosphate is the starting material for all sesquiterpenes and diterpenes.

[3] Two molecules of farnesyl pyrophosphate are converted to **squalene, a C_{30} triterpene.** Squalene is the starting material for all triterpenes and steroids.

The biological formation of geranyl pyrophosphate from the two five-carbon pyrophosphates involves two steps: **nucleophilic attack** on the allylic pyrophosphate, followed by **loss of a proton,** as shown in Mechanism 26.1.

MECHANISM 26.1

Biological Formation of Geranyl Pyrophosphate

Step [1] Nucleophilic attack on the allylic pyrophosphate

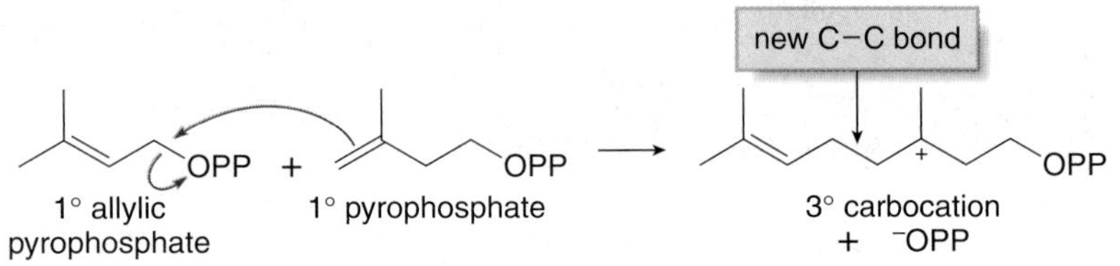

◆ **Step [1]** Although both C_5 pyrophosphates have a leaving group on a 1° carbon, the **allylic pyrophosphate** is more reactive towards nucleophilic attack. Nucleophilic attack followed by loss of this pyrophosphate forms a new C−C bond and a 3° carbocation.

Step [2] Loss of a proton

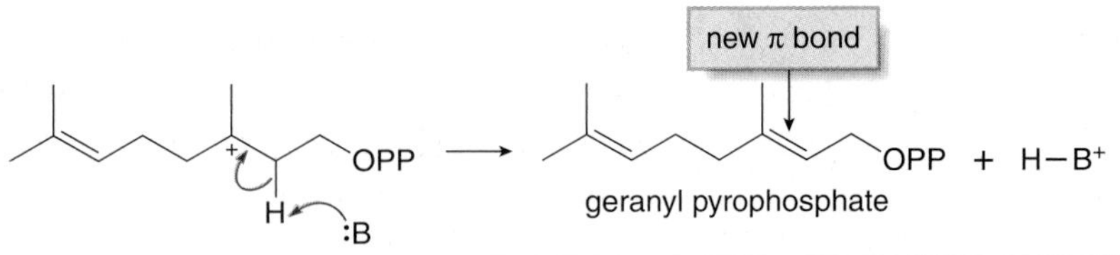

◆ **Step [2]** Loss of a proton forms geranyl pyrophosphate.

The biological conversion of geranyl pyrophosphate to farnesyl pyrophosphate involves the same two steps: **nucleophilic attack** on the allylic pyrophosphate, followed by **loss of a proton** (Mechanism 26.2).

MECHANISM 26.2

Biological Formation of Farnesyl Pyrophosphate

Step [1] Nucleophilic attack on the allylic pyrophosphate

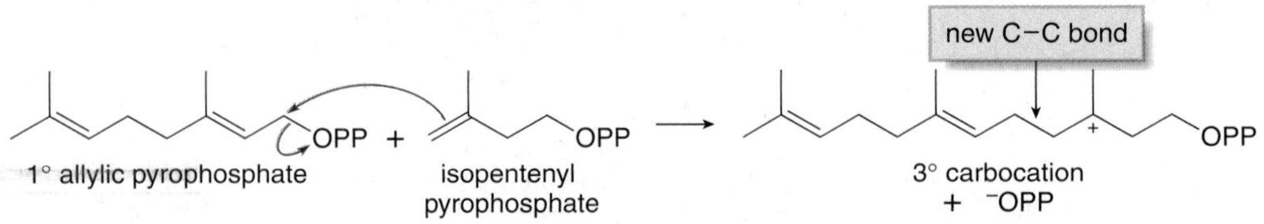

◆ **Step [1]** Nucleophilic attack of the isopentenyl pyrophosphate on the allylic pyrophosphate yields a 3° carbocation. This reaction forms a new carbon–carbon σ bond.

Step [2] Loss of a proton

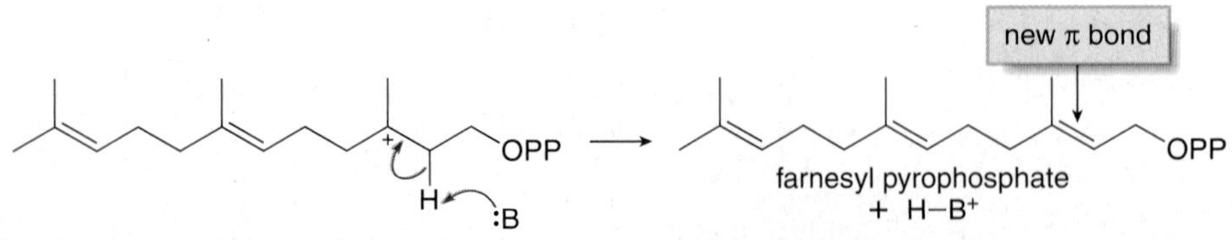

◆ **Step [2]** Loss of a proton forms a new π bond and farnesyl pyrophosphate.

Two molecules of farnesyl pyrophosphate react to form squalene, from which all other triterpenes and steroids are synthesized.

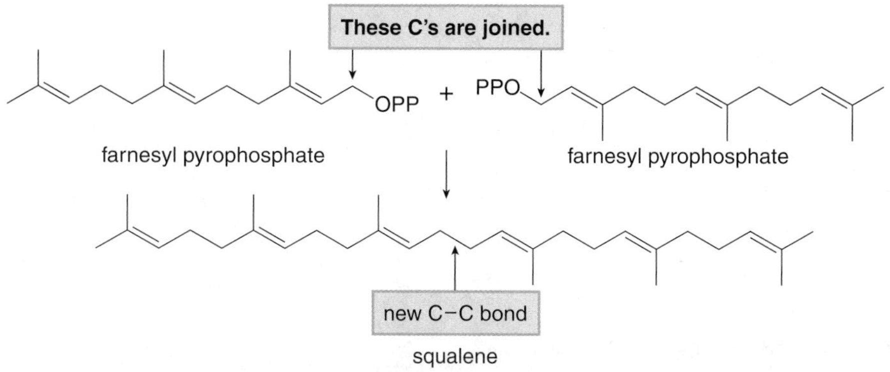

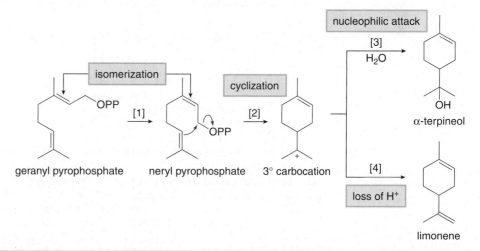

Aqueous hydrolysis of geranyl and farnesyl pyrophosphates forms the monoterpene geraniol and the sesquiterpene farnesol, respectively.

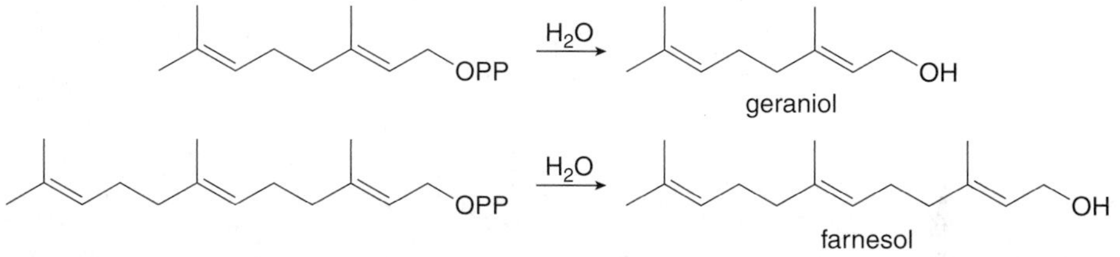

All other terpenes are biologically derived from geranyl and farnesyl pyrophosphates by a series of reactions. Cyclic compounds are formed by intramolecular reactions involving nucleophilic attack of π bonds on intermediate carbocations.

In the synthesis of α-terpineol or limonene, for example, geranyl pyrophosphate isomerizes at one double bond to form neryl pyrophosphate, a stereoisomer (Step [1] in the following reaction sequence). Neryl pyrophosphate then cyclizes to a 3° carbocation by intramolecular attack (Step [2]). Nucleophilic attack of water on this carbocation yields α-terpineol (Step [3]) or loss of a proton yields limonene (Step [4]). Both products are cyclic monoterpenes.

PROBLEM 26.12   Write a stepwise mechanism for the following reaction.

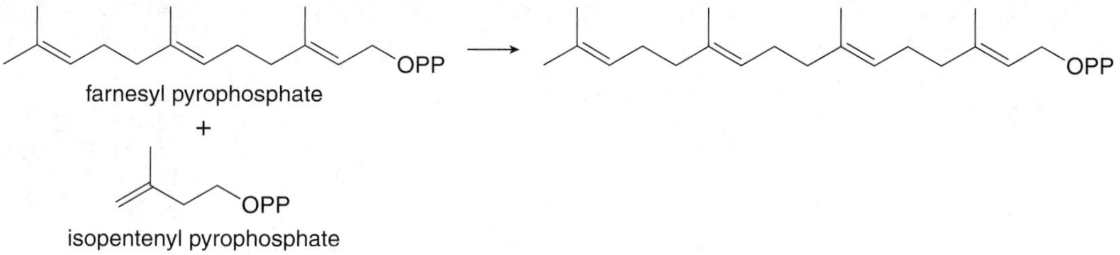

PROBLEM 26.13   Draw a stepwise mechanism for the conversion of geranyl pyrophosphate to α-terpinene.

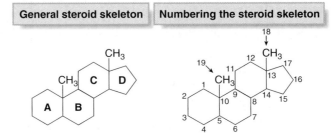

α-terpinene

26.8   Steroids

The steroids are a group of tetracyclic lipids, many of which are biologically active.

26.8A   Steroid Structure

Steroids are composed of three six-membered rings and one five-membered ring, joined together as drawn. Many steroids also contain two methyl groups, called **angular methyl groups,** at the two ring junctions indicated. The steroid rings are lettered **A, B, C,** and **D,** and the 17 ring carbons are numbered as shown. The two angular methyl groups are numbered C18 and C19.

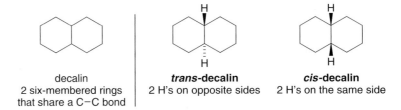

Whenever two rings are fused together, the substituents at the ring fusion can be arranged cis or trans. To see more easily why this is true, consider **decalin,** which consists of two six-membered rings fused together. *trans*-**Decalin** has the two hydrogen atoms at the ring fusion on opposite sides, whereas *cis*-**decalin** has them on the same side.

decalin	*trans*-decalin	*cis*-decalin
2 six-membered rings that share a C–C bond	2 H's on opposite sides	2 H's on the same side

Three-dimensional structures of these molecules show how different these two possible arrangements actually are. The two rings of *trans*-decalin lie roughly in the same plane, whereas the two rings of *cis*-decalin are almost perpendicular to each other. **The trans arrangement is lower in energy and therefore more stable.**

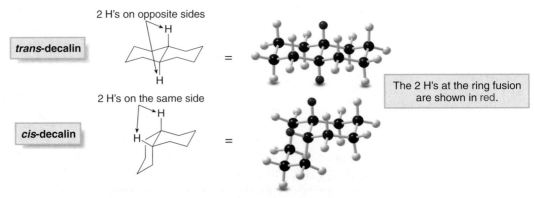

trans-decalin   2 H's on opposite sides

cis-decalin   2 H's on the same side

The 2 H's at the ring fusion are shown in red.

Figure 26.10 The three-dimensional structure of the steroid nucleus

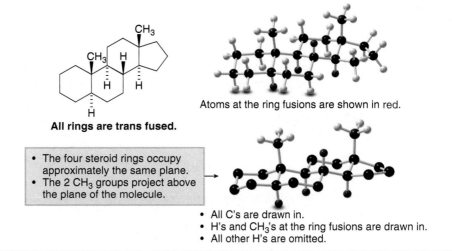

All rings are trans fused.

- The four steroid rings occupy approximately the same plane.
- The 2 CH$_3$ groups project above the plane of the molecule.

Atoms at the ring fusions are shown in red.

- All C's are drawn in.
- H's and CH$_3$'s at the ring fusions are drawn in.
- All other H's are omitted.

In steroids, each ring fusion could theoretically have the cis or trans configuration, but, by far the most common arrangement is all trans. Because of this, **all four rings of the steroid skeleton lie in the same plane,** and the ring system is fairly rigid. The two angular methyl groups are oriented perpendicular to the plane of the molecule. These methyl groups make one side of the steroid skeleton significantly more hindered than the other, as shown in Figure 26.10.

Although steroids have the same fused-ring arrangement of carbon atoms, they differ in the identity and location of the substituents attached to that skeleton.

26.8B Cholesterol

Cholesterol, the chapter-opening molecule, has the tetracyclic carbon skeleton characteristic of steroids. It also has eight stereogenic carbons (seven on rings and one on a side chain), so there are $2^8 = 256$ possible stereoisomers. In nature, however, only the following stereoisomer exists:

> Cholesterol has also been discussed in Sections 3.5C and 4.15. The role of cholesterol in plaque formation and atherosclerosis was discussed in Section 22.17.

[*denotes a stereogenic center]

cholesterol

> Konrad Bloch and Feodor Lynen shared the 1964 Nobel Prize in Physiology or Medicine for unraveling the complex transformation of squalene to cholesterol.

Cholesterol is essential to life because it forms an important component of cell membranes and is the starting material for the synthesis of all other steroids. Humans do not have to ingest cholesterol, because it is synthesized in the liver and then transported to other tissues through the bloodstream. Because cholesterol has only one polar OH group and many nonpolar C−C and C−H bonds, it is insoluble in water (and, thus, in the aqueous medium of the blood).

Cholesterol is synthesized in the body from squalene, a C$_{30}$ triterpene that is itself prepared from smaller terpenes, as discussed in Section 26.7B. Because the biosynthesis of all terpenes begins with acetyl CoA, every one of the 27 carbon atoms of cholesterol comes from the same two-carbon precursor. The major steps in the conversion of squalene to cholesterol are given in Figure 26.11.

The conversion of squalene to cholesterol consists of five different parts:

[1] **Epoxidation** of squalene with an enzyme, squalene epoxidase, gives squalene oxide, which contains a single epoxide on one of the six double bonds.

[2] **Cyclization** of squalene oxide yields a carbocation, called the protosterol cation. This reaction results in the formation of four new C−C bonds and the tetracyclic ring system.

Figure 26.11 The biosynthesis of cholesterol

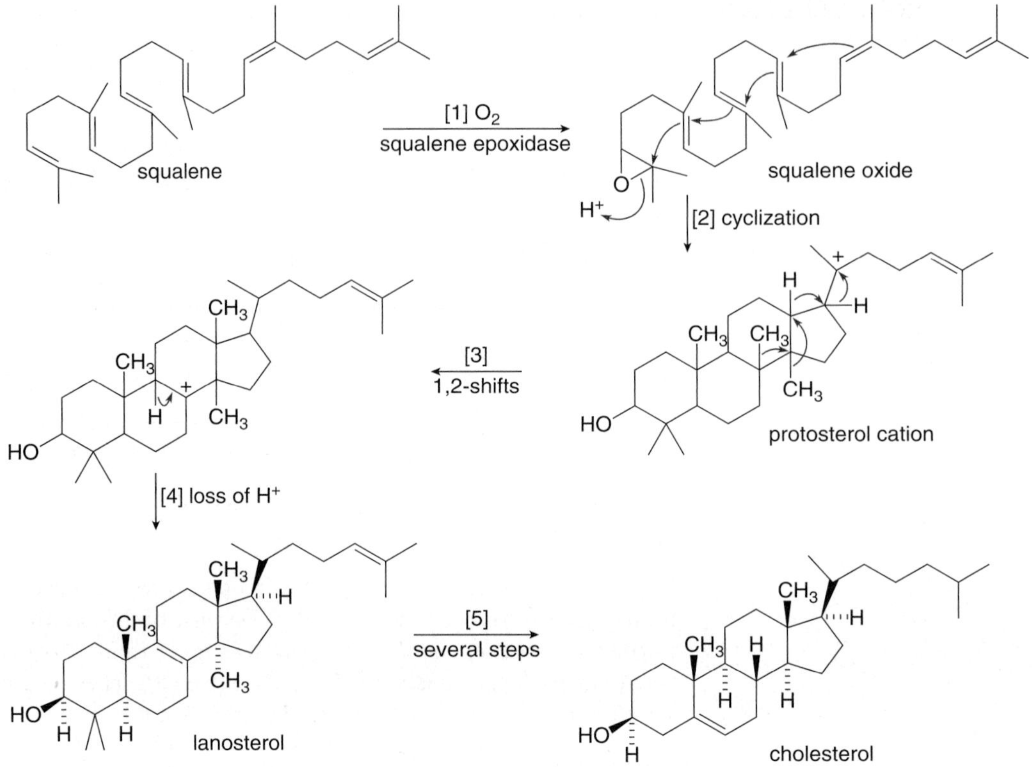

[3] **The protosterol carbocation rearranges** by a series of 1,2-shifts of either a hydrogen or methyl group to form another 3° carbocation.

[4] **Loss of a proton** gives an alkene called **lanosterol.** Although lanosterol has seven stereogenic centers, a single stereoisomer is formed.

[5] Lanosterol is then converted to cholesterol by a multistep process that results in removal of three methyl groups.

Several drugs are now available to reduce the level of cholesterol in the bloodstream. These compounds act by blocking the biosynthesis of cholesterol at its very early stages. Two examples include atorvastatin (Lipitor) and simvastatin (Zocor), whose structures appear in Figure 26.12.

PROBLEM 26.14 Draw the enantiomer and any two diastereomers of cholesterol. Does the OH group of cholesterol occupy an axial or equatorial position?

Figure 26.12 Two cholesterol-lowering drugs

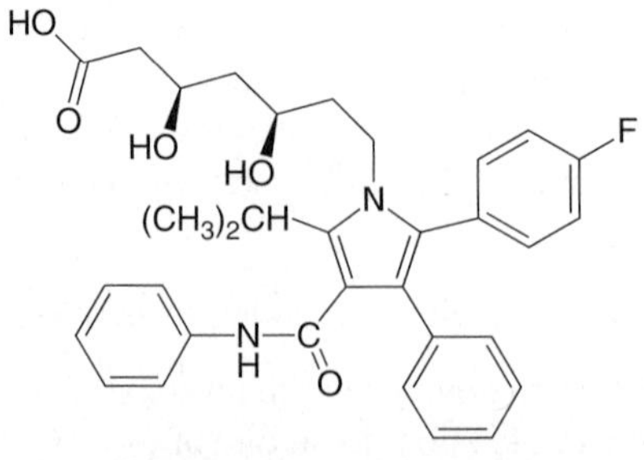

Generic name: atorvastatin
Trade name: Lipitor

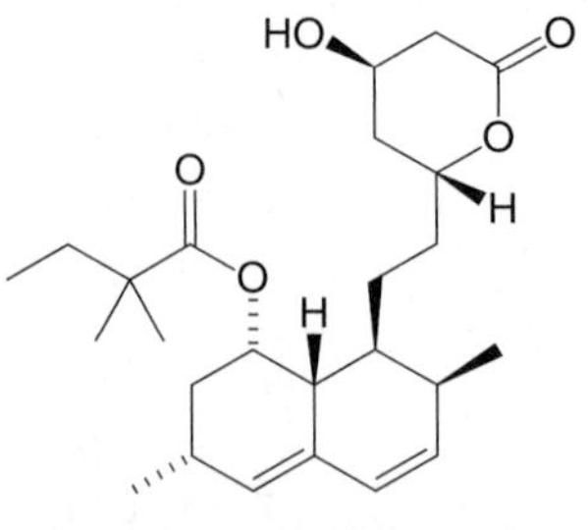

Generic name: simvastatin
Trade name: Zocor

PROBLEM 26.15 Treatment of cholesterol with mCPBA results in formation of a single epoxide **A,** with the stereochemistry drawn. Why isn't the isomeric epoxide **B** formed to any extent?

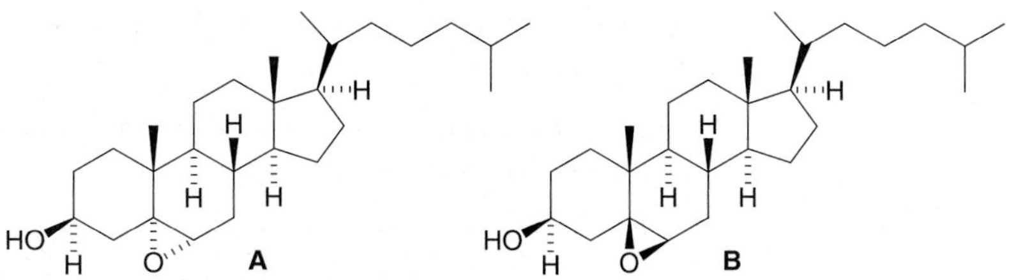

26.8C Other Steroids

Many other important steroids are hormones secreted by the endocrine glands. Two important classes are the **sex hormones** and the **adrenal cortical steroids.**

There are two types of female sex hormones, **estrogens** and **progestins.** The male sex hormones are called **androgens.** The most important members of each hormone type are given in Table 26.6.

TABLE 26.6   The Female and Male Sex Hormones

Structure	Properties
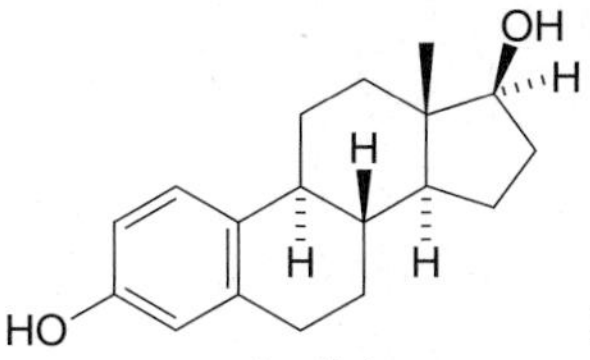 estradiol 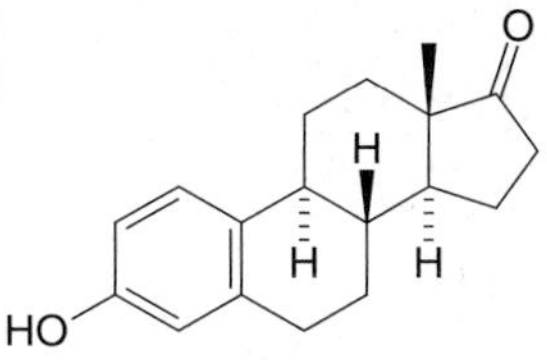 estrone	• **Estradiol** and **estrone** are estrogens synthesized in the ovaries. They control the development of secondary sex characteristics in females and regulate the menstrual cycle.
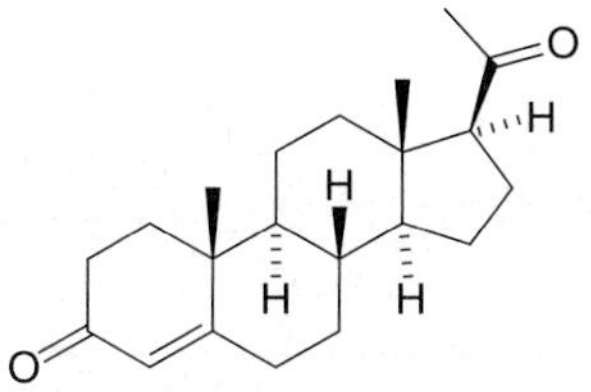 progesterone	• **Progesterone** is often called the "pregnancy hormone." It is responsible for the preparation of the uterus for implantation of a fertilized egg.
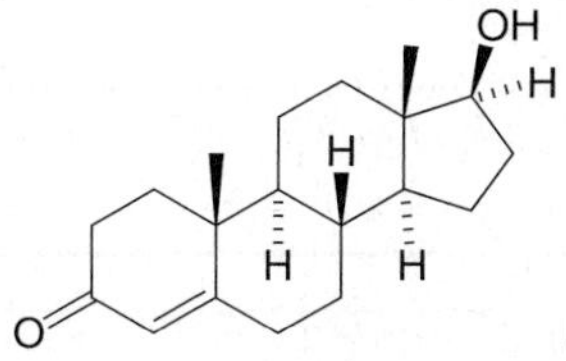 testosterone 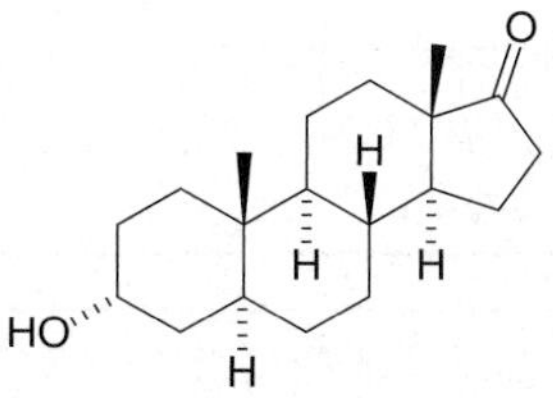 androsterone	• **Testosterone** and **androsterone** are androgens synthesized in the testes. They control the development of secondary sex characteristics in males.

Synthetic analogues of these steroids have found important uses, such as in oral contraceptives, first mentioned in Section 11.4.

Synthetic oral contraceptives

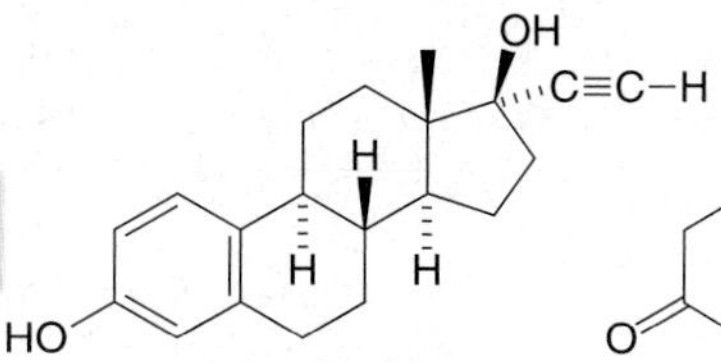

ethynylestradiol 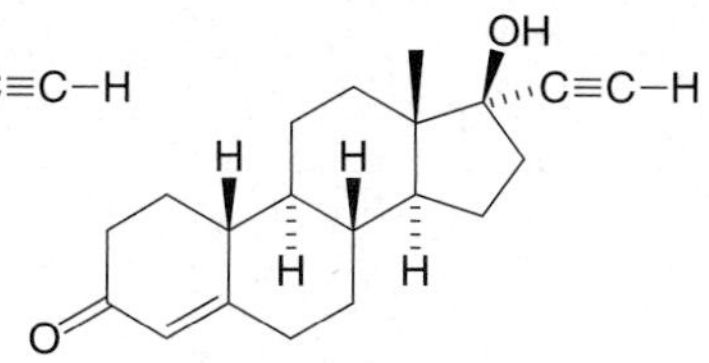   norethindrone

Synthetic androgen analogues, called **anabolic steroids,** promote muscle growth. They were first developed to help individuals whose muscles had atrophied from lack of use following surgery. They have since come to be used by athletes and body builders, although their use is not permitted in competitive sports. Many physical and psychological problems result from their prolonged use.

Anabolic steroids, such as **stanozolol** and **nandrolone,** have the same effect on the body as testosterone, but they are more stable, so they are not metabolized as quickly.

Anabolic steroids

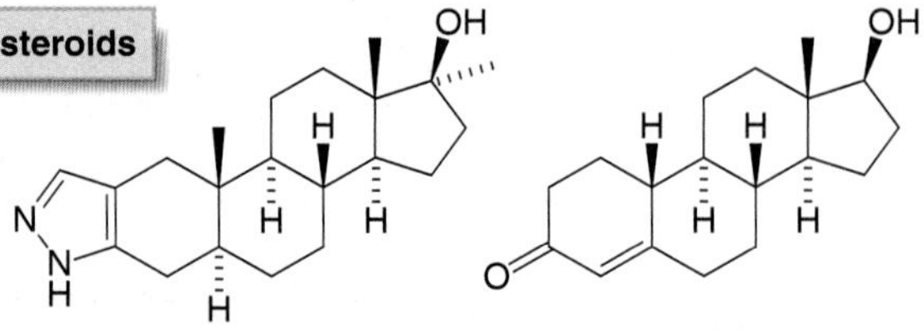

stanozolol   nandrolone

A second group of steroid hormones includes the **adrenal cortical steroids.** Three examples of these hormones are **cortisone, cortisol,** and **aldosterone.** All of these compounds are synthesized in the outer layer of the adrenal gland. Cortisone and cortisol serve as anti-inflammatory agents and they also regulate carbohydrate metabolism. Aldosterone regulates blood pressure and volume by controlling the concentration of Na^+ and K^+ in body fluids.

Three adrenal cortical steroids

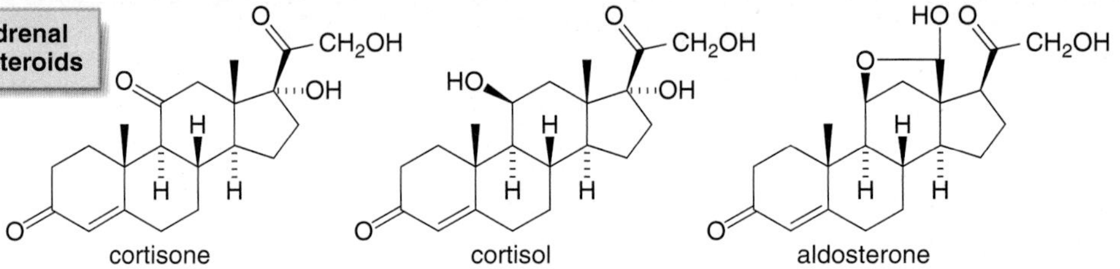

cortisone   cortisol   aldosterone

26.9 Key Concepts—Lipids

Hydrolyzable Lipids

[1] **Waxes** (26.2)—Esters formed from a long-chain alcohol and a long-chain carboxylic acid.

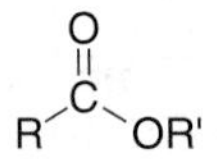

R, R' = long chains of C's

[2] **Triacylglycerols** (26.3)—Triesters of glycerol with three fatty acids.

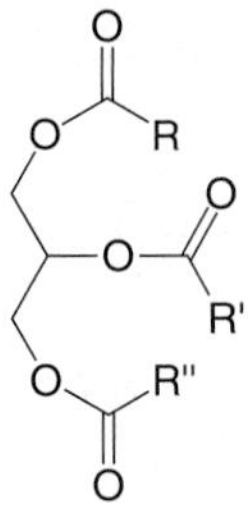

R, R', R" = alkyl groups with 11–19 C's

[3] **Phospholipids** (26.4)

[a] Phosphatidylethanolamine (cephalin)   [b] Phosphatidylcholine (lecithin)   [c] Sphingomyelin

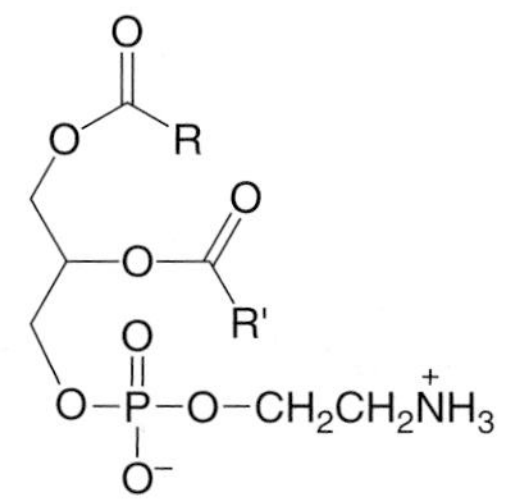

R, R' = long carbon chain

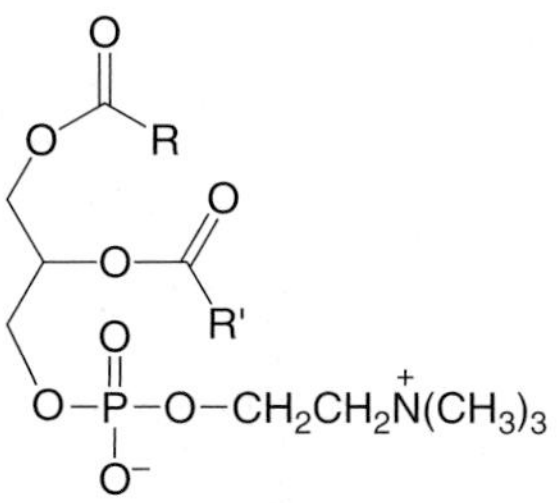

R, R' = long carbon chain

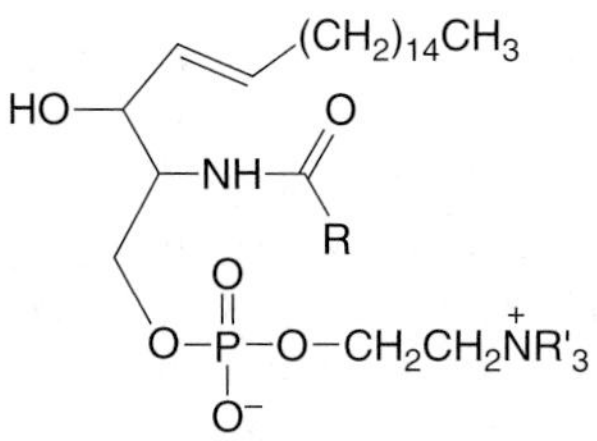

R = long carbon chain
R' = H or CH_3

Nonhydrolyzable Lipids

[1] **Fat-soluble vitamins** (26.5)—Vitamins A, D, E, and K.

[2] **Eicosanoids** (26.6)—Compounds containing 20 C's derived from arachidonic acid. There are four types: prostaglandins, thromboxanes, prostacyclins, and leukotrienes.

[3] **Terpenes** (26.7)—Lipids composed of repeating 5 C units called isoprene units.

Isoprene unit	Types of terpenes			
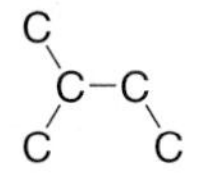 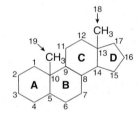	[1] monoterpene	10 C's	[4] sesterterpene	25 C's
	[2] sesquiterpene	15 C's	[5] triterpene	30 C's
	[3] diterpene	20 C's	[6] tetraterpene	40 C's

[4] **Steroids** (26.8)—Tetracyclic lipids composed of three six-membered and one five-membered ring.

Problems

Waxes, Triacylglycerols, and Phospholipids

26.16 One component of lanolin, the wax that coats sheep's wool, is derived from cholesterol and stearic acid. Draw its structure, including the correct stereochemistry at all stereogenic centers.

26.17 Draw all possible constitutional isomers of a triacylglycerol formed from one mole each of palmitic, oleic, and linoleic acids. Locate the tetrahedral stereogenic centers in each constitutional isomer.

26.18 What is the structure of an optically inactive triacylglycerol that yields two moles of oleic acid and one mole of palmitic acid when hydrolyzed in aqueous acid?

26.19 Triacylglycerol **L** yields compound **M** when treated with excess H_2, Pd. Ozonolysis of **L** ([1] O_3; [2] $(CH_3)_2S$) affords compounds **N–P**. What is the structure of **L**?

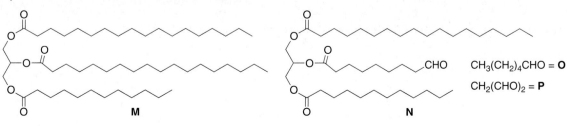

$CH_3(CH_2)_4CHO = O$

$CH_2(CHO)_2 = P$

M   N

26.20 Draw the structure of the following phospholipids:
   a. A cephalin formed from two molecules of stearic acid.
   b. A sphingomyelin formed from palmitic acid.

Prostaglandins

26.21 A difficult problem in the synthesis of $PGF_{2\alpha}$ is the introduction of the OH group at C15 in the desired configuration.
   a. Label this stereogenic center as R or S.
   b. A well known synthesis of $PGF_{2\alpha}$ involves reaction of **A** with $Zn(BH_4)_2$, a metal hydride reagent similar in reactivity to $NaBH_4$, to form two isomeric products, **B** and **C**. Draw their structures and indicate their stereochemical relationship.
   c. Suggest a reagent to convert **A** to the single stereoisomer **X**.

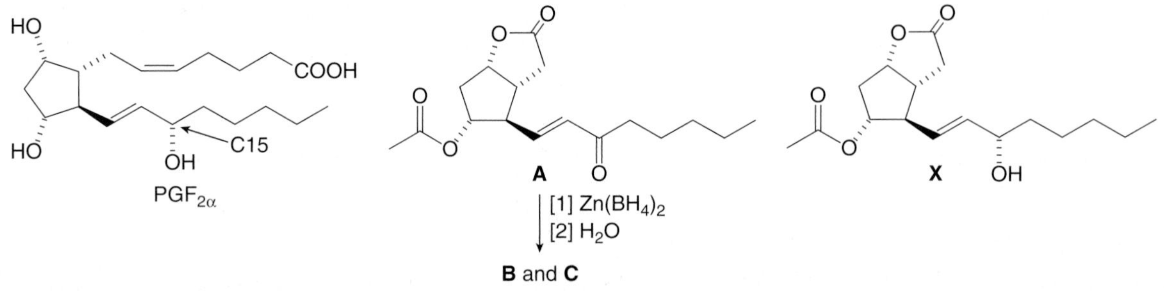

Terpenes

26.22 Locate the isoprene units in each compound.

a. 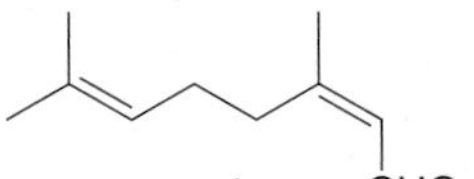 neral   CHO
b. 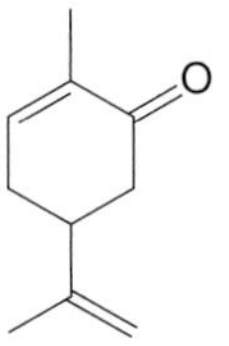 carvone
c. 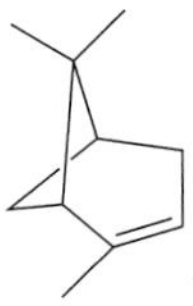 α-pinene
d. 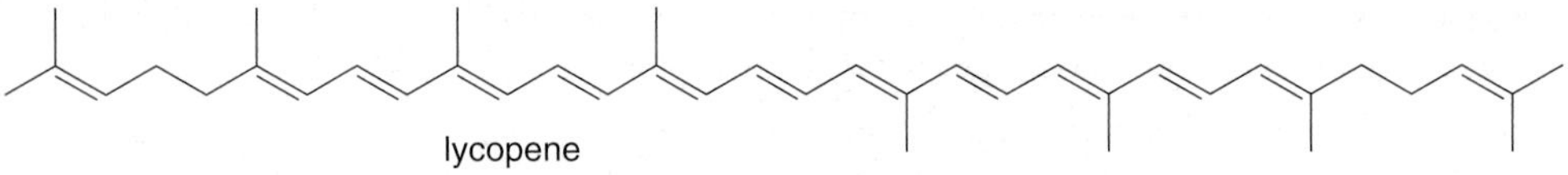 lycopene
e. 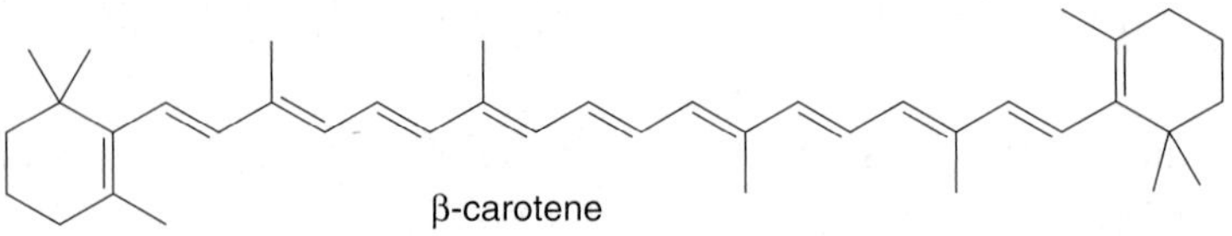 β-carotene
f. humulene
g. patchouli alcohol
h. periplanone B
i. dextropimaric acid COOH
j. β-amyrin

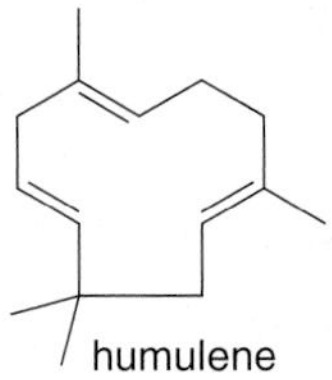

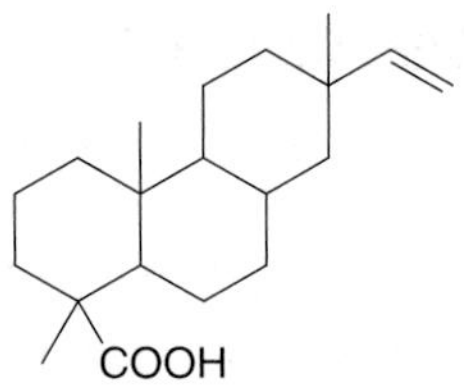

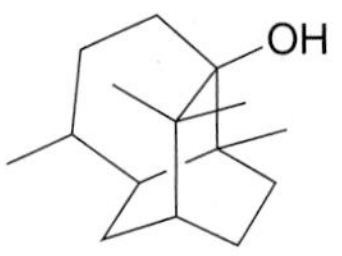

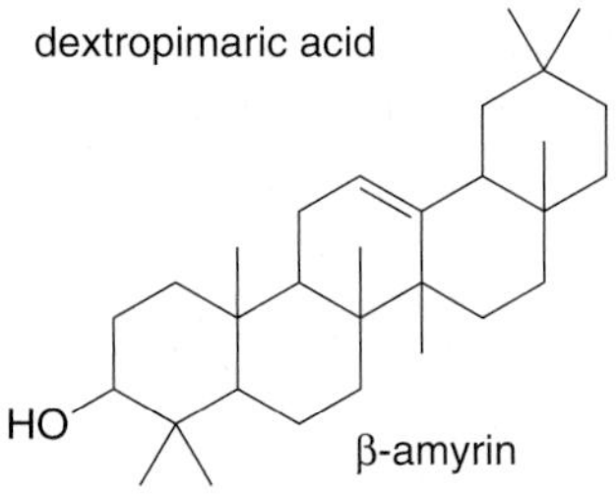

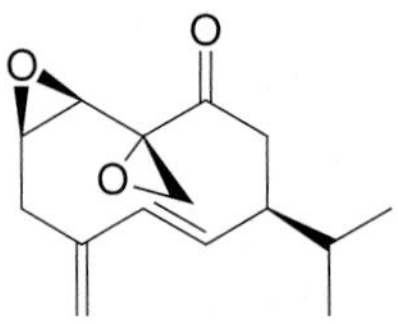

26.23 Classify each terpene in Problem 26.22 (e.g., as a monoterpene, sesquiterpene, etc.).

26.24 An isoprene unit can be thought of as having a head and a tail. The "head" of the isoprene unit is located at the end of the chain nearest the branch point, and the "tail" is located at the end of the carbon chain furthest from the branch point. Most isoprene units are connected together in a "head-to-tail" fashion, as illustrated. For both lycopene (Problem 26.22), and squalene (Figure 26.9), decide which isoprene units are connected in a head-to-tail fashion and which are not.

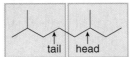

head   tail   tail   head

These two isoprene units are
connected in a head-to-tail fashion.

26.25 Draw a stepwise mechanism for the following reaction.

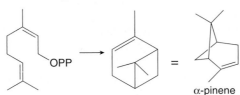

α-pinene

26.26 The biosynthesis of lanosterol from squalene has intrigued chemists since its discovery. It is now possible, for example, to synthesize polycyclic compounds from acyclic or monocyclic precursors by reactions that form several C−C bonds in a single reaction mixture.
  a. Draw a stepwise mechanism for the following reaction.
  b. Show how **X** can be converted to 16,17-dehydroprogesterone. (*Hint:* See Figure 24.4 for a related conversion.)

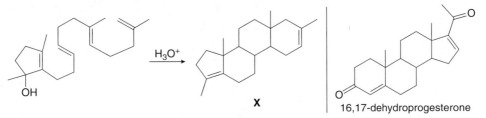

H_3O^+

X   16,17-dehydroprogesterone

Steroids

26.27 Draw three-dimensional structures for each decalin derivative.

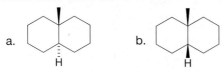

a.   b.

26.28 Draw three-dimensional structures for each alcohol. Label the OH groups as occupying axial or equatorial positions.

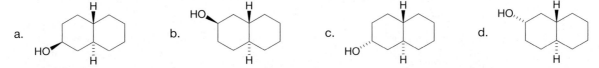

a.   b.   c.   d.

26.29 Axial alcohols are oxidized faster than equatorial alcohols by PCC and other Cr^{6+} oxidants. Which OH group in each compound is oxidized faster?

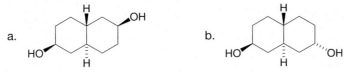

a.   b.

26.30 Draw a three-dimensional representation for androsterone.

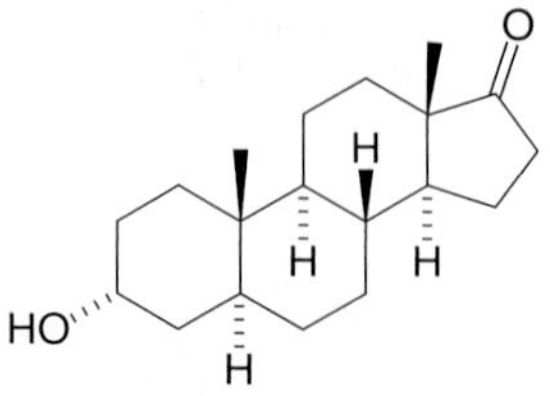

androsterone

26.31 a. Draw a three-dimensional structure for the following steroid.
   b. What is the structure of the single stereoisomer formed by reduction of this ketone with H_2, Pd? Explain why only one stereoisomer is formed.

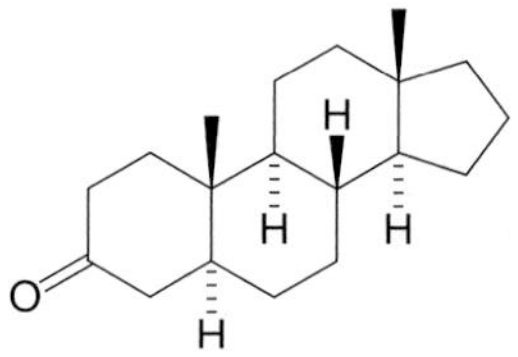

26.32 Draw the products formed when cholesterol is treated with each reagent. Indicate the stereochemistry around any stereogenic centers in the product.
   a. CH_3COCl
   b. H_2, Pd-C
   c. PCC
   d. stearic acid, H^+
   e. [1] $BH_3 \cdot THF$; [2] H_2O_2, ^-OH

Challenge Problems

26.33 Draw a stepwise mechanism for the following reaction.

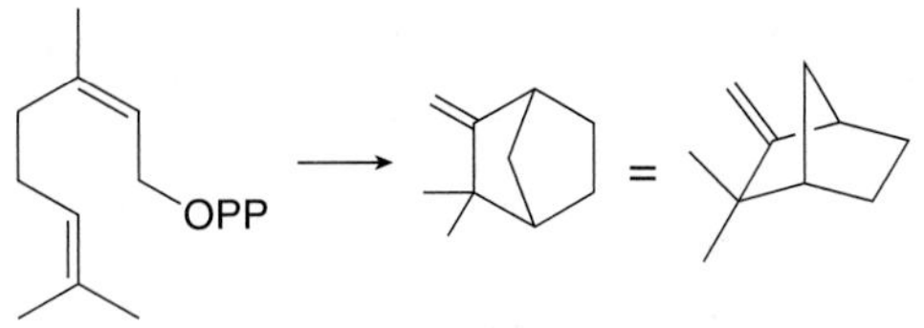

26.34 Draw a stepwise mechanism for the following reaction.

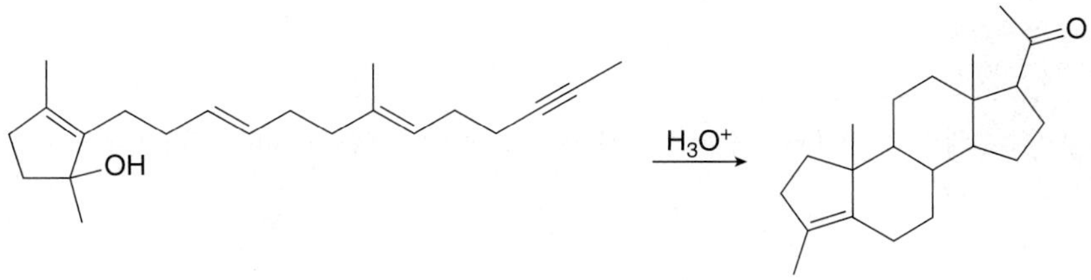

CHAPTER

27

Carbohydrates

Lactose, a carbohydrate formed from two simple sugars, glucose and galactose, is the principal sugar in dairy products. Many individuals, mainly of Asian and African descent, lack adequate amounts of the enzyme necessary to digest and absorb lactose. This condition, lactose intolerance, is associated with abdominal cramping and recurrent diarrhea, and is precipitated by the ingestion of milk and dairy products. In Chapter 27, we learn about the structure, synthesis, and properties of carbohydrates like lactose.

We now turn our attention to **carbohydrates,** the largest group of organic molecules in nature, comprising ~50% of the earth's biomass.

Carbohydrates can be simple or complex, having as few as three or as many as thousands of carbon atoms. They form the support structure of plants and tree trunks. They can be covalently bonded to lipids, forming glycolipids, or to proteins, forming glycoproteins. Carbohydrates on cell surfaces play an important role in cell recognition.

The chemistry of carbohydrates can be complex. Unlike lipids, which contain rings or chains of unfunctionalized carbon atoms, carbohydrates contain many functional groups, which can react with each other and with other reagents. Fortunately, however, almost all of the properties and reactions of carbohydrates can be understood by applying the basic principles of organic chemistry discussed in previous chapters.

27.1   Introduction

Carbohydrates were given their name because their molecular formula could be written as $C_n(H_2O)_n$, making them **hydrates of carbon.**

Carbohydrates such as glucose and cellulose were discussed in Sections 5.1, 6.4, and 21.17.

Although the metabolism of lipids provides more energy per gram than the metabolism of carbohydrates, glucose is the preferred source when a burst of energy is needed during exercise. Glucose is water soluble, so it can be quickly and easily transported through the bloodstream to the tissues.

Carbohydrates, commonly referred to as sugars and starches, are polyhydroxy aldehydes and ketones, or compounds that can be hydrolyzed to them. The cellulose in plant stems and tree trunks and the chitin in the exoskeletons of arthropods and mollusks are all complex carbohydrates. Four examples are shown in Figure 27.1. They include not only glucose and cellulose, but also doxorubicin (an anticancer drug) and 2'-deoxyadenosine 5'-monophosphate (a nucleotide base from DNA), both of which have a carbohydrate moiety as part of a larger molecule.

Carbohydrates are storehouses of chemical energy. They are synthesized in green plants and algae by **photosynthesis,** a process that uses the energy from the sun to convert carbon dioxide and water into glucose and oxygen. This energy is released when glucose is metabolized. The oxidation of glucose is a multistep process that forms carbon dioxide, water, and a great deal of energy (Section 6.4).

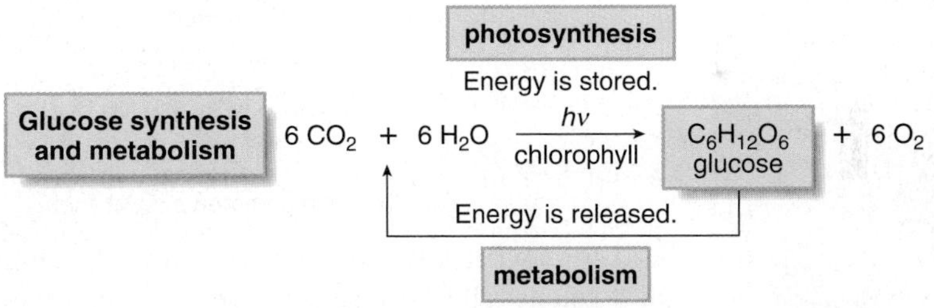

$$6\ CO_2\ +\ 6\ H_2O\ \xrightarrow[\text{chlorophyll}]{hv}\ C_6H_{12}O_6\ +\ 6\ O_2$$

Figure 27.1   Some examples of carbohydrates

These compounds illustrate the structural diversity of carbohydrates. **Glucose** is the most common simple sugar, whereas **cellulose,** which comprises wood, plant stems, and grass, is the most common carbohydrate in the plant world. **Doxorubicin,** an anticancer drug that has a carbohydrate ring as part of its structure, has been used in the treatment of leukemia, Hodgkin's disease, and cancers of the breast, bladder, and ovaries. **2'-Deoxyadenosine 5'-monophosphate** is one of the four nucleotides that form DNA.

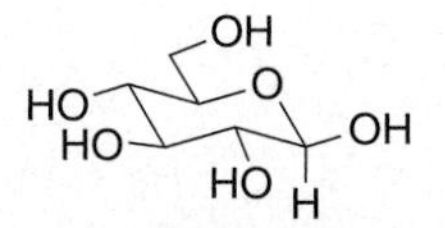

β-D-glucose
most common simple carbohydrate

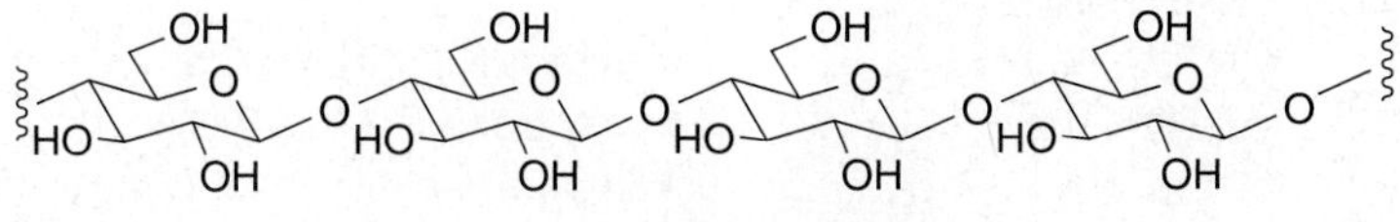

cellulose
main component of wood

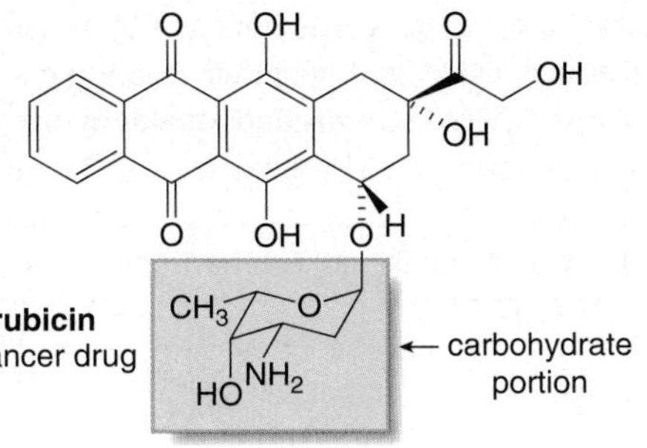

doxorubicin
an anticancer drug

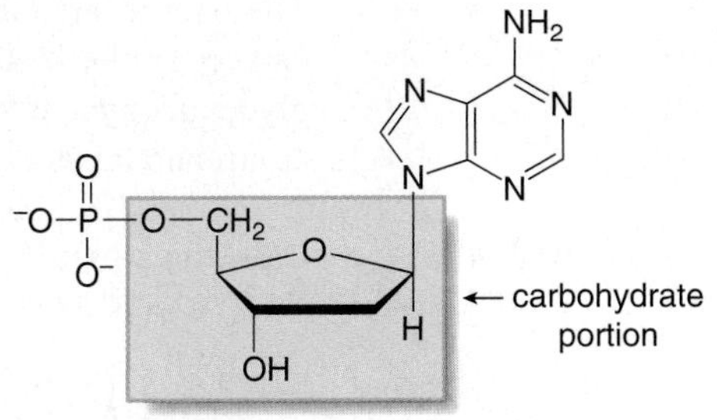

2'-deoxyadenosine 5'-monophosphate
a nucleotide component of DNA

27.2   Monosaccharides

The word saccharide comes from the Latin word *saccharum* meaning *sugar.*

The simplest carbohydrates are called **monosaccharides** or **simple sugars. Monosaccharides have three to seven carbon atoms** in a chain, with a **carbonyl group** at either the terminal carbon (C1) or the carbon adjacent to it (C2). In most carbohydrates, each of the remaining carbon atoms has a **hydroxy group.** Monosaccharides are usually drawn vertically, with the carbonyl group at the top.

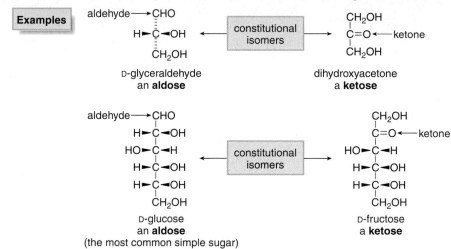

- ◆ Monosaccharides with an aldehyde carbonyl group at C1 are called aldoses.
- ◆ Monosaccharides with a ketone carbonyl group at C2 are called ketoses.

D-Fructose is almost twice as sweet as normal table sugar (sucrose) with about the same number of calories per gram. "Lite" food products use only half as much fructose as sucrose for the same level of sweetness, and so they have fewer calories.

Several examples of simple carbohydrates are shown. D-Glyceraldehyde and dihydroxyacetone have the same molecular formula, so they are **constitutional isomers,** as are D-glucose and D-fructose.

Dihydroxyacetone is the active ingredient in many artificial tanning agents.

aldehyde ⟶ CHO
H ►C◄ OH
CH₂OH
D-glyceraldehyde
an **aldose**

constitutional isomers

CH₂OH
C=O ◄ ketone
CH₂OH
dihydroxyacetone
a **ketose**

aldehyde ⟶ CHO
H ►C◄ OH
HO ►C◄ H
H ►C◄ OH
H ►C◄ OH
CH₂OH
D-glucose
an **aldose**
(the most common simple sugar)

constitutional isomers

CH₂OH
C=O ◄ ketone
HO ►C◄ H
H ►C◄ OH
H ►C◄ OH
CH₂OH
D-fructose
a **ketose**

All carbohydrates have common names. The simplest aldehyde, glyceraldehyde, and the simplest ketone, dihydroxyacetone, are the only monosaccharides whose names do not end in the suffix **-ose.** (The prefix "D-" is explained in Section 27.2C.)

A monosaccharide is called:

- ◆ a triose if it has 3 C's;
- ◆ a tetrose if it has 4 C's;
- ◆ a pentose if it has 5 C's;
- ◆ a hexose if it has 6 C's, and so forth.

These terms are then combined with the words *aldose* and *hexose* to indicate both the number of carbon atoms in the monosaccharide and whether it contains an aldehyde or ketone. Thus, glyceraldehyde is an aldotriose (three C atoms and an aldehyde), glucose is an aldohexose (six C atoms and an aldehyde), and fructose is a ketohexose (six C atoms and a ketone).

PROBLEM 27.1   Draw the structure of (a) a ketotetrose; (b) an aldopentose; (c) an aldotetrose.

27.2A   Fischer Projection Formulas

A striking feature of carbohydrate structure is the presence of stereogenic centers. **All carbohydrates except for dihydroxyacetone contain one or more stereogenic centers.**

The simplest aldehyde, glyceraldehyde, has one stereogenic center, so there are two possible **enantiomers.** Only the enantiomer with the *R* configuration occurs naturally.

Two different representations for each enantiomer of glyceraldehyde

(*R*)-glyceraldehyde
naturally occurring enantiomer

(*S*)-glyceraldehyde

The stereogenic centers in sugars are often depicted following a different convention than is usually seen for other stereogenic centers. Instead of drawing a tetrahedron with two bonds in the plane, one in front of the plane, and one behind it, the **tetrahedron is tipped so that horizontal bonds come forward (drawn on wedges) and vertical bonds go behind (on dashed lines).** This structure is then abbreviated by a **cross formula,** also called a **Fischer projection formula.** In a Fischer projection formula:

◆ A carbon atom is located at the intersection of the two lines of the cross.
◆ The horizontal bonds come forward, on wedges.
◆ The vertical bonds go back, on dashed lines.
◆ The aldehyde or ketone carbonyl is put at or near the top.

Using a Fischer projection formula, (*R*)-glyceraldehyde becomes:

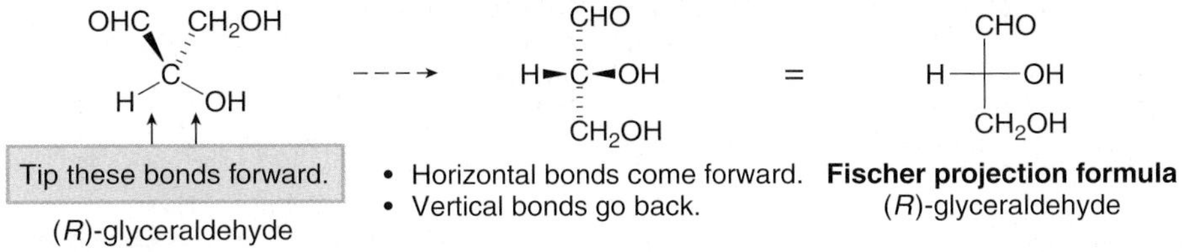

(*R*)-glyceraldehyde

Tip these bonds forward.

• Horizontal bonds come forward.
• Vertical bonds go back.

Fischer projection formula
(*R*)-glyceraldehyde

Do not rotate a Fischer projection formula in the plane of the page, because you might inadvertently convert a compound into its enantiomer. When using Fischer projections it is usually best to convert them to structures with wedges and dashes, and then manipulate them. Although a Fischer projection formula can be used for the stereogenic center in any compound, it is most commonly used for monosaccharides.

SAMPLE PROBLEM 27.1   Convert each compound to a Fischer projection formula.

a.

b.

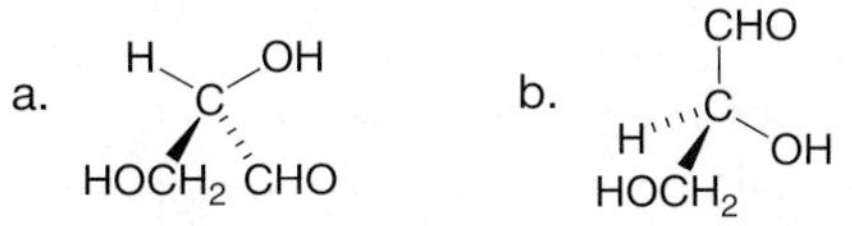

SOLUTION

Rotate and re-draw each molecule to place the horizontal bonds in front of the plane and the vertical bonds behind the plane. Then use a cross to represent the stereogenic center.

a.   re-draw

b.   re-draw

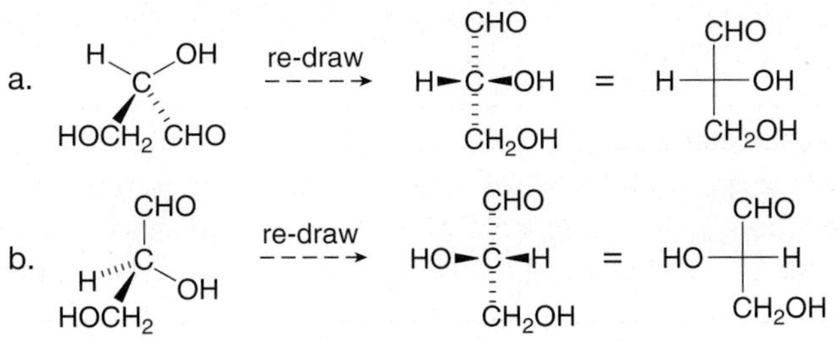

PROBLEM 27.2   Draw each stereogenic center using a Fischer projection formula.

a.   b.   c.   d.

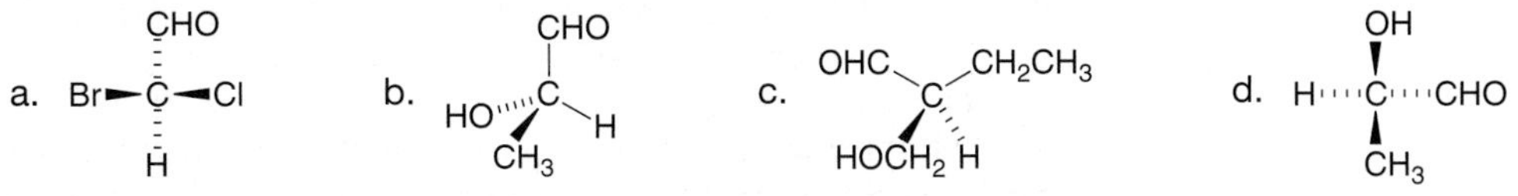

R,S designations can be assigned to any stereogenic center drawn as a Fischer projection formula in the following manner:

[1] Assign priorities (1 → 4) to the four groups bonded to the stereogenic center using the rules detailed in Section 5.6.

[2] When the lowest priority group occupies a vertical bond—that is, it projects *behind* the plane on a dashed line—tracing a circle in the clockwise direction (from priority group 1 → 2 → 3) gives the *R* configuration. Tracing a circle in the counterclockwise direction gives the *S* configuration.

[3] When the lowest priority occupies a horizontal bond—that is, it projects *in front* of the plane on a wedge—reverse the answer obtained in Step [2] to designate the configuration.

SAMPLE PROBLEM 27.2   Re-draw each Fischer projection formula using wedges and dashes for the stereogenic center, and label the center as *R* or *S*.

a. Br——CH₃ (CH₂OH top, H bottom)

b. Cl——H (CHO top, CH₃ bottom)

SOLUTION

For each molecule:

[1] Convert the Fischer projection formula to a representation with wedges and dashes.
[2] Assign priorities (Section 5.6).
[3] Determine *R* or *S* in the usual manner. Reverse the answer if priority group [4] is oriented forward (on a wedge).

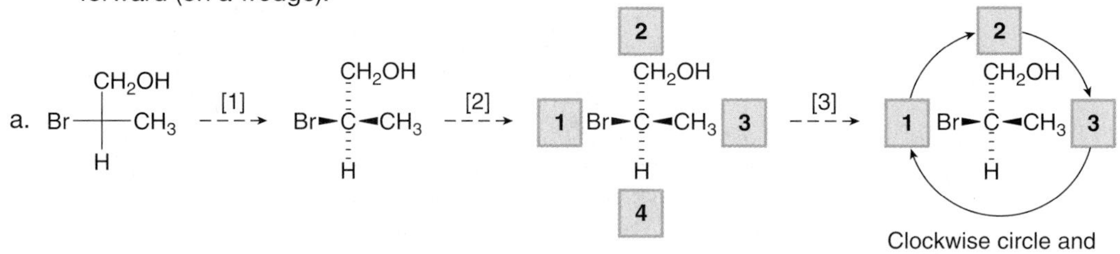

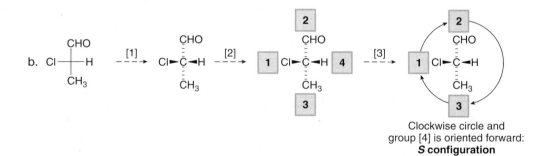

Clockwise circle and group [4] is oriented behind: **R configuration**

Clockwise circle and group [4] is oriented forward: **S configuration**

PROBLEM 27.3   Label each stereogenic center as *R* or *S*.

a. Cl——CH₂Br (CH₂NH₂ top, H bottom)

b. Cl——H (CHO top, CH₂NH₂ bottom)

c. Cl——H (CHO top, CH₂OH bottom)

d. Cl——CH₂Br (COOH top, H bottom)

27.2B Monosaccharides with More than One Stereogenic Center

The number of possible stereoisomers of a monosaccharide increases exponentially with the number of stereogenic centers present. **An aldohexose has four stereogenic centers, and so it has $2^4 = 16$ possible stereoisomers,** or eight pairs of enantiomers.

General structure
of an aldohexose

$$
\begin{array}{c}
\text{CHO} \\
\text{H-C*-OH} \\
\text{H-C*-OH} \\
\text{H-C*-OH} \\
\text{H-C*-OH} \\
\text{CH}_2\text{OH}
\end{array}
$$

4 stereogenic centers
$2^4 = 16$ possible stereoisomers

[* denotes a stereogenic center.]

Fischer projection formulas are also used for compounds like aldohexoses that contain several stereogenic centers. In this case, the molecule is drawn with a vertical carbon skeleton and the stereogenic centers are stacked one above another. Using this convention, **all horizontal bonds project forward (on wedges).** Thus, each bond is drawn in the less stable eclipsed conformation.

Two representations
for D-glucose

$$
\begin{array}{c}
\text{CHO} \\
\text{H}{\blacktriangleright}\text{C}{\blacktriangleleft}\text{OH} \\
\text{HO}{\blacktriangleright}\text{C}{\blacktriangleleft}\text{H} \\
\text{H}{\blacktriangleright}\text{C}{\blacktriangleleft}\text{OH} \\
\text{H}{\blacktriangleright}\text{C}{\blacktriangleleft}\text{OH} \\
\text{CH}_2\text{OH}
\end{array}
\quad = \quad
\begin{array}{c}
\text{CHO} \\
\text{H}{-}{-}\text{OH} \\
\text{HO}{-}{-}\text{H} \\
\text{H}{-}{-}\text{OH} \\
\text{H}{-}{-}\text{OH} \\
\text{CH}_2\text{OH}
\end{array}
$$

D-glucose   Fischer projection
All horizontal bonds are drawn
as **wedges.**

PROBLEM 27.4   Assign *R,S* designations to each stereogenic center in glucose.

27.2C   D and L Monosaccharides

Although the prefixes *R* and *S* can be used to designate the configuration of stereogenic centers in monosaccharides, an older system of nomenclature uses the prefixes D- and L-, instead. Naturally occurring glyceraldehyde with the *R* configuration is called the **D-isomer.** Its enantiomer, (*S*)-glyceraldehyde, is called the **L-isomer.**

Fischer projections for the
enantiomers of glyceraldehyde

$$
\begin{array}{c}
\text{CHO} \\
\text{H}{-}{-}\text{OH} \\
\text{CH}_2\text{OH}
\end{array}
\qquad
\begin{array}{c}
\text{CHO} \\
\text{HO}{-}{-}\text{H} \\
\text{CH}_2\text{OH}
\end{array}
$$

(*R*)-glyceraldehyde   (*S*)-glyceraldehyde
D-glyceraldehyde   L-glyceraldehyde

The letters D and L are used to label all monosaccharides, even those with multiple stereogenic centers. **The configuration of the stereogenic center *furthest* from the carbonyl group determines whether a monosaccharide is D- or L-.**

◆ A D-sugar has the OH group on the stereogenic center furthest from the carbonyl on the right in a Fischer projection (like D-glyceraldehyde).
◆ An L-sugar has the OH group on the stereogenic center furthest from the carbonyl on the left in a Fischer projection (like L-glyceraldehyde).

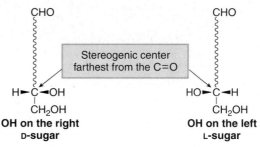

Stereogenic center
farthest from the C=O

OH on the right
D-sugar

OH on the left
L-sugar

Glucose and all other naturally occurring sugars are D-sugars. L-Glucose, a compound that does not occur in nature, is the enantiomer of D-glucose. L-Glucose has the opposite configuration at every stereogenic center.

The two designations, D and *d*, refer to very different phenomena. The "D" designates the configuration around a stereogenic center. In a D monosaccharide, the OH group on the stereogenic center furthest from the carbonyl group is on the right in a Fischer projection. The "*d*," on the other hand, is an abbreviation for "dextrorotatory"; that is, a *d*-compound rotates the plane of polarized light in the clockwise direction. A D-sugar may be dextrorotatory or it may be levorotatory (that is, it may rotate plane-polarized light in a clockwise or counterclockwise direction). There is no direct correlation between D and *d* or L and *l*.

CHO
H——OH
HO——H
H——OH
H——OH
CH₂OH
D-glucose
naturally occurring enantiomer

CHO
HO——H
H——OH
HO——H
HO——H
CH₂OH
L-glucose

> Every stereogenic center has the opposite configuration.

PROBLEM 27.5 How many stereogenic centers are present in each type of monosaccharide: (a) an aldotetrose; (b) a ketohexose?

PROBLEM 27.6 (a) Label compounds **A**, **B**, and **C** as D- or L-sugars. (b) How are compounds **A** and **B** related? **A** and **C**? **B** and **C**? Choose from enantiomers, diastereomers, or constitutional isomers.

CHO
H——OH
H——OH
HO——H
CH₂OH
A

CHO
HO——H
H——OH
HO——H
CH₂OH
B

CHO
HO——H
HO——H
H——OH
CH₂OH
C

27.3 The Family of D-Aldoses

> The common name of each monosaccharide indicates both the number of atoms it contains and the configuration at each of the stereogenic centers. Because the common names are firmly entrenched in the chemical literature, no systematic method has ever been established to name these compounds.

Beginning with D-glyceraldehyde, one may formulate other D-aldoses having four, five, or six carbon atoms by adding carbon atoms (each bonded to H and OH), one at a time, between C1 and C2. Two D-aldotetroses can be formed from D-glyceraldehyde, one with the new OH group on the right and one with the new OH group on the left. Their names are D-erythrose and D-threose: two diastereomers, each with two stereogenic centers.

> D-aldotetroses

CHO
H—*—OH
H—*—OH ← D-sugar
CH₂OH
D-erythrose

CHO   [* denotes a stereogenic center.]
HO—*—H
H—*—OH ← D-sugar
CH₂OH
D-threose

Because each aldotetrose has two stereogenic centers, there are 2^2 or four possible stereoisomers. D-Erythrose and D-threose are two of them. The other two are their enantiomers, called L-erythrose and L-threose, respectively. The configuration around each stereogenic center is exactly the opposite in its enantiomer. All four stereoisomers of the D-aldotetroses are shown in Figure 27.2.

PROBLEM 27.7 How are D-erythrose and L-threose related? L-Erythrose and L-threose?

To continue forming the family of D-aldoses, we must add another carbon atom (bonded to H and OH) just below the carbonyl of either tetrose. Because there are two D-aldotetroses to begin with, and there are two ways to place the new OH (right or left), there are now four D-aldopentoses: D-ribose, D-arabinose, D-xylose, and D-lyxose. Each aldopentose now has three stereogenic centers, so there are $2^3 = 8$ possible stereoisomers, or four pairs of enantiomers. The D-enantiomer of each pair is shown in Figure 27.3.

Figure 27.2 The four stereoisomeric aldotetroses

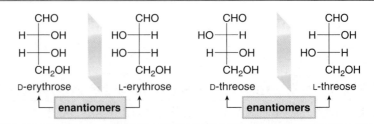

Figure 27.3 The family of D-aldoses having three to six carbon atoms

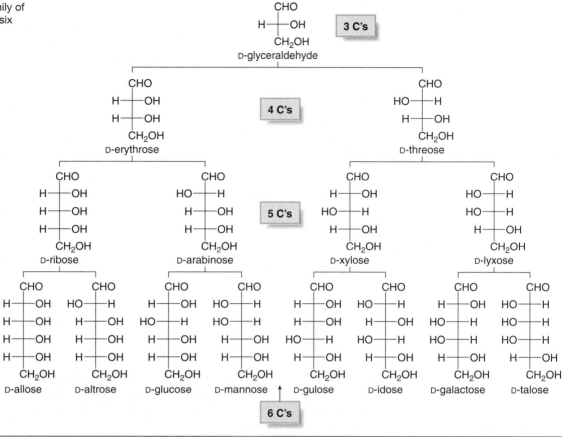

Finally, to form the D-aldohexoses, we must add another carbon atom (bonded to H and OH) just below the carbonyl of all the aldopentoses. Because there are four D-aldopentoses to begin with, and there are two ways to place the new OH (right or left), there are now eight D-aldohexoses. Each aldohexose now has four stereogenic centers, so there are $2^4 = 16$ possible stereoisomers, or eight pairs of enantiomers. Only the D-enantiomer of each pair is shown in Figure 27.3.

The tree of D-aldoses (Figure 27.3) is arranged in pairs of compounds that are bracketed together. Each pair of compounds, such as D-glucose and D-mannose, has the same configuration around all of its stereogenic centers except for one.

◆ **Two diastereomers that differ in the configuration around one stereogenic center only are called *epimers*.**

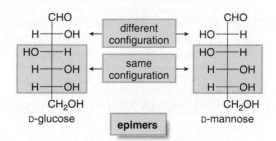

PROBLEM 27.8   How is it possible that D-glucose is dextrorotatory but D-fructose is levorotatory?

PROBLEM 27.9   How many different aldoheptoses are there? How many are D-sugars?

PROBLEM 27.10   Draw two possible epimers of D-erythrose.

Figure 27.4 The family of D-ketoses having three to six carbon atoms

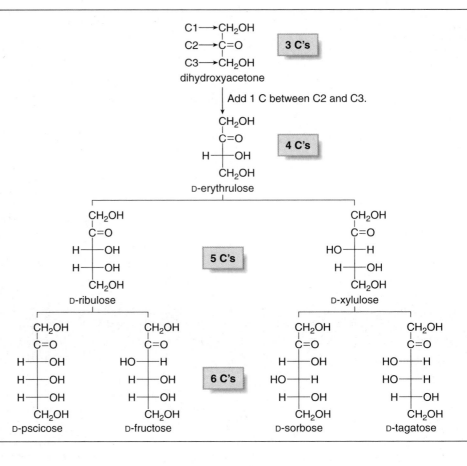

27.4 The Family of D-Ketoses

The family of D-ketoses, shown in Figure 27.4, is formed from dihydroxyacetone by adding a new carbon (bonded to H and OH) between C2 and C3. Having a carbonyl group at C2 decreases the number of stereogenic centers in these monosaccharides, so that there are only four D-ketohexoses. The most common naturally occurring ketose is D-fructose.

PROBLEM 27.11 Referring to the structures in Figures 27.3 and 27.4, classify each pair of compounds as enantiomers, epimers, diastereomers but not epimers, or constitutional isomers of each other.
a. D-allose and L-allose
b. D-altrose and D-gulose
c. D-galactose and D-talose
d. D-mannose and D-fructose
e. D-fructose and D-sorbose
f. L-sorbose and L-tagatose

PROBLEM 27.12 a. Draw the enantiomer of D-fructose.
b. Draw an epimer of D-fructose at C4. What is the name of this compound?
c. Draw an epimer of D-fructose at C5. What is the name of this compound?

27.5 Physical Properties of Monosaccharides

Monosaccharides have the following physical properties:

◆ They are all **sweet tasting,** but their relative sweetness varies a great deal.

◆ They are polar compounds with **high melting points.**

◆ The presence of so many polar functional groups capable of hydrogen bonding makes them **water soluble.**

◆ Unlike most other organic compounds, monosaccharides are so polar that they are **insoluble in organic solvents like diethyl ether.**

27.6   The Cyclic Forms of Monosaccharides

Although the monosaccharides in Figures 27.3 and 27.4 are drawn as acyclic carbonyl compounds containing several hydroxy groups, the hydroxy and carbonyl groups of monosaccharides can undergo intramolecular cyclization reactions to form **hemiacetals** having either five or six atoms in the ring. This process was first discussed in Section 21.16.

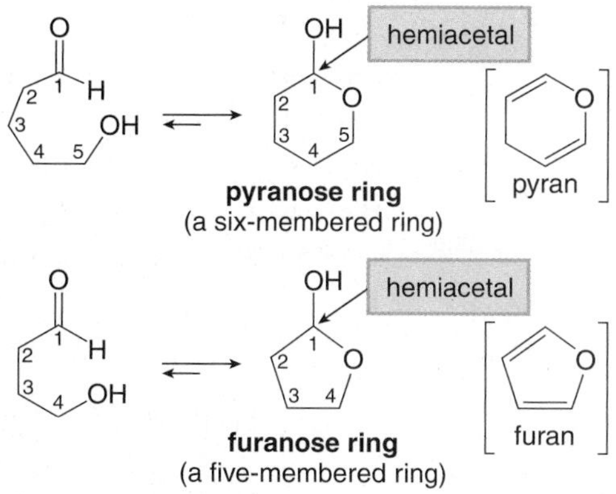

pyranose ring
(a six-membered ring)   [pyran]

furanose ring
(a five-membered ring)   [furan]

◆ A six-membered ring containing an O atom is called a *pyranose* ring.
◆ A five-membered ring containing an O atom is called a *furanose* ring.

Cyclization of a hydroxy carbonyl compound always forms a stereogenic center at the hemiacetal carbon, called the **anomeric carbon.** For a D-sugar, the following terminology is used:

◆ The α anomer has the OH group drawn down.
◆ The β anomer has the OH group drawn up.

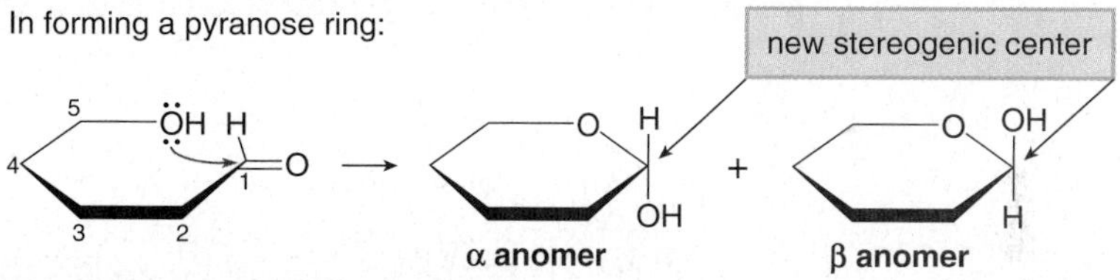

In forming a pyranose ring:

α anomer   β anomer

Cyclization forms the more stable ring size in a given molecule. **The most common monosaccharides, the aldohexoses like glucose, typically form a pyranose ring,** so our discussion begins with forming a cyclic hemiacetal from D-glucose.

27.6A   Drawing Glucose as a Cyclic Hemiacetal

Which of the five OH groups in glucose is at the right distance from the carbonyl group to form a six-membered ring? The **O atom on the stereogenic center furthest from the carbonyl** (C5) is six atoms from the carbonyl carbon, placing it in the proper position for cyclization to form a pyranose ring.

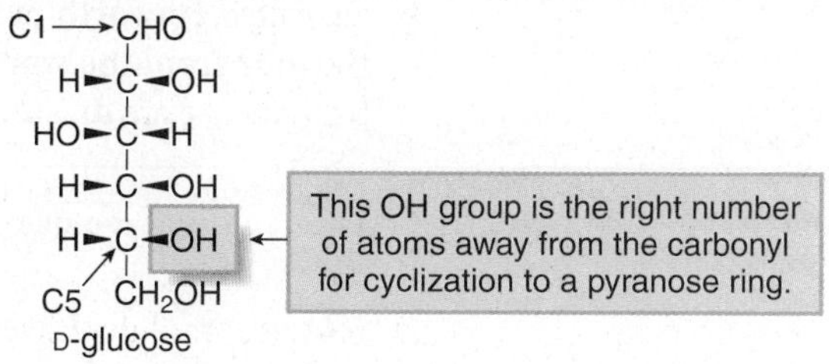

This OH group is the right number of atoms away from the carbonyl for cyclization to a pyranose ring.

D-glucose

To translate the acyclic form of glucose into a cyclic hemiacetal, we must draw the hydroxy aldehyde in a way that suggests the position of the atoms in the new ring, and then draw the ring. **By convention the O atom in the new pyranose ring is drawn in the upper right-hand corner of the six-membered ring.**

Rotating the groups on the bottom stereogenic center in **A** places all six atoms needed for the ring (including the OH) in a vertical line (**B**). Re-drawing this representation as a Fischer projection makes the structure appear less cluttered (**C**). Twisting this structure and rotating it 90° forms **D**. Structures **A–D** are four different ways of drawing the same acyclic structure of D-glucose.

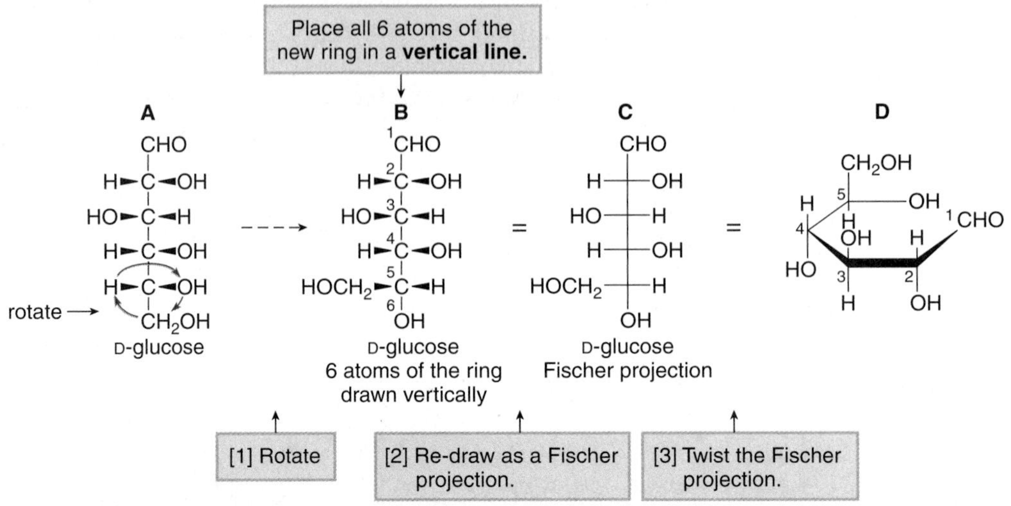

We are now set to draw the cyclic hemiacetal formed by nucleophilic attack of the OH group on C5 on the aldehyde carbonyl. Because cyclization creates a new stereogenic center, there are **two cyclic forms of D-glucose, an α anomer and a β anomer.** All the original stereogenic centers maintain their configuration in both of the products formed.

◆ The α anomer is called α-D-glucose, or α-D-glucopyranose (to emphasize the six-membered ring).
◆ The β anomer is called β-D-glucose, or β-D-glucopyranose (to emphasize the six-membered ring).

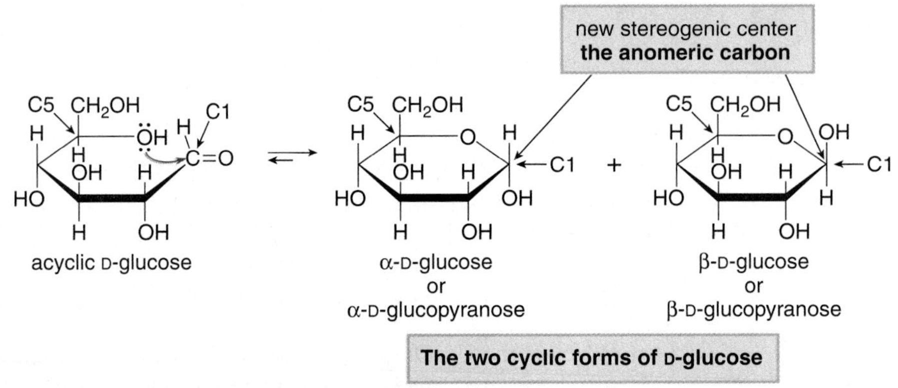

These flat, six-membered rings used to represent the cyclic hemiacetals of glucose and other sugars are called **Haworth projections.** The cyclic forms of glucose now have **five stereogenic centers, the four from the starting hydroxy aldehyde and the new anomeric carbon.** α-D-Glucose and β-D-glucose are **diastereomers,** because only the anomeric carbon has a different configuration.

Figure 27.5   The three forms of glucose

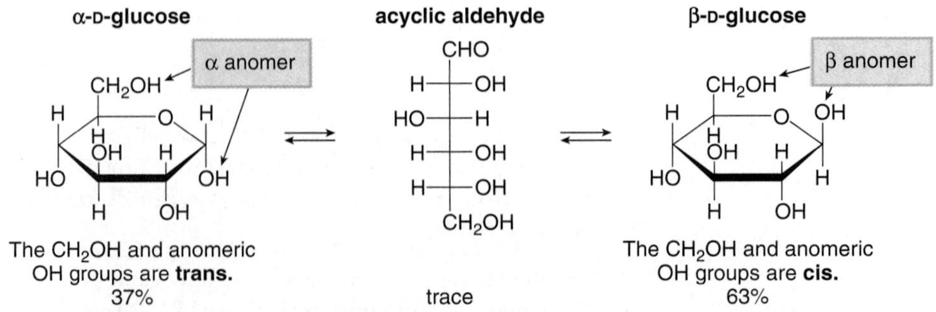

The α **anomer** has the anomeric OH group and the CH_2OH group **trans**. The β **anomer** has the anomeric OH group and the CH_2OH group **cis**.

The mechanism for this transformation is exactly the same as the mechanism that converts a hydroxy aldehyde to a cyclic hemiacetal (Mechanism 21.11). The acyclic aldehyde and two cyclic hemiacetals are all in equilibrium. Each cyclic hemiacetal can be isolated and crystallized separately, but when any one compound is placed in solution, an equilibrium mixture of all three forms results. This process is called **mutarotation.** At equilibrium, the mixture has 37% of the α anomer, 63% of the β anomer, and only trace amounts of the acyclic hydroxy aldehyde, as shown in Figure 27.5.

27.6B Haworth Projections

To convert an acyclic monosaccharide to a Haworth projection, follow a stepwise procedure.

How To	Draw a Haworth Projection from an Acyclic Aldohexose

Example **Convert D-mannose to a Haworth projection.**

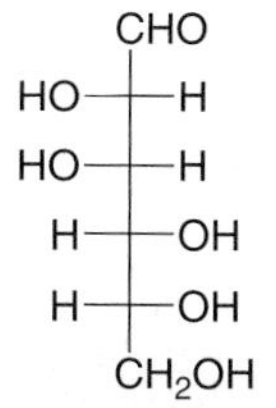

D-mannose

Step [1] **Place the O atom in the upper right corner of a hexagon, and add the CH_2OH group on the first carbon counterclockwise from the O atom.**

- For **D-sugars,** the CH_2OH group is drawn **up.** For **L-sugars,** the CH_2OH group is drawn **down.**

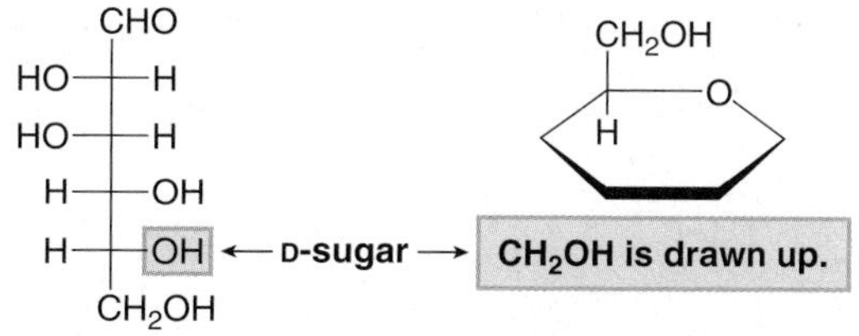

Step [2] **Place the anomeric carbon on the first carbon clockwise from the O atom.**

- For an α **anomer,** the **OH** is drawn **down** in a D-sugar.
- For a β **anomer,** the **OH** is drawn **up** in a D-sugar.

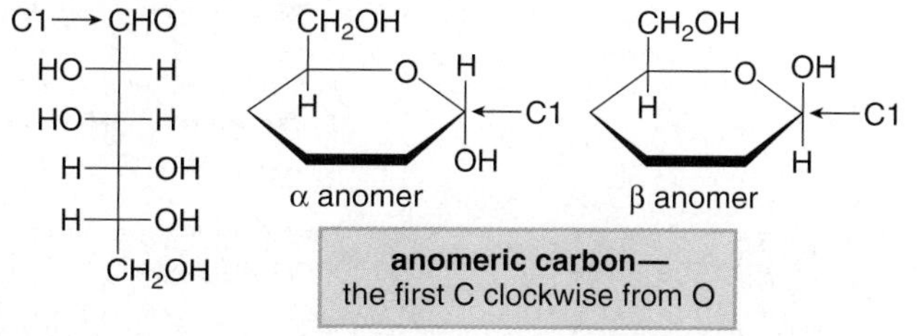

- Remember: The carbonyl carbon becomes the anomeric carbon (a new stereogenic center).

Step [3] **Add the substituents at the three remaining stereogenic centers clockwise around the ring.**

- The substituents on the **right side** of the Fischer projection are drawn **down.**
- The substituents on the **left** are drawn **up.**

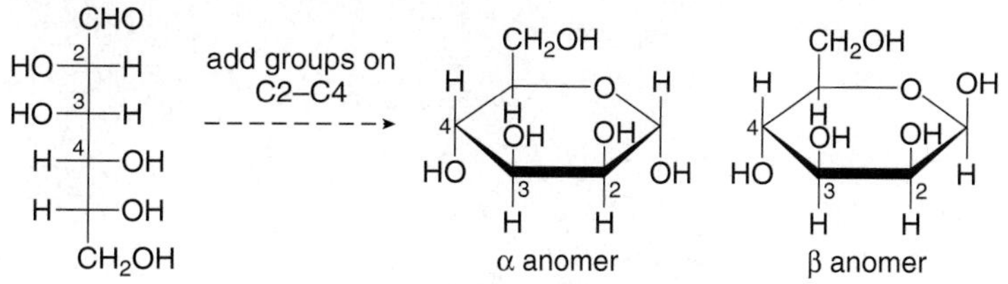

PROBLEM 27.13   Convert each aldohexose to the indicated anomer using a Haworth projection.

a. Draw the α anomer of:

```
      CHO
   H ──┼── OH
   H ──┼── OH
   H ──┼── OH
   H ──┼── OH
      CH₂OH
```

b. Draw the α anomer of:

```
      CHO
  HO ──┼── H
  HO ──┼── H
   H ──┼── OH
  HO ──┼── H
      CH₂OH
```

c. Draw the β anomer of:

```
      CHO
  HO ──┼── H
   H ──┼── OH
   H ──┼── OH
   H ──┼── OH
      CH₂OH
```

Sample Problem 27.3 shows how to convert a Haworth projection back to the acyclic form of a monosaccharide.

SAMPLE PROBLEM 27.3   Convert the following Haworth projection to the acyclic form of the aldohexose.

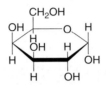

SOLUTION

To convert the substituents to the acyclic form, start at the pyranose O atom, and work in a *counterclockwise* fashion around the ring, and from bottom-to-top along the chain.

[1] Draw the carbon skeleton, placing the CHO on the top and the CH₂OH on the bottom.

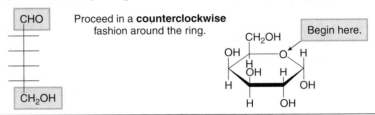

[2] Classify the sugar as D- or L-.
   • The CH₂OH is drawn **up,** so it is a **D-sugar.**
   • A D-sugar has the OH group on the bottom stereogenic center on the **right.**

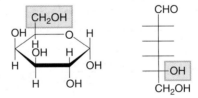

[3] Add the three other stereogenic centers.
   • **Up** substituents go on the **left.**
   • **Down** substituents go on the **right.**

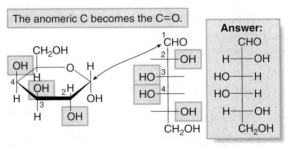

It doesn't matter whether the hemiacetal is the α or β anomer, because both anomers give the same hydroxy aldehyde.

PROBLEM 27.14   Convert each Haworth projection to its acyclic form.

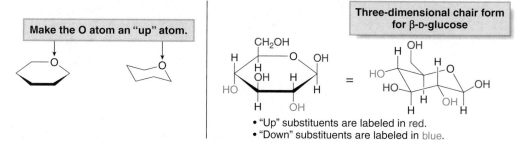

27.6C Three-Dimensional Representations for D-Glucose

Because the chair form of a six-membered ring gives the truest picture of its three-dimensional shape, we must learn to convert Haworth projections into chair forms.

To convert a Haworth projection to a chair form:

◆ Draw the pyranose ring with the O atom as an "up" atom.
◆ The "up" substituents in a Haworth projection become the "up" bonds (either axial or equatorial) on a given carbon atom on a puckered six-membered ring.
◆ The "down" substituents in a Haworth projection become the "down" bonds (either axial or equatorial) on a given carbon atom on a puckered six-membered ring.

As a result, the three-dimensional chair form of β-D-glucose is drawn in the following manner:

Make the O atom an "up" atom.

Three-dimensional chair form for β-D-glucose

• "Up" substituents are labeled in red.
• "Down" substituents are labeled in blue.

Glucose has all substituents larger than a hydrogen atom in the more roomy equatorial positions, making it the most stable and thus most prevalent monosaccharide. The β anomer is the major isomer at equilibrium, moreover, because the hemiacetal OH group is in the equatorial position, too. Figure 27.6 shows both anomers of D-glucose drawn as chair conformations.

PROBLEM 27.15   Convert each Haworth projection in Problem 27.14 to a three-dimensional representation using a chair pyranose ring.

27.6D Furanoses

Certain monosaccharides—notably aldopentoses and ketohexoses—form furanose rings, *not* pyranose rings, in solution. The same principles apply to drawing these structures as for drawing pyranose rings, except the ring size is one atom smaller.

Figure 27.6 Three-dimensional representations for both anomers of D-glucose

α anomer   β anomer

Honey was the first and most popular sweetening agent until it was replaced by sugar (from sugarcane) in modern times. Honey is a mixture consisting largely of D-fructose and D-glucose.

♦ Cyclization always forms a new stereogenic center at the anomeric carbon, so two different anomers are possible. For a D-sugar, the OH group is drawn down in the α anomer and up in the β anomer.
♦ Use the same drawing conventions for adding substituents to the five-membered ring. With D-sugars, the CH₂OH group is drawn up.

With D-ribose, the OH group used to form the five-membered furanose ring is located on C4. Cyclization yields two anomers at the new stereogenic center, which are called **α-D-ribofuranose** and **β-D-ribofuranose**.

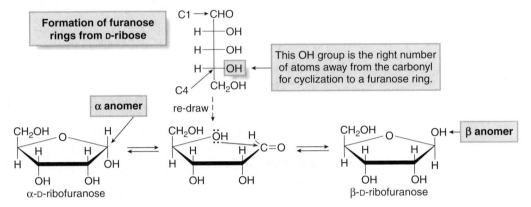

The same procedure can be used to draw the furanose form of D-fructose, the most common ketohexose. Because the carbonyl group is at C2 (instead of C1, as in the aldoses), the OH group at C5 reacts to form the hemiacetal in the five-membered ring. Two anomers are formed.

Formation of furanose rings from D-fructose

This OH group is the right number of atoms away from the carbonyl group for cyclization to a furanose ring.

re-draw

α-D-fructofuranose   α anomer   β anomer   β-D-fructofuranose

---

PROBLEM 27.16   Aldotetroses exist in the furanose form. Draw both anomers of D-erythrose.

27.7 Glycosides

Because monosaccharides exist in solution in an equilibrium between acyclic and cyclic forms, they undergo three types of reactions:

♦ **Reaction of the hemiacetal**
♦ **Reaction of the hydroxy groups**
♦ **Reaction of the carbonyl group**

Le Châtelier's principle was first discussed in Section 9.8.

Even though the acyclic form of a monosaccharide may be present in only trace amounts, the equilibrium can be tipped in its favor by Le Châtelier's principle. Suppose, for example, that the carbonyl group of the acyclic form reacts with a reagent, thus depleting its equilibrium concentration. The equilibrium will then shift to compensate for the loss, thus producing more of the acyclic form, which can react further.

Note, too, that monosaccharides have two different types of OH groups. Most are "regular" alcohols, and as such, undergo reactions characteristic of alcohols. The anomeric OH group, on the other hand, is part of a hemiacetal, giving it added reactivity.

27.7A   Glycoside Formation

Treatment of a monosaccharide with an alcohol and HCl converts the hemiacetal into an acetal called a **glycoside.** For example, treatment of α-D-glucose with CH_3OH and HCl forms two glycosides that are diastereomers at the acetal carbon. The α and β labels are assigned in the same way as anomers: with a D-sugar, an α glycoside has the new OR group (OCH_3 group in this example) down, and a β glycoside has the new OR group up.

> **Keep in mind the difference between a hemiacetal and an acetal:**
>
> OH
> |
> R–C–OR
> |
> H
>
> **hemiacetal**
> - one OH group
> - one OR group
>
> OR
> |
> R–C–OR
> |
> H
>
> **acetal**
> - two OR groups

> α-D-glucose →(CH_3OH / HCl)→ α glycoside + β glycoside
>
> **Only the hemiacetal OH reacts.**

Mechanism 27.1 explains why a single anomer forms two glycosides. The reaction proceeds by way of a **planar carbocation,** which undergoes nucleophilic attack from two different directions to give a mixture of diastereomers. Because both α- and β-D-glucose form the same planar carbocation, each yields the same mixture of two glycosides.

MECHANISM 27.1

Glycoside Formation

Steps [1]–[2]   Protonation and loss of the leaving group

β-D-glucose   →[1]← (H–Cl)   + Cl⁻   →[2]←   loss of H_2O + H_2O

resonance-stabilized cation

◆ Protonation of the hemiacetal OH group, followed by **loss of H_2O,** forms a resonance-stabilized cation (Steps [1] and [2]).

Steps [3]–[4]   Nucleophilic attack and deprotonation

planar carbocation   or [3]

$CH_3\ddot{O}H$ above → ... →[4] (Cl⁻)→   β glycoside + HCl

$CH_3\ddot{O}H$ below → ... →[4]→   α glycoside

◆ **Nucleophilic attack of CH_3OH** on the planar carbocation occurs from both sides to yield α and β glycosides after loss of a proton (Steps [3] and [4]).

The mechanism also explains why only the hemiacetal OH group reacts. Protonation of the hemiacetal OH, followed by loss of H_2O, forms a resonance-stabilized carbocation. A resonance-stabilized carbocation is not formed by loss of H_2O from any other OH group.

Unlike cyclic hemiacetals, **glycosides are acetals, and so they do not undergo mutarotation.** When a single glycoside is dissolved in H_2O, it is *not* converted to an equilibrium mixture of α and β glycosides.

◆ Glycosides are acetals with an alkoxy group (OR) bonded to the anomeric carbon.

PROBLEM 27.17   What glycosides are formed when each monosaccharide is treated with CH_3CH_2OH, HCl: (a) β-D-mannose; (b) α-D-gulose; (c) β-D-fructose?

27.7B   Glycoside Hydrolysis

Because glycosides are acetals, **they are hydrolyzed with acid and water to cyclic hemiacetals and a molecule of alcohol.** A mixture of two anomers is formed from a single glycoside. For example, treatment of methyl α-D-glucopyranoside with aqueous acid forms a mixture of α- and β-D-glucose and methanol.

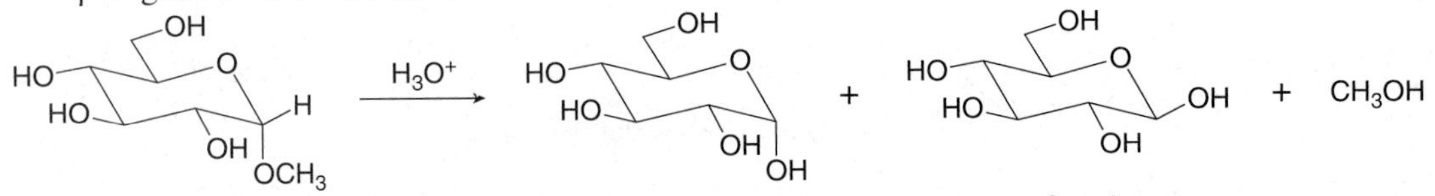

methyl α-D-glucopyranoside   α-D-glucose   β-D-glucose

The mechanism for glycoside hydrolysis is just the reverse of glycoside formation. It involves two parts: **formation of a planar carbocation,** followed by **nucleophilic attack of H_2O to form anomeric hemiacetals,** as shown in Mechanism 27.2.

MECHANISM 27.2

Glycoside Hydrolysis

Steps [1]–[2]   Protonation and loss of the leaving group

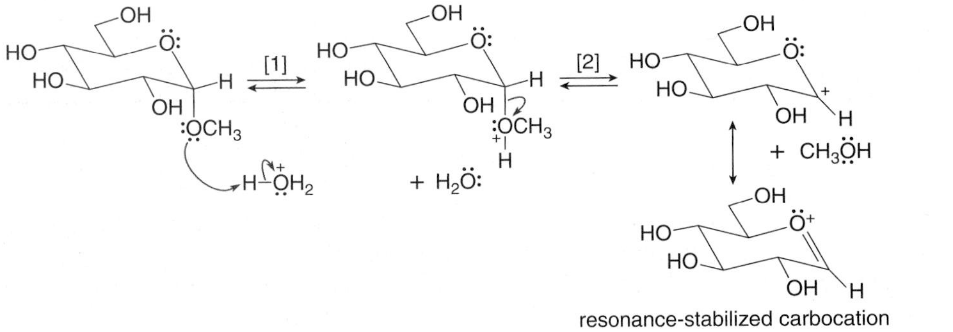

◆ Protonation of the acetal OCH_3 group, followed by **loss of CH_3OH,** forms a resonance-stabilized cation (Steps [1] and [2]).

resonance-stabilized carbocation

Steps [3]–[4]   Nucleophilic attack and deprotonation

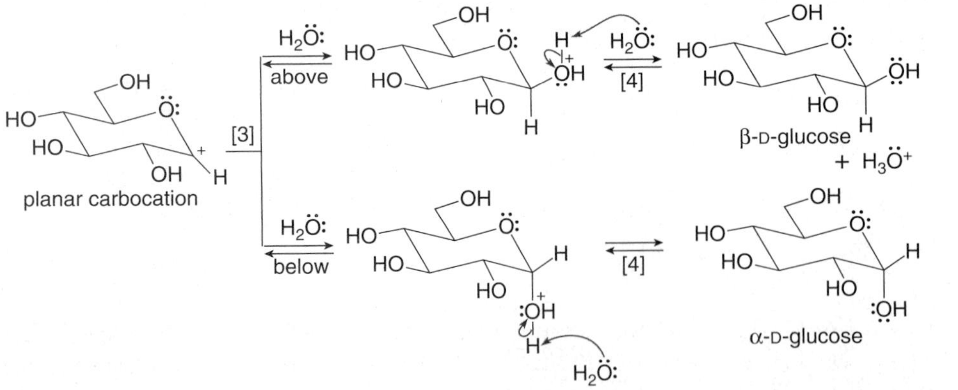

◆ **Nucleophilic attack of H_2O** on the planar carbocation occurs from both sides to yield α and β anomers after loss of a proton (Steps [3] and [4]).

PROBLEM 27.18   Draw a stepwise mechanism for the following reaction.

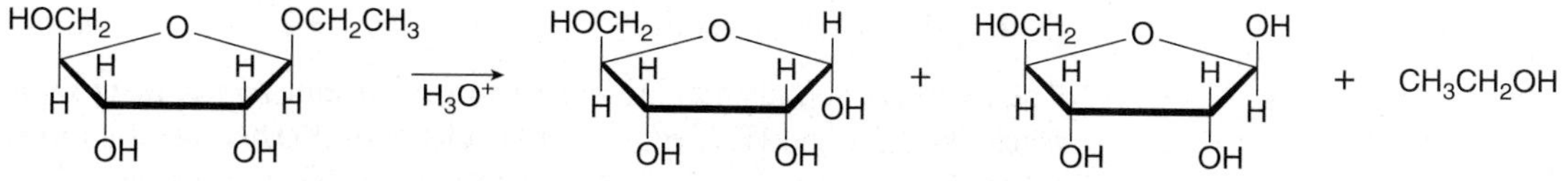

27.7C   Naturally Occurring Glycosides

Salicin, indican, and **erythromycin A** are three naturally occurring compounds that contain a glycoside as part of their structures. Salicin and indican are both synthesized in plants, whereas erythromycin A is an antibiotic produced by a bacterium, *Streptomyces erythreus,* and is used to treat a variety of infections. It is believed that the role of the sugar ring is to increase the water solubility of these compounds.

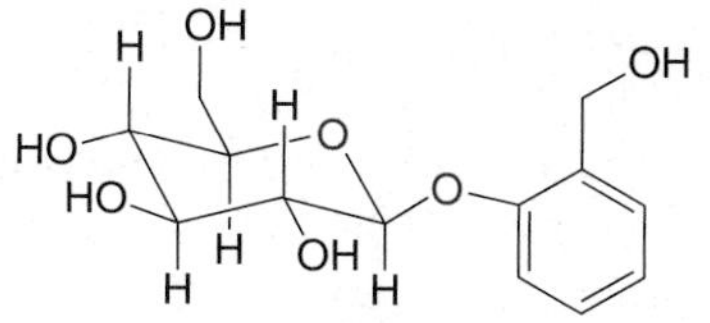

salicin
analgesic from willow bark

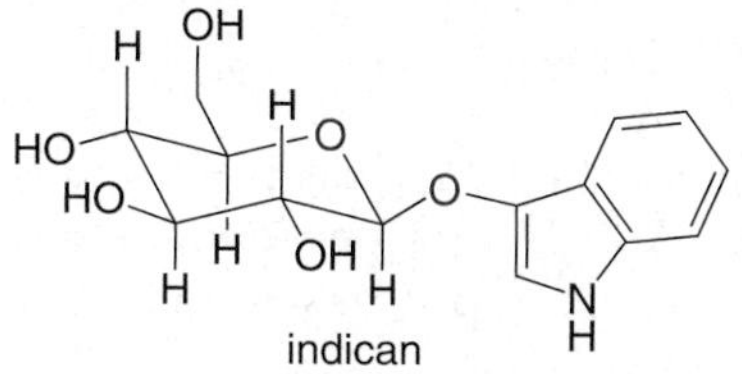

indican
precursor of the blue pigment indigo

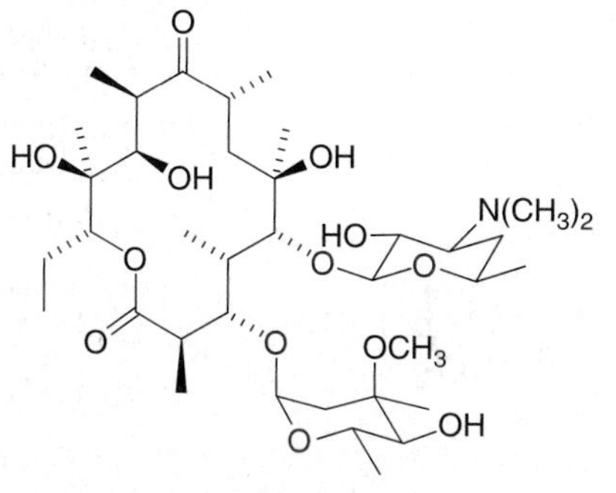

erythromycin A
antibiotic
[The O atoms that are part of the glycosides are drawn in red.]

Glycosides are common in nature. All disaccharides and polysaccharides are formed by joining monosaccharides together with glycosidic linkages. These compounds are discussed in detail beginning in Section 27.12.

PROBLEM 27.19   The alcohol or phenol formed from hydrolysis of a glycoside is called an **aglycon**. What monosaccharide and aglycon are formed by acidic hydrolysis of salicin and indican?

27.8   Reactions of Monosaccharides at the OH Groups

Because monosaccharides contain OH groups, they undergo reactions typical of alcohols—that is, they are converted to **ethers** and **esters.** Because the cyclic hemiacetal form of a monosaccharide contains an OH group, this form of a monosaccharide must be drawn as the starting material for any reaction that occurs at an OH group.

All OH groups of a cyclic monosaccharide are converted to ethers by treatment with base and an alkyl halide. For example, α-D-glucose reacts with silver(I) oxide (Ag_2O, a base) and excess CH_3I to form a pentamethyl ether.

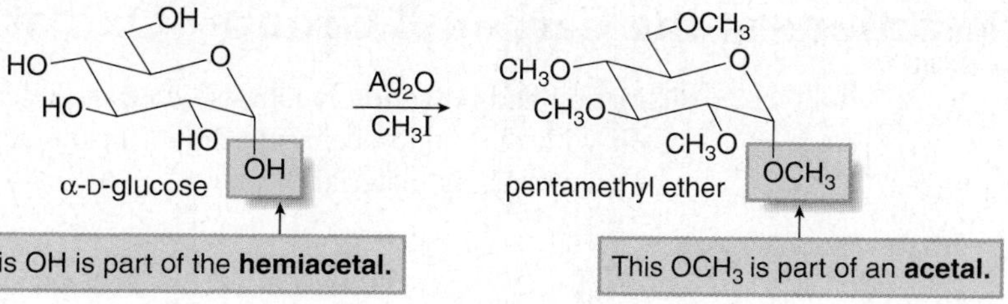

α-D-glucose

This OH is part of the **hemiacetal.**

pentamethyl ether

This OCH_3 is part of an **acetal.**

Ag_2O removes a proton from each alcohol, forming an alkoxide (RO^-), which then reacts with CH_3I in an S_N2 reaction. Because no C–O bonds are broken, the configuration of all substituents in the starting material is **retained,** forming a single product.

The product contains two different types of ether bonds. There are four "regular" ethers formed from the "regular" hydroxyls. The new ether from the hemiacetal is now part of an acetal—that is, a **glycoside.**

The four ether bonds that are *not* part of the acetal do not react with any reagents except strong acids like HBr and HI (Section 9.14). **The acetal ether, on the other hand, is hydrolyzed with aqueous acid** (Section 27.7B). Aqueous hydrolysis of a single glycoside (like the pentamethyl ether of α-D-glucose) yields both anomers of the product monosaccharide.

Only the **acetal** ether bond reacts.

Two anomers are formed.

- "Regular" ethers are shown in blue.
- The acetal ether is shown in red.

The OH groups of monosaccharides can also be converted to esters. For example, treatment of β-D-glucose with either acetic anhydride or acetyl chloride in the presence of pyridine (a base), converts all OH groups into acetate esters.

β-D-glucose

All OH groups react.

acetyl = **Ac**

AcCl

Ac₂O

It is cumbersome and tedious to draw in all the atoms of the esters. To simplify the process, the abbreviation **Ac** can be used for the acetyl group, **CH₃C=O.** The esterification of β-D-glucose can then be written as follows:

β-D-glucose

Monosaccharides are so polar that they are insoluble in common organic solvents, making them difficult to isolate and use in organic reactions. Monosaccharide derivatives that have five ether or ester groups in place of the OH groups, however, are readily soluble in organic solvents.

PROBLEM 27.20   Draw the products formed when β-D-galactose is treated with each reagent.
a. $Ag_2O + CH_3I$
b. $NaH + C_6H_5CH_2Cl$
c. The product in (b), then H_3O^+
d. Ac_2O + pyridine
e. C_6H_5COCl + pyridine
f. The product in (c), then C_6H_5COCl + pyridine

27.9   Reactions at the Carbonyl Group—Oxidation and Reduction

Oxidation and reduction reactions occur at the carbonyl group of monosaccharides, so they all begin with the monosaccharide drawn in the acyclic form. We will confine our discussion to aldoses as starting materials.

27.9A   Reduction of the Carbonyl Group

Like other aldehydes, the **carbonyl group of an aldose is reduced to a 1° alcohol using NaBH₄.** This alcohol is called an **alditol.** For example, reduction of D-glucose with NaBH₄ in CH₃OH yields glucitol (also called sorbitol).

Glucitol occurs naturally in some fruits and berries. It is sometimes used as a substitute for sucrose (table sugar). With six polar OH groups capable of hydrogen bonding, glucitol is readily hydrated. It is used as an additive to prevent certain foods from drying out.

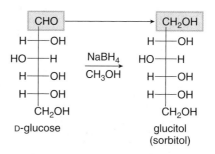

27.9B Oxidation of Aldoses

Aldoses contain 1° and 2° alcohols and an aldehyde, all of which are oxidizable functional groups. Two different types of oxidation reactions are particularly useful—oxidation of the aldehyde to a carboxylic acid (an **aldonic acid**) and oxidation of both the aldehyde and the 1° alcohol to a diacid (an **aldaric acid**).

CHO   COOH   COOH

        [O]

       ─────→   or

CH₂OH   CH₂OH   COOH

aldose   aldonic acid   aldaric acid

---

[1]   **Oxidation of the aldehyde to a carboxylic acid**

The aldehyde carbonyl is the most easily oxidized functional group in an aldose, and so a variety of reagents oxidize it to a carboxy group, forming an **aldonic acid.**

Three reagents used for this process produce a characteristic color change because the oxidizing agent is reduced to a colored product that is easily visible. As described in Section 20.8, **Tollens reagent** oxidizes aldehydes to carboxylic acids using Ag_2O in NH_4OH, and forms a mirror of Ag as a by-product. **Benedict's** and **Fehling's reagents** use a blue Cu^{2+} salt as an oxidizing agent, which is reduced to Cu_2O, a brick-red solid. Unfortunately, none of these reagents gives a high yield of aldonic acid. When the aldonic acid is needed to carry on to other reactions, $Br_2 + H_2O$ is used as the oxidizing agent.

CHO   COOH
H──OH Ag_2O, NH_4OH H──OH   Ag
HO──H   or   HO──H   or
H──OH Cu^{2+} ────→   H──OH   + Cu_2O
H──OH   or   H──OH   or
CH₂OH Br_2, H_2O CH₂OH Br^-
D-glucose   D-gluconic acid

◆ Any carbohydrate that exists as a *hemiacetal* is in equilibrium with a small amount of acyclic aldehyde, so it is oxidized to an aldonic acid.
◆ Glycosides are acetals, not hemiacetals, so they are *not* oxidized to aldonic acids.

Carbohydrates that can be oxidized with Tollens, Benedict's, or Fehling's reagent are called **reducing sugars.** Those that do not react with these reagents are called **nonreducing sugars.** Figure 27.7 shows examples of reducing and nonreducing sugars.

Figure 27.7 Examples of reducing and nonreducing sugars

- Carbohydrates containing a hemiacetal are in equilibrium with an acyclic aldehyde, making them reducing sugars.
- Glycosides are acetals, so they are *not* in equilibrium with any acyclic aldehyde, making them nonreducing sugars.

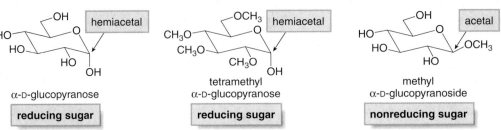

α-D-glucopyranose
reducing sugar

tetramethyl
α-D-glucopyranose
reducing sugar

methyl
α-D-glucopyranoside
nonreducing sugar

PROBLEM 27.21   Classify each compound as a reducing or nonreducing sugar.

a.

b.

c.

d.

lactose

---

[2]   Oxidation of both the aldehyde and 1° alcohol to a diacid

Both the aldehyde and 1° alcohol of an aldose are oxidized to carboxy groups by treatment with warm nitric acid, forming an **aldaric acid.** Under these conditions, D-glucose is converted to D-glucaric acid.

$$
\begin{array}{cc}
\text{CHO} & \\
\text{H} \text{—OH} & \\
\text{HO} \text{—H} & \xrightarrow[\text{H}_2\text{O}]{\text{HNO}_3} \\
\text{H} \text{—OH} & \\
\text{H} \text{—OH} & \\
\text{CH}_2\text{OH} &
\end{array}
\qquad
\begin{array}{c}
\text{COOH} \\
\text{H} \text{—OH} \\
\text{HO} \text{—H} \\
\text{H} \text{—OH} \\
\text{H} \text{—OH} \\
\text{COOH}
\end{array}
$$

D-glucose

D-glucaric acid
an aldaric acid

Because aldaric acids have identical functional groups on both terminal carbons, some aldaric acids contain a plane of symmetry, making them achiral molecules. For example, oxidation of D-allose forms an achiral, optically inactive aldaric acid. This contrasts with D-glucaric acid formed from glucose, which has no plane of symmetry, and is thus still optically active.

The oxidation of aldoses to aldaric acids has been used to determine the relative orientation of the OH groups in monosaccharides, as shown in Section 27.10C.

No plane of symmetry

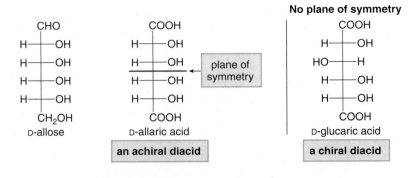

D-allose

D-allaric acid
an achiral diacid

plane of symmetry

D-glucaric acid
a chiral diacid

PROBLEM 27.22   Draw the products formed when D-arabinose is treated with each reagent: (a) Ag_2O, NH_4OH; (b) Br_2, H_2O; (c) HNO_3, H_2O.

PROBLEM 27.23   Which aldoses are oxidized to optically inactive aldaric acids: (a) D-erythrose; (b) D-lyxose; (c) D-galactose?

27.10   Reactions at the Carbonyl Group—Adding or Removing One Carbon Atom

Two common procedures in carbohydrate chemistry result in adding or removing one carbon atom from the skeleton of an aldose. The **Wohl degradation** shortens an aldose chain by one carbon, whereas the **Kiliani–Fischer synthesis** lengthens it by one. Both reactions involve cyanohydrins as intermediates. Recall from Section 21.9 that cyanohydrins are formed from aldehydes by addition of the elements of HCN. Cyanohydrins can also be re-converted to carbonyl compounds by treatment with base.

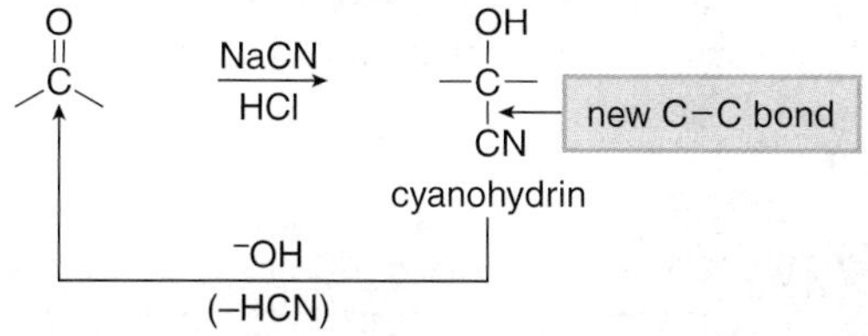

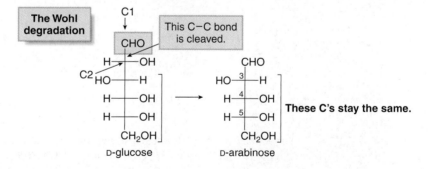

- Forming a cyanohydrin adds one carbon to a carbonyl group.
- Re-converting a cyanohydrin to a carbonyl compound removes one carbon.

27.10A   The Wohl Degradation

The Wohl degradation is a stepwise procedure that shortens the length of an aldose chain by cleavage of the C1–C2 bond. As a result, an aldohexose is converted to an aldopentose having the same configuration at its bottom three stereogenic centers (C3–C5). For example, the Wohl degradation converts D-glucose into D-arabinose.

The Wohl degradation consists of three steps, illustrated here beginning with D-glucose.

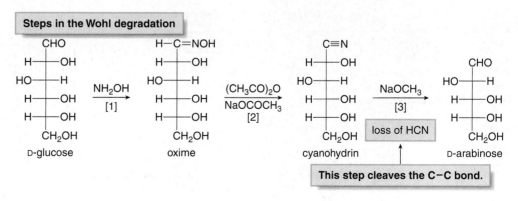

[1] Treatment of D-glucose with hydroxylamine (NH_2OH) forms an **oxime** by nucleophilic addition. This reaction is analogous to the formation of imines discussed in Section 21.11.

[2] Dehydration of the oxime to a nitrile occurs with acetic anhydride and sodium acetate. The nitrile product is a cyanohydrin.

[3] **Treatment of the cyanohydrin with base results in loss of the elements of HCN to form an aldehyde having one fewer carbon.**

The Wohl degradation converts a stereogenic center at C2 in the original aldose to an sp^2 hybridized C=O. As a result, a pair of aldoses that are epimeric at C2, such as D-galactose and D-talose, yield the same aldose (D-lyxose, in this case) upon Wohl degradation.

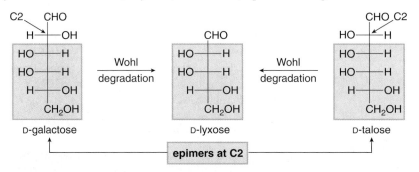

PROBLEM 27.24   What two aldoses yield D-ribose on Wohl degradation?

27.10B   The Kiliani–Fischer Synthesis

The Kiliani–Fischer synthesis lengthens a carbohydrate chain by adding one carbon to the aldehyde end of an aldose, thus forming a new stereogenic center at C2 of the product. The product consists of epimers that differ only in their configuration about the one new stereogenic center. For example, the Kiliani–Fischer synthesis converts D-arabinose into a mixture of D-glucose and D-mannose.

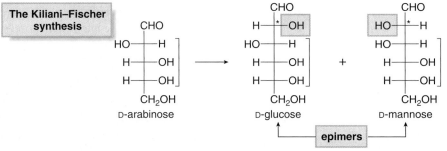

[* denotes the new stereogenic center.]

The Kiliani–Fischer synthesis, shown here beginning with D-arabinose, consists of three steps. "Squiggly" lines are meant to indicate that two different stereoisomers are formed at the new stereogenic center. As with the Wohl degradation, the key intermediate is a cyanohydrin.

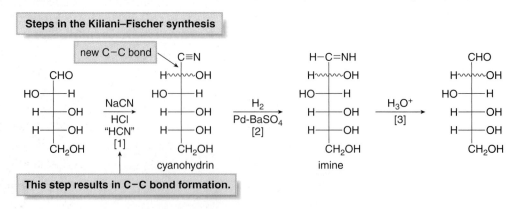

[1]   Treating an aldose with NaCN and HCl adds the elements of HCN to the carbonyl group, forming a **cyanohydrin** and a new carbon–carbon bond. Because the sp^2 hybridized carbonyl carbon is converted to an sp^3 hybridized carbon with four different groups, a new stereogenic center is formed in this step.

[2]   Reduction of the nitrile with H_2 and Pd-BaSO$_4$, a poisoned Pd catalyst, forms an **imine.**

[3]   Hydrolysis of the imine with aqueous acid forms an aldehyde that has one more carbon than the aldose that began the sequence.

Note that the **Wohl degradation and the Kiliani–Fischer synthesis are conceptually opposite transformations.**

- ◆ The Wohl degradation *removes* a carbon atom from the aldehyde end of an aldose. Two aldoses that are epimers at C2 form the same product.
- ◆ The Kiliani–Fischer synthesis *adds* a carbon to the aldehyde end of an aldose, forming two epimers at C2.

PROBLEM 27.25   What aldoses are formed when the following aldoses are subjected to the Kiliani–Fischer synthesis: (a) D-threose; (b) D-ribose; (c) D-galactose?

27.10C   Determining the Structure of an Unknown Monosaccharide

The reactions in Sections 27.9–27.10 can be used to determine the structure of an unknown monosaccharide, as shown in Sample Problem 27.4.

SAMPLE PROBLEM 27.4   A D-aldopentose **A** is oxidized to an optically inactive aldaric acid with HNO$_3$. **A** is formed by the Kiliani–Fischer synthesis of a D-aldotetrose **B**, which is also oxidized to an optically inactive aldaric acid with HNO$_3$. What are the structures of **A** and **B**?

SOLUTION

Use each fact to determine the relative orientation of the OH groups in the D-aldopentose.

Fact [1]   **A D-aldopentose A is oxidized to an optically *inactive* aldaric acid with HNO$_3$.**

An optically inactive aldaric acid must contain a **plane of symmetry.** There are only two ways to arrange the OH groups in a five-carbon D-aldaric acid, for this to be the case. Thus, only two structures are possible for **A,** labeled **A'** and **A".**

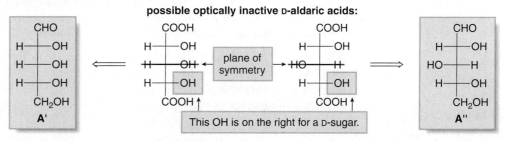

possible optically inactive D-aldaric acids:

This OH is on the right for a D-sugar.

Fact [2]   **A is formed by the Kiliani–Fischer synthesis from a D-aldotetrose B.**

A' and **A"** are each prepared from a D-aldotetrose (**B'** and **B"**) that has the same configuration at the bottom two stereogenic centers.

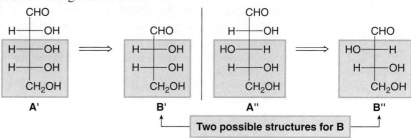

Two possible structures for B

Fact [3]	The D-aldotetrose is oxidized to an optically *inactive* aldaric acid upon treatment with HNO_3.

Only the aldaric acid from **B'** has a plane of symmetry, making it optically inactive. Thus, **B'** is the correct structure for the D-aldotetrose **B**, and therefore **A'** is the structure of the D-aldopentose **A**.

PROBLEM 27.26

D-Aldopentose **A** is oxidized to an optically inactive aldaric acid. On Wohl degradation, **A** forms an aldotetrose **B** that is oxidized to an optically active aldaric acid. What are the structures of **A** and **B**?

27.11 The Fischer Proof of the Structure of Glucose

> The Fischer proof is remarkable because it was done at a time when determining melting points and optical rotations were the most sophisticated techniques available to the chemist.

Both Fischer projections and the Kiliani–Fischer synthesis are named after **Emil Fischer,** a noted chemist of the late nineteenth and early twentieth centuries, who received the Nobel Prize in Chemistry in 1902 for his work in carbohydrate chemistry. Fischer's most elegant work is the subject of Section 27.11.

In 1891, only 10 years after the tetrahedral structure of carbon was proposed, Fischer determined the *relative* configuration of the four stereogenic centers in naturally occurring (+)-glucose. This body of work is called the **Fischer proof** of the structure of glucose.

Because glucose has four stereogenic centers, there are $2^4 = 16$ possible stereoisomers, or eight pairs of enantiomers. In 1891, there was no way to determine the *absolute* configuration of (+)-glucose—that is, the *exact* three-dimensional arrangement of the four stereogenic centers. Because there was no way to distinguish between enantiomers, Fischer could only determine the *relative* **arrangement of the OH groups** *to each other.*

> In 1951, the technique of X-ray crystallography confirmed that (+)-glucose had the D configuration, as assumed by Fischer more than 50 years earlier.

Because of this, Fischer began with an assumption. He assumed that naturally occurring glucose had the **D configuration**—namely, that the OH group on the stereogenic center furthest from the aldehyde was oriented on the *right* in a Fischer projection. He then set out to determine the orientation of all other OH groups relative to it. Thus, the Fischer proof determined which of the eight D-aldohexoses was (+)-glucose.

The strategy used by Fischer is similar to that used in Sample Problem 27.4, in which the structure of an aldopentose is determined by piecing together different facts. The Fischer proof is much more complicated, though, because the relative orientation of more stereogenic centers had to be determined.

The reasoning behind the Fischer proof is easier to follow if the eight possible D-aldohexoses are arranged in pairs of epimers at C2. These compounds are labeled **1–8** in Figure 27.8. When organized in this way, each pair of epimers would also be formed as the products of a Kiliani–Fischer synthesis beginning with a particular D-aldopentose (lettered **A–D** in Figure 27.8).

To follow the steps in the Fischer proof, we must determine what information can be obtained from each experimental result.

Figure 27.8   The D-aldopentoses and D-aldohexoses needed to illustrate the Fischer proof

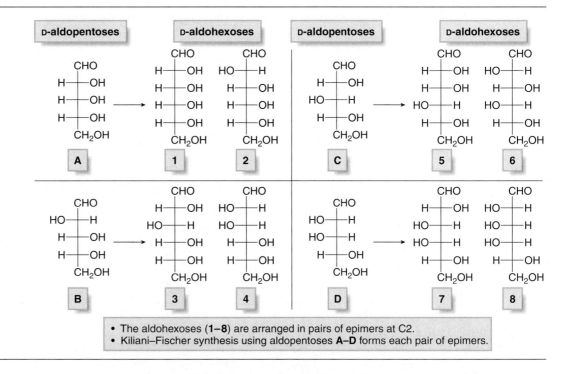

- The aldohexoses (**1–8**) are arranged in pairs of epimers at C2.
- Kiliani–Fischer synthesis using aldopentoses **A–D** forms each pair of epimers.

Fact [1]   **Kiliani–Fischer synthesis of arabinose, an aldopentose, forms glucose and mannose.**

Because the Kiliani–Fischer synthesis forms two epimers at C2, glucose and mannose have the same configurations at three stereogenic centers (C3–C5), but opposite configurations at C2. Thus, glucose and mannose are either **1** and **2**, **3** and **4**, **5** and **6**, or **7** and **8**.

Glucose and mannose are epimers at C2.

CHO → (C2 CHO, H—OH) ←three identical stereogenic centers→ (C2 CHO, HO—H)

CH₂OH   CH₂OH   CH₂OH

Fact [2]   **Glucose and mannose are both oxidized to optically active aldaric acids.**

Because an optically active aldaric acid does not have a plane of symmetry, any aldohexose that forms an aldaric acid with a plane of symmetry can be eliminated.

1 (CHO / H—OH / H—OH / H—OH / H—OH / CH₂OH) →[HNO₃, H₂O]→ (COOH / H—OH / H—OH / H—OH / H—OH / COOH) ←plane of symmetry
optically inactive aldaric acid

7 (CHO / H—OH / HO—H / HO—H / H—OH / CH₂OH) →[HNO₃, H₂O]→ (COOH / H—OH / HO—H / HO—H / H—OH / COOH) ←plane of symmetry
optically inactive aldaric acid

◆ Thus, aldohexoses **1** (and therefore its epimer **2**), and **7** (and its epimer **8**) can be eliminated, so that glucose and mannose are one of two pairs of epimers: either **3** and **4**, or **5** and **6**.

Fact [3]   **Arabinose is oxidized to an optically active aldaric acid.**

We can now narrow the possible structures for arabinose, the aldopentose that forms glucose and mannose from the Kiliani–Fischer synthesis (Fact [1]).

◆ Because glucose and mannose are epimers **3** and **4** or **5** and **6** (Fact [2]), arabinose must be either **B** or **C.**

◆ Because oxidation of **C** forms an optically *inactive* aldaric acid, arabinose must have structure **B,** an aldopentose that gives an optically active aldaric acid.

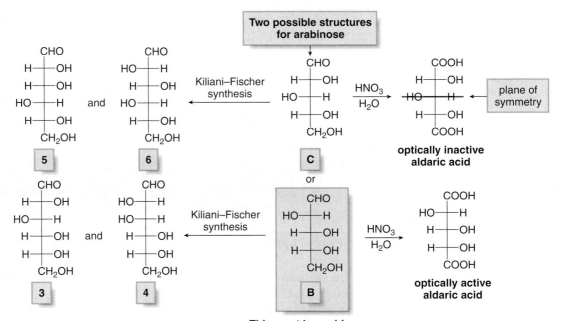

◆ Thus, glucose and mannose are structures **3** and **4,** but, with the given information, it is not possible to decide which is glucose and which is mannose.

Fact [4]   **When the functional groups on the two end carbons of glucose are interchanged, glucose is converted to a different aldohexose.**

To determine whether glucose had structure **3** or **4,** a method was devised to interchange the two functional groups at the ends of an aldohexose. The CHO at C1 was converted to CH_2OH, and the CH_2OH at C6 was converted to CHO.

$$C1 \longrightarrow CHO \longrightarrow CH_2OH$$

$$C6 \longrightarrow CH_2OH \longrightarrow CHO$$

The results of this process are different for compounds **3** and **4.** Compound **4** gives a compound that is identical to itself, whereas compound **3** gives a compound that is different from itself.

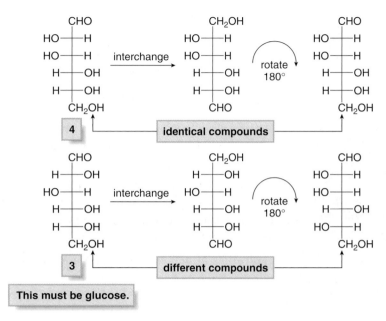

Because glucose gives a different aldohexose after the two end groups are interchanged, it must have structure **3,** and mannose must have structure **4. The proof is complete.**

PROBLEM 27.27   Besides D-mannose, only one other D-aldohexose yields itself when the CHO and CH$_2$OH groups on the end carbon atoms are interchanged. What is the name and structure of this D-aldohexose?

PROBLEM 27.28   A D-aldohexose **A** is formed from an aldopentose **B** by the Kiliani–Fischer synthesis. Reduction of **A** with NaBH$_4$ forms an optically inactive alditol. Oxidation of **B** forms an optically active aldaric acid. What are the structures of **A** and **B**?

27.12   Disaccharides

Disaccharides contain two monosaccharides joined together by a glycosidic linkage. The **general features of a disaccharide** include the following:

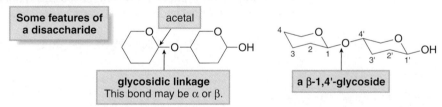

[1] Two monosaccharide rings may be five- or six-membered, but six-membered rings are much more common. The two rings are connected by an O atom that is part of an acetal, called a **glycosidic linkage,** which may be oriented α or β.

[2] The **glycoside is formed from the anomeric carbon of one monosaccharide and any OH group on the other monosaccharide.** All disaccharides have **one acetal,** together with either a hemiacetal or another acetal.

[3] With pyranose rings, the carbon atoms in each ring are numbered beginning with the anomeric carbon. Primes (') are used to designate the atoms in the ring that does not contain the glycoside. The most common disaccharides are formed from the **4' OH group.**

The three most abundant disaccharides are **maltose, lactose,** and **sucrose.**

27.12A   Maltose

Maltose, a disaccharide formed by the hydrolysis of starch, is found in germinated grains such as barley. Maltose contains two glucose units joined together by an α-1,4'-glycoside bond.

Maltose gets its name from malt, the liquid obtained from barley and other cereal grains.

α glycoside bond

maltose

hemiacetal
(β anomer)

Because one glucose ring of maltose still contains a hemiacetal, it exists as a mixture of α and β anomers. Only the β anomer is shown. Maltose exhibits two properties of all carbohydrates that contain a hemiacetal: it undergoes **mutarotation,** and it reacts with oxidizing agents, making it a **reducing sugar.**

Hydrolysis of maltose forms two molecules of glucose. The **C1−O** bond is cleaved in this process, and a mixture of glucose anomers forms. The mechanism for this hydrolysis is exactly the same as the mechanism for glycoside hydrolysis in Section 27.7B.

C1

This bond is cleaved.

α-D-glucose   β-D-glucose

PROBLEM 27.29   Draw a stepwise mechanism for the acid-catalyzed hydrolysis of maltose to two molecules of glucose.

PROBLEM 27.30   Draw the α anomer of maltose. What products are formed on hydrolysis of this form of maltose?

27.12B   Lactose

Milk contains the disaccharide lactose.

As noted in the chapter opener, **lactose** is the principal disaccharide found in milk from both humans and cows. Unlike many mono- and disaccharides, lactose is not appreciably sweet. Lactose consists of **one galactose** and **one glucose unit,** joined by a **β-1,4'-glycoside bond** from the anomeric carbon of galactose to the 4' carbon of glucose.

β glycoside bond

lactose

hemiacetal
(β anomer)

Like maltose, lactose also contains a hemiacetal, so it exists as a mixture of α and β anomers. The β anomer is drawn. Lactose undergoes **mutarotation,** and it reacts with oxidizing agents, making it a **reducing sugar.**

Lactose is digested in the body by first cleaving the β-1,4'-glycoside bond using the enzyme *lactase.* Individuals who are lactose intolerant no longer produce lactase, and so lactose passes through the digestive system unchanged, causing abdominal cramps and diarrhea.

PROBLEM 27.31   Cellobiose, a disaccharide obtained by the hydrolysis of cellulose, is composed of two glucose units joined together in a β-1,4'-glycoside bond. What is the structure of cellobiose?

Figure 27.9   Sucrose

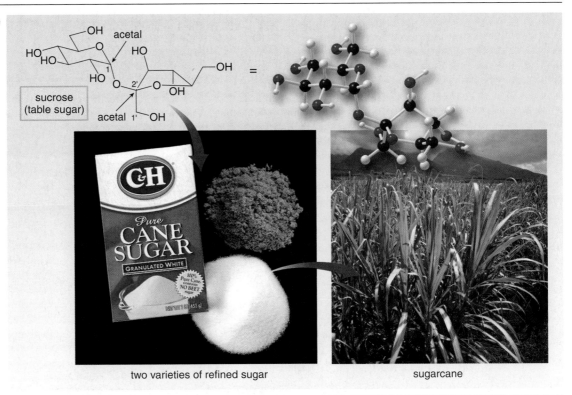

two varieties of refined sugar   sugarcane

27.12C   Sucrose

Sucrose, the disaccharide found in sugarcane and used as table sugar (Figure 27.9), is the most common disaccharide in nature. It contains **one glucose unit** and **one fructose unit.**

The structure of sucrose has several features that make it different from maltose and lactose. First of all, sucrose contains one six-membered ring (glucose) and one five-membered ring (fructose), whereas both maltose and lactose contain two six-membered rings. In sucrose the six-membered glucose ring is joined by an α-glycosidic bond to the 2' carbon of a fructofuranose ring. The numbering in a fructofuranose is different from the numbering in a pyranose ring. The anomeric carbon is now designated as C2, so the anomeric carbons of the glucose and fructose rings are both used to form the glycosidic linkage.

As a result, **sucrose contains two acetals but no hemiacetal.** Sucrose, therefore, is a **nonreducing sugar** and **it does not undergo mutarotation.**

27.13   Polysaccharides

Polysaccharides contain three or more monosaccharides joined together. The three most prevalent polysaccharides in nature are **cellulose, starch,** and **glycogen,** each of which consists of repeating glucose units joined by different glycosidic bonds.

27.13A   Cellulose

The structure of cellulose was previously discussed in Section 5.1.

Cellulose is found in the cell walls of nearly all plants, where it gives support and rigidity to wood and plant stems. Cotton is essentially pure cellulose.

cellulose

The β-1,4'-glycosidic bonds are shown in red.

Ball-and-stick models showing the three-dimensional structures of cellulose and starch were given in Figure 5.2.

Cellulose is an unbranched polymer composed of repeating glucose units joined in a β-1,4'-glycosidic linkage. The β-glycosidic linkage creates long linear chains of cellulose molecules that stack in sheets, creating an extensive three-dimensional array. A network of intermolecular hydrogen bonds between the chains and sheets means that only the few OH groups on the surface are available to hydrogen bond to water, making this very polar compound water insoluble.

Cellulose acetate, a cellulose derivative, is made by treating cellulose with acetic anhydride and sulfuric acid. The resulting product has acetate esters in place of every OH group. Cellulose acetate is spun into fibers that are used for fabrics called *acetates,* which have a deep luster and satin appearance.

cellulose

Ac$_2$O, H$_2$SO$_4$

cellulose acetate

Cellulose can be hydrolyzed to glucose by cleaving all of the β-glycosidic bonds, yielding both anomers of glucose.

Hydrolysis cleaves all the β-glycosidic bonds, shown in red.

β-D-glucose   α-D-glucose

A **β-glycosidase** is the general name of an enzyme that hydrolyzes a β-glycoside linkage.

In cells, the hydrolysis of cellulose is accomplished by an enzyme called a **β-glucosidase,** which cleaves all the β-glycoside bonds formed from glucose. Humans do not possess this enzyme, and therefore cannot digest cellulose. Ruminant animals, on the other hand, such as cattle, deer, and camels, have bacteria containing a β-glucosidase in their digestive systems, so they can derive nutritional benefit from eating grass and leaves.

27.13B   Starch

Starch is the main carbohydrate found in the seeds and roots of plants. Corn, rice, wheat, and potatoes are common foods that contain a great deal of starch.

Starch is a polymer composed of repeating glucose units joined in α-glycosidic linkages. Both starch and cellulose are polymers of glucose, but starch contains α glycoside bonds, whereas cellulose contains β glycoside bonds. The two common forms of starch are **amylose** and **amylopectin.**

amylose
(the linear form of starch)

The α-1,4'-glycoside bonds are shown in red.

amylopectin
(the branched form of starch)

The α-1,6'-glycoside bond is shown in red.

Amylose, which comprises about 20% of starch molecules, has an unbranched skeleton of glucose molecules with **α-1,4'-glycoside bonds.** Because of this linkage, an amylose chain adopts a helical arrangement, giving it a very different three-dimensional shape from the linear chains of cellulose. Amylose was first described in Section 5.1.

Amylopectin, which comprises about 80% of starch molecules, likewise consists of a backbone of glucose units joined in **α-glycosidic bonds,** but it also contains considerable branching along the chain. The linear linkages of amylopectin are formed by **α-1,4'-glycoside bonds,** similar to amylose. The branches are linked to the chain with **α-1,6'-glycosidic linkages.**

Both forms of starch are water soluble. Because the OH groups in these starch molecules are not buried in a three-dimensional network, they are more available for hydrogen bonding with water molecules, leading to greater water solubility than cellulose has.

The ability of amylopectin to form branched polymers is a unique feature of carbohydrates. Other types of polymers in the cell, such as the proteins discussed in Chapter 28, occur in nature only as linear molecules.

Both amylose and amylopectin are hydrolyzed to glucose with cleavage of the glycosidic bonds. The human digestive system has the necessary **α-glucosidase** enzymes needed to catalyze this process. Bread and pasta made from wheat flour, rice, and corn tortillas are all sources of starch that are readily digested.

> **α-Glycosidase** is the general name of an enzyme that hydrolyzes an α-glycoside linkage.

27.13C Glycogen

Glycogen is the major form in which polysaccharides are stored in animals. Glycogen, a polymer of glucose containing **α-glycosidic bonds,** has a branched structure similar to amylopectin, but the branching is much more extensive.

Glycogen is stored principally in the liver and muscle. When glucose is needed for energy in the cell, glucose units are hydrolyzed from the ends of the glycogen polymer, and then further metabolized with the release of energy. Because glycogen has a highly branched structure, there are many glucose units at the ends of the branches that can be cleaved whenever the body needs them.

PROBLEM 27.32 Draw the structure of: (a) a polysaccharide formed by joining D-mannose units in β-1,4'-glycosidic linkages; (b) a polysaccharide formed by joining D-glucose units in α-1,6'-glycosidic linkages. The polysaccharide in (b) is dextran, a component of dental plaque.

27.14 Other Important Sugars and Their Derivatives

Many other examples of simple and complex carbohydrates with useful properties exist in the biological world. In Section 27.14, we examine some carbohydrates that contain nitrogen atoms.

27.14A Amino Sugars and Related Compounds

Amino sugars contain an NH_2 group instead of an OH group at a non-anomeric carbon. The most common amino sugar in nature, **D-glucosamine,** is formally derived from D-glucose by replacing the OH at C2 with NH_2. Although it is not classified as a drug, and therefore not regulated by the Food and Drug Administration, glucosamine is available in many over-the-counter treatments for osteoarthritis. Glucosamine is thought to promote the repair of deteriorating cartilage in joints.

Acetylation of glucosamine with acetyl CoA (Section 22.17) forms **N-acetyl-D-glucosamine,** abbreviated as **NAG. Chitin,** the second most abundant carbohydrate polymer, is a polysaccharide

The rigidity of a crab shell is due to chitin, a high molecular weight carbohydrate molecule. Chitin-based coatings have found several commercial applications, such as extending the shelf life of fruits. Processing plants now convert the shells of crabs, lobsters, and shrimp to chitin and various derivatives for use in many consumer products.

formed from NAG units joined together in **β-1,4'-glycosidic linkages.** Chitin is identical in structure to cellulose, except that each OH group at C2 is now replaced by $NHCOCH_3$. The exoskeletons of lobsters, crabs, and shrimp are composed of chitin. Like those of cellulose, chitin chains are held together by an extensive network of hydrogen bonds, forming water-insoluble sheets.

chitin
a polysaccharide composed of NAG units

> The β-1,4'-glycosidic bonds are shown in red.

Several trisaccharides containing amino sugars are potent antibiotics used in the treatment of certain severe and recurrent bacterial infections. These compounds, such as tobramycin and amikacin, are called **aminoglycoside antibiotics.**

tobramycin

amikacin

PROBLEM 27.33   Treating chitin with H_2O, $^-$OH hydrolyzes its amide linkages, forming a compound called chitosan. What is the structure of chitosan? Chitosan has been used in shampoos, fibers for sutures, and wound dressings.

27.14B   N-Glycosides

N-**Glycosides** are formed when a monosaccharide is reacted with an amine in the presence of mild acid (Reactions [1] and [2]).

[1]

β-D-glucopyranose $\xrightarrow[\text{mild H}^+]{CH_3CH_2NH_2}$ α-*N*-glycoside   +   β-*N*-glycoside

[2]

α-D-ribofuranose $\xrightarrow{\text{mild H}^+}$

The mechanism of *N*-glycoside formation is analogous to the mechanism for glycoside formation, and both anomers of the *N*-glycoside are formed as products.

PROBLEM 27.34   Draw the products of each reaction.

a. $\xrightarrow[\text{mild H}^+]{CH_3NH_2}$

b. $\xrightarrow[\text{mild H}^+]{C_6H_5NH_2}$

PROBLEM 27.35 Draw a stepwise mechanism for the conversion of β-D-glucose to both anomers of *N*-ethyl glucopyranoside, the reaction shown in Reaction [1].

The prefix *deoxy* means *without oxygen*.

The *N*-glycosides of two sugars, **D-ribose** and **2-deoxy-D-ribose,** are especially noteworthy, because they form the building blocks of RNA and DNA, respectively. 2-Deoxyribose is so named because it lacks an OH group at C2 of ribose.

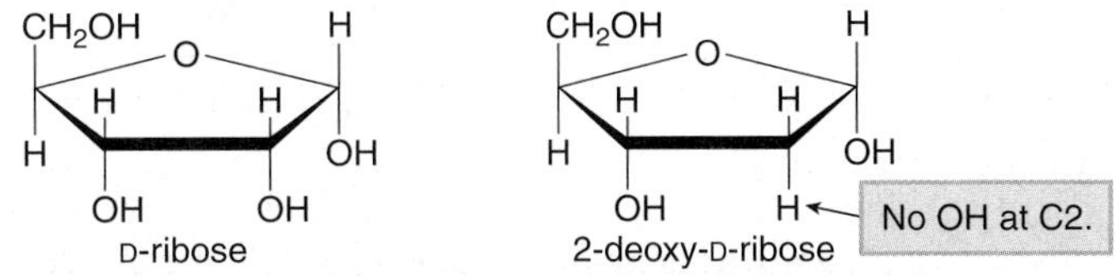

D-ribose

2-deoxy-D-ribose

No OH at C2.

◆ Reaction of D-ribose with certain amine heterocycles forms *N*-glycosides called **ribonucleosides**.
◆ This same reaction of 2-deoxy-D-ribose forms **deoxyribonucleosides**.

An example of a **ribonucleoside** and a **deoxyribonucleoside** are drawn. These *N*-glycosides have the β orientation. Numbering in the sugar ring begins at the anomeric carbon (1'), and proceeds in a clockwise fashion around the ring.

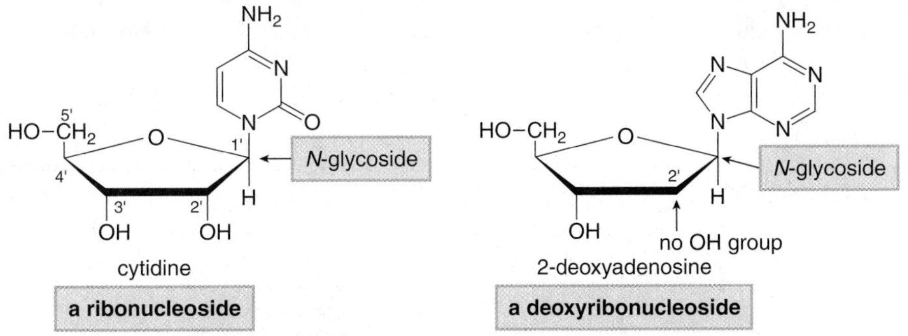

cytidine
a ribonucleoside

2-deoxyadenosine
a deoxyribonucleoside

Only five common nitrogen heterocycles are used to form these nucleosides. Three compounds have one ring, and are derived from a nitrogen heterocycle called **pyrimidine.** Two are bicyclic, and are derived from a nitrogen heterocycle called **purine.** These five amines are referred to as *bases.* Each base is designated by a one-letter abbreviation, as shown in the names and structures drawn. Note that uracil (U) occurs only in ribonucleosides and thymine (T) occurs only in deoxyribonucleosides.

◆ Each nucleoside has two parts, a sugar and a base, joined together by a β *N*-glycosidic linkage.

The five nitrogen heterocycles used to form nucleosides

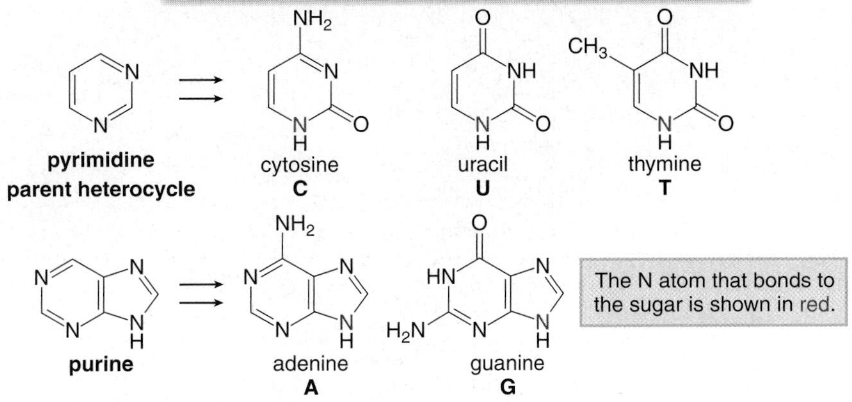

pyrimidine
parent heterocycle

cytosine
C

uracil
U

thymine
T

purine

adenine
A

guanine
G

The N atom that bonds to the sugar is shown in red.

When one OH group of the sugar nucleus is bonded to a phosphate, the derivatives are called **ribonucleotides** and **deoxyribonucleotides**.

cytidine monophosphate

a ribonucleotide

deoxyadenosine monophosphate

a deoxyribonucleotide

◆ Ribonucleotides are the building blocks of the polymer ribonucleic acid, or RNA, the messenger molecules that convert genetic information into proteins.
◆ Deoxyribonucleotides are the building blocks of the polymer deoxyribonucleic acid, or DNA, the molecules that are responsible for the storage of all genetic information.

Short segments of both RNA and DNA are shown in Figure. 27.10.

Much more can be said about DNA and RNA, but the discussion will conclude here because Chapter 27 focuses on carbohydrates, not nucleic acids. Note the central role of the sugar moiety in both RNA and DNA. The sugar residues are bonded to two phosphate groups, thus connecting the chain of RNA or DNA together. The sugar residues are also bonded to the nitrogen base via the anomeric carbon.

DNA backbone

Base

O—CH₂

The sugar ring is central to the structure of DNA and RNA.

DNA backbone

Figure 27.10 Short segments of RNA and DNA

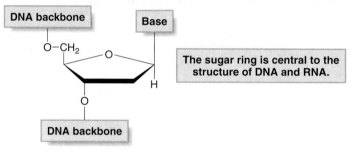

Ribonucleic acid RNA

adenine

cytosine

guanine

uracil

Deoxyribonucleic acid DNA

thymine

guanine

cytosine

adenine

PROBLEM 27.36 Draw the structures of the nucleosides formed from each of the following components: (a) ribose + uracil; (b) 2-deoxyribose + guanine

27.15 Key Concepts—Carbohydrates

Important Terms

- **Aldose** A monosaccharide containing an aldehyde (27.2)
- **Ketose** A monosaccharide containing a ketone (27.2)
- **D-sugar** A monosaccharide with the OH bonded to the stereogenic center furthest from the carbonyl group drawn on the right in the Fischer projection (27.2C)
- **Epimers** Two diastereomers that differ in configuration around one stereogenic center only (27.3)
- **Anomers** Monosaccharides that differ in configuration at the hemiacetal OH group (27.6)
- **Glycoside** An acetal derived from a monosaccharide hemiacetal (27.7)

Acyclic, Haworth, and 3-D Representations for D-Glucose (27.6)

Haworth projection

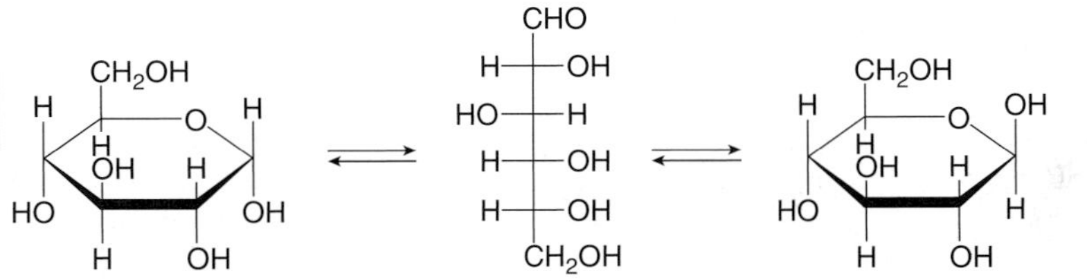

α anomer **acyclic form** **β anomer**

3-D representation

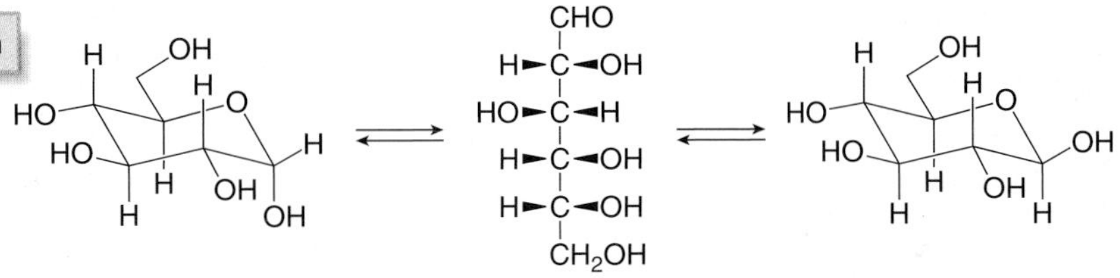

Reactions of Monosaccharides Involving the Hemiacetal

[1] Glycoside formation (27.7A)

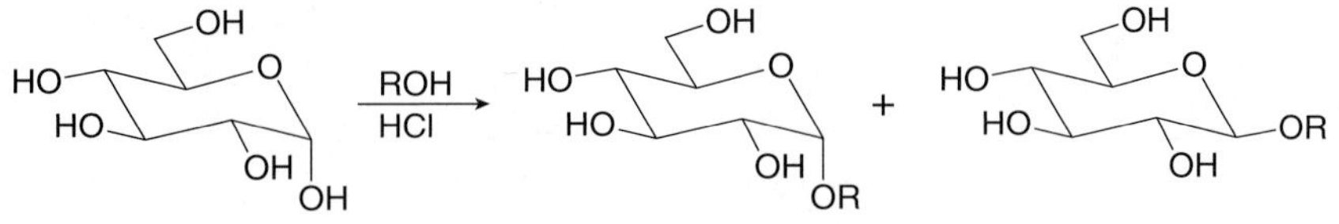

α-D-glucose α glycoside β glycoside

- Only the hemiacetal OH reacts.
- A mixture of α and β glycosides forms.

[2] Glycoside hydrolysis (27.7B)

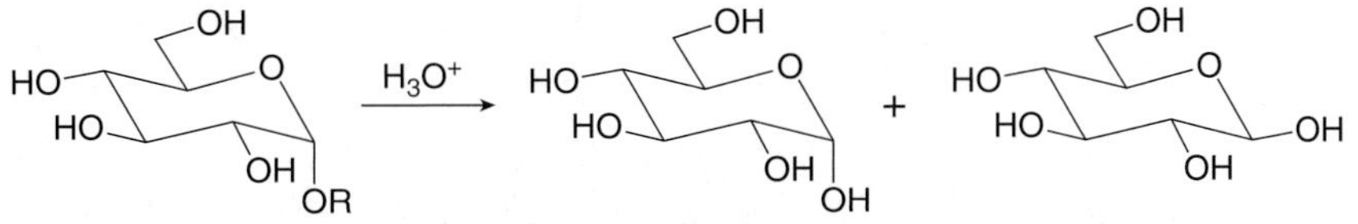

α anomer + ROH β anomer

- A mixture of α and β anomers forms.

Reactions of Monosaccharides at the OH Groups

[1] Ether formation (27.8)

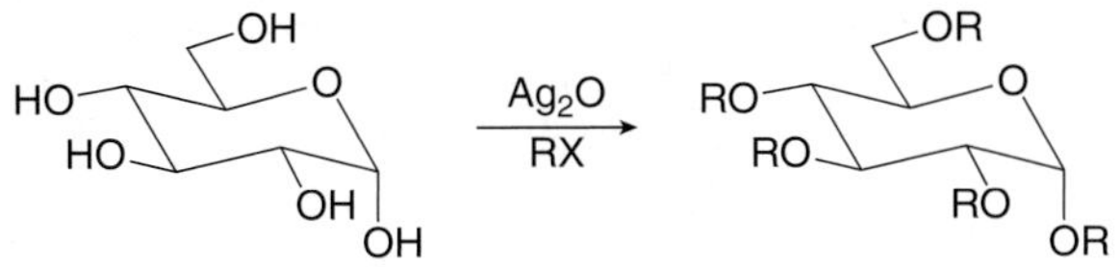

- All OH groups react.
- The stereochemistry at all stereogenic centers is retained.

[2] Ester formation (27.8)

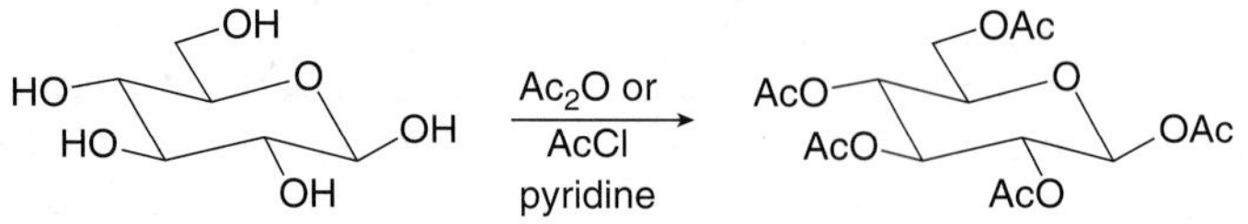

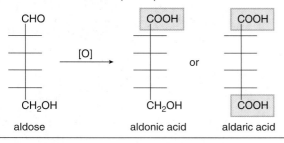

- All OH groups react.
- The stereochemistry at all stereogenic centers is retained.

Reactions of Monosaccharides at the Carbonyl Group

[1] Oxidation of aldoses (27.9B)

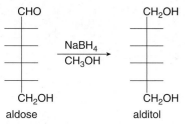

- Aldonic acids are formed using:
  - Ag_2O, NH_4OH
  - Cu^{2+}
  - Br_2, H_2O
- Aldaric acids are formed with HNO_3, H_2O.

[2] Reduction of aldoses to alditols (27.9A)

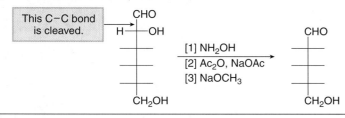

[3] Wohl degradation (27.10A)

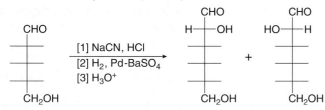

- The C1 – C2 bond is cleaved to shorten an aldose chain by one carbon.
- The stereochemistry at all other stereogenic centers is retained.
- Two epimers at C2 form the same product.

[4] Kiliani–Fischer synthesis (27.10B)

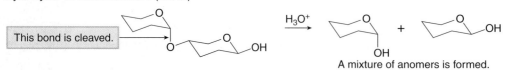

- One carbon is added to the aldehyde end of an aldose.
- Two epimers at C2 are formed.

Other Reactions

[1] Hydrolysis of disaccharides (27.12)

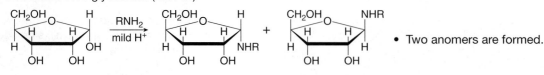

A mixture of anomers is formed.

[2] Formation of N-glycosides (27.14B)

- Two anomers are formed.

Problems

Fischer Projections

27.37 Classify each compound as identical to **A** or its enantiomer.

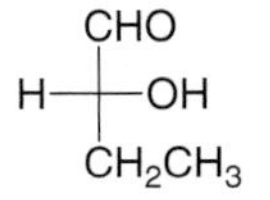

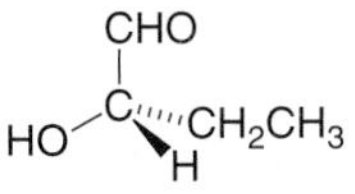

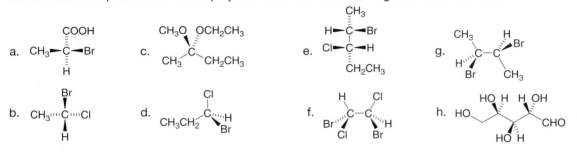

27.38 Convert each compound to a Fischer projection and label each stereogenic center as *R* or *S*.

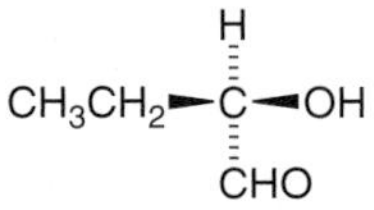

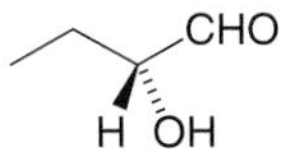

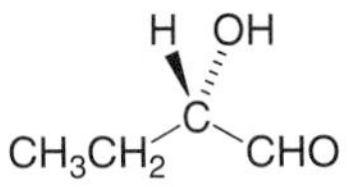

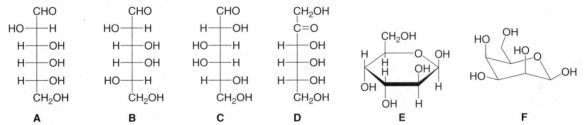

Monosaccharide Structure and Stereochemistry

27.39 Which D-aldopentoses have the *R* configuration at C2?

27.40 Draw the C4 epimer of D-xylose and name the monosaccharide.

27.41 For D-arabinose:
a. Draw its enantiomer.
b. Draw an epimer at C3.
c. Draw a diastereomer that is not an epimer.
d. Draw a constitutional isomer that still contains a carbonyl group.

27.42 Consider the following six compounds (**A–F**).

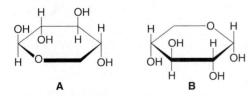

How are the two compounds in each pair related? Choose from enantiomers, epimers, diastereomers but not epimers, constitutional isomers, and identical compounds.
a. **A** and **B**   b. **A** and **C**   c. **B** and **C**   d. **A** and **D**   e. **E** and **F**

27.43 Consider the monosaccharide aldopentoses **A** and **B,** drawn below.

a. Which of the following terms describe **A** and **B:** epimers, anomers, enantiomers, diastereomers, and reducing sugars?
b. Draw the acyclic form of both **A** and **B,** and name each compound.

27.44 Draw a Haworth projection for each compound using the structures in Figures 27.3 and 27.4.
a. β-D-talopyranose   b. β-D-mannopyranose   c. α-D-galactopyranose   d. α-D-ribofuranose   e. α-D-tagatofuranose

27.45 Draw both pyranose anomers of each aldohexose using a three-dimensional representation using a chair pyranose. Label each anomer as α or β.

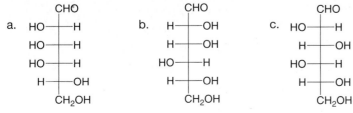

27.46 Convert each cyclic monosaccharide into its acyclic form.

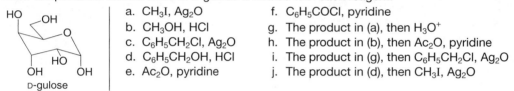

27.47 D-Arabinose can exist in both pyranose and furanose forms.
a. Draw the α and β anomers of D-arabinofuranose.
b. Draw the α and β anomers of D-arabinopyranose.

27.48 The most stable conformation of the pyranose ring of most D-aldohexoses places the largest group, CH₂OH, in the equatorial position. An exception to this is the aldohexose D-idose. Draw the two possible chair conformers of either the α or β anomer of D-idose. Explain why the more stable conformer has the CH₂OH group in the axial position.

Monosaccharide Reactions

27.49 Draw the products formed when α-D-gulose is treated with each reagent.

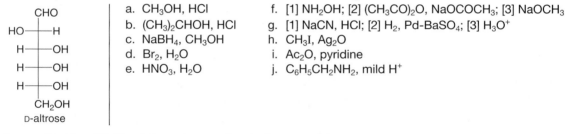

D-gulose

a. CH₃I, Ag₂O
b. CH₃OH, HCl
c. C₆H₅CH₂Cl, Ag₂O
d. C₆H₅CH₂OH, HCl
e. Ac₂O, pyridine

f. C₆H₅COCl, pyridine
g. The product in (a), then H₃O⁺
h. The product in (b), then Ac₂O, pyridine
i. The product in (g), then C₆H₅CH₂Cl, Ag₂O
j. The product in (d), then CH₃I, Ag₂O

27.50 Draw the products formed when D-altrose is treated with each reagent.

CHO
HO──H
H──OH
H──OH
H──OH
CH₂OH
D-altrose

a. CH₃OH, HCl
b. (CH₃)₂CHOH, HCl
c. NaBH₄, CH₃OH
d. Br₂, H₂O
e. HNO₃, H₂O

f. [1] NH₂OH; [2] (CH₃CO)₂O, NaOCOCH₃; [3] NaOCH₃
g. [1] NaCN, HCl; [2] H₂, Pd-BaSO₄; [3] H₃O⁺
h. CH₃I, Ag₂O
i. Ac₂O, pyridine
j. C₆H₅CH₂NH₂, mild H⁺

27.51 Answer Problem 27.50 using D-xylose as the starting material.

27.52 What two aldohexoses yield D-arabinose upon Wohl degradation?

27.53 What products are formed when each compound is subjected to a Kiliani–Fischer synthesis?

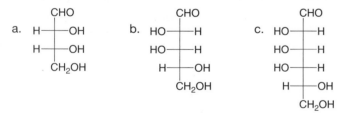

27.54 How would you convert D-glucose into each compound? More than one step is required.

a.
CH₃CH₂O
CH₃CH₂O──
OCH₂CH₃
O
OCH₃
OCH₂CH₃
+ α anomer

b.
CH₂OCH₃
H──OCH₃
CH₃O──H
H──OCH₃
H──OCH₃
CH₂OCH₃

c.
COOH
H──OAc
AcO──H
H──OAc
H──OAc
CH₂OAc

27.55 Which D-aldopentoses are reduced to optically inactive alditols using NaBH₄, CH₃OH?

27.56 What products are formed when each compound is treated with aqueous acid?

a. 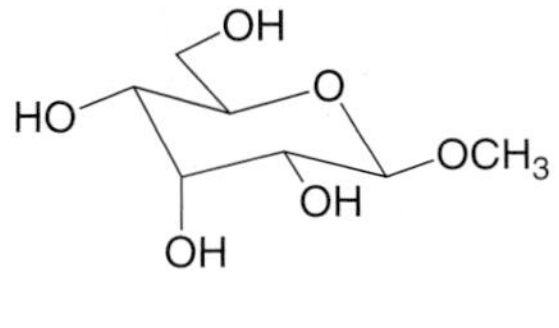

b. 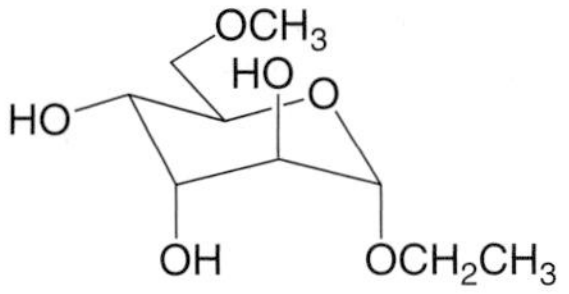

c. 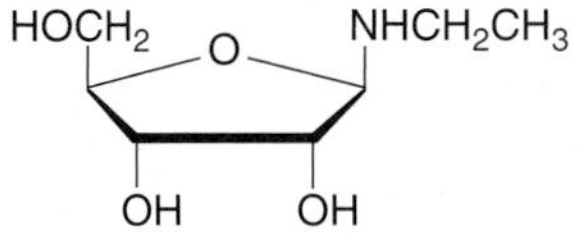

Mechanisms

27.57 Draw a stepwise mechanism for the acid-catalyzed interconversion of two glucose anomers by mutarotation.

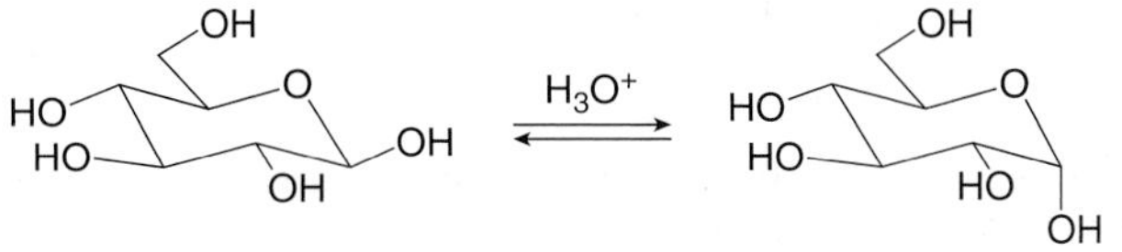

27.58 Draw a stepwise mechanism for the following reaction.

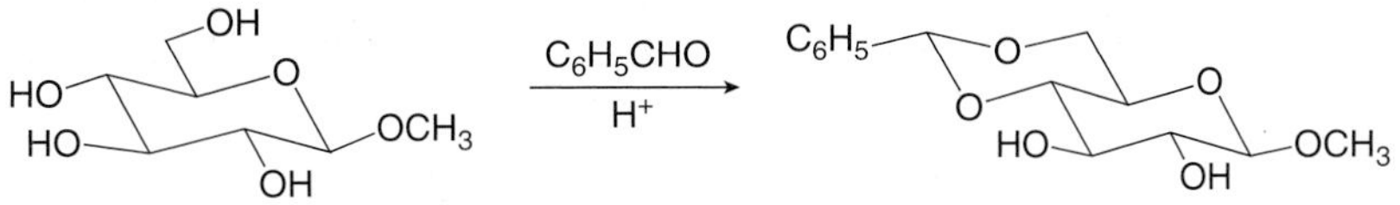

27.59 Draw a stepwise mechanism for each hydrolysis reaction.

a. 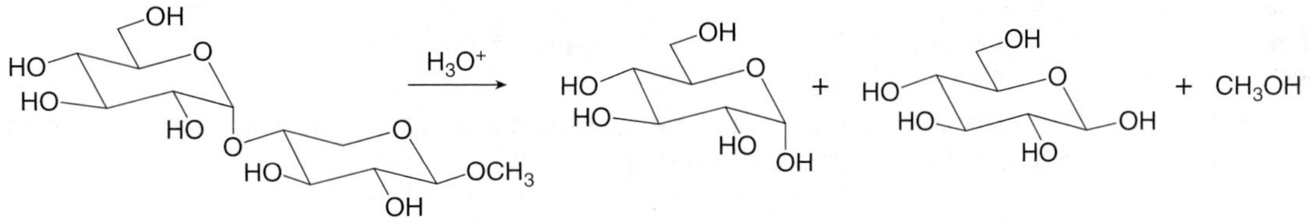

b. 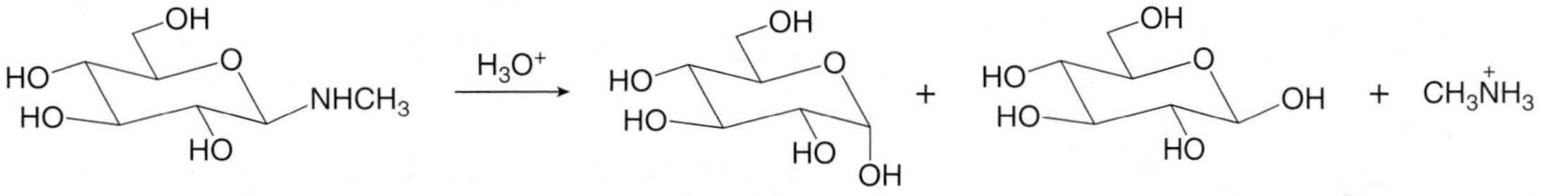

27.60 The following isomerization reaction, drawn using D-glucose as starting material, occurs with all aldohexoses in the presence of base. Draw a stepwise mechanism that illustrates how each compound is formed.

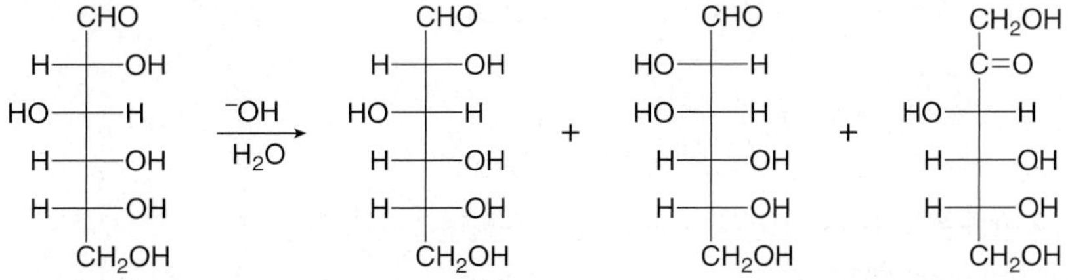

D-glucose

(recovered starting material)

Identifying Monosaccharides

27.61 A 2-ketohexose is reduced with $NaBH_4$ in CH_3OH to form a mixture of D-galactitol and D-talitol. What is the structure of the 2-ketohexose?

27.62 Which D-aldopentose is oxidized to an optically active aldaric acid and undergoes the Wohl degradation to yield a D-aldotetrose that is oxidized to an optically active aldaric acid?

27.63 What other D-aldopentose forms the same alditol as D-arabinose when reduced with $NaBH_4$ in CH_3OH?

27.64 Identify compounds **A–D.** A D-aldopentose **A** is oxidized with HNO_3 to an optically inactive aldaric acid **B. A** undergoes the Kiliani–Fischer synthesis to yield **C** and **D. C** is oxidized to an optically active aldaric acid. **D** is oxidized to an optically inactive aldaric acid.

27.65 A D-aldopentose **A** is reduced to an optically active alditol. Upon Kiliani–Fischer synthesis, **A** is converted to two D-aldohexoses, **B** and **C. B** is oxidized to an optically inactive aldaric acid. **C** is oxidized to an optically active aldaric acid. What are the structures of **A–C?**

27.66 A D-aldohexose **A** is reduced to an optically active alditol **B** using NaBH$_4$ in CH$_3$OH. **A** is converted by Wohl degradation to an aldopentose **C**, which is reduced to an optically inactive alditol **D**. **C** is converted by Wohl degradation to aldotetrose **E**, which is oxidized to an optically active aldaric acid **F**. When the two ends of aldohexose **A** are interconverted, a different aldohexose **G** is obtained. What are the structures of **A–G**?

Disaccharides and Polysaccharides

27.67 Identify the lettered compounds in the following reactions.

a.

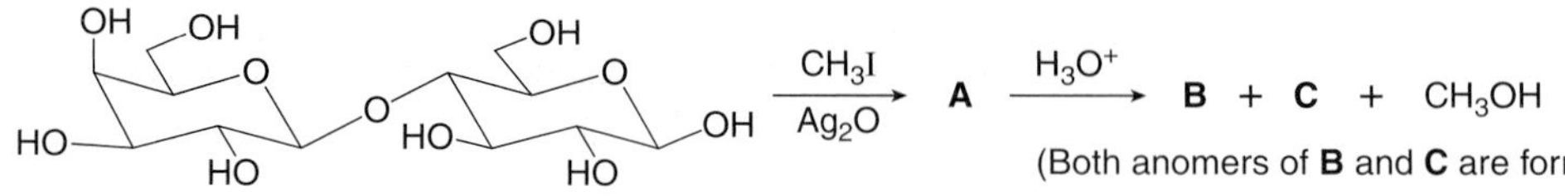

$$\xrightarrow[\text{Ag}_2\text{O}]{\text{CH}_3\text{I}} \mathbf{A} \xrightarrow{\text{H}_3\text{O}^+} \mathbf{B} + \mathbf{C} + \text{CH}_3\text{OH}$$

(Both anomers of **B** and **C** are formed.)

b.

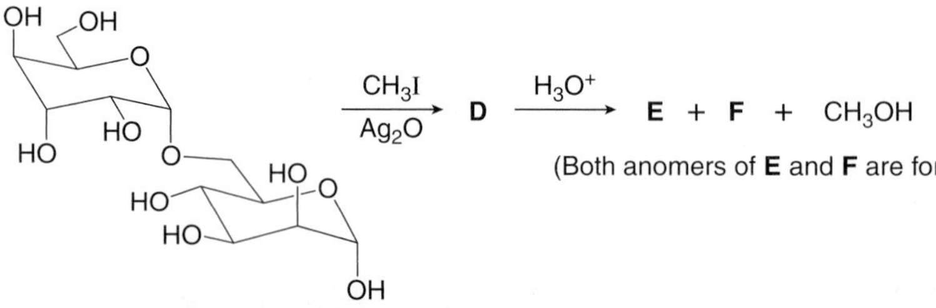

$$\xrightarrow[\text{Ag}_2\text{O}]{\text{CH}_3\text{I}} \mathbf{D} \xrightarrow{\text{H}_3\text{O}^+} \mathbf{E} + \mathbf{F} + \text{CH}_3\text{OH}$$

(Both anomers of **E** and **F** are formed.)

27.68 For each disaccharide in Problem 27.67:
  a. Identify the glycosidic linkage.
  b. Classify the glycosidic bond as α or β and use numbers to designate its location.
  c. Classify each disaccharide as reducing or nonreducing.

27.69 Consider the tetrasaccharide stachyose drawn below. Stachyose is found in white jasmine, soybeans, and lentils. Because humans cannot digest it, its consumption causes flatulence.

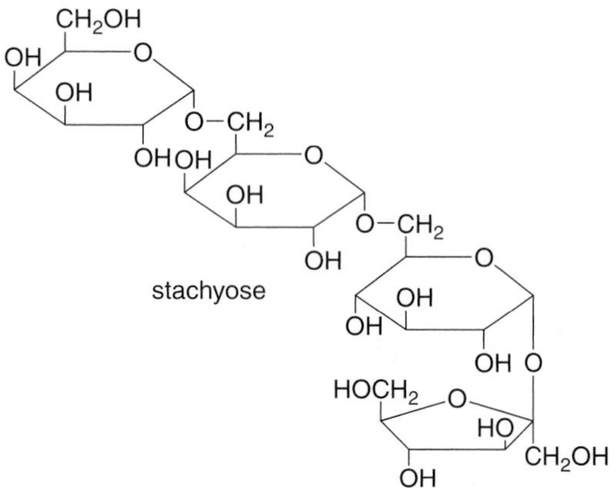

stachyose

  a. Label all glycoside bonds.
  b. Classify each glycosidic linkage as α or β and use numbers to designate its location between two rings (e.g., β-1,4').
  c. What products are formed when stachyose is hydrolyzed with H$_3$O$^+$?
  d. Is stachyose a reducing sugar?
  e. What product is formed when stachyose is treated with excess CH$_3$I, Ag$_2$O?
  f. What products are formed when the product in (e) is treated with H$_3$O$^+$?

27.70 Deduce the structure of the disaccharide isomaltose from the following data.
  [1] Hydrolysis yields D-glucose exclusively.
  [2] Isomaltose is cleaved with α-glycosidase enzymes.
  [3] Isomaltose is a reducing sugar.
  [4] Methylation with excess CH$_3$I, Ag$_2$O and then hydrolysis with H$_3$O$^+$ forms two products:

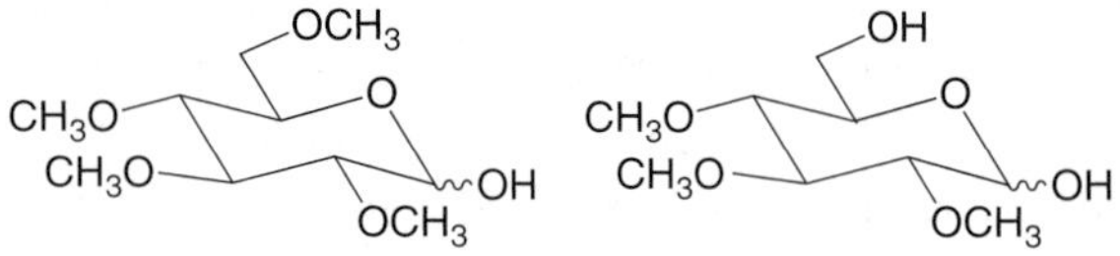

(Both anomers are present.)

27.71 Deduce the structure of the disaccharide trehalose from the following data. Trehalose is the "blood sugar" of the insect world. It is found in bacterial spores, fungi, and many insects whose natural environment has large variations in temperature.

[1] Hydrolysis yields D-glucose exclusively.

[2] Trehalose is hydrolyzed by α-glycosidase enzymes.

[3] Trehalose is a nonreducing sugar.

[4] Methylation with excess CH_3I, Ag_2O, followed by hydrolysis with H_3O^+, forms only one product:

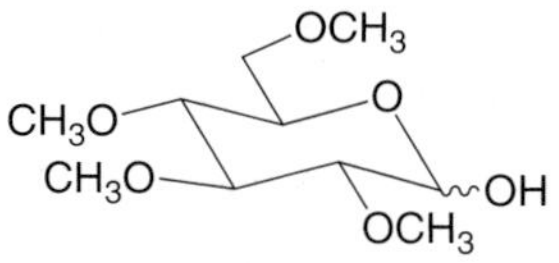

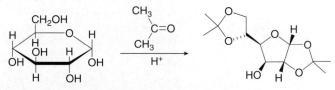

(both anomers)

27.72 Deduce the structure of the following disaccharide using the given information.

[1] It is hydrolyzed to D-glucose and D-galactose.

[2] It is cleaved with α-glycosidase enzymes.

[3] It is a reducing sugar.

[4] Methylation with excess CH_3I, Ag_2O, followed by hydrolysis with H_3O^+, forms two products:

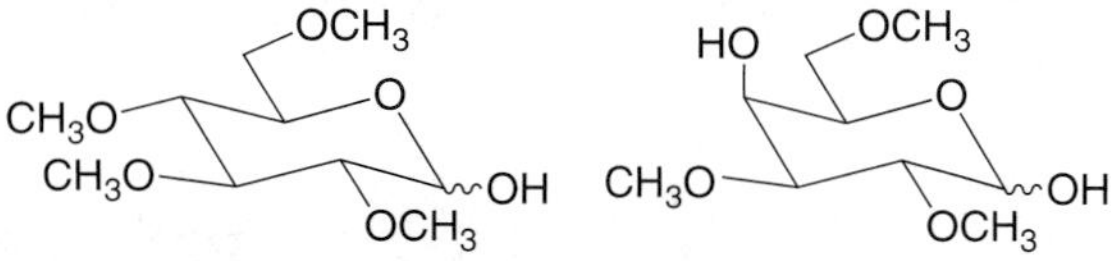

(Both anomers are present.)

27.73 Draw the structure of each of the following compounds:

a. A polysaccharide formed by joining D-glucosamine in α-1,6'-glycosidic linkages.

b. A disaccharide formed by joining D-mannose and D-glucose in a β-1,4'-glycosidic linkage using mannose's anomeric carbon.

c. An α-N-glycoside formed from D-arabinose and $C_6H_5CH_2NH_2$.

d. A ribonucleoside formed from D-ribose and thymine.

Challenge Problems

27.74 Draw a stepwise mechanism for the following reaction.

27.75 Deduce the structure of the trisaccharide (**X**) from the given information.

[1] Methylation with excess CH_3I, Ag_2O and then hydrolysis with H_3O^+ forms three products.

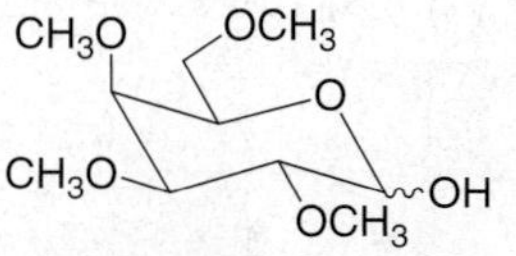

2,3,4,6-tetra-O-methyl-D-galactose 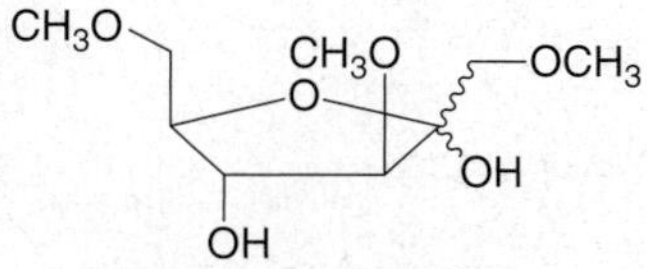   1,3,6-tri-O-methyl-D-fructose 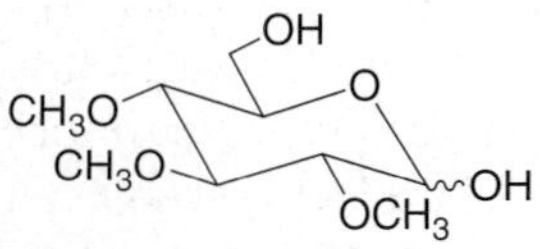   2,3,4-tri-O-methyl-D-glucose

(Both anomers of each compound are formed.)

[2] **X** is cleaved with a β-glycosidase enzyme to give a disaccharide and D-galactose.

[3] **X** is cleaved with an α-glycosidase enzyme to give a disaccharide and D-fructose.

Amino Acids and Proteins

Aspartame (trade name: **Nutrasweet**) is a dipeptide formed by joining two naturally occurring amino acids, aspartic acid and phenylalanine. Since the discovery in 1965 that it is 180 times sweeter than sucrose, common table sugar, it has been used as a substitute sweetener. Aspartame is now widely used in foods and beverages and is marketed to consumers desiring to reduce their caloric intake. In Chapter 28, we discuss the properties of proteins and peptides as well as the amino acids from which they are synthesized.

O f the four major groups of biomolecules—lipids, carbohydrates, nucleic acids, and proteins—proteins have the widest array of functions. **Keratin** and **collagen,** for example, are part of a large group of structural proteins that form long insoluble fibers, giving strength and support to tissues. Hair, horns, hooves, and fingernails are all made up of keratin. **Collagen** is found in bone, connective tissue, tendons, and cartilage. **Enzymes** are proteins that catalyze and regulate all aspects of cellular function. **Membrane proteins** transport small organic molecules and ions across cell membranes. **Insulin,** the hormone that regulates blood glucose levels, **fibrinogen** and **thrombin,** which form blood clots, and **hemoglobin,** which transports oxygen from the lungs to tissues, are all proteins.

In Chapter 28 we discuss proteins and their primary components, the amino acids.

28.1   Amino Acids

Amino acids were first discussed in Section 19.14.

Naturally occurring amino acids have an amino group (NH_2) bonded to the α carbon of a carboxy group (COOH), and so they are called **α-amino acids.**

◆ **All proteins are polyamides formed by joining amino acids together.**

Amino acid

COOH
|
H_2N-C-H
|
R α carbon
α-amino acid

Portion of a protein molecule

28.1A   General Features of α-Amino Acids

The 20 amino acids that occur naturally in proteins differ in the identity of the R group bonded to the α carbon. The R group is called the **side chain** of the amino acid.

The simplest amino acid, called glycine, has R = H. **All other amino acids (R ≠ H) have a stereogenic center on the α carbon.** As is true for monosaccharides, the prefixes D and L are used to designate the configuration at the stereogenic center of amino acids. Common, naturally occurring amino acids are called **L-amino acids.** Their enantiomers, D-amino acids, are rarely found in nature. These general structures are shown in Figure 28.1. According to R,S designations, all L-amino acids except cysteine have the *S* **configuration.**

All amino acids have common names. These names can be represented by either a one-letter or a three-letter abbreviation.

Figure 28.2 is a listing of the 20 naturally occurring amino acids, together with their abbreviations. Note the variability in the R groups. A side chain can be a simple alkyl group, or it can have additional functional groups such as OH, SH, COOH, or NH_2.

Figure 28.1   The general features of an α-amino acid

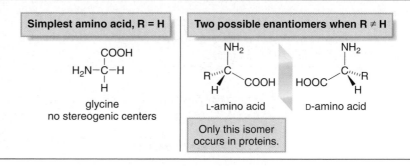

Simplest amino acid, R = H

COOH
|
H_2N-C-H
|
H
glycine
no stereogenic centers

Two possible enantiomers when R ≠ H

L-amino acid   D-amino acid

Only this isomer occurs in proteins.

Figure 28.2   The 20 naturally occurring amino acids

Neutral amino acids

Name	Structure	Abbreviations	Name	Structure	Abbreviations
Glycine	$\underset{\underset{H}{|}}{\overset{\overset{COOH}{|}}{H-C-NH_2}}$	Gly G	Cysteine	$\underset{\underset{H}{|}}{\overset{\overset{COOH}{|}}{HSCH_2-C-NH_2}}$	Cys C
Alanine	$\underset{\underset{H}{|}}{\overset{\overset{COOH}{|}}{CH_3-C-NH_2}}$	Ala A	Methionine*	$\underset{\underset{H}{|}}{\overset{\overset{COOH}{|}}{CH_3SCH_2CH_2-C-NH_2}}$	Met M
Valine*	$\underset{\underset{H}{|}}{\overset{\overset{COOH}{|}}{(CH_3)_2CH-C-NH_2}}$	Val V	Asparagine	$\underset{\underset{H}{|}}{\overset{\overset{COOH}{|}}{H_2NCOCH_2-C-NH_2}}$	Asn N
Leucine*	$\underset{\underset{H}{|}}{\overset{\overset{COOH}{|}}{(CH_3)_2CHCH_2-C-NH_2}}$	Leu L	Glutamine	$\underset{\underset{H}{|}}{\overset{\overset{COOH}{|}}{H_2NCOCH_2CH_2-C-NH_2}}$	Gln Q
Isoleucine*	$\underset{\underset{H}{|}}{\overset{\overset{CH_3\ COOH}{|\ \ \ |}}{CH_3CH_2CH-C-NH_2}}$	Ile I	Phenylalanine*	$\underset{\underset{H}{|}}{\overset{\overset{COOH}{|}}{C_6H_5-CH_2-C-NH_2}}$	Phe F
Serine	$\underset{\underset{H}{|}}{\overset{\overset{COOH}{|}}{HOCH_2-C-NH_2}}$	Ser S	Tyrosine	$\underset{\underset{H}{|}}{\overset{\overset{COOH}{|}}{HO-C_6H_4-CH_2-C-NH_2}}$	Tyr Y
Threonine*	$\underset{\underset{H}{|}}{\overset{\overset{HO\ COOH}{|\ \ \ |}}{CH_3CH-C-NH_2}}$	Thr T	Tryptophan*	$\underset{\underset{H}{|}}{\overset{\overset{COOH}{|}}{CH_2-C-NH_2}}$ (indole)	Trp W
Proline	(pyrrolidine)—COOH	Pro P			

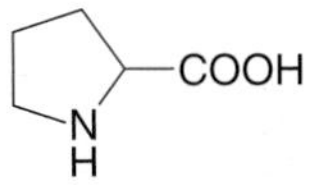

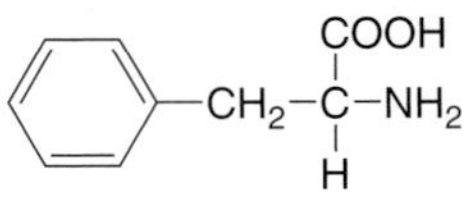

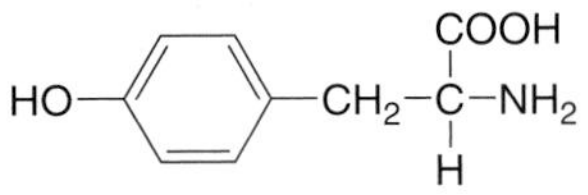

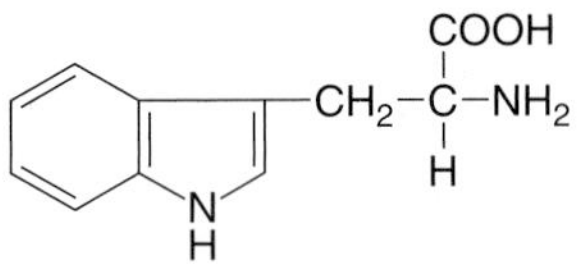

Acidic amino acids   ## Basic amino acids

Name	Structure	Abbreviations	Name	Structure	Abbreviations
Aspartic acid	$\underset{\underset{H}{|}}{\overset{\overset{COOH}{|}}{HO_2CCH_2-C-NH_2}}$	Asp D	Lysine*	$\underset{\underset{H}{|}}{\overset{\overset{COOH}{|}}{NH_2CH_2CH_2CH_2CH_2-C-NH_2}}$	Lys K
Glutamic acid	$\underset{\underset{H}{|}}{\overset{\overset{COOH}{|}}{HO_2CCH_2CH_2-C-NH_2}}$	Glu E	Histidine*	$\underset{\underset{H}{|}}{\overset{\overset{COOH}{|}}{(imidazole)-CH_2-C-NH_2}}$	His H
			Arginine*	$\underset{\underset{H}{|}}{\overset{\overset{NH\ \ \ \ \ \ \ \ \ \ \ \ \ \ \ COOH}{||\ \ \ \ \ \ \ \ \ \ \ \ \ \ |}}{H_2N-C-N-CH_2CH_2CH_2-C-NH_2}}$	Arg R

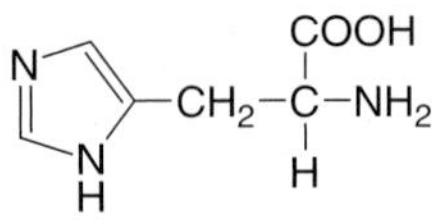

Essential amino acids are labeled with an asterisk (*).

- ◆ Amino acids with an additional COOH group in the side chain are called acidic amino acids.
- ◆ Those with an additional basic N atom in the side chain are called basic amino acids.
- ◆ All others are neutral amino acids.

Look closely at the structures of proline, isoleucine, and threonine.

◆ **All amino acids are 1° amines except for proline,** which has its N atom in a five-membered ring, making it a **2° amine.**

◆ **Isoleucine** and **threonine** contain an additional stereogenic center at the β carbon, so there are four possible stereoisomers, only one of which is naturally occurring.

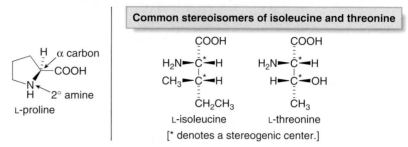

Common stereoisomers of isoleucine and threonine

L-proline   L-isoleucine   L-threonine

[* denotes a stereogenic center.]

Humans can synthesize only 10 of these 20 amino acids. The remaining 10 are called **essential amino acids** because they must be obtained from the diet. These are labeled with an asterisk in Figure 28.2.

PROBLEM 28.1 | Draw the other three stereoisomers of L-isoleucine, and label the stereogenic centers as *R* or *S*.

28.1B Acid–Base Behavior

Recall from Section 19.14B that an amino acid has both an acidic and a basic functional group, so proton transfer forms a salt called a **zwitterion.**

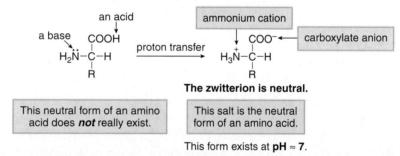

The zwitterion is neutral.

This neutral form of an amino acid does **not** really exist.

This salt is the neutral form of an amino acid.

This form exists at **pH ≈ 7.**

◆ **Amino acids do not exist to any appreciable extent as uncharged neutral compounds. They exist as salts, giving them high melting points and making them water soluble.**

Amino acids exist in different charged forms, as shown in Figure 28.3, depending on the pH of the aqueous solution in which they are dissolved. For neutral amino acids, the overall charge is +1, 0, or –1. Only at pH ~7 does the zwitterionic form exist.

The −COOH and −NH$_3^+$ groups of an amino acid are ionizable, because they can lose a proton in aqueous solution. As a result, they have different pK_a values. The pK_a of the −COOH group is typically ~2, whereas that of the −NH$_3^+$ group is ~9, as shown in Table 28.1.

Some amino acids, such as aspartic acid and lysine, have acidic or basic side chains. These additional ionizable groups complicate somewhat the acid–base behavior of these amino acids. Table 28.1 lists the pK_a values for these acidic and basic side chains as well.

Figure 28.3 How the charge of a neutral amino acid depends on the pH

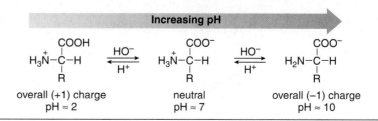

Increasing pH

overall (+1) charge
pH ≈ 2

neutral
pH ≈ 7

overall (−1) charge
pH ≈ 10

Table 28.1 also lists the isoelectric points (pI) for all of the amino acids. Recall from Section 19.14C that the **isoelectric point is the pH at which an amino acid exists primarily in its neutral form,** and that it can be calculated from the average of the pK_a values of the α-COOH and α-NH$_3^+$ groups (for neutral amino acids only).

TABLE 28.1	pK_a Values for the Ionizable Functional Groups of an α-Amino Acid			
Amino acid	**α-COOH**	**α-NH$_3^+$**	**Side chain**	**pI**
Alanine	2.35	9.87	—	6.11
Arginine	2.01	9.04	12.48	10.76
Asparagine	2.02	8.80	—	5.41
Aspartic acid	2.10	9.82	3.86	2.98
Cysteine	2.05	10.25	8.00	5.02
Glutamic acid	2.10	9.47	4.07	3.08
Glutamine	2.17	9.13	—	5.65
Glycine	2.35	9.78	—	6.06
Histidine	1.77	9.18	6.10	7.64
Isoleucine	2.32	9.76	—	6.04
Leucine	2.33	9.74	—	6.04
Lysine	2.18	8.95	10.53	9.74
Methionine	2.28	9.21	—	5.74
Phenylalanine	2.58	9.24	—	5.91
Proline	2.00	10.00	—	6.30
Serine	2.21	9.15	—	5.68
Threonine	2.09	9.10	—	5.60
Tryptophan	2.38	9.39	—	5.88
Tyrosine	2.20	9.11	10.07	5.63
Valine	2.29	9.72	—	6.00

PROBLEM 28.2 What form exists at the isoelectric point of each of the following amino acids: (a) valine; (b) leucine; (c) proline; (d) glutamic acid?

PROBLEM 28.3 Explain why the pK_a of the $-$NH$_3^+$ group of an α-amino acid is lower than the pK_a of the ammonium ion derived from a 1° amine (RNH$_3^+$). For example the pK_a of the $-$NH$_3^+$ group of alanine is 9.7 but the pK_a of CH$_3$NH$_3^+$ is 10.63.

28.2 Synthesis of Amino Acids

Amino acids can be prepared in a variety of ways in the laboratory. Three methods are described, each of which is based on reactions learned in previous chapters.

28.2A S$_N$2 Reaction of α-Halo Acids with NH$_3$

The most direct way to synthesize an α-amino acid is by **S$_N$2 reaction of an α-halo carboxylic acid with a large excess of NH$_3$.**

General reaction

$$\text{R}-\underset{\underset{\text{Br}}{|}}{\text{CHCOOH}} \xrightarrow[\text{(large excess)}]{\text{NH}_3} \text{R}-\underset{\underset{\text{NH}_2}{|}}{\text{CHCOO}^-}\text{NH}_4^+ \ + \ \text{NH}_4^+\text{Br}^-$$

Example

$$(\text{CH}_3)_2\text{CH}-\underset{\underset{\text{Br}}{|}}{\text{CHCOOH}} \xrightarrow[\text{(large excess)}]{\text{NH}_3} \underset{\text{S}_N2}{} \quad (\text{CH}_3)_2\text{CH}-\underset{\underset{\text{NH}_2}{|}}{\text{CHCOO}^-}\text{NH}_4^+ \ + \ \text{NH}_4^+\text{Br}^-$$

valine

Although the alkylation of ammonia with simple alkyl halides does not generally afford high yields of 1° amines (Section 25.7A), this reaction using α-halo carboxylic acids does form the desired amino acids in good yields. In this case, the amino group in the product is both less basic and more sterically crowded than other 1° amines, so that a single alkylation occurs and the desired amino acid is obtained.

PROBLEM 28.4   What α-halo carbonyl compound is needed to synthesize each amino acid: (a) glycine; (b) isoleucine; (c) phenylalanine?

28.2B Alkylation of a Diethyl Malonate Derivative

The second method for preparing amino acids is based on the malonic ester synthesis. Recall from Section 23.9 that this synthesis converts diethyl malonate to a carboxylic acid with a new alkyl group on its α carbon atom.

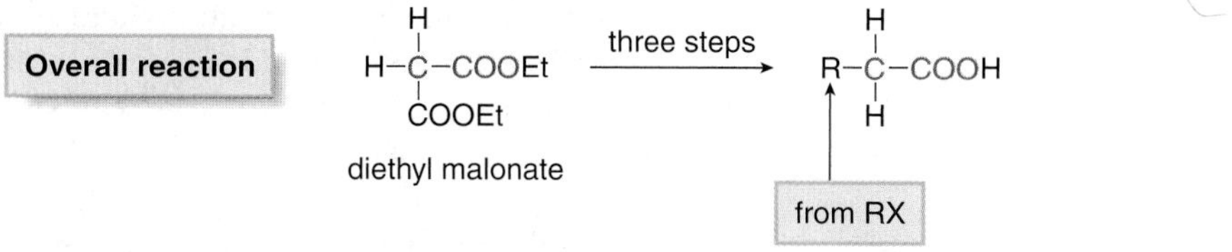

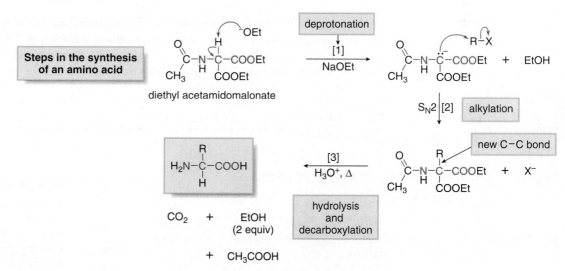

This reaction can be adapted to the synthesis of α-amino acids by using a commercially available derivative of diethyl malonate as starting material. This compound, **diethyl acetamidomalonate,** has a nitrogen atom on the α carbon, which ultimately becomes the NH$_2$ group on the α carbon of the amino acid.

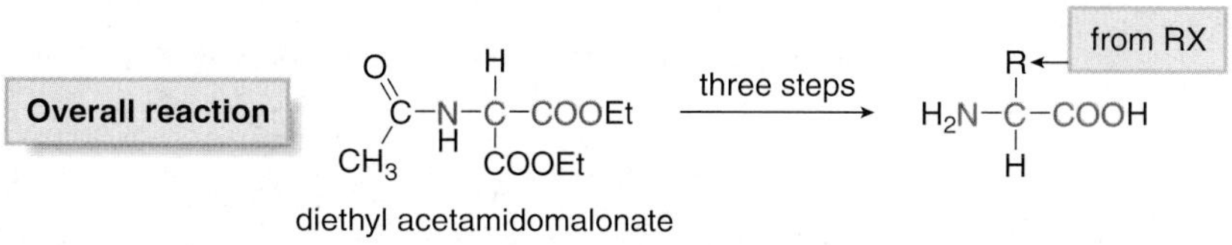

The malonic ester synthesis consists of three steps, and so does this variation to prepare an amino acid.

[1] **Deprotonation** of diethyl acetamidomalonate with NaOEt forms an enolate by removal of the acidic proton between the two carbonyl groups.

[2] **Alkylation** of the enolate with an unhindered alkyl halide (usually CH$_3$X or RCH$_2$X) forms a substitution product with a new R group on the α carbon.

[3] Heating the alkylation product with aqueous acid results in **hydrolysis** of both esters and the amide, followed by **decarboxylation** to form the amino acid.

Phenylalanine, for example, can be synthesized as follows:

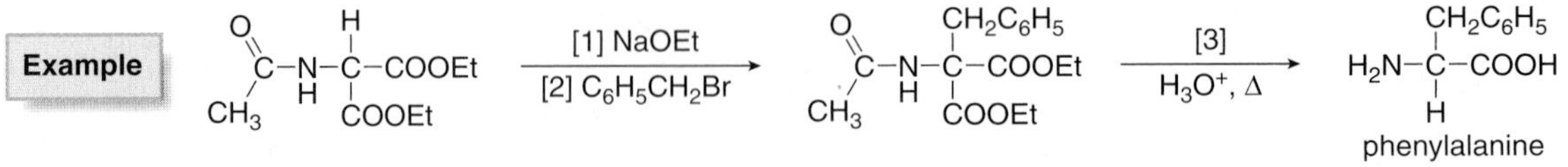

Example

$$CH_3-\underset{\underset{CH_3}{|}}{\overset{\overset{O}{||}}{C}}-\underset{H}{\overset{H}{N}}-\underset{\underset{COOEt}{|}}{C}-COOEt \xrightarrow[\text{[2] } C_6H_5CH_2Br]{\text{[1] NaOEt}} CH_3-\underset{\underset{CH_3}{|}}{\overset{\overset{O}{||}}{C}}-\underset{H}{\overset{\overset{CH_2C_6H_5}{|}}{N}}-\underset{\underset{COOEt}{|}}{C}-COOEt \xrightarrow[H_3O^+, \Delta]{\text{[3]}} H_2N-\underset{H}{\overset{\overset{CH_2C_6H_5}{|}}{C}}-COOH$$

phenylalanine

PROBLEM 28.5 The enolate derived from diethyl acetamidomalonate is treated with each of the following alkyl halides. After hydrolysis and decarboxylation, what amino acid is formed: (a) CH_3I; (b) $(CH_3)_2CHCH_2Cl$; (c) $CH_3CH_2CH(CH_3)Br$?

PROBLEM 28.6 What amino acid is formed when $CH_3CONHCH(CO_2Et)_2$ is treated with the following series of reagents: [1] NaOEt; [2] $CH_2=O$; [3] H_3O^+, Δ?

28.2C Strecker Synthesis

The third method, the **Strecker amino acid synthesis,** converts an aldehyde into an amino acid by a two-step sequence that adds one carbon atom to the aldehyde carbonyl. Treating an aldehyde with NH_4Cl and NaCN first forms an **α-amino nitrile,** which can then be hydrolyzed in aqueous acid to an amino acid.

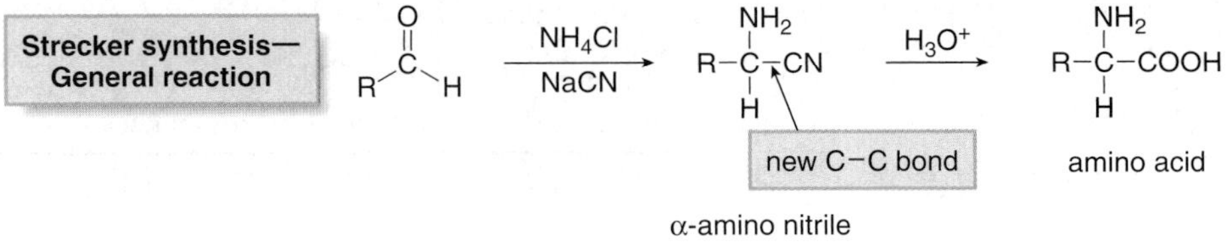

Strecker synthesis— General reaction

$$R-\overset{\overset{O}{||}}{C}-H \xrightarrow[\text{NaCN}]{NH_4Cl} R-\underset{\underset{H}{|}}{\overset{\overset{NH_2}{|}}{C}}-CN \xrightarrow{H_3O^+} R-\underset{\underset{H}{|}}{\overset{\overset{NH_2}{|}}{C}}-COOH$$

new C–C bond   α-amino nitrile   amino acid

The Strecker synthesis of alanine, for example, is as follows:

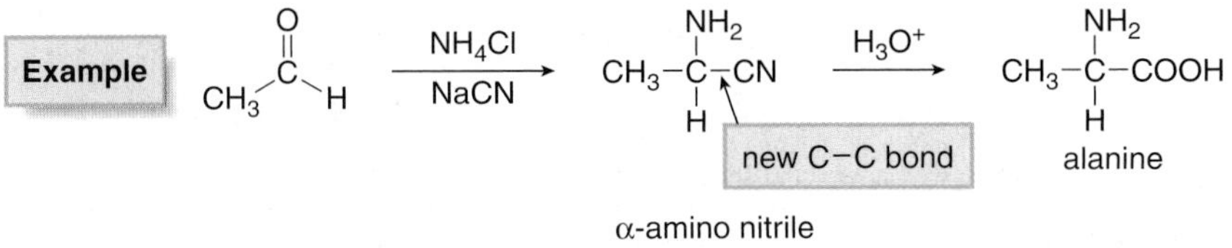

Example

$$CH_3-\overset{\overset{O}{||}}{C}-H \xrightarrow[\text{NaCN}]{NH_4Cl} CH_3-\underset{\underset{H}{|}}{\overset{\overset{NH_2}{|}}{C}}-CN \xrightarrow{H_3O^+} CH_3-\underset{\underset{H}{|}}{\overset{\overset{NH_2}{|}}{C}}-COOH$$

new C–C bond   α-amino nitrile   alanine

Mechanism 28.1 for the formation of the α-amino nitrile from an aldehyde (the first step in the Strecker synthesis) consists of two parts: **nucleophilic addition of NH_3** to form an imine, followed by **addition of cyanide** to the C=N bond. Both parts are related to earlier mechanisms involving imines (Section 21.11) and cyanohydrins (Section 21.9).

MECHANISM 28.1

Formation of an α-Amino Nitrile

Part [1] Nucleophilic attack of NH_3 to form an imine

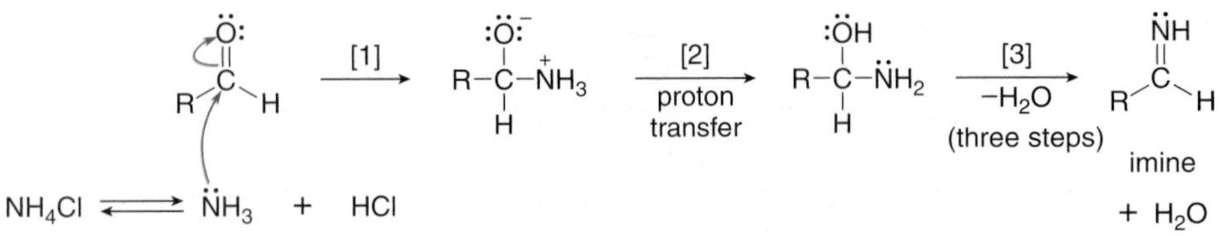

$$R-\overset{\overset{\ddot{O}:}{||}}{C}-H \xrightarrow{\text{[1]}} R-\underset{\underset{H}{|}}{\overset{\overset{:\ddot{O}:^-}{|}}{C}}-\overset{+}{N}H_3 \xrightarrow[\text{transfer}]{\text{[2]}\atop\text{proton}} R-\underset{\underset{H}{|}}{\overset{\overset{:\ddot{O}H}{|}}{C}}-\ddot{N}H_2 \xrightarrow[\text{(three steps)}]{\text{[3]}\atop -H_2O} R-\overset{\overset{\ddot{N}H}{||}}{C}-H$$

$$NH_4Cl \rightleftharpoons \ddot{N}H_3 + HCl \qquad\qquad\qquad\qquad\qquad \text{imine} + H_2O$$

◆ **Part [1]** Nucleophilic attack of NH_3 followed by proton transfer and loss of H_2O forms an imine. Loss of H_2O occurs by the same three-step process outlined in Mechanism 21.5.

Part [2] Nucleophilic attack of $^-$CN to form an α-amino nitrile

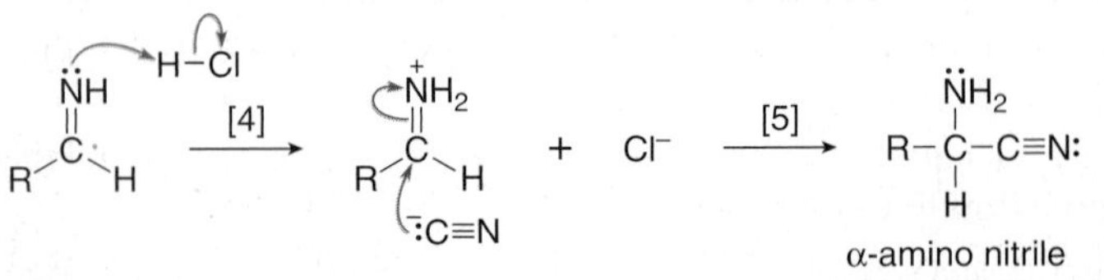

$$R-\overset{\overset{\ddot{N}H}{||}}{C}-H \xrightarrow{\text{[4]}} R-\underset{\underset{:C\equiv N}{|}}{\overset{\overset{+}{N}H_2}{C}}-H + Cl^- \xrightarrow{\text{[5]}} R-\underset{\underset{H}{|}}{\overset{\overset{\ddot{N}H_2}{|}}{C}}-C\equiv N:$$

H–Cl

α-amino nitrile

◆ **Part [2]** Protonation of the imine followed by nucleophilic attack of $^-$CN gives the α-amino nitrile.

Figure 28.4 The synthesis of methionine by three different methods

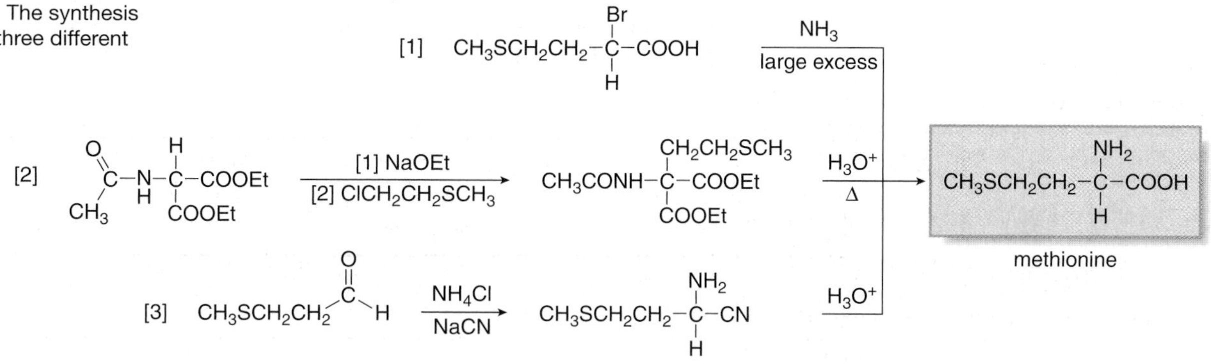
[1]
$CH_3SCH_2CH_2-C(Br)(H)-COOH$
NH_3
large excess
[2]
[1] NaOEt
[2] $ClCH_2CH_2SCH_3$
$CH_3CONH-C(CH_2CH_2SCH_3)(COOEt)-COOEt$
H_3O^+
Δ
[3]
NH_4Cl
NaCN
$CH_3SCH_2CH_2-C(NH_2)(H)-CN$
H_3O^+
$CH_3SCH_2CH_2-C(NH_2)(H)-COOH$
methionine

Three methods of amino acid synthesis:
[1] S_N2 reaction using an α-halo carboxylic acid
[2] Alkylation of diethyl acetamidomalonate
[3] Strecker synthesis

The details of the second step of the Strecker synthesis, the hydrolysis of a nitrile (RCN) to a carboxylic acid (RCOOH), have already been presented in Section 22.18A.

PROBLEM 28.7 What aldehyde is needed to synthesize each amino acid by the Strecker synthesis: (a) valine; (b) leucine; (c) phenylalanine?

28.2D Summary of Methods

Figure 28.4 shows how the amino acid methionine can be prepared by all three methods.

PROBLEM 28.8 Draw the products of each reaction.

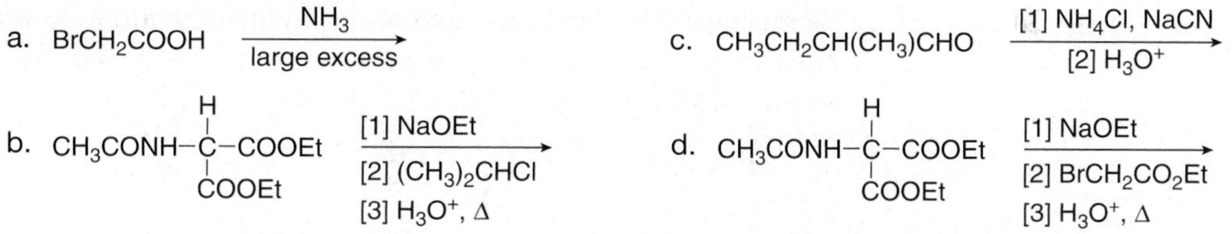
a. $BrCH_2COOH$
NH_3
large excess
b. $CH_3CONH-C(H)(COOEt)-COOEt$
[1] NaOEt
[2] $(CH_3)_2CHCl$
[3] H_3O^+, Δ
c. $CH_3CH_2CH(CH_3)CHO$
[1] NH_4Cl, NaCN
[2] H_3O^+
d. $CH_3CONH-C(H)(COOEt)-COOEt$
[1] NaOEt
[2] $BrCH_2CO_2Et$
[3] H_3O^+, Δ

28.3 Separation of Amino Acids

No matter which of the preceding methods is used to synthesize an amino acid, all three yield a racemic mixture. Naturally occurring amino acids exist as a single enantiomer, however, so the two enantiomers obtained must be separated if they are to be used in biological applications. This is not an easy task. Two enantiomers have the same physical properties, so they cannot be separated by common physical methods, such as distillation or chromatography. Moreover, they react in the same way with achiral reagents, so they cannot be separated by chemical reactions either.

Nonetheless, strategies have been devised to separate two enantiomers using physical separation techniques and chemical reactions. We will examine two different strategies in Section 28.3. Then, in Section 28.4, we will discuss a method that affords optically active amino acids without the need for separation.

- **The separation of a racemic mixture into its component enantiomers is called *resolution*. Thus, a racemic mixture is *resolved* into its component enantiomers.**

Figure 28.5   Resolution of a racemic mixture by converting it to a mixture of diastereomers

Enantiomers A and B can be separated by reaction with a single enantiomer of a chiral reagent, Y. The process of resolution requires three steps:

[1] Reaction of enantiomers **A** and **B** with **Y** forms two diastereomers, **AY** and **BY**.

[2] Diastereomers **AY** and **BY** have different physical properties, so they can be separated by physical methods such as fractional distillation or crystallization.

[3] **AY** and **BY** are then re-converted to **A** and **B** by a chemical reaction. The two enantiomers **A** and **B** are now separated from each other, and resolution is complete.

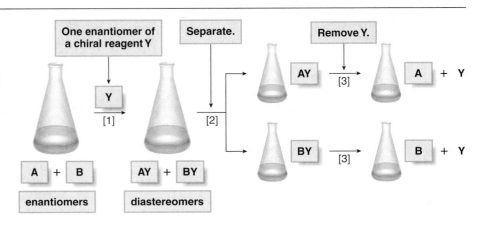

28.3A   Resolution of Amino Acids

The oldest, and perhaps still the most widely used method to separate enantiomers exploits the following fact: **enantiomers have the *same* physical properties, but diastereomers have *different* physical properties.** Thus, a racemic mixture can be resolved using the following general strategy.

[1] **Convert a pair of enantiomers into a pair of diastereomers,** which are now separable because they have different melting points and boiling points.

[2] **Separate the diastereomers.**

[3] **Re-convert each diastereomer into the original enantiomer,** now separated from the other.

This general three-step process is illustrated in Figure 28.5.

To resolve a racemic mixture of amino acids such as (*R*)- and (*S*)-alanine, the racemate is first treated with acetic anhydride to form **N-acetyl amino acids.** Each of these amides contains one stereogenic center and they are still enantiomers, so they are *still inseparable.*

Both enantiomers of *N*-acetyl alanine have a free carboxy group that can react with an amine in an acid–base reaction. **If a chiral amine is used, such as (*R*)-α-methylbenzylamine, the two salts formed are diastereomers, *not* enantiomers.** Diastereomers can be physically separated from each other, so the compound that converts enantiomers into diastereomers is called a **resolving agent.** Either enantiomer of the resolving agent can be used.

(*R*)-α-methylbenzylamine

a resolving agent

How To   Use (*R*)-α-Methylbenzylamine to Resolve a Racemic Mixture of Amino Acids

Step [1]   **React both enantiomers with the *R* isomer of the chiral amine.**

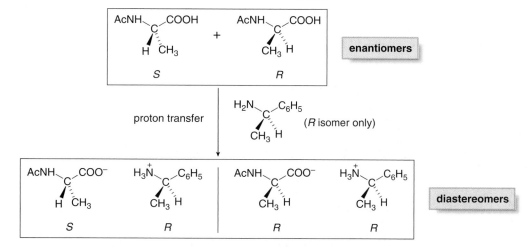

These salts have the *same* configuration around one stereogenic center, but the *opposite* configuration about the other stereogenic center.

Step [2]   **Separate the diastereomers.**

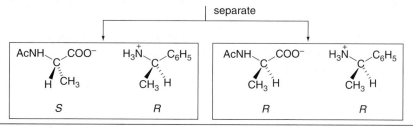

Step [3]   **Regenerate the amino acid by hydrolysis of the amide.**

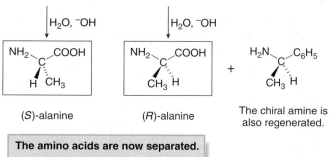

(*S*)-alanine   (*R*)-alanine   The chiral amine is
                                          also regenerated.

The amino acids are now separated.

Step [1] is just an acid–base reaction in which the racemic mixture of *N*-acetyl alanines reacts with the same enantiomer of the resolving agent, in this case (*R*)-methylbenzylamine. The salts that form are **diastereomers, *not* enantiomers,** because they have the same configuration about one stereogenic center, but opposite configurations about the other stereogenic center.

In **Step [2]**, the diastereomers are separated by some physical technique, such as crystallization or distillation.

In **Step [3]**, the amides can be hydrolyzed with aqueous base to regenerate the amino acids. The amino acids are now separated from each other. The optical activity of the amino acids can be measured and compared to their known rotations to determine the purity of each enantiomer.

PROBLEM 28.9   Which of the following amines can be used to resolve a racemic mixture of amino acids?

a. $C_6H_5CH_2CH_2NH_2$

b.

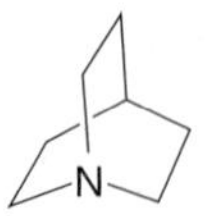

c.

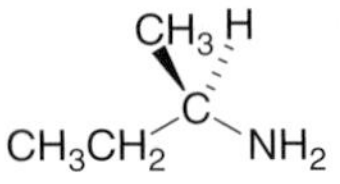

d.

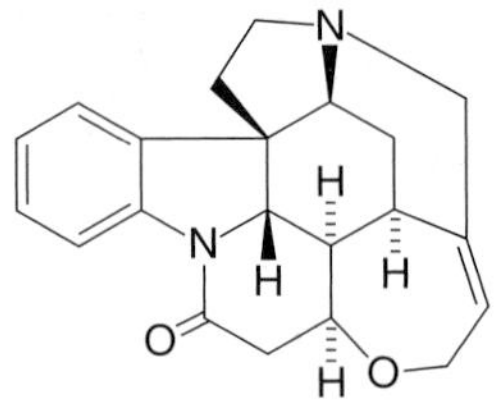

strychnine
(a powerful poison)

PROBLEM 28.10   Write out a stepwise sequence that shows how a racemic mixture of leucine enantiomers can be resolved into optically active amino acids using (R)-α-methylbenzylamine.

28.3B   Kinetic Resolution of Amino Acids Using Enzymes

A second strategy used to separate amino acids is based on the fact that two enantiomers react differently with chiral reagents. An **enzyme** is typically used as the chiral reagent.

To illustrate this strategy, we begin again with the two enantiomers of *N*-acetyl alanine, which were prepared by treating a racemic mixture of (*R*)- and (*S*)-alanine with acetic anhydride (Section 28.3A). A group of enzymes called **acylases** hydrolyze amide bonds, such as those found in *N*-acetyl alanine, but only for amides of L-amino acids. Thus, when a racemic mixture of *N*-acetyl alanines is treated with an acylase, only the amide of L-alanine (the *S* stereoisomer) is hydrolyzed to generate L-alanine, whereas the amide of D-alanine (the *R* stereoisomer) is untouched. The reaction mixture now consists of one amino acid and one *N*-acetyl amino acid. Because they have different functional groups with different physical properties, they can be physically separated.

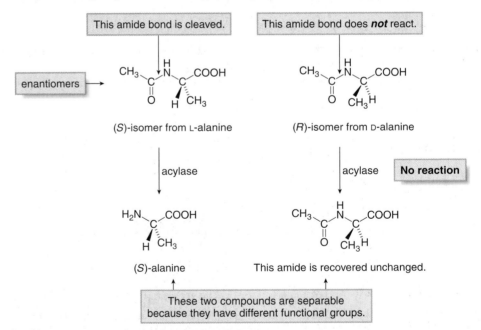

♦ **Separation of two enantiomers by a chemical reaction that selectively occurs for only one of the enantiomers is called *kinetic resolution*.**

PROBLEM 28.11   Draw the organic products formed in the following reaction.

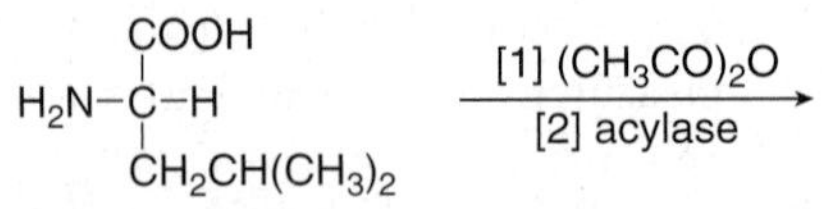

$$
\begin{array}{c}
\text{COOH} \\
| \\
H_2N-\overset{\displaystyle |}{C}-H \\
| \\
CH_2CH(CH_3)_2
\end{array}
\quad
\xrightarrow[\text{[2] acylase}]{\text{[1] }(CH_3CO)_2O}
$$

(mixture of enantiomers)

28.4 Enantioselective Synthesis of Amino Acids

Although the two methods introduced in Section 28.3 for resolving racemic mixtures of amino acids make enantiomerically pure amino acids available for further research, half of the reaction product is useless because it has the undesired configuration. Moreover, each of these procedures is costly and time-consuming.

If we use a chiral reagent to synthesize an amino acid, however, it is possible to favor the formation of the desired enantiomer over the other, without having to resort to a resolution. For example, single enantiomers of amino acids have been prepared by using **enantioselective (or asymmetric) hydrogenation reactions.** The success of this approach depends on finding a chiral catalyst, in much the same way that a chiral catalyst is used for the Sharpless asymmetric epoxidation (Section 12.14).

The necessary starting material is an alkene. Addition of H_2 to the double bond forms an *N*-acetyl amino acid with a new stereogenic center on the α carbon to the carboxy group. With proper choice of a chiral catalyst, the naturally occurring *S* configuration can be obtained as product.

Several chiral catalysts with complex structures have now been developed for this purpose. Many contain **rhodium** as the metal, complexed to a chiral molecule containing one or more phosphorus atoms. One example, abbreviated simply as **Rh***, is drawn below.

This catalyst is synthesized from a rhodium salt and a phosphorus compound, 2,2'-bis(diphenylphosphino)-1,1'-binaphthyl (**BINAP**). It is the BINAP moiety (Figure 28.6) that makes the catalyst chiral.

BINAP is one of a small number of molecules that is chiral even though it has no tetrahedral stereogenic centers. Its shape makes it a chiral molecule. The two naphthalene rings of the

Ryoji Noyori shared the 2001 Nobel Prize in Chemistry for developing methods for asymmetric hydrogenation reactions using the chiral BINAP catalyst.

Figure 28.6 The structure of BINAP

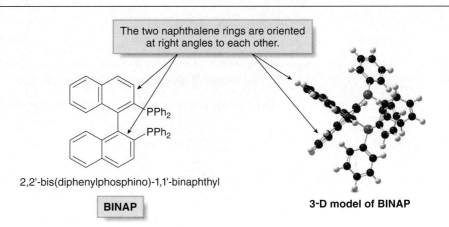

The two naphthalene rings are oriented at right angles to each other.

2,2'-bis(diphenylphosphino)-1,1'-binaphthyl

BINAP

3-D model of BINAP

Twistoflex and helicene (Section 17.5) are two more aromatic compounds whose shape makes them chiral.

BINAP molecule are oriented at almost 90° to each other to minimize steric interactions between the hydrogen atoms on adjacent rings. This rigid three-dimensional shape makes it nonsuperimposable on its mirror image, and thus it is a chiral compound.

The following graphic shows how enantioselective hydrogenation can be used to synthesize a single stereoisomer of phenylalanine. Treating alkene **A** with H_2 and the chiral rhodium catalyst Rh* forms the *S* isomer of *N*-acetyl phenylalanine in 100% *ee*. Hydrolysis of the acetyl group on nitrogen then yields a single enantiomer of phenylalanine.

Example

S enantiomer
100% *ee*

hydrolysis

(*S*)-phenylalanine

enantioselective hydrogenation

PROBLEM 28.12   What alkene is needed to synthesize each amino acid by an enantioselective hydrogenation reaction using H_2 and Rh*: (a) alanine; (b) leucine; (c) glutamine?

28.5   Peptides

When amino acids are joined together by amide bonds, they form larger molecules called **peptides** and **proteins.**

◆ A *dipeptide* has two amino acids joined together by *one* amide bond.
◆ A *tripeptide* has three amino acids joined together by *two* amide bonds.

Dipeptide   **Tripeptide**

Two amino acids joined together.   Three amino acids joined together.

[Amide bonds are drawn in red.]

Polypeptides and **proteins** both have many amino acids joined together in long linear chains, but the term **protein** is usually reserved for polymers of more than 40 amino acids.

◆ The amide bonds in peptides and proteins are called *peptide bonds.*
◆ The individual amino acids are called *amino acid residues.*

28.5A   Simple Peptides

To form a dipeptide, the amino group of one amino acid forms an amide bond with the carboxy group of another amino acid. Because each amino acid has both an amino group and a carboxy group, **two different dipeptides can be formed.** This is illustrated with alanine and cysteine.

[1]   **The COOH group of alanine can combine with the NH$_2$ group of cysteine.**

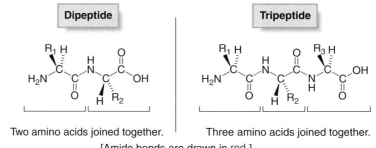

alanine   cysteine   Ala | Cys

Ala–Cys

peptide bond

[2] The COOH group of cysteine can combine with the NH₂ group of alanine.

cysteine   alanine   Cys | Ala   Cys–Ala

peptide bond

These compounds are **constitutional isomers** of each other. Both have a free amino group at one end of their chains and a free carboxy group at the other.

> ◆ The amino acid with the free amino group is called the *N-terminal amino acid.*
> ◆ The amino acid with the free carboxy group is called the *C-terminal amino acid.*

By convention, **the N-terminal amino acid is always written at the left end of the chain and the C-terminal amino acid at the right.** The peptide can be abbreviated by writing the one- or three-letter symbols for the amino acids in the chain from the N-terminal to the C-terminal end. Thus, Ala–Cys has alanine at the N-terminal end and cysteine at the C-terminal end, whereas Cys–Ala has cysteine at the N-terminal end and alanine at the C-terminal end. Sample Problem 28.1 shows how this convention applies to a tripeptide.

SAMPLE PROBLEM 28.1

Draw the structure of the following tripeptide, and label its N-terminal and C-terminal amino acids: Ala–Gly–Ser.

SOLUTION

Draw the structures of the amino acids in order from **left to right,** placing the COOH of one amino acid *next* to the NH₂ group of the adjacent amino acid. Always draw the **NH₂ group on the *left*** and the **COOH group on the *right.*** Then, join adjacent COOH and NH₂ groups together in amide bonds to form the tripeptide.

Make amide bonds here.

Ala   Gly   Ser   N-terminal amino acid   C-terminal amino acid

tripeptide **Ala–Gly–Ser**
[The new peptide bonds are drawn in red.]

The N-terminal amino acid is **alanine,** and the C-terminal amino acid is **serine.**

The tripeptide in Sample Problem 28.1 has one N-terminal amino acid, one C-terminal amino acid, and two peptide bonds.

> ◆ No matter how many amino acid residues are present, there is only *one* N-terminal amino acid and *one* C-terminal amino acid.
> ◆ For *n* amino acids in the chain, the number of amide bonds is *n* − 1.

PROBLEM 28.13

Draw the structure of each peptide. Label the N-terminal and C-terminal amino acids and all amide bonds.

a. Val–Glu   b. Gly–His–Leu   c. Phe–Ile–Tyr–Ile

PROBLEM 28.14   Name each peptide using the three letter abbreviations for the names of the component amino acids.

a.

b.

PROBLEM 28.15   How many different tripeptides can be formed from three different amino acids?

28.5B   The Peptide Bond

The carbonyl carbon of an amide is *sp²* **hybridized** and has **trigonal planar** geometry. A second resonance structure can be drawn that delocalizes the nonbonded electron pair on the N atom. Amides are more resonance stabilized than other acyl compounds, so the resonance structure having the C=N makes a significant contribution to the hybrid.

two resonance structures for the peptide bond

Resonance stabilization has important consequences. Rotation about the C–N bond is restricted because it has partial double bond character. As a result, there are two possible conformations.

> **Two conformations of the peptide bond**

s-trans ⇌ *s*-cis

> Recall from Section 16.6 that 1,3-butadiene can also exist as *s*-cis and *s*-trans conformers. In 1,3-butadiene, the *s*-cis conformer has the two double bonds on the same side of the single bond (dihedral angle = 0°), whereas the *s*-trans conformer has them on opposite sides (dihedral angle = 180°).

- The *s*-trans conformation has the two R groups oriented on *opposite* sides of the C–N bond.
- The *s*-cis conformation has the two R groups oriented on the *same* side of the C–N bond.
- The *s*-trans conformation of a peptide bond is typically more stable than the *s*-cis, because the *s*-trans has the two bulky R groups located further from each other.

A second consequence of resonance stabilization is that **all six atoms involved in the peptide bond lie in the same plane.** All bond angles are ~120° and the C=O and N–H bonds are oriented 180° from each other.

> The planar geometry of the peptide bond is analogous to the planar geometry of ethylene (or any other alkene), where the double bond between *sp²* hybridized carbon atoms makes all of the bond angles ~120° and puts all six atoms in the same plane.

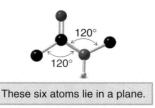

120°
120°

> These six atoms lie in a plane.

The structure of a tetrapeptide illustrates the results of these effects in a long peptide chain.

◆ The *s*-trans arrangement makes a long chain with a zigzag arrangement.
◆ In each peptide bond, the N–H and C=O bonds lie parallel and at 180° with respect to each other.

A tetrapeptide

PROBLEM 28.16   Draw the *s*-cis and *s*-trans conformers for the dipeptide formed from two glycine molecules.

28.5C   Interesting Peptides

Even relatively simple peptides can have important biological functions. **Bradykinin,** for example, is a peptide hormone composed of nine amino acids. It stimulates smooth muscle contraction, dilates blood vessels, and causes pain. Bradykinin is a component of bee venom.

Arg–Pro–Pro–Gly–Phe–Ser–Pro–Phe–Arg
bradykinin

Oxytocin and **vasopressin** are nonapeptide hormones, too. Their sequences are identical except for two amino acids, yet this is enough to give them very different biological activities. Oxytocin induces labor by stimulating the contraction of uterine muscles, and it stimulates the flow of milk in nursing mothers. Vasopressin, on the other hand, controls blood pressure by regulating smooth muscle contraction. The N-terminal amino acid in both hormones is a cysteine residue, and the C-terminal residue is glycine. Instead of a free carboxy group, both peptides have an NH_2 group in place of OH, so this is indicated with the additional NH_2 group drawn at the end of the chain.

oxytocin   vasopressin

The structure of both peptides includes a **disulfide bond,** a form of covalent bonding in which the –SH groups from two cysteine residues are oxidized to form a sulfur–sulfur bond. In oxytocin and vasopressin, the disulfide bonds make the peptides cyclic. Three-dimensional structures of oxytocin and vasopressin are shown in Figure 28.7.

$$2 \quad R-S-H \quad \xrightarrow{[O]} \quad R-S-S-R$$

thiol   disulfide bond

Aspartame's sweetness was discovered in 1965 when a chemist working with it in the lab licked his dirty fingers.

The artificial sweetener **aspartame** is the methyl ester of the dipeptide Asp–Phe. This synthetic peptide is 180 times sweeter (on a gram-for-gram basis) than sucrose (common table sugar). Both of the amino acids in aspartame have the naturally occurring L-configuration. If the D-amino acid is substituted for either Asp or Phe, the resulting compound tastes bitter.

Figure 28.7 Three-dimensional structures of oxytocin and vasopressin

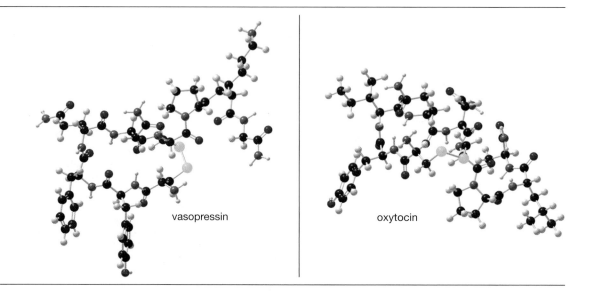

vasopressin

oxytocin

aspartame
the methyl ester of Asp−Phe
a synthetic artificial sweetener

PROBLEM 28.17   Draw the structure of leu–enkephalin, a pentapeptide that acts as an analgesic and opiate, and has the following sequence: Tyr–Gly–Gly–Phe–Leu. (The structure of a related peptide, met-enkephalin, appeared in Section 22.6B.)

PROBLEM 28.18   Glutathione, a powerful antioxidant that destroys harmful oxidizing agents in cells, is composed of glutamic acid, cysteine, and glycine, and has the following structure:

glutathione

a. What product is formed when glutathione reacts with an oxidizing agent?
b. What is unusual about the peptide bond between glutamic acid and cysteine?

28.6 Peptide Synthesis

The synthesis of a specific dipeptide, such as Ala–Gly from alanine and glycine, is complicated because both amino acids have two functional groups. As a result, four products—namely, Ala–Ala, Ala–Gly, Gly–Gly, and Gly–Ala—are possible.

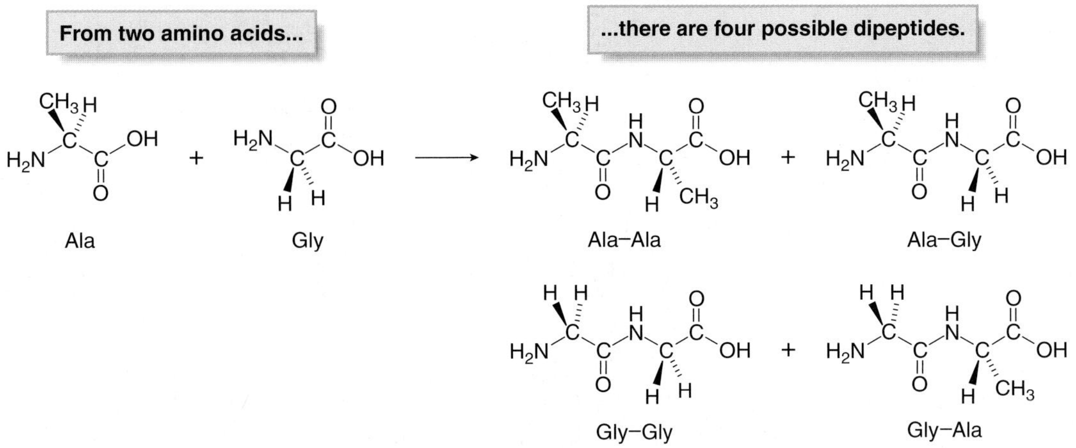

From two amino acids...

...there are four possible dipeptides.

Ala + Gly → Ala–Ala + Ala–Gly + Gly–Gly + Gly–Ala

How do we selectively join the COOH group of alanine with the NH$_2$ group of glycine?

◆ Protect the functional groups that we don't want to react, and then form the amide bond.

How To Synthesize a Dipeptide from Two Amino Acids

Example

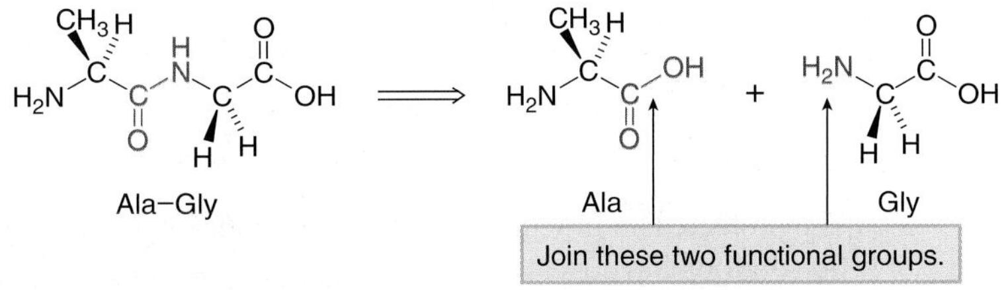

Ala–Gly ⟹ Ala + Gly

Join these two functional groups.

Step [1] **Protect the NH$_2$ group of alanine.**

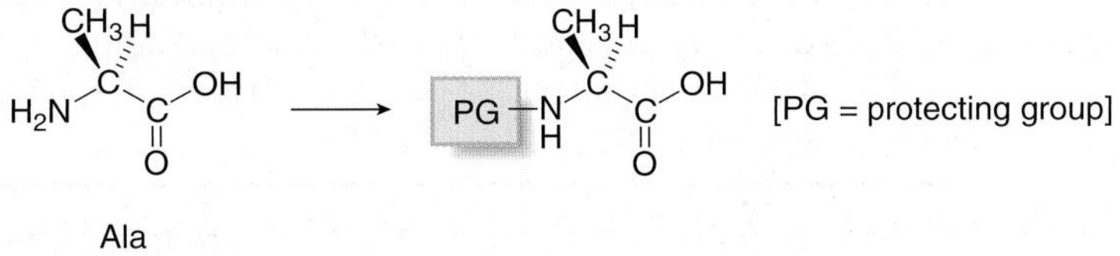

Ala → [PG = protecting group]

Step [2] **Protect the COOH group of glycine.**

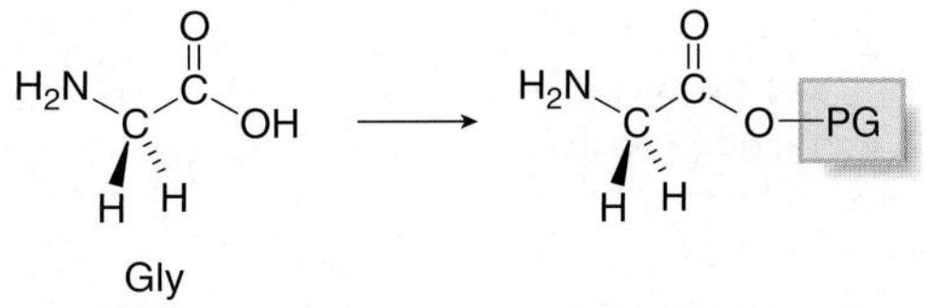

Gly

Step [3] **Form the amide bond with DCC.**

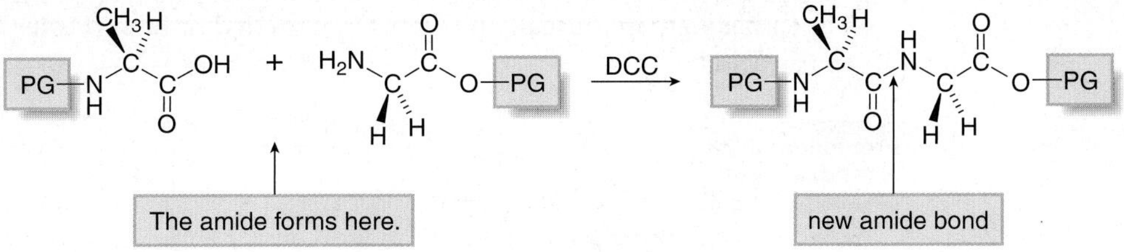

The amide forms here.   new amide bond

Dicyclohexylcarbodiimide (**DCC**) is a reagent commonly used to form amide bonds (see Section 22.10D). DCC makes the OH group of the carboxylic acid a better leaving group, thus **activating the carboxy group toward nucleophilic attack.**

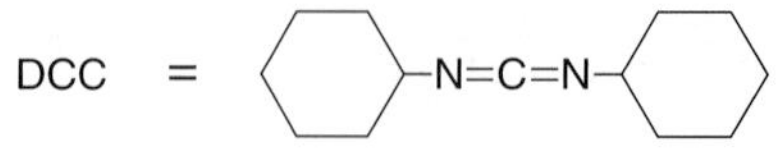

DCC =

dicyclohexylcarbodiimide

Step [4] **Remove one or both protecting groups.**

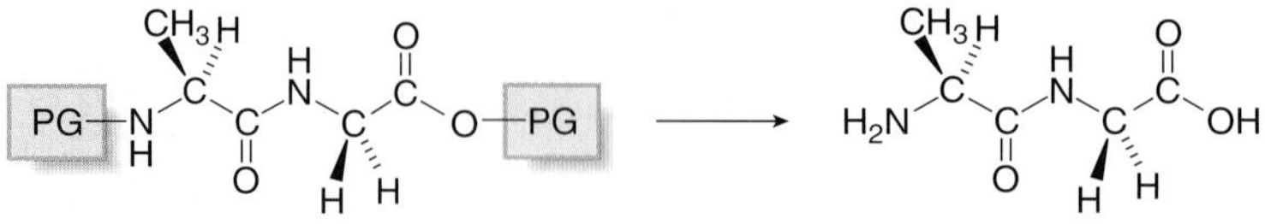

Ala–Gly

A widely used amino protecting group is called the ***tert*-butoxycarbonyl protecting group,** abbreviated as **BOC.** It is formed by reacting the amino acid with di-*tert*-butyl dicarbonate in a nucleophilic acyl substitution reaction.

Protection of an amino group

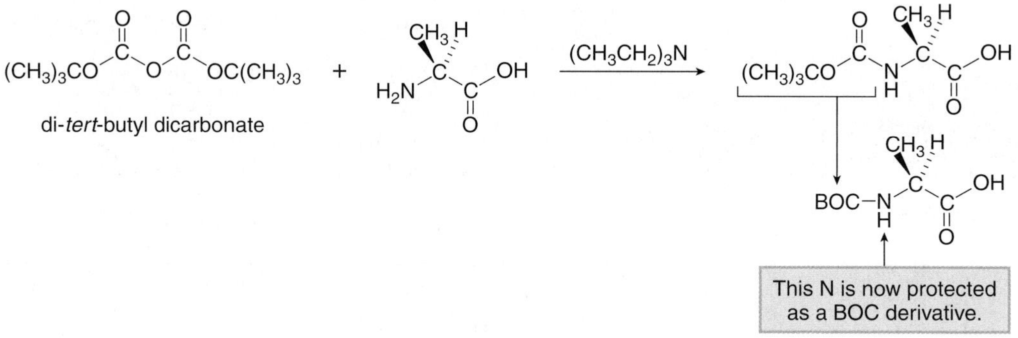

di-*tert*-butyl dicarbonate

This N is now protected as a BOC derivative.

The protected amino group is no longer a nucleophilic amine. It is part of a **carbamate,** a functional group having a carbonyl group bonded to both an oxygen and a nitrogen atom.

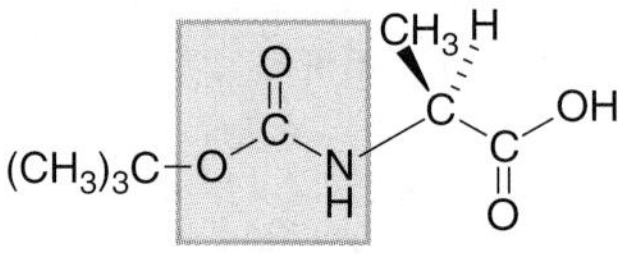

carbamate

To be a useful protecting group, the BOC group must be removable under reaction conditions that do not affect other functional groups in the molecule. It can be removed with an acid such as **trifluoroacetic acid, HCl,** or **HBr.**

Removal of the BOC protecting group

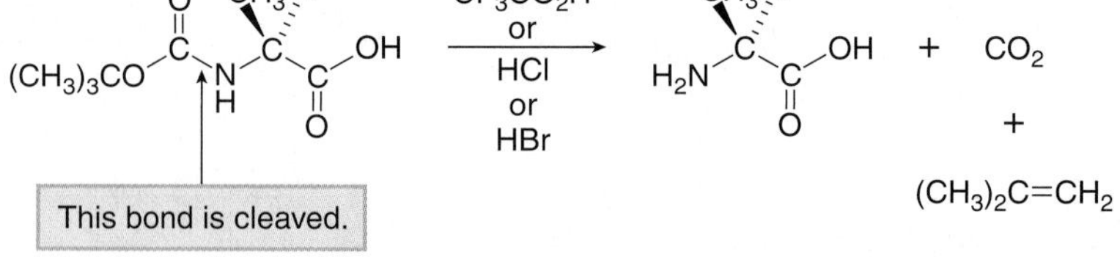

This bond is cleaved.

The carboxy group is usually protected as a **methyl** or **benzyl ester** by reaction with an alcohol and an acid.

Protection of the carboxy group

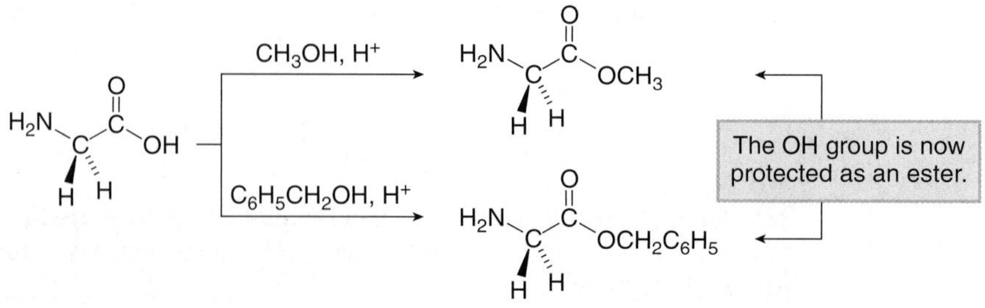

The OH group is now protected as an ester.

These esters are usually removed by hydrolysis with aqueous base.

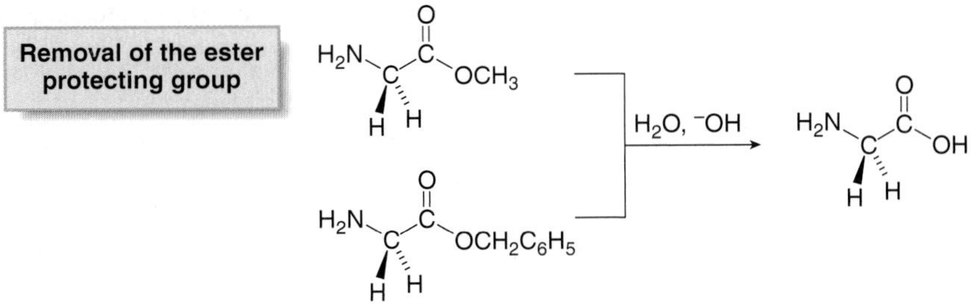

One advantage of using a benzyl ester for protection is that it can also be removed with H_2 in the presence of a Pd catalyst. This process is called **hydrogenolysis.** These conditions are especially mild, because they avoid the use of either acid or base. Benzyl esters can also be removed with HBr in acetic acid.

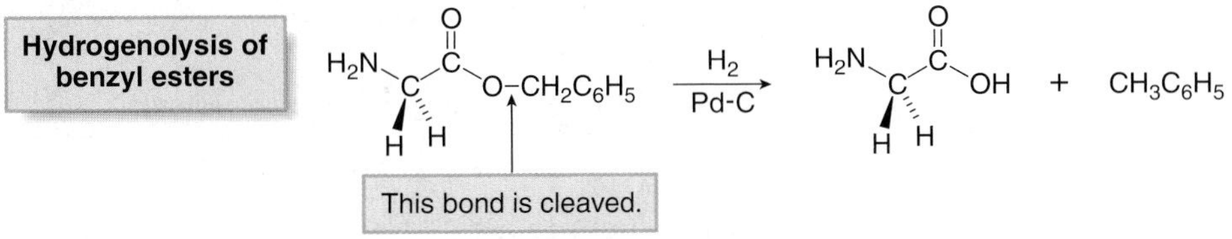

The specific reactions needed to synthesize the dipeptide Ala–Gly are illustrated in Sample Problem 28.2.

SAMPLE PROBLEM 28.2 Draw out the steps in the synthesis of the dipeptide Ala–Gly.

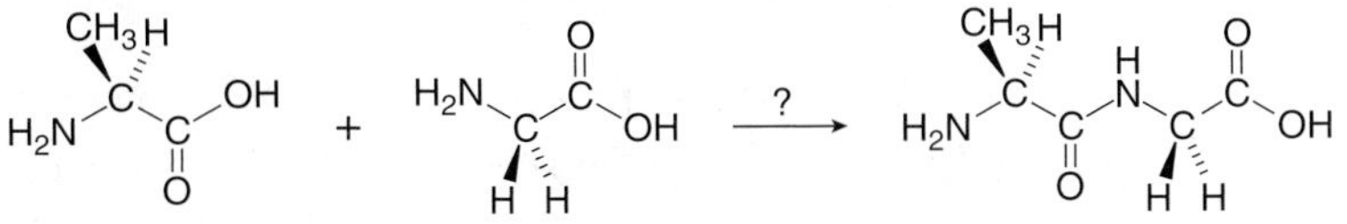

Ala Gly Ala–Gly

SOLUTION

Step [1] Protect the NH_2 group of alanine using a BOC group.

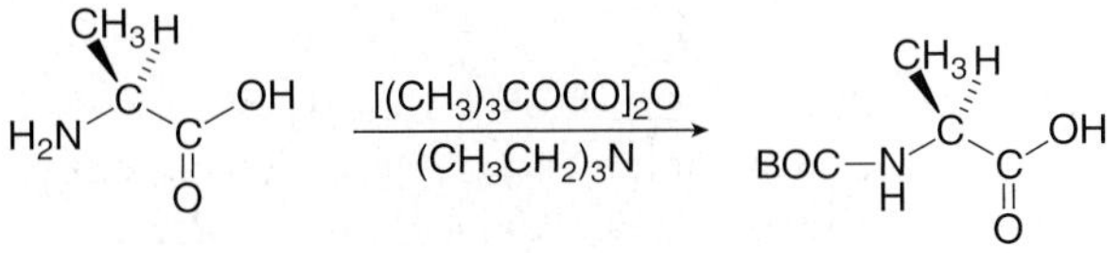

Ala

Step [2] Protect the COOH group of glycine as a benzyl ester.

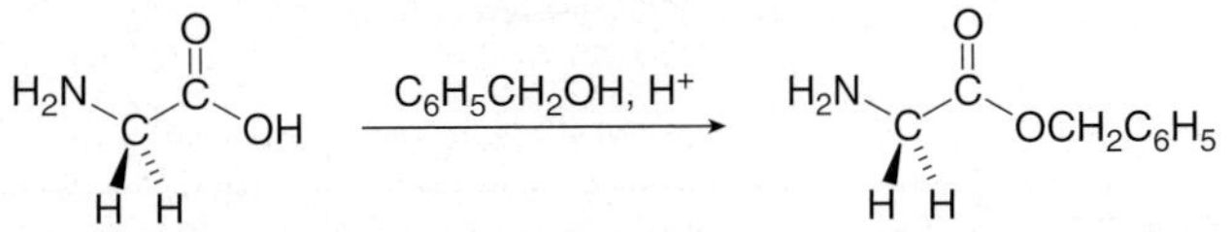

Gly

Step [3] Form the amide bond with DCC.

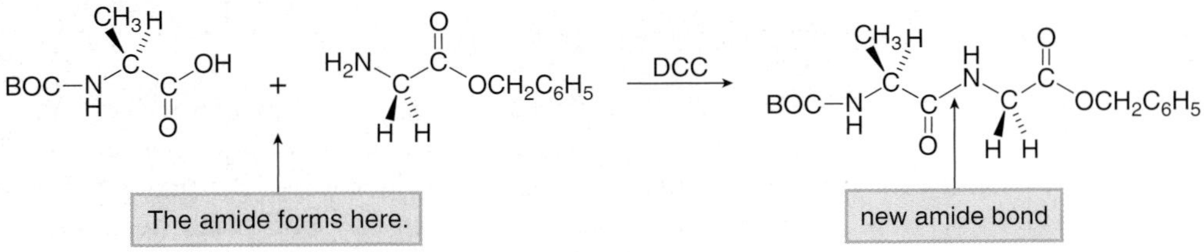

Step [4]   Remove one or both protecting groups.

The protecting groups can be removed in a stepwise fashion, or in a single reaction.

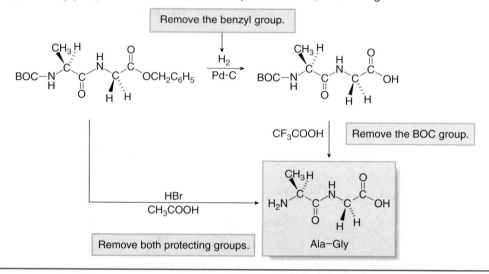

This method can be applied to the synthesis of tripeptides and even larger polypeptides. After the protected dipeptide is prepared in Step [3], only one of the protecting groups is removed, and this dipeptide is coupled to a third amino acid with one of its functional groups protected, as illustrated in the following equations.

PROBLEM 28.19   Devise a synthesis of each peptide from amino acid starting materials: (a) Leu–Val; (b) Ala–Ile–Gly; (c) Ala–Gly–Ala–Gly.

28.7   Automated Peptide Synthesis

The method described in Section 28.6 works well for the synthesis of small peptides. It is extremely time-consuming to synthesize larger proteins by this strategy, however, because each step requires isolation and purification of the product. The synthesis of larger polypeptides is usually accomplished by using the **solid phase technique** originally developed by R. Bruce Merrifield of Rockefeller University.

Development of the solid phase technique earned Merrifield the 1984 Nobel Prize in Chemistry and has made possible the synthesis of many polypeptides and proteins.

In the Merrifield method an amino acid is attached to an **insoluble polymer.** Amino acids are sequentially added, one at a time, thereby forming successive peptide bonds. Because impurities and by-products are not attached to the polymer chain, they are removed simply by washing them away with a solvent at each stage of the synthesis.

The polymer typically used is a **polystyrene derivative** that contains $-CH_2Cl$ groups bonded to some of the benzene rings in the polymer chain. The Cl atoms serve as handles that allow attachment of amino acids to the chain.

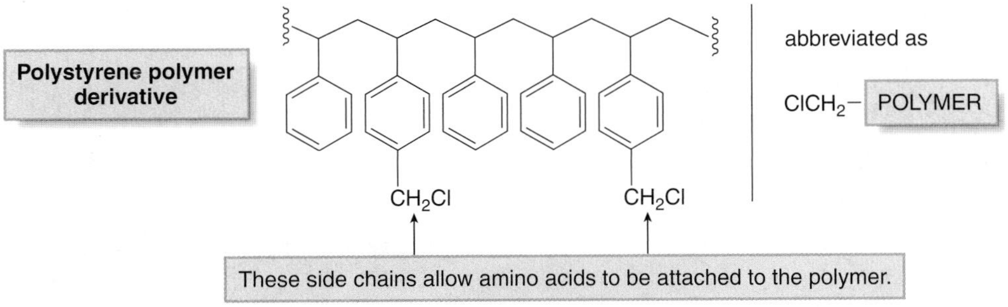

Polystyrene polymer derivative

abbreviated as

$ClCH_2-$ POLYMER

These side chains allow amino acids to be attached to the polymer.

A BOC-protected amino acid can easily be attached to the polymer at its carboxy group by an S_N2 reaction.

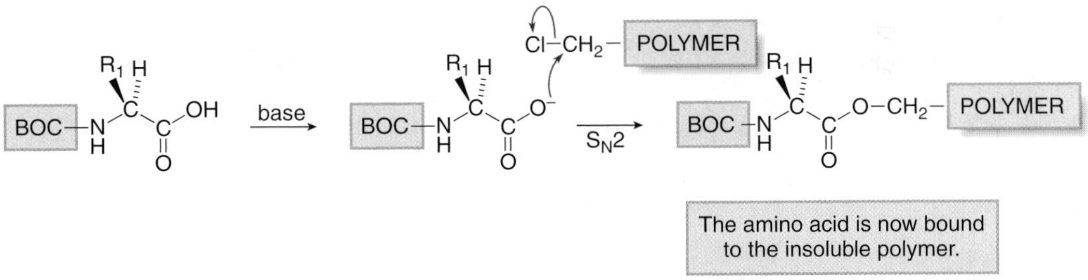

The amino acid is now bound to the insoluble polymer.

Once the first amino acid is bound to the polymer, additional amino acids can be added sequentially. The steps of the solid phase peptide synthesis technique are illustrated in the accompanying scheme. In the last step, HF not only cleaves the polypeptide chain from the polymer, but also removes the BOC-protecting group.

Steps in the Merrifield Solid Phase Technique for Peptide Synthesis

Step [1]

Attach a BOC-protected amino acid to the polymer.

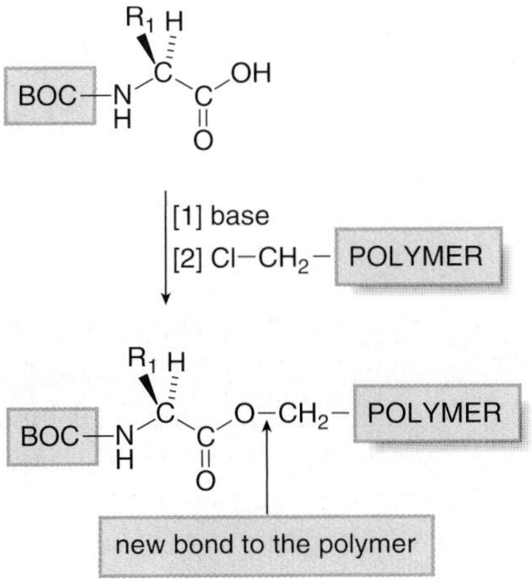

[1] base
[2] Cl$-CH_2-$ POLYMER

new bond to the polymer

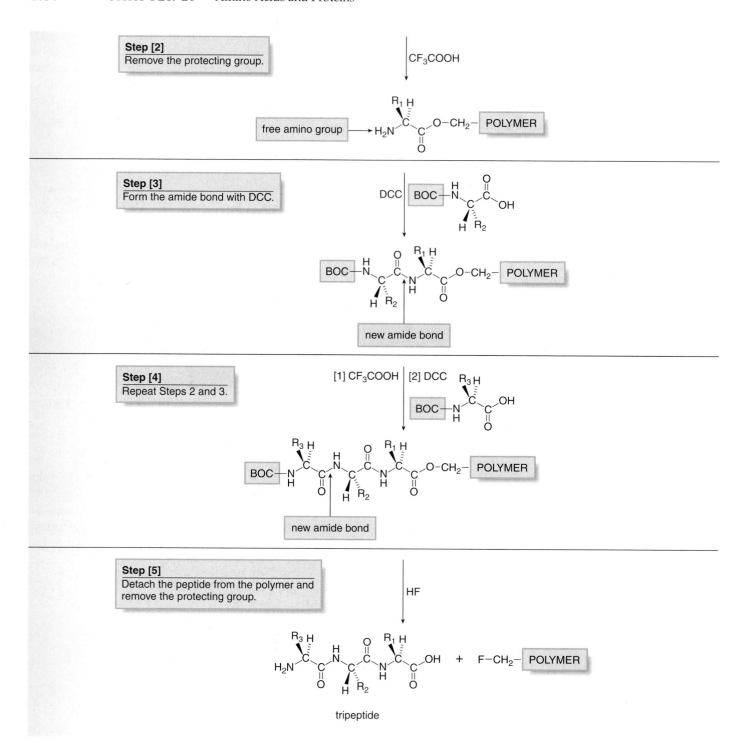

Step [2]
Remove the protecting group.
CF3COOH
free amino group
POLYMER
Step [3]
Form the amide bond with DCC.
DCC
BOC
new amide bond
Step [4]
Repeat Steps 2 and 3.
[1] CF3COOH [2] DCC
new amide bond
Step [5]
Detach the peptide from the polymer and remove the protecting group.
HF
tripeptide

The Merrifield method has now been completely automated, so it is possible to purchase peptide synthesizers that automatically carry out all of the above operations and form polypeptides in high yield in a matter of hours, days, or weeks, depending on the length of the chain of the desired product. The instrument is pictured in Figure 28.8. For example, the protein ribonuclease, which contains 128 amino acids, has been prepared by this technique in an overall yield of 17%. This remarkable synthesis involved 369 separate reactions, and thus the yield of each individual reaction was > 99%.

PROBLEM 28.20 Outline the steps needed to synthesize the tetrapeptide Ala–Leu–Ile–Gly using the Merrifield technique.

28.8 Protein Structure

Now that you have learned some of the chemistry of amino acids, it's time to study proteins, the large polymers of amino acids that are responsible for so much of the structure and function of all living cells. We begin with a discussion of the **primary, secondary, tertiary, and quaternary structure** of proteins.

28.8A Primary Structure

The *primary structure* of proteins is the particular sequence of amino acids that is joined together by peptide bonds. The most important element of this primary structure is the **amide bond.**

♦ Rotation around the amide C−N bond is *restricted* because of electron delocalization, and the *s*-trans conformer is the more stable arrangement.
♦ In each peptide bond, the N−H and C=O bonds are directed 180° from each other.

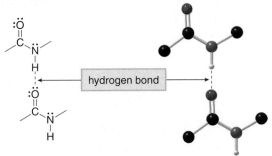

two amide bonds in a peptide chain

restricted rotation

Although rotation about the amide bonds is restricted, **rotation about the other σ bonds in the protein backbone is not.** As a result, the peptide chain can twist and bend into a variety of different arrangements that constitute the secondary structure of the protein.

28.8B Secondary Structure

The three-dimensional conformations of localized regions of a protein are called its secondary structure. These regions arise due to hydrogen bonding between the N−H proton of one amide and C=O oxygen of another. Two arrangements that are particularly stable are called the **α-helix** and the **β-pleated sheet.**

hydrogen bond

Figure 28.8 Automated peptide synthesizer

α-Helix

The **α-helix** forms when a peptide chain twists into a right-handed or clockwise spiral, as shown in Figure 28.9. Four important features of the α-helix are as follows:

[1] **Each turn of the helix has 3.6 amino acids.**
[2] **The N−H and C=O bonds point along the axis of the helix.** All C=O bonds point in one direction, and all N−H bonds point in the opposite direction.
[3] **The C=O group of one amino acid is hydrogen bonded to an N−H group four amino acid residues further along the chain.** Thus, hydrogen bonding occurs between two amino acids *in the same chain.* Note, too, that the hydrogen bonds are parallel to the axis of the helix.
[4] **The R groups of the amino acids extend outward** from the core of the helix.

Figure 28.9 Two different illustrations of the α-helix

[a] The right-handed α-helix

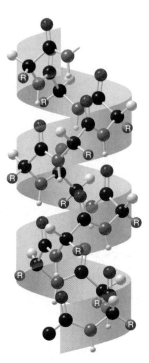

All atoms of the α-helix are drawn in this representation. All C=O bonds are pointing up and all N−H bonds are pointing down.

[b] The backbone of the α-helix

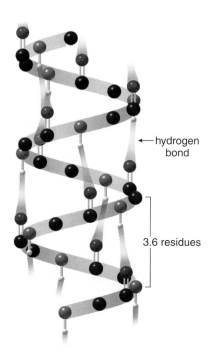

←hydrogen bond

3.6 residues

Only the peptide backbone is drawn in this representation. The hydrogen bonds between the C=O and N−H of amino acids four residues away from each other are shown.

An α-helix can only form if there is rotation about the bonds at the α carbon of the amide carbonyl group, and not all amino acids can do this. For example, proline, the amino acid whose nitrogen atom forms part of a five-membered ring, is more rigid than other amino acids, and its C_α−N bond cannot rotate the necessary amount. Additionally, it has no N−H proton with which to form an intramolecular hydrogen bond to stabilize the helix. Thus, proline cannot be part of an α-helix.

Both the myosin in muscle and α-keratin in hair are proteins composed almost entirely of α-helices.

β-Pleated Sheet

The **β-pleated sheet** secondary structure forms when two or more peptide chains, called **strands,** line up side-by-side, as shown in Figure 28.10. All β-pleated sheets have the following characteristics:

[1] **The C=O and N−H bonds lie in the plane of the sheet.**
[2] **Hydrogen bonding often occurs between the N−H and C=O groups of nearby amino acid residues.**
[3] The **R groups are oriented above and below the plane** of the sheet, and alternate from one side to the other along a given strand.

The β-pleated sheet arrangement most commonly occurs with amino acids with small R groups, like alanine and glycine. With larger R groups steric interactions prevent the chains from getting close together and so the sheet cannot be stabilized by hydrogen bonding.

The peptide strands of β-pleated sheets can actually be oriented in two different ways, as shown in Figure 28.11.

Figure 28.10 Three-dimensional structure of the β-pleated sheet

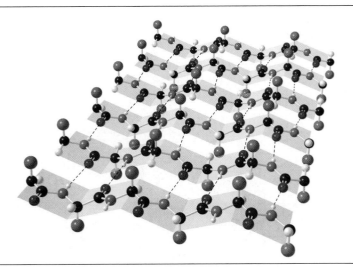

- The β-pleated sheet consists of extended strands of the peptide chains held together by hydrogen bonding. The C=O and N−H bonds lie in the plane of the sheet, and the R groups (shown as orange balls) alternate above and below the plane.

Figure 28.11 The parallel and antiparallel forms of the β-pleated sheet

Parallel β-pleated sheet	**Antiparallel β-pleated sheet**
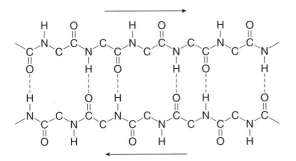	

The two peptide chains are arranged in the same direction. Hydrogen bonds occur between N−H and C=O bonds in adjacent chains.

[Note: R groups on the carbon chain are omitted for clarity.]

The two peptide chains are arranged in opposite directions. Hydrogen bonding between the N−H and C=O groups still holds the two chains together.

◆ In a *parallel* β-pleated sheet, the strands run in the *same* direction from the N- to C-terminal amino acid.
◆ In an *antiparallel* β-pleated sheet, the strands run in the *opposite* direction.

Most proteins have regions of α-helix and β-pleated sheet, in addition to other regions that cannot be characterized by either of these arrangements. Shorthand symbols are often used to indicate regions of a protein that have α-helix or β-pleated sheet. A **flat helical ribbon** is used for the α-helix, and a **flat wide arrow** is used for the β-pleated sheet.

α-helix shorthand   β-pleated sheet shorthand

Spider dragline silk is a strong yet elastic protein because it has regions of β-pleated sheet and regions of α-helix (Figure 28.12). α-Helical regions impart elasticity to the silk because the peptide chain is twisted (not fully extended), so it can stretch. β-Pleated sheet regions are almost fully extended, so they can't be stretched further, but their highly ordered three-dimensional structure imparts strength to the silk. Thus, spider silk suits the spider by comprising both types of secondary structure with beneficial properties.

Figure 28.12 Different regions of secondary structure in spider silk

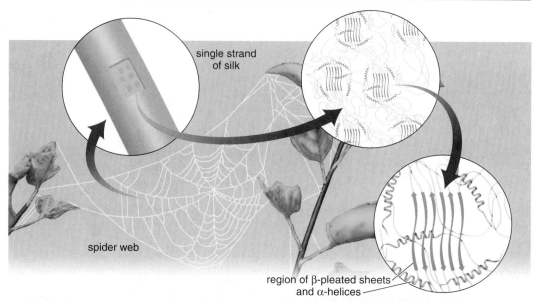

single strand of silk

spider web

region of β-pleated sheets and α-helices

Spider silk has regions of α-helix and β-pleated sheet that make it both strong and elastic. The green coils represent the α-helical regions, and the purple arrows represent the β-pleated sheet regions. The broad yellow line represents other areas of the protein that are neither α-helix nor β-pleated sheet.

PROBLEM 28.21   Suggest a reason why antiparallel β-pleated sheets are generally more stable than parallel β-pleated sheets.

PROBLEM 28.22   Consider two molecules of a tetrapeptide composed only of alanine residues. Draw the hydrogen bonding interactions that result when these two peptides adopt a parallel β-pleated sheet arrangement. Answer this same question for the antiparallel β-pleated sheet arrangement.

28.8C   Tertiary and Quaternary Structure

The three-dimensional shape adopted by the entire peptide chain is called its tertiary structure. A peptide generally folds into a conformation that maximizes its stability. In the aqueous environment of the cell, proteins often fold in such a way as to maximize the number of polar and charged groups on their outer surface, to maximize the dipole–dipole and hydrogen bonding interactions with water. This generally places most of the nonpolar side chains in the interior of the protein, where van der Waals interactions between these hydrophobic groups help stabilize the molecule, too.

In addition, polar functional groups hydrogen bond with each other (not just water), and amino acids with charged side chains like $-COO^-$ and $-NH_3^+$ can stabilize tertiary structure by electrostatic interactions.

Finally, **disulfide bonds are the only covalent bonds that stabilize tertiary structure.** As previously mentioned, these strong bonds form by oxidation of two cysteine residues either on the same polypeptide chain or another polypeptide chain of the same protein.

Disulfide bonds can form in two different ways.

Between two SH groups on the same chain. | Between two SH groups on different chains.

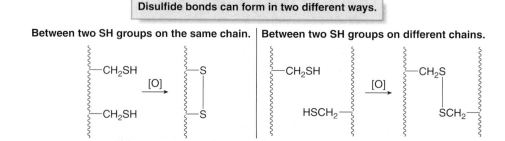

Figure 28.13 Insulin

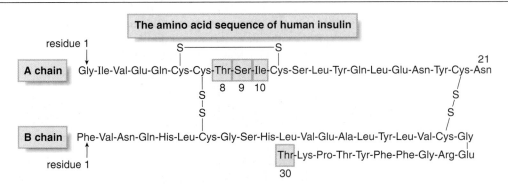

| The amino acid sequence of human insulin |

Insulin is a small protein consisting of two polypeptide chains (designated as the **A** and **B** chains) held together by two disulfide bonds. An additional disulfide bond joins two cysteine residues within the **A** chain.

Synthesized by groups of cells in the pancreas called the islets of Langerhans, insulin is the protein that regulates the levels of glucose in the blood. Insufficiency of insulin results in diabetes. Many of the abnormalities associated with this disease can be controlled by the injection of insulin. Until the recent availability of human insulin through genetic engineering techniques, all insulin used by diabetics was obtained from pigs and cattle. The amino acid sequences of these insulin proteins is slightly different from that of human insulin. Pig insulin differs in one amino acid only, whereas bovine insulin has three different amino acids. This is shown in the accompanying table.

	Chain A			Chain B
Position of residue →	8	9	10	30
Human insulin	Thr	Ser	Ile	Thr
Pig insulin	Thr	Ser	Ile	Ala
Bovine insulin	Ala	Ser	Val	Ala

Figure 28.14 The stabilizing interactions in secondary and tertiary protein structure

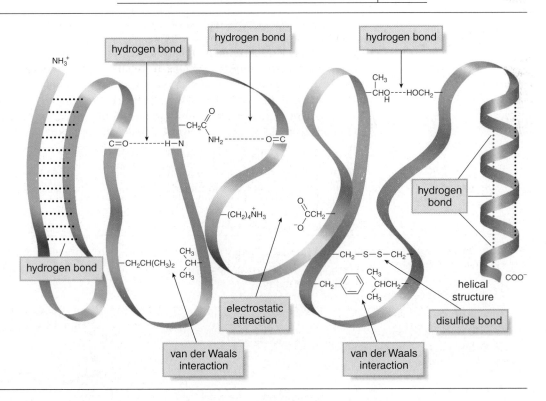

The nonapeptides **oxytocin** and **vasopressin** (Section 28.5C) contain intramolecular disulfide bonds. **Insulin,** on the other hand, consists of two separate polypeptide chains (**A** and **B**) that are covalently linked by two intermolecular disulfide bonds, as shown in Figure 28.13. The **A** chain, which also has an intramolecular disulfide bond, has 21 amino acid residues, whereas the **B** chain has 30.

Figure 28.14 schematically illustrates the many different kinds of intramolecular forces that stabilize the secondary and tertiary structures of polypeptide chains.

The shape adopted when two or more folded polypeptide chains aggregate into one protein complex is called the **quaternary structure** of the protein. Each individual polypeptide chain is called a **subunit** of the overall protein. **Hemoglobin,** for example, consists of two α and two β subunits held together by intermolecular forces in a compact three-dimensional shape. The unique function of hemoglobin is possible only when all four subunits are together.

PROBLEM 28.23   What types of stabilizing interactions exist between each of the following pairs of amino acids?
a. Ser and Tyr   b. Val and Leu   c. Two Phe residues

PROBLEM 28.24   The fibroin proteins found in silk fibers consist of large regions of β-pleated sheets stacked one on top of another. The polypeptide sequence in these regions has glycines at every other residue. Explain how this allows the β-pleated sheets to stack on top of each other.

28.9   Important Proteins

Proteins are generally classified according to their three-dimensional shapes.

◆ **Fibrous proteins** are composed of long linear polypeptide chains that are bundled together to form rods or sheets. These proteins are insoluble in water and serve structural roles, giving strength and protection to tissues and cells.

◆ **Globular proteins** are coiled into compact shapes with hydrophilic outer surfaces that make them water soluble. Enzymes and transport proteins are globular to make them soluble in the blood and other aqueous environments in cells.

28.9A   α-Keratins

α-Keratins are the proteins found in hair, hooves, nails, skin, and wool. They are composed almost exclusively of long sections of α-helix units, having large numbers of alanine and leucine residues. Because these nonpolar amino acids extend outward from the α-helix, these proteins are very water insoluble. Two α-keratin helices coil around each other, forming a structure called **a supercoil** or **superhelix.** These, in turn, form larger and larger bundles of fibers, ultimately forming a strand of hair, as shown schematically in Figure 28.15.

α-Keratins also have a number of cysteine residues, and because of this, disulfide bonds are formed between adjacent helices. The number of disulfide bridges determines the strength of the material. Claws, horns, and fingernails have extensive networks of disulfide bonds, making them extremely hard.

Figure 28.15 Anatomy of a hair—It begins with α-keratin.

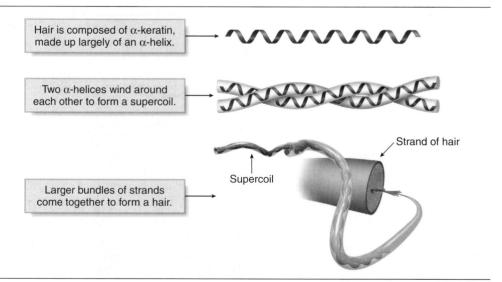

Hair is composed of α-keratin, made up largely of an α-helix.

Two α-helices wind around each other to form a supercoil.

Larger bundles of strands come together to form a hair.

Supercoil

Strand of hair

Figure 28.16 The chemistry of a "permanent"—Making straight hair curly

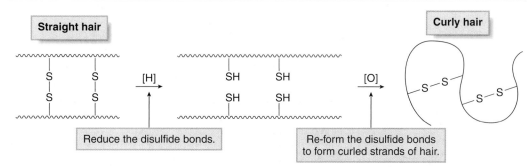

To make straight hair curly, the disulfide bonds holding the α-helical chains together are cleaved by reduction. This forms free thiol groups (–SH). The hair is turned around curlers and then an oxidizing agent is applied. This re-forms the disulfide bonds in the hair, but between different thiol groups, now giving it a curly appearance.

Straight hair can be made curly by cleaving the disulfide bonds in α-keratin, then rearranging and re-forming them, as shown schematically in Figure 28.16. First, the disulfide bonds in the straight hair are reduced to thiol groups, so the bundles of α-keratin chains are no longer held in their specific "straight" orientation. Then, the hair is wrapped around curlers and treated with an oxidizing agent that converts the thiol groups back to disulfide bonds, now with twists and turns in the keratin backbone. This makes the hair look curly and is the chemical basis for a "permanent."

28.9B Collagen

Collagen, the most abundant protein in vertebrates, is found in connective tissues such as bone, cartilage, tendons, teeth, and blood vessels. Glycine and proline account for a large fraction of its amino acid residues, whereas cysteine accounts for very little. Because of the high proline content, it cannot form a right-handed α-helix. Instead, it forms an elongated left-handed helix, and then three of these helices wind around each other to form a right-handed **superhelix** or **triple helix.** The side chain of glycine is only a hydrogen atom, so the high glycine content allows the collagen superhelices to lie compactly next to each other, thus stabilizing the superhelices via hydrogen bonding. Two views of the collagen superhelix are shown in Figure 28.17.

Figure 28.17 Two different representations for the triple helix of collagen

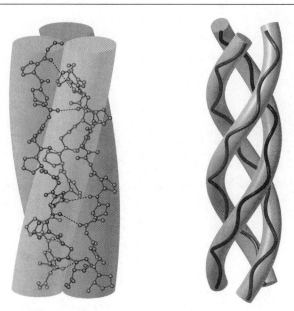

- In collagen, three polypeptide chains having an unusual left-handed helix wind around each other in a right-handed triple helix. The high content of small glycine residues allows the chains to lie close to each other, permitting hydrogen bonding between the chains.

28.9C   Hemoglobin and Myoglobin

Hemoglobin and **myoglobin,** two globular proteins, are called **conjugated proteins** because they are composed of a protein unit and a nonprotein molecule called a **prosthetic group.** The prosthetic group in hemoglobin and myoglobin is **heme,** a complex organic compound containing the Fe^{2+} ion complexed with a nitrogen heterocycle called a **porphyrin.** The Fe^{2+} ion of hemoglobin and myoglobin binds oxygen in the blood. Hemoglobin, which is present in red blood cells, transports oxygen to wherever it is needed in the body, whereas myoglobin stores oxygen in tissues.

heme

Myoglobin has 153 amino acid residues in a single polypeptide chain. It has eight separate α-helical sections that fold back on one another, with the prosthetic heme group held in a cavity inside the polypeptide. Most of the polar residues are found on the outside of the protein so that they can interact with the water solvent. Spaces in the interior of the protein are filled with nonpolar amino acids. Myoglobin gives cardiac muscle its characteristic red color. Whales have a particularly high myoglobin concentration in their muscles. It serves as an oxygen reservoir for the whale while it is submerged for long periods.

Hemoglobin consists of four polypeptide chains (two α subunits and two β subunits), each of which carries a heme unit. Hemoglobin has more nonpolar amino acids than myoglobin. When each subunit is folded, some of these remain on the surface. The van der Waals attraction between these hydrophobic groups is what stabilizes the quaternary structure of the four subunits.

Carbon monoxide is poisonous because it binds to the Fe^{2+} of hemoglobin more strongly than does oxygen. Hemoglobin complexed with CO cannot carry O_2 from the lungs to the tissues. Without O_2 in the tissues for metabolism, cells cannot function, so they die.

The properties of all proteins depend on their three-dimensional shape, and their shape depends on their primary structure—that is, their amino acid sequence. This is particularly well exemplified by comparing normal hemoglobin with **sickle cell hemoglobin,** a mutant variation in which a single amino acid of both β subunits is changed from glutamic acid to valine. The replacement of one acidic amino acid (Glu) with one nonpolar amino acid (Val) changes the shape of hemoglobin, which has profound effects on its function. Deoxygenated red blood cells with sickle cell hemoglobin become elongated and crescent shaped, and they are unusually fragile. As a result, they do not flow easily through capillaries, causing pain and inflammation, and they break open easily, leading to severe anemia and organ damage. The end result is often a painful and premature death.

This disease, called **sickle cell anemia,** is found almost exclusively among people originating from central and western Africa, where malaria is an enormous health problem. Sickle cell hemoglobin results from a genetic mutation in the DNA sequence that is responsible for the synthesis of hemoglobin. Individuals who inherit this mutation from both parents develop sickle cell anemia, whereas those who inherit it from only one parent are said to have the sickle cell trait. They do not develop sickle cell anemia and they are more resistant to malaria than individuals without the mutation. This apparently accounts for this detrimental gene being passed on from generation to generation.

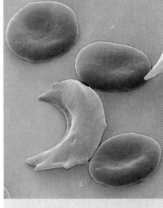

When red blood cells take on a "sickled" shape in persons with sickle cell disease, they occlude capillaries (causing organ injury) and they break easily (leading to profound anemia). This devastating illness results from the change of a single amino acid in hemoglobin. Note the single sickled cell surrounded by three red cells with normal morphology.

28.10   Key Concepts—Amino Acids and Proteins

Synthesis of Amino Acids (28.2)

[1] From α-halo carboxylic acids by S_N2 reaction

$$R-\underset{\underset{Br}{|}}{C}HCOOH \xrightarrow[\substack{\text{(large excess)} \\ S_N2}]{NH_3} R-\underset{\underset{NH_2}{|}}{C}HCOO^-NH_4^+ + NH_4^+Br^-$$

[2] By alkylation of diethyl acetamidomalonate

$$\underset{CH_3}{\overset{O}{\underset{||}{C}}}-\underset{H}{\overset{H}{\underset{|}{N}}}-\underset{COOEt}{\overset{H}{\underset{|}{C}}}-COOEt \xrightarrow[\substack{[2]\ RX \\ [3]\ H_3O^+,\ \Delta}]{[1]\ NaOEt} H_2N-\underset{\underset{H}{|}}{\overset{R}{\overset{|}{C}}}-COOH$$

- Alkylation occurs with unhindered alkyl halides—that is, CH_3X and RCH_2X.

[3] Strecker synthesis

$$\underset{R}{\overset{O}{\underset{}{C}}}\underset{H}{\overset{||}{}} \xrightarrow[NaCN]{NH_4Cl} R-\underset{\underset{H}{|}}{\overset{NH_2}{\overset{|}{C}}}-CN \xrightarrow{H_3O^+} R-\underset{\underset{H}{|}}{\overset{NH_2}{\overset{|}{C}}}-COOH$$

α-amino nitrile

Preparation of Optically Active Amino Acids

[1] Resolution of enantiomers by forming diastereomers (28.3A)
- Convert a racemic mixture of amino acids into a racemic mixture of *N*-acetyl amino acids [(*S*)- and (*R*)-CH₃CONHCH(R)COOH].
- Treat the enantiomers with a chiral amine to form a mixture of diastereomers.
- Separate the diastereomers.
- Regenerate the amino acids by protonation of the carboxylate salt and hydrolysis of the *N*-acetyl group.

[2] Kinetic resolution using enzymes (28.3B)

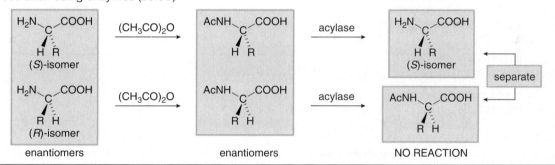

[3] By enantioselective hydrogenation (28.4)

$$\underset{H}{\overset{R}{\underset{}{}}}\underset{}{\overset{NHAc}{C=C}}\underset{COOH}{} \xrightarrow[Rh^*]{H_2} AcNH\underset{\overset{|}{C}}{\overset{COOH}{}}\underset{H\ \ CH_2R}{} \xrightarrow{H_2O,\ ^-OH} H_2N\underset{\overset{|}{C}}{\overset{COOH}{}}\underset{H\ \ CH_2R}{}$$

S enantiomer   *S* amino acid

Rh* = chiral Rh hydrogenation catalyst

Adding and Removing Protecting Groups for Amino Acids (28.6)

[1] Protection of an amino group as a BOC derivative

$$H_2N\underset{}{\overset{R\ \ H}{\overset{|}{C}}}\underset{CO_2H}{} \xrightarrow[(CH_3CH_2)_3N]{[(CH_3)_3COCO]_2O} BOC-\underset{H}{\overset{}{N}}\underset{}{\overset{R\ \ H}{\overset{|}{C}}}\underset{CO_2H}{}$$

[2] Deprotection of a BOC-protected amino acid

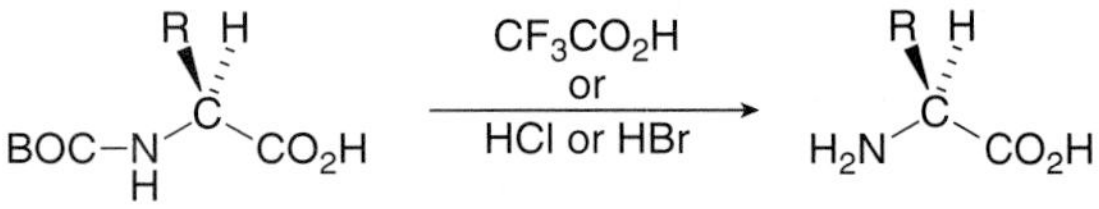

[3] Protection of a carboxy group as an ester

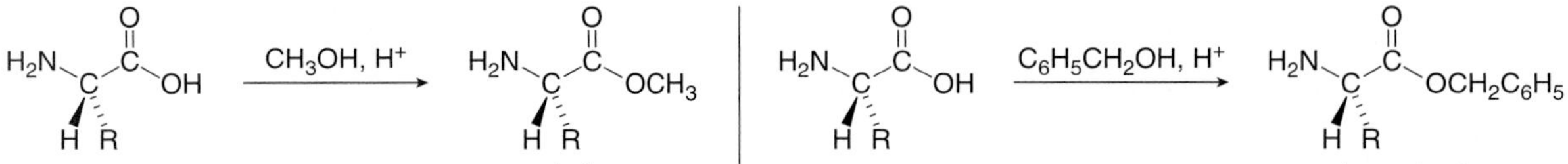

methyl ester   benzyl ester

[4] Deprotection of an ester group

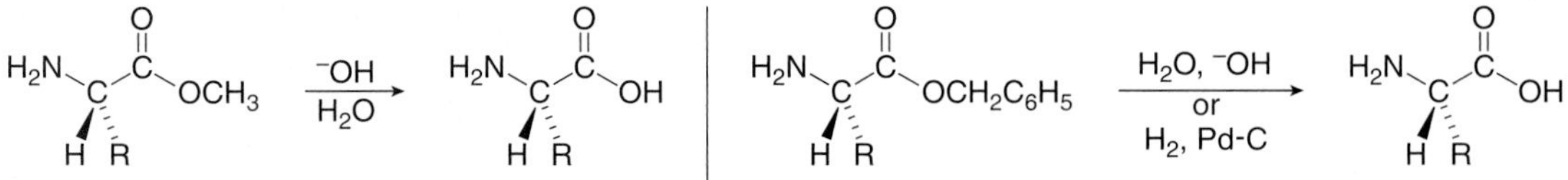

methyl ester   benzyl ester

Synthesis of Dipeptides (28.6)

[1] Amide formation with DCC

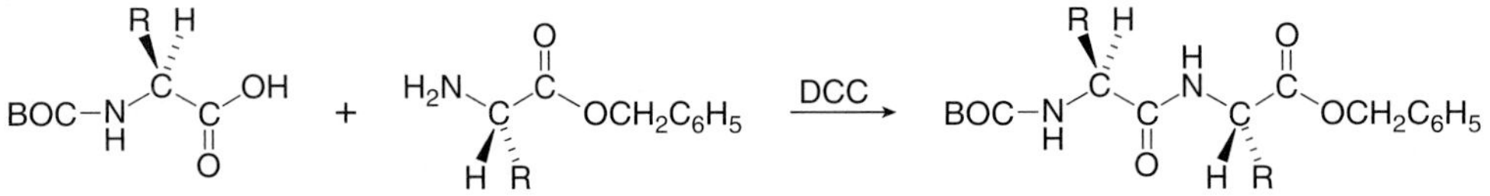

[2] Four steps are needed to synthesize a dipeptide:
  [a] **Protect** the amino group of one amino acid with a BOC group.
  [b] **Protect** the carboxy group of the second amino acid as an ester.
  [c] Form the amide bond with **DCC.**
  [d] **Remove both protecting groups** in one or two reactions.

Summary of the Merrifield Method of Peptide Synthesis (28.7)

[1] Attach a BOC-protected amino acid to a polymer derived from polystyrene.
[2] Remove the BOC protecting group.
[3] Form the amide bond with a second BOC-protected amino acid by using DCC.
[4] Repeat steps [2] and [3].
[5] Detach the peptide from the polymer and remove the protecting group.

Problems

Amino Acids

28.25 Explain why L-alanine has the S configuration but L-cysteine has the R configuration.

28.26

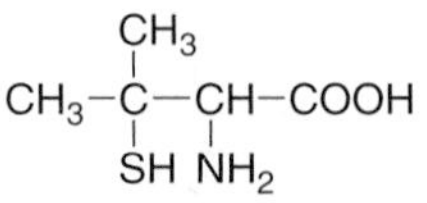

penicillamine

  a. (S)-Penicillamine, an amino acid that does not occur in proteins, is used as a copper chelating agent to treat Wilson's disease, an inherited defect in copper metabolism. (R)-Penicillamine is toxic, sometimes causing blindness. Draw the structures of (R)- and (S)-penicillamine. Which of these compounds is L-penicillamine?
  b. What disulfide is formed from oxidation of L-penicillamine?

28.27 Explain why amino acids are insoluble in diethyl ether but N-acetyl amino acids are soluble.

28.28 Histidine is classified as a basic amino acid because one of the N atoms in its five-membered ring is readily protonated by acid. Which N atom in histidine is protonated and why?

28.29 Tryptophan is not classified as a basic amino acid even though it has a heterocycle containing a nitrogen atom. Why is the N atom in the five-membered ring of tryptophan not readily protonated by acid?

28.30 What is the structure of each amino acid at its isoelectric point: (a) alanine; (b) methionine; (c) aspartic acid; (d) lysine?

28.31 To calculate the isoelectric point of amino acids having other ionizable functional groups, we must also take into account the pK_a of the additional functional group in the side chain.

For an acidic amino acid (one with an additional acidic OH group):

$$pI = \frac{pK_a\ (\alpha\text{-COOH}) + pK_a\ (\text{second COOH})}{2}$$

For a basic amino acid (one with an additional basic NH group):

$$pI = \frac{pK_a\ (\alpha\text{-NH}_3{}^+) + pK_a\ (\text{side chain NH})}{2}$$

a. Indicate which pK_a values must be used to calculate the pI of each of the following amino acids: [1] glutamic acid; [2] lysine; [3] arginine.
b. In general, how does the pI of an acidic amino acid compare to that of a neutral amino acid?
c. In general, how does the pI of a basic amino acid compare to the pI of a neutral amino acid?

28.32 What is the predominant form of each of the following amino acids at pH = 1? What is the overall charge on the amino acid at this pH? (a) threonine; (b) methionine; (c) aspartic acid; (d) arginine

28.33 What is the predominant form of each of the following amino acids at pH = 11? What is the overall charge on the amino acid? (a) valine; (b) proline; (c) glutamic acid; (d) lysine

Synthesis and Reactions of Amino Acids

28.34 Draw the organic product formed when the amino acid leucine is treated with each reagent.
a. CH_3OH, H^+
b. CH_3COCl, pyridine
c. $C_6H_5CH_2OH$, H^+
d. Ac_2O, pyridine
e. HCl (1 equiv)
f. NaOH (1 equiv)
g. C_6H_5COCl, pyridine
h. $[(CH_3)_3COCO]_2O$
i. The product in (d), then $NH_2CH_2COOCH_3$ + DCC
j. The product in (h), then $NH_2CH_2COOCH_3$ + DCC

28.35 Answer Problem 28.34 using phenylalanine as a starting material.

28.36 Draw the organic products formed in each reaction.

a. 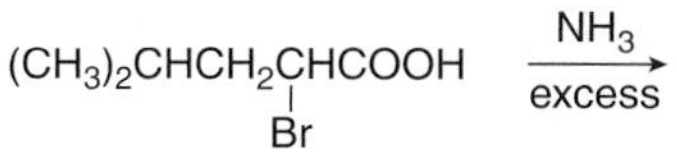 $(CH_3)_2CHCH_2CHCOOH$ (with Br on the carbon) $\xrightarrow[\text{excess}]{NH_3}$

b. 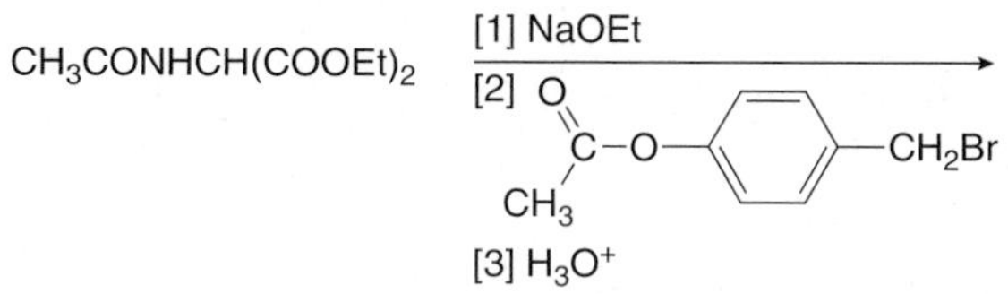 $CH_3CONHCH(COOEt)_2$ $\xrightarrow{[1]\ NaOEt}$ [2] (structure with $CH_3\overset{O}{C}-O-\text{(benzene ring)}-CH_2Br$) [3] H_3O^+

c. (structure: aldehyde chain $H-C(=O)-\cdots-N(H)-C(=O)-CH_3$) $\xrightarrow{[1]\ NH_4Cl,\ NaCN}$ [2] H_3O^+

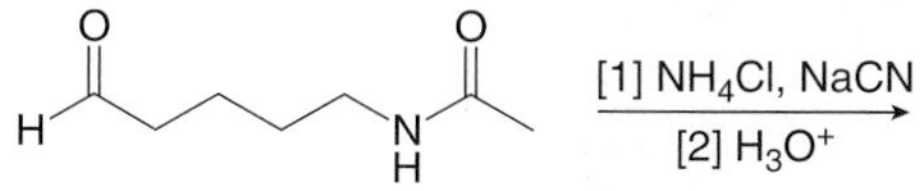

d. 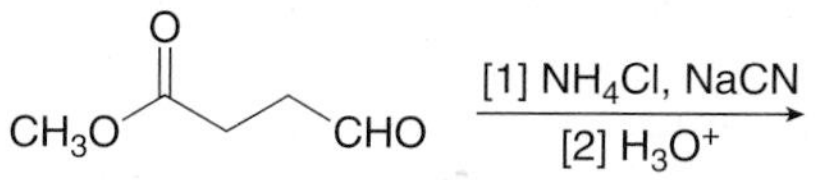 $CH_3O-C(=O)-CH_2CH_2-CHO$ $\xrightarrow{[1]\ NH_4Cl,\ NaCN}$ [2] H_3O^+

e. $CH_3CONHCH(COOEt)_2$ $\xrightarrow{[1]\ NaOEt}$ [2] $ClCH_2CH_2CH_2CH_2NHAc$ [3] H_3O^+

28.37 What alkyl halide is needed to synthesize each amino acid from diethyl acetamidomalonate: (a) Asn; (b) His; (c) Trp?

28.38 Devise a synthesis of threonine from diethyl acetamidomalonate.

28.39 Devise a synthesis of each amino acid from acetaldehyde (CH_3CHO): (a) glycine; (b) alanine.

28.40 Devise a synthesis of each amino acid from 3-methylbutanal [$(CH_3)_2CHCH_2CHO$]: (a) valine; (b) leucine.

28.41 Identify the lettered intermediates in the following reaction scheme. This is an alternative method to synthesize amino acids, based on the Gabriel synthesis of 1° amines (Section 25.7A).

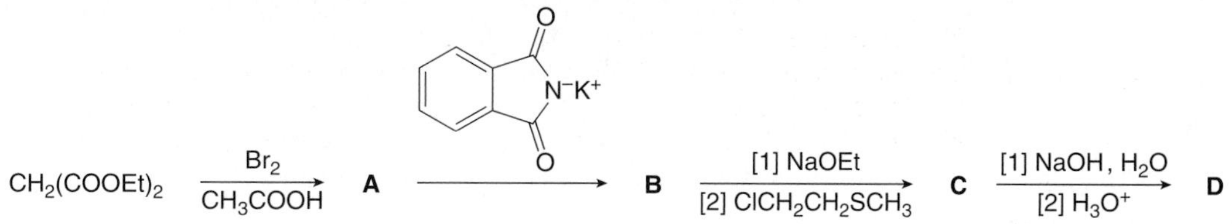

(phthalimide N^-K^+ structure)

$CH_2(COOEt)_2$ $\xrightarrow[CH_3COOH]{Br_2}$ **A** $\longrightarrow$ **B** $\xrightarrow[[2]\ ClCH_2CH_2SCH_3]{[1]\ NaOEt}$ **C** $\xrightarrow[[2]\ H_3O^+]{[1]\ NaOH,\ H_2O}$ **D**

28.42 Glutamic acid is synthesized by the following reaction sequence. Draw a stepwise mechanism for Steps [1]–[3].

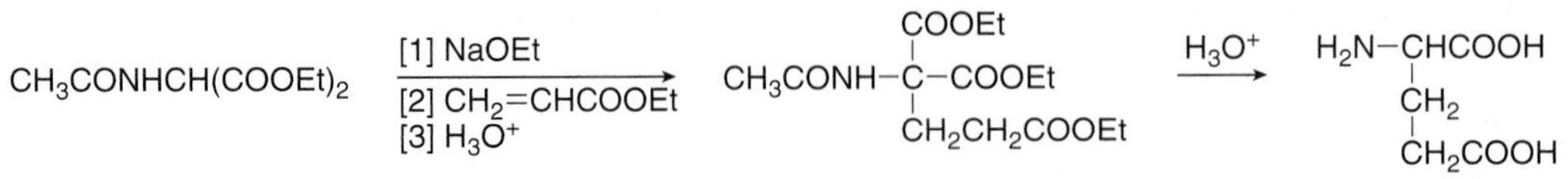

$$CH_3CONHCH(COOEt)_2 \xrightarrow[\substack{[2]\ CH_2=CHCOOEt \\ [3]\ H_3O^+}]{[1]\ NaOEt}\ CH_3CONH-\underset{\underset{CH_2CH_2COOEt}{|}}{\overset{\overset{COOEt}{|}}{C}}-COOEt \xrightarrow{H_3O^+} \underset{\underset{CH_2COOH}{|}}{\underset{\underset{CH_2}{|}}{H_2N-CHCOOH}}$$

glutamic acid

Resolution; The Synthesis of Chiral Amino Acids

28.43 Write out a scheme for the resolution of the two enantiomers of racemic lactic acid [$CH_3CH(OH)COOH$] using (*R*)-α-methylbenzylamine as resolving agent.

28.44 Another strategy used to resolve amino acids involves converting the carboxy group to an ester and then using a *chiral carboxylic acid* to carry out an acid–base reaction at the free amino group. The general plan is drawn below using (*R*)-mandelic acid as resolving agent. Using a racemic mixture of alanine enantiomers and (*R*)-mandelic acid as resolving agent, write out the steps showing how a resolution process would occur.

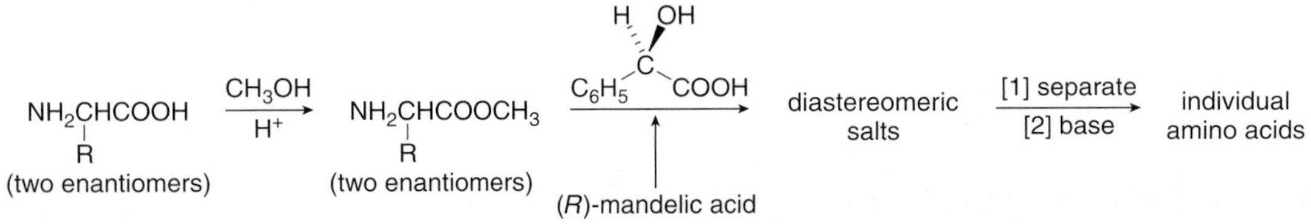

$$\underset{\substack{| \\ R \\ \text{(two enantiomers)}}}{NH_2CHCOOH} \xrightarrow[H^+]{CH_3OH} \underset{\substack{| \\ R \\ \text{(two enantiomers)}}}{NH_2CHCOOCH_3} \xrightarrow{\underset{\underset{\text{(R)-mandelic acid}}{\uparrow}}{}} \underset{\text{salts}}{\text{diastereomeric}} \xrightarrow[\text{[2] base}]{\text{[1] separate}} \underset{\text{amino acids}}{\text{individual}}$$

(*R*)-mandelic acid: C_6H_5 with H, OH, COOH

28.45 Brucine is a poisonous alkaloid obtained from *Strychnos nux vomica,* a tree that grows in India, Sri Lanka, and northern Australia. Write out a resolution scheme similar to the one given in Section 28.3A, which shows how a racemic mixture of phenylalanine can be resolved using brucine.

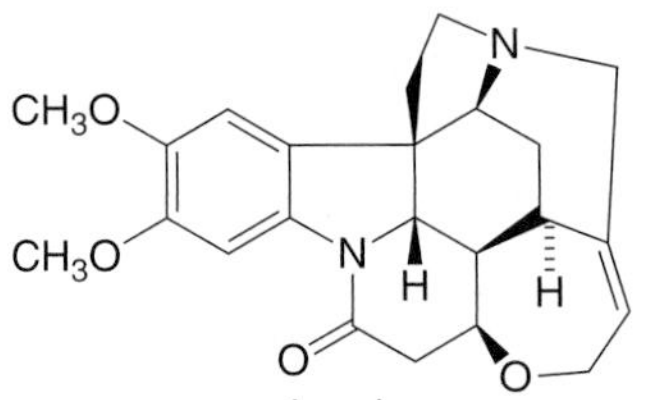

brucine

28.46 Draw the organic products formed in each reaction.

a. 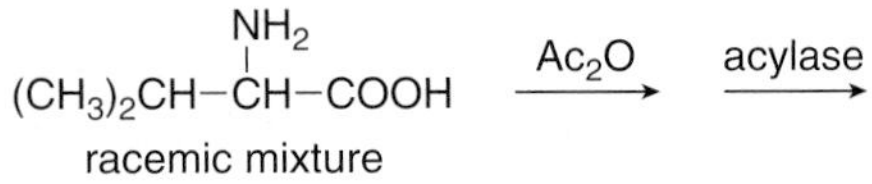 $(CH_3)_2CH-\underset{\underset{NH_2}{|}}{CH}-COOH$ (racemic mixture) $\xrightarrow{Ac_2O} \xrightarrow{acylase}$

b. 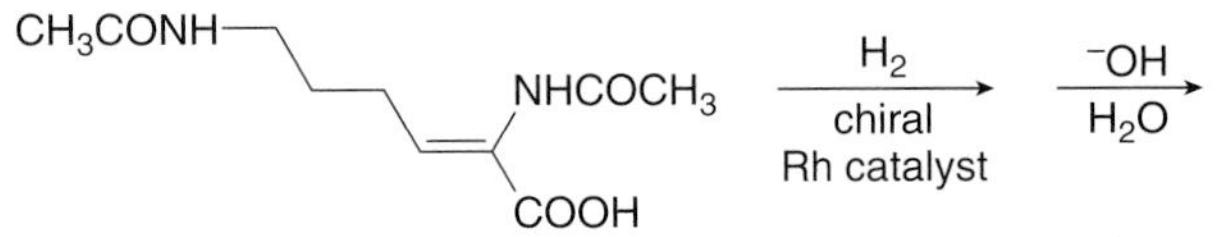 $\xrightarrow[\substack{\text{chiral} \\ \text{Rh catalyst}}]{H_2} \xrightarrow[H_2O]{^-OH}$

c. 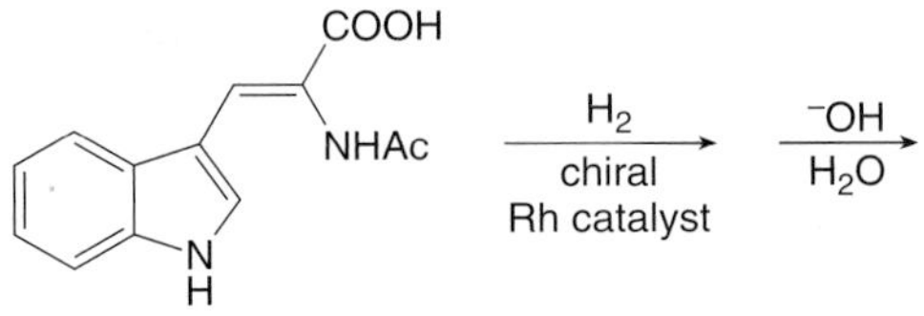 $\xrightarrow[\substack{\text{chiral} \\ \text{Rh catalyst}}]{H_2} \xrightarrow[H_2O]{^-OH}$

Peptide Structure

28.47 Draw the structure for each peptide: (a) Phe–Ala; (b) Gly–Gln; (c) Lys–Gly; (d) Arg–His.

28.48 For each tetrapeptide [1] Ala–Gln–Cys–Ser; [2] Asp–Arg–Val–Tyr:
   a. Name the peptide using one-letter abbreviations.
   b. Draw the structure.
   c. Label all amide bonds.
   d. Label the N-terminal and C-terminal amino acids.

28.49 Name each peptide using both the three-letter and one-letter abbreviations of the component amino acids.

a.

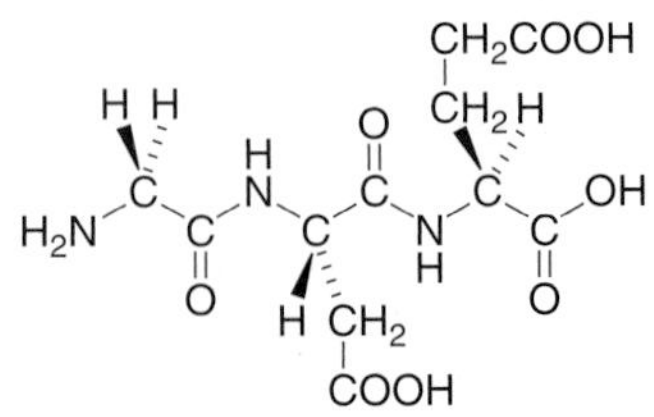

b.

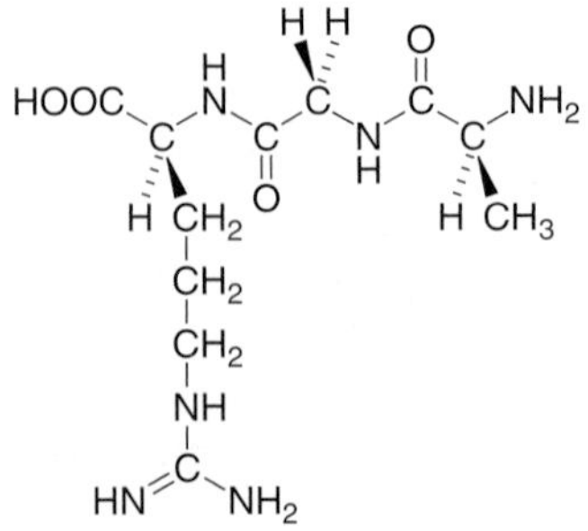

28.50 Explain why a peptide C−N bond is stronger than an ester C−O bond.

28.51 Draw the *s*-trans and *s*-cis conformers of the peptide bond in the dipeptide Ala–Ala.

Peptide Synthesis

28.52 Draw all the products formed in the following reaction.

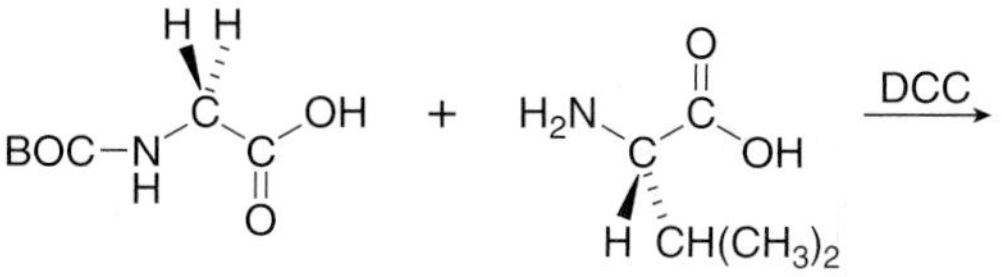

28.53 Draw the organic products formed in each reaction.

a. 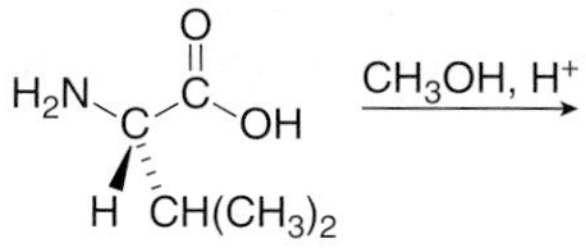 CH_3OH, H^+

b. 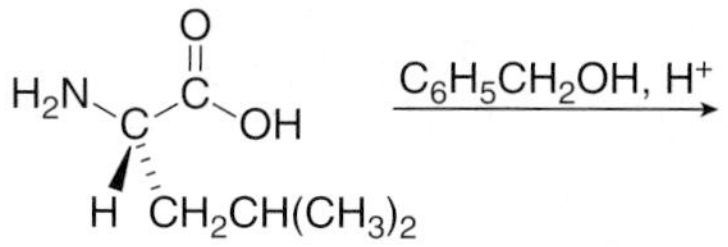 $C_6H_5CH_2OH, H^+$

c. NH_2CH_2COOH $\xrightarrow[\;(CH_3CH_2)_3N\;]{[(CH_3)_3COCO]_2O}$

d. product in (b) + product in (c) $\xrightarrow{DCC}$

e. 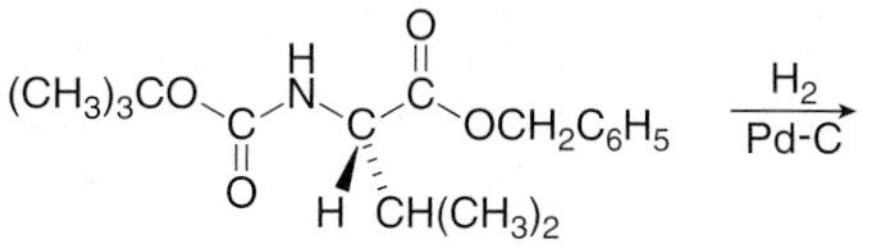 $\xrightarrow{H_2}{Pd-C}$

f. starting material in (e) $\xrightarrow[\;CH_3COOH\;]{HBr}$

g. product in (e) $\xrightarrow{CF_3COOH}$

28.54 BOC protecting groups convert an NH_2 group into a carbamate. Draw all reasonable resonance structures for a carbamate.

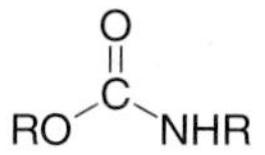

carbamate

28.55 Draw all the steps in the synthesis of each peptide from individual amino acids: (a) Gly–Ala; (b) Phe–Leu; (c) Ile–Ala–Phe.

28.56 Write out the steps for the synthesis of each peptide using the Merrifield method: (a) Ala–Leu–Phe–Phe; (b) Phe–Gly–Ala–Ile.

28.57 An amino acid [RCH(NH₂)COOH] can readily be converted to an *N*-acetyl amino acid [RCH(NHCOCH₃)COOH] using acetic anhydride. Why can't this acetyl group be used as an amino protecting group, in place of the BOC group, for peptide synthesis?

28.58 Another method to form a peptide bond involves a two-step process:
[1] Conversion of a BOC-protected amino acid to a *p*-nitrophenyl ester.
[2] Reaction of the *p*-nitrophenyl ester with an amino acid ester.

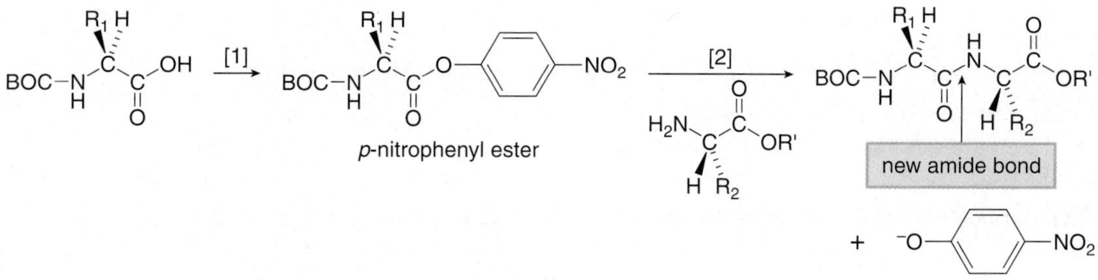

a. Why does a *p*-nitrophenyl ester "activate" the carboxy group of the first amino acid to amide formation?
b. Would a *p*-methoxyphenyl ester perform the same function? Why or why not?

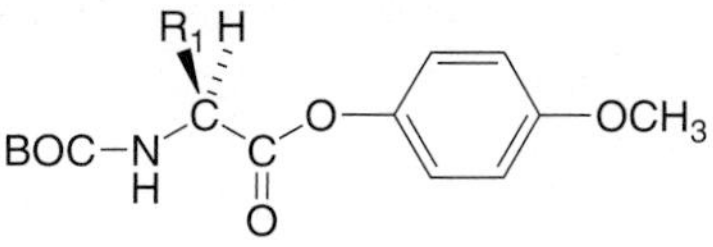

p-methoxyphenyl ester

28.59 An alternative to the BOC protecting group in peptide synthesis is the (9-fluorenyl)methoxycarbonyl group, usually called the FMOC group.

a. Draw the mechanism for the reaction that adds an FMOC group to an amino acid.

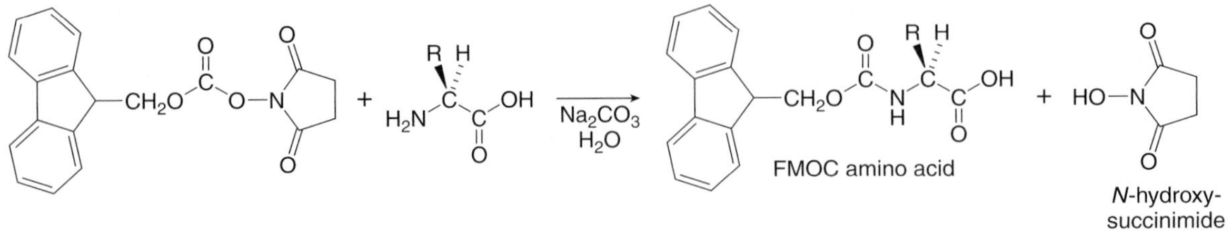

FMOC amino acid

N-hydroxy-succinimide

b. Draw the mechanism for the reaction that removes an FMOC group from an amino acid under the following conditions:

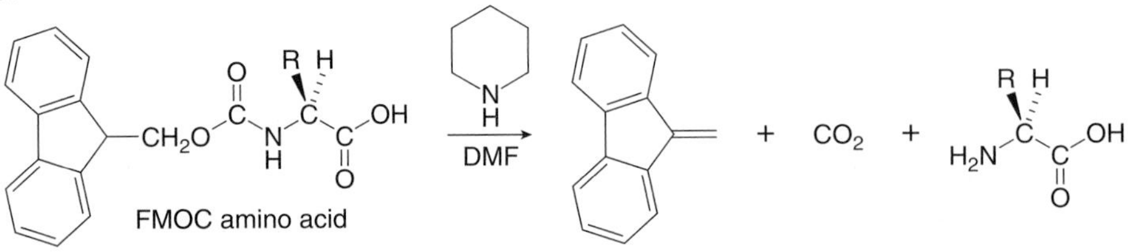

FMOC amino acid

Proteins

28.60 Which of the following amino acids are typically found in the interior of a globular protein, and which are typically found on the surface: (a) phenylalanine; (b) aspartic acid; (c) lysine; (d) isoleucine; (e) arginine; (f) glutamic acid?

28.61 After the peptide chain of collagen has been formed, many of the proline residues are hydroxylated on one of the ring carbon atoms. Why is this process important for the triple helix of collagen?

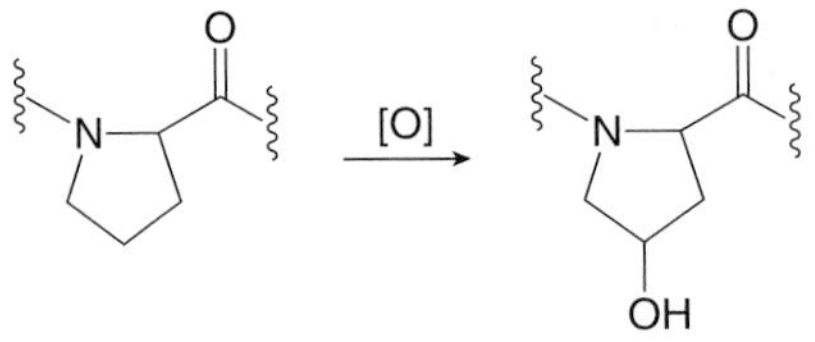

Appendix A

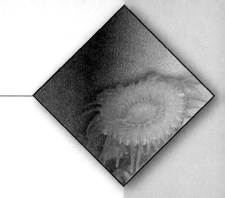

Common Abbreviations, Arrows, and Symbols

Abbreviations

Ac	acetyl, CH_3CO-
BBN	9-borabicyclo[3.3.1]nonane
BINAP	2,2'-bis(diphenylphosphino)-1,1'-binaphthyl
BOC	*tert*-butoxycarbonyl, $(CH_3)_3COCO-$
bp	boiling point
Bu	butyl, $CH_3CH_2CH_2CH_2-$
CBS reagent	Corey–Bakshi–Shibata reagent
DBN	1,5-diazabicyclo[4.3.0]non-5-ene
DBU	1,8-diazabicyclo[5.4.0]undec-7-ene
DCC	dicyclohexylcarbodiimide
DET	diethyl tartrate
DIBAL-H	diisobutylaluminum hydride, $[(CH_3)_2CHCH_2]_2AlH$
DMF	dimethylformamide, $HCON(CH_3)_2$
DMSO	dimethyl sulfoxide, $(CH_3)_2S=O$
ee	enantiomeric excess
Et	ethyl, CH_3CH_2-
FGI	functional group interconversion
HMPA	hexamethylphosphoramide, $[(CH_3)_2N]_3P=O$
HOMO	highest occupied molecular orbital
IR	infrared
LAH	lithium aluminum hydride, $LiAlH_4$
LDA	lithium diisopropylamide, $LiN[CH(CH_3)_2]_2$
LUMO	lowest unoccupied molecular orbital
m-	meta
mCPBA	*m*-chloroperoxybenzoic acid
Me	methyl, CH_3-
MO	molecular orbital
mp	melting point
MS	mass spectrometry
MW	molecular weight
NBS	*N*-bromosuccinimide
NMO	*N*-methylmorpholine *N*-oxide
NMR	nuclear magnetic resonance
o-	ortho
p-	para
PCC	pyridinium chlorochromate
Ph	phenyl, C_6H_5-
ppm	parts per million
Pr	propyl, $CH_3CH_2CH_2-$
TBDMS	*tert*-butyldimethylsilyl
THF	tetrahydrofuran
TMS	tetramethylsilane, $(CH_3)_4Si$
UV	ultraviolet

Arrows

$\longrightarrow$	reaction arrow
$\rightleftharpoons$	equilibrium arrows
$\longleftrightarrow$	double-headed arrow, used between resonance structures
 $\frown$	full-headed curved arrow, showing the movement of an electron pair
 $\frown$	half-headed curved arrow (fishhook), showing the movement of an electron
$\Longrightarrow$	retrosynthetic arrow
 $\xrightarrow{\times}$	no reaction

Symbols

$\longmapsto$	dipole
$h\nu$	light
Δ	heat
δ^+	partial positive charge
δ^-	partial negative charge
λ	wavelength
ν	frequency
$\tilde{\nu}$	wavenumber
HA	Brønsted–Lowry acid
B:	Brønsted–Lowry base
:Nu$^-$	nucleophile
E$^+$	electrophile
X	halogen
◀	bond oriented forward
⁞⁞⁞⁞	bond oriented behind
- - -	partial bond
[ ]‡	transition state
[O]	oxidation
[H]	reduction

Appendix B

pK_a Values for Selected Compounds

Compound	pK_a	Compound	pK_a
HI	−10	C₆H₅—COOH (benzoic acid)	4.2
HBr	−9		
H_2SO_4	−9	CH_3—C₆H₄—COOH	4.3
CH_3—C(=$^+$OH)—CH_3	−7.3		
		CH_3O—C₆H₄—COOH	4.5
CH_3—C₆H₄—SO_3H	−7		
		C₆H₅—$\overset{+}{N}H_3$	4.6
HCl	−7		
$[(CH_3)_2OH]^+$	−3.8		
$[CH_3OH_2]^+$	−2.5	CH_3COOH	4.8
H_3O^+	−1.7	$(CH_3)_3CCOOH$	5.0
CH_3SO_3H	−1.2	CH_3—C₆H₄—$\overset{+}{N}H_3$	5.1
CH_3—C(=$^+$OH)—NH_2	0.0		
		pyridinium ($C_5H_5\overset{+}{N}H$)	5.2
CF_3COOH	0.2		
CCl_3COOH	0.64		
O_2N—C₆H₄—$\overset{+}{N}H_3$	1.0	CH_3O—C₆H₄—$\overset{+}{N}H_3$	5.3
$Cl_2CHCOOH$	1.3	H_2CO_3	6.4
H_3PO_4	2.1	H_2S	7.0
FCH_2COOH	2.7		
$ClCH_2COOH$	2.8	O_2N—C₆H₄—OH	7.1
$BrCH_2COOH$	2.9		
ICH_2COOH	3.2	C₆H₅—SH	7.8
HF	3.2		
O_2N—C₆H₄—COOH	3.4	CH_3—CO—CH—CO—CH_3	8.9
HCOOH	3.8		
Br—C₆H₄—$\overset{+}{N}H_3$	3.9	$HC{\equiv}N$	9.1
		Cl—C₆H₄—OH	9.4
Br—C₆H₄—COOH	4.0	NH_4^+	9.4

Compound	pKa	Compound	pKa
$H_3\overset{+}{N}CH_2COO^-$	9.8	CH_3OH	15.5
phenol–OH	10.0	H_2O	15.7
		CH_3CH_2OH	16.0
		CH_3CONH_2	16
CH_3–(phenol)–OH	10.2	CH_3CHO	17
		$(CH_3)_3COH$	18
HCO_3^-	10.2	$(CH_3)_2C{=}O$	19.2
CH_3NO_2	10.2	$CH_3CO_2CH_2CH_3$	24.5
		$HC{\equiv}CH$	25
NH_2–(phenol)–OH	10.3	$CH_3C{\equiv}N$	25
		$CHCl_3$	25
CH_3CH_2SH	10.5	$CH_3CON(CH_3)_2$	30
$[(CH_3)_3NH]^+$	10.6	H_2	35
		NH_3	38
CH₃CO–CH–COOEt (H)	10.7	CH_3NH_2	40
$[CH_3NH_3]^+$	10.7	benzene–CH_3	41
cyclohexyl–$\overset{+}{N}H_3$	10.7	benzene–H	43
$[(CH_3)_2NH_2]^+$	10.7		
CF_3CH_2OH	12.4	$CH_2{=}CHCH_3$	43
		$CH_2{=}CH_2$	44
EtO₂C–CH₂–CO₂Et (diethyl malonate)	13.3	cyclopropyl–H	46
		CH_4	50
cyclopentadiene–H	15	CH_3CH_3	50

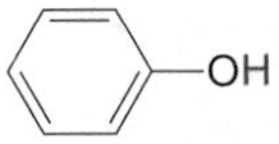

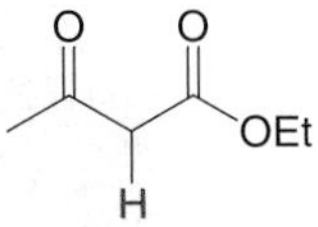

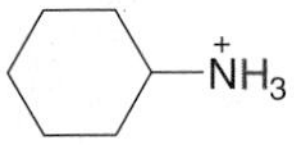

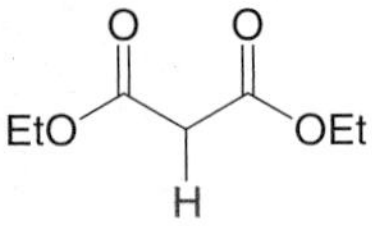

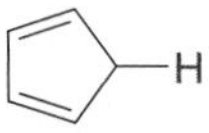

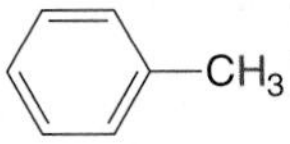

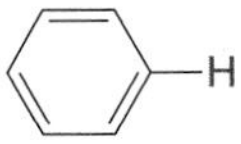

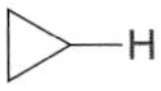

Appendix C

Bond Dissociation Energies for Some Common Bonds [A−B → A• + •B]

Bond	$\Delta H°$ kcal/mol	(kJ/mol)
H−Z bonds		
H−F	136	(569)
H−Cl	103	(431)
H−Br	88	(368)
H−I	71	(297)
H−OH	119	(498)
Z−Z bonds		
H−H	104	(435)
F−F	38	(159)
Cl−Cl	58	(242)
Br−Br	46	(192)
I−I	36	(151)
HO−OH	51	(213)
R−H bonds		
CH_3-H	104	(435)
CH_3CH_2-H	98	(410)
$CH_3CH_2CH_2-H$	98	(410)
$(CH_3)_2CH-H$	95	(397)
$(CH_3)_3C-H$	91	(381)
$CH_2=CH-H$	104	(435)
$HC\equiv C-H$	125	(523)
$CH_2=CHCH_2-H$	87	(364)
C_6H_5-H	110	(460)
$C_6H_5CH_2-H$	85	(356)
R−R bonds		
CH_3-CH_3	88	(368)
$CH_3-CH_2CH_3$	85	(356)
$CH_3-CH=CH_2$	92	(385)
$CH_3-C\equiv CH$	117	(489)

Bond	$\Delta H°$ kcal/mol	(kJ/mol)
R−X bonds		
CH_3-F	109	(456)
CH_3-Cl	84	(351)
CH_3-Br	70	(293)
CH_3-I	56	(234)
CH_3CH_2-F	107	(448)
CH_3CH_2-Cl	81	(339)
CH_3CH_2-Br	68	(285)
CH_3CH_2-I	53	(222)
$(CH_3)_2CH-F$	106	(444)
$(CH_3)_2CH-Cl$	80	(335)
$(CH_3)_2CH-Br$	68	(285)
$(CH_3)_2CH-I$	53	(222)
$(CH_3)_3C-F$	106	(444)
$(CH_3)_3C-Cl$	79	(331)
$(CH_3)_3C-Br$	65	(272)
$(CH_3)_3C-I$	50	(209)
R−OH bonds		
CH_3-OH	91	(381)
CH_3CH_2-OH	91	(381)
$CH_3CH_2CH_2-OH$	91	(381)
$(CH_3)_2CH-OH$	91	(381)
$(CH_3)_3C-OH$	91	(381)

Appendix D

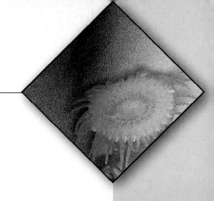

Characteristic IR Absorption Frequencies

Bond	Functional group	Wavenumber (cm^{-1})	Comment
O–H			
	• ROH	3600–3200	broad, strong
	• RCOOH	3500–2500	very broad, strong
N–H			
	• RNH$_2$	3500–3300	two peaks
	• R$_2$NH	3500–3300	one peak
	• RCONH$_2$, RCONHR	3400–3200	one or two peaks; N–H bending also observed at 1640 cm^{-1}
C–H			
	• C$_{sp}$–H	3300	sharp, often strong
	• C$_{sp^2}$–H	3150–3000	medium
	• C$_{sp^3}$–H	3000–2850	strong
	• C$_{sp^2}$–H of RCHO	2830–2700	one or two peaks
C≡C		2250	medium
C≡N		2250	medium
C=O			strong
	• RCOCl	1800	
	• (RCO)$_2$O	1800, 1760	two peaks
	• RCOOR	1745–1735	increasing $\tilde{\nu}$ with decreasing ring size
	• RCHO	1730	
	• R$_2$CO	1715	increasing $\tilde{\nu}$ with decreasing ring size
	• R$_2$CO, conjugated	1680	
	• RCOOH	1710	
	• RCONH$_2$, RCONHR, RCONR$_2$	1680–1630	increasing $\tilde{\nu}$ with decreasing ring size
C=C			
	• Alkene	1650	medium
	• Arene	1600, 1500	medium
C=N		1650	medium

Appendix E

Characteristic ^{1}H NMR Absorptions

Compound type	Chemical shift (ppm)
Alcohol	
R—O—H	1–5
R—C(H)—O—	3.4–4.0
Aldehyde	
R—C(=O)—H	9–10
Alkane	0.9–2.0
RCH$_3$	~0.9
R$_2$CH$_2$	~1.3
R$_3$CH	~1.7
Alkene	
C=C—H sp^2 C–H	4.5–6.0
C=C—C—H allylic sp^3 C–H	1.5–2.5
Alkyl halide	
R—C(H)—F	4.0–4.5
R—C(H)—Cl	3.0–4.0
R—C(H)—Br	2.7–4.0
R—C(H)—I	2.2–4.0

Compound type	Chemical shift (ppm)
Alkyne	
—C≡C–H	~2.5
Amide	
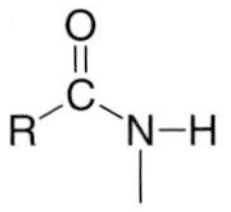	7.5–8.5
Amine	
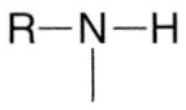 R–N–H	0.5–5.0
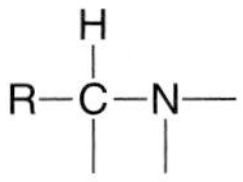	2.3–3.0
Aromatic compound	
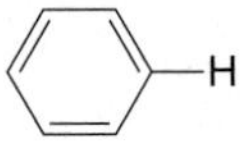 sp^2 C–H	6.5–8
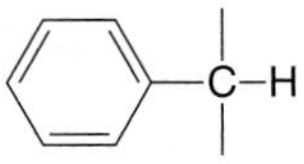 benzylic sp^3 C–H	1.5–2.5
Carbonyl compound	
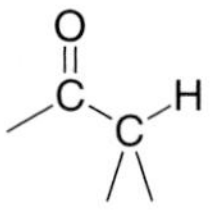 sp^3 C–H on the α carbon	2.0–2.5
Carboxylic acid	
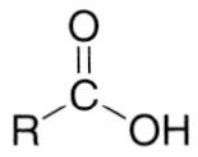	10–12
Ether	
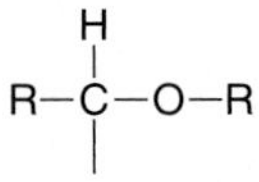	3.4–4.0

Appendix F

General Types of Organic Reactions

Substitution Reactions

[1] Nucleophilic substitution at an sp^3 hybridized carbon atom

[a] Alkyl halides (Chapter 7)   R—X   +   :Nu⁻   ⟶   R—Nu   +   X:⁻
                                              nucleophile

[b] Alcohols (Section 9.11)   R—OH   +   HX   ⟶   R—X   + H_2O

[c] Ethers (Section 9.14)   R—OR'   +   HX   ⟶   R—X   +   R'—X   + H_2O
                                            X = Br or I

[d] Epoxides (Section 9.15)

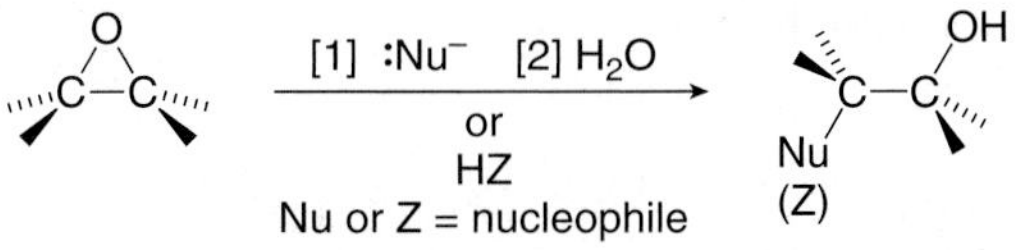

$$\underset{\text{C}}{\overset{\text{O}}{\triangle}}\text{C} \xrightarrow[\substack{\text{or} \\ \text{HZ} \\ \text{Nu or Z = nucleophile}}]{\text{[1] :Nu}^- \quad \text{[2] } H_2O} \underset{\text{Nu}}{\overset{\text{OH}}{\text{C—C}}} \text{(Z)}$$

[2] Nucleophilic acyl substitution at an sp^2 hybridized carbon atom

Carboxylic acids and their
derivatives (Chapter 22)

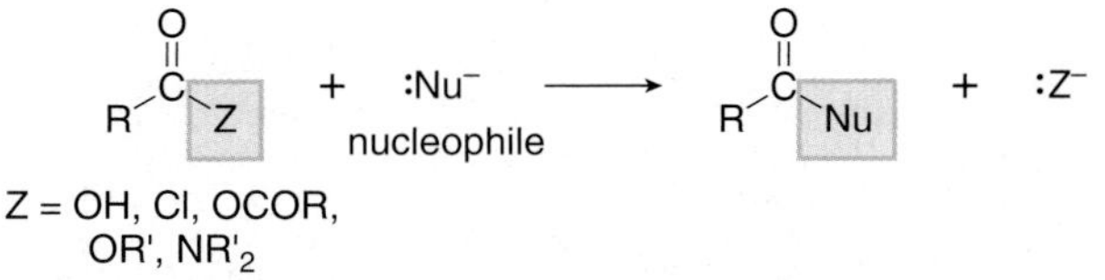

$$\underset{R}{\overset{O}{\|}}\text{C}_{Z} \quad + \quad :\text{Nu}^- \quad \longrightarrow \quad \underset{R}{\overset{O}{\|}}\text{C}_{\text{Nu}} \quad + \quad :\text{Z}^-$$
                                   nucleophile

Z = OH, Cl, OCOR,
    OR', NR'_2

[3] Radical substitution at an sp^3 hybridized C—H bond

Alkanes (Section 13.3)

R—H   + X_2 $\xrightarrow{h\nu \text{ or } \Delta}$ R—X   +   HX

[4] Electrophilic aromatic substitution

Aromatic compounds
(Chapter 18)

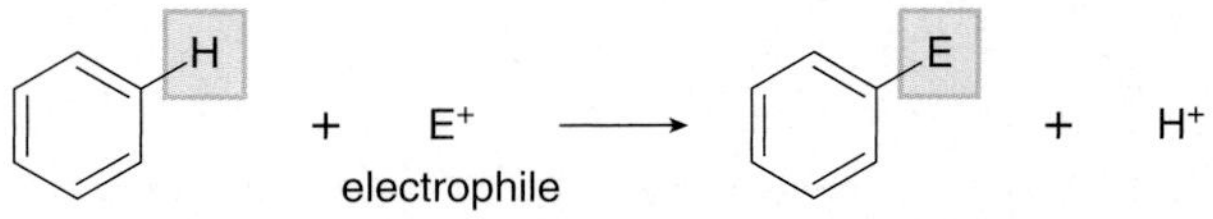

[benzene]—H   +   E⁺   ⟶   [benzene]—E   +   H⁺
                electrophile

Elimination Reactions

β Elimination at an sp^3 hybridized carbon atom

[a] Alkyl halides
(Chapter 8)

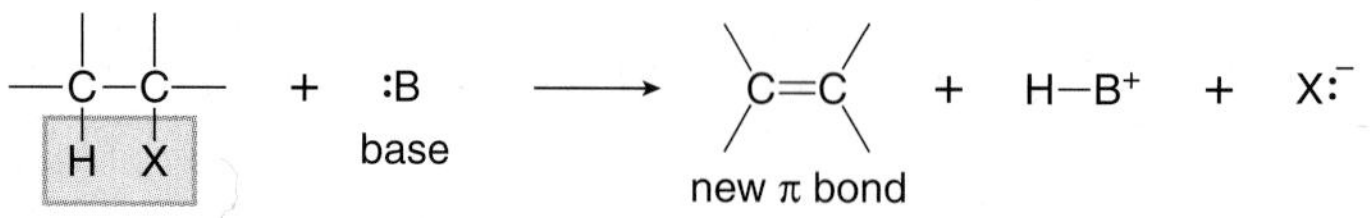

$$
-\underset{\underset{\boxed{\text{H}}}{|}}{\overset{|}{\text{C}}}-\underset{\underset{\boxed{\text{X}}}{|}}{\overset{|}{\text{C}}}- \;+\; :\!\text{B} \;\longrightarrow\; \overset{}{\text{C}}\!=\!\overset{}{\text{C}} \;+\; \text{H}-\text{B}^+ \;+\; \text{X}\!:^-
$$

base   new π bond

[b] Alcohols
(Section 9.8)

$$
-\underset{\underset{\boxed{\text{H}}}{|}}{\overset{|}{\text{C}}}-\underset{\underset{\boxed{\text{OH}}}{|}}{\overset{|}{\text{C}}}- \;\xrightarrow{\text{HA}}\; \overset{}{\text{C}}\!=\!\overset{}{\text{C}} \;+\; \text{H}_2\text{O}
$$

new π bond

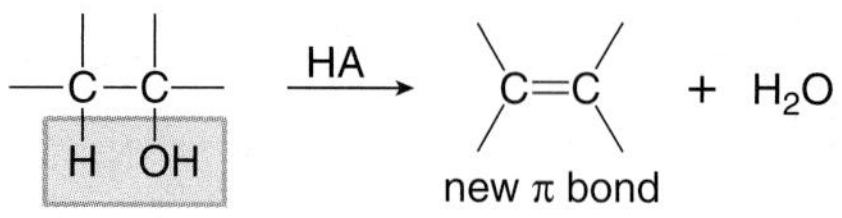

Addition Reactions

[1] Electrophilic addition to carbon–carbon multiple bonds

[a] Alkenes
(Chapter 10)

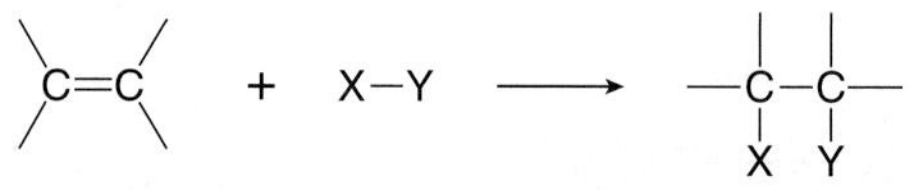

$$
\text{C}\!=\!\text{C} \;+\; \text{X}-\text{Y} \;\longrightarrow\; -\underset{\underset{\text{X}}{|}}{\overset{|}{\text{C}}}-\underset{\underset{\text{Y}}{|}}{\overset{|}{\text{C}}}-
$$

[b] Alkynes
(Section 11.6)

$$
-\text{C}\!\equiv\!\text{C}- \;+\; \text{X}-\text{Y} \;\longrightarrow\; -\underset{\underset{\text{X}}{|}}{\overset{\overset{\text{X}}{|}}{\text{C}}}-\underset{\underset{\text{Y}}{|}}{\overset{\overset{\text{Y}}{|}}{\text{C}}}-
$$

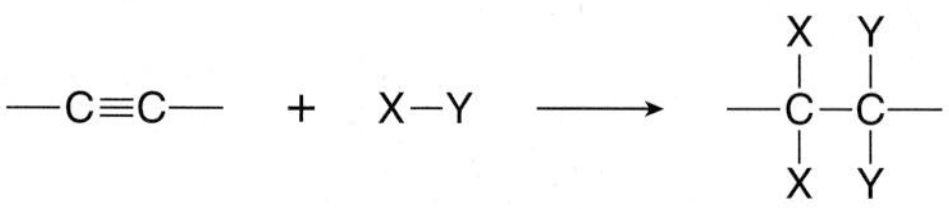

[2] Nucleophilic addition to carbon–oxygen multiple bonds

Aldehydes and ketones
(Chapter 21)

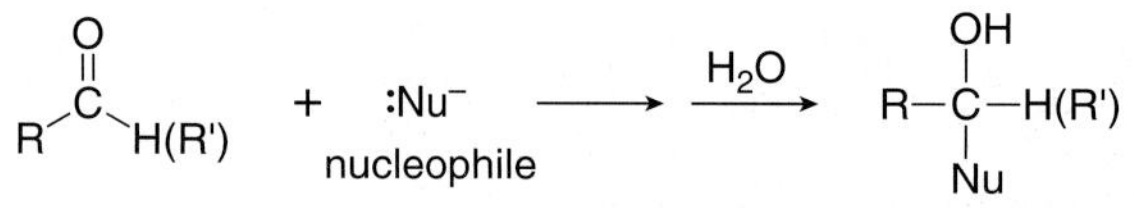

$$
\underset{\text{R}}{\overset{\overset{\displaystyle\text{O}}{\|}}{\text{C}}}\!-\!\text{H(R')} \;+\; :\!\text{Nu}^- \;\xrightarrow{\text{H}_2\text{O}}\; \text{R}-\underset{\underset{\text{Nu}}{|}}{\overset{\overset{\text{OH}}{|}}{\text{C}}}\!-\!\text{H(R')}
$$

nucleophile

Appendix G

How to Synthesize Particular Functional Groups

Acetals
◆ Reaction of an aldehyde or ketone with two equivalents of an alcohol (21.14)

Acid chlorides
◆ Reaction of a carboxylic acid with thionyl chloride (22.10)

Alcohols
◆ Nucleophilic substitution of an alkyl halide with $^-$OH or H_2O (9.6)
◆ Hydration of an alkene (10.12)
◆ Hydroboration–oxidation of an alkene (10.16)
◆ Reduction of an epoxide with $LiAlH_4$ (12.6)
◆ Reduction of an aldehyde or ketone (20.4)
◆ Hydrogenation of an α,β-unsaturated carbonyl compound with H_2 + Pd-C (20.4C)
◆ Enantioselective reduction of an aldehyde or ketone with the chiral CBS reagent (20.6)
◆ Reduction of an acid chloride with $LiAlH_4$ (20.7)
◆ Reduction of an ester with $LiAlH_4$ (20.7)
◆ Reduction of a carboxylic acid with $LiAlH_4$ (20.7)
◆ Reaction of an aldehyde or ketone with a Grignard or organolithium reagent (20.10)
◆ Reaction of an acid chloride with a Grignard or organolithium reagent (20.13)
◆ Reaction of an ester with a Grignard or organolithium reagent (20.13)
◆ Reaction of an organometallic reagent with an epoxide (20.14B)

Aldehydes
◆ Hydroboration–oxidation of a terminal alkyne (11.10)
◆ Oxidation of a 1° alcohol with PCC (12.12)
◆ Oxidative cleavage of an alkene with O_3 followed by Zn or $(CH_3)_2S$ (12.10)
◆ Reduction of an acid chloride with $LiAlH[OC(CH_3)_3]_3$ (20.7)
◆ Reduction of an ester with DIBAL-H (20.7)
◆ Hydrolysis of an acetal (21.14B)
◆ Hydrolysis of an imine or enamine (21.12)
◆ Reduction of a nitrile (22.18B)

Alkanes
◆ Catalytic hydrogenation of an alkene with H_2 + Pd-C (12.3)
◆ Catalytic hydrogenation of an alkyne with two equivalents of H_2 + Pd-C (12.5A)
◆ Reduction of an alkyl halide with $LiAlH_4$ (12.6)
◆ Reduction of a ketone to a methylene group (CH_2)—the Wolff–Kishner or Clemmensen reaction (18.14B)
◆ Protonation of an organometallic reagent with H_2O, ROH, or acid (20.9)

Alkenes

◆ Dehydrohalogenation of an alkyl halide with base (8.3)

◆ Dehydration of an alcohol with acid (9.8)

◆ Dehydration of an alcohol using $POCl_3$ and pyridine (9.10)

◆ β Elimination of an alkyl tosylate with base (9.13)

◆ Catalytic hydrogenation of an alkyne with H_2 + Lindlar catalyst to form a cis alkene (12.5B)

◆ Dissolving metal reduction of an alkyne with Na, NH_3 to form a trans alkene (12.5C)

◆ Wittig reaction (21.10)

◆ Hofmann elimination of an amine (25.12)

Alkyl halides

◆ Reaction of an alcohol with HX (9.11)

◆ Reaction of an alcohol with $SOCl_2$ or PBr_3 (9.12)

◆ Cleavage of an ether with HBr or HI (9.14)

◆ Hydrohalogenation of an alkene with HX (10.9)

◆ Halogenation of an alkene with X_2 (10.13)

◆ Hydrohalogenation of an alkyne with two equivalents of HX (11.7)

◆ Halogenation of an alkyne with two equivalents of X_2 (11.8)

◆ Radical halogenation of an alkane (13.3)

◆ Radical halogenation at an allylic carbon (13.10)

◆ Radical addition of HBr to an alkene (13.13)

◆ Electrophilic addition of HX to a 1,3-diene (16.10)

◆ Radical halogenation of an alkyl benzene (18.13)

◆ Halogenation α to a carbonyl group (23.7)

Alkynes

◆ Dehydrohalogenation of an alkyl dihalide with base (11.5)

◆ S_N2 reaction of an alkyl halide with an acetylide anion, $^-C\equiv CR$ (11.11)

Amides

◆ Reaction of an acid chloride with NH_3 or an amine (22.8)

◆ Reaction of an anhydride with NH_3 or an amine (22.9)

◆ Reaction of a carboxylic acid with NH_3 or an amine and DCC (22.10)

◆ Reaction of an ester with NH_3 or an amine (22.11)

Amines

◆ Reduction of a nitro group (18.14C)

◆ Reduction of an amide with $LiAlH_4$ (20.7B)

◆ Reduction of a nitrile (22.18B)

◆ S_N2 reaction using NH_3 or an amine (25.7A)

◆ Gabriel synthesis (25.7A)

◆ Reductive amination of an aldehyde or ketone (25.7C)

Amino acids

◆ S_N2 reaction of an α-halo carboxylic acid with excess NH_3 (28.2A)

◆ Alkylation of diethyl acetamidomalonate (28.2B)

◆ Strecker synthesis (28.2C)

◆ Enantioselective hydrogenation using a chiral catalyst (28.4)

Anhydrides
◆ Reaction of an acid chloride with a carboxylate anion (22.8)
◆ Dehydration of a dicarboxylic acid (22.10)

Aryl halides
◆ Halogenation of benzene with X_2 + FeX_3 (18.3)
◆ Reaction of a diazonium salt with CuCl, CuBr, HBF_4, NaI, or KI (25.14A)

Carboxylic acids
◆ Oxidative cleavage of an alkyne with ozone (12.11)
◆ Oxidation of a 1° alcohol with CrO_3 (or a similar Cr^{6+} reagent), H_2O, H_2SO_4 (12.12B)
◆ Oxidation of an alkyl benzene with $KMnO_4$ (18.14A)
◆ Oxidation of an aldehyde (20.8)
◆ Reaction of a Grignard reagent with CO_2 (20.14A)
◆ Hydrolysis of a cyanohydrin (21.9)
◆ Hydrolysis of an acid chloride (22.8)
◆ Hydrolysis of an anhydride (22.9)
◆ Hydrolysis of an ester (22.11)
◆ Hydrolysis of an amide (22.13)
◆ Hydrolysis of a nitrile (22.18A)
◆ Malonic ester synthesis (23.9)

Cyanohydrins
◆ Addition of HCN to an aldehyde or ketone (21.9)

1,2-Diols
◆ Anti dihydroxylation of an alkene with a peroxyacid, followed by ring opening with ⁻OH or H_2O (12.9A)
◆ Syn dihydroxylation of an alkene with $KMnO_4$ or OsO_4 (12.9B)

Enamines
◆ Reaction of an aldehyde or ketone with a 2° amine (21.12)

Epoxides
◆ Intramolecular S_N2 reaction of a halohydrin using base (9.6)
◆ Epoxidation of an alkene with mCPBA (12.8)
◆ Enantioselective epoxidation of an allylic alcohol with the Sharpless reagent (12.14)

Esters
◆ S_N2 reaction of an alkyl halide with a carboxylate anion, $RCOO^-$ (7.19)
◆ Reaction of an acid chloride with an alcohol (22.8)
◆ Reaction of an anhydride with an alcohol (22.9)
◆ Fischer esterification of a carboxylic acid with an alcohol (22.10)

Ethers
◆ Williamson ether synthesis—S_N2 reaction of an alkyl halide with an alkoxide, ⁻OR (9.6)
◆ Reaction of an alkyl tosylate with an alkoxide, ⁻OR (9.13)
◆ Addition of an alcohol to an alkene in the presence of acid (10.12)

Halohydrins
◆ Reaction of an epoxide with HX (9.15)
◆ Addition of X and OH to an alkene (10.15)

Imine
◆ Reaction of an aldehyde or ketone with a 1° amine (21.11)

Ketones

◆ Hydration of an alkyne with H_2O, H_2SO_4, and $HgSO_4$ (11.9)

◆ Oxidative cleavage of an alkene with O_3 followed by Zn or $(CH_3)_2S$ (12.10)

◆ Oxidation of a 2° alcohol with any Cr^{6+} reagent (12.12)

◆ Friedel–Crafts acylation (18.5)

◆ Reaction of an acid chloride with an organocuprate reagent (20.13)

◆ Hydrolysis of an imine or enamine (21.12)

◆ Hydrolysis of an acetal (21.14B)

◆ Reaction of a nitrile with a Grignard or organolithium reagent (22.18C)

◆ Acetoacetic ester synthesis (23.10)

Nitriles

◆ S_N2 reaction of an alkyl halide with NaCN (7.19, 22.18)

◆ Reaction of an aryl diazonium salt with CuCN (25.14A)

Phenols

◆ Reaction of an aryl diazonium salt with H_2O (25.14A)

Appendix H

Reactions that Form Carbon–Carbon Bonds

Section	Reaction
11.11A	S_N2 reaction of an alkyl halide with an acetylide anion, $^-C\equiv CR$
11.11B	Opening of an epoxide ring with an acetylide anion, $^-C\equiv CR$
13.14	Radical polymerization of an alkene
16.12	Diels–Alder reaction
18.5	Friedel–Crafts alkylation
18.5	Friedel–Crafts acylation
20.10	Reaction of an aldehyde or ketone with a Grignard or organolithium reagent
20.13A	Reaction of an acid chloride with a Grignard or organolithium reagent
20.13A	Reaction of an ester with a Grignard or organolithium reagent
20.13B	Reaction of an acid chloride with an organocuprate reagent
20.14A	Reaction of a Grignard reagent with CO_2
20.14B	Reaction of an epoxide with an organometallic reagent
20.15	Reaction of an α,β-unsaturated carbonyl compound with an organocuprate reagent
21.9	Cyanohydrin formation
21.10	Wittig reaction to form an alkene
22.18	S_N2 reaction of an alkyl halide with NaCN
22.18C	Reaction of a nitrile with a Grignard or organolithium reagent
23.8	Direct enolate alkylation using LDA and an alkyl halide
23.9	Malonic ester synthesis to form an α-substituted carboxylic acid
23.10	Acetoacetic ester synthesis to form an α-substituted ketone
24.1	Aldol reaction to form a β-hydroxy carbonyl compound or an α,β-unsaturated carbonyl compound
24.2	Crossed aldol reaction
24.3	Directed aldol reaction
24.5	Claisen reaction to form a β-keto ester
24.6	Crossed Claisen reaction to form a β-dicarbonyl compound
24.7	Dieckmann reaction to form a five- or six-membered ring
24.8	Michael reaction to form a 1,5-dicarbonyl compound
24.9	Robinson annulation to form a 2-cyclohexenone
27.10B	Kiliani–Fischer synthesis of an aldose
28.2B	Alkylation of diethyl acetamidomalonate to form an amino acid
28.2C	Strecker synthesis of an amino acid

Glossary

A

Acetal (Section 21.14): An organic compound having two alkoxy groups bonded to the same carbon atom. Acetals are used as protecting groups for aldehydes and ketones.

Acetoacetic ester synthesis (Section 23.10): A stepwise method that converts ethyl acetoacetate into a ketone having one or two alkyl groups on the α carbon.

Acetylation (Section 22.9): A reaction that results in the transfer of an acetyl group (CH_3CO-) from one atom to another.

Acetyl coenzyme A (Section 22.17): A biochemical thioester that acts as an acetylating reagent. Acetyl coenzyme A is often referred to as acetyl CoA.

Acetyl group (Section 21.2E): A substituent having the structure $-COCH_3$.

Acetylide anion (Sections 11.11, 20.9B): An anion and nucleophile formed by treating a terminal alkyne with a strong base. Acetylide anions have the general structure $R-C\equiv C^-$.

Achiral (Section 5.3): Having the property of being superimposable upon its mirror image. An achiral object is not chiral.

Acid chloride (Sections 20.1, 22.1): A compound having the general structure RCOCl.

Acidity constant (Section 2.3): A value that represents the strength of an acid, equal to $[H_3O^+][A:^-]/[HA]$. The acidity constant is represented by the symbol K_a. The larger the K_a, the stronger the acid.

Acid strength (Section 2.3): A measure of the tendency of an acid to donate a proton. The more readily a compound donates a proton, the stronger the acid.

Active site (Section 6.11): The region of an enzyme that binds the substrate. The enzymatic transformation of the substrate occurs at the active site.

Acyclic alkane (Section 4.1): A compound with the general formula C_nH_{2n+2}. Acyclic alkanes are also called saturated hydrocarbons because they contain the maximum number of hydrogen atoms per carbon.

Acylation (Sections 18.5A, 22.17): A reaction that transfers an acyl group from one atom to another.

Acyl chloride (Section 18.5A): A compound having the general structure RCOCl. Acyl chlorides are also called acid chlorides.

Acyl group (Section 18.5A): A substituent having the general structure $-C(=O)R$.

Acylium ion (Section 18.5B): A positively charged electrophile having the general structure $(R-C=O)^+$, formed when the Lewis acid $AlCl_3$ ionizes the carbon–halogen bond of an acid chloride.

Acyl transfer reaction (Section 22.17): A reaction that results in the transfer of an acyl group from one atom to another.

1,2-Addition (Sections 16.10, 20.15): An addition reaction to a conjugated system that adds groups across two adjacent atoms.

1,2-Addition product (Section 16.10): The product that results from adding groups across two adjacent atoms of a conjugated system.

1,4-Addition (Sections 16.10, 20.15): An addition reaction that adds groups to the atoms in the 1 and 4 positions of a conjugated system. 1,4-Addition is also called conjugate addition.

1,4-Addition product (Section 16.10): The product that results from adding groups to the atoms in the 1 and 4 positions of a conjugated system.

Addition reaction (Sections 6.2C, 10.8): A reaction in which elements are added to a starting material. In an addition reaction, a π bond is broken and two σ bonds are formed.

Adrenal cortical steroids (Section 26.8C): A class of steroids synthesized in the adrenal gland that regulates certain biological functions and serves as anti-inflammatory agents.

Aglycon (Section 27.7C): The alcohol formed from hydrolysis of a glycoside.

Alcohol (Section 9.1): An organic compound that contains a hydroxy group (OH group) bonded to an sp^3 hybridized carbon atom.

Aldaric acid (Section 27.9B): The dicarboxylic acid formed by the oxidation of the aldehyde and the primary alcohol of an aldose.

Aldehyde (Section 11.10): An organic compound containing a carbonyl group with a hydrogen atom bonded to the C=O carbon atom.

Alditol (Section 27.9A): The organic compound formed by the reduction of the aldehyde of an aldose to a primary alcohol.

Aldol condensation (Section 21.4B): An aldol reaction in which the initially formed β-hydroxy carbonyl compound loses water by dehydration.

Aldol reaction (Section 24.1A): A reaction in which two molecules of an aldehyde or ketone react with each other in the presence of base to form a β-hydroxy carbonyl compound.

Aldonic acid (Section 27.9B): The organic compound formed by the oxidation of the aldehyde of an aldose to a carboxylic acid.

Aldose (Section 27.2): A monosaccharide comprised of a polyhydroxy aldehyde.

Aliphatic (Section 3.2A): A compound or portion of a compound made up of carbon–carbon σ and π bonds but not aromatic bonds.

Alkaloid (Section 25.6A): A naturally occurring compound containing a basic nitrogen, which is isolated from a plant source.

Alkane (Section 4.1): An aliphatic hydrocarbon having only C–C and C–H σ bonds.

Alkene (Section 8.2A): A hydrocarbon that contains a carbon–carbon double bond.

Alkoxide (Sections 8.1, 9.6): An anionic oxygen nucleophile (base) formed by deprotonating an alcohol with a base. Alkoxides have the structure RO^-.

Alkoxy group (Section 9.3B): A substituent containing an alkyl group bonded to an oxygen (RO group).

Alkylation (Section 23.8): A reaction that transfers an alkyl group from one atom to another.

Alkyl group (Section 4.4A): A group formed by removing one hydrogen from an alkane. Alkyl groups are named by replacing the suffix *-ane* of the parent alkane with *-yl*.

Alkyl halide (Section 7.1): An organic molecule containing a halogen atom bonded to an sp^3 hybridized carbon atom. Alkyl halides have the general molecular formula $C_nH_{2n+1}X$.

1,2-Alkyl shift (Section 9.9): A rearrangement of a less stable carbocation to a more stable carbocation by the shift of an alkyl group from one carbon atom to an adjacent carbon atom.

Alkyl tosylate (Section 9.13): An organic compound having the general structure $ROSO_2C_6H_4CH_3$. Alkyl tosylates are also called tosylates and are abbreviated as ROTs.

Alkyne (Section 8.10): A hydrocarbon that contains a carbon–carbon triple bond.

Allyl carbocation (Section 16.1B): A carbocation that has a positive charge on the atom adjacent to a carbon–carbon double bond. An allyl carbocation is resonance stabilized.

Allyl group (Section 10.3C): A substituent having the structure $-CH_2-CH=CH_2$.

Allylic bromination (Section 13.10A): A radical substitution reaction in which a bromine is substituted on the carbon adjacent to a carbon–carbon double bond.

Allylic carbon (Section 13.10): A carbon atom adjacent to the carbon atom of a carbon–carbon double bond.

Allylic halide (Section 7.1): An organic molecule containing a halogen atom bonded to the carbon atom adjacent to a carbon–carbon double bond.

Allylic substitution (Section 13.10A): A substitution reaction in which a new group is substituted on the carbon adjacent to a carbon–carbon double bond.

Allyl radical (Section 13.10): A radical that has an unpaired electron on the carbon adjacent to a carbon–carbon double bond. An allyl radical is resonance stabilized.

Alpha (α) carbon (Sections 8.1, 19.2B): In an elimination reaction, the carbon that is bonded to the leaving group. In a carbonyl compound, the carbon that is bonded to the carbonyl carbon.

Ambident nucleophile (Section 23.3C): A nucleophile that has two reactive sites.

Amide (Sections 20.1, 22.1): A class of organic compound having the general structure $RCONR'_2$ where R' is either H or alkyl.

Amide base (Sections 8.10, 23.3B): A nitrogen-containing base formed by the deprotonation of an amine.

Amine (Sections 21.11, 25.1): A basic organic nitrogen compound having the general structure RNH_2, R_2NH, or R_3N. An amine has a nonbonded pair of electrons on the nitrogen atom.

α-Amino acid (Sections 19.14A, 28.1): An amino acid having the amino group on the α carbon of a carboxylic acid. α-Amino acids are the building blocks of proteins.

Amino acid residue (Section 28.5): The individual amino acids in peptides and proteins.

Amino group (Section 25.3D): A nitrogen containing substituent having the structure $-NH_2$.

α-Amino nitrile (Section 28.2C): An organic compound having an amino group α to a nitrile.

Amino sugar (Section 27.14A): A carbohydrate that contains an NH_2 group instead of a hydroxy group at a non-anomeric carbon.

Anabolic steroids (Section 26.8C): A class of synthetic androgen analogues designed to promote muscle growth.

Androgens (Section 26.8C): A class of steroidal male sex hormones responsible for development of secondary sexual characteristics in males.

Angle strain (Section 4.11): An increase in the energy of a molecule resulting when the bond angles of the sp^3 hybridized atoms deviate from the optimum tetrahedral angle of $109.5°$.

Angular methyl group (Section 26.8A): A methyl group located at the ring junction of two fused rings of the steroid skeleton.

Anhydride (Section 22.1): An organic compound having two carbonyl groups joined by a single oxygen atom.

Aniline (Section 25.3C): An organic compound having an amine nitrogen bonded to a benzene ring.

Anion (Section 1.2): A negatively charged ion that results from a neutral atom gaining one or more electrons.

Annulation (Section 24.9): A reaction that results in a new ring.

Annulene (Section 17.8A): A hydrocarbon containing a single ring with alternating double and single bonds.

α-Anomer (Section 27.6): The stereoisomer of a cyclic D monosaccharide that has the hydroxy group on the anomeric carbon drawn down. The α anomer has the anomeric OH group and the CH_2OH group trans.

β-Anomer (Section 27.6): The stereoisomer of a cyclic D monosaccharide in which the hydroxy group on the anomeric carbon is drawn up. The β-anomer has the anomeric OH and the CH_2OH groups cis.

Anomeric carbon (Section 27.6): The stereogenic center at the hemiacetal carbon of a furanose or a pyranose ring.

Anti addition (Section 10.8): An addition in which the two parts of the reagent are added from opposite sides.

Antiaromatic compound (Section 17.7): An organic compound that is cyclic, planar, completely conjugated, and has $4n$ π electrons.

Antibonding molecular orbital (Section 17.9A): A high-energy molecular orbital formed when two atomic orbitals of opposite phase overlap.

Anti conformation (Section 4.10): A staggered conformation in which the two larger groups on adjacent carbon atoms have a dihedral angle of $180°$.

Anti dihydroxylation (Section 12.9): An oxidation reaction that involves the addition of two hydroxy groups to opposite faces of a double bond.

Antioxidant (Section 13.12): A compound that stops an oxidation from occurring.

Anti periplanar (Section 8.8A): In an elimination reaction, a geometry where the β hydrogen and the leaving group are on opposite sides of the molecule.

Aromatic compound (Section 17.1): A planar unsaturated cyclic organic compound that has p orbitals on all ring atoms and a total of $4n + 2$ π electrons in the orbitals.

Aromatic heterocycle (Section 17.8C): An aromatic ring containing at least one heteroatom within the aromatic ring.

Aryl group (Section 17.3D): A substituent formed by removing one hydrogen atom from an aromatic ring.

Aryl halide (Sections 7.1, 18.3): An organic molecule containing a halogen atom bonded to an aromatic ring.

Asymmetric carbon (Section 5.3): A carbon atom that is bonded to four different groups. An asymmetric carbon is also called a stereogenic carbon, a chiral center, or a chirality center.

Asymmetric reaction (Sections 12.14, 20.6A, 28.4): A reaction that converts an achiral starting material into predominantly one enantiomer.

Atomic number (Section 1.1): The number of protons in the nucleus of an element.

Atomic weight (Section 1.1): The weighted average of the mass of all isotopes of a particular element. The atomic weight is reported in atomic mass units (amu).

Axial bonds (Section 4.12A): Bonds located above or below and perpendicular to the plane of the chair conformation of cyclohexane. There are three axial bonds that point upwards (on the up carbons) and three axial bonds that point downwards (on the down carbons).

Azo compound (Section 25.15): An organic compound having a nitrogen–nitrogen double bond.

B

Backside attack (Section 7.11C): Approach of a nucleophile from the side opposite the leaving group.

Barrier to rotation (Section 4.10): The energy difference between the lowest and highest energy conformations of a molecule.

Base peak (Section 14.1): The peak in the mass spectrum having the greatest abundance value.

Basicity (Section 7.8): A measure of how readily an atom donates its electron pair to a proton.

Benedict's reagent (Section 27.9B): A reagent for oxidizing aldehydes to carboxylic acids using a Cu^{2+} salt as an oxidizing agent. Brick-red Cu_2O is produced as a side product.

Benzoyl group (Section 21.2E): A substituent having the structure $-COC_6H_5$.

Benzyl group (Section 17.3D): A substituent containing a benzene ring bonded to a CH_2 group ($C_6H_5CH_2-$).

Benzylic halide (Sections 7.1, 18.13): An aromatic compound with a halogen atom bonded to a carbon that is bonded to a benzene ring.

Beta (β) carbon (Sections 8.1, 19.2B): In an elimination reaction, the carbon adjacent to the carbon that is bonded to the leaving group. In a carbonyl compound, the carbon located two carbons from the carbonyl carbon.

Betaine (Section 21.10B): An unstable intermediate in the Wittig reaction.

Bimolecular reaction (Sections 6.9B, 7.10, 7.13A): A reaction in which the concentration of both reactants affects the reaction rate and both terms appear in the rate equation. In a bimolecular reaction, two reactants are involved in the only step or the rate-determining step.

Biomolecule (Sections 3.3, 26.1): An organic compound found in a biological system.

Boat conformation of cyclohexane (Section 4.12B): An unstable conformation adopted by cyclohexane that resembles a boat. The instability of the boat conformation results from torsional strain and steric strain. The boat conformation of cyclohexane is 7 kcal/mol less stable than the chair conformation.

Boiling point (Section 3.5A): The temperature at which molecules in the liquid phase are converted to the gaseous phase. Molecules with stronger intermolecular forces have higher boiling points. Boiling point is abbreviated as bp.

Bond dissociation energy (Section 6.4): The amount of energy needed to homolytically cleave a covalent bond.

Bonding (Section 1.2): The joining of two atoms in a stable arrangement. Bonding is a favorable process that leads to lowered energy and increased stability.

Bonding molecular orbital (Section 17.9A): A low energy molecular orbital formed when two atomic orbitals of similar phase overlap.

Bond length (Section 1.6A): The average distance between the centers of two bonded nuclei. Bond lengths are typically reported in angstroms.

Branched-chain alkane (Section 4.1A): An acyclic alkane that has alkyl substituents bonded to the parent carbon chain.

Bridged ring system (Section 16.13D): A bicyclic ring system in which the two rings share non-adjacent carbon atoms.

Bromination (Sections 10.13, 13.6, 18.3): The reaction of an organic compound with bromine.

Bromohydrin (Section 10.15): A compound having a bromine and a hydroxy group on adjacent carbon atoms.

Brønsted–Lowry acid (Section 2.1): A compound that is a proton donor. A Brønsted–Lowry acid must contain a hydrogen atom.

Brønsted–Lowry base (Section 2.1): A compound that is a proton acceptor. A Brønsted–Lowry base must be able to form a bond to a proton. The base must contain an available electron pair.

C

^{13}C NMR spectroscopy (Section 15.1): A form of nuclear magnetic resonance spectroscopy used to determine the type of carbon atoms in a molecule.

Cahn–Ingold–Prelog system of nomenclature (Section 5.6): The system of designating a stereogenic center as either *R* or *S* according to the arrangement of the four groups attached to the center.

Carbamate (Section 28.6): A functional group containing a carbonyl group bonded to both an oxygen and a nitrogen atom.

Carbanion (Section 2.5D): An ionic species with a negative charge on a carbon atom.

Carbinolamine (Section 21.7B): An unstable intermediate having a hydroxy group and an amine group on the same carbon. A carbinolamine is formed during the addition of an amine to a carbonyl group.

Carbocation (Section 7.13C): A positively charged carbon atom. A carbocation is sp^2 hybridized and trigonal planar, and contains a vacant *p* orbital.

Carbocation rearrangement (Section 9.9): Rearrangement of a less stable carbocation to a more stable carbocation by a shift of a hydrogen atom or an alkyl group.

Carbohydrate (Sections 21.17, 27.1): A polyhydroxy aldehyde or ketone or a compound that can be hydrolyzed to a polyhydroxy aldehyde or ketone. Carbohydrates are also called starches or sugars.

Carbon backbone (Section 3.1): The $C-C$ and $C-H$ σ bond framework that makes up the skeleton of an organic molecule.

Carbon NMR spectroscopy (Section 15.1): A form of nuclear magnetic resonance spectroscopy used to determine the type of carbon atoms in a molecule.

Carbonyl group (Sections 3.2C, 11.9, 20.1): A functional group that contains a carbon–oxygen double bond ($C=O$). The polar carbon–oxygen bond makes the carbonyl carbon electrophilic.

Carboxy group (Section 19.1): An organic functional group having the structure COOH.

Carboxylate anion (Section 19.2C): An anion having the general structure $RCOO^-$, formed by deprotonating a carboxylic acid with a Brønsted–Lowry base.

Carboxylation (Section 20.14): The reaction of an organometallic reagent with carbon dioxide to form a carboxylic acid after protonation.

Carboxylic acid (Section 19.1): An acidic organic compound having the general structure RCOOH.

Carboxylic acid derivatives (Section 20.1): A class of organic compounds including acid chlorides, anhydrides, esters, and amides that can be synthesized from carboxylic acids.

Catalyst (Section 6.10): A substance that speeds up the rate of a reaction, but is recovered unchanged at the end of the reaction and does not appear in the product.

Catalytic hydrogenation (Section 12.3): A reduction reaction involving the addition of H_2 to a π bond in the presence of a metal catalyst.

Cation (Section 1.2): A positively charged ion that results from a neutral atom losing one or more electrons.

CBS reagent (Section 20.6A): A chiral reducing agent formed by reacting an oxazaborolidine with BH_3. CBS reagents predictably give one enantiomer as the major product of ketone reduction.

Cell membrane (Section 3.8A): The structure that separates the aqueous interior cytoplasm of the cell from the aqueous exterior environment. Cell membranes are composed of a lipid bilayer.

Cephalin (Section 26.4A): A phosphoacylglycerol in which the phosphodiester alkyl group is $-CH_2CH_2NH_3^+$. Cephalins are also called phosphatidylethanolamines.

Chain mechanism (Section 13.4A): A reaction mechanism that involves repeating steps.

Chair conformation of cyclohexane (Section 4.12A): A stable conformation adopted by cyclohexane that resembles a chair. The stability of the chair conformation results from the elimination of angle strain (all $C-C-C$ bond angles are $109.5°$) and torsional strain (all groups on adjacent carbon atoms are staggered).

Chemical shift (Section 15.B): The position of an absorption peak on the x-axis in an NMR spectrum relative to an internal standard such as tetramethylsilane.

Chiral (Section 5.3): Having the property of not being superimposable upon its mirror image.

Chirality center (Section 5.3): A carbon atom bonded to four different groups. A chirality center is also called a chiral center, a stereogenic center, and an asymmetric center.

Chiral molecule (Section 5.3): A molecule that is not superimposable upon its mirror image.

Chiral reducing agent (Section 20.6A): A reducing reagent that selectively forms one enantiomer.

Chlorination (Sections 10.13, 13.5, 18.3): The reaction of an organic compound with chlorine.

Chlorofluorocarbons (Section 7.4): Synthetic alkyl halides having the general molecular formula CF_xCl_{4-x}. Chlorofluorocarbons were used as refrigerants and aerosol propellants and contribute to the destruction of the ozone layer. Chlorofluorocarbons are abbreviated as "CFCs."

Chlorohydrin (Section 10.15): A compound having a chlorine and a hydroxy group on adjacent carbon atoms.

Chromate ester (Section 12.12A): An intermediate in the chromium mediated oxidation of an alcohol having the general structure $R-O-CrO_3H$.

s-Cis (Sections 16.6, 28.5B): The conformer of a 1,3-diene that has the two double bonds on the same side of the single bond that joins them.

Cis isomer (Sections 4.13B, 8.3B): An isomer of a ring or double bond that has two groups on the same side of the ring or double bond.

Claisen reaction (Section 24.5): A reaction in which two molecules of an ester react with each other in the presence of base to form a β-keto ester.

Clemmensen reduction (Section 18.14B): A method to reduce aryl ketones to alkyl benzenes using zinc and mercury in the presence of a strong acid.

Coenzyme (Section 12.13): A compound that acts with an enzyme to carry out a biochemical process. Coenzymes typically have simpler structures that the corresponding enzyme.

Combustion (Section 4.14B): An oxidation–reduction reaction, in which an alkane or other organic compound reacts with oxygen to form carbon dioxide and water, releasing energy.

Common name (Section 4.6): The name of a molecule that was adopted prior to and therefore does not follow the IUPAC system of nomenclature. Many common organic structures are still called by their common names.

Compound (Section 1.2): The structure that results when two or more elements are joined together in a stable arrangement.

Concerted reaction (Sections 6.3, 7.11B): A reaction in which all bond forming and bond breaking occurs in one-step.

Condensation polymer (Section 22.16A): A polymer formed by a polymerization process in which small molecules, such as water or HCl, are eliminated during the synthesis.

Condensation reaction (Section 21.4B): A reaction in which a small molecule, often water, is eliminated during the reaction process.

Condensed structure (Section 1.7A): A shorthand representation of the structure of an organic compound in which all atoms are drawn in but bonds and lone pairs are usually omitted. Parentheses are used to denote similar groups bonded to the same atom.

Configuration (Section 5.2): A structure that has a particular three-dimensional arrangement of atoms.

Conformations (Section 4.9): The different arrangements of atoms that are interconverted by rotation about single bonds.

Conformer (Section 4.9): A particular conformation of a molecule showing the arrangement of the atoms in three-dimensions.

Conjugate acid (Section 2.2): The compound that results when a base gains a proton in a proton transfer reaction.

Conjugate addition (Sections 16.10, 20.15): An addition reaction that adds groups to the atoms in the 1 and 4 positions of a conjugated system. Conjugate addition is also called 1,4-addition.

Conjugate base (Section 2.2): The compound that results when an acid loses a proton in a proton transfer reaction.

Conjugated dienes (Section 16.1A): A compound that contains two carbon–carbon double bonds joined by a single σ bond. Pi (π) electrons are delocalized over both double bonds. Conjugated dienes are also called 1,3-dienes.

Conjugated protein (Section 28.9C): A structure composed of a protein unit and a non-protein molecule.

Conjugation (Section 16.1): A bonding pattern that occurs whenever p orbitals are located on three or more adjacent atoms.

Constitutional isomers (Sections 1.4A, 4.1A, 5.2): Two compounds that have the same molecular formula but differ in the way the atoms are connected to each other. Constitutional isomers are also called structural isomers.

Core electrons (Section 1.1): The electrons in the inner shells of orbitals. Core electrons are not usually involved in the chemistry of a particular element.

Counterion (Section 2.1): An ion that does not take part in a reaction and is opposite in charge to the ion that does take part in the reaction. A counterion is also called a spectator ion.

Coupling (Section 15.6B): A spin–spin splitting interaction between two nearby nonequivalent protons.

Coupling constant (Section 15.6A): The frequency difference, measured in Hz, between the peaks in a split NMR signal.

Coupling reaction (Section 25.15): A reaction that forms a bond between two discrete molecules.

Covalent bond (Section 1.2): A bond that results from the sharing of electrons between two nuclei. Electrons are shared so that atoms can attain a full valence shell of electrons. A covalent bond is a two-electron bond.

Crossed aldol reaction (Section 24.2): An aldol reaction in which the two reacting carbonyl compounds are different. A crossed aldol reaction is also called a mixed aldol reaction.

Crossed Claisen reaction (Section 24.6): A Claisen reaction in which the two reacting esters are different.

Crown ether (Section 3.8B): A cyclic ether containing multiple oxygen atoms. Crown ethers bind specific cations depending on the size of their central cavity.

Curved arrow notation (Section 1.5A): A convention that shows how electron position differs between two resonance forms or how electrons move in a reaction mechanism. The curved arrow shows the movement of an electron pair. The tail of the arrow begins at the electron pair and the head points to where the electron pair moves.

Cyanide anion (Section 21.9A): A nucleophilic anion having the structure $^-C \equiv N$.

Cyano group (Section 22.1): A functional group consisting of a carbon–nitrogen triple bond ($C \equiv N$).

Cyanohydrin (Section 21.9): An organic compound having a hydroxy group and a cyano group on the same carbon. A cyanohydrin results from the addition of HCN across the carbonyl of an aldehyde or a ketone.

Cyclo- (Section 4.5): A prefix in the IUPAC name of a molecule that indicates a cyclic structure.

Cycloalkane (Sections 4.1, 4.2): A compound that contains carbons joined in one or more rings. Cycloalkanes with one ring have the general formula C_nH_{2n}.

D

D Sugar (Section 27.2C): A sugar with the hydroxy group on the stereogenic center furthest from the carbonyl on the right side in the Fischer projection formula.

Deactivated catalyst (Section 12.5B): A hydrogenation catalyst with reduced activity. Deactivated catalysts are also called poisoned catalysts.

***cis*-Decalin** (Section 26.8A): Two fused six-membered rings having the hydrogen atoms at the ring fusion on the same side of the rings.

***trans*-Decalin** (Section 26.8A): Two fused six-membered rings having the hydrogen atoms at the ring fusion on opposite sides of the rings.

Decarboxylation (Section 23.9A): A reaction that results in the elimination or loss of carbon dioxide through cleavage of a carbon–carbon bond.

Degenerate orbitals (Section 17.9B): Orbitals (either atomic or molecular) having the same energy.

Degree of unsaturation (Section 10.2): The number of rings or π bonds a molecule contains. The number of degrees of unsaturation compares the number of hydrogens in a compound to that of a saturated hydrocarbon containing the same number of carbons.

Dehydration (Sections 9.8, 22.10B): A reaction that results in the loss of the elements of water from the reaction components.

Dehydrohalogenation (Section 8.1): An elimination reaction in which the elements of hydrogen and halogen are lost from a starting material.

Delta (δ) scale (Section 15.1B): A common scale of chemical shifts used in NMR spectroscopy.

Deoxy (Section 27.14B): A prefix that means without oxygen.

Deoxyribonucleoside (Section 27.14B): An *N*-glycoside formed by the reaction of D-2-deoxyribose with certain amine heterocycles.

Deoxyribonucleotide (Section 27.14B): A DNA building block having a deoxyribose and either a purine or pyrimidine base joined together by an *N*-glycosidic linkage and a phosphate bonded to a hydroxy group of the sugar nucleus.

Deprotection (Section 20.12): The reaction that removes a protecting group, regenerating a functional group.

Deshielding effects (Section 15.3A): An effect in NMR caused by a decrease in electron density that increases the strength of the magnetic field felt by the nucleus. Deshielding shifts an absorption downfield.

Dextrorotatory (Section 5.12A): Rotating plane-polarized light in the clockwise direction. The rotation is labeled *d* or (+).

1,3-Diacid (Section 23.9A): A compound containing two carboxylic acids separated by a single carbon atom. 1,3-Diacids are also called β-diacids.

Dialkylamides (Section 23.3B): An amide base with two alkyl groups attached to the nitrogen, having the general structure $^-NR_2$.

Diastereomers (Section 5.7): Compounds that are stereoisomers but are not mirror images of each other. Diastereomers will have the same *R,S* designation for at least one stereogenic center and the opposite *R,S* designation for at least one of the other stereogenic centers.

Diastereotopic protons (Section 15.2C): Two hydrogen atoms on the same carbon such that substitution of either hydrogen with Z forms diastereomers. The two hydrogen atoms are not equivalent and give two NMR signals.

1,3-Diaxial interaction (Section 4.13A): A steric interaction between two axial substituents of the chair form of cyclohexane. Larger axial substituents create unfavorable 1,3-diaxial interactions, destabilizing a cyclohexane conformer.

Diazonium salt (Section 25.13A): An ionic salt having the general structure $(R-N \equiv N)^+Cl^-$.

Diazotization reaction (Section 25.13A): A reaction that converts primary alkylamines and arylamines to diazonium salts.

1,3-Dicarbonyl compound (Section 23.2): A compound containing two carbonyl groups separated by a single carbon atom.

1,4-Dicarbonyl compound (Section 24.4): A dicarbonyl compound in which the carbonyl groups are separated by three single bonds. 1,4-Dicarbonyl compounds can undergo intramolecular reactions to form five-membered rings.

1,5-Dicarbonyl compound (Section 24.4): A dicarbonyl compound in which the carbonyl groups are separated by four single bonds. 1,5-Dicarbonyl compounds can undergo intramolecular reactions to form six-membered rings.

Dieckmann reaction (Section 24.7): An intramolecular Claisen reaction of a diester to form a ring, typically a five- or six-membered ring.

Diels–Alder reaction (Section 16.12): An addition reaction between a 1,3-diene and a dienophile to form a cyclohexene ring.

Diene (Section 10.3A): A hydrocarbon that contains two carbon–carbon double bonds.

1,3-Diene (Section 16.1A): A compound containing two carbon–carbon double bonds joined by a single σ bond. Pi (π) electrons are delocalized over both double bonds. 1,3-Dienes are also called conjugated dienes

Dienophile (Section 16.12): The alkene component in a Diels–Alder reaction that reacts with the 1,3-diene.

Dihedral angle (Section 4.9): The angle that separates a bond on one atom from a bond on an adjacent atom.

Dihydroxylation (Section 12.9): An oxidation reaction in which two hydroxy groups are added to a double bond to form a 1,2-diol.

Diol (Section 9.3A): An organic compound possessing two hydroxy groups. Diols are also called glycols.

Dipeptide (Section 28.5): Two amino acids joined together by one amide bond.

Dipole (Section 1.11): A separation of electronic charge.

Dipole–dipole interactions (Section 3.4A): An attractive intermolecular interaction between the permanent dipoles of polar molecules. The dipoles of adjacent molecules align so that the partial positive and partial negative charges are in close proximity.

Directed aldol reaction (Section 24.3): A crossed aldol reaction in which the enolate of one carbonyl compound is formed followed by addition of the second carbonyl compound.

Disaccharide (Section 27.12): A carbohydrate containing two monosaccharide units joined together by a glycosidic linkage.

Dissolving metal reduction (Section 12.2): A reduction reaction using alkali metals as a source of electrons and liquid ammonia as a source of protons.

Disubstituted alkene (Section 8.2A): An alkene that has two alkyl groups and two hydrogens bonded to the carbons of the double bond.

Disulfide (Section 28.5C): A compound that contains a sulfur–sulfur bond often formed between the side chain of two cysteine residues.

Diterpene (Section 26.7A): A terpene that contains 20 carbons and four isoprene units.

Doublet (Section 15.6): An NMR signal that is split into two peaks of equal area, caused by one nearby nonequivalent proton.

Doublet of doublets (Section 15.8): A splitting pattern of four peaks observed when a signal is split by two different nonequivalent protons.

Downfield shift (Section 15.1B): In an NMR spectrum, a term used to describe the relative location of an absorption peak. A downfield shift means the peak is shifted to the left in the spectrum to higher chemical shift on the δ scale.

E

E,Z System of nomenclature (Section 10.3B): A system for unambiguously naming alkene stereoisomers.

E1 mechanism (Sections 8.3, 8.6): An elimination mechanism that goes by a two-step process involving a carbocation intermediate. E1 is an abbreviation for "Elimination Unimolecular."

E1cB mechanism (Section 24.1B): A two-step elimination mechanism that goes by a carbanion intermediate. E1cB stands for "Elimination Unimolecular, Conjugate Base."

E2 mechanism (Sections 8.3, 8.4): An elimination mechanism that goes by a one-step concerted process, in which both reactants are involved in the transition state. E2 is an abbreviation for "Elimination Bimolecular."

Eclipsed conformation (Section 4.9): A conformation of a molecule where the bonds on one carbon are directly aligned with the bonds on the adjacent carbon.

Eicosanoids (Section 26.6): A group of biologically active compounds containing 20 carbon atoms derived from arachidonic acid.

Electromagnetic radiation (Section 14.4): Radiant energy having dual properties of both waves and particles.

Electromagnetic spectrum (Section 14.4): The complete range of electromagnetic radiation, arbitrarily divided into different regions.

Electron-donating inductive effect (Section 7.14A): An inductive effect in which an electropositive atom or polarizable group donates electron density through σ bonds to another atom.

Electronegativity (Section 1.11): A measure of an atom's attraction for electrons in a bond. Electronegativity indicates how much a particular atom "wants" electrons.

Electron-withdrawing inductive effect (Sections 2.5, 7.14A): An inductive effect in which a nearby electronegative atom pulls electron density towards itself through σ bonds.

Electrophile (Section 2.8): An electron deficient compound that can accept a pair of electrons from an electron rich compound, forming a covalent bond. Lewis acids are electrophiles.

Electrophilic addition reaction (Section 10.9): An addition reaction in which the first step of the mechanism involves addition of the electrophilic end of the reagent to a carbon–carbon double bond.

Electrophilic aromatic substitution (Section 18.1): A characteristic reaction of benzene in which a hydrogen atom on the ring is replaced by an electrophile.

Electrospray ionization (Section 14.3B): A method for ionizing large biomolecules in a mass spectrometer. Electrospray ionization is abbreviated as "ESI."

Electrostatic potential map (Section 1.11): A color-coded map that illustrates the distribution of electron density in a molecule. Electron-rich regions are indicated in red and electron-deficient regions are indicated in blue. Regions of intermediate electron density are shown in orange, yellow, and green.

β Elimination (Section 8.1): An elimination reaction involving the loss of elements from two adjacent atoms.

Elimination reaction (Sections 6.2B, 8.1): A chemical reaction in which elements of the starting material are "lost" and a π bond is formed. In an elimination reaction, two σ bonds are broken and a π bond is formed between adjacent atoms.

Enamine (Section 21.12): An organic compound having an amine nitrogen atom bonded to a carbon–carbon double bond.

Enantiomeric excess (Section 5.12D): A measurement of how much one enantiomer is present in excess of the racemic mixture. Enantiomeric excess is denoted by the symbol *ee*. Enantiomeric excess is also called optical purity; *ee* = % of one enantiomer – % of the other enantiomer

Enantiomers (Section 5.3): Stereoisomers that are mirror images but not superimposable upon each other. Enantiomers have the exact opposite *R,S* designation at every stereogenic center.

Enantioselective reaction (Sections 12.14, 20.6A, 28.4): A reaction that affords predominantly or exclusively one enantiomer. Enantioselective reactions are also called asymmetric reactions.

Enantiotopic protons (Section 15.2C): Two hydrogen atoms on the same carbon such that substitution of either hydrogen with Z forms enantiomers. The two hydrogen atoms are equivalent and give a single NMR signal.

Endo position (Section 16.13D): A position of a substituent on a bridged bicyclic compound in which the substituent is closer to the longer bridge that joins the two carbons common to both rings.

Endothermic reaction (Section 6.4): A reaction in which the energy of the products is higher than the energy of the reactants. In an endothermic reaction, energy is absorbed and the $\Delta H°$ is a positive value.

Energy diagram (Section 6.7): A schematic representation of the energy changes that take place as reactants are converted to products. An energy diagram indicates how readily a reaction proceeds, how many steps are involved, and how the energy of the reactants, products, and intermediates compares.

Energy of activation (Section 6.7): The energy difference between the transition state and the starting material. The energy of activation is the minimum amount of energy needed to break bonds in the reactants. The energy of activation is denoted by the symbol E_a.

Enolate (Sections 20.15, 23.3): A resonance-stabilized anion formed when a base removes an α-hydrogen from the α-carbon to a carbonyl group.

Enol tautomer (Sections 9.1, 11.9, 20.15): An organic compound having a hydroxy group bonded to a carbon–carbon double bond. The enol tautomer is in equilibrium with the keto tautomer.

Enthalpy change (Section 6.4): The energy absorbed or released in a reaction. Enthalpy change is symbolized by $\Delta H°$ and is also called the heat of reaction.

Entropy (Section 6.6): A measure of the randomness in a system. The more freedom of motion or the more disorder present, the higher the entropy. Entropy is denoted by the symbol $S°$.

Entropy change (Section 6.6): The change in the amount of disorder between reactants and products in a reaction. The entropy change is denoted by the symbol $\Delta S°$. $\Delta S° = S°_{\text{products}} - S°_{\text{reactants}}$.

Enzyme (Section 6.11): A biochemical catalyst composed of at least one chain of amino acids held together in a very specific three-dimensional shape.

Enzyme-substrate complex (Section 6.11): A structure having a substrate bonded to the active site of an enzyme.

Epoxidation (Section 12.7): An oxidation reaction that involves the addition of a single oxygen atom to an alkene to form an epoxide.

Epoxide (Section 9.1): A cyclic ether having the oxygen atom as part of a three-membered ring. Epoxides are also called oxiranes.

Equatorial bonds (Section 4.12A): Bonds located in the plane of the chair conformation of cyclohexane (around the equator). There are three equatorial bonds that point slightly upwards (on the down carbons) and three equatorial bonds that point slightly downward (on the up carbons).

Equilibrium constant (Section 6.5A): A mathematical expression that relates the amount of starting material and product at equilibrium. An equilibrium constant is denoted by the symbol K_{eq}. $K_{\text{eq}} = [\text{products}]/[\text{starting materials}]$.

Equivalent protons (Section 15.2A): Protons in an NMR spectrum that have the same NMR signal and do not split each other's signal.

Essential oil (Section 26.7): A class of terpenes isolated from plant sources by distillation.

Ester (Sections 20.1, 22.1): A class of organic compound having the general structure RCOOR'.

Esterification (Section 22.10C): A reaction that converts a carboxylic acid or a derivative of a carboxylic acid to an ester.

Estrogens (Section 26.8C): A class of steroidal female sex hormones that regulates the menstrual cycle and controls the development of secondary sexual characteristics in females.

Ether (Section 9.1): A functional group having two alkyl groups bonded to the same oxygen.

Ethynyl group (Section 11.2): An alkynyl substituent having the structure $-C\equiv C-H$.

Excited state (Sections 1.8B, 16.15A): A high-energy electronic state in which one or more electrons has been promoted, by absorption of energy, to a higher energy orbital.

Exo position (Section 16.13D): A position of a substituent on a bridged bicyclic compound in which the substituent is closer to the shorter bridge that joins the two carbons common to both rings.

Exothermic reaction (Section 6.4): A reaction in which the energy of the products is lower than the energy of the reactants. In an exothermic reaction, energy is released and the $\Delta H°$ is a negative value.

Extraction (Section 19.12): A laboratory method to separate and purify a mixture of compounds using solubility differences and acid–base principles.

F

Fat (Sections 10.6B, 26.3): A triacylglycerol typically isolated from animal sources that is solid at room temperature and composed of fatty acid side chains with a high degree of saturation.

Fat-soluble vitamin (Section 26.5): A group of nonpolar vitamins that is part of the lipid class of molecules.

Fatty acid (Sections 10.6A, 19.6): A long-chain carboxylic acid having between 12 and 20 carbon atoms.

Fehling's reagent (Section 27.9B): A reagent for oxidizing aldehydes to carboxylic acids using a Cu^{2+} salt as an oxidizing agent. Brick-red Cu_2O is produced as a side product.

Fibrous proteins (Section 28.9): Long linear polypeptide chains that are bundled together to form rods or sheets.

Fingerprint region (Section 14.5B): The region in an IR spectrum at < 1500 cm^{-1}. The region often contains a complex set of peaks and is unique for every compound.

First-order kinetics (Section 6.9B): The order of a rate equation in which there is only one concentration term (raised to the first power).

First-order rate equation (Section 7.10): A rate equation in which the reaction rate depends on the concentration of only one reactant.

Fischer esterification (Section 22.10C): An acid-catalyzed esterification reaction between a carboxylic acid and an alcohol to form an ester.

Fischer projection formula (Section 27.2A): A method for representing stereogenic centers with the stereogenic carbon at the intersection of vertical and horizontal lines. Fischer projections are also called cross formulas.

Fishhook (Section 6.3B): A half-headed curved arrow used in a reaction mechanism to denote the movement of a single electron.

Flagpole hydrogens (Section 4.12B): Hydrogens in the boat conformation of cyclohexane that are on either end of the "boat" and are forced into close proximity of each other.

Formal charge (Section 1.3C): The electronic charge assigned to individual atoms in a Lewis structure. The formal charge is calculated by subtracting an atom's unshared electrons and half of its shared electrons from the number of valence electrons that a neutral atom would possess.

Formyl group (Section 21.2E): A substituent having the structure $-CHO$.

Four-centered transition state (Section 10.16): A transition state that involves four atoms.

Fragment (Section 14.1): Radicals and cations formed by the decomposition of the molecular ion in a mass spectrometer.

Freons (Section 13.9): Chlorofluorocarbons consisting of simple halogen-containing organic compounds that were once commonly used as refrigerants.

Frequency (Section 14.4): The number of waves passing a point per unit time. Frequency is reported in cycles per second (s^{-1}), which is also called hertz (Hz). Frequency is abbreviated with the Greek letter nu (ν).

Friedel–Crafts acylation (Section 18.5A): An electrophilic aromatic substitution reaction in which benzene reacts with an acid chloride in the presence of a Lewis acid to give a ketone.

Friedel–Crafts alkylation (Section 18.5A): An electrophilic aromatic substitution reaction in which benzene reacts with an alkyl halide in the presence of a Lewis acid to give an alkyl benzene.

Frontside attack (Section 7.11C): Approach of a nucleophile from the same side as the leaving group.

Full-headed curved arrow (Section 6.3B): An arrow used in a reaction mechanism to denote the movement of a pair of electrons.

Functional group (Section 3.1): An atom or group of atoms with characteristic chemical and physical properties. The functional group is the reactive part of the molecule.

Functional group interconversion (Section 11.12): A reaction that converts one functional group into another. Functional group interconversion is abbreviated as "FGI."

Functional group region (Section 14.5B): The region in an IR spectrum at ≥ 1500 cm^{-1}. Common functional groups show one or two peaks in this region, at a characteristic frequency.

Furanose (Section 27.6): A cyclic five-membered ring of a monosaccharide containing an oxygen atom.

Fused ring system (Section 16.13D): A bicyclic ring system in which the two rings share one bond and two adjacent atoms.

G

Gabriel synthesis (Section 25.7A): A two-step method that converts an alkyl halide into a primary amine using a nucleophile derived from phthalimide.

Gas chromatography (Section 14.3B): An analytical technique that separates the components of a mixture based on their boiling points and the rate at which their vapors travel down a column.

Gauche conformation (Section 4.10): A staggered conformation in which the two larger groups on adjacent carbon atoms have a dihedral angle of 60°.

Gauss (Section 15.1A): A unit used to measure the strength of a magnetic field. Gauss is denoted with the symbol "G."

GC–MS (Section 14.3B): An analytical instrument that combines a gas chromatograph (GC) and a mass spectrometer (MS) in sequence.

Gem-diol (Section 21.13): An organic compound having two hydroxy groups bonded to the same carbon atom. Gem-diols are also called hydrates.

Geminal dihalide (Section 8.10): An organic compound that has two halogen atoms on the same carbon atom.

Gibbs free energy (Section 6.5A): The free energy of a molecule. Gibbs free energy is denoted by the symbol $G°$.

Gibbs free energy change (Section 6.5A): The overall energy difference between reactants and products. The Gibbs free energy change is denoted by the symbol $\Delta G°$. $\Delta G° = G°_{products} - G°_{reactants}$.

Globular proteins (Section 28.9): Polypeptide chains that are coiled into compact shapes with hydrophilic outer surfaces that make them water soluble.

Glycol (Section 9.3A): An organic compound possessing two hydroxy groups. Glycols are also called diols.

α-Glycosidase (Section 27.13B): An enzyme that can hydrolyze α-glycosidic linkages.

Glycoside (Section 27.7A): A monosaccharide in which the hemiacetal has been converted to an acetal with an alkoxy group bonded to the anomeric carbon.

N-Glycoside (Section 27.14B): A monosaccharide containing a nitrogen bonded to the anomeric carbon.

Glycosidic linkage (Section 27.12): An acetal linkage formed between an OH group on one monosaccharide and the anomeric carbon on a second monosaccharide.

Grignard reagent (Section 20.9): An organometallic reagent having the general structure $R-Mg-X$.

Ground state (Sections 1.8B, 16.15A): The lowest energy arrangement of electrons for an atom.

Group number (Section 1.1): The number above a particular column in the periodic table. Elements having the same group number have similar electronic and chemical properties. Group numbers are represented either by an Arabic (1 to 8) or Roman (I to VII) numeral followed by the letter A or B. The group number of a second-row element is equal to the number of valence electrons in that element.

Guest molecule (Section 9.5B): A small molecule that can bind to a larger "host" molecule.

H

^{1}H NMR spectroscopy (Section 15.1): A form of nuclear magnetic resonance spectroscopy used to determine the number and type of hydrogen atoms in a molecule. ^{1}H NMR is also called "proton NMR spectroscopy."

Half-headed curved arrow (Section 6.3B): An arrow used in a reaction mechanism to denote the movement of a single electron. A half-headed curved arrow is also called a fishhook.

α-Halo aldehyde or ketone (Section 23.7): An aldehyde or ketone with a halogen atom bonded to the α carbon.

Haloform reaction (Section 23.7B): A halogenation reaction of a methyl ketone with excess halogen, that results in formation of a carboxylate anion and CHX_3 (haloform).

Halogenation (Sections 10.13, 13.3, 18.3): The reaction of an organic compound with a halogen.

Halohydrin (Sections 9.6, 10.15): A compound that has a hydroxy group and a halogen atom on adjacent carbon atoms.

Halonium ion (Section 10.13): A positively charged halogen atom. A bridged halonium ion contains a three-membered ring, and is formed in the addition of a halogen (X_2) to an alkene.

Hammond postulate (Section 7.15): A postulate that states that the transition state of a reaction resembles the structure of the species (reactant or product) to which it is closer in energy.

Haworth projection (Section 27.6A): A representation of the cyclic form of a monosaccharide in which the ring is drawn flat.

Head-to-tail polymerization (Section 13.14B): A mechanism of radical polymerization in which the more substituted radical of the growing polymer chain always adds to the less substituted end of the new monomer.

Heat of hydrogenation (Section 12.3A): The $\Delta H°$ of a catalytic hydrogenation reaction equal to the amount of energy released by hydrogenating a π bond.

Heat of reaction (Section 6.4): The energy absorbed or released in a reaction. Heat of reaction is symbolized by $\Delta H°$ and is also called the change in enthalpy.

α-Helix (Section 28.8B): A secondary structure formed when a peptide chain twists into a right-handed or clockwise spiral.

Heme (Section 28.9C): A complex organic compound containing an Fe^{2+} ion complexed with a porphyrin.

Hemiacetal (Section 21.14A): A compound that contains an alkoxy group and a hydroxy group bonded to the same carbon atom.

Hertz (Section 14.4): A unit of frequency measuring the number of waves passing a point per second.

Heteroatom (Section 3.1): An atom in a molecule other than carbon or hydrogen. Common heteroatoms in organic chemistry are nitrogen, oxygen, sulfur, phosphorus, and the halogens.

Heterocycle (Section 9.3B): A cyclic organic compound containing a heteroatom as part of the ring.

Heterolysis (Section 6.3A): The breaking of a covalent bond by unequally dividing the electrons between the two atoms in the bond. Heterolysis generates charged intermediates. Heterolysis is also called heterolytic cleavage.

Hexose (Section 27.2): A monosaccharide containing six carbons.

High density lipoprotein (Section 22.17): Lipoprotein particles known as "good cholesterol" that transfer cholesterol from the tissues back to the liver. High density lipoprotein is abbreviated as "HDL."

Highest occupied molecular orbital (Section 17.9B): The molecular orbital with the highest energy that also contains electrons.

The highest occupied molecular orbital is abbreviated as the HOMO.

High-resolution mass spectrometer (Section 14.3A): A mass spectrometer that can measure mass-to-charge ratios to four or more decimal places. High-resolution mass spectra are used to determine the molecular formula of a compound.

Hofmann elimination (Section 25.12): An E2 elimination reaction that converts an amine into a quaternary ammonium salt as leaving group. The Hofmann elimination gives the less substituted alkene as the major product.

Homologous series (Section 4.1B): A group of compounds that differ by only a CH_2 group in the chain.

Homolysis (Section 6.3A): The breaking of a covalent bond by equally dividing the electrons between the two atoms in the bond. Homolysis generates uncharged radical intermediates. Homolysis is also called homolytic cleavage.

Hooke's law (Section 14.6): A physical law that can be used to calculate the frequency of a bond vibration from the strength of the bond and the masses of the atoms attached to it.

Host–guest complex (Section 9.5B): The complex that is formed when a small "guest" molecule binds to a larger "host" molecule.

Host molecule (Section 9.5B): A large molecule that can bind a smaller "guest" molecule.

Hückel's rule (Section 17.7): A principle that states for a compound to be aromatic it must be cyclic, planar, completely conjugated, and have $4n + 2$ π electrons.

Hybridization (Section 1.8B): The mathematical combination of two or more atomic orbitals (having different shapes) to form the same number of hybrid orbitals (all having the same shape).

Hybrid orbital (Section 1.8B): A new orbital that results from the mathematical combination of two or more atomic orbitals. The hybrid orbital is intermediate in energy compared to the atomic orbitals that were combined to form it.

Hydrate (Sections 12.12B, 21.13): An organic compound having two hydroxy groups on the same carbon atom. Hydrates are also called gem-diols.

Hydration (Sections 10.12, 21.9A): Addition of the elements of water to a molecule.

Hydride (Section 12.2): A negatively charged hydrogen ion ($H:^-$).

1,2-Hydride shift (Section 9.9): A rearrangement of a less stable carbocation to a more stable carbocation by the shift of a hydrogen atom from one carbon atom to an adjacent carbon atom.

Hydroboration (Section 10.16): The addition of the elements of borane (BH_3) to an alkene or alkyne. When combined with a subsequent oxidation step, the two steps add water to a π bond.

Hydrocarbon (Sections 3.2A, 4.1): An organic compound made up of only the elements of carbon and hydrogen.

Hydrogen bonding (Section 3.4A): An attractive intermolecular interaction that occurs when a hydrogen atom bonded to an O, N, or F, is electrostatically attracted to a lone pair of electrons on an O, N, or F atom in an adjacent molecule.

Hydrogenolysis (Section 28.6): A reaction that cleaves a σ bond using H_2 in the presence of a metal catalyst.

α-Hydrogens (Section 23.1): The hydrogen atoms on the carbon bonded to the carbonyl carbon atom (the α-carbon).

Hydrohalogenation (Section 10.9): An electrophilic addition of hydrogen halide (HX) to an alkene or alkyne.

Hydrolysis (Section 21.9A): A cleavage reaction that results in the loss of water from a molecule.

Hydrolyzable lipid (Section 26.1): A lipid that can be cleaved into smaller molecules by hydrolysis with water.

Hydroperoxide (Section 13.11): An organic compound having the general structure $R-O-O-H$.

Hydrophilic (Section 3.5C): Attracted to water. The polar portion of a molecule that interacts with the polar water molecules is hydrophilic.

Hydrophobic (Section 3.5C): Not attracted to water. The nonpolar portion of a molecule that is not attracted to the polar water molecules is hydrophobic.

β-Hydroxy carbonyl compound (Section 24.1A): An organic compound having a hydroxy group on the carbon β to the carbonyl group.

Hydroxy group (Section 9.1): The OH functional group.

Hyperconjugation (Section 7.14B): The overlap of an empty p orbital with an adjacent σ bond.

I

Imide (Section 25.7A): An organic compound having a nitrogen atom between two carbonyl groups.

Imine (Sections 21.7B, 21.11A): An organic compound with the general structure $R_2C=NR'$. Imines are also called Schiff bases.

Iminium ion (Section 21.11A): A resonance-stabilized cation having the general structure $(R_2C=NR'_2)^+$, where $R'=H$ or alkyl.

Inductive effect (Sections 2.5B, 7.14A): The pull of electron density through σ bonds caused by electronegativity differences of atoms.

Infrared spectroscopy (Section 14.5): An analytical technique used to identify the functional groups in an organic molecule based on their absorption of electromagnetic radiation in the infrared region. Infrared spectroscopy is abbreviated as IR spectroscopy.

Initiation (Section 13.4A): The initial step in a radical chain reaction in which two radicals are formed by homolysis of a σ bond.

Inscribed polygon method (Section 17.10): A method to predict the relative energies of cyclic, completely conjugated compounds to determine which molecular orbitals are filled or empty. The inscribed polygon is also called a Frost circle.

In situ (Section 9.12): Occurring directly in the reaction mixture.

Integral (Section 15.5): The stepped curve printed on a ^{1}H NMR spectrum.

Integration (Section 15.5): The area under an NMR signal that is proportional to the number of absorbing nuclei that give rise to the signal.

Intermolecular forces (Section 3.4): The types of interactions that exist between molecules. Functional groups determine the type and strength of these forces. Intermolecular forces are also called noncovalent interactions or nonbonded interactions.

Internal alkene (Section 10.1): An alkene that has at least one carbon atom bonded to each end of the double bond.

Internal alkyne (Section 11.1): An alkyne that has one carbon atom bonded to each end of the triple bond.

Intramolecular aldol reaction (Section 24.4): An aldol reaction in which a dicarbonyl compound undergoes reaction to form a five- or six-membered ring.

Inversion of configuration (Section 7.11C): The opposite relative stereochemistry of a stereogenic center in the starting material and product of a chemical reaction. In a nucleophilic substitution reaction, inversion results when the nucleophile and leaving group are in the opposite position relative to the three other groups on carbon.

Iodoform test (Section 23.7B): A test for the presence of methyl ketones, indicated by the formation of the yellow precipitate, CHI_3, via the haloform reaction.

Ionic bond (Section 1.2): A bond that results from the transfer of electrons from one element to another. Ionic bonds result from strong electrostatic interactions between ions with opposite charges. The transfer of electrons forms stable salts composed of cations and anions.

Ionophore (Section 3.8B): An organic molecule that can form a complex with cations so they may be transported across a cell membrane. Ionophores have a hydrophobic exterior and a hydrophilic central cavity that complexes the cation.

IR inactive (Section 14.6): A type of vibration for a symmetrical bond that does not absorb electromagnetic radiation in the infrared region.

Isoelectric point (Sections 19.14C, 28.1A): The pH at which an amino acid exists primarily in its neutral zwitterionic form. Isoelectric point is abbreviated as pI.

Isolated dienes (Section 16.1A): A compound containing two carbon–carbon double bonds joined by more than one σ bond.

Isomers (Sections 1.4A, 4.1A, 5.1): Two different chemical compounds that have the same molecular formula.

Isoprene unit (Section 26.7): A five-carbon unit with four carbons in a row and a one-carbon branch on one of the middle carbons.

Isotope (Section 1.1): Two or more atoms of the same element having the same number of protons in the nucleus but a different number of neutrons. Isotopes have the same atomic number but different mass numbers.

IUPAC (Section 4.3): The International Union of Pure and Applied Chemistry, an international organization of chemists.

IUPAC System of nomenclature (Section 4.3): A systematic method for naming compounds developed by the International Union of Pure and Applied Chemistry.

K

K_a (Section 2.3): The symbol that represents the acidity constant and equal to $[H_3O^+][A:^-]/[HA]$. The larger the K_a, the stronger the acid.

K_{eq} (Section 2.3): The equilibrium constant, equal to the product of the product concentrations divided by the product of the reactant concentrations.

Kekulé structures (Section 17.1): Two equilibrating structures for benzene. Each structure contains a six-membered ring and three π bonds alternating with σ bonds around the ring.

β-Keto ester (Section 23.10): An organic compound containing a ketone carbonyl on the carbon β to the ester carbonyl group.

Ketone (Section 11.9): An organic compound having a carbonyl with two alkyl groups bonded to the $C=O$ carbon atom.

Ketose (Section 27.2): A monosaccharide comprised of a polyhydroxy ketone.

Keto tautomer (Section 11.9): A tautomer of a ketone that has a $C=O$ and a hydrogen bonded to the α-carbon. The keto tautomer is in equilibrium with the enol tautomer.

Kiliani–Fischer synthesis (Section 27.10B): A reaction that lengthens the carbon chain of an aldose by adding one carbon to the end.

Kinetic enolate (Section 23.4): The enolate that is formed the fastest. When two enolates are possible, the kinetic enolate is the less substituted enolate.

Kinetic product (Section 16.11): In a reaction that can give more than one product, the product that is formed the fastest.

Kinetic resolution (Section 28.3B): The separation of two enantiomers by a chemical reaction that selectively occurs for only one of the enantiomers.

Kinetics (Section 6.5): A study that describes the rate of a chemical reaction.

L

L Sugar (Section 27.2C): A sugar with the hydroxy group on the stereogenic center furthest from the carbonyl on the left side in the Fischer projection formula.

Lactam (Section 22.1): A cyclic amide in which the carbonyl carbon–nitrogen σ bond is part of the ring.

β-Lactam (Sections 14.7, 22.1): A cyclic amide in which the carbonyl carbon–nitrogen σ bond is part of a four-membered ring.

Lactol (Section 21.16): A cyclic hemiacetal.

Lactone (Section 22.1): A cyclic ester in which the carbonyl carbon–oxygen σ bond is part of the ring.

Le Châtelier's principle (Section 9.8D): The principle that a system at equilibrium will react to counteract any disturbance to the equilibrium.

Leaving group (Section 7.6): An atom or group of atoms (Z) that is able to accept the electron density of the C−Z bond during a substitution or elimination reaction.

Leaving group ability (Section 7.7): A measure of how readily a leaving group (Z) can accept the electron density of the C−Z bond during a substitution or elimination reaction.

Lecithin (Section 26.4A): A phosphoacylglycerol in which the phosphodiester alkyl group is $-CH_2CH_2N(CH_3)_3^+$. Lecithins are also called phosphatidylcholines.

Leukotriene (Section 9.16): An unstable and potent biomolecule synthesized in cells by the oxidation of arachidonic acid. Leukotrienes are responsible for biological conditions such as asthma.

Levorotatory (Section 5.12A): Rotating plane-polarized light in the counterclockwise direction. The rotation is labeled l or (−).

Lewis acid (Section 2.8): An electron pair acceptor.

Lewis acid–base reaction (Section 2.8): A reaction that results when a Lewis base donates an electron pair to a Lewis acid.

Lewis base (Section 2.8): An electron pair donor.

Lewis structure (Section 1.3): A representation of a molecule that displays the position of covalent bonds and nonbonding electrons. In Lewis structures, valence electrons are represented by dots and a two-electron covalent bond is represented by a solid line. Lewis structures are also called electron dot structures.

"Like dissolves like" (Section 3.5C): The principle that compounds dissolve in solvents having similar kinds of intermolecular forces; i.e., polar compounds dissolve in polar solvents and nonpolar compounds dissolve in nonpolar solvents.

Lindlar catalyst (Section 12.5B): A catalyst for the catalytic hydrogenation of an alkyne that affords a cis alkene product. The Lindlar catalyst is Pd adsorbed onto $CaCO_3$ with lead(II) acetate and quinoline.

Lipid (Sections 4.15, 26.1): A biomolecule with a large number of carbon–carbon and carbon–hydrogen σ bonds that is soluble in organic solvents and insoluble in water.

Lipid bilayer (Sections 3.8A, 26.4A): A bilayer formed when phospholipids are mixed with water. The ionic heads of the phospholipids are oriented on the outside and the nonpolar tails on the inside.

Lone pair of electrons (Section 1.2): A pair of valence electrons that is not involved in a covalent bond. The lone pair of electrons is not shared with another atom in the molecule. Lone pairs are also called unshared or nonbonded pairs of electrons.

Low density lipoprotein (Section 22.17): Lipoprotein particles known as "bad cholesterol," which transfer cholesterol from the liver to the tissues. Low density lipoprotein is abbreviated as "LDL."

Lowest unoccupied molecular orbital (Section 17.9B): The molecular orbital with the lowest energy that does not contain electrons. The lowest unoccupied molecular orbital is abbreviated as the LUMO.

M

M peak (Section 14.1): The peak in the mass spectrum that corresponds to the mass of the molecular ion. The M peak is also called the molecular ion peak or the parent peak.

M + 1 peak (Section 14.1): The peak in the mass spectrum that corresponds to the mass of the molecular ion plus one. The M + 1 peak is caused by the presence of isotopes that increase the mass of the molecular ion.

M + 2 peak (Section 14.2): The peak in the mass spectrum that corresponds to the mass of the molecular ion plus two. The M + 2 peak is caused by the presence of isotopes typically of a chlorine or a bromine atom.

Macrocyclic lactone (Section 22.6A): A cyclic ester contained in a large ring. Macrocyclic lactones are also called macrolides.

Macrolide (Section 22.6A): A cyclic ester contained in a large ring. Macrolides are also called macrocyclic lactones.

Magnetic resonance imaging (MRI) (Section 15.12): A form of NMR spectroscopy used in medicine.

Malonic ester synthesis (Section 23.9A): A stepwise method that converts diethyl malonate into a carboxylic acid having one or two alkyl groups on the α carbon.

Markovnikov's rule (Section 10.10): The rule that states in the addition of HX to an unsymmetrical alkene, the H atom bonds to the less substituted carbon atom.

Mass number (Section 1.1): The total number of protons and neutrons in the nucleus of a particular atom.

Mass spectrometer (Section 14.1): An analytical instrument in which organic compounds are ionized, fragmented, and separated in a magnetic field to afford a mass spectrum.

Mass spectrometry (Section 14.1): An analytical technique used for measuring the molecular weight and determining the molecular formula of an organic molecule.

Mass spectrum (Section 14.1): A plot of the amount of each cation (its relative abundance) versus its mass-to-charge ratio.

Mass-to-charge ratio (Section 14.1): A ratio of the mass to the charge of a molecular ion or fragment. Mass-to-charge ratio is abbreviated as m/z.

Megahertz (Section 15.1A): A unit of frequency used in NMR spectroscopy for the frequency of the RF radiation.

Melting point (Section 3.5B): The temperature at which molecules in the solid phase are converted to the liquid phase. Molecules with stronger intermolecular forces and higher symmetry have higher melting points. Melting point is abbreviated as mp.

Merrifield method (Section 28.7): A method for synthesizing polypeptides using insoluble polymer supports.

Meso compound (Section 5.8): An achiral compound that contains two or more tetrahedral stereogenic centers and a plane of symmetry.

Meta director (Section 18.7): A substituent on a benzene ring that directs a new group to the meta position during electrophilic aromatic substitution.

Meta isomer (Section 17.3B): A disubstituted benzene ring in which two substituents are on carbon atoms separated by a single ring carbon (i.e., a 1,3-disubstituted benzene). Meta substitution is abbreviated as m-.

Metal hydride reagent (Section 12.2): A reagent containing a polar metal–hydrogen bond that places a partial negative charge on the hydrogen and acts as a source of hydride ions.

Methylation (Section 7.12): A reaction in which a CH_3 group is transferred from one compound to another.

Methylene group (Sections 4.1B, 10.3C): A CH_2 group bonded to a carbon chain ($-CH_2-$) or part of a double bond ($CH_2=$).

1,2-Methyl shift (Section 9.9): A rearrangement of a less stable carbocation to a more stable carbocation by the shift of a methyl group from one carbon atom to an adjacent carbon atom.

Micelles (Section 3.7): Spherical droplets formed by soap molecules having the ionic heads on the surface and the nonpolar tails packed together in the interior. Grease and oil dissolve in the interior nonpolar region.

Michael acceptor (Section 24.8): The α,β-unsaturated carbonyl compound in a Michael reaction.

Michael reaction (Section 24.8): A reaction in which a resonance-stabilized carbanion (usually an enolate) adds to the β carbon of an α,β-unsaturated carbonyl compound.

Miscible (Section 3.5C): The physical characteristic that two liquids can form solutions in all proportions with each other. Ethanol and water are miscible; oil and water are not miscible.

Mixed aldol reaction (Section 24.2): An aldol reaction between two different carbonyl compounds. A mixed aldol reaction is also called a crossed aldol reaction.

Mixed anhydrides (Section 22.1): An anhydride with two different alkyl groups bonded to the carbonyl carbon atoms.

Molecular ion (Section 14.1): The radical cation having the general structure $M^{+\bullet}$, formed by the removal of an electron from an organic molecule. The molecular ion is also called the parent ion.

Molecular orbital (Section 17.9A): An orbital that results from the mathematical combination of atomic orbitals from bonding atoms.

Molecular orbital theory (Section 17.9A): A theory that describes bonds as the mathematical combination of atomic orbitals to form a new set of orbitals called molecular orbitals. Molecular orbital theory is also called MO theory.

Molecular recognition (Section 9.5B): The ability of a host molecule to recognize and bind specific guest molecules.

Molecule (Section 1.2): A compound containing two or more atoms bonded together with covalent bonds.

Monohalogenation (Section 13.3): A halogenation reaction that involves the substitution of a single H by X.

Monomers (Sections 3.4B, 13.14): Small organic compounds that can be covalently bonded to each other (polymerized) in a repeating pattern.

Monosaccharide (Section 27.2): A simple sugar having three to seven carbon atoms.

Monosubstituted alkene (Section 8.2A): An alkene that has one alkyl group and three hydrogens bonded to the carbons of the double bond.

Monoterpene (Section 26.7A): A terpene that contains 10 carbons and two isoprene units.

Multiplet (Section 15.6C): An NMR signal that is split into more than seven peaks.

Mutarotation (Section 27.6A): The process by which a pure anomer of a monosaccharide equilibrates to a mixture of both anomers when placed in solution.

N

$n + 1$ rule (Section 15.6C): The rule that an NMR signal for a proton with n nearby nonequivalent protons will be split into $n + 1$ peaks.

Natural product (Section 7.19): A compound that is isolated from a natural source.

Newman projection (Section 4.9): An end-on representation of the conformation of a molecule. The Newman projection is a graphic that shows the three groups bonded to each carbon atom in a particular C−C bond, as well as the dihedral angle that separates the groups on each carbon.

Nitration (Section 18.4): An electrophilic aromatic substitution reaction in which benzene reacts with a nitronium ion to give a nitrobenzene.

Nitrile (Sections 22.1, 22.18): An organic compound having the general structure $R-C\equiv N$.

Nitronium ion (Section 18.4): An electrophile having the structure $^+NO_2$, formed by protonation of HNO_3 followed by loss of water.

N-Nitrosamine (Sections 7.16, 25.13B): An organic compound having the general structure $R_2N-N=O$. Nitrosamines are formed by the reaction of a secondary amine with a nitrosonium ion.

Nitrosonium ion (Section 25.13): An electrophilic cation having the structure ^+NO.

NMR peak (Section 15.6A): The individual absorptions in a split NMR signal due to nonequivalent nearby protons.

NMR signal (Section 15.6A): The entire absorption due to a particular kind of proton in an NMR spectrum.

NMR spectrometer (Section 15.1A): An analytical instrument that measures the absorption of RF radiation by certain atomic nuclei when placed in a strong magnetic field.

Nonbonded pair of electrons (Section 1.2): A pair of valence electrons that is not involved in a covalent bond. The nonbonded pair of electrons is not shared with another atom in the molecule. Nonbonded electrons are also called unshared or lone pairs of electrons.

Nonbonding molecular orbital (Section 17.10): A molecular orbital having the same energy as the atomic orbitals that formed it.

Nonequivalent protons (Section 15.6C): Protons in different magnetic environments that give rise to different NMR signals. If nonequivalent protons are on the same or adjacent carbons, they will split each other's signal into peaks.

Nonhydrolyzable lipid (Section 26.1): A lipid that cannot be cleaved into smaller units by aqueous hydrolysis.

Nonnucleophilic bases (Section 7.8B): Bases that are poor nucleophiles due to steric hindrance resulting from the presence of bulky groups.

Nonpolar bond (Section 1.11): A covalent bond in which the electrons are equally shared between the two atoms.

Nonpolar molecule (Section 1.12): A molecule that has no net dipole. A nonpolar molecule has either no polar bonds or multiple polar bonds whose dipoles cancel.

Nonreducing sugar (Section 27.9B): A carbohydrate that cannot be oxidized by Tollens, Benedict's or Fehling's reagent.

Normal alkane (Section 4.1A): An acyclic alkane that has all of its carbons in a row. A normal alkane is an "n-alkane" or a straight-chain alkane.

Nuclear magnetic resonance spectroscopy (Section 15.1): A powerful analytical tool that can help identify the carbon and hydrogen framework of an organic molecule.

Nucleophile (Sections 2.8, 7.6): An electron rich compound that donates a pair of electrons to an electron deficient compound, forming a covalent bond. Lewis bases are nucleophiles.

Nucleophilic acyl substitution (Sections 20.2B, 22.1): Substitution of a leaving group by a nucleophile at a carbonyl carbon.

Nucleophilic addition (Section 20.2A): Addition of a nucleophile to the electrophilic carbon of a carbonyl group followed by protonation of the oxygen.

Nucleophilicity (Section 7.8A): A measure of how readily an atom donates an electron pair to other atoms.

Nucleophilic substitution (Section 7.6): A reaction in which a nucleophile replaces the leaving group in a molecule.

Nucleoside (Section 27.14B): A biomolecule having a sugar and either a purine or pyrimidine base joined together by an N-glycosidic linkage.

Nucleotide (Section 27.14B): A biomolecule having a sugar and either a purine or pyrimidine base joined together by an N-glycosidic linkage and a phosphate bonded to a hydroxy group of the sugar nucleus.

Nylon (Section 22.16A): A condensation polymer consisting of many amide bonds between diamines and dicarboxylic acids. Nylons are polyamides.

O

Observed rotation (Section 5.12A): The angle that a sample of an optically active compound rotates plane-polarized light. The angle is denoted by the symbol α and is measured in degrees (°).

Octet rule (Section 1.2): The general rule governing the bonding process for second-row elements. Through bonding, second-row elements attain a complete outer shell of eight valence electrons. By attaining eight valence electrons, second-row elements attain a stable noble gas configuration of electrons.

Oil (Sections 10.6B, 26.3): A triacylglycerol typically isolated from vegetable sources that is liquid at room temperature and composed of fatty acid side chains with a high degree of unsaturation.

Olefin (Section 10.1): An alkene. An organic compound possessing a carbon–carbon double bond.

Open arrow (Section 11.12): An arrow used to indicate a retrosynthetic step working from a product backwards to a starting material.

Optically active (Section 5.12A): Able to rotate the plane of plane-polarized light as it passes through a solution of a compound.

Optically inactive (Section 5.12A): Not able to rotate the plane of plane-polarized light as it passes through a solution of a compound.

Optical purity (Section 5.12D): A measurement of how much one enantiomer is present in excess of the racemic mixture. Optical purity is also called enantiomeric excess and is denoted by the symbol ee; ee = % of one enantiomer − % of the other enantiomer.

Orbital (Section 1.1): A region of space around the nucleus of an atom that is high in electron density. There are four different kinds of orbitals, called s, p, d, and f.

Order of a rate equation (Section 6.9B): The sum of the exponents of the concentration terms in the rate equation of a reaction.

Organic synthesis (Section 7.19): The systematic preparation of an organic compound from readily available starting materials by one or more steps.

Organoborane (Section 10.16): A compound that contains a carbon–boron bond. Organoboranes have the general structure RBH_2, R_2BH, or R_3B.

Organocopper reagent (Section 20.9): An organometallic reagent having the general structure R_2CuLi. Organocopper reagents are also called organocuprates.

Organolithium reagent (Section 20.9): An organometallic reagent having the general structure $R-Li$.

Organomagnesium reagent (Section 20.9): An organometallic reagent having the general structure $R-Mg-X$. Organomagnesium reagents are also called Grignard reagents.

Organometallic reagent (Section 20.9): A reagent that contains a carbon atom bonded to a metal.

Organophosphorus reagent (Section 21.10A): A reagent that contains a carbon–phosphorus bond.

Ortho director (Section 18.7): A substituent on a benzene ring that directs a new group to the ortho position during electrophilic aromatic substitution.

Ortho isomer (Section 17.3B): A disubstituted benzene ring in which two substituents are on adjacent carbon atoms (i.e., a 1,2-disubstituted benzene). Ortho substitution is abbreviated as *o*-.

Oxaphosphetane (Section 21.10B): An intermediate in the Wittig reaction consisting of a four-membered ring containing a phosphorus–oxygen bond.

Oxazaborolidine (Section 20.6A): A heterocycle possessing a boron, a nitrogen, and an oxygen. An oxazaborolidine can be used to form a chiral reducing agent.

Oxidation (Sections 4.14A, 12.1): A process that results in a compound losing electrons. For organic compounds, oxidation results in an increase in the number of $C-Z$ bonds or a decrease in the number of $C-H$ bonds.

Oxidative cleavage (Section 12.10): An oxidation reaction that breaks both the σ and π bonds of a multiple bond to form two oxidized products.

Oxime (Section 27.10A): An organic compound having the general structure $R_2C=NOH$.

Oxirane (Section 9.1): A cyclic ether having the oxygen atom as part of a three-membered ring. Oxiranes are also called epoxides.

Ozone (Section 13.9): An elemental form of oxygen having the structure O_3, formed in the upper atmosphere by reaction of oxygen molecules with oxygen atoms.

Ozonolysis (Section 12.10): An oxidative cleavage reaction in which a multiple bond reacts with ozone (O_3) as the oxidant.

P

Para director (Section 18.7): A substituent on a benzene ring that directs a new group to the para position during electrophilic aromatic substitution.

Para isomer (Section 17.3B): A disubstituted benzene ring in which two substituents are on carbon atoms that are separated by two ring carbons (i.e., a 1,4-disubstituted benzene). Para substitution is abbreviated as *p*-.

Parent ion (Section 14.1): The radical cation having the general structure $M^{+\bullet}$, formed by the removal of an electron from an organic molecule. The parent ion is also called the molecular ion.

Parent name (Section 4.4): The portion of the IUPAC name of an organic compound that indicates the number of carbons in the longest continuous chain in the molecule.

Pentose (Section 27.2): A monosaccharide containing five carbons.

Peptide bond (Section 28.5): The amide bond in peptides and proteins.

Peptides (Sections 22.6B, 28.5): Low molecular weight polymers of less than 40 amino acids joined together by amide linkages.

Percent *s*-character (Section 1.10B): The fraction of a hybrid orbital due to the *s* orbital used to form it. As the percent *s*-character increases, a bond becomes shorter and stronger.

Percent transmittance (Section 14.5B): A measure of how much electromagnetic radiation passes through a sample of a compound and how much is absorbed.

Peroxide (Section 13.2): A reactive organic compound with the general structure $RO-OR$. Peroxides are used as radical initiators by homolysis of the weak $O-O$ bond.

Peroxyacid (Section 12.7): An oxidizing agent having the general structure RCO_3H.

Peroxy radical (Section 13.11): A radical formed by the reaction of oxygen with an alkyl or allylic radical. Peroxy radicals have the general structure $R-O-O\bullet$.

Petroleum (Section 4.7): A fossil fuel containing a complex mixture of compounds, primarily hydrocarbons containing one to forty carbon atoms.

Phenol (Sections 9.1, 13.12): An organic compound that contains a hydroxy group (OH group) bonded to a benzene ring.

Phenyl group (Section 17.3D): A group formed by removal of one hydrogen from benzene, abbreviated as C_6H_5- or $Ph-$.

Pheromone (Section 4.1): A chemical substance used for communication in an animal or insect species.

Phosphatidylcholine (Section 26.4A): A phosphoacylglycerol in which the phosphodiester alkyl group is $-CH_2CH_2N(CH_3)_3^+$. Phosphatidylcholines are also called lecithins.

Phosphatidylethanolamine (Section 26.4A): A phosphoacylglycerol in which the phosphodiester alkyl group is $-CH_2CH_2NH_3^+$. Phosphatidylethanolamines are also called cephalins.

Phosphoacylglycerols (Section 26.4A): A lipid having a glycerol backbone with two of the hydroxy groups esterified with fatty acids and the third hydroxy group as part of a phosphodiester.

Phosphodiester (Section 26.4): A functional group having the general formula $ROPO_2OR'$ formed by replacing two of the H atoms in phosphoric acid with alkyl groups.

Phospholipid (Section 3.8A, 26.4): A hydrolyzable lipid that contains a phosphorus atom.

Phosphonium salt (Section 21.10A): An organophosphorus reagent with a positively charged phosphorus and a suitable counterion; for example, $R_4P^+X^-$. Phosphonium salts are converted to ylides upon treatment with a strong base.

Phosporane (Section 21.10A): A phosphorus ylide; for example, $Ph_3P=CR_2$.

Photon (Section 14.4): A particle of electromagnetic radiation.

Pi (π) bond (Section 1.9B): A bond formed by side-by-side overlap of two *p* orbitals where electron density is not concentrated on the axis joining the two nuclei. Pi (π) bonds are generally weaker than σ bonds.

pK_a (Section 2.3): A logarithmic scale of acid strength. $pK_a = -\log K_a$. The smaller the pK_a, the stronger the acid.

Plane of symmetry (Section 5.3): A mirror plane that cuts a molecule in half, so that one half of the molecule is the mirror reflection of the other half.

Plane-polarized light (Section 5.12A): Light that has an electric vector that oscillates in a single plane. Plane-polarized light arises from passing ordinary light through a polarizer. Plane-polarized light is also called polarized light.

β-Pleated sheet (Section 28.8B): A secondary structure formed when two or more peptide chains line up side by side.

Poisoned catalyst (Section 12.5B): A hydrogenation catalyst with reduced activity that allows selective reactions to occur. Poisoned catalysts are also called deactivated catalysts. The Lindlar catalyst is a poisoned Pd catalyst that converts alkynes to cis alkenes.

Polar aprotic solvent (Section 7.8C): A polar solvent that is incapable of intermolecular hydrogen bonding because it does not contain an $O-H$ or $N-H$ bond.

Polar bond (Section 1.11): A covalent bond in which the electrons are unequally shared between the two atoms. Unequal sharing of electrons results from bonding between atoms of different electronegativity values, usually with a difference of ≥ 0.5 units. A polar bond is also called a polar covalent bond.

Polarimeter (Section 5.12A): An instrument that measures the degree that an organic compound rotates plane-polarized light.

Polarity (Section 1.11): A characteristic that results from a dipole. The polarity of a bond is indicated by an arrow with the head of the arrow pointing toward the negative end of the dipole and the tail with a perpendicular line through it at the positive end of the dipole. The polarity of a bond can also be indicated by the symbols δ^+ and δ^-.

Polarizability (Section 3.4A): A measure of how the electron cloud around an atom responds to changes in its electronic environment.

Polarized light (Section 5.12A): Light that has an electric vector that oscillates in a single plane. Polarized light arises from passing ordinary light through a polarizer. Polarized light is also called plane-polarized light.

Polar molecule (Section 1.12): A molecule that has a net dipole. A polar molecule has either one polar bond or multiple polar bonds whose dipoles reinforce.

Polar protic solvent (Section 7.8C): A polar solvent that is capable of intermolecular hydrogen bonding because it contains an $O-H$ or $N-H$ bond.

Polar reaction (Section 6.3B): A reaction in which a nucleophile reacts with an electrophile.

Polycyclic aromatic hydrocarbon (Sections 9.17, 17.5): An aromatic hydrocarbon containing two or more benzene rings that share carbon–carbon bonds. Polycyclic aromatic hydrocarbons are abbreviated as "PAHs."

Polyene (Section 16.7): An organic compound that contains three or more double bonds.

Polyester (Section 22.16B): A condensation polymer consisting of many ester bonds between diols and dicarboxylic acids.

Polyether (Section 9.5B): An organic compound that contains two or more ether linkages.

Polymer (Sections 3.4B, 13.14): A large organic molecule composed of smaller monomer units covalently bonded to each other in a repeating pattern.

Polymerization (Section 13.14A): The chemical process that joins together monomers to make polymers.

Polysaccharide (Section 27.13): A carbohydrate containing three or more monosaccharide units joined together by glycosidic linkages.

Porphyrin (Section 28.9C): A nitrogen-containing heterocyclic structure that can complex metal ions.

Primary alcohol (Section 9.1): An alcohol having the general structure RCH_2OH. Primary alcohols are denoted as 1° alcohols.

Primary alkyl halide (Section 7.1): An alkyl halide having the general structure RCH_2X. Primary alkyl halides are denoted as 1° alkyl halides.

Primary amide (Section 22.1): An amide in which the nitrogen bonded to the carbonyl carbon has two hydrogens attached directly to it. Primary amides ($RCONH_2$) are denoted as 1° amides.

Primary amine (Sections 21.11, 25.1): An amine containing a nitrogen with one R group and two hydrogens attached directly to it. A primary amine (RNH_2) is denoted as a 1° amine.

Primary carbocation (Section 7.14): A carbocation in which the charged carbon has one alkyl group and two hydrogens attached to it. A primary carbocation (RCH_2^+) is denoted as a 1° carbocation.

Primary carbon (Section 4.1A): A carbon atom that is bonded to one other carbon atom. Primary carbons are abbreviated as "1° C."

Primary hydrogen (Section 4.1A): A hydrogen that is attached to a carbon bonded to one other carbon atom (i.e., a primary carbon). Primary hydrogens are abbreviated as "1° H."

Primary protein structure (Section 28.8A): The particular sequence of amino acids joined together by peptide bonds.

Primary radical (Section 13.1): A reactive intermediate containing a radical on a carbon with one R group and two hydrogens attached directly to it. Primary radicals ($RCH_2•$) are denoted as 1° radicals.

Progestins (Section 26.8C): A class of steroidal female sex hormones responsible for preparation of the uterus for egg implantation.

Propagation (Section 13.4A): The middle part of a radical chain reaction in which a radical reacts with another reactant to form a radical as one of the products. Propagation repeats until a termination step occurs.

Prostacyclin (Section 26.6): A type of eicosanoid that acts as a vasodilator that inhibits platelet aggregation.

Prostaglandin (Section 4.15): A class of lipids containing 20 carbons, a five-membered ring, and a COOH group. Prostaglandins possess a wide range of biological activities.

Prosthetic group (Section 28.9C): The non-protein unit of a conjugated protein.

Protecting group (Section 20.12): A blocking group that renders a reactive functional group unreactive so that it does not interfere with another reaction. The protecting group must be easy to put on and easily removed.

Protection (Section 20.12): The reaction that blocks a reactive functional group with a protecting group.

Proteins (Sections 22.6B, 28.5): High molecular weight polymers of 40 or more amino acids joined together by amide linkages.

Proton (Section 2.1): A positively charged hydrogen ion (H^+).

Proton NMR spectroscopy (Section 15.1): A form of nuclear magnetic resonance spectroscopy used to determine the number and type of hydrogen atoms in a molecule.

Proton transfer reaction (Section 2.2): A reaction that results in the transfer of a proton from an acid to a base. Proton transfer reactions are also called Brønsted–Lowry acid–base reactions.

Puckered conformation (Section 4.11): A conformation adopted by cycloalkanes where the atoms in the rings pucker out of planarity to reduce both angle strain and torsional strain.

Purine (Section 27.14B): A bicyclic aromatic heterocycle having two nitrogens in each of the rings.

Pyranose (Section 27.6): A cyclic six-membered ring of a monosaccharide containing an oxygen atom.

Pyrimidine (Section 27.14B): A six-membered aromatic heterocycle having two nitrogens in the ring.

Pyrophosphate (Section 26.7B): A good leaving group that is often used in biological systems. Pyrophosphate is abbreviated as OPP.

Q

Quantum (Section 14.4): The discrete amount of energy associated with a particle of electromagnetic radiation (i.e., a photon).

Quartet (Section 15.6C): An NMR signal that is split into four peaks having a relative area of 1:3:3:1, caused by three nearby nonequivalent protons.

Quaternary ammonium salt (Section 25.1): An organic compound containing a positively charged nitrogen with four σ bonds for example, $R_4N^+X^-$.

Quaternary carbon (Section 4.1A): A carbon atom that is bonded to four other carbon atoms. Quaternary carbons are abbreviated as "4° C."

Quaternary protein structure (Section 28.8C): The shape adopted when two or more folded polypeptide chains aggregate into one protein complex.

Quintet (Section 15.6C): An NMR signal that is split into five peaks caused by four nearby nonequivalent protons.

R

R,S System of nomenclature (Section 5.6): A system of nomenclature used to distinguish the stereochemistry at a tetrahedral stereogenic center. The system involves assigning a priority to each group connected to the stereogenic center. *R* indicates a clockwise orientation of the three highest priority groups and *S* indicates a counterclockwise orientation of the three highest groups. The system is also called the Cahn–Ingold–Prelog system.

Racemic mixture (Section 5.12B): A mixture of two enantiomers in equal amounts. A racemic mixture is optically inactive. A racemic mixture is also called a racemate.

Racemization (Section 7.13C): The formation of equal amounts of two enantiomeric products from an enantiomerically pure starting material.

Radical (Sections 6.3B, 13.1): A reactive intermediate with a single unpaired electron, formed by homolysis of a covalent bond.

Radical anion (Section 12.5C): A reactive intermediate containing both a negative charge and an unpaired electron.

Radical cation (Section 14.1): A species formed in a mass spectrometer by the bombardment of a molecule with an electron beam. The species has an unpaired electron and a positive charge.

Radical inhibitor (Section 13.2): A compound that prevents radical reactions from occurring. Radical inhibitors are also called radical scavengers.

Radical initiator (Section 13.2): A compound that contains an especially weak bond that serves as a source of radicals.

Radical polymerization (Section 13.14B): A radical chain reaction involving the polymerization of alkene monomers by adding a radical to a π bond.

Radical reaction (Section 6.3B): A reaction that involves a radical as a reactive intermediate.

Radical scavenger (Section 13.2): A compound that prevents radical reactions from occurring. Radical scavengers are also called radical inhibitors.

Rate constant (Section 6.9B): A constant that is a fundamental characteristic of a reaction. The rate constant is a complex mathematical term that takes into account the dependence of a reaction rate on temperature and the energy of activation. The rate constant is denoted by the symbol k.

Rate-determining step (Section 6.8): In a multistep reaction mechanism, the step with the highest energy transition state. The rate-determining step determines how fast the reaction proceeds (i.e., the reaction rate).

Rate equation (Section 6.9B): An equation that shows the relationship between the rate of a reaction and the concentration of the reactants. The rate equation depends on the mechanism of the reaction. The rate equation is also called the rate law.

Rate law (Section 6.9B): An equation that shows the relationship between the rate of a reaction and the concentration of the reactants. The rate law depends on the mechanism of the reaction. The rate law is also called the rate equation.

Reaction arrow (Section 6.1): An arrow in a reaction equation indicating that a chemical reaction occurs. The reaction arrow is drawn between the starting material and product.

Reaction coordinate (Section 6.7): A representation of the progress of a reaction as it proceeds from reactant to product. The reaction coordinate is the *x*-axis in an energy diagram.

Reaction mechanism (Section 6.3): A detailed description of how bonds are broken and formed as a starting material is converted to a product.

Reactive intermediate (Sections 6.3, 10.18): A high-energy unstable intermediate formed during the conversion of a stable starting material to a stable product.

Reactivity–selectivity principle (Section 13.6): The chemical principle that less reactive reagents are generally more selective, typically yielding one major product.

Reagent (Section 6.1): The chemical substance with which an organic compound reacts. The reagent is drawn on either the reactant side of the reaction arrow or above the reaction arrow.

Reciprocal centimeter (Section 14.5A): The unit for wavenumber, which is used to report frequency in IR spectroscopy.

Reducing sugar (Section 27.9B): A carbohydrate that can be oxidized by Tollens, Benedict's, or Fehling's reagent.

Reduction (Sections 4.14A, 12.1): A process that results in a compound gaining electrons. For organic compounds, reduction results in a decrease in the number of $C-Z$ bonds or an increase in the number of $C-H$ bonds.

Reductive amination (Section 25.7C): A two-step method that converts aldehydes and ketones into amines.

Regioselective reaction (Section 8.5): A reaction that yields predominantly or exclusively one constitutional isomer when more than one constitutional isomer is possible.

Resolution (Section 28.3): The separation of a racemic mixture into its component enantiomers.

Resonance (Section 15.1A): In NMR spectroscopy, when an atomic nucleus absorbs RF radiation and "spin flips" to a higher energy state.

Resonance hybrid (Sections 1.5C, 16.4): A structure that is a weighted composite of all possible resonance structures. The resonance hybrid shows the delocalization of electron density due to the different location of electrons in individual resonance structures.

Resonance structures (Sections 1.5, 16.2): Two or more structures of a molecule that differ in the placement of π bonds and nonbonded electrons. The placement of atoms and σ bonds stays the same.

Retention of configuration (Section 7.11C): The same relative stereochemistry of a stereogenic center in the reactant and the product of a chemical reaction. Retention of configuration typically results from a frontside attack at a stereogenic center.

Retention time (Section 14.3B): The length of time required for a component of a mixture to travel through a chromatography column.

Retro Diels–Alder reaction (Section 16.14B): A reaction that is the reverse of a Diels–Alder reaction in which a cyclohexene is cleaved to give a 1,3-diene and an alkene.

Retrosynthetic analysis (Section 10.18): Working backwards from a product to determine the starting material from which it is made.

RF radiation (Section 15.1A): Radiation in the radiofrequency region of the electromagnetic spectrum, characterized by long wavelength and low frequency and energy.

Ribonucleoside (Section 27.14B): An *N*-glycoside formed by the reaction of D-ribose with certain amine heterocycles.

Ribonucleotide (Section 27.14B): An RNA building block having a ribose and either a purine or pyrimidine base joined together by an *N*-glycosidic linkage and a phosphate bonded to a hydroxy group of the sugar nucleus.

Ring current (Section 15.4): A circulation of π electrons in an aromatic ring caused by the presence of an external magnetic field.

Ring flip (Section 4.12B): A two-step process in which one chair conformation of cyclohexane interconverts with a second chair conformation.

Robinson annulation (Section 24.9): A ring forming reaction that combines a Michael reaction with an intramolecular aldol reaction to form a 2-cyclohexenone.

Rule of endo addition (Section 16.13D): The rule that in a Diels–Alder reaction, the endo product is preferred.

S

Sandmeyer reaction (Section 25.14A): A reaction between an aryl diazonium salt and a copper(I) halide to form an aryl halide.

Saponification (Section 22.11B): Basic hydrolysis of an ester to form an alcohol and a carboxylate anion.

Saturated fatty acid (Section 10.6A): A fatty acid having no carbon–carbon double bonds in its long hydrocarbon chain.

Saturated hydrocarbon (Section 4.1): A compound that contains only $C-C$ and $C-H$ σ bonds and no rings, thus having the maximum number of hydrogen atoms per carbon.

Schiff base (Section 21.11A): An organic compound having general structure $R_2C=NR'$. A Schiff base is also called an imine.

Secondary alcohol (Section 9.1): An alcohol having the general structure R_2CHOH. Secondary alcohols are denoted as 2° alcohols.

Secondary alkyl halide (Section 7.1): An alkyl halide having the general structure R_2CHX. Secondary alkyl halides are denoted as 2° alkyl halides.

Secondary amide (Section 22.1): An amide in which the nitrogen bonded to the carbonyl carbon has one hydrogen and one alkyl group attached directly to it. Secondary amides (RCONHR') are denoted as 2° amides.

Secondary amine (Sections 21.11, 25.1): An amine containing a nitrogen with two R groups and one hydrogen attached directly to it. Secondary amines (R_2NH) are denoted as 2° amines.

Secondary carbocation (Section 7.14): A carbocation in which the charged carbon has two alkyl groups and one hydrogen attached to it. Secondary carbocations (R_2CH^+) are denoted as 2° carbocations.

Secondary carbon (Section 4.1A): A carbon atom that is bonded to two other carbon atoms. Secondary carbons are abbreviated as "2° C."

Secondary hydrogen (Section 4.1A): A hydrogen that is attached to a carbon bonded to two other carbon atoms (i.e., a secondary carbon). Secondary hydrogens are abbreviated as "2° H."

Secondary protein structure (Section 28.8B): The three-dimensional conformations of localized regions of a protein.

Secondary radical (Section 13.1): A reactive intermediate containing a radical on a carbon with two R groups and one hydrogen attached directly to it. Secondary radicals ($R_2CH\cdot$) are denoted as 2° radicals.

Second-order kinetics (Section 6.9B): The order of a rate equation in which the sum of the exponents is two.

Second-order rate equation (Section 7.10): A rate equation in which the reaction rate depends on the concentration of two reactants.

Separatory funnel (Section 19.12): A piece of laboratory glassware used for extractions.

Septet (Section 15.6C): An NMR signal that is split into seven peaks caused by six nearby nonequivalent protons.

Sesquiterpene (Section 26.7A): A terpene that contains 15 carbons and three isoprene units.

Sesterterpene (Section 26.7A): A terpene that contains 25 carbons and five isoprene units.

Sextet (Section 15.6C): An NMR signal that is split into six peaks caused by five nearby nonequivalent protons.

Sharpless asymmetric epoxidation (Section 12.14): An enantioselective oxidation reaction that converts the double bond of an allylic alcohol to a predictable enantiomerically enriched epoxide.

Sharpless reagent (Section 12.14): The reagent used in the Sharpless asymmetric epoxidation. The Sharpless reagent consists of *tert*-butyl hydroperoxide, a titanium catalyst, and one enantiomer of diethyl tartrate.

Shielding effects (Section 15.3A): An effect in NMR caused by small induced magnetic fields of electrons in the opposite direction to the applied magnetic field. Shielding decreases the strength of the magnetic field felt by the nucleus and shifts an absorption upfield.

1,2-Shift (Section 9.9): A rearrangement of a less stable carbocation to a more stable carbocation by the shift of a hydrogen atom or an alkyl group from one carbon atom to an adjacent carbon atom.

Sigma (σ) bond (Section 1.8A): A cylindrically symmetrical bond that concentrates the electron density on the axis that joins two nuclei. All single bonds are σ bonds.

Silyl ether (Section 20.12): A common protecting group for an alcohol in which the $O-H$ bond is replaced by an $O-Si$ bond.

Singlet (Section 15.6A): An NMR signal that occurs as a single peak.

Skeletal structure (Section 1.7B): A shorthand representation of the structure of an organic compound in which carbon atoms and hydrogen atoms bonded to them are omitted. All heteroatoms and hydrogens bonded to them are drawn in. Carbon atoms are assumed to be at the junction of any two lines or at the end of a line.

S_N1 mechanism (Sections 7.10, 7.13): A nucleophilic substitution mechanism that goes by a two-step process involving a carbocation intermediate. S_N1 is an abbreviation for "Substitution Nucleophilic Unimolecular."

S_N2 mechanism (Sections 7.10, 7.11): A nucleophilic substitution mechanism that goes by a one-step concerted process, where both reactants are involved in the transition state. S_N2 is an abbreviation for "Substitution Nucleophilic Bimolecular."

Soap (Section 22.12B): The carboxylate salts of long-chain fatty acids prepared by the basic hydrolysis or saponification of a triacylglycerol.

Solubility (Section 3.5C): A measure of the extent to which a compound dissolves in a liquid.

Solute (Section 3.5C): The compound that is dissolved in a liquid solvent.

Solvent (Section 3.5C): The liquid component into which the solute is dissolved.

Specific rotation (Section 5.12C): A standardized physical constant for the amount that a chiral compound rotates plane-polarized light. Specific rotation is denoted by the symbol $[\alpha]$ and defined using a specific sample tube length (l, in dm), concentration (c in g/mL), temperature (25°C) and wavelength (589 nm). $[\alpha] = \alpha / (l \times c)$

Spectator ion (Section 2.1): An ion that does not take part in a reaction and is opposite in charge to the ion that does take part in a reaction. A spectator ion is also called a counterion.

Spectroscopy (Section 14.1): An analytical method using the interaction of electromagnetic radiation with molecules to determine molecular structure.

Sphingomyelin (Section 26.4B): A hydrolyzable phospholipid derived from sphingosine.

Spin flip (Section 15.1A): In NMR spectroscopy, when an atomic nucleus absorbs RF radiation and its magnetic field flips relative to the external magnetic field.

Spin–spin splitting (Section 15.6): Splitting of an NMR signal into peaks caused by nonequivalent protons on the same carbon or adjacent carbons.

Spiro ring system (Section 23.8D): A compound having two rings that share a single carbon atom.

Staggered conformation (Section 4.9): A conformation of a molecule in which the bonds on one carbon bisect the R−C−R bond angle on the adjacent carbon.

Stereochemistry (Sections 4.9, 5.1): The study of the three-dimensional structure of molecules.

Stereogenic center (Section 5.3): A site in a molecule at which the interchange of two groups forms a stereoisomer. A carbon bonded to four different groups is a tetrahedral stereogenic center. A stereogenic center is also called a chirality center, a chiral center, or an asymmetric center.

Stereoisomers (Sections 4.13B, 5.1): Two compounds that have the same molecular formula and same atom connectivity but differ in the way the atoms are oriented in space and cannot be interconverted by rotation around single bonds.

Stereoselective reaction (Section 8.5): A reaction that yields predominantly or exclusively one stereoisomer when two or more stereoisomers are possible.

Stereospecific reaction (Section 10.14): A reaction in which each of two specific stereoisomers of a starting material yields a particular stereoisomer of a product.

Steric effects (Section 7.8B): Energy effects that arise due to two atoms not being able to occupy the same space.

Steric hindrance (Section 7.8B): A decrease in reactivity resulting from the presence of bulky groups at the site of a reaction.

Steric strain (Section 4.10): An increase in energy resulting when atoms in a molecule are forced too close to one another.

Steroid (Sections 16.14C, 26.8): A nonhydrolyzable tetracyclic lipid composed of three six-membered rings and one five-membered ring.

Straight-chain alkane (Section 4.1A): An acyclic alkane that has all of its carbons in a row. Straight-chain alkanes are also called normal alkanes.

Strecker amino acid synthesis (Section 28.2C): A reaction that converts an aldehyde into an amino acid by way of an α-amino nitrile.

Structural isomers (Sections 4.1A, 5.2): Two compounds that have the same molecular formula but differ in the way the atoms are connected to each other. Structural isomers are also called constitutional isomers.

Substituent (Section 4.4): A group or branch attached to the longest continuous chain of carbons in an organic molecule.

Substitution reaction (Section 6.2A): A reaction in which an atom or a group of atoms is replaced by another atom or group of atoms. Substitution reactions involve σ bonds: one σ bond breaks and another is formed at the same atom.

Substrate (Section 6.11): An organic molecule that is transformed by the action of an enzyme.

Sulfa drug (Section 25.17): A sulfur-containing organic compound that acts as an antibiotic.

Sulfonate anion (Section 19.13): An anion having the general structure RSO_3^-, formed by deprotonating a sulfonic acid with a Brønsted–Lowry base.

Sulfonation (Section 18.4): An electrophilic aromatic substitution reaction in which benzene reacts with $^+SO_3H$ to give a benzenesulfonic acid.

Sulfonic acid (Section 19.13): An acidic organic compound having the general structure RSO_3H.

Symmetrical anhydride (Section 22.1): An anhydride that has two identical alkyl groups bonded to the carbonyl carbon atoms.

Symmetrical ether (Section 9.1): An ether with two identical alkyl groups groups bonded to the oxygen.

Syn addition (Section 10.8): An addition in which the two parts of the reagent are added from the same side.

Syn dihydroxylation (Section 12.9): An oxidation reaction that involves the addition of two hydroxy groups to the same face of a double bond.

Syn periplanar (Section 8.8): In an elimination reaction, a geometry in which the β hydrogen and the leaving group are on the same side of the molecule.

Synthetic intermediate (Section 10.18): A stable compound that is the product of one step and the starting material of another step in a multistep synthesis.

Systematic name (Section 4.3): The name of a molecule indicating the compound's chemical structure. The systematic name is also called the IUPAC name.

T

Target compound (Section 11.12): The final product of a synthetic scheme.

Tautomerization (Sections 11.9, 23.2A): The process of converting one tautomer into another.

Tautomers (Section 11.9): Constitutional isomers that differ in the location of a double bond and a hydrogen atom.

Terminal alkene (Section 10.1): An alkene that has the double bond at the end of the carbon chain.

Terminal alkyne (Section 11.1): An alkyne that has the triple bond at the end of the carbon chain.

C-Terminal amino acid (Section 28.5A): The amino acid at the end of a peptide chain with a free carboxy group.

N-Terminal amino acid (Section 28.5A): The amino acid at the end of a peptide chain with a free amino group.

Termination (Section 13.4A): The final step of a radical chain reaction where two radicals combine to form a stable bond. Removing the radicals from the reaction mixture without generating any new radicals stops the reaction.

Terpene (Section 26.7): A lipid composed of repeating five-carbon isoprene units.

Tertiary alcohol (Section 9.1): An alcohol having the general structure R_3COH. Tertiary alcohols are denoted as 3° alcohols.

Tertiary alkyl halide (Section 7.1): An alkyl halide having the general structure R_3CX. Tertiary alkyl halides are denoted as 3° alkyl halides.

Tertiary amide (Section 22.1): An amide in which the nitrogen bonded to the carbonyl carbon has two alkyl groups attached directly to it. Tertiary amides ($RCONR'_2$) are denoted as 3° amides.

Tertiary amine (Sections 21.11, 25.1): An amine containing a nitrogen with three R groups and no hydrogens attached directly to it. Tertiary amines (R_3N) are denoted as 3° amines.

Tertiary carbocation (Section 7.14): A carbocation in which the charged carbon has three alkyl groups and no hydrogens attached to it. Tertiary carbocations (R_3C^+) are denoted as 3° carbocations.

Tertiary carbon (Section 4.1A): A carbon atom that is bonded to three other carbon atoms. Tertiary carbons are abbreviated as "3° C."

Tertiary hydrogen (Section 4.1A): A hydrogen that is attached to a carbon bonded to three other carbon atoms (i.e., a tertiary carbon). Tertiary hydrogens are abbreviated as "3° H."

Tertiary protein structure (Section 28.8C): The three-dimensional shape adopted by the entire peptide chain.

Tertiary radical (Section 13.1): A reactive intermediate containing a radical on a carbon with three R groups attached directly to it. Tertiary radicals ($R_3C\cdot$) are denoted as 3° radicals.

Tesla (Section 15.1A): A unit used to measure the strength of a magnetic field. Tesla is denoted with the symbol "T."

Tetramethylsilane (Section 15.1B): An internal standard used as a reference in ^{1}H NMR spectroscopy. The tetramethylsilane reference peak occurs at 0 ppm on the δ scale.

Tetrasubstituted alkene (Section 8.2A): An alkene that has four alkyl groups and no hydrogens bonded to the carbons of the double bond.

Tetraterpene (Section 26.7A): A terpene that contains 40 carbons and eight isoprene units.

Tetrose (Section 27.2): A monosaccharide containing four carbons.

Thermodynamic enolate (Section 23.4): The enolate that is lower in energy. When two enolates are possible, the thermodynamic enolate is the more substituted enolate.

Thermodynamic product (Section 16.11): In a reaction that can give more than one product, the product that predominates at equilibrium.

Thermodynamics (Section 6.5): A study of the energy and equilibrium of a chemical reaction.

Thioester (Section 22.17): An organic compound with the general structure RCOSR'.

Thromboxane (Section 26.6): A type of eicosanoid that acts as a vasoconstrictor that triggers blood platelet aggregation.

Tollens reagent (Sections 20.8, 27.9B): A reagent that oxidizes aldehydes, and consists of silver(I) oxide in aqueous ammonium hydroxide.

Tollens test (Section 20.8): A test for detecting an aldehyde in a molecule. A positive Tollens test forms a silver mirror on the walls of the reaction flask.

***p*-Toluenesulfonate** (Section 9.13): A very good leaving group having the general structure $CH_3C_6H_5SO_3^-$, and abbreviated as TsO^-. Compounds containing a *p*-toluenesulfonate leaving group are called alkyl tosylates and are abbreviated ROTs.

Torsional energy (Section 4.9): The energy difference between the staggered and eclipsed conformers of a molecule.

Torsional strain (Section 4.9): An increase in the energy of a molecule caused by eclipsing interactions between groups attached to adjacent carbon atoms.

Tosylate (Section 9.13): An organic compound having the general structure $ROSO_2C_6H_4CH_3$, formed from an alcohol. The tosylate group is a very good leaving group. Tosylates are also called alkyl tosylates and are abbreviated ROTs.

***s*-Trans** (Sections 16.6, 28.5B): The conformer of a 1,3-diene that has the two double bonds on opposite sides of the single bond that joins them.

Trans diaxial (Section 8.8B): In an elimination reaction of a cyclohexane, a geometry in which the β-hydrogen and the leaving group are trans with both in the axial position.

Trans isomer (Sections 4.13B, 8.3B): An isomer of a ring or double bond that has two groups on opposite sides of the ring or double bond.

Transition state (Section 6.7): An unstable energy maximum as a chemical reaction proceeds from reactants to products. The transition state is at the top of an energy "hill" and can never be isolated.

Triacylglycerol (Sections 10.6, 22.12A, 26.3): A lipid consisting of the triester of glycerol with three long-chain fatty acids. Triacylglycerols are the lipids that comprise animal fats and vegetable oils. Triacylglycerols are also called triglycerides.

Triose (Section 27.2): A monosaccharide containing three carbons.

Triplet (Section 15.6): An NMR signal that is split into three peaks having a relative area of 1:2:1, caused by two nearby nonequivalent protons.

Trisubstituted alkene (Section 8.2A): An alkene that has three alkyl groups and one hydrogen bonded to the carbons of the double bond.

Triterpene (Section 26.7A): A terpene that contains 30 carbons and six isoprene units.

U

Ultraviolet (UV) light (Section 16.15): Electromagnetic radiation with a wavelength from 200–400 nm.

Unimolecular reaction (Sections 6.9B, 7.10, 7.13A): A reaction where the concentration of only one reactant determines the reaction rate and only one term appears in the rate equation. A unimolecular reaction has only one reactant involved in the rate-determining step.

α,β-Unsaturated carbonyl compound (Section 20.15): A conjugated organic compound containing a carbonyl group and a carbon–carbon double bond separated by a single σ bond.

Unsaturated fatty acid (Section 10.6A): A fatty acid having one or more carbon–carbon double bonds in its hydrocarbon chain. In natural fatty acids, the double bonds have the Z configuration.

Unsaturated hydrocarbon (Section 10.2): A hydrocarbon that has fewer than the maximum number of hydrogen atoms per carbon atom. Hydrocarbons with π bonds or rings are unsaturated.

Unsymmetrical ether (Section 9.1): An ether in which the two alkyl groups bonded to the oxygen are different.

Upfield shift (Section 15.1B): In an NMR spectrum, a term used to describe the relative location of an absorption peak. An upfield shift means a peak is shifted to the right in the spectrum to lower chemical shift.

V

Valence bond theory (Section 17.9A): A theory that describes covalent bonding as the overlap of two atomic orbitals with the electron pair in the resulting bond being shared by both atoms.

Valence electrons (Section 1.1): The electrons in the outermost shell of orbitals. Valence electrons determine the properties of a given element. Valence electrons are more loosely held than the core electrons and thus participate in chemical reactions.

van der Waals forces (Section 3.4A): Very weak intermolecular interactions caused by momentary changes in electron density in molecules. The changes in electron density cause temporary dipoles, which are attracted to temporary dipoles in adjacent molecules. van der Waals forces are also called London forces.

Vicinal dihalide (Section 8.10): A compound that has two halogen atoms on adjacent carbon atoms.

Vinyl group (Section 10.3C): An alkene substituent having the structure $-CH=CH_2$.

Vinyl halide (Section 7.1): A molecule containing a halogen atom bonded to the sp^2 hybridized carbon of a carbon–carbon double bond.

Vitamins (Sections 3.6, 26.5): Organic compounds needed in small amounts by biological systems for normal cell function.

VSEPR (Section 1.6B): Valence Shell Electron Pair Repulsion theory. A theory that determines the three-dimensional shape of a molecule by the number of groups surrounding a central atom. The most stable arrangement keeps the groups as far away from each other as possible.

W

Walden inversion (Section 7.11C): The inversion of a stereogenic center involved in an S_N2 reaction.

Wavelength (Section 14.4): The distance from one point of a wave to the same point on the adjacent wave. In electromagnetic radiation, the wavelength is abbreviated with the Greek letter lambda (λ).

Wavenumber (Section 14.5A): A unit for the frequency of electromagnetic radiation that is inversely proportional to wavelength. Wavenumber is reported in reciprocal centimeters (cm^{-1}). Wavenumber is used to report frequency in IR spectroscopy.

Wax (Sections 4.15, 26.2): A hydrolyzable lipid consisting of an ester formed from a high molecular weight alcohol and a fatty acid.

Williamson ether synthesis (Section 9.6): A method for preparing ethers by reacting an alkoxide ion with a methyl or primary alkyl halide.

Wittig reaction (Section 21.10): A reaction of a carbonyl group and an organophosphorus reagent that forms an alkene.

Wittig reagent (Section 21.10A): An organophosphorus reagent containing a positively charged phosphorus bonded to a negatively charged carbon. Wittig reagents have the general structure $Ph_3P{=}CR'_2$.

Wohl degradation (Section 27.10A): A reaction that shortens the carbon chain of an aldose by removing one carbon from the end.

Wolff–Kishner reduction (Section 18.14B): A method to reduce aryl ketones to alkyl benzenes using hydrazine (NH_2NH_2) and strong base (KOH).

Y

Ylide (Section 21.10A): A chemical species that contains two oppositely charged atoms bonded to each other, and both atoms have octets of electrons.

Z

Zaitsev rule (Section 8.5): In a β elimination reaction, a rule that states that the major product is the alkene with the most substituted double bond.

Zwitterion (Sections 19.14B, 28.1B): A neutral compound that contains both a positive and negative charge.

Credits

Prologue

Page 2 (pills): © Eyewire EP046/Getty; **p. 2 (biker):** © Corbis/Vol. 49; **p. 2 (fabric):** © Randy Faris/Corbis; **p. 2 (paints):** © Photodisc Blue/Getty; **p. 3 (both):** © Royalty-Free/Corbis; **fig. 1a-b:** Courtesy Dr. S. Arthur Reed, Professor Emeritus, Zoology Department, University of Hawaii, Honolulu, Hawaii.

Chapter 1

Opener: © Scott T. Smith/Corbis; **p. 11:** © Rachel Epstein/PhotoEdit; **p. 18 (top):** Courtesy Wilmer Stratton; **p. 18 (bottom):** © The McGraw-Hill Companies, Inc./Elite Images; **p. 28:** Courtesy John Somerville; **p. 31:** © The McGraw-Hill Companies, Inc./Elite Images; **p. 37:** © Corbis/R-F Website; **p. 38:** © PhotoDisc Website/Getty.

Chapter 2

Opener: © The McGraw-Hill Companies, Inc./Elite Images; **p. 72:** © National Geographic/Getty Images.

Chapter 3

Opener: © Jacqui Hurst/Corbis; **p. 85:** Courtesy Ed Czerwinksi, Ontario Ministry of Natural Resources; **p. 90 (top):** © Andrew Syred/Photo Researchers, Inc.; **p. 90 (bottom):** © Mark Moffett/Minden Pictures.

Chapter 4

Opener: © Samuel Ashfield/Photo Researchers, Inc.; **p. 115 (top):** © God of Insects; **p. 115 (bottom):** © Douglas Peebles/Corbis; **p. 120, fig 4.5a, p. 150:** © PhotoDisc Website/Getty.

Chapter 5

Opener: © The McGraw-Hill Companies, Inc./Elite Images; **fig 5.1 (starch):** © Jim Erickson/Corbis; **fig. 5.1 (wheat):** © Vol. 1 PhotoDisc/Getty; **fig. 5.1 (cotton):** © Corbis/R-F Website; **fig. 5.1 (fabric):** Courtesy Daniel C. Smith; **p. 160 (both):** Courtesy Daniel C. Smith; **p. 162:** Courtesy Harry G. Lee, Jacksonville Shell Club; **p. 164:** © Henriette Kress; **p. 166:** © PhotoDisc Website/Getty; **fig 5.13 (left):** © The McGraw-Hill Companies, Inc./Elite Images; **fig 5.13 (right):** © David Sieren/Visuals Unlimited.

Chapter 6

Opener: © Jim Erickson/Corbis.

Chapter 7

Opener: © Corbis/R-F Website; **p. 225 (top):** Courtesy David L. Ballantine, Department of Marine Sciences, University of Puerto Rico; **p. 225 (bottom):** *Time Magazine* June 30,
1947 in the collection of Beverly Naidus, www.artsforchange.org; **p. 253:** © The McGraw-Hill Companies, Inc./Elite Images.

Chapter 8

Opener: © Corbis/R-F Website.

Chapter 9

Opener: Courtesy Peter J. S. Franks, Scripps Institution of Oceanography; **fig 9.3 (left):** © Food Pix/Getty Images; **fig. 9.3 (right):** © eStock Photo; **p. 310 (top):** Courtesy Boston Medical Library in the Francis A. Countway Library of Medicine; **p. 310 (bottom):** Courtesy Florida Marine Research Institute; **p. 322:** © Henriette Kress.

Chapter 10

Opener: © Jim Foster/Corbis; **fig 10.5 (carrots):** © Roy Morsch/Corbis; **fig. 10.5 (citrus):** © FoodPix/Getty Images; **fig. 10.5 (apples & ginger):** © PhotoDisc Website/Getty; **p. 367:** © Richard Megna/Fundamental Photographs, NYC; **fig 10.17:** © Henriette Kress.

Chapter 11

Opener: © Ray Ellis/Photo Researchers, Inc.; **fig 11.1:** © Arachnokulture; **p. 403:** Michael G. Moye.

Chapter 12

Opener: Courtesy John Ghent, USDA Forest Service; **p. 418:** © The McGraw-Hill Companies, Inc./Elite Images; **p. 421:** © Richard Megna/Fundamental Photographs, NYC; **p. 426:** USDA APHIS PPQ Archives, www.forestryimages.org; **fig 12.10:** © The McGraw-Hill Companies, Inc./Elite Images; **p. 437:** © PhotoDisc/Getty Images.

Chapter 13

Opener: © PhotoAlto/eStock Photo; **p. 459 (top):** © Richard Megna/Fundamental Photographs, NYC; **p. 459 (bottom):** SVS, TOMS, NASA; **p. 466:** © The McGraw-Hill Companies/Elite Images; **p. 468 (both):** © Richard Megna/Fundamental Photographs, NYC; **Table 13.2 (all):** © The McGraw-Hill Companies, Inc./John Thoeming, photographer.

Chapter 14

Opener: © Ted Kinsman/Photo Researchers, Inc. Spectra courtesy of the Chemistry Department at Rutgers University.

Chapter 15

Page 529: © The McGraw-Hill Companies, Inc./Elite Images; **fig 15.15 (left):** James Rosato, RT (R) (MR) (CT) MRI Applications Specialist Brigham and Women's Hospital,

Boston MA; **fig. 15.15 (right):** With permission, J. Hornak, Basics of MRI, www.cis.rit.edu. Spectra courtesy of the Chemistry Department at Rutgers University.

Chapter 16

Opener: © Royalty-Free/Corbis; **p. 551:** © Franz-Marc Frei/Corbis; **fig. 16.8:** © Stephen Frink/Corbis; **p. 567:** © Scott Camazine/Photo Researchers, Inc.

Chapter 17

Opener: Photo by Forest & Kim Starr; **fig. 17.3 (left):** © David Young-Wolff/PhotoEdit; **fig. 17.3 (right):** © Corbis; **p. 591:** Courtesy Daniel C. Smith; **p. 599 (soccer ball):** © Getty Images; **p. 599 (left):** © Charles O'Rear/Corbis; **p. 599 (right):** © Lester V. Bergman/Corbis. Spectra courtesy of the Chemistry Department at Rutgers University.

Chapter 18

Opener: Courtesy Department of Plant Pathology, North Dakota State University; **fig. 18.3:** © Time Life Pictures/Getty Images. Spectra courtesy of the Chemistry Department at Rutgers University.

Chapter 19

Opener (babies): © Beranger/Photo Researchers, Inc. **Opener (inset):** © David Young-Wolff/PhotoEdit; **p. 657:** © Ted Nelson/Dembinsky Photo Assoc; **p. 658 (top):** © Royalty-Free/Corbis; **p. 658 (bottom):** © Niels Sloth; **p. 672:** © John A. Raineri. Spectra courtesy of the Chemistry Department at Rutgers University.

Chapter 20

Opener: © Bill Scifres; **p. 690:** © M. K. Ranjitsinh/Photo Researchers, Inc.; **p. 693 (top):** © PhotoDisc/Getty Images; **p. 693 (bottom):** © Vol. 110/Getty Images; **p. 696:** © Digital Vision/Getty Images; **p. 698:** © The McGraw-Hill Companies, Inc./Joe Franek, photographer. Spectra courtesy of the Chemistry Department at Rutgers University.

Chapter 21

Opener: © The McGraw-Hill Companies, Inc.; **fig. 21.6 (lemon grass):** © Light Photographic; **fig. 21.6 (cinnamon):** © Photo Researchers, Inc.; **fig. 21.6 (almonds):** © Photo Alto/eStock Photo; **fig. 21.6 (bleu cheese):** © eStock Photo; **p. 763:** © Ernest Beat Basel. Spectra courtesy of the Chemistry Department at Rutgers University.

Chapter 22

Opener: © Andrew McClenaghan/Photo Researchers, Inc.; **p. 788:** © Gary Retherford/Photo Researchers, Inc.; **p. 794 (top):** © The McGraw-Hill Companies, Inc./Joe Franek, photographer; **p. 794 (bottom):** Photo by Scott Bauer, USDA; **p. 806:** © Mark A. Johnson/Corbis; **p. 810:** Courtesy DuPont; **p. 811:** © Mark A. Johnson/Corbis. Spectra courtesy of the Chemistry Department at Rutgers University.

Chapter 23

Opener: Breast Cancer 1998: © United States Postal Service. Displayed with permission. All rights reserved. Written authorization from the Postal Service is required to use, reproduce, post, transmit, distribute, or publicly display these images; **pp. 856 & 863:** © The McGraw-Hill Companies, Inc./Joe Franek, photographer; **p. 863:** © The McGraw-Hill Companies, Inc./Joe Franek, photographer; **p. 866:** © Sergio Piumatti; **p. 867:** Anita Cooper Courtesy The Vetiver Network; **p. 873:** © Margareta Taute/Visuals Unlimited.

Chapter 24

Opener (both): Courtesy Richard M. Amasino; **p. 888:** © William Weber/Visuals Unlimited.

Chapter 25

Opener: © Science & Society Picture Library; **p. 917 (coffee):** © Inga Spence/Visuals Unlimited; **p. 917 (tobacco):** Taxi/Getty; **p. 917 (hemlock):** © Ken Lucas/Visuals Unlimited; **p. 946 (indigo plant):** © Kirsten Soderlind/Corbis; **p. 946 (snail shell):** © SuperStock; **p. 946 (madder):** Photograph by Paul Redfearn/Ozarks Regional Herbarium/Southwest Missouri State University. Spectra courtesy of the Chemistry Department at Rutgers University.

Chapter 26

Opener: © Andrew Brookes/Corbis; **p. 962:** © Taxi/Getty; **p. 964 (trees):** © Taxi/Getty; **p. 964 (oils):** © The McGraw-Hill Companies, Inc./Elite Images; **p. 964 (herring):** © Rick Price/Corbis; **p. 973 (bottom):** © Vol. 18/PhotoDisc/Getty; **p. 974:** Coral Reef Research Foundation.

Chapter 27

Opener: © Japack Company/Corbis; **p. 990 (top):** © Michael Newman/PhotoEdit; **p. 990 (bottom):** © The McGraw-Hill Companies, Inc./Elite Images; **p. 1002:** © Michelle Garrett/Corbis; **p. 1016 (both):** © The McGraw-Hill Companies, Inc.; **fig. 27.9 (left):** © The McGraw-Hill Companies, Inc./Elite Images; **(right):** © David Muench/Corbis; **p. 1020:** © Lawson Wood/Corbis.

Chapter 28

Opener: © Michael Dalton/Fundamental Photographs, NYC; **fig. 28.8:** Courtesy Protein Technologies, Inc.; **p. 1063:** © Eye of Science/Photo Researchers, Inc.

Index

Page numbers followed by an "f" indicate figures; "t" indicates tabular material.